Frontiers in Surface and Interface Science

Frontiers in Surface and Interface Science

Editors:

Charles B. Duke
Webster, NY, USA

E. Ward Plummer
Knoxville, TN, USA

2002
ELSEVIER

Amsterdam - Boston - London - New York - Oxford - Paris
San Diego - San Francisco - Singapore - Sydney - Tokyo

ELSEVIER SCIENCE B.V.
Sara Burgerhartstraat 25
P.O. Box 211, 1000 AE Amsterdam, The Netherlands

© 2002 Elsevier Science B.V. All rights reserved.

This work is protected under copyright by Elsevier Science, and the following terms and conditions apply to its use:

Photocopying
Single photocopies of single chapters may be made for personal use as allowed by national copyright laws. Permission of the Publisher and payment of a fee is required for all other photocopying, including multiple or systematic copying, copying for advertising or promotional purposes, resale, and all forms of document delivery. Special rates are available for educational institutions that wish to make photocopies for non-profit educational classroom use.

Permissions may be sought directly from Elsevier Science Global Rights Department, PO Box 800, Oxford OX5 1DX, UK; phone: (+44) 1865 843830, fax: (+44) 1865 853333, e-mail: permissions@elsevier.co.uk. You may also contact Global Rights directly through Elsevier's home page (http://www.elsevier.com), by selecting 'Obtaining Permissions'.

In the USA, users may clear permissions and make payments through the Copyright Clearance Center, Inc., 222 Rosewood Drive, Danvers, MA 01923, USA; phone: (+1) (978) 7508400, fax: (+1) (978) 7504744, and in the UK through the Copyright Licensing Agency Rapid Clearance Service (CLARCS), 90 Tottenham Court Road, London W1P 0LP, UK; phone: (+44) 207 631 5555; fax: (+44) 207 631 5500. Other countries may have a local reprographic rights agency for payments.

Derivative Works
Tables of contents may be reproduced for internal circulation, but permission of Elsevier Science is required for external resale or distribution of such material. Permission of the Publisher is required for all other derivative works, including compilations and translations.

Electronic Storage or Usage
Permission of the Publisher is required to store or use electronically any material contained in this work, including any chapter or part of a chapter.

Except as outlined above, no part of this work may be reproduced, stored in a retrieval system or transmitted in any form or by any means, electronic, mechanical, photocopying, recording or otherwise, without prior written permission of the Publisher. Address permissions requests to: Elsevier Science Global Rights Department, at the mail, fax and e-mail addresses noted above.

Notice
No responsibility is assumed by the Publisher for any injury and/or damage to persons or property as a matter of products liability, negligence or otherwise, or from any use or operation of any methods, products, instructions or ideas contained in the material herein. Because of rapid advances in the medical sciences, in particular, independent verification of diagnoses and drug dosages should be made.

First edition 2002
Reprinted from Surface Science, Vol. 500 (2002) Nos. 1–3

Library of Congress Cataloging in Publication Data
A catalog record from the Library of Congress has been applied for.

```
British Library Cataloguing in Publication Data

Frontiers in surface and interface science
   1.Surfaces (Physics)
   I.Duke, Charles B. II.Plummer, E. Ward
   530.4'17
```

ISBN: 0-444-51041-9

∞ The paper used in this publication meets the requirements of ANSI/NISO Z39.48-1992 (Permanence of Paper). Printed in The Netherlands.

Cover illustration: From article *The role of surface science in bioengineered materials*, by Matthew Tirrell, Efrosini Kokkoli and Markus Biesalski (pp. 61-83, fig. 10). The fluid mosaic model describes the essential features of the biological membrane. It is a mosaic because it is a structure that includes different kinds of macromolecules such as phospholipids, proteins, cholesterol, glycoproteins (proteins that are covalently bonded to carbohydrates) and glycolipids (lipids that are covalently bonded to monosaccharides or polysaccharides). A lipid bilayer forms the framework of the membrane with proteins randomly embedded in it and it is a two-dimensional fluid, or liquid crystal, in which the hydrophobic integral components of lipids and membrane proteins are constrained within the plane of the membrane, but are free to diffuse laterally.

Editors' Foreword

Frontiers in Surface and Interface Science

This volume is a collection of articles about the future of surface and interface science in the next millennium. Its intended audience is first year graduate students in a wide variety of scientific and engineering disciplines (e.g., chemistry, physics, molecular and mathematical biology, electrical engineering, mechanical engineering, materials science and engineering, and chemical engineering). Its intent is to express, in simple language, the exciting prospects at the frontiers of surface and interface science at the turn of the millennium.

The astonishing advances in surface science during its first thirty years has been documented elsewhere [1]. At this point in time, we are able to determine the composition, structure, and dynamics of relatively simple crystalline surfaces carefully prepared in a vacuum or electrochemical cell on scales ranging from atomic to macroscopic by a variety of experimental techniques. The emphasis is now changing to the applications of these techniques to more complex systems (including biological materials and buried interfaces), to a wide variety of technologically important interfaces, and to increasingly sophisticated studies of reduced dimensional phenomena that occur at surfaces and interfaces. Theoretical modeling is increasing in power and sophistication to the point that phenomena on multiple space and time scales can be simulated with accuracy, opening new vistas for the design of chemical and physical processes of particular interest in the chemical, biotechnology and electronics industries. Thus, we are reaching a point of confluence of modeling and experimentation that is opening revolutionary opportunities at the intersection of physical science, biological science and practical technology for use in engineering design.

Surface and interface science is expanding in scope as well. The articles on, e.g., nuclei, stars, cosmic dust, biological surfaces, DNA computation, enzyme catalysis, and soft surfaces attest to a discipline which is extending its reach far beyond its historical domains.

The articles in this volume were commissioned in part by the *Surface Science* advisory editorial board whose members suggested topics and authors that they believed to be at the cutting edge of modern surface and interface science. In part the editors used their own knowledge and judgment to commission articles. Finally, publication in this volume was opened to anyone who submitted before September 1999 an abstract that was manifestly compatible with the instructions to authors for the special volume. From these a few were selected. A booklet of abstracts was published and circulated in October 1999 defining the scope of the proposed volume. A few additional invitations subsequently were extended to fill gaps in the topical coverage and authorship.

The editors participated actively in the refereeing and editing of submitted manuscripts, especially with regard to making the articles accessible to a target audience that was not familiar with the technical jargon of surface science and its many subdisciplines. Many contributions have been rewritten as many as four times before they were deemed suitable for the intended audience. Thus, this is a volume of carefully crafted accounts of modern research in surface and interface science by currently active workers. It is not a volume of hastily written pieces by famous researchers of interest at best to specialists in their own subfields. The authors of the articles in this volume worked hard to make them clear and interesting to a general audience. The editors thank them each and every one for their exceptional efforts.

The net result of this effort is a volume in which promising frontiers in surface and interface science are expressed in simple language accessible to anyone with an undergraduate college degree in the physical or biological sciences, materials science, or electrical, materials, mechanical, and biomedical engineering.

PII: S0039-6028(02)01240-2

Some of these articles are really gems: Pieces from which you can glean the essential elements of the topic in pictorial diagrams and carefully crafted examples. One of the articles even affords the use of video clips ("movies") from the electronic version of this volume. Most of them afford a generous supply of technical references to give the interested reader an entrée into the technical literature. Not all of the articles meet all these standards, but a fair fraction do, reflecting enormous efforts on the part of the authors. You can find them easily by skimming the book. If you cannot understand an article at the end of the first two or three pages, abandon it and try another. Almost any first year grad student can find five or ten articles that will really interest you, and give you the flavor of the excitement of the field. This is a book to enjoy: In the airport, on the plane, at the beach, sipping a glass or cup of your favorite beverage, ...

One author is no longer with us. Prof. J.M. (Mayo) Greenberg died on November 29, 2001. His article "Cosmic dust and our origins" is perhaps the final contribution of his life. It is a fitting memorial: a reminder to us all of the long path that life has traveled before reaching the human species.

Publication of this volume marks Charlie Duke's retirement as the editor of *Surface Science* in January of 2002. Copies are being made available free of charge to the over 4,000 active referees of *Surface Science* as a token of gratitude for their contributions to the success of the journal during his ten years as editor. He thanks you one and all for your efforts on behalf of *Surface Science*, leaving this volume as an indication to the field's practitioners, sponsors, and students of the enormous accomplishments during the past forty years and the promise of the next millennium.

Charles B. Duke
Webster, NY, August 2001

E. Ward Plummer
Knoxville TN, August 2001

[1] C.B. Duke, ed., Surface Science: The First Thirty Years, Surf. Sci. 299/300 (1994).

Contents

Editors' Foreword ... vii

Contents ... ix

Introduction: Surface and Interface Science in the New Millennium

Surfaces: a playground for physics with broken symmetry in reduced dimensionality
    E.W. Plummer, Ismail, R. Matzdorf, A.V. Melechko, J.P. Pierce and J. Zhang ... 1

Biomedical surface science: Foundations to frontiers
    D.G. Castner and B.D. Ratner ... 28

The role of surface science in bioengineered materials
    M. Tirrell, E. Kokkoli and M. Biesalski ... 61

Fundamental Phenomena at Surfaces and Interfaces

Electronic transport at semiconductor surfaces—from point-contact transistor to micro-four-point probes
    S. Hasegawa and F. Grey ... 84

Dynamical phenomena including many body effects at metal surfaces
    W.A. Diño, H. Kasai and A. Okiji ... 105

Statistical thermodynamics of soft surfaces
    S.A. Safran ... 127

Solved and unsolved problems in surface structure determination
    D.P. Woodruff ... 147

Magnetism in low dimensionality
    S.D. Bader ... 172

New Materials via Surface Processing

Molecular beam epitaxy
    J.R. Arthur ... 189

Carbon nanotubes: opportunities and challenges
    H. Dai ... 218

Clusters as new materials
    W. Eberhardt ... 242

Clusters and islands on oxides: from catalysis via electronics and magnetism to optics
    H.-J. Freund ... 271

Tailoring magnetism in artificially structured materials: the new frontier
    J. Shen and J. Kirschner ... 300

Frontiers in the Modeling of Surface Processes

Modeling the full monty: baring the nature of surfaces across time and space
    F. Starrost and E.A. Carter ... 323

PII: S0039-6028(02)01241-4

The virtual chemistry lab for reactions at surfaces: Is it possible? Will it be useful?
  A. Groß   347

Catalysis and corrosion: the theoretical surface-science context
  C. Stampfl, M.V. Ganduglia-Pirovano, K. Reuter and M. Scheffler   368

Surface Dynamics, Growth and Etching

Atomic description of elementary surface processes: diffusion and dynamics
  F. Rosei and R. Rosei   395

Fidgety particles on surfaces: how do they jump, walk, group, and settle in virgin areas?
  A.G. Naumovets and Z. Zhang   414

Epitaxy: the motion picture
  P. Finnie and Y. Homma   437

Understanding crystal growth in vacuum and beyond
  E. Vlieg   458

Real time chemical dynamics at surfaces
  M. Bonn, A.W. Kleyn and G.J. Kroes   475

The growth and modification of materials via ion–surface processing
  L. Hanley and S.B. Sinnott   500

Sputtering: the material erosion tool
  M.V. Ramana Murty   523

Surface Science Tools and their Applications

Probing buried interfaces with non-linear optical spectroscopy
  C.T. Williams and D.A. Beattie   545

Frontiers in infrared spectroscopy at surfaces and interfaces
  C.J. Hirschmugl   577

Surface science done at third generation synchrotron radiation facilities
  S. Ferrer and Y. Petroff   605

Low temperature surface chemistry and nanostructures
  G.B. Sergeev and T.I. Shabatina   628

The Surface Science of Biomaterials and Processes

Biological surface science
  B. Kasemo   656

The surface science of enzymes
  T.H. Rod and J.K. Nørskov   678

Computation with DNA on surfaces
  S.D. Gillmor, P.P. Rugheimer and M.G. Lagally   699

Influence of Surface Science on other Disciplines

An atomistic view of electrochemistry
  D.M. Kolb   722

Surface science and the atomic-scale origins of friction: what once was old is new again
    J. Krim   741

The surfaces of compact systems: from nuclei to stars
    R.A. Broglia   759

Cosmic dust and our origins
    J.M. Greenberg   793

It's a dusty Universe: surface science in space
    D.A. Williams and E. Herbst   823

Far-out surface science: radiation-induced surface processes in the solar system
    T.E. Madey, R.E. Johnson and T.M. Orlando   838

Influence of Surface Science on Technology

The surface science of semiconductor processing: gate oxides in the ever-shrinking transistor
    M.K. Weldon, K.T. Queeney, J. Eng Jr., K. Raghavachari and Y.J. Chabal   859

Organic functionalization of group IV semiconductor surfaces: principles, examples, applications, and prospects
    S.F. Bent   879

Surfaces and interfaces in polymer-based electronics
    M. Fahlman and W.R. Salaneck   904

Role of surface science in catalysis
    J.H. Sinfelt   923

The surface chemistry of catalysis: new challenges ahead
    F. Zaera   947

Impact of surface science on the understanding of kinetics of heterogeneous catalytic reactions
    V.P. Zhdanov   966

Role of surface and interface science in chemical vapor deposition diamond technology
    L.K. Bigelow and M.P. D'Evelyn   986

The surface science of xerography
    C.B. Duke, J. Noolandi and T. Thieret   1005

Practical surfaces: beyond the wheel
    S.S. Badesha and J.A. Swift   1024

Author index   1042

Subject index   1044

Materials index   1051

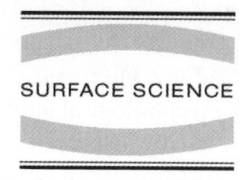

Surfaces: a playground for physics with broken symmetry in reduced dimensionality

E.W. Plummer [a,b,*], Ismail [a,b], R. Matzdorf [c], A.V. Melechko [a,b], J.P. Pierce [a,b], Jiandi Zhang [d]

[a] *Department of Physics and Astronomy, The University of Tennessee, 401 A.H. Nielsen Physics Building, Knoxville, TN 37996-1200, USA*
[b] *Oak Ridge National Laboratory (ORNL), Oak Ridge, TN 37831-6057, USA*
[c] *Universität Würzburg, Experimentelle Physik IV, Am Hubland, 97074 Würzburg, Germany*
[d] *Department of Physics, Florida International University, Miami, FL 33199, USA*

Received 4 January 2001; accepted for publication 11 July 2001

Abstract

With our crystal ball in front of us, we attempt to articulate the opportunities and challenges for a surface physicist in the beginning of the new millennium. The challenge is quite clear: to use the unique environment of a surface or interface to do fascinating physics, while taking full advantage of the skills the community has developed over the last 30 years. The opportunities appear to be endless! In this age of Nanotechnology where the promise is to shape the world atom by atom, leading to the next industrial revolution [Nanotechnology: shaping the world atom by atom, National Science and Technology Council, Committee on Technology, 1999], surface science should be at the very forefront of both technological and scientific advances. The smaller objects become, the more important their surfaces become. In this article we focus on the role of a surface physicist in the emergence of nanoscale collective phenomena in complex materials. © 2001 Elsevier Science B.V. All rights reserved.

Keywords: Surface structure, morphology, roughness, and topography; Surface defects; Magnetic phenomena (cyclotron resonance, phase transitions, etc.)

1. Introduction

As we enter the new millennium the international science agenda is focused on nanoscale science and technology [1,2]. We begin the "age of designer materials," when complex materials are designed to have desired properties, with both basic and technological applications. Some present day examples include artificially patterned structures, forever-smaller integrated circuits, magnetic storage devices, composite materials, polymer blends, doped transition-metal oxides, self-assembling nanostructures, molecular electronics, etc. This volume is intended to guide a surface or interface scientist into this new world, spanning all of the basic disciplines (chemistry, physics,

---

* Corresponding author. Address: Department of Physics and Astronomy, The University of Tennessee, 401 A.H. Nielsen Physics Building, Knoxville, TN 37996-1200, USA. Tel.: +1-865-974-3055; fax: +1-865-974-3895.
E-mail address: eplummer@utk.edu (E.W. Plummer).

biology, etc.), and extending to the exciting areas of technological applications. This article concerns a narrow aspect of surface science, specifically the prospects for a surface scientist in the world of condensed matter physics (CMP). There are other exciting, more global or cosmological aspects of surface science covered in this volume by Broglia (*The surface of compact systems: from nuclei to stars*) [3], Greenberg (*Cosmic dust and other origins*) [4], Herbst and Williams (*It's a dusty universe: surface science in space*) [5], and Madey et al. (*Far-out surface science: radiation-induced surface processes in the solar system*) [6].

2. Themes in modern condensed matter physics

In this section we briefly outline the general themes in CMP as we enter the new millennium. The idea is to understand where CMP is going so that a surface physicist can engage, participate, and contribute. There are three general thematic areas that have emerged in the last several years, which will guide the general direction of CMP in the next decade.

(i) nanoscale science and technology,
(ii) complex systems,
(iii) functional materials.

Worldwide, *Nano-scale Science, Nano Engineering, and Nano Technology* have become the buzzwords for discussing the future of materials sciences. It seems that 'nanocenters' are appearing on university and government campuses at a faster rate than daycare centers. The obvious reason for this interest is stated in the Department of Energy (DOE) report, Nanoscale Science, Engineering and Technology Research Direction [2]. "The reason that nanoscale materials and structures are so interesting is that size constraints often produce qualitatively new behavior. We now understand, in a general way, that when the sample size, grain size, or domain size becomes comparable with the specific physical length scale such as the mean free path, the domain size in ferromagnets or ferroelectrics, the coherence length of phonons, or the correlation length of a collective ground state as in superconductivity, then the corresponding physical phenomenon will be strongly affected. Although such changes in behavior can be the dominant effects in nanoscale structures, we still have remarkably little experience or intuition for the expected phenomena and their practical implication, except for electronic systems. The physics, chemistry and biology of phenomena occurring in nanoscale systems is effectively a new subject with its own set of physical principles, theoretical descriptions, and experimental techniques, which we are only in the process of discovering. Thus, there is an urgent need for broadly based investigations of the physical phenomena associated with confined systems, especially in materials and structural contexts where the implications are not at all well understood."

The DOE, Office of Science Workshop on Science for the 21st Century discussed the many facets of *complex systems*. This report [7] outlines the opportunities and challenges for the future. "As we look to the next century, we find science and technology at yet another threshold: the study of simplicity will give way to the study of *complexity* as the unifying theme."

The new millennium will take us into the world of complexity. Here, simple structures interact to create new phenomena and assemble themselves into devices. Here also, complicated structures can be designed atom by atom for desired characteristics. With new tools, new understanding, and a developing convergence of the disciplines of physics, chemistry, materials science, and biology, we can build on our 20th century successes and begin to ask and solve questions that were, until the 21st century, the stuff of science fiction.

Complexity can take many forms, so the identification of five emerging themes by the DOE workshop participants is useful as an organizational guide. The five themes are

- *Collective phenomena*—Can we achieve an understanding of collective phenomena to create materials with novel, useful properties?
- *Materials by design*—Can we design materials having predictable, and yet often unusual properties?

- *Functional systems*—Can we design and construct multicomponent molecular devices and machines?
- *Nature's mastery*—Can we harness, control, or mimic the exquisite complexity of nature to create new materials that repair themselves, respond to their environment, and perhaps even evolve?
- *New tools*—Can we develop the characterization instruments and the theory to help us probe and exploit this world of complexity?

In our opinion, the theme of Functional Systems is important enough to be broadened in content and listed as one of the three main thematic areas of CMP; Functional Materials.

For a physicist an interesting material has to do something. It has to be a *Functional Material*. It does not make any sense to fabricate a nanoscale material that is complex yet does nothing when its environment is changed. In physiological terms, we would like to design a material that is "schizophrenic" in that it develops a multitude of different personalities that appear in response to a prescribed stimulus. In scientific terms we want a material that has many and varied competing ground states so that an external stimulus (heat, pressure, electric field, magnetic field, etc.) can be used to switch its 'personality.' In the remaining portion of this section we illustrate that there are families of materials that allow you to "have your cake and eat it at the same time." Their properties are not only dictated by what occurs on a nanoscale, but they are also complex and wildly schizophrenic.

Before proceeding, it is important to understand that underlying all three of the thematic areas of modern CMP is our ability to synthesize, fabricate, or process advanced materials. The materials community must develop special tools to synthesize [8, p. 135] and characterize complex systems in order to make rapid progress. This is an area in which the surface science community can excel; growing and characterizing novel nanophase material structures. We are experts in this endeavor!

Although there are many aspects of complexity, we begin by talking about complex structures, and designer materials that exhibit nanoscale behavior, either because of *nature's mastery*, man's design, or a combination of both. Fig. 1 schematically illustrates four different forms of complex materials; (a) soft or biomaterials, (b) transition-metal oxides (TMOs), (c) artificially fabricated structures, and (d) designed quantum objects. Castner/Ratner (*Biomedical surface science: foundations to frontiers*) [9], Fahlman/Salaneck (*Surfaces and interfaces in polymer-based electronics applications*) [10], Kasemo (*Biological surface science*) [11], Norskov and Rod (*The surface science of Enzymes*) [12], Safran (*Statistical thermodynamics of soft surfaces*) [13], and Tirrell et al. (*The role of surface science in bioengineered materials*) [14], all describe the role of surface science in soft materials in their articles in this volume. Bader (*Magnetism in low dimensionality*) [15], Shen/Kirschner (*Tailoring magnetism in artificially structured materials: the new frontier*) [16], and Arthur (*Molecular beam epitaxy*) [17] describe phenomena in two dimensional artificially structured materials like those shown in Fig. 1(c). Dai (*Carbon nanotubes: growth, properties and applications in surface science*) [18], Freund (*Clusters and islands on oxides: from catalysis via electronics and magnetism to optics*) [19], Eberhardt (*Clusters as new materials*) [20], and Naumovets/Zhang (*Fidgety particles on surfaces* [21]. *How do they jump, walk, group, and settle in virgin areas?*) all describe artificially structured quantum objects on a surface as depicted in Fig. 1(d). In this article we focus on the new physics that can occur at the surfaces of chemically or electronically complex materials such as shown in Fig. 1(b).

The high-temperature superconductors offer examples of complexity, reduced dimensionality, and broken symmetry. Here, we use the term "reduced dimensionality" to denote systems in two, one, and zero dimensions. The dimensionality is dictated by the size in a certain direction of the object compared to an appropriate length scale for the phenomena being discussed. If one dimension of the sample is small compared to this length parameter, then the object is two dimensional (2D). Fig. 2 is an artist's view of the high-T_c material $HgBa_2Ca_2Cu_3O_{8+x}$ [7,22]. This material has one of the highest superconducting temperatures

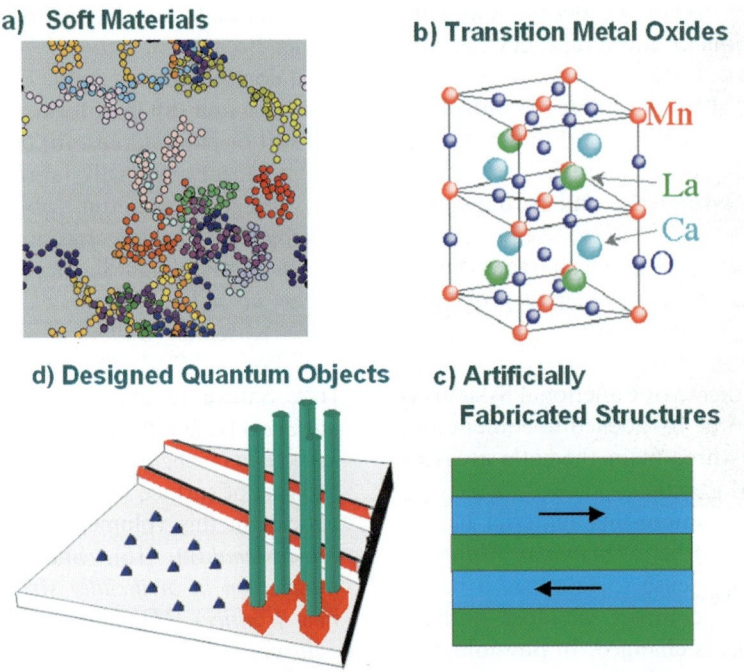

Fig. 1. Schematic illustrations of different important types of complex materials that will be encountered in this special issue of Surface Science. (a) Soft materials, like polymers and biomaterials, are made of interconnected chains of many kinds of molecules. Condensed matter physicists are only beginning to learn how to exploit the unique properties of these immensely complex systems. (b) In ordered materials that are made of many elements, like the TMOs, complex chemical composition leads to highly tunable electronic, magnetic, and structural properties. This figure shows the structure of the perovskite $La_{1-x}Ca_xMnO_3$, where x can range from 0 to 1. (c) 2D complex sandwich structures made of magnetic (blue) and nonmagnetic (green) layers have electronic properties that change drastically in the presence of a magnetic field. These materials have already proven useful in magnetic data reading/writing devices. (d) It has only recently become possible to fabricate quantum objects like nanowires (red) and nanocolumns (green) and quantum dots (blue). These objects are built on length scales which are similar to or smaller than the characteristic lengths over which many fundamental electronic and magnetic interactions occur, which should lead to exotic behavior.

recorded (133.5 K), when optimally doped with oxygen ($x \sim 0.14$) [22]. Doping generally refers to the addition of a small amount of an element that will give or take charge from the host lattice. Conventional wisdom says that the superconductivity occurs in the central CuO_2 plane (shaded with green in the figure), so this is really a 2D system. The parent compound ($x = 0$) is not a superconductor but when excess O is incorporated into the Hg–O lattice it pulls charge out of the CuO_2 plane (called hole doping, because O takes charge from the CuO_2 plane leaving a hole) and drives the material into the superconducting state. Doping with O breaks the translational symmetry of the crystal, creating nanoscale inhomogeneities (see the example in Section 3.3), that may be intimately related to the origin of superconductivity in these materials. A comparison of the coherence length in the superconducting phase dramatically illustrates the difference between these complex oxides and ordinary superconductors. The easiest way to define the coherence length is as the minimum spatial extent of a transition layer between the normal and superconducting states. For a normal superconductor like Al ($T_c = 1.14$ K) the coherence length is isotropic and very long ($\sim 10^4$ Å), while for this material it is anisotropic, and much shorter; ~ 10 Å perpendicular to the CuO_2 plane and ~ 50 Å in the plane. The perpendicular coherence length is less than the 15.86 Å lattice spacing. The inherent 2D-layered nature of the high-T_c materials, coupled with the short inter-

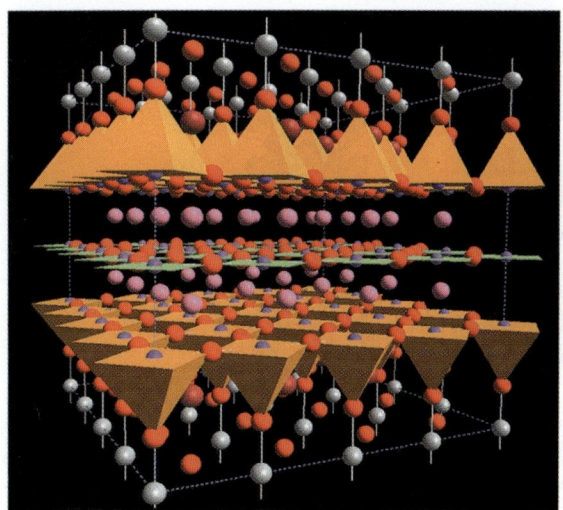

Fig. 2. Artistic view of the crystal structure of the high-temperature superconductor $HgBa_2Ca_2Cu_3O_{8+x}$ [7], showing highly 2D (layered) character. The color code is: silver for Hg, purple for Cu, red for O, pink for Ca and dark red for Ba. The orange pyramids present the appropriate portion of CuO_6 octahedra. The lattice spacing in the vertical direction is 15.86 and 3.86 Å in the plane.

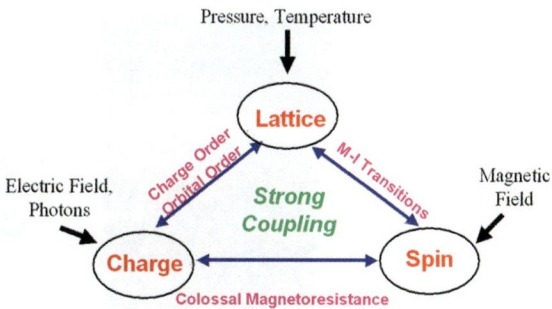

Fig. 3. Artistic representation of a highly "functional material." The electronic, spin and lattice degrees of freedom are tightly coupled, leading to dramatic responses to external probes such as temperature, pressure, and electric or magnetic fields. For example, the application of a magnetic field can shift the M–I and magnetic transition to a different temperature, thus creating a drastic change of the resistivity (i.e., CMR behavior). The application of a light beam (electric field) can melt the charge/orbital ordered insulating state (described later) creating a ferromagnetic metal.

planar coherence length, makes them amenable to investigation using conventional surface probes such as angle-resolved photoemission and scanning tunneling microscopy (STM). There is always the prospect that a clever surface scientist can engineer the growth of a unique superconducting thin film, and perhaps even fabricate one that has a transition temperature above room temperature.

High-temperature superconductors are only one class of a set of materials usually referred to as 'highly correlated electron systems.' A special issue of Science (*correlated electron systems*), April 21, 2000, was devoted to this subject [23]. The commonality these materials have is the dominant role played by electron–electron interaction that goes far beyond our conventional wisdom. No current textbook theory can explain the exotic behavior of these materials [24]. Examples of these highly correlated electron systems are TMOs, heavy fermion metals, organic charge transfer compounds, and 1D and 2D electron gas systems [25]. The TMOs are characterized by a strong coupling between the structural, electronic, and magnetic properties leading to richness in their physical properties, which can be tuned by doping. In this case the word doping is really not correct because the properties are actually tuned by chemical substitution. For example, divalent Ca is substituted for trivalent La in Fig. 1(b). "With this strong correlation, even a small change in an external parameter can have a drastic effect on the system's ground state, changing the electronic and/or magnetic phase of the material from one ground state to another" [24]. Fig. 3 illustrates schematically the effect of this close coupling of spin, charge and the lattice degrees of freedom. In the science compass section of the special issue of Science, Birgeneau and Kaster comment, "Clearly, highly correlated electron systems present us with profound new problems that almost certainly will represent deep and formidable challenges well into this new century [25]."

To illustrate the range of dimensionality in the TMOs, Fig. 4 displays the structure of the Ruddlesden–Popper (RP) series described by the general formula $(A_{1-x}B_x)_{n+1}M_nO_{3n+1}$. In these compounds A is usually a trivalent rare-earth element, B is a divalent alkaline-earth element, and M is a transition-metal ion that can have several different valences. The structure of the RP phases is composed of n consecutive perovskite layers

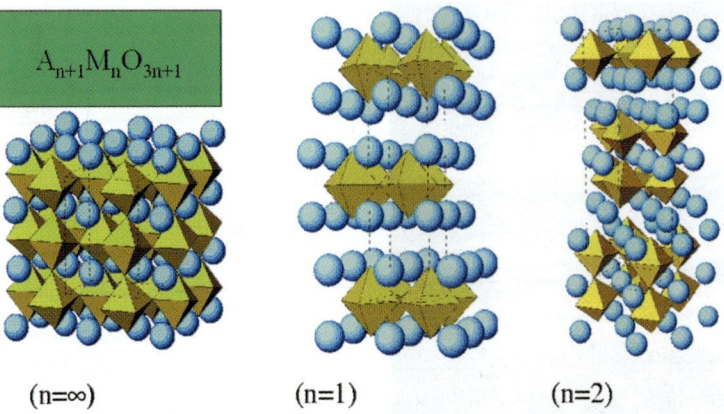

Fig. 4. Artistic view of three members of the RP series ($A_{n+1}M_nO_{3n+1}$) for $n = 1$, 2, and ∞. The A ions are the large-blue spheres with the MO_6 octahedra displayed as the light-green solid object. The structural properties of the $n = 1$ and $n = 2$ phases are clearly low dimensional and are expected to lead to highly anisotropic physical properties, while the $n = \infty$ structure is 3D.

(AMO_3) alternating with rock salt layers so that their formula can be represented by (AO)-$(AMO_3)_n$, where n represents the number of connected layers of vertex sharing MO_6 octahedra (green octahedra in Fig. 4). $n = 1$ is a 2D-layered material with only one layer of octahedra. $n = 2$ has two connected layers of octahedra. As n increases the materials become more 3D in character with $n = \infty$ being the perovskite structure. As an example of the close coupling in these materials consider the perovsike $La_{1-x}Ca_xMnO_3$ shown in Fig. 1(b). The ground state of the parent compound $LaMnO_3$ is an antiferromagnetic (adjacent spins pointing in opposite directions) insulator, but as the concentration of Ca is increased the ground state dramatically changes. Looking at the formal valence of this compound leads to the following representation, $La^{3+}_{1-x}Ca^{2+}_xMn^{3+}_{1-x}Mn^{4+}_xO^{2-}_3$. Increasing the Ca concentration changes the average valence of the Mn, leading to an array of different ground states which are described as ferromagnetic-ordered (FM) metal, charge-ordered insulator (explained later), canted-antiferromagnetic insulator, etc. [26].

When an external stimulus is applied, these materials exhibit quite remarkable changes (Fig. 3) because the energies separating the ground state from quite different excited state configurations are small. Fig. 5 illustrates several commonly observed cooperative phenomena. In Fig. 5(a) a coupled magnetic–electronic transition occurs with changing temperature in the perovskites ($A_{1-x}B_x$-MO_3), for a doping of $x \sim 0.3$. The ground state is a FM metal that changes to a paramagnetic (PM) insulator as the temperature is raised. The signature of a metal is a resistivity that increases as the temperature rises, while the resistivity of a semiconductor or insulator decreases with increasing temperature. The transition temperature for this coupled transition depends on the size and electronic properties of the divalent and trivalent components, ranging from ~ 60 K for $Pr_{0.7}Ca_{0.3}$-MO_3 to ~ 370 K for $La_{0.7}Ca_{0.2}MO_3$. The transition temperature for a given material depends on the strength of the applied magnetic field, moving to lower temperature as the field increases. This produces what is known as colossal magnetoresistance (CMR), as shown in Fig. 5(b). Magnetoresistance is the change in the resistance $\Delta R/R$ of a material in the presence of an applied magnetic field. This is normally on the order of a few percent. The phrase giant magnetoresistance (GMR) is used to describe the magnetoresistance in artificially layered metal films (see Fig. 1(c)) and can be as large as 100% (see Section 3.1). CMR in the TMOs can be as large as 10^5. When light, either from X-rays or a laser, is incident on some of the TMOs, a startling transition occurs from an antiferromagnetic insulator to a ferromagnetic metal, as depicted in Fig. 5(c) [27–29].

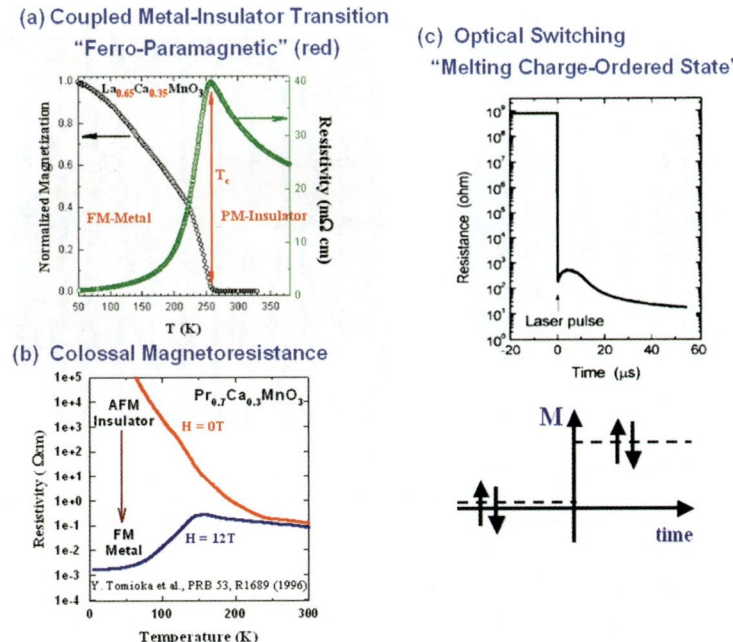

Fig. 5. Illustration of a few of the macroscopic collective phenomena seen in TMOs. (a) A coupled FM to PM phase transition (based on the measurement of magnetization vs. temperature) and a M–I transition (based on the change of resistivity vs. temperature) in $La_{0.65}Ca_{0.35}MnO_3$. (b) CMR, i.e., a drastic decrease in resistivity by applying an external magnetic field is observed in $Pr_{0.7}Ca_{0.3}MnO_3$. The change in resistivity with applied field is a function of the sample temperature, going to zero at room temperature. (c) Optical switching in $Pr_{0.7}Ca_{0.3}MnO_3$, where the physical properties change dramatically with the exposure to light. In this case the material can switch its charge-ordered insulating state with an antiferromagnetic spin structure (nearest neighbor spins are antiparallel with lower magnetization) to its charge-disordered metallic state with a ferromagnetic spin structure (nearest neighbor spins are parallel with higher magnetization). The change in resistivity when the laser pulse hits the sample is shown in the top and the spin alignment below, both as a function of time with $t = 0$ at the time when the laser was pulsed [26–29].

Creating a surface by cleaving a TMO single crystal is a controlled way to disturb the coupled system (spin, charge, and lattice) by breaking the symmetry without changing the stoichiometry. The unique environment at the surface could produce new phenomena, while providing a fresh approach to the study of the spin, charge, and lattice coupling in these complex materials (see Section 3.3). Furthermore, the influence of surfaces and interfaces on thin-film properties is of technological interest for the design of TMO devices.

There is a paradigm shift occurring in CMP, from detailed studies of model systems to embracing the importance of complexity. Associated with complexity is the emergence of new phenomena and the realization that controlled synthesis is the enabler for their discovery and understanding. Discovery of new phenomena requires exploring the frontiers of complexity in all of its forms. One such new phenomenon is electronic nanoscale phase separation in these highly correlated systems [30–38]. In a conventional solid, the ion cores create the symmetry of the solid and the electronic charge density has the same symmetry. In some highly correlated systems there can be static and dynamic variations in the charge density that does not follow the symmetry of the ion cores. The best-known observation of this phenomenon is the striped phases in the high-T_c superconductors reported by Tranquada et al. [30] (see Fig. 6(a)), which has lead to a virtual zoo of predictions on new nanophase separations in TMOs [32–38]. There seems to be a general trend in these systems toward an intrinsic nanophase separation close to the phase boundaries in

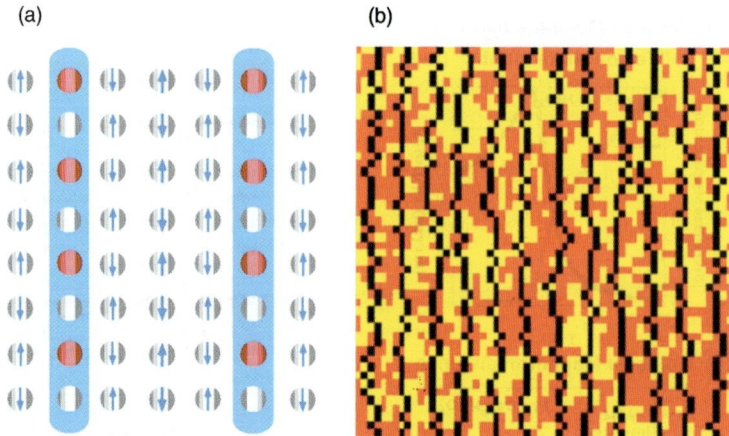

Fig. 6. (a) Theorists view [32] of the static stripe phase of the high-T_c superconductor $La_{01.475}Nd_{0.4}Sr_{0.125}CuO_4$ [30]. Here, the charge is largely confined to the channels shaded in blue. The average charge density along the stripe of +e per two sites is indicated by alternating red and silver circles. Blue arrows indicate magnitude of the magnetic moment on sites containing spins. The strip is an antiphase boundary for the antiferromagnetic order; in the absence of the stripe, the first and third column from the left would have the same spin orientation, not an opposite one. The oxygen ions are not shown. (b) Theoretical image of fluctuating quantum stripes (lines of black dots) moving through a quantum antiferromagnet (red/yellow background) from an imaginary time slice of a quantum Monte Carlo simulation. In the classical limit the background would be, say, uniformly red, corresponding to a simple antiferromagnet. Here the yellow patches represent spin fluctuations [38]. In both (a) and (b) the length scale is in nanometers.

temperature, doping concentration, or magnetic field. Although there is no direct proof that striped phases dictates the properties of high-T_c materials [27], there is growing evidence that dynamic mixed nanophases (see Fig. 6(b)) are an integral part of the CMR behavior [34]. The length scales of the phase separated regions are in the nanometer range, involving fluctuations of charge, spin, and orbital orientation [23,31,32]. Current examples of direct observations of electronic inhomogeneities on the surfaces of these highly correlated electron systems are included in Section 3.3.

Fig. 7 is a picture representing orbital, charge, and spin ordering that occurs in perovskites [23,26,31]. This is a specific example for the quarter-doped sample ($x = 0.25$) $A_{0.75}B_{0.25}MnO_3$, in which there is a 3/1 ratio of Mn^{3+} to Mn^{4+} caused by doping. In principle, the divalent ions (B) are randomly distributed on the blue ball lattice in Fig. 4, but local or long-range charge ordering of the Mn ions can occur at low-temperatures. This is shown in Fig. 7 by having each of the Mn^{4+} ions shown as light and dark red balls surrounded by six Mn^{3+} ions (with light and dark green orbital lobes). Orbital ordering occurs when the occupied orbitals in the Mn^{3+} ions align themselves with respect to the Mn^{4+} as shown by the green orbital. Spin ordering occurs with ferromagnetic coupling between the spin $-3/2$ Mn^{4+} and the spin-2 Mn^{3+} nearest neighbors, as shown in the right-hand panel of Fig. 7. The spin, orbital, or charge ordering can be static or dynamic, short range (light green) or long range depending upon the details of the coupling in the system.

Given the high sensitivity of these materials to their local environment, it is likely that in some cases their surfaces could behave totally different from the bulk. It takes very little imagination on the part of a surface scientist to envision charge-, orbital- or spin-ordered surface states on these materials, or the inverse in which the bulk is a charge-ordered antiferromagnetic insulator and the surface is a ferromagnetically ordered metal. An example, which illustrates the possibilities, is described in Section 3.3.

We have learned that one key characteristic of complex materials is that it is possible to create spectacular, new phenomena simply by adding new components to a given material. This was illustrated by the discovery of high-temperature

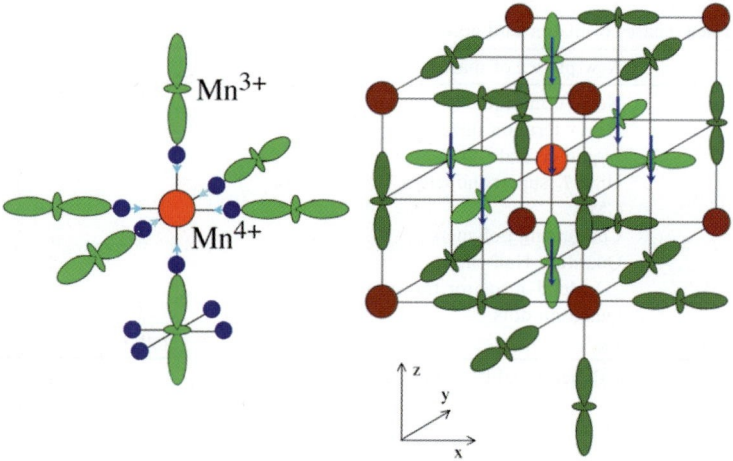

Fig. 7. Schematic representation of the charge, spin and orbital ordering in a perovskite $A_{0.75}B_{0.25}MnO_3$ where A represents trivalent rare-earth ions and B sites are occupied divalent alkaline-earth doped ions [31]. Doping divalent ions creates a Mn mixed-valent configuration, with 25% Mn^{4+} (light/dark red balls) and 75% Mn^{3+} (light/dark green lobes representing the electron orbitals), exhibiting spin, charge, and orbital ordering below a critical temperature. The basic buiding block for the charge/orbital ordering is shown in the cluster in the left panel, a Mn^{4+} ion with spin (3/2) is surrounded by six short bond length oxygen (O) atoms (purple circles) and six Mn^{3+} ions (green orbitals) with spin 2 ferromagnetically coupled to the central Mn^{4+}. The fourth 3d electrons in the Mn^{3+} ions occupy the lobes as drawn, all pointing toward the central Mn^{4+} ion and to local long Mn^{3+}–O bond lengths. The right panel shows the charge/orbital ordered cluster embedded in the host material. It is the charge and orbital ordering which results in the peculiar magnetic structure, in this case with ferromagnetic clusters which can be antiferromagnetically ordered with respect to each other in the solid.

superconductivity, CMR, and conducting polymers. In the future, new phenomena will be discovered by artificially growing complex structures out-of-equilibrium as depicted in Fig. 1(c) and (d). The objective will be to control the *coupling* of cooperative phenomena, such as, the coupling between magnetic ordering and a structural transition, or between magnetism and electrical conductivity or superconductivity. What we have learned is that complex or composite systems do not behave simply as linear combinations of the properties of the parent materials. In a much earlier time, Sir Arthur Eddington aptly and eloquently described this situation in saying

We used to think if we knew one, we knew two, because one and one are two. We are finding that we must learn a great deal more about 'and'.

For complex materials, we must understand the way in which "and" governs all of the factors that combine to produce dramatic new phenomena. One thing that we have learned is that complexity is a Fountain of Youth. Pushing the *frontiers* of chemically and structurally complex materials will always cause new phenomena to emerge. Consequently, the subjects of complexity and cooperative phenomena are *fundamental*, long-term interests of science and society. Some time ago Phil Anderson presciently observed that,

...at each new level of complexity, entirely new properties appear, and the understanding of these behaviors requires research which I think is as fundamental in its nature as any other [39]

A survey of new phases of matter and recent Nobel Prizes clearly illustrates the trend. Table 1 was taken from the 1999 APS centennial presentation by Piers Coleman, titled "An Unfinished Revolution." The table has been updated with recent discoveries and awards. The 2000 Nobel

Table 1
List of recently discovered new phases of matter and accompanying Nobel Prizes

Nobel Prize Year	New phases of matter	Dimensionality
	Non-Fermi liquid	3D
	Heavy fermion metals	
1991	Liquid crystals	
1987	High-T_c superconductors	
	CMR	
2000	Conducting polymers	
1985, 1998	Quantum Hall effect	
	2D Metal–insulator transition	
	Quantum dots	
1996	C_{60}	0D

On the right is a rough measure of the dimensionality of the system.

Prize in physics to Alferov and Kroemer for developing semiconductor heterostructures used in high-speed and opto-electronics should also be included. The message is quite clear. If you want to discover a new phase of matter or win a Nobel Prize you need to embrace complex materials which exhibit collective phenomena. It should also be apparent that for devices to function in the future world of Nanotechnology, coupling of cooperative phenomena is required.

The opportunity, for those who seize it, and the risk, for those who ignore it, is that

Whoever controls the materials controls the science and the technology,

i.e., progress is materials driven. Part of this claim, that whomever controls the materials controls the science, can be substantiated by looking at the list of the 1000 most cited physicists in the period 1981 to June 1997 [40]. Of the top 10 most cited, seven were condensed matter physicists. Six of these seven make materials, ranging from superconductors to conducting polymers, QED.

We are entering a "New World of Designer Materials," designed for scientific and technological impact. There are tremendous opportunities for surface physicists, if we become engaged! These highly correlated electron systems, the questions associated with them, and the importance of nanoscale phenomena seem to be designed for a surface physicist. We have the tools and the experience to make a real impact.

3. Examples: surface physics in the world of contemporary condensed matter physics

In this section, we attempt to illustrate ways for a surface scientist to *engage* in contemporary CMP. Generally, there are three obvious paths: (1) take our surface systems and use them to illustrate the physics relevant to questions being asked by the materials physics community (map our problems onto theirs); (2) use our techniques, both experimental and theoretical, to study new materials relevant to the materials physics community; and (3) fabricate new materials with exotic behavior that attract this community.

Our challenge to you, the reader, is to construct your own version of this section. Ask yourself, "Where are the forefronts of my discipline?" Once you have this list then ask, "What can I do as a surface scientist to have an impact?" If you are a beginning graduate student looking for a thesis advisor, ask each potential advisor how he or she plans to have an impact on science or technology. If he or she does not have an answer, look for another advisor.

3.1. Magnetism in reduced dimensionality

Magnetism in reduced dimensionality illustrates every point that we emphasized in the previous sections. It is by its very nature a cooperative phenomenon involving broken symmetry. The 1990s have been referred to as the decade of magnetism in two dimensions. The important advances in this field in the last decade are due to science-driven fabrication of complex multilayered structures. Our community can be proud of its contribution to the advancement of this field. According to Bader (in Ref. [15] this volume), "The development and adaptation of surface-science techniques are responsible for enabling surface magnetism to flourish."

There are many landmark papers in this field but the discovery of GMR in 1988 [41] spurred major advances in research and development. Within a decade of the discovery of GMR, the phenomenon was adapted to a new generation of high-performance read heads in magnetic hard disks. The simplest device based on GMR is the spin valve developed by IBM and shown in Fig. 8(a). In its simplest form it can be thought of as a sandwich—a thin film of nonmagnetic material (Cu in this case) between two thin films of magnetic material (Co). The key concept is associated with the flow of electron with a spin orientation through this Co/Cu/Co sandwich as depicted in Fig. 8(b) and (c). An electron moves through a magnetic layer with relative ease when its spin is parallel to the magnetization of the layer and with relative difficulty when its spin is opposite to the magnetization direction. A magnetic material that is, for instance, magnetized in the "up" direction can be thought of as having an excess of spin-up electrons, or in the language of the condensed matter physicist, a larger density of states associated with spin up than with the spin "down" electrons. The bottom line is that spin-up electrons move through the up aligned magnetic film easier than do the spin down electrons, so the resistance is different in the two cases. When no external magnetic field is applied to the device, as in Fig. 8(b), the magnetic layers are antialigned and both spin-up and spin-down electrons are scattered equally as they try to move across the two

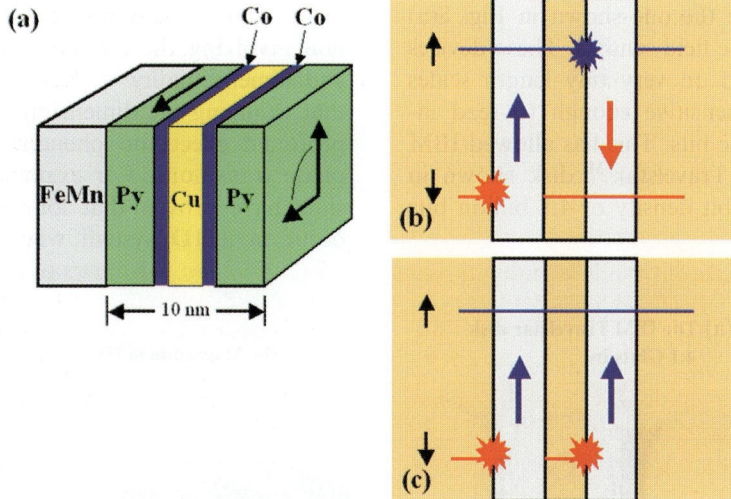

Fig. 8. GMR devices work by exploiting not only the charge of the electrons that pass through them, but also their spin. The simplest device is shown in (a), where the magnetic moment of the green permalloy (Py) layer that is furthest to the left is pinned in one direction by the neighboring FeMn layer. The magnetic moment of the Py layer that is furthest to the right responds to the magnetic field created by the magnetized domain of the "bit" of information that is written on the hard drive spinning underneath it. Passing a test current through the device immediately identifies whether the moments of the two permalloy layers of the read head are aligned (low resistance) or antialigned (high resistance), thus reading the information written on the disc. The heart of the device consists of a nonmagnetic metal film sandwiched between two magnetic layers. In this case Cu between Co layers. As shown in (b), when the magnetic moments of the magnetic layers (large red and blue arrows) are antialigned, spin up and spin down electrons (small black arrows) have an equally difficult journey through the device. When the magnetic moments of the layers are aligned as shown in (c), spin up electrons can easily pass through the device and the resistance decreases. This is the GMR effect.

magnetic layers. When a field is applied, as in Fig. 8(c), it is like opening a valve—spin-up electrons are allowed through, resulting in a large decrease in the resistance, i.e. magnetoresistance. The important length scale in this case is the spin diffusion length, which is the average distance that an electron can travel in a metal before it experiences a spin flip collision. All of the film thickness in the GMR devices must be less than the spin diffusion length for them to act as 2D systems. At room temperature, the spin diffusion lengths are roughly 40 and 20 nm in cobalt and copper, respectively.

In the Co/Cu/Co sandwich shown in Fig. 8(a) the differences in the spin-aligned (c) and spin-antialigned electrical conductivity is only ∼20% because of the finite difference in the spin-up and spin-down density of states in Co. The dream is to fabricate a material where the spin-down density of states is zero and the spin-up density finite. This type of material is called "half-metallic," and would exhibit an infinite magnetoresistance, making an ideal switch [42].

GMR devices like the one shown in Fig. 8(a) are excellent magnetic field sensors. These devices can be manufactured on very tiny length scales (∼10 nm) and are sensitive enough to read incredibly tiny magnetic bits. This has allowed IBM to manufacture the Travelstar™ disk shown in Fig. 9(a) that has a bit density of 4.1 billion bits per square inch.

The future for a surface scientist interested in magnetism is undoubtedly in designed magnetic systems in one or zero dimensions. Fig. 9(b) shows a marble model of a 1D magnetic wire array. Surface scientists can lead the way in research on magnetism in low-dimensional systems. Stepped surfaces can form ideal templates for 1D magnetic nanowires, as illustrated in Fig. 1(d). Novel growth techniques can be used to prepare zero-dimensional magnetic quantum dots (Fig. 1(d)). It should be possible to investigate in a systematic way the dimensional dependence of magnetism [15,16]. The objective is to control or to engineer the coupling in these reduced dimensional systems. For example, you could envision using more complex materials like the TMOs so that an electric field would couple to the magnetic behavior, as displayed in Fig. 5(c). There are two excellent articles in this special issue on the future of this area, by Bader [15], and by Shen/Kirschner [16].

3.2. Phase transitions in reduced dimensionality

A phase transition is a cooperative phenomenon involving the concepts of broken symmetry and dimensionality. It has long been recognized that changing the dimensionality of a system has profound effects on phenomena associated with phase transitions. For example Landau and Lifshitz have proven that long-range order cannot occur in a 1D system whose elements interact

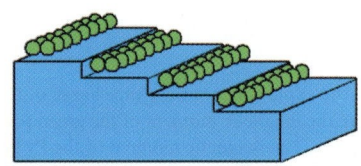

Fig. 9. (a) A photo of the Travelstar™ disc, manufactured by IBM. The disc spins as if it were on a record player and the information on it is read by a tiny GMR read head placed at the end of the "needle" of the record player. GMR read heads, like the one shown in the previous figure, can be manufactured on length scales on the order of 10 nm. The small size and high sensitivity of these devices have led to increased performance of magnetic information storage media. The disc shown above has an information storage density of 4.1 billion bits per square inch. (b) A marble model of a 1D magnetic nanowire array. By miscutting substrate surfaces to form atomic staircases, it is possible to grow parallel arrays of these nanowires using the stepped surface as a template [16].

through finite-range forces [43]. It is also anticipated that defects will play a larger role in phase transitions in low-dimensional systems [44], though microscopic evidence has been limited. This is an area where we as surface scientists should be able to have an impact on the materials community using our systems and our tools.

Historically there have been many investigations of 2D surface phase transitions. One area where immense progress has occurred is in the area of weakly bound inert gas atoms on surfaces [45]. One of the most famous, or at least the most studied, clean surface phase transition is one that occurs on Si(0 0 1) [46–52]. Since 1978 there have been over 4000 papers published about this surface, in large part due to its technological importance. At room temperature the surface reconstructs to eliminate the unbonded electrons created by breaking the Si–Si bonds to form a surface. The reconstructed surface consists of rows of dimers as shown in Fig 10(b), but when the surface is cooled to ∼100 K a new structure appears which is composed of asymmetric dimers (called buckled dimers) as shown in Fig. 10(a) [50]. An electron diffraction study found that the phase transition took place over a fairly broad temperature range and was consequently labeled a second order, order–disorder transition. The buckled dimer configuration shown in Fig. 10(a) melts into the symmetric arrangement as the temperature increases. A second order phase transition is one in which the order parameter that characterizes the existence of one phase changes continuously with temperature. This seemed to be a solved problem, with the ground state being the buckled dimer, which underwent an order–disorder transition with increasing temperature. Ab initio theory confirmed that the buckled dimer was indeed the ground state.

Truth is eternal—until a new truth replaces it
Morton F. Spears, "Capacitance Theory of Gravity"—1991

There was a warning issued in 1994 that a small number of defects could dramatically change the nature of this phase transition. On the eve of the new millennium this consistent textbook story on the surface structure of clean Si(0 0 1) was turned on its head. Low-temperature STM (down to 5 K) revealed that the true ground state of this surface is a symmetric dimer, leading to speculation that antiferromagnetic ordering was involved in the ground state [51,52]. This provides an important lesson to all surface physicists—lower the temperature!

The STM is a magnificent tool to study what the atoms are doing in real space as a phase transition occurs, especially in one or two dimensions.

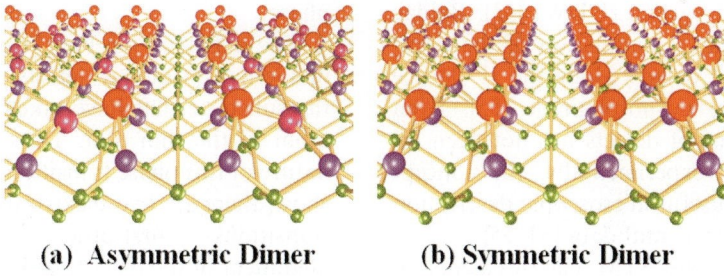

(a) Asymmetric Dimer   (b) Symmetric Dimer

Fig. 10. Stick and ball model of the two reconstructed surface structures seen on the reconstructed surface of Si(0 0 1). When a Si(0 0 1) surface is created by cutting the solid into two pieces each exposed Si atom has lost two out of four of its nearest neighbors, leaving these surface atoms with two broken (dangling) bonds. The surface reconstructs to eliminate as many dangling bonds as possible. On this surface the reconstruction results in rows of dimers as indicated by the red atoms in (b). If the red atoms were removed in this figure the purple Si atoms would be in the positions dictated by the bulk structure. At room temperature the dimer rows are symmetric (b) but when the temperature is lowered to ∼100 K a transition occurs to an asymmetric or buckled configuration shown in (a). In each dimer the bright red atom is up and the dark red atom down with adjacent dimer pairs in a row or across rows being out of phase. Recent STM measurements at ∼5 K have shown that the true ground state is the symmetric dimer configuration shown in (b) [51,52] (courtesy of T. Uda and K. Terakura).

Here we show that an STM study of the role of defects in a 2D phase transition can change the way scientists think about this subject, i.e. a paradigm shift. It is generally recognized that defects can play a critical role in phase transitions, especially in reduced dimensionality [43]. Mobile defects and the onset of their spatial long-range ordering are often invoked as a microscopic explanation of some experimental observations in low-temperature phase transitions [53–55]. For example, Mutka, in his article on the influence of defects and impurities on charge density waves (CDWs) comments [53]:

Defects in CDW compounds are inevitable, and their influence on the CDW phenomena is more a rule than an exception—The strong connection to defects influences the whole physics of CDWs with consequences that are manifest in a wide space and time scale from microscopic to macroscopic... This is the reason why microstructural characterization of the CDW is of primary importance...

A CDW is an electronic perturbation of a solid that lowers the total energy by the formation of an electron CDW with a different periodicity than the solid [56,57], i.e. broken symmetry. A CDW is usually accompanied by a small lattice distortion. Mutka speculates on the possibility of metastable defect configurations and to the concept of defect-density waves (DDW) [53]. Baldae has hypothesized that a modulation of the occupation probability of defects along the 1D lattice occurs in potassium cyano-platinide [55]. Even though the alignment of defects has not been directly demonstrated experimentally, the concept has proven to be a useful microscopic description for a variety of experimental data [53–55].

Before discussing the STM measurements of defect mediated 2D phase transitions the system being investigated needs to be introduced. Fig. 11(a) shows a marble model of what is referred to as the α phase of Sn on Ge(1 1 1) [58]. The Sn atoms (light blue) are positioned on top of three Ge atoms (purple with red bonds) in the first layer of the (1 1 1) surface. The red lines show the basic building block (called unit cell) for the room temperature structure labeled $\sqrt{3} \times \sqrt{3}$ and the low-temperature CDW structure labeled (3×3). The important observation is that in the room temperature phase there is only one Sn atom in a unit cell, which means that every Sn atom is identical. But in the CDW phase there are three Sn atoms in each unit building block. In Fig. 11(b) are the original STM images [58] showing the changes that are observed when the temperature is lowered. The RT STM images shown at the top (small inserts) exhibit the same symmetry in the filled and empty state images, indicating that the STM image is reproducing the geometric structure shown in Fig. 11(a). When the temperature is lowered the images are complimentary. The empty state image has two out of the three Sn atoms in the (3×3) unit cell bright forming a honeycomb structure. In the filled state image, the other atom in the unit cell is bright forming a hexagonal lattice. This complimentarity is the signature of a charge density wave. A simple interpretation is that two atoms in the unit cell are positively charged ($+q/2$ seen in empty state images) and one negatively charged ($-q$ seen in the filled state image). There is a lattice distortion accompanying the CDW, one Sn atom in the unit cell moves up and the other two move down (∼0.3 Å vertical buckling). One can take three waves (120° apart) with the wavelength associated with the (3×3) structure in Fig. 11(a), sum their amplitude and square (charge density) to reproduce the images of the CDW phase.

The red arrows in Fig. 11 indicate the positions of Ge substitutional defects, i.e. Ge atoms from the substrate that have taken the place on an Sn atom in the thin film. These defects control the nature of this phase transition. As we subsequently show, there are two inextricable intertwined phase transitions—a first-order transition involving the alignment of the defects and a second-order CDW condensation. A clarification of the definition of first- and second-order phase transitions is presented later.

The ground state of this system is the (3×3) CDW structure, but at temperatures above the transition temperature Ge substitutional defects stabilize damped CDW-like waves in their vicinity. Experimentally it was realized that these damped

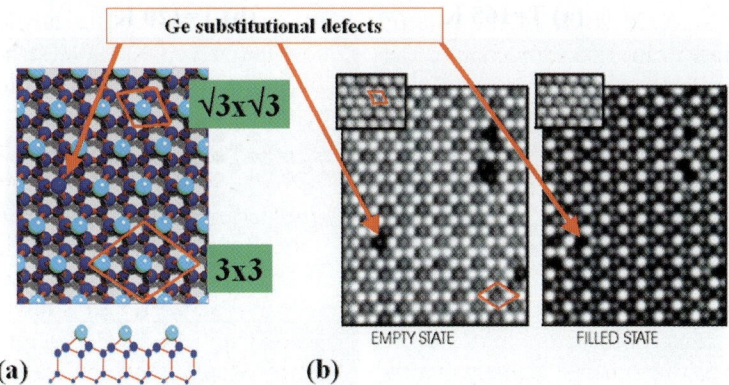

Fig. 11. Structural model and STM images of the symmetry lowering phase transition, from $(\sqrt{3} \times \sqrt{3})R30°$ to (3×3) structure observed in a thin metallic film of Sn on Ge(1 1 1). (a) A marble model of the room temperature structure of this system, where the Sn atoms are shown as light blue balls and Ge atoms are indicated as dark blue. The drawing at the bottom shows a side view of the surface. At room temperature, the Sn atoms form a structure with respect to the bulk truncated Ge structure denoted by $(\sqrt{3} \times \sqrt{3})R30°$. This notation means that the unit cell or basic building block of this structure (shown in red at the top) contains one Sn atom and three surface Ge atoms. When the temperature is lowered there is a transition to a new structure with a 3×3 unit cell shown at the bottom right of (a). The new unit cell triples in size compared to the $\sqrt{3} \times \sqrt{3}$ unit cell and now contains three inequivalent Sn atoms. (b) The STM observations above and below the phase transition temperature are shown in these two sets of images, labeled 'empty' and 'filled' state, which corresponds to the voltage bias between the sample and the STM tip. In 'filled' state images the tip is positive with respect to the sample and electrons tunnel from occupied states (filled) in the sample to the tip. In 'empty' state images the voltage on the tip is negative and electrons tunnel from the tip into unoccupied (empty) states in the sample. The two small images in the upper left hand corners are the room temperature images, and the big STM images are at low-temperature. The Sn atoms appear equivalent in both filled and empty state STM images taken at room temperature. At low temperature there are two bright and one dark atoms per unit cell (seen unit cell in empty state image) observed in the empty state image, thus indicating lowering of the symmetry. The filled state image has one bright and two dark Sn atoms in each unit cell. The complimentarity of the filled and empty state images at low-temperature is a signature of the CDW formation. Arrows in (a) and (b) mark the position of Ge substitutional defects that affects the CDW formation [58].

CDWs were temperature dependent and could be fitted with a simple exponential function $\exp[-|\mathbf{r} - \mathbf{r}_n|/l(T)]$, where $l(T)$ is the decay length of the CDW away from the defect located at $\mathbf{r}_n$ [59,60]. Fig. 12(a) shows a simulation of the damped CDW, stabilized by a single Ge substitutional defect. The defect creates a honeycomb local disturbance (in a filled state STM image). As the temperature decreases, the range of the stabilized CDW extends, shown in the simulation in Fig. 12(b) which illustrates the interference in the stabilized CDWs from two defects. The interference produces the hexagonal lattice seen in the STM filled state images (Fig. 11(b)). This is the first direct indication that defects are important in the creation of the CDW [59]. A single defect would create a honeycomb lattice seen in the filled state STM images, while multiple defects create a hexagonal lattice.

Many STM images were fitted with the simulations depicted in Fig. 12. The positions of the defects in the real image were transferred to the simulation and the magnitude of $l(T)$ adjusted until interference patterns in the experimental and theoretical images agreed. Fig. 13(a) shows the measured inverse decay length $1/l(T)$ as a function of temperature for Sn/Ge(1 1 1) [59]. The length parameter diverges at $T_1 = 70$ K, leading to the prediction that the CDW phase transition from a single defect would occur at this temperature T_1. Fig. 13(b) shows the (3×3) diffraction intensity calculated from the simulated images, or measured with low energy electron diffraction (LEED) [61] as a function of temperature. This is an appropriate order parameter for this phase transition. An order parameter is used to distinguish one phase from the other. It is by definition zero in the high-temperature phase, and saturates in the

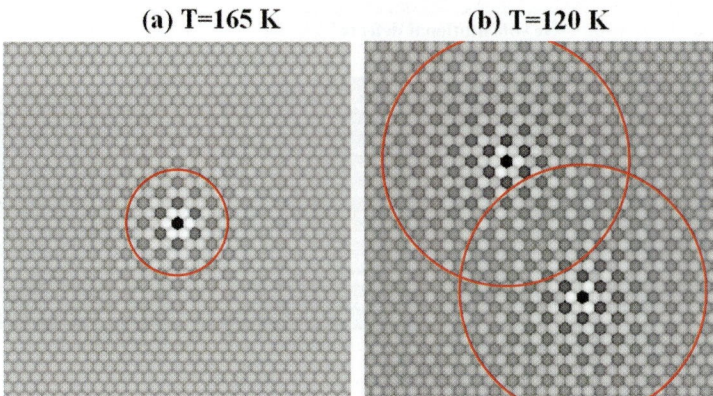

Fig. 12. Simulation of a defect-induced CDW. A defect can produce a periodic oscillation of charge in its vicinity. The extent of these damped density oscillations depends upon the temperature, increasing in range as the temperature is lowered. The simulated STM image for a single Ge defect (a) shows damped CDWs at a temperature equivalent of $T = 160$ K. The characteristic decay length of this wave is shown by the red circle and is $l = 25$ Å. When there is more than one defect present, the resultant charge perturbation is a complex interference pattern. The simulation (b) is a result of interference of waves from two defects at a temperature $T = 120$ K, with a decay length shown by the red circles of $l = 50$ Å [59].

low-temperature phase. Since this order parameter is continuous, this is a second-order phase transition.

There is another transition that occurs in this system. At a temperature $T_2 < 120$ K the decay length is longer than the average defect–defect spacing l_{av} indicated in Fig. 13(a). The (3×3) unit cell in Fig. 11(a) shows that there are three Sn atoms in each cell. This means that there are three possible CDW domains depending on which of the three atoms is negatively charged. Since Ge substitutional atoms do not like to sit on a charge maximum site, a random distribution of Ge defects would have, on average, 1/3 on unfavorable sites. A careful statistical counting of the defects in the STM images has proven that they move as the temperature is lowered [59,60]. The measured result is shown in Fig. 13(c), indicating an abrupt alignment of the defects at ~120 K. Fig. 13(d) shows the results of a Monte Carlo simulation of the defect alignment P_C. P_C is the order parameter for the defect alignment phase transition.

There are two defect density dependent interlocked phase transitions in this system. A second-order CDW transition characterized by the order parameter of the (3×3) diffraction intensity in Fig. 1(b) and a first-order DDW characterized by the defect alignment probability function P_C (Fig. 13(d)). A phase transition is first order if the order parameter is discontinuous. Both transitions depend intimately upon the defect density.

Undoubtedly, this is just the beginning of STM studies of defect mediated phase transitions in one and two dimensions. We can look forward to STM images of DDWs, and new microscopic theories of defect-mediated phase transitions in reduced dimensionality. When there are only a few "wrong" atoms we talk about defects. Understanding the interaction of defects with the host system and with each other is the first step to a microscopic picture of binary systems. On a more global scale the real space images obtained with the STM compliment more established diffraction experiments (momentum space), and contribute a new perspective to the understanding of collective phenomenon in complex–inhomogeneous materials.

The example given here of defect alignment falls in the family of what is now called self-assembly, or self-organization. Self-assembly refers to systems that, under the correct conditions, will organize themselves into a desirable state. Often the final structures are periodic on the nanoscale and consequently are of special importance to the emerging field of nanotechnology.

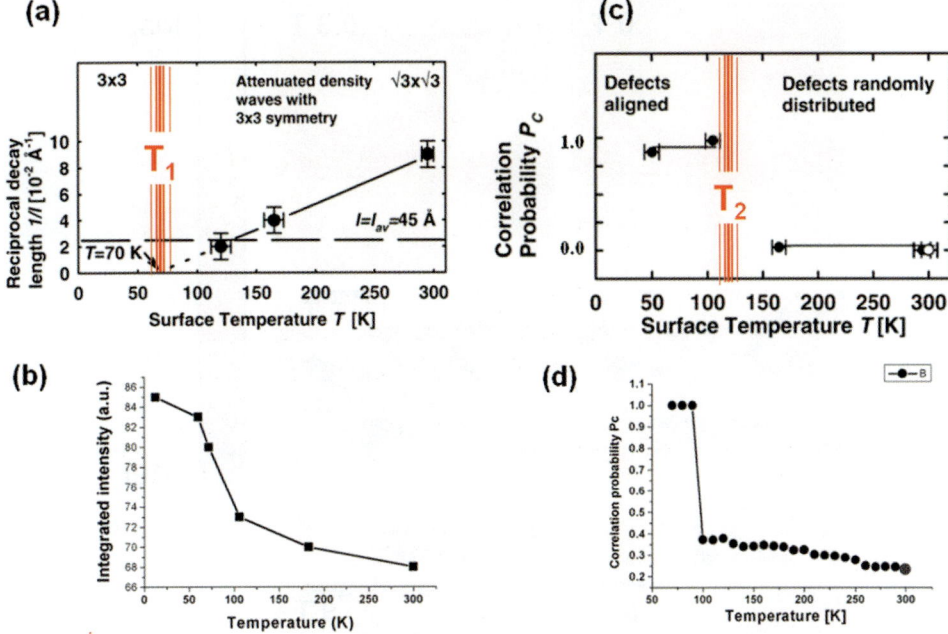

Fig. 13. Representation of the two intertwined phase transitions associated with the system described in Fig. 11, 1/3 of a monolayer of Sn on Ge(1 1 1). (a) The measured temperature dependence of the inverse decay length, $1/l(T)$, of the CDWs induced by substitutional Ge defects in Sn overlayer. Extrapolation can be used to determine the temperature at which the CDW transition would occur in the absence of defects (T_1). The horizontal line marked by l_{av} denotes the average spacing between Ge defects in the Sn overlayer. The order parameter of CDW transition, integrated intensity of (3×3) spot, is shown in (b). In the defect populated system it never becomes zero. Due to the CDW-mediated defect–defect interaction a CDW transition shown in (a) and (b) is accompanied by defect-ordering transition. A probability function P_C is defined as 0 if there is an equal distribution of Ge substitutional atoms on all three possible CDW lattices and is equal one if the Ge atoms in a single CDW domain are all located on the two lattices with charge minimum. The measured order parameter of defect alignment, the correlation probability P_C vs. temperature is shown in (c), indicating an abrupt transition at a temperature T_2. (d) The defect alignment has been simulated using a simple electrostatic model for the interaction of a defect with all of the defect-induced CDWs in the system. The simulated order parameter P_C for the defect alignment transition is shown in (d) [59–61].

3.3. The surfaces of complex highly correlated materials

In this section we illustrate how surface science techniques, both experimental and theoretical, are ideal for uncovering the fascinating physics taking place at the surfaces of highly correlated materials such as TMOs. Three examples will be given. (1) A study of electronic nanophase separation in the CMR materials [62], as discussed in Section 1; (2) the discovery of microscopic spatial electronic inhomogeneity in the superconducting gap in a high-T_c sample [63], and; (3) the identification of surface stabilized structural and magnetic phases in a spin triplet paired TMO superconductor [64].

The phenomenon of electronic nanophase separation is at the heart of the basic physics of CMR materials. In 1999 Fäth et al. used a variable temperature STM equipped with a high magnetic field to study the spatial electronic properties of the CMR materials $La_{0.7}Ca_{0.3}MnO_3$ [62]. Fig. 5(a) displays the coupled metal–insulator (M–I)—ferro–paramagnetic transition in the same material but with a slightly higher doping level, $x = 0.35$. The transition temperature is ∼240 K when there is no applied magnetic field and it decreases with increasing magnetic field. Fig. 14 shows a series of spatial images (STM) of this material at a temperature slightly below the zero field transition temperature (Fig. 5(a)) as a

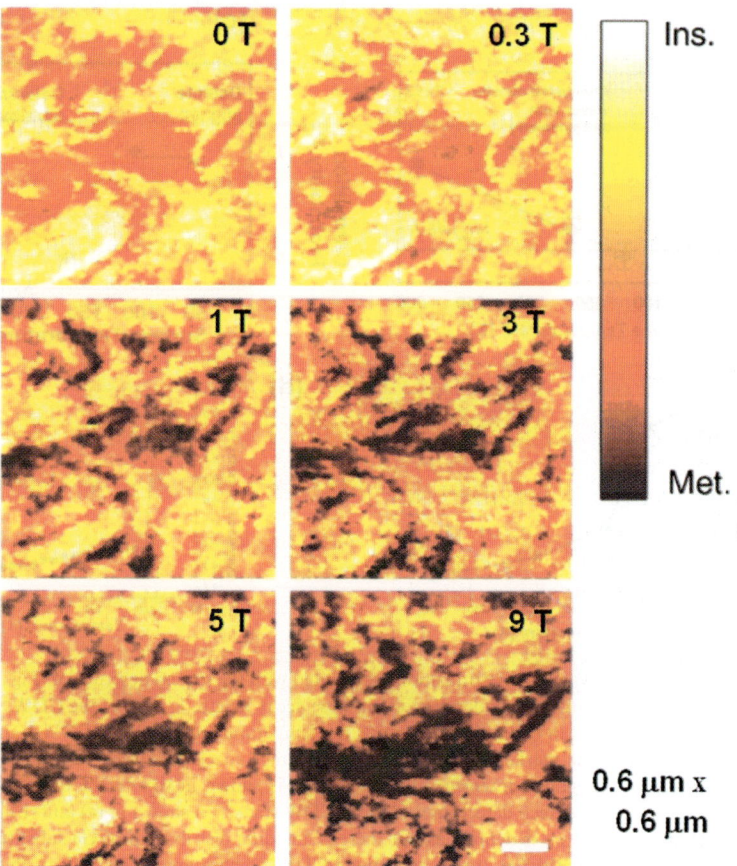

Fig. 14. STM spectroscopic images (0.61×0.61 μm^2) from the surface of La$_{0.7}$Ca$_{0.3}$MnO$_3$ taken just below the M–I transition temperature T_c (240 K), showing the evolution of local electronic structure under the influence of an external magnetic field [61]. The phrase spectroscopic image refers to the measurement procedure, where images were taken by measuring the spatial differential conductance dI/dV (I is the tunneling current and V the applied voltage) at a bias voltage of 3 V. The authors have demonstrated that spectroscopic images at this bias voltage are sensitive to the local metallic behavior. The bar on the right indicates the insulating vs. metal character. The magnetic field is increased from zero to 9 T. It is found that microscopic electronic inhomogeneities appear in this material and that applying a magnetic field enhances the size of the metallic domains until they are interconnected, i.e., a percolation behavior.

function of applied magnetic field. The authors have shown that the differential conductance dI/dV (I is the tunneling current and V the applied voltage) at a voltage of 3 V can be used as a direct measure of the metallic character. The dark regions in the images are metallic. It is easy to see that there are electronic inhomogeneities on the 20 nm to several 100 nm scale, with the coexistence of insulating, metallic, and intermediate regions. The observed nanophase separation is not as well behaved as is predicted theoretically and shown in Fig. 6. As the magnetic field is increased the dark regions (metallic) grow at the expense of the light (insulating) regions. When metallic regions (dark colour) connect on a macroscopic scale the sample becomes a metal. This phenomenon is called percolation.

The STM is an excellent tool for monitoring the surfaces of these spatially inhomogeneous systems as they approach the phase transition temperature or the percolation threshold. It is likely that surface scientists, working in this field over the next decade, will be among the first to understand the processes that drive and control the exotic properties of these materials. There is a tremendous opportunity for a surface scientist who knows how to prepare, modify, and carefully characterize the surfaces or interfaces of these TMOs.

The second example is an exciting measurement of the spatial inhomogenieties in the superconducting gap for $Bi_2Sr_2CaCu_2O_{8+x}$ created by the broken symmetry associated with the excess O doping (x) in these samples [63]. In a conventional metal, excess doping might be expected to create local inhomogeneities in the structure but the electronic structure usually remains homogeneous due to the short-range screening in metals. As will be shown in this example, this is apparently not the case for the high-T_c superconductor $Bi_2Sr_2CaCu_2O_{8+x}$. The structure of this material and its cleavage plane (BiO) are shown in Fig. 15.

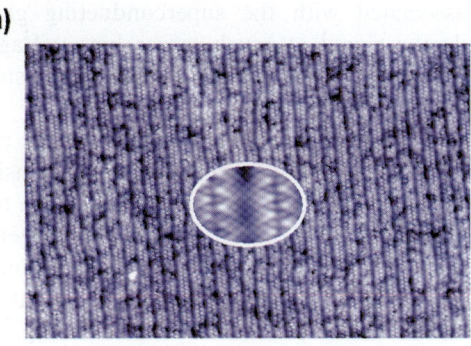

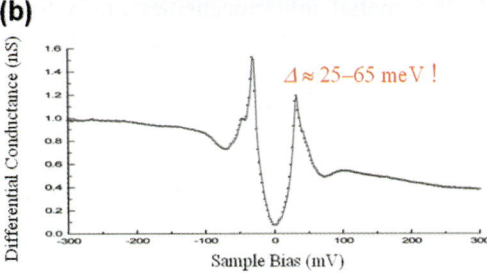

Fig. 16. (a) STM constant-current image at 4.2 K of $Bi_2Sr_2CaCu_2O_{8-x}$ taken at a tunneling current of 100 pA and a voltage of 100 mV [62]. The size of the picture is 80×125 nm^2. The center inset is a zoom-in image (5×) showing the detailed lattice structure on the surface. (b) The differential conductance (dI/dV) vs. sample bias, showing the electronic density of states including the superconducting energy gap of the local position of the sample. The sharp peaks around zero bias are associated the "quasiparticle" density of states and superconducting energy gap (Δ).

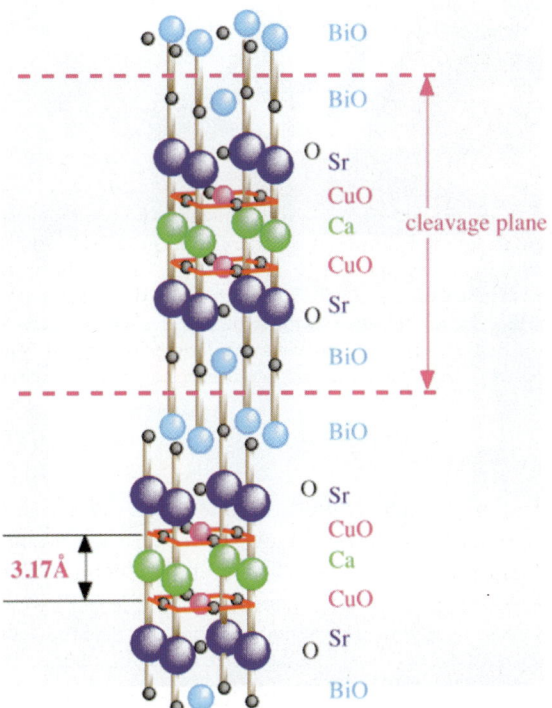

Fig. 15. Structural model of $Bi_2Sr_2CaCu_2O_{8-x}$. The cleavage plane is marked by the arrow and the color code of the atoms is indicated. Notice that the CuO superconducting plane is the third plane from the surface which is the BiO plane.

Superconductivity occurs in the CuO planes which are the third and fifth planes below the surface. The optimized superconducting transition temperature is ~90 K for a doping level of $x \sim 0.16$. Fig. 16(a) shows a high-resolution constant current topograpic image of the surface showing the structure of the BiO plane. The center insert is an enlargement by a factor of five. Fig. 16(b) is a differential conductance (dI/dV) plot for a specific region on the sample. These tunnel junctions (tip to sample) are not ohmic so the resistance and the conductance are a function of voltage. A conductance plot of $I(V)/V$ is not a straight line and differential conductance dI/dV plot-like that in Fig. 16(b) displays structure related to the electronic properties of the local position on the sample. The sharp structure around zero bias voltage is associated with the superconducting gap Δ, while the signal at much larger bias voltages is related to the local density of states at the surface [63].

These authors found that the magnitude of the superconducting gap as well as the local density of states were dependent on the position of the tip on the surface, i.e. spatial electronic inhomogeneity. In a *tour de force*, Pan was able to map out the spatial distribution of the magnitude of the superconducting gap, which is shown in Fig. 17(a) [63]. The spatial inhomogeneities are a few nanometers in extent. Fig. 17(b) shows a histogram of the observed superconducting gap size, showing an average value of 40 meV, the value usually observed in macroscopic measurements. But there is a variation of ~40 meV in the gap size depending on the position of the STM tip. There is a spatial correlation between the local density of states and the superconducting gap. The higher the local density of states the smaller the superconducting gap. Both exhibit Gaussian-like distributions, with a short autocorrelation length of ~14 Å. This length scale is surprising because the in-plane superconducting coherence length is believed to be ~25 Å [63].

The origin of the length scale of the electronic inhomogeneities in the density of states and the superconducting gap seems to be directly related to the density of excess oxygen x. For optimal

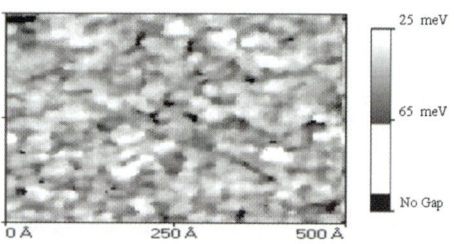

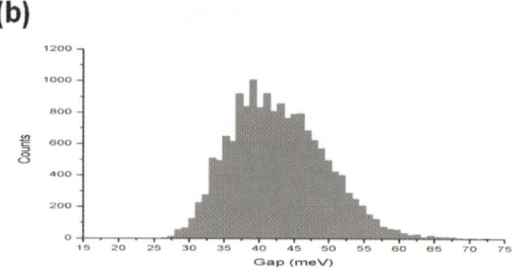

Fig. 17. (a) The spatial distribution map of the measured superconducting energy gap Δ from the surface of $Bi_2Sr_2CaCu_2O_{8-x}$ at 4.2 K, revealing the spatial electronic inhomogeneity on nanometer-length scale. (b) A histogram of the spatial distribution of gap sizes [63] showing a large spatial variation of the gap with a average value of ~40 meV which is usually observed in macroscopic.

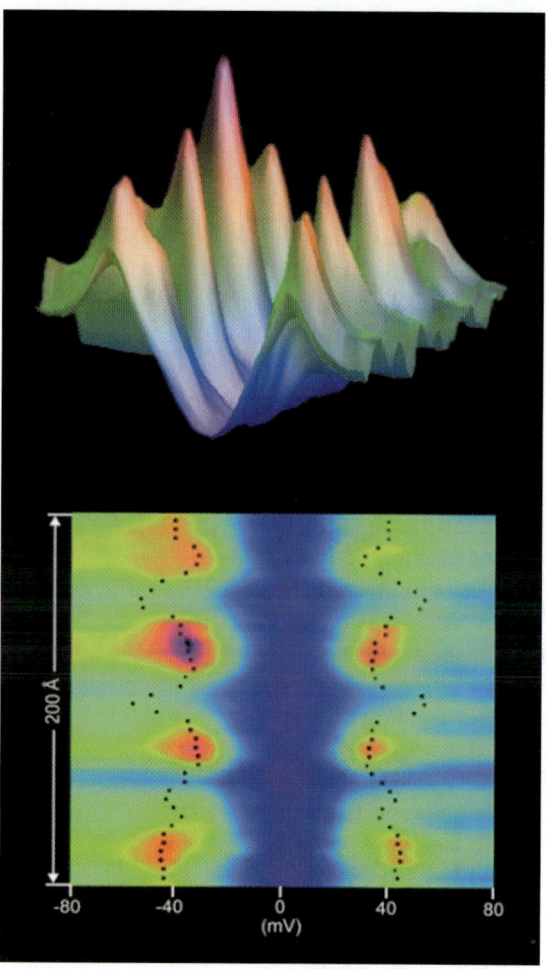

Fig. 18. 2D (bottom) and 3D (top) plots of the spatial dependence of the differential conductance showing the microscopic inhomogeneity in the magnitude of the superconducting gap in $Bi_2Sr_2CaCu_2O_{8-x}$ [63]. The dots in the bottom panel trace out the superconducting band edges.

doping of $x = 0.16$ the average inter-dopant spacing is ~13.5 Å. Fig. 18 is a 2D and 3D drawing of the spatial dependence of the differential conductance in this sample. The speculation is that the local density of states away from the gap is believed to be a good measure of the oxygen concentration [46]. When the STM tip is close to an oxygen atom (dopant) the density of states at high bias voltages is low and the superconducting gap is large, as displayed in Fig. 18.

This study of the spatially dependent electronic properties of a high-T_c superconductor raises many questions and suggests many new experiments. What creates the inhomogeneous superconducting gap? Is it the broken symmetry created by the excess doping of oxygen x? Is this a general property of all high-T_c materials? If so, is this a general and fundamental property of high-T_c superconductors? Is this the first picture of inhomogeneity-induced superconductivity [37]? Will these and future measurements lead to a deeper understanding of high-T_c superconductivity? Is this a surface affect? Is the oxygen concentration at the surface the same as in the bulk? Again, we stress that here is a tremendous opportunity for a surface physicist to have an impact on an important area of CMP.

The last example illustrates the stabilization at the surface of low energy excited states (structural and magnetic) in the bulk of the unconventional superconductor Sr_2RuO_4 [64–67]. Sr_2RuO_4 is the only known layered perovskite without copper that exhibits superconductivity ($T_c \sim 1.5$ K) [65]. It is also a superconductor without additional oxygen doping. It has attracted much attention because it shows so-called spin-triplet pairing symmetry with a p-wave order parameter [66,67]. The pairing symmetry is determined by total spin (spin singlet or triplet) and angular momentum channels (s-wave, p-wave, and so forth) characterizing the Cooper pairs. Pairs of fermions must have antisymmetric wave functions under particle interchange. For a Cooper pair, this requirement implies a relationship between the orbital and spin character. Orbital wave functions with even values of the orbital quantum number ($l = 0, 2, \ldots$) are even under particle interchange and therefore require odd symmetry in the spin wave function, i.e. spin singlets with one spin up and one down. High-T_c cuprates, which are antiferromagnets rendered superconducting by doping, are now known to have spin-singlet pairing in a d-wave ($l = 2$) orbital channel. In Sr_2RuO_4, experiment and theory suggests that the orbital wave function in the Cooper pair has p-character ($l = 1$) such that the spin wave function is required to be triplet (spins aligned). Obviously, spin-triplet pairing favors ferromagnetic spin fluctuation in contrast to that in cuprates where the spin-singlet pairing is associated with antiferromagnetic fluctuation.

The bulk of Sr_2RuO_4 has the RP structure for $n = 1$ shown in Fig. 4, with a nonmagnetic ground state. However, this ground state is close to structural and magnetic instabilities. A common characteristic of this class of materials, is that the energy difference between different structural/magnetic phases is very small. Creating a surface by cleaving a single crystal is a controlled way to disturb the coupled system by breaking the symmetry without changing the stoichiometry. With this example we illustrate that this unique environment at the surface could produce new phenomena, while providing a fresh approach to the study of the spin-charge-lattice coupling in these complex materials.

Fig. 19(b) shows a large scale STM image of a surface cleaved inside a vacuum system and transferred to an STM stage [64]. There are large flat terraces with an extension up to 10 μm. All step heights are integral multiples of half the unit cell (6.4 Å) shown in Fig. 19(a). Experimental electron diffraction determinations of the surface structure and ab initio calculations prove that the surface is the SrO plane, as expected from looking at Fig. 4 ($n = 1$). The enlarged STM image, with atomic resolution, shown in Fig. 19(c) as well as LEED patterns shown in Fig. 20(a) indicate that the surface is not bulk truncated but reconstructed. As in Fig. 11 for Sn/Ge(1 1 1), we have drawn in the basic 2D structure (unit cell) in Fig. 19(c). The reduction in symmetry at the surface caused by the reconstruction creates fractional order diffraction spots in the LEED pattern shown in Fig. 20(a), marked by the white arrows. The measured and calculated surface structure

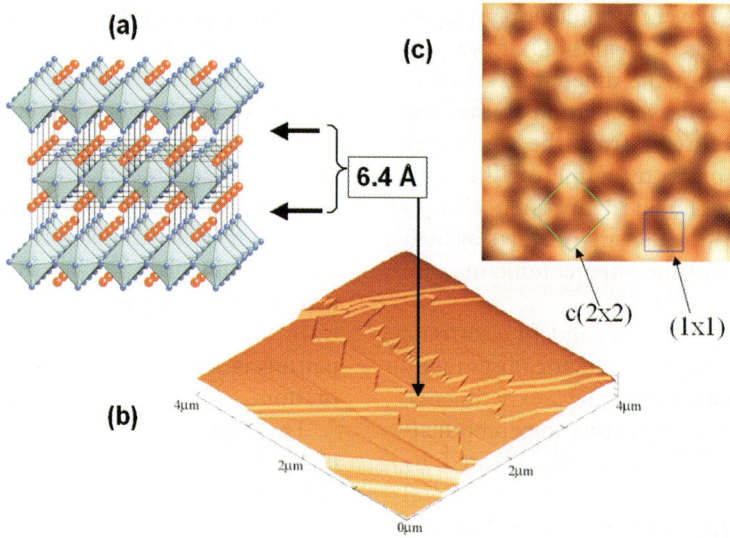

Fig. 19. (a) Structural model of Sr_2RuO_4 (red: strontium, blue: oxygen, green: ruthenium in the center of the octahedron). The dark arrow indicate the cleavage planes and the spacing between these planes. (b) STM image of a 4×4 μm² surface area (room temperature with +0.5 bias voltage) showing extremely large terraces and steps. (c) Atomically resolved STM image of 26×12 Å² containing 7×7 strontium sites. Theoretical calculation of the local density of states showed that the bright spots correspond to the strontium site ($T = 300$ K, sample bias voltage -0.75 V). The bulk (1×1) (blue) and surface $c(2 \times 2)$ reconstructed unit cell (green) are indicated [64].

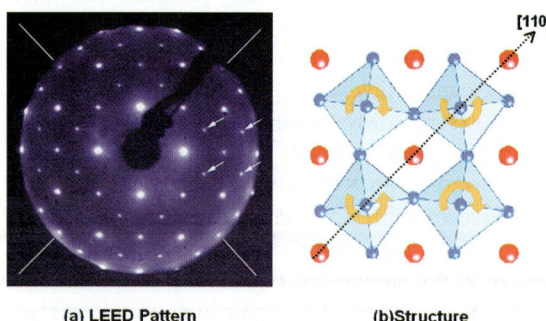

Fig. 20. (a) LEED pattern showing fractional spots (marked by arrows) created to the surface reconstruction. There are missing fractional order spots in the LEED pattern (indicated by white lines) that are important in determining the symmetry of surface structure [64]. Electron energy $E = 195$ eV, temperature $T = 80$ K. (b) Ball model of the surface structure with rotated octahedra (top view). The direction of the rotation is shown by the yellow arrows. The direction of propagation of the soft-bulk phonon is phonon mode is in the diagonal direction [1 1 0].

is shown in Fig. 20(b). The octahedra in the surface plane are rotated alternating clock- and anticlockwise around the surface normal direction. This rotation is exactly the low energy vibrational (phonon) mode in the bulk, so the surface has stabilized a geometric instability present in the bulk.

Experimentally, the octahedra rotation is $9 \pm 3°$. First-principles calculations of the ground state surface structure confirm the octahedra rotation. The optimized structure for a nonmagnetic surface is a surface layer with octahedra rotated by 6.5°. This reconstruction is driven by compressive strain in the RuO_2 layers. According to theory, the reconstruction enhances ferromagnetic ordering, which in turn stabilizes the distortion further and increases the rotation angle to 9°. The ground state of the surface is, according to theory, ferromagnetically ordered, driven by the structural distortion of a rotation of the octahedra. It is worth mentioning that there is a ferromagnetic fluction in the bulk. Broken symmetry at the surface stabilizes not only the lattice but also a magnetic structural instability that existed in the bulk.

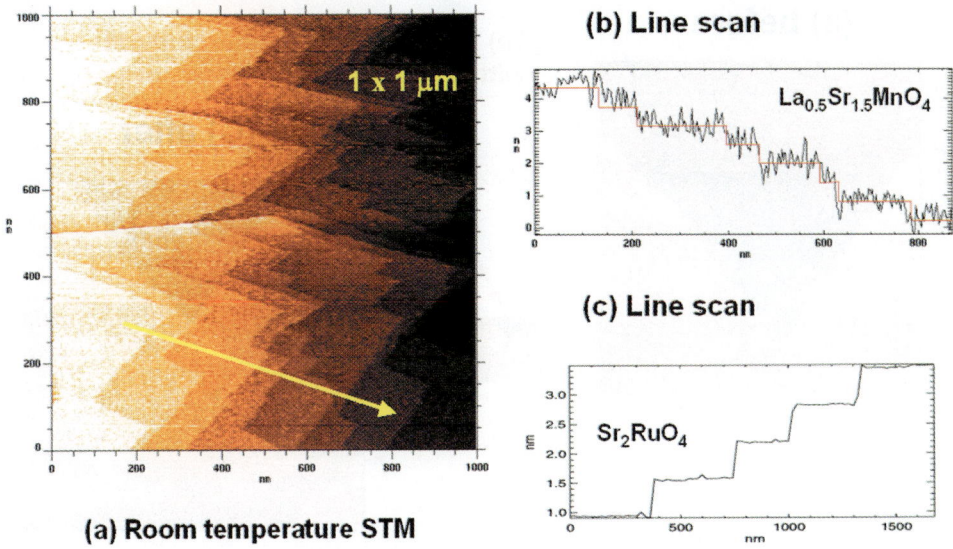

Fig. 21. (a) Large area STM image of $La_{0.5}Sr_{1.5}MnO_4$ cleaved in the vacuum and imaged at room temperature with a bias voltage of 2.5 V. (b) STM constant current height scan along the yellow line shown in (a). (c) Line scan for a freshly cleaved sample of Sr_2RuO_4 like the one shown in Fig. 18. Though both materials have the same bulk structure, the surface roughness is drastically different, caused by the electronic nanophase inhomogeneities occurring in $La_{0.5}Sr_{1.5}MnO_4$ but not in Sr_2RuO_4.

This investigation of the surface of Sr_2RuO_4 illustrates that existing surface structural and electronic probes can be used on these complex materials. It opens up many exciting prospects, relevant to the bulk and surface properties of layered TMOs. For example, there is the possibility of the coexistence of FM and superconductivity in layered superconductors [68] or at the surface of high-T_c materials [69]. The challenge for the surface physicist is to develop techniques capable of structural, magnetic and electronic measurements on that same surface, or ideally at the same place on the surface as a function of temperature and magnetic field.

3.4. Nano-scale patterning of the surface of complex transition-metal oxides

The final example of this paper illustrates the possibility of utilizing the close coupling of the spin, charge, and lattice degrees of freedom in complex TMOs to artificially pattern the surface on a nanoscale. The hint that this might work was presented in Fig. 5(c) where it was shown that a light beam could convert an antiferromagnetic-ordered insulating phase of a TMO into a ferromagnetic-ordered metallic phase. If you can accomplish this switching with a laser beam then you should be able to achieve the same conversion with the current from an STM tip.

We have investigated the layered manganite $La_{0.5}Sr_{1.5}MnO_4$ which has the same basic crystal structure as Sr_2RuO_4 ($n = 1$ in Fig. 4). In contrast to Strontium Ruthenate, this material is a charge-ordered insulator with a charge ordering transition at 220 K and it is antiferromagnetically ordered below 110 K [70]. Fig. 21(a) shows a large scale STM image of a cleaved sample of this material at room temperature. It is easy to see the step edges and terraces as in the equivalent image for Sr_2RuO_4 displayed in Fig. 19. The striking difference is that the material is electronically very rough. A line scan across the sample (yellow line in (a)) is shown in (b) where you can see that the modulations are almost as big as the step height. This is caused by electronic nanophase separation in this material, similar to what was shown for $La_{0.7}Ca_{0.3}MnO_3$ in Fig. 14, except that in this case

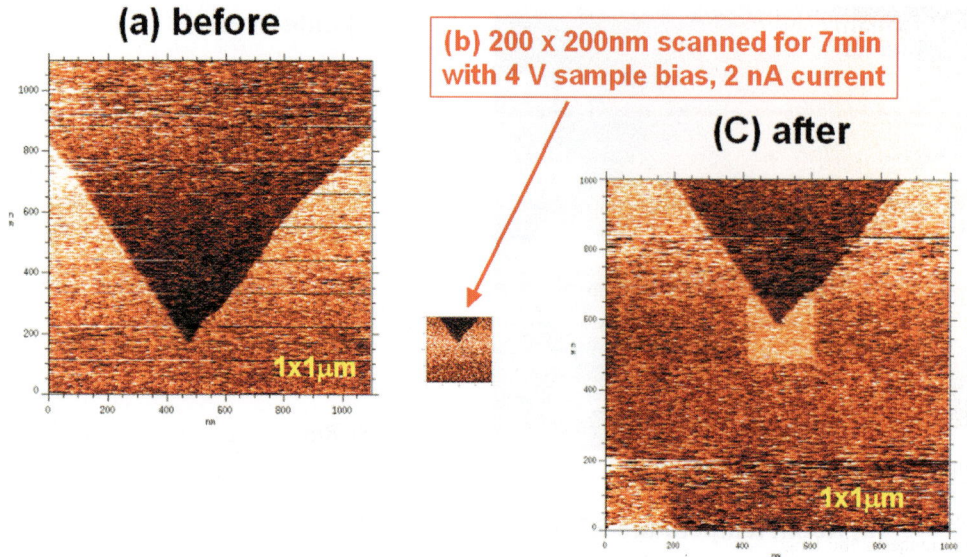

Fig. 22. Panel (a): STM image of a 1100×1100 nm^2 area of La$_{0.5}$Sr$_{1.5}$MnO$_4$ vacuum cleaved surface scanned at room temperature with +2.5 V bias voltage. The triangular structure is the image of the step. (b) After taking the image (a), a 200×200 nm^2 square area centered around the step-corner was scanned for 7 min with 4.0 V bias voltage. (c) 1000×1000 nm^2 area centered around the same step corner imaged after the procedure described above in (b). The excessively scanned area of 200×200 nm^2 looks brighter than the rest of the terraces, indicating a better conductivity of the surface after scanning for some minutes with elevated bias voltage.

the length scale associated with the nanophases is much smaller, ~20 nm. For comparison, Fig. 21(c) shows a similar line scan of Sr$_2$RuO$_4$, where the roughness is caused only by the atomic corrugation.

Fig. 22 illustrates what happens if the tunneling current density is increased dramatically. You write! Panel (a) is a large area scan with a bias voltage of 2.5 V. After this scan was taken, a smaller area (200×200 nm) was scanned with an increased voltage of 4 V at 2 nA current for approximately 7 min. Fig. 22(c) displays a subsequent broad area scan showing that the area scanned in (b) is much brighter, indicating that it is more metallic. It seems possible to write on this surface creating nanoscale ferromagnetic-ordered metal patches! Generally speaking, it should be possible to destroy the bulk charge ordered state (insulating and antiferromagnetic-ordered states) at the surface, creating a ferromagnetic-ordered metal. There is no obvious reason why this conversion cannot be patterned with an electron or photon beam.

4. Conclusion

What an exciting prospect: Surface scientists can change the way the materials community thinks. All we have to do is *engage*.

Many a false Step is Made by Standing Still—Chinese fortune cookie

You miss all of the shots you don't take—Wayne Gretzky

If the surface physics community engages in contemporary materials physics, we can look forward to many decades of exciting discoveries. The sky is the limit, or perhaps more appropriately, the cosmos is the limit. All physicists search for universality that is the ability to map their observations or theories onto a wide variety of problems of very different length scales. Fig. 23 shows that the STM images of the electronic inhomogeneities at the surface of La$_{0.5}$Sr$_{1.5}$MnO$_4$ (top right) map onto spatial inhomogeneities seen in UV emission

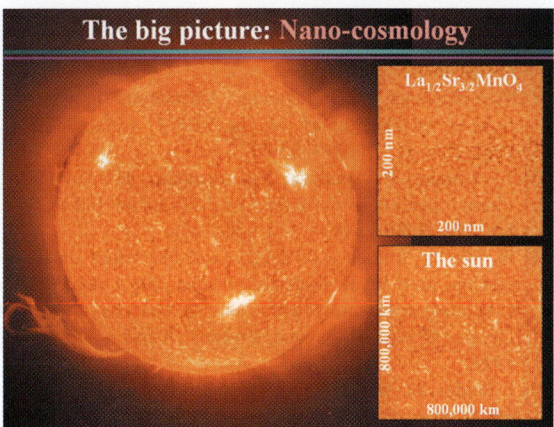

Fig. 23. Nanocosmology, comparing the electronic spatial variation for $La_{0.5}Sr_{1.5}MnO_4$ with the spatial variation in UV emission from the surface of the sun. There are 15 orders of magnitude difference in the length scales. The sun image is a courtesy of SOHO/EIT consortium. SOHO is a project of international cooperation between ESA and NASA.

image of the surface of the sun. This is true universality, since the difference in length scale is ∼15 orders of magnitude. We are truly living in the world of nano-cosmology. Our contribution has come full circle, back to the introduction, where we introduced Broglia's article in this special issue of Surface Science titled "The surface of compact systems: from nuclei to stars".

References

[1] Nanotechnology: Shaping the World Atom by Atom, National Science and Technology Council, Committee on Technology, 1999 http://www.nano.gov/press.htm.
[2] D.H. Lowndes (Ed.), Nanoscale Science, Engineering and Technology Research Directions. Basic Energy Sciences Program, US Department of Energy, ORNL, 1999.
[3] R.A. Broglia, The surfaces of compact systems: from nuclei to stars, Surf. Sci. 500 (2002) 759.
[4] J.M. Greenberg, Cosmic dust and our origins, Surf. Sci. 500 (2002) 793.
[5] D.A. Williams, E. Herbst, It's a dusty Universe: surface science in space, Surf. Sci. 500 (2002) 823.
[6] T.E. Madey, R.E. Johnson, T.M. Orlando, Far-out surface science: radiation-induced surface processes in the solar system, Surf. Sci. 500 (2002) 838.
[7] Complex Systems: Science for the 21st Century, Office of Science Workshop, US Department of Energy, March 5–6, 1999, available at http://www.er.doe.gov/production/bes/complexsystems.htm.
[8] Condensed-Matter and Materials Physics: Basic Research for Tomorrow's Technology, Committee on Condensed-Matter and Materials Physics, National Research Council, National Academy Press, Washington, DC, 1999, p. 10.
[9] D.G. Castner, B.D. Ratner, Biomedical surface science: Foundations to frontiers, Surf. Sci. 500 (2002) 28.
[10] M. Fahlman, W.R. Salaneck, Surfaces and interfaces in polymer-based electronics, Surf. Sci. 500 (2002) 904.
[11] B. Kasemo, Biological surface science, Surf. Sci. 500 (2002) 656.
[12] T.H. Rod, J.K. Nørskov, The surface science of enzymes, Surf. Sci. 500 (2002) 678.
[13] S.A. Safran, Statistical thermodynamics of soft surfaces, Surf. Sci. 500 (2002) 127.
[14] M. Tirrell, E. Kokkoli, M. Biesalski, The role of surface science in bioengineered materials, Surf. Sci. 500 (2002) 61.
[15] S.D. Bader, Magnetism in low dimensionality, Surf. Sci. 500 (2002) 172.
[16] J. Shen, J. Kirschner, Tailoring magnetism in artificially structured materials: the new frontier, Surf. Sci. 500 (2002) 300.
[17] J.R. Arthur, Molecular beam epitaxy, Surf. Sci. 500 (2002) 189.
[18] H. Dai, Carbon nanotubes: opportunities and challenges, Surf. Sci. 500 (2002) 218.
[19] H.-J. Freund, Clusters and islands on oxides: from catalysis via electronics and magnetism to optics, Surf. Sci. 500 (2002) 271.
[20] W. Eberhardt, Clusters as new materials, Surf. Sci. 500 (2002) 242.
[21] A.G. Naumovets, Z. Zhang, Fidgety particles on surfaces: how do they jump, walk, group, and settle in virgin areas? Surf. Sci. 500 (2002) 414.
[22] A. Schilling, M. Cantani, J.D. Guo, H.R. Ott, Superconductivity above 130 K in the Hg–Ba–Ca–Cu–O system, Nature 363 (1993) 56.
[23] Correlated Electron Systems, Science 288 (2000).
[24] I.S. Osborn, Science 288 (2000) 438.
[25] R.J. Birgeneau, M.A. Kastner, Frontier physics with correlated electrons, Science 288 (2000) 437.
[26] M. Imada, A. Fujimori, Y. Tokura, Metal–insulator transitions, Rev. Mod. Phys. 70 (1998) 1039–1263;
Y. Tokura, N. Nagaosa, Orbital physics in transition-metal oxides, Science 288 (2000) 462.
[27] V. Kiryukhin, D. Casa, J.P. Hill, B. Keimer, A. Viliante, Y. Tomioka, Y. Tokura, An X-ray-induced insulator-metal transition in a magnetoresistive manganite, Nature 386 (1997) 813.
[28] M. Fiebig, K. Miyano, Y. Tomioka, Y. Tokura, Visualization of the local insulator–metal transition in $Pr_{0.7}Ca_{0.3}MnO_3$, Science 280 (1998) 1925.
[29] H. Oshima, K. Miyano, Y. Konishi, M. Kawasaki, Y. Tokura, Switching behavior of epitaxial perovskite manganite thin films, Appl. Phys. Lett. 75 (1999) 1473.

[30] J. Tranquada, B.J. Sternlieb, J.D. Axe, Y. Nakamura, S. Ushida, Nature 375 (1995) 338;
J. Tranquada, P. Wochner, D.J. Buttery, Phys. Rev. Lett. 78 (1997) 2133.

[31] Workshop on Soft X-ray Science in the Next Millennium: The Future of Photon-In/Photon-Out Experiments, Pikeville, Tennessee, March 15–18, 2000.

[32] J. Orenstein, A.J. Millis, Science 288 (2000) 468.

[33] J. Zaanen, Self-organized one dimensionality, Science 286 (1999) 251.

[34] A. Moreo, S. Yunoki, E. Dagotto, Phase separation scenario for manganese oxides and related materials, Science 283 (1999) 2034.

[35] S.A. Kivelson, E. Fradkin, V.J. Emery, Electronic liquid-crystal phases of a doped Mott insulator, Nature 393 (1998) 550.

[36] P. Littlewood, Phases of resistance, Science 399 (1999) 529.

[37] J. Eroles, G. Ortiz, A.V. Balatsky, A.R. Bishop, Inhomogeneity-induced superconductivity? Europhys. Lett. 50 (2000) 540.

[38] J. Zaanen, Stripes, defeat the Fermi liquid, Nature 404 (2000) 714.

[39] P.W. Anderson, More is different-broken symmetry and nature of hierarchical structure, Science 177 (1972) 393.

[40] 1000 Most Cited Physicists, 1981 to June 1997 (out of over 5,000 examined) http://www.sst.nrel.gov. Current Contents, Research Department, Institute for Scientific Information (ISI).

[41] M.N. Baibich, J.M. Broto, A. Fert, F. Nguyen Van Dau, F. Petroff, P. Eitenne, G. Creuzet, A. Friederich, J. Chazelas, Giant magnetoresistance of $(001)Fe/(001)$ Cr magnetic superlattices, Phys. Rev. Lett. 61 (1988) 2472.

[42] W.E. Pickett, J.S. Moodera, Phys. Today 54 (2001) 39.

[43] D.L. Landau, E.M. Lifshitz, Statistical Physics, Pergamon Press, New York, 1970.

[44] D.R. Nelson, Defect-mediated phase transitions, in: C. Doumb, J.L. Lebowitz (Eds.), Phase Transitions and Critical Phenomena, Academic Press, London, 1983, pp. 1–99.

[45] L.W. Bruch, M.W. Cole, E. Zaremba, Physical Adsorption: Forces and Phenomena, Oxford University Press, Oxford, 1997.

[46] T. Tabata, T. Aruga, Y. Murata, Order–disorder transition on $Si(001)$-C(4×2) to (2×1), Surf. Sci. 179 (1987) L63–L70.

[47] J.D. Chadi, Atomic and electronic-structures of reconstructed $Si(001)$ surfaces, Phys. Rev. Lett. 43 (1979) 43.

[48] R.M. Tromp, R.J. Hamers, J.E. Demuth, $Si(001)$ dimer structure observed with scanning tunneling microscopy, Phys. Rev. Lett. 55 (1985) 1303;
R.J. Hamers, R.M. Tromp, J.E. Demuth, Scanning tunneling microscopy of $Si(001)$, Phys. Rev. B 34 (1986) 5343.

[49] K. Inoue, Y. Morikawa, K. Terakura, M. Nakayama, Order–disorder phase-transition on the $Si(001)$ surface—critical role of dimer defects, Phys. Rev. B 49 (1994) 14774.

[50] R.A. Wolkow, Direct observation of an increase in buckled dimers on $Si(001)$ at low-temperature, Phys. Rev. Lett. 68 (1992) 2636.

[51] T. Yokoyama, K. Takayanagi, Anomalous flipping motions of buckled dimers on the $Si(001)$ surface at 5 K, Phys. Rev. B 61 (2000) R5078.

[52] Y. Kondo, T. Amakusa, M. Iwatsuki, H. Tokumoto, Phase transition of the $Si(001)$ surface below 100 K, Surf. Sci. 453 (2000) L318.

[53] H. Mutka, Influence of defects and impurities on CDW systems, in: F.W. Boswell, J.C. Bennet (Eds.), Advances in Crystallographic and Microstructural Analysis of Charged Density Waves Modulated Crystals, Kluwer, Dordrecht, 1999.

[54] I. Baldea, M. Badescu, Quasi-regular impurity distribution driven by a charge-density-wave, Phys. Rev. B 48 (1993) 8619.

[55] I. Baldea, M. Apostol, Modulated-impuity mechansism of pinning in Kcp, J. Phys. C 18 (1985) 6135.

[56] A.W. Overhauser, Exchange and correlation instabilities of simple metals, Phys. Rev. 167 (1968) 691.

[57] G. Gruner, Density Waves in Solids, first ed., Addison-Wesley, Reading MA, 1994.

[58] J.M. Carpinelli, H.H. Weitering, M. Bartkowiak, R. Stumpf, E.W. Plummer, Surface charge ordering transition: α phase of $Sn/Ge(111)$, Phys. Rev. Lett. 79 (1997) 2859.

[59] A.V. Melechko, J. Braun, H.H. Weitering, E.W. Plummer, Two-dimensional phase transition mediated by extrinsic defects, Phys. Rev. Lett. 83 (1999) 999.

[60] A.V. Melechko, J. Braun, H.H. Weitering, E.W. Plummer, The role of defects in two-dimensional phase transitions: an STM study of the $Sn/Ge(111)$ system, Phys. Rev. B 61 (2000) 2235.

[61] H.H. Weitering, A. Melesko, J.M. Carpinelli, J. Zhang, M. Bartkowiak, E.W. Plummer, Defect-mediated condensation of a charge density wave, Science 285 (1999) 2107–2110.

[62] M. Fäth, S. Freisem, A.A. Menovsky, Y. Tomioka, J. Aarts, J.A. Mydosh, Science 285 (1999) 1540.

[63] S.H. Pan, J.P. O'Neal, R.L. Badzey, C. Chamon, H. Ding, J.R. Englebrecht, Z. Wang, H. Eisaki, S. Uchida, A.K. Gukpta, K.-W. Ng, E.W. Hudson, K.M. Lang, J.C. Davis, A microscopic picture of the electronic inhomogeneity in high-T_c superconductor $Bi_2Sr_2CaCu_2O_{8+x}$, in press.

[64] R. Matzdorf, Z. Fang, Ismail, J. Zhang, T. Kimura, Y. Tokura, K. Terakura, E.W. Plummer, Ferromagnetism stabilized by lattice distortion at the surface of the p-wave superconductor Sr_2RuO_4, Science 289 (2000) 746.

[65] Y. Maeno, H. Hashimoto, K. Yoshida, S. Nishizaki, T. Fujita, J.G. Bednorz, F. Lichtenberg, Superconductivity in a layered perovskite without copper, Nature 372 (1994) 532.

[66] T.M. Rice, M. Sigrist, Sr_2RuO_4: an electronic analogue of ^{3}He? J. Phys.: Condens. Matter 7 (1995) L643.

[67] I.I. Mazin, D.J. Singh, Competitions in layered ruthenates: ferromagnetism versus antiferromagnetism and triplet versus singlet pairing, Phys. Rev. Lett. 82 (1999) 4324.

[68] W.E. Pickett, R. Weht, A.B. Shick, Superconductivity in ferromagnetic $RuSr_2GdCu_2O_8$, Phys. Rev. Lett. 83 (1999) 3713.

[69] D.E. Pugel, M.B. Salamon, M.B. Weissman, L.H. Greene, Direct dection of spontaneously broken time-reversal symmetry at surfaces of superconducting $Yba_2Cu_3O_{7-\delta}$, in press.

[70] B.J. Sternlieb, J.P. Hill, U.C. Wildgruber, et al., Charge and magnetic order in $La_{0.5}Sr_{1.5}MnO_4$, Phys. Rev. Lett. 76 (1996) 2169.

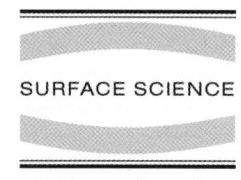

Biomedical surface science: Foundations to frontiers

David G. Castner *, Buddy D. Ratner *

National ESCA and Surface Analysis Center for Biomedical Problems, University of Washington Engineered Biomaterials, Departments of Chemical Engineering and Bioengineering, University of Washington, Seattle, WA 98195, USA

Received 1 December 2000; accepted for publication 23 July 2001

Abstract

Surfaces play a vial role in biology and medicine with most biological reactions occurring at surfaces and interfaces. The foundations, evolution, and impact of biomedical surface science are discussed. In the 19th century, the first observations were made that surfaces control biological reactions. The advancements in surface science instrumentation that have occurred in the past quarter of a century have significantly increased our ability to characterize the surface composition and molecular structure of biomaterials. Similar advancements have occurred in material science and molecular biology. The combination of these advances have allowed the development of the biological model for surface science, where the ultimate goal is to gain a detailed understanding of how the surface properties of a material control the biological reactivity of a cell interacting with that surface. Numerous examples show that the surface properties of a material are directly related to in vitro biological performance such as protein adsorption and cell growth. The challenge is to fully develop the biological model for surface science in the highly complex and interactive in vivo biological environment. Examples of state-of-the-art biomedical surface science studies on surface chemical state imaging, molecular recognition surfaces, adsorbed protein films, and hydrated surfaces are presented. Future directions and opportunities for surface scientists working in biomedical research include exploiting biological knowledge, biomimetics, precision immobilization, self-assembly, nanofabrication, smart surfaces, and control of non-specific reactions. © 2001 Elsevier Science B.V. All rights reserved.

Keywords: Atomic force microscopy; Photoelectron spectroscopy; Secondary ion mass spectroscopy; Self-assembly; Biological compounds; Biological molecules – proteins; Solid–liquid interfaces

---

* Corresponding authors. Tel.: +1-206-5438094; fax: +1-206-5433778 (D.G. Castner). Tel.: +1-206-6851005; fax: +1-206-6169763 (B.D. Ratner).
  E-mail addresses: castner@nb.engr.washington.edu (D.G. Castner), ratner@uweb.engr.washington.edu (B.D. Ratner).

0039-6028/01/$ - see front matter © 2001 Elsevier Science B.V. All rights reserved.
PII: S0039-6028(01)01587-4

1. Introduction

1.1. The rationale and organization for this review

Through the first half of the 20th century physics dominated intellectual thought and discovery in the western world. With Watson and Crick's seminal 1953 paper outlining the structure of DNA, molecular biology quickly assumed a leadership role in the growth, discovery and nucleation of ideas. The surface science model, so successful in catalysis and microelectronics, will find facile partnership with modern biology ideas—the outcome will be revolutionary 21st century technologies.

This article will first trace the roots of biosurface ideas. Then the contemporary surface science model will be described along with its transformation into a biosurface model (Section 2). Surface science methods will next be presented in the context of an area poised for growth, chemical state imaging (Section 3). Biomaterials and surface science have always had a close relationship so traditional and new biomaterials ideas relating to surfaces will be discussed in Section 4. Protein films have also played an essential role in the understanding of biology at surfaces and Section 5. will summarize developments enhancing our abilities to analyze these films. Finally, state-of-the-art ideas, materials and trends that will have profound implications for surface science, technology and medicine are presented in the final section.

1.2. Surfaces in biology and medicine

Biological reactions are frequently described as occurring in the solution phase, for example, the reaction of a soluble enzyme with its substrate. In fact, most reactions in biology occur, not in solution, but at interfaces. Typical interfaces of biological importance include the cell surface/synthetic biomaterial (see Fig. 1), extracellular matrix (ECM)/biomolecule, ECM/cell, hydrated tissue/air (lung) and mineral/protein (bone).

Why would nature evolve molecular systems that exploit surfaces? We can surmise that for nature to do its work efficiently, surfaces offered the following advantages. Surfaces provide high accessibility for reaction. The low energy barrier to mobility in the plane of the surface can be used to facilitate complex reactions (clustering, conformational changes, exposure and burial in membranes). Epitaxy-like phenomena can be readily exploited at surfaces. High surface area geometries can be created to enhance reaction turnover rates. Unique organic microenvironments can enhance specific affinities and reactions. Self assembly in the plane of the interface can be used to orient and space molecules with precision. Surface energy minimization can orient specific structures to interfaces. Molecular recognition, a manifestation of both geometry and chemistry, is readily implemented at surfaces.

Beyond nature's use of surfaces, surface concepts have been adopted in medical and biological technology. Consider implant biomaterials, blood oxygenators, hemodialysis, affinity chromatography, surface diagnostics, cell culture surfaces and biosensors as examples of surface technology applied to biological problems. These applications have been largely driven by early observations that surfaces control biological reactions. Three areas, in particular, have been influential in advancing biological applications for surface science: chromatographic separations, blood compatibility and cell culture. A related, important realization was that proteins in aqueous solution rapidly adsorb as monolayers on surfaces. The thickness of the adsorbed protein monolayer (1–10 nm) is, fortuitously, in the range where most of our surface analysis instrumentation yields optimal results.

At this point it is useful to emphasize two ideas. First, surfaces are critically important to nearly all aspects of biology and biological technology. Second, the rules that govern biological surface phenomena are no different from the rules that govern the reactions at a silicon(1 0 0) surface. There is a limited set of physical laws governing this universe, and even the diversity seen in biology is constrained by these fundamental laws.

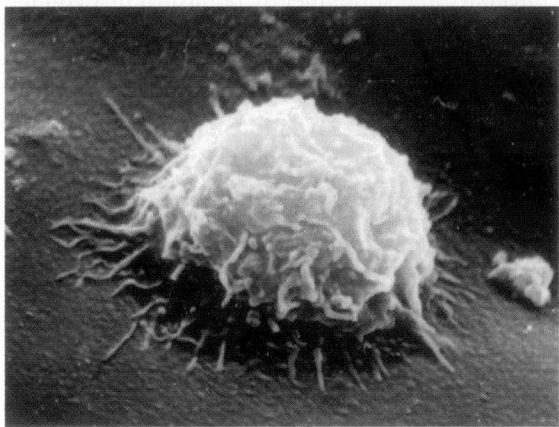

Fig. 1. A scanning electron micrograph of a myoblast cell interacting with a synthetic surface. The cell is approximately 10 μm in length.

What are the differences between the surface science practiced on a platinum catalyst and the surface science practiced on a biological specimen? First, biological systems are fragile in two ways: energetic surface probes can damage organocarbon-based molecules and those molecules can be chemically and morphologically altered by the abuse they suffer with storage and preparation for surface analysis (for example, conformational changes in proteins). Second, the molecules of biology are immensely more complex than the inorganic systems used for semiconductors and catalysts (Fig. 2). Third, biological systems only function normally in aqueous media, a condition challenging for ultrahigh vacuum (UHV) surface analysis equipment. Fourth, many important biological processes occur at relatively deeply buried interfaces. Fifth, most biological surface science specimens are irregularly shaped insulators. For these reasons, special care, understanding and methodologies must be used in the analysis of biological specimens.

1.3. The evolution of biomedical surface science

The importance of surfaces has been empirically appreciated through all history as attested to by early texts such as the 1250 AD manuscript, *De Proprietatibus Rerum* (The Properties of Things), that outlined surface preparations to achieve metal–metal bonding. In the early 18th century, experimentalists such as Dobereiner and Faraday described surface-induced catalysis. By the 19th century, J. Willard Gibbs gave us a useful thermodynamics for surfaces. This early history did not directly acknowledge the relevance of these governing principles to biology. In fact, with Wohler's synthesis of urea (1828), the realization dawned that biology was not a province of matter stemming only from earlier biology, but rather a branch of chemistry, and by inference obedient to the physical laws that govern the properties of matter.

Surface science ultimately relevant to biology was developing in other contexts. Benjamin Franklin's observations on oil films on lakes foreshadowed 19th century studies such as those by Agnes Pockels [1881] on surfactant film surface tension. Pockels' work clearly hinted at modern assembled organic/biological materials.

It was probably the early colloid chemists who, in the late nineteenth century, first speculated on the relationships between interfacial phenomena and the structure and organization of living systems, especially proteins and single cells (see, for example, "Proteins and the Theory of Colloidal Behavior," by J. Loeb, 1922). Many early investigators such as Faraday, Zsigmondy and Tyndall studied gold colloids, dust, bacteria particles or milk. The colloid state allowed some of the earliest appreciation of the concepts of self-assembly and minimization of interfacial energy.

In the first few decades of the twentieth century, the work of Irving Langmuir set the foundation for the surface science we have today. Some of his major contributions included methods for generating high vacuums, thin film deposition, quantitative theories of adsorption and coining the name "plasma" for the ionized gaseous state. Of particular relevance for this article is the development by Langmuir's technician, Katherine Blodgett, of molecular assembled films at an air–water interface that could be transferred in a compressed, monolayer state to a solid surface [1]. The analogy between these structures and the lipid bilayer surrounding living cells was clear and the films themselves suggested possibilities as a basic research tool and for novel applications. In 1946, in a paper presented at a convocation at Princeton University and subsequently published in a volume entitled "Molecular Films, The Cyclotron and the New Biology," Langmuir speculated on the significance of these films for biology. The Langmuir–Blodgett invention made possible experimental systems later used by Zisman to explore fundamental aspects of organic interfaces [2] and early, exciting biomedical applications of these structures by Ringsdorf [3].

The modern characterization of organic surfaces, precursors to biomaterials, started with researchers such as Fowkes [5], Good [4] and Zisman [2] of the Naval Research labs. Zisman's investigations addressed both the preparation of surfaces and the quantitative contact angle measurements to approximate surface energy (critical surface tension). These ideas were expanded upon by

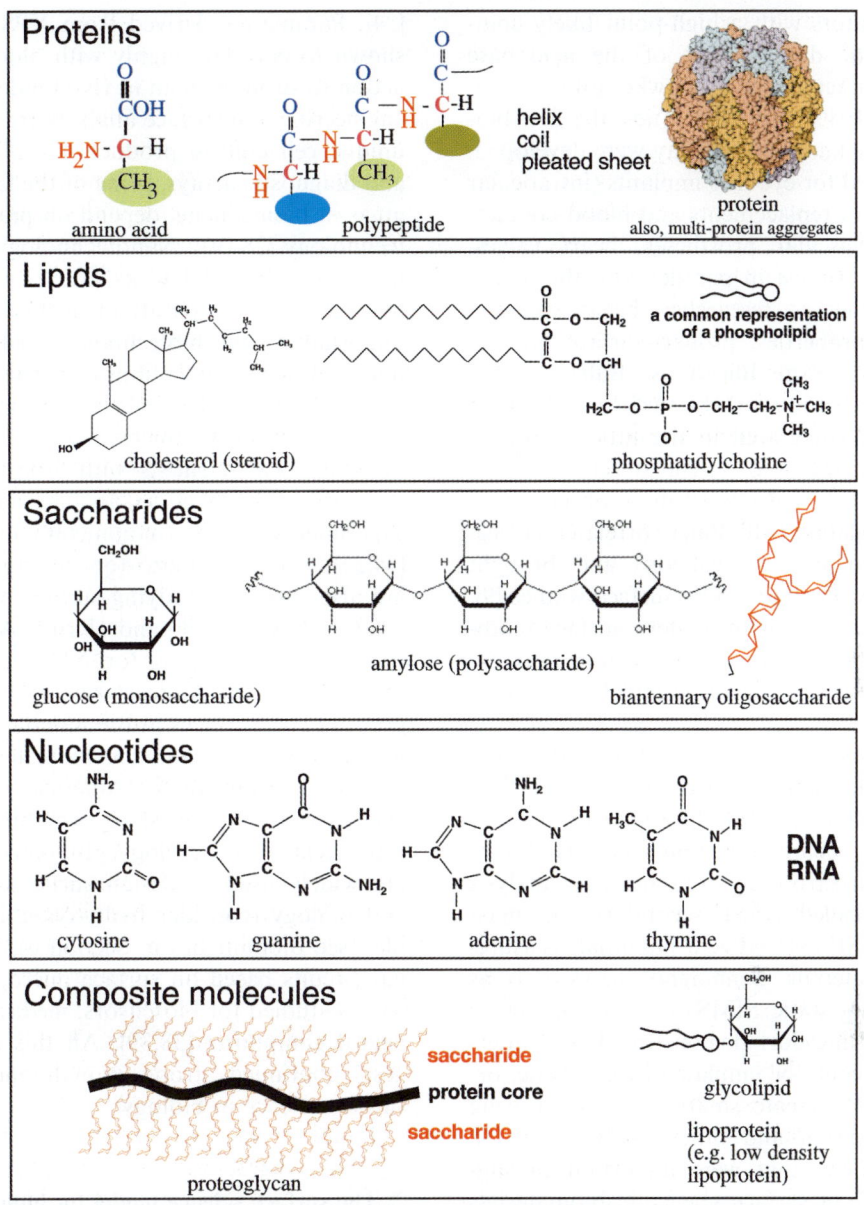

Fig. 2. The complexity of biological molecules of the type that can self-assemble into biological systems is illustrated in this figure. Semiconductor systems might involve the elements Si, O, Au, Cu, Al, Ga, As, B, P, N and perhaps a few other elements. Biology uses a substantial part of the periodic table but mostly C, O, N, P and S. The complexity arises from the many highly specific combinations with which these can be connected. For example, the 20 amino acids (complex molecules in themselves each with a symmetry-driven optical sense) making up the polypeptide chains can be ordered in almost infinite combinations. Specific arrangements yield secondary structures (random coil, helix, β pleated sheet), which then fold in a tertiary structures which then can aggregate into quaternary structures. A typical cell may have 15,000 proteins (about 2000 of those proteins are present in over 50,000 copies). These proteins are organized with sugars, nucleotides and lipids into the elegant supramolecular structure, the cell. And, the sugars, nucleotides and lipids themselves are each remarkably complex considering their structural possibilities and diversity. Protein (rubisco) image by David S. Goodsell, the Scripps Research Institute.

many investigators with a high-point likely coinciding with the development of the acid–base concept of contact angles by Fowkes [6].

In the late 1940s and early fifties, the first biomaterials as we know them today were developed. These were used for eye lens implants (intraocular lenses), hip joint replacements and blood contacting devices (vascular prostheses, heart valves, hemodialysis). Almost in parallel with the invention of these devices, researchers began studying their surface properties, protein–surface interactions and surface modifications. Bull explored protein adsorption to synthetic surfaces starting in the 1930s and going well in the fifties. Vroman made ellipsometric and visual observations of protein adsorption and related those observations to blood coagulation [7,8]. Baier correlated critical surface tension with bioreactions and brought surface infrared methods to biosurface studies [9]. Hoffman led the way with modern surface modification methods and demonstrated how surfaces could be engineered to give desired bioresponses [10].

Andrade [11], building on the foundations developed by Siegbahn in Sweden [12] and then Clark in the UK [13,14], brought a range of modern surface characterization methods including electron spectroscopy for chemical analysis (ESCA, also called XPS), secondary ion mass spectrometry (SIMS) and zeta potential measurement to biomaterials. Benninghoven, as early as 1977, was using static SIMS to study biological systems [15]. Ringsdorf demonstrated that interfacial biology could be emulated by synthetic, organic systems to create smart systems for drug delivery, biosensing and other applications [3]. Some key reviews that were important in suggesting the role of surface science in biomaterials are cited here [11,16–23].

1.4. The impact of biological surface science

Since the introduction of modern surface methods to study and modify materials and surfaces of biological interest, contemporary surface science has had considerable impact on biology and medicine. Surface criteria were used in a pass/fail test for commercial blood vessel substitutes [24]. Parameters derived from XPS spectra were shown to correlate highly with blood platelet reaction to surfaces in an in vivo model [25]. Surface engineering and surface analysis are used to create unique cell culture products [26,27]. Gene chips and diagnostic arrays, some of the fastest growing areas in biomedicine, depend on precision surface technology [28–30]. Nanotechnology exploits surface and interfacial ideas [31,32]. Column chromatography, a separations technology critically important to the biotechnology industry, is built upon surface modifications of particulate supports [33]. Commercial surface plasmon resonance (SPR) instruments are readily available that permit biological scientists with little formal understanding of surfaces to do precision adsorption experiments in the nanometer thickness range [34,35]. Similarly, piezo (quartz crystal) balances are also used for studying surface biointeractions [36,37]. The Food and Drug Administration (FDA) routinely calls for XPS data to qualify medical devices. At least two thriving technical groups (The Surfaces in Biomaterials Foundation, www.surfaces.org, and the Biomaterial Interfaces Technical Group of the American Vacuum Society, www.avs.org) exist to promote surface concepts related to biology. Biomimetic strategies are widely used to design surfaces for medicine and biology (consider hydroxyapatite, for example—see Biomimetics in Section 6). Living neuronal circuits based on surface micropatterning are widely studied for biosensors, medical devices and neural computing [38,39]. All this evidence suggests continued strong growth and impact for surface science in biology.

2. The surface science model for biology

In many research fields such as catalysis and microelectronics, the combination of well-defined model surfaces (e.g., metallic single crystals) with sophisticated surface analysis techniques has resulted in a detailed understanding of the role that surface structure and chemistry play. An important aspect of the success of the surface science model has been the ability to reduce a complex process (e.g., refining of petroleum crude oil) into a set of elementary steps, which then can be studied at a fundamental level with surface analysis techniques [40]. Then the results from the fundamental studies on each of these elementary steps are recombined to provide a description and understanding of the entire process. For example, petroleum reforming uses dual functional catalysts to increase product quality (e.g., gasoline octane) via dehydrogenation, hydrogenation, cyclization, and isomerization reactions. By using different single crystal surfaces of platinum with well-defined surface modifications, reactions such as the dehydrogenation of cyclohexane to benzene can be systematically studied (see Fig. 3, top) [41]. These studies have produced fundamental thermodynamic and kinetic information about the adsorption, desorption, and surface intermediate species involved in dehydrocyclization. This information can then be combined with corresponding information from the other reforming reactions to provide an overall description of the reforming process and the role surface structure and chemistry play in that process. The success of the surface science model for catalysis is due to the availability of detailed surface characterization and catalyst reactivity results that can be correlated.

The surface science model for biology and medicine is not as fully developed as the surface science model for catalysis. The ultimate goal for the biological surface science model would be to provide an understanding of how the surface chemistry and structure of a material can be used to control the biological reactivity of a cell interacting with that surface (see Fig. 3, bottom). To accomplish this goal requires understanding the cell reactivity and characterizing a complex, protein-covered surface. The increased complexity and highly interactive nature of the biological environment, relative to catalytic processes, has

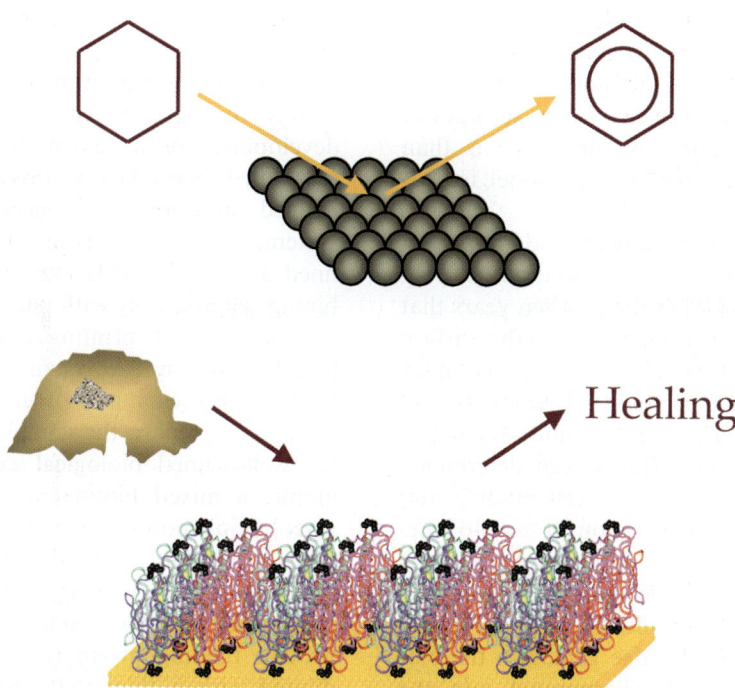

Fig. 3. The surface science model for catalysis showing the dehydrogenation of cyclohexane to benzene (top). To achieve the surface science model for biology an understanding of how the surface chemistry and molecular structure controls the biological reactivity of a cell interacting with a protein-covered surface to produce a normal healing reaction is needed (bottom).

made it difficult to isolate and study the reactivity of cells and biomolecules.

All cells communicate by the release and detection of signaling agents (cytokines) through a network made up of multiple and interactive signaling pathways [42]. The state of a cell (shape, structure, biological activity, etc.) depends on the signals it receives from its biological environment. For example, platelets normally circulate in the bloodstream in a passive state. Upon vascular injury a signaling cascade is initiated that activates the platelets for their role in healing the vascular injury. For this reason, whole blood must be treated (e.g., heparinized) when removed from the body to keep it from coagulating. Thus, the highly interactive nature of the biological environment makes it challenging to isolate a cell or biomolecule for meaningful, fundamental studies. Still, the reductionist approach to biology has been highly successful in isolating components and describing their operation. Similarly, to increase the understanding surface analysis can bring to biology, the complexity of the biological environment must be reduced where possible by employing well-defined model systems. However, to realize the surface science model for biology, the biological reactivity and the corresponding surface analysis studies will require the use of more complex systems than those used to develop the surface model for catalysis.

Significant advances have been made in molecular and cellular biology [43], material science [44], and surface analysis [45] in the past ten years that now make it possible to characterize the surface chemistry and structure of increasingly complex materials and study the biological reactivity and interactions of cells. The structure and function of many biological receptors have been determined along with their mechanism of cell binding and activation. With the combination of state-of-the-art patterning techniques and novel synthetic strategies it is now possible to prepare complex, organic surfaces with well-defined structures and chemistry [46]. The knowledge gained in the biological studies has been incorporated into the synthesis strategies for these surfaces. For example, the RGD (arginine–glycine–aspartic acid) amino acid sequence in proteins is known to play a key role in cell adhesion, so surfaces with short peptide sequences containing the RGD motif have been prepared [47]. RGD surfaces are useful and interesting models, but they only weakly emulate the multifunctionality of proteins such as fibronectin with over 2500 amino acids that contain this RGD sequence. Fibronectin comprises just part of the complex molecular structure of the natural attachment and activation substrate for cells. This example illustrates both the complexity of the problem and the surface related approaches to systematically examine aspects of this complexity (the reductionist approach).

To address the complexity of biology at surfaces, powerful surface tools are needed. Both hardware and software enhancements have been made in surface analysis techniques. The spatial resolution, energy resolution, mass resolution, sensitivity, etc. have improved. New data analysis techniques such as multivariate statistical analysis have been introduced. These combined advances in the biological, materials science, and surface analysis research fields provide the promise of realizing the surface science model for biology in the 21st century.

A major advance in materials fabrication technology during the last 10 years has been the development of self-assembly methods [48]. Self-assembled monolayers (SAMs) provide well-defined structures and chemistries that can be systematically varied (Fig. 4). Also, spatially defined arrays of SAMs can be prepared by combining self assembly with patterning methods such as microcontact printing and photolithography [49]. In addition, SAMs can be used to immobilize peptides, proteins, and other biomolecules to the surface to prepare the complex surfaces required for well-defined biological experiments. For example, a mixed biotinylated thiol/oligo-ethylene glycol thiol monolayer can be assembled onto a gold surface [50]. Since protein molecules are significantly larger in size than the thiol molecules in the SAM, the thiol molecule that contains the protein binding group (e.g., biotin) is typically diluted with a thiol that resists protein binding (e.g., oligo-ethylene glycol). Once the mixed biotinylated/oligo-ethylene glycol SAM is prepared, then the protein streptavidin is bound to it.

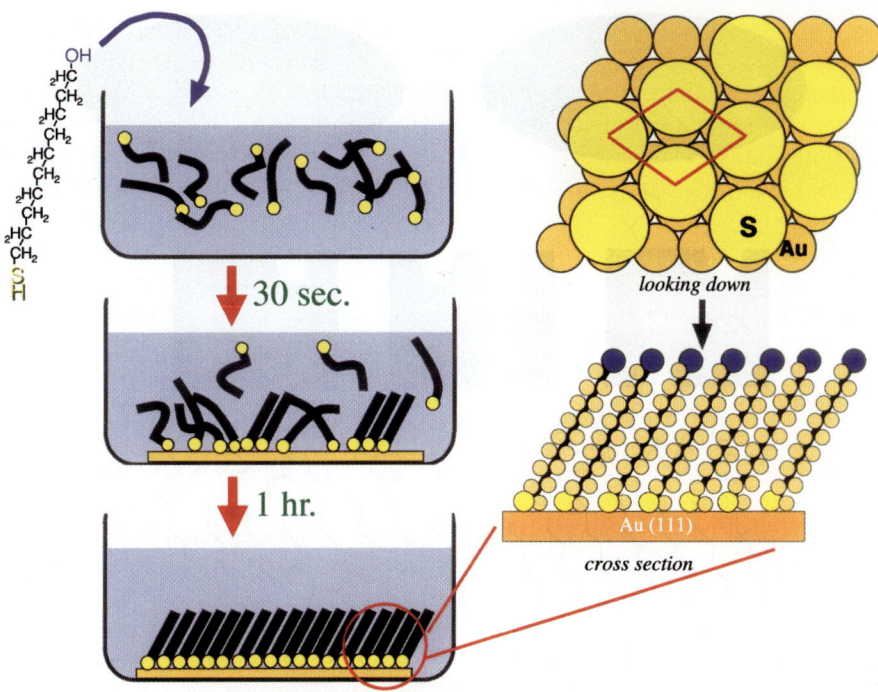

Fig. 4. The self-assembly process. An *n*-alkane thiol is added to an ethanol solution (0.001 M). A gold(1 1 1) surface is immersed in the solution and the self-assembled structure rapidly evolves. A properly assembled monolayer on gold(1 1 1) typically exhibits a $(\sqrt{3} \times \sqrt{3})R30°$ lattice.

Since streptavidin has binding pockets on opposite sides of the molecule, it can be used as a linker to bind other biotinylated molecules. A cartoon depicting the complex, multicomponent organic surfaces that can be prepared using the techniques described in this paragraph is shown in Fig. 5. Similarly, complex surfaces with immobilized peptides and proteins can be prepared using surface functionalization and polymer chemistry methods [51].

SAMs provide the model organic surfaces for use in developing the biological surface science model much the same way that metal single crystal surfaces earlier provided the model surfaces to develop the catalysis surface science model. However, organic surfaces are more fragile than the metallic single crystals since they can degrade when exposed to typical experimental and analysis conditions (elevated temperatures, X-rays, electron beams, etc.) [52]. Thus, the primary surface science techniques used for characterization of organic surfaces are XPS, static SIMS, and scanning probe microscopy (SPM) since, when used with care, they provide detailed information about the composition and molecular structure of organic surfaces without causing extensive degradation of the samples. Additional techniques such as Fourier transform infrared (FTIR), Raman, sum frequency generation (SFG), and high-resolution electron energy loss (HREELS) can be used to obtain surface vibrational spectra with relatively little, if any, damage.

Since it has been difficult to find model biomolecules that represent the reactivity of the entire complex biological environment, advancements in surface analysis techniques must be made so that increasing complex surfaces and processes can be fully characterized. A layer of adsorbed protein mediates the interaction of cells with a biomaterial when that biomaterial is placed in the biological environment. Also, future multilayered engineered surface-biology constructs may have a synthetic material, a hydrogel-like support, a SAM, a variety of tethering "hooks" and a number of oriented,

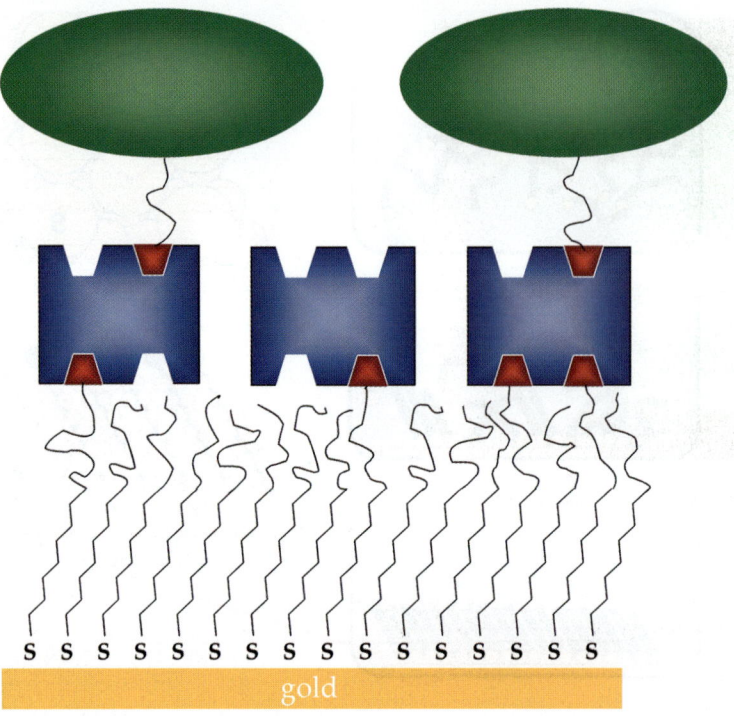

Fig. 5. A cartoon depicting how biomolecules (green) functionalized with biotin groups (red) can be selectively immobilized onto a gold surface using a streptavidin linker (blue) bound to a mixed biotinylated thiol/ethylene glycol thiol self-assembled monolayer.

organized biomolecules. Thus, it is especially important to develop surface analysis methods that can fully characterize these complex layers containing proteins and other biomolecules. An example using static SIMS with multivariate analysis to characterize adsorbed protein films is described in Section 5.

3. Surface chemical state imaging

3.1. Surface chemical state objectives

Prior to the last 5–10 years, spectroscopic analysis dominated biomedical surface science studies. The early biomaterials (silicones, polyurethanes, metals, etc.) typically were laterally homogeneous, so there was not a significant need for spatially resolved surface analysis methods. With the recent, rapid growth of methods for preparing spatially well-defined materials, the focus of biomedical surface science is now on high spatial resolution surface chemical state analysis in the x–y plane. The important objectives for developing spatially resolved spectroscopic analysis techniques are identification of all surface species present, identification of all chemically distinct regions, quantitative determination of the surface composition in each region, and optimization of spatial resolution. The driving forces for developing biomedical surface chemical state imaging techniques are addressed below.

A central goal of modern bioengineering is the development of biomaterial surfaces that direct the biological healing response [53]. These novel surfaces are envisioned to have a well-defined array of biorecognition sites designed to interact specifically with cells since many of the important functions of cells depend on the arrangement of molecules at their surfaces. Fig. 6 shows how the spatially resolved chemistry of a surface controls the shape and structure of a cultured rat bone cell [54]. New developments in surface analysis techniques are required to provide detailed surface

100 microns

Fig. 6. An F-actin stained rat bone cell cultured on a 75 × 75 μm square region of an amino-silane functionalized quartz surface. The area around this square is functionalized with a non-fouling acrylamide/ethylene glycol copolymer. The F-actin stain shows how the surface chemistry controls the shape and cytoskeleton structure of the cultured rat bone cell. See Ref. [54] for further details.

Table 1
Complementary surface chemical state imaging with XPS, static ToF SIMS, SPM, and NEXAFS

Technique	Current strength	Current research focus
XPS	Quantification	Spatial resolution
Static ToF SIMS	Molecular structure	Image analysis
SPM	Spatial resolution	Chemical specificity
NEXAFS	Chemical specificity	Spatial resolution

chemical state information at high spatial resolution for mapping out the presentation of these biorecognition sites. Improving the ability of surface analysis techniques to characterize and understand the composition, molecular structure, orientation, and spatial resolution of surface species will provide the biomedical research community with the tools and information needed to develop novel biomedical devices.

3.2. Surface chemical state imaging techniques

XPS, ToF SIMS, SPM and near edge X-ray absorption fine structure (NEXAFS) each has its own strengths and weaknesses with respect to generating surface chemical state information at high spatial resolution, but together they provide a powerful set of complementary techniques (see Table 1). For example, XPS and ToF SIMS can be used to improve the level of chemical state information obtainable with SPM, while SPM can be used to improve the spatial resolution obtainable with XPS and ToF SIMS.

3.3. Scanning probe microscopy

SPM is the biological surface science technique that provides the highest spatial resolution [55,56]. Depending on the sample being analyzed, individual atoms can be imaged with SPM. However, the inherent chemical specificity of SPM techniques is limited. For biomedical surface science this limitation can be overcome with proper functionalization of the probe tip. Immobilizing a biomolecule onto the tip in an active state allows surfaces to be interrogated at high spatial and temporal resolution while quantifying the binding force between the tip-immobilized biomolecule and the surface [57] (Fig. 7). The SPM techniques can map surface features and measure intermolecular forces at x, y, and z spatial resolutions of 1, 1, and 0.1 Å, respectively. Additionally, rapid advances in SPM technology have resulted in the development of several different modes of operation for image generation (magnetic, electric, electrochemical, thermal, viscoelastic, frictional, adhesive, etc.).

The increasing use of SPM techniques for investigating biological problems is primarily focused in the areas of structural identification and interfacial biophysical phenomena. SPM can provide molecular resolution images of proteins, DNA, lipids and carbohydrates [58]. Cellular structure also may be identified, and some recent work has examined dynamic responses of cells to environmental stimuli [59].

The quantification of biophysical phenomena (e.g., biotin–streptavidin binding) offers exciting opportunities for developing biorecognition SPM imaging methods [60]. Most of the initial research has focused on chemical specificity, studying the

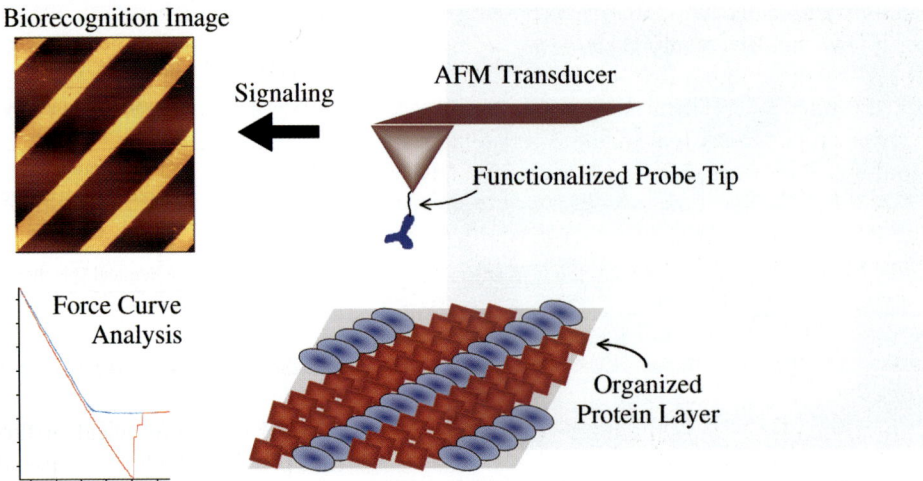

Fig. 7. A cartoon showing elements involved in biorecognition SPM imaging. A probe tip is functionalized to specifically interact with a protein on the sample surface. From analysis of the force curves taken as the tip is scanned across the sample an image of the patterned protein surface is generated.

interaction of different chemical groups on a well-defined surface with a functionalized SPM tip [61]. SPM studies typically use force curve spectroscopy, lateral force imaging, or phase shift imaging to interrogate the tip–sample binding phenomena [62].

The force curve spectroscopic method measures the interaction force at discrete points in the image. Lateral or friction force is a scanning technique that generates an image from differential lateral twisting of the cantilever. The magnetic mode is one of most recently developed phase shift imaging methods [63]. It is especially well-suited for biorecognition imaging since it can be used in liquids and the cantilever oscillation frequency can be tuned to a value appropriate for the biological binding event being examined.

The preparation of functionalized tips is the key to successful biorecognition with SPM. A significant amount of research activity is focused on the preparation and characterization of tips functionalized with well-defined chemical and biological species [64]. For biomolecules, the goal is to immobilize them in a well-defined orientation while maintaining their full biological activity and selectivity. One strategy for maintaining the activity of a biomolecule is to tether it to the tip with a poly(ethylene glycol) (PEG) spacer [65]. The PEG tether allows the biomolecule sufficient mobility so it can selectively bind to surface biorecognition sites. It is also important that the density of immobilized biomolecules be low enough so only a single binding event is detected for each localized force curve measurement. Due to their small size, it is challenging to directly characterize functionalized SPM tips.

The most commonly used SPM method for biological recognition imaging has been to acquire an array of force curves. The biological affinity information is contained in the retraction portion of the force curve [60]. Unfortunately none of methods used to date to extract the biological affinity force from the total adhesion force provide unambiguous, quantitative results. Thus, development of an imaging mode that is based on measurement of quantitative chemical and biological recognition forces would represent a significant advancement in SPM biorecognition capabilities.

3.4. *X-ray photoelectron spectroscopy and near X-ray absorption fine structure*

In contrast to SPM, the spatial resolution of surface science techniques using X-rays (XPS, NEXAFS, etc.) is limited. This is because of the difficulties associated with focusing X-rays. Until recently the smallest sample area that could be analyzed with a laboratory XPS system was 150×150 µm^2 [66]. Through a combination of improvements in X-ray focusing and lens/analyzer technology, laboratory XPS systems are now available that do real time imaging at spatial resolutions less than 10 µm. Although this spatial resolution is still orders of magnitude higher than the spatial resolutions obtained with SPM, XPS has a significant advantage in terms of its ability to quantify the surface composition of a sample. The recent construction of low-emittance synchrotrons such as the Advanced Light Source at the Lawrence Berkeley Laboratory has resulted in significant improvements in the spatial resolution obtainable with XPS and NEXAFS. Currently spatial resolutions in the 20–40 nm range are achievable using zone plate focusing elements. With further improvements in the microfabrication methods used to make the zone plates, even higher spatial resolutions should be achievable.

The theory of XPS and its application for identifying and quantifying surface chemical species is well developed [67]. Thus, the frontier for biomedical XPS is to improve its spatial resolution. However, a major challenge will be to improve the spatial resolution without introducing significant X-ray induced sample degradation. For large area analysis of organic and biological materials with monochromatized XPS, sample degradation is typically not a concern. However, as the X-ray beam is focused into increasingly smaller areas, the X-ray brightness (photons per unit area) increases and as does the potential for sample degradation. Thus, when changing from low spatial resolution analysis to high spatial resolution analysis, XPS can shift from a "non-destructive" to a "destructive" technique. To increase the spatial resolution of XPS while maintaining non-destructive analysis conditions will require advancements in methods for acquiring images with lower X-ray doses (rastering of the X-ray beam, improved sensitivity of the analyzer and detector, etc.). This need is especially critical for the high X-ray brightness achievable in micro-XPS and NEXAFS experiments done using synchrotron radiation. Fragile organic and biological samples can be completely destroyed under those analysis conditions.

3.5. Static time-of-flight secondary ion mass spectrometry

The spatial resolution of AES and ToF SIMS falls between that of SPM and laboratory XPS instruments. Most biological samples are readily degraded by the high-energy electron source used in AES, so AES is not widely used in biomedical surface science studies. Static ToF SIMS generates a mass spectrum from the outer 1–2 nm of sample, providing detailed information about the molecular structure of organic and biological materials [68]. With liquid metal sources, ToF SIMS images can be acquired at spatial resolutions down to 0.1 µm, which is sufficient for cellular resolution (1–100 µm). The mass resolution obtainable at a given spatial resolution also must be considered. The mass resolution of a Cs$^+$ source focused to 5 µm is degraded to the point that peaks 1 amu apart are barely resolved in the low mass range. This results in a loss of information for samples that have more than one peak at a given nominal mass. The Ga$^+$ source can be operated at full mass resolution ($M/\Delta M > 8000$ for conducting samples) at spot sizes down to 1 µm. This combination of high-spatial resolution and high-mass resolution is essential for imaging complex biological samples. Thus, static ToF SIMS has the capability to produce detailed molecular structural information at high spatial resolutions, making it a valuable technique for biomedical surface analysis [69]. However, many challenges need to be addressed before the full power of imaging ToF SIMS can be realized. These challenges include (1) analyzing large data sets, (2) images with low signal-to-noise ratios, (3) chemical species identification typically requires the use of several peaks and (4) distinguishing topography and chemistry effects.

With modern ToF SIMS instrumentation, a vast amount of data can be collected in a relatively short period of time. A mass spectrum can be acquired for each pixel in a 256×256 static ToF SIMS image (Fig. 8). Thus a total of 65,536 mass spectra, each containing several hundred

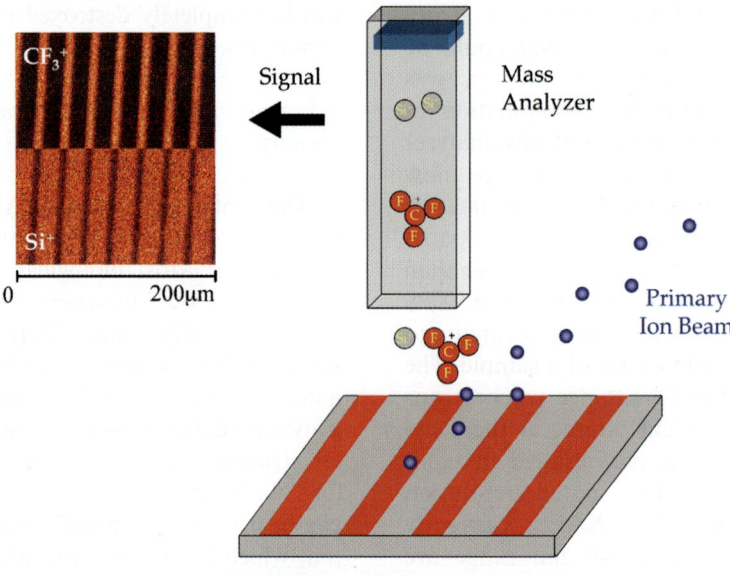

Fig. 8. A cartoon of the static ToF SIMS imaging process. A primary ion beam is scanned across the surface, which results in the ejection of secondary ions (Si^+, CF_3^+, etc.). The secondary ions are then mass analyzed to generate an image of the patterned fluorocarbon stripes on a silicon surface.

secondary ion peaks, needs to be analyzed for each image. Using the traditional ToF SIMS approach of examining images from selected individual mass peaks or the average spectrum from a selected subsection of the image can become time consuming. Even more importantly, a large portion of the acquired data is ignored and all the information present in a static ToF SIMS image is not used. Thus, improved image processing methods that use all of the data present in a static ToF SIMS image are needed to ensure all the surface chemical species are identified and quantified [69]. By examining all of the data, a set of peaks or combinations of peaks can be identified that increase the signal-to-noise ratio, improve the contrast between the chemical constituents, enhance the distinction of topographical features from chemical features, and make it easier to obtain the pure component spectra. Recognizing patterns and relationships in a set of hundreds or thousands of measured variables is a formidable task for any researcher using traditional data reduction methods. Thus, image processing methods need to be developed for ToF SIMS that permit a more efficient use of all the data in an image. Some of these methods are described below.

3.6. Image analysis methods

The analysis of static ToF SIMS images can be addressed in a three-step process. First, the raw data is denoised. Images acquired in the static mode (i.e., low ion dose) at high spatial resolution typically have low signal-to-noise ratios. Many peaks in the ToF SIMS spectra may only have a few counts per pixel. Thus, noise reduction methods that allow weak signals to be identified are important to use in the first step of imaging processing. Methods such as wavelet, median, and boxcar filtering are typically used to denoise images.

The second step is to identify the chemical species present in the sample. If one knows, a priori, what components are present in the sample this step is straightforward. This is usually not the case, so a method for extracting this information from the spectra is required. The fact that multiple peaks in the spectra can be associated with each chemical species adds complexity to this identification. Multivariate analysis techniques such as principle component analysis (PCA) can be use to identify the chemical species present and to compress the amount of data to be analyzed [70]. It also provides information about which combination of peaks in the static ToF SIMS spectrum originate from the same chemical component. This is important since static ToF SIMS imaging is typically limited by low signal-to-noise. The counts per pixel in the image can be increased significantly by combining the counts from several spectral peaks. However, selecting the wrong combination of peaks will decrease the contrast between chemical components in the image. Thus, correct and complete identification of the chemical species present along with their characteristic mass spectrum is essential to successfully completing the third step, image construction. If each chemical component has one unique mass fragment with strong intensity then the image reconstruction is straightforward. The relative intensities of the unique mass fragments show directly how the chemical species are distributed spatially on the surface (Fig. 8). When this is not the case, then images can be constructed directly from the PCA results using the appropriate combinations of peaks to identify the spatial locations of the various chemical components. Image segmentation algorithms such as region growing can also be used for image construction. These methods are particularly useful for images with a large number of surface phases [71].

4. Biocompatibility, biomaterials and molecular biorecognition surfaces

Medical implant materials, loosely referred to as biomaterials, have played a pivotal role in bringing surface concepts to biology. The ability of these materials to save human lives and the significant economic implications of medical devices have spurred many avenues of research, including surface-biology models. This section will clarify some of the concepts surrounding modern biomaterials, especially with reference to surfaces.

A word central to biomaterials, that distinguishes them from other materials, is biocompatibility. However, biocompatibility is poorly defined. Some properties that have been suggested to correlate with biocompatibility include surface energy, negative charges, hydrogels, heparin, titanium, phosphatidyl choline, polysulfones, etc. Many of the attempts to correlate materials properties with biocompatibility invoke surface properties. Yet, to this day, there are no clear rules that can be used to design a material for biocompatibility—good evidence that we do not yet understand biocompatibility. What is biocompatibility and what route might we take exploiting surfaces to obtain a precise definition of biocompatibility?

Millions of medical devices are implanted into humans each year with reasonable levels of success (Table 2). The FDA and other regulatory agencies "stamp" our medical devices as biocompatible. So, why is this word poorly defined? Consider the following two ideas. First, smooth materials that do not leach biologically reactive substances will heal in the body in a manner now considered biocompatible. Are all non-leaching materials equally biocompatible irrespective of surface properties? Second, the body reacts similarly to nearly all materials that we call biocompatible and walls them off in an avascular, tough, collagenous bag, roughly 50–200 μm thick (Fig. 9). This reaction is

Table 2
Medical implants used in the United States

Device	Number/year	Biomaterial
Intraocular lens	2,700,000	PMMA
Contact lens	30,000,000	Silicone acrylate
Vascular graft	250,000	PTFE, PET
Hip and knee prostheses	500,000	Titanium, PE
Cathether	200,000,000	Silicone, Teflon
Heart valve	80,000	Treated pig valve
Stent (cardiovascular)	>1,000,000	Stainless steel
Breast implant	192,000	Silicone
Dental implant	300,000	Titanium
Pacemaker	130,000	Polyurethane
Renal dialyzer	16,000,000	Cellulose
Left ventricular assist devices	>100,000[a]	Polyurethane

[a] Since inception.

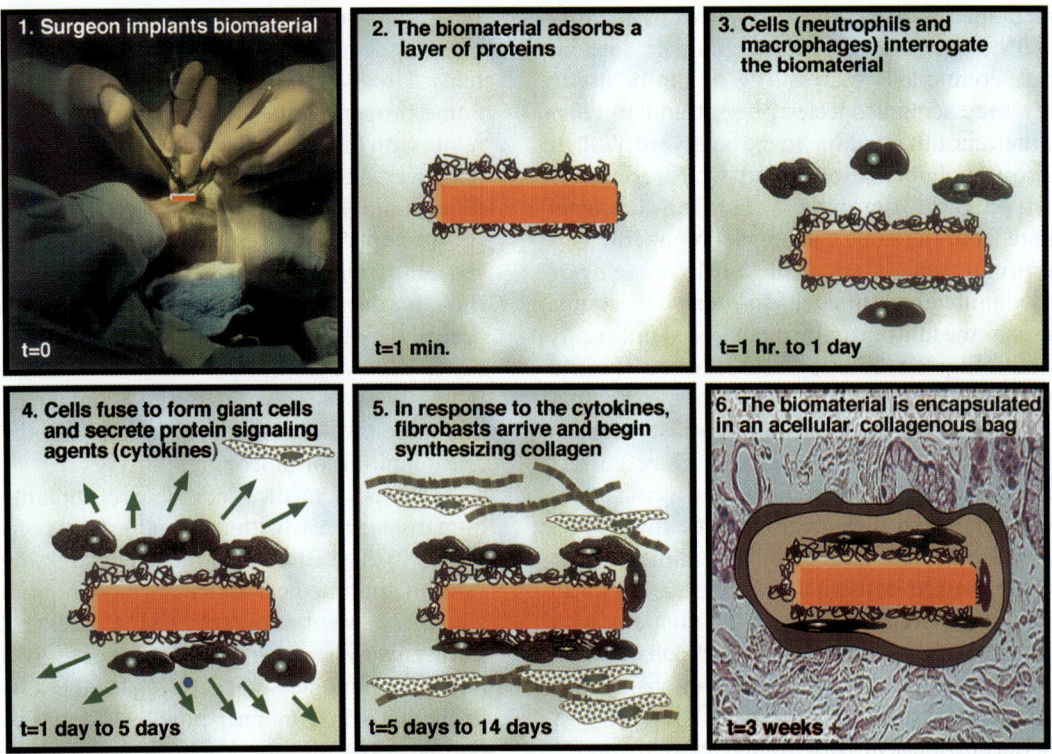

Fig. 9. The foreign body reaction is the normal reaction of a higher organism to an implanted synthetic material and is schematically illustrated here. (1) A surgeon implants a biomaterial in a surgical site (an injury). (2) Quickly, the implant adsorbs a layer of proteins, the normal process for a solid surface in biological fluids. (3) Cells (neutrophils and then macrophages) interrogate and attack the "invader," i.e., the biomaterial. (4) When the macrophages find they cannot digest the implant, they fuse into giant cells to engulf the object. However, it is too large to completely ingest. The giant cells send out chemical messengers (cytokines) to call in other cells. (5) Fibroblast cells arrive and begin synthesizing collagen. (6) The end stage of the reaction has the implant completely encased in an acellular, avascular collagen bag. There are macrophages between the collagen sac and the implant.

referred to as the foreign body reaction. Surprisingly, the accepted regulatory definition of biocompatibility revolves around this reaction of the body to rid itself of "biocompatible" biomaterials.

What are the concerns with today's biomaterials and how they heal? Uncontrolled biological encapsulation directly confounds the performance of many implanted devices. Consider, for example, implant electrodes, drug delivery systems, and breast implants. The presence of this capsule seriously degrades their performance by preventing intimate contact between device and tissue. The reaction associated with this foreign body response (long term, low level inflammation and macrophage activation) may also inhibit the luminal healing of vascular grafts, trigger capsular opacification found with intraocular lenses, lead to the extrusion of percutaneous devices, exacerbate device calcification, induce contact lens discomfort and generally lead to complications and less than desirable outcomes associated with today's medical devices. In contrast, our body has an excellent capacity to heal wounds and injuries with healthy, vascularized tissue. Could this normal healing be wrong? Why do "biocompatible" implants shut off normal wound healing? We already know how to get devices to heal with a foreign body capsule. So, what's next? Can we go beyond this aberrant healing? These questions and comments require clarification and justification.

Given a list of 10 common materials used in medicine (for example, gold, polyurethane, silicone rubber, polytetrafluoroethylene (PTFE), polyethylene (PE), poly(methyl methacrylate) (PMMA), poly(2-hydroxyethyl methacrylate) (PHEMA), poly(ethylene terephthalate) (PET), titanium, alumina) materials that are hydrophilic, hydrophobic, hard, soft, polymeric, ceramic and metallic are represented. Yet, after one month implantation in mammals, they are all found to heal essentially identically. On the other hand, each material will be found, in vitro, to adsorb different proteins, and to show substantially different cell attachment and cell growth behavior. The origin of this striking difference between in vivo and in vitro represents one of the pervasive problems in biomaterials science.

The commonality among the ten materials in the previous paragraph is that they adsorb a complex, non-specific layer of proteins. Each will have a different protein mixture at its surface, but all the materials will quickly acquire a layer that contains many proteins (possibly comprised of 200 or more proteins) in many states of orientation and denaturation [72]. Nature *never* uses such non-specific layers—nature's use of proteins as signaling agents comes from one (or a few) specific proteins in fixed conformations and orientations so they optimally deliver signals. A hypothesis has been developed suggesting that the body views this non-physiologic proteinaceous layer as something with which it has no experience and reacts to it as an unrecognized foreign invader that must be walled off [53,73]. One of the authors (BDR) refers to these non-specific layers as "the enemy." For progress to be made, we must go beyond this ill-controlled reaction, i.e., defeat the enemy. Hence, surfaces must be developed that control the conformation and orientation of proteins with precision so that the body will specifically recognize them.

In a normal wound, the macrophage cell responsible for "orchestrating" wound healing is activated. In the presence of an uncomplicated wound, the macrophage turns on the pathways leading to normal healing by first cleaning up the wound site and then secreting the appropriate cytokine messenger molecules. These soluble messengers activate processes in the cell types needed for healing (fibroblast, keratinocyte, osteoblast, etc.).

The surfaces of today's biomaterials, if present in the wound site, turn this normal healing process off. The macrophages adhere to the biomaterial. They do not recognize it and spread on its surface as they try to phagocytose it. They cannot digest or engulf this large mass, so, to increase their effectiveness, they fuse together to form multinucleated giant cells. Of course, these cells still cannot engulf a macroscopic medical device. The giant cells signal to the body that there is a large foreign body to be walled off. The fibroblasts arrive and generate the collagen capsule, most likely guided by the macrophages. Although there is not complete consensus on how the body reacts to implanted synthetic materials, most experts would agree with the general outline described above.

There are a number of steps that must be taken to realize "biomaterials that heal." First, a serious investment must be made in the study of the basic biology of normal wound healing, in contrast to wound healing with a biomaterial present. This basic study will tell us what molecular and cellular pathways to turn on and what pathways to turn off. Second, the non-specific adsorption of proteins and other biomolecules must be inhibited. Finally, the surfaces of biomaterials should be synthesized to present to the body the same signaling groups as the surface of a clean, fresh wound.

This hypothesis on healing and the foreign body reaction opens many opportunities for surface scientists. Indeed, the basic biology studies are best left to the biological researchers. However, once the biological discoveries are made, the ability to inhibit non-specific interactive events on surfaces and the intellectual challenges of delivering the specific biological signals opens exciting frontiers for surface scientists.

Many strategies to inhibit non-specific protein adsorption (non-fouling surfaces) have been developed [74–82] (Table 3). How resistant to protein pickup can such surfaces be made? Why are they resistant to protein adsorption? How long can they remain resistant to protein fouling? Can they be functionalized with organic groups permitting the immobilization of active biomolecules on a bland background? These questions drive research in this area. A number of recent issues of *Journal of Biomaterials Science: Polymer Edition* (Volume 11,

Table 3
Strategies to achieve protein-resistant (non-fouling) surfaces

Surface strategy	Comments	Reference
PEG[a]	Effective but dependent on chain density at the surface; damaged by oxidants	[74]
PEG-like surfaces by plasma deposition	Applicable for the treatment of many substrates and geometries; highly effective	[76]
PEG oligomers in self-assembled monolayers	Highly effective; applicable for precision molecularly engineered surfaces; durability to elevated temperature is low	[80]
PEG-containing surfactants adsorbed to the surface	A simple method for achieving non-fouling surfaces; durability may be low and high surface densities are hard to reach	[78]
PEG blocks in other polymers coated on the surface	May provide a relatively low density of surface PEG groups	[77]
Saccharides	Nature's route to non-fouling surfaces; some successes but much territory remains to explored	[148,149]
Choline headgroups (phosphatidyl choline)	Has shown good success in many applications	[79]
Hydrogen bond acceptors	Possibly, this principle imparts non-fouling properties to PEG surfaces; this is leading to new discoveries of surface functional groups for non-fouling	[134]
Adsorbed protein layer	A pre-adsorbed protein layer resists further adsorption of proteins; this approach, long used by biologists, is easy to implement but of low durability	[150]
Hydrogels, in general	PEG is in this class; many other hydrogels have shown non-fouling behavior	[151]

[a] Also called poly(ethylene oxide) (PEO).

2000) have focused on these points. As an example, surfaces made by the RF-plasma deposition of tetraethyleneglycol dimethylether (tetraglyme) have been explored in our group. These surfaces have been characterized by modern surface techniques [83] revealing a crosslinked PEG-like structure and have been shown to have extremely low protein pickup (Fig. 10).

Within the University of Washington Engineered Biomaterials (UWEB) program (a National Science Foundation Engineering Research Center), key molecules that turn on and off normal healing have been explored [84,85]. Can these molecules retain their effectiveness when bound to surfaces? What strategies might be used to immobilize them in a precise manner?

Surfaces that interact with precision with biological systems will be complex—multicomponent, multilayer, orientated, patterned. Given the complexity of the molecular structures that make up the individual biomolecules comprising these surfaces, fabrication and characterization of such surfaces will push the skills of surface scientists to their limits.

In the future, tissue engineering (coupled with truly biocompatible scaffolds), stem cell technology, control of regeneration and the knowledge of the human genome will completely change the way we work with biomaterials and medical devices [86–91]. But, before these revolutionary technologies replace today's biomaterials, we still probably have 30 years during which biomaterials as we know them today will be of increasing importance. Thus, there is strong impetus to evolve the surface strategies needed to control biological interactions.

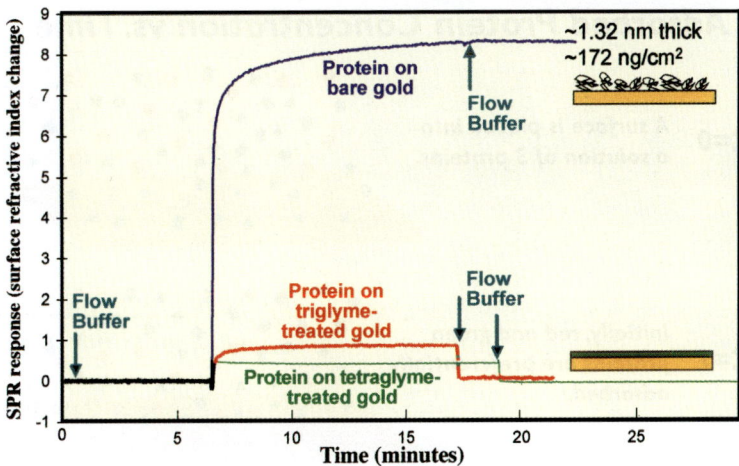

Fig. 10. A SPR experiment on protein (bovine serum albumin, 1 mg/ml) adsorption to three surfaces. Protein adsorption is rapid and high on a bare gold SPR element (blue). When buffer is flowed through the system at about 18 min, essentially no protein desorbs from the gold. If the gold is treated in an RF plasma under the vapors of triglyme (red) or tetraglyme (green), a treatment that deposits a tightly bound poly(ethylene glycol)-like layer, little protein is noted to adsorb. Protein that does adsorb is washed away when the buffer flow commences. (Data of Mar et al. see Ref. [76] for further details).

5. Characterization of complex biological surfaces

5.1. Adsorbed protein films

The adsorption of proteins onto a biomaterial surface from the surrounding fluid phase is rapid, with the surface properties of the biomaterial determining the type, amount, and conformation of the adsorbed proteins [92]. The composition of the adsorbed protein layer (i.e., the type and concentration of the proteins present in the adsorbed film) can differ from the fluid phase composition and can change with time adsorbed. This is shown schematically in Fig. 11 using three different proteins (red, green and blue). Initially the surface concentration of "red" and "green" proteins is higher than their solution concentration. With time the red and green proteins are displaced from the surface by the "blue" protein. In addition to the time-dependent compositional changes, each absorbed protein can undergo conformational and orientational changes, as shown schematically in Fig. 12. Upon adsorption, a protein can retain the conformation or structure it has in the biological environment or it may conformationally change in response to local environments. The nature of the surface strongly influences the composition and recognizability of the adsorbed protein layer, which in turn affects the subsequent cellular interactions. Thus, to understand the biological response to a material, especially in vitro, one must fully understand the nature of the adsorbed protein film that forms on that material.

The extremely high analytical sensitivity of static ToF SIMS, its sampling depth of 1–2 nm, and the molecular information it provides about the chemistry of the adsorbed protein film and the substrate offer the potential to use static ToF SIMS to gain a detailed understanding of the composition, conformation and orientation of adsorbed proteins. For an intact protein adsorbed in its native conformation, the static ToF SIMS spectrum will represent only the amino acids present on the surface of the protein since most proteins have dimensions between 4 and 10 nm, which is significantly larger than the static ToF SIMS sampling depth. For proteins with a heterogeneous distribution of amino acids across the three dimensional domain of the protein molecule, the relative intensities of the amino acid fragments detected in the SIMS spectrum will be sensitive to the orientation of the adsorbed protein and its degree of conformational alteration [93]. As a protein adjusts to the surface and changes its

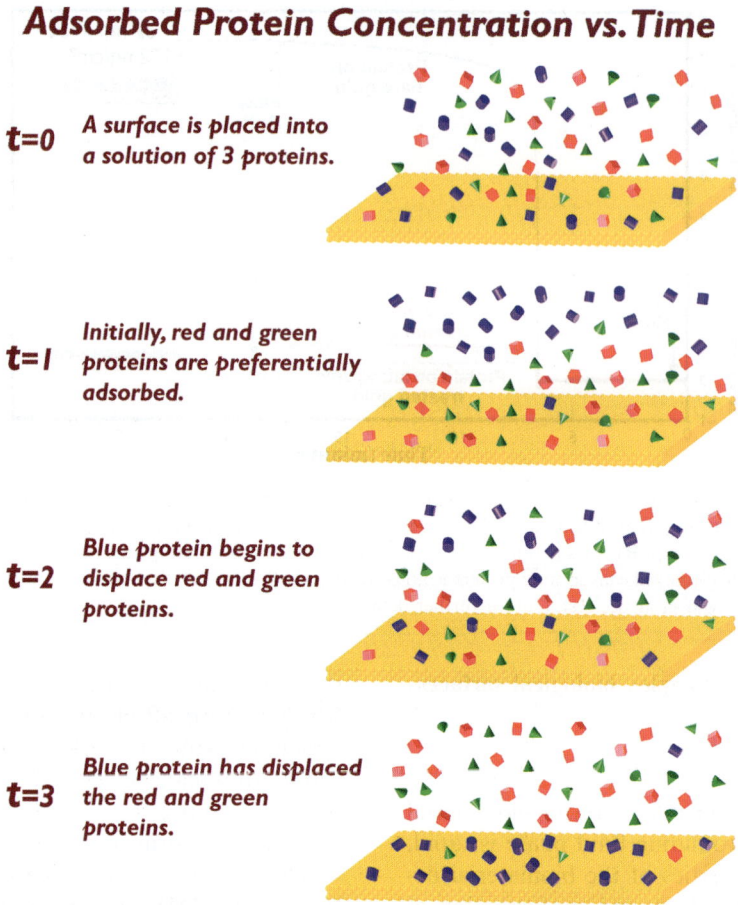

Fig. 11. A surface placed in a protein mixture will be covered with a layer of adsorbed proteins in a matter of seconds. The concentration of proteins in the adsorbed film is typically different from their solution concentration and can change with time. In this example, initially the "red" and "green" proteins are preferentially adsorbed. With increasing adsorption time (minutes to hours) the red and green proteins are displaced by the "blue" protein.

conformation or orientation, new regions of the protein with different amino acid compositions will be exposed to the static SIMS beam. This means that static ToF SIMS has the potential to provide a microscopic, chemical glimpse of changes in the protein conformation (see Fig. 13). Furthermore, during the denaturation process, the ratio of bare substrate to protein may change as the protein unfolds and spreads over the surface.

The use of static ToF SIMS for characterizing adsorbed protein films has shown the potential to probe protein conformation, assess surface coverage, measure protein concentration with extreme analytical sensitivity, map protein distributions and identify different proteins (see below). In addition, it can measure contamination and analyze synthetic substrates and binding chemistries. Thus, static ToF SIMS can make important contributions to biomaterials development. In vitro, static ToF SIMS should play an important role in the development of cell culture surfaces, biosensors, protein and DNA diagnostic arrays, immunoassays, non-fouling surfaces and chromatographic supports. In vivo, static ToF SIMS will be used to characterize surfaces engineered with specifically immobilized signal molecules, measure uncontrolled fouling, and relate surface structures to blood interactions. An example of the power of

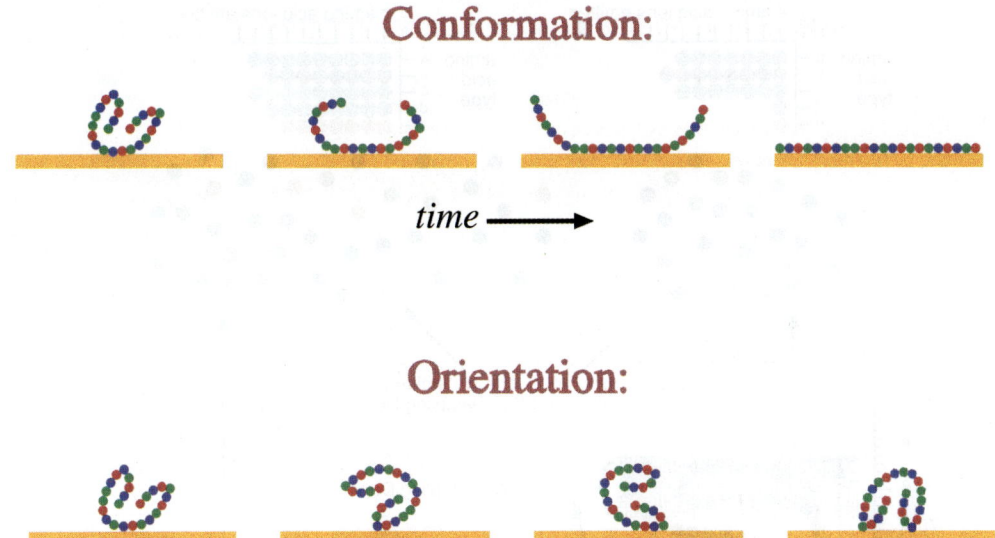

Fig. 12. The conformation and orientation of adsorbed proteins depend on adsorption conditions and surface properties. The top schematic shows a protein denaturing with increasing adsorption time. The bottom schematic shows a protein adsorbing to the surface in different orientations.

static ToF SIMS combined with multivariate analysis for analyzing the complex protein films formed on surfaces, so important for the many applications mentioned, will now be presented.

Static ToF SIMS spectra of adsorbed protein films are complex and contain peaks from all of the amino acids. Since the same 20 amino acids are present in all proteins, it is the relative intensities of the amino acid fragments in the ToF SIMS spectra that contain the information needed to identify adsorbed proteins. To do this identification efficiently for a large number of proteins requires the use of a pattern recognition method such as PCA [70]. PCA is an unsupervised classification method that can be used to reduce the dimensionality of the complex static ToF SIMS protein spectra, making it straightforward to identify proteins and also to develop an understanding why static ToF SIMS can make this identification. For example, static ToF SIMS with PCA has successfully identified 13 different proteins adsorbed onto mica from pure protein solutions [94]. It was determined that PCA was distinguishing the adsorbed proteins based on their different bulk amino acid compositions. Static SIMS with PCA is also able to distinguish albumin proteins from different species (human, cow, pig, chicken, and turkey). In addition to identification of protein type, the combination of PCA and static ToF SIMS can be used to quantify the amount of adsorbed protein present in mixed films [94]. To date, it has been successfully applied to binary protein mixtures. A challenge for the future will be to determine how much the complexity of the protein mixture can be increased (i.e., how many proteins can be present in the adsorbed film) while still retaining the ability to identify and quantify all of the proteins. It has been shown that using a PCA model built from the pure protein data set, it is possible to draw qualitative conclusions about the protein composition of films adsorbed from 1% bovine plasma [94]. It was found that initially the protein film is enriched in fibrinogen. With increasing adsorption time the fibrinogen concentration of the film decreases.

In addition to determining the composition of an adsorbed protein layer, it would also be desirable to determine the accessibility and location of binding sites on a protein molecule. For example, the aggregation of fibrin proceeds by the staggered overlap of the fibrin molecules, where the central portion of one molecule interacts with the terminal

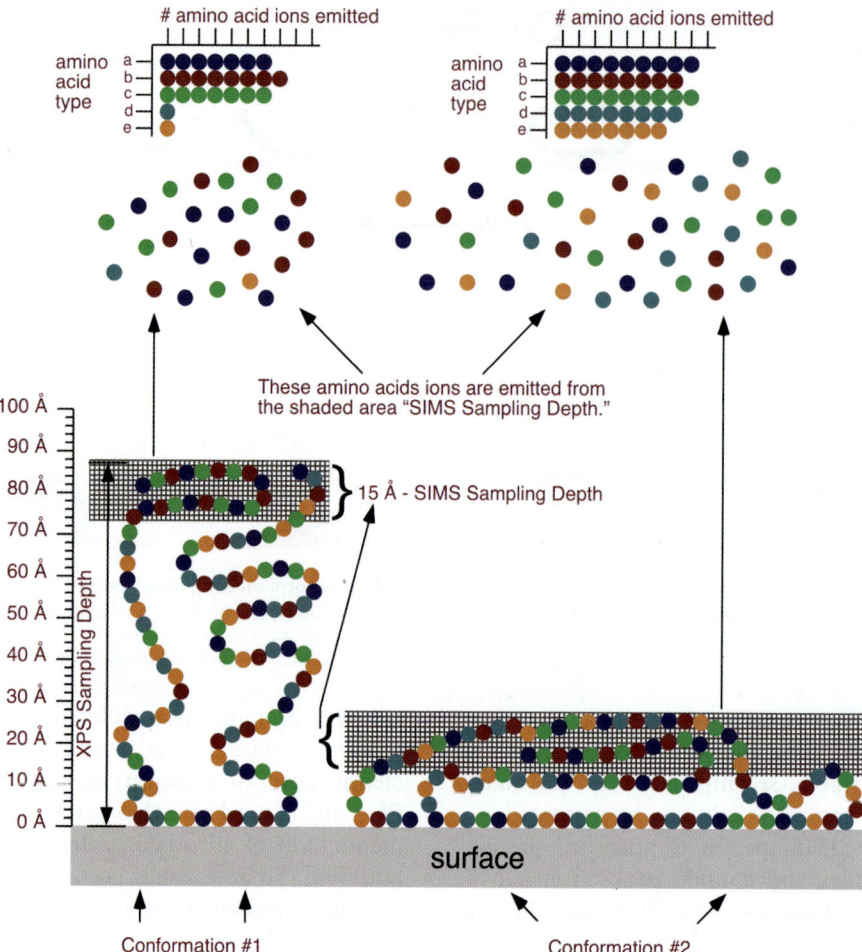

Fig. 13. A schematic showing the sensitivity of static ToF SIMS to protein conformation. Both proteins have the same bulk amino acid composition, but since the amino acid composition is not uniform across the protein molecule, conformation #1 will produce a different intensity pattern of static ToF SIMS fragments than conformation #2.

regions of neighboring molecules. Peptides are known to bind to these polymerization sites in fibrin [95], so peptides containing ^{13}C, ^{15}N, or F labels can be used as a probe for fibrin polymerization sites. Labeled peptides are necessary to generate unique SIMS amino acid fragments from the peptides that can be distinguished from the large number of unlabeled amino acid fragments that would originate from the protein. This approach should be generally applicable to protein binding reactions since the fibrin polymerization process is prototypical of the binding between specific, localized peptidic sites on two protein molecules.

5.2. Hydrated surfaces

Hydrated surfaces are challenging for UHV surface science, but normal and relevant for biology. Methods applicable to the solid–aqueous interface, and therefore relevant to biological surface studies include contact angle measurements, second harmonic generation (SFG), Brewster angle microscopy, X-ray reflectivity, SPM, frozen-hydrated UHV techniques (XPS, SIMS, etc.), environmental scanning electron microscopy, attenuated total reflectance IR (subtract out the water signal), ellipsometry, SPR and neutron scattering. Most of these methods work in aqueous environments without appreciable interference from the water. Unique aspects associated with UHV biological surface science are addressed below.

UHV XPS studies of wet surfaces have been performed using frozen, hydrated specimens for electrochemistry [96] and biomaterials [97,98]. In biological and biomaterial XPS studies, a specific protocol has been adopted based on the following rationale. The wet sample, when frozen under atmospheric pressure conditions, will adsorb a layer of adventitious contaminants. After freezing, if the temperature is kept below −160 °C, water will not sublime from the sample. Once the sample temperature is raised above −100 °C, water will quickly sublime from the frozen ice layer "blasting off" the contaminant layer. The objective is to stop the ice sublimation when an ice layer approximately 1–2 nm thick remains so XPS can see through the ice layer to analyze the frozen, hydrated sample. To stop the ice sublimation and stabilize the thin ice overlayer, the sample temperature is rapidly lowered to −160 °C. In this way, a hydrated, frozen specimen can be studied under UHV.

The specific University of Washington protocol for UHV frozen-hydrated analysis is as follows. First, the preparation chamber stage and the analytical chamber stage in the XPS instrument are cooled to −160 °C while the specimen to be examined is hydrated by placing a drop of water on its surface. Second, the hydrated sample is cooled below −100 °C in the preparation chamber under an atmosphere of dry, purified nitrogen gas. Third, the preparation chamber is pumped down to UHV while keeping the sample temperature well below −100 °C. Fourth, the sample temperature is now raised to typically −90 °C (above the sublimation point of ice but below the polymer glass transition temperature). The sample temperature is quickly lowered below −120 °C to stop the sublimation process when a 2 nm ice overlayer remains. Fifth, the sample is quickly moved from the preparation chamber to the pre-chilled analytical chamber stage at −160 °C and XPS data is acquired, frequently at multiple photoemission take-off angles. Finally, XPS data from the "dehydrated" sample is obtained by returning the sample to the preparation chamber and bringing it to room temperature to liberate the remaining frozen water. The dehydrated sample is then returned to the analytical chamber at −160 °C for analysis. A Surface Science Instruments (SSI) X-probe XPS instrument was adapted for these cryogenic studies by adding liquid nitrogen cooled stages to the entry/preparation chamber and to the analytical chamber. Further details have been described elsewhere [98].

An example illustrating the powerful influence of the sample environment on the surface chemistry is taken from the literature [97,98]. In these studies a silicone elastomer film covalently grafted with the hydrophilic polymer PHEMA was examined both in the frozen, hydrated state and in the dehydrated state. In the frozen, hydrated state, a spectrum resembling PHEMA was observed. In the dehydrated state, the spectrum had the characteristics of silicone rubber. The data suggest that when the sample is wet, the hydrophilic PHEMA chains dominate the surface to reduce the interfacial energy between solid and water. When in air, the hydrophobic silicone chains dominate the specimen, reducing interfacial energy. Thus, for a sample of this type with polar and non-polar moieties, and considerable polymer chain mobility, it must be studied dry and wet to fully characterize the surface.

5.3. Photon in/photon out techniques

Although the standard electron and ion based surface science techniques (XPS, static SIMS, etc.) provide excellent surface sensitivity and detailed information about composition and molecular structure, they cannot be used to directly examine a biomaterial surface in an aqueous environment. The strong interactions of electrons and ions with materials provide surface sensitivity. However, these strong interactions also require that these probes be used in an UHV environment since low-energy electrons and ions cannot penetrate the aqueous-based biological environment surrounding an implanted biomaterial.

The previous section discussed how the aqueous environment can be simulated using a frozen, hydrated method, allowing the structure of the hydrated surface to be examined under UHV conditions. Although this method provides information about the composition and structure of hydrated surfaces, it would be preferable to directly examine the surface of biomaterials in aqueous environments. Then, changes in surface composition and structure could be monitored in real time.

One method for investigating the properties of the solid (biomaterial)–liquid (biological environment) interface is to use photon in/photon out techniques since photons have longer mean free paths than low-energy electrons and ions. However, in many cases the surface sensitivity of the technique is compromised. For example, fluorescent X-rays can be detected instead of electrons in NEXAFS experiments. This allows NEXAFS experiments to be done at significantly higher pressures, but also results in the sampling depth increasing from 5 nm (electrons) to 200 nm (soft X-rays) [99]. If the surface species under investigation has a unique spectroscopic feature (e.g., adsorption peak) then the surface specificity can be regained since the signal from the surface species only contributes to that particular spectroscopic feature.

Another method for regaining surface specificity is to use total external reflection experimental conditions (grazing incident and reflection angles) [100]. This can be done for photon energies from the IR to hard X-ray regions, but requires large, flat surfaces. However, if the bulk of the material contains similar species as the surface and a suitably flat surface cannot be prepared, then most photon in/photon out techniques will not provide the needed surface sensitivity to characterize the hydrated biomaterial surface.

One optical technique that can directly examine the structure of the solid–liquid interface is SFG [101]. SFG is a second order non-linear optical process where a pulsed visible laser beam (ω_{vis}) is overlapped with a tunable, pulsed IR laser beam (ω_{ir}) to generate a signal at the sum frequency (ω_{sum}). Emission of the sum frequency light does not occur for the bulk phase of most materials. However, the symmetry of the bulk phase is broken at a surface or interfaces, so surface species do produce sum frequency signals. Thus, SFG provides both surface sensitivity and direct interrogation of the structure at the solid–liquid interface.

The SFG intensity plotted versus the frequency of the IR laser provides a vibrational spectrum of the surface species with submonolayer sensitivity. By using different polarization conditions (e.g., s-polarized sum frequency, s-polarized visible, and p-polarized IR) the orientation of surface species can be determined. Typically, it takes a few minutes to acquire a SFG spectrum over a few hundred wave number range, so by monitoring the changes in a given spectral region (e.g., C–H stretch), time dependent changes in the surface structure can be monitored with a resolution of minutes. The time-dependent restructuring of polymeric materials (migration of end groups, copolymer components, etc.) that occur upon hydration and dehydration have been determined with SFG [102].

Future opportunities for improving the SFG technique include decreasing the time resolution from minutes to seconds and expanding the vibration frequency range of the SFG spectrum. Typically SFG spectra are acquired in the range from 2500 to 3600 cm^{-1}. Expanding that range to cover 1000–4000 cm^{-1} would significantly increase the number of vibrational bands that can be accessed with SFG, thereby expanding the molecular structure information that can be determined with SFG. For example, expanding the range below 2000 cm^{-1} would allow the structure of adsorbed proteins to be determined using the amide bands. The major limitation of SFG to date has been the fact that the concentration of the surface species detected by SFG cannot be quantified. However, SFG used in combination with frozen-hydrated XPS and static ToF SIMS may provide a method for overcoming this limitation.

There are other photon in/photon out techniques that can provide information about the thickness (ellipsometry) and amount (SPR) of a deposited species, but the chemical composition and molecular structure information provided by these techniques is limited. Both ellipsometry and SPR detect changes in the refractive index, which only provides indirect information about chemical and biological species. Thus, these techniques need to be used in combination with other techniques such as XPS and static SIMS which can provide direct information about surface composition and structure.

6. Future directions of biomedical surface science

Opportunities are plentiful in biomedical surface science. Start-up businesses based on biological surface science abound. New discoveries in biology beg for application. Surface analysis instrumentation steadily improves. Self-assembly ideas have made the routine synthesis of nearly perfect organic surfaces a reality. What paths might one take to get involved at the cutting edge of this exciting field?

A few themes will dominate the future of biomedical surface science: learning from biology, biomimetics, precision immobilization, self assembly, nanofabrication, control of non-specific reactions, and smart surfaces. We will briefly address each of these.

6.1. Biological knowledge

Biological systems use surfaces with precision. The analysis of these surfaces and their emulation (biomimetics) represents an important avenue to improved, functional biosurfaces for basic research and technological applications. Examples will be given based on cell membranes, biominerals, and the extracellular matrix.

Fig. 14 highlights the basic components of a cell membrane. This precision supramolecular structure is much more than a protective barrier. It inhibits non-specific interactions, recognizes specific ligands, performs enzymatic (catalytic) functions, pumps ions and reconfigures its topography and geometry. Surface researchers are now attempting to model cell surfaces with supported lipid bilayer membranes [103–107]. These surface assemblies lightly tether a lipid bilayer film onto a hydrated, hydrogel support conferring mobility and order. Such synthetic structures are roughly analogous to the elegant supramolecular structure pictured in Fig. 14.

Nature has its own equivalent of "single crystal surface science." Inorganic crystalline structures are used throughout biology. Complex calcium phosphate crystalline phases comprise roughly two thirds of bone. Calcium carbonate crystals form into otolith structures in the inner ear responsible for balance. Mollusks synthesize calcium carbonate-based nacre (mother of pearl). Diatoms extract silicon from the ocean to make their silica skeletons. Interestingly, these crystals rarely exist in isolation. Most commonly, they are closely complexed with organic components. In fact, there are many aspects of this process suggestive of epitaxy and molecular recognition.

For much of the history of biology, the material between cells, the extracellular matrix, was thought to be uninteresting, amorphous filler. In recent years, it has been found to be organized into precise structures that control many functions central to life. Along with a mechanical function, the ECM has important roles in cell adhesion, migration, proliferation and differentiation. As an example of the elegant reactions that occur on ECM, consider protein binding to hyaluronic acid. This polysaccharide, which can have molecular weights up to 25,000,000 and forms an amorphous gel, binds specifically to a 30 amino acid peptide sequence with a molecular weight of approximately 3000. This is an elegant surface interaction engineered over approximately 7.5 nm of linear surface.

Based upon these examples of complex biological surfaces, we can begin to perceive the challenge. It is fourfold: (1) What is the biological significance of these structures (functionality and mechanism)? This is the challenge of biological discovery. (2) How can we characterize surface structures given their oriented multilayer organization and remarkable chemical complexity, mobility and fragility? (3) How can we bring the surface science model of structure and reactivity into congruence with the biological model of surface functionality? (4) How can we emulate nature's elegance and create biomimetic surfaces?

6.2. Biomimetics

Since nature uses surfaces with precision, there is clear justification in emulating nature's methods

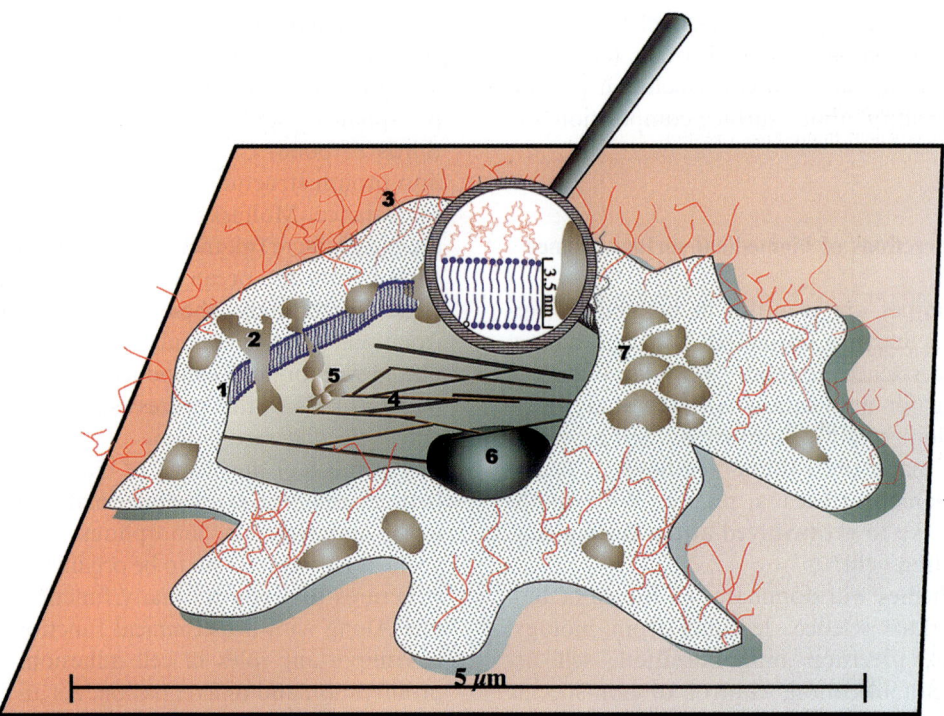

Fig. 14. The basic components of a cell membrane: (1) the lipid bilayer membrane, (2) an embedded protein through the lipid bilayer membrane, (3) saccharide chains on the surface of the cell, (4) the cell cytoskeleton, linked to a transmembrane protein through (5) a series of smaller proteins (talin, vincullin, etc), (6) the cell nucleus, (7) proteins exposed at the cell surface.

to make more functional, controllable surfaces. A variety of approaches have been developed to copy nature's way of doing things. Many of these address pharmacological strategies, which have only limited relevance to this article. The efforts directed to materials that are stronger, lighter, tougher, cheaper, cleaner to manufacture and biodegradable largely focus on bulk properties. However, these are generally multiphase, composite materials. The nature of the interface plays a key role in the ultimate properties. For example, nacre, the material lining an abalone shell, is a mechanically tough substance. The calcium carbonate that makes up 95% of nacre is a brittle mineral. The 5% protein dispersed between mineral "bricks" is sufficient to confer strength and toughness to the composite [108]. The interfacial interactions between the protein component and the carbonate platelets are central to explaining the substantial enhancement in mechanical properties. Other examples of biomimetic strategies involving surfaces and interfaces include synthetic mussel adhesive [109], nanopits for protein recognition [110], supported lipid bilayer membranes to mimic natural cell membranes (discussed in the previous section) and hydroxyapatite surfaces for bone incorporation [111]. Further examples of biomimetic surface approaches are presented in each of the following sections.

6.3. Precision immobilization

A surface skill used in nature with elegance and precision is the ability to order and organize complex molecules at surfaces. Precision immobilization typically aims to copy nature's way of organizing molecules and is thus an example of a biomimetic strategy. Such ordering permits biomolecular signals to be delivered with great precision. Biomolecules used in precision immo-

Table 4
Methods to immobilize active biomolecules to surfaces

Method	Comment
Non-specific adsorption	Little control is afforded of protein orientation or activity; low durability
Non-specific covalent immobilization	Little control is afforded of protein orientation or activity
Immobilization on an antibody surface	Using monoclonal antibodies, protein orientation can be controlled
HIS tags	Histidine sequences (HIS tags) can be specifically engineered into proteins for attachment and orientation
Biotin/streptavidin	A flexible strategy for tightly fixing protein to surfaces; in vivo biological reaction to streptavidin is a concern
Crystallized protein layers	Useful only in limited cases
Immobilization to a template structure	An evolving field
Biomimetic recognition sites	An evolving field
Incorporation in a supported bilayer	As an emulation of the cell membrane this has the possibility to stabilize fragile proteins
Nucleotide conjugation/hybridization	Many possibilities are being explored
Electrostatic	A non-specific approach to immobilizing proteins when the protein has an isoelectric point higher or lower than seven and a surface has a positive or negative charge

bilization strategies include proteins, lipids, polypeptides, polynucleotides and polysaccharides.

Possibilities for surface immobilization of biomolecules are suggested in Table 4. Two books that overview this field are cited here [112,113]. The degree of specificity (precision) in immobilization ranges from relatively low to extremely high. The characteristics of successful precision engineered biorecognition surfaces include the presence of one receptor site, an appropriate surface density of those sites, controlled orientation of the sites, some molecular mobility to enhance "docking," and stability (of the biomolecular conformation and the film integrity). The ability to inhibit non-specific reactions (in particular, protein adsorption) is also essential to succeed at emulating nature's surface signal delivery strategy. The ultimate goals in surface immobilization of biomolecules are high activity (functionality) and specificity.

6.4. Self-assembly

Self-assembly can be used to create bulk materials or ordered surfaces. Molecular mobility allows complex, often flexible molecules sufficient time and geometric opportunity to associate and assume their lowest energy state, the crystal. Self-assembled surfaces prepared from organic molecules are valuable as models (analogous to the metallic single crystal models) for exploring biological-like, hierarchical systems and also present possibilities to nanofabricate real surfaces for technological applications. SAMS were discussed earlier in this review. However, they offer so many opportunities that further elaboration is useful.

Three- and two-dimensional self-assembly are well known. Two-dimensional self-assembly is more relevant for this surface article. The commonality in systems that show 2-D self-assembly are a relatively simple molecular geometry, a driving force for interacting with a smooth surface, a lateral interactive force between molecules to stabilize them in the crystal and a chemical group that forms the outermost surface of these systems. The scientific roots of this area of study lie in the Langmuir–Blodgett deposition of lipids and surfactants [114]. The discovery in 1983 of thiol assembly on gold [115] (see Fig. 4) launched an explosion of publications and new discoveries. Self-assembly of complex organic structures on solid surfaces has been observed for phospholipids [116], silanes [117], n-alkyl thiols [115,118,119], porphyrins [120], nucleotide bases [121], hydrocarbons [122], proteins [123,124], and many other organic structures. An example from nature of the self-assembly of proteins on the surface of a bacterium is shown in Fig. 15 (also see Ref. [123]). Recent developments in the self-assembly of multilayer systems of polyions is also interesting in this context [125].

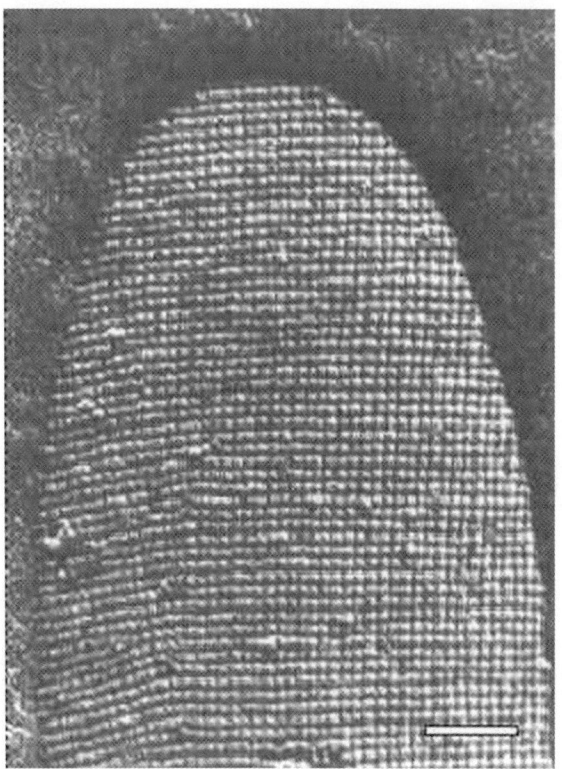

Fig. 15. A TEM image of a bacterial cell with an ordered S-layer protein array with square lattice symmetry. Bar = 100 nm (used with permission of Prof. U. Sleytr, http://www.boku.ac.at/zuf/sxl9.htm).

6.5. Nanofabrication

Nanofabrication, nanotechnology and nanoscience represent growth areas in research and development. However, nature has been using nanofabrication ideas since the beginnings of biological evolution. For example, the coordinated workings of receptor and enzyme mechanisms in the cell membrane suggest clever, nanoscale machines [126]. Topics already addressed here such as precision immobilization, self-assembly and biomimetics are all examples of nanofabrication. Thinking at the nanoscale does open interesting possibilities for synthesis of new structures and the interface of biology and materials. Particularly interesting work has been done using dendrimers (tree and star-like polymers) [127], rotoxanes [128] and DNA [129,130] as building components for creating nanostructures. Tools that the surface scientist can apply for nanofabrication are the atomic force microscope [32,131] and electron beam lithography.

6.6. Control of non-specific reactions

The subject of protein resistant (non-fouling) surfaces has been addressed earlier in this article. How such surfaces function, how to optimize them and how to use them in technology remains an important frontier area in biointerfaces. Non-fouling surfaces will be important for biomaterials, biosensors, medical diagnostics, heat exchangers, ship bottoms and food processing plants. Improved biofouling-resistant surfaces will become a reality when we have an enhanced understanding of why such surfaces function as they do. Theories focussing on polymer chain excluded volume, polymer chain entropy, water structure and hydrogen bond acceptors have been put forth [132–136].

6.7. Smart surfaces

The term "smart material" has been used to describe materials that go through rapid phase transitions with a small change in environmental conditions leading to a useful physical property change. Many examples of such materials have been produced—this is a branch of chemistry demonstrating creativity and promise [137–144]. When coupled to enzymes or other specific biological receptors, the smart materials are made smarter still [144,145]. Surfaces that undergo rapid shifts in surface properties with small external changes open many possibilities and present a frontier area for biosurface science [146,147].

7. Conclusions

Biomedical surface science will contribute to both fundamental knowledge and technology. From the basic science perspective, surface science models will assist in the understanding how nature does its work–chemical and biological models are insufficient to provide this knowledge without invoking the special properties that characterize the surface state. Technology will clearly benefit from a surface science model of biology in more functional medical implants, improved biosensors, chip-based neuronal computing, precision medical diagnostics, barnacle-resistant ships, finer biomolecule separations, biosynthetic production of plastics and chemicals and interfacially engineered biocomposite materials. Areas such as nanotechnology and smart materials hold untold promise and will certainly partner with biomedical surface science to implement novel technologies and new discoveries. Advances in surface analysis methodology will be central to these developments in basic science and technology.

Acknowledgements

The authors acknowledge generous support from the National ESCA and Surface Analysis Center for Biomedical Problems (NESAC/Bio, funded by NIH grant RR-01296 from the National Center for Research Resources) and the UWEB Engineering Research Center (funded by grant EEC-9529161 from the NSF) during the preparation of this manuscript and for some of the experimental work described herein. We also would like to thank our colleagues for many stimulating discussions over the years regarding the development of biomedical surface science.

References

[1] K.B. Blodgett, Films built by depositing successive monomolecular layers on a solid surface, J. Am. Chem. Soc. 57 (1935) 1007–1022.
[2] W.A. Zisman, Relation of the equilibrium contact angle to liquid and solid constitution, in: F.M. Fowkes (Ed.), Contact Angle, Wettability and Adhesion, ACS Advances in Chemistry Series, vol. 43, American Chemical Society, Washington, DC, 1964, pp. 1–51.
[3] H. Ringsdorf, B. Schlarb, J. Venzmer, Molecular architecture and function of polymeric oriented systems: models for the study of organization, surface recognition, and dynamics of biomembranes, Angew. Chem. Int. Ed. Engl. 27 (1988) 113–158.
[4] L.A. Girafalco, R.J. Good, A theory for the estimation of surface and interfacial energies. I. Derivation and application to interfacial tension, J. Phys. Chem. 61 (1957) 904–909.
[5] F.M. Fowkes, Additivity of intermolecular forces at interfaces. I. Determination of the contribution to surface and interfacial tensions of dispersion forces in various liquids, J. Phys. Chem. 67 (1963) 2538–2541.
[6] F.M. Fowkes, Donor–acceptor interactions at interfaces, in: L.H. Lee (Ed.), Adhesion and Adsorption of Polymers Part A, Plenum Publishing, New York, 1980, pp. 43–52.
[7] L. Vroman, Effects of hydrophobic surfaces upon blood coagulation, Thrombos. Diathes. Haemorrh. 10 (1964) 455–493.
[8] L. Vroman, A.L. Adams, Findings with the recording ellipsometer suggesting rapid exchange of specific plasma proteins at liquid/solid interfaces, Surf. Sci. 16 (1969) 438–446.
[9] R.E. Baier, A.E. Meyer, J.R. Natiella, R.R. Natiella, J.M. Carter, Surface properties determine bioadhesive outcomes: methods and results, J. Biomed. Mater. Res. 18 (1984) 337–355.
[10] A.S. Hoffman, Principles governing biomolecule interactions at foreign interfaces, J. Biomed. Mater. Res. Symp. 5 (1974) 77–83.
[11] J.D. Andrade, Surface Chemistry and Physics, in: Surface and Interfacial Aspects of Biomedical Polymers, vol. 1, Plenum Press, New York, 1985.
[12] K. Siegbahn, Electron spectroscopy for atoms, molecules, and condensed matter, Science 217 (1982) 111–121.
[13] D.T. Clark, Some experimental and theoretical aspects of structure, bonding and reactivity of organic and polymeric systems as revealed by ESCA, Phys Scripta 16 (1977) 307–328.
[14] D.T. Clark, J. Peeling, L. Colling, An experimental and theoretical investigation of the core level spectra of a series of amino acids, dipeptides and polypeptides, Biochim. Biophys. Acta 453 (1976) 533–545.
[15] A. Benninghoven, W. Sichtermann, Secondary ion mass spectrometry: a new analytical technique for biologically important compounds, Organic Mass Spectrom 12 (1977) 595–597.
[16] B.D. Ratner, Surface characterization of materials for blood contact applications, ACS Adv. Chem. Ser. 199 (1982) 9–23.
[17] J.D. Andrade, Surface analysis of materials for medical devices and diagnostic products, MD and DI 22–33 (1980).
[18] B.D. Ratner, B.J. McElroy, Electron spectroscopy for chemical analysis: applications in the biomedical sciences, in: R.M. Gendreau (Ed.), Spectroscopy in the Biomedical Sciences, CRC Press, Boca Raton, Fl, 1986, pp. 107–140.
[19] B. Kasemo, J. Lausmaa, Surface science aspects on inorganic biomaterials, CRC Crit. Rev. Biocompat. 2 (1986) 335–380.
[20] B.D. Ratner, A.B. Johnston, T.J. Lenk, Surface properties of biomaterials, in: J.G. Webster (Ed.), Encyclopedia of Medical Devices and Instrumentation, vol. 1, Wiley, New York, 1988, pp. 366–381.

[21] B.D. Ratner, Characterization of biomaterial surfaces, Cardiovasc. Pathol. (Suppl 2) (1993) 87S–100S.
[22] R.E. Baier, A.E. Meyer, Biosurface chemistry for fun and profit, Chemtech. 16 (1986) 178–185.
[23] B.D. Ratner, S.C. Porter, Surfaces in biology and biomaterials: description and characterization, in: J.L. Brash, P.W. Wojciechowski (Eds.), Interfacial Phenomena and Bioproducts, Marcel Dekker, New York, 1996, pp. 57–83.
[24] R.I. Shapiro, F.B. Cerra, J. Hoffman, R. Baier, Surface chemical features and patency characteristics of chronic human umbilical vein arteriovenous fistulas, Surg. Forum 29 (1978) 229–231.
[25] S.R. Hanson, L.A. Harker, B.D. Ratner, A.S. Hoffman, In vivo evaluation of artificial surfaces with a nonhuman primate model of arterial thrombosis, J. Lab. Clin. Med. 95 (1980) 289–304.
[26] J.A. Chinn, T.A. Horbett, B.D. Ratner, Laboratory preparation of plasticware to support cell culture: surface modification by radio frequency glow discharge deposition of organic vapors, J. Tissue Cult. Meth. 16 (1994) 155–159.
[27] C.D. Tidwell, S.I. Ertel, B.D. Ratner, B. Tarasevich, S. Atre, D. Allara, Endothelial cell growth and protein adsorption on terminally functionalized, self assembled monolayers of alkanethiolates on gold, Langmuir 13 (1997) 3404–3413.
[28] M. Eggers, M. Hogan, R.K. Reich, J. Lamture, D. Ehrlich, M. Hollis, B. Kosicki, T. Powdrill, K. Beattie, S. Smith, R. Varma, R. Gangadharan, A. Mallik, B. Burke, D. Wallace, A microchip for quantitative detection of molecules utilizing luminescent and radioisotope reporter groups, BioTechniques 17 (1994) 516–525.
[29] K. Douglas, G. Devaud, N.A. Clark, Transfer of biologically derived nanometer-scale patterns to smooth substrates, Science 257 (1992) 642–644.
[30] A.C. Pease, D. Solas, E.J. Sullivan, M.T. Cronin, C.P. Holmes, S.P.A. Fodor, Light-generated oligonucleotide arrays for rapid DNA sequence analysis, Proc. Natl. Acad. Sci. USA 91 (1994) 5022–5026.
[31] K.E. Drexler, Molecular tip arrays for molecular imaging and nanofabrication, J. Vac. Sci. Technol. B 9 (1991) 1394–1397.
[32] T. Boland, E.E. Johnston, A. Huber, B.D. Ratner, Recognition and nanolithography with the atomic force microscope, in: B.D. Ratner, V.V. Tsukruk (Eds.), Scanning Probe Microscopy of Polymers, vol. 694, American Chemical Society, Washington, DC, 1998, pp. 342–350.
[33] J. Nawrocki, B. Buszewski, Influence of silica surface chemistry and structure on the properties, structure and coverage of alkyl-bonded phases for high-performance liquid chromatography, J. Chromatogr. 449 (1998) 1–24.
[34] J. Davies, Surface plasmon resonance—the technique and its applications to biomaterial processes, Nanobio. 3 (1994) 5–16.
[35] B. Liedberg, I. Lundstrom, E. Stenberg, Principles of biosensing with an extended coupling matrix and surface plasmon resonance, Sens. Actuators 11 (1993) 63–72.
[36] C. Fredriksson, S. Kihlman, M. Rodahl, B. Kasemo, The piezoelectric quartz crystal mass and dissipation sensor: A means of studying cell adhesion, Langmuir 14 (1998) 248–251.
[37] F. Patolsky, A. Lichtenstien, J. Willner, Amplified microgravimetric quartz-crystal-microbalance assay of DNA using oligonucleotide-functionalized liposomes or biotinylated liposomes, J. Am. Chem. Soc. 122 (2000) 418–419.
[38] J.J. Hickman, S.K. Bhatia, J.N. Quong, P. Shoen, D.A. Stenger, C.J. Pike, C.W. Cotman, Rational pattern design for in vitro cellular networks using surface photochemistry, J. Vac. Sci. Technol. A 12 (1994) 607–616.
[39] M. Matsuzawa, K. Umemura, D. Beyer, K. Sugioka, W. Knoll, Micropatterning of neurons using organic substrates in culture, Thin Solid Films 305 (1997) 74–79.
[40] G.A. Somorjai, Introduction to surface chemistry and catalysis, Wiley, New York, 1994.
[41] X. Su, Y.R. Shen, G.A. Somorjai, The surface chemistry of 1,3-cyclohexadiene and 1,4-cyclohexadiene on Pt(1 1 1) studied by surface vibrational spectroscopy with sum frequency generation, Chem. Phys. Lett. 280 (1997) 302–307.
[42] J.A. Hunt, P.J. McLaughlin, B.F. Flanagan, Techniques to investigate cellular and molecular interaction in the host response to implanted biomaterials, Biomaterials 18 (1997) 1449–1459.
[43] C.A. Parent, P.N. Devreotes, A cell's sense of direction, Science 284 (1999) 765–770.
[44] A.P. Alivisatos, P.F. Barbara, A.W. Castelman, J. Chang, D.A. Dixon, M.L. Klein, G.L. McLendon, J.S. Miller, M.A. Ratner, P.J. Rossky, S.I. Stupp, M.E. Thomsen, From molecules to materials: current trends and future directions, Adv. Mater. 10 (1998) 1297–1336.
[45] B.D. Ratner, Advances in the analysis of surfaces of biomedical interest, Surf. Interf. Anal. 23 (1995) 521–528.
[46] A.S. Blawas, W.M. Reichert, Protein patterning, Biomaterials 19 (1998) 595–609.
[47] M.C. Porté-Durrieu, C. Labrugére, F. Villars, F. Lefebvre, S. Dutoya, A. Guette, L. Bordenave, C. Baquey, Development of RGD peptides grafted onto silica surfaces: XPS characterization and human endothelial cell interactions, J. Biomed. Mater. Res. 46 (1999) 368–375.
[48] G.M. Whitesides, G.S. Ferguson, D. Allara, D. Scherson, L. Speaker, A. Ulman, Organized molecular assemblies, Crit. Rev. Surf. Chem. 3 (1993) 49–65.
[49] N.L. Jeon, R.G. Nuzzo, Y. Xia, M. Mrksich, G.M. Whitesides, Patterned self-assembled monolayers formed by microcontact printing direct selective metalization by chemical vapor deposition on planar and nonplanar substrates, Langmuir 11 (1995) 3024–3026.
[50] L. Häussling, H. Ringsdorf, F.-J. Schmitt, W. Knoll, Biotin-functionalized self-assembled monolayers on gold:

surface plasmon optical studies of specific recognition reactions, Langmuir 7 (1991) 1837–1840.

[51] L. Zhang, C.A. Booth, P. Stroeve, Phosphatidylserine/cholesterol bilayers supported on a polycation alkylthiol layer pair, J. Coll. Interf. Sci. 228 (2000) 82–89.

[52] B.D. Ratner, D.G. Castner, Advances in XPS instrumentation and methodology: Instrument evaluation and new techniques with special reference to biomedical studies, Coll. Surf. B: Biointerfaces 2 (1994) 333–346.

[53] B.D. Ratner, The engineering of biomaterials exhibiting recognition and specificity, J. Mol. Rec. 9 (1997) 617–625.

[54] C.H. Thomas, J.B. Lhoest, D.G. Castner, C.D. McFarland, K.E. Healy, Surface designed to control the projected area and shape of individual cells, J. Biomech. Eng. 121 (1999) 40–48.

[55] S. Kasas, N.H. Thomson, B.L. Smith, P.K. Hansma, J. Miklossy, H.G. Hansma, Biological applications of the AFM: from single molecules to organs, Int. J. Imaging Syst. Technol. 8 (1997) 151–161.

[56] C.F. Quate, Scanning probes as a lithography tool for nanostructures, Surf. Sci. 386 (1997) 259–264.

[57] E.-L. Florin, V.T. Moy, H.E. Gaub, Adhesion forces between individual ligand-receptor pairs, Science 264 (1994) 415–417.

[58] L.A. Bottomley, Scanning probe microscopy, Anal. Chem. 70 (1998) 425R–475R.

[59] C. Rotsch, M. Radmacher, Drug-induced changes of cytoskeletal structure and mechanics in fibroblasts: an atomic force microscopy study, Biophys. J. 78 (2000) 520–535.

[60] T. Boland, B.D. Ratner, Direct measurement by atomic force microscopy of hydrogen bonding in DNA nucleotide bases, Proc. Natl. Acad. Sci. USA 92 (1995) 5297–5301.

[61] C.D. Frisbie, L.F. Rozsnyai, A. Noy, M.S. Wrighton, C.M. Lieber, Functional group imaging by chemical force microscopy, Science 265 (1994) 2071–2074.

[62] B.D. Ratner, V. Tsukruk, Scanning probe microscopy of polymers, vol. 694, American Chemical Society, Washington, DC, 1998.

[63] W. Han, S.M. Lindsay, T. Jing, A magnetically driven oscillating probe microscope for operation in liquids, Appl. Phys. Lett. 78 (1996) 4111–4113.

[64] G.R. Harper, S.S. Davis, M.C. Davies, M.E. Norman, T.F. Tadros, D.C. Taylor, M.P. Irving, J.A. Waters, J.F. Watts, Influence of surface coverage with poly (ethylene oxide) on attachment of sterically stabilized microspheres to rat kupffer cells in vitro, Biomaterials 16 (1995) 427–439.

[65] W. Baumgartner, P. Hinterdorfer, H. Schindler, Data analysis of interaction forces measured with atomic force microscope, Ultramicroscopy 82 (2000) 85–95.

[66] R.L. Chaney, Recent developments in spatially resolved ESCA, Surf. Interface Anal. 10 (1987) 36–47.

[67] B.D. Ratner, D.G. Castner, Electron spectroscopy for chemical analysis, in: J.C. Vickerman (Ed.), Surface Analysis—The Principal Techniques, Wiley, Chichester, UK, 1997, pp. 43–98.

[68] A. Benninghoven, Surface analysis by secondary ion mass spectrometry (SIMS), Surf. Sci. 299/300 (1994) 246–260.

[69] A. Willse, B.J. Tyler, Multivariate methods for TOF-SIMS imaging, in: G. Gillen, R. Lareau, J. Bennett, F. Stevie (Eds.), SIMS XI, Wiley, Chichester, UK, 1998, pp. 843–846.

[70] S. Wold, K. Esbensen, P. Geladi, Principal component analysis, Chemometrics and Intelligent Lab. Syst. 2 (1987) 37–52.

[71] S.D. Bohmig, B.M. Reichl, Segmentation and scatter diagram analysis of scanning Auger images, a critical comparison of results, Fresenius' J. Anal. Chem. 346 (1993) 223–226.

[72] T.A. Horbett, Principles underlying the role of adsorbed plasma proteins in blood interactions with foreign materials, Cardiovasc. Pathol. 2 (1993) 137S–148S.

[73] B.D. Ratner, Molecular design strategies for biomaterials that heal, Macromol. Symp. 130 (1998) 327–335.

[74] E.W. Merrill, Poly(ethylene oxide) blood contact, in: J.M. Harris (Ed.), Poly(ethylene glycol) chemistry: Biotechnical and biomedical applications, Plenum Press, New York, 1992, pp. 199–220.

[75] N.B. Holland, Y. Qiu, M. Ruegsegger, R.E. Marchant, Biomimetic engineering of non-adhesive glycocalyx-like surfaces using oligosaccharide surfactant polymers, Nature 392 (1998) 799–801.

[76] M.N. Mar, B.D. Ratner, S.S. Yee, An intrinsically protein-resistant surface plasmon resonance biosensor based upon a RF-plasma-deposited thin film, Sensors and Actuators B 54 (1999) 125–131.

[77] G.L. Kenausis, J. Voros, D.L. Elbert, N. Huang, R. Hofer, L. Ruiz-Taylor, M. Textor, J.A. Hubbell, N.D. Spencer, Poly(L-lysine)-g-poly(ethylene glycol) layers on metal oxide surfaces: attachment mechanism and effects of polymer architecture on resistance to protein adsorption, J. Phys. Chem. B 104 (2000) 3298–3309.

[78] J.-T. Li, K.D. Caldwell, Plasma protein interactions with Pluronic™-treated colloids, Coll. Surf. B: Biointerfaces 7 (1996) 9–22.

[79] Y. Iwasaki, S.-I. Sawada, N. Nakabayashi, G. Khang, H.B. Lee, K. Ishihara, The effect of the chemical structure of the phospholipid polymer on fibronectin adsorption and fibroblast adhesion on the gradient phospholipid surface, Biomaterials 20 (1999) 2185–2191.

[80] K.L. Prime, G.M. Whitesides, Adsorption of proteins onto surfaces containing end-attached oligo(ethylene oxide): a model system using self-assembled monolayers, J. Am. Chem. Soc. 115 (1993) 10714–10721.

[81] P. Harder, M. Grunze, R. Dahint, G.M. Whitesides, P.E. Laibinis, Molecular conformation in oligo(ethylene glycol)-terminated self-assembled monolayers on gold and silver surfaces determines their ability to resist protein adsorption, J. Phys. Chem. B 102 (1998) 426–436.

[82] A.J. Pertsin, M. Grunze, I.A. Garbuzova, Low-energy configurations of methoxy triethylene glycol terminated

alkanethiol self-assembled monolayers and their relevance to protein adsorption, J. Phys. Chem. B 102 (1998) 4918–4926.

[83] G.P. Lopez, B.D. Ratner, Molecular adsorption and the chemistry of plasma-deposited thin organic films: Deposition of oligomers of ethylene glycol, Plasmas and Polymers 1 (1996) 127–151.

[84] T.R. Kyriakides, K.J. Leach, A.S. Hoffman, B.D. Ratner, P. Bornstein, Mice that lack the angiogenesis inhibitor, thrombospondin 2, mount an altered foreign body reaction characterized by increased vascularity, Proc. Natl. Acad. Sci. USA 96 (1999) 1–6.

[85] M. Malyankar, M. Scatena, K.L. Suchland, T.J. Yun, E.A. Clark, C.M. Giachelli, Osteoprotegerin is an $\alpha_v\beta_3$-induced, NF-κB-dependent survival factor endothelial cells, J. Biol. Chem. 275 (2000) 20959–20962.

[86] R. Langer, J.P. Vacanti, Artificial organs, Sci. Am. 273 (1995) 130–133.

[87] G.K. Naughton, W.R. Tolbert, T.M. Grillot, Emerging developments in tissue engineering and cell technology, Tissue Eng 1 (1995) 211–219.

[88] Y. Kuboki, M. Sasaki, A. Saito, H. Takita, H. Kato, Regeneration of periodontal ligament and cementum by BMP-applied tissue engineering, Eur. J. Oral Sci. 106 (1998) 197–203.

[89] D.J. Prockop, Marrow stromal cells as stem cells for nonhematopoietic tissues, Science 276 (1997) 71.

[90] L. Rowen, G. Mahairas, L. Hood, Sequencing the human genome, Science 278 (1997) 605–607.

[91] G. Vogel, Harnessing the power of stem cells, Science 283 (1999) 1432–1434.

[92] T.A. Horbett, L.A. Klumb, Cell culturing: Surface aspects and considerations, in: J.L. Brash, P.W. Wojciechowski (Eds.), Interfacial Phenomena and Bioproducts, Marcel Dekker, New York, 1996, pp. 351–445.

[93] C.D. Tidwell, D.G. Castner, S.L. Golledge, B.D. Ratner, K. Meyer, B. Hagenhoff, A. Benninghoven, Static ToF SIMS and XPS characterization of adsorbed albumin and fibronectin films, Surf. Interface Anal. 31 (2001) 724–733.

[94] M.S. Wagner, D.G. Castner, Characterization of adsorbed protein films by ToF SIMS with PCA, Langmuir 17 (2001) 4649–4660.

[95] R.F. Doolittle, Fibrinogen and fibrin, Sci. Am. 245 (6) (1981) 126–135.

[96] F.T. Wagner, T.E. Moylan, Modeling the aqueous-metal interface in ultrahigh vacuum via cryogenic coadsorption, ACS Symp. Ser. 378 (1988) 65–82.

[97] B.D. Ratner, P.K. Weathersby, A.S. Hoffman, M.A. Kelly, L.H. Scharpen, Radiation-grafted hydrogels for biomaterial applications as studied by the ESCA technique, J. Appl. Polym. Sci. 22 (1978) 643–664.

[98] K.B. Lewis, B.D. Ratner, Observation of surface restructuring of polymers using ESCA, J. Coll. Interf. Sci. 159 (1993) 77–85.

[99] D.A. Fischer, J. Colbert, J.L. Gland, Ultrasoft (C,N,O) X-ray fluorescence detection: proportional counters, focusing multilayer mirrors, and scattered light systematics, Rev. Sci. Instrum. 60 (1989) 1596–1602.

[100] T.A. Roberts, K.E. Gray, Total-reflection X-ray fluorescence spectroscopy for in situ, real-time analysis of growing films, MRS Bulletin 20 (1995) 43–46.

[101] Y.R. Shen, Surface properties probed by second-harmonic and sum-frequency generation, Nature 337 (1989) 519–525.

[102] D.H. Gracias, Z. Chen, Y.R. Shen, G.A. Somorjal, Molecular characterization of polymer and polymer blend surfaces combined sum frequency generation surface vibrational spectroscopy and scanning force microscopy studies, Acc. Chem. Res. 32 (1999) 930–940.

[103] J. Salafsky, J.T. Groves, S.G. Boxer, Architecture and function of membrane proteins in planar supported bilayers: a study with photosynthetic reaction centers, Biochemistry 35 (1996) 14773–14781.

[104] H.T. Tien, A. Ottova-Leitmannova, Membrane biophysics as viewed from experimental bilayer lipid membranes, Elsiever, Amsterdam, New York, 2000.

[105] E. Gizeli, M. Liley, C.R. Lowe, H. Vogel, Antibody binding to a functionalized supported lipid layer: A direct acoustic immunosensor, Anal. Chem. 69 (1997) 4808–4813.

[106] A. Plant, Supported hybrid bilayer membranes as rugged cell membrane mimics, Langmuir 15 (1999) 5128–5135.

[107] I. Reviakine, A. Brisson, Formation of supported phospholipid bilayers from unilamellar vesicles investigated by AFM, Langmuir 16 (2000) 1806–1815.

[108] A. Sellinger, P.M. Weiss, A. Nguyen, Y. Lu, R.A. Assink, W. Gong, C.J. Brinker, Continuous self-assembly of organic-inorganic nanocomposite coatings that mimic nacre, Nature 394 (1998) 256–260.

[109] M.P. Olivieri, R.E. Loomis, A.E. Meyer, R.E. Baier, Surface characterization of mussel adhesive protein films, J. Adhes. Sci. Tech. 4 (1990) 197–204.

[110] H. Shi, W.-B. Tsai, S. Ferrari, B.D. Ratner, Template imprinted nanostructural surfaces for protein recognition, Nature 398 (1998) 593–597.

[111] G. Daculsi, J.-M. Bouler, R.Z. LeGeros, Adaptive crystal formation in normal and pathological calcifications in synthetic calcium phosphate and related biomaterials, Int. Rev. Cytol. 172 (1997) 129–191.

[112] T. Cass, F.S. Ligler, Immobilized biomolecules in Analysis. A practical approach, Oxford University Press, Oxford, 1997.

[113] G.T. Hermanson, Bioconjugate techniques, Academic Press, New York, 1996.

[114] H. Kuhn, Functionalized monolayer assembly manipulation, Thin Solid Films 99 (1983) 1–16.

[115] R.G. Nuzzo, D.L. Allara, Adsorption of bifunctional organic disulfides on gold surfaces, J. Am. Chem. Soc. 105 (1983) 4481–4483.

[116] A.S. Rudolph, Biomaterial biotechnology using self-assembled lipid microstructures, J. Cellular Biochem. 56 (1994) 183–187.

[117] M. Pomerantz, A. Segmuller, L. Netzer, J. Sagiv, Coverage of Si substrates by self-assembling monolayers and multilayers as measured by IR, wettability and X-ray diffraction, Thin Solid Films 132 (1985) 153–162.

[118] A. Ulman, J.E. Eilers, N. Tillman, Packing and molecular orientation of alkanethiol monolayers on gold surfaces, Langmuir 5 (1989) 1147–1152.

[119] C.D. Bain, G.M. Whitesides, Molecular-level control over surface order in self-assembled monolayer films of thiols on gold, Science 240 (1988) 62–63.

[120] M. Kunitake, N. Batina, K. Itaya, Self-organized porphyrin array on iodine-modified Au(111) in electrolyte solutions: In situ scanning tunneling microscopy study, Langmuir 11 (1995) 2337–2340.

[121] T. Boland, B.D. Ratner, Two dimensional assembly of purines and pyrimidines on Au(111), Langmuir 10 (1994) 3845–3852.

[122] A. Wawkuschewski, H.-J. Cantow, S.N. Magonov, Scanning tunneling microscopy of alkane adsorbates at the liquid/graphite interface, Langmuir 9 (1993) 2778–2781.

[123] D. Pum, U.B. Sleytr, Monomolecular reassembly of a crystalline bacterial cell surface layer (S-layer) on untreated and modified silicon surfaces, Supramol. Sci. 2 (1995) 193–197.

[124] W. Frey, W.R. Schief, V. Vogel, Two-dimensional crystallization of streptavidin studied by quantitative brewster angle microscopy, Langmuir 12 (1996) 1312–1320.

[125] S. Dante, R. Advincula, C.W. Frank, P. Stroeve, Photoisomerization of polyionic layer-by-layer films containing azobenzene, Langmuir 15 (2000) 193–201.

[126] S. Miyamoto, H. Teramoto, O.A. Coso, J.S. Gutkind, P.D. Burbelo, S.K. Akiyama, K.M. Yamada, Integrin function: Molecular hierarchies of cytoskeletal and signaling molecules, J. Cell Biol. 131 (1995) 791–805.

[127] D.A. Tomalia, Starburst™/cascade dendrimers: fundamental building blocks for a new nanoscopic chemistry set, Aldrichimica Acta 26 (1993) 91–101.

[128] P. Laitenberger, C.G. Claessens, L. Kuipers, F.M. Raymo, R.E. Palmer, J.F. Stoddart, Building supramolecular nanostructures on surfaces: the influence of the substrate, Chem. Phys. Lett. 279 (1997) 209–214.

[129] M.J. Heller, R.H. Tullis, Self-organizing molecular photonic structures based on functionalized synthetic nucleic acid (DNA) polymers, Nanotechnology 2 (1991) 165–171.

[130] T.H. Labean, H. Yan, J. Kopatsch, F. Liu, E. Winfree, J.H. Reif, N.C. Seeman, Construction, analysis, ligation and self-assembly of DNA triple crossover complexes, J. Am. Chem. Soc. 122 (2000) 1848–1860.

[131] C.B. Ross, L. Sun, R.M. Crooks, Scanning probe lithography. 1. Scanning tunneling microscope induced lithography of self-assembled n-alkanethiol monolayer resists, Langmuir 9 (1993) 632–636.

[132] S.I. Jeon, J.H. Lee, J.D. Andrade, P.G. DeGennes, Protein-surface interactions in the presence of polyethylene oxide, J. Coll. Interf. Sci. 142 (1991) 149–158.

[133] S.J. Lee, K. Park, Protein interaction with surfaces: Separation distance-dependent interaction energies, J. Vac. Sci. Technol. A 12 (1994) 2949–2955.

[134] R.G. Chapman, E. Ostuni, S. Takayama, E. Homlin, L. Yan, G.M. Whitesides, Surveying for surfaces that resist the adsorption of proteins, J. Am. Chem. Soc. 122 (2000) 1848–1860.

[135] E.A. Vogler, Structure and reactivity of water at biomaterial surfaces, Adv. Coll. Interf. Sci. 74 (1998) 69–117.

[136] A.J. Pertsin, M. Grunze, Computer simulation of water near the surface of oligo(ethylene glycol)-terminated alkanethiol self-assembled monolayers, Langmuir 16 (2000) 8829–8841.

[137] J.P. Chen, A.S. Hoffman, Polymer-protein conjugates. II. Affinity precipitation of human IgG by poly(N-isopropyl acrylamide)—protein A conjugate, Biomaterials 11 (1990) 631–634.

[138] H. Iwata, I. Hirata, Y. Ikada, Atomic force microscopic analysis of a porous membrane with pH-sensitive molecular valves, Macromolecules 31 (1998) 3671–3678.

[139] T. Okano, Y.H. Bae, H. Jacobs, S.W. Kim, Thermally on-off switching polymers for drug permeation and release, J. Controlled Release 11 (1990) 255–265.

[140] L.R. Brown, E.R. Edelman, F. Fischel-Ghodsian, R. Langer, Characterization of glucose-mediated insulin release from implantable polymers, J. Pharm. Sci. 85 (1996) 1341–1345.

[141] D.H. Carey, G.S. Ferguson, A smart surface: entropic control of composition at polymer/water interface, J. Am. Chem. Soc. 118 (1996) 9780–9781.

[142] K. Fujimoto, C. Iwasaki, C. Arai, M. Kuwako, E. Yasugi, Control of cell death by the smart polymeric vehicle, Biomacromolecules 1 (2000) 515–518.

[143] O.-S. Jung, Y.J. Kim, Y.-A. Lee, J.K. Park, H.K. Chae, Smart molecular helical springs as tunable receptors, J. Am. Chem. Soc. 122 (2000) 9921–9925.

[144] A.S. Hoffman, P.S. Stayton, V. Bulmus, G. Chen, C. Chueng, A. Chilkoti, Z. Ding, R. Fong, C.A. Lackey, C.J. Long, M. Miura, J.E. Morris, N. Murthy, Y. Nabeshima, T.G. Park, O.W. Press, T. Shimoboji, S. Shoemaker, H.J. Yang, N. Monji, R.C. Nowinski, C.A. Cole, J.H. Priest, J.M. Harris, K. Nakamae, T. Nishino, T. Miyata, Really smart bioconjugates of smart polymers and receptor proteins, J. Biomed. Mater. Res. 52 (2000) 577–586.

[145] P.S. Stayton, T. Shimboji, C. Long, A. Chilkoti, G. Chen, J.M. Harris, A.S. Hoffman, Control of protein-ligand recognition using a stimuli-responsive polymer, Nature 378 (1995) 472–474.

[146] M. Sisido, M. Harada, K. Kawashima, H. Ebato, Y. Okahata, Photoswitchable peptide antigens on solid surfaces, Biopolymers 47 (1998) 159–165.

[147] Y.V. Pan, R.A. Wesley, R. Luginbuhl, D.D. Denton, B.D. Ratner, Plasma polymerized n-isopropylacrylamide: synthesis and characterization of a smart thermally responsive coating, Biomacromolecules 2 (2001) 32–36.

[148] L. Dai, H.A.W. St John, J. Bi, P. Zientek, R.C. Chatelier, H.J. Griesser, Biomedical coatings by the covalent

immobilization of polysaccharides on gas-plasma activated polymer surfaces, Surf. Interf. Anal. 29 (2000) 46–55.
[149] M. Morra, C. Cassinelli, Surface studies on a model cell resistant system, Langmuir 15 (1999) 4658–4663.
[150] L.A. Cantarero, J.E. Butler, J.W. Osborne, The adsorptive characteristics of proteins for polystyrene and their significance in solid-phase immunoassays, Anal. Biochem. 105 (1980) 375–382.
[151] G.P. Lopez, B.D. Ratner, R.J. Rapoza, T.A. Horbett, Plasma deposition of ultrathin films of poly(2-hydroxyethyl methacrylate): Surface analysis and protein adsorption measurements, Macromolecules 26 (1993) 3247–3253.

 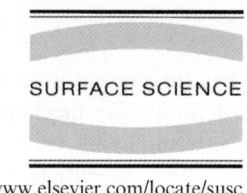

The role of surface science in bioengineered materials

Matthew Tirrell [*], Efrosini Kokkoli [1], Markus Biesalski [2]

Department of Chemical Engineering and Materials Research Laboratory, University of California at Santa Barbara, Santa Barbara, CA 93106, USA

Received 7 August 2000; accepted for publication 22 May 2001

Abstract

Materials employed in biomedical technology are increasingly being designed to have specific, desirable biological interactions with their surroundings, rather than the older common practice of trying to adapt traditional materials to biomedical applications. Moreover, materials scientists are also increasingly deriving new lessons from naturally occurring materials (from mollusk shells to soft animal tissue) about useful composition–structure property relationships that might be mimicked with synthetic materials. Together, these two areas of effort constitute what we may call bioengineered materials. It is possible to set down a reasonably thorough set of characteristics that bioengineered materials have in common. Among these characteristics we discuss the following: self-assembly, bioengineered materials often rely on information content built into structural molecules to determine the order and organization of the material; hierarchical structure, in most bioengineered materials several different length scales of structure are essential and are formed spontaneously and simultaneously via self-assembly; precision synthesis, fundamental to biological material structures is the idea of macromolecules constructed in a precise manner; templating, ordered structures in bioengineered materials are often propagated from one element or set of instructions, to another; specific and non-specific interactions, the forces involved in holding biomaterials structures together. In the future, a carefully selected combination of this set of characteristics will enable us to bioengineer surfaces that are capable to direct and control a desired biological response. Eventually, such bioengineered surfaces will become important tools to comprehend and analyze how materials interact in nature. © 2001 Elsevier Science B.V. All rights reserved.

Keywords: Adhesion; Biological compounds; Chemisorption; Self-assembly; Surface chemical reaction

1. Introduction

Materials communicate with their environment through their interfaces. Both, the kind and the strength of such "communications" are determined by the interfacial properties of the material. The last 50 years has witnessed dramatic progress in the ability to comprehend and to characterize these types of communications, significantly advancing the fields of materials and surface science. During the same time, cell and molecular biology

---

[*] Corresponding author. Address: Office of the Dean of Engineering, College of Engineering, University of California at Santa Barbara, Santa Barbara, CA 93106-5130, USA. Tel.: +1-805-8933141; fax: +1-805-8938124.
E-mail address: tirrell@engineering.ucsb.edu (M. Tirrell).
[1] Present address: Department of Chemical Engineering, University of Massachusetts at Amherst, Amherst, MA 01003, USA.
[2] Present address: Institute for Microsystem Technology (IMTEK), University of Freiburg, Georges Koehler Allee 103, D-79110 Freiburg, Germany.

has undergone a revolution in the understanding of molecular mechanisms and signaling cascades between living cells and their environment. For example, scientists have uncovered an array of mechanisms that control how cells communicate with their environment and, in particular, with proteins and other cells that compose the surrounding commonly known as the extra-cellular matrix (ECM) [1–7]. As a consequence, any serious attempt to engineer a successful biomaterial must merge the knowledge of materials surface science and cell and molecular biology and will allow us to establish and accurately control the interfacial interactions needed for biospecificity.

2. Defining the character of bioengineered materials

2.1. Biomedical materials

A number of materials have been used in medicine ranging from metals and ceramics for dental and orthopedic applications to "soft materials" for cardiovascular (clinical surgery on heart and blood vessels) and plastic surgery. For example, polymers like polymethyl methacrylate, polyethylene terephthalate, and polyamides (e.g. Nylon®) have long been used in surgery. Together with polytetrafluoro ethylene, polyurethanes, polypropylene and polyvinyl chloride they form an important class of "soft" materials in biomedical technology. A number of outstanding articles, reviews and books cover the present challenges and future aspects of biomedical materials, especially, in the field of polymeric biomaterials [8–15].

The question of "what defines the character of a biomaterial that is used in medical surgery?" is directly related to the questions "what is the response of the host to the material?" and "what happens to the material in such an aggressive environment?". More simply, "does the living (human) body accept the material or not?". Anderson et al. [16] describe these features in a textbook on materials used in medicine. Briefly, if a biomedical material is implanted into a living tissue, a cascade of host reactions occur at the interface between the tissue and the material, known as an inflammatory response [16–18]. This inflammatory response ends in an "encapsulation" of the material, which means the development of a scar tissue surrounding the biomaterial [16].

The first step in this response is the interaction of proteins and blood cells from body fluids with the surface of the biomaterial. As a consequence, the adsorption of proteins at the surface of the biomaterial is an important issue in its design. A large number of physicochemical and biochemical studies have been carried out so far on the fundamentals and the applications of proteins adsorbed to a materials surface [19–27].

Proteins typically adsorb to the surface of a biomaterial in a non-specific way. Here the term "non-specific" means that the proteins are only "physically" attached to the surface. We discuss the forces involved in these interactions between the surface and the proteins, such as electrostatic and hydrophobic forces later.

An example where the adsorption of proteins to a surface can lead to serious problems is schematically shown in Fig. 1. After proteins have adsorbed to the material, which for example could be an artificial blood vessel or any other material in contact with the mammalian blood stream a cascade of complex events takes place, which is shown in a very simplified fashion in Fig. 1: small disk-shaped cells called "platelets" adhere to the interface. Among other signaling molecules released after adsorbing to the surface, the platelets release the protein "thrombin", which directs the formation of a fibrin matrix. This matrix stabilizes the adsorbed cells. As a consequence the adhesion of proteins and cells to a surface can form huge aggregates ("thrombus"). Finally, such a thrombus can cause vascular obstruction at the place of its formation, or, even more severe, if the process takes place in a blood vessel in vivo and the blood current removes the thrombus from the materials surface the clot can block an artery ("embolism"). Hence, an important goal in the design of biocompatible materials is to create surfaces that minimize non-specific interactions with biological material such as proteins and (blood) cells. Interesting candidates to fulfill such requirements are, for example, surfaces modified with polymer hydrogels [8,10,12,20–23,28–30] or polymer brushes [24–27]. In Fig. 2 planar surfaces modified with

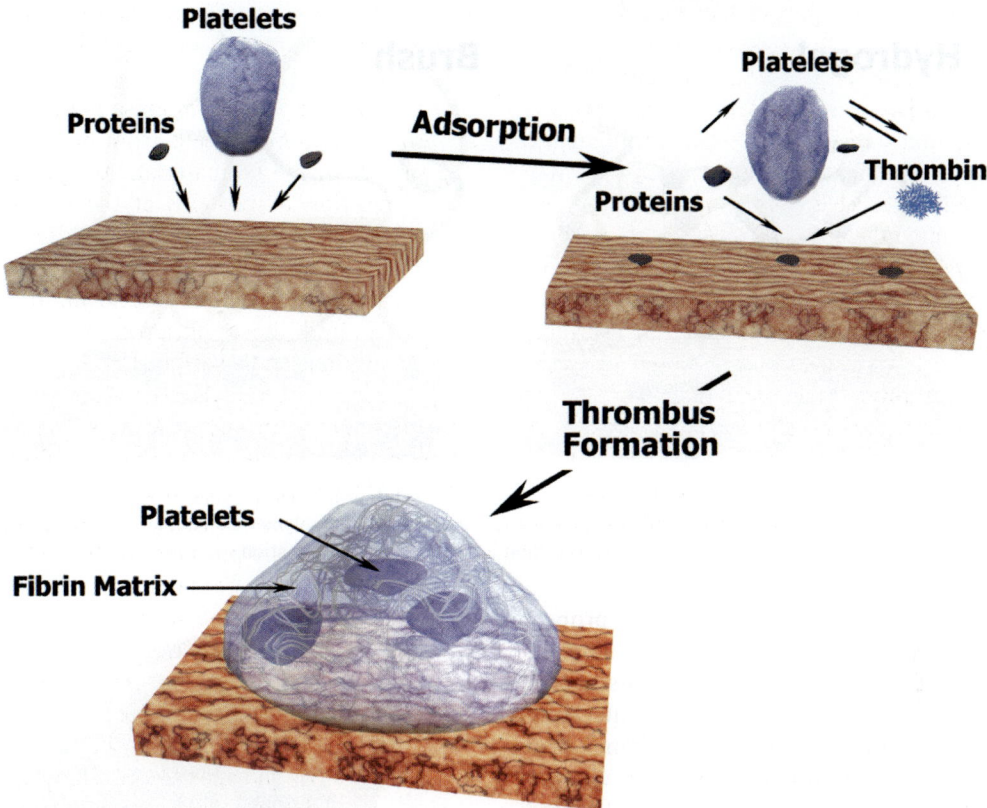

Fig. 1. Formation of a clot of small blood cells and proteins adsorbed to a surface ("thrombus formation"). The scheme is not drawn to scale and the formation of the "thrombus" is shown in a very simplified fashion.

such polymer layers are shown in a schematic fashion.

Polymer hydrogels (Fig. 2, left) are either physically or covalently crosslinked polymer chains forming a highly water-swellable polymer network layer. Once the hydrogel is swollen in an aqueous environment a displacement of the water molecules from the polymer layer is thermodynamically unfavorable. Thus, the strong interactions between water molecules and the polymer prevent attractive interactions between proteins and the hydrogel. As a result proteins approaching a material, which is covered with a polymer hydrogel are repelled from the surface.

The self-association of water molecules in a three-dimensional network driven by hydrogen bonds between the molecules is an important issue of materials in contact with biological fluids.

Vogler [31] describes how biological responses to a biomaterial such as the adsorption/repulsion of proteins or the formation of a thrombus can be discussed in terms of differences in the water layer structures, and the reactivity of water at surfaces. For a detailed insight the reader is referred to this review article and references therein [31].

Polymer brushes are linear polymer chains terminally anchored to a solid surface (Fig. 2, right). Such polymeric monolayers play an important role in a wide field of colloid stabilization, tribology, lubrication, and rheology [32–36]. If the distance between the anchoring points of the surface-grafted chains is small, interchain correlations occur and the tethered chains are stretched away from the surface leading to a "brush"-like conformation. Polymer brushes consisting of polyethylene glycol (PEG) have opened a wide door in

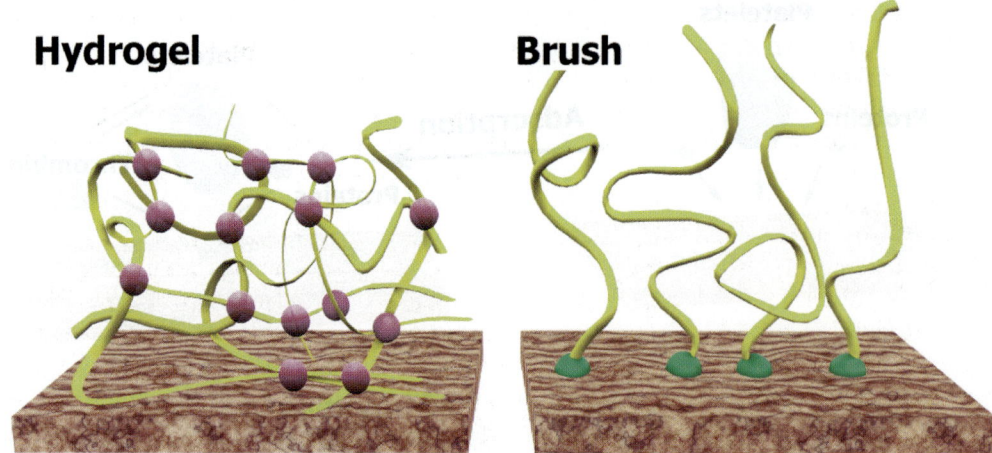

Fig. 2. A polymer hydrogel (left) and a polymer brush (right) adsorbed at a solid surface. The polymer hydrogel consists of crosslinked water-swellable polymer chains. The polymer brush represents terminally anchored polymer chains that form a monolayer at the surface. At high anchoring densities these polymer chains exhibit a "brush"-like conformation.

biomaterials research due to the suitable properties of the polymer. PEG is a water-swellable, non-toxic and biocompatible polymer and its use in biotechnology is widely reviewed in the literature [22,29,37–39]. With water-swellable polymer brushes, the repulsion to protein adsorption relies on factors like the grafting density of the tethered chains and the surface–polymer interactions [24]. Later we discuss some of the various synthetic efforts that have explored the use and modification of hydrogel and brush covered surfaces to establish biocompatible surfaces with good mechanical properties.

Another approach to establish bio-, and especially, blood-compatible ("non-thrombogenic") surfaces is to directly seed cells onto the surface of a material, analogous to the natural lining of blood vessels by endothelial cells (Fig. 3) [40]. Up to now, mid-size and large diameter artificial blood vessels ("vascular grafts") coated with endothelial cells are well established in surgery [41–43]. Such vascular grafts remain excellent for more than 10 years after implantation [42, 43]. Surface-hydrophilized polytetrafluoro ethylene (PTFE, Teflon®) or polyethylene terephthalate (PET, Dacron®) have been reported to be suitable substrate candidates establishing an endothelial cell layer [42]. Although some promising approaches exist, the fabrication of small caliber

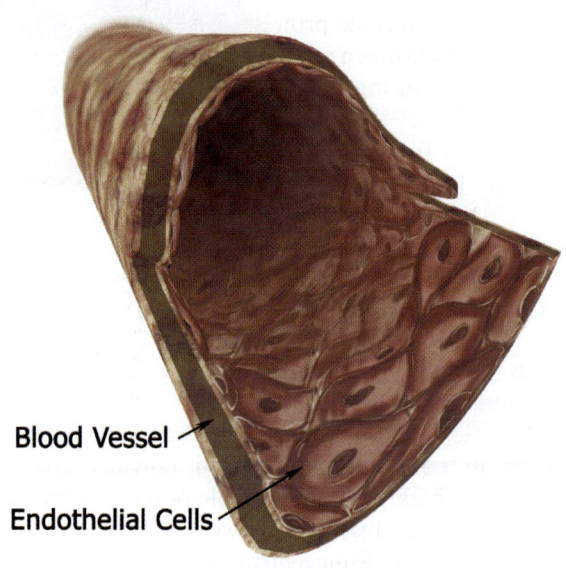

Fig. 3. Schematic picture of a human blood vessel. Among others one characteristic of the blood vessels is the natural lining by endothelial cells.

grafts (diameter < 6 mm) is still problematic but necessary in the reconstruction of small (peripheral) arteries [42]. Typically, such a seeding of endothelial cells requires an indirect approach where distinct proteins that mediate adhesion, spreading and proliferation of cells in vivo are "pre-adsorbed" to the material. It is known that the strong adhesion of the cells to such surfaces is due to the binding of small ligands ("adhesion motifs", found in ECM proteins) to receptor molecules located within the cellular membrane. Consequently, a more direct approach to control the interaction of biomaterials with living cells is to create surfaces that explore distinct adhesion motifs from the ECM at the surface and, therefore, are capable of "mimicking" the natural environment of a living cell. We discuss the character of such "biomimetic materials" in the following section.

2.2. Biomimetic materials

A common theme in engineering cell and tissue behavior at device surfaces is to modify the material to interact selectively with a specific cell type through biomolecular recognition events. The cell surface has a variety of receptors that bind with other cells or specific proteins, which compose the environment (known as the ECM) surrounding the cells. A promising approach is the biomimetic modification of the material in which peptides (a sequence of two or more amino acids joined by a chemical bond between the carbonyl group of the first amino acid and the amino group of the second amino acid) containing the adhesion domains of the ECM proteins are attached to the base material. Amino acids are often designated by either a three-letter abbreviation or a one-letter symbol to facilitate concise communication (Table 1).

The central hypothesis of biomimetic surface engineering is that peptides that mimic part of the ECM affect cell attachment to the material, and that surfaces or three-dimensional matrices modified with these active peptides can induce tissue formation conforming to the cell type seeded on the material (Fig. 4). Therefore extensive research over the last decade has been performed on the incorporation of adhesion promoting oligopeptides into biomaterial surfaces [8,12,13,44–46]. Since identification of the RGD peptide sequence as mediating the attachment of cells to several plasma and ECM proteins, including fibronectin, vitronectin, type I collagen, osteopontin and bone sialoprotein (BSP) [1], researchers have been depositing RGD-containing peptides on biomaterials to promote cell attachment. Its ability to bind a variety of cells through ligand–receptor interactions makes RGD an exceptionally useful sequence for incorporating onto biomaterial surfaces [47].

RGD peptides that have not been designed to bind certain integrins (adhesion receptors on the cell surface with two subunits, designated α (alpha) and β (beta)) mimic a number of adhesion proteins and bind more than one receptor [5]. The main two design strategies to increase integrin specificity have been the chemical synthesis of cyclic peptides to provide conformational constrain and selected sequences flanking the RGD to give increasing affinity and selectivity [12,48,49]. The affinity of these peptides however is relatively low compared to the ECM proteins. The GRGDSP which is derived from the cell attachment site of fibronectin, is 1000 times less effective in cell attachment assays than fibronectin itself [50]. An advantage, though, of employing short bioactive peptides rather than the complete parent glycoprotein is selectivity for targeted cell types. GRGDSP, for example, is quite specific in its activity for the fibronectin receptor and changes as small as the replacement of the aspartic acid (D) with a glutamic acid (E) reduce the activity 100-fold or more [50].

There are several important fundamental issues about how biomimetic surfaces can display specificity and binding affinity. One important parameter is the conformation of the amino acids. Human melanoma (cancer) cells spread on looped RGD biomimetic surfaces in a concentration-dependent manner, spread indiscriminately on carboxyl-coupled RGD, and did not spread on amino-coupled RGD surfaces (Fig. 5) [51]. The surface density of RGD peptides has been demonstrated to elicit different cellular responses [51,52]. A surface density of 1×10^{-15} mol/cm^2 for GRGDY covalently grafted to the surface of otherwise poorly adhesive glass substrate was sufficient to promote fibroblast cell (common cell type found in connective tissue) spreading but focal contact formation (small region on the surface of the fibroblast that is anchored to the substrate and is mediated by clusters of integrin receptors) was observed only at concentrations of 10×10^{-15} mol/cm^2 and higher [52]. These measurements provide a benchmark for the design of

Table 1
Amino acids

Amino acids	Three-letter abbreviation	One-letter symbol	R-group
Alanine	Ala	A	$-CH_3$
Arginine	Arg	R	
Asparagine	Asn	N	$-CH_2-CO-NH_2$
Aspartic acid	Asp	D	$-CH_2-COOH$
Cysteine	Cys	C	$-CH_2-SH$
Glutamine	Gln	Q	$-(CH_2)_2-CO-NH_2$
Glutamic acid	Glu	E	$-(CH_2)_2-COOH$
Glycine	Gly	G	$-H$
Histidine	His	H	
Isoleucine	Ile	I	
Leucine	Leu	L	$-CH_2-CH-(CH_3)_2$
Lysine	Lys	K	$-(CH_2)_4-NH_2$
Methionine	Met	M	$-(CH_2)_2-S-CH_3$
Phenylalanine	Phe	F	
Proline	Pro	P	
Serine	Ser	S	$-CH_2-OH$
Threonine	Thr	T	
Tryptophan	Trp	W	
Tyrosine	Tyr	Y	
Valine	Val	V	$-CH-(CH_3)_2$

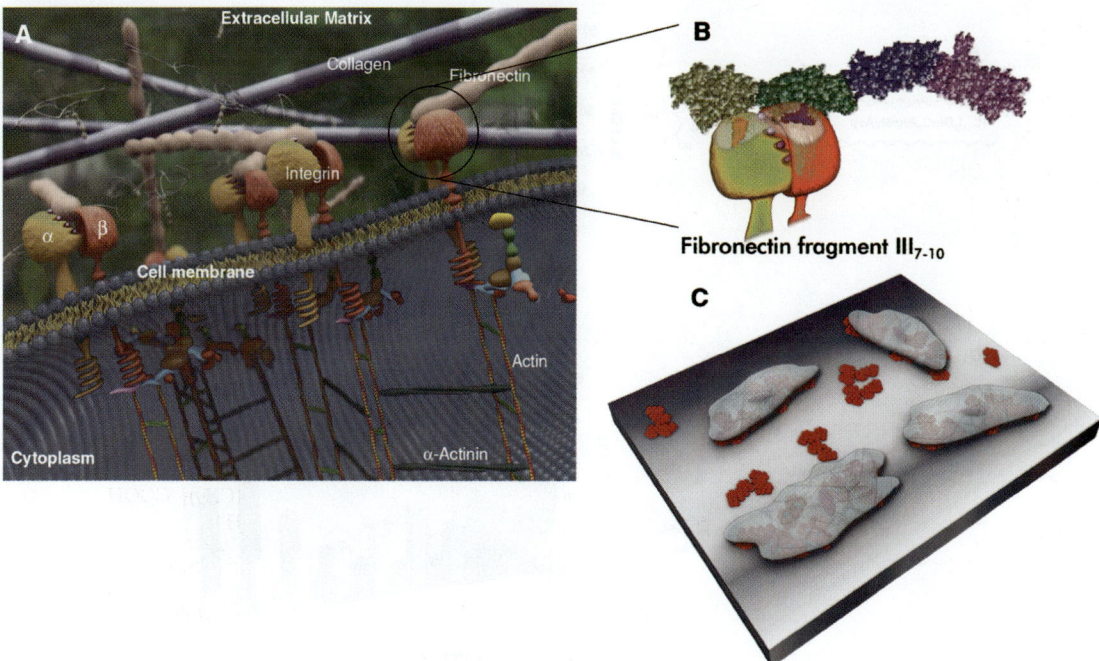

Fig. 4. (A) Integrins are composed of two subunits, α and β, and they hold a cell in place by attaching at one end to proteins of the ECM (and in specific to the cell binding domain of the protein), like fibronectin, and at the other end to the cytoskeleton, the structural framework of the cell. They connect to the cytoskeleton through a highly organized aggregate of proteins, such as actin filaments, α-actinin and others. (B) Integrins bind to the adhesion domains of the ECM proteins by recognizing specific sites of that domain. A model for the recognition of the fibronectin GRGDS site and PHSRN synergy site by an integrin is shown here. The four repeats of the fibronectin fragment III_{7-10} are shown in different colors: purple for III_7 to yellow for III_{10}. The synergy site PHSRN in III_9 repeat is marked in blue and the GRGDS in III_{10} in red. The minimal amino acid sequence RGD in fibronectin and other cell adhesion proteins is the most important recognition site for about half of all known integrins. The synergy site is also required for high affinity binding. (C) The central hypothesis of the biomimetic surface engineering is that short peptides that mimic part of the ECM proteins (like RGD, shown in red) are recognized by integrins via "specific interactions" and can affect cell attachment to the material. Cells adhere and spread on RGD biomimetic surfaces and surfaces modified with these active peptides can therefore induce tissue formation conforming to the cell type seeded on the material.

practical peptide biomimetic biomaterials, as they indicate a minimum RGD density of 10×10^{-15} mol/cm^2, corresponding to a spacing of about 140 nm between peptide ligands. Another parameter that can influence cell adhesion is peptide clustering at the nanoscale level [53]. Clustering of YGRGD ligand significantly reduced the average ligand density required to support fibroblast cell migration whereas non-clustered ligands supported cell attachment but did not promote full spreading [53]. Cell behavior also can be influenced by creating a biofunctional surface in which the accessibility of the ligand is used as the control parameter. Therefore selective masking of a peptide ligand by PEG chains of varying length on a biomimetic surface is another new method of controlling the surface bioactivity [54].

Even though the most common cell-binding domain which has been used extensively as a candidate peptide to enhance cell adhesion onto biomaterial surfaces is the RGD sequence, other non-RGD-containing cell-binding domains exist [47], such as YIGSR and IKVAV in laminin [55], REDV and LDV in fibronectin [56], DGEA in collagen I, and various heparin-binding domains [57]. Certain studies have demonstrated that a more "complete" cell response (e.g. cell attachment, spreading, focal contact formation and

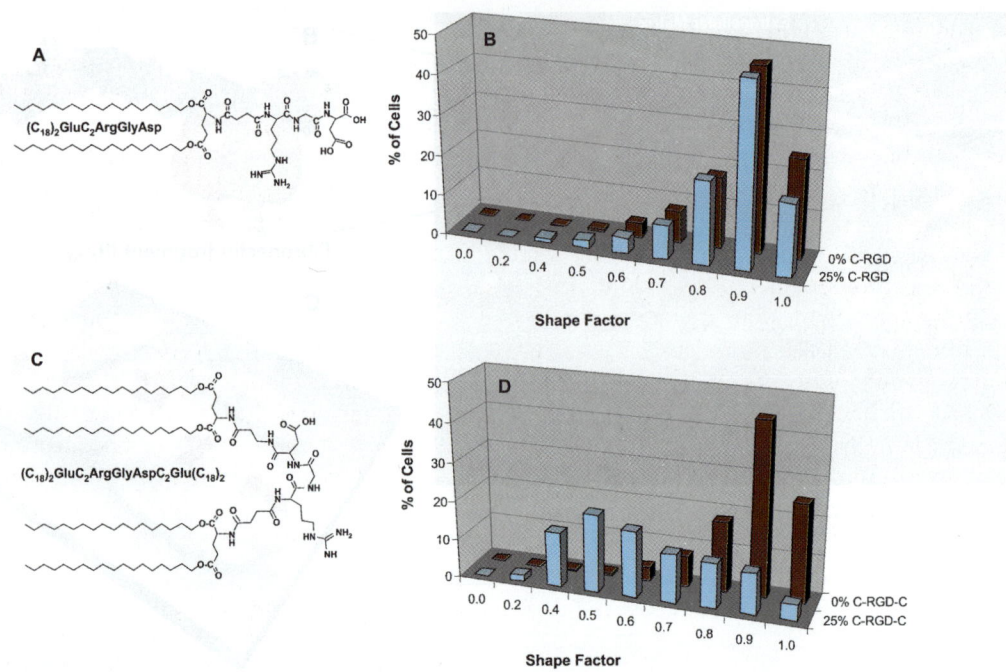

Fig. 5. (A) Chemical structure of the amino-coupled RGD peptide amphiphile, $(C_{18})_2$–Glu–C_2–RGD. (B) Shape factor histogram measured for the spreading of M14#5 human melanoma cells on monolayers containing molar mixtures of the methyl ester amphiphile $(C_{18})_2$–Glu–C_2–COOCH$_3$ and the peptide amphiphile $(C_{18})_2$–Glu–C_2–RGD. (C) Chemical structure of the looped RGD peptide amphiphile, $(C_{18})_2$–Glu–C_2–RGD–C_2–Glu–$(C_{18})_2$. (D) Shape factor histogram measured for the spreading of M14#5 human melanoma cells on monolayers containing molar mixtures of the methyl ester amphiphile $(C_{18})_2$–Glu–C_2–COOCH$_3$ and the peptide amphiphile $(C_{18})_2$–Glu–C_2–RGD–C_2–Glu–$(C_{18})_2$. Molar percentages are given for RGD. Cell shape factor is calculated as a ratio of area to perimeter squared, such that a round (i.e., non-spread) cell has a shape factor of one. As shape factor goes to zero, cells are assumed to be increasingly more spread.

organized cytoskeletal assembly) was obtained by providing the cell with both the cell-binding (RGD containing) and heparin-binding domains of fibronectin [57] or BSP [58,59].

The central objective of biomimetic surface engineering for bone regeneration is to examine whether a monolayer of different active peptides can affect bone cell adhesion and preferentially induce mineralization of the synthesized ECM on modified surfaces. This objective was recently tested by designing model biomimetic surfaces containing different ratios of both RGD and heparin-binding domains of BSP capable of controlling mammalian cell behavior [58,59]. These results demonstrate that utilizing peptide sequences incorporating both cell- and heparin-adhesive motifs can enhance the degree of cell surface interactions and influence the long-term formation of mineralized ECM in vitro. Events leading to the final integration of an implant into bone take place at the tissue–implant interface. The development of this interface is complex and involves numerous factors that include not only implant-related factors, such as, shape, topography and surface chemistry, but also mechanical loading, surgical technique and bone quantity and quality [60].

Another kind of biomimetic materials is a class of functional polymers termed carboxymethyl benzylamide polysaccharides. These functionalized polysaccharides are soluble and can be tailored in arrays of various copolymers with different statistical distributions of chemical groups. Depending on the chemical nature of the functional groups and their relative proportions, these polysaccharides can reveal binding affinities for proteins that range from non-specific to highly specific. Such polysaccharides contain chemical groups that mimic natural ligands with biospecific sites. A recent review describes the properties of this family of dextran (polysaccharides that yield glucose units on hydrolysis) derivatives that are endowed with specific interactions with respect to coagulation proteins, complement enzymes, viral glycoproteins, growth factors and cell receptors [61].

Significant progress has been made in incorporating adhesion peptides into practical biomaterials that might be used as implants and different methods, including self-assembly, precision synthesis and templating, have been used to fix the peptide in place on surfaces. While the classification of some approaches into one of these three classes is somewhat arbitrary, these classes do provide a reasonable framework for considering common features of bioengineered materials, as described in the next section.

3. Specific common features of bioengineered materials

3.1. Self-assembly

One of the great challenges for materials science is the creation of supramolecular materials in which the constituent units are highly regular nanostructures. Self-assembly of large oligomers into supramolecular arrangements has the potential for creating such nanostructures [62]. Supramolecular assemblies such as amphiphilic (having both hydrophobic and hydrophilic regions) self-assembled, multilayered, and nanostructured systems have been used successfully to modify biomaterials [8]. These systems are not covalently linked, but can build two- and three-dimensional structures based on non-covalent interactions.

The cell membrane is an example of a self-assembled system. Simplified models of cell membranes have been the subject of intense study. Solid-supported membranes are being used increasingly as model systems for fundamental biophysical research, for biosensors, and for the design of phantom cells exhibiting well-defined adhesive properties and receptor densities [63]. Several strategies for the assembly of biomembranes onto various solid supports have been reviewed [63–68]. The major techniques are direct vesicle fusion [69], the Langmuir–Blodgett (LB) technique [70] and molecular self-assembly from dilute organic solutions [71]. Three types of supported membranes can be assembled [72]: (a) supported lipid bilayer membranes prepared by vesicle spreading or LB deposition at hydrophilic surfaces, (b) covalently tethered self-assembled monolayers (SAMs) of functionalized lipid derivatives, prepared by coadsorption of membrane-forming components and (c) asymmetric or hybrid bilayer membranes, composed of an outer lipid layer and an inner SAM. The most natural membrane bilayer environment is the first approach. When designing a model system, it is desirable to have supports that enable the lateral mobility of the membrane components. Different strategies are to separate the lipid bilayer from the solid support by an ultrathin layer of water [73,74], a crystalline bacterial surface layer proteins [75] or a soft cushion composed by a water-swellable hydrophilic polymer or polyelectrolyte film [63,67,76]. A neutron reflectometry study has recently evaluated the structures resulting from different methods of preparing polymer-cushioned lipid bilayers and showed that the initial conditions of the polymer layer are a critical factor for the successful formation of desired structures, i.e., a continuous bilayer atop a hydrated polymer layer [77].

Amphiphilic self-assembly relevant to the production of devices and artificial implantable organs has been examined using LB films. Using a multiblock copolymer of PEG and bisphenol A, LB films were formed and transferred to hydrophobized glass. The platelet adhesion on LB films was no different from that on glass; however, fewer hepatocytes (epithelial cells constituting the major cell type in the liver) adhered to the LB films, indicating the presence of PEG on the surface [78]. In another study, an amphiphile, with two alkyl chains terminated with acrylic groups attached to a sulfate group was used to form LB films on polyethylene. Three layers were deposited on the surface and polymerized, producing a highly sulfated hydrophilic surface. The LB films were tested for adhesion of platelets and the number of well-spread platelets was reduced to about six times less than with polyethylene [79]. These studies indicate the possibility of using LB films to produce dense, uniform coatings on biomedical polymers.

Another method of creating novel biomaterials with distinct protein-like structures is to directly synthesize the peptide onto a template that induces secondary (conformation of a protein or peptide with respect to nearest-neighbor amino acids, i.e., α helices and β sheets) and tertiary (complex 3-D form of a macromolecule) structures and either serves directly as a biomaterial or is compatible with biomaterial surfaces. Ideally, one wants to create a system by which synthetic linear peptide chains self-assemble into desirable secondary and tertiary structures. Novel "peptide-amphiphiles" were synthesized to serve this purpose, whereby a peptide "headgroup" incorporates a cell binding sequence and has the propensity to form a distinct structural element, while a hydrophobic "tail" serves to align the peptide strands and induce secondary and tertiary structure formation as well as providing a hydrophobic surface for self-association and/or interaction with other surfaces (Fig. 6) [80–83]. The peptide headgroup incorporated a collagen-model cell binding sequence that promoted melanoma cell spreading [84]. Both the molar concentration and the molecular packing of the collagen-like peptide-amphiphile played an important role in mediating the cellular response. In addition to promoting adhesion and cell spreading, collagen peptide amphiphiles also induce cell signal-transduction (a process by which a cell converts an extra-cellular signal into a response) [85]. Other examples of bioactive peptide-amphiphiles have been documented. A series of peptides were found for example to activate murine spleen cells [86]. Combining the peptide sequence SFLLRN, that is recognized by the human thrombin receptor, with a phospholipid tail produced a peptide-amphiphile that promoted platelet aggregation only one order of magnitude smaller than the peptide itself [87]. Although at an early stage of development, peptide-amphiphiles appear to be quite promising reagents for surface modification applications.

SAMs formed on the adsorption of long-chain alkyl thiols to the surface of gold (Fig. 7) or alkyl silanes to hydroxylated surfaces are well ordered organic surfaces that permit control over the properties of the interface at the molecular scale. The ability to present molecules, peptides and proteins at the interface makes SAMs especially useful for fundamental studies of protein adsorption and cell adhesion [88]. Mixed SAMs that presented an RGD peptide in an inert, non-adsorbing interface of oligoethylene glycol moieties provided a surface that promoted endothelial cell attachment and spreading at relative RGD mole fractions of 0.001, while preventing the remodeling of the substrate (within hours of plating cells onto substrates the cells deposit new ECM proteins) [89]. An improved understanding of molecular surface determinants required for adhesion-dependant cell growth and proliferation using ultrathin, highly organized functionalized SAMs indicated that well-controlled culture surfaces modulate differential cell adhesion, spreading and growth through modulations of the amounts and conformations of adsorbed ECM molecules [90].

3.2. Hierarchical structures

Tissues, designed to serve specific functions in the human body are among the most advanced structural composite materials known made of macromolecular building blocks. Hierarchical structures in biocomposite systems such as in bone (Fig. 8), teeth, and collagenous connective tissue have many scales or levels, have highly specific interactions between these levels, and have the architecture to accommodate a complex spectrum of property requirements [94]. Bioengineered materials used as implants for hierarchical tissues, as in the case of implants for bone formation, demonstrate several different length scales of structure and are formed spontaneously and simultaneously via self-assembly.

Hydroxyapatite (HA) is an effective, commercially available bone implant material because it contains an interconnected network of channels appropriate for bone cell ingrowth. HA also stimulates bone growth without immune response complications [95,96] and small particles of HA may be biodegradable and stimulate bone ingrowth as they dissolve, depending on the initial composition [97]. Various HA/collagen composites

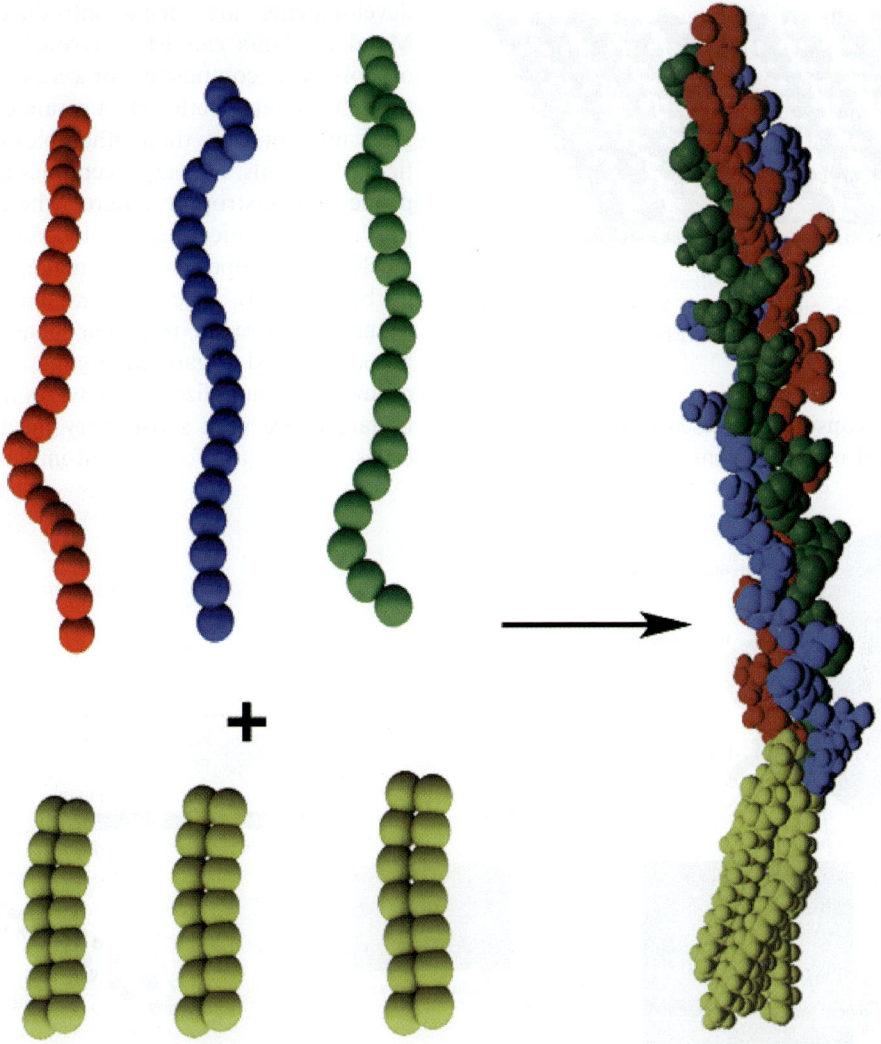

Fig. 6. Structure of novel peptide-amphiphiles incorporating triple-helical protein-like molecular architecture. Long-chain dialkyl ester lipid tails are connected to linear peptide chains. The tails associate by hydrophobic interactions, inducing and/or stabilizing the 3-D structure of the peptide headgroup. Triple-helical (shown here) and α-helical protein-like molecular architecture is stabilized in the peptide-amphiphile.

are developed as potential biomaterials for bone substitutions due to their compositional analogy to bone [98–100]. HA/collagen composites are biodegradable [101] and good matrices for bone cell attachment and proliferation [102] as well as new bone formation [103].

Recently, aragonite polymorph of $CaCO_3$ in gastropod abalone nacre was converted to HA within its ordered, interpenetrating organic matrix [104]. HA-converted nacre may have biomedically useful properties. Flat, *ab*-face pieces of both nacre [105] and HA [106] induce bone formation by osteoblasts (bone-forming cells) in vitro and bivalve nacre can be fashioned into a strong, long lasting dental root prosthesis [107]. Therefore a material with the nacre ultrastructure and HA composition may be useful for either in vivo or in vitro bone growth studies.

Fig. 7. A schematic of SAM (*n*-alkanethiol $CH_3(CH_2)_n SH$ molecules) formation on a Au(111) sample. Color code: Au—yellow, S—red, C—blue and H—white.

There is still considerable research activity in the area of surgical metal implant materials and new developments are being introduced regularly. Metal implants can have porous outer coatings of polymers, composites, or ceramics. Our body though tolerates better most ceramics used in surgical implantation than other biomaterials, particularly metals. However, ceramics are brittle and prone to catastrophic failure, whereas, naturally produced ceramic organic biocomposites possess mechanical properties that make them suitable as biomaterials. The mollusc shell is one such example, where an inorganic phase grows onto a charged and organized organic template. This natural biomineralization provides a calcium carbonate phase with satisfactory mechanical properties appropriate for surgical implantation [108].

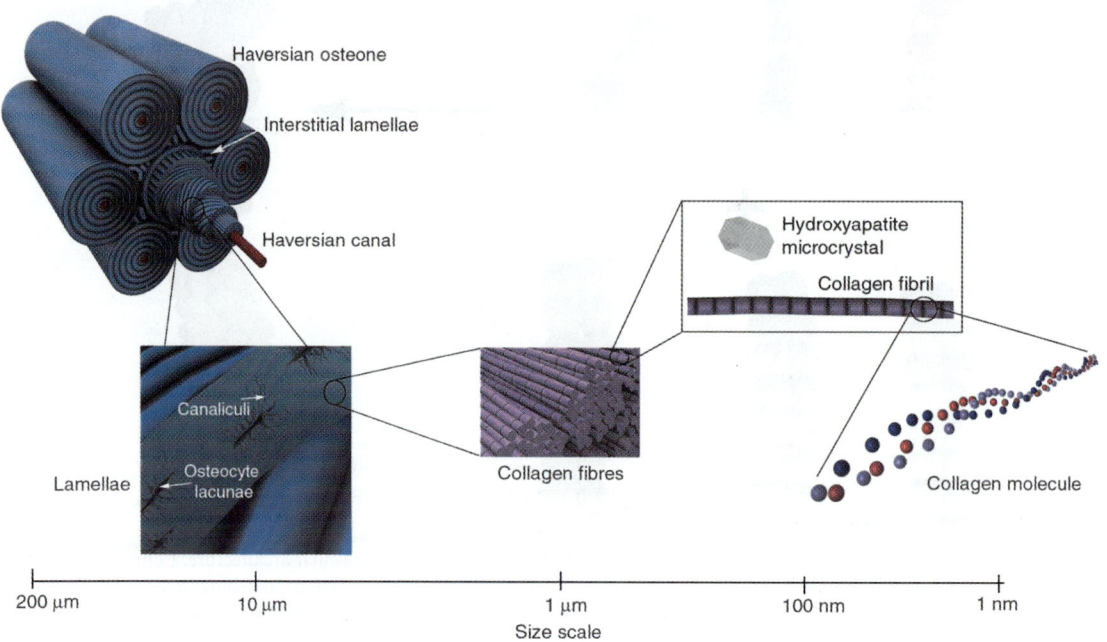

Fig. 8. Human bone is an hierarchical composite of different constituents which are employed in the construction of bone tissue of two varieties, namely woven and lamellar bone (shown here). These structural components are the protein scaffold, dahllite crystals, also known as carbonate hydroxyapatite [$Ca_5(PO_4,CO_3)_3(OH)$], and osteocytes (bone cells) each of which occupies its own cavity, or lacunae, within the matrix, with fine channels (canaliculi) radiating from each lacunae [91–93]. The protein scaffold consists mainly of collagenous fibers that are composed of much finer collagen fibrils (80–100 nm in diameter) carrying the distinctive periodic cross-banding of about 68 nm. Each fibril is made up of packed triple-helical collagen molecules (three polypeptide chains, about 1000 amino-acids long, wound together in a triple helix). The triple-helical molecules are all parallel, but their ends are separated by holes of about 35 nm and the neighboring triple-helical molecules are offset or staggered by 68 nm. The HA crystals (on average $50 \times 25 \times 2$ nm^3) are plate-shaped and are organized in layers that traverse across the fibrils. In lamellar bone, the collagen fibers are organized into unit layers of sheets, the lamellae (a few microns thick). The lamellae are arranged like skins of onion concentrically around a central vascular "Haversian" canal, blood vessels (the Haversian canal with its associated concentric lamellae constitute a "Haversian system" or "osteone"), and as "interstitial lamellae", interposed between the Haversian systems. In lamellae the osteocyte lacunae are laid out in a regular, orderly pattern.

LB films with methyl and calcium carboxylate functional groups have been used as organic templates for calcium phosphate crystallization. The calcium carboxylate functional group controls calcium phosphate crystallization by a biomimetic approach [108]. This study strongly suggests that this approach has the potential for producing high performance ceramics comparable with those made by mollusc shells.

3.3. Precision synthesis

Based on the knowledge of the structure motifs involved in intercellular signaling and cell adhesion, such as proteins containing adhesive peptide sequences as described above, scientists have tried to create and modify surfaces of biomaterials with such short peptide sequences in various synthetic pathways. The challenge of this strategy is to precisely construct surfaces that prevent non-specific interactions with biological material and, furthermore, are capable of possessing and controlling only receptor-mediated cell responses to the modified surface, thus providing a synthetic ECM.

Surfaces modified with a large number of different functional polymers, such as polymer hydrogels and polymer brushes have been synthetically modified to mimic the ECM by the attachment of specific cell adhesion sites to the polymer surface. Both synthetically and naturally derived polymers have been used for this purpose. An example for a naturally derived polymer hydrogel that is widely used in cell encapsulation, cell transplantation and tissue engineering, is the polysaccharide alginate [109]. Precise modification of alginates was shown to fulfill several requirements for the establishment of successful interplay between living cells and surfaces modified with the hydrogel [109–111]. In one example, Rowley et al. [111] have used organic wet chemistry, in particular, carbodiimide chemistry, to covalently graft short RGD-containing peptide sequences onto a crosslinked alginate hydrogel. Surfaces prepared with these functionalized hydrogels promote the adhesion, and the proliferation of cells (mouse skeletal myoblast cells) as a function of the surface density of bioactive peptide sequences. The amount of RGD incorporated can be controlled by parameters of the chemical grafting reaction [111].

As noted in Section 2.1, PEG is extensively used in constructing biocompatible surface coatings. A large number of surface modifications with PEG containing specific cell-recognition sites were reported, where PEG layers are applied in various conformations on surfaces including crosslinked PEG hydrogels [112,113], surface-immobilized PEG brushes [54], star-like PEG molecules [53], and PEG covered lattices [114]. Hern and Hubbell [113] established photo-polymerized crosslinked networks of PEG diacrylate that are functionalized with GRGDS and YRGDS sequences by the attachment of an acrylate unit onto the free amine of the short peptide and copolymerization of this peptide monomer with the PEG diacrylate monomer. Both, the concentration of active peptide sequence and the distance of the active site from the hydrogel-modified surface determine the adhesion and spreading of human foreskin fibroblast cells [113].

In another approach star-like PEG tethers modified with different amounts of RGD peptides were attached to glass surfaces, which had been previously covered with a crosslinked PEG hydrogel [53]. The motility of mouse fibroblast cells on such surfaces was investigated as a function of the spatial distribution of the active sites and the amount of active sites applied per PEG-star assembly. The observations demonstrate that the motility of the cells, which can be taken as a measure for the strength of the cell–surface interactions, may be regulated by the variation of this spatial distribution [115,116].

Using surfaces modified with PEG brushes it was found that the grafting density of the surface-attached polymer chains is the main factor that determines the non-adhesiveness of the prepared systems [24–26]. Generally, polymer brushes at solid surfaces can be prepared in two different pathways: Either by physisorption of blockcopolymers or by chemisorption, which means by grafting of polymer chains "to" or "from" the surface. Both methods are described in books, articles, and references therein [36,117,118]. Finally, further examples, in which PEG or other uncharged polymer hydrogel modified surfaces such as polyacryl amide hydrogels or biodegradable polylactide based hydrogels used to induce specific interactions with living cells, are reviewed in the literature [8,10].

Beside hydrogels and polymer brushes, the broad field of silane- and thiol-chemistry has been used in synthetic surface modification to achieve active communication of surfaces with living cells. Organic silane compounds can be reacted easily with glass surfaces. Since these compounds can be obtained with a large variety of functionalities they are suitable candidates to establish functionalized surfaces, including bioactive surfaces [58,59,89, 119]. In one example, Rezania and Healy [59] have modified aminosilane-functionalized glass surfaces with short RGD containing peptides and peptide sequences originating from heparin-binding domains (FHRRIKA). The short peptide sequences were linked via a thiol group present on a cysteine residue of the peptide to the organic layer at the surface. The adhesion, and mineralization of osteoblasts (cells that aid the growth and development of bones) was investigated as a function of the surface peptide composition. The results suggest that surfaces modified with peptide sequences containing both cell- and heparin-adhesive sites are suitable candidates to control the degree of cell–surface interactions. In a similar approach Porte-Durrieu et al. [119] used substrates functionalized with a maleimide to anchor adhesive peptide motifs. Massia and Hubbel [52] prepared alkyl siloxane-modified glass surfaces with the active peptide GRGDY, showing that even a large spacing between the peptide groups of 440 nm is enough to induce spreading and 140 nm to observe focal contact and stress fiber formation of foreskin fibroblasts [52]. Because this approach relies on a random chemical attachment of the peptide with the alkylated surface the characterization and analysis of the peptide surface concentration may be difficult.

Another technique, in which the "precise synthesis" is a tool to generate bioactive surface coatings, is the preparation and the use of surface layers consisting of peptide-amphiphiles [54,80–85]. These molecules easily self-assemble into ordered layers that have been shown to promote cell adhesion and spreading. The construction of such peptide-amphiphiles was established by a combination of wet organic chemistry and peptide solid phase synthesis [80–85]. RGD- and looped RGD-modified lipids have been synthesized as well as lipids containing a peptide headgroup derived from collagen (IVH1-sequence). Mixed layers of such peptide-amphiphiles with "inert" lipids, which have not been modified with a peptide headgroup, were shown to promote the adhesion and spreading of human melanoma and endothelial cells [54,85]. Due to the defined surfaces generated, the analysis of the peptide surface concentration was easy to achieve and very accurate. However, a high concentration of approximately 50 mol% peptide-amphiphiles within the mixed lipid layers was necessary to obtain optimum adhesion and spreading characteristics of human melanoma cells [54].

3.4. Templating

A successful approach in the bioengineering of materials requires not only surfaces that are capable of specifically communicating with a living environment, but that also are able to define, induce and control a distinct cellular architecture and outgrowth. Different approaches exist to achieve this sophistication in the control of cell/surface interactions where the spatial distribution of bioactive sites are restricted to micron-scale patterns.

It is known that living cells respond both to chemical as well as topographical cues. A number of articles, reviews, and books exist that describe and discuss recent advances in this challenging field of engineering biomaterials [120–125]. The first observations of cells responding to certain topographies of a substrate were obtained almost a century ago [126]. Up to now, scientists are still controversially discussing whether the "self-alignment of cells" relies on the topography of the material, which means the cell is "contact-guided", or whether the fabrication of such topographies actually produces a "chemical pattern" guiding the cells into distinct arrays. On the other hand, it may be possible that cells are aligned at surface topographies unintentionally formed by the pattern of a chemical adhesive track.

Despite this ongoing discussion both topographically and/or chemically patterned surfaces have shown large potentials to control the spatial distribution of cells on solid substrates. Among others techniques micro-contact printing (µCP) [89, 120, 127,128], micro-fluidic/micro-molding [129–131] or photo-lithography [121,132–136] are typically used to imprint both, distinct chemical and/or topographical patterns on planar surfaces.

Lithographic methods define patterns in radiation-sensitive materials. The lateral spacing of the pattern is determined by the source of radiation used. A beam of light (e.g. UV light) is suitable for µm-resolution, whereas for smaller lateral scales it is necessary to use electron beam lithography. Photo-lithographic techniques have been used recently to create micro-patterns on solid surfaces that can be subsequently chemically modified to control the interactions of such surfaces with biological material. Examples include surfaces used in cell alignment [132–134] and bioassays [135,136].

Spargo et al. [132] have studied the adhesion, spreading and proliferation of endothelial cells on patterned glass surfaces that were coated with SAMs of N-(2-aminoethyl)-3-aminopropyl trimethoxysilane (EDA) or EDA subsequently treated with physically adsorbed ECM components (fibronectin and heparin sulfate). The surface pattern on the glass substrates was generated by UV light (ArF excimer laser; $\lambda = 193$ nm) ablation of a pre-adsorbed film of EDA with a lateral spacing between the ridges of 100 µm. Cell-adhesion experiments were carried out on glass, EDA-modified glass, and substrates where ECM components, such as fibronectin were adsorbed to EDA regions. The authors showed that endothelial cells respond to the surface chemical pattern of the substrate (chemical composition or hydrophilicity), and that further proliferation and cord formation of these cells along a distinct pattern can be enhanced by the addition of growth factors [132]. In a chemically similar approach Hickman et al. [133] were able to control the growth of hippocampal neurons by spatial dictation of the cell adhesion to restricted areas that have been previously modified with SAMs of EDA. Ruehe et al. [134] used lithographically patterned polymer brushes that were covalently attached to planar solid substrates to direct the growth of neuronal cells.

Another approach where chemically modified surfaces can be used to pattern cells on solid substrates is µCP [120,127,136]. This technology relays on the formation of SAMs, which are "inked" on a polymer (PDMS) stamp and subsequently "printed" onto various planar solid substrates. Zhang et al. [127] have synthesized oligopeptides containing a cell-adhesive motif $(RADS)_{2-3}$ at the N-terminus followed by an oligoalanine linker and a cysteine residue at the C-terminus. The oligopeptides were attached to gold layers on glass surfaces via the thiol group of the cysteine. The peptides were mixed with short PEG molecules, preventing non-specific interactions of biological material with the prepared surfaces. Bovine aortic endothelial cells were grown to a distinct spatial distribution on the patterned surfaces. Due to the simplicity of this type of approach the µCP of small peptides to planar surfaces provides a playground for scientists to address important questions of "how cells react to external stimuli?" or "how cells communicate with each other?".

Another similar, non-lithographic method, to chemically pattern a solid surface, is the micro-fluidic patterning technique [130,131]. This approach requires a treatment of a partially protected hydrophobic polymer PDMS stamp with oxygen plasma [130]. Non-protected areas of the PDMS stamp are hydrophilized by this treatment. Afterwards the PDMS mold is transferred onto a substrate and aqueous solutions containing biomaterials can be brought into contact with the surface of the substrate on restricted areas via capillary forces.

Patel et al. [130] modified a polystyrene substrate with a biodegradable polylactide–polyethylene glycol (PLA–PEG) copolymer presenting biotin moieties at the surface. Biotin is a small molecule that acts as a growth factor in many cells and is well known because of its high affinity binding to the protein Avidin (a protein originating from egg white). The biotin–avidin binding is considered to be irreversible. A partially hydrophilized PDMS stamp was molded onto this surface and ligands containing the cell-adhesion motif GRGDS linked to biotinylated-PLA–PEG copolymer were attached to the surface via a biotin–avidin coupling. With this technique the authors were able to spatially control the adhesion and spreading of endothelial and PC12 nerve cells. Folch and Toner [131] used the same technique to pattern various polymer surfaces, but first injected a protein solution into the micro-capillaries, from which adhesion factors were allowed to adsorb on the surface. Subsequently, fibroblast cells were adhered to the protein-patterned surfaces.

Beside photo-lithographical and µCP methods more sophisticated techniques, such as sulfur bearing templates attached to gold-sputtered patterns on polymer surfaces [137] or the implantation of metal-ions into polymer films [138] exist to create suitable spatially patterned surfaces for the alignment of cells onto materials. Saneinejad and Shoichet [137] reported on the spatial alignment of hippocampal neuronal cells on patterned surfaces that were obtained by sputtering thin gold layers through an electron microscopy grid onto polychlorotrifluoro ethylene–polyethylene glycol (PCTFE–PEG) copolymer films. Subsequently, cell-adhesive motifs were attached via cysteine residues to the metal layers. The authors show that Hippocampal neuronal cells were adherent on areas containing the active peptide sequences, but did not adhere on the PCTFE–PEG polymer film [137].

Walboomers et al. [139] and Britland et al. [140] have shown that cells can be "contact-guided" along deep grooves on a patterned surface without any chemical treatments to the surface itself. In an interesting article, Britland et al. [140] modified glass substrates with both chemical and topographical cues. The chemical adhesive tracks (laminin, an ECM protein carrying the RGD peptide sequence) were patterned both parallel and perpendicular to the topographical grooves as shown in Fig. 9. With this kind of approach the authors show two interesting things. Cells orient along chemical adhesive track, if the grooves on the substrate are less than 500 nm in depth. However, with deeper grooves the contact-guidance by the topography dominates the cellular behavior (alignment).

In conclusion, contact-guidance and chemical cues are balanced to direct the orientation of cells

Fig. 9. Modified glass substrates with both chemical (blue lines) and topographical cues as used by Britland et al. [140] to investigate parameters that guide living cells to form a distinct pattern on the surface.

in a specific spatial distribution on a surface of a biomaterial. This mutual dependence on both, the topography of the material and chemical cues at the interface provide another means to direct cell behavior and growth at a biomaterials surface.

3.5. Specific and non-specific forces

The term "specific interactions" is used to refer to the involvement of biochemically specific receptor-ligand bonds in cell adhesion. Receptor-ligand recognition is often considered a lock-and-key mechanism. In many cases, it is believed that adhesion would not occur without these interactions, and thus the expression of receptors on a cell and/or the modulation of receptor affinity or number with time serves to control the type of surface and cells with which it interacts [141–143]. "Non-specific interactions" are defined as interactions between a cell and surface that do not provide receptors, and can increase or decrease the overall strength of the interaction. The non-specific forces for cell–cell and cell–surface adhesion are electrostatic, steric, van der Waals and hydrophobic forces [141–143].

Electrostatic forces, which may be attractive or repulsive, occur between charged molecules (or between charged segments of large molecules). The surface of a cell consists of a lipid bilayer con-

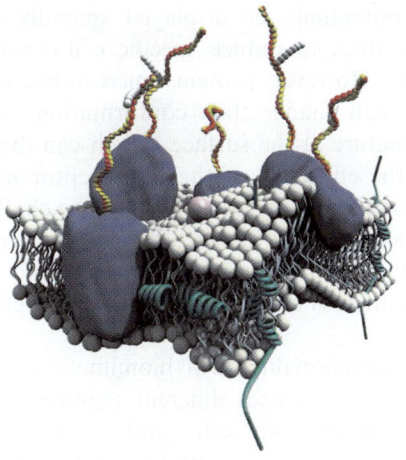

Fig. 10. The fluid mosaic model describes the essential features of the biological membrane. It is a mosaic because it is a structure that includes different kinds of macromolecules such as phospholipids, proteins, cholesterol, glycoproteins (proteins that are covalently bonded to carbohydrates) and glycolipids (lipids that are covalently bonded to monosaccharides or polysaccharides). A lipid bilayer forms the framework of the membrane with proteins randomly embedded in it and it is a two-dimensional fluid, or liquid crystal, in which the hydrophobic integral components of lipids and membrane proteins are constrained within the plane of the membrane, but are free to diffuse laterally.

taining receptors and other embedded molecules as well as the glycocalyx (charged macromolecules, typically of polysaccharide origin) (Fig. 10). The glycocalyx is of the order of 100 Å thick and is negatively charged due to the numerous sialic acid residues present [141]. The electrostatic contribution of the cell–surface interaction depends on the surface chemistry and can be positive or negative.

The glycocalyx consists of polymers in a hydrated environment. As molecules of polymeric size approach each other, the layers overlap and some of the water of hydration is pushed out. The osmotic tendency of the water to return, together with the steric compression of the polymer chains, results in a repulsive force. Distortion of the population of the chain configurations from its most random or maximum-entropy distribution leads to an elastic response tending to restore maximally random configurations [142]. These forces are generally referred to as steric interactions. They dominate at small separation distances and act to prevent significant interpenetration of the glycocalices on two cells.

Attractive van der Waals forces result from the charge interactions of polarizable molecules, including molecules with no net charge, on the cell surface and in the solvent. The hydrophobic interaction has been the subject of much recent debate [144,145]. The strong and long-range attraction found between macroscopic hydrophobic surfaces occupies a unique place in surface science and undoubtedly plays a role in cell adhesion phenomena [146].

The structural integrity of proteins and supramolecular assemblies depends critically on the strength [147,148] and spatial distribution [149] of the non-specific forces on a surface. Bongrand and Bell calculated the magnitudes of the non-specific forces for the case of cell–cell adhesion [150]. They demonstrated that at small separation distances, repulsive potentials dominate the cell–cell interaction, which diminish at separation distances of the order of 200 Å (much beyond glycocalyx interpenetration). Cell–cell and cell–surface adhesion is observed at separation distances of 100–300 Å, so these repulsive forces are important to consider in a treatment of cell adhesion [141]. As the separation distance increases, attractive forces increase the likelihood of cell–cell adhesion as they only weakly overcome the electrostatic and steric repulsion [141].

The non-specific forces provide only a weak attractive force. In order to strengthen adhesive interactions as well as provide specificity, receptors must play a role. In a theoretical discussion of the specific adhesion of cells to cells, Bell has calculated the equilibrium strength of a receptor-ligand bond in the case of an antigen–antibody to be 120 pN/bond from the relation $f_c = E_0/r_0$, where f_c is the force required to break the bond, E_0 is the free energy of bond formation and r_0 is the range of the bond potential energy minimum [151]. A related approach to analyze the strength of a receptor-ligand bond is the kinetic that was introduced by Bell [151] and is based on the kinetic theory of the strength of solids [152]. According to the kinetic point of view $f_c = 40$ pN/bond [151]. The kinetic approach gives a similar but lower estimate than the equilibrium approach, because it does not require all bonds to break simultaneously. In fact, both experiments [153] and simulations [154] suggest that receptor-ligand pairs typically break in multiple steps. These bonds appear to be strong or weak depending on how fast they are loaded [155], and their strength of adhesion at interfaces appears to be quite dynamic with time- and loading rate-dependent properties [156].

Receptors are the ultimate "smart" materials that respond to their environment. They can respond both structurally and functionally when they are occupied by ligands [157]. Their tendency to bind ligands, and hence their function, can be altered by changes in temperature, pH, or ionic strength. Furthermore, when occupied by ligands, they can generate or participate in the generation of signals (biochemical reactions such as phosphorylation and acetylation) that in turn might change the function of the receptor or the function of molecules (other receptors or intracellular molecules) that interact with the adhesion receptor [142].

4. Perspectives in design and application of bioengineered materials

In the previous sections we discuss the role of the surface science in bioengineering materials. Although much progress has been made so far, many questions in the design and the application of bioengineered materials remain open, thus this is a challenging and broad field of interdisciplinary research.

The concept of designing biomimetic materials that combine synthetic materials with cell-recognition sites is attractive. Within these kinds of hybrid materials one future challenge is to provide surfaces with enhanced mechanical strength and/or degradation properties, as well as a specific biological activities.

There are still difficulties in targeting specific cell types among a population of many different cells, and in inducing function and architecture in multiple cells when they interact with a biomimetic material surface. A promising approach could be the patterning of proteins, where several cell-adhesion ligands with highly specific recognition could potentially be displayed spatially only in certain areas to induce specific cell-organization schemes. However, proteins micro-fabricated at a surface can change their conformation, i.e., they can denature at the surface, which can drastically lower the efficiency of the cell-receptor mediated response to a living environment. The challenge in this case is to provide surfaces on which proteins exploring distinct ligands for the specific interactions with cells can adsorb without changing their activity.

Multifunctionality in a biomimetic engineered material that utilizes different peptide sequences incorporating both cell- and heparin-adhesive motifs as well as polypeptide growth factors that regulate a variety of cell functions are highly desirable. Self-assembly of peptide-amphiphiles is a novel method of engineering bioactive, conformationally constrained peptides onto biomaterial surfaces with distinct protein-like structures. Thus, their ability to self-assemble into well-defined and intricate structures and their applicability in incorporating any kind of peptide head group makes them a promising tool towards the design of a biomimetic multipeptide surface.

By using materials that self-assemble into ordered hierarchical structures it might be possible to induce new levels of ordering in cell-biomaterial responses. The overall shape of a hierarchical three-dimensional matrix determines the gross shape and size of the tissue; the μm size and structure of the matrix pores coordinates cell invasion and growth; and the surface chemistry of the material controls the adhesion and spreading of cells. Advances in surface science and nanotechnology should allow the synthesis of such materials with desirable properties.

Biomineralization is another important area. Biomineralized tissues such as bones and teeth are complex multifunctional composites that are of fundamental importance in medicine and health care. There are important implications of biomineralization research for new advances in materials surface science. For example, the use of biomineralization proteins and their synthetic analogues for the control of crystal properties and hierarchical organization is an area of growing interest. Such biomolecules are likely to be used as templates for the fabrication of inorganic systems that could serve as biomimetic structures for more conventional uses in biomaterials engineering.

Future design of engineering composite materials may therefore be influenced by the complex structures and multifunctions of biological soft and hard tissues. These biological materials include soft tissues such as mucus and cartilage; and hard tissues such as teeth and mollusk shells. We are now learning how to produce more complex materials inspired by nature. We must learn much more about the relationships between structure and physical properties of natural, hierarchically complex surfaces. Perhaps more important and far more reaching are the inspirations we can get from the study of biological tissues. Such an inspiration may prove invaluable in the design of new advanced bioengineered materials.

Bioengineered materials are important for future applications in directing and controlling a certain receptor-mediated host response. Moreover, bioengineered surfaces are useful tools in molecular/cell biology and pathology in exploring and understanding how materials communicate in nature.

Acknowledgements

We thank Peter Allen at the Office of the Dean of Engineering—Publication Department for valuable assistance with the graphic design of the figures. Financial support by the Materials Research Laboratory Program of the National Science Foundation under Award DMR00-80034 is gratefully acknowledged. M.B. likes to thank the German National Science Foundation (DFG) for a fellowship within the "Emmy-Noether-Program".

References

[1] M.D. Pierschbacher, E. Ruoslahti, Cell attachment activity of fibronectin can be duplicated by small synthetic fragments of the molecule, Nature 309 (1984) 30.
[2] M.D. Pierschbacher, E. Ruoslahti, Variants of the cell recognition site of fibronectin that retain attachment-promoting activity, Proc. Natl. Acad. Sci. USA 81 (1984) 5985.
[3] E. Ruoslahti, M.D. Pierschbacher, New perspectives in cell-adhesion—RGD and integrins, Science 238 (1987) 491.
[4] B.D. Ratner, in: B.D. Ratner (Ed.), Biomaterials—an Introduction to Materials in Medicine, Academic Press, San Diego, 1996.
[5] M.J. Humphries, The molecular basis and specificity of integrin ligand interactions, J. Cell Sci. 97 (1990) 585.
[6] F.G. Giancotti, E. Ruoslahti, Transduction—integrin signaling, Science 285 (1999) 1028.
[7] F.G. Giancotti, Complexity and specificity of integrin signaling, Nature Cell Biol. 2 (2000) E13.
[8] D.L. Elbert, J.A. Hubbell, Surface treatments of polymers for biocompatibility, Annu. Rev. Mater. Sci. 26 (1996) 365.
[9] D. Klee, H. Hoecker, Polymers for biomedical applications: improvement of the interface compatibility, Adv. Polym. Sci. 149 (1999) 1.
[10] L.G. Griffith, Polymeric biomaterials, Acta Mater. 48 (2000) 263.
[11] R. Langer, J.P. Vacanti, Tissue engineering, Science 260 (1993) 920.
[12] J.A. Hubbell, Bioactive biomaterials, Curr. Opin. Biotechnol. 10 (1999) 123.
[13] J.A. Hubbell, Biomaterials in tissue engineering, Bio-Technology 13 (1995) 565.
[14] N.A. Peppas, R. Langer, New challenges in biomaterials, Science 263 (1994) 1715.
[15] D.F. Williams (Ed.), Definitions in Biomaterials, Progress in Biomedical Engineering, vol. 4, Elsevier, Amsterdam, 1987.
[16] J.M. Anderson et al., in: B.D. Ratner, et al. (Eds.), Biomaterials—an Introduction into Materials in Medicine, Academic Press, San Diego, 1996, p. 165.
[17] R.A.F. Clark, P.M. Henson (Eds.), The Molecular and Cellular Biology of Wound Repair, second ed., Plenum Press, New York, 1996.
[18] J.I. Gallin, R. Snyderman (Eds.), D.T. Feraon, B.F. Haynes, C. Nathan (assoc. Eds.), Inflammation: Basic Principals and Clinical Correlates, third ed., Lippincott Williams and Williams, Philadelphia, 1999.
[19] J.L. Brash, T.A. Horbett (Eds.), Proteins at Interfaces: Physicochemical and Biochemical Studies, Adv. Symp. Ser. 343, Am. Chem. Soc., Washington, 1987.
[20] T.A. Horbett, J.L. Brash (Eds.), Proteins at Interfaces II: Fundamentals and Application, Adv. Symp. Ser. 602, Am. Chem. Soc., Washington, 1995.
[21] R. Kunz, C. Anders, L. Heinrich, K. Gersonde, Investigation into the mechanism of bacterial adhesion to hydrogel-coated surfaces, J. Mater. Sci. Mater. Med. 10 (1999) 649.
[22] C.R. Jenney, J.M. Anderson, Effects of surface-coupled polyethylene oxide in human macrophage adhesion and foreign body giant cell formation in vitro, J. Biomed. Mater. Res. 44 (1999) 206.
[23] D.L. Hern, J.A. Hubbell, Incorporation of adhesion peptides into nonadhesive hydrogels usefull for tissue resurfacing, J. Biomed. Mater. Res. 39 (1998) 266.

[24] I. Szleifer, M.A. Carignano, Macrom. Tethered polymer layers: phase transitions and reduction of protein adsorption, Rapid Commun. 21 (2000) 423.

[25] Z. Xu, R.E. Marchant, Adsorption of plasma proteins on polyethylene oxide-modified lipid bilayers studied by total internal reflection fluorescence, Biomaterials 21 (2000) 1075.

[26] Z.H. Yang, J.A. Galloway, H.U. Yu, Protein interactions with poly(ethylene glycol) self-assembled monolayers on glass substrates: diffusion and adsorption, Langmuir 15 (1999) 8405.

[27] T. Lehmann, J. Ruehe, Polyethyloxazoline monolayers for polymer supported biomembrane models, Macrom. Symp. 142 (1999) 1.

[28] (a) S.I. Jeon, J.H. Lee, J.D. Andrade, P.G. De Gennes, Protein surface interactions in the presence of poly(ethylene oxide), 1. Simplified theory, J. Coll. Interf. Sci. 142 (1991) 149;
(b) S.I. Jeon, J.D. Andrade, Protein surface interactions in the presence of poly(ethylene oxide), 2. Effect of protein size, J. Coll. Interf. Sci. 142 (1991) 159.

[29] J.D. Andrade, V. Hlady, S.I. Jeon, Poly(ethylene oxide) and protein resitance—principles, problems, and possibilities, Adv. Chem. Ser. 248 (1996) 51.

[30] N.A. Peppas, Y. Huang, M. Torres-Lugo, J.H. Ward, J. Zhang, Physicochemical foundations and structural design of hydrogels in medicine and biology, Annu. Rev. Biomed. Eng. 2 (2000) 9.

[31] E.A. Vogler, Structure and reactivity of water at biomaterial surfaces, Adv. Coll. Int. Sci. 74 (1998) 69.

[32] D.H. Napper, Steric Stabilization of Colloidal Dispersions, Academic Press, London, 1983.

[33] J. Klein, E. Kumacheva, Confinement-induced phase transitions in simple liquids, Science 269 (1995) 816.

[34] J. Klein, Shear, friction, and lubrication forces between polymer-bearing surfaces, Annu. Rev. Mater. Sci. 26 (1996) 581.

[35] R.S. Parnas, Y. Cohen, A terminally anchored polymer chain in shear-flow—self-consistent velocity and segment density profiles, Rheol. Acta 33 (1994) 485.

[36] G.J. Fleer, M.A. Cohen-Stuart, J.M.H.M. Scheutjens, T. Cosgrove, B. Vincent (Eds.), Polymers at Interfaces, Chapman and Hall, London, 1993.

[37] J.M. Harris (Ed.), Poly(ethylene glycol) Chemistry, Plenum Press, New York, 1992.

[38] M. Amiji, K. Park, Surface-modification of polymeric biomaterials with poly(ethylene oxide)—a steric repulsion approach, J. Biomater. Sci. Polym. Ed. 4 (1993) 217.

[39] G.R. Llanos, M.V. Sefton, Does polyethylene oxide posses a low thrombogenicity, J. Biomater. Sci. Polym. Ed. 4 (1993) 381.

[40] C.J. Kirkpatrick, M. Otto, T. van Kooten, V. Krump, J. Kriegsmann, F. Bittinger, Endothelial cell cultures as a tool in biomaterial research, J. Mater. Sci. Mater. Med. 10 (1999) 589.

[41] S.E. Greenwald, C.L. Berry, Improving vascular grafts: the importance of mechanical and haemodynamic properties, J. Pathol. 190 (2000) 292.

[42] G.W. Bos, A.A. Poot, T. Beugeling, W.G. van Aken, J. Feijen, Small-diameter vascular graft prostheses: current status, Arch. Physiol. Biochem. 106 (1998) 100.

[43] M. Deutsch, J. Meinhart, T. Fischlein, P. Preiss, P. Zilla, Clinical autologous in vitro endothelialization of infrainguinal ePTFE grafts in 100 patients: a 9-year experience, Surgery 126 (1999) 847.

[44] Y. Ikada, Surface modifications of polymers for medical applications, Biomaterials 15 (1994) 725.

[45] B.-S. Kim, D.J. Mooney, Development of biocompatible synthetic extracellular matrices for tissue engineering, Trends Biotechnol. 16 (1998) 224.

[46] A.K. Dillow, M. Tirrell, Targeted cellular adhesion at biomaterial surfaces, Curr. Opin. Solid State Mater. Sci. 3 (1998) 252.

[47] E. Ruoslahti, RGD and other recognition sequences for integrins, Annu. Rev. Cell Dev. Biol. 12 (1996) 697.

[48] M.D. Pierschbacher, E. Ruoslahti, Influence of the stereochemistry of the sequence ARG–GLY–ASP–XAA on binding specificity in cell adhesion, J. Biol. Chem. 262 (1987) 17294.

[49] S. Cheng, W.S. Craig, D. Mullen, J.F. Tschopp, D. Dixon, M.D. Pierschbacher, Design and synthesis of novel cyclic RGD-containing peptides as highly potent and selective alpha(IIB)beta(3) integrin antagonists, J. Med. Chem. 37 (1994) 1.

[50] A. Hautanen, J. Gailit, D.M. Mann, E. Ruoslahti, Effects of modifications of the RGD sequence and its context on recognition by the fibronectin receptor, J. Biol. Chem. 264 (1989) 1437.

[51] T. Pakalns, K.L. Haverstick, G.B. Fields, J.B. McCarthy, D.L. Mooradian, M. Tirrell, Cellular recognition of synthetic peptide amphiphiles in self-assembled monolayer films, Biomaterials 20 (1999) 2265.

[52] S.P. Massia, J.A. Hubbell, An RGD spacing of 400 nm is sufficient for integrin alpha-V-beta-3-mediated fibroblast spreading and 140 nm for focal contact and stress fiber formation, J. Cell Biol. 114 (1991) 1089.

[53] G. Maheshwari, G. Brown, D.A. Lauffenburger, A. Wells, L.G. Griffith, Cell adhesion and motility depend on nanoscale RGD clustering, J. Cell Sci. 113 (2000) 1677.

[54] Y. Dori, H. Bianco-Peled, S.K. Satija, G.B. Fields, J.B. McCarthy, M. Tirrell, Ligand accessibility as means to control to bioactive bilayer membranes, J. Biomed. Mater. Res. 50 (2000) 75.

[55] S. Vukicevic, F.P. Luyten, H.K. Kleinman, A.H. Reddi, Differentiation of canalicular cell processes in bone cells by basement membrane matrix components: regulation by discrete domains of laminin, Cell 63 (1990) 437.

[56] A.P. Mould, A. Komoriya, K.M. Yamada, M.J. Humphries, The CS5 peptide is a second site in the IIICS region of fibronectin recognized by the integrin $\alpha_4\beta_1$. Inhibition of $\alpha_4\beta_1$ function by RGD peptide homologous, J. Biol. Chem. 266 (1991) 3579.

[57] B.A. Dalton, C.D. McFarland, P.A. Underwood, J.G. Steele, Role of the heparin-binding domain of fibronectin

in attachment and spreading of human bone-derived cells, J. Cell Sci. 108 (1995) 2083.
[58] K.E. Healy, A. Rezania, R.A. Stile, Designing biomaterials to direct biological responses, Ann. N.Y. Acad. Sci. 875 (1999) 24.
[59] A. Rezania, K.E. Healy, The detachment strength and morphology of bone cells contacting materials modified with a peptide sequence found within bone sialoprotein, Biotechnol. Prog. 15 (1999) 19.
[60] D.A. Puleo, A. Nanci, Understanding and controlling the bone–implant interface, Biomaterials 20 (1999) 2311.
[61] D. Logeart-Avramoglou, J. Jozefonvicz, Carboxymethyl benzylamide sulfonate dextrans (CMDBS), a family of biospecific polymers endowed with numerous biological properties: a review, J. Biomed. Mater. Res. 48 (1999) 578.
[62] S.I. Stupp, V. LeBonheur, K. Walker, L.S. Li, K.E. Huggins, M. Keser, A. Amstutz, Supramolecular materials: self-organized nanostructures, Science 276 (1997) 384.
[63] E. Sackmann, Supported membranes: scientific and practical applications, Science 271 (1996) 43.
[64] C.A. Steinem, A. Janshoff, W.P. Ulrich, M. Sieber, H.J. Galla, Impedance analysis of supported lipid bilayer membranes: a scrutiny of different preparation techniques, Biochim. Biophys. Acta-Biomembr. 1279 (1996) 169.
[65] G. Puu, I. Gustafson, Planar lipid bilayers on solid supports from liposomes: factors of importance for kinetics and stability, Biochim. Biophys. Acta 1327 (1997) 149.
[66] A.L. Plant, Supported hybrid bilayer membranes as rugged cell membrane mimics, Langmuir 15 (1999) 5128.
[67] E. Sackmann, M. Tanaka, Supported membranes on soft polymer cushions: fabrication, characterization and applications, Trends Biotechnol. 18 (2000) 58.
[68] S.G. Boxer, Molecular transport and organization in supported lipid membranes, Curr. Opin. Chem. Biol. 4 (2000) 704.
[69] R.G. Horn, Direct measurement of the force between two lipid bilayers and the observation of their fusion, Biochim. Biophys. Acta 778 (1984) 224.
[70] L.K. Tamm, H.M. McConnell, Supported phospholipid bilayers, Biophys. J. 47 (1986) 105.
[71] C. Duschl, M. Liley, H. Lang, A. Ghandi, S.M. Zakeeruddin, H. Stahlberg, A. Nemetz, W. Knoll, H. Vogel, Sulphur-bearing lipids for the covalent attachment of supported lipid bilayers to gold surfaces: a detailed characterization and analysis, Mater. Sci. Eng. C 4 (1996) 7.
[72] A.N. Parikh, J.D. Beers, A.P. Shreve, B.I. Swanson, Infrared spectroscopic characterization of lipid-alkylsiloxane hybrid bilayer membranes at oxide substrates, Langmuir 15 (1999) 5369.
[73] E.A. Evans, E. Sackmann, Translational and rotational drag coefficients for a disk moving in a liquid membrane-associated with a rigid substrate, J. Fluid. Mech. 194 (1988) 553.
[74] J.T. Groves, N. Ulman, S.G. Boxer, Micropatterning fluid lipid bilayers on solid supports, Science 275 (1997) 651.
[75] E. Györvary, B. Wetzer, U.B. Sleytr, A. Sinner, A. Offenhäusser, W. Knoll, Lateral Diffusion of lipids in silane-, dextran-, and S-layer-supported mono- and bilayers, Langmuir 15 (1999) 1337.
[76] J. Majewski, J.Y. Wong, C.K. Park, M. Seitz, J.N. Israelachvili, G.S. Smith, Structural studies of polymer-cushioned lipid bilayers, Biophys. J. 76 (1998) 2363.
[77] J.Y. Wong, J. Majewski, M. Seitz, C.K. Park, J.N. Israelachvili, G.S. Smith, Polymer-cushioned bilayers. I. A structural study of various preparation methods using neutron reflectometry, Biophys. J. 77 (1999) 1445.
[78] C.S. Cho, T. Kotaka, T. Akaike, Cell-adhesion onto block copolymer langmuir-blodgett-films, J. Biomed. Mater. Res. 27 (1993) 199.
[79] M. Uchida, T. Tanizaki, T. Oda, T. Kajiyama, Control of surface chemical-structure and functional property of langmuir-blodgett-film composed of new polymerizable amphiphile with a sodium-sulfonate, Macromolecules 24 (1991) 3238.
[80] P. Berndt, G.B. Fields, M. Tirrell, Synthetic lipidation of peptides and amino-acids-monolayer structure and properties, J. Am. Chem. Soc. 117 (1995) 9515.
[81] Y.-C. Yu, P. Berndt, M. Tirrell, G.B. Fields, Self-assembling amphiphiles for construction of protein molecular architecture, J. Am. Chem. Soc. 118 (1996) 12515.
[82] Y.-C. Yu, M. Tirrell, G.B. Fields, Minimal lipidation stabilizes protein-like molecular architecture, J. Am. Chem. Soc. 120 (1998) 9979.
[83] Y.-C. Yu, V. Roontga, V.A. Daragan, K.H. Mayo, M. Tirrell, G.B. Fields, Structure and dynamics of peptide-amphiphiles incorporating triple-helical protein-like molecular architecture, Biochemistry 38 (1999) 1659.
[84] Y.-C. Yu, T. Pakalns, Y. Dori, J.B. McCarthy, M. Tirrell, G.B. Fields, Construction of biologically active protein molecular architecture using self-assembling peptide-amphiphiles, Meth. Enzymol. 289 (1997) 571.
[85] G.B. Fields, J.L. Lauer, Y. Dori, P. Forns, Y.-C. Yu, M. Tirrell, Proteinlike molecular architecture: Biomaterial applications for inducing cellular receptor binding and signal transduction, Biopolymers 47 (1998) 143.
[86] W. Prass, H. Ringsdorf, W. Bessler, K.H. Wiesmuller, G. Jung, Lipopeptides of the n-terminus of escherichia-coli lipoprotein—synthesis, mitogenicity and properties in monolayer experiments, Biochim. Biophys. Acta 900 (1987) 116.
[87] T.M. Winger, P.J. Ludovice, E.L. Chaikof, Lipopeptide conjugates: biomolecular building blocks for receptor activating membrane-mimetic structures, Biomaterials 17 (1996) 437.
[88] M. Mrksich, G.M. Whitesides, Using self-assembled monolayers to understand the interactions of man-made surfaces with proteins and cells, Annu. Rev. Biophys. Biomol. Struct. 25 (1996) 55.
[89] C. Roberts, C.S. Chen, M. Mrksich, V. Martichonok, D.E. Ingber, G.M. Whitesides, Using mixed self-assembled monolayers presenting RGD and (EG)(3)OH groups to characterize long-term attachment of bovine capillary

endothelial cells to surfaces, J. Am. Chem. Soc. 120 (1998) 6548.

[90] K.B. McClary, T. Ugarova, D.W. Grainger, Modulating fibroblast adhesion spreading and proliferation using self-assembled monolayer films of alkylthiolates on gold, J. Biomed. Mater. Res. 50 (2000) 428.

[91] N.M. Hancox, Biology of Bone, Cambridge University Press, London, 1972.

[92] S. Weiner, W. Traub, Bone structure: from ångstroms to microns, FASEB J. 6 (1992) 879.

[93] S. Weiner, H.D. Wagner, The material bone: structure–mechanical function realtions, Annu. Rev. Mater. Sci. 28 (1998) 271.

[94] M. Sarikaya, I.A. Aksay (Eds.), Biomimetics: Design and Processing of Materials, AIP Press, Woodbury, NY, 1995.

[95] E. White, E.C. Shors, Biomaterial aspects of interpore-200 porous hydroxyapatite, Dent. Clin. North Am. 30 (1986) 49.

[96] H. Ohgushi, M. Okumura, T. Yoshikawa, K. Inoue, N. Senpuku, S. Tamai, E.C. Shors, Bone-formation process in porous calcium-carbonate and hydroxyapatite, J. Biomed. Mater. Res. 26 (1992) 885.

[97] C.A. van Blitterswijk, J.J. Grote, W. Kuypers, W.T.H. Daems, K. de Groot, Macropore tissue ingrowth: a quantitative and qualitative study on hydroxyapatite ceramic, Biomaterials 1 (1986) 137.

[98] K.I. Clarke, S.E. Graves, A.T.-C. Wong, J.T. Triffitt, M.J.O. Francis, J.T. Czernuszka, Investigation into the formation and mechanical properties of a bioactive material based on collagen and calcium phosphate, J. Mater. Sci. Mater. Med. 4 (1993) 107.

[99] R.Z. Wang, F.Z. Cui, H.B. Lu, H.B. Wen, C.L. Ma, H.D. Li, Synthesis of nanophase hydrohyapatite/collagen composite, J. Mater. Sci. Lett. 14 (1995) 490.

[100] A.C. Lawson, J.T. Czernuszka, Collagen-calcium phosphate composites, Proc. Instn. Mech. Eng. 212H (1998) 413.

[101] F.Z. Cui, C. Du, X.W. Su, X.D. Zhu, N.M. Zhao, Biodegradation of a nano-hydroxyapatite/collagen composite by peritoneal monocyte-macrophages, Cells Mater. 6 (1996) 31.

[102] C. Du, F.Z. Cui, X.D. Zhu, K. de Groot, Three-dimensional nano-HAp/collagen matrix loading with osteogenic cells in organ culture, J. Biomed. Mater. Res. 44 (1999) 407.

[103] C. Du, F.Z. Cui, Q.L. Feng, X.D. Zhu, K. de Groot, Tissue response to nano-hydroxyapatite/collagen composite implants in marrow cavity, J. Biomed. Mater. Res. 42 (1998) 540.

[104] C.M. Zaremba, D.E. Morse, S. Mann, P.K. Hansma, G.D. Stucky, Aragonite-hydroxyapatite conversion in gastropod abalone nacre, Chem. Mater. 10 (1998) 3813.

[105] E. Lopez, B. Vidal, S. Berland, S. Camprasse, G. Camprasse, C. Silve, Demonstration of the capacity of nacre to induce bone-formation by human osteoblasts maintained invitro, Tissue Cell 24 (1992) 667.

[106] J.-M. Sautier, J.-R. Nefussi, N. Forest, Ultrastructural-study of bone formation on synthetic hydroxyapatite in osteoblast cultures, Cells Mater. 1 (1991) 209.

[107] E. Lopez, B. Vidal, S. Berland, S. Camprasse, G. Camprasse, C. Silve, Demonstration of the capacity of nacre to induce bone-formation by human osteoblasts maintained invitro, Tissue Cell 24 (1992) 667.

[108] N. Costa, P.M. Maquis, Biomimetic processing of calcium phosphate coating, Med. Eng. Phys. 20 (1998) 602.

[109] N.A. Peppas, in: Hydrogels in Medicine and Pharmacy, vol. 1, CRC Press, Boca Raton, 1986.

[110] A.C. Jen, M.C. Wake, A.G. Mikos, Review: hydrogels for cell immobilization, Biotechnol. Bioeng. 50 (1996) 357.

[111] J.A. Rowley, G. Madlambayan, D.J. Mooney, Alginate hydrogels as synthetic extracellular matrix materials, Biomaterials 20 (1999) 45.

[112] A.S. Sawhney, C.P. Pathak, J.A. Hubbell, Bioerodible hydrogels based on photopolymerized poly(ethylene glycol)-co-poly(alpha-hydroxy acid) diacrylate macromers, Macromolecules 26 (1993) 581.

[113] D.L. Hern, J.A. Hubbell, Incorporation of adhesion peptides into nonadhesive hydrogels useful for tissue resurfacing, J. Biomed. Mater. Res. 39 (1998) 266.

[114] P. Banerjee, D.J. Irvine, A.M. Mayes, L.G. Griffith, Polymer latexes for cell-resistant and cell-interactive surfaces, J. Biomed. Mater. Res. 50 (2000) 331.

[115] D.A. Lauffenburger, A.F. Horwitz, Cell migration: a physically integrated molecular process, Cell 84 (1996) 359.

[116] S.P. Palecek, J.C. Loftus, M.H. Ginsberg, D.A. Lauffenburger, A.F. Horwitz, Integrin-ligand binding properties govern cell migration speed through cell-substratum adhesiveness, Nature 385 (1997) 537.

[117] O. Prucker, J. Ruhe, Synthesis of poly(styrene) monolayers attached to high surface area silica gels through self-assembled monolayers of azo initiators, Macromolecules 31 (1998) 592.

[118] M. Biesalski, J. Ruhe, Preparation and characterization of a polyelectrolyte monolayer covalently attached to a planar solid surface, Macromolecules 32 (1999) 2309.

[119] M.C. Porte-Durrieu, C. Labrugere, F. Villars, F. Lefebvre, S. Dutoya, A. Guette, L. Bordenave, Development of RGD peptides grafted onto silica surfaces: XPS characterization and human endothelial cell interactions, J. Biomed. Mater. Res. 46 (1999) 368.

[120] C.S. Chen, M. Mrksich, S. Huang, G.M. Whitesides, D.E. Ingber, Geometric control of cell life and death, Science 276 (1997) 1425.

[121] J. Voldman, M.L. Gray, M.A. Schmidt, Microfabrication in biology and medicine, Annu. Rev. Biomed. Mater. 1 (1999) 401.

[122] H.C. Hoch, L.W. Jelinski, H.G. Craighead (Eds.), Nanofabrication and Biosystems: Integrating Materials Science Engineer Cambridge University Press, Cambridge University Press, London, 1996.

[123] A.S.G. Curtis, C.D. Wilkinson, Reaction of cells to topography, J. Biomat. Sci. Polym. Ed. 9 (1998) 1313.

[124] C.D.W. Wilkinson, A.S.G. Curtis, J. Crossan, Nanofabrication in cellular engineering, J. Vac. Sci. Technol. B 16 (1998) 3132.

[125] A. Curtis, C. Wilkinson, Topographical control of cells, Biomaterials 18 (1997) 1573.

[126] R.G. Harrison, On the stereotropism of embryonic cells, Science 34 (1911) 279.

[127] S.G. Zhang, L. Yan, M. Altmann, M. Laessle, H. Nugent, F. Frankel, D.A. Lauffenburger, G.M. Whitesides, A. Rich, Biological surface engineering: a simple system for cell pattern formation, Biomaterials 20 (1999) 1213.

[128] R.S. Kane, S. Takayama, E. Ostuni, D.E. Ingber, G.M. Whitesides, Patterning proteins and cells using soft lithography, Biomaterials 20 (1999) 2363.

[129] M. Mrksich, C.S. Chen, Y. Xia, L.E. Dike, D.E. Ingber, G.M. Whitesides, Controlling cell attachment on contoured surfaces with self-assembled monolayers of alkanethiolates on gold, Proc. Natl. Acad. Sci. USA 93 (1996) 10775.

[130] N. Patel, R. Padera, G.H.W. Sanders, S.M. Cannizzaro, M.C. Davies, R. Langer, C.J. Roberts, S.J.B. Tendler, P.M. Williams, K.M. Shakesheff, Spatially controlled cell engineering on biodegradable polymer surfaces, FASEB J. 12 (1998) 1447.

[131] A. Folch, M. Toner, Cellular micropatterns on biocompatible materials, Biotechnol. Prog. 14 (1998) 388.

[132] B.J. Spargo, M.A. Testoff, T.B. Nielsen, D.A. Stenger, J.J. Hickman, A.S. Rudolph, Spatially controlled adhesion spreading and differentiation of endothelial-cells on self-assembled molecular monolayers, Proc. Natl. Acad. Sci. USA 91 (1994) 11070.

[133] J.J. Hickman, S.K. Bhatia, J.N. Quong, D.A. Stenger, C.J. Pike, C.W. Cotman, Rational pattern design for in-vitro cellular networks using surface photochemistry, J. Vac. Sci. Technol. A 12 (1994) 607.

[134] J. Ruhe, R. Yano, J.S. Koberle, W. Knoll, A. Offenhaeusser, Tailoring of surfaces with ultrathin polymer films for survival and growth of neurons in culture, J. Biomater. Sci. Polym. Ed. 10 (1999) 859.

[135] C. Sproessler, M. Denyer, S. Britland, W. Knoll, A. Offenhaeusser, Electrical recordings from rat cardiac muscle cells using field-effect transistors, Phys. Rev. E 60 (1999) 2171.

[136] M.C.T. Denyer, M. Riehle, J. Hayashi, M. Scholl, C. Sproessler, S.T. Britland, A. Offenhaeusser, W. Knoll, Bioassay development: the implications of cardiac myocyte motility, in vitro, In Vitro Cell. Dev. Biol. Animal 35 (1999) 352.

[137] S. Saneinejad, M.S. Shoichet, Patterned poly(chlorotrifluoroethylene) guides primary nerve cell adhesion and neurite, J. Biomed. Mater. Res. 50 (2000) 465.

[138] J.S. Lee, M. Kaibara, M. Iwaki, H. Sasabe, Y. Suzuki, M. Kusakabe, Selective adhesion and proliferation of cells on ion-implanted polymer domains, Biomaterials 14 (1993) 958.

[139] X.F. Walboomers, W. Monaghan, A.S.G. Curtis, J.A. Jansen, Attachment of fibroblasts on smooth and microgrooved polystyrene, J. Biomed. Mat. Res. 46 (1999) 212.

[140] S. Britland, C. Perridge, M. Denyer, H. Morgan, A. Curtis, C. Wilkinson, Morphogenetic guidance cues can interact synergistically and hierachically in steering nerve cell growth, Exp. Biol. Online 1 (1996) 2.

[141] D.A. Lauffenburger, J.L. Linderman, Receptors: Models for Binding, Trafficking, and Signaling, first ed., Oxford, New York, 1993.

[142] D.A. Hammer, M. Tirrell, Biological adhesion at interfaces, Annu. Rev. Mater. Sci. 26 (1996) 651.

[143] P.F. Luckham, P.G. Hartley, Interactions between biosurfaces, Adv. Coll. Interf. Sci. 49 (1994) 341.

[144] E. Kokkoli, C.F. Zukoski, Interactions between hydrophobic self-assembled monolayers. Effect of salt and the chemical potential of water on adhesion, Langmuir 14 (1998) 1189, and references therein.

[145] E. Kokkoli, C.F. Zukoski, Effect of solvents on interactions between hydrophobic self-assembled monolayers, J. Coll. Interf. Sci. 209 (1999) 60, and references therein.

[146] C.J. van Oss, Interfacial Forces in Aqueous Media, first ed., Marcel Dekker, New York, 1994.

[147] J. Schneider, P. Berndt, K. Haverstick, S. Kumar, S. Chiruvolu, M. Tirrell, Force and adhesion measurements between hydrogen-bonded layers of glycine-functionalized amphiphiles, J. Am. Chem. Soc. 120 (1998) 3508.

[148] J. Schneider, Y. Dori, M. Tirrell, R. Sharma, Effect of substrate anchoring on the mechanical strength of Langmuir–Blodgett bilayers, Thin Solid Films 327–329 (1998) 772.

[149] E. Kokkoli, C.F. Zukoski, Surface pattern recognition by a colloidal particle, Langmuir 17 (2001) 369.

[150] P. Bongrand, G.I. Bell, in: A.S. Perelson, C. DeLisi, F.W. Wiegel (Eds.), Cell Surface Dynamics: Concepts and Models, Marcel Dekker, New York, 1984, p. 459.

[151] G.I. Bell, Models for specific adhesion of cells to cells, Science 200 (1978) 618.

[152] S.N. Zhurkov, Kinetic concept of the strength of solids, Int. J. Fract. Mech. 1 (1965) 311.

[153] E.-L. Florin, V.T. Moy, H.E. Gaub, Adhesion forces between individual ligand-receptor pairs, Science 264 (1994) 415.

[154] R. Vijayendran, D. Hammer, D. Leckband, Simulations of the adhesion between molecularly bonded surfaces in direct force measurements, J. Chem. Phys. 108 (1998) 7783.

[155] R. Merkel, P. Nassoy, A. Leung, K. Ritchie, E. Evans, Energy landscapes of receptor-ligand bonds explored with dynamic force spectroscopy, Nature 397 (1999) 50.

[156] E. Evans, K. Ritchie, Dynamic strength of molecular adhesion bonds, Biophys. J. 72 (1997) 1541.

[157] M.J. Humphries, Integrin activation: the link between ligand binding and signal transduction, Curr. Opin. Cell Biol. 8 (1996) 632.

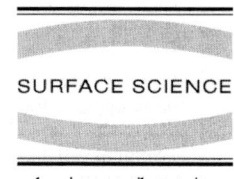

Surface Science 500 (2002) 84–104

www.elsevier.com/locate/susc

Electronic transport at semiconductor surfaces—from point-contact transistor to micro-four-point probes

Shuji Hasegawa [a,b,*], François Grey [c]

[a] *Department of Physics, School of Sciences, University of Tokyo, 7-3-1 Hongo, Bunkyo-ku, Tokyo 113-0033, Japan*
[b] *Core Research for Evolutional Science and Technology, The Japan Science and Technology Corporation, Kawaguchi Center Building, Hon-cho 4-1-8, Kawaguchi, Saitama 332-0012, Japan*
[c] *MIC, Technical University of Denmark, Bldg. 345 East, DK-2800 Lyngby, Denmark*

Received 27 July 2000; accepted for publication 24 April 2001

Abstract

The electrical properties of semiconductor surfaces have played a decisive role in one of the most important discoveries of the last century, transistors. In the 1940s, the concept of surface states—new electron energy levels characteristic of the surface atoms—was instrumental in the fabrication of the first point-contact transistors, and led to the successful fabrication of field-effect transistors. However, to this day, one property of semiconductor surface states remains poorly understood, both theoretically and experimentally. That is the conduction of electrons or holes directly through the surface states. Since these states are restricted to a region only a few atom layers thick at a crystal surface, any signal from them might be swamped by conduction through the underlying bulk semiconductor crystal, as well as greatly perturbed by steps and other defects at the surface. Yet recent results show that this type of conduction is measurable using new types of experimental probes, such as the multi-tip scanning tunnelling microscope and the micro-four-point probe. The resulting electronic transport properties are intriguing, and suggest that semiconductor surfaces should be considered in their own right as a new class of electronic nanomaterials because the surface states have their own characters different from the underlying bulk states. As microelectronic devices shrink even further, and surface-to-volume ratios increase, surfaces will play an increasingly important role. These new nanomaterials could be crucial in the design of electronic devices in the coming decades, and also could become a platform for studying the physics of a new family of low-dimensional electron systems on nanometre scales. © 2001 Elsevier Science B.V. All rights reserved.

Keywords: Electrical transport measurements; Scanning tunneling microscopy; Surface electrical transport (surface conductivity, surface recombination, etc.); Surface electronic phenomena (work function, surface potential, surface states, etc.); Silicon; Silver

1. Introduction

In many ways, the modern science of semiconductor surfaces was born along with the invention of the solid-state transistor in 1947 by William Shockley, John Bardeen and Walter Brattain. Prior to this date, work by Shockley and others to

---

[*] Corresponding author. Address: Department of Physics, School of Sciences, University of Tokyo, 7-3-1 Hongo, Bunkyo-ku, Tokyo 113-0033, Japan. Tel.: +81-3-5841-4167; fax: +81-3-5841-4167.

E-mail address: shuji@surface.phys.s.u-tokyo.ac.jp (S. Hasegawa).

0039-6028/01/$ - see front matter © 2001 Elsevier Science B.V. All rights reserved.
PII: S0039-6028(01)01533-3

make a field-effect transistor had failed. The driving idea behind the earlier attempts was simple; use a gate electrode, as shown in Fig. 1, to induce excess charges in a semiconductor, which travel through it as carriers of current flow, resulting in an increase in conductivity [1]. This kind of device already had been outlined in a prescient patent by Julius Lillienfeld in 1929. By changing the gate voltage, the conductance through the semiconductor could be varied. The gate was thus a tap that could control a flow of electrons—the prerequisite for amplification.

Alas, semiconductors resisted all attempts to fashion them into such a device. It took the analytical mind of Bardeen to realize what was going wrong. Bardeen postulated that there were charged electronic states at the semiconductor surface, which were screening out the electric field induced by the gate [2]. In other words, the induced charges are trapped in the surface states, becoming immobile. Thus the charges in the bulk of the semiconductor hardly felt the effect of the gate, and the current flow could be modified only minimally. This was a great insight, because not only did it explain a known phenomenon, but also it suggested ways to mitigate it. Together with Brattain, the experimental genius behind the first transistor, Bardeen began searching for ways to mitigate the screening effect. The result was the first point-contact type transistor, as shown in Fig. 2, in which two metal needles with a small spacing were brought into contact to a germanium crystal surface [3]. A current was injected into the crystal through the first needle (emitter), and electric potential distribution near the point contact was investigated with the second needle (collector) as a probe, as a function of the voltage applied to the germanium crystal via the base electrode.

Brattain and Bardeen discovered amplification during such measurements; a voltage signal fed through the emitter is transmitted to the collector as a larger voltage signal. The degree of amplification depends on the voltage of the base electrode. In one of the great ironies of technological history, what the researchers had made was not a field-effect transistor at all, but rather a precursor to modern bipolar transistors [4], where the surface states play a less important role, rather minority carriers injected into the bulk states are crucial for the transistor action. Nevertheless, when field-effect transistors were finally fabricated successfully some years later, it was due to Bardeen's insight. In particular, a key breakthrough was finding a surface where the number of surface states could be drastically reduced, greatly diminishing the shielding effect. This surface (or interface) was between silicon and silicon oxide. Interfaces involving the silicon(1 0 0) crystallographic surface showed the lowest density of the states, and for this reason, this crystallographic surface is still the most popular for modern electronic chip fabrication.

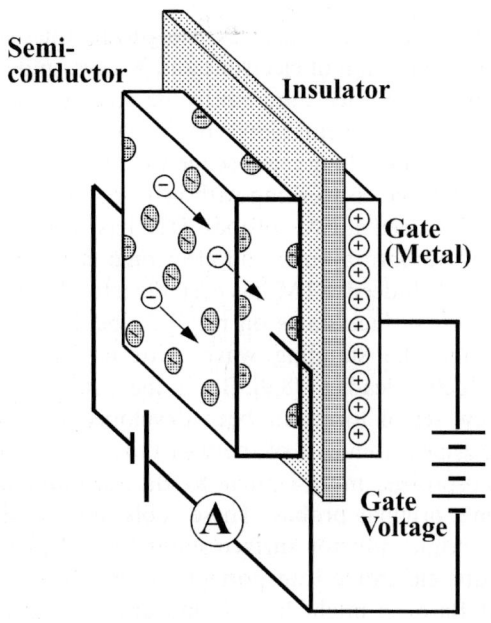

Fig. 1. Schematic of a field-effect transistor [1]. When a voltage is applied between a metal gate and a semiconductor, separated by a thin insulator, excess charges, negative ones on the semiconductor side, for example, are induced because of a capacitance of the structure. This should effectively vary the electrical conductance of the semiconductor, due to the change in number of carriers (electrons) flowing through it. Thus the gate voltage controls the current flow through the semiconductor, the basis of amplification. The field effect is diminished if there are surface states at the semiconductor–insulator interface, because they screen the electric field due to the gate by capturing the induced charges. Thus minimizing the amount of electronic states at the interface is a key requirement for field-effect transistor action.

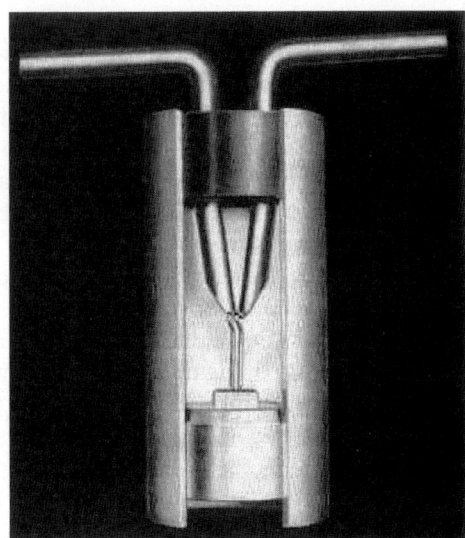

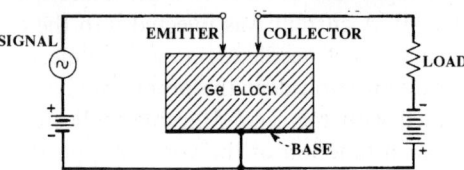

Fig. 2. A point-contact transistor of the type invented by Brattain and Bardeen [3]. The cutaway model shows the semiconductor at the center, below the two point contacts. Typical dimensions of the semiconductor were a couple of millimetres. The casing around the device protected the contacts from changes in ambient conditions that could affect the contacts. Below is a schematic diagram of the electrical circuit for the transistor action.

By the standards of modern surface science, the surface of the Germanium crystal that Brattain and Bardeen used in their pioneering experiments was dirty and ill-defined, since the experiments were carried out in air. Bardeen's idealized picture of surface states probably never applied to such surfaces. But today, under ultra-high vacuum (UHV) conditions in which the number of residual gas molecules is less than in the outer space, it is possible to tailor the crystallographic structure of semiconductor surfaces accurately, for example, by dosing the surface with small amounts (around a monolayer) of specific materials under precise control. As a result, a variety of *surface superstructures*—peculiar, but regular atomic arrangements of the topmost atomic layers on surfaces, a kind of two-dimensional (2D) crystals—are created [5]. There is an intimate and profound relation between the crystallographic structure of a material and its electronic properties. This is well known for three-dimensional (3D) crystals (consider insulating diamond and conducting graphite, both of which are consisted of carbon atoms). It is also true of the 2D periodic structures at crystal surfaces.

What would Brattain and Bardeen have discovered with their two-point probe, had they had access to the well-defined surface states that can be produced today? Oddly enough, although their experiments are over 50 years old, there has been a long hiatus in surface science, during which comparatively little research was carried out on the electronic transport properties of surface states. In the same period, much has been learned about the atomic structure of surfaces, using techniques such as the diffraction of electrons and X-rays, and also about the surface-state electronic structure using primarily photoemission spectroscopy [6]. The scanning tunnelling microscope (STM), which can be compared to working with just one of Brattain and Bardeen's two probes [7], has helped to elucidate both atomic and electronic structure of surfaces. Indeed, STM has even detected transport through surface states on metals, albeit indirectly through the standing wave patterns that such conduction leads to [8,9]. But in the last few years, a new set of tools has been developed for more direct measurements of the transport, microscopic multi-probes, more similar to Brattain and Bardeen's original probes. These tools are revealing that semiconductor surface states have their own unique electronic transport properties. In a sense that Bardeen might have found amusing, history has come full circle.

In the next section, general descriptions of the interest in the electronic structures and electrical conduction at semiconductor surfaces are given. In Section 3, model systems on a Si(1 1 1) surface are shown to give a flavour of a huge variety of surface superstructures and surface states on a silicon crystal. Section 4 is devoted to results of surface-state transport obtained with recently developed probes. These are just an opening of the entrance to a rich field of study on the transport properties of surface states.

2. Surface-state bands and electrical conduction

2.1. What is a semiconductor surface state? And what is so special about it?

At the surface of a crystal, the atomic bonding geometry in the bulk is no longer necessarily the minimum free energy configuration. Atoms at the surface lack neighbours *above* them. For semiconductors such as silicon and germanium, the bonds are highly directional (covalent nature). So-called *dangling bonds*, unpaired bonds pointing in the direction of missing neighbours, are very reactive. These bonds naturally want to pair up with other unsaturated bonds on other surface atoms. The result can be a quite drastic change in the atomic arrangement in only one or two atomic layers on a crystal surface, leading to a lower free energy, but also a completely new periodic structure. This is a *surface reconstruction*, result of which is a *surface superstructure*. For example, the 7×7 superstructure of the (1 1 1) clean surface of a silicon crystal refers to a structure with a unit cell that is 49 times larger than that of the ideal surface, as described later in detail [10]. This is a resultant atomic arrangement of the minimum free energy at the clean surface, leads to only 19 dangling bonds, much fewer than on an unreconstructed surface. Furthermore, the same surface can go through a variety of complete structural rearrangements on absorption of only a fraction of a monolayer of foreign atoms to attain the lowest surface free energy [5]. These are *adsorbate-induced surface superstructures* in which atomic arrangements are in general quite different from those of the corresponding bulk alloys and compounds.

The periodic nature of a crystal leads to energy bands in which electrons occur: a valence band, which is practically full of electrons, and a conduction band, which is almost empty. Between them, a forbidden energy gap opens up (Fig. 3). The physics behind this is similar to diffraction of photons of X-rays in crystals; at some wavelengths and energies, a travelling wave is forbidden. In a map of wave energy against inverse wavelength (*k*-vector) of electrons (*E–k* space, or dispersion relations), the regions where travelling waves are allowed have the shape of energy bands (see Fig. 3(c)). When the periodic structure at the surface of a crystal is different from the bulk of the crystal, the allowed energy bands for electrons on the surface atomic layers may be different from those in the bulk. This is *surface-state bands*.

Sometimes an electron energy level in the allowed surface-state band is at the same time a level where bulk electrons are forbidden. For semiconductors, such surface-state bands become particularly important. The surface-state bands can alter the behaviour of the so-called *surface space–charge layer* just under the surface, by attracting excess negative or positive charges into the surface states, which then has to be compensated by an equal and opposite charges accumulated in the space–charge layer, typically about 1 µm thick. This charging of the surface effectively alters the positions of the bulk energy bands in the vicinity of the surface, an effect called *band bending* (see Fig. 3(a) and (b)). In this way surface states are responsible for the phenomena of charge carrier accumulation, depletion and inversion near a semiconductor surface. Without going into detail here, it is worth noting that such effects have led to 2D electron gases (2DEG) in such space–charge layers, where the quantum Hall effect, field-effect transistor action, and a rich variety of other electronic transport properties have been found [11].

In this article, however, the focus is on the conduction properties of the surface-state bands themselves, not of the space–charge layer. An obvious effect is that, for example, if a surface-state band crosses the Fermi level—the energy level above which the electron states are empty and below which they are full at very low temperatures—which lies in the bulk-band gap of the semiconductor (a metallic character), the surface-state band could be conducting. So the surface has completely different electronic transport properties from the bulk crystal. As we illustrate in the following sections, this is exactly what is seen in some cases.

2.2. Electrical conduction at semiconductor surfaces

The most common way to measure the electrical conductivity is to use a four-point probe, as shown in Fig. 4(a). A current is made flow through the

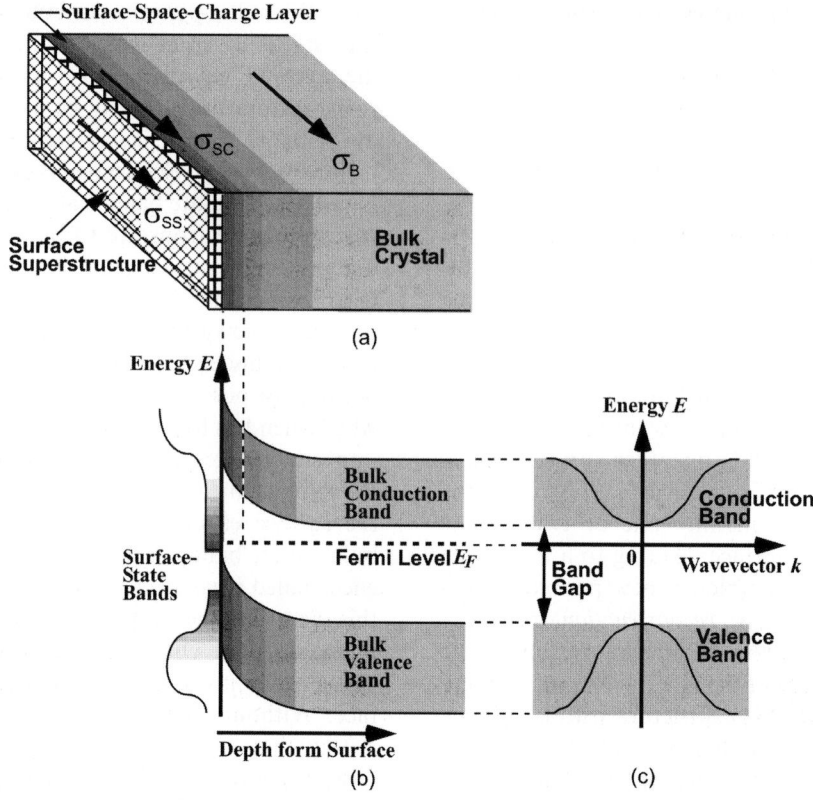

Fig. 3. Panel (a): Schematics of a semiconductor crystal, showing three channels for current flow; bulk states in the interior crystal (its conductivity is σ_B), surface-space–charge layer near the surface (σ_{SC}), and surface-state bands inherent in the surface superstructure at the topmost layers (σ_{SS}). Panel (b): An energy diagram near the surface, showing the bulk band structure, band bending beneath the surface, and surface-state bands. Panel (c): A band dispersion diagram, showing a relation between energy and wavevector of electrons in a crystal structure.

outer pair of probes, and a voltage drop is measured across the inner pair of probes, using a voltmetre with such large impedance that it draws hardly any current. As a result, the measured voltage drop V is predominantly that occurring across the semiconductor surface due to the current I flowing through the specimen. The four-point-probe resistance R of the sample is then obtained by $R = V/I$ (with a geometrical correction factor depending on the shape of specimen and probe spacing).

If we look more closely at the situation in Fig. 4(a), where the probes have macroscopic spacing (i.e. much larger than the thickness of the space–charge layer) in most cases, the current flows through three parallel channels [12]:

1. surface-state bands on the topmost atomic layers (when well-ordered surface superstructures are formed),
2. surface-space–charge layer due to band bending (in which conductivity can be considerably larger or smaller than in the bulk of the crystal in some cases),
3. unperturbed bulk bands inside the crystal (independent of the surface conditions and treatments).

The conductivity measured by a four-point probe method contains contributions from all of the three channels. In general it is very difficult to separate the contributions from each channel. For measurements made in air, without surface

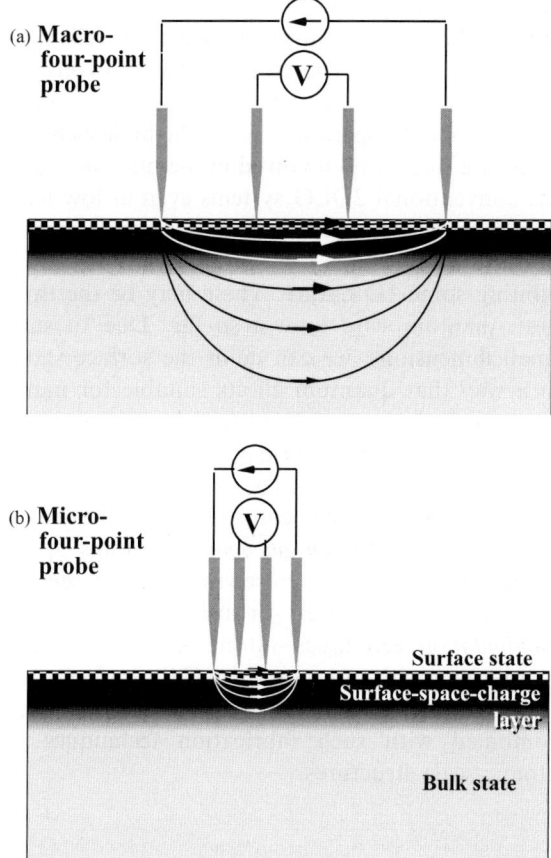

Fig. 4. Schematic diagram of a linear four-point probe measurement. A constant current source injects current through the left outermost probe, which is collected in the right outermost one. The potential difference measured across the inner probes, with a suitably high-impedance voltmetre, is then dependent only on the voltage drop across the specimen surface, eliminating contact resistance phenomena. Panel (a): Macroscopic probes in which the probe spacing is much larger than the thickness of the surface space–charge layer, and so most of the current passes through the bulk of the crystal. Panel (b): Microscopic probes, with schematic illustrations of current paths near a semiconductor surface, penetrating only a distance comparable to the spacing between the probes.

superstructures on the crystal, the measured resistance is normally interpreted as bulk resistance only. Possible contributions from the surface space–charge layer and surface-state bands are ignored. This is usually a good approximation, since on macroscopic scales, the surface-layer contributions are relatively small because a large fraction of the current tends to flow through the interior of the crystal, far from the surface, as illustrated in Fig. 4(a). Despite this drawback, it has been possible in careful measurements with macroscopic probes to detect the effects of conductance through the surface space–charge layer, and in special cases also conductance directly through surface-state bands. As described in Section 4, these have been measured in UHV with crystals having well-ordered surface superstructures [13].

In the early days of surface science, a number of techniques were reported for measuring the surface conductivity of a semiconductor. In general, the data were interpreted in terms of conductivity in the surface space–charge layer, due to band bending, rather than the surface-state conductivity, because the latter was assumed (without much basis) to be negligible. Brattain and Bardeen [14] found that the band bending could be varied in a reproducible way by exposing the surface to different gaseous ambients (Brattain–Bardeen cycle). Since their experiments were not carried out with clean semiconductor surfaces in UHV environment, the gases adsorb randomly on the outer surface of the oxide film on the semiconductor in a form of ions, and electric fields due to these ions penetrate into the space–charge region across the oxide layer. The conductivity actually changed by varying the band bending due to the gas exposures. From the conductivity changes, reversibly, the amounts of band bending were deduced [15].

Changes in surface conductivity accompanied by conversions of surface superstructures in a well-controlled manner in UHV were first measured by Mönch and co-workers on clean Si and Ge crystals [16]. Their results showed a strong correlation between the structural conversion and change in surface conductivity, from which they could deduce a change in band bending under the surface. The band bending depends on the charge in the surface states, which redistributes due to the reconstruction of the surface atomic structure. In these pioneering experiments, the surface-state conductivity was not taken into consideration, because it was assumed to be negligibly small.

As described in Section 4, the surface-state electrical conduction has been confirmed first by a macroscopic four-point probe method through a

phenomena of carrier doping into a surface-state band [17], followed by other surfaces with conductive surface-state bands of metallic nature [13].

Another approach to detect and study the surface-state electronic transport is miniaturizing the four-point probes. By reducing the probe spacing to a distance similar to the space–charge layer thickness, as shown in Fig. 4(b), a larger fraction of the current flows near the surface because the current penetrates to an extent similar to the probe spacing. This results in a more surface-sensitive measurement than with the macroscopic four-point probes shown in Fig. 4(a) [18]. The actual current distribution in the crystal may not be as simple as illustrated in Fig. 4, because of a possible Schottky barrier between the surface states and the underlying bulk states [19] or a possible pn-junction between the surface-space–charge layer and the interior bulk. But, this simple picture appears to be at least qualitatively true, as demonstrated in recent measurements [20–23]. Two types of such microscopic four-point probes are introduced in Section 4.

The surface-state bands provide a new class of reduced-dimension electron systems, which may have the following new intrinsic features when compared with the other low-dimensional electron systems. Three forces described below drive the study of the transport properties of surface states.

(1) Variety: More than 300 kinds of surface superstructures are found on silicon crystals with foreign adsorbates of around monolayer coverages [5]. Their surface-state bands are known to show a rich variety of characters, some of which are introduced in Section 3. We can play with them to study almost all kinds of low-dimensional phenomena just on silicon surfaces. Furthermore, of course, more huge variety of surface states is also known on surfaces of other materials.

(2) Scales: As described in Section 3, some surface-state bands have a nearly 2D free-electron character, which is quite similar to the conventional 2DEG systems formed at the space–charge layers (Fig. 5). One of the main differences between them, however, is *thickness*. The surface-state systems are much thinner, around a monatomic thick, compared with a thickness of around 10 nm for the conventional 2DEG. This results in much larger energy spacing between the quantized energy levels in the confinement potential wells normal to the surface. Therefore, we have to consider only the lowest energy level for the surface states even at room temperature, while the higher-energy levels are frequently involved in the phenomena at the conventional 2DEG systems even at low temperatures. Some other surface-state bands consist of 1D atomic chains only a few atoms wide, exhibiting some 1D nature. These may be the thinnest quantum wire known so far. Due to such small dimensions, we can tailor the surface states in a way that quantum effects suitable for nanometre-scale devices can be expected even for metals having short Fermi wavelengths.

(3) Controllability: The structures on surfaces can be precisely controlled and fabricated using *self-organization phenomena*—atoms spontaneously rearrange themselves to make regular surface superstructures, or using *atom/molecule manipulation* techniques—atoms are made arrange one by one with *atom tweezers* [24]. Surface-state transport may be controlled in a novel way when combined with such fabrication techniques of atomic-scale structures.

3. Silicon(1 1 1): a case study

To get a more quantitative picture of surface states of silicon crystals, one of the most popular materials in surface science, consider surface-state bands using the schematic illustration given in Fig. 6. When a Si atom makes a bond with a neighbouring atom, the energy level splits into bonding and anti-bonding states. The valence electrons are accommodated in the bonding state, so that the anti-bonding state is empty. Since in a crystal many atoms make bonds to each other to arrange themselves periodically, these energy levels are broadened to make bands, valence band and conduction band, respectively. These are electronic states in a bulk crystal (see also Fig. 3).

But on the surface, there are the dangling bonds, which have an energy level located between the bonding and anti-bonding states, or, within an energy gap of the bulk band structure. Actually, the dangling-bond state on a clean Si(1 1 1) surface

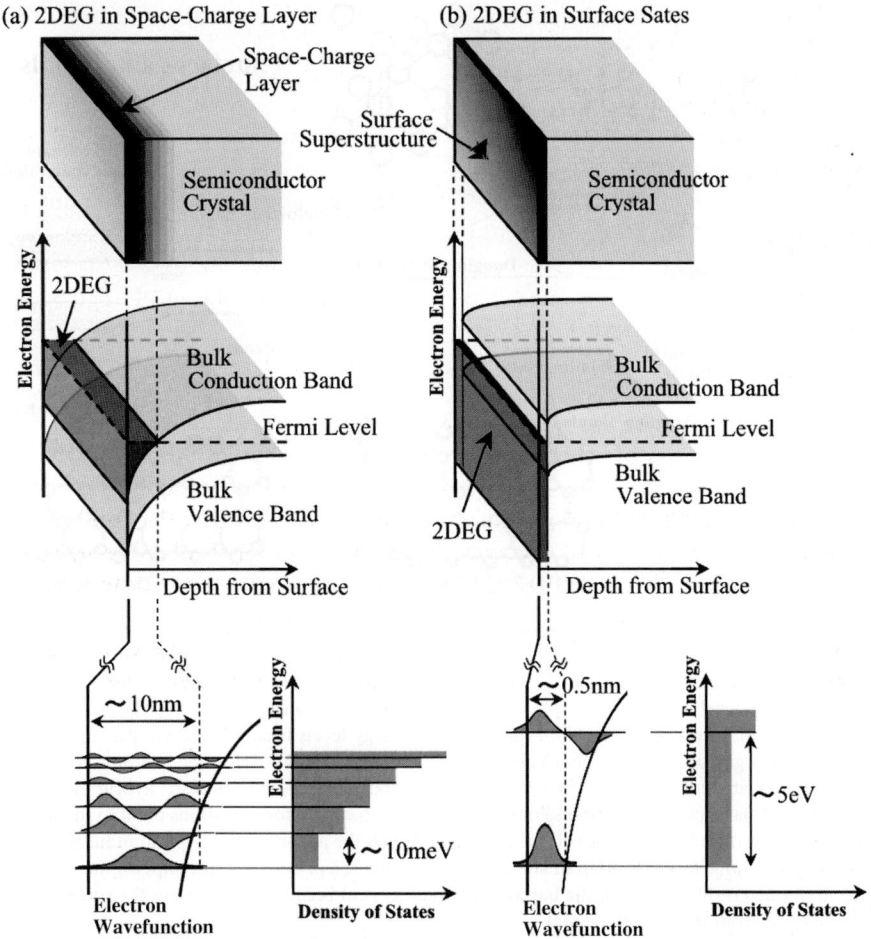

Fig. 5. Comparison of 2DEG systems between (a) at a surface space–charge layer and (b) at surface states. Upper panels: real-space illustrations showing where the 2DEGs are formed. Middle panels: energy band diagrams showing the electronic states of the 2DEGs. The former 2DEG (a) is formed in bulk state bands extending over 10 nm into the crystal, while the latter 2DEG (b), of which thickness is around 0.5 nm, is formed at the topmost few atomic layers on the crystal surface. Lower panels: potential wells and wavefunctions of the 2DEGs, and also their density of states. The spacing of energy levels is much larger in (b) than in (a) due to the much narrower well width.

is known to lie around the middle of the band gap [25,26]. But when foreign atoms bond with the topmost Si atoms, the dangling-bond state is again split into bonding and anti-bonding states. When, furthermore, the adsorbates induce a surface superstructure to generate overlaps between atomic orbitals of neighbouring sites, the bonding and anti-bonding states are broadened into bands. These are *surface-state bands*, inherent in the surface superstructure, which have their own character, independent of the bulk electronic states. To give a hint of the great variety of surface-state band structures that can arise on silicon, we give three examples on a Si(1 1 1) crystal surface.

3.1. Si(1 1 1)–√3 × √3-Ag: a two-dimensional metal

The Si(1 1 1)–√3 × √3-Ag surface superstructure is formed by depositing one monolayer (ML) of Ag atoms on a Si(1 1 1) surface at temperatures higher than 250 °C. One ML means an atom

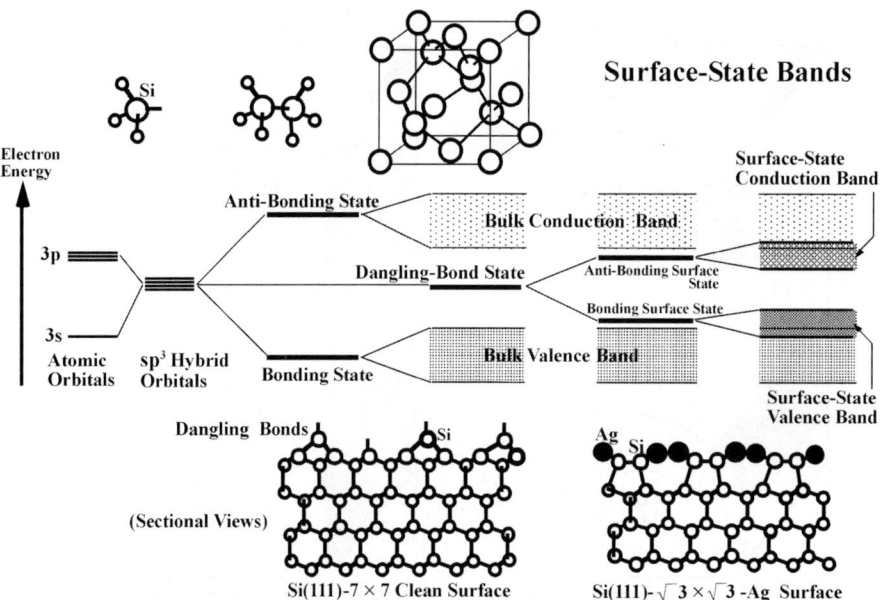

Fig. 6. A schematic illustration of energy diagram of atoms, molecules, and also in bulk and at surface of a silicon crystal. The valence electrons of an isolated Si atom are in 3p and 3s atomic orbitals. These orbitals are 'hybridized' into sp³ orbitals when the atoms arrange in tetrahedral structures like in a diamond-lattice crystal and in SiH₄ molecules. When the atoms make bonds with the neighbouring atoms, the energy level of the orbitals splits into anti-bonding and bonding states. Furthermore, in the crystal the atomic orbitals overlap with each other between the neighbouring sites, the energy levels broaden into bands, conduction band and valence band, between which an energy gap opens. This is an electronic state in the bulk crystal of Si. The Si atoms on the topmost surface layer of the crystal, however, have dangling bonds, of which energy level is similar to the unpaired sp³ hybrid orbitals, locating within the band gap. The dangling-bond state split into anti-bonding and bonding states when foreign atoms bond to the topmost Si atoms. These states are created only in the surface layer, surface electronic states. When the adsorbate adsorption induces a surface superstructure, the surface states become bands due to the overlap of surface states among the periodic atomic sites on the surface. These are surface-state bands. They have characters different from the bands in the bulk crystal because the atoms in the surface superstructure arrange in a way completely different from in the bulk crystal.

density equal to that of the topmost layer of Si(1 1 1) face, 7.8×10^{14} atoms/cm². The atomic arrangement of this surface is known to be a so-called honeycomb-chained triangle (HCT) structure as shown in Fig. 7(a) [27–29], in which each Ag atom makes ionic covalent bond with a substrate Si atom, leaving no dangling bonds, resulting in an exceptionally stable surface.

This surface has a characteristic surface-state band that is revealed by a technique called angle-resolved photoemission spectroscopy (ARPES) [30,31]. The surface-state band is parabolic and crosses the Fermi level (E_F), as indicated by S_1 band in Fig. 7(b). A parabolic band is characteristic of a 2D *free-electron-like state*, in other words, a state in which the electrons behave like free travelling waves along the surface. Furthermore, this band is partially filled with electrons, a requisite of a metal, and confined only in the topmost layer of Ag and Si atoms. Therefore we can call this surface a 2D metal.

Such a 2D free-electron state can be visualized in low-temperature STM images in the form of so-called *electron standing waves* (more technically *energy-resolved Friedel oscillations* [32]). Such standing waves were first observed on metal surfaces independently by Eigler's group [8] and Avouris's group [9], and extended by several others [33–36]. They are essentialy the consequence of electron travelling waves bouncing off steps and other defect structures, leading to a self-interference pattern that changes the probability for electrons tunnelling from the tip into the sample surface. Although not a direct measurement of

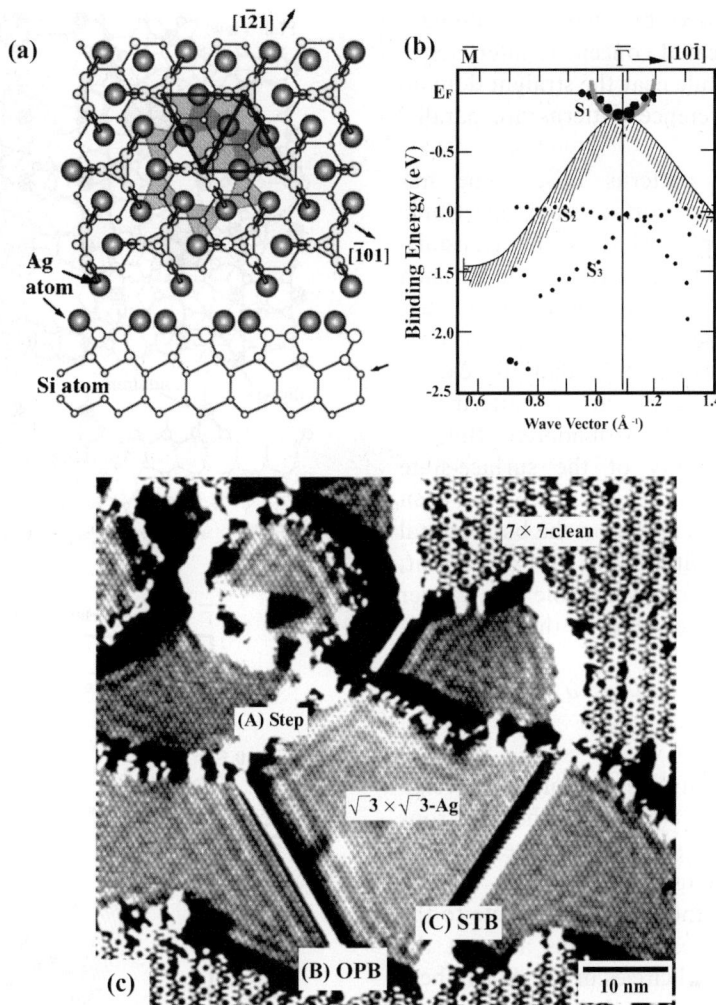

Fig. 7. Panel (a): Schematic illustrations of atomic arrangements (upper panel: plan view, lower panel: sectional view) of the Si(1 1 1)–$\sqrt{3} \times \sqrt{3}$-Ag surface superstructure (a HCT model [27]). Panel (b): Its band dispersion diagram showing surface-state bands, constructed by angle-resolved photoemission spectroscopy measurements [57]. The shallowest state S_1 has a parabolic shape crossing the Fermi level E_F, indicating a nearly free-electron character. The curve with hatching shows a projected region of the bulk bands of a silicon crystal. $\overline{\Gamma}$ and $\overline{M}$ indicate symmetric points in the $\sqrt{3} \times \sqrt{3}$ surface Brillouin zone. Panel (c): An empty-state STM image of this surface (partially the 7 × 7 domains remain) taken at 6 K [32]. 'OPB' is an out-of phase domain boundary at which a mismatch in the $\sqrt{3} \times \sqrt{3}$-periodicity occurs between the adjacent domains. 'STB' is a 'surface twin boundary' at which domains in twin relation in structure meet [31]. Ripples are observed near step edges (A) and OPBs (B), while no observable ripples at STB (C). This means that the steps and OPBs act as potential barriers for surface-state electrons to 'reflect' the electron waves, resulting in an interference between the incident and reflected waves to form the standing wave patterns.

surface transport, they provide indirect proof of the free-electron nature of the surface-state electrons.

Fig. 7(c) shows an STM image of the $\sqrt{3} \times \sqrt{3}$-Ag surface taken at 6K (though the 7 × 7 clean domains partially remain, because of a Ag coverage smaller than 1 ML). In the $\sqrt{3} \times \sqrt{3}$-Ag domains, fine periodic corrugations are seen, corresponding to the periodicity of this superstructure. Additionally, one can see standing wave patterns superimposed near step edges (A) and domain boundaries (B). In a small domain on the upper right, surrounded by steps and domain boundaries, a complicated concentric interference pattern is observed, while near the straight domain boundaries, the interference patterns are parallel to the boundaries.

These interference patterns raise some important questions for surface-state electronic transport: For example, what is the transmission coefficient of electron wavefunction at such boundaries? This governs an important parameter, the mobility of surface-state carriers. Certainly from Fig. 7(c), one can deduce that the carrier mobility must be lowered by carrier scattering by the step edges and domain boundaries. But, by how much? The mobility of the surface-state electrons on the $\sqrt{3} \times \sqrt{3}$-Ag surface has been measured to be lower than that in the bulk crystal by two orders of magnitude [17]. But at present, there is no consensus on what role the carrier scattering from step edges plays in this difference.

3.2. The Si(111) 7 × 7 surface: a poor metal or Mott insulator?

For the 7 × 7 clean surface without any foreign atoms adsorbed (Fig. 8(a)) [10], the electrons which contribute to the surface state are those on the remaining dangling-bonds on the surface atoms, which are also those that contribute to the STM images. Since these Si atoms are not the nearest neighbours (filled circles in Fig. 8(a)), there is very little overlap between their electron wavefunction. The degree of overlap controls how the curvature of the surface-state band in E–k space. On the $\sqrt{3} \times \sqrt{3}$-Ag surface the overlap is large, hence the parabolic shape of the surface-state band S_1 (Fig. 7(b)). On the 7 × 7 surface the small overlap leads to an almost flat surface state (nearly no variation of energy with k-vector), as shown by S_1 band in Fig. 8(b).

As a result of this difference, electrons on the 7 × 7 surface behave much more as isolated states than as travelling waves. This is the reason why there are no observable standing waves on the 7 × 7 domains in Fig. 7(c), even though the surface state is very close to the Fermi level, suggesting a metallic behaviour. The actual position of S_1 band in Fig. 8(b) is disputed; if it crosses the Fermi level,

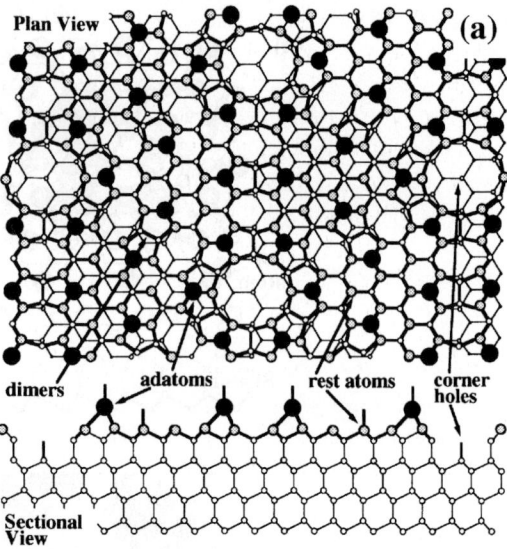

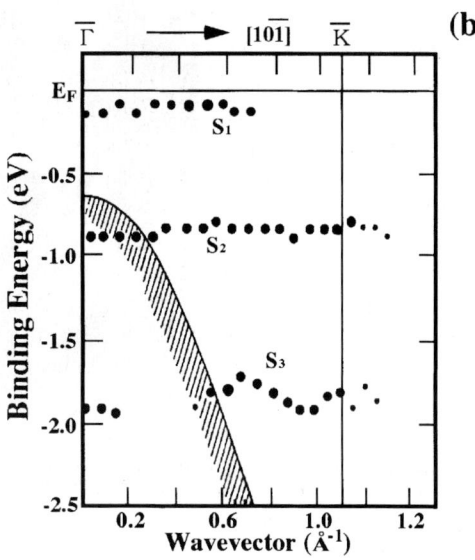

Fig. 8. Si(111)-7 × 7 clean surface. Panel (a): Schematics of the atomic arrangement. Every circles indicate Si atoms, but have different names. Filled circles: adatoms. Shaded circles: rest atoms. The 7 × 7 unit cell is a lozenge having the corner at 'corner holes'. Panel (b): Its band dispersion diagram, constructed by angle-resolved photoemission spectroscopy [13]. The shallowest state S_1 shows an almost flat band, indicating its localized character. This is the dangling-bond surface state of the topmost Si atoms (filled circles in (a)). The curve with hatching shows a projected region of the bulk bands of a silicon crystal. $\overline{\Gamma}$ and $\overline{K}$ indicate symmetric points in the 1 × 1 surface Brillouin zone.

then strictly the surface is metallic [37]. Otherwise an energy gap opens up around the Fermi level, or Mott insulator [38]. Temperature dependent measurements of electrical conductivity through this state, as yet unavailable, could help to distinguish between these two possibilities. Because of this localized nature, the surface-state conductivity is measured to be four orders of magnitude lower than that of the $\sqrt{3} \times \sqrt{3}$-Ag surface, as discussed further on in this article [19].

3.3. Si(1 1 1)–4 × 1-In: a one-dimensional metal

The final example, Si(1 1 1)–4 × 1-In surface superstructure is formed by depositing 1 ML of In atoms on a Si(1 1 1) surface at around 300 °C. Its atomic arrangement is still a matter of debate [39], but one model is shown in Fig. 9(a) [40,41]. According to this model, indium atoms arrange in four lines, between which a zigzag Si chain runs. As one can imagine from this atomic arrangement, its surface electronic state is highly anisotropic. From photoemission spectroscopies [42,43], it is metallic along the In chains, while insulating in the perpendicular direction. In other words, the electrons are mobile along the In chains, while they can hop only with considerable energy to neighbouring chains. This is the recipe for a 1D metal.

The properties of a 1D metal have been appreciated since the early days of quantum mechanics, because solving Schrödinger equation is much easier in 1D than in 2D or 3D. Easier it may be, a 1D metal is inherently unstable. It spontaneously develops some sort of periodic lattice distortion known as a Peierls instability [44]. A charge-density wave (CDW)—a spatially periodic modulation in electron density—accompanies with the distortion to lower the electronic energy.

Fig. 9(b) is an STM image of the 4 × 1-In surface taken at 70 K, showing stripes corresponding to the In chains in four lines given in Fig. 9(a). Along each stripe, one can clearly see the modulations with the double periodicity [45]. The ripples are not so clear on some stripes, due to fluctuations of the structure. A range of other surface science techniques has confirmed that this is a Peierls state, and the ripples are 1D CDW accompanied with a lattice distortion [45,46] (al-

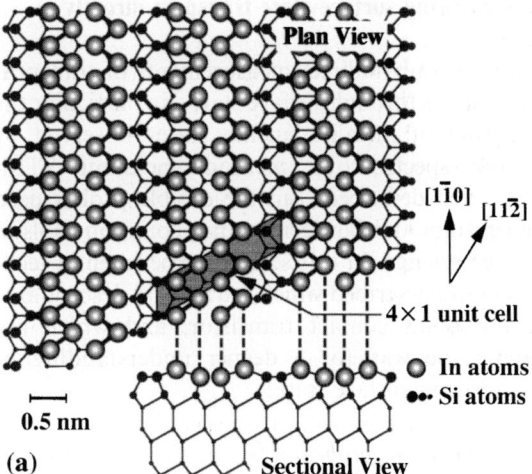

Fig. 9. Panel (a): A model of atomic arrangement of the Si(1 1 1)–4 × 1-In surface [40]. Shaded big circles are indium atoms, and the others are Si atoms. Panel (b): Its empty-state STM image at 70 K [45]. A 4 × 1 unit cell is indicated by a parallelogram. Each stripe corresponds to the four lines of In atoms shown in (a). Modulations of the double periodicity are seen along the stripes, which is due to charge-density waves.

though the latest study casts a doubt against this picture [47]). The surface-state transport must be influenced by this Peierls transition. Although such transitions are known to occur at quasi-1D bulk materials [48,49], the 4 × 1-In surface was the first surface state to show this phenomenon.

4. Measuring surface-state transport directly

So far, only indirect measurements of transport of electrons through surface states, in particular the signature of metallic surface states observed by various spectroscopic methods including STM, have been described. But the double probes used by Brattain and Bardeen are not common tools in surface science. In this section, experimental techniques are described which are natural descendants of the point contact transistor, and which are paving the way for a deeper understanding of transport in surface states.

4.1. 'One-point' probes

As described in Section 2, four-point probes provide the most accurate and direct way of measuring surface transport properties. Nevertheless, much has been learned about the transport with a 'one-point' probe—the STM tip. So some of the outstanding results obtained so far with this technique are considered in this subsection.

Hasegawa(Y), Lyo, and Avouris [50] suggested that the electrical conduction through the dangling-bond surface state on the Si(1 1 1)-7 × 7 is detected as an excess leakage current at a nanometre-scale point contact between the STM tip and the silicon surface. As the radius r of the contact area is reduced to nanometre scale, the current flowing through the metal/semiconductor interface at the contact is reduced in proportion to r^2, while the leakage current through the periphery of the contact area is proportional to r. Therefore the leakage dominates at nanometre scale contacts. The current flows along the surface under such a condition, not directly into the crystal, which is illustrated in the inset of Fig. 10. This is a reason one expects to obtain the surface-state conductivity by this method.

Fig. 10 shows the conductance between the tip and specimen, measured as a function of tip displacement towards the clean Si(1 1 1)-7 × 7 surface, together with the data for a clean Si(1 0 0) and Au surfaces for comparison [50]. In a tunnelling regime, where the tip–sample distance is larger (0–0.5 nm range in the horizontal axis), the conductance increases exponentially with decrease of

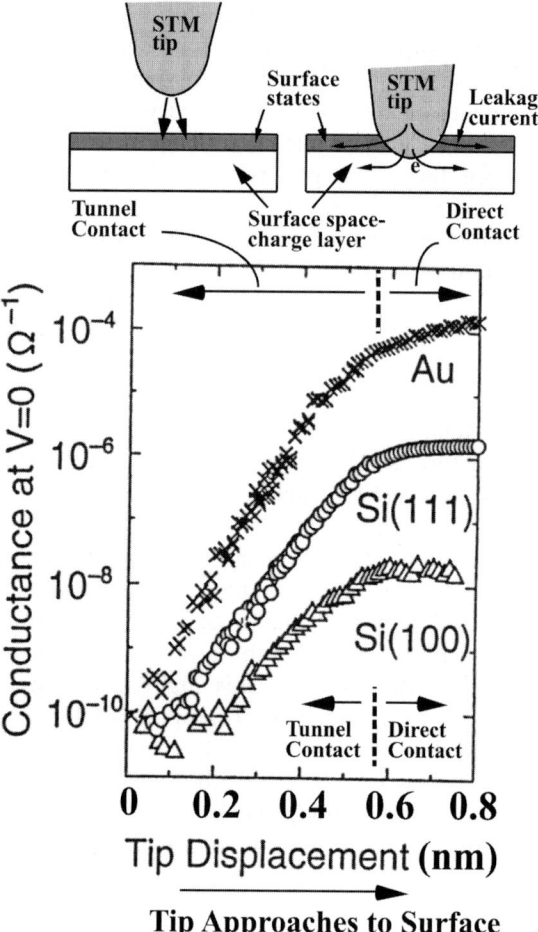

Fig. 10. Conductance between a STM tip and specimen surfaces, as a function of tip displacement toward the Si(1 1 1)-7 × 7 surface, Si(1 0 0)-2 × 1 surface, and polycrystalline Au surface, respectively [50]. The upper panels illustrate current flows at tunnel regime (left) and at direct contact (right), respectively.

the tip–sample distance (forward displacement of the tip). When the tip comes contact into the sample surfaces, the conductances show saturation; the saturated values are different depending on the surfaces. The 7 × 7 surface shows a larger conductance than the Si(1 0 0) surface, whereas its value is found to decrease by adsorption of oxygen onto the 7 × 7 surface, presumably because oxygen saturates the dangling bonds and reduces the density of surface state. These measurements suggest that excess leakage conductance via the point contact between the STM tip and the clean 7 × 7 surface is due to current paths parallel to the surface, in other words, the surface-state conductance through the dangling-bond surface state. Its conductance is estimated to be around 10^{-6} Ω^{-1}.

Another approach to measuring semiconductor surface conductance with STM relies on fabricating nanoscale structures on the surface using the STM [19]. Heike et al. first fabricated thin insulating trenches on the 7 × 7 clean surface, by applying a relatively high bias voltage with a high tunnelling current in the STM. After that, they observed the structured surface in a conventional STM mode. Fig. 11(a) and (b) shows a half-closed tape-shaped pattern, surrounded by the insulating trench. The apparent height of the surface area surrounded by the trench is lower (darker) (by approximately 0.2 nm) in the empty-state images (b) as the STM tip approaches to the closed-end of the tape, while the brightness of the whole tape area is observed to be flat in the filled-state images (a).

This result is interpreted as follows: The electrons tunnelling from the tip to the sample flow along the tape through the surface state of only the region restricted by the trench, when the tip is positioned inside the tape (see the top panel in Fig. 11). This is because a Schottky barrier between the surface state and the bulk state forces the current to travel along the surface for a while before leaking into the bulk state. Then, a voltage drop occurs along the tape due to a finite resistance of the surface state, as measured in Fig. 11(c). This is a kind of scanning tunnelling potentiometry [51–53]. By comparing the measured voltage drop along the tape with the calculated one, the conductance of the dangling-bond surface state on the 7 × 7 surface was deduced to be 8.7×10^{-9} Ω^{-1}.

This value is much smaller than that obtained by the point-contact method of Hasegawa(Y) et al. mentioned above. Heike et al. [19] suggest that this discrepancy arises because the method of Hasegawa(Y) et al. involves the conductance through the surface space–charge layer as well as through the surface states. It is postulated that tunnel contact may be a much more effective way to inject carriers into the surface-state bands, whereas direct contact between the probe and sample surface

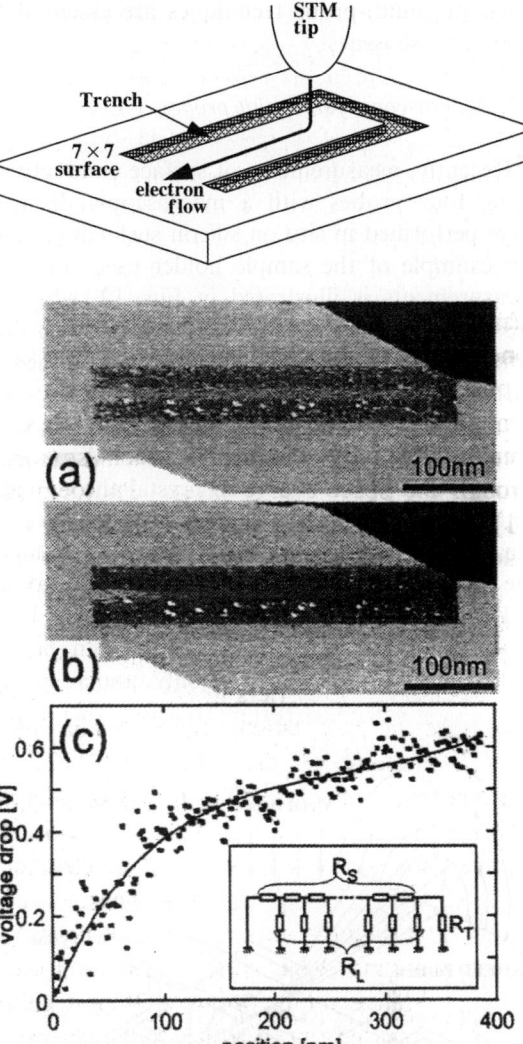

Fig. 11. Top Panel: Schematic diagram of preferred direction of current flow along an artificial nanostructure on a Si(1 1 1)-7 × 7 clean surface. STM images of (a) the filled-state and (b) empty-state on this surface [19]. In (a), the area surrounded by the trench looks flat, while in (b) the area looks darker with approaching the right end of the area. In (c) the voltage drop measured along the tape-shaped area in (b) is shown, together with a simulation of the voltage drop calculated using a simple series resistance model (calculation is solid line, as modeled in inset).

injects carriers into both of the bulk and surface bands. However, because all 'one-point' measurements are susceptible to resistance effects at the point of contact, which are difficult to control or calculate, multi-probe techniques are essential to resolve these issues.

4.2. Macroscopic four-point probes

Recently, measurements of surface conductivity using four probes with a macroscopic distance were performed in situ on silicon surfaces [13,54]. An example of the sample holder used for such measurements is illustrated in Fig. 12, which is designed for simultaneous metal deposition and reflection high energy diffraction RHEED observations, to prepare well-defined surface superstructures on the whole are of surface [54]. After preparing the superstructures, a small direct current is fed into the specimen crystal through the end clamp electrodes, and the voltage drop is measured between a pair of thin Ta-wire contacts, the spacing of which is usually as large as about 5–10 mm.

Surface-state electrical conduction can be detected in this configuration only under special conditions, when small amounts (around 0.01 ML) of atoms of monovalent metals (noble and alkali metals) are deposited on top of the Si(1 1 1)–$\sqrt{3} \times \sqrt{3}$-Ag surface. These adatoms are found to enhance the surface conductivity [55]. From photoemission measurements to reveal the changes in surface electronic states, it has been shown that the adatoms donate their valence electrons into the 2D-free-electron-like surface-state band S_1 of the $\sqrt{3} \times \sqrt{3}$-Ag substrate (Fig. 7(b)), thus enhancing the surface-state conductivity [17,55]. This is quite similar to in bulk crystals of semiconductor in which impurity atoms act as donors or acceptors to vary the carrier concentrations and conductivity.

With increasing the coverage of the adatoms up to around 0.15 ML, 2D islands nucleate and arrange to make a new ordered structure with a $\sqrt{21} \times \sqrt{21}$ periodicity [13,56]. The $\sqrt{21} \times \sqrt{21}$ superstructures made by monovalent-metal-atom adsorptions on the Si(1 1 1)–$\sqrt{3} \times \sqrt{3}$-Ag surface have high surface conductivities. This is due to new dispersive metallic surface-state bands that are created in the superstructures, while the surface-space–charge-layer conductance is suppressed [57,58]. In these cases, the surface-state electrical conduction is so large that it can be detected experimentally using the macroscopic four-probe method. The conductances for these structures are on the order of $10^{-4} \, \Omega^{-1}$, which is higher than that of the 7×7 clean surface by about four orders of magnitude.

4.3. Microscopic four-point probes

From the expectation of enhanced surface sensitivity by reducing the probe spacings as shown in Fig. 4, micro-four-point probes with probe spacings down to 2 µm have recently been developed using silicon-based micro-fabrication technology [20], one of which is shown in Fig. 13. Such microscopic probes also enable mapping of the local conductivity distribution on material surfaces [59].

The micro-four-point probes have been used in UHV to measure the surface conductances of the Si(1 1 1)-7×7 clean and $\sqrt{3} \times \sqrt{3}$-Ag surfaces. Fig. 14(a) is a scanning electron micrograph showing a micro-four-point probe of 8 µm spacing, contacting to a silicon surface for the electrical

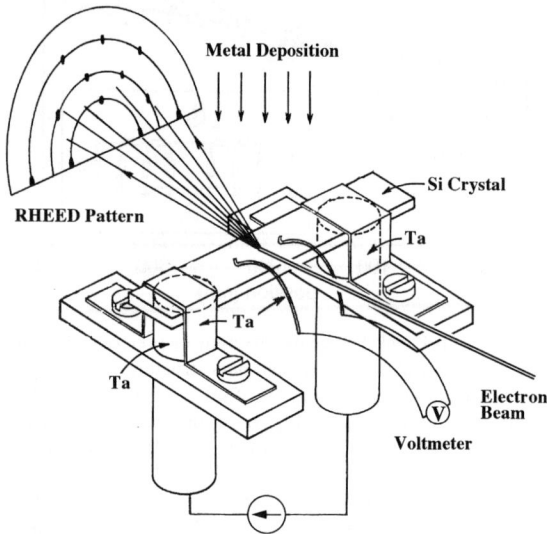

Fig. 12. A sample holder for in-situ electrical conductivity measurements in UHV using a "four-probe" method [54]. The inner probe spacing is typically several millimetres. After the surface superstructures are confirmed by using RHEED, the electron beam is always turned off during the subsequent conductivity measurements. A small current is fed through the clamps at both ends, and the voltage drop is measured by a pair of Ta wire contacts.

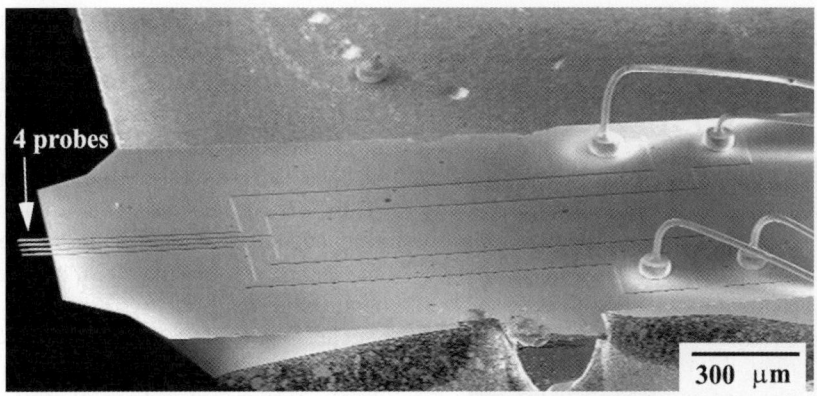

Fig. 13. A SEM image of a silicon chip with a micro-four-point-probe array projecting over the chip edge. The probes are made of silicon oxide, coated with metal. Bonding pads on the chip facilitate electrical connections to the probes. An underetching technique ensures that the probes are not shorted when a metallic layer is evaporated over the whole chip to make the probes conducting. The whole chip is 6×4 mm^2 and mounted on a ceramic holder for convenience [20].

measurements. The resistance measured in this way on the $\sqrt{3} \times \sqrt{3}$-Ag surface was smaller than that for the 7×7 clean surface by about two orders of magnitudes [22]. This should be compared with the result obtained by macroscopic four probes of about 10 mm probe spacing, where the difference of resistance between the two surfaces was as small as about 10% [60]. These results imply that reducing the probe spacing makes the measurements more surface sensitive, as expected from Fig. 4.

From a quantitative analysis, the extremely high conductance of the $\sqrt{3} \times \sqrt{3}$-Ag surface compared with that of the 7×7 surface cannot be explained by the surface-space–charge-layer conductivity. Rather it must be attributed to the surface-state band inherent in the $\sqrt{3} \times \sqrt{3}$-Ag superstructure [20]. Although the surface-state conductivity on this surface had been detected by macroscopic four-probe measurements as a carrier-doping effect into the surface-state band, as mentioned before, micro-four-point probes enabled the direct detection of the contribution of the surface-state conductance [20,22].

Atomic steps on the surface scatter the surface-state electrons as seen as standing waves in Fig. 7(c). This should cause additional electrical resistance at steps, which has been measured with the micro-four-point probe. A technique to control the step configuration on the surface recently has been developed [61], which has been utilized to obtain almost step-free terraces as wide as the probe spacing, as shown in Fig. 14(b). The surface exhibits flat terraces of about 10 μm wide and step bunches of about 2 μm wide where around 300 monatomic steps are accumulated. It is wholly covered homogeneously by the 7×7 superstructure or $\sqrt{3} \times \sqrt{3}$-Ag superstructure due to 1 ML Ag deposition.

By observing the probes and the sample surface by scanning electron microscopy in situ, the probes can be positioned on a large flat terrace, or positioned across a step bunch running between the inner pair of probes. Fig. 14(a) illustrates the latter case. In this way, the surface area under measurement can be selected by shifting the probe position, so that the influence of atomic steps upon the surface conductivity can be intentionally avoided or included. It has been found then that the resistance measured across a step bunch is much larger than that measured on a step-free terrace. Although this result is intuitively reasonable, it is the first direct transport measurement, confirming that atomic steps on a surface causes an additional resistance [22].

4.4. Multi-tip STM

In order to measure the local conductivity in smaller areas down to nanometre scales, several groups have constructed multi-tip STMs in which

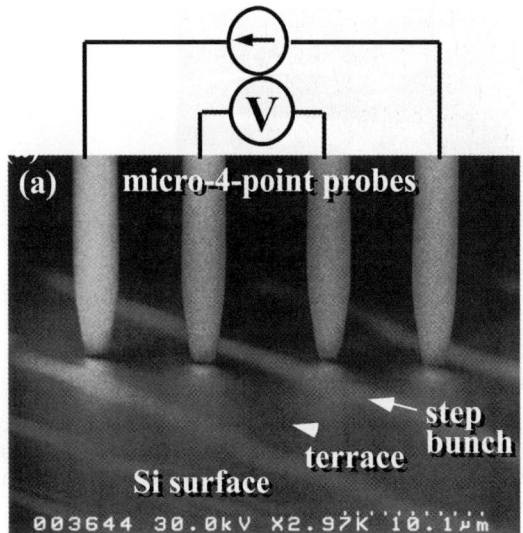

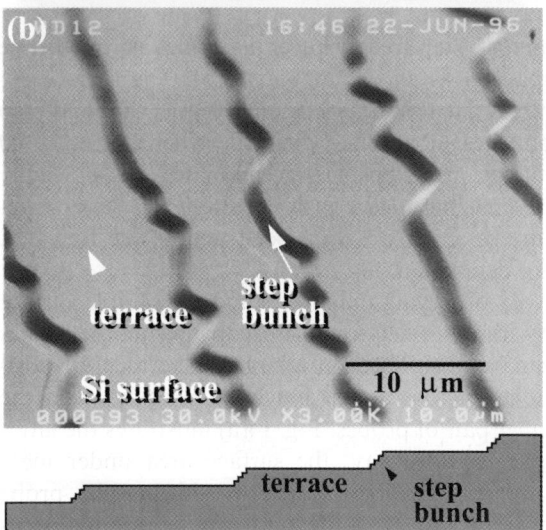

Fig. 14. Panel (a): Scanning electron micrographs showing the micro-four-point probe contacting to a silicon surface for measuring the conductivity in UHV. Slightly brighter bands on the sample surface are step bunches, and wider darker bands are terraces [21]. Panel (b): Grazing-incidence scanning electron micrographs of a Si(1 1 1) specimen surface, with step bunching as illustrated at the bottom. The surface is wholly covered by the $\sqrt{3} \times \sqrt{3}$-Ag surface superstructure with 1 ML Ag.

the tips are utilized as nanoprobes to measure electric properties. The technical challenge of contacting independent probes is considerably greater than using the monolithic micro-four point probe described above. The advantages are that each probe can be controlled much more accurately with variable probe spacings, and smaller probe spacings could be achievable. Prospects for such multi-tip STMs are the measurement of electrical conductivity in a single domain of a surface superstructure, and the study of ballistic transport as well as other quantum phenomena that occur only on the nanometre scale. The usefulness of double-tip STM has been discussed in Ref. [62].

The first trial to make such an STM was done with electrically isolated two tips mounted on a single scanning head, so that the probe spacing could not be changed [63]. A machine with independently driven double tips in UHV was later constructed by Aono et al. [64], in which the tips can be brought together as close as about 100 nm. The macroscopic radius of each tip determines the minimum distance attainable between the two tips. Such tips can be used as an emitter and collector, like in the point-contact transistor, with much better control on much more well-defined surfaces.

Fig. 15 shows SEM images of four STM tips that are independently driven on a single stage in UHV [23]. With such a device, the surface conductivities were measured as a function of probe spacing ranging from 1 mm to 1 μm [23]. The results were in accord with the expectation from Fig. 4. The probe-spacing dependences of the measured resistance for the 7×7 clean surface and the $\sqrt{3} \times \sqrt{3}$-Ag surface were quite different from each other, indicating quite different ways of current flowing; the current flows dominantly only at the surface for the latter superstructure, while the bulk conductivity is dominant for the former superstructure. Such measurements lead to clarifying the influence of surface defects and intrinsic nature of surface-state transport in a more detailed way.

5. Concluding remarks

In the present paper, by taking the silver- and indium-covered Si(1 1 1) surfaces as well as the clean surface as examples, various phenomena revealed in the surface-state bands and electronic transport phenomena are described. Some of the readers may think that it is neither surprising nor

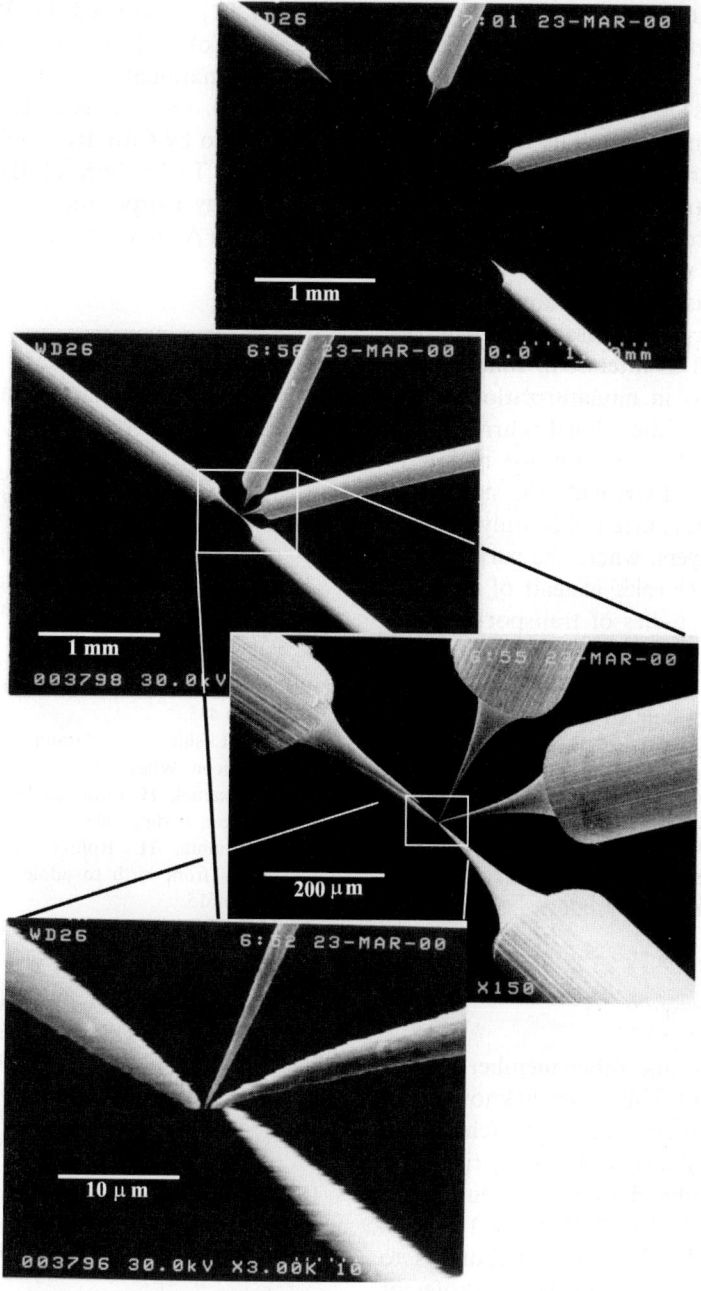

Fig. 15. SEM images of four STM tips, independently driven in UHV [23]. The tips are separated as far as about 1 mm in the top panel, while they are brought together to within a distance less than 1 μm spacing in the other panels.

new to have metallic electronic states and enhanced conductivity on these surfaces because 'metal' atoms are deposited on the surface. But such considerations are too naive. The deposited metal atoms do not exhibit their own properties expected from their bulk crystals. The metal atoms do not make metallic bonds with each other to produce conduction electrons like in their bulks, rather make bonds with the substrate Si atoms to form characteristic surface electronic states. This is the main point for the properties of surface phases as new 'nanomaterials', which are completely different from those of bulk materials of composite atoms. The atomic arrangements on the surface and the resulting electronic states directly characterize the transport properties.

The electronic transport properties of surface-state bands are increasingly important subjects in the near future because of necessity for semiconductor devices as well as interest in fundamental physics. With progress in miniaturization of microelectronics devices, the signal currents are forced to flow only near the surface region of semiconductor crystals. Eventually the signals are expected to process with current flow only through one or two atomic layers, where the surface electronic states play main roles instead of the bulk states. Therefore the studies of transport properties of surfaces lead to an important branch of nanotechnogy when coupled with the fabrication technology for atomic-scale structures on surfaces. For that, the microscopic multi-probes described here are indispensably important. Thus it is tempting to conclude that the descendants of the point-contact transistor are still having a major impact on surface science—and microelectronics—more than half a century later.

Acknowledgements

Dr. Tadaaki Nagao and other members of the group in University of Tokyo are acknowledged for their supportive collaboration. Dr. Ichiro Shiraki of University of Tokyo and Drs. Christian L. Petersen, Peter Bøggild, Torben M. Hansen at MIC (Technical University of Denmark) are acknowledged for their dedicated contributions to micro-four-point probe measurements. Professor Martin Henzler at Hannover University gave us valuable advice on surface-state conduction. The preparation of this manuscript, as well as some of the work described here was supported in part by Grants-in-Aid from the Ministry of Education, Science, Culture, and Sports of Japan, especially through that for Creative Basic Research (no. 09NP1201) conducted by Professor Katsumichi Yagi of Tokyo Institute of Technology, and also for International Collaboration Program (no. 11694059). One of us (S.H.) has also been supported also by Core Research for Evolutional Science and Technology of the Japan Science and Technology Corporation conducted by Professor Masakazu Aono of Osaka University and RIKEN.

References

[1] W. Shockley, G.L. Pearson, Modulation of conductance of thin films of semiconductors by surface charges, Phys. Rev. 74 (1948) 232.
[2] J. Bardeen, Surface states and rectification at a metal–semiconductor contact, Phys. Rev. 71 (1947) 717.
[3] J. Bardeen, W.H. Brattain, The transistor a semiconductor triode, Phys. Rev. 74 (1948) 230;
For a popular introduction to this discovery, see M. Riordan, L. Hoddeson, Crystal Fire, Norton Publishers, 1997.
[4] W. Shockley, The theory of p–n junctions in semiconductors and p–n junction transistors, Bell Sys. Tech. J. 28 (1949) 435.
[5] V.G. Lifshits, A.A. Saranin, A.V. Zotov, Surface Phases on Silicon, Wiley, Chichester, 1994.
[6] For example H. Lüth, Surfaces and Interfaces of Solids, Springer, Berlin, 1993.
[7] G. Binnig, H. Rohrer, Scanning tunnelling microscopy—from birth to adolescence, Rev. Mod. Phys. 59 (1987) 615.
[8] M.F. Crommie, C.P. Lutz, D.M. Eigler, E.L. Heller, Waves on a metal surface and quantum corrals, Surf. Rev. Lett. 2 (1995) 127.
[9] Ph. Avouris, I.-W. Lyo, R.E. Walkup, Y. Hasegawa, Real space imaging of electron scattering phenomena at metal surfaces, J. Vac. Sci. Technol. B 12 (1994) 1447.
[10] K. Takayanagi, Y. Tanishiro, S. Takahashi, M. Takahashi, Structure analysis of Si(1 1 1)-7 × 7 reconstructed surface by transmission electron diffraction, Surf. Sci. 164 (1985) 367.
[11] For example S. Datta, Electronic Transport in Mesoscopic Systems, Cambridge, New York, 1995.
[12] M. Henzler, Electronic transport at surfaces, in: J.M. Blakely (Ed.), Surface Physics of Materials I, Academic Press, New York, 1975, p. 241.
[13] For review S. Hasegawa, X. Tong, S. Takeda, N. Sato, T. Nagao, Structures and electronic transport on silicon surfaces, Prog. Surf. Sci. 60 (1999) 89;
S. Hasegawa, Surface-state bands on silicon as electron systems in reduced dimensions at atomic scales, J. Phys. C: Condens. Matter 12 (2000) R463.
[14] W.H. Brattain, J. Bardeen, Surface properties of germanium, Bell Syst. Tech. J. 32 (1953) 1.

[15] J. Bardeen, S.R. Morrison, Surface barriers and surface conductance, Physica 20 (1954) 873.
[16] W. Mönch, On the correlation of geometrical structure and electronic properties at clean semiconductor surfaces, Surf. Sci. 63 (1977) 79.
[17] Y. Nakajima, S. Takeda, T. Nagao, S. Hasegawa, X. Tong, Surface electrical conduction due to carrier doping into a surface-state band on Si(1 1 1)-$\sqrt{3} \times \sqrt{3}$-Ag, Phys. Rev. B 56 (1997) 6782.
[18] S. Hasegawa, Atomic imaging of macroscopic surface conductivity, Curr Opinion Solid State Mater. Sci. 4 (1999) 429.
[19] S. Heike, S. Watanabe, Y. Wada, T. Hashizume, Electron conduction through surface states of the Si(1 1 1)-(7 × 7) surface, Phys. Rev. Lett. 81 (1998) 890.
[20] C.L. Petersen, F. Grey, I. Shiraki, S. Hasegawa, Microfour-point probe for studying electronic transport through surface states, Appl. Phys. Lett. 77 (2000) 3782.
Microfour-point probes are now commercially available; http://www.capres.com.
[21] I. Shiraki, C.L. Petersen, P. Bøggild, T.M. Hansen, T. Nagao, F. Grey, S. Hasegawa, Microfour-point probes in a UHV-scanning electron microscope for in-situ surface conductivity measurements, Surf. Rev. Lett. 7 (2000) 533.
[22] I. Shiraki, C.L. Petersen, P. Bøggild, T.M. Hansen, T. Nagao, F. Grey, S. Hasegawa, Surface electronic transport measured with microfour-point probes, Nature, submitted for publication.
[23] I. Shiraki, F. Tanabe, R. Hobara, T. Nagao, S. Hasegawa, Independently driven four-tip probes for conductivity measurements in ultrahigh vacuum, Surf. Sci., 493 (2001) 633.
[24] D.M. Eigler, E.K. Schweizer, Positioning single atoms with a scanning tunnelling microscope, Nature 344 (1990) 524–526;
J.A. Stroscio, D.M. Eigler, Atomic and molecular manipulation with the scanning tunnelling microscope, Science 254 (1991) 1319.
[25] F.J. Himpsel, G. Hollinger, R.A. Pollak, Determination of the Fermi-level pinning position at Si(1 1 1) surfaces, Phys. Rev. B 28 (1983) 7014.
[26] J. Viernow, M. Henzler, W.L. O'Brien, F.K. Men, F.M. Leibsle, D.Y. Petrovykh, J.L. Lin, F.J. Himpsel, Unoccupied surface states on Si(1 1 1)$\sqrt{3} \times \sqrt{3}$-Ag, Phys. Rev. B 57 (1998) 2321.
[27] T. Takahashi, S. Nakatani, N. Okamoto, T. Ishikawa, S. Kikuta, Study on the Si(1 1 1)$\sqrt{3} \times \sqrt{3}$-Ag surface structure by X-ray diffraction, Jpn. J. Appl. Phys. 27 (1988) L753;
T. Takahashi, S. Nakatani, N. Okamoto, T. Ishikawa, S. Kikuta, A study of the Si(1 1 1)$\sqrt{3} \times \sqrt{3}$-Ag surface by transmission X-ray diffraction and X-ray diffraction topography, Surf. Sci. 242 (1991) 54;
T. Takahashi, S. Nakatani, Refinement of the-Si(1 1 1)$\sqrt{3} \times \sqrt{3}$-Ag structure by surface X-ray diffraction, Surf. Sci. 282 (1993) 17, and references therein.

[28] M. Katayama, R.S. Williams, M. Kato, E. Nomura, M. Aono, Structure analysis of the Si(1 1 1)$\sqrt{3} \times \sqrt{3}R30°$-Ag surface, Phys. Rev. Lett. 66 (1991) 2762.
[29] E. Vlieg, E. Fontes, J.R. Patel, Structure analysis of Si(1 1 1)-($\sqrt{3} \times \sqrt{3}$)$R30°$/Ag using X-ray standing waves, Phys. Rev. B 43 (1991) 7185.
[30] L.S.O. Johansson, E. Landemark, C.J. Karlsson, R.I.G. Uhrberg, Fermi-level pinning and surface-state band structure of the Si(1 1 1)/($\sqrt{3} \times \sqrt{3}$)$R30°$Ag surface, Phys. Rev. Lett. 63 (1989) 2092;
L.S.O. Johansson, E. Landemark, C.J. Karlsson, R.I.G. Uhrberg, Structure of the ($\sqrt{3} \times \sqrt{3}$)$R30°$ Ag/Si(1 1 1) surface from first-principles calculations, Phys. Rev. Lett. 69 (1992) 2451.
[31] S. Hasegawa, N. Sato, I. Shiraki, C.L. Petersen, P. Boggild, T.M. Hansen, T. Nagao, F. Grey, Surface-state bands on silicon—Si(1 1 1)-$\sqrt{3} \times \sqrt{3}$-Ag surface superstructure, Jpn. J. Appl. Phys. 39 (2000) 3815.
[32] N. Sato, S. Takeda, T. Nagao, S. Hasegawa, Electron standing waves on the Si(1 1 1)-$\sqrt{3} \times \sqrt{3}$-Ag surface, Phys. Rev. B 59 (1999) 2035.
[33] T. Yokoyama, M. Okamoto, K. Takayanagi, Electron waves in the π^*-surface band of the Si(0 0 1) surface, Phys. Rev. Lett. 81 (1998) 3423;
T. Yokoyama, K. Takayanagi, Size quantization of surface-state electrons on the Si(0 0 1) surface, Phys. Rev. B 59 (1999) 12232.
[34] L. Li, W.-D. Schneider, R. Berndt, S. Crampin, Electron confinement to nanoscale Ag islands on Ag(1 1 1): a quantitative study, Phys. Rev. Lett. 80 (1998) 3332.
[35] D. Fujita, K. Amemiya, T. Yakabe, H. Nejoh, Anisotropic standing-wave formation on an Au(1 1 1)-(23×$\sqrt{3}$) reconstructed surface, Phys. Rev. Lett. 78 (1997) 3904.
[36] L. Petersen, P.T. Sprunger, Ph. Hofmann, E. Lægsgaard, B.G. Briner, M. Doering, H.-P. Rust, A.M. Bradshaw, F. Besenbacher, E.W. Plummer, Direct imaging of the two-dimensional Fermi contour: Fourier-transform STM, Phys. Rev. B 57 (1998) R6858.
[37] R. Losio, K.N. Altmann, F.J. Himpsel, Fermi surface of Si(1 1 1)7 × 7, Phys. Rev. B 61 (2000) 10845.
[38] F. Flores, A. Levy Yeyati, J. Ortega, Metal–insulator transition in the Si(1 1 1)-(7 × 7) surface, Surf. Rev. Lett. 4 (1997) 281.
[39] A.A. Saranin, V.G. Lifshits, M. Katayama, K. Oura, Composition and surface structure of quantum chains on a In/Si(1 1 1) surface, Jpn. J. Appl. Phys. 39 (2000) L306.
[40] O. Bunk, G. Falkenberg, J.H. Zeysing, L. Lottermoser, R.L. Johnson, M. Nielsen, F. Berg-Rasmussen, J. Baker, R. Feidenhans'l, Structure determination of the indium-induced Si(1 1 1)-(4 × 1) reconstruction by surface X-ray diffraction, Phys. Rev. B 59 (1999) 12228.
[41] J. Nakamura, S. Watanabe, M. Aono, Anisotropic electronic structure of the Si(1 1 1)-(4 × 1)-In surface, Phys. Rev. B 63 (2001) 193307.
[42] T. Abukawa, M. Sasaki, F. Hisamatsu, T. Goto, T. Kinoshita, A. Kakizaki, S. Kono, Surface electronic structure of a single-domain Si(1 1 1)4 × 1-In surface: a

[43] I.G. Hill, A.B. McLean, Metallicity of In chains on Si(1 1 1), Phys. Rev. B 56 (1997) 15725.
[44] R.E. Peierls, Quantum Theory of Solids, Clarendon Press, Oxford, 1964.
[45] H.W. Yeom, S. Takeda, E. Rotenberg, I. Matsuda, K. Horikoshi, J. Schaefer, C.M. Lee, S.D. Kevan, T. Ohta, T. Nagao, S. Hasegawa, Instability and charge density wave of metallic quantum chains on a silicon surface, Phys. Rev. Lett. 82 (1999) 4898.
[46] O. Gallus, Th. Pillo, M. Hengsberger, P. Segovia, Y. Baer, A system with a complex phase transition: Indium chains on Si(1 1 1), Eur. Phys. J. B 20 (2001) 313.
[47] C. Kumpf, O. Bunk, J.H. Zeysing, Y. Su, M. Nielsen, R.L. Johnson, R. Feidenhans'l, K. Bechgaard, Low-temperature structure of indium quantum chains on silicon, Phys. Rev. Lett. 85 (2000) 4916.
[48] T. Nishiguchi, M. Kageshima, N. Ara-Kato, A. Kawazu, Behavior of charge density waves in a one-dimensional organic conductor visualized by scanning tunneling microscopy, Phys. Rev. Lett. 81 (1998) 3187.
[49] M. Ishida, T. Mori, H. Shigekawa, Surface charge-density wave on the one-dimensional organic conductor β-(BEDT-TTF)$_2$PF$_6$, Phys. Rev. Lett. 83 (1999) 596.
[50] Y. Hasegawa, I.-W. Lyo, Ph. Avouris, Measurement of surface state conductance using STM point contacts, Surf. Sci. 357/358 (1996) 32.
[51] J.R. Kirtley, S. Washburn, M.J. Brady, Direct measurement of potential steps at grain boundaries in the presence of current flow, Phys. Rev. Lett. 60 (1988) 1546.
[52] A.D. Kent, I. Maggio-Aprile, Ph. Niedermann, O. Fischer, Direct measurements of the effects of inhomogeneities on the normal-state transport properties of YbaCuO thin films, Phys. Rev. B 39 (1989) 12363.
[53] P. Muralt, GaAs pn junction studied by scanning tunneling potentiometry, Appl. Phys. Lett. 49 (1986) 1441;
P. Muralt, H. Meier, D.W. Pohl, H.W.M. Salemink, Scanning tunneling microscopy and potentiometry on a semiconductor heterojunction, Appl. Phys. Lett. 50 (1987) 1352.
[54] S. Hasegawa, S. Ino, Surface structures and conductance at epitaxial growths of Ag and Au on the Si(1 1 1) surface, Phys. Rev. Lett. 68 (1992) 1192.
[55] Y. Nakajima, G. Uchida, T. Nagao, S. Hasegawa, Two-dimensional adatom gas on the Si(1 1 1)-$\sqrt{3} \times \sqrt{3}$-Ag surface detected through changes in electrical conduction, Phys. Rev. B 54 (1996) 14134.
[56] S. Hasegawa, K. Tsuchie, K. Toriyama, X. Tong, T. Nagao, Surface electronic transport on silicon: donor- and acceptor-type adsorbates on Si(1 1 1)-$\sqrt{3} \times \sqrt{3}$-Ag substrate, Appl. Surf. Sci. 162/163 (2000) 42.
[57] X. Tong, C.-S. Jiang, S. Hasegawa, Electronic structure of the Si(1 1 1)-$\sqrt{21} \times \sqrt{21}$-(Ag + Au) surface, Phys. Rev. B 57 (1998) 9015.
[58] X. Tong, C.-S. Jiang, K. Horikoshi, S. Hasegawa, Surface-state electrical conduction on the Si(1 1 1)-$\sqrt{3} \times \sqrt{3}$-Ag surface with noble-metal adatoms, Surf. Sci. 449 (2000) 125.
[59] P. Bøggild, T.M. Hanssen, O. Kuhn, F. Grey, T. Junno, L. Montelius, Scanning nanoscale multiprobes for conductivity measurements, Rev. Sci. Instrum. 71 (2000) 2781–2783;
P. Bøggild, F. Grey, T. Hanssenkam, D.R. Greve, T. Bjornholm, Direct measurement of the microscale conductivity of conjugated polymer monolayers, Adv. Mater. 12 (2000) 947.
[60] C.-S. Jiang, S. Hasegawa, S. Ino, Surface conductivity for Au or Ag on Si(1 1 1), Phys. Rev. B 54 (1996) 10389.
[61] T. Ogino, Self-organization of nanostructures on Si wafers using surface structure control, Surf. Sci. 386 (1997) 137.
[62] Q. Niu, M.C. Chang, C.K. Shih, Double-tip scanning tunneling microscope for surface analysis, Phys. Rev. B 51 (1995) 5502.
[63] S. Tsukamoto, B. Siu, N. Nakagiri, Twin-probe scanning tunneling microscope, Rev. Sci. Instrum. 62 (1991) 1767.
[64] M. Aono, C.-S. Jiang, T. Nakayama, T. Okuda, S. Qiao, M. Sakurai, C. Thirstrup, Z.-H. Wu, The present and future of nano-lithography using scanning probes—how to measure the properties of nano-lithographed surfaces, J. Surf. Sci. Soc. Jpn. 9 (1998) 698, in Japanese.

Dynamical phenomena including many body effects at metal surfaces

Wilson Agerico Diño [a,b,*], Hideaki Kasai [a], Ayao Okiji [c]

[a] *Department of Applied Physics, Osaka University, 2-1 Yamadaoka, Suita, Osaka 565-0871, Japan*
[b] *Institute of Industrial Science, The University of Tokyo, Meguro-ku, Tokyo 153-8505, Japan*
[c] *Wakayama National College of Technology, Gobō, Wakayama 644-0023, Japan*

Received 5 July 2000; accepted for publication 25 May 2001

Abstract

In this contribution, we give a brief survey of some elementary many body processes observable at surfaces. We begin the survey by discussing how the electron ground state would behave and look like in real space at surfaces. Next, we discuss how these electrons behave when they are perturbed by external fields characterized by ultrashort time scales. We follow this with a discussion of how the dynamics of electrons would then affect the motion of adsorbates on surfaces. Finally, we cite some possible technological applications utilizing this knowledge. We also discuss possible trends or directions of scientific research in this century. © 2001 Elsevier Science B.V. All rights reserved.

Keywords: Many body and quasi-particle theories; Electron density, excitation spectra calculations; Desorption induced by electronic transitions (DIET); Surface electrical transport (surface conductivity, surface recombination, etc.); Alloys; Metallic surfaces; Conductivity

1. Introduction

A recent and continuing trend in surface science is towards an understanding of the dynamical nature of surfaces. The basic rationale underlying this trend is as follows: Let us consider, e.g., how dynamical processes proceed around us. A slight change (e.g., induced by light irradiation) in the electronic states making up a solid surface interacting with atoms or molecules gives rise to a slight change in the position of atoms and molecules, which in turn initiates, further, a marked change in the electronic states. Eventually, the coupled electronic and atomic processes proceed to the final state on the surface. This may be properly regarded as characteristic of all dynamical processes occurring on surfaces, including chemical reactions. It is also a characteristic of all dynamical processes occurring around us, starting from those involving life forms and extending to other complex systems. Thus, in order to deepen our understanding of the dynamics of complex systems, it is necessary to clarify the fundamental mechanism behind the dynamics of each of the elementary processes. The most important breakthrough we can achieve towards a thorough understanding of

---

* Corresponding author. Tel.: +81-3-5452-6132; fax: +81-3-5452-6159.
  E-mail addresses: wilson@dyn.ap.eng.osaka-u.ac.jp, wilson@iis.u-tokyo.ac.jp (W.A. Diño).

the dynamics occurring around us would come from an accurate understanding of the elementary processes, in which mass, charge, and energy transport play important roles.

With recent advances in experimental techniques in the field of surface science, we are now at the stage where we can prepare well characterized solid surfaces, which, in a sense, serve as *playgrounds* for scientists. The solid surface provides us with a stage to study the dynamics of complex systems, where state transitions of the corresponding electron system are closely connected with the changes of atomic and molecular motion on various scales of magnitude with respect to time, space, and energy. For this reason, it is not only interesting but also necessary and important to study the excited as well as the ground states of the electron system constituting the system of a surface interacting with atoms and molecules.

We begin our brief survey by considering the dynamics of surface electrons as the temperature is lowered to 0 K (the electron ground state!) in Section 2. We then consider the dynamics of electrons perturbed by external fields characterized by ultrashort time scales in Section 3. We then continue by considering how the dynamics of atoms or molecules adsorbed on surfaces are affected by the ultrafast electron dynamics in Section 4.

2. A physical glimpse of how electrons at surfaces look as $T \to 0$ K

2.1. The Kondo effect and the resistance minimum phenomenon—a brief history

The electric resistance is related to the amount of electrons (back) scattered from defects and the vibrations of the atoms (phonons), which effectively hinders the motion of the electrons through the crystal. As the temperature of a material (e.g., a pure metal) is lowered, the vibrations of the atoms (phonons) that make up the material become small, and it becomes easier for the electrons to travel through the material. Thus, usually, the electric resistance of a material decreases as the temperature is lowered. However, due to static defects (which are not affected by the temperature) in the material, the resistance saturates as the temperature is further lowered below some critical temperature (≈ 10 K). The value of the low-temperature resistance depends on the number of defects in the material. Adding defects increases the value of this "residual resistance" without changing the character of the temperature dependence. However, the temperature dependence of the resistance changes dramatically when magnetic atoms, e.g., Fe, are added. Rather than saturating, the electric resistance increases as the temperature is lowered further.

The observation of a resistance minimum in some metals, as a function of temperature T, one manifestation of the effect of magnetic impurities, has been known as early as the 1930s. Meissner and Voigt [1] discovered that instead of decreasing monotonically as one would expect [2], the electric resistance R of Mg, as with other samples they have tested, has a shallow minimum occurring at a low temperature T_K (O(10 K)), now widely known as the Kondo temperature. The Mg sample showed a higher electric resistance at the temperature $T = 1$ K, than resistance at $T = 4$ K, i.e., $R(T = 1 \text{ K}) > R(T = 4 \text{ K})$. Following Meissner and Voigt, de Boer and coworkers [3] found that the characteristic R vs. T curves of several other samples also showed minima. Why?, When? and How? does this "resistance minimum phenomenon" occur? These were probably the questions occupying the minds of the condensed matter physicists at that time. Later, it was found that the occurrence of an electric resistance minimum was due to the presence of magnetic impurities. Typical examples are when trace amounts of Fe, Mn and Co are found in the samples of Cu, Ag, and Au. (For more examples, we refer the readers to, e.g., Refs. [4–8].) However, because resistance minima were observed even for minute quantities (e.g., in units of parts per million) of magnetic impurities, it was difficult to prepare samples and perform experiments that would, without a trace of doubt, establish the presence of the magnetic impurities as the origin of the resistance minimum observed. To make matters even worse, metals with nonmagnetic impurities like Sn also exhibited a resistance minimum. Sarachik et al. [9] found the solution. By measuring the electric resistance of samples prepared by Clogston et al. [10] they found that there was a one-to-one correspondence between the existence of a localized magnetic moment and the existence of a resistance minimum. The weak magnetic impurity concentration dependence of the temperature of the minimum resistance and the scaling of the low-temperature resistance with impurity concentration indicated that this phenomenon arose from the interaction of the sea of conduction electrons (mobile electrons that behave like a sea—Fermi sea—that fills the entire sample) with isolated magnetic impurities, and not from the impurity–impurity interactions [4,5]. A typical example of this so-called *resistance minimum phenomenon* is shown in Fig. 1. (For a more thorough review of the experimental facts that were available until 1964, we refer the readers to the paper by van den Berg [12].)

The mystery enshrouding the resistance minimum phenomenon persisted for over 30 years. A very significant advance in the theory of magnetic impurities was an explanation of this phenomenon by Kondo in 1964 [13], 30 years after it had been observed experimentally. As inferred above, the electric resistance is related to the amount of back scattering from defects, which hinders the motion of the electrons through the sample crystal. Usually, one can readily calculate the probability with which an electron will be scattered when the defects are small. For larger defects, one could usually perform the calculation using perturbation theory—an iterative process in which the answer is usually written as a sum of a series of smaller and smaller terms. However, in 1964, Kondo, when he considered the scattering from a magnetic impurity that interacts with the spins of the conduction electrons, found that the second term in the perturbation calculation could be larger than the first. Kondo showed, using perturbation theory, that many body effects due to dynamical interactions and scattering of the conduction electrons at the Fermi level (all the states in the Fermi sea with energies below the so-called Fermi level are occupied, while the higher energy states are empty) by a single localized magnetic impurity with an associated local spin moment leads to a $-\ln T$ contribution to the resistance. The $-\ln T$ term increases at low temperatures and when this term is included with the phonon contribution (scattering of the conduction electrons due to the vibrations of atoms—phonons—in the sample) to the resistance, it is sufficient to explain the observed resistance minimum (Fig. 2).

The occurrence of a resistance minimum is connected with the existence of localized magnetic moments on the impurity atoms. As mentioned above, the criterion for a localized moment is a term in the magnetic susceptibility inversely proportional to the temperature, with a coefficient proportional to the density of magnetic atoms as predicted by (Pierre) Curie's law [6–8]. The moment of the free atom is determined by Hund's rules [6–8], which in turn are based on considerations of the intra-atomic Coulomb (and, to a lesser degree, spin–orbit) interactions. Where a resistance minimum is found, there is inevitably a local moment. Kondo has shown that the anomalous high scattering probability of magnetic atoms at low temperatures is a consequence of the dynamic nature of the scattering by the exchange

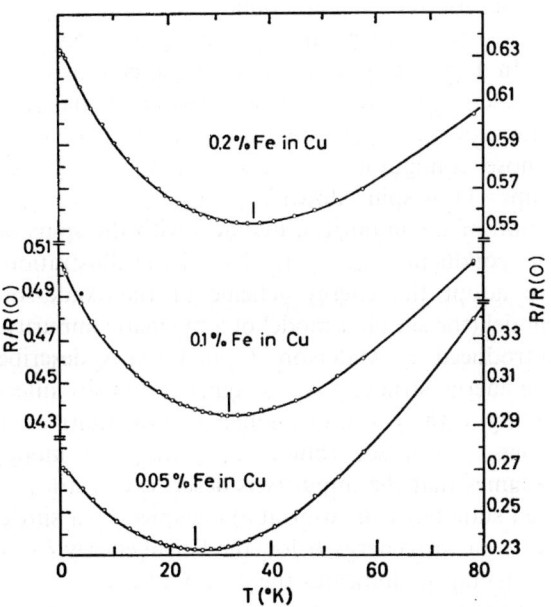

Fig. 1. The resistance minimum as a function of temperature for Cu with different concentrations of iron as impurity. $R(0)$ is the corresponding resistance at $T \approx 273$ K. The resistance minimum phenomenon is clearly evident. Note that the position of the minimum depends on the concentration of iron. From Ref. [11].

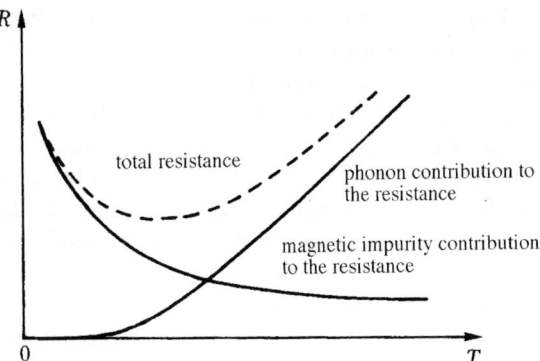

Fig. 2. Temperature T dependence of the resistance R associated with a low concentration of magnetic impurities in a metal. *Phonon contribution to the resistance*: Electric resistance is related to the amount of back scattering from defects and the vibrations of the atoms (phonons), which hinders the motion of the electrons through the crystal. The electric resistance of a pure metal usually drops as its temperature is lowered. This is because the vibrations of the atom (phonons) that impede the electrons get smaller with decreasing temperature. However, the resistance saturates as the temperature is lowered below a certain critical temperature (≈ 10 K) due to static defects in the material. The value of the low-temperature resistance depends on the number of defects in the material. Adding defects increases the value of this "saturation resistance" but the character of the temperature dependence remains the same. *Magnetic impurity contribution to the resistance*: The temperature dependence of the resistance changes dramatically when magnetic atoms, e.g., Fe, are added. Rather than saturating, the electric resistance increases as the temperature is lowered further, due to additional scattering of the electrons by the magnetic impurity through the interaction of their intrinsic angular momentum or "spin". The so-called Kondo temperature—roughly speaking the temperature at which the resistance starts to increase again—completely determines the low-temperature electronic properties of the material.

coupling and of the sharpness of the Fermi surface at low temperatures. Hence the solution of a long-standing puzzle!

2.2. The electron ground state is a singlet!

However, there were difficulties with the theory. Note that because $\ln T$ terms diverge as $T \to 0$, Kondo's perturbation calculations could not be valid at low temperatures. Although Kondo's theory correctly describes the observed upturn of the resistance at low temperatures, it also makes the unphysical prediction that the resistance will be infinite at even lower temperatures. It turns out that Kondo's result is correct only above a certain temperature, which became known as the Kondo temperature T_K. A more comprehensive theory was needed to explain questions such as 'What would be the actual low-temperature ($T \to 0$) behavior of the electron systems showing resistance minima?' That is 'What would be the ground state of the electron system of a dilute magnetic alloy?' The search for such a theory opened a new and fascinating field of research for many body condensed matter physics, and became known as the *Kondo problem* [14–20].

The results of the perturbation calculations of Yosida [21] and Okiji [22], who showed that the results of Yosida's perturbation calculation [21] persist even after considering higher order terms in the perturbation calculation, for the susceptibility suggest that the localized spin falls into a *many-body singlet ground state* (Fig. 3). Furthermore, they concluded that the reason why earlier investigators, who also performed perturbation calculations, got unphysical results—the logarithmic divergence—was because they disregarded the existence of the singlet bound state and, instead, started with degenerate localized spin states.

In Fig. 3 we illustrate what happens when a magnetic impurity (with an associated magnetic moment/spin, analogous to having a bar magnet whose configuration is described by being spin "up" ($\uparrow$) or spin "down" ($\downarrow$), localized at the position of the impurity), interacts with the spins of the conduction electrons. To aid the illustration, we adopt the energy scheme of the Anderson model (the simplest model of a magnetic impurity, introduced by Anderson [34] in 1961) to describe the energy structure of a magnetic impurity interacting with a sea of mobile (conduction) electrons—Fermi sea (blue). The Anderson model assumes that the impurity is described by a localized state (e.g., the d-orbital) occupied by a single electron with energy below the Fermi energy E_F of the metal. E_d indicates the energy of the localized state of the magnetic impurity when it is occupied by a single electron of either spin $\uparrow$ or spin $\downarrow$ (red). Occupancy of the localized state by another electron of opposite spin results in an increase in the corresponding energy level by U, the Coulomb energy. Removing an electron from the localized

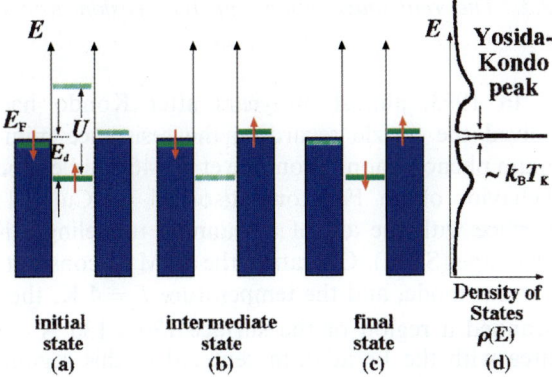

Fig. 3. Energy scheme of the Anderson model (the simplest model of a magnetic impurity, introduced by Anderson [34] in 1961) describing the energy structure of a magnetic impurity interacting with a sea of mobile (conduction) electrons—Fermi sea (blue). The Anderson model assumes that the impurity is described by a localized state (e.g., the d-orbital) occupied by a single electron with energy below the Fermi energy E_F of the metal. E_d indicates the energy of the localized state of the magnetic impurity when it is occupied by a single electron of either spin ↑ or spin ↓ (red). Occupancy of the localized state by another electron of opposite spin results in an increase in the corresponding energy level by U, the Coulomb energy. Removing an electron from the localized state requires at least $|E_d|$. Suppose that a spin ↑ electron is initially occupying the energy level E_d (a). It may tunnel out of the impurity site to briefly occupy a classically forbidden "intermediate state" (b) outside of the impurity. According to Heisenberg's uncertainty principle, within the time scale of $\approx h/|E_d|$, another electron must tunnel from the Fermi sea back towards the impurity (c). Within the time scale $h/|E_d|$, another electron must tunnel from the Fermi sea back towards the impurity (c). However, the spin of this electron may point in the opposite direction. Thus, the initial and final states of the impurity may have different spins. This spin "flipping" or spin exchange qualitatively changes the energy spectrum of the system (d). The Kondo effect is thus produced by a combination of such events taken together, and this results in the appearance of an extra resonance level/a new state (the singlet bound state) at the Fermi energy—known as the Yosida–Kondo resonance (Yosida–Kondo peak) [21,22] (d).

state requires at least $|E_d|$. Although it is classically forbidden to take an electron from the localized impurity state without feeding energy into the system, in quantum mechanics, the Heisenberg uncertainty principle allows such a configuration to exist for a very short time ($\approx h/|E_d|$, where h is Planck's constant). Thus, the spin ↑ electron initially occupying the energy level E_d (Fig. 3(a)), may tunnel out of the impurity site to briefly oc-cupy a classically forbidden "intermediate state" (Fig. 3(b)) outside of the impurity. Within the time scale $h/|E_d|$, another electron must tunnel from the Fermi sea back towards the impurity (Fig. 3(c)). However, the spin of this electron may point in the opposite direction. Thus, the initial and final states of the impurity may have different spins. This spin "flipping" or spin exchange qualitatively changes the energy spectrum of the system (Fig. 3(d)). The Kondo effect is thus produced by a combination of such events taken together, and this results in the appearance of an extra resonance level/a new state (the singlet bound state) at the Fermi energy—known as the Yosida–Kondo resonance (Yosida singlet state) [21,22] (Fig. 3(d)). Thus, we can see that, taken as a whole (the system consisting of a magnetic impurity interacting with the conduction electrons), the localized spin of the magnetic impurity does not survive upon interaction with the spins of the conduction electrons, at very low temperature—the low-temperature spin of the system is neither spin ↑ nor spin ↓, as assumed by earlier investigators. We now know, from hindsight, that the low-temperature increase in the resistance is an indication of the existence of this new state. Such a resonance state is very effective at scattering electrons near the Fermi level, and since the same electrons are responsible for the low-temperature conductivity of a metal, the strong scattering greatly contributes to the electric resistance.

From then on, starting with the ground state properties and resistance near absolute zero temperature, several physical quantities were clarified with more detailed studies [23–25]. Starting with Wilson's [14] solution, which was valid for all temperature range, one after the other, exact solutions [16,26,27] to the two basic models used to study dilute magnetic alloys, viz., the s–d model [4,19,21,22,28–33] and the Anderson model (Fig. 3(a)) [5,34–38], were discovered in the early 1980s. These studies confirmed the generality of earlier perturbation theory results, and further resolved fundamental problems related to the Kondo effect relevant to a single magnetic impurity. Since then, numerous systematic and extensive studies [18,39–42] of a variety of physical properties, e.g., magnetism and superconductivity, and of heavy fermion systems were made, as an extension of the Kondo impurity system. Other fields of condensed matter physics, especially, the studies of high transition temperature T_c superconductivity have greatly benefited from the results of these studies.

For more information on researches carried out in magnetism, we refer the readers to the articles by Bader [43], Goedkoop [44], and Shen and Kirschner [45].

One important thing to note about the Yosida–Kondo resonance (Yosida singlet state) is that it is always "on resonance", as it is fixed at the Fermi level. Although the system may start with the energy of the localized state of the magnetic impurity (E_d in Fig. 3) located very far from the Fermi level, dynamical interaction between the conduction electrons and the localized spin at the magnetic impurity changes the energy spectrum of the system so that it is always on resonance. The only requirement for the Yosida–Kondo resonance (Yosida singlet state) to appear is that the system should be cooled to temperatures sufficiently below the Kondo temperarture T_K.

Because many electrons are involved in the spin–flip processes (as described above and in Fig. 3) to build the Yosida–Kondo resonance, the Yosida–Kondo resonance state is a many body singlet bound state. Also note that each of the electrons that has previously interacted with the same magnetic impurity (the so-called Kondo cloud) contains information about the same magnetic impurity, each electron effectively contains information about the other electrons (the electrons making up the Kondo cloud interact with each other via the localized spin at the magnetic impurity). Thus, we could also say that the Kondo effect is a strongly correlated, many body phenomenon/effect.

The holy grail for research on the Kondo effect is the quest for a possible means of measuring and controlling the Kondo cloud. Another equally important quest is to understand how such a many body quantum state evolves in time? In the next section, we will discuss how Surface Science made another important contribution to this quest, by allowing us to visually observe the Yosida–Kondo peak, and of course, the accompanying Kondo cloud.

2.3. The real-space image of the Yosida–Kondo peak

In 1993, almost 30 years after Kondo had solved the puzzle regarding the resistance minimum phenomenon, Crommie et al. [46] studied the behavior of an Fe atom adsorbed on Cu(1 1 1) surface with the aid of a scanning tunneling microscope (STM). Operating the STM at constant-current mode, and the temperature $T = 4$ K, they scanned a region of the surface 130×130 Å^2 in area with the Fe adatom centered in this region. The corresponding STM image of the region showed a central main peak structure at the position of the Fe adatom, and concentric standing waves around it (cf. Fig. 4(a)). These concentric standing waves arise as a result of the interaction/interference between the conduction electrons scattered by the Fe adatom. The height of the standing wave is distinguishable to within 0.01 Å, which gives us an idea of the high degree of resolution attainable with STMs! By modeling the Fe as a cylindrically symmetric scattering potential, they were able to derive an expression for the change in the local density of states (LDOS) at the Fermi level E_F surrounding the Fe atom, and reproduce the standing wave structure observed *far from the Fe* (with surface lateral distances $r > 20$ Å from Fe) in the STM image (cf. Fig. 4(b)). In addition to this, they also measured the corresponding density of states distribution (dI/dV vs. V curve).

Since the Fe/Cu(1 1 1) system is a *classic* example of a dilute magnetic alloy showing the Kondo effect, these data, as well as the *sharp peak structure* observed at the position of the Fe, motivated Kasai and coworkers [47–49] to propose observing the Kondo effect, in particular the Yosida–Kondo peak (Yosida singlet state) in real space, with the aid of the STM. A few years later, independent [50] of the works of Kasai and coworkers [47–49], Manoharan et al. [51] succeeded in obtaining a real-space image of the Yosida–Kondo peak for the system Co/Cu(1 1 1) (cf. Fig. 1(c) of Ref. [51]), using the same procedure suggested earlier by Kasai and coworkers [47–49] (Fig. 6).

Madhavan et al. [52] and Li et al. [53] measured the corresponding differential conductance dI/dV vs. bias voltage V spectra for a magnetic atom on

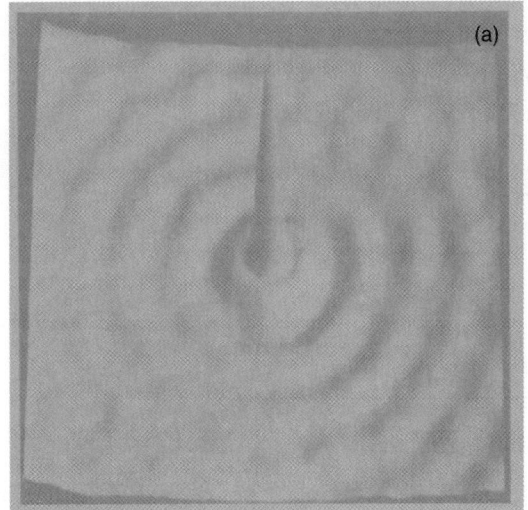

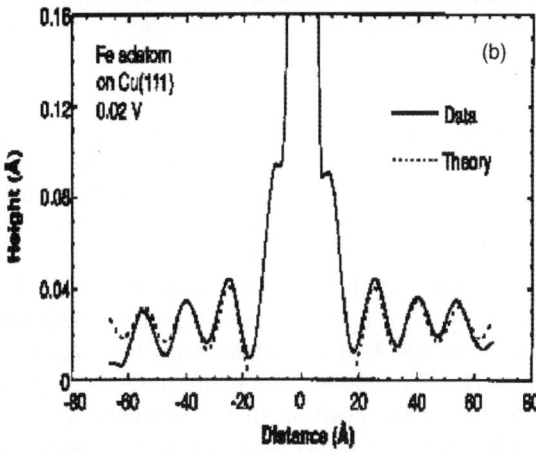

Fig. 4. (a) Constant-current 130 × 130 Å² image of a single Fe adatom on a Cu(1 1 1) surface (acquired with a bias voltage (of the sample with respect to the STM tip) of $V = 0.02$ V, and current $I = 1.0$ nA). The apparent height of the Fe adatom is 0.9 Å. The LDOS at the Fermi level E_F surrounding the Fe adatom are marked by concentric rings (circular standing waves centered at the position of the Fe adatom), a result of the interaction/interference between the conduction electrons scattered by the Fe adatom. (b) Solid line: cross-sectional slice through the center of the Fe adatom image in (a). By modeling the Fe adatom as a cylindrically symmetric scattering potential, at low energies, the change in the LDOS at distances r far from the Fe adatom ($r \geqslant \pm 20$ Å) can be approximated as $\propto (1/kr)[\cos^2(kr - (\pi/4) + \delta_0) - \cos^2(kr - (\pi/4))]$, $k = [2m^*E/\hbar^2]^{1/2}$. m^* is the effective mass of a surface-state electron (0.38m_e). E is the energy measured from the surface-state band edge. Dashed line: obtained by fitting the experimental data (solid line) to the expression above, with $\delta_0 = -80° \pm 5°$ for the phase shift of the $l = 0$ scattered partial wave. From Ref. [46].

a metal surface—Co/Au(1 1 1) and Ce/Ag(1 1 1), respectively. Their results show that near the magnetic atom, the corresponding dI/dV vs. V curve shows either an asymmetric or a symmetric dip structure about the Fermi level ($V = 0$). Analysis of their data suggests that the energy width of the observed dip structure is directly related to the Kondo temperature of the corresponding system (cf., e.g., Fig. 5 [52]).

For more details on the utilization of the STM and its main achievements, we refer the readers to the articles by Plummer et al. [54], Crommie [55], Hasegawa and Grey [56], and Ho [57], and also Refs. [58–60].

In Fig. 6 we show the calculated results [47–49], for the spatial distribution of the tunneling current $I(\mathbf{R}) = I(x, y, z)$ for the case when the STM tip is placed at a distance $z = 4.0$ Å from the surface, and scanned along the surface. In the vicinity of the Fe atom ($r \leqslant 5$ Å, Fig. 6(a)), we can see a narrow peak structure, the Yosida–Kondo peak, which grows large with decreasing temperature. This can be interpreted as due to the relaxation of the electron system to the ground state (Yosida singlet state), as $T \to 0$, and the narrow peak is the spatial manifestation of the ground state wave function. In the region $r > 5$ Å (Fig. 6(b)) we can observe some oscillations (Friedel oscillations) in $I(\mathbf{R})$. (Note that the ordinates of Fig. 6(a) and (b) differ by three orders of magnitude.) We can see that there are some regions of r where $I(\mathbf{R})$ increases in amplitude as $T \to 0$, and some regions of r where $I(\mathbf{R})$ does not vary much with T.

In Fig. 7 we show the calculated results [47,49] for the variation in STM tip height from the surface, when we operate the STM at constant-current mode and scan along the surface. The qualitative features of the calculated results are comparable with experimental results for the STM tip height variations (cf. Fig. 4(b)) [46]. The peak height of ≈ 0.5 Å is of the same order as that of experimental results [46]. Similarly, the standing wave has an amplitude variation of ($\approx 0.02 \pm 0.01$ Å) and a period (≈ 15 Å) that resembles the experimental data. We note that the period of the standing wave is closely related to the Fermi wave number (corresponding to the above-mentioned Fermi level) k_F.

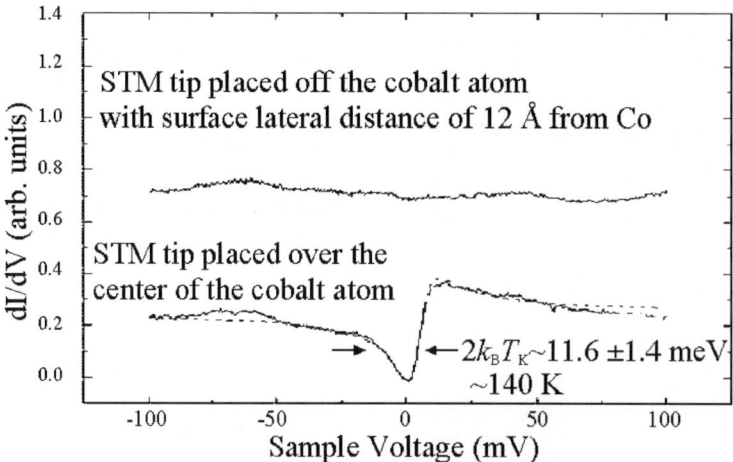

Fig. 5. A pair of dI/dV spectra taken with the STM tip held over a single Co atom (acquired with a bias voltage (of the sample with respect to the STM tip) of $V = 0.1$ V, and current $I = 1.0$ nA) and over the nearby bare Au surface (a constant slope has been substracted from both curves, and they have been shifted vertically). The feature identified as a Yosida–Kondo resonance appears over the Co atom (the ratio of the amplitude of the resonance feature to the overall conductivity is 0.3). From Ref. [52].

Thus, as mentioned earlier, the surface has provided us with a playground to study the Kondo effect, in particular, the Yosida–Kondo resonance (Yosida singlet state) and the accompanying Kondo cloud. Because of the existence of conduction electrons localized at the surface (surface electron state, Fig. 8) and the ability of surface scientists to probe the electrons at surface with an STM, we are now at the stage where we could measure and visualize the Kondo cloud. With further developments like the ability to probe the surface with the STM under varying temperatures and magnetic fields (cf. further discussion below regarding the Kondo temperature), we will soon be able to understand how such a many body quantum state evolves in time?

2.4. A few words on the Kondo temperature T_K

At present, as far as we know, there is no way one can predict what the Kondo temperature T_K would be for a particular system. However, as discussed in Section 2.3 (for more details also cf., e.g., Refs. [47–49]), this should be measurable from the dI/dV spectra. The width of the resonance at/near the Fermi level is $\approx 2k_B T_K$. However, if one does not have a reproducible dI/dV spectrum, one could still experimentally estimate the value of T_K from the temperature dependent I vs. $r(=(x^2+y^2)^{1/2})$ curve. For systems with low T_K, e.g., Fe/Cu(1 1 1) (<20 K) and Co/Au(1 1 1) ($\approx$55 K), the magnitude of the peak structure at the position of the magnetic adatom would decrease by 50% upon an increase in temperature from 0 to 100 K. For systems with high T_K, e.g., Ce/Ag(1 1 1) ($\approx$250 K), the magnitude of the peak structure at the position of the magnetic adatom would decrease by no more than 10% upon an increase in temperature from 0 to 100 K. Thus, one would have reached T_K when the peak structure is $\approx$75% of that at 0 K. Alternatively, one could monitor the width of the dI/dV spectra and determine how the resonance width changes with temperature T. For temperatures $T < T_K$, the resonance width slowly increases ($\propto T^2$) as you sweep the T from $T \ll T_K$ up to T_K. Upon crossing T_K, one observes an abrupt change in the resonance width. Further increases in T causes the resonance to become negligible. Another way of verifying that we are actually seeing the real-space image of the Yosida–Kondo peak is to study how it changes upon the application of an external magnetic field [63], which is also experimentally feasible [64,65].

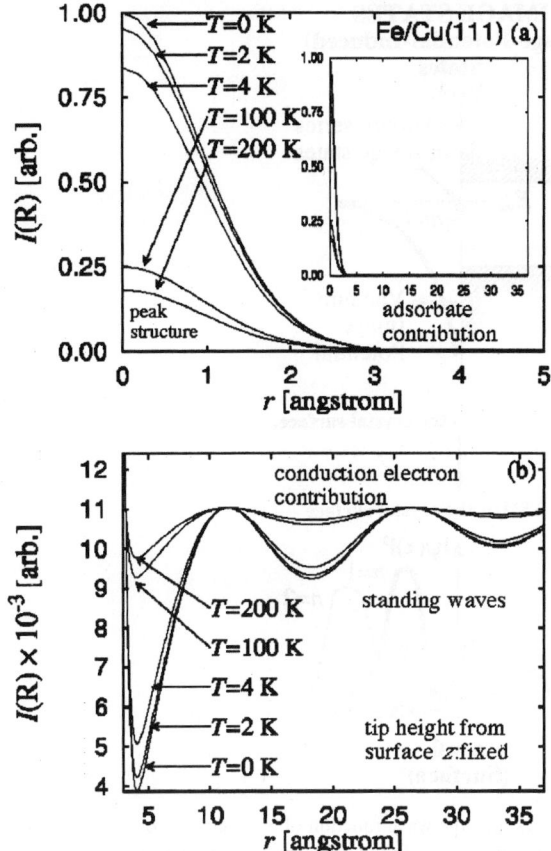

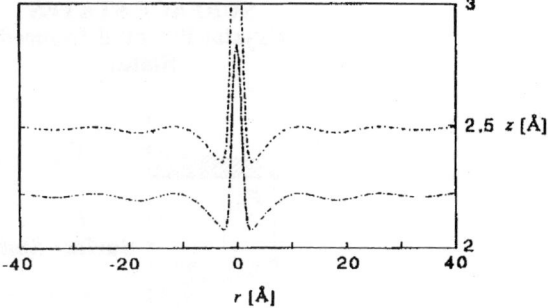

Fig. 6. Simulated tunnel current $I(\mathbf{R}) = I(x, y, z)$ (arbitrary units) as a function of the distance $r = (x^2 + y^2)^{1/2}$ (Å) of the STM tip from Fe (which is located at $(x, y, z) = (0, 0, 0)$) and T dependence (a) for $r = 0–5$ Å, (b) for $r = 4–35$ Å. The STM tip is positioned at a distance $z = 4.0$ Å from the surface. From Ref. [47].

Fig. 7. Simulated STM tip height variation at constant-current mode (corresponds to the spatial distribution of the density of states) and $T = 4$ K. The slight difference in peak width as compared to experimental results can be attributed to the omission of the effects of the structure of the STM tip (cf. Fig. 4 [46]). Inclusion of the effects of the STM tip structure [61] and the continuum of bulk-derived states [62] should give the correct peak width. From Ref. [47].

Thus, we see the manifestation of the Kondo effect in the corresponding STM image of a magnetic atom on a metal surface, or the corresponding spectra, via the electron tunneling between the STM tip and the localized orbitals of the magnetic atom, and between the STM tip and the states of metal conduction electrons scattered by the magnetic impurity atom. We hope that with further studies, we can gain a better understanding of how a magnetic atom behaves on a metal surface. With this new ability to *observe the Kondo effect in real space*, one of our long-standing dreams for doing research on dilute magnetic alloys may come true.

3. Ultrafast electron dynamics studied by time-resolved two-photon photoemission spectroscopy

3.1. Femtosecond time-resolved two-photon photoemission spectroscopy

How do electrons in the metal surface respond when perturbed with external fields on ultrashort time scales? This is a fundamental question, since ultrafast dynamical processes involving electrons affect chemical reactions and optical responses at the surfaces of metals. Let us consider, for example, photon-stimulated desorption. Initially, we photoexcite the electrons of the metal surface at/on which atoms and molecules (adparticles) are adsorbed. Subsequently, the adparticles gain enough kinetic energy to desorb from the surface because of energy and charge transfers between the metal and the adparticles due to electronic transitions [73,74]. Since the excited electrons relax in a femtosecond (10^{-15} s) time scale, however, it is difficult to ascertain the dynamics of electrons in such ultrafast surface processes. Fortunately, recent progress in femtosecond laser technology has revolutionized the study of ultrafast dynamics. An effective experimental method for studying ultrafast electron dynamics is time-resolved two-photon photoemission (TR2PPE) spectroscopy.

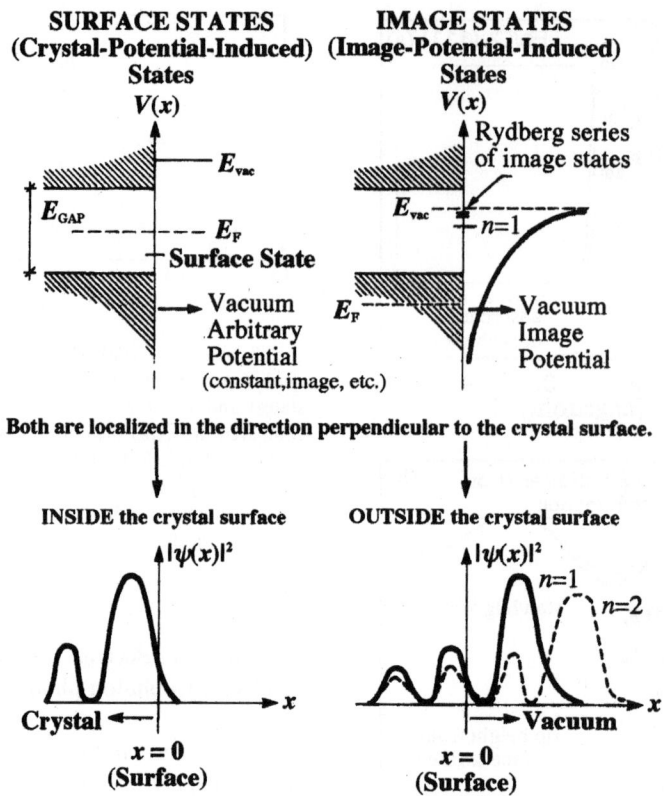

Fig. 8. Inside a perfectly periodic crystal, the electrons occupy Bloch states, i.e., the wave function of an electron moving in a crystal lattice, described by a periodic potential, has the form of a plane wave, modulated by a function which has the same periodicity as the periodic potential. Because of the breaking of the crystal three-dimensional symmetry, the surface of a solid introduces strong modifications into the electronic structure. These modifications come in the form of states appearing inside an electron energy band or between electron energy bands (inside electron energy band gaps). The wave functions (lower left panel) that represent these electrons decay strongly on both sides of the surface (bulk and vacuum). The electron states localized near the vicinity of solid surfaces are generally called surface states. Surface states located inside an electron energy band, called surface resonances, are sometimes distinguished from those located in band gaps, called proper surface states. The amplitude of the wave functions of the former decay exponentially into the bulk. The amplitude of the wave functions of the latter are still fairly localized in the surface region but can, however, leak into the bulk (persist throughout the crystal). It should be noted that the existence of surface states are influenced by the existence of impurities and defects at the surface, and they also exist near interfaces or point contact between two solids. Historically, surface states are distinguished into two, depending on how surface states are modeled. Tamm states (after Tamm [66,67]) occur near band edges and are due to the difference in the potential at the surface and in the bulk. Shockley states (after Shockley [68]) occur within hybridizational band gaps far from band edges and require a multi-band model. For completeness, we should also mention the image states (right panel) which are mainly located outside the surface. They refer to a Rydberg series (lower right panel) of states generated by the Coulombic tail of the image potential field associated with the electrostatic image force experienced by an electron outside a metal surface [69]. For more details, we refer the readers to the review articles by Davison and Levine [70], Echenique and Pendry [69], Osgood and Wang [71] and in the book by Davison and Stęślicka [72].

A schematic diagram of TR2PPE spectroscopy is given in Fig. 9(a), the metal surface is irradiated with pump and probe laser pulses, and then the intensity of the photoelectron emitted from the surface is measured as a function of the photoelectron energy and pump–probe delay time (TR2PPE spectrum). An energy diagram of a two-photon photoemission (2PPE) process is shown in Fig. 9(b). First a pump photon excites an electron from an initial state below the Fermi level into an

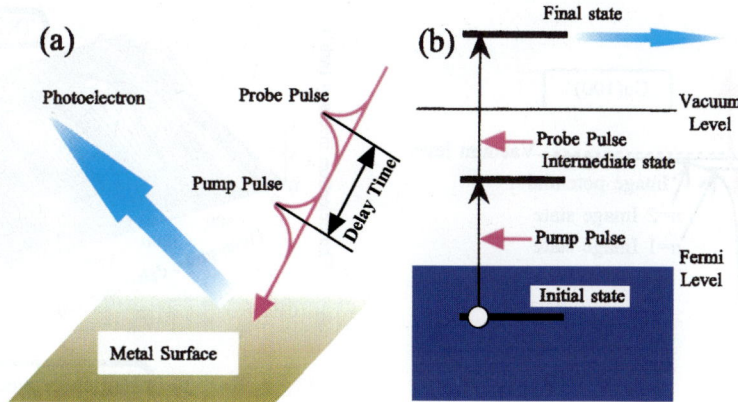

Fig. 9. (a) Schematic picture of TR2PPE spectroscopy. (b) Schematic diagram of the 2PPE process which may be described by a three level model. First an electron in an initial state is excited into an intermediate state by a pump photon and subsequently the electron is excited into a final state by a probe photon.

intermediate state above the Fermi level (hence a hole is excited/created at the initial state level). The temporal evolution of the electron density in the intermediate state can be affected by electron hopping between the substrate metal and the adparticle and various scattering processes, e.g., electron–electron and electron–phonon scattering and electron scattering at impurities and defects. As a result, the electron density decreases and the electron system loses coherence (phase information). Next a probe photon excites the electron from the intermediate state into a final state above the vacuum level, and hence the electron is likely to be emitted from the surface. This process can be described by, e.g., the optical Bloch equations or the Liouville–von Neumann equations [75].

3.2. Photoinduced dynamics of electrons in localized states at surfaces

When an electron moves in front (immediately outside) of a clean metal surface (Fig. 8), the motion of the electron is affected by the image potential which accounts for an attractive force towards the surface due to the dielectric response of the other electrons in the vicinity of the surface. As a result, the electron can be trapped between the image potential and the surface, which acts like a hard wall barrier if there exists a projected band gap at the surface. Hence, localized electron states called *image states* may exist in front of the surface, which form a Rydberg-like series below the vacuum level E_V (Fig. 10). Because of their accessibility, image states provide a useful system to study ultrafast electron dynamics both theoretically and experimentally.

In clean Cu(1 0 0), the vacuum level is located at the center of the projected band gap, and image states with various quantum number n exist just below the vacuum level (the upper panel of Fig. 10). The wave functions of these states spread several nanometers out of the surface, and the lifetimes of these states are of the order of $O(10–1000$ fs$)$ long [76]. Irradiating the surface with a femtosecond laser pulse can result in excitation of an electron wave packet oscillating in front of the surface (quantum beats), when the full width at half maximum (FWHM) of the laser pulse is shorter than 100 fs (the FWHM of the energy spectrum is larger than 13 meV and hence more than two image states are included in this energy range), as 2PPE processes via various image states interfere with each other. Sakai et al. [77] theoretically demonstrated a procedure for analyzing the corresponding TR2PPE spectrum to aid in the understanding of the dynamics of electrons, which provides information on the temporal evolution of the phase of the probability amplitude, and thereby on the motion of the electron wave packets in the *space domain* (Fig. 11).

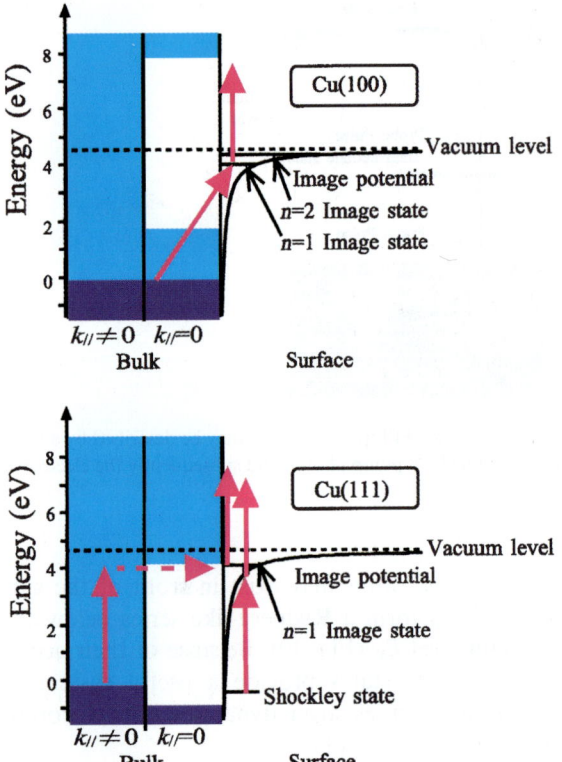
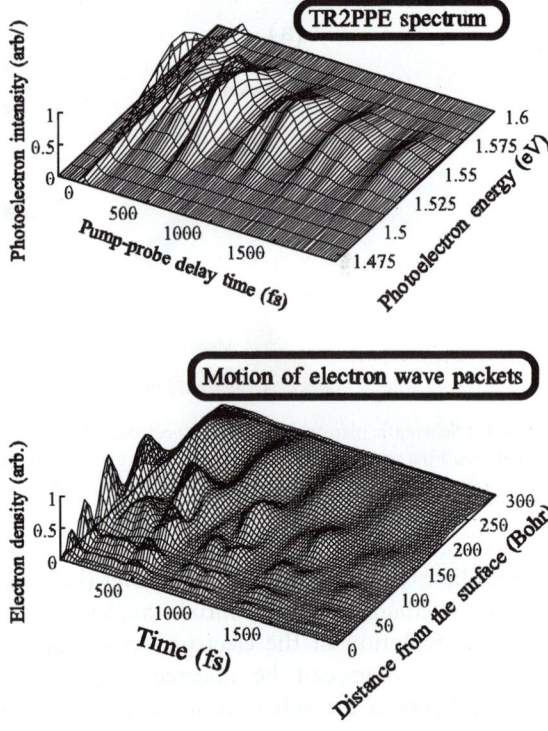

Fig. 10. Energy structure of the electron system of Cu(1 0 0) and Cu(1 1 1) and 2PPE processes (normal emission). $k_\parallel$ gives the component of the electron wave vector parallel to the surface. Dark and light blue areas denote occupied and unoccupied bulk states. Horizontal lines in the right side denote surface localized states, i.e., the image states and the Shockley state. In Cu(1 0 0), image states with quantum numbers $n \geqslant 3$ exist within the energy region between the energy level of the $n = 2$ image state and the vacuum level. Solid purple arrows denote excitation by UV and visible laser pulses, and broken arrow denotes electron transition into the image state by scattering due to Coulomb interactions between electrons. The energy level E_n and wave function $\psi_n(z)$ of an image state with quantum number n, respectively, can be expressed approximately by simple formulae $E_n = E_{vac} - 0.85 \text{ eV} \times (n+a)^{-2}$ and $\psi_n(z) \propto \exp(-z/n^2)L_n(z)$. Here, E_{vac} gives the vacuum level, z gives the distance of the electron from the surface, a gives the quantum defect due to the difference between the effective potential for an electron in the vicinity of the surface and the classical image potential $-e^2/4z$, and $L_n(z)$ are Laguerre polynomials. From Ref. [71].

Fig. 11. Simulations [77] of the TR2PPE spectrum of Cu(1 0 0) (upper panel) and the electron density in the wave packets of the image states with quantum numbers 4, 5 and 6, $\sum_{n=4}^{6}\sum_{n'=4}^{6}\psi_{n'}^*(z)\langle n'|\rho(t)|n\rangle\psi_n(z)$, as a function of the time (the pump pulse intensity reaches a maximum at 0 fs) and the distance from the surface. Here, $|n\rangle$ stands for the eigenvector of the image state with quantum number n and $\rho(t)$ stands for the density matrix in the Schrödinger representation. The origin of the energy is the vacuum level. The pump and probe photon energies are 4.7 and 1.57 eV and their durations are 95 fs. The TR2PPE spectrum shown in the upper panel provides information on the temporal change of the phase, which controls the motion of the wave packets shown in the lower panel. For example, the delay times when the photoelectron intensity in the upper panel reaches maxima ($t_d \simeq 0$, 500 and 1500 fs) correspond to the times when the electron density in the vicinity of the surface in the lower panel reaches maxima.

In clean Cu(1 1 1), the vacuum level is located just above the top of the projected band gap (lower panel of Fig. 10). As a consequence, image states with associated quantum numbers $n \geqslant 2$ are in resonance with the bulk states and the peak structure due to the states in the 2PPE spectrum becomes indistinct, and the $n = 1$ image state has a lifetime ≈ 10 fs [82]. Moreover, a localized state called *Shockley state* exists at the surface [72,78,79] The $n = 1$ image state of Cu(1 1 1) is an attractive target for researchers of ultrafast electron dynamics in metals, because the 2PPE process via this state is well defined, although complicated.

The 2PPE (normal emission) from Cu(1 1 1) occurs via the following two processes—(i) Process (SS), where an electron in the Shockley state with $k_\parallel = 0$ ($k_\parallel$ stands for the component of the wave vector parallel to the surface) below the Fermi level is first excited into the image state with $k_\parallel = 0$ by a pump photon, and then, subsequently, it is further excited into a free electron state above the vacuum level. (ii) Process (IS), where an electron in a state with $k_\parallel \neq 0$ is first excited by a pump photon (hence a hole with $k_\parallel \neq 0$ is excited), and then, subsequently, electron transfer into the image state with $k_\parallel = 0$ occurs accompanied by scattering of the photoexcited electron or the hole by the cold Fermi sea, and finally the electron in the image state is excited by the probe photon. Process (SS) accounts for the peak in the 2PPE energy spectrum at an energy position obtained by adding the pump and probe photon energies to the energy level of the Shockley state, while process (IS) accounts for the peak at an energy position obtained by adding the probe photon energy to the energy level of the image state.

When the pump photon energy is not in resonance with the energy region between the image and Shockley states, relaxation of the electron excited from the Shockley state into the image state occurs in a time shorter than the intrinsic lifetime of the image state. As a result, the correlation trace at the SS peak is not affected by lifetimes of either electrons or holes. On the other hand, the correlation trace at the IS peak is affected by the lifetimes of the electrons and holes involved in the 2PPE process. Sakaue et al. [80] theoretically demonstrated that the effect of Coulomb interactions between electrons, which is important in the understanding of the TR2PPE spectrum of bulk states [81], is equally as important in the understanding of the TR2PPE spectrum of the image state (Fig. 12).

In Fig. 12, the increase in the lifetime of the electron in the image state can be explained by the reduction in the scattering probability of the electron by other electrons in the bulk, since the wave function of the image state is pushed away from the surface by Xe adsorption [83]. Here, what should be emphasized is that the lifetime of the photoexcited electron or the hole is longer in the

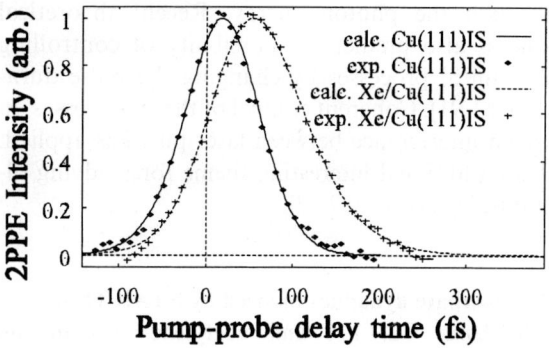

Fig. 12. Simulations [80] of the cross-correlation trace at the IS peak in the TR2PPE spectrum of Cu(1 1 1). Solid and broken curves show the results for a lifetime of the electron in the image state $\tau_{I0} = 7$ fs and a lifetime of the photoexcited electron or the hole $\tau_{\mu j} = 3$ fs, and those for $\tau_{I0} = 30$ fs and $\tau_{\mu j} = 7$ fs, respectively. Here, the nonresonant transitions are neglected for simplicity. The pulse duration is obtained to be 59 fs by fitting the simulation correlation trace to the experimental data for the IS peak [82,83] considering the difference between the pump photon energy and excitation energies from the Shockley state to the image state. We see that the solid and dashed curves agree well with the experimental data for the clean Cu(1 1 1) surface (diamonds) [82] and those of the Xe/Cu(1 1 1) surface (crosses) [83].

experiment on the Xe/Cu(1 1 1) surface than that on the clean Cu(1 1 1) surface. The results show that the pump photon with a lower energy excites an electron and a hole into states with energies closer to the Fermi level, which have longer lifetimes as understood qualitatively from the Fermi liquid theory [84].

3.3. Optical control of dynamical processes at surfaces

Extensive studies of ultrafast electron dynamics in metals by TR2PPE spectroscopy [86–90] and the theoretical analyses [91] of the experimental results obtained are gradually clarifying the mechanisms of the optical response of electrons in bulk states and localized states at metal surfaces. Possible next targets of these studies will be the development of methods to control dynamical processes involving atomic and molecular motions on surfaces by femtosecond laser pulses. Information like that presented in the previous section may serve as helpful hints for controlling dynamical processes through the photon energy. Recent theoretical studies also predict the possibility of controlling dynamical processes by changing the pulse duration [81,91]. Coherent control of these processes, in which interference between laser pulses is applied, is an additional interesting theme for studying ultrafast dynamics [87,92].

4. Adsorbate dynamics induced by STMCO desorption from Cu(1 1 1) and acetylene rotation on Cu(0 0 1)

Continuing on and developing over the question we asked earlier in the beginning of Section 3, we address the question how adsorbates (atoms/molecules on a solid surface) would respond, in turn, when the corresponding electron system is perturbed by some external fields and forced to make a transition from the ground state of the electron system to an excited state? What physical mechanism governs the motion induced? In what follows, we will consider the STM-induced desorption of a single CO from Cu(1 1 1) and acetylene rotation on Cu(0 0 1), and discuss the mechanism behind the corresponding adsorbate motion induced from a microscopic point of view.

4.1. CO desorption from Cu(1 1 1)

In 1998, by combining the atomic length-scale spatial resolution of tunneling spectroscopy (TS) and the femtosecond time-scale temporal resolution of 2PPE, Bartels et al. [85] were able to observe the motion of a single CO induced by *the tunneling of a single electron* from the STM tip to the metal substrate. The bonding of CO to Cu(1 1 1) is described in Fig. 13. The electron tunnels into the CO $2\pi^*$ orbital, thereby destabilizing the bonding of the CO to Cu(1 1 1) as indicated in Fig. 14. Bartels et al. [85] also determined the lifetime of the tunneling electron in the CO $2\pi^*$ orbital to be 0.8–5 fs.

Hasegawa et al. [96] introduced a microscopic model to describe the physical mechanism leading to the above-mentioned STM-induced motion of the CO, which we present below. For simplicity, let us consider only the center-of-mass (CM) motion

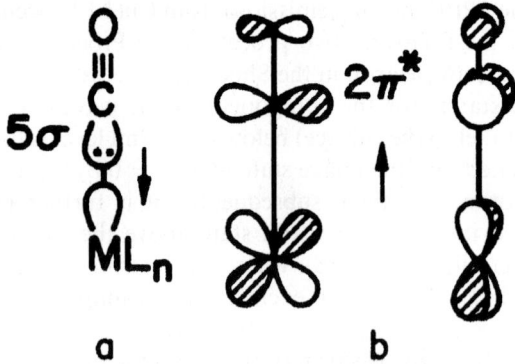

Fig. 13. Adsorption model of CO chemisorbed on a metal surface. A trace of the bonding in the chemisorbed CO reveals that the $2\pi^*$ interaction with the surface d_π is responsible for a good part of the bonding. (a) Forward donation from the carbonyl lone pair 5σ to some appropriate hybrid on a partner metal fragment. (b) Back donation involving the $2\pi^*$ of CO and a d_π orbital, xz, yz of the metal. Shading corresponds to a positive phase of the wave function, and no shading corresponds to a negative phase of the wave function. Alternatively, shading may also mean a wave function with a positive sign, and no shading means the same wave function with a negative sign. From Refs. [93–95].

of CO in the direction normal to the Cu(1 1 1), and neglect the hindered rotational motion and other internal degrees-of-freedom (DOF), e.g., C–O bond stretching, etc., and also the effect of the presence of the STM tip. We further assume that the motion of CO along the direction normal to the Cu(1 1 1) corresponds to the motion of its CM along a Morse potential, whose structure is determined by the ground state energy of the CO/Cu(1 1 1) electron system. Considering a CO adsorbed on the surface (Fig. 15(a)), Hasegawa et al. [96] started with an initial state where a single electron, supposedly originating from the STM tip, is initially occupying the $2\pi^*$ orbital of the adsorbed CO. This electron initially *already* occupying the $2\pi^*$ orbital of the adsorbed CO then, subsequently, transfers/tunnels to the metal substrate (Fig. 15(b)). Because electron hopping/transfer between the $2\pi^*$ orbital of the adsorbed CO and the metal orbital depends on the distance between the CO and the metal surface, the tunneling of the electron from the $2\pi^*$ orbital of the adsorbed CO to the metal substrate may induce the excitation of different CO-surface vibrational

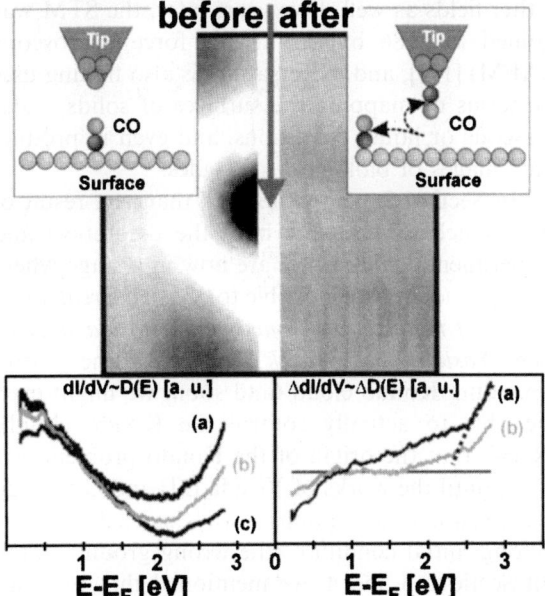

Fig. 14. Center: STM image of a single CO molecule on Cu(1 1 1) at 1.5 nA and 2 V. The left side of the 25 × 25 Å^2 area, including the center of the CO molecule, is shown in the left panel. Throughout the experiment, the temperature is maintained at 15 K. The tip is subsequently positioned at the center of the image (top right sketch), and then the bias voltage is switched to 2.7 V for 3.8 s. During this time, an abrupt change in the simultaneously recorded tunneling current I and LDOS of the sample $D(E) \approx dI/dV$ (in atomic units, measured as functions of the positive sample bias voltage $V \approx E - E_F$) occurs, which can be interpreted as a transfer of the CO molecule away from its initial position to another adsorption site nearby (including the tip apex) (top left sketch), and as the scanning is resumed, the right side of the 25 × 25 Å^2 area turns out dark, as shown in the right panel. Three dI/dV spectra (bottom left diagram) are obtained by performing TS (a) at the center of the CO molecule, (c) at the edge of the image, and (b) in between—instead of inducing a transfer. The difference spectra $\Delta D(E) \approx \Delta dI/dV$ approximately reveal the CO-induced LDOS (bottom right diagram), which shows a rise above ≈ 1 V followed by a steep increase starting at ≈ 2.4 V (dashed line). From Ref. [90].

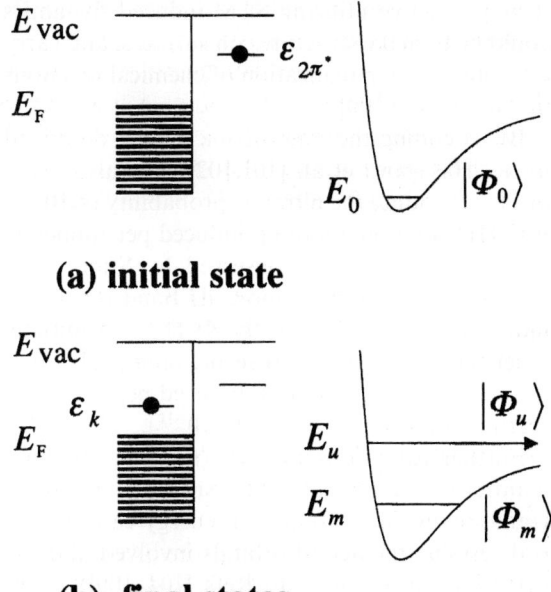

Fig. 15. Schematic energy diagram of the initial and final states of the CO/Cu(1 1 1) system. (a) The initial state configuration. The Fermi energy E_F is below the vacuum level E_{vac}. The CO $2\pi^*$ orbital, with corresponding energy $E_{2\pi^*}$, is initially occupied by a single electron and the system is in the ground state CO-surface stretching vibration $|\Phi_0\rangle$. (b) The final states configuration. The single electron, initially occupying the CO $2\pi^*$ orbital, subsequently transfers/tunnels to the metal $|k\rangle$ orbital, with corresponding energy ϵ_k, and the system is excited either into the mth excited state of the CO-surface stretching vibration $|\Phi_m\rangle$ or an unbound state of the CO-surface stretching vibration $|\Phi_u\rangle$.

stretching modes. When the CO-surface stretching mode in the neutral CO-surface potential is excited to a state with a corresponding energy greater than the adsorption energy of CO on Cu(1 1 1), CO desorption/relocation from/on Cu(1 1 1) occurs, otherwise, the CO stays in place.

With the microscopic model described above, Hasegawa et al. [96] were able to reproduce the low probability of CO desorption induced per tunneling electron observed experimentally ($\approx 10^{-11}$), as well as the experimentally observed isotope effects. Moreover, Hasegawa et al. [96] gave an estimation of what the mean translational energy of CO would be upon desorption—≈ 10 meV. This value is very small compared to the mean translational energy of ≈ 2 eV for a H$^+$ desorbing (induced by electron-beam bombardment) from a W surface [97], and a mean translational energy of ≈ 100 meV for NO desorbing (induced by light irradiation) from a Pt surface [98]. It seems to give us an indication of the energy necessary/relevant/appropriate for manipulating individual adsorbates on surfaces by STM. And also an indication of how effective utilizing STM-induced dynamics would be to make structures on surfaces, and carry out a step-by-step realization of chemical reactions on surfaces [99,100].

By extending the microscopic model described above, Hasegawa et al. [101,102] were also again able to reproduce the ultralow probability ($\approx 10^{-12}$) of C_2HD rotational motion induced per tunneling electron experimentally observed by Stipe et al. [103], for the case when the C–H bond is initially under the STM tip. With the C–D bond initially under the STM tip, the corresponding probability of C_2HD rotational motion induced per tunneling electron is one order higher ($\approx 10^{-11}$).

Further extensions of these models, to attain quantitative agreement with experimental results would involve the actual total energy calculations to determine the actual orbitals involved and the activation barriers (cf., e.g., Refs. [104–109]). From these calculations, one could, e.g., calculate the STM image of the C_2H_2 adsorbed on Cu(1 0 0) [109], and distinguish the STM manipulation mode—pulling, sliding, or pushing—from the feedback loop signal [107].

5. Outlook

Significant contributions from many researchers in the field of condensed matter physics, starting with the observation of a resistance minimum in some metals, in the early 1930s; the recognition that this minimum was a manifestation of many body effects, i.e., the Kondo effect, which results from the interaction between a local magnetic moment associated with a spin and the conduction electrons of the host metal in dilute magnetic alloys; and our present understanding of this physical phenomenon—not only enriched the field in terms of content, the achievements also served as inspirations to other fields of research. The invention of the STM by Binnig and Rohrer in 1981 [110], also had a similar effect. Not only did it contribute to a clearer understanding of the structure of clean and adsorbate-covered surfaces, it has also become a necessary tool in the investigation and understanding of the physical properties of (associated with) atom-size regions, and in other fields as well. Five years later, the STM was joined in 1986 by the atomic force microscope (AFM) [111], and its versatility is also finding uses in terms of mapping the surfaces of solids under gaseous or liquid conditions, and even in producing images of biological molecules.

In Section 2, we pointed out that as a result of recent achievements by both the theoretical and experimental studies, we are now in a stage where it might actually be possible to observe *a real-space image of the electron ground state (Yosida singlet)*, the *Yosida–Kondo peak*, including the corresponding Kondo cloud (and soon, we might even be able to actually control the Kondo cloud). Recall that the origin of the Kondo problem was that, until the works of Yosida [21] and Okiji [22], perturbation calculations always started with the wrong initial condition (the wrong ground state). In Sections 3 and 4, we mentioned that by combining the temporal (femtosecond time-scale) resolution of TR2PPE spectroscopy and the spatial (atomic length-scale) resolution of STM, together with scanning tunneling spectroscopy, we have now within our means a powerful tool to induce and observe ultrafast dynamics of electrons, atoms, and molecules, and also the ability to manipulate and construct different atomic structures on the surface (as discussed in Sections 3 and 4). Numerous achievements have been made with the STM, as reported in the literature. Some examples are the ability to observe the surface charge density distribution; the observation and study of atomic or molecular adsorbate–surface systems, steps, defects and vacancies; atomic-scale manipulations, nanolithography, nanofabrication (construction of quantum corrals [46,51], atom-size bridges [112–115], etc.); STM-induced dynamics (desorption from surface, rotation or dissociation on surface); and luminescence. Even biology has found a use for STMs, e.g., DNA sequencing with an STM. All these achievements, regardless of the scientific field of research involved, were made possible by having an actual image of a natural phenomenon that occurs in the atomic or molecular level. Thus, it would be desirable to have a deeper understanding of the physical mechanism behind each STM image, and STM related phenomena—e.g., How is quantum transport of electrons through an atom-size bridge affected by the shape of the constriction [116]? To what [117] can we attribute the deviation of the conductance through an atom-size point contacts made from STM tips [112] from the universal value of $G_0 = 2e^2/h$? How do many body effects (electron–electron scattering) manifest in the corresponding conductance through an atom-size point contacts made from STM tips [118–121]?

So what does the future have in store for us [122]? Regardless of what (popular) keyword [123]—quantum generators, universal replicators, space drive, etc.—may come to mind, complete control of matter at the atomic level would be necessary. Thus, one answer to our earlier question would be Nanotech [124], where materials to be studied will measure in the range of 1–100 nm (for reference, a Cu measures 0.25 nm across). As Plummer [54] would say "Whoever controls the materials controls the science and technology". So what will the future materials be? What will replace the silicon or the silicon-based computers? Optical computers? Understanding ultrafast dynamics (Section 3) will surely help in the development of this technology. Molecular computers [125], where individual molecules act like switches, wires and even memory elements; where instead of high current or no current representing bits 0 and 1, we use, the orientation [126] of the molecule with respect to the surface? We would surely benefit from an understanding of adsorbate dynamics (Section 4). How about DNA computers [127,128], or quantum computers, where we use the internal DOF of individual atoms? Even increasing the capacities of magnetic data storage devices [129,130] would require knowledge of how magnetic particles interact with each other when jammed together in a very small region (Kondo lattice, cf., e.g., Ref. [20]).

Obviously, controlling things on the atomic scale is not a real big problem. It is only a matter of time before we perfect the technique, be it by using an STM, an AFM, a laser tweezer, or some combination of these. Who knows, some more ingenious method might even be discovered along the way. However, being able to observe the atoms that make up surfaces, or being able to manipulate individual atoms, by itself, will not lead to technological application. The next big question would be 'How do we make these atoms and molecules come together to build large scale functional devices/materials?' How can we take advantage of the inherent products/properties of naturally occurring elementary processes around us to create functional structures?

Nature, most probably, has all the answers. We, living things, for example, are built by natural elementary processes that operate at infinitesimally small scales, and yet functional structures like whales and redwoods exist. Thus, it is not really suprising to observe that recent developments in large scale integration of new devices are getting their hints from processes found in living organisms. It would not be far into the future when we will be able to custom design processes (reactions) much *like those occurring in nature*. However, we must remember that every time we take away something from nature, nature always has a means of exacting her toll on us, in turn. Thus, the big challenge would be 'How do we continue to improve our way of life and still maintain our balance with nature?' How should we design processes, such that by-products would not be the cause of a new disease, the extinction of a species of organism, or the degradation of our environment? Most likely, we would have to find means of designing processes based on naturally occurring processes that are not only user friendly, but also nature (environment) friendly.

In any event, we again find ourselves in the earlier stages of a new quest with many unanswered questions. However, we believe that further significant contributions and new discoveries are on the way. In a recent article, Diño et al. [126] mentioned the significance of understanding the mechanisms behind the dynamics of atom/molecule–surface reactions. We hope that with this article we have shown that the dynamical processes involving the surface electrons of an adsorbate–surface system are equally as interesting and important. As a result of the dynamics of the surface electrons, we now have a means with which we could *actually observe* a physical phenomena that would otherwise not be possible in the bulk. As the saying goes, *a picture says more than a thousand words*. Furthermore, coupled with recent developments in computation techniques for doing first principles electron structure calculations, we now have within our means a powerful tool for systematic analysis of the fundamental mechanism behind dynamical processes occurring at surfaces. And we believe that in order to gain a full understanding of physical phenomena occurring around us in nature, investigations involving surface physical properties such as those presented here are necessary and important.

A few words of explanation on references: In covering subjects like the Many Body Effects, for which there is a vast literature, the problem of which references to cite is an especially difficult one. We have found no good solution to this problem, but we have adopted the following guiding principle—The papers have been selected as the ones most closely related to the particular subject we are discussing, or the ones most familiar to us. Inevitably, much interesting related work will appear to have been overlooked (we sincerely apologize to any contributors to the field who feel that this applied to their work). We do, however, strongly encourage the readers to use the cited references as "seed" references. Used in this way, together with the reference list of the cited review articles, they should be able to build up a comprehensive list to the whole field or any part of it which is of particular interest to them.

Acknowledgements

We gratefully acknowledge the tremendous effort devoted by our collaborators Dr. Hiroshi Nakanishi, Dr. Mamoru Sakaue, and Dr. Kazuhiko Hasegawa in helping us prepare this manuscript, and for sharing with us their opinions on how to prepare this manuscript. We would also like to acknowledge useful discussions with Dr. Atsushi Fukui and Dr. Yoshio Miura. We would also like to thank Prof. Mike Crommie, Dr. Donald Eigler, Prof. Wilson Ho, Prof. Wolf-Dieter Schneider, and Prof. Ludwig Bartels, who were always willing to explain what are and can be obtained in their experiments, and also for sending us some of the original figures that appear in this article. And last, but not the least, special thanks to Dr. Charles Duke and Prof. Ward Plummer, who initiated and devoted a lot of their time into making this special issue of Surface Science happen. Our original works cited here are partly supported by the Ministry of Education, Culture, Sports and Science and Technology of Japan through Grants-in-Aid for Creative Basic Research (no. 09NP1201), COE Research (no. 10CE2004), and Scientific Research (no. 11640375) programs, and by the Japan Science and Technology Corporation (JST), through their Research and Development Applying Advanced Computational Science and Technology program. Some of the calculations in our works cited here were done using the computer facilities of the Japan Science and Technology Corporation (JST), the Institute of Solid State Physics (ISSP) Super Computer Center (University of Tokyo), the Yukawa Institute (Kyoto University, the Center of Computational Physics (Tsukuba University), and the Cybermedia Center (Osaka University). W.A.D. gratefully acknowledges Fellowship grant from the Japanese Society for the Promotion of Science (JSPS), under their Postdoctoral Fellowship Program for Foreign Researchers, and financial support from the Inoue Foundation for Science (Inoue Research Award for Young Scientists).

References

[1] W. Meissner, G. Voigt, Messungen mit Hilfe von flüssigem Helium XI: Widerstand der reinen Metalle in tiefen Temperaturen (Meaurements with liquid helium XI: resistance of pure metals at low temperatures) Ann. Phys. 7 (1930) 761, 892 (in Germany).
[2] J.M. Ziman, Electrons and Phonons: The Theory of Transport Phenomena in Solids, Clarendon Press, Oxford, 1963.
[3] W.J. de Haas, J. de Boer, G.J. van den Berg, The electrical resistance of gold, copper, and lead at low temperatures, Physica 1 (1934) 1115.
[4] J. Kondo, Theory of dilute magnetic alloys, in: F. Seitz, D. Turnbull, H. Ehrenreich (Eds.), Solid State Physics, vol. 23, Academic Press, New York, 1969, p. 183.
[5] A.J. Heeger, Localized moments and nonmoments in metals: the Kondo effect, in: F. Seitz, D. Turnbull, H. Ehrenreich (Eds.), Solid State Physics, vol. 23, Academic Press, New York, 1969, p. 283.
[6] N.W. Ashcroft, N.D. Mermin, Solid State Physics, Saunders College Publications, Philadelphia, 1976.

[7] W.A. Harrison, Solid State Theory, Dover Publications, New York, 1979.
[8] C. Kittel, Introduction to Solid State Physics, seventh ed., Wiley, New York, 1996.
[9] M. Sarachik, E. Corenzwit, L.D. Longinotti, Resistivity of Mo–Nb and Mo–Re alloys containing 1% Fe, Phys. Rev. A 135 (1964) 1041.
[10] A.M. Clogston, B. Matthias, M. Peter, H.J. Williams, E. Corenzwit, R.C. Sherwood, Local magnetic moment associated with an iron atom dissolved in various transition metal alloys, Phys. Rev. 125 (1962) 541.
[11] J.P. Franck, F.D. Manchester, D.L. Martin, The specific heat of pure copper and of some dilute copper + iron alloys showing a minimum in the electrical resistance at low temperatures, Proc. R. Soc. (London) A 263 (1961) 494.
[12] G.J. van den Berg, Anomalies in dilute metallic solutions of transition elements, Prog. Low Temp. Phys. 4 (1964) 194.
[13] J. Kondo, Resistance minimum in dilute magnetic alloys, Prog. Theor. Phys. 32 (1964) 37.
[14] K.G. Wilson, The renormalization group: critical phenomena and the Kondo problem, Rev. Mod. Phys. 47 (1975) 773.
[15] J. Kondo, The Physics of Dilute Magnetic Alloys, Syokabo, Tokyo, 1983.
[16] N. Andrei, K. Furuya, J. Lowenstein, Solution of the Kondo problem, Rev. Mod. Phys. 55 (1983) 331.
[17] J. Kondo, Two-level systems in metals, in: J. Kondo, A. Yoshimori (Eds.), Fermi Surface Effect, Springer Series in Solid-State Science, vol. 77, Springer, Berlin, 1988, p. 1.
[18] A.C. Hewson, The Kondo Problem to Heavy Fermions, Cambridge University Press, Cambridge, MA, 1993.
[19] K. Yosida, in: Theory of Magnetism, Springer Series in Solid-State Science, vol. 122, Springer, Berlin, 1996.
[20] D.L. Cox, A. Zawadowski, Exotic Kondo Effects in Metals, Taylor and Francis, London, 1999.
[21] K. Yosida, Bound state due to the s–d exchange interaction, Phys. Rev. 147 (1966) 223.
[22] A. Okiji, Bound state due to the s–d exchange interaction—effect of higher order perturbations, Prog. Theor. Phys. 36 (1966) 712.
[23] S.D. Silverstein, C.B. Duke, Greens's function derivation of the low equation for the scattering amplitude in dilute magnetic alloy systems, Phys. Rev. Lett. 18 (1967) 695–698.
[24] S.D. Silverstein, C.B. Duke, Theory of s–d scattering in dilute magnetic alloys. I. Perturbation theory and the derivation of the low equation, Phys. Rev. 161 (1967) 456–469.
[25] C.B. Duke, S.D. Silverstein, Theory of s–d scattering in dilute magnetic alloys. II. Derivation and solution of linear vertex equation, Phys. Rev. 161 (1967) 470–477.
[26] A.M. Tsvelick, P.B. Wiegmann, Exact results in the theory of magnetic alloys, Adv. Phys. 32 (1983) 453.
[27] A. Okiji, Bethe Ansatz treatment of the Anderson model for a single impurity, in: J. Kondo, A. Yoshimori (Eds.), Fermi Surface Effect, Springer Series in Solid-State Science, vol. 77, Springer, Berlin, 1988, p. 63.
[28] C. Zener, Interaction between the d shells in the transition metals, Phys. Rev. 81 (1951) 440.
[29] M.A. Ruderman, C. Kittel, Indirect exchange coupling of nuclear magnetic moments by conduction electrons, Phys. Rev. 96 (1954) 99.
[30] K. Yosida, Magnetic properties of Cu–Mn alloys, Phys. Rev. 106 (1957) 893.
[31] T. Kasuya, A theory of metallic ferro- and antiferromagnetism on Zener's model, Prog. Theor. Phys. 16 (1956) 45.
[32] T. Kasuya, s–d and s–f interaction and rare earth metals, in: G.T. Rado, H. Suhl (Eds.), Magnetism, vol. IIB, Academic Press, New York, 1966, p. 215.
[33] K. Yosida, Anomalous electrical resistivity and magnetoresistance due to an s–d interaction in Cu–Mn alloys, Phys. Rev. 107 (1957) 396.
[34] P.W. Anderson, Localized magnetic states in metals, Phys. Rev. 124 (1961) 41.
[35] K. Yosida, K. Yamada, Perturbation expansion for the Anderson model, Prog. Theor. Phys. 46 (Suppl.) (1970) 244.
[36] K. Yamada, Perturbation expansion for the Anderson model. II, Prog. Theor. Phys. 53 (1975) 970.
[37] K. Yosida, K. Yamada, Perturbation expansion for the Anderson model. III, Prog. Theor. Phys. 53 (1975) 1286.
[38] K. Yamada, Perturbation expansion for the Anderson model. IV, Prog. Theor. Phys. 54 (1975) 316.
[39] T. Kasuya, T. Saso (Eds.), Theory of heavy fermions and valence fluctuations, Springer Series in Solid-State Science, vol. 62, Springer, Berlin, 1985.
[40] A. Sumiyama, Y. Oda, H. Nagano, Y. Ōnuki, K. Shibutani, T. Komatsubara, Coherent Kondo state in a dense Kondo substance: $Ce_xLa_{1-x}Cu_6$, J. Phys. Soc. Jpn. 55 (1986) 1294.
[41] A. Yoshimori, H. Kasai, Theory of dense Kondo system, J. Magn. Magn. Mater. 31–34 (1983) 475.
[42] A. Yoshimori, H. Kasai, Heavy electrons in alloys, Solid State Commun. 58 (1986) 259.
[43] S.D. Bader, Magnetism in low dimensionality, Surf. Sci. 500 (2002) 172.
[44] J.B. Goedkoop, http://www.science.uva.nl/research/cmp/goedkoop/.
[45] J. Shen, J. Kirschner, Tailoring magnetism in artificially structured materials: the new frontier, Surf. Sci. 500 (2002) 300.
[46] M.F. Crommie, C.P. Lutz, D.M. Eigler, Confinement of electrons to quantum corral on a metal surface, Science 262 (1993) 218.
[47] T. Kawasaka, H. Kasai, A. Okiji, Adsorbate-induced charge density oscillations and many body effects in STM image, Phys. Lett. A 250 (1998) 403.
[48] T. Kawasaka, H. Kasai, W.A. Diño, A. Okiji, Spatial and spectroscopic profiles of the Kondo resonance for magnetic atoms on metal surfaces, J. Appl. Phys. 86 (1999) 6970.

[49] H. Kasai, W.A. Diño, A. Okiji, Behavior of a magnetic atom on a metal surface—real space image of the Kondo effect, J. Electron Spectrosc. Rel. Phenom. 109 (2000) 63.
[50] D.M. Eigler, private communication.
[51] H.C. Manoharan, C.P. Lutz, D.M. Eigler, Quantum mirages formed by coherent projection of electronic structure, Nature 403 (2000) 512.
[52] V. Madhavan, W. Chen, T. Jamneala, M.F. Crommie, N.S. Wingreen, Tunneling into a single magnetic atom: spectroscopic evidence of the Kondo resonance, Science 280 (1998) 567.
[53] J.T. Li, W.-D. Schneider, R. Berndt, B. Delley, Kondo scattering observed at a single magnetic impurity, Phys. Rev. Lett. 80 (1998) 2893.
[54] E.W. Plummer, Ismail, R. Matzdorf, A.V. Melechko, J.P. Pierce, J. Zhang, Surfaces: a playground for physics with broken symmetry in reduced dimensionality, Surf. Sci. 500 (2002) 1.
[55] M. Crommie, http://physics.berkeley.edu/research/crommie/.
[56] S. Hasegawa, F. Grey, Electron transport—from point-contact transistor to micro-four-point probes, Surf. Sci. 500 (2002) 84.
[57] W. Ho, http://www.ps.uci.edu/physics/wilsonho/wilsonho.html.
[58] H.-J. Güntherodt, R. Wiesendanger (Eds.), Scanning Tunneling Microscopy I, second ed., Springer Series in Surface Science, vol. 20, Springer, Berlin, 1994.
[59] R. Wiesendanger, H.-J. Güntherodt (Eds.), Scanning tunneling microscopy II, second ed., Springer Series in Surface Science, vol. 28, Springer, Berlin, 1995.
[60] C. Bai, in: Scanning Tunneling Microscopy and its Application, Springer Series in Surface Science, vol. 32, Springer, Berlin, 1995.
[61] M. Tsukada, N. Shima, Theory of electronic processes of scanning tunneling microscopy, Phys. Soc. Jpn. 56 (1987) 2875.
[62] S. Crampin, Surface states as probes of buried impurities, J. Phys.: Condens. Matter 6 (1994) L613.
[63] W.A. Diño, K. Imoto, H. Kasai, A. Okiji, Spatial and spectroscopic profiles of the Kondo resonance for a single magnetic atom on a metal surface under an external magnetic field, Jpn. J. Appl. Phys. 39 (2000) 4359.
[64] R. Wiesendanger, M. Bode, R. Drombowski, M. Getzlaff, M. Morgenstern, C. Wittneven, Local electronic properties in the presence of internal and external magnetic fields studied by variable-temperature scanning tunneling spectroscopy, J. Appl. Phys. 37 (1998) 3769.
[65] K. Sueoka, K. Mukasa, K. Hayakawa, Possibility of observing spin-polarized tunneling current using scanning tunneling microscope with optically pumped GaAs, J. Appl. Phys. 32 (1993) 2989.
[66] I.E. Tamm, Über eine mögliche Art der Elektronenbindung an Kristalloberflächen (On a possible type of electron binding at crystal surfaces), Phys. Z. Sowjet. 1 (1932) 733.
[67] I.E. Tamm, Über eine mögliche Art der Elektronenbindung an Kristalloberflächen (On a possible type of electron binding at crystal surfaces), Z. Physik 76 (1932) 849.
[68] W.B. Shockley, On the surface states associated with a periodic potential, Phys. Rev. 56 (1939) 317.
[69] P.M. Echenique, J.B. Pendry, Theory of image states at metal surfaces, Prog. Surf. Sci. 32 (1989) 111–172.
[70] S.G. Davison, J.D. Levine, Surface states, in: H. Ehrenreich, F. Seitz, D. Turnbull (Eds.), Solid State Physics, vol. 525, Academic Press, New York, 1970, p. 1.
[71] R.M. Osgood Jr., X. Wang, Image state on single-crystal metal surfaces, in: H. Ehrenreich, F. Spaepen (Eds.), Solid State Physics, vol. 51, Academic Press, San Diego, Chestnut Hill, 1998.
[72] S.G. Davison, M. Stęślicka, Basic Theory of Surface States, Oxford Science Publications, New York, 1992.
[73] M. Mizuno, H. Kasai, A. Okiji, A model calculation for photo-stimulated desorption of molecules adsorbed on metal surface, Surf. Sci. 310 (1994) 273.
[74] H. Tsuchiura, H. Kasai, A. Okiji, A model calculation for photo-stimulated desorption, J. Phys. Soc. Jpn. 66 (1997) 2805.
[75] F. Abelés (Ed.), Optical Properties of Solids, North-Holland, Amsterdam, 1972.
[76] U. Höfer, I.L. Shumay, Ch. Reuß, U. Thomann, W. Wallauer, Th. Fauster, Time-resolved coherent photoelectron spectroscopy of quantized electronic states on metal surfaces, Science 277 (1997) 1480.
[77] T. Sakai, M. Sakaue, H. Kasai, A. Okiji, Theory of dynamics of electron wave packets in time-resolved two-photon photoemission via image states, Appl. Surf. Sci. 169-170 (2001) 57.
[78] A. Zangwill, Physics at Surfaces, Cambridge University Press, New York, 1988.
[79] M.C. Desjonquères, D. Spanjaard, Concepts in Surface Physics, second ed., Springer, Berlin, 1996.
[80] M. Sakaue, H. Kasai, A. Okiji, Theory of time-resolved two-photon photoemission from Cu(1 1 1): effect of Coulomb interactions among electrons, Appl. Surf. Sci. 169–170 (2001) 68.
[81] M. Sakaue, H. Kasai, A. Okiji, Theory of time-resolved two-photon photoemission from a metal surface: the effect of Coulomb interactions between electrons, J. Phys. Soc. Jpn. 69 (2000) 160.
[82] T. Hertel, E. Knoesel, M. Wolf, G. Ertl, Ultrafast electron dynamics at Cu(1 1 1): response of an electron gas to optical excitation, Phys. Rev. Lett. 76 (1996) 535.
[83] M. Wolf, E. Knoesel, T. Hertel, Ultrafast dynamics of electrons in image-potential states on clean and Xe-covered Cu(1 1 1), Phys. Rev. B 54 (1996) R5295.
[84] A.A. Abrikosov, L.P. Gor'kov, I.E. Dzyaloshinskii, Quantum field theoretical methods in statistical physics, Pergamon Press, Oxford, New York, 1965, translated to English by D.E. Brown, edited by D. ter Haar.
[85] L. Bartels, G. Meyer, K.-H. Rieder, D. Velic, E. Knoesel, A. Hotzel, M. Wolf, G. Ertl, Dynamics of electron-

induced manipulation of individual CO molecules on Cu(1 1 1), Phys. Rev. Lett. 80 (1998) 2004.
[86] M. Bauer, S. Pawlik, M. Aeschlimann, Decay dynamics of photoexcited alkali chemisorbates: real-time investigations in the femtosecond regime, Phys. Rev. B 60 (1999) 5016.
[87] S. Ogawa, H. Nagano, H. Petek, Optical decoherence and quantum beats in Cs/Cu(1 1 1), Surf. Sci. 427–428 (1999) 34.
[88] H. Petek, M.J. Weida, H. Nagano, S. Ogawa, Real-time observation of adsorbate atom motion above a metal surface, Science 288 (2000) 1402.
[89] M. Bauer, S. Pawlik, M. Aeschlimann, Femtosecond lifetime investigations of excited adsorbate states: atomic oxygen on Cu(1 1 1), Surf. Sci. 377–379 (1997) 350.
[90] A. Hotzel, K. Ishioka, E. Knoesel, M. Wolf, G. Ertl, Can we control lifetimes of electronic states at surfaces by adsorbate resonances? Chem. Phys. Lett. 285 (1998) 271.
[91] M. Sakaue, H. Kasai, A. Okiji, The effect of interband excitations on time-resolved two-photon photoemission via a localized state at a metal surface, J. Phys. Soc. Jpn. 67 (1998) 2058.
[92] H. Petek, S. Ogawa, Femtosecond time-resolved two-photon photoemission studies of electron dynamics in metals, Prog. Surf. Sci. 56 (1997) 239.
[93] R. Hoffmann, A chemical and theoretical way to look at bonding on surfaces, Rev. Mod. Phys. 60 (1988) 601–628.
[94] R. Hoffmann, How chemistry and physics meet in the solid state, Angew. Chem. Int. Ed. Engl. 26 (1987) 846–879.
[95] J.L. Whitten, H. Yang, Theory of chemisorption and reactions on metal surfaces, Surf. Sci. Rep. 218 (1996) 55–124.
[96] K. Hasegawa, H. Kasai, W.A. Diño, A. Okiji, Dynamics of STM-induced CO desorption from Cu(1 1 1), Phys. Soc. Jpn. 67 (1998) 4018;
K. Hasegawa, H. Kasai, W.A. Diño, A. Okiji, A microscopic theory of STM-induced CO desorption from Cu(1 1 1), Surf. Sci. 438 (1999) 283.
[97] M. Nishijima, F.M. Propst, Electron-impact desorption of ions from polycrystalline tungsten, Phys. Rev. B 2 (1970) 2368.
[98] K. Fukutani, A. Peremans, K. Mase, Y. Murata, Photodesorption of NO from Pt(0 0 1) at $\lambda = 193$, 248, and 352 nm, Phys. Rev. B 47 (1993) 4007;
K. Fukutani, A. Peremans, K. Mase, Y. Murata, Photostimulated desorption of NO from a Pt(0 0 1) surface, Surf. Sci. 283 (1993) 158.
[99] H.J. Lee, W. Ho, Single bond formation and characterization with a scanning tunneling microscope, Science 286 (1999) 1719–1722.
[100] S.-W. Hla, L. Bartels, G. Meyer, K.-H. Rieder, Inducing all steps of a chemical reaction with the scanning tunneling microscope tip: towards single molecule engineering, Phys. Rev. Lett. 85 (2000) 2777–2780.

[101] K. Hasegawa, W.A. Diño, H. Kasai, A. Okiji, Dynamics of STM-induced acetylene rotation on Cu(1 0 0), Surf. Sci. 454–456 (2000) 1052–1057.
[102] K. Hasegawa, H. Kasai, W.A. Diño, A. Okiji, Adsorbate dynamics induced by STM, Appl. Surf. Sci. 169–170 (2001) 25.
[103] B.C. Stipe, M.A. Rezaei, W. Ho, Coupling of vibrational excitation to the rotational motion of a single adsorbed molecule, Phys. Rev. Lett. 81 (1998) 1263.
[104] K. Stokbro, B.Y.-K. Hu, C. Thirstrup, X.C. Xie, First-principles theory of inelastic currents in a scanning tunneling microscope, Phys. Rev. B 58 (1998) 8038–8041.
[105] U. Kürpick, T.S. Rahman, Tip induced motion of adatoms on metal surfaces, Phys. Rev. Lett. 83 (1999) 2765–2768.
[106] S. Corbel, J. Cerdá, P. Sautet, Ab initio calculations of scanning tunneling microscopy images within a scattering formalism, Phys. Rev. B 60 (1999) 1989–1999.
[107] X. Bouju, C. Joachim, C. Girard, Single-atom motion during a lateral STM manipulation, Phys. Rev. B 59 (1999) R7845–R7848.
[108] N. Mingo, K. Makoshi, Excitation of vibrational modes of adsorbates with the scanning tunneling microscope: many-orbital theory, Surf. Sci. 438 (1999) 261–270.
[109] N. Mingo, K. Makoshi, Calculation of the inelastic scanning tunneling image of acetylene on Cu(1 0 0), Phys. Rev. Lett. 84 (2000) 3964.
[110] G. Binnig, H. Rohrer, Scanning tunnelling microscopy, Helv. Phys. Acta 55 (1982) 726.
[111] G. Binnig, C.F. Quate, Atomic force microscope, Phys. Rev. Lett. 56 (1986) 930.
[112] L. Olesen, E. Lægsgaard, I. Stensgaard, F. Besenbacher, J. Schiøtz, P. Stoltze, K.W. Jacobsen, J.K. Nørskov, Quantized conductance in an atom-sized point contact, Phys. Rev. Lett. 72 (1994) 2251.
[113] F. Komori, K. Nakatsuji, Quantized conductance through atomic-sized iron contacts at 4.2 K, Phys. Soc. Jpn. 68 (1999) 3786.
[114] H. Ohnishi, Y. Kondo, K. Takayanagi, Quantized conductance through individual rows of suspended gold atoms, Nature 395 (1998) 780.
[115] Y. Kondo, K. Takayanagi, Gold nanobridge stabilized by surface structure, Phys. Rev. Lett. 79 (1997) 3455.
[116] H. Kasai, K. Mitsutake, A. Okiji, Effects of confining geometry on ballistic transport in quantum wires, Phys. Soc. Jpn. 60 (1991) 1679.
[117] H. Kasai, T. Kakuda, A. Okij, Conductance quantization in a atom-sized contact between an STM-tip and a metal surface, Surf. Sci. 363 (1996) 428.
[118] Y. Kawahito, H. Kasai, H. Nakanishi, A. Okiji, Effects of intra-site Coulomb interaction on quantized conductance in a quantum wire between an STM-tip and a metal surface, Surf. Sci. 409 (1998) L709.
[119] Y. Kawahito, H. Kasai, H. Nakanishi, A. Okiji, Quantum transport in an atom-sized bridge between a metal surface and a tip of a scanning tunneling microscope at finite temperatures, J. Appl. Phys. 85 (1999) 947.

[120] H. Nakanishi, H. Kasai, A. Okiji, Effects of the intra-site Coulomb interaction on electron transport in an atom bridge, Appl. Surf. Sci. 169/170 (2001) 73.

[121] Y. Morigaki, H. Nakanishi, H. Kasai, A. Okiji, Quantized conductance in an atom-size bridge made from magnetic materials, J. Appl. Phys. 88 (2000) 2682.

[122] An interesting series of articles on how our work and society could be in the 21st century can be found online at http://www.time.com/time/reports/v21/home.html.

[123] A.C. Clarke, Beyond 2001, Reader's Digest (Asia Edition) February (2001) 100.

[124] C. Macilwain, Nanotech thinks big, Nature 405 (2000) 730.

[125] M.A. Reed, J.M. Tour, Computing with MOLECULES, Sci. Am. June (2000) 69.

[126] W.A. Diño, H. Kasai, A. Okiji, Orientational effects in dissociative adsorption/associative desorption dynamics of $H_2(D_2)$ on Cu and Pd, Prog. Surf. Sci. 63 (2000) 63.

[127] K. Sakamoto, H. Gouzu, K. Komiya, D. Kiga, S. Yokoyama, T. Yokomori, M. Hagiya, Molecular computation by DNA hairpin formation, Science 288 (2000) 1223.

[128] A. Cho, Hairpins trigger an automatic solution, Science 288 (2000) 1152.

[129] S. Sun, C.B. Murray, D. Weller, L. Folks, A. Moser, Monodisperse FePt nanoparticles and ferromagnetic FePt nanocrystal superlattices, Science 287 (2000) 1989.

[130] R.F. Service, Nanocrystal may give boost to data storage, Science 287 (2000) 1902.

Statistical thermodynamics of soft surfaces

S.A. Safran *

Department of Materials and Interfaces, Weizmann Institute of Science, Rehovot 76100, Israel

Received 27 June 2000; accepted for publication 27 April 2001

Abstract

We review the continuum, statistical thermodynamics of surfaces and interfaces in soft matter where both the energy and entropy of the surface are comparable. These systems include complex fluids that are dominated by either surface tension or the interfacial curvature, such as: fluid and solid interfaces, colloidal dispersions, macromolecular solutions, membranes, and other self-assembling aggregates such as micelles, vesicles, and microemulsions. The primary focus is on the theoretical concepts, their universality, and the role of fluctuations and inhomogeneities with connections to relevant experimental systems. © 2001 Elsevier Science B.V. All rights reserved.

Keywords: Equilibrium thermodynamics and statistical mechanics; Bending of surfaces; Self-assembly; Surface structure, morphology, roughness, and topography; Surface tension; Surface thermodynamics (including phase transitions)

1. Introduction

1.1. What is soft matter?

During the last half-century, the growth of the electronics industry drove the development of the surface science of "hard" materials such as crystals. In the next several decades, the coming of age of biotechnology and the chemical industries will provide the impetus for the quantitative understanding of the properties of the surfaces and interfaces of "soft" matter such as liquid crystals, colloidal dispersions, macromolecular solutions and gels, and self-assembling, amphiphilic systems such as membranes. While hard materials are characterized by global or local atomic-scale order

* Tel.: +972-8934-3362; fax: +972-8934-4138.
E-mail address: sam.safran@weizmann.ac.il (S.A. Safran).

and by interaction energies that are generally much greater than room temperature (we shall often measure energy scales in terms of room temperature) soft matter is characterized by relatively weaker interactions that promote only partial order on length scales that are much larger than those of a single molecule. For example, polymers in solution can be described by disordered, blob-like structures where the typical blob size can be hundreds of Angstroms [1]. Water–surfactant–oil dispersions can exhibit disordered sponge-like structures consisting of domains of water in oil with the surfactant monolayers at the water–oil interfaces [2,3]. The domains sizes can be a hundred times the size of an individual soap molecule. Even those soft materials [4] that show solid-like properties, such as gels or colloidal crystals, have mesh or lattice constants that are in the range of hundreds to thousands of Angstrom range. The weak interactions in such matter result in shear elastic constants that can be several orders of magnitude weaker than those of crystals. The weak, non-covalent nature of the interactions [5] in these systems are often competitive with the entropy of the system; this leads to a rich variety of structures and phases as the temperature or composition is varied. The surfaces and interfaces in soft materials show a similar sensitivity.

While much of the surface science of hard matter focuses on the atomic-scale structure, the study of soft surfaces involves the large-scale nature of the interface and especially its thermal fluctuations. The intellectual challenges include the following questions: on what length scale is the surface or interface flat? Does the surface retain its integrity as a nominally flat, atomically thin, sheet whose volume is negligible compared with that of the two, macroscopic bulk phases that it separates? Or, does the interface develop a macroscopic amount of undulations so that one can speak of a thermodynamic amount of internal surface within two comingled bulk phases? In the latter case, what are the sizes, shapes, and phase behavior of the system of interface? The nature of soft materials is such that an analysis [2,4] of their surface and interfacial properties [5] is important for the understanding of even their bulk behavior.

1.2. Soft matter systems

We begin with some examples of soft matter systems that are of both intellectual and practical interest. The sketches shown below depict these structured fluids contained in "test tubes"; note, however, that the system scale can range from tens to thousands of Angstroms, as explained in what follows.

(1) *Liquid crystals:* are phases of anisotropic (e.g., rod-like) molecules that show orientational order, but not necessarily positional (translational) order (see Fig. 1). Typical properties of interest include: cooperative phase transitions towards oriented and partially ordered states as a function of temperature, orientational response to weak electric and magnetic fields, partial ordering into layered states with in-plane fluidity (smectic state). Applications include their use in displays where the orientational sensitivity to small external fields

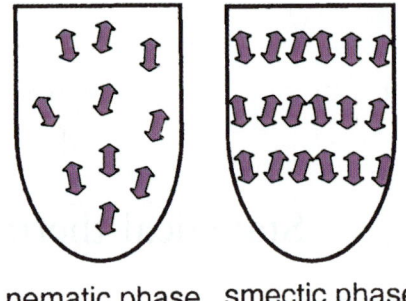

Fig. 1. Liquid crystal phases.

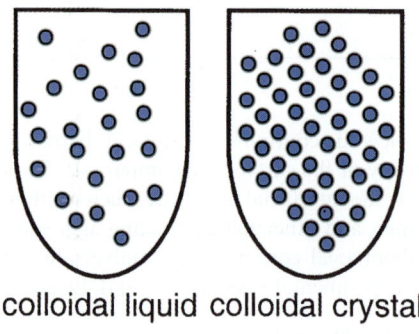

Fig. 2. Colloidal liquids and crystals.

is used to modulate their optical properties as well as uses in the chemical industry of the anisotropic viscosity of smectic phases.

(2) *Colloidal dispersions:* consist of a molecular solvent and solid or liquid particles that are much larger than a molecular size (see Fig. 2). The properties of interest include the tendency for the dispersion to be stable or to phase separate into a particle-rich phase and a nearly pure solvent phase, the existence of ordered, large scale crystals with lattice constants that can be of the order of optical wavelengths and which melt under very small shear forces, and changes in viscosity that can be controlled by concentration. Colloidal dispersions are used in paints, inks, drug encapsulation; lysosomes are a type of liquid dispersion that are found in biological cells and are used for digestion.

(3) *Macromolecular solutions:* are composed of long-chain molecules in solution. Beyond some length scale, usually several nanometers (but in

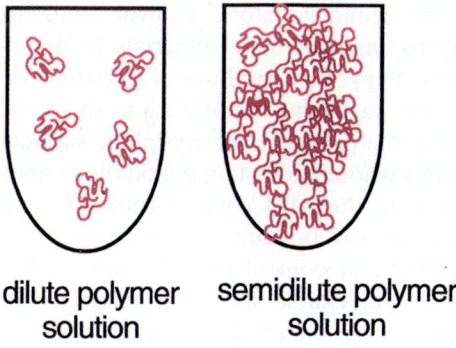

Fig. 3. Polymer solutions.

double stranded DNA, 50 nm), the chains are flexible and their conformations are dominated by their entropy in solution (see Fig. 3). The questions of interest include the scaling of the size of a chain as a function of its stretched length (or molecular weight), the dependence of the solution properties such as the turbidity and viscosity on polymer concentration, and the cooperative nature of the adsorption of polymers at interfaces. In the chemical industry, polymers are used as additives, lubricants and in the formation of plastics. In biological systems, linear macromolecules with a variety of conformational structures abound; these include proteins that show intricate folding patterns due to different types of molecular interactions of the monomers in the chains, long sugar molecules, as well as DNA and RNA.

(4) *Gels:* are composed of long-chain molecules that are either chemically or physically interconnected by junctions (see Fig. 4). If the density of these junctions exceeds a certain critical value, known as the gelation point, the system shows a solid-like response to external shear; this is in contrast to a fluid—even a very viscous one like a polymer solution—that shows no shear response in the low frequency (long time) regime. The study of the large-scale structure of gels (the typical mesh size can be in the hundreds of Angstrom range) focuses on the gelation transition as a function of concentration of junction moieties, the swelling of the mesh by solvents, inhomogeneities and fluctuations within the bulk gel, and the stability of the shear and yield moduli. Applications range from diapers (where the large swelling of the gel by the addition of solvent is enabled by the unfolding or entropic stretching of the polymer coils in the gel network) to the use of actin gels by most biological cells; these gels give the cell its elasticity and shape integrity upon deformation by surfaces or intercellular contacts.

(5) *Self-assembling amphiphilic systems:* are composed of molecules with both hydrophobic (water hating) and hydrophilic (water loving) parts that cause these molecules to be interfacially active. The systems include some alcohols (that covalently combine a hydrocarbon chain and a polar, water-loving OH group), surfactants or soap-like molecules (that have hydrocarbon chains bonded to charged groups such as $(Na)^+(SO_3)^-$ or to non-polar, but still hydrophilic ester groups), and lipids (that typically consist of two hydrocarbon chains bonded to hydrophilic groups). These molecules reduce the surface tension of water or oil systems and in some cases allow for the stabilization of phases with a macroscopic amount of internal interface. Three-component systems of water, oil and an amphiphile that form an equilibrium dispersion of oil in water (or water in oil) domains with most of the amphiphile located at the internal interfaces are called microemulsions (see Fig. 5). A similar phenomenon occurs in two-component mixtures (e.g., amphiphile and water) where two layers of amphiphile separate either water or oil domains; these bilayer membranes can then fold into a variety of shapes including spherical vesicles or convoluted, sponge-like phases. The properties that are studied include the changes of large-scale structure and phase behavior

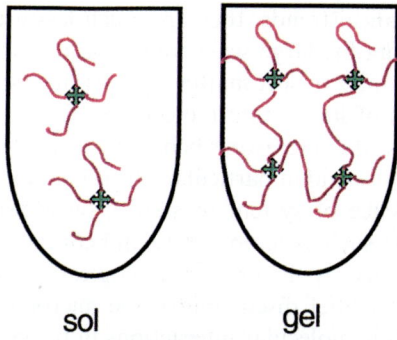

Fig. 4. Gel phases.

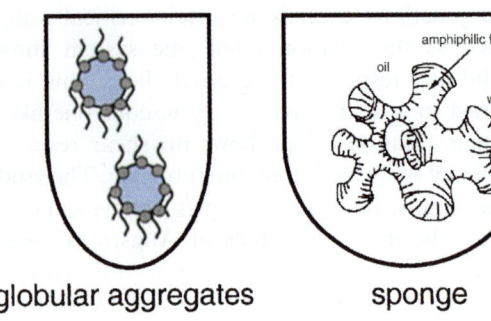

Fig. 5. Self-assembly of amphiphiles.

as a function of concentration, temperature and salinity, the viscosity, and the interfacial tension of these systems that can sometimes show ultra-low values, 5–6 orders of magnitude lower than typical water/oil tensions. Self-assembly is used by the chemical industry in dispersion, encapsulation, and as oil-spill cleaning agents. In the biotechnology industry, they are being studied in order to engineer smart drug encapsulation schemes. Nature has found many uses for self-assembly in biological systems ranging from the formation of the cell and nuclear-pore membranes that are relatively simple in shape, to the endoplasmic reticulum that shows a complex, sponge-like structure. The variety of structures observed in amphiphilic self-assembly are also found in the block copolymers where one part of the polymer may be hydrophilic and the other block hydrophobic. Block copolymer self-assembly gives rise to various surface microstructures that are used in nanolithography and template patterning.

1.3. Challenges

The efforts of the last twenty years to quantify the physical properties of synthetic, soft matter systems such as polymers, colloids and surfactants, as well as the current emphasis on biomaterials is motivated by the need to engineer these materials to a high level of sophistication. The intense efforts to understand the materials properties of crystalline systems in the previous century paved the way to predictability and control—one knows how even small temperature changes or impurity concentrations will change the electronic conductivity of some semiconductors; control leads to the ability to engineer these materials to permit the dazzling array of capabilities of electronic devices today. Of particular importance in these applications are the properties of crystalline surfaces and thin films whose understanding on the atomic level has formed the hard core of traditional surface science. Similar fundamental understanding, predictability and control of soft matter and their surfaces will enable the 21st century vision of bioengineering, smart drugs, cellular level medicine such as gene therapy, and replacement tissue and organism parts to become a reality.

In this article, we present a pedagogical review of the concepts and applications of statistical thermodynamics of interfaces (the boundary between two condensed phases) and surfaces (the boundary between a condensed phase and vapor or vacuum) in soft matter [2,4] with an emphasis on their large-scale behavior. The general approach is to first predict the average, or mean-field interfacial shape, and then to consider the role of fluctuations in possibly destabilizing the average interfacial configuration. A "coarse grained" approach is taken whereby many microscopic details are "lumped" into a small number of phenomenological constants; often the precise values of the "elastic constants" that enter these phenomenological models can be extracted from experiment. An alternative approach, beyond the scope of this article, is to focus on the atomic or molecular interactions that give rise to these elastic constants. While these treatments, which are most often numerical and involve large scale computer simulations, can provide more accurate estimates of the atomic or molecular scale behavior, important for predicting the chemical trends, they are much less useful for describing the large scale configurations and thermodynamics of soft matter systems that consist of millions of interacting molecular units. In almost all cases, the interesting behavior of the system at large scales and in particular, the predictions of the dependence of system properties on physical parameters such as temperature and concentration is best treated by the phenomenological approach.

After a brief discussion of the microscopic origins of the molecular interactions in the system we relate the interfacial or surface tension to the structure of the interface and the tendency of the system to separate into two coexisting phases separated by the interface. The tension can be reduced by the presence of surfactant molecules and this important, practical property can be simply understood in terms of the entropy of the amphiphilic monolayer. Thermal fluctuations are particularly important in the context of surfaces and interfaces due to their reduced dimensionality and their role in the roughening transition of crystal surfaces and in tension-induced morphological instabilities of soft columnar structures is explored. We then move from tension dominated systems to self-assembling interfaces where the energetics of the interface is determined by its curvature [6]. After reviewing the microscopic origins of this curvature or bending energy, we show how thermal fluctuations can destabilize simple interfacial shapes to give the variety of structures observed in self-assembling systems. Entropic interactions that can stabilize locally fluctuating, but globally, one-dimensionally ordered structures in lamellar systems are discussed. Finally, we close with a survey of current and future prospects that include the role of long-range electrostatic interactions, the coupling of linear macromolecules and membranes, and the coupling of tension, shear, and curvature elasticity in determining the interfacial properties of soft matter.

2. Surfaces and interfaces

2.1. Geometry

In order to predict the equilibrium structure, fluctuations and stability of surfaces and interfaces, we have to first specify the surface configuration (position, area, curvature) and then minimize the free energy as a function of this configuration. We shall see that in some cases, the surface free energy is proportional to the local surface area while in others it is related to the local curvatures.

Simple surfaces that do not double back along themselves can be described by the surface height, $h(x,y)$, above a fixed plane (see Fig. 6). The height can vary with the position along the plane, here taken to be the x–y plane (whose normal is the unit

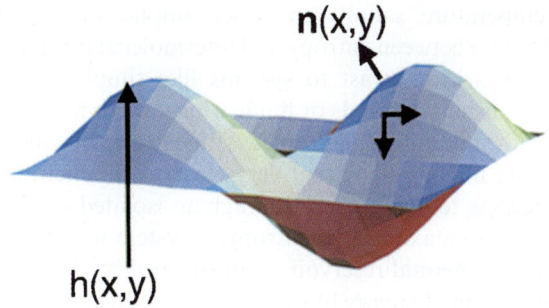

Fig. 6. Surface parameterized by its height above the x–y plane, $h(x,y)$ and showing the normal unit vector $\hat{n}(x,y)$ and the surface tangents.

vector $\hat{z}$), and the surface normal deviates from a constant unit vector in the $\hat{z}$ direction, as shown in Fig. 6. For surfaces that are nearly planar, the excess area, above and beyond the area of the base plane, is proportional [2] to the square of the gradient of $h(x,y)$. The surface curvature is related to the second derivative of the height function with respect to x and y; a flat wedge, which has no curvature, has a finite first derivative but a vanishing second derivative. One can show that there are two invariants of the curvature—i.e. two quantities whose values do not depend on the reference frame in which they are measured [2]: the mean curvature, proportional to the average of the xx and yy second derivatives of $h(x,y)$, and the Gaussian curvature, proportional to the product of the xx and yy second derivatives less the square of the xy second derivative. The Gaussian curvature is a measure of the saddle-like nature of the surface. For simple geometries, such as a sphere, the mean curvature is the inverse of the radius, R, and the Gaussian curvature is positive and equal to R^{-2}. For a cylinder, the curvature along the axis is zero so that the mean curvature is $R/2$ (where R is the cylinder radius) and the Gaussian curvature is zero. For a torus, the two curvatures have opposite sign so that the Gaussian curvature is negative.

2.2. Interface and surfaces: statistical thermodynamics

The many changes that occur in the structure and phase behavior of soft matter as the composition, temperature, salinity are varied, implies a delicate balance between entropy and intermolecular forces. This is in contrast to systems like simple three-dimensional crystals or fluids which are dominated by the intermolecular forces, with entropy playing a more minor role except very close to the melting/freezing temperature. Although an isolated system tends to maximize its entropy, a system in contact with a thermal reservoir is constrained by the conservation of energy of the system and the reservoir. This implies [7] that the system will adapt a configuration that will minimize its free energy, F, given by the difference between the internal energy and the entropy of the system. Because of the opposite signs of the energy and entropy in the definition of F, *maximizing* the entropy (with the constraint of conservation of the total energy) is equivalent to *minimizing* the free energy. Thus, having more states available (a configuration which allows more randomness) lowers the free energy as does having a lower internal energy. At zero temperature, the criterion for minimizing F and the internal energy are identical and at very high temperatures where the entropy dominates, the criterion for minimizing F and maximizing the entropy coincide. While we shall define the internal energy to include the surface and curvature energies and the entropy to account for fluctuations of the surface about its average configuration, it is worthwhile to note that in the coarse-grained view of a surface, even this internal energy can be temperature dependent. That is because the coarse graining process that allows one to use the large scale, continuum description of a surface, averages over small scale, molecular conformations that are favored by a balance of both energy and entropy.

The statistical thermodynamics of surfaces are often treated by first determining the surface configuration that minimizes the internal energy. The effect of thermal fluctuations are then accounted for by minimizing the free energy with respect to all possible surface configurations (e.g., all possible functions of the surface height, $h(x,y)$). We shall see quite generally, that the most probable fluctuations are those that modulate the interface configuration over long-length scales; these fluctuations have the smallest internal energy cost. Since the probability is proportional to the Boltzmann factor which is a decreasing exponential function of the internal energy divided by the temperature, this increases the probability of such a state.

An important measure of the fluctuations of the interface is the correlation function that describes its height fluctuations. For example, the statistical mean-square average of the difference between the height of the surface at the origin $x = y = 0$ and the height at some other value of (x,y) is written $\langle (h(0) - h(x,y))^2 \rangle$ where the angle brackets signify a thermal average; if this quantity is small, this means that the surface is nearly flat—if it is large, it can signify the onset of surface roughening and wandering. One can show that the thermal fluctuations are related to the behavior of the interface under local pressures. In particular, the correlation function is a measure of how the interface position at one point is affected by changes in the local pressure at another.

3. Tension dominated systems

3.1. Interfacial tension: reduction in soft matter

We now proceed to examine the properties of soft surfaces that are dominated by their interfacial tension. Tension dominated systems often show an interesting variety of behaviors ranging from the breakup of a column of fluid into spherical droplets, with its modern analogies in tubular, biological cells to the intricate patterns observed when complex fluids wet complex surfaces. To understand these phenomena, it is useful to first understand the physical origin of the interfacial tension or energy. For systems where the interfacial position is well defined, one expects that the energy cost to create interface is proportional to the amount of interface; the energy per unit area is known as the interfacial tension, γ, and for systems with stable interfaces, this quantity is positive. The tension resists increases in the interfacial area; systems with zero or negative tension have unstable interfaces since the free energy is reduced if the interfacial area is increased.

The additional free energy per unit area to remove molecules from the bulk and create interface between two coexisting phases is known as the surface or interfacial tension, γ, which has the units of energy per unit area (erg/cm^2), or force per unit length (dynes/cm). Both the energy and the entropy of the molecules are different at surfaces or interfaces and at the very least, the molecule loses entropy by being confined to the interface instead of occupying all its possible positions in the bulk. A mechanical interpretation of the surface tension is that it is the two-dimensional analogue of the pressure; it is the force per unit length parallel to the interface. Typical values of interfacial tensions are water-vapor $\gamma \approx 73$ dynes/cm, water-oil $\gamma \approx 57$ dynes/cm, mercury-water $\gamma \approx 415$ dynes/cm. However, some surfactant systems show interfacial tensions that are several of orders of magnitude lower than these values [3]. The interfacial tension arises from the free energy cost of mixing, in the interfacial region, the two types of molecules that would prefer to be in two separate bulk phases (or a dense phase of molecules with its vapor in the case of fluid/vapor equilibrium). The fact that the bulk system prefers to be phase separated under some conditions, is due to the microscopic interactions in the system. In fluid systems, the van der Waals interactions between molecules usually leads to an attraction between like species. The intermolecular interactions thus favor phase separation which is opposed only by the entropy of mixing. Thus, at low enough temperatures, where the entropic contribution to the free energy is smaller than the enthalpic, demixing terms, the system phase separates into two coexisting phases with an interface. Near the interface, however, the two phases are, of necessity, mixed; molecules of one type are adjacent to molecules of the other type and this results in a free energy cost that translates into the interfacial tension. In the case of fluid/vapor equilibrium, the dense phase is adjacent to the vapor phase; molecules in the dense phase located at this boundary have fewer neighbors and thus a higher free energy.

Phase separation and the resulting formation of interfaces is particularly important in soft materials such as those used in modern chemical technologies (paints, cleaners, dispersants, smart drug delivery) or those found in biological systems (membranes and vesicles with several lipid components, proteins found in membranes) that often consist of multicomponent systems. A common, universal question important in all these systems, whether they be solid (as in alloys or semiconductors), liquid (paints, cleaners, emulsions) or biological is to what extent a given composition of several components remains homogeneously distributed and to what extent one or more of the components separates into a distinct, macroscopic phase that coexists, even in equilibrium, with another macroscopic phase with an entirely different composition. We are all familiar with the demixing of water and oil which leads to phase separation (coexistence of a phase that is almost pure water with one that is almost pure oil) and with the complete mixing of some solutes in water such as salt. But even salt will phase separate from water and form crystals at high concentrations or low enough temperatures. As we explained above, the interfacial tension of a phase separated system is very much related to the thermodynamics of the two coexisting phases.

The thermodynamic stability of the phase separated state implies that the bulk thermodynamic potential (when studied as a function of composition) has two equivalent minima; the two phases, whose compositions are given by these minima, have equal pressures and chemical potentials [2]. The thermodynamic potential, G, is a function of the local composition, and can be derived from a microscopic model that includes both the molecular interactions and the entropy. In general, the thermodynamic potential per unit area, for a one-dimensional composition variation that describes the two phases and their interface is given by Ref. [2] the sum of a bulk term that determines the thermodynamic state of each of the phases and an inhomogeneous term, proportional to the square of the gradient of the local composition. The energy cost of having gradients in the system, such as those that exist near interfaces, is related to the molecular scale energy divided by a molecular size. Near the interface, where the two different species are forced to be in close contact, the molecular attraction of each species to its own kind is lost—hence the dependence of the gradient term on the interaction energy. At high temperatures, the bulk term shows only one minimum representing the composition of the homogeneous phase while for low temperatures, the minimization of the bulk term indicates that there are two coexisting phases with different compositions.

The free energy, including the gradient term, can be minimized to find the spatially varying composition profile. The profile is constant far from the interface at $z = 0$ and is approximately linear in the interfacial region, with a width proportional to the bulk correlation length for composition fluctuations. Near the critical temperature of the phase transition $T = T_c$, the correlation length, that characterizes the scale of fluctuations in the bulk as well as the interface width, diverges. When the equation for this profile is inserted into the free energy and the contribution of the bulk phases is subtracted off, one can find the free energy cost per unit area, γ, of the interface. This quantity is known as the surface tension, which in a simple mean-field model [2] vanishes as $(T - T_c)^{3/2}$ since the two coexisting phases become identical as the temperature approaches the critical temperature.

Soft interfaces are often characterized by low interfacial tension. This allows the interface to deform with a relatively small energy cost and is important in applications such as tertiary oil recovery where one attempts to expel microscopic oil droplets found in the crevices of rock, by external pressure; to escape from the pore network in the rocks, the drops must deform and this costs interfacial energy. A significant reduction of this tension is accomplished by the addition of a third component whose free energy is lowered when it resides on the surface. Examples of such agents are amphiphiles that are molecules with both polar and non-polar groups joined together by chemical bonds. Anionic surfactants have a counterion that is soluble in a polar liquid (such as water) and is positively charged (e.g., $R_n OSO_3 Na^+ SO_3^-$ where the Na^+ is water soluble and R_n signifies the hydrocarbon chain). Cationic surfactants have negative counterions (e.g., Br_2^-) and non-ionic surfactants can have a dipolar group, but no ion in solution (e.g., OCH_2CH_2OH).

While the interfacial energy between two coexisting phases opposes the formation of additional surface, the addition of a surfactant that is not soluble in the solution promotes the formation of more surface area. Having a larger interface makes the surfactant monolayer at the interface more dilute and increases its entropy; the surfactant exerts a lateral pressure on the interface that tends to expand it and this reduces the net interfacial tension. One can estimate the magnitude of this effect by considering a monolayer of an ideal gas on the surface of a fluid with a fixed area. The surface energy consists of the sum of the bare surface energy of the fluid and the entropy of the ideal gas of the surfactant whose area density is, σ. We assume here that there are no molecules in the bulk solvent. The net surface tension in the system of fluid and surfactant is given by the change in the free energy as the amount of interfacial area is varied and one finds [2] that $\gamma = \gamma_0 - k_B T \sigma$; each surfactant molecule lowers the bare interfacial free energy, γ_0, by an amount $k_B T$. This reduction in the tension is due to the fact that as the interfacial area is increased, the translational entropy of the surface active component is increased (for a fixed number of the surface active molecules). This increase in entropy lowers the system free energy and thus reduces γ from its bare value.

3.2. Fluctuations and instabilities of soft interfaces

In thermal equilibrium, the surfaces of fluids and even solids are not perfectly flat; neither are the interfaces between two different phases. Thermal fluctuations roughen the regions between two materials; how strong these effects are depends on dimensionality (i.e., if the interface is a line or a plane) and on temperature. For fluid interfaces, two-dimensional systems with one-dimensional interfaces have a mean-square roughness of the interface that varies linearly with the system size; the corresponding quantity in three-dimensional systems is logarithmic with the system size. For solids, the periodic potential tends to decrease the roughness of the interface; however, above the roughening transition temperature, which is typically some fraction of the melting temperature, the fluctuations of the position of the interface resemble those of a liquid. We shall thus focus next on the effect of thermal fluctuations on the shape of interfaces; these effects are opposed by the interfacial tension and are thus the most pronounced for soft interfaces where this tension is low. In addition to the undulations of nearly flat interfaces, there are other interesting and technologically important instabilities of interfaces. These include the tension driven breakup of cylindrical structures into spheres and the spontaneous nucleation and growth of holes even in flat, thin films.

3.2.1. Thermal fluctuations

In order to predict the effects of thermal fluctuations [7] on the shapes of interfaces and surfaces, one first needs an expression for the free energy of an interface with a shape that is not necessarily flat. Phenomenologically, one can write that the free energy is the product of the interfacial tension, γ, and the area. It can be shown that approximating the interfacial free energy as the product of the tension and the area is correct for systems where the spatial variation of the interface position occurs over length scales much larger than the typical width of the interface. It is only in this limit—frequently, the important limit in many physical situations where the interface is well defined to within a molecular scale—that one can study the long-wavelength undulations of the interface.

In order to determine to what extent the thermal motion of the interface causes deviations from the flat state, we calculate the thermal average of the mean square of the local height difference, $\langle (h(0) - h(x,y))^2 \rangle$ which turns out to be proportional to the logarithm of the distance between the origin and the point (x, y). This means that fluctuations *increase* as the distance between the two points along the surface increases, implying that the interface is not asymptotically flat. However, in two dimensions, the deviation from flatness is only logarithmic in the distance between the two points. For example, two points, one cm apart have an rms fluctuation in their heights of about 10 Å. For one-dimensional interfaces, applicable to the boundaries between different domains of a surface monolayer [8], the fluctuation effects are much larger and the correlation function diverges linearly in the distance; one dimensional systems are much more sensitive to the effects of thermal fluctuations. In addition, for vertical surfaces, gravity will further suppress these undulations [2]. However, since this effect scales inversely with the interfacial tension, soft surfaces, such as surfactant covered interfaces, will show considerably larger undulation effects. These can be measured via capillary wave experiments [9] that involve [2] the Fourier decomposition of the real-space fluctuations discussed above. This is one way that the interfacial tension can be measured and it is particularly important in determining how surfactants can lower the value of γ.

Recent scattering experiments [10] have probed the applicability of the surface tension theory outlined above to the liquid/vapor interface, down to the nanometer regime. They find that the assumption of an interfacial free energy, that is the product of the interfacial area and a constant interfacial tension, begins to break down in the ten nanometer range. In this range, the interface becomes *softer* than one would predict from the constant surface-tension model. The effect is rather large, with a maximum softening by more than a factor of two at a length scale of a few nanometers. Most theories that attempt to improve on the simple model predict corrections that tend to *increase* the local surface tension for small distance scales and result in a less rough interface (see the references in Ref. [10]) at small wavelengths. A possible resolution to this experimental and conceptual puzzle may lie in the presence of van der Waals attractive interactions between the bumps on the surface that result from the thermal undulations; this may favor the presence of fluctuations and hence locally soften the interface [11].

Another important manifestation of the thermal undulations of an interface occurs for the crystal–vapor surface, particularly near the melting transition. Below a characteristic temperature which is some fraction of the melting temperature, the crystal interface is faceted and well defined; above this roughening temperature, T_R, [12], the interface is smooth and show thermal fluctuations that are characteristic of the liquid phase. The transition between these two states is known as the roughening transition [2,4]. Above T_R, the interactions that are responsible for crystallinity have a thermal average that is equal to zero while below T_R, the thermal average of the crystal potential is non-zero and suppresses the interface fluctuations. At high temperatures, the resulting crystal face will be smooth and not faceted. At low temperatures, the lattice potential results in stepwise growth of the interface and the formation of facets.

3.2.2. Instabilities of soft interfaces

However, there are interface geometries where even a simple interface shape—such as a cylindrical tube of one phase in another or the surface of a cylinder of fluid or solid—is intrinsically unstable because there are other geometries of lower surface areas and hence lower interfacial free energies. A simple energetic/thermodynamic argument can be used to show that the free energy of a cylinder with undulations whose wavelength exceeds a critical value proportional to the cylinder radius, is lower than that of the perfect cylinder [2]. These undulations eventually lead to the Rayleigh instability of the cylinder into spheres—which naturally have a lower surface/volume ratio.

This instability is well known in the case of cylindrical columns of fluid. For the case of the stability of tubes composed of solid matter, relevant to cylindrical inclusions in a matrix, there is an additional restoring force that suppresses this instability that arises from the shear elastic energy of the cylinder. The instability is not opposed by the compressional elastic energy (which is present in both liquid and solid systems) since it arises even if the total volume is perfectly conserved (i.e., the limit of an incompressible system). However, the undulations of a cylinder that locally change the cylindrical radius, cause a shear deformation and this opposes the instability. This additional restoring force [13] does not vanish for long-wavelength undulations and one finds that for cylinder radii larger than a critical value $R_c \sim \gamma/E$ (where E is the shear modulus and γ is the surface tension), the instability is completely suppressed by the shear modulus. In most hard systems, the shear modulus scales as a typical bond energy divided by a molecular volume, while γ scales as a typical bond energy divided by a molecular area; the critical radius $R_c \sim \gamma/E$ is thus on the order of a molecular size and cylinders with radii much larger than molecular sizes should show no undulation instability. However, in soft systems, such as a very dilute gel, one might expect that the shear forces scale as $k_B T$ divided by the mesh size cubed. Since the mesh size, ζ, can be much larger than the molecular scale, the shear modulus may be much weaker than the surface tension which tends to scale as $k_B T/a^2$—namely, the temperature divided by the square of a molecular size, a. The critical radius scales as $R_c \sim \zeta^3/a^2$ and can be much larger than a molecular scale. Undulations of cylinders smaller than this critical radius, which is much larger than any molecular scale, will destabilize the cylindrical structure at long wavelengths. Recent experiments [13] on the instabilities of biological cells with cylindrical protrusions caused by their adhesion to surfaces, have noted the observed instability of the cylinders and their breakup into a string of pearls. The wavelength of the instability (that determines the distance between the pearls) is controlled by the shear modulus of the cytoskeletal, actin gel contained in the bulk of the cell. In the experiments (see Fig. 7), the shear modulus of the gel is weakened by a drug which depolymerizes the actin and the experiments measured the change in the wavelength of the pearling instability as a function of the drug concentration, proportional to the change in the shear modulus of the gel.

While the Rayleigh instability of a cylindrical surface is driven by the surface energy, the flat surface of a thin film can become unstable under the influence of van der Waals forces. Eventually, this can lead to the formation of holes in the film that was initially spread to a uniform thickness on the substrate. In this case, the surface tension provides a restoring force which stabilizes short-wavelength undulations. The destabilizing force comes from an attractive van der Waals interac-

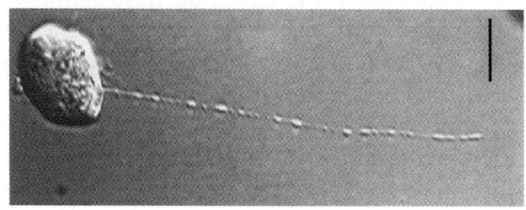

Fig. 7. Rayleigh instability in a cylindrical section of a fibroblast cell under tension. Here, the instability is inhibited by shear modulus of the internal, gel like cytoskeleton. For details, see Ref. [13]. Picture courtesy of E. Moses.

tion between the thin film and the substrate. Understanding the mechanism and dynamics of hole formation is important in the design of thin films that are to remain stable under a variety of thermal, compositional and chemical conditions. The critical wavelength, below which the surface tension is ineffective in stabilizing the film against these undulations, is determined by a balance of the two energies and there is a characteristic wavelength, above which, the film is unstable. These instabilities, that eventually lead to hole formation and film rupture have been studied in polymeric systems [14] where the time and length scales can be large due to the large polymer molecular weight; in some cases, the instability can be prevented [15] by the presence of a dense polymer layer, but the mechanism behind this restabilization is still unclear.

4. Curvature dominated systems: self-assembly

In self-assembling matter the energy cost for creating interfaces effectively vanishes, resulting in the formation of complex structures with internal interfaces; in this case, the properties of the system will be determined by the geometrical nature of the interface and not by its area. While this is hardly relevant for two-component, phase separated materials (except at the critical point where many other properties of the system are anomalous), multi-component systems, where one of the components is interfacially active, can be at effectively zero interfacial tension. An important example of recent experimental and theoretical interest [3] are the fascinating and useful variety of shapes, sizes, and forms found in the self-assembly of amphiphiles to form micelles (amphiphilic aggregates in a single solvent such as water), membranes, vesicles (closed, bilayer structures formed by amphiphiles in a single solvent), and microemulsions (three-component dispersions of water, oil, and surfactant). Often, the domain sizes are much larger than molecular scales and one can speak of an interfacial film. In these systems, the amount of internal interface is (for fairly insoluble amphiphiles) determined by the amount of amphiphile in the dispersion; the system is self-regulating (hence the term self-assembly)—the amount of interface formed is determined by the minimization of the free energy per unit area, f, of the dispersion with respect to the packing area per amphiphile, Σ, and not by external constraints. Thus, to zeroth order, $\partial f/\partial \Sigma = \partial(fA)/\partial A = 0$ if we assume that the total area of interface, A, is equal to the product of the number of amphiphiles and Σ. It is in this sense, that these systems are at effectively zero interfacial tension so that their properties are controlled by their relatively weak curvature energy, interactions, and entropy as discussed below.

The interfacially active, amphiphilic molecules that form fluid membranes have both polar and non-polar parts that are covalently bonded. Because of the hydrophobic interactions, these molecules tend to form monolayer films at polar–non-polar (e.g., water–oil) interfaces with the polar part of the molecule solvated in the water and the hydrocarbon part of the molecule in the oil. In this case, the properties of the film are, in general, not symmetric with respect to the interface. In a single solvent (e.g., water), these molecules tend to form bilayers where the hydrocarbon parts of each monolayer are aggregated in the middle of the bilayer to reduce the contact between the water and the non-polar parts of the molecule. When composed of a single species, the properties of such bilayer films are symmetric with respect to their two sides. Lipid molecules are surfactant-like entities that generally have a polar head group and a double-chained hydrocarbon tail. They are important in biological applications. Another system that shows amphiphilic properties are block copolymers, which consist of two immiscible polymers joined together by a covalent bond. If the two polymers are water soluble and oil soluble, respectively, these blocks are directly analogous to amphiphiles and thus form monolayer films at water–oil interfaces. Block copolymers can also form films at the interfaces between the two (immiscible) homopolymers that comprise the block. Such molecules are useful as compatibilizers of the two homopolymers and can be used to produce stable dispersions of one polymer in the other. This allows the formation of composite materials with particular properties which can be optimized by the dispersion of the two types of polymers. In the absence of the compatibilizer, the homopolymers would phase separate in equilibrium—i.e., the dispersion would be unstable.

4.1. Curvature elasticity of fluid membranes

4.1.1. Deformation modes

Although fluid membranes can be composed of many different types of chemical and molecular species, their behavior (shapes, fluctuations, thermodynamics) can be understood from a unified point of view that considers the free energy of deformation of the membrane. If the membrane were constrained to lie in a plane, the only relevant energy would be the compression of the molecules—i.e., a change in the average area per molecule. However, since the membrane can also deform in the normal direction (out of the plane), there is an additional set of "modes" describing the conformations of the film. These out-of-plane deformations are known as bending or curvature modes and the free energy associated with such modes is known as the curvature free energy [16].

4.1.2. Saturated interfaces and compressions

We consider a fluid, monolayer, membrane at a water–oil interface in equilibrium with a dilute solution of amphiphiles in the water and oil. In general, there is an equilibrium between those amphiphiles adsorbed at the interface and those in the bulk solution. For extremely small volume fractions of amphiphile, the amphiphiles will preferentially stay in solution due to their higher entropy of mixing with the solvent; the interface will have a relatively small number of amphiphiles adsorbed per unit area. However, this is not the case when the amphiphilic molecules are strongly insoluble in either solvent due to the unfavorable interactions of the polar groups with hydrocarbon solvents and of the hydrocarbon groups with polar solvents. The large energy cost of keeping these molecules in solution overcomes their entropy of mixing and at even moderately small volume fractions (which in practice can be very low $\sim 10^{-4}$ or less for amphiphiles which strongly prefer the interface), the free energy cost for remaining in solution is too high and the amphiphiles will tend to accumulate at the interface.

As one increases the volume fraction of amphiphiles in the solution, more and more would go to the interface and the area per molecule, Σ, on the interface would decrease. However, the molecules cannot pack at infinite density at the interface. In the case where there exists a minimum in the packing energy of the flat interface at a value of $\Sigma = \Sigma_0$, the system will keep adding amphiphiles to the flat interface until Σ is reduced to a value close to Σ_0. If even more molecules are added to the system, instead of decreasing Σ further and thus increasing the free energy (since $\Sigma = \Sigma_0$ is a minimum), the amphiphiles will maintain their packing at $\Sigma \approx \Sigma_0$ and accommodate the extra molecules by creating *more* interface (e.g., by rippling the flat interface or by incorporating oil into the water with the additional molecules located at the extra interface that is thus generated). When this happens, one says that the interface is saturated; instead of changing the packing area, the system accommodates more amphiphiles by making more interface under the condition of minimizing the free energy with respect to Σ. Of course, the interface may then have some curvature and the actual value of Σ may depend on the curvature [2].

In general, one must consider the chemical potential of a molecule at the interface and in the solution. The equality of the two chemical potentials is the criterion for equilibrium and hence determines the area per molecule on the interface. When the amount of interface is fixed, as in the case of a single water–oil interface, this equality fixes Σ. However, when the amount of interface can vary to minimize the free energy, Σ is determined by minimizing the interfacial free energy per molecule; the chemical potential then determines the number of interfaces that exist in the system as well as the (small) volume fraction of amphiphile that is not incorporated in these interfaces. However, the properties of each individual interface are determined to a first approximation by the minimization of the local free energy of the film.

A detailed study of the thermodynamics [6] shows that there is a critical volume fraction of amphiphile, ϕ^* (typically of order 10^{-6}–10^{-4} for good amphiphiles) above which there are many interfaces in the system; the amount of amphiphile

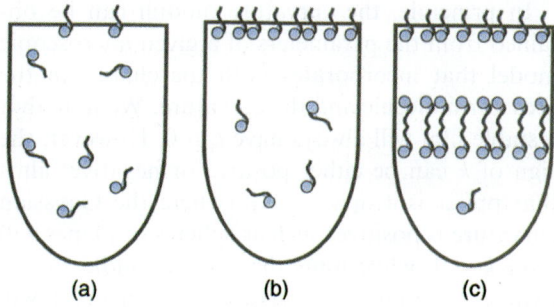

Fig. 8. (a) Dilute dispersion of amphiphiles in water. The entropy dominates and they are randomly dispersed. Some amphiphile covers the surface; this reduces the surface tension at the expense of the entropy of solution. (b) At higher concentrations, a complete surface monolayer is formed. This monolayer coexists with amphiphiles in solution. The monolayer shown here is saturated and no further amphiphiles can be added. (c) As more amphiphile is added to the system in (b), they form self-assembled aggregates in solution, such as the bilayer pictured here.

not incorporated into these interfaces is small and remains approximately constant as the overall amount of amphiphile, ϕ, is increased (see Fig. 8). We therefore consider the simple case of amphiphiles that are strongly surface active (strongly insoluble in the bulk) so that at even very small volume fractions of amphiphile ($\phi^* \ll \phi \ll 1$), there are many interfaces (e.g., vesicles, microemulsions) in the system in equilibrium. In this approximation, the fraction of amphiphiles in solution is very small and their volume fraction is approximately constant. The properties of the system are obtained by focusing on the properties of the interfaces. When, in addition, the interactions between these interfaces and their translational entropy can be neglected compared with the local deformation energies of the films, one can first minimize these local deformation energies to find the size and shape of the interfaces and then take into account the entropic and interaction effects as higher order corrections to the shape determined by the curvature energy defined previously.

First consider a locally flat, isolated interface. Saturation occurs when the interfacial free energy achieves a minimum: $\partial f / \partial \Sigma = 0$ where f is the free energy per molecule for a flat layer and Σ is the area per molecule. The free energy per molecule is minimized when $\Sigma = \Sigma_0$. The optimal value of the area per molecule arises from a balance of terms such as the entropy, and the interfacial-tension terms or attractions. The entropy favors a large area per molecule—because of the larger number of center of mass positions and chain conformations—while the interfacial-tension terms (e.g., contact of the hydrocarbon chains with the water) and attractions favor a small value of Σ. Of course, there can be deviations in the area per molecule from this minimum; however, these deformations are typically of higher energy than the curvature deformations; a membrane can change its shape or size with a much lower free energy cost than that required to compress or expand it. It is important to remember therefore, that for insoluble amphiphiles, it is the saturation of the interface and the minimization of the area per molecule that permits the usual surface-tension term to be neglected; the derivative $\partial f / \partial \Sigma = 0$. The surface tension is no longer relevant since the molecules adjust their area to optimize the free energy and it is therefore the curvature energy that mainly determines the properties of the film.

4.1.3. Curvature energy

We now consider a membrane as a curved interface that is described by its mean, H, and Gaussian, K, curvatures as described above. These quantities are invariant to rotations or other reparameterizations of the coordinate system used to describe the membrane. To describe the novel, large-scale structures observed in these systems and to characterize the low-energy deformations that are most strongly influenced by thermal fluctuations, we consider radii of curvature whose length scales are much larger than molecular sizes. Specific molecular models for the free energy of the curved interface are mentioned in the following section; what is important is that all these models yield an expression for the energy to bend an interface whose form is identical with that given by the phenomenological curvature elastic energy. The curvature energy is written as an expansion of the free energy of the interface in powers of the curvatures, keeping terms up to quadratic order in the mean curvature. This expansion is physically meaningful because we are interested in radii of curvature that are much larger than molecular sizes (e.g. the thickness of the bilayer) so there is a natural small parameter in the problem, the product of the amphiphile layer thickness and the curvature. In biological systems, the curvatures are of order of microns, while the thickness of a bilayer is typically 60 Å. In chemical systems such as microemulsions, the curvatures can range from hundreds to thousands of inverse Å and in units of the molecular size, are again small.

The most general form [16] of the curvature free energy, f_c, per unit area, up to quadratic order in the mean and Gaussian curvatures, contains one term proportional to the Gaussian curvature and another proportional to the square of the deviation of the mean curvature from its preferred or spontaneous curvature c_0. This is a type of "Hooke's law" for the curvature (which is in general, a very complex function of the surface shape) and states that deviations of the mean curvature from c_0 are associated with an energy cost (analogous to the "spring constant" in the energy for a stretched spring) known as the bending or curvature modulus, k. The parameter $\bar{k}$, known as the saddle-splay modulus, measures the energy cost of saddle-like deformations (where the two curvatures can have opposite sign) relative to isotropic, sphere-like deformations (where the two curvatures have the same sign).

The parameter c_0 is the spontaneous curvature that describes the tendency of the surfactant film to bend toward either the water ($c_0 < 0$ by convention) or the oil ($c_0 > 0$). It arises—in the absence of long-range interactions—to arise from the competition between the packing areas of the polar head and hydrocarbon tail of the amphiphilic molecules. If the interactions between the polar heads (as mediated through the intervening water and electrolyte) favor a smaller packing area than that dictated by the tail-oil-tail interactions, the surfactant film will tend to curve so that the heads (and the water) are on the "inside" of the interface. The bending moduli, k and $\bar{k}$, arise from the elastic constants determined by the head-head and tail-tail interactions. It is expected that these moduli will be strongly dependent on the surfactant chain length but only weakly dependent on the head-head interaction strength.

In principle, the curvature moduli can be obtained from the parameters of a given microscopic model that incorporates both the change in the area per molecule and the curvature. We note that a stable film will always have $k > 0$. However, the sign of $\bar{k}$ can be either positive or negative; films that prefer isotropic shapes (where the Gaussian curvature is positive) such as spheres or planes will have $\bar{k} < 0$, while films that prefer saddle shapes (where the Gaussian curvature is negative) will have $\bar{k} > 0$.

A simple example of the role of the curvature energy in stabilizing various structures can be found in microemulsions that are dispersions of water in oil or oil in water stabilized by surfactant monolayers at the interface. For simplicity, we first consider a system where the saddle splay modulus is zero. Sometimes, these dispersions consist of spherical domains and the lowest bending energy state is that of a drop whose radius is equal to the inverse of the spontaneous curvature $R = 1/c_0$. If, by adding more of the inner-phase solvent, one tries to increase the drop size, the system fails to emulsify the additional water or oil and it remains in a separate, excess phase that coexists with the drops of the optimal size $R = 1/c_0$. This simple type of phase separation combines with entropically driven effects to produce a complex equilibrium that results in three-phase coexistence. The complete progression of structures in these systems from simple spheres, to cylinders, to interconnected networks that lead to bicontinuous, sponge-like structures have only recently been observed (see Fig. 9) and analyzed [17]. The competition between the compositional constraints in these systems (e.g., the volume to surface ratios dictated by the relative amounts of water and amphiphile) that determine a compositional length [3] scale, and the bending energy that favors structures with curvature c_0, is responsible for the wide variety of structures that are possible in both biological and synthetic self-assembling systems [3,17–19].

It is important to note that the bending energy, in contrast to the interfacial tension is not an extensive quantity; it is not proportional to the area of the membrane. For example, the energy to bend a membrane with zero spontaneous curvature into

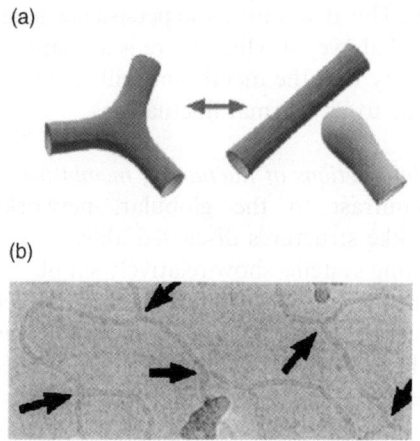

Fig. 9. (a) Theoretically predicted junction structure that leads to large scale networks in self-assembling surfactant systems governed by spontaneous curvature. (b) Experimentally observed network structure. From Tlusty and coworkers [17]. Micrograph courtesy of A. Grosswaser and I. Talmon.

a sphere is proportional to the modulus k and is independent of the size of the sphere! The physical reason for this is that although larger spheres have more molecules that are deformed, the bending angle per molecule is smaller; these two effects exactly compensate each other. A result of the non-extensivity of the curvature energy is the fact that self-assembling systems with no external surface tension are particularly soft and are strongly affected by thermal fluctuations.

4.1.4. Microscopic models

As mentioned above, the phenomenological bending energy can be derived from a variety of microscopic models of amphiphiles at interfaces, at least for deformations whose length scale is much larger than the interface thickness. The simplest of these models considers the hydrocarbon chains of the amphiphilic molecules as simple, polymeric springs whose length is determined by the incompressibility of the system; the incompressibility translates into a volume constraint on the curved film, whose thickness is then a function of the curvature [6]. The bending moduli are proportional to the spring constants of the chain and the third power of the amphiphile film thickness; this is also the result of continuum elastic theory for bent plates [20]. The curvature modulus $k > 0$, while the saddle-splay modulus $\bar{k} < 0$, signifying that spherical topologies (as opposed to saddle-shaped structures) are preferred. The chain-spring model also predicts that the spontaneous curvature is given by the difference between the optimal packing volume of the chains and the heads. With v_0 as the volume per incompressible surfactant molecule, Σ_0, the optimal area per head (determined by electrostatic and steric interactions) and ℓ_s as the optimal chain length (the unstretched spring length in this model) one finds that the spontaneous curvature is proportional to the difference $(v_0 - \ell_s \Sigma_0)$. This model endows c_0 with a simple physical meaning: When the imposed head area, Σ_0 is larger than the optimal area, v_0/ℓ_s, dictated by the chain packing, the preferred curvature is negative; the system prefers to pack with the heads on the "outside".

This simple spring model implies that the dependence of the curvature modulus on the chain length is non-linear. Relatively small changes in the number of hydrocarbon units in the amphiphilic tail can result in large changes in the bending stiffness of the membrane [3,21]. For example, consistent with these predictions, experimental observations found that an increase in the hydrocarbon chain length from C_8 to C_{12} results in an increase of the elastic constant by a factor of 3.6 [22,23].

4.2. Fluctuations of membranes

4.2.1. Thermal undulations

As mentioned above, the non-extensive nature of the curvature energy makes self-assembling systems particularly soft and amplifies the effects of thermal fluctuations compared with systems dominated by their interfacial tension. In particular, the height fluctuations (e.g., as measured by the difference in height at two different points in the membrane) promoted by thermal fluctuations and resisted only by the curvature energy, are qualitatively larger than in systems where the restoring force is the surface tension. In contrast to the logarithmic divergence discussed above, height fluctuations in self-assembled membranes diverge quadratically with the distance between the two points on the surface.

The relative flatness or curvature of the surface is determined by the changes in the surface normal as one proceeds along the surface. One can define a characteristic distance, denoted as the persistence length of the membrane, ξ_k, as the scale at which the deviations of the normal correlations are of order unity and one finds that $\xi_k \sim a \exp[4\pi k/3k_B T]$. Thus, relatively small changes in the stiffness of the membrane with respect to bending fluctuations can make large differences in the persistence length. On length scales smaller than ξ_k the membrane is flat while on larger length scales it performs a "random walk", of course constrained by its self-avoidance. The random nature of the membrane fluctuations at large length scales has been used to explain [24] the behavior of "balanced" microemulsions (where the spontaneous curvature is zero) and other amphiphilic systems that show sponge-like, bicontinuous structures [3]. Recent experiments [17] have shown, however, that interconnected structures can also be observed in systems with finite spontaneous curvatures that form cylindrical networks and the transition [25] from curvature dominated to fluctuation dominated structures remains to be resolved.

The long-wavelength fluctuations of interfaces and membranes effectively reduce the bending energy since the interface spontaneously undergoes a certain amount of bending due to the increased entropy of the disordered structure—to what extent this occurs is of course a balance of the bending energy and the thermal forces. Thus, for any given bend imposed on the system, there is some probability that the interface will have already been bent in a somewhat similar configuration due to the thermal fluctuations. Entropy thus increases the probability of spontaneous bends and can be thought of as effectively reducing the bending energy.

Theoretical treatments of this effect are reviewed in Refs. [3,6] and have also been recently studied using computer simulations [26]. The effective bending modulus, k_r, of the system at a length scale ρ, is reduced from its bare value, k, by a term proportional to $k_B T \log[\rho/a]$, where a is the molecular size. One can identify the length scale at which there is no longer any bending resistance ($k_r = 0$). This determines the persistence length, ξ_k, discussed above, at which there is a relatively high probability that the membrane will spontaneously bend due to its thermal fluctuations.

4.2.2. Interactions of fluctuating membranes

In contrast to the globular, network, and sponge-like structures discussed above, some self-assembling systems show relatively simple lamellar structures where the membranes form a stack with a typical intermembrane distance, d. However, even these seemingly ordered phases reflect the "softness" of the system and the importance of thermal fluctuations. In some systems, the average spacing between membranes can be large—thousands of Å, while in others, the average spacing has a maximum value of only several tens of Å; trying to swell the system further by the addition of solvent results in phase separation of a membrane phase and excess solvent. Obviously, the "unbound" systems that can be swollen to large distances have only negligible attractive interactions, while the tightly bound systems are affected by the van der Waals attractions, that tend to decrease as $1/d^4$ as the spacing is increased. While charged systems at low salt concentrations can have strong electrostatic repulsions that can also stabilize phases with large values of d, all membrane stacks are subject to a universal repulsive interaction arising from entropic restriction of the thermal fluctuations of the membranes. This interaction is long range, scaling as $1/d^2$ and is sometimes termed the "Helfrich" interaction [27]. This mechanism is an example of how entropy can sometimes stabilize structures resulting in relatively long-range order; the ordered structure allows more fluctuations to an individual sheet than the disordered one.

The physical origin of this entropic repulsion is the fact that, the membranes in a stack collide and lose entropy in these collisions. This can be understood [2] if one considers the case of hard-core repulsions between membranes; the excluded volume of the neighboring membranes limits the configuration of any given membrane, thus reducing its entropy. The entropic limitation implies that the free energy per membrane of a stack of membranes must be greater than that of a single, free film. Effectively, this results in a repulsion which ultimately causes the system to order in a somewhat periodic array. Stiffer membranes show a smaller repulsion since their thermal undulations are suppressed by their bending stiffness. Both experiment and theory show [28] that the one dimensional order in this array is less perfect than that of an ordinary crystal due to the fluctuations of membranes which are extended objects whose fluctuations away from perfect periodicity are much larger than those of single atoms in an ordinary crystal.

5. Outstanding problems

The competition between the repulsive, "Helfrich" interaction that arises from the thermal undulations of membranes and the attractive interactions related to the van der Waals forces has been predicted [29] to give rise to a novel phase transition. For small attractions or high temperatures, the entropic repulsions should dominate and the membranes are expected to be in an unbound state—the lamellar structure should swell as additional solvent that is presented to the system. For larger attractive strengths or lower temperatures, the membranes should be bound (to either each other or a substrate) with a preferred intermembrane distance—any additional solvent presented to the system remains in an excess solvent phase. The continuous or discontinuous nature of this transition and its experimental manifestations is still being investigated [30].

Looking forward to the future, one can guess that one area of both conceptual and practical importance involves the coupling between the interface structure, fluctuations, phase behavior and the internal degrees of freedom (such as composition and charge) of interfaces and membranes. These questions are of particular importance to applications in biomaterials science and cell biology since the cell membrane contains not only a mixture of lipids but also various proteins and even macromolecules. All these components interact within the membrane and the spatial dependence of their composition (i.e., whether are they uniformly distributed within the membrane or are clustered due to interactions) as well as the synergism between the local composition and the membrane structure are important in order to predict and control the properties of these systems. The coupling of the curvature elasticity and the differences in the local composition in the two leaflets of a membrane bilayer can stabilize spherical vesicles [31]; in single component systems, these can normally be obtained only by non-equilibrium methods in which the equilibrium, lamellar structure is broken.

A very different consequence of the interactions between membranes and inclusions is in the distribution of proteins in membranes; the curvature deformations induced by these inclusions results in an effective protein–protein interaction whose nature, attractive or repulsive, is responsible for either the ordering, clustering or dispersion of the proteins [32]. In addition, biological membranes are coupled to the cytoskeleton which provides the cell with shear rigidity [13] and shape integrity. This coupling, however, is not microscopic—the membrane and cytoskeleton are connected by linker molecules that are relatively far apart—and one might expect that the local properties of the membrane may be dramatically different from its large scale properties where it acts in consort with the much more rigid, cytoskeletal gel [33].

Biological materials are often highly charged. At small distances or low salt concentrations, the long-range Coulomb interactions can strongly modify the usually local properties of interfaces or membranes such as the tension or curvature elastic modulus. The charge is a self-regulating degree of freedom and predictions of these effects are subtle. A current problem [34] of practical importance concerns the intercalation of DNA into oppositely charged membranes. The application that drives this research is the possibility of delivering DNA to the cell nucleus to substitute for or repair damaged genes. However, since both the DNA macromolecule and the cell membrane are negatively charged (and neutralized by positive counterions in solution), there is a repulsive barrier that must be overcome. This problem can be solved by intercalating the DNA into a positively charged, artificial, lipid membrane (see Fig. 10). The resulting electrostatic attraction between the membrane

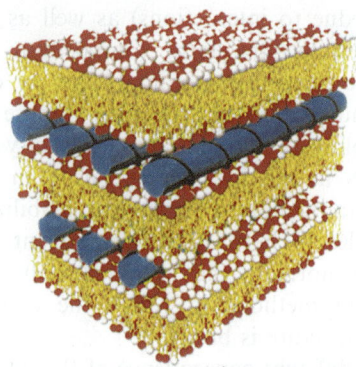

Fig. 10. Sketch of intercalation of DNA (rods) in oppositely charged, bilayer membranes. Courtesy of C. Safinya.

and the DNA allows the membrane counterions to increase their entropy by migrating to the solution while the attraction between the artificial lipid and the biological cell membranes allows the intercalated DNA to enter the cytoplasm. The physical questions of interest include an understanding of the optimal spacing and ordering of DNA in these systems and how the membrane deformations induced by the intercalation process affect the system. Again, the focus here is on the coupling of the electrostatic and compositional (i.e., DNA packing density and ordering) degrees of freedom that are coupled to the membrane structure.

While the present review has focused on the equilibrium properties of soft interfaces, many practical applications that involve the processing of these materials, require a fundamental understanding of the dynamics of these systems. A recent collection of reviews is presented in Ref. [35]. Some of the work focuses on the approach to equilibrium of self-assembled structures [36] and how various ordered structures develop. Other research probes the nature of the dynamical fluctuations of systems in equilibrium and asks how the correlation between fluctuations decay in time to their equilibrium values. While the dynamical modes [37] of such systems are simple functions of the system viscosity, bending modulus, and wavelength, recent experiment and theory have shown that the time correlation functions do not show a simple exponential decay [38]. However, the approach to or fluctuations about equilibrium are not necessarily the most dramatic dynamical signature of soft interfaces. Because these systems most often occur in the fluid state, they are particularly sensitive to external shear. Understanding their shear response is of particular importance because the processing of these materials involves the application of forces that result in various shear deformations. Shear can cause strong deformations of the interface or membrane and even induce phase transitions between different morphologies. The different local structures can sometimes coexist with each other in a type of shear-induced phase separation. Equilibrium thermodynamics and statistical mechanics (and their associated experimental techniques) provide fundamental information about structure and phase behavior; it is extremely difficult to deduce these properties from rheological measurements alone. However, while the equilibrium properties provide the underpinning, it is clear that future research in the area of soft interfaces and membranes will also focus on the novel effects induced by shear and other non-equilibrium external fields. The practical applications certainly warrant these directions and the conceptual goal of formulating a theory of structure and phase behavior for steady-state systems that are far from equilibrium is one that should challenge soft matter science in the coming years.

Acknowledgements

The author acknowledges useful discussions with T. Tlusty, U. Schwarz, D. Lukatsky, and A. Zilman.

References

[1] P.G. de Gennes, Scaling Concepts in Polymer Physics, Cornell, Ithaca, NY, 1979.
[2] S.A. Safran, Statistical Thermodynamics of Surfaces, Interfaces, and Membranes, Addison-Wesley, NY, 1994.
[3] W. Gelbart, D. Roux, A. Ben-Shaul (Eds.), Micelles, Membranes, Microemulsions, and Monolayers, Springer, NY, 1994.

[4] P.M. Chaikin, T.C. Lubensky, Principles of Condensed Matter Physics, Cambridge Press, Cambridge, 1995.
[5] J. Israelachvili, Intermolecular and Surface Forces, Academic Press, NY, 1992.
[6] S.A. Safran, Curvature elasticity of thin films, Adv. Phys. 48 (1999) 395.
[7] L.D. Landau, E.M. Lifshitz, Statistical Physics, third ed., Pergamon, New York, 1980.
[8] D. Andelman, F. Brochard, C. Knobler, F. Rondolez, Structures and phase transitions in Langmuir monolayers, in: W. Gelbart, A. Ben-Shaul, D. Roux (Eds.), Micelles, Membranes, Microemulsions and Monolayers, Springer, NY, 1994, p. 559.
[9] D. Langevin, J. Meunier, Interfacial tension: theory and experiment, in: W. Gelbart, A. Ben-Shaul, D. Roux (Eds.), Micelles, Membranes, Microemulsions and Monolayers, Springer, NY, 1994, p. 485.
[10] C. Fradin et al., Reduction in the surface energy of liquid interfaces at short length scales, Nature 403 (2000) 871.
[11] K.R. Mecke, S. Dietrich, Effective Hamiltonian for liquid-vapor interfaces, Phys. Rev. E 59 (1999) 6766.
[12] J.D. Weeks, The roughening transition, in: T. Riste (Ed.), Ordering in Strongly Fluctuating Condensed Matter Systems, Plenum Press, NY, 1980, p. 293;
A.C. Levi, M. Kotrla, Theory and simulation of crystal growth, J. Phys. Condens. Matter 9 (1997) 299.
[13] R. Bar-Ziv, T. Tlusty, E. Moses, S.A. Safran, A. Bershadsky, in: Pearling in cells: a clue to understanding cell shape, Proceedings National Academy of Sciences 96 (1999) 10140.
[14] G. Reiter et al., Thin film instability induced by long-range forces, Langmuir 15 (1999) 2551.
[15] R. Yerushalmi-Rozen, J. Klein, L.J. Fetters, Suppression of rupture in thin, nonwetting liquid-films, Science 263 (1994) 793;
S.A. Safran, J. Klein, Surface instability of viscoelastic thin films, J. de Physique II 3 (1993) 749.
[16] W. Helfrich, Elastic properties of lipid bilayers: theory and possible experiments, Z. Naturf. (c) 28 (1973) 693.
[17] A. Bernheim-Groswasser, T. Tlusty, S.A. Safran, Y. Talmon, Direct observation of phase separation in microemulsion networks, Langmuir 15 (1999) 5448;
T. Tlusty, S.A. Safran, R. Strey, Topology, phase instabilities and wetting of microemulsion networks, Phys. Rev. Lett. 84 (2000) 1244.
[18] U.S. Schwarz, G. Gompper, Stability of bicontinuous cubic phases in ternary amphiphilic systems, J. Chem. Phys. 112 (2000) 3792;
Z.-G. Wang, S.A. Safran, Curvature elasticity of ordered bicontinuous structures, Europhys. Lett. 11 (1990) 425.
[19] W. Helfrich, J. de Physique I 37 (1976) 1335–1345;
see also the volume by R. Lipowsky, E. Sackmann (Eds.), Structure and Dynamics of Membranes, North Holland, Amsterdam, 1995.
[20] L.D. Landau, E.M. Lifshitz, Theory of Elasticity, third ed., Pergamon, Oxford, 1986.
[21] I. Szleifer, D. Kramer, A. Ben-Shaul, W.H. Gelbart, S.A. Safran, Molecular theory of curvature elasticity in surfactant films, J. Chem. Phys. 92 (1990) 6800.
[22] M. Gradzielski, D. Langevin, B. Farago, Experimental investigation of the structure of nonionic microemulsions and their relation to the bending elasticity of the amphiphilic film, Phys. Rev. E 53 (1996) 3900;
M. Gradzielski, Bending constants of surfactant layers, Curr. Opin. Colloid Interface Sci. 5 (1998) 478.
[23] R.S. Cantor, Nonionic diblock copolymers as surfactants between immiscible solvents, Macromolecules 14 (1981) 1186;
L. Leibler, Emulsifying effects of block copolymers in incompatible polymer blends, Makromol. Chem., Macromol. Symp. 16 (1988) 1;
Z.-G. Wang, S.A. Safran, Curvature elasticity of diblock copolymer monolayers, J. Chem. Phys. 94 (1991) 679;
S.T. Milner, T.A. Witten, Bending moduli of polymer surfactant interfaces, J. Phys. Paris 49 (1988) 1951.
[24] K. Palmer, D. Morse, Statistical mechanics of microemulsions: Droplet phases and macroscopic interfaces, J. Chem. Phys. 105 (1996) 11147;
P.G. de Gennes, C. Taupin, Microemulsions and the flexibility of oil/water interfaces, J. Phys. Chem. 69 (1978) 2984.
[25] T. Sottmann, R. Strey, Ultralow interfacial tensions in water-n-alkane-surfactant systems, J. Chem. Phys. 106 (1997) 8606.
[26] G. Gompper, D.M. Kroll, Fluctuations of polymerized, fluid and hexatic membranes: continuum models and simulations, Curr. Opin. Colloid Surf. Sci. 2 (1997) 373.
[27] W. Helfrich, Steric interactions of fluid membranes in multilamellar systems, Z. Naturforsch. 33a (1977) 305.
[28] C.R. Safinya, D. Roux, G.S. Smith, S.K. Sinha, P. Dimon, N.A. Clark, A.M. Bellocq, Steric interactions in a model multimembrane system: A synchrotron X-ray study, Phys. Rev. Lett. 57 (1986) 2718.
[29] R. Lipowsky, Generic Interactions of Flexible Membranes, in Ref. [19], p. 521.
[30] M. Vogel, C. Munster, W. Fenzl, T. Salditt, Thermal unbinding of highly oriented phospholipid membranes, Phys. Rev. Lett. 84 (2000) 390.
[31] H.T. Jung, B. Coldren, J.A. Zasadzinski, D.J. Iampietro, E.W. Kaler, The origins of stability of spontaneous vesicles, PNAS 98 (2001) 1353;
E.W. Kaler, A.K. Murthy, B.E. Rodriguez, J.A. Zasadzinski, Spontaneous vesicle formation in aqueous mixtures of single-tailed surfactants, Science 245 (1989) 1371;
S.A. Safran, P. Pincus, D. Andelman, Theory of spontaneous vesicle formation in surfactant mixtures, Science 248 (1990) 354.
[32] S. May, A. Ben-Shaul, A molecular model for lipid-mediated interaction between proteins in membranes, Phys. Chem. Chem. Phys. 2 (2000) 4494;
P. Sens, S.A. Safran, Inclusions induced phase separation in mixed lipid film, Eur. Phys. J. A 1 (2000) 237;

S.L. Keller, S.M. Bezrukov, S.M. Gruner, M.W. Tate, I. Vodyanoy, V.A. Parsegian, Probability of alamethicin conductance states varies with nonlamellar tendency of bilayer phospholipids, Biophys. J 65 (1993) 23;

N. Dan, S.A. Safran, The effect of non-bilayer forming lipids on the structure of transmembrane proteins, Biophys. J. 76 (1998) 1410;

R. Bruinsma, P. Pincus, Protein aggregation in membranes, Curr. Opin. Solid State Mater. Sci. 3 (1996) 401;

N. Dan, P. Pincus, S.A. Safran, Membrane-induced interactions between inclusions, Langmuir 9 (1993) 2768.

[33] Physical Basis of Self-Organization and Function of Membranes: Physics of Vesicles, in Ref. [19], p. 213.

[34] I. Koltover, T. Salditt, J.O. Radler, C.R. Safinya, An inverted hexagonal phase of cationic liposome-DNA complexes related to DNA release and delivery, Science 281 (5373) (1998) 78;

D. Harries, S. May, W.M. Gelbart, A. Ben-Shaul, Structure stability and thermodynamics of lamellar DNA-lipid complexes, Biophys. J. 75 (1998) 159.

[35] Dynamic aspects of colloids and interfaces in Current Opinion in Colloid and Interface Science, vol. 3, 1998.

[36] Z.-G. Wang, Kinetics and anisotropic fluctuations in weakly ordered block copolymers, Curr. Opin. Colloid Interface Sci. 3 (1998) 428.

[37] S.T. Milner, S.A. Safran, Dynamical fluctuations of droplet microemulsions and vesicles, Phys. Rev. A 36 (1987) 437.

[38] A.G. Zilman, R. Granek, Undulations and dynamic structure factor of membranes, Phys. Rev. Lett. 77 (1996) 4788;

E. Freyssingeas, D. Roux, Quasi-elastic light scattering study of highly swollen lamellar and sponge phases, J. de Phys. France II 7 (1997) 913.

Solved and unsolved problems in surface structure determination

D.P. Woodruff *

Department of Physics, University of Warwick, Coventry CV4 7AL, UK

Received 1 June 2000; accepted for publication 30 August 2000

Abstract

The range of surface structural problems of interest in understanding the physics and chemistry of solid surfaces is reviewed with reference to the available methods and their strengths and limitations. Key challenges being addressed currently concern the achievable precision of measurements and their physical and chemical significance, and the complexity of the problems which may be solved. Past progress and future problems are illustrated with a series of examples ranging from the relaxation of simple clean metal surfaces through complex semiconductor reconstructions to large molecular adsorption and coadsorption systems and complex adsorbate-induced reconstructions. The strengths and limitations of scanning tunnelling microscopy as a complement to true quantitative structural methods are discussed, as is the role of chemical state specificity and elemental specificity in solving complex molecular adsorbate systems. © 2001 Elsevier Science B.V. All rights reserved.

Keywords: Surface structure, morphology, roughness, and topography; Chemisorption; Surface relaxation and reconstruction

1. Introduction

The structure of a surface is commonly regarded as the starting point for understanding a wide range of surface phenomena. For example, the structure influences the electronic structure of surfaces and interfaces, the latter being potentially of special importance in semiconductor devices. Similarly, many heterogeneous catalytic processes are described as 'structure sensitive', meaning that the reaction 'turnover' rate depends on the morphology of the surface, while catalytic chemists frequently describe this in terms of 'the active sites', implying special local geometries on the catalyst surface. To address this need for quantitative surface structural information a great variety of specialised techniques have been developed, and a huge amount of progress has been made in the last 30 years or so. There remain, however, many unsolved problems. Some of these are well within the bounds of current methods and will doubtless be resolved with time. Others are at the limits of these methods, either because of limited precision or because of inadequate specificity in the study of more complex problems. Some future developments to overcome these limitations can be anticipated, but there are also problems for which the current prospects for solution seem poor.

* Tel.: +44-2476-523378; fax: +44-2476-692016.
  E-mail address: d.p.woodruff@warwick.ac.uk (D.P. Woodruff).

The objective of this review is to illustrate both the past successes and the future challenges in surface structure determination. The material is mostly organised according to the type of problem of interest. Surface structural problems which have been and are being studied fall into three broad categories: clean surface relaxations and reconstructions, adsorption site determinations, and adsorbate-induced reconstructions. A further somewhat distinct issue concerns the structure of subsurface interfaces; this could be regarded as a special form of adsorption system, in that many interface investigations involve in situ growth of overlayers and the monitoring of the structure at different stages, but ultimately the determination of the structure at the interface after it has been buried presents a very special set of experimental problems. In general, what has been, and will be achieved, is closely related to the methods available. Understanding the future limitations, in particular, requires an appreciation of some of the details of these methods. Before discussing specific examples of the different classes of surface structures we therefore consider some general issues concerning the techniques.

2. The techniques

Indirect information concerning surface structure can be obtained from a variety of surface electronic and vibrational spectroscopies but this information will not be considered here. Past experience has identified areas where long-standing 'conventional wisdom' based on such methods has been found to be fundamentally flawed, signalling a clear need for caution in the future. By contrast, there is a significant armoury of methods which provide specific quantitative structural information. In general these are either based on diffraction (and thus exploit long-range order) or are essentially local in character. A detailed presentation of these techniques is clearly beyond the scope of this review, but it is important to identify the key physical principles and their strengths and limitations.

The main diffraction methods are low energy electron diffraction (LEED) [1–3] and surface X-ray diffraction (SXRD) [4,5]. LEED was the first quantitative surface structural method to be developed, a key reason for it being the method that has so far led to the largest number of published structures [6], and it remains something of a benchmark against which new methods can be evaluated. LEED exploits the very strong interaction, both elastic and inelastic, of low energy (typically 30–300 eV) electrons with atoms, which ensures the experiment is surface specific. The very strong elastic scattering cross-sections, however, also mean that multiple scattering processes are important, so structure determination can only be achieved by comparing the measured LEED intensities with the results of theoretical simulations for a series of trial structures. This process is computationally intensive, but with the increasing availability of low-cost high-powered computers, this is a diminishing problem. However, it still places some limits on the size of the surface unit mesh that can be addressed. LEED typically involves electron incidence at, or close to, perpendicular to the surface, while the diffracted beam intensities are also usually measured well away from grazing emission. This leads to diffracted intensities which are more sensitive to atomic positions perpendicular to the surface than parallel to it.

By contrast, SXRD exploits the very weak elastic scattering of X-rays by atoms which allows X-ray diffraction to be used routinely to determine the structure of bulk solids.

Surface specificity is achieved either by concentrating on measurements in regions of momentum transfer space for which there is no coherent bulk scattering (such as fractional-order diffraction beams associated only with a surface unit mesh which is larger than that of the substrate) or by using very grazing incidence. The simplest SXRD experiments involve measuring diffracted beam intensities at fixed momentum transfer perpendicular to the surface, when the technique is *only* sensitive to atomic positions parallel to the surface. In this form the method is thus fundamentally different in its structural information from LEED. However, SXRD experiments can also include measurements of the intensity of diffracted beams as a function of the momentum transfer perpendicular to the surface (known as reciprocal lattice 'rod scans' in SXRD or I–E (intensity–energy) spectra in LEED), and in this case the information content of the two methods is far more similar. SXRD is typically more instrumentally-demanding than LEED, but the theoretical analysis is simpler because the weak scattering typically allows neglect of multiple scattering. For this reason SXRD has proved particularly attractive for studies of relatively large surface mesh structures for which LEED may prove computationally intractable.

The two most widely used local structural probes are photoelectron diffraction [7–10] and surface extended X-ray absorption fine structure (SEXAFS) [11–16]. These two methods exploit the same underlying physics as LEED, but instead of using an external electron source, they use a photoelectron ejected from a core level of the atoms (usually adsorbate atoms) whose local geometry is sought. Fig. 1 shows schematically the (single) scattering contributions involved in LEED, photoelectron diffraction and SEXAFS. In photoelectron diffraction the components of the outgoing photoelectron wavefield which are elastically scattered by the surrounding atoms interfere coherently with the component directly emitted to the photoelectron detector placed in a specific direction outside the crystal. In SEXAFS this interference occurs at the core of the emitter atom itself, leading to a modification of the local wavefield amplitude which in turn influences the X-ray absorption

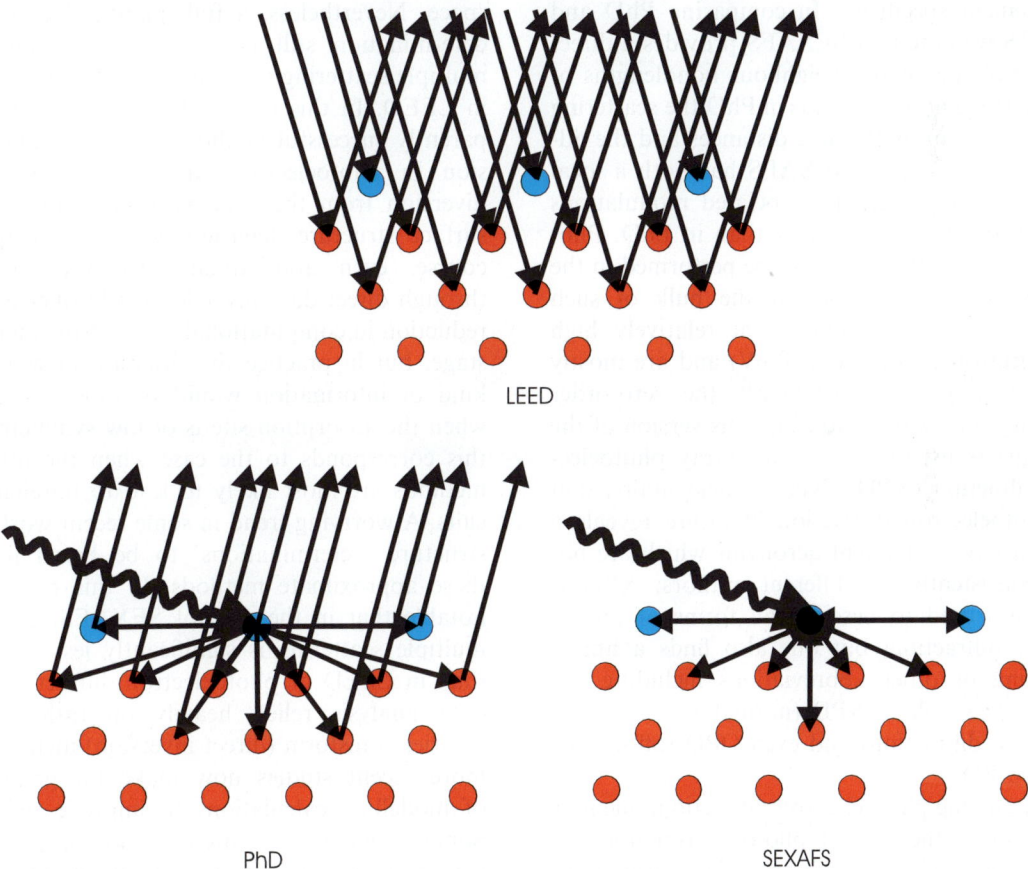

Fig. 1. Schematic comparing the electron single-scattering paths contributing to the surface structural information in LEED, photoelectron diffraction and SEXAFS.

probability. In a SEXAFS experiment one scans the photon energy through an adsorbate core level absorption edge and detects the modulations in photoabsorption cross-section as the photoelectron energy (and hence wavelength) varies and the 180° backscattering paths shift in and out of phase. Scanned-energy mode photoelectron diffraction (referred to here as PhD but also sometimes known as ARPEFS—angle-resolved photoemission fine structure) measures similar modulations in the photoemission signal as a function of photoelectron energy, and is thus closely related to the LEED I–E measurement. A key difference, however, is that the electron source in PhD is the adsorbate atom, so *all* the interfering scattering pathlengths are influenced by the position of this atom. The use of photoemission from the adatom core level also provides the method with element specificity. In comparing PhD and SEXAFS we note that SEXAFS provides a direct measure of the nearest-neighbour bondlengths of the absorbing atom, whereas in PhD the scattering paths depend on both these distances and the adsorption site. Because SEXAFS has both a local emitter *and* absorber, its associated modulations are typically 10 times smaller than in PhD. Photoelectron diffraction can also be performed in the angle-scan mode, but so far the bulk of such studies have been conducted at relatively high photoelectron energies (>500 eV) and are mostly designed to exploit particularly the zero-order (zero angle) forward scattering; this version of the technique is usually known as X-ray photoelectron diffraction (XPD). Notice, incidentally, that the photoelectron diffraction literature reveals a great variety of different acronyms which are not used consistently by different authors; XPD is sometimes used to describe all forms of photoelectron diffraction, but one also finds a liberal sprinkling of other abbreviations including PD and PED as well as NPD (normal-emission photoelectron diffraction) and even OPD (off-normal emission PD).

In discussing photoelectron diffraction, one may also mention the role of photoelectron holography, a topic which has been rather fashionable in the last decade. In terms of physical processes, photoelectron holography is simply a different name for photoelectron diffraction, but the modified vocabulary is used to stress the fact that a complete angular distribution measurement at fixed energy is effectively a photoelectron hologram with the directly emitted photoelectron wave being the reference wave and the elastically scattered components the object wave. Of course, this description raises the idea that this hologram may be directly inverted to produce the real-space structure, and a whole series of algorithms have been explored aimed at achieving this inversion computationally. What has emerged from a range of such attempts is that reasonable inversions which provide an *approximate* picture of the real-space structure are possible, but only if one collects angular distribution data over a substantial range of different photoelectron energies to provide an adequate sampling of electron momentum scattering space. Nevertheless, a full quantitative structure determination still needs proper trial-and-error multiple scattering modelling, similar to that used in LEED. In this regard the development of apparently successful methods of direct data inversion in photoelectron diffraction is mainly a diversion from the primary task of quantitative surface structure determination. In principle, of course, even approximate structures obtained through direct data inversion could offer a major reduction in computational effort at the modelling stage, but in practice the situation in which this kind of information would be most valuable is when the adsorption site is of low symmetry, and this corresponds to the case when the inversion methods are most likely to lead to unreliable results. A worrying trend in some recent work is for structure 'determinations' to be based *only* on these approximate methods. In this regard it is notable that in the case of SEXAFS, for which multiple scattering is significantly less important than in LEED or photoelectron diffraction, early data analysis relied heavily on rather simple Fourier transform (direct inversion) methods, but more recent studies now make far greater use of modelling calculations to improve reliability. Some recent work seems to be taking photoelectron diffraction in the opposite direction!

One further method providing local structural information (but actually based on diffraction in the substrate crystal) is X-ray standing waves (XSW) [17,18]. In this technique an X-ray Bragg reflection is established in the bulk solid (which requires long-range order in the substrate) and one exploits the characteristic X-ray absorption lineshape in a specific elemental species (typically an adsorbate) as the X-ray wavelength or incidence angle is scanned through the Bragg condition and the X-ray standing wave, produced by interference between the incident and diffracted beam shifts through the absorber. Recently, most XSW surface structure experiments have been performed at close to normal incidence to the Bragg scatterer planes (NIXSW) which offers some advantages in the range of materials to which the method can be applied. Like photoelectron diffraction and SEXAFS, this is an element-specific method due to the characteristic absorption edge energy.

The final group of quantitative surface structural methods which have been applied to quite a number of problems are based on ion scattering [19,20]. These techniques generally exploit the effects of elastic scattering 'shadowing' of one atom by another (either on the incident or emerging trajectories) in the near-surface region. Strictly, such effects are local, but in general most of the experiments conducted using ion scattering do rely on a reasonable degree of long-range order at the surface. The main differences in the different ion scattering methods relate to their energies. Low energy ion scattering (LEIS—less than about 10 keV) has strong elastic scattering which leads to wide shadow cones and intrinsically high surface specificity. At higher energies (high—HEIS—is typically 1 MeV or greater, while medium energy ion scattering (MEIS) uses 50–400 keV) the shadow cones are much narrower; in this case surface specificity can be achieved by using specific (shadowing) crystallographic incidence directions, but one also has the possibility to probe the subsurface region. The narrow shadow cones also offer the potential for greater structural precision.

Although not a quantitative surface structural probe, scanning tunnelling microscopy (STM) is now widely exploited in surface structural studies [21]. It is therefore important to put this method in context. We should first recall the underlying physics. STM exploits the fact that if two conducting materials are brought into very close (atomic scale) proximity, electrons can tunnel across the vacuum gap between them. This tunnelling current depends exponentially on the size of the vacuum gap. By scanning a conducting tip over a surface, adjusting its height to maintain a constant tunnelling current, one can therefore obtain an atomic-scale contour of the surface. This contour is, of course, electronic in origin. Insofar as the tunnelling current depends on the overlap of the exponential tails of the tip and surface wavefunctions, the height contour followed by the tip is essentially one of constant (very low) electron density outside the surface. This charge contour might be expected to be furthest from the surface directly above surface atoms, but it is clear that this is not necessarily the case, especially when the surface layer comprises two or more different elements, or when the valence charge occupies relatively local and directional orbitals, as in covalently-bonded solids. However, atomic-resolution STM images are extremely beautiful and their interpretation in terms of atomic asperities is seductively intuitive. Unfortunately, there is ample evidence that this kind of interpretation can be extremely misleading. As a simple example of this, there is now considerable evidence that chemisorbed oxygen atoms on surface commonly (but not invariably) appear in the STM image as hollows rather than asperities. STM can certainly help to provide plausible structural models (a key problem in most of the quantitative methods described above, which rely on 'guessing' the basic structural models and then testing them against the experimental data). Even more importantly, as a real-space microscopy, STM can provide information on the inhomogeneity of a surface and the way the surface behaves during structural transformations; these are key ingredients in helping to understand surface structures and their formation.

One final comment on methods of structure determination concerns the role of theoretical calculations of surface structure. The development of computer codes based on density functional theory (DFT) using the local density approximation, and the further application of the generalised gradient approximation has led to a remarkable level of success in establishing the minimum energy structure of surfaces, as judged by comparison with experimental results [22]. Such calculations have correctly determined the adsorption sites of atomic and molecular adsorbates on metal, semiconductor and oxide surfaces, and have similarly determined clean surface reconstructions and reproduced key features of epitaxial growth systems. Bearing in mind the increasing availability of low-cost high-speed computing remarked upon earlier, one might ask whether *experimental* surface structure determination is still necessary! Apart from the serious philosophical issues raised by this suggestion, it is clear that despite the remarkable advances achieved by these new computational codes, the time to dispense with experiments has not arrived. Significant discrepancies between theory and experiment still do exist, some of which greatly exceed the expected precision. It is also notable that so far almost all such calculations determine the total energy of the different structures in a rigid lattice, excluding all effects of finite temperature. Advances in this area are occurring, but there is still much to overcome. In this regard one might also mention the role of metastable surface structural phases; experimental studies, especially of more complex surfaces, are increasingly revealing the coexistence of more than one surface structural phase which may be in thermodynamic equilibrium, or of phases which transform irreversibly upon heating, implying the existence of structures with very similar total energy but separated by energy barriers of different heights. Evidently effective modelling of these situations also places increasing demands on the precision of the theoretical calculations.

3. Clean surface structures

3.1. Reconstruction and relaxation

In discussing the various types of surface structure which have been, and are being studied, it is helpful to classify them according to the nature of the modification of the surface structure relative to an ideally-terminated bulk structure, and according to the type of material which also bears on the dominant nature of the interatomic bonding within the solid and at its surface. For a clean surface, the simplest effect which can occur as a result of the presence of the surface (the termination of the bulk solid) is that the outermost layer spacings perpendicular to the surface may be modified. This is commonly referred to as *relaxation*. More significant modifications, typically involving atom movements parallel to the surface, generally lead to an increase in the the size of the surface unit mesh (and thus to 'extra' diffracted beams from the surface), are generally referred to as *surface reconstruction*. Notice that in many simple cases the enlarged periodicity can be referred to in the Wood notation as simply $(n \times m)$ where n and m are the integral factors defining the ratio of the size of the substrate and surface meshes in the directions of the vectors defining these unit nets. Such reconstructions were first recognised at clean semiconductor surfaces and may be seen as a consequence of the dominant directional covalent bonding in these materials, but they are now also known to occur at some clean metal surfaces and are quite a common feature of adsorbate-covered surfaces, the adsorbate causing a reconstruction in the underlying substrate surface. In the following sections some of the main problems related to these different types of surface structure are summarised, organised roughly in order of increasing complexity.

3.2. Unreconstructed metal surfaces

The structure of clean elemental solid surfaces presents problems of variable complexity. Superficially, at least, the simplest problems of this type concern surfaces which retain the same periodicity parallel to the surface as that of the underlying solid and so, in most cases, are regarded as not 'reconstructed'. Nevertheless, the exact positions of the atoms in the near-surface layers do differ from those expected in an ideally-terminated bulk structure. The great majority of metal surfaces fall into this category. The subtle movements of the surface atoms relative to the bulk-termination positions is attributable to the effect of the changing valence charge density at the metal–vacuum interface.

In a bulk metal one regards the average valence electron density associated with each atom as occupying an approximately spherical region (actually a Wigner–Seitz cell) around the atom, so if one then creates a surface by cutting away half of the bulk, surface atoms will have regions between them which are devoid of charge (where the electrons were associated with an atom which has been removed) whereas there will be charge above the surface atoms in the near-vacuum region. Clearly one expects this electron charge to be redistributed at the surface to produce a much smoother charge variation parallel to the surface in the near-vacuum region. In the simplest picture of metallic bonding the atoms in the bulk seek to achieve an optimal value of the surrounding charge density, and at the interface with the vacuum the average charge density at the surface atoms is reduced due to this electron smoothing; there is therefore generally a contraction of the outermost layer spacing to compensate for this. The effect is largest for open-packed (and hence more corrugated) surfaces, such as fcc (1 1 0), where its magnitude can be 10% or more of the bulk spacing, and is smallest for the closest-packed surfaces, such as fcc(1 1 1), where it may be only 1–2% and indeed there is evidence for both contraction and expansion. More subtly, there is a periodic variation in the outermost layer spacings, with fluctuations between contraction and expansion, which damps as one moves to deeper layers; this can also be understood in a qualitative fashion as a consequence of the same surface charge-density smoothing.

These effects have been observed in quite a large number of metal surfaces, especially using LEED and are also understood in general terms theoretically [23]; however, there appear to be significant discrepancies between experimental and theoretical values in some cases [24]. The origin of this discrepancy is unclear. One issue is certainly the precision available in the experimental methods; even in the most favourable cases for which a precision of ± 0.02 Å is probably realistic, the effects of interest can be only of comparable size. Greater and more consistent experimental precision here would certainly be helpful in identifying possible weaknesses of even the best density-functional theory calculations and helping to provide an accurate description of this fundamental phenomenon. A related topic which has attracted some interest recently is the issue of anomalous thermal expansion coefficients in these near-surface layer spacings which have been investigated both experimentally and theoretically [25–27]. The great majority of current DFT calculations take no account of finite temperatures, so the advances being made to understand this physical problem could open up a range of new issues to theoretical investigation.

3.3. Semiconductor surface reconstructions

An early challenge for surface structure determination was to gain a quantitative (indeed, initially, even a qualitative) understanding of the nature of clean surface *reconstructions*, first recognised to occur at the surfaces of Si and Ge. For some years surface reconstruction was believed to be a characteristic *only* of covalently-bonded semiconductor surfaces, at which the unsaturated directional 'dangling bonds' provide a driving force for rearrangement of the surface atomic positions. The classic problems here are presented by the clean surface structures of silicon. Quite early qualitative LEED observations [28] established that on the (1 0 0) surface one sees a doubling of the surface periodicity to form a (2×1) structure, whereas on Si(1 1 1) the equilibrium structure is represented by a (7×7) mesh, thus having an area 49 times larger than that expected for the ideally-terminated solid. Solving the Si(1 1 1)(7 × 7) surface structure was for many years a kind of 'holy grail' in surface structure determination, because the large surface net size and thus the large number of atomic positions to be determined clearly represents a very significant challenge for any technique.

By contrast, the Si(1 0 0)(2 × 1) structure appeared far more straight-forward. Even prior to quantitative structural studies of this surface it was proposed that the reconstruction involved pairing of surface Si atoms into dimers parallel to the surface in order to reduce the number of dangling bonds at the surface (two per surface atom for an ideally-terminated solid, one per surface atom after producing singly-bonded dimers), and this view

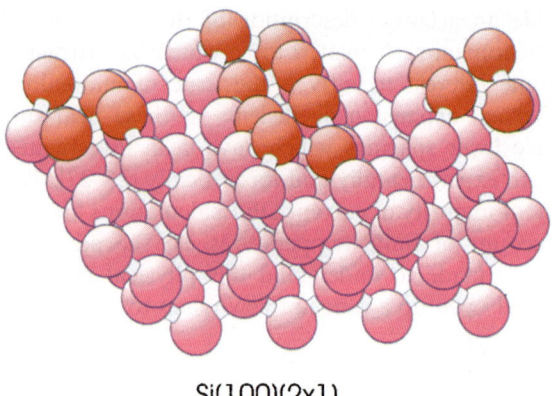

Fig. 2. Schematic diagram in an oblique view of the Si(1 0 0)(2 × 1) room temperature structure, showing random distribution of the asymmetric dimers which are dynamically flipping. For clarity the Si dimer atoms are distinguished in colour (red) from those of the underlying substrate (pink).

has proved to be essentially correct. We now know, however, that there are important subtleties in this system. In fact the surface Si dimers are not parallel to the surface but are asymmetric (Fig. 2). The (2 × 1) periodicity seen at room temperature involves random disordering of these dimers which are dynamically flipping between the two inequivalent orientations [29]. The clearest evidence for this asymmetry first came from electronic spectroscopy rather than true structural probes, while the situation was confused by early STM images which appeared to show symmetric dimers; the resolution of this apparent conflict relates to the sampling time-scales of the different techniques, the electronic states being probed on a time-scale short compared with the flipping, while the STM imaging time-scale is comparably long. At low temperature the flipping is suppressed and ordering of alternately oriented asymmetric dimers leads to larger surface mesh structures, the static asymmetric dimers then being seen in STM [30]. Because the room temperature (2 × 1) structure is strictly *not* a long-range ordered phase, but has randomly ordered asymmetric dimers placed on a (2 × 1) mesh, some of the detailed quantitative surface structural determinations need to be treated with caution. In particular, most mainstream diffraction analyses (including even a quite recent and careful LEED study [31] conducted at low temperature) have assumed true (2 × 1) ordering, and as such must incorporate incorrect descriptions of the local multiple electron scattering which will introduce systematic errors. The key parameters, however, such as the existence of asymmetric dimers and the approximate tilt angle, are consistent across a broad range of techniques including local probes. Thus, while it is still questionable whether there is a complete quantitative analysis which takes full account of all the complexities, the main qualitative features and much of the quantitative detail of the Si(1 0 0) reconstruction phases are no longer controversial.

As remarked above, the Si(1 1 1)(7 × 7) surface structure presents a very difficult problem for all structural probes. Even before one can consider trying to determine the exact positions of the very many near-surface atoms in the unit mesh, one needs a reliable *structural model* for the origin of the (7 × 7) periodicity. In the absence of any hard quantitative information this issue of the correct structural model was hotly debated during the late 1970s and early 1980s, with many possibilities being aired. The breakthrough in solving this problem came from (high-energy) transmission electron diffraction, a technique well out of the mainstream of surface structural methods, by Takayanagi et al., the original solution being reported in 1982 (see Ref. [32]). Their model is referred to as the DAS model because its key ingredients are dimers, adatoms, and a stacking fault over one half of the surface unit mesh. This structure is shown schematically in the lower part of Fig. 3. It is clearly rather complex, but the key features are the structural modifications undertaken in order to reduce the number of dangling surface bonds; these include not only dimerisation (as on Si(1 0 0)) but also adatoms bonding to three surface Si atoms. In fact this particular structure is often cited as an example of the early success of STM in solving complex surface structures, but it is probably fairer to say that the STM results helped the DAS model to gain widespread acceptance.

The upper part of Fig. 3 shows a typical STM image of this surface, and displays just 12 asperities per unit mesh which might reasonably be

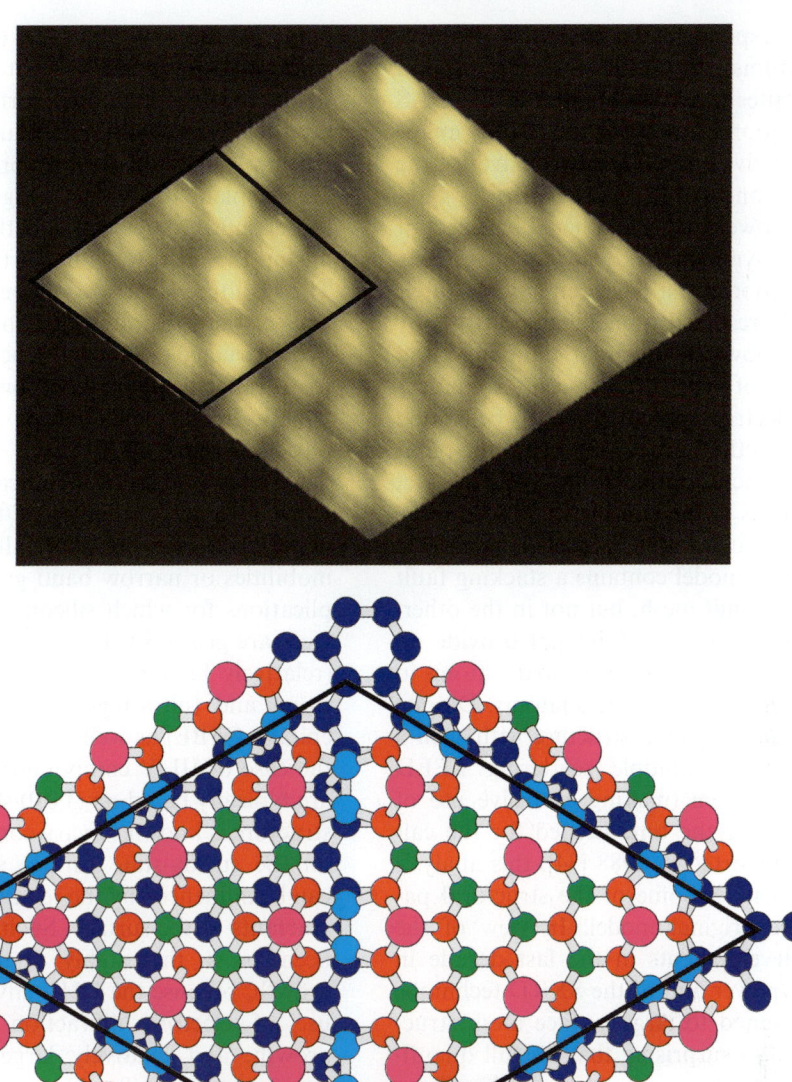

Fig. 3. STM image of the Si(1 1 1)(7 × 7) structure is shown at the top, covering a region of four surface unit meshes (the surface unit mesh is denoted by the bold lines on the left). Below is shown a schematic diagram, in plan view, of the DAS model of this surface; the bold lines again show the surface unit mesh but for clarity the model shows some of the atoms in the edges of adjacent surface unit meshes. In this diagram the adatoms imaged as the asperities in STM are shown as large pink spheres, while the dimerised Si atoms are shown as pale blue. The red spheres show un-dimerised Si atoms in this same layer. The Si atoms in the layer below are shown green, while those in deeper layers are dark blue. Notice that in the right-hand (unfaulted) half of the unit mesh these lower atoms lie directly below those in the outermost two layers.

assumed to correspond to the positions of outermost layer Si atoms. Notice that these correspond to the adatom sites (large pink balls) of the DAS model. In addition, however, the STM image shows some inequivalence of the two halves of the unit mesh, with one group of six asperities typically appearing lower than the other. This effect of reduced image symmetry is dependent on the tunnelling conditions, and has been studied in elegant atomically-resolved tunnelling spectroscopy studies which allow the different dangling-bond electronic states of this surface seen in conventional surface electron spectroscopies to be related to specific sites on the surface [33]. From the point of view of the structural model, however, the fact that the two halves of the unit mesh are seen to be electronically inequivalent is consistent with the fact that the DAS model contains a stacking fault in one half of the unit mesh, but not in the other. Evidently, however, the STM images provide no information on the locations of the many atoms in the surface unit mesh below the adatoms. Despite the complexity of treating a structure with such a large unit mesh, a multiple scattering LEED analysis of this system (making extensive use of symmetry to reduce the time needed for the calculations) was reported in 1988 [34]; this analysis allowed refinement of some of the structural parameters of the original model. In view of the considerable developments in the last decade in computational power and of the SXRD technique (which is well suited to large surface mesh structure), it is perhaps surprising that no full quantitative analysis using newer methods has been performed to check and refine the details of this structure, although in view of the large number of diffracted beams which need to be measured and the large number of structural parameters to be optimised, it remains a very challenging problem.

While the reconstructions of elemental semiconductors, notably Si because of its dominance as a device material, but also of Ge and even α-Sn, have attracted attention as the (superficially) simplest problems, many compound semiconductor surfaces have also been investigated. The (1 1 0) natural cleavage faces of III–V compounds, notably GaAs and other materials having the bulk zinc blende structure, provided the simplest starting point. In the bulk solid the (1 1 0) atomic planes comprise equal numbers of atoms of each element, so the ideally-terminated surface structure would involve these atoms in a coplanar geometry. In fact the in-plane anion–cation bond rotates out of the surface plane to produce a significant distortion, a rather different response to the problem of minimising the energy of the surface dangling bonds. The resulting structure retains the (1×1) periodicity of the substrate but is normally classified as 'reconstructed'. Similar structure determinations for a range of II–VI compounds also reveal systematic trends which can be reconciled by simple physical arguments.

Of course, there is significant technological interest in quite a number of these semiconductor materials for devices, especially where high carrier mobilities or narrow band gaps offer possible applications for which silicon is ill-suited. Such devices are grown by a variety of methods including (relatively high pressure) chemical vapour deposition and (ultra-high vacuum) molecular beam epitaxy (MBE) as well as hybrid intermediates. In the case of III–V compounds, in particular, these are typically based on (1 0 0) slices, and the surface structure is known to vary considerably depending on the steady-state surface stoichiometry which can be influenced by the relative arrival rates of the reactants at the surface. Such surface phases have been most thoroughly characterised in MBE growth, because the UHV environment allows the use of electron diffraction monitoring during growth, and relatively large surface unit mesh structures (e.g. $c(8 \times 2)$) are seen. Rather few of these structures have been analysed by true quantitative methods. The reason for this seems to be threefold. Firstly, large surface mesh structures present the intrinsic problems of complexity mentioned above; there are a large number of possible structural models and each of these has a large number of atoms per unit mesh, the coordinates of which must be located. A second technical problem is that some of these structures, of real interest in understanding the kinetics of the epitaxial growth, are only stable under growth conditions, and the range of structural methods applicable under these conditions is limited. SXRD is potentially the most attractive in that X-rays can easily be used even at relatively high pressures and may often be used in a grazing-in grazing-out geometry which does not interfere with the growth equipment. On the other hand, combining synchrotron radiation and in situ growth is very demanding experimentally and has been used relatively little. A final factor, however, is that the development of an interest in these structural problems has coincided with the development of commercial STM instruments, and much of the attention has switched to the use of STM as the structural probe. Unfortunately, STM in isolation is able to offer only a range of speculative structural models which are increasingly regarded as solved 'surface structures'. This is an area in which more true quantitative surface structure determination is really needed.

3.4. Metal surface reconstructions

Unlike the covalent bonding which dominates elemental semiconductors and plays a major role in compound semiconductors, metallic bonding is non-directional in character, so a simple metal surface lacks the localised dangling bonds which are considered to drive the reconstruction of semiconductor surfaces. For many years, therefore, it was generally believed that clean metal surfaces do not reconstruct. In fact it transpires that this is not true, and there are a small number of metal surfaces which do show significant reconstruction. In some ways the behaviour of the Au(1 1 1) surface is simplest to understand, although obtaining a quantitative experimental description of the structure is certainly not simple. Like other fcc (1 1 1) surfaces, the ideally-terminated surface structure of Au(1 1 1) involves a hexagonally close-packed layer of Au atoms, yet the clean surface actually reconstructs to form an outermost layer which is also hexagonally close-packed but with an interatomic spacing approximately 4% smaller. The driving force for this transformation is, at least in part, the same as that which gives rise to the contraction of the outermost layer spacing at unreconstructed metal surfaces; namely, the atoms rearrange themselves to raise the local valence charge density back to a value more like that in the bulk, to compensate for the spill-out of charge into the vacuum. Contracting the outermost layer spacing is generally insufficient to achieve this, so the atoms would like to reduce their interatomic spacing parallel to the surface. Of course, the surface atoms see a corrugation in the potential attracting them to the substrate which has the periodicity of the underlying bulk, so in general surface atoms do not move parallel to the surface but remain in a state of tensile surface stress [35]. In the case of Au(1 1 1) this stress in the unreconstructed surface is evidently too large, and the reconstruction occurs. The small mismatch of the surface and substrate interatomic spacings leads to a very large mesh size coincidence lattice, with periodic faulting to avoid surface Au atoms being forced into the most unfavourable sites; the resultant surface is commonly referred to as having the herring-bone reconstruction, an allusion to the appearance of STM images of this surface phase. The most quantitative description of this large surface mesh structure was achieved by SXRD. The reconstruction of Au(1 0 0) to form a similar hexagonally close-packed outermost layer can be rationalised in the same way.

The tendency of several fcc (1 1 0) faces, including that of Au(1 1 0), to form a (1 × 2) missing row structure has also been attributed to a similar mechanism. In these structures alternate close-packed ⟨0 1 1⟩ rows of atoms in the outermost layer are missing, which clearly leads to a *more* strongly corrugated surface, in apparent contradiction to the idea of increased surface atomic density. On the other hand, the faces exposed on the 'hill and valley' faces are then of {1 1 1} character, so whereas the unreconstructed (1 1 0) surface is a rather open and corrugated surface on an atomic scale, leading to marked reduction in average valence charge density after smoothing, the reconstructed surface presents {1 1 1} 'nano-facets' which are locally close-packed, and which can also become even closer-packed by some outer layer contraction. Thus, although on a microscopic (or more correctly, nanoscopic) scale, the missing rows make the surface rougher, on an atomistic scale the surface becomes locally smoother (see Fig. 4).

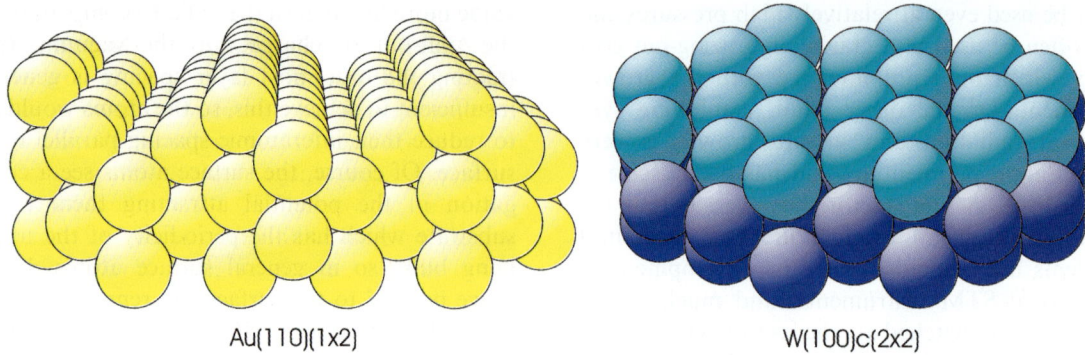

Fig. 4. Oblique views of models of the Au(1 1 0)(1 × 2) clean surface 'missing row' structure and the W(1 0 0)c(2 × 2) 'zig-zag' displacive reconstruction. In the case of the W surface the outermost layer atoms are shown in a different shade of blue for clarity.

While these reconstructions of certain fcc surfaces can be rationalised (albeit not necessarily with complete accuracy) in terms of these simple closest-packing metallic bonding arguments most relevant to free-electron-like (s–p metal) materials, the (1 0 0) surfaces of bcc W and Mo have also been found to undergo 'displacive' reconstructions in which the atomic density of the outermost layer is unchanged from that of the bulk termination, but there is significant movement of the atoms within this layer parallel to the surface producing on W(1 0 0) 'zig-zag' chains of W atoms which create a reconstruction with a c(2 × 2) periodicity (Fig. 4). Although these effects can be reproduced theoretically, the underlying explanation appears to be significantly more complex, and seems to be generally agreed to be related to the high density of d-electron states at the Fermi levels in these materials [23].

3.5. Other material surfaces

It is probably a reflection of the perceived importance of understanding semiconductor surfaces and interfaces for device applications, and metal surfaces in relation to their role in heterogeneous catalysis, which has meant that the great majority of surface structural studies have concentrated on these two classes of materials. Although there has also been a longstanding appreciation of the importance of oxide surfaces (including their role in heterogeneous catalysis) these materials have received far less attention until recently. Of course, a problem with oxides is that they are commonly insulating and thus can be difficult to study by techniques based on electron scattering (including LEED) or electron emission due to problems of sample charging. There has also been doubt as to how to prepare the necessary well-characterised surfaces. Recently the number of groups studying oxide surfaces has increased significantly, in part triggered by the appreciation that some oxides can be grown as quite good quality epitaxial films on metallic substrates with thicknesses small enough to overcome problems of electron charging. With this expansion in general interest has come quite an explosion of structural studies, especially by SXRD [36], of both epitaxial films and of the surfaces of bulk single crystals. Controversies are emerging in this area which are as yet unresolved [37]. One concerns the structure of polar surfaces such as the (1 1 1) surface of oxides having the rock salt structure (e.g. NiO, MgO) which might be expected to be intrinsically unstable, while other (and related) issues concern the extent to which structures really are characteristic of clean surfaces (a key issue being whether OH rather than O termination may be present due to interaction with H_2O). One can certainly anticipate further expansion of surface structural studies on oxide surfaces. Until recently there was a dearth of experimental data with which to establish the systematics of phenomena like surface layer rumpling of non-polar surfaces and the extent of intrinsic surface reconstructions. Such data will also confront the results of theoretical total energy calculations and allow these theoretical studies to advance.

4. Adsorbate overlayer structures: simple atoms to large molecules

The simplest adsorbate structural problems are those of atomic adsorbates forming long-range-ordered phases in which there is just one adsorbed atom per surface unit mesh. All the adsorbed atoms occupy identical sites with respect to both the substrate atoms and the surrounding adsorbate atoms, and on metal surfaces the adsorbed atoms almost always occupy the highest coordination 'hollow' sites which the next layer of substrate atoms would occupy if the solid was continued into the vacuum. Indeed, this effect is even found on fcc (1 1 1) surfaces where there are two inequivalent hollow sites on the surface, both having three neighbours in the top layer, but one of these has a single second layer atoms directly below, while the other has three equivalent substrate neighbours in the second layer and one directly below in the third layer. These two sites are referred to respectively at hcp and fcc hollow sites by analogy with the close-packed layer stacking sequence in these two bulk structural phases. Clearly the energy difference in bonding to these two sites must be small, yet the fcc site is favoured in almost all atomic adsorbate systems on these surfaces. These basic facts were established quite early in the development of surface structural techniques, but later studies have shown that even in these very simple cases the adsorbed atoms can introduce local distortions of the substrate; for example the substrate atoms in the second layer below the adsorbate generally have a different layer spacing to those which do not lie below the adsorbed atoms. These are commonly quite subtle effects, but remind us that surfaces are not rigid chequer-boards on which the adsorbates sit, but rather that the adsorption process causes some modification of both the adsorbate and the substrate. On semiconductor surfaces, of course, one can have more directional bonding which can lead to low coordination sites being occupied.

In this general group of 'simple' adsorbate systems, however, perhaps the greatest surprise has come from investigations of alkali and inert gas atoms adsorbed on some close-packed (fcc (1 1 1) and hcp (0 0 0 1)) metal surfaces [38]. While one would anticipate that these adsorbate species would bond in a non-directional fashion and thus occupy the maximally-coordinated hollow sites, there are a significant number of cases in which *atop* sites are found to be preferred. At first sight this results seems quite extraordinary, yet it has also been reproduced by DFT calculations. In order to understand how this can come about, one needs to recognise that the effective radii of alkali atoms, in particular, are very large, so that on a close-packed surface if one simply rolls a hard-sphere alkali atom over a hard-sphere close-packed substrate, the corrugation amplitude is rather small. For example, if one takes a K atom with a metallic radius of 2.27 Å on a Ni(1 1 1) surface, the highest Ni–K layer spacing would occur in the atop site with a value of 3.51 Å and the lowest value corresponding to hollow site adsorption would be 3.21 Å, yielding a corrugation of only 0.30 Å. Of course, this simple hard-sphere picture is not strictly appropriate, and the effective radius of the adsorbed alkali atom actually depends on the site and the nature of the bonding; moreover, the results of this argument would still favour hollow site occupation, but it does shows how 'flat' the substrate surface may appear, suggesting that the atop site may not be as unfavourable as one may first expect. What is also revealed by both the DFT calculations and the experimental structure determinations in several of these systems is that the outermost substrate layer actually rumples so that the substrate atom directly below the alkali is some 0.1 Å deeper in the surface than the surrounding six next-nearest neighbours (Fig. 5); the atop site which is strictly onefold coordinated then moves towards a state of sevenfold coordination with the six surrounding surface atoms moving closer to the adsorbate (a situation to be compared with the threefold coordination of the hollow sites). Indeed, the behaviour of the superficially simple problem of alkali atom adsorption on metal surfaces has also led to equally surprising surface alloy formation in some systems as we will discuss in the following section.

While atomic adsorbates on metal surfaces are generally expected to form non-directional adsorbate–substrate bonds and thus favour high coordination sites, the same is not true of many

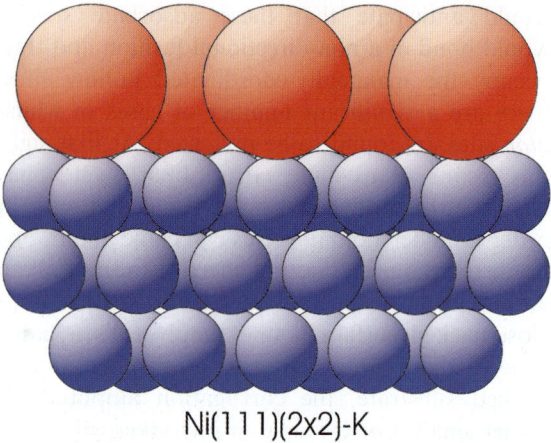

Fig. 5. Schematic side view of the Ni(1 1 1)(2 × 2)–K structure. The radii of both the Ni and K atoms have been reduced by about 10% relative to their metallic values in drawing this model to reduce overlap.

molecular adsorbates, and especially those of interest for their modified chemical reactivity in the adsorbed state which forms the basis of heterogeneous catalysis. The classic example here is CO adsorption. CO/surface reactions are important because CO is a feedstock in the chemical industry and also pollution control demands that CO should be removed from the exhaust gases of automobile engines by catalytic oxidation to CO_2. CO is well-known as a ligand in metal coordination chemistry, bonding via the C atom to one, two or more metal atoms. There are direct surface analogues of this situation, and on metal surfaces of different surface orientation, different metals or different coverage, CO is found to bond atop, in bridging sites or in higher-coordinated hollow sites. The behaviour is complex and the detailed systematics are not understood; on some surfaces one may see atop adsorption at low coverage with displacement to higher coordination at higher coverage, whereas on other surfaces (even of the same substrate material) the converse is true. In view of the huge number of studies devoted to CO adsorption on metal surfaces, there are remarkably few quantitative structure determinations for this adsorbate, most adsorption site assignments being based on vibrational spectral fingerprinting. In particular, building on the body of evidence from metal carbonyl analogues, the C–O stretching frequency was used as an indicator of CO–metal coordination, generally agreed frequency ranges being believed to correspond to one, two and higher-fold coordination [39]. In the early 1990s a few proper quantitative structural studies showed that some well-established conventional wisdom based on these assignments was false, and these results have led to greater caution in the application of this approach. Fig. 6 shows models of the first of these structures to be resolved, that of the Ni(1 1 1)c(4 × 2)–CO surface phase. On the basis of vibrational spectroscopy all the CO molecules were believed to occupy local bridging sites, and the associated model is shown on the left in Fig. 6. The model is attractive because although the 0.5 ML coverage (1 ML or one monolayer corresponds to one adsorbate molecule per surface layer substrate atom) known to be found for this phase implies that there are two (inequivalent) CO molecules per primitive unit mesh (and thus four CO molecules per c(4 × 2) mesh), this bridging model places all the CO adsorbates in *locally* identical sites, and on a primitive hexagonal CO sub-mesh. The two inequivalent sites arise because these sites are only twofold symmetric and are related by the threefold (120°) rotation operation of the substrate. Despite the attractiveness of the model, there is now convincing evidence first from SEXAFS [40], then PhD [41] and finally quantitative LEED [42] that all the CO adsorbate occupy hollow sites, but with equal occupation of the two inequivalent 'hcp' and 'fcc' hollows above second and third layer Ni atoms respectively (Fig. 6).

While this discovery has necessarily caused a re-evaluation of local adsorption site 'determination' using vibrational spectroscopy, most assignments of CO adsorption sites based on the value of the C–O stretching frequency which have been checked by proper quantitative structural methods have been found to be correct, yet the small number of incorrect assignments certainly highlights the need for caution in spectral fingerprinting. Nevertheless, a more recent investigation of this same adsorption system by another method of spectral fingerprinting, that of core level photoelectron binding energy shifts, indicates that even the hollow sites model of Fig. 6 may not be wholly

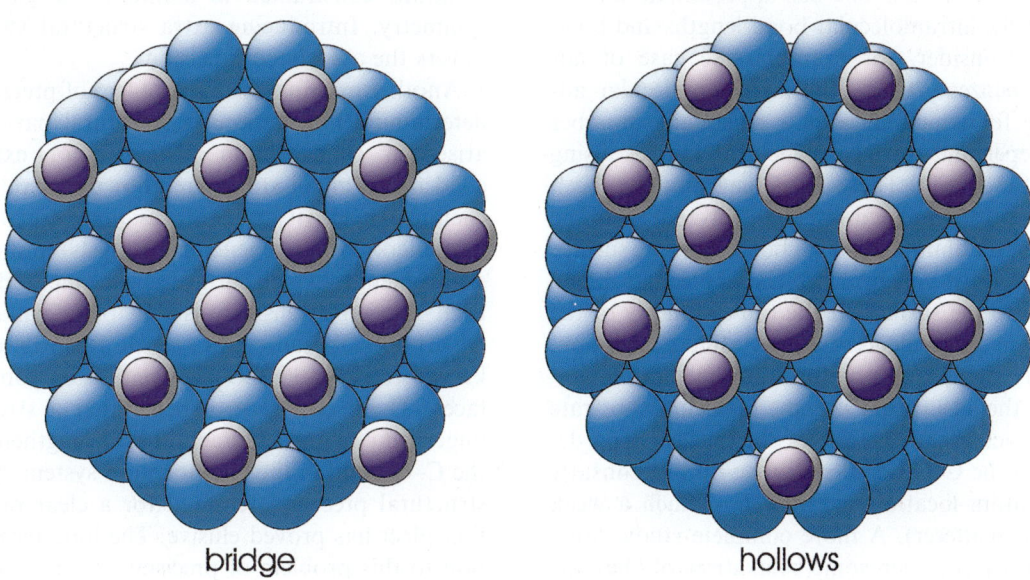

Fig. 6. Schematic plan views of the Ni(1 1 1)c(4 × 2)–CO structural models; the bridging model on the left was long held to be correct, but quantitative structural studies have shown the correct model is that based on occupation of the two inequivalent hollows.

correct as the surface is heated from around 100 K to room temperature. These data show transfer of photoemission intensity into a differently-shifted peak which has been interpreted as indicative of partial occupation of lower-coordination (notably onefold) adsorption sites [43]. The underlying implication is that the energetic differences between the different local adsorption sites is sufficiently small that thermal activation can lead to partial occupation of the less favourable sites. Especially with the increasing availability of high-flux high-resolution synchrotron radiation beamlines on third-general sources one can anticipate a significant growth in the application of this kind of high-resolution X-ray photoelectron spectroscopy as a fingerprint of possible local geometry changes. It remains to be seen how reliable this form of spectral fingerprinting will prove to be as an indicator of local adsorption sites. However, by combining a true quantitative structural method which can be monitored with core level photoemission intensities (such as photoelectron diffraction or X-ray standing waves) with this 'chemical state' (or adsorption site) sensitivity of the core level shifts, one may provide true site-specific tests of this question.

As in other branches of surface science an important trend in surface structure determination is to problems of increasing complexity, broadening the base of data but also allowing investigations of increasing relevance to some of the underlying technological problems to which surface science relates. In the case of 'simple' (unreconstructed) adsorbate structures there are three such areas of complexity, large molecules, coadsorption structures, and surfaces involving multiple adsorption sites as is commonly associated with large surface mesh periodic structures. A problem common to all of these is that of the large number of structural variables to be determined and the associated problem of uniqueness of the solution. Precision is also an important issue for many of these problems.

The specific problem of molecular adsorbates illustrates both the achievements and the problems quite well. Once one moves beyond simple diatomics such as CO, the number of structural variables increases markedly. In effect one has the three Cartesian coordinates of each atom within the molecule, although the fact that the adsorbate *is* a molecule means one has approximate knowledge of the intramolecular bond lengths and bond angles. Consider, for example, the case of adsorbed benzene, C_6H_6, the largest molecular adsorbate for which there is a reasonable number of independent surface structural studies using LEED and PhD. These investigations have been significantly simplified by the fact that benzene appears to almost always bond to the surface through its π-system with the molecular plane parallel to the surface. If the full symmetry of the gas-phase molecule is retained after adsorption, the only significant structural parameters remaining are the adsorption site, the molecule–substrate layer spacing and the size of the benzene ring defined by the C–C bondlength (we do not consider the H atom locations because H is such a weak electron scatterer). A more complete study, however, must consider some relaxation of the gas-phase symmetry. For example (especially in the case of adsorption on threefold symmetric surfaces), does the molecule undergo the in-plane Kekulé distortion with alternating long and short C–C bonds? Are there any out-of plane distortions to 'boat' or 'chair' conformations? One must also consider at least some possible molecular tilt (with different components in different azimuths) and any local substrate relaxation. All of these aspects can, and have been, addressed in recent structural studies. The starting question in structural studies of this kind is the adsorption site, but the more subtle questions of adsorption-induced molecular distortions are of considerable potential importance; much of the motivation for this work is to understand the modified chemical reactivity of adsorbed species, and one may expect subtle structural modifications to accompany the electronic changes which control the chemical processes. Unfortunately, the structural studies conducted so far generally fail to answer these questions with sufficient precision to be chemically significant. Typically the best-fit solution *do* show distortions, but these are almost always of such a magnitude as to fall within the estimated limits of precision. The most consistent picture in this case is that there almost certainly is some expansion of the C–C bonds induced by adsorption, although this result is typically most significant in structural solutions constrained to maintain the gas-phase symmetry. Introducing extra structural variables lowers the associated precision.

Another example of this problem of precision in determining intramolecular structural parameters arises in coadsorption systems. For example, coadsorbed alkali atoms are known to modify the chemical reactivity of surfaces (and are used as 'modifiers' in heterogeneous catalysts) and a classic problem to which surface science methods have been applied extensively is that of CO/alkali coadsorption, because the alkali adsorbates are known to promote CO dissociation on some surfaces. There is a general belief that a structural fingerprint of this interaction is a lengthening of the C–O bond in the coadsorption system, but the structural precision required for a clear proof of this effect has proved elusive. The long-term solution to this problem of precision is far from clear and presents a major challenge for the future of surface structure determination.

The relatively small number of determinations of the structure of adsorbed molecular species more complex than benzene provides some indication of the future of this field. In order to overcome the problem of the very large number of associated structural variables there are two main routes forward. One is to try to separate out the variables by finding a method of determining them in an essentially independent fashion; the other is to impose constraints. The former solution is the one offered by element-specific structural probes such as PhD, SEXAFS and XSW. For an adsorbed molecular species containing atoms of several different elements, these techniques allow one to obtain data on the structural surroundings (PhD and SEXAFS) or atomic position relative to the substrate (XSW) for each elemental species independently. Of course, in the case of the photoelectron scattering methods (PhD and SEXAFS), intramolecular scattering contributes to the data, so the information is not wholly independent of the location of the other atoms within the molecule, but often the substrate scattering dominates and at least a first-order local geometry can be established independently. An example of this approach is that adsorption of the simplest amino acid, glycine, on Cu(1 1 0); strictly the adsorbate is a glycinate species (NH$_2$CH$_2$COO–) which is glycine with the carboxylic acid (–COOH) H atom removed by interaction with the Cu surface. The adsorption geometry of this system is shown in Fig. 7. Independent O 1s and N 1s PhD data were used to show that the species adopts a geometry which implies bridge-bonding through the two O atoms of the carboxylate group and near-atop bonding of the N atom [44]. In fact this structure determination did not locate the positions of the C atoms; there are two inequivalent C atoms so C 1s PhD data comprise an incoherent sum of the data from each of these atoms, and analysis is significantly less straight-forward and reliable.

The solution to this problem is to exploit the 'chemical' shift in the 1s photoelectron binding energies which must exist between the two inequivalent C atoms, as discussed above. Separate PhD spectra for the two components provides a means to determine their local geometries in the same, largely independent, fashion as was used for

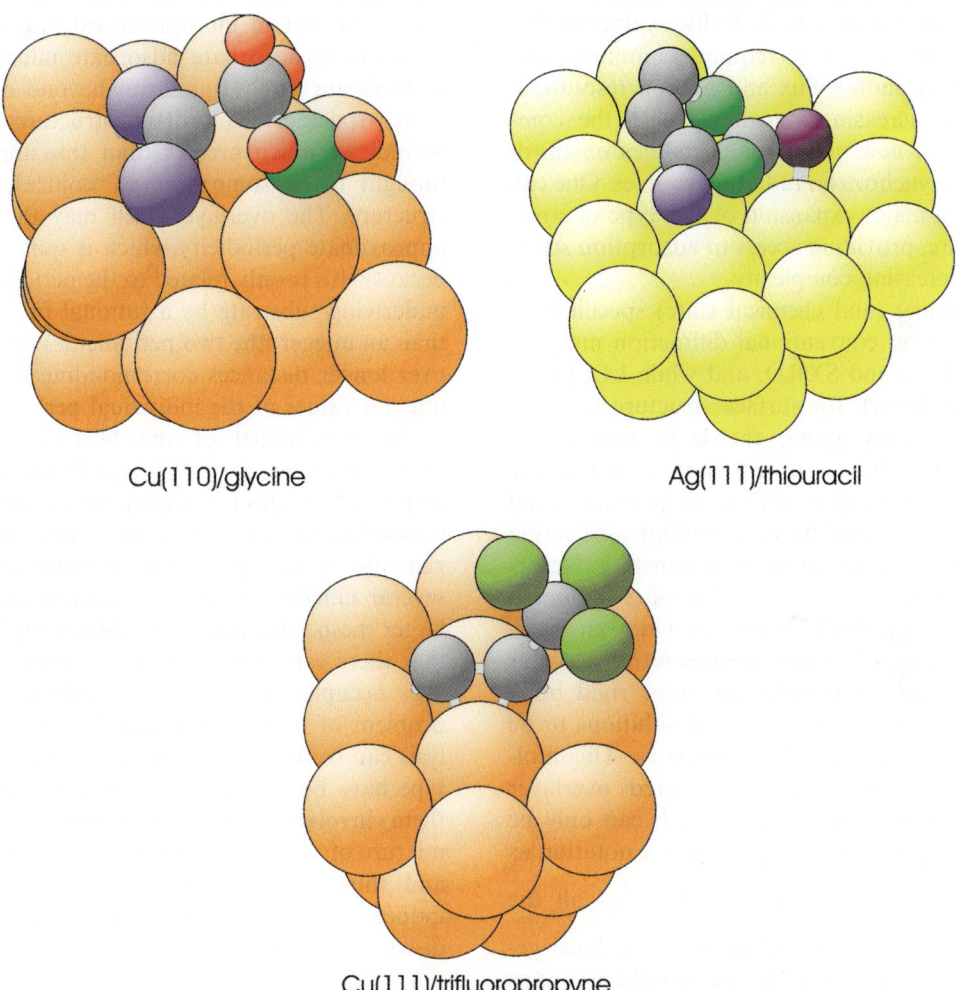

Fig. 7. Schematic oblique views of the local adsorption geometries found for glycine on Cu(1 1 0), trifluoropropyne on Cu(1 1 1) and thiouracil on Ag(1 1 1). In all cases the C atoms are shown grey, O atoms blue and N atoms dark green; in thiouracil the S atom is coloured purple while in trifluoropropyne the F atoms are shown as pale green. None of these structure determinations were able to locate the H atoms in the molecules, but the locations of the H atoms derived from a total energy calculation are included in the case of glycine coloured red.

the O and N atoms. This 'chemical-state-specific' structural information is available from any true quantitative structural technique which can be monitored by core level photoemission, and has been demonstrated using both PhD and NIXSW. A very recent chemical-shift PhD study of this kind [45] concerns propyne (CH_3–CH≡CH) on Cu(1 1 1); the methyl and acetylenic C atoms also show a rather small C 1s chemical shift, but by using the trifluoropropyne species (CF_3–CH≡CH) a large separation could be achieved and the structure determined rather fully to have the conformation shown in Fig. 7. Fully exploiting this chemical-state specificity requires simultaneously high incident photon flux and photon resolution, but this is increasingly available with the commissioning of new undulator beamlines on third-generation synchrotron radiation sources. One can anticipate a major expansion of this type of study in the future, providing access to adsorption structures of increasing complexity.

This element (and chemical state) specificity is not available in conventional diffraction methods such as LEED and SXRD, and while LEED has been a benchmark for surface structure determination for many years, the large number of structural variables is a real barrier in tackling complex problems such as those of large adsorbed molecules, especially because multiple scattering can lead to coupling of these parameters. Nevertheless, a recent study by LEED of 2-thiouracil ($C_4H_4N_2OS$) on Ag(1 1 1) [46] provides an interesting example of how constraining intramolecular structural parameters (referred to as 'rigid body constraints') may allow structural solutions to be obtained for such complex adsorbates. This molecule does form a complex ordered overlayer structure having a unit mesh, which can only be described by the more complex matrix notation as $\begin{vmatrix} 8 & 6 \\ -1 & 2 \end{vmatrix}$ containing four molecules, but all the molecules are found to adopt the same local geometry with the species bound via the S atoms near bridging sites (see Fig. 7). Notice that while one of the interesting aspects of molecular adsorption is the extent to which the adsorption modifies the intramolecular conformation, we do not expect such effects to be large in parts of a molecule far removed from the substrate. Nevertheless, as in this example, some molecular distortions can be investigated in a second stage of structural optimisation. Indeed, in three-dimensional X-ray crystallography, which now is used routinely to tackle (bulk) molecular structures of enormous complexity, constraints of this general kind are imposed commonly on the solutions.

Another form of complexity which may limit one's ability to determine surface structures is that of large unit mesh phases, a problem commonly related to that of multiple co-existing adsorption sites. Direct adsorbate–adsorbate interactions on surfaces are rarely significant beyond one or two lattice spacings, so ordered structures having a periodicity much greater than this are generally thought to be some kind of 'coincidence mesh' structure. The overlayer itself may then have an approximate periodicity which is quite small, but because this is only related to the periodicity of the underlying substrate by a rational fraction rather than an integer, the two periodicities coincide only over longer distances corresponding to (different) integral values of the individual periodicities. The bridge site model of the Ni(1 1 1)c(4 × 2)–CO structure shown in Fig. 6 is a rather special example of this effect in which the inequivalent CO molecules occupy locally equivalent sites; more generally a coincidence mesh structure involves several different local site registries and a much larger mesh. One inevitable consequence of this is that there can be many different local adsorption sites occupied relative to the substrate, and the problem of obtaining a complete structural analysis can be severe. A number of structures of this type have been tackled successfully, but many of them involve surface layers which comprise a mixture of adsorbate and substrate atomic species and thus fall into the category of adsorbate-induced substrate reconstructions which will be discussed in the following section.

5. Adsorbate-induced surface reconstruction

While we now know that adsorbates invariably cause local modification of the structure of the underlying surface, there are many examples in which they introduce substantial reconstruction. They may also introduce 'unreconstruction'—i.e. the presence of adsorbates may lift a clean surface reconstruction. In the case of a covalently-bonded semiconductor surface, for example, adsorbates may saturate the 'dangling bonds' which would occur on an ideally-terminated surface and thus obviate the need for reconstruction. A simple example is the Si(1 0 0) surface (Fig. 2). There is now evidence that quite a number of species may adsorb onto the remaining single dangling bands of the dimerised surface Si atoms, and this bonding appears to largely remove the asymmetry of the dimer; examples are H_2O and NH_3 adsorption when it is believed that a hydrogen atom bonds to one end of the dimer and the remaining OH or NH_2 bonds to the other end. Exposing the clean Si(1 0 0)(2 × 1) surface to atomic hydrogen, however, can lead first to H atoms adsorbing at both ends of the dimer, but with higher exposure the dimers break to allow an unreconstructed (1 × 1) surface to occur with each Si atom at the surface bonded to two H atoms. This kind of adsorbate-induced unreconstruction also occurs on some metal surfaces. One such example is the missing-row Pt(1 1 0)(1 × 2) surface structure (cf. Fig. 2), which reverts to a more ideally-terminated (1 × 1) surface in the presence of certain adsorbates, such as CO. If one conducts a simple catalytic reaction over this surface, such as the oxidation of CO (an important ingredient of the role of the automobile exhaust catalyst), the nature of the surface reconstruction depends on the relative coverage of the O and CO reactants and this phase transformation has been considered to be a factor in determining the oscillations in time and position on the surface (so-called spatio-temporal patterns) involved in this reaction.

The phenomenon of adsorbate-induced reconstruction of metal surfaces which do not reconstruct when clean is also rather widespread, and is potentially important if one is to gain a fundamental understanding of heterogeneous catalysis. Although some of the broad aspects of the reactivity to metal surfaces and their relative utility as catalysts can be derived from an understanding of the electronic structure of the metals and their clean surfaces, the surface under reaction conditions may be quite different. There will almost certainly be a steady-state coverage of some of the reactant molecules or reaction intermediates, and these may induce structural modifications to the surface which in turn can influence the local electronic and chemical properties. For example, nickel surfaces are known to catalyse the methanation reaction in which CO reacts with H_2 to produce methane, CH_4, and the surface under reaction conditions is known to have a partial coverage of atomic carbon. In fact atomic C is known to cause reconstruction of all three low index faces ((1 0 0), (1 1 0) and (1 1 1)) of nickel, and the behaviour on the (1 0 0) surface is especially well-characterised. Fig. 8 shows a comparison of the known surface structures of Ni(1 0 0) in the presence of 0.5 ML of atomic C and with the same coverage of atomic oxygen. The oxygen adsorption structure is a typical case of a c(2 × 2) adsorption phase, in which the O atoms occupy alternate fourfold coordinated hollow sites on the surface with only relatively subtle distortions of the underlying Ni layers perpendicular to the surface. Although C adsorbate atoms occupy essentially the same sites, they induce a so-called 'clock' reconstruction of the Ni surface in which the four Ni atoms in the outermost layer surrounding each C adsorbate atom are displaced outwards but rotate around the C atoms in alternate clockwise and counter-clockwise fashion to produce an interlocking set of distortions which leads to a (2 × 2) rather than c(2 × 2) periodicity. This arises because the Ni atom rotations around the centre C atom have the opposite sign to those at the corners and so are no longer equivalent. While the equilibrium geometry of this reconstructed phase is now well-established by a variety of methods, the exact origin of the phase transition is still debated. Initial ideas centred on the fact that the reconstruction enlarges the hollow occupied by the C atom, allowing this atom to occupy a site deep in the surface layer (almost coplanar) and to bond to the second layer Ni atom directly below. However, quite a large lateral clock distortion (0.41 Å) actually enlarges the hollow by only 0.04 Å. It is also well-established that C (like many adsorbates, especially electronegative ones) produces a

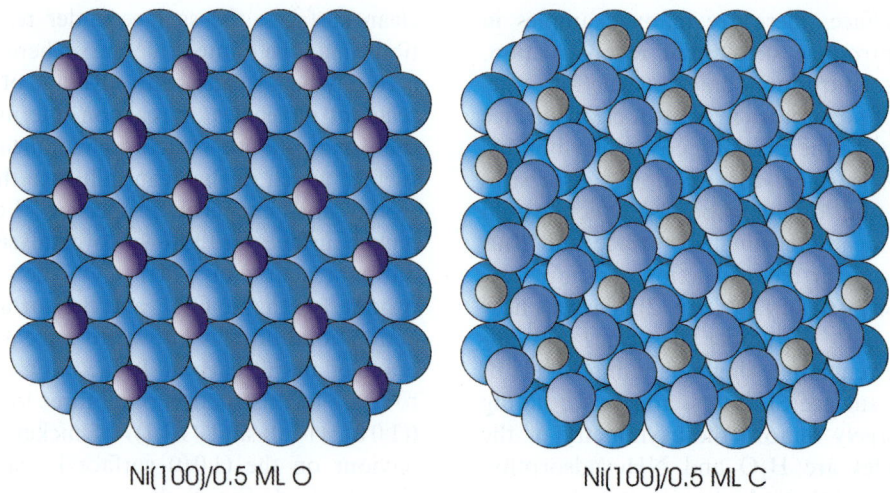

Fig. 8. Plan views of the half-monolayer adsorption structures formed on Ni(1 0 0) by atomic O and by atomic C. In the case of the C 'clock-reconstructed' phase the outermost layer Ni atoms are shown as slightly smaller spheres in a different colour for clarity.

significant compressive stress in the surface, and the reconstruction does halt the rise in this stress as the coverage increases when it nucleates at about 0.3 ML. In this regard some insight may be gained from a study of the local adsorption geometry of the C atoms at lower coverage below that required for the reconstruction. An STM study of this state suggested that a local radial expansion of the Ni atoms around the hollow site occurs, interpreted as indicative of direct Ni–C repulsion. More recently, however, a true quantitative structural study using the PhD technique of this low coverage state has shown that no such distortion occurs (to a precision of 0.03 Å) and that it seems more probable that the reconstruction is driven by adsorbate-induced Ni–Ni repulsion [47].

This study highlights another potentially important but demanding future direction for surface structural studies, namely the determination of *local* structural parameters for adsorbates at low coverage. The problem is important because many surface chemical processes do occur at low (steady-state) coverage and subtle adsorbate-induced changes in the surrounding surface (which is less constrained than at high coverage in an ordered adsorption phase) may influence the reactivity. It is demanding because the problem cannot be addressed by some of the most powerful methods which rely on long-range order, and because low coverage inevitably means weaker signals in whatever technique is used. This Ni/C example, however, does show that existing methods (in this case exploiting a new third-generation synchrotron radiation source) are becoming available.

In many of the more complex (large unit mesh) adsorbate-induced metal surface reconstructions, it has been commonly thought that the driving force is the formation of local surface compounds, effectively very thin films of an epitaxial bulk compound. For example, atomic sulphur is known to produce a range of complex reconstructions on surfaces of Cu and Ni, and a commonly-held view was that many of these were probably sulphide phases involving one or two layers of S and reconstructed substrate atoms. The fact that these are large unit mesh structures does make them quite challenging to solve, so many of them remain controversial. Large surface mesh structures are not only computationally demanding by methods such as LEED, but also involve large numbers of structural parameters and are far less constrained in terms of possible structural models. This results in problems of uniqueness and, as in any method based on trial and error comparison with the results of model calculations, the solutions are limited by the imagination of the researchers who must include the true structure in the set of models which are tested. Nevertheless, there is now considerable evidence that the simple surface sulphide interpretation is not generally correct for the phases formed on Cu and Ni, and may even not be correct for any of these particular structures. In the case of the Cu surfaces, in particular, there is evidence from SXRD that a key building-block of these reconstructions is a quartet of Cu adatoms with a S atom in the resulting local hollow site.

Another class of adsorbate-induced metal surface reconstructions for which there is growing evidence is the case of a surface which, in the presence of an adsorbate, reconstructs to form a surface layer having essentially the same structure as that of this adsorbate on another crystal face of the same substrate. The group of systems studied which illustrate this behaviour are the so-called pseudo-(1 0 0) reconstructions [48] of (1 1 1) and (1 1 0) faces of Ni and Cu which can occur in the presence of adsorbed C, N and in some cases S. In these systems the (1 0 0) surfaces typically adopt a simple $c(2 \times 2)$ hollow-site overlayer as in the Ni(1 0 0)/O case of Fig. 8, or the clock-distorted form of this structure for Ni(1 0 0)/C and Ni(1 0 0)/N. It appears that these surface phases have sufficiently low energy that the (1 1 1) or (1 1 0) face may reconstruct to a slightly distorted form of these same structures, despite the additional energy cost of the interface with the substrate of different symmetry. There is a whole range of subtle variations to be found in these systems. For example, while S on Ni(1 0 0) forms a simple $c(2 \times 2)$ overlayer, the pseudo-(1 0 0) reconstruction formed by S on Ni(1 1 1) includes a kind of clock reconstruction. On the same Ni(1 1 1) surface the pseudo-(1 0 0) phase formed by C appears to have the same clock reconstruction which characterises the Ni(1 0 0)/C phase itself, but on Cu(1 1 1) the pseudo-(1 0 0) phase formed by N adsorption is not clock reconstructed.

The fact that so many of the adsorbate-induced surface reconstructions do produce large surface mesh ordered phases means that there are still only a small number of these structures which have been fully explored, and an even smaller number for which there is a full consensus achieved from several different methods. Gaining a good understanding of the underlying systematic trends needs a larger body of hard structural data than currently exists. In this regard, some comments on the role of STM are certainly appropriate. STM has important strengths and weaknesses in the study of adsorbate-induced reconstructions. The weaknesses surround the fact that one cannot be sure that all atomic-sized asperities in an STM image do correspond to atomic positions (and certainly not to exact atomic positions), and it is even more difficult to know the identity of imaged atoms in a surface comprising two or more elemental species. All of this reflects the fact that STM is a probe of valence electron charge rather than atom core positions.

Despite these problems of interpretation, STM can be extremely valuable in these problems. At the simplest level, one sometimes finds that several different surface reconstructions can coexist, and while it is difficult to separate these using conventional structural techniques which average over large (several mm^2) of the surface, STM allows one to image the coexistent domains of the different phases separately. This situation occurs, for example, in the case of S adsorption on Cu(1 1 1). STM can also be extremely valuable in identifying local periodicities within a long-range ordered structure of large surface mesh. As remarked earlier, large surface mesh structures do not generally arise because of very long-range interactions between widely-separated atoms, but occur due to the local mismatch of short-range periodicities in the substrate and surface. A true diffraction method sees the large mesh of the shared periodicity, but the far more local view offered by STM allows the much shorter sub-periodicity of the surface to be identified, greatly helping in constructing plausible structural models.

Perhaps an even more important aspect of STM, however, is its ability to study the surface dynamics associated with structural transformations. An early example of this is the case of the Cu(1 1 0)-(2×1) structure formed by oxygen adsorption. Prior to the development of STM there had been a number of structural studies of this phase which had concluded that the equilibrium structure comprises O atoms bonded in 'long-bridge' sites mid-way between adjacent Cu atoms in the open ⟨1 0 0⟩ directions on the surface, but with alternate Cu atoms in the close-packed ⟨1 1 0⟩ surface rows missing. In this structure, however, as in the Au(1 1 0)(1 × 2) clean surface structure with alternate Au atoms missing in ⟨1 0 0⟩ rows (Fig. 2), there was skepticism that the large amount of metal atom transport needed on the surface could occur at room temperature. A key question was, 'where do the missing row atoms go?' STM studies of Cu(1 1 0) during the uptake of atomic O provided an elegant answer. Cu–O–Cu atoms chains grow out over the surface from step edges as the structure forms. Thus, these experiments show that Cu atoms are very mobile on the surface at room temperature, that the structure formed is strictly an 'added row' rather than 'missing row' phase (although these are equivalent in the equilibrium structure) and that the step edges act as a source of Cu atoms. These results thus reinforce the original structural interpretation, but also provide new insight into the associated mechanisms of the phase transition.

One further aspect of adsorbate-induced surface reconstruction which was touched on earlier in this review is that of surface alloying. Clearly, if one adsorbs onto a surface of material A atoms of a second species B which are known to form bulk alloys, one can anticipate that if the surface is allowed to equilibrate, B will dissolve into A, but the thermodynamics may favour the retention of a significant surface concentration of B which may intermix with A in the surface layer. While there are many examples of this behaviour, there are also quite a number of examples in which intermixing in the outermost layer (or even outermost layers) occurs even when A and B are immiscible in the bulk. Specific examples of this include the alkali metals adsorbed on Al surfaces, as described earlier. Detailed studies of this kind of intermixing of the surface layers during epitaxial growth are revealing complex situations in which both alloying and dealloying appear to occur in the same system as a function of coverage. So far many of these more complex systems have been studied only by STM and not by true quantitative structural methods, yet they appear to expose a rich seam of problems for fuller future characterisation.

6. New problems: buried interfaces and beyond

The discussion above concentrates on mainstream UHV studies of clean and adsorbate-covered surfaces, especially of metals and semiconductors, but also including oxides. While understanding the structure of clean surfaces is a fundamental issue of surface physics, most of the adsorbate studies are driven by a need to understand aspects of surface chemistry, whether this is in relation to heterogeneous catalysts or semiconductor device processing. Much progress has been made in these areas, but one can also define key problems for the future. How can we improve experimental precision to provide stronger challenges to basic theory and to produce more chemically significant information about structural implications of surface molecular bonding? Can we gain sufficient sensitivity with local methods to study adsorbates at low coverages including investigations of the role of surface defects, such as steps and kinks at vicinal surfaces? What are the limitations placed on increased complexity in terms of large unit mesh ordered phases, larger molecular adsorbates, and more complex coadsorption systems? Increased use of local structural probes with both elemental and chemical-state specificity should allow progress in this final area, while new third-generation synchrotron radiation sources should allow progress in low coverage investigations. Establishing the limits of structural precision needs more careful comparisons of different methods applied to a few key model structural problems. One further general topic in the area of UHV solid–vacuum interface studies is the role of elevated temperature. The relatively subtle effect of anomalous surface thermal expansion coefficients at Ag(1 1 1) has already been mentioned, as has the issue of how to include fully the effect of finite temperatures in DFT theoretical calculations. It is also worth noting, however, that elevated temperature studies are not without their difficulties experimentally, yet can be important in obtaining quantitative structure information about thermally-activated surface phase transitions. High temperatures lead to enhanced diffuse scattering and strong Debye–Waller attenuation in conventional diffraction methods, significantly degrading the quality of the data. Enhanced thermal vibrations similarly degrade the quality of structural data from local electron and ion scattering techniques. Elevated temperature phase transitions thus *can* be studied, but typically can only be addressed with reduced precision.

Beyond UHV solid–vacuum interfaces are other classes of surface structural problems which are so far unexplored or which have been investigated only in a limited fashion. One obvious problem of this kind is 'buried interfaces'. Most obviously, one may consider solid–solid interfaces which may be close to a free surface and may have been produced by in situ epitaxial growth. Until a few years ago the problems of obvious interest would be associated with semiconductor heterojunctions, semiconductor–metal interfaces and semiconductor–oxide interfaces. More recently there has been growing interest in metal–metal and metal–oxide interfaces in relation to novel magnetic multilayer structures and heterogeneous catalysts. The problem here is that most surface science methods have been developed to provide information which is surface-specific and thus is intrinsically insensitive to sub-surface phenomena. There are, however, methods in the surface structure armoury which are actually not implicitly surface specific, but can be made surface specific under appropriate conditions, notably SXRD and MEIS (and HEIS). These methods can be adapted to study shallowly-buried interfaces, and there have been notable successes in the application of both methods to such problems, although the topic remains in its infancy.

A more general form of this problem is the study of any solid interface which does not involve UHV as the second phase, for example solid–liquid (including solid–electrolyte) and solid–gas interfaces. In these cases the only true quantitative structural method that has been used is SXRD. X-rays (of sufficiently high energy) can penetrate not only gases at elevated pressures, but also thin (aqueous) liquid films, and there have been a number of structural studies of solid–electrolyte interfaces by this method, while very recently SXRD has been used to study the structure of an adsorbed CO layer on Ni [49] in the presence of an elevated pressure of the gas, revealing a structural phase not previously observed under more typical non-equilibrium UHV surface science conditions. This is potentially an important development, because there are often arguments about the so-called 'pressure gap' in UHV studies of surface chemistry, the idea being that at the high pressures relevant to practical heterogeneous catalysis the surface phenomena can sometimes be fundamentally different. Optical methods (notably infrared reflection-absorption but also sum-frequency generation) provide a means of spectroscopic characterisation of these elevated pressure interfaces, but SXRD opens the way to true structural studies under these conditions. Of course, STM and related scanning probe microscopies SPM—(notable atomic force microscopy) also offer a way to obtain structural information at both these types of interface, albeit with the problems of interpretation discussed earlier. SPM methods have been used with some success to obtain atomic-resolution information on solid–liquid interfaces although as yet there appear to have been no comparable reports of gas–surface interface studies.

Acknowledgements

The author is pleased to acknowledge the Engineering and Physical Sciences Research Council for the award of a senior research fellowship.

References

[1] J.B. Pendry, Low Energy Electron Diffraction, Academic, New York, 1974.
[2] M.A. VanHove, W.H. Weinberg, C.-M. Chan, Low-energy Electron Diffraction Experiment Theory and Surface Structure, Springer, Berlin, 1986.
[3] K. Heinz, LEED and DLEED as modern tools for quantitative surface structure determination, Rep. Prog. Phys. 58 (1995) 637.
[4] R. Feidenhans'l, Surface structure determination by X-ray diffraction, Surf. Sci. Rep. 10 (1989) 105.
[5] I.K. Robinson, D.J. Tweet, Surface X-ray diffraction, Rep. Prog. Phys. 55 (1992) 599.
[6] P.R. Watson, M.A. Van Hove, K. Hermann, NIST Surface Structure Database, version 3.0, NIST Standard Reference Database 42; NIST, Gaithersburg, 1999.

[7] C.S. Fadley, Synchrotron radiation research, in: R.Z. Bachrach (Ed.), Advances in Surface and Interface Science and Techniques, vol. 1, Plenum, New York, 1992, p. 422.
[8] D.P. Woodruff, Photoelectron and Auger electron diffraction, Surf. Sci. 300 (1994) 183.
[9] J.J. Barton, C.C. Bahr, S.W. Robey, Z. Hussain, E. Umbach, D.A. Shirley, Adsorbate geometry determination by measurement and analysis of angle-resolved-photoemission extended-fine-structure data, Phys. Rev. B 34 (1986) 3807.
[10] D.P. Woodruff, A.M. Bradshaw, Adsorbate structure determination on surfaces using photoelectron diffraction, Rep. Prog. Phys. 57 (1994) 1029.
[11] J. Stöhr, in: R. Prins, D.C. Koeningsberger (Eds.), X-ray Absorption, Principles, Techniques, Applications of EXAFS, SEXAFS and XANES, Wiley, New York, 1988, p. 443.
[12] P.H. Citrin, X-ray absorption spectroscopy applied to surface structure – SEXAFS and NEXAFS, Surf. Sci. 300 (1994) 199.
[13] J. Haase, Surface structure determinations using X-ray absorption spectrscopy, J. Chem. Soc. Faraday Trans. 92 (1996) 1653.
[14] D. Arvanitis, K. Baberschke, Adsorbate substrate bonding and dynamics as determined by SEXAFS, J Electron Spectrosc. Rel. Phenom. 75 (1995) 149.
[15] J.E. Rowe, Synchrotron radiation research, in: R.Z. Bachrach (Ed.), Advances in Surface and Interface Science, Techniques, vol. 1, Plenum, New York, 1992, p. 117.
[16] D.P. Woodruff, Fine structure in ionisation cross sections and applications to surface science, Rep. Prog. Phys. 49 (1986) 683.
[17] J. Zegenhagen, Surface structure determination with X-ray standing waves, Surf. Sci. Rep. 18 (1993) 199.
[18] D.P. Woodruff, Normal incidence X-ray standing wave determination of adsorbate structures, Prog. Surf. Sci. 57 (1998) 1.
[19] H. Niehus, W. Heiland, E. Taglauer, Low-energy ion-scattering at surfaces, Surf. Sci. Rep. 17 (1993) 213.
[20] J.F. van der Veen, Ion beam crystallography of surfaces and interfaces, Surf. Sci. Rep. 5 (1985) 199.
[21] R. Wiesendanger, Scanning Probe Microscopy and Spectroscopy Methods and Applications, Cambridge University Press, Cambridge, 1994.
[22] M. Scheffler, C. Stampfl, Theory of adsorption on metal substrates, in: K. Horn, M. Scheffler (Eds.), Handbook of Surface Structure, Electronic Structure, vol. 2, Elsevier, Amsterdam, 1999.
[23] J.E. Inglesfield, Reconstructions and relaxations at metal surfaces, Prog. Surf. Sci. 20 (1985) 105.
[24] P.J. Feibelman, Disagreement between experimental and theoretical metal surface relaxations, Surf. Sci. 360 (1996) 297.
[25] P. Statiris, H.C. Lu, T. Gustafsson, Temperature-dependent sign reversal of the surface contraction of Ag(1 1 1), Phys. Rev. Lett. 72 (1994) 3574.
[26] K. Pohl, J.H. Cho, K. Terakura, M. Scheffler, E.W. Plummer, Anomalously large thermal expansion at the (0 0 0 1) surface of beryllium without observable interlayer anharmonicity, Phys. Rev. Lett. 80 (1998) 2853.
[27] J. Xie, S. de Gironcoli, S. Baroni, M. Scheffler, Temperature-dependent surface relaxations of Ag(1 1 1), Phys. Rev. B 59 (1999) 970.
[28] J.J. Lander, J. Morrison, Low-energy electron diffraction study of silicon surface structures, J. Chem. Phys. 37 (1962) 729.
[29] J. Dabrowski, M. Scheffler, Self-consistent study of the electronic and structural properties of the Si(1 0 0)(2 × 1) surface, Appl. Surf. Sci. 56–58 (1992) 15.
[30] R.A. Wolkow, Direct observation of an increase in buckled dimers on Si(1 0 0) at low temperature, Phys. Rev. Lett. 68 (1992) 2636.
[31] H. Over, J. Wasserfall, W. Ranke, C. Ambiatello, R. Sawitzki, D. Wolf, W. Moritz, Surface atomic geometry of Si(0 0 1)(2 × 1): a low energy electron diffraction structure analysis, Phys. Rev. B 55 (1997) 4731.
[32] K. Takayanagi, Y. Yanishiro, M. Takahashi, S. Takahashi, Structural analysis of Si(1 1 1)(7 × 7) by UHV transmission electron diffraction and microscopy, J. Vac. Sci. Technol. A 3 (1985) 1502.
[33] R.J. Hamers, R.M. Tromp, J.E. Demuth, Surface electronic structure of Si(1 1 1)(7 × 7) resolved in real space, Phys. Rev. Lett. 56 (1986) 1972.
[34] S.Y. Tong, H. Huang, C.M. Wei, W.E. Packard, F.K. Men, G. Glander, M.B. Webb, Low energy electron diffraction analysis of the Si(1 1 1)(7 × 7) structure, J. Vac. Sci. Technol. A 6 (1988) 615.
[35] H. Ibach, The role of surface stress in reconstruction, epitaxial growth and the stabilization of mesoscopic structures, Surf. Sci. Rep. 29 (1997) 193.
[36] G. Renaud, Oxide surfaces and metal/oxide interfaces studied by grazing incidence X-ray scattering, Surf. Sci. Rep. 32 (1998) 1.
[37] M.A. Van Hove, Challenges of structure determination of compound surfaces, Prog. Surf. Sci. 64 (2000) 157.
[38] R.D. Diehl, R. McGrath, Current progress in understanding alkali metal adsorption on metal surfaces, J. Phys.: Condens. Matter 9 (1997) 951.
[39] N. Sheppard, N.T. Nguyen, The vibrational spectra of carbon monoxide chemisorbed on the surfaces of metal catalysts—a suggested scheme of interpretation, Adv. IR Raman Spectrosc. 5 (1978) 67.
[40] L. Becker, S. Aminpirooz, B. Hillert, M. Pedio, J. Haase, D.L. Adams, Threefold-coordinated hollow adsorption site for Ni(1 1 1)c(4 × 2)-CO – a surface-extended X-ray-absorption fine-structure study, Phys. Rev. B 47 (1993) 9710.
[41] K.-M. Schindler, Ph. Hofmann, K.-U. Weiss, R. Dippel, V. Fritzsche, A.M. Bradshaw, D.P. Woodruff, M.E. Davila, M.C. Asensio, J.C. Conesa, A.R. Gonzalez-Elipe, Is the frequency of the internal mode of an adsorbed diatomic molecule a reliable guide to its local adsorption site? J. Electron Spectrosc. Rel. Phenom. 64/65 (1993) 75.

[42] L.D. Mapledoram, M.P. Bessent, A.D. Wander, D.A. King, An automated tensor LEED analysis of the Ni(1 1 1)c(4 × 2)-CO structure, Chem. Phys. Lett. 228 (1994) 527.

[43] G. Held, J. Schuler, W. Sklarek, H.-P. Steinrück, Determination of adsorption sites of pure and coadsorbed CO on Ni(1 1 1) by high resolution X-ray photoelectron spectroscopy, Surf. Sci. 398 (1998) 154.

[44] N.A. Booth, D.P. Woodruff, O. Schaff, T. Gießel, R. Lindsay, P. Baumgartel, A.M. Bradshaw, Determination of the local structure of glycine adsorbed on Cu(1 1 0), Surf. Sci. 397 (1998) 258.

[45] R.L. Toomes, R. Lindsay, P. Baumgärtel, R. Terborg, J.-T. Hoeft, A. Koebbel, O. Schaff, M. Polcik, J. Robinson, D.P. Woodruff, A.M. Bradshaw, R.M. Lambert, Structure determination of propyne and 3,3,3-trifluoropropyne on Cu(1 1 1), J. Chem. Phys. 112 (2000) 7591.

[46] M. Deschauer, W. Moritz, to be published.

[47] R. Terborg, J.T. Hoeft, M. Polcik, R. Lindsay, O. Schaff, A.M. Bradshaw, R.L. Toomes, N.A. Booth, D.P. Woodruff, E. Rotenberg, J. Denlinger, The coverage dependence of the local structure of C on Ni(1 0 0): a structural precursor to adsorbate-induced reconstruction, Surf. Sci. 446 (2000) 301.

[48] D.P. Woodruff, Adsorbate-induced reconstruction of surfaces; an atomistic alternative to microscopic faceting? J. Phys.: Condens. Matter 6 (1994) 6067.

[49] K.F. Peters, C.J. Walker, P. Steadman, O. Robach, H. Isern, S. Ferrer, Adsorption of carbon monoxide on Ni(1 1 0) above atmospheric pressure investigated with surface X-ray diffraction, Phys. Rev. Lett. 86 (2001) 5325.

Magnetism in low dimensionality

S.D. Bader *

Materials Science Division 223, Argonne National Laboratory, Argonne, IL 60439, USA

Received 30 November 2000; accepted for publication 17 May 2001

Abstract

The collective creativity of those working in the field of surface magnetism has stimulated an impressive range of advances. Once wary, theorists are now eager to enter the field. The present article attempts to take a snapshot of where the field has been, with an eye to the more speculative issue of where it is going. Selective examples are used to highlight three general areas of interest: (i) characterization techniques, (ii) materials properties, and (iii) theoretical/simulational advances. Emerging directions are identified and discussed, including laterally confined nanomagnetism and spintronics. © 2001 Published by Elsevier Science B.V.

Keywords: Magnetic films; Magnetic surfaces; Magnetic interfaces; Magnetic phenomena (cyclotron resonance, phase transitions, etc.); Iron; Cobalt

1. Introduction

Remarkable progress has been made in the field of surface magnetism in the last few decades. In this article we will examine some of the major achievements in order to identify emerging prospects for the future. As we enter a new millennium there is a spirit of passing the gauntlet to the next generation of scientists. Of course that new generation will want to define its own directions. But historical perspective is always valuable.

At the outset the scope is defined as encompassing research on ultrathin films, interfaces, and on the surface properties of bulk crystals. There is also a broad range of bulk materials whose intrinsic properties and crystallographic structures qualify them as low dimensional. We will mention only the naturally layered manganites as such an example.

1.1. Looking back

Since no one can predict the future, the best gauge of future progress is to examine the evolution of the field from its past. The deep past in the area of surface magnetism takes us back to the late 1960s and early 1970s. Two significant milestones can be highlighted. One is in the papers by Lieberman et al. [1] that announced the appearance of magnetic *dead* layers at the surface. Dead layers refer to atomic layers that carry no magnetic moment and hence are non-magnetic even though the underlying bulk material is magnetic. The paradox in this historic work is that the message was totally misleading. Lieberman's layers were indeed dead, but the cause was not intrinsic; instead, as was later realized, uncontrolled contaminants induced

---

* Tel.: +1-630-252-4960; fax: +1-630-252-9595.

the state. Surface contamination is a perennial problem for the entire field of surface science. The art of how to prepare and preserve well-defined surfaces gave birth to the scientific study of surfaces. The success of the field of surface science revolved around the emergence of equipment and techniques that could assure the cleanliness of the surface. Two pivotal techniques paved the way: low energy electron diffraction (LEED) and Auger electron spectroscopy. These techniques monitor the crystallographic and chemical integrity of the surface, respectively, and are described elsewhere in the present volume. Both now have electron spin-polarized adaptations meant specifically for the study of magnetic signals emanating from the near surface region [2]. The lesson to be learned from Lieberman's pioneering study that remains valid to this day is that the surface properties of magnetic materials can be as experimentally challenging and subtle as those in any other area of surface science. An additional lesson is that being misleading can be just as stimulating as being correct *if* you are creative in the process. Presently it is well documented that surfaces can in fact be magnetically alive, *and* that adsorbates (i.e. contaminants) can enhance as well as quench surface magnetism [3].

A second milestone, and one that remains unassailed, is by Palmberg et al. [4]. They examined NiO via LEED in 1968. NiO(1 0 0) is an anti-ferromagnet (AF); hence its magnetic unit cell is twice the size of its chemical unit cell. Since a diffraction experiment probes reciprocal space, the unit-cell doubling in real space gives rise to half-order diffraction beams. They identified half-order beams and monitored their disappearance above the magnetic ordering temperature (the Néel temperature). This demonstrated that the beams originated from the magnetic order. Hence the study provided the first LEED experiment in coherent scattering from an ordered magnetic spin system. This study stands out for its cleverness in that it was possible to probe surface magnetism without using exotic instrumentation. Today by contrast many specialized techniques are available to probe magnetic surfaces. An example of a state-of-the-art approach is in the imaging of the antiferromagnetic structure at surface defects of NiO(1 0 0) by means of photoelectron emission microscopy with 50 nm lateral resolution [5]. In this experiment, synchrotron radiation was used as the soft X-ray source, and magnetic linear dichroism with electron yield detection served as the method for generating the magnetic contrast at the surface. The level of sophistication of the instrumentation is staggering compared to that available to Palmberg in his earlier study. Fig. 1 shows related images of the magnetic domain structure of an antiferromagnetic film of LaFeO$_3$ covered by a ferromagnetic film of Co [6]. The magnetic contrast for the ferromagnetic Co was obtained by means of magnetic circular dichroism (MCD). The correlation between the domains of the two magnetic structures is possible because of the elemental sensitivity of the imaging technique to probe Fe and Co separately. The understanding of such correlations is an important and active area of research associated with a topic known as exchange bias that we will consider in Section 4 as part of the discussion of magnetic coupling phenomena.

This highlights the importance given to the development of such advanced instrumentation in the national agenda. Major research funding is being invested in this manner because the expectation is that there will be a clear payoff, not only in basic-science breakthroughs, but also in solving important technological problems. But, one might ask, what are the anticipated hot spots? The hot spots in the applications arena today are associated with the needs of the magnetic recording industry, as we will see.

1.2. Looking ahead

Traditional surface science has been largely preoccupied with epitaxial monolayers. This pristine two-dimensional (2D) platform with characteristic defects—i.e. steps and kinks—has served as a proving ground for modeling phenomena such as catalytic activity and understanding structural reconstructions and relaxations. As we look ahead the next global frontier appears to be in the area of lateral patterning. This represents the quest for lower dimensionality in the progression from the bulk, to surfaces, to quantum wires and dots. The

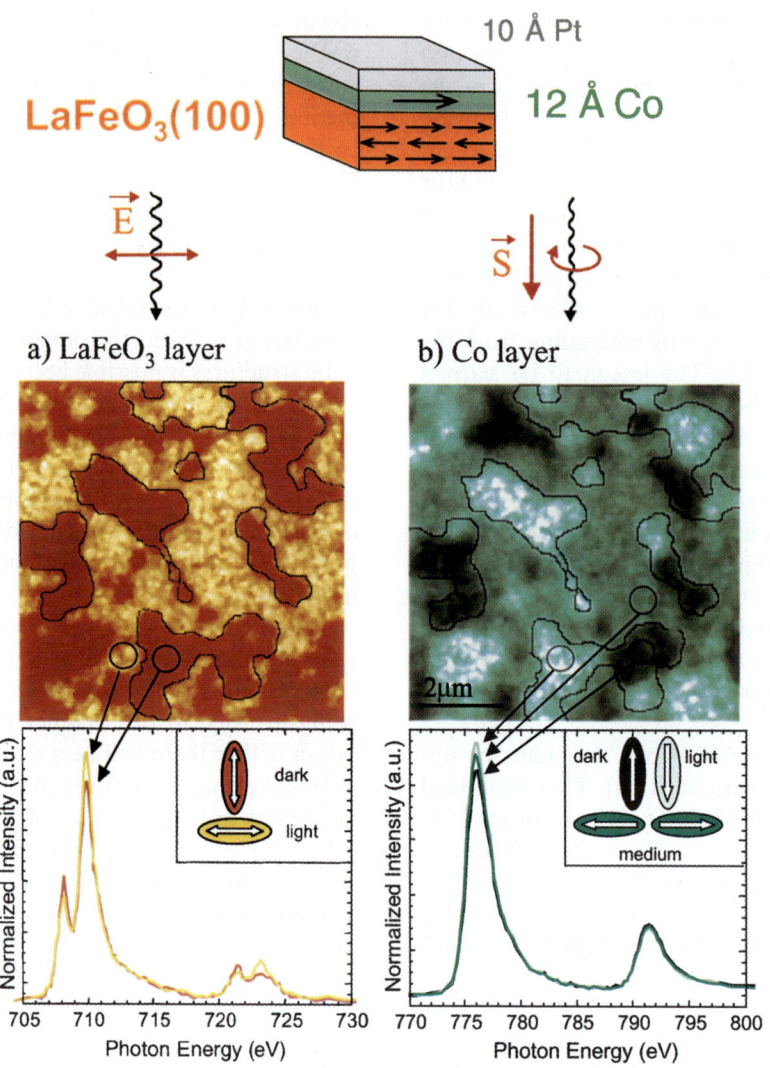

Fig. 1. Element-sensitive magnetic domain imaging of approximately 10×10 μm² region of ferromagnetic Co on antiferromagnetic LaFeO₃ by means of synchrotron X-ray magnetic circular dichroism (right) and magnetic linear dichroism (left), respectively, from Ref. [6]. A schematic of the structure of the layering of the sample is shown at the top, below which is indicated the directions of the photon momentum and E-vector in each of the two cases. At the bottom are representative dichroism spectra for a core level of Fe (left) and Co (right) from the indicated regions of the micrographs, with insets that indicate the color coding of magnetization orientations. The ferromagnetic domains correlate with the antiferromagnetic domains, confirming the pinning of the former by the latter.

focus of surface magnetism research started out analogously and appears to be on the same course as its "parent" field of surface science.

The defining moment for the field of surface magnetism came with the advent of giant magnetoresistance (GMR) [7]. GMR is a phenomenon in layered magnetic structures associated with enhanced sensitivity of the electrical resistivity to external magnetic fields. Within about 10 years of its discovery GMR was adapted to a new generation of high-performance read heads in commercial magnetic hard disk systems. GMR has thus been identified as a crowning achievement of nanotechnology. GMR now stands as one of a number of newly identified types of magnetic exchange couplings across buried interfaces. We introduce GMR because of its pivotal role in defining the direction and focus of the field of surface magnetism. GMR melted the boundary—possibly forevermore—between pure and applied magnetism research. The era in which basic researchers eschewed applied research is gone. It is now quite trendy to work on applied systems and phenomena relevant to technology.

Nanoscale devices are themselves rapidly becoming the laboratory of the future. The quest is for lower dimensional nanostructures, to progress from layers to stripes and dots, to simulate one-dimension and "zero"-dimension, respectively. The features can be in periodic arrays to form lateral superlattices with their own intriguing diffraction properties. They can be self-assembled or fabricated via high-tech methods utilizing lithographic techniques developed originally for semiconductor devices [8]. From this perspective surface magnetism appears to be developing within the wake created by the semiconductor miniaturization revolution. But, take note, the situation might be considerably more insidious. The hidden ambition is for nanoscale magnetic devices to overtake and replace semiconductors. Will the field emerge from the shadows and displace semiconductor technology? We cannot know. But it is realistic to see attempts to develop *magnetic electronics* and hybrid semiconductor-magnetic devices [9]. Maybe magnetism will at least get to share the limelight. The code word for the impending revolution is *spintronics* because magnetic devices utilize not only the charge of the electron, but its *spin* as well. Just as semiconductor solid-state electronics emerged to replace the preceding vacuum-tube era, the dream is that spintronics will blossom into a dominant player in the future.

Today magnetic films are already a symbiotic part of the information revolution because of the role of hard disc data storage systems in computer technology. Hard disc media are magnetic and, as mentioned above, the read heads utilize magneto-resistive sensing. The 0s and 1s of binary lingo are represented by "spin-up" and "spin-down" orientations of the magnetic bits. As the head flies over the medium, the binary code translates into streams of low- and high-resistance states. The pace of development is such that every year the areal density of magnetic storage doubles. Toward the latter part of the year 2000, commercial products have densities as high as about 20 Gbits/in.^{2} and 56 Gbits/in.^{2} has been demonstrated [10] in the laboratory. The roadmap for development anticipates that the present trend can continue to about 300 Gbits/in.^{2} Then a change of technology will be needed to progress from today's continuous, longitudinal media to another format, such as vertical and/or discrete media. The term *continuous* refers to media for which the bits are written onto magnetic films, while *discrete* refers to media for which the bits are written onto patterned magnetic arrays such that each element of the array represents a bit. Longitudinal and vertical media refer to the orientation of the spin as either in the plane or perpendicular to the plane of the media, respectively.

Magnetic discs have a grip on information storage because, among other reasons, they are *nonvolatile* (the information is retained when the power source is removed). This differentiates them from conventional semiconductor-based technologies. In the future this property will be exploited to configure a new generation of magnetic random access memory (MRAM) to replace semiconductor-based RAM. Fig. 2 shows a schematic of an MRAM array element. An obvious advantage will be to eliminate the boot-up process when power is first applied to one's laptop. Another advantage could be in the lower power consumed by MRAM because the memory would not need to be refreshed periodically as does present-day RAM. In the distant future one can dream of programmable magnetic logic elements configured to form magnetic central processor units, the brains of the computer. In such a scenario the CPU could be

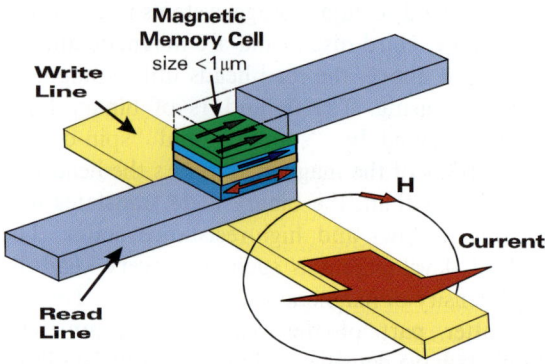

Fig. 2. Schematic of MRAM array element, from Ref. [2]. Binary coding corresponds to the magnetizations of the two ferromagnetic (blue) layers being parallel or antiparallel. The top ferromagnetic is the reference layer. Its magnetization orientation is fixed because it is pinned by the antiferromagnetic (green) layer above it. The bottom ferromagnetic layer is written onto by pulsing current through the orthogonal read–read lines to create magnetic fields, as shown for the write line. The element is read magnetoresistively. If the spacer layer between the two ferromagnetic layers is an insulator, then the spin-dependent transport is governed by the tunneling magnetoresistance, and if the spacer is a non-ferromagnetic metal, then the transport is governed by the giant magnetoresistance (GMR) effect. In either case, parallel orientations of the ferromagnetic layers yield a low-resistance state (binary 0) and antiparallel orientations yield a high-resistance state (binary 1).

reprogrammed on the fly so that the architecture of the machine optimally matches the sub-task at hand. Hence, there is a conceptual framework for far-reaching innovations based on magnetic technologies that are today still in their infancy.

Opportunities exist to harness the methodologies of surface science to open new vistas in the world of technology. However, before this can happen, the fundamental underpinnings of the field have to be mastered so that engineers are comfortable that each stage of miniaturization can be controlled without triggering changes in the principles that govern the operation of the device. What we are witnessing today is a veritable feeding frenzy on the possibility that technology will further embrace magnetic devices. This is increasingly an area where the training of new graduate students will take place and where new grants will be awarded.

2. Where we are

Now that we have acquired something of a sense of where the field might be going, let us see where we are. Magnetic nanostructures, like other nanostructures, are probed with electron beams, synchrotron radiation, lasers, scanning probes, etc.; often however, specialized adaptations are utilized to acquire sensitivity to the magnetic spin arrangement. Representative examples will be used to illustrate this point rather than categorizing the full plethora of approaches.

2.1. Electrons

Electron beam methods can be properly highlighted using the example of SEMPA—scanning electron microscopy with polarization analysis. A spin detector is used to analyze the electron spin polarization of the sample being probed. There are many types of such detectors, but they are all based on a left–right (Mott) scattering asymmetry of the spin up–down electrons due to the spin–orbit interaction. Detailed descriptions of the method appear in the literature [11]. Magnetic domain structure at the surface can readily be imaged via SEMPA. The most dramatic experiment was the original observation of oscillatory magnetic coupling of two Fe layers across a Cr spacer layer of nanometer thickness. The pioneering experiment was recently recounted in a superb review [11]. An Fe(0 0 1) whisker (a fraction of a millimeter wide) was used in the experiment because it provides a quite perfect, atomically flat surface. The whisker has two opposing magnetic domains along its length, unlike a compass needle that consists of only one domain. A Cr layer of increasing thickness along the length of the whisker forms a wedge-like structure that separates the whisker from a second Fe layer. The interaction across the metallic Cr spacer mediates a magnetic coupling between the two Fe layers that alternates between parallel and anti-parallel alignment of their respective magnetizations. The SEMPA image of this new oscillatory coupling phenomenon (as shown in Fig. 3) provided such a dramatic illustration of the effect at the time that it rapidly became an icon to symbolize the new era of nanomagnetism.

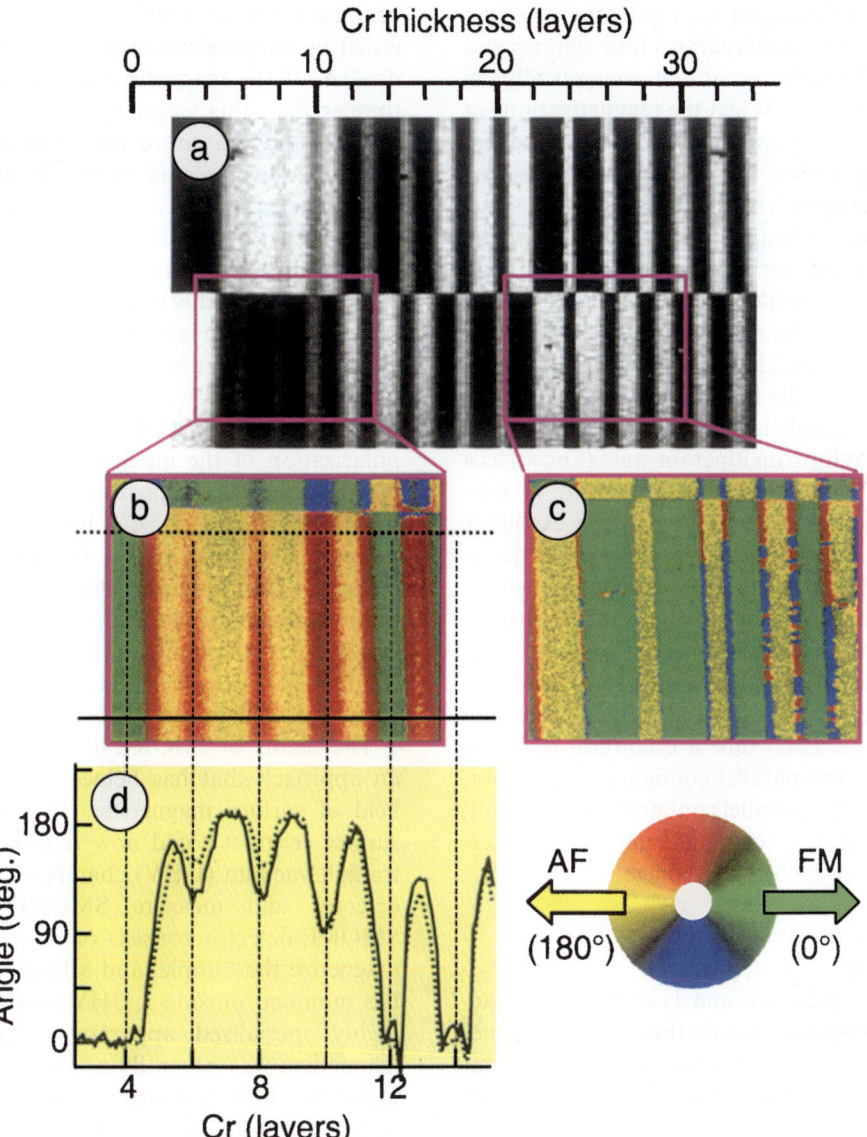

Fig. 3. SEMPA imaging of oscillatory interlayer exchange coupling of Fe film across Cr wedge grown on an Fe whisker, from Ref. [11]. The whisker is a fraction of a millimeter wide and possesses two oppositely oriented magnetic domains running along its length, shown as the two horizontal stripes in (a). The checkerboard pattern in (a) shows that the top Fe layer has its magnetization oriented alternately parallel or antiparallel to the magnetization of the whisker below it. Enlarged regions are shown in (b) and (c) with magnetization orientation color coded according to the calibration wheel at the bottom. These enlargements and the line scan rendition in (d) show that the magnetization in the transition regions can be orthogonally oriented. Such checkerboard patterns have become an icon of the oscillatory coupling associated with the GMR effect because they provided the first direct view of the coupling.

Fe/Cr/Fe is a prototypical GMR material. The oscillatory coupling is understood as arising from magnetic quantum-well phenomena, as summarized in recent reviews [12,13]. As the Cr thickens, the coupling alternates to generate the checkerboard pattern shown in the figure. The parallel and anti-parallel configurations give rise to the low- and high-resistance states, respectively, as alluded to earlier. For a ferromagnet the electronic structure of the conduction band separates into spin-up and spin-down sub-bands, separated energetically by an exchange splitting. When the magnetizations of two Fe layers are aligned in parallel, one of the two spin sub-bands shunts the current across the trilayer structure, giving rise to a low-resistance state. However, when the magnetizations of the two Fe layers are opposed, carriers from either sub-band encounter strong spin-dependent scattering in their attempt to traverse the heterostructure, resulting in a high-resistance state. In the language of quantum wells, when the two Fe layers are spin aligned, one of the spin sub-bands is matched in energy across the interfaces to form an itinerant state (known as a Block state). The other spin sub-band is mismatched with the Cr states due to the exchange splitting. The mismatched Cr state becomes quantum confined. The position of the confined states relative to the Fermi energy E_F changes with Cr thickness. When a confined state is at E_F, the resultant high density of states triggers an instability that is removed by anti-aligning the magnetization of the Fe layers. Each time a quantum-well state crosses E_F the anti-parallel configuration is stabilized, otherwise the parallel configuration wins out; hence the oscillatory coupling and the checkerboard pattern in the SEMPA image.

2.2. Photons

Synchrotron radiation and laser beams are accessible to magnetism mainly through the magneto-optic interaction based also on the spin–orbit interaction. The magneto-optic Kerr effect (MOKE) is a standard technique that originated well over 100 years ago in the work of John Kerr, an early follower of Michael Faraday. MOKE can be used to trace out magnetic hysteresis loops of ferromagnets. Moog and Bader [14] first adapted it in 1985 to the study of surface magnetism and coined the acronym SMOKE to emphasize its utility in probing epitaxial magnetic *surface* layers, as is recounted in a recent review [15]. The technique can readily detect ultrathin films undergoing the magnetic phase transition as a function of temperature with sensitivity such that 2D magnetic critical exponents can be precisely measured. The probing depth of visible light in a metal is governed by its penetration depth, and is a few hundred Å. If the magnetic overlayer is significantly thinner than this length scale, it is said to be ultrathin, and the entire film thickness is probed. The monolayer regime is clearly in the ultrathin limit, hence the utility of SMOKE.

In MOKE (and SMOKE) linearly polarized light has its polarization rotated and it becomes elliptical upon reflection from a magnetized surface. The rotation is clockwise or counterclockwise depending on if the magnetization of the target sample is up or down. The reflected light is passed through a polarizer that is almost crossed with the polarization of the incident light. As the applied magnetic field is swept through a hysteresis loop, the rotation changes and hence the intensity of reflected light that passes through the polarizer changes. This intensity change is monitored to provide a signal that is proportional to the magnetization of the sample. This is the essence of the technique. More detailed descriptions can be found in the literature [15].

The beauty of SMOKE lies in its simplicity. It is an approach that has helped to democratize the field of surface magnetism. This is because any surface scientist could now retrofit his or her ultrahigh vacuum (UHV) chamber with a SMOKE detector and measure SMOKE signals. The SMOKE detector consists of an electromagnet to magnetize the sample, and a laser and simple optics mounted outside a UHV window. Previously highly specialized apparatus to probe surface magnetism was generally needed which tended to make the field a rather exclusive club. Just as SEMPA has helped to *visualize* surface magnetic phenomena (in zero applied magnetic field), SMOKE hysteresis loops have helped to make such phenomena more tangible in complementary ways. SMOKE is routinely used in systematic searches for the existence of magnetic overlayer phenomena, and in determining the easy direction of magnetization. Spin reorientation phenomena in ultrathin films are especially amenable to SMOKE analysis because as the sample reorients its magnetic easy axis from in-plane to perpendicular, longitudinal to polar SMOKE geometries, respectively, can be used to track the signal.

Today a whole family of MOKE techniques is being used in surface magnetism research. These include MOKE imaging, which is limited in lateral resolution by the wavelength of the light; near-field, scanning magneto-optic microscopy, which can transcend the diffraction limit; second-harmonic MOKE (SHMOKE) [16] to enhance surface and interface sensitivity; and XMOKE [2] which extends the technique to the short wavelengths of X-rays. Diffraction MOKE (DMOKE) [17] is used to probe lateral superlattices whose array spacing is comparable to the wavelength of the incident light. Synchrotron techniques are increasingly emerging as powerful probes of surface magnetism. Linear dichroic spectromicroscopy of AFs was cited in the introduction and in Fig. 1 [5,6]. MCD is also routinely used at soft and hard X-ray sources also to provide element specific signals, as in Fig. 1b [18]. Spin-polarized photoemission is another technique that especially benefits from the photon brightness available at third-generation synchrotron facilities, i.e. those equipped with wigglers and undulators to enhance the characteristics of the beam [2].

3. Materials frontiers

The first task is to establish the existence of magnetism in reduced dimensionality. A recent review by Elmers [19] is dedicated to monolayer magnetism in epitaxial overlayers. Techniques such as SMOKE, highlighted above, play an important role in identifying surface magnetism.

3.1. Fe/Cu(1 0 0)

The magnetic phase diagram of Fe/Cu(1 0 0) provides an instructive case in point. Iron is of course the prototypical ferromagnet. But ordinary iron that we are familiar with in the three-dimensional world crystallizes in the body-centered cubic (bcc) phase. A face-centered cubic (fcc) phase exists at elevated temperatures in the bulk phase diagram and can be epitaxially stabilized on Cu(1 0 0) with only slight tetragonal distortion. The actual magnetic phase diagram for ultrathin Fe/Cu(1 0 0) films grown as a function of Fe thickness and substrate temperature during deposition is quite nuanced. There were four discernible phases at last count [20]. These include face-centered tetragonal (fct) phases whose easy axes of magnetization are either perpendicular or in-plane (separated by a spin-reorientation transition), a relaxed fcc phase that supports antiferromagnetic coupling of the top Fe layers [21], and the ordinary bcc phase as the Fe thickness is increased. The Fe/Cu(1 0 0) system has provided an active playing field and training ground for theoreticians and experimentalists alike. Since it is so close to multiple instabilities in its magnetic and structural properties, the details of the synthetic pathway determine the types of metastability that get quenched into the system. In the future, laterally confined Fe/Cu(1 0 0) will probably continue to engage and perplex a new generation of researchers.

The Fe/Cu(1 0 0) system illustrates the interplay between structure and magnetism. Within the fct phase there is a spin-reorientation transition governed by the magnetic energy competition between a perpendicular surface anisotropy and an in-plane shape anisotropy [22]. The surface anisotropy is due to the spin–orbit interaction and can be calculated from the band structure, while the shape anisotropy acts to minimize the dipolar energy of the system. It is the shape anisotropy, for example, that dictates that the magnetization of a compass needle is along its length.

3.2. Step-induced anisotropy

One of the more recent directions in the tailoring of magnetic anisotropies is in the use of substrates with regular arrays of steps to induce new anisotropies and spin reorientations as a function of temperature and/or thickness [15]. To create steps the substrate is cut at an angle from a low Miller-index face. For thermodynamically stable cuts that do not facet, the surface consists of low Miller-index terraces separated by atomic steps. The step morphology can be characterized by conventional surface science methods (i.e. LEED and STM). The step density is determined by the vicinal angle of the cut. A clever innovation is the introduction of curved substrates to give rise to a

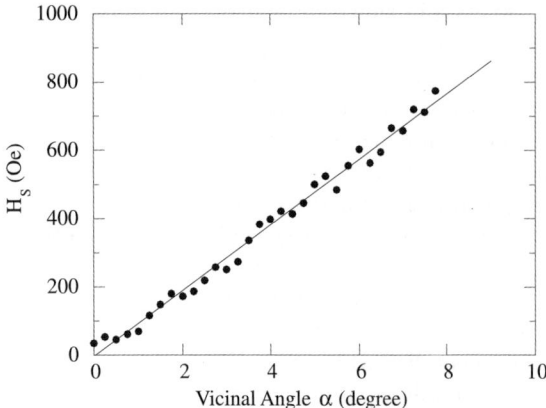

Fig. 4. SMOKE-derived data, from Ref. [23], for step-induced magnetic anisotropy as a function of vicinal angle across a 1.8 monolayer Fe film grown on a curved Pd(100) substrate to provide a continuous distribution in the step density. The field H_S provides a measure of the anisotropy strength. The data show that the magnetic anisotropy increases with step density. This demonstrates a way to tailor the magnetic anisotropy by adjusting the step density.

continuous gradient in the step density. Then a focused laser beam, for example, can scan the curved surface in much the same manner as wedged samples are probed, but in this case the anisotropy is determine as a function of the step density rather than as a function of the wedge thickness. Fig. 4 shows the variation in magnetic anisotropy from such a curved sample [23]. A message to take from this is that as the regime of lateral confinement becomes prime territory for the exploration of new properties, the steps and edges that surround the confined regions might be anticipated to play as important a role as in the case cited above.

3.3. Scaling

Within a symmetry class, the dimensionality determines the scaling properties in the critical region around a second-order phase transition. Three symmetry systems of note are the Ising, XY, and Heisenberg, with one, two and three spin degrees of freedom, respectively. In the Ising class the spins orient *up* or *down*, in XY the spins are confined within a plane, and in the Heisenberg class the spins are isotropic in all three directions.

The scaling exponents have been extensively explored theoretically in different dimensionalities. For the magnetic phase transition the magnetization serves as the order parameter of the system [24]. Most interestingly, in 2D for isotropic and XY systems there is no long-range order (LRO) at finite temperature. Fortunately the various anisotropies in the systems under consideration break the symmetry and impart to the system an Ising-like transition. Ising exponents, most extensively explored for the magnetization exponent, are reported in Elmer's review [19]. The Ising model is best envisaged for perpendicular easy axes, and indeed these systems exhibit 2D Ising exponents and have broad critical regions characteristic of 2D nearest-neighbor exchange interactions. However, for magnetic overlayers with in-plane easy axes, unusual exponent values have been observed that were subsequently attributed to a *finite-size* 2D XY model [25]. Finite size itself destroys LRO and rounds the transition, so the exponent has a formally different character than for systems for which the LRO is preserved. There is always a brisk interplay between theorists and experimentalists in the area of scaling. The theorists are necessarily precise in their statements and mathematical expressions, while the experimentalists are left to deal with the issues imposed by the real world. There is always the question of how closely any real system can be made to simulate one of the ideal theoretical constructs. In this regard it is heartening that theorists have been exploring realms that include finite-size effects and even random anisotropies.

4. Magnetic couplings

The great discovery and development of GMR has certainly brought to the forefront the importance of magnetic couplings. GMR involves the coupling of two ferromagnetic layers across an intervening metallic spacer. There are many other possible spacers to explore, and each provides interesting properties and materials challenges. Tunneling magnetoresistance (TMR) involves structures with insulating spacers. These are now candidate systems for the creation of magnetic

RAM, as was mentioned above. A recent TMR review covers theory and experiment alike [26]. Semiconducting interfaces and their interfaces with magnetic semiconductors offer possible integration of magnetic and semiconducting devices. A recent review highlights intriguing fast spin dynamics that have been probed in GaAs/MnAs quantum-well superlattices [27].

Another area is to interleave superconductors (S) and ferromagnets (F) [28]. It is well known that magnetism tends to preclude superconductivity because the magnetic spins break up the singlet-state Cooper pairs of conventional superconductivity. The competition in such S/F systems gives rise to new types of superconductivity (i.e. theoretically predicted π-phase phenomena, whereby the superconducting order parameter couples across the intervening ferromagnet either in-phase or π apart in an oscillatory fashion as a function of the thickness of the ferromagnet). Conversely for GMR systems, if the metallic spacer is cooled below its superconducting transition it is anticipated from theoretical considerations that the oscillatory magnetic coupling will be dampened [29]. This is because there is a competition between Cooper pairing and establishing the static spin density wave needed to communicate the magnetic coupling.

4.1. Exchange bias

Another type of coupling is embodied in the *exchange biasing effect* that occurs between a ferromagnet and an AF [30,31]. The bias is set by cooling the AF through its Neel temperature in the presence of an external magnetic field that aligns the magnetization of the ferromagnet. This acts to pin the ferromagnet to the AF, so that when a ferromagnetic hysteresis loop is swept out, the loop is *biased* in the direction of the pinning. The bias field is defined as the offset of the loop from zero field. It provides a measure of the coupling strength. Exchange bias has become of more than academic interest because of its use to set the *reference* layer in GMR, TMR and MRAM devices. The issue is that experiment generally finds that the magnitude of the coupling strength is say 10^1–10^2 times weaker than anticipated from the theoretical interfacial coupling strength. This has

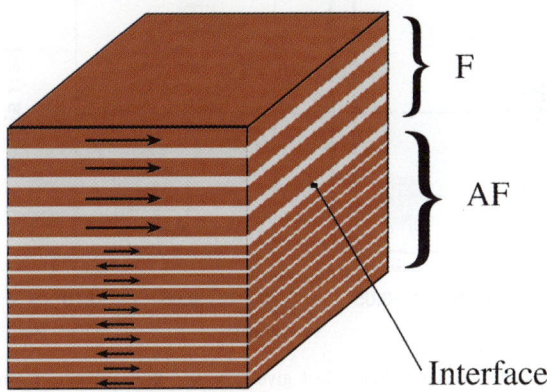

Fig. 5. Schematic of epitaxial Fe/Cr(2 1 1) double superlattice with uniaxial in-plane magnetic easy axes, as shown by the arrows, based on Ref. [32]. The lower stack has the adjacent Fe layers antiferromagnetically (AF) coupled, while the interlayer magnetic coupling of the Fe layers in the upper stack is ferromagnetically (F) coupled. The interfacial Cr layer defines a F–AF junction for the study of exchange bias phenomena.

led to a variety of ideas involving spin configurations that lower the switching field. The root of the problem is that in real systems there is spin frustration at the F/AF interface due to roughness and intermixing. In a recent paper Fe/Cr GMR *double* superlattices are used to create the ferro- and antiferromagnetic layers, as shown schematically in Fig. 5 [32]. The interface hence is epitaxial; there is no lattice mismatch, but only a change of Cr thickness across the two superlattices. Interfacial roughness is less important here than for atomic AFs because the relevant length scale is the Fe/Cr bilayer thickness rather than an atomic layer spacing. In this case the bias can be field set, rather than thermally set. Fig. 6 shows an example of an offset sub-loop measured both by means of the Kerr effect and magnetometry utilizing a superconducting quantum interference device. This system yields bias field values that are in quantitative accord with the origin theoretical idea proposed by Meiklejohn and Bean [33] over 40 years ago! Hence, we project that in the future we can look forward to better controlled interfaces, better characterized interfaces, and artificial methods to alter length scales of a problem in order to properly test ideas and hypotheses.

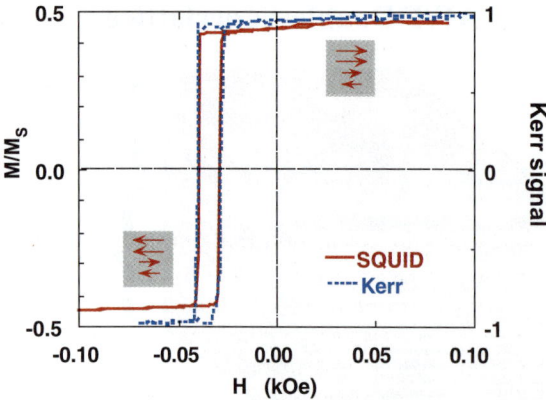

Fig. 6. Exchange bias effect gives rise to offset subloop from Fe/Cr(2 1 1) double superlattice sample, based on Ref. [32]. The normalized magnetization is plotted vs. applied magnetic field H, based on measurements taken utilizing either a superconducting quantum interference device (SQUID) or magneto-optic (Kerr) detection. The spin configurations are schematically indicated for two F layers coupled to an AF layer. Since only the ferromagnetically coupled Fe layers switch their magnetization orientation (and not the AF layer), the measured signal is nominally half (0.5) the full signal possible, as indicated by the left abscissa labeling.

4.2. Exchange spring magnets

In contrast to exchange biasing, another important interfacial pinning effect occurs in *exchange spring magnets* [34]. These are formed between hard and soft ferromagnets (i.e. ferromagnetic heterostructures with high and low values of the coercivity, respectively). Spring magnets provide a nanotech strategy to realize stronger permanent magnets than are presently commercially available. Strong permanent magnets should have the property that their hysteresis loops encompass large areas. To make the loops wide requires high magnetic anisotropies, and to make the loops tall requires high magnetization per unit volume. These two properties tend to work against each other. For example, rare-earth elements tend to have high anisotropies associated with their f orbitals, but because they have large atomic radii (c.f. compared to Fe) their magnetization per unit volume is low. In the spring concept a hard, rare-earth containing phase is interleaved or nanodispersed with a magnetically soft material with high magnetization (i.e. Fe or Co). The hard material

pins the soft material at the interface. Spin frustration is not an important impediment, as it is at F/AF interfaces, because both phases can orient their magnetization in the same direction, even in the presence of interfacial roughness. The key constraint to yield a high and wide loop is that the thickness of the soft material be less than that of a magnetic domain wall. Otherwise the part of the soft layer that is distant from the interface will develop a spin structure that can twist in the direction of an external magnetic field that is applied in the reverse direction. This twist can then facilitate the magnetization reversal and degrade the desirable property of interest. Such interfacial pinning and twisting away from the interface is depicted schematically in Fig. 7. The term *spring magnet* refers to the ability of the twist structure to *spring* back to full remanence when the reverse field is removed. For ordinary ferromagnets such a reversal tends to lead to a sub-loop with reduced remanence. This exchange spring hardening approach has been estimated to give rise in principle

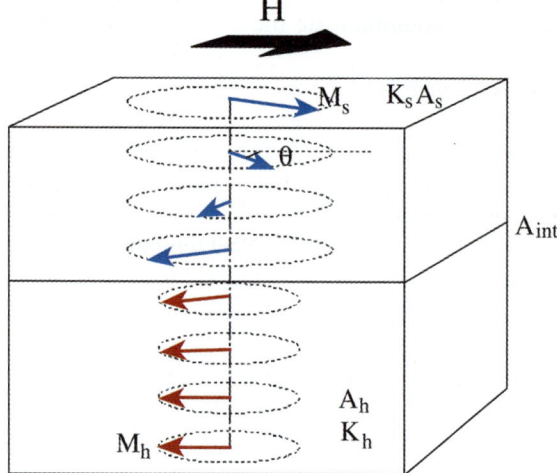

Fig. 7. Schematic of canted magnetic structure of a hard (h)/soft (s) ferromagnetic spring magnet, as in Ref. [34]. The red (blue) arrows denote atomic layer-dependent magnetization for the hard (soft) stack, where θ is a cant angle, and A is the exchange, K the anisotropy, and M the magnetization of the indexed layer. The subscript "int" denotes "interface". The stack is originally magnetized in the direction opposite to that of the applied magnetic field H.

to magnets that are two-to-three times as strong as existing commercial magnets. The commercial material of choice today is NdFeB. Its discovery in 1983 led to notable increases in efficiency and reduction in size and weight of electric motors and devices like audio speakers that rely on permanent magnets. In recent years incremental improvements in the magnetic microstructure of NdFeB continue to be made. But it is anticipated that spring magnets will dramatically impact device development in the future. It remains to be seen if spring magnets become the next great achievement of nanotechnology.

5. Conceptual frontiers

Theoretical and simulational advances are too vast to be inclusively surveyed, so highlights will be chosen and comprehensive reviews cited. Some areas have already been touched upon, such as the 2D Ising model. Onsager's solution of this model serves as one of the cornerstones of modern physics and has been adapted to describe phenomena far removed from the field of magnetism. The Mermin–Wagner theorem proves that thermal fluctuations destroy LRO at finite temperature in 2D isotropic and XY systems. Kosterlitz–Thouless theory describes the orientational order that develops in 2D XY systems. An infinitesimal anisotropy is all that is needed, however, to trigger LRO at finite temperature. Pokrovsky's recent review deals with 2D magnetic phase transitions [24].

5.1. 2D spin reorientation transition

Surface magnetic anisotropies can now be described from first principles using relativistic band structure calculations [35]. The perpendicular surface anisotropy can compete with the shape anisotropy to give rise to the 2D *spin reorientation* transition, as already mentioned. In the vicinity of this transition, as the dominant anisotropies tend to cancel, a rich magnetic phase diagram is predicted with intermediate phases that possess orientational order, analogous to that found in liquid crystals [24]. These magnetic phases are just starting to be uncovered experimentally; much careful experimental work remains to be done to unravel this gem of a problem with its rich interdisciplinary overtones.

5.2. Surface spin-flop transition

A fascinating transition that was predicted theoretically decades ago but only uncovered experimentally in recent years is the *surface spin-flop* transition [36]. By way of contrast, the bulk spin-flop transition is well known to occur in AFs in the presence of a magnetic field. The requirement for its occurrence is that the AF has a uniaxial anisotropy and that the magnetic field is applied along this direction. As the field is increased, once it exceeds the anisotropy that holds the sub-lattice magnetizations in place, the magnetizations flop by 90° and then also cant somewhat in the direction of the field. (For a quick lesson in *flip-flop* terminology: a spin *flip* is a 180° event, while a spin *flop* is 90° minus the cant angle.) Consider now AFs with spin structures that consist of ferromagnetic sheets that alternate in direction from one layer to the next. In this case the surface layer is less strongly coupled than the bulk because it is only coupled to one sheet—the one below it. The theoretical prediction was that the surface would flop at a lower field than the bulk. But for years the prediction was left to languish because of the need for atomically flat surfaces and magnetic probes with atomic sensitivity. The breakthrough involved a clever utilization of antiferromagnetic Fe/Cr(2 1 1) GMR superlattices that possess uniaxial, in-plane magnetic anisotropy. Here atomic scale roughness can be tolerated because the magnetic length scale again is the Fe/Cr bilayer thickness, which is many atomic layers thick, as in the double superlattice exchange bias structures discussed above.

A very interesting result is that superlattices with even and odd numbers of bilayers behave differently with respect to the surface spin-flop transition. For even numbers of bilayers, the top and bottom Fe layers are magnetically inequivalent, while for the odd case, these two terminal Fe layer magnetizations point in the same direction. What happens is that for *odd*-layered stacks, the

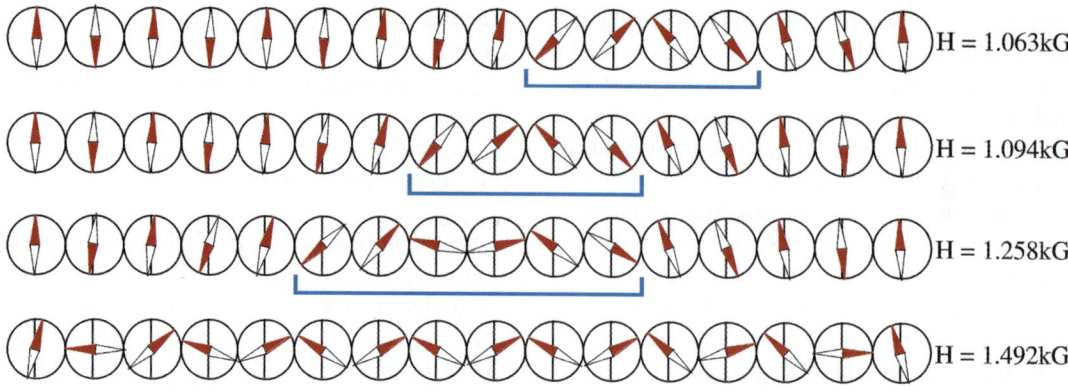

Fig. 8. Schematic spin structure for the inhomogeneous propagation of the surface spin flop transition as the magnetic field increases, based on Ref. [36]. The compass arrows denote the layer-dependent magnetization direction of Fe layers within an antiferromagnetically coupled Fe/Cr(2 1 1) superlattice with an even number of Fe layers within the stack (so that the terminal Fe layers are magnetically inequivalent). The vertical lines indicate the easy axis (zero-field magnetization axis). In the top panel, the field denoted for the simulation is just above the surface spin-flop transition; in subsequent panels the front (bracketed in blue) moves into the interior from the near surface region until, in the bottom panel, the bulk spin-flop field is just exceeded.

spin flop transition is homogeneous throughout the stack, but the presence of the surface causes it to occur at a lower field than the bulk spin flop transition. However, for *even*-layered stacks the transition is inhomogeneous. What this means is that the near-surface region of one side of the stack (either the top or bottom) flops at a lower field than the bulk. Then the transition front progressively moves into the stack as the field is raised, as shown schematically in Fig. 8, until the bulk is fully transformed at the bulk flop field. In each case, for even or odd number of bilayers, the flop is followed by a gradual rotation of the magnetization to a state of parallel spin alignment as the field increases toward saturation (as in the closing of scissors). Operationally the transition is subtle enough that it is more readily detected via magnetic susceptibility measurements (the field derivative of the magnetization) rather than in the magnetization itself. Another point is that if the stack is thicker than an optical penetration depth (a few hundred Ångstrom), then there is an interesting asymmetry in the hysteresis when measured magneto-optically. In one field-sweep direction both surface and bulk flops are discernable at their characteristic field values, but in the other sweep direction only the bulk transition is detected. This is because in the latter case, the surface transition is seeded from the buried surface and the light does not get to it. There is still great interest and controversy to fully understand the phase diagram and dynamics of this elementary phase transition [37].

5.3. Micromagnetics

Another area of current interest is in micromagnetics. This represents classical physics but because of the quest for smaller bits in magnetic recording and smaller device structures in general, the domain pattern and stability and switching mechanisms are of central importance [38]. Great progress is being made to simulate stable feature sizes and shapes, and to image them and compare to the simulations. This also requires an understanding of dynamical effects because stray fields generated during the switching process can be large enough to irreversibly alter the domain structures [39]. Remember, in device applications one is striving for a domain structure that can withstand more than 10^{10} reversal cycles. The art of designing stable domain structures is rapidly becoming a science due to the advent of high-performance computer simulations. There are a number of valuable sources of both basic and hands-on information about modern micromagnetics. The reader is urged to examine the book by Aharoni [40] and visit a full-service website [41] that even provides standard problems and codes.

6. Emerging opportunities in nanomagnetism and spintronics

Throughout this article the area of laterally confined structures has been highlighted as a major new direction. Activity is already well underway in many areas. These include the definition of small structures using photolithographies and electron and focused ion beam techniques. Another approach involves electrochemical filling of nonmagnetic nanopores [42], i.e. polymers, buckytubes, Si or alumina honeycombs, etc. with magnetic atoms. Also, a variety of self-assembly approaches are being formulated and pursued, from the decoration of step edges to form quantum stripes [43] and even lateral superlattices, to the organic synthesis of nanoarrays utilizing diblock copolymer building blocks [44]. The field of magnetic nanostructures is fully embracing the breadth of metallurgical, chemical and physical creativity. Most elegant in the regime of artificial structuring is the atom-by-atom assembly made possible by STM manipulation of individual atoms to form quantum corrals and, most recently, *phantom ellipse* structures that utilize concepts in focusing acoustics and wave mechanics at the ultimate limits of miniaturization [45]. Scanning probes are also being used as *quantum quills*, or dipping pens, to write small structures with organic ink [46]. Small structures are also being stamped or embossed onto substrates. Innovative synthesis is leading the way to new generations of magnetic nanostructures.

6.1. Molecular magnets

Parallel to this development is the exploration of the field of *molecular magnets* [47,48]. Here bulk crystals are probed but the magnetism is largely confined within discrete molecular entities within the unit cell. The most familiar example is of Mn_{12} acetate, which exhibits quantum tunneling at low temperatures within the Mn_{12} clusters. The tunneling signature consists of discrete steps in the magnetic hysteresis curve. Such large-scale molecular structures might be utilized in emerging paradigms for quantum computing. The role of magnetism has recently been outlined in an intriguing review [49]. That naturally occurring bulk materials can exhibit low-dimensional behavior calls to mind a rich array of potential systems to explore for nanoscale phenomena, from spinoidal decompositions in metallurgy to nanophase separation in the colossal magnetoresistive (CMR) manganites.

6.2. Naturally layered manganites

The manganites offer a richness of competing physical effects that give rise to an intriguing metal–insulator transition at the Curie temperature [50]. Just above T_C an external magnetic field can coax the material from the insulating-to-metallic state, hence the CMR effect. The lower the T_C value, the larger the effect because the resistivities of metallic and insulating phases diverge from each other on cooling. The manganites are being considered for possible future applications [51]. Presently they provide an important intellectual bridge in unlocking the secrets of complex oxides that extend to those of the high-temperature cuprate superconductors [52]. The role of characterizing and understanding short-range order, including on the nanoscale, is fundamental here. Within the manganite CMR family of materials are the naturally layered manganites [53]. These structures are related to layered cuprate and ruthanate materials. Their attraction is that they introduce an extreme anisotropy, if not a lower dimensionality that alters the various competing physical effects and causes dramatically different behavior within the *ab* plane as compared to across the *c*-axis. The structures belong to a Ruddlesdon–Popper series and posses manganite octahedra separated by SrO rock-salt blocking layers. A representative chemical formula is $SrO(La_{1-x}Sr_xMnO_3)_n$, where x is the Sr doping level, and n is the number of layers. (see Fig. 9). Thus, there could be tunneling conductance across the *c*-axis, while within the *ab* plane the conduction is

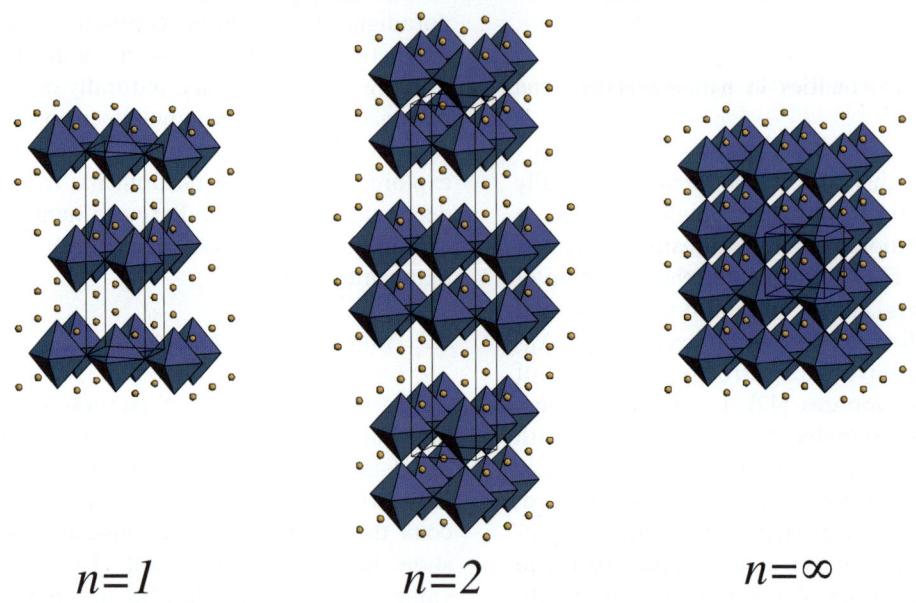

Fig. 9. Schematic of the Ruddlesden–Popper series for manganite oxides doped with Sr, as in Ref. [53]. This structure supports colossal magnetoresistance (CMR) for a certain doping range. The structure on the right represents a bulk (infinite) perovskite while the other two represent naturally layered manganites with $n = 1$ and 2 perovskite layers separated by SrO rock salt layers. The perovskite layers consist of corner-sharing oxygen octahedra (purple) with Mn ions inside. Actual crystallographic distortions of the octahedra are not shown. The La(Sr) ion positions are indicated by small yellow circles. A unit cell is indicated by black line drawings in each case. The chemical formula for the series is denoted at the top in two different ways.

analogous to that of the bulk perovskite manganite structures. The naturally layered manganites should remain a favored playground for the study of competing spin, charge and orbital effects in complex oxides.

7. Summary

The field of low dimensional magnetism is vibrantly alive and spawning new research directions in the exploration of basic and applied phenomena. This article attempts to catalogue some of these developments and point the motivated reader toward more comprehensive sources of information. A major resource utilized in the reference listing is the special issue of the *Journal of Magnetism and Magnetic Materials*, volume 200, entitled "Magnetism Beyond 2000". The reader is encouraged to peruse the 46 review articles in that volume. While the future of technology remains unpredictable, the depth of challenge should provide gratifying intellectual stimulation to a new generation of researchers. This is so because as the physical limits of miniaturization are approached, the role of surface science should continue to be heightened.

Acknowledgements

I thank the many close colleagues with whom I have had the honor to work in this field. This work was supported by the US Department of Energy, Division of Basic Energy Sciences—Material Sciences under contract no. W-31-109-ENG-38.

References

[1] L.N. Leiberman, D.R. Fredkin, H.B. Shore, Two dimensional "ferromagnetism" in iron, Phys. Rev. Lett. 22 (1969) 539; L.N. Leiberman, J. Clinton, D.M. Edwards, J. Mathon, "Dead" layers in ferromagnetic transition metals, Phys. Rev. Lett. 25 (1970) 232.

[2] J.B. Kortright, D.D. Awschalom, J. Stöhr, S.D. Bader, Y.U. Idzerda, S.S.P. Parkin, I.K. Schuller, H.-C. Siegmann, Research frontiers in magnetic materials at soft X-ray synchrotron radiation facilities, J. Magn. Magn. Mater. 207 (1999) 7.

[3] P.D. Johnson, A. Clarke, N.B. Brooks, S.L. Hulbert, B. Sinkovic, N.V. Smith, Exchange-split adsorbate bands: the role of substrate hybridization, Phys. Rev. Lett. 61 (1988) 2257.

[4] P.W. Palmberg, R.E. DeWames, L.A. Vredevoe, Direct observation of coherent exchange scattering by low-energy electron diffraction from antiferromagnetic NiO, Phys. Rev. Lett. 21 (1968) 682.

[5] J. Stöhr, A. Scholl, T.J. Regan, S. Anders, J. Lüning, M.R. Scheinfein, H.A. Padmore, R.L. White, Images of the antiferromagnetic structure of a NiO(100) surface by means of X-ray magnetic linear dichroism spectromicroscopy, Phys. Rev. Lett. 83 (1999) 1862.

[6] F. Nolting, A. Scholl, J. Stöhr, J.W. Seo, J. Fompeyrine, H. Siegwart, J.-P. Locquet, S. Anders, J. Lüning, E.E. Fullerton, M.F. Toney, M.R. Scheinfein, H.A. Padmore, Direct observation of the alignment of ferromagnetic spins by antiferromagnetic spins, Nature 405 (2000) 767.

[7] M.N. Baibich, J.M. Broto, A. Fert, F.N. Van Dau, F. Petroff, P. Etienne, G. Creuzet, A. Friederich, J. Chazelas, Giant magnetoresistance of (001)Fe/(001)Cr magnetic superlattices, Phys. Rev. Lett. 61 (1988) 2472.

[8] M. Madou, Fundamentals of Microfabrication, CRC Press, Boca Raton, 1997.

[9] G.A. Prinz, Magnetoelectronics applications, J. Magn. Magn. Mater. 200 (1999) 57.

[10] E.N. Abarra, A. Inomata, H. Sato, I. Okamoto, Y. Mizoshita, Longitudinal magnetic recording media with thermal stabilization layers, Appl. Phys. Lett. 77 (2000) 2581.

[11] D.T. Pierce, J. Unguris, R.J. Celotta, M.D. Stiles, Effect of roughness, frustration, and antiferromagnetic order on magnetic coupling of Fe/Cr multilayers, J. Magn. Magn. Mater. 200 (1999) 290.

[12] M.D. Stiles, Interlayer exchange coupling, J. Magn. Magn. Mater. 200 (1999) 322.

[13] F.J. Himpsel, K.N. Altmann, G.J. Mankey, J.E. Ortega, D.Y. Petrovykh, Electronic states in magnetic nanostructures, J. Magn. Magn. Mater. 200 (1999) 456.

[14] E.R. Moog, S.D. Bader, SMOKE signals from ferromagnetic monolayers: p(1 × 1) Fe/Au(100), Superlatt. Microstruct. 1 (1985) 543.

[15] Z.Q. Qiu, S.D. Bader, Surface magneto-optic Kerr effect (SMOKE), J. Magn. Magn. Mater. 200 (1999) 664.

[16] K.H. Bennemann, Theory for nonlinear magnetooptics in metals, J. Magn. Magn. Mater. 200 (1999) 679.

[17] I. Guedes, N. Zaluzek, M. Grimsditch, V. Metlushko, P. Vavassori, B. Ilic, P. Neuzil, R. Kumar, Magnetization of negative magnetic arrays: elliptical holes on a square lattice, Phys. Rev. B 62 (2000) 11719.

[18] J. Stöhr, Exploring the microscopic origin of magnetic anisotropies with X-ray magnetic circular dichroism (XMCD) spectroscopy, J. Magn. Magn. Mater. 200 (1999) 470.

[19] H.-J. Elmers, Ferromagnetic monolayers, Int. J. Mod. Phys. B 9 (1995) 3115.

[20] D. Li, M. Freitag, J. Pearson, Z.Q. Qiu, S.D. Bader, Magnetic phases of ultrathin Fe grown on Cu(100) as epitaxial wedges, Phys. Rev. Lett. 72 (1994) 3112.

[21] R.E. Camley, D. Li, Theoretical calculation of magnetic properties of ultrathin Fe films on Cu(100), Phys. Rev. Lett. 84 (2000) 4709.

[22] S.D. Bader, D. Li, Z.Q. Qiu, Magnetic and structural instabilities of ultrathin Fe(100) wedges, J. Appl. Phys. 76 (1994) 6425.

[23] H.J. Choi, R.K. Kawakami, E.J. Escorcia-Aparicio, Z.Q. Qiu, J. Pearson, J.S. Jiang, D. Li, S.D. Bader, Curie temperature enhancement and induced Pd magnetic moments for ultrathin Fe films grown on stepped Pd(001), Phys. Rev. Lett. 82 (1999) 1947.

[24] V.L. Pokrovsky, Two-dimensional magnetic phase transitions, J. Magn. Magn. Mater. 200 (1999) 515.

[25] S.T. Bramwell, P.C.W. Holdsworth, Magnetization and universal sub-critical behaviour in two-dimensional XY magnets, J. Phys. Condens. Matter 5 (1993) L53.

[26] J.S. Moodera, G. Mathon, Spin polarized tunneling in ferromagnetic junctions, J. Magn. Magn. Mater. 200 (1999) 248.

[27] D.D. Awschalom, N. Samarth, Spin dynamics and quantum transport in magnetic semiconductor quantum structures, J. Magn. Magn. Mater. 200 (1999) 130.

[28] C.L. Chien, D.H. Reich, Proximity effects in superconducting-magnetic multilayers, J. Magn. Magn. Mater. 200 (1999) 83.

[29] C.A.R. Sá de Melo, Magnetic exchange coupling in ferromagnet/superconductor/ferromagnet multilayers, Phys. Rev. Lett. 79 (1997) 1933.

[30] A.E. Berkowitz, K. Takano, Exchange anisotropy, J. Magn. Magn. Mater. 200 (1999) 552.

[31] J. Nogués, I.K. Schuller, Exchange Bias, J. Magn. Magn. Mater. 192 (1999) 203.

[32] J.S. Jiang, G.P. Felcher, A. Inomata, R. Goyette, C. Nelson, S.D. Bader, Exchange-bias effect in Fe/Cr(211) double superlattice structures, Phys. Rev. B 61 (2000) 9653.

[33] W.H. Meiklejohn, C.P. Bean, New magnetic anisotropy, Phys. Rev. 105 (1957) 904.

[34] E.E. Fullerton, J.S. Jiang, S.D. Bader, Hard/soft magnetic heterostructures: model exchange-spring magnets, J. Magn. Magn. Mater. 200 (1999) 392.

[35] R. Wu, A.J. Freeman, Spin–orbit induced magnetic phenomena in bulk metals and their surfaces and interfaces, J. Magn. Magn. Mater. 200 (1999) 498.

[36] R.W. Wang, D.L. Mills, E.E. Fullerton, J.E. Mattson, S.D. Bader, Surface spin-flop transition in Fe/Cr(211) superlattices; experiment and theory, Phys. Rev. Lett. 72 (1994) 920.

[37] C. Micheletti, R.B. Griffiths, J.M. Yeomans, Surface spin-flop and discommensuration transitions in antiferromagnets, Phys. Rev. B 59 (1999) 6239.

[38] D. Weller, A. Moser, Thermal effect limits in ultrahigh-density magnetic recording, IEEE Trans. Magn. 35 (1999) 4423.

[39] M.R. McCartney, R.E. Dunin-Borkowski, M.R. Scheinfein, D.J. Smith, S. Gider, S.S.P. Parkin, Origin of magnetization decay in spin-dependent tunnel junctions, Science 286 (1999) 1337.

[40] A. Aharoni, Introduction to the Theory of Ferromagnetism, Clarendon Press, Oxford, 1996.

[41] http://www.ctcms.nist.gov/~rdm/mumag.org.html.

[42] A. Fert, L. Piraux, Magnetic nanowires, J. Magn. Magn. Mater. 200 (1999) 338.

[43] F.J. Himpsel, J.E. Ortega, G.J. Mankey, R.F. Willis, Magnetic nanostructures, Adv. Phys. 47 (1998) 511.

[44] F.S. Bates, G.H. Fredrickson, Block copolymers—designer soft materials, Phys. Today 52 (2) (1999) 32; M.W. Matsen, F.S. Bates, Block copolymer microstructures in the intermediate-segregation regime, J. Chem. Phys. 106 (1997) 2436.

[45] H.C. Manoharan, C.P. Lutz, D.M. Eigler, Quantum mirages formed by coherent projection of electronic structure, Nature 403 (2000) 512.

[46] R.D. Piner, J. Zhu, F. Xu, S. Hong, C.A. Mirkin, "Dip Pen" nanolithography, Science 283 (1999) 661.

[47] A. Caneschi, D. Gatteschi, C. Sangregorio, R. Sessoli, L. Sorace, A. Cornia, M.A. Novak, C. Paulsen, W. Wernsdorfer, The molecular approach to nanoscale magnetism, J. Magn. Magn. Mater. 200 (1999) 182.

[48] B. Barbara, L. Thomas, F. Lionti, I. Chiorescu, A. Sulpice, Macroscopic quantum tunneling in molecular magnets, J. Magn. Magn. Mater. 200 (1999) 167.

[49] D.P. DiVincenzo, D. Loss, Quantum computers and quantum coherence, J. Magn. Magn. Mater. 200 (1999) 202.

[50] Y. Tokura, Y. Tomioka, Colossal magnetoresistive manganites, J. Magn. Magn. Mater. 200 (1999) 1.

[51] A. Gupta, J.Z. Sun, Spin-polarized transport and magnetoresistance in magnetic oxides, J. Magn. Magn. Mater. 200 (1999) 24.

[52] A.M. Goldman, V. Vas'ko, P. Kraus, K. Nikolaev, V.A. Larkin, Cuprate/manganite heterostructures, J. Magn. Magn. Mater. 200 (1999) 69.

[53] S.D. Bader, D.J. Miller, J.F. Mitchell, R.M. Osgood III, J.S. Jiang, Role of intergrowths in the properties of naturally layered manganite single crystals, J. Appl. Phys. 83 (1998) 6385.

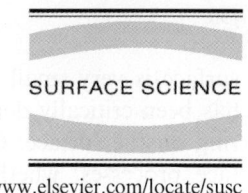

Molecular beam epitaxy

John R. Arthur

Department of Electrical and Computer Engineering, Oregon State University, Corvallis, OR 97331, USA

Received 6 September 2000; accepted for publication 16 March 2001

Abstract

Molecular beam epitaxy (MBE) is a process for growing thin, epitaxial films of a wide variety of materials, ranging from oxides to semiconductors to metals. It was first applied to the growth of compound semiconductors. That is still the most common usage, in large part because of the high technological value of such materials to the electronics industry. In this process beams of atoms or molecules in an ultra-high vacuum environment are incident upon a heated crystal that has previously been processed to produce a nearly atomically clean surface. The arriving constituent atoms form a crystalline layer in registry with the substrate, i.e., an epitaxial film. These films are remarkable because the composition can be rapidly changed, producing crystalline interfaces that are almost atomically abrupt. Thus, it has been possible to produce a large range of unique structures, including quantum well devices, superlattices, lasers, etc., all of which benefit from the precise control of composition during growth. Because of the cleanliness of the growth environment and because of the precise control over composition, MBE structures closely approximate the idealized models used in solid state theory.

This discussion is intended as an introduction to the concept and the experimental procedures used in MBE growth. The refinement of experimental procedures has been the key to the successful fabrication of electronically significant devices, which in turn has generated the widespread interest in the MBE as a research tool. MBE experiments have provided a wealth of new information bearing on the general mechanisms involved in epitaxial growth, since many of the phenomena initially observed during MBE have since been repeated using other crystal growth processes. We also summarize the general types of layered structures that have contributed to the rapid expansion of interest in MBE and its various offshoots. Finally we consider some of the problems that remain in the growth of heteroepitaxial structures, specifically, the problem of mismatch in lattice constant between layers and between layer and substrate. The discussion is phenomenological, not theoretical; MBE has been primarily an experimental approach based on simple concepts. © 2001 Elsevier Science B.V. All rights reserved.

Keywords: Models of surface kinetics; Molecular beam epitaxy; Reflection high-energy electron diffraction (RHEED); Epitaxy; Single crystal surfaces; Heterojunctions; Quantum wells; Superlattices

1. Introduction

As scientists have learned how the properties of materials depend on their microstructure, there have been ever increasing efforts to design the structure to produce the desired behavior. While the fraction of the atoms in a solid that lie on the

E-mail address: jread@proaxis.com (J.R. Arthur).

surface is very small, the formation of that solid has been critically dependent on processes occurring on the surface. Crystal growth involves surface processes, whether it takes place from the vapor or from a melt. The better we can understand these processes, the more we can control the manner in which the solid is formed and its subsequent properties.

The invention of the transistor and the beginning of the computer age had an enormous impact on the science of materials. The new semiconducting devices were critically dependent on the availability of very perfect and extremely pure semiconductor crystals. The economic importance of the semiconductor revolution quickly stimulated researchers to develop a variety of new methods of growing crystals in order to produce the purity and perfection demanded by the new devices. One approach was to use a slice of semiconductor as the seed on which to deposit additional material in the form of a thin film in order to obtain electrical properties in the film that were superior to those of the starting substrate material. If the film had a crystalline structure that was ordered with respect to that of the underlying substrate, it was described as "epitaxial". Epitaxial films could be grown on a substrate of the same material, in which case the film was "homoepitaxial"; alternatively, if grown on a substrate of a different material, the film was "heteroepitaxial".

Epitaxial films of semiconductors have continued to play a major role in device processing because they can be produced with electrical properties different from those of the substrate, either higher purity, or fewer defects or with a different concentration of electrically active impurities as desired. By depositing a sequence of epitaxial layers with specific properties, specialized device structures can be realized without the need for processing steps involving the diffusion of impurities to produce doped layers.

Many semiconductor materials can be grown epitaxially by allowing a suitable mixture of gaseous vapors containing the constituent elements to react with a heated seed or substrate crystal, a process known as vapor phase epitaxy (VPE). Alternatively, placing a seed crystal wafer in contact with a liquid solution saturated with the semiconductor constituents can be used to grow an epitaxial layer by liquid phase epitaxy (LPE) as the solution is very slowly cooled. Each of these methods has advantages, e.g. VPE is a relatively rapid method of film growth which is readily scaled to manufacturing volume, and LPE produces relatively pure films; however each has disadvantages as well, e.g. VPE takes place at relatively high temperatures which can enhance bulk diffusion, and LPE does not produce films of uniform thickness.

Recently I listened to a radio program on the Public Broadcasting Network in which a group of experts presented brief statements describing their work in "Nanotechnology". While the presentations left me a bit impatient because of the heavy emphasis on the more flamboyant future possibilities that research may provide, there was some brief, but interesting, discussion of the impact on materials that could be fabricated "one atom at a time". It struck me that for more than thirty years, some of us have been doing this, in one dimension at least, by the process known as molecular beam epitaxy (MBE). And in fact, the resulting materials have indeed opened new doors in physics, chemistry, material science and electronics.

MBE, as the name suggests, uses localized beams of atoms or molecules in an ultra-high vacuum (UHV) environment to provide a source of the constituents to the growing surface of a substrate crystal. The beams impinge on the crystal kept at a moderately elevated temperature that provides sufficient thermal energy to the arriving atoms for them to migrate over the surface to lattice sites. The UHV environment minimizes contamination of the growing surface. In the UHV environment, the beam atoms and molecules travel in nearly collision-free paths until arriving either at the substrate or else at chilled walls of the chamber where they condense and are thus effectively removed from the system. When a shutter is interposed in a beam, that beam is effectively turned off almost instantly. These features make it possible to grow the films very slowly without contamination, and, most importantly, to change the composition of the arriving atom stream very abruptly; in fact, the composition of the flux can be changed in times much shorter than that needed to grow a single atom layer of the film. Very simplistically, MBE growth might be likened to "spray painting" the substrate crystal with layers of atoms, changing the composition or impurity level in each layer until a desired structure is obtained. In this sense MBE is nearly the ideal approach to material preparation since the composition can be tailored, layer-by-layer. We shall see, however, that much more is involved, and that unraveling the details of growth has added much to our understanding of surface processes. In fact, it may be argued that the greatest value of MBE is the insight into crystal growth that it continues to provide.

2. Historical background

The growth of semiconductor thin films from the vapor has a long history; however prior to the 1970s these films were not structurally equivalent to bulk material and thus were of little use from a device standpoint. One particular problem for compound semiconductor films had to do with the very different vapor pressures of the pure constituents, as much as two orders of magnitude for Ga and As at temperatures useful for film growth. Thermal evaporation of separate As and Ga sources would require impossibly precise temperature control to produce equal vapor pressures, and thus equal arrival rates of the constituent atoms at the substrate. Collins et al. [1] used two crucibles containing Ga and Sb to evaporate films of GaSb onto a glass substrate; their method was based on the concept that due to the angular decrease in the flux of each element away from the centerline of its crucible, there would be a location somewhere between the two crucibles where the flux ratio would produce a stoichiometric film. The resulting films were unfortunately highly polycrystalline, so that little could be determined about actual composition. Earlier, Günther attempted to provide the proper vapor ratio by separate control of the temperatures of nonmetal, metal and substrate [2]. Again, the resulting films were not well ordered and so were of small interest to the device engineers.

The reader today may find it difficult to realize how difficult it was to obtain information about the condition of the substrate, the composition of the vacuum, and the crystallinity of films grown in those early days (1950s–1960s). Usually it was necessary to carry out post-growth electron diffraction studies to determine whether the films were crystalline or not, and often the results took several days to obtain—long after the growth procedure had been all but forgotten! However a revolution in surface analysis took place in the late 60s with the introduction of small mass spectrometers, Auger electron spectroscopy and compact electron diffraction equipment. In fact, the discovery of the MBE process came about not with the intent of finding a new method of crystal growth, but rather as a study of surface–vapor interactions with a new, compact mass spectrometer [3]. In 1968, John LePore and I were studying the reflection of pulsed molecular beams of Ga and As_2 from GaAs surfaces in UHV in order to measure the energy of adsorption. It became clear that the presence of a layer of Ga greatly increased the bond strength of As adsorbed on the Ga compared to the energy of As on GaAs. This suggested that the vapor pressure/arrival rate of different species might not be controlling the composition, since excess As appeared to be quickly desorbed from the clean surface at temperatures above 300 °C. In fact, simply by maintaining a slight overpressure of As and/or P to insure complete reaction of Ga, we were able to obtain epitaxial growth of GaAs and GaP at around 500 °C [4]. This temperature was significantly lower than was needed for VPE, and thus MBE seemed to hold some promise as a low temperature alternative to the current epitaxial techniques.

In 1969 A.Y. Cho published landmark results reporting the first in situ observations of the MBE growth process using high energy electron diffraction [5]. This structural analysis capability proved to be crucial for characterizing MBE epitaxy because it provided an instantaneous feedback on the influence of growth conditions on film structure. Cho demonstrated that MBE growth could produce atomically flat, ordered layers; thus these studies marked the beginning of the use of MBE for practical device fabrication. Cho went on to publish a number of key papers during this period among which he demonstrated how to incorporate doping impurities [6], and ultimately, how to fabricate GaAs–Al$_x$Ga$_{1-x}$As laser structures, which at the time was considered the crucial test for a III–V materials process [7].

There were other important advances that quickly followed. In 1973 Chang et al. [8] reported on the growth of a superlattice structure consisting of alternating layers of GaAs and AlGaAs. Later, they were able to observe evidence in the transverse conductance of these structures that indicated the resonant tunneling between the GaAs layers predicted by their theory [9]. At the same time, Dingle et al. observed structure in the optical absorption spectrum of superlattice structures associated with quantized energy levels due to carrier confinement in one dimension [10]. Thus in the short period of about five years, MBE moved from being a novel but very uncertain film growth method to becoming an established research tool in which the general directions for future work were well established. Since then MBE research has escalated rapidly, with the publication of numerous books, monographs and conference proceedings that summarize the extensive body of work [11–16].

3. Experimental methods

MBE is an experimental approach to epitaxial film growth, which has emphasized including the modern tools of surface analysis to obtain a real time analysis of the surface and its environment. It has also been a very demanding art in which the economic stakes have been high. As a result there has been a great deal of experimental innovation that frequently has been of high value in other disciplines. In this discussion of the experimental aspects of MBE we will discuss the working components of MBE systems and indicate their evolution as the details of the growth process have become clearer from the happy union of surface analysis with crystal growth.

Fig. 1 shows a schematic front view of a basic MBE growth chamber. A thin, crystalline substrate wafer is mounted on a heater block such that it can be brought to face the source ovens used to evaporate the constituent atoms or molecules. Mechanical shutters driven from outside the vacuum chamber are used to switch the beams on and off. Because of the extensive use of chilled walls surrounding the source ovens and the substrate, the beams make essentially a single pass through the chamber before condensing on the cold walls, and the background pressure in the system remains very low. This preserves the purity of the growing film and at the same time allows the reflection high energy electron diffraction (RHEED) gun to operate without damage from corrosive reaction with residual vapors. The RHEED system provides a diffraction pattern on a phosphor-coated window that is indicative of the ordering of the substrate surface. Thus the observer can immediately see the effect on film crystallinity due to changes in the growth conditions, e.g., exposing the surface to the source beams or changing beam intensity or substrate temperature. This is the essential MBE system, very similar to the ones in use in the 70s. For many research applications this type of system is completely adequate, for example, where relatively small substrates are used and where sample throughput is not a big issue. However, for modern semiconductor device fabrication, throughput is very important and large substrate wafers are used so that modifications must be made to improve deposition uniformity and to minimize the downtime.

3.1. Vacuum considerations

The essence of the MBE concept is that the growth surface is kept clean by the UHV; thus the vacuum environment surrounding the growing crystal must be kept as low as possible to avoid contamination that might affect electrical properties, film morphology and even whether or not epitaxial growth takes place. To put this into perspective, consider that the number n of gas atoms impinging on unit area of surface in unit time is

$$\frac{dn}{dt} = \frac{P}{\sqrt{2\pi mkT}} \text{ cm}^{-2}\text{s}^{-1} \qquad (1)$$

P is the gas pressure, m is the atomic mass, k is the Boltzmann constant and T is the absolute

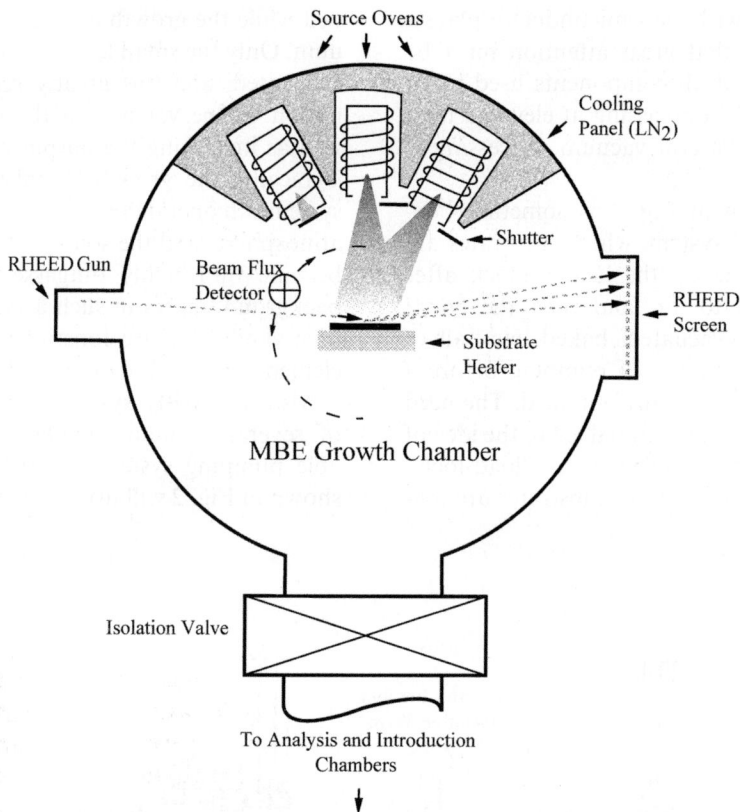

Fig. 1. Top view of a simple MBE chamber showing the essential growth sources, shutters, beam flux detector and the RHEED system for monitoring structure during growth.

temperature. For example, if the vacuum pressure P is measured in Torr and if the atomic mass m is converted to molecular weight M in g, then Eq. (1) becomes

$$\frac{dn}{dt} = 3.5 \times 10^{22} \frac{P}{\sqrt{MT}} \text{ cm}^{-2}\text{s}^{-1} \quad (1a)$$

For a typical residual gas molecular weight of about 40 g and a temperature of 25 °C, the rate of gas arrival is $3.2 \times 10^{20} P$ (Torr), and for a pressure of 10^{-6} Torr the arrival rate is 3.2×10^{14} cm^{-2} s^{-1}. The number of atoms in a cm^2 of the cube face of Si is also about 3.2×10^{14}, thus the arrival rate in a second at 10^{-6} Torr is nearly equal to the number of atoms in a cm^2 of surface. Hence in one second a complete monolayer of residual gas arrives at the surface, (although all the arriving atoms may not react or "stick" on the surface). The sobering implication of Eq. (1) is that even at a good UHV of 10^{-10} Torr, a clean surface will become badly contaminated with *reactive* background gas in just a few hours. We emphasize "reactive" because, fortunately, many semiconductor surfaces are relatively unreactive with common background gases, and thus can be preserved in the clean state significantly longer.

However, the purity constraint for MBE growth is even more severe. If it is desirable to maintain the background impurity level at, say, one part per million, then clearly the contamination of each surface layer must be kept at that level. MBE growth is relatively slow, typically about 1 ML/s, thus to keep the arrival rate of background species at one part per million would require a pressure of 10^{-12} Torr! For many specialized semiconductor devices, the impurity level must be much less than 10^{-6}, thus it is perhaps surprising that such low levels can be obtained. While the slow reaction of many residual gases with the semiconductor plays a part, it is also true that great attention must be given to the design of all components used in an MBE system to avoid outgassing at elevated temperatures, and that careful vacuum processing is critical.

The system shown in Fig. 1 is sometimes described as a "batch" system, where substrates are loaded one at a time on the heater block after opening the system to air, and then the entire growth chamber is evacuated, baked, etc., all of which requires many hours of pumping before a suitable vacuum level is finally obtained. The need for greater throughput of material led to the idea of introducing substrates through a "load-lock" chamber in which one or more substrates are loaded while the growth chamber remains under vacuum. Only the small loading chamber need then be evacuated, and this greatly reduces the contamination of the vacuum in the growth chamber as well as increasing the output of processed wafers. Typically, the growth chamber in a load-locked system can operate for months without exposure to atmosphere, and the source ovens, shutters, etc. all become thoroughly outgassed; the predominant background gases in such a system are simply the high vapor pressure constituents, e.g. the group V elements in the case of III–V MBE.

Modern MBE systems will normally consist of several vacuum chambers, each with a suitable pumping system. A fairly typical system as shown in Fig. 2 will usually consist of at least three

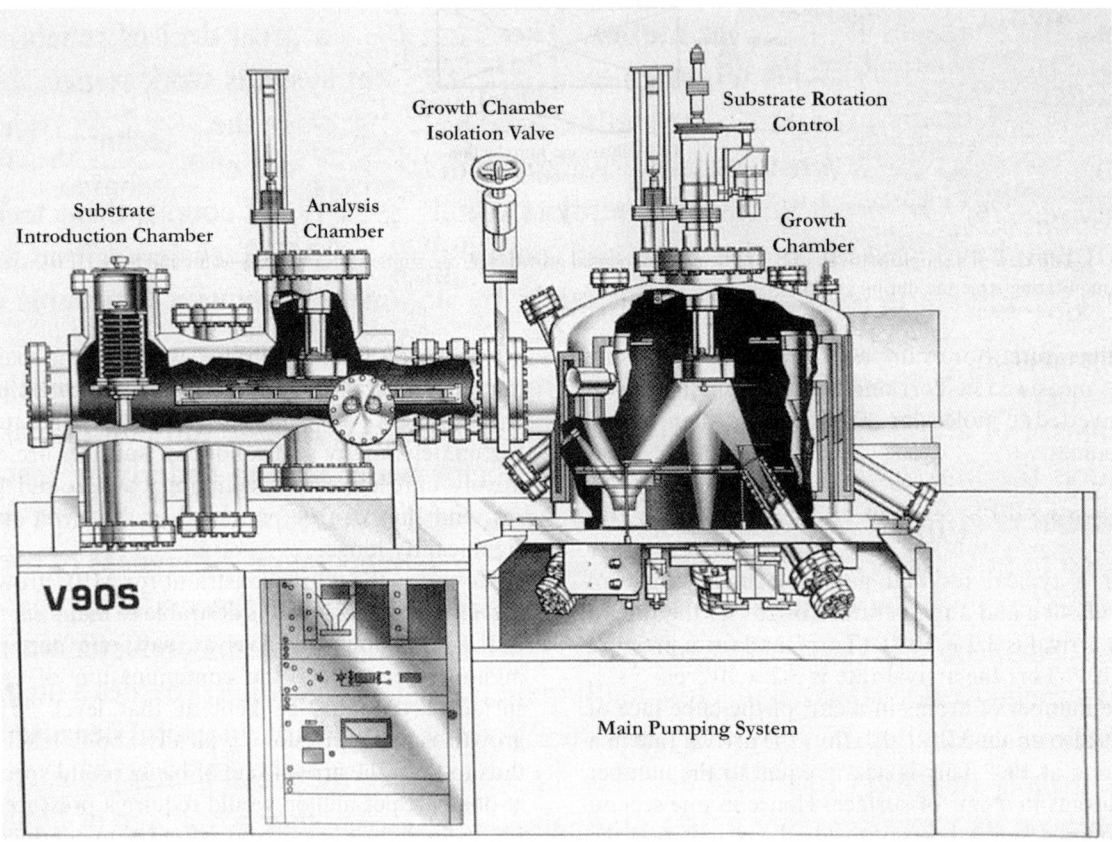

Fig. 2. Side cutaway view of a commercial MBE system with a substrate introduction chamber on the left, analysis chamber, and growth chamber on the right. Substrate transport occurs on trolley connecting the chambers (courtesy of Vacuum Generators, Ltd.).

independently pumped chambers, although there can be many variations on the general theme. A vacuum interlock or introduction chamber is used to allow substrates to be installed on a transport device in a small volume which is pumped down prior to opening the valve to the rest of the system to reduce the amount of air load introduced into the main vacuum. The introduction chamber often will have provision for heating the substrates prior to introduction into the main vacuum system in order to outgas the substrates and their holders. A second chamber is often used for additional substrate preparation and surface characterization, using tools such as AES and/or XPS. An essential feature of MBE from the very beginning has been the presence of surface analysis tools in the systems. The UHV environment and the limited spatial extent of the atom beams make it possible to include some tools, typically electron diffraction and mass spectroscopy, in the immediate growth area to provide a real-time measure of the nature of the growing surface and its environment. In the earlier batch systems an Auger spectrometer was frequently included, but current systems typically include a separate analysis chamber in which to examine the condition of the surface before and after growth. The third chamber is used for the actual growth and usually can be isolated from the rest of the system during both growth and substrate introduction in order to (1) avoid the contamination of the analysis chamber and the surface probes with vapors produced during growth and (2) minimize vacuum contamination of the growth chamber during pressure bursts when the intro chamber is opened to the main vacuum. Additional growth chambers are sometimes included in order to carry out the epitaxy of additional layers where the constituents of the different layers may be incompatible. For example, it is not a good idea to grow II–VI and III–V semiconductors in a single chamber since the components of one are dopants in the other, and there are usually significant residual pressures from the more volatile species remaining after growth.

Load-locked systems require a means for transporting substrates from the loading chamber to the growth chamber, and the development of simple and reliable means for moving substrates and coupling them to holders in UHV has been very useful both to MBE technology and to surface science in general. Sample transport from chamber to chamber has been accomplished in a variety of innovative ways. The requirements are to have means for selecting a wafer from a multiple wafer stack in the introduction chamber, then to move that wafer into another chamber and to dock it securely onto a heater/manipulator, all without mishap or contamination of the wafer by dust particles in the system. The bearings providing smooth transport or rotation must not seize in the UHV environment, and there must be positive control over the sample position, particularly when docking and undocking from the chamber manipulators. These are difficult requirements, and have been met by using magnetically coupled transport rods, by trolley systems with rotary linkage to the external world (as in Fig. 2), or by bellows sealed rods providing extended linear motion. There has been a great deal of reliability testing by MBE researchers, to an extent that present day systems work remarkably well, regardless of the particular method used. The success of these techniques have made it possible to construct large systems with many growth and analysis chambers which can even be used by multiple operators simultaneously. Of course, these techniques have added to the cost and complexity of MBE systems, but the economic consequences of MBE research have been sufficient justification.

Modern production MBE systems are designed to maximize throughput for commercial fabrication of MBE devices. High production rates are achieved by using industry standard, large area substrates, and by growing on multiple wafers simultaneously. This is achieved by mounting the large substrate wafers on a rotating platen that can move under multiple source ovens. This increases the size of the system and the growth chamber in particular, and the larger dimensions then demand increased output from the sources to maintain the flux at the substrates. Obviously there are many additional considerations in the scaling up of such a system, which are beyond the scope of the present discussion.

3.2. Growth chamber details

The growth chamber is where the critical part of the process occurs and it typically contains the following essential components for MBE growth:

1. source ovens,
2. beam shutters and actuating mechanism,
3. substrate heater and sample docking mechanism,
4. in situ growth characterization tools,
5. mass spectrometer and/or separate beam flux monitor,
6. cryopanels to act as cryopumps and to condense unused beam flux.

Normally, an isolation valve is used to close off the growth chamber except when substrates are loaded or removed. All of these components must be designed to minimize the outgassing of impurities, particularly those components that become heated by either the source ovens or by radiation from the substrate. For example, molybdenum metal typically contains a small amount of sulfur impurity, which can evaporate at elevated temperatures; furthermore S is a donor dopant in III–V materials. Thus Mo is undesirable for use where there is a possibility of it getting too hot. Ta sheet or foil is much less prone to produce volatile impurities and is preferred.

The source ovens typically are mounted so that each is surrounded by a chilled panel (usually filled with liquid N_2), which acts to reduce radiative heating of the chamber. A large variety of source oven designs have been used depending on the temperature needed to evaporate the source material and on whether the material is a major or a minor constituent. Fig. 3 shows a fairly standard type of source oven. The crucible containing the source is normally a high grade of pyrolytic boron nitride (PBN) since much experimentation has shown this material to be least reactive with a wide range of source materials, including highly reactive elements such as Al and Ga. The conical shape has been shown to reduce the focused "beaming" of the evaporating flux as well as maintaining a more constant angular dependence as the contents of the crucible is depleted, i.e. a lengthy tubular region

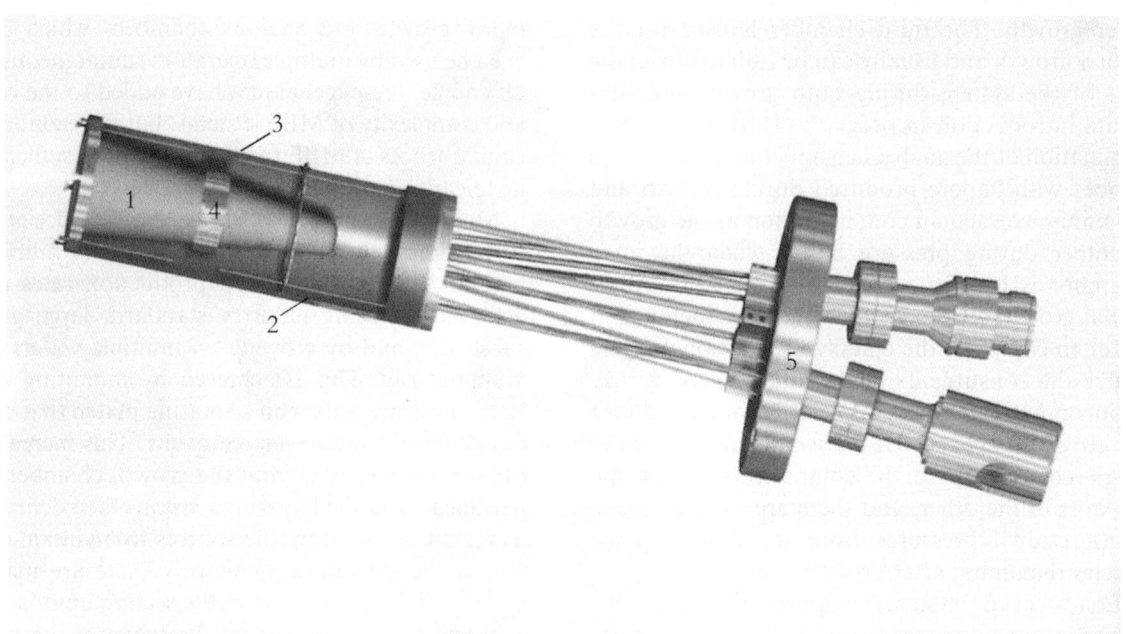

Fig. 3. Cutaway view of an MBE thermal effusion furnace: (1) pyrolytic BN crucible (2) resistive heater filament, (3) metal foil radiation shields, (4) wrap-around thermocouple and (5) mounting flange (courtesy of EPI, Inc.).

above the charge increases the beaming of the flux along the tube axis, leading to poorer uniformity in the deposited layer. The crucible itself is surrounded by several layers of Ta foil that serve as a radiation shield to improve the power efficiency of the source and to reduce the heating of the surrounding cryopanel and loss of LN_2. The temperature of the crucible is measured with a thermocouple attached to a Ta belt wrapped around the crucible. The thermocouple is normally a Type C (W/Re 5/26%) since those alloys are refractory, relatively clean, and resistant to reaction with vapors from the crucible.

The angular distribution of flux from a crucible is an important issue in terms of the thickness uniformity of an MBE film. A simple calculation provides a first order approximation. The intensity of the flux radiating out from a point source clearly is proportional to $1/r$, the distance of the source from the substrate. For a substrate perpendicular to the axis of the crucible a distance r_0 away, the distance r at any point off the axis of the crucible is equal to $r_0 \cos \theta$, where θ is the angle between the crucible axis and the line to the point. However a unit area of the substrate intersects a solid angle which is also proportional to $\cos \theta$; and finally, if the substrate is not perpendicular to the source, the area of the source viewed by the substrate is again proportional to $\cos \theta$. Thus the flux should vary approximately as $\cos^3 \theta$. It is clear that there is a fairly rapid decrease in flux with angle away from the centerline of the source. One can minimize the variation either by using small substrates or by moving the substrate far from the source, but this reduces the overall intensity, i.e., the deposition rate. Fortunately there is another solution that we will discuss below.

The oven shown in Fig. 3 is useful for a wide range of elements, e.g. In, Ga, Al, etc., but high vapor pressure materials require modifications. For example, growth of III–V films requires that an excess pressure of Group V element be used, thus a large volume of source material is needed to minimize refills (and consequent contamination of the growth chamber). Furthermore, it has been found that fewer defects in the film occur when the V element vapor, normally composed of tetramers, e.g. As_4, are thermally cracked to dimers. This can be accomplished in a tube containing a heated filament on which the As_4 molecules impinge en route to the substrate. Fig. 4 shows a source oven for high vapor pressure materials equipped with a cracking filament. The container for the source material has a large volume that is located on the exterior of the growth chamber while the heated source tube (containing the cracker section) extends through a flange into the high vacuum. A valve separates the source container from the source tube to allow precise regulation of the beam flux, and also to allow the source container to be closed off from the growth chamber during bakeout and when the growth chamber is opened to air. This prevents excessive deposits of high vapor pressure material in the growth chamber. P_4 deposits, in particular, can ignite when exposed to air. It is also very important to prevent oxidation of the source material, since the presence of oxides in the beam flux can lead to structural defects in the film.

This extended, but certainly incomplete, explanation for the oven design is intended to convey the idea that a very large amount of both analysis and trial and error have been expended on designing the ovens to produce clean, reproducible sources of atoms or molecules. Fortunately for the experimenter, a large assortment of source ovens and other components are now available from commercial sources [17], and these have been thoroughly tested and specifically optimized for a wide range of elements and compound sources.

The beam shutters are simply plates that can be interposed across a beam to prevent the flux from reaching the substrate. However, even with these simple components, some important design features have been implemented after lengthy experience. Originally, the shutters were Ta plates positioned to pass rather closely in front of a source crucible to cut off the beam as effectively as possible. Experiments have shown that this configuration is not optimum because the Ta sheet will reflect radiation back into the source crucible, causing its temperature distribution to increase. Then when the shutter is opened, there will be an initially higher flux transient due to the hotter average temperature in the source. By positioning the shutters farther from the source ovens, angling

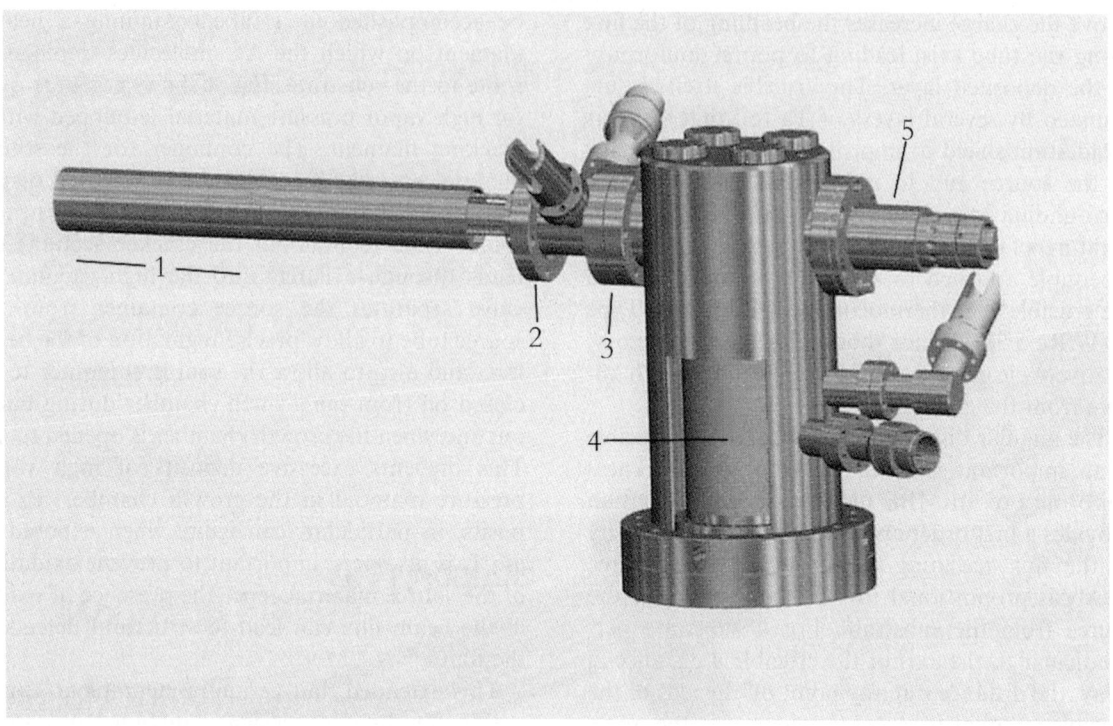

Fig. 4. Cutaway view of valved oven for high-vapor-pressure elements (e.g., As_4, P_4, etc.): (1) internal hot zone for molecular cracking of tetramers to dimers, (2) mounting flange, (3) valve seat for isolating charge, (4) externally heated large volume PBN insert crucible and (5) controlled leak valve stem (courtesy of EPI, Inc.).

them, and constructing them from PBN, there is less radiation reflected and consequently the shutter transient is greatly reduced. In the original configuration with closely mounted metal shutters, my students have observed transients in the metal beam flux of as much as 20% or more with a decay time of 30 s or so. Thus the effect is non-trivial, at least when very thin layers are grown. Again, the point to be made is that the strength of MBE is the control over growth conditions, and yet some very subtle effects must be taken into account in order to achieve that control.

Substrate mounting for epitaxy is important because quite often the precise control of temperature during growth can be critical. It is surprisingly difficult to make good thermal contact between a semiconductor wafer and an underlying metal heater. One successful approach that has been used since the early work on MBE is to bond the semiconductor wafer to a metal heater plate using a low melting metal such as In which provides a liquid thermal contact at the growth temperature. Indium is particularly useful because it is usually relatively insoluble in either the substrate or the metal heater plate, and because the vapor pressure is fairly low up to about 600 °C. Another advantage of the liquid metal bonding is that the surface tension of the liquid provides fairly strain free adhesion of the substrate to the heater so that clamping can be kept to a minimum; as a result, the epitaxial layers after growth do not show the thermal strain-induced slip lines that appear near the points of attachment on substrates which are firmly clamped to the heater unit. Finally, the metal bonding approach is very useful for odd shaped substrates, where no special provision need be taken for the geometry. There are problems with metal bonding, however. For higher growth temperatures the In will vaporize and may contaminate the growth surface; In induced defects can frequently be observed near the edges of In-mounted wafers. Also, if the wafers are subsequently to be processed in a semiconductor fabrication facility, the In metal must be thoroughly removed from the back of the wafer before further processing steps to avoid contamination of the process line.

To avoid the use of In backing, the use of sample holders with radiative coupling between substrate and heater has become quite popular in recent years. The substrates, which must be of exact dimensions to fit a particular holder, are placed in an open ring, which then is locked into position in close proximity to a ribbon or wire heater wound so as to closely fill the area behind the wafer. Radiation from the heater that is more energetic than the band gap of the semiconductor wafer is absorbed by the wafer, and heats it rapidly because of the low thermal mass. To avoid temperature gradients due to hot spots directly over the radiant elements, a thermal diffusing plate made from sapphire or BN is often placed between the heater and the substrate to even out the radiation pattern. Aside from the advantage of not needing to clean the back of the substrate after growth, another advantage of this method of substrate heating is that the smaller thermal mass of heater and substrate allows faster changes in temperature.

Measurement of the substrate temperature is surprisingly difficult to do. Accurate temperature measurement with a thermocouple requires a good thermal contact between the thermocouple and the substrate material. Most semiconductor MBE systems are arranged so as to allow rotation of the substrate during growth (see next section); thus it is not feasible to bond a thermocouple directly to either the substrate or the heater. A non-contacting stationary thermocouple can be mounted directly behind the substrate to measure the radiant flux from the substrate, but this arrangement provides only a relative measure of temperature changes so accurate temperature calibration is necessary. To calibrate substrate temperatures, many investigators use an infrared optical pyrometer in order to be able to measure temperatures in the range 400–600 °C. If semiconducting substrates are used, the bandpass of the pyrometer must be chosen to be an energy window centered above the bandgap of the semiconductor; otherwise the pyrometer simply observes the radiation from the substrate heater element that is transmitted by the substrate. Further complicating matters are the changes in total emissivity that are produced by the growth of an epitaxial layer. Coating of the optical window through which the pyrometer observes the substrate can also cause problems [18,19]. Single temperature point calibration is possible using the melting of In or Sn or other metal eutectic structures attached to the surface of the substrate, or by RHEED observations of surface phase transitions, however these methods are obviously not as accurate as is possible with a plot of thermocouple reading vs. substrate temperature covering the entire temperature range of interest.

We mentioned above the problem of obtaining a uniform beam flux across a large substrate wafer. There is an additional aspect to consider, which is that in most instances more than one beam source is involved in the growth. For example, in Fig. 5, a Ga source is directed at a large substrate from one side of the substrate and an In source is directed at the other, with the purpose of depositing a binary alloy film of, say, $In_{0.5}Ga_{0.5}As$. It is clear from the figure that the angular distributions from the two sources change the ratio of In to Ga across the wafer. To avoid having a concentration gradient, the wafer can be rotated continuously throughout the growth. Since typical growth rates are ~ 1 ML/s the rotation rate should be fairly rapid to avoid the formation of a periodic or "superlattice" structure (see below). Needless to say, rotation of the substrate creates some design problems because the heater structure under the substrate generally needs to remain stationary (slip ring electrical contacts are not practical for the amount of power needed), and because of the tendency for hot bearings to freeze up in UHV. However, commercial MBE systems have solved these difficulties and are generally reliable.

3.3. Growth characterization and rate monitors

From the very beginning, MBE has benefited enormously by the inclusion of analytic tools that provide real time information on the topography of the surface, the condition of the vacuum and the

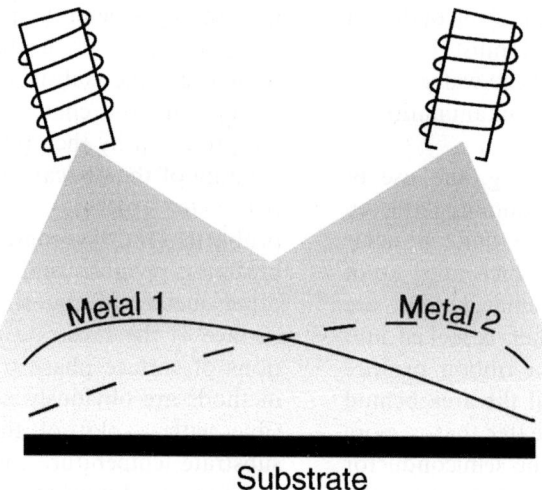

Fig. 5. Diagram showing the coverage distribution on a non-rotating substrate for dual opposing metal source crucibles.

precise growth rate. While the earliest MBE systems used a single chamber for both analysis and growth, the modern systems limit the analysis in the growth chamber to only those probes which provide real time information about the growth process, such as RHEED [5,20]. It is also useful to have some means for measuring the flux of atoms from the source ovens, since the growth rate may not be a good indication of the flux. The simplest way of determining beam flux is with an ionizing detector such as an ion gauge or mass spectrometer that can be placed directly in the molecular beam. While an ion gauge is compact and can easily be positioned in the beam, a mass spectrometer has the big advantage of distinguishing between various vapor species, and, in addition, can be used to detect problems with the vacuum. For beams that condense on cooled surfaces, such as Si, accurate flux measurements can be made with a quartz crystal microbalance; however uncertainties in the condensation coefficient as a function of the elevated temperature of the substrate make some other calibration methods essential [21]. Both types of equipment are commercially available with valves and shutters to minimize unwanted deposits. Atomic absorption by the beams provides a direct measure of the beam flux but requires some more complicated hardware [22]. Various other optical techniques have been used to observe or infer the growth rate of films, including infrared reflectometry [23,24] and ellipsometry [25,26]. However, growth rate measurements are most easily obtained from RHEED observations, which we describe below.

Undoubtedly the single most important analytical tool for the film grower has been the RHEED system for real time observation of the crystal structure of the growing film. Most MBE systems today either include an electron gun and phosphor screen for displaying RHEED patterns while the film is growing, or a low energy electron diffraction (LEED) system for viewing the structure before and after growth. The great advantage of RHEED is that the geometry allows the system to operate while the substrate is exposed to the molecular beams, and thus one can obtain real-time structural information. The appearance of a RHEED pattern not only shows when the surface oxide is removed (since the oxide is amorphous and gives rise to a diffuse diffraction pattern) but also shows the improvement in the surface ordering that occurs with the subsequent annealing. Clean semiconductor surfaces are reconstructed into a geometric configuration that minimizes the energy of electrons in localized bonds at the surface; this reconstruction is evident in the RHEED patterns as diffraction features positioned between the bulk, or "integral order" diffraction spots. The fractional order beams indicate that the surface unit cell is a multiple of the bulk spacing, e.g., the well-known Si(1 1 1) 7 × 7 structure. The presence of these "fractional order features" provides a qualitative measure of the long range ordering of the surface.

The most remarkable application of RHEED information, however, has come from the inference of the mechanism of film growth obtained from the time dependence of the intensity of the diffraction features. I have a vivid memory of a dark, stormy night in Minnesota (it was indeed during a blizzard) when L. Curtis Shannon and I were trying out our new MBE machine. We turned off the room lights to improve the view of the RHEED screen; because of the storm the room was almost completely dark. When Shannon opened the shutter controlling the Ga oven to begin the growth sequence, we observed with amazement that the diffraction spots were pulsing in brightness. As we continued the growth, the diffraction beams became steady, but after turning off the Ga beam for a period of time, then restarting the growth, the same behavior occurred. We were stunned, with no idea of an explanation. In the days that followed a steady stream of visitors came in to view our strange phenomenon, but no one could explain it. Unknown to us at the time, two other groups, one in England [27], the other a few miles away from us at the University of Minnesota [28], had come up with a very interesting explanation for this amazing behavior. On a properly cleaned and annealed surface that has been smoothed by the growth of a few tens of monolayers of material and annealed to improve the ordering, growth occurred mainly by the addition of new atoms/molecules to the edges of monolayer steps. (A monolayer on the (1 0 0) face of a crystal of GaAs is taken to be a bilayer of Ga plus As.) This is a mechanism in which the growth of each atom layer is largely completed before the next begins, i.e. a layer-by-layer growth mechanism [29]. Fig. 6 shows a sketch of the intensity of the specular RHEED beam as a function of time,

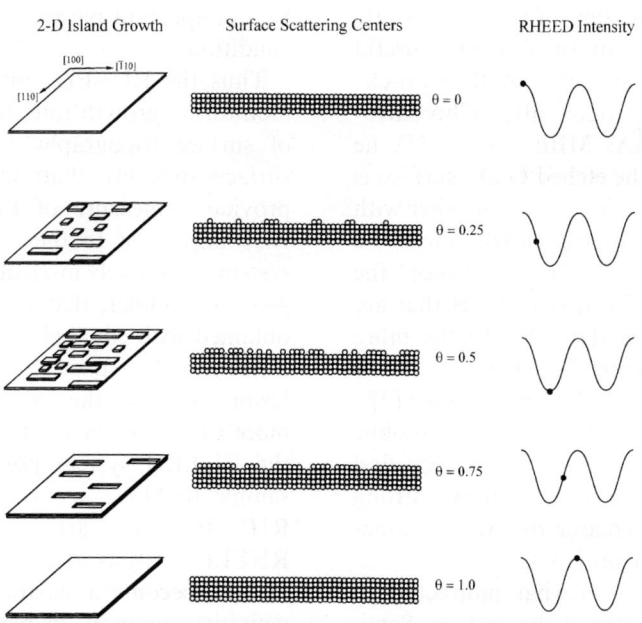

Fig. 6. Schematic diagram of the correlation of surface coverage of 2-D clusters with idealized RHEED oscillations.

beside two views of the condition of the surface. The arriving atoms first nucleate in 2-D islands if steps are not present on the smooth surface. Subsequently, arriving atoms can migrate to the existing step edges to complete the monolayer and return the surface to a smooth condition. Thus the surface cycles between smooth and atomically rough, with a period corresponding to the time to complete a monolayer of growth. The rougher surface causes more diffuse scattering of the RHEED beam, leading to a lower intensity of the diffracted beams. The correspondence between the RHEED oscillation period and the monolayer growth rate was clearly established by empirical measurements of film thickness after growth [27,28]. Thus the RHEED oscillations provide a precise method of measuring growth rates in real time. While RHEED oscillations were first observed in the growth of GaAs layers by MBE, a large body of literature now exists demonstrating the same effect in other materials including metals [30,31], and by a variety of growth processes including various types of chemical beam epitaxy (CBE), atomic layer epitaxy (ALE), and chemical vapor deposition (CVD). The general growth mechanism that produces oscillations is clearly widespread for a large variety of materials.

Besides providing a measure of the growth rate of a film, the RHEED pattern also gives useful information about the geometry of the surface, including the roughness. Soon after Cho introduced RHEED into GaAs MBE systems [5], he also demonstrated that the etched GaAs surface is relatively rough, but becomes much smoother with growth [11]. RHEED patterns clearly show the improvement of surface geometry, since the smoother surface has diffraction features that are streaked normal to the surface due to the more 2-D character of an atomically flat surface. The improvement of the surface is a direct result of the fact that the arriving atoms/molecules are mobile on the heated substrate and predominantly find bonding sites at step edges. Thus there is a strong tendency for terraces to enlarge by the accumulation of material at the edges [13,32].

RHEED also serves (somewhat indirectly) to identify the chemical nature of the surface. Semiconductor surfaces typically reconstruct in order to minimize dangling bond densities at the surface. In the case of compound semiconductors, where the crystals are often composed of alternating layers of metal and nonmetal atoms, the low index planes can either be predominantly metallic or nonmetallic. This is evident in the diffraction pattern since the reconstruction changes as the surface composition changes. A well known example is the GaAs(1 0 0) surface, where the As-rich reconstruction, which is observed during the growth of GaAs under an excess of As, is described as the 2×4 structure, while the Ga-rich structure which forms at a lower As to Ga flux ratio is described as a 4×2 structure. There are a number of other structures which have been observed on this particular surface, which depend on the growth temperature and the ratio of As to Ga in the arriving flux [33]; however the transition from one structure to another, such as from the principal 2×4 As-rich structure to the Ga-rich structure, can be observed (from the RHEED pattern) to occur very rapidly (<1 s) when the As beam is interrupted during MBE growth (at rates of ~ 1 ML/s), indicating that the transition is the result of submonolayer changes in surface concentrations. The point we wish to make is that the surface structure is quite sensitive to the stoichiometry of the top layer, and this can be an important parameter in determining growth conditions.

Thus the RHEED system provides a tool for monitoring growth rate, for a qualitative measure of surface topography and for monitoring the surface structure that, in certain instances, can provide a measure of the surface composition. Also important to the user is that the RHEED system is relatively insensitive to the ambient in the growth chamber, that is, RHEED images can be obtained with equal clarity either while beams are incident on the substrate or when growth has been terminated and the substrate is cooled; furthermore the geometry is such that there is no part of the RHEED system positioned in front of the sample to block access to the surface since the RHEED beam arrives at a glancing angle. RHEED analysis of crystal growth dynamics has, in fact, become a separate field of study, and a definitive summary of recent work has been published by Braun [34].

There are a few potential problems with the use of RHEED in a deposition system. The phosphor screen used to display the diffraction pattern can gradually become coated with the deposited material, but it is not too difficult to clean the glass window and replace the phosphor. Glass windows are presently available commercially which can be kept at an elevated temperature to reduce the condensation of film material. For film growth where film purity and crystalline perfection must be optimized, it is undesirable to flood the growing surface with energetic electrons because beam-induced cracking of residual gases can take place, thus it is generally preferable not to use the RHEED for an extended time on the actual surface of the growing film. Many crystal growers will have a small auxiliary substrate that can be used to establish the beam flux and growth rate prior to actually growing on the large substrate intended for device fabrication.

An alternative to RHEED for growth rate analysis is the use of a thickness measurement that is not dependent on the angular position of the substrate, and which will not cause degradation of the film. Optical methods that do not depend on the angular position of the substrate are becoming more popular for the measurement of deposition rates. One of these techniques relies on the interference in IR reflectivity when a film is deposited on a substrate with different dielectric constant. In the simplest version, the IR pyrometer observing the substrate at constant temperature measures the change in emission as the film grows [23,35]. This method requires that the film be relatively transparent to the optical band detected by the pyrometer. It also requires that the pyrometer window be kept free from deposits that would reduce the optical intensity. A clever way to avoid deposits on the window is to use a Si surface as a mirror to reflect the optical emission from the substrate and yet avoid a direct path to the window for atoms/molecules desorbing from the substrate [18]. The reflectivity of the Si seems to be little affected by a thin film of deposited material from the growth, and the longer indirect path to the window prevents coatings from forming. Heating the optical windows is also an effective way in which to keep them free of coatings [22].

Other workers have used spectral ellipsometry to obtain real time information about surface composition, optical properties and growth rate of films in an MBE system [36,37]. Spectral ellipsometry measures the reflectivity of the perpendicular and parallel polarized components of a light beam over a range of wavelengths in order to determine the complex index of refraction and layer thickness of a transparent film on a reflective substrate. The index of refraction can be related to the composition using the effective medium approximation; thus spectral emissivity provides significantly different information than does RHEED. Furthermore, provided a substrate holder which can rotate precisely in the plane parallel to the substrate surface is used, it is possible to obtain this information while the substrate rotates to maintain uniform growth over a large area. One limitation of this method is again the need to prevent condensation on the optical windows.

Maracas et al. [37] used spectral ellipsometry to monitor beam flux ratios. The RHEED oscillation technique for measuring growth rate (described above) is primarily an indication of the metal flux, since the non-metal, As in the case of GaAs, is provided in excess. Thus the growth rate is limited by the metal flux. The As flux can be measured only approximately using the mass spectrometer. However, after depositing a known amount of Ga on GaAs (determined by the growth rate), these investigators then determined the length of time required for the As beam to convert the Ga into epitaxial GaAs based on measurements of the dielectric function as it returned to the values for GaAs. There were several experimental challenges that were overcome, a principal problem being the design of a substrate manipulator with sufficient mechanical stability to allow measurements during growth when the substrate was heated.

Even though the vapors in the chamber used for epitaxial growth can produce deposits of the film constituents which may form insulating films on the electron optics of analytical tools such as AES, there is always a need for additional tools in the growth chamber to obtain more real time information about growth. Chambers et al. [38] describe a system that uses the RHEED beam to excite Auger electrons from the growing film. The electrons are analyzed in a small, high throughput spectrometer which does not block the substrate from the molecular beams. The system described by Chambers et al. is used to grow epitaxial oxide films and contains electron beam heated sources to provide metal beams (Mo and Cr) and an electron cyclotron resonance (ECR) source to provide O atoms. The system is also provided with quartz crystal oscillator monitors to measure beam fluxes; however the in situ AES capability showed clearly that the CRO monitors gave an incorrect measure of film composition because not all of the incident flux was incorporated into the growing film on the 750 °C substrate. The AES analysis has also been crucial in determining the composition of mixed metal oxides during growth. The system contains a separate analysis chamber with both XPS and X-ray diffraction capability. The message from these experimenters is quite clear: as the material systems become more complex, more in situ analytic capability becomes essential. Because of the UHV environment, many of the new tools of surface physics are easily accommodated within the growth chamber for this purpose.

4. General mechanism of molecular beam epitaxy growth

The information provided by these techniques has been absolutely essential in understanding how to optimize the growth process and has given surface scientists some remarkable insight into the nature of the dynamics of the surface of a growing crystal. The in situ studies of surface structure by electron diffraction have led to an understanding of the dynamic motion of steps on a growing surface, while the use of tools that unveil the details of the atomic structure of the surface, i.e., the scanning electron microprobe, has shown that MBE growth produces extremely well ordered surfaces on many materials. This makes it possible to study the detailed structure of such surfaces without the complications resulting from contamination or the loss of constituents due to the preparation process. Furthermore the experimenter has great control over the precise conditions of the surface, i.e., the surface chemical composition can be altered at will and the effect on the arrangement of surface atoms can be studied. In this section we will describe in rather general terms the prevailing wisdom about growth mechanism of MBE films. Tsao has reviewed the theory of this basic growth model in considerable detail [33]. A number of elegant STM studies on Si surfaces exposed to Si vapor have corroborated most of the elements of this general scheme [39,40].

The etched substrate surface after the thermal removal of surface oxide is typically rough on an atomic scale as shown by a spotty RHEED pattern and by TEM observations that indicate rough features as much as 10 nm above the surrounding flat areas. The degree of roughness is very much a function of the polishing treatment and the subsequent annealing in UHV. Once epitaxial growth begins, however, the surface rapidly becomes much smoother and the RHEED pattern shows this smoothing by developing streaked diffracted features [20]. This smoothing of the surface was predicted by Frank and van der Merwe [41] based on a model that involved the migration of arriving atoms/molecules over flat terraces on a rough surface with incorporation into the lattice at step edges. Wider terraces have a larger collection area for vapor species and thus have bounding step edges that advance more rapidly than those bounding narrow terraces. The consequence of this step growth is that terraces tend to become similar in size, the smaller terraces disappear and the surface becomes smoother. Eventually the surface evolves to a nearly uniform array of terraces, as can be seen in the STM image of a Si(1 0 0) surface in Fig. 7 [42]. An interesting feature of the Si(1 0 0) step edges is the way in which the edges alternate between relatively smooth and relatively ragged. This has to do with the manner in which the Si surface atoms reconstruct by forming dimer chains to minimize the number of unpaired electrons at the surface; the smoothness of the step edges depends on whether the reconstructed surface has the Si dimer chains lying parallel to or perpendicular to the step [43]. However, the main feature of the surface to consider is the nearly perfect flatness of the terraces.

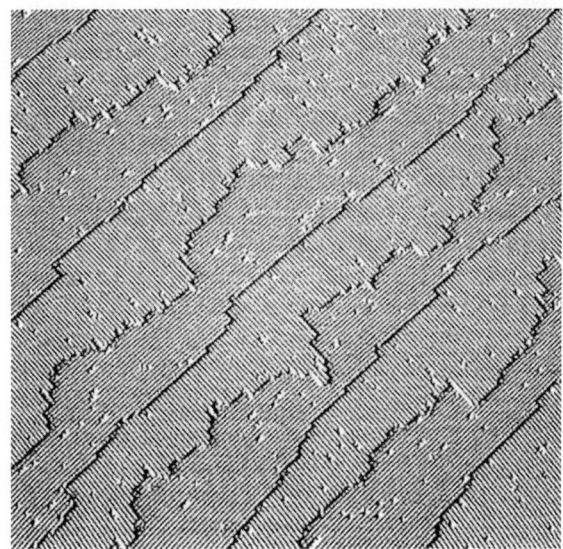

Fig. 7. Scanning tunneling microscope image of a Si surface, ~0.3° off (1 0 0) orientation showing the type A steps (Si dimers parallel to steps) and type B steps (Si dimers perpendicular to steps). Uppermost part of the surface is at lower right, with downward tilt to upper left. Scale is ~110 nm square (courtesy of Prof. Max Lagally).

Once the surface becomes atomically smooth, with only an array of monolayer steps as shown in the figure, further growth can proceed in either of two modes, depending on the nature of the surface and the mean free path of atoms/molecules on the surface. If the terrace width is comparable to or less than the diffusion length of atoms, then under normal growth conditions it is possible for the atoms arriving on the surface to diffuse to step edges, and growth occurs by the steady growth of steps, which advance across the surface, described as the "step flow" regime. Step flow will occur on surfaces similar to that shown in Fig. 7 that are slightly off the principal plane axis, i.e., vicinal surfaces on which the terraces are not too wide, at temperatures, which are high enough to provide good surface mobility. Aoki and coworkers have shown that the difficult problem of producing flat interfaces in the strained growth of $In_xGa_{1-x}As/In_yAl_{1-y}As$ grown on InP can be solved by using "step flow" growth on the (4 1 1) surface [44]. On the low index (1 0 0) surface, these authors observed that the wider terrace width does not permit step flow growth and the interfaces are much rougher, as indicated by broadening of the photoluminescence peak.

The other growth mode occurs when the terraces are wider than the diffusion length. In this instance 2-D nucleation occurs on the terraces, which leads to periodic roughening and smoothing of the surface as each monolayer fills in again. This is the situation that produces the dramatic oscillations in the RHEED intensities, discussed above. It is not surprising that increasing the surface mobility (by, for example, increasing the temperature) can lead to a transition from the 2-D nucleation mode to the step flow mode, which is indicated by a disappearance of the RHEED oscillations as the temperature increases. Neave et al. have used measurements of this type to infer the kinetic parameters and diffusion length of Ga atoms on vicinal surfaces whose terrace length can be calculated from the crystal orientation [45]. Alternatively, lowering the temperature can cause the onset of RHEED oscillations, at least until the surface mobility decreases so much that atoms are no longer able to form large islands. Clearly, surface mobility is a key element in determining growth mode.

5. Beam flux and stoichiometry

An important consideration is the control of the composition of the deposited film. The material properties of semiconductors are particularly sensitive to deviations from stoichiometry, where vacancies of one or the other of the components may form electrically active centers. Clearly this would be undesirable, and yet we have previously indicated that precise control of the temperature of the ovens providing the constituent vapor flux is difficult to achieve.

(A) Elemental growth: The simplest growth system is one with only a single component, and Si, because of its technological importance, has been studied more than any other material system. In the case of the growth of a single element, compositional control is not an issue. What is necessary is a source of Si vapor sufficiently intense to provide a growth rate much larger than the

contamination rate from residual impurities. While solid Si has been occasionally used as a Si vapor source, the vapor pressure is so low at the melting point, $\sim 4 \times 10^{-4}$ Torr, that the deposition rate is too low to deposit more than a few monolayers. Therefore typically in Si MBE the Si vapor is derived from an electron beam heated source crucible, to obtain a large enough flux at the substrate. Thus the term "beam" is used very loosely, since the Si atoms simply evaporate radially outward from the hot zone of the evaporator. Because of the low vapor pressure of Si at the typical growth temperature of ~ 600 °C, the condensation rate of the Si vapor is near unity with very little reevaporation.

The vapor pressure of Ge is approximately two orders of magnitude greater than that of Si, but is too low at the melting point to use a subliming source. It is possible to derive a sufficient flux of Ge from a molten charge in a heated crucible to grow an elemental film at very slow growth rates, but it is much better to use an electron beam heater for the source to avoid contamination due to slow deposition. Again, the condensation on the substrate is essentially unity. In any event, stoichiometry is not an issue for a single component material.

(B) Compound growth: When the films have more than one constituent, the problem of compositional control becomes an issue. We can consider three possible scenarios, depending on the vapor pressure of the constituents.

In the first case, if both constituents have low vapor pressure, so that the adsorbed species have a long surface lifetime before reevaporation, then the problem for the crystal grower is simply to control the flux of both kinds of atoms/molecules to produce the desired ratio in the film. In other words, the film composition is determined by the arrival rate of atoms at the surface. An example of this is the alloy Si_xGe_{1-x} where there is complete miscibility in the solid phase and reevaporation of either element is minimal. The challenge then lies in precise temperature control of the sources. As indicated above, one-degree temperature change produces on the order of 2% variation in beam flux in sources using the evaporated elements. However, a small deviation from the desired ratio of Si to Ge is unimportant, since Si and Ge are incorporated on equivalent lattice sites. A variation of the ratio of Si to Ga does not lead to vacancy formation and thus does not have a major effect on the electrical properties.

The second possibility is one in which one of the constituents has a higher vapor pressure as an element than in the compound. For example, elemental As has a vapor pressure at the typical GaAs growth temperature of 550 °C that is roughly seven orders of magnitude larger than the vapor pressure of As in equilibrium with GaAs. As we have indicated above, a (nearly) stoichiometric film can be obtained by supplying an excess flux of As (typically as As_2 molecules) to react with all of the arriving Ga; the unreacted As will be reevaporated from the surface. (We are so far ignoring the defect structure of the film; the beam flux ratio of As to Ga can have a significant influence on the concentration of vacancies of either species.)

The third possibility is that both of the constituents have higher vapor pressures as elements than in the compound. This is true for most of the II–VI compounds. For these materials, the first monolayer of either component can adsorb on the substrate, but subsequent layers will be less strongly bound due to the weaker elemental bond, and at elevated temperatures will simply desorb. By exposing the substrate alternately to each beam long enough to deposit successive monolayers of each component, stoichiometry is achieved because only a monolayer will stick on each exposure, and each monolayer reacts with the preceding one. This process has been termed ALE [46,47]. ALE is useful in producing a film consisting of alternating layers of the constituents, e.g., the ideal configuration of the zincblende and wurtzite structures in the $\langle 1\,0\,0 \rangle$ direction consisting of alternating layers of each species. ALE has the advantage that uniform coverage is not as dependent on surface migration as is MBE, since the components arrive at adsorption sites from the vapor or by rapid migration from a physisorbed layer. As a result, ALE is particularly useful in providing uniform coverage over non-planar surfaces (although in this case it is often true that the substrate is not a single crystal and so the film may not be epitaxial).

The ALE approach can also be used to grow materials of the second type described above. For example, GaAs can be grown by exposing the substrate alternatively to first Ga and then As beams separately. Since the shutters on a standard MBE system can usually be programmed to operate automatically, the timing can be adjusted to any desired cycling of beams. It has been observed that the surface mobility of adsorbed Ga atoms is significantly greater when As is not incident; thus if the shutter timing is such that Ga begins arriving in the absence of an As flux, there is greater smoothing of the surface. This type of growth has been termed migration-enhanced epitaxy (MEE) [48], and is reported to improve epitaxial growth at low temperatures, where the reduced mobility of the metal atoms normally leads to roughening of the surface. It may be particularly useful when growing a heterostructure where the different metal species may have widely differing surface mobilities at a fixed growth temperature. This would normally lead to a roughening of the interface as soon as the lower mobility metal is deposited.

A clever variant of this approach is to phase-lock the shutter operation to the RHEED oscillation period to begin a heterointerface only at the time when the maximum in the RHEED intensity indicates that optimum smoothness of the growing surface has been achieved. For example, in very narrow quantum wells, the variation in well width produced by disorder in a single layer of atoms can produce noticeable degradation in the optical properties such as increasing the width and reducing the intensity of the main recombination peak in the photoluminescent spectrum. In Fig. 8 we show schematically the structure of a quantum well device with interfaces formed when the surfaces are relatively rough and when the surfaces are maximally smooth. Superlattices grown by phase-locked epitaxy (PLE) have shown significant improvement in optical and structural properties over superlattices grown with shutter operation at a random phase of the surface molecular coverage [12,49].

6. Surfactant assisted growth

We have emphasized the role of surface migration in improving the crystalline quality of

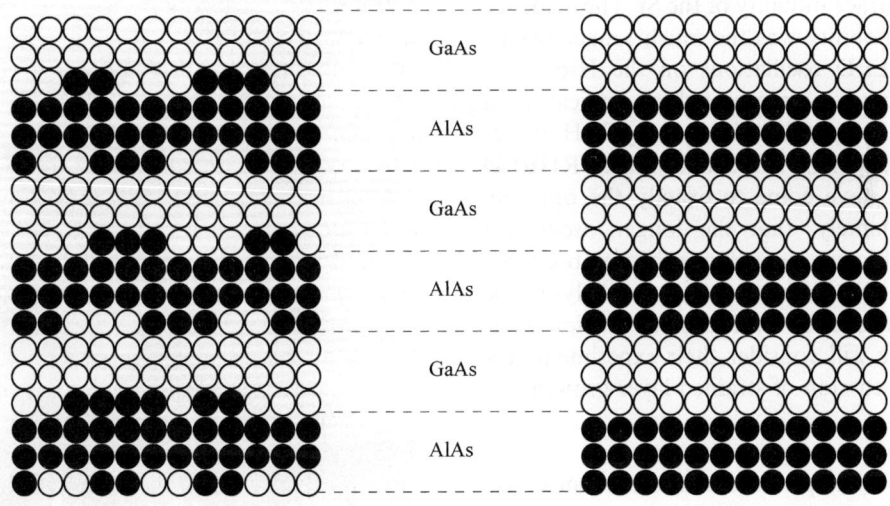

Fig. 8. Cross-sectional diagram of superlattices grown with random shutter timing and with shutters timed to change at completion of monolayer growth, i.e., PLE.

epitaxial films. If the surface species are able to migrate to step edges, a planar growth interface is maintained. However, if the temperature is too low, mobility may not be great enough to achieve a thermodynamically stable structure, and smooth growth may not result. Recently several authors have explored the use of impurities described as "surfactants" to improve surface mobility; these are materials that act to weaken the bonding of the arriving constituent atoms to the substrate, thus enhancing their surface mobility. The impurities themselves are almost entirely rejected from the growing film. For example, Okada and Harris have found that irradiating the surface of GaAs during growth with atomic H permits the growth temperature to be lowered from 580–600 °C to as low as 330 °C with no loss of structural quality of the epitaxial film [50]. There is much that is not clear about the process, e.g., how much H is adsorbed and how does it leave the growing film? Presumably the H film is segregated to the surface as the film grows with little incorporation into the film. Nevertheless the H (derived from a tungsten filament heated in H_2 gas) clearly makes a significant difference in the growth kinetics. Hydrogen is also used as a surfactant in the homoepitaxial growth of Si [51]; however these authors believe that H on Si *reduces* the surface mobility of the Si. The evidence for reduced mobility was the much longer persistence of RHEED oscillations on the H-covered surface, which was interpreted as indicating a shorter mean free path for the arriving Si. However Pillai et al. [52]; have observed similar RHEED behavior and have come to exactly the opposite conclusion, namely, that the longer oscillation persistence indicates a *greater* surface mobility because the surface layers are more likely to be completed without 2-D nucleation of the next layer occurring. We will consider their experiment further in the discussion of strained layer growth.

7. Heteroepitaxy, superlattices and quantum wells

One of the most amazing and important consequences of the 2-D growth by MBE is the ability to produce structures consisting of alternating thin layers of two different semiconductor materials with similar crystal lattice constants. If this stack of layers is grown in such a way that there is a periodicity to the structure, it is described as a "superlattice" since the period of the structure is larger than the lattice spacing in the crystal. The fact that such structures can be grown routinely with individual layers as thin as a few monolayers, and with the total number of layers essentially limited only by the grower's patience says a great deal about the growth process. Cross-section transmission electron micrographs clearly show that the perfection of the superlattice is as good at the end of the growth as it is at the beginning, if not better! Fig. 9 shows a repeated stack of four monolayers of GaAs on four layers of AlAs, with the growth continuing from the bottom to the top. In order for the interfaces to improve, the starting surface must have been almost atomically smooth, with only a regular array of monatomic steps, as in Fig. 7. Furthermore, this nearly ideal surface must have been preserved after the completion of each layer, otherwise the surface irregularities would quickly become magnified to an extent that the interfaces would lose their definition as the layers

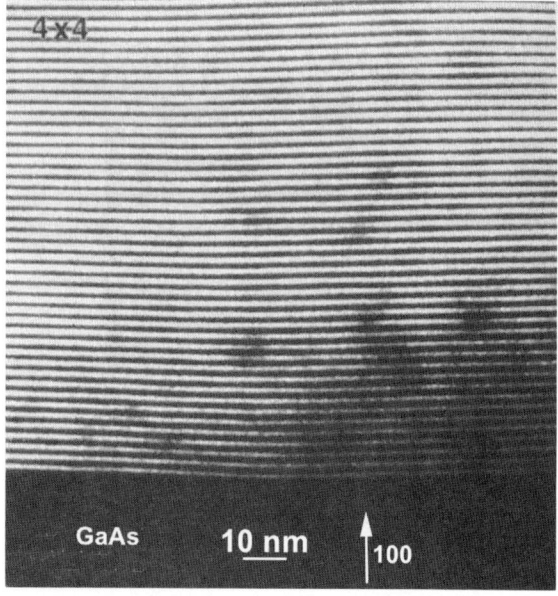

Fig. 9. Cross-sectional transmission micrograph of a GaAs/AlAs superlattice with four layers of each material in a period (courtesy of Lucent Technologies, Inc.).

accumulate. While GaAs and AlAs are an important example of materials with nearly the same lattice constant, this is, of course, only rarely true. For other III–V compounds it is more likely that the superlattice structure will be strained, unless efforts are made to use ternary or quaternary alloys, e.g., $In_xGa_{1-x}As$, that are lattice-matched to each other and/or to the substrate. We discuss strained layers in the next section.

"Quantum wells" are produced by depositing a narrower band gap semiconductor between layers of a different material, with a wider bandgap than the well material. If the thickness of the narrow bandgap center well is much less than the wavelength of electrons in the material, in the range from a few to tens of atom layers, the electron energy levels are discrete in the well, i.e., "quantized" and the structure is described as a quantum well. A structure comprised of a periodic sequence of quantum wells is a superlattice. Quantum wells and superlattices are of great interest because the electronic and optical properties are quite different from bulk semiconductor material. Many of the early theoretical predictions of solid state physics have been precisely confirmed in these devices that so closely approximate the ideal models used in the theory. For example, the energy levels of electrons in a quantum well no longer form a continuum of states as they do in the conduction band of a bulk semiconductor, but rather they are "quantized" into discrete levels by the fact that the quantum well has one dimension, perpendicular to the surface, which is of the order of the wavelength of electrons in the film. This can lead to a number of strange and sometimes useful properties. Instead of transmitting light of energy less than the band gap and absorbing all wavelengths above the band gap, as in a bulk semiconductor, a quantum well absorbs light only at energies corresponding to the energy difference between the valence band and the discrete levels in the quantum well. This leads to an absorption spectrum with a series of peaks at the edge just below the continuum. Of course, it would be difficult to measure the minute optical absorption of a single quantum well, so the experiment is carried out on a stack of many wells, which are so reproducible to make that all are the same width and hence show the same spectrum. This, in fact, was the historic experiment carried out by Dingle et al. that was mentioned in the discussion of the history of MBE [10].

Another technologically significant phenomenon observed with quantum wells is the greatly increased electron mobility that occurs, particularly at low temperatures. When the surrounding wide bandgap barrier layers are doped with donor impurities while the quantum well itself is undoped, the electrons produced from the ionized donors drift into the quantum well, i.e. spill over into a region of lower energy, where transport parallel to the layer can occur without scattering from the ionized donors. In effect, the free carriers have been separated physically from the ionized donors by confining or "modulating" the doping to only the barrier regions. This produces a much higher electron mobility than would occur in bulk material doped to produce the same electron density. Field effect transistors, which utilize modulated doping, are known as high electron mobility transistors, or HEMTs; they are also described as modulated doped FETs, or MODFETs [53]. Fig. 10 shows a graphical representation of the conduction band in a MODFET. Because of the separation of positive and negative charges, an electric field is produced which shows up as curvature in the conduction band edge at the boundary between the barrier and well. In order to better separate the electrons from their parent ions, a small, undoped spacer layer of AlGaAs is normally included in the structure and is indicated in the diagram.

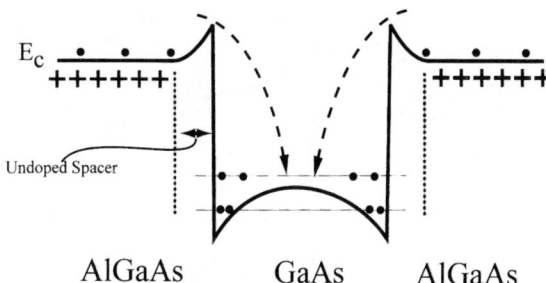

Fig. 10. Schematic diagram of the conduction band in a AlGaAs/GaAs MODFET. Ionized donors represented by (+), electrons represented by (•). Diagram includes an undoped spacer in the AlGaAs to improve the physical separation of charges to reduce scattering.

Much of the early work on superlattices and quantum wells was based on the material system GaAs–Ga$_x$Al$_{1-x}$As because the Al content in the alloy increases the bandgap and yet the two binary compounds have nearly the same lattice constant. The MBE process, at least in principle, makes possible the growth of selected ternary and quaternary III–V alloys, since the composition of the film can be controlled by the relative beam fluxes. Fig. 11 shows the lattice constant vs. bandgap for III–V binary materials and their ternary alloys. It is evident that the lattice constant varies widely for the various alloys. If one is restricted to binary compound substrates, the mismatch between ternary film and binary substrate can be substantial. Thus the technology for growing strained layers is extremely important in order to have access to the full range of materials and properties offered by the III–V alloy family, as well as other families, notably, Si–Ge.

8. Strained layer epitaxy

The subject of strain in epitaxial layers is extremely important because of the obvious fact that most often heteroepitaxial layers are grown on a substrate with a different lattice constant. Clear evidence for the strongly ordering nature of a clean substrate is given by observations that the initial growth of a film mismatched to the substrate often occurs with the film adopting the two-dimensional spacing of the substrate, i.e., the film growth is "pseudomorphic" with the substrate. Fig. 12 illustrates the difference between commensurate, or lattice-matched growth, pseudomorphic growth with uniaxial distortion of the film, and relaxed, incommensurate growth. Depending on the amount of mismatch, the distortion of pseudomorphic growth causes increasing strain in the film to an extent that relaxation eventually occurs with the generation of misfit dislocations in the plane of the interface. Frequently the relaxation is catastrophic with extensive slip, and disruption of the planarity of the surface. A number of calculations have been made to determine the critical thickness of the epilayer at which the strain is sufficient to cause the generation of dislocations. The original theory of Matthews and Blakesley was a straightforward mechanical energy balance, and provides a reasonable fit to more recent experimental data [54]. Fig. 13 is a plot of the variation of critical thickness vs. percent mismatch for the system of Si$_x$Ge$_{1-x}$ on Si. The figure shows that there are definite limitations on the thickness of pseudomorphic growth. It should be noted that the exact value of critical thickness depends on the elastic constants for the film material; thus this is not a universal curve. People and Bean [55] have discussed the more detailed considerations for strained layer growth and have modified the curves somewhat, although in general, the simpler theory provides a good fit to the data.

There are various ways to get around the mismatch problem. As indicated in Fig. 13, as long as the film thickness is below the critical thickness, a planar film can be grown on a mismatched substrate. Strain effects will, however, alter the band structure of the film, which may be an important consideration for subsequent experiments. The traditional method of dealing with mismatch between film and substrate has been to grow a graded buffer layer, where varying the composition from that of the substrate to that of the film gradually alters the lattice constant of the film.

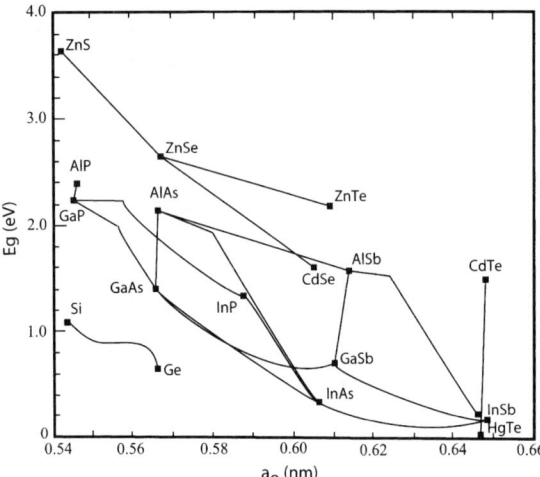

Fig. 11. Plot of band gap vs. lattice constant for elemental, III–V and II–VI semiconductors. The solid squares represent the values for the indicated binary compounds; the lines show the band-gap values for the intermediate ternary alloys.

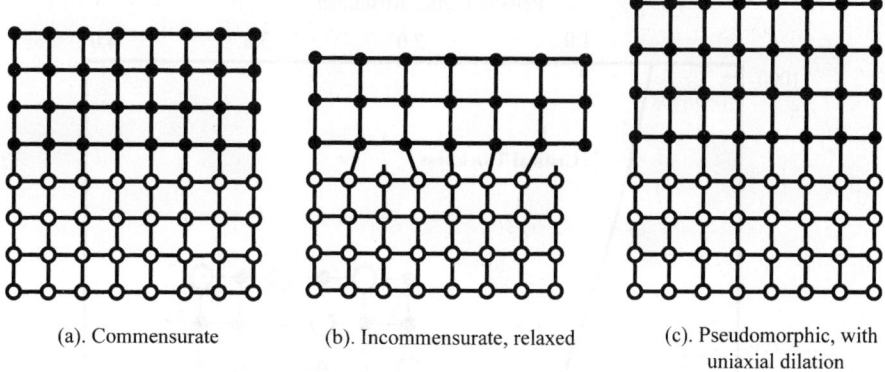

(a). Commensurate   (b). Incommensurate, relaxed   (c). Pseudomorphic, with uniaxial dilation

Fig. 12. Cross-section schematic of atomic arrangement in various modes of epitaxial growth: (a) lattice-matched, commensurate growth; (b) lattice-mismatched, relaxed growth; (c) lattice-mismatched, strained pseudomorphic growth.

MBE is a particularly effective method to do this because the relative beam fluxes of constituents can be varied in a very controlled fashion; thus it is easy to begin by growing e.g., GaAs and then to direct a gradually increasing In flux toward the substrate by a programmed temperature sequence of the In source to eventually deposit a film of $In_xGa_{1-x}As$ where x can be as large as 0.3 or greater for buffer thicknesses of the order of 0.5 μm. (InP is normally used as a substrate for larger values of x.) Graded buffers, however, do not eliminate stress; they merely distribute it over a larger volume of material, and threading dislocations are often observed in TEM studies of such structures.

Another approach is to introduce a strained superlattice buffer, where alternating layers are strained either in tension or compression. This has the advantage that dislocations are often constrained to lie parallel to the interface, and thus do not extend into the final layer of interest. The growth of strained superlattices with the alternate layers in compression and tension is an active research area that is much too broad for the scope of this chapter. Ref. [56] provides an excellent presentation of the current status of research in this field [56].

There is an additional problem with strained layers, which is that strain can be relieved by the rejection from the bulk of the species causing strain, i.e., there can be strain-enhanced diffusion to the surface with the accumulation on the surface of the atoms causing strain. For example, InGaAs quantum wells grown with GaAs barriers on a GaAs substrate show a skewing of the In spatial distribution toward the surface as a result of such strain-enhanced outdiffusion and excess evaporation of the In [57]. Not only can this occur with the layer constituents, but doping impurities can also be rejected or segregated to the surface if they introduce sufficient strain into the lattice. This effect can be useful in certain instances (see the discussion below on the use of surfactants to facilitate growth), as well as being a problem.

While in some circumstances the structure of the substrate may force the epilayer into a metastable pseudomorphic configuration, at sufficiently high temperatures the surface layer is quite mobile, and as a result will attain a thermodynamically stable structure that depends on the energetics of surface and interface and also on the lattice strain. This may lead to the formation of islands of the epitaxial material, either on the clean substrate (which is described as Volmer–Weber growth [58]) or after the growth of a monolayer or so (Stranski–Krastanov growth [59]) instead of a smooth surface by the desired layer-by-layer (Frank–van der Merwe [40]) growth. Several investigators [60,61] have shown that the presence of an intermediate layer, a surfactant, can reduce the tendency for island formation and at the same time can reduce the temperature needed for surface mobility. In the growth of Si/Ge heterostructures, As and Sb have proven useful for this purpose. Arsenic in particular was

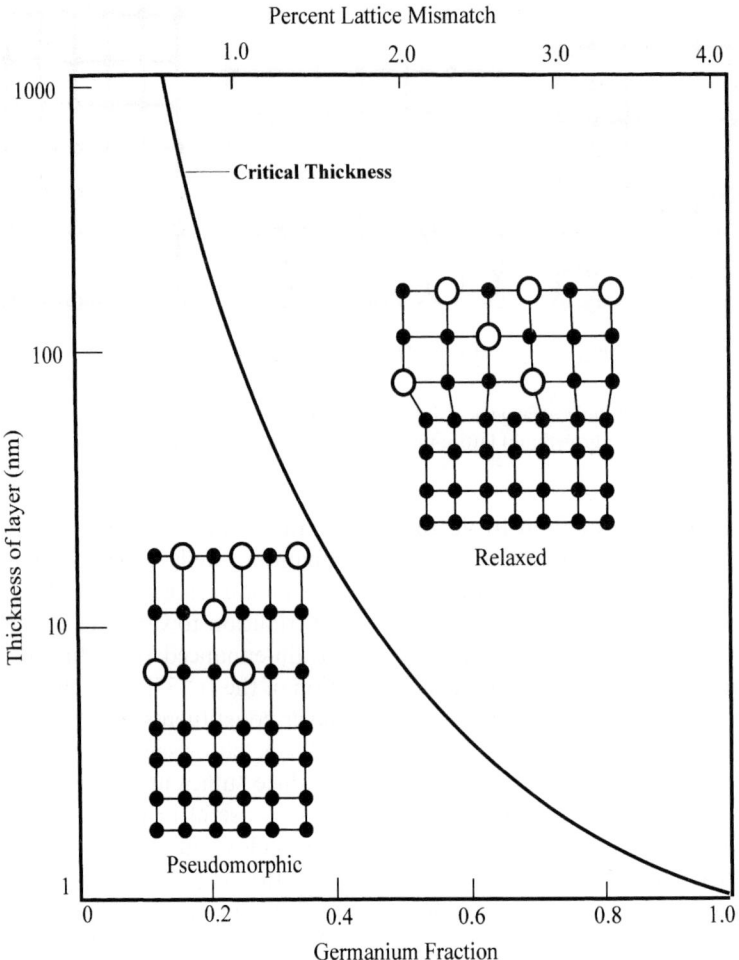

Fig. 13. Transition between pseudomorphic and relaxed growth modes for Si_xGe_{1-x} alloy films deposited on Si. The solid line is based on the calculations by Matthews and Blakesley [54].

shown to promote the growth of flat Ge layers on Si, and to prevent the incorporation of Ge into the growing Si layer. In both cases only monolayer amounts of the surfactant are required, since the As and Sb are strongly segregated to the growth surface. An As layer in particular permits the growth of very thick layers of Ge on Si, since strain is relieved not by misfit dislocations but by stacking fault arrays ("V defects") which do not destroy the planarity of the surface [60]. Unfortunately, both As and Sb are doping impurities in the Group IV elements, and sufficient amounts of the surfactants are incorporated into the growing films to alter the doping.

Sn does not act as a dopant in Si/Ge, since it is isoelectronic, i.e., has the same electronic configuration as Si and Ge, and it has also been found to facilitate the growth of Ge layers on Si and to minimize the segregation of Ge into the subsequent Si layer [62]. It is strongly segregated to the surface of the growing film so that only monolayer quantities are needed. There is a major difficulty with Sn-mediated Ge growth that is that the Ge layer thickness cannot exceed about 4 monolayers to avoid Ge island formation, i.e., the growth mode is clearly Stranski–Krastanov. Strained layers of $In_xGa_{1-x}As$ on GaAs with $x = 0.3$–0.4 have been grown using Bi as a surfactant [52]. The Bi was co-deposited along with the constituents, yet was undetectable by X-ray diffraction on the completed film. Atomic force microscopy after growth showed island formation both with and without the Bi, yet the islands were much larger when the Bi was included, and, as was mentioned in the previous section, RHEED showed enhanced surface mobility when Bi was included. While these films were not the perfect planar structures to be wished for, there was certainly improvement in the film morphology with the use of a surfactant. Clearly, the use of surfactant-mediated growth is a very interesting research area that is relatively unexplored.

Recent reports of the use of compliant substrates have indicated another alternative to dealing with strained growth [63,64]. Rather than attempting growth on a rigid substrate that produces stress in the growing film, if growth occurs on a thin film substrate that is able to deform to match the growing layer, it should be much easier to grow thicker films that do not relax by dislocation generation. GaAs and AlAs have only a small mismatch ($\sim$0.15%) and have been favored for electronic devices because of the bandgap offset and because of the ease in growing thin layers of either material on the other. In addition, AlAs is chemically much more reactive than GaAs, which makes it possible to detach very thin films of GaAs from an AlAs substrate (or more likely, from a thin layer of AlAs on a GaAs substrate) by chemical etching. Thus it is possible to grow a very thin film of GaAs on a AlAs substrate or layer, bond the GaAs to a second substrate by a flexible adhesive such as liquid In and subsequently detach the thin GaAs from the original substrate by chemical etching in dilute HCl. The resulting structure consists of a very thin film of GaAs bonded by a compliant material to a supporting structure. The GaAs is thin enough and only weakly attached to the supporting structure so that it can deform to match the structure of an epitaxial film of different lattice constant deposited on it. This approach has proven very promising in the growth of InGaAs, where layers with a thickness as much as eight times the critical thickness have been grown without generating significant misfit dislocations. While the method is certainly novel and may be quite useful in certain instances where very small areas of substrate are needed, it does not seem to be the solution for mass production of electronic devices. The problems of handling large area compliant substrates appear formidable.

9. Quantum boxes

We have discussed quantum wells in one dimension. Why not extend the concept to three dimensions? The flippant response to that question is that it is considerably more difficult to do! MBE offers tremendous control over composition normal to the crystal surface, however control across the surface is not easily accomplished. The molecular beams used in MBE are very diffuse, and since they consist of neutral atoms, there is no way to focus and steer them electrically. Some 3-D control has been demonstrated using in situ shadow masks; however, because the beams originate from large area sources, and because multiple sources are required, it is difficult to use shadow masks for the definition of small features. Standard lithography following the growth of quantum wells has been used, as has the growth of wells on a lithographically patterned substrate. Both approaches are limited by the size of features that can be defined by optical lithography—at present typically one or two orders of magnitude larger than the width of quantum well structures.

Another new approach to quantum boxes takes advantage of the clustering of metal clusters on the surface of a III–V substrate. Whenever excess Group III metal is present on the surface of a III–V substrate at temperatures above the melting point of the metal, the metal film forms a very uniform array of small droplets, with the droplet size proportional to the amount of metal present. This has often been a problem in attempting to grow III–V materials under metal-rich conditions, where it is easy to get metal droplets on the surface and thus destroy the planarity. Now, however, a number of authors are using this general scheme to great advantage in producing quantum dots. Stinz et al. were able to produce InAs quantum dots surrounded by strained InGaAs by inserting a metal-rich step in the growth sequence, which caused the growth mode to shift from Frank–van der Merwe to Stranski–Krastanov long enough to produce the dots [65]. This was followed by encapsulation with first InGaAs and subsequently GaAs. The resulting dots were approximately 80 Å in diameter. From a device standpoint, the results were promising because the photoluminescense produced by the dots was close to the technologically important 1.3 μm wavelength; the material is therefore a good candidate to produce 1.3 μm laser diodes. After annealing at high temperatures above 600 °C, there was a blue shift in the luminescence and a large reduction in intensity, suggesting that the quantum wells were gradually merging with the surrounding matrix, due to bulk diffusion. Understanding the processes leading to the formation and dissolution of quantum dots will clearly be a useful and active research topic.

10. Summary and conclusions

The discussion above has barely scratched the surface of the very active research currently in progress in MBE. We have attempted to give just a bit of the flavor of the wide diversity of research that is based on MBE growth. Of course, since its initial use in the growth of compound semiconductors, one of the principal driving forces behind MBE research has been the industrial need for specialized semiconductor devices, such as optoelectronic devices and high-speed transistors. These devices take advantage of the almost atomically abrupt interfaces between different materials comprising the crystal that are produced by the low growth temperatures and the ability to switch the growth beams rapidly. The ability to grow layers of different compositions with differing bandgaps and with precisely controllable thicknesses has made possible many novel devices such as HEMTS and multiple quantum well (MQW) devices which had previously not been attempted because the necessary control over the device structure was not available. Now that the fabrication of these devices has been demonstrated using MBE, there are other growth techniques, particularly the method known as organo-metallic chemical vapor deposition (OMCVD or MOCVD), which have matured and which are more suitable for mass production than is the relatively slow method of MBE. Many of the new devices are now also fabricated using these faster techniques; however it was the demonstration of the properties of these structures from MBE that provided the incentive to improve the other techniques.

MBE has also aided in understanding fundamental solid-state physics. Thin films comprised of layers of differing semiconductors with precise boundaries and geometries approach the ideal models long used by theorists. As a result, some dramatic demonstrations of quantum effects in solids have been observed using MBE structures. The MQW structures have shown the effect of spatial confinement in one (or more) dimensions on the electronic properties, where a continuous band of states is split into discrete quantum states, with remarkable agreement with previous theory. Resonant quantum tunneling devices in which carriers are transported across potential barriers at specific resonant energies have been studied and have been used as extremely high frequency oscillators. Thus the ability to produce prototypical film structures that are close to the idealized models amenable to theory has added significantly to our understanding of semiconductor crystals.

These new materials and devices have opened a significant window on the nature of thin films, their surfaces, and the dynamics of their growth. Studies of thin films in the past suffered from the non-reproducible nature of the films and surfaces. The new methods for growth and characterization have allowed the widely reproducible creation of large area, single crystal films with customized characteristics. This has certainly had much to do with the present excitement and interest in the field of surface science. While it is certainly true that other growth techniques have been modified and improved to facilitate the production of many of the structures and devices described above, MBE has been the research tool that has lead the way in demonstrating how to make novel film structures. This has occurred because of the unique combination of (a) ultra-clean vacuum conditions, (b) low growth temperatures, (c) precisely controllable sources of the film constituents and (d) in situ surface analysis tools that are all characteristic of the MBE approach.

In the future, MBE should continue as an important research tool for materials preparation. It seems likely that there will be an ever-stronger marriage of the MBE technique with more sophisticated analysis tools. The combination of MBE with the various scanning probe microscopies has been a union of great productivity, since MBE is able to supply the highly ordered surfaces that are most informative. There is an increasing movement toward combining MBE as the surface conditioning tool in surface experiments, e.g., those involving synchrotron radiation studies, where MBE equipment is incorporated directly into the beam line. More complex films, such as oxide superconductors, are being prepared by MBE. It is evident that more complex material systems will require new sources for the constituents and dopants, such as focused ion sources to allow selected area doping, and a variety of chemical vapor sources to allow the growth of novel materials. Industrial use of MBE will involve more automated systems with capability for handling larger wafers in keeping with industry standards.

Finally, from a personal perspective, MBE has come a very long way from the first films grown in a glass tabletop UHV system on sub cm-sized substrates, to today's massive production systems, capable of growth on multiple six-inch (and larger!) wafers. Even research systems have become significantly more complex, as more analysis capability has been incorporated. It has been awe inspiring to see the huge increase in research interest and in sophistication of equipment and experiment over the years. But I still feel a profound excitement when I come upon a working MBE system in a darkened lab and view the RHEED monitor as it cycles up and down, indicating the atomic layers adding to a newborn crystal.

References

[1] R.J. Collins, F.W. Reynolds, G.R. Stilwell, Electrical and optical properties of GaSb films, Phys. Rev. 98 (1955) 227.
[2] K.G. Günther, Evaporated layers of semiconducting III–V compounds, Naturwissenschaften 45 (1958) 415.
[3] J.R. Arthur, Interaction of Ga and As_2 molecular beams with GaAs surfaces, J. Appl. Phys. 39 (1968);
J.R. Arthur, Interaction of As_2, P_2 and Bi molecular beams with GaAs and GaP($\bar{1}\bar{1}\bar{1}$) surfaces, in: G.A. Somerjai (Ed.), The Structure and Chemistry of Solid Surfaces, Wiley, New York, 1969, p. 46.
[4] J.R. Arthur, J.J. LePore, GaAs, GaP and $GaAs_xP_{1-x}$ epitaxial films grown by molecular beam depositions, J. Vac. Sci. Technol. 6 (1969) 545.
[5] A.Y. Cho, Morphology of epitaxial growth of GaAs by a molecular beam method: the observation of surface structures, J. Appl. Phys. 41 (1970) 782;
GaAs epitaxy by a molecular beam method: observation of surface structure on the (001) face, J. Appl. Phys. 42 (1971) 2074.
[6] A.Y. Cho, I. Hayashi, Epitaxy of silicon doped gallium arsenide by molecular beam method, Metall. Trans. 2 (1971) 777;
A.Y. Cho, I. Hayashi, P–N junction formation during molecular beam epitaxy of Ge-doped GaAs, J. Appl. Phys. 42 (1971) 4422.
[7] A.Y. Cho, H.C. Casey, GaAs–Al_xGa_{1-x}As double-heterostructure lasers prepared by molecular beam epitaxy, Appl. Phys. Lett. 25 (1974) 288.
[8] L.L. Chang, L. Esaki, W.E. Howard, R. Ludeke, The growth of a GaAs–GaAlAs superlattice, J. Vac. Sci. Technol. 10 (1973) 11.
[9] L.L. Chang, L. Esaki, R. Tsu, Resonant tunneling in semiconductor double barriers, Appl. Phys. Lett. 24 (1974) 593.
[10] R. Dingle, W. Wiegmann, C.H. Henry, Quantum states of confined particles in very thin Al_xGa_{1-x}As–GaAs–Al_xGa_{1-x}As heterostructures, Phys. Rev. Lett. 33 (1974) 827.
[11] A.Y. Cho, J.R. Arthur, Molecular beam epitaxy, Prog. Solid-State Chem. 10 (1975) 157.
[12] M.A. Sitter, H. Sitter, Molecular beam epitaxy, Fundamentals and Current Status, second ed., Springer, Berlin, 1996.
[13] J.Y. Tsao, Materials Fundamentals of Molecular Beam Epitaxy, Academic Press, Boston, 1993.
[14] B.A. Joyce, Kinetic and surface aspects of MBE, in: L.L. Chang, K. Ploog (Eds.), Molecular Beam Epitaxy and Heterostructures, Martinus Nijhoff Publishers, Dordrecht, 1985, p. 37.
[15] A.Y. Cho, Growth and properties of III–V semiconductors by molecular beam epitaxy, in: L.L. Chang, K. Ploog (Eds.), Molecular Beam Epitaxy and Heterostructures, Martinus Nijhoff Publishers, Dordrecht, 1985, p. 191.
[16] C.W. Tu, J.S. Harris Jr. (Eds.), Molecular Beam Epitaxy, North Holland, Amsterdam, 1990.
[17] See, for example, EPI Product Guide, EPI MBE Products Group, St. Paul, MN 55127.
[18] A.J. Springthorpe, A. Majeed, Epitaxial growth rate measurements during molecular beam epitaxy, J. Vac. Sci. Technol. B 8 (1990) 266.
[19] S. Strite, M. Kamp, H.P. Meier, Reliable substrate temperature measurements for high temperature AlGaAs

molecular-beam epitaxy growth, J. Vac. Sci. Technol. B 13 (1995) 290.
[20] M.G. Lagally, D.E. Savage, M.C. Tringides, Diffraction from disordered surfaces: an overview, in: P.K. Larsen, P.J. Dobson (Eds.), Reflection High-Energy Electron Diffraction and Reflection Electron Imaging of Surfaces, Plenum Press, New York, 1988, p. 139.
[21] P. Link, G. Grobbel, M. Worz, S. Bauer, H. Berger, W. Gebhardt, J.J. Paggel, K. Horn, Photoelectron spectroscopy from atomic layer epitaxially grown ZnTe/ZnSe heterostructures, J. Vac. Sci. Technol. A 13 (1995) 11.
[22] P. Pinsukanjana, A. Jackson, J. Tofte, K. Maranowski, S. Campbell, J. English, S. Chalmers, L. Coldren, A. Gossard, Real-time simultaneous optical-based flux monitoring of Al, Ga and In using atomic absorption for molecular beam epitaxy, J. Vac. Sci. Technol. B 14 (1996) 2147.
[23] A.J. Springthorpe, T.P. Humphreys, A. Majeed, W.T. Moore, In situ growth rate measurements during molecular beam epitaxy using an optical pyrometer, Appl. Phys. Lett. 55 (1989) 2138.
[24] R.N. Sacks, R.M. Sieg, S.A. Ringel, Investigation of the accuracy of pyrometric interferometry in determining $Al_xGa_{1-x}As$ growth rates and compositions, J. Vac. Sci. Technol. B 14 (1996) 2157.
[25] D.E. Aspnes, W.E. Quinn, S. Gregory, Application of ellipsometry to crystal growth by organometallic molecular beam epitaxy, Appl. Phys. Lett. 56 (1990) 2569.
[26] D.E. Aspnes, Real-time optical diagnostics for epitaxial growth, Surf. Sci. 307 (1995) 1017.
[27] J.J. Harris, B.A. Joyce, P.J. Dobson, Oscillations in the surface structure of Sn-doped GaAs during growth by MBE, Surf. Sci. 103 (1981) L90;
C.E.C. Wood, RED intensity oscillations during growth by MBE, Surf. Sci. 108 (1981) L441.
[28] J.M. Van Hove, C.S. Lent, P.R. Pukite, P.I. Cohen, Damped oscillations in reflection high energy electron diffraction during GaAs MBE, J. Vac. Sci. Technol. B 1 (1983) 741.
[29] B.A. Joyce, J.H. Neave, J. Zhang, P.J. Dobson, RHEED intensity oscillations during MBE growth of III–V compounds an overview, in: P.K. Larsen, P.J. Dobson (Eds.), Reflection High-Energy Electron Diffraction and Reflection Electron Imaging of Surfaces, Plenum Press, New York, 1988, p. 397.
[30] D.E. Chambliss, K.E. Johnson, Using scanning tunneling microscopy to understand diffraction oscillations: Fe growth on Cu(100), J. Vac. Sci. Technol. A 13 (1995) 1522.
[31] C.M. Gilmore, J.A. Sprague, A molecular dynamics study of transient processes during deposition on (001) metal surfaces, J. Vac. Sci. Technol. A 13 (1995) 1160.
[32] M.A. Herman, H. Sitter, Molecular Beam Epitaxy, Fundamentals and Current Status, second ed., Springer, Berlin, 1996.
[33] C.T. Foxon, B.A. Joyce, Interaction kinetics of As_2 and Ga on (100) GaAs surfaces, Surf. Sci. 64 (1977) 293.

[34] W. Braun, in: Applied RHEED—Reflection High Energy Electron Diffraction During Crystal Growth, Springer Tracts in Modern Physics, vol. 154, Springer, Berlin, 1999.
[35] K.A. Bertness, Smart pyrometry for combined sample temperature and reflectance measurements in molecular-beam epitaxy, J. Vac. Sci. Technol. B 18 (2000) 1426.
[36] W.M. Duncan, M.J. Bevan, H.D. Shih, Real-time diagnostics of II–VI molecular beam epitaxy by spectral ellipsometry, J. Vac. Sci. Technol. A 15 (1997) 216.
[37] G.N. Maracas, C.H. Kuo, S. Anand, R. Droopad, G.R.L. Sohie, T. Levola, Ellipsometry for III–V epitaxial growth diagnostics, J. Vac. Sci. Technol. A 13 (1995) 727.
[38] S.A. Chambers, T.T. Tran, T.A. Hileman, Auger electron spectroscopy as a real-time compositional probe in molecular beam epitaxy, J. Vac. Sci. Technol. A 13 (1995) 83.
[39] R.J. Hamers, U.K. Kohler, J.E. Demuth, Epitaxial growth of silicon on Si(001) by scanning tunneling microscopy, J. Vac. Sci. Technol. A 8 (1990) 195.
[40] Y.W. Mo, B.S. Swartzentruber, R. Kariotis, M.B. Webb, M.G. Lagally, Growth and equilibrium structures in the epitaxy of Si on Si(001), Phys. Rev. Lett. 63 (1989) 2393.
[41] F.C. Frank, J.H. van der Merwe, One-dimensional dislocations I Static theory, Proc. Roy. Soc. A 198 (1949) 205;
F.C. Frank, J.H. van der Merwe, One-dimensional dislocations. II. Misfitting monolayers and oriented overgrowth, Proc. Roy. Soc. 200 (1949) 125.
[42] B.S. Swartzentruber, Y.W. Mo, R. Kariotis, M.G. Lagally, M.B. Webb, Direct determination of step and kink energies on vicinal Si(001), Phys. Rev. Lett. 65 (1990) 1913.
[43] R. Becker, R. Wolkow, Semiconductor surfaces, in: J.A. Stroscio, W.J. Kaiser (Eds.), Scanning Tunneling Microscopy, Academic Press, Boston, 1993, Chapter 5.
[44] T. Aoki, T. Kitada, S. Shimomura, S. Hiyamizu, Superflat interfaces in pseudomorphic $In_{0.72}Ga_{0.28}As/In_{0.52}Al_{0.48}As$ quantum wells grown on (411) InP substrates by molecular beam epitaxy, J. Vac. Sci. Technol. B 18 (2000) 1598.
[45] J.H. Neave, P.J. Dobson, B.A. Joyce, J. Zhang, Reflection high energy electron diffraction oscillations from vicinal surfaces—a new approach to surface diffusion measurements, Appl. Phys. Lett. 47 (1985) 100.
[46] T. Suntola, J. Hyvaerinen, Atomic layer epitaxy, Ann. Rev. Mater. Sci. 15 (1985) 177.
[47] C.H.L. Goodman, M.V. Pessa, Atomic layer epitaxy, J. Appl. Phys. 60 (1986) R65.
[48] Y. Horikoshi, M. Kawashima, N. Kobayashi, Optical investigation of GaAs growth process in molecular beam epitaxy and migration—enhanced epitaxy, J. Cryst. Growth 111 (1991) 200.
[49] T. Yao, H. Nakahara, H. Matuhata, Y. Okada, Fabrication of AlAs/Al/AlAs heterostructures by molecular beam epitaxy and migration enhanced epitaxy, J. Cryst. Growth 111 (1991) 221.
[50] Y. Okada, J.S. Harris Jr., Basic analysis of atomic-scale growth mechanisms for molecular beam epitaxy of GaAs using atomic hydrogen as a surfactant, J. Vac. Sci. Technol. B 14 (1996) 1725.

[51] K. Sakamoto, H. Matsuhata, K. Miki, T. Sakamoto, Adsorption and desorption of atomic hydrogen on Si(0 0 1) and its effects on Si MBE, J. Cryst. Growth 157 (1995) 295.

[52] M.R. Pillai, S.-S. Kim, S.T. Ho, S.A. Barnett, Growth of $In_xGa_{1-x}As$/GaAs heterostructures using Bi as a surfactant, J. Vac. Sci. Technol. B 18 (2000) 1232.

[53] H. Morkoc, The HEMT: a superfast transistor, IEEE Spectrum 21 (1984) 28.

[54] J.H. Matthews, A.E. Blakesley, Defects in epitaxial multilayers. I. Misfit dislocations, J. Cryst. Growth 27 (1974) 118.

[55] R. People, J.C. Bean, Calculation of critical layer thickness versus lattice mismatch for Ge_xSi_{1-x}/Si strained layer heterostructures, Appl. Phys. Lett. 47 (1986) 322.

[56] T.P. Piersall (Ed.), Strained-Layer Superlattices: Physics, Academic Press, Boston, 1991;
T.P. Piersall (Ed.), Strained-Layer Superlattices: Materials Science and Technology, Academic Press, Boston, 1991.

[57] J.-P. Reithmaier, H. Riechert, H. Schlotterer, Indium desorption during MBE growth of strained InGaAs layers, J. Cryst. Growth 111 (1991) 407.

[58] M. Volmer, A. Weber, Keimbildung in übersättigten Gebilden, Z. Phys. Chem. 119 (1926) 277.

[59] I.N. Stranski, L. Krastanov, Zur Theorie der orientierten Ausscheidung von ionenkristallen aufeinander, Ber Akademie der Wissenschaften und der LiteraturMathematisch-Naturwissenschaftliche Klasse 146 (1939) 797.

[60] M. Copel, M.C. Reuter, M. Horn, von Hoegen, R.M. Tromp, Influence of surfactants in Ge and Si epitaxy on Si(0 0 1), Phys. Rev. B 42 (1990) 11682;
F.K. LeGoues, M. Copel, R.M. Tromp, Microstructure and strain relief of Ge films grown layer by layer on Si(0 0 1), Phys. Rev. B 42 (1990) 11690.

[61] F.K. LeGoues, M. Copel, R.M. Tromp, Novel strain-induced defect in thin molecular-beam-epitaxy layers, Phys. Rev. Lett. 63 (1989) 1826.

[62] X.W. Lin, Z. Liliental-Weber, J. Washburn, E.R. Weber, A. Sasaki, A. Wakahara, T. Hasegawa, Ge/Si heterostructures grown by Sn-surfactant-mediated molecular beam epitaxy, J. Vac. Sci. Technol. B 13 (1995) 1805.

[63] C. Carter-Coman, A.S. Brown, R. Bicknall-Tassius, N.M. Jokerst, M. Allen, Strain-modulated epitaxy: a flexible approach to 3D band structure engineering without surface patterning, Appl. Phys. Lett. 69 (1996) 257.

[64] C. Carter-Coman, R. Bicknall-Tassius, A.S. Brown, N.M. Jokerst, Analysis of $In_{0.07}Ga_{0.93}As$ layers on GaAs compliant substrates by double crystal X-ray diffraction, Appl. Phys. Lett. 70 (1997) 1754.

[65] A. Stintz, G.T. Liu, A.L. Gray, R. Spillers, S.M. Delgado, K.J. Malloy, Characterization of InAs quantum dots in strained $In_xGa_{1-x}As$ quantum wells, J. Vac. Sci. Technol. B 18 (2000) 1496.

 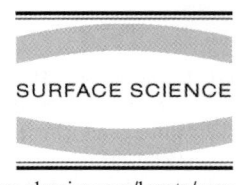

Carbon nanotubes: opportunities and challenges

Hongjie Dai *

Department of Chemistry, Stanford University, Stanford, CA 94305, USA

Received 14 August 2000; accepted for publication 11 July 2001

Abstract

Carbon nanotubes are graphene sheets rolled-up into cylinders with diameters as small as one nanometer. Extensive work carried out worldwide in recent years has revealed the intriguing electrical and mechanical properties of these novel molecular scale wires. It is now well established that carbon nanotubes are ideal model systems for studying the physics in one-dimensional solids and have significant potential as building blocks for various practical nanoscale devices. Nanotubes have been shown to be useful for miniaturized electronic, mechanical, electromechanical, chemical and scanning probe devices and materials for macroscopic composites. Progress in nanotube growth has facilitated the fundamental study and applications of nanotubes. Gaining control over challenging nanotube growth issues is critical to the future advancement of nanotube science and technology, and is being actively pursued by researchers. © 2001 Elsevier Science B.V. All rights reserved.

Keywords: Carbon; Chemical vapor deposition; Electrical transport measurements; Quantum effects; Chemisorption

1. Introduction

A new form of carbon, buckministerfullerene C_{60} was discovered in 1985 by a team headed by Smalley, Kroto and coworkers [1], and led to the Nobel Prize in chemistry in 1997. C_{60} is a soccer ball-like molecule made of pure carbon atoms bonded in hexagon and pentagon configurations. Besides diamond, graphite and C_{60}, quasi one-dimensional nanotube is another form of carbon first reported by Iijima in 1991 when he discovered multi-walled carbon nanotubes (MWNT) in carbon-soot made by an arc-discharge method [2]. About two years later, he made the observation of single-walled nanotubes (SWNTs) [3]. Since then, nanotubes have captured the attention of researchers worldwide. A significant amount of work has been done in the past decade to reveal the unique structural, electrical, mechanical, electromechanical and chemical properties of carbon nanotubes and to explore what might be the key applications of these novel materials.

A SWNT is a graphene sheet (Fig. 1) rolled-over into a cylinder with typical diameter on the order of 1.4 nm (Fig. 3a–d), similar to that of a C_{60} bucky-ball. A MWNT consists of concentric cylinders with an interlayer spacing of 3.4 Å and a diameter typically on the order of 10–20 nm (Fig. 2a–c). The lengths of the two types of tubes can be up to hundreds of microns or even centimeters. A SWNT is a molecular scale wire that has two key structural parameters. By folding a graphene sheet into a

* Fax: +1-650-725-0259.
 E-mail address: hdai1@stanford.edu (H. Dai).

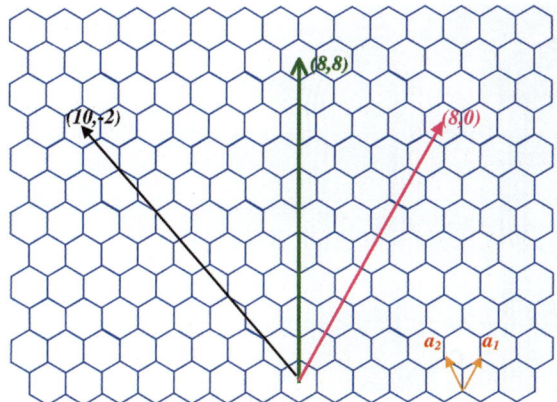

Fig. 1. Schematic honeycomb structure of a graphene sheet. Carbon atoms are at the vertices. SWNTs can be formed by folding the sheet along lattice vectors. The two basis vectors a_1 and a_2, and several examples of the lattice vectors are shown.

cylinder so that the beginning and end of a (m,n) lattice vector in the graphene plane join together (Fig. 1), one obtains an (m,n) nanotube. The (m,n) indices determine the diameter of the nanotube, and also the so-called 'chirality'. (m,m) tubes are 'arm-chair' tubes, since the atoms around the circumference are in an arm-chair pattern (Fig. 3a). $(m,0)$ nanotubes are termed 'zigzag' in view of the atomic configuration along the circumference (Fig. 3b and c). The other types of nanotubes are chiral, with the rows of hexagons spiraling along the nanotube axes (Fig. 3d).

In terms of mechanical properties, nanotubes are among the strongest and most resilient materials known to exist in nature. A nanotube has a Young's modulus of 1.2 TPa and tensile strength about a hundred times higher than steel and can tolerate large strains before mechanical failure [4]. The electrical properties of nanotubes depend sensitively on the (m,n) indices and therefore on the diameter and chirality. A SWNT can be either a metal, semiconductor or small-gap semiconductor depending on the (m,n) structural parameters [5]. For a graphene sheet, the conduction and valence bands touch each other at the six corner points of the first Broulloin zone. These states are filled with electrons that have the highest energy (Fermi's energy). A graphene sheet is therefore semimetallic with a zero band gap. The electronic states of an infinitely long nanotube are parallel lines in **k** space, continuous along the tube axis and quantized along the circumference (Fig. 3a–d). For (m,m) arm-chair tubes, there are always states crossing the corner points of the first Brillouin zone (Fig. 3a), suggesting that arm-chair tubes should always be metallic. For (m,n) nanotubes with $m-n \neq 3 \times$ integer, the electronic states (lines) miss the corner points (Fig. 3c and d) and the nanotubes are semiconducting. The energy gap scales with the tube diameter as $1/d$ and is on the order of 0.5 eV for a SWNT with typical diameter $d = 1.4$ nm. For $m-n = 3 \times$ integer, certain electronic states of the nanotube land on the corner points of the first Broulloin zone (Fig. 3b). These types of tubes would be semimetals but become small-gap semiconductors (band gap scales with $1/d^2 \sim 10$ meV for $d \sim 1.4$ nm) due to a curvature induced orbital rehybridization effect [6]. The extreme sensitivity of electronic property on structural parameters is unique for carbon nanotubes. This uniqueness leads to rich physical phenomena in nanotube systems, and poses a significant challenge to chemical synthesis in terms of controlling the nanotube diameter and chirality.

The intriguing properties of carbon nanotubes have led to an explosion of research efforts worldwide [7]. Understanding these properties and exploring their potential applications have been a main driving force in this area. Theoretical and experimental work has been focusing on the relationship between nanotube atomic structures and electronic structures, transport properties, electron–electron and electron–phonon interaction effects. Extensive effort has been made to investigate the mechanical properties of nanotubes, including their Young's modulus, tensile strength, failure processes and mechanisms. It has also been an important fundamental question regarding how mechanical deformation in a nanotube affects its electrical properties.

Thus far, nanotubes have been utilized individually or as an ensemble to build functional device prototypes, as has been demonstrated by many research groups. Ensembles of nanotubes have been used for field emission based flat-panel displays, composite materials with improved mechanical properties and electromechanical actuators. Bulk

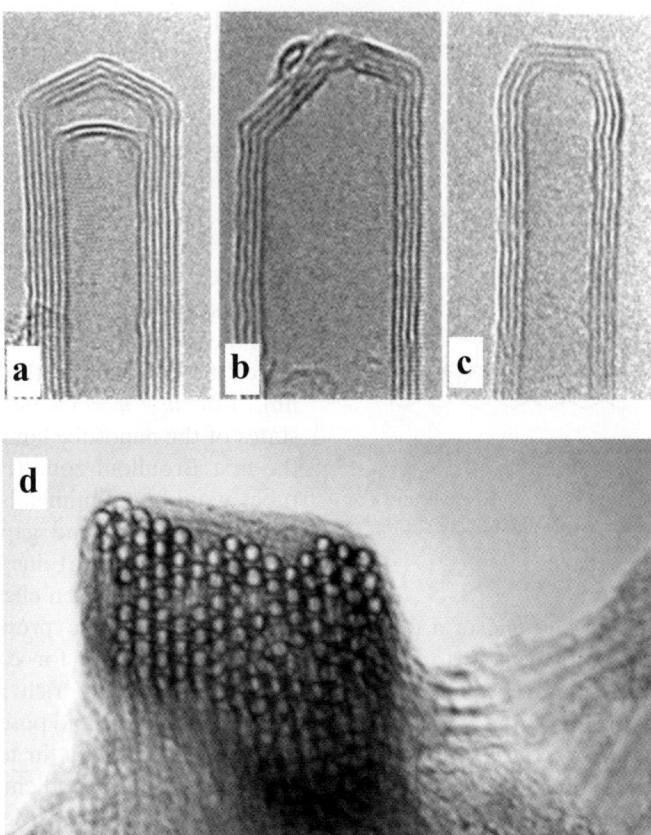

Fig. 2. Viewing the invisible: carbon nanotubes imaged by transmission electron microscopy (TEM). TEM is a technique used in the discovery of both MWNTs and SWNTs. (a–c) TEM images of MWNTs with closed caps (courtesy of S. Iijima). The parallel lines are the cross-sections of the sidewalls of concentric cylinders. Diameters of MWNTs are typically on the order of 10–20 nm. (d) TEM image of the cross-section of a bundle of SWNTs (courtesy of R. Smalley). Each circle represents the cross-section of a SWNT with diameter ∼1.4 nm.

quantities of nanotubes have also been suggested as high-capacity hydrogen storage media. Individual nanotubes have been used for field emission sources, tips for scanning probe microscopy, nanotweezers and chemical sensors. Nanotubes are also promising as the central elements for future miniaturized electronic devices.

The success in nanotube growth has led to the wide availability of nanotube materials, which is a main catalyst behind the recent leaps-and-bounds in basis physics studies and applications of nanotubes [7]. The full potential of nanotubes for applications will not be realized until the growth of nanotubes can be further optimized and controlled.

Real-world applications of nanotubes require either large quantities of bulk materials or device integration in a scale-up fashion. For applications such as composites and hydrogen storage, it is desired to obtain high quality nanotubes at the kilogram or ton level using growth methods that are simple, efficient and inexpensive. For devices such as nanotube based electronics, scale-up will unavoidably rely on self-assembly techniques or controlled growth strategies on surfaces combined with microfabrication techniques. Significant work has been carried out to tackle these issues. Nevertheless, many challenges remain in the nanotube growth area. An efficient growth approach to

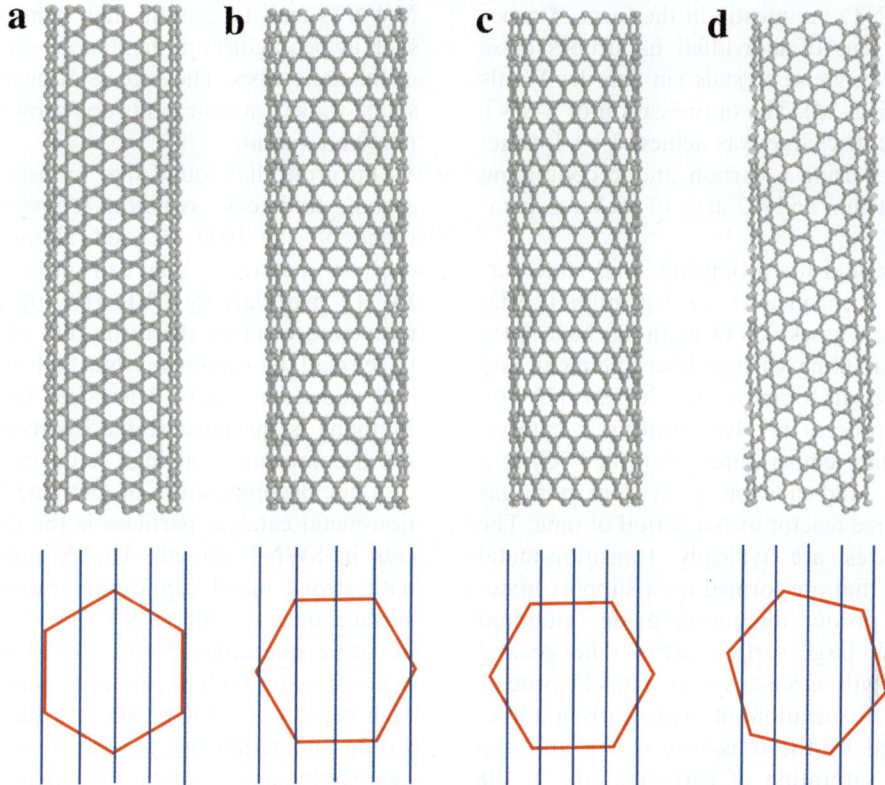

Fig. 3. Schematic structures of SWNTs and how they determine the electronic properties of the nanotubes. (a) A (10,10) arm-chair nanotube. Bottom panel: the hexagon represents the first Broulloin zone of a graphene sheet in reciprocal space. The vertical lines represent the electronic states of the nanotube. The center-line crosses two corners of the hexagon, resulting in a metallic nanotube. (b) A (12, 0) zigzag nanotube. The electronic states cross the hexagon corners, but a small band gap can develop due to the curvature of the nanotube. (c) The (14, 0) zigzag tube is semiconducting because the states on the vertical lines miss the corner points of the hexagon. (d) A (7, 16) tube is semiconducting. This figure illustrates the extreme sensitivity of nanotube electronic structures to the diameter and chirality of nanotubes.

structurally perfect nanotubes at large scales is not yet at hand. Growing defect-free nanotubes continuously to macroscopic lengths is still difficult. Also, it is desired to gain exquisite control over nanotube growth on surfaces and to obtain large-scale ordered nanowire structures. Finally, there is a seemingly formidable task of controlling the chirality of SWNTs during growth.

2. Nanotube growth

Arc-discharge and laser ablation methods for the growth of nanotubes have been actively pursued in the past 10 years. In 1992, a breakthrough in MWNT growth by arc-discharge was first by Ebbesen and Ajayan who achieved growth and purification of high quality MWNTs at the gram level [8]. The synthesized MWNTs have lengths on the order of 10 μm and diameters in the range of 5–30 nm. The nanotubes are typically bound together by strong van der Waals interactions and form tight bundles. For the growth of single-walled tubes, a metal catalyst is needed in the arc-discharge system. The first success in producing substantial amounts of SWNTs by arc-discharge was achieved by Bethune and coworkers in 1993 [9]. The growth of high quality SWNTs at the 1–10 g scale was achieved by Smalley and coworkers using a laser ablation (laser oven) method [10]. The produced SWNTs are mostly in the form of ropes consisting of tens of individual nanotubes close packed into hexagonal crystals via van der Waals interactions (Fig. 2d). The optimization of SWNT growth by arc-discharge was achieved by Journet and coworkers using a carbon anode containing 1.0 at.% of yttrium and 4.2 at.% of nickel as catalysts [11].

Another method for producing solid state carbon materials is chemical vapor deposition (CVD) of hydrocarbon gases. CVD methods have been successful in making carbon fiber, filament and nanotube materials for more than 20 years [12–16]. The growth process involves heating a catalyst material to high temperatures (500–1000 °C) in a tube furnace, and flowing a hydrocarbon gas through the tube reactor over a period of time. The catalytic species are typically transition-metal nanoparticles that are formed on a support material such as porous aluminum oxide (alumina) materials with large surface areas. The general nanotube growth mechanism in a CVD process involves the dissociation of hydrocarbon molecules catalyzed by the transition metal, and dissolution and saturation of carbon atoms in the metal nanoparticle. The precipitation of carbon from the saturated metal particle leads to the formation of tubular carbon solids in an sp^2 structure. Tubule formation is favored over other forms of carbon such as graphitic sheets with open edges. This is due to that a tube contains no dangling bonds and therefore is in a low energy form. For MWNT growth, most of the CVD methods employ ethylene or acetylene as the carbon feedstock and the growth temperature is typically in the range of 550–750 °C. Iron, nickel or cobalt nanoparticles are often used as catalysts. The rationale for choosing these metals as catalysts for CVD growth of nanotubes lies in the phase diagrams for the metals and carbon. At high temperatures, carbon has a finite solubility in these metals, which leads to the formation of metal–carbon solid state solutions and therefore to the aforementioned growth mechanism.

CVD methods have great potential for scaled-up nanotube materials synthesis. Multi-walled nanotubes have already been produced commercially on the kilogram to ton level. Nevertheless, these MWNTs tend to contain high densities of defects such as pentagons and heptagons on the sidewalls of the nanotubes. The success of nearly defect-free single-walled carbon nanotube growth by CVD is relatively recent.

Our group has found that by using methane as carbon feedstock, reaction temperatures in the range of 850–1000 °C and alumina supported catalyst materials, one can grow high quality SWNT materials by CVD [17–20]. High growth temperatures favor the formation of SWNTs that have small diameters and thus high strain energies, and allow for nearly defect-free tube structures. Methane is the most stable hydrocarbon against self-decomposition at high temperatures, and the catalytic decomposition of methane by the transition-metal catalyst particles is the dominant process in SWNT growth. Highly porous catalysts with strong metal–support interactions and high surface areas produce SWNTs with better quality at larger quantities. Strong metal–support interaction means high metal dispersion and thus a high density of catalytic sites on the support. The strong interaction can prevent metal species from aggregating and forming unwanted large particles that could yield to graphitic particles or defective multi-walled tubular structures in CVD. Large surface area and a highly porous structure of the catalyst increase the yield of SWNTs, because of a high density of catalytic sites owing to the former and rapid diffusion and efficient supply of carbon feedstock to the catalytic sites owing to the latter.

Liu and coworkers made significant progress recently in obtaining an excellent catalyst for methane CVD growth of SWNTs [21]. The catalyst preparation involved supercritical drying at high pressure and temperature conditions. The catalyst exhibited ultra-high surface area and porous structure as preserved by the supercritical drying condition under which destructive capillary forces are non-existent. The catalyst yielded abundant SWNTs in CVD. In general, understanding the chemistry involved in catalyst preparation and nanotube growth should lead to further breakthroughs towards scale-up of the growth of perfect SWNT materials.

CVD of methane for the growth of bulk amounts of SWNTs has been investigated by Rao and coworkers using catalysts based on mixed oxide spinels [22]. Good quality and quantity nanotubes were obtained with nanoparticles of a Fe–Co alloy. Colomer and coworkers recently reported the growth of bulk quantities of SWNTs by CVD of methane using a cobalt catalyst supported on magnesium oxide materials [23].

It has been demonstrated that catalytic growth of SWNTs can be achieved by CVD of several types of hydrocarbons and certain carbon-containing molecules. A CVD approach to SWNTs was developed by Smalley and coworkers who used supported catalysts, ethylene as carbon feedstock and growth temperature around 800 °C [24]. In this case, low partial-pressure ethylene was employed in order to reduce amorphous carbon formation due to the self-decomposition of ethylene at the high growth temperature.

Catalyst particles can be generated from the gas phase in situ for the growth of SWNTs by CVD. Cheng and coworkers reported a method that employs benzene as the carbon feedstock, hydrogen as the carrier gas, and ferrocene as the catalyst precursor for SWNT growth [25]. In this method, ferrocene is vaporized and carried into a reaction tube by benzene and hydrogen gases. The reaction tube is heated at 1100–1200 °C. The vaporized ferrocene decomposes in the reactor, which leads to the formation of iron particles that can catalyze the growth of SWNTs. More recently, the Smalley group developed a gas phase catalytic process to grow bulk quantities of SWNTs [26]. Carbon monoxide (CO) is used as the carbon feedstock and the growth temperature is in the range of 800–1200 °C. Catalytic particles for SWNT growth are generated in situ by thermal decomposition of iron pentacarbonyl in a reactor heated to the growth temperature. CO provides the carbon feedstock for the growth of nanotubes off the iron catalyst particles.

CVD and gas phase catalytic growth of nanotubes have been actively pursued in recent years because of their significant potential for materials scale-up. With continued effort in this area, it is expected that large quantities (up to tons) of high quality nanotube materials can be produced in the near future.

3. Controlled nanotube growth by chemical vapor deposition

With arc-discharge or laser ablation techniques, only tangled nanotubes mixed randomly with various impurities are obtainable. Recent research activities in CVD nanotube growth have also been sparked by the idea that aligned and ordered nanotube structures can be grown on surfaces with control [17]. Xie et al. have grown aligned MWNTs out of the pores of mesoporous silica [27,28] by a CVD approach. The catalyst used is iron oxide particles embedded in the pores of silica. The carbon feedstock is 9% acetylene in nitrogen at an overall 180 Torr pressure, and the growth temperature is 600 °C. Remarkably, nanotubes with lengths up to milimeters are made (Fig. 4a) [28]. Ren and coworkers have been able to grow large-diameter MWNTs forming oriented 'forests' (Fig. 4b) on glass substrates [29]. Plasma-assisted CVD at 660 °C is used in this work with nickel as the catalyst, and acetylene as the carbon feedstock. Fan and coworkers have obtained ordered MWNT structures by CVD on catalytically patterned substrates [30–32]. MWNTs self-assemble into aligned structures during CVD growth. Squared iron patterns on porous silicon substrates are employed for the growth. Regularly positioned arrays of nanotube towers are grown on the substrate (Fig. 4c and d). The nanotube towers exhibit very sharp edges and corners with no nanotubes branching away from the blocks. The MWNTs within each block are well aligned along the direction perpendicular to the substrate surface [30]. The mechanism of nanotube self-orientation involves the nanotube base-growth mode substrates [30]. During CVD growth, the outmost walls of the nanotubes interact with their neighbors via van der Waals forces to form a rigid bundle, which allows the nanotubes to self-orient and grow perpendicular to the substrate.

It is challenging to grow SWNTs into structures with well controlled orientations. Nevertheless, suspended SWNT networks with directionality on substrates containing lithographically patterned silicon pillars have been grown. Contact printing is used to transfer catalyst materials onto the tops of

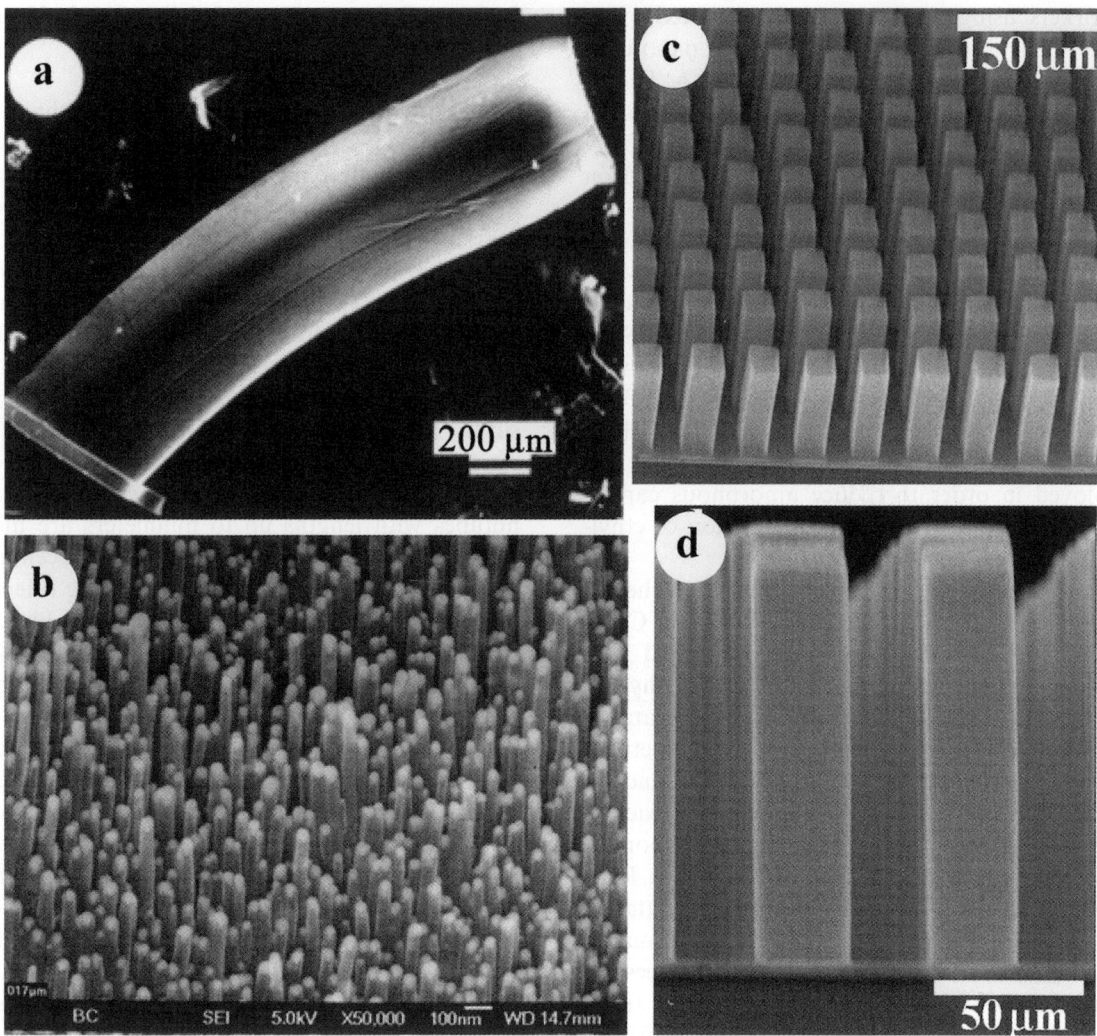

Fig. 4. Aligned MWNT structures grown by CVD methods on various substrates. These structures illustrate that ordered nanotubes can be obtained by direct chemical synthesis approaches. (a) An ultra-long nanotube bundle (courtesy of Dr. S. Xie). The slightly curved cylindrical object (diameter ∼600 μm, length ∼2 mm) is a large bundle of many aligned nanotubes along the length of the cylinder. (b) An oriented MWNT forest grown on a glass substrate (courtesy of Dr. Z. Ren). Each rod-like object in the image is a MWNT with diameter on the order of tens of nanometers. The nanotubes are grown on a glass substrate (not shown in the image) and are oriented perpendicular to the substrate. (c) SEM image of self-oriented MWNT arrays (Fan et al.). Each tower-like structure is formed by many closely packed MWNT. Nanotubes in each tower are oriented perpendicular to the substrate. (d) Microscopic "twin towers": SEM image of the side view of the towers.

pillars selectively. CVD of methane using the substrates leads to suspended SWNTs forming nearly ordered networks with the nanotube orientations directed by the pattern of the pillars (Fig. 5) [33,34]. Nanotubes are nucleated on the tops of the pillars and lengthen from there as they grow. The methane flow keeps the nanotubes floating and waving in the 'wind' since the flow velocity near the bottom surface is substantially lower than that at the level of the tower tops. This prevents

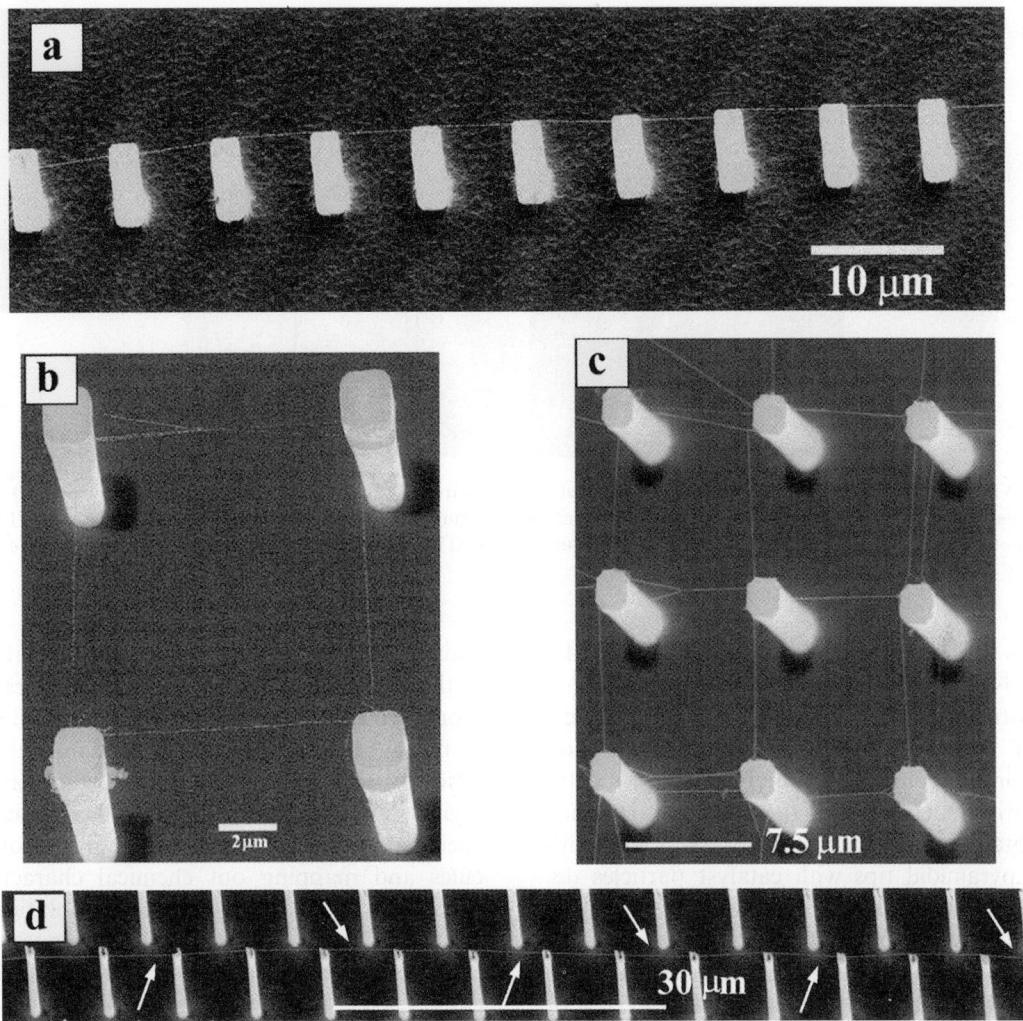

Fig. 5. 'Self-directed' growth of suspended SWNTs. (a) SEM image of suspended SWNT 'power-line' grown on a row of silicon pillars (bright post-like objects). The line-like structures bridging the posts are SWNTs. (b) A square of suspended SWNT. (c) A square network of suspended SWNTs. (d) A long SWNT bridging many silicon pillars. These results demonstrate that ordered networks of SWNTs can be obtained by self-assembled nanotube growth, which could be utilized for building interesting circuits/devices of nanotubes.

the SWNTs from being caught by the bottom surface. The nearby towers on the other hand provide fixation points for the growing tubes. If the waving SWNTs contact adjacent towers, the tube-tower van der Waals interactions will catch the nanotubes and hold them aloft. Single-wall nanotubes that are as long as 0.2 mm (200 μm) in length have been grown (Fig. 5d) [34].

An important aspect of nanotube growth is related to carbon nanotubes as advanced probe tips for scanning probe microscopy (SPM). Nanotubes present ideal characteristics for enhancing the capabilities of SPM in imaging, manipulation and nanolithography. This is owing to the sharpness of the nanotubes, their high aspect ratios, high mechanical stiffness and resilience, and tunable chemical characteristics [35–43]. Nanotube AFM tips have been previously obtained by attaching MWNTs [35,36] and SWNT bundles [40] to the sides of silicon pyramidal tips. Recently, CVD

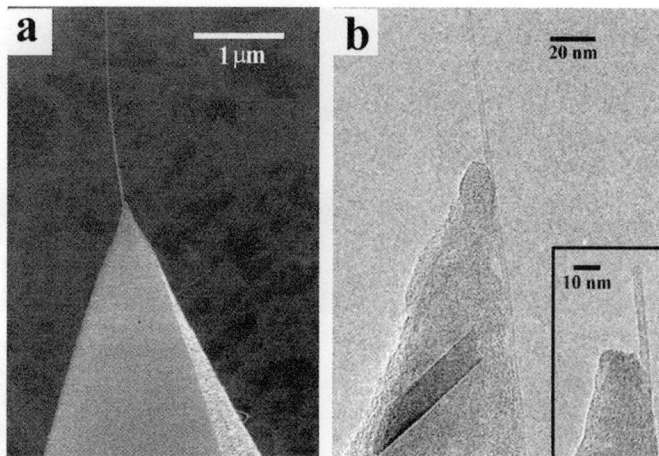

Fig. 6. SWNTs as tips for atomic force microscopy. Nanotubes can be directly grown on conventional silicon tips by CVD methods. (a) SEM image of a SWNT grown on a conical Si tip. The line-like structure extending out of the Si tip is a nanotube. (b) TEM image of the nanotube tip before and after (inset) the length of the tube is reduced by discharging. The nanotube tip has a diameter of only ∼2 nm, highly desirable for high resolution AFM imaging.

methods were used to directly grow MWNTs [41] and SWNTs [42,43] on silicon tips for the synthesis of nanotube AFM tips. Hafner and Lieber grew MWNT probe tips by the CVD of ethylene on silicon probes containing straight holes with catalyst particles deposited at the bottom [41]. They also obtained SWNT probe tips by CVD with silicon pyramidal tips with catalyst particles deposited on the sides of the silicon tips [42]. Our group has grown SWNT tips using dip coating of silicon pyramids in a liquid phase catalyst [33] followed by CVD of methane [43]. In the case of SWNT tips, the orientations of the nanotubes (individual or small bundles, Fig. 6a and b) are maintained by van der Waals interactions between the nanotube and the pyramid surface. The SWNTs extending from silicon pyramid tips typically range 1–20 μm in length beyond the pyramid tip. The nanotubes are shortened to ∼30–100 nm in order to obtain rigid AFM probe tips (e.g., Fig. 6b inset) needed for probing [43].

Wong and coworkers obtained chemically functionalized carbon nanotube probe tips by opening the ends of nanotubes and terminate the ends with oxygen containing groups such as –COOH [36]. Nanotube tips with a variety of functional chemical groups at the ends were obtained. Biological molecules were also attached to the ends of nanotube tips by reacting functional groups at the ends of nanotubes with reactive sites on the biomolecules. These nanotube tips were shown to be not only superior in obtaining high lateral resolution in real-space imaging of surfaces and biological molecules, but also promising for measuring chemical forces between small numbers of molecules and mapping out chemical characteristics of surfaces with 1 nm spatial resolution [36]. Nanotube tips were also ideal for nanoscale surface modification and scanning probe lithography [38,43]. MWNT tips were used to fabricate ∼10–20 nm wide oxide lines on silicon surfaces. SWNT tips were used to fabricate oxide structures with feature size below 10 nm [43]. Nanotubes tips had an overall longer life-time in both imaging and nanofabrication processes than silicon tips owing to their excellent mechanical strength and resilience.

4. Probing the properties of individual nanotube molecular wires

4.1. Correlation between atomic structures and electronic properties

Nanotubes are ideal for studying structure–property relations for atomically well-defined one-dimensional materials. Elegant work has been carried out by Lieber and Dekker's group who used scanning tunneling microscopy (STM) to directly reveal the real-space atomic structures of SWNTs with various chiralities. At the same time, the electronic structures of the nanotubes are probed and then correlated with the structural properties [44,45]. In STM, electrons tunnel between a sharp metal tip biased against a conducting substrate (in this case nanotubes deposited on metal surfaces). The tunnel current is proportional to the local electronic density of states of the nanotube. Real-space images reflecting the atomic corrugation of the nanotubes can be obtained (Fig. 7a). Local electronic density of states can be directly measured by current vs. bias-voltage characteristics recorded with the tip positioned above a specific point of the nanotube [46] (Fig. 7b). STM and spectroscopy are powerful in correlating microscopic structures with electronic properties in solid state systems, and this is certainly the case for nanotubes. Excellent agreement is found with the main theoretical predictions. That is, arm-chair SWNTs are metallic with a finite density of states at the Fermi level, and semiconducting tubes exhibit energy gaps with zero density of states inside the gap. As expected, the energy gaps of semiconducting SWNTs scale with $1/d$ where d is the diameter of the nanotube. STM studies have also probed interesting quantum phenomena associated with the finite lengths of SWNTs. Venema et al. observed electron standing waves along the axis of short nanotubes by direct STM imaging. The electron density oscillates periodically along the nanotube with wavelength $\approx 2L/n$ (L is the tube length, n the integer). This is a quantum confinement effect as electrons behave like 'particles-in-a-box' with the box being a 1D wire in this case. Odom et al. observed single electron charging effects for electron tunneling from an STM tip to short nanotubes deposited on substrates [47].

4.2. Electron transport in nanotubes

Integrating individual nanotubes into addressable structures is important to elucidate the electron transport properties of nanotubes. Dekker and coworkers made electrical contacts to indi-

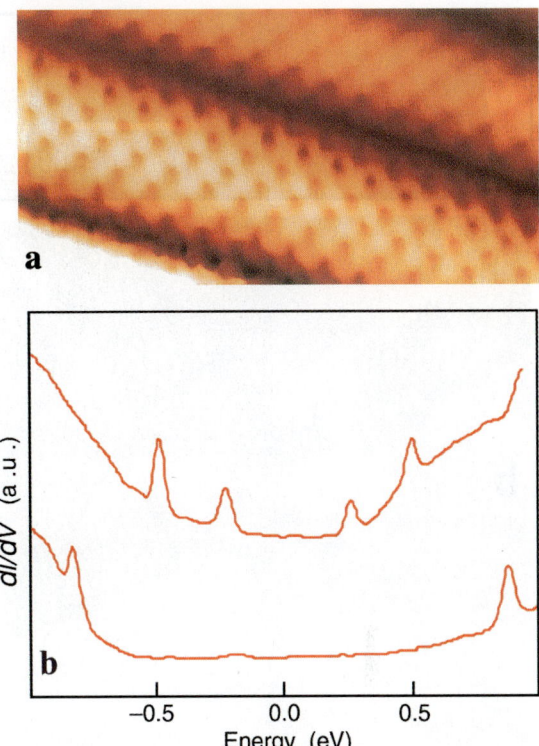

Fig. 7. Probing the real-space atomic structures of SWNTs and correlating structures with electronic properties (courtesy of Dr. C. Lieber). (a) Atomically resolved (dots) STM image of a (12,−2) semiconducting tube (center region between the two darker lines) in a bundle. The dots are associated with the atomic corrugations. (b) Local density of states $(dI/dV)/(I/V)$ for a metallic (bottom curve) and semiconducting (top curve) SWNT respectively. The two curves are shifted artificially for clarity. For the semiconducting nanotube (top curve), the density of state in the energy range of −0.2 to 0.2 eV is zero, corresponding to a semiconducting band gap. For the metallic nanotube (bottom curve), the density of states is non-zero at all energies without a band gap. Density of states is defined as the number of electronic states as a function of electronic energy.

vidual SWNTs grown by laser ablation by depositing SWNTs from liquid suspensions onto predefined electrodes [48]. McEuen and coworkers used AFM to locate SWNTs deposited on a substrate and then placed metal electrodes to contact the nanotubes [49–51]. Our group developed a growth and integration method to contact SWNTs grown from patterned catalyst islands. Large numbers of individually addressable nanotubes are obtained this way (Fig. 8a) [17,31,32,52–58].

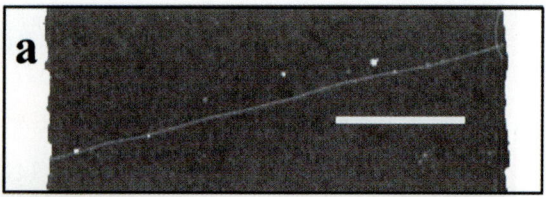
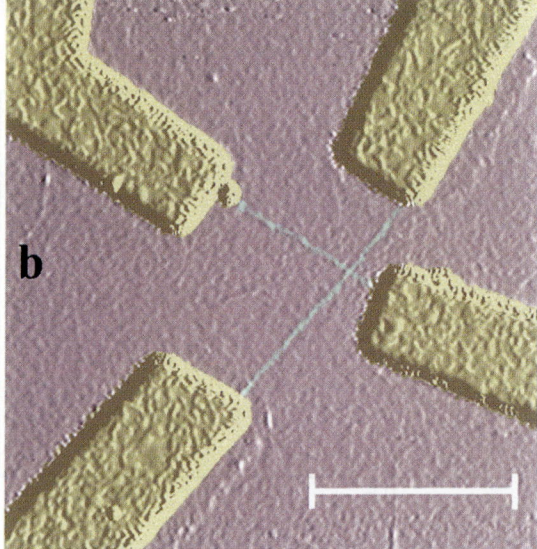
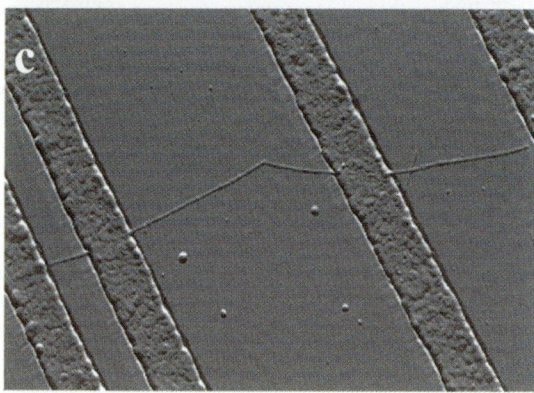

Fig. 8. Connecting SWNTs to the macroscopic world for electrical measurements. (a) AFM image of a SWNT contacted by metal electrodes (bright regions at the left and right of the image). (b) AFM image of two crossing nanotubes each connected to two metal electrodes (courtesy of Dr. P. McEuen). (c) AFM image of a nanotube heterojunction formed by a metallic tube connected to a semiconducting tube (courtesy of Drs. C. Dekker and Z. Yao).

Complex SWNT electrical devices have also been obtained by various approaches. These include intra-tube heterojunctions containing sharp kinks at the junctions [59] (Fig. 8c), SWNT crosses on substrates [57,60] (Fig. 8b) and cross-structures of suspended SWNTs [61].

The ability to connect to nanotubes from the macroscopic world has allowed the investigation of electrical transport properties of individual nanotubes. Generally, a SWNT is identified to be metallic when its room temperature electrical conductance is insensitive to gate voltages applied by an electrode hundred of nanometers away from the tube. The role of the gate voltage is to electrostatically change the chemical potential or Fermi energy of the nanotube. For a metallic tube, the shift in Fermi energy does not significantly change the density of states at the Fermi level. Therefore no significant change in the conductance is expected. An intriguing question has been under what conditions metallic SWNTs behave as ballistic wires. For a metallic SWNT, the Fermi level crosses two energy bands. Therefore, two channels and four electronic states (due to spin up and down degeneracy) should contribute to transport. In the absence of electron scattering and ideal electrical contacts to the nanotube, the conductance of the system should be $4e^2/h = 2G_0$ [62], where $G_0 = 2e^2/h$ is the conductance quantum. That is, each channel should contribute G_0 to the conductance of the system. The corresponding resistance for ballistic transport in a metallic SWNT should be 6.45 kΩ, one half of the resistance quantum $R_Q = h/2e^2$.

Tans [48] and Bockrath [49] carried out the earliest transport measurements of individual metallic SWNTs. These nanotube samples have high resistance on the order of megaohms mainly due to poor metal–tube contacts. Interestingly, nanotubes were found to act as quantum wires. Quantum phenomena observed in these samples include single-electron charging and resonance tunneling through discrete energy states of nanotubes arising from quantum confinement along the length of the tubes [48,49]. Single-electron charging or Coulomb blockade [63] occurs when two conditions are met by a small metallic dot (in this case a tube) connected to two electrodes. The first is that the connections are weak characterized by contact resistance larger than R_Q. The second is that the dot is sufficiently small, therefore the capacitance is small and the energy needed for adding an electron to the system e^2/C is larger than thermal energy K_BT. The capacitance of a nanotube scales linearly with the inverse of its length. For a micron long nanotube, the single-electron charging energy is sufficiently high for the Coulomb blockade to be observed at liquid helium temperature 4.2 K [48,49]. In the Coulomb blockade regime, transport occurs via individual electrons. Discrete energy levels exist for electrons in a nanotube with finite length, the physics of which is well described by the particle-in-a-box picture with the box being a 1D wire with finite length. When a discrete energy state is aligned with the Fermi level of the system, resonance tunneling occurs [48,49]. Nanotubes are ideal systems to study quantum effects at relatively high temperatures due to their small size, relatively large Coulomb charging energy and energy level spacing for sub-micron long nanotubes.

Single electron charging related phenomena are observed in SWNT samples with high contact resistance. With improved metal–tube contacts, the intrinsic electrical properties of nanotubes can be elucidated. Low contact resistance has been obtained by the growth and integration approach involving titanium electrodes evaporated onto SWNTs [52]. Metallic SWNTs with several microns length between electrodes exhibit two-terminal resistance on the order of tens to hundreds of kΩ. The resistance of a 3 μm long SWNT can be as low as 12 kΩ (Fig. 9a). The linear resistance of the sample decreases as the temperature is decreased, and a slight upturn is observed below 30 K in the resistance vs. temperature curve. Coulomb blockade is not observed because the contact resistance is less than R_Q. The decrease of resistance as temperature decreases is due to reduced phonon scattering at low temperatures [64,65]. The upturn in resistance at low temperature is believed to be due to electron localization [64,65] in 1D systems. Localization is essentially a quantum interference phenomenon that the electron wave function decays exponentially from a point due to a random potential distribution caused by disorder. Localization effects have also been observed by Avouris and coworkers in SWNT rings [66].

Yao and Dekker obtained low resistance metallic SWNT samples by placing SWNTs on top of

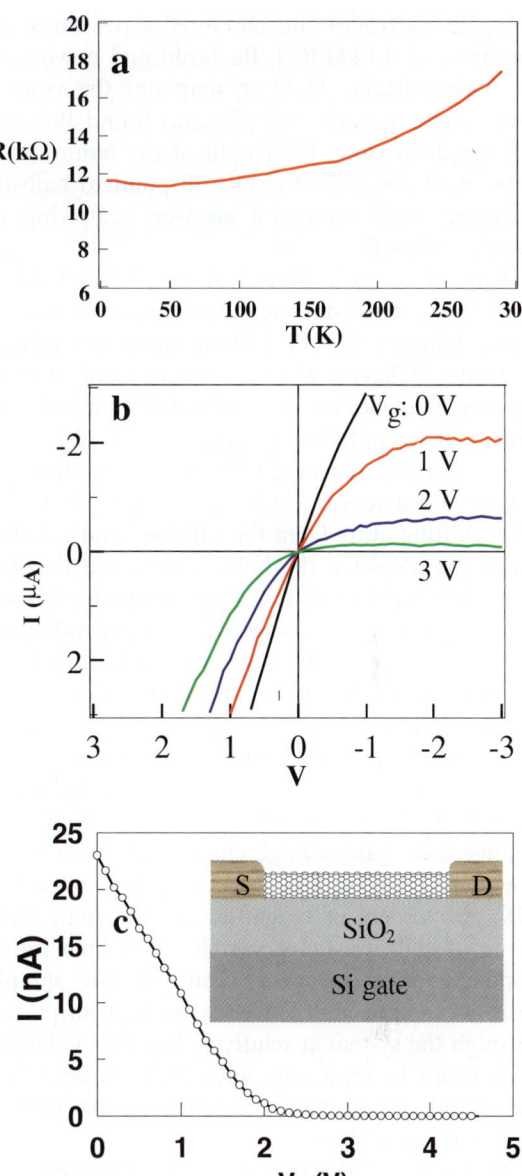

Fig. 9. Electrical properties of individual SWNTs. (a) Resistance as a function of temperature for a metallic SWNT contacted by two Ti electrodes. (b) I–V curves for a semiconducting SWNT under various gate voltages V_g. The gate voltage is applied to the sample substrate as illustrated in the inset below. (c) I vs. V_g characteristics for the nanotube. The inset shows the schematic sample configuration. The metal-contacted nanotube device with source (S) and drain (D) electrodes is fabricated on a 500 nm thick silicon oxide surface. The silicon substrate underneath the silicon oxide is highly doped and electrically conducting. A gate voltage can be applied to this substrate to capacitively couple to the nanotube.

flat gold electrodes and measured a resistance on the order of 15 kΩ [67]. Bachtold and coworkers used electrostatic AFM to map out the voltage drop across metallic SWNTs and found that the voltage drop along the length of the nanotube is very small [68]. These results all point to ballistic transport with minimum electron scattering in metallic SWNTs.

Tans et al. [69], Martel et al. [70] and Zhou et al. [54] carried out measurements of individual semiconducting SWNTs. These nanotubes exhibit transistor behavior at room temperature, that is, their conductance can be changed dramatically (by orders of magnitude) by gate voltages (Fig. 9b). The nanotubes appeared to be doped with holes as p-type. Positive gate voltages caused the Fermi level shifting away from the valence band into the band gap, depleted the holes and turned the system into insulating states. p-type transistor behavior was observed consistently in semiconducting SWNTs by a number of groups. Hole doping of SWNTs is attributed to adsorbed oxygen [71]. With good metal–tube contacts, low resistance semiconducting SWNT samples on the order of hundreds of kiloohms were obtained by Zhou et al. [54]. The transconductance (ratio of current change over gate-voltage change) of these semiconducting tube samples can be up to ∼200 nA/V [54], two orders of magnitude higher than measured in high resistance samples. The high transconductance is a direct result of low sample resistance, since high currents can be transported through the system at relatively low bias voltages. This result is important since high transconductance and voltage gain is essential to the performance of transistors.

McEuen and coworkers studied the electrical properties of crosses of SWNTs [60]. Electron transport from metallic to metallic, metallic to semiconducting, and semiconducting to semiconducting tubes were investigated. The conductance of the metal–metal tube junction was found to be ∼0.1 e^2/h, suggesting appreciable electron scattering at the crosses. Metallic and semiconducting tubes were found to form Schottky junctions at the crosses. A Schottky junction is typically formed at the interface between a bulk semiconductor and a pure metal. The difference in workfunction between the two materials causes energy band bending in the semiconductor side. Characteristic of a Schottky junction is rectifying behavior in the current vs. voltage curves [60]. Current flows across the junction only when the metal side is negatively biased. Tombler et al. studied SWNT crosses and used scanning probes as gates to identify the metallic or semiconducting nature of the crossing nanotubes [57].

Chico et al. proposed theoretically that by joining a metallic and semiconducting SWNT with topological defects, such as pentagons and heptagons, novel intramolecular devices can be obtained [72]. Yao et al. investigated intra-tube devices experimentally and observed a sharp kink on an individual SWNT (Fig. 8c) [59]. Electron transport across the kink exhibited rectifying behavior (Fig. 10). The system was suggested to contain a metallic tube joining a semiconducting tube with a Schottky junction formed at the kink. This result demonstrated that extremely small electrical devices are obtainable on an individual molecular wire.

4.3. Chemical doping

Chemical doping effects to the electrical properties of SWNTs have been investigated by several groups. SWNT doping with electron withdrawing

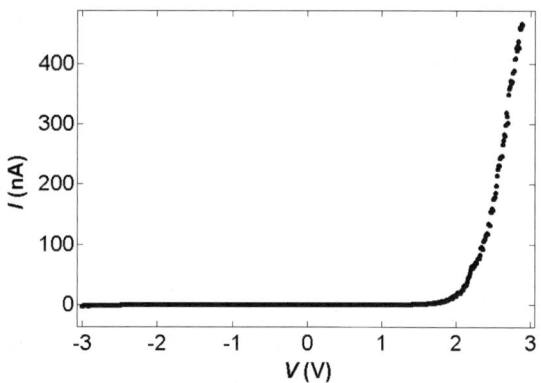

Fig. 10. SWNT heterojunction as intra-molecular rectifier: the world's smallest rectifying junction. The rectifying current vs. voltage characteristics was observed with a SWNT containing a sharp kink in the structure (Fig. 8c).

(Br$_2$, I$_2$) and donating species (K, Cs) were first carried out with bulk SWNT mats by Lee et al. and Grigorian et al. [73,74]. Bundles of SWNTs were also studied [75,76]. With bulk mats, chemical doping can lead to orders of magnitude reduction in the resistance of bulk SWNT materials due to the increase in the average of hole or electron carriers in the nanotubes. The doping species absorb on to nanotube surfaces and intercalate into the empty space of bundles of SWNTs [73,74].

It is necessary to investigate individual SWNTs since doping effects in metallic and semiconducting SWNTs can be drastically different. Bockrath et al. carried out potassium doping of a semiconducting SWNT at room temperature. A transition from p-type to n-type was observed as K atoms absorb onto the outer surface of the nanotube and donate electrons to the tube [76]. Similar phenomenon was observed by Kong et al. [77]. Fig. 11a shows the evolution of the conductance as a function of time during K-doping for a diameter ~2.5 nm and length = 0.4 µm SWNT contacted by Ti electrodes. The conductance decreased dramatically by five orders of magnitude and then recovered. This corresponds to the nanotube evolving from p-type to intrinsic and then to n-type semiconducting states under progressive K-doping.

More can be learned about K-doping of single tubes by characterizing the doped nanotubes at low temperatures. Temperature dependent measurements of K-doped SWNTs reveal Coulomb blockade [63] oscillations at high temperatures up to 160 K (Fig. 11b). The Coulomb oscillation peaks in gate voltage (V_g) is periodic and evenly spaced at $\Delta V_g \sim 0.45$ V in regime I of Fig. 11b, where each peak corresponds to charging and transporting one electron through the system. In regime II, the oscillations are periodic with $\Delta V_g \sim 0.45$ V at high temperatures but some of the peaks split into doublets at low temperatures ($T < 110$ K). In regime III, periodic oscillations are observed only at $T < 90$ K with a small period of $\Delta V_g \sim 0.15$ V. These peaks show no splitting down to 20 K. The change of ΔV_g by ~0.45 V causes one electron addition or removal from the dot, the gate capacitance is therefore $C_g = e/\Delta V_g = 0.35$ aF. The total capacitance of the system is estimated to be $C_\Sigma = 3.72$ aF, corresponding to a charging energy of $U = e^2/C_\Sigma = 43$ meV. For charging energy $U \sim 50$ meV, the size of the quantum dot seen in regime I appears smaller than the tube length $L = 0.4$ µm between the edges of metal electrodes. The dot in regime I has an effective length of $L_{\text{eff}} \sim 0.1$–0.15 µm, about one third of the actual geometry of the nanotube. This phenomenon is attributed by an inhomogeneous K-doping profile along the nanotube, as a result of the randomness of K-evaporation onto the sample. That is, an inhomogeneous effect can lead to a high temperature SET due to the formation of a single quantum dot with its size smaller than the geometrical tube length. This can occur when the two barriers for the dot are located within the length of the tube while the rest of the system is well coupled to the metal contacts. This suggests that to obtain high temperature nanotube SETs reliably, it is desired to obtain samples with 10–50 nm tube lengths and have great chemical homogeneity along the lengths.

Single-electron transistors operating at room temperatures have been pursued actively with Si and metal systems [78–82]. Nanotubes are potential candidates for high performance SETs. With nanotubes, room temperature single-electron charging will require a tube 'dot' with length on the order of 10–50 nm, so that charging energy can be $U \geqslant 100$ meV $\sim 4 K_B T$ (300 K). Room temperature nanotube SETs could be realized in the near future.

4.4. Nanotube electromechanical properties

The question of how mechanical deformation affects the electrical properties of carbon nanotubes has been under theoretical and experimental studies due to the potential application of nanotubes for nanoscale electro-mechanical (NEMs) devices. The effects of mechanical deformation to the electrical properties of nanotubes were studied theoretically by several groups. Nardelli and coworkers modeled the bending of an arm-chair metallic SWNT and calculated the electrical conductance vs. bending angle [83–85]. The nanotube conductance was found to change only slightly under small bending angles. Rochefort et al. carried out simulations of nanotube bending and found that at larger bending angles ~45°, the electrical conductance of a metallic SWNT can be lowered by up to

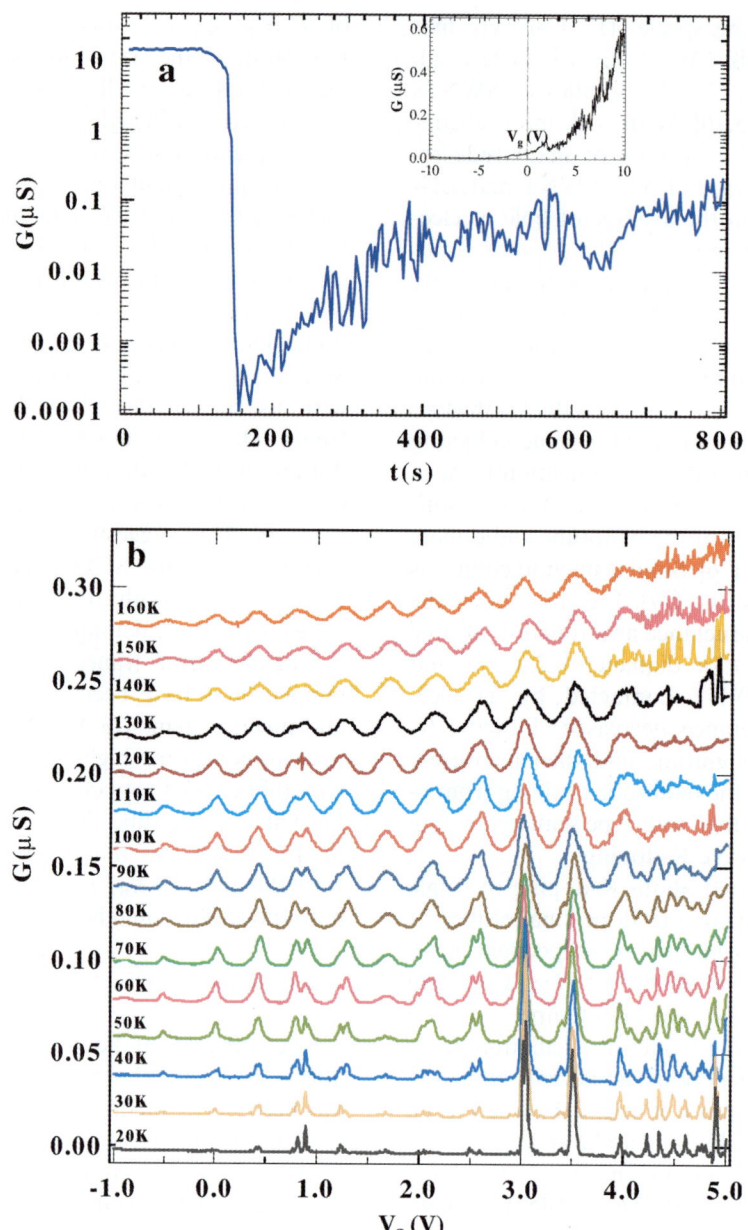

Fig. 11. Elucidating how chemical doping affects the electrical properties of individual SWNTs. (a) Conductance vs. time for a semiconducting SWNT during K doping. The drastic conductance decrease to a nearly insulating state followed by a recovery corresponds to transitions from a p-type (hole rich) to intrinsic to n-type (electron rich) semicondutor, due to electron donation from adsorbed K atoms. Inset: conductance vs. V_g curve at room temperature showing n-type behavior after K-doping. (b) Conductance vs. V_g curves for the doped SWNT at various temperatures showing Coulomb oscillations corresponding to single-electron charging effect. The K-doped nanotube behaves as a single-electron transistor functioning at relatively high temperatures.

10-fold. The conductance decrease was attributed to a σ–π hybridization effect, i.e., some sp³ bonding characteristics developed at the bent due to increased curvature under high bending angles

[84,85]. The sp³ atomic structure causes the electrons to be localized and is responsible for the reduced nanotube conductance.

Experimental investigation of the electromechanical properties of nanotubes can be carried out with suspended nanotubes, as shown by Tombler et al. [58]. A nanotube can be grown from patterned catalyst sites across pre-fabricated trenches on SiO_2/Si substrates [58]. This leads to an individual SWNT that is partially suspended over the trenches (Fig. 12a). The suspended part of the nanotube can be manipulated with an AFM tip, while the resistance of the sample is being monitored (Fig. 12b). The nanotube conductance decreases each time the AFM tip pushes the nanotube down, but recovers as the tip retracts (Fig. 12c). The full reversibility of the nanotube electrical conductance upon tip retraction suggests that the metal–tube contacts are not affected each time when the tip deflects the suspended part of the nanotube. The observed change in sample conductance is entirely due to the mechanical deformation of the SWNT caused by the pushing tip.

The conductance is found to decrease by a factor of 2 at ∼5° bending angle (strain ∼0.3%), but decreases more dramatically by two orders of magnitude at a bending angle ∼14° (strain ∼3%, Fig. 12d) [58]. Wu and coworkers have carried out order-N non-orthogonal tight-binding molecular-dynamics simulations of a tip deflecting a metallic (5,5) SWNT, with the tip modeled by a short and stiff (5,5) SWNT cap [58,86]. At relatively small bending angles, the nanotube is found to retain sp² bonding throughout its structure, but exhibits significant bond distortion for the atoms in the region near the tip. As tip-pushing and bending proceed, the nanotube structure progressively evolves and larger structural changes occur in the nanotube region in the vicinity of the tip. At a 15° bending angle, the average number of bonds per atom in this region is found to increase to ∼3.6, suggesting the appearance of sp³-bonded atoms (marked in red in Fig. 12e). This causes a significant decrease in the local π-electron density as revealed by electronic structure calculations. Since the π-electrons are delocalized and responsible for electrical conduction, a drastic reduction in the π-electron density is responsible for the significant decrease in conductance. Simulations find that the large local sp³ deformation is highly energetic, and its appearance is entirely due to the forcing tip. The structure is found to fully reverse to sp² upon moving the tip away in the simulation. The combined experimental and theoretical study leads to an in-depth understanding of nanotube electromechanical properties, and suggests that SWNTs could serve as reversible electro-mechanical transducers that are potentially useful for NEMs devices.

5. Surface science in nanotubes

5.1. Metal–nanotube interactions

The interactions between nanotubes and various metals are important to low resistance ohmic contacts to nanotubes and other issues such as forming metal or superconducting nanowires on nanotube templates [87–89]. Experimental and theoretical investigations of the interactions between metals and various carbon materials have been carried out previously. It is known that in three-dimensional bulk materials, different metals exhibit different interactions with carbon. The ability for transition metals to bond with carbon atoms increases with the number of unfilled d-orbitals. Metals such as Al, Au and Pd have no d-vacancies and negligible affinity for carbon. Metals with few d-vacancies such as Ni, Fe and Co exhibit finite solubility for carbon in certain temperature ranges. 3d and 4d metals with many d-vacancies such as Ti and Nb can form strong chemical bonds with carbon and thus highly stable carbide compounds. It is generally believed that the interactions between deposited metals and a graphite basal plane are weak [90–99]. The interactions are suggested to be through van der Waals forces and do not involve chemical bond formations between the metal and carbon atoms in the graphite basal plane. For C_{60} [100–102], it has been shown that certain metals interact much more strongly with C_{60} than graphite including Ti and Ni [100,102]. Deposition of Ti on C_{60} can lead to the formation of Ti–C carbide bonds as revealed by X-ray photoelectron spectroscopy [100].

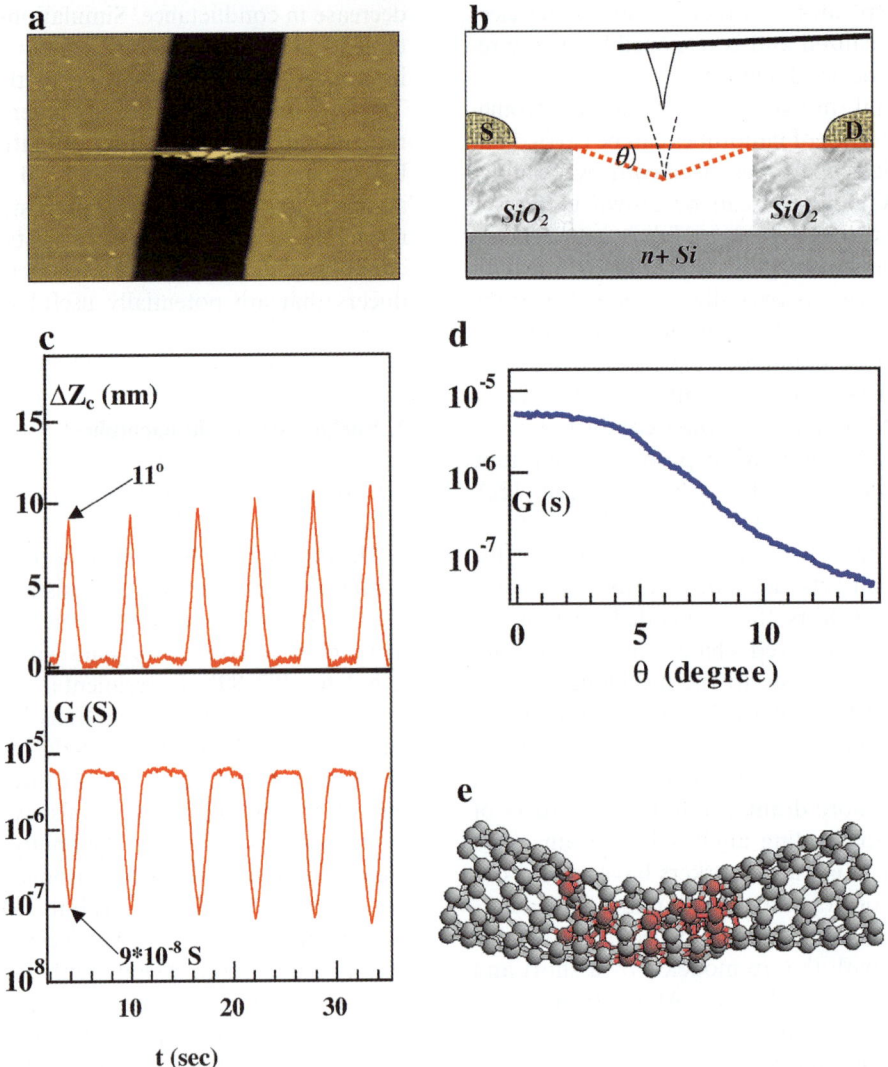

Fig. 12. Elucidating how mechanical deformation affects the electrical properties of SWNTs: electromechanical properties. (a) AFM image of a suspended SWNT over a trench. (b) Experimental scheme for measuring the electromechanical property of the nanotube. The suspended nanotube is pushed downwards by an AFM tip, while the deflection of the cantilever and the conductance of the nanotube sample across the source (S) and drain (D) electrodes are monitored. (c) Upper panel: cantilever deflection (ΔZ_c) vs. time during repeated pushing-down of the suspended nanotube by the AFM tip. The cantilever deflection signal can be used to compute the vertical deflection and bending angle (θ) of the nanotube. Lower panel: Conductance (G, in unit of Ω^{-1}) of the nanotube vs. time recorded simultaneously as the deflection signal. (d) Nanotube conductance vs. bending angle obtained by analyzing the data in (c). (e) Atomic configuration revealed by molecular dynamics simulations (MD) at a bending angle of 15° manipulated by a simulated tip. The red atoms are highlighted because they are bonded to four other atoms (sp^3-like bonding) instead of three atoms as in the unperturbed form of the nanotube. Such structural distortion leads to energy barriers to electron transport and is mainly responsible for the reduced electrical conductance of the nanotube system.

A nanotube differs from a graphene sheet in that the cylindrical sidewall of a nanotube is curved and is in a non-planar sp^2 bonding configuration.

A nanotube differs from C_{60} in dimensionality, radius and that C_{60} contains pentagons in its structure, whereas the sidewall of a nanotube contains exclusively hexagons. C_{60} therefore is expected to be more reactive with metals than a nanotube because of the pentagons in the structure. A recent theoretical study by Menon et al. shed some light into metal–tube interactions [103]. They used a tight-binding molecular-dynamics method to calculate bonding configurations of Ni with the sidewall atoms on a SWNT, and compared the results with Ni bonding on a graphene sheet. Covalent bonding characteristics of Ni (on certain sites) with carbon atoms on the nanotube were identified from the calculations. The interaction was found to be stronger than ionic-like (charge transfer) bonding with a graphene sheet. The strong Ni-SWNT interaction was attributed to curvature-induced rehybridization of carbon sp^2 orbitals with the Ni d-orbital [103].

An experimental investigation of metal–tube interaction was carried out recently by Zhang et al. by studying the structures of various metals formed on suspended SWNTs deposited by electron beam evaporation [104]. 5 nm of Ti, Ni, Pd, Au, Al, and Pb are evaporated to coat suspended SWNTs and drastically different coating results are observed (Fig. 13). Ti forms continuous nanowires on suspended SWNTs (Fig. 13a). Ni and Pd form quasi-continuous and uniform coatings on SWNTs (Fig. 13b and c respectively), with the wire structures occasionally disconnected along the lengths. Au, Al, and Pb deposited on the nanotubes form disconnected crystalline particles, leaving sections of the nanotubes free of metal coating (Fig. 13d, e and f respectively). The Au particles decorating SWNTs are up to ~60 nm wide, much larger than the deposited film thickness. This indicates that Au atoms deposited on the tubes have migrated and merged together to form large particles.

The continuous and uniform Ti coating on nanotubes suggests a high nucleation density and strong Ti–SWNT interaction. On the contrary, the Au and Al coatings are highly discontinuous with a very low nucleation density due to weak Au, Al–SWNT interactions. A low Au-tube binding energy points to a low activation barrier for adatom diffusion. Rapid motion of Au atoms on the nanotube sidewall can cause the atoms or even small clusters to merge into isolated large particles [98]. Ti atoms deposited on nanotubes exhibit the highest condensation/sticking coefficient among Ni and other metals. The Ti–SWNT interaction should be stronger than that for Ni–SWNT and could involve covalent bonding due to the high affinity of Ti for carbide formation and the curvature induced rehybridization effect. The intimate Ti–SWNT interaction and the fact that Ti–C is highly conducting are consistent with the result that low resistance ohmic electrical contacts to individual SWNTs can be reliably made by depositing Ti electrodes onto nanotubes [52]. It was found previously that Ti also makes ohmic contacts to doped diamonds. In this case, as deposited Ti forms carbide bonds with the very surface of diamond, and ohmic contacts are obtained by thermal annealing that leads to a layer of carbide at the Ti–diamond interface [105].

5.2. Interactions with small molecules: nanotube chemical sensors

How molecules interact with carbon nanotubes and affect their physical properties is of fundamental interest, and may have important implications to their applications. The electrical properties of metallic SWNTs are relatively insensitive to their chemical environment and interactions with other species. For an example, potassium doping of a metallic SWNT does not affect its conductance significantly [106], because the band shift due to charge-transfer interactions does not change the density of states at the Fermi level for a metallic tube. Semiconducting SWNTs, on the other hand, are very sensitive to K-doping, and can change from p-type to n-type accompanied by orders of magnitude change in conductance as discussed earlier. Semiconducting SWNTs are also very sensitive to gas molecule adsorption, exhibiting significant changes in their electrical conductance.

Colins et al. [71] have studied molecular oxygen adsorption on carbon nanotubes. The electrical properties of nanotubes are found to be highly sensitive to oxygen adsorption effects. In a vacuum chamber, the conductance of bulk SWNTs increases when oxygen is introduced into the system and recovers when vacuum is restored. Full desorption of oxygen can occur when heating the

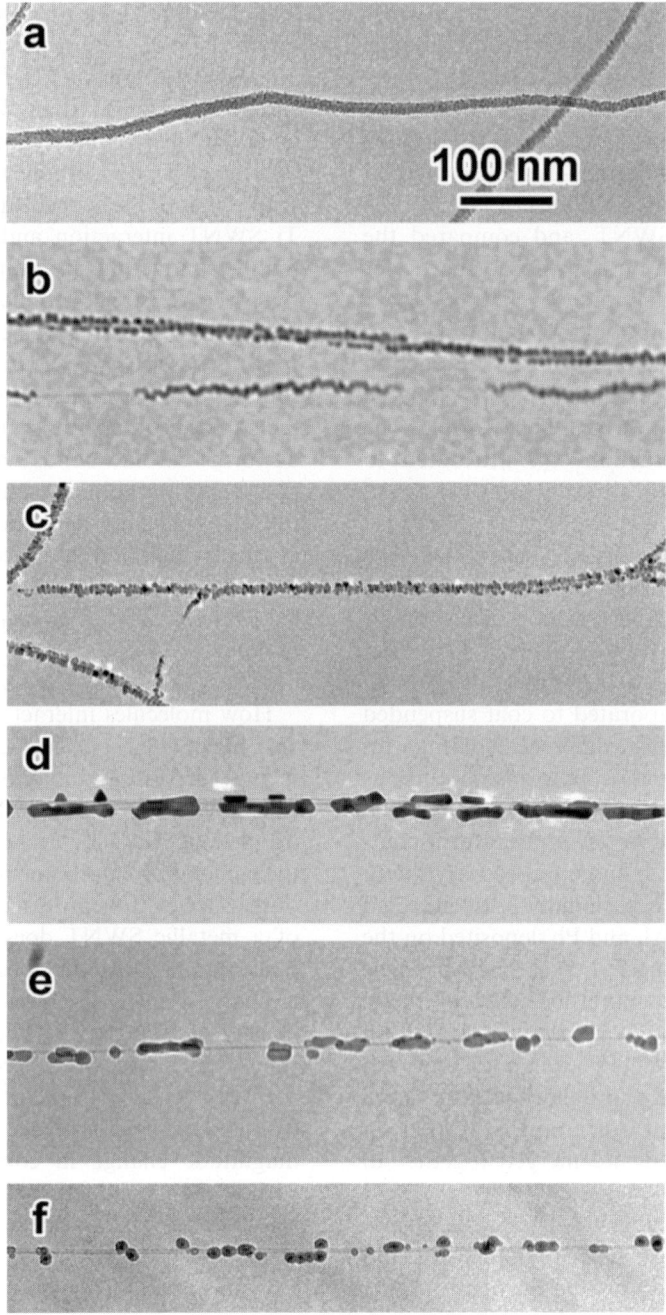

Fig. 13. TEM study of metal–SWNT interactions. The images show the structures of various types of metal deposited onto suspended SWNTs. (a) 5 nm Ti coating on a suspended SWNT. Clearly, Ti atoms deposit uniformly on the SWNT. (b–f) Structures of 5 nm Ni, Pd, Fe, Au, Al and Pb coatings respectively on suspended SWNTs. These metals tend to form discrete particles of various sizes when deposited on SWNTs due to weak metal–tube interactions.

nanotubes to high temperatures in vacuum. Cohen's group carried out density-functional theory (DFT) calculations on oxygen–SWNT complexes and found that oxygen adsorbs on a nanotube with a binding energy of approximately 0.25 eV [107]. An oxygen molecule withdraws about 0.1 electron from the nanotube, causing hole-doping to the nanotube.

Kong et al. found that nanotubes can be used for miniature chemical sensors to detect small concentrations of gas molecules with high sensitivity at room temperature [55]. Chemical sensors based on individual or ensembles of SWNT can detect chemicals such as nitrous oxide (NO_2) and ammonia (NH_3) [55]. For a semiconducting single-wall nanotube exposed to 200 ppm of NO_2, it was found that the electrical conductance can increase by up to three orders of magnitude in a few seconds (Fig. 14a). On the other hand, exposure to 2% NH_3 caused the conductance to decrease by up to two orders of magnitude (Fig. 14b). As a general comparison, conventional solid state sensors for NO_2 and NH_3 typically operate at temperatures over 400 °C, and conducting polymers provide only limited sensitivity. Sensors made from SWNT have high sensitivity and a fast response time at room temperature, which are important advantages for sensing applications.

DFT calculations carried out by Cho and coworkers have revealed that an NO_2 molecule can bind to a semiconducting SWNT with a binding energy of ∼0.9 eV (18.4 kcal/mol) [55]. This suggests that the nature of the molecule–tube interaction is strong physisorption and approaching the chemisorption regime. The oxidizing NO_2 molecule withdraws about one-tenth of an electron charge from the nanotube. The charge transfer leads to increased hole carriers and enhanced conductance for the p-type nanotube. The interaction between NH_3 and a SWNT is physisorption in nature. NH_3 is a Lewis base that can donate a small amount of electrons to nanotubes and therefore reduce the hole-carriers.

Modification of nanotubes by chemical or physical means should represent a viable way to enable the development of highly sensitive and selective chemical sensors useful for practical applications. For an example, semiconducting SWNTs coated

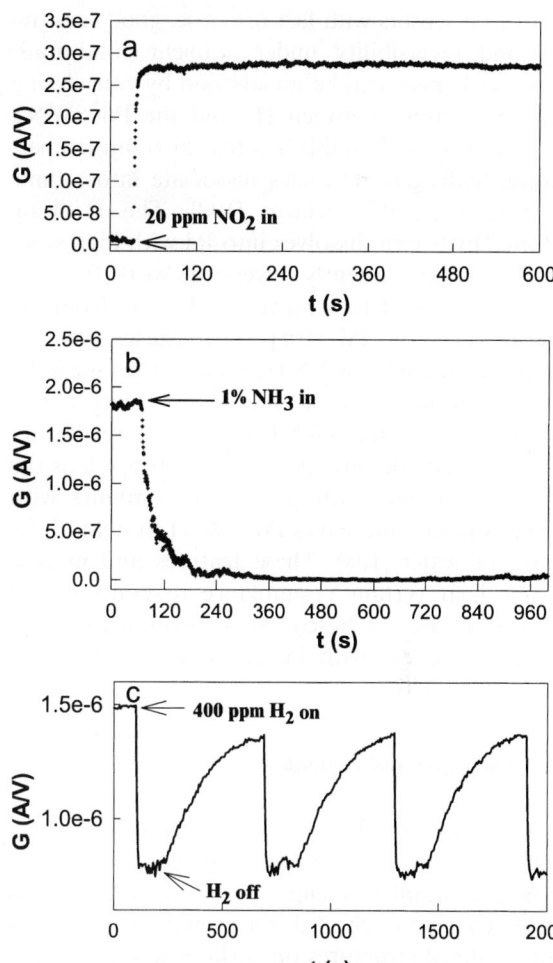

Fig. 14. Elucidating the interactions between molecules and SWNTs and exploring carbon nanotubes for advanced miniature chemical sensors. (a) Conductance response of an individual semiconducting SWNT to 200 ppm of NO_2 in argon. (b) Conductance response of an individual SWNT to 1% NH_3 in argon. (c) Conductance of a Pd-decorated semiconducting SWNT vs. time in a flow of air with 400 ppm of H_2 on and off over several cycles. These results illustrate the potential of nanotube based chemical sensors.

with a non-continuous thin layer of palladium metal become extremely sensitive to molecular hydrogen. The conductance of a palladium-coated semiconducting single-wall nanotube decreases rapidly upon exposure to a flow of air mixed with 400 ppm of hydrogen, and reverses when the hydrogen flow is turned off (Fig. 14c). This clearly shows that palladium modified SWNTs are excellent hydrogen sensors with fast response, good sensitivity and reversibility under ambient conditions. The mechanism can be understood by considering the interactions between H_2 and the Pd–SWNT system. It is well established that at room temperature, hydrogen molecules dissociate into atomic hydrogen on Pd surfaces [108]. The resulting atomic hydrogen dissolves into Pd with high solubility and consequently lowers the work function (energy required to remove an electron from the Fermi level) of Pd [108]. This causes electron transfer from Pd to SWNTs, which lowers the hole-carriers in the nanotube and thus its conductance.

The Pd-modified SWNT sensors are readily reversible when the hydrogen flow is stopped, as the dissolved atomic hydrogen in Pd combines with oxygen in air and leaves Pd–SWNT system in the form of water [108]. These features and mechanisms bear certain resemblance to conventional hydrogen sensors based on semiconductor field-effect transistors with Pd gates, as pioneered by Lundstrom [109].

6. Conclusion and outlook

Carbon nanotubes present significant opportunities to basic science and nanotechnology, and pose significant challenge for future work in this field. The approach of direct growth of nanowires into ordered structures on surfaces is a promising route to approach nanoscale problems and create novel molecular-scale devices with advanced electrical, electromechanical and chemical functions. Gaining further control in nanotube growth will continue to open new possibilities in basic science and real-world applications. It is an ultimate goal for growth to produce defect-free nanotubes at the ton level by simple and efficient methods, gain control over the nanotube chirality and diameter, and be able to direct the growth of a semiconducting or metallic nanowire from and to any desired sites.

Acknowledgements

We are grateful to Drs. C. Lieber, S. Iijima, P. McEuen, C. Dekker, Z. Yao, S. Xie and Z. Ren for providing some of the data shown in this article. Work carried out at Stanford were in collaboration with Professors C. Quate, S. Fan, S. Y. Wu, C. S. Jayanthi, K. Cho, S. Manalis and C. Marcus, supported by National Science Foundation, DARPA/ONR, a Packard Fellowship, a Terman Fellowship, Semiconductor Research Corporation/Motorola Co., Semiconductor Research Corporation/Semetech, ABB Group Ltd., the National Nanofabrication Users Network at Stanford, Stanford Center for Materials Research, a Packard Fellowship for Science and Engineering, a Terman Fellowship, the Camile Henry-Dreyfus Foundation and the American Chemical Society.

References

[1] H.W. Kroto, J.R. Heath, S.C. O'Brien, R.F. Curl, R.E. Smalley, C_{60}: Buckminsterfullerene, Nature 318 (1985) 162–163.
[2] S. Iijima, Helical microtubules of graphitic carbon, Nature 354 (1991) 56–58.
[3] S. Iijima, T. Ichihashi, Single-shell carbon nanotubes of 1-nm diameter, Nature 363 (1993) 603–605.
[4] B.I. Yakobson, R.E. Smalley, Fullerene nanotubes: $C_{1,000,000}$ and beyond, Am. Sci. 85 (1997) 324–337.
[5] M.S. Dresselhaus, G. Dresselhaus, P.C. Eklund, Science of Fullerenes and Carbon Nanotubes, Academic Press, San Diego, 1996.
[6] N. Hamada, S. Sawada, A. Oshiyama, New one-dimensional conductors: graphitic microtubules, Phys. Rev. Lett. 68 (1992) 1579–1581.
[7] M.S. Dresselhaus, G. Dresselhaus, P. Avouris (Eds.), Carbon Nanotubes, Springer, Berlin, 2001.
[8] T.W. Ebbesen, P.M. Ajayan, Large-scale synthesis of carbon nanotubes, Nature 358 (1992) 220–222.
[9] D.S. Bethune, C.H. Kiang, M. DeVries, G. Gorman, R. Savoy, J. Vazquez, R. Beyers, Cobalt-catalysed growth of carbon nanotubes with single-atomic-layer walls, Nature 363 (1993) 605–607.
[10] A. Thess, R. Lee, P. Nikolaev, H. Dai, P. Petit, J. Robert, C. Xu, Y. Lee, S. Kim, A. Rinzler, D. Colbert, G. Scuseria, D. Tomanek, J. Fischer, R. Smalley, Crystalline ropes of metallic carbon nanotubes, Science 273 (1996) 483–487.
[11] C. Journet, W. Maser, P. Bernier, A. Loiseau, M. Delachapelle, S. Lefrant, P. Deniard, R. Lee, J. Fischer, Large-scale production of single-walled carbon nanotubes by the electric-arc technique, Nature 388 (1997) 756–758.
[12] G.G. Tibbetts, M.G. Devour, E.J. Rodda, An adsorption-diffusion isotherm and its application to the growth of carbon filaments on iron catalyst particles, Carbon 25 (1987) 367–375.

[13] G.G. Tibbetts, Why are carbon filaments tubular? J. Cryst. Growth 66 (1984) 632–638.
[14] R.T.K. Baker, Catalytic growth of carbon filaments, Carbon 27 (1989) 315–323.
[15] H.G. Tennent, Hyperion Catalysis International, Inc. US patent #4663230, USA, 1987.
[16] M. Endo, Grow carbon fibers in the vapor phase, Chemtech 18 (1988) 568–576.
[17] H. Dai, Growth and characterization of carbon nanotubes, in: M.S. Dresselhaus, G. Dresselhaus, P. Avouris (Eds.), Carbon Nanotubes, Springer, Berlin, 2001, pp. 29–53.
[18] J. Kong, A.M. Cassell, H. Dai, Chemical vapor deposition of methane for single-walled carbon nanotubes, Chem. Phys. Lett. 292 (1998) 567–574.
[19] J. Kong, H. Soh, A. Cassell, C.F. Quate, H. Dai, Synthesis of individual single-walled carbon nanotubes on paterned silicon wafers, Nature 395 (1998) 878.
[20] A. Cassell, J. Raymakers, J. Kong, H. Dai, Large scale single-walled nanotubes by CVD synthesis, J. Phys. Chem. 103 (1999) 6484–6492.
[21] M. Su, B. Zheng, J. Liu, A scalable CVD method for the synthesis of single-walled carbon nanotubes with high catalyst productivity, Chem. Phys. Lett. 322 (2000) 321–326.
[22] E. Flahaut, A. Govindaraj, A. Peigney, C. Laurent, C.N. Rao, Synthesis of single-walled carbon nanotubes using binary (Fe, Co, Ni) alloy nanoparticles prepared in situ by the reduction of oxide solid solutions, Chem. Phys. Lett. 300 (1999) 236–242.
[23] J.-F. Colomer, C. Stephan, S. Lefrant, G. Tendeloo, I. Willems, Z. Kónya, A. Fonseca, C. Laurent, J. Nagy, Large-scale synthesis of single-wall carbon nanotubes by catalytic chemical vapor deposition (CCVD) method, Chem. Phys. Lett. 317 (2000) 83–89.
[24] J. Hafner, M. Bronikowski, B. Azamian, P. Nikolaev, D. Colbert, R. Smalley, Catalytic growth of single-wall carbon nanotubes from metal particles, Chem. Phys. Lett. 296 (1998) 195–202.
[25] H. Cheng, F. Li, G. Su, H. Pan, M. Dresselhaus, Large-scale and low-cost synthesis of single-walled carbon nanotubes by the catalytic pyrolysis of hydrocarbons, Appl. Phys. Lett. 72 (1998) 3282–3284.
[26] P. Nikolaev, M. Bronikowski, R. Bradley, F. Rohmund, D. Colbert, K. Smith, R. Smalley, Gas-phase catalytic growth of single-walled carbon nanotubes from carbon monoxide, Chem. Phys. Lett. 313 (1999) 91–97.
[27] W.Z. Li, S. Xie, L. Qian, B. Chang, B. Zou, W. Zhou, R. Zhao, G. Wang, Large-scale synthesis of aligned carbon nanotubes, Science 274 (1996) 1701–1703.
[28] Z. Pan, S.S. Xie, B. Chang, C. Wang, Very long carbon nanotubes, Nature 394 (1998) 631–632.
[29] Z.F. Ren, Z.P. Huang, J.W. Xu, J.H. Wang, Synthesis of large arrays of well-aligned carbon nanotubes on glass, Science 282 (1998) 1105–1107.
[30] S. Fan, M. Chapline, N. Franklin, T. Tombler, A. Cassell, H. Dai, Self-oriented regular arrays of carbon nanotubes and their field emission properties, Science 283 (1999) 512–514.
[31] H. Dai, J. Kong, C. Zhou, N. Franklin, T. Tombler, A. Cassell, S. Fan, M. Chapline, Controlled chemical routes to nanotube architectures, physics and devices, J. Phys. Chem. 103 (1999) 11246–11255.
[32] H. Dai, Controlling nanotube growth, Phys. World 13 (2000) 43–47.
[33] A. Cassell, N. Franklin, T. Tombler, E. Chan, J. Han, H. Dai, Directed growth of free-standing single-walled carbon nanotubes, J. Am. Chem. Soc. 121 (1999) 7975–7976.
[34] N. Franklin, H. Dai, An enhance chemical vapor deposition method to extensive single-walled nanotube networks with directionality, Adv. Mater. 12 (2000) 890.
[35] H. Dai, J. Hafner, A. Rinzler, D. Colbert, R. Smalley, Nanotubes for nanoprobes, Nature 384 (1996) 147–150.
[36] S. Wong, E. Joselevich, A. Woolley, C. Cheung, C. Lieber, Covalently functionalized nanotubes as nanometrel-sized probes in chemistry and biology, Nature 394 (1998) 52–55.
[37] S. Wong, J. Harper, M. Lansbury, C.M. Lieber, Carbon nanotube tips: high-resolution probes for imaging biological systems, J. Am. Chem. Soc. 120 (1998) 603–604.
[38] H. Dai, N. Franklin, J. Han, Exploiting the properties of carbon nanotubes for nanolithography, Appl. Phys. Lett. 73 (1998) 1508–1510.
[39] S. Wong, A. Wooley, E. Joselevich, C. Lieber, Functionalization of carbon nanotube AFM probes using tip-activated gases, Chem. Phys. Lett. 306 (1999) 219–225.
[40] S. Wong, A. Woolley, E. Joselevich, C. Cheung, C. Lieber, Covalently functionalized single-walled carbon nanotube probe tips for chemical force microscopy, J. Am. Chem. Soc. 120 (1998) 8557–8558.
[41] J. Hafner, C. Cheung, C. Lieber, Growth of nanotubes for probe microscopy tips, Nature 398 (1999) 761–762.
[42] J. Hafner, C. Cheung, C.M. Lieber, Direct growth of single-walled carbon nanotube scanning probe microscopy tips, J. Am. Chem. Soc. 121 (1999) 9750–9751.
[43] E.B. Cooper, S. Manalis, H. Fang, H. Dai, K. Matsumoto, S. Minne, T. Hunt, C.F. Quate, Terabit-per-square-inch data storage with the atomic force microscope, Appl. Phys. Lett. 29 (1999) 3566–3568.
[44] T. Odom, J. Huang, P. Kim, C.M. Lieber, Atomic structure and electronic properties of single-walled carbon nanotubes, Nature 391 (1998) 62–64.
[45] J.W.G. Wildoer, L.C. Venema, A.G. Rinzler, R.E. Smalley, C. Dekker, Electronic structure of atomically resolved carbon nanotubes, Nature 391 (1997) 59.
[46] M. Ouyang, J.-L. Huang, C.M. Lieber, Harvard University, unpublished results.
[47] T.W. Odom, J.L. Huang, P. Kim, C.M. Lieber, Structure and electronic properties of carbon nanotubes, J. Phys. Chem. 104 (2000) 2794–2809.

[48] S.J. Tans, M. Devoret, H. Dai, A. Thess, R. Smalley, L. Geerligs, C. Dekker, Individual single-wall carbon nanotubes as quantum wires, Nature 386 (1997) 474–477.

[49] M. Bockrath, D. Cobden, P. McEuen, N. Chopra, A. Zettl, A. Thess, R. Smalley, Single-electron transport in ropes of carbon nanotubes, Science 275 (1997) 1922–1925.

[50] D. Cobden, M. Bockrath, N. Chopra, A. Zettle, P. McEuen, A. Rinzler, R. Smalley, Spin splitting and even–odd effects in carbon nanotubes, Phys. Rev. Lett. 81 (1998) 681–684.

[51] J. Nygard, D.H. Cobden, M. Bockrath, P.L. McEuen, P.E. Lindelof, Electrical transport measurements on single-walled carbon nanotubes, Appl. Phys. A 69 (1999) 297–304.

[52] H. Soh et al., Integrated nanotube circuits: controlled growth and ohmic contacting of single-walled carbon nanotubes, Appl. Phys. Lett. 75 (1999) 627–629.

[53] A. Morpurgo, J. Kong, C. Marcus, H. Dai, Gate controlled superconducting proximity effect in nanotubes, Science 286 (1999) 263–265.

[54] C. Zhou, J. Kong, H. Dai, Electrical measurements of individual semiconducting single-walled nanotubes of various diameters, Appl. Phys. Lett. 76 (1999) 1597.

[55] J. Kong, N. Franklin, C. Zhou, M. Chapline, S. Peng, K. Cho, H. Dai, Nanotube molecular wires as chemical sensors, Science 287 (2000) 622–625.

[56] C. Zhou, J. Kong, H. Dai, Intrinsic electrical properties of single-walled carbon nanotubes with small band gaps, Phys. Rev. Lett. 84 (2000) 5604.

[57] T. Tombler, C. Zhou, J. Kong, H. Dai, Gating individual nanotubes and crosses with scanning probes, Appl. Phys. Lett. 76 (2000) 2412.

[58] T. Tombler, C. Zhou, L. Alexeyev, J. Kong, H. Dai, L. Liu, C. Jayanthi, M. Tang, S. Wu, Reversible nanotube electro-mechanical characteristics under local probe manipulation, Nature 405 (2000) 769.

[59] Z. Yao, H.W.C. Postma, L. Balents, C. Dekker, Carbon nanotube intramolecular junctions, Nature 402 (1999) 273–276.

[60] M.S. Fuhrer, J. Nygård, L. Shih, M. Forero, Y. Yoon, M. Mazzoni, H. Joon, J. Choi, S. Louie, A. Zettl, P. McEuen, Crossed nanotube junctions, Science 288 (2000) 494–497.

[61] T. Rueckes, K. Kim, E. Joselevich, G. Tseng, C. Cheung, C. Lieber, Carbon nanotube-based nonvolatile random access memory for molecular computing, Science 289 (2000) 94–97.

[62] C.T. White, T.N. Todorov, Carbon nanotubes as long ballistic conductors, Nature 393 (1998) 240.

[63] H. Grabert, M.H. Devoret (Eds.), Single Charge Tunneling, Plenum Press, New York, 1992.

[64] J.E. Fischer, H. Dai, A. Thess, R. Lee, N. Hanjani, D. De Haas, R. Smalley, Metallic resistivity in crystalline ropes of single-wall carbon nanotubes, Phy. Rev. B-Condens. Matter 55 (1997) R4921–R4924.

[65] C.L. Kane, E. Mele, R. Lee, J. Fischer, P. Petit, H. Dai, A. Thess, R. Smalley, A. Verschueren, A. Tans, C. Dekker, Temperature dependent resistivity of single wall carbon nanotubes, Euro. Phys. Lett. 6 (1998) 683–688.

[66] H.R. Shea, R. Martel, M. Avouris, Electrical transport in rings of single-wall nanotubes: one-dimensional localization, Phys. Rev. Lett. 84 (2000) 4441.

[67] Z. Yao, C. Dekker, personal communication.

[68] A. Bachtold, M. Fuhrer, S. Plyasunov, M. Forero, E. Anderson, A. Zettl, P. McEuen, Scanned probe microscopy of electronic transport in carbon nanotubes, Phys. Rev. Lett. 84 (2000) 6082–6085.

[69] S. Tans, A. Verschueren, C. Dekker, Room-temperature transistor based on a single carbon nanotube, Nature 393 (1998) 49–52.

[70] R. Martel, T. Schmidt, H.R. Shea, T. Hertel, P. Avouris, Single- and multi-wall carbon nanotube field-effect transistors, Appl. Phys. Lett. 73 (1998) 2447–2449.

[71] P.G. Collins, K. Bradley, M. Ishigami, A. Zettl, Extreme oxygen sensitivity of electronic properties of carbon nanotubes, Science 287 (2000) 1801–1804.

[72] L. Chico, V.H. Crespi, L.X. Benedict, S.G. Louie, M.L. Cohen, Pure carbon nanoscale devices—nanotube heterojunctions, Phys. Rev. Lett. 76 (1996) 971–974.

[73] R.S. Lee, H.J. Kim, J.E. Fischer, A. Thess, R.E. Smalley, Conductivity enhancement in single-wall carbon nanotube bundles doped with K and Br, Nature 388 (1997) 255–257.

[74] L. Grigorian, K. Williams, S. Fang, G. Sumanasekera, A. Loper, E. Dickey, S. Pennycook, P. Eklund, Reversible intercalation of charged iodine chains into carbon nanotube ropes, Phys. Rev. Lett. 80 (1998) 5560–5563.

[75] R.S. Lee, H. Kim, J. Fischer, J. Lefebvre, M. Radosavljevi, J. Hone, A. Johnson, Transport properties of a potassium-doped single-wall carbon nanotube rope, Phys. Rev. B 61 (2000) 4526–4529.

[76] M. Bockrath, J. Hone, A. Zettl, P. McEuen, A. Rinzler, R. Smalley, Chemical doping of individual semiconducting carbon-nanotube ropes, Phys. Rev. B 61 (2000) R10606–R10608.

[77] J. Kong, C. Zhou, E. Yenilmez, H. Dai, Alkaline metal doped n-type semiconducting nanotubes as quantum dots, Appl. Phys. Lett. 77 (2000) 3977.

[78] H. Ishikuro, T. Fujii, T. Saraya, G. Hashiguchi, T. Hiramoto, T. Ikoma, Coulomb blockade oscillations at room temperature in a Si quantum wire metal-oxide-semiconductor field-effect transistor fabricated by anisotropic etching on a silicon-on-insulator substrate, Appl. Phys. Lett. 68 (1996) 3585–3587.

[79] H. Ishikuro, T. Hiramoto, Quantum mechanical effects in silicon quantum dot in a single-electron transistor, Appl. Phys. Lett. 71 (1997) 3691–3693.

[80] L. Zhuang, L. Guo, S.Y. Chou, Silicon single-electron quantum-dot transistor switch operating at room temperature, Appl. Phys. Lett. 72 (1998) 1205–1207.

[81] K. Matsumoto, M. Ishii, K. Segawa, Y. Oka, B. Vartanian, J. Harris, Room temperature operation of a single

electron transistor made by the scanning tunneling microscope nanooxidation process for the TiO$_x$/Ti system, Appl. Phys. Lett. 68 (1996) 34–36.

[82] Y.A. Pashkin, Y. Nakamura, J.S. Tsai, Room-temperature Al single-electron transistor made by electron-beam lithography, Appl. Phys. Lett. 76 (2000) 2256–2258.

[83] M. Nardelli, J. Bernholc, Mechanical deformations and coherent transport in carbon nanotubes, Phys. Rev. B 60 (1998) R16338–16341.

[84] A. Rochefort, D. Salahub, P. Avouris, The effect of structural distortions on the electronic structure of carbon nanotubes, Chem. Phys. Lett. 297 (1998) 45–50.

[85] A. Rochefort, F. Lesage, D. Salhub, P. Avouris, Conductance of distorted carbon nanotubes, 1999 xxx.lanl.gov/cond-mat/9904083.

[86] L. Liu, C. Jayanthi, M. Tang, S. Wu, T. Tombler, C. Zhou, L. Alexeyev, J. Kong, H. Dai, Controllable reversibility of an transition of a sp^2 to sp^3 single wall nanotube under the manipulation of an AFM tip: a nanoscale electromechanical switch? Phys. Rev. Lett. 84 (2000) 4950.

[87] H. Dai, E.W. Wong, Y.Z. Lu, F. Shoushan, C.M. Lieber, Synthesis and characterization of carbide nanorods, Nature 375 (1995) 769–772.

[88] W.Q. Han, S.S. Fan, Q.Q. Li, Y.D. Hu, Synthesis of gallium nitride nanorods through a carbon nanotube-confined reaction, Science 277 (1997) 1287–1289.

[89] A. Bezryadin, C.N. Lau, M. Tinkham, Quantum suppression of superconductivity in ultrathin nanowires, Nature 404 (2000) 971.

[90] Q. Ma, A. Rosenberg, Interaction of Ti with the (0001) surface of highly oriented pyrolytic graphite, Phys. Rev. B 60 (1999) 2827–2832.

[91] M. Baumer, J. Libuda, H. Freund, The temperature dependent growth mode of nickel on the basal plane of graphite, Surf. Sci. 327 (1995) 321–329.

[92] P. Kruger, A. Rakotomahevitra, J. Parlebas, C. Demangeat, Magnetism of epitaxial 3d-transition-metal monolayer on graphite, Phys. Rev. B (1998) 5276–5280.

[93] S.S. Peng, B.R. Cooper, Y.G. Hao, Modeling of adsorption of iron impurities at the surface of graphite, Philo. Mag. 73 (1996) 611–617.

[94] D. Tomanek, W. Zhong, Palladium–graphite interaction potentials based on first-principles calculations, Phys. Rev. B 43 (1991) 12623–12625.

[95] Q. Ma, R. Rosenberg, Interaction of Al clusters with the (0001) surface of highly oriented pyrolytic graphite, Surf. Sci. 391 (1997) L1224–1229.

[96] I. Moulett, Ab-initio molecular-dynamics study of the interaction of aluminum clusters on a graphite surface, Surf. Sci. 333 (1995) 697–702.

[97] E. Ganz, K. Sattler, J. Clarke, Scanning tunneling microscopy of Cu, Ag, Au and Al adatoms small clusters, and islands on graphite, Surf. Sci. 219 (1989) 33–67.

[98] L. Barfotti, P. Jensen, A. Hoareau, M. Treilleux, B. Cabaud, A. Perez, F. Aires, Diffusion and aggregation of large antimony and gold clusters deposited on graphite, Surf. Sci. 367 (1996) 276.

[99] J. Arthur, A. Cho, Adsorption and desorption kinetics of Cu and Au on (0001) graphite, Surf. Sci. 36 (1973) 641–660.

[100] T.R. Ohno, Y. Chen, S. Harvey, G. Kroll, J. Weaver, R. Haufler, R. Smalley, C_{60} bonding and energy-level alignment on metal and semiconductor surfaces, Phys. Rev. B 44 (1991) 13747–13755.

[101] V. Vijayakrishnan, A. Santra, R. Seshadri, R. Nagarajan, T. Pradeep, C. Rao, A comparative study of the interaction of nickel clusters with buckministerfullerene C_{60} and graphite, Surf. Sci. 262 (1992) L87–90.

[102] E. Parks, K. Kerns, S.R.B. Winter, Adsorption of C_{60} on nickel clusters at high temperatures, Phys. Rev. B 59 (1999) 13431.

[103] M. Menon, A. Andriotis, G. Froudakis, Curvature dependence of metal catalyst atom interaction with carbon nanotubes walls, Chem. Phys. Lett. 320 (2000) 425–434.

[104] Y. Zhang, N. Franklin, R. Chen, H. Dai, A metal coating study of suspended carbon nanotubes and its implications to metal-tube interactions, Chem. Phys. Lett. 331 (2000) 35–41.

[105] T. Tachibana, B. Williams, J. Glass, Correlation of the electrical properties of metal contacts on diamond films with the chemical nature of the metal–diamond surface. II. Titanium contacts: a carbide forming metal, Phys. Rev. B 45 (1992) 11975–11981.

[106] J. Kong, H. Dai, unpublished results.

[107] S. Jhi, S.G. Louie, M.L. Cohen, Electronic properties of oxidized carbon nanotubes, Phys. Rev. Lett. 85 (2000) 1710.

[108] A. Mandelis, C. Christofides, Physics Chemistry and Technology of Solid State Gas Sensor Devices, Wiley, New York, 1993.

[109] I. Lundstrom, M. Shivaraman, C. Svensson, Chemical reactions on palladium surfaces studied with Pd-MOS structures, Surf. Sci. 64 (1977) 497.

Clusters as new materials

W. Eberhardt *

IFF, Forschungszentrum Jülich GmbH, 52425 Jülich, Germany

Received 20 September 2000; accepted for publication 25 May 2001

Abstract

Over the past two decades methods have been developed to produce clusters with an exactly determined number of atoms. Due to their finite size these small particles have totally different structures and 'materials properties' than their bulk crystalline counterparts. Even more, these properties sometimes change drastically whenever a single atom is added to or removed from the cluster. This opens the pathway for a whole new world of tailor made materials in the future. In this article we describe the present state of the knowledge of the properties of clusters of atoms which in their bulk form conventional metals or semiconductors. The questions addressed include the development of the electronic structure as a function of cluster size and for example what remains of the 'metallic' properties of the bulk solid in these very small clusters. Technological advances are expected using clusters on a specific support material in the areas of catalysis, magnetic storage media or electronic materials, and even solids assembled totally from clusters. Examples from each of these fields will be discussed in the context of this article. © 2001 Elsevier Science B.V. All rights reserved.

Keywords: Clusters; Catalysis; Chemisorption; Magnetic phenomena (cyclotron resonance, phase transitions, etc.); Quantum effects; Fullerenes; Photoelectron spectroscopy

1. Introduction

The term 'cluster' identifies aggregates of various kinds and clusters have become a quite familiar term in everyday's life. Pre-school children know about clusters in their cereal, whereas materials scientists are thinking of aggregates containing anywhere between 2 and about 10^4 atoms as a cluster. An astrophysicist may call an aggregation of stars a cluster.

Some clusters are just weakly bound aggregates of condensed molecules. The more interesting clusters however are assembled from atoms, where the bulk solid is a metal or semiconductor. These clusters have distinctly different properties than the corresponding bulk solid. In chemistry it is known since a long time that these clusters or small particles exhibit different catalytic properties than their bulk counterparts. Accordingly, a lot of work in chemistry and chemical engineering is devoted to the production of ensembles of particles with 'special' properties by narrowing down the size distribution or by establishing preparative procedures to fabricate tailor made catalysts with well defined properties.

The special properties of clusters have been exploited by mankind for many centuries. The colored glass used already by the ancient Romans

---

* Present address: BESSY GmbH, Albert Einstein Str. 15, D-12489 Berlin, Germany.
  E-mail address: w.eberhard@fz-juelich.de (W. Eberhardt).

Fig. 1. Roman glass cup (Licurgus cup) manufactured in the 4th century. The glass contains small clusters, which cause the characteristic optical properties. It appears red in transmission and pale green in reflected light.

(see Fig. 1) or the fantastically colored glass of the church windows from the Middle Ages owe their color to finely dispersed silver, copper, or gold particles (clusters) embedded by special fabrication procedures into the glass.

Clusters come in unusual shapes, which most of the time reflect structures that do not resemble parts of the crystal structure of the corresponding solid materials. The most well known example is the soccer ball shaped C_{60} cluster [1] (Fig. 2), which, in addition, also forms the first true cluster material when condensed into a solid bound not by chemical interaction but only by weak electrostatic forces. This solid is nevertheless stable at room temperature and it has totally different properties than diamond or graphite, even though all three materials consist exclusively of carbon atoms. Graphite is a semimetal, whereas diamond is a wide-band gap semiconductor with an energy gap between the occupied and unoccupied electronic states of about 5.5 eV. The fullerenes on the other hand condense into a molecular solid with an onset of the electronic excitations around 1.5–2 eV. Nevertheless, the fullerene solids can be doped to be metallic and even become superconducting at transition temperatures exceeding 30 K. Next to the so called high T_c superconducting oxides these are the superconductors with the highest transition temperatures known so far.

Not only carbon clusters have unique geometry and properties. A few more 'unexpected' structures are presented in Fig. 3. Fig. 3a shows the structure of a sixfold ring of Au atoms, which is the ground state of this cluster when it is negatively charged by one electron. This structure was initially inferred from the detection of the characteristic vibrational (breathing) mode, where the whole ring expands and contracts, in investigations by the Smalley group [2]. This planar ring which resembles the well known structure of benzene is far from the fcc solid state crystallographic structure of Au in its crystalline form.

Fig. 3b shows a pentagonal bipyramid, containing seven atoms. This is the ground state of a 7

Fig. 2. Cover of Nature magazine announcing the discovery of C_{60} [1].

atom Si cluster, as derived from electron spectroscopy data [3]. This interpretation was also confirmed by Raman spectroscopy after depositing these clusters into a rare gas matrix [4]. The Si atoms at the two apexes are fivefold coordinated, again having a ground state structure, which is not found in the extended solid. The lowest energy optical excitation of this cluster is observed at 1.5 eV hardly larger than in crystalline Si [5]. With increasing cluster size this value should approximate the value of the bulk band gap. This demonstrates that the electronic structure is also quite unique and cannot be extrapolated by the standard quantum confinement models using bulk crystalline Si as the starting point.

Fig. 3c shows a 13 atom icosahedral structure, which is the ground state geometry for 13 Al atoms [6]. These fivefold symmetric icosahedral structures also are not found in crystalline solids, since no space filling fivefold symmetric structure with translational periodicity exists. (Quasicrystals exhibit a fivefold symmetry, but do not have a strict translational periodicity.) Apart from its unusual shape, theory and experiment also show that this Al-cluster is magnetic, while bulk aluminum certainly does not display magnetic behavior.

This small collection of established cluster structures is shown here to convey the message that clusters may not be viewed as small chunks of the corresponding well known extended crystalline phases. For all systems where the bulk is a conventional semiconductor or metal, the corresponding clusters are truly new materials where both the structural and also the electronic properties are individually different. Depending on the number of atoms, each cluster has its own individual properties. This makes any extrapolation from one cluster-size to another quite uncertain, especially when dealing with particles containing less than 100 atoms. This also means that in theoretical descriptions each cluster has to be calculated at its own individual molecular level and

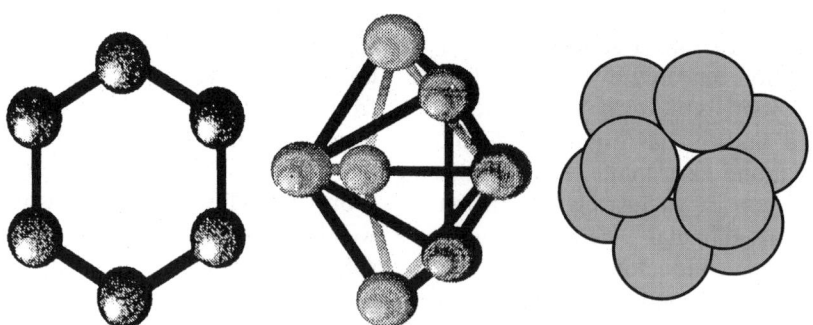

Fig. 3. Structure of Au_6 [2], Si_7 [3], and Al_{13} [6].

that structural freedom and optimization is a crucial ingredient for these calculations. Nevertheless common features and trends can still be observed in some cases, as we see below.

So far it was implied that the number of atoms in a cluster uniquely defines its structure and properties. This is not always true. In some cases, especially in covalently bound systems such as for example C and Si, structural isomers have been observed to exist [7,8]. This is not only a question of the total energy difference for the various isomers, but also the growth kinetics and the barriers for transformation from one structure to another play an important role in this respect [9,10].

In order to make use of these unique properties, the clusters have to be individually (mass) selected and preserved by depositing them into a matrix or by landing them 'softly' onto a surface. This is the next challenging step in cluster physics and a few groups worldwide have started on this endeavor. Of course the surface or the matrix will also interact with the clusters to the extent that their structural and electronic properties are modified. In catalysis this is referred to as 'metal support interaction'. But it is not only in chemistry or catalysis, where these unique properties might be exploitable technologically. Quantum dot lasers are another example, where the properties of quantized structures, generally however of a much larger scale, are exploited. Another technological vision concerns the single electron transistor, where a cluster on an oxide barrier changes the electrostatic potential by transfer of a single electron into or out of its quantized states sufficiently to switch the current in the channel of a field effect transistor. Last but certainly not least in magnetic recording and for magnetic sensors clusters also might offer unique properties or advantages in performance.

The cluster and surface science communities join forces in the attempt to produce unique cluster defined materials. In surface science various methods have been explored to grow clusters on surfaces by self aggregation. Some of the examples exploit the decoration of steps to produce one-dimensional quantum wires. However, only with few exceptions do these methods really produce unique and as well defined homogeneous clusters.

2. 'Metallic' properties of clusters

What constitutes a metal or 'metallic' behavior? This is a favorite exam question and the answers can be manifold. Certainly one characteristic feature is that there is no energy gap in the electronic excitation spectrum and accordingly the electrons contribute to the specific heat of the system. The energy spectrum of the electronic states can be readily investigated by photoemission spectroscopy of negatively charged species (anions), as we show later. All of the clusters assembled from atoms of the group I (alkali metals) and group Ib (noble metals) of the periodic system exhibit a discrete electronic excitation spectrum due to their small and finite size and in the solid state language they exhibit a 'band gap'.

Another definition of a metal hinges on the delocalization of the electrons throughout the system, and again another one is the increase of resistivity with increasing temperature at moderate temperatures due to the dominant contribution of electron–phonon scattering to the overall electrical conductivity. These definitions are not at all readily verified experimentally. Even for bulk systems the verification of the delocalized nature of the electrons is at best indirect. For nanoscale systems and especially clusters this is even more difficult. To my knowledge there exists presently only one true four probe measurement of the resistive properties of a cluster, in this case a carbon nanotube [11].

The delocalized nature of the electronic wave function in clusters was already postulated from the specific abundancies and stabilities observed in the mass spectra of alkali clusters. Knight and his collaborators [12] noted that alkali clusters having 2, 8, 18, 20, 40 valence electrons in total are more prominent in the mass spectra and concluded these clusters are especially stable because these electron configurations correspond to closed shell configurations in an electronic system, where all the valence electrons are put into one spherically symmetric potential well, resembling the clusters as a whole. In such a potential the eigenstates of a delocalized electronic system should be arranged in shells with well defined quantum numbers and degeneracies. The electronic states and the level structure for spherically symmetric particles are schematically indicated in Fig. 4 for a system containing 40 electrons [13]. The electronic states are classified by the quantum numbers nl. These quantum numbers correspond to the number of radial (n) and spherical (l) nodes of the delocalized wave function, whereby the spectroscopic notation s,p,d,f,... is used for the spherical quantum number l. In this framework a level sequence of 1s(2), 1p(6), 1d(10), 2s(2), 1f(14), 2p(6)... is predicted. Here the numbers in the brackets denote the degeneracy of the corresponding shell. Obviously, if the spherical symmetry of the particle is broken, the degeneracies of these shells are lifted.

We have to add here a few sentences to clarify the meaning of delocalized electronic states. Obviously in a covalently bound semiconductor, such as Si, the valence electrons are also delocalized throughout the solid as described by the appropriate Bloch functions. Similarly the electrons in C_{60} are also delocalized over the cluster as a whole as shown in Fig. 5, where the charge density of C_{60} is plotted [14,15]. However, one obvious conclusion from this figure is the electron charge density is most prominently concentrated along the bonds directed from each C-atom to its neighbors. This constitutes the covalent bonding character of the electrons. In a delocalized metallic system on the other hand, ideally the valence electron charge density would be smeared out completely, independent of the exact positions of the nuclei. This is where the description of the cluster by one common, even though not necessarily spherically symmetric potential well gives adequate results. Accordingly when referring to delocalized metallic states, we imply that these electrons are not reflecting the local bond axes or directions between the individual atoms within the cluster.

The proof that in alkali, Cu, and Ag clusters the electronic states are arranged in electronic shells as predicted by this shell model, which is also used in nuclear physics to describe the protons and neutrons and their excited states in the core, was one of the first objects of our own investigations [16,17]. Such a verification in turn gives strong evidence to the delocalized metallic nature of the valence electrons in these clusters. Our experimental setup for photoelectron spectroscopy on mass selected cluster anions is shown schematically in Fig. 6 [18]. Mass selected clusters are produced by a (conventional) laser vaporization of a solid target or by a pulsed arc cluster ion source (PACIS). The laser or the electric arc vaporizes the chosen material. The atoms and ions of the plasma are condensed into small aggregates in a high pressure inert gas atmosphere. Subsequently, a molecular beam is formed through adiabatic expansion, containing a distribution of non-interacting particles. The negative cluster ions within this beam are accelerated and deflected in a pulsed Wiley McLaren type time of flight setup. Photoemission (photodetachment) of electrons from cluster anions of a selected mass only is achieved by adjusting the time of the detachment laser pulse to the time of arrival of clusters of a certain mass at the source region of the magnetic bottle type time-of-flight electron spectrometer. In order to achieve an overall resolution of better than 10 meV, the Doppler broadening has to be reduced

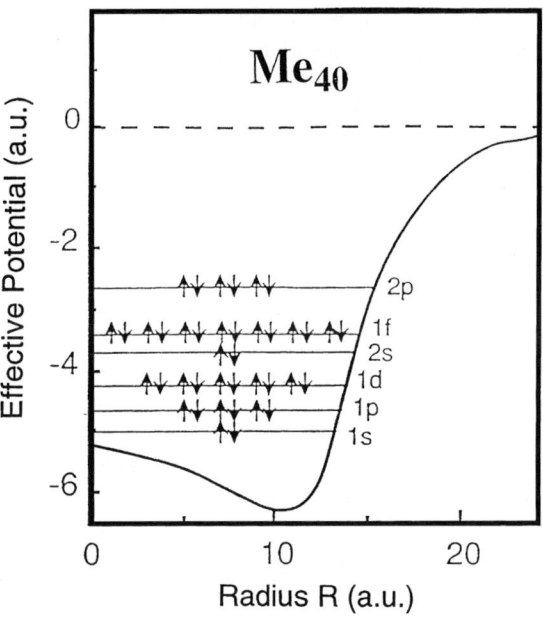

Fig. 4. Shell structure of delocalized orbitals in a spherical potential of the shape as indicated. The energy levels are characterized by the quantum numbers nl, where n denotes the radial nodes and l the angular nodes of the wave functions. For l the spectroscopic notation (s,p,d,f...) is used.

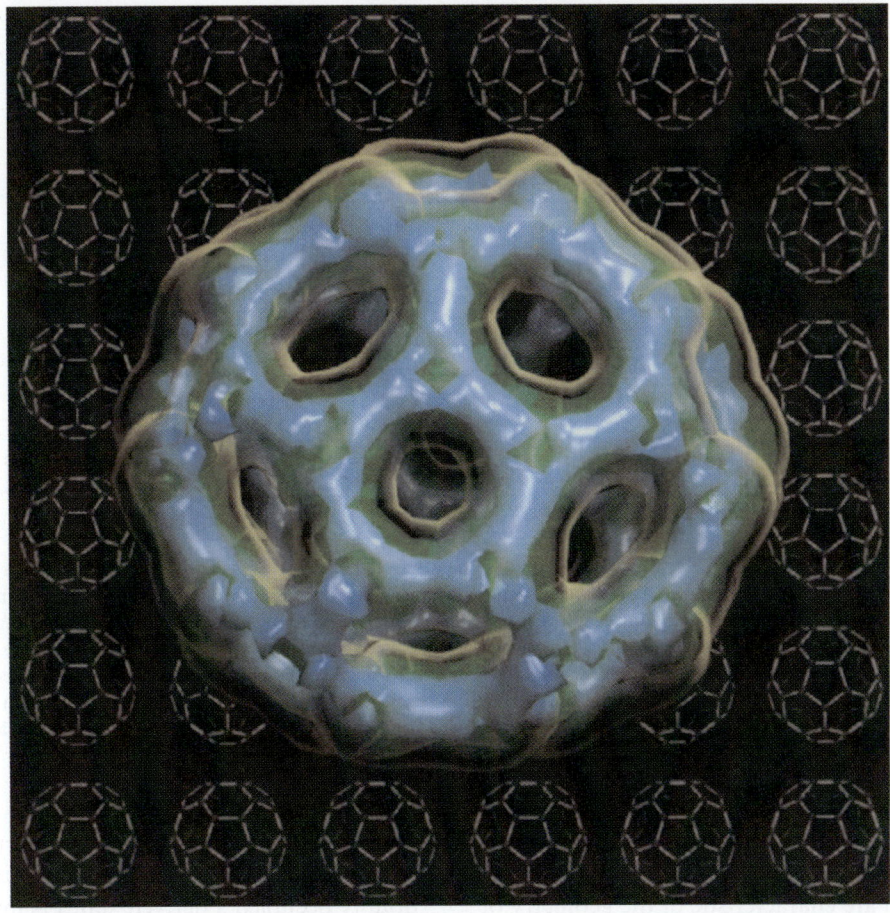

Fig. 5. Charge density of C_{60} [14] showing the concentration of the charge along the bond direction in covalent bonding [15].

by decelerating the clusters prior to the photodetachment process.

Mass selecting the clusters prior to the photoemission process is an essential step to ensure the correct assignments. This is quite costly, since it decreases the target density for the photoemission experiment by about three orders of magnitude. The photoemission process in small clusters frequently results in the production of fragments. Accordingly an experiment, where for example the photoelectrons and the ionic (or neutral) species are detected in coincidence does not allow for an unambiguous assignment.

At low resolution the photoelectron spectra allow the determination of the electronic structure. The spectra show the various electronic transitions from the ground state of the anion into the ground and excited state configurations of the neutral system. In cases where the ground state geometries exhibit large differences, the electronic configurations in the appropriate geometry of the anion have to be taken into account. Examples for this can be found for the metal trimers, where often the ground state of the anion has a linear configuration, whereas the neutral ground state exhibits a geometry close to an equilateral triangle.

High resolution spectra additionally may exhibit fine structure due to the excitation of vibrational quanta of the neutral cluster, which are excited in the photoemission process. The relative intensities of the vibrational sidebands are determined by the change in geometry between the anion ground

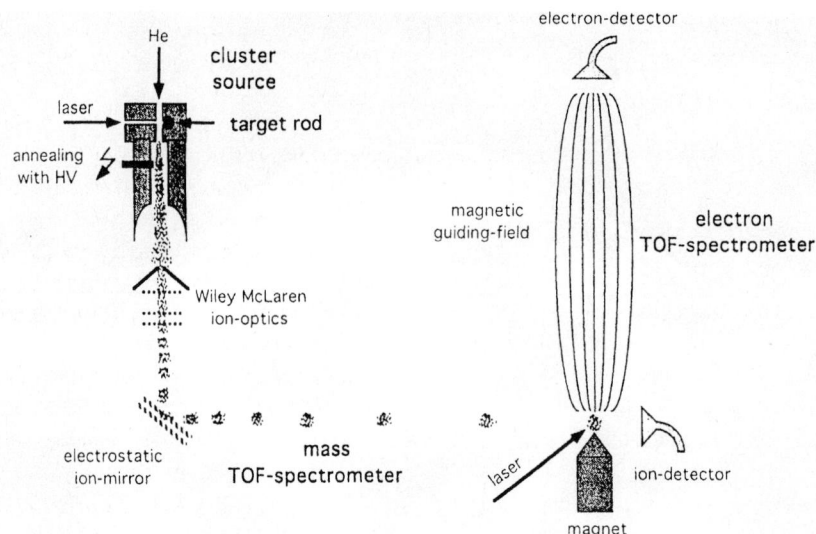

Fig. 6. Experimental setup for photoelectron spectroscopy of mass selected cluster anions [18]. The clusters are produced by laser vaporization of the target rod material and subsequent condensation within in a high pressure He atmosphere. In some cases an additional electric discharge is used to heat (anneal) the clusters. The cluster/He mixture is expanded into vacuum forming a molecular beam of adiabatically cooled non-interacting particles. The negatively charged cluster anions present in this beam are accelerated to a fixed kinetic energy by a pulsed electric field and deflected by the ion optics into the electron spectrometer. At fixed kinetic energy clusters of different mass can be distinguished by their velocity and arrive at different times in the source region of the electron spectrometer. Adjusting the timing of the ionization laser, electrons from clusters of a specific selected size (mass) are emitted. The kinetic energy of the emitted electrons is analyzed by a time-of flight electron spectrometer.

state and the selected electronic configuration of the neutral cluster. The vibrational frequencies thus observed are characteristic of the geometry of the particle.

A series of photoelectron (photodetachment) spectra for Cu_1^- to Cu_8^- clusters is shown in Fig. 7 [19]. In these spectra, taken at 3.6 eV photon energy, already the features of the electronic shell model are dominating the emission characteristics of the states, which derive their character largely from the atomic 4s states. Obviously the Cu 3d electrons remain very localized and do not participate in the delocalized 'metallic' states. Their binding energy is increasing rapidly [20], such that within the binding energy range shown in Fig. 7 only the spectrum of the negatively charged atom exhibits two lines between 2.5 and 3 eV binding energy, which are attributed to the Cu 3d emission. This energy separation corresponds exactly to the spin–orbit splitting of the $3d_{5/2}$ and $3d_{3/2}$ final hole state. The strong feature at a binding energy of about 1.2 eV is attributed to emission from the doubly occupied 4s level. This spectrum thus also demonstrates that the emission cross section from doubly occupied s derived states is much larger at these low photon energies than the emission from the atomic d states, which have a degeneracy of 6 and 4, respectively.

The spectra of the larger clusters are dominated by the even/odd alternation of the vertical detachment energy (VDE), i.e., the binding energy of the outermost electron of the negatively charged cluster. The VDE of the cluster anion differs from the electron affinity of the neutral cluster in cases where the geometry is different, due to the fact that the electronic transitions are orders of magnitude faster than the nuclear motion. Therefore the electronic states are observed in the geometry of the anion (Born Oppenheimer approximation).

According to the observed alternation of binding energies, all the even numbered clusters, which are containing an odd number of electrons since they are negatively charged, have low VDEs and all the odd numbered clusters have higher VDEs.

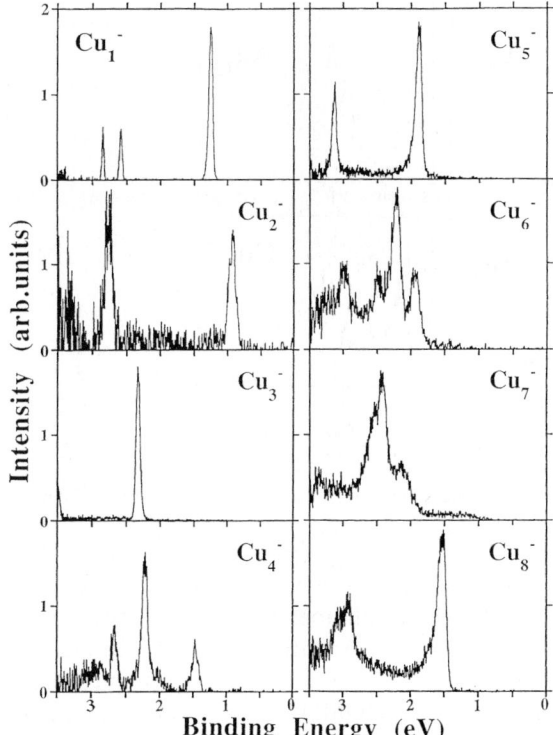

Fig. 7. Photoelectron spectra of negatively charged Cu_n^- clusters [16,19] excited with a photon energy of $hv = 3.5$ eV. In the most simplified interpretation each one of the observed lines corresponds to the removal of a particular electron from the cluster. The difference in binding energies reflects the shell structure of the electronic states in the cluster.

This pattern exactly matches the even/odd alternation in the abundancies of these clusters in the mass spectra mentioned earlier. The clusters, which have generally higher electron binding energies, are more stable and accordingly more abundant in the mass spectra. This general 'rule' is not only valid for 'metallic' clusters but can also be applied to covalently bound species and even to fullerenes.

As all rules the correlations can be shown for quite a few individual cluster systems, but a general proof cannot be established. This is related to the fact that, due to double counting of the interaction energies, the total energy of any given system cannot be measured by adding the single particle binding energies of all the valence electrons observed in a photoelectron spectrum.

Following this brief discussion directed at a more advanced audience, we continue with the interpretation of the photoelectron spectra of the Cu anions. In the negatively charged atom the s-shell is already completely filled. Accordingly, Cu_2^- has three electrons in the delocalized 4s-derived states and these are occupying two shells. The smaller peak in the spectrum is attributed to the emission from the p-symmetry-shell, whereas the emission of the s-shell gives rise to two peaks, separated in energy by the singlet–triplet splitting of the final state. This splitting results from the fact that the two unpaired electrons in the neutral system can arrange their spin either parallel or antiparallel with each other. Only the lower energy peak, attributed to the triplet final state, is seen in Fig. 7. At larger excitation energies, however, the corresponding singlet state is observed as well [20]. The spectra of Cu_4^- and Cu_6^- clearly exhibit this splitting in the second emission feature. The singlet triplet splitting decreases with increasing cluster size as expected for more delocalized electronic states.

Within the admittedly crude approximation of the shell model the interpretation of the spectra is straightforward. For Cu_7^- the p shell of the cluster is completely filled, since this cluster contains eight delocalized electrons. Accordingly, the next larger cluster, which now contains an additional d-symmetry delocalized electron, exhibits a clearly split off sharp peak at about 1.6 eV binding energy. For even larger clusters this trend can be followed up to the closing of the d-shell at altogether 18 valence electrons, which is observed for Cu_{17}^-.

Additional confidence in this assignment and the existence of delocalized electronic states that are described by a shell model is gained from a comparison of clusters containing atoms from the same group within the periodic table. For Cu these are Ag and Au clusters. In such a comparison the alkali's should be included as well since all these elements have one outer s-electron, which in the solid forms the metallic bond. In Fig. 8 [19] we show some spectra of Ag and Na clusters which may be directly compared with the spectra of the Cu clusters. Na has a much smaller bandwidth than Cu but contains no atomic d-electrons at all. Accordingly the binding energy scale has been

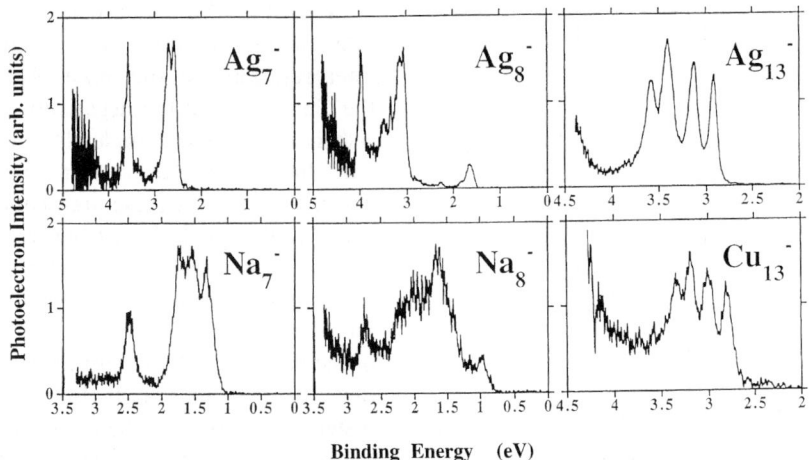

Fig. 8. Photoelectron spectra of Na_n^-, Ag_n^-, and Cu_n^- clusters demonstrating the similarity of the electronic structure of clusters consisting of the elements of the group I and Ib of the periodic table.

stretched by a factor of 2 in Fig. 8 for the Na clusters. Ag on the other hand, has 10 d-electrons in the atom just as Cu does, but the d-states of Ag in the solid have twice the binding energy, measured relative to the top of the occupied valence states (E_F), as the Cu d-states. Thus any hybridization between the atomic s- and d-states is much less likely to occur for Ag, than for Cu clusters. Actually, whereas we observed some indications of s–d hybridization in the Cu clusters [16], which were later confirmed by detailed calculations [23], we were not able to identify anything comparable in the spectra of Ag clusters.

Comparing the spectra of the 7 and 8 atom clusters of Cu, Ag, and Na shown in Figs. 7 and 8, we notice a remarkable similarity. The Ag_7^- spectrum shows the emission of three non-degenerate 1p delocalized states, whereas for the Na_7^- the equivalent 1p state emission is observed between 1 and 2 eV binding energy and the 1s emission is observed at 2.5 eV binding energy. All this points to a full occupation of the 1p shell for the 7 atom cluster-anions. For the 8 atom clusters an additional electron has been added, which now occupies a new state. Within the shell model this is the 1d state. Also the spectra of Cu_{13}^- and Ag_{13}^- (Fig. 8) are very similar, now predominantly showing emission from 1d delocalized states within the range of binding energies displayed here.

From these photoemission data the lowest energy electronic excitation energy of the clusters, the HOMO–LUMO gap (highest occupied molecular orbital (HOMO) and lowest unoccupied molecular orbital (LUMO)) as it is called in a molecular notation, can be unambiguously determined. In the solid state analog this corresponds to the band gap of the system. For any closed shell neutral system, which corresponds to the even numbered clusters, this gap is directly measured in the spectra. The energy separation of the lowest binding energy state, which in the final state configuration corresponds to the ground state of the neutral cluster, and the energy difference to the next higher state is the HOMO–LUMO gap. Most likely this next higher state is a triplet state. Then this transition energy does not correspond to the optical gap since the creation of the triplet state from the singlet ground state is forbidden because it involves a spinflip transition. The values of the HOMO–LUMO gap energies thus derived for Cu and Ag clusters are shown in Fig. 9 [19]. Typically these values are in the 1 eV range. Apart from a slow general trend towards smaller values, there is fairly little systematic evolution in these energy gaps. Moreover, all small clusters investigated here are far from having a continuous density of states near the top of the occupied electronic states. In the solid state definition these clusters are all

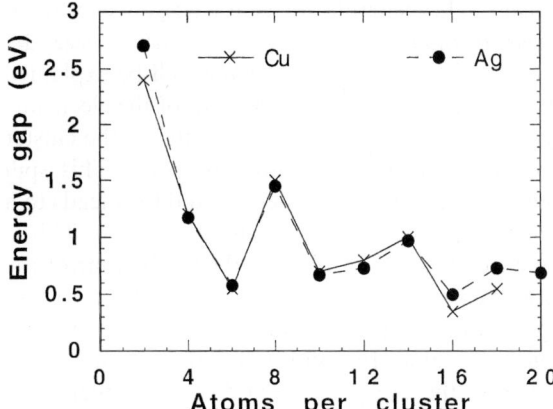

Fig. 9. HOMO–LUMO gap of Cu_n and Ag_n clusters [19] as function of the cluster size n. The values are directly extracted from the photoelectron spectra as described in the text.

semiconducting. From low resolution photoemission data Smalley and coworkers [20] estimated that the density of states of the solid is reasonably reached once the cluster contains 400–500 atoms. This observation holds for all metals and their clusters, where the s electrons define the top of the electronic states near the Fermi level of the solid. Obviously open d-shell systems do not exhibit such a behavior.

For small Ag-clusters containing less than 10 atoms there are quantum-chemical calculations (HF-CI) available, which allow for a more accurate assignment of the states observed in these spectra [21]. With a few exceptions the calculated states agree quite well with the transitions observed in our detachment spectra. Moreover, in all these cases, the symmetry of the calculated CI states corresponds to the symmetry of the state based upon our assignment within the shell model. This is certainly quite an encouragement for carrying out an assignment of the cluster spectra on the basis of this simple, back-of-the-envelope, shell model. A similar comparison has been carried out by Andreoni [22], who found a large overlap between the electronic state wave functions of Na-clusters, calculated by first principles, and the wave functions calculated in the shell model.

Obviously the complete picture of the electronic states can be used to derive the geometry of the clusters by comparing the electronic structure of the same size cluster, calculated for various geometric configurations, with our spectra. Following the publication of our Cu spectra, Car and coworkers were encouraged to calculate these particles using a simulated annealing LDA calculation [23]. In general they confirmed our conclusions and assignments of the electronic states. These calculations also included the d-electrons and accordingly they were able to refine some of the aspects by pointing out the contributions due to s–d hybridization in the electronic structure for some of the clusters, which is well beyond the scope of the shell model presented above. Additionally they obtained the structure of these clusters and could give the ground state geometries for clusters containing up to nine atoms. Calculations for even larger clusters including transition metals with open d-shells are at the limit of present day computational capabilities.

The exact geometries and structures of (transition) metal clusters having more than 10 atoms are therefore still undetermined. This is an opportunity for research using a free electron laser (FEL). These novel sources in the VUV and X-ray range, several of which are currently in the planning or under construction, carry the promise of high power pulses, tunable in photon energy from the UV to the X-ray range and comparable in photon intensities to conventional laser sources. With these sources structural determinations for free mass selected clusters by structure sensitive spectroscopic techniques such as EXAFS are within range. Furthermore core level photoemission will also help to elucidate the structures similar to the studies of core level binding energy shifts on reconstructed surfaces.

2.1. Plasmons and collective excitations

In metals the electrons can display density modulations which are coherent over many atomic sites. These plasmons or collective electron excitations are responsible for the characteristic 'metallic' appearance. The corresponding excitations exist also in clusters and their distinctive differences due to the small size have been exploited 'technologically' for many centuries. The beautiful colored glass windows of medieval church windows and the intricate glassware of ancient times owe their color to finely dispersed Au and Ag particles (clusters) embedded into the glass matrix. Even though this process has been used by skilled craftsmen for centuries, the physics of it was first explained by Gustav Mie [24] at the beginning of the last century, when he developed the electrodynamic treatment and the theory describing the optical properties of small metal particles.

It is quite difficult to measure the absorption spectrum of a cluster in gas phase. Nevertheless early on the plasmon resonances were explored by depletion spectroscopy, where the absorption is detected by depleting or 'bleaching' a certain cluster size in the mass spectra using a tunable laser to interact with the cluster beam. The underlying assumption is that the absorption of a photon by a cluster leads to fragmentation and/or ionization and thus the original mass abundance is altered. To obtain photoabsorption cross-sections this way is a very difficult task, since both the cluster intensity and also the intensity of the tunable laser have to be determined accurately.

Mostly the early data concentrate on the collective 'plasmon' resonances of alkali- and Ag-clusters. In general the characteristic energy of these resonances shifts, as the particle size decreases. These shifts could quite well be explained by the Mie theory if the 'spill-out' of the electronic wave functions beyond the potential of the cluster core was taken into account properly. This spill out reduces the electron density and thus leads to a red-shift of the plasmon excitation energy. This is an effect superimposed onto the predictions of the Mie theory, where the plasmon energy of a small spherical particle is reduced by $1/\sqrt{3}$ compared to the bulk value. In contrast to the data for alkali clusters, the Ag-cluster plasmon frequencies exhibit a strong blue shift with decreasing particle size. The explanation for this effect is given by Liebsch as due to a reduced s–d interaction in the outer regions of the clusters, which proportionally increases as the particle size decreases. Similar effects are also observed on the surface of a Ag crystal [25].

Additionally, for some clusters, multiple resonances are observed. This is nicely illustrated by the spectra of Ag_n^+ clusters taken by Tiggesbäumler et al. [26] shown in Fig. 10. Already classical electrodynamics teaches that the polarizability of

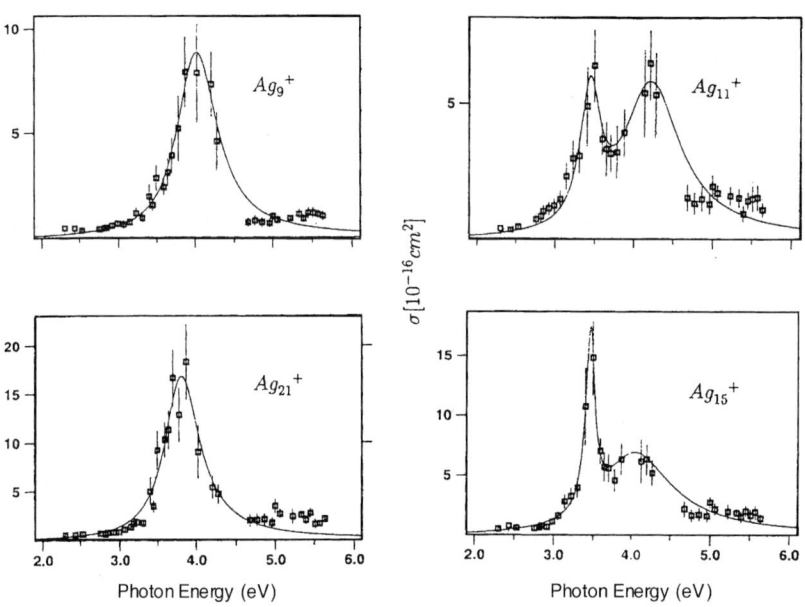

Fig. 10. Pseudo 'absorption' spectra of Ag_n^+ clusters measured by depletion spectroscopy [26]. The spectra are totally dominated by resonances corresponding to plasmon excitations, which show a pronounced splitting for spherically non-symmetric particles.

an ellipsoidal metallic object differs depending on the orientation of the electric field relative to the axes of the ellipsoid. Ag_9^+ and Ag_{21}^+ are nearly spherically symmetric particles and accordingly display one dominant 'plasmon' resonance. On the other hand, Ag_{11}^+ and Ag_{14}^+ are strongly deviating from the spherical shape, according to the predictions of the shell model, Ag_{11}^+ with 10 valence electrons is predicated to have a prolate (cigar like) shape, whereas Ag_{15}^+ with 14 valence electrons is supposed to have an oblate shape. Based on these simple models, two different resonance frequencies are expected for the electrical field being oriented along the short and long axes respectively. Accordingly, the two peak spectra and their relative intensities can be rationalized quite well.

To date some of the best data of the absorption spectra in the optical and UV region of clusters are available for Ag_n clusters mass selected and deposited into a rare gas matrix [27]. These spectra are shown in Fig. 11. It is interesting to note that in the size range of clusters having between 10 and 20 atoms, these spectra also exhibit similar structures to the ones observed for the isolated species in gas phase. The multiple peak structure indicates that the clusters as deposited in the matrix most likely have quite similar shapes and geometries as the isolated clusters. Consequently one is led to conclude that the deposition process and the landing are sufficiently soft and thus do not alter the geometry of the particles significantly.

The collective excitations in Ag particles offer some very interesting possibilities for the manipulation of the size distribution of particles deposited on surfaces. This is demonstrated in the experiments concerning the deposition and agglomeration of Ag clusters on quartz substrates by Stietz and Träger [28]. Following deposition of atoms from an oven with the substrate at room temperature, the intrinsic particle growth process leads to a fairly broad distribution of cluster sizes as indicated in the top of Fig. 12. The optical absorption spectrum of this ensemble of clusters is shown in Fig. 13. The pronounced two peak structure of the absorption spectrum is attributed in general terms to the ellipsoidal (island) shape of the agglomerated particles [29]. Irradiating these clusters subsequently with laser light of 532 and

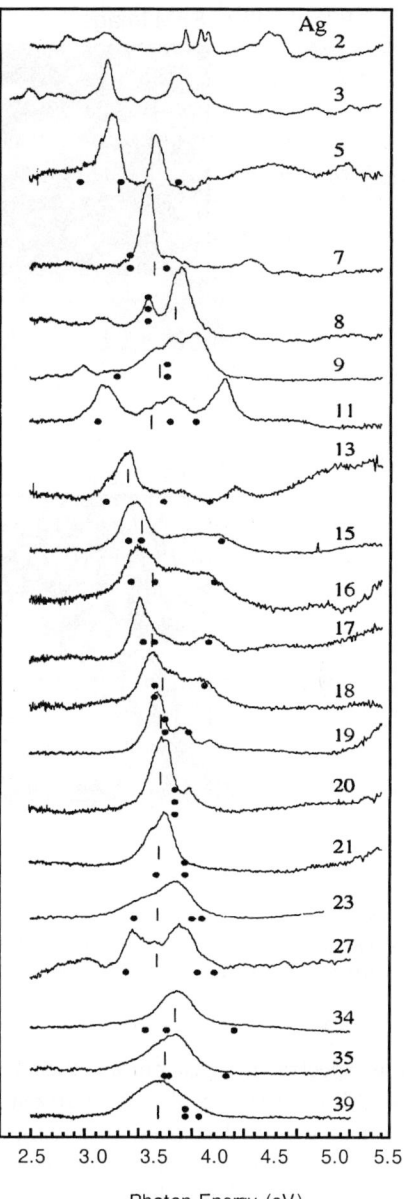

Fig. 11. Absorption spectra of Ag_n clusters deposited into a solid rare gas matrix. The spectra are quite similar to the spectra of the isolated clusters in gas phase [27].

355 nm wavelength, produces a much sharper spectrum as shown in Fig. 13c. This corresponds to a narrowing of the size distribution of the clusters as verified by scanning force microscopy in the bottom of the previous Fig. 12. Similar to the

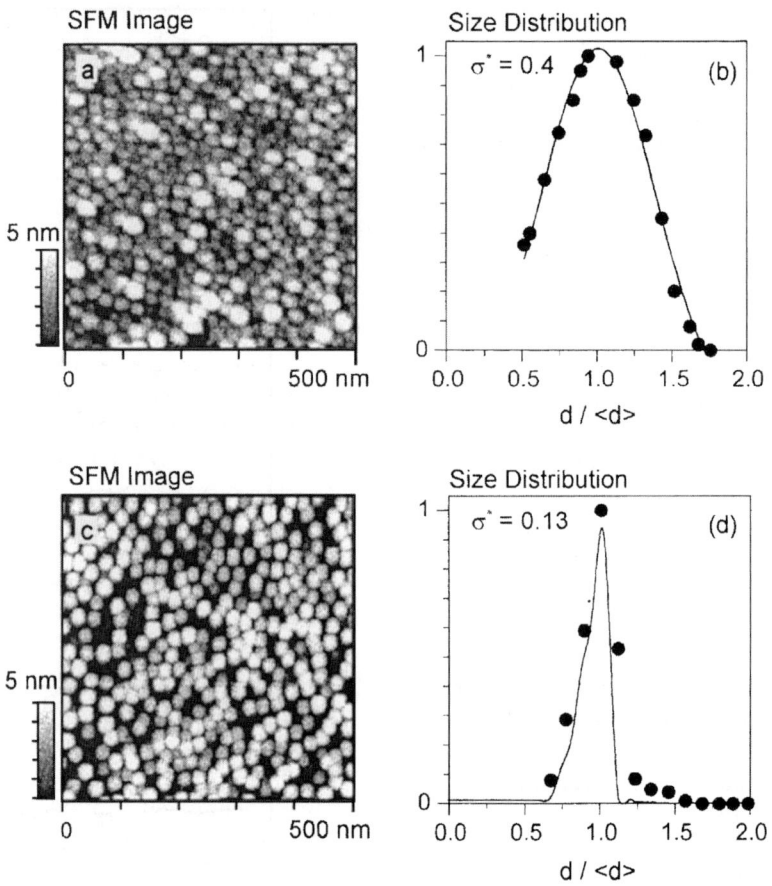

Fig. 12. Scanning force microscopy (SFM) images and size distributions of Ag clusters grown by evaporation on a MgO substrate. The top image and distribution shows the clusters as deposited whereas the bottom image and distribution is taken after laser treatment, aimed at narrowing down the distribution of sizes [28].

depletion spectra in gas phase, the deposited clusters, at least on an insulating quartz substrate, are fragmented and destroyed following the absorption of a photon. Certain size clusters, which absorb the laser photons during the irradiation process, are bleached from the ensemble of clusters. Since both wavelengths smaller and larger than the absorption maximum are used, both the clusters smaller than the average value as well as the largest clusters are destroyed preferentially and sequentially. Thus the size distribution of the clusters is substantially narrowed. This is a unique demonstration of the controlled modification of the distribution of clusters on surfaces by optical methods. Especially for clusters of free electron metals this method seems generally applicable, since these clusters all exhibit pronounced plasmon resonances which shift more or less continuously with size.

3. Magnetic properties of clusters

Most atoms are magnetic. According to Hund's rule, the preferred state of any electronic open shell system is a high spin configuration. Due to the combination of orbital and spin moments, the total magnetic moments of atoms are quite large. In the transition to the solid state system, the orbital moments are largely quenched, since the

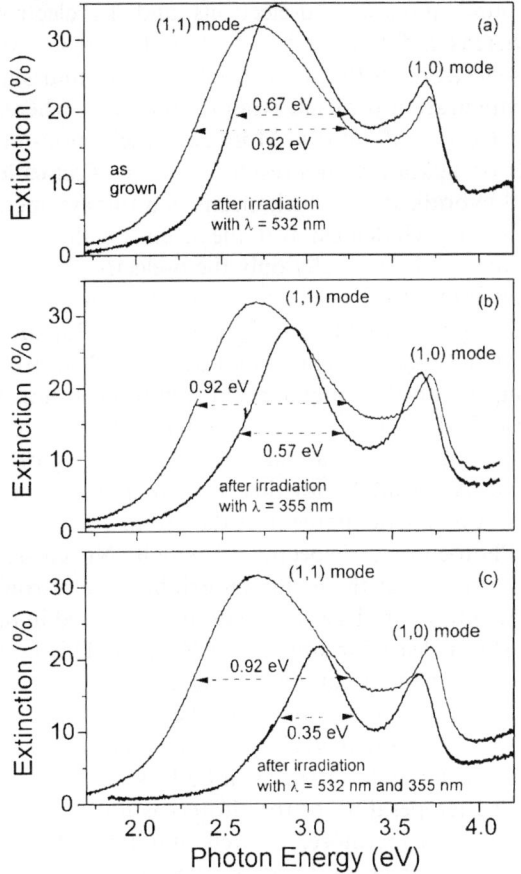

Fig. 13. Optical absorption spectrum of Ag clusters grown by evaporation on a MgO substrate [28]. After irradiation with two different wavelengths (522 and 355 nm) the absorption spectrum is significantly modified corresponding to the much narrower size distribution of the clusters as shown in the previous Fig. 12.

orbital momentum is no longer a good quantum number for an electron in a delocalized electronic state of the solid. Moreover, for most systems even the spins do not form a long range ordered state. Only a temporally and spatially long range order of the atomic moments results in the macroscopic magnetization of the corresponding solid.

As a general phenomenon it is observed that, starting from the extended bulk system, the surfaces of the itinerant transition metal magnets exhibit an increase in the magnetic moment due to the reduction of the coordination number of the surface atoms. Additionally, in thin film systems approaching the thickness of only a few monolayers the Curie temperature decreases [30]. Simple magnetostatics predicts the preferred magnetization of thin film systems to be in the plane of the film. This is the same reason why conventional rod shaped magnets are most stable when magnetized along the long axis. For some thin film systems however, a transition to a perpendicular magnetic anisotropy has been reported [31]. This is especially interesting for magnetic recording, since it allows for a higher recording density.

Considering these two limiting cases, the single or few atom system and the thin films, clusters hold the promise of some interesting properties for magnetic data storage. If the clusters are small enough, then for energy reasons only a single magnetic domain exists whenever the particle is magnetically ordered. The questions arise, however, at what temperature and for which timeframe this configuration is stable. Will the clusters exhibit a stable magnetization at room temperature like a ferromagnet or will they exhibit (super) paramagnetic behavior, which is only stable in an external magnetic field. This magnetic behavior of the system is determined by an intricate competition between thermal fluctuations and the total energy of a ferromagnetically ordered state.

Due to the technological background, the investigation of magnetic properties of clusters has received considerable attention early on. The magnetic properties of mass selected clusters in a molecular beam were measured by observing their deflection in the asymmetric field of a Stern–Gerlach magnet. Whenever such a deflection is observed, this indicates the presence of a magnetic moment, at least while the cluster is subjected to the external magnetic field. Temperature variation of the clusters in the beam results in a variation of the observed moments reasonably well described by the Langevin equation. Accordingly the clusters behave just as superparamagnetic species with moments considerably larger than the bulk values. These moments are shown as a function of cluster size in Fig. 14 for Ni, Co, and Fe clusters [32]. Magnetic moments larger than $1\mu_B$/atom for Ni or $2\mu_B$/atom for Co, as they are reported here, are larger than the corresponding values of the bulk solids. This can be explained either by a change in

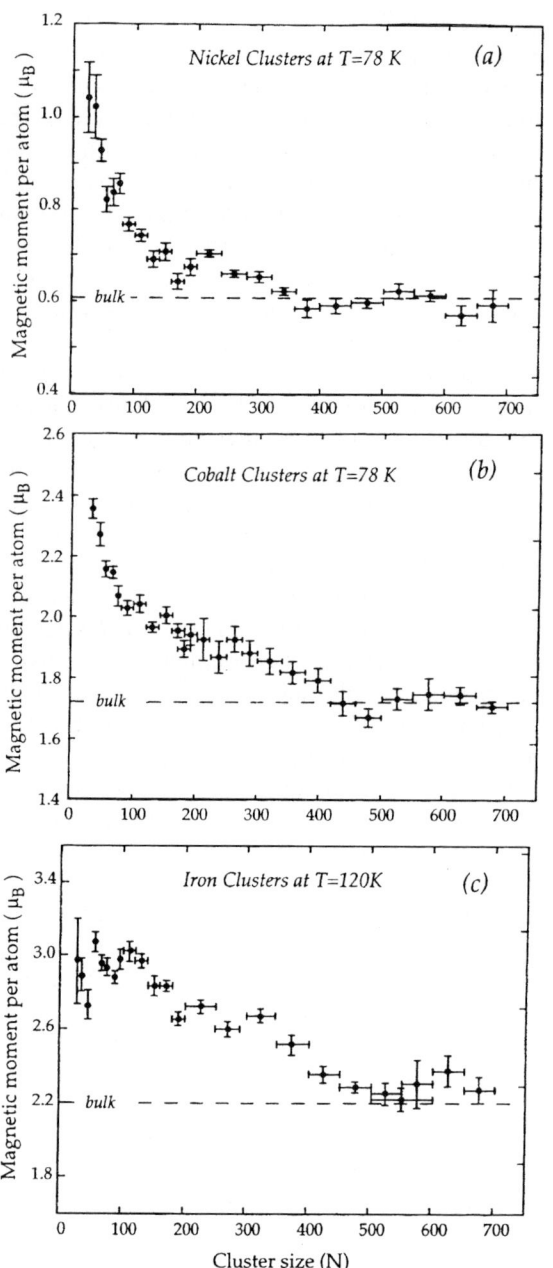

Fig. 14. Magnetic moment/atom of small transition metal clusters measured by deflection of a cluster beam in the inhomogeneous field of a Stern–Gerlach magnet [32].

s and d-band occupation numbers or by the fact that the orbital angular momentum is contributing substantially to the total magnetic moment. The valence electron configuration of a Ni atom in the ground state is 9 d-electrons and 1s electron, whereas a Co atom has only 8 d-electrons and 1 s-electron. In the solid the average ground state configuration for each atom is about d^9s, whereas for Co it is about d^8s. More accurately, non-integer occupation values result due to an effect called s–d hybridization, which means that this symmetry is not as well defined as in the atom. Furthermore, in the solid essentially only the d-electrons (holes) contribute to the magnetic moment which accordingly is about $1\mu_B$/atom for Ni or $2\mu_B$/atom for Co. In the atom however both s and d electrons couple with their angular and spin momentum, as described by the various coupling schemes, to a total angular momentum of the atomic ground state. Associated with this angular momentum is also a magnetic moment of the atom.

Photoemission spectra of small Ni clusters confirmed that these clusters exhibit an electronic structure where 1 s electron/atom is observed in the cluster at least for sizes up to Ni_7 [33]. This large spin moment in conjunction with the orbital angular moment accounts for the large moments observed for small Ni clusters [34], which are up to twice as large per atom as in bulk Ni.

The temperature of the clusters is not very well defined. Furthermore, it is well established that the various degrees of freedom; rotation, vibration, translation, and the electronic excitation are generally not equilibrated in a cluster beam. Thus the analysis of the magnetic deflection data observed in the Stern–Gerlach experiments on clusters in a molecular beam is quite complicated, since the rotation of the free clusters has to be taken into account as well. This quite complicated analysis is consistent with the conclusion that the transition metal clusters are superparamagnetic. The moments of the transition metal clusters are not coupled to the coordinate system of the clusters but rather rotate freely and follow the direction of the external applied field, independent of the actual rotational motion of the cluster itself. It is needless to point out that in terms of magnetic data storage such a behavior is not useful. However, we expect a transition to occur at a certain temperature, called blocking temperature T_B, when the magnetic anisotropy becomes larger than the thermal fluctuations. Under these conditions, the moments lock into a fixed registry with the crystalline lattice. Studies on Co islands grown by self aggregation on the Au (1 1 1) surface were carried out to determine this blocking temperature as a function of the cluster size [35].

For rare earth clusters the observations indicate different magnetic properties [36]. Gd clusters exhibit a locked moment behavior, where the magnetic moment has a fixed orientation relative to the cluster coordinates and rotates in space as the cluster undergoes rotational motion. This alignment with the lattice depends on the vibrational temperature of the cluster. As the vibrational temperature increases, the alignment becomes progressively less perfect and eventually the cluster will turn paramagnetic. This transition depends strongly on the cluster size. Another interesting observation is that in the Gd clusters the magnetic moment is drastically lower than in Gd thin films or solids, which is interpreted by the authors as indication of a non-collinear or antiferromagnetic alignment of the moments of the individual atoms within the cluster [36].

In general, the studies of the magnetism of free clusters in a molecular beam are very difficult not only from the experimental viewpoint, but also as far as the interpretation is concerned. Recently there have been considerable advances in the controlled fabrication of magnetic quantum structures by self aggregation on surfaces. The state of this field is excellently described in the review by Himpsel et al. on "Magnetic nanostructures" [37]. More recently step decoration was successfully used to grow monatomic Fe or Co wires on stepped surfaces. The combination Fe on vicinal Cu(1 1 1) [38] and Co on Pt(9 9 7) [39] so far yielded the best results. Fig. 15 shows an STM image of Co wires grown on the stepped Pt(9 9 7) surface at a coverage of 0.07 ML of Co. Upon carefully chosen growth conditions, on this surface a coverage of 0.1 ML Co corresponds to extremely well ordered monatomic Co wires decorating the steps. As characterized by STM and He atom scattering experiments even a second row of Co atoms can be attached to form nearly perfect wires along the steps, two atoms wide [39].

Magnetic circular dichroism studies of these wires demonstrate that they are superparamag-

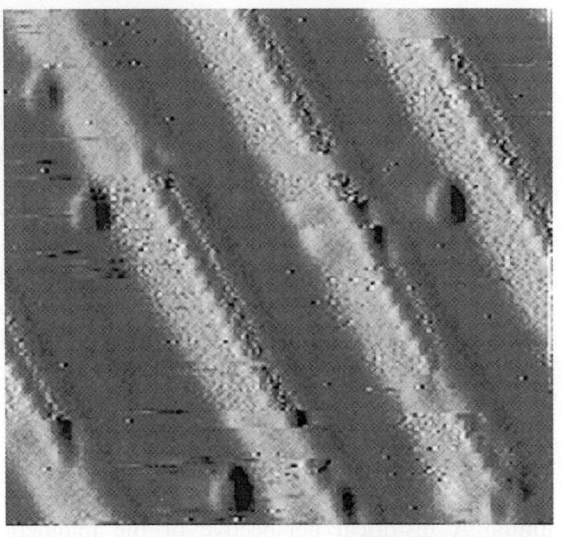

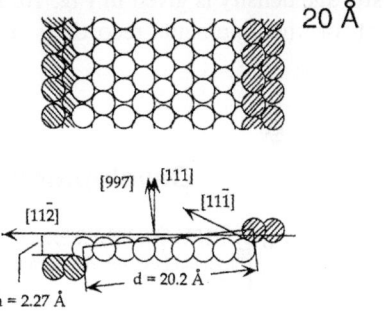

Fig. 15. STM image of the Pt(997) surface covered with 0.07 monolayers of Co at a substrate temperature of 250 K, showing the decoration of the step edges by monatomic Co wires [39]. The geometry of the Pt(9 9 7) surface with terraces nine atoms wide by one atom high is shown at the bottom.

netic down to temperatures around 10 K [40]. Magnetic circular dichroism is the difference in the absorption of circularly polarized light, when an alignment of the magnetization parallel and antiparallel to the photon polarization is compared. This difference signal is proportional to the magnetic moment of the system under investigation [41,42]. Without an external field there is no remnant magnetization observed for the nanowires. Applying the well-known sum rules for CMXD [41,42], the magnetic moments can be estimated. These magnetic moments are substantially larger than for the fcc Co solid or even at the Co surface. Analogous to the proposed behavior of the clusters, the ratio between the orbital moment and the spin moment shows a considerable increase. Furthermore, there is a considerable anisotropy observed. The magnetization prefers to be oriented perpendicular to the direction of the wire. Along the wire, magnetic alignment can also be induced, but the magnetization saturates at a value substantially lower than in the perpendicular direction [40]. Because of the technological impact on magnetic data storage, the area of controlled deposition of clusters and the controlled growth of clusters or islands of magnetic systems and their characterization will be an important field of research in the future. These systems need to be explored and specifically designed as the magnetic storage density is pushed to higher and higher limits. A historical perspective of the magnetic data storage density is given in Fig. 16. In terms of interest of the magnetic recording industry the investigation of the magnetic properties of clusters boils down to the question whether we can find new materials that exhibit superior magnetic properties in the form of clusters or cluster assembled systems. For certain types of magnetic recording media, mostly tapes, it is a well established technology to use magnetic particles of a fairly uniform size embedded into some polymeric 'binder'.

Presently, the highest density recording media have bit sizes as short as 100 nm measured along the track which has a width of about 300 nm. This limit is partially determined by the sensitivity and positioning accuracy of the read/write head, but an important factor is also the data storage lifetime. At room temperature a bit lifetime of 10 years is the accepted standard of the recording industry. New materials engineered on the nanometer scale to stabilize and utilize the large magnetic moments

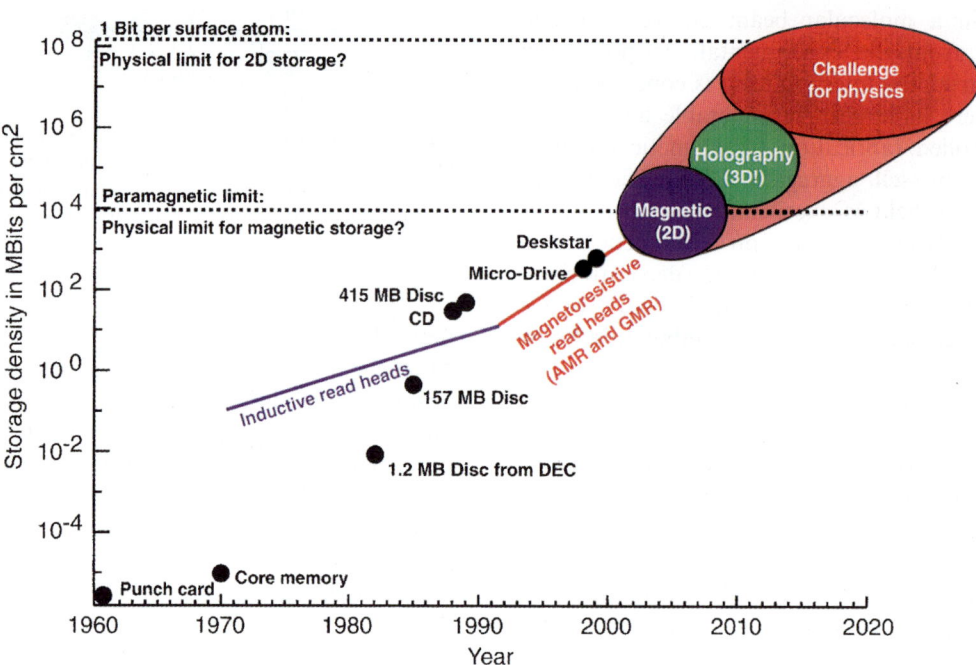

Fig. 16. Development of magnetic data storage density over the last decades and projections into the future. Holographic storage media and paper (punch) cards are included for comparison. The paramagnetic limit is approximately indicated. The atomic density of a crystalline surface corresponds to the physical limit for two-dimensional storage.

present in small clusters may open the pathway to further advances of the state of the art of the magnetic data storage.

3.1. Fullerenes, the third state of solid carbon

The soccer ball shaped C_{60} cluster is certainly the most well known cluster. Kroto, Smalley, and Curl were awarded the Nobel Prize in Chemistry in 1996 for the discovery of this cluster and for proposing this unique structure [1]. At the time, this required a lot of 'chemical intuition', since all they had as 'hard data' was the indication of a cluster consisting of 60 carbon atoms and being somewhat more stable than the neighboring mass clusters which contained 58 or 62 carbon atoms. Conventional structural techniques only could be applied after sufficient quantities of this cluster were available from the arc discharge production developed by Kraetschmer [43] about five years after the initial discovery. NMR and later on X-ray and STM studies eventually were applied to verify the structural model.

This opened the road to the first unique cluster assembled material. Upon condensation from the vapor phase C_{60} molecules, and also other larger fullerenes, form single crystalline solids, which are bound by electrostatic polarization (van der Waals) forces. Due to their mass and large polarizability, the sublimation temperatures of these solids are well above 400 °C. These solids consist purely of carbon and are generally classified as 'semiconducting', exhibiting the onset of electronic excitations in the 1–2 eV range. Thus this third form of solid carbon has distinctly different properties than graphite, which is a semimetal, or the wide band gap (5.5 eV) transparent diamond. Even more fascinating, C_{60}, and so far only C_{60}, can be (electron-)doped using (earth-)alkali atoms to a metallic state, which even is superconducting with transition temperatures above 30 K. Thus the fullerenes, as a class of materials, are second only to the superconducting oxides as far as the superconducting transition temperature is concerned. Contrary to the oxides, where the superconductivity is strongly anisotropic and confined to the Cu–O planes, the fullerides form truly three-dimensional superconductors. Additionally, the superconductivity of the fullerenes can be consistently explained by the standard BCS-mechanism [44]. This conclusion followed from an experimental determination of the electron–phonon coupling constants by photoemission of C_{60}^- in gas phase [44]. Even though a considerable effort has been devoted to explore these systems, there is still a substantial potential that other fullerenes or different dopants may lead to a superconductor exhibiting an even higher transition temperature. Recently for hole doped C_{60} a superconducting transition temperature of 52 K was reported [74].

Fullerenes are only under certain growth conditions the most prominent structures found in the distribution of carbon clusters. This is demonstrated in Fig. 17 [45]. These are cluster mass

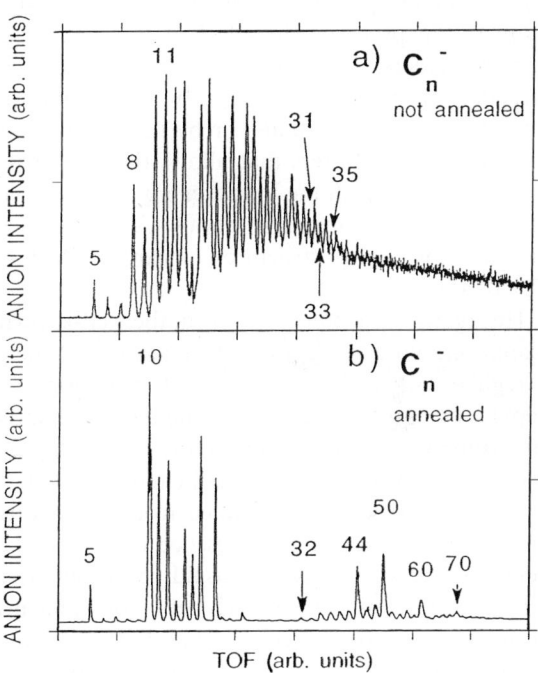

Fig. 17. Time-of flight (TOF) mass spectra of carbon clusters produced by laser vaporization of a carbon rod under different growth conditions. The difference between the mass distribution shown in the top and bottom panel is that for the latter one the clusters were additionally annealed by burning a low energy electric discharge in the high-pressure zone prior to the adiabatic expansion [45]. The bottom spectrum reflects the characteristic abundancies of the fullerenes.

spectra produced by a laser vaporization source. The top spectrum shows a distribution where clusters with even and odd numbers are approximately equally abundant. As explained below, from topological considerations fullerenes all have even numbers of carbon atoms. Accordingly under the growth conditions present to produce the top mass spectrum fullerenes may be present but they are definitely not the dominating structures. This is completely different in the cluster distribution shown at the bottom of Fig 17. Now in the mass range above 30 atoms only even numbered clusters are present and they exhibit the characteristic distribution where C_{44}, C_{50}, C_{60}, and C_{70} are more abundant than the other clusters. The difference in the production conditions between the two mass spectra are due to the fact that in order to obtain the distribution in the bottom panel of Fig. 17 the clusters were annealed by a low energy glow discharge following the laser vaporization and agglomeration process. The effect of this discharge can be understood either as a preferential destruction of less stable clusters, by fragmentation or evaporation of C_2 units, or by the addition of energy to these clusters that allows them to cross the barriers to reach the lowest energy most stable configurations. Starting at about C_{30}, fullerenes are the most stable structures of carbon clusters [7,9,10,45].

Hexagons of carbon atoms form the well-known planar graphite structure, when assembled into a regular lattice. Incorporation of pentagons or heptagons into the planar graphite structure adds curvature to it. According to the rules already formulated by the famous mathematician Euler, closed cage structures assembled from pentagons and hexagons require the presence of exactly 12 pentagons. Accordingly C_{20} would be the smallest fullerene consisting of only 12 pentagons and no hexagon. C_{60} with 12 pentagons and 20 hexagons is the smallest fullerene where all pentagons are completely isolated and surrounded by hexagons only. This is certainly a unique structure and structures with isolated pentagons are considered more favorably, since the pentagons add the curvature to the structure and therefore bring some stress into the system. This is referred to as the isolated pentagon rule (IPR). There are other isomers of C_{60} possible, but none of these has isolated pentagons.

Even though fullerenes smaller than C_{60} can be produced by the standard cluster production methods, i.e. laser ablation or electric arc discharge followed by condensation in a high pressure inert gas atmosphere, only C_{60}, C_{70} and larger fullerenes could so far be successfully extracted by solvents from the soot produced according to the Krätschmer–Huffmann method [43]. This leads to speculations about the stability or reactivity of the fullerenes. It is supposed that in the extraction and separation process the smaller fullerenes and also C_{62}, C_{64}, C_{66}, and C_{68} are unstable against either reaction–agglomeration or fragmentation. Whether this is really the explanation, this is one of the so far unresolved questions in fullerene research.

The question of stability of the smaller fullerenes against either fragmentation or reaction has been addressed by a theoretical study combined with photoemission of the mass separated cluster anions [10]. As suggested, by 'chemical intuition' the clusters having the largest HOMO–LUMO gap are also found by theory to be especially stable. The correlation of the HOMO–LUMO gaps and the theoretically determined stabilities is shown in Fig. 18. The gap values are determined as

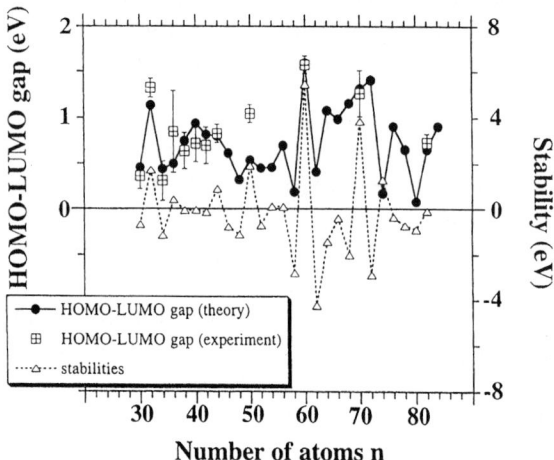

Fig. 18. Experimental and theoretical values of the HOMO–LUMO gap of the fullerenes as a function of cluster size [10]. Also shown is the stability S of the clusters defined by the energy difference relative to the next higher and lower fullerene $[S(n) = E(n-2) + E(n+2) - 2E(n)]$.

described above for the metal clusters. The theoretical stability is more difficult to define. Here the energy differences are plotted showing the binding energies per atom in comparison with the next smaller and next larger fullerene. This is a realistic approach for the cluster growth in gas phase, since we know that, for example, fullerenes shrink by (successive) evaporation of C_2 dimers [46]. On the other hand, the process of joining two fullerenes and forming a much larger structure is not taken into account here. From this graph C_{32}, C_{44}, C_{50}, C_{60}, and C_{70} are expected to form especially stable clusters and cluster materials.

Recently a group from Berkeley has claimed to have isolated C_{36} and formed a new solid consisting of these molecules [47]. While they show X-ray scattering and scanning tunneling spectroscopy data in support of their claim, others claim that the NMR data are not in agreement with the proposed structure [48]. Furthermore, desorption mass spectra never showed the presence of clean C_{36}, only a peak at a mass corresponding to $C_{36}H_4$ was found [47].

The most intriguing questions in fullerene research, however, is why only C_{60} can be doped to exhibit metallic properties and this is observed only in a fairly narrow doping range at a stoichiometry around A_3C_{60}, where A is an alkali atom intercalated into the fullerene solid by standard intercalation procedures. For the sake of being complete we have to add here that doping with alkaline earths or a mixed alkali-alkaline earth doping of C_{60} also produces metallic and superconducting compounds, Ba_4C_{60} being one example of these [49]. In these compounds the second conduction band (t_{1g}) is partially filled with electrons from the dopant, even though the charge transfer might not be complete. Because of the extremely high electron affinity of C_{60} 'chemical' or charge transfer doping has only resulted in electron doped C_{60}. Recently hole doping for C_{60} was accomplished by subjecting C_{60} to electric fields in a special FET geometry and a superconducting transition temperature of 52 K was reported [74]. For the sake of simplicity we have omitted these compounds in the following discussions. C_{70}, on the other hand, which at a rough glance has a very similar electronic structure, cannot be doped to exhibit metallic conductivity at any doping level and neither can the other fullerenes so far be doped to a metallic state.

The most simplistic approach to understand this doping is by a straightforward charge transfer from the alkali atom into the first unoccupied band of the C_{60} solid, which has t_{1u} character in the molecule and is threefold degenerate. Accordingly one would expect that any doping level resulting in less than six electrons in this manifold of states will be metallic. Since this is not the case, obviously this system is much more complicated than this oversimplified rigid band doping picture suggests.

The next higher level approach to explain this behavior takes electron correlations into account. Essentially the electronic states of C_{60} are of molecular nature and the van der Waals bonding of the solid results only in a small dispersion and increase of the bandwidth W by a few tenths of an eV usually around 0.5–0.6 eV [50,51]. The electron electron Coulomb repulsion on the other hand is quite strong, because of the small size of the molecule and amounts to a value about 1.6 eV [52]. Whenever the bandwidth is smaller than the electron electron repulsion, charge transfer from one C_{60} unit to its neighbor is prohibited and the system is insulating. Following these criteria the alkali doped C_{60} solids should be so called Mott–Hubbard insulators and now we are faced with the opposite question: how can it happen that A_3C_{60} is a metal and even more a superconductor? Early on Lof et al. [52] estimated that stoichiometric A_3C_{60} would be a Mott–Hubbard insulator with a band gap of 0.7 eV and they proposed that the metallic phase originates from off-stoichiometry phases where either electrons or holes are doped into this phase similar to the hole doping of high T_c superconductors. Later Gunnarsson and coworkers [51], however, pointed out that in N-fold degenerate bands, the hopping is increased approximately proportional to $\sqrt{N}$ and thus the Mott–Hubbard gap is substantially reduced and the transition into the Mott–Hubbard insulating state scales not as U/W but roughly with $U/W\sqrt{N}$.

Another factor to be taken into account is that any charge transfer into the LUMO of C_{60} will result in a deformation of the molecular structure.

Associated with this is a Jahn–Teller splitting of the initially degenerate t_{1u} states into three individual bands, which are at least partially overlapping in energy since the Jahn–Teller splitting is of the same order of magnitude as the solid state band dispersion. The occupation of these bands and the magnetic configuration will be determined by the relative order of the Jahn–Teller energy and the multiplet splitting or the exchange energy, which results in Hund's rule. These parameters will determine whether the ground state is a high spin 'magnetic' metal or a low spin (non-magnetic) insulator. Furthermore the modification of the electronic structure by the JT-effect also will result in a change of the Coulomb repulsion energy, which could be different for the different bands [53].

All this obviously points to a very delicate balance of the various electron electron and electron phonon interactions. The present state of our knowledge is that several factors are conspiring favorably to cause a metallic ground state for A_3C_{60} whereas all other dopant stages or any of the other fullerenes as host materials are Mott–Hubbard insulators [51].

This 'coincidence' as strange or accidental as it may be, immediately points to one of the challenges of the future. Can we produce other fullerenes that are superconductors with potentially even higher transition temperatures? Alkali doping has been extensively investigated [49], but an alternate route might be via endohedral doping. Even though quite a few endohedral fullerenes could be produced and isolated, any of the higher endohedral fullerenes are semiconducting. Examples foremost include $RE@C_{82}$ [RE = rare earth], which is one of the most prominent and widely studied classes, since these endohedrals are available in large quantities following extraction from fullerene soot and chromatographic separation [54].

Nominally La should be able to transfer three valence electrons to the cage. The other rare earths also have occupied 4f states, but thereby the question is open in which form these are involved in a charge transfer. Even for La, however, the charge transfer is not a simple electron donation into the unoccupied states of the fullerene cage.

Resonant photoemission of $La@C_{82}$ has been used to demonstrate that there is a sizeable hybridization of the electronic states between the endohedral La atom and some of the valence orbitals of the fullerene cage [55]. Interestingly the structures that are most strongly enhanced are not the topmost orbitals as can be seen from Fig. 19, but rather some valence orbitals in the binding energy range between 2 and 4 eV. This is in agreement with calculations of the electronic structure, which also give the largest hybridization with orbitals in this binding energy range [56].

While all higher endohedral fullerenes investigated so far exhibit a band gap, the question remains whether C_{60} doped endohedrally with metal atoms might be metallic. La and the rare earths are especially promising candidates, since the external doping by alkali atoms results in a metallic state when three electrons are doped into each C_{60} molecule as discussed extensively above. Recently a Japanese group has produced and isolated [57] various endohedral compounds of C_{60}, but again the UV–VIS absorption spectrum of $Eu@C_{60}$, for example, exhibited a small red-shift compared to C_{60} but not a metallic behavior.

We have produced $Ce@C_{60}$ and $La@C_{60}$ as isolated clusters deposited onto a graphite substrate [58]. This was done in our laboratory using a specifically designed cluster source, which is coupled to a magnetic mass separator. At the output of this mass separator clusters of a specific mass can be deposited with adjustable energy onto a substrate. While $Ce@C_{60}$ displays a gap of about 0.3 eV, $La@C_{60}$ shows metallic behavior as observed by $I(V)$ curves taken locally with an STM whose tip was placed directly above the cluster [58]. So far, however, we have not been able to produce sufficient material to deposit a solid film and to check for superconductivity.

3.2. Surface properties of clusters

So far very little is known about the differences between surfaces and bulk properties of clusters. Naturally for small particles almost every atom is a surface atom and only with increasing size will there be atoms in the bulk interior. For icosahedral structures, which are the preferred structures

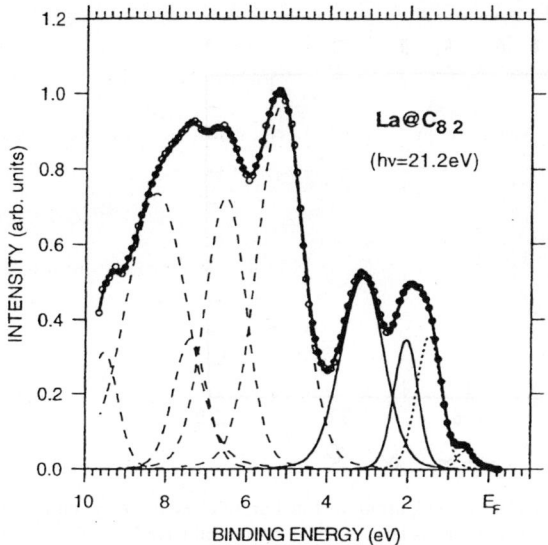

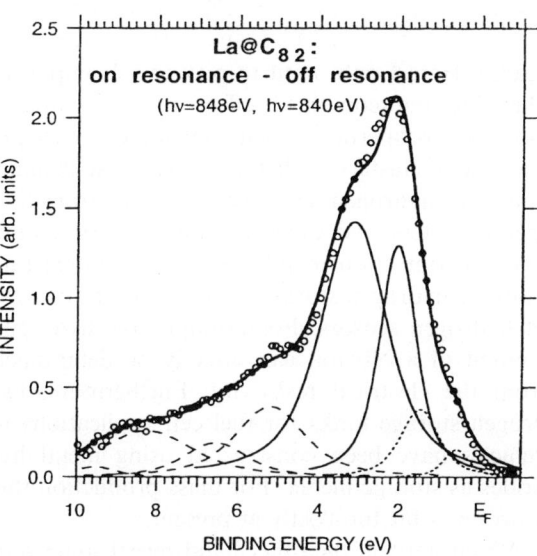

Fig. 19. Valence band photoemission of LaC$_{82}$ taken with low energy (21.2 eV) photons (top) in comparison with a difference spectrum emphasizing the resonant valence band emission (bottom) [55]. The largest resonance effects are observed for the states between 1 and 4 eV binding energy. Also shown is a deconvolution of the spectra into individual line (Voigt) profiles under the condition that the center positions of the lines are unchanged between the top and bottom panel.

and the ratio of surface to bulk atoms are listed also. Since these are close packed structures, for any real system this is the largest bulk to surface atom ratio achievable. Most likely the real clusters will have more surface atoms than the numbers given in Fig. 20. The best example for this is the fullerenes, where for any structure established so far all the atoms are located on the surface and, except for endohedral dopants, there are no interior bulk atoms.

Additionally, many industrial catalysts contain small clusters of metal atoms, which form the essential centers where the chemical reactions take place. One reason for this high dispersion is obviously due to the fact, that, especially if expensive metals such as Pt are involved, the total amount of material should be kept low and used most economically. The other reason, however, is that small clusters exhibit different reactivities, or what is often more important selectivity, than the corresponding bulk material. This is illustrated in the 'classical' plot by Masson [59], shown as Fig. 21, where the hydrogenation efficiency of ethylene by Pt clusters on Al_2O_3 is plotted as a function of cluster size. Solid surfaces exhibit a small but finite activity and the atom is almost totally unreactive, whereas the largest activities are reported for an ensemble of small clusters. The question remains, whether the largely enhanced activity is due to one particular size cluster, which is just present within the ensemble at different concentrations, or whether it is due to clusters within a particular small range of sizes.

As a next step one might ask the question of whether clusters are more catalytically active because they have more surface sites or special sites for the reaction to occur. It is also known for some chemical reactions that steps or kinks on the otherwise perfect flat surface are the actual reaction sites. In clusters there are obviously many more of these sites available than on the almost perfect surface of a single crystal. The question, however, is deeper to the extent that the steps and kinks not only allow for different adsorption geometries, but they also have distinctly different electronic properties. The ultimate response to these questions, however, obviously is that the structure and geometry on one hand and the

of some cluster systems, these numbers can be given exactly. These structures are shown in Fig. 20 and the exact numbers of atoms in each cluster

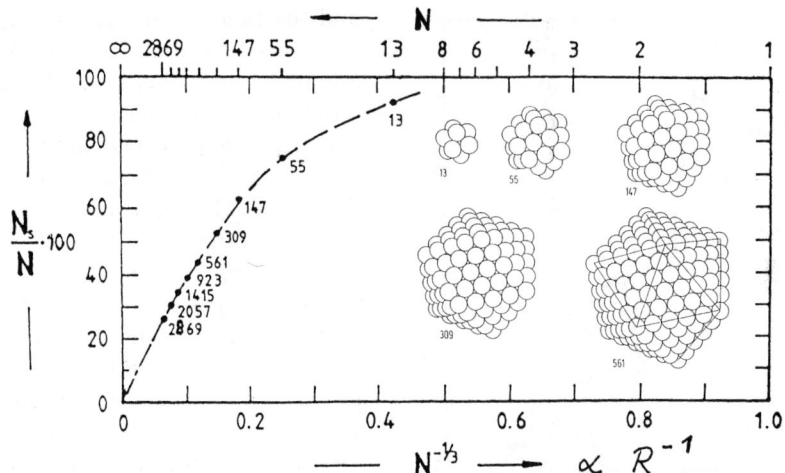

Fig. 20. Structures of closed shell icosahedral clusters and the ratio of surface to bulk atoms as function of clusters size N, plotted as total number of atoms N (top scale) and $1/R \sim N^{-1/3}$ (bottom scale). Even clusters as large as 2000 atoms still have 30% of all atoms located in the outermost surface shell.

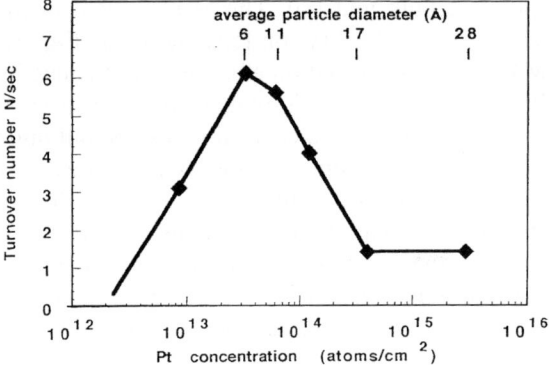

Fig. 21. Efficiency of ethylene hydrogenation by ensembles of Pt clusters produced by evaporation and agglomeration of Pt onto Al_2O_3 [59]. The cluster size distributions and average particle diameters were determined by TEM and correlated with the total number of atoms deposited.

electronic structure of these particles on the other hand are intimately related and their effect on the catalytic reactivity cannot be independently determined. Nevertheless in the literature frequently chemisorption properties of clusters are discussed predominantly stressing either the electronic or the geometric factors.

For these general reasons chemisorption studies on mass resolved clusters in a molecular beam were started early on in the field of cluster research. Here I only want to pick two examples to elucidate the combined effect of steric and/or electronic properties, which contribute to the special role of clusters in chemistry. The first example concerns hydrogen chemisorption on transition metal clusters and especially on Nb. As a bulk material transition metal hydrides are interesting both for hydrogen storage as well as in the context of hydrogen sensors. For example, the hydrogen content of a Nb foil can directly be determined from the electrical resistivity. Furthermore, hydrogen storage tanks for fuel cell applications in vehicles have been constructed using metal hydrides as storage media. For mass production this however is far too costly at present.

Nb clusters on the other hand reveal some very interesting properties when reaction with hydrogen (H_2 or D_2) is investigated. First of all, their reactivity is quite high and saturation is reached at about a hydrogen/metal ratio of 1.3 [60]. Furthermore, there are some clusters, Nb_8, Nb_{10}, and Nb_{16}, which exhibit a drastically reduced reactivity for the reaction with and chemisorption of H_2 (D_2). This cannot readily be explained by the standard models relating the IP's of the clusters with the chemical reactivity. Detailed studies of the electronic structure of these clusters have shown, that Nb_8, Nb_{10}, and Nb_{16} are the only Nb clusters in this size range that exhibit a HOMO–LUMO gap of about 0.5 eV in the electronic states of the neutral species [61]. All other Nb clusters have an open shell structure and a vanishing electronic gap. The correlation between a closed shell electronic structure, as evidenced by a filled d-band of a transition metal surface or the gap of the clusters, and low H_2 chemisorption rate was originally pointed out by Harris and Andersson, when the differences between H_2 chemisorption on Ni and Cu were discussed [62]. In cases where the substrate has a closed electronic shell, the total charge cannot react flexibly to the approaching H_2. Since the molecule also is surrounded by a large s-electron cloud it bounces off the electronically inflexible surface and gets repelled by the surface long before it comes within reach of the d-electrons, which would help to dissociate the molecule. Similarly clusters with a band gap cannot readily accommodate any additional charge and are therefore less reactive. Thus the peculiar reactivity of the Nb clusters is probably related to an electronic structure effect, even though as we pointed out above, the geometry and the electronic structure of these clusters are intimately related [61].

The second example concerns the CO chemisorption on transition metal clusters. Transition metal carbonyls can be prepared as saturated species by chemical means and are available in large quantities. Accordingly they have been the center of attraction for many experimental studies [63]. However, as far as the saturated species are concerned they exhibit chemisorption properties disappointingly similar to single crystalline surfaces. This is evidenced by the fact, that in most cases the CO stretch frequency, which can be taken as a sensitive indicator of the bond activation is with values above 2000 cm^{-1} quite similar to the surface and even the gas phase isolated molecule values. For reactive chemisorption and an activated CO bond a large shift to lower frequencies to values below 1600 cm^{-1} can be expected. According to this criterion the small clusters at the core of the transition metal carbonyls only rarely exhibit any special catalytic properties. This is, however, a much too simplistic approach. As the coverage of the surface or on the metal cores of the carbonyls increases there is obviously a competition for the backbonding charge into the 2π molecular orbital complex. Moreover neighboring CO molecules exhibit a dipole interaction which also results in a frequency shift. Exactly as on surfaces [64], as the coverage increases, the interaction and bond activation for each individual molecule is reduced and the frequency shifts back to values closer to the isolated molecule.

Experiments on clusters allow to study unsaturated carbonyls, which are not available as stable compounds and the systematics with increasing coverage. Accordingly CO chemisorption on clusters in a molecular beam has attracted the attention of various groups. A fairly extensive compilation of these studies is given in Ref. [65]. Essentially all of these used mass spectroscopy to determine the saturation behavior and reactivity. Detailed studies of the electronic structure of the saturated and unsaturated carbonyls were only performed recently [66]. Some of these high resolution data are shown in Fig. 22 for Ni(CO)$_3$. Interestingly not only the CO stretch frequency could be resolved, but also the carbon metal stretch frequency of Ni(CO)$_3$ is observed in the band at 3.5 eV binding energy. This second band does not show any features indicating that the CO

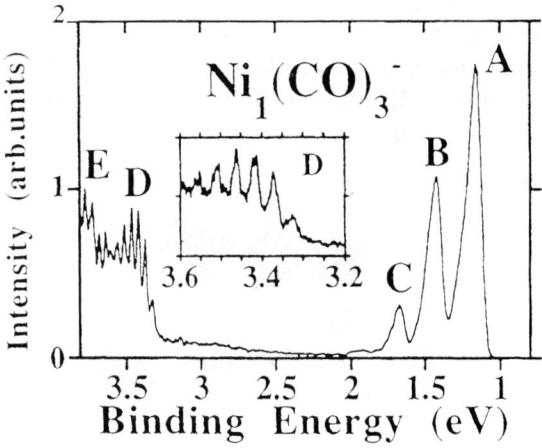

Fig. 22. High-resolution photoemission spectra of Ni(CO)$_3^-$ taken with laser excitation at $h\nu = 4$ eV photon energy. The finestructure spectrum is due to vibrational sidebands corresponding to the excitation of the CO stretch frequency (2090 ± 80 cm^{-1}, peaks A B C) and the CO–metal stretch frequency of 380 ± 30 cm^{-1} (peak D) [66].

bond is affected upon the removal of this electron. For symmetry reasons therefore this band is attributed to a $d_z\sigma$ bonding or antibonding orbital. In terms of CO chemisorption studies on surfaces, this type of bonding orbital has not been observed or identified before, even though simple symmetry arguments unambiguously point to the existence of such a feature.

These studies of the electronic structure of saturated and unsaturated transition metal carbonyls also provide an answer to another intriguing question on transition metal carbonyl chemistry. This relates to how many carbonyls can be attached to a given metal core and whether the saturation is dominated by steric hindrance or electronic effects. The available data clearly pointed towards an electronic effect. For example the Fe (or Ru) trimer, which has a triangular geometry, saturates with 12 CO ligands as shown schematically in Fig. 23, whereas the corresponding Pt trimer saturates when only six CO ligands are chemisorbed. This Pt carbonyl cluster is planar, with all 6 CO's chemisorbed in the plane of the Pt_3 triangle. Three of the CO's are bonded directly to a Pt atom and the other three are chemisorbed at bridge positions between two Pt atoms. Obviously this planar structure of $Pt_3(CO)_6$, offers ample room for additional CO ligands on top of or below the metal atoms, but this space does not get occupied. Another example, which needs an explanation is that Ni_4 saturates with 10 CO ligands [65], whereas Pt_4 which is isoelectronic in terms of the valence shell, saturates with only eight CO ligands [67].

In chemistry these phenomena have been 'explained' or rationalized by the electron counting rules [65]. However, as any 'rule' in physics this is lacking a rigorous theoretical proof. The high resolution photoemission studies mentioned above now provide evidence that the saturation is due to an electronic effect. Saturation in the Pt carbonyls coincides with the complete filling of the valence electronic shell in the neutral cluster. For the saturated neutral clusters a band gap of several tenths of an eV is observed. Thus indirectly the basic concept of the electron counting rules is confirmed to the extent that upon saturation a closed shell electronic configuration has been reached.

4. Outlook: mass selected deposited clusters

The pathway to applications for electronics or catalysis leads via deposition of clusters onto suitable substrates to better defined cluster based materials. Compared to the methods used for the production of today's catalysts; this combines a better definition of the particle sizes with the hope to improve the performance of the system. Initial experiments with mass selected clusters deposited with monatomic size resolution onto a solid substrate have been performed. Monatomic mass resolution alone is not sufficient, but also fragmentation upon impact onto the solid or agglomeration by diffusion after deposition have to be prevented. The concept of 'soft landing' was developed to keep the clusters intact upon deposition onto a substrate. This can be controlled readily by adjusting the voltages of the ion beam deposition. In some experiments a rare gas buffer layer has been used additionally to serve as a soft cushion for the clusters impacting on the surface. Furthermore, following the initial landing, diffusion of the atoms and clusters on the surface has to be suppressed by control of the substrate temperature, in order to prevent agglomeration into larger particles.

For different applications high internal cluster energies and large impact energies may be desired. Thus for example novel coatings can be produced

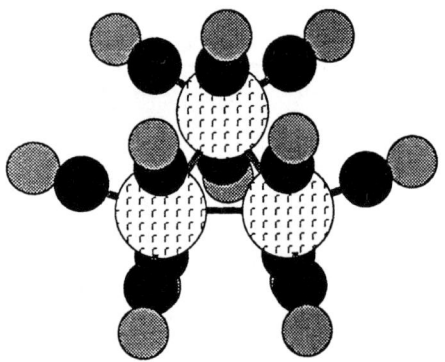

Fig. 23. Geometry of the saturated $Ru(CO)_{12}$ or $Fe(CO)_{12}$ carbonyl cluster.

exhibiting a superior adhesion to the substrate, due to a large intermixing upon impact of the clusters. Specific examples include metal deposition on Teflon or glass and also TiN coatings for biocompatible surfaces of artificial joints and other medical implants [68].

The first experiments on mass resolved deposited clusters by Fayet and coworkers [69] examined the sensitivity of Ag clusters to light, to explore details about the photographic process. They found that the smallest clusters reacting to light exposure were Ag_4 particles. Later on Pt clusters could be deposited with monatomic mass resolution and characterized by photoelectron spectroscopy [70]. Recently a series of very interesting chemistry experiments on mass resolved deposited Pt clusters were carried out by Heiz and coworkers [71,72]. This was achieved using a laser vaporization source for the production of clusters rather than the sputter source used earlier. The sputter source, even though it offers higher intensities, is more limiting since it does not include a cluster growth process. Accordingly the development of a high intensity laser vaporization source by Heiz was an important breakthrough for this field. Fig. 24 shows the oxidation efficiency for CO conversion into CO_2 by mass selected Pt clusters deposited onto a MgO(1 0 0) substrate. The smallest clusters are not very reactive, but starting around 15 Pt atoms/cluster, the reactivity exhibits a sharp increase. The future prospect of this field is very well described in the review article about "Nano-assembled model catalysts" by Heiz and Schneider [72].

5. Summary

I hope to have been able to convey some of the excitement that has carried us in studies on clusters in a molecular beam. Applications in chemistry or information technology, which will follow once an interesting system of technological relevance has been discovered and characterized, will require a much higher rate of production. In the pursuit of these goals a two prong approach might be the most successful tactics. Characterization of these new materials will always require the best possible specificity or selectivity for the production of these samples. Since there will be generally only minute quantities of samples available, novel photon sources such as a FEL will lead to a substantial improvement in terms of the characterization possibilities using methods like EXAFS, scattering, or photoelectron spectroscopy. On the other hand, the actual applications might not require such a high definition. This might open the route to alternate sources or means of deposition such as STM nanofabrication or even deposition by electrochemical means in a modified STM setup [73].

Thin film systems, where the fabrication and composition of the materials is controlled along one coordinate in three-dimensional space form the basis of today's multi billion $ industry in information technology. Analogously, nanofabrication of tailor made materials, where all three dimensions of the system are precisely controlled in order to achieve novel properties, will be a large and profitable endeavor in the future. I am convinced that the 'bottom-up' approach, starting from well defined clusters in the direction to new materials will be a very valuable and profitable route, not only in terms of scientific accomplishments

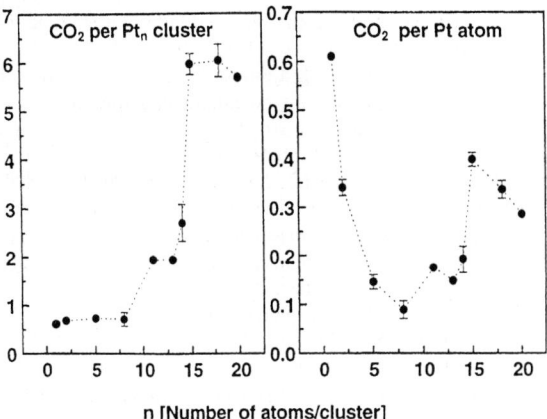

Fig. 24. Efficiency of the catalytic oxidation of CO by mass selected Pt clusters deposited onto a MgO(1 0 0) substrate as a function of the cluster size. The reactivity is abruptly increasing for clusters containing 10 and more atoms (U. Heiz, A. Sanchez, S. Abbet, W.D. Schneider, ISSPIC 9, 0.19 Lausanne (1998)).

and understanding but also in terms of technological advances and achievements.

Acknowledgements

I would like to thank the past and present members of the 'cluster physics' group at the IFF for their many extremely valuable contributions and discussions. Foremost among those are G. Ganteför, B. Kessler, M. Neeb, and P.S. Bechthold, but especially I also would like to thank the students C.Y. Cha, H. Handschuh, H. Kietzmann, R. Klingeler, G. Lüttgens, M. Maus, J. Morenzin, N. Pontius, and G. Schulze Icking-Konert, who have contributed a great deal during their thesis work.

References

[1] H.W. Kroto, J.R. Heath, S.C. O'Brien, R.F. Curl, R.E. Smalley, C_{60}: Buckminsterfullerene, Nature 318 (1985) 162.
[2] K.J. Taylor, C. Jin, J. Conceicao, L.S. Wang, O. Chesnovsky, B.R. Johnson, P.J. Nordlander, R.E. Smalley, Vibrational autodetachment spectroscopy of Au_6^-: image-charge-bound states of a gold ring, J. Chem. Phys. 93 (1990) 7515.
[3] E.C. Honea, A. Ogura, C.A. Murray, K. Raghavachari, W.O. Sprenger, M.F. Jarrold, W.L. Brown, Raman spectra of size-selected silicon clusters and comparison with calculated structures, Nature 366 (1993) 42.
[4] E.C. Honea, A. Ogura, D.R. Peale, C. Felix, C.A. Murray, K. Raghavachari, W.O. Sprenger, M.F. Jarrold, W.L. Brown, Structures and coalescence behavior of size-selected silicon nanoclusters studied by surface-plasman-polarization enhanced Raman spectroscopy, J. Chem. Phys. 110 (1999) 12161.
[5] M. Maus, G. Ganteför, W. Eberhardt, The electronic structure and the band gap of nano-sized Si particles: competition between quantum confinement and surface reconstruction, Appl. Phys. A 70 (2000) 535.
[6] U. Röthlisperger, W. Andreoni, P. Giannozzi, Thirteen-atom clusters: equilibrium geometries, structural transformations, and trends in Na, Mg, Al, and Si, J. Chem. Phys. 96 (1992) 1248.
[7] G. van Helden, M.T. Hsu, N. Gotts, M.T. Bowers, Carbon cluster cautions with up to 84 atoms: structures, formation, mechanism, and reactivity, J. Phys. Chem. 97 (1993) 8182.
[8] K.M. Ho, A.A. Shvartsburg, B. Pan, Z.Y. Lu, C.Z. Wang, J.G. Wacker, J.L. Fai, M.F. Jarrold, Structures of medium-sized silicon clusters, Nature 392 (1998) 582.
[9] R.O. Jones, G. Seifert, Structure and bonding in carbon clusters C_{14} to C_{24}: chains, rings, bowls, plates, and cages, Phys. Rev. Lett. 79 (1997) 443.
[10] H. Kietzmann, R. Rochow, G. Ganteför, W. Eberhardt, K. Vietze, G. Seifert, P.W. Fowler, Small fullerenes: evidence for the high stability of C_{32}, Phys. Rev. Lett. 81 (1998) 5378.
[11] A. Bezryadin, A.R.M. Verschueren, S.J. Tans, C. Dekker, Multiprobe transport experiments on individual single-wall carbon nanotubes, Phys. Rev. Lett. 80 (1998) 4036.
[12] W. Knight, K. Clemenger, W.A. de Heer, W.A. Saunders, M.Y. Chou, M.L. Cohen, Electronic shell structure and abundances of sodium clusters, Phys. Rev. Lett. 52 (1984) 2141.
[13] W. de Heer, The physics of simple metal clusters: experimental aspects and simple models, Rev. Mod. Phys. 65 (1993) 611.
[14] Calculated electron density of C_{60} [15] and structure as it was published as cover of Science 254, December 21 (1991).
[15] The calculations were performed by J. Bernholc, Q.M. Zhang, J.Y. Yi, C. Brabec, T. Palmer.
[16] C.Y. Cha, G. Ganteför, W. Eberhardt, Photoelectron spectroscopy of Cu_n^- clusters: comparison with jellium model predictions, J. Chem. Phys. 99 (1993) 6308.
[17] H. Handschuh, C.Y. Cha, P.S. Bechthold, G. Ganteför, W. Eberhardt, Electronic shells or molecular orbitals: photoelectron spectra of Ag_n^- clusters, J. Chem. Phys. 102 (1995) 6406.
[18] H. Handschuh, G. Ganteför, W. Eberhardt, Vibrational spectroscopy of clusters using a magnetic bottle electron spectrometer, Rev. Sci. Instrum. 66 (1995) 3838.
[19] G. Ganteför, C.Y. Cha, H. Handschuh, G. Schulze Icking Konert, B. Kessler, O. Gunnarsson, W. Eberhardt, Electronic and geometric structure of small mass selected clusters, J. Electr. Spectrosc. 76 (1995) 37.
[20] O. Cheshnovsky, K.J. Taylor, J. Conceicao, R.E. Smalley, Ultraviolet photoelectron spectra of mass-selected copper clusters: evolution of the 3d band, Phys. Rev. Lett. 64 (1990) 1785.
[21] V. Bonacic-Koutecky, L. Cespiva, P. Fantucci, J. Pittner, J. Koutecky, Effective core potential-configuration interaction study of electronic structure and geometry of small anionic Ag_n clusters: predictions and interpretation of photodetachment spectra, J. Chem. Phys. 100 (1994) 490.
[22] U. Röthlisberger, W. Andreoni, Structural and electronic properties of sodium micro-clusters ($n = 2$–20) at low and high temperatures: new insights from ab initio molecular dynamics studies, J. Chem. Phys. 94 (1991) 8129.
[23] C. Massobrio, A. Pasquarello, R. Car, First principles study of photoelectron spectra of Cu_n^- clusters, Rev. Lett. 75 (1995) 2104.
[24] G. Mie, Beiträge zur Optik früher Medien, speziell kolloidaler Metallösungen, Annalen d. Physik 25 (1908) 377.
[25] J. Tiggesbäumler, L. Köller, K.H. Meiwes-Broer, A. Liebsch, Blue shift of the Mie plasma frequency in Ag clusters and particles, Phys. Rev. A 48 (1993) 1749.
[26] J. Tiggesbäumler, L. Köller, H.O. Lutz, K.H. Meiwes-Broer, Giant resonances in silver-cluster photofragmentation, Chem. Phys. Lett. 190 (1992) 42.

[27] W. Harbich, Soft landing of size-selected clusters in chemically inert substrates, Philos. Mag. B 79 (1999) 1307.

[28] F. Stietz, F. Träger, Surface plasmons in nanoclusters elementary electronic excitations and their applications, Philos. Mag. B 79 (1999) 1281.

[29] M. Vollmer, U. Kreibig, Optical Properties of Metal Clusters, Springer Series in Materials Sciences, Berlin, vol. 24, 1995.

[30] F. Huang, M.T. Kief, G.J. Mankey, R.F. Willis, Magnetism in the few-monolayers limit: a surface magneto-optic Kerr-effect study of the magnetic behavior of ultrathin films of Co, Ni, and Co–Ni alloys on Cu(100) and Cu(111), Phys. Rev. B 49 (1994) 3962.

[31] B. Schulz, K. Baberschke, Crossover from in-plane to perpendicular magnetization in ultrathin Ni/Cu(001) films, Phys. Rev. B 50 (1994) 13467.

[32] I.M.L. Billas, A. Chatelain, W.A. deHeer, Magnetism from the atom to the bulk in iron, cobalt, and nickel clusters, Science 265 (1994) 1682.

[33] G. Ganteför, W. Eberhardt, Localization of 3d- and 4d-electrons in small clusters: the "roots" of magnetism, Phys. Rev. Lett. 76 (1996) 4975.

[34] S.E. Apsel, J.W. Ennert, J. Deng, L.A. Bloomfield, Surface-enhanced magnetism in nickel clusters, Phys. Rev. Lett. 76 (1996) 1441.

[35] H.A. Dürr, S.S. Dhesi, E. Dudzik, D. Knabben, G. van der Laan, J.B. Goedkoop, F.U. Hillebrecht, Spin and orbital magnetization in self-assembled Co clusters on Au(111), Phys. Rev. B 59 (1999) R701.

[36] D.C. Douglass, A.J. Cox, J.P. Bucher, L.A. Bloomfield, Magnetic properties of free cobalt and gadolinium clusters, Phys. Rev. B 47 (1993) 12874.

[37] F.J. Himpsel, J.E. Ortega, G.J. Mankey, R.F. Willis, Magnetic nanostructures, Adv. Phys. 47 (1998) 511.

[38] J. Shen, M. Klaua, P. Ohresser, H. Jenniches, J. Barthel, Ch.V. Mohan, J. Kirschner, Structural and magnetic phase transitions of Fe on stepped Cu(111), Phys. Rev. B 56 (1997) 11134.

[39] P. Gambardella, M. Blanc, L. Bürgi, K. Kuhnke, K. Kern, Co growth on Pt(997): from monatomic chains to monolayer completion, Surf. Sci. 449 (2000) 93.

[40] A. Dallmeyer, K. Maiti, M.C. Malagoli, C. Carbone, W. Eberhardt, P. Gambardella, K. Kern, Magnetism of one-dimensional Co quantum wires, in press.

[41] B.T. Thole, P. Carra, F. Sette, G. van der Laan, X-ray circular dichroism as a probe of orbital magnetization, Phys. Rev. Lett. 68 (1992) 1943.

[42] C.T. Chen, Y.U. Idzerda, H.J. Lin, N.V. Smith, G. Meigs, E. Chaban, G.H. Ho, E. Pellegrin, F. Sette, Experimental confirmation of the X-ray magnetic circular dichroism sum rules for iron and cobalt, Phys. Rev. Lett. 75 (1995) 152.

[43] W. Krätschmer, L.D. Lamb, K. Fostiropoulos, D.R. Huffman, Solid C_{60}: a new form of carbon, Nature 347 (1990) 354.

[44] O. Gunnarsson, H. Handschuh, P.S. Bechthold, B. Kessler, G. Ganteför, W. Eberhardt, Photoemission Spectra of C_{60}^-: electron–phonon coupling, Jahn–Teller-effect, and superconductivity in the fullerides, Phys. Rev. Lett. 74 (1995) 1875.

[45] H. Handschuh, G. Ganteför, B. Kessler, P.S. Bechthold, W. Eberhardt, Stable configurations of carbon clusters: chains, rings, and fullerenes, Phys. Rev. Lett. 74 (1995) 1095.

[46] F.D. Weiss, S.C. O'Brien, J.L. Elkind, R.F. Curl, R.E. Smalley, Fourier transform ion cyclotron resonance studies of H_2 chemisorption on niobium cluster caution, J. Am. Chem. Soc. 110 (1988) 4464.

[47] E. Piscoti, T. Yarger, A. Zettl, C_{36}, a new carbon solid, Nature 393 (1998) 771.

[48] G. Seifert, private communication.

[49] A.F. Hebard, Superconductivity in doped fullerenes, Physics Today 45 (November) (1992) 26;
A.R. Kortan, N. Kopylov, S. Glarum, E.N. Gyorgy, A.P. Ramirez, R.M. Fleming, O. Zhou, F.A. Thiel, P.L. Thervot, R.C. Haddon, Superconductivity at 8.4 K in calcium-doped C_{60}, Nature 355 (1992) 529.

[50] O. Gunnarsson, E. Koch, R.M. Martin, Mott–Hubbard insulators for systems with orbital degeneracy, Phys. Rev. B 56 (1997) 1146.

[51] O. Gunnarsson, S.C. Erwin, E. Koch, R.M. Martin, Role of alkali atoms in A_4C_{60}, Phys. Rev. B 57 (1998) 2159.

[52] R.W. Lof, M.A. van Veenendaal, B. Koopmans, H.T. Jonkman, G.A. Sawatzky, Band gap, excitons, and Coulomb interaction in solid C_{60}, Phys. Rev. Lett. 68 (1992) 3924.

[53] M. Fabrizio, E. Tosatti, Nonmagnetic molecular Jahn–Teller Mott insulators, Phys. Rev. 55 (1997) 13465;
A. Auerbach, N. Manini, E. Tosatti, Electron–vibron interactions in charged fullerenes. I. Berry phases, Phys. Rev. B 49 (1994) 12998.

[54] H. Shinohara, Endohedral metallofullerenes, Rep. Prog. Phys. 63 (2000) 843.

[55] B. Kessler, A. Bringer, S. Cramm, C. Schlebusch, W. Eberhardt, Y. Suzuki, Y. Achiba, F. Esch, M. Barnaba, D. Cocco, Evidence for incomplete charge transfer and La-derived states in the valence bands of endohedrally doped La@C_{82}, Phys. Rev. Lett. 79 (1997) 2289.

[56] D.M. Poirier, M. Knupfer, J.H. Weaver, W. Andreoni, K. Laasonen, M. Parrinello, D.S. Bethune, K. Kikuchi, Y. Achiba, Electronic and geometric structure of La@C_{82} and C_{82}: theory and experiment, Phys. Rev. B 49 (1994) 17403.

[57] T. Inoue, Y. Kubuzono, S. Kashino, Y. Takabayashi, K. Fujitaka, M. Hida, M. Inoue, T. Kanbara, S. Emura, T. Uruga, Electronic structure of Eu@C_{60} studied by XANES and UV–VIS absorption spectra, Chem. Phys. Lett. 316 (2000) 381.

[58] R. Klingeler, G. Kann, I. Wirth, S. Eisebitt, P.S. Bechthold, M. Neeb, W. Eberhardt, La@C_{60}: a metallic endohedral fullerene, J. Chem. Phys. 115 (2001) 7215.

[59] A. Masson, B. Bellamy, Y. Hadj Romdhane, M. Che, H. Roulet, G. Dufour, Intrinsic size effect of platinum particles supported on plasma-grown amorphous alumina in the hydrogenation of ethylene, Surf. Sci. 173 (1986) 479.

[60] J.L. Elkind, F.D. Weiss, J.M. Alford, R.T. Laakaonen, R.E. Smalley, Fourier transform ion cyclotron resonance studies of H_2 chemisorption on niobium cluster cautions, J. Chem. Phys. 88 (1988) 5215.

[61] H. Kietzmann, J. Morenzin, P.S. Bechthold, G. Ganteför, W. Eberhardt, Photoelectron spectra of Nb_n^--Clusters: correlation between electronic structure and hydrogen chemisorption, J. Chem. Phys. 109 (1998) 2275.

[62] J. Harris, S. Andersson, H_2 dissociation at metal surfaces, Phys. Rev. Lett 55 (1985) 1583.

[63] E.W. Plummer, W.R. Salaneck, J.S. Miller, Photoelectron spectra of transition-metal carbonyl complexes: comparison with the spectra of adsorbed CO, Phys. Rev. B 18 (1978) 1673.

[64] L. Surnev, Z. Xu, J. Yates, IRAS study of the adsorption of CO on Ni(1 1 1): interrelation between various bonding modes of chemisorbed CO, Surf. Sci. 201 (1988) 1.

[65] S. Vajda, T. Leisner, S. Wolf, L. Wöste, Reactions of size-selected metal cluster ions, Philos. Mag. B 79 (1999) 1353.

[66] G. Schulze Icking-Konert, H. Handschuh, G. Ganteför, W. Eberhardt, Bonding of CO to metal particles: photoelectron spectra of $Ni_n(CO)_m^-$ and $Pt_n(CO)_m^-$ clusters, Phys. Rev. Lett. 76 (1996) 1047; Int. J. Mass Spectr. Ion Phys. 159 (1996) 81.

[67] P.A. Hintz, K.M. Ervin, Nickel group cluster anion reactions with carbon monoxide: rate coefficients and chemisorption efficiency, J. Chem. Phys. 100 (1994) 5715.

[68] H. Haberland, Z. Insepov, M. Moseler, Molecular-dynamics simulation of thin-film growth by energetic cluster impact, Phys. Rev. B 51 (1995) 11061.

[69] P. Fayet, F. Granzer, G. Hegenbart, E. Moisar, B. Pischel, L. Wöste, Latent-image Generation by deposition of monodispersed silver clusters, Phys. Rev. Lett 55 (1985) 3002.

[70] W. Eberhardt, P. Fayet, D.M. Cox, Z. Fu, A. Kaldor, R. Sherwood, D. Sondericker, Photoemission from mass-selected monodispersed Pt clusters, Phys. Rev. Lett. 64 (1990) 780.

[71] U. Heiz, A. Sanchez, S. Abbet, W.D. Schneider, The reactivity of gold and platinum metals in their cluster phase, Eur. Phys. D 9 (1999) 35.

[72] U. Heiz, W.D. Schneider, Nanoassembled model catalysts, J. Phys. D Appl. Phys. 33 (2000) R85.

[73] D.M. Kolb, An atomistic view of electrochemistry, Surf. Sci. 500 (2002) 722.

[74] J.H. Schön, C. Kloc, B. Batlogg, Superconductivity at 52 K in hole-doped C_{60}, Nature 408 (2000) 549.

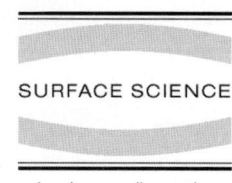

Clusters and islands on oxides: from catalysis via electronics and magnetism to optics

Hans-Joachim Freund *

Fritz-Haber-Institut der Max-Planck-Gesellschaft, Faradayweg 4-6, D-14195 Berlin-Dahlem, Germany

Received 5 July 2000; accepted for publication 27 April 2001

Abstract

The study of metal deposits on oxides represents a field of wide interest with respect to applications as well as to basic science. The state of the art of the field is reviewed on the basis of examples from various research groups. An attempt is made to define and discuss a series of new experiments that could be undertaken to answer some key questions in the field. © 2001 Elsevier Science B.V. All rights reserved.

Keywords: Models of surface chemical reactions; Catalysis; Surface diffusion; Aluminum oxide; Clusters; Insulating films

1. Introduction

Why in the world would anyone be interested in a rather specific subject like this? There are many reasons why you should be interested and those range from the industrial importance of catalysis to the beauty of the pre-Roman art of making stained glasses:

(1) Think about your car and the pollution control in the exhaust system. Fig. 1 shows a schematic diagram with a typical exhaust catalyst in its housing [1]. The catalyst consists of a monolithic backbone covered internally with a washcoat made of mainly alumina but also ceria and zirconia, which itself is mesoporous and holds the small metal particles, often platinum or rhodium. An electron microscope allows us to take a close look at the morphology of the catalyst at the nanometer scale. In order to be active, the metal particles have to be of a few nanometer in diameter and also the support has to be treated in the right way. To a certain extent the preparation is an art, some call it even "black magic". A full understanding of the microscopic processes occurring at the surface of the particles or at the interface between particle and support, however, is unfortunately lacking. We have to realize that catalysis in connection with pollution control—the specific example chosen here—does only utilize a small fraction of the world market for solid catalysts. Human welfare is considerably depending on automotive, petroleum and other industries which constitute a market of $100 billion per year and growing rapidly. Given the situation, it is clear that we eventually must achieve a good understanding of the processes. Interestingly, even though the problem is strongly connected to applications, there is a lot of fundamental insight that has to be gained.

* Tel.: +49-30-84134102; fax: +49-30-84134101.
E-mail address: freund@fhi-berlin.mpg.de (H.-J. Freund).

0039-6028/01/$ - see front matter © 2001 Elsevier Science B.V. All rights reserved.
PII: S0039-6028(01)01543-6

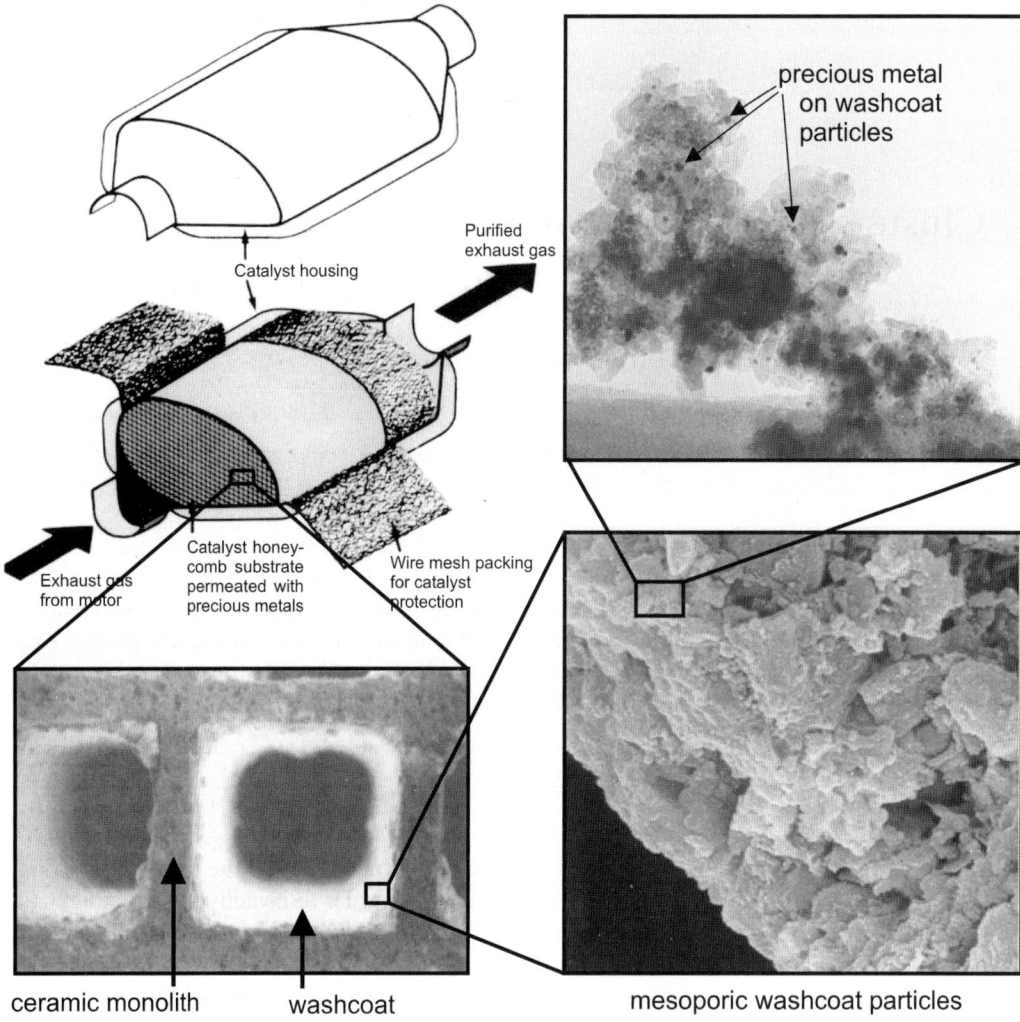

Fig. 1. Schematic representation of the car exhaust catalyst in its housing. Transmission electron micrographs with increasing resolution show the various constituing ceramic and metallic materials in their morphology. Adapted from Ref. [1].

(2) Think about the problem how one could create an artificial nose! Sensors [2–4] allow a computer to smell via "communication with a chemical reaction". An example is shown in Fig. 2. A schematic representation of a device which is called a metal-oxide-semiconductor-field-effect-transistor (MOSFET) is shown (Fig. 2a). In such a device a thin metal film is separated from a Si-crystal through an isolating layer of SiO_2. The idea is to modulate the conductance of a small semiconductor slice by means of an electric field perpendicular to the semiconductor surface. A positive charge on the metal layer induces a negative charge in the semiconductor resulting in a change in the lateral conduction. The charging of the metal layer critically depends on its morphology and can be influenced in a characteristic way by adsorbing gases onto it. These changes upon adsorption allow the MOSFET to "smell", but the details of the elementary steps are not fully understood. The actual device, which was developed about half a century after the initial idea, looks more like the one

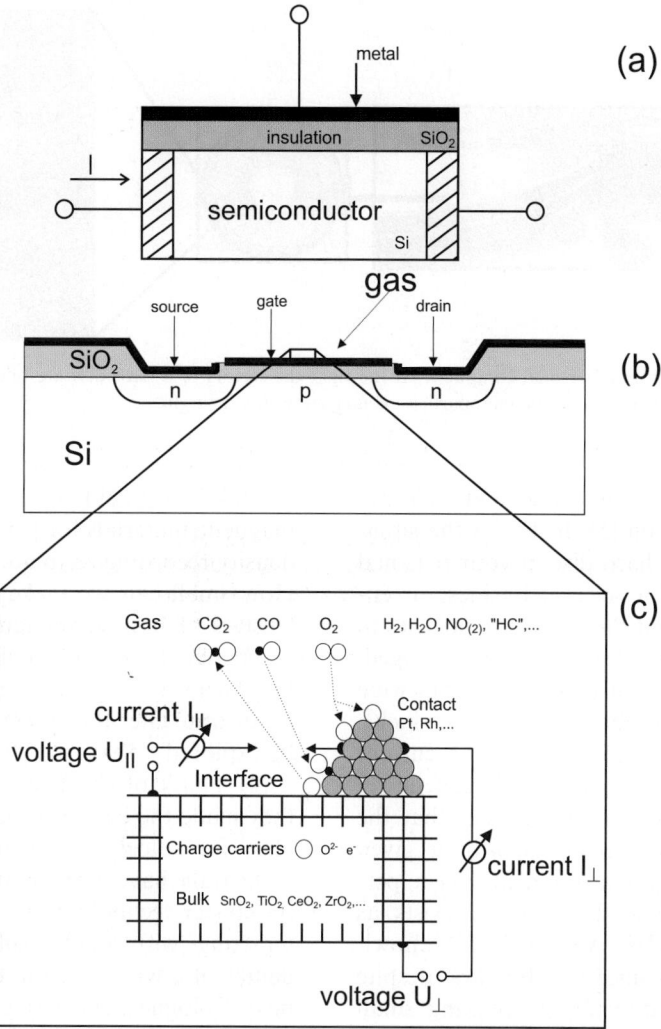

Fig. 2. (a) Schematic representation of the principle of a metal oxide field effect transistor (MOSFET). (b, c) Schematic representation of the design of a MOSFET, and representation of the morphology and the adsorption processes at the metal–oxide interface (2c adapted from Ref. [2]).

schematically shown in Fig. 2b. A n/p/n-transistor is generated via local doping of the semiconductor. It is possible to shape the oxide film and the metal layer as indicated, thus forming a source-gate-drain-structure. It is now easy to envision that the performance, durability and chemical sensitivity of such a device depends heavily on the microscopic structure of the metal layer (see Fig. 2c). The control of the structure of the metal overlayer film inturn depends on our understanding of the elementary steps in nucleation and growth of metal islands and their coalescence to form the film. Cluster formation is an intermediate step in this process, in fact a rather important one. In the series of elementary steps governing the shape size and distribution of islands cluster formation is crucial. The changes in conductance, i.e. the ability and sensitivity to "smell", via the interaction with a gas phase, depends largely on the shape and size of islands, exposure of facets, and other more complex factors, such as co-adsorbates, contaminants etc. of the film. These properties need to be investigated.

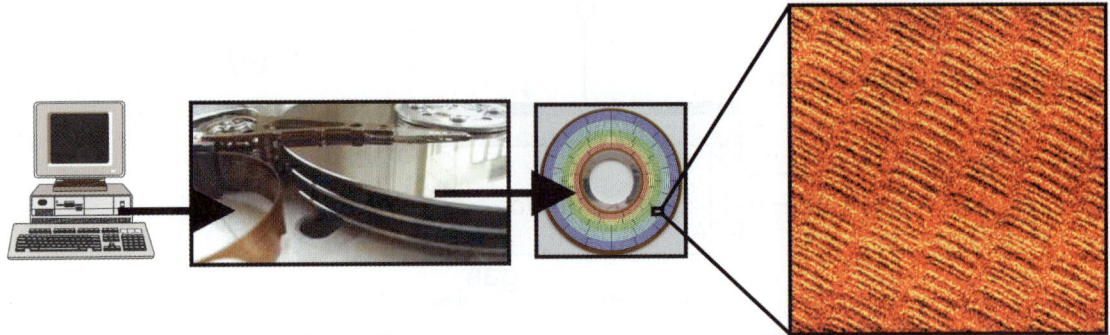

Fig. 3. Picture of a hard disk in a personal computer. At the right the sectors of a hard disc are schematically represented and in addition a small area consisting of magnetic nanoparticles is imaged on the far right.

(3) Think of the way magnetic materials are used to store information [5]! In Fig. 3 the situation is illustrated for a hard disc in your personal computer, be it used for science, business or entertainment. On the right the magnetic particles in one sector of the hard disc have been imaged. These particles are 650 nm long and 50 nm wide for a storage density of about 20 Gbit per inch^2. Their size can go down to 150×15 nm^2. A current goal is to make them even smaller. Magnetic recording industries make use of the fact that the ferromagnetic state of a material with a given orientation of the magnetic moment has a permanent magnetization if the material exists as nanometer-sized particles. Although this knowledge has been around since 40–50 years, despite intense activities, the difficulty in making small enough particles of good quality has slowed down the advancement of this applied field. Only recently, this difficulty in making small particles has been overcome and new experimental techniques have been developed. Conceptually, the ferromagnetic state of bulk metals is a surprisingly rarely observed property if we consider that most atoms have non-zero magnetic moments or spin. Apparently, the formation of metallic bonds leads to non-magnetic bulk metal. It appears quite natural to ask how the spin systems of clusters evolve as a function of particle size. Do we understand this in detail? The answer is probably No! Therefore we need deeper insight into the magnetic behavior of nanoparticles, which is potentially important for the development of new fundamental theories of magnetism and in modeling new magnetic materials for permanent magnets or high density recording as to above. Questions such as i) How small can we make a ferromagnet? ii) Can clusters of non-ferromagnetic materials be ferromagnetic? If yes, how many atoms do we need? iii) In which way does the interaction of a cluster with a substrate alter the magnetic properties? Can now be explored.

(4) Think of the Romans! Do you know how they made the beautiful stained glasses or how the glass for colossal, colorful windows in medieval cathedrals have been manufactured? It is based on cluster technology, i.e., the use of clusters interacting with oxidic substrates! Fig. 4 shows details of a window from the Altenberg Cathedral near Cologne. The red color is caused by gold particles embedded in the glass matrix and the matrix is an amorphous silica–alumina mixture. Since the work of the physicist A. Mie, published in "Annalen der Physik" in 1908 [6], it is understood that absorption of light in a collective excitation of electrons on a sphere of metal—we now call it a plasmon excitation—is the cause of the color. When the particles become smaller and smaller in size the electrons start to "realize" that they are confined in space and then the optical properties become size dependent in a way that has not been predicted by classical Mie-theory but is a consequence of quantum mechanics: If you put an electron in a "box", i.e. a potential well and the dimensions of the "box" reach atomic dimensions, e.g. 10 Å or so, then the states

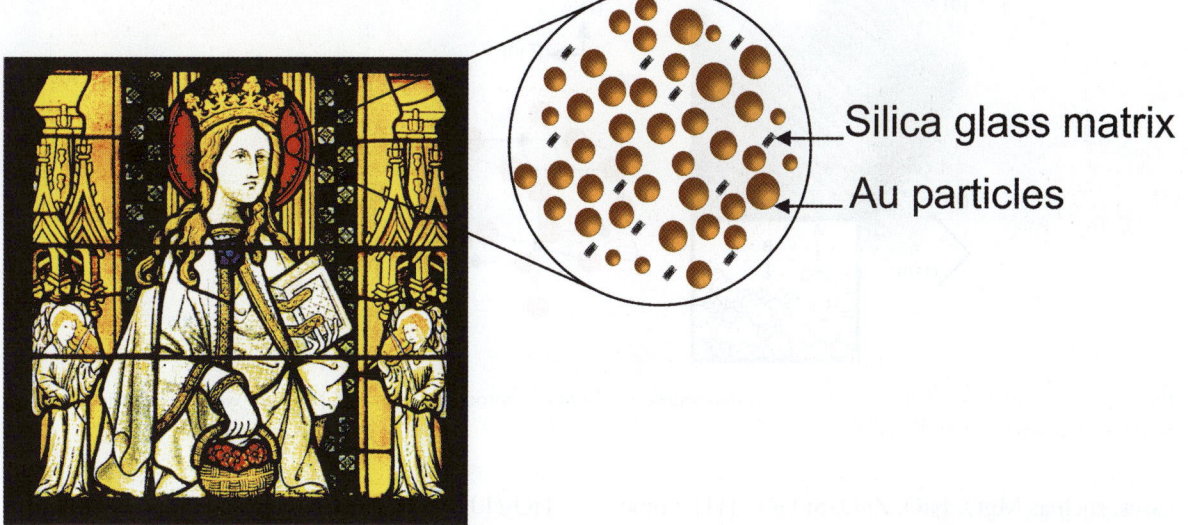

Fig. 4. Detail of a window in the Altenberg Cathedral (near Cologne). The red colored stained glass consists of small gold colloid particles residing in a glass matrix.

of these electrons in the box are not continuous but quantized. The energetic separation between the quantized levels are now depending on the size and therefore the energies for electron excitation will depend upon size, and, as it turns out, also on shape. This opens possibilities for manipulation of optical properties. If one, for example, asymmetrically stretches a solidifying glass the particles assume a certain shape which not only influences the optical absorption energy but also its polarization. Optical polarization filters [7] can be produced. While these filters depend on the linear optical properties of the material, also the non-linear optical responses are changed if high light intensities, e.g. from laser sources, are used. Keep in mind that the "matrix" surrounding the particle, in our case the glass, has an influence on the optical properties of the small particles so that there are many parameters that can be manipulated in order to design new materials with unexpected optical properties.

The chosen examples all have a connection to everyday life, including information transfer and storage, environmental pollution control, arts and even entertainment. It should be obvious therefore that though the topic is specific it has wide implications, and:

You must be interested!

In the following we discuss a variety of case studies and then in the final chapter speculate about things to do and where the field is going.

2. Where do we stand?

This question is answered in two steps. The first concerns the insulating substrate. How well do we understand its structure and properties? This is of importance to understand any modification to the substrate by deposition of additional material. In the second step, we deal with similar questions for the metal deposits, sometimes also called aggregates or clusters.

2.1. The oxide support

Let us start with a look at the oxide supports and answer the question: How do you make oxide surfaces? The preparation of a clean oxide surface is a rather difficult task. Several strategies have been followed [8–10].

The most straightforward strategy is in situ cleavage under ultrahigh-vacuum conditions. This, however, only leads to good results in certain

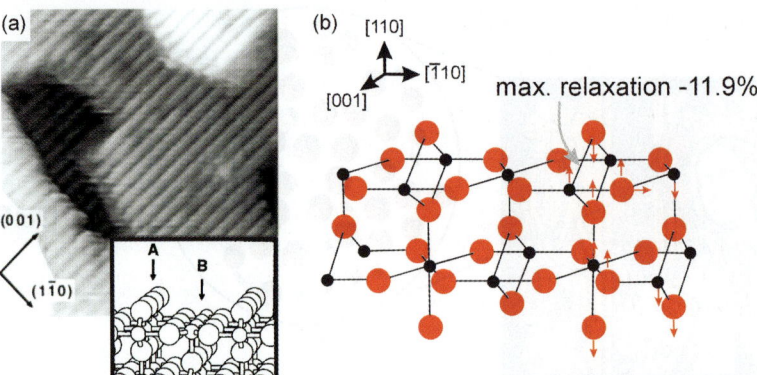

Fig. 5. Structure of TiO$_2$(1 1 0) (1 × 1) surface as determined via STM (a, reproduced from Ref. [16]) and via grazing incidence X-ray scattering (b, adapted from Ref. [23]).

cases, such as MgO, NiO, ZnO, SrTiO$_3$, [11]. Some interesting materials such as Al$_2$O$_3$, SiO$_2$, TiO$_2$, etc. are hard to cleave, i.e., they tend to form rough surfaces upon cleave [10]. A general disadvantage of cleaved bulk single crystal insulators with respect to experimental investigations is their low thermal and electrical conductivity. An alternative way of bulk single crystal surface preparation is ex situ cutting and polishing followed by an in situ treatment by sputtering and subsequent annealing in oxygen. Through such a process a sufficient number of defects is created in the near surface region and in the bulk to support conductivity of the material. This leads to a situation where electron spectroscopies as well as scanning tunneling microscopy (STM) can be applied to elucidate the electronic and geometric structure of the system [10].

Single crystalline oxide surfaces also may be prepared via the growth of thin oxide films on single crystal metal supports [8,9,12]. To such systems all surface science tools can be applied directly. If the oxide film is supposed to represent the bulk situation special care has to be taken in the control of film thickness because the film should represent the situation in the bulk. Also, if adsorption and reactivity studies are intended the continuity of the film has to be guaranteed. There are several examples in the literature where this has been achieved [12–14]. Probably, the best-studied clean oxide surfaces are TiO$_2$(1 0 0) and TiO$_2$(1 1 0) [8,10,15]. A STM image of the clean (1 × 1) TiO$_2$(1 1 0) surface taken by Diebold et al. [16] is shown in Fig. 5. One of the first atomically resolved images of this surface was reported by Thornton and co-workers [17,18]. The inset shows a ball and stick model of the surface. There is now accumulating evidence from theoretical modeling of the tunneling conditions, but also from adsorbate studies using molecules which are assumed to bind to the exposed Ti sites, that the bright rows represent Ti atoms. Iwasawa and co-workers [19–22] have successfully used formic acid in such a study and showed in line with the theoretical predictions, and counter intuitive with respect to topological arguments, that the Ti-ions are imaged as bright lines and the oxygen rows as dark lines. Taking the resolvable interatomic distances within the surface layer the values correspond to the structure of the stoichiometric (1 1 0) surface [23,24]. Interatomic distances normal to the surface, however, are substantially different from the bulk values as revealed by X-ray scattering experiments [23]. The top layer sixfold coordinated Ti atoms move outward and the fivefold-coordinated Ti atoms inward. This leads to a rumpling of 0.3 ± 0.1 Å. The rumpling repeats itself in the second layer down with an amplitude of about half of that in the top layer. Bond length variations range from 11.9% contraction to 9.3% expansion. These strong relaxations are not atypical for oxide surfaces and had been theoretically predicted [25–27].

Recently, the RuO$_2$(1 1 0) surface, which is isostructural with the TiO$_2$(1 1 0) surface, has been characterized via STM and low energy electron diffraction (LEED) as a means to determine structures [28,29]. It appears that in this case the contrast in the STM images has been reversed as compared with the $TiO_2(110)$ surface. The oxygen rows on the $RuO_2(110)$ surface are protruding while the ruthenium rows appear dark. In this case, CO adsorption has been used to show this, i.e. CO resides in the dark Ru rows. The structural relaxations as documented in the LEED studies are similar to the $TiO_2(110)$ surface.

There are several experimental results [30–33] where relaxations are particularly pronounced [25–27] basically corroborating the theoretical predictions although the quantitative agreement is not always good [34–37]. Specifically, the (0001) surfaces of materials such as Al_2O_3 [34,35], Cr_2O_3 [36] and Fe_2O_3 [37] have been studied with X-ray diffraction, quantitative LEED as well as with STM and theoretical methods. Fig. 6 shows the results of structural determinations for the three related systems $Al_2O_3(0001)$, $Cr_2O_3(0001)$ and $Fe_2O_3(0001)$ as addressed above. In all cases a stable structure in UHV is the metal ion terminated surface retaining only half of the number of metal ions in the surface as compared to a full buckled layer of metal ions within the bulk. The interlayer distances are strongly relaxed down to several layers below the surface. The perturbation of the structure due to the presence of the surface in oxides is considerably more pronounced than in metals, where the interlayer relaxations are typically of the order of a few percent [38]. The absence of the screening charge in a dielectric material such as an oxide contributes to this effect considerably. It has recently been pointed out [39] that oxide structures may not be as rigid as one might think based on the relatively high energy needed to excite lattice vibrations in the bulk.

Bulk oxide stoichiometries depend strongly on oxygen pressure, a fact that has been recognized for a long time [40]. So do oxide surface structures and stoichiometries [37]. In fact, if a Fe_2O_3 single crystalline film is grown in low oxygen pressure, the surface is metal terminated while growth under higher oxygen pressures leads to a complete oxygen termination [37]. Calculations by the Scheffler group [37] have shown, that a strong rearrangement of the electron distribution as well as relaxation between the layers leads to stabilization of the system. STM images by Weiss and co-workers [37] corroborate the coexistence of oxygen and iron terminated layers and thus indicate that stabilization must occur.

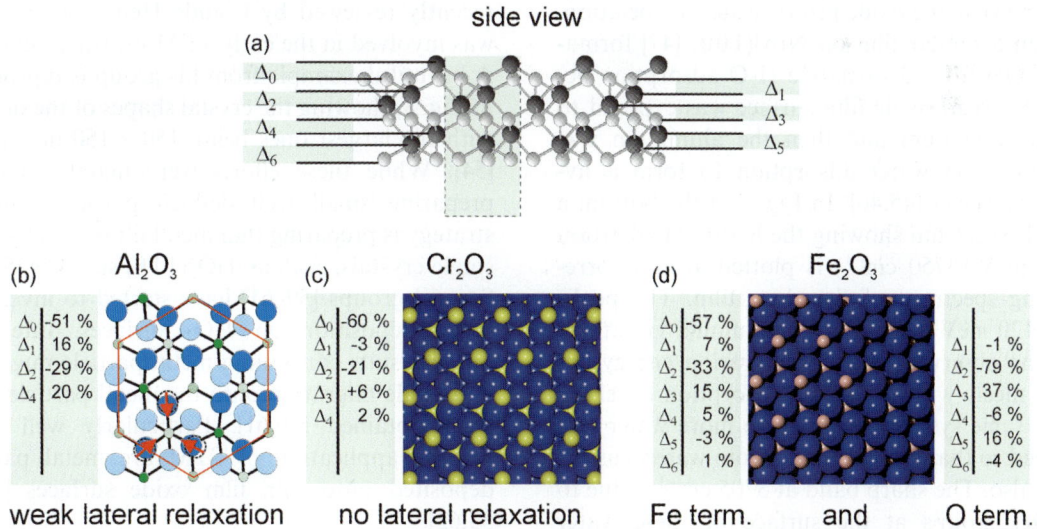

Fig. 6. (a) Experimental data on the structure of corundum-type depolarized (0001) surfaces (side and top views). (b) Adapted from Ref. [24], (c) Refs. [31,32], and (d) Ref. [37].

Another important stabilization mechanism for oxygen terminated surfaces proceeds via chemical means. Charge reduction can occur by replacing oxygen at the surface by hydroxyls. On the basis of energetic considerations [41], real crystallites must be terminated partly by polar surfaces, the charges of which are reduced by surface OH groups. The experimental confirmation was delivered much later [42,43]. For Al_2O_3 surfaces with oxygen termination it was shown recently by theoretical methods that OH termination also leads to the most stable surfaces [44]. Since hydrogen is difficult to detect with structural methods [24], vibrational spectroscopies are well suited to study this aspect. In fact, hydroxyl groups may be used to modify the chemical nature of oxide surface which in turn, has consequences for the adsorption of further material such as metal deposits [45,46]. We show in Fig. 7 the results of such a hydroxylation as measured with vibrational spectroscopies. Vibrational spectra can be measured either by infrared absorption after reflection of infrared light from the surface and recording the spectra with an interferometer (Fourier-transform infrared spectroscopy, FTIR) or by scattering electrons from the surface and measuring the loss of energy due to excitation of vibrations (electron energy loss spectroscopy, HREELS) [45,46]. In the case of a thin alumina film on NiAl(1 1 0) it was impossible to hydroxylate the oxide just by water dissociation, while on a similar film on NiAl(1 0 0) [47] formation of OH from dissociative H_2O adsorption occurs. The clean oxide film surface was exposed to metallic aluminum and then the aluminum was hydrolyzed via water adsorption to form a hydroxyl overlayer [45,46]. In Fig. 7 at the bottom a HREEL-spectrum showing the hydroxyl vibration at 465 meV (3750 cm^{-1}) is plotted atop a corresponding spectrum of the clean film. The peaks below 120 meV are due to the alumina vibrations [48]. The observed hydroxyl vibration energy coincides nicely with the FTIR absorption observed for the same system. In this case more water was adsorbed so that a broad band from water clusters is seen also. The sharp band at 3705 cm^{-1} is due to free OH groups at the surface of these water clusters [49], as they are known from the surface of ice. In fact, if a thick ice film is grown on the alumina film this particular vibration is observed (see Fig. 7). In comparison with literature data [50] it is now possible to assign the hydroxyl loss on the alumina surface. According to a review article by Knözinger and Ratnasamy [50] an OH-vibration at 3750 cm^{-1} is characteristic of hydroxyls bridging aluminum ions both in octahedral, or one in an octahedral and one in a tetrahedral site. On alumina films grown on a different NiAl substrate other types of OH species may be formed [47].

2.2. The metal particle–oxide system

So far, we have been considering the clean oxide surface and its reactivity. In the following, we consider the modification of the oxide surface by deposition of metals.

Over the last years several strategies have been followed [51]. Small metal particles have been deposited onto oxide bulk single crystal surfaces, particularly MgO, and characterized by transmission electron microscopy (TEM). A transmission electron micrograph is produced by transmitting electrons with an energy of several hundred (typically 200–400 keV) kiloelectronvolts through a sample using the contrast produced by the electron density in the system for imaging. Helmut Poppa has been the pioneer in the field of imaging small metal particles [52]. Contributions to it have been recently reviewed by Claude Henry, who himself was involved in the early TEM measurements [53]. A beautiful example from his group is reproduced in Fig. 8, showing the crystal shapes of the deposits with the largest ones being 150 × 150 nm^2 in size [54]. While these efforts were mainly aimed at preparing small well defined particles, another strategy is preparing thin metal films on bulk oxide single crystals, such as TiO_2(1 1 0) surfaces [55–58]. Several groups [59–61] have started to investigate metal deposition on TiO_2 surfaces. Interesting initial results concerning metal particle migration, and oxide migration onto the metal particles have been obtained [60,61]. Particularly well suited for the application of STM are metal particles deposited onto thin film oxide surfaces [12,13,53,62].

Often, well-ordered alumina films have been used as substrates. In Fig. 9 we show the result of a

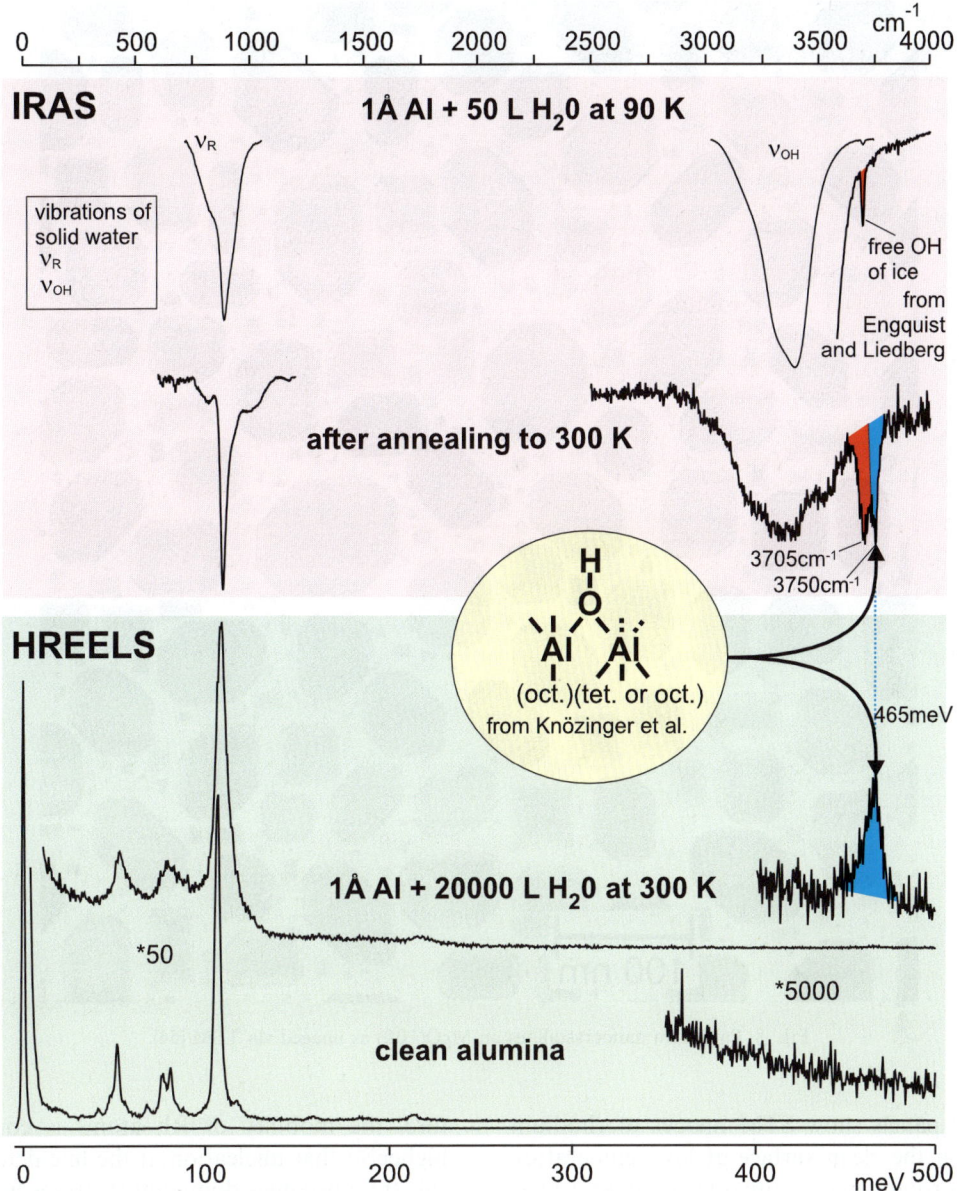

Fig. 7. Fourier transform IR spectra (IRAS) and electron energy loss spectra (HREELS) of a clean and an OH(+H_2O)-covered alumina film.

STM study from our laboratory. The upper left panel (a) shows the clean alumina surface as imaged by a scanning tunneling microscope [63]. The surface is well ordered and there are several kinds of defects on the surface. One of them are reflection domain boundaries between the two growth directions of $Al_2O_3(0001)$ on the NiAl(110) surface [48]. There are anti-phase domain boundaries within the reflection domains, and, in addition, there are point defects which are not resolved in the images. The image does not change dramatically after hydroxylating the film [45]. The

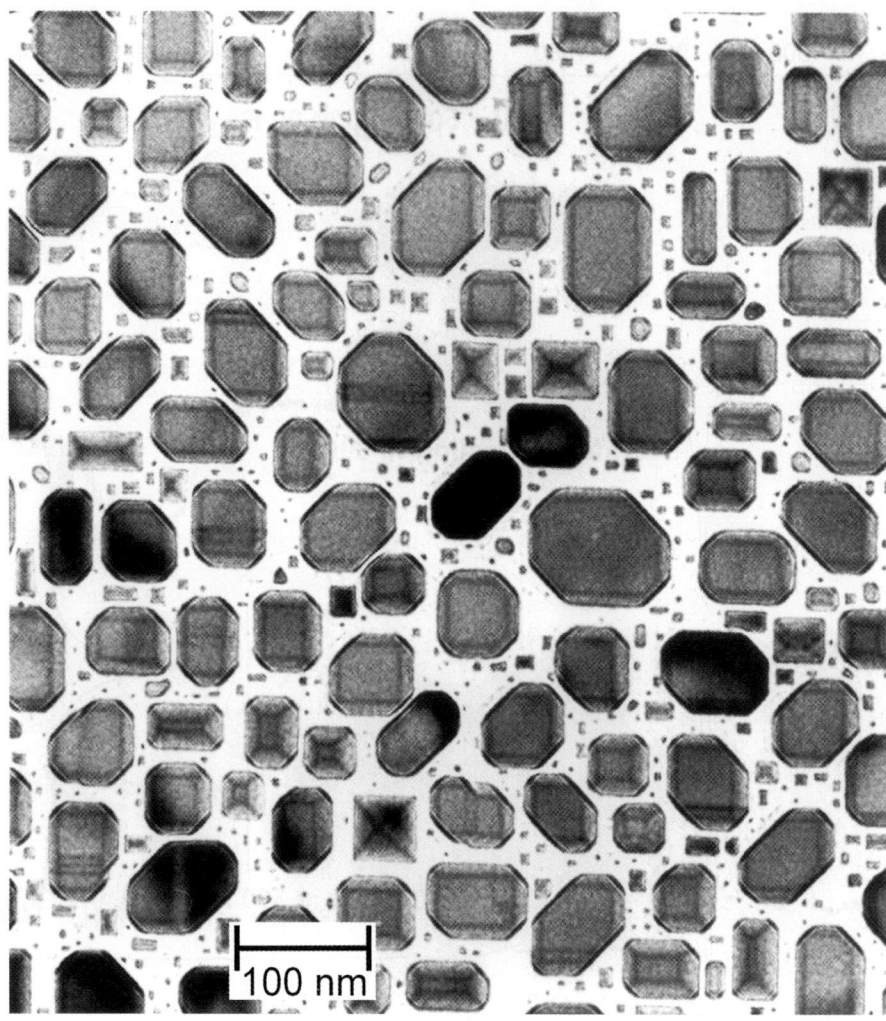

Fig. 8. Palladium nanocrystallites on MgO(100) as imaged via TEM [54].

additional panels show STM images of rhodium deposits on the clean surface at low temperature (b), and at room temperature (c) [64,65], as well as an image after deposition of Rh at room temperature on a hydroxylated substrate (d) [66]. The amount deposited onto the hydroxylated surface is equivalent to the amount deposited onto the clean alumina surface at room temperature. Upon vapor deposition of Rh at low temperature (the protrusions shown in Fig. 9b), small particles nucleate on the point defects of the substrate and a narrow distribution of sizes of particles is generated. If the deposition of Rh is performed at room temperature, the mobility of Rh atoms is considerably higher so that nucleation at the line defects of the substrate becomes dominant (features line up with the light lines in Fig. 9c). Consequently, all the material nucleates on steps, reflection domain and anti-phase domain boundaries. The particles have a relatively uniform size, in turn depending on the amount of material deposited. If the same amount of material is deposited onto a hydroxylated surface, the particles (the protrusions shown in Fig. 9d) are considerably smaller and distributed across the entire surface, i.e. a much higher metal dispersion is obtained [45].

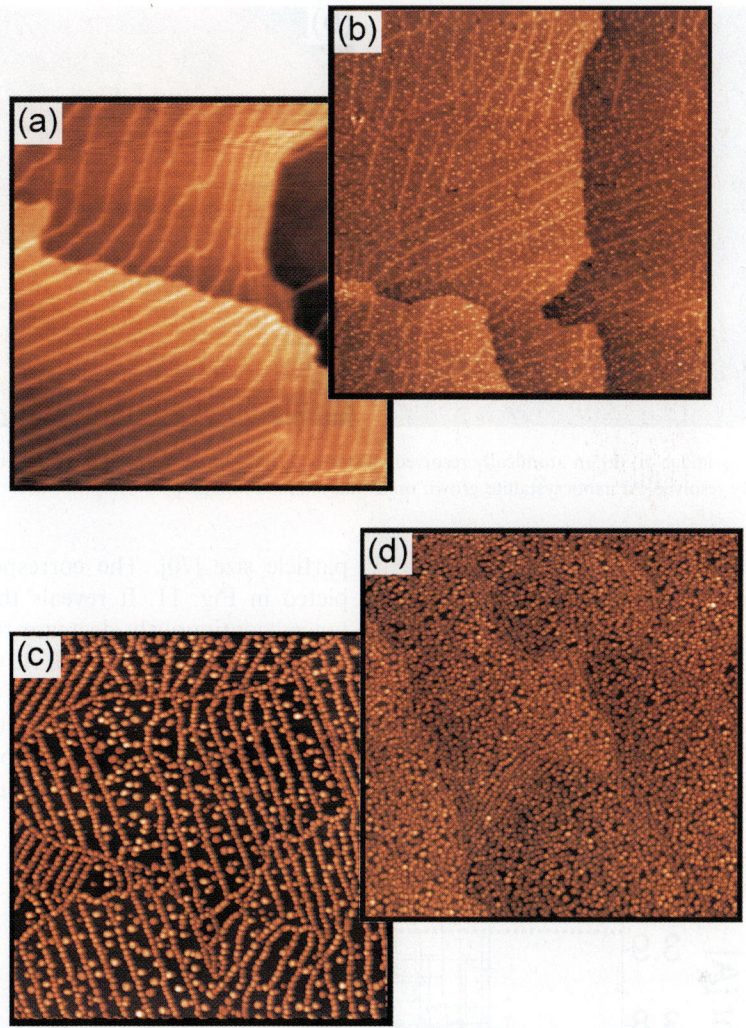

Fig. 9. Scanning tunneling images (3000×3000 Å^2, Al$_2$O$_3$/NiAl(1 1 0), $U_{tip} = 8$ V, $I = 0.8$ nA): (a) Clean alumina film, (b) after deposition of 0.1 Å of Rh at 90 K, (c) after deposition of 2 Å of Rh at 300 K, and (d) after deposition of 2 Å of Rh at 300 K on hydroxylated substrate onto the pre-hydroxylated alumina film.

The sintering process is an interesting subject. Research on this process is just beginning [64]. A more basic process is metal atom diffusion on oxide substrates.

Diffusion studies [67] could profit from atomic resolution, once it is obtained routinely for deposited aggregates on oxide surfaces. While for clean TiO$_2$ surfaces and a few other oxide substrates atomic resolution may be obtained routinely, there are few studies on deposited metal particles where atomic resolution has been reported [68]. An image of Pd metal clusters on MoS$_2$ is shown in Fig. 10a and exhibits 27 metal atoms in the cluster. A joint effort between Fleming Besenbacher and our group [69] has led to atomically resolved images of Pd aggregates deposited on a thin alumina film. Fig. 10b shows such an image of an aggregate of about 50 Å in width. The particle is crystalline and exposes on its top a (1 1 1) facet. Also, on the side, (1 1 1) facets, typical for a cuboctahedral particle, can be discerned.

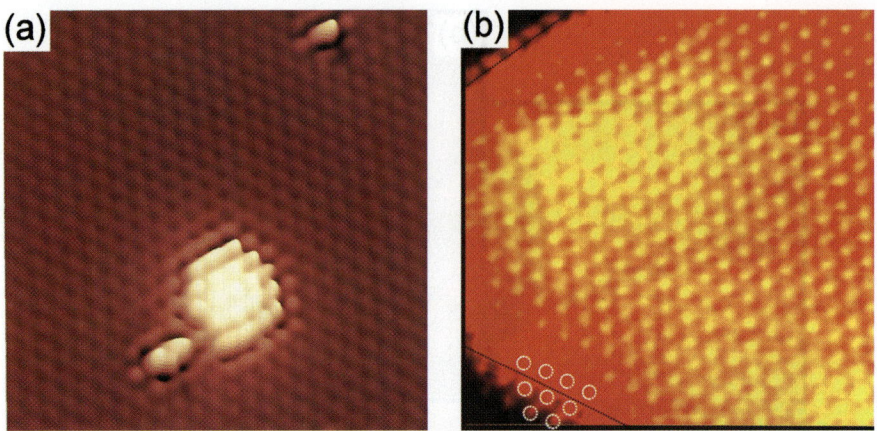

Fig. 10. Scanning tunneling image of: (a) an atomically resolved cluster of 27 Pd atoms arranged in two layers on a MoS$_2$ substrate [68], and (b) an atomically resolved Pd nanocrystallite grown on a thin alumina film [69].

While STM reveals the surface structure of deposited particles, their internal structure, in particular as a function of size, is not easily accessible through STM. In this connection, TEM studies on the same model systems help [70]. On the basis of numerous high resolution transmission electron microscopy (HRTEM) images, it has been possible to calculate the lattice constants as a function of particle size [70]. The corresponding plot is depicted in Fig. 11. It reveals that the atomic distances continuously decrease to 90% of the bulk value at a cluster size of 10 Å. On the other hand, the lattice constant approaches the Pt bulk value at a diameter of 30 Å. This effect also has been detected for Ta and Pd clusters on the thin alumina film, but seems to be less pronounced in these cases

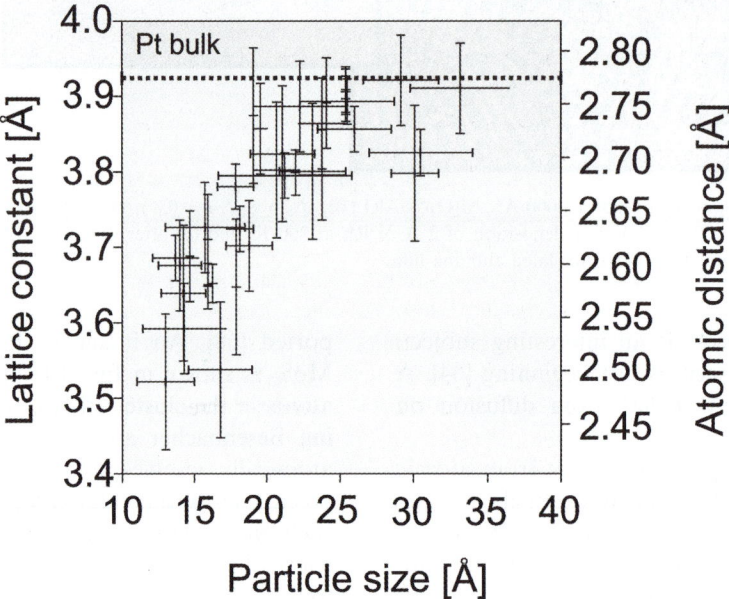

Fig. 11. Lattice constants and interatomic distances of Pt particles grown on Al$_2$O$_3$/NiAl(1 1 0) as a function of their size (the horizontal bars represent the difference of the widths and the lengths of the clusters, while the vertical bars are error bars).

[71,72]. Variations of interatomic distances as a function of particle size are known from calculations on isolated clusters and have occasionally been reported for deposits [73].

The deposits discussed so far were prepared with the intention to keep the size distribution narrow. The lateral distribution of aggregates on the surface, however, has not been an issue. If we consider reacting systems, interdiffusion of species between the particles, i.e. spillover processes, may become an important issue. Therefore, it may be desirable to control not only particle size and morphology but also interparticle distances. Based on electron beam lithography, Rupprechter et al. [74,75] have reported the preparation of two-dimensional arrays of Pt particles deposited onto amorphous SiO_2 layers. Particles of 25–40 nm average size could be produced as shown in Fig. 12. The image reveals an average height of 20 nm of these particles. In these studies [76–80] the average size is an order of magnitude larger than the particles imaged in Fig. 10.

The electronic structure of deposited metal aggregates reflects to a certain extent the geometric structure and vice versa.

Starting from an atomic level diagram, Fig. 13 shows how such a level diagram develops when more and more metal atoms agglomerate and finally form a solid with a periodic lattice. Upon formation of an aggregate from equivalent atoms, the atomic levels split into cluster orbitals. The splittings are characteristic of the interatomic interactions. Depending on the interaction strength, the split levels derived from a given atomic orbital start to energetically overlap with levels derived from other atomic orbitals. As long as the system has molecular character, there is an energy gap left between occupied and unoccupied levels. This is in contrast to the situation encountered for an infinite periodic metallic solid as presented on the right hand side of the figure, where no longer a gap between occupied and unoccupied levels exists. It is not hard to envision that, as we enlarge the number of atoms in an agglomerate, the gap between occupied and unoccupied orbitals effectively vanishes. This is the case if the energy gap decreases to a value close to the thermal energy in the system kT.

The question arises: How many atoms are necessary to induce a transition from an insulator to a metallic cluster? Reports in the literature claim numbers ranging from 20 to several hundred atoms in this respect [72,81–92]. One interesting extrapolation deduced from spectroscopic measurements of the gap of inorganic carbonyl cluster compounds containing a transition metal kernel and CO molecules as a ligand sphere as a function of the size of the metal kernel is shown in Fig. 14 [83]. It suggests that 70 atoms are sufficient to close the gap. A study from the author's laboratory on CO covered Pd and Rh clusters [86,92] yields a comparable value. In those cases where larger values have been obtained, the metals were Cu, Ag, Au, Al or alkali metals [81,82,93]. It is likely that the specific electronic structure of metals has an influence on the exact value.

Experiments on electronic structure so far have dealt with ensembles of clusters and relied upon the preparation of ensembles with narrow size distributions. Recording current–voltage curves in a scanning tunneling microscope for a given position (this procedure is called scanning tunneling spectroscopy), enables the investigation of single clusters, e.g., aggregates deposited on oxides [94]. Fig. 15 shows typical current–voltage curves for some aggregate sizes, i.e. Au on $TiO_2(1\,1\,0)$ [95]. While the large particles do not exhibit a plateau near $I = V = 0$, the smaller clusters do show the behavior expected for a system with a gap.

The electronic structure of deposited aggregates has also been probed via optical response [96–98]. Fig. 16 shows the optical absorption as well as the atomic force microscopy (AFM) image of an ensemble of small Ag clusters on mica [97]. The two absorption bands are associated with the optical excitation of a surface plasmon, i.e., a collective excitation of the electrons on a sphere, which corresponds to the so called Mie plasmon [6] mentioned in the introduction. There are two bands because the three possible oscillatory directions in a sphere no longer lead to the same plasmon energy for a free sphere deposited on a substrate. The oscillation perpendicular to the surface appears at higher energy than the two equal-energy oscillations within the surface plane [96]. This is illustrated in Fig. 17 where the dipole

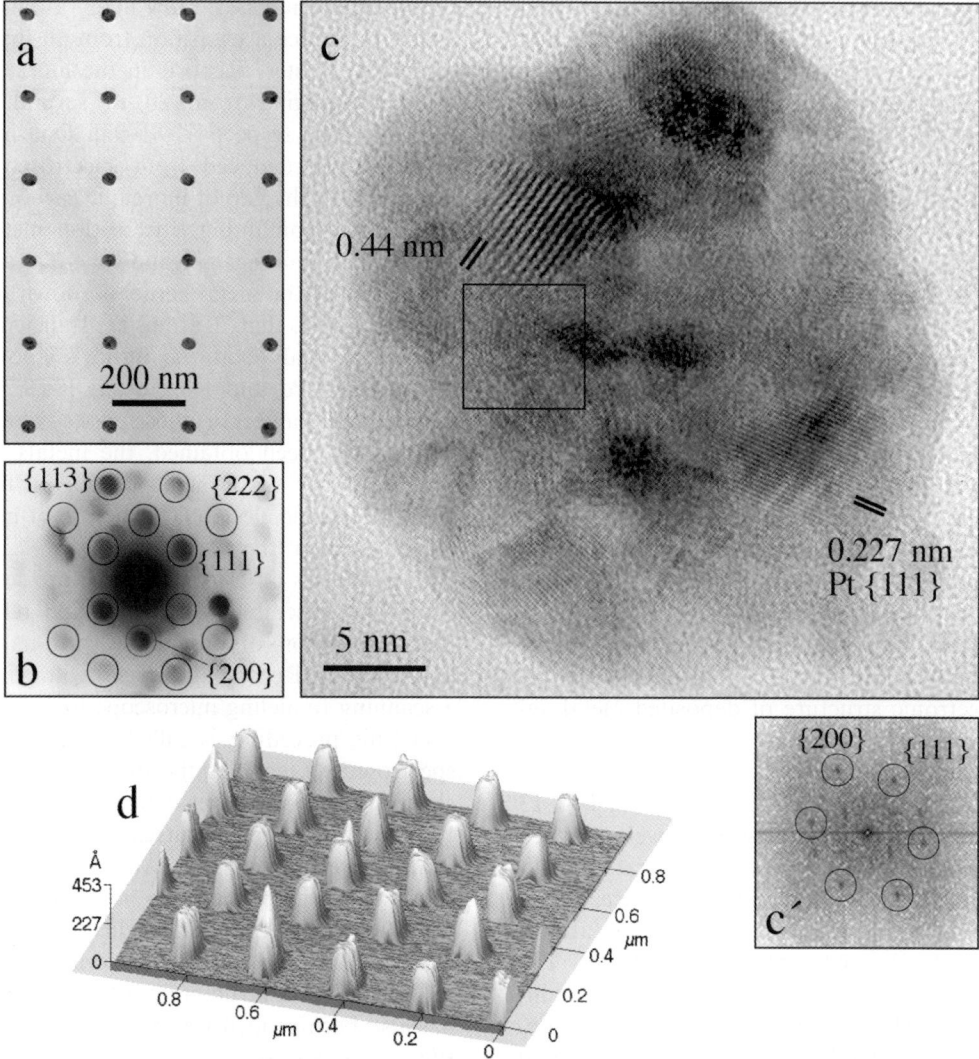

Fig. 12. (a) Transmission electron micrograph of a platinum nanoparticle array on SiO$_2$ (mean particle diameter 40 nm; interparticle distance 200 nm), (b) Microdiffraction pattern of an individual platinum particle showing its polycrystallinity (spots originating from a (1 1 0)-oriented crystalline grain within the polycrystalline platinum particle are marked by circles), (c) HRTEM micrograph and (c′) fast Fourier transform of a 25-nm platinum model catalyst particle. (d) AFM image of a platinum nanocluster array after several reaction-cleaning cycles [74].

in a sphere is indicated together with its image dipole in the substrate. The perpendicular dipoles couple to form a large dipole moment harder to excite (blue shift), while the parallel dipole couple to form a reduced dipole moment easier to excite (red shift). The widths of the bands depend on the size and the shape distributions of the clusters. Since there is a stronger variation in lateral shape than in height the blue shifted band is wider. The widths are therefore inhomogeneous, i.e., each cluster exhibits its own shift and the experiment measures the sum of these. Experiments on either a monodisperse cluster ensemble of single shape or experiments on individual clusters would be needed to investigate the homogeneous widths. Such experiments have been recently reported by using a

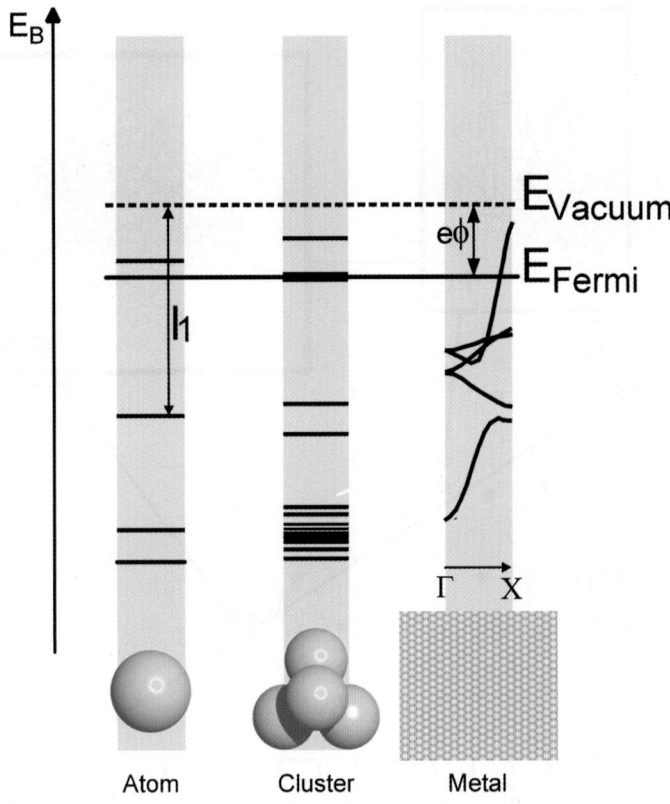

Fig. 13. Diagram illustrating the transition from an atom to a metal (E_B, binding energy; I_1, first ionization energy; e: electron charge; ϕ: work function; Γ, X: symmetry points in the Brillouin zone).

scanning tunneling device [99]. Schematically the setup is shown in Fig. 18a [100]. The tip is used to inject electrons into individual Ag clusters, in this case deposited on alumina for excitation. Then the light emitted from the clusters upon radiative decay is measured via a spectrometer outside the vacuum chamber. Fig. 18b shows the fluorescence spectra as a function of size referring to the specific clusters in the STM image, which occurs blurred because it was taken at high tunneling voltage necessary for excitation. A better representation of the size distribution of the Ag clusters is imaged in the second inset in Fig. 18b although even in this case one has to take account of the fact that due to tip convolution the actual size is considerably smaller than the imaged one. The peak shows a pronounced blue shift as a function of size consistent with observations on cluster ensembles of varying size. In this context it is interesting to look at the line widths of the resonance as a function of size. This is plotted in Fig. 18c. The line width is smallest for the larger clusters, i.e. 0.15 eV, and increases to 0.3 eV for the smallest ones studied. We consider this to be the homogeneous line width. The fact that it changes following an inverse cluster radius reveals the influence of the cluster surface becoming more important for smaller systems as a channel to deactivate the excited state through electron-surface scattering without generating radiation.

In the introduction we referred to the interaction of species from the gas phase with the deposited clusters. This is an important issue in catalysis as well as in understanding sensors.

An advantageous technique to expose a cluster to a gas and then re-establish ultrahigh vacuum is FTIR because it provides the resolution to differentiate between various adsorbed species. Again,

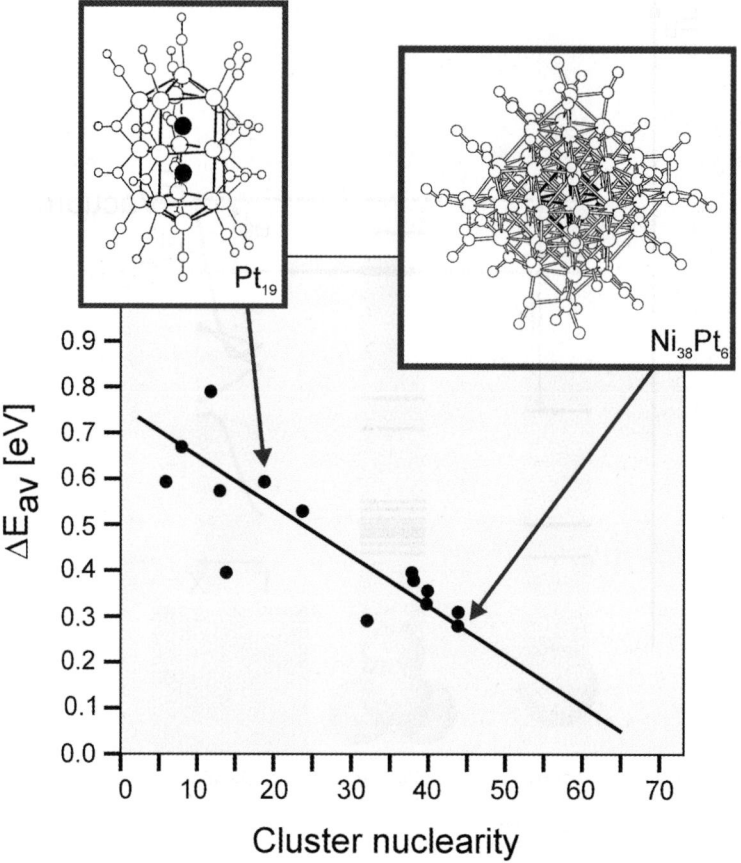

Fig. 14. Electronic excitation of lowest energy for several cluster compounds as a function of the number of metal atoms in the cluster (ΔE_{av} is the energy gap between occupied and unoccupied electronic states for cluster compounds). Reproduced with permission from de Biani et al. [83].

the thin film based systems are particularly well suited since the metallic support of the oxide films acts as a mirror at infrared frequencies. It is, however, also possible to perform such experiments on surfaces of bulk dielectrics as shown by the Hayden group [101,102].

Wayne Goodman and his group have published an interesting study of CO adsorption on Pd aggregates on Al_2O_3 films [94]. The results have been interpreted as characteristic for the adsorption of CO on different facets of the small crystalline aggregates. Although this interpretation does not take into account adsorption on the various defect sites of the aggregates [86], as pointed out in a more recent study [103], the data are indicative of the potential of this technique for the study of size dependent absorption phenomena. The presence of adsorbed molecules can change the morphology of deposited particles because in the presence of adsorbates molecular species may be found that detach themselves from a cluster and move across the surface. Such phenomena are interesting with respect to redispersing metal on a surface. For example, a catalyst could have been deactivated by cluster agglomeration. This process can be reversed to a certain extent by the formation of mobile species which can re-nucleate small metal particles when treated properly.

The infrared spectrum taken from a Rh deposit prepared and saturated with CO at 90 K is displayed in Fig. 19 (second spectrum) [104]. The most prominent feature in the stretching region of

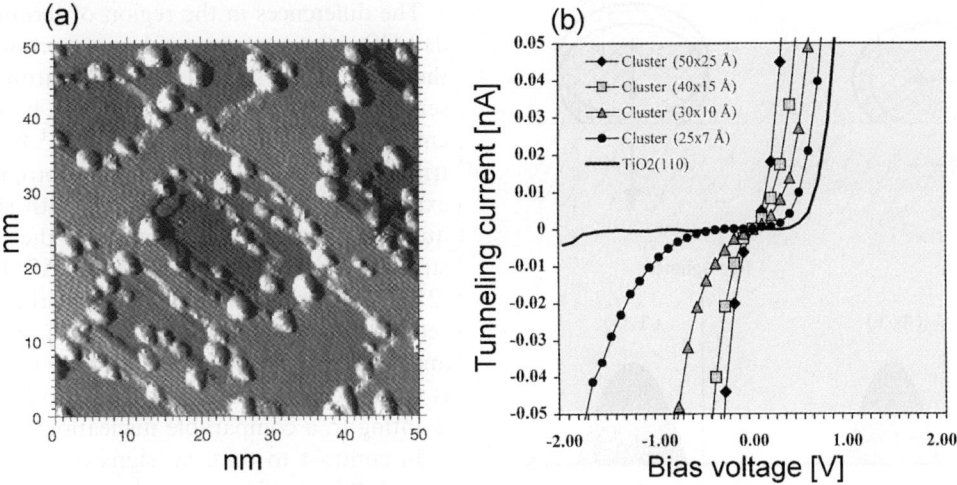

Fig. 15. Current–voltage (b) recorded for Au clusters of various sizes deposited onto a TiO$_2$(110) surface. A typical STM picture of the system is shown in (a). (Adapted from Ref. [95].)

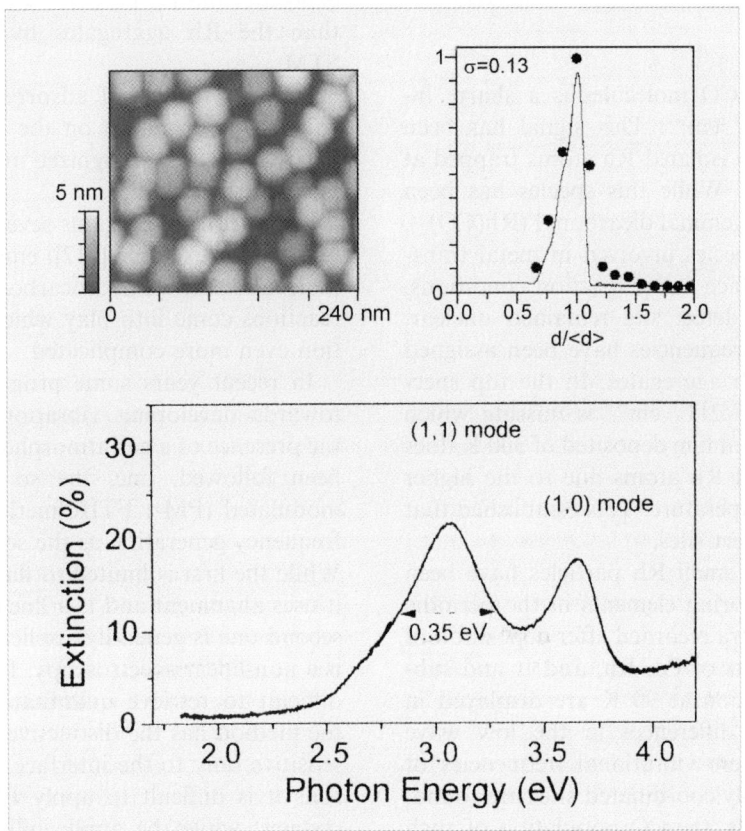

Fig. 16. Extinction spectra of small silver particles in the range of 2 eV < E(photon) < 4 eV. The insets contain: (upper left panel) an AFM image of the particle distribution and (upper right panel) the normalized size distribution of the particles.

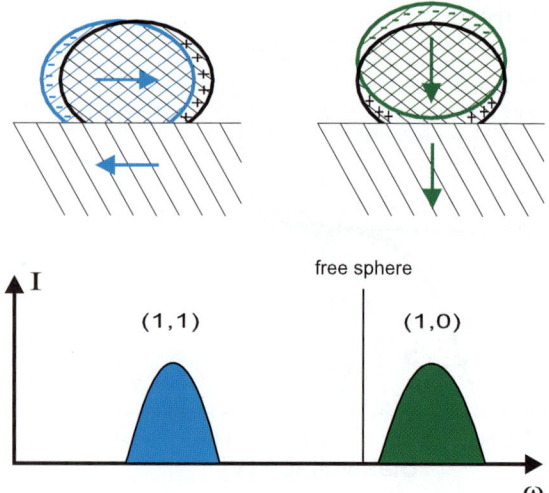

Fig. 17. Schematic representation of the surface plasmon excitations for ellipsoids attached to a solid substrate. The modes with parallel and perpendicular excitation dipole are indicated and the resulting spectrum is schematically indicated below.

terminally bonded CO molecules is a sharp, intense band at 2117 cm^{-1}. This signal has been shown to arise from isolated Rh atoms trapped at oxide defects [104]. While this species has been identified to be the geminal dicarbonyl (Rh(CO)$_2$), known to be the species involved in metal transport across the surface under reaction conditions, the nature of the defect site remained unclear. Features at lower frequencies have been assigned to molecules on Rh aggregates. In the top spectrum the feature of 2117 cm^{-1} is missing which means that a rhodium film deposited of 300 K does not contain isolated Rh atoms due to the higher mobility at this temperature. This established that Rh(CO)$_2$ sits in defect sites.

These studies on small Rh particles have been extended to neighboring elements in the periodic table. Infrared spectra recorded after deposition of comparable amounts of Pd, Rh, and Ir and subsequent CO saturation at 90 K are displayed in Fig. 20. We note differences in the low wave number region, where vibrational frequencies of molecules in multiply coordinated sites are found. As on single crystals, the CO population of such sites is highest on Pd [105,106], while no such CO is observed on Ir [107,108].

The differences in the region of terminally bonded CO, however, are much more pronounced. In the case of Ir, several distinct features are observed. In analogy to the Rh(CO)$_2$ band at 2117 cm^{-1}, the sharp signal at 2107 cm^{-1} has been attributed to Ir(CO)$_2$ species via isotopic mixture experiments (not shown). Bands with similar frequencies have been assigned to the symmetric stretch of Ir$^+$(CO)$_2$ on technical Ir/Al$_2$O$_3$ catalysts (2107–2090 cm^{-1}) [109] and on the iridium-loaded zeolite H-ZSM-5 (2104 cm^{-1}) [110]. The appearance of a number of bands at lower wave number is reminiscent of the 90 K Rh deposits (Fig. 20), pointing to a comparable nucleation behavior.

In contrast to that, no signs of atomically dispersed Pd or of structurally well-defined aggregates are observed. Indeed, the infrared spectrum is rather similar to that observed on much larger, disordered Pd aggregates [111]. At the same metal exposure, the Pd particles are found to be larger than the Rh aggregates by room temperature STM.

Infrared spectra of adsorbed CO thus provide valuable information on the size of metal nanoparticles, as long recognized in the catalysis related literature.

The literature contains several adsorption studies, (see for example [112]) employing other probe molecules such as hydrocarbons but in these cases reactions come into play which renders the situation even more complicated.

In recent years some progress has been made towards developing vibrational spectroscopy in the presence of a gas atmosphere. Two routes have been followed, one, the so called polarization modulated (PM-) FTIR method [113], and sum-frequency generation as the second one [114–116]. While the first is limited to flat substrates because it uses alignment and is a linear spectroscopy, the second one is generally applicable in principle but is a non-linear spectroscopy. In the latter case it is difficult to retrieve quantitative information but the method has the distinctive advantage of being sensitive only to the interface. For the above reasons it is difficult to apply PM-FTIR to cluster systems, while the applicability of the latter has just been demonstrated. On the basis of these methods one may study whether the ideas devel-

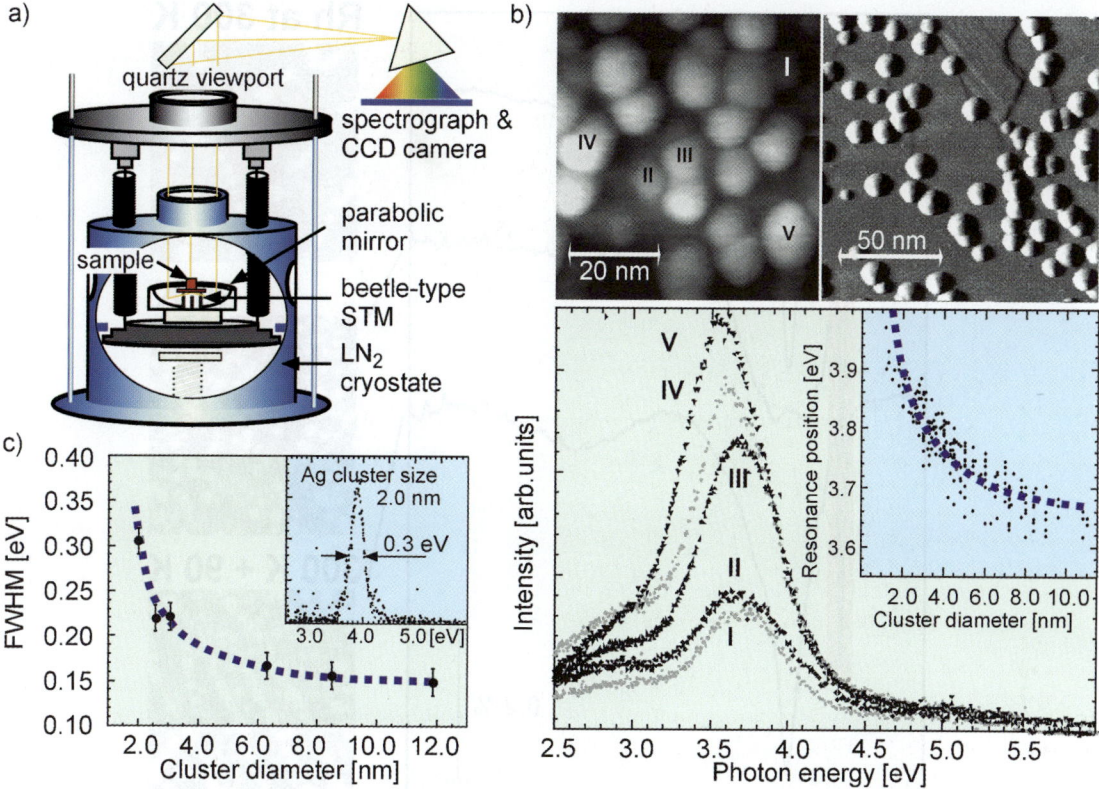

Fig. 18. (a) Schematic diagram of the experimental setup for the photon emission scanning tunneling microscope, (b) photon emission spectra as a function of particle size. The corresponding particles are marked in the upper left panel. The upper right panel shows a topological image of a typical cluster covered area. The size dependence of the resonance position of the plasmon excitation is shown in the inset on the left. (c) Line widths of the observed plasmon excitation as a function of particle size.

oped through studies under ultrahigh vacuum conditions can be extended to ambient pressures. Initial results look encouraging [116].

Vibrational spectroscopy on individual clusters has not been reported yet, but vibrational spectroscopy has been performed on carbon monoxide molecules bound to mass-selected deposited clusters. Heiz and coworkers [117–119] have mass selected in a quadrupole mass spectrometer Pt and Au clusters and deposited them onto a thin MgO film grown on a Mo substrate [13]. The assumption is that the clusters when deposited stay as deposited and the ensemble remains monodisperse. So far no one has demonstrated this by scanning probe microscopies although this should be possible. Fig. 21 shows FTIR spectra of CO adsorbed on mass-selectively deposited Pt_8 and Pt_{20} [119]. While the small Pt_8 cluster only exhibits bridging sites, the larger Pt_{20} cluster also shows a band typical for threefold hollow sites on Pt single crystals. An interesting result has been obtained on deposited Au clusters, which are, as mentioned before, also interesting research objects for low temperature CO oxidation. CO adsorbed on Au_8 clusters exhibit an infrared spectrum even at or above room temperature [120], indicating rather strong bonding. Fig. 21 contains the corresponding spectrum showing an unusually high CO stretching frequency. This is a result very different from CO adsorption on Au single crystal surfaces and larger Au particles [121]. The result could be important to understand why CO can be oxidized to CO_2 at very low temperature on gold catalysts!

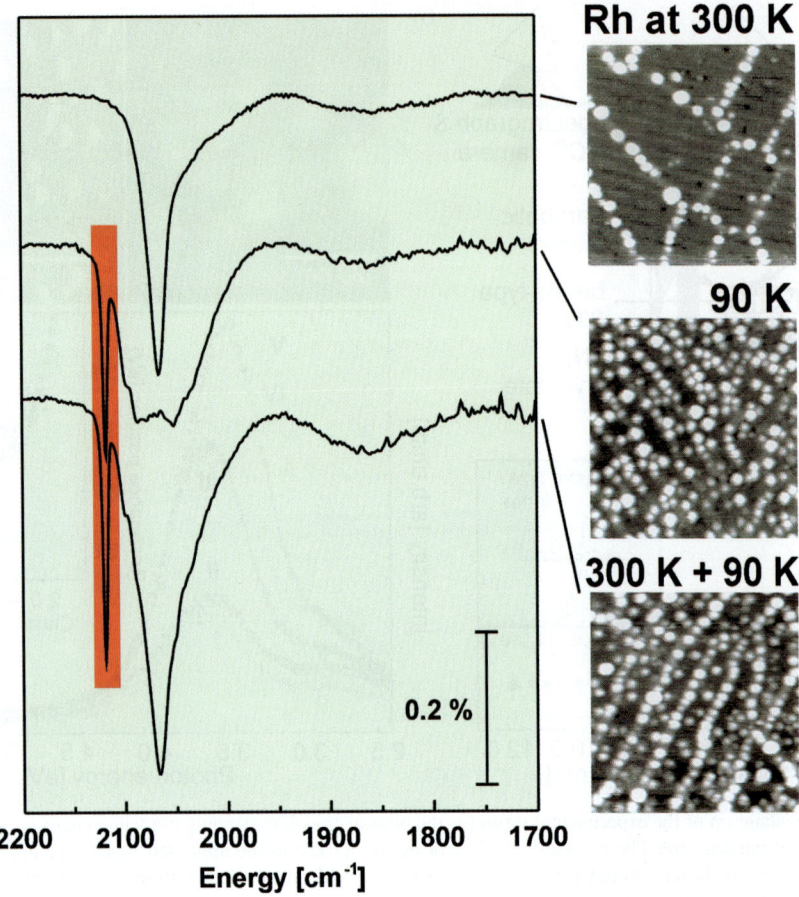

Fig. 19. Infrared spectra recorded after CO saturation of Rh deposits at 90 K, along with corresponding room temperature STM images (500 × 500 Å²). Top: 0.057 ML Rh deposited at 300 K, middle: 0.057 ML Rh deposited at 90 K and bottom: 0.057 ML Rh deposited at 300 K, followed by the same exposure at 90 K.

The examples reviewed in this section indicate that cluster systems are complex and our knowledge on deposited clusters on insulating surfaces has increased enormously over the last ten years. Still we are only at the beginning of a crude understanding of the properties of aggregates and of the processes involved in their formation. There are many fascinating experiments one can think of that could be done on the basis of the knowledge so far accumulated. Some of these, which are to a large extent based on speculations, are considered in the next section in order to motivate more interest and activity in this field as suggested in the first line of the introduction.

3. Which are open questions and how could they be answered?

It is evident from the previous sections that the field is in its infancy and there are still many interesting questions unanswered. In the following we will examine a series of these questions and speculate about solutions and experiments to do. Even though we primarily discuss experimental aspects, the field can only develop and prosper through a concerted effort in experiment and theory. In particular the modern simulation methods are required to provide insight that cannot be gained through experiment alone.

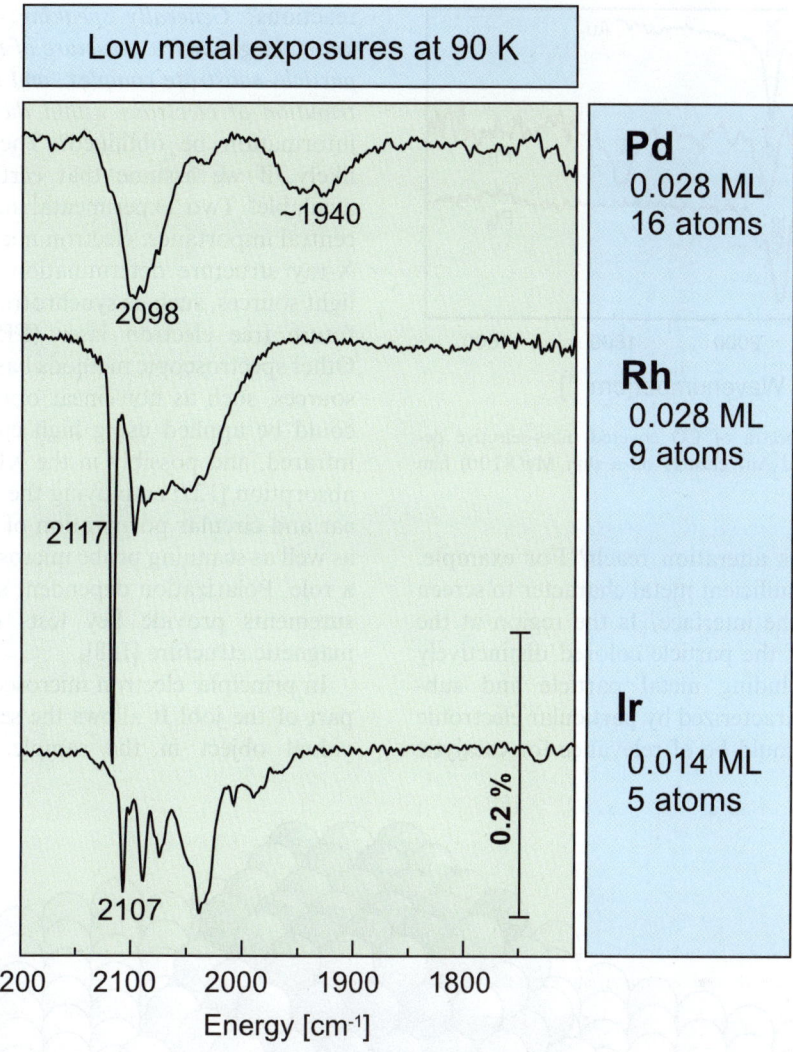

Fig. 20. Infrared spectra of Pd, Ir, and Rh deposited at 90 K and saturated with CO at the same temperature.

Before we become more specific let us identify areas of interest. Obvious ones are:

- Structure (geometric and electronic, including magnetic) at the atomic level of the metal particle–oxide interface under ambient conditions
- Control of structure and morphology
- Chemistry at the atomic level on the cluster
- Ultra fast dynamics and coherent control in and on clusters

There is information both on the oxide as the substrate as well as on the deposited metal particles. Although this may not yet be sufficient, there is very much less known on the metal particle–oxide interface [122]. Some of the important issues, however, are connected with this knowledge which renders the investigation of the metal particle–oxide interface an important one. Consider, as indicated in Fig. 22, a particle as an idealized cubo octahedron. Would it not be interesting to know whether the structure of the support underneath the particle is the same as the uncovered substrate? How is the electronic and geometric structure of the metal atoms in the aggregate in direct proximity to the substrate altered and how deep into the

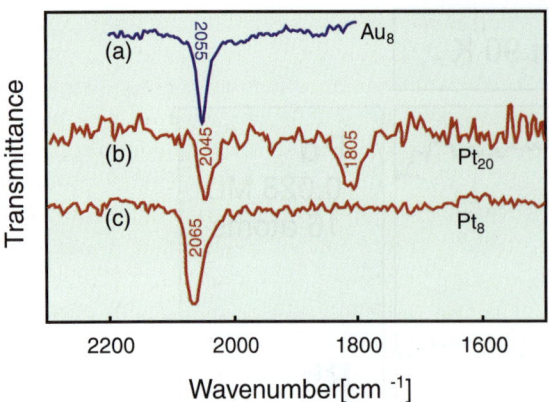

Fig. 21. Infrared spectra of CO covered mass-selective deposited Pt_8, Pt_{20} and Au_8 clusters on a thin MgO(100) film [119,120].

reactions? *Generally speaking, we would like to know the geometric structure of the entire deposited particle–substrate complex, and in addition the distribution of electrons within the system.* Can this information be obtained? The answer is: Very likely, if we assume that certain tools become available! Two experimental methods will be of central importance: electron microscopy [123], and X-ray structure determination with very intense light sources, such as synchrotrons [124] and in the future free electron laser (FEL) sources [125]. Other spectroscopic methods based on intense light sources, such as non-linear optical methods [126] could be applied using high energies in the UV, infrared, and possibly in the XUV regions. X-ray absorption [127] employing the high degree of linear and circular polarization of synchrotron light, as well as scanning probe microscopy can also play a role. Polarization dependent spectroscopic measurements provide key tests of electronic and magnetic structure [128].

In principle, electron microscopy can do a good part of the job! It allows the selection of an individual object in the sample, determines local

particle does this alteration reach? For example, has the particle sufficient metal character to screen the changes at the interface? Is the region at the circumference of the particle colored distinctively in Fig. 22, including metal particle and substrate atoms, characterized by particular electronic properties that could be of relevance for catalytic

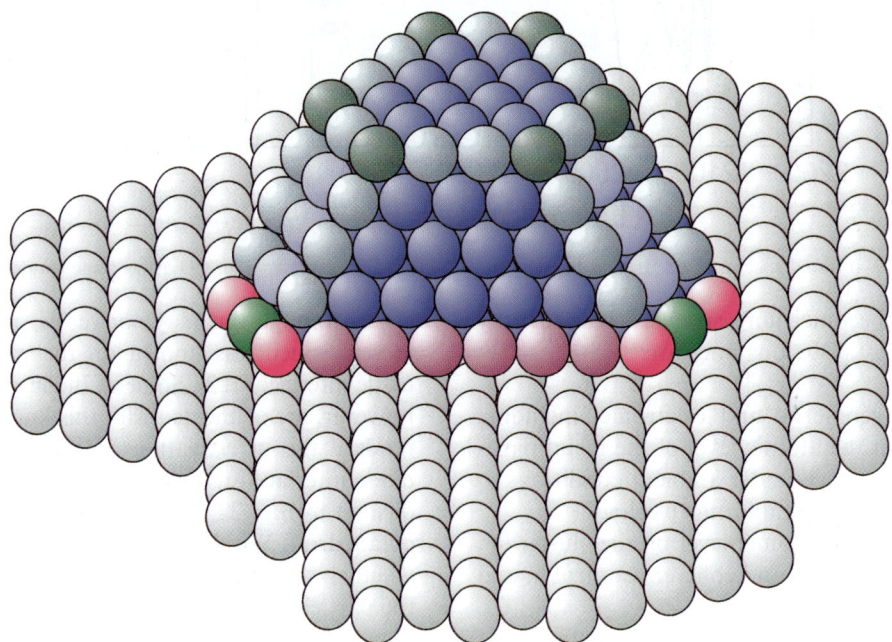

Fig. 22. Schematic representation of a cubo octahedral metal cluster on an insulating substrate. See also text. Chemically different sites on the cluster surface are colored differently.

structure with atomic resolution, and can be applied to cross-sectioned samples, so that we can image directly the interface. But: Electron microscopy is destructive, and is difficult to do under the influence of gases. It has chemical identification power, though limited, and may not be the method of choice to investigate electronic structure. Even today it is particularly difficult to image surfaces at atomic resolution with electron microscopy. For the purpose of surface imaging scanning probe microscopy is the method of choice. Photon based methods are by definition less destructive than electron based methods. Moreover, they can be used in the presence of a gas atmosphere, although care has to be exercised in this case as well. If it is possible to prepare uniform arrays of islands which are crystalline, —and some progress has already been demonstrated—then X-ray scattering and X-ray standing waves [129] are the promising tools to investigate the structure of deposited particles even under ambient conditions. If one had coherent X-rays from an X-ray laser [130] the investigation of structure and dynamics under ambient conditions would be in reach. With the advances recently made to shift the wavelength of coherent light into the XUV range by building free electron lasers, for example, the one at HASYLAB in Hamburg, a new generation of structure related experiments, including holography, would become available. The entire field of non-linear optical techniques, currently used in the lower frequency regimes could be applied at high energies opening up the possibility to selectively study interfaces similar to what is done at present in vibrational spectroscopy. These techniques could be used to study the metal particle–oxide interface. Photon based microscopy can benefit largely from the advent of high brilliance sources. There are several projects under way already to push the limits of lateral resolution to below 10 Å [131–133]. Also, X-ray optics is progressing with X-ray microscopy pushing the limits of resolution. In combinations with scanning probe microscopy, performed also under ambient conditions—a technology which is just being tested—experimental techniques would be available allowing us to answer the above question: Yes!

As has been indicated above, and alluded to on several occasions in the previous chapter, *it is of utmost importance to be able to control structure and morphology of cluster formation by understanding the elementary steps in the processes*, such as metal diffusion and aggregate migration on insulators. A good understanding of these processes would help to improve theoretical modeling studies. It is not known in detail how a single metal atom diffuses across an insulator surface. How does this process compare with metal-on-metal diffusion, where experimental studies have been successfully undertaken [134]? How does a cluster migrate on an oxide surface, and how does the concerted movement of atoms across the surface occur? Can a cluster dig itself into the substrate? Such questions need to be answered experimentally. The answers to these questions are intimately connected with knowing the real, i.e., defect-filled structure of the oxide substrate in the surface region. Scanning probe microscopy can play a decisive role in unraveling the number and distribution of such defect sites as well as their mobility. Whether these techniques, perhaps used in an inelastic tunneling mode, can be employed is an open question at present. It is conceivable to try to observe the vibrations at these local sites, similar to the observation of vibrational modes reported recently [135]. Definitely, photoelectron spectroscopy and vibrational spectroscopy are tools that can be used. The former can be combined with microscopy so that certain areas of the sample could be selected. Vibrations at such defects will have rather low frequencies so that traditional infrared techniques may be difficult to apply. There is certainly a chance for synchrotron based IR techniques [136]. If Raman spectroscopy could be developed into an even more sensitive technique to allow the study of species on single crystalline sample surfaces at low concentration, this would be an ideal tool [137,138]. There is also room for field ion microscopy since it has been shown to be applicable to study diffusion of individual metal atoms on oxide films.

Yet another aspect that has to be considered, when the area of structural control is discussed, is chemical modification of the substrate and self organization of the metal deposit. It is very important for the development of the field that preparative techniques are established, and tested. That the study of crystal growth remains a field of intense study to guarantee the availability of samples of highest quality. The examples discussed in the previous chapter illustrated how chemical modification can change the distribution, the sintering properties, and also the electronic structure of particles. This topic is at the heart of catalysis.

The third area addressed at the beginning of this chapter, namely how to study chemistry on clusters is another topic connected with of the examples mentioned in the introduction. Of course, again, of particular importance is this area for catalysis. *We would like to aim at a truly size dependent understanding of the chemistry on individual clusters.* This is meant in the broadest sense, namely understanding the making of a surface-to-atom or -molecule bond, breaking it, and reacting two species with each other, whereby the reaction may be triggered through thermal energy or other means, for example photochemically. Through a combination of tunneling microscopy and tunneling spectroscopy this is possible, and experiments are actually under way. There is no reason why the single molecule experiments via inelastic tunneling spectroscopy performed on single crystal metal surfaces could not be repeated on deposited clusters. By controlling the site of the molecule on the aggregate, i.e., on the terraces (Fig. 22), on the edges or on the corners one could study properties not only cluster specific but also site specific. In fact, it is to be expected that on small clusters edges or corners have a different influence than on large clusters, even if the overall symmetries are similar. It was demonstrated in the previous sections that the response of a cluster to light depends on the particle size. A consequence of this is that photochemistry depends on particle size, a fact that has already been demonstrated [139]. Still, however, there is a lot to be done in this area. While experiments on individual clusters have a certain appeal, experiments on ensembles of well defined size should not be forgotten. Molecular beam experiments can be decisive to provide quantitative information on kinetic parameters, so important in applications [140–142]. The big advantage is that all structural parameters are controlled simultaneously at the atomic level. In this arena the future has already begun.

One of the important questions in chemistry in general is: How does a molecular bond break [143–145]? This happens on the time scale of a molecular vibration, i.e., a picosecond. However, since energy dissipation via electron–electron interaction happens on the even shorter time scale of a few femtoseconds, *it is necessary to perform ultra fast spectroscopic experiments if we want to understand the dynamics of making and breaking of bonds on clusters at the atomic level.* Such experiments have been performed in the gas phase [146], and there are experiments on single crystal metal surfaces [147] but next to nothing is known for aggregates on insulating substrates. On the basis of what has been discussed we can expect that there are interesting size dependent effects. Again, the sample quality plays an important role to be able to perform ensemble averaging experiments. It is possible to probe such processes by performing photoelectron spectroscopy after absorption of two time delayed photons [147]. Here the photoelectron is emitted after absorption of two photons, whose pulse widths are of the order of femtoseconds. The delay between the absorption events of the photons can be chosen and varied, which in turn defines the time resolution. Since the detected particle is an electron photoelectron microscopy could be used to try to detect these processes even with spatial resolution. If it were possible to bring down the spatial resolution to the size of one cluster one could do ultra short experiments on individual clusters. Another route is to use an STM tip as an electron detector to bring down the spatial resolution [148]. In fact feasibility studies in this area are under way. In order to perform such experiments in the presence of a gas atmosphere electron based experiments may not be the optimal choice. Optical microscopy using ultra-small glass fiber tips to excite and/or record a spectrum (so-called scanning near field optical microscopy, SNOM), if developed further, may be an option, although it cannot be foreseen that with this method a single cluster can be selected out of a relatively dense cluster arrangement [149]. Usually, SNOM assumes that the investigated sample area contains only a few species. It is clear, however, that ultra fast spectroscopy combined with atomic resolution of reacting systems will be one of the most fascinating areas in chemical physics in these systems.

We have briefly discussed in the last section some open questions and strategies to possibly solve them. This catalogue of questions and answers can be extended almost infinitely. The specific areas touched here are considered to be opening up avenues to perform fascinating basic research that might be directly linked to applications.

4. Final remark

The attempt has been made to answer the simple question, why is the study of metal aggregates on oxides, described in this paper, an interesting subject. I have tried to put the field in a perspective that one can enter through applications from everyday life. The examples were chosen to document the current state of the art, and at the same time demonstrate that still a lot has to be achieved. In the third section we explore a whole variety of new, demanding experiments, which were chosen such, that in all likelihood they could be done given that the necessary tools are available and the scientific community receives continuous support, and, most importantly, there is a sufficient number of young talented scientists who are willing to take up a challenge.

Acknowledgements

This article has been written while I was staying as a chercheur associé at the Centre de Recherche sur les Mécanismes de la Croissance Cristalline (CRMC^2), a laboratory of CNRS in Luminy, France. I would like to thank my host, Claude Henry, for many fruitful discussions. Also I would like to thank the organizations sponsoring our research: Max-Planck-Society, Ministry of Education and Research of the Federal Republic of Germany, Deutsche Forschungsgemeinschaft, German Israeli Foundation, New Energy and Industrial Technology Development Organization (NEDO) of Japan, DAAD, Humboldt Foundation as well as Synetix (a member of the ICI group). I am indepted to my collaborators whose names appear in some of the references. For their help with this article I would like to thank Heiko Hamann, Yvonne Gründer, and Helmut Kuhlenbeck.

References

[1] G. Ertl, H. Knözinger, J. Weitkamp (Eds.), Handbook of Heterogeneous Catalysis, vol. 4, Wiley-VCH Verlagsgesellschaft mbH, Weinheim, 1997.

[2] Sensors update H. Baltes, W. Gospel, J. Hesse (Eds.), Sensor Technology Applications—Markets, vol. 1, VCH Verlagsgesellschaft mbH, Weinheim, 1996.

[3] H. Ulmer, J. Mitrovics, G. Noetzel, U. Weimar, W. Goepel, Odours and flavours identified with hybrid modular sensor systems, Sens. Actuators B 43 (1997) 24.

[4] I. Lundstroem, Why bother about gas-sensitive field-effect devices? Sens. Actuators A 56 (1996) 75.

[5] First Toyota Workshop on Magnetism and Magnetic Materials for High Density Information Storage, Toyota Institute of Technology, 1997; in J. Magn. Magn. Mater., vol. 175, Elsevier, Brussels, Belgium.

[6] A. Mie, Beiträge zur optik trüber medien, speziell kolloidaler metalllösungen, Ann. Phys. 25 (1908) 377.

[7] P. Chakraborty, Metal nanoclusters in glasses as nonlinear photonic materials, J. Mat. Sci. 33 (1998) 2235.

[8] H.-J. Freund, E. Umbach (Eds.), Adsorption on Ordered Surfaces of Ionic Solids and Thin Films, Springer Series in Surface Sciences, vol. 33, Springer, Heidelberg, 1993.

[9] G.H. Vurens, M. Salmeron, G.A. Somorjai, The preparation of thin ordered transition metal oxide films on metal single crystals for surface science studies, Progr. Surf. Sci. 32 (1989) 333.

[10] V.E. Henrich, P.A. Cox, The Surface Science of Metal Oxides, Cambridge University Press, Cambridge, 1994.

[11] H.-J. Freund, H. Kuhlenbeck, V. Staemmler, Oxide surfaces, Rep. Progr. Phys. 59 (1996) 283, Review.

[12] H.-J. Freund, Adsorption of gases on complex solid surfaces, Angew. Chem. Int. Ed. Engl. 36 (1997) 452, Review.

[13] D.W. Goodman, Model catalysts—from extended single crystals to supported particles, Surf. Rev. Lett. 2 (1995) 9.

[14] P.L.J. Gunter, J.W.H. Niemantsverdriet, F.H. Ribeiro, G.A. Somorjai, Surface science approach to modeling supported catalysts, Catal. Rev. Sci. Eng. 39 (1997) 77.

[15] D.A. Bonnell, Scanning tunneling microscopy and spectroscopy of oxide surfaces, Progr. Surf. Sci. 57 (1998) 187.

[16] U. Diebold, J.F. Anderson, K.-O. Ng, D. Vanderbilt, Evidence for the tunneling site on transition-metal oxides: TiO$_2$(1 1 0), Phys. Rev. Lett. 77 (1996) 1322.

[17] P.W. Murray, F.M. Leibsle, C.A. Muryn, H.J. Fisher, C.F.J. Flipse, G. Thornton, Interrelationship of structural

elements on $TiO_2(100)$-(1×3), Phys. Rev. Lett. 72 (1994) 689.
[18] P.W. Murray, N.G. Condon, G. Thornton, Effect of stoichiometry on the structure of $TiO_2(110)$, Phys. Rev. B 51 (1995) 10989.
[19] H. Onishi, Y. Iwasawa, Reconstruction of $TiO_2(110)$ surface: STM study with atomic-scale resolution, Surf. Sci. 313 (1994) L783.
[20] H. Onishi, Y. Iwasawa, STM-imaging of formate intermediates adsorbed on a $TiO_2(110)$ surface, Chem. Phys. Lett. 226 (1994) 111.
[21] H. Onishi, K. Fukui, Y. Iwasawa, Atomic-scale surface structures of $TiO_2(110)$ determined by scanning tunneling microscopy—a new surface-limited phase of titanium oxide, Bull. Chem. Soc. Jpn. 68 (1995) 2447.
[22] H. Onishi, Y. Iwasawa, Dynamic visualization of a metal-oxide-surface/gas-phase reaction: time-resolved observation by scanning tunneling microscopy at 800 K, Phys. Rev. Lett. 76 (1996) 791.
[23] G. Charlton, P.B. Howes, C.L. Nicklin, P. Steadman, J.S.G. Taylor, C.A. Muryn, S.P. Harte, J. Mercer, R. McGrath, D. Norman, T.S. Turner, G. Thornton, Relaxation of $TiO_2(110)$-(1×1) using surface X-ray diffraction, Phys. Rev. Lett. 78 (1997) 495.
[24] G. Renaud, Oxide surfaces and metal/oxide interfaces studied by grazing incidence X-ray scattering, Surf. Sci. Rep. 32 (1998) 1.
[25] P.W. Tasker, Adv. Ceram. 10 (1984) 176.
[26] W.C. Mackrodt, R.J. Davey, S.N. Black, R. Docherty, The morphology of α-Al_2O_3 and α-Fe_2O_3: the importance of surface relaxation, J. Cryst. Growth 80 (1987) 441.
[27] C. Noguera, Physics and Chemistry at Oxide Surfaces, Cambridge University Press, Cambridge, 1996.
[28] H. Over, Y.D. Kim, A.P. Seitsonen, S. Wendt, E. Lundgren, M. Schmid, P. Varga, A. Morgante, G. Ertl, Atomic-scale structure and catalytic reactivity of the $RuO_2(110)$ surface, Science 287 (2000) 1474.
[29] Y.D. Kim, A.P. Seitsonen, M. Schwegmann, H. Over, Epitaxial Growth of $RuO_2(100)$ on $Ru(10\bar{1}0)$: Surface Structure and Other Properties, J. Phys. Chem. B 105 (2001) 2205–2211.
[30] P. Guénard, G. Renaud, A. Barbier, M. Gautier-Soyer, Determination of the α-$Al_2O_3(0001)$ surface relaxation and termination by measurements of crystal truncation rods, Surf. Rev. Lett. 5 (1998) 321.
[31] F. Rohr, M. Bäumer, H.-J. Freund, J.A. Mejias, V. Staemmler, S. Müller, L. Hammer, K. Heinz, Strong relaxations the $Cr_2O_3(0001)$ surface as determined via low-energy electron diffraction and molecular dynamics simulations, Surf. Sci. 372 (1997) L291.
[32] F. Rohr, M. Bäumer, H.-J. Freund, J.A. Mejias, V. Staemmler, S. Müller, L. Hammer, K. Heinz, Erratum: Strong relaxations a the $Cr_2O_3(0001)$ surface as determined via low-energy electron diffraction and molecular dynamics simulations, Surf. Sci. 372 (1997) L291, Surf. Sci. 389 (1997) 391.

[33] W. Weiss, Structure and composition of thin epitaxial iron oxide films grown onto $Pt(111)$, Surf. Sci. 377–379 (1997) 943.
[34] I. Manassidis, A. De Vita, M.J. Gillan, Structure of the (0001) surface of α-Al_2O_3 from first principles calculations, Surf. Sci. 285 (1993) L517.
[35] I. Manassidis, M.J. Gillan, Structure and energetics of alumina surfaces calculated from first principles, J. Am. Ceram. Soc. 77 (1994) 335.
[36] C. Rehbein, N.M. Harrison, A. Wander, The structure of the α-$Cr_2O_3(0001)$ surface: an ab initio total energy study, Phys. Rev. B 54 (1996) 14066.
[37] X.-G. Wang, W. Weiss, S.K. Shaikhoutdinov, M. Ritter, M. Petersen, F. Wagner, R. Schlögl, M. Scheffler, The hematite α-$Fe2O3(0001)$ surface: evidence for domains of distinct chemistry, Phys. Rev. Lett. 81 (1998) 1038.
[38] K. Heinz, Geometrical and chemical restructuring of clean metal surfaces as retrieved by LEED, Surf. Sci. 299–300 (1994) 433.
[39] N.M. Harrison, X.-G. Wang, J. Muscat, M. Scheffler, The influence of soft vibrational modes on our understanding of oxide surface structure, Faraday Disc. 114 (1999) 305.
[40] H. Schmalzried, Chemical Kinetics of Solids, Verlag Chemie, Weinheim, 1995.
[41] J.G. Fripiat, A.A. Lucas, J.M. André, E.G. Derouane, On the stability of polar surface planes of macroscopic ionic crystals, Chem. Phys. 21 (1977) 101.
[42] D. Cappus, C. Xu, D. Ehrlich, B. Dillmann, C.A. Ventrice Jr., K. Al-Shamery, H. Kuhlenbeck, H.-J. Freund, Hydroxyl groups on oxide surfaces: $NiO(100)$, $NiO(111)$, and $Cr_2O_3(111)$, Chem. Phys 177 (1993) 533.
[43] F. Rohr, K. Wirth, J. Libuda, D. Cappus, M. Bäumer, H.-J. Freund, Hydroxyl driven reconstruction of the polar $NiO(111)$ surface, Surf. Sci. 315 (1994) L977.
[44] X.-G. Wang, A. Chaka, M. Scheffler, Effect of the environment on α-$Al_2O_3(0001)$, Phys. Rev. Lett. 84 (2000) 3650.
[45] J. Libuda, M. Frank, A. Sandell, S. Andersson, P.A. Brühwiler, M. Bäumer, N. Mårtensson, H.-J. Freund, Interaction of rhodium with hydroxylated alumina model substrates, Surf. Sci. 384 (1997) 106.
[46] M. Heemeier, M. Frank, J. Libuda, K. Wolter, H. Kuhlenbeck, M. Bäumer, H.-J. Freund, The influence of OH groups on the growth of Rhodium on alumina: a model study, Catal. Lett. 68 (2000) 19.
[47] M.M. Ivey, H.C. Allen, A. Avoyan, K.A. Martin, J.C. Hemminger, Dimerization of 1,3-butadiene on highly characterized hydroxylated surfaces of ultrathin films of gamma-Al_2O_3, J. Am. Chem. Soc. 120 (42) (1998) 10980.
[48] R.M. Jaeger, H. Kuhlenbeck, H.-J. Freund, M. Wuttig, W. Hoffmann, R. Franchy, H. Ibach, Formation of a well-ordered aluminium oxide overlayer by oxidation of $NiA(110)$, Surf. Sci. 259 (1991) 235.
[49] I. Engquist, B. Liedberg, D_2O ice on controlled wettability selfassembled alkanethiolate monolayers—Cluster formation and substrate–adsorbate interaction, J. Phys. Chem. 100 (1996) 20089.

[50] H. Knözinger, P. Ratnasamy, Catalytic aluminas: surface models and characterization of surface sites, Catal. Rev. Sci. Eng. 17 (1978) 31.
[51] C.B. Duke (Ed.), Surface Science: The First Thirty Years, vol. 299/300, Elsevier, Amsterdam, 1994.
[52] H. Poppa, Nucleation growth, and, TEM analysis of metal particles and clusters deposited in UHV, Catal. Rev. Sci. Eng. 35 (1993) 359.
[53] C.R. Henry, Surface studies of supported model catalysts, Surf. Sci. Rep. 31 (1998) 231.
[54] C. Goyhenex, C.R. Henry, J. Urban, In-situ measurements of the lattice parameter of supported palladium clusters, Phil. Mag. A 69 (1994) 1073.
[55] M.-C. Wu, P.J. Møller, Studies of the electronic structure of, Surf. Sci. 224 (1989) 250.
[56] P.J. Møller, M.-C. Wu, Surface geometrical structure and incommensurate growth: ultrathin Cu films on $TiO_2(110)$, Surf. Sci. 224 (1989) 265.
[57] M.C. Wu, P.J. Møller, Direct and substrate-phonon-assisted electronic transitions in $Cu/TiO_2(110)$ observed by AREELS, Surf. Sci. 235 (1990) 228.
[58] P.J. Møller, J. Nerlov, Ultrathin films of Cu on $ZnO(11-20)$: growth and electronic structure, Surf. Sci. 307–309 (1993) 591.
[59] P.W. Murray, J. Shen, N.G. Condon, S.J. Pang, G. Thornton, STM study of PD growth on $TiO_2(100)$-(1×3), Surf. Sci. 380 (1997) L455.
[60] P. Stone, R.A. Bennett, M. Bowker, Reactive re-oxidation of reduced $TiO_2(110)$ surfaces demonstrated by high temperature STM movies, New J. Phys. 1 (1998–99) 8.
[61] U. Diebold, Structure(s) and reactivity of the $TiO_2(110)$ surface, Proc. of 1st Int. Workshop on Oxide Surfaces, Elmau, Germany 1999.
[62] C.T. Campbell, Ultrathin metal films and particles on oxide surfaces: structural, electronic and chemisorption properties, Surf. Sci. Rep. 27 (1997) 1.
[63] J. Libuda, F. Winkelmann, M. Bäumer, H.-J. Freund, T. Bertrams, H. Neddermeyer, K. Müller, Structure and defects of an ordered alumina film on NiAl(110), Surf. Sci. 318 (1994) 61.
[64] M. Bäumer, H.-J. Freund, Metal deposits on well-ordered oxide films, Progr. Surf. Sci. 61 (1999) 127.
[65] S. Stempel, M. Bäumer, H.-J. Freund, STM studies of rhodium deposits on an ordered alumina film—resolution and tip effects, Surf. Sci. 402–404 (1998) 424.
[66] M. Heemeier, Ph.D. thesis, Freie Universität Berlin, in preparation.
[67] N. Ernst, B. Duncombe, G. Bozdech, M. Naschitzki, H.-J. Freund, Field ion microscopy of platinum adatoms deposited on a thin Al_2O_3 film on NiAl(110), Ultramicroscopy 79 (1999) 231.
[68] A. Piednoir, E. Perrot, S. Granjeaud, A. Humbert, C. Chapon, C.R. Henry, Atomic resolution on small three-dimensional metal clusters by STM, Surf. Sci. 391 (1997) 19.
[69] K.H. Hansen, T. Worren, S. Stempel, E. Lægsgaard, M. Bäumer, H.-J. Freund, F. Besenbacher, I. Stensgaard, Palladium nanocrystals on Al_2O_3: structure and adhesion energy, Phys. Rev. Lett. 83 (1999) 4120.
[70] M. Klimenkov, S. Nepijko, H. Kuhlenbeck, M. Bäumer, R. Schlögl, H.-J. Freund, The structure of Pt-aggregates on a supported thin aluminum oxide film in comparison with unsupported alumina—a transmission electron microscopy study, Surf. Sci. 391 (1997) 27.
[71] S.A. Nepijko, M. Klimenkov, H. Kuhlenbeck, D. Zemlyanov, D. Herein, R. Schlögl, H.-J. Freund, TEM study of tantalum clusters on Al_2O_3/NiAl(110), Surf. Sci. 412/413 (1998) 192.
[72] S.A. Nepijko, M. Klimenkov, M. Adelt, H. Kuhlenbeck, R. Schlögl, H.-J. Freund, Structural investigation of palladium clusters on γ-Al_2O_3(111)/NiAl(110) with transmission electron microscopy, Langmuir 15 (1999) 5309.
[73] C. Solliard, M. Fluëli, Surface stress and size effect on the lattice parameter in small particles of gold and platinum, Surf. Sci. 156 (1985) 487.
[74] G. Rupprechter, A.S. Eppler, G.A. Somorjai, A new model catalyst studied by HREM and AFM: Platinum nanoparticle arrays on silica fabricated by electron beam lithography, in: H.A.C. Benavides, M.J. Yacamàn (Eds.), Electron Microscopy, vol. II, Institute of Physics Publishing, Bristol, 1998, p. 369.
[75] A.S. Eppler, G. Rupprechter, L. Guczi, G.A. Somorjai, Model catalysts fabricated using electron beam lithography and pulsed laser deposition, J. Phys. Chem. B 101 (1997) 9973.
[76] K. Wong, S. Johansson, B. Kasemo, Nanofabricated model catalysts, Faraday Disc. 105 (1996) 237.
[77] P.W. Jacobs, F.H. Ribeiro, G.A. Somorjai, S.J. Wind, New model catalysts—uniform platinum cluster arrays produced by electron beam lithography, Catal. Lett. 37 (1996) 131.
[78] W. Gotschy, K. Vonmetz, A. Leitner, F.R. Aussenegg, Thin films by regular patterns of metal nanoparticles—tailoring the optical properties by nanodesign, Appl. Phys. B. Laser and Optics 63 (1996) 381.
[79] F. Cerrina, C. Marrian, A path to nanolithography, MRS Bull. 21 (1996) 56.
[80] P.B. Fischer, S.Y. Chou, 10 nm electron beam lithography and sub-50 nm overlay using a modified scanning electron microscope, Appl. Phys. Lett. 62 (1993) 2989.
[81] G. Schmid (Ed.), Clusters and Colloids From Theory to Applications, VCH Verlagsgesellschaft mbH, Weinheim, 1994.
[82] W. Ekardt, Metal Clusters, Wiley, Chichester, 1999.
[83] F.F. deBiani, C. Femoni, M.C. Iapalucci, G. Longoni, P. Zanello, A. Ceriotti, Redox behavior of $[H_{6-n}Ni_{38}Pt_6(CO)_{48}]^{n-}$ ($n = 4$–6) anions: a series of metal carbonyl clusters displaying electron-sink features, Inorg. Chem. 38 (1999) 3721.
[84] B.F.G. Johnson, Transition Metal Clusters, Wiley, Chichester, 1980.
[85] G.K. Wertheim, S.B. DiCenzo, D.N.E. Buchanan, Noble and transition metal clusters: the d bands of silver and palladium, Phys. Rev. B 33 (1986) 5384.

[86] A. Sandell, J. Libuda, P. Brühwiler, S. Andersson, A. Maxwell, M. Bäumer, N. Mårtensson, H.-J. Freund, Electron spectroscopy studies of small deposited metal particles, J. Electron Spectrosc. Rel. Phenom. 76 (1995) 301.

[87] R. Unwin, A.M. Bradshaw, Photoelectron spectroscopy of palladium particle arrays on a carbon substrate, Chem. Phys. Lett. 58 (1978) 58.

[88] G.K. Wertheim, Electronic structure of metal clusters, Z. Phys. D 12 (1989) 319.

[89] S.-T. Lee, G. Apai, M.G. Mason, R. Benbow, Z. Hurych, Evolution of band structure in gold clusters as studied by photoemission, Phys. Rev. B 23 (1981) 505.

[90] V. De Gouveia, B. Bellamy, Y. Hadj Romdhane, A. Mason, M. Che, Electronic effect induced by variation of size for Pd clusters in 1,3-butadiene hydrogenation, Z. Phys. D 12 (1989) 587.

[91] C. Kuhrt, M. Harsdorff, Photoemission and electron microscopy of small supported palladium clusters, Surf. Sci. 245 (1991) 173.

[92] A. Sandell, J. Libuda, P.A. Brühwiler, S. Andersson, A.J. Maxwell, M. Bäumer, N. Mårtensson, H.-J. Freund, Interaction of CO with Pd clusters supported on a thin alumina film, J. Vac. Sci. Technol. A 14 (1996) 1546.

[93] M. DeCrescenzi, M. Diociaiuti, L. Lozzi, P. Picozzi, S. Santucci, Surface electron-energy-loss fine-structure investigation on the local structure of copper clusters on graphite, Phys. Rev. B 35 (1987) 5997.

[94] D.R. Rainer, D.W. Goodman, in: R.M. Lambert, G. Pacchioni (Eds.), Chemisorption and Reactivity on Supported Clusters and Thin Films Series E, vol. 331, Kluwer, Dordrecht, 1997, p. 27.

[95] M. Valden, X. Lai, D.W. Goodman, Onset of catalytic activity of gold clusters on titania with the appearance of nonmetallic properties, Science 281 (1998) 1647.

[96] U. Kreibig, M. Vollmer (Eds.), Optical Properties of Metal Clusters, Springer Series in Materials Science, vol. 25, Springer, Berlin, 1995.

[97] F. Stietz, F. Träger, Surface plasmons in nanoclusters: elementary electronic excitations and their applications, Phil. Mag. B 79 (1999) 1281.

[98] K.-P. Charlé, L. König, S. Nepijko, I. Rabin, W. Schulze, The surface plasmon resonance of free and embedded Ag-Clusters in the size range $1,5$ nm $< D < 30$ nm, Crystal Res. Technol. 33 (1998) 1085.

[99] N. Nilius, N. Ernst, H.-J. Freund, Photon emission spectroscopy of individual oxide-supported silver clusters in the STM, Phys. Rev. Lett. 84 (2000) 3994.

[100] J.K. Gimzewski, B. Reihl, J.K. Sass, R.R. Schlittler, Photon emission experiment with the scanning tunneling microscope, J. Microsc. 152 (1988) 325.

[101] J. Evans, B. Hayden, F. Mosselman, A. Murray, Adsorbate induced phase changes of rhodium on $TiO_2(110)$, Surf. Sci. 279 (1992) L159.

[102] J. Evans, B. Hayden, F. Mosselman, A. Murray, The chemistry of rhodium on $TiO_2(110)$ deposited by MOCVD of $[Rh(CO)_2Cl]_2$ and MVD, Surf. Sci. 301 (1994) 61.

[103] M. Frank, R. Kühnemuth, M. Bäumer, H.-J. Freund, Vibrational spectroscopy of CO adsorbed on supported ultra-small transition metal particles and single metal atoms, Surf. Sci. 454–456 (2000) 968.

[104] M. Frank, R. Kühnemuth, M. Bäumer, H.-J. Freund, Oxide-supported Rh particle structure probed with carbon monoxide, Surf. Sci. 427/428 (1999) 288.

[105] P. Uvdal, P.-A. Karlsson, C. Nyberg, S. Andersson, N.V. Richardson, On the structure of dense CO overlayers, Surf. Sci. 202 (1988) 167.

[106] T. Giessel, O. Schaff, C.J. Hirschmugl, V. Fernandez, K.M. Schindler, A. Theobald, S. Bao, R. Lindsay, W. Berndt, A.M. Bradshaw, C. Baddeley, A.F. Lee, R.M. Lambert, D.P. Woodruff, A photoelectron diffraction study of ordered structures in the chemisorption system Pd(111)-CO, Surf. Sci. 406 (1998) 90.

[107] G. Kisters, J.G. Chen, S. Lehwald, H. Ibach, Adsorption of CO on the unreconstructed and reconstructed Ir(100) surface, Surf. Sci. 245 (1991) 65.

[108] J. Lauterbach, R.W. Boyle, M. Schick, W.J. Mitchell, B. Meng, W.H. Weinberg, The adsorption of CO on Ir(111) investigated with FT-IRAS, Surf. Sci. 350 (1996) 32.

[109] F. Solymosi, È. Novàk, A. Molnàr, Infrared spectroscopic study on CO-induced structural changes of iridium on an alumina support, Phys. Chem. 94 (1990) 7250.

[110] T.W. Voskobojnikov, E.S. Shpiro, H. Landmesser, N.I. Jaeger, G. Schulz-Ekloff, Redox and carbonylation chemistry of iridium species in the channels of H-ZSM-5 zeolite, J. Mol. Catal. A-Chem. 104 (1996) 299.

[111] K. Wolter, O. Seiferth, H. Kuhlenbeck, M. Bäumer, H.-J. Freund, Infrared spectroscopic investigation of CO adsorbed on Pd aggregates deposited on an alumina model support, Surf. Sci. 399 (1998) 190.

[112] C. DeLaCruz, N. Sheppard, In situ IR spectroscopic study of the adsorption and dehydrogenation of ethene on a platinum-on-silica catalyst between 100 and 294 K, J. Chem. Soc. Faraday Trans. 93 (1997) 3569.

[113] G.A. Beitel, A. Laskov, H. Oosterbeek, E.W. Kuipers, Polarization modulation infrared reflection absorption spectroscopy of CO adsorption on Co(0001) under a high-pressure regime, J. Phys. Chem. 100 (1996) 12494.

[114] G.A. Somorjai, G. Rupprechter, Molecular studies of catalytic reactions on crystal surfaces at high pressures and high temperatures by infrared-visible sum frequency generation (SFG) surface vibrational spectroscopy, J. Phys. Chem. B 103 (1999) 1623.

[115] H. Härle, A. Lehnert, U. Metka, H.R. Volpp, L. Willms, J. Wolfrum, In situ detection and surface coverage measurements of CO during CO oxidation on polycrystalline platinum using sum frequency generation, Appl. Phys. B 68 (1999) 567.

[116] T. Dellwig, G. Rupprechter, H. Unterhalt, H.-J. Freund, Bridging the pressure and materials gaps: High pressure sum frequency generation study on supported Pd nanoparticles, Phys. Rev. Lett. 85 (2000) 776.

[117] U. Heiz, W.-D. Schneider, in: K.H. Meiwes-Broer (Ed.), Cluster Surface Interaction, Springer, Berlin, 1999.

[118] F. Vanolli, U. Heiz, W.-D. Schneider, Vibrational coupling of CO adsorbed on monodispersed Ni_{11} clusters supported on magnesia, Chem. Phys. Lett. 277 (1997) 527.

[119] U. Heiz, A. Sanchez, S. Abbet, W.-D. Schneider, Catalytic oxidation of carbon monoxide on monodispersed platinum clusters: each atom counts, J. Am. Chem. Soc. 121 (1999) 3214.

[120] U. Heiz, W.-D. Schneider, Nanoassembled model catalysts, J. Phys. D: Appl. Phys. 33 (2000) 85.

[121] U. Heiz, A. Sanchez, S. Abbet, W.D. Schneider, The reactivity of gold and platinum metals in their cluster phase, Eur. Phys. J. D 9 (1999) 35.

[122] Y. Iwasawa (Ed.), X-ray Absorption Fine Structure for Catalysts and Surfaces, Series on Synchrotron Radiation Techniques and Applications, vol. 2, World Scientific, Singapore, 1996.

[123] L. Reimer, Transmission Electron Microscopy: Physics of Image Formation and Microanalysis, Springer, Berlin, 1997.

[124] Seventh Annual Workshop–European Synchrotron Light Sources, BESSY, Berlin, 1999.

[125] Das Jahrbuch des Forschungszentrums, DESY, Hamburg, 2000.

[126] Y.R. Shen, The Principles of Nonlinear Optics, Wiley, New York, 1984.

[127] J. Stöhr, NEXAFS spectroscopy, Springer Series in Surface Sciences, vol. 25, Springer, Berlin, 1992.

[128] W. Kuch, R. Fromter, J. Gilles, D. Hartmann, C. Ziethen, C.M. Schneider, G. Schonhense, W. Swiech, J. Kirschner, Element-selective magnetic imaging in exchange-coupled systems by magnetic photoemission microscopy, Surf. Rev. Lett. 5 (1998) 1241.

[129] J. Falta, D. Bahr, G. Materlik, B.H. Muller, M. Horn-Von-Hoegen, X-ray characterization of buried delta layers, Surf. Rev. Lett. 5 (1998) 145.

[130] DESY Forschungs- und Entwicklungsprogramm, DESY, Hamburg, 1999–2000.

[131] R. Wichtendahl, R. Fink, H. Kuhlenbeck, D. Preikszas, H. Rose, R. Spehr, P. Hartel, W. Engel, R. Schlögl, H.-J. Freund, A.M. Bradshaw, G. Lilienkamp, T. Schmidt, E. Bauer, G. Benner, E. Umbach, SMART: an aberration-corrected XPEEM/LEEM with energy filter, Surf. Rev. Lett. 5 (1998) 1249.

[132] H. Ade, W. Yang, S.L. English, J. Hartman, R.F. Davis, R.J. Nemanich, V.N. Litvinenko, L.V. Pinayev, Y. Wu, J.M.J. Madey, A free electron laser—photoemission electron microscope system (FEL-PEEM), Surf. Rev. Lett. 5 (1998) 1257.

[133] T. Duden, E. Bauer, Spin-polarized low energy electron microscopy, Surf. Rev. Lett. 5 (1998) 1213.

[134] T.R. Linderoth, S. Horch, E. Laegsgaard, I. Stensgaard, F. Besenbacher, Dynamics of Pt adatoms and dimers on Pt(1 1 0)-(1 × 2) observed directly by STM, Surf. Sci. 402–404 (1998) 308.

[135] B.C. Stipe, M.A. Rezaei, W. Ho, Single-molecule vibrational spectroscopy and microscopy, Science 280 (1998) 1732.

[136] F.M. Hoffmann, G.P. Williams, Synchrotron radiation in the far infrared: infrared reflection absorption spectroscopy, in: W. Eberhardt (Ed.), Applications of Synchrotron Radiation—High Resolution Studies of Molecules and Molecular Adsorbates on Surfaces, Springer Series in Surface Sciences, vol. 35, Springer, Berlin, 1995, p. 263.

[137] A. Campion, Raman spectroscopy in vibrational spectroscopy of molecules on surfaces, in: J.T. Yates, T.E. Madey (Eds.), Methods of Surface Characterization, vol. 1, Plenum, New York, 1987, p. 345.

[138] Y.T. Chua, P.C. Stairs, A novel fluidized bed technique for measuring UV Raman spectra of catalysts and adsorbates, J. Catal. 196 (2000) 66.

[139] K. Watanabe, Y. Matsumoto, M. Kampling, K. Al-Shamery, H.-J. Freund, Photochemistry of methane on Pd/Al_2O_3 model catalyst: control of photochemistry on transition metal surfaces, Angew. Chem. Int. Ed. 38 (1999) 2192.

[140] C. Becker, C.R. Henry, Cluster size dependent kinetics for the oxidation of CO on a Pd/MgO(1 0 0) model catalyst, Surf. Sci. 352 (1996) 457.

[141] I. Meusel, J. Hoffmann, J. Hartmann, M. Heemeier, M. Bäumer, J. Libuda, H.-J. Freund, The interaction of oxygen with alumina-supported palladium particles, Catal. Lett. 71 (2001) 5.

[142] V.P. Zhdanov, B. Kasemo, Simulations of the reaction kinetics on nanometer supported particles, Surf. Sci. Rep. 39 (2000) 25.

[143] H. Eyring, J. Phys. Chem. 3 (1935) 107.

[144] M.G. Evans, M. Polanyi, Trans. Faraday Soc. 31 (1935) 875.

[145] M.G. Evans, M. Polanyi, Trans. Faraday Soc. 33 (1937) 448.

[146] A.H. Zewail (Ed.), Femtochemistry—Ultrafast Dynamics of the Chemical Bond, vols. 1 and 2, World Scientific Publishing, Singapore, 1994.

[147] H. Petek, S. Ogawa, Femtosecond time-resolved two-photon photoemission studies of electron dynamics in metals, Progr. Surf. Sci. 56 (1998) 239.

[148] M.J. Feldstein, P. Vöhringer, W. Wang, N.F. Scherer, Femtosecond optical spectroscopy and scanning probe microscopy, J. Phys. Chem. 100 (1996) 4739.

[149] T. Klar, M. Perner, S. Grosse, G. von Plessen, W. Spirkl, J. Feldmann, Surface-plasmon resonances in single metallic nanoparticles, Phys. Rev. Lett. 80 (1998) 4249.

 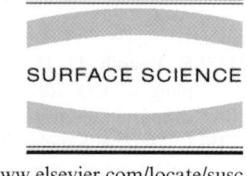

Tailoring magnetism in artificially structured materials: the new frontier

J. Shen [a,*], J. Kirschner [b]

[a] *Solid State Division, Oak Ridge National Laboratory, Oak Ridge, TN 37831-6057, USA*
[b] *Max Planck Institut für Mikrostrukturphysik, Halle 06120, Germany*

Received 30 June 2000; accepted for publication 11 May 2001

Abstract

The current standard of electronic devices and data storage media has reached a level such that magnetic materials have to be fabricated on a nanometer scale. In particular, the emerging concept of spintronics, which is based on fact that current carriers have not only charge but also spin, requires the assembling of nanometer-sized magnetic structures with desired magnetic properties. It is this background that motivates scientists and engineers to attempt to grow and characterize magnetic objects at smaller and smaller length scales, from 2D films and multilayers to 1D wires and eventually to 0D dots. In this article, some of the most significant progress in recent years in the effort of growing artificially structured magnetic materials are reviewed. The new structural and magnetic properties of these materials are discussed, with an emphasis on the correlation between structure and magnetism, which also serves as guidance for improving their magnetic properties. The emerging emphasis is on converting the existing knowledge into growing and studying low-dimensional complex materials, which promise to have considerably higher "tuning" ability for desired properties. © 2001 Elsevier Science B.V. All rights reserved.

Keywords: Magnetic films; Magnetic interfaces; Metal–metal magnetic heterostructures; Epitaxy; Magnetic phenomena (cyclotron resonance, phase transitions, etc.); Electrical transport (conductivity, resistivity, mobility, etc.)

1. Introduction

In modern society, applications of magnetic materials can be found almost everywhere. From the familiar small magnets that hang our children's drawings on the refrigerators, to the almost as familiar industrial transformer cores, motor generators and microwave guides, our daily life is intimately tied to magnetism and magnetic materials. The most noteworthy impact of magnetism occurs via information transport and data storage devices, which mostly consist of artificially structured magnetic materials. Here artificially structured materials (ASM) refer to the materials which are either made into reduced dimensions such as two-dimensional (2D) ultrathin films, one-dimensional (1D) wires and zero-dimensional (0D) dots, or assemblies of these low-dimensional structures such as multilayers, wire and dot arrays. Magnetic ASM have been utilized in technologies with a large and growing economic impact. The magnetic

* Corresponding author. Tel.: +1-865-2414828.
E-mail address: Shenj@ornl.gov (J. Shen).

recording industry alone exceeds $100 billion in annual sales. It is continuously pushing the technical envelope as evidenced by the fact that over the past decade, the density of information storage has been increasing at a compound annual rate exceeding 60% per year while the cost of storing information on a computer hard drive has been decreasing at the same rate [1].

This progress has been made possible by a series of scientific and technological advances, mostly marked by the synthesis of artificially structured magnetic materials. Consider, as an example, the read head of a disk drive. The technological challenge in this case is to sense the ever smaller magnetization of each "bit", whose size keeps decreasing with increasing density. The traditional inductive read sensors could no longer be scaled to the rapidly decreasing sizes required and a more sensitive sensor was clearly needed. In 1988, a phenomenon called giant magnetoresistance (GMR) effect was first observed by German and French scientists independently in a metallic multilayer system consisting of stacked thin films of magnetic and nonmagnetic materials [2,3]. Fig. 1 demonstrates the principle of GMR effect using a schematic view of a ferromagnetic Co/nonmagnetic Cu trilayer. Here both Co and Cu contain electrons with opposite spin directions, which are defined as "spin up" and "spin down". The difference is that the number of spin up and spin down electrons are equal in nonmagnetic Cu and unequal in ferromagnetic Co. If a voltage is applied between the two Co layers, both "up" and "down" electrons will sense the voltage and try to travel from one Co to the other Co layer through the Cu layer as well as the two Co/Cu interfaces. However, the electronic structures of Co and Cu are such that the up electrons of Co and Cu have matching potentials, and the Co down electrons are in a considerably lower potential than that of Cu. This means that the Co up electrons will have no difficulty to pass through the first Co/Cu interface and reach the second Co/Cu interface, while the Co down electrons will mostly be scattered back at the first Co/Cu interface. At the second Co/Cu interface, the Co up electrons can only pass through if there are "empty space" available for up electrons in the second Co layer.

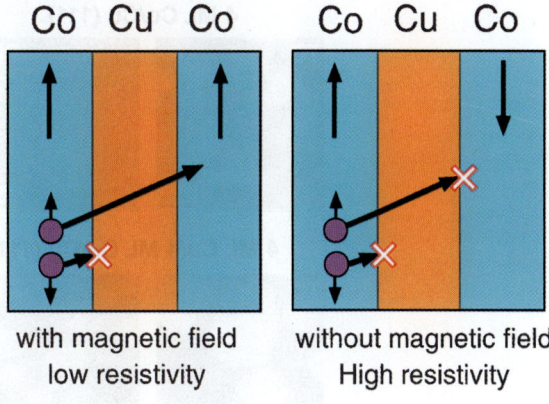

Fig. 1. Schematic view of a Co/Cu/Co trilayer which exhibits GMR effect. When a bias voltage is applied between Co(I) and Co(II) layers, the spin up and spin down electrons in Co(I) layer form two independent current channels to travel across the trilayer. Due to the potential barrier between the Co and Cu electrons, the spin down electrons are scattered back at the Co(I)/Cu interface independent of the relative magnetization orientations between the two Co layers. The spin up electrons in Co(I) have much less difficulty to reach Co(II) when the magnetization of the two Co layers are parallel as compared to the antiparallel case. As a result, the resistance of the trilayer is minimum (maximum) when the magnetization of the two Co layers are parallel (antiparallel).

From Co electronic structure we know that such "empty space" is created when the magnetization of the two Co layers are parallel. As a result, the electrical resistance of such a Co/Cu trilayer system reduces to the minimum when the magnetization of the two Co layers are parallel, and increases to maximum when the magnetization of the two Co layers are antiparallel. The change of the resistance with respect to the maximum resistance is often referred as magnetoresistance. The word "giant" is used for the multilayer system because its magnetoresistance is typically around 10–100%, which is one to two orders of magnitude larger than the conventional magnetoresistance in bulk ferromagnetic metals.

The key to have a large GMR effect is to make sure the two Co layers are absolutely antiparallel in the absence of an external field. This relies on an interaction called indirect exchange coupling between the two Co layers. The indirect exchange coupling has a oscillatory behavior which favors either parallel (ferromagnetic coupling) or

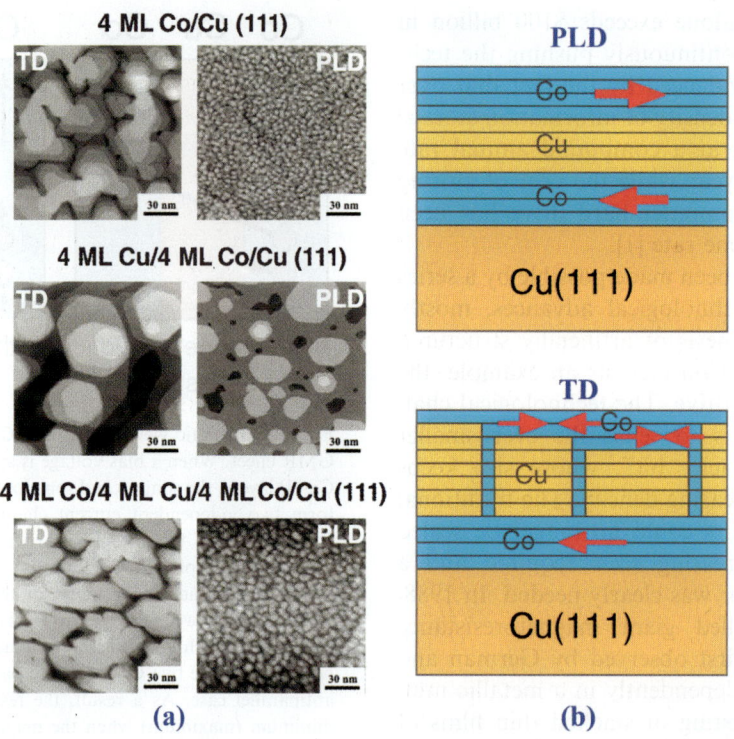

Fig. 2. (a) Surface morphology of a Co/Cu/Co trilayer grown on Cu(1 1 1) using MBE (left) and PLD. The images were taken as snapshots after the completion of each layer by STM. The roughness of the TD trilayer is much higher than that of the PLD trilayer. The Cu spacer layer does not completely separate the two Co layers in the TD trilayer. (b) Schematic views of the TD and PLD trilayer shown in (a). Magnetic measurements confirm that only the PLD trilayer has realized a complete antiferromagnetic coupling between the two Co layers.

antiparallel alignment (antiferromagnetic coupling) depending on the Cu interlayer thickness. For a complete antiferromagnetic coupling, the Cu interlayer has to be uniformly thick in space. In Fig. 2 we again use the example of Co/Cu trilayer to demonstrate that synthesis plays a key role to achieve desired magnetic properties of a GMR device. On a single crystal Cu with (1 1 1) surface orientation, Co/Cu/Co trilayers were grown by two techniques, i.e. molecular beam epitaxy (MBE or TD) and pulsed laser deposition (PLD). The major difference between these two growth techniques is the deposition rate, with PLD rate (during each pulse) being six orders of magnitude larger than that of the TD. From the surface topography images taken by scanning tunneling microscope (STM), as shown in Fig. 2(a), the TD grown Co/Cu trilayer is so rough that the Cu interlayer does not completely separates the two Co layers. The PLD trilayer, however, grows in a nearly ideal layer-by-layer manner and thus form a high quality trilayer with sharp Co/Cu interfaces. Consequently, the two Co layers are completely antiparallel aligned in the PLD trilayer, and only partially antiparallel aligned in the TD trilayer, as schematically shown in Fig. 2(b) [4]. This example clearly tells us that synthesis can make a real difference in magnetic properties of ASM.

Artificially structured magnetic materials are also critical to the future progress of the computer and electronics industries as they confront the fundamental barriers that stand in the way of continued exponential growth in integration density and circuit speed. Recently, a new breed of electronic devices has emerged, promising major breakthroughs in these areas. These new devices are called spintronic or magnetoelectronic devices. They utilize the fact that the current carriers (either electrons or holes) in electronic devices carry not only charge but also spin [5]. An illustrative example is the so-called magnetic random access memory (MRAM). The MRAM, based on the GMR effect or spin-dependent tunneling effect, has the advantage of being nonvolatile, yet with faster speed, higher density and lower cost, compared to the current dynamic random access memory (DRAM) devices, which are based on charged capacitors. Several major recording industries are working intensively on the MRAM projects and it is highly hoped that in the next few years PCs will all be "instant-on" computers since the MRAM, with its nonvolatile nature, does not require any booting time!

As we show in Figs. 1 and 2, the promise of these technologies can only be attained to the extent that we learn how to fabricate high quality magnetic thin films, multilayers or even smaller objects. This has not always been an easy task despite the rapidly improved techniques and knowledge of thin film growth. The major difficulty comes from the fact that there are many parameters that influence growth quality, because the ASM are often grown under conditions that are far from thermodynamic equilibrium. The quality of thin films depends on both the thermodynamics and growth kinetics. The presence of substrate often adds to the complexity.

To gain a fundamental understanding of the magnetic behavior of the ASM is another great challenge. This is because these materials often have a very different dimensionality, symmetry, or crystallographic structure compared to the traditional bulk magnets which we have known for a long time. For example, the thin films or multilayers are usually 2D, and sometimes they can have a crystallographic structure which does not exist in bulk at similar temperatures. As we will see later, by properly controlling the growth, 1D wires or 0D dots can also be fabricated. For years theoretician have speculated that reduced dimensionality will strongly affect the magnetism of magnetic materials [6]. Now with real samples in hand, scientists can finally study what kind of new magnetic behavior they would show. With this background, since the early 1980s, there have been a large number of research groups in the world working on magnetism of low-dimensional materials, and the observation of the GMR effect is one of the examples reflecting the significance and success of the research. With the ever improved knowledge of low-dimensional magnetism, particularly the correlation between the magnetic properties, electronic properties and structural properties, we are achieving an ability to design and to fabricate low-dimensional materials with desired magnetic properties, or, as summarized in our title, to tailor magnetism in ASM.

In this paper, we review some of the most important progress and successes in this part of the magnetism research in the past two decades. This, of course, cannot be done without addressing the synthesis of these ASM. In Section 2 we will discuss current knowledge about how to grow 2D films, 1D wires and 0D dots. While these low-dimensional objects remain to be ASM themselves, they are also building blocks of more complex artificial materials. The magnetic properties of these building blocks are reviewed in Section 3, and some general conclusions or comments on the correlation between structure and magnetism of these systems are presented. In Section 4 we address the fabrication of more complex artificial structures composed of these building blocks, and demonstrate how to achieve desired magnetic properties by strategically stacking them. The final section, Section 5, we project the future of designer magnets in various research and application directions.

2. Growth of low-dimensional magnets

2.1. 2D: epitaxial growth of ultrathin films

From magnetism point of view, the word "ultrathin" refers to the thickness regime within which the ferromagnetic exchange coupling strength is large enough to hold the magnetic moments parallel across the film thickness. The ultrathin limit, when expressed in film thickness, is typically of the order of 20–30 atomic layers for most magnetic thin film systems [7]. Because of this low thickness limit, small surface and interface roughness can have significant influence on the magnetic properties of the films. For this reason, it is highly desired to grow the films in a layer-by-layer manner.

The growth mode of the magnetic films, mostly 3d elements of Fe, Co, and Ni or their alloys, is determined by a number of key parameters. These include the lattice mismatch between the substrate and the film, the substrate temperature during growth, the growth rate, and the surface free energies of both the substrate and the film. Bauer has given a simple free energy argument, which includes a comparison of the surface free energy of the bare substrate (which exists before film growth) and the sum of the free energies of the film surface and film/substrate interface (which are created after film growth), to predict the growth mode of films [8]. The general trend is that the higher the surface free energy of the substrate compared to that of the film, the more the films tend to wet the substrate. Three typical growth modes have been classified, they are layer-by-layer growth (Frank van de Merve growth), 3D island growth (Vollmer–Webber growth), or monolayer growth followed by 3D island growth (Stranski–Kastanov growth).

Although Bauer's free energy argument is plausible and physically transparent, it is a rule for growth under equilibrium condition. In real world, more than often the films are grown under a condition that the substrate temperature is too low and (or) the deposition rate is too high for thin film systems to reach its thermodynamic equilibrium. The growth kinetics plays an important and sometimes a dominant role which generally makes the islands denser and smaller. This, provided the lattice mismatch between the film and the substrate is not too large, will assist some of the thin films to grow layer-by-layer which is otherwise not favored from the free energy consideration. Such kinetics-limited growth is particularly important for growing 2D magnetic films, since the magnetic elements such as Fe, Co and Ni all tend to have a higher surface free energy compared to the noble metals like copper, silver and gold, which are often used as templates.

Experimentally the most reliable way to verify the layer-by-layer growth mode is using electron diffraction technique such as reflection high energy electron diffraction (RHEED) or medium energy electron diffraction (MEED). In real experiments the diffraction geometry is set up in a way where the reflected electron beams from neighboring terraces separated by monolayer high steps diffract constructively (in-phase) or destructively (out-of-phase). Thus, the reflected beam intensity is directly related to the layer distribution of the films, and will oscillate with a periodicity of 1 ML. For example, in out-of-phase condition, the beam intensity will reach a minimum when the surface layer is half filled, and a maximum when the surface layer is fully filled. Fig. 3 shows the MEED intensity oscillations of Fe films grown on Cu(1 0 0) substrate at room temperature. As indicated in the figure, above 2 ML the MEED intensity starts to oscillate layer by layer. The lack of oscillation for the first layer is primarily due to the poor first layer growth [9]. The corresponding surface morphologies of the films at the arrow-marked points (3

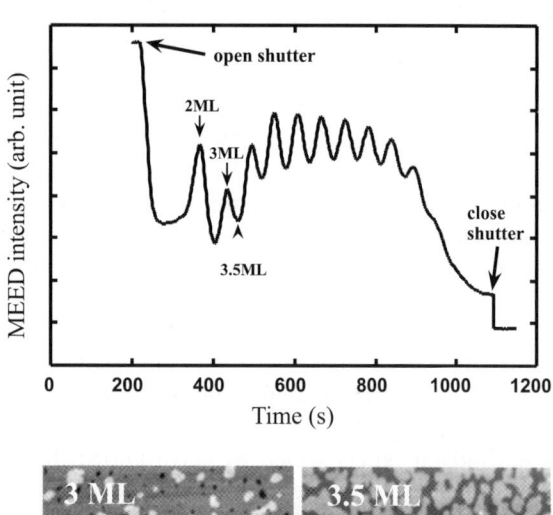

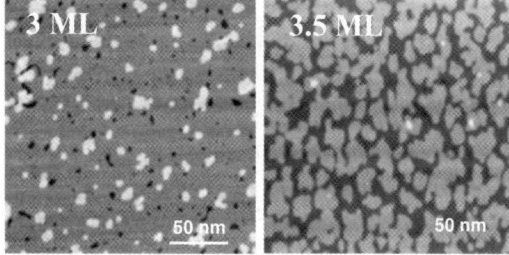

Fig. 3. Growth of Fe ultrathin films on Cu(1 0 0). Above 2 ML, Fe films grow layer by layer on Cu(1 0 0), as indicated by the periodic MEED oscillations (upper) and smooth STM surface morphology (lower).

and 3.5 ML) are shown by STM images below the MEED curve. Apparently the film grows layer by layer from 2 to 11 ML, as proved by both the MEED oscillations and the STM morphology, despite the fact that Fe has higher surface free energy than copper.

However, there do exist some magnetic thin film systems where layer-by-layer growth is extremely hard to achieve under usual growth conditions because of a rather large island edge diffusion barrier which hinders the necessary interlayer diffusion for the 2D growth. For example, Co has a 3D island growth mode on a Cu(1 1 1) substrate no matter what substrate temperature is used. The rough morphology of Co/Cu(1 1 1) films appears to be the main reason that MBE-grown {Co/Cu(1 1 1)} multilayers do not show the expected GMR effect [10,11]. To reduce the island edge diffusion barrier, one could either use surfactants such as Pb [12] or minimize the island size by ultrafast deposition [13], and in both cases good layer-by-layer growth of Co/Cu(1 1 1) films has obtained.

2.2. 1D: parallel atomic chains or stripes along the substrate steps

Magnetic 1D chains or stripes are distinguished from conventional magnetic nanowires by the fact that the former have a limited height and width which are less than few nanometers, while the latter can range from few to several hundred nanometers. As we pointed out earlier, for a ferromagnetic metal several nanometer is the length scale beyond which the individual magnetic moments are no longer all aligned in one direction. So far the growth of 1D magnetic chains or stripes has been mostly done using substrates with parallel aligned steps. The presence of steps can strongly alter the atomistic process of adatom diffusion. At a step edge the diffusion barrier tends to be higher than that on the terrace, hindering adatoms to cross over the steps. The local stress field around the step edges is also different from that on the regular terrace. In some cases the diffusion across the step edges is greatly suppressed resulting in a step decoration effect, i.e. deposited atoms stay in close vicinity to step edges. If the diffusion along the step edges is sufficiently active, 1D chains or stripes will be formed along the step edges. Fig. 4 shows the STM images of 1.0 ML Fe grown on flat (a) and vicinal with 1.2° miscut (b) Cu(1 1 1) substrates. On the flat substrate, Fe forms triangular-shaped islands with double- or triple-layer in height. On the vicinal surface, almost all Fe atoms are decorated along the step edges and form parallel stripes. The Fe stripes are typically 4 nm wide, 0.4 nm high and micrometer long along the step edges. The height and the width of the Fe stripes can be modified by

Fig. 4. Surface morphology of 1.0 ML Fe on flat (left) and vicinal (right) Cu(1 1 1) substrate. Scanning area is 200 × 200 nm². The Fe nanowires formed on the vicinal substrate are about 5 nm wide and 0.4 nm high.

changing the total dosage of Fe. Recently, 1D atomic chains of Co have been successfully grown on a high density stepped Pt(9 9 7) surface [14].

The step decoration effect is only applicable to a limited number of systems. If the substrates have a higher surface free energy than that of the film, a more general method can be used to grow 1D wires. This method is called step flow growth. The substrate should still be miscut to introduce parallel steps on the surface. The magnetic materials will then be deposited onto the substrate at a temperature which is high enough for the step-flow growth. If the dosage is a small fraction of a monolayer, parallel aligned stripes will be formed along the steps. 1D Fe stripes have been formed in this way on W(1 1 0) substrates [15].

While it is common to take advantage of step effect, there are alternative ways to grow 1D chains or stripes. The noteworthy ones include the formation of alternative 1D stripes of magnetic Co (or Fe) and nonmagnetic Ag on Mo(1 1 0) substrate [16], and deposition on grating NaCl(1 1 0) template at a glancing angle with respect to the surface [17]. Recently, it was observed that deposition at glancing angle on a flat substrate breaks the symmetry of islands [18], and can yield elongated islands with 1D appearance when the incident angle is close to 90° off surface normal at low temperatures [19].

2.3. 0D: ordered quantum dot arrays

Traditionally there are two ways to grow arrayed quantum dots (QDs), one is "serial" and the other is "parallel". The most typical serial way is using an STM tip to assist the deposition of magnetic clusters onto a chosen substrate. By precise control of the tip position, the QDs can be prepared one by one into any desired pattern [20,21]. Fig. 5 shows the morphology of Fe QD arrays prepared by STM-assisted chemical vapor deposition. After iron pentacarbonyl [Fe(CO)] gas is introduced into the UHV chamber, the STM tip had been moved into the location where the Fe QD is wished to be placed. A voltage pulse of about 10–20 V is applied between the tip and the sample. The local electric field dissociates the iron pentacarbonyl molecules. Iron clusters are depos-

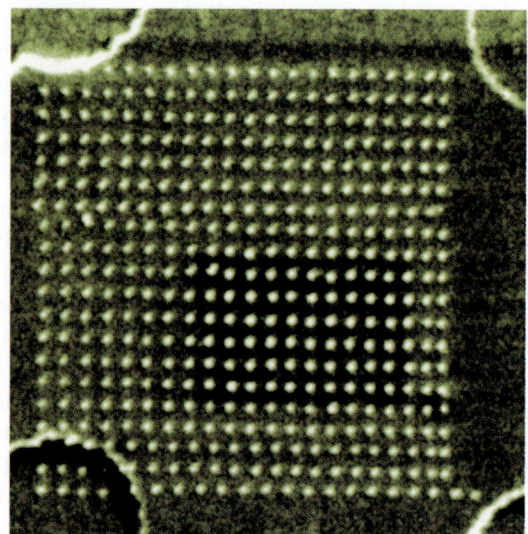

Fig. 5. Scanning electron microscope images (4.5×4.5 μm^2) of square Fe dot arrays grown by STM-assisted CVD onto a substrate of two-dimensional electron system (GaAs–Ga$_{0.7}$Al$_{0.3}$As) (image copied from Ref. [139]).

ited underneath the tip. The procedure is then repeated at all desired locations on the surface. A patterned QD array can be formed in this way.

The major disadvantage of the serial method is its slowness. Although the time scale appears to be acceptable for laboratory research, it is virtually impossible to use such a method in manufacturing applications, where mass production of QD arrays needs to be done in a reasonably short time. Therefore, much of the recent research effort has gone to the parallel approach—self assembling the QD arrays. So far most of the self assembling of QD arrays have been done based on periodic stress effect. For example, in some thin film systems dislocations form a network with a particular symmetry. If growing magnetic material on top of such kind of films, the dislocations often repel adsorbed atoms diffusing over the surface, and so they serve as templates for the confined nucleation of nanostructures from adatoms. Fig. 6 shows the QD array of Fe nucleated on the dislocation network of a Cu bilayer on Pt(1 1 1) at 250 K [22]. The lateral dimension of the Fe islands and their distance are as small as about 2 nm, which means a density of the order of 10 Terabit/in.^{2}. Similar

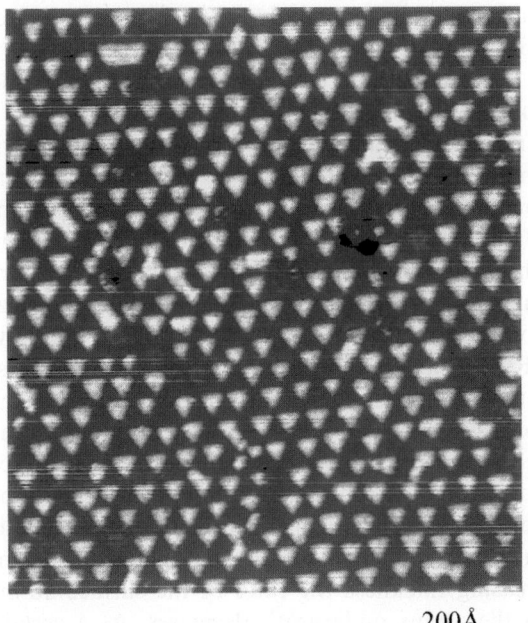

Fig. 6. STM image of a periodic array of Fe islands nucleated on the dislocation network of a Cu bilayer on Pt(1 1 1) at 250 K (image copied from Ref. [22]).

densities of magnetic dots has also been grown on a Au(1 1 1) $(23 \times \sqrt{3})$ reconstructed surface [23,24]. Teichert et al. have recently shown that it is possible to grow magnetic dot arrays by depositing magnetic materials at an glancing angle onto a semiconductor substrate with existing arrayed nanostructures, and an areal density of $0.25 \times 10^{12}/\text{in.}^2$ has been achieved [25].

The limitation of the parallel approach is that it is system specific and depends on whatever mother nature has offered to us. The key to creating a more general means to grow QD arrays is to eliminate the substrate effect. In the past, scientists have tried to use a cluster beam source to directly land QDs on substrates [26]. Another interesting approach is to adsorb inert gas elements (Xe, for example) on the substrate prior to deposition [27]. The adsorption temperature is kept low enough to have a solid phase on top of the substrate. The magnetic elements are subsequently deposited onto the inert buffer layer at cold temperature. The adatoms will have a very high mobility to easily form QDs on the buffer layer. The system is then gently warmed up to evaporate the buffer layer away, and the QDs will softly land on the real substrate. Very recently, it was suggested that the QDs will form a triangular lattice if they are electrically charged before the landing process [28].

3. Magnetism of low-dimensional materials

3.1. 2D ultrathin films

In the middle of the 1960s, Mermin and Wagner predicted that a 2D system would not exhibit long-range magnetic order at finite temperatures if the system is magnetically isotropic [29]. This prediction appeared to be confirmed by early experimental results on Ni thin films, where magnetic "dead layers" were claimed at the surface [30]. Although these results are now known to be an artifact due to the contamination in the film preparation process, they stimulated great interest in studying magnetic properties of surfaces and thin films. The follow-up experiments on various ultrathin films have provided overwhelming evidence to show that ferromagnetic order does exist even in monolayer films. The long-range order is now understood to be stabilized by either magnetic uniaxial anisotropy—an energy term which favors all magnetic moments aligned along one particular direction [31], or long-range dipolar interaction—an interaction between the magnetic moments similar to attraction or repulsion between two ordinary magnets [32]. Both the uniaxial anisotropy and the dipolar interaction suppress the long-wave length spin fluctuations at finite temperatures, which in turn stabilize the ferromagnetic order.

The existence of ferromagnetic order in ultrathin films obviously is great news for scientists who are interested in low-dimensional magnetism. In the past two decades, magnetic ultrathin films have been extensively studied and some striking differences between 2D ultrathin films and 3D bulk, as shown below, have been observed and understood. For ultrathin films, while the reduced dimensionality and the surface effect are the main reasons for their new magnetic properties, growth-linked parameters such as roughness and stress can also significantly affect the magnetic properties. In the following, we mention a number of important and general magnetic properties associated with 2D ultrathin films.

3.1.1. Surface magnetic anisotropy and spin reorientation

In general the magnetization direction of 2D ultrathin films tends to be parallel to the surface plane. This is because for a slab-shaped ultrathin film, if all the magnetic moments are aligned parallel to the film plane, there are no exposed magnetic free poles (much similar to the south and north poles of a permanent magnet) throughout the sample except at the very ends of the film. The number of magnetic free poles increases when the moments are aligned along any other directions. By having the least exposed magnetic free poles, the magnetostatic energy reaches the minimum for the system. Therefore, if there are no any other energy terms involved, the magnetization direction of an ultrathin film always lies in the film plane. This tendency towards in-plane magnetization, since it is originated from the slab shape of the film, is termed as shape anisotropy. Besides the shape anisotropy, there exist other energy terms that favor certain orientations of the magnetization. For example, for a magnet with crystallographic symmetry, there exists an energy term called magnetocrystalline anisotropy, which favors magnetization direction along certain major crystal symmetry axes. At the surface of a magnet, as first pointed out by Néel [33], due to the missing neighbor atoms, the magnetocrystalline anisotropy is generally one to two orders of magnitude larger than that of the bulk. This new energy term is often referred to as surface anisotropy, which may favor magnetization direction either parallel or perpendicular to the film plane depending on the system. In case the surface anisotropy favors perpendicular magnetization and is large enough to overcome the in-plane shape anisotropy, the magnetization direction of the film will be perpendicular to the film plane. Such perpendicular magnetization has been observed in numerous magnetic ultrathin film and multilayer systems [34–36], which have attracted great attention because of their potential use as high density recording materials.

The competition between the surface, shape and some other anisotropy energies gives rise to a striking magnetic phenomenon called spin reorientation (SRT) which involves 90° rotation of the magnetization direction from perpendicular to in-plane or vice versa. Experimentally it has been observed that the SRT can be driven by film thickness, temperature [37,38], and film chemical composition [39]. As an example, we explain how SRT occurs via increasing film thickness in a system with perpendicular surface anisotropy. At low thickness, the ratio of the surface atoms to the atoms underneath is large enough that the surface anisotropy wins the competition against the shape anisotropy. The magnetization of the system is thus perpendicular to the film plane. With increasing thickness the ratio of the surface atoms to the atoms underneath decreases. At a certain critical thickness, the surface anisotropy becomes equal to the shape anisotropy. Above this thickness, the shape anisotropy dominates and the system becomes in-plane magnetized. In most systems, the SRT is from perpendicular (at low thickness or low temperature) to in-plane (at high thickness or high temperature), and the critical thickness or critical temperature defines the point where the surface anisotropy just compensates the shape anisotropy. However, reversed SRT from in-plane to perpendicular has also been observed in Ni/Cu(1 0 0) system [40,41], whose origin was attributed to the involvement of an additional anisotropy energy other than the surface and shape anisotropy. For interested readers, the description of the reversed SRT is referred to the original references listed above.

From a thermodynamic point of view SRT represents a phase transition [42]. Consider the perpendicular component of the magnetization, or the angle of the magnetization with respect to the film plane, their change with increasing thickness or temperature can be of first order, when it proceeds via abrupt rotation of magnetization, or of second order when the magnetization rotation is continuous. In real systems the former is characterized by the coexistence of perpendicular and in-plane magnetic domains [43,44], and the latter is featured with a single domain with magnetization direction pointing at an angle between 0° and 90° with respect to the film plane. A recent study on ultrathin Fe/Gd(0 0 0 1) films has shown that both types of SRT occur in two steps with increasing temperature [45]. Comprehensive theoretical work has been carried out by Millev et al. [46].

3.1.2. Enhanced magnetic moment

The enhancement of the magnetic moment in a 2D system can be generally understood as a result of reduced nearest neighbor atoms. Take Fe as an example, the magnetic moment of an Fe atom is $4.0\mu_B$ in isolated form (no nearest neighbors), and is reduced to $2.2\mu_B$ in bulk crystal (eight nearest neighbors). A 2D Fe film, which has an intermediate number of nearest neighbors, should thus have a moment between that of the bulk and a free atom. Theoretical calculation of a free-standing Fe(1 0 0) single-atomic-layer (monolayer) thick film has indeed revealed a moment of $3.07\mu_B$ [47]. An Fe film has to be supported by a substrate. Since the electronic hybridization between film and substrate often counteract with the reduced dimensionality, the best substrates are those whose electronic structure is clearly separated from that of the substrate. In this respect Fe/MgO(1 0 0) serves as a good candidate since the Fe and Mg electronic structures are well separated. Indeed, the calculation performed on monolayer Fe on MgO(1 0 0) substrate yields magnetic moment of $3.04\mu_B$ [46], which is very close to that of the free-standing Fe(1 0 0) monolayer. Experimental measurements of magnetic moments of 2D films are generally very challenging, although torsion magnetometry and polarized neutron reflection have been applied successfully. Enhanced magnetic moments have been observed in ultrathin film systems such as Fe/W(1 1 0) [48] and Fe/Ag(1 0 0) [49].

As seen in Fig. 3, even for a thin film system with good layer-by-layer growth, the exposed surface layer is not ideally flat due to the presence of some islands. Since a reduced number of nearest neighbors tend to increase the magnetic moment of surface atoms, the moment of magnetic atoms at the edges of islands would be even larger because their nearest neighbors are further reduced. Albrecht et al. have compared the magnetic moments of Fe/W(1 1 0) films with rough surface (more step edge atoms) and smooth surface (less step edge atoms), and concluded that the moment enhancement at the step edge has a upper limit of about $1.1\mu_B$ [50]. Similar results have been obtained by nonlinear magneto-optical Kerr effect measurements on Co/Cu(1 0 0) films [51], where one monolayer oscillation period of the surface magnetization was observed due to variation of roughness with the same periodicity.

3.1.3. Curie temperature and critical behavior

The temperature dependence of magnetic order also depends upon the dimensionality of a magnet. The Curie temperature, which separates the ferromagnetic and paramagnetic phases, is determined by the strength of magnetic anisotropy and dipolar interaction because the aforementioned Mermin–Wagner theorem shows that isotropic exchange interaction gives no long-range order. Because the strength of both magnetic anisotropy and dipolar interaction are often smaller than the strength of exchange coupling, the Curie temperature is, not surprisingly, reduced for a 2D film. However, as pointed out by Erickson and Mills [52], the Curie temperature of the films increases with thickness and will be close to that of the bulk when the films reach several monolayer thickness. The thickness dependence of the Curie temperature is most properly expressed by finite-size scaling theories [53], which were backed by various experimental results on ultrathin films of Fe [54], Co [55], and Ni [56].

The temperature dependence of spontaneous magnetization near the Curie temperature is called the critical behavior, which is often characterized by the exponent of a phenomenological power-law fitting of the magnetization vs. temperature data. The smaller the power-law exponent, the faster the magnetization falls to zero when approaching the Curie temperature. The amplitude of the power-law exponent depends on the anisotropy and the dimensionality of the magnetic system, generally the lower the dimensionality, the smaller the power-law exponent. Huang et al. [57] have reported the values of the power-law exponents for various ultrathin films and their cross-over from 2D to 3D when increasing the film thickness.

3.1.4. Correlation between structure and magnetism

The magnetic properties and the structural properties of 2D ultrathin films are strongly correlated. Magnetic quantities such as magnetic moment, magnetic anisotropy and Curie temperature are closely linked to structural parameters such as lattice constant, strain, roughness etc. A good example to illustrate the magnetic moment–lattice constant correlation is the fcc Fe ultrathin films on Cu(1 1 1) substrate. Fig. 7 shows the thickness dependence of magnetic moment of Fe/Cu(1 1 1) films measured by the magneto-optical Kerr effect. The linear slope for films above 3 ML is about three times smaller than that of the films below 3 ML, indicating that the magnetic moment of the Fe films drops by a factor of 3 at about 3 ML or above. Detailed structural analysis by RHEED shows that the lattice constant of the films happens to become smaller also at 3 ML or above, which is the likely origin of the reduction of the moment [58].

The magnetic anisotropy in 2D films is affected by a number of structural parameters including the strain, roughness and intermixing between film and substrate. Epitaxially grown films, if they have a lattice mismatch with the substrate, their lattice constant will be stretched or compressed to match the substrate lattice in the lateral directions, which will result in a contraction or an expansion of the film lattice in the vertical direction (tetragonal distortion), respectively. This tetragonal distortion essentially breaks the crystallographic symmetry of the films. As a result, the strain-induced magnetic anisotropy shows up similarly to surface anisotropy. The strain-induced anisotropy can contribute significantly to the magnetization orientation in films, depending on both the sign and the amplitude of the lattice mismatch between the film and the substrate. We use an unusual SRT in Ni/Cu(1 0 0) ultrathin films to demonstrate the importance of the strain-induced anisotropy. As mentioned, the conventional SRT in ultrathin films involves 90° spin rotation from perpendicular to in-plane direction with growing film thickness. The Ni/Cu(1 0 0) films, however, have an interesting reversed SRT which goes from in-plane to perpendicular. It is now understood that the Ni/Cu(1 0 0) films have an in-plane surface anisotropy and a perpendicular strain-induced anisotropy [59]. The contribution of surface anisotropy is inversely proportional to the film thickness. Therefore above a critical thickness the perpendicular strain-induced anisotropy takes over despite of the in-plane shape anisotropy. This explains the reversed SRT and shows how significant the strain-induced anisotropy can be in an ultrathin film system.

The surface roughness has its influence on magnetic anisotropy by changing the effective surface symmetry or shape of the films. The surface symmetry is broken at step edges, whose density is directly associated with the surface roughness. Using a simple model, Bruno has argued that the existence of the step edge atoms would cause the reduction of the effective surface anisotropy [60]. The surface roughness also reduces the in-plane

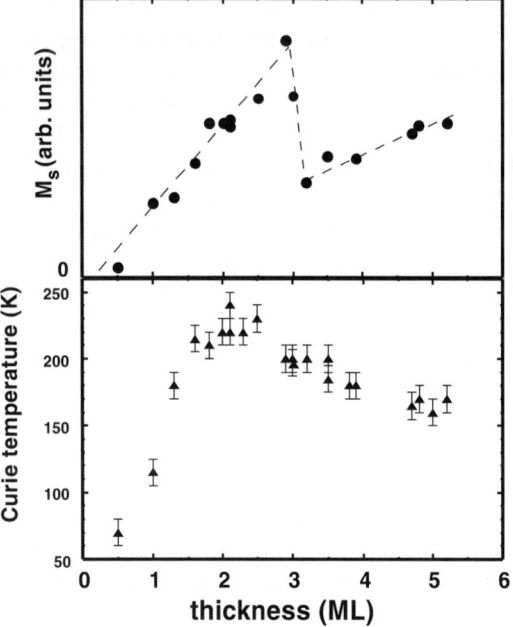

Fig. 7. Saturation magnetization along the surface normal (upper panel) and Curie temperature (lower panel) as a function of the thickness of the Fe/Cu(1 1 1) films prepared by PLD M_s increases linearly with thickness reaching its maximum at about 3 ML, and then drops to a smaller value. At higher thickness M_s again linearly increases but with a smaller slope. The change of the slope reflects a magnetic phase transition from a high moment ferromagnetic phase to a low moment ferromagnetic phase. The Curie temperature reaches a maximum at 2 ML, and then gradually decreases with increasing thickness.

shape anisotropy [61]. In an extreme case, the shape anisotropy could favor perpendicular magnetization if the roughness is high enough to show columnar islands [62]. For ordinary 2D films, the effect of roughness is generally too small to change the direction of magnetization, although it does appear to have some influence on the magnetization direction at the critical thickness of the SRT where other anisotropies of the film cancel out [63].

In some magnetic ultrathin film systems, the correlation between structure and magnetism can be complex. No other system shows such complexity better than the Fe/Cu(1 0 0) ultrathin film system. Initially this system attracted attention because at room temperature Fe grows on Cu in fcc structure [64], a high temperature phase which has been predicted to have rich magnetic phases by theories [65–67]. Early experimental results indeed indicated signs that more than one magnetic state exists in this system [68–70]. It was later found out, as seen in Fig. 8, that the system can be divided into three regions based on their magnetic states [71]. In region I (below 4 ML), the films have a uniform high moment (~2.5μ_B) ferromagnetic phase as shown by the linear increase of magnetic moments as a function of thickness and proved by other experimental measurement [72,73]. The high moment ferromagnetic phase was understood to be linked with increased atomic volume of Fe due to the tetragonal expansion [74]. In region II (4–11 ML), the magnetic moment of the film is nearly a factor of two smaller than that of the 4 ML film, and kept nearly constant for all thicknesses in region II. More careful experimental measurements indicates that the moment of the film actually oscillates with increasing thickness. The current understanding of such interesting magnetic behavior is that the top one or two layers of the films have a high-moment ferromagnetic phase, and the layers underneath have some sort of antiferromagnetic structure [75,76]. The structural origin of the coexistence of the ferromagnetic and antiferromagnetic phases have been discussed mainly based on the fact that the films undergo a structural transition from fct (tetragonally distorted fcc) to fcc for all layers except the top two in region II, although the real structure of the films is more complex due to a fcc to bcc Martensitic structural phase trans-

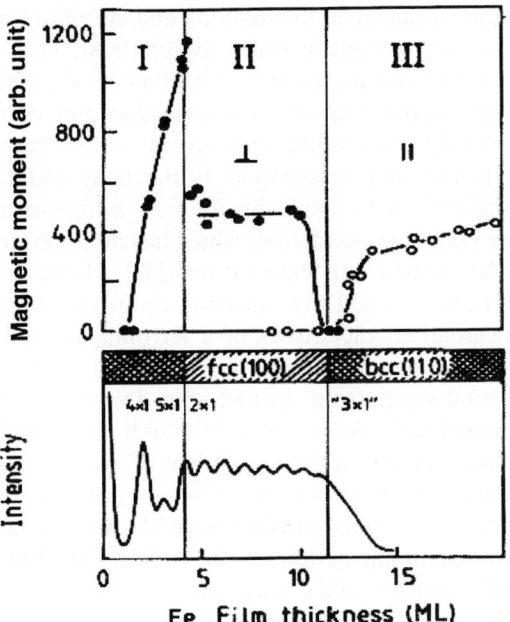

Fig. 8. Correlation between magnetism, structure and growth of Fe/Cu(1 0 0) ultrathin films. In the upper portion the measured magnetic moment at $T = 0$ K are depicted as a function of film thickness. Filled and open circles represent data recorded when external field is applied perpendicular ($\perp$) and parallel ($\|$) to the film plane, respectively. In the first two regions the magnetization is perpendicular to the film surface and parallel in region III. In the lower portion the intensity of the (0 0) MEED beam is depicted during Fe deposition showing the three different regions. The corresponding surface reconstructures for each region (4×1, 5×1 in region I, 2×1 in region II, and 3×1 in region III) are denoted as well. Between 4 and 11 ML in region II, the film is in fcc structure with (1 0 0) orientation (fcc(1 0 0)). Above 11 ML the films transform from the fcc(1 0 0) structure to a bcc structure (bcc(1 1 0)). (Figure copied from Ref. [140]).

formation [77,78]. In region III (>11 ML), the fcc to bcc structural transformation completes and as a result, the bcc films become again uniformly magnetized with an easy magnetization being parallel to the film plane. The precise location of the transition regions on the thickness scale is in addition influenced by adsorbates, e.g. carbon monoxide [79] and hydrogen [80].

3.2. 1D stripes

As mentioned, in two dimensions the magnetic anisotropy or even dipolar interaction can suppress magnetic fluctuations and stabilize long-range ferromagnetic order. By contrast, in a 1D magnetic system, theoretical arguments show that even a highly anisotropic system does not yield a zero-field equilibrium spontaneous magnetization [81]. The prototype model is the Ising chain of localized spins or spin blocks with nearest-neighbor exchange interaction, which has no long-range order at nonzero temperatures [82]. However, it was also realized that an Ising chain could show the usual characteristics of a ferromagnet at low temperatures, because of the separation of the two ground states (spin up and spin down) and that thermal activation is needed for spin rotation [83]. Experimentally a large number of work on 1D magnetism were performed on organic or inorganic compounds with chains of 3d ions. The large intra-chain interaction (on the order of 10–100 K) and a much weaker inter-chain interaction (typically 100 or 1000 times smaller) have made these compounds quasi-1D magnetic systems. These compound materials have provided a large playground to study some fundamental magnetism issues such as 1D magnetic excitations and correlation between magnetic order and chain–chain interactions [84–86]. In this article we discuss 1D stripes formed by 3d magnetic metals such as Fe, Co and Ni, which are potentially important for future spin electronic devices.

So far only a limited amount of work has been done on 1D magnetic stripes. The magnetic measurements were pioneered by Elmers et al. on Fe stripes grown on W(1 1 0) substrate [87]. These Fe stripes showed a strong in-plane anisotropy perpendicular to the stripe axis and a magnetic behavior similar to that of a 2D film with uniaxial anisotropy, despite their quasi-1D appearance. It was later argued by the same authors, based on a Monte Carlo simulation, that a inter-stripe dipolar interaction has helped to stabilize the long-range order, which was consequently termed as "dipolar superferromagnetism" [88]. Fig. 9 shows the morphology (a) and the sharp magnetic phase transition (b) of the Fe/W(1 1 0) stripes. Magnetic domain imaging has been recently performed on such Fe stripes [89].

Distinctly different magnetic behavior has been observed in 1D Fe stripes grown on a Cu(1 1 1)

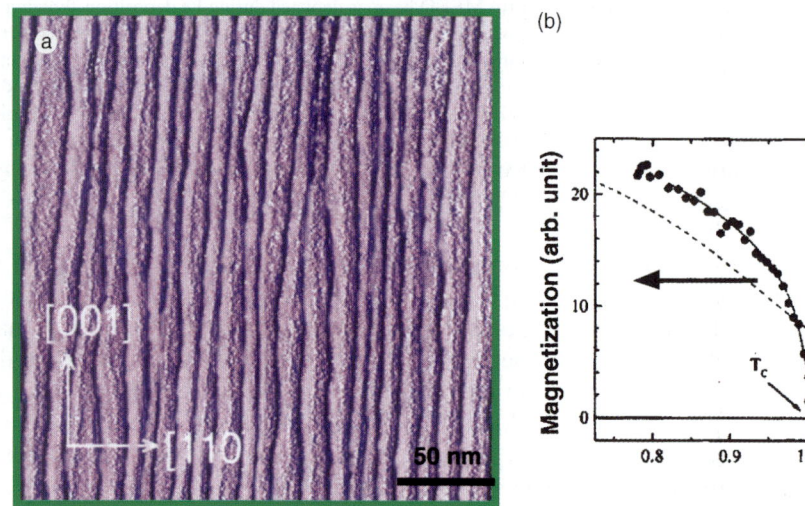

Fig. 9. (a) Differentiated STM image (250 × 250 nm²) of a vicinal W(1 1 0) surface with steps along [0 0 1], covered by 0.8 ML of Fe. Note that the lighter colored stripes adjacent to the step edges are the Fe 1D stripes. (b) Remanent magnetization (●) and saturation field (○) as a function of temperature ($T_c = 179$ K). The full line represents a fit by a power law of exponent $\beta = 0.32$, convoluted with a Gaussian distribution of mean value $T_c = 179$ K and width of $\Delta T_c = 1.7$ K. The dashed line shows the same function but with a broader distribution of width $\Delta T_c = 15$ K, as would be expected for noninteracting stripes. (Figure copied from Ref. [88]).

substrate [90]. Since the Fe/Cu(1 1 1) stripes have an easy magnetization axis perpendicular to the substrate surface, dipolar interaction between the wires would not favor ferromagnetic ordering in this case. Although the wires do show hysteresis loops at low temperatures, their magnetization, in the absence of external field, is not stable with respect to time. The differences between the Fe/W(1 1 0) and the Fe/Cu(1 1 1) stripes indicate that the interaction between the stripes can affect significantly the overall magnetism of 1D stripes.

Besides magnetic ordering, several other magnetic properties of 1D wires have been discussed in the literature. A theoretical calculation has shown very interesting magnetic anisotropic behavior for 1D chains [91]. Not only does the magnetic anisotropy energy appear to be one order of magnitude larger than that of the 2D film, the anisotropy direction could even oscillate from along the chain to perpendicular to the chain depending on the chain width. For thicker wires (on the order of micrometer scale in width), the presence of domain walls appear to be significant for both magnetization reversal [92] and the transport properties [93] of the wires.

3.3. 0D particles and dot arrays

A small magnetic particle is a quasi-0D object which has different magnetic properties compared to their 2D or 3D counterparts. With decreasing size, magnetic particles first become single domain when the gain in magnetostatic energy (by forming domains) is overcompensated by the energy required to form a domain wall. As a rule of thumb, this occurs when the size is smaller than the 3D domain wall width (0.1–1 µm). At even smaller size, the particle approaches the superparamagnetic limit (10–20 nm) where the thermal energy activates the spontaneous spin flip at room temperature. For such nanometer-sized particles, spin flip can occur even at very low temperatures (<10 K), at which the thermal activation may be completely suppressed, by means of magnetic quantum tunneling [94–97]. The stability of the magnetization of a particle with respect to magnetization direction becomes one of the major concerns for applications in high-density magnetic recording.

3.3.1. Isolated particles

The equilibrium of isolated single-domain particles at nonzero temperatures is generally described by well-established superparamagnetic theory [98–101]. Under this theory, the magnetic moments within each magnetic particle move coherently and can be treated as a single giant moment. The giant magnetic moment is coupled to magnetic anisotropy, which provides several equivalent equilibrium states depending on the actual type of the anisotropy. For an assembly of such kind of particles, the giant moments will have a balanced distribution in those equivalent states, leaving zero net magnetization of the system. An important concept associated with superparamagnetism is the blocking temperature, below which the thermal energy is not large enough to flip the giant moment within the measuring time. An operational definition of blocking temperature includes at least two requirements. First, above this temperature the magnetization curve must show no hysteresis loop. Second, the magnetization curves taken at different temperatures above the blocking temperature must superimpose when plotted against the magnetic field normalized by temperature after correction for the temperature dependence of the spontaneous magnetization. The particles can be considered as ferromagnetic particles below the blocking temperature, which is determined by the competition between the magnetic anisotropy barrier and thermal activation.

One of the characteristic features of a 0D particle is its large surface to volume ratio. It has been speculated that the large surface to volume ratio would give rise to enhanced magnetic moment and anisotropy. Early experiments on isolated Fe [102] and Co [103] particles embedded in nonmagnetic metals indicated only small changes of 0 K saturation magnetization as a function of the particle size. The measured results, however, are not necessarily connected with the intrinsic properties of 0D free clusters, since the electronic hybridization at particle/host interface could largely affect the surface magnetic moment. In the 1990s, there is growing interest of re-examining the magnetic moment of 0D clusters. These are free clusters made by cluster beam sources and their magnetic moments were measured by a conventional

Stern–Gerlach apparatus. Enhanced magnetic moments were reported for Fe [104], Co [105] and Ni [106] free clusters which typically contain several ten to several hundred atoms. Even Rh, which is nonmagnetic in the bulk, exhibits a sizable moment in cluster form [107]. In most cases the magnetic moment oscillates with the number of atoms contained in the cluster, reaching maxima or minima for open or close geometric shell, respectively. Since the open and close shells represent the largest and the smallest surface/volume ratio, respectively, this is a clear demonstration of the significance of the surface effect on the moment enhancement [108].

3.3.2. Magnetic dot arrays

While the fundamental 0D magnetic behavior can be addressed in studying the magnetic properties of isolated particles or free clusters, magnetic dot arrays attract the most attention due to their potential application in high density recording media. Various methods have been introduced to study the magnetization reversal process of the arrayed particles under external field and changing temperature [109,110]. When the dot arrays are dense enough, the interaction between the dots appears to have its influence on the magnetic properties. For example, the simplest dipolar interaction was found to increase the blocking temperature by effectively increasing the energy barrier between adjacent low energy states [111,112].

Since high blocking temperature is crucial for converting magnetic dot arrays into magnetic recording devices, it is necessary to increase the anisotropic barrier of the dots by all possible means. Fruchart et al. have recently further extended the idea of Co/Au(1 1 1) dot arrays to grow arrayed Co pillars on Au(1 1 1) substrate [113]. The principle is to deposit sequentially a fraction x of an atomic layer of Co and $1 - x$ atomic layers of Au on the Au(1 1 1) substrate, which yielded arrayed pillars with 2:1 vertical aspect ratio. Because of the large number of Co atoms per pillar, the Co pillars have a much higher superparamagnetic blocking temperature (300 K) than that of the double layer-high Co/Au(1 1 1) dots (∼20 K). In another approach, Sun et al. have used a chemical process to grow arrayed FePt dots, and were able to increase the blocking temperature from 20 to 30 K to higher than room temperature by thermal annealing. These authors have shown that the dot arrays support the magnetic reversal transition after a magnetic read–write process had been performed.

Besides magnetic recording, magnetic dot arrays have also been studied in terms magneto-optical response [114], vortex pinning in superconducting materials [115], and even possible application in magnetic refrigeration materials [116].

4. Tailoring magnetism in artificially structured materials

As shown above, the 2D, 1D and 0D objects all have interesting magnetic properties. The accumulated knowledge of their structural and electronic origins opens the route to modify their magnetic properties in a controllable and desired way. With such a capability to produce magnet design principles, low-dimensional magnetism has found itself right on the frontier of modern science and technology.

4.1. Spin engineering

The orientation of magnetization is one of the key features for a magnet. Magnetic properties of 2D magnetic thin films are often closely tied to their actual magnetization direction. If the magnetization direction can be oriented into any desired direction with high accuracy, i.e. spin engineering, magnetic thin films can be used with much higher efficiency in data storage and sensor devices.

Since the magnetization direction depends on the sign of the effective magnetic anisotropy constant, it is natural to control the spin direction by modifying the magnetic anisotropy. So far most attempts emphasized the modification of the surface anisotropy, which can be done by means of capping films with typically a fraction of a monolayer of magnetic [117] or nonmagnetic layers [118], or gas adsorption on the film surface [119]. In both cases the surface anisotropy constant can be changed controllably by controlling the coverage of the capping layers or absorbing gases. For example, Webber et al. have reported that three hundredths of a monolayer of copper capping layer is sufficient to rotate spins of a 20 ML thick Co/Cu(1 0 0) film by 90° [118]. This means that one nonmagnetic Cu atom can switch the spin direction of 500 Co atoms. This seemingly large effect of Cu atoms is understood as a result of Cu atoms preferentially locate at the step edges of Co surface, which affects the magnetic anisotropy in a most efficient way. The spin direction of the Co film can be finely tuned by further adding copper atoms, which eventually comes back to its original direction after 1 ML of copper has been deposited. Hope et al. have combined the effects of CO adsorption and Cu capping on Co/Cu(1 1 0) films, which make it possible to direct rotation of spins in a fine step between two perpendicular directions of [1 −1 0] and [0 0 1] [119].

4.2. Spin electronic devices

The field of spin electronics has been growing dramatically in recent years. The central idea of spin electronics is centered around the fact that conduction electrons (or holes) carry not only charge, but also spin. According to Mott's observation of electrical conduction in a ferromagnetic material [120], the current is carried by two distinct channels which consist of electrons with antiparallel and nonequal spins. Depending on the magnetization direction, electrons of one spin polarization are more heavily scattered than those with the other polarization. The recognition of this distinction is the key which promises to unlock a new generation of spin electronics devices whose operation relies on differential manipulation of independent families of current carriers with opposite spin polarization.

The simplest demonstration of a spin electronic device is a magnetic multilayer system which shows GMR. Due to the differential spin scattering at the ferromagnetic/nonmagnetic interface [121], the electrical resistance of a GMR multilayer is the smallest (largest) when two magnetic layers are parallel (antiparallel) magnetized. The change of the magnetoresistance can be more than 100% [122], making the GMR multilayers an excellent candidate to construct magnetic field sensors. In real applications, instead of using the normal GMR multilayers, a multilayer system called spin valve has been introduced to increase the field sensitivity. Fig. 10 shows the fundamental structure of a spin valve multilayer. A nonmagnetic (Cu) layer separates two ferromagnetic layers. The magnetization of one of the ferromagnetic layers (Py) is pinned by an uni-directional anisotropy, which is provided by the uni-directional exchange bias between a ferromagnetic/antiferromagnetic (Py/FeMn) interface. The other ferromagnetic layer is free to rotate under a small external field. The resistance of this device varies depending on the external field direction, i.e. a field sensor. The GMR spin valve so far has already been successfully used in commercial magnetic disk readheads as shown in Fig. 10, and is being considered as one of the top candidates for building nonvolatile MRAM.

An important concept associated with spin-dependent transport is called "spin accumulation", which is simply an excess of the number density of up-spins over down-spins and can, thus, be viewed as a divergence in the electrochemical potentials of the two spin channels. Using this effect, Johnson [123] has constructed a three-terminal spin transistor, which includes a nonmagnetic base layer sandwiched by a ferromagnetic emitter layer and a ferromagnetic collector layer. The spin accumulation effect makes the base layer to have a different chemical potential than the collector layer, and will either pump the current into or out of the collector, depending on the magnetization direction of the collector layer. Later a more sophisticated spin valve transistor, which uses two Si layers as emitter and collector, and a GMR multilayer as the base, also, has been demonstrated [124].

Datta and Das have proposed another spin electronic device called spin-injection high electron mobility transistor [125]. In their design, as seen in Fig. 11, the spin-polarized electrons are injected from a ferromagnetic metal pad into a 2D electron (2DEG) gas layer, which would be an inversion layer formed at the heterojunction between InAlAs and InGaAs. The 2DEG layer provides a high mobility channel, free of spin flip scattering. The spin-polarized electrons, when passing through the

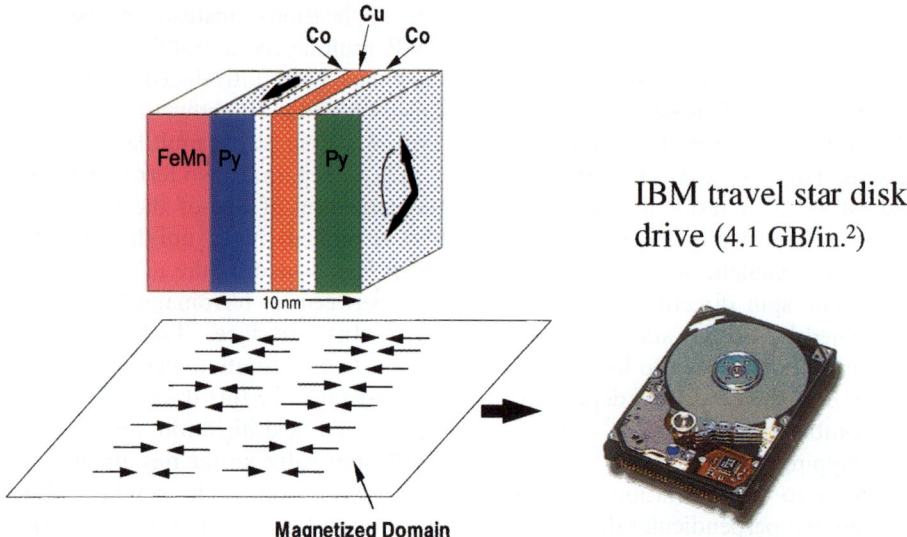

Fig. 10. GMR read sensor. Two magnetic layers (here each layer consists of a $Ni_{0.81}Fe_{0.19}$ alloy (Py) plus a few angstroms of Co) are separated by a Cu spacer layer. The moments of one of the magnetic layers is pinned by an FeMn antiferromagnetic layer. The other layer is free to rotate under the influence of the fringing fields from the magnetized domains on the spinning disk. As the magnetic moments of the free layer rotate, the resistance of the film changes allowing the information recorded in the magnetized domains to be read.

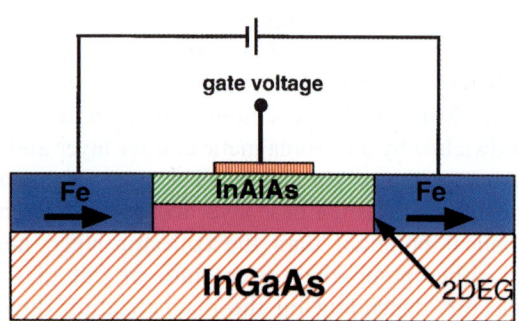

Fig. 11. Spin-polarized field-effect transistor scheme. A biased voltage applied between two ferromagnetic Fe pads (parallel magnetization) drives a spin-polarized electronic current through an interface layer (pink) between InAlAs and InGaAs. This interface layer is called a 2D electron gas layer, which does not flip electron spin directions of the current carriers. The spin direction of the current carriers can be tuned, however, by a gate voltage as shown in the figure. Similar to what has been described in Fig. 1, the resistivity varies depending on the relative orientation between the spin direction of the current carriers and the Fe pad (on the right hand side). A proof-of-principle experiment has been done by Hammar et al. (see Ref. [126]).

2DEG, would sense an effective magnetic field due to the electric field between the heterojunction, and hence will precess. This precession turns the spin direction somewhat off the magnetization direction of the spin injector, with the offset angle tunable by an external bias voltage applied to the heterojunction. Another ferromagnetic metal pad will be used as a detector located at the other end of the 2DEG layer. In this way the current passing through the 2DEG can be modulated by the bias voltage applied to the heterojunction, which is virtually a spin-polarized field-effect transistor. The most challenging part to realize this device is to successfully inject spin-polarized electrons from the ferromagnetic metal into the semiconductor, and good progress has been made recently [126].

4.3. Artificially ordered alloys

Low-dimensional magnetic alloys are attractive because these materials have, via composition, an additional adjustable parameter for tailoring materials with desired properties. The development of thin film growth techniques in ultrahigh vacuum has made it possible to fabricate artificial ordered alloys which may or may not exist in bulk equilibrium. New alloy phases can be achieved in thin film growth due to factors such as enhanced surface diffusion, symmetry and strain influence by the substrate, and the low dimensionality effect. For example, it has been demonstrated that the strain effect could result in compositionally modulated alloy structures in bimetallic system which are immiscible in bulk [127,128].

The fact that 2D films can be made atomically smooth is particularly useful to grow artificial layered alloys, by alternatively stacking nanometer-thick layers of magnetic and nonmagnetic metals. Striking magnetic properties such as perpendicular GMR [129] in Fe/Cr multilayers and large magneto-optical effect [130] in Fe/Pt alloys have been reported. The ultimate limit of such kind of layered alloys are the monoatomic stacked $L1_0$ structure, which exists in nature for some bulk alloys such as FePt. In bulk synthesis the disadvantage is that one has to live with the phase diagram nature has provided, so only a very limited kinds of magnetic $L1_0$ alloys can be produced. Using epitaxial growth, a variety of $L1_0$ alloys, which do not exist in nature, can be artificially made by stacking monoatomic metal layers alternatively. Magnetic $L1_0$ alloys such as Fe/Cu [131] and Fe/Au [132] have been successfully fabricated in this way.

5. Outlook

The emergence of low-dimensional magnetic materials in the last two decades of the 20th century has already had a tremendous impact on modern technology. Moving into the 21st century, major efforts will be spent to make devices based on these artificial magnetic materials even smaller in size and function at a faster speed. Meanwhile scientists will keep looking for new low-dimensional magnetic properties and constructing better artificial magnetic materials. With this in mind, we foresee that the following directions will likely generate new excitement.

5.1. Magnetic imaging with atomic resolution

Since the dimensions of magnetic materials is being made smaller and smaller, the current nanometer resolution of magnetic imaging techniques are becoming more and more dissatisfying for research. For most people spin-polarized STM (SPSTM) holds the most promise to solve this problem. The operating principle of SPSTM is close to that of the STM. In both cases a sharp tip is scanning across the surface and record the magnetic (SPSTM) or topographic (STM) signal. Since STM is well know to have atomic spatial resolution, SPSTM is highly expected to show similar capability. The major challenges for developing the SPSTM include the very small signal and the tip–sample interaction. The spin-polarized tunneling current is usually several percent or less of the total tunneling current, and thus can be easily hidden by the noise. Adapting the lock-in technique appears to be able to improve the signal-to-noise ratio, and some initial success has been achieved [133]. The tip–sample interaction is due to the high magnetic stray field induced by the magnetic tip, which is on the order of about 0.1 T. Under such a large field, the local magnetic moment direction will be disturbed or even flipped for many magnetic thin films. In order to suppress the tip–sample interaction, instead of using ferromagnetic tips, people have tried other kinds of tips such as nonmagnetic tips coated with a magnetic thin film [134], or a GaAs tip pumped by circularly polarized light [135,136]. These tips seem to affect the sample considerably less, and have been applied to image magnetic domain structures of ultrathin films. There are also discussions about using antiferromagnetic or superconducting materials as a STM tip probe, and a new concept of using a two-terminal nonmagnetic tip [137]. Despite the technical difficulties, the SPSTM research is currently being hotly pursued. A first report on atomic resolution has been given recently [88].

5.2. New artificially structured magnetic materials

Searching for new magnetic materials with a desired function has always been a major topic in the history of magnetism research. In most cases, a single magnetic element falls short to simultaneously match all aspects of technical demands. As a result, these days the best performed hard and soft magnetic materials are alloys or compounds

[138]. The traditional bulk synthesis of magnetic alloys or compounds are limited by the bulk phase diagram. The ASM, constructed layer by layer, row by row, or even atom by atom, can be made of elements which are immiscible in the bulk. For this reason the ASM could have a much better tunability. In addition, the ASM are low-dimensional materials whose size can be made small and their response time fast. Following the trend of modern technology to make devices smaller and faster, ASM is perfectly suited for the future development of magnetic and electronic devices. For example, spin electronic devices are anticipated to become a new generation of electronic devices in the near future [5]. The new materials also will cross-fertilize fundamental research. As suggested by history, new exciting physics are often linked to new materials.

Materials science will be a major thrust for technology in the 21st century. Magnetic materials will play a major role. Artificially structured magnetic materials are a new frontier of fundamental and applied research in magnetism.

Acknowledgements

Oak Ridge National Laboratory is managed by UT-Battelle, LLC, for the US Department of Energy under contract DE-AC05-00OR22725.

References

[1] D.G. McKendrick, R.F. Doner, S. Haggard, From Silicon Valley to Singapore—Location and Competitive Advantage in the Hard Disk Drive Industry, Stanford University Press, Stanford, CA, 2000.
[2] M.N. Baibich, J.M. Broto, A. Fert, F. Nguyen Van Dau, F. Petroff, P. Etienne, G. Creuzet, A. Friederich, J. Chazelas, Giant magnetoresistance of (001)Fe/(001)Cr magnetic superlattices, Phys. Rev. Lett. 61 (1988) 2472.
[3] G. Binasch, P. Grunberg, F. Saurenbach, W. Zinn, Enhanced magnetoresistance in layered magnetic-structures with antiferromagnetic interlayer exchange, Phys. Rev. B 39 (1989) 4828.
[4] M. Zheng, J. Shen, Ch.V. Mohan, P. Ohresser, J. Barthel, J. Kirschner, Suppression of the fcc-hcp stacking fault in Co/Cu(111) ultrathin films by pulsed laser deposition, Appl. Phys. Lett. 74 (1999) 425.
[5] G. Prinz, K. Hathaway, Magnetoelectronics, Phys. Today 48 (1995) 24.
[6] L. Valenta, Theorie der spontanen Magnetisierung, Phys. Status. Solidi 2 (1962) 112.
[7] B. Heinrich, J.A.C. Bland, Ultrathin Magnetic Structures I, Spinger, Berlin, 1994.
[8] E. Bauer, Phänomenologische Theorie der Kristallabscheidung an Oberflächen, Z. Kristallogr. 110 (1958) 372.
[9] D.D. Chambliss, K.E. Johnson, Using STM to understand diffraction oscillations for Fe growth on Cu(100), Surf. Sci. 313 (1994) 215.
[10] M.T. Johnson, R. Coehoorn, J.J. de Vries, N.W.E. McGee, J. aan de Stegge, P.J.H. Bloemen, Orientational dependence of the oscillatory exchange interaction in Co/Cu/Co, Phys. Rev. Lett. 69 (1992) 969.
[11] J. Camarero, T. Graf, J.J. de Miguel, R. Miranda, W. Kuch, M. Zharnikov, A. Dittschar, C.M. Schneider, J. Kirschner, Surfactant-mediated modification of the magnetic properties of Co/Cu(111) thin films and superlattices, Phys. Rev. Lett. 76 (1996) 4428.
[12] J.J. de Miguel, J. Camarero, J. de la Figuera, J.E. Prieto, R. Miranda, Tailoring epitaxial growth of low-dimensional magnetic heterostructures by means of surfactants, in: Z. Zhang, M.G. Lagally (Eds.), Morphological Organization Eptixial Growth and Removal, World Scientific, Singapore, 1998, p. 367.
[13] M. Zheng, J. Shen, J. Barthel, P. Ohresser, Ch.V. Mohan, J. Kirschner, Growth, structure and magnetic properties of Co ultrathin films on Cu(111) by pulsed laser deposition, J. Phys.-Condens. Matter 12 (2000) 783.
[14] P. Gambardella, M. Blanc, L. Bürgi, K. Kuhnke, K. Kern, Co growth on Pt(997): from monatomic chains to monolayer completion, Surf. Sci. 449 (2000) 93.
[15] H.J. Elmers, J. Hauschild, H. Höche, U. Gradmann, H. Bethge, D. Heuer, U. Köhler, Submonolayer magnetism of Fe(110) on W(110): finite width scaling of stripes and percolation between islands, Phys. Rev. Lett. 73 (1994) 898.
[16] E.D. Tober, R.F.C. Farrow, R.F. Marks, G. Witte, K. Kalki, D.D. Chambliss, Self-assembled lateral multilayers from thin film alloys of immiscible metals, Phys. Rev. Lett. 81 (1998) 1897.
[17] A. Sugawara, T. Coyle, G.G. Hembree, M.R. Scheinfein, Self-organized Fe nanowire arrays prepared by shadow deposition on NaCl(110), Appl. Phys. Lett. 70 (1997) 1043.
[18] S. van Dijken, L.C. Jorritsma, B. Poelsema, Steering-enhanced roughening during metal deposition at grazing incidence, Phys. Rev. Lett. 82 (1999) 4038.
[19] J. Zhong, E. Wang, Q. Niu, Z.Y. Zhang, Morphological symmetry breaking during epitaxial growth at grazing incidence, Phys. Rev. Lett. 84 (2000) 3895.
[20] D.D. Awschalom, M.A. McCord, G. Grinstein, Observation of macroscopic spin phenomena in nanometer-scale magnets, Phys. Rev. Lett. 65 (1990) 783.
[21] A.D. Kent, T.M. Shaw, S. von Molnar, D.D. Awschalom, Growth of high-aspect-ratio nanometer-scale magnets

with chemical-vapor-deposition and scanning-tunneling-microscopy, Science 262 (1993) 1249.
[22] H. Brune, M. Giovannini, K. Bromann, K. Kern, Self-organized growth of nanostructure arrays on strain-relief patterns, Nature 394 (1998) 451.
[23] D.D. Chambliss, R.J. Wilson, S. Chiang, Nucleation of ordered Ni island arrays on Au(1 1 1) by surface-lattice discolations, Phys. Rev. Lett. 66 (1991) 1721.
[24] B. Voigtlander, G. Meyer, N.M. Amer, Epitaxial growth of thin magnetic cobalt films on Au(1 1 1) studied by scanning tunneling microscopy, Phys. Rev. B 44 (1991) 10354.
[25] C. Teichert, J. Barthel, H.P. Oepen, J. Kirschner, Fabrication of nanomagnet arrays by shadow depostion on self-organized semiconductor substrates, Appl. Phys. Lett. 74 (1999) 588.
[26] K. Bromann, C. Felix, H. Brune, W. Harbich, R. Monot, J. Buttet, K. Kern, Controlled deposition of size-selected silver nanoclusters, Science 274 (1996) 956.
[27] L. Huang, S.J. Chey, J.H. Weaver, Buffer-layer-assisted growth of nanocrystals: Ag–Xe–Si(1 1 1), Phys. Rev. Lett. 80 (1998) 4095.
[28] Z.Y. Zhang, J.F. Wendelken, private communication.
[29] N.D. Mermin, H. Wagner, Absence of ferromagnetism or antiferromagnetism in one- or two-dimensional isotropic Heisenberg models, Phys. Rev. Lett. 17 (1966) 1133.
[30] L.N. Liebermann, J. Clinton, D.M. Edwards, J. Mathon, "Dead" layers in ferromagnetic transition metals, Phys. Rev. Lett. 25 (1970) 4.
[31] M. Bander, D.L. Mills, Ferromagnetism of ultrathin films, Phys. Rev. B 38 (1988) 12015.
[32] Y. Yafet, J. Kwo, E.M. Gyorgy, Dipole–dipole interactions and two-dimensional magnetism, Phys. Rev. B 33 (1986) 6519.
[33] L. Néel, Anisotropie magnetique superficielle et surstructures D'orientation, J. de Phys. et le Rad. 15 (1954) 225.
[34] U. Gradmann, J. Müller, Flat ferromagnetic, epitaxial 48Ni/52Fe(1 1 1) films of few atomic layers, Phys. Status Solidi 27 (1968) 313.
[35] P.F. Carcia, A.D. Meinhaldt, A. Suna, Perpendicular magnetic-anisotropy in Pd/Co thin-film layered structures, Appl. Phys. Lett. 47 (1985) 178.
[36] B.N. Engel, C.D. England, R.A. Van Leeuwen, M.H. Wiedmann, C.M. Falco, Interface magnetic anisotropy in epitaxial superlattices, Phys. Rev. Lett. 67 (1991) 1910.
[37] D.P. Pappas, K.-P. Kamper, H. Hopster, Reversible transition between perpendicular and in-plane magnetization in ultrathin films, Phys. Rev. Lett. 64 (1990) 3179.
[38] Z.Q. Qiu, J. Pearson, S.D. Bader, Asymmetry of the spin reorientation transition in ultrathin Fe films and wedges grown on Ag(1 0 0), Phys. Rev. Lett. 70 (1993) 1006.
[39] A. Dittschar, M. Zharnikov, W. Kuch, M.-T. Lin, C.M. Schneider, J. Kirschner, Composition-driven spin-reorientation transition in ferromagnetic alloy films, Phys. Rev. B 57 (1998) R3209.

[40] B. Schulz, K. Baberschke, Crossover from in-plane to perpendicular magnetization in ultrathin Ni/Cu(0 0 1) films, Phys. Rev. B 50 (1994) 13467.
[41] M. Farle, W. Platow, A.N. Anisimov, P. Poulopoulos, K. Baberschke, Anomalous reorientation phase transition of the magnetization in fct Ni/Cu(0 0 1), Phys. Rev. B 56 (1997) 5100.
[42] D. Pescia, V.L. Pokrovsky, Perpendicular versus in-plane magnetization in a 2D Heisenberg monolayer at finite temperatures, Phys. Rev. Lett. 65 (1990) 2599.
[43] M. Speckmann, H.P. Oepen, H. Ibach, Magnetic domain structures in ultrathin Co/Au(1 1 1): on the influence of film morphology, Phys. Rev. Lett. 75 (1995) 2035.
[44] H.P. Oepen, M. Speckmann, Y. Millev, J. Kirschner, Unified approach to thickness-driven magnetic reorientation transitions, Phys. Rev. B 55 (1997) 2752.
[45] C.S. Arnold, D.P. Pappas, A.P. Popov, Second- and first-order phase transitions in the magnetic reorientation of ultrathin Fe on Gd, Phys. Rev. Lett. 83 (1999) 3305.
[46] Y.T. Millev, H.P. Oepen, J. Kirschner, Influence of external field on spin reorientation transitions in uniaxial ferromagnets. I. General analysis for bulk and thin-film systems, Phys. Rev. B 57 (1998) 5837;
Y.T. Millev, H.P. Oepen, J. Kirschner, Influence of external fields on spin reorientation transitions in uniaxial ferromagnets. II. Ultrathin ferromagnetic films, Phys. Rev. B 57 (1998) 5848.
[47] C. Li, A.J. Freeman, Giant monolayer magnetization of Fe on MgO: a nearly ideal two-dimensional magnetic system, Phys. Rev. B 43 (1991) 780.
[48] U. Gradmann, M. Przybylski, H.J. Elmers, G. Liu, Ferromagnetism in the thermodynamically stable monolayer Fe(1 1 0) on W(1 1 0), coated by Ag, Appl. Phys. A 49 (1989) 563.
[49] J.A.C. Bland, R.D. Bateson, B. Heinrich, Z. Celinski, H.J. Lauter, Spin-polarized neutron reflection studies of ultrathin magnetic-films, J. Magn. Magn. Mater. 104 (1992) 1909.
[50] M. Albrecht, U. Gradmann, T. Furubayashi, W.A. Harrison, Magnetic-moments in rough Fe surfaces, Europhys. Lett. 20 (1992) 65.
[51] Q.Y. Jin, H. Regensburger, R. Vollmer, J. Kirschner, Periodic oscillations of the surface magnetization during the growth of Co films on Cu(0 0 1), Phys. Rev. Lett. 80 (1998) 4056.
[52] R.P. Erickson, D.L. Mills, Anisotropy-driven long-range order in ultrathin ferromagnetic films, Phys. Rev. B 43 (1991) 11527.
[53] A.E. Ferdinand, M.E. Fisher, Bounded and inhomogeneous Ising models. I. Specific-heat anomaly of a finite lattice, Phys. Rev. 185 (1969) 832;
G.A.T. Allan, Critical temperature of Ising lattice films, Phys. Rev. B 1 (1970) 352;
M.E. Fisher, M.N. Barber, Scaling theory for finite-size effects in the critical region, Phys. Rev. Lett. 28 (1972) 1516;

[54] K. Binder, P.C. Hohenberg, Surface effects on magnetic phase-transitions, Phys. Rev. B 9 (1974) 2194.
[54] M. Stampanoni, A. Vaterlaus, M. Aeschlimann, F. Meier, Magnetism of epitaxial bcc iron on Ag(001) observed by spin-polarized photoemission, Phys. Rev. Lett. 59 (1987) 2483.
[55] C.M. Schneider, P. Bressler, P. Schuster, J. Kirschner, J.J. de Miguel, R. Miranda, Curie temperature of ultrathin films of fcc-cobalt epitaxially grown on atomically flat Cu(100) surfaces, Phys. Rev. Lett. 64 (1990) 1059.
[56] R. Bergholz, U. Gradmann, Structure and magnetism of oligatomic Ni(111)-films on Re(0001), J. Magn. Magn. Mater. 45 (1984) 389.
[57] F. Huang, M.T. Kief, G.J. Mankey, R.F. Willis, Magnetism in the few-monolayers limit: a surface magneto-optic Kerr-effect study of the magnetic behavior of ultrathin films of Co, Ni, and Co–Ni alloys on Cu(100) and Cu(111), Phys. Rev. B 49 (1994) 3962.
[58] P. Ohresser, J. Shen, J. Barthel, M. Zheng, Ch.V. Mohan, M. Klaua, J. Kirschner, Growth, structure and magnetism of fcc Fe ultrathin films on Cu(111) by pulsed laser deposition, Phys. Rev. B 59 (1999) 3696.
[59] M. Farle, B. Mirwald-Schulz, A.N. Anisimov, W. Platow, K. Baberschke, Higher-order magnetic anisotropies and the nature of the spin-reorientation transition in face-centered-tetragonal Ni(001)/Cu(001), Phys. Rev. B 55 (1997) 3708.
[60] P. Bruno, Magnetic surface anisotropy of cobalt and surface roughness effects within Neel's model, J. Phys. F: Phys. 18 (1988) 1291.
[61] P. Bruno, Dipolar magnetic surface anisotropy in ferromagnetic thin films with interfacial roughness, J. Appl. Phys. 64 (1988) 3153.
[62] Y. Yafet, E.M. Gyorgy, L.R. Walker, Directional dependence of the demagnetizing energy in imperfect thin magnetic-films, J. Appl. Phys. 60 (1986) 4236.
[63] M. Zheng, J. Shen, P. Ohresser, Ch.V. Mohan, M. Klaua, J. Barthel, J. Kirschner, Influence of growth temperature on the spin reorientation of Ni/Cu(100) ultrathin films, J. Appl. Phys. 85 (1999) 5060.
[64] U. Gradmann, P. Tillmanns, Supersaturation and mode of growth for Fe films on Cu(111)—experimental-study using LEED and AES, Phys. Status Solidi (a) 44 (1977) 539.
[65] C.S. Wang, B.M. Klein, H. Krakauer, Theory of magnetic and structural ordering in iron, Phys. Rev. Lett. 54 (1985) 1852.
[66] F.J. Pinski, J. Staunton, B.L. Gyorffy, D.D. Johnson, G.M. Stocks, Ferromagnetism versus antiferromagnetism in face-centered-cubic iron, Phys. Rev. Lett. 56 (1986) 2096.
[67] V.L. Moruzzi, P.M. Marcus, K. Schwarz, P. Mohn, Ferromagnetic phases of bcc and fcc Fe, Co, and Ni, Phys. Rev. B 34 (1986) 1784.
[68] A. Amiri Hezaveh, G. Jennings, D. Pescia, R.F. Willis, K. Prince, M. Surman, A. Bradshaw, Quenching of exchange splitting in face-centered cubic Fe observed by angle resolved photoemission, Solid State Commun. 57 (1986) 329.
[69] M.F. Onellion, C.L. Fu, M.A. Thompson, J.L. Erskine, A.J. Freeman, Electronic structure and properties of epitaxial Fe on Cu(100): theory and experiment, Phys. Rev. B 33 (1986) 7322.
[70] P.A. Montano, G.W. Fernando, B.R. Cooper, E.R. Moog, H.M. Naik, S.D. Bader, Y.C. Lee, Y.N. Darici, H. Min, J. Marcano, Two magnetically different, closely lying states of fcc iron grown on copper (100), Phys. Rev. Lett. 59 (1987) 1041.
[71] J. Thomassen, F. May, B. Feldmann, M. Wuttig, H. Ibach, Magnetic live surface layers in Fe/Cu(100), Phys. Rev. Lett. 69 (1992) 3831.
[72] D.J. Keavney, D.F. Storm, J.W. Freeland, I.L. Grigorov, J.C. Walker, Site-specific Mössbauer evidence of structure-induced magnetic phase transition in fcc Fe(100) thin films, Phys. Rev. Lett. 74 (1995) 4531.
[73] R.D. Ellerbrock, A. Fuest, A. Schatz, W. Keune, R.A. Brand, Mössbauer effect study of magnetism and structure of fcc-like Fe(001) films on Cu(001), Phys. Rev. Lett. 74 (1995) 3053.
[74] S. Müller, P. Bayer, C. Reichel, K. Heinz, B. Feldmann, Z. Zillgen, M. Wuttig, Structural instability of ferromagnetic fcc Fe films on Cu(100), Phys. Rev. Lett. 74 (1995) 765.
[75] D. Li, M. Freitag, J. Pearson, Z.Q. Qiu, S.D. Bader, Magnetic phases of ultrathin Fe grown on Cu(100) as epitaxial wedges, Phys. Rev. Lett. 72 (1994) 3112.
[76] Ch. Würsch, C.H. Back, L. Bürgi, U. Ramsperger, A. Vaterlaus, U. Maier, D. Pescia, P. Politi, M.G. Pini, A. Rettori, Direct observation of antiferromagnetic phase transition in fcc Fe films, Phys. Rev. B 55 (1997) 5643.
[77] J. Giergiel, J. Kirschner, J. Landgraf, J. Shen, J. Woltersdorf, Stages of structural transformation in iron thin film growth on copper (100), Surf. Sci. 310 (1994) 1.
[78] K. Kalki, D.D. Chambliss, K.E. Johnson, R.J. Wilson, S. Chiang, Evidence for martensitic fcc–bcc transition of thin Fe films on Cu(100), Phys. Rev. B 48 (1993) 18344.
[79] A. Kirilyuk, J. Giergiel, J. Shen, M. Straub, J. Kirschner, Growth of stabilized gamma—Fe films and their magnetic properties, Phys. Rev. B 54 (1996) 1050.
[80] R. Vollmer, J. Kirschner, Influence of H_2 adsorption on magnetic properties of Fe films on Cu(001), Phys. Rev. B 61 (2000) 4146.
[81] G.F. Newell, E.W. Montroll, On the theory of the Ising models of ferromagnetism, Revs. Modern Phys. 25 (159) (1953) 353.
[82] E. Ising, Beitrag zur theorie des Ferromagnetismus, Z. Phys. 31 (1925) 253.
[83] I.S. Jacobs, C.P. Bean, Fine particles, thin films and exchange anisotropy, in: G.T. Rado, H. Suhl (Eds.), Magnetism, vol. III, Academic Press, New York, London, 1963.
[84] M. Hase, I. Terasaki, K. Uchinokura, Observation of the spin-Peierls transition in linear Cu^{2+} (spin-1/2) chains in

an inorganic compound $CuGeO_3$, Phys. Rev. Lett. 70 (1993) 3651.
[85] F.D.M. Haldane, Non-linear field-theory of large-spin Heisenberg anti-ferromagnets—semi-classically quantized solitons of the one-dimensional easy-axis Neel state, Phys. Rev. Lett. 50 (1983) 1153.
[86] K. Katsumata, H. Hori, T. Takeuchi, M. Date, A. Yamagishi, J.P. Renard, Magnetization process of an $S = 1$ linear-chain Heisenberg antiferromagnet, Phys. Rev. Lett. 63 (1989) 86.
[87] H.J. Elmers, J. Hauschild, H. Höche, U. Gradmann, H. Bethge, D. Heuer, U. Köhler, Submonolayer magnetism of Fe(1 1 0) on W(1 1 0): finite width scaling of stripes and percolation between islands, Phys. Rev. Lett. 73 (1994) 898.
[88] J. Hauschild, H.J. Elmers, U. Gradmann, Dipolar superferromagnetism in monolayer nanostripes of Fe(1 1 0) on vicinal W(1 1 0) surfaces, Phys. Rev. B 57 (1998) R677.
[89] O. Pietzsch, A. Kubetzka, M. Bode, R. Wiesendanger, Real-space observation of dipolar antiferromagnetism in magnetic nanowires by spin-polarized scanning tunneling spectroscopy, Phys. Rev. Lett. 84 (2000) 5212.
[90] J. Shen, R. Skomski, M. Klaua, H. Jenniches, S. Sundar Manoharan, J. Kirschner, Magnetism in one dimension: Fe on Cu(1 1 1), Phys. Rev. B 56 (1997) 2340.
[91] J. Dorantes-Dávila, G.M. Pastor, Magnetic anisotropy of one-dimensional nanostructures of transition metals, Phys. Rev. Lett. 81 (1998) 208.
[92] W. Wernsdorfer, B. Doudin, D. Mailly, K. Hasselbach, A. Benoit, J. Meier, J.-Ph. Ansermet, B. Barbara, Nucleation of magnetization reversal in individual nanosized nickel wires, Phys. Rev. Lett. 77 (1996) 1873.
[93] U. Ruediger, J. Yu, S. Zhang, A.D. Kent, S.S.P. Parkin, Negative domain wall contribution to the resistivity of microfabricated Fe wires, Phys. Rev. Lett. 80 (1998) 5639.
[94] D.D. Awschalom, M.A. McCord, G. Grinstein, Observation of macroscopic spin phenomena in nanometer-scale magnets, Phys. Rev. Lett. 65 (1990) 783.
[95] L. Thomas, F. Lionti, R. Ballou, D. Gatteschi, R. Sessoli, B. Barbara, Macroscopic quantum tunnelling of magnetization in a single crystal of nanomagnets, Nature 383 (1996) 145.
[96] E.M. Chudnovsky, L. Gunther, Quantum tunneling of magnetization in small ferromagnetic particles, Phys. Rev. Lett. 60 (1988) 661;
E.M. Chudnovsky, Macroscopic quantum tunneling of the magnetic-moment, J. Appl. Phys. 73 (1993) 6697.
[97] C. Sangregorio, T. Ohm, C. Paulsen, R. Sessoli, D. Gatteschi, Quantum tunneling of the magnetization in an iron cluster nanomagnet, Phys. Rev. Lett. 78 (1997) 4645.
[98] E.C. Stoner, E.P. Wohlfarth, A mechanism of magnetic hysteresis in heterogeneous alloys, Philos. Trans. R. Soc. London A 240 (1948) 599.
[99] L. Néel, Theorie du trainage magnetique des ferromagnetiques en grains fins avec applications aux terres cuites, Ann. Geophys. 5 (1949) 99.

[100] W.F. Brown, Thermal fluctuation of a single domain particle, Phys. Rev. 130 (1963) 1677.
[101] A. Aharoni, I. Eisenstein, Theoretical relaxation-times of large superparamagnetic particles with cubic anisotropy, Phys. Rev. B 11 (1975) 514.
[102] J.J. Becker, Magnetic method for the measurement of precipitate particle sizes in a Cu–Co alloy, Trans. Am. Inst. Mining. Met. Petrol. Eng. 209 (1957) 59;
C.P. Bean, J.D. Livingston, D.S. Rodbell, The anisotropy of very small Co particles, J. Phys. et le Radium 20 (1959) 298.
[103] F.E. Luborsky, P.E. Lawrence, Saturation magnetization and size of iron particles less than 100 Å in diameter, J. Appl. Phys. (Suppl.) 32 (1961) 231S;
F.E. Luborsky, Loss of exchange coupling in the surface layers of ferromagnetic particles, J. Appl. Phys. 29 (1958) 309.
[104] I.M.L. Billas, A. Chatelain, W.A. de Heer, Magnetism from the atom to the bulk in iron, cobalt, and nickel clusters, Science 265 (1994) 1682.
[105] J.P. Bucher, D.C. Douglass, L.A. Bloomfield, Magnetic properties of free cobalt clusters, Phys. Rev. Lett. 66 (1991) 3052.
[106] S.E. Apsel, J.W. Emmert, J. Deng, L.A. Bloomfield, Surface-enhanced magnetism in nickel clusters, Phys. Rev. Lett. 76 (1996) 1441.
[107] A.J. Cox, J.G. Louderback, L.A. Bloomfield, Experimental observation of magnetism in rhodium clusters, Phys. Rev. Lett. 71 (1993) 923.
[108] C.P. Bean, J.D. Livingston, Superparamagnetism, J. Appl. Phys. (Suppl.) 30 (1959) 120S.
[109] S. Wirth, S. von Molnar, M. Field, D.D. Awschalom, Magnetism of nanometer-scale iron particles arrays, J. Appl. Phys. 85 (1999) 5249.
[110] R.P. Cowburn, D.K. Koltsov, A.O. Adeyeye, M.E. Welland, D.M. Tricker, Single-domain circular nanomagnets, Phys. Rev. Lett. 83 (1999) 1042.
[111] S. Mørup, E. Tronc, Superparamagnetic relaxation of weakly interacting particles, Phys. Rev. Lett. 72 (1994) 3278.
[112] J. Garcia-Otero, M. Porto, J. Rivas, A. Bunde, Influence of dipolar interaction on magnetic properties of ultrafine ferromagnetic particles, Phys. Rev. Lett. 84 (2000) 167.
[113] O. Fruchart, M. Klaua, J. Barthel, J. Kirschner, Self-organized growth of nanosized vertical magnetic Co pilars on Au(1 1 1), Phys. Rev. Lett. 83 (1999) 2769.
[114] H. Takeshita, Y. Suzuki, H. Akinaga, W. Mizutani, K. Tanaka, T. Katayama, A. Itoh, Magneto-optical response of nanoscaled cobalt dots array, Appl. Phys. Lett. 68 (3040) 1996.
[115] J.I. Martin, M. Vélez, A. Hoffmann, I.K. Schuller, J.L. Vicent, Artificially induced reconfiguration of the Vortex lattice by arrays of magnetic dots, Phys. Rev. Lett. 83 (1022) 1999.
[116] R.D. McMichael, R.D. Shull, L.J. Swartzendruber, L.H. Bennett, R.E. Watson, Magnetocaloric effect in superparamagnets, J. Magn. Magn. Mater. 111 (1992) 29.

[117] J. Shen, A.K. Swan, J.F. Wendelken, Determination of critical thickness of spin orientation in metastable magnetic ultrathin films, Appl. Phys. Lett. 75 (1999) 2987.
[118] W. Webber, C.H. Back, A. Bischof, D. Pescia, R. Allenspach, Magnetic switching in cobalt films by adsorption of copper, Nature 374 (1995) 788.
[119] S. Hope, E. Gu, B. Choi, J.A.C. Bland, Spin engineering in ultrathin Cu/Co/Cu(110), Phys. Rev. Lett. 80 (1998) 1750.
[120] N.F. Mott, The electrical conductivity of transition metals, Proc. R. Soc. A 153 (1936) 699.
[121] S.S.P. Parkin, Origin of enhanced magnetoresistance of magnetic multilayers: spin-dependent scattering from magnetic interface states, Phys. Rev. Lett. 71 (1993) 1641.
[122] S.S.P. Parkin, 100% GMR in (110) Co/Cu Multilayers, unpublished work.
[123] M. Johnson, Bipolar spin switch, Science 260 (1993) 320.
[124] D.J. Monsma, J.C. Lodder, Th.J.A. Popma, B. Dieny, Perpendicular hot electron spin-valve effect in a new magnetic field sensor: the spin-valve transistor, Phys. Rev. Lett. 74 (1995) 5260.
[125] S. Datta, B. Das, Electronic analog of the electrooptic modulator, Appl. Phys. Lett. 56 (1990) 665.
[126] P.R. Hammar, B.R. Bennett, M.J. Yang, M. Johnson, Observation of spin injection at a ferromagnet-semiconductor interface, Phys. Rev. Lett. 83 (1999) 203.
[127] R.Q. Hwang, Chemically induced step edge diffusion barriers: dendritic growth in 2D alloys, Phys. Rev. Lett. 76 (1996) 4757.
[128] E.D. Tober, R.F.C. Farrow, R.F. Marks, G. Witte, K. Kalki, D.D. Chambliss, Self-assembled lateral multilayers from thin film alloys of immiscible metals, Phys. Rev. Lett. 81 (1998) 1897.
[129] M.A.M. Gijs, S.K.J. Lenczowski, J.B. Giesbers, Perpendicular giant magnetoresistance of microstructured Fe/Cr magnetic multilayers from 4.2 to 300 K, Phys. Rev. Lett. 70 (1993) 3343.
[130] S. Mitani, K. Takanashi, M. Sano, H. Fujimori, A. Osawa, H. Nakajima, Perpendicular magnetic-anisotropy and magnetooptical Kerr rotation in FePt(001) monoatomic multilayers, J. Magn. Magn. Mater. 148 (1995) 163.
[131] S.S. Manoharan, M. Klaua, J. Shen, J. Barthel, H. Jenniches, J. Kirschner, Artificially ordered Fe–Cu alloy supperlattices on Cu(100): I. Studies on the structural and magnetic properties, Phys. Rev. B 58 (1998) 8549;
W. Kuch, M. Salvietti, X.Y. Gao, M.-T. Lin, M. Klaua, J. Barthel, Ch.V. Mohan, J. Kirshchner, Artificially ordered Fe–Cu alloy supperlattices on Cu(100): II. Spin-resolved electronic properties and magnetic dichroism, Phys. Rev. B 58 (1998) 8556.
[132] K. Takanashi, S. Mitani, M. Sano, H. Fujimori, H. Nakajima, A. Osawa, Artificial fabrication of an L10-type ordered FeAu alloy by alternative monatomic deposition, Appl. Phys. Lett. 67 (1995) 1016.
[133] W. Wulfhekel, J. Kirschner, Spin-polarized scanning tunneling microscopy on ferromagnets, Appl. Phys. Lett. 75 (1999) 1944.
[134] M. Bode, R. Pascal, R. Wiesendanger, Scanning tunneling spectroscopy of Fe/W(110) using iron covered probe tips, J. Vac. Sci. Technol. A 15 (1997) 1285.
[135] S.F. Alvarado, P. Renaud, Observation of spin-polarized-electron tunneling from a ferromagnet into GaAs, Phys. Rev. Lett. 68 (1992) 1387;
S. Alvarado, Tunneling potential barrier dependence of electron spin polarization, Phys. Rev. Lett. 75 (1995) 513.
[136] Y. Suzuki, W. Naghan, K. Tanaka, Magnetic domains of cobalt ultrathin films observed with a scanning tunneling microscope using optically pumped GaAs tips, Appl. Phys. Lett. 71 (1997) 3153.
[137] P. Bruno, Magnetic scanning tunneling microscopy with a two-terminal nonmagnetic tip, Phys. Rev. Lett. 79 (1997) 4593.
[138] S. Chikazumi, Physics of Ferromagnetism, second ed., Clarendon Press, Oxford, 1997.
[139] S. Wirth, S. von Molnár, Hall cross size scaling and its application to measurements on nanometer-size iron particle arrays, Appl. Phys. Lett. 76 (2000) 3283.
[140] M. Wuttig, B. Feldmann, T. Flores, The correlation between structure and magnetism for ultrathin films and surface alloys, Surf. Sci. 331–333 (1995) 659.

 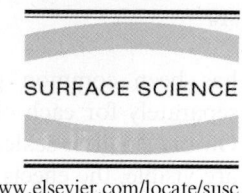

Modeling the full monty: baring the nature of surfaces across time and space

Frank Starrost, Emily A. Carter *

Department of Chemistry and Biochemistry, University of California, Box 951569, 405 Hilgard Avenue, Los Angeles, CA 90095-1569, USA

Received 10 July 2000; accepted for publication 5 May 2001

Abstract

Computational materials research has made great strides in recent years in the description of the science of surfaces and interfaces. So far, however, the approaches generally have been limited to fixed resolutions of time and space. In effect, each group of scientists has chosen its specific scale for the "road maps" used by them to investigate surfaces and interfaces, some focusing only on highly resolved "city maps", others considering the big picture of the "countrywide" view. So, just as in the planning for improvements of, e.g., a nation's infrastructure, future progress in the field requires limitations of the models to single length and time scales to be overcome. Recently, the first steps have been taken to set up multi-scale modeling techniques, often involving collaborations of chemists, physicists, and engineers. Here, it is our aim to present a representative survey of these techniques. In particular, we discuss informed continuum approaches, the quasicontinuum method, the kinetic Monte Carlo technique and accelerated molecular dynamics simulation. We show where they have been used to date and outline where their future application holds promise. © 2001 Elsevier Science B.V. All rights reserved.

Keywords: Solid–gas interfaces; Catalysis; Growth; Etching; Ab initio quantum chemical methods and calculations; Density functional calculations; Molecular dynamics; Monte Carlo simulations

1. Introduction

Picture a microscope where, with a twist of a knob, one sees first a smooth solid surface, then, zooming in, one notices cracks in the solid, or perhaps dust particles, or, in a liquid droplet, some cells. Zooming in further, suppose our microscope sees what is on the surface or in the crack or in the cells, in the form of islands or steps or molecules, comprised of atoms, which are, in turn, composed of electrons and nuclei. We have just described a multi-scale picture of an object of interest, where the "scale" here is size or "length".

Over the course of the last century, improvements in instrumentation have made it possible to observe all these scales. The smaller length scales, right down to the atom on a surface, have become accessible only recently, for example, through the invention of the scanning tunneling microscope [1,2]. For the interpretation of most of the observations, companion calculations are necessary. It

---

* Corresponding author. Tel.: +1-310-206-5118; fax: +1-310-267-0319.

E-mail address: eac@chem.ucla.edu (E.A. Carter).

has been common practice to deduce theories separately for each of the length scales observed. On the smallest scales, where electrons and nuclei are visible, the effects of quantum mechanics have been taken into account; on larger scales classical and statistical mechanics (with the quantum effects hidden in empirical parameters) was generally sufficient to explain the experiment. Progress in computational materials science in the last two decades made possible ever more precise predictions of properties on these separate length scales. With the beginning of the new millenium, the borders between theories (and theorists) describing different scales are beginning to dissolve, and a more unified look at materials science is beginning to emerge. Chemists, physicists and engineers working in the field of theoretical materials science find it advantageous (or even necessary) to collaborate, and find that connections can and should be made across scales, from big to small, so we can know how small affects big, and vice versa. This new "multi-scale" approach sharpens the theoretical understanding of the interconnectedness of processes observed by our "microscope" on the different length scales and is referred to as multi-scale modeling. Wherever necessary, observables such as energies are calculated by treating individual atoms with quantum mechanics. Wherever possible, large numbers of atoms are treated with coarser techniques, e.g., by taking average values to determine their behavior as a subunit of the whole system. In between these descriptions, there can be 10 orders of magnitude from the atomic length scale (10^{-10} m) to the chunk of steel length scale (1 m). To describe certain properties of some materials, it may be necessary to consider at the same time the tiny realms of quantum physics and quantum chemistry and the macroscopic partitions of materials considered in, for example, finite element methods of mechanical engineering [3].

In the same way, in the time domain, single ultra-short events, such as vibrational motion or the breaking of a bond at a surface, should be treated at least partially by quantum mechanics. For example, the evolution of a moderately sized system (typically tens of atoms) can be described using the "ab initio" molecular dynamics approach (integrating Newton's classical equations of motion using the forces calculated including quantum mechanical effects) for extremely short time periods, currently extending to picoseconds at most [4]. So while this approach can describe how ultra-short events unfold, it is insufficient to describe processes happening on time scales of seconds or more, such as the growth of a material by the adsorption (sticking to the surface) of atoms and molecules from the gas phase. To satisfy these aims, it is necessary to use coarser methods that, in their more advanced realizations, make use of the quantum mechanical results for the description of single events, that comprise the overall phenomenon.

Many chemical processes happen at surfaces and interfaces, where different materials in different phases (solid, liquid, gaseous) come together. When the different length scales are taken into account, more realistic and complete descriptions can be achieved for a wide range of behavior. These include surface phase transitions and reconstructions (structural reorganizations), surface and interface strain, dislocation (a long-range defect, such as a missing row of atoms) formation, materials growth and etching, and chemically induced materials failure, e.g., embrittlement or corrosion, in which a material is attacked at a surface and weakened so that it may crack under stress. In the time domain, rates can vary by ten powers of ten or more, e.g., in materials growth and surface melting, in adsorption and desorption (release from the surface), in biological processes, and in catalytic reactions or cooperative phenomena such as oscillating reactions.

It is our intention to give a description of the general concepts of current multi-scale modeling techniques and then to focus on their application to the investigation of surfaces and interfaces. We give examples and describe results that have been and can only be obtained by multi-scale modeling. We do not attempt to provide details on every method. The techniques discussed are intended to illustrate the basic ideas behind multi-scale modeling: informed continuum methods, where macroscopic material constants are determined from lower length scales (including from quantum mechanics); the quasicontinuum method, which links atomistic and continuum approaches on-the-fly;

the kinetic Monte Carlo technique, which allows one to reach arbitrarily long time scales (at the expense of full determinism); and accelerated molecular dynamics techniques, which simulate the evolution of many-particle systems by making rare events happen at a faster pace. We close with some suggestions for future research.

2. Theory of the multi-scale modeling of surfaces

Obviously, all real-life objects have surfaces. However, for the determination of many material properties, it is often assumed that the material is infinite in all directions. (Additionally it is often assumed that it is comprised of perfectly periodic units (see the unit cell of a body-centered cubic (bcc) crystal shown in Fig. 1), which neglects all the real-life defects in solids such as missing atoms, impurities, incorrectly placed atoms, cracks, dislocations, grain boundaries, etc.) In many objects, the assumption of infinite expanse makes sense when observed on the atomic level: In a macroscopic chunk of material, almost all the atoms are far away from the surface. The distance from the surface is so large that it no longer influences the local region around the inner atoms: For all practical purposes they "see" an infinite solid.

Indeed, a cube of 1 cm edge length contains about 10^{23} atoms, while there are only $\sim 10^{16}$ atoms on its surface. Although the latter certainly is a large number, a factor of 10 million separates the two numbers—and the volume-to-surface ratio becomes larger as the size of the body increases. Thus, in many applications the surface can be neglected.

Things are different, however, in other instances. On modern computer chips, so much functionality has been crowded onto so tiny a space that wires are only a few hundred or thousand atoms in diameter—and miniaturization is expected to continue. In fact, in nanometer-size (1 nm = 10^{-9} m; e.g., the diameter of a hydrogen atom is 0.1 nm) structures, e.g., nanoporous materials such as zeolites, which often are used as "molecular sieves", the number of surface atoms can be comparable to or even exceed the number of bulk atoms. Surface effects dominate the behavior of these materials, so their correct description becomes crucial. Surface effects also are important when it comes to interactions of materials with the outside world. These interactions usually take place at or near surfaces and can occur through electromagnetic radiation (light) or single atoms, molecules, liquids or other solids. The area of materials growth and etching, important, for example, in the design of nanoscale wires and microscale motors, requires consideration

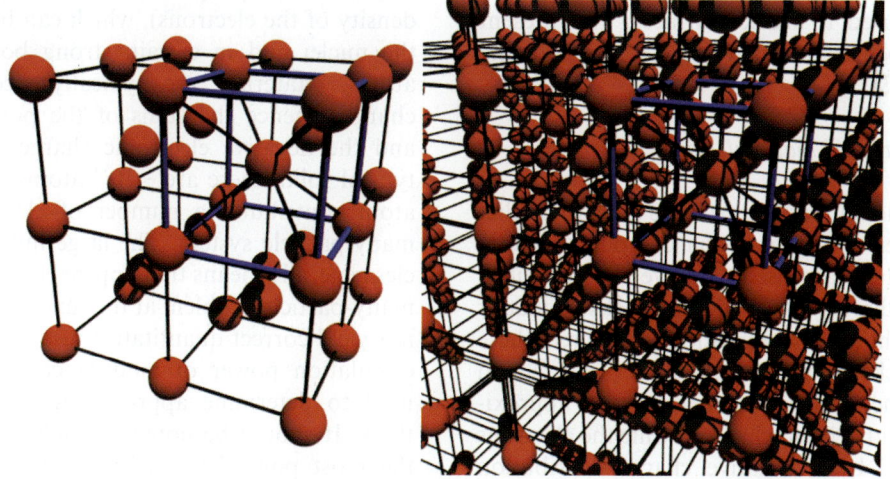

Fig. 1. Left: The unit cell of a body-centered cubic (bcc) lattice (blue) and seven other unit cells (grey). The atoms are denoted by red spheres. Right: A larger representation of a bcc crystal. Elements that crystallize in this structure include chromium (Cr), iron (Fe) and sodium (Na).

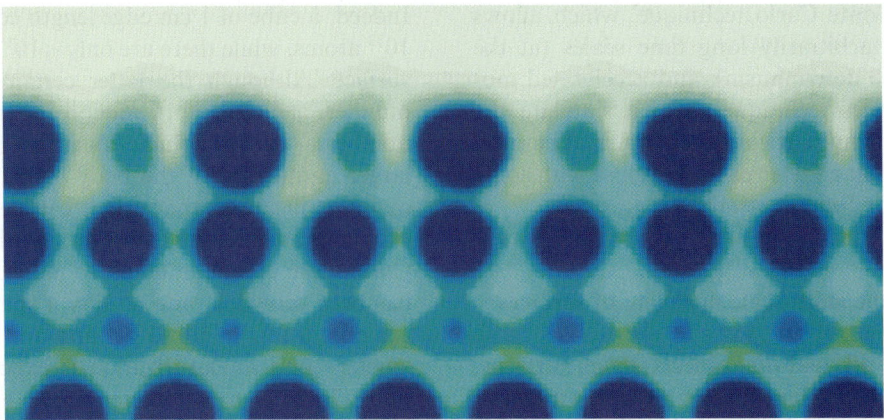

Fig. 2. The electron charge density at a surface and an interface: a monolayer film of Al_2O_3 on Ni. The darker the color, the higher is the probability of finding an electron at that position. The white area on the top indicates vacuum; the top layer is alumina (Al_2O_3) with the O atoms (big blue regions) close to the cross-section, the Al atoms further away (smaller, lighter blue regions). The lower three layers are composed of nickel (Ni) atoms, the middle layer being out of the paper plane. (Figure courtesy of E.A. Jarvis, from a density functional theory calculation [91].)

of the surface by definition. In catalytic reactions, the catalyst is a material that accelerates the rate of a chemical reaction without being consumed or essentially altered. In the case of heterogeneous catalysis, the reaction takes place at the surface and the bulk is irrelevant.

Computational science has developed a number of methods to determine theoretically the properties of solids and their surfaces. We next give a general introduction to these methods, describe the approaches used to model surfaces, and present the theory behind relevant multi-scale modeling techniques.

2.1. Computational materials science

Computational materials science aims to predict the properties of materials via computer simulation. Computers are used to solve the equations that govern the way the building blocks of matter interact. At the highest resolution considered, these building blocks are atoms composed of nuclei and electrons. The nuclei are generally well approximated as positive point charges, while the electrons are described by a negative charge distribution spread out in space. The extremely small mass of the electrons makes it necessary to describe them using quantum mechanics. In the quantum world, speed and position of low mass particles cannot be known exactly at the same time, as expressed in Heisenberg's uncertainty principle. One can determine, however, the probability for the electrons to be at a specific point in space. This quantity is called the charge density, and it can be measured experimentally by X-ray diffraction. Fig. 2 shows the charge density for a cross-section of a metal–ceramic interface and a crystal surface. The dark areas indicate a high electron density (the charge density of the electrons), which can be found near the nuclei and in certain strong bonds between atoms. Materials are typically electrically uncharged, hence the sums of the positive nuclear and the negative electronic charges cancel. In a typical solid, there are $\sim 10^{23}$ atoms/cm^3 and each atom contributes a number of electrons to this many-particle system. The large number of particles generally means that approaches to solving the many-particle problem at the fundamental level do not yield correct quantitative results. Yet, the vast calculation power of modern computers can be used to determine approximate numerical solutions. It should be noted, though, that even with the most powerful computers and the most advanced techniques, an explicit, exact description of the many-particle electron system is not possible. Clever approximations have been invented to deal with this problem. We next outline an important subset of the methods developed to describe properties of solids, surfaces and interfaces, and their interaction with the environment: electronic density functional theory and quantum chemistry for describing electrons and nuclei, molecular dynamics and Monte Carlo "atomistic" methods, and continuum theory of macroscopic solids.

Within computational materials science, three disciplines intersect: those of computational solid state physics, theoretical chemistry, and mechanical engineering. Each field is wide by itself and the boundaries are sometimes fuzzy. The following definitions give an idea of the primary contributions of each of the disciplines to materials science. Theoretical solid state physics is concerned with rendering as accurately as possible the geometric and the electronic structure of a given solid. The geometric structure is given by the positions of the atoms, the electronic structure by the energies electrons have in the solid and their wave function. (The wave function completely describes the electrons' behavior. Among other properties, it provides the charge density.) In theoretical surface chemistry, the main aim is to model atomic and molecular adsorption, desorption, diffusion (movement along and in the surface) and reactions occurring at solid surfaces. Important aspects are identifying the ways a certain reaction can happen (its mechanism) and the relative likelihood for each of these pathways to be traversed (reaction rates). In treating atoms or molecules interacting with crystals, it becomes necessary to treat non-periodic entities within or near periodic ones, each of which are described differently. While physicists often tend to use methods that are developed to treat periodic solids, chemists tend to employ techniques that are optimized to describe molecules. Mechanical engineers, on the other hand, usually deal with questions of stability and durability of materials. Traditionally, they have ignored the atomic structure of solids. Rather, they consider average mechanical properties to determine how large structures of material, from bridges to airplane wings, may behave under stress. Recently, however, the calculations have begun to include atomic scale modeling. One example of this is the quasicontinuum method, discussed later.

Theoretical predictions are most convincing when they are calculated with as few empirically determined parameters as possible. In materials theory, the only parameter that must be supplied is the makeup of the solid to be studied, i.e., which elements comprise the material and, preferably, in which approximate geometry they are arranged. A method that takes only these external data as input and computes the geometric and electronic structure based on the laws of quantum mechanics is called an ab initio, or first-principles, method. These methods are usually costly computationally and currently are restricted to the treatment of crystals made up of unit cells of at most a few thousand (and more typically, $\sim$10–100) atoms, corresponding to a cube of $\sim 10^{-20}$ cm^3 volume. Unit cells are groups of atoms that are treated as the building blocks of the crystal; the unit cells are repeated identically through space to create the atomic structure of the crystalline solid (see Fig. 1). Significantly less burdensome computationally are so-called empirical or semi-empirical methods. Empirical methods employ parameters based on experimental values; semi-empirical techniques are based on results from both experiment and ab initio approaches.

One successful ab initio method for the determination of the electronic and the geometric structure of materials is density functional theory [5]. Within density functional theory, the problem of finding a description for all the electrons in the solid is reduced to the problem of determining the correct electron density. Hohenberg and Kohn showed that all the ground state properties of a system of electrons can be determined from its charge density alone (where "ground state" refers to the lowest energy state the system can be in). This is an enormous simplification, since the electron density is a function of merely three position variables, such as x, y and z, whereas the wave function describing the electrons themselves depends on the coordinates of all the electrons. The total energy of all the electrons is the sum of the kinetic energy (the energy of motion) of all the electrons, the Coulomb energy due to their attraction to the positive nuclei, the Coulomb energy due to the repulsion of the negative electrons amongst themselves, and the non-classical, exchange-correlation energy (a term that lowers the energy due to the electrons' tendency to avoid each other). The local density approximation for the last contribution yields a convenient set of single-electron equations to determine the electron density [6]. In a number of cases, an approximation to the exchange-correlation energy that not only depends on the electron density (as in the local density approximation) but also on its gradient has been found to improve results (generalized gradient approximation [7]). In the single-electron approximation, the wave function is determined for a single electron moving subject to a potential due to the nuclei and the other electrons. Density functional theory has certain limitations, particularly when it comes to calculating not the ground state but states with higher energies, or excited states. However, the list of successful applications of density functional theory within the local density approximation (and generalized gradient approximation) to condensed matter phenomena is long and impressive. Unfortunately, conventional density functional theory algorithms become very slow and cumbersome when the unit cells become large. The computational time required for the calculation of the electron energies scales with the number of electrons, n, to the third power. While one might expect that a system twice as large should only cost twice as much to calculate, a density functional theory calculation would be eight times as costly (2^3). Physically, one might expect the cost of models to scale linearly with size of the system. Algorithms which achieve a linear dependence on n are dubbed "linear scaling," or $O(n)$ (pronounced "order n"), methods [8,9].

In quantum chemistry, the system of electrons within each atom or molecule is described by a many-electron wave function. There are several stages of approximation and different approaches for the way these wave functions are constructed from single-electron solutions ("orbitals"). The typical starting point is Hartree–Fock theory [10], where the wave function, a product of orbitals, is constructed to obey certain fundamental laws of quantum mechanics, namely the Pauli exclusion principle, which prohibits electrons (of the same spin) from occupying the same region of space. In any system, such as an atom with a nucleus and electrons, nature demands a minimum total energy, which involves a compromise between kinetic energy and potential energy. (The potential energy describes the energy stored in a system, usually depending on the relative position of, e.g., electrons or atoms. The term "potential" is used for the potential energy given as a function of spatial coordinates.) The electron–electron Coulomb repulsion energy is always positive, and therefore nature acts to minimize it, so long as the kinetic energy does not rise too steeply as a result. Nature reduces electron–electron repulsion by correlating the motion of the electrons so as to keep them as far apart from each other as possible at all times. Thus, the many-particle wave function should contain electron correlation. This is done in the so-called configuration interaction approach. Other techniques that model the correlation of electronic motion and the reduction of the Coulomb repulsion include many-body perturbation theory and the coupled cluster method (see, e.g. Ref. [10]). Modern quantum chemistry methods can yield accurate results, but the scaling of the computation with the number of electrons prohibits their use for more than tens of atoms. Recent advances have reduced the scaling of some of these methods to near linear, but their overall cost is still high [11,12].

The molecular dynamics technique allows one to follow the evolution of a collection of atoms through time. Based on the velocities of the atoms and the forces between them, the trajectories (time-evolving positions) of all the atoms are calculated. Systems of interest for surface chemistry are atoms and molecules near surfaces in such processes as scattering, catalytic reactions, growth, and etching. Fig. 3 shows the elementary processes which happen when molecules interact with surfaces: adsorption, dissociation, diffusion, reaction, and desorption. In surface simulations, the atoms inside the solid, far away from the surface, are close to their equilibrium positions (i.e., the positions the atoms adopt after they have moved according to the forces acting on them and the forces have balanced out). However, dramatic reconfigurations of the surface atoms may happen. In molecular dynamics, the electrons are usually not

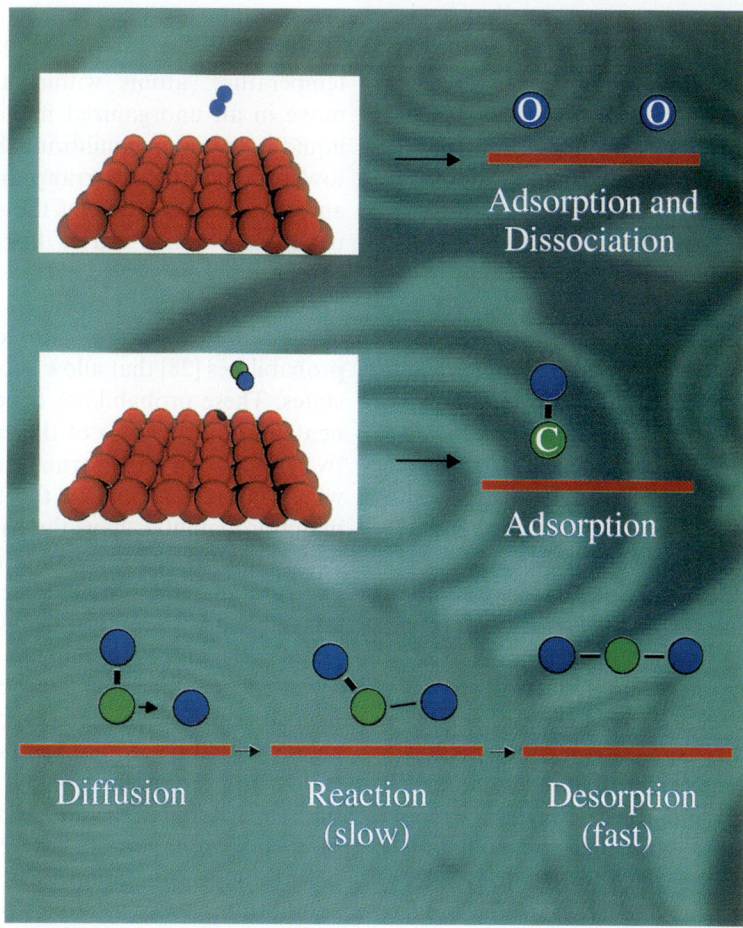

Fig. 3. The fundamental steps of the catalytic oxidation of carbon monoxide (CO) on a platinum surface. The two upper panels show oxygen being molecularly adsorbed and dissociating easily on the surface. Middle panels: CO is molecularly adsorbed. In the lower three panels, CO diffuses on the surface and meets an oxygen atom (left), CO reacts slowly with the oxygen atoms (middle), but the resulting carbon dioxide desorbs quickly from the surface (right). This reaction generates complex patterns of the oxygen and CO concentrations on the platinum surface. In the spatial pattern displayed in the background (whose width corresponds to about 0.2 mm) the dark areas are covered by oxygen, the light areas by CO (courtesy of G. Ertl [93]).

explicitly considered for the interaction among atoms. Molecular dynamics techniques still take the individual atoms into account, but typically via an empirical analytic interaction potential, which allows forces to be evaluated cheaply. In this way up to millions of atoms can be followed, especially when parallel processing is used to split up the solid into subsystems [13]. By contrast, the more accurate ab initio molecular dynamics accounts for the relaxation of the electrons as the atoms move, and thus is significantly slower [4,14,15].

An example of the predictive power of the molecular dynamics technique for surface chemistry is given by ab initio-derived molecular dynamics calculations (a modest multi-scale model, where parameters describing the atomic interactions ($\sim 10^{-9}$ m) were obtained not from experiment but from quantum mechanical ($\sim 10^{-10}$ m) calculations) of the chemisorption of fluorine atoms and molecules on a silicon surface. A number of predictions made about the chemisorption processes [16–18], some of them in direct contravention of contemporary chemical thinking, were confirmed in later experiments [19–23]. In particular, these simulations characterized quantitatively two reaction pathways (atom abstraction by the surface and dissociative adsorption) as a function of how fast the molecules hit the surface and of how the molecules are oriented as they impinge on the surface [17]. Defying conventional wisdom, the less exothermic (heat-releasing) pathway of atom abstraction dominates for slow molecules, while dissociative adsorption is preferred only by fast molecules (experimental confirmation reported in Refs. [20,22]). The molecular dynamics simulations [17] predicted that abstraction and dissociation go through a common intermediate, whereby dissociation occurs stepwise through abstraction followed by release of a halogen atom that either goes into the gas phase or quickly finds a nearby Si atom on which to bind (confirmed experimentally in Ref. [21]). Conventional dissociative adsorption is thought to happen with both atoms of a diatomic molecule simultaneously attaching to a surface, but in this case the "dangling bonds" (localized single electrons) of Si alter the reaction mechanism. These molecular dynamics simulations [18] further showed that when halogen atoms are released into the gas phase, they come off at an angle with respect to the surface normal and rather slowly, losing considerable energy to the solid (measured in Ref. [23]). Finally, molecular dynamics simulations of high-pressure etching [16] showed that local heating of the surface, due to strong chemical bond formation, disorders the Si crystal, which was subsequently confirmed by experiment [19]. On a practical note, the theory culminated [18] in a suggestion for a two-cycle etching process; a related process is now used commercially [24].

Monte Carlo methods are frequently used to simulate physical events based on their *probabilities* (related to either thermal equilibrium or kinetic rates). The name is derived from the famous casino in the Mediterranean seaport of Monaco and points to the use of random numbers, as in the outcome of spinning a roulette wheel. One use of Monte Carlo methods is the study of equilibrium properties of matter (a system in equilibrium remains in that state unless acted upon by some external influence), e.g., the equilibrium structures of surfaces and interfaces [25–27]. At a given temperature, atoms within a condensed phase move in an unorganized manner. If the solid or liquid is not in an equilibrium structure, it evolves towards one. The rearrangements of the atoms, and thus the evolution of the system, can be simulated by Monte Carlo. For the move toward equilibrium, the system evolves not only simply to lower energy states (which, of course, occurs) but changes states according to so-called Boltzmann probabilities [28] that allow access to higher energy states. These probabilities depend strongly (exponentially) on the ratio of the energy difference between states to the thermal energy, the latter of which is proportional to the temperature. Boltzmann probabilities have the characteristic that the probability of changing state increases with temperature and, that the more energy required, the less likely it is that the system can provide this energy for the process to happen. If the system is in a lower energy state after a process occurs, it is less likely to leave it than the former state, because more energy would be required to escape. However, the Boltzmann probabilities allow the system to eventually rearrange itself into another state, that may be even more favorable energetically. After a sufficiently long simulation time, the system reaches equilibrium, where a stable structure is achieved.

An example for the description of *non*-equilibrium systems by Monte Carlo is the growth of films on surfaces [29]. Because actual film growth occurs much more slowly (one layer per second) than one can simulate with, e.g., molecular dynamics (typical time scales of nanoseconds instead of seconds), one means to model film growth is by "equilibrating" the system between each adsorption of an atom, using Monte Carlo. In particular, atoms can be deposited randomly on the surface, changing the state of the system. The evolution is again modeled by Monte Carlo using Boltzmann probabilities for the rearrangement of atoms on or near the surface. If the equilibration "time" between depositions of atoms is sufficiently large, the current system, real or simulated, may reach an equilibrium state which is destroyed by the next deposition. The process is repeated until the desired film thickness is reached.

In engineering, a common approximation is to ignore the discrete nature of solids and to assume a continuity of structure. Every volume of the continuum solid, at all length scales, is assumed to exhibit the properties observed in the macroscopic material. For example, if a certain coefficient of elasticity is measured for a rod of steel of 1 m length, this coefficient is assumed be valid for all portions of the rod, be they 1 cm long or 1 μm = 10^{-6} m. (Linear elasticity means in the case of a spring that the counteracting force is proportional to the extension away from equilibrium.) At the atomic and molecular length scale, however, this assumption no longer holds. For many applications, though, the equations of continuum mechanics can be used safely to describe deformations of solids. In continuum mechanics, few material parameters (e.g., elastic constants) are needed to describe how the solid reacts to applied stress or pressure.

In mechanical engineering and other fields, material behavior is frequently simulated by finite element methods, which are numerical techniques for obtaining the response of physical systems to external influences. In these methods, a solid is divided into little space-filling polyhedra of finite volume whose corners are called nodes. In the two-dimensional example shown in Fig. 4(b), these polyhedra are triangles which cover the square piece of material. A cut is used to simulate a crack whose evolution is studied by simulating forces on the material. The finite elements are thought of as continuum particles [30], representing a large number of atoms (e.g., $\sim 10^6$). During the finite element simulation, the value of an unknown function (such as the displacement of each of the nodes covering the piece of material when considering the response of, e.g., a steel beam to forces, or a temperature distribution in a simulation of, e.g., an engine part in a gas turbine) is determined at each node by minimizing the energy over the whole body, taking into account material parameters and the sharing of the boundaries of the finite elements. At the same time, the number of nodes can be adjusted to account for spatially varying forces that may act on the body, possibly leading to its deformation. Sophisticated algorithms have been invented to refine adaptively the

(a)
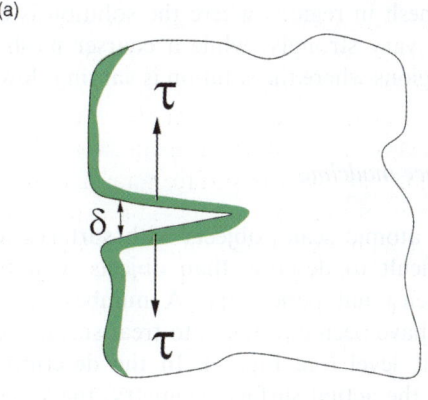

(b)
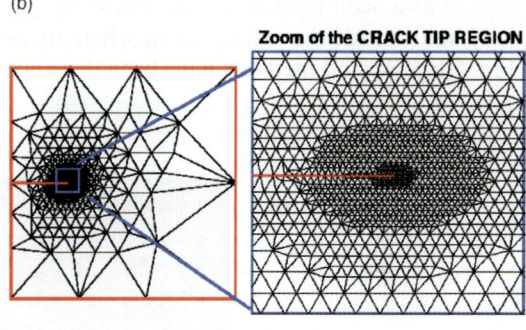

(c)
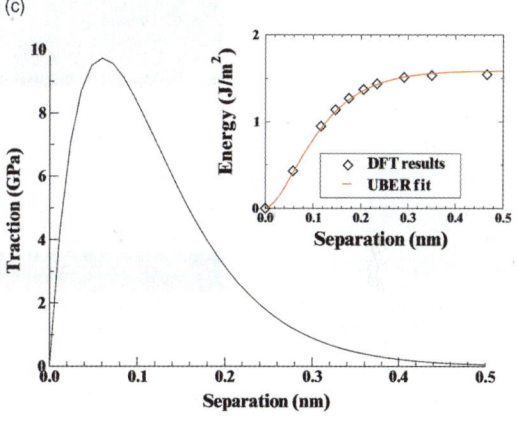

Fig. 4. Panel (a): A schematic of the forming of a crack. Applied is a traction τ (applied force per unit area) and the separation δ is the distance between the surfaces. Bulk aluminum (Al, white) oxidizes to alumina (Al_2O_3, green) at the surface. Panel (b): A finite element/cohesive zone model in which results derived from density functional theory are used to describe crack formation (courtesy of M. Ortiz). Panel (c): The calculated traction is shown as a function of the separation, as derived from a universal binding energy relation (UBER) [94] fitted to density functional theory-generalized gradient approximation results for unrelaxed surfaces (inset) [71] (courtesy of R.L. Hayes).

element mesh in regions where the solution is expected to vary strongly, while a coarser mesh is kept in regions where the solution is varying slowly [31].

2.2. Surface modeling

At the atomic scale, objects with surfaces are more difficult to describe than objects with full three-dimensional periodicity. A number of approaches have been developed to treat surfaces on the atomic level (see Fig. 5). In the description closest to the actual surface geometry, the system is treated as a semi-infinite crystal where the periodicity holds inside the solid (shown schematically in Fig. 5(b)). Several methods have been developed to calculate the electronic properties of the semi-infinite crystal (e.g., Refs. [32–34]).

In other approaches, the surface is treated as a so-called slab. In the "stand-alone" geometry (Fig. 5(c)) this is a single film of several layers of atoms with vacuum on both sides (see, e.g., Ref. [35]). The slab geometry necessarily leads to the existence of two surfaces, both of which can be taken into account, or the "bottom" layer of the slab can be fixed to the bulk structure.

A large number of electronic structure methods have been designed with three-dimensional periodicity in mind. This allows the use of so-called fast Fourier transform routines [36] to evaluate certain properties (such as the classical electron–electron Coulomb repulsion). Three-dimensional

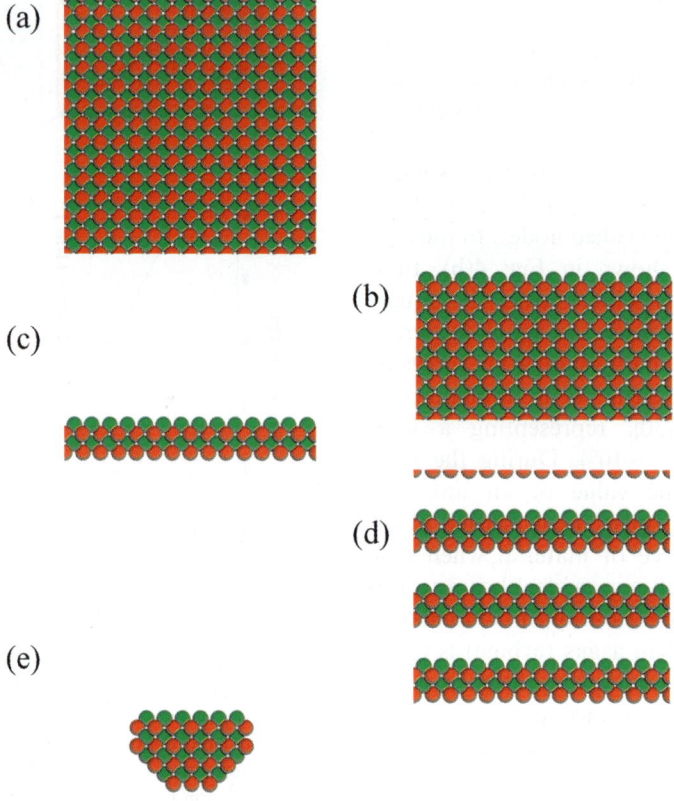

Fig. 5. Several geometries used to simulate surfaces of crystals, shown here for the zincblende crystal, suppressing the dimension perpendicular to the page: Panel (a): bulk, panel (b): half-space, panel (c): single slab, panel (d): super-cell slab, and panel (e): cluster geometries. (The geometries shown are only schematic; generally, in panels (c) and (d) the slabs have to be thicker, as do the vacuum regions in panel (d); also the cluster in panel (e) consists generally of a smaller number of atoms.)

periodicity also makes possible the use of plane-waves ($\exp[i\mathbf{k} \cdot \mathbf{r}]$) rather than more complicated functions, to construct the wave function. Due to the periodicity of the crystal, the wave functions fulfill a mathematical condition, called Bloch's theorem, which plane waves account for by default. To mimic a surface while retaining periodicity in the third dimension, the "super-cell" slab approach can be used: Here, an infinite number of parallel slabs separated by vacuum layers is simulated (Fig. 5(d)). The computational cost is reduced by using thin slabs and thin vacuum layers. The slab has to be thick enough to simulate the bulk crystal beneath the surface, and both the slab and the vacuum layer have to be sufficiently thick to prevent the surfaces from interacting with each other across either layer—they must not "know" of the existence of the other surfaces. In practice, the unit cells employed are chosen to be as small as possible in the directions parallel to the surface as well. The size of the cell in this lateral dimension is dictated by, e.g., the coverage of adsorbates on the surface.

An alternative approach in common use with quantum chemistry techniques, is to use a cluster to model the surface (Fig. 5(e)). In the cluster model, only a small number of surface and subsurface atoms are treated, while the rest of the solid is neglected. For this smaller number of atoms, accurate results may be obtained using, e.g., configuration interaction techniques of quantum chemistry (see, e.g., Ref. [37]). The challenge is to minimize the influence of the artificial surfaces created by the finite size of the cluster [38].

A compromise between super-cell slab and finite cluster models is to use an embedding model. In embedding methods, a cluster of atoms is treated separately from a larger number of background atoms in which they are "embedded" [38]. More accurate but computationally more costly schemes have been developed where the background atoms are treated by super-cell slab density functional theory and the embedded atoms are described by high-accuracy quantum chemistry calculations [39, 40]. Embedding techniques make accurate treatments of ground and excited states of defects in crystals or adsorbed molecules on surfaces possible.

2.3. Multi-scale modeling techniques

Multi-scale modeling involves at least two techniques, operating at different resolutions (degrees of detail) in space and/or in time. Results of the higher resolution simulation are fed into the lower resolution method. In some but not all techniques, there is a feedback from the coarser-grained, or low resolution, method to the finer-grained, or high resolution, one. The feedback changes the environment at the small-scale level and leads to adapted results at this level. The calculations at the different length scales are intimately related. When there is no feedback from the coarser level, the calculations are essentially decoupled (e.g., in the kinetic Monte Carlo method) and the calculation can be done by two separate programs. In the following, several techniques are illustrated to which the bridging of length and/or time scales is central.

The first two classes of methods are based on a finite element representation of the solid. The first class to be discussed does not use feedback from the coarser level. Methods in this class are called informed, or physics-based, finite element methods. The parameters governing the response of the material to certain interactions have been obtained by a higher resolution theory, e.g., molecular dynamics techniques based on an empirical description of the atomic interactions or from measured bulk properties. An example is the use of elastic constants to describe how a piece of material deforms due to a small force. For each finite element, the appropriate material properties are taken or interpolated from a database constructed from higher resolution theory or experiment. Then the finite element algorithm determines the result for the whole solid, based on the values obtained for each of the elements.

The quasicontinuum method is a finite element technique with feedback [31]. It was initially developed to analyze the structure of crystal defects [41]. To do this, a finite element mesh is set up for the solid, with the nodes corresponding to "representative atoms". These atoms represent a large number of atoms far away from the defect, where the variation from the ideal crystal structure is small, whereas they describe single atoms close to the defect. The positions of other atoms are arrived at by interpolation. The total energy of the crystal is then derived from the energy contributed by each representative atom weighted by the number of atoms it represents. More accurate results are obtained if this is done in an ab initio fashion, for example, by applying density functional theory. When many atoms are involved, however, the exact quantum mechanical treatment of density functional theory has been too demanding computationally so that generally more efficient, yet less accurate, methods have been used to date, e.g., empirical interatomic potentials [41], or a semi-empirical quantum mechanical method such as a tight-binding technique [42]. The physical properties for elements in the higher resolution region are obtained by (approximately) treating the actual atomic structure. For consistency, the values in the coarse-grained region are generally computed with the same interatomic description as in the fine-grained region. One difficulty is the construction of a correct description of the boundary between the two regions. In particular, the questions of how to transport heat or mass through the boundaries have not been solved.

The concurrent coupling of length scales scheme [43,44] is an alternative to the quasicontinuum method (see Fig. 6), where tight binding is used to treat the region of the solid where highest resolution is necessary (e.g., if chemical bonds are made or broken in this region, a quantum mechanical treatment is desirable). Fig. 6 shows this region in the lowest panel. The tight binding region is coupled to a larger region described by molecular dynamics with empirical potentials (the middle panel in Fig. 6), which in turn is connected to the largest region, the rest of the solid (the upper panel in Fig. 6). It is treated by a finite element method, assuming a continuum approximation of the solid in that region. By contrast, the quasicontinuum approach treats the whole solid by a finite element method, where the properties in the elements are derived from one single method for both the representative atoms and the explicit atoms on the finest scale. Also, unlike the quasicontinuum method, the concurrent coupling of length scales technique does not offer the possibility to adapt the finite element grid after it has been set up. The

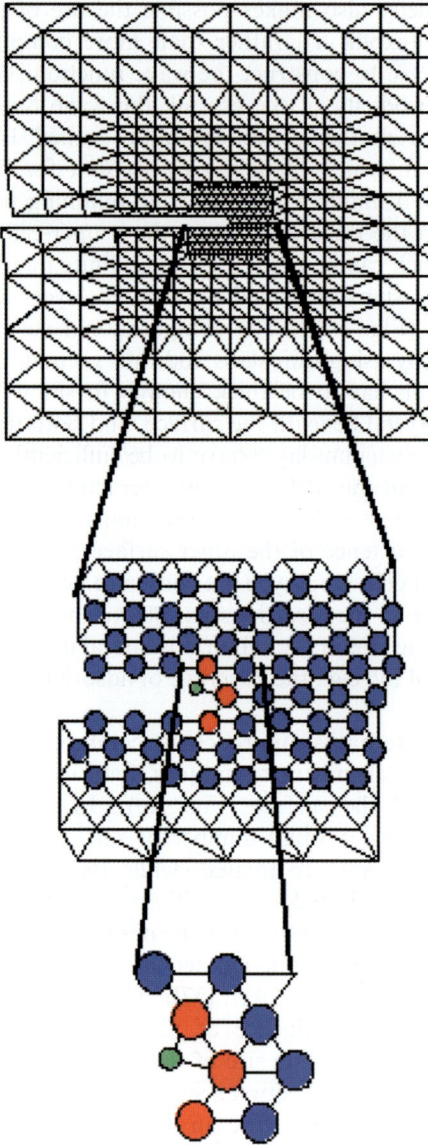

Fig. 6. The figure shows the resolution at different length scales as used in the concurrent coupling of length scales method. The largest length scale is treated by a finite element grid using the continuum approximation of the solid. Increasing the resolution, the individual atoms are considered by molecular dynamics with empirical potentials. Quantum mechanical effects are taken into account at the highest resolution using a tight-binding method. In the concurrent coupling of length scales method, the mesh refinement is defined once at the beginning. In the quasicontinuum method, the whole solid is covered by the finite element mesh, which can be adaptively refined during the simulation to atomic resolution in the region near the crack tip (courtesy of W.A. Curtin).

concurrent coupling of length scales method ensures that the total energy is conserved, however, while this is not guaranteed in the quasicontinuum approach. In the concurrent coupling of length scales method, three different algorithms are at work. They are connected at the region boundaries by appropriate "handshaking", making sure forces are correctly coupled across region boundaries. It is the boundary where atomic resolution and the continuum approximation meet that poses the greatest problems. To improve the situation, a replacement for the finite element part has been suggested: coarse-grained molecular dynamics [44], a scale-dependent generalization of finite elements which passes smoothly into molecular dynamics as the mesh is reduced to atomic spacing. Finally, use of tight binding in the concurrent coupling of length scales strategy makes the modeling of metals difficult in many cases; the authors of this hybrid scheme suggest that the quantum mechanical heart might be replaced by another rather inexpensive quantum mechanical method such as kinetic energy density functional techniques [45].

A different method of bridging length, but more importantly time, scales is the kinetic Monte Carlo technique. This method provides the possibility to study the evolution of a system for longer times and with more participating species than is possible by traditional molecular dynamics. Kinetic Monte Carlo is a variant of ordinary Monte Carlo, where processes have a probability of execution governed not by the Boltzmann probabilities but by ratios of rate constants that quantify how often processes happen. In some cases, elementary processes can be isolated in experiments and their rate constants measured. Alternatively, the rate constants are obtained from ab initio or from more approximate, empirical potential calculations, or left as phenomenological input variables. Once the relevant processes have been found and their rate constants obtained—not a trivial undertaking—the Monte Carlo part can begin. Some implementations (e.g. Refs. [46,47]) start with the random selection of a species and choose at random a process that is allowed for that species. Then a random number between 0 and 1 is selected by the computer and is compared to the rate constant of the process divided by the largest rate constant, the reference rate constant. If this rate constant ratio is greater than the random number, the species reacts, otherwise it is left unchanged. This means that "faster" processes (which have larger rate constants) happen more often. After N attempted moves, where N is the number of species, the clock is advanced. In some implementations, the time step is the inverse of the reference rate constant, corresponding to the time needed to execute the fastest process [46]. In others the time step is drawn from a specially adapted pool of time steps [48]. Another implementation of kinetic Monte Carlo [49] is based on the so-called n-fold way algorithm [50], where the probability and selection of a process to occur is given by its rate constant divided by the sum of rate constants, the "total" rate constant. Here, an event happens at every time step while, in the implementations mentioned earlier, the execution is conditional on the comparison of the rate constant ratio to a random number.

Kinetic Monte Carlo can simulate arbitrarily long times, depending on the rate constants involved. It works most efficiently, however, when the processes simulated happen on similar time scales. In particular, processes that are very fast (where the energy barrier, which has to be surmounted by the system to move from one state to another, is almost zero) in comparison to the others can be problematic, since then the individual time step is very short. This means that the simulation of longer time intervals is costly. To avoid this problem, one can have all species undergo the very fast processes according to their equilibrium probabilities before attempting a traditional kinetic Monte Carlo move. The very fast processes are assumed to be executed instantaneously, i.e., without incrementing the clock, and can be treated in a computationally cheaper manner [47]. In all implementations of kinetic Monte Carlo, the explicit simulation of unproductive events (i.e., events, that ultimately leave the system unchanged) is omitted, increasing its efficiency compared to molecular dynamics. Also, it is easy to change the concentration of species and to study the effect of such changes. Kinetic Monte Carlo also can predict the kinetic order, i.e., how the rate at which a reaction happens depends on the concentration of the reactants (for an example, see Ref. [47]).

Kinetic Monte Carlo can only give an accurate rendering if all the relevant processes are known in advance, since they are inputs to the simulation. If important processes are left out, the simulation can produce qualitatively incorrect results. For example, Jónsson and co-workers found that the shape of platinum islands in simulated growth on a platinum surface can depend dramatically on the inclusion or exclusion of a single process [51]. An example of a counter-intuitive process that might have been left out of a kinetic Monte Carlo simulation is how metal atoms can diffuse on metal surfaces. It was discovered only recently that on some surfaces diffusion occurs not primarily via repeated hops on top of the surface as illustrated in Fig. 7(a), but rather by a concerted displacement in which the adatom replaces an atom in the surface layer, which in turn moves to the surface, reminiscent of the butterfly swimming style [52–54] (as displayed in Fig. 7(b)).

Another factor limiting kinetic Monte Carlo is its assumption of a complete decoupling of processes. Like any Monte Carlo scheme, the kinetic Monte Carlo method is stochastic (random or probabilistic) by definition. Also, the way a state is reached is not taken into account and thus cannot influence the evolution of a system. The molecular dynamics method, on the other hand, is deterministic: any configuration of the system at any time step is connected by Newton's laws to the configuration at prior and subsequent times. Molecular dynamics methods also give access to properties of the system that can only be determined from a trajectory. Among these are so-called time-correlation functions, from which, e.g., vibrational spectra can be derived that reveal the motion of an atom or molecule on a surface. As another example, energy and angular distributions of colliding molecules can be predicted from gas-surface scattering events [17,18]. However, since the evolution time step is limited by the fastest motion in the system, typically an atomic vibration ($\sim 10^{-12}$ s), it only can simulate events happening on very short time scales. Rare events (those with a high energy barrier) generally do not occur during these short time intervals. If the objective is to study a rare event, though in principle it is not necessary to know the processes a priori, it is required in practice because the simulation usually cannot be run long enough for the rare event to happen unprompted. In particular, transition states—those unstable states which, if perturbed slightly, will evolve into either the initial or the final state—can be located a priori. An example of a transition state is denoted in Fig. 8(a) by a cross. One can follow the rare event of crossing a transition state by starting there, and following the molecular dynamics trajectory forwards and backwards in time to construct the entire trajectory [55,56]. A further drawback is that generally the quality of the calculation is not sufficient to study chemical reactions, with a few exceptions [4]. However, if the quality is improved by using ab initio calculations for the forces, the method is considerably slowed down, and only a few trajectories may be sampled, leading to poor statistics.

Several methods for treating longer times using variants of molecular dynamics have been developed. One can retain the advantages of traditional molecular dynamics as long as one makes

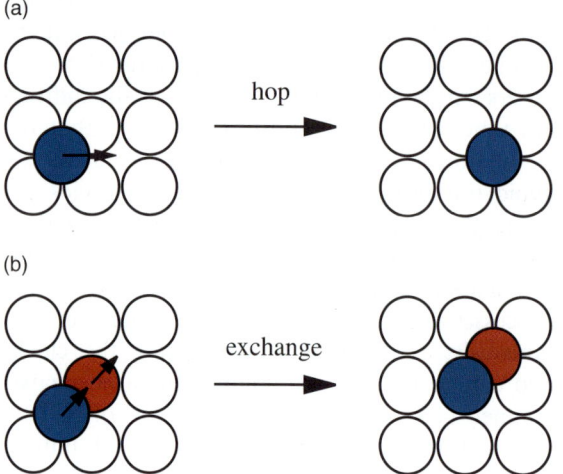

Fig. 7. Schematic illustration of two mechanisms of diffusion on a metal surface. To model diffusion in a kinetic Monte Carlo simulation, all processes likely to happen must be included in the model. The hop process shown in panel (a) is easily imagined. Some metals prefer, however, a cooperative exchange mechanism (shown in panel (b)) over the hop process [52]. (Figure after A.F. Voter, private communication.)

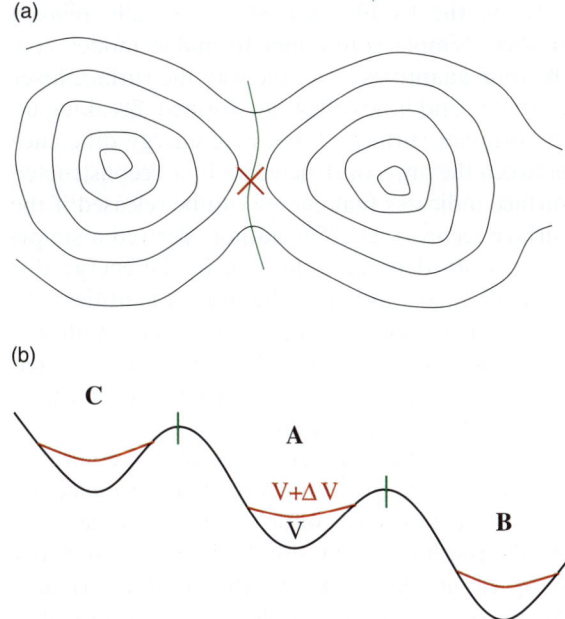

Fig. 8. Schematic representation of potential basins separated by energy barriers (the "watersheds" are denoted by green lines). Panel (a) shows a "map" view with contour lines indicating lines with the same potential height V. The transition state is the point through which the transition from one basin to the next requires the least energy. It is indicated by a red cross. In accelerated molecular dynamics [58,59], a bias potential ΔV is added to the potential V (see panel (b)) to make the potential basins shallower, leading to an extension of the time that can be simulated. (After A.F. Voter, private communication.)

the additional assumption that the system obeys transition state theory (TST) [57]. TST assumes that barrier crossings are uncorrelated and that no recrossings of barriers occur. One can then consider the evolution of a system as a sequence of infrequent transitions from one potential basin to another. (A potential basin is a region in space where the potential energy is lower than in the surrounding region, as shown in Fig. 8(a)). A system in one of these basins spends a lot of time meandering around, or vibrating, in one state, going up the potential hills in different directions, then down again, until it at some point climbs over a hill to cross through a transition state into another basin. One idea to speed up the method is to shorten the time spent in the basins. To get meaningful results, this quickening of the pace has to be done keeping the relative probabilities of different escape routes unchanged with respect to the original system. Several approaches achieve this by artificially shallowing the potential basins [58,59], most recently the hyper-molecular dynamics method [60–62]. As depicted in Fig. 8(b), a bias potential is added to the deep potential wells, keeping the transition state potential energy unchanged. This modification preserves the ratio of the TST escape rates from basin A to any two adjacent states, since the energy barrier is lowered by the same amount for both transitions. This means that the escape to a given adjacent state (say, B) happens with the correct relative probability. The boosted time can be calculated from the time elapsed during the trajectory occurring on the biased potential surface. In cases where empirical potentials determine the forces (ab initio molecular dynamics is currently out of reach), this has led to a speed-up of up to four orders of magnitude. However, in more complicated systems, the determination of the bias potential can be cumbersome [63]. An approach which avoids the difficulty of determining the bias potential is the temperature-accelerated dynamics method [63].

3. Multi-scale modeling in surface science

A multi-scale approach to the modeling of materials is necessary whenever elementary processes taking place in a small region or a short time affect behavior on larger scales in a manner not describable by interpolation. A class of examples that does not require multi-scale modeling is the simulation of a material before and after a complete change of phase, e.g., from solid to liquid due to an increase in temperature or from one crystal structure to another due to a change in pressure. The properties of the uniform phases can be determined from the investigation of small unit cells, usually neglecting the imperfections of a real material. However, when the aim is to understand the phase transition *itself*, a look at different length scales simultaneously is a must. Atoms rearrange themselves in their immediate neighborhood, zones of different phases form within the material, complex breaking and forming of bonds—chemistry—happens at the zone boundaries. The zones expand, needing a description at a higher length scale, while the elementary processes continue to happen at the atomic length scale. We next give examples of current and future applications in surface science that demand the use of multi-scale methods.

3.1. Surface modeling across length and time scales

Conventional single-length scale modeling can be performed at the atomic or a higher length scale. Extrapolating material properties from atomic results obtained with assumption of perfect periodicity neglects the defects and irregularities present in a real crystal. This atomic-scale-only approach is as limiting as a coarser model that includes empirical input or, worse, fitting parameters. While the latter includes large-scale effects that cannot be described on the atomic scale, an understanding of the basic reasons for the phenomenon investigated is lacking.

3.1.1. Bridging length scales

We now examine the contribution to the understanding of surface phenomena that can be achieved by combining the low-resolution continuum model with higher resolution input. Our first example is the reconstruction of crystal surfaces that occurs when a crystal is cleanly cut into pieces. "Reconstruction" refers to the rearrangement of atoms at the surface to lower their energy (bonds are broken when the surface is created, raising the energy of the surface atoms). Many theoretical approaches have been and are being used to describe surface reconstruction [64]. A multi-scale description is particularly important in cases where the reconstructed pattern (the surface unit cell) is large compared to the interatomic distance. Cammarata used an informed continuum model to describe the reconstruction of metal surfaces of two orientations [65]. Four properties of the newly created (i.e., unreconstructed) surface, among them surface tension and surface stress, were obtained from empirical embedded atom method [66] or density functional theory calculations (depending on the metal). A continuum analysis yielded an estimate for the energy difference between the freshly cleaved and a fully relaxed surface. Simple continuum formulae model how the four quantities affect the way the surface layer contracts and how defects are formed, breaking up the original surface. A negative energy difference between the unreconstructed and the reconstructed surface indicates that energy can be released if the surface reconstructs. The authors derived a simple criterion to determine the sign of the energy difference, involving only the four quantities obtained at a higher resolution and one additional parameter. They evaluated this criterion for ten metals and can correctly predict (in most cases) which surfaces reconstruct.

The description of crystal growth requires a knowledge not only of the way atoms or molecules react on the surface (on a short time scale, with atomic resolution) but also of the way islands develop (long time scale, rougher resolution) [67]. Arrays of small islands, called quantum dots, that confine electronic motion in all three spatial directions, are being considered for a variety of applications in the electronics industry. The shape and the stability of quantum dots have been investigated in a multi-scale approach. For a quantum dot to form (rather than a film), the gain from elastic relaxation energy has to be larger than the cost of increased surface energy (the three-dimensional island has a greater surface than an equivalent two-dimensional film). The relaxation and surface energies also are the main terms in the total energy of a quantum dot. In the calculations of Pehlke et al., surface reconstructions and surface energies for different semiconductor surfaces have been calculated by density functional theory [68–70]. The surfaces are shown in the upper panel of Fig. 9. The elastic energy and the strain field (the local distortion of the lattice) can be accurately described by elasticity theory (if the island has more than one thousand atoms, as the authors find). The equilibrium shape of the island is then calculated within a finite element framework by determining the minimum of the total energy of the island with respect to its shape. Each finite element represents a part of the island and the substrate in which the atoms are similarly arranged. Taking the strain into account by modeling all the atoms would be expensive computationally,

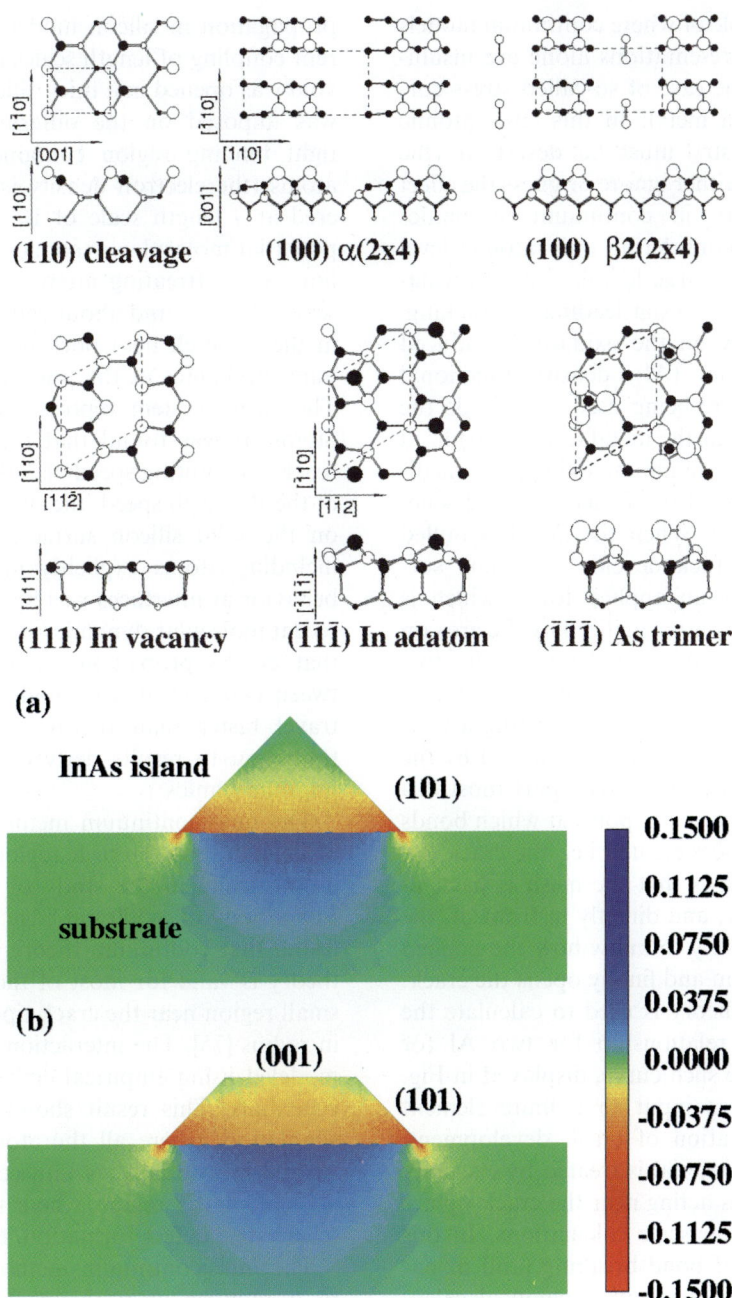

Fig. 9. Island shape determined by a multi-scale method combining ab initio calculations of surface energies and stresses with an elasticity theory description of long-range strain fields and strain relaxations. The top panel shows the surface models that were used in the quantum mechanical calculations. The lower panel shows the strain (a measure of the distortion of the crystal) for two islands of different shape and the underlying substrates. (Figure from Ref. [69], courtesy of N. Moll.)

and moreover unnecessary, since long-range strain is well described by elasticity theory. The lower panel of Fig. 9 shows the color-coded strain of the island calculated with the finite element method.

As another example of where continuum models alone or atomic representations alone are insufficient, we consider the case of so-called stress corrosion cracking of a metal. In this case, atomic level surface chemistry must be described (the usual tendency is to either ignore or guess the effect of surface chemistry in continuum mechanics models). However, considering only atomic level chemistry ignores the large length scale deformations and defects in a crystal leading to cracking. In an ongoing study on the oxidatively induced fracture of aluminum (Al), density functional theory calculations are being used to provide the forces necessary to tear the metal (and its oxidized surface) apart [71]. Aluminum oxidizes immediately when it is exposed to air, as indicated schematically in Fig. 4(a). When the metal is pulled (under tension), a crack develops, forming surfaces. The pulling is an applied force, which is called traction when it is normalized by the area on which it is felt. The distance between the two surfaces of the crack is called the separation δ. Fig. 4(b) shows a finite element mesh covering a rectangular body which has a crack, indicated by the red line drawn into the body. The part most difficult to describe is the actual point at which bonds are broken and surfaces created i.e., the crack tip. The zoom image shows that the mesh is particularly fine there. Near, and directly in front of, the crack tip, it is necessary to know how the applied traction induces strain and finally opens the crack. Density functional theory is used to calculate the traction–separation relationship for two Al (or Al_2O_3) surfaces. One such curve, displayed in Fig. 4(c), may be used as input to a finite element method-based simulation of crack development. The interior of the material is treated by elasticity theory, but the forces acting near the crack tip are taken from the atomic-scale calculations. In this way, the chemistry of bond breaking (and oxidation) is described by quantum-mechanical calculations, while the crack formation in the solid is treated by a continuum model, called the cohesive zone model [72,73]. This approach also can be used to study grain boundary embrittlement by impurities (vide infra).

To get a feeling for the dimensions and numbers involved, let us take a look at the study of crack propagation in silicon modeled using the concurrent coupling of length scales method [43]. Here, a void was opened in a thick silicon slab and a strain was imposed on the outermost boundary. The tight binding region contained several hundred atoms (the electron density is (implicitly) considered at a length scale of 10^{-10} m), the empirical potential molecular dynamics region about a million atoms (treating atomistic physics at the nm scale, 10^{-9} m), and about 250,000 nodes were used in the finite element portion of the solid (continuum mechanics at the µm scale, 10^{-6} m) [43,44]. The entire system represented about 11 million atoms. It was found that the propagating crack reached a limiting speed of 2770 m/s, equal to 85% of the Rayleigh speed, the speed that sound travels on the solid silicon surface. The importance of including atoms explicitly in the description of behavior at interfaces was highlighted further in a recent molecular dynamics study [74], which found that cracks propagating along an interface between two crystals can, under certain conditions, travel faster than the Rayleigh speed, in contradiction to results derived purely from continuum mechanics.

The quasicontinuum method has been used to investigate how large a region has to be treated atomistically in the study of crack propagation. For a body of ~500 nm edge length, Miller et al. found that continuum theory and linear elasticity theory is valid for most of the solid, except for a small region near the crack tip of only about 1 nm in radius [75]. The interaction between atoms was modeled using empirical embedded atom method potentials. This result shows that a brute force simulation, where all the atoms in the body are treated individually, is unnecessary except within ~1 nm of the region where the transformations (chemistry, defect formation, etc.) are occurring.

The quasicontinuum method has been applied to a number of problems researching the mechanical stability of crystals. One study investigated what happens when cracks encounter grain boundaries in solids [75]. Grain boundaries are found where small fragments of crystal of the same material come together but the crystal lattices do not match exactly. Depending on the situation, the grain boundary can stop (blunt) the crack or the crack can reach the boundary and then branch, running along the boundary leading to a separation of the grains. Surface chemistry can be important here, when impurities segregate to the grain boundary. These may make the material more brittle. Again, density functional theory combined with continuum theory is needed to address both length scales; such work is in progress [76].

As shown above, we have taken the first steps toward accurate modeling of surface and interface phenomena varying in length scale by powers of ten. We next consider current methods to bridge the time scale gap between atomic motions (10^{-12} s) and materials growth (seconds or minutes).

3.1.2. Across time scales

In this arena, the main surface science problem that has attracted attention is crystal growth. Of particular interest is how to choose the process conditions to achieve specific structures. The atoms comprising newly grown layers often come from the gas phase and tend to be involved in complex interactions with the surface atoms. Kinetic Monte Carlo is a natural method to choose, since it can include atomic scale processes while overcoming the long time scales for growth. One recent application of kinetic Monte Carlo, which illustrates its versatility, used rate constants derived from embedded atom method potentials to examine nickel (Ni) growth in the presence of hydrogen (H) [77]. Here, the elementary processes exhibit time constants (the inverse of their rate constants) ranging from $\sim 10^{-9}$ s for an isolated H hop to $\sim 10^{7}$ s for the breaking of a Ni–Ni bond, spanning 16 orders of magnitude from the extremely fast to the very slow. There is no other method available that can span such an impressive spread in the characteristic times for various processes. The presence of small amounts of H was predicted to increase the average Ni island size, helping the Ni atoms to gather together more effectively than they would on their own. Additionally, kinetic Monte Carlo has been used in connection with rate constants from density functional theory to investigate the growth of Al on Al [49], specific chemical reactions in diamond growth (in connection with a tight binding method) [78], and the nucleation of silver (Ag) islands on platinum [79].

In desorption of hydrogen molecules from silicon surfaces, the high accuracy of ab initio quantum chemistry calculations was exploited to calculate the rate constants for kinetic Monte Carlo [4,37,47]. This model included many surface processes, e.g., H atom diffusion, defect migration, monohydride to dihydride interconversion, etc. While the kinetics were found to be first order—in agreement with experiment—for all coverages, temperatures, and independent of the presence of steps or defects, the overall desorption rate constant was quite sensitive to the concentration of surface vacancies (missing atoms). This helped rationalize the factor of 16 range in measured desorption rate constants, as perhaps due to surface preparation differences which result in variable concentrations of vacancies. To derive these results from a molecular dynamics investigation would have been prohibitively expensive due to the range of time scales involved (from very fast adatom surface diffusion to slow molecule desorption).

As an alternative to kinetic Monte Carlo, accelerated molecular dynamics can be used with empirical potentials to explore long time scales. For example, Voter has used his hyperdynamics method with embedded atom method potentials to study the diffusion of a 10-atom Ag cluster on a Ag surface at room temperature [60]. In a simulated time of 220 μs, he observed three distinct mechanisms affecting cluster motion on the surface. The temperature-accelerated dynamics approach with embedded atom method potentials has been used to study film ripening (the evolution of an initially random surface until it settles into a stable configuration) on a copper (Cu) surface [63]. A simulation of 14 adatoms on a Cu slab with the three top layers free to move (182 moving atoms) was run for a simulated time of *several hours*. An ordinary molecular dynamics calculation with the same computational work would have only reached a time on the μs scale. The simulation showed that a compact island shape was reached after 17 min, which remained stable for hours. 48 different states were visited and about 3000 transitions occurred. Such details would not have been observed on the shorter, μs, time scale, illustrating the power and need for accelerated molecular dynamics.

3.2. Twenty-first century modeling

Surface science has matured as a field to the point that it is now found in many arenas beyond chemistry and physics, including geology, dentistry, prosthetic medicine, and sensor technology. Likewise, the field of computational sciences has grown to the point where methods are no longer routinely used in isolation. The complex problems of the 21st century demand a battery of approaches. The maturing of both surface science and computational methods allows us to dream of understanding—and then using our knowledge to solve—important problems, where surfaces and interfaces abound. They include: inorganic–inorganic interfaces and inorganic–biological interfaces found below the Earth's surface [80], in the human body [81], and in the products of corrosion [82], as well as fluid–solid interfaces important for both gas and liquid sensors [83]. The action at these surfaces involves multiple players on multiple length and time scales. For example, bone regrowth involves transport of calcium ions (atomic scale) and adhesion of cells (100 μm) via anchor proteins (<1 μm) [81]. Gas sensors involve adsorption of molecules and transport of, e.g., oxygen, while the sensor measurements may involve mechanical deformations or electrical conductivity on the microscale [83].

In the 21st century, we expect to see multi-scale modeling become a powerful partner with experiment to discern atomic scale effects on macroscopic properties. The future is expected to bring the ability to carry out, e.g., accelerated molecular dynamics and quasicontinuum calculations with forces derived not from empirical potentials but from accurate quantum calculations. Indeed the density functional theory-quasicontinuum coupling, via orbital-free density functional theory [45], is already under development [84]. More accurate means to couple different scales should appear (and other methods already exist that we have omitted in the interest of brevity!). In the end, this should allow us to simulate surface chemistry in real time and to determine its influence on the macroworld. In what follows, we give a few examples of where we can expect multi-scale modeling to impact surface science in the near future.

A classic surface chemical reaction is the oxidation of carbon monoxide (CO) by oxygen (O_2) as catalyzed by platinum surfaces. This reaction takes place daily as part of exhaust cleanup by the catalytic converter in automobiles. While the kinetics and thermodynamics has been understood for at least two decades [85], more recently fascinating kinetic oscillations have been uncovered, which are a beautiful example of atomic scale events affecting much larger length scale phenomena. In particular, complex concentration patterns emerge on the surfaces in space and time. In some regions of the surface, the dominant species is adsorbed oxygen atoms, while in others CO prevails. A variety of striking patterns have been observed, for which several mechanisms have been postulated depending on the temperature, the amount of CO and O_2 available, and the platinum surface used [85,86]. The basic origin of these patterns is the catalytic oxidation of the adsorbed CO to carbon dioxide (CO_2) (see Fig. 3). The patterns of the surface concentrations of CO and O, that can be observed in time, are fundamentally due to: (i) CO "poisoning", where high coverages of CO inhibit the dissociative adsorption of oxygen and (ii) the slow conversion rate of CO and O to CO_2. These two effects favor a high CO coverage. The adsorption of CO changes the structure of the platinum surface, however, altering at times the adsorption rate for oxygen. This allows an oxygen layer to be established, restoring the original surface structure [87]. By contrast, an oxygen layer, once formed, does not stop CO from being adsorbed on the surface. CO thus slowly returns and expels the oxygen layer. Spatial patterns on the micron scale develop when the reactions occur at different times in different areas on the surface. So far, kinetic models of these oscillating reactions embodied rate constants obtained from experiment [86]. The local time dependence of the concentrations is obtained through (differential) equations but long-range effects are neglected, in particular, the long-range surface reconstructions. A multi-scale description might start with ab initio calculation of the rate constants needed for a kinetic Monte Carlo simulation. Furthermore, it would go beyond the purely local description of the concentrations, taking into account size and shape of the areas (on the micron scale) where one or the other species is dominant. A two-dimensional finite element method could be combined with the kinetic Monte Carlo method to keep track of these areas. The same method could keep track of the subsurface oxygen, which is formed from the surface species, and thought to alter the surface structure [86]. This approach could yield a unified description of this complex system that may reveal more about the interplay of chemistry and structure across length scales.

An example with important industrial implications is the development of thermal barrier coatings to protect metal engine parts (e.g., in turbine engines) from the high operating temperatures (up to 1700 K) and oxidizing environment of the combustion gas mixture [88]. Such coatings could be used more generally to protect any material from harsh environments. While initial density functional theory studies characterized the strength of interfaces in thermal barrier coatings [89–91], accelerated molecular dynamics and kinetic Monte Carlo studies could help determine the interaction of gases with the thermal barrier coating, the importance of ion interdiffusion at interfaces, and the effect of heat on the mechanical stability of the coating–engine interface. Linking the density functional theory results to continuum descriptions of fracture is important for a comprehensive understanding of how and why the thermal barrier coatings fail and ultimately to the design of coatings with long lifetimes.

A promising area of high-tech research is the development of microelectromechanical systems. These are small devices of an overall size of a few millimeters, with feature sizes of a few μm. The devices have electrical and mechanical components that are combined into one unit and are designed to act as, e.g., miniature sensors, actuators, valves, or gears. In these miniaturized devices, surface effects are extremely important in every normal or unintentional contact during their operation [92]. Surface roughness needs to be minimized in order to reduce wear in moving parts. In a micromotor, roughness can lead to localized electric field concentrations and therefore premature breakdown. Additionally, at small length scales, the effect of friction on operation is much more dramatic than in larger elements. Exposure of microelectromechanical systems to humidity increases the friction. Several materials have been examined as structural materials or as thin film coatings to reduce friction and wear and to improve reliability and lifetime. Because of the optimization of silicon fabrication technology, these devices are typically made from silicon, followed by deposition of other materials as needed. Modeling both the production and coating procedures by density functional theory combined with accelerated molecular dynamics and kinetic Monte Carlo could help select the optimal processing strategy. Theoretical tests of proposed designs—commonly used for this purpose at larger length scales—could be used to model the element in its entirety (on the 10^{-3} m length scale), incorporating atomic-level friction (on the 10^{-10} m length scale of the electron density) as needed. Such multi-scale simulations offer a cheap possibility to customize the microelectromechanical systems for very different fields of operations. Simulations of microelectromechanical sensors for gas detection could reveal the atomic level reasons for measured changes in conductivity [83], which at this point are elusive. This is critical to development of sensors that not only are sensitive to species at the parts per billion (ppb) level and stable over a long time, but also exhibit the desired selectivity. Valuable information could be gleaned by linking the atomic scale adsorption processes to large length scale ion mobility measurements, so that sensors could be built that can discriminate between a garden-variety pesticide and a chemical warfare agent.

4. Conclusions

Multi-scale modeling methods are the key to deciphering some of the most intricate problems of surface science. The problems of materials degradation and synthesis, heterogeneous catalysis, gas sensing, liquid–solid interfaces in the body (e.g., cell adhesion) are far too complex on several different length and time scales to be treated by one method alone. For example, using atomic resolution, even coupled with an explicit treatment of quantum mechanics, is not an option to describe the structures that are present in modern day steels at the length scale of microns. Even if all atoms could be treated individually, such a simulation would be extraordinarily wasteful. It makes much more sense to try to describe regions where atoms behave similarly on a coarser scale, and to reserve atomic resolution for those areas where large deviations from the original structure are present. Multi-scale modeling promises to contribute to a wide variety of industrial applications including coating technology and gas microsensors. Multiscale simulations also can help solve fundamental questions of chemistry, such as the complex processes involved in the seemingly simple oxidation of carbon monoxide, that give rise to the spatiotemporal patterns displayed in Fig. 3. Many problems are treated now at a single length scale simply for lack of a better description. They will soon be within the reach of current and future multi-scale techniques. The methods discussed here, as well as ones anticipated that link accurate quantum energies and forces with higher length scale dynamics, are expected to become full-fledged partners with experiment in revealing the naked truth about surface and interface phenomena in the years to come.

Acknowledgements

This work has been funded by the DoD-MURI program, administered by the Air Force Office of Scientific Research. We wish to thank R.L. Hayes, E.A. Jarvis, Dr. N. Moll, Dr. E. Pehlke, Prof. W.A. Curtin, Prof. G. Ertl, and Prof. M. Ortiz for several illustrations.

References

[1] G. Binnig, H. Rohrer, C. Gereber, E. Weibel, Tunneling through a controllable vacuum gap, Appl. Phys. Lett. 40 (1982) 178.
[2] N.D. Spencer, Scanning probe microscopies, in: J.H. Moore, N.D. Spencer (Eds.), Encyclopedia of Chemical Physics and Physical Chemistry, vol. 2, IOP Publishing, 2001, p. 1467.
[3] J.N. Reddy, An Introduction to the Finite Element Method, McGraw-Hill, New York, 1984.
[4] M.R. Radeke, E.A. Carter, Ab initio dynamics of surface chemistry, Annu. Rev. Phys. Chem. 48 (1997) 243.
[5] P. Hohenberg, W. Kohn, Inhomogeneous electron gas, Phys. Rev. 136 (1964) B864.
[6] W. Kohn, L.J. Sham, Self-consistent equations including exchange and correlation effects, Phys. Rev. 140 (1965) A1133.
[7] J.P. Perdew, K. Burke, M. Ernzerhof, Generalized gradient approximation made simple, Phys. Rev. Lett. 77 (1996) 3865.
[8] G. Galli, Large-scale electronic structure calculations using linear scaling methods, Phys. Status Solidi B 217 (2000) 231.
[9] S.C. Watson, E.A. Carter, Linear scaling parallel algorithm for the first principles treatment of metals, Comp. Phys. Commun. 128 (2000) 67.
[10] A. Szabo, N.S. Ostlund, Modern Quantum Chemistry: Introduction to the Advanced Electronic Structure Theory, Dover Publications, Mineola, NY, 1996.
[11] M. Schütz, G. Hetzer, H.-J. Werner, Low-order scaling local electron correlation methods. I. Linear scaling local MP2, J. Chem. Phys. 111 (1999) 5691.
[12] G.E. Scuseria, P.Y. Ayala, Linear scaling coupled cluster and perturbation theories in the atomic orbital basis, J. Chem. Phys. 111 (1999) 8330.
[13] A. Omeltchenko, M.E. Bachlechner, A. Nakano, R.K. Kalia, P. Vashishta, I. Ebbsjö, A. Madhukar, P. Messina, Stress domains in $Si(111)/a$-Si_3N_4 nanopixel: ten-million-atom molecular dynamics simulations on parallel computers, Phys. Rev. Lett. 84 (2000) 318.
[14] K. Schwarz, E. Nusterer, P. Margl, P.E. Blöchl, Ab initio molecular dynamics calculations to study catalysis, Int. J. Quant. Chem. 61 (1997) 369.
[15] M. Parrinello, Simulating complex systems without adjustable parameters, Comput. Sci. Eng. 2 (2000) 22.
[16] P.C. Weakliem, C.J. Wu, E.A. Carter, First-principles-derived dynamics of a surface reaction: fluorine etching of $Si(100)$, Phys. Rev. Lett. 69 (1992) 200;
P.C. Weakliem, C.J. Wu, E.A. Carter, First-principles-derived dynamics of a surface reaction: fluorine etching of $Si(100)$, Phys. Rev. Lett. 69 (1992) 1475E.
[17] L.E. Carter, S. Khodabandeh, P.C. Weakliem, E.A. Carter, First-principles-derived dynamics of F_2 reactive scattering on $Si(100)$-(2×1), J. Chem. Phys. 100 (1994) 2277.
[18] L.E. Carter, E.A. Carter, Ab initio-derived dynamics for F_2 reactions with partially fluorinated $Si(100)$ surfaces: translational activation as a possible etching tool, J. Phys. Chem. 100 (1996) 873.
[19] C.W. Lo, P.R. Varekamp, D.K. Shuh, T.D. Durbin, V. Chakarian, J.A. Yarmoff, Substrate disorder induced by a surface chemical reaction: the fluorine–silicon interaction, Surf. Sci. 292 (1993) 171.

[20] J.A. Jensen, C. Yan, A.C. Kummel, Energy dependence of abstractive versus dissociative chemisorption of fluorine molecules on the silicon (111)-(7 × 7) surface, Science 267 (1995) 493.

[21] J.A. Jensen, C. Wan, A.C. Kummel, Direct chemisorption site selectivity for molecular halogens on the Si(111)-(7 × 7) surface, Phys. Rev. Lett. 76 (1996) 1388.

[22] E.R. Behringer, H.C. Flaum, D.J.D. Sullivan, D.P. Masson, E.J. Lanzendorf, A.C. Kummel, Effect of incident translational energy and surface temperature on the sticking probability of and on sticking probability of F_2 and O_2 on Si(100)-(2 × 1) and Si(111)-(7 × 7), J. Phys. Chem. 99 (1995) 12863.

[23] M.R. Tate, D. Gosalvez-Blanco, D.P. Pullman, A.A. Tsekouras, Y.L. Li, J.J. Yang, K.B. Laughlin, S.C. Eckman, M.F. Bertino, S.T. Ceyer, Fluorine atom abstraction by Si(100). I. Experimental, J. Chem. Phys. 111 (1999) 3679.

[24] A.A. Ayón, R. Braff, C.C. Lin, H.H. Sawin, M.A. Schmidt, Characterization of a time multiplexed inductively coupled plasma etcher, J. Electrochem. Soc. 146 (1999) 339.

[25] P.C. Weakliem, E.A. Carter, Surface and bulk equilibrium structures of silicon–germanium alloys from Monte Carlo simulations, Phys. Rev. B 45 (1992) 13458.

[26] L.E. Carter, P.C. Weakliem, E.A. Carter, Temperature and composition dependent structures of Si_xGe_{1-x}/Si and Si_xGe_{1-x}/Ge superlattices, J. Vac. Sci. Technol. A 11 (1993) 2059.

[27] M.R. Radeke, E.A. Carter, Interfacial strain-enhanced reconstruction of Au multilayer films on Rh(100), Phys. Rev. B 51 (1995) 4388.

[28] J.P. Valleau, S.G. Whittington, A guide to Monte Carlo for statistical mechanics, in: B.T. Berne (Ed.), Statistical Mechanics A. Modern Theoretical Chemistry, vol. 5, Plenum, New York, 1977, pp. 137–194.

[29] S. Ozawa, Y. Sasajima, D.W. Heermann, Monte Carlo simulations of film growth, Thin Solid Films 272 (1996) 172.

[30] J.H. Weiner, Statistical Mechanics of Elasticity, Wiley, New York, 1983.

[31] V.B. Shenoy, R. Miller, E.B. Tadmor, D. Rodney, R. Phillips, M. Ortiz, An adaptive finite element approach to atomic-scale mechanics—the quasicontinuum method, J. Mech. Phys. Solids 47 (1999) 611.

[32] J.A. Appelbaum, D.R. Hamann, Self-consistent electronic structure of solid surfaces, Phys. Rev. B 6 (1972) 2166.

[33] P.J. Feibelman, Force and total-energy calculations for a spatially compact adsorbate on an extended, metallic crystal surface, Phys. Rev. B 35 (1987) 2626.

[34] J.E. Inglesfield, G.A. Benesh, Surface electronic structure: embedded self-consistent calculations, Phys. Rev. B 37 (1988) 6682.

[35] H. Krakauer, M. Posternak, A.J. Freeman, Linearized augmented plane-wave method for the electronic band structure of thin films, Phys. Rev. B 19 (1979) 1706.

[36] W.H. Press, S.A. Teukolsky, W.T. Vetterling, B.P. Flannery, Numerical Recipes in C: The Art of Scientific Computing, second ed., Cambridge University Press, Cambridge, 1992.

[37] M.R. Radeke, E.A. Carter, A dynamically and kinetically consistent mechanism for H_2 adsorption/desorption from Si(100)-(2 × 1), Phys. Rev. B 54 (1996) 11803.

[38] J.L. Whitten, H. Yang, Theory of chemisorption and reactions on metal surfaces, Surf. Sci. Rep. 24 (1996) 59.

[39] N. Govind, Y.A. Wang, E.A. Carter, Electronic-structure calculations by first-principles density-based embedding of explicitly correlated systems, J. Chem. Phys. 110 (1999) 7677.

[40] T. Klüner, N. Govind, Y.A. Wang, E.A. Carter, Prediction of electronic excited states of adsorbates on metal surfaces from first principles, Phys. Rev. Lett. 86 (2001) 5954.

[41] E.B. Tadmor, M. Ortiz, R. Phillips, Quasicontinuum analysis of defects in solids, Phil. Mag. A 3 (1996) 1529.

[42] E.B. Tadmor, G.S. Smith, N. Bernstein, E. Kaxiras, Mixed finite element and atomistic formulation for complex crystals, Phys. Rev. B 59 (1999) 235.

[43] J.Q. Broughton, F.F. Abraham, N. Bernstein, E. Kaxiras, Concurrent coupling of length scales: methodology and application, Phys. Rev. B 60 (1999) 2391.

[44] R.E. Rudd, J.Q. Broughton, Concurrent coupling of length scales in solid state systems, Phys. Status Solidi B 217 (2000) 251.

[45] Y.A. Wang, E.A. Carter, Orbital-free kinetic-energy density functional theory, in: S.D. Schwartz (Ed.), Theoretical Methods in Condensed Phase Chemistry, Kluwer, Boston, 2000, pp. 118–184.

[46] Y.-T. Lu, H. Metiu, Growth kinetics simulations of the Al–Ga self-organization on GaAs(100) stepped surfaces, Surf. Sci. 245 (1991) 150.

[47] M.R. Radeke, E.A. Carter, Ab initio derived kinetic Monte Carlo model of H_2 desorption from Si(100)-(2 × 1), Phys. Rev. B 55 (1997) 4649.

[48] K.A. Fichthorn, W.H. Weinberg, Theoretical foundations of dynamical Monte Carlo simulations, J. Chem. Phys. 95 (1991) 1090.

[49] P. Ruggerone, A. Kley, M. Scheffler, Bridging the length and time scales: from ab initio electronic structure calculations to macroscopic proportions, Comments Cond. Mat. Phys. 18 (1998) 261.

[50] A.B. Bortz, M.H. Kalos, J.L. Lebowitz, A new algorithm for Monte Carlo simulation of Ising spin systems, J. Comput. Phys. 17 (1975) 10.

[51] G.S. Sun, M. Villarba, H. Jónsson, private communication.

[52] D.W. Bassett, P.R. Webber, Diffusion of single adatoms of platinum, iridium and gold on platinum surfaces, Surf. Sci. 70 (1978) 520.

[53] J.D. Wrigley, G. Ehrlich, Surface diffusion by an atomic exchange mechanism, Phys. Rev. Lett. 44 (1980) 661.

[54] G.L. Kellogg, P.J. Feibelman, Surface self-diffusion on Pt(001) by an atomic exchange mechanism, Phys. Rev. Lett. 64 (1990) 3143.

[55] C.H. Bennett, Exact defect calculations in model substances, in: A.S. Nowick, J.J. Burton (Eds.), Diffusion in

Solids: Recent Developments, Academic Press, New York, 1975, pp. 73–113.

[56] D. Chandler, Statistical mechanics of isomerization dynamics in liquids and the transition state approximation, J. Chem. Phys. 68 (1978) 2959.

[57] D.G. Truhlar, W.L. Hase, J.T. Hynes, Current status of transition-state theory, J. Phys. Chem. 87 (1983) 2664.

[58] E.K. Grimmelmann, J.C. Tully, E. Helfrand, Molecular dynamics of infrequent events: thermal desorption of xenon from a platinum surface, J. Chem. Phys. 74 (1981) 5300.

[59] A.P.J. Jansen, Compensating hamiltonian method for chemical reaction dynamics: Xe desorption from Pd(100), J. Chem. Phys. 94 (1991) 8444.

[60] A.F. Voter, A method for accelerating the molecular dynamics simulation of infrequent events, J. Chem. Phys. 106 (1997) 4665.

[61] A.F. Voter, Hyperdynamics: accelerated molecular dynamics of infrequent events, Phys. Rev. Lett. 78 (1997) 3908.

[62] A.F. Voter, Parallel replica method for dynamics of infrequent events, Phys. Rev. B 57 (1998) R13985.

[63] M.R. Sørensen, A.F. Voter, Temperature-accelerated dynamics for simulation of infrequent events, J. Chem. Phys. 112 (2000) 9599.

[64] G.P. Srivastava, Theory of semiconductor surface reconstruction, Rep. Prog. Phys. 60 (1997) 561.

[65] R.C. Cammarata, Continuum model for surface reconstruction in (111) and (100) oriented surfaces of fcc metals, Surf. Sci. 279 (1992) 341.

[66] M.S. Daw, M.I. Baskes, Semiempirical, quantum mechanical calculation of hydrogen embrittlement in metals, Phys. Rev. Lett. 50 (1983) 1285.

[67] M. Tomellini, M. Fanfoni, Kinetic theory of cluster impingement in film growth at solid interfaces: interplay between island density and adatom lifetime, Surf. Sci. 450 (2000) L267.

[68] E. Pehlke, N. Moll, A. Kley, M. Scheffler, Shape and stability of quantum dots, Appl. Phys. A 65 (1997) 525.

[69] N. Moll, M. Scheffler, E. Pehlke, Influence of surface stress on the equilibrium shape of strained quantum dots, Phys. Rev. B 58 (1998) 4566.

[70] Q.K.K. Liu, N. Moll, M. Scheffler, E. Pehlke, Equilibrium shapes and energies of coherent strained InP islands, Phys. Rev. B 60 (1999) 17008.

[71] R.L. Hayes, E.A. Jarvis, E.A. Carter, O. Nguyen, M. Ortiz, in press.

[72] A. Needleman, A continuum model for void nucleation by inclusion debonding, J. Appl. Mech., ASME 54 (1987) 525.

[73] J.R. Rice, Dislocation nucleation from a crack tip: an analysis based on the Peierls concept, J. Mech. Phys. Sol. 40 (1992) 239.

[74] F.F. Abraham, H. Gao, How fast can cracks propagate? Phys. Rev. Lett. 84 (2000) 3113.

[75] R. Miller, E.B. Tadmor, R. Phillips, M. Ortiz, Quasicontinuum simulation of fracture at the atomic scale, Modell. Simul. Mater. Sci. Eng. 6 (1998) 607.

[76] P. Maragakis, I. Park, R.L. Hayes, E.A.A. Jarvis, O. Nguyen, E. Kaxiras, E.A. Carter, M. Ortiz, work in progress.

[77] K. Haug, N.-K.N. Do, Kinetic Monte Carlo study of the effect of hydrogen on the two-dimensional epitaxial growth of Ni(100), Phys. Rev. B 60 (1999) 11095.

[78] I.I. Oleinik, D.G. Pettifor, A.P. Sutton, J.E. Butler, Theoretical study of chemical reaction on CVD diamond surfaces, Diamond Rel. Mater. 9 (2000) 241.

[79] K.A. Fichthorn, M. Scheffler, Island nucleation in thin-film epitaxy: a first-principles investigation, Phys. Rev. Lett. 84 (2000) 5371.

[80] G.E. Brown Jr. et al., Metal oxide surfaces and their interactions with aqueous solutions and microbial organisms, Chem. Rev. 99 (1999) 77.

[81] F.H. Jones, Teeth and bones: applications of surface science to dental materials and related biomaterials, Surf. Sci. Rep. 42 (2001) 75.

[82] Special issue on corrosion science, MRS Bulletin, July 1999.

[83] S. Semancik, R. Cavicchi, Kinetically controlled chemical sensing using micromachined structures, Acc. Chem. Res. 31 (1998) 279.

[84] R.L. Hayes, M. Fago, F. Starrost, P.A. Madden, M. Ortiz, E.A. Carter, work in progress.

[85] C.T. Campbell, G. Ertl, H. Kuipers, J. Segner, A molecular beam study of the catalytic oxidation of CO on a Pt(111) surface, J. Chem. Phys. (1980) 5862.

[86] A. von Oertzen, H.H. Rotermund, A.S. Mikhailov, G. Ertl, Standing wave patterns in the CO oxidation reaction on a Pt(110) surface: experiments and modeling, J. Phys. Chem. B 104 (2000) 3155.

[87] M.P. Cox, G. Ertl, R. Imbihl, J. Rustig, Non-equilibrium surface phase transitions during the catalytic oxidation of CO on Pt(100), Surf. Sci. 134 (1983) L517.

[88] A. Christensen, E.A.A. Jarvis, E.A. Carter, Atomic-level properties of thermal barrier coatings: characterization of metal–ceramic interfaces, in: R.A. Dressler, C. Ng (Eds.), Chemical Dynamics in Extreme Environments, Advanced Series in Physical Chemistry, vol. 11, World Scientific, Singapore, 2001, p. 490.

[89] A. Christensen, E.A. Carter, First principles characterization of a hetero-ceramic interface: $ZrO_2(001)$ deposited on an α-$Al_2O_3(1\bar{1}02)$ substrate, Phys. Rev. B 63 (2000) 16968.

[90] A. Christensen, E.A. Carter, Adhesion of ultrathin $ZrO_2(111)$ films on Ni(111) from first principles, J. Chem. Phys. 114 (2001) 5816.

[91] E.A. Jarvis, A. Christensen, E.A. Carter, Weak bonding of alumina coatings on Ni(111), Surf. Sci. 487 (2001) 55.

[92] Z. Rymuza, Control tribological and mechanical properties of MEMS surfaces. Part 1: Critical review, Microsyst. Technol. 5 (1999) 173.

[93] S. Nettsheim, A. von Oertzen, H.H. Rotermund, G. Ertl, Reaction diffusion patterns in the catalytic co-oxidation on Pt(110), J. Chem. Phys. 98 (1993) 9977.

[94] J.H. Rose, J.R. Smith, J. Ferrante, Universal features of bonding in metals, Phys. Rev. B 28 (1983) 1835.

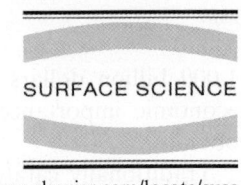

The virtual chemistry lab for reactions at surfaces: Is it possible? Will it be useful?

Axel Groß *

Physik-Department T30, Technische Universität München, James-Franck Str., D-85747 Garching, Germany

Received 5 July 2000; accepted for publication 27 April 2001

Abstract

Ab initio total-energy calculations based on electronic structure theory have tremendously enlarged our knowledge about the geometrical and electronic structure of clean and adsorbate-covered low-index surfaces and reactions on these surfaces. In technological applications, however, extended flat surfaces are very rarely used. Hence the applicability of the theoretical results for the technological surfaces are indeed questionable. In this review I will reflect on the question whether ab initio calculations of reactions at surfaces can contribute to the development of, e.g., better catalysts. Simulations alone will not be able to lead to new products but it will be demonstrated that they can contribute enormously to the development process. Thus the virtual chemistry lab is indeed possible and helpful. © 2001 Elsevier Science B.V. All rights reserved.

Keywords: Ab initio quantum chemical methods and calculations; Atomistic dynamics; Computer simulations; Density functional calculations; Models of surface chemical reactions; Catalysis; Sticking; Surface chemical reaction

1. Introduction

It has always been the goal of theoretical surface science to understand the fundamental principles that govern the geometric and electronic structure of surfaces and the processes occurring on these surfaces like gas-surface scattering, reactions at surfaces or growth of surface layers [1]. Processes on surfaces play an enormously important technological role. We are all surrounded by the effects of these processes on surfaces in our daily life. Some are rather obvious to us like rust and corrosion. These are reactions that we would rather like to avoid. Less obvious are surface reactions that are indeed very advantageous. Many chemical reactions are promoted tremendously if they take place on a surface that acts as a catalyst. Actually most reactions employed in the chemical industry are performed in the presence of a catalyst. Catalysts are not only used to increase the output of a chemical reaction but also to convert hazardous waste into less harmful products. The most prominent example is the car exhaust catalyst. The enormous technological relevance of catalysts is also reflected by the large worldwide demand of catalysts which, e.g., amounted to 6.4 billion dollars in 1996. More than 90% of the chemical manufacturing processes employed throughout the world utilize catalysts in one form or another [2]. And since in total chemicals worth more than

---

* Tel.: +49-89-289-12355; fax: +49-89-289-12296.
  E-mail address: agross@ph.tum.de (A. Groß).

1,000 billion dollars are produced annually, the economic importance of catalyst development is obvious.

Traditionally catalysts have been designed and improved by a trial and error approach. For example, before the iron-based catalyst for the ammonia synthesis was found, about 10,000 different catalysts had been tried by A. Mittasch in the first decade of the century [3]. This trial and error approach is still employed nowadays. A modern version of this approach is used in Combinatorial Chemistry [4–6] where the search is automated so that a huge number of different systems can be checked within a short period of time.

Now the question arises whether theoretical surface science can contribute to technological advances in the field of processes at surfaces. In the last decade there had been a tremendous progress as far as the accurate and reliable theoretical description of the geometric and electronic structure of surfaces and the interaction of atoms and molecules with surfaces is concerned. This progress was caused partially by the improvement in computer power, but also to a large extent by the improvement of programs based on ab initio electronic structure calculations [7–11] and other methods that use information coming from electronic total-energy calculations as an input.

The information obtained by electronic structure calculations is complete and detailed enough in order to construct whole potential energy surfaces (PESs) which allows to perform dynamical simulations of the reactions. The great advantage of such calculations compared to the experiment is that the calculations can shed light on the underlying electronic factors that determine bond making and breaking processes at surfaces; in addition, one can follow the microscopic steps of a reaction while in the experiment usually only the initial reactants and final products can be determined. Using some examples I will illustrate what has been learned in recent years by modern computational simulations about reactions at surfaces.

There is one basic question for any scientific research with the claim to contribute to the technological progress. Is it better to study and discover fundamental principles and mechanisms that will in the long run be beneficial for the overall progress, or should one concentrate on the development of particular products? In my opinion there is no definite answer to this question. Often the lines are even blurred between the two approaches so that this clear distinction between a more basic and a more applied approach cannot be made. But I will show that ab initio based calculations of properties and processes at surfaces have not only tremendously enlarged our knowledge about basic mechanisms, but that they are also now accepted as a very valuable tool for the development of particular products.

Quantum chemistry methods based on Hartree–Fock theory have been an integral part of research and development in the chemical and pharmaceutical industry for more than 20 years. These methods that are based on representations of the electronic many-body wave function have also significantly contributed to our understanding of processes at surfaces (see, e.g., Refs. [12,13]). Unfortunately, wave function based methods are limited to rather small systems. Electronic structure algorithms based on density functional theory (DFT) can be applied to larger systems. The theoretical chemistry community has long been reluctant to accept DFT methods because they considered them as not accurate enough, but now DFT becomes more and more accepted due to some fundamental developments in the formulation of DFT algorithms that greatly improved the accuracy of the DFT calculations. And now even material science related industries such as catalyst manufacturers and semiconductor industries begin to realize the potential of electronic structure calculations.

The theoretical description of reactions is often much less expensive than carrying out many different experiments. Hence besides satisfying the curiosity of a scientist there is actually an economic need for the quantitative description of reaction pathways at surfaces. The computational tools are now reliable enough for a sufficiently accurate determination of chemical interactions at surfaces for many systems. I will show that the surface science approach is not only an intellectual endeavor, but that even industrially used catalysts have been developed by a combination of theoretical and experimental surface science techniques.

In this review I will first introduce basic theoretical concepts necessary to treat processes at surfaces. Although electronic structure theory calculations based on Hartree–Fock theory continue to contribute to the field of theoretical surface science, I will concentrate on DFT because of its widespread use in the theoretical treatment of surface processes. I will give some examples of successful applications of ab initio calculations in the fields of both basic and applied research. The list of examples does not claim at all to be complete. The selection is certainly biased through my personal view, but I am convinced that the studies presented will show that a virtual chemistry lab of reactions at surfaces is indeed possible and helpful.

However, we are still far away from the development of technological products solely on a theoretical basis. It is even rather likely that this will never become possible. It is fair to say that most quantitative simulations of reactions at surfaces are still limited to rather simple systems like molecular dissociation on close-packed metal surfaces. There is plenty of room for the theoretical description of more complicated systems like reactions on structured substrates. The determination of the reaction pathways for these more complicated reactions and their dynamical simulation even may require the improvement of the theoretical tools. I will sketch the theoretical challenges that are still lying ahead for a quantitative description of reaction pathways, but also the opportunities that open up once these problems have been solved.

2. Brief sketch of electronic structure theory

The theoretical treatment of reactions at surfaces requires in principle the solution of the quantum mechanical many-body problem including all electronic and ionic degrees of freedom. In practice, this solution is not possible in closed form. However, in many chemical reactions the electrons follow the motion of the nuclei almost adiabatically due to the large mass difference. We will concentrate on reactions of this type in this review, i.e., we will assume that the Born–Oppenheimer approximation [14] is valid. This means that we can break up the theoretical description of a chemical reaction. First the electronic problem for a particular configuration of the nuclei will be solved, and only then the motion of the nuclei has to be considered.

Still the solution of the electronic problem represents a formidable task. Of central importance for the progress in the theoretical description of processes at surfaces are the recent developments in electronic structure theory. There are two main different approaches to electronic structure calculations, based on either Hartree–Fock theory or DFT. Both approaches are called first-principles or ab initio theories because in principle they do not require any empirical parameter, i.e. they are derived *from first principles*.

Electronic structure calculations based on Hartree–Fock theory determine the many-body wave function by directly solving the Schrödinger equation. They are also called wave function based methods. As far as these methods are concerned, I refer to the excellent "non-mathematical" review by Head-Gordon [7] and references therein. Wave function based methods usually describe the electrons by a localized basis set derived from atomic orbitals. Within this localized description the surface is modeled by a finite cluster. Since the computational effort scales very unfavorably with the system size in wave function based methods, these calculations are limited to a rather small number of atoms, typically about 10–20. This is often not sufficient to model an extended substrate. Cluster calculations can still yield qualitative trends, but they are "often best used for explanatory rather than predictive purposes" [7].

Electronic structure calculations based on DFT allow the treatment of larger systems. Since I focus on calculations using DFT, I will briefly sketch some fundamentals of DFT. As the name DFT already suggests, the electron density is the basic quantity in DFT [15]. The Hohenberg–Kohn theorem [16] states that the ground state expectation value of the total energy is a unique functional of the ground state electron density. Even more important is the existence of a variational principle for the energy functional, namely that the exact ground state density and energy can be determined by the minimization of this energy functional.

This still sounds rather technical. But instead of using the many-body quantum wave function which depends on many coordinates now we only have a function of three coordinates which has to be varied. In practice, however, no direct variation of the density is performed. The density is rather expressed as a sum over single-particle states. These states are obtained by self-consistently solving effective single-particle equations, the Kohn–Sham equations [17], where the quantum mechanical many-body effects are contained in the so-called exchange-correlation potential. In principle the solution of the single-particle equation still requires the diagonalization of a rather large matrix. This diagonalization is avoided in almost all modern DFT algorithms by using the fact that the diagonalization can be regarded as a minimization problem for which many efficient algorithms exist.

The exchange-correlation potential is an universal functional of the electron density, i.e., it does not depend on any particular system. Unfortunately this non-local functional is not known; probably it is even impossible to determine it exactly in a closed form. What is known is the exchange-correlation potential for the homogeneous electron gas, i.e. for a system with a constant electron density [18]. In the so-called local density approximation (LDA), the exchange-correlation potential for the homogeneous electron gas is also used for non-homogeneous situations. At any point this local potential is used for the corresponding density, ignoring the non-locality of the true potential. Surprisingly, for many solid state systems the LDA yields rather satisfying results in good agreement with experiment. This is still not fully understood but probably due to a cancellation of opposing errors in the LDA. For chemical reactions in the gas phase and at surfaces the LDA results are not sufficiently accurate. That is the reason why many theoretical chemists were rather reluctant to use DFT for a long time. Only with the advent of exchange-correlation functionals in the generalized gradient approximation (GGA) [19–23] this situation has changed. In the GGA the gradient of the density is also included in the exchange-correlation functional, but the dependence on the gradient is modified in such a way as to satisfy important sum rules. DFT calculations in the GGA arrive at chemical accuracy (error $\leqslant 0.1$ eV) for many chemical reactions. Still there are important exceptions where the GGA also does not yield sufficient accuracy. The development of more accurate expressions for the exchange-correlation functional is certainly not finished yet.

In most molecules and materials the chemical interaction is governed by the valence electrons while the core electrons are hardly involved in the binding process. This fact is used in the pseudopotential concept [24] in which the influence of the core electrons on the other electrons is represented by an effective potential, the pseudopotential. This approach reduces the number of electrons dramatically that have to be taken into account explicitly in the calculations. Most modern DFT studies presented in this review employ the pseudopotential concept, and many large-scale computations would be impossible without the use of pseudopotentials.

There are very efficient DFT algorithms based on a plane-wave expansion of the Kohn–Sham single-particle states. These programs often originate from solid-state applications. However, they usually require a three-dimensional periodicity of the considered system. In the so-called supercell approach surfaces are modeled by periodically repeated slabs with a sufficient vacuum layer between them in order to avoid any interaction between the slabs. Furthermore, the slabs have to be thick enough to reproduce the correct electronic structure. Such a supercell geometry is shown in Fig. 1. In this approach the extended nature of the surfaces in the lateral direction is taken into account. Using this technique, modern efficient DFT algorithms can treat more than 100 [25] or even 300 [36] atoms per supercell.

3. The potential energy surface

In Section 2 we have described how the electronic structure problem can be solved for a particular configuration of the nuclei. To describe a chemical reaction, we need to know the potential energy of the system as a function of the coordi-

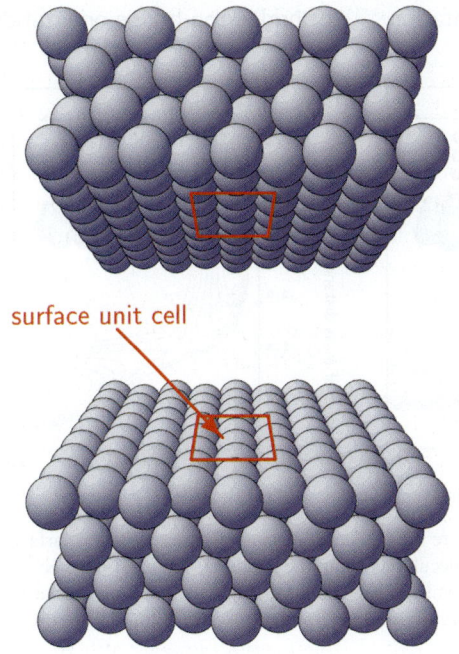

Fig. 1. Supercell approach to model surfaces: the surface is represented by a periodic stack of slabs with a sufficient layer of vacuum between them. The surface unit cell is indicated by the boxes on the surface.

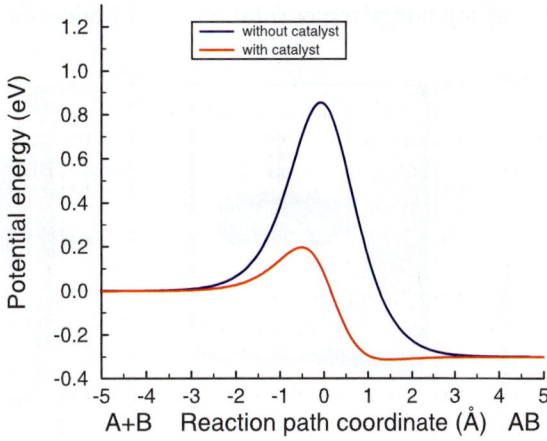

Fig. 2. Schematic drawing of the energetics along the reaction path for the reaction $A + B \rightarrow AB$ without and in the presence of a catalyst.

nates of the reactants. This defines the PES. The PES is the central quantity in any theoretical description because all chemical processes are governed by the underlying PES.

One important concept in the discussion of chemical reactions is the reaction path which is usually defined as the minimum energy path combining the region of reactants with the region of products in the PES. A plot of the potential energy along this path reveals, e.g., the minimum energy barrier hindering a reaction. This concept is usually very helpful for understanding the thermodynamics and kinetics of a certain reaction. Furthermore, qualitative concepts can thus be explained. This is demonstrated in Fig. 2 where the concept of catalysis is illustrated for a certain reaction $A + B \rightarrow AB$. This reaction might be hindered in the gas phase by a large barrier. In the presence of the catalyst the barrier is strongly reduced so that the reaction proceeds much more efficiently.

However, simple one-dimensional representations of the energetics are often oversimplifying the reaction mechanism and can be rather misleading. For example, Fig. 2 suggests that the presence of a catalyst simply leads to a reduction of the reaction barrier, nothing else. However, most often the reduction of the barrier is achieved through some additional intermediate reaction steps like the adsorption of the reactants on the catalyst. In a pictorial way one could say that the presence of the catalyst makes a detour in the multidimensional PES possible that has smaller barriers than the direct route.

Hence for a proper discussion of a chemical reaction the multidimensionality of the relevant PES has to be taken into account. One example of the multidimensionality of the PES is shown in Fig. 3. These two-dimensional cuts through the PES are derived from DFT calculation of the adsorption of O_2/Pt(1 1 1) by Eichler and Hafner [27]. The notation above the figures characterizes the dissociation path. For example, top–bridge–top orientation means that the center-of-mass is fixed over the bridge site and the two O atoms are oriented towards the top sites. The dissociation path is also illustrated in the insets in Fig. 3. These figures do not, however, correspond to the number of atoms considered in the calculations. As mentioned in the last section most DFT algorithms used nowadays for surface science studies employ a plane-wave basis set which requires a

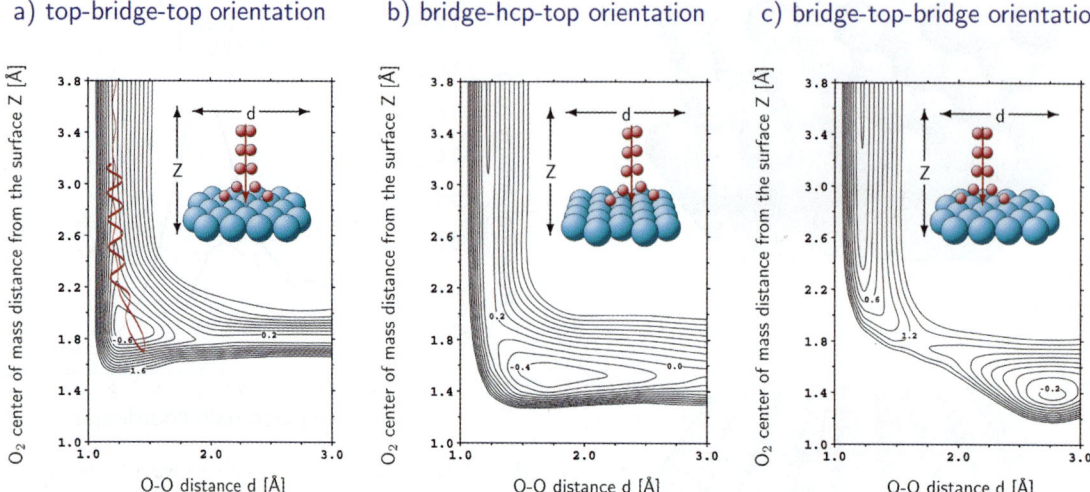

Fig. 3. PES of the dissociation of O_2/Pt(1 1 1). The coordinates in the figure are the O_2 center-of-mass distance from the surface Z and the O–O interatomic distance d. The insets illustrate the orientation of the molecular axis and the lateral O_2 center-of-mass coordinates. For the notation above the figures see the text. Energies are in eV per O_2 molecule. The contour spacing is 0.2 eV. The contour plots are derived from DFT calculations by Eichler and Hafner [27]. In (a) a trajectory of an O_2 molecule scattered at Pt(1 1 1) is also plotted.

three-dimensional periodicity of the calculated structures. In supercell calculations typically slabs are used that are between three and 10 layers thick while the periodic surface unit cell contains usually less than 10 atoms.

Even for a rather simple reaction like molecular adsorption and dissociation on a close-packed metal surface, the PES shows a strong dependence on the lateral coordinates of the molecular center-of-mass. While molecular adsorption above the bridge and the three-fold hollow sites is non-activated, above the top site the molecular adsorption is hindered by a barrier of more than 1.2 eV. What is not illustrated in Fig. 3 is the anisotropy of the PES. In the upright orientation the O_2 molecule cannot adsorb, i.e., the PES is repulsive.

Although the PES is an abstract object, the movement of the corresponding reactants on a PES can be understood and analyzed in very simple terms. An illustration is shown in Fig. 3a where also a trajectory of an O_2 molecule is plotted. This molecule is initially non-vibrating, but after the scattering process it is vibrationally excited. This can be inferred from the oscillating trajectory which corresponds to the molecular vibration. Furthermore, it is rather obvious that if the molecule is already initially vibrating, i.e., if it is oscillating back and forth in the d-direction, then it will be easier for the molecule "to make it around the curve" and enter the dissociation channel. This is called vibrationally enhanced dissociation [28] and was already realized 30 years ago [29].

One note of caution should be added. Total-energy calculations only yield a finite number of potential energies. Hence plotted PESs like in Fig. 3 always involve an interpolation scheme between the actual calculated points. If the grid of ab initio points is rather sparse then large regions of the PES are actually determined by the interpolation scheme and not by the actual calculations. Sometimes different interpolation schemes can give surprisingly varying results. In principle always the grid of the calculated ab initio points and the interpolation scheme should be reported, however, this is hardly ever done. The PES in Fig. 3 has actually been calculated by a tight-binding interpolation scheme [30], and the tick marks at the axes correspond to the chosen grid.

If many degrees of freedom are involved, PESs are often too complex to be mapped out in detail. Furthermore, for kinetic simulations not the whole

PES is of interest but rather the minimum energy barrier hindering a certain transition or reaction. In such a situation the saddle points in the multi-dimensional PES yield the relevant information. Once these saddle points are located, transition rates can be approximated via transition state theory [31]. There is no unique way to find a saddle point between two minima of the PES. However, powerful methods have been developed for climbing up the PES from minima to saddle points (see, e.g., Refs. [32,33]). These methods are often already part of standard ab initio electronic structure packages.

Another complimentary way of illustrating a reaction path obtained by ab initio calculations is to plot some atomic positions along this path. This is shown for the case of the CO oxidation on Ru(0 0 0 1) in Fig. 4 [34]. The sequence of snapshots shows how the CO axis is tilted and the atoms are lifted during the oxidation process. However, these snapshots do not yield any information about the energetics along the reaction path.

For a deeper understanding of a particular reaction the knowledge of the PES alone is usually not sufficient. One should rather analyze the electronic and geometric factors that lead to the specific topology of a PES. Fortunately total energy calculations also yield information on the electronic density and structure. Charge density plots are a good tool to analyze the interaction and hybridization of the chemical bonds between the reactants. We will show an example in Section 4.

Furthermore, an analysis of the density of states (DOS) for different configurations is also very helpful in understanding the reasons for the specific topology of a PES, for example, why and where energetic barriers in bond breaking or making processes are formed. However, the interpretation of a DOS evaluated from the Kohn–Sham eigenvalues of DFT calculations does not rest on firm theoretical grounds. The Kohn–Sham one-particle energies enter the theory as Lagrange multipliers ensuring the normalization of the wave functions; thus they correspond to quasiparticles with no specific physical meaning except for the highest occupied state [15]. Still it is almost always taken for granted that the Kohn–Sham eigenenergies can be interpreted, apart from a rigid shift, as the correct electronic one-particle energies. This is justified by the success since the Kohn–Sham eigenenergy spectrum indeed very often gives meaningful physical results, as will be shown later in this review.

4. Applications of first-principles calculations to reactions at surfaces

In this section I will illustrate the contribution that first-principles calculations have made to our understanding of reactions and processes at surfaces. First I will focus on fundamental studies in the field of theoretical surface science. Then I will show that nowadays first-principles studies, in particular DFT calculations, are advanced enough to even contribute to the development of technological products.

First I will review the situation in the early nineties and report on some of the first spectacular results of ab initio studies of hydrogen dissociation on clean flat surfaces. I will then proceed to some very recent examples of fundamental studies in which surface imperfections like steps and adatoms are considered. These studies represent the cutting edge of modern research. They show how far applications of electronic structure calculations can go, and at the same time they demonstrate what is just possible with modern algorithms and computers. Hence these examples can be regarded

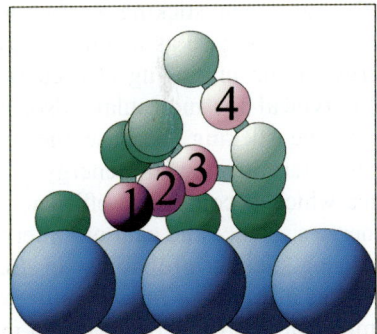

Fig. 4. Atomic positions along the reaction path of the CO oxidation on Ru(0 0 0 1). The blue, purple and green circles denote the Ru, C and O atoms, respectively (from Ref. [34]).

as an outlook; they also indicate the direction further ab initio investigations will take.

4.1. Basic research

In the early 1990s the computer power and the development of efficient electronic structure algorithms had been improved enough so that for example hundreds of DFT calculations with a few tens of atoms in a repeated supercell structure (see Fig. 1) became feasible on workstations. This made the calculation of the PES of simple reactions like the dissociative adsorption of hydrogen on metal surfaces possible [35–40]. In these calculations the surface is represented by a repeated slab structure separated by a sufficiently wide region of vacuum between the surfaces so that there is no interaction between the slabs. The first ab initio studies of molecular dissociation at surfaces mainly focused on hydrogen dissociation, the main reason being that the interaction of hydrogen with surfaces can be well-studied experimentally as well as theoretically [28,41]. This established the role of hydrogen adsorption as a model system where a fruitful interaction between theory and experiment has been possible.

One of the surprising results of these studies was that the PES of hydrogen dissociation on close-packed metal surfaces is highly corrugated. Before, it was anticipated that the PES depends only very weakly on the location of the molecule in the surface unit cell since the electron density in front of metal surfaces is rather smeared out [42,43]. This picture of a flat, structureless surface was seemingly confirmed by the experimentally found normal energy scaling of the sticking probability [44–46], i.e., the parallel component of the incident kinetic energy seemed to have no influence on the sticking. However, the bond-breaking process during the dissociative adsorption involves the hybridization of chemical bonds which has a rather localized nature.

The results of the ab initio calculations motivated new efforts in the dynamical simulation of processes at surfaces. These simulations have to be done quantum mechanically due to the light mass of hydrogen. For the dissociation of hydrogen on close-packed metal surfaces the recoil of the substrate atoms is usually negligible due to the large mass mismatch between the impinging hydrogen molecule and the metal atoms. Hence the hydrogen dissociative adsorption can be simulated within a six-dimensional PES corresponding to the six molecular degrees of freedom and fixed substrate atoms. Taking into account all hydrogen degrees of freedom quantum mechanically in a dynamics calculations was until recently still considered to be computationally impractical [47,48]. In the first studies the dynamics calculations were still performed in a restricted geometry, but the PES used in these simulations had features derived from electronic structure calculations [49–51]. One of the first successes of dynamics calculations based on a PES derived from ab initio calculations was the reconciliation of normal energy scaling with strong corrugation. Darling and Holloway showed [49] that for a special form of corrugation termed "balanced" corrugation [28] where the higher barriers are further away from the surface than the lower ones the effects of the corrugation for non-normal incidence of the impinging molecular beam effectively cancel. These features are indeed present in the calculated PES of H_2/Cu [35,36].

In 1995, the first dynamical study of hydrogen dissociative adsorption was performed in which all hydrogen degrees of freedom were treated quantum mechanically [52] using a time-independent coupled-channel scheme [53]. This study not only represented a technical achievement, but it also led to a new qualitative picture of hydrogen dissociation at reactive transition metal surfaces. In these systems the hydrogen sticking probability often shows an initial decrease as a function of the kinetic energy of the impinging molecules. Such a behavior is typical for molecular adsorption, i.e. non-dissociative sticking, because molecular adsorption is determined by the energy transfer to the surface which becomes less efficient at higher kinetic energies. Hence the canonic interpretation of a decreasing sticking probability in *dissociative adsorption* was a *precursor model*: before dissociation the molecule has to be trapped in a molecular adsorption state, the precursor to dissociation, and it is the trapping probability that determines the overall sticking probability.

The system chosen for the first six-dimensional quantum dynamical study, $H_2/Pd(100)$, indeed showed an initially decreasing sticking probability in the experiment [54]. However, the ab initio PES [37] on which the dynamical study was based did not exhibit any indication of a molecular precursor state. Still the dynamical calculations reproduced the experimentally observed trend of the sticking probability as a function of the initial kinetic energy of the hydrogen molecules.

The advantage of a computer simulation compared to an experiment is that the simulation is performed under well-defined conditions and can be analyzed at any point of the simulation. This analysis showed that the initially decreasing sticking probability is caused by a dynamical process which had been proposed before [55] but whose efficiency had been grossly underestimated: dynamical steering. This process can only be understood if one takes the multidimensionality of the PES into account. The PES of $H_2/Pd(100)$ shows purely attractive paths toward dissociative adsorption, but the majority of reaction paths for different molecular orientation and impact points exhibits energetic barriers hindering the dissociation. This coexistence of activated and non-activated paths is similar to the situation in the system $O_2/Pt(111)$ plotted in Fig. 3.

At very low kinetic energies the particles are so slow that they can be very efficiently steered to a favorable configuration for dissociation. This leads to a very high dissociation probability. Since this mechanism becomes less effective at higher kinetic energies, the reaction probability decreases. This scenario is illustrated in Fig. 5 where snapshots of two classical molecular dynamics runs of $H_2/Pd(100)$ on the PES based upon the ab initio results [37] are shown. The trajectories have the same initial conditions except for the kinetic energy. The initially non-rotating molecules approach the surface in an almost upright orientation in which the interaction with the surface is purely repulsive. However, in Fig. 5(a) the molecule is so slow ($E_{kin} = 0.01$ eV) that the torque acting upon the molecule is able to turn the molecule in a parallel orientation so that the molecule can directly dissociate. If the molecule is much faster as in Fig. 5(b) ($E_{kin} = 0.12$ eV), the rotation to the favorable parallel orientation cannot be completed, the molecule hits the repulsive wall of the potential and is scattered back to the gas phase.

The efficiency of the steering effect was also confirmed in an independent theoretical study [56]. Furthermore, predictions of the dependence of the reaction probability on the molecular rotation and

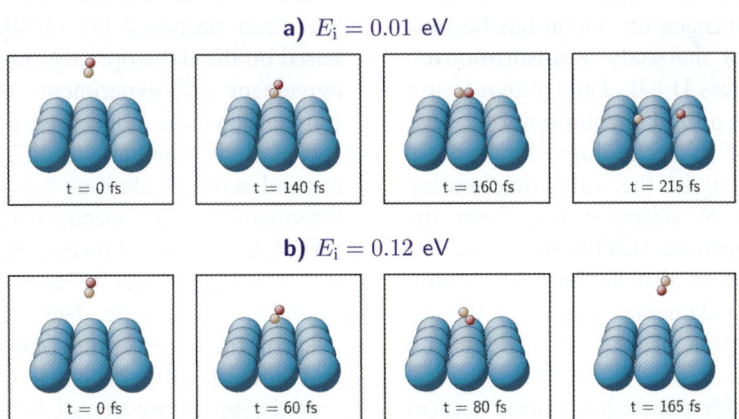

Fig. 5. Illustration of the steering effect. The figures show snapshots of classical molecular dynamics runs of $H_2/Pd(100)$ on a PES based upon ab initio results. Both trajectories have the same initial conditions except for the kinetic energy which is 0.01 eV in (a) and 0.12 eV in (b). Initially the molecules are not rotating. The slower molecule is turned to a parallel orientation which is favorable for dissociation while the faster molecule hits the repulsive wall of the potential before the re-orientation is completed and is scattered back into the gas phase.

orientation based on the ab initio dynamics calculations [52] have later been confirmed experimentally [57–59]. In the meantime, hydrogen dissociation in the activated systems $H_2/Cu(1\,0\,0)$ [60,61] and $H_2/(2 \times 2)S/Pd(1\,0\,0)$ [62] has also been treated with six-dimensional quantum dynamics calculations on PESs derived from DFT calculations.

Calculations of the reaction dynamics based upon PESs derived from DFT calculations are not only important for a deeper understanding of the reaction mechanism. In fact they also serve as a check of the accuracy of the DFT calculations. The information obtained from static total energies is often not sufficient to really judge the quality of the calculated PES. In the experiment the PES is never directly measured but the consequences of the PES on reaction rates and probabilities. For a true comparison with experiment also reaction rates and probabilities have to be derived from the calculated PESs. For a more detailed discussion of the theoretical treatment of reaction dynamics see the contribution by Bonn et al. in this volume [63].

In contrast to close-packed metal surfaces, semiconductor surfaces show a strong surface rearrangement upon hydrogen adsorption. This is caused by the covalent nature of the bonding in semiconductors where an additional chemisorbed adsorbate strongly perturbs the bonding situation of the substrate. Hydrogen on silicon has become the model system for the study of adsorption on semiconductor surfaces [1,64]. This system is not only interesting from a fundamental point of view, but also because of its technological relevance. Hydrogen desorption is the rate determining step in the growth of silicon wafers from the chemical vapor deposition (CVD) of silane. We will address this system also in the next section. Here we focus on fundamental aspects of the adsorption and desorption process of hydrogen on Si(1 0 0).

The so-called barrier puzzle has caused a great interest for this system: While the sticking coefficient of molecular hydrogen on Si surfaces is very small [65,66] indicating a high barrier to adsorption, the low mean kinetic energy of desorbed molecules [67] suggests a small adsorption barrier. This puzzle was assumed to be caused by the strong surface rearrangement of Si upon hydrogen adsorption [67]: The hydrogen molecules impinging on the Si substrate from the gas phase typically encounter a Si configuration which is unfavorable for dissociation, while desorbing hydrogen molecules leave the surface from a rearranged Si configuration with a low barrier.

It was immediately realized that the large influence of the dissociation barrier on the substrate configuration should cause a strong surface temperature dependence of the hydrogen dissociation probability on Si [68]. This predicted strong dependence was indeed confirmed experimentally [66,69,70]. Still it turned out later that the solution of the barrier puzzle involves a much more complex scenario taking into account surface structure and coverage effects, as will be shown below.

As for the ab initio calculations, $H_2/Si(1\,0\,0)$ is a system that initially was studied extensively by quantum chemical methods in which the extended substrate was modeled by a finite cluster [71–73]. Due to the localized nature of the covalent bonds in semiconductors it was believed that the cluster description for a Si surface might be appropriate. These cluster studies could not reproduce the experimentally observed activation energy for H_2 adsorption from $Si(1\,0\,0)$ for a clean surface. Therefore defect-mediated desorption mechanisms had been proposed [71,73,74]. DFT calculations based on the slab approach, however, were in good agreement with experiment, as far as the desorption energy was concerned [75–77]. There were speculations whether the difference between cluster and slab calculations was due to the different treatment of the electron exchange-correlation [78]. A later study showed [79] that one has to use rather large clusters to appropriately model the extended Si substrate. The slab approach is more efficient for representing semiconductor surfaces and consequently becomes more and more accepted [80]. However, all GGA functionals used nowadays have limitations as far as their accuracy is concerned [78]. Hence the development of improved exchange-correlation functionals still represents a very active area of research (see below).

The situation was not only confusing from the theoretical side, but also experiments of the adsorption of H_2/Si(1 0 0) showed large quantitative differences [65,70]. However, slowly a consistent picture of the interaction of hydrogen with Si(1 0 0) is emerging that is able to explain even the sometimes seemingly contradictory results [81,82]. This progress in the understanding has been achieved in a close collaboration between experiment and electronic structure calculations. It was realized that it is very important to determine the exact surface structure. At surface imperfections like steps the reactivity of a surface can be extremely altered. Indeed it was found experimentally on vicinal Si(1 0 0) surfaces that the sticking coefficient at steps is up to six orders of magnitude higher than on the flat terraces [83]. This finding was supported by DFT studies which showed that non-activated dissociation of H_2 on the so-called rebonded D_B steps on Si(1 0 0) is possible [25,83], while on the flat Si(1 0 0) terraces the dissociative adsorption is hindered by a barrier of 0.4 eV [77].

Indeed adsorbates can have a similar effect on the dissociation probability as steps since the electronic structure of the dangling bonds is perturbed in a similar way by both steps and adsorbates [81]. Recent scanning tunneling microscope experiments demonstrated that predosing the Si(1 0 0) surface by *atomic* hydrogen creates active sites at which the H_2 adsorption is considerably facilitated [84]. Actually the predosing of atomic hydrogen makes the adsorption of H_2 in an *interdimer* configuration possible, an adsorption configuration that is energetically unfavorable on the clean Si(1 0 0) surface compared to the *intradimer* pathway [85] (recent DFT calculations using large surface unit cells to avoid elastic strain, however, obtain the opposite result [81]).

Electronic structure calculations cannot only yield information on the energetics of a particular configuration or reaction pathway, but they can also provide an analysis of the underlying electronic structure that contributes to the PES. This is illustrated in Fig. 6 for the H_2 interdimer adsorption path on the Si(1 0 0) surface where one additional H atom acts as spectator. In order to understand the influence of this additional atom, charge density isosurfaces for single bands are

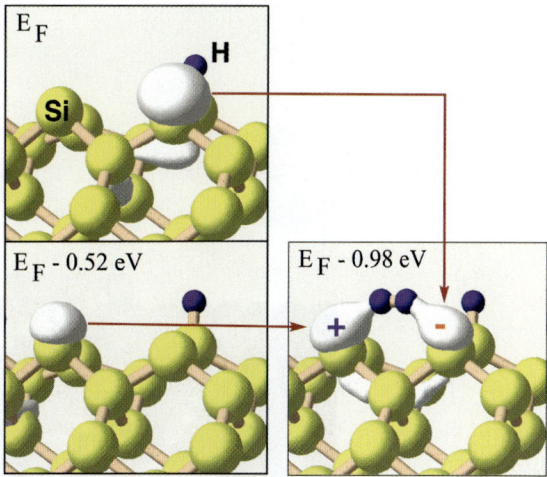

Fig. 6. Charge density isosurfaces of single bands illustrating the H_2 dissociation mechanism on a Si(1 0 0) surface along the interdimer reaction path where one adsorbed H atom on one of the Si dimers acts as a spectator. Left panel: clean surface before H_2 adsorption; right panel: at the transition state geometry (courtesy of E. Pehlke).

plotted. The isosurfaces correspond to a density of 0.005 electrons/bohr^3, and the states are characterized by their eigenenergies.

The left panel shows the dangling bond states of the interdimer Si atoms before the adsorption of the H_2 molecule. These states are at the Fermi energy and 0.5 eV below the Fermi energy. The right panel of Fig. 6 exhibits the charge distribution of the state 1.0 eV below the Fermi energy at the transition state to dissociation where the H–H bond breaks. The phase relation between both dangling bonds is indicated by the + and − sign. By a representation like this the chemical nature of bonds that are broken respectively formed during a dissociation process at a surface can be nicely analyzed.

Not only on semiconductor surfaces, but also on metal surfaces, surface imperfections can modify the reactivity enormously. Some of these surface imperfections are illustrated in Fig. 7. This scenario corresponds to a snapshot of different molecules in front of a close-packed surface with a straight step. In addition, atoms of another sort are incorporated substitutionally into the surface. These additional atoms may form an ordered alloy

Fig. 7. Schematic scenario of different kinds of molecules in front of a surface with surface imperfections like steps or substitutionally incorporated adsorbates (courtesy of B. Hammer).

structure (see the dark atoms on the upper terrace in Fig. 7).

While most surface science studies deal with well-characterized low-index surface planes, real catalysts are usually highly dispersed as small particles in order to increase the active surface area. This is commonly called the "structure gap". The catalyst is often microcrystalline, the surface is not well ordered but consists of microfacets with a lot of surface imperfections. Thus, ab initio investigations of active sites at non-perfect surfaces are already rather close to applied research since they contribute to close the structure gap.

Later in this review I will give an example of the effect of alloying on the activity of a catalyst. Here I focus on the influence of steps. A DFT study of the NO dissociation at corrugated Ru(0 0 0 1) [86] has demonstrated the tremendous effect of steps on the reactivity at surfaces. Atoms at steps have a lower coordination number than at terraces. At transition metal surfaces this leads to an upward shift of the metal d states and causes a stronger bonding to NO-2π orbitals [86]. Thus the NO molecular chemisorption is stabilized at the Ru steps with respect to the flat Ru surface by about 0.4 eV. Even much more dramatic is the effect of the Ru steps on the NO dissociation. While the dissociation of NO on flat Ru(0 0 0 1) is hindered by a barrier of 1.3 eV, on a corrugated surface with steps along the [1 0 $\bar{1}$ 0] direction the barrier is reduced to 0.15 eV. Assuming an Arrhenius behavior of the reaction, this corresponds to an increase in the reaction rate by 20 orders of magnitude at room temperature. At a surface with even a small amount of steps the dissociation on the steps will dominate the overall reactivity. Consequently, the diffusion to the step which is hindered by energy barriers of about 0.5 eV will be the rate limiting step.

Note that the calculations reported in Ref. [86] did not correspond to the geometry shown in Fig. 7. The terraces used in these calculations were actually only four atom rows wide. This is, however, sufficient to differentiate between dissociation on the terraces and the steps. Electronic structure calculations do not only yield numbers, they also help to understand the particular reason for this strong reduction in the barrier height. The energy of the barrier can be decomposed into an interaction energy between the two reactants N and O and the bonding energy of the reactants alone. By this decomposition it was shown that it is mainly the increase in the bonding energy of the atomic fragments that causes the dramatic reduction in barrier height [86]. Additionally, at the steps the dissociating atoms do not share nearest neighbor surface atoms. This reduces the repulsion between the reaction products and further reduces the dissociation barrier.

In recent years the focus of ab initio studies of reactions on surfaces has shifted from hydrogen dissociation to reactions involving oxygen [34,87–90]. These important class of reactions is technologically very important, for example in the context of the car exhaust catalyst. Since these reactions are covered by another contribution in this volume [91], I will not discuss them here any further.

4.2. Applied research

The examples shown in this review so far have been devoted to the study of fundamental processes at surfaces. The number of atoms that were considered in these studies per periodic cell were well below 100. Unfortunately, almost no technical device exhibits perfect structures on a large length scale. Therefore the question arises whether electronic structure calculations for periodic structures have any relevance for the development of real products?

Indeed there are already companies that prosper by applying combinatorial technologies to discover and develop new materials for the chemical and electronic industry [6]. In this approach a vast number of microscale quantities of compounds is created and automatically tested. However, the preparation and performance of new materials can often not be tested with microscale quantities. For example, a new lubricant still has to be tested under realistic conditions in an engine. An excellent discussion of this issue has been given by Wimmer et al. [3]. They demonstrate that for the development of a catalyst that contains five elements one could end up with about 50,000 combinations. If one additionally considers different preparation methods and performance conditions, easily about one million experiments have to be carried out. This is a formidable task, even with modern automated synthesis and screening methods. Hence computational methods that at least noticeably reduce the number of experiments which have to be performed can have a sizable economic impact. Thus the goal of the computations is not to create a new product theoretically from the scratch, but much more humble, namely "to add more relevant information" [3] for the research and development process.

The authors of the cited article [3] were actually employed by a company that offers computational software; hence they might be slightly biased as far as the quality and impact of their products is concerned. Still they give very convincing examples of the contribution that calculations can make (although they are faced with the problem of confidentiality that always comes up when industrial products are involved). The first example they give is concerned with an iron-oxide based catalyst. In this catalyst Cr had to be replaced because under production conditions it was oxidized to Cr(VI) which is rather poisonous. In electronic structure calculations the influence of Cr on the oxidation state of the Fe atoms in the iron oxide catalyst material was studied. This helped to establish a ranking among possible Cr substitutes that were then tested in experiment thus providing an efficient strategy for the testing.

Another example of applied research using ab initio methods was actually performed by scientists employed by the electronic industry [92]. The materials for electronic devices are rather well characterized which is a great advantage for the application of electronic structure methods. As we have already learned in the last section, hydrogen desorption is often the rate limiting step in the CVD growth of semiconductor devices. Currently there has been growing interest in SiGe alloys as a material for high speed electronic devices. It has been observed experimentally that already small amounts of Ge incorporated into a Si surface significantly enhance the deposition during CVD growth. This is caused by the fact that the presence of Ge facilitates the desorption of hydrogen which then creates additional surface sites for deposition.

Hydrogen on pure Ge is less strongly bound than on pure Si. However, it was not clear whether the presence of Ge in the SiGe alloys globally lowers the barriers for hydrogen desorption or whether hydrogen desorbs locally from the Ge sites so that first hydrogen diffusion to these sites is required. Using clusters with 35 Si and/or Ge atoms, barriers for hydrogen diffusion and desorption were determined by DFT calculations [92]. It was found that the barriers for diffusion from Si to Ge are indeed smaller than the barriers for hydrogen desorption from the Si sites. Thus the surface diffusion on mixed SiGe surfaces leads to an enhanced desorption via Ge surface sites. The calculated barrier heights were then implemented into a kinetic reactor model [92]. The computed growth rates were in rather good agreement with experiments. The authors of this article conclude that ab initio electronic structure calculations "provide access to data where measurements would be much more time consuming, expensive, and sometimes basically not possible".

Ab initio calculations have also been performed in the industry in order to determine the electronic structure of compact low pressure discharge fluorescent lamps [93]. The fluorescent lamp cathode surfaces consist of tungsten because of its high melting point. It is usually coated by some alkaline earth-metal oxide in order to reduce the work function. The emitter material has to exhibit a good thermal stability due to the high operation temperatures of 1000–1500 K. Because of these severe constraints, it is very important to understand the oxide–substrate interaction for further cathode development.

The researchers at the company that manufactures the fluorescent lamps chose the BaO/W(001) system to study this interaction. They performed DFT slab calculations in the c(2 × 2) geometry using a five layers thick tungsten slab. This geometry is shown in Fig. 8, the box in the top view of Fig. 8(B) indicates the surface unit cell.

This study demonstrates the ability of DFT calculations for analyzing the electronic structure of a particular system. The DOS for the clean and BaO covered tungsten surface is plotted in Fig. 8(C). As already mentioned in Section 2, there is no firm theoretical justification for interpreting the Kohn–Sham eigenvalues as the electronic one-particle energies that can be measured for example by photoemission studies. However, the peaks in the Kohn–Sham DOS of the clean surface in Fig. 8(C) at -0.8 and -4.2 eV can be identified with surface resonances obtained in photoemission experiments [94]. An agreement like this lends credibility to the analysis of the DOS from DFT calculations.

As Fig. 8(C) shows, these surface resonances are strongly perturbed by the presence of BaO. A state-resolved analysis of the DOS shows [93] that the large increase in the DOS between -5 and -3 eV is mainly due to oxygen 2p states that interact with the W surface resonance at -4.2 eV. Furthermore, the Ba d states which are unoccupied in the free Ba atom are shifted down and partially occupied. They strongly hybridize with the W d states leading to a covalent bond. This transfers electronic charge into the interface between the adlayer and the substrate atoms and induces a dipole moment that causes the reduction of the work function. At the same time it stabilizes the surface–adsorbate bonding leading to the high thermal stability of the BaO coating. Insights like these in the microscopic origins of the work function reduction and high thermal stability of adlayers will be very helpful for the future cathode materials development.

Finally I like to turn to a celebrated success story [95] which is one of the first examples where a close collaboration between fundamental academic research, both experimental and theoretical, and industrial development has led to the design of a successful catalyst for the steam-reforming process. This impressive success will be reviewed elsewhere in this volume [96]. Besides, just recently a very detailed review about this subject has been published [97]. Here I will focus on the contribution of electronic structure calculations [95,98] to the design of the catalyst.

In the steam-reforming process, hydrocarbon molecules (mainly CH_4) and water are converted into H_2 and CO. This is a very important process since it is the first step for several large scale chemical processes as ammonia synthesis, methanol production or reactions that need H_2 [97]. The catalysts usually used for this reaction are based

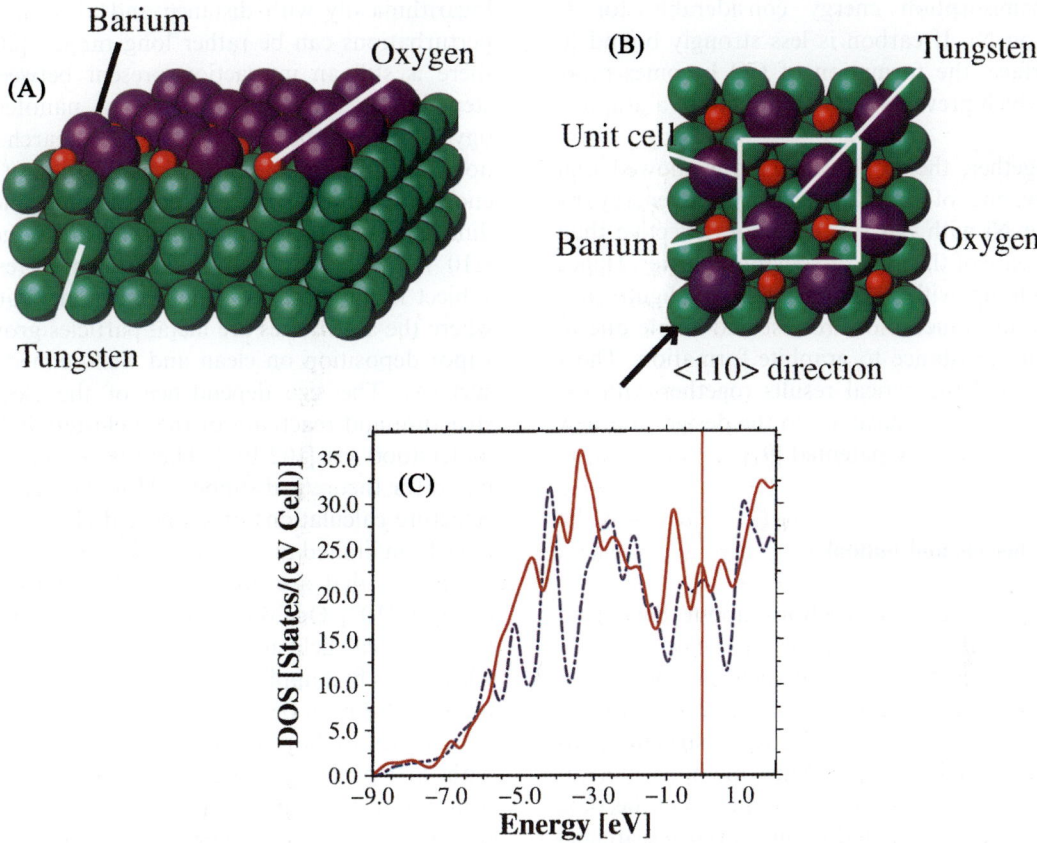

Fig. 8. Atomic geometry and electronic structure of c(2 × 2)-BaO/W(0 0 1): (A) Side view and (B) top view of the atomic geometry used in DFT calculations [93]. The box in (B) indicates the surface unit cell. (C) Calculated DOS of the clean (dashed blue line) and the BaO covered (solid red line) W(0 0 1) surface. The vertical red line indicates the Fermi energy (after [93]).

on Ni. However, during the catalyzed reaction also an unwanted by-product, namely graphite, is formed. A graphite overlayer on the Ni surface leads to a poisoning of the reaction, i.e., it lowers the activity of the catalyst. Such poisoning processes are very costly since they reduce the time the catalyst can be used and require a more frequent maintenance of the reactor unit in the chemical plant.

One way of changing the reactivity of metal surfaces is to modify their chemical composition by alloying them with other metals. Some metals that are immiscible in the bulk may still be able to form alloys at the surface. Au and Ni is such a system. The rate limiting process in the steam-reforming process on Ni is the dissociation of CH_4 into CH_3 and H. DFT calculations by Kratzer et al. showed that this process is hindered by a relatively high barrier of 1.1 eV on Ni(1 1 1) [98]. If a Ni atom on the (1 1 1) surface has one or two Au atoms as neighbors, this barrier is even increased by 165 and 330 meV, respectively. Due to the fact that Au is a noble metal, the CH_4 dissociation barrier over the Au atom is even much higher [95]. An analysis of the calculated electronic structure revealed that the presence of neighboring Au atoms leads to a downshift of the d states at the Ni atom which reduces the reactivity at the Ni atoms [99]. Hence alloying a Ni surface with Au atoms leads to a reduced activity of the catalyst. However, DFT calculations also demonstrated [95] that the presence of the Au atoms lowers the chemisorption energy considerably for C atoms on Ni. If carbon is less strongly bound to the surface, the formation of CO becomes more likely which prevents the building up of a graphite layer.

Altogether, the DFT calculations showed that the lowering of the C chemisorption energy by alloying Ni with Au is much more effective than the increase of the CH_4 dissociation barrier. Hence one ends up with a catalyst that is slightly less reactive but much more robust and stable due to its higher resistance to graphite formation. These fundamental theoretical results together with experimental studies have led to the design of a new catalyst that is now patented [97].

5. Conclusions and outlook

The two preceding sections about basic and applied research have been devoted to examples of successful theoretical surface science studies. They have indicated the deep insight that can be gained into surface processes by electronic structure calculations based on DFT. Such methods are not only used by academic researchers but become more and more accepted by industrial researchers as a valuable and versatile tool in the research and development process.

However, not all materials and systems can be addressed with the same level of accuracy by electronic structure methods; there are still severe problems and strong limitations associated, in particular with DFT calculations. In the following I will sketch some of the problems. These problems represent very challenging tasks where still a lot of work is to do. But they also offer many opportunities for ambitious projects that might be very rewarding.

As already mentioned in Section 3, usually well below 100 atoms are considered either in the periodic supercells or in the finite clusters representing surfaces in electronic structure calculations. Consequently, most studies are still concerned with highly ordered structures. For example, in calculations of stepped surfaces the distance between the steps is usually two to four atom rows [86,100]. Since the stress field of line defects vanishes only logarithmically with distance and also electronic perturbations can be rather long ranged [26,101], there is still an interaction present between the steps in these studies. Furthermore, nanotechnology is one of the most exciting research fields nowadays. But the supercells are still not large enough to model "real" nanostructures since the dimensions of the surface unit cell are in general <10 Å. One very active experimental research subject is the study of supported model catalysts where the active sites are metal particles grown by vapor deposition on clean and well-defined oxide surfaces. The size dependence of the electronic structure and reactivity of these clusters is hardly understood yet [102,103]. There is certainly a demand for theoretical support. However, electronic structure calculations of supported clusters so far have been limited to very small clusters of less than ten atoms that are not really relevant for metal catalysts [102]. One very demanding, but important technological field is the development of sensors where only first applications of electronic structure theory have been performed [104].

In order to be able to address larger systems with first-principles calculations one can either wait for faster computers or try to improve the algorithms. One example of a very successful improvement is the development of ultrasoft pseudopotentials by Vanderbilt [105] which made it possible to use much smaller energy cutoffs, i.e. much smaller plane-wave basis sets, in the supercell calculations. Many DFT studies including transition metals which have rather localized d states actually use these ultrasoft pseudopotentials [27,86,88,100].

Another possibility to make the calculations faster is to run them in a massively parallel fashion on many processors. Usually the communication, i.e. the exchange of data, between different processors is the computational bottleneck in massively parallel implementations. In this context so-called O(N) (order N) methods are rather well suited [106]. These methods take advantage of the localization properties of the fundamental interactions in materials [107]. Thus they are able to exhibit linear scaling with respect to the size of the system. Due to the locality of the algorithm only little communication is needed between the processors which treat each some localized region of the system. However, O(N) applications so far have mainly used tight binding [108] or semiempirical methods. This is due to the fact that O(N) methods which use DFT still have certain shortcomings that require some algorithmic progress [106]. Another way to address larger systems is to use multiscale modeling techniques in which different aspects of a certain system are treated within different levels of microscopic accuracy. Such multiscale approaches are presented elsewhere in this volume [91,109].

Progress is also still needed as far as the exchange-correlation functionals used in DFT calculations are concerned. While quantum chemists using DFT prefer exchange-correlation functionals that are fitted to a number of reference reactions in the gas phase [19,20], physicists rather rely on functionals that are derived without any adjustment of parameters [21,22]. However, Hammer et al. [23] have shown that within the construction of the widely used PBE functional [22] there is a certain freedom in the interpolation scheme. Results obtained with different GGA functionals that follow the same construction can have discrepancies by up to 0.5 eV as far as chemisorption and dissociation energies are concerned [23]. This is certainly a very unsatisfying situation. There are attempts to improve the accuracy of the functionals, for example by constructing so-called meta-GGAs [110]. But they require a second order gradient expansion which makes them computationally less efficient so that they have not found a wide-spread recognition. Particularly problematic for present GGA functionals are oxygen systems. For example, the binding energy of O_2 comes out wrong in DFT pseudopotential calculations by more than 0.4 eV [23].

All examples of ab initio studies presented in this review so far were devoted to electronic ground state properties; consequently the Born–Oppenheimer approximation was used for dynamical simulations. But very interesting physics and chemistry is concerned with processes involving electronic excitations [111]. Ordinary DFT methods, for example, are not well suited to address electronically excited states. Here wave function based calculations are the method of choice at the moment [112]. One very promising approach to treat excited states is based on time-dependent DFT [113], other attempts employ the so-called GW approximation [114,115]. However, the algorithms based on these approaches are still not mature enough to allow a complete detailed description of reactions at surfaces with electronic excitations. This means that there is plenty of room for further research.

A problem closely related to the treatment of excited states is the fact that band gaps are grossly underestimated in DFT calculations based on LDA. In the LDA, the Si band gap is only half the experimental value while Ge even turns out to be metallic instead of being a semiconductor. This is a severe limitation for, e.g., the study of optoelectronic materials where an accurate description of optical excitation energies is needed [3]. There have been several attempts to tackle this problem like the screened-exchange LDA approach [116,117] or pseudopotentials incorporating the so-called self-interaction correction [118]. These approaches are often only well suited for particular problems and not universally applicable.

An important materials class that has not been discussed so far are the actinides and lanthinides where the f-shell is in most cases only partially filled so that the f electrons are essential for understanding most properties of these elements. Because of the strong electronic localization in these heavy elements, which is associated with high electronic kinetic energies, these systems have to be treated relativistically. This strong localization also renders the use of pseudopotentials very ineffective. One method to treat these systems with strong electron correlation is the $LDA + U$ method where the non-local and energy dependent self energy is approximated by a non-local screened Coulomb potential [119]. This method is still a mean-field approximation, hence it is not really suited for strongly correlated metals and often only gives qualitative information.

This list of problems associated with electronic structure algorithms is certainly not complete. These obstacles have prevented the use of ab initio methods for certain material classes. But alternative methods are either not accurate or not efficient enough. However, once any of these particular problems is solved, an almost "virgin" research field for calculations and simulations opens up with all its exciting opportunities. The examples presented in this review show what kind of contribution electronic structure theory can make to a research field. A similar impact is to be expected in other research fields. There is every reason to believe that the theoretical study of processes on surfaces based on ab initio electronic structure methods will also prosper in the future. It will provide further exciting insights into fundamental mechanisms but it will also become a very valuable tool in the research and development process.

Acknowledgements

Scientific work is almost never done alone. I have profited enormously from the close collaborations and fruitful discussions with, among others, Wilhelm Brenig, Matthias Scheffler, Bjørk Hammer, Steffen Wilke, Eckhard Pehlke, Dimitrios Papaconstantopoulos, Peter Kratzer, Andreas Eichler, and Jürgen Hafner. I like to thank them all for the insights they have shared with me. As for this article, I like to thank in particular Bjørk Hammer, Eckhard Pehlke and Jürgen Almanstötter for providing me with nice illustrations. Furthermore, I am grateful to again Eckhard Pehlke and Markus Lischka for a careful proofreading of the manuscript.

References

[1] A. Groß, Reactions at surfaces studied by ab initio dynamics calculations, Surf. Sci. Rep. 32 (1998) 291.
[2] J.M. Thomas, W.J. Thomas, Principles and Practice of Heterogeneous Catalysis, VCH, Weinheim, 1997.
[3] E. Wimmer, J.-R. Hill, P. Gravil, W. Wolf, Industrial Use of Electronic Structure Methods, Newsletter 34 of the Psi-k Network, 1999, http://psi-k.dl.ac.uk/psi-k/highlights.html.
[4] R. Schlögl, Combinatorial chemistry in heterogeneous catalysis: a new scientific approach or the "king's new clothes", Angew. Chem. Int. Ed. 37 (1998) 2333.
[5] J.M. Thomas, Design, synthesis, and in situ characterization of new solid catalysts, Angew. Chem. Int. Ed. 38 (1999) 3589.
[6] P. Cong, R.D. Doolen, Q. Fan, D.M. Giaquinta, S. Guan, E.W. McFarland, D.M. Poojary, K. Self, H.W. Turner, W.H. Weinberg, High-throughput synthesis and screening of combinatorial heterogeneous catalyst libraries, Angew. Chem. Int. Ed. 38 (1999) 483.
[7] M. Head-Gordon, Quantum chemistry and molecular processes, J. Phys. Chem. 100 (1996) 13213.
[8] P. Blaha, K. Schwarz, P. Sorantin, S.B. Trickey, Full-potential, linearized augmented plane wave programs for crystalline systems, Comput. Phys. Commun. 59 (1990) 399.
[9] M.C. Payne, M.P. Teter, D.C. Allan, T.A. Arias, J.D. Joannopoulos, Iterative minimization techniques for ab initio total-energy calculations: molecular dynamics and conjugate gradients, Rev. Mod. Phys. 64 (1992) 1045.
[10] G. Kresse, J. Furthmüller, Efficiency of ab-initio total energy calculations for metals and semiconductors using a plane-wave basis set, Comp. Mat. Sci. 6 (1996) 15.
[11] M. Bockstedte, A. Kley, J. Neugebauer, M. Scheffler, Density-functional theory calculations for poly-atomic systems: electronic structure, static and elastic properties and ab initio molecular dynamics, Comput. Phys. Commun. 107 (1997) 187.
[12] G.W. Trucks, K. Raghavachari, G.S. Higashi, Y.J. Chabal, Mechanism of HF etching of silicon surfaces: a theoretical understanding of hydrogen passivation, Phys. Rev. Lett. 65 (1990) 504.
[13] J.L. Whitten, H. Yang, Theory of chemisorption and reactions on metal surfaces, Surf. Sci. Rep. 24 (1996) 55.
[14] M. Born, J.R. Oppenheimer, Zur Quantentheorie der Molekeln, Ann. Phys. 84 (1927) 457.
[15] R.M. Dreizler, E.K.U. Gross, Density Functional Theory, Springer, Berlin, 1990.
[16] P. Hohenberg, W. Kohn, Inhomogeneous electron gas, Phys. Rev. 136 (1964) B864.
[17] W. Kohn, L.J. Sham, Self-consistent equations including exchange and correlation effects, Phys. Rev. 140 (1965) A1133.
[18] D.M. Ceperley, B.J. Alder, Ground state of the electron gas by a stochastic method, Phys. Rev. Lett. 45 (1980) 566.
[19] A.D. Becke, Density-functional exchange-energy approximation with correct asymptotic behavior, Phys. Rev. A 38 (1988) 3098.
[20] C. Lee, W. Yang, R.G. Parr, Development of the Colle-Salvetti correlation-energy formula into a functional of the electron density, Phys. Rev. B 37 (1988) 785.
[21] J.P. Perdew, J.A. Chevary, S.H. Vosko, K.A. Jackson, M.R. Pederson, D.J. Singh, C. Fiolhias, Atoms, molecules, solids, and surfaces: Applications of the generalized gradient approximation for exchange and correlation, Phys. Rev. B 46 (1992) 6671.
[22] J.P. Perdew, K. Burke, M. Ernzerhoff, Generalized gradient approximation made simple, Phys. Rev. Lett. 77 (1996) 3865.
[23] B. Hammer, L.B. Hansen, J.K. Nørskov, Improved adsorption energetics within density-functional theory using revised Perdew–Burke–Ernzerhoff functionals, Phys. Rev. B 59 (1999) 7413.

[24] M. Fuchs, M. Scheffler, Ab initio pseudopotentials for electronic structure calculations of poly-atomic systems using density-functional theory, Comput. Phys. Commun. 119 (1999) 67.

[25] E. Pehlke, P. Kratzer, Density-functional study of hydrogen chemisorption on vicinal Si(001) surfaces, Phys. Rev. B 59 (1999) 2790.

[26] A. Bogicevic, S. Ovesson, P. Hyldgaard, B.I. Lundqvist, H. Brune, D.R. Jennison, Nature, strength, and consequences of indirect adsorbate interactions on metals, Phys. Rev. Lett. 85 (2000) 1910.

[27] A. Eichler, J. Hafner, Molecular precursors in the dissociative adsorption of O_2 on Pt(111), Phys. Rev. Lett. 79 (1997) 4481.

[28] G.R. Darling, S. Holloway, The dissociation of diatomic molecules at surfaces, Rep. Prog. Phys. 58 (1995) 1595.

[29] J.C. Polanyi, W.H. Wong, Location of energy barriers. I. Effect on the dynamics of reactions A + BC, J. Chem. Phys. 51 (1969) 1439.

[30] A. Groß, M. Scheffler, M.J. Mehl, D.A. Papaconstantopoulos, Ab initio based tight-binding Hamiltonian for the dissociation of molecules at surfaces, Phys. Rev. Lett. 82 (1999) 1209.

[31] P. Hänggi, P. Talkner, M. Borkovec, Reaction-rate theory: fifty years after Kramers, Rev. Mod. Phys. 62 (1990) 251.

[32] G. Mills, H. Jónsson, G.K. Schenter, Reversible work transition state theory—application to dissociative adsorption of hydrogen, Surf. Sci. 324 (1995) 305.

[33] G. Henkelman, H. Jónsson, A dimer method for finding saddle points on high dimensional potential surfaces using only first derivatives, J. Chem. Phys. 111 (1999) 7010.

[34] C. Stampfl, M. Scheffler, Density-functional theory study of the catalytic oxidation of CO over transition metal surfaces, Surf. Sci. 433–435 (1999) 119.

[35] B. Hammer, M. Scheffler, K.W. Jacobsen, J.K. Nørskov, Multidimensional potential energy surface for H_2 dissociation over Cu(111), Phys. Rev. Lett. 73 (1994) 1400.

[36] J.A. White, D.M. Bird, M.C. Payne, I. Stich, Surface corrugation in the dissociative adsorption of H_2 on Cu(100), Phys. Rev. Lett. 73 (1994) 1404.

[37] S. Wilke, M. Scheffler, Potential-energy surface for H_2 dissociation over Pd(100), Phys. Rev. B 53 (1996) 4296.

[38] A. Eichler, G. Kresse, J. Hafner, Quantum steering effects in the dissociative adsorption of H_2 on Rh(100), Phys. Rev. Lett. 77 (1996) 1119.

[39] C. Stampfl, M. Scheffler, Anomalous behavior of Ru for catalytic oxidation: a theoretical study of the catalytic reaction CO + 1/2 O_2 to CO_2, Phys. Rev. Lett. 78 (1997) 1500.

[40] G. Wiesenekker, G.J. Kroes, E.J. Baerends, An analytical six-dimensional potential energy surface for dissociation of molecular hydrogen on Cu(100), J. Chem. Phys. 104 (1996) 7344.

[41] G.J. Kroes, Six-dimensional quantum dynamics of dissociative chemisorption of H_2 on metal surfaces, Prog. Surf. Sci. 60 (1999) 1.

[42] J.R. Chelikowsky, M. Schlüter, S.G. Louie, M.L. Cohen, Self-consistent pseudopotential calculation for the (111) surface of aluminum, Solid State Commun. 17 (1975) 1103.

[43] P.J. Feibelman, D.R. Hamann, Electronic structure of a 'poisoned' transition-metal surface, Phys. Rev. Lett. 52 (1984) 61.

[44] H.A. Michelsen, D.J. Auerbach, A critical examination of data on the dissociative adsorption and associative desorption of hydrogen at copper surfaces, J. Chem. Phys. 94 (1991) 7502.

[45] C.T. Rettner, D.J. Auerbach, H.A. Michelsen, Role of vibrational and translational energy in the activated dissociative adsorption of D_2 on Cu(111), Phys. Rev. Lett. 68 (1992) 1164.

[46] K.D. Rendulic, A. Winkler, Adsorption and desorption dynamics as seen through molecular beam techniques, Surf. Sci. 299/300 (1994) 261.

[47] A.E. DePristo, A. Kara, Molecule-surface scattering and reaction dynamics, Adv. Chem. Phys. 77 (1991) 163.

[48] S. Kumar, B. Jackson, Dissociative adsorption of H_2 on Cu(110): a mixed quantum-classical study, J. Chem. Phys. 100 (1994) 5956.

[49] G.R. Darling, S. Holloway, The role of parallel momentum in the dissociative adsorption of H_2 at highly corrugated surfaces, Surf. Sci. 304 (1994) L461.

[50] A. Groß, B. Hammer, M. Scheffler, W. Brenig, High-dimensional quantum dynamics of adsorption and desorption of H_2 at Cu(111), Phys. Rev. Lett. 73 (1994) 3121.

[51] A. Groß, The role of lateral surface corrugation for the quantum dynamics of dissociative adsorption and associative desorption, J. Chem. Phys. 102 (1995) 5045.

[52] A. Groß, S. Wilke, M. Scheffler, Six-dimensional quantum dynamics of adsorption and desorption of H_2 at Pd(100): steering and steric effects, Phys. Rev. Lett. 75 (1995) 2718.

[53] W. Brenig, T. Brunner, A. Groß, R. Russ, Numerically stable solution of coupled channel equations: the local reflection matrix, Z. Phys. B 93 (1993) 91.

[54] K.D. Rendulic, G. Anger, A. Winkler, Wide range nozzle beam adsorption data for the systems H_2/nickel and H_2/Pd(100), Surf. Sci. 208 (1989) 404.

[55] D.A. King, Kinetics of adsorption, desorption, and migration at single-crystal metal surfaces, CRC Crit. Rev. Solid State Mater. Sci. 7 (1978) 167.

[56] M. Kay, G.R. Darling, S. Holloway, J.A. White, D.M. Bird, Steering effects in non-activated adsorption, Chem. Phys. Lett. 245 (1995) 311.

[57] M. Beutl, M. Riedler, K.D. Rendulic, Strong rotational effects in the adsorption dynamics of H_2/Pd(111): evidence for dynamical steering, Chem. Phys. Lett. 247 (1995) 249.

[58] D. Wetzig, R. Dopheide, M. Rutkowski, R. David, H. Zacharias, Rotational alignment in associative desorption of D_2 ($v = 0$ and 1) from Pd(100), Phys. Rev. Lett. 76 (1996) 463.

[59] M. Gostein, G.O. Sitz, Rotational state-resolved sticking coefficients for H_2 on Pd(111): testing dynamical steering in dissociative adsorption, J. Chem. Phys. 106 (1997) 7378.

[60] G.J. Kroes, E.J. Baerends, R.C. Mowrey, Six-dimensional quantum dynamics of dissociative chemisorption of ($v=0, j=0$) H_2 on Cu(100), Phys. Rev. Lett. 78 (1997) 3583.

[61] D.A. McCormack, G.J. Kroes, R.A. Olsen, E.J. Baerends, R.C. Mowrey, Rotational effects on vibrational excitation of H_2 on Cu(100), Phys. Rev. Lett. 82 (1999) 1410.

[62] A. Groß, C.-M. Wei, M. Scheffler, Poisoning of hydrogen dissociation at Pd(100) by adsorbed sulfur studied by ab initio quantum dynamics and ab initio molecular dynamics, Surf. Sci. 416 (1998) L1095.

[63] M. Bonn, A.W. Kleyn, G.J. Kroes, Real time chemical dynamics at surfaces, Surf. Sci. 500 (2002) 475.

[64] K.W. Kolasinski, Dynamics of hydrogen interactions with Si(100) and Si(111) surfaces, Int. J. Mod. Phys. B9 (1995) 2753.

[65] K.W. Kolasinski, W. Nessler, K.-H. Bornscheuer, E. Hasselbrink, Beam investigations of D_2 adsorption on Si(100): on the importance of lattice excitations in the reaction dynamics, J. Chem. Phys. 101 (1994) 7082.

[66] P. Bratu, K.L. Kompa, U. Höfer, Optical second-harmonic investigations of H_2 and D_2 adsorption on Si(100)2 × 1: the surface temperature dependence of the sticking coefficient, Chem. Phys. Lett. 251 (1996) 1.

[67] K.W. Kolasinski, W. Nessler, A. de Meijere, E. Hasselbrink, Hydrogen adsorption on and desorption from Si: considerations on the applicability of detailed balance, Phys. Rev. Lett. 72 (1994) 1356.

[68] W. Brenig, A. Groß, R. Russ, Detailed balance and phonon assisted sticking in adsorption and desorption of H_2/Si, Z. Phys. B 96 (1994) 231.

[69] P. Bratu, U. Höfer, Phonon-assisted sticking of molecular hydrogen on Si(111)-(7 × 7), Phys. Rev. Lett. 74 (1995) 1625.

[70] P. Bratu, W. Brenig, A. Groß, M. Hartmann, U. Höfer, P. Kratzer, R. Russ, Reaction dynamics of molecular hydrogen on silicon surfaces, Phys. Rev. B 54 (1996) 5978.

[71] C.J. Wu, I.V. Ionova, E.A. Carter, Ab initio H_2 desorption pathways for H/Si(100): the role of $SiH_{2(a)}$, Surf. Sci. 295 (1993) 64.

[72] Z. Jing, G. Lucovsky, J.L. Whitten, Mechanism of H_2 desorption from monohydride Si(100)-2 × 1H, Surf. Sci. 296 (1993) L33.

[73] P. Nachtigall, K.D. Jordan, C. Sosa, Theoretical study of the mechanism of recombinative hydrogen desorption from the monohydride phase of Si(100): the role of defect migration, J. Chem. Phys. 101 (1994) 8073.

[74] Z. Jing, J.L. Whitten, Multiconfiguration self-consistent-field treatment of H_2 desorption from Si(100)-2 × 1H, J. Chem. Phys. 102 (1995) 3867.

[75] P. Kratzer, B. Hammer, J.K. Nørskov, The coupling between adsorption dynamics and the surface structure: H_2 on Si(100), Chem. Phys. Lett. 229 (1994) 645.

[76] E. Pehlke, M. Scheffler, Theory of adsorption and desorption of H_2/Si(001), Phys. Rev. Lett. 74 (1995) 952.

[77] A. Groß, M. Bockstedte, M. Scheffler, Ab initio molecular dynamics study of the desorption of D_2 from Si(100), Phys. Rev. Lett. 79 (1997) 701.

[78] P. Nachtigall, K.D. Jordan, A. Smith, H. Jonsson, Investigation of the reliability of density functional methods: reaction and activation energies for Si–Si bond cleavage and H_2 elimination from silanes, J. Chem. Phys. 104 (1996) 148.

[79] E. Penev, P. Kratzer, M. Scheffler, Effect of the cluster size in modeling the H_2 desorption and dissociative adsorption on Si(001), J. Chem. Phys. 110 (1999) 3986.

[80] T.-C. Shen, J.A. Steckel, K.D. Jordan, Electron-stimulated bond rearrangements on the H/Si(100)-3 × 1 surface, Surf. Sci. 446 (2000) 211.

[81] E. Pehlke, Highly reactive dissociative adsorption of hydrogen molecules on partially H-covered Si(001) surfaces: a density-functional study, Phys. Rev. B 62 (2000) 12932.

[82] F.M. Zimmermann, X. Pan, Interaction of H_2 with Si(001)-(2 × 1): solution of the barrier puzzle, Phys. Rev. Lett. 85 (2000) 618.

[83] P. Kratzer, E. Pehlke, M. Scheffler, M.B. Raschke, U. Höfer, Highly site-specific H_2 adsorption on vicinal Si(001) surfaces, Phys. Rev. Lett. 81 (1998) 5596.

[84] A. Biedermann, E. Knoesel, Z. Hu, T.F. Heinz, Dissociative adsorption of H_2 on Si(100) induced by atomic H, Phys. Rev. Lett. 83 (1999) 1810.

[85] A. Vittadini, A. Selloni, Density functional study of H_2 desorption from monohydride and dihydride Si(100) surfaces, Chem. Phys. Lett. 235 (1995) 334.

[86] B. Hammer, Bond activation at monatomic steps: NO dissociation at corrugated Ru(0001), Phys. Rev. Lett. 83 (1999) 3681.

[87] A. Alavi, P.J. Hu, T. Deutsch, P.L. Silvestrelli, J. Hutter, CO oxidation on Pt(111)—an ab initio density functional theory study, Phys. Rev. Lett. 80 (1998) 3650.

[88] A. Eichler, J. Hafner, Reaction channels for the catalytic oxidation of CO on Pt(111), Surf. Sci. 435 (1999) 58.

[89] M.V. Ganduglia-Pirovano, M. Scheffler, Structural and electronic properties of chemisorbed oxygen on Rh(111), Phys. Rev. B 59 (1999) 15533.

[90] T. Sasaki, T. Ohno, Calculations of the potential-energy surface for dissociation process of O_2 on the Al(111) surface, Phys. Rev. B 60 (1999) 7824.

[91] C. Stampfl, M.V. Ganduglia-Pirovano, K. Reuter, M. Scheffler, Catalysis and corrosion: The theoretical surface-science context, Surf. Sci. 500 (2002) 368.

[92] M. Hierlemann, C. Werner, A. Spitzer, Equipment simulation of SiGe heteroepitaxy—model validation by ab initio calculations of surface diffusion processes, J. Vac. Sci. Technol. B 15 (1997) 935.

[93] J. Almanstötter, T. Fries, B. Eberhard, Electronic structure of fluorescent lamp cathode surfaces: BaO/W(001), J. Appl. Phys. 86 (1999) 325.

[94] S.-L. Weng, E.W. Plummer, T. Gustafsson, Experimental and theoretical study of the surface resonances on the (100) faces of W and Mo, Phys. Rev. B 18 (1978) 1718.

[95] F. Besenbacher, I. Chorkendorff, B.S. Clausen, B. Hammer, A.M. Molenbroek, J.K. Nørskov, I. Stensgaard,

Design of a surface alloy catalyst for steam reforming, Science 279 (1998) 1913.
[96] F. Rosei, R. Rosei, Atomic description of elementary surface processes: diffusion and dynamics, Surf. Sci. 500 (2002) 395.
[97] J.H. Larsen, I. Chorkendorff, From fundamental studies of reactivity on single crystals to the design of catalysts, Surf. Sci. Rep. 35 (1999) 165.
[98] P. Kratzer, B. Hammer, J.K. Nørskov, A theoretical study of CH_4 dissociation on pure and gold-alloyed Ni(111) surfaces, J. Chem. Phys. 105 (1996) 5595.
[99] B. Hammer, J.K. Nørskov, Why gold is the noblest of all the metals, Nature 376 (1995) 238.
[100] P.J. Feibelman, Interlayer self-diffusion on stepped Pt(111), Phys. Rev. Lett. 81 (1998) 168.
[101] M.F. Crommie, C.P. Lutz, D.M. Eigler, Imaging standing waves in a two-dimensional electron gas, Nature 363 (1993) 524.
[102] C.R. Henry, Surface studies of supported model catalysts, Surf. Sci. Rep. 31 (1998) 235.
[103] K.H. Hansen, S. Stempel, E. Laegsgaard, M. Bäumer, H.-J. Freund, F. Besenbacher, I. Stensgaard, Palladium nanocrystals on Al_2O_3: structure and adhesion energy, Phys. Rev. Lett. 83 (1999) 4120.
[104] T.S. Rantala, I.T. Rantala, V. Lantto, Computational studies for the interpretation of gas response of SnO_2(110) surface, Sens. Actuat. B 65 (2000) 375.
[105] D. Vanderbilt, Soft self-consistent pseudopotentials in a generalized eigenvalue formalism, Phys. Rev. B 41 (1990) 7892.
[106] S. Goedecker, Linear scaling electronic structure methods, Rev. Mod. Phys. 71 (1999) 1085.
[107] W. Kohn, Density functional and density matrix method scaling linearly with the number of atoms, Phys. Rev. Lett. 76 (1996) 3168.
[108] S. Goedecker, L. Colombo, Efficient linear scaling algorithm for tight-binding molecular dynamics, Phys. Rev. Lett. 73 (1994) 122.
[109] F. Starrost, E.A. Carter, Modeling the full monty: baring the nature of surfaces across time and space, Surf. Sci. 500 (2002) 323.
[110] J.P. Perdew, S. Kurth, A. Zupan, P. Blaha, Accurate density functional with correct formal properties: a step beyond the generalized gradient approximation, Phys. Rev. Lett. 82 (1999) 2544.
[111] M. Wolf, Femtosecond dynamics of electronic excitations at metal surfaces, Surf. Sci. 377 (1997) 343.
[112] T. Klüner, H.-J. Freund, V. Staemmler, R. Kosloff, Theoretical investigation of laser induced desorption of small molecules from oxide surfaces: A first principles study, Phys. Rev. Lett. 80 (1998) 5208.
[113] E. Runge, E.K.U. Gross, Density-functional theory for time-dependent systems, Phys. Rev. Lett. 52 (1984) 997.
[114] A. Schindlmayr, T.J. Pollehn, R.W. Godby, Spectra and total energies from self-consistent many-body perturbation theory, Phys. Rev. B 58 (1998) 12684.
[115] M. Rohlfing, P. Krüger, J. Pollmann, Role of semicore d electrons in quasiparticle band-structure calculations, Phys. Rev. B 57 (1998) 6485.
[116] D.M. Bylander, L. Kleinman, Good semiconductor band gaps with a modified local-density approximation, Phys. Rev. B 41 (1990) 7868.
[117] S. Picozzi, A. Continenza, R. Asahi, W. Mannstadt, A.J. Freeman, W. Wolf, E. Wimmer, C.B. Geller, Volume and composition dependence of direct and indirect band gaps in ordered ternary III–V semiconductor compounds: A screened-exchange LDA study, Phys. Rev. B 61 (2000) 4677.
[118] D. Vogel, P. Krüger, J. Pollmann, Self-interaction and relaxation-corrected pseudopotentials for II–VI semiconductors, Phys. Rev. B 54 (1996) 5495.
[119] V.I. Anisimov, F. Aryasetiawan, A.I. Lichtenstein, First-principles calculations of the electronic structure and spectra of strongly correlated systems: the $LDA + U$ method, J. Phys.: Condens. Matter 9 (1997) 767.

Catalysis and corrosion: the theoretical surface-science context

Catherine Stampfl [a,b,*], M. Veronica Ganduglia-Pirovano [a], Karsten Reuter [a], Matthias Scheffler [a]

[a] *Fritz-Haber-Institut der Max-Planck-Gesellschaft, Faradayweg 4-6, D-14195 Berlin-Dahlem, Germany*
[b] *Department of Physics and Astronomy, Northwestern University, 2145 Sheridan Road, Evanston, IL 60208-3112, USA*

Received 11 September 2000; accepted for publication 21 May 2001

Abstract

NNumerous experiments in ultra-high vacuum as well as ($T = 0$ K, $p = 0$) theoretical studies on surfaces have been performed over the last decades in order to gain a better understanding of the mechanisms, which, for example, underlie the phenomena of catalysis and corrosion. Often the results achieved this way cannot be extrapolated directly to the technologically relevant situation of finite temperature and high pressure. Accordingly, modern surface science has realized that bridging the so-called pressure gap (getting out of the vacuum) is the inevitable way to go. Of similar importance are studies in which the temperature is changed systematically (warming up and cooling down). Both aspects are being taken into account in recent experiments and ab initio calculations.

In this paper we stress that there is still much to learn and important questions to be answered concerning the complex atomic and molecular processes which occur at surfaces and actuate catalysis and corrosion, although significant advances in this exciting field have been made over the years. We demonstrate how synergetic effects between theory and experiment are leading to the next step, which is the development of simple concepts and understanding of the different modes of the interaction of chemisorbed species with surfaces. To a large extent this is being made possible by recent developments in theoretical methodology, which allow to extend the ab initio (i.e., starting from the self-consistent electronic structure) approach to poly-atomic complexes with 10,000 and more atoms, time scales of seconds, and involved statistics (e.g., ab initio molecular dynamics with 10,000 and more trajectories). In this paper we will

1. sketch recent density–functional theory based hybrid methods, which bridge the length and time scales from those of electron orbitals to meso- and macroscopic proportions, and
2. present some key results on properties of surfaces, demonstrating their role in corrosion and heterogeneous catalysis. In particular we discuss
   - the influence of the ambient gas phase on the surface structure and stoichiometry,
   - adsorbate phase transitions and thermal desorption, and
   - the role of atoms' dynamics and statistics for the surface chemical reactivity.

© 2001 Elsevier Science B.V. All rights reserved.

---

*Corresponding author. Address: Department of Physics and Astronomy, Northwestern University, 2145 Sheridan Road, Evanston, IL 60208-3112, USA. Tel.: +1-8474918637; fax: +1-8474676857.
E-mail address: cathy@venus.phys.nwu.edu (C. Stampfl).

0039-6028/01/$ - see front matter © 2001 Elsevier Science B.V. All rights reserved.
PII: S0039-6028(01)01551-5

Keywords: Density functional calculations; Non-equilibrium thermodynamics and statistical mechanics; Catalysis; Corrosion; Oxidation; Surface chemical reaction; Surface thermodynamics (including phase transitions); Ruthenium

1. Introduction

1.1. The cutting edge

Surfaces are the cutting edge of material sciences, i.e., a surface is the place where molecules from the gas phase or a liquid come into contact with a material, and where chemical bonds of these approaching molecules may be cut and new bonds formed. If we look around, everything we see is surfaces, and to understand the properties of materials, to understand how materials can be produced, how they can be grown, why and how they corrode or rust, and to how to protect them, for example, how to make the surface hard, and to understand how catalysis works and how it may be improved,—for all this, one has to understand surfaces.

The processes which occur at surfaces play a critical role in the manufacture and performance of advanced materials. Examples are: electronic, magnetic, and optical devices, sensors, catalysts, and hard coatings. In this paper we will mention only very few examples, i.e., focus on molecular surface processes which rule catalysis and corrosion. We note, however, that the same concepts and methodology also apply to the modeling of dopant profiles, surface segregation, crystal growth, and more.

Obviously, a better knowledge of surface- and interface-physics and chemistry is vital to the way we live, i.e. a better knowledge is necessary to support and advance the high technology which very much determines our life style, and it is needed in order to protect the environment. It is evident that there is still much to learn about the intricate molecular and atomic processes that occur at surfaces. The reason for this lack of understanding is largely due to the length and time scales involved (see Fig. 1). Molecular processes at surfaces proceed on a length scale of 0.1 nm (1 nm = 10^{-9} m), electrons move and adjust to perturbations in the femto- (10^{-15}) second time

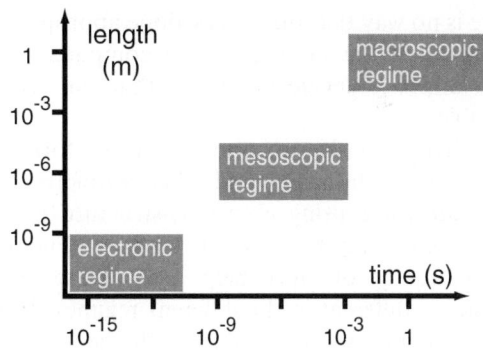

Fig. 1. Schematic presentation of the time and length scales relevant to catalysis and corrosion: The molecular processes, which rule the behavior of a system, proceed in the "electronic regime", but the observable effects only develop after meso- and macro-scopic lengths and times.

scale, and atoms vibrate on a time scale of a pico- (10^{-12}) second. However, for surface phase transitions, corrosion, crystal growth, or catalysis, the relevant time scale is of the order of microseconds or even seconds. For example, a single dissociation event takes some femtoseconds, and diffusion over a distance of 10 nm may take a time of the order of several picoseconds. Thus, the combination of the various molecular processes which take part in a catalytic reaction (see Section 1.3) may take about a nanosecond. Clearly, the realization of the surface phenomena mentioned above involve very many such reactions and consequently the "whole concert" takes much longer. Therefore in order to evaluate turn-over rates, for example, one needs to perform a statistical average over many such concerted events, which implies that in total, a theoretical simulation must span a time period of several microseconds. For corrosion (as well as for crystal growth), the typical time period is even longer, i.e., here we are dealing with a phase transition or detachment of atoms which proceed at a speed of about one atomic layer per minute. Thus, we have to go to the meso- or even macro-scopic regime (see Fig. 1). Everything is determined by the lower left box in Fig. 1, but how systems evolve and behave is described in the regimes of the other two boxes. The main problem is in the time scale, where we have to bridge 12 orders of magnitude: from 10^{-12} s to seconds. Obviously there is no way that this can be done appropriately by simply using more powerful computers. Not even in 100 years do we expect that this will be possible.

In fact, even if it were possible to perform calculations for the meso- and macroscopic regimes by brute force using electronic-structure theory, i.e., by enlarging the lower left box of Fig. 1, it would not be of much help. The nature of the physics is different in the different regimes: In the lower left box we are dealing with the nature of the chemical bond, and this describes the behavior of the electrons and the interactions between atoms and molecules. However, the other regions are governed by the electronically determined microscopic parameters plus the laws of thermodynamics and statistical mechanics. The electronic regime tells us what can happen. However, whether a process found in this regime is actually relevant for the full concert of possible molecular processes will be decided by the physics in the other regimes. If one likes to understand what it is that determines real-life phenomena, one has to combine (hopefully seamlessly) electronic-structure theory with a proper treatment of statistical mechanics. And this is what we will emphasize and describe in this paper.

The physics of the lower left box of Fig. 1, i.e. the electronic regime and molecular processes, is best described by density-functional theory (DFT) [1–3]. DFT has developed into the most successful and wide-spread approach for accurately calculating and predicting various physical properties of a wide range of materials systems including molecules, bulk solids, and surfaces (see e.g. Ref. [4] and other contributions in that volume). It is called a "first-principles" theory, which in Latin reads "ab initio". For practical calculations it means that everything is evaluated by starting from a self-consistent electronic-structure calculation.

In order to describe the properties and performance of "real materials", as for example those involved in catalysis and corrosion, one has to go beyond the electronic-structure calculations. For this second step, four different methods have been developed, which apply to different situations:

- If it is necessary to follow the motion of the atoms in detail, ab initio molecular dynamics is used [5,6]. Here ab initio refers to the fact that the forces acting on the nuclei are calculated by DFT. And then the atoms are allowed to move, being accelerated by the ab initio forces as described by Newton's laws [7,8]. For atoms heavier than hydrogen, the nuclear motion is indeed classical—within the required accuracy—and thus, the ab initio molecular dynamics treatment is appropriate [6].
- For systems which involve hydrogen, an ab initio quantum dynamics approach has been developed. Here a Schrödinger equation is also solved for the nuclei, where the potential energy entering the Hamiltonian is the DFT total energy [6,9–11]. Thus, zero-point vibrations and tunneling are allowed. In Section 3.3 we give an example of what can be learnt from such studies. Other examples are given in the paper by Groß in this volume.
- Often processes are close to thermodynamic equilibrium. Then it is e.g. important to consider the possibility of exchanging atoms with the environment. When it comes to a description of surface segregation, adsorbate phase transitions as a function of temperature, and even as a function of heating (time dependent change of temperature), a lattice-gas Hamiltonian approach is appropriate (e.g. Ref. [12]), which enables us to evaluate the partition function, and, of course, from the partition function all thermodynamic potentials can be obtained. In Section 2 we give two examples of thermodynamic equilibrium and/or close to thermodynamic equilibrium studies relevant to the context of this paper.
- For systems which are not in thermodynamic equilibrium, but which are determined by the dynamics of the atoms, and when the time scale is beyond that possible in the direct molecular dynamics calculations, the appropriate method

is the kinetic Monte Carlo (kMC) approach [13]. This employs statistical sampling of the various adparticle surface processes and brings us into the time regime of seconds and minutes. Corrosion, crystal growth, and also catalysis are examples for which this approach will be useful. In Section 4 we describe the state-of-the-art of this method.

One may argue that using a different methodology for different types of problems lacks elegance and beauty. Admittedly this is true, i.e., the choice of methodology is done pragmatically. It is taken because we like to obtain understanding. In fact, we believe that "one theory for everything" is an irrational fiction. What is needed is understanding, as we have advocated before [4], and this requires knowledge about various areas: electronic-structure theory, condensed matter physics, material science, chemistry, computer science, thermodynamics, and statistical mechanics. In other words, this field is very interdisciplinary in nature, which, besides curiosity, makes it appealing and fascinating.

In the present paper, we concentrate on metal and metal-oxide surfaces, and do not e.g. discuss zeolites, the use of which as catalysts represents an area of its own. Zeolites have properties that may not undergo a significant change at high pressures and temperatures, instead, other factors play a role in their activity such as the geometry of the cages, pores, and channels. Hence, this material class lies outside the emphasis of the present paper and we refer the interested reader to the contribution by Sinfelt in this volume. The remainder of this paper is organized as follows. In Section 1.2 we summarize the main concepts and definitions behind heterogeneous catalysis and corrosion. Then in Section 1.3 we extend this description by discussing the various molecular processes, which happen at surfaces and rule, though in a concerted and statistical manner, catalysis and corrosion. In Section 2 we describe concepts (and results) for a surface under realistic conditions, i.e., when it is in contact with a realistic atmosphere at finite temperature and pressure. In Section 3 we discuss the asymptotic interaction of atoms and molecules at, and with, surfaces (reactivity theory); it describes the nature of the chemical bond at the transition state of a surface chemical reaction (Section 3.2), and it emphasizes the importance of the dynamics and statistics for a proper account of surface reactivity (Section 3.3). We discuss in Section 3.4 the importance of an optimum adparticle–substrate bondstrength in relation to its reactivity and describe examples of how it can be modified either by adjusting the coverage or through electronic excitation. Then in Section 4 we describe how the various individual molecular processes could be combined and the statistics taken into account. That is, we show how the full concert of a sustained catalytic process as well as corrosion could be treated. This section describes work in progress and is pointing towards future theoretical work. In this way it contains the outlook and concludes the paper.

1.2. Concepts and definitions of heterogeneous catalysis and corrosion

A "catalyst" is a substance that enhances the rate at which a certain chemical is produced, possibly it also reduces the rate of another competing reaction. And a catalyst does this without being consumed in the process. Then "catalysis" is the phenomenon of a catalyst in action. A heterogeneous catalytic reaction is a concert of processes whereby the word "heterogeneous" refers to the situation that the net reaction occurs when reactants and catalyst are present in different phases. The catalytic reaction between adsorbed particles exhibits a free-energy barrier which is lower than that of the chemical reaction between the reactants in the gas phase. The result is that the reaction is kinetically accelerated. We refer to Ref. [14] for a comprehensive publication on heterogeneous catalysis. Fig. 2 gives a schematic example of a gas phase reaction (left) and how it is changed by a catalyst (right). Catalytic reactions at surfaces have traditionally been classified as different types: a so-called "Langmuir–Hinshelwood" reaction is where the reactants are both adsorbed on the surface prior to reaction; this is the most usual case in heterogeneous catalysis. An "Eley–Rideal" reaction mechanism is where the chemical reaction takes place between a gas phase particle (e.g. CO)

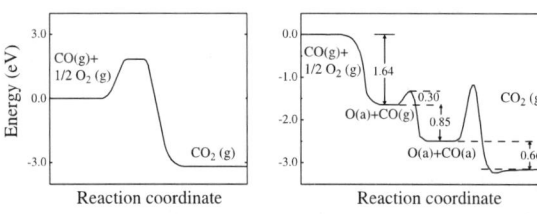

Fig. 2. Schematic illustration of the energy diagram of a simple gas phase reaction, using the CO oxidation as an example (left) and the corresponding reaction over a catalyst (right). In this reaction, the CO molecule interacts with an O atom, either from the O_2 molecule (left) or with an adsorbed O atom (right), to form carbon dioxide, CO_2. The rate for crossing energy barriers is $\Gamma = \Gamma_0 \exp(-\Delta E/k_B T)$, where Γ_0 is the attempt frequency, ΔE the height of the energy barrier, k_B is Boltzmann's constant and T is the temperature. Note that the energy scales of the left and right figures differ by a factor of two. Obviously, the catalyzed reaction is much more efficient. Labels (g) and (a) denote "gas phase" and "adsorbate". (Right figure from Ref. [16].)

which scatters at an adsorbed particle (e.g. O). Thus, the reaction product (CO_2) is formed during a scattering event. This reaction mechanism is usually assumed to be un- or less-likely because the number of gas phase particles hitting the surface per site per second, at a given temperature and pressure is typically low compared to the attempt frequency of two particles already adsorbed on the surface in the case of the Langmuir–Hinshelwood process. The classification of these two types of reaction mechanisms was originally put forward by Langmuir [15]. Finally, for oxidation catalysis over metal oxides, a frequently considered assumption is that the required oxygen is taken out of the oxide material, which is called the "Mars–van Krevelen" mechanism.

In industry, catalytic substances (such as transition metals) are typically utilized as a highly dispersed catalyst on a "support material". This is for a number of reasons: firstly, to increase the available surface area for reaction; support materials should thus have a high surface area, be durable, and inexpensive. Second, the interaction of the dispersed catalyst with the support can lead to an enhanced chemical efficiency compared to reaction over the catalytic material alone, as for example, in the case of some catalytic nanoparticles. Finally, transition metals are expensive and limited in supply; already in 1991, 87% of the world demand of Rh went to catalysts [17].

The nature of the surface, which is exposed to the reactants, is crucial. However, the knowledge about the surface, as it develops under the actual catalytic conditions, is typically uncertain, and we believe that many of the "established" ideas and concepts are incomplete, and some may be wrong. Two examples are briefly mentioned to illustrate this point. Oxidation catalysis at a ruthenium catalyst probably does not take place at the Ru metal, initially introduced as "the catalytic material". Instead, under normal (or high) pressure and high temperature conditions, the Ru metal undergoes a phase transition into a mesoscopically structured system with areas of very different behavior, which most likely can be characterized as oxygen adsorbate regions, RuO_2 oxide regions, and other (strained) Ru oxides [18–20] (see also the discussion in Section 4). We believe that the co-existence of the different regions is crucial for understanding catalysis. But we also note that this is not an accepted or widely shared view, and "standard" surface-science studies on single crystals are often discussed in terms of their direct significance for catalysis. The other example is the dehydrogenation of ethylbenzene ($C_6H_5CH_2CH_3$) to form styrene ($C_6H_5CH=CH_2$) using potassium promoted iron-oxide catalysts. Recent work surprisingly shows that after the "induction period" (see below) ethylbenzene apparently does not interact with the iron oxide and thus the catalytic activity is in fact not related to iron oxide; rather the catalyst is a "new material" that is created during initial stages of the interaction of ethylbenzene with the iron-oxide surface: It seems that the catalytically active material (for styrene formation) is a certain type of carbon [21]. In relation to the nature of surfaces under actual catalytic conditions, we mention that industrial catalysts typically require an induction period, i.e. a time to get into efficient action (sometimes of hours, or days) before performance reaches steady state, which indicates that the native catalyst material initially undergoes significant changes in its surface structure and/or composition. We believe that most (maybe all) catalysts which exhibit an induction period may undergo such dramatic

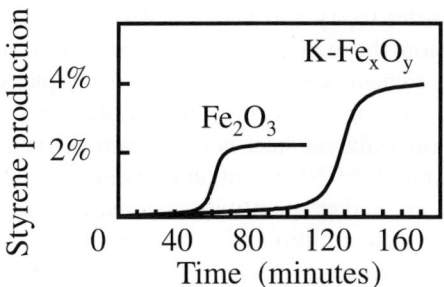

Fig. 3. Induction period of styrene production from ethylbenzene (EB) using hematite (Fe_2O_3) and a potassium-promoted iron-oxide catalyst. The plot is schematic, but based on data of Ranke and coworkers [21,22], who used a 1 cm² sized model-catalyst under near-industrial conditions ($T = 870$ K, $p_{EB} = 3$ mbar, and the ratio of EB to water steam is 1:10).

changes and we expect many more surprises, of similar significance and impact as that of the above mentioned "promoted iron-oxide catalyst". Fig. 3 illustrates this point for styrene production which shows the required induction period where an iron-oxide catalyst is used. A similar, but longer induction period occurs for the K-promoted iron-oxide catalyst, which results in a higher formation rate of styrene. (We note that in Fig. 3 the rate is low because the size of the catalyst used in this study is small (1 cm²). Industrial reactors consist of catalyst-filled tubes that are several meters long, and then the rate is about 50%.)

The behavior of a catalyst under ultra-high vacuum (UHV) conditions (10^{-13}–10^{-9} atm) and low temperatures (e.g. room temperature and below) may be different from the behavior at high pressures and temperatures, and this is called the "pressure gap". In this respect, one of the main concerns in catalysis research has been whether it is possible to extrapolate between the UHV conditions and single crystal substrates of surface science, to the high pressures and temperatures, and the often quite complex structures of industrial catalysts under reaction conditions. This deviation between the single crystals and "working" catalyst materials, has been coined the "material gap". Because the native catalyst material may undergo significant changes when under actual operating conditions, the material and pressure gaps are typically linked (see e.g., the contribution by Zaera in this volume). Another aspect is that in real catalysis the support material may play a more significant and different role than often assumed, and this support material, as well as the interface between nanoparticles and the support, may be substantially different under realistic catalytic conditions than in hitherto performed UHV studies.

We also note that industrial catalysis involves more aspects than those mentioned above and in Section 1.3. For example, there is "selectivity", which means that only the desired reaction should take place, and competing reactions yielding unwanted products are suppressed. Sometimes this is the main problem; for example, in ammonia synthesis the dissociation of nitrogen is the rate-limiting step, and for the cracking of large hydrocarbon molecules into smaller ones, which is a crucial process in the petroleum industry, the catalyst must actuate scission of the carbon–carbon bonds. Also the buffering of intermediate chemical products is important, as is the self-maintenance of the catalyst. And often it is important that no poisonous by-products are released. Furthermore, the role and mechanisms of promoters and poisoners are very important but are rarely well understood. In order to enhance the performance of a catalyst, various atomic species are sometimes added to the catalyst. These additives are called modifiers and if it is a "good modifier", that is, if it brings about an improvement in the performance of the catalyst, then it is termed a "promoter". Conversely, a "bad modifier" is referred to as a "poison". Usually additives are deliberately introduced but sometimes they arise unintentionally through contaminents. The economic importance of catalysis for creating desired substances is significant; but perhaps even more important is that catalysis is necessary for preserving the environment. The catalytic converter to clean the automotive exhaust (i.e. convert CO to CO_2) is just one example.

Corrosion has a similar economic importance, but, in contrast to catalysis, it is unwanted. Corrosion can take place in different environments such as in the atmosphere, in solution, and in soil. Most metals are thermodynamically unstable in aqueous environments, and their corrosion in water or an electrolytic solution, made by dissolving a salt, acid, or base in water, is an electrochemical process where the basic mechanism involves removal of valence electrons from the metal atoms which become ions that move into the solution, and then often more stable compounds such as iron oxide (e.g., rust) may form. This basic mechanism is called an "anodic reaction", and for the corrosion process to proceed there must be a corresponding "cathodic reaction" that accepts the electrons. We discuss this in more detail in Section 1.3. Corrosion of a metal brings the metal to its thermodynamic ground state under the existing environmental conditions. Fig. 4 illustrates this point by giving the energetics of the formation of an iron-oxide layer on iron due to the interaction of O_2 with Fe. Often metal oxides are insulators, but RuO_2, for example, is metallic with a conductivity comparable to that of copper [23]. Obviously, corrosion compounds start to form on the surface as thin films, and if they are hard and impenetrable, and if they adhere well to the material, the progress of corrosion will stop (or be significantly slowed down) after a typical thickness of the order of 2–5 nm. This is then known as passivation. For example, a reactive metal like iron, which corrodes at half a millimeter per year in salt water, corrodes a thousand times slower in environments that allow the formation of a passive film.

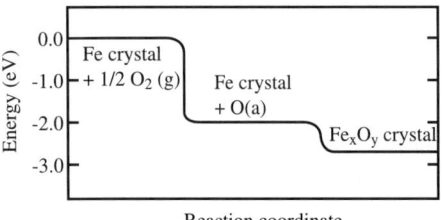

Fig. 4. Schematic energy diagram of metal corrosion, for the example of O_2 at an Fe surface, that illustrates the point that corrosion processes lower the energy of the metal/environment system as far as possible, through the formation of more stable structures: In an O_2 atmosphere the metal oxide (Fe_xO_y) is thermodynamically more stable than the metal, but typically, the oxide layer formed at the surface prevents the phase transition being completed. The "reaction coordinate" on the horizontal axis represents, in a general way, the atomic positions of the atoms involved in the reaction.

In order to try and help protect the metal from corrosion through passive film formation, an alloying element can be added, for example, chromium to iron to form stainless steel. (A most common stainless steel is comprised of iron with 18% Cr and 8% Ni, but other elements may also be added in smaller quantities such as Mo, Se.) Apparently, the addition of chromium has the result that the surface-oxide film formed is more resistant to breakdown. We point out that as yet there is not a good quantitative understanding of how chromium makes stainless steel stainless, but it is thought that the Cr in Fe segregates to the surface and covers the surface with chromium oxide. Chromium and aluminum form the most stable and protective of such films. The films that form on copper and iron as a result of corrosion are known as tarnish and rust, respectively. The corrosion resistance of a passive film is determined by its ionic and electronic transport properties, which are largely determined by the film's crystallographic structure, defects, and its microstructure. In general, the atomic structure of such oxide films is still poorly understood. We like to note, however, that a recent in situ X-ray diffraction study of the passive film on iron that forms in aqueous solution, which has been an outstanding question for decades, concluded that it is a highly defective nanocrystalline spinel oxide with a high concentration of vacancies and interstitials [24,25]. Clearly, there is much quantitative and interesting work to be done in this field.

Like catalysis, understanding of the various molecular processes involved in corrosion encompasses concepts from a variety of disciplines, including thermodynamics, electrochemistry, physical metallurgy, condensed matter, and statistical mechanics. The adsorption of reactants on the surface (atoms or molecules from the environment) represents the first elementary step in corrosion (and in a catalytic reaction cycle). Thus, understanding the chemisorption step (bond breaking or weakening in the reactant and the making of new bonds to the surface) has been (and still is) one of the prime goals. The main aspects that have been addressed by theoretical surface scientists, in particular, are the atomic and electronic-structure, the binding energies of atoms and molecules on metal surfaces and the nature of the bonding, activation energies, and diffusion at the surface and from the surface into the bulk. In addition, factors such as temperature, gas phase pressure, etc., namely, the chemical and thermal contact with the environment can play a decisive role in the behavior of the system so that it is also desirable to take it into account. Furthermore, often interesting phenomena occur far away from equilibrium so that in order to understand such processes, it can be necessary to go beyond ground state and equilibrium properties. We will discuss some of these aspects in this paper, and some are still waiting to be treated properly (see Section 4).

1.3. Molecular processes relevant to heterogeneous catalysis and corrosion

In this section we describe important molecular processes which are part of corrosion and heterogeneous catalysis and which will be drawn upon in subsequent sections in discussing selected examples of recent studies. Understanding of these individual processes is crucial, but we also emphasize that in the end what counts is how they act together, and it is the statistical sampling that determines which of these processes are really relevant. The main molecular processes are:

- the adsorption of atoms and molecules (often dissociatively) on a solid surface from the gas or liquid phase,
- diffusion on the surface,
- diffusion of adsorbates into the subsurface region and of substrate metal atoms (or ions) through a surface-oxide layer (if the latter is present),
- interactions between different adparticles at the surface,
- chemical reaction of different adparticles,
- desorption of the reaction product.

In Fig. 5 we illustrate some of these molecular processes, referring to the interaction of Ru with O_2: Molecules present in the gas phase may bond to the surface in a molecular form or may dissociate into atoms depending on a number of factors such as the temperature and pressure at which adsorption takes place, surface structure, and the extent to which the surface is covered by adsorbate

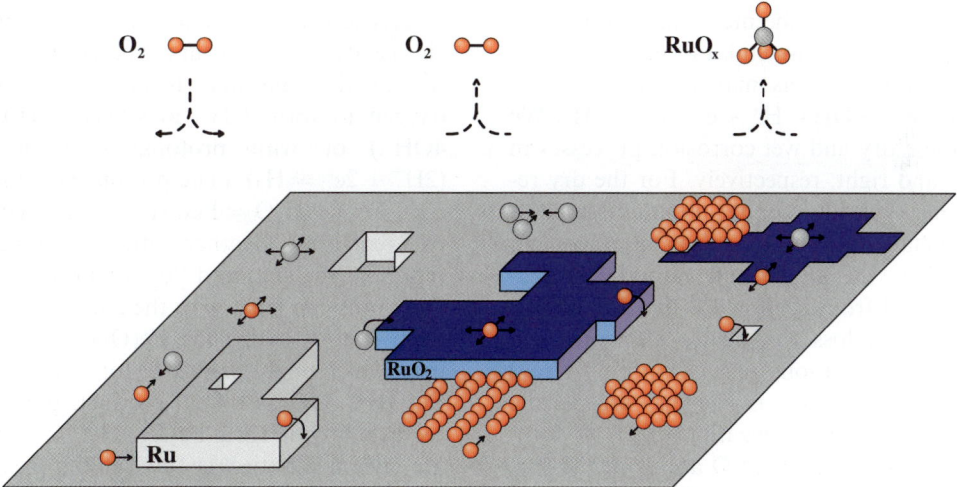

Fig. 5. Schematic illustration of some microscopic atomic and molecular processes that take place at surfaces and which are relevant for heterogeneous catalysis and corrosion. The example substrate here is ruthenium in an oxygen environment. For example, oxygen molecules, O_2, may dissociatively adsorb on the surface (indicated by the down and outward arrows), the O atoms may diffuse on the surface (as indicated by small circles with arrows), and may associatively desorb into the gas phase as O_2 (indicated by the upward and inward arrows). Alternatively they may react with the Ru atoms of the substrate to form islands of ruthenium dioxide, RuO_2, or may desorb as RuO_x molecules.

species. These adparticles then diffuse around on the surface and may cluster together forming islands, or they may react with other adsorbed species to form a molecular product that desorbs from the surface, or to form a new phase that stays on the surface, e.g. an oxide. Furthermore, surface segregation of metal cations may occur. In heterogeneous catalysis and corrosion, almost always one of the approaching molecules dissociates and it is the availability of bonding sites on the substrate that affords and favors the catalytic or corrosion process.

With respect to corrosion in solution, we elaborate a little further below on some of the typical reactions and processes that take place and compare with the case of corrosion in a pure gas environment. Here the examples are iron in water (which contains oxygen) and in a pure oxygen atmosphere. Corrosion reactions that do not involve water or aqueous solutions are often called "dry" corrosion reactions and those that do are often called "wet" corrosion reactions. Firstly, we point out that corrosion can be viewed as a heterogeneous chemical reaction at a metal/non-metal interface in which one of the reactants is the metal itself, and the non-metal reactants are the environment. One of the reaction products will always be an oxidized form of the metal and the other a reduced form of the non-metal. For example, non-metal reduction reactions may be $O_2 + 2me^- \rightleftharpoons 2O^{m-}$, $OH + e^- \rightarrow OH^-$, $H^+ + e^- \rightarrow (1/2)H_2$. We illustrate these dry and wet corrosion processes in Fig. 6, left and right, respectively. For the dry reaction, which typically requires elevated temperatures, the metal/oxide and oxide/gas interfaces can be regarded as the anode and cathode, respectively. We recall from Section 1.2 that the anodic reaction involves loss of valence electrons and creation of metal cations (e.g. $Fe \rightleftharpoons Fe^{z+} + ze^-$) and the cathodic reaction is one that accepts the electrons; here it is the ionization of O_2 to form adsorbed, negatively charged O atoms at the surface (e.g., $(1/2)O_2 + 2e^- \rightleftharpoons O^{2-}$). The chemisorbed O atoms cause a high electric field so the system tries to lower it by either "pulling" metal ions from the metal or "pushing" oxygen ions into the metal lattice (segregation and diffusion). This mass transport enables the formation of the corrosion

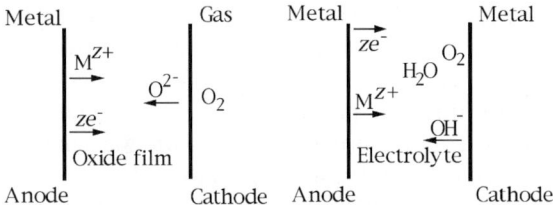

Fig. 6. Schematic diagram illustrating the interaction of a metal with dry, O_2, (left) and wet, e.g. $O_2 + H_2O$, (right) environments, and some of the mechanisms that give rise to the formation of the passive oxide layer and the effective anodes and cathodes. In the former case the metal atoms lose valence electrons (ze^-) and become cations (M^{z+}). This is the anodic reaction and can be regarded as occurring at the metal/oxide interface. The electrons are accepted by oxygen to form negatively charged O atoms at the surface; this is called the cathodic reaction and can be thought of as taking place at the oxide/gas interface. In the wet case (right), the anodic reaction is the same except that the metal ions diffuse into the electrolyte. The electrons lost from the metal atoms may react with the oxygen and water to form, e.g., OH^- which may react with the cations to form the oxide product.

product, i.e. the metal oxide formation, which as we showed in Fig. 4 for the example of iron oxide, has a lower energy than that of clean Fe in such an environment. In the wet case, the anodic reaction is the same, except that the metal ions diffuse into the solution. This sets up an electric potential difference between the metal and the solution and charge separation occurs. The electrons lost from the metal atoms may react with the water and oxygen to form OH^- ions ($O_2 + 2H_2O + 4e^- \rightleftharpoons 4OH^-$) or with protons to form hydrogen ($2H^+ + 2e^- \rightleftharpoons H_2$). (The protons being formed via e.g. $Fe^{2+} + H_2O \rightleftharpoons FeOH^+ + H^+$.) These latter two reactions, together with the anodic reaction, represent the standard "dissolution process". The OH^- ions can react with the metal cations, e.g., to form ferrous hydroxide $Fe(OH)_2$ for the case of iron, which might oxidize to form ferric hydroxide, $Fe(OH)_3$, and as a final product, the well known reddish brown rust, $Fe_2O_3 \cdot H_2O$ of which there are several varieties (i.e. hydrated iron, where each molecule of iron oxide is chemically bound to one or more water molecules). If there is insufficient oxygen, Fe_3O_4 or α-Fe_2O_3 may form [26].

In Fig. 7 we illustrate in more detail some of these microscopic steps that occur in the building-up of the films. The film structure of Fe in an O_2

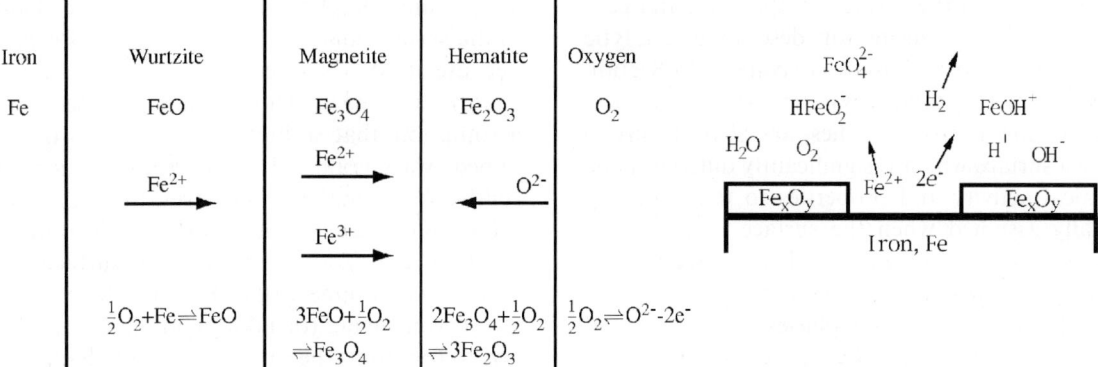

Fig. 7. Illustration of the formation of iron oxide in (left) an oxygen gas environment (for temperatures greater than about 570 °C) showing the different states of oxidation of the film (FeO, Fe_3O_4, Fe_2O_3) and (right) in the vicinity of an imperfection or pore at an iron surface which is immersed in an aqueous environment containing water and oxygen. In the latter, reactions at the pore (bare Fe surface) present either anodic or dissolution processes, while those on the oxide surface are cathodic in nature. Besides water, H_2O, and oxygen, O_2, metal ions, Fe^{2+} and electrons, $2e^-$, some of the molecular reaction products are illustrated. (Various solid compounds also form as mentioned in the text, but are not explicitly shown for simplicity.) The left figure is a simplified form of that given in Ref. [27].

environment is believed to consist of oxide layers of different stoichiometries: initially, at sufficiently high temperatures (>570 °C) a (wurtzite) FeO layer forms on the surface, which, with further oxidation, leads to growth of Fe_3O_4 and Fe_2O_3 (see Fig. 7, left). We refer to Refs. [28,29] for a discussion of theories of the initial oxidation of metals and some recent results. In the wet environment (Fig. 7, right) we illustrate how quite naturally the corrosion processes can occur at an imperfection in the surface (e.g. a pit or pore exposing clean Fe). A local anode and cathode are formed where the bare Fe acts as the former and the developing oxide as the latter. Some of the characteristic reactions and reaction products are shown. In the present paper our subsequent discussions relate more to the case of dry corrosion processes.

2. Surfaces in or close to thermodynamic equilibrium

When a surface is in contact with a realistic environment, e.g. an atmosphere containing oxygen and water, atoms or molecules from the environment can adsorb on the surface and/or atoms from the surface can be released into the environment. What is typically called a stable surface structure is in fact a statistical average over adsorption and desorption processes. The surface composition and geometry which is assumed in thermodynamic equilibrium depends on the type of environment, e.g. the partial pressures of O_2 and H_2O and temperature. Considering, for example, iron oxide in thermodynamic equilibrium, the energy to take an oxygen atom out of the bulk, or out of the surface, or out of a gas phase O_2 molecule, or out of a gas phase H_2O molecule is the same. This equality, as exemplified for the oxygen atom, is in fact the definition of thermodynamic equilibrium, and if the chemical potentials of a certain species in different coexisting phases were not the same, the system would change, e.g. by altering the surface geometry or composition, until thermodynamic equilibrium is reached. This process can be slow and kinetically hindered, and corrosion is one example where kinetic hindrance stops oxidation if a passive film forms and has attained a certain thickness. Thus, under these circumstances the corrosion compound which forms on a metal surface is not in thermodynamic equilibrium with the metal bulk. However, the surface (at least at not too low temperatures) is likely to be in thermodynamic equilibrium with the ambient atmosphere.

Knowledge of the surface composition and geometry is a prerequisite for describing catalytic processes and corrosion or passivation. In Section 2.1 we show that (and why) UHV surface science, and also low pressure studies, are likely to investigate a surface which is significantly different in its chemical activity and properties to that which is actually assumed when the surface is in contact with a realistic atmosphere. Then in Section 2.2 we describe how the temperature and pressure dependence of adsorbate phases can be treated theoretically, with periodic as well as disordered surface structures. As an example of this approach we present an ab initio calculation of temperature programmed desorption, appropriate for not too fast heating rates so that the surface structure can always assume the thermal equilibrium surface geometry.

2.1. The role of the ambient atmosphere

For metal oxides [30] (and this includes the oxide film which develops at a metal surface under realistic conditions) the understanding of the surface terminations (composition and atomic geometry) is still very shallow. The reason is that electron scattering and spectroscopy techniques and scanning tunneling microscopy (STM) are hampered by the insulating nature of the material. Also theoretical studies for metal oxides are very demanding because they have to deal with a rather open structure, oxygen with very localized wave functions, huge atomic relaxations, big supercells, and often, as for example in iron oxide, 3d electrons and magnetism. In this section we describe concepts and some recent theoretical work on atomistic thermodynamics of metal oxides in contact with a realistic liquid or gas phase. We start with α-Fe_2O_3 (hematite) in an oxygen atmosphere, and at the end we add some words about α-Al_2O_3 (corundum, or sapphire) and the changes which occur, when in addition to O_2 also hydrogen (either in the form of H_2 or water) is present in the environment. While Fe_2O_3 may be a possible catalyst, Al_2O_3 is a frequently used support material.

Before 1998, theoretical work had concluded that the most stable surface structure of α-Fe_2O_3 was metal terminated, with one Fe atom in the surface unit cell. A termination with three O atoms in the surface unit cell (denoted, O_3) was excluded because it was believed to yield an electrostatic energy divergence. This argument, based on the assumption that a bulk-like geometry is maintained, was supported by some calculations. Recently, Wang et al. [31], however, through detailed DFT–GGA calculations allowing full atomic relaxation and possible exchange of surface atoms with the environment (the O_2 atmosphere acting as a reservoir giving (or taking) any amount of oxygen to (or from) the sample, without changing the temperature or pressure) showed that this is not correct. In particular, by calculating the surface energy for various hypothetical surface terminations as a function of the oxygen chemical potential, which is uniquely determined by the environment (see e.g. Refs. [31,32]), the energetically most favorable structures could be derived for certain conditions. In Fig. 8 we show these results: While under oxygen poor conditions, it can

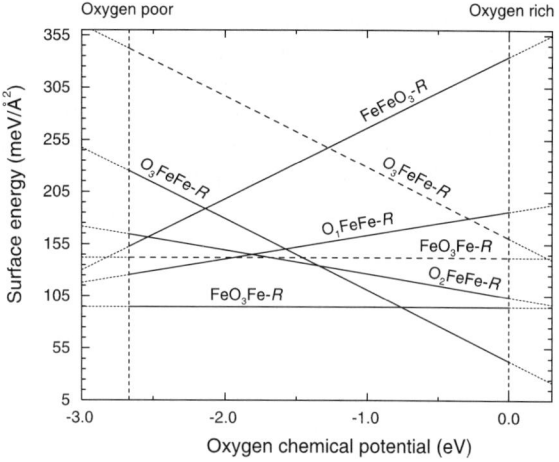

Fig. 8. Surface energies of different $Fe_2O_3(0001)$ surface terminations. The lines are labeled by the sequence of layers from the surface toward the bulk, and $-R$ stands for the bulk-like continuation. The zero of the chemical potential is set to $(1/2)E_{O_2}$, the total energy per oxygen atom of molecular O_2 at $(T = 0$ K). The allowed range of the oxygen chemical potential is indicated by the vertical dashed lines, where the left one corresponds to strongly Fe-rich (i.e., oxygen-poor) conditions, and the right one corresponds to strongly oxygen-rich conditions (i.e., high oxygen gas pressure). Full lines show results for relaxed geometries, and dashed lines give for comparison results for unrelaxed surfaces (after Ref. [31]).

be seen that the most stable surface is indeed the mentioned iron terminated one, however, at higher oxygen pressure, the O_3 terminated structure surprisingly becomes more stable. In fact, the crossing point of the surface energies of these two surface configurations is roughly at room temperature and 1 atm, which implies, that this previously rejected surface geometry may certainly play a significant role. Considering the surface relaxations involved, they are found to be huge (changes of interlayer spacings can be as large as 80% of the bulk value), and the surface electronic-structure, as well as the magnetism are very different to that of a truncated bulk geometry. This may be already guessed from Fig. 8 because the difference between the surface energies of the unrelaxed (dashed lines) and relaxed (full lines) structures is significant (for details see Ref. [31]).

Analogous studies for Al_2O_3 [33] showed that its (0 0 0 1) surface, when in contact with an O_2 environment, is always Al terminated, in the whole range of accessible oxygen chemical potentials. This difference to the iron oxide is related to the fact that transition metals have more flexible valence states than Al. Interestingly, when in addition to oxygen, also H_2 and H_2O were considered as part of the environment, the surface of aluminum oxide was found to be the O_3 terminated one, where, however, each oxygen also binds an H atom. Thus, we may also say that this surface is OH terminated.

These are just two examples (Fe_2O_3 and Al_2O_3) which show that under realistic conditions surfaces can be very different to those often studied in UHV. The difference can be dramatic, and the electronic properties and performance can have little resemblance to the low pressure results. We like to emphasize this point by also stating that surfaces in general (as compared to the bulk or even to the bulk terminated surface) and in particular as they develop under realistic conditions, represent a new material.

For corrosion that results in passivation of the surface, thermodynamic equilibrium between the film and the substrate is not achieved; the reason that the film stops growing is because of kinetic hindrance. Nevertheless, the film may have a sufficient thickness in order to obtain a rough estimate of the metal-rich (oxygen poor) boundary of the surface structures and the compositions which may develop.

2.2. Surface phase transitions and thermal desorption

For surfaces that are in, or close to, thermodynamic equilibrium and which depend on the atmospheric environment and temperature, such as adsorbate phases which may initiate corrosion and surface phase transitions of a catalyst under high temperature and pressure conditions, and/or to describe surface segregation, an appropriate and general approach to describe the structure and to analyze adatom–adatom interactions as a function of temperature and pressure is a lattice-gas (or Ising) model [34,35]. This approach is quite powerful, and e.g., also enables us to evaluate medium range interactions between different adatoms. Furthermore, the Hamiltonian need not be restricted to the description of adsorbates on a surface; it can, for example, be extended to account for subsurface species, describing their mutual interactions, as well as their interactions with on-surface adsorbates.

This approach starts by defining possible sites for the adatoms, and that this is possible is the main assumption of the approach. Then, in a second step it is determined whether or not these sites are occupied, and there is no restriction that this occupation should be dense or sparse, or periodic, or disordered. From the lattice-gas Hamiltonian one can evaluate the partition function and thus obtain the thermodynamic properties of the system. Furthermore, through combination with rate equations, a description of the kinetics (if it proceeds close to thermodynamic equilibrium) can be obtained, such as thermal desorption.

The lattice-gas Hamiltonian is built-up from the energy of an isolated adatom at a surface including its partition function accounting for vibration perpendicular, and frustrated translation parallel, to the surface. Interactions from possible neighboring adsorbates are then added, for example, two-body or pair-wise interaction energy contributions due to adatoms at nearest neighbor lattice sites, at second neighbor sites, and third neighbor sites etc. Also, three-body (or trio) interactions that account for modifications because the interaction between two adsorbed atoms is changed, when a third adatom is close by can be included, as can higher-order interactions. The number and type of interactions (and atomic species) can be increased at will until it is judged that the Hamiltonian is sufficiently accurate. These quantities can all be obtained from DFT calculations: Specifically, the individual adatom–adatom interactions can be deduced from calculations with different supercells and different adatom geometries by expressing the adatom adsorption energy in terms of its interactions with neighboring adsorbates. When done for the different structures, this yields a system of equations that can be solved simultaneously. Direct calculation of the individual adatom–adatom interactions would not be readily accessible because of too high computational costs.

In what follows, we give an example of this approach for the system of oxygen at Ru(0001) where the temperature-programmed desorption of O_2 is calculated in the coverage regime from low coverage to a full monolayer [36]. Temperature-programmed desorption is one of the most widely used experimental techniques for studying the binding energies of adsorbed species. In this experiment one prepares an adsorbate layer of a given initial coverage at a given temperature and measures the desorption rate of the particular species as a function of increasing temperature. The rate depends exponentially upon the negative of the activation energy barrier for desorption, which in turn is also a function of the coverage. The prefactor to the exponential involves a sticking probability (as will be discussed in Section 3.3). The lattice-gas Hamiltonian constructed in this study [36] used two-body interactions up to third neighbor distances and three types of three-body (trio) interactions. Also two types of adsorption sites were included, namely, the hcp- and fcc-hollow site, and also interactions between the O atoms in these two types of hollow sites were included up to third neighbor distances. The calculated interaction parameters agree well with those determined recently from STM studies [37]. Quantitative information on adsorbate–adsorbate interactions have also been obtained for nitrogen on Fe(100) from STM investigations [38].

In Fig. 9 we show the theoretical temperature programmed thermal desorption spectra for O_2, which is compared to recent experimental data [39]. It can be seen that the theoretical spectra agree well with the experiment, in particular the shift of the peak maxima to lower temperatures for higher initial coverages which is due to repulsive adsorbate–adsorbate interactions. The structure (e.g. the various shoulders) in the calculated TPD spectra is due to the formation of ordered phases.

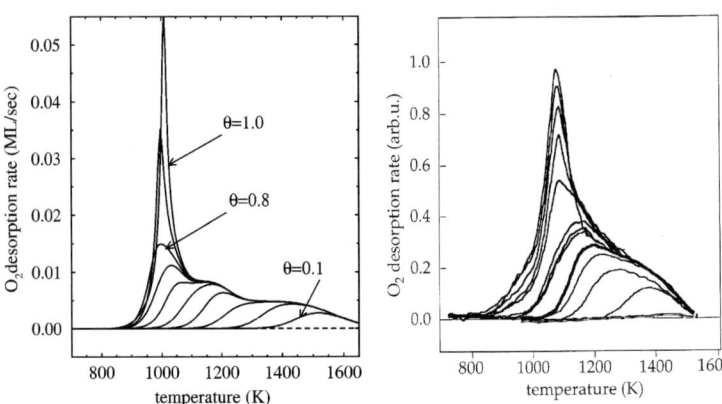

Fig. 9. Theoretical (left panel) and experimental (right panel) temperature programmed desorption curves (heating rate of 6 K/s). Each curve shows the rate of oxygen molecules that desorb from the Ru(0001) surface as a function of temperature when the system is prepared with a given initial oxygen coverage. The initial coverages are reflected by the area under the curve. For the theoretical results initial coverages are 0.1–1.0 ML in steps of 0.1; the experimental results also span this coverage region (from Ref. [36]).

The presence of such phases can be predicted by calculation of the isosteric heat of adsorption (the energy which an O_2 molecule gains by dissociatively adsorbing on the surface) as a function of coverage, which was also presented in Ref. [36]. From the calculated isosteric heat of adsorption, stable structures of O on Ru(0 0 0 1) were identified for coverages 1/4, 1/2, 3/4, and 1 ML which correspond to each of the ordered phases that form in nature, namely, (2×2)-O, (2×1)-O, (2×2)-3O, and (1×1)-O [36]. We note that similar first-principles based approaches have successfully been used for studying bulk phase diagrams [40,41]. Also, recently the lattice-gas Hamiltonian, together with the kMC approach has been used to study the diffusion and island nucleation at metal surfaces [42].

3. Reactivity theory

In the catalytic synthesis of ammonia, for example, it is the dissociative adsorption of N_2, which largely determines the turn-over rate (see e.g., the contribution by Sinfelt in this volume), and for various examples of oxidation catalysis it can be the dissociative adsorption of O_2. In "catalytic cracking" where large hydrocarbon molecules are decomposed into smaller ones, it is necessary that the catalyst breaks the carbon–carbon bonds. And also for corrosion it is necessary that molecules from the environment dissociate in order to take part in the formation of the corrosion compound. In general, the ability of a surface to break bonds of an approaching molecule is referred to as the surface reactivity. And this aspect will be discussed in the following three subsections. We also mention that it will often be important to specify the surface reaction to which the "surface reactivity" refers (e.g. H_2 dissociation and O_2 dissociation can be quite different at the same surface).

In gas phase chemistry "reactivity indices", based on the analysis of the electronic structure of reactants before they interact, have been useful. In Section 3.1 we give a brief summary of these ideas, but then we will stress (in Section 3.2) that the interaction of molecules at surfaces is typically quite strong and therefore will often fall outside the applicability range of "reactivity indices". And in Section 3.3 we discuss the importance of the statistics of the dissociation dynamics, demonstrating this point with an example showing that at low temperature palladium has the same surface reactivity as rhodium, although from the electronic structure Pd should be much less reactive.

3.1. The static, asymptotic behavior (reactivity indices)

The description of the early stages of a chemical reaction, i.e. when the interaction between the reactants is still weak, may be characterized by the so-called reactivity indices. These are given by the changes in the electronic structure of a reactant (before the reaction takes place) as stimulated by an external perturbation. It was Fukui [43,44], who taking notice of the principal role played by the valence electrons, studying condensed aromatic hydrocarbons, found an almost perfect agreement between the actual position of electrophilic attack and the site of large electron density. In the search for a quantitative correlation between reactivity and electronic configuration, the concept of frontier orbitals was established: for the case of reaction with an electrophilic reagent or electron acceptor, the position that is most susceptible to attack is that of the highest occupied (molecular) orbital (HOMO). For the case of reaction with a nucleophilic reagent or electron donor, the lowest unoccupied (molecular) orbital (LUMO) plays the principal role. When two molecules approach each other, the reaction will start with a nuclear configuration that favors best HOMO–LUMO overlap. Thus, reactivity indices of the frontier type identify sites in a molecule where new chemical bonds are likely to form.

Pearson [45] introduced the electronic "softness", the magnitude of the change in the electronic structure of a molecule due to a change of the number of electrons in the system, as a measure of reactivity. Species are then classified as "soft" if only a small energy is required to change their electronic configuration, i.e., if the valence electrons are easily distorted, polarized, removed, or added. A "hard" species has the opposite properties, holding its valence electrons more tightly [45,46]. The utility of these concepts is based on the so-called hard and soft–acid base (HSAB) principle formulated by Pearson [45] which states that when two reactants interact, either hard–hard or soft–soft interactions are preferred. In the case of poly-atomic or extended systems, the HSAB principle is used in a local version: the soft (hard) parts of one reactant prefer to interact with soft (hard) parts of the other. An example of a soft–soft interaction is CO adsorption on a metal surface. It involves electron donation from the molecule, which is considered to be the base, to the metal substrate which may be regarded as the acid; hence it can be called an acid–base reaction. Soft–soft interactions typically involve covalent-like bond formation and can be described as "frontier controlled" since the reactivity is essentially determined by the frontier orbitals. A hard–hard reaction occurs predominantly between highly charged species that are difficult to polarize or ionize, where the interaction is electrostatic in nature involving very little charge transfer. Interactions of this type are called "charge controlled". An example of a hard base (donor) is OH^- and a hard acid (acceptor) is, for example, Mg^{2+}. From these considerations the soft–soft and hard–hard interaction preference can be understood; e.g. soft bases will not be able to achieve a strong covalent bond with a hard acid which has a low tendency to accept electrons, nor can a strong electrostatic bonding be achieved due to the small or negligible charge of the soft base.

Parr and Yang [47] have expressed these ideas within DFT and defined the "local softness" in terms of the change in the electron density due to a change of the number of electrons of the system. Analogously, for metals, the local softness is expressed in terms of the response of the electron density to a change in the electron chemical potential. More precisely, it was shown that the local softness is the electron density of states at the Fermi level screened by a response function closely related to the static dielectric function [48]. It is intuitively clear that for both cases (semiconductors and metals) this is closely related to the HOMO and LUMO concept mentioned earlier.

For extended systems molecular orbitals and levels are not well defined and models of chemical reactions at surfaces were formulated using basis sets of localized atomic-like orbitals and the projected density of states (see e.g. Refs. [49–52]). Wilke et al. introduced the "local isoelectronic reactivity" [53], which is closely related to the local polarizability of valence electrons induced by an atom or molecule starting to chemically interact with the surface. The spatial variation of Wilke's function essentially gives the density of the LUMO (represented by regions of positive Wilke density) and HOMO (represented by regions of negative Wilke density) [53]. With similar limitations as those from which the general Fukui theory suffers, Wilke's representation gives a rationalization of the initial preference of a reaction pathway.

As already mentioned, chemical reactivity theory can give quantitative information only about the early stages of a chemical interaction, because in most cases the actual reaction will be determined by the non-linear response of the electron density to a local perturbation, and it will be governed by many states, not just the states at the Fermi energy. Despite this restriction, the local softness, in connection with the HSAB principle, has become an important predictive tool in analyzing reactions between molecules (see e.g. Ref. [54]).

3.2. At the transition state

For strong overlap between the HOMO and LUMO of reactants, the concept of reactivity indices breaks down. For dissociation and other reactions at metal surfaces we believe that this warning applies frequently, and the main reason for this difference between gas phase chemistry and the chemistry at metal surfaces is that in the interaction of a molecule with a surface, we are dealing with very unequal partners. The approaching molecule is able to translate and rotate, but the substrate is not (besides small relaxations of surface atoms). Furthermore, a metal substrate has a nearly infinite number of electrons, which implies that the position of the highest occupied and lowest unoccupied energy level is fixed. But for the molecule, the states can shift significantly, and if the highest occupied level is only fractionally filled, this level will not remain unchanged, but it will be subject to electron transfer and adjust to the substrate Fermi level. Also, it has been found from studies of the dissociation of H_2 at metal surfaces, through accurate first-principles calculations of high dimensional potential-energy surfaces (PESs), that the lowest energy barrier (when present) is actually very close to the surface and at a H–H distance which is significantly stretched (by about 100%) compared to the free molecule. Thus, the H–H bond is nearly broken when the molecule has reached the top of the energy barrier and the analysis showed that the differences between the behavior of H_2 at different metal surfaces should be described in a covalent picture. Earlier attempts, which applied a description in terms of Pauli repulsion, and/or frontier orbitals of the unperturbed non-interacting constituents do not account properly for the character and strength of the interaction. A more detailed discussion of these aspects is given in Ref. [4].

In Fig. 10 we summarize the view developed by Hammer et al. [52,55–57] in their analysis of H_2 at Cu(1 1 1) and H_2 at NiAl(1 1 0) (see also the earlier study by Hjelmberg and Lundqvist for H_2 at jellium [58]). At the transition state the interaction of the molecule with the surface has already produced a clear splitting into states which are bonding between the molecule and the substrate and ones which are antibonding. Assuming a substrate from the middle of the transition-metal series (e.g. Ru or Rh) implies that the low energy resonances, which are σ_g and σ_u derived, are filled with electrons. These states are bonding with respect to the molecule–substrate interaction, and thus their filling implies an attraction of the molecule to the surface. But the filling of the σ_u resonance also implies a weakening of the H–H bond. Thus, when the substrate Fermi level is in the middle of the d-band, we understand that molecules are strongly attracted to the surface and at the same time the molecular bond is broken. On the other hand, when the substrate Fermi level is well above the d-band, as for a noble metal, also the states which are antibonding with respect to the molecule–surface interaction become filled (the high energy DOS in panels (c) and (d) of Fig. 10).

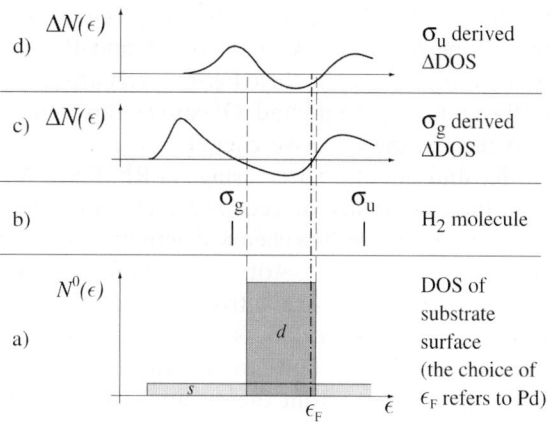

Fig. 10. Schematic description of the interaction of H_2 at the transition state toward dissociative adsorption at transition-metal surfaces. The bottom panel (a) shows the density of states for a transition metal before adsorption, and panel (b) shows the energy levels of a free H_2 molecule: the bonding state $[\phi_{\sigma_g} = \phi_{H1s}(R_1) + \phi_{H1s}(R_2)]$ and the antibonding state $[\phi_{\sigma_u} = \phi_{H1s}(R_1) - \phi_{H1s}(R_2)]$. The σ_g-level is filled with two electrons and the σ_u-level is empty. The interaction between the H_2 σ_g-level and the substrate s- and d-bands gives rise to a broadening and the formation of an antibonding level at about the upper edge of the d-band and a bonding level below the d-band—see panel (c). Panel (d) shows that the interaction between the H_2 σ_u-level with the substrate s- and d-bands gives rise to a broadening and the formation of a bonding level (at about the lower edge of the d-band) and an antibonding level (above the d-band). From Ref. [4].

This implies that the net interaction between the molecule and the substrate is repulsive. This can be thought of as a so-called four-electron two-orbital interaction, in which the stabilization due to the occupied bonding orbital is weaker than the destabilization due to the occupation of the antibonding orbital. This is typically described as a consequence of Pauli repulsion (see e.g. Refs. [50,59]). Thus, an energy barrier is built up which hinders the dissociation.

Calculations for the dissociation of H_2 at Rh, Pd, and Ag surfaces (i.e. for substrates from the left to the right in the periodic table) found that for Rh most pathways are not hindered, for Pd most pathways are hindered, and for Ag there is always an energy barrier hindering the dissociated adsorption [60]. From Fig. 10 it is clear that and why Ag is chemically rather inert (its Fermi level is about 3 eV above the top of the d-band); but its left neighbor, Pd, is chemically more active. And the left neighbors of Pd, namely Rh and Ru, are even more so, because their Fermi levels are closer to the middle of the d-band. Of course, this trend is also seen in the cohesive energies, which is large for Ru and smoothly decreasing via Rh, Pd to Ag. It is the availability of occupied and unoccupied d-states which rules the chemical activity and bond strength. In the same spirit, Toulhoat [61], studying the strength of the nitrogen–metal bond in transition-metal nitrides as a function of the number of d-electrons, found that the bond strength decreases from the group VIA elements (configuration $d^5 s^1$ or $d^4 s^2$) with increasing number of valence electrons, being lowest for the noble metals (e.g. Cu), where the d-band is well below the Fermi level. He suggests that such ab initio derived bond lengths provide practical guides for research in catalysis design. Clearly, earlier work using the Andersen–Grimley–Newns Hamiltonian, or, equivalently, Nørskov's d-band center formulation, work on the same principle. In this context we also mention the early work of Pettifor on the development of our understanding of the nature of bonding of transition metals [62].

Obviously, the situation described by Fig. 10 falls outside the range of validity of the concepts discussed in Section 3.1, i.e., the position of the transition state and the subsequent dissociation are determined by strong covalent interactions. However, the general trend of the reactivity of various transition metal surfaces described above, is apparently also predicted by the reactivity indices. In the next subsection, we will see that in a full, dynamic description of H_2 dissociation as a measure of reactivity, a different conclusion will result. We will also see in Section 3.4 that an increased bond strength does not imply that the system is a good catalyst, but (besides other aspects) the bond strength should be intermediate.

3.3. The role of dynamics and statistics

To obtain a full description of surface reactivity, i.e. of the probability of dissociation, it is necessary to calculate the dynamics of the molecules approaching the surface. This subsection shows, that sometimes dynamical effects can be significant and differences from a theory that just considers reactivity indices, noticeable. A good quality theory of the dissociation dynamics was first developed by Gross et al. [6,9] and Kroes et al. [10], who took the high dimensionality of the PES into account and, studying H_2 dissociation, also treated the hydrogen nuclei as quantum particles. By comparison with a classical treatment of the dynamics, Gross and Scheffler [6] showed which quantum effects are important (mainly zero-point vibrations, and only little tunneling).

In Fig. 11 we display the sticking probability, S, for two different substrates, obtained by solving the Schrödinger equations for the nuclei. The sticking probability is the probability that an incoming H_2 molecule dissociates and that the atoms then adsorb at the surface. One could also say that $(1 - S)$ is the probability that an incoming H_2 molecule gets reflected back into the vacuum. One unexpected and surprising result of Fig. 11 is that for low kinetic energies ($E_i \lesssim 0.05$ eV) the sticking probabilities for the Pd and Rh substrates are very

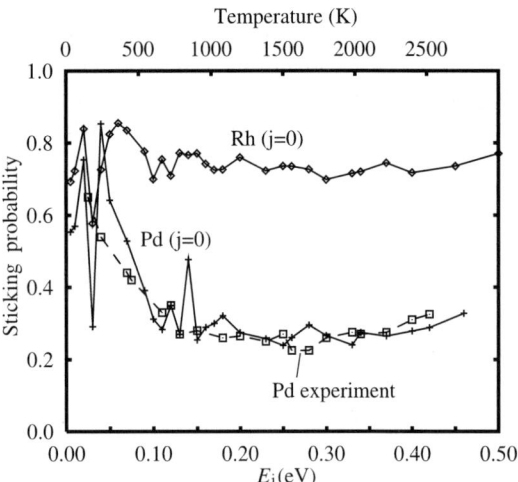

Fig. 11. Initial sticking probability (the probability that an incoming H_2 molecule dissociates and that the atoms adsorb at the surface) versus kinetic energy for an H_2 beam under normal incidence on a clean Rh(0 0 1) and Pd(0 0 1) surface. The H_2 molecules are in their rotational and vibrational ground state. Theoretical results are from Eichler et al. [60], and the experimental data are from Rendulic et al. [63]. We also mention that for H_2 at Ag(0 0 1) the calculations give a sticking probability of zero in the whole range of energies E_i shown in the figure.

similar. In view of the discussion of reactivity indices (Section 3.1) and the discussion of a strong covalent bond formation (Section 3.2), this is not understandable because at the Fermi level Rh has more occupied, and in particular, more empty d-states than Pd. Thus, Rh should be more reactive than Pd. Nevertheless, both substrates give a sticking probability as high as 75% for low temperature ($T \lesssim 250$ K). While for Rh the sticking probability always remains high, it decreases for Pd to about 25% when the temperature of the H_2 beam is increased. This is found in the theory as well as in the experimental data. The analysis of the H_2 dynamics revealed that the effect is best described by the word "steering", which means the same as on the road: If one is going slowly, one will make it well along a curvy street. However, when one is going too fast one is pulled out of the curve. Thus, molecules which are approaching slowly will be able to follow a pathway along curvy valleys of this high dimensional potential-energy landscape, and find the way toward the point where they can dissociate without an energy barrier. They will dissociate even if their initial orientation was unfavorable (e.g. perpendicular to the surface) because they are steered toward a more favorable transition-state geometry. On the other hand, fast molecules will not be able to make it around all the curves. They will bump against an energy barrier and be reflected back into the gas phase.

Thus, the behavior we see for the Pd substrate which increases the sticking to nearly 75% at low energies of the H_2 beam, is truly a dynamical effect. In a static picture the high reactivity of Pd cannot be understood. The efficiency of steering depends on the speed of the incoming molecule and on the shape of the PES. Therefore, to evaluate the sticking probability, which we consider a good measure of the surface reactivity, it is important to consider all degrees of freedom and the dynamics of the nuclei. In addition, it is necessary to include good statistics; a corresponding calculation for the results shown in Fig. 11, but calculated using molecular dynamics, would require more than 100,000 trajectories in order to obtain an adequate description. Obviously, as much as "steering" is important to understand the high reactivity of Pd at low E_i, for other systems, which on the grounds of the electronic-structure alone may be expected to exhibit a high reactivity, an "anti-steering" may occur, which drives approaching molecules not toward the best transition state, but against an energy barrier.

3.4. Weakening the adatom–surface bonds

In Sections 3.1–3.3 we dealt with "surface reactivity" in terms of the dissociation and/or adsorption of an approaching molecule. However, there is an additional aspect which determines whether or not a surface is possibly useful in a catalytic cycle: the adsorbed atom or molecule should not be bound too strongly, because otherwise there would be little or no reason for the next steps, namely diffusion on the surface in order to find a partner, and reaction toward an intermediate or the final reaction product. That is, a good catalyst is one that can efficiently dissociate molecules but not bind the adsorbates too strongly. In this section we will discuss two examples which highlight this point and show the importance of adsorbate–surface bonds being sufficiently weak, and how this can be achieved. Both examples refer to oxygen at ruthenium, with some reference to the CO oxidation.

From the above sections, one may expect that Ru, being situated in the middle of the transition-metal series, will exhibit a good reactivity. However, in standard UHV experiments its activity towards CO oxidation is very low [64]. This is because although Ru effectively dissociates molecular oxygen, it binds the oxygen atoms too strongly under these conditions [16,65]. In what follows we will see how this scenario can be improved.

In standard UHV experiments and exposing a Ru(0 0 0 1) surface to O_2, the saturation coverage has been reported to be approximately half a monolayer [66,67]. Recent studies employing NO_2 or very high O_2 exposures have shown, however, that Ru(0 0 0 1) can support higher coverages, for example, ordered structures with $\Theta = 0.75$ [68–70] and with $\Theta = 1$ [71], as had initially been predicted by DFT–GGA calculations [72].

Furthermore it was predicted that subsurface adsorption is possible but will occur only after completion of the monolayer structure at elevated temperatures ($T \approx 600$ K) [39,72–75]. Thus, the (apparent) oxygen saturation coverage noted above for low (or room) temperature UHV conditions and typical exposure by O_2, is solely due to kinetic hindering of the O_2 dissociation. This is part of the "pressure gap" problem, but as shown by Stampfl and Scheffler [72], if the reason is understood, it is possible to bridge the pressure gap.

In Fig. 12 it can be seen that the adsorption energy of oxygen at Ru(0001) markedly decreases when the coverage increases. The reason can be well understood in terms of the repulsive interaction between the partially ionized O adatoms. For low coverage the adsorption energy is high and oxygen is satisfying its bonding needs well. Thus, this oxygen will not be keen on undergoing another chemical reaction (leading to the mentioned low CO_2 turnover rates in UHV [64]). The rationalization behind this statement is that energy barriers for diffusion and chemical reactions roughly scale with the energy of the initial state. Thus, even if a well bound initial state is less favorable than the end product of a surface chemical reaction, the energy barrier to reach the product is probably high. In relation to this, Alavi et al. [76] proposed on the basis of first-principles calculations, that the weakening of the O–Pt bond is a significant contribution to the energy activation barrier for CO_2 formation over the Pt(111) surface. Similar calculations were performed by Eichler and Hafner [77], where in addition, reaction of CO with molecular oxygen was investigated. We also note that in Fig. 9 calculated and measured TPD spectra for the O/Ru adsorbate system were shown, and one sees that oxygen from the high-coverage phase (i.e., where it is more weakly bound) leaves the surface at significantly lower temperature ($T \approx 900$ K) than the oxygen from the low-coverage phase ($T \approx 1300$ K). In the 1 ML phase, the oxygen binding to the surface is still, however, sufficiently strong to disfavor a reaction with a CO molecule as indicated by DFT calculations [65].

From Fig. 12 it can be seen that for structures involving both on-surface and subsurface oxygen, the average bond strength is even weaker than for on-surface O alone. In view of this, the high-coverage phases of O/Ru(0001) should be substantially more active towards CO oxidation than the lower coverage UHV phases. This is indeed the case as found by experiments performed using high O_2 gas pressures [78]. Adding to this, experimental work of Böttcher and Niehus [74], in agreement with recent calculations [75,79], pointed out that at high oxygen pressure Ru can accomodate much more than a monolayer (cf. Fig. 12). In fact, Böttcher et al. report an uptake corresponding to more than 30 ML and that this "loading" of the Ru with oxygen gives rise to a significant structural destabilization and the formation of oxide crystallites at the surface, which were shown to be RuO_2 [19]. This surface phase transition is nothing else than the dry corrosion mentioned in Section 1.3, and we see here an example where catalysis and corrosion meet: Böttcher et al. [18,80] and Over et al. [19] emphasize that the true catalyst, which is operative when the high activity of Ru for oxidation catalysis is discussed, is not the originally introduced Ru metal (which binds O too strongly) but the material which comes into being when Ru metal is in an oxygen atmosphere at high pressure and high temperature; that is, the predominant catalytic activity arises due to the presence of RuO_2. We will come back to this point in Section 4.

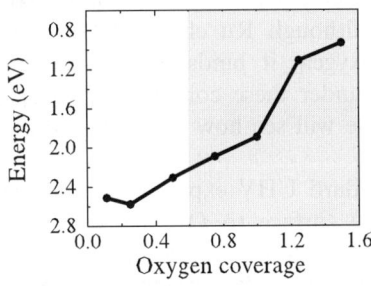

Fig. 12. Average adsorption energy of oxygen on Ru(0001) for various coverages, with respect to 1/2 O_2. The coverage range $\Theta > 0.6$ is marked in gray to indicate that here O_2 dissociation is kinetically hindered. For coverages $\Theta > 1$ oxygen atoms also occupy subsurface sites [75] and a significant decrease in the average adsorption energy can be noticed.

Our second example highlighting the importance of weak oxygen–ruthenium bonds for increased catalytic activity, addresses the recent work of Bonn et al. [81]. Here, the weakening of the bonds is induced by laser-light excitation. The laser light is mainly absorbed in the substrate, where it gives rise to a high density of hot electrons ($T_{el} \approx 6000$ K). With a time delay, corresponding to some atomic vibrations (about 1–2 ps), the electron and phonon temperatures equilibrate, and then the temperature of both, phonons and electrons, is about 2000 K. Using 110 fs short laser pulses ($\hbar\omega = 1.5$ eV), Bonn et al. could show that CO_2 formation is enhanced during the initial time period, i.e., when the electrons are hot, but atomic vibrations not yet significant. Thus, electronic excitation enhances the desired reaction, a phenomenon known as photochemistry of a surface chemical reaction. With a noticeable time delay, CO was found to desorb. Obviously, this is a thermal desorption process, i.e. driven by the atomic vibrations. DFT calculations, evaluating the electron free energy, explained that the photochemistry of the CO_2 formation is largely driven by a weakening of the O–Ru bond, i.e., for high electron temperatures the oxygen binding energy is significantly reduced (see Fig. 13). This is due to the presence of an O–Ru antibonding state just above the Fermi energy (see the shaded part of the density of states in Fig. 13). Hence we see that weakening of the strong O–Ru bond can either be realized by modifying the catalyst's surface structure, in our example by increasing the O concentration, or by externally stimulating the chemical reaction. In relation to the latter, recently it was demonstrated using (third generation) synchrotron radiation, that site-selective targeting and breaking of individual bonds, even for identical atoms which are only in slightly different chemical environments could be achieved [82].

The importance of an optimum bond strength was further illustrated in recent studies by Toulhoat et al. [83] and Raybaud et al. [84]. Here the hydrodesulfurization process was considered, for which transition-metal sulfides (TMS) are the only class of materials that act as good catalysts (crude oil contains many organosulfur compounds and one wants to abstract the sulfur, producing H_2S

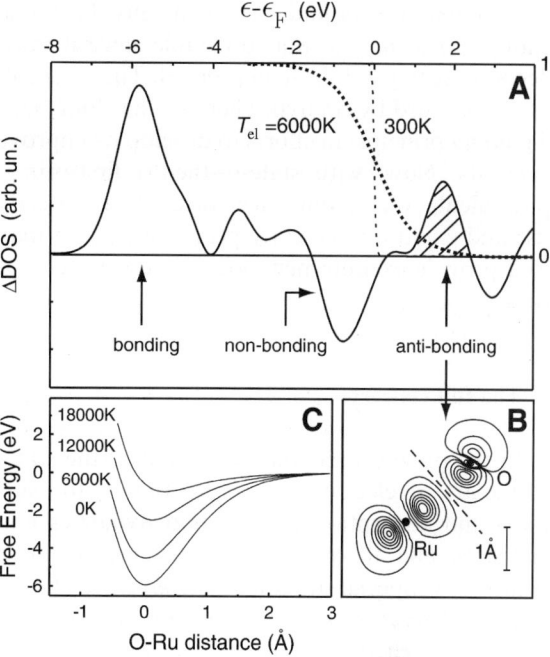

Fig. 13. Results of DFT–GGA calculations for (2×1)-O on Ru $(0\,0\,0\,1)$. (A) The change in the density of states (ΔDOS) upon adsorption. A bonding state appears well below the Fermi level and an antibonding state at 1.7 eV above. The Fermi functions $f(T_{el}, \epsilon - \epsilon_F)$ at $T_{el} = 300$ and 6000 K (referred to the right ordinate axis) demonstrate that with increasing temperature the antibonding level becomes populated. (B) Electron density of the antibonding state at $\epsilon_F + 1.7$ eV. (C) Free energy as a function of O–Ru distance and electronic temperature. The bond is strongly weakened as the electron temperature increases (after Ref. [81]).

and hydrocarbons). Toulhoat et al. showed that if the TM–S bond strength (i.e. here, the cohesive energy per metal–sulfur bond) is calculated from first-principles, one observes the so-called "volcano" curve [14,85], i.e., it exhibits a maximum in reactivity for an intermediate bond strength (here for ruthenium sulfide), or other measure of the stability of the reaction intermediate. In relation to this, Raybaud et al., studying in particular MoS_2, showed how such an optimum bond strength and increased activity could also be achieved through an appropriate mixture of selected atomic species to the TMS. The typical volcano-shaped curves were first proposed by Balandin [85]. They reflect the Sabatier principle which emphasized the intermediate compound, in that it must be stable enough to form, but not too stable since it must decompose to yield the final product. This concept, as put forward by catalytic chemists, has long been a guiding principle in efforts to develop or improve catalysts. Now, with state-of-the-art first-principles calculations at our finger-tips, we can directly use and explore these early principles, which may well provide exciting new possibilities in the near future.

4. The full concert

In the previous sections we described the state-of-the-art of electronic-structure theory for surfaces and adsorption, dissociation dynamics and statistics, and thermodynamics of surface phase transitions. Special attention was given to the influence of temperature and pressure. The various aspects discussed will all play a role in modeling and understanding a sustained catalytic reaction, or the time and structural evolution of the growth of a corrosion film on a metal surface. The new aspect in this sentence, which will be emphasized in the present section, is in the words "sustained" and "evolution", and this implies the necessity to describe time scales of the order of milliseconds or more, and length scales of 10 nm or more. Thus, an adequate theory of catalysis and corrosion should describe the atomistic level, i.e. a time period of a few femtoseconds, the formation of small islands (microseconds), as well as the evolution of mesoscopic and macroscopic structures (tenths of seconds). In Table 1 we list different theoretical approaches which address different time and length scales. These techniques should be regarded as complementary to each other rather than as alternatives, and possibly they should be combined. We will discuss the latter in this section.

Previous and present "academic" research is mainly concerned with isolated molecular processes. While this is indeed important, we also note that it is not sufficient for a realistic modeling and trustworthy understanding of catalysis or the evolution of a corrosion film. In addition to the knowledge of possibly relevant, individual molecular processes, it is in fact crucial to know if they can, and how they will, "play together". Let us make this point more clear, again through the example of the oxidation of carbon monoxide at Ru(0 0 0 1). At high gas pressures and temperatures, a Ru surface will at some stage start to transform into an oxide. Hence, the CO oxidation reaction could either take place on patches of this newly formed oxide, or occur on those parts of the surface that are still unchanged. The surface-science approach to this problem was hitherto to focus on perfect surfaces of one or the other. However, the crucial point for catalysis, is to understand how, and how often, reaction to CO_2 is actually realized; and after reaction and desorption events have taken place, how, and how fast they are built up again. Thus while it is useful, and even necessary, to study the mentioned isolated surface reactions, whether or not they ultimately play a significant role in "the full concert" of various molecular processes, that must play together in a sustained catalytic reaction, remains to be seen. We note in passing that the processes of CO oxidation have been studied experimentally as well as by DFT calculations but we only refer here

Table 1
The time and length scales handled by different theoretical approaches to study chemical reactions, the evolution of new structures, and crystal growth

	Type of information	Time scale	Length scale
DFT	Microscopic	–	$\lesssim 10^3$ atoms
Ab initio molecular dynamics	Microscopic	$t \lesssim 100$ ps	$\lesssim 10^2$ atoms
Semi-empirical molecular dynamics	Microscopic	$t \lesssim 10$ ns	$\lesssim 10^3$ atoms
kMC	Microscopic to mesoscopic	1 ps $\lesssim t \lesssim$ 1 h	$\lesssim 1$ μm
Rate equations	Averaged	0.1 s $\lesssim t \lesssim \infty$	All
Continuum equations	Macroscopic	1 s $\lesssim t \lesssim \infty$	$\gtrsim 10$ nm

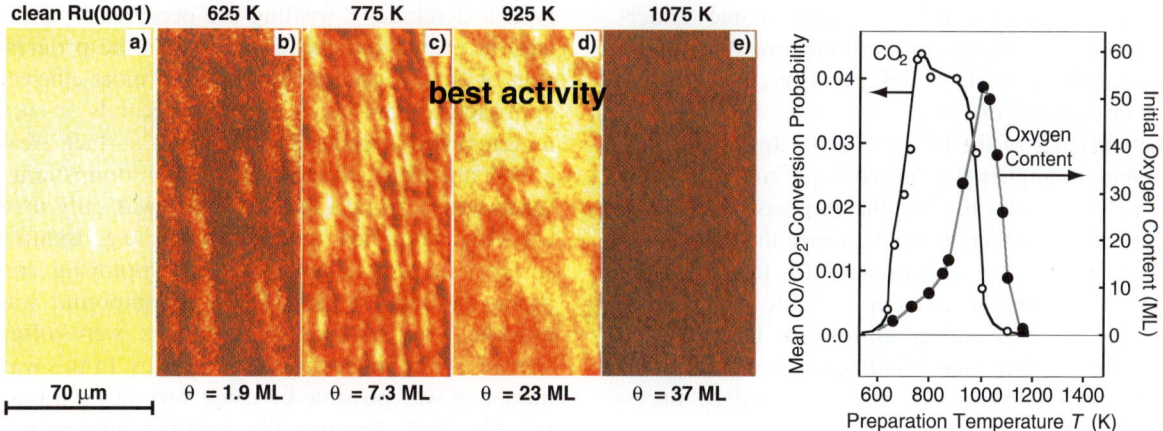

Fig. 14. Oxidation of Ru(0001) and corresponding CO → CO_2 formation rate. Left: PEEM images of the Ru(0001) surface, when clean and when exposed to an O_2 atmosphere of 1 mbar for 1 s at different temperatures (noted at the top of the figure). The resulting uptake of oxygen by Ru (noted at the bottom of the figure) was determined by temperature programmed desorption (see Ref. [20] for details). Regions with different shades correspond to materials with different workfunctions, Φ: White regions (highest photoelectron energy) have $\Phi = 5.3$ eV, and black regions (no photoelectron emission) have $\Phi > 6$ eV. Right: Rate of CO_2 formation over the differently prepared Ru + O systems.

to the book chapter by Scheffler and Stampfl [4] and a recent publication by Kim et al. [86] for details.

In Fig. 14 we show recent photoemission electron microscopy (PEEM), temperature programmed desorption, and surface reactivity experiments for the CO oxidation reaction over Ru surfaces prepared with different and systematically controlled concentrations of oxygen in the surface region [20]. These results demonstrate that depending on the pressure and temperature, different phases coexist on the surface and this coexistence apparently markedly affects the reactivity. The highest reactivity was found for conditions where a significant amount of O is stored on and in the surface, and where quite differently behaving materials (cf. the bright and dark domains in Fig. 14) coexist on a mesoscopic scale. We note that although Ru behaves quite differently to other transition metals (see Ref. [4] for details), the important issue, likely to be relevant to all realistic catalytic processes, is that *complexity* plays a more significant role than typically assumed—correspondingly, this aspect must also apply to theoretical descriptions. In this respect we envisage that for relatively "simple" catalytic reactions such as CO oxidation over transition metals, understanding and a good quality theoretical treatment will be achieved soon, e.g. in less than 10 years. In what follows we will explain what we believe should be the next steps in modeling and achieving understanding of catalysis. Analogously the modeling of corrosion may proceed.

Surface science, catalysis research, and the science of crystal growth (which includes the kinetics of formation of the corrosion compound) have to undergo (and are already undergoing) a noticeable change. Present day methods to develop new catalysts or to improve materials properties are still mainly done by trial and error. And the theoretical modeling of technologically relevant chemical processes so far proceeds by employing phenomenological methods (e.g. rate equations or hydrodynamic theories) together with effective parameters. Typically the latter have a very limited physical meaning and are not transferable, e.g. to situations with significantly different temperature, pressure, or material composition. This may be called "modeling without microscopic understanding". Using this term does not imply that it is bad or not useful. In fact, so far it has been very useful, but now there is clearly a need for improvement. In particular for basic research, it is clear that the next step, i.e. the development of a theoretical approach "with microscopic understanding", is what should be done now. We note in passing that combinatorial chemistry, appreciating the importance of complexity, attempts to improve the efficiency of the trial and error approach. This seemingly implies that microscopic understanding is no longer relevant, but this impression is in fact incorrect. In order to create a meaningful catalog of 10,000 or more experiments, i.e. for constructing a meaningful "descriptor" for the various studied systems and conditions so that directed searches (rather than blind scans) can be carried out, microscopic understanding is indeed necessary.

In the history of surface science we have worked our way down many orders of magnitude (both experimentally and theoretically) from the rate equations of the beginning of this century to the microscopic description of individual elementary processes, and have gained much understanding. Now, in the new century, we can begin to work our way back up, introducing some controlled complexity with a microscopic basis, and attempting to assemble these processes in a unified theoretical description of, for example, a sustained catalytic reaction or the evolution or a corroding surface. One promising approach that we see to do this is outlined below.

A single molecular process can be studied either by performing a molecular dynamics simulation or by using transition-state theory. Both approaches typically will give the same result. Problems with the latter may occur when the energy barrier is comparable to the thermal energy and correlated effects cannot be neglected. With respect to the former, presently (and the next years) due to computer limitations, it is unlikely that it will be possible to perform MD calculations with good statistics for surface reactions, that is, for the "whole concert" of reactions involved in describing surface phenomena. Thus, typically transition-state theory is the preferred approach (calculating energy barriers and prefactors by DFT). When there are competing processes (and typically there are), a statistical sampling becomes crucial. Handling of the statistics is the essence of the kMC method [13]. It describes the time evolution of chemical reactions and/or growth in terms of "the typical dynamics", treating the occurrence of the various events statistically as well as the progression of time. Consequently it is the most efficient approach to study long time (e.g. seconds or even minutes) and large length scales (e.g. surface areas of 10^4 nm^2) but still able to provide atomistic information. Up until recently kMC was only used for very simplified model systems, e.g. treating surfaces as simple-cubic lattices, employing empirical parameters, and, when a compound was modeled, the two different species were often treated as one single "effective" species. In this way modeling was possible by using very few parameters. In fact, because the possible microscopic processes and the associated energy barriers were unknown (and this situation still holds for many systems), there was no way to do it differently.

The key idea in kMC is that the microscopic processes are described by rates. Thus, the first step is the analysis and identification of all possibly relevant processes and determination of the associated rate. Once the rates are known, the various processes and their interplay can be handled by standard methods from statistical physics. The two basic quantities describing the rate of a process $\Gamma = \Gamma_0 \exp(-\Delta E/k_B T)$, are the attempt frequency Γ_0 and the difference ΔE, of the total energy with the particle at the minimum and at the saddle point of the potential energy curve along a reaction path of the process. T is the temperature and k_B the Boltzmann constant. A typical kMC cycle consists of, (i) from inspection of all atoms of the system, determine the atomic processes that are possible and calculate the rates, (ii) generate two random numbers, (iii) advance the system (i.e. move an atom) according to the atom and process selected by the first random number, (iv) increment the clock according to the rates and the second random number.

Ruggerone et al. [87] and Ratsch et al. [88] were the first to employ energy barriers calculated by DFT in the kMC approach for studying nucleation and growth. In a subsequent study, also prefactors were calculated by DFT [89]. For compounds, where the interplay of surface reconstruction, diffusion, dissociation, and two chemically different species also play a role, the description becomes more complicated as evident in the recent ab initio kMC study of Kratzer and Scheffler [90,91]. This study describes the deposition of $As_2 + Ga$ on a reconstructed GaAs surface in order to study the temperature and pressure dependence of diffusion, dissociation, island nucleation, and growth. The treatment of catalysis could proceed analogously, but in view of the above mentioned mesoscopically structured surface, which contains different materials of very different properties, the corresponding modeling will be even more involved. In this context we also note the recent work by Hansen and Neurock [92] who performed Monte Carlo calculations based on a mixed scheme of first-principles calculations and semi-empirical models, to study nitric oxide decomposition on Rh(1 0 0) and ethylene hydrogenation over Pd(1 0 0). The work by Kratzer and Scheffler [90,91] showed that "modeling with microscopic understanding" requires taking into account more than 30 processes. This is a significantly higher number than that used for semi-empirical studies where one gets along with about five "effective" processes. Still, the risk that a relevant process may be overlooked in a DFT–kMC calculation certainly exists. For example, Ovesson et al. [93] recently showed that diffusion rates at island corners are noticeably different to what was assumed in the earlier work [87,88], and this had important consequences for the island shapes, that develop under growth conditions. The description and inclusion of all relevant and important processes is one of the challenges that will have to be faced in order to perform and elaborate upon such a description.

Over the last decades, years, and months much has been learned about the chemistry at surfaces, which is often noticeably different to the chemistry between gas phase molecules: Gas phase chemistry is well established, but surface-science chemistry is still in a very early stage and full of surprises. We illustrated this through two examples that revealed that wide-held expectations and understanding were incorrect, namely, that it is not the Ru metal which is the main catalyst for CO oxidation at high pressure and temperature, nor is it the metal oxide Fe_2O_3 that is the main catalyst for styrene production under industrial conditions, but it is a new material which is produced during the induction period. And it is the environment that controls composition and structure, and the detailed material composition seems to be essential. With this in mind, we may even speculate that also for other oxidation catalysts the initially introduced metal does not just adsorb oxygen, but the catalytically active material contains subsurface oxygen as well as surface-oxide phases, some of them may even be unknown to date, as they may not exist under UHV conditions. From the theoretical side, we see the need for more accurate electronic-structure calculations (as well as molecular dynamics) for surface processes, but the big next step will be the combining of electronic-structure theory with dynamics and statistical mechanics of catalysis and corrosion on realistic time and length scales, at realistic temperatures and pressures. We have described how we believe this goal can be achieved, but also note that many difficulties are to be expected when this is really formulated and programmed. However, we are convinced that the knowledge gained along the way will be significant and consequential.

References

[1] P. Hohenberg, W. Kohn, Inhomogeneous electron gas, Phys. Rev. B 136 (1964) 864.
[2] R.M. Dreizler, E.K.U. Gross, Density Functional Theory, Springer, Berlin, 1990.
[3] W. Kohn, L. Sham, Self-consistent equations including exchange and correlation effects, Phys. Rev. A 140 (1965) 1133.
[4] M. Scheffler, C. Stampfl, Theory of adsorption on metal substrates, in: K. Horn, M. Scheffler (Eds.), Handbook of Surface Science, vol. 2: Electronic Structure, Elsevier, Amsterdam, 2000.
[5] A. Gross, M. Bockstedte, M. Scheffler, Ab initio molecular dynamics study of the desorption of D_2 from Si(1 0 0), Phys. Rev. Lett. 79 (1997) 701.
[6] A. Gross, M. Scheffler, Ab initio quantum and molecular dynamics of the dissociative adsorption of hydrogen on Pd(1 0 0), Phys. Rev. B 57 (1998) 2493.
[7] M.P. Allen, D.J. Tildesley, Computer Simulation of Liquids, Oxford University Press, Oxford, 1997.
[8] D. Frenkel, B. Smit, Understanding Molecular Simulation, Academic Press, San Diego, 1996.
[9] A. Gross, S. Wilke, M. Scheffler, Six-dimensional quantum dynamics of adsorption and desorption of H_2 at Pd(1 0 0): steering and steric effects, Phys. Rev. Lett. 75 (1995) 2718.

[10] G.J. Kroes, E.J. Baerends, R.C. Mowrey, Six-dimensional quantum dynamics of dissociative chemisorption of ($v = 0$, $j = 0$) H_2 on Cu(100), Phys. Rev. Lett. 78 (1997) 3583; Erratum: Phys. Rev. Lett. 81 (1998) 4781.

[11] G.J. Kroes, Six-dimensional quantum dynamics of dissociative chemisorption of H_2 on metal surfaces, Prog. Surf. Sci. 60 (1999) 1.

[12] H.J. Kreuzer, S.H. Payne, Theoretical approaches to the kinetics of adsorption, desorption and reactions at surfaces, in: M. Borowko (Ed.), Computational Methods in Surface and Colloid, Marcel Dekker, New York, 2000.

[13] K. Binder, M.H. Kalos, Monte Carlo methods in statistical physics, in: K. Binder (Ed.), Springer Topics in Current Physics, second ed., Springer, Berlin, 1986, p. 225.

[14] G. Ertl, H. Knözinger, J. Weitkamp (Eds.), Handbook on Heterogeneous Catalysis, Wiley, New York, 1997.

[15] I. Langmuir, Chemical reactions on surfaces, Trans. Faraday Soc. 17 (1921) 607.

[16] C. Stampfl, M. Scheffler, Density functional theory study of the catalytic oxidation of CO over transition metal surfaces, Surf. Sci. 433 (1999) 119.

[17] M. Shelef, G.W. Graham, Why rhodium in automotive three-way catalysts?, Catal. Rev. Sci. Eng. 36 (1994) 433.

[18] A. Böttcher, H. Conrad, H. Niehus, Characterization of oxygen phases created during oxidation of Ru(0001), J. Chem. Phys. 112 (2000) 4779.

[19] H. Over, Y.D. Kim, A.P. Seitsonen, S. Wendt, E. Lundgren, M. Schmid, P. Varga, A. Morgante, G. Ertl, Atomic-scale structure and catalytic reactivity of the $RuO_2(110)$ surface, Science 287 (2000) 1474.

[20] A. Böttcher, B. Krenzer, H. Conrad, H. Niehus, Mesoscopic-scale pattern formation induced by oxidation of Ru(0001), Surf. Sci. Lett. 466 (2000) L811;
B. Krenzer, in: Ph.D. Thesis, Humbolt University, Berlin, 2000, p. L811.

[21] C. Kuhrs, Y. Arita, W. Weiss, W. Ranke, R. Schlögl, Understanding heterogeneous catalysis on an atomic scale: a combined surface science and reactivity investigation for the dehydrogenation of ethylbenzene over iron oxide catalysts, Topics Catal. 14 (2001) 111.

[22] W. Ranke, private communication.

[23] W.D. Ryden, A.W. Lawson, C.C. Sartain, Electrical transport properties of IrO_2 and RuO_2, Phys. Rev. B 1 (1970) 1494.

[24] M.P. Ryan, M.F. Toney, A.J. Davenport, L.J. Oblonsky, In situ X-ray diffraction studies of passive oxide films, MRS Bull. 24 (1999) 29.

[25] M.F. Toney, A.J. Davenport, L.J. Oblonsky, M.P. Ryan, C.M. Vitus, Atomic structure of the passive oxide film formed on iron, Phys. Rev. Lett. 79 (1997) 4282.

[26] L.L. Shreir (Ed.), Corrosion, vol. 1, Newnes-Butterworths, London, Boston, 1976, Sec. 3, p. 4.

[27] K. Hauffe, Oxidation of Metals, Plenum Press, New York, 1965, p. 285.

[28] A.T. Fromhold, in: Theory of Metal Oxidation, vols. I and II, North-Holland Publishing, Amsterdam, 1976.

[29] S.J. Roosendaal, A.M. Vredenberg, F.H.P.M. Habraken, Oxidation of iron: the relation between oxidation kinetics and oxide electronic structure, Phys. Rev. Lett. 84 (2000) 3366.

[30] V.E. Henrich, P.A. Cox, The Surface Science of Metal Oxides, Cambridge University Press, Cambridge, 1994.

[31] X.-G. Wang, W. Weiss, Sh.K. Shaikhutdinov, M. Ritter, M. Petersen, F. Wagner, R. Schlögl, M. Scheffler, The hematite (α-Fe_2O_3)(0001) surface: evidence for domains of distinct chemistry, Phys. Rev. Lett. 81 (1998) 1038.

[32] K. Reuter, M. Scheffler, Composition, structure and stability of $RuO_2(110)$ as a function of oxygen pressure, submitted to Phys. Rev. B.

[33] X.-G. Wang, A. Chaka, M. Scheffler, Effect of the environment on α-Al_2O_3(0001) surface structures, Phys. Rev. Lett. 84 (2000) 3650.

[34] D. de Fontaine, Statics and Dynamics of Alloy Phase Transformations, NATO ASI Series, Plenum Press, New York, 1994.

[35] H.J. Kreuzer, S.H. Payne, in: W. Rudziński, W.A. Steele, G. Zgrablich (Eds.), Equilibria and Dynamics of Gas Adsorption on Heterogeneous Solid Surfaces, vol. 104, Elsevier, Amsterdam, 1997.

[36] C. Stampfl, H.J. Kreuzer, S.H. Payne, H. Pfnür, M. Scheffler, First-principles theory of surface thermodynamics and kinetics, Phys. Rev. Lett. 83 (1999) 2993.

[37] S. Renisch, R. Schuster, J. Wintterlin, G. Ertl, Dynamics of adatom motion under the influence of mutual interactions: O/Ru(0001), Phys. Rev. Lett. 82 (1999) 3839.

[38] L. Österlund, M.Ø. Pedersen, I. Stensgaard, E. Lægsgaard, F. Besenbacher, Quantitative determination of adsorbate–adsorbate interactions, Phys. Rev. Lett. 83 (1999) 4812.

[39] A. Böttcher, H. Niehus, S. Schwegmann, H. Over, G. Ertl, CO oxidation reaction over oxygen-rich Ru(0001) surfaces, J. Chem. Phys. 101 (1997) 11185.

[40] A. Zunger, in: P.E.A. Turchi, A. Gonis (Eds.), Statics and Dynamics of Alloy Phase Transformations, NATO ASI Series, Plenum Press, New York, 1994, p. 361.

[41] D.J. Liu, T.L. Einstein, P.A. Sterne, L.T. Wille, Phase diagram of a two-dimensional lattice-gas model of oxygen ordering in $YBa_2Cu_3O_z$, Phys. Rev. B 52 (1995) 9784.

[42] K. Fichthorn, M. Scheffler, Island nucleation in thin-film epitaxy: a first-principles investigation, Phys. Rev. Lett. 84 (2000) 5371.

[43] K. Fukui, T. Yonezawa, H. Shingu, Molecular orbital theory of orientation in aromatic, heteriaromatic, and other conjugated molecules, J. Chem. Phys. 20 (1952) 722.

[44] K. Fukui, Role of Frontier orbitals in chemical reactions, Science 218 (1982) 747.

[45] R.G. Pearson, Hard and soft acids and bases, J. Am. Chem. Soc. 85 (1963) 3533;
R.G. Pearson, Acid and bases, Science 151 (1966) 172.

[46] R.G. Parr, R.G. Pearson, Absolute hardness: companion parameter to absolute electronegativity, J. Am. Chem. Soc. 105 (1983) 7512.

[47] R.G. Parr, W. Yang, Density Functional Theory of Atoms and Molecules, Oxford University, New York, 1989.

[48] M.H. Cohen, M.V. Ganduglia-Pirovano, J. Kudrnovský, Electronic and nuclear reactivity, J. Chem. Phys. 101 (1994) 8988.
[49] L. Pauling, The Nature of the Chemical Bond, Cornell University Press, Ithaca, NY, 1960.
[50] R. Hoffmann, Solids and Surfaces: A Chemist's View of Bonding in Extended Strutures, VHC Verlag, Weinheim, Germany, 1988.
[51] M.H. Cohen, M.V. Ganduglia-Pirovano, J. Kudrnovský, Orbital symmetry, reactivity, and transition metal surface chemistry, Phys. Rev. Lett. 72 (1994) 3222.
[52] B. Hammer, M. Scheffler, Local chemical reactivity of a metal alloy surface, Phys. Rev. Lett. 74 (1995) 3487.
[53] S. Wilke, M.H. Cohen, M. Scheffler, Local isoelectronic reactivity of solid surfaces, Phys. Rev. Lett. 77 (1996) 1560.
[54] M.H. Cohen, Strengthening the foundations of chemical reactivity theory, in: R. Nalewajski (Ed.), Topics of Current Chemistry, Springer, Heidelberg, 1996.
[55] B. Hammer, M. Scheffler, K.W. Jacobsen, J.K. Nørskov, Multidimensional potential energy surface for H_2 dissociation over Cu(111), Phys. Rev. Lett. 73 (1994) 1400.
[56] B. Hammer, J.K. Nørskov, Why gold is the noblest of all the metals, Nature 376 (1995) 238.
[57] B. Hammer, J.K. Nørskov, Electronic factors determining the reactivity of metal surfaces, Surf. Sci. 343 (1995) 211.
[58] H. Hjelmberg, B.I. Lundqvist, Self-consistent calculation of molecular chemisorption on metals, Phys. Scr. 20 (1979) 192.
[59] J.K. Burdett, Chemical Bonds: A Dialog, Wiley, Chichester, 1997.
[60] A. Eichler, J. Hafner, A. Groß, M. Scheffler, Trends in the chemical reactivity of surfaces studied by ab initio quantum-dynamics calculations, Phys. Rev. B 59 (1999) 13297; A. Eichler, J. Hafner, A. Groß, M. Scheffler, Rotational effects in the dissociation of H_2 on metal surfaces studied by ab initio quantum-dynamics calculations, Chem. Phys. Lett. 311 (1999) 1.
[61] H. Toulhoat, private communication of submitted work.
[62] D. Pettifor, A quantum-mechanical critique of the Miedema rules for alloy formation, Solid State Phys. 40 (1987) 43; D. Pettifor, in: Bonding and Structure of Molecules and Solids, Clarendon Press, Oxford University Press, New York, 1995, p. 43.
[63] K.D. Rendulic, G. Anger, A. Winkler, Wide range nozzle beam adsorption data for the systems H_2/Ni and H_2/Pd(100), Surf. Sci. 208 (1989) 404.
[64] T. Engel, G. Ertl, in: D.A. King, D.P. Woodruff (Eds.), The Chemical Physics of Solid Surfaces and Heterogeneous Catalysis, vol. 4, Elsevier, Amsterdam, 1982.
[65] C. Stampfl, M. Scheffler, Anomalous behavior of Ru for catalytic oxidation: a theoretical study of the catalytic reaction $CO + \frac{1}{2}O_2 \rightarrow CO_2$, Phys. Rev. Lett. 78 (1997) 1500.
[66] H. Pfnür, D. Held, M. Lindroos, D. Menzel, Oxygen induced reconstruction of a close-packed surface: a LEED IV study on Ru(001)-p(2×1)O, Surf. Sci. 220 (1989) 43.

[67] L. Surnev, G. Rangelov, G. Bliznakov, Interaction of oxygen with a Ru(001) surface, Surf. Sci. 159 (1985) 299.
[68] K.L. Kostov, M. Gsell, P. Jakob, T. Moritz, W. Widdra, D. Menzel, Observation of a novel high density 3O(2 × 2) structure on Ru(001), Surf. Sci. 394 (1997) L138.
[69] Y.D. Kim, S. Wendt, S. Schwegmann, H. Over, G. Ertl, Structural analyses of the pure and cesiated Ru(0001)-(2 × 2)-3O phase, Surf. Sci. 418 (1998) 267.
[70] M. Gsell, M. Stichler, P. Jacob, D. Menzel, Formation and geometry of a high-coverage oxygen adlayer on Ru(001), the p(2 × 2)-3O phase, Israel J. Chem 38 (1998) 339.
[71] C. Stampfl, S. Schwegmann, H. Over, M. Scheffler, G. Ertl, Structure and stability of a high-coverage (1 × 1) oxygen phase on Ru(0001), Phys. Rev. Lett. 77 (1996) 3371.
[72] C. Stampfl, M. Scheffler, Theoretical study of O adlayers on Ru(0001), Phys. Rev. B 54 (1996) 2868.
[73] W.J. Mitchell, W.H. Weinberg, Interaction of NO_2 with Ru(001): formation and decomposition of RuO_x layers, J. Chem. Phys. 104 (1996) 9127.
[74] A. Böttcher, H. Niehus, Formation of subsurface oxygen on Ru(0001), J. Chem. Phys. 110 (1999) 3186.
[75] C. Stampfl, H.J. Kreuzer, S.H. Payne, M. Scheffler, Challenges in predictive calculations of processes at surfaces: surface thermodynamics and catalytic reactions, Appl. Phys. A 69 (1999) 471.
[76] A. Alavi, P. Hu, T. Deutsch, P.L. Silvestrelli, J. Hutter, CO oxidation on Pt(111): an ab initio density functional theory study, Phys. Rev. Lett. 80 (1998) 3650.
[77] A. Eichler, J. Hafner, Reaction channels for the catalytic oxidation of CO on Pt(111), Phys. Rev. B 59 (1999) 5960.
[78] C.H.F. Peden, D.W. Goodman, Kinetics of CO oxidation over Ru(0001), J. Phys. Chem. 90 (1986) 1360.
[79] K. Reuter, C. Stampfl, M.V. Ganduglia-Pirovano, M. Scheffler, Atomistic description of oxide formation on metal surfaces: the example of Ruthenium, submitted to Chem. Phys. Lett.; Metastable precursors during the oxidation of the Ru(0001) surface, submitted to Phys. Rev. B.
[80] A. Böttcher, M. Rogozia, H. Niehus, H. Over, G. Ertl, Transient experiments on CO_2 formation by the CO oxidation reaction over oxygen-rich Ru(0001) surfaces, J. Chem. Phys. 103 (1999) 6267.
[81] M. Bonn, S. Funk, Ch. Hess, D.N. Denzler, C. Stampfl, M. Scheffler, M. Wolf, G. Ertl, Phonon- versus electron-mediated desorption and oxidation of CO on Ru(0001), Science 285 (1999) 1042.
[82] R. Romberg, N. Heckmair, S.P. Frigo, A. Ogurtsov, D. Menzel, P. Feulner, Atom-selective bond breaking in a chemisorbed homonuclear molecule induced by core excitation: N_2/Ru(001), Phys. Rev. Lett. 84 (2000) 374.
[83] H. Toulhoat, P. Raybaud, S. Kasztelan, G. Kresse, J. Hafner, Transition metals to sulfur binding energies relationship to catalytic activities in HDS: back to Sabatier with first principles calculations, Catal. Today 50 (1999) 629; J. Phys.: Condens. Matter 9 (1997) 11085.
[84] P. Raybaud, J. Hafner, G. Kresse, S. Kasztelan, H. Toulhoat, Structure, energetics, and electronic properties

of the surface of a promoted MoS_2 catalyst: an ab initio local density functional study, J. Catal. 190 (2000) 128.

[85] A.A. Balandin, Modern state of the multiplet theory of heterogeneous catalysis, Adv. Catal. Rel. Subj. 19 (1969) 1.

[86] Y.D. Kim, H. Over, G. Krabbes, G. Ertl, Identification of RuO_2 as the active phase in CO oxidation on oxygen-rich ruthenium surfaces, Top. Catal. 14 (2001) 95.

[87] P. Ruggerone, C. Ratsch, M. Scheffler, Density-functional theory of epitaxial growth of metals, in: D.A. King, D.P. Woodruff (Eds.), Growth and Properties of Ultrathin Epitaxial Layers, The Chemical Physics of Solid Surfaces, vol. 8, Elsevier Science, Amsterdam, 1997, pp. 490–544.

[88] C. Ratsch, P. Ruggerone, M. Scheffler, Study of strain and temperature dependence of metal epitaxy, in: Z. Zhang, M.G. Lagally (Eds.), Morphological Organization in Epitaxial Growth and Removal, vol. 14, World Scientific, Singapore, 1998, pp. 3–29.

[89] C. Ratsch, M. Scheffler, Density-functional theory calculations of hopping rates of surface diffusion, Phys. Rev. B 58 (1998) 13163.

[90] P. Kratzer, M. Scheffler, Reaction-limited island nucleation in molecular beam epitaxy of compound semiconductors, Phys. Rev. Lett. 88 (2002), in press.

[91] P. Kratzer, M. Scheffler, Molecular modeling of surfaces from first principles. Computing in Science & Engineering, in press, to appear in November 2001.

[92] E.W. Hansen, M. Neurock, Modeling surface kinetics with first-principles based molecular simulation, Chem. Eng. Sci. 54 (1999) 3411.

[93] S. Ovesson, A. Bogicevic, B.I. Lundqvist, Origin of compact triangular islands in metal-on-metal growth, Phys. Rev. Lett. 83 (1999) 2608.

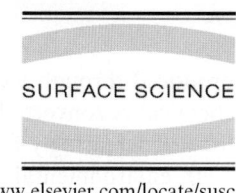

Atomic description of elementary surface processes: diffusion and dynamics

Federico Rosei [a,b], Renzo Rosei [c,d,*]

[a] *Institute of Physics and Astronomy and Center for Atomic-Scale Materials Physics, University of Århus, 8000 Århus C, Denmark*
[b] *Unità INFM and Dipartimento di Fisica, Università di Roma "Tor Vergata", Via della Ricerca Scientifica n1, 00133 Roma, Italy*
[c] *Dipartimento di Fisica, Università di Trieste, 34127 Trieste, Italy*
[d] *Laboratorio TASC—INFM, Basovizza, 34012 Trieste, Italy*

Received 5 July 2000; accepted for publication 11 June 2001

Abstract

A large fraction of processes which are at the foundation of our technological society involve physical and chemical properties of surfaces. Catalytic reactions and semiconductor devices production are two of the most important ones.

This paper describes a sample of some of the most relevant surface science experiments which have been recently performed, in order to understand elementary surface processes of model catalytic reactions and in semiconductor technology at the atomic level. The focus is on experiments performed with scanning tunneling microscopy and atomic force microscopy which have represented, in some cases, real breakthroughs in our understanding of these phenomena.

We then present an overview of possible experimental technique developments that can be foreseen for the future and that may give us a more in-depth understanding of the elementary processes which form the basis of important complex surface phenomena. Finally, some of the challenging tasks that lie ahead for surface scientists and the collateral opportunities are discussed. © 2001 Elsevier Science B.V. All rights reserved.

Keywords: Scanning tunneling microscopy; Chemisorption; Growth; Epitaxy; Surface chemical reaction; Surface diffusion

1. Introduction

The last few atomic layers of a solid constitute its interface with the environment; on this interface a great number of molecular and atomic processes continuously take place in a quasi two-dimensional (2D) world.

These processes form the basis of some of the most important technologies on which our modern society rests for its prosperity and economic development. Information technology and telecommunications would not have been possible without the progress in device fabrication, which has required profound knowledge of semiconductor surface technology.

Most of the modern industrial chemical production is based on catalytic reactions which are just sequences of elementary processes, taking place on the surface of a suitable catalyst. Every year, more than 150 million tons of ammonia are

---

[*] Corresponding author. Address: Dipartimento di Fisica Università di Trieste, 34127 Trieste, Italy. Fax: +39-4037-58565.
E-mail address: renzo.rosei@elettra.trieste.it (R. Rosei).

produced through a catalytic process; most of this chemical is converted into much needed fertilizers for an increasingly hungry world.

In environmental chemistry, catalytic processes are of vital importance for breaking down pollutants such as automotive exhausts and industrial emissions.

The atomic/molecular surface processes play also a key role in other economically important phenomena like corrosion, metal embrittlement, friction and lubrication, just to name a few.

The interest in deepening our understanding of the phenomena which take place at surfaces is therefore very strong. A more profound knowledge holds the promise of optimizing the production rate and the selectivity of a catalyst or the physical properties and size of semiconductor devices and could be the key for mastering other promising technologies, like hydrogen storage and fuel cell development.

2. Surface structure and elementary surface phenomena

The ensemble of surface phenomena taking place at the interface between a solid and a gas phase shows usually a bewildering complexity. For this reason surface scientists have mainly confined themselves so far to the study of the well characterized surfaces which result from cutting or cleaving a mono-crystalline sample along one of the planes with lowest Miller indices. In studying a single crystal surface, one tries to obtain a suitable sample on which to conduct experiments, with a well known geometry and with a definite chemical identity.

Fig. 1a shows a picture of a clean (1 1 1) surface of aluminum as determined by scanning tunneling microscopy(e) (STM) [1]. Even though the atoms in the last plane are not exactly located in the position one would expect by an ideal "cut" through a (1 1 1) plane of an aluminum crystal (they are relaxed inwardly by a few % of the lattice parameter), still the picture may leave the impression that a surface is a well behaved sort of checkerboard: just a simple playground for the foreign atoms and molecules we may want to de-

Fig. 1. Structure of surfaces as seen by STM: (a) clean Al surface (adapted from Ref. [1], with permission); (b) Large scale topograph of a Si(1 0 0) surface (from Ref. [2]. Copyright—1993 by Oxford University Press, Inc. used by permission of Oxford Univ. Press); (c) Image of the reconstructed Pt(1 0 0) surface (adapted from Ref. [3], with permission). The inset shows the arrangement of atoms in a non-reconstructed (1 0 0) surface.

posit on it. This is actually not so. Fig. 1b shows for instance the (1 0 0) surface of Si as determined by STM measurements [2]. Large irregular terraces with the nominal orientation are present but also a number of random steps and kinks are clearly visible. Fig. 1c gives another illustrative example. It shows a Pt(1 0 0) surface again as imaged by STM [3]. The inset in Fig. 1c shows the ideal positions that surface atoms of a (1 0 0) surface should have when the platinum single crystal has been cut (a square array). Obviously, the surface atoms do not occupy such ideal positions but rather form a very elaborate pattern. The surface in this case is "reconstructed". A number of small pits are visible both in Fig. 1b and c, which are signaling the vacancy of several atoms in the topmost layer.

Even at the outset therefore, we are confronted with the difficulty that also "model" surfaces, prepared with the purpose of reducing the level of complexity, do present a relatively large number of defects and/or other anomalies.

When a surface interacts with an external flux of atoms and/or molecules, many phenomena may take place; some of the most important ones are sketched in Fig. 2.

Process no. 1 shows a diatomic molecule which impinges on a surface. On hitting the surface it has a finite probability p of sticking to the surface and a probability $1-p$ of bouncing back in the gas phase. These molecules often form a chemical bond with the surface and choose a site which maximizes the binding energy. However, they do not reside long in the site where they have landed: if the surface temperature is high enough, they start migrating, jumping from one site to the next in a random way (see Process no. 2 in Fig. 2) giving rise to a 2D diffusion process.

The chemical bond that the molecule has formed with the surface weakens considerably the intramolecular bond so that the molecule may end up dissociating into its constituents; these, in turn, stick to the surface with strong chemical bonds (Process no. 3). Also, atoms originating from the dissociation process are mobile on the surface and during their 2D random walk, they may encounter another chemisorbed particle (atom or molecule) with which they might combine forming a new molecule, which may in turn leave the surface (Process no. 4).

A surface however is not a simple inert checkerboard. Upon adsorption of a foreign atom or molecule, it undergoes a local deformation; the atoms of the surface, while forming new chemical bonds with host particles, modify the structure of the bonds with the underlying bulk. When potassium is adsorbed on an Al(1 1 1) surface, for instance, it induces a rumpling of the first layer such that the Al atoms directly beneath the K atoms

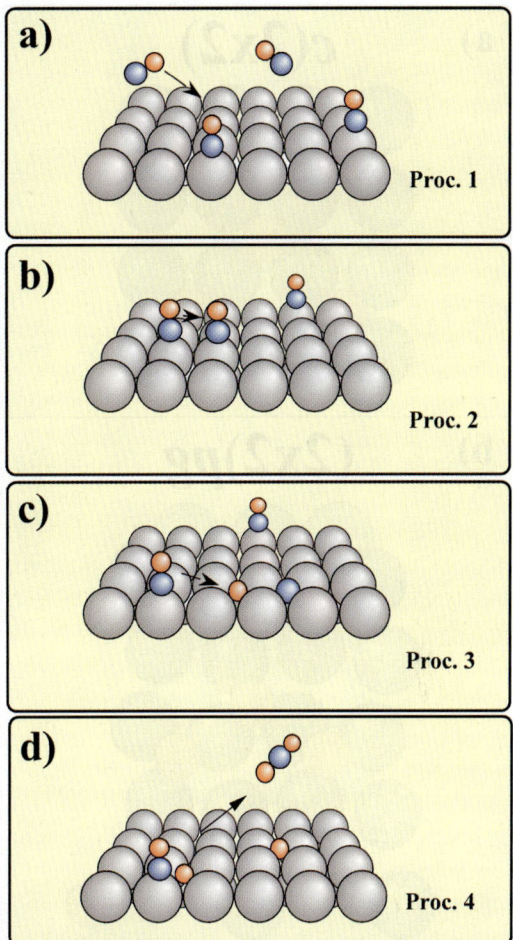

Fig. 2. Sketch of a number of elementary surface processes taking place on a surface exposed to an external atmosphere of reacting gases. Panel (a) shows the process of the adsorption of a diatomic molecule; Process no. 2 in panel (b) shows an elementary step of diffusion, while in panel (c) a process of dissociation is sketched. Finally panel (d) shows the reaction step of combination between two ad-particles, producing a molecule which is released in the gas phase.

are displaced towards the bulk by as much as 0.25 Å [4].

For stronger interactions the adsorbate may cause reconstruction of the underlying surface, which can range from a simple distortion of the elementary cell, up to a complete rearrangement of the surface atomic geometry.

An example of the first type of reconstruction is shown in Fig. 3. While Fig. 3a shows a c(2 × 2)

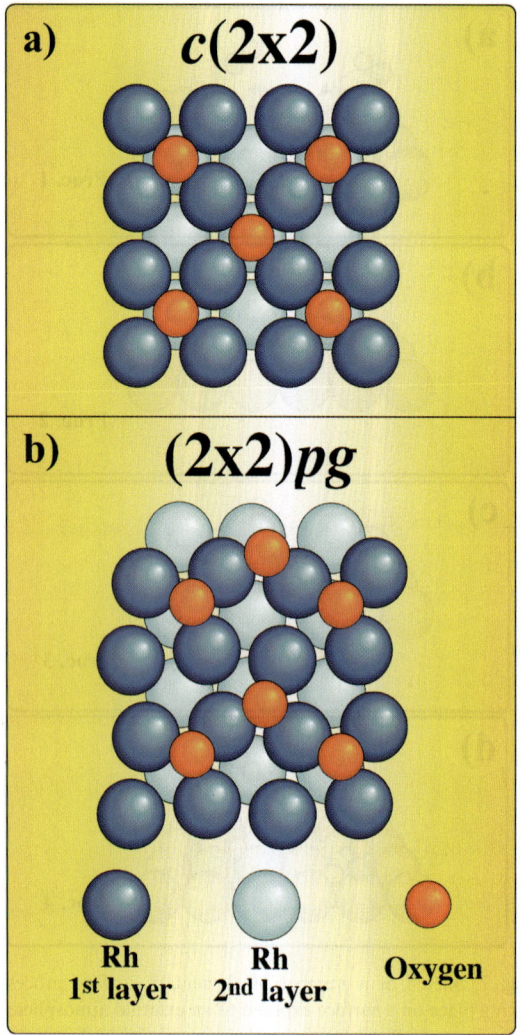

Fig. 3. Adsorbate overlayers on a (1 0 0) f.c.c. metal surface at 0.5 ML coverage: (a) regular c(2 × 2) structure; (b) (2 × 2) pg structure formed by oxygen on Rh(1 0 0).

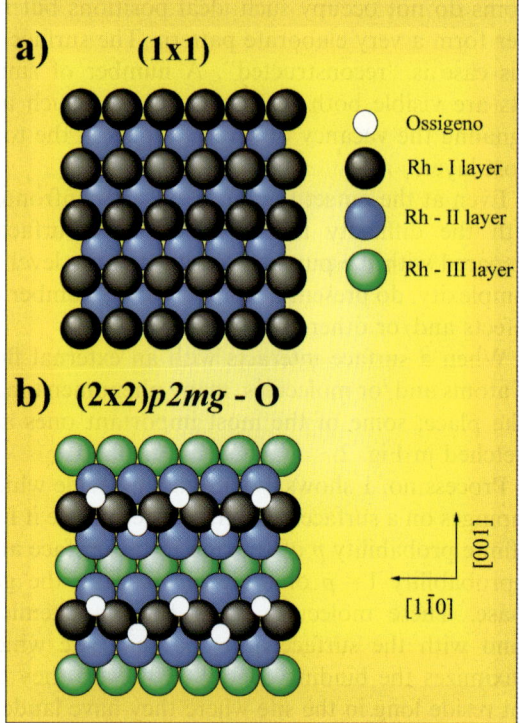

Fig. 4. (a) Structure of a clean Rh(1 1 0) surface; (b) change of structure upon adsorption of 0.5 ML of oxygen. The morphology of the reconstructed (2 × 2) p2mg surface implies a major process of mass transport during formation.

structure, which is the most common formed by an adsorbate on a (1 0 0) surface of a face centered cubic crystal at a coverage of 0.5 ML, the distortion induced by oxygen adsorption on Rh(1 0 0) is clearly visible in Fig. 3b. When the coverage approaches 0.5 ML (mono-layers, ML), the elementary cell elongates from the original square to a rhombus and the oxygen sits in one of the corners forcing a new surface symmetry [5,6].

Fig. 4a shows the structure of a Rh(1 1 0) surface while Fig. 4b shows the same surface after 0.5 ML adsorption of oxygen [7]. A considerable mass transport of rhodium atoms takes place in this case and gives rise to a (2 × 2) "missing row" reconstruction, characterized by depressions three atomic steps deep.

Surface reconstruction, whether spontaneous or induced by an adsorbate, is an important phenomenon since it implies, besides a change in the geometry, a strong change of the electronic structure and of the chemical reactivity of the surface.

In order to obtain an understanding of complex surface processes like catalytic reactions or the formation of metal–semiconductor interfaces, we need to know which are the various species that participate in the process, how they are bound to the surface and what are the activation energies of the single elementary steps they go through.

The detection, identification and characterization of these elementary processes, however, is a very challenging task: we are dealing with a physical system which consists merely of a 2D phase and, therefore, extremely surface-sensitive experimental tools are needed.

Ideally, in order to fully describe the behavior of atoms and molecules on a surface, we would require insight at atomic length scale and at the characteristic times of elementary step evolution (i.e. sub-picosecond time resolution).

3. Selected experimental results

In order to address these challenging tasks, in the last 30 years or so, surface scientists have developed an arsenal of different experimental techniques [8]. In the following, we describe some of the exciting results which have been obtained in recent years with two of these techniques, STM [9] and atomic force microscopy(e) (AFM) [10].

Structural information at the required spatial resolution has been obtained by these scanning probe techniques, which, however, still have poor time resolution and a limited "chemical" sensitivity, although a number of groups have demonstrated the possibility of acquiring STM images displaying "chemical contrast", mostly in the case of metallic alloys [11–14].

The STM is generally capable of acquiring atom-resolved images of the uppermost surface layer of a conducting solid. Beautiful and dramatic images of the atomic arrangements have been obtained for a rather large number of metal and semiconductor surfaces. Most spectacular in the history of surface science has been the visualization of the geometric structure of the (7×7) reconstruction of the Si(1 1 1) surface [15]. This complex reconstruction had proven to be one of the most intriguing problems in surface science. Due to the fact that this particular reconstruction extends to a few layers underneath the surface, the complete determination of this structure had to wait until the detailed transmission electron diffraction measurements of Takayanagi et al. [16].

Recently, some very impressive STM experiments have been performed in Ertl's group in Berlin, Germany, and in Besenbacher's group in Århus, Denmark, for characterizing the diffusion processes of adparticles on surfaces (Process no. 2 in Fig. 2b).

Previous macroscopic experiments for determining adsorbate diffusion constants have encountered severe difficulties. Even a well prepared single crystal surface in fact, exhibits defects (mostly atomic steps, kinks and impurities). Hopping energies at sites with different local structure may widely differ and macroscopic experiments can only probe "effective" diffusion parameters averaged on many different elementary hopping processes.

By using newly developed fast STMs, capable of acquiring up to 20 images per second (typically 100 Å × 100 Å wide) while still conserving atomic resolution, these groups have been able to follow the dynamic behavior of individual host atoms on several metal surfaces. The method they have implemented consists in taking a long sequence of pictures of the same region of the surface and record those atoms which have moved by one lattice parameter from one frame to the following one.

Counting the atoms that have not moved yields the probability P that atoms are still found on their original site at time t. If we can assume that the jumps are statistically independent, then $P(t) = \exp(-t/\tau)$, where τ is the mean time an adatom spends on its adsorption site. The jumping rate v is just the inverse of τ.

Fig. 5 shows for example, an image obtained from an STM movie of nitrogen atoms adsorbed on a Fe(1 0 0) surface [17]. In the inset of the figure, open circles mark the original positions of four N atoms and it can be seen that in the time lapse between two pictures, three atoms have moved by one lattice parameter.

By performing the experiment at several different temperatures, it is found, as expected, that the elementary jump rate follows the law of an activated process, i.e.:

$$v = v_0 \exp(-E_d/k_B T)$$

where E_d is the energy barrier the atoms have to surmount, v_0 is a pre-exponential factor generally referred to as "attempt frequency" and k_B is the Boltzmann constant.

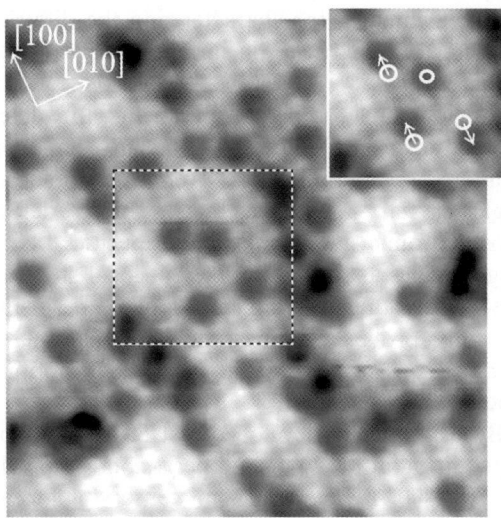

Fig. 5. STM image obtained from an STM movie of nitrogen atoms adsorbed on a Fe(1 0 0) surface at 300 K. N atoms are imaged as dark spots. The inset shows an area (marked by a dashed rectangle) 62 s later. Open circles in the inset mark the original position of four isolated N adatoms (from Ref. [17], with permission).

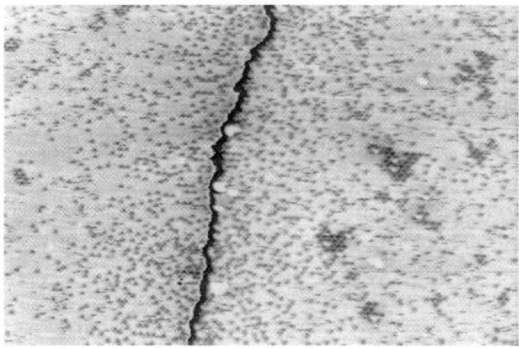

Fig. 6. STM image of a Ru(0 0 0 1) surface after the dissociative adsorption of 0.3 L of NO at 315 K, showing two terraces separated by a monatomic step (black almost vertical stripe). The black dots are N atoms. Their concentration profile signals that NO molecules have dissociated at the step (reprinted with permission from Ref. [22]. Copyright—1996 American Association for the Advancement of Science).

The values of the pre-exponential v_0 and of the energy barrier have now been determined for several systems [17–20].

These accomplishments must be considered as a real breakthrough: having isolated a single elementary step of a surface process, this becomes now amenable of a meaningful comparison between theory and experiment, which is the only foundation for advancing our understanding. A comprehensive review of diffusion phenomena of adsorbates on metal surfaces has recently appeared in the literature [21].

Concerning the use of STM in understanding the elementary steps of a catalytic reaction, the last four years or so have witnessed some really spectacular results. Zambelli et al. [22] for instance, addressed Process no. 3 in Fig. 2c in the case of dissociative chemisorption of NO (a decisive step in the catalytic reduction of nitric oxides), on the (0 0 0 1) surface of ruthenium, which is known to be the most selective catalyst for this reaction.

They exposed the surface at room temperature to a small dose of NO (0.3 L) and recorded an STM topography 0.5 h after exposure. The results are shown in Fig. 6. The ragged line which crosses the figure almost vertically is a monatomic step (these defects are always present even on the best prepared surfaces). A large number of small dark dots is visible on both sides of the step and are identified as N atoms originating from dissociated NO molecules. The small clusters at larger distances from the step are islands of oxygen atoms. These identifications were made on the basis of previous experiments with pure N or O layers in which it had been found that O atoms tend to cluster in islands with a (2 × 2) structure, while N atoms remain randomly distributed [23]. The NO molecules at 300 K and at low coverages are extremely mobile and are not detectable by STM. The picture which emerges from the distribution of the product atoms is very intriguing. The NO molecules, once adsorbed, diffuse rapidly across the terraces of the surface until they meet a step. Here, they apparently have a very high probability of dissociating. After dissociation, O atoms move relatively rapidly away from the steps. Nitrogen atoms move more slowly so that the image shows a diffusion profile of N atoms with its origin at the step. This interpretation has been later confirmed by detailed density functional theory calculations [24].

Both theory and experiment demonstrate clearly that the Ru(0 0 0 1) surface is not uniformly active for the process of dissociation of NO, but, rather, that the steps are highly preferential sites. This notion of special "active sites" on surfaces is an old one in catalysis; however, their exact nature and the mechanism by which they act, have remained elusive for a long time.

It is interesting to note that both sides of the step are covered with a comparable number of N atoms. Observing that a N atom does not cross the step upon dissociation, this means that NO molecules approaching the step, both from the lower and from the upper adjoining terrace, dissociate with about the same probability. From this observation Zambelli et al. conclude that the reaction takes place at the frontmost metal atoms at the step.

Besides confirming directly the concept of "active site" in catalysis, this experiment demonstrates that the kinetics is influenced by surface structure at sub-nanometer level and confirms that "macroscopic" experiments can only measure an overall reactivity which results from weighted contributions from various surface structure elements at the microscopic level.

The elementary step by which a molecule and an atom adsorbed on a surface react and their end products is released in the gas phase (Process no. 4 in Fig. 2d) is well exemplified by the oxidation of CO, for instance on Pt(1 1 1).

This is one of the most studied model-like catalytic reaction also because it represents a key step for the removal of CO from exhaust gases.

Despite the detailed knowledge of many aspects of the CO and O reaction, the usual macroscopic experiments have not been able to determine a mechanistic model completely consistent with all the features of the measured kinetics.

In a very elegant experiment, Wintterlin et al. [25] have recently observed this reaction with STM and were able to atomically resolve the reactants, monitor their reactions as a function of time and derive a quantitative rate equation, based solely on the statistics of the atomic process.

The experiment was performed as a titration, by first covering the sample with a submonolayer of oxygen and by exposing it afterwards to a constant CO pressure.

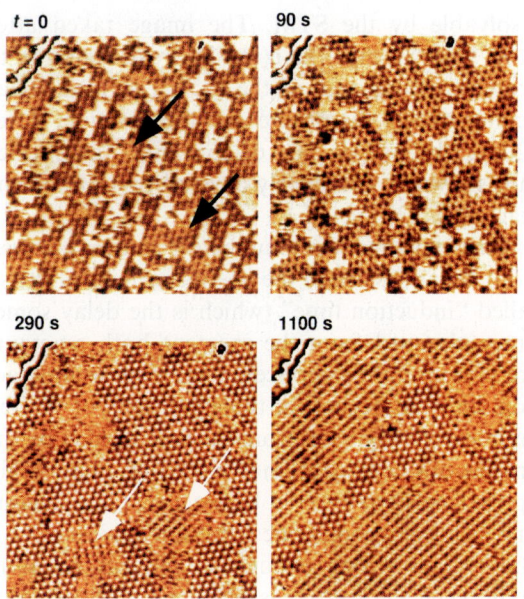

Fig. 7. Series of STM images, recorded during reaction of adsorbed oxygen atoms with co-adsorbed CO molecules at 247 K, from the same area of a Pt(1 1 1) crystal (reprinted with permission from Ref. [25]. Copyright—1997 American Association for the Advancement of Science). Black arrows in the $t = 0$ image mark two of the (2×2) oxygen islands. The white arrows, after 290 s, shows the formation of some $c(4 \times 2)$ CO islands.

As shown in Fig. 7, after preparation (time $t = 0$), the oxygen atoms (imaged as dark dots) form small irregular islands with the known (2×2) structure (two of them are indicated by black arrows). The white patches represent bare platinum areas. Adsorption of CO on this surface produces (after 90 s) a compression of the oxygen layer (the dark dot covered areas are now larger). The total area covered by oxygen has not changed at this time and the CO molecules, that are certainly present on the clean Pt areas between oxygen islands, are not visible because of their high mobility.

After 290 s of CO exposure, the oxygen islands start to shrink, signaling the beginning of the reaction. The areas not covered by oxygen show now (see white arrows) a $c(4 \times 2)$ structure, distinctive of near saturation CO adsorption. By this time in fact, the CO molecules are closely packed and therefore practically immobilized, becoming resolvable by the STM. The image taken after 1100 s shows that the c(4 × 2) areas continue to grow at the expense of the (2 × 2) oxygen areas, demonstrating the progress of the reaction. After 2200 s the oxygen has disappeared, the surface (not shown) is completely covered by CO and the reaction is completed.

This experiment is illuminating in many ways:

(i) it clarifies one of the possible origins of the so called "induction time" (which is the delay sometimes observed from the moment both reactants are present on the surface, to the time when reaction starts to occur). In the conditions of the experiment, it originates from the time needed by the process of CO adsorption from the gas phase to fill up the areas not covered by oxygen. This suggests that at low coverages, the repulsion between CO_{ads} and O_{ads} is strong enough to prevent reaction. Only when the surface is filled up with reactants and the CO islands push against the border of O islands, the reaction begins;

(ii) in the conditions of the experiment the reactants were clearly not distributed randomly but were forming separate domains and the reaction was only taking place at the boundaries between the (2 × 2) oxygen islands and the c(4 × 2) CO domains.

The STM images were also amenable of quantitative treatment and the authors evaluated the CO_2 production rate by determining the (2 × 2) oxygen fraction as a function of time. Repeating the process at different temperatures, it was then possible to evaluate the kinetic parameters of the reaction.

This work is a beautiful and so far unique example of how, by observing elementary steps of catalytic phenomena at the atomic scale, one can derive in a unique way the kinetics of the process directly from the microscopic mechanism.

4. Semiconductor elementary surface processes and new device fabrication

We have presented so far examples of elementary surface processes which are mainly relevant to the modeling of simple catalytic reactions. Some of the best catalysts are transition metals like platinum or rhodium and the examples have been chosen accordingly.

In this class of phenomena, the surface which catalyzes the transformation of reactants into the reacted species returns to its original conditions after the reaction has taken place. When completed, the reaction, at least in principle, leaves the surface essentially untouched and unmodified.

On semiconductors, on the other hand, the elementary phenomena which are of highest interest are those aimed at modifying the surface, with the purpose of device fabrication. In the following, we present some of the most relevant processes which take place during deposition reactions in which one or more layers of a host semiconductor species are deposited on a semiconductor surface.

Remarkably, in the last two decades or so, the quest for fabricating new materials with desired electronic and microscopic properties down to the atomic scale has had a tremendous impact on research on semiconductor surfaces and on the development of new micro and optoelectronic devices. One very important method which has been developed for this purpose is epitaxial growth.

In principle, three epitaxial growth modes have been studied and classified:

1. layer by layer growth in lattice matched systems (Frank–Van der Merwe);
2. the island growth mode (Volmer–Weber);
3. the layer by layer, then island growth mode (Stranski–Krastanov, SK) in lattice mismatched systems.

One typical example of semiconductor on semiconductor growth which we would like to discuss here is layer by layer growth followed by island formation, which takes place in the SK growth mode in heteroepitaxy [26–28]. The strain caused by the difference in lattice parameters between the materials employed, naturally leads to island formation after the completion of a thin (a few monolayers thick) layer, the so called 2D "wetting layer", on top of a template substrate. At the early stage of island formation, the islands can be free of defects (coherent), whereas during further growth defects and dislocations are introduced.

Strain relaxation and island formation in lattice mismatched semiconductor systems in the SK growth mode have been studied in great detail in the (InGa)As/GaAs and Si/SiGe material systems. Strain relaxation in this stage is believed to occur due to the elastic deformation of the substrate and at the free edges of the islands, and possibly to intermixing between the atoms of the constituent species; the elastic energy reduction outweighs the cost of additional surface energy due to island formation to form a lower energy configuration than that of the planar strained film. It has also been shown that coherent islands are energetically favored over dislocated islands below a critical size with a distinct coherent-to-incoherent transition for continued growth or ripening.

In the last decade or so quantum dots (QDs) have become particularly interesting, both from a fundamental and a technological point of view. The goal is therefore to fabricate these QDs, i.e. 3D structures which are so small that the charge carriers inside them find themselves confined in atomic-like energy levels. The possibilities of future applications are quite obvious: controlling the size and spatial distribution of the islands would enable to tailor the energy levels to one's needs, leading for example to the fabrication of completely new optoelectronic devices.

In the area of compound semiconductors such as InGaAs/GaAs and InAs/GaAs, self-organized QDs have been intensively studied and successively employed to build lasers with a low threshold current. In this context, size uniformity of the dots is the crucial aspect.

In the course of further control of the formation of (InGa)As islands in the SK growth mode it is desirable to achieve vertical ordering as well as lateral ordering towards the formation of 3D arrays of 'dots'. Vertical alignment of dots in the growth direction during the subsequent deposition of multiple layers of islands separated by thin spacer layers has been observed by several groups in the (InGa)As/GaAs and Si/SiGe systems and is attributed to the presence of non-uniform strain fields during the growth of successive layers.

This vertical alignment of the islands during the growth of successive layers has been reported several times [29], and it is probably due to the formation of a region of tensile stress above an island, which favors the positioning of islands on top of it, leading to column stacking of islands.

On the other hand, heteroepitaxial Ge/Si alloys seem to be promising materials for both microelectronic [30] and optoelectronic [31] applications: due to the fact that Si still rules the semiconductor market, the possibility of integrating micro and optoelectronics on the same Si chip is one of the main objectives in this field of research.

The main questions to be pursued in this context are: how the strain leads to island formation, and the role of Ge–Si intermixing; both questions have been addressed in a series of recent papers [30–35]. Besides, it is not yet possible to grow islands of desired size, shape and density, partly because the experimental results available today are lacking detailed knowledge of the SK growth dynamics, particularly on the atomic scale.

Again, in order to gain complete insight on the details of these growth processes, fast atomic-length-scale monitoring is necessary.

5. Change of the reconstruction in the Ge/Si system

As was discussed earlier, it is well known that in many lattice mismatched heteroepitaxial systems the SK growth mode is observed; this consists initially of the formation of a 2D strained layer. During the growth of thicker layers, the increasing misfit strain leads to a transition to different growth morphologies: the formation of coherent, dislocation-free 3D islands or the formation of misfit dislocations are efficient ways of relieving the misfit strain, at least partially. Besides this transition to different growth morphologies during the later stages of growth, even during the formation of the initial 2D layer, the surface reconstruction may change in order to accommodate for the increasing strain energy.

For example, in the case of Ge epitaxy on Si(0 0 1), a $(2 \times N)$ reconstruction is observed on the 2D strained layer (the wetting layer); this reconstruction consists of a periodic array of missing dimers of the (2×1) dimer reconstruction: every Nth dimer of the (2×1) reconstruction is missing.

Additionally, the reconstruction periodicity (N) is dependent on the stoichiometry of the wetting layer, i.e. the amount of Si/Ge intermixing [36]. (The wetting layer, which has been introduced previously, is a layer a few MLs thick: after the completion of it, layer by layer growth is not favorable any more, and therefore after the critical thickness is reached the growth proceeds by island nucleation.)

In the case of Ge deposited on Si(1 1 1) on the other hand, the reconstruction gradually changes from the typical 7 × 7 reconstruction of the Si(1 1 1) surface to the 5 × 5 of the Ge–Si wetting layer. This is an interesting phenomenon: it means that the reconstruction is still hexagonal, but with a different degree of symmetry.

After deposition of 0.45 ML of Ge the 7 × 7 reconstruction is maintained; this has been observed by STM and was also confirmed by the RHEED pattern, that can probe a larger and deeper region of the surface. The absence of islands on the terraces and of reconstructed (2 × N) areas of Ge suggests the diffusion of Ge atoms into the Si substrate. As proposed in the case of sub-monolayer deposition of Ge on Si(0 0 1) surfaces, Ge might exchange places with Si and the mobile Si might diffuse to the substrate step edges (displacive adsorption).

When the Ge coverage exceeds 3 ML [32,33], several 3D islands appear at random locations on the surface. At the beginning they have a triangular shape (truncated triangular pyramids, or tetrahedra), average lateral dimensions 500–1000 Å and are about 1 ML high. It has been suggested that this morphological transition occurs when the free energy of the islands is lower than that of the strained layer. At 9 ML coverage two kinds of islands are visible: small, tall islands (typically 1800 Å wide × 100 Å high); large and ripened flat islands (average dimensions 3500 Å wide × 25 Å high). It is argued that, after nucleation, the islands grow vertically up to a critical value at which the strain energy introduces a morphological transition with the possible formation of dislocations. Subsequently the islands grow laterally, apparently drawing material from the top or collapsing together with other islands. This process is called ripening and leads to the formation of flat,

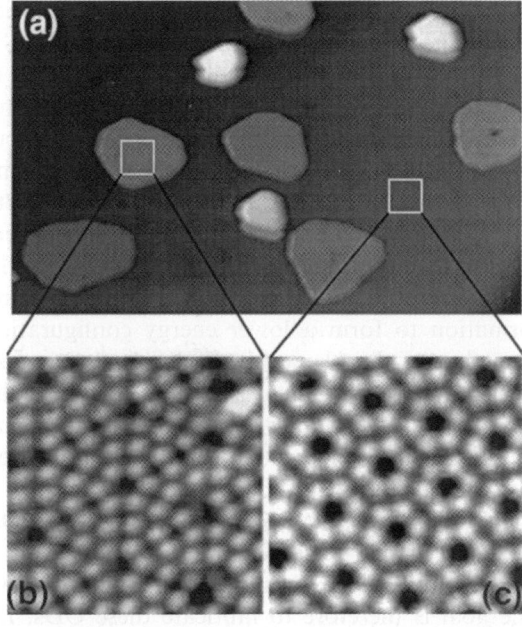

Fig. 8. STM images of a Ge–Si(1 1 1) surface after deposition of 9 ML of Ge. Two different kinds of islands are visible in (a): tall and triangular shaped (white) and low and rounded (gray). Magnifications 80 × 80 Å^2 (b) on the island's top and (c) on the substrate display the 7 × 7 structure of the former and the 5 × 5 of the latter (from Ref. [32]).

round islands. Some of these islands display a hole at their center, whereas other ones display a depletion surrounding them. These features should be related to strain and intermixing [37–40]. Enlarged STM shots on the top of the islands (both for strained and ripened islands) and on the substrate display the 7 × 7 structure of the former and the 5 × 5 reconstruction of the latter (see Fig. 8).

6. Results for intermixing, local spectroscopy

Recently, Boscherini et al. [34,35] have provided direct evidence of Si–Ge intermixing in self-organized strained and unstrained Ge QDs grown on Si(1 0 0) and Si(1 1 1) surfaces, also determining a quantitative measurement of the average composition. This was done by exploiting Ge K-edge X-ray absorption spectroscopy (XAFS); the measurements were taken at the ESRF in Grenoble

(France). The XAFS technique is a powerful tool for the study of the local chemical composition of solids and surfaces, its usefulness resting on the fact that it is a local spectroscopy, and therefore it does not require a sample exhibiting long range order.

Previously [32,33,41] several STM studies had argued the occurrence of some degree of intermixing in this heteroepitaxial system, but no conclusive answer had been found.

The results reported in Refs. [34,35] have provided a direct answer to the question of intermixing in these systems; the authors discuss their observations in terms of the energetics of island formation. It is argued that the important energy terms are in fact the strain energy, the surface energy of epilayer and substrate and, in the case of relaxed or partially relaxed layers, the dislocation formation energy. Intermixing can have an important role in decreasing the strain energy; in fact, the total strain energy of an ensemble of dots is given by a volume integral in which the elements of the strain tensor appear quadratically. The magnitude of the strain tensor elements depends linearly on the difference between the local free lattice parameter (in each direction) and that imposed by the SK growth on the substrate. Since the free lattice parameter can be reduced by intermixing it follows that diffusion of Si into the Ge dots will reduce the strain energy per atom.

In principle it could seem strange that such a high level of intermixing should occur at the Ge–Si interface, at the relatively low temperatures that have been used for growth: the bulk interdiffusion coefficient of Ge in Si has been estimated [42] to have an activation energy of 3.93 eV, which would lead to an extremely slow interdiffusion rate.

In a series of recent papers [43,44], Bottomley has proposed new theoretical insight on the formation of semiconductor QDs, with particular interest in the phenomenon of intermixing.

His main, quite radical, idea is the following: he proposes that the large heteroepitaxial stress present on semiconductor surfaces during heteroepitaxy (in particular for InAs on GaAs(0 0 1), but also the case of Ge on Si(0 0 1) is briefly discussed in his papers) causes InAs to melt when deposited on GaAs(0 0 1) at about 770 K. This naturally leads to mixing with the substrate in order to obtain a local minimum of the Gibbs free energy of the liquid phase. The liquid phase clearly facilitates mass transport, leading to QD formation. This is a phenomenon which is strictly related to surface properties, and it has no bulk analogue.

His calculations predict that island formation occurs after about 2 ML of liquid InAs material accumulate in order to minimize the surface tension, but without effectively reducing the net coordination of the liquid phase atoms.

When applied to the Ge/Si(0 0 1) system for typical growth temperatures such as about 970 K (about 700 °C), the calculations again predict a liquid phase, and he expects that in the limit of equilibrium thermodynamics the molten Ge should mix with the Si substrate to yield an alloy of approximate composition $Si_{0.5}Ge_{0.5}$.

In principle, if one writes down the expression of the thermodynamic force acting on a Si and on a Ge atom at opposite sides of the Ge/Si interface, one would find that they have opposite signs, meaning that Si wants to diffuse into the Ge layers, and vice versa.

Clearly hydrostatic pressure has the effect of reducing the melting point of both Ge and Si; recent works [45,46] measure an incremental film stress of up to 5 GPa for Ge grown on Si(0 0 1) and up to 7 GPa for Ge on Si(1 1 1), while theoretical calculations [47] also predict an isotropic stress in the range 6–9 GPa for Ge on Si(1 1 1) and about 3 GPa for Ge grown on Si(0 0 1) (the stress is larger under the islands, smaller in island-free regions). Therefore the intuition that the compressed Ge should become more mobile (even without actually melting) is qualitatively justified. This means that the process of atomic exchange (which is always present, but too slow at the relatively low temperatures used for growth) can be strongly accelerated in the Ge layers which are compressed, but not into the Si layers which are expanded.

We observe here that we are not in the presence of hydrostatic pressure, but of planar stress. One can however expect that uniaxial stress should always favor fusion, whatever its sign, either compressive or tensile; in fact both cases imply an increase in elastic energy, which is obviously absent in the liquid phase.

7. Self-organization of nanostructures versus lithographic techniques

Performance and integration density of large-scale-integrated circuits are growing fast as time passes, despite the limitations that were predicted just a few decades ago. At the same time, new techniques for nanofabrication which do not make use of conventional lithography are attracting interest and attention worldwide, for the purpose of processing the new generation of micro and optoelectronic devices. This is due to the fact that optical lithography is presently not capable of fabricating structures with lateral resolution smaller than a few hundreds of nanometers, whereas new device effects using single electron and other quantum effects require structure control of just a few Å.

One recent potential breakthrough in nanofabrication is represented by the development of techniques which employ scanning probe microscopies [48]. Indeed, as was shown several years ago by Eigler and co-workers [48], the STM is a surface-science tool which is both capable of fabricating and studying nanostructures on the atomic scale; one example is the manipulation of single atoms on a surface using the STM. The main problem with these techniques is the fact that, due to their serial nature, they cannot be applied directly to the large-scale production of integrated devices.

In order to achieve wafer-scale control, many authors [49] have been trying to develop self-organization techniques; the idea here is simply to "let nature do it herself": nanostructures are formed by spontaneous growth, without the use of lithographic techniques.

It is generally believed that self-assembling processes have the disadvantage of poor controllability of size uniformity and arrangement of positions; however, the improvement of control would obviously lead to a breakthrough, because in this way nanostructures could be simultaneously developed on the whole wafer surface.

A recent review article by Ogino et al. [50] has shown that self-organization can potentially be exploited to produce uniform and artificially designed nanostructure patterns, provided that critical surface features such as reconstruction, atomic steps and the phase boundaries of reconstructed domains can be controlled.

Once surface control is achieved, these critical features may be used as templates for controlled growth of hetero-nanostructures. This is especially

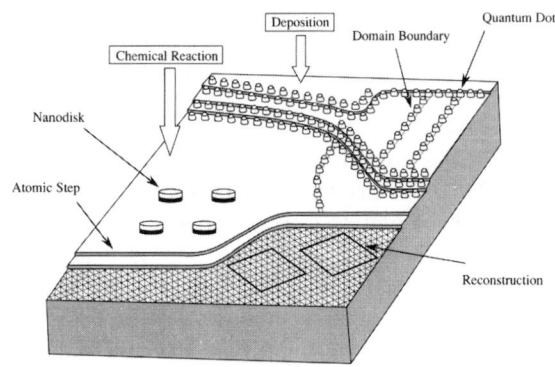

Fig. 9. The idea of nanofabrication based on surface structure control. Nanowires and QDs are self-organized on Si surfaces by controlling reconstruction, atomic steps, and reconstructed domain boundaries, which are to be used as nanofabrication templates (from Ref. [50], with permission).

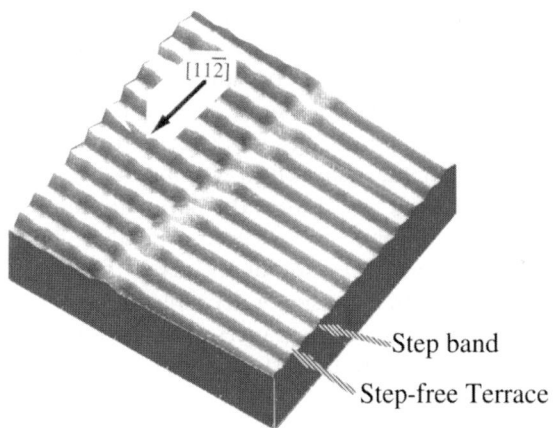

Fig. 10. AFM image of the step bunches that are formed on a Si(1 1 1) surface. The density of steps on the reconstructed surface depends critically on the flashing conditions. The Si substrate is misoriented toward the ⟨1 1 2̄⟩ direction by 1.5° from the exact (1 1 1) surface. The patterns used to control the step arrangement were hole arrays with a pith of 3.5 μm (right region) and 4.0 μm (left region) (from Ref. [50], with permission).

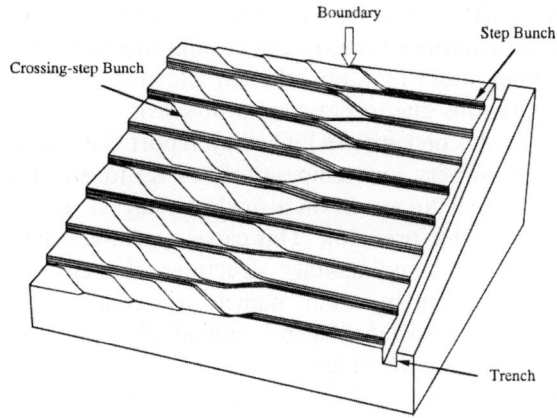

Fig. 11. Schematic of the step network formed as shown in Fig. 10. The step network in the left region consists of main step bunches steps crossing the terraces, whereas that in the right region exhibits only main bunches. This difference is due to the boundary conditions (from Ref. [50], with permission).

true in the case of Si surfaces, since their structure can be controlled down to the atomic level [50] (see Figs. 9–11).

Fig. 9 shows the key idea of nanofabrication: the approach is based on surface structure control. Nanowires and QDs are self-organized on Si surfaces by controlling reconstruction, atomic steps, and reconstructed domain boundaries as selective templates for nanofabrication.

Fig. 10 displays a Si(1 1 1) surface with a regular step arrangement and regular spacing of terraces between two consecutive steps; the substrate is misoriented toward the $\langle 11\bar{2}\rangle$ direction by 1.5° with respect to the exact (1 1 1) surface. Fig. 11 shows a schematic of the step bunching network and cross step bunching on a Si(1 1 1) surface, referred to the surface shown in Fig. 10.

The idea here is to control the formation of steps and the width of the terraces, for the purpose of using the terraces as templates for the growth of nanostructures. If one wants to envision the production of optoelectronic devices using new materials based on Si substrates, it is essential to achieve an alignment of the islands on the surface. This can be obtained in principle by controlling the stress on the surface, for example by patterned implantation of gas (such as Ar) layers buried deep within the samples [51] leading to regions of enhanced/reduced stress on the topmost layers of the surface. It has in fact already been shown that the control of surface stress leads to a higher concentration of adatoms on tensile stress regions [51].

The process is divided in "elementary" steps, as follows:

1. The first step is surface structure control, down to the atomic level; this is possible in principle for Si(1 1 1) surfaces; so far the Si(0 0 1) surface is the most widely used but there are many good reasons for trying to employ the Si(1 1 1) surface. For example, recent studies have shown how it is possible to create large terraces on this surface [52,53].
2. The second stage is the fabrication of semiconductor hetero-nanostructures whose formation sites are determined in advance, by controlling surface structure.
3. Finally, electronic devices require insulating layers and conducting layers in addition to the semiconductor nanostructures: the third stage is the fabrication of alternate semiconductor/insulator/conductor nanostructures.

The basic idea of this approach is the compatibility of atomically controlled nanofabrication and wafer scale integration. In current Si-MOS (MOS: metal oxide semiconductor) technology, the Si(0 0 1) surface is exclusively used, mainly for two reasons: because it is more easily oxidized, therefore allowing easy lithographic patterning, and because its Si/SiO$_2$ interface exhibits a lower density of defects compared to other surfaces.

However, in trying to control self-organizing nanostructures, the Si(1 1 1) surface has the advantage that an atomically flat surface and a well-defined step-terrace structure can be easily obtained, by a combination of thermal and chemical treatment.

8. Nanofabrication and atomic-scale manipulation using scanning probe microscopies

In a visionary speech entitled "There's plenty of room at the bottom", about four decades ago

Richard Feynman speculated on the fascinating properties of matter made by arranging "the atoms one by one the way we want them" [54].

Since the mid 1970s, before the invention of the STM, low dimensional confinement structures called quantum wells could be made by using growth techniques such as molecular beam epitaxy and metal-organic chemical vapor deposition. High-energy barriers at both sides of the quantum well confine the charge carriers into a 2D layer, giving experimental proof of the particle-in-a-box quantization "textbook problem", in the thin epitaxial direction. Though these techniques have atomic control in the epitaxial growth direction, they do not allow us to artificially structure materials in the remaining two lateral dimensions: only this achievement would create total quantum confinement.

A variety of techniques that achieve lower dimensional structures—quantum wires (1D) and QDs (0D)—have been attempted with varying degrees of success. Combining 2D epitaxial technology with nanometer-scale lithographic techniques, it has been possible to carve out thin lines and dots. The quantum confinement of the electrons in these structures is reflected in the energy states of the wire or dot, measured by optical or electronic techniques. A final demonstration of the quantum confinement would be the spatial mapping of the standing-wave patterns of electrons inside the structure; until recently, however, the electron wavefunctions of low-dimensional structures could only be derived by inference.

Indeed, the idea of manipulating matter on the atomic scale seemed to be just a fascinating dream—or even an unreachable goal—just a couple of decades ago. The underlying fascination may have been motivated simply by scientific and technological opportunities, or from curiosity about the consequences of being able to 'play' with atoms, placing them in particular locations on a surface, using them as 'building blocks' to construct just about anything on a chosen substrate.

In a series of outstanding papers published in the last decade (Refs. [55–57] just to mention a few) Eigler and co-workers have shown that the advances reached in STM have made this prospect a reality: single atoms and molecules may be manipulated in a variety of ways by using an STM, and therefore structures can be built to a particular design, atom by atom (though allowing for some limitations due to practical problems).

In the first article [55], they report the use of STM at very low temperatures (4 K) to position individual Xe atoms on a single-crystal Ni surface with atomic precision. This capacity enables them to fabricate rudimentary structures of their own design, atom by atom, namely to actually *write words* on surfaces using individual atoms for constructing the letters.

They have used the fact that an STM tip may exert a force on an atom or molecule adsorbed on a surface; by adjusting the position and the tunneling parameters (in this case the voltage) of the tip it is possible to tune the magnitude and the direction of the force, so that the STM tip can push or drag an atom or molecule across a surface, with the adsorbate still remaining bound to the surface.

The decision to study Xe on Ni(1 1 0) was not casual. Rather, it was dictated by the necessity that the corrugation in the surface potential be sufficiently large for the Xe atoms to be imaged without inadvertently moving them, yet sufficiently small that enough lateral force could be exerted to move the Xe atoms across the surface. The experiments were performed using an STM contained in a UHV system cooled to 4 K. The main reason for using low temperatures is to prevent the Xe atoms from thermally diffusing across the surface after they have been positioned in the desired location.

In order to move a single atom, they begin by using the microscope in a normal (non-perturbative) imaging mode, by which they locate the atom which is to be moved and decide where to move it; then scanning is stopped and the tip is placed directly above the atom which is to be moved. The successive step is to increase the interaction between tip and atom, by lowering the tip towards the atom. Finally, the tip is moved to the target destination on the surface, dragging the Xe atom along with it. At this point the tip is again withdrawn, by reducing the tunneling current to the value used for imaging. This effectively terminates the attraction between the Xe atom and the tip,

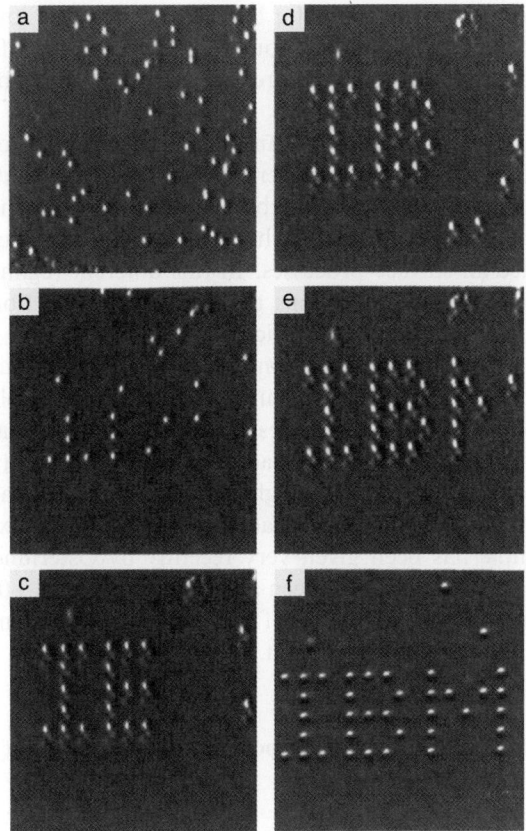

Fig. 12. A sequence of STM images taken during the construction of a patterned array of xenon atoms on a nickel (1 1 0) surface. The atomic structure of the nickel surface is not resolved. The ⟨1 1 0⟩ direction of the surface runs vertically. (a) The surface after xenon dosing, (b)–(f) various stages during the construction of the word IBM. Each letter is 50 Å from top to bottom (from Ref. [55], with permission).

leaving the Xe atoms bound to the surface at the desired location.

Fig. 12a shows the surface after Xe dosing. Each Xe atom appears as a 1.6 Å high bump on the surface, initially positioned at random locations. Fig. 12b–f shows various stages of the surface during the construction of the word "IBM". Each letter is 50 Å from top to bottom. At this gap impedance the interaction of the Xe atom with the tip is sufficiently weak to leave the Xe atom essentially unperturbed during the imaging process. The exact "periodicity" of the Xe atom spacing is derived from the crystalline structure of the underlying Ni(1 1 0) surface.

Creating or synthesizing 'custom' made structures is perhaps one of the most exciting applications of manipulation; it should be possible in this way to create totally new phases and structures of matter that are not normally accessible in nature: after creating these new structures, they can be investigated with the STM in the normal imaging and spectroscopy modes.

The main problem here, from a practical point of view, is the serial nature of the manipulations, versus parallel processes like lithography, which are therefore much faster and economically advantageous.

The STM has also been applied to induce chemical transformations on surfaces on the atomic scale by several groups of researchers [58]. Quite astonishing results have been recently obtained by Ho's group in Cornell. They were able for instance, to use the STM to break a C–H bond in a single acetylene molecule adsorbed on a Cu(0 0 1) surface [59]; they further proceeded to dehydrogenate the ethinil species to form dicarbon (CC) [59]. A comprehensive review of these fascinating results appears in this issue of Surface Science [60].

It has been shown that besides the breaking of selected molecular bonds, the STM can also induce a single bond formation. Lee et al. [61], have been able to form Fe(CO) and Fe(CO)$_2$ molecules starting from Fe atoms and CO molecules adsorbed on an Ag(1 1 0) surface. More recently, Rieder's group [62] has demonstrated the feasibility of inducing all the steps of a surface chemical reaction by using an STM tip. They were able in fact to synthesize biphenil molecules starting from iodobenzene adsorbed on a Cu(1 1 1) surface. These groups have therefore demonstrated the concrete possibility of controlling chemistry at the spatial limit of individual atoms and molecules and have provided new insights into the nature of chemical bonds and reactivity.

The ability to manipulate matter with atomic scale precision suggests that single molecular synthesis is possible with the STM; there are several motivations for such synthesis. One could study how the environment of the reactants affects surface reactions, or how the conformation of reactants affects reaction barriers. If one could perform such studies with an AFM it would be possible to study the forces between reactants as a function of conformation. Manipulation combined with force measurement seems particularly useful. For example, if one could measure the force on an atom during the sliding process, then it would be possible to map the potential between the atom and the surface. It should be possible to use the AFM to image and manipulate metal atoms on an insulating substrate, which would open the door to the study of electron transport in extremely small structures.

9. Perspectives for the future

One of the limiting factors towards our goal of a deeper understanding of surface phenomena is still the rather poor time resolution that most of the surface science techniques have. In practice, the time window in which we can observe phenomena is presently limited between a few milliseconds to a few thousand seconds. Extending this range to shorter times would be extremely beneficial to our investigation. On one hand, it would offer the possibility of detecting new intermediate species in kinetic processes which would be critical for a more in-depth understanding of the basic phenomena. As a second major benefit, higher time resolution would allow to explore processes at higher temperatures where, as a general rule, they tend to become much faster.

In recent years there have been several efforts to build ever faster STM and AFM instruments. While the time for the acquisition of a single frame (typically 50 Å × 50 Å) for a commercial instrument is of the order of several seconds, precision laboratory prototypes have been built which acquire up to 10–20 images per second, while still retaining atomic resolution [63–65].

In perspective, it is important to recall however that elementary dynamic phenomena at the atomic level take place on a surface at the femtosecond time scale, and even the recent impressive experimental advances (and foreseeable future ones) are missing this range by many orders of magnitude.

The second obstacle is the so called "pressure gap". While most of the surface science model studies carried out so far have been using very low pressures (less than about 10^{-6} mbar), real life processes are run at pressures many orders of magnitude higher.

Higher pressures mean very high coverage of the reactants. Many experiments have shown that, as the coverage of adsorbed species increases, the lateral interactions which develop strongly influence all surface processes. In this different regime all the elementary steps and the overall reaction rate may change dramatically.

It will be therefore compulsory to improve our present experimental techniques in order to be able to probe surface processes in higher environmental pressures. STM has already shown the capability of performing in situ high pressure studies of the atomic structure of surfaces. An outstanding example has been given very recently by Österlund et al. [66], who have been able to image a Cu(1 1 0) surface up to pressure of 1 bar of H_2, still maintaining atomic resolution.

10. Challenges and opportunities for a surface scientist

One of the problems which has hampered our progress so far, lies in the fact that experimental techniques, for the lack of space or time resolution (or both), usually observe an "ensemble" of different elementary steps and measure therefore parameters which are averaged out over several processes.

A great challenge for surface scientists is going to be the possibility of singling out and characterizing every step at the atomic level separately. This will require a substantial improvement of present techniques and probably, the development of new ones. The effort will be extremely exciting however, since it may also lead to the detection of new important "co-operative" many-body effects of which we had, so far, only occasional glimpses.

The goal of learning to design the atomic geometry of surfaces in order to confer them specific properties, is going to be another exciting challenge for the future. This endeavor would allow, for instance, to optimize the production rate and selectivity of catalysts or the physical properties and size of semiconductor devices and would therefore be extremely rewarding from a practical point of view.

A rather impressive step towards the general goal of designing a catalyst from "first principles" has already been obtained by the combined effort of the theoretical group of Nørskov and the experimental group of Besenbacher. In a series of papers, they have presented a beautiful example in which the full step has been taken from atomistic surface science studies of a model system to the design of a new alloy catalyst for the steam-reforming catalytic reaction [67]. This is just an example which shows how a combined effort and a strong interplay between theory and experiment has produced significant advances in a very demanding field. The collaboration between theory and experiment is pursued by many surface science groups around the world, often producing very fruitful results. An even stronger interplay between theory and experiment would be very rewarding and should be further pursued and encouraged.

Finally, an extremely challenging task which is going to bring a lot of new opportunities for the future, is to enter the arena of bridging the gap between surface model systems and real surfaces, with the ultimate aim of achieving a fast technology transfer between research laboratories and high-tech industries.

Although surface science is a very demanding field of research, both in terms of experimental and theoretical resources, in its short history very remarkable results have been obtained in understanding surface phenomena at the atomic level, and new challenges are awaiting us in the near future in this exciting interdisciplinary field which has never been more lively.

Acknowledgements

The authors are indebted to F. Besenbacher for continuous support, and for making Ref. [66] available prior to publication, and to E. Tosatti for fruitful discussions. We are grateful to N. Motta for providing the original figures from Refs. [32,33] and to T. Ogino for providing the original figures of Ref. [50]. We would like to acknowledge Giovanni Comelli, Kim Hansen, Eric Ganz, Friso Van der Veen, Tommaso Zambelli for a critical reading of the manuscript. We are also grateful to S. Rosei for an early reading criticism of the manuscript, to G. Risicato for helping prepare the final layout and to C. Africh for preparing the figures. One of us (F.R.) would like to thank the Physics Department of the University of Rome "Tor Vergata" and NTT Basic Research Laboratories (Atsugi, Kanagawa, Japan) for hospitality.

References

[1] J. Wintterlin, J. Wiechers, H. Brune, T. Gritsch, T. Hofer, R.J. Behm, Atomic-resolution imaging of close-packed metal surfaces by scanning tunneling microscopy, Phys. Rev. Lett. 62 (1989) 59.

[2] C.J. Chen, Introduction to Scanning Tunneling Microscopy, Oxford University Press, New York, Oxford, 1993.

[3] A. Borg, A.M. Hilmen, E. Bergene, STM studies of clean, CO and O_2 exposed Pt(100)-hex-R0.7°, Surf. Sci. 306 (1994) 10.

[4] C. Stampfl, M. Scheffler, H. Over, J. Burchhardt, M. Nielsen, D.L. Adams, M. Moritz, Identification of stable and metastable adsorption sites of K adsorbed on Al(111), Phys. Rev. Lett. 69 (1992) 1532.

[5] A. Baraldi, J. Cerdá, J.A. Martin-Gago, G. Comelli, S. Lizzit, G. Paolucci, R. Rosei, Oxygen induced reconstruction of the Rh(100) surface: general tendency towards threefold oxygen adsorption site on Rh surfaces, Phys. Rev. Lett. 82 (1999) 4874.

[6] D. Alfè, S. de Gironcoli, S. Baroni, The reconstruction of Rh(100) upon oxygen adsorption, Surf. Sci. 410 (1998) 151.

[7] P.W. Murray, F.M. Leibsle, Y. Li, Q. Guo, M. Bowker, G. Thornton, U.R. Dhanak, K.C. Prince, R. Rosei, Scanning tunneling microscopy study of the oxygen-induced reconstruction of Rh(110), Phys. Rev. B 47 (1993) 12976.

[8] D.P. Woodruff, T.A. Delchar, Modern Techniques of Surface Science, Cambridge University Press, New York, 1986.

[9] H. Rohrer, Scanning tunneling microscopy: a surface science tool and beyond, Surf. Sci. 299/300 (1994) 956.

[10] C.F. Quate, The AFM as a tool for surface imaging, Surf. Sci. 299/300 (1994) 980.

[11] M. Schmidt, H. Stadler, P. Varga, Direct observation of surface chemical order by scanning tunneling microscopy, Phys. Rev. Lett. 70 (1993) 1441.

[12] P.T. Wouda, B.E. Nieuwenhuys, M. Schmid, P. Varga, Chemically resolved STM on a Pt Rh(100) surface, Surf. Sci. 359 (1996) 17.

[13] P.M. Holmblad, J.H. Larsen, I. Chorkendorff, L.P. Nielsen, F. Besenbacher, I. Stensgaard, E. Laegsgard, P. Kratzer, B. Hammer, J.K. Nørskov, Designing surface alloy with specific active sites, Catal. Lett. 40 (1996) 131.

[14] E. Ganz, I.S. Hwang, F. Xiong, S.K. Theiss, J. Golovchenko, Growth and morphology of Pb on Si(1 1 1), Surf. Sci. 257 (1991) 259.

[15] G. Binnig, H. Rohrer, Ch. Gerber, E. Weibel, 7 × 7 reconstruction on Si(1 1 1) resolved in real space, Phys. Rev. Lett. 50 (1983) 120.

[16] K. Takayanagi, Y. Tanishiro, S. Takahashi, M. Takahashi, Structure analysis of Si(1 1 1)–7 × 7 reconstructed surface by transmission electron diffraction, Surf. Sci. 164 (1985) 367.

[17] M.Ø. Pedersen, L. Österlund, J.J. Mortensen, M. Mavrikakis, L.B. Hansen, I. Stensgaard, E. Laegsgaard, J.R. Nørskov, F. Besenbacher, Diffusion of N adatoms on the Fe(1 0 0) surface, Phys. Rev. Lett. 84 (2000) 4898.

[18] J. Wintterlin, R. Schuster, G. Ertl, Existence of a "Hot" atom mechanism for the dissociation of O_2 on Pt(1 1 1), Phys. Rev. Lett. 77 (1996) 123.

[19] T. Zambelli, J. Trost, J. Wintterlin, G. Ertl, Diffusion and atomic hopping of N atoms on Ru(0 0 0 1) studied by scanning tunneling microscopy, Phys. Rev. Lett. 76 (1996) 795.

[20] S. Renish, R. Schuster, J. Wintterlin, G. Ertl, Dynamics of adatom motion under the influence of mutual interactions: O/Ru(0 0 0 1), Phys. Rev. Lett. 82 (1999) 3839.

[21] J.V. Barth, Transport of adsorbates at metal surfaces: from thermal migration to hot precursors, Surf. Sci. Rep. 40 (2000) 75.

[22] T. Zambelli, J. Wintterlin, J. Trost, G. Ertl, Identification of the "active sites" of a surface-catalyzed reaction, Science 273 (1996) 1688.

[23] J. Trost, T. Zambelli, J. Wintterlin, G. Ertl, Adsorbate–adsorbate interactions from statistical analysis of STM images: N/Ru(0 0 0 1), Phys. Rev. B 54 (1996) 17850.

[24] B. Hammer, Steps: NO dissociation at corrugated bond activation at monatomic Ru(0 0 0 1), Phys. Rev. Lett. 83 (1999) 3681.

[25] J. Wintterlin, S. Völkening, T.V.W. Janssens, T. Zambelli, G. Ertl, Atomic and macroscopic reaction rates of a surface-catalyzed reaction, Science 278 (1997) 1931.

[26] J.M. Moison, F. Houzay, F. Bathe, L. Leprince, E. Andrè, O. Vatel, Self-organized growth of regular nanometer-scale InAs dots on GaAs, Appl. Phys. Lett. 64 (1994) 196.

[27] R. Noetzel, J. Temmyo, H. Kamada, T. Furuta, T. Tamamura, Strong photoluminescence emission at room temperature of strained InGaAs quantum disks (200–30 nm diameter) self-organized on GaAs(3 1 1) B substrates, Appl. Phys. Lett. 65 (1994) 457.

[28] R. Noetzel, self-organized growth of quantum-dot structures, Semicond. Sci. Tech. 11 (1998) 1365.

[29] C.J. Huang, D.Z. Li, B.W. Cheng, J.Z. Wu, Q.M. Wang, Oblique alignment of columns of self-organized Ge–Si(0 0 1) islands in multilayer structure, Appl. Phys. Lett. 77 (2000) 2852, and references therein.

[30] L.J. Schowalter, MRS Bull. Mater. Res. Soc. 21 (1996) 18.

[31] T.P. Pearsall, Mater. Sci. Engin. B 9 (1991) 225.

[32] N. Motta, A. Sgarlata, R. Calarco, Q. Nguyen, J. Castro Cal, F. Patella, A. Balzarotti, M. De Crescenzi, Growth of Ge–Si(1 1 1) epitaxial layers: intermixing, strain and island formation, Surf. Sci. 406 (1998) 254.

[33] N. Motta, A. Sgarlata, R. Calarco, Q. Nguyen, J. Castro Cal, P. Prosposito, A. Balzarotti, M. De Crescenzi, Scanning tunneling microscopy studies of Ge/Si films on Si(1 1 1): from layer by layer to quantum dots, J. Vac. Sci. Technol. B 16 (1998) 1555.

[34] F. Boscherini, G. Capellini, L. Di Gaspare, F. Rosei, N. Motta, S. Mobilio, Ge–Si intermixing in Ge quantum dots on Si(0 0 1) and Si(1 1 1), Appl. Phys. Lett. 76 (2000) 672.

[35] F. Rosei, N. Motta, A. Sgarlata, G. Capellini, F. Boscherini, Formation of the wetting layer in Ge/Si(1 1 1) studied by STM and XAFS, Thin Solid Films 369 (2000) 29.

[36] B. Voigtlander, M. Kästner, Evolution of the strain relaxation in a Ge layer on Si(0 0 1) by reconstruction and intermixing, Phys. Rev. B 60 (1999) R5121.

[37] W. Seifert, N. Carlsson, J. Johansson, M. Pistol, L. Samuelson, In situ growth of nano-structures by metalorganic vapour phase epitaxy, J. Cryst. Growth 170 (1997) 39.

[38] G. Capellini, N. Motta, A. Sgarlata, R. Calarco, Evolution of strained Ge islands grown on Si(1 1 1): a scanning probe microscopy study, Solid State Commun. 112 (1999) 145.

[39] F. Boscherini, G. Capellini, L. Di Gaspare, M. De Seta, F. Rosei, A. Sgarlata, N. Motta, S. Mobilio, Ge–Si intermixing in Ge quantum dots on Si, Thin Solid Films 380 (2000) 173.

[40] A. Sgarlata, F. Rosei, M. Fanfoni, N. Motta, A. Balzarotti, STM/AFM study of Ge quantum dots grown on Si(1 1 1), IEEE Proc. 11th Intl. Semicond. Insulating Mater. Confer., 2000.

[41] T. Fukuda, Random adatom heights in Ge/Si(1 1 1)-5 × 5 surfaces, Surf. Sci. 351 (1996) 103.

[42] K. Nakajima, A. Konishi, K. Kimura, Direct observation of intermixing at Ge/Si(0 0 1) interfaces by high-resolution Rutherford backscattering spectroscopy, Phys. Rev. Lett. 83 (1999) 1802.

[43] D.J. Bottomley, The physical origin of InAs quantum dots on GaAs (0 0 1), Appl. Phys. Lett. 72 (1998) 783.

[44] D.J. Bottomley, Formation and shape of InAs nanoparticles on GaAs surfaces: fundamental thermodynamics, Jpn. J. Appl. Phys. 39 (2000) 4604.

[45] G. Wedler, J. Walz, T. Hesjedal, E. Chilla, R. Koch, Stress and relief of misfit strain of Ge/Si(0 0 1), Phys. Rev. Lett. 80 (1998) 2382.

[46] J. Walz, A. Greuer, G. Wedler, T. Hesjedal, E. Chilla, R. Koch, Stress and relief of misfit strain of Ge/Si(1 1 1), Appl. Phys. Lett. 73 (1998) 2579.

[47] F. Rosei, P. Raiteri, On the composition of Ge quantum dots on low index Si surfaces, Thin Solid Films, submitted for publication.

[48] D.M. Eigler, C.P. Lutz, W.E. Rudge, An atomic switch realized with the scanning tunneling microscope, Nature 352 (1991) 600.

[49] F. Liu, M.G. Lagally, Self-organized nanoscale structures in Si/Ge films, Surf. Sci. 386 (1997) 169.

[50] T. Ogino, H. Hibino, Y. Homma, Y. Kobayashi, K. Prabhakaran, K. Sumitomo, H. Omi, Fabrication and integration of nanostructures on Si surfaces, Acc. Chem. Res. 32 (1999) 447.

[51] M. Gsell, P. Jacob, D. Menzel, Effect of substrate strain on adsorption, Science 280 (1998) 717.

[52] K. Miki, H. Tokumoto, Step control on silicon surfaces by electric fields, Nanotechnology 3 (1992) 142.

[53] S. Stoyanov, J.J. Métois, V. Tonchev, Current induced bunches of steps on the Si(1 1 1) surface – a key to measuring the temperature dependence of the step interaction coefficient, Surf. Sci. 465 (2000) 227.

[54] R.P. Feynman, There is plenty of room at the bottom, Paper presented at the American Physical Society Annual Meeting, 29 December 1959; reprinted in: H.D. Gilbert (Ed.), Miniaturization, Reinhold, NY, 1961 (an internet link to this famous speech may be found within the website of zyvex.com, http://www.zyvex.com/nanotech/feynman.html).

[55] D.M. Eigler, E.K. Schweizer, Positioning single atoms with a scanning tunneling microscope, Nature 344 (1990) 524.

[56] J.A. Stroscio, D.M. Eigler, Atomic and molecular manipulation with the scanning tunneling microscope, Science 254 (1991) 1319.

[57] A.F. Crommie, C.P. Lutz, D.M. Eigler, Imaging standing waves in a two-dimensional electron gas, Nature 363 (1993) 524.

[58] W. Ho, Inducing and viewing bond selected chemistry with tunneling electrons, Acc. Chem. Res. 31 (1998) 567, and references therein.

[59] L. Lauhon, W. Ho, Control and characterization of a multistep unimolecular reaction, Phys. Rev. Lett. 84 (2000) 1527.

[60] W. Ho, http://www.ps.uci.edu/physics/wilsonho/wilsonho.html.

[61] H.J. Lee, W. Ho, Single-bond formation and characterization with a scanning tunneling microscope, Science 286 (1999) 1719.

[62] S.-W. Hla, L. Bartels, G. Meyer, K.-H. Rieder, Towards single molecule engineering, Phys. Rev. Lett. 85 (2000) 2777.

[63] L. Kuipers, M.S. Hoogeman, J.W.M. Frenken, Step dynamics on Au(1 1 0) studied with a high-temperature, high-speed scanning tunneling microscope, Phys. Rev. Lett. 71 (1993) 3517.

[64] J. Wintterlin, J. Trost, S. Renisch, R. Schuster, T. Zambelli, G. Ertl, Real-time STM observations of atomic equilibrium fluctuations in an adsorbate system: 0/Ru(0 0 1), Surf. Sci. 394 (1997) 159.

[65] F. Besenbacher, Scanning tunneling microscopy studies of metal surfaces, Rep. Prog. Phys. 59 (1996) 1737, and references therein.

[66] L. Österlund, P.B. Rasmussen, P. Thorstrup, E. Laegsgaard, I. Steensgaard, F. Besenbacher, Bridging the pressure gap in surface science at the atomic level, Phys. Rev. Lett 86 (2001) 460.

[67] F. Besenbacher, I. Chorkendorff, B.S. Clansen, B. Hammer, A.M. Molenbroek, J.K. Nørskov, I. Stensgaard, Design of a surface alloy catalyst for steam reforming, Science 279 (1998) 1913, and references therein.

Fidgety particles on surfaces: how do they jump, walk, group, and settle in virgin areas?

A.G. Naumovets [a,*], Zhenyu Zhang [b,c]

[a] *Institute of Physics, National Academy of Sciences of Ukraine, 46 Prospect Nauki, UA-03028, Kiev 28, Ukraine*
[b] *Solid State Division, Oak Ridge National Laboratory, Oak Ridge, TN 37831-6032, USA*
[c] *Department of Physics, University of Tennessee, Knoxville, TN 37996, USA*

Received 29 June 2000; accepted for publication 14 May 2001

Abstract

This article is aimed at describing the history and current status of surface diffusion, its vital importance in diverse surface phenomena and numerous industrial applications, the recent significant progress in its understanding, and, in particular, many unsettled questions concerning surface diffusion mechanisms. From the fundamental point of view, the interest in surface diffusion and, at the same time, the difficulties in its understanding, stem primarily from the collective (many-body) nature of this phenomenon. On the practical side, the importance of the knowledge of the physics behind surface diffusion is becoming increasingly crucial in the coming era of nanotechnology, which will bring an unprecedented level of miniaturization in electronics and a broad employment of various nanomaterials. © 2001 Elsevier Science B.V. All rights reserved.

Keywords: Surface diffusion; Models of surface kinetics; Scanning tunneling microscopy; Field ion microscopy; Surface thermodynamics (including phase transitions); Adatoms; Clusters; Single crystal surfaces

1. Introduction

1.1. Historical remarks

Suppose we are able to discern individual atoms on a surface and to trace their movements. Then the surface will be seen not as a lifeless frozen landscape, but rather as an extremely animated arena resembling a square filled with fidgety children, who jump spontaneously up and down, run around, collide or dance with each other, arrange themselves into quaint patterns or try to hide away from the crowd. Such a picture, in many aspects, is a close analogue of surface diffusion.

The mobility of substances on surfaces has many manifestations. The most common one is wetting of surfaces and spreading of various films, as seen in diverse capillary processes, spreading of oily films over water surfaces, etc. Such wetting phenomena had been observed with the naked eye already by the ancients and widely used for ages. Nobody was too surprised by such observations, because one was so accustomed to the fluidity of liquids. In contrast, the mobility of surface atoms on *solids* was inferred by Volmer and Estermann only in 1921 from their growth studies of mercury

---

[*] Corresponding author. Fax: +380-44-265-1589.
E-mail address: naumov@iop.kiev.ua (A.G. Naumovets).

crystals from the vapor phase [1]. They found that Hg crystals grew in some directions much faster (by as many as three orders of magnitude) than could be expected due to direct condensation from the vapor phase.

The classical differential equations for diffusion were derived by A. Fick in 1855, well before the discovery of surface diffusion. These equations, known as Fick's laws, are in principle applicable to both volume and surface diffusion. However, for several decades studies of surface diffusion were far less numerous than volume diffusion studies, which were strongly stimulated by practical needs of metallurgy and other important "volume" technologies. Nevertheless, the pioneers and veterans of surface diffusion investigations—I. Langmuir, J.A. Becker, E.W. Müller, R. Gomer, M. Drechsler, C. Herring, W.W. Mullins, Ya. Geguzin, I. Sokolskaya, and others—had already established prior to the 1960s a number of important regularities in surface diffusion. Example aspects included distinct roles of substrate atomic structures, strong dependence on adsorbate coverage, different diffusion mechanisms for submonolayer and multilayer deposits, and effects of high electric fields in diffusion kinetics. This stage of investigations was summarized in reviews [2–6]. However, a real breakthrough in surface diffusion studies, as in surface science in general, took place only in the 1960s, when ultra-high vacuum, macroscopic single crystals and various highly sensitive surface characterization methods became broadly accessible. Nowadays we can investigate surface diffusion in detail under very diverse conditions—from observation of random walks of individual atoms to recording surface diffusion over macroscopic distances. No less important is the fact that an ever-increasing number of new problems, both fundamental and practical, call for deeper insight into the underlying physics of surface diffusion.

1.2. Why so much interest in surface diffusion?

The stimuli for surface diffusion studies are manifold. Any surface physical or chemical process which involves displacement of surface atoms or molecules (and there are very few surface processes that do not) naturally depends on surface diffusion in an important way. Here is an inclusive list of fields and industries which are inherently tied to surface diffusion, and can therefore greatly benefit from any major advance in its understanding: crystal growth (including epitaxial growth of single crystal films widely used in microelectronics); heterogeneous catalysis (without which the modern chemical industry could hardly be imagined); wetting, spreading and capillarity (it is not an overstatement to say that life would be impossible without such phenomena); tribology and development of new lubricants; corrosion protection; electrochemistry; and sintering technology (which, in particular, is the basis of powder metallurgy).

For a given system of size R, the time necessary for the occurrence of some compositional or morphological changes (desirable or troublesome) caused by surface diffusion varies as R^2. Therefore, the role of surface diffusion in influencing the performance characteristics and long-term stability of functional elements in electronics and other technologies grows rapidly with progressing device miniaturization. For this reason, surface diffusion is critical in such modern technologies as growth of nanostructures, fabrication of quantum well, wire and dot devices, modern semiconductor lasers, magnetic multilayer recording heads based on giant magnetoresistance, etc. When we create nanostructures and nanodevices, the elemental building blocks are actually still individual atoms and molecules. We must therefore understand how they can be moved and move themselves, and how strongly they stick to each other and to the substrates. Otherwise, we may not be able to realize a desired construction, or, once created, such a construction may rapidly stop functioning due, for example, to electromigration, just like the drifting and scattering of a sand pile under the bursts of wind.

Thus, surface diffusion is actually an interdisciplinary problem of immense importance. As a rough quantitative indicator, a survey of recent publications in the journal of Surface Science reveals that about one-fifth of these publications is connected, in one way or another, with problems of surface diffusion. Our aim is to give the readers a flavor of what is going on in surface diffusion studies now and what topics in this field are likely to become increasingly important in the foreseeable future.

The organization of this article is as follows. Section 1.3 gives a simplified view of surface diffusion as a start for further treatment. Section 2 presents a selection of current results, which give a feeling for topics studied today and illustrate the importance of basic knowledge of surface diffusion. In Section 3 we discuss in more detail some problems that are waiting for further investigation. The final section presents a summary.

1.3. Surface diffusion simplified

The initial views on surface diffusion were rather naive or oversimplified. First, the substrate was considered as a purely static participant of the diffusion process, defining the "scenery", viz. the corrugation and symmetry of a frozen potential energy landscape felt by the diffusing atoms or molecules. These particles were assumed to jump (due to thermal excitations) over the static potential energy barriers set up by the substrate (see Fig. 1). Correspondingly, the calculations of the activation energy of surface diffusion were based on the primitive assumptions about a static substrate and additive interactions of diffusing species with the substrate atoms. In fact, the first atomic-scale investigations of surface diffusion, made by Ehrlich and Hudda in 1966 using the field ion microscopy (FIM), showed that such a static-substrate picture failed to predict even the qualitative correspondance between diffusion activation energies and crystal surfaces with different Miller indices [7].

More recent observations revealed that the role of the substrate is not that passive at all. Instead, the diffusion jumps normally result from concerted movements of the diffusing particles and their substrate neighbors. In some cases, diffusion can also proceed via the exchange mechanism, where a diffusing adatom pushes out a substrate atom and takes its position while the former substrate atom starts to diffuse [8–12].

The second simplification was related to the situation when many particles are diffusing at a time. The particles were assumed to be non-interacting (independent). For such a case, the solutions of the diffusion equations are quite simple and universal [14], but this idealized situation has little in common with reality either. In fact, the particles on surfaces interact with each other by manifold and non-additive lateral forces, which can be attractive, repulsive or oscillatory, short or long ranged, and, generally, anisotropic [15–19].

Therefore, surface diffusion is really highly cooperative in nature. This endows it with rich physical content, making its investigation challenging and fascinating at the same time.

2. Examples of present-day knowledge

2.1. Diffusion of single adatoms

When an atom is adsorbed onto the surface of a substrate, it is called an adatom, and its energetically favorable location is called an adsorption site. For thermally activated diffusion of such an adatom, the surface diffusion coefficient, D, can be written as $D = (1/4)a^2 k_s$, where a is the hopping distance between two neighboring sites (see Fig. 1), and the total rate for the adatom to hop out of a given site, k_s, can further be written as $k_s \propto \exp\{-V_s/k_B T\}$, with V_s expressing the potential energy barrier the adatom must overcome in hopping from site to site, T the substrate temperature, and k_B the Boltzmann constant.

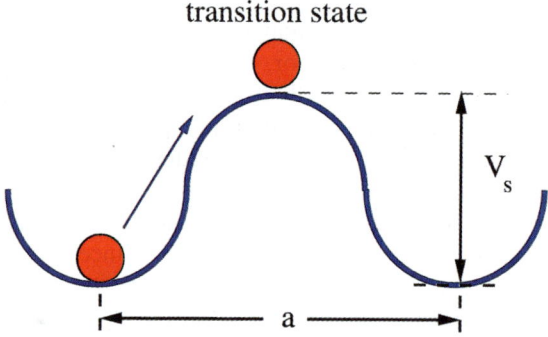

Fig. 1. Schematic representation of the potential energy barrier, V_s, against the diffusion of an atom adsorbed on a surface site (the well). The average site-to-site hopping distance is measured by a.

The simplest type of surface diffusion is tracer diffusion, when an adatom makes thermally activated hops without feeling the presence of any other diffusing objects. Such a regime can be realized when only one adatom is placed onto an otherwise empty terrace, as done in typical FIM experiments; it is also applicable to cases where the adatom density is very low and their mutual interaction is negligible, as in the earliest stages of epitaxial growth.

Studies of tracer diffusion can be most readily carried out using atomic-resolution real-space techniques such as the FIM and the scanning tunneling microscopy (STM). The FIM, invented by Müller in the early 1950s [20], provides an image of the protruding atoms at the apex of a sharp needle, commonly referred to as a tip. Such an image is obtained by applying a high positive potential to the tip in the presence of a background gas, called the imaging gas. The high electric field ionizes the imaging gas atoms, and accelerates them to an image detector where image spots are formed. As first demonstrated by Ehrlich and Hudda [7], the FIM allows one to view an individual atom and track its motion as it moves across a well-defined surface of the tip. Since then, the FIM has proven to be a powerful technique for tracer diffusion studies, leading to the discovery of a wealth of important concepts in this field [11–13]. The main limitation of the FIM is its relatively narrow range of applicability, so far only to a selected number of noble and transition metals. This limitation is inherently tied to the fact that the FIM essentially produces images of the tip, whereas not too many elements and compounds can be explored in the tip configuration and under the high field imaging conditions.

Unlike the FIM, in STM the tip scans over the surface and records the tunneling current between the tip and the surface atoms. The bias potential inducing the tunneling current is typically on the order of a few volts or smaller, much weaker than what is required in FIM. The variation of the current can be converted into images of the surface, and the high (atomic-scale) resolution is due to the tunneling nature of the current, characterized by its exponential dependence on the tip–surface separation. Since the invention of the STM [21], at least four conceptually different modes of operation have been pursued to use the STM for tracer diffusion studies. The first is to monitor the fluctuations in the tunneling current caused by the passing of adatoms underneath an STM tip [22,23], in analogy to field emission microscopy [24]. The second is to count the density of islands formed by the diffusing adatoms, and deduce the tracer mobility by invoking classical nucleation theory [25]. The physical basis of this approach is as follows: At a given deposition rate, the faster an adatom diffuses, the longer it travels before running into an existing island (or other surface defect) and being captured, thereby reducing its chance to run into another adatom and nucleate a new island, leading to fewer islands. This approach, first demonstrated by Mo et al. [26], has received the widest attention among the four so far. The third approach is by direct imaging of the displacements of the diffusing species induced by thermal annealing, in analogy to the working principle of the FIM [27]. The fourth, and highly appealing, method is by atom tracking, where the STM tip automatically follows the position of the diffusing species [28]. Each of these approaches has its merits and limitations, but collectively, the power of the STM in surface diffusion measurement is clearly superior to the FIM. In STM, it is the surfaces and the features on the surfaces that are being imaged; therefore, one can essentially use the same tip to image a wide variety of systems, as long as the latter are conductive under the bias potential between the tip and the substrate.

On the theoretical side, with the recent significant advances in high-performance computing, tracer diffusion, in particular the potential energy barriers involved, can now be routinely studied using first-principles approaches for both metal and semiconductor systems [29,30]. Such studies not only allow quantitative comparisons of the activation energies between experiment and theory, but often also shed new light on the microscopic diffusion mechanisms involved. As a compelling example, it was first predicted by Feibelman using first-principles calculations [29], and also confirmed experimentally [31,32], that adatom diffusion on some fcc(1 0 0) metal surfaces can proceed via the place exchange mechanism [29],

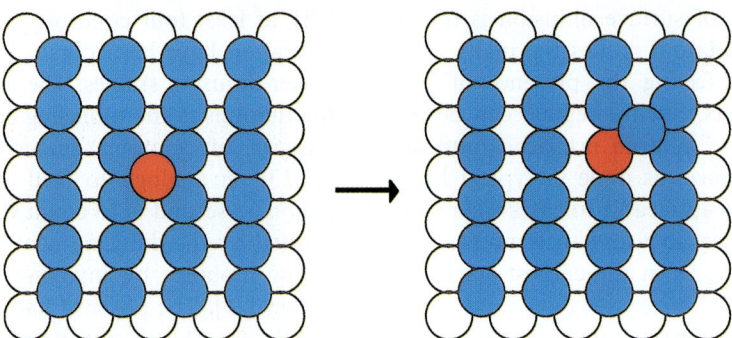

Fig. 2. Schematic representation of the place exchange mechanism for cross-channel diffusion of an adatom on an fcc(1 1 0) surface. The diffusing adatom (red) kicks out an atom in the wall (blue), and takes its position; the latter emerges in a neighbor channel and continues the diffusion process [8,33].

similar to what had been observed earlier on the fcc(1 1 0) surfaces [8,33]. Such place exchange mechanisms are schematically illustrated in Figs. 2 and 3, showing vividly the many-body and cooperative nature of the diffusion of even single adatoms.

If an adatom diffuses solely via the place exchange mechanism on a fcc(1 0 0) surface, it can only visit a subset of the lattice sites. These visited and unvisited sites, of equal population, each form a (2 × 2) mesh with sides parallel to the [1 0 0] and [0 1 0] directions of the substrate lattice (see Fig. 4a). As a signature of the place exchange mechanism, this intriguing feature has indeed been exposed in the self-diffusion measurements on Pt(1 0 0) [31] (see also Fig. 4b) and Ir(1 0 0) [32].

Precisely why the place exchange mechanism is preferred on some of the relatively smooth fcc(1 0 0) surfaces is still controversial, and the underlying reason may also vary from system to system [29, 34–36]. For self-diffusion on Al(1 0 0), Feibelman used the local chemical bonding argument to rationalize his finding: The exchange mechanism preserves a bond between the two participating Al atoms at the transition state, effectively lowering

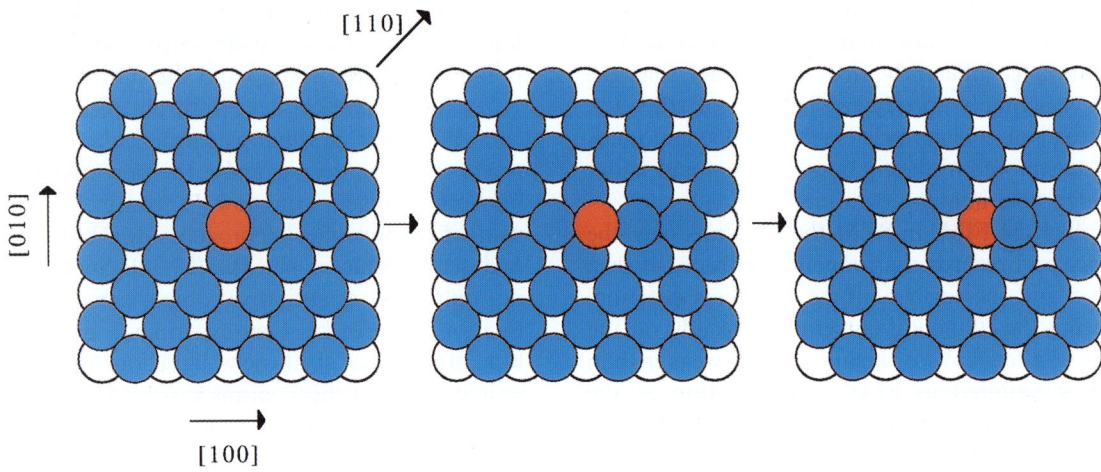

Fig. 3. Schematic representation of the place exchange mechanism for adatom diffusion on an fcc(1 0 0) surface. The diffusing adatom (red) kicks out an atom from the top layer of the substrate (blue), and takes its position; the latter emerges in a next nearest neighboring site (see also Fig. 4) and continues the diffusion process [12,29,31,32].

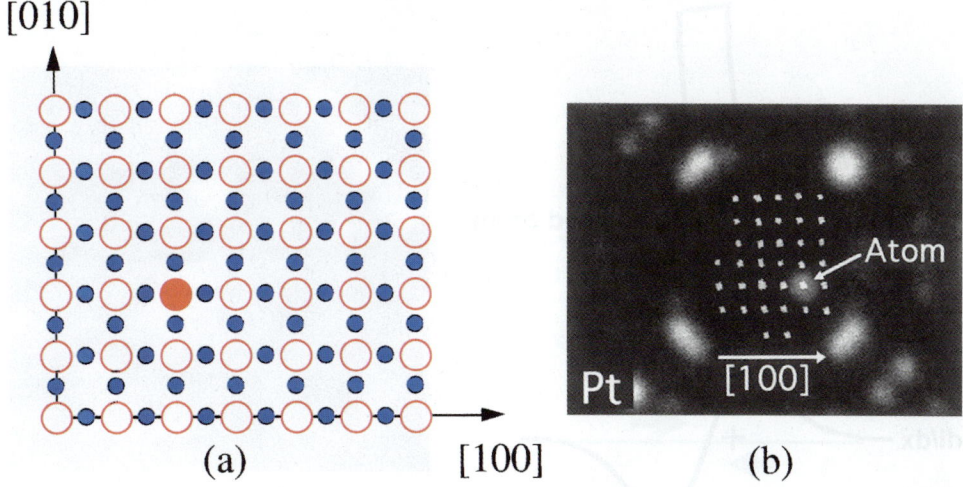

Fig. 4. (a) Accessible sites (empty red circles) to an adatom (filled red) diffusing on an fcc(1 0 0) surface via the place exchange mechanism shown in Fig. 2. The smaller filled circles are the top-layer atoms of the substrate. (b) Sites visited by a diffusing Pt adatom on Pt(1 0 0), as mapped out by Kellogg using the field ion microscopy [31] (courtesy of G.L. Kellogg). Similar results were also obtained by Chen and Tsong for Ir self-diffusion on the (1 0 0) surface [32].

the activation energy barrier [29]. Alternatively, it has been suggested that the tensile surface stress [34,35], or the substrate relaxation energy around the diffusing adatom [36], can also play a crucial role in activating the place exchange process for self-diffusion on fcc(1 0 0) surfaces.

2.2. *Diffusion of atom clusters and islands: from individual atom jumps to reptation, gliding, caterpillar motion, leapfrog games, and rolling stones*

Let us next consider more closely the case of diffusion when the average adatom concentration on the substrate is low—say, about 0.1 of a monolayer, which is typical, e.g., for conditions of thin film growth. Recent studies have shown that lateral interactions of adatoms impart surface diffusion a collective character and can dramatically affect surface diffusivity even at such low coverages. In particular, the presence of lateral attraction entails the formation of atom clusters (oligomers) and, at the later stages of overlayer condensation and ripening, of larger islands. The more diverse degrees of freedom of adatom clusters will naturally increase the number and complexity of their diffusion pathways, and a great deal of effort has been devoted to the elucidation of such complexities over the decades, with some examples given below.

The most systematic experimental studies of diffusion kinetics for clusters containing different but precisely known number of atoms have been carried out for metal systems using the FIM [11–13,37–41]. In recent years, an increasing number of investigations of cluster mobility (for both metal and semiconductor systems) have also been made using the STM [27,28,42–49]. To date, an intriguing variety of mechanisms have been proposed to describe the migration of clusters that are relatively small in size. Fig. 5 shows an example for the diffusion of a Si dimer adsorbed on top of the Si(1 0 0) surface, as observed using the atom-tracking method [28]. Here the adsorbed dimer, or addimer, is found to rotate between two orthogonal orientations as it diffuses along a fast track defined on top of a dimer row of the reconstructed surface [28,42,50]. A selected number of other examples are collectively presented in Fig. 6. They include mechanisms that can be readily envisioned, such as the sequential displacement of individual atoms (one atom at a time, see Fig. 6a); reptation, or successive shearing of some compact blocks (Fig. 6b); and gliding (moving a cluster as a whole, see Fig. 6c). Other examples are more exotic in

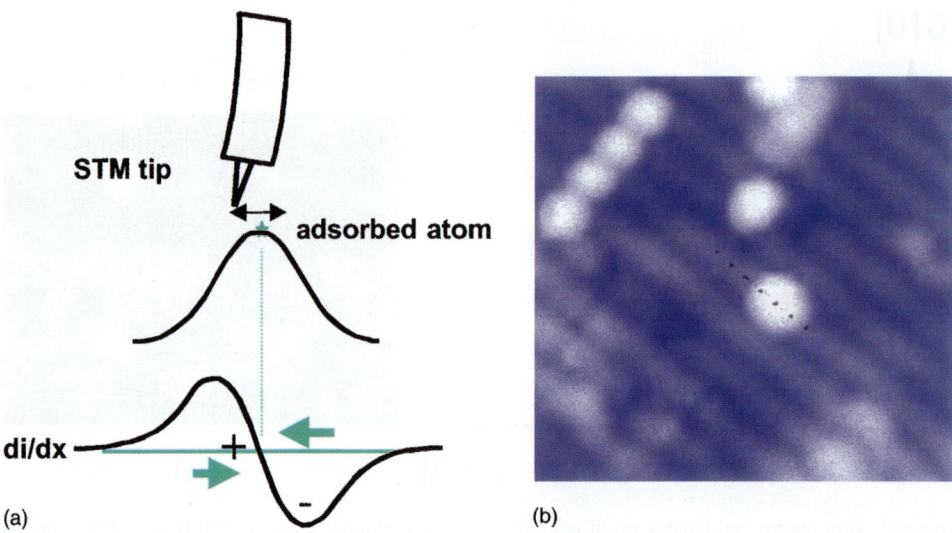

Fig. 5. (a) Schematic representation of the atom-tracking method for surface diffusion measurement, in which the STM tip follows the motion of the adatom through a feedback of the di/dx signal. (b) The diffusion track (the dark dashed line) of a Si dimer adsorbed on top of a dimer row of the reconstructed Si(1 0 0) surface as imaged using the method shown in (a). Also shown in the STM image are several other features (the bright ones) adsorbed on the surface [28] (courtesy of B.S. Swartzentruber).

nature, such as the leapfrog mechanism (Fig. 6f), dislocation (soliton) diffusion (Fig. 6g), and rolling of three-dimensional clusters (Fig. 6h). Let us consider these mechanisms more closely.

Perhaps the simplest scenario for cluster diffusion is to shift one atom ahead at a time, as, e.g., was suggested by Kellogg in his FIM studies of the size dependence of the mobility of compact Rh clusters containing from 2 to 12 atoms on Rh(1 0 0) [38]. Subsequently, empirical-potential total energy calculations revealed that the self-diffusion of compact clusters consisting of less than 9 atoms on the fcc(1 0 0) surfaces may involve successive shearing of dimers belonging to the cluster [51]. The two different diffusion mechanisms, sequential motion of individual atoms and dimer shearing, are contrasted in Fig. 6a and b for the case of a tetramer. The dimer shearing mechanism is preferred because it does not require simultaneous breaking of two nearest-neighbor bonds in the transition state, similar to the adatom diffusion via place exchange [29].

Molecular dynamics simulations performed recently by Chirita et al. [52] revealed that the shearing mechanism of cluster diffusion can be even more diverse and widespread than formerly recognized. Namely, not only dimers, but also larger groups of atoms ("subcluster units") can shear simultaneously. Chirita et al. [52] used the term *reptation* to describe this mechanism, since such a motion of clusters resembles the reptation of side winding snakes (*Crotalus cerastes* in Latin). The term *reptation* was introduced earlier by de Gennes [53] in a different context, to describe the displacement of molecules in polymers. Chirita et al. have predicted that for Pt$_n$ clusters with $n < 7$, the reptation mechanism should be as probable as the concerted gliding of the whole cluster [52]. However, the reptation becomes preferred at $n > 7$. In clusters with sizes up to ∼20 atoms, this mechanism is implemented through successive concerted shearing of subcluster blocks ("reptons") which can contain from 2 to 10 atoms. The reptation mechanism may also be important in diffusion and shape evolution of fractal (randomly ramified) and dendritic (symmetrically branched) islands.

The self-diffusion of Ir$_n$ and Pt$_n$ clusters on the (1 1 1) surfaces, with n ranging from 2 to 7, has been extensively investigated experimentally by Ehrlich et al. [11,37,39–41]. Overall, a general

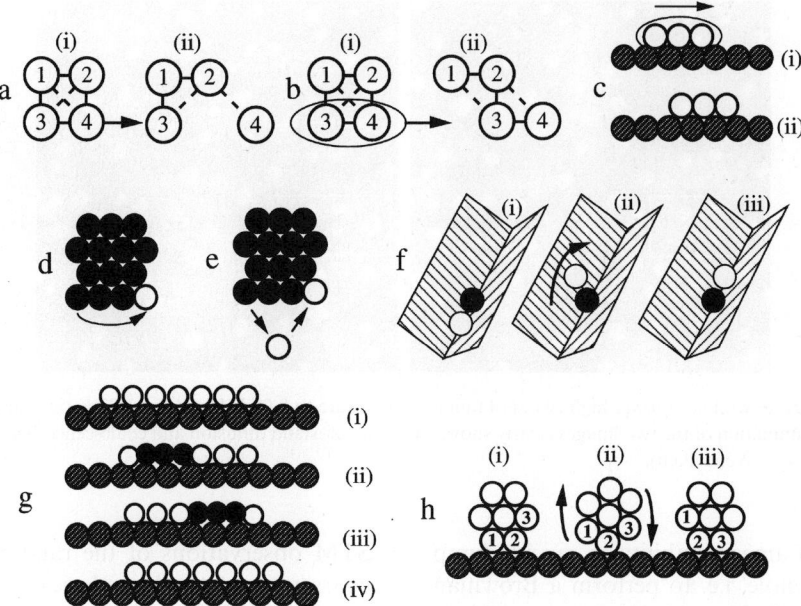

Fig. 6. Schematics of some diffusion mechanisms for clusters and atomic islands. (a) Sequential displacement of individual atoms for the diffusion of a tetramer on fcc(1 0 0). (b) The dimer shearing mechanism for the diffusion (and dissociation) of compact clusters of sizes less than 9 atoms on fcc(1 0 0). (c) The gliding mechanism: the cluster glides as a whole from one position to the next. (d) The edge diffusion mechanism: the motion of a loose adatom along the island edge causes a shift of the center of mass of the island. (e) The evaporation + condensation mechanism: the interchange of adatoms between different islands (and steps) results in the Brownian motion of the islands, accompanied by coarsening of the islands. (f) Diffusion of a dimer in an atomic channel (shown schematically as a trough) by the leapfrog mechanism. (g) The dislocation (soliton) mechanism: adatoms in the dislocation (soliton) are shown as black balls. Note the displacement of the atoms from (i) to (ii). (h) Schematic of the rolling mechanism.

trend for a decrease in the diffusivity has been found with increasing n. The decrease is monotonic for Pt_n clusters but is non-monotonic (with a rather sharp local minimum at $n = 4$) for Ir_n clusters. The experimental data suggest that "closed shell" (*magic*) clusters, having a symmetrical compact shape, should have peculiar energetics and diffuse in a different way than non-compact ones. Amazingly, the addition or subtraction of a single atom to or from the compact configuration changes dramatically the cluster diffusion parameters. On the basis of these findings it is inferred that the compact Ir_7 and Ir_{19} clusters may diffuse by a collective gliding mechanism (see Fig. 6c), whereas non-compact clusters seem to move by edge diffusion (Fig. 6d).

Sometimes, for example on the fcc(1 1 0) and even on some fcc(1 0 0) surfaces, clusters may assume linear (chain-like) configurations. As a rule, linear oligomers were found to diffuse by successive shear displacements of atoms in the chain (atom-by-atom motion, similar to the process shown in Fig. 6a). However, theoretical calculations [54] predicted that dimers adsorbed at the bottom of atomic channels, formed due to surface reconstruction, will either make a concerted jump as a whole or move by a *leapfrog* mechanism. In the latter case, one adatom temporarily leaves the bottom and climbs the channel wall to leapfrog the other adatom remaining at the bottom (Fig. 6f). This mechanism was found to be even more probable for longer adatom chains, and experimental evidence has been provided for linear Pt clusters on the reconstructed Pt(1 1 0)-1 × 2 surface [49].

A long discussion in the literature is concerned with the collective gliding and dislocation mechanisms of cluster motion. Studies in the 1960s revealed that under some conditions the clusters

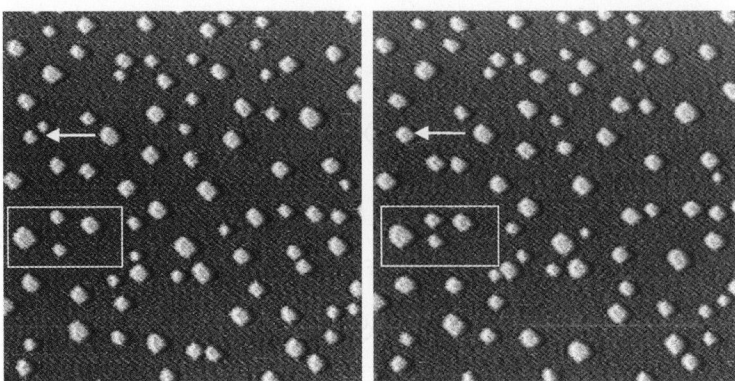

Fig. 7. A Cu(1 0 0) surface with monolayer high islands 4 min after exposure to 0.08 ML of Cu at 297 K (left) and then 8 min and 50 s later (right). Close examination of the two images clearly shows examples of island diffusion and coalescence. The images are 200 × 200 nm^2 [46] (courtesy of J.F. Wendelken).

(islands) show an amazing ability to quickly jump and rotate as a whole, i.e. to perform a Brownian motion on the substrate (for a history of this problem, see reviews [55,56] and references therein). The possibility for an island to rotate and/or to shift with activation energies comparable to that of the diffusion of an individual adatom is attributed mainly to a "counterweight effect". In the collective motion of the island, some atoms climb up the potential maxima while others simultaneously slide down to the bottom of the potential wells, the net activation barrier being rather low. A similar effect occurs when a misfit dislocation (in other terms, a soliton or a domain wall) arises and runs across the island [55,57,58] (Fig. 6g). It also has an analogue in fauna: the motion of a caterpillar by displacement of the hump along its trunk (see also Section 2.3 about the soliton diffusion mechanism in continuous dense overlayers). According to computer simulations [58], the dislocation diffusion mechanism should dominate over the edge diffusion and gliding in rather large compact islands ($20 < n < 100$) [58]. The dislocation diffusion mechanism was predicted to be particularly efficient in magic islands, which contain a misfit dislocation in their lowest energy configuration. For example, a 61-atom island should have a much higher mobility at room temperature than the 37- and 91-atom islands [58].

However, different conclusions about the diffusion mechanisms have been drawn from recent STM observations of the random migration and coarsening of the boundaries of large adatom or vacancy clusters (with $n > 100$) [44–46]. An example for the case of Cu islands on Cu(1 0 0) is shown in Fig. 7 [46]. The mobility of such large clusters is determined mainly by two competing mechanisms: adatom edge diffusion (Fig. 6d), and evaporation and condensation of adatoms (Fig. 6e) [44–46,59]. By analyzing both the island mobility and decay, Pai et al. [46] have shown that the self-diffusion of large Ag and Cu clusters on the (1 0 0) surfaces is dominated by the edge diffusion mechanism at room temperature. At higher temperatures, the migration and coarsening of large Cu islands on Cu(1 0 0) can also be caused by the evaporation and condensation of vacancies [60].

So far all our discussions have been restricted to diffusion of two-dimensional (2D) (or monatomic-layer high) islands. Recent molecular dynamics simulation studies of the diffusion of 3D clusters on lattice-mismatched substrates have also resulted in the identifications of several new intriguing mechanisms [61–64]. One specific example is the diffusion of a 3D cluster containing 13 atoms, having an icosahedral shape, on a fcc(1 1 1) surface [64]. The diffusion proceeds rather fast, and the cluster mobility is close to that of an individual adatom. The diffusion rate is the highest when there is an optimal structural mismatch between the cluster and the substrate, which induces a stress at the interface and lowers the diffusion barrier. The cluster is found to move as a whole by rolling (Fig. 6h) rather than by single adatom jumps (Fig. 6a). In the elementary step of the cluster rolling, an atom is receding from the surface while another one is approaching it, and two other atoms located at the icosahedron edge preserve their positions. Because of such a counterbalance effect, the rolling diffusion barrier was found to be by a factor of about 3 lower than the barrier against cluster translation.

Though far from inclusive, the results presented above do demonstrate a fascinating diversity of the diffusion mechanisms of 2D and 3D clusters which are important entities in crystal growth, thin film technologies, catalysis, etc.: from apparently "simple" successive jumps of individual atoms to collective movements resembling gliding, motion of reptiles and caterpillars, and rolling stones. Such a broad spectrum of observations, and the corresponding strong discordance in the literature about their interpretations, will continue to challenge the wisdom of the surface science community.

2.3. Surface diffusion and phase transitions: inseparable partners

Up to now we have considered the situations typical for low average coverages ($\theta < 0.1$, with θ defined as the ratio of the concentration of adsorbed particles to the concentration of substrate surface atoms), and with *attractive* lateral interactions between the diffusing species. In these regimes and under quasi-equilibrium conditions, the clusters and islands coexist with a dilute phase (2D gas or, more often, 2D-lattice gas), occupying a major part of the surface area.

Next we move on to the "medium and high" coverage regimes (from $\theta \geqslant 0.1$ up to 1 monolayer and above). Under such conditions, diffusion proceeds within 2D *condensed* phases which can further be ordered if the diffusion temperature is below the order–disorder transition temperature. Under repulsive interaction, ordered (but dilute) phases can also form even at coverages as low as ~ 0.1 of a monolayer [19,65]. These dilute phases are followed by a series of denser ones at higher coverages, up to a close-packed monolayer; then higher monolayers can be built up (see, e.g., Refs. [55,66,67] for a general review of phase transitions in overlayers).

One can represent the diffusion flux, J, of interacting particles in the form of the first Fick law

$$J = -D(\theta)\,\text{grad}\,\theta. \tag{1}$$

It is important to stress that in this case, contrary to diffusion of non-interacting particles, the coefficient D, termed *chemical* ("*collective*") *diffusivity*, depends upon the coverage. D is determined, on the one hand, by the frequency and the length of the jumps executed by the diffusing particles (*the kinetic factor*) and, on the other hand, by the coverage derivative of the chemical potential (*the thermodynamic factor*), which accounts for the lateral interaction [14,68–70]. Recall that the chemical potential is equal to the change in the free energy when a particle is added to a system. The dependence of the chemical potential on concentration, temperature and other external parameters is known to govern the overall picture of the phase transitions in the system (the *phase diagram*) [66,67]. Therefore, phase transitions and diffusion kinetics are closely interrelated.

To give an idea about the strength of the phase effects on surface diffusion, let us consider a specific example: diffusion of dysprosium on a Mo(1 1 2) surface, with the Dy atoms occupying one-half of the surface initially (Fig. 8a) [71]. Such a deposit can be set up with the aid of a mask that shields a half of the sample from the adsorbate flux coming from an evaporator. The adsorbate distribution over the surface can be recorded by any appropriate technique which is able to determine the coverage locally (the data depicted in Fig. 8a were obtained by local work function measurements). The sample was heated at a temperature for various times to cause the spreading of the initial deposit.

Fig. 8a shows that the shape of the Dy coverage profiles experimentally recorded in the diffusion zone differs radically from the symmetrical featureless shape of the profiles theoretically predicted for diffusion of non-interacting particles [14]. In the experimental profiles, one sees clear-cut segments having substantially different gradients of the Dy coverage. Intuition suggests that the flat θ terraces should correspond to θ intervals where

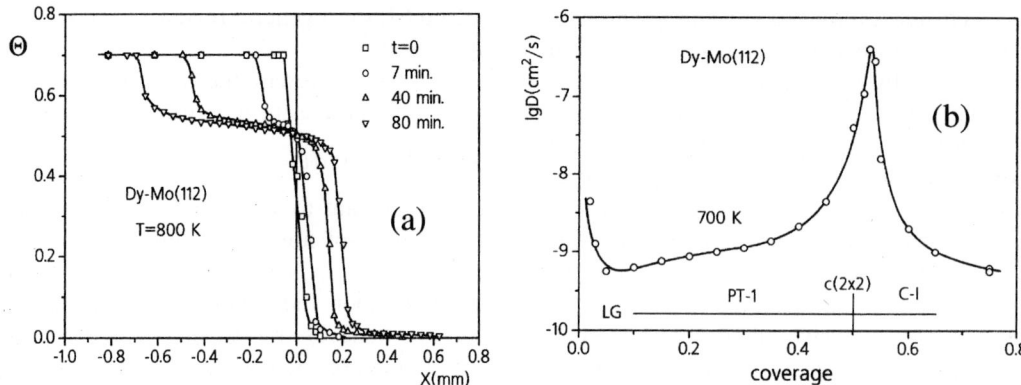

Fig. 8. (a) Time evolution of coverage profiles for Dy diffusion on Mo(1 1 2). This surface has a strongly anisotropic (channeled) atomic corrugation. The edge of the initial profile ($t = 0$) is perpendicular to the channels, so the profiles relate to Dy diffusion along the channels. (b) Diffusivity of Dy on Mo(1 1 2) as a function of coverage. A section of the phase diagram of Dy on Mo(1 1 2) is shown at the coverage axis. LG: lattice gas; PT-1: first-order phase transition to the c(2×2) structure; C-I: commensurate–incommensurate phase transition.

the coverage differences can be leveled off quickly, i.e. the diffusivity is high. In contrast, the steep sections of the θ profiles should correspond to θ intervals with low diffusivity. These qualitative considerations are confirmed in more rigorous theoretical analysis, and the Boltzmann–Matano method allows extraction of the coverage-dependent diffusivity from the experimental coverage profiles [14].

Fig. 8b shows the results of such an evaluation of the profiles shown in Fig. 8a. First, there is a very strong (by as much as three orders of magnitude!) and non-monotonic variation of the diffusivity with the coverage. Second, there is a clear correlation between the variation in D and the phase transitions in the overlayer. A section of the phase diagram of the Dy/Mo(1 1 2) system, documented in separate experiments by low-energy electron diffraction [72], is shown in Fig. 8b along the θ axis. As θ increases, there occurs a first-order transition from a lattice gas, observed at $\theta < 0.1$, to the c(2×2) commensurate phase corresponding to $\theta = 0.5$. An adsorbate phase is called commensurate if its lattice periods are multiples of the substrate periods (in other words, such an overlayer conforms to the substrate atomic corrugations). At $\theta > 0.5$, the c(2×2) phase is replaced by an incommensurate phase via a continuous compression of the overlayer along the atomic channels. In an incommensurate phase, the adatoms generally "ignore" the substrate atomic corrugations.

It is seen that D has *high* values at the low coverages and at $\theta \approx 0.53$. At low coverages, diffusion is most probably affected by jumps of rather mobile individual Dy adatoms and Dy oligomers. The coverage $\theta \approx 0.53$ corresponds to an early stage of the commensurate–incommensurate transition characteristic of the emergence of incommensurate walls separating commensurate c(2×2) domains. The incommensurate walls were shown to have properties of solitons (*solitary waves*, depicted schematically by the black balls in Fig. 6g) that can move through the commensurate phase conserving their configuration [57]. Alternatively, the soliton configuration is called dislocation, as more commonly accepted in description of the diffusion of islands (see Section 2.2). The high peak observed in the Dy diffusivity at $\theta \approx 0.53$ as well as the similar effects found for a number of other systems suggest that the solitons should have a high mobility. At coverages exceeding the range for the commensurate phase, each soliton contains an extra row of adatoms and for this reason its motion provides the mass transport across the commensurate phase [57]. Diffusion of the solitons, which represent quasi-linear units in a 2D overlayer, proceeds through their random walks executed by generation of kinks in the solitons. As soon as a moving soliton reaches the low-concentration boundary of the commensurate phase, it advances the boundary by one lattice period of the commensurate overlayer. New solitons are being generated at the high-concentration boundary of the commensurate phase through implantation of interstitial adatoms into this phase (the adatoms are strongly attracted to the substrate!). It should be emphasized again that soliton motion represents a clearly pronounced collective ("relay") process. Experimental data corroborating the existence of the soliton diffusion mechanism were also obtained in the investigations of the cross-correlation function of density fluctuations of potassium atoms adsorbed on a tungsten surface within two small areas which were separated by a mesoscopic ($\sim$10 nm) distance [73].

The extensive range of *low* diffusivity at medium coverages corresponds to the region of the first-order phase transition in the overlayer. This is just what should be expected in the presence of attractive forces causing the formation of a 2D condensate. It is easy to understand that lateral attraction should reduce the driving force of diffusion connected with the concentration gradient and therefore slow down diffusion in comparison with the case of non-interacting particles (in contrast, lateral repulsion increases this driving force and enhances the diffusion rate) [74].

The coverage dependence of the diffusivity, which originates from the lateral interaction of diffusing particles, results in a process of dynamical self-organization of the diffusion zone. Suppose the diffusion temperature is below the order–disorder transition temperature in the overlayer, so the diffusing species, under local equilibrium conditions, can form a series of ordered structures in the diffusion zone. Obviously, one cannot speak about a strict equilibrium between the phases in the diffusion zone, e.g. from the standpoint of the phase rule. This zone is an evolving "object" which emerges in the transition from an initial (non-equilibrium) to a final (equilibrium) particle distribution over the surface. The symmetry and lattice parameters of the 2D phases depend both upon the characteristics of the lateral interaction and upon the symmetry and parameters of the potential corrugation of the substrate. In turn, the surface diffusion parameters (the diffusivity, the activation energy, and the pre-exponential factor in the Arrhenius equation) in each point of the diffusion zone depend not only on the substrate properties, but also on the local structure of the diffusing overlayer itself and its interaction potentials. Any chosen area in the diffusion zone receives particles from the adjacent area with a higher chemical potential and at the same time feeds with particles the adjacent area with a lower chemical potential. The quasi-continuity of the diffusion transport is attained by self-fitting the concentration gradient to the throughput (diffusivity) of each particular phase emerging in the diffusion zone. The result of such a dynamical self-organization of the diffusion zone is that the larger the area occupied in the diffusion zone by an adsorbate phase corresponding to a particular coverage and diffusion temperature, the higher is the diffusivity peculiar to this phase. One can find an analogy between surface diffusion in a concentration gradient and some processes in the human society, e.g. in pioneering new territories. The pioneers settle new lands, take up there the most attractive places, self-organize, create an infrastructure (roads, communications etc.) and thus facilitate a further flux of newcomers.

Experimental investigations of phase effects in surface diffusion, an example of which has been considered above, are very laborious. This is the reason why so far the experimental work trudges behind the theory and simulations of such effects. On the other hand, theory and simulations are able to treat only some of the simplest model systems.

For example, there are rather numerous investigations, both analytical and simulational, that purposely treat surface diffusion kinetics in the region of second order (continuous) transitions within 2D systems with various symmetry and interaction potentials [69,72,75–77]. The coverage dependence of D in these transitions is found to be rather sensitive to the system parameters. Some combinations of interaction potentials give sharp extrema in D at particular coverages. Generally, the simulations are very useful because they allow the disentanglement of the contributions of various physical factors to surface diffusion kinetics and the elucidation of the diffusion mechanisms.

An interesting and important problem, both experimental and theoretical, is surface diffusion in the regions of the first-order phase transitions in the diffusing overlayer where it consists of two coexisting phases, i.e. has a patchy structure. Under equilibrium (i.e. in a "ripened" system), the two phases have the same chemical potential. However, the systems that we consider are non-equilibrated at the beginning of the diffusion process. The spatial distribution of the diffusate and its evolution in the coexistence region of the phase diagram will depend on an array of factors: θ, T, line tension at the phase boundary, distribution and nature of surface defects, etc. Thus the local minimization of the surface free energy will proceed with contributions from interaction energy of adatoms with each other and with defects, elastic strain energy induced by the overlayer in the substrate, as well as from the configurational entropy due to dispersion of the coexisting phases and/or to possible formation of a rough (in particular fractal) boundary between the phases. In general, since the overlayer becomes heterogeneous in the first-order transition region and each of the phases has its own diffusion characteristics, macroscopic experiments can only give an effective diffusivity which is an average of D's peculiar to coexisting phases [78]. Experimental observations (see e.g. Fig. 8b) show that mass transport in the first-order transition regions is comparatively slow [71]. This result is usually attributed to the necessity of detaching the particles from the condensed phase.

The above examples clearly show that surface diffusion of interacting particles and surface phase transitions are actually inseparable phenomena: mobility of the particles determines the kinetics of phase transitions whereas the properties of the emerging phases affect the adatom mobility. Undoubtedly, studies of the phase effects in surface diffusion will remain a fertile research area in the predictable future because they involve intricate questions from non-equilibrium phenomena, many-body problems and non-linear dynamics.

3. Problems and outlooks

In the previous section, we have surveyed a selection of results typical for the present state of the art in surface diffusion. Of course, the scope of problems in this field is much more diverse and broad. In the remaining space of this paper we will touch some of them both to illustrate what we want to learn more about surface diffusion processes and to discuss how we could control them to gain technological advantages.

3.1. Impact of defects: anomalous surface diffusion

A saying goes that studies are the learning of rules whereas experience is the learning of exceptions. Recent investigations have revealed a number of situations where diffusion regularities are *anomalous*, i.e. deviate from the classic law [79,80]. This law predicts the mean-square displacement of diffusing tracer species to be proportional to the diffusion time: $\langle r^2 \rangle \propto t$, as this is the case for instance for Brownian motion (see e.g. Ref. [81] for an introductory review). The possible reasons for the deviation can be geometrical or energetic disorder of the substrate as well as some peculiarities of atomic movements and diffusion mechanisms. Actually, all substrates prepared with the aid of present-day techniques contain only rather limited areas ($\leqslant 0.1$ to 1 µm in size) which are perfect, i.e. free of steps, various point defects and impurities, grain boundaries, etc. Therefore, strictly speaking, any surface diffusion process extending over larger distances occurs in a disordered medium. It is clear that diffusion in disordered media attracts particular interest from the practical standpoint. This problem has been addressed in much detail for a number of special cases, first of all for percolating media [80,82,83]. It has been predicted that there should exist a hierarchy of diffusion times in such media corresponding to different diffusion regularities. This hierarchy is connected with the existence of distinct length scales in the structure of percolating systems. The easy paths for diffusion are termed *links*, which intersect in *nodes*. Some links can be cut in *dead ends*. Multiconnected regions in the links are named *blobs*. The most important is the length of the links between adjacent nodes, which is termed the correlation length, ξ. The diffusion obeys the classic law $\langle r^2 \rangle \propto t$, on the one hand, at the very early stage when a particle hops in the close vicinity to its initial position and, on the other hand, at the late stage when $\langle r^2 \rangle$ is much larger than ξ. For the case when $\langle r^2 \rangle$ is smaller than ξ, the diffusion is predicted to be anomalous and obeys the dependence

$$\langle r^2 \rangle \propto t^{2/d_w}, \qquad (2)$$

where d_w is named the anomalous diffusion exponent. For a number of 2D models d_w as found to lie in the range 2.69–2.87 [83]. This is the so-called sublinear ("dispersive") anomalous diffusion. The slowing down of the diffusion in this regime is due to complex structure of the links and to the existence of the dead ends in the percolation network. The asymptotic classical regime is achieved when the structure is averaged over large ($\gg \xi$) distances [79,82,84]. On the other hand, superlinear (enhanced) anomalous diffusion with $d_w < 2$ is found in systems where diffusion coexists with convection flow, in layered and porous systems as well as in the case of turbulent media [81]. Generally, experimental evidence on anomalous surface diffusion is so far rather scarce (see, e.g., Ref. [79] for a review), and this important and interesting field still awaits future investigations.

It should be expected that linear defects such as steps or grain boundaries can play a more essential part in diffusion than point defects. Atomic steps have received much attention in recent years also because of their remarkable role in crystal growth. It was shown that steps can induce a strong diffusion anisotropy in many cases, the diffusion being considerably faster along the steps than across them [85–87]. The impact of various interaction potentials of diffusing particles with steps on the diffusion process was simulated by Uebing and Gomer [88]. Their results demonstrated a variety of possible diffusion scenarios on stepped surfaces. The impact of point defects can be also diverse, but generally it is less pronounced because the diffusing particles can simply go around such defects [84].

As we see, the presence of surface defects can strongly diversify the diffusion processes.

3.2. Surface mobility under high electric fields and high current densities

The problem of surface diffusion in high electric fields is not new. A considerable amount of related data was amassed with the aid of field emission and field ion microscopes during the previous decades [3,9]. However, the current elaboration of the methods of manipulating single surface atoms and molecules with the aid of the STM has inspired renewed interests in this topic. One well-known and beautiful example showing the capability of using the STM to induce adatom motion and achieve atomic-scale manipulation is the quantum corral shown in Fig. 9, assembled by Crommie et al. using 48 Fe atoms on Cu(1 1 1) [89]. It is foreseeable that with such levels of precision in atom manipulation and assembly, important new science will likely be gained regarding electron motion at the nanoscale, and such advances in basic research will in turn likely lead to breakthroughs in technological applications.

Because the external fields of 10^7 to 10^8 V/cm applied to adatoms are comparable in magnitude to those of the intra-atomic fields, their effect on the diffusion kinetics of the adatoms can be quite substantial. For instance, the activation energy of surface diffusion can change in the electric field by tens of percent (see e.g. Ref. [9] for a review). If the field is inhomogeneous, there occurs in addition a drift of surface atoms into or out of the region of the highest field strength. This effect underlies the method of sliding the adatoms and molecules along the surface with an STM tip [89,90].

The effects of high electric fields on surface diffusion kinetics are usually treated in terms of the polarizability and permanent dipole moment of the adsorption bond. If the field effect can be ascribed solely to the polarization of the adparticle by the external electric field, the particle should always be pulled into the region of the highest field, independently of the sign of the field. In the STM configuration, when a tip is located in a close proximity to the surface, an atom adsorbed immediately under the tip feels an additional potential well caused by the van der Waals forces [91]. However, if the adsorption bond has a substantial permanent dipole moment, the adparticles can be

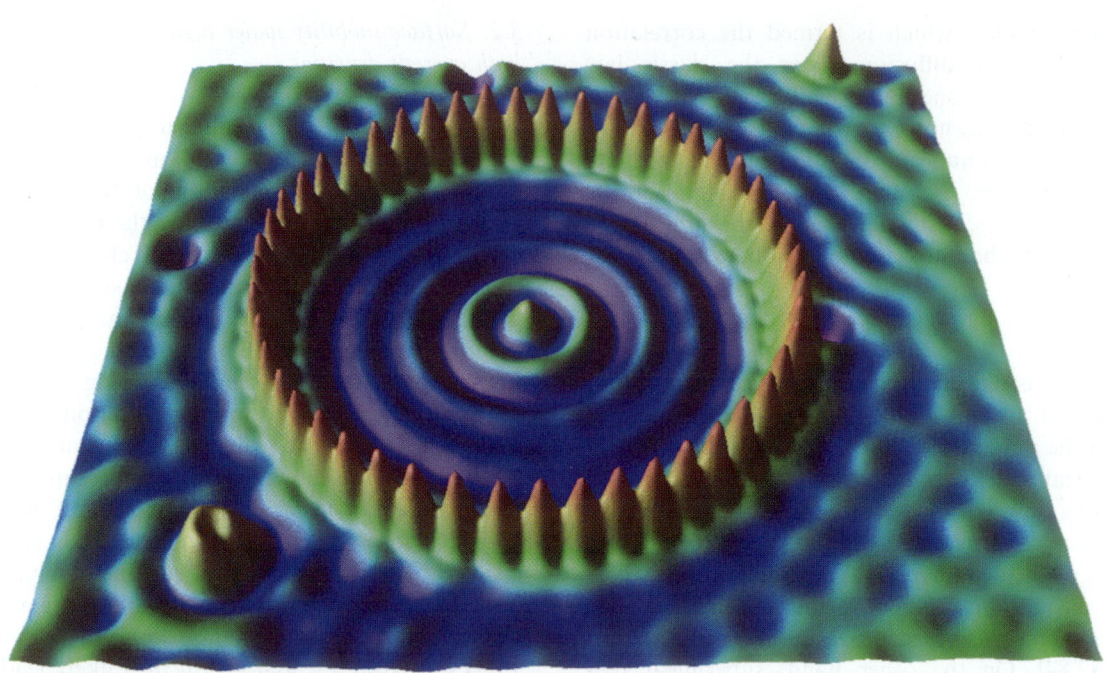

Fig. 9. An STM image of the quantum corral, consisting of 48 iron atoms on Cu(1 1 1). The image size is 200×200 nm^2 [89] (courtesy of IBM Almaden Research Center).

either pulled into the region of the highest field or pushed away from it depending on the relative orientation of the electric field and the dipole moment, determined by the chemical nature of the adsorbate and substrate [9].

It should be noted, however, that the polarizability of an adsorbed particle, which is used to evaluate the driving force of the field drift, can substantially differ from its polarizability in the free space. This is caused by the fact that the adsorbed particle is actually a part of the solid, and the external field sensed by it can be screened over a very short distance (shorter than the atom diameter in the case of metals). Furthermore, the effective polarizability of the surface particle should depend on its position: a protruding particle is more exposed to the external field than the particle built into a dense atomic plane. Similar considerations pertain also to the adatom dipole moment "felt" by the field. This problem calls for a closer theoretical inspection.

In STM experiments, one more powerful factor is added to the purely electrostatic effects considered above: a high-density current that passes through the adparticle. Even if the tunneling current is kept at a level of 1 nA, its density within an adatom amounts to $\sim 10^6$ A/cm^2. Electrons passing through the adparticle can excite its vibrations by the inelastic resonance tunneling [91,92]. The distribution of the excited vibrational levels was shown to correspond approximately to the Maxwell–Boltzmann statistics. The resulting vibrational temperature is determined by the competition between the excitation rate and the vibrational damping. The estimations made in Ref. [91] showed that a tunneling current of 100 nA can increase the vibrational temperature of Xe adatoms on Ni by a factor of 1.5 (from 4 to 6 K). For hydrogen adatoms, the vibrational damping is dramatically reduced because of the strong difference between the hydrogen atom mass and the mass of the atoms of typical substrates. This leads to a much stronger vibrational heating of H adatoms in inelastic electron tunneling: a tunneling current of ~ 1 nA through a single Si–H bond increases the vibrational temperature of H adatoms from 300 to about 600 K [91]. The excited vibrations, in turn, can promote diffusion, chemical reactions (both dissociative and associative), desorption and atom transfer between the tip and the surface [93]. In particular, the above results show that field emission measurements, under certain conditions, may induce some mobility of adatoms (especially of hydrogen adatoms) [92].

The regularities of surface mobility under extreme conditions discussed in this section deserve exploration in much more detail. Our present knowledge about the electric field effects in surface diffusion is, at best, at a semi-quantitative level.

3.3. Mobility under bombardment: migration induced by electronic transitions

Thermal excitation is not the only way to induce mobility of surface particles. A similar effect can be achieved by electron or photon irradiation. The phenomena of electron and photon stimulated *desorption* are well known and have been investigated in much detail for decades [94]. The potential barriers for surface diffusion are lower by about an order of magnitude than those for desorption. Correspondingly, it can be expected that electron and photon stimulated surface mobility should be observed in experiments more frequently than desorption. This, however, is not the case due to a number of experimental complications. The main one is the necessity to experiment at sufficiently low temperatures when the obscuring effect of thermally activated surface diffusion is effectively suppressed. Furthermore, a technique must be employed to sense the displacements of surface particles induced by irradiation.

Up to now, the most systematic studies of electron stimulated surface diffusion have been carried out using LEED [95]. The experiment is initiated with a completely ordered overlayer. For this state one records the intensity of the LEED superstructure beams, which measures the degree of the long-range order. Then the overlayer is irradiated by an electron beam. The irradiation induces some mobility of the adparticles by displacing some of them from their "regular" positions on the substrate, thereby producing a number of defects in the ordered overlayer. This diminishes the degree of long-range order, which is evaluated from the intensity of the superstructure electron beams measured after the irradiation. Such data can be used to determine the effective cross-section of the creation of the defects (actually of the adparticle displacement) due to electron bombardment [95].

The energy dependence of the effective cross-sections suggests that the electron-stimulated displacement of adparticles is initiated by the excitation of electron shells. When the valence or inner electron shells of the adatom become excited by the incident electron, the adatom starts to move to a new equilibrium position with respect to the substrate. In this motion it can acquire a kinetic energy sufficient to overcome the barrier separating the adjacent adsorption sites (or even the desorption barrier). It should be realized that due to complex 3D character of the potential energy barrier the kinetic energy can rather readily be transferred from one degree of freedom to another, e.g. from motion normal to the surface to motion along the surface. As the energy of incident electrons is increased, different possible "channels" are sequentially opened in this process, each of them having its own threshold energy. A possibility also exists that the mobility can be initiated not by the excitation of the adparticle itself, but by the excitation of an adjacent substrate atom. Then the energy accumulated in it can be transferred to the adparticle by an interatomic Auger transition [94].

It is intriguing that no energy threshold in the electron-stimulated mobility has been found for hydrogen overlayers on the W and Mo(1 1 0) surfaces [95]. It means that even electrons having zero kinetic energy with respect to the sample can displace the hydrogen adatoms while recombining with the surface. It should be kept in mind, however, that such an electron possesses a potential energy equal to the work function ($\sim$4–5 eV for most solids). In principle, this energy when transformed into the energy of adatom motion along the surface can be sufficient to induce a displacement of the adatom. The transformation can proceed via an intermediate negative ion state, or due to direct excitation of H adatom vibrations by the slow incident electron [92]. Recently, the negative ion resonance excitation was found to induce the mobility of benzoate molecules on Cu(1 1 0) [96]. This mechanism was proposed as a possibility to manipulate the molecules with STM.

The above observations, although still quite few in number, show that surface mobility induced by incident electrons or photons should be a rather widely occurring effect [97,98]. It can play a significant role as a starter of various surface modification processes. Because electronic transitions in adsorbate and substrate atoms can be initiated not only by electrons, but also by photons, it is appropriate to speak, in a more general way and by analogy with DIET (*desorption* induced by electronic transitions), about MIET: *migration* induced by electronic transitions. It is anticipated that MIET will receive more attention in the nearest future both due to its manifold physical scenarios and due to its practical potentials (such as promoting various processes where surface diffusion is the rate-limiting stage).

3.4. Diffusion from three-dimensional deposits. Spreading droplets

There is an interesting special case of surface diffusion in the concentration gradient which is observed when the coverage in the initial deposit exceeds one monolayer (it can be termed as surface diffusion from 3D objects or deposits). It is not a rare situation that the cohesion within the bulk adsorbate is weaker than the adhesion between the adsorbate and the substrate. Under such conditions, the adatom mobility within the first monolayer is usually lower than in the subsequent monolayers and diffusion proceeds via the mechanism in which the uppermost atoms slide down to the substrate and form a "slippery" bedding for the following adatoms. The result is that one observes the spreading of the *first* monolayer which occurs through diffusion in the *second and following* monolayers. The adatoms within the first monolayer itself are actually immobile. The mechanism resembles a carpet unrolling over the floor or a situation when wind scatters a pile of sand over a softened asphalt.

This mechanism was first suggested by Gomer to explain low-temperature diffusion of gas multilayers on metals in which the first monolayer is chemisorbed and the following monolayers are physisorbed [3,10]. The same behavior was found for a number of metal-on-metal systems in which the first monolayer is strongly adsorbed and less mobile than the next ones. However, for some systems of this type the first monolayer proves to be more mobile despite the fact that the heat of adsorption in it is higher than in the second monolayer [9]. This seems to be caused by the strong repulsive interaction within the close packed first monolayer.

In general, the surface diffusion process reveals very graphically that dynamical properties of the substance in a multilayer film depend essentially on the distance from the interface. For example, spreading microdroplets of non-volatile oily liquids were shown to assume stepped-pyramid shapes ("stratified droplets") [99]. Each successive step corresponds to a monomolecular layer, and its diffusion parameters are dependent on the layer number. Such observations suggest that some new interesting effects can be discovered yet at the front edge of spreading liquids.

3.5. This open, non-equilibrium, and non-linear world...

In most "ordinary" cases it goes usually without saying that linear diffusion equations are applicable to describe the processes under study. Thus the concentration (chemical potential) gradients are assumed to be small. It should be realized, however, that if the diffusion coefficient in Eq. (1) is coverage dependent, the equation generally becomes non-linear. This important feature, plus the non-equilibrium character of the (chemical) diffusion, plus an "open" configuration of the diffusion zone (there is a flux of particles into it and out of it) make this zone a possible arena for emergence of various dissipative structures. Such structures which appear in catalytic reactions in the presence of surface diffusion are considered in the article by Eiswirth and Ertl dedicated to this topic and published in this volume (see also Ref. [100]).

There seem to exist also a number of other potential situations in surface diffusion where Prigogine's concept of dissipative structures may be fruitful. Let us recall as an example the step bunching effect on stepped surfaces which is induced by surface electromigration [101,102]. The phenomenon, which attracts much practical interest as a method to control surface morphology, has all signs of a self-organization with formation of dissipative structures [103]. It is expected that non-linear effects will be one of the hot areas in surface diffusion.

3.6. From adatom diffusivity to control of nanostructures

As presented in Section 1, many fundamentally important areas and vital industrial technologies are based on surface diffusivity of matter and will naturally benefit from better understanding of the rich variety of surface diffusion phenomena. One particularly dynamic area over the last decade and expectedly in the foreseeable future is the use of knowledge on surface diffusion to fabricate desired stable nanostructures of technological significance. Examples of such low-dimensional artificial structures include ultra-thin films and magnetic multilayers, metallic quantum wires and carbon nanotubes, and semiconductor or magnetic quantum dots. The overall trend for miniaturization is driven by the requirement that the de Broglie wavelength of electrons in such structures is comparable to the dimensions of the systems even at room temperature in order to make such structures possible elements of quantum devices. Because nanometer-scale structures naturally demand atomic-scale precision, it is even more stringent to control the fabrication process of such structures. In turn, such degrees of precision control can be achieved only after a complete understanding of the various intricate atomic processes involved in the fabrication process.

An extensive coverage on recent progress and standing challenging issues regarding adatom mobility and the corresponding morphological evolution in growth of low-dimensional structures is beyond the scope of the present article. We refer the readers to several recent reviews [104,105], and the articles on epitaxial growth by Comsa and others in this volume, for detailed discussion of these problems. In the following, we highlight two important concepts as general rules of thumb in controlling the morphological evolution of epitaxial growth.

(a) *Step-edge barrier effect in multilayer growth.* When an atom lands on top of an island and attempts to descend to a lower layer from the step edge, it will typically encounter a higher potential energy barrier at the edge than the barrier for diffusion in the central region of the island [7,106]. Such a reflective wall surrounding the island edge will hinder the descent of the atop adatom, thus increasing the chance of nucleation and growth of a higher layer on top of the island because of the landing of subsequent atoms onto the same island. In 1990, Villain showed for the first time that if the step-edge barrier (also commonly known as the Ehrlich–Schwoebel barrier) is effective in hindering interlayer adatom transport, then an effective uphill current for mass transport exists, leading to unstable growth and the evolution of rough films [107]. Since then, many research activities have evolved around this concept, ranging from quantitative determination and possible modification of the step-edge barriers in various systems to scaling properties of the evolving films [108]. As a rule of thumb, any manipulation of growth kinetics or modification of interface thermodynamics that can weaken the step-edge barrier effect and thus enhance interlayer diffusion will naturally lead to the formation of smoother films.

(b) *Island-corner barrier effect in submonolayer growth.* In the earliest stages of film growth, only monolayer-high islands are formed on the substrate, and the roughness of such ultra-thin films is better measured by the compactness of the 2D islands. When an adatom reaching an island edge from the terrace attempts to reach a neighboring edge by crossing the island corner defined by the two edges, it typically will also need to overcome a higher potential energy barrier than that for diffusion along the straight edge. As in the case of step-edge barrier, the existence of a higher potential energy barrier for island corner crossing is due to reduced coordination of the adatom at the island corner site. It was first postulated [104], and later established on more firm grounds [109], that an island must develop a non-compact (fractal or dendritic) shape if adatoms reaching one edge of the island cannot effectively cross the island corner

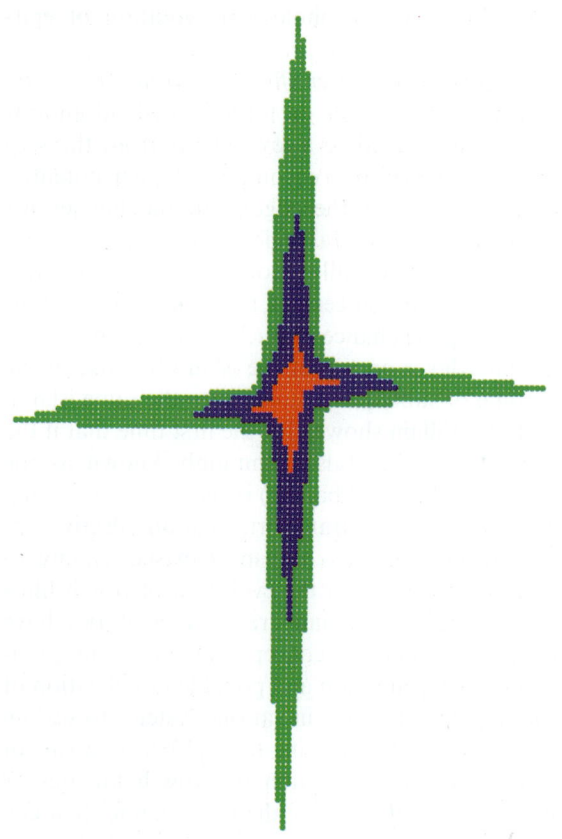

Fig. 10. A dendritic island formed on a square lattice with infinitely frequent edge diffusion but zero corner crossing of the adatoms, illustrating the island-corner barrier effect [104,109]. The number of atoms contained is 100, 400, and 1000 in the red, blue, and green zone, respectively. The fractal dimension of the island is about 1.45.

sites to reach the neighboring edges. This conclusion is true irrespective of how fast the adatom can diffuse along the island edge. One example illustrating this conclusion is shown in Fig. 10, showing the growth of a dendritic island on a square lattice with an infinitely frequent diffusion of adatoms along the straight edges of the island but with a zero rate for corner crossing [109]. As a rule of thumb, any enhancement of island corner crossing will always lead to the formation of more compact islands.

The size and density distributions as well as the shapes of the islands formed in the submonolayer growth regime may greatly influence the morphological evolution of the films in the multilayer growth regime. Therefore, the existence of the additional potential energy barrier at the island corners may also directly influence the evolution of the films in the multilayer growth regime, as demonstrated in two recent studies [110,111].

4. Conclusions

The body of results obtained up to now in surface diffusion studies is rich and spectacular. Does it mean that this scientific field has been worked out? Not at all.

Actually some surface diffusion mechanisms are understood only roughly, particularly in the case where the lateral interaction between the diffusing species is strong. There are such poorly explored phenomena as long jumps of the particles and situations when rather large islands seem to move as a whole (gliding). The effect of so-called transient mobility remains to be unraveled, too. We still understand very vaguely what happens, on the atomic or molecular level, at the front edge of a liquid drop or film spreading over a solid surface. There are also interesting unresolved questions concerning surface electromigration and surface diffusion stimulated by electron or photon irradiation. It is highly desirable to continue the investigation of quantum effects in surface diffusion of hydrogen and its isotopes [93,112], first of all using techniques with an atomic resolution. This is prompted not only by the interesting physics involved in the phenomenon, but also by the importance of understanding dynamics of hydrogen overlayers for the problems of catalysis.

The problem of interrelation between the surface diffusion kinetics and surface phase transitions is also very intricate and has barely been touched by investigators. For instance, an interesting case is surface diffusion in the region of the first-order phase transition when the diffusing overlayer becomes heterogeneous. Generally, the available data on the link between surface diffusion and phase transitions are still too few in number and fragmentary to allow extraction of reliable regularities. There remain rather broad gaps and controversies between theory and experiment in surface diffusion studies. In particular, one of the puzzles is the values of the pre-exponential factor in the Arrhenius expression of the diffusivity, which are found to span many orders of magnitude in different systems and under different conditions [113].

A highly important problem is surface diffusion on heterogeneous (disordered) substrates, in particular those containing defects of various types (steps, impurities, grain boundaries). In fact we mean here all *real* surfaces. One is faced with such complicated situations as percolation in the concentration gradient, fractal structures of the diffusion front, and anomalous diffusion characterized by a non-linear dependence of the mean-square diffusive displacement on time. In general, non-linearities in surface diffusion remain so far almost unexplored. The same can be said about diffusion in mixed (coadsorbed) overlayers, in spite of the importance of this case for catalysis, motion of surface reaction fronts, etc. Another highly important issue, in which surface physics comes into close contact with biology [114], is surface diffusion of large and complex-shaped organic molecules [115–117].

There is no doubt that all these problems, and many further ones that will emerge in the course of their investigation, will make surface diffusion an attractive research area which will give interesting and valuable physical results. It is important to note that the arsenal of experimental tools applicable to study surface diffusion has recently been supplemented with such powerful techniques as variable-temperature STM, the atom-tracking method using STM, scanning contact-potential microscopy, and new optical methods.

From the practical standpoint, it is evident that a deeper understanding of surface diffusion on an atomic level is basic to the most important nanofabrication processes of today's and even more so of tomorrow's age, as well as to further development of surface materials science in all of its important branches.

Acknowledgements

We gratefully thank Drs. Gary Kellogg, Brian Swartzentruber, John Wendelken, and Michael Crommie for providing us the FIM and STM images used in Figs. 4, 5, 7, and 9, respectively, and Drs. Riccardo Ferrando and John Wendelken for their critical readings of this paper. We also thank many of our collaborators, past and present, for their contributions. The preparation of this paper has been partially supported by the Ministry of Ukraine for Education and Science (grant # 2.4/776), by Oak Ridge National Laboratory, managed by UT-Battelle, LLC, for the U.S. Department of Energy under contract DE-AC05-00OR22725, and by the U.S. National Science Foundation (grant # DMR-0071893).

References

[1] M. Volmer, I. Estermann, Über den Mechanismus der Molekülabscheidung an Kristallen, Z. Phys. 7 (1921) 13.
[2] R.H. Good Jr., E.W. Müller, Field emission, in: S. Flügge (Ed.), Handbuch der Physik 21, Springer, Berlin, 1956, p. 176.
[3] R. Gomer, Field Emission and Field Ionization, Harvard University Press, Cambridge, 1961 (Chapter 4).
[4] J.M. Blakely, Surface diffusion, in: B. Chalmers (Ed.), Progress in Materials Science 10, Pergamon, Oxford, 1963, p. 395.
[5] Ya.Ye. Geguzin (Ed.), Poverkhnostnaya Diffuzıya i Rastekaniye (Surface Diffusion and Spreading), Nauka, Moscow, 1969, in Russian.
[6] G. Ehrlich, Diffusion in surface layers, CRC Crit. Rev. Solid State A. Mater. Sci. 10 (1982) 391.
[7] G. Ehrlich, F.G. Hudda, Atomic view of surface self-diffusion: tungsten on tungsten, J. Chem. Phys. 44 (1966) 1039.
[8] D.W. Bassett, P.R. Webber, Diffusion of single adatoms of platinum, iridium and gold on platinum surfaces, Surf. Sci. 70 (1978) 520.
[9] A.G. Naumovets, Y.S. Vedula, Surface diffusion of adsorbates, Surf. Sci. Rep. 4 (1985) 365.
[10] R. Gomer, Diffusion of adsorbates on metal surfaces, Rep. Progr. Phys. 53 (1990) 917.
[11] G. Ehrlich, Diffusion of individual adatoms, Surf. Sci. 299/300 (1994) 628.
[12] G. Kellogg, Field ion microscope studies of single-atom surface diffusion and cluster nucleation on metal surfaces, Surf. Sci. Rep. 21 (1994) 1.
[13] T.T. Tsong, Atomic, molecular and cluster dynamics on flat and stepped surfaces, Progr. Surf. Sci. 64 (2000) 199.
[14] J. Philibert, Atom Movements. Diffusion and Mass Transport in Solids, Les Editions de Physique, Les Ulis, 1991, p. 11.
[15] T.T. Tsong, Field-ion microscopic observations of indirect interaction between adatoms on metal surfaces, Phys. Rev. Lett. 31 (1973) 1207.

[16] K.H. Lau, W. Kohn, Indirect long-range oscillatory interaction between adsorbed atoms, Surf. Sci. 75 (1978) 68.

[17] O.M. Braun, V.K. Medvedev, Interactions between particles adsorbed on metal surfaces, Sov. Phys. Uspekhi 32 (1989) 328.

[18] T.L. Einstein, Interactions between adsorbate particles, in: W.N. Unertl (Ed.), Handbook of Surface Science, Elsevier, Amsterdam, 1996, p. 577.

[19] A.G. Naumovets, Two-dimensional phase transitions in alkali-metal adlayers, in: D.A. King, D.P. Woodruff (Eds.), The Chemical Physics of Solid Surfaces, Elsevier, Amsterdam, 1994, p. 163.

[20] E.W. Müller, T.T. Tsong, Field Ion Microscopy, Principles and Applications, Elsevier, New York, 1969.

[21] G. Binnig, H. Rohrer, C. Gerber, E. Weibel, Surface studies by scanning tunneling microscopy, Phys. Rev. Lett. 49 (1982) 57.

[22] G. Binnig, H. Fuchs, E. Stoll, Surface diffusion of oxygen atoms individually observed by STM, Surf. Sci. 169 (1985) L295.

[23] M.L. Lozano, M.C. Tringides, Surface-diffusion measurements from STM tunneling current fluctuations, Europhys. Lett. 30 (1995) 537.

[24] R. Gomer, Current fluctuations from small regions of adsorbate covered field emitters—method for determining diffusion coefficients on single-crystal planes, Surf. Sci. 38 (1973) 373.

[25] J.A. Venables, Rate equation approaches to thin-film nucleation kinetics, Phil. Mag. 27 (1973) 697.

[26] Y.W. Mo, J. Kleiner, M.B. Webb, M.G. Lagally, Activation energy for surface diffusion of Si on Si(001): a scanning-tunneling-microscopy study, Phys. Rev. Lett. 66 (1991) 1998.

[27] E. Ganz, S.K. Theiss, I.S. Hwang, J. Golovchenko, Direct measurement of diffusion by hot tunneling microscopy—activation energy, anisotropy, and long jumps, Phys. Rev. Lett. 68 (1992) 1567;
Y.W. Mo, Direct determination of surface diffusion by displacement distribution measurement with scanning tunneling microscopy, Phys. Rev. Lett. 71 (1993) 2923.

[28] B.S. Swartzentruber, Direct measurement of surface diffusion using atom-tracking scanning tunneling microscopy, Phys. Rev. Lett. 76 (1996) 459.

[29] P.J. Feibelman, Diffusion path for an Al adatom on Al(001), Phys. Rev. Lett. 65 (1990) 1990.

[30] G. Brocks, P.J. Kelly, R. Car, Binding and diffusion of a Si adatom on the Si(100) surface, Phys. Rev. Lett. 66 (1991) 1729.

[31] G.L. Kellogg, P.J. Feibelman, Surface self-diffusion on Pt(001) by an atomic exchange mechanism, Phys. Rev. Lett. 64 (1990) 3143.

[32] C.L. Chen, T.T. Tsong, Displacement distribution and atomic jump direction in diffusion of Ir atoms on the Ir(001), Phys. Rev. Lett. 64 (1990) 3147.

[33] J.D. Wrigley, G. Ehrlich, Surface diffusion by an atomic exchange mechanism, Phys. Rev. Lett. 44 (1980) 661.

[34] G.L. Kellogg, A.F. Wright, M.S. Daw, Surface diffusion and adatom-induced substrate relaxations of Pt, Pd, and Ni atoms on Pt(001), J. Vac. Sci. Technol. A 9 (1991) 1757.

[35] B.D. Yu, M. Scheffler, Physical origin of exchange diffusion on fcc (100) metal surfaces, Phys. Rev. B 56 (1997) R15569.

[36] P.J. Feibelman, R. Stumpf, Adsorption-induced lattice relaxation and diffusion by concerted substitution, Phys. Rev. B 59 (1999) 5892.

[37] S.C. Wang, G. Ehrlich, Structure stability and surface diffusion of clusters—Ir_x on Ir(111), Surf. Sci. 239 (1990) 301.

[38] G.L. Kellogg, Oscillatory behavior in the size dependence of cluster mobility on metal surfaces: Rh on Rh(100), Phys. Rev. Lett. 73 (1994) 1833.

[39] S.C. Wang, G. Ehrlich, Diffusion of large surface clusters: direct observations on Ir(111), Phys. Rev. Lett. 79 (1997) 4234.

[40] S.C. Wang, U. Kurpick, G. Ehrlich, Surface diffusion of compact and other clusters: Ir_x on Ir(111), Phys. Rev. Lett. 81 (1998) 4923.

[41] K. Kyuno, G. Ehrlich, Diffusion and dissociation of Pt clusters on Pt(111), Surf. Sci. 437 (1999) 29.

[42] Z.Y. Zhang, F. Wu, H.J.W. Zandvliet, B. Poelsema, H. Metiu, M.G. Lagally, Energetics and dynamics of Si ad-dimers on Si(001), Phys. Rev. Lett. 74 (1995) 3644.

[43] B. Borovsky, M. Krueger, E. Ganz, Diffusion of the silicon dimer on Si(001): new possibilities at 450 K, Phys. Rev. Lett. 78 (1997) 4229.

[44] J.M. Wen, S.L. Chang, J.W. Burnett, J.W. Evans, P.A. Thiel, Diffusion of large 2-dimensional Ag clusters on Ag(100), Phys. Rev. Lett. 73 (1994) 2591.

[45] K. Morgenstern, G. Rosenfeld, B. Poelsema, G. Comsa, Brownian motion of vacancy islands on Ag(111), Phys. Rev. Lett. 74 (1995) 2058.

[46] W.W. Pai, A.K. Swan, Z.Y. Zhang, J.F. Wendelken, Island diffusion and coarsening on metal (100) surfaces, Phys. Rev. Lett. 79 (1997) 3210.

[47] B.G. Briner, M. Doering, H.P. Rust, A.M. Bradshaw, Microscopic molecular diffusion enhanced by adsorbate interactions, Science 278 (1997) 257.

[48] T.R. Linderoth, S. Horch, E. Laegsgaard, I. Stensgaard, F. Besenbacher, Surface diffusion of Pt on Pt(110): Arrhenius behavior of long jumps, Phys. Rev. Lett. 78 (1997) 4978.

[49] T.R. Linderoth, S. Horch, L. Petersen, S. Helveg, E. Laegsgaard, I. Stensgaard, F. Besenbacher, Novel mechanism for diffusion of one-dimensional clusters: Pt/Pt(110)-(1 × 2), Phys. Rev. Lett. 82 (1999) 1494.

[50] P.J. Bedrossian, Si binding and nucleation on Si(100), Phys. Rev. Lett. 74 (1995) 3648.

[51] Z.P. Shi, Z.Y. Zhang, A.K. Swan, J.F. Wendelken, Dimer shearing as a novel mechanism for cluster diffusion and dissociation on metal (100) surfaces, Phys. Rev. Lett. 76 (1996) 4927.

[52] V. Chirita, E.P. Munger, J.E. Greene, J.-E. Sundgren, Reptation: a mechanism for cluster migration on (1 1 1) face-centered-cubic metal surfaces, Surf. Sci. 436 (1999) L641.

[53] P.-G. de Gennes, Scaling Concepts in Polymer Physics, Cornell University Press, Ithaca, 1979 (Chapter 8).

[54] F. Montalenti, R. Ferrando, Leapfrog diffusion mechanism for one-dimensional chains on missing-row reconstructed surfaces, Phys. Rev. Lett. 82 (1999) 1498.

[55] R. Kern, G. Le Lay, J.J. Metois, Basic mechanisms in the early stages of epitaxy, in: E. Kaldis (Ed.), Current Topics in Materials Science, vol. 3, North-Holland, Amsterdam, 1979, p. 131.

[56] I. Markov, S. Stoyanov, Mechanisms of epitaxial growth, Contemp. Phys. 28 (1987) 267.

[57] I.F. Lyuksyutov, A.G. Naumovets, V. Pokrovsky, Two-Dimensional Crystals, Academic Press, Boston, 1992.

[58] J.C. Hamilton, Magic size effects for heteroepitaxial island diffusion, Phys. Rev. Lett. 77 (1996) 885.

[59] S.V. Khare, N.C. Bartelt, T.L. Einstein, Diffusion of monolayer adatom and vacancy clusters—Langevin analysis and Monte Carlo simulations of their Brownian motion, Phys. Rev. Lett. 75 (1995) 2148.

[60] J.B. Hannon, C. Klunker, M. Giesen, H. Ibach, N.C. Bartelt, J.C. Hamilton, Surface self-diffusion by vacancy motion: island ripening on Cu(0 0 1), Phys. Rev. Lett. 79 (1997) 2506.

[61] P. Deltour, J.L. Barrat, P. Jensen, Fast diffusion of a Lennard-Jones cluster on a crystalline surface, Phys. Rev. Lett. 78 (1997) 4597.

[62] P. Jensen, Growth of nanostructures by cluster deposition: experiments and simple models, Rev. Mod. Phys. 71 (1999) 1695.

[63] W.D. Luedtke, U. Landman, Slip diffusion and Levy flights of an adsorbed gold nanocluster, Phys. Rev. Lett. 83 (1999) 1702.

[64] D.Y. Sun, X.G. Gong, Cluster on the fcc(1 1 1) surface: structure, stability and diffusion, Surf. Sci. 445 (2000) 41.

[65] R.D. Diehl, R. McGrath, Structural studies of alkali metal adsorption and coadsorption on metal surfaces, Surf. Sci. Rep. 23 (1996) 43.

[66] B.N.J. Persson, Ordered structures and phase transitions in adsorbed layers, Surf. Sci. Rep. 15 (1992) 1.

[67] L.D. Roelofs, Phase transitions and kinetics of ordering, in: W.N. Unertl (Ed.), Handbook of Surface Science, vol. 1, Elsevier, Amsterdam, 1996, p. 713.

[68] D.A. Reed, G. Ehrlich, Surface diffusion, atomic jump rates and thermodynamics, Surf. Sci. 102 (1981) 588.

[69] A.V. Myshlyavtsev, A.A. Stepanov, C. Uebing, V.P. Zhdanov, Surface diffusion and continuous phase transitions in adsorbed overlayers, Phys. Rev. B 52 (1995) 5977.

[70] A. Danani, R. Ferrando, E. Scalas, M. Torri, Lattice-gas theory of collective diffusion in adsorbed layers, Int. J. Mod. Phys. B 11 (1997) 2217.

[71] A.T. Loburets, A.G. Naumovets, Yu.S. Vedula, Diffusion of dysprosium on the (1 1 2) surface of molybdenum, Surf. Sci. 399 (1998) 297.

[72] F.M. Gonchar, V.K. Medvedev, T.P. Smereka, G.V. Babkin, Adsorption of dysprosium on the (1 1 2) face of a molybdenum crystal, Sov. Phys. Sol. State 32 (1990) 1092.

[73] Ch. Kleint, Evidence for soliton dynamics in alkali adlayers on transition metals, Solid State Phenom. 12 (1990) 1.

[74] M. Bowker, D.A. King, Adsorbate diffusion on single crystal surfaces, I and II, Surf. Sci. 71 (1978) 583; 72 (1978) 208.

[75] C. Uebing, V.P. Zhdanov, Surface diffusion near the points corresponding to continuous phase transitions, J. Chem. Phys. 109 (1998) 3197.

[76] A. Danani, R. Ferrando, E. Scalas, M. Torri, Collective surface diffusion on triangular and square interacting lattice gases, Surf. Sci. 409 (1998) 117.

[77] A.T. Loburets, A.G. Naumovets, Yu.S. Vedula, Surface diffusion and phase transitions in atomic overlayers, in: M.C. Tringides (Ed.), Surface Diffusion: Atomistic and Collective Processes, Plenum, New York, 1997, p. 509.

[78] T. Ala-Nissila, S.C. Ying, Theory of classical surface diffusion, Progr. Surf. Sci. 39 (1992) 227.

[79] R. Kutner, A. Pekalski, K. Sznajd-Weron (Eds.), Anomalous Diffusion. From Basics to Applications, Springer, Berlin, 1999.

[80] W. Steele, Computer simulation of surface diffusion in adsorbed layers, in: W. Rudzinski, W.A. Steele, G. Zgrablich (Eds.), Equilbria and Dynamics of Gas Adsorption on Heterogeneous Solid Surfaces, Elsevier, Amsterdam, 1997, p. 451.

[81] J. Klafter, M.F. Schlesinger, G. Zumofen, Beyond Brownian motion, Phys. Today 49 (1996) 33;
R. Metzler, J. Klafter, The random walk's guide to anomalous diffusion: a fractional dynamics approach, Phys. Rep. 339 (2000) 1.

[82] J.-F. Gouyet, Diffusion in heterogeneos materials, in: A.L. Laskar, et al. (Eds.), Diffusion in Materials, Kluwer, Dordrecht, 1990, p. 155.

[83] T. Nakayama, K. Yakubo, R. Orbach, Dynamical properties of fractal networks: scaling, numerical calculations, and physical realizations, Rev. Mod. Phys. 66 (1994) 381.

[84] E. Arapaki, P. Argyrakis, I. Avramov, A. Milchev, Anomalous diffusion in disordered lattices: effect of bias, in: R. Kutner, A. Pekalski, K. Sznajd-Weron (Eds.), Anomalous Diffusion. From Basics to Applications, Springer, Berlin, 1999, p. 83.

[85] H. Wagner, Transport of adsorbed species: correlations with concentration and step structure, in: V.T. Binh (Ed.), Surface Mobilities on Solid Materials, Plenum, New York, 1983, p. 161.

[86] M. Snabl, M. Ondrejcek, V. Chab, Z. Chvoj, W. Stenzel, H. Conrad, A.M. Bradshaw, Surface diffusion of K on Pd{1 1 1}: coverage dependence of the diffusion coefficient

[87] J.W. Ma, X.D. Xiao, M.M.T. Loy, Experimental study of surface diffusion rate enhancement along steps: CO on Pt(1 1 1), Surf. Sci. 436 (1999) L661.

[88] C. Uebing, R. Gomer, Monte Carlo study of diffusion on stepped surfaces, I, II and III, Surf. Sci. 306 (1994) 419, 427; 317 (1995) 165.

[89] M.F. Crommie, C.P. Lutz, D.M. Eigler, Confinement of electrons to quantum corrals on a metal surface, Science 262 (1993) 218.

[90] J.A. Stroscio, D.M. Eigler, Atomic and molecular manipulation with the scanning tunneling microscope, Science 254 (1991) 1319.

[91] R.E. Walkup, D.M. Newns, Ph. Avouris, Vibrational heating and atom transfer with the STM, J. Electron. Spectrosc. Rel. Phenom. 64/65 (1993) 523.

[92] O.M. Braun, E.A. Pashitskii, Vibrational excitation and surface diffusion of hydrogen atoms on W, Phys. Chem. Mech. Surf. 3 (1985) 1989.

[93] X.D. Zhu, A. Lee, A. Wong, U. Linke, Surface diffusion of hydrogen on Ni(1 0 0): an experimental observation of quantum tunneling diffusion, Phys. Rev. Lett. 68 (1992) 1862.

[94] R.D. Ramsier, J.T. Yates Jr., Electron-stimulated desorption: principles and applications, Surf. Sci. Rep. 12 (1991) 243.

[95] A.G. Fedorus, E.V. Klimenko, A.G. Naumovets, E.M. Zasimovich, I.N. Zasimovich, Electron-stimulated mobility of adsorbed particles, Nucl. Instr. Meth. B 101 (1995) 207.

[96] B.G. Frederick, Q. Chen, F.M. Leibsle, S.S. Dhesi, N.V. Richardson, Electron-stimulated disordering in c(8 × 2) benzoate/Cu(1 1 0): a combined STM, LEED and HREELS study, Surf. Sci. 394 (1997) 26.

[97] H.J. Ernst, F. Charra, L. Douillard, Interband electronic excitation-assisted atomic-scale restructuring of metal surfaces by nanosecond pulsed laser light, Science 279 (1998) 679.

[98] K. Nakayama, J.H. Weaver, Electron-stimulated modification of Si surfaces, Phys. Rev. Lett. 82 (1999) 980.

[99] N. Fraysse, M.P. Valignat, A.M. Cazabat, F. Heslot, P. Levinson, The spreading of layered microdroplets, J. Colloid Interf. Sci. 158 (1993) 23.

[100] R. Imbihl, G. Ertl, Oscillatory kinetics in heterogeneous catalysis, Chem. Rev. 95 (1995) 697.

[101] H. Yasunaga, A. Natori, Electromigration on semiconductor surfaces, Surf. Sci. Rep. 15 (1992) 205.

[102] Y.N. Yang, E.S. Fu, E.D. Williams, An STM study of current-induced step bunching on Si(1 1 1), Surf. Sci. 356 (1996) 101.

[103] H. Emmerich, C. Misbah, K. Kassner, T. Ihle, Vicinal surfaces: growth structures close to instability and far beyond, J. Phys. Condens. Matter 11 (1999) 9985.

[104] Z.Y. Zhang, M.G. Lagally, Atomistic processes in the early stages of thin-film growth, Science 276 (1997) 377.

[105] H. Brune, Microscopic view of epitaxial metal growth: nucleation and aggregation, Surf. Sci. Rep. 31 (1998) 121.

[106] R.L. Schwoebel, E.J. Shipsey, J. Appl. Phys. 37 (1966) 3682.

[107] J. Villain, Continuum models of crystal growth from atomic beams with and without desorption, J. Phys. I 1 (1991) 19.

[108] Z.Y. Zhang, M.G. Lagally (Eds.), Morphological Organization in Epitaxial Growth and Removal, Series on Directions in Condensed Matter Physics, vol. 14, World Scientific, Singapore, 1998.

[109] J.X. Zhong, T.J. Zhang, Z.Y. Zhang, M.G. Lagally, Island-corner barrier effect in two-dimensional pattern formation at surfaces, Phys. Rev. B 63 (2001) 3403.

[110] O. Pierre-Louis, M.R. D'Orsogna, T.L. Einstein, Edge diffusion during growth: the kink Ehrlich–Schwoebel effect and resulting instabilities, Phys. Rev. Lett. 82 (1999) 3661.

[111] M.V.R. Murty, B.H. Cooper, Instability in molecular beam epitaxy due to fast edge diffusion and corner diffusion barriers, Phys. Rev. Lett. 83 (1999) 352.

[112] R. DiFoggio, R. Gomer, Diffusion of hydrogen and deuterium on the (1 1 0) plane of tungsten, Phys. Rev. B 25 (1982) 3490.

[113] X.R. Wang, X.D. Xiao, Z.Y. Zhang, Apparent anomalous prefactor enhancement for surface diffusion due to surface defects (to be published).

[114] B. Kasemo, Biological surface science, Current Opin. Sol. State Mater. Sci. 3 (1998) 451.

[115] J. Weckesser, J.V. Barth, K. Kern, Direct observation of surface diffusion of large organic molecules at metal surfaces: PVBA on Pd(1 1 0), J. Chem Phys. 110 (1999) 5351.

[116] R.D. Tilton, Mobility of biomolecules at interfaces, in: M. Malmsten (Ed.), Biopolymers at Interfaces, Marcel Dekker, New York, 1998, p. 363.

[117] V.P. Zhdanov, B. Kasemo, Monte Carlo simulation of diffusion of adsorbed proteins, Proteins Struct. Funct. Genetics 39 (2000) 76.

Epitaxy: the motion picture

Paul Finnie [a,*], Yoshikazu Homma [b,1]

[a] *Institute for Microstructural Sciences, National Research Council of Canada, Building M-50, Montreal Road, Ottawa, Ont., Canada K1A OR6*
[b] *NTT Basic Research Laboratories, 3-1 Morinosato-Wakamiya, Atsugi-Shi, Kanagawa 243-0198, Japan*

Received 5 July 2000; accepted for publication 7 December 2000

Abstract

The engineering of many modern electronic devices demands control over a crystal down to the thickness of a single layer of atoms—and future demands will be even more challenging. Such control is achieved by the method of crystal growth known as epitaxy, and that makes this method the subject of intense study. More than that, recent advances are revolutionizing our knowledge of how surfaces grow. In fact, growing surfaces show a beautifully rich variety of phenomena, many of which are only now beginning to be uncovered. In the past few years many surface imaging techniques have been used to give us a close look at how crystals grow—*while they are growing*. The purpose of this article will be to illustrate some of the ways real surfaces grow and change as revealed by some of the latest in situ microscopic imaging technologies.

It is often said that crystal growth is more of an art than a science. Here we will show that it is emphatically both. Crown Copyright © 2001 Published by Elsevier Science B.V. All rights reserved.

Keywords: Epitaxy; Growth; Molecular beam epitaxy; Electron microscopy; Scanning tunneling microscopy

1. Overview

1.1. What is epitaxy?

Epitaxy is a kind of crystal growth. Specifically, it is the ordered growth of one crystalline layer upon a pre-existing crystalline surface [1]. There are many kinds of epitaxy, the main difference being the supply of source atoms: the source could be a molecular beam, gas, liquid or even an amorphous solid layer. Distinguishing epitaxy from other forms of crystal growth is the substrate, which is a single crystal, usually in the form of a thin, smooth, finally polished wafer. A very high-quality thin layer can often be grown.

This article will deal mainly with molecular beam epitaxy (MBE), in which the source material comes from a unidirectional beam, and the ambient pressure is very low [1,2]. In an industrial context, MBE is only one of several common epitaxial technologies [2–4]. However, from the surface science perspective it is one of the most important. This is because the high vacuum

---

☆ Movies available on the Elsevier Website: http://www.elsevier.com/inca/homepage/sak/susc/movies/.
* Corresponding author. Fax: +613-990-0202.
E-mail addresses: paul.finnie@nrc.ca (P. Finnie), homma@will.brl.ntt.co.jp (Y. Homma).
[1] Fax: +81-46-240-4718.

conditions enable the use of many well-established surface analysis techniques during the growth process.

On a microscopic scale, the epitaxial growth process is truly remarkable. In MBE, source atoms arrive randomly on the surface. At first these atoms are only loosely bound to the surface—although they cannot escape, they remain mobile and migrate. Almost miraculously, these atoms are assembled into a nearly perfect three-dimensional lattice. This process is driven by the minimization of energy that follows when the atoms increase their atomic coordination and become bound to the crystal. The assembly of individual atoms into the crystalline lattice takes place almost entirely on the surface. In the past few years microscopy has revealed the intricacy and the beauty of this process.

1.2. The value of epitaxy

In epitaxy, process parameters such as substrate temperature, growth rate, and surface preparation can be controlled very precisely. In situ analysis techniques can be applied, particularly during growth under vacuum conditions. The combination of process control and process monitoring make the epitaxial technique ideal for fundamental scientific studies of surface growth. This combination explains why previously impossible complex, engineered structures of extremely high quality are routinely grown by epitaxy.

In terms of technology, the quality of the grown materials and versatility of the technique has enabled the fabrication of a wide variety of solid-state devices impossible to make by any other technique. These devices are revolutionizing optics and electronics from transistors to light bulbs, from lasers to sensors. They are valuable—amounting to markets measured in billions of dollars per year. This alone drives research and development. It is impossible to do justice to the engineering and economic value of epitaxy in this short article. Here we simply touch on a few specific highlights.

The most important application of epitaxy, in general, is the growth of *heterostructures*. A heterostructure is a crystal in which the chemical composition is changed while the crystalline lattice remains largely unchanged. Epitaxy enables the fabrication of abrupt interfaces between layers of different compositions. In particular, it is often useful to embed thin layers of different materials in a larger crystal. Because different materials have different electronic energy levels, by changing the composition it is possible to sculpt the electronic structure. The freedom to shape the electronic structure makes for new and faster electronic devices, and in many material systems it opens the door to a myriad of optoelectronic applications, in which light is generated, detected or manipulated.

Epitaxial materials are used today in cell phones, where the design advantages provided by high-quality energy band engineered structures are used for high speed transistors [5]. Light emitting diodes (LEDs) are made epitaxially—in particular, solid state laser diodes. Laser diodes are used in laser printers, compact disk players, and digital videodisk players, as the sources for signals on fiber optic cables in telecommunications, as laser pointers and countless other applications [6]. Recent breakthroughs in the epitaxy of blue and blue–green LEDs, together with older red and yellow LEDs mean that display applications are emerging, from traffic signals to stadium-sized video screens [7]. The combination of colors can be used to make bright, energy efficient white light, and if the cost comes down it could mean that the light bulb may soon follow the typewriter and the phonograph into history [8]. As light detectors, epitaxial layers can be used to monitor pollution, measure temperatures, and monitor chemical processes. From CD players to traffic lights to the networks of the world wide web, epitaxial layers are everywhere.

1.3. Motivation for microscopy

It is natural, though not necessarily easy in practice, to use microscopy to study epitaxial growth. From the earliest experiments, optical and electron microscopes have been used to characterize grown structures. Gross morphological defects in the surface can be seen using an optical microscope and transmission electron microscopes (TEMs) are often able to discern interfaces between substrates and grown layers, or between layers of different compositions. However, to really see what is going on, it is much better to obtain images while the sample is still in the growth chamber—if possible, while growing.

Microscopic imaging techniques are needed now more than ever, as the size of grown structures decreases to the nanometer scale and below. It is essential to monitor length scales comparable to the size of the intended structures, and many techniques can resolve the nanometer scales important to present and future devices. Moreover, in situ techniques are particularly useful because they enable rapid or even instant feedback—this can amount to savings of hours or days spent on analysis for every sample that is grown.

From a purely academic perspective, in situ imaging technologies are valuable because they allow old theoretical models of crystal growth to be tested thoroughly, deconstructed where necessary and modified or even reformulated. Early models may have worked well, but with some exceptions, they only had to explain the properties of the crystal after growth, given some initial conditions. But all theories, including the older ones, make concrete predictions for the dynamics of growth. Now it is possible to observe structures second by second, minute by minute, and any reasonably complete theory must explain *everything*.

1.4. New imaging technologies

For the most part, it is not the imaging technologies themselves that are new, but rather it is the use of these imaging technologies in the context of growth that is new. While almost any analysis technique can be used to map a surface and generate an image, there are certain specific approaches that have proved especially useful. They will be briefly introduced below.

At least two broad classes of imaging technologies can be discerned. One class is the scanning probe microscopies (SPM). These can be traced back to the scanning tunneling microscope (STM) [9]. In STM, a very sharp point is raster scanned across a surface. The current that is able to traverse the gap between the tip and the sample is highly dependent on the size of this gap. The current can be measured at each point in the raster scan, or the tip can be raised or lowered to keep the current constant. Either way, a map of the surface is obtained—and these maps can have resolution that is better than even the size of a single atom. More generally, almost any local probe can be scanned across the surface like this to generate a map, and so there is almost an infinite variety of possible SPM.

The second broad class of imaging technologies is electron microscopy [10–12]. Instead of using photons as in conventional optical microscopes, electrons can be used for imaging. This overcomes the micron-scale resolution limits of optical microscopes due to the wavelength of light. The resolution limit in electron microscopy is, in principle, well below an Angstrom. In electron microscopes, electromagnetic lenses and mirrors are used rather than optical ones. Some techniques rely on beams of electrons, which, like the probes in SPM techniques, are raster scanned to generate an image. Others act much more like optical microscopes, generating the entire image simultaneously.

In classical surface science, electron diffraction has been used extensively to determine surface structure. A number of the imaging techniques in use now have evolved from earlier electron diffraction experiments.

1.5. Why now?

In this volume, there are many examples of how surface science has contributed to the development of electronic technology. Arguably, the main reason why imaging technologies are only now having an impact on the surface science of epitaxial growth is the converse. Computers recently became able to handle the amount of data that is required to store and manipulate images and movies. Somewhat earlier, videotape became widely available and it was possible to record, replay and analyze imaging experiments. Some imaging technologies are quite new (e.g. atomic force microscopy), while some are old (e.g. TEM). All can be used to make movie data now because the data processing equipment is relatively easy to use and available relatively cheaply in the form of consumer electronics.

The overall picture of surface growth is changing rapidly in large part because imaging techniques generate such a wealth of data. Moreover, the data tends to be in a form with which we are naturally familiar. For example, an STM generated image of a surface does have a quite different meaning than a conventional image derived from reflected light, but nonetheless it is more intuitive for us to interpret than a reciprocal space diffraction pattern of the same surface. In no way should this be construed as a criticism of the more traditional non-imaging surface science techniques. Indeed, some argue against becoming too dependent on imaging data [13]. Certainly, there are pitfalls which may lead to the misinterpretation of data, some of which will be touched upon in this article. In general, however, imaging technology is helping greatly to advance surface science.

2. Epitaxy: examples

2.1. Adatoms

An STM can provide an incredibly detailed look at epitaxial growth when restricted to the conditions in which it works best: stable, non-rotating substrates, very slow growth rates, and low temperatures. It should be cautioned that although STM-based studies show the initial phases of epitaxy, in some cases the conditions are very different from those used for conventional growth. Typically, in an industrial context, vastly higher fluxes (with correspondingly higher growth rates) and somewhat higher substrate temperatures are used. Fortunately, more and more the STM is used to study practically relevant growth conditions. Regardless, even when the conditions are chosen to obtain optimum image quality rather than optimum material quality, the STM remains an excellent tool for use in trying to understand the "earliest phase" of epitaxy.

In the earliest phase of MBE, atoms evaporated from a source strike the surface and, if it is cold enough, stay there. Atomic forces and the effects of thermal energy are roughly balanced such that the newly arrived atoms remain mobile on the surface. For a time, these atoms cannot escape the surface, but they can avoid being trapped permanently. They diffuse across the surface, but ultimately they are either absorbed into the crystal or else they desorb from it. Such unsettled atoms are called adatoms [14]. (Note, however, that outside the sub-field of epitaxial growth, many surface scientists reserve the word "adatom" to refer to very stable structures.)

A particularly simple example of adatom movement and consequent growth is shown in Fig. 1 [15]. This is a platinum surface, technically called Pt(1 1 0)-(1 × 2) to classify its structure. This surface is very simple, with atoms lining up to form troughs (dark) separated by raised rows of atoms (light). In ultrahigh vacuum conditions, after the preparation of a flat surface, additional platinum atoms were deposited, and then the sample was transferred to an STM chamber. The deposited atoms were unable to become permanently incorporated into the crystal. By heating the sample these adatoms become mobile and it is possible to observe how they move and become incorporated. Fig. 1 shows the diffusion of adatoms, which for this particular surface is essentially one dimensional.

A fundamental step in crystal growth can be seen several times in this movie. The single adatoms diffuse up and down the valleys. On occasion these solitary adatoms strike a neighboring adatom. The two react with one another and create an "island" of just two atoms (called a "dimer"). The island is much less mobile than the individual atoms were. This process will continue, with more and more adatoms joining the pair. The islands are not static—atoms can escape and the islands can move, albeit much more slowly than adatoms. Although this one-dimensional situation in which adatoms and islands are confined to one trough is a very special case, accumulation of adatoms at islands such as these is one of the fundamental processes by which many crystals grow.

Already, imaging has provided a great deal of insight. However, it is worth highlighting a pitfall. A fundamental problem of any imaging technique is that it takes a finite time to acquire the image.

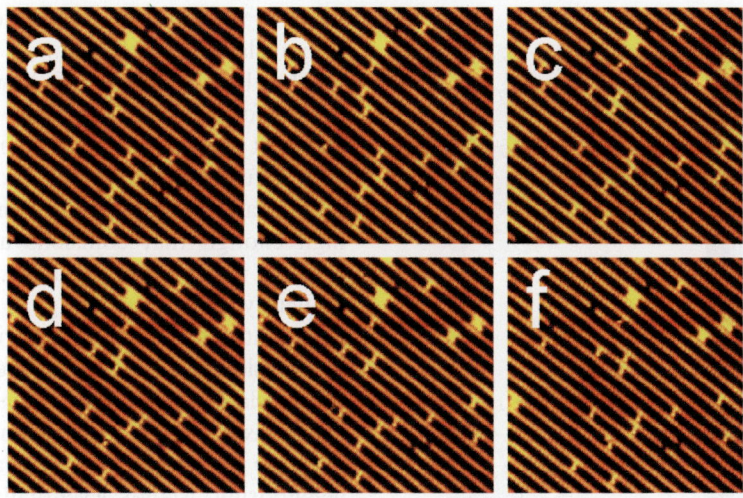

Fig. 1. The motion of adatoms. This STM movie shows platinum atoms diffusing through valleys on a platinum surface. Occasionally these adatoms collide and coalesce into larger islands. The figure shows a time series of still images taken from the movie. The frames are labeled a, b, c, d, e, f in sequence. An equal time interval separates each frame. (For movie, see: http://www.elsevier.com/inca/homepage/sak/susc/movies/, reproduced with permission from Flemming Besenbacher, University of Aarhus.)

Rather than having data that specifies the location of adatoms at all times, instead we have data that tells what is present at a given location at certain *discrete* times. The interval between those times is unknown. This is a well-known problem in many kinds of experiments: the sampling problem. It is especially severe in surface imaging. Atoms can move very quickly—hopping processes can occur in femtoseconds, while an STM image may take minutes to acquire. It is essential to keep this effect in mind!

Fortunately, even events that cannot be seen directly can be inferred with the help of statistical models. This very movie (Fig. 1) was used to show that adatoms sometimes hop one step at a time, but a fraction of the time they skip the nearest site and move immediately to a second site [15,16].

So far the focus has been on adatoms. Now let us look more closely at islands, which are the growing seeds of the new crystalline layer. As the islands grow larger it might be expected that they would become locked in place. However, it is found experimentally that these islands can move around—a lot [17]. This appears to be contrary to the islanding model, and would seem very energetically unfavored. Imaging technologies can also help to unveil the mechanism of the motion. It seems that there are always surprises in store when it becomes possible to actually see the surface!

In Fig. 2, selected frames are shown from a movie of the motion of a one-dimensional island, along with a schematic illustration of the process. This experiment, backed up by modeling [18] shows that it is not the entire collection of atoms in the island that are moving. First, an adatom disappears from the left-hand side of the island. The island brightens, and then an adatom attaches to the right-hand side. Rather than all the atoms constituting the island moving, what is happening is that occasionally an atom detaches from the edge of the island, and reverts to adatom status. It leapfrogs over the island, sticking to the other side. This might create the impression that the entire island has moved, and indeed its center of mass has, but this has only required the displacement of a single atom. Most of the atoms making up the islands are essentially immobilized, as expected, but this leapfrog effect shows that the island edges remain very much a dynamic part of the surface.

Leapfrogging is only one very special case illustrating the interaction of islands with adatoms. More generally, atoms may be ejected from

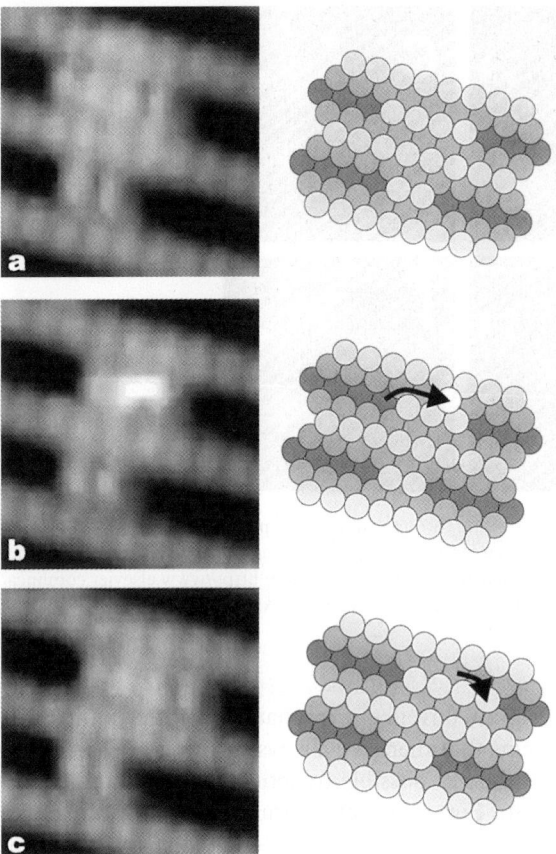

Fig. 2. Adatoms leapfrogging over islands. Frames from an STM movie (a, b, and c) show an island made up of four platinum atoms. An atom from the left-hand side of the island escapes, leapfrogging over the island (b), reattaching at the right side (c). The leapfrogging is accompanied by a brightening of the island. The leapfrogging effect creates the illusion that the entire island has moved. A schematic diagram illustrates the process (right). (Originally published in Ref. [17]. Used by permission.)

islands, as well as being incorporated into islands. They may detach and reattach, or attach and then detach, or they may even glide along the periphery of the island. Thus the center of mass of an island may be displaced even though most atoms in the island stay fixed.

2.2. Two-dimensional islands

The above effectively one-dimensional situation is somewhat unusual. More commonly, adatoms move in the plane of the surface, and two-dimensional islands often result.

The basic picture of how two-dimensional islands form and grow is easily understood. Atoms arrive at the surface and become adatoms, which then diffuse in a random walk type process across the surface. Wherever the density of adatoms is high enough that collisions are likely, adatoms coalesce into miniscule islands. These islands subsequently grow by acting as seeds and adding new adatoms at their peripheries. Since there is a line tension associated with the stepped perimeter of the island—neglecting other effects for the time being—island expansion is energetically favored over the nucleation of new islands. The edges of neighboring islands eventually collide and the islands themselves coalesce into a complete monolayer. The process repeats, "layer-by-layer". This is the first of the three main classes of growth modes, which has been labelled "Frank–van der Merwe" or layer-by-layer growth [19,20]. This growth mode is very common when the grown layer has the same composition as the substrate.

In modeling the initial stages of layer-by-layer growth, a useful simplification in many cases is to consider the island edge to be a smooth, rounded curve. Individual adatoms can be ignored, and instead a locally averaged adatom density describes the surface at any given point. This density satisfies a diffusion equation, with boundary conditions at the island edges. The island grows smoothly and continuously as it accretes adatoms from its surroundings. In stark contrast to this often used, simplified model, STM experiments of silicon epitaxy on the Si(1 1 1) surface revealed that two-dimensional islands do form and grow, but that they take a triangular shape—and the island sizes are quantized! [21].

This quantization may seem mysterious at first, but movies help to determine the physical origin. In Fig. 3, islands grow quickly by adding a row of material on to an existing straight edge. An island grows into a triangle bounded by slow growth planes. The triangular island symmetry originates in the triangular crystalline symmetry of the underlying surface. The quantized sizes are simple but surprising consequences of the discrete spacing between slow growth planes.

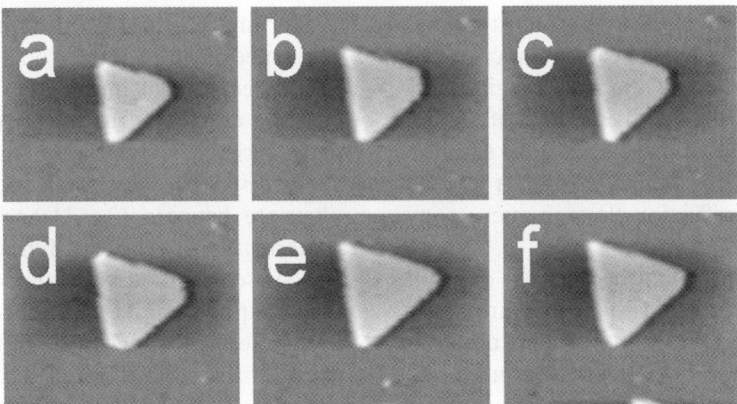

Fig. 3. The growth of a "magic" island. This STM movie shows a two-dimensional silicon island growing rapidly, row-by-row, pausing at magic or quantized island sizes. The figure shows a time series of still images taken from the movie. The frames are labeled a, b, c, d, e, f in sequence. An equal time interval separates each frame. This and other growth movies are available on the world wide web at http://www.fz-juelich.de/video/voigtlaender/. (The movie can also be found on the Elsevier site: http://www.elsevier.com/inca/homepage/sak/susc/movies/, reproduced with permission from Bert Voigtländer, Forschungszentrum Jülich.)

Interestingly, the rows separating successive quantized triangles are not a single atom wide. To explain this it is necessary to understand the Si(1 1 1) surface in detail. At temperatures below about 850 °C, the Si(1 1 1) surface is known to form the so-called (7 × 7) structure, which is made up of adjacent equilateral triangles of "faulted" and "unfaulted" types [22]. These triangles have the same surface area, but it takes more energy to transform a "faulted" triangle into the bulk crystal structure than it does to transform the "unfaulted" triangle.

The quantized triangular islands are bounded by faulted triangles only. For the island to grow, the (7 × 7) structure on the substrate surface must be destroyed, and this takes more energy for the faulted triangles. However, once a single faulted triangle has been destroyed, growth on adjacent areas can proceed very rapidly. Growth proceeds triangle-by-triangle, and this is why rows are more than a single atom wide.

This effect of quantized islands was, for the most part, unexpected, but once discovered, imaging techniques enabled a straightforward explanation. Again, the real surface growth process has turned out to be much more intricate than anticipated on the basis of earlier models. Dynamical imaging plays a powerful role in unveiling the physical process.

2.3. Moving islands

Like one-dimensional islands, two-dimensional islands interact dynamically with their surroundings. Atoms escape from step edges, hopping up and diffusing over the islands or hopping back into the adatom "sea" surrounding the islands. Surprisingly, on the Si(1 0 0) surface,[2] islands can be driven in any direction with the controlled application of an electric bias [23]. This will be explained in terms of the interaction of adatoms with electric fields and the influence of the surface structure. First, the imaging technique used to obtain this surprising result is briefly introduced.

Very generally, the development of surface science has relied heavily on diffraction. The periodicity of the two-dimensional lattice on the surface causes a diffraction pattern in a beam of electrons. This pattern can be used to deduce the surface structure. Measuring the intensity of a particular diffraction spot as the beam is moved across a surface, it is possible to construct a map of the surface structure. One such method, in which the

---

[2] People who grow samples epitaxially for industry often use a slightly different naming convention than many surface scientists. While surface scientists refer to the (0 0 1) surface, growers tend to refer to the (1 0 0) surface.

diffraction of reflected electrons is monitored, is called scanning reflection electron microscopy (SREM). Even better for dynamical imaging purposes, by using a more diffuse beam, the diffraction from a large area can be monitored simultaneously. This is the principle behind a related technique known as reflection electron microscopy (REM).

An important point is that because the electron beam grazes the surface, REM images are squashed along the vertical direction, so, for example, a circle appears like a very flattened, oblong oval. This geometrical distortion may be a drawback when it comes to interpreting an image, but it means that instruments involved in imaging are well separated from MBE sources and monitoring equipment, which are typically located as close to the substrate normal as possible. This is a very real practical consideration for MBE since the entire area in front of the sample is typically crowded, with today's machines literally bristling with deposition and monitoring equipment. The "foreshortening" effect should be kept in mind when viewing any REM image.

Turning the discussion from the apparatus to the experimental sample, on the Si(1 0 0) surface, atoms reconstruct into a (2×1) structure, giving rise to a (2×1) diffraction pattern. However, with each successive monolayer the surface structure rotates by 90°. As a result, successive monolayers alternate between (2×1) and (1×2) diffraction patterns. REM relies on diffraction to produce an image so that, under appropriate conditions, (2×1) appears dark, while (1×2) appears light. It is therefore possible to see monolayer islands because of the change in reflected beam intensity.

In Fig. 4, REM has been used to show the effect of electric bias on the motion of two-dimensional islands. In this experiment, the flux exactly balances desorption, so that the island is at a kind of dynamical equilibrium with its surroundings. In the figure, (1×2) islands (white) move to the left relative to landmarks on the surface (arrows), towards the positive terminal, while (2×1) islands (black) move opposite, towards the negative terminal.

The reasons are subtle, and the explanation is still perhaps somewhat controversial. First, ad-

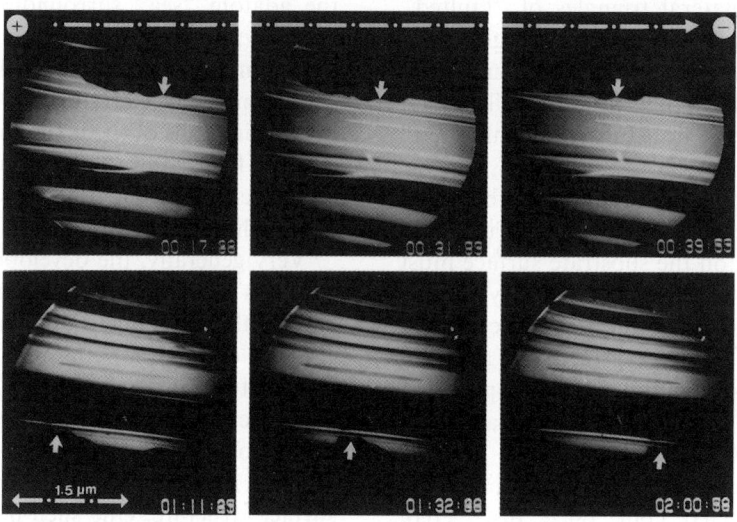

Fig. 4. Driven islands. This series of images from REM movies shows how an island moves when an electrical bias is applied to the substrate. The left-hand side is positively biased (+), and the right-hand side is negatively biased (−), so current flows from left to right (top arrow). The top row of three frames shows the motion of a white (1×2) island with respect to a fixed surface imperfection (down arrows). The bottom row of three frames shows the motion of a dark (2×1) island with respect to another fixed surface imperfection (up arrows). The time scale at the bottom of each image is in minutes, seconds and hundredths of seconds. The marker at bottom left represents 1.5 μm in the horizontal direction, however distances in the vertical direction are highly foreshortened. (Originally published in Ref. [23]. Used by permission.)

atoms are thought to have an effectively positive charge on Si(1 0 0), so they move preferentially toward the negative terminal (right). Second, the speed at which adatoms diffuse is very anisotropic on Si(1 0 0). They move rapidly left-to-right on the (1 × 2) surface (light), but slowly in the orthogonal direction.

The island motion can be understood considering the limiting case in which diffusion only happens at an appreciable rate left-to-right on the light (1 × 2) surface. The light (1 × 2) islands move to the right because atoms detach from the left-hand side, hop up and undergo biased diffusion to the right, over the island, to be incorporated at the right-hand side. On the other hand, the dark (2 × 1) islands move to the left because adatoms on the surrounding light (1 × 2) terrace are moving primarily left to right. Adatoms accumulate on the left-hand side of the island, while at the same time atoms escape easily from the right-hand side.

Remarkably, this experiment shows islands can be driven in any desired direction by the application of an appropriate electric field. The imaging technique has provided evidence that adatoms carry an effective charge, shown the importance of surface structure on surface dynamics, and highlighted the complex interactions between adatoms and islands.

2.4. Strained islanding

Up to this point the grown material has been the same as the substrate material. However, it is more useful to grow layers of differing compositions. Industrially, this is the main use of epitaxy. For example, thin layers of aluminum gallium arsenide (AlGaAs) in gallium arsenide (GaAs) are used to make lasers for CD players and printers [24]. Layers of indium gallium arsenide phosphide (InGaAsP) in indium phosphide (InP) are used for components in fiber optic networks [25], and layers of silicon–germanium in silicon enable transistors to be made on silicon chips which are fast enough for present day cell phone applications [26].

When the growing layer is different from the substrate there is almost always at least some strain. This is because atoms in the substrate crystal have a given characteristic interatomic spacing, while the atoms in the growth material (or epilayer) have another, different characteristic spacing. The epilayer would ordinarily crystallize with its own lattice constant, but it must also be matched to the substrate lattice constant at the interface. The difference in lattice constants is extremely small for some material combinations (e.g. AlAs on GaAs), and very large for others (e.g. InAs on GaAs). The competition between differing lattice constants gives rise to a rich variety of structures during growth.

For germanium grown on Si(1 0 0), if fewer than three atomic layers are grown, the germanium adjusts to the silicon lattice constant. The cost in strain energy of conforming to the pre-existing lattice structure is less than the energy cost of relaxing to the natural germanium lattice constant—which would necessarily involve the creation of defects with their own energy cost. For slightly more than three layers, the germanium is on the verge of relaxing to its own preferred lattice spacing—and this makes the growth interesting.

Above three layers, the strained Ge layer relaxes by spontaneously forming three-dimensional islands. Many materials grow in this so-called Stranski–Krastanov growth mode for which the islands form atop a continuous "wetting" layer; many other materials follow the Volmer–Weber growth mode where the islands form without a continuous layer [19,20,27]. The balance between strain energy, the surface energies of the epilayer and substrate, and the interface energy determines which growth mode is energetically favored. Layer-by-layer (Frank–van der Merwe) growth is favored for most unstrained materials and even for strained materials if the tension (free energy per unit area) associated with the substrate surface exceeds the sum of the tension associated with the exposed surface of the epilayer and the tension at the interface between substrate and the epilayer. Island growth is favored in strained systems since much of the strain energy can be relieved by allowing the material in the islands to relax towards the equilibrium lattice spacing, even at the cost of additional surface energy. Whether islands grow in the Stranski–Krastanov or Volmer–Weber mode depends on the details of the energy balance, involving the surface energies of the epilayer and the interface energy between the substrate and the epilayer.

Although these modes have been well known for a very long time, until recently three-dimensional islands had been considered a nuisance, causing roughness in an otherwise ideal surface. Interest in producing nanometer-sized devices lead to a complete reinterpretation: rather than being a undesirable, the islands were to be used as structures by themselves [28]. Imaging played a substantial role in this radical change of perspective.

When examined in detail with various imaging techniques it was found that, surprisingly, germanium islands did not fit into the classical Stranski–Krastanov mode. Instead, bimodal distributions of island sizes were seen [29,30]. STM imaging studies revealed that the different sizes corresponded to different shapes: smaller islands are "pyramids" and larger islands are more rounded "domes" [31]. Movies are helping to uncover how islands grow and change shape in this complex growth mode [32–35].

Fig. 5a shows a TEM movie of germanium islands on a silicon substrate. The image is formed by passing electrons through a thinned region of a silicon (0 0 1) wafer. Growth occurs during observation as digermane gas (Ge_2H_6) is supplied, decomposes as it strikes the heated substrate. The digermane decomposes as it strikes the heated substrate, leaving germanium which diffuses across the surface until it is incorporated into an island. The strain field around each island locally changes the scattering of the electrons by the specimen. Thus the TEM image does not show the actual shape of the islands, but rather it produces a kind of map of the strain field around each island, and the size of the strain field can be related to the island size.

Most striking is the smooth growth of each island and the continuous "coarsening" process which takes place during deposition: some islands grow while others shrink and disappear. After about 30 s two different sizes of islands can be seen, one smaller and one larger. The smaller ones are the "pyramids" and the larger ones are "domes". Ultimately the smaller islands disappear, but the larger ones continue to grow. The graph in Fig. 5b summarizes the movie data.

The differences between the two island shapes are more apparent in low energy electron microscopy (LEEM) movies. Certainly one of the most versatile and promising of imaging techniques, and is derived from low energy electron diffraction (LEED), a classical technique for surface analysis, LEEM is very sensitive to surface structure. Like TEM, LEEM can generate an entire image at one time, so it is well suited to movie making. A review of the LEEM imaging technique and its history are given in Refs. [12,36–38].

Fig. 6a shows a LEEM movie of the growth of silicon–germanium (SiGe) islands on silicon. An alloy of silicon and germanium, SiGe is emerging as an important material for band gap engineering on silicon [39]. When growing on silicon substrates, the more silicon included in the SiGe, the lower the strain, so the driving force for islanding is weaker. Thus, although the same two shapes form for SiGe as for pure Ge, SiGe islands tend to form later, and grow larger, before the shape change occurs. While smaller pure germanium islands cannot be clearly resolved with present day LEEM, the shape change for these larger SiGe islands can be resolved directly.

The LEEM is sensitive to the particular facet planes which bounds the island, with different facets differing in brightness. In the movie, the same island coarsening and growth can be seen as before, but the pyramids can now be seen converting slowly into domes by the gradual addition of extra facets around the edges. The asymmetric transitional shapes convert to domes on cooling, and imaging during growth is the only way to see this phenomenon. The island size evolution is summarized in the graph (Fig. 6b).

The above evolution in island sizes and shapes is a specific manifestation of an old, very general principle known as Ostwald ripening [40–42]. Initially, because atoms are arriving so quickly at the surface, many small islands are nucleated. This is a metastable state, however, and given enough time, the system will evolve in such a way as to minimize energy. Neglecting anisotropies in surface tension, it is energetically more favorable, given a fixed amount of material, to form islands that are as large as possible. This is because minimizing the total surface area minimizes the contribution of

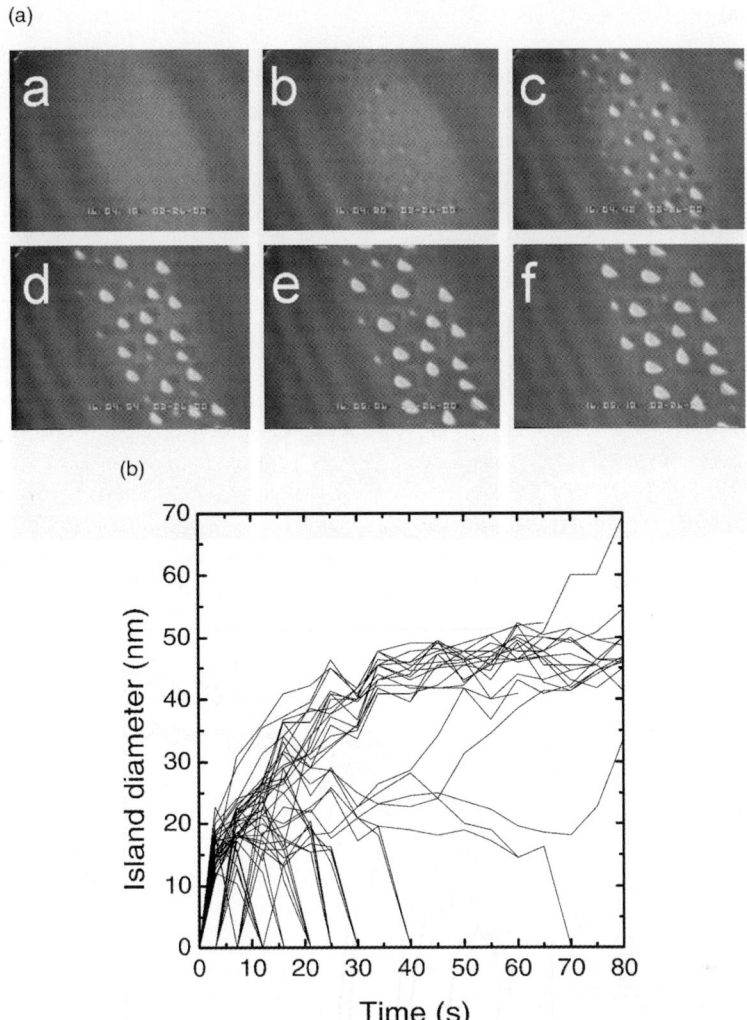

Fig. 5. Strained islands I: germanium on silicon. (a) This TEM movie shows the nucleation and growth of germanium islands on a silicon surface. Islands appear as variation in contrast due to the local strain field, which modifies the electron scattering. Island nucleation, growth and coarsening are visible. The field of view is 800 nm and the growth conditions are 650+/−25 C and 5×10^{-7} Torr digermane. The figure shows a time series of still images taken from the movie. The frames are labeled a, b, c, d, e, f in sequence. An equal time interval separates each frame. (For movie, see: http://www.elsevier.com/inca/homepage/sak/susc/movies/, reproduced with permission from Frances Ross, IBM Research.) (b) This graph of the TEM data shows the size of each island a function of time, highlighting the differing evolution of the two types of island. One type of island continues to grow while the other type ultimately disappears. (Orginally published in Ref. [34]. Used by permission.)

surface tension to the total energy of the system. Large islands are thermodynamically driven to grow at the expense of smaller ones. Atoms escape preferentially from the small islands diffuse across the surface, and incorporate preferentially into the large islands.

Anisotropy in surface tension is responsible for the different island types, which represent different metastable states. The more stable island type "ripens" while the less stable type "withers". A model which includes both the free energy of each type of island and Ostwald ripening can explain

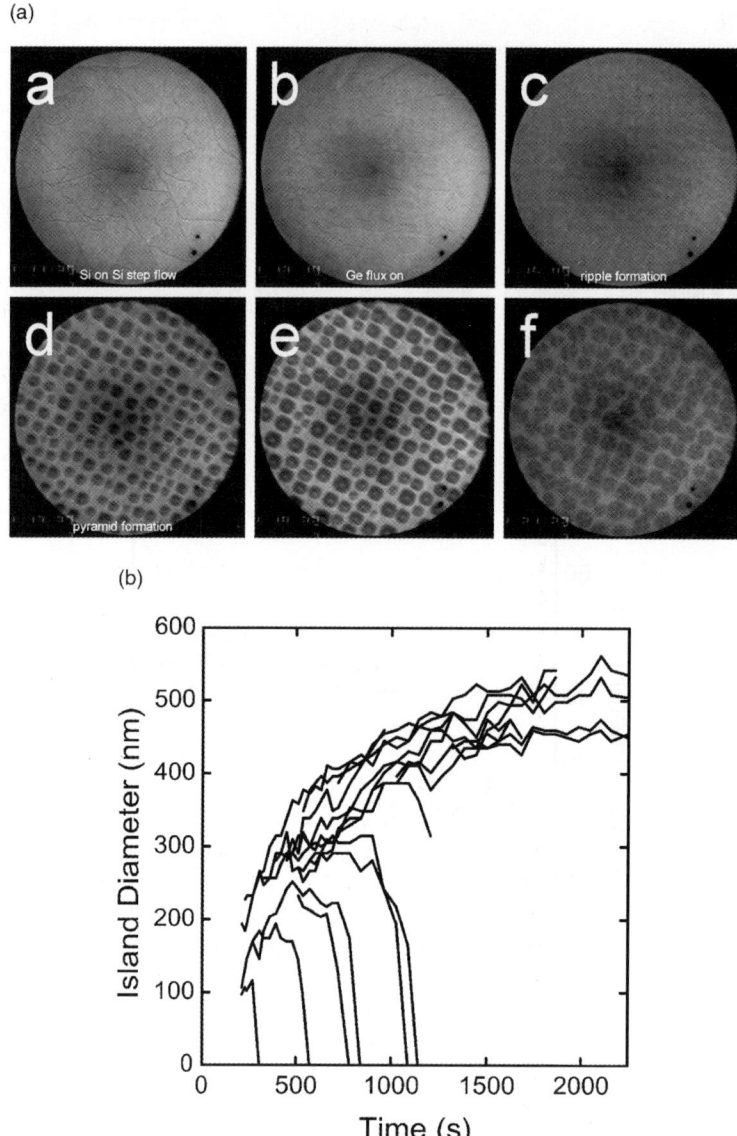

Fig. 6. Strained islands II: silicon germanium on silicon. (a) This LEEM movie shows the nucleation and growth of silicon germanium islands on a silicon (1 0 0) surface. A different image intensity is obtained for different surface structures. Initially pyramid shaped islands form and as they grow larger they slowly convert into domes by the gradual addition of extra facets around the edges. The field of view is 4 μm in diameter. The temperature is 690 ± 25 °C and the germanium content is 30%. The video has been sped up from real time by a factor of 30. The figure shows a time series of still images taken from the movie. The frames are labeled a, b, c, d, e, f in sequence. An equal time interval separates each frame. (For movie, see: http://www.elsevier.com/inca/homepage/sak/susc/movies/. Reproduced with permission from Frances Ross, IBM Research.) (b) A graph of the LEEM data showing the island size as a function of time. The disppearance of some pyramids and the slow conversion of the others into domes can be seen. (Originally published in Ref. [32]. Used by permission.)

the development of individual islands and the bi-modal distribution of island sizes, and can predict which growth conditions will lead to uniform island sizes, and can predict which growth conditions will lead to uniform island sizes and shapes, as desired for the fabrication of electronic devices. Direct visualization of strained island growth in real time has clarified the mechanism of this complex process.

2.5. Step flow

Even when islands are not present, a real crystalline surface is not perfectly flat. Typically, atomic layer high steps separate rows of otherwise flat terraces. On the best substrates commercially available for epitaxy, the terraces may be anywhere from tens of nanometers to microns wide. By intentionally cutting and polishing a crystal surface at an angle slightly tilted away from perfect alignment with the natural crystalline planes of the substrate, a staircase of parallel steps is produced. Crystal growers often find that the intentional introduction of arrays of steps in this way results in better quality epitaxial material. Substrates with such step arrangements said to be "vicinal". On vicinal substrates, there is usually no island formation, and growth occurs by another major class of growth modes: step flow.

In step flow growth, atoms arrive on the surface and diffuse until they reach an atomic step where they stick due to the increased atomic coordination at the step edge. Steps advance as they incorporate adatoms, and in this way the crystal grows. Islands do not form because the average adatom is able to reach a step edge in a shorter time than it is likely to collide with another adatom.

It is not hard to see that there should be a transition between the island mode and the step flow mode. For example, if the flux—that is, the deposition rate—is increased the number of adatoms on the surface will increase and, just like people in a crowd, there is in increased probability that they will bump into one another. When adatoms collide they tend to form the nucleus of a new island. Thus, step flow growth should occur at lower fluxes, but at high fluxes island nucleation will occur.

Step spacing also has an impact on the growth mode. Suppose there is no desorption and adatoms remain mobile until they reach a step or nucleate an island. For a perfectly flat substrate, there would be no pre-existing steps, and growth could only occur by islanding. If terraces are wide relative to the characteristic adatom diffusion length, adatoms are more likely to nucleate islands than incorporate into the steps. At the opposite limit, if the steps are close together (as compared to this diffusion length) adatoms will almost always reach the steps before forming islands. There is therefore a characteristic step spacing at which a transition between island nucleation and step flow occurs.

The transition between step flow growth and islanding can be seen clearly in an experiment using a specially prepared substrate in REM images (Fig. 7) [43–46]. Latyshev et al. exploited the

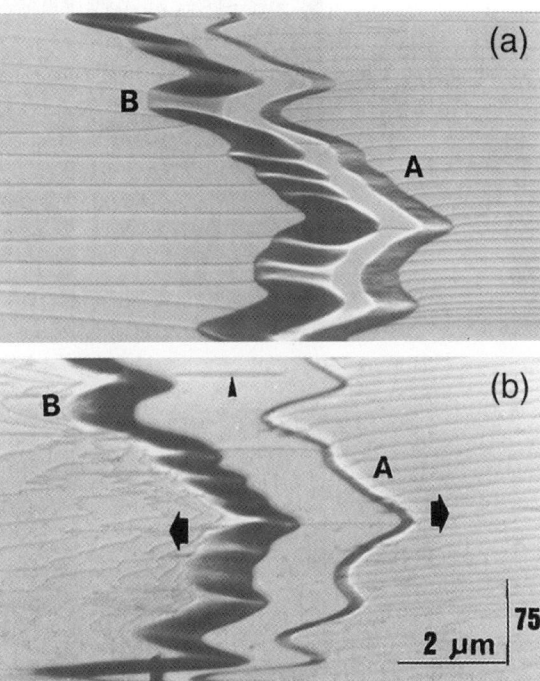

Fig. 7. Step flow to islanding transition. These REM images show a ridge, created when atomic steps bunch due to the influence of an applied electric field. The sides of the ridge are dark (labeled "A" and "B") and individual atomic steps can be seen on either side. Before growth (a) the ridge is flat and so is devoid of atomic steps. When the growth starts the ridge expands (as indicated by the horizontal arrows in (b)). Once the flat top exceeds a critical width islands (vertical arrow) are nucleated near the center. These islands subsequently expand. (Originally published in Ref. [44]. Used by permission.)

tendency for steps to rearrange when heated by an electrical current to prepare an atomically flat plateau bounded by bands of many atomic steps. When a flux is supplied, adatoms attach to the steps, so the bands advance, and as a result the plateau expands. This expansion continues uneventfully until the plateau exceeds the critical width. Suddenly, islands nucleate at the center of the plateau and these islands subsequently grow. Thus an imaging technique has been used to show that such "phase transitions" occur in epitaxial growth.

In Fig. 8, REM movies illustrate both growth modes for steps on the (1 1 1) face of silicon. This experiment was conducted at much higher temperatures than those of Fig. 3, high enough that the (7×7) structure does not form and thus "magic" islands do not form either. Instead, in Fig. 8(a), steps advance continuously. In contrast, Fig. 8(b) shows the coalescence of large islands into a complete film, and the subsequent nucleation of new islands. This is the ideal Frank–van der Merwe growth mode. The difference between (a) and (b) is a change in temperature, which has a

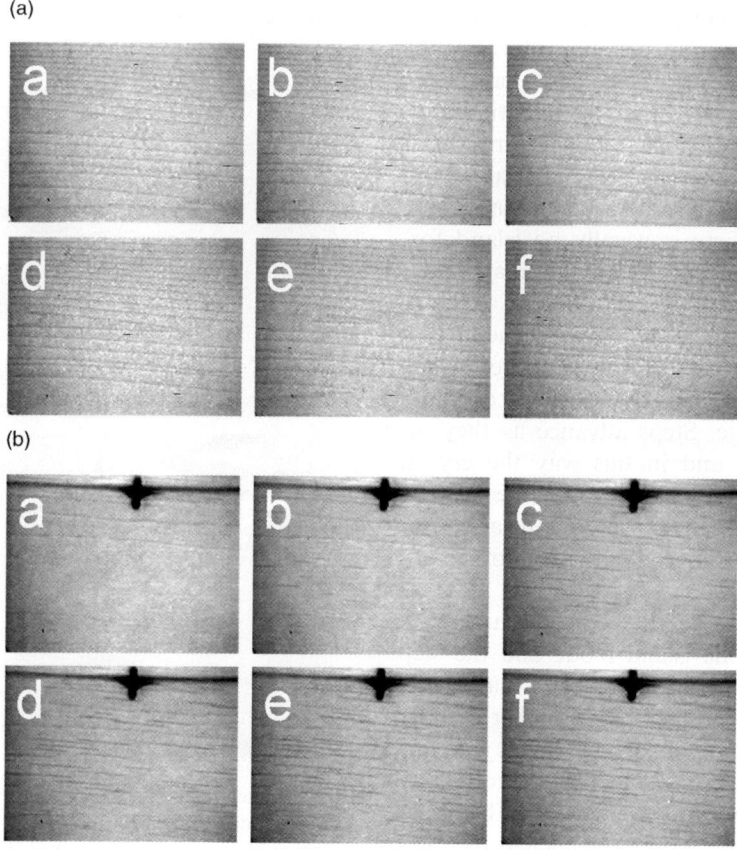

Fig. 8. Step flow vs. island nucleation. These REM movies show growth by step flow and growth by two-dimensional island nucleation. As always for REM the field of view distorted: although the image shows several microns in the horizontal direction, it is distorted to approximately 100 μm along the vertical. In (a) steps advance smoothly without nucleating islands. This is the ideal step flow growth mode. In (b) large islands coalesce into a complete layer. Islands nucleate on this flat surface, and these islands expand, only to coalesce and repeat the process again and again. Both figures show a time series of still images taken from the movie. The frames are labeled a, b, c, d, e, f in sequence. An equal time interval separates each frame. (Both movies can be seen at: http://www.elsevier.com/inca/homepage/sak/susc/movies/, reproduced with permission from A.V. Latyshev, SB RAS.)

strong effect on the diffusion of adatoms, changing their characteristic diffusion length—that is, how far they can diffuse on average, before desorbing or being immobilized. The critical spacing between steps required for nucleation to occur is comparable to this diffusion length. In (a) the diffusion length is very long and so there is no new nucleation, while in (b) the diffusion length is shorter, comparable to the step spacing, and therefore island nucleation is observed.

2.6. Shape instability

Islanding and step flow have both long been recognized, but they are not the only modes of epitaxial growth. It was more recently recognized that there are at least two distinct modes of step flow [47]. In order to explain the reason for this, it is worth looking at the properties of steps in more detail.

First, just as a real surface is seldom, if ever, perfectly flat, a real step is seldom, if ever, perfectly straight. Steps typically wander back and forth on a very small scale. This is particularly true during growth as the step incorporates new adatoms. Even if a step was perfectly straight initially, to grow it must fluctuate in shape somewhat.

The shape of a surface is fundamentally related to surface tension. A flat surface is very often a minimum energy surface, and the surface tension can be defined as the energetic change associated with distorting the surface. Likewise, by analogy with two-dimensional surfaces, a straight step tends to be the minimum energy configuration. The energy cost associated with distorting the step shape is defined as the line tension, or more colloquially, the step stiffness.

Consider a shape that is locally distorted—in particular consider the effect of a bulge outward in the step down direction (Fig. 9). Adatoms on the lower step can reach the bulge very easily—meaning that the bulge tends to grow, making the step more and more distorted. However, adatoms on the upper terrace must travel further to reach the bulge, which is less likely on average, and this therefore tends to cause the bulge to decay. Additionally, a bulge will tend to decay in order to minimize the step curvature and thereby minimize the impact the step stiffness on the overall energy balance.

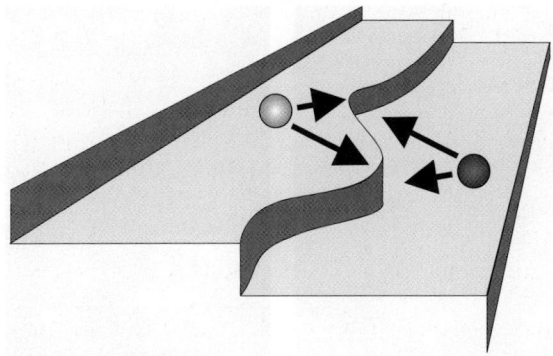

Fig. 9. A distorted step. A single step is shown with a distorted shape. The spheres represent adatoms on the surface. Adatoms on the lower terrace (dark spheres) can reach the bulge quickly. Adatoms on the upper terrace (light sphere) are less likely to reach the bulge.

If the probabilities of incorporation into the step are much lower for adatoms on the upper terrace than for adatoms on the lower terrace, so much so that the step stiffness is overcome, then any small bulge will grow indefinitely, even under normal growth conditions [47]. The step shape is unstable. Alternatively if the upper terrace is substantially smaller than the lower terrace, even if the incorporation probabilities are equal, the reduced terrace size will be manifested by a reduced supply of adatoms arriving at the step from the upper terrace [48]. In this step geometry, too, the step shape is unstable.

Scanning electron microscope images of epitaxial growth have shown the transition between stable and unstable growth [48–50]. For example, if the upper terrace is large enough, steps advance in a stable manner. But if the upper terrace is too small, the adatom supply from it is reduced, and the step begins to wander almost sinusoidally (Fig. 10). The amplitude of the oscillation increases with time. Like the islanding mode, the onset of the transition depends on the flux. In terms of flux, the morphologically unstable phase lies between the stable step flow and islanding regimes.

In practical terms, the danger of unstable growth is that it might roughen the entire surface, and the quality of devices manufactured by epitaxy usually depends on the smoothness of grown interfaces. With imaging technologies this danger is easily avoided.

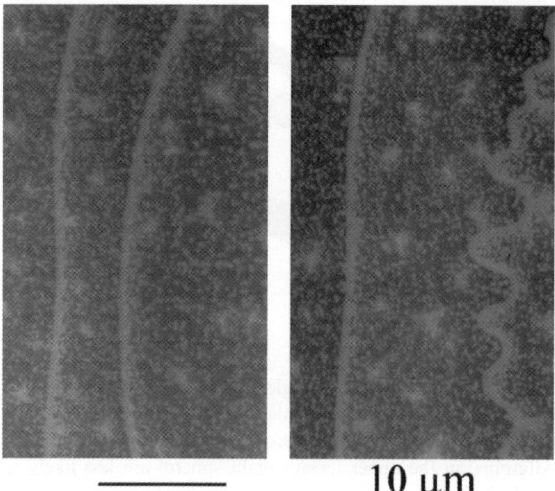

Fig. 10. Shape instability. These SEM images show the same steps before (left) and after (right) epitaxial growth. Both steps move to the right, but the rightmost step, which moved further, develops a shape instability (as evidenced by the sinusoidal wandering of the step).

2.7. Roughness of thicker layers

Just as the step shape may become deformed, the entire two-dimensional surface may become rougher. Any real surface will have some associated roughness, which will evolve. Surface tension tends to flatten the surface, while natural fluctuations in the incoming flux, anisotropies in diffusion probabilities, slow-growth facet formation and many other factors can make it rougher. Imaging techniques make it possible to characterize the roughness and its evolution.

One measure of the roughness of a surface is the root mean square displacement on the surface about its average height. Indeed, this is sometimes considered to be the definition of the term roughness or roughness scale. This is analogous to characterizing a mountain range by the average height of the mountains in the range. Another statistical measure is the lateral spacing between surface features, sometimes called the coarseness or coarseness scale. This is analogous to characterizing a mountain range by the average spacing between its peaks.

Intriguingly, it is found that both these measures of surface roughness often scale as a power law in the growth duration [51]. That is, the roughness scale and the coarseness scale increase to some power of the growth time. This occurs even though the average thickness is simply proportional to the growth duration. The roughness begins with small, closely packed fluctuations in the surface, and develops into large, widely spaced fluctuations.

Imaging techniques used during growth allow the roughness scales to be monitored (Fig. 11) and

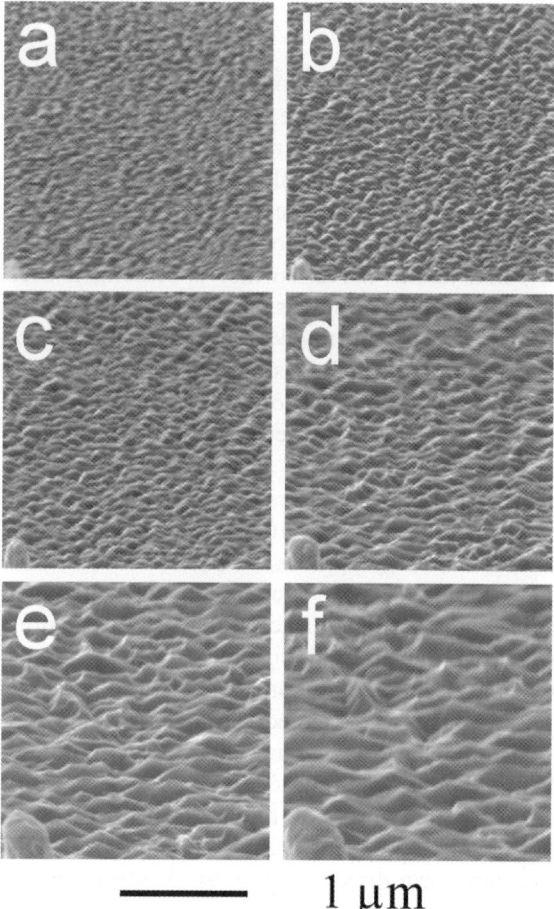

Fig. 11. Roughness scaling. This is a time sequence from an SEM movie showing a gallium arsenide film grown on silicon. The surface roughness begins at very small scales, but the characteristic scale grows as a power law in the growth duration. The images are at various intervals after the initiation of the growth flux (a) 5.1 min (b) 13 min (c) 20 min (d) 45 min (e) 97 min and (f) 169 min. The dust particle at the bottom left of each image serves as a landmark. (Originally published in Ref. [52]. Used by permission.)

the various scaling exponents to be measured [52]. The scaling exponents are useful as a theoretical tool making it possible to eliminate models with different asymptotic behaviors. The fact that power law scaling of roughness scales is found for so many models of surface growth is telling us something fundamental about the nature of growth itself—arguably, this is something not fully understood to this day.

3. Future prospects

3.1. Disclaimer

Long-term prediction of scientific progress is notoriously error prone. Nonetheless, it is worthwhile to look ahead to try to see what lies on the horizon. So, with this disclaimer, bearing in mind the spirit and purpose of this volume, below are some areas where progress seems likely.

3.2. Gaps in theory

There are two radically different approaches to modeling epitaxy. One method relies on an atomistic description of the growth process. Individual atoms (or aggregated blocks representing many atoms) are added to an initial surface where they move and attach according to pre-defined rules. The rules may be quantum mechanical, as is the case in the most ambitious "ab initio" models [53], or simple and phenomenological, as in the much cruder but surprisingly powerful "solid-on-solid (SOS)" models [54]. Generally, a computer is used to keep track of the particles, and the evolution of the surface as a consequence of particle attachment is studied.

Atomistic models can be difficult to handle, and therefore it is often simpler to use a continuum mechanics approach. In this case the underlying atomistic structure of a crystal is ignored (except perhaps that it grows in monolayers) and a continuum equation is used to define the surface at any given time. An evolution equation—a kind of diffusion equation—describes how the surface propagates with time. One of the most successful models of this type is the venerable theory by Burton, Cabrerra, and Frank (BCF) [55,56]. Continuum models like BCF theory can be very helpful for developing a physical intuition about growth.

The atomistic models are advantageous because they can adequately represent fast processes on very small scales, while continuum models are better at explaining averaged, large-scale variations over longer time periods. This highlights a fundamental difficulty for theories of growth—they must successfully bridge an enormous range of parameter space [57]. Length scales of fractions of Angstroms ($<10^{-10}$ m) are important since a slight displacement of atoms implies substantial strain, while is often required that a sample be uniform on length scales greater than centimeters ($>10^{-2}$ m). That is a span of over eight orders of magnitude. Moreover, adatoms can move on time scales of femtoseconds (10^{-15} s), while a micron of growth may take a matter of hours (10^4 s), for a span of nineteen orders of magnitude! Flux rate, pressure and other parameters can also vary over a wide range. Perhaps the abundance of experimental data, together with creative theoretical work will bridge these enormous ranges. It would be a tremendous challenge to obtain a "unified" theoretical method with spans all the ranges of interest. For the foreseeable future it is necessary to cobble together several theories, each applicable in their own ranges. It would, at minimum, be worthwhile to have an explicit prescription for moving between theories and scales.

For a more specific theoretical challenge, let us focus on well-established continuum models of growth. Typically, the boundary conditions for a continuum diffusion equation are set at atomic steps. Usually the boundary conditions are phenomenological, not predicted from first principles by theory. It is seldom known a priori whether the boundary conditions are symmetric on either side of the step, for example. It is seldom clear how barrier heights would change given a change in crystal structure, or moving to a new material. It ought to be possible to determine all these boundary value parameters unambiguously from first principles. It is not yet—no "periodic table" of boundary conditions is yet available.

Experimental feedback should help advance theory in these and other directions. In particular, precision measurements of dynamical phenomena will help to rigorously test any theory. Consider continuum theories. Such theories are "effective theories" in the sense that they are known to fail on some scale. A promising direction might be to push continuum theories to their breaking points by making precision measurements of dynamical parameters for situations where these theories are vulnerable.

3.3. Predictions awaiting verification

The existing theories of crystal growth define a number of control parameters and various dependent properties of the surface. A fundamental property of the surface in many continuum theories is the local adatom density. Any particular theory will describe how this density varies across a surface, and how it depends on the growth parameters. A map of adatom density as a function of position away from a step edge would be a powerful tool to discriminate between models. Remarkably, this has not been obtained, except indirectly, for any material. Even the *average* adatom density is usually unknown.

Another prediction of a number of theoretical works is the possibility of morphological instabilities. As described above, this is one area beginning to be explored by imaging technologies. In the future, detailed studies of instabilities can be expected. The conditions leading to the onset of the instabilities, the evolution of the unstable surface or step, and the underlying mechanism of instability will be explored. Ultimately it should be possible to determine which theoretical model, if any, applies to a given unstable growth regime.

An area where considerable progress is being made is in the physics of step-step interactions. At the simplest level, steps must interact because adatoms move to and from steps and terraces. But strain and entropic effects have also been observed. Thus far this has primarily been done on surfaces at equilibrium [58]. The same interactions may be important during growth, however, and this is an area ready for experimentation.

3.4. Overcoming limitations of microscopy

There will always be room to improve imaging techniques. The profound instrumental challenges in imaging surfaces during growth have been and will continue to be an important frontier for research.

While each imaging technology works well under its own specific operating conditions, these may not be the optimum growth conditions. For example, STM requires high vacuum, and stable conducting substrates. Substrates are often rotated during epitaxy, so they are not stable, insulating substrates may be necessary for some device structures, and most importantly, some epitaxial techniques require atmospheric pressures or even a liquid or solid phase environment. Each individual technique has its own regime of usefulness, but this certainly does not cover all regimes of interest. It will be a continuing challenge to adapt imaging techniques to interesting growth situations.

From quantum physics, any observation must necessarily influence the subject of observation, however slightly. Present day imaging techniques disturb the surface much more than this quantum limit. For example, during epitaxy with an STM, the STM probe casts a "shadow" in the molecular beam, and in addition it locally modifies the potential surface. Electron beams may cause local charging, and may cause changes in the surface by a variety of mechanisms. Imaging should be as non-evasive as possible, and there is a lot of room for progress in this direction.

The flip side of "observation without modification" is to use the imaging apparatus to intentionally modify the surface. For example, there has already been widespread success in using the STM to modify the surface, and much more work of this kind can be expected.

3.5. Practical application of imaging

Until now, obtaining images of surface growth phenomena has been a challenge in itself. There would be great possibilities for progress, however, if imaging were applied to practical engineering epitaxy. There are many areas that imaging could contribute if it became a standard part of the fabrication toolkit. To do so, imaging technologies should be relatively inexpensive, easy to use, and cause negligible degradation in the quality of the grown crystal. Meeting these conditions will be an ongoing challenge for research in instrumentation.

One obvious area where imaging would have an impact would be in nanofabrication. Nanometer-sized quantum dots and quantum wires, as promising as they may be, are far removed from the original intentions. Ideally it would be possible to reproducibly fabricate structures of arbitrary shape and size in all three dimensions, with high-quality interfaces between them. Through rapid feedback and direct, precise characterization, microscopy may bring us closer to this ideal.

Another area where imaging could be helpful is in the control and minimization of surface and interface roughness. As detailed earlier, microscopy has already helped to explore the origin and evolution of surface roughness. This could be extended to the industrial context where it is often critical to ensure that interfaces are very smooth, and where it may occasionally be useful to introduce a controlled roughening. Microscopic mapping of the strain or distortion in a layer could also be useful.

Apart from the structure and morphology, some imaging techniques, such as LEEM and STM, have chemical specificity. That is to say, they can distinguish between atomic species on the surface. This has important implications for practical epitaxial growth. First, the ability to deduce the composition of a layer while it grows would be a valuable capability, since the electronic structure of a layer varies with its composition. Mapping the composition of a layer would be useful, especially in the case of compound semiconductors, which can deviate from stoichiometry. If chemical species can be detected with sufficient sensitivity, measuring the abundance and distribution of trace impurities could potentially be even more useful. This is because such impurities are intentionally introduced to function as dopants in almost every electronic device structure.

There are vast frontiers to explore in the field of epitaxial growth. At the moment we are like navigators with a few well traveled, well charted and very profitable routes, but we still lack a world map. The application of the many microscopic in situ imaging techniques to epitaxy should dramatically expand our scientific understanding, as well as the possibilities for technological innovation.

4. Conclusion

The availability of high resolution, dynamical images of the crystal growth process is revolutionizing the science of epitaxy. From adatom diffusion to island expansion to atomic step flow, epitaxial growth processes are now literally under the microscope. Models are being tested with unprecedented rigor, and new phenomena are being steadily uncovered. In the future, such imaging will continue to expand the frontiers, both scientific and technological, of epitaxial crystal growth.

References

[1] A.Y. Cho, Molecular Beam Epitaxy, AIP Press, Woodbury, NY, 1994.
[2] M.A. Herman, H. Sitter, Molecular Beam Epitaxy: Fundamentals and Current Status, second ed., Spinger, Berlin, 1996.
[3] M. Razeghi, The MOCVD Challenge Volume 1: A Survey of GaInAsP–InP for Photonic and Electronic Applications, Adam Hilger, Bristol, 1989.
[4] M.B. Panish, H. Temkin, Gas Source Molecular Beam Epitaxy: Growth and Properties of Phosphorus Containing III–V Heterostructures, Springer, Berlin, 1993.
[5] S. Entwistle, Trends in the market for GaAs devices, Compound Semiconductor 6, May/June 2000.
[6] E.J. Lerner, Diode lasers light up disks, communications, and printers, Laser Focus World, January 1998.
[7] T. Whitaker, Lighting the Future with LEDs, Compound Semiconductor 7, April 2000.
[8] S.J. Mathews, Back To Basics: III–V Light-Emitting Diodes, Laser Focus World, April 2001.
[9] G. Binnig, H. Rohrer, Scanning tunneling microscopy, IBM J. Res. Develop. 30 (1986) 355.
[10] L. Reimer, Scanning Electron Microscopy, Springer, Berlin, 1985.
[11] P.K. Larsen, P.J. Dobson (Eds.), Reflection High-Energy Electron Diffraction and Reflection Electron Imaging of Surfaces, Plenum Press, NewYork, 1988.
[12] E. Bauer, Low energy electron microscopy, Rep. Prog. Phys. 57 (1994) 895.
[13] G. Comsa, Impact of surface science on epitaxial growth, private communication.

[14] A. Pimpinelli, J. Villain, Physics of Crystal Growth, Cambridge University Press, Cambridge, 1998, p. 4.
[15] T.R. Linderoth, S. Horch, E. Læsgaard, I. Stensgaard, F. Besenbacher, Surface diffusion of Pt on Pt(1 1 0): Arrhenius behavior of long jumps, Phys. Rev. Lett. 78 (1997) 4978.
[16] H.T. Lorensen, J.K. Nørskov, K.W. Jacobsen, Mechanisms of self-diffusion on Pt(110), Phys. Rev. B 60 (1999) R5149.
[17] T.R. Linderoth, S. Horch, L. Petersen, S. Helveg, E. Læsgaard, I. Stensgaard, F. Besenbacher, Novel mechanism for diffusion of one-dimensional clusters: Pt/Pt(1 1 0)-(1 × 2), Phys. Rev. Lett. 82 (1999) 1494.
[18] F. Montalenti, R. Ferrando, Leapfrog diffusion mechanism for one-dimensional chains on missing-row reconstructed surfaces, Phys. Rev. Lett. 82 (1999) 1498.
[19] E. Bauer, Phenomenological theory of crystal precipitation on surfaces I, Zeitschrift fuer Kristallographie 110 (1958) 372, in German.
[20] I.V. Markov, Crystal Growth for Beginners, World Scientific, Singapore, 1996, p. 376.
[21] B. Voigtländer, M. Kästner, P. Smilauer, Magic islands in Si/Si(1 1 1) homoepitaxy, Phys. Rev. Lett. 81 (1998) 858.
[22] K. Takayanagi, Y. Tanishiro, S. Takahashi, M. Takahashi, Structure analysis of Si(1 1 1)-7 × 7 reconstructed surface by transmission electron diffraction, Surf. Sci. 164 (1985) 367.
[23] J.-J. Métois, J.-C. Heyraud, A. Pimpinelli, Steady-state motion of silicon islands driven by a DC current, Surf. Sci. 420 (1999) 250.
[24] D.B. Carlin, Y. Tsunoda, Diode lasers for mass market applications: optical recording and printing, Proc. IEEE 82 (1994) 469.
[25] J. Shibata, T. Kajiwara, Optics and electronics are living together, IEEE Spectrum (February 1989) 34.
[26] R. Dixon, Silicon germanium, Compound Semiconductor 6, January/February 2000.
[27] A. Zangwill, Physics at Surfaces, Cambridge University Press, Cambridge, 1988, p. 431.
[28] D. Leonard, M. Krishnamurthy, C.M. Reaves, S.P. Denbaars, P.M. Petroff, Direct formation of quantum-sized dots from uniform coherent islands of InGaAs on GaAs surfaces, Appl. Phys. Lett. 63 (1993) 3203.
[29] M. Goryll, L. Vescan, K. Schmidt, S. Mesters, H. Lüth, K. Szot, Size distribution of Ge islands grown on Si(0 0 1), Appl. Phys. Lett. 71 (1997) 410.
[30] T.I. Kamins, E.C. Carr, R.S. Williams, S.J. Rosner, Deposition of three-dimensional Ge islands on Si(0 0 1) by chemical vapor deposition at atmospheric and reduced pressures, J. Appl. Phys. 81 (1997) 211.
[31] G. Medeiros-Ribeiro, A.M. Bratkovski, T.I. Kamins, D.A.A. Ohlberg, R.S. Williams, Shape transition of germanium nanocrystals on a silicon (0 0 1) surface from pyramids to domes, Science 279 (1998) 353.
[32] F.M. Ross, R.M. Tromp, M.C. Reuter, Transition states between pyramids and domes during Ge/Si island growth, Science 286 (1999) 1931.
[33] F.M. Ross, P.A. Bennett, R.M. Tromp, J. Tersoff, M. Reuter, Growth kinetics of $CoSi_2$ and Ge islands observed with in situ transmission electron microscopy, Micron 30 (1999) 21.
[34] F.M. Ross, J. Tersoff, R.M. Tromp, Coarsening of self-assembled Ge quantum dots on Si(0 0 1), Phys. Rev. Lett. 80 (1998) 984.
[35] F.M. Ross, J. Tersoff, M. Reuter, F.K. Legoues, R.M. Tromp, In situ transmission electron microscopy observations of the formation of self-assembled Ge islands on Si, Microsc. Res. Tech. 42 (1998) 281.
[36] E. Bauer, J. Electron. Spectrosc. Relat. Phenom., in press.
[37] E. Bauer, LEEM basics, Surf. Rev. Lett. 5 (1998) 1275.
[38] E. Bauer, Surface electron microscopy: the first thirty years, Surf. Sci. 299/300 (1994) 102.
[39] B.S. Meyerson, High speed silicon–germanium electronics, Sci. Am. 270 (1994) 42.
[40] W. Ostwald, Studien uber die bildung und umwandlung fester korper, Zeitschrift fuer physikalische Chemie 22 (1897) 289.
[41] I.V. Markov, Crystal Growth for Beginners, World Scientific, Singapore, 1996, p. 139.
[42] A. Pimpinelli, J. Villain, Physics of Crystal Growth, Cambridge University Press, Cambridge, 1998, p. 138.
[43] A.V. Latyshev, A.B. Krasilnikov, A.L. Aseev, In situ reflection electron microscope observation of two-dimensional nucleation on Si(1 1 1) during epitaxial growth, Thin Solid Films 281–282 (1996) 20.
[44] A.V. Latyshev, A.B. Krasilnikov, A.L. Aseev, Self-diffusion on Si(1 1 1) surfaces, Phys. Rev. B 54 (1996) 2586.
[45] A.V. Latyshev, A.B. Krasilnikov, A.L. Aseev, UHV reflection electron microscopy investigation of the monoatomic steps on the silicon (1 1 1) surface at homo- and heteroepitaxial growth, Thin Solid Films 306 (1997) 205.
[46] A.V. Latyshev, A.L. Aseev, Monatomic steps on silicon surfaces, Physics-Uspekhi. 41 (1998) 1015, translated from Uspekhi Fizicheskii Nauk. 168 (1998) 1117.
[47] G.S. Bales, A. Zangwill, Morphological instability of a terrace edge during step-flow growth, Phys. Rev. B 41 (1990) 5500.
[48] P. Finnie, Y. Homma, Stability–instability transitions in silicon crystal growth, Phys. Rev. Lett. 85 (2000) 3237.
[49] P. Finnie, Y. Homma, In situ observation of instability in step morphology during epitaxy and erosion, Proceedings of the Materials Research Society 2000 Fall Meeting, 2001, p. 10.12.
[50] Y. Homma, P. Finnie, M. Uwaha, Morphological instability of atomic steps observed on Si(1 1 1) surfaces, Surf. Sci. 492 (2001) 125.
[51] A.-L. Barabási, H.E. Stanley, Fractal Concepts in Surface Growth, Cambridge University Press, Cambridge, 1995.
[52] P. Finnie, Y. Homma, Island growth and surface roughness scaling of epitaxial GaAs on Si observed by in situ scanning electron microscopy, Phys. Rev. B 59 (1999) 15240.

[53] E. Kaxiras, Review of atomistic simulations of surface diffusion and growth on semiconductors, Comp. Mater. Sci. 6 (1996) 158.
[54] Y. Saito, Statistical Physics of Crystal Growth, World Scientific, Singapore, 1996.
[55] W.K. Burton, N. Cabrera, F.C. Frank, Philos. Trans. R. Soc. London A 243 (1951) 299.
[56] A. Pimpinelli, J. Villain, Physics of Crystal Growth, Cambridge University Press, Cambridge, 1998, p. 89.
[57] D.D. Vvedensky, Atomistic modeling of epitaxial growth: comparisons between lattice models and experiment, Comput. Mater. Sci. 6 (2) (1996) 182.
[58] H.C. Jeong, E.D. Williams, Steps on surfaces: experiment and theory, Surf. Sci. Rep. 34 (1999) 171.

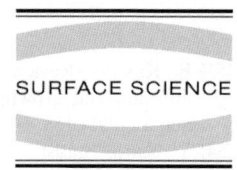

Understanding crystal growth in vacuum and beyond

Elias Vlieg *

NSR-RIM, Department of Solid State Chemistry, University of Nijmegen, Toernooiveld 1, 6525 ED Nijmegen, Netherlands

Received 10 November 2000; accepted for publication 4 May 2001

Abstract

For a detailed understanding of crystal growth, the atomic-scale structure of the growing interface must be known. While such knowledge is available in vacuum environments, this is not the case if the crystal grows from a solution, melt or solid. X-ray diffraction is one of the few techniques that can be applied for this purpose and it is starting to provide information on the structure of both sides of a growing interface. This means that structural details like relaxation and reconstruction on the crystal surface and ordering in the solution can be included in the theoretical description of crystal growth. © 2001 Elsevier Science B.V. All rights reserved.

Keywords: X-ray scattering, diffraction, and reflection; Growth; Molecular beam epitaxy; Surface relaxation and reconstruction; Surface structure, morphology, roughness, and topography; Solid–liquid interfaces

1. Crystal growth

Crystals are grown in many ways, used in a great variety of applications and come in many shapes and sizes. Silicon crystals, for example, form the basic material of the present information age, and the technology to grow large and nearly perfect crystals is well developed. So-called boules of single crystal silicon, with a diameter of up to 12 inches and a length of up to 1 m, are cut into wafers from which the chips that power all computers are manufactured. Thus the importance of these silicon crystals can hardly be overstated, but smaller and less perfect crystals are important as well, since most solid materials exist in a (poly)crystalline state. Rock salt and sugar are two familiar examples, and many people have seen crystallisation when honey forms sugar crystals or when water freezes on a window. The materials that are synthesised or processed in the chemical industry are often obtained in crystalline form. These are typically not single crystals of appreciable size, but consist of small grains forming a powder or aggregate.

Crystallisation of a mixture of many components produces crystals that are purer than the starting material. In industry, crystallisation is therefore frequently used as an energy-efficient way to purify a material. Margarine, for example, is made in this way by cooling a mixture of fats. The size and shape of the crystals are important in many applications and determine physical properties such as strength and solubility. The photographic industry, for example, uses sophisticated processes to obtain AgBr crystals of precisely defined shape and size. While earlier photographic

---

* Tel.: +31-24-365-3070; fax: +31-24-365-3067.
 E-mail address: vlieg@sci.kin.nl (E. Vlieg).

film contained small, block-like grains of AgBr, the crystals produced today are plate-like and make thus a more efficient use of this expensive material. A more mundane case in which the shape of crystals is important, is when a material needs to be filtered. Here needle or plate-like crystals are to be avoided since they tend to block the filters.

In addition to size and shape, the specific crystallographic structure of a crystal is also important. Many substances crystallise in different structures called polymorphs. This is not only true for complicated molecules, but also for such a 'simple' material as carbon, which occurs as both graphite and diamond (and in even more forms when we consider C60 and related crystals!). The control of polymorphism is particularly important in the pharmaceutical industry, where the active ingredient of a medicine is often present in the form of small crystals of a specific polymorph. A different polymorph can completely alter the effect of the medicine, owing to a different dissolution rate. As another example, the taste and feel of chocolate are determined by the polymorphic form of its crystalline compounds.

A different application of crystals is for the structure determination of the molecules it is made of by using X-ray diffraction. This 'X-ray crystallography' was discovered in 1912 and has been the dominant tool for the structure determination of crystals ever since. In recent years, the structure determination of proteins (protein crystallography) has become very important in understanding biological processes. This is largely driven by the pharmaceutical industry that wants to make drugs in a rational way: "drug engineering". One of the limiting steps however, is the growth of protein crystals of sufficient size and quality. As soon as a structure is determined, a crystal becomes obsolete in this case, although storage and injection of proteins in the form of crystals is contemplated for medical applications.

The most prominent features of a single crystal are its facets and the distinct angles between these facets. Natural crystals of quartz or other minerals provide strikingly beautiful examples of these features (Fig. 1). Although many characteristics of crystals were understood for a long time (the famous astronomer Kepler speculated about the in-

Fig. 1. A natural quartz crystal, showing the various facets created during growth (source: www.majestic-quartz.com).

ternal structure of crystals as early as 1611 [1]), only the first X-ray diffraction experiments proved that a crystal consists of a regular arrangement of atoms or molecules. From this knowledge the growth shape ('habit' or 'morphology') of a crystal can be understood: the facets reflect the regular arrangement of the atoms and molecules in a crystal. Here a remarkable range of length scales is covered, because the structure at the atomic level determines the macroscopic crystal shape. In the 1950s, Hartman and Perdok developed theories that predict, starting from the bulk crystallographic structure, the growth shape of a crystal by calculating which facets are the most stable [2]. Note that the growth shape of a crystal generally differs from the equilibrium shape [3], because kinetics plays an important role during growth.

From a surface-science point of view, the assumption in the Hartman–Perdok theory that the structure at the surface is the same as in the bulk of a crystal appears to be somewhat naive, but it is a good starting point that has proven to be quite successful. Most crystals for which habit calculations have been done are not grown in a high-vacuum environment where surface-science techniques are available to obtain details of the surface structure such as relaxation (displacement of the atoms without changing the symmetry) or reconstruction (rearrangement of the atoms with a change in symmetry). This lack of experimental information has stopped the further refinement of the crystal growth theories to include the effect of these factors. Here lies an important challenge.

However, it is not only the crystallographic structure that determines the shape and quality of a crystal. There is a large amount of empirical knowledge about the influence of impurities [4] and the effects of solvents and supersaturation. The first publication about the role of impurities dates from 1783 [5], when Romé de l'Isle reported that the shape of rock salt crystals changes from cubic to octahedral when urea is added to the aqueous growth solution. The macroscopic effects of these parameters must have a microscopic origin, but again, experimental information is lacking.

The cases mentioned so far mainly dealt with the growth of bulk crystals. A completely different area is the deposition of thin films, which is particularly relevant for the manufacture of semiconductor devices and metallic (multi)layers. Here other issues are important, such as alloying and the influence of the difference in the atomic distances between the two materials (lattice mismatch). Thin films are often grown using vacuum or low-pressure techniques like molecular-beam epitaxy (MBE) and chemical-vapour deposition (CVD). These growth techniques are compatible with a number of surface-science techniques. Using the full range of available surface-science techniques, the morphology, dynamics and atomic structure have been determined in many growth systems. For surfaces prepared in vacuum, displacements as small as 0.001 nm (1 pm) of the topmost atoms can be measured, changes in the near-surface composition can be determined and surface defects can be identified.

There is no doubt that the field of bulk crystal growth would benefit enormously from experimental data with a similar level of refinement. As discussed earlier, the main problem lies in the different growth environment, which is often a solution or melt in bulk crystal growth. Here one needs 'interface science', rather than surface science in the classical sense. Two techniques, scanning-probe microscopy (SPM) and surface X-ray diffraction (SXRD), which both have their roots in vacuum surface science, are now starting to provide interface information with similar detail as common in surface science and are thus bridging the gap in understanding between thin film and bulk crystal growth. Some other techniques are possible as well (e.g. in situ electron microscopy [6] or second harmonic generation [7]), but are less widely applicable.

SPM techniques, such as scanning-tunnelling microscopy (STM) and atomic-force microscopy (AFM), have had a great impact on surface science and are very suitable for non-vacuum applications. Since many crystals are non-conducting, AFM is in practice the most common technique in studies of solution growth [8,9]. For a full understanding of crystal growth, a microscopic technique is an absolute necessity, because the local structure on a crystal surface (steps, impurities) is one of the determining factors in growth. In this paper, however, we will restrict the discussion to SXRD. Being a diffraction technique, SXRD provides an averaged picture of a growth system, but at a much higher resolution than AFM. The two techniques are thus complementary and form an ideal combination.

2. Surface X-ray diffraction

A crystal consists of a basic unit (unit cell) of atoms or molecules that is repeated many times in all directions (Fig. 2). It is impossible to see the structure of a crystal directly, as the distances and

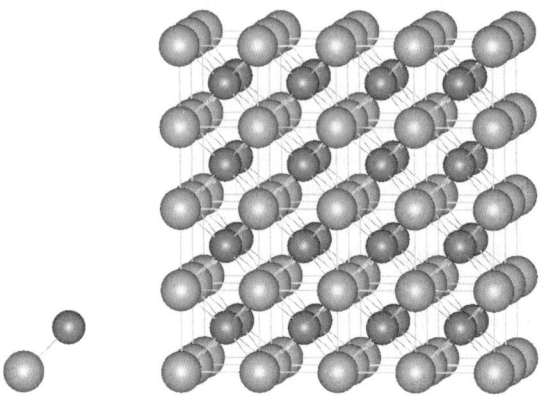

Fig. 2. A schematic of a crystal in which a basic unit (drawn left) is repeated many times in all directions in space. In this example, the basic unit consists of two atoms. The cube that contains these atoms and that is repeated in the crystal is called the unit cell.

sizes of the atoms and molecules are much smaller than the wavelength of visible light. A typical atomic distance is 0.3 nm, while light has a wavelength of approximately 500 nm. X-rays, on the other hand, have a typical wavelength of 0.1 nm and can therefore be used to infer the structure of a crystal, but only in an indirect way through the phenomenon of diffraction.

In order to explain how this works, we can compare the illumination of a crystal by X-rays to the illumination of a handkerchief by a visible laser (Fig. 3). The wavelength of visible light is several thousand times larger than that of X-rays, but the regular distance between the threads in a handkerchief is a similar factor larger than the regular atomic distances in a crystal. The handkerchief experiment is therefore analogous to X-ray diffraction, but is, with the use of a laser pointer, much easier to perform. When inserting a handkerchief into the laser beam, one finds a regular pattern of spots on a screen placed after the handkerchief (Fig. 3a). These spots are called diffraction spots or reflections and occur owing to the regular arrangement of the tiny holes between the threads in the handkerchief. At a diffraction spot, the light from all the holes adds up constructively, yielding a high intensity at that location. The smaller the spacing between the threads, the larger the distance between the diffraction spots: the screen behind the handkerchief therefore displays what is called 'reciprocal space'. The better the threads are ordered (the more expensive the handkerchief!), the sharper the diffraction spots. The exact shape of the holes in the handkerchief may be (theoretically) calculated from the intensity of the diffraction spots. Simply looking with a magnifying lens at the handkerchief (in 'real space') is of course much easier, but for a crystal this is not possible. In that case the width, spacing and intensity of the X-ray diffraction spots are used to determine the structure and ordering of a crystal. The atoms in a crystal are regularly arranged along three dimensions, and therefore reciprocal space of a crystal is also three-dimensional (3D). The X-ray diffraction spots correspond with points in this reciprocal space that are by convention labelled using three so-called diffraction indices $(h\,k\,l)$, where h, k and l are integer numbers.

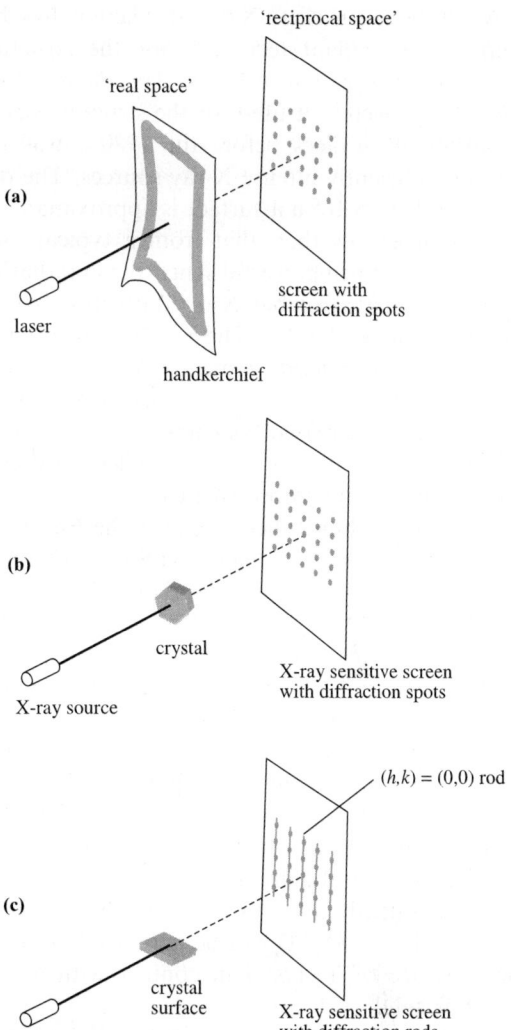

Fig. 3. Diffraction occurs when an object with a periodic structure is illuminated by 'light' with a wavelength somewhat smaller than the periodic distances. The diffraction of laser light by a handkerchief (a) is therefore analogous to the diffraction of X-rays by a crystal (b). On the screen after the object (reciprocal space), diffraction spots are observed from which the 'real space' structure can be derived. By convention, each diffraction spot (reflection) gets a label with three integer indices $(h\,k\,l)$. (c) When X-rays diffract from a crystal with an atomically flat surface, one observes a combination of the diffraction spots from the bulk crystal with diffraction rods originating from the surface. The central rod corresponds to specular reflectivity and is labelled the $(h,k) = (0,0)$ rod. For all the other rods, there is a side-ways reflection component, which has as consequence that these rods are sensitive to the surface structure and ordering in the lateral direction.

As discussed earlier, X-ray diffraction has become the dominant technique for the structure determination of bulk crystals. The main reason why X-rays were not used for the structure determination of surfaces before the 1980's, was the lack of sufficiently intense X-ray sources. The diffracted intensity from a surface is approximately a million times less than that from a typical bulk crystal. Only when powerful synchrotron radiation sources became available, X-ray diffraction from a surface became feasible. The intensity from such sources has increased tremendously in the last decades and the possibilities for using X-rays on surfaces have increased concomitantly. The latest synchrotron sources are of the so-called third generation, providing unprecedented fluxes of hard X-rays. Fig. 4 shows such a source, the European Synchrotron Radiation Facility (ESRF) in Grenoble, France. For most surfaces, damage caused by such powerful X-ray beams is minimal, but for organic materials, this can be a problem.

SXRD was initially used as a technique to study surfaces in vacuum [10]. However, it was quickly realised that X-rays are also powerful in other environments. The penetration power of X-rays is well known when they are used for 'an X-ray' in medical applications. Whereas many surface-science techniques require a vacuum environment and consequently are only suitable for solid–vacuum interfaces, SXRD can be employed to study the interface of a crystal in contact with a gas, liquid or solid.

SXRD is often compared with low-energy electron diffraction (LEED), the most frequently used surface-crystallography technique. The major difference between the two techniques is that low-energy electrons are strongly scattered in a crystal, while X-rays are scattered only very weakly. This means that LEED has to a low penetration depth (high surface sensitivity) and a high yield. However, it is difficult to interpret the scattered intensity, because many electrons are scattered more than once (multiple scattering). X-rays have a high penetration depth, low yield (hence the necessity of synchrotron radiation sources) and the major advantage of a straightforward interpretation of the intensity owing to the absence of multiple scattering.

Fig. 4. An aerial photograph of the European Synchrotron Radiation Facility (ESRF) in Grenoble, France. A synchrotron source consists of a ring-shaped vacuum pipe in which high-energy electrons are circulating. Radiation is produced at each location where the electron path is curved. The ESRF is one of the three hard X-ray synchrotron radiation sources of the third generation that are currently in operation. The other two are located in the USA (APS, in Argonne) and in Japan (Spring-8, in Nishi Harima). (Source: ESRF.)

While the large penetration depth of X-rays is necessary for in situ studies in non-vacuum environments, we need to discuss how this is compatible with surface sensitivity. After all, the surface is only a small part of the total illuminated area. The main trick here is to investigate specific diffraction spots at appropriate places in reciprocal space. This can be understood by comparing Fig. 3b and c. A bulk crystal yields intensity only at sharp spots in reciprocal space. For a crystal that terminates in an atomically flat surface the bulk reflections are connected by tails of diffuse intensity in the direction perpendicular to the surface. (By convention the diffraction index l describes the

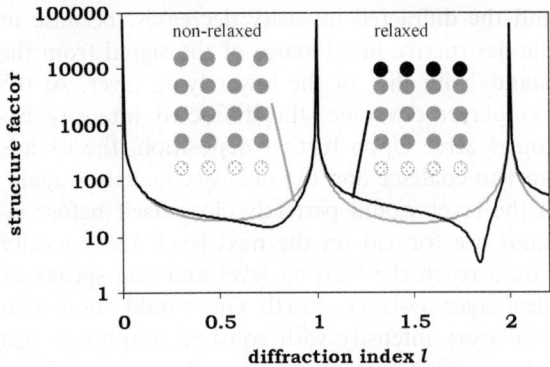

Fig. 5. The variation of the structure factor as a function of the diffraction index l along a diffraction rod for a simple cubic crystal. The measured intensity is proportional to the square of the structure factor. l is the direction perpendicular to the surface. For integer values of l one finds diffraction spots from the bulk crystal with a very high intensity. In between these spots, the structure factor is surface sensitive. This is illustrated by a calculation for a crystal that has the surface atoms at the same positions as in the bulk (grey curve), and one calculation for a crystal in which the topmost layer is relaxed inward by 10% (black curve). The two curves only differ significantly away from the bulk diffraction spots.

out-of-plane direction.) These tails are called crystal-truncation rods [11]. On such a rod the intensity varies continuously as a function of l and reaches a maximum for integer values of l at the bulk spots, while the maximum surface sensitivity is obtained exactly midway between these spots. Fig. 5 shows an example of the intensity changes along a crystal-truncation rod for a simple cubic crystal. When the surface atoms are displaced from their bulk-extrapolated positions (i.e., when surface relaxation occurs), the intensity distribution along the rod changes. It is clear that this change becomes invisible near the diffraction spots from the bulk crystal (at integer values of l) and that one should thus measure in between these spots.

The signal from the surface is weak and it is important to minimise the background scattering from the bulk crystal. Diffraction experiments are therefore typically performed at small incoming or outgoing angles, because then the penetration depth is reduced.

An experimental set-up consists of a sample environment chamber that is coupled to a so-called diffractometer [10]. This is a big sample manipulator that allows accurate positioning and scanning of both the sample and the detector. In this way the intensities of various diffraction rods can be measured. Using standard procedures [12], each measured intensity can be converted into the amplitude of the so-called structure factor for a particular reflection $(h\ k\ l)$. Structure determination using X-ray diffraction revolves around these structure factors. The structure factor contains information about the type of atoms in the crystal, their position and their thermal vibrations. In the case of diffraction from a surface, the structure factor is the sum of the bulk and surface contributions and is sensitive to surface relaxation, reconstruction and roughness [13].

The structure of a crystal surface is typically derived from model calculations [14]. Luckily, model-independent information can also be derived directly from the data using methods developed for bulk X-ray crystallography [10,15].

The integrated intensity of a diffraction spot determines the structure factor and thus contains the crystallographic information. The profile of a diffraction spot contains a second important piece of information, because it depends on the long-range order of the surface. Disordered surfaces give broad diffraction profiles, while well-ordered surfaces give narrow ones [13,16]. Measuring diffraction profiles during growth thus yields information about the evolution of the surface morphology.

3. The experimental state-of-the-art

3.1. In situ studies of epitaxial growth

Two common methods to grow an epitaxial film are MBE and CVD. (A film is called epitaxial if its orientation mimics that of the substrate.) The growth process can be accurately followed using diffraction since the reflected signal is sensitive to surface roughness. Typically, the specular beam is used for this. This is the central diffraction rod in Fig. 3c and is labelled as the $(h,k) = (0,0)$ rod. Maximum sensitivity to surface roughness is obtained for points on this rod exactly in between two bulk peaks. In this case the signal from one layer is exactly out of phase with that of the underlying one (destructive interference).

Fig. 6 explains the behaviour of the diffracted signal for the various growth modes that can occur as a function of the substrate temperature. At high substrate temperatures (Fig. 6a), the deposited atoms are very mobile and diffuse over the surface until a step edge is encountered. They attach to this edge and are incorporated in the growing terrace when more atoms arrive. This regime is called step-flow growth. Since the roughness on the surface does not change, the diffracted X-ray intensity remains constant. For lower substrate temperatures (Fig. 6b), the atoms do not reach a step edge, but encounter other deposited atoms with which they form islands (two-dimensional (2D) island nucleation). Initially the islands grow and the diffracted intensity decreases, because of the destructive interference of the signal from the islands with that of the lower lying layer. At 0.5 monolayer coverage, the diffracted intensity becomes zero. Upon further deposition, the islands start to coalesce and the intensity increases again. If the layer would perfectly close itself before islands are formed on the next level, the intensity would reach the starting level and one speaks of ideal layer-by-layer growth. One would observe an oscillatory intensity with constant amplitude and with a period corresponding to the growth of one monolayer. In practice, this ideal behaviour does not occur and damped oscillations are found instead. Finally, at sufficiently low temperatures, the atoms are largely immobile and the surface quickly becomes rough (Fig. 6c). During this rough growth, no oscillations are observed and the signal decays rapidly.

The behaviour illustrated in Fig. 6 was indeed observed in the first X-ray study of a growing crystal [17], in which Ge was deposited on Ge(1 1 1) using MBE. X-ray oscillations with monolayer period were measured, similar to the oscillations found in such growth systems using reflection high-energy electron diffraction (RHEED) [18]. In observing the growth mode or speed, X-rays offer no advantages over RHEED. In fact, X-rays are a far more elaborate and expensive method for this. However, in interpreting the scattered intensity and deriving surface crystallographic information, X-rays offer distinct advantages. In the case of Ge(1 1 1) it was found that low temperature growth does not proceed via the full double-layers of the crystal structure, but that partial layers form as well [17].

In non-vacuum growth environments, RHEED cannot be used. Using a special geometry, however, Rijnders et al. [19] were able to use RHEED during pulsed laser deposition of oxide films at a pressure of 0.5 mbar. For higher pressures this is not an option, and therefore, Fuoss et al. designed a set-up in which metal–organic chemical vapour deposition (MOCVD) could be combined with X-ray diffraction [20]. Typical system pressures during growth are 100 mbar. Fig. 7 shows a recent and beautiful example of the observed growth modes during MOCVD growth of GaN [21]. All the

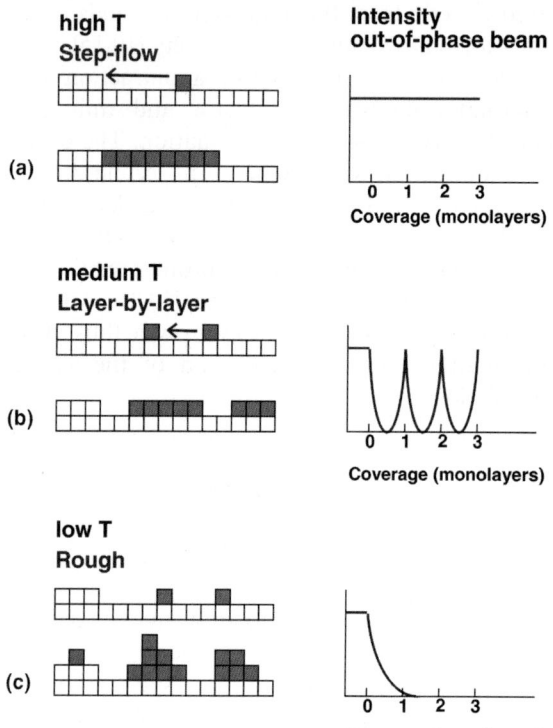

Fig. 6. A schematic diagram of the growth modes during homoepitaxial growth of a crystal as a function of the substrate temperature. On the right, the corresponding diffracted signal is shown for a reflection for which the scattering amplitudes of consecutive layers are exactly out-of-phase. For decreasing temperatures, the growth modes vary from (a) step-flow, (b) layer-by-layer, to (c) rough.

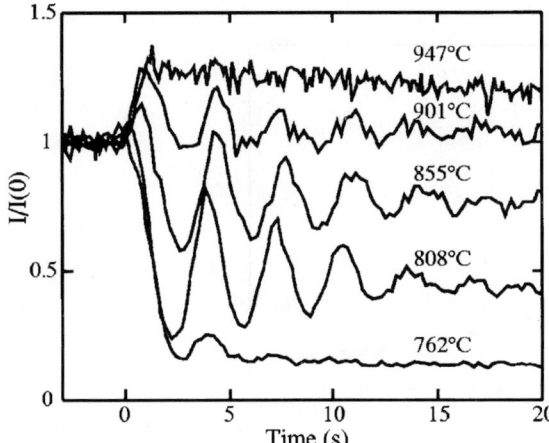

Fig. 7. The diffracted intensity of a surface sensitive reflection during the growth of GaN using metal–organic chemical vapour deposition. The observations for the different substrate temperatures closely follow the schematic behaviour shown in Fig. 6. The oscillation amplitude decays because during layer-by-layer growth the surface roughness is steadily increasing (from Ref. [21], with permission).

growth modes that were schematically shown in Fig. 6 are clearly visible as a function of the growth temperature.

A special type of layer-by-layer growth was observed by Bennett et al. [22] at a buried interface. While depositing palladium on Si(1 1 1) they concluded from the observed oscillatory X-ray signal, that the silicon substrate below the palladium film was consumed layer-by-layer during the room-temperature formation of palladium silicide. This experiment fully exploited the penetrating power of the X-rays.

By choosing an appropriate location in reciprocal space, information about particular features of a crystal is obtainable. We will illustrate this with the case of the homoepitaxial growth of Ag(1 1 1) in the presence of Sb [23,24]. Using the specularly reflected beam, it was found that clean Ag(1 1 1) does not exhibit a layer-by-layer growth mode. Adding submonolayers of Sb as a 'surfactant', did induce this growth mode, as evidenced by the observed X-ray oscillations [23]. This striking effect can be explained by the Ehrlich–Schwoebel barrier (island-edge diffusion barrier) where atoms landing on top of an island are prevented from descending to the lower terrace on the time scale of the growth process. The Sb lowers this barrier and thus induces the layer-by-layer growth mode in Ag(1 1 1). These observations used the specular beam that is insensitive to the lateral ordering of the growing layer.

In order to determine the lateral ordering, a diffraction rod with lateral reflection component has to be measured, i.e. a reflection with $(h,k) \neq (0,0)$ (Fig. 3c). Fig. 8a shows the intensity of the $(h,k,l) = (0,1,0.3)$ reflection during Ag deposition after an initial 0.3 monolayer of Sb was deposited. For normal crystal growth, the intensity oscillations should behave as shown in Fig. 6b, with a decaying amplitude never exceeding the initial intensity. The observations shown in Fig. 8a are very different from this and indicate that the stacking of the top layer is different from that of the bulk crystal. At a coverage of 1/3 monolayer of Sb, the system reconstructs with the Sb atoms occupying substitutional sites and with a top layer that has the wrong stacking [25]. Fig. 8b illustrates that during growth, major rearrangements take place. The Sb segregates to the top layer, while the lower lying Ag atoms return to their correct crystallographic positions. One thus observes effectively a floating stacking fault [26]. For an Sb coverage of less than 1/3 monolayer, this stacking fault does not occur.

Seeing the floating stacking fault requires a penetration depth of only a few layers. This is always possible with X-rays, but difficult (or impossible) with many other probes. When stacking faults occur deeper in a crystal, X-rays soon become the only option for in situ observations. In the case of Cu(1 1 1) growth, for example, small twin crystallites buried under tens of layers were observed [27]. Here it was discovered that another surfactant, indium, prevented the formation of these twin crystallites.

During growth, not only is the total roughness important, but also the lateral ordering of the deposited material. Such ordering may for example occur between the islands during layer-by-layer growth. These islands are typically separated by a distance comparable to the diffusion length of the atoms at the given deposition rate and substrate temperature. If such a preferred island–island distance indeed occurs, the diffraction profile will

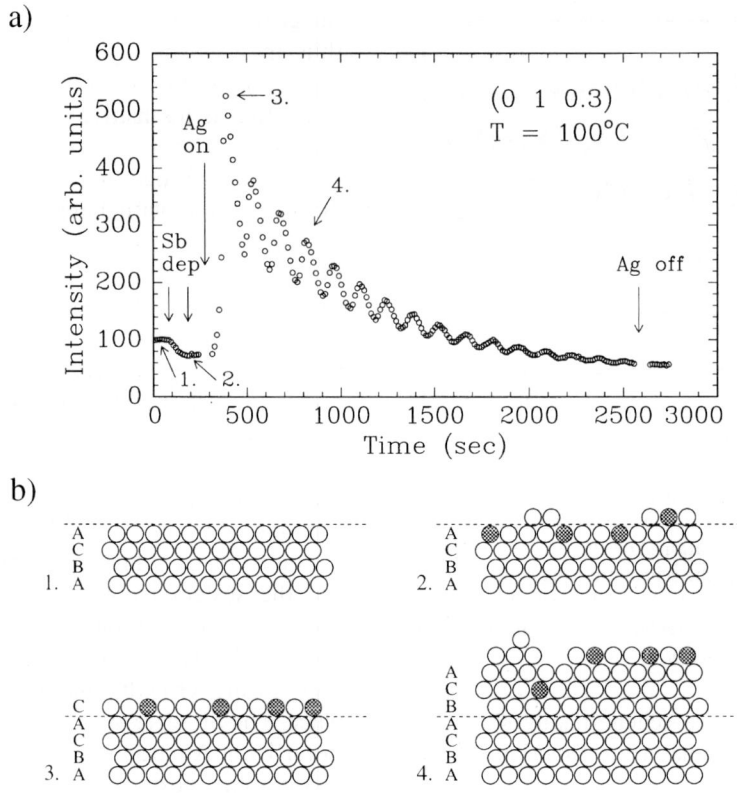

Fig. 8. (a) The intensity of a reflection with an in-plane diffraction component, $(h,k,l) = (0,1,0.3)$, which is sensitive to the in-plane positions of the growing layers. A deposition of 0.3 monolayer Sb is followed by 16 monolayers of Ag at 100 °C. The numbers indicate the situations depicted in (b). (b) Schematic side view of the surface during different stages of deposition. Open circles represent Ag atoms and filled circles represent Sb atoms. On the clean Ag(1 1 1) surface with the ABC stacking of an fcc crystal (1), 0.3 monolayer of Sb is deposited (2). After a total deposition of one monolayer Ag (3), all top layer atoms have the wrong stacking (C instead of B). After continued deposition (4), the surface atoms are partly correctly stacked and partly wrongly stacked. The (buried) starting surface has returned to the correct fcc stacking (from Ref. [26]).

develop shoulders. By measuring the diffraction profile, this and other types of order can be determined. Fuoss et al. [28] were the first to measure such profiles during vapour-phase epitaxial growth of GaAs.

3.2. Solution growth

We now turn our attention to the growth of bulk crystals from a solution. As discussed earlier, in this area atomic-scale information about the surface structure is almost completely lacking. This is definitely a 'niche market' for X-ray diffraction, which can be used to obtain the in situ surface structure. Additional and important information like step heights and roughness can be gained from SPM and optical microscopy, but crystallographic information appears to be the exclusive realm for X-ray diffraction for this type of systems.

Measuring data from a crystal immersed in a solution is far more difficult than the corresponding vacuum experiments. The X-rays need to penetrate the liquid, both on the way in and towards the detector. This leads to an attenuation of the intensity that can be minimised by using hard X-rays (with an energy typically above 15 keV) and by reducing the travel path through the liquid. The latter is achieved by using a foil to produce a thin film in reflection geometry, or by using transmission geometry with small dimensions [29,30]. These principles are illustrated in Fig. 9a and b,

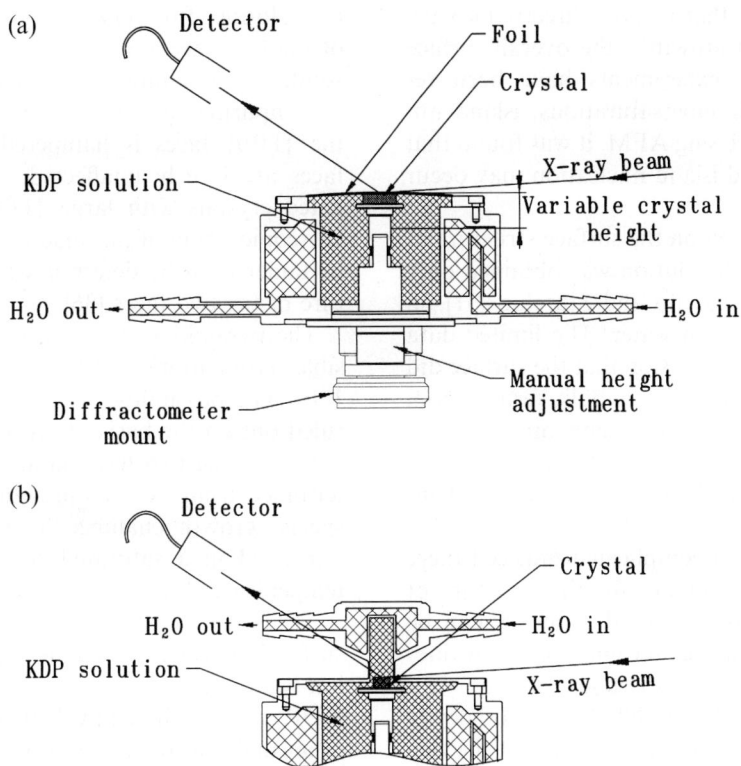

Fig. 9. Two possible scattering geometries for in situ studies of a solid–liquid interface. This is illustrated for the case of KDP crystals in a saturated solution that are kept at a specific temperature by a water bath. (a) Reflection geometry, in which a thin layer of liquid is generated by pressing a foil against the crystal. (b) Transmission geometry, which is suitable for experiments during growth (from Ref. [30]).

respectively. A second and more problematic effect of the liquid is that it scatters X-rays more or less isotropically and thus generates a large background. In vacuum, signal-to-background ratios are easily above 10, but in solution this may decrease by a factor of 100–1000 [30], depending on the solution and crystal used. For this reason, high quality crystals with narrow reflection profiles are preferred when performing such in situ SXRD measurements.

The growth behaviour in a solution differs considerably from that during the MBE or CVD growth shown in Fig. 6. This is owing to a considerable difference in the typical supersaturation used. The supersaturation in a solution is given by the ratio of the concentration of the growth units during growth and that during equilibrium. A low supersaturation indicates that growth occurs close to equilibrium. For MBE and CVD, where growth occurs from a vapour, the supersaturation can be several hundred percent, while in solution growth, supersaturation values below 10% are normal. At such low supersaturations it is difficult to form islands of the type sketched in Fig. 6b, because their creation requires energy (nucleation barrier). The contribution of island nucleation to the growth is then negligible compared with step flow growth. The observable layer-by-layer growth during MBE occurs only at high supersaturations and thus is not typical for solution growth. No X-ray growth oscillations have therefore been observed so far during solution growth. Note that what is called 'step flow' in the vacuum growth community, is called 'layer growth' in the solution growth community. This layer growth mode does not lead to strong variations in the diffracted signal. The quantity that is most directly measurable during solution growth is the overall surface roughness, but few experiments have been reported [30]. At high supersaturations, island nucleation does occur. Using AFM, it was found that step flow growth and island nucleation may occur simultaneously [8].

The first information on the surface structure of a crystal in its growth solution was obtained only five years ago by Chiarello and Sturchio [31] on cleaved calcite ($10\bar{1}4$) in water. The limited data set only allowed the conclusion that the surface did not differ significantly from a bulk termination. More elaborate experiments using more intense synchrotron sources have since led to much improved data sets and therefore more detailed information [32].

For a crystal with a complicated unit cell there are often several possibilities for the structure of the surface. The Hartman–Perdok theory can be used to find these various terminations, but does not provide simple means to choose the most favourable one. Using X-ray diffraction, Gidalevitz et al. [33] were able to determine which of the two alternative terminations occurs on β-alanine and α-glycine (0 1 0) surfaces.

As an example of a solution growth study, we will now discuss KDP (KH_2PO_4, potassium dihydrogen-phosphate) crystals in some detail. Few systems have received more attention in solution growth than KDP [34]; one could call it the 'fruit fly' of crystal growth. Initially, its ferroelectric properties were exploited, but now its main application is as a non-linear optical material for laser applications (frequency conversion and optical switching). The National Ignition Facility (NIF), currently under construction in the USA for studies of an energy production method called inertial confinement fusion, will contain hundreds of high-quality KDP crystals with a size of up to 40 cm and thickness of approximately 1 cm.

KDP crystals are grown from aqueous solution. The Hartman–Perdok theory predicts that the pyramidal {1 0 1} and the prismatic {1 0 0} faces are the most stable [35] and these are indeed the faces observed when such crystals are grown. Fig. 10a shows the typical habit of a KDP crystal. The exact shape is highly dependent upon the purity of the solution. For very pure growth conditions, all of the faces are approximately the same size. For solutions containing small traces of trivalent metal ion impurities like Fe^{3+} and Cr^{3+}, the growth of the {1 0 0} faces is hampered, while the {1 0 1} faces are largely unaffected. This leads to elongated crystals with large {1 0 0} facets. The microscopic origin of this macroscopic behaviour can be understood by determining the interface structure of the two faces [36].

The pyramidal {1 0 1} faces can have two possible terminations: a K^+ or a $H_2PO_4^-$ layer (Fig. 10b). The possibility of a mixed termination was ruled out on the basis of AFM measurements [8]. Which of the two terminations occurs can be determined using X-ray diffraction. This requires a special growth chamber in which the crystal is immersed in a saturated solution at a constant temperature (Fig. 9).

The X-ray measurements require a flat and clean surface. In vacuum, this is typically achieved by ion-beam sputter and annealing cycles. For the KDP crystals, the equivalent procedure is etching in an undersaturated solution ("bathing or showering"). This leads to surfaces that are sufficiently flat to measure the diffraction rod intensity over the entire range between the bulk reflections. The results of the measurements performed at the ESRF are shown in Fig. 11, where the (1,0) diffraction rod data are shown as open circles. The dashed and dotted curves are the calculated rod profiles for bulk $H_2PO_4^-$ and K^+ termination, respectively. The K^+ termination clearly describes the data much better. In this case, the data is of sufficient quality to also include relaxation of the two outermost layers in the fitting procedure. It was found that the K^+ ions relax outwards by an amount of 0.10 ± 0.05 Å and the $H_2PO_4^-$ groups by 0.04 ± 0.05 Å. A similar analysis was done for the {1 0 0} surface, where one termination was expected and found. This surface consists of alternating K^+ and $H_2PO_4^-$ groups.

Having established the atomic structure of both faces, we can now understand the effect of the positive metal ion impurities. The pyramidal face has only K^+ ions on the surface of the crystal, and will repel the positive metal impurities. The growth of this face is therefore unaffected by these impu-

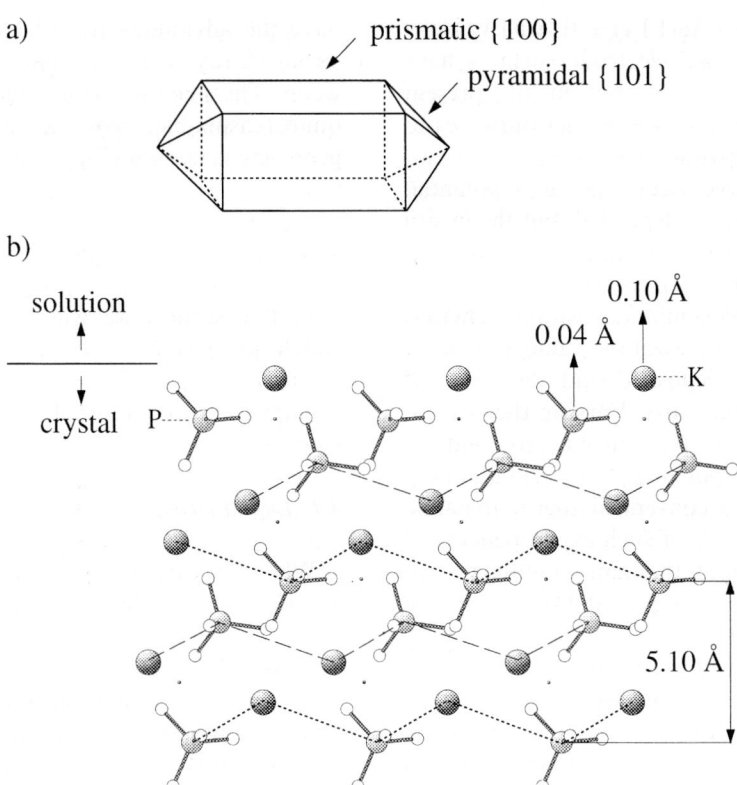

Fig. 10. (a) Growth habit of a KDP crystal with the prismatic and pyramidal faces indicated. In solutions containing metal impurities the growth of the prismatic faces is hindered, but the pyramidal faces are hardly affected. This results in elongated crystals. (b) Schematic side view of the pyramidal face, KDP(1 0 1). From the bulk crystallographic structure one expects this face to either terminate in a potassium layer (dotted curve) or a phosphate layer (dashed curve). X-ray diffraction data show that the first possibility occurs. The outward relaxation of the top layers is shown as well (from Ref. [36]).

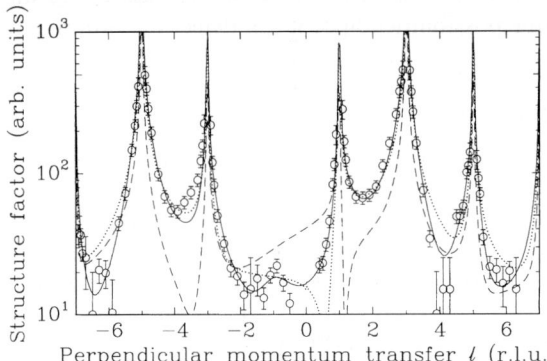

Fig. 11. Structure factor amplitudes along the (1,0) diffraction rod for KDP(1 0 1) as a function of the diffraction index l. The dotted line is a calculation for a bulk K^+-terminated surface, the dashed curve for a $H_2PO_4^-$-terminated one. The solid line is the best fit, starting from a K^+-terminated layer and allowing for perpendicular relaxation [36].

rities. The prismatic face, on the other hand, has both positive and negative ions at the surface, so that the impurities can adsorb easily and block the growth. This leads to the observed elongated crystal shape in impure solutions.

3.3. Electrochemistry

The structure of the solid–liquid interface in electrochemical systems has been the subject of many investigations. Even though many results have been obtained by transferring such interfaces to vacuum and then applying surface-science techniques [37], it is clear that the in situ structure determination using X-ray diffraction offers many benefits. The first X-ray study of this type was done by Samant et al. [38] on the deposition of lead monolayers onto Ag(1 1 1) and Au(1 1 1) surfaces. Since then, 'clean' electrode surfaces have also been investigated [39], but in the present context of crystal growth, we restrict ourselves to studies involving deposition.

At sufficiently large electrochemical potential thick metal layers can be deposited, but the in situ experiments reported so far have only dealt with the 2D layers that are formed when smaller voltages are used (underpotential deposition). There is a finite voltage range between the value at which a single monolayer is deposited and the value at which bulk deposition starts. Varying the electrochemical potential over this range corresponds to varying the chemical potential of the 2D film [40]. Thus the voltage is a convenient thermodynamic variable. As an example of such experiments consider the deposition of thallium monolayers on Ag(1 1 1) as studied by Toney et al. [41]. The lattice constant of the adsorbed thallium layer is different from that of the substrate and the layer is slightly rotated as well. By measuring the position (in reciprocal space) of the diffraction rods, the nearest-neighbour distance of the thallium atoms can be determined as a function of the electrochemical potential (Fig. 12). This distance was found to decrease with the applied voltage. From the data shown in Fig. 12b the compressibility of the 2D layer can be derived. Only with SXRD have such measurements become possible.

Compared to the solution growth experiments discussed earlier, the electrochemistry experiments have the advantage that the substrate is often a strong X-ray scatterer while the liquid is mainly water. The signal-to-background ratio is therefore quite reasonable, provided the proper scattering geometry is chosen [29].

4. Prospects for the future

In this section we will discuss three areas in which some results have already been obtained, but that are expected to grow in importance owing to improvements in sample preparation and X-ray sources.

4.1. Liquid structure

When a crystal grows from a solution or melt, the actual growth occurs at the solid–liquid interface. For a proper understanding of the atomic-scale processes that occur, knowledge of both the crystal surface and the liquid structure near the crystal interface is important. The liquid may influence the growth and the resulting crystal shape in various ways. The interaction of the liquid with the crystal may change the relative stability of the various growth faces, the growth rate can change [42] and concentration gradients may exist. Impurities may also be present in the liquid and pre-ordering of the growth units may occur at the interface.

In this area, X-ray diffraction again offers unique possibilities since it can detect the ordering in

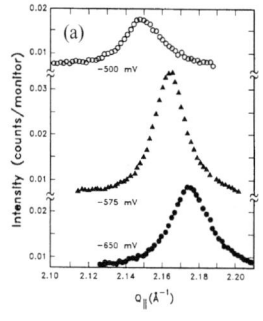
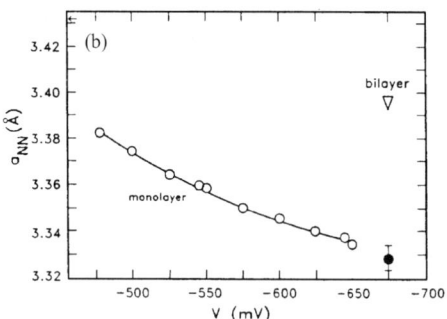

Fig. 12. (a) A diffraction peak position of a thallium film on a silver (1 1 1) substrate as a function of the electrochemical potential. For more negative voltages, the peak shifts to larger values in reciprocal space, indicating a contraction in real space. (b) The thallium–thallium nearest-neighbour distance as function of the electrochemical potential. From such a curve the compressibility of the film can be determined (from Ref. [41], with permission).

the liquid without disturbing it. Solid–liquid interfaces are also of fundamental interest, and a number of theoretical studies have shown that a liquid in contact with a crystal should show layering in the direction perpendicular to the surface [43]. Such layering over a small distance has been observed for a number of systems [44], but much remains to be discovered.

Layering is the ordering component in the direction perpendicular to the surface, but lateral components should exist as well. The lateral ordering is expected to be much weaker. The liquid near the interface 'feels' the two-dimensionally periodic potential of the substrate [45]. This means that the liquid will contribute (weakly) to the substrate diffraction rods. By measuring these rods with high accuracy, the full ordering characteristics of the liquid near the crystalline interface can be derived. However, these effects are weak and the signal-to-background ratio in an experiment is often poor due to the presence of the liquid. The thinner the liquid, the easier the experiment. The most detailed information of this type has been obtained on a simple model system: a quasi-liquid monolayer of Pb on Ge(1 1 1) [46]. Recently, results from thicker liquid films in contact with KDP have been obtained [47]. These experiments require all the X-ray intensity and stability available at the current generation of synchrotron radiation sources.

4.2. Two-dimensional crystals

We remarked earlier that protein crystallography is the main technique used to unravel the structure of proteins. For this, 3D crystals are used. There are many proteins, however, that are hard to crystallise. In these cases crystallisation in the form of 2D membranes may be an alternative [48,49]. Many organic molecules form ordered layers at the interface between air and the solution containing these molecules [50]. This ordering allows the structure of these molecules to be studied, although the resolution depends strongly on the level of ordering. Since the lateral ordering is often poor, the majority of the experiments have so far used only the specular beam to determine the layer structure in the direction perpendicular to the surface.

The use of these 2D crystals for protein crystallography requires an improvement in the quality of the crystals [48]. Currently, most samples consist of many domains that are rotationally disordered with respect to each other [51]. Improvements can be expected in two areas. Firstly, crystallisation methods need to be developed that lead to larger domains. Secondly, improvements in X-ray sources and beam lines will facilitate these measurements. If smaller sample areas still yield sufficiently strong signals, the demands on the domain size are equally reduced. Similar developments have helped 3D protein crystallography tremendously in the past, as presently the crystals can be much smaller than before.

While it seems hard to imagine that the resolution for 2D crystals will become the same as for 3D crystals, useful information is already now being obtained. The situation is, to some extent, similar to the fully developed crystallography of thin layers absorbed on a substrate, but that type of film is easier to study because the substrate induces a high order in the absorbed layer and the structure of such layers is far simpler than that of a protein.

4.3. Coherent X-rays

The applications discussed so far were all based on the use of incoherent X-rays. However, a number of techniques based on coherent X-ray beams are also used. Typically, an incoherent source is made coherent by selecting a small part of the beam using a pinhole of micrometer sized dimension [52]. As a consequence the beam intensity is strongly reduced and only the most powerful synchrotron beam lines yield a sufficiently intense coherent beam. In case of coherent radiation, the diffracted intensity becomes sensitive to the absolute position of the objects contributing to the signal. Instead of a smooth profile, one observes a speckle pattern [52]. The dynamics of a sample can be determined by measuring the changes in this speckle pattern as function of time, a technique called X-ray photon-correlation spectroscopy [53].

Such coherent sources are potentially useful when studying crystal growth. We discussed earlier that the diffracted X-ray intensity from an incoherent beam is constant for step-flow growth (Fig. 6a). However, for a coherent beam this is not true. During step-flow growth the positions of the steps change and thus a different speckle pattern will be generated. In an ideal case, the step pattern after one monolayer will be the same as the starting one, and thus the speckle pattern should change periodically over a one monolayer period. This type of experiment is very difficult, because now the weak surface signal is combined with the weak coherent beam. Step-flow growth has therefore not been observed so far. However, in a related experiment Robinson et al. were able to infer the morphology of Si during the etching of the native oxide [54].

Even using powerful synchrotron sources, the use of coherent X-rays for crystal growth investigations is hampered by the low fluxes currently available. About 10 years from now a new generation of sources is expected to become available, the X-ray free-electron laser (XFEL) [55]. Such a source will produce fully coherent and very fast X-ray pulses with a peak intensity that is expected to be at least a million times higher than that currently available. This will open entirely new directions of research. One possibility is the observation of the nucleation stage of crystallisation. In standard nucleation theory, a smooth size distribution of small crystallites is assumed to be present up to a critical size. It could well be that in reality the size distribution is not so smooth, but that certain sizes are preferred. In case of 2D nucleation on metal surfaces, such 'magic clusters' have been observed using STM [56]. The subcritical nuclei may also have a shape or structure that deviates substantially from that of bulk crystals. Observing the nucleation stage of crystals is not possible with today's X-ray sources, but the X-ray free-electron laser holds promise to cover this entirely new territory.

5. Conclusions

The examples discussed in this paper demonstrate the type of structural information that can be obtained using in situ surface X-ray diffraction. Compared to conventional surface science, the applications to non-vacuum environments are still relatively new and much remains to be investigated. New X-ray sources will enable completely new experiments, but the powerful synchrotron sources available today are sufficiently powerful to enlarge this field of 'surface science under water'. All the phenomena observed on surfaces in vacuum can be found on crystal surfaces in a non-vacuum environment: relaxation, reconstruction, adsorbates, kinetic roughening and thermodynamic roughening. These effects are interesting in their own right, but their role on the growth behaviour of crystals is particularly important.

The liquid side of the interface adds a whole range of new parameters whose role also needs to be explored. What is the ordering behaviour of the various components in the solution? What are their concentration gradients? How can this affect the nucleation of a particular polymorph? How will the surface structure vary depending on the solvent used? These questions are also relevant for other fields, such as (wet) etching and lubrication. It is an exciting idea that these questions can now finally be addressed using X-ray diffraction. The future holds undoubtedly many surprises for the crystal growth field.

Acknowledgements

The work described here was done together with and/or discussed with many colleagues, of whom I would like to thank in particular J. Arsic, P. Bennema, J. De Yoreo, W.J.P. van Enckevort, S. Ferrer, M. Lohmeier, J.E. Macdonald, H. Meekes, C. Norris, M.F. Reedijk, I.K. Robinson, J.M.C. Thornton, J.F. van der Veen, H.A. van der Vegt and S.A. de Vries. I thank J. Loh and D. van Rijsingen for a critical reading of the manuscript.

References

[1] J. Keppler, Strena seu de nive sexangula (G. Tampach, Frankfurt/Main, 1611).
[2] P. Hartman, W.G. Perdok, On the relations between structure and morphology of crystals I, Acta Cryst. 8 (1955) 49;
P. Bennema, in: D.T.J. Hurle (Ed.), Handbook of Crystal Growth, vol. 1B, 1993, p. 477.

[3] C. Herring, Some theorems on the free energies of crystal surfaces, Phys. Rev. 82 (1951) 87.

[4] J. Nyvlt, J. Ulrich, Admixtures in Crystallization, VCH, Weinheim, 1995.

[5] J.B. Romé de l'Isle, Cristallographie (Imprimerie de Monsieur, Paris, 1783), vol. 1, p. 379.

[6] S. Arai, S. Tsukimoto, S. Muto, H. Saka, Direct observation of the atomic structure in a solid–liquid interface, Microsc. Microanal. 6 (2000) 358.

[7] R. Srinivasan, I.I. Suni, Electroless deposition of Au onto Si(1 1 1) studied by surface second harmonic generation, Surf. Sci. 408 (1998) L698.

[8] J.J. De Yoreo, T.A. Land, B. Dair, Growth morphology of vicinal hillocks on the {1 0 1} face of KH_2PO_4: from stepflow to layer-by-layer growth, Phys. Rev. Lett. 73 (1994) 838.

[9] M. Plomp, W.J.P. van Enckevort, E. Vlieg, Etching and surface termination of $K_2Cr_2O_7$ {0 0 1} faces observed using in situ atomic force microscopy, J. Cryst. Growth 216 (2000) 413.

[10] R. Feidenhans'l, Surface structure determination by X-ray diffraction, Surf. Sci. Rep. 10 (1989) 105;
I.K. Robinson, in: G.S. Brown, D.E. Moncton (Eds.), Handbook on Synchrotron Radiation, vol. 3, 1991, p. 221.

[11] I.K. Robinson, Crystal truncation rods and surface roughness, Phys. Rev. B 33 (1986) 3830;
S.R. Andrews, R.A. Cowley, Scattering of X-rays from crystal surfaces, J. Phys. C: Solid State Phys. 18 (1985) 6427.

[12] E. Vlieg, Integrated intensities using a six-circle surface X-ray diffractometer, J. Appl. Cryst. 30 (1997) 532.

[13] E. Vlieg, J.F. van der Veen, S.J. Gurman, C. Norris, J.E. Macdonald, X-ray diffraction from rough, relaxed and reconstructed surfaces, Surf. Sci. 210 (1989) 301.

[14] E. Vlieg, ROD, a program for surface crystallography, J. Appl. Cryst. 33 (2000) 401.

[15] I.K. Robinson, D.J. Tweet, Surface X-ray diffraction, Rep. Prog. Phys. 55 (1992) 599.

[16] C.S. Lent, P.I. Cohen, Diffraction from stepped surfaces—I. Reversible surfaces, Surf. Sci. 139 (1984) 121;
S.K. Sinha, E.B. Sirota, S. Garoff, H.B. Stanley, X-ray and neutron-scattering from rough surfaces, Phys. Rev. B 38 (1988) 2297.

[17] E. Vlieg, A.W. Denier van der Gon, J.F. van der Veen, J.E. Macdonald, C. Norris, Surface X-ray scattering during crystal growth: Ge on Ge(1 1 1), Phys. Rev. Lett. 61 (1988) 2241.

[18] J.H. Neave, B.A. Joyce, P.J. Dobson, N. Norton, Dynamics of film growth of GaAs by MBE from RHEED observations, Appl. Phys. A 31 (1983) 1;
J.M. van Hove, C.S. Lent, P.R. Pukite, P.I. Cohen, Damped oscillations in reflection high energy electron diffraction during GaAs MBE, J. Vac. Sci. Technol. B 1 (1983) 741.

[19] A.J.H.M. Rijnders, G. Koster, D.H.A. Blank, H. Rogalla, In situ monitoring during pulsed laser deposition of complex oxides using reflection high energy electron diffraction under high oxygen pressure, Appl. Phys. Lett. 70 (1997) 1888.

[20] P.H. Fuoss, D.W. Kisker, G. Renaud, K.L. Tokuda, S. Brennan, J.L. Kahn, Atomic nature of organometallic-vapor-phase-epitaxial growth, Phys. Rev. Lett. 63 (1989) 2389.

[21] G.B. Stephenson, J.A. Eastman, C. Thompson, O. Auciello, L.J. Thompson, A. Munkholm, P. Fini, S.P. Den Baars, J.S. Speck, Observation of growth modes during metal–organic chemical vapor deposition of GaN, Appl. Phys. Lett. 74 (1999) 3326.

[22] S.L. Bennett, B. DeVries, I.K. Robinson, P.J. Eng, Layerwise reaction at a buried interface, Phys. Rev. Lett. 69 (1992) 2539.

[23] H.A. van der Vegt, H.M. van Pinxteren, M. Lohmeier, E. Vlieg, J.M.C. Thornton, Surfactant induced layer-by-layer growth of Ag on Ag(1 1 1), Phys. Rev. Lett. 68 (1992) 3335.

[24] J. Vrijmoeth, H.A. van der Vegt, J.A. Meyer, E. Vlieg, R.J. Behm, Surfactant-induced layer-by-layer growth of Ag on Ag(1 1 1): origins and side-effects, Phys. Rev. Lett. 72 (1994) 3843.

[25] P. Bailey, T.C.Q. Noakes, D.P. Woodruff, A medium energy ion scattering study of the structure of Sb overlayers on Cu(1 1 1), Surf. Sci. 426 (1999) 358;
S.A. de Vries, W.J. Huisman, P. Goedtkindt, M.J. Zwanenburg, S.L. Bennett, I.K. Robinson, E. Vlieg, Surface atomic structure of the $(\sqrt{3} \times \sqrt{3})R30°$-Sb reconstructions of Ag(1 1 1) and Cu(1 1 1), Surf. Sci. 414 (1998) 159.

[26] S.A. de Vries, W.J. Huisman, P. Goedtkindt, M.J. Zwanenburg, S.L. Bennett, E. Vlieg, Floating stacking fault during homoepitaxial growth of Ag(1 1 1), Phys. Rev. Lett. 81 (1998) 381.

[27] H.A. van der Vegt, J. Alvarez, X. Torrelles, S. Ferrer, E. Vlieg, In-induced layer-by-layer growth and suppression of twin formation in the homoepitaxial growth of Cu(1 1 1), Phys. Rev. B 52 (1995) 17443.

[28] P.H. Fuoss, D.W. Kisker, F.J. Lamelas, G.B. Stephenson, P. Imperatori, S. Brennan, Time-resolved X-ray scattering studies of layer-by-layer epitaxial growth, Phys. Rev. Lett. 69 (1992) 2791.

[29] H. You, C.A. Melendres, Z. Nagy, V.A. Maroni, W. Yun, R.M. Yonco, X-ray reflectivity study of the copper–water interface in a transmission geometry under in situ electrochemical control, Phys. Rev. B 45 (1992) 11288.

[30] S.A. de Vries, P. Goedtkindt, W.J. Huisman, M.J. Zwanenburg, R. Feidenhans'l, S.L. Bennett, D.-M. Smilgies, A. Stierle, J.J. De Yoreo, W.J.P. van Enckevort, P. Bennema, E. Vlieg, X-ray diffraction studies of potassium dihydrogen phosphate (KDP) crystal surfaces, J. Cryst. Growth 205 (1999) 202.

[31] R.P. Chiarello, N.C. Sturchio, The calcite (1 0 1 4) cleavage surface in water: early results of a crystal truncation rod study, Geochim. Cosmochim. Acta 59 (1995) 4557.

[32] P. Fenter, P. Geissbühler, E. DiMasi, G. Srajer, L.B. Sorensen, N.C. Sturchio, Surface speciation of calcite observed in situ by high-resolution X-ray reflectivity, Geochim. Cosmochim. Acta 64 (2000) 1221.

[33] D. Gidalevitz, R. Feidenhans'l, S. Matlis, D.-M. Smilgies, M.J. Christensen, L. Leiserowitz, Monitoring in situ growth and dissolution of molecular crystals: towards determination of the growth units, Angew. Chem. Int. Ed. Engl. 36 (1997) 955.

[34] L.N. Rashkovich, KDP-Family Single Crystals, Adam Hilger, Bristol, 1991.

[35] P. Hartman, The morphology of zircon and potassium dihydrogen phosphate in relation to the crystal structure, Acta Cryst. 9 (1956) 721;
B. Dam, P. Bennema, W.J.P. van Enckevort, The mechanism of tapering on KDP-type crystals, J. Cryst. Growth 74 (1986) 118.

[36] S.A. de Vries, P. Goedtkindt, S.L. Bennett, W.J. Huisman, M.J. Zwanenburg, D.-M. Smilgies, J.J. De Yoreo, W.J.P. van Enckevort, P. Bennema, E. Vlieg, Surface atomic structure of KDP crystals in aqueous solution: an explanation of the growth shape, Phys. Rev. Lett. 80 (1998) 2229.

[37] E.M. Stuve, N. Kizhakevariam, Chemistry and physics of the liquid/solid interface: a surface science perspective, J. Vac. Sci. Technol. A 11 (1993) 2217.

[38] M.G. Samant, M.F. Toney, G.L. Borges, L. Blum, O.R. Melroy, Grazing incidence X-ray diffraction of lead monolayers at a silver (1 1 1) and gold (1 1 1) electrode/electrolyte interface, J. Phys. Chem. 92 (1988) 220.

[39] B.M. Ocko, J. Wang, A. Davenport, H. Isaacs, In situ X-ray reflectivity and diffraction studies of the Au(1 0 0) reconstruction in an electrochemical cell, Phys. Rev. Lett. 65 (1990) 1466.

[40] O.R. Melroy, M.F. Toney, G.L. Borges, M.G. Samant, J.B. Kortright, P.N. Ross, L. Blum, Two-dimensional compressibility of electrochemically adsorbed lead on silver (1 1 1), Phys. Rev. B 38 (1988) 10962.

[41] M.F. Toney, J.G. Gordon, M.G. Samant, G.L. Borges, O.R. Melroy, D. Yee, L.B. Sorensen, Underpotentially deposited thallium on silver (1 1 1) by in situ surface X-ray scattering, Phys. Rev. B 45 (1992) 9362.

[42] X.Y. Liu, E.S. Boek, W.J. Briels, P. Bennema, Prediction of crystal growth morphology based on structural analysis of the solid–fluid interface, Nature 374 (1995) 342.

[43] W.A. Curtin, Density-functional theory of the solid–liquid interface, Phys. Rev. Lett. 59 (1987) 1228.

[44] M.F. Toney, J.N. Howard, J. Richer, G.L. Borges, J.G. Gordon, O.R. Melroy, D.G. Wiesler, D. Yee, L.B. Sorensen, Voltage-dependent ordering of water molecules at an electrodo–electrolyte interface, Nature 368 (1994) 444;
B.M. Ocko, Smectic-layer growth at solid interfaces, Phys. Rev. Lett. 64 (1990) 2160;
W.J. Huisman, J.F. Peters, M.J. Zwanenburg, S.A. de Vries, T.E. Derry, D.L. Abernathy, J.F. van der Veen, Layering of a liquid metal in contact with a hard wall, Nature 390 (1997) 379.

[45] M.F. Toney, J.G. Gordon, M.G. Samant, G.L. Borges, O.R. Melroy, L.S. Kau, D.G. Wiesler, D. Yee, L.B. Sorensen, Surface X-ray scattering measurments of the substrate-induced spatial modulation of an incommensurate adsorbed monolayer, Phys. Rev. B 42 (1990) 5594.

[46] F. Grey, R. Feidenhans'l, J.S. Pedersen, M. Nielsen, R.L. Johnson, Pb/Ge(1 1 1)1 × 1: an anisotropic two-dimensional liquid, Phys. Rev. B 41 (1990) 9519;
S.A. de Vries, P. Goedtkindt, P. Steadman, E. Vlieg, Phase transition of a Pb monolayer on Ge(1 1 1), Phys. Rev. B 59 (1999) 13301.

[47] M. Reedijk, J. Arsic, F.F.A. Hollander, S.A. de Vries, E. Vlieg, to be published.

[48] S.A.W. Verclas, P.B. Howes, K. Kjaer, A. Wurlitzer, M. Weygand, G. Büldt, N.A. Dencher, M. Lösche, X-ray diffraction from a single layer of purple membrane at the air/water interface, J. Mol. Biol. 287 (1999) 837;
C. Vénien-Bryan, P.F. Lenne, C. Zakri, A. Renault, A. Brisson, J.F. Legrand, B. Berge, Characterization of the growth of 2D protein crystals on a lipid monolayer by ellipsometry and rigidity measurements coupled to electron microscopy, Biophys. J. 74 (1998) 2649.

[49] D. Gidalevitz, Z. Huang, S.A. Rice, Urease and hexadecylamine-urease films at the air–water interface: an X-ray reflection and grazing incidence X-ray diffraction study, Biophys. J. 76 (1999) 2797.

[50] J. Als-Nielsen, D. Jacquemain, K. Kjaer, F. Leveiller, M. Lahav, L. Leiserowitz, Principles and applications of grazing incidence X-ray and neutron scattering from ordered molecular monolayers at the air–water interface, Phys. Rep. 246 (1994) 252.

[51] I. Weissbuch, P.N.W. Baxter, I. Kuzmenko, H. Cohen, S. Cohen, K. Kjaer, P.B. Howes, J. Als-Nielsen, J.M. Lehn, L. Leiserowitz, M. Lahav, Oriented crystalline monolayers and bilayers of 2 × 2 silver (I) grid architectures at the air–solution interface: their assembly and crystal structure elucidation, Chem. Eur. J. 6 (2000) 725.

[52] M. Sutton, S.G.J. Mochrie, T. Greytak, S.E. Nagler, L.E. Berman, G.A. Held, G.B. Stephenson, Observation of speckle by diffraction with coherent X-rays, Nature 352 (1991) 608.

[53] S. Brauer, G.B. Stephenson, M. Sutton, R. Brüning, E. Dufresne, S.G.J. Mochrie, G. Grübel, J. Als-Nielsen, D.L. Abernathy, X-ray intensity fluctuation spectroscopy observations of critical dynamics in Fe_3Al, Phys. Rev. Lett. 74 (1995) 2010.

[54] I.K. Robinson, J.L. Libbert, I.A. Vartanyants, J.A. Pitney, D.-M. Smilgies, D.L. Abernathy, G. Grübel, Coherent X-ray diffraction imaging of silicon oxide growth, Phys. Rev. B 60 (1999) 9965.

[55] R. Brinkmann, G. Materlik, J. Rossbach, J.R. Schneider, B.H. Wiik, An X-ray FEL laboratory as part of a linear collilder design, Nucl. Instrum. Meth. Phys. Res., Sect. A 393 (1997) 86.

[56] G. Rosenfeld, A.F. Becker, B. Poelsema, L. Verheij, G. Comsa, Magic clusters in 2 dimensions, Phys. Rev. Lett. 69 (1992) 917.

 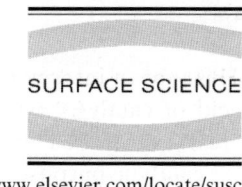

Real time chemical dynamics at surfaces

M. Bonn, A.W. Kleyn *, G.J. Kroes

Leiden Institute of Chemistry, Leiden University, Gorlaeus Laboratories, P.O. Box 9502, 2300 RA Leiden, Netherlands

Received 13 July 2000; accepted for publication 10 July 2001

Abstract

It is a major goal in surface science to make movies of molecules on surfaces, in which the reaction of the molecules on the surface can be followed on a femtosecond time scale, with sub-nanometer resolution. By moving the actors (the molecules) to precisely determined positions on the stage (the surface) at some well-defined moment in time, and subsequently making a space- and time-resolved documentary of what happens next, we would be able to understand the reactive interactions between molecules on surfaces in the greatest possible detail. This would enable us to set the stage and bring together the actors in such a way as to produce the chemical outcomes our society needs, by improving existing catalysts and designing novel catalysts, and by engineering novel reactions on surfaces. Any future director of such movies needs to know which techniques (i.e., which theoretical and experimental methods) hold promise for movie making, what has been done with these techniques, and what can be done with appropriate extensions. The methods we discuss are: (i) the time-dependent wave packet method, which is a theoretical method for simulating molecule–surface reactions with sub-nanometer resolution on a femtosecond time scale, (ii) molecular beam experiments, which allow detailed investigation of the molecule–surface interaction at a molecular level, and (iii) time-resolved laser pump–probe experiments, which allow reactions to be studied with femtosecond resolution. In particular, we discuss (i) theoretical studies of the dissociation reaction of hydrogen on metal surfaces, the reactive system presently understood at the greatest level of detail, (ii) the reactive and non-reactive scattering of heavy diatomics (NO,CO) from metal surfaces, and (iii) the competition between reaction of coadsorbed CO with O and desorption of CO, again on a metal surface. We examine possibilities to extend these methods to make movies at the desired level of detail. We also discuss which reactions are likely to provide good material for plots of movies that will be exciting for future generations of surface scientists. © 2001 Elsevier Science B.V. All rights reserved.

Keywords: Atomistic dynamics; Laser methods; Molecule–solid reactions; Chemisorption; Desorption induced by photon stimulation; Photochemistry; Second harmonic generation; Surface chemical reaction

1. Introduction

Over the past decades surface science has constructed an impressive database on the structure of solid surfaces, with respect to both the chemical composition and the electronic and geometric structures. Most of this information pertains to the static equilibrium properties of the surface. Questions regarding the dynamics of chemical bond rearrangement on these surfaces determining, for example, how these surface structures are actually formed, have been studied to a much lesser extent.

---

* Corresponding author.
 E-mail address: a.kleyn@chem.leidenuniv.nl (A.W. Kleyn).

0039-6028/01/$ - see front matter © 2001 Elsevier Science B.V. All rights reserved.
PII: S0039-6028(01)01550-3

This type of information is vital in for instance the field of catalysis, where recent progress has been great: ever more complex materials are being used for catalytic purposes and novel catalysts are being synthesised. However, in many cases this progress is dictated by experience, intuition and trial and error. For many applications, the understanding of the mechanism behind the catalytic reaction is unavailable and cannot be obtained by the synthesis of novel catalysts. A full atomic scale picture of reaction dynamics at surfaces is desirable so that existing processes can be optimised and new and better catalysts can be designed. Especially for applications in the automotive industry the demands on the catalysts are increasing to such an extent that true 'atomic engineering' is becoming a necessity to fully control the catalytic process.

There are several routes to a detailed microscopic understanding of the dynamics of surface reactions. It is the goal of this article to sketch the state-of-the-art in three techniques that can provide detailed insights into chemical dynamics at surfaces: quantum dynamics calculations, molecular beam scattering experiments and femtosecond laser experiments. As we will show, the combination of these complementary techniques can yield information with sub-nanometre resolution on a femtosecond timescale.

To study the reaction of molecules on surfaces in detail, it is important to consider the different degrees of freedom the system has. The reacting molecule's rotational, vibrational and translational degrees of freedom will affect its reactivity as it hits the surface. In addition, the surface has lattice vibrations (phonons) and electronic degrees of freedom. Both the lattice vibrations and the electrons can act as efficient energy absorbers or energy sources in a chemical reaction.

The key concept in chemical reaction dynamics in general is the potential energy surface (PES). The PES determines the electronic energy of the system as a function of the different degrees of freedom of the system mentioned above. For instance, for the dissociation of a di-atomic molecule like H_2 on a surface two pivotal coordinates are the distance between the molecule and the surface and the inter-atomic distance within the molecule. For a more complete description additional coordinates such as the angle of the molecular axis with respect to the surface can be taken into account. An example of the PES for hydrogen dissociating on a surface is given in the left part of Fig. 1. This shows a typical "elbow-potential", where in the upper left of the potential molecular hydrogen is far from the surface, and in the lower right two hydrogen atoms are stuck to the surface ("chemisorbed"). A molecule approaching the surface with enough translational energy will move from the "molecule + surface" potential well (upper left corner) to the "atoms + surface" well (lower right corner). In doing so, it will favor following the minimum energy path connecting reactants and products. As the molecule moves towards the surface, one observes that the H–H bond length increases. Next, the molecule has to cross an energetic barrier (which may be viewed as a "mountain pass"), which separates the "molecule + surface" part of the PES from the "atoms + surface" part. The system is in the so-called transition state when it resides on this barrier. For surface reactions in which the electronic state of the system does not change, sufficiently detailed knowledge of the PES leads to a complete understanding of the microscopic reaction dynamics, since the PES characterises the complete reaction process for such reactions. The potential energy surface determines the motion of reactants in the multi-dimensional space of the atomic coordinates, and therefore the rate at which the reaction occurs. It determines the energetics of the reaction as well as lifetimes of reaction intermediates and transition states.

The first technique that we will describe, molecular level modelling of the dynamics of surface reactions, has proven to be a powerful tool to unravel the mechanisms of molecule–surface reactions [1,2]. In the most simple approach, the PES is used to predict the motion of the reactants, either through the Newtonian dynamics or even through the quantum-mechanical molecular motion on this potential. In this way information can be obtained on molecular distance- and timescales. Theorists are presently capable of making 'real time movies' of how reactants move through a transition state to a final product. These movies are conceptually very useful, since they demon-

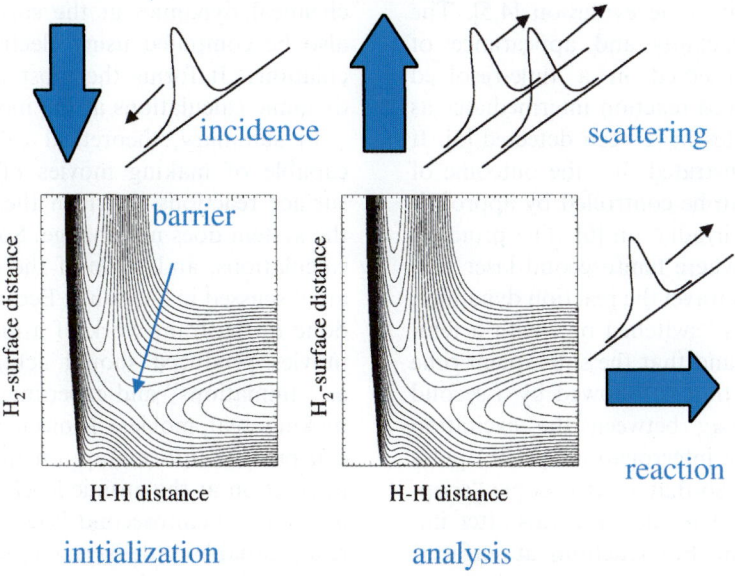

Fig. 1. Panel illustrating the generic form of a potential energy surface describing the dissociative chemisorption of a diatomic molecule, like H_2, on a surface. The shape of the PES is explained in the text. The figure also illustrates the first and final stage of a time-dependent wave packet calculation (see also text of Section 2). In the first, initialization stage, the initial wave packet is set up (left). In the final, analysis stage, the scattered and reacted parts of the wave function are analyzed (right). For further details, see the text.

strate how chemists should envision the progress of chemical reactions, and provide insights into transition states and/or bottlenecks of chemical reactions [3]. Theoretical simulations can be used for the interpretation of existing experimental information, but also for predicting new trends in reactivity and can lead to new concepts concerning reaction dynamics.

Experiments, on the other hand, also have the capacity to reveal new trends in reactivity and to discover new phenomena concerning reaction dynamics. In addition, experiments provide the ultimate testing ground of any theoretical prediction. What is more, new experimental discoveries can demonstrate to theorists what important elements their theories should contain, and what kind of simplifications are appropriate. Experiments should *not* be done at the level of overall rates, because very different microscopic dynamics can fortuitously lead to the same overall 'macroscopic' kinetics. Rather, experiments should investigate the surface reaction dynamics directly at the molecular level. In this way, theory and experiment can complement each other ideally, and benefit from the mutual feedback.

What is the ideal experiment? Such an experiment would involve the experimental analogue of a "movie camera" making snapshots of chemical reactions while molecules on the surface pass through the transition state and other reaction intermediates. The ideal experimental set-up would have the ability to resolve individual atoms as chemical bonds are rearranging in time. This means a spatial resolution on the order of a fraction of a nanometer (1 nm = 10^{-9} m) and a time resolution of tens of femtoseconds (1 fs = 10^{-15} s). The latter time is the vibrational period of a simple diatomic molecule like O_2 or NO. To reach this goal ultrafast spectroscopic tools are required that can distinguish individual atoms or molecules. Such tools are not yet available.

However, as far as reaction time scales are concerned, it has become possible to follow reaction dynamics 'real-time' with the recent advent of ultra-short laser pulses: Ultra-short laser pulses can be used to initiate a chemical reaction and subsequently probe its time evolution [4,5]. The disappearance of reactants and appearance of products can be observed in a time-resolved manner, and short-lived reaction intermediates as well as transition states have been detected [3]. It has even been demonstrated that the outcome of chemical reactions can be controlled by appropriate short-pulse laser irradiation [6]. The principle behind experiments where femtosecond laser pulses are employed to unravel the reaction dynamics, is that the reaction is "switched on" with a femtosecond laser pulse and that the subsequent time evolution of the reaction is followed by a second pulse. The time delay between the excitation (pump) pulse and the interrogation (probe) pulse can be varied at will, so that spectroscopic "snapshots" can be made at regular intervals after initiation of the reaction. For reactions at surfaces, there are a few complicating factors such as the low optical density of one layer of molecules on the surface, making spectroscopic detection experimentally challenging. Nonetheless, these femtosecond spectroscopic techniques have recently been transferred to the study of surface chemical dynamics [7–9], but the spatial resolution remains limited to ~1 μm, determined by the wavelength of light.

Scanning tunnelling microscopy (STM) allows the study of surfaces with a resolution of fractions of nanometres. This technique is well documented in the surface science literature, see also other chapters of this volume. However, the time resolution of the method is limited to microseconds at the very best.

Although ultrafast spectroscopy can provide important insights into the dynamical evolution of a reactive system, viz. the dynamics of the molecular motion, it is not particularly well suited for determination of potential energies surfaces. A method that does provide information on the molecular-level interaction parameters that define the potential energy surface for a molecule approaching the surface is the method of molecular beam scattering. By colliding molecular beams of reactants with a surface, information about the sub-nanometre PES describing the molecule–surface interaction can be obtained from the scattering pattern of the molecules. As already mentioned, the PES directly determines the chemical dynamics at the surface. The PES can also be computed using electronic structure calculations. It forms the most important input to dynamic calculations at the molecular level.

In summary, theoretical calculations are now capable of making movies of simple molecule–surface reactions in which the electronic state of the system does not change. Some results of these calculations, and some of the methods employed are discussed in Section 2. Because calculations are done in the framework of a model of reality, the movies produced should actually be considered as "animations" and experiments are required for making real movies. Molecular beam scattering can provide information on the molecule–surface interaction at the atomic level, and is discussed in Section 3. Femtosecond laser experiments are already capable of delivering the time-resolution for moviemaking, and are discussed in Section 4. In Section 5 we discuss prospects for the future.

2. Theoretical calculations on dissociative chemisorption of H_2 on metal surfaces

During the last decade, theoretical efforts have focused on what is perhaps the most simple molecule–surface reaction that can be imagined, the dissociation of H_2 on metal surfaces. Living in an age where the human genome will soon be determined completely, one might think that this is an easy problem, but it is not! While much progress has been made on the dynamics of dissociation of H_2 on surfaces, it is still not possible to determine the barrier to such reactions theoretically with "chemical accuracy" (4 kJ/mol) [10–12]. In this respect calculations regarding gas–surface reaction dynamics differ greatly from gas phase reaction dynamics calculations, where this accuracy has been achieved for three-atom reactions involving light atoms [13]. One challenge taken up by theoretical surface scientists is to try to model the dissociation of H_2 on metal surfaces as accurately as possible, using a bottom-up ("first-principles") approach borrowed from physics. Because theorists can benchmark their tools on them, and because these are the systems that can be investigated in the greatest possible detail, H_2–metal surface dissociation reactions will continue to be studied during the first decades of the 21st century.

As depicted in Fig. 2, a variety of events can occur if an H_2 molecule collides with a metal surface, and these events may be probed using two types of experiments. To understand these possibilities, a short introduction to molecular quantum mechanics can be useful. Quantum mechanically, a molecule can exist in different vibrational and rotational states, which are associated with vibrational motion of the atoms relative to one another, and the rotation of the molecule, respectively. A definite energy is associated with each state, and the molecule can only take on the specific energy values associated with these states (the vibrational and rotational energies of the molecules are "quantised"). Vibrational states are labelled by the vibrational quantum number v, and rotational states by the rotational quantum numbers j (defining the magnitude of the angular momentum associated with the molecule's rotation, which is a vector) and m_j (defining the projection of the angular momentum vector on an axis which is normal to the surface). To a good approximation, the vibrational energy of a molecule is equal to $(v + 1/2) \times h$ times the vibrational frequency of the molecule, h being Planck's constant. Similarly, to a good approximation the rotational energy of a molecule is proportional to $j \times (j + 1)$. In other words, high values of v and j correspond to high vibrational and rotational energies, respectively.

A hydrogen molecule approaching the surface at a collision energy E in a particular initial vibrational and rotational state (panel B of Fig. 2) can either react (panel C) or it can scatter back into the gas phase (panel D). For sufficiently low collision energies (up to about 1.5 eV or 150 kJ/mol), due to energetic constraints reaction will result in the dissociated atoms remaining on the surface (this is called dissociative chemisorption). This process can be investigated using molecular beam experiments (as depicted schematically in the left part of panel A and discussed more fully in Section 3), in which the collision energy of the molecule can be well controlled [14,15]. If one assumes that "microreversibility" holds (i.e., if one assumes that the rate of the reaction equals the rate of the reverse process), dissociative chemisorption can also be investigated using experiments on "associative desorption" [15–17]. In this process, two atoms recombine at the surface to form a molecule

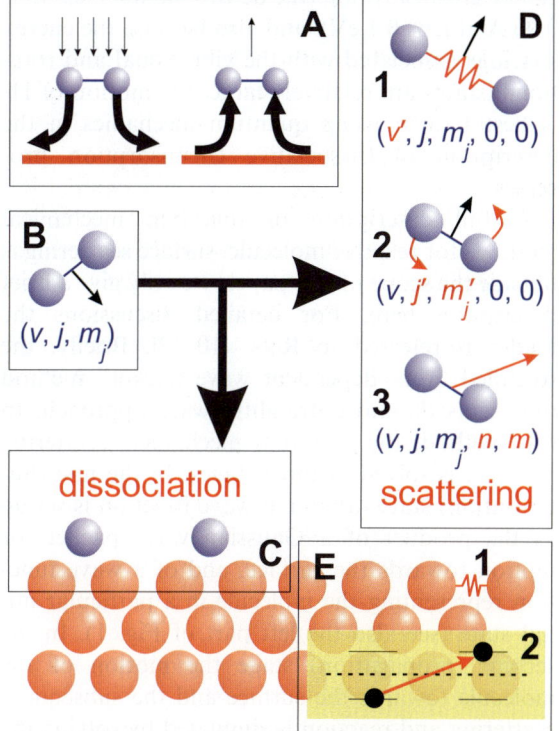

Fig. 2. Experimental methods to study reaction and scattering are shown, as well as the dynamical outcomes of a collision of a diatomic molecule with a reactive surface which can be probed in these experiments. According to microreversibility, the measured distribution of translational desorption energies E in an associative desorption experiment (panel A, right) should reflect the dissociation probability measured for the collision energy E in a molecular beam experiment (A, left). This is because associative desorption, in which two atoms recombine at the surface to form a molecule desorbing with translational energy E, is the reverse process of dissociative chemisorption at collision energy E. The energy required for associative desorption is provided by the temperature of the surface. A molecule approaching the surface under specific initial conditions defined by quantum numbers v and (j, m_j) (B, see also text) may react (C) or scatter back to the gas phase (D). In the latter case, it can change its vibrational state to v' (D1), its rotational state to (j', m'_j) (D2), or it can diffract (see text for explanation D3). Both reaction and scattering can occur in coincidence with excitations of the surface lattice vibrations (E1) and with electronic excitations (E2). For further explanations, see the text.

desorbing with translational energy E (depicted in the right of panel A).

As stated before, instead of reacting the molecule can also scatter back to the gas phase (panel D of Fig. 2). In scattering, the molecule can change its vibrational state (to v', this is called vibrationally inelastic scattering, panel D1), its rotational state (to (j', m'_j); rotationally inelastofic scattering, panel D2), and it can gain additional momentum associated with translational motion parallel to the surface (panel D3). On an ideal, rigid surface with perfect translational symmetry, changes in the molecule's parallel translational momentum are quantised (molecular diffraction), and changes in the molecules parallel translation can be defined through the quantum numbers for diffraction n and m. Both scattering and chemisorption can occur in coincidence with excitation of the surface lattice vibrations (called phonons, panel E1) and with excitations of electrons from one level in the conduction band to another level in that band (these excitations are called electron–hole pair excitations, panel E2).

Reaction can be studied using molecular beam and associative desorption experiments [15–17]. Molecular beam experiments can also be employed to analyze non-reactive scattering. Molecular beam and associative desorption experiments have yielded information on the influence of translational energy and incidence angle on reaction. In the most advanced associative desorption experiments, the influence of translational energy, the molecule's vibrational and rotational state, and molecular orientation on reaction are determined simultaneously [18]. In the most advanced molecular beam experiments, lasers are used to overpopulate a particular vibrationally excited, rotational level in the H_2 incident beam to determine state-to-state probabilities for rotationally elastic and inelastic scattering as well as for rovibrationally inelastic scattering (in the latter process, both the vibrational and the rotational state of the molecule change) [19–22]. The most complete set of data has been obtained for the reaction of H_2 and D_2 on Cu(1 1 1) [15].

The extent to which molecular translational motion is "quantum-like" can be inferred from the de Broglie wavelength, which is the wavelength associated with translational motion of the molecule at a specific translational energy. The lighter a molecule, the larger its de Broglie wavelength, and the more quantum-like the motion will be. Because for H_2 this wavelength can take on considerable values for collision energies of interest to dissociative chemisorption (the de Broglie wavelength is 0.06 Å at $E = 0.1$ eV), and also because the energy spacings associated with the vibrational and rotational states are relatively large, the motion of H_2 is best treated using quantum mechanics in the description of dissociative chemisorption processes.

A full description of quantum mechanical methods for reactive molecule–surface scattering is outside the scope of this paper; we only give a brief description here. For detailed discussions the reader is referred to Refs. [10–12]. Briefly, the so-called "time-dependent wave packet" method represents the most straightforward approach. In this method, the quantum mechanical scattering problem is solved in three stages. In the first (initialisation) stage, an initial wave function is set up as the product of a Gaussian wave packet for motion towards the surface, and of a wave function representing the molecule in a particular initial state (see also the left part of Fig. 1). In the second (propagation) stage the motion of the molecule towards the surface and the subsequent scattering and reaction is simulated by solving the quantum mechanical equations for nuclear motion. This stage of the calculation requires as input the PES, which describes the interaction of the molecule with the surface. Such potentials can now be computed reasonably accurately using density functional theory, a method which is not discussed here because it is explained in detail in the contributions of Gross and of Scheffler and coworkers to the present volume. Finally, in the third (analysis) stage probabilities of the molecule scattering back into a particular final state are computed by analysing the reflected part of the wave function, and reaction probabilities are computed by calculating the flux through a dividing surface positioned in the product region of the potential (see right part of Fig. 1).

An advantage of the time-dependent wave packet method is that it allows one to follow the dynamics of the reaction on the time scale in which reaction takes place, i.e., with femtosecond resolution. An example of snapshots from a movie of $v = 0$ H$_2$ reacting on the top site of Cu(1 0 0) is shown in Fig. 3 [23]. The wave function at $t = 51$ fs represents molecules approaching the surface at various collision energies (the initialisation stage referred to above). Many of these molecules are seen to react at $t = 88$ fs. The third and fourth frame ($t = 102$ and 131 fs) show molecules scattering back, and the "nodal pattern" indicates that many of the reflected molecules have been excited to the $v = 1$ vibrationally excited state in the collision (the $v = 1$ vibrational wave function exhibits two maxima, and is zero ("has a node") in between these maxima).

The earliest quantum mechanical calculation of dissociative chemisorption was a two-dimensional (2D) dynamics calculation, in which the influence on reaction of the molecule's vibration and its motion towards the surface was investigated [24]. The motion was described using an 'elbow-type' PES, as depicted in Figs. 1 and 3. This type of analysis was soon followed by studies of higher dimensionality (3D and 4D) which also described either rotations or translational motion parallel to the surface; see for instance Refs. [10–12]. These low-dimensional (2D–4D) calculations were able to explain several experimentally observed trends. For instance, vibrationally exciting the molecule prior to the collision with the surface enhances the reaction on copper because the barrier for that system is "late" (occurs at an extended H–H distance, see also Fig. 1), allowing vibrational energy to flow into the reaction co-ordinate. [25,26].

Whereas these quantum simulations were able to explain several experimentally observed trends, classical trajectory calculations showed that for obtaining quantitatively accurate results it should be necessary to perform six-dimensional (6D) calculations, in which motion in all six molecular degrees of freedom is treated explicitly [27]. Excitations of the surface electrons and surface lattice vibrations (panel E of Fig. 2) are still neglected in such a treatment, but a full description is given of dissociation and scattering (panels C and D in Fig. 2).

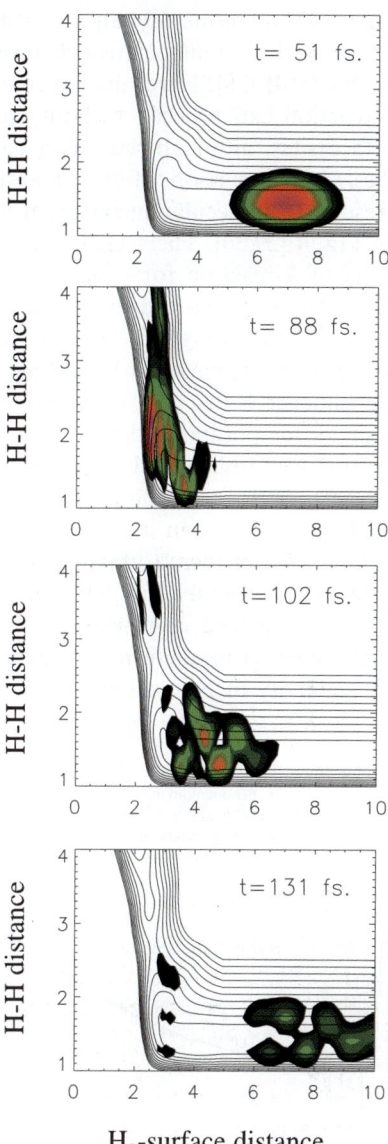

Fig. 3. Snapshots from a 2D wave packet calculation on the dissociation of $v = 0$ H$_2$ above the top site of Cu(1 0 0) [23]. The wave packet is seen moving on the PES, which is depicted in a contour plot, and exhibits a late barrier to reaction (occurring at a value $r = 2.7a_0$ of the H–H distance, which is considerably larger than the equilibrium bond distance of H$_2$, which is 1.4 a_0). In the frame where the time $t = 102$ fs, the back scattered portion contains fast travelling molecules in the vibrational ground state $v = 0$ (at large Z) and slower molecules in the vibrationally excited $v = 1$ state (at smaller Z). The frame for $t = 131$ fs shows that some molecules remain adsorbed on the surface for quite a long time, in trapped states called resonances.

The first 6D quantum dynamical study addressed the unactivated dissociative chemisorption of H_2 on Pd(100) [28]. Here, unactivated means that the reaction can proceed without barrier at some surface sites, although there is a barrier at other sites. A comparison of the quantum dynamics results to molecular beam experiments is shown in Fig. 4 [29,30]. The system considered is an example of a reaction for which the reaction probability first decreases and then increases with increasing collision energy [30–32]. An important question for such systems has been whether this energy dependence is due to a mechanism involving trapping or temporary residence in a molecular chemisorption well (this would allow the molecule time to react), or to a mechanism in which the repulsive forces acting on the incoming molecules are able to steer them to favourable reactive orientations at sufficiently low collision energies ("steering"). Gross and co-workers have shown that it is the steering mechanism that explains the increased reactivity of H_2 on Pd(100) at low energies [28]. On the basis of their results, they also discussed how the predicted mechanism could be confirmed experimentally, namely by showing that at low energies the reaction is not as efficient for rotating molecules as for non-rotating molecules (because non-rotating H_2 is more easily steered into a favourable reactive configuration). Subsequently, their results were confirmed for the closely related $H_2 + Pd(111)$ system [33]. For a more detailed discussion of steering, we refer to the discussion accompanying Fig. 4 of the contribution of Gross to this volume.

The first 6D quantum dynamical calculations on activated dissociation (in which the dissociation is activated for each surface site) were performed for $H_2 + Cu(100)$. The 6D quantum results are compared with a fit to molecular beam and associative desorption experiments in Fig. 5 [34–36]. This comparison and the comparison for $H_2 + Pd(100)$ reveal a problem with the theory. The agreement between theory and experiment is reasonably good, but not excellent. For instance, for $H_2 + Cu(100)$ the theoretical curve is shifted upwards relative to the experimental curve by about 0.1 eV, suggesting that the barriers in the PES that was used to describe the molecule–surface

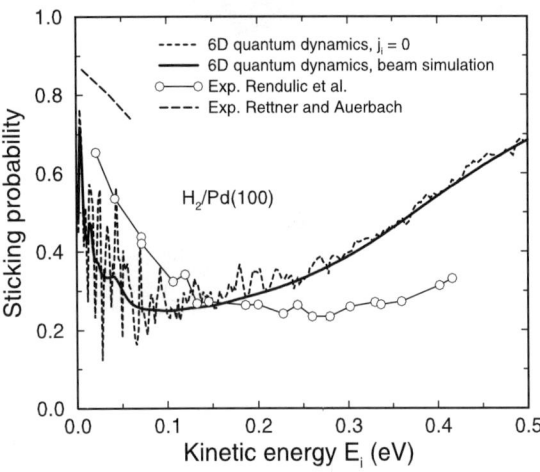

Fig. 4. The probability for dissociation of H_2 on Pd(100) is plotted as a function of normal incidence energy. 6D quantum dynamics results are given for H_2 incident in its vibrational and rotational ground state, and for H_2 incident a distribution over initial vibrational and rotational states that is representative of molecular beam experiments. Results from molecular beam experiments are also shown, for normal incidence and for an incident angle of 15°. Reprinted from Surface Science Reports 32, A. Gross, Reactions at surfaces studied by ab initio dynamics calculations, pp. 291–340. Copyright (1998), with permission from Elsevier Science [102].

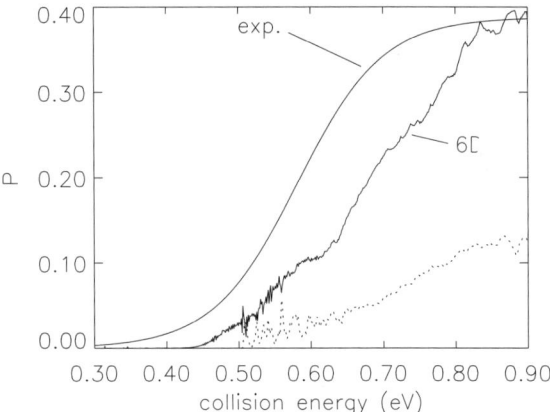

Fig. 5. The probability for dissociation of H_2 incident on Cu(100) in its $(v = 0, j = 0)$ vibrational and rotational ground state. The dissociative chemisorption probability obtained from 6D quantum dynamics calculations is plotted as a function of normal incidence energy, and compared to the experimentally fitted reaction probability curve. Also shown is the calculated probability for vibrational excitation $P(v = 0, j = 0 \to v = 1)$ in scattering (dotted line) (from Refs. [34,35]).

interaction are too high on average by about that amount.

The potentials employed in the 6D calculations on $H_2 + Pd(100)$ and $H_2 + Cu(100)$ were taken from a density functional approach, which is currently the best method available for molecules interacting with metal surfaces [37,38], and is discussed in some detail in the contributions of Gross and of Scheffler and coworkers to the present volume. A important ingredient in this approach is the use of the so-called generalised gradient approximation [39]. This approximation is related to a problem that is fundamental to density functional theory, which holds that the ground state electronic energy of a system is uniquely determined by its electronic density, but has so far not provided a unique description to compute this density, and the electronic energy from it.

This means that improvement of the accuracy of quantum dynamical calculations on molecule–surface reactions depends crucially on fundamental research aimed at devising better density functionals (which relate electronic densities to energies), or alternative electronic structure methods which perform better than density functional theory.

In the future, techniques will be needed that allow for the visualisation of the reaction dynamics in spite of the high dimensionality of the calculations, to overcome the black box character of the calculations that have been done so far. One possible way forward is presented by methods called "flux analysis methods". Fig. 6 shows a specific application of such an analysis method in a 6D quantum dynamics calculation on dissociation of H_2 on $Cu(100)$ [40]. In this example, the flux

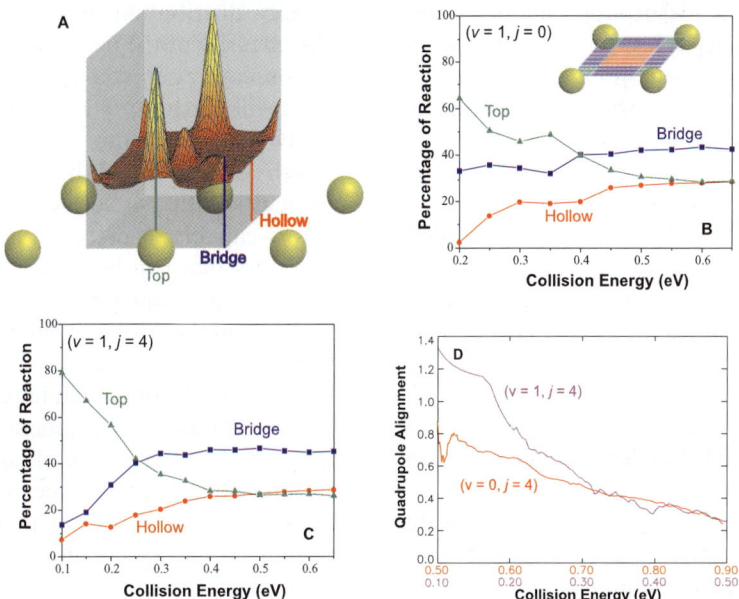

Fig. 6. Computed site-specific reaction probabilities for $v = 1$ states, for reaction of H_2 on $Cu(100)$ (A–C). The reaction probability flux density—which, integrated over the surface unit cell, yields the reaction probability—is shown as a function of surface site for the vibrationally excited ($v = 1, j = 0$) incident state at a collision energy of 0.2 eV (A). A strong preference for reaction at the top site is found, which is also reflected by the probability for reaction in the vicinity of this site being higher than for the bridge and hollow sites, at low energies (B). The site-specific reaction probabilities shown were obtained by integrating the reaction probability flux density over pre-assigned areas of the surface unit cell (inset to B). At low energies top site reaction is also preferred for the rotationally excited $j = 4$ state within the vibrationally excited $v = 1$ level(C). The difference in preferred reaction site between $v = 1$ and $v = 0$ (which reacts at the bridge site [40]) could be confirmed experimentally by showing that the rotational quadrupole alignments measured for H_2 associatively desorbing in the ($v = 1, j = 4$) state are higher than for H_2 desorbing in ($v = 0, j = 4$), at desorption (or collision) energies where the reaction probabilities are low (a few percent, as occurs between 0.1 and 0.2 eV for $v = 1$, and between 0.5 and 0.6 eV for $v = 0$) (D) (from Ref. [40]).

analysis method allowed the theorists to predict, with sub-nanometre resolution, that vibrationally excited ($v = 1$) H_2 will react on top of a surface Cu atom, i.e. at the top site, which is surprising because the barrier to reaction is higher (by 0.13 eV) at the top site than at the bridge site (between two copper atoms), where ($v = 0$) H_2 preferentially reacts. According to the theorists, the prediction can be confirmed by performing a velocity resolved associative desorption experiment in which the alignment of the rotational angular momentum of H_2 desorbing in the ($v = 0, j = 4$) and ($v = 1, j = 4$) states is measured. This alignment (or, more precisely, the rotational quadrupole alignment, which takes on positive values for molecular rotation in a plane parallel to the surface) will usually be high, because only those molecules which rotate in a plane more or less parallel to the surface can react. According to the theorists, especially at desorption energies where the reaction probability is low (a few percent), the alignment for the $v = 1$ state should be even higher than for the $v = 0$ state, because the molecule–surface interaction constrains reaction at the top site, at which $v = 1$ H_2 reacts, to parallel orientations even more than at other sites [40].

In a second possible way forward, sensible projection techniques could be employed to make meaningful movies of high dimensional calculations, similar to the movie showing results of a 2D calculation already presented (Fig. 3).

3. Interactions of simple heavier diatomic molecules

The advantage of studying hydrogen is that the system is relatively simple due to the large mass mismatch between the incoming reacting molecule and the surface atoms. Due to the limited number of electrons involved in the molecular bond detailed first principle calculations are feasible and PES for 6D calculations can be computed. In addition, scattering and adsorption can be studied theoretically using quantum mechanical dynamics methods. Even for simple heavier diatomics like O_2, CO or NO this is no longer true. PES can be calculated, but quantum mechanical dynamical calculations are too complex due to the larger number of degrees of freedom, because energy exchange with the lattice cannot be excluded for these heavier molecules. Therefore, most dynamical calculations have been carried out using classical trajectory methods.

Instead of discussing results of classical trajectory calculations on reaction of heavy diatomic molecules on surfaces, we will use this example to illustrate the use of the molecular beam scattering method. As discussed in the introduction, this method can be used for determining parameters that specify the molecule–surface interaction at a molecular level.

A schematic of a possible molecular beam set-up is shown in Fig. 7. Briefly, a collimated stream of molecules with the same velocity is aimed at a surface in ultra-high vacuum. The gas at typically atmospheric pressure expands through a small hole into vacuum, yielding a beam with a narrow velocity distribution and a variable energy [41]. The intensity of the beam at the crystal is typically a monolayer per second. The molecules in the beam will partly be reflected from the surface, and partly absorbed onto it. The reflection pattern (which is determined by the interaction of the molecules with the surface) is monitored by detectors capable of detecting individual molecules. Detectors that can be rotated around the surface are mass spectrometers with electron impact ionisation, and these are used in most cases. However, photo-ionisation of scattered molecules can often also be used to obtain state-resolved detection of heavier molecules. This generally requires lasers as light sources. The degree of adsorption (the sticking probability) can be measured by monitoring the residual gas pressure in the chamber while beaming at the crystal under study, and comparing to the case of an inert crystal [42,43] onto which no adsorption occurs. In case of a large sticking probability the pressure will be low, since most molecules will remain on the surface.

Some of the physical pictures that arise from the analysis of molecular beam scattering patterns are shown in Fig. 8 which represents the situation for a molecule that hits a smooth and inert surface. Because the surface is smooth the component of the momentum along the surface ($\mathbf{p}_\parallel$) is conserved as for light reflecting from a mirror. In the case of

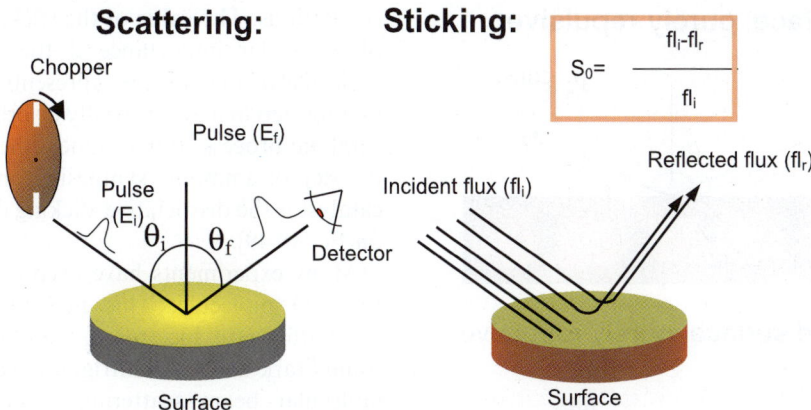

Fig. 7. Panel illustrating a beam surface scattering experiment (left) and a measurement of sticking coefficients using molecular beams (right). For details, please see the text.

He scattering from close packed metal surfaces this picture may be appropriate. Molecules like CO, NO or O_2, which have masses of the same order of magnitude as the mass of the surface atoms, can transfer (or gain) perpendicular momentum ($\mathbf{p}_\perp$) to (from) the surface. This implies that the molecule can heat (or cool) the surface. This situation can be represented by the scattering of an ellipsoid (the molecule) from a set of flat cubes, each of which has a mass corresponding to that of a surface atom. The scattering pattern from such a surface can be computed easily if quantities like the mass ratio, the translational energy of the molecule and the surface temperature are known. The width of the resulting angular distribution $\Delta\theta$ is rather narrow if the energy of the molecule is higher than thermal. Although this picture clearly is an oversimplification of reality, it nevertheless describes the interaction between thermal noble gas atoms and closed packed metal surfaces quite well [44,45].

Another limiting case occurs when the surface can be thought of as individual atoms, represented by little balls in Fig. 8B. In this case the velocity parallel to the surface is not conserved, and the scattering can be seen as the reflection of light from a rippled mirror, just as can be seen in case of a sunset over a quiet sea. In this case, the width of the resulting angular distribution $\Delta\theta$ is broader if the energy of the molecule is higher than thermal. This situation is typically encountered for the scattering of fast ions from surfaces, but can already be observed for O_2 impinging on a Ag(1 1 1) surface molecules at energies of about 100 kJ/mol [46,47].

In both of these cases no attractive force between molecule and surface was taken into consideration. If one does consider such a force, the situation of Fig. 8C is obtained. Here 'glue' representing a attractive, chemical force is added, causing the molecules to chemisorb to the surface. The attractive force is of longer range than the repulsive force and deflects the molecule towards the surface, where it collides with respect to an effective local normal. In order to chemisorb, all translational energy of the molecule has to be absorbed by the surface. Since the surface is extended and has many degrees of freedom, this is often possible.

Molecular beams can be used to study absorption as well as scattering processes (see also Fig. 2). When molecules are reflected from the surface, this leads to a pressure rise in the ultra-high vacuum chamber surrounding the crystal [42,43]. When the molecules are adsorbed (or pumped) by the crystal this pressure rise is absent. By measuring the pressure rise in the ultra-high vacuum chamber when a beam of molecules impinges on the surface, the sticking coefficient S_0 of the surface can be measured for various conditions of the incident beam. This sticking coefficient is a quantity that can also be computed using molecular dynamics

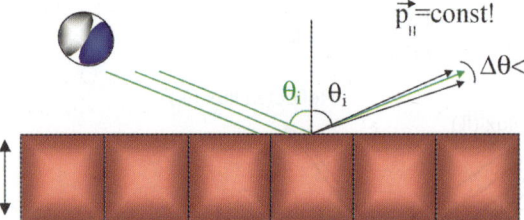

A: Flat surface, purely repulsive

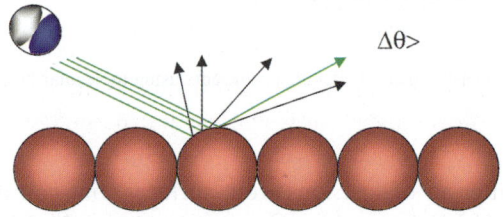

B: Corrugated surface, purely repulsive

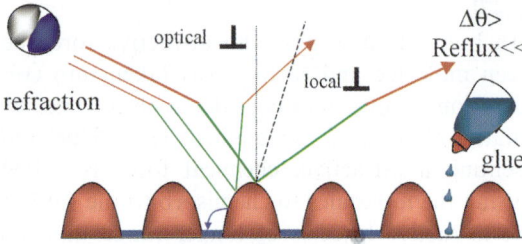

C: Corrugated surface + attractive well

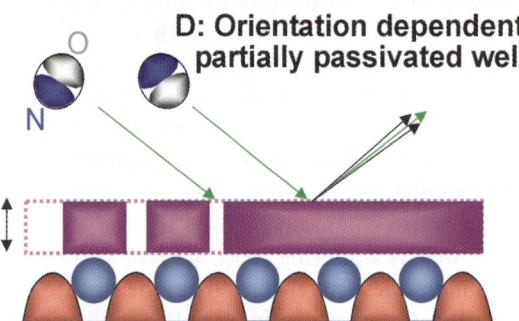

D: Orientation dependent partially passivated well

Fig. 8. Schematic diagrams of prototypes of gas–surface interactions as can be probed by molecular beams, presented as side views of the surface atoms or cubes. (A) Molecular scattering in which parallel momentum is conserved and the surface is represented by hard cubes. (B) Molecular scattering from individual surface atoms. (C) Molecular scattering in the presence of a strong chemisorption well. (D) Molecular scattering for a partially passivated surface, containing specific sites where chemisorption is possible. Note that in this case the interaction is also strongly orientation dependent.

simulations. The value of the sticking coefficient is of particular importance if the sticking of the molecule to the surface represents the rate determining step in an industrially or otherwise relevant catalytic process. For instance, the rate determining step of ammonia synthesis from N_2 on an Fe-catalyst is the dissociative sticking coefficient of N_2 on Fe [48,49].

Many experiments have been carried out using the NO molecule. This molecule has a fairly straightforward spectroscopy and can be oriented using Stark fields. An intriguing recent example of molecular beam scattering experiments and its connection to theory is the scattering of NO or CO from clean and H-covered Ru(0 0 0 1) [50,51]. At the clean surface the angular distribution of scattered particles is very broad, as shown in Fig. 9. This points to a picture of a collision of the molecule with at an atom-like surface (Fig. 8B). In addition, the sticking coefficient is almost unity, implying that most molecules stick to the surface through the action of an attractive chemical force that seems to be uniform over the surface (Fig. 8C),

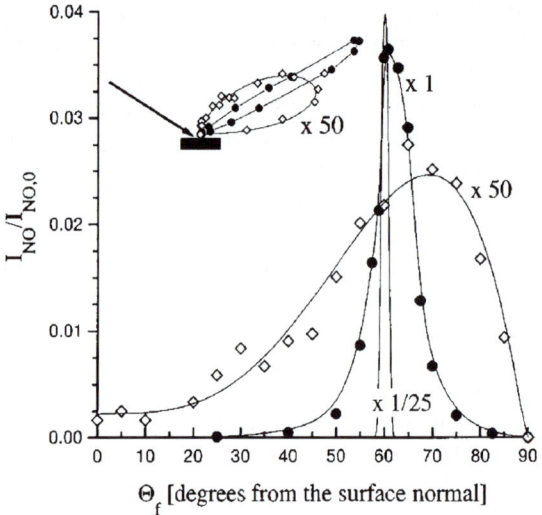

Fig. 9. Angular distributions measured for scattering of NO with a translational energy of about 160 kJ/mol from clean (◇) and H-covered (●) Ru(0 0 0 1). The sharp distribution around the specular angle of 60° indicates the angular width of the primary beam. Note the remarkable decrease of the angular width in case of H-coverage. Hydrogen turns the surface into a molecular mirror (from Ref. [50]).

with few 'bumps' and lots of glue. It is well known that Ru binds CO and NO very strongly. The metal atoms at the surface are lacking their neighbours (low co-ordination) and have electrons available to form chemical bonds with reactive species from the gas phase. Recent calculations from our group confirm this picture [52].

The scattering of NO from a Ru surface becomes more interesting if the surface is pre-covered by hydrogen atoms. The scattering pattern changes dramatically, as also shown in Fig. 9. The reflected molecules appear in a very narrow cone around the specular direction: the hydrogen atoms have passivated the Ru surface, and turned it into a molecular mirror. One could perhaps rationalise this by invoking that the H-atoms have formed chemical bonds with the available Ru electrons, in analogy to the case of silicon surfaces. These Si surfaces in vacuum have molecular orbitals available (dangling bonds directed away from the surface) that can actively form chemical bonds. Once chemical bonds are formed, the surface becomes completely inert. This also occurs for diamond surfaces, which are the hardest of all. On metals, however, such a picture is harder to imagine since the electrons are delocalised. Nevertheless, to our great surprise H-covered Ru does still bind NO and CO in an activated process [50–52]. This means that our surface is adsorbing molecules in one part of the unit cell, while elsewhere in the same unit cell the surface is like a mirror. Thus the character of the force field changes dramatically over distances of less than a tenth of a nanometre! The physical picture of the interaction that emerges is one of a mirror with very localised—chemically active—holes in it, as shown in Fig. 8D.

Density functional theory enables one to study the origin of these phenomena. It turns out that H makes the Ru surface repulsive, with the exception of a single spot, on top of a Ru atom [52]. If CO approaches this site it feels the repulsion by the three hydrogen atoms surrounding the Ru-atom. However, it is possible to 'push these aside' to form a chemical bond with the Ru atom. In order to form the bond the Ru atoms moves up out of the surface plane by about 0.06 nm. This is an activated process, that leads to a weak chemical bond when the surface is allowed to relax; i.e., to adapt its geometry to form the strongest chemical bond. The orientation of the CO is very important in this process. The above is only possible when the molecule approaches the surface with the C-end towards the surface. When the O-end is in front, no chemical bond is formed whatsoever. Experimentally, this can be studied for NO molecules. Using the Stark effect and hexapole lens focusing a state selected beam of NO molecules can be prepared in which NO is oriented either with the O-end or with the N-end pointing towards the sample surface [53]. Such experiments show that there is a very strong orientation dependence of the interaction [54,55]. This is very reasonable, because it is known that NO and CO always bind through their N- or C-end. Such experiments with oriented beams also indicate that in many cases molecules reorient themselves before they form a chemical bond. In other words, NO oriented with the O-end down can reorient itself on a femtosecond timescale, so that it can form a bond with the reactive N-end [56]. This phenomenon is also known as 'steering' and was discussed in Section 2 in the context of the H_2–Pd(1 0 0) system [28, 33,57]. Like the experiments, theoretical calculations have found steering to be important for heavy molecules such as NO colliding with surfaces [58]. The experiments and their analysis shows that the bonding of the NO molecule is very orientation dependent. In Fig. 8D this is indicated by the fact that for the O-end down molecules no binding sites or holes exist.

From the examples given above it is clear that is it possible to obtain a detailed picture of the atomic scale PES and the interaction dynamics from the analysis of molecular beam scattering experiments. The examples refer to the direct formation of bonds, or to a direct scattering or reflection event.

4. Studies with femtosecond pulses

It has been a long-standing goal in surface science to real-time monitor chemical reactions on surfaces. Achievement of this goal allows us to answer important questions concerning the reaction mechanism, such as: where does the energy for the re-arrangement of chemical bonds come from? How fast does this energy flow? Therefore, mechanisms and rates of energy transfer at surfaces, in particular between the adsorbates and the substrate, have been the subject of ongoing experimental and theoretical efforts. The substrate has electronic and phonon degrees of freedom from which energy can be transferred into the vibrational and electronic states of the adsorbate, and vice versa. Knowledge of the time-scales and mechanisms of energy transfer at surfaces and the resulting rate of reaction is essential for an integral understanding of reactions on surfaces.

The recent advent of lasers producing femtosecond light pulses has made such experiments possible. The principle behind experiments in which femtosecond laser pulses are employed to unravel the reaction dynamics is illustrated in Fig. 10. The reaction is "switched on" with a femtosecond laser pulse, and the subsequent time-evolution of the reaction is followed by a second pulse. The time delay between the excitation (pump) pulse and the interrogation (probe) pulse can be varied at will, so that spectroscopic "snapshots" can be made at regular intervals after initiation of the reaction. For reactions at surfaces, there are a few complicating factors such as the low optical density of one layer of molecules on the surface, making spectroscopic detection difficult. However, techniques for this detection are now available, as illustrated by Fig. 10.

Real-time observations of the damping of molecular vibrations of molecules adsorbed on metal surfaces constitute a fine example of energy transfer from adsorbed molecules to surface degrees of freedom [59–61]. Vibrational relaxation of high-frequency modes (such as the C–O stretch vibration) typically occurs on a timescale of a few picoseconds by creation of electron–hole pairs in the metal. The reverse pathway, energy flow from the surface to the adsorbate, has also been observed in a time-resolved manner by exciting the surface with a short optical pulse and monitoring the effect on an adsorbate vibration [62–66].

Time-resolved investigations of chemical surface reactions started with so-called two-pulse correlation measurements, in which the integrated reaction yield, e.g. gas-phase NO after desorption from Pd(1 1 1), is monitored as a function of the delay between two ultra-short laser pulses [67]. This type of experiment provides important information on the mechanism behind the chemistry—in particular on the lifetime of the excitation responsible for the chemistry. Shortly following that, a pump–probe scheme was employed, in which the pump pulse initiated the chemistry, and the time-delayed probe pulse monitored the effect of the pump pulse. In this way, the desorption of CO from Cu

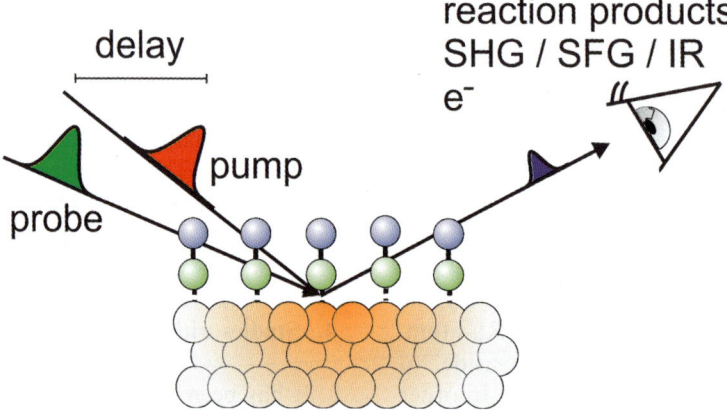

Fig. 10. Illustration of femtosecond laser pump–probe experiments to investigate the dynamics of reactions in surfaces. Two pulses are incident on the surface, where the first initiates the chemistry (pump pulse) at $t = 0$, so that the temporal evolution of the reaction can be investigated by a second, probe pulse at time $t = t_{\text{delay}}$. Detection possibilities include detection of reflected light (IR), its SHG, or the sum frequency of two probe fields (SFG) [101], as well as detection of reaction products or photo-electrons.

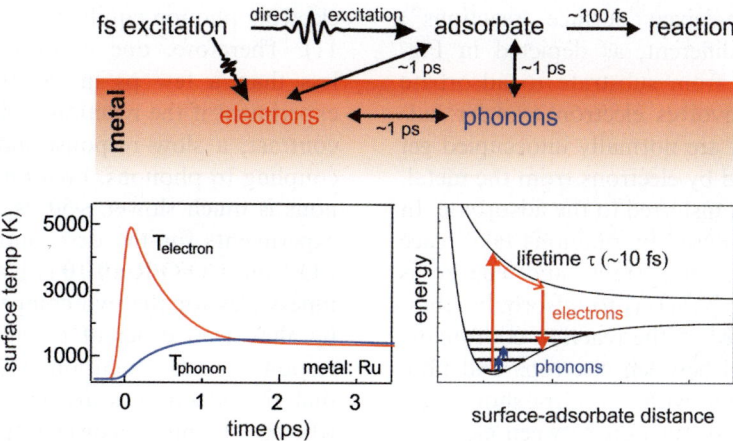

Fig. 11. Mechanisms and time-scales of surface chemistry following femtosecond laser excitation. In the direct excitation pathway, an electronic transition in the adsorbate–substrate bond is accomplished directly by the excitation pulse. In the two substrate-mediated mechanisms, the chemistry can be triggered by coupling of the adsorbate to the electrons or the surface lattice vibrations ("phonons"). Initially, the energy is deposited in the metal electrons, which reach a very high transient temperature (lower left panel), after which "electron–phonon coupling" causes equilibration between electrons and phonons. The lower right panel illustrates the fundamental difference between "electron-mediated" and "phonon-mediated" chemistry: Electron-mediated chemistry involves the transition of an electron from the substrate to the adsorbate, where it will reside for a certain time. Upon relaxation to the ground state, the system will have acquired vibrational energy that can be used for the reaction. In contrast, phonon-mediated chemistry takes place in the electronic ground state, where the vibrational ladder is 'climbed' in consecutive steps.

(1 1 1) was the first surface reaction to be monitored real-time, by means of time-resolved second harmonic generation (SHG) [68]. From this work, it was concluded that desorption occurs very rapidly, within ∼325 fs.

We will discuss here two recent examples of time-resolved spectroscopic investigations of surface chemical dynamics: the oxidation and desorption of CO on Ru(0 0 0 1), and the real-time observation of a Cs atom moving away from a Cu(1 1 1) surface [69–71].

Since in this type of investigation, an ultra-short laser pulse is used to "switch on" the reaction, it is helpful to consider the various reaction pathways upon femtosecond excitation. Generally, chemical changes (e.g. desorption or oxidation) of an adsorbate on a metal surface due to excitation with laser pulses can be either direct or substrate-mediated. The direct process involves the optical excitation of an electronic state of the adsorbate, so that the photon-energy is directly inserted into the reactant, as observed for the Cs/Cu(1 1 1) system discussed below [70]. More common (especially if the laser pulses are short and the wavelength is not in the UV) is a substrate-mediated process, as depicted in Fig. 11, in which the metal electrons play a key role: in the excitation process initially the electrons near the metal surface are excited by the optical field (the optical penetration depth of light into metals is typically ∼10 nm). Thermalisation within the "electron gas" to a hot, equilibrated electron distribution occurs rapidly by "electron–electron scattering". Due to the very small heat capacity of the electron gas, peak electronic temperatures of thousands of Kelvin above the equilibrium melting point can be reached. Cooling of the surface electrons occurs by diffusion of hot electrons into the bulk and by simultaneous heat transfer to the surface lattice vibrations ("phonons") through "electron–phonon coupling", typically on a picosecond time scale. The time dependence of calculated electron and phonon temperatures at the Ru(0 0 0 1) surface after excitation with a femtosecond pulse is also depicted in Fig. 11. A surface reaction can subsequently be triggered either by coupling of the adsorbate to these hot electrons or by its coupling to phonons.

Surface reactions driven by exciting surface-lattice vibrations ("phonon-driven surface reactions") and reactions driven by exciting the surface electrons ("electron-driven surface reactions") are fundamentally different, as depicted in Fig. 11. Energy transfer from substrate to adsorbate through electrons involves electron transfer: adsorbate orbitals that are normally unoccupied get transiently populated by electrons from the metal, by which energy is transferred to the adsorbate. In contrast, reactions caused by phonons take place in the electronic ground state, and energy is transferred from the phonons to adsorbate vibrations that are coupled to the reaction coordinate. The non-equilibrium between electrons and phonons upon excitation with an ultra-short laser pulse allows for the distinction between electron- and phonon-induced processes, which is not possible with conventional heating. Summarizing, there are essentially three pathways from femtosecond laser excitation to chemical reaction: (i) via direct adsorbate excitation, or ((ii) and (iii)) through the substrate, more specifically (ii) via its phonons or (iii) via its electrons. Below, we will discuss one example of each pathway.

It was recently demonstrated that, on a ruthenium surface on which carbon monoxide and oxygen are co-adsorbed, oxidation of carbon monoxide could be initiated by an ultra-short laser pulse, alongside desorption of CO [69]. It is interesting to note that this reaction does not proceed thermally under ultra-high vacuum (UHV) conditions: upon conventional heating, only CO is released from the surface; no CO_2 is formed. The laser induces novel, non-thermal chemistry, which allowed the authors to unravel the time scales and mechanisms of both the desorption and the oxidation surface reactions. It was found that the oxidation of CO is caused by coupling to electrons, whereas the desorption of CO is caused by coupling to phonons.

The predominant reaction path (phonon- or electron-driven) can be determined by means of a "two-pulse correlation measurement" [67]. In such an experiment, the photo-reaction yield is measured as a function of the delay between two pulses of equal intensity. Because of the fast electron–phonon coupling, high electronic temperatures resulting from the combined effect of the two pulses can only be reached when the two pulses are separated by a time-span that does not exceed the electron–phonon equilibration time of ~ 1 ps (Fig. 11). Therefore, one expects a fast response of less than a few ps in the two-pulse-correlation experiment if the reaction is electron-mediated. In contrast, a slow response indicates predominant coupling to phonons, because the cooling of phonons is much slower ~ 50 ps. The results of such experiments for the desorption and oxidation of CO from CO/O/Ru(0 0 0 1) have revealed that the time-scales for the two competing processes differ by almost an order of magnitude. The ultra-fast response for the oxidation of 3 ps FWHM suggests that this reaction is driven by the hot electrons, whereas the much slower response for desorption of 20 ps implies coupling to phonons.

That the oxidation reaction is electron-mediated and the desorption reaction phonon-mediated, was verified by an investigation into isotope-effects for the two reactions: For the desorption no isotope effect was observed, whereas for the oxidation a very large effect was seen. When the ruthenium surface is pre-covered with a 50/50 mixture of ^{16}O and ^{18}O, CO_2 containing the light O-isotope is preferentially formed (yield ratio $^{16}OCO/^{18}OCO = 2.2$). This isotope effect can be rationalised in terms of multiple electronic transitions: the energetic metal electrons can be scattered into an unoccupied orbital of the oxygen atom, as illustrated in Fig. 11 [72]. The resulting short-lived, negatively charged oxygen will favour a Ru–O distance different from that of the neutral state, so that the O-atom is accelerated with respect to the surface. The electron residence time is equal for both isotopes, and thus the lighter isotope will traverse a larger distance than the heavier isotope, because it is accelerated faster. This means that when the electron transfers back to the substrate, leaving the oxygen in its electronic ground state, the light isotope is further from its equilibrium position and has therefore acquired more vibrational energy which is available for surmounting the reaction barrier. Hence the larger yield for the lighter isotope.

Besides confirming that the oxidation reaction is electron-mediated, the O-isotope effect also demonstrated that activation of the Ru–O bond by electrons is the rate-determining step in the oxidation reaction: as soon as the adsorbed oxygen atoms are activated by the hot substrate electrons (the amount of energy uptake being mass-dependent), the reaction can take place. Due to the high coverage, O and CO are in close proximity on the surface (∼3.5 Å) so that after activation of the O, the required additional decrease in O–CO distance is readily realised by the thermally activated lateral motion of CO.

Important reaction parameters such as activation energies and reaction rates could be extracted from the data with the use of appropriate models [73,74]. The adsorbate–electron and adsorbate–phonon coupling strengths, as well as reaction activation barriers were fitted to the data. For the oxidation, an activation energy of 1.8 eV was observed, and an adsorbate–electron coupling time of 500 fs. For the desorption, the 0.8 eV activation energy obtained from the thermal desorption measurements was used (since the desorption is triggered by coupling to phonons), and an adsorbate-phonon coupling time of 1 ps was observed. The different activation mechanisms for the two competing reactions (the desorption is phonon-driven, whereas the oxidation is electron-mediated) implies that the temporal evolution of the two reactions is different. This is illustrated in Fig. 12, which depicts the rate of CO_2 formation and of CO desorption after excitation with the laser pulse. The oxidation reaction requires the presence of hot electrons, so that its rate peaks at 0.7 ps after

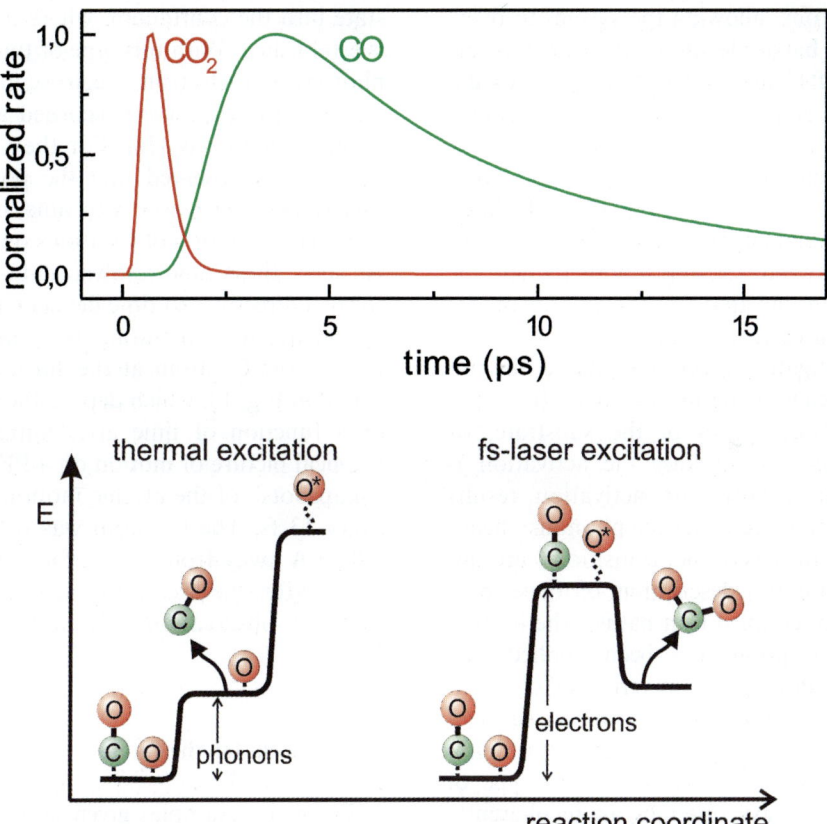

Fig. 12. Upper panel: calculated rates of formation for the oxidation of CO driven by the hot electrons, and for the phonon-driven desorption of CO after excitation with a femtosecond laser pulse at $t = 0$ ps. The formation of CO_2 is very rapid and almost complete before the desorption of CO commences. Lower panel: schematic representation of the observation that only desorption occurs under conventional heating (left): the CO has desorbed before the oxygen is activated (O*). Under laser excitation (right), the high activation barrier (1.8 eV) of oxygen is overcome by coupling to the hot electrons, so that CO_2 is formed.

excitation, whereas the desorption is much slower and peaks around 4 ps after excitation. It is evident that the oxidation is almost complete when desorption sets in. It is this separation of time scales that enables the oxidation to take place upon femtosecond excitation, whereas it is not possible normally. This is illustrated in the lower panel of Fig. 12. In order for oxidation to occur, the Ru–O bond needs to be activated, after which the reaction with CO takes place. However, the energy required for CO desorption (0.8 eV) is considerably lower than the barrier for CO_2-formation (1.8 eV), so that attempts to drive the reactions thermally fail, because CO is desorbed before the oxygen is activated to react. In contrast, laser excitation enables the oxidation through activation of the Ru–O bond by the laser-heated ruthenium electrons, allowing the system to overcome the 1.8 eV barrier leading to CO_2, before the phonons are sufficiently hot to cause CO to desorb. Thus, the competing process of desorption is outpaced. In a recent density functional theory study of CO oxidation on Ru(0 0 0 1) an activation barrier of 1.4 eV was reported [75]. This calculated value is the minimal-energy barrier, and the slight discrepancy (between 1.4 and 1.8 eV) might be understood by noting that in the experiment not all pathways will be that of least energy.

For the investigation mentioned above, the initiation of the reaction occurs incoherently either through the phonon bath of the substrate, or through its electrons. Because the activation is substrate-mediated, adsorbate activation results that is incoherent, since either the phonons "heat" the adsorbate, or electronic transitions are involved. Therefore the description of these processes in terms of coupled heat baths—the friction model type description—has been applied successfully, but a description in terms of coherent wave packet dynamics on potential energy surfaces (transient occupation of excited states) is inappropriate. For the same reason, in this type of experiment one always observes an incoherently averaged system response. This drawback has been circumvented recently with time-resolved two-photon-photoemission spectroscopy of Cs on Cu(1 1 1), which has an excited electronic state that can be directly, resonantly populated with an ultra-short laser pulse. What is more, this electronic state exhibits an unusually long lifetime, allowing very efficient energy transfer from electronic to nuclear degrees of freedom.

In these experiments, Petek et al. covered a Cu(1 1 1) surface with ~0.1 monolayer of Cs [70]. Upon Cs adsorption, an occupied Cu–Cs bonding orbital appear, as well as unoccupied anti-bonding Cu–Cs orbitals, separated by ~3.1 eV, as illustrated in Fig. 13. Electrons can be photo-excited from the ground state to the repulsive excited state with an ultra short (pump) pulse, after which the wave packet (the atomic probability distribution) motion on the upper PES can be monitored real-time time by photo-emission of the electron from the upper PES into the continuum: a second—probe—photon ejects the electron from this state into the continuum, where its kinetic energy is established. With this type of time resolved two-photon photoelectron spectroscopy (2PPE), electron energy spectra are recorded as a function of pump–probe delay [76]. For the Cs/Cu(1 1 1) system, it was observed that the population in the excited state decreases with time, indicative of the dissociative motion of Cs atoms on the upper PES. As such, these time resolved data could be transformed directly into time dependent motion of the adsorbate, demonstrating the time-resolved motion of the Cs atom at the surface. This is illustrated in Fig. 13, which depicts the Cu–Cs distance as a function of time after initial excitation. A classical picture of motion on a PES emerged, and "snapshots" of the atomic motion could be taken every 13 fs. The Cs atom was observed to move ~0.35 Å away from the surface during the initial 160 fs after the excitation, amounting to the first real-time observation of atomic motion at a surface.

5. Prospects for the future

From the examples given above it is clear that the study of chemical dynamics at surfaces is in an extremely interesting phase. First of all, theory is getting to such a level of sophistication that it can strongly interact with experiment. Secondly, the experiments are succeeding at obtaining lateral

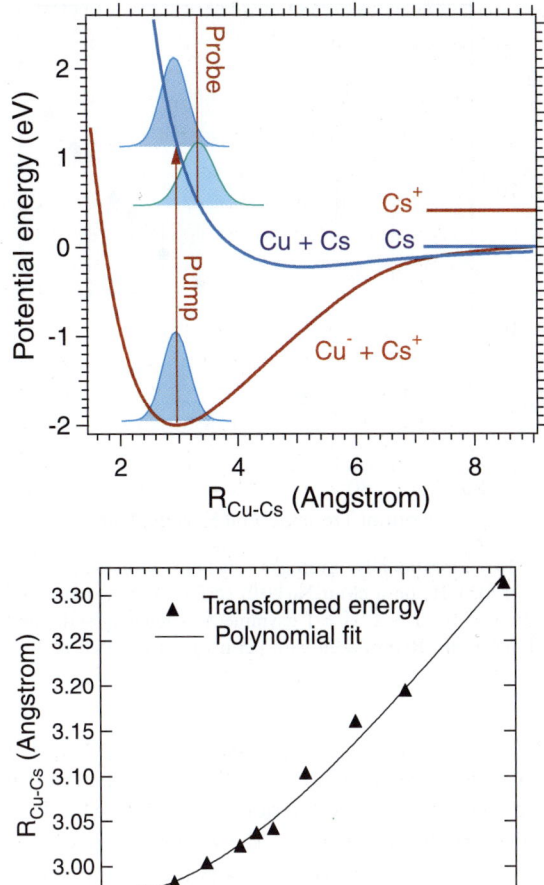

Fig. 13. Upper panel: Schematic PES for Cs adsorbed on Cu: the energy of adsorption of Cs^+ is 1.9 eV, at a Cu–Cs distance of 2.97 Å. The excited state potential, in which an electron has been transferred from the surface to the Cs atom, lies 3.1 eV higher. Lower panel: time-resolved movement of the Cs atomic wave packet on the upper PES, as recorded with time-resolved 2PPE. Reprinted with permission from [H. Petek, M.J. Weida, H. Nagano, and S. Ogawa, Real-time observation of adsorbate atom motion above a metal surface, Science 288 (2000) 1402.] Copyright 2000 American Association for the Advancement of Science.

Although the overall description of the interaction between H_2 and metal surfaces is extremely good compared to almost any other system, quantitative discrepancies remain between theory and experiment. The most likely source of this disagreement lies in the limitations of the electronic structure approach currently available for computing molecule–surface interactions, which is density functional theory. As already discussed fundamental research is needed to obtain functionals which can predict reaction barrier heights with the required accuracy (4 kJ/mol or 0.04 eV).

Other fundamental questions concerning dissociative chemisorption of H_2 on metal surfaces that remain to be explored are the influence of excitations of the surface electrons and of surface lattice vibrations. Billing has claimed that electronic excitations have an important influence on dissociation of H_2 on Cu(1 0 0), but his model has lead to predictions that have not yet been confirmed experimentally [77]. Other approaches for dealing with electronic excitations have been developed by Baer and Kosloff, but have not been tested yet on dissociative chemisorption problems [78,79]. In addition, the influence of "resonances" (metastable states in which H_2 is trapped on the surface) needs to be explored. These "resonances" were found on H_2–metal surface reactions like the reaction of H_2 on Pd(1 0 0) [28,29,80,81]. A more detailed discussion of these issues can be found in Ref. [12].

Problems for which the solution will come in reach during the next decade include the dissociation of H_2 at stepped surfaces and at defects. The time-dependent wave packet method discussed in Section 2 should allow this process to be studied with femtosecond, sub-nanometre resolution. Questions that can then most likely be answered include: under what conditions is dissociation at steps or defects dominant? How big does a terrace have to be for dissociation on the terrace to be like dissociation on the perfect corresponding low index surface?

Another problem that will be tackled theoretically during the next century is the dissociative chemisorption of H-containing polyatomic molecules, such as methane and higher alkanes. Polyatomic molecules pose qualitatively new problems,

and time resolutions which are sufficient to study chemical dynamics at surfaces with sub-nanometre resolution on a femtosecond time scale. To be more specific, we will give an outlook for some of the areas already discussed in Section 2.

such as: if the reaction can be promoted by exciting one (and only one) of the normal modes with one quantum of vibration, which mode would be the best one to choose? Electronic structure theory for molecule–surface interactions has already progressed to the point where it can be used to guide experimentalists in designing new catalysts for conversion of such molecules [82]. On the other hand, the dynamical mechanisms of methane dissociation are still highly controversial, as also illustrated by recent experiments on methane dissociation [83–88]. The experimental evidence concerning the key question that we just posed (which vibrational modes promote dissociation best when excited prior to the collision with the surface) is inconclusive. Some experimentalists have argued that the deformation or bending modes should be most important [83], while others have argued that excitation of the stretching modes should be most effective at least at high collision energies [86].

A new development that is highly relevant to the question of which vibrational mode is best excited to promote reaction is that it is now possible to selectively excite a vibration in methane using an infrared laser, and to monitor the effect this has on the reactivity in a molecular beam experiment [87]. Experiments on $CH_4 + Ni(100)$ were able to show that the excitation of the asymmetric stretching mode with one quantum leads to a large increase in the reactivity, although not enough to explain the large effect of aselectively raising the molecules' vibrational temperature observed in seeded beam experiments on the same system (see Fig. 14) [89]. The present technique only allows the excitation of the asymmetric modes, so that quantum dynamics calculations on the influence of the symmetric modes could provide vital additional information.

Another key question is what causes the increase of the reaction probability with decreasing collision energy seen at low energies in some experiments. In the case of $CH_4 + Pt\{110\}$ (1×2), this finding was attributed to a dynamical steering mechanism, in which the repulsive forces of the potential steer the molecule towards a favourable reactive configuration at low energies [86]. Mullins and co-workers attributed a similar observation for $CH_4 + Ir(110)$ to a different mechanism in

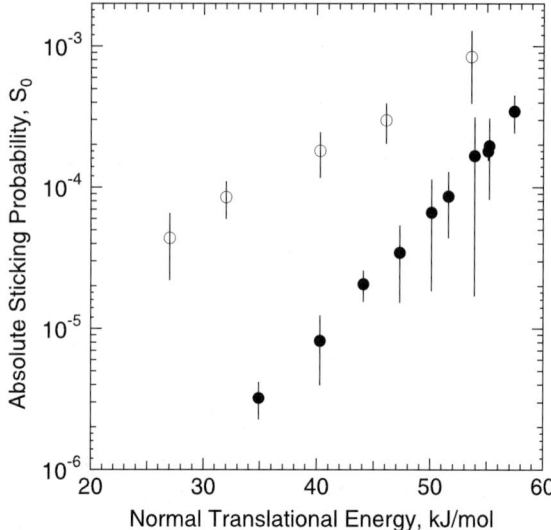

Fig. 14. Vibrationally averaged sticking probabilities are shown for CH_4 on a clean $Ni(100)$ surface with (○) laser excitation of the $J = 2$, $v_3 = 1$ asymmetric stretching state, and without it (●). Reproduced with permission from Ref. [87].

which the molecule becomes temporarily trapped in a molecular adsorption well, allowing it more time to overcome the barrier to reaction [88]. This is just another example of the many controversies surrounding the reaction of polyatomic molecules on metal surfaces. Similar questions surround the dissociation of higher alkanes on metal surfaces [90,91]. Theory will play a very important role in answering these questions.

A theoretical method which holds great promise for dissociative chemisorption of polyatomic molecules is the multi-configuration time-dependent Hartree method [92]. In this method the time-dependent wave function is expanded in time-dependent basis functions, so that the wave function can move with the system, leading to more favourable scaling properties. Recent examples of high-dimensional calculations with this method include a 10D study of inelastic scattering (but not reaction) of CH_4 from $Ni(111)$ [93] and a 24D study of the dynamics of relaxation in pyrazine [94]. The high-dimensionality achieved in these calculations clearly suggests the usefulness of the method for dissociative chemisorption of polyatomic molecules.

Also for heavier molecules experiments with quantum state specificity will become more and more common. The experiments by the Utz group using the novel methods of laser-preparation of molecular CH_4 beams have already been mentioned [87]. The groups of Auerbach and Wodtke have managed to prepare molecular beams of NO in highly excited vibrational states [95,96]. An example is shown in Fig. 15. Here the kinetic energy and surface temperature dependence of the excitation and de-excitation rate of $NO(v=2)$ is shown [95,96]. Both rates increase with increasing energy. But de-excitation shows no dependence on surface temperature, whereas the excitation shows a very strong one. Analysis of the data shows that the surface is the main sink or source of vibrational energy. Mechanisms involving energy transfer to and from the surface electrons have been invoked. However, according to Huang et al. no electron mediated process put forward to date is capable of explaining simultaneously the vibrational excitation and de-excitation observed in Fig. 15. Clearly, coupling to the electron–hole pair continuum of the surface is a process important for chemical dynamics at surfaces. This process is far from being fully understood.

Another new direction is the study of reactions involving complex surfaces, for instance those having relevance to atmospheric chemistry. The dynamics of the interaction of HCl and H_2O–ice surfaces is a new field of study to which the tools and knowledge developed for dynamics at surfaces can be applied [97].

Femtosecond laser technology will soon allow to record molecular spectra with femtosecond resolution. Recently, the use of femtosecond time-resolved sum-frequency generation to monitor an internal vibrational mode (the C–O stretch vibration of adsorbed CO) while desorption of CO was occurring, was reported [98]. These and other experiments [99] demonstrate the possibility of monitoring surface reactions real-time through vibrational spectroscopy. An advantage of time-resolved sum-frequency generation is that it enables one to take adsorbate-specific "snapshots" of the C–O stretch vibration while desorption is occurring. The most important strength of the new technique is its very high chemical sensitivity: the probe pulse is in the infrared, so that changes in the *vibrational state* of molecules adsorbed on surfaces due to the reaction are investigated. Chemical bonds are essentially springs that vibrate with a particular eigenfrequency. This frequency is determined specifically by the strength of the chemical bond, and the mass of the atoms involved. The beauty of this technique therefore lies in the fact that, as chemical bond re-arrangement takes place on the surface, the vibrations provide direct access to the chemical process. An alternative experimental scheme to perform time-resolved sum-frequency generation has recently been introduced [100], testifying to the promise this technique holds for time resolving surface reactions. This will allow one to really observe bond breaking and bond making at surfaces for the first time.

In summary, theory can be used to produce animations of molecule–surface reactions, with sub-nanometre and femtosecond time-resolution. Theory now allows for realistic calculations on dissociative chemisorption of H_2 on metal surfaces, but quantitative differences with experiment remain due to problems with the accuracy of existing density functionals. Also, the role of the

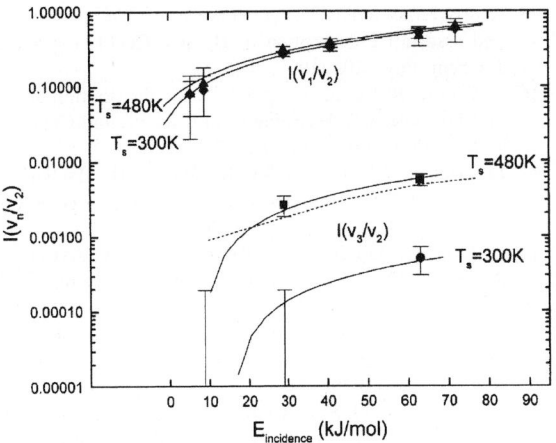

Fig. 15. Incidence kinetic energy dependence of the excitation and de-excitation probability for $NO(v=2)$ scattering from Au(1 1 1) for two surface temperatures. De-excitation ($v = 2 \rightarrow 1$): triangles and diamonds. Excitation ($v = 2 \rightarrow 3$): squares and circles. Lines are explained in the original reference (from Ref. [95]).

surface electrons has not yet been clarified. The extension of theoretical investigations to irregular (stepped) surfaces and polyatomic molecules will become possible in the next few decades and are expected to address many questions of physical interest. Such extensions will be highly useful to the interpretation of molecular beam scattering experiments which are increasingly carried out at the full state-to-state level, also for heavier molecules like NO and CO. Also for these molecules the role of electronic excitations is highly uncertain and in need of further investigation. Time-resolved experiments can now unravel molecule–surface reactions with the time-resolution needed to follow the breaking of bonds, and will yield much useful information on the role of surface lattice vibrations and electrons in surface chemical reactions. In this area, new techniques (such as time-resolved sum-frequency generation) hold great promise. However, an experimental technique that yields both sub-nanometre spatial resolution *and* femtosecond time-resolution has yet to be developed. In short, there is much to do for young researchers interested in the art of making movies of surface chemical reactions.

Acknowledgements

We are grateful to M. Wolf and H. Petek for sharing their results, to A. Gross, D.J. Auerbach, and A.L. Utz for providing us with original figures from their work, and to D.A. McCormack and M.F. Somers for providing some of the art work. Some parts of the work described are part of the research program of FOM and are supported financially by NWO. Part of the research has been supported by the KNAW.

References

[1] R.A. VanSanten, J.W. Niemantsverdriet, Chemical kinetics and catalysis, in: M.V. Twigg, M.S. Spencer, Plenum Press, New York and London, 1995.
[2] C.T. Rettner, D.J. Auerbach, J.C. Tully, A.W. Kleyn, Chemical dynamics at the gas–surface interface, J. Phys. Chem. 100 (1996) 13021–13033.
[3] J.C. Polanyi, A.H. Zewail, Direct observation of the transition state, Acc. Chem. Res. 28 (1995) 119–132.
[4] A.H. Zewail, Femtochemistry: recent progress in studies of dynamics and control of reactions and their transition states, J. Phys. Chem. 100 (1996) 12701.
[5] A.H. Zewail, Femtochemistry: atomic scale dynamics of the chemical bond, J. Phys. Chem. A 104 (2000) 5660.
[6] A. Assion, T. Baumert, M. Bergt, T, Brixner, B. Kiefer, V. Seyfried, M. Strehle, G. Gerber, Control of chemical reactions by feedback-optimized phase-shaped femtosecond laser pulses, Science 282 (1998) 919.
[7] R.R. Cavanagh, T.A. Germer, E.J. Heilweil, J.C. Stephenson, time-resolved measurements of substrate to adsorbate energy transfer, Faraday Discuss. 96 (1993) 235–243.
[8] R.R. Cavanagh, D.S. King, J.C. Stephenson, T.F. Heinz, Dynamics of nonthermal reactions—femtosecond surface chemistry, J. Phys. Chem. 97 (1993) 786–798.
[9] H.-L. Dai, W. Ho (Eds.), Laser Spectroscopy and Photochemistry on Metal Surfaces, World Scientific, Singapore, 1995.
[10] G.R. Darling, S. Holloway, The dissociation of diatomic molecules at surfaces, Rep. Prog. Phys. 58 (1995) 1595.
[11] A. Gross, Reactions, at surfaces studied by ab initio dynamics calculations, Surf. Sci. Rep. 32 (1998) 291.
[12] G.J. Kroes, Six-dimensional quantum dynamics of dissociative chemisorption of H_2 on metal surfaces, Prog. Surf. Sci. 60 (1999) 1.
[13] D.C. Clary, Quantum theory of chemical reaction dynamics, Science 279 (1998) 1879.
[14] K.D. Rendulic, A. Winkler, Adsorption and desorption dynamics as seen through molecular beam techniques, Surf. Sci. 299/300 (1994) 261.
[15] C.T. Rettner, H.A. Michelsen, D.J. Auerbach, Quantum-state-specific dynamics of the dissociative adsorption and associative desorption of H_2 at a Cu(1 1 1) surface, J. Chem. Phys. 102 (1995) 4625.
[16] G. Comsa, R. David, The purely "fast" distribution of H_2 and D_2 molecules desorbing from Cu(1 0 0) and Cu(1 1 1) surfaces, Surf. Sci. 117 (1982) 77.
[17] D. Wetzig, M. Rutkowski, R. David, H. Zacharias, Rotational corrugation in associative desorption of D_2 from Cu(1 1 1), Europhys. Lett. 36 (1996) 31.
[18] H. Hou, S.J. Gulding, C.T. Rettner, A.M. Wodtke, D.J. Auerbach, The stereodynamics of a gas–surface reaction, Science 277 (1997) 80.
[19] M. Gostein, H. Parhikhteh, G.O. Sitz, Survival probability of H_2 ($v = 1, J = 1$) scattered from Cu(1 1 0), Phys. Rev. Lett. 75 (1995) 342.
[20] M. Gostein, G.O. Sitz, Rotational state-resolved sticking coefficients for H_2 on Pd(1 1 1): testing dynamical steering in dissociative adsorption, J. Chem. Phys. 106 (1997) 7378.
[21] M. Gostein, E. Watts, G.O. Sitz, Vibrational relaxation of H_2 ($v = 1, J = 1$) on Pd(1 1 1), Phys. Rev. Lett. 79 (1997) 2891.

[22] A. Hodgson, P. Samson, A. Wight, C. Cottrell, Rotational excitation and vibrational relaxation of H_2 ($v = 1$, $J = 0$) scattered from Cu(1 1 1), Phys. Rev. Lett. 78 (1997) 963.

[23] G.J. Kroes, G. Wiesenekker, E.J. Baerends, R.C. Mowrey, D. Neuhauser, Dissociative chemisorption of H_2 on Cu(1 0 0): a four-dimensional study of the effect of parallel translational motion on the reaction dynamics, J. Chem. Phys. 105 (1996) 5979.

[24] B. Jackson, H. Metiu, The dynamics of H_2 dissociation on Ni(1 0 0): a quantum mechanical study of a restricted two-dimensional model, J. Chem. Phys. 86 (1987) 1026.

[25] M.R. Hand, S. Holloway, The scattering of H_2 and D_2 from Cu(1 0 0): vibrationally assisted dissociative adsorption, Surf. Sci. 211–212 (1989) 940.

[26] J. Harris, On vibrationally-assisted dissociation of H_2 at metal surfaces, Surf. Sci. 221 (1989) 335.

[27] C. Engdahl, U. Nielsen, Hydrogen dissociation on copper: importance of dimensionality in calculations of the sticking coefficient, J. Chem. Phys. 98 (1993) 4223.

[28] A. Gross, S. Wilke, M. Scheffler, Six-dimensional quantum dynamics of adsorption and desorption of H_2 at Pd(1 0 0): steering and steric effects, Phys. Rev. Lett. 75 (1995) 2718.

[29] C.T. Rettner, D.J. Auerbach, Search for oscillations in the translational energy dependence of the dissociation of H_2 on Pd(1 0 0), Chem. Phys. Lett. 253 (1996) 236.

[30] K.D. Rendulic, G. Anger, A. Winkler, Wide range nozzle beam adsorption data for the systems H_2/nickel and H_2/Pd(1 0 0), Surf. Sci. 208 (1989) 404.

[31] D.A. Butler, B.E. Hayden, The indirect channel to hydrogen dissociation on W(1 0 0)-c(2 × 2) Cu. Evidence for a dynamical precursor, Chem. Phys. Lett. 232 (1995) 542.

[32] A.T. Pasteur, S.J. Dixon-Warren, Q. Ge, D.A. King, Dynamics of hydrogen dissociation on Pt(1 0 0): steering, screening, and thermal roughening effects, J. Chem. Phys. 106 (1997) 8896.

[33] M. Beutl, M. Riedler, K.D. Rendulic, Strong rotational effects in the adsorption dynamics of H_2/Pd(1 1 1): evidence for dynamical steering, Chem. Phys. Lett. 247 (1995) 249.

[34] G.J. Kroes, E.J. Baerends, R.C. Mowrey, Six-dimensional quantum dynamics of dissociative chemisorption of ($v = 0$, $j = 0$) H_2 on Cu(1 0 0), Phys. Rev. Lett. 78 (1997) 3583.

[35] G.J. Kroes, E.J. Baerends, R.C. Mowrey, Erratum: six-dimensional quantum dynamics of dissociative chemisorption of ($v = 0$, $j = 0$) H_2 on Cu(1 0 0), Phys. Rev. Lett. 81 (1998) 4781.

[36] H.A. Michelsen, D.J. Auerbach, A critical examination of data on the dissociative adsorption and associative desorption of hydrogen at copper surfaces, J. Chem. Phys. 94 (1991) 7502.

[37] S. Wilke, M. Scheffler, Potential-energy surfaces for H_2 dissociation over Pd(1 0 0), Phys. Rev. B 53 (1996) 4926.

[38] G. Wiesenekker, G.J. Kroes, E.J. Baerends, An analytical six-dimensional potential energy surface for dissociation of molecular hydrogen on Cu(1 0 0), J. Chem. Phys. 104 (1996) 7344.

[39] J.P. Perdew, J.A. Chevary, S.H. Vosko, K.A. Jackson, M.R. Pederson, D.J. Singh, C. Fiolhais, Atoms, molecules, solids, and surfaces: applications of the generalized gradient approximation for exchange and correlation, Phys. Rev. B 46 (1992) 6671.

[40] D.A. McCormack, G.J. Kroes, R.A. Olsen, J.A. Groeneveld, J.N.P.V. Stralen, E.J. Baerends, R.C. Mowrey, Molecular knife throwing: aiming for dissociation at specific surface sites through state-selection, Chem. Phys. Lett. 328 (2000) 317.

[41] G. Scoles (Ed.), Atomic and Molecular Beams Methods I, Oxford University Press, Oxford, 1988.

[42] D.A. King, M.G. Wells, Surf. Sci. 29 (1972) 454.

[43] B. Berenbak, D.A. Butler, B. Riedmüller, D.C. Papageorgopoulos, S. Stolte, A.W. Kleyn, Sticking probability measurement in a reactive system, Surf. Sci. 414 (1998) 271–278.

[44] R.M. Logan, in: M. Green (Ed.), Solid State Surface Science, Marcel Dekker, New York, 1973, pp. 1–103.

[45] R.J.W.E. Lahaye, A.W. Kleyn, S. Stolte, S. Holloway, The scattering of Ar from Ag(1 1 1): a molecular dynamics study, Surf. Sci. 338 (1995) 169–182.

[46] A. Raukema, R.J. Dirksen, A.W. Kleyn, Probing the (dual) repulsive wall in the interaction of O_2, N_2 and Ar with the Ag(1 1 1) surface, J. Chem. Phys. 103 (1995) 6217–6231.

[47] A.W. Kleyn, D.A. Butler, A. Raukema, Dynamics of the interaction of O_2 with silver surfaces, Surf. Sci. 363 (1996) 29–41.

[48] J.J. Mortensen, L.B. Hansen, B. Hammer, J.K. Norskov, Nitrogen adsorption and dissociation on Fe(1 1 1), J. Catal. 182 (1999) 479–488.

[49] C.T. Rettner, H. Stein, Effect of translational energy on the chemisorption of N_2 on Fe(1 1 1): activated dissociation via a precursor state, Phys. Rev. Lett. 59 (1987) 2768–2771.

[50] D.A. Butler, B. Berenbak, S. Stolte, A.W. Kleyn, Elastic scattering in a reactive environment: NO on Ru(0 0 0 1)-(1 × 1)H, Phys. Rev. Lett. 78 (1997) 4556–4653.

[51] B. Berenbak, B. Riedmuller, D.A. Butler, C.T. Rettner, D.J. Auerbach, S. Stolte, A.W. Kleyn, Molecular beam study on interaction dynamics in a reactive system: NO on bare Ru(0 0 0 1), Phys. Chem. Chem. Phys. 2 (2000) 919–923.

[52] B. Riedmüller, I.M. Ciobîca, D.C. Papageorgopoulos, F. Frechard, B. Berenbak, A.W. Kleyn, R.A. van Santen, CO adsorption on hydrogen saturated Ru(0 0 0 1), J. Chem. Phys. 115 (2001) 5244–5251.

[53] M.G. Tenner, E.W. Kuipers, W.Y. Langhout, A.W. Kleyn, G. Nicolasen, S. Stolte, Molecular beam apparatus to study interactions of oriented NO and surfaces, Surf. Sci. 236 (1990) 151–168.

[54] A.W. Kleyn, Non-reactive orientations of molecules at surfaces, Prog. Surf. Sci. 54 (1997) 407–420.

[55] U. Heinzmann, S. Holloway, A.W. Kleyn, R.E. Palmer, K.J. Snowdon, Orientation in molecule-surface interaction, J. Phys. Condens. Matter 8 (1996) 3245–3269.

[56] R.J.W.E. Lahaye, S. Stolte, S. Holloway, A.W. Kleyn, Orientation and energy dependence of NO scattering from Pt(1 1 1), J. Chem. Phys. 104 (1996) 8301–8311.

[57] M. Kay, G.R. Darling, S. Holloway, J.A. White, D.M. Bird, Steering effects in non-activated adsorption, Chem. Phys. Lett. 245 (1995) 311–318.

[58] D. Lemoine, T. Duhoo, Quantum study of oriented NO scattering from Ag(1 1 1): orientational steering and effects of surface corrugation, Chem. Phys. 238 (1998) 58.

[59] P. Guyot-Sionnest, Phys. Rev. Lett. 64 (1990) 2156.

[60] J.D. Beckerle, R.R. Cavanagh, M.P. Casassa, E.J. Heilweil, J.C. Stephenson, Subpicosecond transient infrared spectroscopy of adsorbates—vibrational dynamics of CO/Pt(1 1 1), J. Chem. Phys. 95 (1991) 5403–5418.

[61] M. Morin, N.J. Levinos, A.L. Harris, Vibrational energy transfer of CO/Cu(1 0 0): nonadiabatic vibration/electron coupling, J. Chem. Phys. 96 (1992) 3950–3956.

[62] T.A. Germer, J.C. Stephenson, E.J. Heilweil, R.R. Cavanagh, Hot carrier excitation of adlayers—time-resolved measurement of adsorbate-lattice coupling, Phys. Rev. Lett. 71 (1993) 3327–3330.

[63] T.A. Germer, J.C. Stephenson, E.J. Heilweil, R.R. Cavanagh, Picosecond time-resolved adsorbate response to substrate heating: spectroscopy and dynamics of CO/Cu(1 0 0), J. Chem. Phys. 101 (1994) 1704–1716.

[64] M. Headgordon, J.C. Tully, Molecular dynamics with electronic frictions, J. Chem. Phys. 103 (1995) 10137–10145.

[65] J.P. Culver, M. Li, L.G. Jahn, R.M. Hochstrasser, A.G. Yodh, Vibrational response of surface adsorbates to femtosecond substrate heating, Chem. Phys. Lett. 214 (1993) 431–437.

[66] A. Bandara, J. Kubota, K. Onda, A. Wada, S.S. Kano, K. Domen, C. Hirose, Time-resolved SFG study of the vibrational excitation of adsorbed CO on Ni(1 1 1) and NiO(1 1 1) surfaces under the irradiation of UV and visible photons, Surf. Sci. 428 (1999) 331–336.

[67] F. Budde, T.F. Heinz, M.M.T. Loy, J.A. Misewich, F. de Rougemont, H. Zacharias, Femtosecond time-resolved measurement of desorption, Phys. Rev. Lett. 66 (1991) 3024.

[68] J.A. Prybylla, H.W.K. Tom, G.D. Aumiller, Femtosecond time-resolved surface reaction: desorption of CO from Cu(1 1 1) in <325 fs, Phys. Rev. Lett. 68 (1992) 503.

[69] M. Bonn, S. Funk, C. Hess, D.N. Denzler, C. Stampfl, M. Scheffler, M. Wolf, G. Ertl, Phonon- versus electron-mediated desorption and oxidation of CO on Ru(0 0 0 1), Science 285 (1999) 1042–1045.

[70] H. Petek, M.J. Weida, H. Nagano, S. Ogawa, Real-time observation of adsorbate atom motion above a metal surface, Science 288 (2000) 1402–1404.

[71] S. Funk, M. Bonn, D.N. Denzler, C. Hess, M. Wolf, G. Ertl, Desorption of CO from Ru(0 0 1) induced by near-infrared femtosecond laser pulses, J. Chem. Phys. 112 (2000) 9888.

[72] J.A. Misewich, T.F. Heinz, D.M. Newns, Desorption induced by multiple electronic transitions, Phys. Rev. Lett. 68 (1992) 3737.

[73] M. Brandbyge, P. Hedegard, T.F. Heinz, J.A. Misewich, D.M. Newns, Electronically driven adsorbate excitation mechanism in femtosecond-pulse laser desorption, Phys. Rev. B—Condens. Matter 52 (1995) 6042–6056.

[74] L.M. Struck, L.J. Richter, S.A. Buntin, R.R. Cavanagh, J.C. Stephenson, Femtosecond laser-induced desorption of CO from Cu(1 0 0): comparison of theory and experiment, Phys. Rev. Lett. 77 (1996) 4576–4579.

[75] C.J. Zhang, P. Hu, A. Alavi, A density functional theory study of CO oxidation on Ru(0 0 0 1) at low coverage, J. Chem. Phys. 112 (2000) 10564.

[76] H. Petek, S. Ogawa, Femtosecond time-resolved two-photon photoemission studies of electron dynamics in metals, Prog. Surf. Sci. 56 (1998) 239–310.

[77] G.D. Billing, Electron–hole pair excitation in molecule–surface collisions, J. Chem. Phys. 112 (2000) 335.

[78] R. Baer, R. Kosloff, Quantum dissipative dynamics of adsorbates near metal surfaces: a surrogate Hamiltonian theory applied to hydrogen on nickel, J. Chem. Phys. 106 (1997) 8862.

[79] R. Baer, R. Kosloff, Hydrogen transport in nickel(1 1 1), Phys. Rev. B 55 (1997) 10952.

[80] A. Gross, M. Scheffler, Reply to Comment on "Six-dimensional quantum dynamics of adsorption and desorption of H_2 at Pd(1 0 0): steering and steric effects", Phys. Rev. Lett. 77 (1996) 405.

[81] C.T. Rettner, D.J. Auerbach, Comment on "Six-dimensional quantum dynamics of adsorption and desorption of H_2 at Pd(1 0 0): steering and steric effects", Phys. Rev. Lett. 77 (1996) 404.

[82] F. Besenbacher, I. Chorkendorff, B.S. Clausen, B. Hammer, A.M. Molenbroek, J.K. Nørskov, I. Stensgaard, Design of a surface alloy catalyst for steam reforming, Science 279 (1998) 1913.

[83] S.T. Ceyer, New mechanisms for chemistry at surfaces, Science 249 (1990) 133.

[84] J. Harris, J. Simon, A.C. Luntz, C.B. Mullins, C.T. Rettner, Thermally assisted tunneling: CH_4 dissociation on Pt(1 1 1), Phys. Rev. Lett. 67 (1991) 652.

[85] V.A. Ukraintsev, I. Harrison, A statistical model for activated dissociative adsorption: application to methane dissociation on Pt(1 1 1), J. Chem. Phys. 101 (1994) 1564.

[86] A.V. Walker, D.A. King, Dynamics of the dissociative adsorption of methane on Pt{1 1 0} (1 × 2), Phys. Rev. Lett. 82 (1999) 5156.

[87] L.B.F. Juurlink, P.R. McCabe, R.R. Smith, C.L. DiCologero, A.L. Utz, Eigenstate-resolved studies of gas–surface reactivity: CH_4 (v_3) dissociation on Ni(1 0 0), Phys. Rev. Lett. 83 (1999) 868.

[88] D.C. Seets, M.C. Wheeler, C.B. Mullins, Mechanism of the dissociative chemisorption of methane over Ir(1 1 0): trapping mediated or direct? Chem. Phys. Lett. 266 (1997) 431.

[89] P.M. Holmblad, J. Wambach, I. Chorkendorff, Molecular beam study of dissociative sticking of methane on Ni(1 0 0), J. Chem. Phys. 102 (1995) 8255.

[90] A.V. Hamza, R.J. Madix, The activation of alkanes on Ni(1 0 0), Surf. Sci. 179 (1987) 25.

[91] D. Kelly, W.H. Weinberg, Isotope effects in trapping-mediated chemisorption of ethane and propane on Ir(1 1 0), J. Chem. Phys. 106 (1996) 3789.

[92] U. Manthe, H.-D. Meyer, L.S. Cederbaum, Wave-packet dynamics within the multiconfiguration Hartree framework: general aspects and applications to NOCl, J. Chem. Phys. 97 (1992) 3199.

[93] R. Milot, A.P.J. Jansen, Ten-dimensional wave packet simulations of methane scattering, J. Chem. Phys. 109 (1998) 1966.

[94] G.A. Worth, H.D. Meyer, L.S. Cederbaum, Relaxation of a system with a conical intersection coupled to a bath: a benchmark 24-dimensional wave packet study treating the environment explicitly, J. Chem. Phys. 109 (1998) 3518.

[95] Y. Huang, A.M. Wodtke, H. Hou, C.T. Rettner, D.J. Auerbach, Observation of vibrational excitation and deexcitation for $NO(v = 2)$ scattering from Au(1 1 1): evidence for electron–hole-pair mediated energy transfer, Phys. Rev. Lett. 84 (2000) 2985–2988.

[96] Y.H. Huang, C.T. Rettner, D.J. Auerbach, A.M. Wodtke, Vibrational promotion of electron transfer, Science 290 (2000) 111.

[97] A. Al Halabi, A.W. Kleyn, G.J. Kroes, New predictions on the sticking of HCl to ice at hyperthermal energies, Chem. Phys. Lett. 307 (1999) 505–510.

[98] M. Bonn, C. Hess, S. Funk, J.H. Miners, B.N.J. Persson, M. Wolf, G. Ertl, Femtosecond surface vibrational spectroscopy of CO adsorbed on Ru(0 0 1) during desorption, Phys. Rev. Lett. 84 (2000) 4653–4656.

[99] A. Bandara, J. Kubota, K. Onda, A. Wada, S.S. Kano, K. Domen, C. Hirose, Short-lived reactive intermediate in the decomposition of formate on NiO(1 1 1) surface observed by picosecond temperature jump, J. Phys. Chem. B 102 (1998) 5951–5954.

[100] J.A. McGuire, W. Beck, X. Wei, Y.R. Shen, Fourier-transform sum-frequency surface vibrational spectroscopy with femtosecond pulses, Opt. Lett. 24 (1999) 1877.

[101] Y.R. Shen, Surface-properties probed by 2nd-harmonic and sum-frequency generation, Nature 337 (1989) 519.

[102] A. Gross, Reactions at surfaces studied by ab initio dynamics calculations, Surf. Sci. Rep. 32 (1998) 291–340.

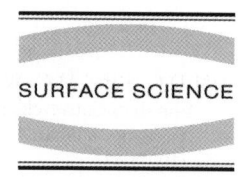

The growth and modification of materials via ion–surface processing

Luke Hanley [a,*], Susan B. Sinnott [b]

[a] *Department of Chemistry, The University of Illinois at Chicago, Chicago, IL 60607-7061, USA*
[b] *Department of Materials Science and Engineering, The University of Florida, 100 Rhines Hall, Gainesville, FL 32611, USA*

Received 12 July 2000; accepted for publication 31 January 2001

Abstract

A wide variety of gas phase ions with kinetic energies from 1–10^7 eV increasingly are being used for the growth and modification of state-of-the-art material interfaces. Ions can be used to deposit thin films; expose fresh interfaces by sputtering; grow mixed interface layers from ions, ambient neutrals, and/or surface atoms; modify the phases of interfaces; dope trace elements into interface regions; impart specific chemical functionalities to a surface; toughen materials; and create micron- and nanometer-scale interface structures. Several examples are developed which demonstrate the variety of technologically important interface modification that is possible with gas phase ions. These examples have been selected to demonstrate how the choice of the ion and its kinetic energy controls modification and deposition for several different materials. Examples are drawn from experiments, computer simulations, fundamental research, and active technological applications. Finally, a list of research areas is provided for which ion–surface modification promises considerable scientific and technological advances in the new millennium. © 2001 Elsevier Science B.V. All rights reserved.

Keywords: Ion–solid interactions; Growth; Ion etching; Ion bombardment; Ion implantation; Sputtering; Computer simulations; Molecular dynamics

1. Overview and motivation

The development of new materials continues to be one of the ongoing technological and scientific revolutions at the dawn of the 21st century. Interface science is at the forefront of this revolution because surfaces of materials dictate many of their most important properties. Continuing progress in many technological applications requires the engineering of interfaces with a wide range of chemical, physical, and/or morphological characteristics.

A wide variety of gas phase ions with kinetic energies from 1–10^7 eV increasingly are being used for the growth and modification of state-of-the-art material interfaces. Ions can be used to deposit thin films; expose fresh interfaces by sputtering; grow mixed interface layers from ions, ambient neutrals, and/or surface atoms; modify the phases of interfaces; dope trace elements into interface

---

* Corresponding author. Tel.: +1-3129960945; fax: +1-3129960431.
E-mail addresses: lhanley@uic.edu (L. Hanley), sinnott@mse.ufl.edu (S.B. Sinnott).

0039-6028/01/$ - see front matter © 2001 Elsevier Science B.V. All rights reserved.
PII: S0039-6028(01)01528-X

regions; impart specific chemical functionalities to a surface; toughen materials; and create micron- and nanometer-scale interface structures. Ion-induced processes allow the engineering of interfaces with specific wettability, hardness, resistance to corrosion, optical parameters, electronic functionality, dimensionality, and/or biocompatibility. Technologically important ion–surface processes include the production of hard coatings for machine tools and computer disk drives; biocompatible surfaces for tissue culture, contact lenses, medical implants, and other biomedical devices; antireflective and protective optical coatings; polymer films for packaging and adhesives; combinatorial arrays; and chemical sensors. Current or potential steps of microelectronics manufacturing that involve ion–surface modification include reactive ion etching, ion beam repair of photolithography masks, doping of semiconductors, and development of commercially viable sub-100 nm ion projection lithography.

Its versatility puts ion–surface modification at the center of a wide range of fundamental research. Not only can interface properties be adjusted experimentally via ion–surface interaction, but experimental data can be supplemented by an array of computational methods that accurately model those interactions. Ion–surface collisions also play an important role in other interface modification methods that might be considered unrelated at first glance, including plasma processing and pulsed laser deposition. However, the collision between a specific ion of a given kinetic energy with a surface can be readily modeled by computer simulations. By contrast, plasmas, sputter plumes, and laser plumes involve multiple particles with broad energy distributions that can only be simulated by correspondingly more complex computer models. Few interface modification methods possess computer modeling tools as powerful and accurate as those afforded to ion–surface modification.

Unlike neutrals, ions can be easily accelerated to any kinetic energy between 1 and 10^7 eV. The amount of energy imparted to a surface or buried interface region of a material is readily controlled by varying this kinetic energy. The chemical and physical processes that are induced in the surface by ion impact also are controlled by this kinetic energy [1–8]. A specific terminology has arisen to define approximate ion energy ranges: ion kinetic energies <1 eV are known as thermal, 1–500 eV are hyperthermal, 0.5–10 keV are low energy, 10–500 keV are medium energy and >0.5 MeV are high energy.

Fig. 1 displays how ion–surface processes vary with kinetic energy. Thermal ions can physisorb or chemisorb on the surface, dissociate and chemisorb to the surface (activated dissociative chemisorption), or simply scatter off the surface with some loss in their kinetic energy (direct inelastic scattering). Hyperthermal ions additionally can abstract atoms from surfaces (abstraction reactions) or dissociate and scatter if they are polyatomic (dissociative scattering). The ions also may react with the substrate atoms to create altogether new chemical species. These new chemical species may be volatile or readily sputtered, leading to surface etching (reactive ion etching). The ions may directly react with the surface atoms to form new species (i.e., oxide/nitride growth). Low energy ions can implant into the surface to form an alloy or doped material (ion implantation).

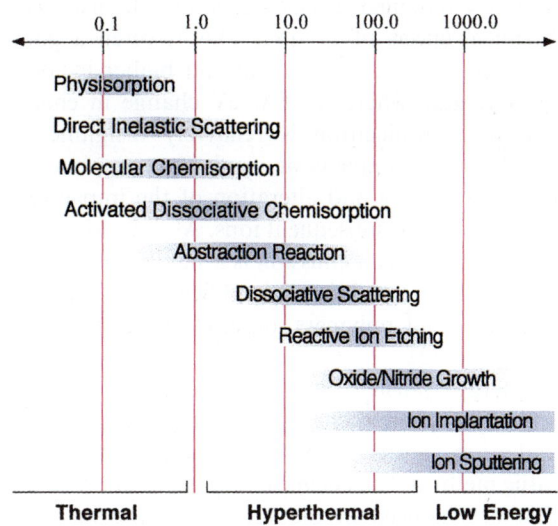

Fig. 1. Fundamental molecule/surface processes: Each shaded region designates the ion kinetic energy range over which the associated process is studied. Adapted from Ref. [150].

Hyperthermal and low energy ions deposit their kinetic energy solely into the surface without affecting the bulk region. This energy deposition causes sputtering of surface atoms as ions or neutrals, annealing of the surface, mixing of surface atoms, creation of unique surface topologies, and defect formation. Low energy ions also can induce secondary collisions of substrate atoms which cause collision cascades and result in energy transfer far from the initial impact point. The effects of low energy ion–surface impacts have been extensively studied by researchers in the field of secondary ion mass spectrometry, where the sputtered ions are used for surface chemical analysis [9,10]. The effects of the impacting ion migrate further into the bulk region of the substrate and away from the impact point with increasing ion energy. As the ion energy increases into the medium and high energy range, the ion–surface interaction moves from predominantly a nuclear–nuclear collision between projectile and target to one that is primarily electronic in nature [11].

A variety of charge transfer processes occur for ion–surface collisions across the full range of incident ion energies. These are not covered in this paper due to their complexity and considerable diversity [5]. It is often assumed that ions simply neutralize at a surface, with little additional effect upon the modified surface beyond the ballistic and chemical effects of the ion. This is often a good assumption for hyperthermal and higher incident ion energies, where the few eV change in energy due to neutralization is relatively insignificant. However, ion impacts at insulating surfaces can lead to charging and alteration of the impact energies or flux of subsequent ions. Also, ion-induced secondary electron emission is a significant process that can cause additional modification of the surface region of polymers, organic films, and other materials [12].

Energy transfer, interface chemical reactions, and other phenomena synergistically combine to make ion–surface modification an extremely versatile method. The chemical and physical processes that are induced in the surface by the ion also are influenced by its chemical identity. A beam of ions of a given chemical identity can be readily prepared by mass selection for subsequent surface interaction. If the ion is polyatomic and impacts with sufficiently low energy, it may attach intact and thereby transfer its chemical functionality to the interface. However, polyatomic ions with kinetic energies far in excess of those required to completely fragment interact differently with a surface than their energy scaled atomic constituents [10,13]. A range of interface chemistries and structures can be imparted through the selection of a projectile ion for its elemental identity, chemical structure, and number of constituent atoms. This modification can be selected for the interface while preserving the native properties of the bulk substrate.

The incredible flexibility of ion–surface modification has been implemented in a wide range of techniques and industrial tools. These are described schematically in Fig. 2. Ion beam synthesis, ion beam deposition, ion beam sputtering, ion beam sputter deposition, ion beam assisted deposition, dual ion beam sputtering, magnetron sputtering, electron beam ion assisted deposition, chemical assisted ion beam etching, reactive ion beam etching, pulsed laser deposition, plasma processing, plasma immersion, ion implantation, intense pulsed ion beams, ionized plasma vapor deposition, physical vapor deposition, and filtered arc deposition all employ ions—at least in part—to modify material interfaces (see Fig. 2 for definitions of these terms) [1,2,6,14–20]. The role of energetic ions is often critical in films grown by these various processes. Ion beam tools are also widely used in semiconductor processing [21] and are under consideration for sub-100 nm lithography [22,23].

Computationally, ion–surface interactions are studied using a variety of methods. One popular computational method is molecular dynamics, an atomistic approach where Newton's second law of motion is solved quantitatively to determine the response of atoms to the forces that are applied to them [24]. The strengths of this method include its ability to model both dynamical and non-equilibrium events. Its main weakness is that it is limited to short time scales on the order of a few picoseconds or nanoseconds. However, this does not prevent molecular dynamics from being used to study ion–surface interactions as the time scales of

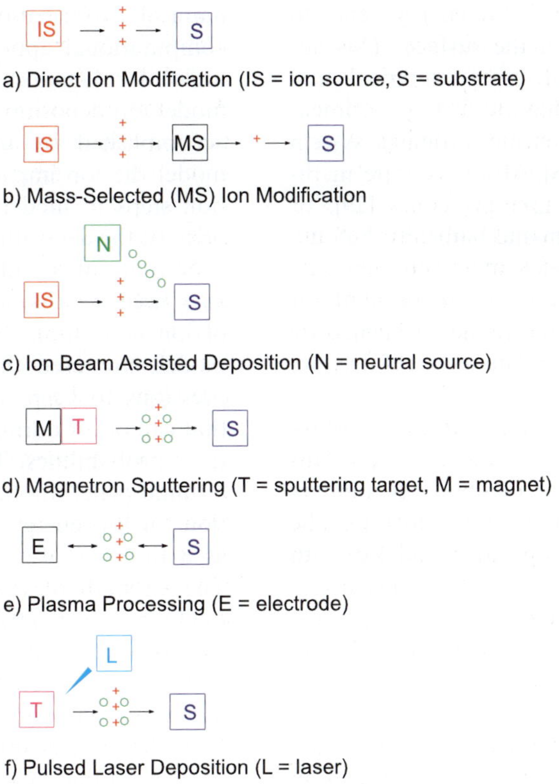

a) Direct Ion Modification (IS = ion source, S = substrate)

b) Mass-Selected (MS) Ion Modification

c) Ion Beam Assisted Deposition (N = neutral source)

d) Magnetron Sputtering (T = sputtering target, M = magnet)

e) Plasma Processing (E = electrode)

f) Pulsed Laser Deposition (L = laser)

Fig. 2. Schematic diagrams of several general experimental configurations in which ion–surface modification is either central or plays an important role. (a) Direct ion modification involves simply an ion source (IS) to deposit ions (+) on a substrate (S). (b) Mass-selected ion modification filters out ions of a single mass-to-charge ratio for deposition on the surface. Ion beam synthesis, ion beam deposition, ion beam sputtering, ion beam sputter deposition, reactive ion beam etching, and dual ion beam sputtering are all varieties of direct or mass-selected ion modification. (c) Ion beam assisted deposition simultaneously adds a source of neutral species (o), to deposit additional material or provide a reagent for ion-induced chemistry. Two variants on this method are electron beam ion assisted deposition and chemical assisted ion beam etching. (d) Magnetron sputtering uses a magnetically confined discharge to sputter ions and neutrals from a target (T) onto S. The discharge here and in plasma processing is established by a direct or alternating voltage difference applied between T and S. (e) Plasma processing uses the gas feed into the chamber to establish the discharge, with (hopefully) no sputtering of the electrode. Plasma immersion ion implantation, intense pulsed ion beams, ionized plasma vapor deposition, physical vapor deposition and filtered arc deposition are all types of plasma processing that utilize different electrode geometries (i.e., separate second electrode used instead of substrate) and/or electrical waveforms for generating the plasma. (f) Pulsed laser deposition employs a pulsed laser (L) to ablate a target material and thereby eject a plume of neutrals and ions for deposition onto S. All types of ion–surface modification typically occur in a vacuum system with an operating pressure of $\leqslant 10^{-4}$ torr and an input of a feed gas for the ion gun (or discharge) into the chamber (not shown). The ion energy is defined by the voltage difference between the ion source and substrate (not shown). Amongst the wide variety of ion sources used are electron impact, discharges, thermionic, plasmas, supersonic expansion, laser ablation and combinations thereof (often with magnetic confinement) [6]. Configurations (a) through (f) are often supplemented by one or more experimental probes to structurally and/or chemically analyze the interface region following ion modification.

these collisions are typically on the order of a few picoseconds.

Modeling ion–surface interactions with atomistic simulations such as molecular dynamics requires careful consideration of the method or potential used to calculate the energies, and therefore the forces, acting on both the ions and the surfaces. The potential used depends on the nature of the ion's interaction with the surface, which, in turn, depends significantly on the ion's energy. For example, ions with several eV energy tend to chemisorb or physisorb on the surface. This adsorption can be modeled using empirical and semiempirical potentials that are fit to experimental data and are used to determine atomistic system energies and forces [25–28]. More accurate methods are developed from first principles [29], or model the atoms quantum mechanically but use empirical methods to model interatomic interactions [30]. These same methods can be used for impacts at higher energies where the ions penetrate and disrupt the surface and/or dissociate if they are polyatomic.

At incident energies of around 10 keV, the binary collision approximation is usually used. This method assumes a series of two-body encounters between the ion and atoms in the substrate. The elastic scattering interactions are modeled with repulsive interatomic potentials while inelastic interactions are modeled with potentials that assume electronic excitation has occurred. While the binary collision approximation has found success in modeling high energy processes, recent simulations suggest that many-body methods are more exact [31]. Finally, electronic stopping is dominant in collisions at energies of about 1 MeV, requiring the use of methods that either treat the electrons in the system explicitly or are derived from such methods [32].

Ion–surface interactions also are modeled with atomistic Monte Carlo simulation methods [6,11,33]. The most popular of the computer codes that implement these methods for ion–surface collisions is known as TRIM (transport of ions in matter), although a variety of related computer codes have evolved from the original TRIM [34]. Monte Carlo simulation methods compare the energy of a proposed new configuration to that of the original configuration using the same methods to calculate energies listed above. If the new configuration is lower in energy it is accepted automatically. However, if the new configuration is higher in energy a random number is used to determine whether or not to accept it. The strength of this method is that it relatively quickly reaches the equilibrium geometry for the system under study, such as a thin film that has been deposited from incident ions. Its weakness is that no dynamical information is available. Sometimes computational approaches use a combination of molecular dynamics and Monte Carlo methods to model ion deposition of thin films. This is usually accomplished by using molecular dynamics to model the ion impacts and Monte Carlo simulation steps to model the relaxation of the system prior to the deposition of the next ion in the beam.

Non-atomistic phenomenological models are also used to characterize the cumulative effects of ion deposition. For instance, semiquantitative models are used that rely on mathematical expressions to keep track of the various processes that can occur during ion–surface interactions and their probabilities. These models are used to determine penetration, adhesion, and defect formation for low energy ion bombardment of surfaces as a function of the reaction conditions [35–37]. The strength of this approach is that after the model has been applied and the results compared to experimental data, information on which ion–surface interactions are most important in producing a given thin film is available. Its weakness is that the model might represent an unphysical conclusion that nevertheless matches experimental data. Additional mathematical models focus on sputtering rather than thin-film growth [7,9].

2. Examples

Several examples are used to demonstrate the variety of technologically important interface modification that is possible with gas phase ions. These examples have been selected to demonstrate how the choice of the ion and its kinetic energy controls modification and deposition for several different materials. However, the wide breadth of the field dictates that these few examples do not fully cover the entire range of ion–surface modification.

2.1. Thin-film growth and chemical modification

2.1.1. Hard carbon nitride film growth
Ion–surface collisions play an important role in the production of mechanically hard films [38]. Theoretical calculations predict that a specific

phase of carbon nitride, β-C_3N_4, is harder than diamond and displays similarly high thermal conductivity [39]. The many potential applications for such a material have encouraged the materials science community to synthesize either β-C_3N_4 or some other variety of hard carbon nitride film. Some of the attempts to produce β-C_3N_4 films utilize mass-selected hyperthermal or low energy ions by codeposition of C^+ and N^+, implantation of N^+ or N_2^+ into graphite or diamond, or codeposition of C_2^- and CN^- [40–44]. The resulting carbon nitride films typically display differing morphologies, mixed phases, and varying C/N ratios. Analytical modeling of the ion beam growth of carbon nitride provides insight about the competition between the processes that lead to thin-film formation—deposition, implantation, diffusion, and densification—and the sputtering and damage to the film from the energetic ions [36]. Magnetron sputtering (see Fig. 2) of graphite targets with nitrogen/argon gas mixtures also may be applied to grow β-C_3N_4 films [40,45,46]. Films grown from magnetron sputtering with a substrate bias of −300 eV are both hardest and smoothest, although they are not of the desired β-C_3N_4 phase [46]. Since the substrate bias causes the acceleration of ions, it is clear that a hyperthermal ion energy is important to growth of high quality carbon nitride films. Magnetron sputtering is now used by many manufacturers to grow hard carbon nitride films on hard disk drive surfaces, improving hard disk lifetimes by reducing wear from the read/write heads [44,46]. Formation of high quality, single phase, crystalline β-C_3N_4 remains a controversial topic, although recent advances are moving closer to this goal [47]. Nevertheless, the commercial success of hard carbon nitride films in hard disk drives remains an important vindication of ion-based film growth methods.

2.1.2. Metal thin films

The controlled growth of metallic thin films is important for the production of a variety of electronic devices such as transistors. Because the metal needs to cover areas on the order of a few micrometers, conventional deposition techniques, such as electroplating, cannot be used. The preferred method of deposition is currently magnetron sputtering. However, this method is not precise enough to be used for contacts on the nanometer scale, which are becoming increasingly important as the size of electronic devices continues to decrease. Two methods that show promise for the production of metallic films with the needed precision are metal ion beam deposition and metal cluster beam deposition.

The quality of metal films produced through metal ion deposition depends on several factors, as indicated schematically in Fig. 3 [5]. The most important is the incident energy which is usually on the order of several eV. Other important factors

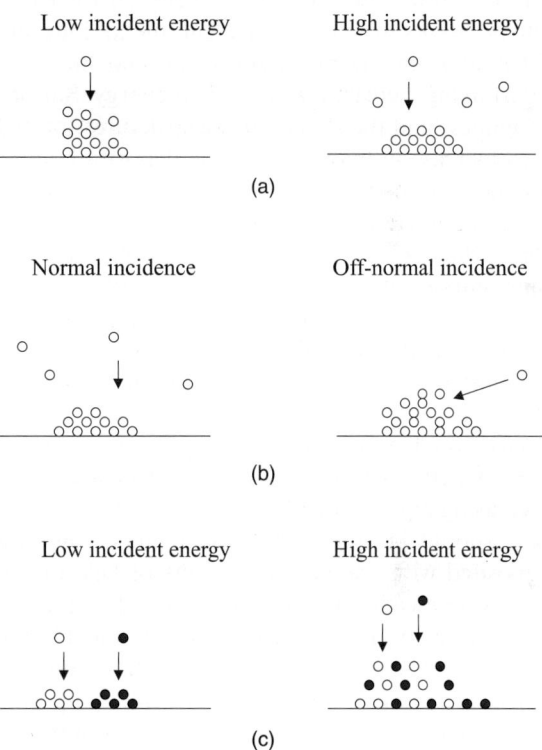

Fig. 3. Schematic showing the various processes that can occur for metal ion deposition. Panel (a) shows how deposition at low incident energy leads to the formation of rough thin films while deposition at higher incident energies yields smoother films but also leads to an increase in sputtering. Panel (b) shows how deposition at normal incidence leads to the formation of smoother films than deposition at an off-normal angle but that the amount of sputtering is higher at normal incidence. Panel (c) compares the deposition of mixed films at low and high incident energies. The degree of mixing and smoothness of the film increases with energy.

are the crystal structure of the surface, the incident angle, and the relative densities of the incident ion and the surface atoms. As the incident energy increases, the deposited ions have enough energy to preferentially diffuse such that layer-by-layer growth is achieved if this is thermodynamically preferred for the system in question. In contrast, very low energies generally yield thin films with significant surface roughness [48]. Combined molecular dynamics and Monte Carlo simulations provide details about the diffusion and coalescence of islands during deposition to form single layers and how these processes change with variations in reaction conditions [49]. For example, the simulations show how intermixing between disparate layers occurs when the incident metal ions are different from the metal substrate, and how this intermixing increases with incident energy. Surface roughness and the shapes of the structures formed also increase in magnitude if the deposition angle increases from the normal [50]. This is due to long-range interactions between the incident ions and the surface, called steering, that become increasingly important at large angles. Steering causes the ions to deposit preferentially on top of islands rather than at island edges, which accounts for the roughness of the surface. Sputtering of the surface is a side effect of metal ion deposition. As expected, the amount of sputtering increases as the ion's kinetic energy goes up but decreases with increasing angles from the surface normal [51].

It should be noted that when metal ions are deposited with energies in the tens of keV range, collision cascades are created within the material that cause local melting [52]. When the molten metal cools, nanometer-scale amorphous regions and defect clusters form within the bulk [53]. This is especially apparent when heavier ions impact a less dense metal substrate. Interstitial point defects and islands can also result [54]. Thus, these high energy bombardments create unique nanostructures and defects near the surface with unusual properties.

The deposition of metallic thin films from clusters is well studied because of the unique properties of cluster beams compared to atomic ion beams [55]. These include the large concentration of atoms and energy that is deposited in a localized region of the surface that facilitates thin-film growth and at the same time causes increased surface damage and sputtering. However, except under extreme conditions, clusters do not penetrate as deeply as single atoms because of their large size [56]. Thus, they have the ability to modify selectively a shallower region of the surface than single ion beams. Despite these benefits, there are experimental problems producing clusters in a reliable manner and verifying their size distribution [57].

One dependable method for depositing metal clusters is with a combination of a magnetron sputter discharge and a gas aggregation source [58]. Deposition conditions include relatively low temperatures of about 500 K and high kinetic energies of 10 eV/atom for clusters of several thousand atoms [59–61]. Thin films grown using this method are smoother than comparable films formed by evaporated deposition and have a dense "mirror-like" finish and good adhesion to a variety of substrates. This process is modeled with molecular dynamics simulations. The top four panels of Fig. 4 contains snapshots from a simulation where a 2000 atom Cu cluster is deposited on a Cu surface with 10 keV of energy [62]. Immediately after impact, a temperature of several 1000 K and a pressure of 80 GPa is produced at the contact area. This pressure pulse is the driving force for a shear process that forms a surface crater. However, the uphill shear motion is impeded by the additional surface atoms so the crater rim formation is suppressed. In the opposite direction there is no such obstacle which allows the rim to slip downhill. The high pressures and temperatures cause an alloying of the substrate with the deposited material leading to a strong adhesion of the coating.

In contrast, highly disordered, porous structures are formed if lower incident energies are used. An example of such a film is shown in the bottom panel of Fig. 4. In this case, a beam of Cu clusters containing 1000 atoms each are deposited on a Cu surface with 10 eV of energy (leading to so-called soft landings) in molecular dynamics simulations [63]. The resulting film exhibits dendritic growth with high roughness, low density and poor adhesion to the substrate. These predictions agree with

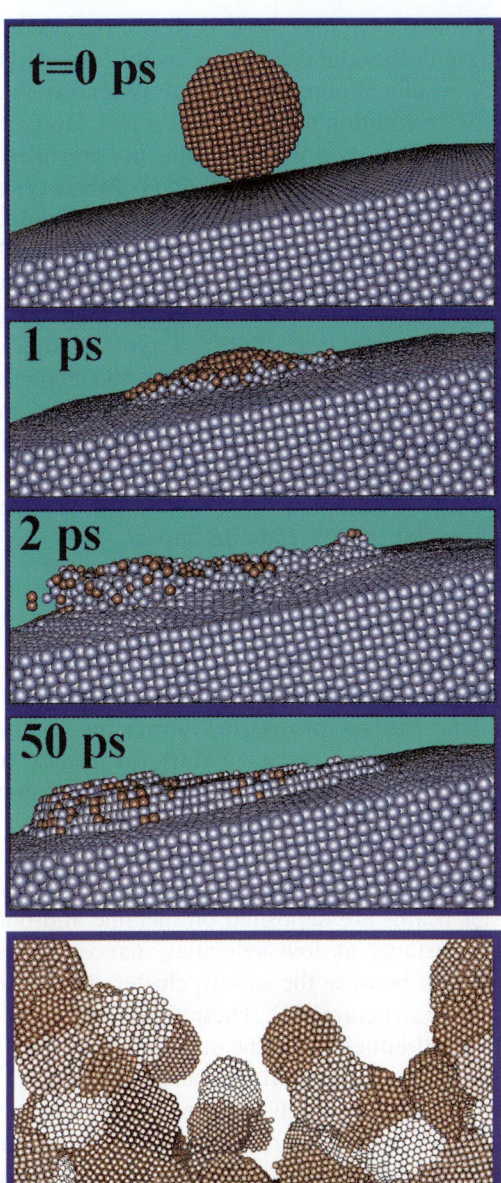

Fig. 4. The top four panels show snapshots from a molecular dynamics simulation of a 2000 atom Cu cluster (red spheres) impinging onto a tilted Cu(0 0 1) surface (blue spheres) at 10 keV of energy. Each panel is labeled with the time from initial contact of the cluster with the surface. The bottom panel shows the final snapshot from a molecular dynamics simulation of a beam of 1000 atom Cu clusters with 10 eV kinetic energy depositing onto a Cu(0 0 1) surface. Note how the quality and character of the films change dramatically with incident cluster energy. Figure used with permission of Elsevier [62,63].

the results of experiments that use laser vaporization to produce smaller (5–10 atoms/cluster) charged and neutral metal clusters that impact with thermal or hyperthermal energies. These experiments reveal that the clusters keep their shape after deposition giving rise to unusual nanostructured films [64].

Atomistic simulations [65–68] shed light on additional ways in which changes in the reaction conditions lead to fundamentally different thin films. For example, they show that smaller clusters (of a few atoms) are more likely to embed themselves into the surface while larger clusters (of more than 50 atoms) are more likely to spread across the surface. In the case of deposition on a surface composed of a different material, they provide information on the way in which the outcome depends on the relative cohesive energies, chemical properties, and atomic masses of the clusters and surfaces. For example, Al does not penetrate Ni because of the mass difference but instead spreads epitaxially, while Ni clusters easily penetrate and embed themselves into Al surfaces. Finally, at high enough impact energies of a few eV/atom, simulations have shown that excitation of electrons is more likely to occur during the early stage of the cluster contact with the surface when the cluster is highly compressed. These core-excited atoms decay by Auger emission following scattering off the surface [32].

2.1.3. Organic thin-film deposition and chemical modification

Hyperthermal atomic ions can be used to chemically modify surfaces or deposit thin films [2]. Small polyatomic ions can induce selective chemical modification of polymer and semiconductor surfaces [69,70]. Certain polyatomic ions land as intact species upon fluorocarbon surfaces at ~10 eV collision energies [71,72]. Polymeric films for applications in optoelectronics can even be grown from organic ion sources [73].

Chemical modification by polyatomic, organic ions can be quite selective, with strong dependence upon the ion structure and kinetic energy [74,75]. Fig. 5 displays the chemical composition of fluorocarbon films grown on a polystyrene surfaces from 25–100 eV CF_3^+ and $C_3F_5^+$ ions. Each ion

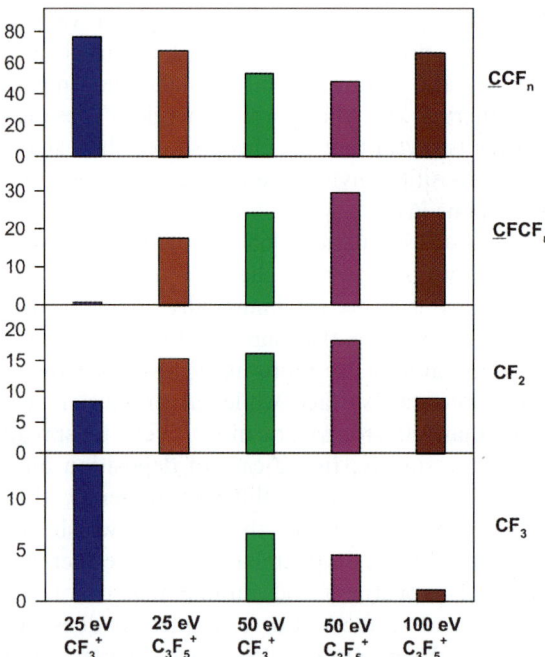

Fig. 5. Chemical composition of polystyrene thin films modified by 25–100 eV CF_3^+ and $C_3F_5^+$ ion beams, prepared with ion doses equivalent to 1.5×10^{16} F atoms/cm^2. The percentages of each fluorocarbon component are normalized by removal of the non-fluorinated component. These fluorocarbon films are several nm thick and 30–60% fluorine, with the remaining composition carbon. Figure adapted from Ref. [75].

leads to different polystyrene film chemistry, when compared at either similar total ion energy or energy/atom. These ions form a distribution of different fluorocarbon functional groups in amounts dependent upon the incident ion energy, structure, and fluence (Fig. 5). Both ions deposit mostly intact upon the surface at 25 eV, with $C_3F_5^+$ maintaining a structure similar to the proposed $CF_2=CFCF_2^+$ gas phase structure [74,75]. However, both ions undergo some polymerization upon deposition. Fragmentation of the ions increases as the ion energies are increased to 50 eV. Both ions form covalent bonds with the polystyrene surface at all energies. Finally, organic film growth from hyperthermal ions is actually a balance between growth and etching processes [8].

Further information on hyperthermal $C_3F_5^+$ deposition is obtained by molecular dynamics simulations on the analogous hydrocarbon system, $C_3H_5^+$ on polystyrene. Fig. 6 shows the effect of changes in the ionic structure, in this case $C_3H_5^+$, on the results of impacts with polystyrene at 50 eV [74,75]. In addition to the ground state $CH_2–C^+H–CH_2$ structure, two excited states are considered, $CH_3–C^+=CH_2$ and $CH_3–CH–C^+H$. Panels (a)–(c) in Fig. 6 show representative dissociation products for these isomers in snapshots from the simulations while panel (d) summarizes some of the important dissociation products that remain on the surface as a function of isomer type. Significant differences are predicted, such as the enhanced production of methyl radicals for isomers that already contain CH_3 structures. The simulations reaffirm that the ground state isomer is the most stable and predict that the $CH_3–CH=C^+H$ isomer is the easiest to dissociate on impact.

2.1.4. Organic cluster ion deposition

The deposition of carbon or hydrocarbon clusters is used to grow several types of films from polymeric to diamond-like carbon. Such films could be used as protective, lubricating coatings for computer hard disks, coatings for medical implants, or as dielectrics. As in the case of metal cluster deposition, the incident energy plays a crucial role in determining the structure and properties of the film. For example, when carbon cluster beams are deposited on metallic and polymeric surfaces at low velocities, nanostructured films form because the carbon clusters tend to retain their structure [76]. These films are found to have good adhesion to the substrate and may be used in electrochemical applications because of their stability and high surface area. In contrast, when carbon clusters ranging from tens to hundreds of atoms are deposited at thermal energies, similar nanostructured diamond-like films form that are only metastable with low adhesion to the substrate [77].

Atomistic simulations are used to study the chemistry that occurs within the cluster following organic cluster–surface impacts [78]. For instance, simulations describe the deposition of pure carbon clusters ranging in size from a few tens of atoms to a few hundreds of atoms, where the larger clusters are fullerenes. The clusters are deposited with energies of 0.1–1 eV/atom. The simulations reveal

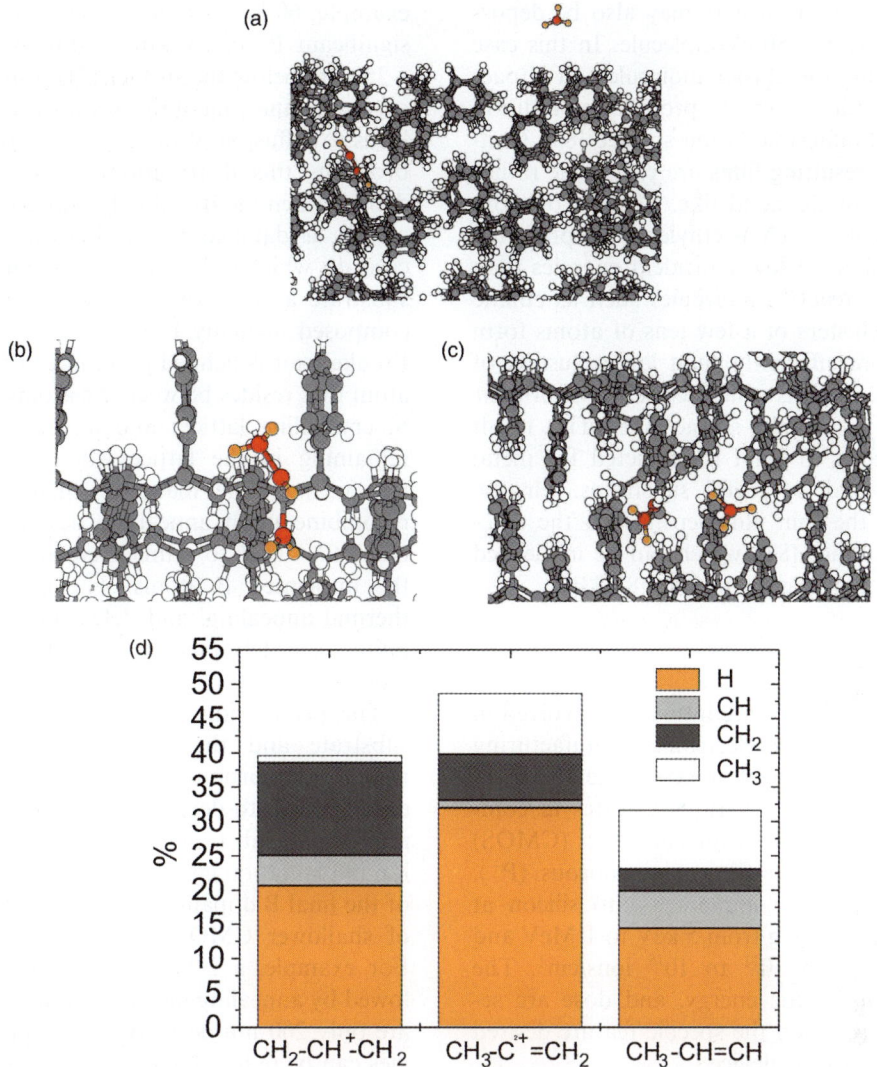

Fig. 6. The top three panels show snapshots from molecular dynamics simulations of three isomers of $C_3H_5^+$ deposited on polystyrene at incident energies of 50 eV. The bottom panel shows a statistical representation of some of the results. Panel (a) is a representative snapshot from the deposition of CH_3–CH–C^+H, panel (b) is a representative snapshot from the deposition of CH_2–C^+H–CH_2, and panel (c) is a representative snapshot from the deposition of CH_3–C^+–CH_2. Note how the ion fragments change with changes in the isomer structure. Panel (d) shows the percentage of the indicated products that remain bonded to or embedded in the polystyrene surface after deposition has fragmented the original ion. The results in panel (d) are averaged over 40 trajectories for each isomer to provide a statistical representation of possible outcomes. Figures adapted from Refs. [74,75].

that cluster size distribution plays a crucial role [79]. Films are homogeneous, graphitic, and relatively high density when grown with a bimodal distribution of cluster sizes and a larger fraction of small clusters compared to large clusters. Films are more porous and lower density with a random inhomogeneous, graphitic structure when a bimodal distribution is used with more large clusters than small clusters. Films become more dense and uniform as the clusters' energy increases—in agreement with experimental results—when a unimodal distribution of clusters is used.

Organic molecular clusters may also be deposited at energies of 5–80 eV/molecule. In this case collisions among the cluster molecules on impact play an important role in promoting addition chemistry and adhesion to the surface [80]. Consequently, the resulting films are polymeric rather than graphitic or diamond like. Clusters of more reactive molecules, such as ethylene, are predicted to form thin films at lower incident energies than clusters of less reactive molecules such as ethane [80]. Smaller clusters of a few tens of atoms form thin films more efficiently than large clusters of several hundred atoms which scatter most of their products away from the surface [81]. This result contrasts directly to what is predicted for metal clusters deposited on metal substrates. Finally, nucleation of the thin film depends on the reactivity of the surface [82], which can be influenced by the presence of a growing film [82,83].

2.2. Doping

Ion–surface collisions are intimately involved in several steps of the microelectronics manufacturing processes [84]. One crucial process is the doping of silicon wafers to create various features in complementary metal-oxide-semiconductor (CMOS) transistors [21]. Boron (B^+), phosphorous (P^+), and arsenic (As^+) are implanted into silicon at kinetic energies ranging from 5 keV to 1 MeV and doses ranging from 10^{11} to 10^{15} ions/cm^2. The dopant ion, its kinetic energy, and dose are selected depending upon the specific feature desired in the microelectronic device.

The need to shrink the dimensions of integrated circuits is progressing rapidly. The gate length of a CMOS transistor constructed on a silicon-based integrated circuit, 2 μm in 1975 and 0.35 μm in 1997, is now targeted to drop below 100 nm within the next few years [21]. Successful construction of these sub-100 nm CMOS transistors requires that B-doped regions be confined to similarly sized volumes of the Si substrate.

2.2.1. B doping of Si with low energy B^+

The keV to MeV energies of B^+ ion beams commonly used to dope Si substrates are expected to be too high to achieve sub-100 nm features. For example, 60 keV B^+ implantation in Si leads to a significant B concentration ranging from ~40 to ~260 nm below the surface [21]. Furthermore, the required annealing of the Si substrate subsequently leads to diffusion of the implanted B atoms which broadens this distribution out to an ~400 nm depth. When the B^+ initially impacts the surface, it creates a damage region known as a collision cascade, which is due to the secondary collisions of substrate atoms. This collision cascade region is composed of many Frenkel pairs, where a single Frenkel pair is defined as an interstitial atom (a Si atom that resides between the atoms in the regular Si crystalline lattice) and a vacancy (the space remaining in the lattice when that Si atom is removed). While many of these Frenkel pairs recombine upon annealing, defective regions remain that lead to damage-enhanced diffusion of the implanted B atoms. The ion implantation, thermal annealing, and defect production can be quite accurately predicted by Monte Carlo simulations.

The penetration depth of the ion into the Si substrate and the substrate defects are both roughly proportional to the ion energy: higher ion energies lead to deeper B penetration and greater numbers of substrate defects. Therefore, lower B^+ ion implantation energies reduce the spatial extent of the final B dopant profile and allow production of shallower CMOS transistor gates [21,85–87]. For example, a 2 keV B^+ ion implantation followed by annealing produces B-doped regions that are only 200 nm deep. B^+ ion implantation energies can be reduced again to 100–1000 eV, further narrowing the preannealing depth profile to below 100 nm. However, damage-enhanced diffusion that occurs when the substrate is subsequently annealed becomes the limiting factor in the final B depth distributions for <1 keV ions. Various annealing and preamorphization strategies have been considered to overcome the problem of damage-enhanced diffusion during annealing.

2.2.2. B doping of Si with low energy $B_{10}H_{14}^+$

Another method for producing ultrashallow junctions in Si is B doping with $B_{10}H_{14}^+$ [56,88]. One advantage of using $B_{10}H_{14}^+$ for B doping is the polyatomic-surface collision dynamics reduce the implantation depth of the B individual atoms into the Si substrate. Although coulombic repulsion between ions becomes significant and begins to limit beam current at lower ion energies [21], the 10 B atoms per charge for a beam of $B_{10}H_{14}^+$ allows an order of magnitude greater flux of B atoms compared with a B^+ at similar ion energy and current. Furthermore, the energy per atom for $B_{10}H_{14}^+$ is only 9% of that for B^+ at similar absolute ion energy.

Molecular dynamics simulations on the impact of 4 keV $B_{10}H_{14}^+$ ions with Si [89] reveal that 90% of the preannealed, implanted B lies within 5 nm of the surface, although 10% of the B atoms are channeled to depths > 9 nm. The interstitial atoms and vacancies induced by 4 keV $B_{10}H_{14}^+$ ion impact are confined to the uppermost 5 nm of the surface. Fig. 7 displays the outcome for the impact of a single 4 keV $B_{10}H_{14}^+$ at the Si surface. Panel (a) of Fig. 7 displays the slight swelling and crater induced by the $B_{10}H_{14}^+$ impact. The same collision event at a 11.5 nm square Si crystal is shown from a different perspective in panel (b) to display the spatial distribution of the B atoms (orange balls), H atoms from $B_{10}H_{14}^+$ (green balls), and displaced surface Si atoms (purple balls). Panel (c) of Fig. 7 displays the same perspective as in panel (b), but displays only the interstitial Si atoms (purple balls). These results reveal that 4 keV $B_{10}H_{14}^+$ implantation in Si induces very shallow preannealed B doping and defect distributions. Experiments confirm the basic findings of the simulations [56,88,90], but other factors must be considered for commercial implementation [90].

2.3. Etching and lithography

2.3.1. Etching and growth of SiO on Si

Etching and lithography are two other processes in which ion–surface modification plays a central role [21–23,91]. The controlled growth of silicon oxide thin films from oxygen plasmas or ion beams is a method available for microelectronics manufacturing [5,84]. This thin-film growth process is balanced between deposition and etching processes. Fundamental reaction mechanisms inherent in these processes can be explored with a monoenergetic ion beam under ultrahigh vacuum

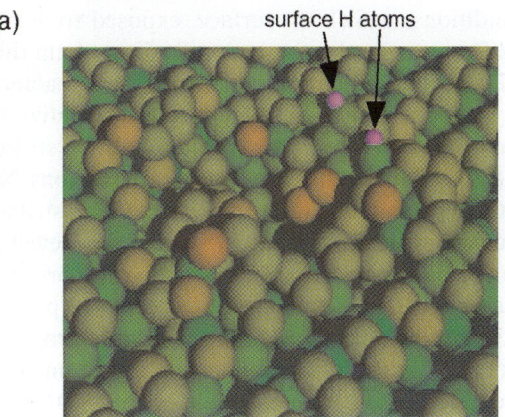

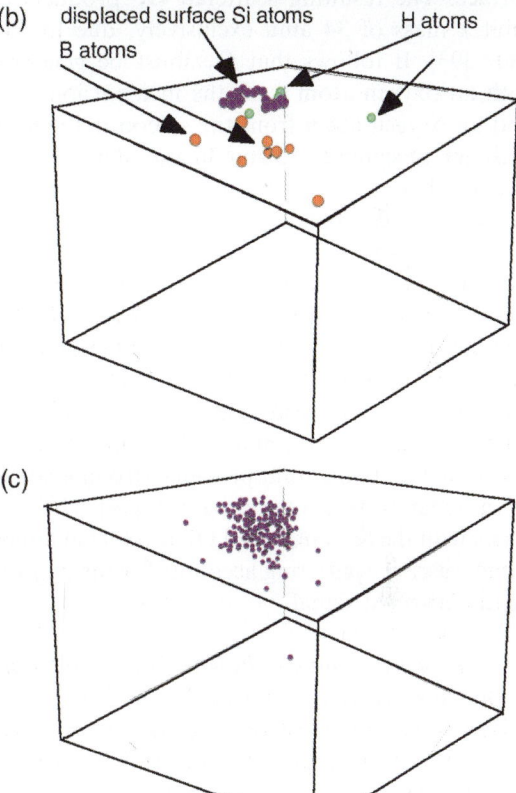

Fig. 7. Panel (a) shows the damage of a Si(1 0 0) crystal after impact of 4 keV $B_{10}H_{14}^+$. Note the surface swelling adjacent to a small crater. Panel (b) shows the crystal drawn from a different perspective where only the Si adatoms (purple balls) and the implanted atoms are shown (orange balls for B and green balls for H). Panel (c) displays the distribution of the interstitial Si atoms (purple balls). In panels (b) and (c), the crystal surface is 11.5 nm square. Figure used by permission of the American Institute of Physics [89].

conditions. A Si(1 0 0) surface, exposed to 40–200 eV O$^+$ ions, develops a thin oxide film ~4 nm thick [92]. Further oxidation of the surface is balanced by erosion of the oxide layer during reactive O$^+$ ion bombardment. For example, 70 eV O$^+$ striking the SiO$_x$ film produces various scattered ions: Si$^+$, SiO$^+$, O$^-$, and O$_2^-$. The product ion yields strongly increase with O$^+$ dose as the oxide film develops. Isotopic labeling experiments define the origin of each oxygen atom within the nascent O$_2^-$ product [93]. An isotopically pure Si^{16}O$_x$ film is grown on Si(1 0 0) using a mass-filtered ^{16}O$^+$ ion beam. Then the ion beam is switched to deliver only ^{18}O$^+$ to the surface. The resulting scattered O$_2^-$ products exhibit a mass of 34 amu exclusively, due to ^{18}O–^{16}O$^-$ [93]. It follows that O$_2^-$ must be generated with an oxygen atom from the incident ion beam and an oxygen atom from the silicon oxide layer. The lack of signal at 32 and 36 amu demonstrated that O$_2^-$ is not formed by sputtering of the oxide film or by recombination of two incident oxygen atoms, respectively. Measurements of the scattered ion angular and kinetic energy distributions (not shown) indicate that the peak of the scattered O$_2^-$ flux lay near the specular angle. Furthermore, the mean kinetic energy of the O$_2^-$ products increased linearly with incident O$^+$ kinetic energy. The correlation between incoming O$^+$ momentum and outgoing O$_2^-$ momentum provides strong evidence for a direct abstraction mechanism [94]. This contrasts with the Si$^+$, SiO$^+$ and O$^-$ product channels, where cascade sputtering accounts for the majority of the observed signal.

2.3.2. Repairing photomasks with focused ion beams

One commercial application for focused ion beams is the repair of photomasks. Photomasks are the templates used for transferring the patterns of microelectronic circuitry from a computer design file onto raw Si wafers via optical lithography [84]. The photomask is a glass plate with Cr features that are transferred onto the glass surface by an electron beam or laser writing process. Each photomask is inspected following production to reveal errors in the writing process. Two types of errors are typical: either extraneous Cr features appear where they do not belong or Cr features are missing where they should appear. These errors are corrected by use of a focused ion beam repair tool [95]. This instrument has a low energy Ga$^+$ beam that is focused to ~100 nm. It also has a secondary ion mass spectrometry imaging system that distinguishes Cr$^+$ sputtered from the Cr features by the incident Ga$^+$ ion beam from Si$^+$ sputtered from the glass. Extraneous Cr features are repaired by sputtering them away with the Ga$^+$ beam until the Cr$^+$ signal is no longer observed. Panel (a) in Fig. 8 displays a photomask with Cr features displayed as blue and glass features as

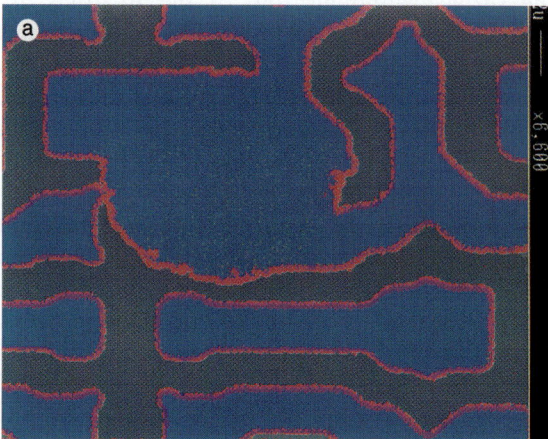

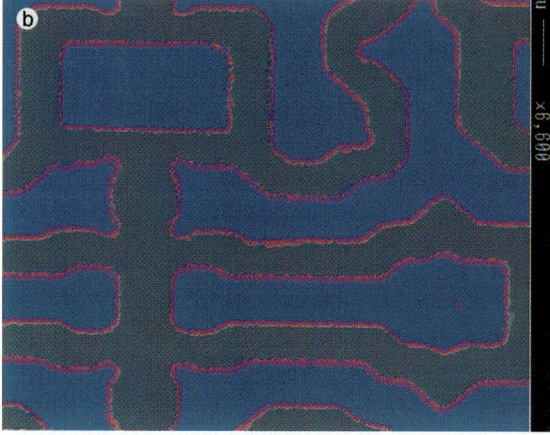

Fig. 8. Panel (a) shows a photomask with Cr features displayed as blue and glass features as either gray or pink when adjacent to Cr (bar at right = 2 μm length). A large blue feature corresponding to extraneous Cr dominates the upper center. Panel (b) shows the same photomask following repair with the Ga$^+$ ion beam: the extraneous Cr feature is now missing and has been replaced with the correct Cr/glass pattern.

either gray or pink when adjacent to Cr (bar at right = 2 μm length). A large blue feature corresponding to extraneous Cr dominates the upper center of panel (a). Panel (b) in Fig. 8 displays the photomask following repair with the Ga^+ ion beam: the extraneous Cr feature is missing and has been replaced with the correct Cr/glass pattern. Missing Cr features also can be repaired by Ga^+ ion beam stimulated deposition of graphitic carbon onto the glass surface. Graphite carbon is formed by decomposition of an adsorbed organic compound by the energetic Ga^+ ions. The organic compound can be pyrene, styrene, indene, or other aromatic hydrocarbon that is introduced as a vapor into the system [96].

2.3.3. Patterning self-assembled monolayers and polymers

Ion beam etching also can be used to pattern organic materials and polymers. 50–140 eV Ar^+ ion beams may be used to pattern organosilane self-assembled monolayers by simply masking a portion of the beam from the monolayer [97]. The ion beam patterns are then transformed into >100 nm thick Ni features by a wet chemical process known as electroless metallization. Poly(methyl methacrylate) and other polymeric photoresists also can be etched by a ∼100 keV broad beam of H^+, H_2^+, or He^+ ions projected through a stencil mask that blocks portions of the beam [22,23]. This technology is the basis for ion projection lithography, which is being considered for sub-100 nm lithography to produce the next generation of microelectronics devices. Alternatively, 300 nm line widths may be created on organosilane self-assembled monolayers by focused beams of 50–280 keV Ga^+, Si^{2+}, Au^+, and Au^{2+} followed by electroless metallization [97]. These line widths readily can be narrowed to <50 nm by better ion beam focusing [21]. Photoresist polymers also can be etched by focused ion beams [98].

2.4. Materials Toughening

2.4.1. Toughening polymers with keV–MeV atomic ion beams

Ion implantation of polymeric materials can produce a hard and wear-resistant surface while maintaining the bulk properties of the polymers [99]. Such changes could allow for polymers to be used in, for example, manufacturing applications where mechanical degradation is currently a problem. These improved properties result from cross-linking at the polymer surface caused by localized structural damage from the ions. Increasing the incident energy increases the penetration depth of the ions but does not increase the amount of cross-linking. For ion energies of several hundreds of keV to several MeV, the modification depth is only a few microns, with the exact depth decreasing as the mass of the incident ion increases. However, ions with higher masses are more efficient at inducing the surface cross-linking than lighter ions. To get a good balance between depth and efficiency of cross-linking, multiple ions (e.g., He^+ and Ar^+ at 1 MeV) can be used [100,101].

Ion implantation has been used successfully to modify several polymers such as polycarbonate, polyimide, polystyrene, and polyetheretherketone. Ion implantation can lead to increases in hardness of about two orders of magnitude and reduction of friction coefficients by about one order of magnitude. Polytetrafluoroethylene and certain other polymers do not respond well to ion bombardment, however, and simply degrade. It is thought [102] that when the incident ions transfer their energy to the surface through ionization or electronic excitation of the substrate atoms, cross-linking results, while direct nuclear collisions that scatter the ions result in surface degradation. Classical molecular dynamics simulations of ion implantation on polyethylene [103,104] predict, however, significant cross-linking near the surface without including electronic effects. They also predict that the ratio of broken C–H to C–C bonds decreases with the increasing mass of the ions and that the total number of bonds broken depends linearly on the incident energy in agreement with experimental observations.

2.4.2. Toughening steel with keV–MeV atomic ion beams

The implantation of N, C, O, Hf, Er, Cr, Ti, and Y ions into steel significantly improves the hardness, toughness and wear properties of these materials [105–111]. In addition, the friction coefficients are decreased [112,113]. These findings have important implications for steel tools for which considerable cost savings could be achieved if the lifetimes of active use of the tools could be extended. The ions are generally implanted at low to medium energies, leading to considerable changes in the structure and microstructure of the surface and near-surface regions of the alloy. The mechanisms by which these various ions modify the mechanical properties of the material are not well understood and appear to depend on the type of ion implanted. For example, there is evidence that carbon implantation leads to the formation of amorphous areas with fine metal carbide grains and graphitic phases that could act as lubricating layers [112,113]. In the case of nitrogen implantation, hard metal nitride formation is thought to be responsible for the increased hardness of steel, while oxygen ion bombardment leads to the formation of oxide layers in the near-surface region of the steel [114]. Studies [114–118] have also found that implanting a metal ion and an oxygen ion either at the same time or sequentially leads to even greater improvement in the mechanical properties of the steel. When one considers the implications of this research just on manufacturing and the lack of fundamental understanding of the mechanisms behind these results, it is clear that this area deserves further study at both the fundamental and applied level.

3. The future

The examples developed in the previous section show that the growth and modification of materials through ion–surface processing can be used to create a wide range of materials for a variety of applications. Nevertheless, the mechanisms by which many of these processes occur remain undetermined. This is especially true for materials that are themselves inherently complex. For example, it is difficult to determine fundamental mechanisms for ion modification of steel, which contains numerous components and whose microstructure and properties can vary widely with changes in processing conditions. It is expected that an improved understanding of these mechanisms should lead to important improvements in manufacturing that are crucial for the processing of the next generation of tools and devices. An indication of research areas for which ion–surface modification promises considerable scientific and technological advances in the new millennium is provided below.

3.1. Formation of nanostructures

Nanometer-scale structured films and materials (nanostructures) are an increasingly active area of research because they promise unique mechanical, electronic, and chemical properties. One area for future advances is increased process control for the doping of electronic materials as the dimensions of electronic devices continue to shrink. A promising new direction is the creation of buried interfaces by doping selective regions of microelectronic circuits with medium energy ion beams. In addition, metallic superlattice structures can exhibit giant magnetoresistance provided they are grown with smooth, defect-free interface structures. Improvements in ion–surface processing could result in the manufacture of superlattice structures with more layers that will have the necessary properties to be used as sensors, disk drives, or memory elements.

One area where nanoscale science has already begun to interact with ion–surface modification is the energetic surface collisions of C_{60}^+ and other fullerene ions. Thermal deposition of buckyball fullerenes (C_{60}) leads the formation of nanostructured thin films where the components of the film retain the electronic structure of the deposited buckyballs [77,119]. Simulations of 0.15 keV C_{60} deposition predict both reactive and elastic collisions [120]. Experiments at similar energies [121] find that 0.15 keV impacts depend greatly on the nature of the substrate being impacted. At 0.5 keV, graphite targets will scatter back some of the incident buckyballs, while metal substrates flatten and dissociate them. At ∼1 keV, smooth, amorphous, diamond-like, hydrogen-free films are produced that contain graphitic crystallites embedded within the amorphous lattice [121]. Atomistic simulations at ∼1 keV show the buckyballs fall apart and the atoms then agglomerate within the surface to form carbon nanostructures [122]. At energies of several keV, simulations predict that a hot zone develops in the area of impact that propagates away in waves [123]. This leads to the formation of protrusions and hillocks that have been experimentally observed [124].

Fullerene–surface collisions are but one means of forming nanostructures by the transfer of intact nanometer-sized species to surfaces. Formation of nanostructures also can be accomplished with metal cluster ion deposition (see Section 2). These methods are being explored to make novel cluster catalyst materials that are stabilized against sintering. Biomolecular ions—formed by methods developed for biomolecular mass spectrometry [125]—can also be collided with surfaces to form nanostructures. For example, 150 keV lysozyme ion impacts create well-defined nm sized defects in graphite [126]. In fact, the shapes of these surface defects are used to determine the gas phase structure of the lysozyme ion. The size of biomolecular ions can range from <0.5 to >10 nm, depending upon their mass and structure: deposition of these ions opens the possibility of nanostructured surfaces with similar (albeit random) dimensionality. More subtle effects might occur at yet lower energies. For example, ion scattering studies of 200–700 amu peptide ions found a strong dependence in ion–surface energy transfer with peptide mass [127]. These unusual energy transfer effects could translate into new phenomena in biomolecular ion–surface modification.

The nanostructure of ion-induced substrate features also is affected by the ion-induced collision cascades that extend into the 10s nm range as the ion energy approaches 10 keV. MeV ion–surface bombardment can transfer a large amount of energy to the substrate, albeit via electronic excitations. These electronic excitations cause collision cascades within the substrate and can lead to the formation of unusual defects and nanostructures with unique properties (e.g., damage tracks extended deep into a substrate) (for example, see Ref. [128] and other articles in that issue).

Simple sputtering of surfaces with low energy atomic ions can lead to regular or irregular nanostructures. Cones can form on surfaces by removal of material around impurities that sputter inefficiently [7]. Tuning the incident ion angle and surface temperature can cause nm-sized regular ripples on a surface formed by competition between erosion and surface atom diffusion [129]. Hyperthermal ion sputtering of graphite followed by reactive etching with oxygen gas forms well-defined nm-scale etch pits [130]. Researchers are increasingly likely to produce unique nanostructures by various combinations of low energy ion sputtering, deposition of sputtered atoms, annealing, and gas–surface reactions. Slow multiply charged ions also can be used to produce nanostructures due to their unique charge transfer-induced sputtering mechanism [131].

Hyperthermal polyatomic ions can also be used to generate nanostructures at interfaces. Fig. 9 is an image from an atomic force microscope (in tapping mode) of a fluorocarbon film grown on a polystyrene surface by 100 and 50 eV $C_3F_5^+$ (see Fig. 5 for chemical analysis of this same surface). Both dimensions of these features vary considerably with ion energy. The polyatomic ion energy, fluence, and identity thereby constitute direct means of controlling the nanostructure on an interface. Nanostructuring of polystyrene is also observed for low energy atomic ion bombardment [132].

Yet another means of nanostructuring interface regions is via focused ion beams (see Section 2). Focused ion beams with beamwidths as narrow as 20 nm are now available and can be employed for lithography via ion exposure of a resist, ion milling, gas-assisted etching, and ion assisted deposition of material [133]. Three-dimensional nanoscale patterning is feasible with focused ion beams [134]. Focused ion beams or ion projection lithography also can be combined with microcontact printing techniques to nanopattern even curved surfaces [135].

3.2. Combinatorial searching of new materials

Combinatorial materials science is rapidly becoming the dominant means by which new materials with favorable properties are discovered [136]. The method involves the production of a two-dimensional array of material samples (library)

3.3. Biointerfaces

Biointerfaces are widely considered one of the important frontiers in surface and interface science (see the article by Matthew Tirrell in this volume). Several ion–surface modification strategies are being actively employed to create surfaces with tailored biological activity by controlling their chemistry, morphology, or mechanical properties. Ion implantation improves corrosion resistance or reduces wear in implanted metal devices [138]. Low energy Mg^+ implantation into alumina improves the growth of bone cells on the alumina surface [139]. Low energy Ar^+ ion bombardment of a polysiloxane polymer surface improves fibroblast cell growth on this polymer [140]. Focused ion beams are used to create nm-scale defects with controlled protein adsorption properties [141]. A significant advantage from a regulatory standpoint is the extremely small amounts of material deposited from the ion beam are often considered unlikely to induce toxic or other unfavorable responses when in contact with human tissue. These facts all indicate that ion–surface modification will play an increasingly important role in the production of biointerfaces.

The deposition of intact biological and synthetic polymers on surfaces shows promise for creating new biointerfaces [125,142–147]. Even large DNA ions have been soft-landed intact on surfaces [144]. These techniques are now being explored for the creation of arrays for use in combinatorial chemistry or chemical sensors. Intense beams of large biomolecular ions can be readily produced, although depending upon experimental conditions the ions so produced can be neat, clustered to solvent molecules, or dissolved in micron-sized solvent droplets. Furthermore, the characteristics of films deposited by these methods have been explored only briefly. For example, these experiments have not applied surface analysis methodologies beyond scanning probe microscopy and mass spectral analysis of the deposited species by redesorption/ionization. Nevertheless, these techniques open up an entirely new class of molecular and biomolecular species for use in ion–surface modification of biointerfaces.

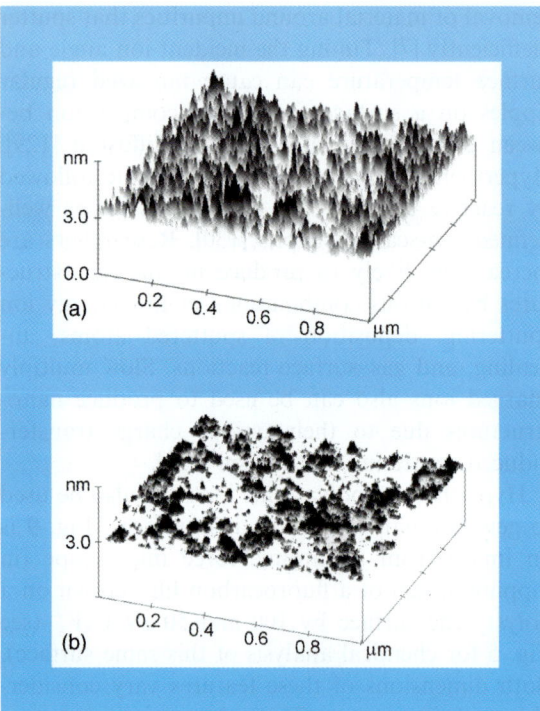

Fig. 9. Panel (a) shows the atomic force microscope image of fluorocarbon film grown on a polystyrene surface by 100 eV $C_3F_5^+$ ions. Panel (b) is the same except the film is formed by 50 eV ions. (see Fig. 5 for chemical analysis of this same surface). Differently shaped features with size ranging from 1.5 to 3.0 nm are observed for both ion-modified surfaces. The roughnesses of the 50 and 100 eV $C_3F_5^+$ ion-modified polystyrene surfaces are 1.8 ± 0.2 and 7.4 ± 0.6 Å (RMS), respectively, indicating that the higher ion energy leads to a rougher surface. By contrast, the roughness of the unmodified polystyrene surface is 1.0 ± 0.2 Å.

with each sample produced under slightly different processing conditions. These libraries are screened for desirable physical properties, then the data used to further refine the processing conditions or the optimal sample selected for use. The entire process is typically highly automated to permit the rapid screening of large numbers of libraries. Libraries also can be produced as a continuous film with smoothly varying processing conditions [137]. Ion beam processing is ideal for combinatorial searching of new materials since it is simple to vary the ion energy, fluence, or ion identity in an automated fashion [151].

3.4. Isotopically pure materials

Mass-selected ion beams are inherently composed of a single isotope and therefore can be employed to create isotopically pure materials interfaces or dopants in materials. Controlling the isotopic distribution of a materials interface promises to affect at least its thermal conductivity and refractive index. Isotopically pure films of ^{28}Si on Si(1 0 0) and ^{107}Ag on Ni(1 0 0) can be grown from the corresponding hyperthermal atomic ions [148,149]. Technical limitations such as low vacuum conditions can complicate the production of isotopically pure materials. Nevertheless, once these experimental difficulties are overcome, a wide range of isotopically pure materials interfaces can begin to be explored in depth.

3.5. Challenges to manufacturing with ion–surface processing

A major challenge for the expansion of ion–surface processing into high volume manufacturing is the relatively high cost associated with these methods. Decreasing the processing cost so that it is the same order of magnitude as paint application, for example, would dramatically increase its use in a variety of industries where it is currently cost prohibitive, such as the automotive industry. However, ion–surface processing possesses a considerable environmental advantage over wet chemical methods in that the former employs no solvents and wastes little reagent. This savings in chemical purchase and disposal costs can sometimes compensate for the added capital costs of ion-modification equipment.

Other details of the manufacturing process also dictate whether ion–surface processing becomes more widely utilized. These include the ability to produce ion beams of sufficient high ion current to reduce processing times and the determination whether mass selection of the ions is always required for a given process. One possible strategy is to use mass-selected ion beams to prototype new materials, then actually manufacture them in bulk by plasma, magnetron sputtering, or other methods already adapted to the industrial scale. Finally, ion–surface modification must still compete with other, often equally powerful technologies. For example, sub-100 nm ion projection lithography is viable, but it must still compete with electron, vacuum ultraviolet, and X-ray lithography for a place in the large-scale production of microelectronic circuits.

Simulations are increasingly being applied to engineering processes of technological importance. In the near future these models may be used in production settings to adjust manufacturing conditions, especially for processes that are somewhat simple, such as sputtering and doping. Atomistic simulations are usually applied to systems that have been simplified somewhat from actual manufacturing conditions. However, as computing power continues to increase and the devices being manufactured continue to decrease in size, it is increasingly useful to apply atomistic approaches to product design.

In conclusion, ion–surface processing promises a rich yield of new, rationally designed materials for technological applications in the new millennium. It also promises to yield many exciting advances in fundamental knowledge regarding the growth of new materials by a combination of experimental and computational approaches.

Acknowledgements

The authors would like to thank the many people who contributed in various ways to the production of this article. Dennis Jacobs and his students at the University of Notre Dame provided Fig. 1. Michael Moseler of the Georgia Institute of Technology and Hellmut Haberland of the Freiburg Materials Research Center provided Fig. 4. Roger Smith of Loughborough University and Roger Webb of University of Surrey provided Fig. 7. Steve Ruatta of Photronics provided Fig. 8. Gerry Zajac of BP Amoco provided scientific assistance and access to the atomic force microscope used to collect the data for Fig. 9. Wayne Rabalais of the University of Houston and Yip-Wah Chung of Northwestern University provided reprints, preprints, and instructive information on isotopically pure and carbon nitride films, respectively. LH's graduate students Muthu Wijesundara and Erick

Fuoco of the University of Illinois at Chicago collected the data in Figs. 5 and 9. SBS's students and postdoctoral research associates Yuan Ji, and Boris Ni ran the molecular dynamics simulations presented in Fig. 6. Finally, the National Science Foundation and the Petroleum Research Fund provided funding for the authors' research efforts in ion–surface collisions (LH: NSF CHE-9986226; SBS: NSF CHE-9708049).

References

[1] J.E. Greene, T. Motooka, J.-E. Sundgren, D. Lubben, S. Gorbatkin, S.A. Barnett, The role of ion/surface interactions and photo-induced reactions during film growth from the vapor phase, Nucl. Instrum. Meth. Phys. Res. B 27 (1987) 226–242.
[2] J.J. Cuomo, S.M. Rossnagel, H.R. Kaufman (Eds.), Handbook of Ion Beam Processing Technology: Principles Deposition Film Modification and Synthesis, Noyes Publications, Park Ridge, 1989.
[3] S.R. Kasi, H. Kang, C.S. Sass, J.W. Rabalais, Inelastic processes in ion–surface collisions, Surf. Sci. Rep. 10 (1989) 1–105.
[4] E. Taglauer, Surface cleaning using sputtering, Appl. Phys. A 51 (1990) 238–251.
[5] J.W. Rabalais (Ed.), Low Energy Ion–Surface Interactions, Wiley, Chichester, 1994.
[6] M. Nastasi, J.W. Mayer, J.K. Hirvonen, Ion–Solid Interactions: Fundamentals and Applications, Cambridge University Press, Cambridge, 1996.
[7] V.S. Smentkowski, Trends in sputtering, Prog. Surf. Sci 64 (2000) 1–58.
[8] W. Jacob, Surface reactions during growth and erosion of hydrocarbon films, Thin Solid Films 326 (1998) 1–42.
[9] A. Benninghoven, F.G. Rudenauer, H.W. Werner, Secondary Ion Mass Spectrometry: Basic Concepts, Instrumental Aspects, Applications and Trends, Wiley, New York, 1987.
[10] L. Hanley, O. Kornienko, E.T. Ada, E. Fuoco, J.L. Trevor, Surface mass spectrometry of molecular species, J. Mass Spectrom. 34 (1999) 705–723.
[11] J.F. Ziegler, J.P. Biersack, U. Littmark, The Stopping and Range of Ions in Solids, Pergamon Press, New York, 1985.
[12] X.-Y. Zhu, Surface photochemistry, Annu. Rev. Phys. Chem. 45 (1994) 113–144.
[13] E.T. Ada, L. Hanley, Comparing hyperthermal molecular and atomic ion sputtering of adsorbates: Xe^+ vs. SF_5^+ on $NH_3/CO/Ni(1\,1\,1)$, Int. J. Mass Spectrom. Ion Process. 174 (1998) 231–244.
[14] W.D. Sproul, Multi-cathode unbalanced magnetron sputtering systems, Surf. Coat. Technol. 49 (1991) 284–289.
[15] W. Ensinger, Plasma immersion ion implantation for metallurgical and semiconductor research and development, Nucl. Instrum. Meth. Phys. Res. B 120 (1996) 270–281.
[16] D.J. Rej, H.A. Davis, J.C. Olson, G.E. Remnev, A.N. Zakoutaev, V.A. Ryzhkov, V.K. Struts, I.F. Isakov, V.A. Shulov, N.A. Nochevnaya, et al., Materials processing with intense pulsed ion beams, J. Vac. Sci. Technol. A 15 (1997) 1089–1097.
[17] S.M. Rossnagel, Directional and ionized physical vapor deposition for microelectronics applications, J. Vac. Sci. Technol. B 16 (1998) 2585–2608.
[18] W.R. Sobie, Ion beam technology for thin film applications, Vacuum and Thinfilm, April 1999 (pp. 36–39).
[19] P.J. Martin, A. Bendavid, H. Takikawa, Ionized plasma vapor deposition and filtered arc deposition: processes, properties, and applications, J. Vac. Sci. Technol. A 17 (1999) 2351–2359.
[20] M.E. Graham, Directions in PVD coatings, in: A. Kumar, Y.-W. Chung, J.J. Moore, J.E. Smugeresky (Eds.), Surface Engineering: Science and Technology I, The Minerals, Metals and Materials Society, 1999.
[21] E. Chanson, S.T. Picraux, J.M. Poate, J.O. Borland, M.I. Current, T. Diaz de la Rubia, D.J. Eaglesham, O.W. Holland, M.E. Law, C.W. Magee, et al., Ion beams in silicon processing and characterization, J. Appl. Phys. 81 (1997) 6513–6561.
[22] J. Melgailis, A.A. Mondelli, I.L. Berry, R. Mohondro, A review of ion projection lithography, J. Vac. Sci. Technol. B 16 (1998) 927–957.
[23] R. Kaesmaier, H. Loschner, G. Stengl, J.C. Wolfe, P. Ruchhoeft, Ion projection lithography: international development program, J. Vac. Sci. Technol. B 17 (1999) 3091–3097.
[24] B.J. Garrison, P.D.S. Kodali, D. Srivastava, Modeling of surface processes as exemplified by hydrocarbon reactions, Chem. Rev. 96 (1996) 1327–1341.
[25] S.M. Foiles, M.I. Baskes, M.S. Daw, Embedded-atom-method functions functions for the FCC metals Cu, Ag, Au, Ni, Pd, Pt, and their alloys, Phys. Rev. B 33 (1986) 7983.
[26] J. Tersoff, New empirical potential for the structural properties of silicon, Phys. Rev. Lett. 56 (1986) 632.
[27] D.W. Brenner, Empirical potential for hydrocarbons for use in simulating the chemical vapor deposition of diamond films, Phys. Rev. B 42 (1990) 9458–9471.
[28] M.P. Allen, D.J. Tildesley, Computer Simulation of Liquids, Oxford University Press, New York, 1987.
[29] R. Carr, M. Parrinello, Unified approach for molecular dynamics and density-functional theory, Phys. Rev. Lett. 55 (1985) 5311.
[30] C.S. Jayanthi, S.Y. Wu, J. Cocks, Z.L.X.N.S. Luo, M. Menon, G. Yang, Order-N method for a nonorthogonal tight-binding Hamiltonian, Phys. Rev. B 57 (1998) 3799.
[31] R.S. Averback, M. Ghaly, Fundamental aspects of defect production in solids, Nucl. Instrum. Meth. Phys. Res. B 127 (1997) 127.

[32] M.H. Shapiro, T.A. Tombrello, Simulation of core excitation during cluster impacts, Phys. Rev. Lett. 68 (1992) 1613.

[33] W. Eckstein, Computer Simulation of Ion–Solid Interactions, Springer, New York, 1991.

[34] J.P. Biersack, L. Haggmark, A Monte Carlo computer program for the transport of energetic ions in amorphous targets, Nucl. Instrum. Meth. Phys. Res. 174 (1980) 257.

[35] K.J. Boyd, D. Marton, J.W. Rabalais, S. Uhlmann, T. Frauenheim, Semiquantitative subplantation model for low energy ion interactions with surfaces. I. Noble gas ion–surface interactions, J. Vac. Sci. Technol. A 16 (1998) 444–454.

[36] K.J. Boyd, D. Marton, J.W. Rabalais, Y. Lifshitz, Semiquantitative subplantation model for low energy ion interactions with surfaces. II. Ion beam deposition of carbon and carbon nitride, J. Vac. Sci. Technol. A 16 (1998) 455–462.

[37] K.J. Boyd, D. Marton, J.W. Rabalais, S. Uhlmann, T. Frauenheim, Semiquantitative subplantation model for low energy ion interactions with surfaces. III. Ion beam homoepitaxy of Si, J. Vac. Sci. Technol. A 16 (1998) 463–471.

[38] W.D. Sproul, New routes in the preparation of mechanically hard films, Science 273 (1996) 889–892.

[39] A.Y. Liu, M.L. Cohen, Structural properties and electronic structure of low-compressibility materials: β-Si_3N_4 and hypothetical β-C_3N_4, Phys. Rev. B 41 (1990) 10727.

[40] D. Marton, K.J. Boyd, A.H. Al-Bayati, S.S. Todorov, J.W. Rabalais, Carbon nitride deposited using energetic species: a two-phase system, Phys. Rev. Lett. 73 (1994) 118.

[41] K.J. Boyd, D. Marton, S.S. Todorov, A.H. Al-Bayati, J. Kulik, R.A. Zuhr, J.W. Rabalais, Formation of C–N thin films by ion beam deposition, J. Vac. Sci. Technol. A 13 (1995) 2110.

[42] C. Ronning, H. Feldermann, R. Merk, H. Hofsass, P. Reinke, J.-U. Thiele, Carbon nitride deposited using energetic species: a review on XPS studies, Phys. Rev. B 58 (1998) 2207–2215.

[43] I. Gouzman, R. Brener, A. Hoffman, Electron spectroscopic study of C–N bond formation by low-energy nitrogen ion implantation of graphite and diamond surfaces, J. Vac. Sci. Technol. A 17 (1999) 411.

[44] N. Tsubouchi, B. Enders, A. Chayahara, A. Kinomura, Optical properties of carbon and carbon nitride films prepared by mass-separated energetic negative carbon and carbon nitrogen ions, J. Vac. Sci. Technol. A 17 (1999) 2384.

[45] B.Z. Zhou, Y.-W. Chung, Recent advances in the synthesis and properties of amorphous and crystalline carbon nitride, J. Chin. Inst. Eng. 21 (1998) 691–700.

[46] W.-C. Chan, B. Zhou, Y.-W. Chung, C.S. Lee, S.T. Lee, Synthesis, composition, surface roughness and mechanical properties of thin nitrogenated carbon films, J. Vac. Sci. Technol. A 16 (1998) 1907–1911.

[47] M.L. Wu, M.U. Guruz, V.P. Dravid, Y.-W. Chung, S. Anders, F.L. Freire Jr., G. Mariotto, Formation of carbon nitride with sp^3-bonded carbon in CN_x/ZrN superlattice coatings, Appl. Phys. Lett. 76 (2000) 2692–2694.

[48] X.W. Zhou, H.N.G. Wadley, Atomistic simulations of the vapor deposition of Ni/Cu/Ni multilayers: the effects of adatom incident energy, J. Appl. Phys. 84 (1998) 2301.

[49] J. Jacobsen, B.H. Cooper, J.P. Sethna, Simulations of energetic beam deposition: from picoseconds to seconds, Phys. Rev. B 58 (1998) 15847.

[50] S. van Dijken, L.C. Jorritsma, B. Poelsema, Grazing-incidence metal deposition: pattern formation and slope selection, Phys. Rev. B 61 (2000) 14047.

[51] C.F. Abrams, D.B. Graves, Cu sputtering and deposition by off-normal, near-threshold Cu^+ bombardment: molecular dynamics simulations, J. Appl. Phys. 86 (1999) 2263.

[52] M. Ghaly, R.S. Averback, The formation of vacancy-type defect clusters by ion bombardment: a problem revisited with molecular dynamics, J. Phys. Chem. Solids 55 (1994) 945–953.

[53] K. Nordlund, M. Ghaly, R.S. Averback, M. Caturla, T. Diaz de la Rubia, J. Tarus, Defect production in collision cascades in elemental semiconductors and fcc metals, Phys. Rev. B 57 (1998) 7556.

[54] K. Nordlund, R.S. Averback, Point defect movement and annealing in collision cascades, Phys. Rev. B 56 (1997) 2321.

[55] N. Toyoda, H. Kitani, J. Matsuo, I. Yamada, Reactive sputtering by SF_6 cluster ion beams, Nucl. Instrum. Meth. Phys. Res. B 121 (1997) 484–488.

[56] D. Takeuchi, N. Shimada, J. Matsuo, I. Yamada, Shallow junction formation by polyatomic cluster ion implantation, Nucl. Instrum. Meth. Phys. Res. B 121 (1997) 345–348.

[57] W.L. Brown, M.F. Jarrold, R.L. McEachern, M. Sosnowski, G. Takaoka, H. Usui, I. Yamada, Ion cluster beam deposition of thin films, Nucl. Instrum. Meth. Phys. Res. B 59/60 (1991) 182.

[58] H. Haberland, M. Karrais, M. Mall, Y. Thurner, Thin films from energetic cluster impact: a feasibility study, J. Vac. Sci. Technol. A 10 (1992) 3266.

[59] H. Haberland, M. Mall, M. Moseler, Y. Qiang, T. Reiners, Y. Thurner, Filling of micron-sized contact holes with copper by energetic cluster impact, J. Vac. Sci. Technol. A 12 (1994) 2925.

[60] H. Haberland, Z. Insepov, M. Moseler, Molecular-dynamics simulation of thin-film growth by energetic cluster impact, Phys. Rev. B 51 (1995) 11061.

[61] H. Haberland, Z. Insepov, M. Karrais, M. Mall, M. Moseler, Y. Thurner, Thin films from energetic cluster impact: experiment and molecular dynamics simulations, Nucl. Instrum. Meth. Phys. Res. B 80/81 (1993) 1320.

[62] M. Moseler, O. Rattunde, J. Nordiek, H. Haberland, On the origin of surface smoothing by energetic cluster impact: molecular dynamics simulation and mesoscopic

modeling, Nucl. Instrum. Meth. Phys. Res. B 164/165 (2000) 522.

[63] M. Moseler, J. Nordiek, O. Rattunde, H. Haberland, Simple models for film growth by energetic cluster impact, Radiat. Eff. 142 (1997) 38.

[64] W. Bouwen, E. Kunnen, K. Temst, P. Thoen, M.J. Van Bael, F. Vanhoutte, H. Weidele, P. Lievens, R.E. Silverans, Characterization of granular Ag films grown by low-energy cluster beam deposition, Thin Solid Films 354 (1999) 87.

[65] H. Hsieh, R.S. Averback, Molecular-dynamics investigations of cluster-beam deposition, Phys. Rev. B 42 (1990) 5365.

[66] H. Hsieh, R.S. Averback, H. Sellers, C.P. Flynn, Molecular-dynamics simulations of collisions between energetic clusters of atoms and metal substrates, Phys. Rev. B 45 (1992) 4417.

[67] R.S. Averback, M. Ghaly, H. Zhu, Cluster–solid interactions: a molecular dynamics investigation, Radiat. Eff. Defects Solids 130–131 (1994) 211.

[68] G. Vandoni, C. Felix, C. Massobrio, Molecular-dynamics study of collision, implantation and fragmentation of Ag_7 on $Pd(100)$, Phys. Rev. B 54 (1996) 1553.

[69] W.M. Lau, R.W.M. Kwok, Engineering surface reactions with polyatomic ions, Int. J. Mass Spectrom. Ion Process. 174 (1998) 245–252.

[70] E.T. Ada, O. Kornienko, L. Hanley, Chemical modification of polystyrene surfaces by low-energy polyatomic ion beams, J. Phys. Chem. B 102 (1998) 3959–3966.

[71] S.A. Miller, H. Luo, S.J. Pachuta, R.G. Cooks, Soft-landing of polyatomic ions at fluorinated self-assembled monolayer surfaces, Science 275 (1997) 1447–1450.

[72] H. Luo, S.A. Miller, R.G. Cooks, S.J. Pachuta, Soft landing of polyatomic ions for selective modification of fluorinated self-assembled monolayer surfaces, Int. J. Mass Spectrom. Ion Process. 174 (1998) 193–217.

[73] H. Usui, Polymeric film deposition by ionization-assisted method for optical and optoelectronic applications, Thin Solid Films 365 (2000) 22–29.

[74] M.B.J. Wijesundara, L. Hanley, B. Ni, S.B. Sinnott, Effects of unique ion chemistry on thin-film growth by plasma–surface interactions, Proc. Nat. Acad. Sci. USA 97 (2000) 23–27.

[75] M.B.J. Wijesundara, Y. Ji, B. Ni, S.B. Sinnott, L. Hanley, Quantifying the effect of polyatomic ion structure on thin-film growth: experiments and molecular dynamics simulations, J. Appl. Phys. 88 (2000) 5004–5016.

[76] L. Diederich, E. Barborini, P. Piseri, A. Podesta, P. Milani, A. Schneuwly, R. Gallay, Supercapacitors based on nanostructured carbon electrodes grown by cluster-beam deposition, Appl. Phys. Lett. 75 (1999) 2662.

[77] V. Paillard, P. Melinon, V. Dupuis, A. Perez, J.P. Perez, G. Guiraud, J. Fornazero, G. Panczer, Production of diamondlike carbon films through random compact cluster stacking, Phys. Rev. B 49 (1994) 11433.

[78] W. Christen, U. Even, Collisional energy loss in cluster surface impact: experimental, model, and simulation studies of some relevant factors, J. Phys. Chem. A 102 (1998) 9420.

[79] D. Donadio, L. Colombo, P. Milani, G. Benedek, Growth of nanostructured carbon films by cluster assembly, Phys. Rev. Lett. 83 (1999) 776.

[80] L. Qi, S.B. Sinnott, Polymerization via cluster–solid surface impacts—molecular dynamics simulations, J. Phys. Chem. B 101 (1997) 6883–6890.

[81] L. Qi, S.B. Sinnott, Effect of cluster size on the reactivity of organic molecular clusters: atomistic simulations, Nucl. Instrum. Meth. Phys. Res. B 140 (1998) 39–46.

[82] L. Qi, W.L. Young, S.B. Sinnott, Effect of surface reactivity on the nucleation of hydrocarbon thin films through molecular-cluster beam deposition, Surf. Sci. 426 (1999) 83–91.

[83] L. Qi, S.B. Sinnott, Generation of 3D hydrocarbon thin films via organic molecular cluster collisions, Surf. Sci. 398 (1998) 195–202.

[84] S.K. Ghandhi, VLSI Fabrication Principles: Silicon and Gallium Arsenide, second ed., Wiley, New York, 1994.

[85] E.J.H. Collart, K. Weemers, D.J. Gravesteijn, J.G.M. van Berkum, Characterization of low-energy (100–10 keV) boron ion implantation, J. Vac. Sci. Technol. B 16 (1998) 280–285.

[86] W.L. Harrington, C.W. Magee, M. Pawlik, D.F. Downey, C.M. Osburn, S.B. Felch, Techniques and applications of secondary ion mass spectrometry and spreading resistance profiling to measure ultrashallow junction implants down to 0.5 keV B and BF_2, J. Vac. Sci. Technol. B 16 (1998) 286–291.

[87] C.L. Hartford, R.J. Hillard, R.G. Mazur, M.A. Foad, High-resolution damage depth profiles of unannealed sub-100 nm B^+ implants in (100) silicon, J. Vac. Sci. Technol. B 16 (1998) 316–319.

[88] M. Sosnowski, The prospects for low energy implantation with large molecular ions—the case of decaborane, Proc. 15th Int. Conf. Appl. Accelerators Res. Ind., Denton, TX, 1998, AIP Conf. Proc., p. 475, 775.

[89] R. Smith, M. Shaw, R.P. Webb, M.A. Foad, Ultrashallow junctions in Si using decaborane? A molecular dynamics simulations study, J. Appl. Phys. 83 (1998) 3148–3152.

[90] M.A. Foad, R. Webb, R. Smith, J. Matsuo, A. Al-Bayati, T. Shen-Wang, T. Cullis, Shallow junction formation by decaborane molecular ion implantation, J. Vac. Sci. Technol. B 18 (2000) 445–449.

[91] J.W. Coburn, Surface-science aspects of plasma-assisted etching, Appl. Phys. A 59 (1994) 451–458.

[92] S.S. Todorov, E.R. Fossum, Growth mechanism of thin oxide films under low-energy oxygen-ion bombardment, J. Vac. Sci. Technol. B 6 (1988) 466.

[93] C.L. Quinteros, T. Tzvetkov, D.C. Jacobs, Eley-Rideal reaction of O^+ with oxidized $Si(100)$, J. Chem. Phys. 113 (2000) 5119.

[94] C.T. Rettner, Dynamics of the direct reaction of hydrogen atoms adsorbed on $Cu(111)$ with hydrogen atoms

incident from the gas phase, Phys. Rev. Lett. 69 (1992) 383.

[95] S.A. Ruatta, E. Smith, A. Yasako, Advanced photomask reconstruction with the Seiko SIR 3000, 17th Annual BACUS Symposium on Photomask Technology Management, Redwood City, CA, 1987.

[96] M.J. Vasile, L.R. Harriot, Focused ion beam stimulated deposition of organic compounds, J. Vac. Sci. Technol. B 7 (1989) 1954–1958.

[97] E.T. Ada, L. Hanley, S. Etchin, J. Melngailis, W.J. Dressick, M.-S. Chen, J.M. Calvert, Ion beam modification and patterning of organosilane self-assembled monolayers, J. Vac. Sci. Technol. B 13 (1995) 2189–2196.

[98] J.S. Huh, M.I. Shepard, J. Melngailis, Focussed ion beam lithography, J. Vac. Sci. Technol. B 9 (1991) 173–175.

[99] E.H. Lee, Ion-beam modification of polymeric materials—fundamental principles and applications, Nucl. Instrum. Meth. Phys. Res. B 151 (1999) 29–41.

[100] E.H. Lee, M.B. Lewis, P.J. Blau, L.K. Mansur, Improved surface properties of polymer materials by multiple ion beam treatment, J. Mater. Res. 6 (1991) 610.

[101] G.R. Rao, E.H. Lee, Effects of sequential He^+ and Ar^+ implantation on surface properties of polymers, J. Mater. Res. 11 (1996) 2661.

[102] E.H. Lee, Y. Lee, W.C. Oliver, L.K. Mansur, Hardness measurements of Ar^+ beam treated polyimide by depth-sensing ultralow load indentation, J. Mater. Res. 8 (1993) 377.

[103] D.W. Brenner, O. Shenderova, C.B. Parker, Ion beam damage of polymer surfaces: insights from molecular-dynamics simulations, Materials Research Society, Pittsburgh, PA, 1997.

[104] K. Beardmore, R. Smith, Ion bombardment of polyethylene, Nucl. Instrum. Meth. Phys. Res. B 102 (1995) 223.

[105] D.M. Ròck, D. Boos, I.G. Brown, Improvement in wear characteristics of steel tools by metal ion implantation, Nucl. Instrum. Meth. Phys. Res. B 80/81 (1993) 233.

[106] R.J. Rodriguez, A.L. Sanz, A.M. Medrano, The search for new applications of ion implantation treatments, Surf. Coat. Technol. 84 (1996) 594.

[107] N.J. Mikkelsen, S.S. Eskildsen, C.A. Straede, N.G. Chechenin, Formation of low friction and wear-resistant carbon coatings on tool steel by 75 keV, high dose carbon ion implantation, Surf. Coat. Technol. 65 (1994) 154.

[108] J. Sasaki, K. Hayashi, K. Sugiama, O. Ichico, Y. Hashiguchi, Implantation of titanium, chromium, yttrium, molybdenum, silver, hafnium, tantalum, tungsten, and platinum ions generated by a metal vapor deposition source into 440C stainless steel, Surf. Coat. Technol. 51 (1992) 166.

[109] F. Alonso, J.L. Viviente, J.I. Onate, B. Torp, B.R. Nelsen, Effects of implantation treatments on micromechanical properties of M2 steel, Nucl. Instrum. Meth. Phys. Res. B 80/81 (1993) 254.

[110] C.A. Straede, N.J. Mikkelsen, Industrial implementation of ion implantation on tools and wear parts based on a strategic approach, Surf. Coat. Technol. 103–104 (1998) 191.

[111] F. Alonso, A. Garcia, J.J. Ugarte, J.L. Viviente, J.I. Onate, P.S. Baranda, C.V. Cooper, Changes in tribological properties of AISI 440C martensitic stainless steel after ion implantation of carbon at very high doses, Surf. Coat. Technol. 83 (1996) 263.

[112] S. Yan, W.J. Zhao, D.M. Ròck, J.M. Xue, Y.G. Wang, Study of tribological properties of high-speed steel implanted by high-dose carbon ions, Surf. Coat. Technol. 103–104 (1998) 348.

[113] F. Jhrling, D.M. Ròck, H. Fuess, Hardness, wear and microstructure of carbon implanted AISI M2 high speed steel, Surf. Coat. Technol. 111 (1999) 111.

[114] P.J. Evens, T. Vilaithong, L.D. Yu, O.R. Monteiro, K.M. Yu, I.G. Brown, Tribological effects of oxygen ion implantation into stainless steel, Nucl. Instrum. Meth. Phys. Res. B 168 (2000) 53.

[115] J. Sasaki, K. Hayashi, K. Sugiyama, O. Ichiko, Y. Hashiguchi, Changes in friction characteristics and microstructure of steel by ion implantation of titanium and additional carbon in various doses (I), Surf. Coat. Technol. 65 (1994) 160.

[116] K. Hayashi, J. Sasaki, K. Sugiyama, O. Ichiko, K. Tanaka, Y. Hashiguchi, Changes in friction characteristics and microstructure of steel by ion implantation of titanium and additional carbon in various doses (II), Surf. Coat. Technol. 65 (1994).

[117] N. Dytlewski, P.J. Evans, I.G. Brown, F. Liu, E.M. Oks, G.Y. Yushkov, Time-of-flight ERDA of dual implanted metals, Nucl. Instrum. Meth. Phys. Res. B 136–138 (1998) 644.

[118] J.I. Othate, F. Alonso, J.L. Viviente, A. Arizaga, A study of dual chromium plus carbon ion implantation into high speed steel, Surf. Coat. Technol. 65 (1994) 165.

[119] V. Paillard, P. Melinon, V. Dupuis, J.P. Perez, A. Perez, B. Champagnon, Diamondlike carbon films obtained by low energy cluster beam deposition: evidence of a memory effect of the properties of free carbon clusters, Phys. Rev. Lett. 71 (1993) 4170.

[120] R.C. Mowrey, D.W. Brenner, B.I. Dunlap, J.W. Mintmire, C.T. White, Simulations of C_{60} collisions with a hydrogen-terminated diamond 1 1 1 surface, J. Chem. Phys. 95 (1991) 7138.

[121] E.E.B. Campbell, I.V. Hertel, Hyperthermal chemistry and cluster collisions, Nucl. Instrum. Meth. Phys. Res. B 112 (1996) 48.

[122] H.-P. Cheng, Cluster–surface collisions: characteristics of Xe_{55}- and C_{20}-Si(1 1 1) surface bombardment, J. Chem. Phys. 111 (1999) 7583.

[123] M. Kerford, R.P. Webb, An investigation of the thermal profiles induced by energetic carbon molecules on a graphite surface, Carbon 37 (1999) 859.

[124] G. Brauchle, S. Richard-Schneider, D. Illig, R.D. Beck, H. Schreiber, M.M. Kappes, STM investigation of

energetic carbon cluster ion penetration depth into HOPG, Nucl. Instrum. Meth. Phys. Res. B 112 (1996) 105.
[125] T. Nohmi, J.B. Fenn, Electrospray mass spectrometry of poly(ethylene glycols) with molecular weights up to five million, J. Am. Chem. Soc. 114 (1992) 3241–3246.
[126] C.T. Reimann, R.A. Sullivan, J. Axelsson, A.P. Quist, S. Altmann, P. Roepstorff, I. Velazquez, O. Tapia, Conformation of highly-charged gas-phase lysozyme revealed by energetic surface imprinting, J. Am. Chem. Soc. 120 (1998) 7608–7616.
[127] L. Hanley, H. Lim, D.G. Schultz, S. Garbis, C. Yu, E.T. Ada, M.B.J. Wijesundara, Energetics, timescales and chemistry of low energy molecular ion–organic surface collisions, Nucl. Instrum. Meth. Phys. Res. B 157 (1999) 174–182.
[128] M.B.H. Breese, Focused MeV ion neams for materials analysis and microfabrication, Mater. Res. Soc. Bull. 25 (2000) 11.
[129] S. Rusponi, G. Costantini, F. Buatier de Mongeot, C. Boragno, U. Valbusa, Patterning a surface in the nanometric scale by ion sputtering, Appl. Phys. Lett. 75 (1999) 3318–3320.
[130] J.R. Hahn, H. Kang, Spatial distribution of defects generated by hyperthermal Ar^+ impact onto graphite, Surf. Sci. 446 (2000) L77–L82.
[131] A. Arnau, F. Aumayr, P.M. Echenique, M. Grether, W. Heiland, J. Limburg, R. Morgenstern, P. Roncin, S. Schippers, R. Schuch, et al., Interaction of slow multicharged ions with solid surfaces, Surf. Sci. Rep. 27 (1997) 113–239.
[132] S. Netcheva, P. Bertrand, Surface topography development of thin polystyrene films under low energy ion irradiation, Nucl. Instrum. Meth. Phys. Res. B 151 (1999) 129–134.
[133] F. Machalett, K. Edinger, L. Ye, J. Melngailis, T. Venkatesan, M. Diegel, K. Steenbeck, Focused-ion beam writing of electrical connections into platinum oxide films, Appl. Phys. Lett. 76 (2000) 3445–3447.
[134] J. Gierak, E. Cambril, M. Schneider, C. David, D. Mailly, J. Flicstein, G. Schmid, Very high-resolution focused ion beam nanolithography improvement: a new three-dimensional pattering capability, J. Vac. Sci. Technol. B 17 (1999) 3132–3136.
[135] P. Ruchhoeft, M. Colburn, B. Choi, H. Nounu, S. Johnson, T. Bailey, S. Damle, M. Stewart, J. Ekerdt, S.V. Sreenivasan, et al., Patterning curved surfaces: template generation by ion imprint proximity lithography and relief transfer by step and flash imprint lithography, J. Vac. Sci. Technol. B 17 (1999) 2965.
[136] B. Jandeleit, D.J. Schaefer, T.S. Powers, H.W. Turner, W.H. Weinberg, Combinatorial materials science and catalysis, Angew. Chem. Int. Ed. 38 (1999) 2494–2532.
[137] R.B. van Dover, L.F. Schneemeyer, R.M. Fleming, Discovery of a useful thin-film dielectric using a composition-spread approach, Nature 392 (1998) 162–164.

[138] B.D. Ratner, Biomaterials science: an interdisciplinary endeavor, in: B.D. Ratner, A.S. Hoffman, F.J. Schoen, J.E. Lemons (Eds.), Biomaterials Science: An Introduction to Materials in Medicine, Academic Press, New York, 1996, pp. 1–8.
[139] H. Zreiqat, P. Evans, C.R. Howlett, The effect of surface chemical modification of bioceramic on phenotype of human bone-derived cells, J. Biomed. Mater. Res. 44 (1999) 389–396.
[140] C. Satriano, G. Marletta, E. Conte, Cell adhesion on low-energy ion beam-irradiated polysiloxane surfaces, Nucl. Instrum. Meth. Phys. Res. B 148 (1999) 1079–1084.
[141] A.A. Bergman, J. Buijs, J. Herbig, D.T. Mathes, J.J. Demarest, C.D. Wilson, C.T. Reimann, R.A. Baragiola, R. Hull, S.O. Oscarsson, Nanometer-scale arrangement of human serum albumin by adsorption on defect arrays created with a finely focused ion beam, Langmuir 14 (1998) 6785–6788.
[142] X. Cheng, D.G.I. Camp, Q. Wu, R. Bakhtiar, D.L. Springer, B.J. Morris, J.E. Bruce, G.A. Anderson, G.G. Edmonds, R.D. Smith, Molecular weight determination of plasmid DNA using electrospray ionization mass spectrometry, Nucl. Acids Res. 24 (1996) 2183–2189.
[143] C.J. Buchko, K.M. Kozloff, A. Sioshansi, K.S. O'Shea, D.C. Martin, Electric field mediated deposition of bioactive polypeptides on neural prosthetic devices, Mat. Res. Soc. Symp. Proc. 414 (1996) 23–28.
[144] B. Feng, D.S. Wunschel, L.P. Tolic, R.D. Smith, ESI-FTICR analysis and recovery of DNA and PCR products through soft-landing, 46th ASMS Conf. Mass Spectrom. Allied Topics, Orlando, FL, 1998.
[145] Y. Xia, R.J. Geiger, M.G. Bartlett, K.L. Busch, Soft-landing of biological compounds using a beam mass spectrometer, 46th ASMS Conf. Mass Spectrom. Allied Topics, Orlando, FL, 1998.
[146] V.N. Morozov, T.Y. Morozova, Electrospray deposition as a method for mass fabrication of mono- and multi-component microarrays of biological and biologically active substances, Anal. Chem. 71 (1999) 3110–3117.
[147] F. Charbonnier, L. Berthelot, C. Rolando, Differentiating between capillary and counter electrode processes during electrospray ionization by opening the short circuit at the collector, Anal. Chem. 71 (1999) 1585–1591.
[148] J.W. Rabalais, A.H. Al Bayati, K.J. Boyd, D. Marton, J. Kulik, Z. Zhang, W.K. Chu, Ion-energy effects in silicon ion-beam epitaxy, Phys. Rev. B 53 (1996) 10781–10792.
[149] S.S. Todorov, H. Bu, K.J. Boyd, J.W. Rabalais, C.M. Gilmore, J.A. Sprague, Ion beam deposition of Ag-107(1 1 1) films on Ni(1 0 0), Surf. Sci. 429 (1999) 63–70.
[150] D.C. Jacobs, The role of internal energy and approach geometry in molecule/surface reactive scattering, J. Phys. Condens. Matter 7 (1995) 1023.
[151] M.B.J. Wijesundara, E. Fuoco, L. Hanley, Preparation of chemical gradient surfaces by hyperthermal polyatomic ion-deposition: A new method for combinatorial materials science, Langmuir 17 (2001) 5721–5726.

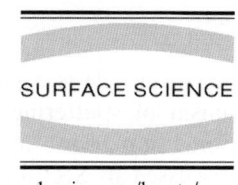

Sputtering: the material erosion tool

M.V. Ramana Murty [*,1]

E2O Communications Inc., 3601 Calle Tecate Road, Suite C, CA 93012, USA

Received 10 July 2000; accepted for publication 4 September 2001

Abstract

The erosion of materials with energetic ions is an indispensable tool in the laboratory and the industry. The ion–solid collision process leading to the ejection of atoms contains fascinating physics and materials science of surfaces. In this article, different aspects of sputtering relevant to applications are discussed. Current research is emphasized and future directions are outlined. © 2001 Elsevier Science B.V. All rights reserved.

Keywords: Sputtering; Sputter deposition; Secondary ion mass spectroscopy; Ion etching; Surface structure, morphology, roughness, and topography

1. Introduction

Imagine sending a high velocity projectile toward a target and one instantly conjures up a picture of light, sound, and debris. Sputtering is a process where the projectile is an atom or molecule and the debris consists of atoms ejected from the target. Scientists and engineers have turned these "microscopic explosions" into a versatile tool for precisely controlling the morphology of a material.

Sputtering finds applications in the industry and the laboratory. It even occurs in nature. The fabrication of a microprocessor, laser or microelectromechanical device involves not one but several steps of ion beam etching [1,2]. The devices are built step by step by defining submicron scale features with lithography and "drilling" them using ion beams. Energetic ions are routinely used to clean surfaces and to deposit thin films. On a planetary scale, sputtering by energetic particles in the solar wind (H^+, He^+) has important implications for the evolution of planetary surfaces [3–5].

The first observation of sputtering was made in the 19th century when Grove found metallic deposits on the glass walls of a discharge tube [6]. The proof that sputtering is caused by the positive ions in the discharge tube was demonstrated by Goldstein when he extracted a beam of canal rays (positive ions) and showed the disappearance of a gold coating on the glass wall facing the beam [6]. Sputtering can be caused with varying efficiency by all kinds of particles including ions, electrons and neutrons. There are a number of phenomena similar to sputtering in that they cause erosion of material, including ablation by an intense photon beam, and blistering caused by occluded gas. In this article, we focus on material erosion by energetic ions.

---

[*] Tel.: +1-818-4662815; fax: +1-818-8789163.
E-mail address: mvrmurty@e2oinc.com (M.V.R. Murty).
[1] Present address: E2O Communications Inc., 3601 Calle Tecate Road, Suite C, Camarillo, CA 93012, USA.

This article is organized as follows. The mechanism of sputtering is the subject of Section 2. Some important applications of sputtering are discussed in Section 3. Section 4 contains a description of nanoscale pattern formation by ion erosion, a subject that has attracted much attention recently. Section 5 contains a discussion of directions for further investigation in sputtering.

Sputtering can be caused by both energetic ions and neutrals, and likewise, the sputtered atoms could be in the ionized, excited or ground state. To simplify description, the bombarding particle is referred to as the ion and the target particles as atoms. Also the terms sputtering, etching and atom ejection are used here interchangeably and refer to the same phenomenon.

2. The sputtering process

The process of ion sputtering may remind one of sandblasting, the erosion of materials with a stream of abrasive particles such as alumina or silicon carbide. Indeed, the variation of the erosion rate with the incident angle of the abrasive particles is analogous to that observed in ion sputtering [7,8]. The analogy ends there, however, since the average kinetic energy per atom in sandblasting (few meV) is much lower than that in sputtering (eV to several keV). Unlike the abrasive particles, ions can penetrate deep into the target, and energy and momentum transfer occurs on a different scale.

Early investigations of sputtering were concerned with predicting the sputter yield—how many target atoms are removed per incident ion? There are of course other observables such as the effects on surface morphology and lattice damage. A detailed investigation of these other effects has had to wait for an understanding of the sputtering process and the development of experimental tools to measure surface structure.

The many-body aspect of sputtering poses a formidable challenge to theorists. Over a century passed from the time of the first observation of sputtering before a theory could explain the general trends of experiments. Computers provide the much needed visualization of the sputtering process and contribute tremendously to our understanding. Experiments continue to be refined and yield data with increasingly better resolution along with more challenges for modeling.

The first explanation of the sputtering process came from Stark [6], who suggested both an evaporation theory and a transport theory. In the evaporation theory, ion bombardment leads to heating of the target and subsequent boiling of atoms from the surface region. An important experimental observation is that ion erosion occurs even in the limit of a small flux, and the sputter yield is independent of the flux. In other words, even a single ion can remove target atoms. Target heating therefore refers to a localized heating caused by the energetic ions. Molecular dynamics simulations have shown that thermal spikes do occur [9]. When crystalline materials are bombarded with keV ions, part of the material in the path of the ions has the radial distribution function of a liquid, consistent with a local heating of the target. However, the sputter yield from such heating is low due to the short-lived nature of the spikes, typically a few picoseconds [9]. There just is not enough time for atoms to boil off from the target.

Better success has been achieved with the transport theory [6,10]. In this theory, one envisages the incident ion losing its energy to the target atoms through a series of collisions as indicated in Fig. 1. The target atoms involved in these collisions, in turn, undergo further collisions thereby setting up collision cascades. Some atoms near the surface receive a sufficient kick in these cascades to leave the target.

The building of particle accelerators and the growing technological applications of ion implantation in the 1950s and 1960s generated an immediate need for a quantitative understanding of the transport of energetic ions in solids. Developments in this field had an enormous impact on the understanding of sputtering. An energetic ion loses energy through collisions with both the nuclei and the electrons as it penetrates the target. The range of the ions can be determined to a good approximation by treating the nuclear stopping and the electronic stopping independently [11]. For all except the light ions (such as H^+, He^+), the nuclear

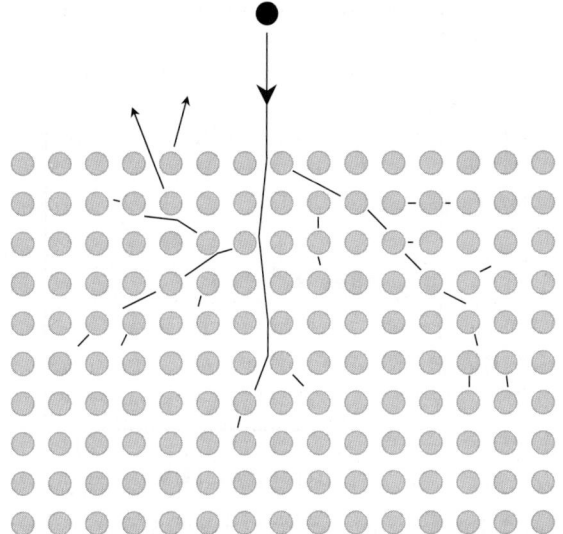

Fig. 1. The sputtering process. The incident ion (black) loses its energy to target atoms (gray) through a series of collisons. In the collision cascades that ensue, some of the atoms receive a sufficient kick to leave the target. The momentum transfer to the ejected atoms occurs through multiple collisions.

energy loss dominates at low (keV) energies. The nuclear energy loss can be calculated by modeling the interaction of the ion and the target nucleus with a (repulsive) screened Coulomb potential [11]. Lindhard, Scharff and Schafer wrote a universal expression for the nuclear stopping cross-section (nuclear energy loss per unit path length per unit atomic density) $S_n(E)$, now called the LSS expression, for any ion–target combination [11]. It has the form shown in Fig. 2. At low energies, $S_n(E)$ increases with energy because the ions get close to the nuclei of the target atoms and undergo strong scattering. The decrease in $S_n(E)$ at high energies is due to less time spent by the (fast) ions near an atom.

The transport theory of sputtering developed by Sigmund [6,10] and Thompson [12] incorporated the ideas of the stopping theory of Lindhard and co-workers. Being a continuum theory, it applies best to amorphous and polycrystalline targets. While the mathematical treatment of the ion transport is sophisticated, the results take a simple form. The sputter yield $Y \approx \alpha N S_n(E)/E_B$, where N is the target atom density, E_B is the binding energy of a surface atom, and α contains material and geometric parameters. The product $NS_n(E)$ represents energy loss per unit length in the target.

Fig. 3 shows the measured sputter yield of Si from noble gas ions at normal incidence [13]. The dependence of the sputter yield on the incident ion energy in Fig. 3 has the same shape as the nuclear energy loss (Fig. 2). Andersen and Bay [14] and Matsunami et al. [15] have collected information on sputter yield for various ion–target combinations. Sputter yields can be quite sensitive to surface contamination and measurements prior to 1970 were not always made under sufficient vacuum conditions. In fitting the experimental data with the transport theory, the surface binding energy E_B often becomes an adjustable parameter [15].

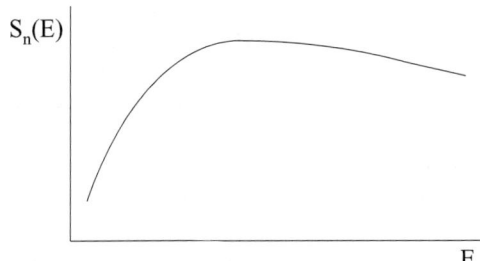

Fig. 2. Nuclear stopping cross-section $S_n(E)$ of a material as a function of the incident particle energy E.

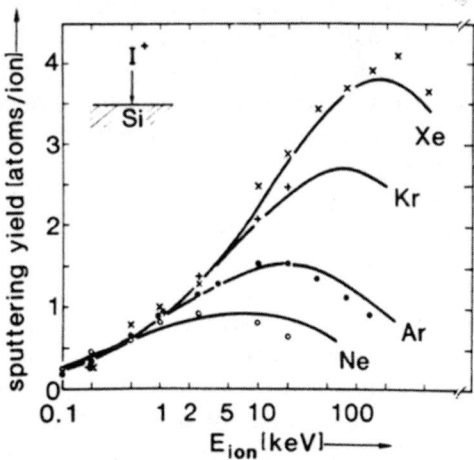

Fig. 3. Sputter yield of Si with various noble gas ions at normal incidence measured as a function of the ion energy (reprinted from Ref. [13], with permission).

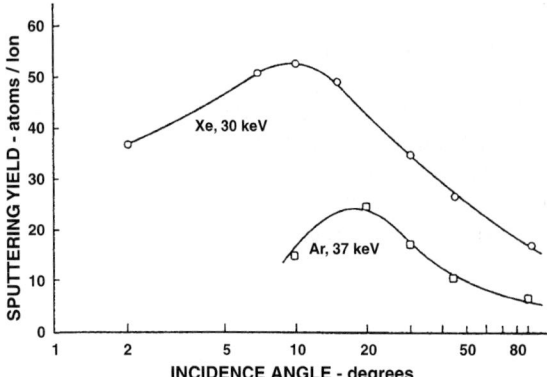

Fig. 4. The dependence of the sputter yield of Cu on the incidence angle for Ar and Xe ions. Note that the incidence angle is measured from the surface plane (from Ref. [16] with permission).

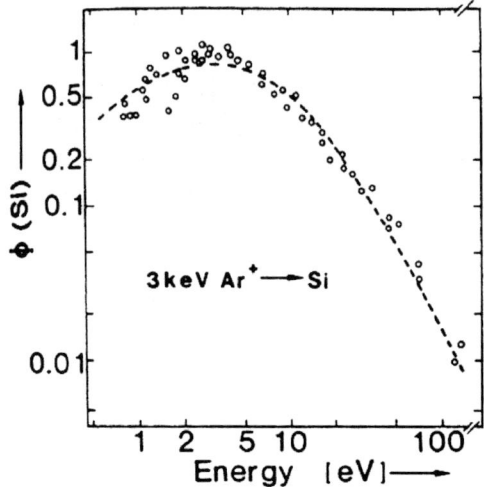

Fig. 5. The energy distribution of sputtered Si atoms during 3 keV Ar$^+$ ion bombardment (reprinted from Ref. [13] with permission).

Fig. 4 shows the variation of the sputter yield of copper with the incidence angle of the ion beam for Ar and Xe [16]. (Note that the incidence angle in Fig. 4 is $90° - \phi$, ϕ being measured from the surface normal.) As the incidence angle is increased, more energy is deposited in the surface region since the ion spends more time near the surface along its path in the solid. A simple projection of the energy deposited by the ion onto the surface indicates that the sputter yield should increase as $(\cos\phi)^{-1}$. A more detailed calculation using the transport theory shows a slightly faster rise than $(\cos\phi)^{-1}$ [10,17]. A faster than $(\cos\phi)^{-1}$ increase is indeed observed in the experiments (including the system in Fig. 4 [6]). Interestingly, the sputter yield decreases at glancing incidence. This is due to an increase in the reflection of the incident ions [18].

What about the sputtered atoms? Fig. 5 shows the energy distribution of Si atoms generated by 3 keV Ar$^+$ ion bombardment of a Si target [13]. Stuart, Wehner and Andersen have measured the energy distribution of sputtered target atoms for a number of elements [19]. The sputtered particle energy distribution is typically peaked at a few eV and has a long tail that goes up to the (order of) incident ion energy. The transport theory predicts a E^{-2} tail for the particle energy distribution [10,12] and is in reasonable agreement with the experimental observations.

Fig. 6 shows sputtered Ag atoms collected on a glass plate from single crystal Ag(1 1 1) and Ag(1 0 0) targets [20]. The deposits clearly show that atom ejection occurs preferentially in the close-packed directions. Such spot patterns get washed out when polycrystalline materials are used and more or less equal ejection in all directions is observed (cosine variation with the polar angle) [21].

While the transport theory describes the general trends of sputtering in amorphous and polycrystalline materials, it still has adjustable parameters. Being a continuum theory, it does not address the inherent inhomogeneity of materials and hence, single crystal sputtering is not described as well. Also the assumption of a purely repulsive force between atoms and the ion is valid only for ion energies above a few keV.

These deficiencies are addressed by computer simulations which provide a direct view of the various events in sputtering. The motion of energetic particles in solids is the subject of some of the earliest molecular dynamics simulations of any kind [22]. All simulations of sputtering conducted to date use classical dynamics. Two types of techniques are popular—the binary collision method and molecular dynamics.

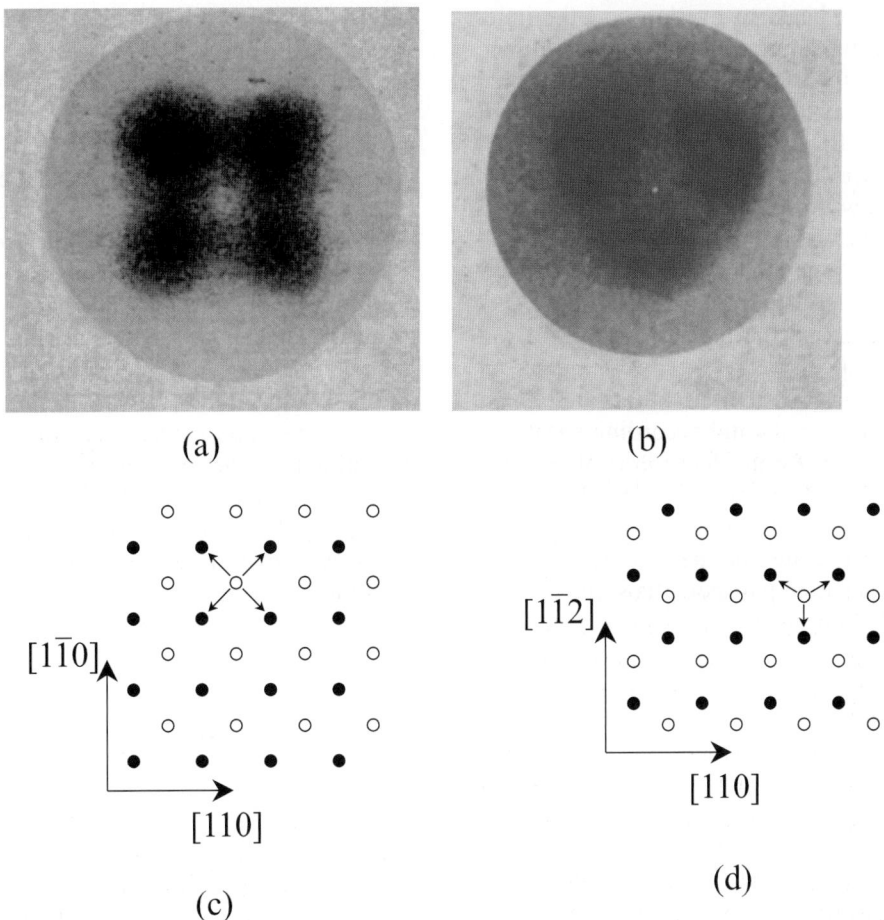

Fig. 6. Ag atoms deposited on a glass plate during Ar$^+$ ion bombardment of single crystal (a) Ag(1 0 0) and (b) Ag(1 1 1). Schematic showing the atoms in the first layer (filled circles) and the second layer (open circles) for the (c) Ag(1 0 0) and (d) Ag(1 1 1) surface. The fourfold and threefold patterns in (a) and (b) occur due to energy channeling along close-packed directions indicated by the arrows in (c) and (d), respectively (adapted from Ref. [20] with permission).

In the binary collision (BC) method, the interaction between the ion (or one of the atoms) and the atoms is treated sequentially. The interaction is modeled by a pair potential, i.e. the potential energy of the two particle system depends only on their separation. Each collision is treated independently and all the energetic particles are followed until they are thermalized. A closed-form expression for scattering due to a central force between two particles [23] makes the BC scheme fast. A popular code of this type, MARLOWE, was developed by Robinson and co-workers [24]. Biersack developed an analytical approximation for the two-body scattering integral further increasing the computing speed. This is implemented in the popular code TRIM.SP [25]. The pair potential is usually taken to be some form of a screened Coulomb potential [26]. A universal potential has been developed [27,28] to model the extensive data on the range of ions in solids and is used in the code TRIM.SP. The universal potential, like the screened Coulomb potential, is purely repulsive.

The binary collision approximation is reasonable for incident ion energies above a few keV, where the individual collision cascades are separated in space. At lower ion energies, the various collision cascades begin to overlap and the independent collision approximation breaks down. Also attractive forces between particles become more important at lower energies. Even for high energy ions, the BC approximation does not provide a good description of the collision cascades as the particles slow down.

Below a few keV, the simultaneous motion of all target atoms and the ion must be included for a proper description of sputtering. Sputtering events typically occur within the first picosecond of ion impact: readily calculable by today's computers. In molecular dynamics simulations, atoms are treated as point particles interacting according to an interatomic potential. An incident ion is placed just outside the range of interaction from the target. Newton's equations of motion are numerically integrated to calculate the particle trajectories. Early simulations used pair potentials due to their simplicity [26]. Many-body interatomic potentials that incorporate more of the material properties have been developed in the last two decades for semiconductors [29,30] and metals [31,32]. Molecular dynamics simulations have reproduced many of the experimental observations on the energy- and angular-distribution of sputtered atoms from single crystal targets [33].

When comparing the results of computer simulations to experiments, it is important to collect good statistics by averaging over many incident ion trajectories. In this sense computer simulations mimic experiments. The target must be chosen to be sufficiently large so that the ion impact is contained within and the effects of reflections from the boundaries are small.

The most important input for molecular dynamics simulations is the interatomic potential. The commonly used interatomic potentials are derived by fitting the equilibrium properties of a material. Whether the potentials give a good description of the collisions where atoms approach a lot closer than their normal separation is an open question.

In the above introduction to the sputtering process many features have been left out. The sputtered flux can consist of dimers, trimers, and larger clusters in addition to monomers. Not all atoms that receive a kick are able to leave the surface and the result is the formation of adatom–vacancy pairs and island nucleation at the surface. Michely and co-workers have investigated adatom generation on single-crystal Pt(1 1 1) during noble gas ion bombardment [34]. The study showed that the ratio of adatom to sputter yield can vary over an order of magnitude for any given ion–target combination. Ion channeling, the steering of the incident ion along the "channels" between columns or planes of atoms, and energy channeling, a series of collisions in a straight line along close-packed directions, are important in single crystals and have a pronounced effect on sputter yield [35]. Chemical reaction between the ion and the target, or induced by the ion impact can lead to an enhanced sputter yield and is referred to as chemical sputtering. In contrast, what has been described above is known as physical sputtering. Our understanding of sputtering becomes less and less when we include these extra features.

3. Applications of sputtering

Applications of sputtering were conceived soon after the phenomenon was discovered. Plucker noted the possibility of thin film coating in 1858 [6]. Today ion erosion has become a routine tool in the fabrication of micro-devices [1,2]. Here we discuss three distinct applications of sputtering—thin film growth, materials characterization, and etching of electronic materials. Other applications include ion milling for preparation of transmission electron microscopy specimens, and focussed ion beams for mask repair and integrated circuit failure detection [1,2].

3.1. Sputter deposition

One of the simplest applications of sputtering is to collect the sputtered atoms on a substrate to form a thin film. A schematic is shown in Fig. 7. The technique, called sputter deposition, is relatively easy to implement and is one of the most widely used methods of film deposition. With sputter deposition one can avoid the high source temperatures often required for molecular beam epitaxy and the potentially hazardous organometallic precursors associated with chemical vapor deposition. Sputter deposition is used for depositing the interconnect material aluminum, diffusion barriers such as TiN and TiW alloys in integrated circuits, and magnetic materials for magnetoresistive read heads in hard disk drives [2,36,37]. Multilayers deposited by sputter deposition often outperform those prepared by evaporation. For example, multilayers composed of alternate high and low atomic number (Z) materials such as Mo/Si and W/B_4C serve as excellent mirrors in the X-ray and deep ultraviolet region of the spectrum. The reflectivity of the mirrors is extremely sensitive to interface roughness. For example, a 2 Å rms roughness at the interfaces in a Mo/Si multilayer with a 20 Å period will reduce the reflectivity by 25%. Here too, sputter deposition yields the best films with measured reflectivities close to the theoretical values for perfect smooth interfaces.

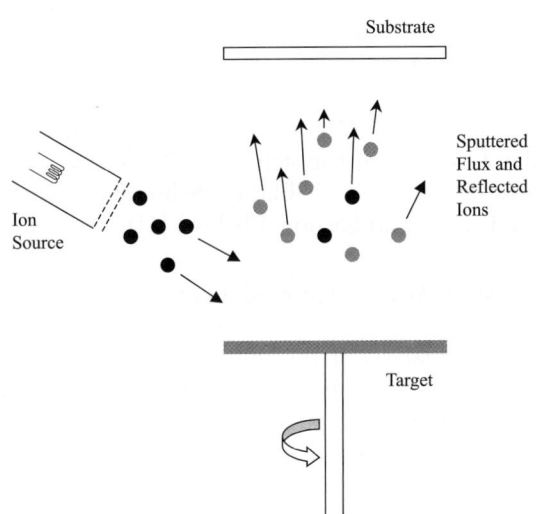

Fig. 7. Schematic of a sputter deposition system. An ion source is pointed toward a target made of the material to be deposited. The sputtered flux is collected on a substrate to form a thin film. The substrate also gets bombarded by reflected ions from the target in this arrangement. The chamber containing the ion source, target and substrate is initially evacuated. It is backfilled with the source gas for the ion gun during operation. Rotation of the target helps to maintain uniform erosion across its surface.

Why are sputter deposited films so different from those deposited by evaporation? The difference lies in the energy of the incident particles. In conventional evaporation the energy of the incident atoms reflects the temperature of the source, typically ≈ 0.1 eV. As discussed in Section 2 above, sputtered atoms have an energy distribution with a peak at few eV and a long tail. The average energy of the incident atoms is not a well-defined quantity in sputter deposition due to the E^{-n} distribution of the particle energies with n close to 2. A change in the incident ion energy on the target can swing the value of $\langle E \rangle$. The most probable energy (the energy at the peak in the energy distribution) serves as a better number for comparison although a complete description requires the knowledge of the entire energy (and angular) distribution. Since the energy of the sputtered atoms is comparable to the bond energies, they can more easily overcome diffusion barriers at the surface, the surmounting of which would otherwise require high temperature.

Sputter deposition can have a dramatic effect on the film microstructure. An example is Si deposition on a dihydride-terminated Si(0 0 1) surface (nominally 2 ML of H on the Si surface). Fig. 8(a) shows a cross-sectional transmission electron microscope image of a Si film deposited by conventional molecular beam epitaxy (MBE) [38]. The film is amorphous from the beginning. Fig. 8(b) shows an image of a sputter deposited Si film on the dihydride-terminated Si(0 0 1) surface [39]. The film is crystalline. Molecular dynamics simulations give insight into the origin of this dramatic difference in the film microstructure. Panels (a) and (b) of Fig. 9 show snapshots of the atoms in the surface layer 1 ps after the impact of a thermal energy (0.2 eV) Si atom and a 4 eV Si atom [38,39]. The Si atom with thermal energy is unable to penetrate the H atoms at the surface. Since there are no good lattice sites for the incident atom, it is more likely to bond to other incident Si atoms (not simulated) resulting in a random network, or amorphous silicon. The Si atom with 4 eV energy penetrates the H layer and goes underneath a surface Si atom leading to the segregation of the SiH_2 unit. The energetic Si atoms are thus able to find lattice sites leading to a crystalline film.

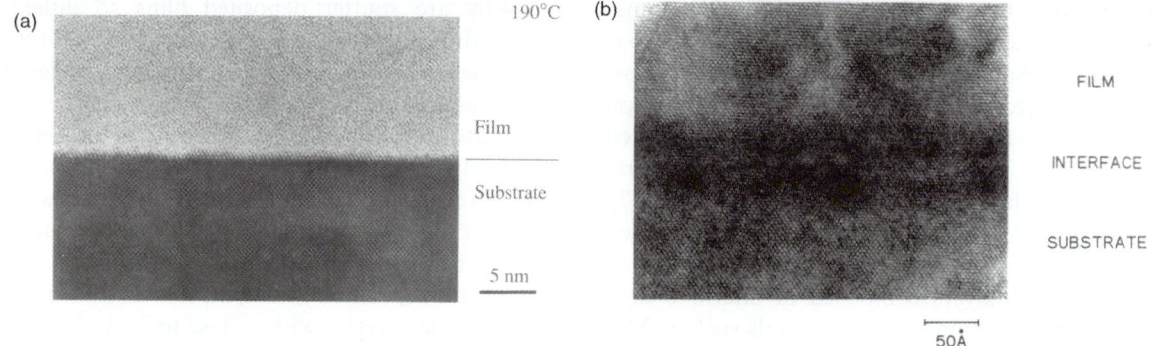

Fig. 8. Cross-sectional transmission electron microscope images along the ⟨1 1 0⟩ direction of Si films deposited on a dihydride-terminated Si(0 0 1) surface. The film is deposited by (a) conventional molecular beam epitaxy at 190 °C, (from Ref. [38]) and (b) by sputter deposition at 300 °C (from Ref. [39] with permission). The film in (a) is amorphous and (b) is crystalline.

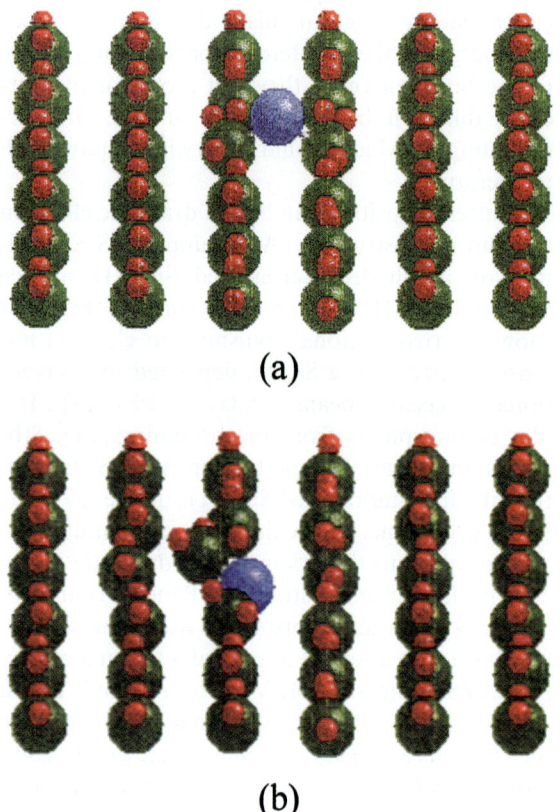

Fig. 9. Snapshots of the surface 1 ps after the deposition of a (a) thermal energy (0.2 eV) Si atom, and (b) 4 eV Si atom on a dihydride-terminated Si(0 0 1) surface. Only the incident Si atom (blue), and top layer Si (green) and H (red) atoms around the point of impact are shown. The thermal energy Si atom is unable to penetrate the H atom layer whereas the 4 eV Si atom substitutes a surface Si atom (from Ref. [38]).

Several other factors make sputter deposition different from conventional MBE. The surface morphology of a film is strongly influenced by two-dimensional island nucleation. Reflected ions from the target can dramatically change the island density [40]. Even if the fraction of reflected ions from the target is low, they can significantly influence the surface kinetics if they possess sufficient energy to nucleate an island when they hit the substrate. A high rate of island nucleation will result in a more rough surface compared to conventional MBE. A more controlled increase in the island nucleation can actually result in smoother films in systems with Ehrlich–Schwoebel barriers (see Fig. 23 and Section 4 below) [41].

3.2. Secondary ion mass spectrometry

Mass analysis of the sputtered atoms, clusters and molecules can be used to identify the chemical composition of the target. In the most commonly used technique, secondary ion mass spectrometry (SIMS), a fraction of the sputtered particles that happen to be ionized is detected using a mass spectrometer. Fig. 10 shows a schematic of the SIMS setup. In practice, SIMS is carried out by rastering the incident beam over a rectangular area (typically 100×100 μm^2). The rastered beam is electronically gated and the sputtered ions are only collected when the incident beam is over the central portion of the sputtered region in order to eliminate edge effects. A mass resolution of at least

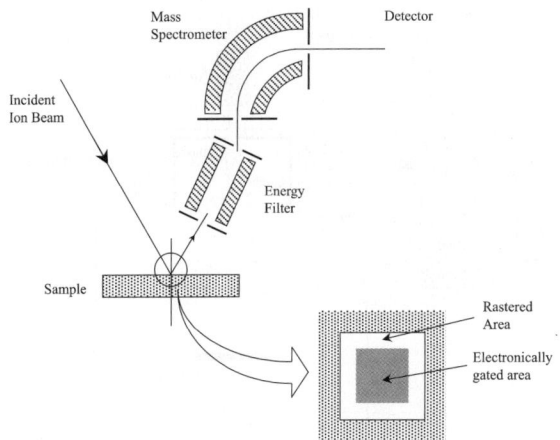

Fig. 10. Schematic of a SIMS system. The incident ion beam is rastered across a small region, typically 100×100 μm^2. A fraction of the sputtered flux that happens to be ionized passes through an energy filter and a mass filter for composition analysis. The detector is electronically gated so that the sputtered flux is only counted when the incident beam is over the center of the rastered area. This eliminates any edge effects.

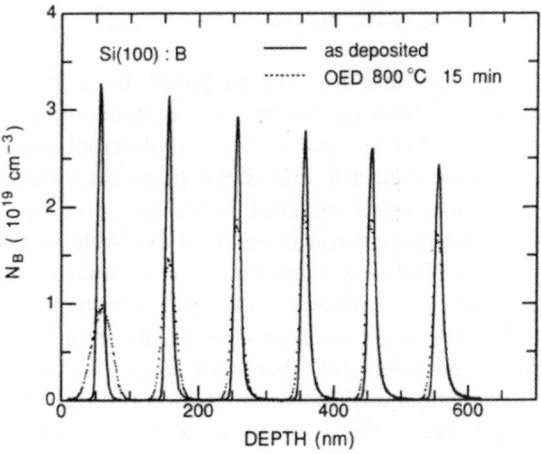

Fig. 11. SIMS depth profile of a Si film with six δ-doped B layers before (solid line) and after (dotted line) oxidation at 800 °C for 15 min. Before oxidation, the six B peaks are sharp (a slight broadening with increasing depth occurs due to the degradation of resolution from intermixing and surface roughness). After oxidation, all B peaks are broadened, with the broadening increasing towards the surface. Oxidation enhanced diffusion (OED) of B atoms occurs due to injection of Si interstitials (from Ref. [43] with permission).

10^{-3} amu is required for distinguishing between species with similar mass, e.g. Si$_2$ from Fe, or OH from NH$_3$. For noble gas ion bombardment, the fraction of sputtered atoms that are ionized is usually small. It is common to use either Cs$^+$ or O$_2^+$ ions as the bombarding species to increase the ionization fraction of the sputtered particles. Oxygen atoms residing at the surface readily accept an electron and increase the fraction of positive ions among the sputtered species. They are commonly used for detecting metal atoms in a target. Likewise, cesium ions readily give up an electron and are commonly used for detecting electronegative elements such as C, N and halogens. Certain elements such as H and Si can both accept and give up an electron, and hence both Cs$^+$ and O$_2^+$ can be used. For quantitative analysis, the fraction of species that are ionized has been experimentally determined for a number of elements embedded in commonly used substrates such as Si and GaAs [42]. SIMS can be sensitive at parts per billion (ppb) level for certain species such as Mg in Si.

Fig. 11 shows a SIMS spectrum of Si with six B δ-doped layers before and after oxidation at 800 °C for 15 min [43]. Each δ-doped layer is a two-dimensional sheet of the dopant (B) atom and the film is grown by molecular beam epitaxy. Oxidation introduces Si interstitials into the Si bulk which in turn cause diffusion of the dopant atoms. Hence, the B layer closest to the surface is the most broadened. Studies of this type have enhanced our knowledge of the diffusion of dopants and self-interstitials in Si, processes of great importance to the semiconductor industry.

The full width at half maximum (FWHM) of $\approx$31 Å for the B layers [43] in the as-deposited film is much wider than the ideal "1 monolayer (ML)." This represents the resolution of the SIMS instrument, primarily limited by two factors: (a) Intermixing—energetic ion bombardment causes intermixing of atoms in different planes due to the collision cascades, and (b) Surface roughness—off-axis ion bombardment often leads to the formation of ripples on the surface (see Section 4 below) and degrades the resolution. One way to reduce ripple formation is to rotate the sample during ion bombardment [44]. Note that the rotation axis must coincide with the axis of the electronically gated area.

3.3. Etching electronic materials

Integrated circuits are prepared in a brick-by-brick fashion by starting with a semiconductor or glass substrate and laying out different semiconductor, dielectric and metal films. At various stages, unwanted material needs to be chipped away and ion erosion is one tool available to the engineer. However, as we know ion bombardment can also cause lattice damage. Yet, the manufacture of the latest microprocessors, flat panel displays, and lasers for fiber optic communications involves the use of ion beam erosion not once but several times. Why is ion beam etching so attractive given the usually challenging specifications on the performance and reliability of devices? The answer lies in the unique advantages it offers including anisotropic etching and profile control, selectivity, and the high purity of the etchant.

Ion beam etching is often referred to as "dry etching" to contrast wet etching using chemical solutions. Compared to solutions, gases can be obtained at higher purities which is an important consideration. Typically, the gases are further purified up to part-per-billon levels using point-of-use purifiers. While wet chemical etching is generally isotropic (with some exceptions such as Si etching with KOH and InP etching with HCl), ion beam etching can be made highly anisotropic, i.e., different etch rates vertically and laterally. This is important for defining trenches and vias at sub-micron dimensions.

A variety of ion sources and reactor configurations have been developed for etching electronic materials [1]. A typical configuration for an etching system is shown in Fig. 12. Ions are generated by striking a discharge between the cathode and anode. In most etching systems, the substrate is exposed not only to the energetic ions but also the various radicals of the gas species formed in the plasma. Consequently the etching process involves both a physical component involving kinetic energy transfer from the ions and a chemical component involving reactions between the ions/radicals and the surface atoms. Physical sputtering dominates at low pressure (<10 mTorr) due to the longer mean free path and is used for anisotropic etching. However, the etching is less selective and

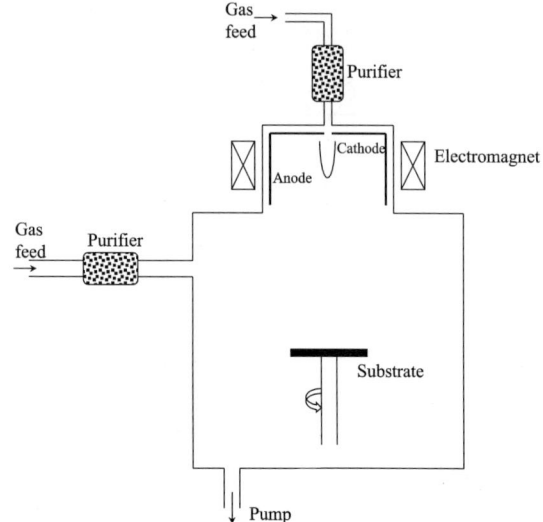

Fig. 12. Schematic of a dc plasma ion source. Gases enter the chamber through a purifier that reduces impurities to parts-per-billion level. A plasma is ignited by applying a voltage between cathode and anode exceeding (typically a few times) the ionization potential of the source gas. The substrate is exposed to various gas radicals, electrons and photons in addition to the ions. Increasing the distance between the substrate and the source, and placing a (negative) grid at the mouth of the source increases the role of ions in the etching process. In the absence of any substrate bias, the ion energy equals the plasma potential. With conducting substrates, the ion energy can be increased by applying a negative substrate bias. Applying an axial magnetic field increases the degree of ionization resulting in a higher flux (and consequently a higher etch rate). Radio-frequency (RF) sources (filamentless) are commonly used for etching insulators. More recently, electron cyclotron resonance (ECR) and inductively coupled plasma (ICP) sources have been developed that allow one to maintain high plasma density at low pressure, 10^{-5}–10^{-2} Torr.

damage is a concern. Chemical etching dominates at higher pressures and gives good selectivity, but is more isotropic.

The materials removed by ion beam etching are many including Si, SiO_2, Si_3N_4, GaAs, metals such as Al and W alloys, and photoresist [1]. Selectivity is achieved by adding various chemical species to the gas stream. For example, a mixture of argon, fluorocarbons (CF_4), and hydrogen is used to remove SiO_2 and gives a high etch rate ratio for SiO_2/Si [45]. When removing a material it is important to know when to stop. In order to remove all material, some amount of overetching is necessary due to the surface roughness. Overetching can cause damage to the underlying layer. Knowing when to stop requires accurate knowledge of the sputter rate and the film thickness. However, sputter rates are largely unknown for the etching chemistries used. Engineers take a systems approach to the problem. The system variables including input power, pressure, gas composition and flow rates, plasma properties (electron temperature, plasma potential) and sample bias are optimized for a high etch rate, selectivity, uniformity across the wafer and minimum residue left behind by the etching process. Real-time feedback systems that monitor the gaseous reaction by-products are used to signal the endpoint for etching [46].

4. Surface morphology evolution during sputtering

A detailed knowledge of surface morphology evolution during sputtering is becoming increasingly important for further progress in a number of applications. As noted above, surface roughness limits the depth resolution in SIMS and makes overetching necessary for the removal of all material during device fabrication. The development of in situ, real-space probes such as scanning tunneling microscopy (STM) [47], and diffraction tools such as electron and X-ray scattering [48–50], and progress in modeling has led to a better understanding of ion erosion in the last decade. Interestingly, these studies have also suggested new ways to create self-organized zero- and one-dimensional nanostructures.

To understand surface morphology evolution, let us first examine some of the possible outcomes of a single ion impact as illustrated in Fig. 13. The simplest outcome is just the removal of one atom from the surface layer. More than one atom may be removed by the incident ion. These atoms could come from the top layer as well as the subsurface region. While some atoms may leave the substrate after an ion impact, others may not have sufficient kinetic energy to overcome the attraction of the surface atoms as they attempt to leave. Such atoms get trapped at the surface leading to the creation of adatom-vacancy pairs. The significant amount of

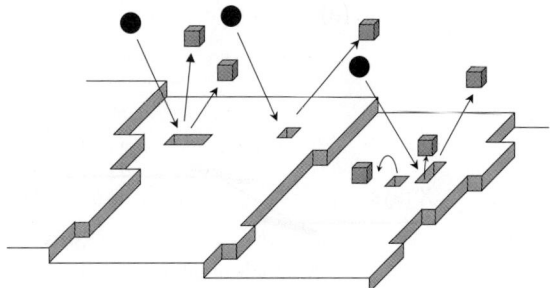

Fig. 13. Some of the possible outcomes of ion bombardment include single vacancy creation, adatom–vacancy pair generation, and island nucleation.

energy deposited by the incident ion can also lead to lattice damage. Ion bombardment can also cause amorphization, and intermixing of atoms from different layers. Semiconductors such as Si and Ge with the diamond structure get amorphized when bombarded at room temperature with Ar^+ ions with energies as low as 50 eV. Higher temperatures are needed to keep these materials crystalline. Close-packed metals such as Cu and Au are more resilient and remain crystalline at room temperature even under 1 keV Ar^+ ion bombardment.

To a first approximation, the sputtering process is the inverse of growth. During sputtering, atoms are removed creating vacancies that subsequently hop around just as atoms deposited during growth. (A neighboring atom hopping into the vacant site constitutes vacancy diffusion.) This analog explains a number of features (but not all) observed on ion eroded surfaces.

Using common diffraction probes such as X-rays, electrons, or He atoms changes in surface structure and morphology can be observed in real-time as the material is being irradiated. A schematic of the X-ray diffraction geometry is shown in Fig. 14. The type of information extracted from diffraction depends on the magnitude and direction of the scattering vector. If the scattering vector $\mathbf{q}$ is normal to the surface ($\alpha = \beta$, $\Delta = 0$), information about surface roughness can be obtained. If $\mathbf{q}$ has an in-plane component, then lateral correlations in the surface morphology can be detected.

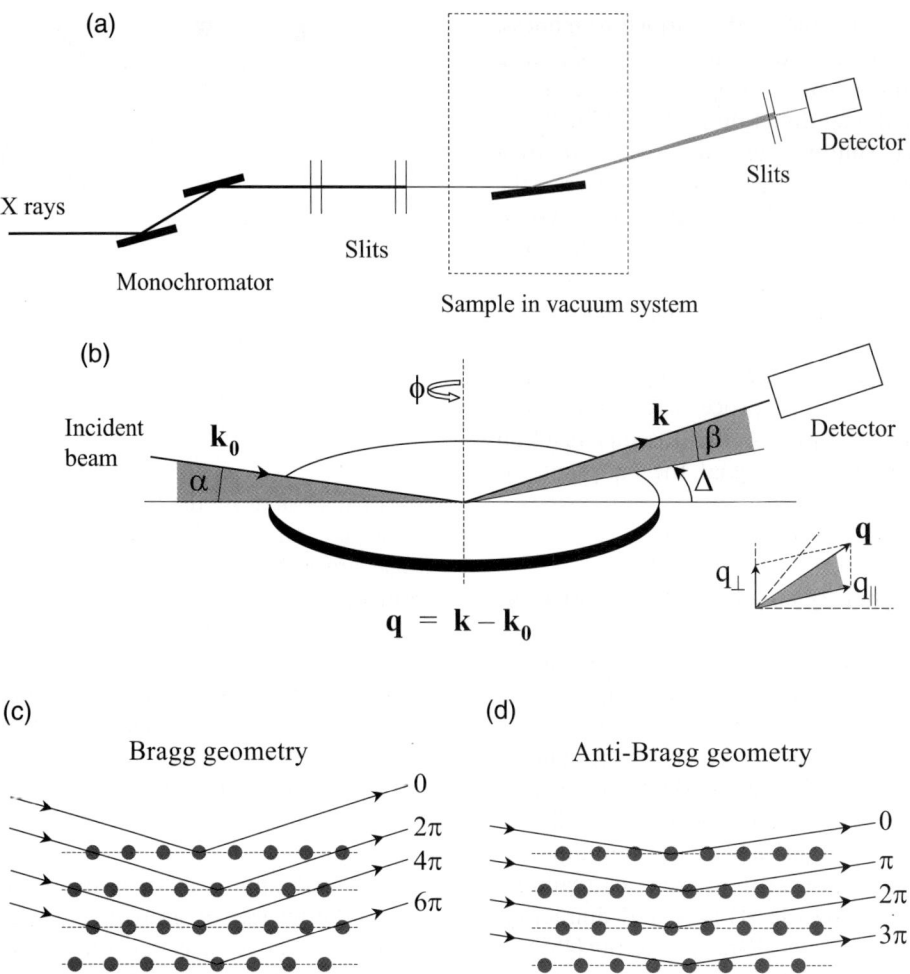

Fig. 14. (a) Schematic of a synchrotron X-ray diffraction experiment. X-rays from the synchrotron pass through a double bounce monochromator (e.g., Si(1 1 1) crystals or W/B$_4$C multilayer mirrors) that selects a specific (user defined) wavelength. The size of the X-ray beam is defined by pairs of horizontal and vertical slits. The X-rays enter and exit the vacuum system through beryllium windows which have low absorption (due to low electron density). (b) Diffraction geometry commonly employed to extract surface morphology information. For single crystals, $q_\perp$ gives information about roughness and $q_\parallel$ about lateral correlations in the surface morphology. In a four circle diffractometer, the angles α, β, Δ and ϕ can be varied independently to access all of reciprocal space. (c) At the Bragg condition, scattering between two successive planes in the crystal is in-phase resulting in a large intensity. (d) At the anti-Bragg condition, scattering between two succesive planes is 180° out-of-phase. The resulting intensity is several orders of magnitude smaller than the intensity at the Bragg condition. For X-ray scattering, a synchrotron source is needed to provide sufficient incident intensity to measure the (small) intensity at the anti-Bragg position. Electrons, on the other hand, are strongly scattered by atoms and significant intensity can be measured at the anti-Bragg position using a simple laboratory source.

The familiar Bragg's law, shown in Fig. 14(c), corresponds to a specular ($\alpha = \beta$) geometry with scattering between two successive planes being in-phase. In surface studies it is useful to adopt the geometry shown in Fig. 14(d) with the incident angle of the probe selected such that scattering from two successive planes of the crystal is 180° out-of-phase. This is sometimes referred to as the anti-Bragg position. If scattering between two successive planes exactly cancels out, why is the scattered intensity finite in this geometry? To see this, take the amplitude of the scattered wavefunction from one plane as A, and the attenuation of the incident beam intensity in the crystal as κ. Then, the amplitude of the scattered wave from the entire crystal [49] is $A(1 - \exp(-\kappa/2) + \exp(-\kappa) - \cdots) = A/2$ as $\kappa \to 0$. There are other ways to arrive at the same result [50]. This argument applies best to X-rays where scattering is weak (small κ). With electrons, the scattering is strong (large κ) and it is more difficult to write an expression for the scattered intensity. However, qualitative conclusions can still be drawn from a measurement of change in the scattered intensity during ion erosion. If a fraction of a monolayer θ is removed from a smooth surface, the scattered wave amplitude becomes $(A/2 - \theta)$. Intensity at the anti-Bragg geometry is thus extremely sensitive to deviations from a smooth surface and removal of as little as 0.01 ML can be detected.

Fig. 15 shows an example of specular beam intensity variation at the anti-Bragg position during 500 eV Ar$^+$ ion bombardment of Au(1 1 1) [51]. X-rays of 11 keV energy (wavelength $\lambda = 1.13$ Å) were used for the experiments. The intensity variation can be understood by referring to the schematic in Fig. 16. Every surface, however well oriented and polished, is misaligned from a low index Miller plane by some amount. The miscut is accommodated by steps (frequently of monolayer height) separated by terraces. Kinks generated by thermal excitation and/or imposed by the miscut cause the step edges to meander. Sputtering creates vacancies and at sufficiently high temperature, the vacancies have sufficient mobility to diffuse all the way up to a step. Once a vacancy has attached to a step, it can diffuse along the step edge to reach kink sites where it is accommodated with even lower energy. Hence, the surface evolves by step retraction and the specular beam intensity remains constant (270 °C). At lower temperatures, vacancies diffuse less rapidly and are more likely to meet other vacancies on the terrace before they reach a step. Such chance meeting of vacancies leads to vacancy island formation. These vacancy islands then grow laterally as other vacancies created both within and outside the islands diffuse and stick to the edges of the vacancy islands. In this way a whole layer is peeled off and the process repeats again for the next layer. The surface thus oscillates

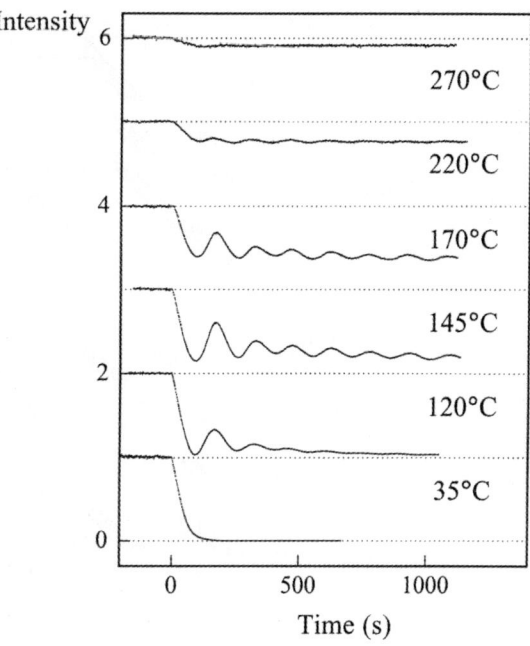

Fig. 15. Evolution of specular beam intensity at the anti-Bragg position (0 0 1.5)$_{hex}$ during 500 eV Ar$^+$ ion irradiation of Au(1 1 1) at different temperatures. The ions were incident at 45° with respect to the surface normal. The data at different temperarures has been offset vertically for clarity (from Ref. [51]).

between rough and smooth at a monolayer level and that shows up as oscillations of the specular beam intensity (120–220 °C). The intensity oscillations occur with a period of (approximately) 1 ML [52,53]. Such layer-by-layer removal has also been observed on Pt(1 1 1), Si(1 0 0), and Si(1 1 1) [54–56]. If the temperature is reduced even further, new vacancy islands are formed in the second layer even before the first layer is completely removed. The surface roughens and the specular beam intensity drops monotonically to zero.

Both layer-by-layer erosion and step retraction are attractive from the point of view of keeping the surface smooth. Layer-by-layer erosion has the advantage that one can count the number of layers removed using a diffraction probe. However, as Fig. 15 shows, the specular beam intensity oscillations are not perfect. By choosing appropriate conditions, over 100 specular beam intensity oscillations have been observed on Au(1 1 1), though

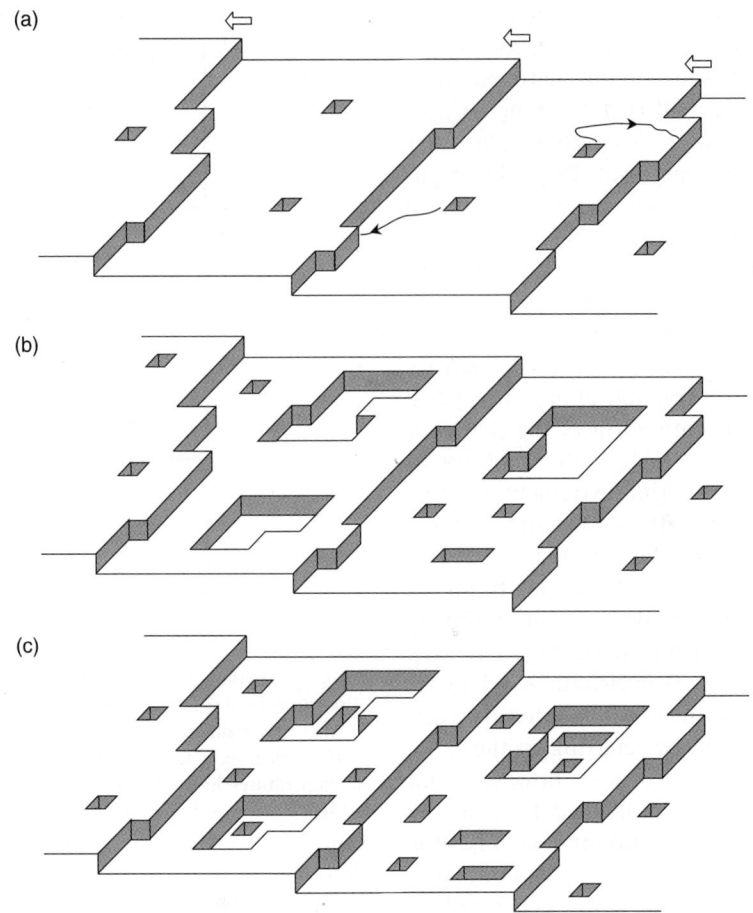

Fig. 16. Schematic of the surface morphology showing evolution by (a) step retraction, (b) two-dimensional island nucleation and growth leading to layer-by-layer removal, and (c) three-dimensional rough erosion.

eventually the oscillations decay away implying that the surface roughens [57]. Step retraction is the best choice for maintaining a smooth surface if one can tolerate the high temperatures. However, note that step retraction can also be unstable leading to step bunching [58,59], and wavy steps [60,61]. Such instabilities show up more quickly when the vicinal angle is high (step spacing is small). Wavy steps and pit formation can be seen in the STM image in Fig. 17 of ion eroded Ag(0 0 1) with ≈0.8° vicinal angle [62]. The time for such instabilities to develop depends strongly on the vicinal angle, and at low miscuts (≈0.1°) several hundreds of nanometers can be removed without pit formation.

Surfaces eventually get rough during ion erosion, so we need to understand how the roughness develops. Fig. 18 shows images of ion eroded GaSb(0 0 1) [63] and SiO_2 [64] surfaces. The most striking aspect of these images is that the features, mounds on GaSb and ripples on SiO_2, are so well organized. The incident ions in these experiments arrive at the surface uncorrelated in time and space. What then causes the atoms or molecules to organize themselves into such regular arrays?

In general, the formation of a regular array of features indicates the instability of the starting smooth surface under ion bombardment. The removal of atoms by ions creates continuous perturbations on the surface morphology. The

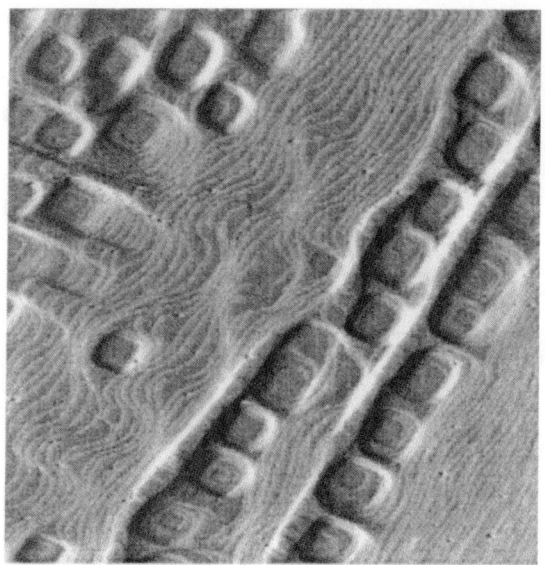

Fig. 17. STM image of ion eroded ≈0.8° vicinal Ag(0 0 1) showing wavy steps and pit formation (reprinted from Ref. [62]. Courtesy of Corrado Boragno).

starting surface orientation is unstable if (at least some of) the perturbations grow with time.

A rippled surface morphology is also observed after sandblasting [65] and was suggested as an analog [66]. However, the period of the ripples is of the order of the incident particle size in sandblasting whereas the ripple period can be an order of magnitude larger than the range of the incident ion during ion erosion. Using the transport theory of sputtering [10], Bradley and Harper [17] showed that ripples are formed due to a curvature dependent sputter yield.

As discussed in Section 2, sputter yield can be related to the amount of energy deposited at the surface. According to the transport theory, the energy distribution in the solid due to an ion impact can be modeled as emerging from a point below the surface. Consider sections of a surface with opposite curvature as shown in Fig. 19 with the ions incident normal to the (average) surface [17]. The energy deposited at point O by ions striking the surface at O is the same as that deposited at O' by ions striking at that point. However, the energy deposited at point O due to an ion striking at point A is greater than that deposited at O' by an ion incident at A', and likewise for B and B'. The rate of erosion is therefore greater at O than O' causing any perturbation to the surface morphology to grow. A quantitative analysis shows that the curvature κ of the surface provides a first order correction to the sputter yield, $\Delta Y \propto \kappa$ [17,67]. Simultaneously surface diffusion (more precisely, a downhill current of atoms as discussed below) acts to smooth out any perturbation caused by erosion. The curvature dependent erosion rate and surface diffusion driven smoothing have different dependences on the wave vector $\mathbf{q}$ of a perturbation. Curvature κ is proportional to $|\mathbf{q}|^2$ whereas surface diffusion driven smoothing is proportional to $|\mathbf{q}|^n$ with $n \approx 3\text{--}5$ [68,69] depending on the microscopic mechanism. The net result is that the wave vector $\mathbf{q}$ that gets selected is the one that grows the fastest. When ions are incident normal to a surface, the two transverse directions (x and y) are equivalent (for an isotropic material) and the resulting pattern consists of mounds and pits.

Under oblique incidence, the symmetry of the two transverse directions (x and y) is broken and a ripple structure is formed. Consider the erosion of a trough and a crest by an ion beam incident at an angle ϕ as shown in Fig. 20. There are two factors to consider here. The energy deposited at point O (O') comes mostly from ions striking near P (P'). From geometrical considerations it is clear that the erosion rate at O is increased relative to O'. Secondly, the number of ions that arrive at P per unit time is less than that at P', resulting in a faster erosion rate at O' compared to O. The former effect dominates near normal incidence whereas the latter effect dominates at grazing incidence (ϕ close to 90°). The net result [17] is the formation of ripples with the wave vector parallel to the surface component of the incident beam direction at small ϕ, and the wave vector perpendicular to the surface component of the beam direction for large ϕ as shown in Fig. 21.

Ripples have been observed on a number of materials including SiO_2, Si(0 0 1), and Ge(0 0 1) [64,70,71]. A rotation of the ripple wave vector

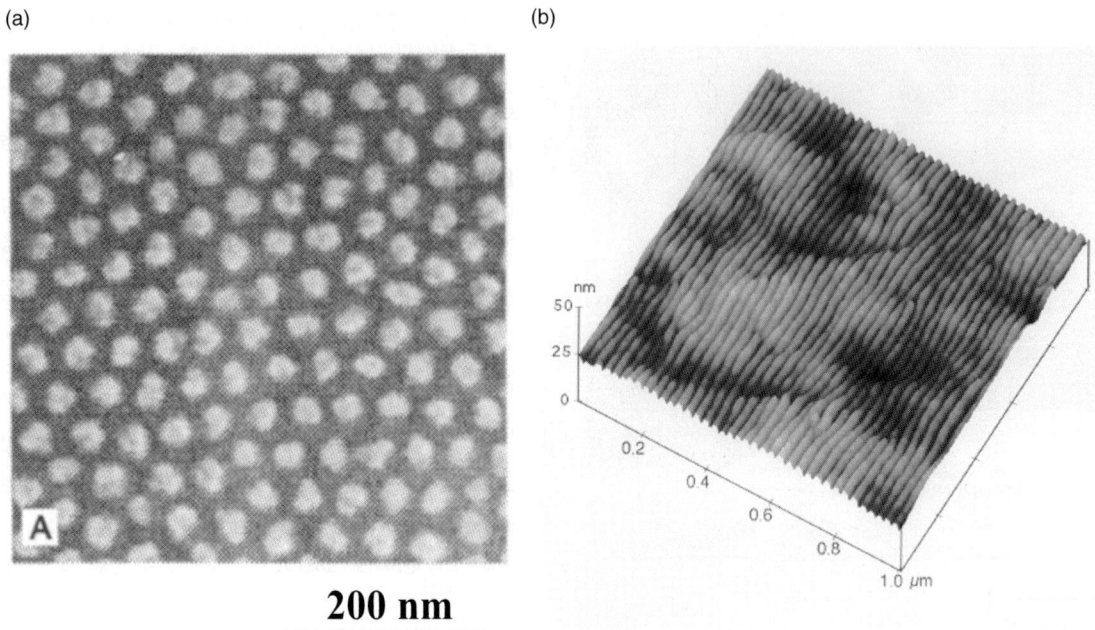

Fig. 18. (a) Scanning electron microscope image of GaSb(0 0 1) irradiated with 420 eV Ar$^+$ ions incident normal to the surface. The average separation between mounds in the two-dimensional array is 45 nm (reprinted with permission from Ref. [63]. Copyright 1999 American Association for the Advancement of Science). (b) Atomic force microscope image of SiO$_2$ after removal of ≈20 nm with 1 keV Xe$^+$ ions incident at 55° with respect to the surface normal. The ripple period is 30 nm and the wave vector is parallel to the surface component of the incident ion beam (from Ref. [64]. Courtesy of Eric Chason).

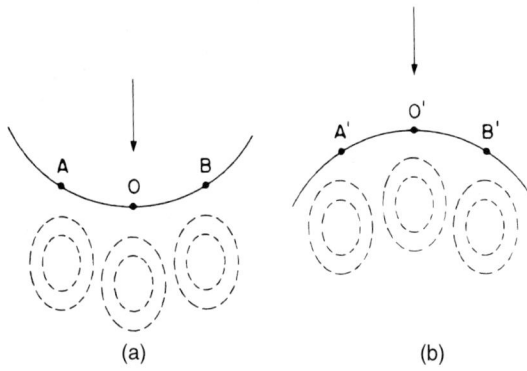

Fig. 19. Contours of constant energy deposition for ions incident at surfaces of opposite curvature according to the transport theory. The ions are incident normal to the average surface. Adding up energy deposited by ions incident at O and neighboring points such as A and B, it is evident that more energy is deposited per unit area at point O compared to O′. The sputter rate is thus expected to be higher at O compared to O′ (from Ref. [17] with permission. Courtesy of Jim Harper).

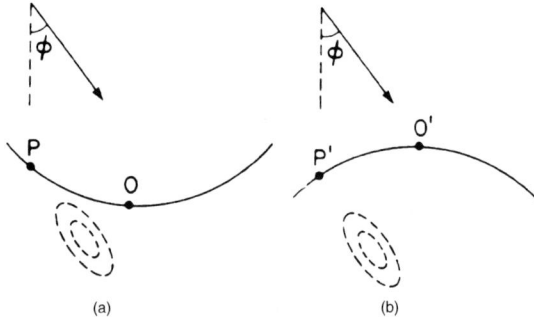

Fig. 20. Same as Fig. 19, except that ions are now incident at an oblique angle with respect to the (average) surface normal. The energy deposited at O by an ion incident at P is higher than the energy deposited at O′ by an ion incident at an equal distance P′. However, the ion flux is higher over the section P′–O′ compared to P–O. The former effect dominates at small ϕ and the latter dominates at large ϕ (grazing incidence) (from Ref. [17] with permission. Courtesy of Jim Harper).

with the incident angle ϕ has been observed on Cu(1 1 0) and graphite (0 0 0 1) [72,73]. The ripple wavelength can be controlled by changing the substrate temperature, ion flux, and the angle of

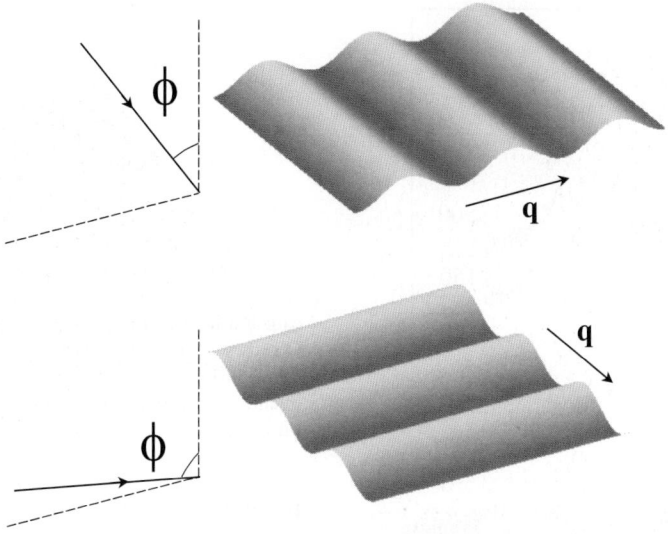

Fig. 21. (a) (top) For small incidence angle ϕ, the ripple wave vector **q** is parallel to the surface component of the incident beam direction. (b) (bottom) Near grazing incidence, the ripple wave vector **q** is perpendicular to the surface component of the incident beam direction.

the incident ion beam. The curvature dependence of the sputter yield results in the selection of a specific wavelength, and the wavelength does not change as the erosion proceeds.

On crystalline materials, the characteristic length scale of morphological features on the surface often increases during ion erosion. Fig. 22(a) shows transverse X-ray scans taken during 500 eV Ar^+ ion erosion of Au(1 1 1) at 35 °C [74]. For the scattering geometry chosen, X-rays scattered between two successive planes are 22° out-of-phase (for $\Delta = 0$) and the angle Δ is scanned (see Fig. 14). The satellite peak indicates regularly spaced features on the surface. Changes in the surface morphology during erosion can be followed in real-time by such measurements. The STM image of ion eroded Au(1 1 1) in Fig. 22(b) shows that the features are mounds of a characteristic size [75]. The characteristic length scale l can be estimated from the location of the satellite peak and Fig. 22(c) shows the variation of l with the amount of material removed for several different bombardment conditions. The length scale increases with a power law. The inset in Fig. 22(c) shows the ratio of the rms roughness w to the length scale l. The nearly constant value of w/l indicates that the features maintain a constant aspect ratio as they grow in size. Coarsening of features during ion erosion also has been observed on Ge(0 0 1), Pt(1 1 1) and Cu(1 1 0) surfaces [76–78].

A different class of microscopic mechanisms involving diffusion bias have been identified [79] that can explain both mound and pit formation, and their coarsening during ion erosion. Fig. 23 shows a schematic of the surface with the activation barriers for several different hops for surface vacancies indicated. The activation barrier for a single vacancy to diffuse on the terrace is ε_d. When the vacancy reaches an upper step edge, it faces a slightly higher barrier ($\varepsilon_d + \varepsilon_s$) to join the step than to hop back (ε_d). The excess barrier ε_s is often called the Ehrlich–Schwoebel barrier [80,58]. When the vacancy reaches a lower step edge, it faces a slightly lower barrier to join the step ($\varepsilon_d - \Delta$) than to hop back (ε_d). Finally, the activation barrier to diffuse along the step edge is ε_e. The presence of any one of the three processes, an Ehrlich–Schwoebel barrier for vacancies [79], a vacancy step attraction [81], or step edge diffusion of vacancies [61,82,83] can give rise to pattern formation. All three of these mechanisms result in a diffusion bias—on average, vacancies hop more often in the downhill direction, i.e. toward the lower step (this is equivalent to atoms making

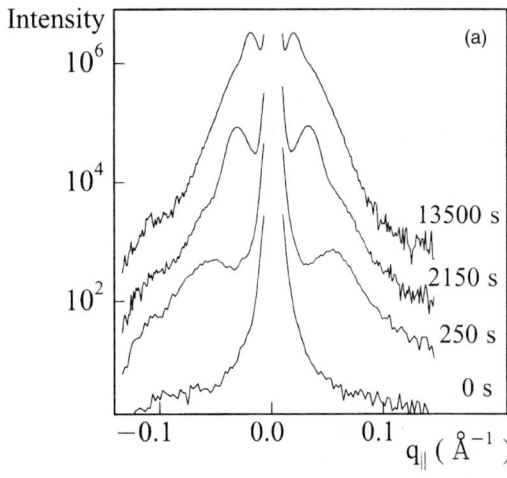

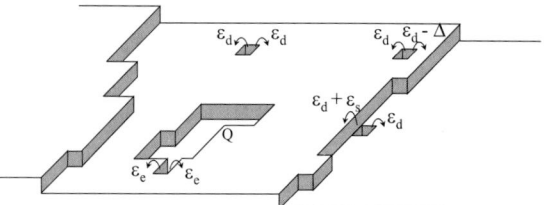

Fig. 23. Schematic of a surface showing three microscopic sources of instability—the Ehrlich–Schwoebel barrier ε_s, the step-vacancy attraction Δ, and step edge diffusion ε_e. For vacancies diffusing along a step edge, outside corner sites such as Q represent the lowest energy sites. All three sources lead to a diffusion bias, i.e. cause vacancies (atoms) to jump more frequently in the downhill (uphill) direction. This leads to mound/pit formation during ion erosion.

more hops in the uphill direction) [79,84]. The symmetry breaking in the surface diffusion is fairly easy to see for the Ehrlich–Schwoebel barrier or the vacancy-step edge attraction. Whenever a vacancy reaches close to a step edge, it is either repelled (upper step, Ehrlich–Schwoebel barrier) or attracted (lower step) leading to more hops in the downhill direction. For vacancies diffusing along a step edge, the lowest energy sites correspond to the outside corner sites such as Q (Fig. 23) since atoms along the step edge achieve the highest coordination. On average, vacancies move toward the outside corners once they join an island [61]. This is the same as the downhill direction. Note that downhill current does not mean that vacancies have to hop from one terrace to the lower terrace, only that they move toward the downhill direction.

The downhill current vanishes at a low index orientation by symmetry (atoms are just as likely

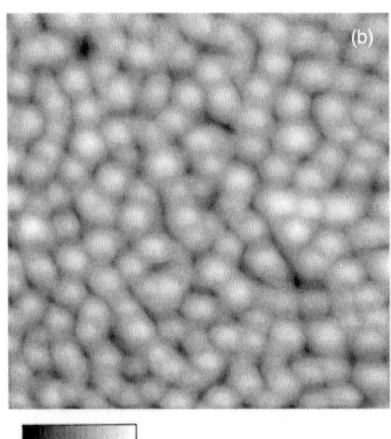

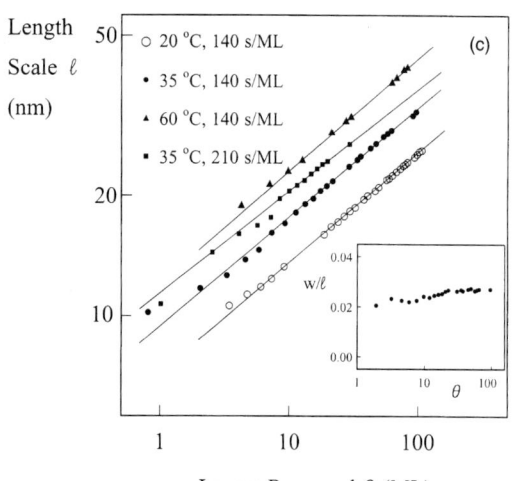

Fig. 22. (a) Transverse X-ray scans taken during 500 eV Ar$^+$ ion irradiation of Au(1 1 1) at 35 °C. The satellite peak indicates the presence of regularly arranged features on the surface. The scans taken at different times have been offset vertically for clarity. (b) STM image (193 × 193 nm^2) of ion irradiated Au(1 1 1) at 25 °C shows that the features on the surface are mounds. (c) The variation of the length scale with time for several ion erosion conditions is consistent with a power law, $l \sim t^{0.27 \pm 0.02}$. Inset: the evolution of the ratio of the rms roughness w to the length scale l. The nearly constant value of w/l indicates that the mounds maintain a constant aspect ratio as they grow in size (panels (a) and (c) from Ref. [74]. Panel (b) from Ref. [75] Courtesy of Aaron Judy).

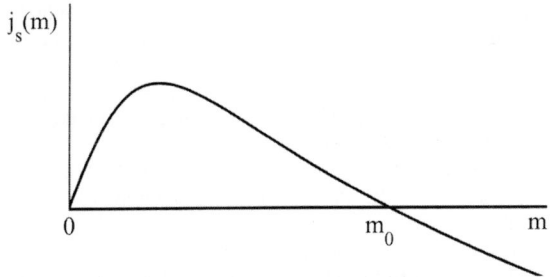

Fig. 24. Variation of the surface diffusion current as a function of slope. The slope m_0 where $j_s = 0$ and $\partial j_s/\partial m < 0$ corresponds to a stable slope and small perturbations from m_0 get damped.

to hop in any direction). Small deviations from low index orientations are usually accommodated by steps of monatomic height, and the presence of a positive ε_s, Δ or step edge diffusion results in a positive vacancy downhill current $j_s(m)$ for these slopes. A typical form [85] for j_s is shown in Fig. 24. Any perturbation created by ion bombardment to a smooth starting surface with slope $0 < m < m_0$ starts to grow. The local slope increases until it reaches a value m_0 where $j_s(m_0) = 0$. Any perturbation from the slope m_0 decays away since $\partial j_s/\partial m < 0$. This is frequently referred to as slope selection [85] in the literature and accounts for the constant aspect ratio of features observed in experiments. Unlike the curvature dependent sputter yield mechanism, the above mechanisms do not select a characteristic length scale. Rather the slope gets selected and the feature size coarsens during ion erosion.

We end the discussion of surface morphology evolution by noting that one does not always get "patterns". For example, rough, self-affine surfaces have been observed on graphite under certain bombardment conditions [86]. Cone-shaped features are sometimes observed on metal surfaces especially in the presence of impurities [87,88].

5. Future directions

Besides sputtering, erosion of a material can be caused in many ways including wet chemical etching, sandblasting, and polishing. For the manufacture of microelectronic devices, ion sputtering or dry etching, has gradually replaced chemical etching in the last two decades due to its many advantages noted above.

As new materials get incorporated into electronic devices, there is a critical need for developing etching chemistries for selective and anisotropic etching combined with high throughput. Some of the new materials include high dielectric constant oxides such as Ta_2O_5, ferroelectric materials such as $PbZr_xTi_{1-x}O_3$ (PZT) and $SrBi_2Ta_2O_9$ (SBT), low-k dielectrics such as F-doped SiO_2, and noble metals such as Pt and Ir [45,89]. A typical reactive ion etching process contains more than 20 reaction products. Modeling such processes to optimize selectivity and minimizing damage will continue to be important. However, it has become clear from computer simulations that small changes in the interatomic potential can result in significantly changes in the sputter yield [33]. It is therefore unlikely that a quantitative knowledge of erosion rates will emerge from the analytical studies, at least to the accuracy desired. Rather, real-time feedback systems that signal the endpoint of etching will continue to be the method of choice. This too is becoming more challenging when the material being removed from exposed etch areas constitutes a small fraction (<1%) of the total surface area [46].

Failure analysis is an important part of the development of new integrated circuits. Focussed ion beam etching is widely used to expose potential areas of device failure such as interconnects for further analysis by electron microscopy. Focussed ion beams can also be used to write features with nanoscale dimensions.

As we reach the limits of conventional optical lithography, there has been a steady drive toward nanotechnology. Self-organized growth, of self-assembly of atoms and molecules into nanoscale dots and wires appears to be a potential successor to optical lithography. Tremendous effort has gone in the last decade toward the fabrication and characterization of nanoscale features using such diverse methods as solution synthesis [90] and strain engineering in epitaxial films [91].

Sputtering can be used to generate highly regular arrays of zero- (dots) and one-dimensional (wires) nanometer scale particles as we saw in

Section 4. The particle size and separation can be adjusted by varying the temperature, ion energy and flux, and the ion irradiation geometry. The underlying microscopic mechanisms are sufficiently general to make nanoscale features of any material. One aspect that is largely unknown at this point is the degree of order in the patterns which differs significantly in the experiments on various materials reported thus far. On GaSb(0 0 1) (Fig. 18(a)), the autocorrelation function shows spots up to sixth order indicating a highly regular array of particles [63]. It is more common to see just the first order spots. It appears that patterns are better ordered on amorphous materials compared to crystalline materials (the surface region of GaSb(0 0 1) is amorphous in Fig. 18 [63]). This may be related to the fact that the curvature dependent sputter yield mechanism selects a specific wavelength and does not cause coarsening (unlike the mechanisms shown in Fig. 23). Where would one use ion sputtering for such an application? The technique is best suited for materials that are resilient to damage, most notably metals.

In the nearly 150 years since sputtering was discovered, it has been put to many uses. The simplicity of the generation and transport of ion beams, and the tremendous flexibility ion erosion offers a device designer will make it a key fabrication process in the new century.

Acknowledgements

This work was supported by the U.S. Department of Energy, BES–DMS, under contract W-31-109-ENG-38, and by the State of Illinois, under HECA.

References

[1] M. Madou, Fundamentals of Microfabrication, CRC Press, Boca Raton, FL, 1997, Chapter 2.
[2] I. Brodie, J.J. Muray, The Physics of Micro/Nano-Fabrication, Plenum, New York, 1992, Chapters 2 and 3.
[3] W.L. Brown, L.J. Lanzerotti, R.E. Johnson, Fast ion bombardment of ices and its astrophysical implications, Science 218 (1982) 525.
[4] C.L. Melcher, D.J. LePoire, B.H. Cooper, T.A. Tombrello, Erosion of frozen sulfur dioxide by ion bombardment: applications to Io, Geophys. Res. Lett. 9 (1982) 1151.
[5] T.E. Madey, R.E. Johnson, T. Orlando, Far-out surface science: radiation induced surface processes in the solar system, Surf. Sci. 500 (2002) 838.
[6] P. Sigmund, Sputtering by ion bombardment. Theoretical concepts, in: R. Behrisch (Ed.), Sputtering by Particle Bombardment I, Topics in Applied Physics, vol. 47, Springer, Berlin, 1981, p. 9.
[7] I. Finnie, Erosion of surfaces by solid particles, Wear 3 (1960) 87.
[8] J.G.A. Bitter, A study of erosion phenomena. Part II, Wear 6 (1963) 169.
[9] T. Diaz de la Rubia, R.S. Averback, R. Benedek, W.E. King, Role of thermal spikes in energetic displacement cascades, Phys. Rev. Lett. 59 (1987) 1930.
[10] P. Sigmund, Theory of sputtering I. Sputtering yield of amorphous and polycrystalline targets, Phys. Rev. 184 (1969) 383; Errata, Phys. Rev. 187 (1969) 768.
[11] M.A. Nastasi, J.W. Mayer, J.K. Hirvonen, Ion–Solid Interactions: Fundamentals and Applications, Cambridge University Press, New York, 1996.
[12] M.W. Thompson, II. The energy spectrum of ejected atoms during the high energy sputtering of gold, Philos. Mag. 18 (1968) 377.
[13] P.C. Zalm, Ion-beam assisted etching of semiconductors, Vacuum 36 (1986) 787.
[14] H.H. Andersen, H.L. Bay, Sputtering yield measurements, in: R. Behrisch (Ed.), Sputtering by Particle Bombardment I. Topics in Applied Physics, vol. 47, Springer, Berlin, 1981, p. 145.
[15] N. Matsunami, Y. Yamamura, Y. Itikawa, N. Itoh, Y. Kazumuta, S. Miyagawa, K. Morita, R. Shimizu, H. Tawara, Energy dependence of the ion-induced sputtering yields of monatomic solids, At. Data Nucl. Data Tables 31 (1984) 1.
[16] K.B. Cheney, E.T. Pitkin, Sputtering at acute incidence, J. Appl. Phys. 36 (1965) 3542.
[17] R.M. Bradley, J.M.E. Harper, Theory of ripple topography induced by ion bombardment, J. Vac. Sci. Technol. A 6 (1988) 2390.
[18] M.J. Witcomb, Prediction of the apex angle of surface cones on ion-bombarded crystalline materials, J. Mater. Sci. 9 (1974) 1227.
[19] R.V. Stuart, G.K. Wehner, G.S. Anderson, Energy distribution of atoms sputtered from polycrystalline metals, J. Appl. Phys. 40 (1969) 803.
[20] G.K. Wehner, Controlled sputtering of metals by low-energy Hg ions, Phys. Rev. 102 (1954) 690.
[21] W.O. Hofer, Angular, energy, and mass distribution of sputtered particles, in: R. Behrisch (Ed.), Sputtering by Particle Bombardment II. Topics in Applied Physics, vol. 52, Springer, Berlin, 1983, p. 15.
[22] J.B. Gibson, A.N. Goland, M. Milgram, G.H. Vineyard, Dynamics of radiation damage, Phys. Rev. 120 (1960) 1229.

[23] H. Goldstein, Classical Mechanics, Addison-Wesley, Reading, MA, 1981, Chapter 3.
[24] M.T. Robinson, I.M. Torrens, Computer simulation of atomic-displacement cascades in solids in the binary collision approximation, Phys. Rev. B 9 (1974) 5008.
[25] J.P. Biersack, W. Eckstein, Sputtering studies with the Monte Carlo program TRIM.SP, Appl. Phys. A 34 (1984) 73.
[26] I.M. Torrens, Interatomic Potentials, Academic, New York, 1972.
[27] J.P. Biersack, L.G. Haggmark, A Monte Carlo computer program for the transport of energetic ions in amorphous targets, Nucl. Instrum. Meth. 174 (1980) 257.
[28] J.F. Ziegler, J.P. Biersack, U. Littmark, The Stopping and Range of Ions in Solids, Pergamon, New York, 1985.
[29] F.H. Stillinger, T.A. Weber, Computer simulation of local order in condensed phases of silicon, Phys. Rev. B 31 (1985) 5262.
[30] J. Tersoff, Empirical interatomic potential for silicon with improved elastic properties, Phys. Rev. B 38 (1988) 9902.
[31] S.M. Foiles, M.I. Baskes, M.S. Daw, Embedded-atom-method functions for the fcc metals Cu, Ag, Au, Ni, Pd, Pt, and their alloys, Phys. Rev. B 33 (1986) 7983.
[32] K.W. Jacobsen, J.K. Norskov, M.J. Puska, Interatomic interactions in the effective medium theory, Phys. Rev. B 35 (1987) 7423.
[33] D.R. Harrison, Application of molecular dynamics simulations to the study of ion-bombarded metal surfaces, Crit. Rev. Solid State Mater. Sci. 14 (1988) S1.
[34] T. Michely, C. Teichert, Adatom yields, sputtering yields, and damage patterns of single-ion impacts on Pt(111), Phys. Rev. B 50 (1994) 11156.
[35] M.T. Robinson, O.S. Oen, The channeling of energetic atoms in crystal lattices, Appl. Phys. Lett. 2 (1963) 30.
[36] D.J. Smith, A.R. Modak, T.A. Rabedeau, S.S.P. Parkin, Growth and structural characterization of highly oriented sputter-deposited (111), (110) and (100) Co/Cu superlattices, Appl. Phys. Lett. 71 (1997) 1480.
[37] G.R. Harp, S.S.P. Parkin, Epitaxial growth of metals by sputter deposition, Thin Solid Films 288 (1996) 315.
[38] M.V.R. Murty, H.A. Atwater, Silicon epitaxy on hydrogen-terminated silicon surfaces using thermal and energetic beams, Surf. Sci. 374 (1997) 283.
[39] T. Ohmi, T. Ichikawa, H. Iwabuchi, T. Shibata, Formation of device-grade epitaxial silicon films at extremely low temperatures by low-energy bias sputtering, J. Appl. Phys. 66 (1989) 4756.
[40] M. Kalff, M. Breeman, M. Morgenstern, T. Michely, G. Comsa, Effect of energetic particles on island formation in sputter deposition, Appl. Phys. Lett. 70 (1997) 182.
[41] G. Rosenfeld, N.N. Lipkin, W. Wulfhekel, J. Kliewer, K. Morgenstern, B. Poelsema, G. Comsa, New concepts for controlled homoepitaxy, Appl. Phys. A 61 (1991) 455.
[42] R.G. Wilson, F.A. Stevie, C.W. Magee, Secondary Ion Mass Spectrometry: A Practical Handbook for Depth Profiling and Bulk Impurity Analysis, Wiley, New York, 1989.

[43] H.-J. Gossmann, C.S. Rafferty, H.S. Luftman, F.C. Unterwald, T. Boone, J.M. Poate, Oxidation enhanced diffusion in Si B-doping superlattices and Si self-interstitial diffusivities, Appl. Phys. Lett. 63 (1993) 639.
[44] R.M. Bradley, E.-H. Cirlin, Theory of improved resolution in depth profiling with sample rotation, Appl. Phys. Lett. 68 (1996) 3722.
[45] I. Morey, A. Asthana, Etch challenges of low-k dielectrics, Solid State Technol. 42 (1999) 71.
[46] P. Biolsi, D. Morvay, L. Drachnik, S. Ellinger, An advanced endpoint detection solution for <1% open areas, Solid State Technol. 39 (1996) 59.
[47] C.J. Chen, Introduction to Scanning Tunneling Microscopy, Oxford University Press, New York, 1993.
[48] M.G. Lagally, Diffraction techniques, in: R.L. Park, M.G. Lagally (Eds.), Solid State Physics: Surfaces. Methods of Experimental Physics, vol. 22, Academic, Orlando, 1985, p. 237.
[49] R. Feidenhans'l, Surface structure determination by X-ray diffraction, Surf. Sci. Rep. 10 (1989) 105.
[50] I.K. Robinson, Crystal truncation rods and surface roughness, Phys. Rev. B 33 (1986) 3830.
[51] M.V.R. Murty, T. Curcic, A. Judy, B.H. Cooper, A.R. Woll, J.D. Brock, S. Kycia, R.L. Headrick, Real-time X-ray scattering study of surface morphology evolution during ion erosion and epitaxial growth of Au(111), Phys. Rev. B 60 (1999) 16956.
[52] M.V.R. Murty, B. Cowles, B.H. Cooper, Surface smoothing during sputtering—vacancy diffusion versus adatom detachment and diffusion, Surf. Sci. 415 (1998) 328.
[53] M.C. Bartelt, J. Evans, Transition to multilayer kinetic roughening for metal (100) homoepitaxy, Phys. Rev. Lett. 75 (1995) 4250.
[54] B. Poelsema, L.K. Verheij, G. Comsa, 'Two-layer' behavior of Pt(111) surface during low-energy Ar$^+$ ion sputtering at high temperatures, Phys. Rev. Lett. 53 (1984) 2500.
[55] P. Bedrossian, J.E. Houston, J.Y. Tsao, E. Chason, S.T. Picraux, Layer-by-layer sputtering and epitaxy of Si(100), Phys. Rev. Lett. 67 (1991) 124.
[56] P. Bedrossian, T. Klitsner, Surface reconstruction in layer-by-layer sputtering of Si(111), Phys. Rev. B 44 (1991) 13783.
[57] M.V. Ramana Murty, A.J. Couture, B.H. Cooper, A.R. Woll, J.D. Brock, R.L. Headrick, Persistent layer-by-layer sputtering of Au(111), J. Appl. Phys. 88 (2000) 597.
[58] R.L. Schwoebel, E.J. Shipsey, Step motion on crystal surfaces, J. Appl. Phys. 37 (1966) 3682.
[59] J.Y. Tsao, Materials Fundamentals of Molecular Beam Epitaxy, Academic, Boston, 1993, Chapter 6.
[60] G.S. Bales, A. Zangwill, Morphological instability of a terrace edge during step-flow growth, Phys. Rev. B 41 (1990) 5500.
[61] M.V. Ramana Murty, B.H. Cooper, Instability in molecular beam epitaxy due to fast edge diffusion and corner diffusion barriers, Phys. Rev. Lett. 83 (1999) 352.

[62] G. Costantini, S. Rusponi, R. Gianotti, C. Boragno, U. Valbusa, Temperature evolution of nanostructures induced by Ar$^+$ sputtering on Ag(001), Surf. Sci. 416 (1998) 245.

[63] S. Fackso, T. Dekorsky, C. Koerdt, C. Trappe, H. Kurz, A. Vogt, H.L. Hartnagel, Formation of ordered nanoscale semiconductor dots by ion sputtering, Science 285 (1999) 1551.

[64] T.M. Mayer, E. Chason, A.J. Howard, Roughening instability and ion-induced viscous relaxation of SiO_2 surfaces, J. Appl. Phys. 76 (1994) 1633.

[65] I. Finnie, Y.H. Kabil, On the formation of surface ripples during erosion, Wear 8 (1965) 60.

[66] G. Carter, M.J. Nobes, F. Paton, J.S. Williams, J.L. Whitton, Ion bombardment induced ripple topography on amorphous solids, Rad. Eff. 33 (1977) 65.

[67] R. Cuerno, A.-L. Barabasi, Dynamic scaling of ion-sputtered surfaces, Phys. Rev. Lett. 74 (1995) 4746.

[68] W.W. Mullins, Solid surface morphologies governed by capillarity, in: N.A. Gjostein, W.D. Robertson (Eds.), Metal Surfaces: Structure, Energetics, and Kinetics, American Society for Metals, Metals Park, OH, 1963, p. 99.

[69] M.V.R. Murty, Morphological stability of nanostructures, Phys. Rev. B 62 (2000) 17004.

[70] J. Erlebacher, M.J. Aziz, E. Chason, M.B. Sinclair, J.A. Floro, Spontaneous pattern formation on ion bombarded Si(001), Phys. Rev. Lett. 82 (1999) 2330.

[71] E. Chason, T.M. Mayer, B.K. Kellerman, D.T. McIlroy, A.J. Howard, Roughening instability and evolution of the Ge(001) surface during ion sputtering, Phys. Rev. Lett. 72 (1994) 3040.

[72] S. Rusponi, G. Costantini, C. Boragno, U. Valbusa, Scaling laws of the ripple morphology on Cu(110), Phys. Rev. Lett. 81 (1998) 4184.

[73] S. Habenicht, W. Bolse, K.P. Lieb, K. Reimann, U. Geyer, Nanometer ripple formation and self-affine roughening of ion-beam-eroded graphite, Phys. Rev. B 60 (1999) R2200.

[74] M.V. Ramana Murty, T. Curcic, A. Judy, B.H. Cooper, A.R. Woll, J.D. Brock, S. Kycia, R.L. Headrick, X-ray scattering study of the surface morphology of Au(111) during Ar$^+$ ion irradiation, Phys. Rev. Lett. 80 (1998) 4713.

[75] A.J. Couture, Surface morphology evolution during ion erosion of metal surfaces: scanning tunneling microscopy studies, Ph.D. Thesis, Cornell University, 1999.

[76] S.J. Chey, J.E. van Nostrand, D.G. Cahill, Surface morphology of Ge(001) during etching by low energy ions, Phys. Rev. B 52 (1995) 16696.

[77] T. Michely, G. Comsa, The scanning tunneling microscope as a means for the investigation of ion bombardment effects on metal surfaces, Nucl. Instr. Meth. B 82 (1993) 207.

[78] S. Rusponi, C. Boragno, U. Valbusa, Ripple structure on Ag(110) surface induced by ion sputtering, Phys. Rev. Lett. 78 (1997) 2795.

[79] J. Villain, Continuum models of crystal growth with and without desorption, J. Phys. I (France) 1 (1991) 19.

[80] G. Ehrlich, F.G. Hudda, Atomic view of surface self-diffusion: tungsten on tungsten, J. Chem. Phys. 44 (1966) 1039.

[81] J.G. Amar, F. Family, Step-adatom attraction as a new mechanism for instability in epitaxial growth, Phys. Rev. Lett. 77 (1996) 4584.

[82] S. Schinzer, M. Kinne, M. Biehl, W. Kinzel, The role of step edge diffusion in epitaxial crystal growth, Surf. Sci. 439 (1999) 191.

[83] O. Pierre-Louis, M.R. D'Orsogna, T.L. Einstein, Edge diffusion during growth: kink Ehrlich–Schwoebel effect and resulting instabilities, Phys. Rev. Lett. 82 (1999) 3661.

[84] J. Krug, M. Plischke, M. Siegert, Surface diffusion currents and universality classes of growth, Phys. Rev. Lett. 70 (1993) 3271.

[85] J.A. Stroscio, D.T. Pierce, M.D. Stiles, A. Zangwill, L.M. Sander, Coarsening of unstable surface features during Fe(001) homoepitaxy, Phys. Rev. Lett. 75 (1995) 446.

[86] E.A. Eklund, R. Bruinsma, J. Rudnick, R.S. Williams, Submicron-scale surface roughening induced by ion bombardment, Phys. Rev. Lett. 67 (1991) 1759.

[87] G.K. Wehner, D.J. Hajicek, Cone formation on metal targets during sputtering, J. Appl. Phys. 42 (1971) 1145.

[88] D.J. Barber, Radiation damage in ion-milled specimens: characteristics, effects and methods of damage limitation, Ultramicroscopy 52 (1993) 101.

[89] S.P. DeOrnellas, A. Cofer, Etching new IC materials for memory devices, Solid State Technol. 41 (1998) 53.

[90] S. Sun, C.B. Murray, D. Weller, L. Folks, A. Moser, Monodisperse FePt nanoparticles and ferromagnetic FePt nanocrystal superlattices, Science 287 (2000) 1989.

[91] Q. Xie, A. Madhukar, P. Chen, N.P. Kobayashi, Vertically self-organized InAs quantum box islands on GaAs(100), Phys. Rev. Lett. 75 (1995) 2542.

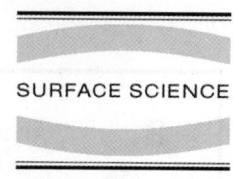

Probing buried interfaces with non-linear optical spectroscopy

Christopher T. Williams [a,*], David A. Beattie [b,1]

[a] *Department of Chemical Engineering, University of South Carolina, Columbia, SC 29208, USA*
[b] *Physical and Theoretical Chemistry Laboratory, University of Oxford, South Parks Road, Oxford OX1 3QZ, UK*

Received 19 July 2000; accepted for publication 10 April 2001

Abstract

The importance of buried interfaces in our everyday lives and in current scientific research is highlighted, along with experimental difficulty associated with studying such systems. We present an overview of the application of second harmonic generation and sum-frequency spectroscopy to the study of buried interfaces. Several examples from the current literature are presented, ranging from chemical and biological, to electrical and magnetic interfaces. The importance of this work in the context of ongoing research in these areas is discussed. Finally, we provide a snapshot of the state of the art in non-linear optical spectroscopy by mentioning several new directions that are likely to have a large impact on future research into the physics and chemistry of buried interfaces. © 2001 Elsevier Science B.V. All rights reserved.

Keywords: Non-linear optical methods; Second harmonic generation; Sum frequency generation; Catalysis; Adsorption kinetics; Tribology; Magnetic phenomena (cyclotron resonance, phase transitions, etc.); Self-assembly

1. Introduction and background

1.1. Buried interfaces everywhere

Interfaces are present throughout our everyday world. They exist either between two different materials or two different phases of the same material. What makes interfaces so interesting is that they exhibit properties and behavior that are very different from the adjacent bulk phases [1]. The thermodynamic constraints enforced by the two-dimensional surface results in the emergence of interesting phenomena. Atoms and molecules are typically present in concentrations different from either of the bulk materials. Groups of atoms and molecules will arrange themselves differently, creating near two-dimensional layers that have varying degrees of geometric order. Indeed, a range of mechanical, electrical, optical, and magnetic properties are altered considerably at surfaces. Given the unique properties of interfaces, perhaps it is not surprising that they play a central role both in nature and within a variety of different technological applications, devices, and industrial processes.

Consider the following three examples of interfaces that we encounter on a regular basis, depicted in Fig. 1. The catalytic converter in automobiles is

---

* Corresponding author. Tel.: +1-803-777-0143; fax: +1-803-777-8265.
E-mail address: willia84@engr.sc.edu (C.T. Williams).
[1] Present address: Ian Wark Research Institute, University of South Australia, Mawson Lakes, Adelaide, SA 5095, Australia.

0039-6028/01/$ - see front matter © 2001 Elsevier Science B.V. All rights reserved.
PII: S0039-6028(01)01536-9

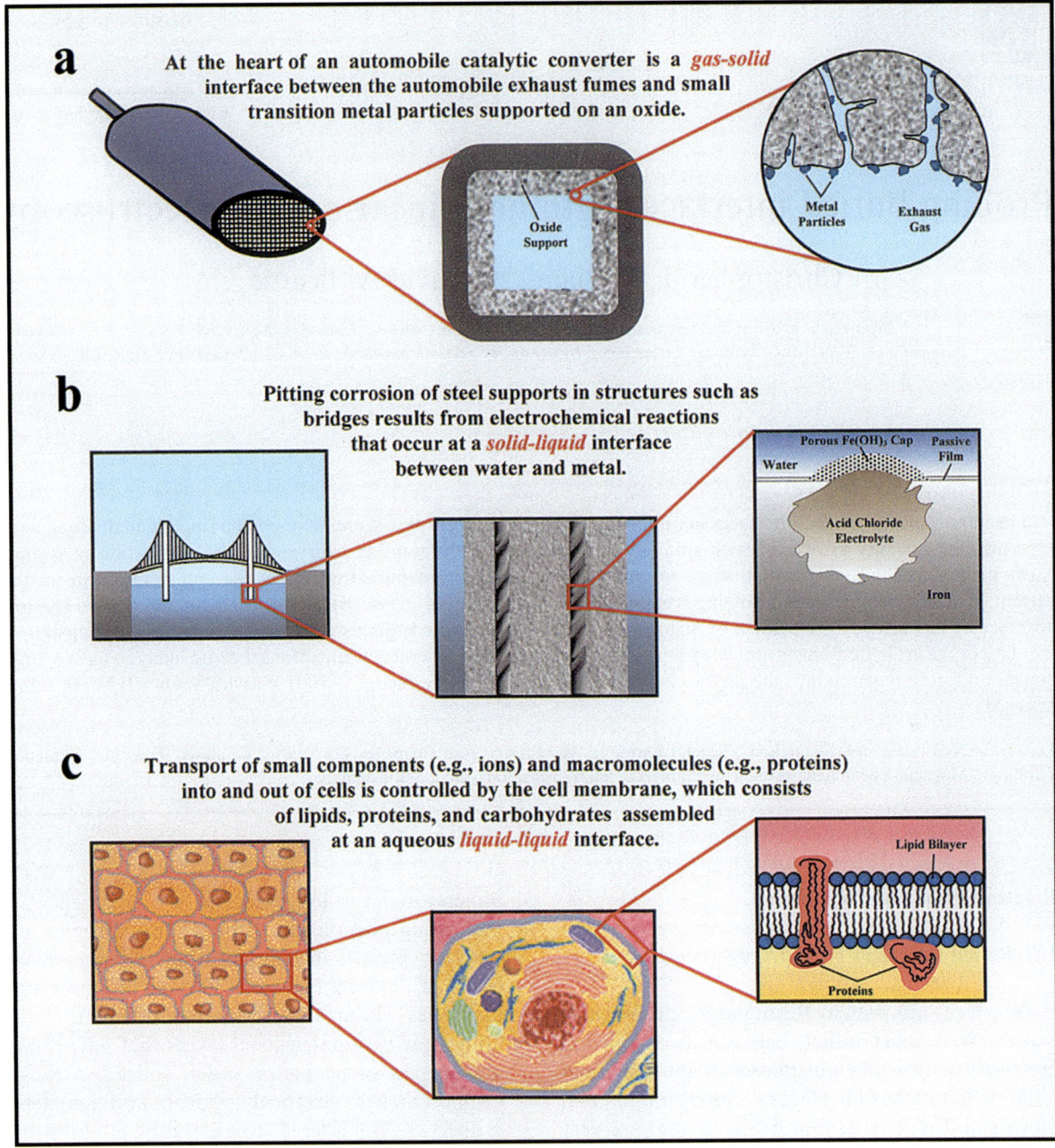

Fig. 1. Buried interfaces everywhere.

used to reduce harmful gas emissions such as carbon monoxide and nitrogen oxide (so-called NO_x) species. At the center of this working device is a type of *solid–gas interface*. In this case, the solid is a catalyst that consists of small transition-metal particles supported on a ceramic oxide surface (Fig. 1(a)). Pollutant gas molecules stick to the catalyst and are oxidized to form carbon dioxide and nitrogen, which are environmentally safe. Other applications in which solid–gas interfaces play an important role include heterogeneous catalytic reactions to produce industrial chemicals (e.g., ethylene oxide, ammonia) and thin film growth during microelectronics processing. *Solid–liquid interfaces* are ubiquitous throughout the environment and in industry, arising in applications ranging from cleaning and lubrication to electrochemical reactions and oil recovery. The corrosion of metals in the presence of water (Fig. 1(b)) is perhaps one of the most destructive (and easily observable!) environmental problems resulting from undesired solid–liquid electrochemical reactions. *Liquid–liquid interfaces* play a central role in the stabilization of emulsions and microemulsions, which are important in applications such as foods, detergents, and paints. They are also very important in biological processes such as transport across cell membranes (Fig. 1(c)). A cell membrane is made up of a lipid bilayer, proteins, and carbohydrates that are assembled at an aqueous liquid–liquid interface. The membrane regulates the transport of both small and large molecules into and out of the cell. Yet there are still more distinct types of interfaces that exist in our world. *Liquid–gas interfaces*, particularly the one between air and water, play a large role in environmental problems such as acid rain and water pollution. Finally, *solid–solid interfaces* also occur in many technological applications, including adhesion and lubrication between materials and in the thin film layers that make up electronic materials, including magnetic storage devices.

In many cases, the desired (or undesired!) action of a chemical, mechanical, or electronic system will depend centrally upon the structure and properties of an interface. A range of conditions will influence what occurs at the interface, including applied electric fields and mechanical agitation, thermodynamic properties such as temperature and pressure, and bulk properties such as composition, polarity, and viscosity. If scientists and engineers can gain a more thorough understanding of how interfacial properties are influenced by these factors (several of which are controllable), they may design new applications and devices that are tailored to specific applications. The study of interfaces is therefore of great interest for both fundamental and technological reasons.

Unfortunately, this endeavor is not often straightforward or easy. To a greater or lesser extent, all of the systems described above and in Fig. 1 can be classified as having "buried" interfaces. The description refers to the fact that the atoms and molecules that make up the interface are also present in (typically) much larger quantities in either (or both) of the bulk phases. To see why this situation is so problematic for surface scientists, let us first consider an interface that is *not* buried. Perhaps the most illustrative example is that of a solid metal surface in an ultrahigh vacuum (UHV) environment ($\approx 10^{-10}$ Torr). It is very straightforward to characterize such a solid–vacuum interface with modern surface analytical techniques, of which there are a large number [2,3]. This is because the UHV environment provides a collision-free path for electrons and photons to reach the surface. When these tiny packages of energy hit the surface they are perturbed in ways that can be detected by modern electrical devices. Thus, the full range of powerful electron- and photon- (i.e., light) based spectroscopic approaches can be readily used to determine surface electronic structure, surface morphology, and the identity of interfacial molecules or atoms. Furthermore, scanning probe microscopic methods such as atomic force microscopy (AFM) and scanning tunneling microscopy (STM) can be used to obtain information regarding the morphology of such a surface down to atomic dimensions (AFM and STM are discussed elsewhere in this volume).

Clearly, different technological applications and devices involve many and varied types of interfaces. In place of the vacuum, a gas, liquid, or solid phase of a given material (or more likely a mixture of materials) is present. The presence of this bulk phase poses significant problems for surface scientists who wish to directly examine these interfaces at a molecular level. One major problem is that the number of surface analytical techniques that are applicable under these conditions is very small. Electron-based techniques are unable to be used outside of a vacuum environment because electrons cannot penetrate the bulk phase. While photon-based techniques are often still applicable, they can also suffer from (often severe) bulk-phase interference effects that occur if the materials of interest are opaque or strongly absorbing at the wavelength of the probe light being used. Finally, probe microscopic methods such as AFM and STM are largely limited to examining solid surfaces in contact with gases or liquids. The second problem to be overcome is that the bulk media contain most (if not all) of the constituents that make up the interface. As a result, the interfacial spectroscopic signal is often mixed in with very similar bulk-phase signals that can be several orders of magnitude larger. This often makes it impossible to extract the desired interfacial spectroscopic information (cf. Fig. 2).

The most widely used spectroscopic approaches for studying buried interfaces in the last two decades have been photon-based techniques. Infrared and Raman spectroscopy have made their marks on our understanding of a wide array of interesting phenomena. Both of these spectroscopic techniques will continue to make a strong impact, especially given the improvements in instrumentation that continue to occur regularly. Nevertheless, such *linear optical* (cf. Fig. 2) approaches are always limited, to varying degrees, in their ability to effectively probe buried interfaces. As a result, laboratories around the world are continuously trying to develop new and more powerful surface analytical approaches for this purpose. One particularly powerful tactic has been to attempt to exploit the *non-linear* optical properties (cf. Fig. 3) of materials and interfaces. The two main non-linear optical techniques that have been developed are known as second harmonic

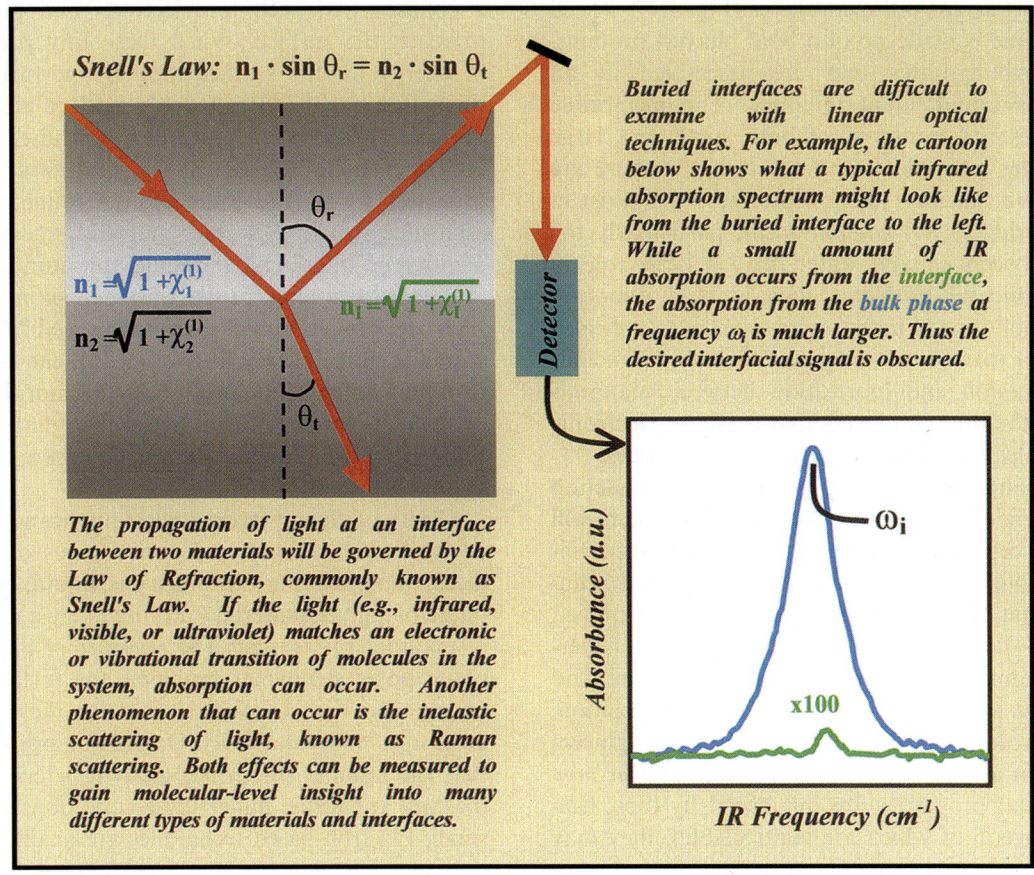

Fig. 2. Linear optics and the case of infrared spectroscopy.

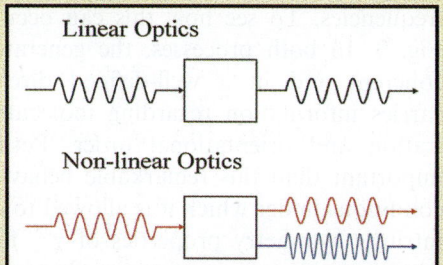

Fig. 3. Non-linear optics and frequency generation.

In an applied electric field, E, the polarization, P, of a system can be written as a power series in E (see box to the right), where $P^{(0)}$ is the static polarization and $\chi^{(i)}$ is the ith order susceptibility. The higher order terms in the expansion are responsible for the non-linear optical effects. It is the second order susceptibility, $\chi^{(2)}$, that is responsible for second harmonic generation (SHG). Sum frequency generation (SFG) will occur if the incident electric fields are at different frequencies. The diagram to the left shows a comparison of linear and non-linear optics in the case of normal transmittance and SHG, respectively, in a non-linear optical crystal. To see how SHG or SFG can arise, consider two incident electromagnetic fields at frequencies ω_1 and ω_2 ($E_1 = E_1(r)\cos\omega_1 t$, $E_2 = E_2(r)\cos\omega_2 t$). The first two terms in the expanded second-order polarization equation (see box below) represent an oscillating polarization at frequencies $2\omega_1$ and $2\omega_2$. The other two terms oscillate at the *sum* and *difference* of the two frequencies. These oscillating polarizations emit light with an intensity proportional to $(P^{(2)})^2$. It is this emitted light that is measured during non-linear optical spectroscopic experiments.

$$P = P^{(0)} + P^{(1)} + P^{(2)} + P^{(3)} + \ldots$$
$$= P^{(0)} + \varepsilon_0\left(\chi^{(1)} \cdot E + \chi^{(2)}{:}EE + \chi^{(3)}{\vdots}EEE + \ldots\right)$$

$$P^{(2)} = \varepsilon_0 \chi^{(2)} : E_1 E_1 + \varepsilon_0 \chi^{(2)} : E_2 E_2 + 2\varepsilon_0 \chi^{(2)} : E_1 E_2$$
$$= \varepsilon_0 \chi^{(2)} : E_1(r)E_1(r)(1 - \cos 2\omega_1 t) + \varepsilon_0 \chi^{(2)} : E_2(r)E_2(r)(1 - \cos 2\omega_2 t)$$
$$+ 2\varepsilon_0 \chi^{(2)} E_1(r)E_2(r)(\cos(\omega_1 + \omega_2)t + \cos(\omega_1 - \omega_2)t)$$

generation (SHG) and sum-frequency spectroscopy (SFS). Their key attribute is the ability to probe both electronic and molecular structure and dynamics at interfaces *while completely avoiding interference from bulk phases*. As a result, these techniques have steadily grown in popularity, especially over the past decade.

1.2. Linear versus non-linear optics

To understand why non-linear optical techniques are so well suited to the study of buried interfaces, it is first helpful to briefly review *linear* optics. Fig. 2 provides an illustrative example of linear optics at the interface between two materials, using infrared light as an example. When a group of molecules at an interface or in a bulk phase are placed in an electromagnetic field, a polarization (which is a dipole moment per unit volume) will be induced. This polarization can be represented mathematically as a power series, consisting of first, second and higher order terms (cf. Fig. 3). If the electric field is of the order of 10^4–10^5 V/m (as is the case, for example, with a continuous wave laser) the higher order terms will be negligible. The polarization will then be directly proportional to a material property known as the linear susceptibility, $\chi^{(1)}$. $\chi^{(1)}$ is in turn related directly to the number of molecules, N, times their polarizability, α, averaged over all molecular orientations in the material or at an interface. The polarizability is essentially a measure of how much the electrons and nuclei within a material will deform when placed in an electric field. The refractive index, n, which is typically used to describe linear optical properties, is a complex number that is related to $\chi^{(1)}$ (see Fig. 2). The real part of the refractive index is what is used to describe propagation of light through various media by the laws of reflection and refraction (i.e., Snell's law). The imaginary part of the refractive index determines the degree to which light will be absorbed by a given material. The theory of linear optics is well established and will not be discussed in any further detail here. The interested reader is directed towards two excellent monographs on the subject [4,5]. As mentioned above, the key obstacle when using linear optical spectroscopic approaches to examine buried interfaces is that it is often difficult (or impossible) to distinguish surface signals from those in the bulk. This problem is illustrated in Fig. 2 by using the example of an infrared absorption measurement.

Now, consider the case where the same group of molecules is exposed to the very large electric fields ($\approx 10^{10}$–10^{12} V/m) that can be generated by a pulsed laser. As shown in Fig. 3, in this case the induced polarization will have contributions from the higher order, or non-linear, terms. The first term that will contribute is called the second-order susceptibility, $\chi^{(2)}$, which is a third rank tensor having 27 elements that is related to the number of molecules times their hyperpolarizability, β, averaged over all molecular orientations in the material or at the interface (cf. Figs. 4 and 5). The hyperpolarizability determines the extent of second-order non-linear deformations of electrons and nuclei when materials are placed in a very strong electric field. While higher order terms can contribute if the electric fields are strong enough, we will limit ourselves in this article to discussion of second-order effects, since these are the most widely used for surface characterization. Several books are available that explore the theoretical basis of non-linear optics and trace the development of the field over the past four decades [6–8].

As described so far there does not appear to be much difference between linear and non-linear optics, other than the extra terms. However, the non-linearity in the polarization results in a surprising occurrence: the frequency of light can be altered! This phenomenon can manifest itself in a number of different ways. In one, the oscillating polarization emits light at twice the frequency of the incident electric field in a process known as SHG. In another, two electric fields having different frequencies mix together to generate light at the sum (SFG) or difference (DFG) of the two frequencies. To see how this can occur, refer to Fig. 3. In both processes, the generated light is coherent and in a well-defined direction, and carries information regarding molecular concentration and orientational order. Perhaps more important than this remarkable behavior are the conditions under which it is allowed to occur. The intrinsic symmetry properties of $\chi^{(2)}$ forbid non-linear processes from occurring in centrosymmetric materials (i.e., those that have an inversion center) such as bulk gases, liquids, and a large number of solids. Thus, while SHG and SFG occur from interfaces (where the bulk symmetry is broken), *no signals arise from the bulk phase*. [2] As a result, spectroscopic approaches based on the measurement of these effects are uniquely suited to probe buried interfaces.

1.3. Second harmonic generation and sum-frequency spectroscopy

SHG was the first non-linear optical technique to be used for the study of interfaces. The initial measurements of bulk non-linear effects from non-centrosymmetric solids were first made in the early 1960s. Around the same time, Bloembergen and co-workers [9,10] developed a theoretical framework to describe SHG and SFG from interfaces. Since the first studies of surfaces in the early 1970s, SHG has evolved into a powerful surface analytical tool for probing buried interfaces. SHG can

---

[2] This is indeed the case for the large number of materials for which the electric dipole approximation holds. However, non-linear optical signals can arise from the centrosymmetric materials if magnetic dipole and/or electric quadrupole contributions are significant [6–8].

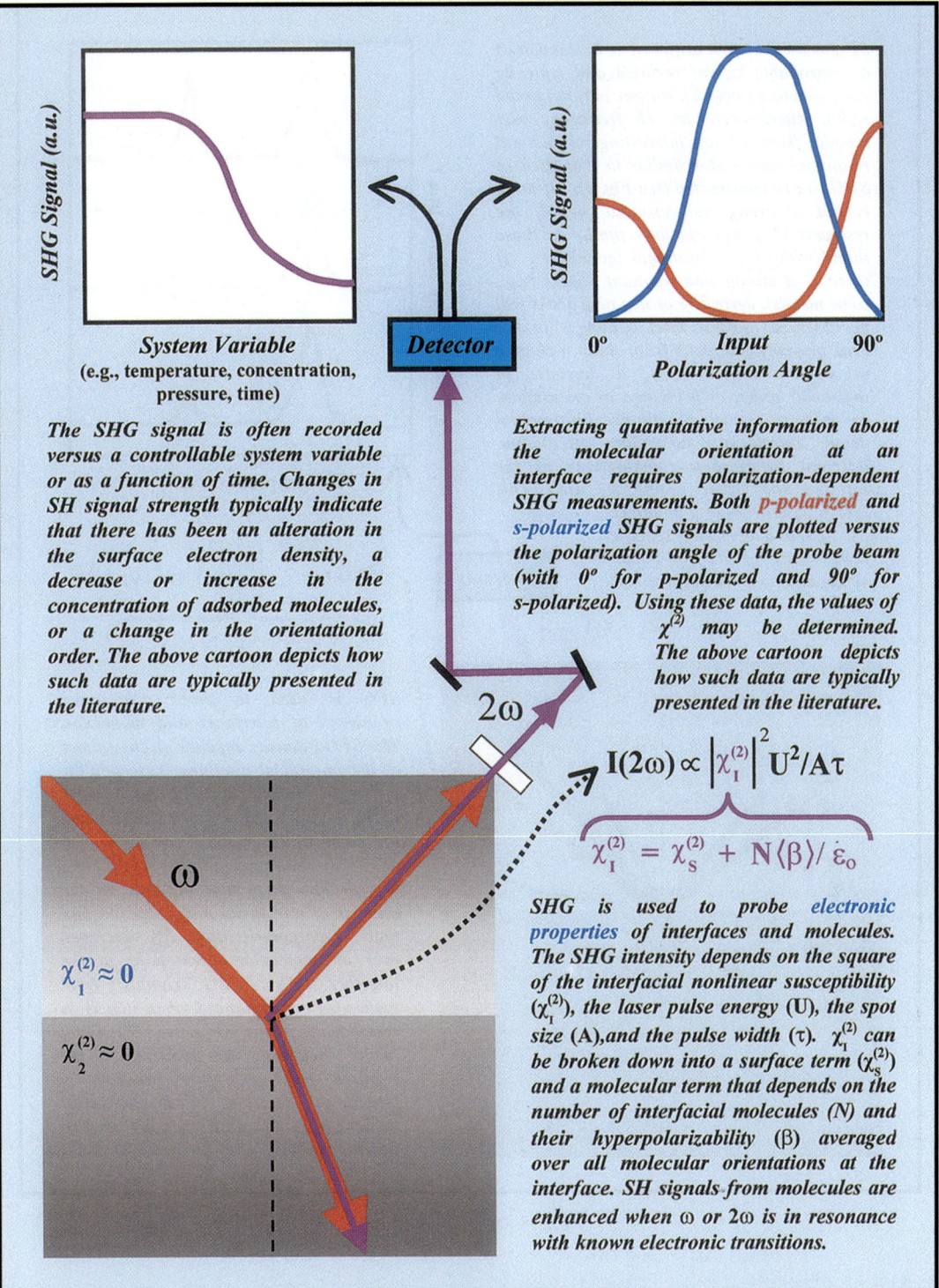

Fig. 4. Overview of SHG.

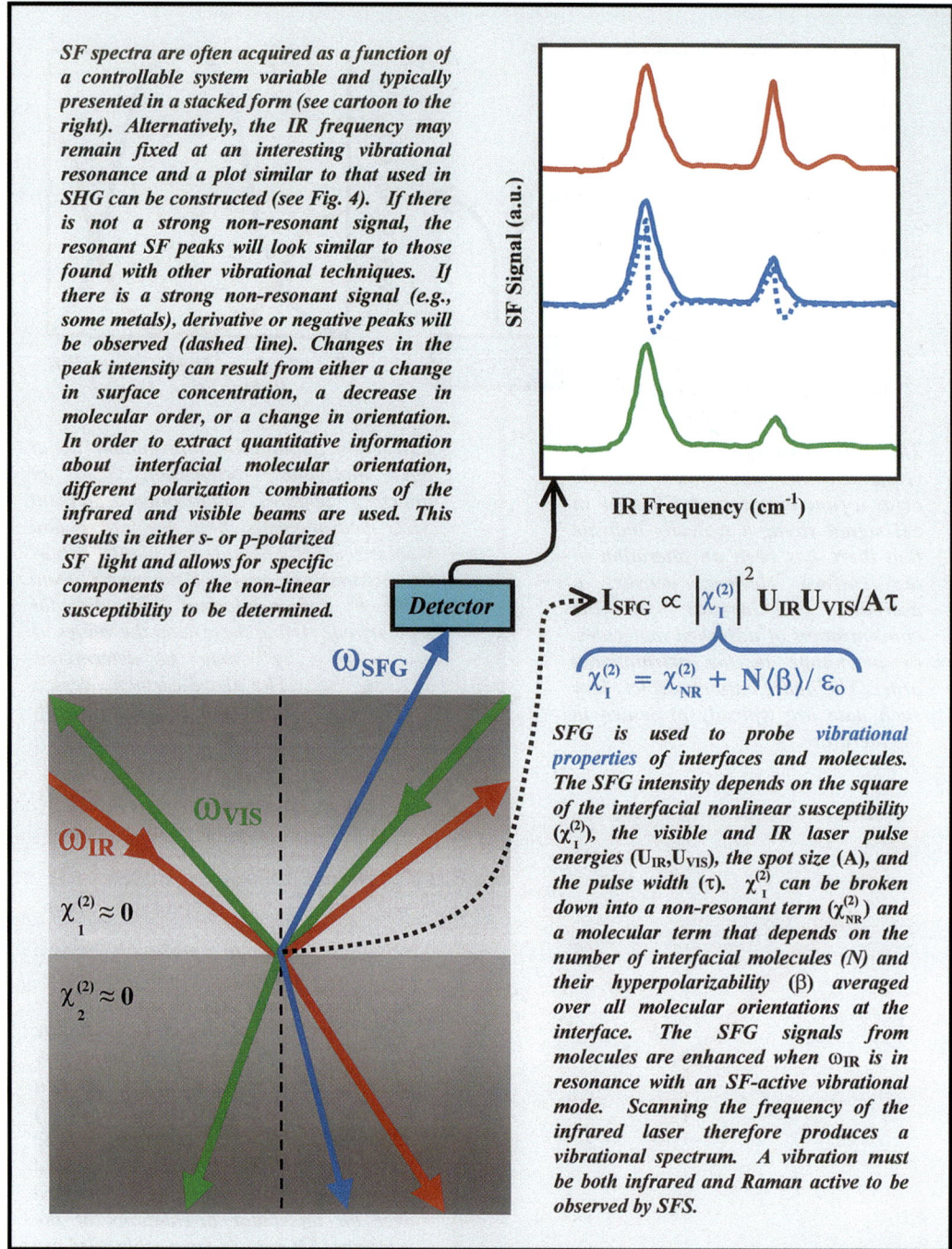

Fig. 5. Overview of SFS.

probe the electronic or magnetic properties of surfaces and molecules. The experimental setup is perhaps the most straightforward of all surface non-linear optical techniques. A pulsed laser is used as the light source, and the beam is polarized and filtered before being sent to the interface. The emitted SHG light is sent through another polarizer and separated from the fundamental light before being detected. Fig. 4 provides a schematic of how the SHG experiment is performed and the type of data that is typically obtained, along with some theoretical aspects. Several excellent review articles are available that deal with both theoretical and experimental aspects of SHG from interfaces [11–17].

It would take over a decade after the development of SHG as a surface probe for SFG to be used in a similar fashion. In 1986, Shen and co-workers obtained the first sum-frequency spectrum from Coumarin dye adsorbed on glass [18]. While experimentally more complicated than SHG, the attraction of SFG is its ability to probe vibrations of molecules at interfaces. Thus, there is considerably more chemical specificity and it is more straightforward to ascertain the presence (and orientation) of specific molecules. For SFG experiments, a pulsed fixed frequency laser (usually in the visible) and a pulsed tunable infrared laser are typically used. Both beams are polarized before reaching the interface and are often focused to some extent. The pulses must be in both spatial and temporal overlap at the interface in order for SFG to occur. This generated light, which will emerge at a different angle from either of the probe beams, is filtered and collected. The key to the technique is that the SFG signal will be resonantly enhanced when the IR laser frequency matches certain vibrational modes of interfacial molecules. Thus, scanning the infrared beam over the desired frequency range produces a vibrational spectrum, which is a series of peaks representing molecular vibrations. Molecular vibrational frequencies are often expressed in the energy unit of cm^{-1}, which is called a wave number. Fig. 5 provides a schematic of how the SFG experiment is performed and the type of data that is typically obtained, along with some theoretical aspects. SFS, as it is often referred to now, has fast become one of the most powerful surface analytical techniques available for probing buried interfaces. The interested reader is directed to several excellent reviews that have been published on this subject [19–24].

2. Current applications

In the following sections we present an overview of current work using non-linear optical spectroscopy to examine buried interfaces. Examples are culled from several research groups around the world, with the aim of providing the reader with a window into this fast growing field of research. The examples have been chosen to cover a wide scope of buried interfaces, ranging from chemical and biological to electrified and magnetic phenomena. However, we stress that no attempt has been made to make this a *comprehensive* review on the subject. Indeed, we have decided to largely leave out discussion of the important liquid–gas interface, although it is considered briefly in the section on interfacial dynamics. In particular, air–water interfaces play a central role in environmental processes (e.g., acid rain and water pollution), as well as in surfactant science and technology. They have been the subjects of many non-linear optical investigations over the years [11,14,15,20,22,24]. Similarly, within each interfacial science area there is much important work that has not been specifically mentioned or cited. The interested reader is directed to the many excellent review articles cited in the previous section for further reading. Furthermore, we limit our discussion largely to SHG and SFS, since these surface non-linear optical techniques are by far the most widely employed. The emphasis is placed on relaying the new physical and chemical insight that has been gained by these approaches rather than the specific experimental data that was obtained.

2.1. Solid–gas interfaces

Solid–gas interfaces are the most straightforward to examine with non-linear optical techniques. The experiment is usually performed using an external reflection geometry for the incoming probe light (i.e., it hits the interface from the gas side). Such an optical arrangement is also often used for the study of liquid–gas interfaces. Figs. 4 and 5 provide examples of what the external reflection geometry looks like. Most of the SHG and SFS studies of gas–solid interfaces have involved the study of adsorption (i.e., the sticking of molecules on a surface), desorption (i.e., when molecules pop off the surface), and reactions occurring on catalyst and semiconductor surfaces. In many cases, previously unavailable information has been obtained regarding the state of the surface under industrially relevant conditions.

In the area of heterogeneous catalysis, the group of Somorjai [23] has been using SFS to examine oxidation, hydrogenation, and dehydrogenation reactions on metal surfaces. One of the most common tactics in heterogeneous catalysis research is to study model catalytic surfaces (usually single-crystal metals) under UHV conditions. As mentioned above, the reason for this is that many surface analytical techniques may be used under these conditions to examine molecular adsorption/desorption and surface reactions. The information obtained from UHV studies is often used in combination with high-pressure kinetic measurements of reaction rates to propose reaction mechanisms [25]. However, the recent results from Somorjai's laboratory using SFS underscore the fact that it can be difficult (and even misleading) to extrapolate UHV behavior to elevated pressure environments.

One of the clearest examples of this has come from studies [23] of CO adsorption and oxidation on a platinum single crystal at ambient gas pressures. CO oxidation is an important reaction in the automobile catalytic converter (cf. Fig. 1(a)). A single-crystal metal surface is one that is well ordered on an atomic scale. For example, Fig. 6 shows temperature-dependent SF spectra that were obtained in situ from Pt(1 1 1) under CO-rich reaction conditions (100 Torr CO, 40 Torr O_2). At 590 K, an SF peak is observed at 2095 cm^{-1}, indicating that CO is adsorbed predominantly on single (i.e., terminal) platinum atom sites. This is one of the adsorbed CO species that has been observed previously in many UHV-based investigations. However, as the temperature was raised to 1100 K, the concentration of this species decreased and a broad SF feature centered at 2050 cm^{-1} was observed to appear, along with a tail extending as low as 1700 cm^{-1}. Somorjai and co-workers attribute these SF features to a combi-

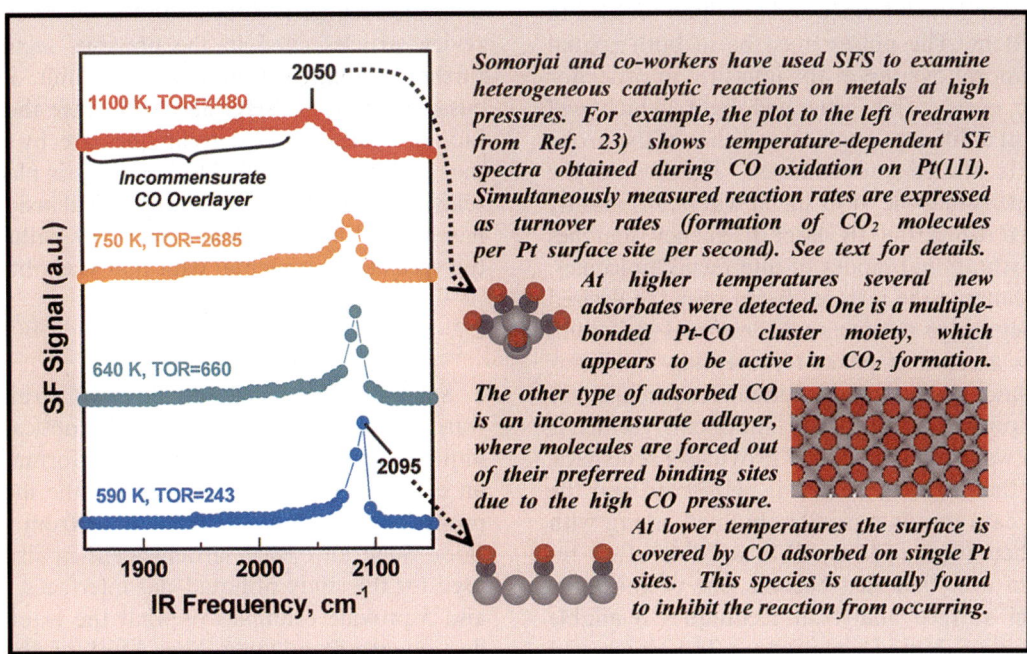

Fig. 6. SFS of high-pressure CO oxidation on Pt(1 1 1). Data reprinted with permission [23]—copyright 1999 American Chemical Society.

nation of multiple-bonded platinum–carbon monoxide clusters and an incommensurate CO adlayer, where the CO molecules are forced out of their preferred terminal configuration (see cartoons in Fig. 6). While there were some minor differences, similar results were also found when using an O_2-rich reaction mixture.

The most interesting characteristic about these new adsorbed Pt–CO cluster species is that they appear to play an active role in the CO oxidation reaction at elevated temperatures. Evidence for this came from examining the dependence of the reaction rate on the surface concentration of this species at 720 K. A linear dependence was found, which suggests that these "cluster"-type CO species may be the intermediate that is oxidized to form CO_2. In contrast, at lower temperatures the reaction rate had a negative dependence on the surface concentration of the terminal bound CO. This adsorbate is therefore not an important intermediate but rather appears to be a reaction inhibitor. Thus, the catalytically interesting surface species in this case could only be observed under the *actual* high-pressure reaction conditions. Several other unexpected phenomena have been revealed through SFS studies of catalysts surfaces under industrially important reaction conditions [23].

2.2. Solid–liquid interfaces

Non-linear optical studies of liquid–solid (and liquid–liquid) interfaces have grown steadily in number over the years. This is due in large part to the fact that SHG and SFG are often the only techniques that can effectively provide molecular-level information about such buried interfacial systems. Two different experimental tactics are typically employed. The first uses external reflection through the liquid phase, similar to the preferred method for linear optical studies of electrochemical systems. Any type of flat surface (i.e., oxides, semiconductors, metals) may be examined with this approach. While SHG experiments in this optical geometry are often straightforward, SFS measurements can be hindered if the tunable infrared beam is strongly absorbed by the liquid. This problem may be minimized through design of a liquid cell that allows the solid surface to be examined under a thin ($\approx$5–50 μm) layer of liquid. The other optical setup that is widely used is to direct the laser beams to the interface through the solid sample, typically above the critical angle for total internal reflection (TIR) [26]. This method has several advantages, including a spectroscopic signal enhancement that arises from increased electric field strengths, an avoidance of absorption of infrared light (in the case of SFS) by the bulk liquid, and none of the mass transfer limitations that can be imposed by a thin layer geometry. Fig. 7 provides a brief overview of TIR and its role in non-linear optical spectroscopy. The TIR approach has largely been used so far to examine adsorption on optically transparent substrates (e.g., SiO_2, Al_2O_3) [24]. However, it has been recently shown that metals can be examined in a similar fashion by depositing them as ultrathin films on such substrates [27]. Indeed, in this fashion the TIR approach should be readily extended to a wide range of new materials.

Electrochemists have long been at the forefront of attempting to use new techniques to examine metal–solution interfaces. Perhaps not surprisingly then, non-linear optical techniques have been used extensively to probe electrochemical interfaces with a view towards obtaining fundamental information regarding adsorption and formation of surface species during electrochemical processes. SHG is a very sensitive probe of the free-electron density at metal electrode surfaces [14]. It has been shown that the surface signal will increase upon adsorption of a molecule (e.g., CO) or atomic species (e.g., H) that increases this density. In contrast, the signal will decrease if this density is lowered upon adsorption, as occurs, for example, with the formation of surface hydroxides or oxides. The major drawback is that such indirect SHG measurements (where the metal surface generates the signal rather than the adsorbed species) lack *chemical* sensitivity. This deficiency can be alleviated to some extent if either the fundamental or SH frequency is in resonance with a known adsorbate electronic transition (cf. Fig. 4).

While the majority of non-linear optical studies of electrode surfaces over the past two decades have employed SHG, the use of SFS has been

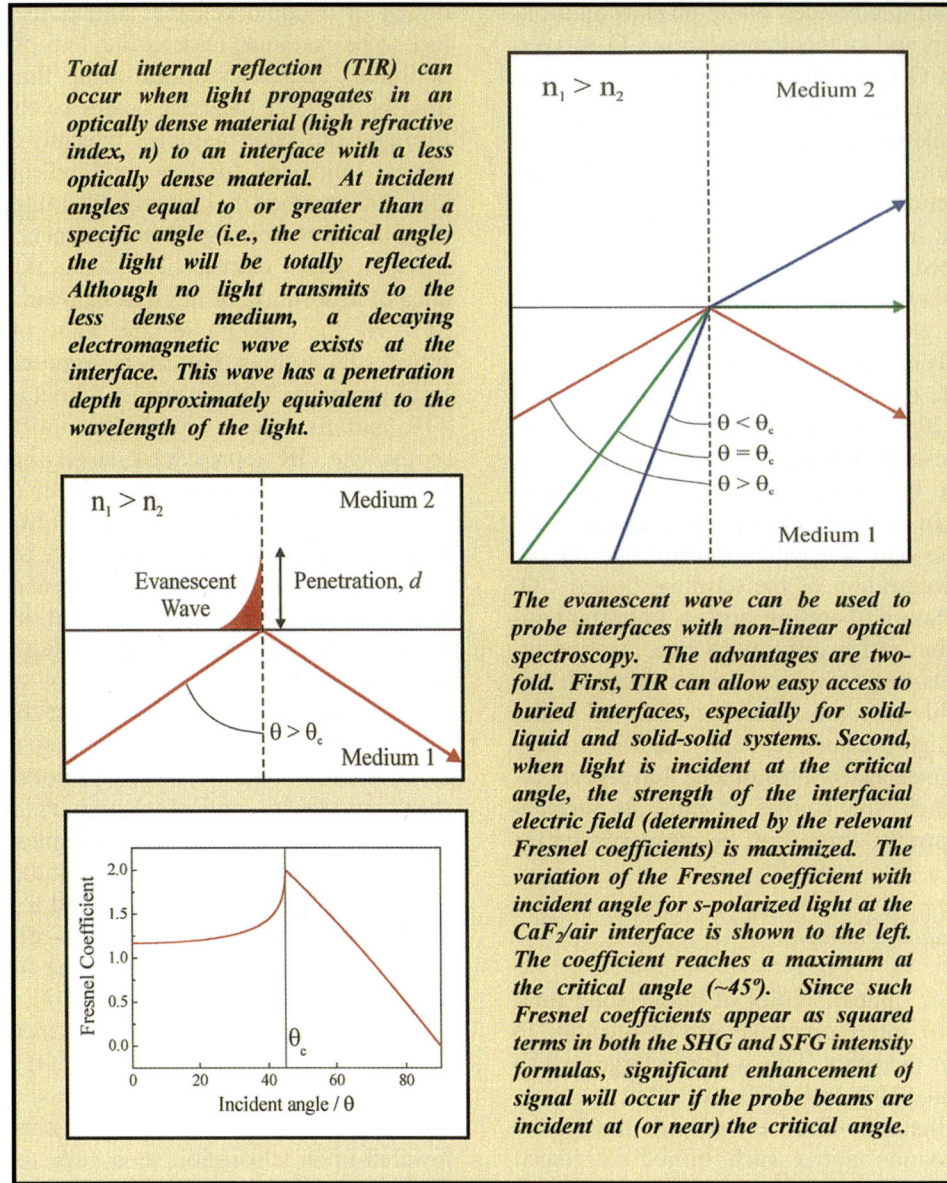

Fig. 7. Advantages of TIR.

steadily increasing in recent years. The ability to identify distinct chemical species on the electrode surface through characteristic molecular vibrations offers a clear advantage over SHG. Tadjeddine and Peremans have been at the forefront of applying SFS to the study of electrochemical systems [21]. They have focused much attention on platinum electrode surfaces, owing to the general importance of this metal in both electrochemical and heterogeneous catalytic processes. One area of interest has been the hydrogen evolution reaction (HER), which is a process by which water is broken down electrochemically to form H_2 molecules in aqueous solutions. The HER is of central importance in the operation of H_2/O_2 fuel cells, in which platinum is one of the major active components. As a result, elucidation of the mechanism of this reaction on platinum is an important goal of electrochemical research around the world.

For example, Tadjeddine and co-worker have studied several different platinum electrode surfaces with SFS in aqueous sulfuric acid solutions under a range of electrode potentials [28,29]. The goal was to examine the surface species that formed at electrode potentials both positive (in the "underpotential" region) and negative (in the "overpotential" region) of the HER potential. Fig. 8 shows one set of potential-dependent SF spectra that was obtained for a Pt(1 1 1) electrode in 0.1 M aqueous sulfuric acid. Two SF resonances were observed in the underpotential region, one at 2020 cm^{-1} and the other at 1945 cm^{-1}. Both peaks are consistent with the Pt–H stretch of adsorbed atomic hydrogen on single platinum atom sites. Studies of other surfaces, including Pt(1 1 0), Pt(1 0 0), and polycrystalline Pt, also revealed similar SF resonances, although the frequencies were slightly different. This indicates that the adsorption of hydrogen is structure sensitive, consistent with previous electrochemical studies of these systems. This was the first time that two distinct Pt–hydrogen vibrations had been observed under these conditions. The interpretation of these two SF resonances is that they arise from the interaction of adsorbed hydrogen with water. It has been shown that the interfacial water concentration increases in the underpotential region at the expense of other solution-phase species. This water will then form hydrogen bonds with adsorbed hydrogen atoms, resulting in shifts of the

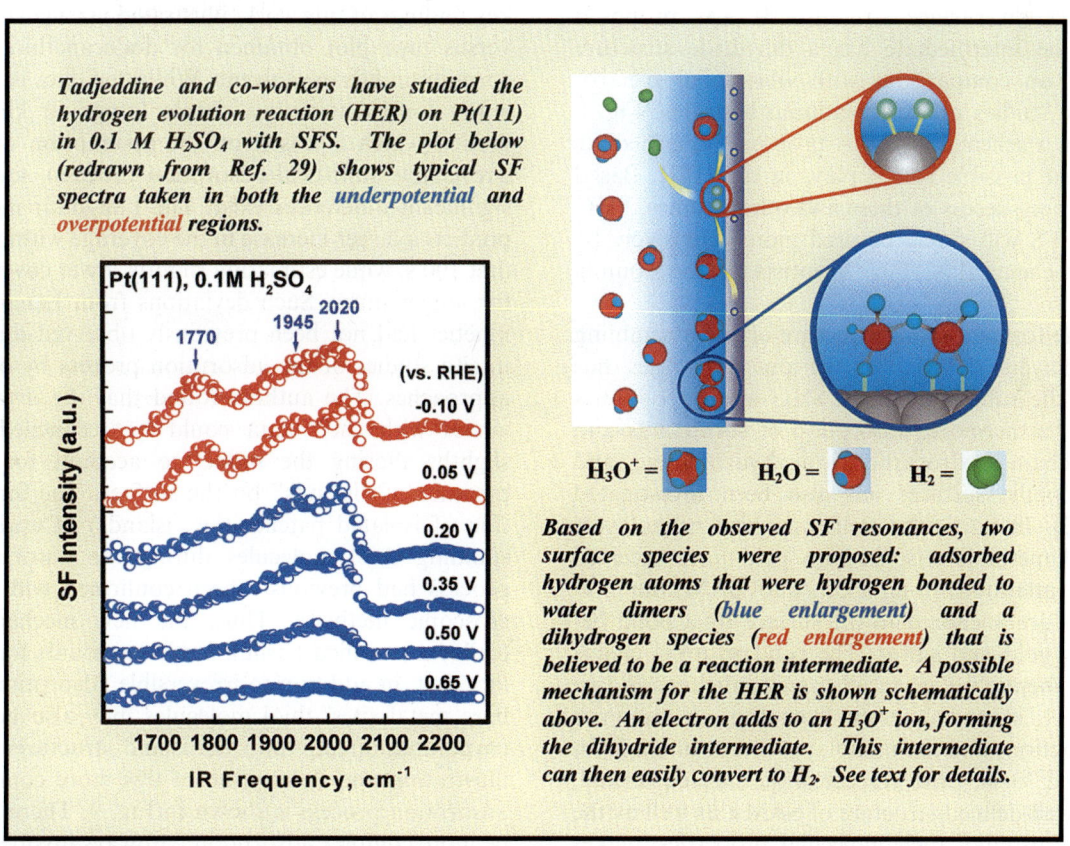

Fig. 8. SFS of Pt(1 1 1) during the H_2 evolution reaction. Data reprinted with permission [29]—copyright 1996 Elsevier.

Pt–H vibrational frequency to lower values. Several different types of interfacial water configurations have been proposed in the literature. The only configuration that could induce the observed double SF resonance is a dimer that interacts (unequally) with three different hydrogens. An illustrative cartoon of this interaction is provided in Fig. 8.

As the potential is lowered into the HER regime, an SF peak appears at 1770 cm^{-1} (Fig. 8). This feature has never been observed before during hydrogen adsorption on Pt surfaces and only appeared during hydrogen evolution. Thus, Tadjeddine and co-workers tentatively assigned the feature to an active intermediate in the HER. Further support for this claim was provided by SFS studies of other platinum electrodes, which revealed that this 1770 cm^{-1} resonance is identical on all Pt surfaces. This appears to be consistent with the well-known independence of the HER kinetics on surface structure. It was proposed that the intermediate has a dihydride structure based on comparisons with vibrational spectroscopic studies of bulk platinum hydrides. Fig. 8 shows a schematic of this proposed intermediate and the possible role it plays in the HER. Based upon the success of these and other studies, SHG and SFS will surely be used more extensively by electrochemical surface scientists in the coming years.

The formation and structure of self-assembling monolayers (SAMs) on various substrates has been the subject of many SHG and SFS studies [19]. Furthermore, adsorption of surfactants and long-chain hydrocarbons on hydrophobic and hydrophilic surfaces has also been investigated [19,22]. In particular, self-assembled monolayers of alkanethiols adsorbed on gold have received much attention. In an alkanethiol SAM, the molecules are anchored to the surface by a bond between gold and sulfur. The pendant alkyl chains align themselves in an all-*trans* configuration (see Fig. 11, second diagram from top) to minimize interactions and the monolayer tilts at an angle of around 30–40° with respect to the surface normal. The well-defined structure of SAMs, as well as the ability to tailor their molecular properties, makes these surfaces useful for fundamental studies of interfacial phenomena such as adsorption, wetting, and tribology. Current and future technological applications of alkanethiol SAMs include their use in biological sensors, in microelectronics applications and "molecular" computing, in synthesizing biomimetic substrates, and as corrosion inhibitors. In order to be able to design new SAM systems for these applications, a detailed understanding of their adsorption and formation kinetics must be achieved. Unfortunately, there are many conflicting findings in the literature regarding such processes.

As part of an effort to resolve some of these issues, Buck and co-workers recently performed a comprehensive in situ SHG investigation of alkanethiol adsorption onto gold [30]. Using the time resolution afforded by the technique, the formation of the monolayer was studied in real time as a function of thiol solution concentration, solvent, and alkane chain length. Fig. 9 reveals one of the key findings of this work, illustrated via a coverage versus time plot obtained for dodecanethiol adsorption in hexane solvent. While the data is generally consistent with a simple Langmuir kinetic model (which is based on the assumption of an irreversible statistical adsorption process), several significant differences exist. The Langmuir model predicts a larger increase in the coverage within the first 100 s, while estimating slightly lower coverage for longer times. Such deviations from Langmuir kinetics had not been previously observed during in situ studies of the adsorption process by other approaches. The authors found that the discrepancies with their data could be reconciled by slightly altering the model to account for the presence of "islands" on the surface. The formation of isolated patches (i.e., islands) of uprightstanding thiol molecules during the adsorption process had previously been confirmed with microscopic methods. Thus, the new mechanism (called a modified Kisliuk model) accounts for the fact that, in addition to irreversible adsorption on bare metal sites, thiol molecules may also be incorporated directly into the island structures. An illustration and description of this more complex adsorption process is shown in Fig. 9. There may be more complex adsorption pathways involved in the early stages of SAM formation that have not

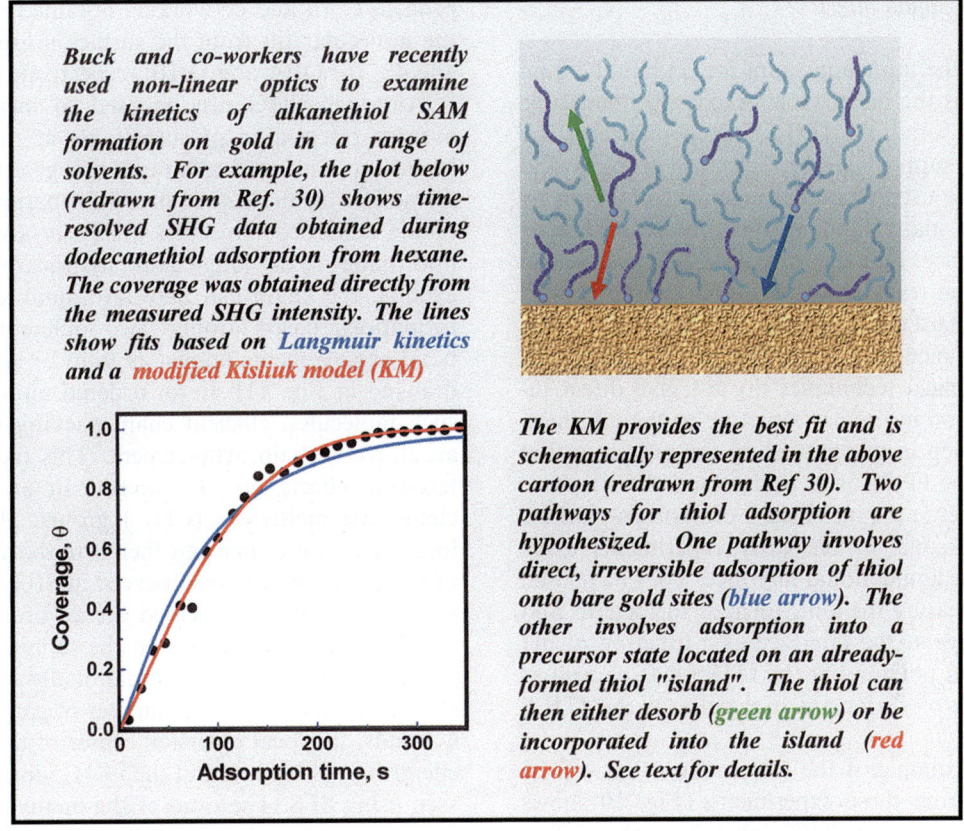

Fig. 9. SHG of SAM formation kinetics. Data reprinted with permission [30]—copyright 1999 American Chemical Society.

been revealed yet. However, SHG has clearly provided new insight into this process and will continue, along with SFS [31], to improve our understanding of these important systems.

We end this section by discussing pure organic liquid/solid interfaces, which are among the most challenging to examine experimentally. Adsorption of organic molecules on solids from aqueous solutions had been studied with considerable success prior to the advent of SHG and SFS. However, it was the development of these approaches that finally allowed direct in situ spectroscopic measurements of adsorption on solids from *pure* organic liquids. Indeed, Hatch et al. [32] obtained the first vibrational spectrum of such an interface for acetonitrile/ZrO_2 in 1992 using SFS. Shen and co-workers have been at the forefront of this area and the reader is directed to their recent comprehensive review of the application of SFS to liquid interfaces [24]. Both SHG and especially SFS experiments of neat liquid/solid interfaces are often run in the TIR mode (cf. Fig. 7), so as to avoid the necessity of passing the probe beams through the liquid phase. So far, studies have been almost completely limited to examination of adsorption and structure of interfacial molecules. However important opportunities exist for studying *reactions* at liquid organic/solid interfaces, for which there is currently no direct in situ spectroscopic information. An important area of research along these lines is the study of heterogeneous catalytic reactions that occur in organic liquids and solvents, which are of interest in the fine chemicals and pharmaceuticals industry. Another area of interest is to examine the surface reactions that occur in alternative energy source devices such as lithium ion batteries and direct methanol fuel cells.

2.3. Liquid–liquid interfaces

One of the most important model liquid–liquid interfaces is the interface between two immiscible electrolyte solutions (ITIES). Such interfaces, by definition, support ion and electron transfer and can thus be used as model systems for the investigation of many fundamental processes in separation science, biology, and electrochemistry. Ion and electron transfer at the ITIES can be probed using standard electrochemical techniques, such as AC impedance and voltammetry [33]. However, electrochemical techniques do not give direct information on molecular structure at the interface. The first step in understanding the processes that occur at the ITIES is to understand the structure of molecules at the interface. Corn and co-workers performed some of the early experiments using SHG at the liquid–liquid interface. They have used SHG to measure the ionic form of surfactants and the coverage at the interface as a function of pH and applied potential at the ITIES [34], as well as orientation of surfactant molecules at the ITIES [35].

As an example of the information that can be obtained from these experiments, Fig. 10 shows the variation of the SHG signal from the surfactant 2-(n-octadecylamino)naphthaline-6-sulfonate (ONS), as a function of applied potential, at the water/1,2-dichloroethane interface. The intensity of the surface SHG can be related to the coverage of ONS at the interface. The authors were able to fit the SHG data to a Frumkin adsorption isotherm from which a number of facts were gleaned. First, the surfactant is present at the interface in its anionic form, and second, the ONS head group is partially embedded in the aqueous phase. The authors also performed another study in which they determined the orientation of two different surfactants, 4-n-octyloxybenzoic acid (OBA) and octyl-4-hydroxybenzoate (OHB), at the water/1,2-dichloroethane interface. To determine orientation, measurements of SHG intensity need to be taken as a function of incoming light polarization angle. The variation of the s- or p-polarized SHG light as a function of the incident pump polarization angle can be analyzed to yield the orientation of molecules at the interface [36]. With this approach, Corn and co-workers obtained values for the molecular tilt from the surface normal of 34° and 40° for OBA and OHB, respectively (Fig. 10).

Although SHG can be used to measure the average orientation of surfactants at the liquid–liquid interface, the technique can give us no information on the degree of conformational order of the adsorbed molecules; SFS *can* give us this information. In long-chain hydrocarbon surfactants, the chain can take any number of conformations based around two different C–C–C bond arrangements known as *trans* or *gauche* (see drawing in Fig. 11). In an ordered monolayer of such molecules, efficient chain packing results in an all-*trans* chain arrangement. This fact has interesting effects on SF spectra. In an all-*trans* chain, the methylene (CH_2–) groups all have a local inversion center and therefore the vibrations of these groups are not observed in SFS. The only vibrations that are observed for an ordered surfactant layer are those due to the terminal methyl group (–CH_3). In a conformationally disordered chain, i.e., one that has a number of gauche C–C–C bonds, the local inversion center of the methylene groups is broken and the –CH_2 vibrations are seen using SFS. The ratio of the methylene peaks to the methyl peaks in a SF spectrum is thus an indicator of conformational order.

The application of SFS to the liquid–liquid interface is not as straightforward as the application of SHG. Absorption of the probe infrared laser by the bulk liquids can seriously diminish the intensity of the SF generated light. One approach taken to circumvent this problem by Richmond and co-workers is to use the water–CCl_4 interface and direct the incident lasers through the CCl_4 liquid. Fig. 11 shows a schematic diagram of the arrangement of the incident lasers. By performing the experiment in this way, the C–H stretching region of the adsorbed molecules (and the O–H stretching region of water) can be probed with SFS. Richmond and co-workers have used this method to investigate the structure of surfactants at the water–CCl_4 interface [37]. The surfactant sodium dodecylsulfate (SDS) was observed to undergo an increase in conformational order as the concentration of SDS in the bulk water solution was increased. This behavior is revealed quite

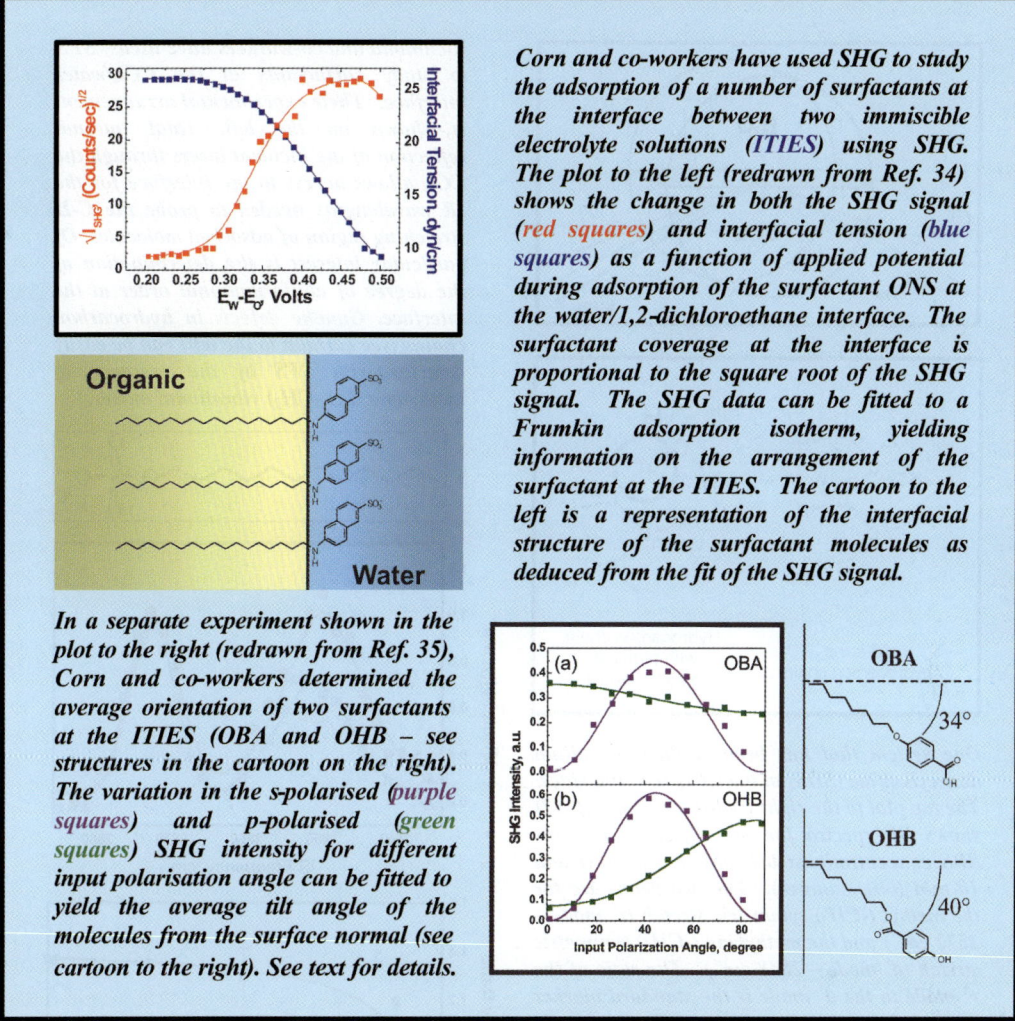

Fig. 10. SHG of surfactants adsorbed at the ITIES. Data reprinted with permission [34,35]—copyright 1993 American Chemical Society, copyright 1993 Elsevier, respectively.

clearly in the plots shown in Fig. 11. The intensity of the d^+ vibrational mode (which arises from a symmetric stretch of the methylene group) of SDS is seen to decrease relative to the r^+ mode (which arises from a symmetric stretch of the methyl group) with increasing concentration of the surfactant in the bulk solution.

2.4. Solid–solid interfaces

Of all the buried interfaces that we have highlighted in this article the solid–solid interface at first glance seems the most impenetrable (and perhaps the least interesting) from the point of view of determining molecular structure. Happily, both of these preconceptions are wrong. SHG and SFS have been used to probe two types of solid–solid interfaces where structure/property relationships are thought to have large effects on the performance of the systems in question. One area of particular interest is the determination of the morphology/magnetic property relationships in magneto-optical recording materials. Most people will be familiar with magneto-optical recording

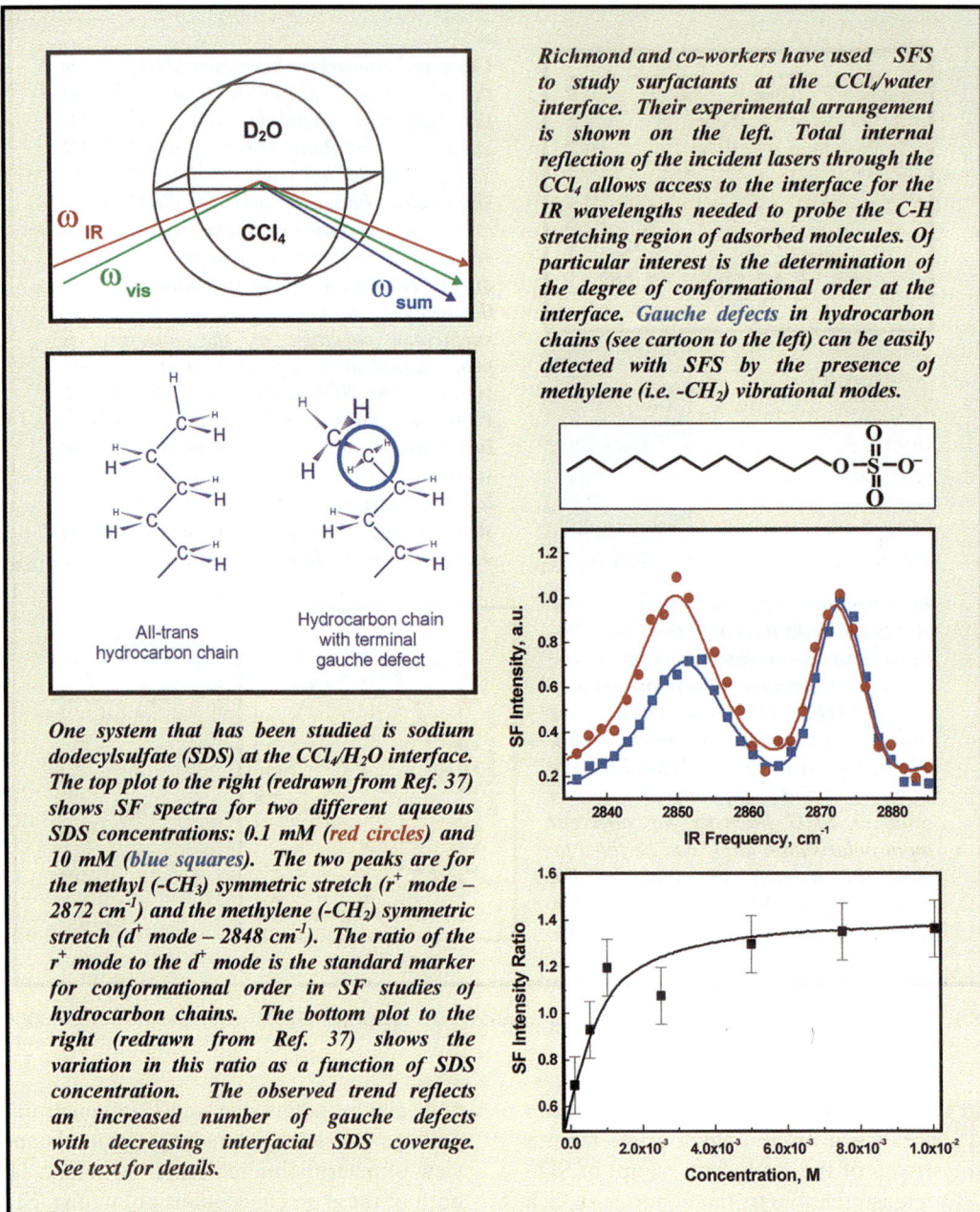

Fig. 11. SFS of surfactants at the liquid–liquid interface. Data reprinted with permission [37]—copyright 1996 American Institute of Physics.

media from the discs used in MiniDisc players. Magneto-optical recording media rely on the properties of light reflected from a magnetic surface; the surface actually rotates the plane of polarization of the incident light. This phenomenon is called the optical Kerr effect. The degree to which the incident light polarization angle is changed upon reflection is related to the magnetization of the surface. By manipulating the direction of the magnetization of the surface, one is able to rotate the polarization angle of the light in different directions. Detection of one or other of the directions from a magnetized region of a surface can be interpreted in a digital device as a 1 or a 0. Multilayer films of cobalt or a cobalt/nickel alloy with layers of platinum are commonly used as magneto-optical recording media. The cobalt layer is normally sandwiched between a thin platinum overlayer and a thicker platinum underlayer. The cobalt layer is the magnetic material and therefore the interface in question is most assuredly buried. Determining the relationship between the interfacial magnetic properties and the interfacial structure in these thin films is important if effective magneto-optical media are to be produced.

The magnetic and morphological properties of the magnetic multilayer films can be probed using techniques based on X-rays. However, these approaches rely upon synchrotron (high-energy X-ray) sources to generate the probe radiation. Unfortunately, such sources are only available at large multiuser facilities. A lab-based technique, rather than one that relies on a central research facility, would be more convenient to study these interfaces. Rasing and co-workers have used magnetization induced SHG (MSHG) to investigate the relationship between interfacial structure and the magnetic properties of the thin film [38,39]. Although the magnetic interfaces are buried, SHG can be used to study them since the light can penetrate through the multilayer (which consists of very thin individual layers, see Fig. 12). A SH signal is generated at every interface of the system (air/Pt, Pt/CoNi, CoNi/Pt) and all the signals are detected during the experiment. While multiple SHG signals complicate the interpretation of the information gained in the experiments, the second-order susceptibility tensor elements ($\chi^{(2)}_{ijk}$) can be extracted by measuring the SHG response as a function of the intermediate CoNi layer thickness. MSHG relies on the fact that certain components of $\chi^{(2)}$ are proportional to the surface magnetization of a material at an interface. The magnitude of these elements is then proportional to the normal SHG response of the interface

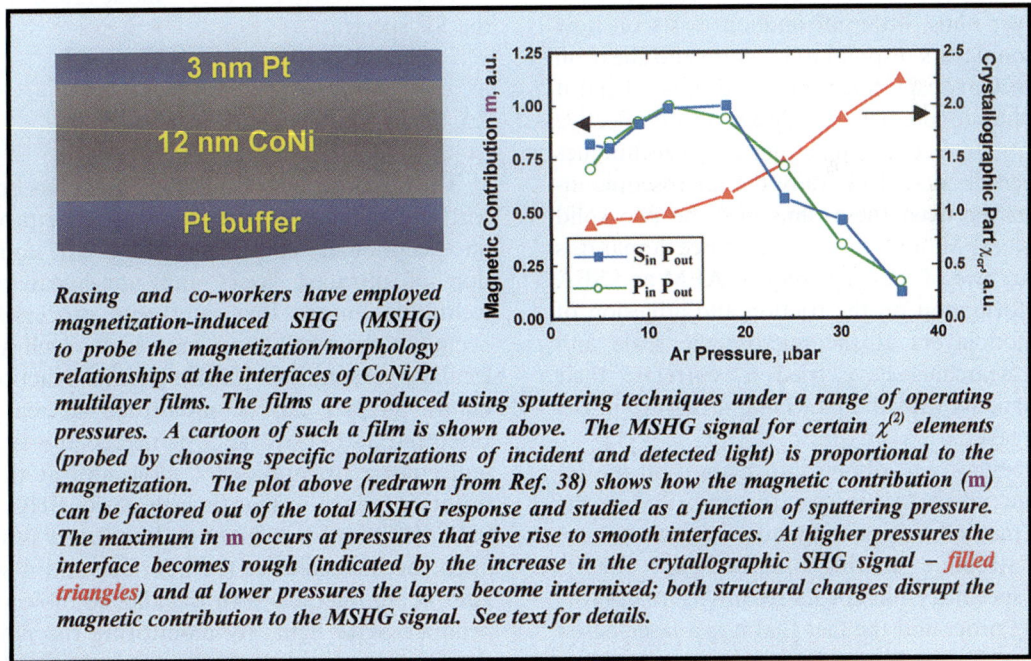

Rasing and co-workers have employed magnetization-induced SHG (MSHG) to probe the magnetization/morphology relationships at the interfaces of CoNi/Pt multilayer films. The films are produced using sputtering techniques under a range of operating pressures. A cartoon of such a film is shown above. The MSHG signal for certain $\chi^{(2)}$ elements (probed by choosing specific polarizations of incident and detected light) is proportional to the magnetization. The plot above (redrawn from Ref. 38) shows how the magnetic contribution (m) can be factored out of the total MSHG response and studied as a function of sputtering pressure. The maximum in m occurs at pressures that give rise to smooth interfaces. At higher pressures the interface becomes rough (indicated by the increase in the crystallographic SHG signal – *filled triangles*) and at lower pressures the layers become intermixed; both structural changes disrupt the magnetic contribution to the MSHG signal. See text for details.

Fig. 12. MSHG of CoNi/Pt multilayer films. Data reprinted with permission [38]—copyright 1998 American Institute of Physics.

and to the magnetization. It is possible to factor out the normal SHG response and to ascertain the magnetic contribution to the SHG light. In their work on CoNi/Pt multilayer films, Rasing and co-workers were able to correlate the interfacial roughness of their films (varied by altering the preparation conditions) with changes in the magnetization of the films. Fig. 12 shows the magnetization changes seen in such CoNi/Pt films. MSHG can also be used to investigate magnetic materials that have uses in other areas of magnetic recording, including giant magneto-resistance, which is a property of the thin films that are used for recording heads in devices.

A more challenging interface to probe experimentally is the lubricated contact between two solid surfaces. The behavior of thin organic films trapped at the solid–solid interface is of great interest to researchers in the field of tribology, which involves the study of friction, lubrication and wear. Organic monolayers are frequently employed to modify the interactions between solid surfaces, either as boundary lubricants or to prevent adhesion of contacting materials. Whereas there is a large body of empirical knowledge concerning the choice and performance of such monolayer films, little information exists on how these monolayers respond to pressure and shear on the molecular scale. Such information is crucial if one wishes to predict and design more effective lubricants. In recent years a number of techniques have been developed to study the microscopic interactions between these films and the two solid surfaces. AFM and the surface force apparatus (SFA) are two of these techniques. AFM and SFA yield information on the friction and adhesion of these monolayers at the microscopic scale and many researchers have tried to correlate their measurements with the molecular structure in the organic layer.

The best way to obtain information on molecular structure is to use spectroscopy. SFS is the perfect method for gaining information on molecular structure at this interface, not for its surface specificity but for its sensitivity to conformational order and the fact that it is a laser-based technique. To study a solid–solid contact with a spectroscopic technique, at least one of the solids in question needs to be transparent to the probe radiation. A technique that uses lasers allows for the use of optical components as the surfaces in contact. Du et al. [40] were the first to apply SFS to the solid–solid interface, however, they were unable to record a vibrational spectrum from their sample. Bain and co-workers [41,42] have adopted a more successful approach that takes advantage of electric field enhancements when the two input lasers are incident at the critical angle for the interface. In their experiments they used a prism and a lens to make the contact interface and directed the input beams through the prism onto the solid–solid contact. The monolayer material used in their experiments was zinc arachidate, a fatty acid salt. The vibrational spectroscopy of these monolayers is well known and good quality monolayers can be produced using straightforward deposition techniques. Fig. 13 shows a schematic diagram of the experimental setup used by Bain and co-workers and one of the SF spectra recorded at the interface. Under the pressures that were applied to the interface (up to 60 MPa), the monolayer retained an ordered structure throughout. This can be seen by the lack of an increase in the d^+ vibrational mode (2850 cm^{-1}) relative to the r^+ mode (2876 cm^{-1}) in the SF spectra.

2.5. Interfacial dynamics

The development of ultrafast lasers has brought with it the ability to monitor molecular processes on exceptionally short time scales. The combination of ultrafast lasers and non-linear optical techniques means that interfacial processes that occur on the picosecond time scale, such as molecular reorientation [15,43–45] and reaction dynamics [46,47] can be monitored in real time. Eisenthal and co-workers pioneered the study of molecular reorientation of molecules at the air–water interface using time-resolved SHG (TR-SHG) [43]. In this technique a laser beam is used to align the molecules at the interface and a second laser beam interacts with the aligned molecules to produce SHG light. By monitoring the polarization dependence of the SHG as a function of delay time between the pump and SHG pulses, one is

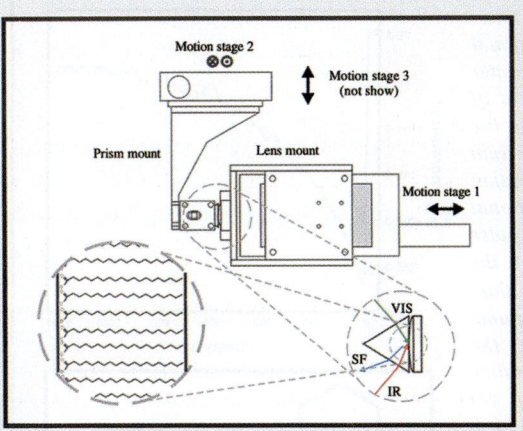

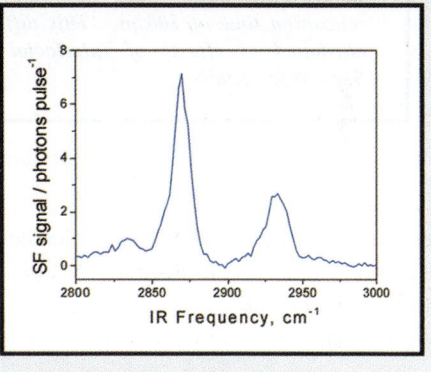

Using SFS, Bain and co-workers have obtained the first vibrational spectrum of an organic monolayer at the solid–solid interface. The schematic diagram on the left illustrates their experimental setup for studying this interface. They used a prism and a lens as the two solid surfaces and computer controlled motion stages (submicron resolution) to move the prism and lens into contact in an accurate and reproducible fashion. The arrangement of the motion stages allows for the study of organic monolayers under both shear and varying pressure. The monolayer chosen for the study was Zn arachidate, a fatty acid salt. The first inset shows the arrangement of the incident and emitted light. The second inset shows a cartoon of the Zn arachidate molecules trapped between the two surfaces.

The plot to the right (redrawn from Ref. 42) shows a SF spectrum of Zn arachidate in contact between a CaF_2 prism and a MgF_2 lens, at a pressure of approximately 60 MPa. The pressure in the contact region can be obtained from measurements of the contact diameter (observed through the lens by viewing Newton's rings from back illumination). The spectrum shows the C–H stretching region with the main peaks being the $-CH_3$ symmetric stretch (r^+ mode - 2876 cm^{-1}) and the Fermi resonance of the $-CH_3$ symmetric stretch with an overtone of a degenerate bending mode (r^+_{FR} - 2936 cm^{-1}). The ratio of the $-CH_2$ symmetric stretching mode (d^+ mode - 2840 cm^{-1}) to the r^+ mode is very low, indicating that the monolayer is well ordered in the contact region, i.e. the majority of the hydrocarbon chains are all-trans (see second inset above). See text for details.

Fig. 13. SFS of organic monolayers at the solid–solid interface. Data reprinted with permission [42]—copyright 2000 Elsevier.

able to determine the time it takes for the molecules at the interface to return to their original orientation. The important result from this kind of work arises from comparing relaxation times in the bulk liquid to those obtained for molecules at the interface. Such a comparison can yield important information on solute–solvent and solvent–solvent interactions.

Eisenthal and co-workers have studied the orientational relaxation of a number of dye molecules at the air–water interface. Fig. 14 shows their results for the dye molecule Coumarin 314 [44]. Using TR-SHG the authors were able to determine a relaxation time of 350 ps compared to 100 ps obtained for the same molecule in bulk water.

The slower reorientation time of the Coumarin molecules at the interface relative to the bulk can be explained by considering the effects of interfacial friction. At the water surface, the water molecules form a more ordered hydrogen bonding structure than in the bulk. Reorientation of Coumarin molecules at the interface relies on the molecule being able to move freely through the water molecules, a process that is harder at the interface than in the bulk. Other molecules have been studied in this way, including Eosin B by Girault and co-workers [45], for which it was found that the relaxation time at the interface was much faster than in the bulk. Since Eosin B forms strong hydrogen bonds with water, the fewer water

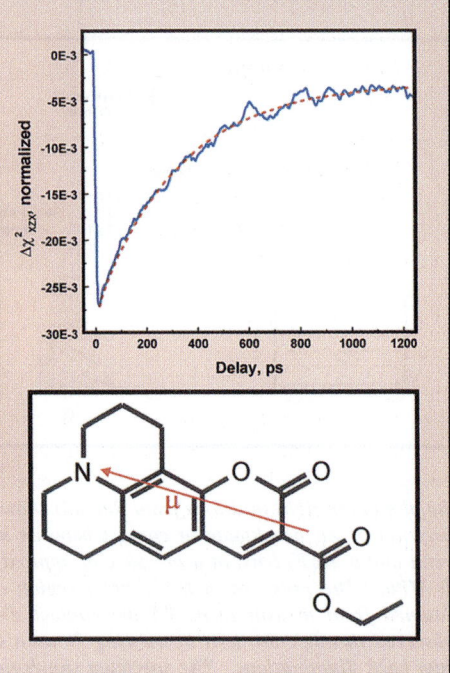

Fig. 14. TR-SHG studies of rotational relaxation. Data reprinted with permission [44]—copyright 1999 American Chemical Society.

molecules surrounding the Eosin B molecules at the interface results in less molecular friction and more rapid reorientation.

3. Future directions

Surface non-linear optics has clearly come a long way over the past two decades. It is our hope that the previous section will spark further interest in this rapidly growing field. While the specific examples reviewed were necessarily limited in number, they were chosen to represent a wide range of interfaces that might be examined by these tactics. Both SHG and SFS have reached sufficient maturity so that they have become accepted techniques within the arsenal of the surface scientist. Nevertheless, improvements in both laser technology and non-linear optical theory promise to lead to further increases in the use of these (and other) non-linear optical techniques in the future.

In the early days of SHG and SFS experimentation, laser technology was the major stumbling block. Indeed, SFS had to await the invention of tunable infrared lasers before it could be attempted successfully! The situation has changed dramatically in the past decade. Tunable laser systems are now available over the full range of pulse widths (i.e., nanoseconds to femtoseconds) and repetition rates, providing a wide selection for the experimentalist. Increases in the tunability of both visible and infrared lasers will continue to enhance the capability of both techniques. In the case of SHG, the majority of studies in the literature involve the use of only one fundamental frequency. This is often chosen to match an electronic transition at the interface in order to increase the signal intensity. The availability of ever more robust (and easily operated!) lasers that can be tuned throughout the entire visible to near infrared spectral regions are enabling *spectroscopic* SHG measurements to be performed more easily and often. Conducted in this fashion, SHG can have much more chemical specificity. Likewise, the extension of wavelength coverage out to the far infrared is necessary if SFS is to be able to probe a wider range of vibrations at interfaces. At the moment the current practical IR wavelength limit for performing SFS with benchtop laser systems is around 12–13 μm (≈ 800 cm^{-1}). It would clearly be desirable to extend this range out to 40 μm (the equivalent of 250 cm^{-1}). At the moment the only way of performing SFS in the far infrared is to employ free-electron lasers [48], which are available at international user facilities in several countries.

In order to correctly interpret the data obtained during non-linear optical experiments, a detailed knowledge of both the surface (i.e., non-resonant) and molecular (i.e., resonant) second-order susceptibilities must be obtained. In the case of surface non-linear response, models have been developed over the years that can account for phenomena observed at metal, semiconductor, and dielectric surfaces with varying degrees of success. Complete understanding of the molecular non-linear response depends on knowledge about the hyperpolarizabilities of electronic and vibrational transitions for individual molecules. More experimentation must be performed to begin to build up a library of molecular hyperpolarizability data. The continued development of more powerful computers with increasing computational speed will have a large impact in this area. For example, reliable ab initio quantum mechanical calculations to predict molecular hyperpolarizabilities should enhance our ability to extract information about interfacial properties (e.g., orientation, conformation) from SHG and SFS data.

In the remainder of this section, we will discuss some important future uses of non-linear optical spectroscopy in surface science. Of course, there will certainly be new developments that we cannot foresee. In addition, several current applications will be taken in new directions. For example, while the study of interfacial dynamics by SHG and SFS has been well established, applications to buried interfaces will surely become more widespread. Indeed, it will be interesting to see how such studies dovetail with the massive efforts being put forth to examine bulk gas- and liquid-phase dynamic processes. We have chosen to focus on applications of SHG and SFS to three areas: the study of small particles, the examination of biological interfaces, and development of non-linear microscopic imaging. The areas have several things in common. First, while their potential has been demonstrated by the pioneering efforts of several research groups, there is much opportunity for emerging surface scientists to make important and lasting contributions in each field. Second, the technology that is required is available now, which should allow new efforts to progress at a very rapid pace. Finally, and perhaps most importantly, all three applications have the potential to dramatically improve our understanding of buried interfacial processes, especially in the area of biological systems.

3.1. Non-linear optical studies of particles

Buried interfaces need not only be those involving planar surfaces. The surfaces of microparticles are as (or even more) important to industrial and biological science as planar surfaces. Some of the most novel experiments using non-linear spectroscopy in the last few years have been those carried out on particles. Although SHG and SFG are forbidden processes from systems that possess inversion symmetry, this restriction only applies on length scales that are shorter than the coherence length of the SHG process. At these shorter lengths, SHG from molecules that are oriented in opposite directions will cancel out (due to a phase mismatch of 180°). However, if molecules are adsorbed on colloidal particles that have at least one dimension larger than the coherence length, the SHG from opposite sides of the particle will not add coherently. Thus, they will not cancel out. Fig. 15 provides an illustration of this interesting phenomenon. Eisenthal and co-workers first demonstrated that SHG can be obtained from the surfaces of microparticles [49]. They showed that malachite green molecules adsorbed on the surface of polystyrene microspheres give rise to very strong SHG. The import of this work lies in the non-invasive nature of SHG as a probe of molecular adsorption. SHG can be used to measure the coverage of molecules on the particles. The only effective method of performing this task without SHG involves centrifugation, which cannot be applied to other important systems such as

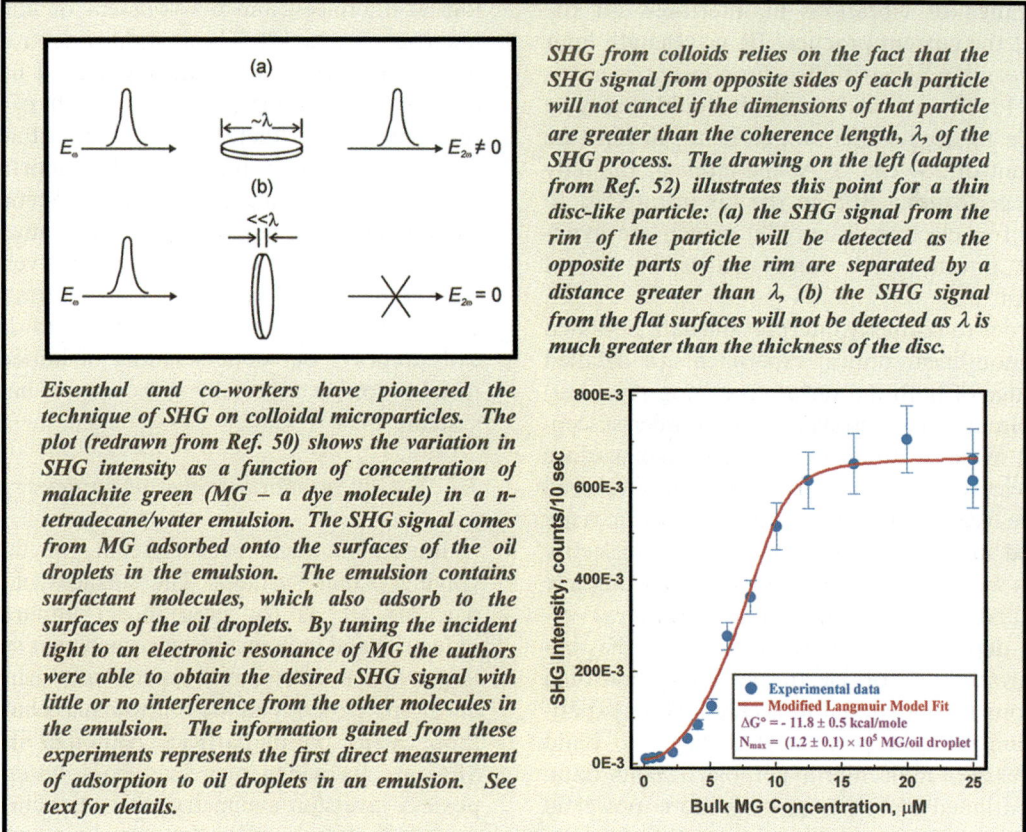

Fig. 15. SHG from colloidal microparticles. Data reprinted with permission [50,52]—copyrights 1998 and 1999, respectively, American Chemical Society.

emulsions. Eisenthal and co-workers have also obtained the first direct measurement of the adsorption of molecules to the surface of oil droplets in a n-tetradecane/water emulsion [50], thus confirming the applicability of SHG to a large variety of systems (see plot in Fig. 15).

Other systems that have been studied by Eisenthal's group include measurement of the surface potential of both polystyrene sulfate microspheres and oil droplets in an emulsion [51]. The surface potential of the microparticles is one of their most important physical parameters, determining many physical and chemical characteristics within the system. The most common method for determining the surface potential of microparticles is electrophoresis, a technique that relies on the motion of the particles under an applied electric field.

SHG is a less perturbative technique for gaining the same information. The surfaces of clay microdiscs have also been probed with SHG [52]. Clays are well-known catalysts both in organic and inorganic chemistry. The surface properties and the electrostatic interactions with adsorbates on the surface of the particles are thought to be very important factors in the chemistry of clays. Eisenthal has been able to probe the organization of water molecules at the surface of the clay particles, thus probing their electrostatic properties. He has also studied the adsorption of 4-(2-pyridylazo)resorcinol to the surface of the clay particles. Finally, the Eisenthal group has also looked at adsorption of molecules to the surface of TiO_2 particles [53]. TiO_2 is a wide band gap semiconductor and colloidal TiO_2 particles have found uses in many areas of photocatalysis. One area of particular interest is the use of TiO_2 in solar energy harvesting. For a semiconductor such as TiO_2, only ultraviolet light can excite electrons from the valence to the conduction band, which is an inefficient use of the available solar energy. However, if one adsorbs an organic molecule to the surface it is possible to form a charge-transfer complex whereby the conduction band of TiO_2 can be populated with electrons from photoexcited adsorbed molecules. Eisenthal and co-workers have observed the adsorption of catechol to the surface of TiO_2 particles and have determined the energy of the charge-transfer band for the catechol/TiO_2 system. Catechol has a broad absorption band centered around 456 nm, a much better region for making use of solar energy.

As can be seen from the quoted examples, SHG of small particles and droplets has the potential to make a large impact in colloid science. However an even greater advance would be to apply SFS to microparticles. As with all the other examples presented in this article, SFS offers chemical specificity and the ability to ascertain conformational order of adsorbed molecules. No research group has yet published a sum-frequency spectrum of molecules adsorbed on microparticles. There is also some promise for using non-linear spectroscopy to look at nanometer-sized particles. For example, Hupp and co-workers have recently demonstrated that *incoherent* SHG, known as hyper-Rayleigh scattering, can be used to probe surfaces of SiO_2 [54] and Au [55] nanoparticles.

3.2. Non-linear optical studies of biological interfaces

Perhaps the most important buried interfaces that we encounter every day reside in our bodies. The membranes of cells (cf. Fig. 1(c)), which consist of lipid bilayers, cholesterol, proteins, and carbohydrates, are responsible for regulating many important biological processes. The transport of components into (e.g., nutrients) and out of (e.g., waste) cells occurs either by diffusion through these membranes or by structural alteration of membrane components. The fluidity and flexibility of the lipid bilayer is ideal for allowing both the growth and division of cells. Proteins in cell membranes are responsible for interacting with the surfaces of other cells in ways that guide the formation of distinct cellular populations (i.e., recognition). Membranes also enable the adhesion of cells to one another, forming connections through the extracellular matrix of proteins outside of cells. Understanding the mechanism by which these membranes interact with water, ions and macromolecules such as proteins is therefore a central goal for cell physiologists. However, biological interfaces are not only limited to living systems. For example, successful application of "DNA on a chip" technology [56,57] for combinatorial biochemical studies and chemically tailored surfaces for use as biotoxin sensors requires a detailed understanding of the interactions of biological components with solid surfaces.

Due to their complexity (and sometimes fragility) biological interfaces are among the most difficult to examine. Their study will likely be greatly influenced by non-linear optical approaches over the coming years [58]. The majority of investigations that have been performed on biological interfaces so far have employed SHG, although experiments with SFS will increase in number. Studies have included the examination of membrane potentials and molecular structure, as well as adsorption properties [58]. So far non-linear optical studies have largely dealt with "model" systems that mimic the behavior of actual biological interfaces. Perhaps the most popular are the so-called hybrid bilayer membranes (HBMs), in which a monolayer of phospholipids is allowed to physically adsorb onto a methyl-terminated monolayer on a solid surface [59]. This forms a rugged interface that has been shown to be similar to actual membranes in terms of structure and stability. HBMs facilitate non-linear optical measurements in either the external reflection or TIR configuration. Notwithstanding these and similar efforts with model biological systems, an important goal is to extend SHG and SFS to real biological systems (e.g., groups of cells). One promising area of research along these lines involves the use of non-linear optical *microscopy*, which will be discussed below in its own section.

Another fairly recent development is the use of non-linear optics to probe the chiral properties of biological interfaces. Chirality indicates that a molecule is not superimposable on its mirror image. Many biological molecules have a chiral structure and this property influences how they interact with biologically active agents (e.g., drugs). Indeed, gaining information about the secondary, chiral structure of proteins is important for understanding how they interact with membranes and surfaces, as well as other molecules. The standard linear optical methods for obtaining such data for proteins in solutions are optical rotary dispersion (ORD), which examines rotation of linearly polarized light, and circular dichroism (CD), which measures absorption of circularly polarized light. While these effects are inherently small, both Raman and infrared spectroscopic versions of CD are used widely to probe bulk solutions. However, when used to examine buried interfaces these approaches suffer from all of the associated problems with linear optics (cf. Fig. 2). But several researchers have shown that non-linear optical spectroscopy can be used very effectively to probe chirality at interfaces. The first demonstration of SHG from adsorbed chiral molecules was by Petralli-Mallow, Hicks and co-workers in 1993 [60]. These researchers [61], as well as the group of Persoons [62], have since contributed importantly to the development of non-linear optical experiments in this area. These relatively new SHG approaches have been given the names SHG-ORD and SHG-CD, in concert with their linear optical analogues.

To see what the future holds for examining chiral surfaces with SHG, we review here some exciting recent work by Hicks and Petralli-Mallow [61]. They studied the adsorption of a model protein, cytochrome c, on three different types of chemically modified surfaces. They were as follows: a hydrophobic surface in the form of a SAM of dodecanethiol on gold, a hydrophilic surface in the form of a SAM of 16-mercaptohexanoic acid on gold, and an HBM surface formed by adsorbing a mixture of lipids onto the hydrophobic SAM. Fig. 16 shows a set of polarization-dependent SHG-CD data for the hydrophobic surface submerged in a buffer solution. The blue circles represent the SHG signal obtained from the hydrophobic SAM, while the red circles are data acquired after cytochrome c adsorption. In the latter case, significant differences in the SHG signals are observed upon switching between right- and left-polarized light, clearly indicating the presence of the protein at the surface. Similar results were obtained for the other two surfaces and can be presented in a more compact form as SHG-CD percentages (defined and tabulated in Fig. 16).

Something unexpected occurred when ascorbic acid was added to the solution. Under such conditions, the proteins in solution will be reduced at the heme (i.e., iron) site, causing a change in chirality. This bulk effect has been observed using linear optical CD spectroscopy. The positive shift in the SHG-CD for cytochrome c adsorbed on both the hydrophobic and hydrophilic surfaces indicates that a similar process is occurring in the adsorbed protein. Interestingly, at the HBM interface there is no positive shift in the SHG-CD percentage. This data suggests that the heme group of adsorbed protein is shielded in some way from reduction by ascorbic acid. Support for this conclusion was provided by additional experiments that showed positive SHG-CD shifts upon exposure of *prereduced* cytochrome c to the lipid surface. It has been proposed that this reduction shielding effect may arise from the proteins being tightly bound or even imbedded in the HBM (see the cartoons in Fig. 16). While more experiments are necessary to fully explain this phenomenon, these initial results bode well for the future of using SHG to probe protein–surface interactions. One exciting possibility is that SFS could be adapted for the study of chiral surfaces. The added chemical sensitivity provided by this vibrational technique would surely be of value in the study mentioned above. One complication is that, unlike SHG, SFS is not forbidden in bulk chiral liquids. As of this writing, we are unaware of anyone having performed SFS experiments on chiral surfaces.

3.3. Non-linear optical microscopy

The desire to elucidate the spatial structure of surfaces and interfaces has driven the development

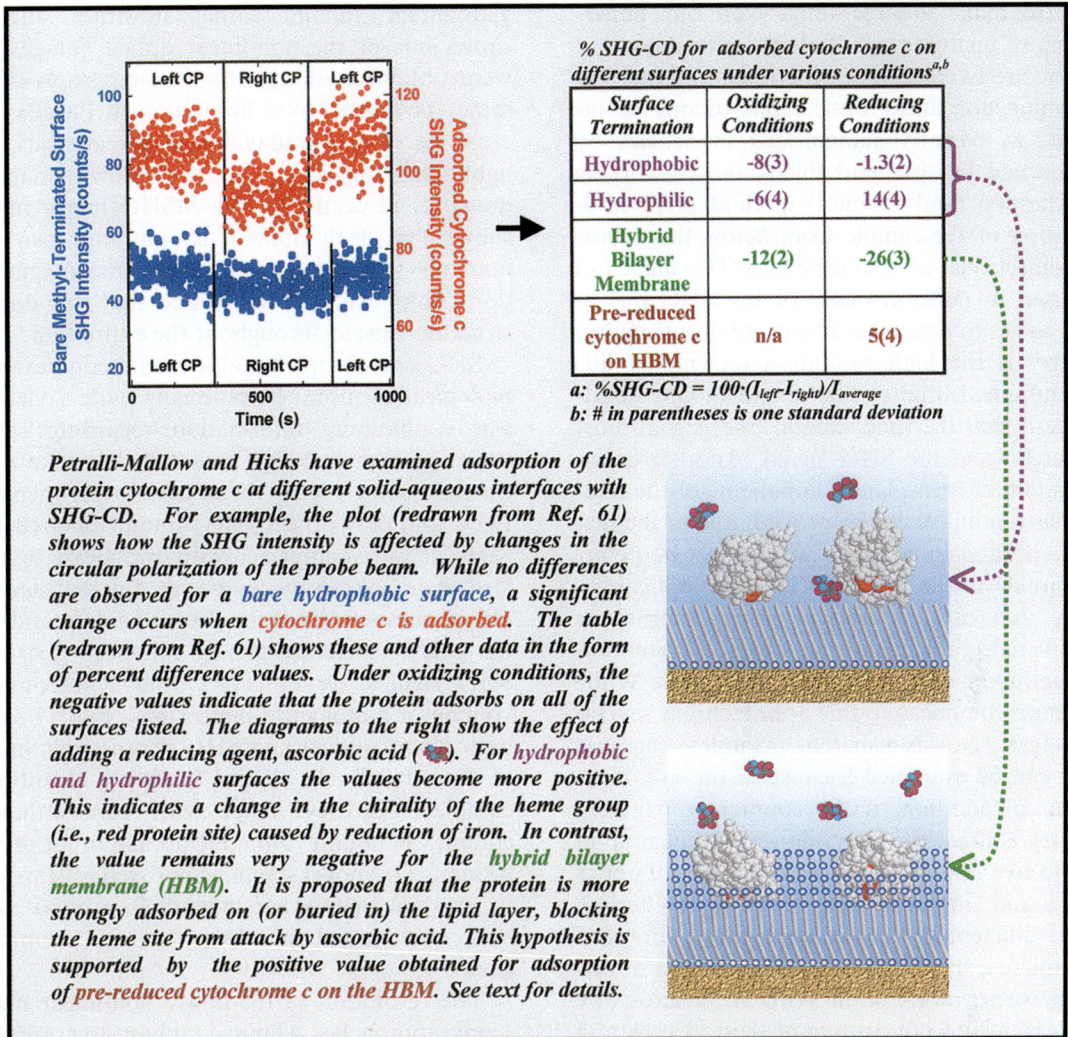

Fig. 16. Chiral SHG of protein adsorption on surfaces. Data reprinted with permission [61, Fig. 4, Table 1]—copyright 1999 Springer.

of many types of optical microscopes. Classic linear optical approaches such as ellipsometric, polarization, and Brewster angle microscopy have been providing information about the optical properties and thickness of thin films and surfaces for years. Fluorescence microscopy is a powerful tool used extensively to examine cells and biological interfaces. Raman and infrared microscopy, with their vibrational "fingerprinting" capability, can produce chemical images of surfaces. The image plane spatial resolution of all these techniques can be on the order of the wavelength of the probe light being used. These approaches are often also combined with a confocal optical configuration, providing superior spatial resolution in the axial direction as well. In the last few years, several research groups have worked on developing various types of non-linear optical microscopes [63,64]. Their motivation is clear: develop a microscope that is capable of probing surface chemical, electrical, and magnetic properties while avoiding interfering signals from bulk phases. The results have been promising (especially for SHG), suggesting that non-linear optical microscopy is poised to make a large impact on our understanding of microscopic interfacial phenomena.

There are two different experimental setups for performing non-linear optical microscopy of interfaces, as recently summarized in articles by Yamada and Lee [63] and Fuchs and co-workers [64]. The first (and simplest) method involves illumination of the sample from below the microscope objective at normal incidence. This tactic has been used to perform many of the SHG microscopy studies to date. The advantage of this optical geometry is the high resolution (essentially diffraction limited) that can be obtained. One disadvantage is that the fundamental laser signal must be filtered from the SHG signal. Another drawback is that only in-plane components of the non-linear susceptibility can be probed, due to the fact that the fundamental beam will always be polarized parallel to the surface. The other optical geometry that can be used is either external or internal reflection from the surface using co-propagating or counter-propagating lasers. While a bit more complicated, this approach has several advantages. Non-transparent samples such as metals can be examined because of the use of reflection. In addition, if the counter-propagating geometry is used, the lasers do not impinge upon the objective, which avoids the possibility of optics damage and simplifies the task of light filtering. Several different optical configurations for non-linear microscopes are shown in Fig. 17. As of this writing, we are aware of only two SFS microscopy investigations [65,66]. In one of these, Fuchs and co-workers examined the structure of ordered monolayer films on fused silica [65].

SHG microscopy has been used to investigate magnetic domain patterns on surfaces, the quality of ordered monolayer films, patterned self-assembled monolayers, and cell membranes. We consider here two particularly beautiful examples of images that have been obtained with this technique. Rasing and co-workers [67] have been using MSHG to examine a range of magnetic surfaces and interfaces. They have demonstrated that magnetic domain patterns on thin film surfaces can be observed using SHG microscopy. One of the first systems that they studied with their microscope was a garnet film grown on a (2 1 0) gadolinium gallium garnet substrate. Fig. 17 shows one of the non-linear optical images that were obtained, along with a micrograph of the same spot taken by a linear optical Faraday microscope. The Faraday image reveals dark and light areas, suggesting "up" and "down" magnetic domains, respectively. The MSHG image further shows that each domain is split into two subdomains that have different in-plane magnetization components. It was found that this domain structure existed throughout the entire film.

SHG microscopy has also been used to examine biological membranes, especially with a view towards obtaining information regarding spatial variation in membrane potential. We take as an example some recent remarkable experiments by Loew and co-workers [68]. Nanometer-sized gold particles were complexed with antibodies that are known to selectively bind with different sites on cell membranes. Since the electric fields at and near the gold particle are greatly enhanced [69], large SHG signals are observed from a surrounding volume of nanometric dimensions. Fig. 17 shows both linear optical and SHG microscopic images of a group of cells stained by this gold–antibody complex. The SHG image clearly reveals that the complex is bound to the membrane in an orderly fashion. This novel staining approach is being used to examine membrane potentials around *single molecules* located at selected cellular membrane sites.

The resolution of the above non-linear microscopic approaches is limited (at best) to around the wavelength of light being used or generated. However, it is possible to overcome this limitation by operating the microscope in what is called the "near field". The concept is relatively simple to picture. A light source is directed through a fiberoptic probe that is tapered down to a fine aperture (say on the order of 100 nm). If this probe is brought in close proximity to a sample surface, the light will interact with the surface before it undergoes diffraction. By scanning the near-field probe over a surface, an optical image may be obtained that has resolution limited only by the distance from the surface and the size of the probe aperture. Near-field microscopy shows great promise for performing a wide range of spectro-

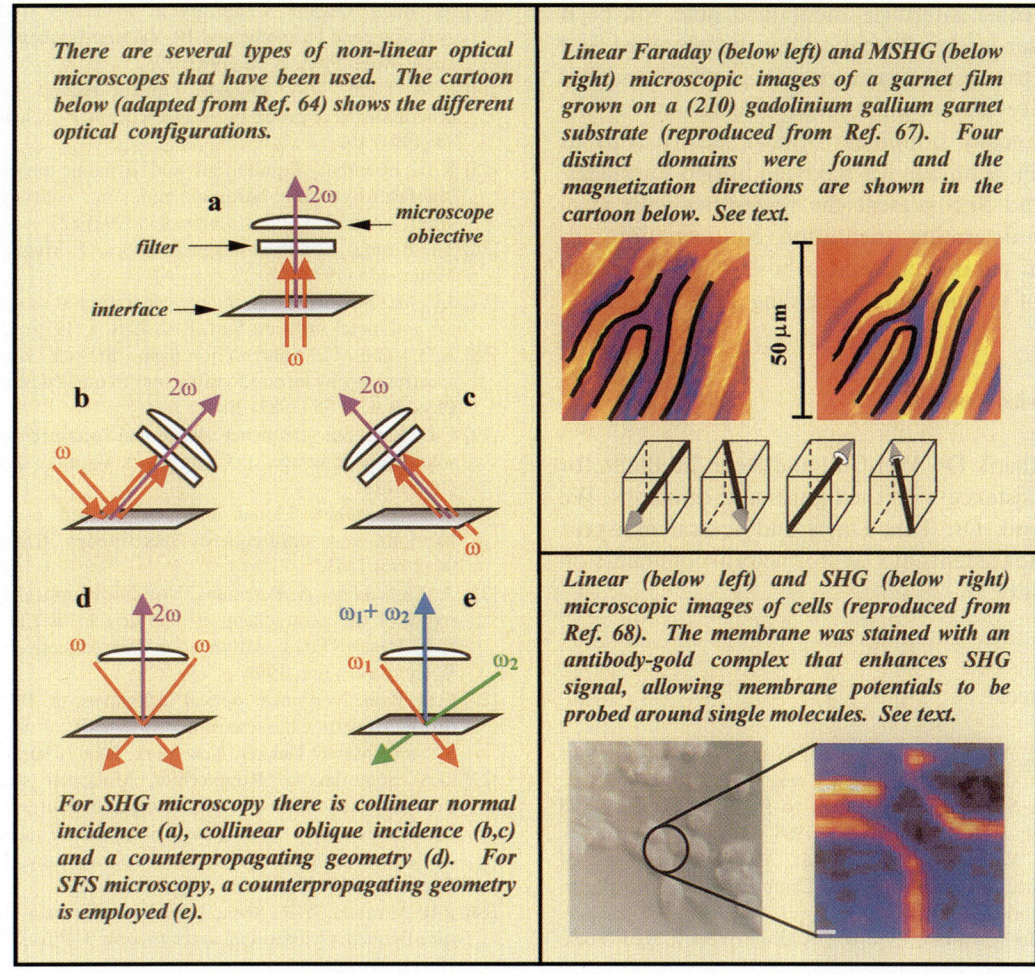

Fig. 17. Non-linear optical microscopy. Data reprinted with permission [64,67,68]—copyright 1998 Elsevier, copyright 1998 American Institute of Physics, copyright 1999 National Academy of Sciences, USA, respectively.

scopic techniques with nanometric resolution. This is because the tapered probe can also be used to *collect* light that is generated at interfaces. Dunn has recently reviewed this exciting and relatively new research area [70]. Fluorescence and Raman spectroscopic microscopy conducted in the near-field mode are being developed as powerful probes of nanoscale interfacial phenomena. Recent near-field SHG investigations have included studies of systems as diverse as rough metal surfaces [71], ordered monolayer films [72], and cell membranes [73]. Again, there have been no reports to date of any near-field SFS investigations. Development of such a nanometric-scale vibrational spectroscopic approach will certainly be a major goal of the non-linear optical spectroscopists over the coming years.

4. Concluding remarks

Buried interfaces represent the last unexplored region of surface science. The paucity of adequate probes of the processes that occur at these interfaces has been the main obstacle preventing rapid progress in this area. The only way that a real understanding will be achieved for the many processes that occur at liquid/liquid, solid/liquid, and all the other interfaces mentioned here will be if molecular-level information can be obtained and related to overall system properties. We hope that in this review we have been able to convey the importance of non-linear optical spectroscopy as the main technique for studying buried interfaces. SHG and SFS present the researcher with ideal tools to determine orientation, conformation, adsorption parameters, and a host of other crucial pieces of information concerning such systems.

Acknowledgements

We thank Dr. Wei Chu and Rene Le Blanc for their assistance with the figures and drawings. We also thank Dr. Tom Davis and several first-year graduate students at USC for helpful comments on readability.

References

[1] A.W. Adamson, A.P. Gast, Physical Chemistry of Surfaces, sixth ed., Wiley, New York, 1997.
[2] G.A. Somorjai, Introduction to Surface Chemistry and Catalysis, Wiley, New York, 1994.
[3] D.P. Woodruff, T.A. Delchar, Modern Techniques of Surface Science, second ed., Cambridge University Press, Cambridge, 1994.
[4] E. Hecht, Optics, third ed., Addison-Wesley, New York, 1998.
[5] M. Born, E. Wolf, Principles of Optics, seventh ed., Cambridge University Press, Cambridge, 1999.
[6] N. Bloembergen, Nonlinear Optics, fourth ed., World Scientific, New Jersey, 1996.
[7] Y.R. Shen, The Principles of Nonlinear Optics, Wiley, New York, 1984.
[8] R.W. Boyd, Nonlinear Optics, Academic Press, New York, 1992.
[9] J.A. Armstrong, N. Bloembergen, J. Ducuing, P.S. Pershan, Interactions between light waves in a nonlinear dielectric, Phys. Rev. 127 (1962) 1918.
[10] N. Bloembergen, P.S. Pershan, Light waves at the boundary of nonlinear media, Phys. Rev. 128 (1962) 606.
[11] G.L. Richmond, J.M. Robinson, V.L. Shannon, Second harmonic generation studies of interfacial structure and dynamics, Prog. Surf. Sci. 28 (1988) 1.
[12] T.F. Heinz, Second-order nonlinear optical effects at surfaces and interfaces, in: H.-E. Ponath, G.I. Stegeman (Eds.), Nonlinear Surface Electromagnetic Phenomena, North Holland, Amsterdam, 1991, p. 353.

[13] M. Buck, Organic Chemistry at interfaces studied by optical second harmonic and IR-Vis sum-frequency generation, Appl. Phys. A 55 (1992) 395.
[14] R.M. Corn, D.A. Higgins, Optical second harmonic generation as a probe of surface chemistry, Chem. Rev. 94 (1994) 107.
[15] K.B. Eisenthal, Equilibrium and dynamic processes at interfaces by second harmonic and sum frequency generation, Annu. Rev. Phys. Chem. 43 (1992) 627.
[16] Th. Rasing, Nonlinear magneto-optics, J. Magn. Magn. Mater. 175 (1997) 35.
[17] J.F. McGilp, Second harmonic generation at semiconductor and metal surfaces, Surf. Rev. Lett. 6 (1999) 529.
[18] X.D. Zhu, H. Suhr, Y.R. Shen, Surface vibrational spectroscopy by infrared-visible sum frequency generation, Phys. Rev. B 35 (1987) 3047.
[19] C.D. Bain, Sum-frequency vibrational spectroscopy of the solid/liquid interface, J. Chem. Soc., Faraday Trans. 91 (1995) 1281.
[20] K.B. Eisenthal, Liquid interfaces probed by second-harmonic and sum-frequency spectroscopy, Chem. Rev. 96 (1996) 1343.
[21] A. Tadjeddine, A. Peremans, Non-linear optical spectroscopy of the electrochemical interface, in: R.J.H. Clark, R.E. Hester (Eds.), Advances in Spectroscopy, vol. 26, Wiley, New York, 1998, p. 159.
[22] C.D. Bain, Non-linear optical techniques, in: B.P. Binks (Ed.), Modern Characterization Methods of Surfactant Systems, Marcel Dekker, New York, 2000 (Chapter 9).
[23] G.A. Somorjai, G. Rupprechter, Molecular studies of catalytic reactions on crystal surfaces at high pressures and high temperatures by infrared-visible sum frequency generation (SFG) surface vibrational spectroscopy, J. Phys. Chem. B 103 (1999) 1623.
[24] P.B. Miranda, Y.R. Shen, Liquid interfaces: a study by sum-frequency vibrational spectroscopy, J. Phys. Chem. B 103 (1999) 3292.
[25] D.W. Goodman, Correlations between surface science models and "real-world" catalysts, J. Phys. Chem. 100 (1996) 13090.
[26] F.M. Mirabella, Internal-reflection spectroscopy, Appl. Spectrosc. Rev. 21 (1985) 45.
[27] C.T. Williams, Y. Yang, C.D. Bain, Total internal reflection sum-frequency spectroscopy: a strategy for studying molecular adsorption on metal surfaces, Langmuir 16 (2000) 2343.
[28] A. Peremans, A. Tadjeddine, Electrochemical deposition of hydrogen on platinum single crystals studied by infrared-visible sum-frequency generation, J. Chem. Phys. 103 (1995) 7197.
[29] A. Tadjeddine, A. Peremans, Vibrational spectroscopy of the electrochemical interface by visible infrared sum frequency generation, J. Electroanal. Chem. 409 (1996) 115.
[30] O. Dannenberger, M. Buck, M. Grunze, Self-assembly of n-alkanethiols: a kinetic study by second harmonic generation, J. Phys. Chem. B 103 (1999) 2202.

[31] M. Himmelhaus, F. Eisert, M. Buck, M. Grunze, Self-assembly of *n*-alkanethiol monolayers: a study by IR-visible sum frequency spectroscopy (SFG), J. Phys. Chem. B 104 (2000) 576.

[32] S.R. Hatch, R.S. Polizzotti, S. Dougal, P. Rabinowitz, Surface vibrational spectroscopy of the vapor/solid and liquid/solid interface of acetonitrile on ZrO_2, Chem. Phys. Lett. 196 (1992) 97.

[33] H.H. Girault, D.J. Schiffrin, Electrochemistry of liquid/liquid interfaces, Electroanal. Chem. 15 (1989) 1.

[34] D.A. Higgins, R.M. Corn, Second harmonic generation studies of adsorption at a liquid–liquid electrochemical interface, J. Phys. Chem. 97 (1993) 489.

[35] D.A. Higgins, R.R. Naujok, R.M. Corn, Second harmonic generation measurements of molecular adsorption and coadsorption at the interface between two immiscible electrolyte solutions, Chem. Phys. Lett. 213 (1993) 485.

[36] D.J. Campbell, D.A. Higgins, R.M. Corn, Molecular second harmonic generation studies of methylene blue chemisorbed onto a sulfur-modified polycrystalline platinum electrode, J. Chem. Phys. 94 (1990) 3681.

[37] J.C. Conboy, M.C. Messmer, G.L. Richmond, Investigation of surfactant conformation and order at the liquid–liquid interface by total internal reflection sum-frequency vibrational spectroscopy, J. Chem. Phys. 100 (1996) 7617.

[38] A. Kirilyuk, Th. Rasing, M.A.M. Haast, J.C. Lodder, Probing structure and magnetism of CoNi/Pt interfaces by nonlinear magneto-optics, Appl. Phys. Lett. 72 (1998) 2331.

[39] Th. Rasing, Nonlinear magneto-optical probing of magnetic interfaces, Appl. Phys. B 68 (1999) 477.

[40] Q. Du, X.-d. Xiao, D. Charych, F. Wolf, P. Frantz, Y.R. Shen, M. Salmeron, Nonlinear-optical studies of monomolecular films under pressure, Phys. Rev. B 51 (1995) 7456.

[41] R. Fraenkel, G.E. Butterworth, C.D. Bain, In situ vibrational spectroscopy of an organic monolayer at the sapphire–quartz interface, J. Am. Chem. Soc. 120 (1998) 203.

[42] D.A. Beattie, S. Haydock, C.D. Bain, A comparative study of confined organic monolayers by Raman scattering and sum-frequency spectroscopy, Vib. Spectrosc. 24 (2000) 109.

[43] A. Castro, E.V. Sitzmann, D. Zhang, K.B. Eisenthal, Rotational relaxation at the air/water interface by time-resolved second harmonic generation, J. Phys. Chem. 95 (1991) 6752.

[44] D. Zimdars, J.I. Dadap, K B. Eisenthal, T.F. Heinz, Anisotropic orientational motion of molecular adsorbates at the air–water interface, J. Phys. Chem. B 103 (1999) 3425.

[45] R. Antoine, A.A. Tamburello-Luca, P. Hebert, P.F. Brevet, H.H. Girault, Picosecond dynamics of Eosin B at the air/water interface by time-resolved second harmonic generation: orientational randomization and rotational relaxation, Chem. Phys. Lett. 288 (1998) 138.

[46] A. Bandara, J. Kubota, K. Onda, A. Wada, K. Domen, C. Hirose, Short-lived reactive formate species on NiO(1 1 1) observed by picosecond temperature jump, Surf. Sci. 433–435 (1999) 83.

[47] K. Domen, A. Bandera, J. Kubota, K. Onda, A. Wada, S.S. Kano, C. Hirose, SFG study of unstable surface species by picosecond pump–probe method, Surf. Sci. 427–428 (1999) 349.

[48] D. Oepts, A.F.G. van der Meer, P.W. van Amersfoort, The free-electron-laser user facility FELIX, Infrared Phys. Technol. 36 (1995) 297.

[49] H. Wang, E.C.Y. Yan, E. Borguet, K.B. Eisenthal, Second harmonic generation from the surface of centrosymmetric particles in bulk solution, Chem. Phys. Lett. 259 (1996) 15.

[50] H. Wang, E.C.Y. Yan, Y. Liu, K.B. Eisenthal, Energetics and population of molecules at microscopic liquid and solid surfaces, J. Phys. Chem. B 102 (1998) 4446.

[51] E.C.Y. Yan, Y. Liu, K.B. Eisenthal, New method for determination of surface potential of microscopic particles by second harmonic generation, J. Phys. Chem. B 102 (1998) 6331.

[52] E.C.Y. Yan, K.B. Eisenthal, Probing the interface of microscopic clay particles in aqueous solution by second harmonic generation, J. Phys. Chem. B 103 (1999) 6056.

[53] Y. Liu, J.I. Dadap, D. Zimdars, K.B. Eisenthal, Study of interfacial charge-transfer complex on TiO_2 particles in aqueous suspension by second harmonic generation, J. Phys. Chem. B 103 (1999) 2480.

[54] F.W. Vance, B.I. Lemon, J.A. Ekhoff, J.T. Hupp, Interrogation of nanoscale silicon dioxide/water interfaces via hyper-Rayleigh scattering, J. Phys. Chem. B 102 (1998) 1845.

[55] F.W. Vance, B.I. Lemon, J.T. Hupp, Enormous hyper-Rayleigh scattering from nanocrystalline gold particle suspensions, J. Phys. Chem. B 102 (1998) 10091.

[56] G. MacBeath, S.L. Schreiber, Printing proteins as microarrays for high-throughput function determination, Science 289 (2000) 1760.

[57] D.D. Shoemaker, E.E. Schadt, C.D. Armour, Y.D. He, P. Garrett-Engele, P.D. McDonagh, P.M. Loer, Experimental annotation of the human genome using microarray technology, Nature 409 (2001) 922.

[58] V. Vogel, What do nonlinear optical techniques have to offer the biosciences? Curr. Opin. Colloid Interface Sci. 1 (1996) 257.

[59] A.L. Plant, Supported hybrid bilayer membranes as rugged cell membrane mimics, Langmuir 15 (1999) 5128.

[60] T. Petralli-Mallow, T.M. Wong, J.D. Byers, H.I. Yee, J.M. Hicks, Circular dichroism spectroscopy at interfaces: a surface second harmonic generation study, J. Phys. Chem. 97 (1993) 1383.

[61] J.M. Hicks, T. Petralli-Mallow, Nonlinear optics of chiral surface systems, Appl. Phys. B 68 (1999) 589.

[62] T. Verbiest, M. Kauranen, A. Persoons, Second-order nonlinear optical properties of chiral thin films, J. Mater. Chem. 9 (1999) 2005.

[63] S. Yamada, I.-Y.S. Lee, Recent progress in analytical SHG spectroscopy, Anal. Sci. 14 (1998) 1045.
[64] M. Flörsheimer, M. Bösch, C. Brillert, M. Wierschem, H. Fuchs, Second harmonic imaging of surface order and symmetry, Thin Solid Films 327–329 (1998) 241.
[65] M. Flörsheimer, C. Brillert, H. Fuchs, Chemical imaging of interfaces by sum frequency microscopy, Langmuir 15 (2000) 5437.
[66] M. Flörsheimer, C. Brillert, H. Fuchs, Chemical imaging of interfaces by sum frequency generation, Mat. Sci. Eng. C-Bio. S. 8–9 (1999) 335.
[67] V. Kirilyuk, A. Kirilyuk, Th. Rasing, A combined nonlinear and linear magneto-optical microscopy, Appl. Phys. Lett. 70 (1997) 28.
[68] G. Peleg, A. Lewis, M. Linial, L.M. Loew, Nonlinear optical measurement of membrane potential around single molecules at selected cellular sites, Proc. Nat. Acad. Sci. USA 96 (1999) 6700.
[69] A. Campion, P. Kambhampati, Surface-enhanced Raman scattering, Chem. Soc. Rev. 27 (1998) 241.
[70] R.C. Dunn, Near-field scanning optical microscopy, Chem. Rev. 99 (1999) 2891.
[71] I.I. Smolyaninov, A.V. Zayats, C.C. Davis, Near-field second harmonic generation from a rough metal surface, Phys. Rev. B 56 (1997) 9290.
[72] S.I. Bozhevolnyi, T. Geisler, Near-field optical spectroscopy of Langmuir–Blodgett films, J. Opt. Soc. Am. A 15 (1998) 2156.
[73] R.D. Schaller, C. Roth, D.H. Raulet, R.J. Saykally, Near-field harmonic imaging of granular membrane structures in natural killer cells, J. Phys. Chem. B 104 (2000) 5217.

Frontiers in infrared spectroscopy at surfaces and interfaces

Carol J. Hirschmugl *

University of Wisconsin-Milwaukee, Laboratory for Surface Studies, Department of Physics, Milwaukee, WI 53211, USA

Received 3 August 2000; accepted for publication 27 April 2001

Abstract

An overview of advances in infrared (IR) spectroscopy is presented. Recent results in several areas of topical interest, including examples in biocompatibility, dielectric films, in situ chemical reactions, and electron relaxation in nanoparticles, are highlighted. Major advances in IR experimental methods include the development of accelerator-based sources of IR light and the application of novel techniques to examine complex systems. These advances, and their role in elucidating crucial insights about surfaces and interfaces, are illustrated by recent work in the literature. After reviewing the current state of the art, promising future directions are discussed. In particular, superb opportunities are expected to develop in a broad range of scientific disciplines, e.g., biology, device engineering, chemistry, and physics. © 2001 Elsevier Science B.V. All rights reserved.

Keywords: Photon absorption spectroscopy; Infrared absorption spectroscopy; Vibrations of adsorbed molecules; Catalysis; Ellipsometry; Electrical transport (conductivity, resistivity, mobility, etc.); Energy dissipation; Biological compounds

1. Introduction

The chemical and physical properties of the surfaces of solid materials and the interfaces between solids and fluids play an overarching role in a host of processes of interest to science and technology. Pertinent problems include reducing the friction between sliding machine parts, growing ultrapure crystals of electronic materials, controlling the binding of chemicals to biological membranes, understanding how charge flows near the surface of a metallic solid, and promoting desired chemical reactions in a chemical plant or automobile exhaust system. The field of surface science strives to understand the fundamental processes occurring at surfaces and interfaces, and, ultimately, to provide a means for controlling and manipulating such systems through this understanding.

Infrared (IR) spectroscopy is an incisive, non-destructive tool for examining interfaces and surfaces, providing valuable knowledge about chemical forces between atoms, vibrational frequencies found in molecules and bulk solids, and the motion of electrons within solids. This tool thereby allows one to determine the flow of energy in matter, or to identify the chemical species present at the surface or interface under investigation, while, in many cases, determining their molecular structure. IR spectroscopy is a mature field, yet it is poised to make contributions to the newer directions embraced by surface scientists in solid–fluid

---

* Tel.: +1-414-229-4474/5748; fax: +1-414-229-5589.
E-mail address: cjhirsch@uwm.edu (C.J. Hirschmugl).

0039-6028/01/$ - see front matter © 2001 Elsevier Science B.V. All rights reserved.
PII: S0039-6028(01)01523-0

interfaces: the ability of IR light to penetrate many forms of matter confers access to vital information even under adverse conditions. For example, IR absorption measurements performed on the electrodes of an electrochemical cell can identify chemical composition and molecular structure occurring at a solid–liquid interface as a function of the applied electric potential [1,2].

Electromagnetic radiation, including IR light, drives the motion of electric charges in matter. If the natural time scale of any oscillations of the charges in a molecule or solid is close to the period of the electromagnetic radiation shining on the system, a condition known as resonance occurs. Like an adult timing his pushes on a child's swing to coincide with the motion of the swing, a driving force having the same frequency as the system's natural frequency efficiently couples to and excites the oscillation. Near resonance, therefore, IR light is efficiently absorbed by the system, allowing the identification of the frequencies of low energy (1–500 meV) excitations found in the system under study. These excitations may involve nuclear motion, such as vibrating molecules, ions, radicals, or chains of atoms in extended materials, or only electrons, as in the promotion of a single electron from one energy state to a higher one or in the excitation of low energy collective oscillations of many electrons.

As a practical matter, IR spectroscopy has found its widest application in identifying the chemical compounds present in an unknown sample by the virtue of frequencies of IR light the sample absorbs. Since the resonance condition occurs over a narrow range of frequencies, which differs for different compounds, the exact frequency of the absorbed light provides a characteristic signature of the molecules, ions, or radicals present in the sample. Extensive gas-phase studies have identified these "fingerprints" for a host of chemical compounds, which can be used in interpreting surface and interface data. For example, the vibrational stretching motion of a triple-bonded CO unit (such as found in CO gas) absorbs IR light at 5.70×10^{13} Hz. Similarly, CO weakly bound to a single atom on a solid surface absorbs IR light at 5.53×10^{13} Hz. The analytical capabilities of IR spectroscopy are invaluable in identifying chemical composition at surfaces and interfaces.

Einstein won a Nobel Prize in Physics (1921) for showing that the energy carried by electromagnetic radiation is directly related to the frequency of its oscillation. Thus, IR spectroscopy allows the determination of the energy of the excitations it probes, and thereby sheds light on the microscopic origin of the excitation. For example, the energies of vibrational excitations provide insight into the interatomic potentials found in molecules, including knowledge about the bond strength. Since the strength of bonding within a molecule is altered upon adsorption (e.g., bonding) to a surface, a sensitive determination of the bond strength can be invaluable in probing surface reaction paths. The oscillations associated with electronic excitations also provide insight into surface processes. It can be determined from IR spectroscopic investigations of a molecule adsorbed on a solid whether vibrational excitations of the adsorbed molecule decay by transferring their energy to bulk atomic vibrations of the solid or to electronic excitations in the solid. In turn, this sheds light on the nature of the bond between the molecule and solid surface.

The origins of IR spectroscopy date back to the 1880s. Recent advances in instrumentation, including the design of spectrometers and detectors and the development of new sources, however, provide the means to enhance significantly the capabilities of this mature field. Thus, more complex interfaces, such as those between fluids and biological or biomimetic materials, or those between the layers of magnetic materials used in constructing computer hard disks, can be accessed. Over the years there has been continual improvement in spectrometers, detectors, window materials, and data processing. Recently, however, there have been major developments in new sources of IR radiation. These sources include radiation emitted from a storage ring (radiation is emitted when swift charged particles are accelerated by a magnetic field), lasers, and free electron lasers (FELs). FEL radiation is generated from swift charged particles that are accelerated several times by a spatially varied magnetic field over a short distance. Thus, several collinear micro-beams are generated and add in-phase to create the resulting beam. All three sources possess two important properties, brightness and pulsed time structures, which are invaluable in studies of topical interest. A high-brightness source emits radiation that originates from a point-like source, and the light is therefore confined to a well defined, small beam (e.g., laser beams). This property is required to measure spatially resolved data with spatial resolution comparable to the wavelength of the light (<1 μm to ten's μms, where 1 μm equals 1 millionth of a meter). Moreover, these bright sources are pulsed in time: they provide their radiation in staccato bursts, lasting anywhere between several femtoseconds to nanoseconds (1 fs equals one quadrillionth of a second and 1 ns equals one billionth of a second). Thus, modern sources allow the possibility to study time-dependent changes in chemical composition, structure, or electronic structure on time scales as short as femtoseconds.

In this paper, the properties of IR spectroscopy and its application to probe fundamental phenomena at solid surfaces and interfaces are reviewed. In Section 2, a survey is presented of the physical phenomena and questions where IR-based measurements (perhaps in combination with complementary measurements) can provide new information is presented. In Section 3, fundamental interactions between IR light and materials are reviewed, together with descriptions of allowed excitations in the 1–500 meV range that couple to incident IR light. In Section 4, selected recent applications are reviewed. These applications were chosen to highlight a wide range of fundamental and applied experiments in a variety of disciplines. Finally, several future applications for which IR is expected to play an important role are presented in Section 5, including discussions of new techniques to obtain spatially resolved and time-dependent chemical and structural information.

2. Modern surface and interface issues

Before discussing the details of IR spectroscopy, two topical areas, reactions at surfaces and materials synthesis and device construction, are introduced. These topics are inherently interdisciplinary and encompass many important issues of current interest. The significant role that IR-based examinations can play is also briefly discussed.

2.1. Reactions at surfaces

Under suitable conditions, e.g., elevated temperatures, high pressures, externally applied potentials, or immersion in acidic or basic liquids, surfaces often induce complex reactions. During a given reaction, the interfacial chemical composition and structure can undergo both reversible and irreversible changes, ultimately leaving the surface in its original state or an altered state, respectively. Although ex situ examinations of the surface allow a thorough understanding of the final condition of the interface, they leave the understanding of the intermediate steps incomplete. In situ investigations of the interface permit direct examinations of the intermediate reaction steps. Examples of important surface and interface reactions include electrochemical reactions, catalytic hydrogenation and oxidation, chemical synthesis, and reactions at biological interfaces.

Surface chemical reactions play an ubiquitous role in both industrial and commonplace processes. Fundamental examinations of ideal surfaces in controlled environments afford scientists the opportunity to discover basic reaction mechanisms and their dependence on interface composition and structure. For instance, investigations of ammonia synthesis over well-characterized Fe single crystals revealed that only specific crystallographic faces could be responsible for catalyzing the reaction using polycrystalline Fe [3]. Ultimately, complex and realistic samples and conditions, which correspond more closely to real chemical applications, can be examined. The results can be compared with control experiments to arrive at a more complete understanding of reaction mechanisms, which in turn can shape future efforts in chemical and materials synthesis.

2.2. Materials synthesis and device construction

Synthesis of novel heterostructures is an extremely important and expanding area of surface and interface science in which IR spectroscopy, a non-destructive probe, can provide important insight about interface and surface dynamics. Sometimes, the structure of the interface can sometimes be ascertained from these results. Technologically relevant and fundamentally interesting examples include quantum engineering, magnetic-thin-film construction, and growth of thin films with superior insulating properties for micro-electronics. IR spectroscopies can play a valuable role in examining the surfaces and interfaces of these manufactured materials through a variety of mechanisms. For some materials, the IR probe couples directly to low energy atomic vibrational excitations in the solid. Other discrete excitations correspond to transitions of an electron between energy levels, like raising a can of peas from the bottom shelf of a kitchen cabinet to a higher shelf. Due to the fact that the regular repetition of atoms in a crystal terminates at a surface or interface, electrons can occupy new, distinct states that do not occur in the repeating structure of the bulk. These new states can couple directly or indirectly to the IR light. For other cases, excitations of low energy vibrational absorptions provide insight into the force that a given atom in a chemical compound exerts on its neighbor.

Nanostructured materials, which have one or more spatial dimensions that are limited to a few trillionths of a meter, constitute an important new class of materials. This length scale is important because it is comparable to the effective electron wavelength in that medium. The accompanying confinement of electrons leads to quantum effects in the flow of heat and electricity, as well as the spacing of the energy levels of electrons (the "excitation spectrum"). These nanostructured materials can behave differently from the familiar metallic or semiconducting solids of macroscopic size. Examples include single-electron transistors, carbon nanotubes, quantum dots, quantum wells, and semiconductor superlattices. Several excellent reviews of these materials have been recently published [4–6].

Nanostructured solids are of interest for the fundamental quantum phenomena they exhibit and for the novel transport devices that can be constructed from them. On the fundamental side, such systems allow the exploration of few-electron systems confined by potentials that mimic those found in textbook problems. Artificial atoms, square wells, and harmonic oscillators can be created with a variable number of occupying particles. The observation of the excitation spectrum of such structures allows tests of our ability to understand and treat electronic interactions in a realistic setting. On the applied side, the ability to control the spatial dimensions allows one to manipulate and even tune the excitation spectrum. This level of control can be used to construct devices that operate at a desired energy scale or respond to a desired external stimulus. For example, the semiconductor laser found in common compact-disc players owes its particular spectral output to the engineered, nm-scale periodicity of alternating layers of materials with contrasting bandgaps. (1 nm is 1 trillionth of a meter)

The drive in the computing industry to make ever more efficient, smaller computers necessitates the development of processes to manufacture high-quality thin-film/substrate interfaces. Non-invasive IR examinations of these interfaces can be invaluable. For instance, a fundamental understanding of SiO_2 growth and the SiO_2/Si interfacial structure becomes more important as the insulating layer is made thinner to allow construction of smaller devices. Non-destructive IR examinations of the interface can detect vibrational signatures of the molecular fragments found there. The signatures reveal the local potential, and can be used to infer interfacial structures at the nanometer scale [7,8]. Alternatively, IR measurements of low energy electronic excitations can provide fundamental insight into ac transport processes. For example, the conductivity in thin metallic films that are used for interconnects in integrated circuits depends on the film thickness, the film/substrate interface, and the circuit frequency. While the measurements described here are at driving frequencies that are several orders of magnitude higher than circuitry in existing computers, future applications are expected to demand ever increasing operating frequencies. In all of these cases, the adsorbate/substrate interfaces become more important as the circuit size decreases, due to the increased surface-to-bulk ratio. As a routine, yet incisive, tool, IR spectroscopy is invaluable for shedding light onto both the structure and dynamics at these technologically useful interfaces.

3. Overview of infrared spectroscopy

Some of the more important aspects of the interplay between light and matter that allow us to probe materials with IR radiation are reviewed in this section. Einstein explained that energy in light waves is not continuous, but is instead limited to a discrete set of values, i.e., is quantized, into small bundles called *photons*. The energy of each photon is $E = hc/\lambda$ where λ is the wavelength of the light, h is Planck's constant and c is the speed of light. Interactions between photon beams and matter offer insight into a host of fundamental processes and properties of materials systems. Upon interaction with matter, the beam may be elastically scattered (e.g., reflected or diffracted), inelastically scattered (e.g., by exciting a vibrational mode), or absorbed. Each of these mechanisms can be exploited using a photon beam of appropriate energy. If none of these interactions is strong, the light may penetrate deeply into the material, or transmit unperturbed through the material. Typical penetration depths for IR radiation range from hundreds of Ås for metals to μms (1 μm is one millionth of a meter) for metallic oxides, to tens of kilometers for certain insulators. The penetration depth depends on the frequency-dependent conductivity of the sample. Buried interfaces can be studied if the materials on at least one side of the interface are thinner than the IR penetration depth, rendering that layer transparent to IR radiation.

3.1. Dielectric function

The *dielectric function*, ε, is a mathematical entity commonly used to describe (as a function of frequency) the interaction between light and matter. This function is closely related to the frequency-dependent index of refraction, $\tilde{n}$, that governs how light is bent by lenses. The index of refraction is separated into two parts, the real part n, which is proportional to the speed of the light as it passes through the material and the imaginary part k, which is a measure of how quickly the light is absorbed by the material. The speed of light in a material, v is equal to c/n, where c is the speed of light in vacuum. A strongly absorbing material that attenuates an incident beam is characterized by a large value of k. The dielectric function ε is also separated into two parts, the real part ε' and the complex part ε''. The dielectric function and the index of refraction are related to each other by $\varepsilon = \tilde{n}^2$.

3.2. IR absorption and energy and momentum conservation

For IR spectroscopy, the process of interest is absorption. IR photons are absorbed by electrons (if an electronic transition to an excited state is possible), or by vibrations that induce *dynamic dipoles*. Dynamic dipoles are oscillations in the density of electrons or electron charge due to atomic motion (the electrons follow the motion of the nuclei). Dynamic dipoles can absorb a photon when the photon's electric field is parallel to the charge oscillation, and when the frequencies of the light and the oscillation are similar (resonance). Dynamic dipoles occur in both molecules and extended systems, and their natural frequencies are related to the masses of the displaced atoms. Vibrational modes are also referred to as *normal modes*, which are defined so that each normal mode is independent of all the other modes. Due to energy and momentum conservation (the momentum of a photon is equal to h/λ or E/c), only photons with energy and momentum matching that of an available excitation may be absorbed; light of higher or lower frequencies cannot couple to the excitation. Sometimes additional momentum is gained by electrons when excited electrons scatter from atomic or ionic displacements or impurities. An absorption spectrum is a plot that shows how well different frequencies of light couple to excitations in the sample. An absorption spectrum is commonly plotted as one of the following (See Fig. 1): (1) Percent transmittance T vs frequency v, which plots the ratio of transmitted intensity to the incident intensity, or (2) absorbance A vs frequency v, which is related to the transmittance by $A = -\log T$. (It is conventional to convert the units for frequency v from Hz (s^{-1}) to wave numbers (cm^{-1}) by dividing v by the speed of light c.) Transmittance curves, shown schematically in panel (a) of Fig. 1, exhibit absorption dips at energies characteristic of excitations available in the material. No additional absorption for a sample compared to its background measurement is manifested as 100% transmittance. Absorbance curves, depicted panel (b) of Fig. 1, exhibit peaks at energies where the sample has absorbed energy from the incident beam. Zero absorbance means no energy has been absorbed.

3.2.1. Molecular vibrations

In an isolated molecule, the potential energy arising from bonding varies with internuclear separation as depicted in Fig. 2 (panel (a)); the minimum in the potential energy curve defines the equilibrium separation of the molecule in its ground state. Due to the constraints imposed by quantum mechanics, only certain vibrational excitations, having discrete energies, may exist. These allowed vibrational energies are represented by the horizontal lines in Fig. 2. The system may be excited to a higher vibrational state by absorbing a photon. Note, that in higher vibrational excited states, the nuclei are displaced more from their equilibrium positions than it is in the energetically lower states.

3.2.2. Bulk vibrational and electronic excitations

In extended, periodic systems, the discrete, allowed vibrational energy states described above broaden into allowed energy bands. Vibrational states in solids, called *phonons*, consist of displacements of chains of atoms. (See Fig. 2, top of panel (b)) Each phonon has a specific wavelength λ associated with it, as depicted in the figure. Also, in extended, periodic systems, electrons are free to move within the solid. As accounted for in quantum mechanics, electrons have a wave nature, and their motion may be described in terms of a propagating wave.

For both phonons and electrons, the wave number k is used to measure how many wavelengths can fit within a unit length and the linear momentum carried by the phonon or electron is given by $hk/2\pi$. Due to the quantum requirement

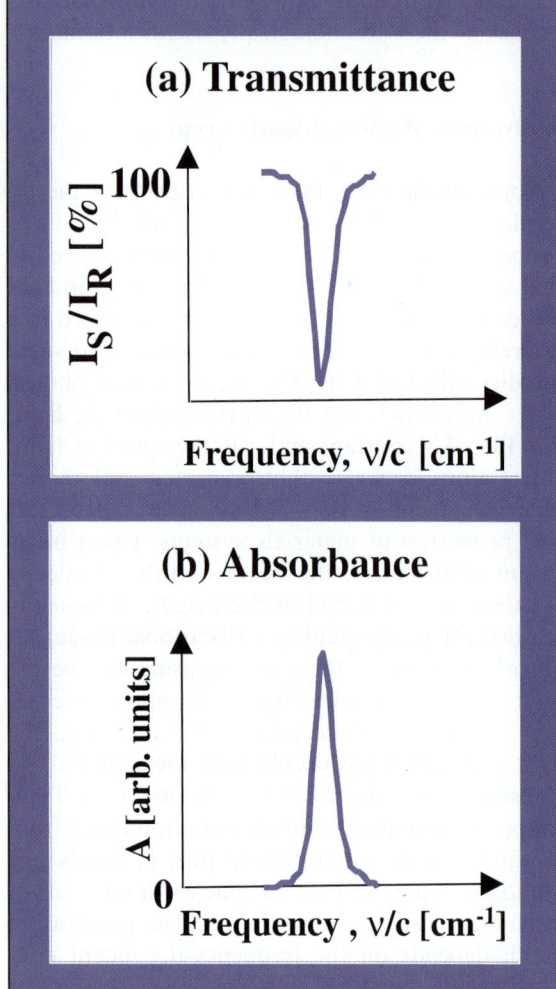

Fig. 1. IR absorption spectra are plotted as either the percentage of transmitted signal (top), or the absorbed signal (bottom), versus frequency v (conventionally, v is divided by c, where c is the speed of light). The transmittance is obtained by dividing a measured sample intensity I_S by a reference intensity I_R. Typically, a reference spectrum is obtained from a clean surface, from a sample held at a certain electrical potential or temperature, or by using a given beam polarization. In these cases, the spectrum of interest, the sample spectrum, may be obtained from an adsorbate-covered surface, from a sample held at a different potential or temperature, or using a different polarization, respectively. In transmittance discrete absorption dips are observed at the energies of vibrational excitations. The absorbed signal A is equal to $-\log(I_S/I_R)$ and absorbance spectra exhibit discrete absorption peaks at the energies of vibrational excitations. Broadband features typically arise from phonons (vibrations within an extended solid) or from electronic excitations.

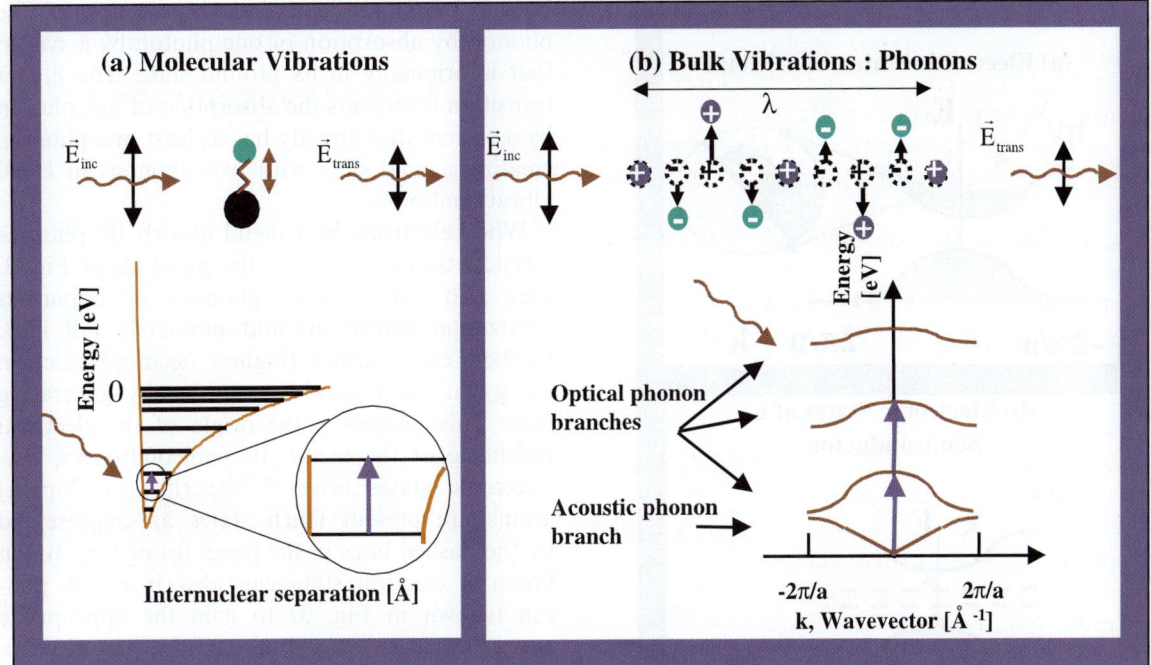

Fig. 2. IR photon probes allow the examination of the oscillations of charge around both vibrating nuclei in molecules and undulating chains of nuclei in extended materials (phonons). A plot of the typical effective potential energy vs internuclear separation for two bonded atoms is shown in the bottom of panel (a). The horizontal lines represent the allowed (discrete) vibrational energy states for a molecule, while the vertical (purple) lines are used to represent light absorption. In extended, periodic systems, the discrete, allowed vibrational energy states described above broaden into allowed energy bands. Vibrational states in solids, called phonons, consist of displacements of chains of atoms and has a wavelength λ, as shown at the top of panel (b) (the displacements of the nuclei are small compared to the distance between atoms). Phonons carry linear momentum, which is proportional to the number of waves per unit length. This quantity is known as the wave number k of the phonon and is equal to 2π divided by the wavelength of the phonon. For example, a phonon with wave number $k = 0.2\pi$ Å^{-1} = $2\pi(0.1)$ Å^{-1} contains 1/10 a wavelength in 1 Å, or a phonon wavelength of 10 Å. The momentum carried by a phonon is given by $hk/2\pi$. Dispersion curves, which plot energy E vs wave number k, are used to represent the allowed vibrational energies of extended periodic systems (bottom of panel (b)). Acoustic phonons correspond to displacements of the nuclei when a sound wave propagates through the sample, and optical phonons correspond to phonons that can be created by absorbing photons. The vibrational transitions excited by the absorption of an IR photon have been indicated in the energy diagrams (purple). Excitations to higher bands are allowed (as shown) but are less likely to occur.

for phonons and electrons in periodic solids, only certain vibration and electron energies are allowed. Dispersion curves, which plot energy E vs wave number k, are used to represent the allowed vibrational energies of extended periodic systems. (See Fig. 2, bottom of panel (b)) Dispersion curves are also used to describe allowed electronic energies. In panel (a) of Fig. 3, dispersion curves (solid lines) for the highest occupied (gray) and lowest unoccupied electronic states in a semiconductor are shown. In panel (b) of Fig. 3, dispersion curves (solid lines) for a metal are shown, with the ground state occupation indicated in gray.

3.2.3. Allowed transitions

Since IR beams have low energy and no mass, their momentum is small compared to that of electrons or vibrations in solids. Therefore, the excited state created after IR absorption has essentially the same momentum as the initial state (i.e., only "vertical transitions" between occupied states and unoccupied states in the dispersion curves are ordinarily allowed).

Figs. 2 and 3 show some typical transitions allowed when IR photons are absorbed. The bottom sketch in panel (a) of Fig. 2 shows a typical transition representing excitation of a vibration in an

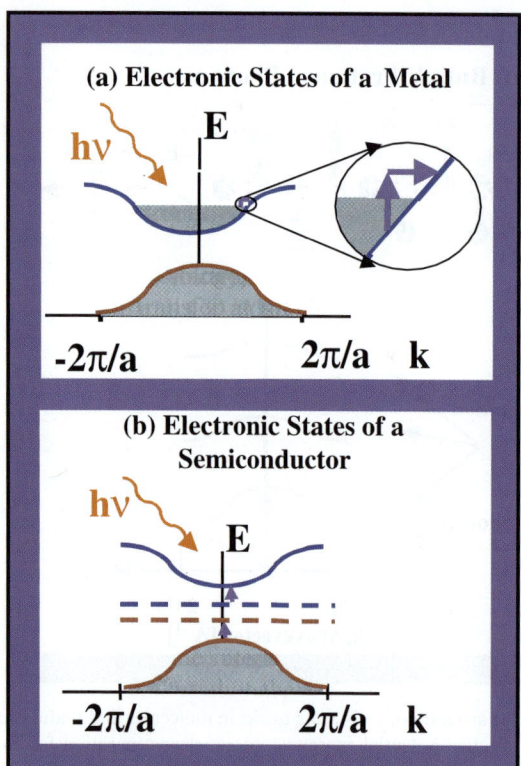

Fig. 3. Energy E vs wave number k dispersion diagrams for bulk metallic (panel (a)) and semiconductor (panel (b)) crystals, with typical electronic transitions (purple arrows) arising from an incident infrared probe depicted. In panel (a), dispersion curves (solid lines) for a metal are shown, with the ground state occupation indicated in gray. In panel (b), dispersion curves (solid lines) for the highest occupied (gray) and lowest unoccupied electronic states in a semiconductor are shown. Since IR photons have negligible momentum, only vertical transitions can be directly excited. The length of the unit cell is denoted by a, and is usually several Å. In panel (a) the electrons in metals close to the highest occupied energy level in a metal can gain momentum by scattering from phonons (horizontal transition) which give rise to the excitations depicted above when coupled with the vertical transition of the photon. In panel (b), for semiconductors, electrons can be excited from the highest occupied bulk level into the lowest impurity state (dashed line) if the impurity states are unoccupied. If the impurity states are occupied, a transition can be excited into the unoccupied bulk levels at higher energy.

isolated molecule. In the bottom sketch of panel (b) of Fig. 2, two typical vertical transitions representing excitations of phonons in a solid are depicted in a set of phonon dispersion curves. The bottom transition represents the creation of one phonon by absorption of one photon by a system that is originally in its ground state. The higher transition represents the absorption of one photon by a system that already has at least one phonon, creating a final state with two phonons in $k = 0$ vibrational modes.

When electrons in a metal absorb IR photons (vertical transition), as in the panel (a) of Fig. 3, they also scatter from phonons or impurities (horizontal transition) and ultimately fall back to the Fermi surface (highest occupied electron energy in the ground state), thereby generating heat. This process is the origin of the electrical resistance of the metal. In semiconductors, new electronic states arise if impurity (or dopant) atoms are present (such states are represented by the dashed lines in the panel (b) of Fig. 3). An impurity electron state can absorb an IR photon (shown in Fig. 3) to gain the appropriate energy to occupy a bulk conduction electron (normally unoccupied) electronic state. Alternatively, an electron in a semiconductor valence band (normally occupied) electronic state can absorb an IR photon to occupy an empty impurity electronic state.

Although IR photons do not carry enough energy to promote the transition of electrons across the energy gaps found in most semiconductors, nanostructured materials, in general, contain an array of electronic excitations that can absorb IR photons. One example, a localized quantum well, is shown in Fig. 4. The limited sizes of these materials result in fundamental energy separations between electronic energy levels of the order of meV to hundreds of meV.

3.3. Properties of a photon probe

Photon probes confer practical advantages. Since they are chargeless they can be used to interrogate insulating materials without complications due to charging. Photon beams are not deflected by strong electric or magnetic fields present in many sample preparation procedures. Moreover, the relatively weak interaction of IR light with some materials under a variety of environmental conditions, ranging from mTorr pres-

3.4. Surface-sensitive geometries

Most of the properties cited above that allow IR beams to reveal fundamental low-energy processes of solids are not specific to surfaces or interfaces. To maximize sensitivity to the surface, particular experimental geometries must be used [9,10]. Several of these are indicated in Fig. 6. The most favorable geometry for a given interface measurement depends on the properties of the materials on either side of the interface.

The grazing-incidence geometry depicted in the top panel of Fig. 6 is ideal to investigate a gas/metal interface, where the material on one side of the interface is transparent (gas) and the other highly reflective (metal). This technique is commonly referred to as infrared reflection absorption spectroscopy (IRAS). Dynamic dipoles with displacements perpendicular to the interface are excited by the polarization component in the incident beam that is perpendicular to the interface, and the excitations are called the *dipole-allowed* modes. The polarization component of the beam parallel to the surface destructively interferes with itself upon reflection, and cannot directly excite vibrations with displacements parallel to the interface, i.e., *dipole-forbidden* modes. The optimal angle of incidence is dependent upon the conductivity of the reflective material, and the energy of the excitation. For example, a broad range of incidence angles, ranging between 78–87° from the sample normal, is acceptable to measure high frequency C–O stretch modes (2000 cm^{-1}) for CO adsorbed on Cu with maximum sensitivity. Since, however, a narrower range of angles is needed to allow the electric field component polarized perpendicular to the surface to couple efficiently to lower frequency C–Cu stretches (associated with the bonding mode) at 300 cm^{-1}, incidence angles between 85–88° are ideal [9].

Investigations of the interface between two highly transparent materials (e.g., gas and an insulator like MgO) can be performed with the geometries indicated in either panel (b) or panel (c) of Fig. 6. Polarization and angle-dependent measurements are useful when using the transmission geometry (b). In general, when unpolarized light is used, the vibrations both parallel and

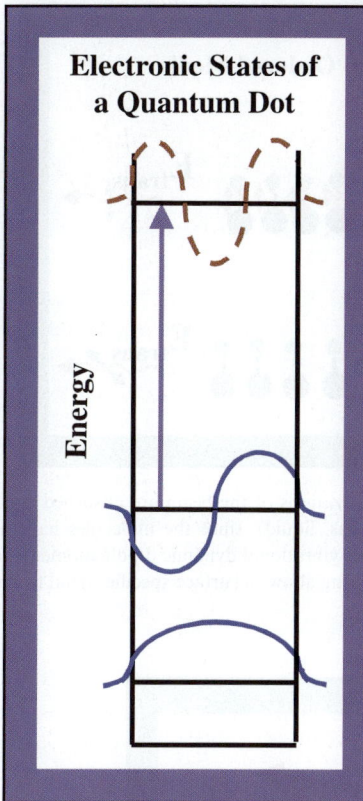

Fig. 4. Schematic of the lowest three discrete energy states of a quantum dot. Quantum structures support discrete electronic energy levels due to quantum confinement; a quantum dot is a 0-dimensional structure. The electronic transition (purple arrow) between the second and third states of the dot is shown; the energy of this transition typically lies in the IR range.

sures (environment with 100,000 fewer gas molecules per unit volume than atmosphere) of reacting gases to liquid cells, provides the opportunity to investigate immersed interfaces. Finally, the polarized nature of light presents another degree of freedom that may be exploited to provide sensitivity to orientational effects. Panel (a) of Fig. 5 illustrates that randomly oriented dynamic dipoles absorb photons of both polarizations of an incident beam. Dynamic dipoles that are oriented with a preferred direction absorb only IR photons that are polarized parallel to the dipole, as indicated in panel (b) of Fig. 5. Since species are often oriented at an interface, this discrimination can prove useful (as discussed in Section 4.3).

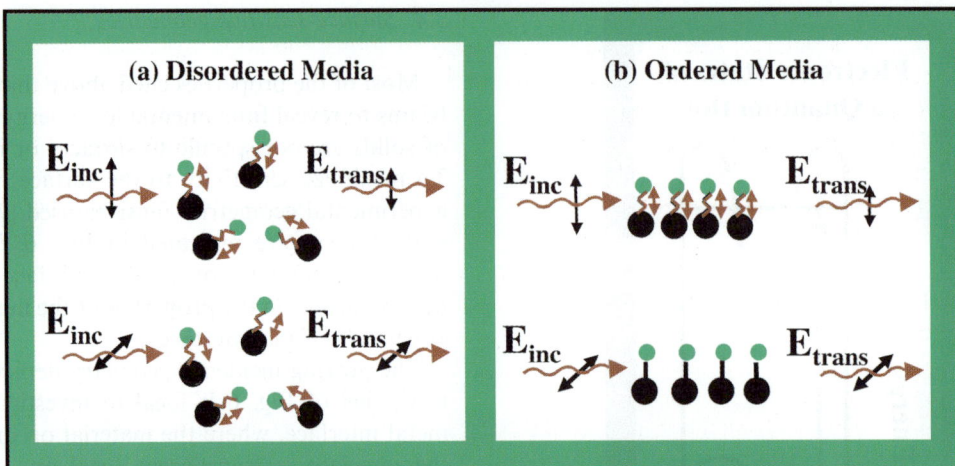

Fig. 5. Polarized light interactions with disordered and ordered media. Both polarizations of the beam are absorbed equally by the vibrational dynamic dipole moments for disordered molecules (panel (a)) (e.g., gas, liquid), since the molecules have all possible orientations. One polarization of the incident beam is preferentially absorbed by the vibrational dynamic dipole moments for ordered molecules (panel (b)) (e.g., molecules adsorbed on surfaces). This sensitivity to ordering allows a surface-specific signal to be recovered even in an elevated absorbing background using polarization modulation.

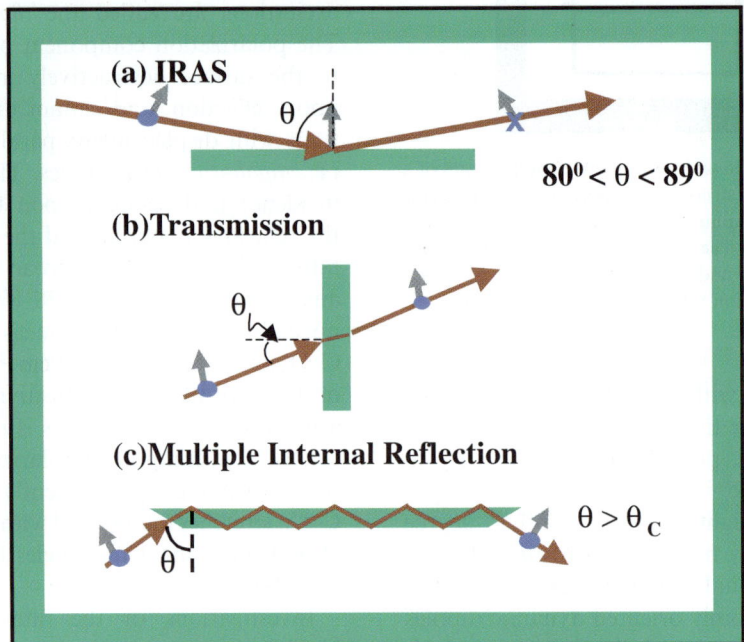

Fig. 6. Typical geometries for surface sensitive IR experiments including the polarization dependence. Geometries for IRAS (panel(a)) transmission geometry at non-normal incidence (panel (b)), and multiple internal reflection geometry (panel (c)). The red arrows represent the propagation direction of the IR beam, the blue circles and crosses represent the polarization direction out of and into the page, respectively, and the gray arrows represent the polarization in the plane of the page. The geometry in panel (a) is typically used for studies at metallic surfaces, where the beam couples only to dynamic dipoles perpendicular to the surface (dipole-allowed modes). Notice the change in the direction of the polarization into and out of the page upon reflection. The geometries depicted in panels (b) and (c) are typically used for surface studies of transparent media, e.g., semiconductors and insulators.

perpendicular to the interface are excited. However, when the light is polarized, selected dynamic dipoles are excited, from which the orientation of the dipoles can be inferred. The total internal reflection (TIR) geometry, indicated in (c), is achieved when the angle of incidence at the second interface is higher than the critical angle for reflection, trapping the photon beam in the solid. In this geometry, both polarization directions couple to vibrational modes at the interface, allowing investigations of vibrations both parallel and perpendicular to the interface. This TIR geometry is also used to investigate the interface between a transparent and absorbing medium (e.g., semiconductor and absorbing liquid). Further discussion of the applications of these geometries can be found in several review articles [11,12].

3.5. Experimental advances in IR sources, detectors and instrumentation

Advances in sample fabrication and preparation and experimental tools are responsible for recent scientific progress and new opportunities in IR examinations of surfaces and interfaces. Improvements in vacuum technology have revolutionized the ability to create and prepare samples of technological and fundamental interest. For example molecular beam epitaxy, and metallorganic chemical vapor deposition (MOCVD) growth techniques have become routine, affording the ability to manufacture novel nanoscale structures.

IR detectors have seen continuous improvement. One class of detector, the bolometer, operates on the principle that the resistance of the detector element is a sensitive function of temperature, and thus responds to IR light. New photoconductive detectors are available with greater spectral response. Many detectors have been developed to operate optimally at liquid nitrogen or liquid helium temperatures, thus reducing the noise from the thermal IR background found even at room temperature. In sum, stable interferometers and advanced detectors have substantially improved the spectral reproducibility over a wide frequency range.

Traditionally, a rod held at high temperature (a "blackbody source") provided the incident IR radiation. New bright sources of IR light, e.g., lasers, synchrotron radiation from electron storage rings [13], and FELs [14], are becoming available, and provide significant advantages for studying interfaces and surfaces compared to traditional blackbody sources. A bright source is necessary both for the grazing-incidence geometry that is required in several classes of surface/interface investigations, and for microscopy applications using the transmission geometry. Storage rings provide stable, broadband sources of light, allowing, in particular, high-quality examinations of low energy, weak absorption signatures. Lasers provide higher-intensity beams (on average 1000 times more photons) than either standard sources or storage rings. However, laser intensities can fluctuate and lasers operate over narrower bandwidths.

Electron storage rings, lasers, and FELs all produce polarized IR beams that are pulsed in time. This contrasts with traditional blackbody sources, which produce continuous, unpolarized IR beams. These properties afford exciting opportunities to investigate low energy, time-dependent phenomena. In particular, storage rings produce pulses between tens of picoseconds (trillionths of a second) to nanoseconds (billionths of a second) long, while FELs and lasers produce pulses as short as femtoseconds. The intensity of the pulses over the duration of the pulse is typically 1000 times larger for the lasers than for storage ring sources. Since excited electronic and vibrational states relax anywhere between femtoseconds to microseconds (millionths of a second) after excitation, pulsed sources provide new tools to investigate directly the relaxation processes for vibrations and electronic excitations at surfaces and interfaces.

4. Highlights of recent results

4.1. Vibrational excitations at interfaces

Surface and interface science spans the range from adsorption of gas molecules onto a solid surface to growth of ultrathin solid films onto a host substrate. In both cases, the vibrational excitations are modified at the interface. Adsorption of a molecule on a surface generates a mode corresponding to a stretch mode of the adsorbate–substrate bond. Moreover, the translational and rotational modes that are unconstrained in the gas phase become hindered when the molecule is tethered to the surface. The vibrational energies of even the internal molecular normal modes (stretching vibrations) are altered by the change in chemical environment. Finally, the molecule may dissociate, resulting in an obvious change of normal modes present.

One might think of comparing ultra-thin films to their bulk counterparts. New modes arise at the interface due to bonds between the host and film atoms, however, and these can reveal how the film is bonded to the substrate. Even the vibrational modes of the thin-film material may differ from those found in bulk material, because phonon frequencies can be affected due to a lack of long-range periodicity in one or more directions.

4.1.1. Adsorbate–substrate binding vibrations using IRAS (vibrations at interfaces)

Typically, lab-based IR examinations of adsorbate/substrate systems emphasize investigations of either the intramolecular modes of adsorbates, such as the C–O stretch of CO molecules adsorbed on metal substrates [15], or the stretch vibration between atomic hydrogen and the substrate. Three experimental factors are responsible for this focus. Firstly, intramolecular modes and hydrogen/substrate modes lie between 1000–3000 cm^{-1}, energies for which laboratory sources provide adequate intensity (the maximum intensity for 1200 K blackbody source is found near 2000 cm^{-1}). Secondly, at these higher frequencies, the optimal angle of incidence (defined from the normal to the surface) for the grazing-incidence geometry ranges between 78–87°. This range is far enough away from strict grazing incidence to allow sufficient intensity to be projected from a blackbody source onto the sample. Thirdly, the intensity of the absorption bands for a monolayer of adsorbates (approximately 10^{15} molecules/cm^2) can be as much as several percent. In contrast, adsorbate/substrate vibrational modes, which are directly related to bonding at the interface, have lower frequencies and stricter experimental requirements. Typically, they are weak absorbers and the optimal angle of incidence is larger, i.e., closer to grazing. This latter restriction, coupled with the reduced photon flux of a blackbody source at these lower energies, significantly limits the number of photons available to probe the adsorbate–substrate bond. This combination of weaker absorption bands and reduced signal renders lab-based investigations of adsorbate/substrate modes difficult at best. The technological advances mentioned in Section 3.1 provide the opportunity to garner direct information about adsorbate–substrate interactions, and open the possibility of investigating the interactions between atomic adsorbates (other than hydrogen) and substrates.

Recently, several groups [16–19] have reported direct observations of adsorbate/substrate vibrational excitations. Bradshaw and coworkers [16] reported the first observations of molecular adsorbate/substrate stretches for CO/Pt(1 1 1). A couple of years later, the first atomic adsorbate/substrate absorption measurements were reported by Greenler and coworkers [17] for O/Ag(1 1 1). More recently, Engstrom and Ryberg [18] have investigated O/Pt(1 1 1), where they have observed an extremely weak absorption band for the adsorbate/substrate stretch. These studies have all used blackbody sources.

Several examinations of adsorbate/substrate modes have exploited a bright storage-ring-based far-IR source. In one example, Hirschmugl et al. [19] studied CO adsorption on single crystal Cu surfaces, where the C atoms bind to individual Cu atoms (atop site) and the CO adsorbs molecularly. For this structure, two absorption bands with dynamic dipoles perpendicular to the Cu surface, the C–O and C–metal stretches, were expected. However, in addition to observing two absorption bands, an unexpected, discrete, anti-absorption feature and concomitant broadband absorption were observed as shown in Fig. 7. A common technique, isotopic substitution, was used to identify the origin of the unexpected features. Recalling that the frequency of vibration is inversely proportional to the square root of the mass of the participating atoms, one can shift the vibrational frequencies by changing the masses of the atoms in the molecules or adatoms. Isotopes are used to

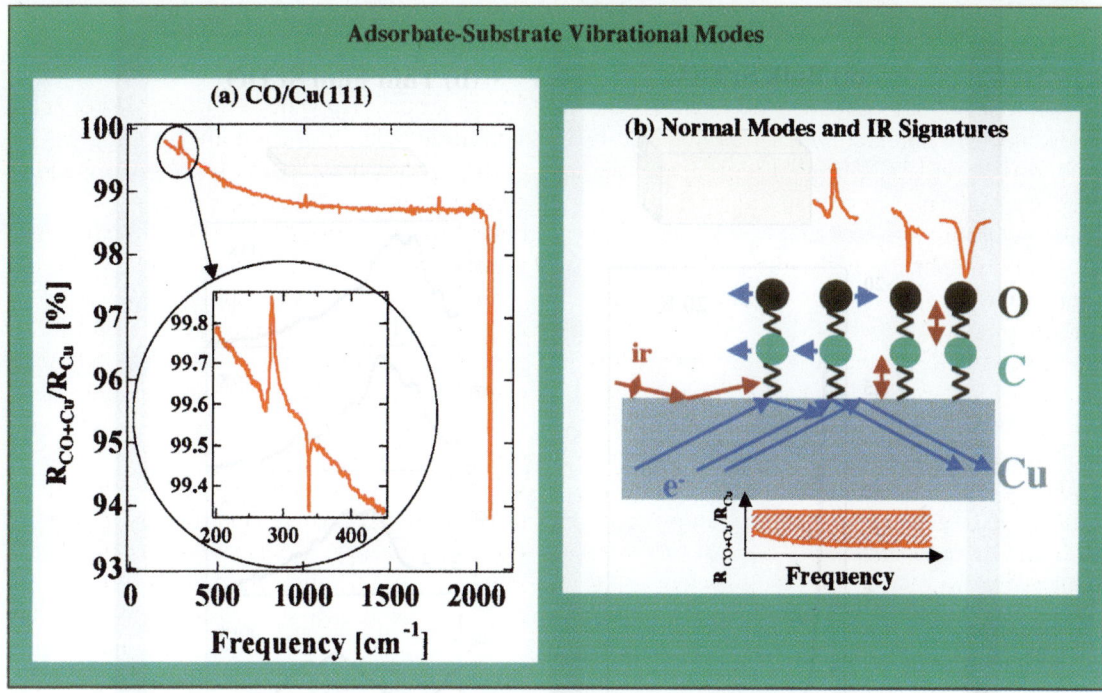

Fig. 7. Normalized IRAS spectrum taken with a storage-ring IR source for CO/Cu(1 1 1) (panel (a)). The origin of the absorption bands is from the C–metal and C–O stretches (displacements perpendicular to the surface), while the origin of the anti-absorption peak is from the hindered rotation (displacement parallel to the surface) (see panel (b)). The origin of the broadband absorption is attributed to substrate electron scattering from the adsorbates. Importantly, both the anti-absorption peak and the broadband absorption appear concomitantly, suggesting that the substrate electrons do not scatter as often from molecules with hindered rotation vibrations as other molecules.

keep the chemical environment unaltered. For example, for CO, one can use ^{13}CO, $^{13}C^{18}O$ or $C^{18}O$. Using these isotopes, the anti-absorption and low-frequency absorption features were attributed to the hindered rotation of the CO molecule (the C and O atoms oscillating parallel to the surface and out of phase) and the C–metal stretch, respectively [19]. Remarkably, the hindered rotation is observed despite the fact that its dynamic dipole is parallel to the surface. Since the electric field of the IR radiation that is parallel to the surface undergoes 180° phase change upon reflection, no electric field parallel to the surface should remain, and the excitation is forbidden by the dipole selection rules. The origin of the unexpected features is explained below. These results have broader implications, since they demonstrate the ability to measure adsorbate–substrate bonding vibrations directly. Thus, measurements of vibrations associated with bonding between heavier atoms, and molecules with surfaces, such as the interaction between oxygen and metals of interest for corrosion, or the interaction between heavy metal atoms and surfaces during growth of a new film on a surface, may be possible.

4.1.2. Phonons in ultra-thin films

Presently, there is enormous interest in identifying and perfecting the growth of dielectric thin films having a large dielectric constant for applications in high-density integrated circuits, since the enhanced dielectric constant should allow the production of more compact capacitor structures. Low energy, dynamical phenomena in dielectrics can play an important role in determining the low frequency dielectric constant. Bulk $SrTiO_3$, a ferroelectric material, possesses an extremely high dielectric constant (20,000) at low temperature,

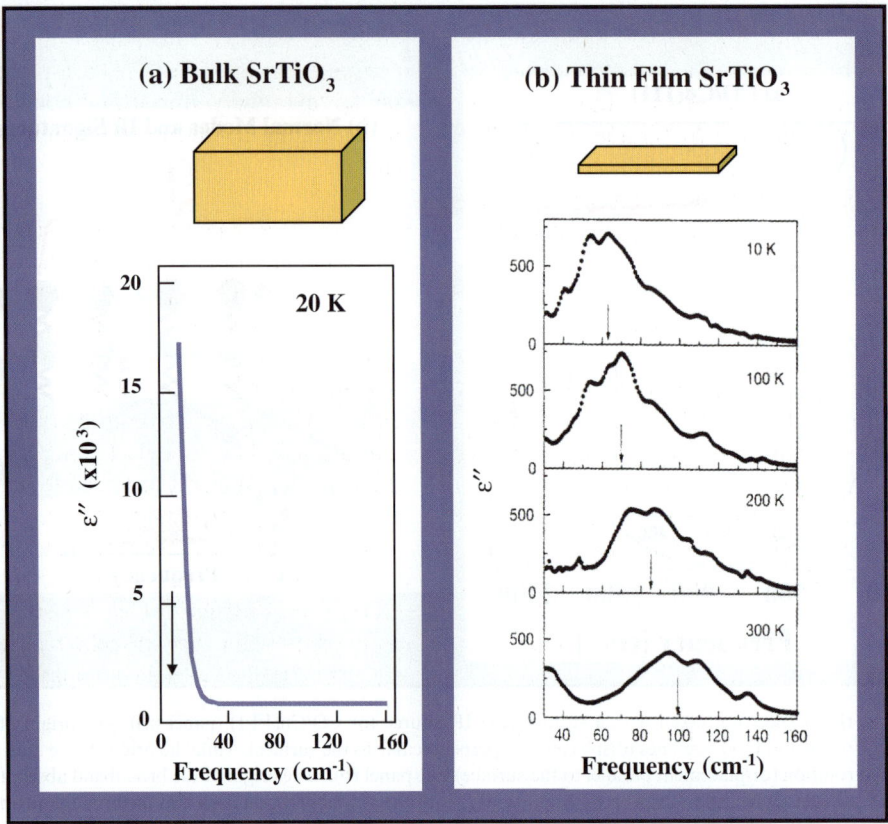

Fig. 8. Imaginary parts of the dielectric constant as a function of energy for a bulk sample (schematic, panel (a)) and for a thin film of SrTiO$_3$ (data, panel (b)). The bulk dielectric function exhibits very high response at low frequencies and temperatures due to a structural distortion and accompanying mode softening. The thin film results, from Sirenko et al., show the low-frequency phonon (soft mode) shifting to lower frequency as a function of cooling. However, the mode never shifts completely to zero frequency, indicating a fundamental limitation in the achievable dielectric constant for SrTiO$_3$ thin films. Therefore thin SrTiO$_3$ films cannot provide the high dielectric constant that bulk materials can provide. Panel (b) is reproduced from Ref. [20] by permission of *Nature*.

several orders of magnitude greater than typical "high" dielectric constants found at room temperature. However, thin SrTiO$_3$ films fail to achieve similarly large low temperature dielectric constants. Recently, an IR examination of the low-frequency response of a thin SrTiO$_3$ film, exhibiting high crystalline quality, was undertaken. Using a storage ring source [20], the far-IR ellipsometric (polarization-dependent) response of the film was measured at grazing incidence, from which the real and imaginary parts of the frequency-dependent dielectric function, ε' and ε'', can be directly determined [21].

Sirenko et al. [20] identified that the static dielectric constant (ε' and ε'' at zero frequency) for SrTiO$_3$ thin films reaches a fundamental limit originating from the behavior of the lowest-frequency phonons. In *bulk* SrTiO$_3$, as the sample is cooled to a critical temperature T_c, lattice instabilities shift the lowest optical phonon mode to even lower frequencies. The frequency of this normal mode shifts to zero at the critical temperature, which leads to a large, increasing ε'' as the frequency approaches zero cm^{-1}, (i.e., dc field), as shown in Fig. 8. This frequency dependence of ε'' is indicative of the material supporting a low frequency, very long wavelength phonon. For the same conditions under which the frequency of the bulk phonon approaches zero, ellipsometric IR measurements of *thin-film* SrTiO$_3$ show a smaller shift in the lowest frequency phonon mode. Since the width of the phonon mode is still finite, and its frequency is not close to zero, the dc dielectric constant is not as large for the thin film as for the bulk material. As suggested by Sirenko et al. [20], this important result could have implications for other ferroelectric materials, and for the "size effect" on ferroelectric properties in polycrystalline ceramics.

4.2. Low energy electronic excitations

4.2.1. Interfacial electronic excitations

The broken periodicity encountered at surfaces and interfaces can support low energy electronic states that are not present in the bulk phase. Recent studies have revealed several interfacial low energy electronic excitations not previously observed. These surface-dominated phenomena have important implications for electron transport and heat generation in small structures of current fundamental and practical interest.

4.2.1.1. Broadband electronic excitations: scattering of free electrons at interfaces. Surprising lineshapes have been observed for IRAS investigations of CO/Cu, including an anti-absorption band and concomitant broadband absorption [22] (see Fig. 7). The broadband absorption originates from substrate electrons absorbing IR radiation when the scatter from impurities (i.e., adsorbates) at the surface (see Fig. 3, panel (a)). The unusual anti-absorption feature (i.e., peak) occurs because the electrons do not scatter off of the adsorbate at this specific frequency. As previously mentioned, this frequency corresponds to the hindered rotation of the adsorbed molecule. Thus, at this frequency, the substrate electrons for an adsorbate-covered surface do not absorb more energy from the incident field than they do for a clean surface. The adsorbate vibration with displacements parallel to the surface resonates with the electron oscillations in the metal. Importantly, the broadband absorption strength is dependent upon the molecular species adsorbed. An extremely weak broadband absorption occurs when formic acid or oxygen is adsorbed on Cu [23] and unusually strong band when ethylene is coadsorbed with CO [24]. The associated electron scattering is responsible for an increase in resistivity in thin Cu films of up to 30% compared to bulk Cu. This dramatic increase may be an extremely important consideration when using Cu as a thin-film interconnect in computer chips.

4.2.1.2. Charge density waves. Materials in which electronic motion is nearly one-dimensional or two-dimensional (2D) sometimes undergo a spontaneous redistribution of charge that lowers the symmetry and energy of the system. This redistribution results in a periodic charge density corrugation, known as a charge density wave (CDW), and also modifies the electronic band structure of the material. While it has been long suspected that the lower symmetry of surfaces and interfaces could support a CDW, only recently there have been claims of direct observations. Evidence for CDW formation in Pb/Ge(1 1 1) was presented in a combined scanning tunneling microscopy and electron energy loss spectroscopy experiment [25]. IRAS also provides a potentially powerful tool to examine surface and interface CDW at high resolution, since the charge redistribution sets up a charge dipole that partially screens an aligned electric field.

Unusual spectral features observed in IRAS [26] for alkali/Cu(1 1 1) interfaces are attributed to a pinned surface CDW, shown in Fig. 9. An anti-absorption band is observed for metallic layers of K and Na adsorbed on Cu, and occurs at approximately the same frequency for both K and Na adsorption. Since there is a negligible shift in frequency with the adsorbate mass, this spectral feature cannot arise from a vibrational excitation of the adsorbate. Moreover, the energy of the excitation is higher than the top of the phonon band in the substrate, and the excitation therefore cannot be attributed to a vibration in the substrate. In contrast, IRAS spectra for insulating overlayers, e.g., oxidized potassium overlayers or sparse alkali layers (below 0.2 ML alkali coverage) on Cu(1 1 1) exhibit no change in IR reflectivity at the frequency of the anti-absorption band. These insulating layers have no free charges that can be transferred to the substrate, but the metallic K and Na layers transfer some of their charge to

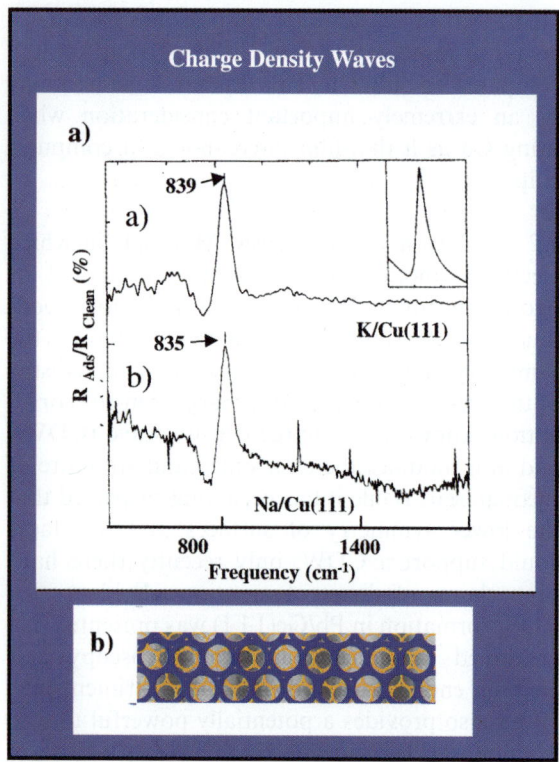

Fig. 9. IRAS spectra for (a) K/Cu(1 1 1), and (b) Na/Cu(1 1 1), showing unexpected anti-absorption features. Theoretical predictions (inset in panel (a)) and experimental results suggest that this feature is due to a CDW at the interface (panel (b)). The CDW is represented by a grayscale variation indicating redistribution of the "free electron charge density" at the interface into a periodic charge density distribution with 15 Å intervals between peak charge densities. The large and small circles represent alkali and Cu atoms, respectively.

the normally unoccupied Cu(1 1 1) surface states. Consequently, the origin of the anti-absorption feature has been attributed to the donated charge forming a pinned charge corrugation (CDW) at the alkali/Cu(1 1 1) interface.

4.2.1.3. Interfacial charge dynamical transfer. Upon molecular adsorption at a surface, the highest occupied molecular orbital of the molecule can hybridize with the substrate electronic states. When the resulting mixed state is partially occupied, and a vibration of the adsorbate modulates the energy of the state, then electron charge redistribution can be induced at the interface, and the charge density oscillates with the vibration. This phenomenon is called interfacial dynamical charge transfer (IDCT) [27]. A combined IR and sum-frequency generation (SFG, which is explained in detail in another article [28] in this volume) experiment has observed the dependence of adsorbate electronic structure on the IDCT. IRAS of C_{60} adsorbed on Ag(1 1 1) reveals an absorption band corresponding to a vibrational mode of C_{60} that cannot absorb IR light for molecular C_{60}. (Such a mode is said to be "IR-inactive") This mode is instead normally "Raman-active," meaning that the mode can lower the frequency of incident visible light by a process known as Raman scattering [27,29]. The unusual IR activity of this mode upon adsorption is attributed to IDCT. A feature at the same energy as in the IR experiment was observed in the SFG experiments, which was attributed to a Raman active vibrational mode of C_{60}. The authors conclude that the IR activity of this mode is due to the existence of IDCT at the C_{60}/Ag(1 1 1) interface.

4.2.2. Electronic excitations in nanostructures

Quantum confinement effects can be realized in materials in which electronic motion exhibits the effective dimensionality of two, one, or zero. A surface state is a familiar example of a 2D electronic system, i.e., electrons in a surface state are free to travel in a plane, but are (exponentially) confined in the transverse dimension. Such states can also be formed at internal interfaces. Moreover, a 2D quantum well may be formed by sandwiching a very thin layer of one material between two other layers with a contrasting band gap. Systems where the electrons are confined to one dimension are typically called quantum wires, and, finally, systems where all three spatial dimensions are nm scale are called quantum dots.

4.2.2.1. Quantum dots. Initially, quantum dots were formed using lithographic techniques to build up buried, nm-scale "dots" one at a time. More recently, arrays of highly uniform dots have been constructed exploiting the growth mode of many lattice-mismatched semiconductors. For example, deposition of InAs on GaAs results in one monolayer (ML) of continuous, lattice-matched (and hence strained) InAs, followed by the formation of small, three-dimensional clusters of strain-relieved InAs (this growth mode is called "Stranski-Krastanov" or layer-plus-islanding). These clusters can be prepared to be highly regular and uniform. The resultant array of dots can then be capped with more GaAs or AlGaAs material, isolating the quantum dots.

If such an array is constructed to be part of a field-effect device such as that found in a simple transistor, then the chemical potential of the electronic system of the dot can be varied by varying the gate voltage. This affords the exciting possibility of tuning the occupancy of the dots. As a result, the spectroscopic and transport properties of dots with various filling factors may be probed. In essence, one can add or subtract electrons to this "artificial atom" at will.

Collective charge density excitations of self-assembled InAs quantum dots embedded in GaAs have been measured with IR photo-absorption spectroscopy as a function of applied magnetic field [30]. The cylindrical dots were approximately 7 nm tall and 20 nm in diameter. Since the incident beam can only be absorbed by electrons that undergo a dipole-allowed transition, and the original occupation of the quantum dot was controlled by the applied gate voltage, it was possible to identify the origin of the electronic transitions. For example, when the gate voltage was tuned such that an electron filled only the lowest normally unoccupied state, only one excitation band was observed, which corresponded to the 1S-to-1P transition (lowest energy state to the next-highest energy state). Ultimately, it was possible to identify which electron states were filled as a function of the externally applied gate voltage by the number of electronic transitions observed in the IR absorption spectrum.

4.3. Chemical and structural identification

4.3.1. In situ observations of reactions
4.3.1.1. Electrochemical cells. Electrochemical cells are used to promote the conversion of electrical energy into chemical energy and vice versa. Electrochemical interfaces are rich with molecular reactions—spontaneous oxidation–reduction reactions that generate electrical current and induced oxidation–reduction reactions that occur in response to an applied voltage. These processes have been studied for nearly two hundred years, with reaction mechanisms being inferred from voltammetry experiments (i.e., monitoring the current induced as a function of the voltage applied). Direct probes of molecules and atoms present at the electrode surfaces have not been available until recently. IR spectroscopy, providing an in situ probe of the constituents adsorbed at the interface, has proved a valuable new tool for understanding reaction pathways in this complex environment. IR spectra acquired at different points in the voltammogram can be compared to unveil the electrochemistry.

In situ IR spectroscopy has been instrumental in identifying hydrogen dynamics at the GaAs(1 0 0)/electrolyte interface. Erne and collaborators [1] have used a TIR geometry (See Fig. 6) to investigate this interface in both acidic and basic environments. In particular, hydrogen binds to As sites at the surface when the GaAs electrode is electron rich. Conversely, when the GaAs electrode is electron poor, the hydrogen adsorbates are replaced by OH species at the As sites. The reader is cautioned that these measurements are sensitive only to *changes* at the interface, and therefore are not sensitive to adsorbate occupation that does not change over the voltammogram cycle. One key IR absorption spectrum that supports the proposed conclusions is shown in Fig. 10.

Weaver [2] has recently compiled an extensive review, comparing a wide variety of interfaces in ultra high vacuum (UHV) environments (environment with one ten-trillionth of the gas molecules per unit volume compared to atmospheric pressure) and equivalent electrochemical interfaces. He concludes that CO and NO molecules adsorb to different sites on metallic surfaces for the two different environments. In the electrochemical case, Weaver suggests that the presence of the water surrounding the adsorbates changes the energetics of the system dramatically, making it more energetically favorable for CO and NO to adsorb to bridge sites. Under UHV conditions, CO and NO are found to adsorb at atop sites.

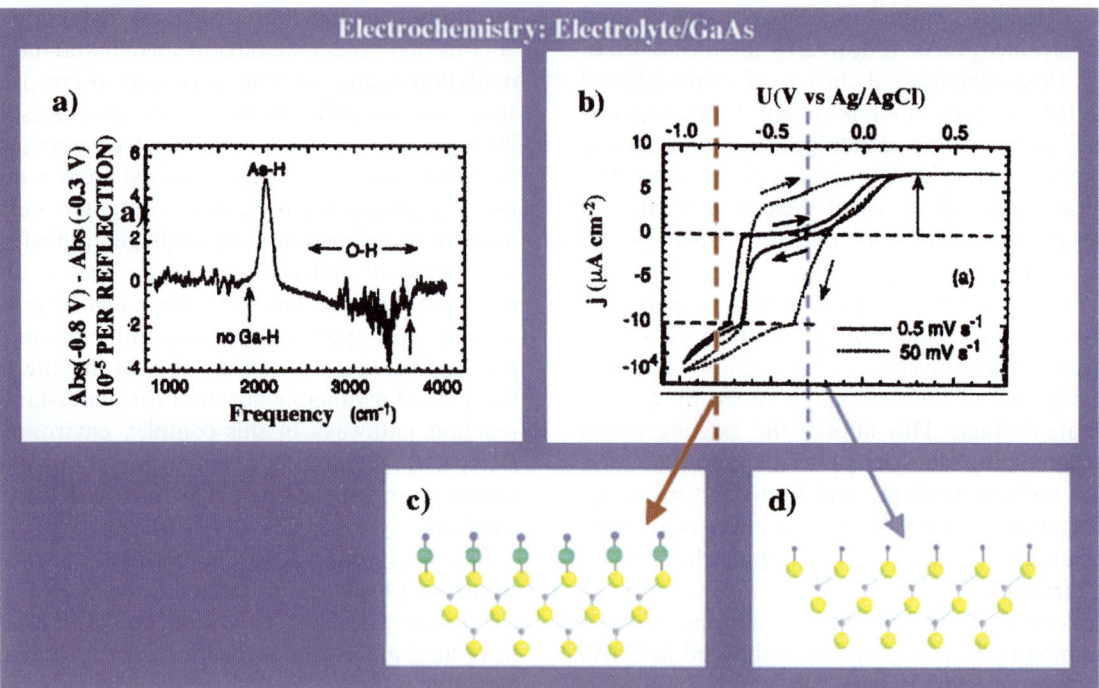

Fig. 10. Difference between the IR absorption spectrum for the electron-rich GaAs electrode (at −0.8 V potential) and a similar spectrum for the electron-poor GaAs electrode (at −0.3 V potential) is used to identify which sites are active during the electrochemical process. In this spectrum, a peak corresponds to a vibrational frequency that absorbs when the potential is −0.3 V, and stops absorbing when the potential is −0.8 V. Anything that remains the same for both conditions will not contribute to this spectrum. Thus H–As bonds (vibrational frequency 2005 cm^{-1}) exist when the potential is −0.3 V (panel (c)), and they are replaced by hydroxyl species when the potential is −0.8 V (panel (d)). Absence of another feature reveals no electroactive H–Ga species. Panels (a) and (b) are reproduced with permission from Ref. [1]. Copyright 1998 by the American Society.

4.3.1.2. Catalysis. Beitel et al. [31] used polarization–modulation IRAS (PM–IRAS) techniques for in situ investigations of a commercially important, high-pressure reaction. The PM–IRAS technique compares the IR absorbance of the two components of the beam having the two possible orthogonal polarizations. Due to the preferential coupling of the IR beam to molecules whose dynamic dipoles are aligned with the electric field, this technique is particularly sensitive to the presence of ordered molecules (uniformly adsorbed on a surface) in a sea of disordered molecules (e.g., a liquid or gaseous environment) (see Fig. 5). Beitel and coworkers have examined a Fisher–Tropsch (FT) reaction at a single crystal Co(0 0 0 1) surface. The FT reaction is an important class of catalytic reaction, from which long-chain hydrocarbon reactants are produced by mixtures of CO and H$_2$. The textbook reaction is represented by:

$$3CO + 5H_2 \rightarrow C_3H_4 + 3H_2O$$

This reaction occurs at elevated temperatures and pressures over group-VIII transition-metal surfaces. Typical reaction temperatures and pressures range between 450–570 K and 5–30 mbar. The authors have shown that Co defect sites play an important role in catalyzing this reaction by investigating both annealed and sputtered surfaces in the presence of H$_2$/CO mixtures (syngas) at a range of pressures and temperatures up to 300 mbar and 490 K. Polarization–modulation is crucial to the success of these experiments because it discriminates against IR absorption signals from the gas, and rejects IR emission signals from the heated sample, allowing small absorption changes at the reaction surface to be measured with phase-locked techniques.

Beitel and coworkers find that CO molecules depopulate defect sites in the presence of syngas as the temperature and gas pressure is raised for both annealed (almost defect free) and sputtered (full of defects) Co surfaces. They hypothesize that the defect sites become populated at high temperatures or pressures with the reaction product, namely, hydrocarbon chains. While there is spectral evidence that the terrace sites are populated by a mixture of CO and H adsorbates at reaction temperature and pressure (see Fig. 11), there is no direct evidence for the presence of hydrocarbon chains under these conditions. However, subsequent X-ray photoelectron spectroscopy (XPS) studies in UHV conditions indicate the presence of several distinct carbon species, including features typically associated with hydrocarbons [31]. The lack of IR spectral signatures for the hydrocarbons at high pressure could be due to one or more of the following reasons. Firstly, if C–C stretches and C–H stretches are oriented parallel to the surface, they are dipole forbidden. Secondly, the hydrocarbon chains occupy defect sites that lack homogeneity, increasing inhomogeneous broadening which produces a broader, weaker IR absorption signal. Thirdly, the C–C stretch and the C–H stretch can exhibit weak dynamic dipole moments in unfavorable cases [32].

4.3.2. Interface structure: devices

Structural information about technologically relevant oxide/metal and oxide/semiconductor interfaces is another area where IR spectroscopy can provide new information. For instance, the vibrational energy of the oxygen–substrate stretch is

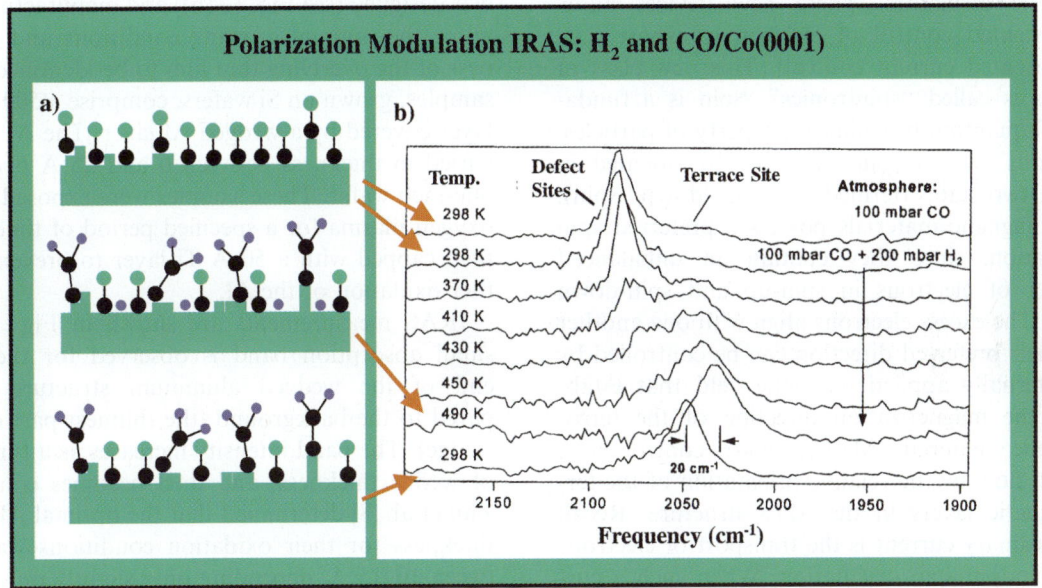

Fig. 11. Sequence of PM–IRAS curves taken for a sputtered Co(0 0 0 1) surface during exposure to mixtures of H_2 and CO, with the temperature and pressure for each spectrum indicated in the figure. Differential PM absorption from the gas-phase signal is zero, since absorptions from both polarizations of the incident IR beam are equal. However, the differential absorption from the adsorbate species is large, since only the p-polarized infrared beam can be absorbed by adsorbate species due to the metal surface-selection rule. The diagrams to the left show the interface structure under different reaction conditions, as concluded from the interpretation of the IR data and subsequent XPS data. Initially, CO occupies a-top sites on the Co at room temperature. Under higher temperature and pressure conditions H starts to occupy some available sites, and hydrocarbons form at the defect sites. Finally, at room temperature and under vacuum conditions the hydrogen desorbs from the Co, while the CO and hydrocarbons remain. Panel (b) is reproduced with permission from Ref. [31].

used as a guide to determine the structure of the interface. For example, Stefanov et al. [33], using a combination of IR absorption measurements and first principles theoretical calculations, have identified that the SiO$_2$/Si interface contains a Si$_2$O$_3$ and Si$_2$O$_5$ (Si epoxide) species. They have studied this system using the transmission geometry, to increase their sensitivity to a wider spectral range, and determined the energetically favorable geometries for Si–O clusters using first principles, gradient-corrected density functional calculations. Subsequently, using the resulting cluster geometries from the first principles calculation to predict vibrational excitations for the different clusters [34], they were able to identify the nature of the clusters at the interface. This work is reviewed further in this volume by Weldon et al. [35].

4.3.2.1. Spintronics. Magnetic tunnel junctions (MTJ) are multilayer sandwiches of two thin film ferromagnetic conductors separated by non-magnetic insulators. These new devices allow spin-oriented control of electron transport (i.e., spin-oriented current control). This new class of devices is called "spintronics". Spin is a fundamental quantum mechanical property of particles. Electrons have a spin that may be oriented in one of two states (termed "up" or "down" spin). Ferromagnetic materials possess a preferred spin orientation, since they contain an unbalanced number of electrons in spin-up and spin-down states. The excess electrons align with one another and their preferred direction can be controlled by an externally applied magnetic field that establishes the magnetization direction of the ferromagnetic material. MTJ provide control over current flow by the relative orientation of the ferromagnetic layers in the MTJ structure. Recall that ordinary current is the transport of electrons with random spin orientation. When such a current encounters a MTJ with ferromagnetic layers whose preferred spin orientations are all aligned, only the electrons with similar spins pass unhindered through the device (low resistance device). If some of the ferromagnetic layers' spin orientations are anti-parallel to the spin orientation of the others, no current can pass (high resistance device). The situation is analogous to light passing through polarizing filters, which can either be aligned (to transmit light) or crossed (to block all light). Since the preferred spin orientation of the layers can be changed by an external field, a MTJ constitutes a switch, which is a fundamental building block of memory circuits. Since the spin states of these devices are set by a magnetic field and then remain fixed, they do not need constant power to retain their status, unlike capacitor-based devices currently used, and therefore represent an attractive alternative for future computer memory.

Improved multilayered structures for magnetic hard drives are another important area in which interface structures can be inferred from a non-invasive IR probe. Zhu et al. [8] have investigated the effect of plasma oxidation on aluminum–cobalt bilayers. In these devices, it is crucial to create a thin insulating surface layer next to a magnetic metallic layer, but the underlying magnetic layer must not be oxidized during the process. A sample whose thickness varied along its length (a "wedge") was designed (see Fig. 12 (b)) and manufactured to allow the optimal growing conditions and thickness of the overlying thin film to be identified. The samples, grown on Si wafers, comprise a 500 Å Co layer covered by a wedged Al layer. The Al layers varied in thickness between 0 and 25 Å over the one-inch wafer. These samples were exposed to an oxygen plasma for a specified period of time, and then capped with a 50 Å Ti layer to prevent further oxidation of the Al.

IRAS measurements are shown in Fig. 12. A small absorption band is observed for the thin end of the wedged aluminum structure compared to the background (the thinnest part of the wedge). The band intensity increases as a function of wedge thickness, and then becomes constant. Zhu et al. [8] determined that the optimal Al layer thickness for their oxidation conditions was between 10–15 Å, depending only slightly on oxidation times between 30 and 600 s. There were clear indications that the Co layer was not oxidized under any of these conditions—suggesting that the barrier for cobalt oxidation is sufficiently high to allow high quality, smooth aluminum oxide/cobalt interfaces to be formed. These conclusions were verified by magneto-resistance measurements, that are sensitive to the presence of an unwanted Al

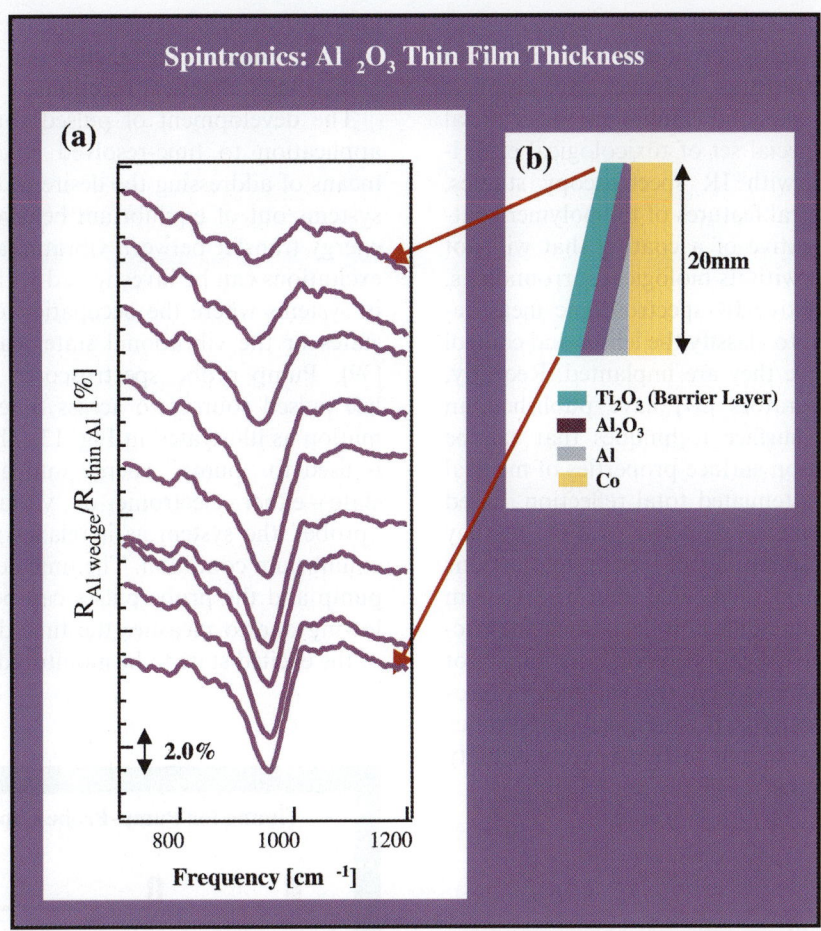

Fig. 12. IR reflection absorption spectra from an Al-wedge/Co multilayer device, oxidized for 480 s in plasma oxygen. A schematic of the sample geometry is shown on the right. An Al wedge (thickness = 5–25 Å) is grown on top of a Co film (500 Å) and then exposed to plasma oxidation. Then the sample is capped with a Ti_2O_3 barrier layer (50 Å) to prevent any additional oxidation of the Al film. Since the Ti_2O_3 film is a uniform thickness and IR transparent, no Ti_2O_3 features are observed in the transmittance as a function of position. The reference scan was taken at the thin end of the Al wedge, and the rest of the spectra were collected at 2 mm intervals along the length of the sample. An IR absorption feature (at $\sim$960 cm^{-1}), identified as an Al–O longitudinal optical phonon, initially shows increasing intensity and shifting frequency as a function of wedge thickness, and then becomes constant for the remainder of the sample. This series of measurements clearly demonstrates that the maximum thickness of the aluminum oxide film in this sample is 12 Å. The lack of a Co–O phonon also suggests that the Co has not oxidized.

metallic layer, or an undesired CoO layer, at the interface, since the electron transport would be affected in either case.

4.3.3. Interface chemistry: biocompatibility

Understanding interfacial reactions between biomaterials (materials that are used for implants and other biomedical purposes) and their biological environments is a key step in gaining fundamental knowledge in the area of biocompatibility. In many cases it is desirable to learn how to modify the surface of proposed biomaterials, without affecting their bulk properties. A chemical understanding of these surfaces can provide insight into the nature of biological reactions, which is required to assess the ultimate compatibility of the proposed materials and their living environments. Polymer coatings are a common surface termination, and can readily be investigated by IR spectroscopy. For example, Afanasyeva and Bruch [36] have investigated polysiloxane polymer and sapphire materials, which are good candidates for artificial lenses. Using a special set of toxicological experiments combined with IR spectroscopy studies, they identify spectral features of the polymer coating that are indicative of a coating that will not adversely interact with its biological surroundings. Thus, non-destructive IR spectroscopic measurements can be used to classify the lenses and control their quality before they are implanted. Recently, Reid and collaborators [37] have published an article reviewing surface techniques that can be used to shed light on surface properties of medical implants. Using attenuated total reflection, based on the TIR geometry depicted in Fig. 6, they measured the IR signatures of biofilm interactions with bacteria, which indicated that the biofilm acted as an adequate diffusion barrier for the bacteria. Other IR investigations reveal the impact of biological environments on the surfaces of proposed biomaterials, which is of great importance to gain an understanding of the biocompatibility of biomedical implants [38].

5. Outlook

5.1. Frontier interdisciplinary techniques: microscopy and time-resolved spectroscopy

Frontier research in many disciplines requires the ability to investigate smaller structures (down to Å spatial resolution) and faster processes (down to femtosecond time resolution). For example, the role that interfaces play in heterogeneous materials, and the identity of the individual constituents of these materials, can be addressed with probes having improved spatial resolution. The development of time-resolved studies is imperative to gain fundamental knowledge about changing systems. Dynamic systems are not only more common than those in equilibrium, but also more important, since all processes of interest are by definition non-equilibrium. For example, the fastest time-resolved studies (fs resolution) focus on examining the fundamental interactions between electron motion and nuclear displacement, which can control macroscopic effects such as diffusion, bonding, dissociation, and chemical reactions.

The development of pulsed sources, and their application to time-resolved studies, provides a means of addressing the desire to understand how systems out of equilibrium behave. For example, energy transfer between vibrational and electronic excitations can be investigated with pulsed sources in systems where the occupation of the electronic states or the vibrational states can be controlled [39]. Pump–probe spectroscopic techniques utilize pulsed sources to access time-resolved information as illustrated in Fig. 13. The pulsed source is used to "pump" energy into a chosen excited state—either electronic or vibrational—and to "probe" the system as it relaxes into its relaxed, equilibrium condition. The time delay between the pump and the probe pulses can be controlled, allowing one to measure the time-dependent decay of the excited state to be monitored. This approach

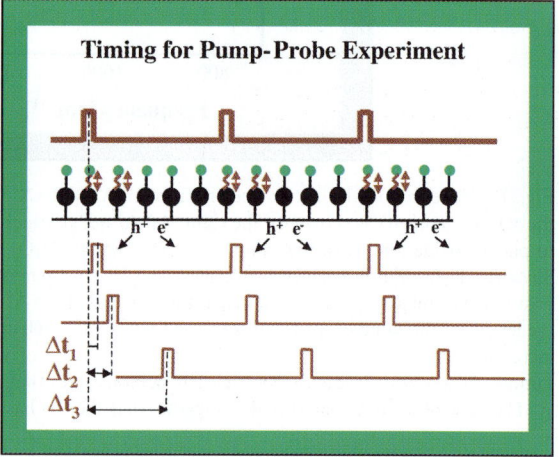

Fig. 13. A schematic of a pump–probe, time-resolved experiment. The top trace represents the arrival of a pump pulse at the surface. The molecules are vibrationally excited at that time as shown below. The rest of the traces represent probing pulses of IR arriving at three different time delays. The pulses arriving at Δt_1, Δt_2, and Δt_3 can probe the system in its excited state, intermediate state, and relaxed state, respectively. The relaxation can occur through electron–hole pair creation (shown) or excitation of a phonon. Direct measurement of the lifetime of a vibrational excitation can shed light on the mechanism of energy dissipation.

directly measures the change in the response of the system during the relaxation process.

5.2. In situ chemical reactions: time-resolved investigations and real-time growth diagnostics

As highlighted above, IR spectroscopy is an extremely useful tool to identify chemistry at surfaces under reaction environments, and is even more powerful with the polarization–modulation technique. One can gain insight into the reaction kinetics at these interfaces by in situ time-resolved measurements. As illustrated by Ruett-Robey and collaborators [40] it is possible to monitor the time dependent IR absorption of diffusing species at a surface by controlling the arrival of molecules at the surface with a pulsed molecular beam and rapidly collecting spectra after each 600 μs pulse. This approach allowed an experimental time resolution of tens of ms. In future experiments, pulsed molecular beam methods could be combined with a step-scan technique to achieve greater time resolution (these methods are discussed in detail by Palmer and coworker [41] and references therein). This technique is alternatively referred to as dynamic IR linear dichroism, and offers the possibility of collecting time-dependent chemical information about the surface reaction. Recently, several IR instrument manufacturers have incorporated control systems to routinely perform step-scan measurements with new spectrometers.

Growth of materials within high-pressure reaction environments, such as MOCVD, is another area where IR spectroscopy is poised to play an important role. MOCVD is the dominant form of thin-film growth used in industry. The process relies on a gas stream to deliver chemical precursors to a heated substrate. The desired compounds are deposited at the growth surface by thermally driven chemical reactions. The ability to obtain in situ IR absorption spectra allows chemical identification at the growth interface, which would provide an invaluable diagnostic. In particular, one could determine the concentrations of constituents at the surface in real time during growth using polarization–modulation techniques, and could tune gas composition or substrate temperature and response.

5.3. IR microspectroscopy at surfaces and interfaces

Advances in IR microspectroscopy have developed the technique into an incisive, routine method to obtain spatially resolved chemical maps for a wide range of samples, including samples of biological [42], geological [43], and industrial interest [44]. IR microspectrometers, microscopes using collinear IR/visible reflective optics coupled to an interferometer, afford the opportunity to optically choose a sample area of interest and to collect an IR spectrum for the same area. Individual spectra are collected through a fixed aperture as a function of sample position. A chemical map is created from these spectra by determining the strength of an absorption band at each measuring position. Different absorption bands are used to create additional chemical maps (see Fig. 14). Recently, several scientists have achieved sub-micron spatial resolution by coupling bright sources of IR light (storage rings [45] and FELs [46]) to IR microspectrometers. It is possible that IR array detectors (several IR detectors arranged in an array to detect spatially resolved signals) can offer alternative methods to achieve high spatial resolution. Efforts are underway to develop near-field IR microspectroscopy to further improve the spatial resolution perhaps to a fraction of the diffraction limit [47,48]. Access to chemical maps of sub-micron spatial resolution can have far-ranging impact in diverse fields in the physical sciences, biological sciences, and technical arenas.

Recent IR microspectroscopy results include spatially resolved chemical maps of living cells [49], and of bone [50]. For example, Jamin and co-workers have identified the presence of two individual nuclei in a cell undergoing division, indicated by the presence of two distinct regions within the cell having high protein/low lipid concentrations as shown in Fig. 14. Miller and collaborators have investigated variations in bone composition of monkeys given ovarectomies to intentionally induce the onset of osteoporosis. They identified clear differences in the mineral-to-protein ratio in the affected bones (compared to undiseased bone). IR microspectroscopy investigations of inherently heterogeneous biomaterials that require biocompatibility could lead to

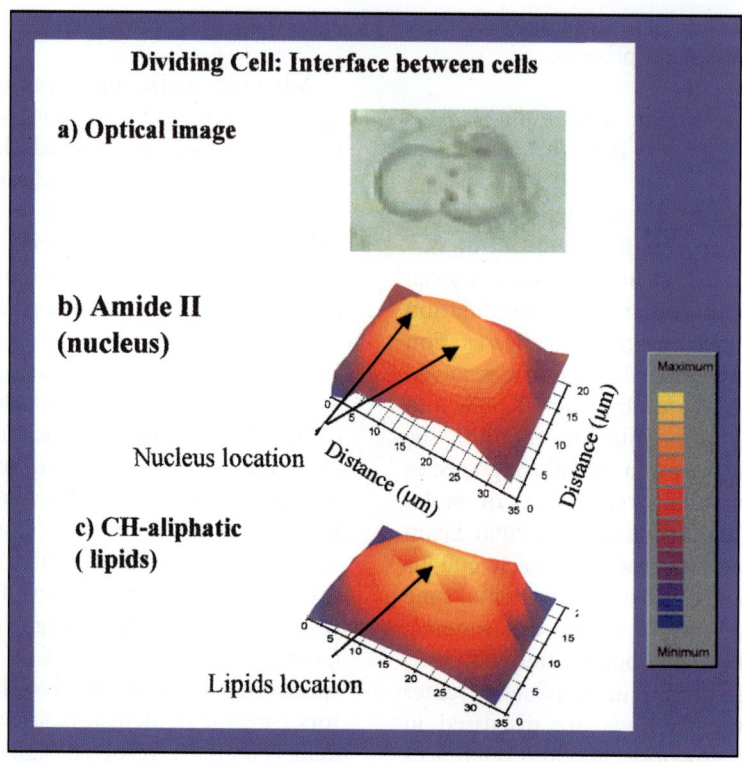

Fig. 14. Optical image and two chemical maps of a dividing cell (35 μm × 20 μm) obtained from IR spectromicroscopy. The chemical maps are derived from the strength of absorption bands from approximately 100 IR transmittance spectra collected over the area of the dividing cell. The maximum in the color scale (yellow) represents the position of the maximum absorption by the cell, while the minimum (blue) represents no absorption by the cell. Strikingly, the intensity map of the amide II absorption band shows two peaks in the center of the two halves of the cell, representing the position of two separate nuclei before the cell division is complete. The intensity map of the C–H stretch absorption bands shows that lipids are concentrated at the contractil ring, where the cleavage furrow is located. The figure is reproduced with permission from Ref. [49]. Copyright (1998) National Academy of Sciences, USA.

valuable, spatially resolved chemical maps of these interfaces. In addition, with the recent advances in near-field IR microscopy using FEL sources it is feasible that IR microspectroscopy will one day provide chemical identification at nanometer spatial resolution for these important research areas.

IR microspectroscopy techniques could potentially be coupled with UHV surface science tools to bring chemical identification at micron length scales to reacting surfaces. Structural contrasts in reaction fronts in UHV have previously been observed with low energy electron microscopy [51], and the chemistry has been identified with in situ IR measurements [52]. To identify the fundamental processes occurring at the reaction front, it would be beneficial to obtain a chemical, spatially resolved map of the reaction front. Moreover, μm scale chemical analysis would also allow surface science to address inhomogeneous surfaces.

IR microspectroscopy can play an important role in future computing and optoelectronic devices based on nanostructures. With computer devices becoming smaller, and with the advent of quantum devices, future computers may incorporate quantum-confinement materials. The resulting nanostructures could have electronic transitions corresponding to IR energies and large IR absorption cross-sections. Furthermore, these transitions can be fairly sensitive to the quality of the structure [53]. An IR microscope with nanoscale spatial resolution could be a useful non-invasive diagnostic tool in the development and processing stages of devices constructed of nanostructures.

5.4. Ultrafast time-resolved techniques to investigate energy relaxation

In the 1980s several efforts focused on pump–IR–probe investigations of adsorbate/substrate interactions, examining the roles of electron–phonon coupling, and electron–hole pair creation at different adsorbate/substrate interfaces. This subject has been reviewed by Cavanagh et al. [54]. In particular, an understanding of the adsorbate vibrational-decay mechanisms was of great interest, since these energy pathways are important in understanding reaction mechanisms. Surface reactions, such as the dissociation or rebonding of admolecules and the bonding between adatoms and the substrate, are governed by the available energy pathways of the surface/adsorbate system. Surface diffusion and molecular reconfiguration are intimately linked to vibrational motion. Moreover, excited adsorbate vibrations can dissipate their energy via other excitations in the system having comparable energy, including other molecular vibrations, bulk vibrations (phonons), and electronic excitations. Vibrational energy transfer at an interface is a fundamental process that controls the dynamics of surface reactions. For example, interfacial electronic structure plays an important role in the process of vibrational energy dissipation [55]. These pioneering pump–IR–probe efforts have elucidated that different energy pathways are active for internal molecular vibrations, depending on the substrate electronic structures. These methods, utilizing new pump and probe sources could be employed to study the relaxation mechanisms for bonding vibrational modes at surfaces.

5.4.1. Quantum dots: lasers

A fundamental understanding of relaxation energy pathways in manufactured structures can be gained by utilizing novel pump–probe techniques. Time-resolved techniques are ideal to examine the lifetimes of a variety of excited states, and to understand the fundamental mechanisms by which the states relax. For example, it has been predicted [56,57] that the excited electronic states of quantum dots might be a source of long-lived (e.g., ns) electronic states, since electron–phonon coupling, which is usually a dominant mechanism for relaxation pathways in bulk semiconductor materials, is expected to be inefficient for quantum dots. This inefficiency originates from the discrete, widely separated electronic states, for which the separation greatly exceeds phonon energies. Many studies [58–61] have nonetheless observed fast decay rates. This relaxation can possibly be explained by an electron–electron Auger mechanism that depends on electron–hole coupling. Recently Guyot-Sionnest et al. [62] have used a novel approach to examine the lifetime of intraband decays, and have thereby separated the electron and hole dynamics

5.4.2. Dynamics at surfaces and interfaces

With the recent development of stable, bright, broadband, pulsed IR sources [13] (storage rings) and bright, high-intensity IR sources [14] (lasers and FELs), opportunities abound for novel developments in time-resolved IR spectroscopy of a wide variety of surface and interface systems. Recently, Carr et al. [63] have synchronized a visible pump laser (Ti–Sapphire) to the time structure of a storage ring [64]. Nahon et al. [65] and Palmer et al. [66] discussed experiments with a UV-FEL synchronized to the time structure of an IR storage-ring source. In these cases, the stability, brightness, and broadband emission of the storage-ring beam are utilized for the probe pulse, while bright, high-intensity UV pump beams are obtained from the FEL. It is possible that an IR-FEL could be housed in the same building as a low-energy storage ring, which would foster IR pump-broadband IR probe surface and interface studies. Although, with present technology, the time resolution in such experiments would be limited to tens of ps, there has been a recent report of the observation of storage ring pulses with fs duration [67]. In addition, recent time-resolved desorption studies by Bonn and collaborators [68] have discovered that CO_2 can be produced at a CO- and O-co-adsorbed Ru surface when the system is excited with fs IR laser pulses. This reaction is initiated by the presence of hot substrate electrons, and is fundamentally different than the CO desorption process mediated by thermal excitation. This pioneering work suggests that a rich area of dynamics can be explored at surfaces and interfaces with the range of accelerator-based pulsed IR beams that are becoming more accessible and reliable.

The dynamics of electron relaxation in nanostructures offers another venue of fast relaxation phenomena that can be explored with the ultrafast pulsed IR probes. The pioneering work by Guyot-Sionnest et al. [62] have shown that surface modifications to quantum dots can profoundly affect electron relaxation and recombination pathways. Smith and collaborators [69] have investigated (with fs time resolution) the time-dependent electron relaxation for metallic nanoparticles, and found a competition in relaxation pathways (surface scattering and electron–phonon coupling) that was dependent upon the size of the particles. Progress in producing novel nanostructured materials and the development of pulsed IR sources provide ample opportunity for further studies of nanoparticles.

6. Synopsis

IR spectroscopy has become a routine and incisive tool to provide chemical-specific investigations, in a wide variety of environments, at surfaces and interfaces. This probe is weakly interacting with many materials, is polarization dependent, and chargeless. These properties are instrumental in its ability to provide insight into many problems. There have been many technical developments, e.g., new measurement techniques, stable interferometers, high quality detectors, and novel, bright, pulsed sources that afford an array of exciting research directions. The impact of IR spectroscopy is felt across a broad ensemble of scientific disciplines, including forefront biology, chemistry, geology, device engineering, and physics.

Acknowledgements

This work was supported in part by the following three grants: NSF-CHE-9984931 and NSF-DMR-9806055 and an Innovative Researcher award from the Research Corporation. The author is grateful for critical readings of the manuscript by Dr. Richard Sorbello, Dr. Carolyn Aita, Mr. Trevor Johnson and Mr. XiaoFeng Hu.

References

[1] B.H. Erne, F. Ozanam, J.-N. Chazlviel, Dynamics of hydrogen adsorption on GaAs electrodes, Phys. Rev. Lett. 80 (1998) 4337.

[2] M.J. Weaver, Binding sites and vibrational frequencies for dilute carbon monoxide and nitric oxide adlayers in electrochemical versus ultrahigh vacuum environments: The role of double-layer solvation, Surf. Sci. 437 (1999) 215.

[3] G. Ertl, M. Huber, S.B. Lee, Z. Paal, M. Weiss, Interactions of nitrogen and hydrogen on iron surfaces, Appl. Surf. Sci. 8 (1981) 373.

[4] F.T. Vasko, A.V. Kuznetsov, Electronic States and Optical Transitions in Semiconductor Heterostructure, Springer, New York, 1999.

[5] B.K. Ridley, Quantum Processes in Semiconductors, Oxford University Press, Oxford, 1999.

[6] M.A. Reed, Quantum dots, Sci. Am. 268 (1993) 118.

[7] B.B. Stefanov, A.B. Gurevich, M. Weldon, K. Raghavachari, Y. Chabal, Silicon epoxide: unexpected intermediate during silicon oxide formation, Phys. Rev. Lett. 81 (1998) 3908.

[8] W. Zhu, C.J. Hirschmugl, A.D Laine, B. Sinkovic, S.S.P. Parkin, Determination of the depth in plasma oxidized Al films as barrier of magnetic tunneling junctions, Appl. Phys. Lett. 3103 (2001).

[9] R.G. Greenler, Design of a reflection absorption experiment for studying the IR spectrum of molecules adsorbed on a metal surface, J. Vac. Sci. Technol. 12 (1975) 1410.

[10] J.E. Butler, V.M. Bermudez, R.L. Rubinovitz, Diode laser spectroscopy of chemical moieties at surfaces, SPIE 669 (1986) 125–127.

[11] Y.J. Chabal, Surface infrared spectroscopy, Surf. Sci. Rep. 8 (1988) 211.

[12] P.W. Bohn, Localized optical phenomena and the characterization of materials interfaces, Ann. Rev. Mater. Sci. 27 (1997) 469.

[13] G.K. Green, Brookhaven National Laboratory Report 50595 (1977);
G.P. Williams, Nucl. Instrum. Methods 195 (1982) 383;
W.D. Duncan, G.P. Williams, Appl. Opt. 22 (1983) 2914.

[14] R. Rohatgi, H.A. Schwettman, T.I. Smith, R.L. Swent, The SCA FEL program in the infrared, visible and ultraviolet, Nucl. Instrum. Methods A 272 (1988) 32;
E.R. Crosson, G.E. James, H.A. Schwettman, T.I. Smith, R.L. Swent, Multi-user operation at an FEL facility, Nucl. Instrum. Methods B 144 (1998) 25.

[15] P. Hollins, J. Pritchard, Surf. Sci. 89 (1979) 486;
B.N.J. Persson, R. Ryberg, Solid State Commun. 36 (1980) 613.
[16] D. Hoge, M. Tushaus, E. Schweizer, A.M. Bradshaw, Chem. Phys. Lett. 151 (1988) 230.
[17] X.-D. Wang, W.T. Tysoe, R.G. Greenler, K. Truskowska, Reflection-absorption infrared spectroscopy study of the adsorption of atomic oxygen on silver, Surf. Sci. 257 (1991) 335;
X.-D. Wang, W.T. Tysoe, R.G. Greenler, K. Truskowska, Reflection–absorption infrared spectroscopy study of the adsorption of dioxygen species on a silver surface, Surf. Sci. 258 (1991) 335.
[18] U. Engstrom, R. Ryberg, Atomic oxygen on a Pt(1 1 1) surface studied by infrared spectroscopy, Phys. Rev. Lett. 82 (1999) 2741.
[19] C.J. Hirschmugl, G.P. Williams, F.M. Hoffmann, Y.J. Chabal, Adsorbate–substrate resonant interactions observed for CO on Cu(1 0 0) in the far-IR, Phys. Rev. Lett. 65 (1990) 480.
[20] A.A. Sirenko, C. Bernhard, A. Golnick, A.M. Clark, J. Hao, W. Si, X.X. Xi, Soft-mode hardening in $SrTiO_3$ thin films, Nature 404 (2000) 374.
[21] A. Roeseler, Infrared Spectroscopy Ellipsometry, Springer, Berlin, 1990.
[22] C.J. Hirschmugl, G.P. Williams, B.N.J. Persson, V.I. Volokitin, Adsorbate vibrational dynamics in the anamolous skin-effect frequency region, Surf. Sci. 317 (1994) L1141.
[23] C.L. Hsu, E.F. McCullen, R.G. Tobin, Evidence for an adsorbate dependent mechanism for surface resistivity: Formic acid, oxygen and CO on Cu(1 0 0), Chem. Phys. Lett. 316 (2000) 336.
[24] M. Hein, P. Dumas, A. Otto, G.P. Williams, CO interaction with Co-adsorbed C_2H_4 on Cu(1 1 1) as revealed by friction with the conduction electrons, Surf. Sci. 465 (2000) 249.
[25] J.M. Carpinelli, H. Weitering, E.W. Plummer, R. Stumpf, Direct observation of a surface charge density wave, Nature 381 (1996) 398.
[26] F. Hoffmann, B.N.J. Persson, W. Walter, D.A. King, C.J. Hirschmugl, G.P. Williams, Antiabsorption resonances in infrared reflectance spectroscopy of alkali-Cu(1 1 1) adsorbate systems: Is the ground state a surface charge density wave condensate?, Phys. Rev. Lett. 72 (1994) 1256.
[27] A. Peremans, Y. Caudano, P.A. Thiry, P. Dumas, W.Q. Zhang, A. LeRille, A. Tadjeddine, Electronic tuning of dynamical charge transfer at an interface: K doping of C_{60}/Ag(1 1 1), Phys. Rev. Lett. 78 (1997) 2999.
[28] C.T. Williams, D.A. Beattie, Probing buried interfaces with non-linear optical spectroscopy, Surf. Sci. 500 (2002) 545.
[29] P. Rudolf, PhD Thesis, University of Namur, Belgium, 1995.
[30] M. Fricke, A. Lorke, J.P. Kotthaus, G. Medeiros-Ribeiro, P.M. Petroff, Shell structure and electron–electron interaction in self-assembled InAs quantum dots, Europhys. Lett. 36 (1996) 197.
[31] G.A. Beitel, C.P.M. de Groot, H. Oosterbeek, J.H. Wilson, A combined insitu PM-RAIRS and kinetic study of single crystal cobalt catalysts under synthesis gas at pressures up to 300 mbar, J. Phys. Chem. B 101 (1997) 4053.
[32] A.M. Bradshaw, E. Schweizer, Infrared reflection absorption spectroscopy of adsorbed molecules, in: R.E. Hester (Ed.), Advances in Spectroscopy: Spectroscopy of Surfaces, Wiley, New York, 1988.
[33] B.B. Stefanov, A.B. Gurevich, M. Weldon, K. Raghavachari, Y. Chabal, Silicon epoxide: Unexpected intermediate during silicon oxide formation, Phys. Rev. Lett. 81 (1998) 3908.
[34] GAUSSIAN 94, M.J. Frisch et al., Gaussian, Pittsburgh PA, 1995.
[35] M.K. Weldon, K.T. Queeney, J. Eng Jr., K. Raghavachari, Y.J. Chabal, The surface science of semiconductor processing: gate oxides in the ever-shrinking transistor, Surf. Sci. 500 (2002) 859.
[36] N.I. Afanasyeva, R.F. Bruch, Biocompatibility of polymer surfaces interacting with living tissue, Surf. Interf. Anal. 27 (1999) 204.
[37] G. Reid, H.J. Busscher, S. Sharma, M.W. Mittelmanm, S. McIntyre, Surface properties of catheters, stents and bacteria associated with urinary tract infections, Surf. Sci. Rep. 21 (1995) 251.
[38] H.M. Kim, Y. Kim, S.J. Park, C. Rey, H.M. Lee, M.J. Glimcher, J.S. Ko, Thin film of low-crystalline calcium phosphate apatite formed at low temperature, Biomaterials 21 (2000) 1129;
T. Chandy, G.S. Das, R.F. Wilson, G.H.R. Rao, Use of plasma glow for surface-engineering biomolecules to enhance blood compatibility of dacron and PTFE vascular prosthesis, Biomaterials 21 (2000) 699.
[39] R.R. Cavanagh, E.J. Heilweil, J.C. Stephenson, Time-resolved measurements of energy transfer at surfaces, Surf. Sci. 299/300 (1994) 643.
[40] J.E. Ruett-Robey, D.J. Doren, Y.J. Chabal, S.B. Christman, CO diffusion on Pt(1 1 1) with time-resolved infrared-pulsed molecular-beam methods—critical tests and analysis, J. Chem. Phys. 93 (1990) 9113.
[41] R.D. Palmer, J.L. Chao, R.M. Dittmar, V.G. Gregoriou, S. Plunkett, Investigation of time-dependent phenomena by use of step-scan FTIR, Appl. Spectrosc. 47 (1993) 1297.
[42] H.J. Humecki, Practical Guide to Infrared Microspectroscopy, Marcel Dekker, New York, 1995.
[43] N. Guilhaumou, P. Dumas, G.L. Carr, G.P. Williams, Synchrotron infrared microspectrometry applied to petrography in micrometer-scale range: fluid chemical analysis and mapping, Appl. Spectrosc. 52 (1998) 1029.
[44] J.-L. Bantignies, G. Fuchs, G.L. Carr, P. Dumas, C. Wilhelm, Industrial applications of accelerator-based infrared sources: analysis using infrared microspectroscopy, SPIE 3153 (1997) 125.
[45] H.Y.N. Holman, D.L. Perry, M.C. Martin, G.M. Lamble, W.R. McKinney, J.C. Hunter-Cevera, Real-time

characterization of biogeochemical reduction of Cr(VI) on Basalt surfaces by SR-FTIR imaging, Geomicrobiology 16 (1999) 307.

[46] F. Glotin, J.-M. Ortega, J.-L. Poncy, F. Tourdes, J.-L. Lefaix, P. Dumas, Frequency dependence of the ablation of pig tissue with an IR-FEL: a microspectroscopic analysis, SPIE 3757 (1999) 113;
A.G. Jeung, S. Errmilli, M.K. Hong, P. Huie, T.I. Smith, Analysis of biological tissue using scanning near field microscopy, SPIE 3153 (1997) 117.

[47] D.B. Talley, L.B. Shaw, J.S. Sanghera, I.D. Aggarwal, A. Cricenti, R. Generosi, M. Luce, G. Margaritondo, J.M. Gilligan, N.H. Tolk, Scanning near field infrared microscopy using chalcogenide fiber tips, Mater. Lett. 42 (2000) 6.

[48] F. Keilmann, B. Knoll, A. Kramer, Long-wave-infrared near-field microscopy, Phys. Status Solidi B 215 (1999) 849.

[49] N. Jamin, P. Dumas, J. Moncuit, W.-H. Fridman, Highly resolved chemical imaging of living cells by using synchrotron infrared microspectrometry, Proc. Nat. Acad. Sci. 95 (1998) 4837.

[50] L.M. Miller, C.S. Carlson, G.L. Carr, G.P. Williams, M.R. Chance, Synchrotron infrared micorspectroscopy as a means of studying the chemical composition of bone: applications to osteoarthritis, SPIE 3153 (1997) 141;
L.M. Miller, J. Tibrewalla, C.S. Carlson, Examination of bone chemical composition in osteoporosis using fluorescence-assisted synchrotron infrared microspectroscopy, Cell Mol. Biol. 46 (2000) 1035–1044.

[51] B. Rausenberger, W. Swiech, W. Engel, A.M. Bradshaw, E. Zeitler, LEEM and selected area LEED studies of reaction front propogation, Surf. Sci. 287 (1993) 235.

[52] J.H. Miners, R. Martin, P. Gardner, R. Nalezinski, A.M. Bradshaw, Monitoring adsorbed species during oscillatory reactions using vibrational spectroscopy, Surf. Sci. 377–379 (1996) 791.

[53] Y. He, Q.S. Zhu, Z.T. Zhong, G.Z. Zhang, J. Xiao, Z.P. Cao, X.H. Sun, H.Z. Yang, Linewidth of the infrared absorption spectra due to bound-to-continuum transition in $GaAs/Al_xGa_{1-x}As$ multiple quantum well structures, Appl. Phys. Lett. 73 (1998) 1131.

[54] R.R. Cavanagh, E.J. Heilweil, J.C. Stephenson, Time-resolved measurements of energy transfer at surfaces, Surf. Sci. 299/300 (1994) 643.

[55] D.C. Langreth, M. Persson, Laser spectroscopy and photochemistry on metal surfaces, in: H.L. Dai, W. Ho (Eds.), Advanced Series in Physical Chemistry, vol. 5, World Scientific, Singapore, 1995, p. 498 and references therein.

[56] U. Bockelmann, G. Bastard, Phonon scattering and energy relaxation in 2-dimensional, one-dimensional and zero-dimensional electron gases, Phys. Rev. B 42 (1990) 8947.

[57] H. Benisty, C.N. Sotomayor-Torres, C. Weisbuch, Intrinsic mechanism for the poor luminescence properties of quantum box systems, Phys. Rev. B 44 (1992) 10945.

[58] G. Wang, S. Fafard, D. Leonard, J.E. Bowers, J.L. Merz, P.M. Petroff, Time-resolved optical characterization of InGaAs/GaAs quantum dots, Appl. Phys. Lett. 64 (1994) 2815.

[59] S. Grosse, J.H.H. Sandmann, G. vonPlessen, J. Feldmann, H. Lipsanen, M. Sopanen, J. Tulkki, J. Ahopelto, Carrier relaxation dynamics in quantum dots: scattering mechanisms and state-filling effects, Phys. Rev. B 55 (1997) 4473.

[60] D. Gammon, E.S. Snow, B.V. Shanabrook, D.S. Katzer, D. Park, Homogeneous linewidths in the optical spectrum of a single gallium arsenide quantum dot, Science 273 (1996) 87.

[61] V. Klimov, D. McBranch, Femtosecond 1P-to-1S electron relaxation in strongly confined semiconductor nanocrystals, Phys. Rev. Lett. 80 (1998) 4028.

[62] P. Guyot-Sionnest, M. Shim, C. Matranga, M. Hines, Intraband relaxation in CdSe quantum dots, Phys. Rev. B 60 (1999) R2181.

[63] G.L. Carr, R.P.S.M. Lobo, C.J. Hirschmugl, J. LaVeigne, D.H. Reitze, D.B. Tanner, Time-resolved spectroscopy using infrared synchrotron pulses, SPIE 3153 (1997) 80.

[64] G.L. Carr, High-resolution microspectroscopy and sub-nanosecond time-resolved spectroscopy with the synchrotron infrared source, Vib. Spectrosc. 19 (1999) 53.

[65] L. Nahon, E. Renault, M.E. Couprie, D. Nutarelli, D. Garzella, M. Billardon, G.L. Carr, G.P. Williams, P. Dumas, Two color experiments combining the UV storage ring free electron laser and the SA5 IR beamline at super-ACO, SPIE 3775 (1999) 145.

[66] R.A. Palmer, G.D. Smith, V.N. Litvinenkl, G.S. Edwards, Fourier transform infrared picosecond time-resolved spectroscopy with a UV free-electron laser pump synchrotron probe, SPIE 3775 (1999) 137.

[67] R.W. Schoenlein, S. Chattopadhyay, H.H.W. Chong, T.E. Glover, P.A. Heimann, C.V. Shank, A.A. Zholents, M.S. Zolotorev, Generation of femtosecond pulses of synchrotron radiation, Science 287 (2000) 2237.

[68] M. Bonn, S. Funk, C. Hess, D.N. Denzler, C. Stampfl, M. Scheffler, M. Wolf, G. Ertl, Phonon-versus electron-mediated desorption and oxidation of CO on Ru(0 0 0 1), Science 285 (1999) 1042.

[69] B.A. Smith, J.Z. Zhang, U. Giebel, G. Schmid, Direct probe of size-dependent electronic relaxation in single-sized Au and nearly monodisperse Pt colloidal nano-particles, Chem. Phys. Lett. 270 (1997) 139.

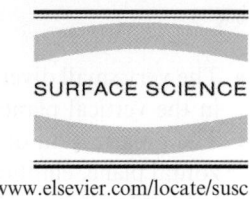

Surface science done at third generation synchrotron radiation facilities

S. Ferrer, Y. Petroff *

European Synchrotron Radiation Facility, 6 rue Jules Horowitz, B.P. 220, F-38043 Grenoble Cedex, France

Received 3 August 2000; accepted for publication 30 July 2001

Abstract

In this paper we present a few examples of surface science done at third generation synchrotron facilities. As explained in the introduction, third generation sources are characterised by a gain in brightness of three or four orders of magnitude. This allows performing experiments which were difficult or impossible before. The first part of the paper is dealing with experiments on magnetic materials and shows how dichroism and surface diffraction can bring new information. In the second part, we discuss two examples related to catalysis: the elementally resolved imaging of chemical waves and the structure of chemisorbs layers on a nickel surface at atmospheric pressure. How do atoms assemble in monatomic liquids? Do they form clusters? This question has been without answer for many years and it is only recently that an X-ray experiment has solved the problem. The fourth part of the paper describes recent results on the electronic properties of high T_c superconductors and heavy fermions, studied by high resolution photoemission. Finally, we present a prospect of a few experiments that could be done in the near future. © 2001 Elsevier Science B.V. All rights reserved.

Keywords: Synchrotron radiation photoelectron spectroscopy; X-ray absorption spectroscopy; X-ray scattering, diffraction, and reflection; Surface chemical reaction; Catalysis; Solid–liquid interfaces; Chemisorption; Magnetic surfaces

1. Introduction

Synchrotron radiation has been used extensively during the last 30 years to investigate the properties of matter. It is produced when charged particles (like electrons) are accelerated by a magnetic field. This happens particularly in circular accelerators, called storage rings, developed initially for high-energy physics. In these machines, highly relativistic electrons travel along a circular path for many hours. The synchrotron radiation is produced by the transverse acceleration due to the magnetic forces in the bending magnets. For production of ultraviolet or soft X-ray radiation, the energy of the electrons vary between 500 MeV and 1 GeV. Hard X-rays are produced with electrons in the range of 2–8 GeV.

Why synchrotron radiation has been so successful? (There are today about 40 synchrotron radiation centers in Asia, Europe, USA and South America.) This is due to the remarkable properties of the radiation:

- The small size of the source (between few mm and few µm).

---

* Corresponding author. Tel.: +33-476882000; fax: +33-476882020.

- The very small divergence of the beam (<1 mrad in the vertical plane).
- The polarization of the light (linear in the horizontal plane, elliptical if observed from above or below the plane).
- The large spectrum emitted (from far infrared to hard X-rays).
- The pulsed structure (20–100 ps) due to the fact that the electrons are grouped in bunches by the rf cavities necessary to compensate for the energy loss due to the emission of synchrotron radiation.

The radiation emitted by a bending magnet can be increased by three or four orders of magnitude by installing in the ring insertion devices called undulators. They consist in an alternate distribution of magnetic fields which force the electrons to do a sinusoidal type of motion which in turn results in an increased emission of radiation. Fig. 1 shows the narrow beam emitted by such a device: the emission being in the X-ray should not be visible but the intense power of the beam ionises the air molecules.

As we have already pointed out, the first storage rings were built for high energy physics studies but at the end of the 1960s, few beamlines were added to extract the synchrotron radiation: they are called "first generation sources". Due to the success of these beamlines, it was decided to build accelerators only for the use of synchrotron radiation: they were mostly based on bending magnets.

The third generation (beginning in the 1990s) is mostly based on insertion devices and is characterized by very high brightness (density of photons in spatial and angular dimensions). The typical brightness is in the range of 10^{18}–10^{20} ph/s/0.1 BW/100 ma/mm^2/mrad^2, which is about 10^9 times more than an X-ray tube.

The impact of these new sources has been immediate in some areas as:

- X-ray inelastic scattering with meV resolution. This is a new field: previously inelastic scattering was done essentially with neutrons, due to lack of resolution with X-rays.
- The possibility to study materials under extremely high pressure (1–3 Mbar) even on light elements like hydrogen.
- The exploitation of the coherence of the beam (small divergence, monochromaticity) allowing to obtain three-dimensional images of non-absorbing materials.

For the studies of surfaces and interfaces, the impact of these new sources was immediate in the case of X-ray surface scattering experiments. It took many more years in the case of photoemission or spectromicroscopy.

This is not a review and due to lack of space, we have chosen very few examples, especially in the case of liquids.

2. Surface magnetism

For a long time magnetic materials were studied with neutrons, due to the strong coupling of the neutron spin to the magnetic moments in the sample. Today, there are about 10 different techniques based on synchrotron radiation to study magnetism (spin-polarised photoemission, dichroism, inelastic scattering, magnetic scattering, nuclear scattering, etc.).

Why this sudden interest? Mostly due to the amazing properties (discovered between 1986 and 1988) of magnetic multilayers: these structures

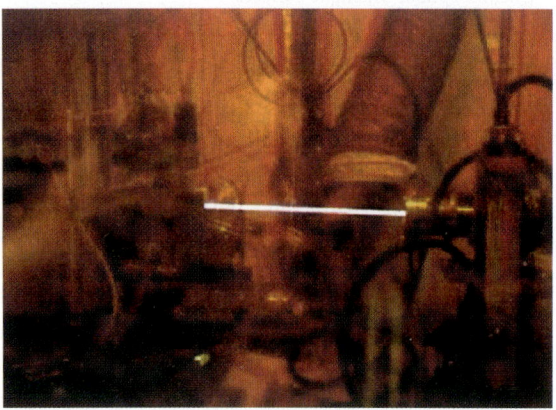

Fig. 1. Photon beam emitted by an insertion device at the ESRF (Grenoble). The X-ray beam is visible due to the ionisation of the air molecules. Notice the small divergence of the beam (courtesy from A. Kvick).

contain alternate layers of a magnetic element, such as iron or cobalt, and a non-magnetic one, such as chromium or copper. The layers are typically between 50 Å and a few angstroms thick.

In 1986, Peter Grünberg's group in Jülich [1] found that, for a given thickness of the non-magnetic chromium layer in a three-layer iron/chromium/iron structure, the magnetisation of the iron layers pointed in opposite directions. It was found later that the alignment (ferro- or antiferromagnetic) depends only on the thickness of the non-magnetic layer (see Fig. 2) [2]. The exchange coupling between the magnetic moments in the layers and the spin of the conduction electrons in the chromium layer produces an oscillation as a function of spacer thickness.

In 1988, Fert's group at the Université Paris-Sud [3] observed that the electrical resistivity of an antiferromagnetically coupled iron/chromium super lattice dropped by a factor of two when a sufficiently high magnetic field was applied (Fig. 2). This effect is called now "giant magnetoresistance". Resistance in metals is caused by electrons being scattered from impurity atoms and lattice imperfections. Giant magnetoresistance is thought to be due to the spin dependence of this scattering: the amount of scattering and hence the resistance depends on whether the electron spins are parallel or antiparallel to the magnetisation.

These two experiments have opened a new field with many industrial applications.

Before presenting a few recent examples of new possibilities, we would like to introduce X-ray magnetic dichroism and magnetic scattering.

X-ray magnetic circular dichroism (XMCD) is a technique for the study of ferromagnetic materials (but also paramagnetic systems polarised by an intense external magnetic field). It consists of the measurement, as a function of photon energy, of the difference in the absorption coefficient for X-rays of opposite circular polarisation, which in many cases of interest can be relatively large and easily observable in the vicinity of an absorption edge (see Fig. 3 for Fe $L_{2,3}$ edge) [4].

XMCD has rapidly gained considerable popularity because of its attractive features, including atomic selectivity, through the choice of the absorption edge, and high sensitivity, as the tech-

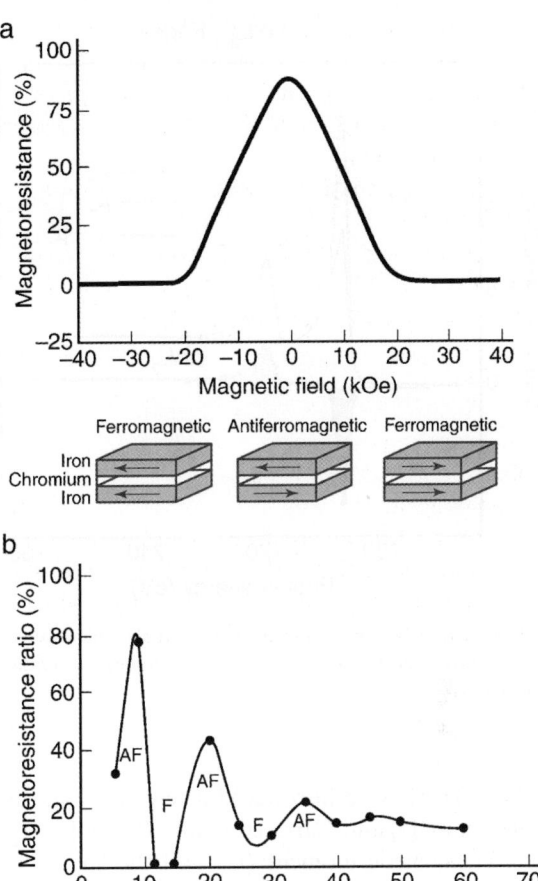

Fig. 2. (a) Magnoresistance curve of an iron–chromium multilayer at 4.2 K. The iron layers are 30 Å thick whilst the chromium layers are 9 Å (there are 30 layers each or iron and chromium). In zero field, the magnetisations of the iron layers are antiparallel and the resistance is at maximum (p_{max}): at high fields the magnetisations are parallel and the resistance is at a minimum (ρ_{min}). The current and the applied field are parallel to the layers. (b) Variation of the magnetoresistance ratio—$100(\rho_{max} - \rho_{min})/\rho_{min}$—of a Co/Cu multilayer as a function of the copper thickness at 4.2 K. The Co layers are 15 Å thick and there are 30 layers each of Co and Cu. Maxima correspond to antiferromagnetic coupling and minima to ferromagnetic coupling (from Ref. [2]).

nique allows the detection of the magnetisation in just a few atomic layers of magnetic material. A further attractive feature of this novel technique is the possibility to extract precise quantitative information on the atomic magnetic moments, thanks to theoretical results known as "sum

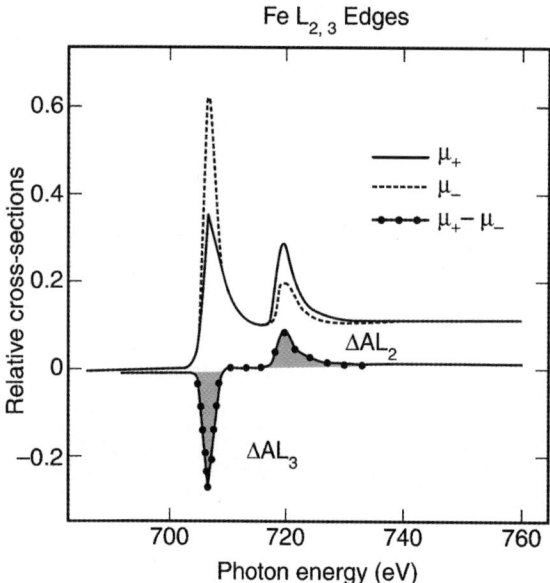

Fig. 3. X-ray absorption and magnetic circular dichroism of iron at the L_2 and L_3 edges taken with two opposite saturation magnetisation (from Ref. [4]).

of magnetic circular dichroism (CDAD) and linear magnetic dichroism (LMDAD) can also give important information [9].

The interaction of X-rays with matter is dominated by the coupling of the electric field with the electrons of the atoms. However, in materials where the atoms exhibit a magnetic moment, the coupling of the magnetic field of the radiation with the magnetic moment of the atoms gives rise to a very small, but observable effect. It was first experimentally demonstrated by Bergevin and Brunel [10]. It is named non-resonant magnetic scattering. The effect is exploited routinely to investigate the magnetic structure of bulk materials. It is, however, too weak to be useful to investigate the magnetism of surfaces with sensitivity of one atomic layer. In 1988, the group of Gibbs at NSLS (Brookhaven) discovered that if the photon energy was such that a magnetic atom could be excited via dipolar or multipolar electric transitions, a new term (called resonant) appeared in the scattering cross section that was sensitive to the magnetic moment of the atoms and that had a magnitude several orders larger than the non-resonant magnetic scattering [11]. Thanks to this resonant effect, surface magnetic scattering experiments with high surface sensitivity have become feasible.

2.1. Images of antiferromagnetic structure of a NiO(100) surface

After the discovery of the giant magnetoresistance [3] it started to be obvious that a technique to study antiferromagnetic films was necessary. The study of antiferromagnetic surfaces and interfaces has been difficult due to the fact that optical X-ray and neutron techniques are mostly bulk-sensitive. This limitation has been overcome by the use of XMLD carried out by means of surface sensitive electron yield detection. In contrast to the XMCD which directly measures the magnetic moment $\langle M \rangle$ through transfer of the X-ray angular momentum vector in the absorption process, XMLD spectroscopy measures the expected value of the square of the magnetic moment $\langle M^2 \rangle$ because linearly polarised photons have only axiality.

rules". These were introduced a few years ago by Thole et al. [5] and Carra et al. [6]. They allow the orbital magnetic moment $\langle L_z \rangle$ and the spin magnetic moment $\langle S_z \rangle$ to be obtained from the dichroism measurements. As for the bulk, they give very good results for the transition metals (Fe, Co, Ni, etc.). It has been shown [5] that, in an atomic model:

$$\langle L_z \rangle = B[\Delta AL_3 + \Delta AL_2] \quad (1)$$

$$\langle S_z \rangle + C\langle T_z \rangle = D[\Delta AL_3 - 2\Delta AL_2] \quad (2)$$

where ΔAL_3 and ΔAL_2 represent the dichroic signal at the L_2 and L_3 edges: B, C, D are simple constants and $\langle T_z \rangle$ is an additional term which vanishes for bulk transition metals [6].

Magnetic circular dichroism, which is sensitive to the expectation value of the local magnetic moment, will vanish for antiferromagnets. However, it has been shown theoretically [7] and experimentally [8] that, using linearly polarised light, a magnetic linear dichroism (X-ray magnetic linear dichroism, XMLD) signal can be observed since it is proportional to $\langle M^2 \rangle$. The angular dependence

By combining XMLD with a photoelectron emission microscope (PEEM) one should be able to image antiferromagnetic domains, similarly to XMCD-PEEM spectroscopy of ferromagnets. The technique offers elemental specificity, surface sensitivity (~2 nm) and a good spatial resolution (20 nm).

An attempt was reported by Spanke et al. [12] but the first unambiguous surface images with antiferromagnetic contrast were presented on NiO(1 0 0) grown on MgO(1 0 0) by Stöhr et al. [13].

The results were obtained at the Advanced Light Source (Berkeley) with a spatial resolution of 50 nm. In order to avoid charging effect, the NiO(1 0 0) sample was coated with a 2 nm Cu layer.

The resonant absorption intensity of linearly polarised radiation is given by [19]:

$$I(\varphi, \theta, T) = a + b(3\cos^2\varphi - 1)\langle Q_{zz}\rangle + c(3\cos^2\theta - 1)\langle M^2\rangle_T + d\Sigma_{ij}\langle \hat{S}_i \cdot \hat{S}_j\rangle_T \quad (3)$$

The constant term is independent of the X-ray polarisation and sample temperature T, the second term expresses the X-ray polarisation dependence due to the presence of a quadrupole moment of charge, $\langle Q_{zz}\rangle$, where φ is the angle of $\vec{E}$ with the crystallographic z axis. This term gives rise to the conventional linear dichroism effect in X-ray absorption [14]. The third term is responsible for the XMLD effect [15]. It depends on the X-ray polarisation through the angle θ between $\vec{E}$ and the magnetic axis A and on temperature through $\langle M^2\rangle_T$. This term vanishes above the Néel temperature. Finally the last term expresses the temperature spin–spin correlation function $\langle S_i \cdot S_j\rangle_T$ [16].

In NiO the charge-only linear dichroism vanishes[1] because of the cubic symmetry ($\langle Q_{zz}\rangle = 0$). However, the alignment of the Ni spins along the antiferromagnetic axis $\vec{A}$ [17] and leads to a pronounced XMLD effect whose angular and temperature dependence has been studied in detail [15,16,21].

In particular, the Ni L_2 edge (~870 eV) exhibits two muliplets peaks of which the lower energy peaks A is larger for $\vec{E} \perp \vec{A}$ and the higher peak B is larger for $\vec{E} \| \vec{A}$ [15].

The peak intensity ratio and its temperature dependence is dominated by the XMLD effect. It can therefore be used as spectroscopic contrast [15] when imaging the magnetic configuration of the antiferromagnetically ordered NiO surface. Fig. 4 shows an antiferromagnetic image with a 40 μm field of view for a 10 nm thick NiO(1 0 0) film. The image is generated by dividing an image acquired at 871.5 eV (peak B) by one recorded at 870.3 eV (peak A). This procedure eliminates topographical errors and produces antiferromagnetic contrast.

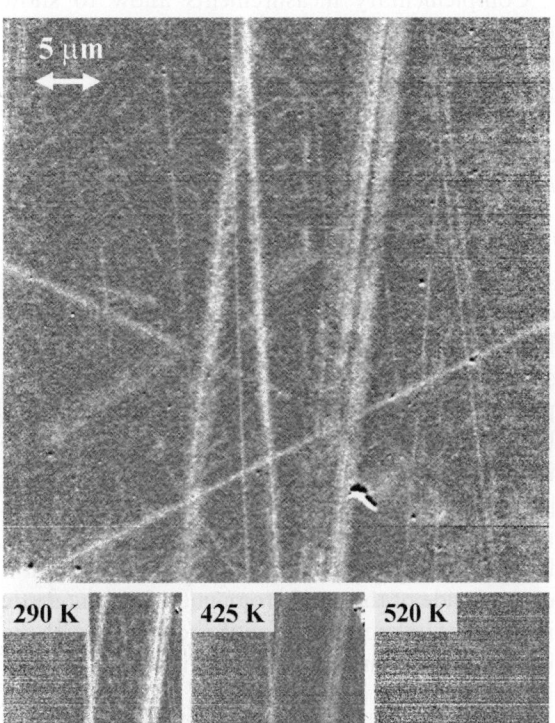

Fig. 4. XMLD antiferromagnetic image of a 10-nm-thick NiO(1 0 0) film on MgO(1 0 0) recorded at room temperature and the temperature dependence of a region within the image (from Ref. [13]).

---

[1] The antiferromagnetic order results in a slight rhombohedral deformation in NiO but the resulting $\langle Q_{zz}\rangle$ term is negligibly small.

The image exhibits straight bright lines with typical width within 400–2000 nm. They are several 100 μm long and the fact that similar structures are also observed by atomic force microscopy indicates that these lines are correlated with the surface. However the contrast in Fig. 4 is of antiferromagnetic and not of topographic nature. This is demonstrated by the temperature dependence of the image shown in Fig. 4. Clearly the contrast disappears as the Néel temperature of the film is approached. The temperature dependence is fully reversible and the XMLD contrast is fully restored upon returning to room temperature. The disappearance of magnetism goes hand in hand with the reduction in image contrast. What is the origin of the stripes? They may be caused by a more in-plane orientation of $\overleftrightarrow{A}$ or by a reduced value of $\langle M^2 \rangle$ associated with individual or an agglomeration of surface defect lines.

Complementary measurements allow to show that these surface line defects have a reduced Néel temperature (455 ± 10 K) relative to the rest of the film surface and the bulk (523 K).

2.2. Observation of the alignment of ferromagnetic by antiferromagnetic spins

After obtaining images of antiferromagnetic surface domains the next step is to examine more complicated systems, as the ferromagnetic–antiferromagnetic interface because the arrangement of spins an interfaces has an important effect on the properties of the material. The directional coupling between the spins in an antiferromagnet and those in an adjacent ferromagnet was discovered in 1956 [18] and referred to as exchange bias. It plays an important role in magnetic device technology but it is still not well understood. The spin orientation on each side of the antiferromagnet–ferromagnetic interface is unknown. The antiferromagnet pins the magnet so it cannot change direction to align itself with an external magnetic field.

Recently Nolting et al. [19] have investigated a thin Co film on top of a 40 nm $LaFeO_3$ film grown on $SrTiO_3(0\,0\,1)$. Spectromicroscopy studies, similar to the ones described in the previous paragraph were carried out using the PEEM facility at the Advanced Light source (Berkeley). The focussed X-rays, whose polarisation could be changed from linear to right or left circular, were incident on the sample at an angle of 30° from the surface and from a 30 μm spot.

The low energy secondary photoelectrons from the sample are imaged by an all-electrostatic photoemission microscope (PEEM) with magnification onto a phosphor screen that is read by a charge-coupled device camera. By tuning the photo energy to either the Fe L-edge near 710 eV or the Co L-edge near 780 eV it is possible to record separate images of the AF $LaFeO_3$ layer and the ferromagnetic Co layer. Linear X-ray polarisation is used to image the microscopic AFM domain structure for $LaFeO_3$ and right or left handed circular polarisation to image the FM Co domain structure.

Fig. 5 shows images of the domain structure in the AFM $LaFeO_3$ and in a 1.2-nm-thick FM Co layer on top of the very same substrate region. The magnetic contrast in the left image arises from AFM domains in $LaFeO_3$ with an in-plane projection of the AFM axis A oriented horizontally (light) and vertically (dark). The four AFM domains in epitaxial $LaFeO_3$ have their AFM axes tilted by 45° from the surface normal with orthogonal in-plane projections. As in the experimental geometry E lies in the surface plane, we cannot distinguish the two domains with collinear in-plane projections and we only observe two of the four AFM domains. Comparison of TEM and PEEM images shows that the AFM domains are seeded by the two epitaxial crystallographic domains [20]. XMLD spectra recorded in the light and dark regions shown underneath, reveal the spectroscopic origin of the AFM contrast. The FM Co image (Fig. 5(b)) exhibits three distinct grey scales, corresponding to FM domains aligned vertically up (black) and down (white) and horizontally left or right (grey). For the experimental geometry used for Fig. 5, corresponding to a vertical X-ray propagation direction (and angular momentum) it is not possible to distinguish left from right horizontally oriented FM domains, but these were resolved by a 90° rotation of the sample (not shown). XMCD spectra recorded for regions with different grey scales (Fig. 5(b)) illustrate the origin of the intensity contrast. Comparison of the in-plane projections of the AFM axis and the FM

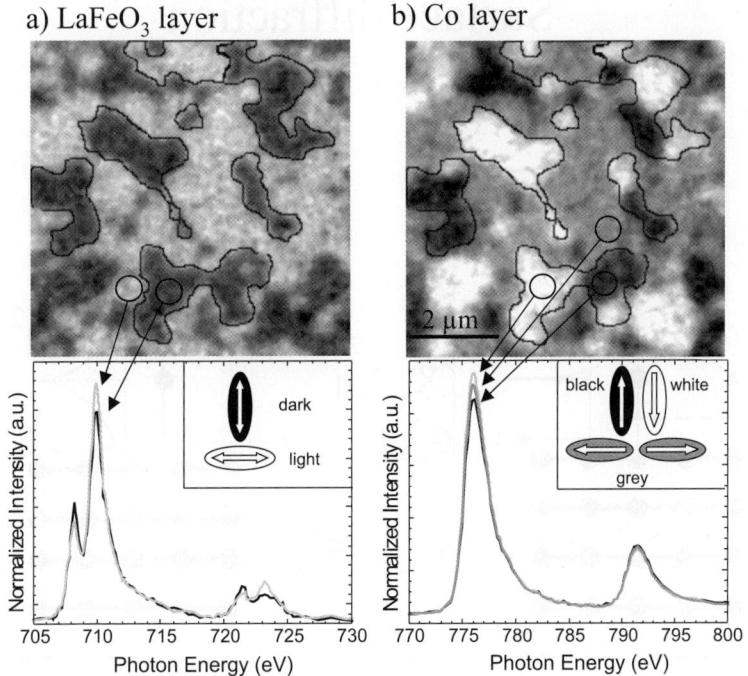

Fig. 5. Images and local spectra from the antiferromagnetic and ferromagnetic layers for 1.2 nm Co on LaFeO$_3$/SrTiO$_3$(0 0 1). (a) Fe L-edge XMLD image, (b) Co L-edge XMCD image. The contrast in the image arises from antiferromagnetic domains in Co(b) with in-phase orientations of the antiferromagnetic axis and ferromagnetic spins as indicated below the images. The spectra shows underneath were recorded in the indicated area and illustrate the origin of the intensity contrast in the PEEM images (from Ref. [19]).

spin directions, illustrated below the images, reveals that the FM Co spins are aligned parallel or antiparallel to the in-plane projection of the AFM axis. The correlation revealed by Fig. 5 is of magnetic rather than crystallographic origin as the Co is polycrystalline. The magnetic alignment of the Co domains, which exhibit an in-plane easy axis, must therefore be caused by a coupling to uncompensated spins at the LaFeO$_3$ surface with an in-plane component parallel to the in-plane projection of the AFM axis. Then the alignment of the ferromagnetic spins in the Co is in fact correlated with the spin orientation in the LaFeO$_3$ antiferromagnetic layer.

2.3. Surface magnetic X-ray diffraction: Co/Pt(1 1 1)

For a 100 years, X-rays have been extensively used to investigate the crystalline structure of matter. The key physical phenomenon which has allowed thousands of crystalline structures to be determined, is the existence of constructive interferences of X-ray waves scattered from crystal planes. The diffraction of the X-rays by crystals is described by the Bragg law that establishes the conditions for such interferences to occur. In Fig. 6, left part, an infinite set of crystal planes is represented as blue lines joining the atoms. If we consider the direction perpendicular to these planes (L axis in the figure), then for specific scattering conditions that satisfy Bragg law, one finds very intense diffracted beams that resemble delta functions as schematised in the figure. The intensity of these Bragg reflections arises from the constructive interference of many (typically several thousand) crystal planes. The information contained in the Bragg reflections concerns average properties of these diffracting planes which constitute the crystal.

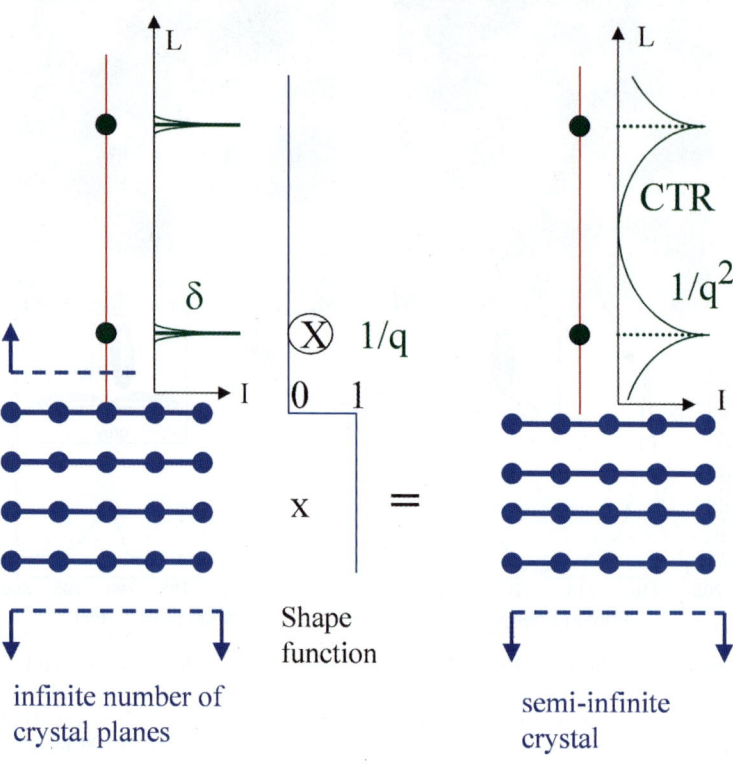

Fig. 6. Principle of surface diffraction. Left side: The diffracted intensity from an infinite set of crystal planes (in blue) in a direction L perpendicular to the planes (red) is very concentrated around the reciprocal space points which satisfy Bragg conditions (green circles). The intensity distribution along L consists of very sharp peaks as indicated in the I versus L plot. Right side: A crystal terminated on a surface may be visualised as the result of multiplying an infinite set of planes by a shape function (middle of the figure) equal to zero outside the surface and to one from the surface to interior of the crystal. In reciprocal space, the above multiplication results in a convolution product which has the effect of transforming the sharply peaked diffracted intensities to continuous intensity distributions which vary slowly with L as indicated in the green curve.

Let us consider the mid-point in the L axis between the two Bragg reflections (circles) of the figure. At this precise location, the scattering conditions are such that a destructive interference occurs between the X-ray waves scattered from two consecutive crystal planes. This causes a very strong cancellation which reduces the diffracted intensity to almost zero. It is precisely this "almost zero" which is the object of surface diffraction. It may be shown that the remaining intensity under destructive interference conditions arises from the topmost atomic planes of the crystal. Thus, by measuring at the destructive interference conditions, one does not sense the bulk of the crystal but only the surface.

In a slightly more precise manner, one may describe a crystal terminated by a surface by multiplying by a step shape function an infinite set of crystal planes as schematised in the right part of the figure. In Fourier space (i.e. in the space where the scattering geometry is described) the simple product transforms in a convolution product as indicated by the symbol $\otimes$. As the Fourier transform of the step function is proportional to $1/q$ where q is the scattering vector (difference of wavevectors of the scattered and incident X-ray beams), the result is that the distribution of the intensity of the scattered X-rays does not look as a set of sharply peaked delta functions as in the left part of the figure. These continuous intensity distributions are usually designated as crystal truncation rods (CTR). They are curves that connect the Bragg reflections. At the minima of the CTR the surface sensitivity is highest as discussed above. In practice however, the intensities of the CTR are sensitive to the surface structure everywhere except at the immediate vicinity of the Bragg reflections.

Surface X-ray diffraction basically consists of measuring the intensities of the CTRs which contain information on the structure of the topmost atomic planes of the crystal (the surface). The structure is normally determined by fitting to the measured CTRs, calculated CTRs obtained from aprioristic surface models. The program in the control and understanding of ultra-thin film growth in the last decade has led to a renaissance in research on magnetism and magnetic materials. Artificially layered materials involving magnetic and non-magnetic elements are the foundation of magnetic recording in today's computers. It is predicted that magnetic storage technology will continue evolving during the next decade or so. It may even happen that the significant development in magnetic technology reaches the level competing with technological areas traditionally dominated by semiconductors. Advances in growth techniques will lead to increasingly more sophisticated nanostructures which will be, in addition, very interesting systems from a fundamental point of view since they will allow basic concepts of magnetic properties at interfaces to be consolidated. These advances in growth techniques have to be accompanied with advances in characterisation tools that match the reduced dimensions and the complexity of the magnetic microstructures. We will briefly describe how surface X-ray scattering is presently contributing to the development of the field.

The interaction of X-rays with matter is usually dominated by the Thomson scattering which results from the coupling of the electric field of the X-rays with the charge of the electrons in the atoms. Moreover, the interaction of X-rays with atoms possessing a magnetic moment may result in a measurable coupling between the magnetic field of the radiation and the magnetic moment of the atoms. This magnetic effect which is much weaker than the charge scattering, was brought to evidence more than 20 years ago and since then it has been used to study bulk magnetic materials. The application of this phenomenon to surface magnetism was, however, rather hopeless even with synchrotron X-ray sources, since surfaces, basically two-dimensional objects, are very diluted systems compared with bulk materials which have three-dimensions. A few years ago it was discovered [11] that under some resonant conditions (the photon energy has to coincide with the energy of some atomic absorption edges), the sensitivity of the X-rays to the magnetic moment of the atoms in resonant condition is enhanced by several orders of magnitude. This important result opened the possibility of performing surface magnetic diffraction experiments aimed at probing the magnetism of a single atomic plane. Furthermore, in a sample with different types of atoms, the enhancement effect has the additional advantage of being selective to only the resonant atoms. In spite of the resonant effect, the magnetic part of the scattered signal is very small compared to the part that originates from the charge scattering. One has therefore to find some 'tricks' to isolate if possible the magnetic contribution from the scattered intensity. The interaction of the X-rays with the atoms is described by the atomic diffusion factor which is a sum of various terms the most important being, by far, the one which corresponds to the Thomson scattering. In addition, there are the magnetically-sensitive terms; the most important one writes (under resonant conditions):

$$nm(\mathbf{e}_f \times \mathbf{e}_i) \cdot \mathbf{m} \qquad (4)$$

where $\mathbf{e}_i$ and $\mathbf{e}_f$ are respectively unit vectors along the polarisation of the incoming and scattered beams, $\mathbf{m}$ is the unit vector along the magnetisation of the surface and nm is a number which describes the strength of the magnetic effect. In atomic units, usually nm is around one. This has to be compared with the strength of the Thomson scattering which is close to the atomic number of the atoms of the sample. As the scattered intensity depends on the square of the sum of the total diffusion factor, it will contain the cross products of the term Eq. (4) with other terms (as the

Thomson scattering amplitude one). The sign of these crossed terms may be reversed if by means of an externally applied magnetic field the magnetisation **m** is reversed. Therefore, the asymmetry ratio

$$R = \frac{I\uparrow - I\downarrow}{I\uparrow + I\downarrow} \quad (5)$$

obtained by measuring the intensity for both magnetisation orientations, will be different from zero since only the odd terms in **m** will survive. Thanks to this, the magnetic part of the scattered intensity may be separated from the charge part.

We will illustrate with a simple example the sensitivity and possibilities of this method.

Co/Pt multilayers are commonly used in magnetic technology since the Pt atoms tend to favour perpendicular magnetic anisotropy thanks to their relatively important spin–orbit coupling. Ferromagnetic Co atoms induce a magnetic moment on the Pt atoms which largely determine the magnetic properties at the interface. Let us consider a film of ∼8 atomic layers of Co deposited on a Pt(1 1 1) substrate. A relevant scientific question is: to what depth do the Pt atoms below the Co film exhibit induced magnetic moment?

To answer that, one first has to tune the photon energy to a Pt adsorption edge where a large resonant effect occurs. The L_3 edge is very suitable since its relatively high ionisation energy (∼11.5 keV) results in a comfortable photon wavelength, close to 1 Å, which is adequate to explore reciprocal space. The magnetic moment of the Pt atoms is proportional to the quantity nm mentioned before (it can be roughly visualised as the number of unpaired electrons in a Pt atom). By measuring the asymmetry R, one has access to nm. In order to extract the values of nm for different Pt planes beneath the Co film, one has to measure, as in any crystallographic determination, the magnitude of R along the direction perpendicular to the surface, i.e. one has to measure how R changes for different values of the perpendicular component of the diffraction vector. Fig. 7 displays the results of such a measurement obtain by Ferrer et al. [21] on the surface diffraction beamline of the ESRF (Grenoble).

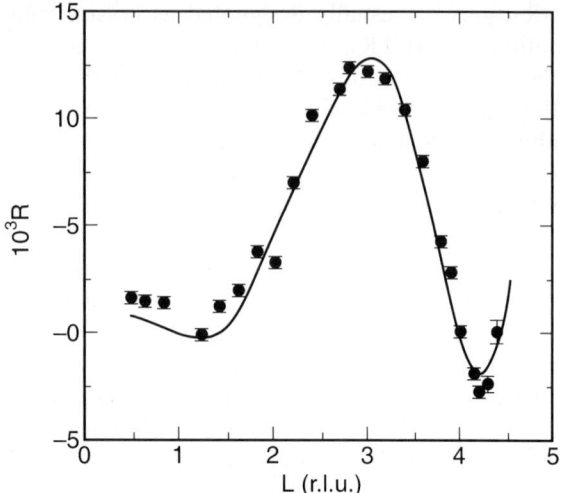

Fig. 7. Dependence of the asymmetry ratio at resonance with the perpendicular momentum transfer for a film of eight atomic layers of Co on the Pt substrate. The continuous curve is a fit to the data that arises from essentially one atomic layer of magnetised Pt at the interface (from Ref. [21]).

The quantity L in the horizontal axis is a reciprocal space co-ordinate that is proportional to the perpendicular momentum transfer. The measurements were done while keeping the parallel momentum transfer constant and equal to one reciprocal lattice unit of the Pt reciprocal lattice. The range of the measurement is about 4 units in L which corresponds to four Brillouin zones in reciprocal space. The maximum measured values of R are about 0.01 which shows that only a small part of the intensity diffracted by the surface is sensitive to magnetism. At $L = 1$ and 4, bulk Bragg conditions in the Pt crystal are fulfilled and therefore at these values R probes the bulk magnetism of the Pt crystal which is measured to be zero as expected since an ultra-thin Co film is not expected to magnetise the bulk of the Pt crystal. Between $L = 1$ and 4 only the topmost atomic planes of the crystal contribute to the diffracted intensity. In this particular example, the analysis of the data is very simple since the measured values of R arise only from the topmost atomic Pt planes in contact with the Co with no interferences from the Co crystal lattice. The continuous curve is the result of a calculation which demonstrates that only the topmost atomic plane of Pt has significant induced moment (nm = 0.5); in the next plane, the induced moment is about 10 times smaller. Note that the results also illustrate the sensitivity of the technique: it is perfectly possible to measure with good statistics the magnetic diffracted signal from a single atomic plane. The abruptness of the induced Pt magnetism is a consequence of the compactness and abruptness of the Co/Pt interface. A similar experiment on a Pt(1 1 0) shows that the magnetisation of the Pt extends at least to three atomic Pt planes. The reasons being the more open atomic structure of the (1 1 0) planes and a less chemically abrupt Co/Pt interface. The combination of the magnetic diffraction technique with ordinary surface diffraction (both techniques may be used in the same sample at the same time) may help to get detailed information on some important topics as for example the relationship between strain and interface magnetism. The high momentum resolution of standard X-ray diffraction (due to the fact that the synchrotron beam sources are very small and are very far away from the sample), allows to observe small changes in lattice spacing (how small depends exclusively on the quality of the sample) allowing precise measurements of interface strain. The corresponding chemically specific magnetic surface diffraction would tell how the magnetic moments and magnetic anisotropies of the atoms under resonant conditions depend on their strain. This would be precious information to design interface structures with some predetermined requisites. As seen in the figure, the error bars of the data points are relatively small (the acquisition time of each data point is typically 10–15 min). This means that more ambitious problems may be investigated. For example stepped surfaces obtained by cutting a low index crystal face at some angle away from the low index planes usually consist in regular arrays of steps separated by identical terraces. They constitute therefore ideal templates for growth of magnetic materials. It is rather well known that steps influence the magnetic anisotropy of the surface. One could envisage determining the relationship between strain and magnetism at the steps atoms to study 1D magnetism.

3. Surface chemistry

Surface chemistry has been another field that has benefited from the advent of third generation synchrotrons. We will give two different examples. Both of them are related to the field of catalysis. Let us keep in mind that the majority of chemical processes that are utilised in chemical technology are catalytic. Among others, transition metals are widely used in the chemical industry as catalysts. One of the major functions of a catalyst consists of accelerating a chemical reaction. Generally speaking, this occurs because the activation barriers that exist between the reactant atoms in the gas phase are largely reduced when the reactants are adsorbed on the surface of the catalyst. Due to this, the speed of the reaction may increase enormously. The microscopic description of the elementary process of a catalytic reaction is an important area of research in surface science due to both technological and basic interest.

The first example concerns the so-called chemical waves and how their study provides key elements for the understanding of the elementary processes of the chemical reaction. The second one concerns the effect of the gas pressure on the surface structure of the catalyst.

3.1. Chemical waves

It was realised several years ago that some simple catalytic surface reactions display instabilities that may result in the formation of surface waves. Let us consider the reaction $NO + H_2 \rightarrow N_2 + H_2O$ occurring on the surface of an Rh(1 1 0) crystal [22]. The reaction proceeds by dissociation of molecularly adsorbed NO into atomic oxygen and nitrogen, and that of molecularly adsorbed H_2 into atomic hydrogen atoms. The main species present at the surface under reaction conditions ($T > 500$ K) are however atomic oxygen and nitrogen. It turns out that for specific values of the pressures of the gas phase species (NO and H_2), the surface is not chemically homogeneous but it displays areas rich in oxygen or in nitrogen. Moreover, these chemical inhomogeneities may display a temporal evolution in the form of chemical waves, i.e. propagating reaction fronts across the surface. The investigation of the dependence of the characteristics of the chemical waves (propagation directions, velocities, etc.) with reaction parameters, may allow the elementary steps of the surface reaction to be determined and therefore the detailed microscopic mechanism.

The laterally varying adsorbate concentrations under reaction conditions have been evidenced during the last few years with PEEM. This method, which yields a spatial resolution around 1 μm, is based on the fact that the yield of photoelectrons depends sensitively on the local work function as one illuminates the sample with photons (usually provided by discharge lamps). Fig. 8 shows a PEEM image of chemical waves in reaction conditions. The white elliptical rings propagate radially away from their centre and correspond to relatively low values of the surface work function which are associated to oxygen-deficient areas. The elliptical and non-circular wave fronts are due to anisotropy of the surface. In the (1 1 0) surface of Rh there are two orthogonal surface directions which are very different. In the [1,−1,0] direction the atoms form atomic rows closely packed whereas along the [0,0,1] direction, two consecutive atoms are separated by $\sqrt{2}$ nearest neighbour distances. As a consequence, surface diffusion of adsorbed atoms is faster in the direction [1,−1,0] giving rise to elliptical wave fronts with elongation in that direction.

A further step in the understanding of this reaction has been made at ELETTRA (Trieste) by

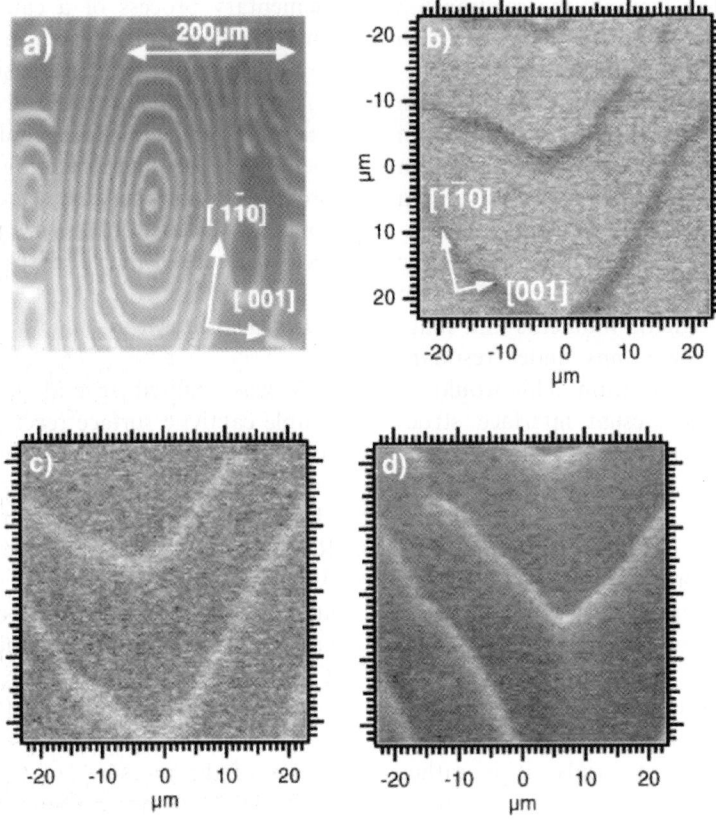

Fig. 8. Pulse propagation in the system Rh(1 1 0)/NO + H_2: (a) PEEM image of target pattern. Experimental conditions: $T = 530$ K, $P_{NO} = 1.7 \times 10^{-7}$ mbar, $P_{H_2} = 7 \times 10^{-2}$ mbar. Elemental maps recorded with SPEM showing the distribution of the O_{1S} (b), N_{1S} (c) and $Rh3d_{5/2}$ (d) intensity during pulse propagation. The pulses propagate in the upward direction. Time per frame: 330 s; photon energy 625.7 eV. Experimental conditions: $T = 530$ K, $P_{NO} = 1.7 \times 10^{-7}$ mbar, $P_{H_2} = 6.4 \times 10^{-7}$ mbar (from Ref. [23]).

Schaak et al. [23] substituting the photon beam from the discharge lamp by a synchrotron beam focused down to 0.15 μm by means of a zone plate. Furthermore, instead of collecting the secondary electrons from the sample, an energy analyser allowed only the photoelectrons arising from some specific chemical levels to be selected. In this way, chemically-specific lateral information of the wave fronts was achieved. Fig. 8(b–d) shows the distribution of the photoelectrons from the 1s level of adsorbed oxygen and nitrogen and that of the photoelectrons arising from the ionisation of the $3d_{5/2}$ electronic level of the Rh atoms. The images were taken while a reaction pulse was propagating. The pulses had a width of 5 μm and propagated at ~0.5 μm/s. The images show that within the pulse there is mainly adsorbed nitrogen making the surface oxygen concentration almost negligible. This essential information was taken as the main experimental fact that was reproduced by integrating a set of differential equations describing the kinetics of several elementary surface reactions. As a result, an improved understanding of the reaction process was achieved. In summary, the different processes are (i) suppose, to start with, that a specific part of the surface is completely covered with oxygen. At some specific surface site where a defect exists, hydrogen adsorbs and combines with the adsorbed oxygen leaving the surface. A reaction front has been created and spreads out. (ii) On the free surface thus created, hydrogen and NO adsorb and dissociate. However, adsorbed oxygen is removed through reaction with hydrogen and therefore nitrogen accumulates on the surface. (iii) On the nitrogen-covered surface, NO may still adsorb and dissociate. The oxygen coverage increases and due to the repulsive interaction between adsorbed nitrogen and oxygen, nitrogen desorbs. The result is a surface rich in oxygen as in the initial state.

The scientific outcome described in the present example is the result of an important development in instrumentation. In addition to the high photon flux required for the experiment, the large lateral resolution achieved by the SPEEM technique is impressive and it is due to the important recent progress in optics development. The key element determining the resolution is the zone plate made with lithographic technology. More progress will certainly be made in the near future since more and more sophisticated zone plates are being built. In only a few years the lateral resolution will probably reach a few hundred angstroms and more detailed information on microscopic processes will be achievable. Coming back to our example, one could, among other things, try to resolve the nature of the surface defects that initiate the chemical wave. If these defects were known, they could perhaps be artificially created. In this case one could modify the reaction rates by acting directly on the surface of the catalyst in a programmed way. This would open absolute new avenues in the field of surface chemistry and catalysis.

3.2. The structure of chemisorbed layers at atmospheric pressures

Since the discovery in the 19th century that some gas molecules adsorbed onto a metal surface are readily converted into other molecules, *heterogeneous catalysis* has achieved tremendous technological, environmental, and commercial importance. Understanding the gas–metal interaction via the relevant adsorbate structures is a primary goal of modern surface science. During the last 20 years, the geometry and chemical bonding of gases adsorbed on metal surfaces has thus been determined and catalogued for more than one thousand systems [24]. Practically all the structural studies have been done under the extreme vacuum conditions of 10^{-8}–10^{-14} bar where appropriate techniques were available. The most widely utilised techniques involve electrons as probing particles. In particular, low energy electron diffraction (LEED) based on the elastic interaction with the adsorbate structure of electron beams with kinetic energies around 100 eV has been very successful. However, due to the strong interaction of electrons with matter, electron based techniques require vacuum environment. Since catalytic reactions are performed at atmospheric pressures, the essential question of the "pressure gap" has been open for many years: "*are the known vacuum structures the ones that actually exist on the surfaces of the catalysts under real reaction conditions?*"

There are fundamental questions which induce to suspect that things may be different between UHV and the atmosphere. The chemical potential of a gas, in the simplistic ideal gas approximation, is $\mu = kT \log p$. The 13 orders of magnitude from 10^{-10} mbar to 1 bar imply a change $\Delta\mu = 13kT$, i.e. more than 300 meV at room temperature. This additional chemical energy may play a decisive role in surface chemistry.

Several investigations on the adsorption of gases at pressures near the atmosphere have been recently performed on metal surfaces of interest in catalysis, by means of suitable probing techniques unaffected by the atmosphere of gas around the samples. Non-linear optical techniques involving the reflection of laser beams [25,26] measure the vibrations of the CO molecules bonded to the metal which provide information on the structure of the adsorbed layer. Also, scanning tunneling microscopy (STM) investigations [27] can determine some structural characteristics of the chemisorbed layer such as symmetry elements and linear dimensions. However, none of the above techniques allows determining the relevant parameters of the structure such as bond-lengths, bond-angles, distortions on the substrate etc.

Surface X-ray diffraction is an important and suitable technique to be considered. Third generation synchrotron sources generate beams of enough brilliance to result in easy measurable diffracted intensities from chemisorbed layers as we will see. Also, the availability of relatively high photon energies ($\sim$20 keV) allows performing X-ray scattering experiments in ambient pressures of one ore several bars with negligible absorption.

Here, we will provide an example of an affirmative answer of the essential question of the pressure gap (at room temperature) and negative one (at elevated temperature) for the archetypal case of CO over Ni(1 1 0). Based on X-ray diffraction measurements, the CO/Ni(1 1 0) structure was determined in situ from 10^{-10} to 2.3 bar CO at 25 °C [28]. Interestingly, the vacuum structure persists unchanged over 10 orders of magnitude in pressure. A subsequent warming to $\sim$130 °C at 2.3 bar then causes a massive restructuring of the Ni surface consisting of the development of microfacets with (1 1 1) orientation and surface strain probably due to carbon dissolution. These results confirm the relevance of vacuum studies to catalysis and offer a glimpse at the complexity of elevated-pressure surface chemistry.

Nickel catalysts are used most frequently to produce methane from carbon monoxide and hydrogen [24]. The reaction is thermally activated, (usually performed from 150–400 °C) carried at or above atmospheric pressure and is thought to involve dissociation of CO adsorbed on the metal surface. In vacuum, adsorbed CO on Ni(1 1 0) forms a 2×1 structure consisting in an ordered zigzag arrangement of tilted molecules on shortbridge sites of the substrate as depicted in Fig. 9. The first column of Table 1 displays the results from the most recent study by LEED [29].

The experiments reported in the following were performed at the surface diffraction beamline at the ESRF with X-rays of wavelength 0.74 Å and a specially designed ultrahigh vacuum (UHV)/high pressure chamber mounted on a high precision diffractometer. The chamber had a cylindrical, X-ray transparent, Be window which allowed the incoming and diffracted beams to reach the sample surface and the detector respectively.

Two independent sets of crystallographic data are collected in situ after exposing a well prepared Ni surface to 10^{-10} bar and 2.3 bar of CO. (The gas

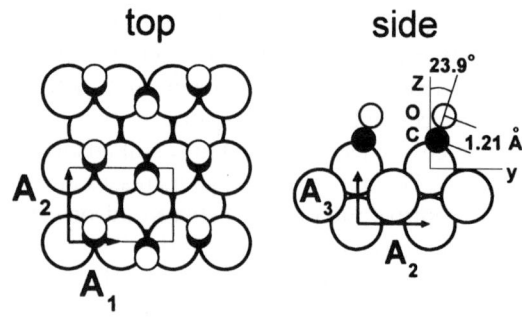

Fig. 9. Top and side views of the CO/Ni(1 1 0) 2×1 structure. A_1, A_2 and A_3 are the lattice vectors used to describe the crystal lattice. $A_1 = A_3 = a\sqrt{2}$ ($a_0 = 3.524$ Å is the lattice constant on Ni) and $A_2 = a_0$. The rectangle shows the 2×1 unit cell. The tilt angle and bond lengths of the adsorbed CO molecules are those obtained from the fit of the data set at 2.3 bars of CO. In reciprocal space, h, k and l are coordinates parallel to A_1, A_2 and A_3 respectively [28].

Table 1
Structural parameters of CO/Ni(1 1 0)(1 × 2)

	Ref. [29]	Peters et al. [28]	
		10^{-10} bar	2.3 bar
Ni–C tilt angle (deg)	20(4)	21.7(5)	21.3(5)
C–O tile angle (deg)	20(4)	23.5(9)	23.9(7)
Ni–C bond length (Å)	1.85(4)	1.84(2)	1.83(2)
C–O bond length (Å)	1.15(7)	1.19(3)	1.21(3)
Ni–Ni expansion (Å)	–	0.052(3)	0.058(2)

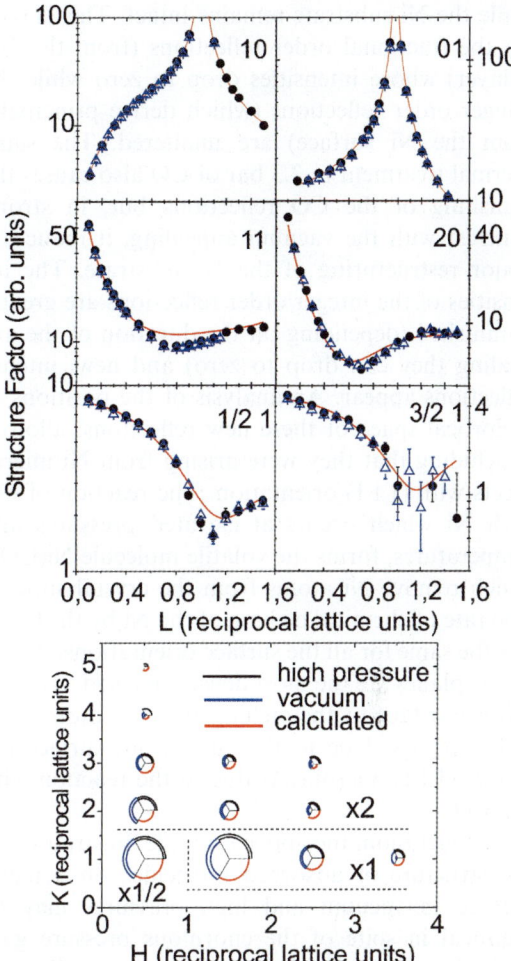

Fig. 10. Crystallographic structure factors of (2×1) CO/Ni(1 1 0) at room temperature (from Ref. [28]). Upper panel: Blacks circles: measured structure factors from four integer rods $[(h,k) = (0,1), (1,1)$ and $(2,0)]$ and two fractional order rods $[(h,k) = (1/2,1), (3/2,1)]$ in 2.3 bars of CO; Blue open triangles: structure factors measured in 10^{-10} bars of CO. Bottom panel: In-plane fractional order structure factors measured at $L = 0.1$. The radius of the black 120° sectors are proportional to the structure factors of the reflections from the structure in 2.3 bar of CO. The error bars are indicated by the two radii. The blue sectors represent the structure factors for the vacuum structure and the red ones are calculated from the fit to the high pressure data.

was kept in the chamber during the measurements.) At crystallographic positions (h, k, l) in reciprocal space, the structure factor was obtained by measuring the integrated intensity of the diffracted beam as the crystal was rocked around the surface normal and by applying the appropriate corrections.

Fig. 10 shows the results. The top panel gives the structure factors of six diffraction rods. Four have integer-valued h and k corresponding to the periodicity of the substrate lattice and two rods have fractional index arising from the CO 2×1 periodicity. The bottom panel displays the structure factors of the fractional reflections at $l \cong 0$. Inspection of the figure shows immediately that the vacuum and high pressure data sets are virtually identical and thus demonstrates that both structures are the same. Crystallographic analysis of the high pressure data results in the continuous red curve in the figure. The best fit model resulted in the structural parameters shown in Table 1. The agreement with the results of Ref. [29] is excellent. In addition, our analysis reveals a slight expansion of the Ni planes that was not detected [29]. The table also shows the results of the analysis of the low pressure data which coincide with these at high pressure.

From the previous experiments we can conclude unambiguously that the equilibrium structure at room temperature of CO on Ni(1 1 0) at 2.3 bars of CO ambient pressure is the same as that obtained under UHV conditions by dosing the Ni(1 1 0) surface at saturation with 10^{-10} bar of CO.

The effect of temperature on both cases is, however, drastically different. Annealing the Ni crystal to ∼130 °C in 10^{-10} bar of CO causes disordering and desorption of the chemisorbed layer while the Ni substrate remains intact. This is seen via the fractional order reflections (from the CO adlayer) whose intensities drop to zero while the integer order reflections (which derive principally from the Ni surface) are unaltered. The same thermal treatment in 2.3 bar of CO also causes the vanishing of the CO reflections but, in strong contrast with the vacuum annealing, it induces a major restructuring of the Ni substrate. The intensities of the integer order reflections are greatly diminished (depending on the duration of the annealing they can drop to zero) and new, intense reflections appear. An analysis of the positions in reciprocal space of these new reflections, allowed concluding that they were arising from Ni microfacets with (1 1 1) orientation. The reaction of CO with Ni which occurs at elevated pressures and temperatures, forms the volatile molecule $Ni(CO)_4$ which remove Ni atoms from the crystal surface. The rate of chemical etching of the Ni by the CO is not the same for all the surface orientations. As the (1 1 1) planes are the most densely packed, they are more resistant to the etching that the more open (1 1 0) planes. Due to this, the (1 1 0) surface develops (1 1 1) microfacets due to the reaction with the CO.

In conclusion, the above example illustrates that the structure of adsorbed molecules on a metal surface in vacuum and high pressures may be identical in spite of the enormous pressure gap, and that, at the same time, completely different surface chemistry may take place at elevated temperatures.

4. Local structure of liquid lead

A perfect crystal is characterized by a long-range order. What about monatomic liquids? The local point symmetry in these liquids has been for a long time an open question. We know that the liquid state exhibits short-range order extending only a few atomic distances given by the liquid correlation length, which is typically in the range of 3–8 Å.

In the 1950s and the 1960s Frank [30] and Bernal [31] demonstrated that the local liquid correlations in these liquids which exhibit centrosymmetric interactions favoring a close-packed local environment, will favor icosahedron building blocks (with 12 atoms). Later on it was suggested that these blocks would break to form fragments (half of icosahedron) exhibiting fivefold symmetry.

If one uses X-ray to study the structure of these clusters in the liquid phase, the scattering experiment will average (because the clusters are randomly oriented) and no information will be obtained. One has to find a trick to align the fragments. This has been done recently by Reichert et al. [32] by trapping them near a Si(0 0 1) surface. The Pb–Si interaction is strong enough to hold the lead clusters against the Si surface without breaking them. These difficult experiments were performed at the European Synchrotron Radiation Facility (Grenoble) and at the Hamburg Synchrotron Radiation Laboratory. The experimental set-up is shown in Fig. 11: it is designated to measure the structure of the clusters at the interface liquid lead–silicon over few Å. A high energy X-ray (71.5 KeV) microbeam (8 μm vertical size) penetrates a cylindrical Si crystal from the side and impinges the Si–Pb interface. For angles below the critical angle of total internal reflection ($\alpha_c = 0.041°$) the incident beam is totally reflected. An evanescent wave is produced within a thin layer but the X-ray field is absent in the bulk liquid. The wall presents only a small lateral periodic perturbation potential, which is incommensurate with the liquid nearest-neighbour distances. Then, there is no wall-induced long-range translational ordering. Fig. 12 shows the results obtained:

(a) By choosing an incidence angle $\alpha_i = 0.032°$ well below the critical angle for total internal reflection, the evanescent X-ray tunnels 55 Å into the liquid Pb. By scanning the 2θ angle (see Fig. 11) one can measure the structure factor of liquid Pb and the curve obtained is typical of bulk-liquid Pb. In particular, the maximum agrees ($q_1 = 2.18$ Å^{-1}) agrees with the know bulk value.

(b) In the second experiment (Fig. 12b), maintaining the 2θ angle around the first diffraction maximum, azimuthal (ϕ) scans are performed. For a bulk liquid, this type of scan shows a constant intensity associated with the random orientation

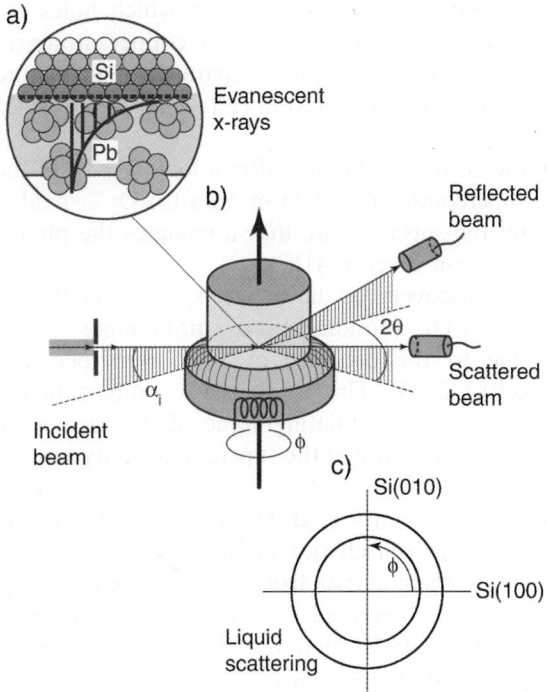

Fig. 11. (a) A well-defined high-energy X-ray microbeam ($E = 71.5$ keV, 8 µm vertical size) penetrates a cylindrical Si solid ($r = 20$ mm) from the side and impinges to the Si–Pb interface. For angles below the critical angle of total internal reflection ($\alpha_c = 0.041°$) the incident beam is totally reflected. The angle of incidence α_i and the position of the interface in the X-ray beam are controlled to an accuracy of 4 µrad and 0.1 µm; (b) In a UHV in situ diffraction chamber the liquid Pb and the Si(0 0 1) surface are cleaned and brought into contact at $T - T_m(P_b) + 10$ K. A 360° Be window gives access to the interface; (c) Top view of the azimuthal rotation ϕ at a fixed momentum transfer q on the liquid scattering (from Ref. [32]).

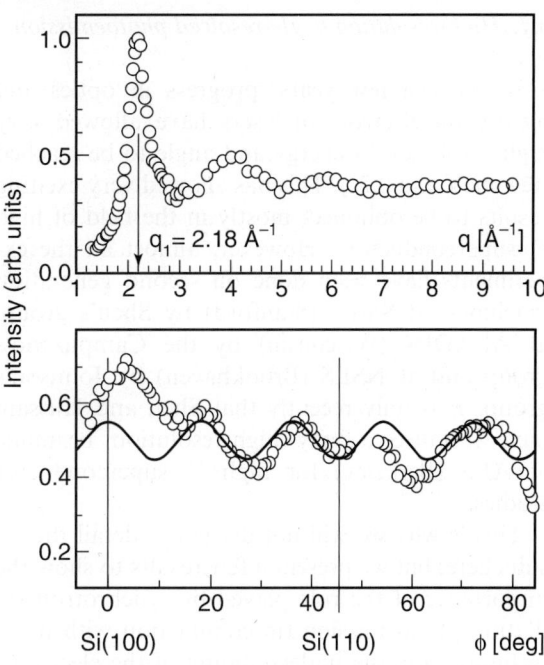

Fig. 12. Scattering at the Pb(liq.)–Si(0 0 1) interface. (a) Pb interfacial liquid structure factor measured in arbitrary units. The evanescent X-ray wave "tunnels" 55 Å deep into the liquid Pb. (b) Azimuthal distribution of the liquid scattering at the first diffraction maximum. The line is a fit of a sinusoidal to the experimental data (circles). The epitaxial relation to the Si substrate is indicated by vertical lines (from Ref. [32]).

of the clusters. Here five intensity maxima are observed with a fixed epitaxial relationship to the substrate. The Pb clusters can be centred about a Si atom (minimum overlap between the Pb and Si atoms) or a hollow site between Si atoms. From energy considerations the first position is favored and this is exactly what it is observed experimentally.

These results are important and could lead to a better understanding of freezing and supercooling (which happens when a substance remains liquid as it is cooled below its freezing point) because clusters with fivefold symmetry may in fact compete with crystallization rather than enhance it.

5. Electronic structures

The electronic structures are mostly studied by photoemission spectroscopy. At low photon energy (20–30 eV) angle-resolved photoemission is used to determine the band structures of solids. At higher photon energy (few hundred eV) one can either exploit the fact that the mean free path of the photoelectron increases with the kinetic energy (going from 5 Å at 50 eV to 15 Å around 1000 eV) and then probe the bulk or use circularly polarised radiation and resonant photoemission to study materials in an antiferromagnetic phase.

5.1. High-resolution angle-resolved photoemission

In the last few years, progress in optics and mostly on electron analysers have allowed very high resolution in energy and angle to be reached (few meV and 0.2°). This has allowed very exciting results to be obtained, mostly in the field of high T_c superconductors. However, almost all the experiments have been done on second generation machines: at SSRL (Stanford) by Shen's group, at ALADIN (Wisconsin) by the Campuzano's group and at NSLS (Brookhaven) by Johnson's group. It is only recently that Shen and Hussain have installed a very high-resolution beamline at ALS (Berkeley) for high T_c superconductor studies.

This is why we will not discuss in detail the results here, but we present a few results to show the importance of the role played by synchrotron radiation photoemission (in conjunction with other techniques) in the understanding of the electronic properties of the high T_c superconductors. The high T_c superconductors like $Bi_2Sr_2CaCu_2O_{8+\delta}$ have a very rich temperature-doping phase diagram. In the undoped case (by varying δ), the copper oxide plans have an odd number of electrons per unit cell and are antiferromagnetic insulators. Band theory predicts that there should be metallic but the strong Coulomb interaction between the localised Cu 3d electrons suppress the charge mobility and produce insulators. At high doping, the high T_c superconductors are metallic and present a band structure whose Fermi surface is consistent with local density band calculations (LDA). We present here the most important results obtained:

- the observation of the opening of a gap in the superconducting phase for an optimally-doped sample by Imer et al. [33];
- the large anistropy of this gap, showing a $d_{x^2-y^2}$ symmetry [34,35];
- the discovery of a pseudogap in the normal state for undoped samples [36,37];
- the observation of a Fermi surface for near optimum doping and the disappearance of part of the Fermi surface for undoped materials [38,39];
- the observation of stripes (in which holes are confined to parallel lines of copper atoms in the Cu–O oxygen plane separated by insulating region without holes) in $(La_{1.28}Nd_{0.6}Sr_{0.2})$ CuO_4 [40];
- the clear observation, after a long search, of the superconducting state in $YBa_2Cu_3O_{7-\delta}$, hidden by the surface state that dominates the photoemission spectra [41];
- the discovery of the bilayer splitting in Bi_2Sr_2-$CaCu_2O_8$ (bonding and antibonding bands due to the presence of two layers per unit cell) [42–44]. This result is very important because it is eliminating some of the theoretical models explaining the superconductivity mechanism;
- finally, Lanzara et al. [45] have argued recently that the self-energy effects in angle-resolved photoemission data from the high T_c compounds could be described in terms of strong-phonon coupling. It is obvious that this interpretation will create hot debates.

5.2. Heavy fermions

Electron correlations are known to play an important role in determining the unusual physical properties of a variety of compounds such as some of the transitions metals, the high T_c superconductors or the heavy fermions. High-resolution photoemission spectroscopy has been intensively used to probe the electronic states of these materials (particularly near the Fermi level). However the experiments until now have been done mostly at low energy (20–120 eV) where the photoelectron mean free path is very short (<5 Å) [46] which means that there are very surface-sensitive. We have already seen in the case of magnetism that the properties in the bulk can be quite different from the surface. This has been nicely illustrated recently by Sekiyama et al. [47] who have studied the Ce compounds $CeRu_2Si_2$ and $CeRu_2$ at two photon energies: 120 eV (probing mostly the surface) and 880 eV (probing mostly the bulk).

In weakly hybridised Ce compounds, where the valence of the Ce ions is close to 3+ f^1, the Ce 4f electrons are localised due to their strong Coulomb interaction. In more strongly hybridised cases, however, the 4f electrons form the so-called Kondo singlet with conductions electrons due to the Kondo effect [48]. The strength of hybridisation is represented by the Kondo temperature T_K, the hybridisation being stronger in compounds with higher T_K. CeRu$_2$Si$_2$ with a $T_K \sim 22$ K is a typical heavy fermion system [49]. It is considered as being at the boundary between the localised and itinerant Ce 4f states. In more strongly hybridised compounds the Ce ions are valence fluctuating between Ce^{3+}(f^1) and Ce^{4+}(f^0) states. In such cases, the 4f electrons may no longer be localised and can be regarded as fairly itinerant. CeRu$_2$ with a T_K of the order of 1000 K, is a typical example of the strongly fluctuating 4f systems [50]. Then the bulk photoemission spectra of CeRu$_2$Si$_2$ and CeRu$_2$ should be different. Why? Because if the valence-band electrons are quite itinerant, the measured valence band spectra are in good agreement with band structure calculations. The spectra of localised Ce 4f states disagree with band structure calculations and are usually explained by the single impurity Anderson model [51] shown schematically in Fig. 13. The ground state is a linear combination of the f^0 and f$^1_{5/2}$ states due to finite hybridisation. There can be several 4f photoelectron final states: f^0 (called the poorly screened state) and the well-screened f$^1_{5/2}$ and f$^1_{7/2}$ states which correspond to a tail of the Kondo peak and its spin orbit partner. The f^0 peak is generally considered to be weak when the hybridisation is strong. Furthermore, it has been shown [52] that the tail of the Kondo peak can be observed as an extremely narrow peak in the vicinity of the Fermi energy (E_F) in the 4f spectra of rather hybridised Ce compounds.

Sekiyama [47] have exploited the high brilliance of Spring 8 to obtain very high-resolution photoemission spectra. The results (Fig. 14) have been obtained at two photon energies: 120 eV corresponding to the Ce 4d–4f resonance and at 880 eV (3d–4f resonance). In the 3d–4f spectrum of CeRu$_2$Si$_2$, there is a sharp peak near E_F and a broad tail ranging from -1 to -5 eV. According to the single impurity Anderson model, the former corresponds to the contribution of both the tail of the Kondo peak (f$^1_{5/2}$) and its spin–orbit partner. The latter is ascribed to the f^0 final state. In the

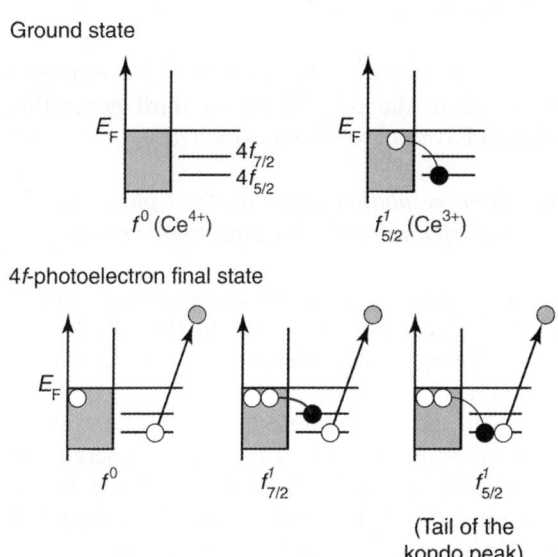

Fig. 13. Schematic picture of the single impurity Anderson model. Filled and open circles represent the occupied electron and the hole, respectively, while shaded circles stand for the photoelectron. E_F denotes the Fermi level.

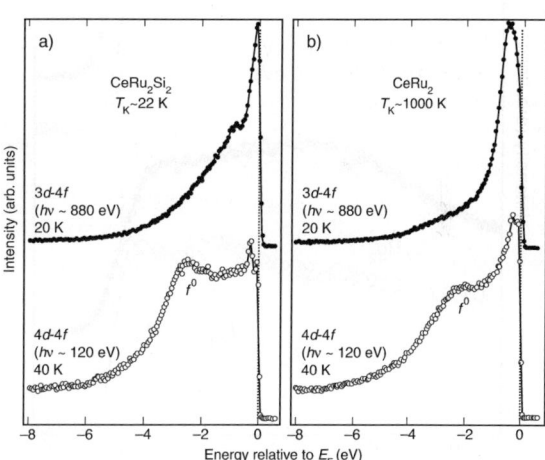

Fig. 14. Ce 4f spectra obtained from Ce 3d–4f and 3d–4f resonance photoemission with energy resolution of 200 and 80 meV, respectively: (a) CeRu$_2$Si$_2$, (b) CeRu$_2$. The Ce 4f contributions are estimated by subtracting the off-resonance spectra taken at $\hbar\omega = 875$ (122) eV for the Ce 3d–4f (4d–4f) resonance. The relative intensity between the 3d–4f and 4d–4f spectra is arbitrary. The surface contribution to the 4f spectra is 55% for the 4d–4f and 15% for the 3d–4f resonance (from Ref. [47]).

4d–4f spectrum on the other hand, the f^0 final state located at -2.5 eV is prominent, reflecting the localised character of the surface 4f states. In the 3d–4f spectrum of $CeRu_2$, the f^0 broad tail is greatly suppressed compared to $CeRu_2Si_2$, indicating that the bulk Ce 4f states are considerably hybridised. These differences reveal that the 4d–4f spectra mainly reflect the surface 4f electronic states. The high resolution spectra near E_F are shown in Fig. 15. We note that the 4f–4f spectral line-shapes of $CeRu_2Si_2$ and $CeRu_2$ are very similar, despite the T_K values of these compounds differing by order of magnitude: there are two comparable structures, located near E_F and near -0.3 eV. However, the 3d–4f spectra of these two compounds are very different. For $CeRu_2Si_2$ there is a prominent peak near E_F and a weak shoulder near -0.3 eV. These two structures originate from the bulk $f^1_{5/2}$ and $f^1_{7/2}$ final states, which means from the tail of the Kondo peak and its spin orbit partner. The observation of such a strong tail is observed for the first time thanks to the photon energy (880 eV) and the very high resolution (100 meV). The two peaks in the 4d–4f spectrum of $CeRu_2Si_2$ show that the hybridisation is weaker at the surface.

The 3d–4f line line-shape of $CeRu_2$ is, however, rather surprising as no structure is seen except for a broad feature centred at 0.5 eV, very different from what can be expected from the single impurity Anderson model. The experimental results that the 3d–4f spectrum shows a rather conventional Fermi cut-off and a broad peak at -0.5 eV cannot be deduced from that model. Therefore bulk $CeRu_2$ presents fairly itinerant 4f bands; this is revealed for the first time.

The main conclusion of this study is that one has to be very careful when comparing data obtained with low photon energy photoemission (very surface sensitive) and techniques that measure the bulk properties (specific heat, etc.).

6. Perspectives

We would like to illustrate by a few examples areas where the possibilities of third generation sources have not yet been exploited.

6.1. High resolution angle resolved photoemission and high spatial resolution (spectromicroscopy)

Spectromicroscopy has been intensively used at various places but essentially in the soft X-rays, people being mostly interested by core level spectroscopy.

It will be very interesting to obtain angle-resolved photoemission at low photon energy (20–30 eV) with high spatial resolution (few μm). Why is that interesting? Before the discovery of the STM, there were a lot of studies of the surface states of the Si (2×1) and Si (7×7) reconstructions. They presented many discrepancies. Examination of the surface by STM some time later showed that for a supposed Si (7×7) surface only a small fraction of

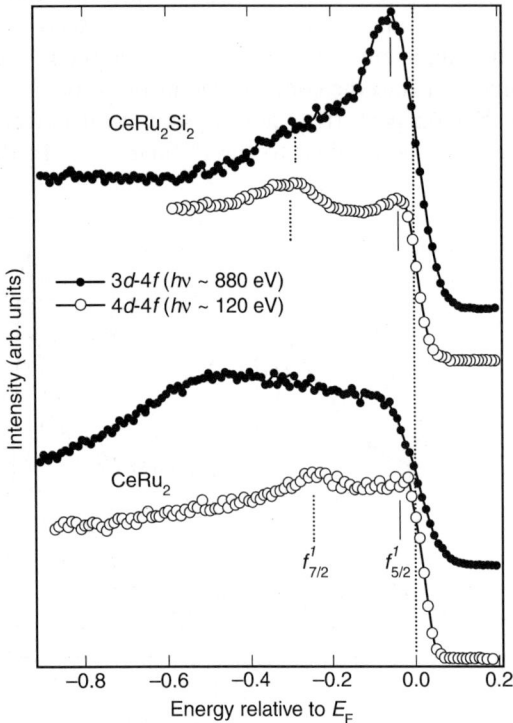

Fig. 15. High resolution Ce 4f spectra in the 3d–4f (resolution 100 meV) and 4d–4f (resolution 50 meV) resonances of $CeRu_2Si_2$ and $CeRu_2$. Here the raw spectra taken at $\hbar\omega = 882.6$ are displayed for the Ce 3d–4f spectra since the Ce 4f resonance enhancement is predominant. The Ce contributions to the 4d–4f. The Ce 4f contributions are obtained by subtracting the off-resonance spectra from the one-resonance ones (from Ref. [47]).

the surface was really a 7 × 7, with a lot of other reconstructions and disorder.

In 1990, Personn et al. [53] did some very interesting experiments on high T_c superconductors, $YBa_2Cu_3O_7$ and $Bi_2Sr_2CaCu_2O_8$. This was done with high resolution energy loss spectroscopy (8 meV) at low temperature (22 K) to observe the superconducting gap.

They were able to have a small electron beam diameter (50 μm) to explore the various regions of the sample. The results showed very strong lateral inhomogeneiticity and a variety of terminations with different properties. Only a small fraction of the surface was superconducting: this is not too surprising because oxygen atoms can be easily removed at the surface (including by the electron beam).

Then a 5 μm spot and 20 meV energy resolution for photoemission experiments could give interesting information on the high T_c but also for other materials.

6.2. Time resolved experiments

Using Laue diffraction from a polychromatic single bunch it is possible today to determine the structure of proteins down to 100 ps (10^{-10} s) resolution.

In the previous sections we showed the important developments in the field of magnetic multilayers. However, for the moment all the experiments have been done with static magnetic fields. Due to the developments of magnetic media technology it will be important to perform experiments at the nanosecond scale. It will be interesting to have an element selective probe for magnetisation switching studies because the dynamical coercivity will be different from the static one and the dynamic magnetic coupling in multilayer systems will also be different.

Time resolved XMCD experiments can be carried out using a pump-probe approach, thanks to a single-bunch time structure associated with a pulsed magnetic (new ns) field generated by a 50 μm coil. This has been done recently by Bonfim et al. [54]. However the next step is to couple this experiment with a PEEM to be able to measure single domains.

6.3. Photoemission electron microscopy with 1–2 nm lateral resolutions

PEEM is limited today by chromatic and spherical aberrations to a resolution of 25–30 nm. With an electron mirror 2 nm theoretical resolution with high transmission should be achieved. New experiments with this kind of resolution are under construction at Bessy II (Berlin) and ALS (Berkeley) opening new possibilities in magnetic and polymer nanostructure research.

6.4. Surface studies under high pressure (few atmospheres)

As pointed out already, industrial catalytic reactions are performed at atmospheric pressure while almost all the experiments on catalysts in the last twenty years have been done under ultrahigh vacuum.

The first surface diffraction experiments have produced very interesting results: they should be followed soon by photoemission experiments planned at ALS by Salmeron et al. [55]. This is clearly opening a new field.

References

[1] P. Grünberg, R. Schreiber, Y. Pang, M.B. Brodsky, H. Sowers, Layered magnetic structures: evidence for antiferromagnetic coupling of Fe layers across Cr interlayers, Phys. Rev. Lett. 57 (1986) 2442.

[2] A. Barthelemy, A. Fert, R. Morel, L. Steren, Giant steps with tiny magnets, Phys. World 7 (1994) 34.

[3] M.N. Baibich, J.M. Broto, A. Fert, F.N. Van Dau, F. Petroff, P. Etienne, G. Creuzet, A. Friedrich, J. Chazelas, Giant magnetoresistance of Fe/Cr magnetic superlattices, Phys. Rev. Lett. 61 (1988) 2472.

[4] C.T. Chen, Y.U. Ydzerda, H.J. Lin, N.V. Smith, G. Meigs, E. Chaban, G.H. Ho, E. Pellegrin, F. Sette, Experimental confirmation of the X-ray magnetic circular dichroism sum rules for iron and cobalt, Phys. Rev. Lett. 75 (1995) 152.

[5] B.T. Thole, P. Carra, F. Sette, G. van der Laan, X-ray circular dichroism as a probe of orbital magnetisation, Phys. Rev. Lett. 68 (1992) 1943.

[6] P. Carra, B.T. Thole, M. Altarelli, X.D. Wang, X-ray circular dichroism and local magnetic fields, Phys. Rev. Lett. 70 (1993) 694.

[7] B.T. Thole, G. van der Laan, G.A. Sawatzky, Strong magnetic dichroism predicted in the $M_{4,5}$ X-ray absorption

spectrum of magnetic rare-earth materials, Phys. Rev. Lett. 55 (1985) 2086.
[8] G. van der Laan, B.T. Thole, G.A. Sawatzky, J.B. Goedkoop, J.C. Fuggle, J.M. Esteva, R. Karnatak, J.P. Remeika, H.A. Dabkowska, Experimental proof of magnetic X-ray dichroism, Phys. Rev. Lett. 34 (1986) 6529.
[9] G. van der Laan, in: E. Beaurepaire, B. Carrière, J.P. Kappler (Eds.), Magnetism and Synchrotron Radiation, Mittelwihr, Les Editions de Physique, 1996, pp. 285–295.
[10] F. de Bergevin, M. Brunel, Observation of magnetic superlattice peaks by X-ray diffraction on an antiferromagnetic NiO crystal, Phys. Lett. A 39 (1972) 141.
[11] D. Gibbs, D.R. Harshman, E.P. Isaacs, D.B. McWhan, D. Mills, C. Vettier, Polarisation and resonance properties of magnetic X-ray scattering in holmium, Phys. Rev. Lett. 61 (1988) 1241.
[12] D. Spanke, V. Solinus, D. Knabben, F.U. Hillebrecht, F. Cicacci, L. Gregoratti, M. Marsi, Evidence for in-plane antiferromagnetic domains in ultrathin NiO films, Phys. Rev. B 58 (1998) 5201.
[13] J. Stöhr, A. Scholl, T.J. Regan, S. Anders, J. Lüning, M.R. Scheinfein, H.A. Padmore, R.L. White, Images of the AF structure of a NiO(100) surface by means of X-ray magnetic linear dichroism spectromicroscopy, Phys. Rev. Lett. 83 (1999) 1862.
[14] J. Stöhr, X-ray magnetic circular dichroism spectroscopy of transition metal thin films, J. Electron. Spectrosc. Relat. Phenom. 75 (1995) 253.
[15] D. Alders, L.H. Tjeng, F.C. Voogt, T. Hibma, G.A. Sawatzky, C.T. Chen, J. Vogel, M. Saachi, S. Iacobucci, Temperature and thickness dependence of magnetic moments in NiO epitaxial films, Phys. Rev. B 57 (1998) 11623.
[16] D. Alders, J. Vogel, C. Levelut, S.D. Peacor, T. Hibmu, M. Sacchi, L.H. Tjeng, C.T. Chen, G. van der Laan, B.T. Thole, G.A. Sawatzky, Europhys. Lett. 32 (1995) 259.
[17] S. Sato, M. Miura, K. Kurosawa, Optical observation of antiferromagnetic s domains in NiO(111) platelets, J. Phys. C 13 (1980) 1513.
[18] W.H. Meiklejohn, C.P. Bean, New magnetic anisotropy, Phys. Rev. 102 (1956) 1413.
[19] F. Nolting, A. Scholl, J. Stöhr, J.W. Seo, J. Fompeyrine, H. Siegwart, J.-P. Locquet, S. Anders, J. Lüning, E.E. Fullerton, M.F. Toney, M.R. Scheinfein, H.A. Padmore, Direct observation of the alignment of ferromagnetic spins by antiferromagnetic spins, Nature 405 (2000) 767.
[20] A. Scholl et al., Observation of antiferromagnetic domains in epitaxial thin films, Science 287 (2000) 1014.
[21] S. Ferrer, J. Alvarez, E. Lundgren, X. Torrelles, P. Fajardo, F. Boscherini, Surface X-ray diffraction from Co/Pt(111) ultra-thin films and alloys: structure and magnetism, Phys. Rev. B 56 (1997) 9848.
[22] F. Meertens, R. Imbihl, Square chemical waves in the catalytic reaction $NO + H_2$ on a rhodium (110) surface, Nature 370 (1994) 124.
[23] A. Schaak, S. Günther, F. Esch, E. Schütz, M. Hinz, M. Marsi, M. Kiskinova, R. Imbihl, Elementally-resolved imaging of dynamic surface processes: chemical waves in the system Rh(110) $NO + H_2$, Phys. Rev. Lett. 83 (1999) 1882.
[24] G.A. Somorjai, Introduction to Surface Chemistry and Catalysis, Wiley, New York, 1994.
[25] K.Y. Kung, P. Chen, F. Wei, Y.R. Shen, G.A. Somorjai, Sum-frequency generation spectroscopic studies of CO adsorption and dissociation on Pt(111) at high pressure and temperature, Surf. Sci. 463 (2000) L627.
[26] T. Dellwig, G. Rupprechter, H. Unterhalt, H.J. Freund, Bridging the pressure and materials gaps: high pressure sum frequency generation study of supported Pd nanoparticles, Phys. Rev. Lett. 85 (2000) 776.
[27] P. Cernota, K. Rider, H.A. Yoon, M. Salmeron, G.A. Somorjai, Dense structure formed by CO on Rh(111) studied by scanning tunnelling microscopy, Surf. Sci. 445 (2000) 249.
[28] K.F. Peters, C.J. Walker, P. Steadman, O. Robach, H. Isern, S. Ferrer, Adsorption of carbon monoxide on Ni(110) above atmospheric pressure investigated with surface X-ray diffraction, Phys. Rev. Lett. 86 (2001) 5325.
[29] C. Zhao, M.A. Passler, A reinvestigation of the surface structure of Ni(110)-(2×1) CO by LEED, Surf. Sci. 320 (1994) 1.
[30] F.C. Frank, Supercooling of liquids, Proc. R. Soc. Lond. A 215 (1952) 43.
[31] J.D. Bernal, Geometry of the structure of monatomic liquids, Nature 185 (1960) 68.
[32] H. Reichert, O. Klein, H. Dosch, M. Denk, V. Honkimäki, T. Lippmann, G. Reiter, Observation of five-fold local symmetry in liquid lead, Nature 408 (2000) 839;
See also E. DiMasi, The search for liquid five fold symmetry, Phys. World 14 (2001) 20.
[33] J.M. Imer, F. Patthey, B. Dardel, W.D. Schneider, Y. Baer, Y. Petroff, A. Zettl, High-resolution photoemission study of the low-energy excitation reflecting the superconducting state of Bi–Sr–Ca–Cu–O single crystals, Phys. Rev. Lett. 62 (1989) 336.
[34] Z.X. Shen, D.S. Dessau, B.O. Wells, D.M. King, W.E. Spicer, A.J. Arko, D. Marshall, L.W. Lombardo, A. Kapitulnik, P. Dickinson, S. Doniach, J. Di Carlo, A.G. Loeser, C.H. Park, Anomalously large gap anisotropy in the a–b plane of $Bi_2Sr_2CaCu_2O_{8+\delta}$, Phys. Rev. Lett. 70 (1993) 1553.
[35] H. Ding, M.R. Norman, J.C. Campuzano, M. Randeria, A.F. Belman, T. Yokoya, T. Takahashi, T. Mochiku, K. Kadowaki, Angle-resolved photoemission spectroscopy study of the superconducting gap anisotropy in $Bi_2Sr_2CaCu_2O_{8+\delta}$, Phys. Rev. B 54 (1996) R9678.
[36] H. Ding, T. Yokoya, J.C. Campuzano, T. Takahashi, M. Randeria, M.R. Norman, T. Mochiku, K. Kadowaki, J. Giapintrakis, Spectroscopic evidence for a pseudo gap in the normal state of underdoped high T_c superconductors, Nature 382 (1996) 51.
[37] A.G. Loeser, Z.X. Shen, D.S. Dessau, D.S. Marshall, C.H. Park, P. Fournier, A. Kapitulnik, Excitation gap in the normal state of underdoped $Bi_2Sr_2CaCu_2O_{8+\delta}$, Science 273 (1996) 325.

[38] D.S. Marshall, D.S. Dessau, A.G. Loeser, C.H. Park, A.Y. Matsura, J.N. Eckstein, I. Bozovic, P. Fournier, A. Kapitulnik, W.E. Spicer, Z.X. Shen, Unconventional electronic structure evolution with hole doping in $Bi_2Sr_2CaCu_2O_{8+\delta}$, Phys. Rev. Lett. 76 (1996) 4841.

[39] A. Damascelli, Z.X Shen, Z. Hussain, Angle-resolved photoemission spectroscopy of the cuprate superconductors, Rev. Mod. Phys., to be published.

[40] X.J. Zhou, P. Bogdanov, S.A. Kellar, T. Noda, H. Eisaki, S. Uchida, Z. Hussain, Z.X. Shen, One dimensional electronic structure and suppression of d-wave node state in $(La_{1.28}Nol_{0.6}Sr_{0.2})CuO_4$, Science 286 (1999) 268.

[41] D.H. Lu, F. Ronning, Z.X. Shen, D.A. Bonn, R. Liang, W.N. Haroly, A.I. Rykov, S. Tajima, Superconducting gap and strong in-plane anisotropy in untwined $YBa_2Cu_3O_{7-\delta}$, Phys. Rev. Lett. 86 (2001) 4370.

[42] D.L. Feng, N.P. Ermitage, D.H. Lu, A. Damascelli, J.P. Hu, P. Bogdanov, A. Lanzara, F. Ronning, K.M. Shen, H. Eisaki, C. Kim, Z.X. Shen, J.I. Shimoyama, K. Kishio, Bilayer splitting in the electronic structure of heavily overdoped $Bi_2Sr_2CaCu_2O_{8+\delta}$, Phys. Rev. Lett. 86 (2001) 550.

[43] Y.D. Chuang, A.D. Gromko, A.F. Fedorov, Y. Aiura, K. Oka, Y. Ando, H. Eisaki, S. Uchida, Doubling of the bands in overdoped $Bi_2Sr_2CaCu_2O_8$ – probably evidence for c-axis bilayer coupling, e-print: cond-mat/0102386, 2001.

[44] P.V. Bogdanov, A. Lanzara, X.J. Zhou, S.A. Kellar, D.L. Feng, E.D. Lu, J.I. Shimoyama, K. Kishio, Z. Hussain, Z.X. Shen, ARPES study of Pb doped $Bi_2Sr_2CaCu_2O_8$ – a new Fermi surface picture, e-print: cond-mat/0005394, 2000.

[45] A. Lanzara, P.V. Bogdanov, X.J. Zhou, S.A. Kellar, D.L. Feng, E.D. Lu, T. Yoshida, H. Eisaki, A. Fujimori, K. Kisho, J.I. Shimyama, T. Noda, S. Uchida, Z. Hussain, Z.X. Shen, Evidence for ubiquitous strong electron–phonon coupling in high temperature superconductors, Nature 412 (2001) 510.

[46] S. Tanuma, C.J. Powell, D.R. Penn, Proposed formula for electron inelastic mean free paths based on calculations for 31 materials, Surf. Sci. 192 (1987) L849.

[47] A. Sekiyama, T. Iwasaki, K. Matsudu, Y. Suitoh, Y. Onuki, S. Suga, Probing bulk states of correlated electron systems by high-resolution resonant photoemission, Nature 403 (2000) 396.

[48] J. Kondo, Due to the magnetic interaction between ions and conduction electrons, Progr. Theor. Phys. 32 (1964) 37.

[49] H. Aoki, S. Uji, A.K. Albessard, Y. Onuki, Observation of heavy electrons in $CeRu_2Si_2$ via the dHvA effect, J. Phys. Soc. Jpn. 61 (1992) 3457.

[50] S.H. Yang, H. Kunigashira, T. Yokoya, A. Chainani, T. Takahashi, H. Takeya, K. Kadowaki, High-resolution photoemission study of $CeRu_2$: the dual character of 4f electrons, Phys. Rev. B 53 (1996) R11946.

[51] O. Gunnarson, K. Schönhammer, Photoemission from Ce compounds: exact model calculations in the limit of large degeneracy, Phys. Rev. Lett. 50 (1983) 604.

[52] F. Pathey, J.M. Imer, W.D. Schneider, H. Beck, Y. Baer, B. Delley, High-resolution photoemission study of the low-energy excitations in 4f-electron systems, Phys. Rev. B 42 (1990) 8864.

[53] B.N. Persson, J.E. Demuth, High-resolution electron-energy-loss study of the surfaces and energy gaps of cleaved high-temperature superconductors, Phys. Rev. B 42 (1990) 8057;
J.E. Demuth, B.N.J. Persson, F. Holtzberg, C.V. Chandrasekhar, Phys. Rev. Lett. 64 (1990) 603.

[54] M. Bonfim, G. Ghiringhelli, F. Montaigne, S. Pizzini, N.B. Brookes, F. Petroff, J. Vogel, J. Camarero, A. Fontaine, Element-selective nanosecond magnetization dynamics in magnetic heterostructures, Phys. Rev. Lett. 86 (2001) 3646.

[55] M. Salmeron, Ambient pressure photoelectron spectroscopy, unpublished.

Low temperature surface chemistry and nanostructures

G.B. Sergeev *, T.I. Shabatina

Department of Chemistry, M.V. Lomonosov Moscow State University, 119899 Moscow, Russia

Received 26 June 2000; accepted for publication 7 May 2001

Abstract

The new scientific field of low temperature surface chemistry, which combines the low temperature chemistry (cryochemistry) and surface chemistry approaches, is reviewed in this paper. One of the most exciting achievements in this field of science is the development of methods to create highly ordered hybrid nanosized structures on different organic and inorganic surfaces and to encapsulate nanosized metal particles in organic and polymer matrices. We consider physical and chemical behaviour for the systems obtained by co-condensation of the components vapours on the surfaces cooled down to 4–10 and 70–100 K. In particular the size effect of both types, the number of atoms in the reactive species structure and the thickness of growing co-condensate film, on the chemical activity of the system is analysed in detail. The effect of the internal mechanical stresses on the growing interfacial co-condensate film formation and on the generation of fast (explosive) spontaneous reactions at low temperatures is discussed. The examples of unusual chemical interactions of metal atoms, clusters and nanosized particles, obtained in co-condensate films on the cooled surfaces under different conditions, are presented. The examples of highly ordered surface and volume hybrid nanostructures formation are analysed. © 2001 Elsevier Science B.V. All rights reserved.

Keywords: Surface energy; Clusters; Surface chemical reaction; Growth

1. Introduction: frontiers in low temperature surface chemistry

In the present paper we concentrate our attention on the following new phenomena produced by combination of low temperature chemistry and surface chemistry (LTSC) approaches. LTSC considers chemical reactions at *low* (−196–173 °C, 77–100 K) and *superlow* (−269–243 °C, 4–30 K) temperatures. It is a new rapidly growing field of chemical and physical sciences. The realisation of chemical reactions at such low temperatures was questionable for a long time due to two main reasons. Most chemical substances are solid at low temperatures. Reagent diffusion is usually necessary for chemical interaction takes place in the condensed phase. The diffusion rate decreases by a factor of 10^5 during the reactive system transition from gas to liquid phase, and it decreases more by a factor of 10^5 during the phase transition to the solid phase. The second reason is connected with the temperature dependence of the chemical reaction rate. According to the Arrhenius equation the rates of the majority of chemical reactions are

* Corresponding author. Tel.: +7-959395442; fax: +7-959390283.
E-mail addresses: gbs@kinet.chem.msu.ru (G.B. Sergeev), tsh@kinet.chem.msu.ru (T.I. Shabatina).

0039-6028/01/$ - see front matter © 2001 Elsevier Science B.V. All rights reserved.
PII: S0039-6028(01)01563-1

negligible at temperatures close to nitrogen boiling point. These phenomena are the theoretical basis of different processes of conservation and stabilisation of goods, food and technical products at low temperatures.

By the investigations begun in the middle of the last (20th) century many new unusual chemical reactions were discovered, taking place with extremely high rates at low and superlow temperatures. These are the reactions of free radicals, polymerisation processes, interactions of halogens and hydrogen halides with double bonds, etc. [1–3].

Chemical reactions at low temperatures are carried out usually by joint or separate condensation of the reagent vapours on cooled surfaces of different cryostats. The interactions between two different substances on the cooled surface or included in thin condensed films usually are initiated by the reagent molecules contacting via the former formation of their unstable complexes of different substances. We distinguish interfacial reactions of gas–solid, liquid–solid and solid–solid types. The first step of these processes occurs as physical and/or chemical adsorption. The accommodation effects, surface diffusion and tunnelling interactions on the surface chemical reactions at low temperatures are very important (see Fig. 1). Nowadays considerable attention is given to the surface structural changes and energy transformations during the reaction due to the development of new surface structure testing and microscopic methods.

LTSC uses various matrix isolation and preparative cryochemistry techniques for carrying out the reactions of atoms and molecules on the cooled surfaces. The formation of hybrid metal–organic supramolecular structures, metal clusters and nanosize metal particles can take place during low temperature co-condensation and during further thermal treatment of the film samples.

One of the exciting achievements of this science is the possibility of the formation of different metal species beginning from small clusters up to nanosized particles, combining hundreds and thousands of atoms. Such structures can be considered as almost fully surface ones, because all their atoms are on-surface or near-surface. Varying the conditions and increasing the particles size we can obtain one-dimensional (1D), 2D and 3D nanostructures. Thus we have the unique possibility to study in situ the effect of surface/volume ratio on physical and chemical properties of chemical systems and their reactivity.

Most interesting in LTSC are the size effects of various sorts. As such effects we consider the effect of metal species size (the number of atoms in structure) on their reactivity in different processes. The existence of such effects leads to reconsideration of many widely used thermodynamic and kinetic concepts. On the other hand, at low temperatures the diffusion processes, mechanic stability and reactivity in the thin condensed films are highly dependent on the sample thickness. Thus LTSC will allow us to reveal the borders between macroscopic and microscopic properties of the sample.

The special attention is given to the processes, which occur in the course of the formation of the reactive film samples during a reagent's condensation on the cooled surface. Just these processes determine mainly the kinetics and the energetics of the chemical interactions. For example, spontaneous fast (explosive) reactions in the growing co-condensed films at low and superlow temperatures lead in some cases to the production of unusual chemical products and to the unusual temperature dependences of the reaction rates. The examples of obtaining multi-component nanostructured films including several metals and organic or inorganic components are considered taking into account different effects via surface interactions.

We especially consider the possibilities of self-organising systems such as liquid crystals and

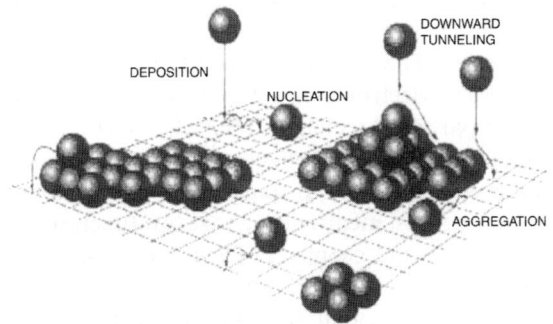

Fig. 1. Schematic of model for metal atom deposition on the surface and condensate film growth.

polymers for the formation of anisotropic nanostructures and discuss their properties.

2. General peculiarities of chemical reagent vapour interactions with cooled surfaces

2.1. Specifics of the samples

The chemistry at low (77 K) and superlow (4 K) temperature enlarges fundamental knowledge about the nature of elementary chemical reactions and opens the new possibilities of producing unusual chemical molecules and complex compounds. The initial reagents are usually in the vapour phase. The preparation of samples at low temperatures is carried out by either separate or joint condensation (co-condensation) of the reagent's vapours on the cooled surface [4,5]. The general scheme of the processes under deposition of diffusing species (metal atoms or molecules) that can aggregate on the cooled surface is presented in Fig. 1. Small metal species can be formed initially. By heating they can migrate on the surface resulting in larger-sized particles formation. Such aggregation/stabilisation reactions are highly dependent on surface temperature, deposition rate (particles flux per surface site and per unit of time). Obviously, processes taking place during interaction of atomic or molecular beams with the surface are very important and determine some specific features of chemical reactions in low temperature deposited films and control the morphology of nanometre-scale surface structures formed. This allows us to consider the LTSC as a special branch of *surface chemistry*.

The technique of sample preparation plays an important role in chemical reactions at low temperatures. Two methods are widely used nowadays [4,5]. The first one is based on the condensation of substances under an excess of inert gas. This is the well known method of matrix isolation. The second one is based on the co-condensation of the reagent's vapours under conditions which exclude interaction in the gas phase (molecular beam deposition). Different variants of these methods are possible: condensation of particles of one type; condensation of particles with an excess of inert substances; layer by layer condensation of two different types of particles; co-condensation of two different types of particles; co-condensation of more than two different types of particles.

It is important to note that in all cases various types of initiation (light, X-ray, high frequency radiation, heat) may be used to promote the interaction. The low temperature reactions are strongly dependent on the nature of the reacting particles and the methods of sample preparation.

Most difficult is it to understand the properties of the reactive low temperature film co-condensates, which arise due to the non-homogeneity and non-equilibrium state of these systems. This leads to at least two consequences. First, the particles in co-condensates are kinetically, thermodynamically and energetically non-equivalent, and second, low temperature co-condensates possess accumulated energy. In connection with this it is necessary to mention some types of the energy excess accumulated in the growing film of low temperature co-condensate: redundant energy of metastable and amorphous phases (typical values 5–20 kJ/mol); energy of donor–acceptor interactions (typical values 4–20 kJ/mol); heat effects of chemical reactions (typical values 100 kJ/mol); stabilisation energy for active particles, such as atoms and radicals (their recombination energy is in the order of the chemical bond energy 100–400 kJ/mol); energy of mechanical stresses (typical values −10 kJ/mol).

2.2. Size effects

The size effects can be conventionally divided into two main groups: inner size effects related to specific changes in surface and volume properties and outer size effects related to external field properties. The first one is connected with the influences of electronic and structural characteristics of the particles, including their chemical activity, bond energy, ionisation potential, melting point and crystal structure. It is important to find out all the properties mentioned above as functions of particle size and geometry. The external size effect can reveal itself as collective electronic or lattice excitations, which in the case of metal particles leads to the appearance of shallow electron plasma resonance in optical spectra of nanosize particles.

It is well established that different physical (phase transitions, optical, magnetic and etc.) properties of small particles depend not only on their sizes [6], but also shapes [7]. Size effects in LTSC occur in two main forms depending on the chemical nature of the system under investigation. One of them is the effect of the reactive species' size (number of structural units—atoms and molecules) on the mechanism and kinetics of their chemical interactions [8,9]. The second is the effect of thickness of low temperature deposited film on its chemical reactivity [10]. The connection of the reactivity with the size parameters is one of the most complex and important problems in modern LTSC. In both cases a small change in size leads to a dramatic effect on chemical behaviour.

One of the most exciting examples is presented in Fig. 2. The metal clusters are known to exhibit high and sometimes unusual reactivity and show a strong variation in their reactivity depending on cluster size. It is demonstrated clearly for niobium clusters (Nb_x) in reaction with alkyl and allyl halides, benzene, cyanides and carbon dioxide using a combination of mass-spectroscopy and supersonic beam technique [9]. The yield of Nb_xBr produced by the reaction of Nb_x clusters ($x = 5$–20) with isopropyl bromide and ethyl bromide (Fig. 2) is the highest for $x = 4$ and decreases by rising $x > 4$ for the first compound due to an increase of steric hindrance of niobium clusters interaction with isopropyl bromide.

Metal atoms and their small aggregates—dimers, trimers, and higher clusters—play an important role in LTSC transformations in the surface co-condensate films. Such metal species include some atoms and are a few nanometers in size (1–10 nm, or 10–100 Å). They possess high chemical reactivity due to non-compensation of chemical bonds for surface atoms. Simple estimations show that the part of surface atoms is equal to 100% for particles 1 nm in size (~12–15 atoms) and about 15% for particles 10 nm in size. Chemical interactions with such highly reactive particles open up new possibilities for the synthesis of new organometallic compounds and composite materials with unique combinations of optical, electric and magnetic properties.

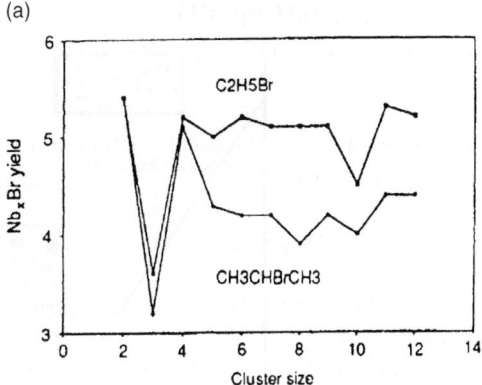

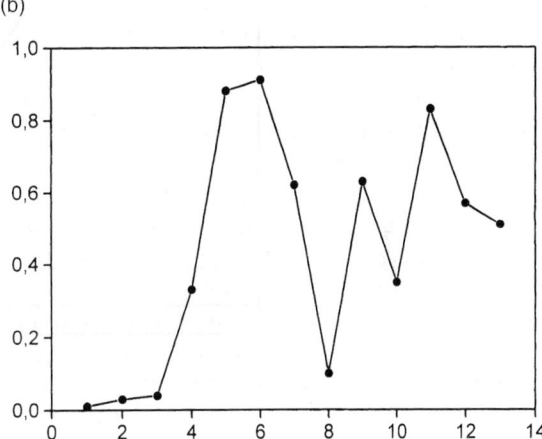

Fig. 2. Size dependence of the reactivity of niobium clusters (Nb_x) to alkyl bromides (a) and to benzene (b), as the number of atoms (x) in niobium clusters was varied as $x = 1$–12 [8].

High reactivity of such species is closely connected with the size effects for nanometer systems. Such effects appear in general when particle size correlates with the characteristic size for a definite system property, for example, magnetic domain or the electron free length. The structural and phase non-homogeneity is characteristic for nanosize systems, and the new co-ordinate—'dispersity'—should be added to classical physical and chemical analysis resulting in diagrams 'composition–structure–property'. Thus, the particle size (the number of atoms in the particle's structure) becomes the active thermodynamic property, determining the system state and its chemical activity. The relative chemical activity of metal species upon their size is presented schematically in Fig. 3.

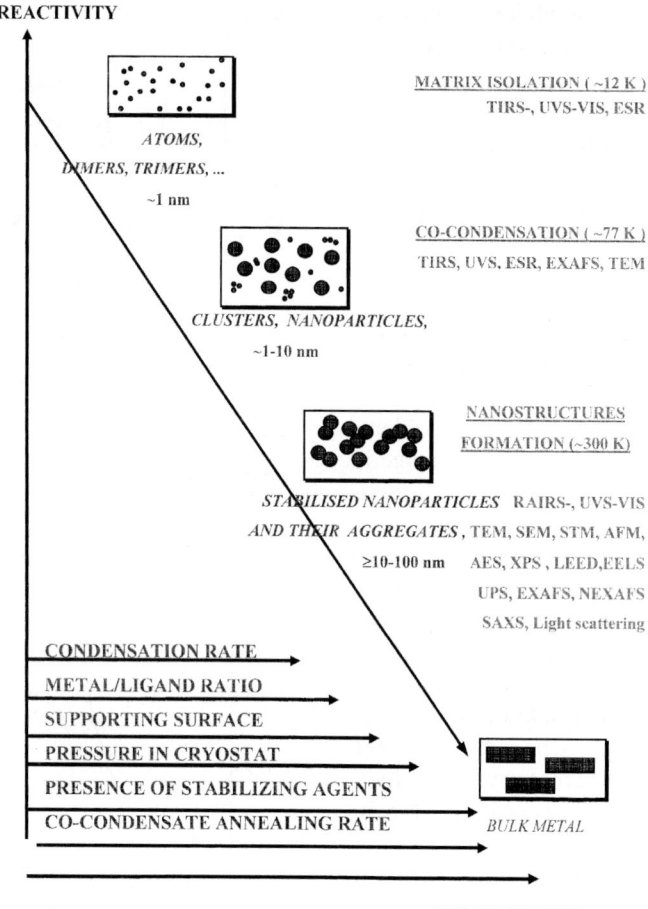

Fig. 3. The general scheme, which describes the relative chemical activity of metal species versus their size. According to this scheme, the chemical activity decreases in the following direction: free metal atoms–oligomeric clusters–nanoparticles–aggregates–bulk metal. The main factors determining the size of metal species formed in the chemically active systems and some experimental techniques used for their characterisation are also indicated: TIRS—transmission infrared spectroscopy; UVS–VIS—ultraviolet and visible electronic spectroscopy; RAIRS—reflection absorbtion infrared spectroscopy; ESR—electron spin resonance spectroscopy; EXAFS—extended X-ray absorption fine structure; NEXAFS—near edge X-ray absorption fine structure; TEM—transmission electron spectroscopy; SEM—scanning electron spectroscopy; AFM—atomic force microscopy; AES—Auger electron spectroscopy; XPS—X-ray photoelectron spectroscopy; UPS—ultraviolet photoelectron spectroscopy; LEED—low energy electron diffraction; EELS—electron energy-loss spectroscopy; SAXS—small angle X-ray scattering; and various sorts of light scattering techniques.

According to this scheme, the chemical activity decreases in the following direction: free metal atoms–oligomeric clusters–nanoparticles–aggregates–bulk metal. The atoms and small metal clusters possess the highest reactivity. Their properties are studied usually using a special technique for their stabilisation in inert matrices at superlow temperatures. Increasing of the reactive specie size leads to decrease of the number of active surface atoms. It is followed by the proper decrease of the system chemical activity per atom going from dimers and trimers to nanosized particles and then to rougher dispersions and bulk metals. It should be mentioned that the plot of chemical activity against reactive species size is not a monotonic one. There should be some extremes due to the higher stability of metal clusters with definite (magic) atom numbers which are specific for individual metals and their combinations. The loss in chemical activity is more significant for growing small particles and less sharp for the final transition from rough dispersions to bulk.

The size of metal particles forming such systems and their reactivity are determined by the combination of different experimental conditions (Fig. 3). The main experimentally controllable factors are the support temperature, the metal/ligand ratio, reagent condensation rate and the rate of sample annealing. It was shown that the lower is the temperature of the support surface, the less are diffusion controlled interactions between reactive particles and the more possible is the formation of high energetic and high reactive species at low temperatures. The metal/ligand ratio highly affect the size of metal particles obtained via low temperature co-condensation. Raising this ratio and annealing of the sample usually lead to increasing of the part of clusters and higher aggregated metal particles. The component condensation rate has a complex effect on the properties of low temperature co-condensate film. The lifetime of highly active species (metal atoms, their dimers or trimers) during co-condensation at the cooled surface is inversely proportional with the condensation rate. It depends on the nature of relaxation processes in the co-condensate system. Intensity of particle beam determines the number of collisions of the molecules with the surface and with each other. The temperature of the surface determines the relaxation processes. Together with the chemical nature of the reagents these factors determine the pathway of processes leading, or not, to the reaction. The processes which occur during the real condensation are more complicated than the given scheme. The experimental techniques usually applied for the study of structure, texture and particle size in co-condensate films are shown also in Fig. 3.

2.3. Defects and mechanical stresses in surface films of low temperature condensates

A number of recent investigations [10,11] allow us to conclude that stresses and defects generated in the growing co-condensate film are very important for the chemical interactions taking place in the system and determine in general the behaviour of low temperature reactive co-condensates.

The results were analysed by the model of rapid explosive processes initiating plastic deformation of the growing co-condensate film up to crack formation and mechanical destruction of the film [10]. The film is forming under the mechanical stresses of deposited layers. The film formation begins as the growth of a separated island nucleus of half-round shape. At this stage only slow surface stresses occur. The growing islands can coalescence and form columnar aggregates packed very close to each other. The atomic forces at the grain borders can cause stretching stresses. By raising the thickness of growing co-condensate film the size of columnar aggregates can increase, which is connected with the increase in columns' diameter and length up to the critical value of the film thickness. Such surface and volume reconstruction leads to decrease of the sample stability. It causes also crack formation, structural transitions and initiates a fast chemical reaction and fast crystallisation of the sample. The general scheme of the processes taking place in film co-condensates growing via low temperature deposition of the reagent vapours is presented in Fig. 4. The arrows in the scheme show the possible ways of mutual transitions of mechanic and chemical energies and realisation of both explosive chemical reaction and explosive crystallisation. The initial co-condensate film is in the amorphous state, which is characterised by the excess of free energy. Internal mechanical stresses cause the film plastic deformation and crack formation when the thickness of the growing co-condensate film reaches the critical value. Plastic deformation causes the increase of molecular mobility and accelerates the steady state chemical reaction and crystallisation processes. Crack formation can initiate a fast (explosive) chemical reaction and fast crystallisation of the amorphous film. Both crack formation and chemical reaction or crystallisation are interconnected by a positive feedback, which can generate a wave explosive chemical reaction or crystallisation during co-condensate film formation. The behaviour of the co-condensate system is determined by the reactive species' size, the degree of

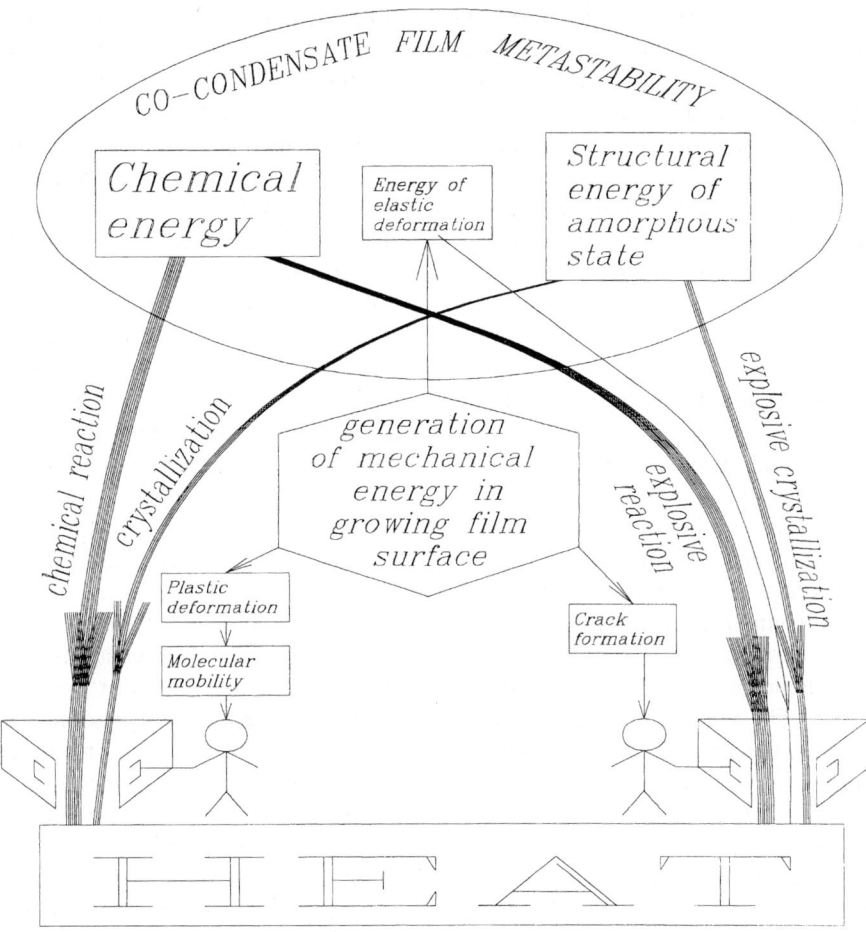

Fig. 4. The scheme of mechanical energy generation in co-condensate film growing on the support surfaces and possible ways of its transformation to the energy of chemical reactions and crystallisation taking place in the system. Accumulation of energy excess in the growing co-condensate films leads to sample plastic deformations, increasing molecular mobility and cracks formation. Rapid explosive chemical processes were initiated by crack formation during the formation of growing co-condensate film up to its tensile strength.

deviation of the system from the equilibrium state and the rate of relaxation processes.

The proposed model explains the dependence of the mechanical stresses in the growing co-condensate film upon the conditions of its formation. By raising the support temperature the molecular mobility increases. This leads to the growth of the existing nucleuses and not to the formation of the new ones. During further film formation the grains grow leading to lowering of the internal stresses due to decreasing of the grain border surface. This fact leads also to decreasing of mechanical stresses and increasing of critical thickness of the sample.

The mechanical stresses and critical film thickness were measured directly for deposited water films using a low temperature strain-measuring setup and low temperature calorimeter [11]. The data obtained confirm the general nature of the processes of spontaneous crystallisation of the co-condensate film sample and of the crack formation, which initiate the crystallisation of amorphous water.

A feature of the modern chemical and physical sciences is the rising interest in non-equilibrium dynamic systems, which possess the internal accumulated energy. Low temperature co-condensates belong to such systems. Generation and accumulation of mechanical energy in the deposited films growing at low temperatures lead both to sample plastic deformations, increase of molecular mobility and crack formation. Thus, slow interactions can cause dramatic effects in low temperature systems. The investigations of low temperature reactions in solid state film co-condensates allow us to understand better these unusual and interesting phenomena.

3. Reactions in low temperature co-condensate films

3.1. General features of chemical reactions at low temperatures

One of the most important achievements of LTSC research is establishing the difference in reactivity for small reactive species: metal atoms, dimers and small clusters ($n = 3$–10).

High chemical activity of metal atoms and small clusters lead the investigators to the use of special equipment [12–15] and a variety of metal clusters have been isolated and examined by matrix isolation spectroscopy [16]. Thus, clusters Al_3, Cu_3, Ag_3 and Au_3 have been isolated and studied by ESR [17], lanthanide metal dimers and higher clusters, Na_n, Ag_n and Au_n ($n = 2$–5), Cr_4 and Cr_5, Mn_2 and Mn_n, Sc_n and Y_n clusters have been studied by different low temperature spectroscopic techniques [18–21]. Mass selection/matrix deposition techniques have been developed recently [22] for spectroscopic study of pure metal clusters. Bimetallic dimers were isolated by two metals co-deposition in an inert matrix [23,24]. A variety of metal atom and cluster coordination saturated and unsaturated complexes with N_2, CO, C_2H_4, C_2H_2 and some other molecules were obtained and spectroscopically characterised using low temperature matrix isolation technique [25–27].

The second method of LTSC is based on the co-condensation of reagent vapours on the cooled surface of special cryostats under conditions which exclude the interaction in the gas phase. The reaction usually takes place in the thin interfacial layer on the surface of frozen bulk co-condensate during the co-condensation of the reagents. By annealing of the sample the reaction can expand into the co-condensate volume due to the molecular mobility rising. Different parallel and sequential stages occur during the annealing of the co-condensate system: aggregation of metal atoms, surface adsorption and chemisorption of ligands, chemical reactions of the components. The scheme in Fig. 5 illustrates the fundamental possibility of using low temperatures to obtain metal clusters, mono- and polynuclear metal complexes and ligand stabilised nanoparticles by co-deposition of metal vapours with different ligands on cooled

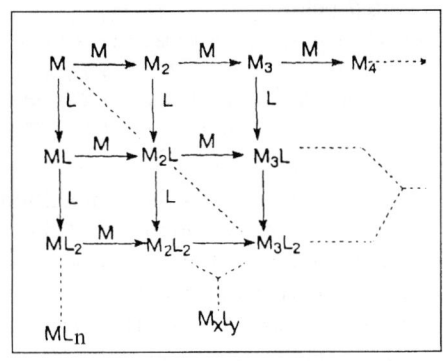

Formation of organometallic compounds and complexes

metal atoms aggregation and formation of metal clusters and nanoparticles

formation of polymolecular associates

Fig. 5. Scheme of chemical processes taking place in metal/ligand co-condensates at low temperatures. Metal and ligand interactions start from atoms and proceed as a number of parallel and sequential reactions lead to aggregation and metal nanoparticles' formation and formation of organometallic compounds and complexes of different nuclearity.

surfaces. Thus, chemical interactions in low temperature co-condensates begin with metal atoms. Upon condensation and during the annealing of the samples the competition of two processes: metal species aggregation and their stabilisation in the matrix, takes place. Input of ligand into the system causes some additional processes leading to the formation of metal clusters of different nuclearity and their stabilisation by the interactions with ligand molecules. Metal atom aggregation and their interaction with ligands occur practically without activation energy. The high reactivity of small metal species is the main difficulty in establishing the relation between size of the particle and its reactivity. There are also problems in producing and isolating the compounds with definite composition. The reactions shown in Fig. 5 are complex multi-factor processes taking place under highly non-equilibrium conditions and for a wide or narrow distribution of the reactive species on their chemical activity. This fact is reflected usually in the kinetics of low temperature processes.

3.2. Reactions of metal atoms and small species

Metal atoms and ligand vapours are simultaneously condensed in vacuum onto a cooled surface. Metal atom aggregation takes place during condensation and by the annealing of the sample and leads to formation of metal dimers, trimers, clusters, and nanoparticles. Thus chemical interactions in low temperature metal–organic co-condensates start at atomic level and involve metal species of different size during annealing of the samples. It is found that metal atoms, clusters and nanoparticles could react by pathways which do not realise at room temperatures. Some unusual reactions of metal atoms and small clusters are given in Ref. [8]. Below are presented some excellent examples.

A systematic study of some transition metal atoms' co-condensates with methane/Ar mixtures at 10 K is presented in Ref. [28]. There is shown the oxidative addition of H–H and C–H bonds in the presence of the electronically excited transition metal atom [29]. The normal alkenes react by C–H oxidative addition producing CH_3FeH, C_2H_5FeH, C_3H_7FeH species [30].

Co-deposition of transition metal atoms with arenes at 77 K results in bis-arene metal. For example, co-condensation of the lanthanide atoms with 1,3,5-tri-*tert*-butylbenzenes gives the thermally stable bis(arene)-lanthanide(0) sandwich complexes for Nd, Tb, Dy, Ho, Er and Lu and thermally unstable complexes for La, Pr, Sm, and highly unstable complexes for Eu, Tm and Yb [31]. In recent works [32] 1,3-diphospha-benzene derivatives are synthesised. The vapours of titanium and vanadium reacted with 1,4-di-*tert*-butylbuta-1,3-diene to form novel compounds $[M(\eta\text{-}tBu_2\text{-}C_2H_4)_2]$, where M = Ti, V [33]. The complexes are found to be very reactive towards a variety of other ligands: phosphines, diazadienes and isonitriles leading to the replacement of diene ligands and forming novel half-sandwich complexes. Co-condensation of V and Cr atoms with tetraphenylsilane ligands give stable sandwich-like compounds [34]. Interesting synthesis of binuclear and trinuclear cluster complexes of Mo, Re, Pd has been performed in Ref. [35].

Another exciting example is the interaction of magnesium with carbon polyhalides. The chemists know that at ambient temperature carbon tetrachloride does not react with bulk magnesium. The situation changes dramatically in magnesium (Mg)–carbon tetrachloride (CCl_4) film co-condensates on cooled surfaces at temperatures close to the temperature of liquid nitrogen boiling (77–80 K). The chemical reactions taking place are presented below:

$$Mg + CCl_4 \longrightarrow \begin{cases} CCl_3MgCl \xrightarrow{H_2O} CHCl_3 \\ CCl_3 \rightarrow C_2Cl_6 \\ CCl_2 \rightarrow C_2Cl_4 \end{cases}$$

The scheme presents the competition of chemical reactions and the formation of different chemical intermediates and products: Grignard's reagent (CCl_3MgCl), tri-chloromethyl radical (CCl_3), dichlorocarbene (CCl_2). The reaction mechanism was studied in detail in Ref. [36]. Different reactivity of magnesium atoms and clusters with alkyl/aryl halides (RX) was shown in Refs. [37,38]. Magnesium atoms initiate a radical reaction resulting in a number of recombination products. Magnesium atoms aggregate in the systems with

Mg/RX ratio in the range 1–100 and yield the Grignard's reagent.

High reactivity of Mg and Ca clusters in comparison to their atoms was shown in Ref. [39]. The relative reactivity of metallic particles and different alkyl halides (RX) is found to be the following: for metals—$Ca_x=Ca_2 > Mg_x=Mg_4 > Mg_3=Mg_2 > Ca > Mg$ and for RX—$CH_3J > CH_3F > CH_3Br > CH_3Cl$. The theoretical calculations [40] also predict the instability and high reactivity of CH_3Mg_2X (X = Br, Cl) particles.

The co-condensation of Na, Mg, Sm with alcohols lead to the competition of two processes [41]. Metal atoms and small clusters at low metal/ligand ratio (~1:500) reduce alcohols to hydrocarbons during co-condensation at 77 K, while larger clusters produce the metal alcoholates by annealing of the system with higher metal/ligand (~1:5) ratio. The reaction scheme is presented below:

$$Sm \xrightarrow[80K]{ROH} RSmOH \xrightarrow[80K]{ROH} RH + (RO)_{1.5}Sm(OH)_{1.5}$$
$$\searrow aggregation$$
$$Sm_n \xrightarrow[80-300K]{ROH} Sm(OR)_3 + H_2$$

Thus, chemical activity of metal species depends upon their size. This fact is confirmed by the results on synthesis and catalytic properties of Sm–hydrocarbon co-condensates [42]. Hexane, hexene-1 and cyclohexene were chosen as model compounds. The catalytic system obtained by low temperature co-condensation of Sm and hexene is more active in hydrogenation of hexene-1 than the catalytic system obtained by low temperature co-condensation of samarium and cyclohexene. The same effect is observed for hydrogenation of cyclohexene using the catalytic system obtained by low temperature co-condensation of Sm and cyclohexene. The effect is called 'matrix memory'. Normal or cyclic hydrocarbon product, formed during the reagent's co-condensation, fixes such a position of samarium atoms in cluster as is suitable for hydrogenation, and determines the high selectivity of the process. The analogy with enzymatic reactions can be found, when enzyme–substrate interactions realise via rearrangement of the active center according to substrate structure.

The results on chemical reactions of atoms, small clusters and nanoparticles of magnesium and samarium with carbon dioxide, ethylene and alkyl halides [43–45] also clearly demonstrated the differences in the reactivity of metal species of different size. The results obtained for the reactions of samarium and magnesium species are summarised in Table 1. The analysis of the data presented in Table 1 shows that samarium atoms and clusters possess the higher redundant energy and reactivity than magnesium clusters in the same reactions. The general feature for both metals lies in the fact that the relative chemical activity for dimers and trimers (small clusters) is higher than for atoms.

Thus, as shown for metal atoms, clusters and nanoparticles interacting with different organic and inorganic substances, develops the original trend of chemistry and chemical material engineering. Knowing the properties of bulk metals it is impossible to predict the chemical behaviour of their atoms, dimers, trimers, small clusters and nanoparticles. More experimental data are needed in order to predict the reactivity of such complex systems. The frontier task is to study how to control the activity and selectivity of metal species interactions depending upon their size from atoms to clusters and polynuclear complexes and solvated nanoparticles.

3.3. Fast spontaneous reactions at low temperatures

The most interesting for LTSC are fast spontaneous reactions proceeding in low temperature co-condensates at a high rate, in some cases explosively, without any external initiation. Such reactions utilise internal redundant energy of co-condensate film.

In this paper we discuss only three reactions of such type. All these reactions have different mechanisms but one common feature—fast, practically explosive rates in the solid phase at low temperatures. The first example is a reaction of halogenation and hydrohalogenation of olefines. This type of reactions is studied in detail in Refs. [5,46]. It is shown that unstable donor–acceptor molecular complexes play a crucial role in their realisation at low temperatures. The correlation between

Table 1
Reactions of magnesium and samarium species in matrices at 10–40 K [44,45]

Metal species	CO_2-ligand	C_2H_4/C_2D_4-ligand	CH_3X (X = Cl, Br)-ligand
Mg	$Mg^+CO_2^-$ by matrix annealing	Cyclic complex $Mg(C_2H_4)_2$ by matrix annealing	CH_3MgX by irradiation ($\lambda \approx 280$ nm)
Mg_{2-4}	$Mg^+CO_2^-$ by co-condensation		CH_3MgX by irradiation ($\lambda > 300$ nm)
Mg_x	$Mg^+CO_2^-$ by matrix annealing		
Sm	$Sm^+CO_2^-$, CO, $SmCO_3$ by matrix annealing	Complexes $Sm(C_2H_4)/(C_2D_4)$ and $Sm(C_2H_4)_2/(C_2D_4)_2$	Formation of CH_4 by co-condensation
Sm_2	$Sm^+CO_2^-$, CO, $SmCO_3$ by co-condensation		
Sm_x			

Table 2
The dependence of reaction rate in 2-methylbutene-2–HCl system (HCl/olefine = 6/1, 80 K)

Complexing degree (%)	Reaction conversion rate during co-condensation (% conversion/h)	Reaction conversion rate during annealing (% conversion/h)
100	39	25
50	22	13
0	0	0

reactivity and the degree of the reagents complexation is nicely illustrated by the results for the hydrogen chloride-2-methylbutene system (see Table 2). The main conclusion from the table is that absence of complexation led to the absence of the reaction conversion.

The explosive reaction in the growing co-condensate films for the magnesium 1,2-dichlorethane system [10] is $Mg + CH_2Cl–CH_2Cl \Rightarrow CH_2=CH_2 + MgCl_2$.

The explosion takes place at a certain critical thickness (L_{cr}), the reaction yield after the explosion is nearly 100%. The reaction is accompanied by fast heat and gas release and mechanical crack of the film. The critical thickness (L_{cr}) depends upon the reagents ratio and support temperature. The critical thickness (L_{cr}) has a minimum, which corresponds to a sample composition with magnesium and di-chloroethane mole/mole ratio 1:1. An increase in the temperature narrows the region of explosions. It is important for this system that the initial sample is in the amorphous state, and it is crystalline after explosion. The reaction occurs only in a part of the co-condensate film. There is a defined interface between the reacted and unreacted parts of the sample. Thus, the self-propagating waves are formed, and the internal energy of mechanical stresses, accumulated in the growing film, initiates explosions. The probability of explosions decreases with the increase of surface temperature and lowering the rate of reagent's condensation. Similar features are found for the reaction of acetyl chloride and di-ethylamine and for reaction of cyclopentadiene with $TiCl_4$ [10,47]. The generation and accumulation of mechanical energy lead to plastic deformation up to the film sample crack formation and initiate both fast reaction and fast crystallisation of the film sample.

4. Hybrid nanostructures formed by low temperature condensation and co-condensation on cooled organic/inorganic surfaces

Different highly ordered surface nanostructures have been obtained by atom/molecular beam condensation on cooled surfaces of different dielectric, semiconductor and metal supports.

4.1. Molecular condensate films

Molecular nitrogen films on the basal-plane surface of graphite are probably the most extensively studied systems. As the result, the "N_2-graphite" phase-diagram as a function of the temperature (T) and surface coverage (n) is known.

In a recent paper [48] the authors use superlow temperatures (12–20 K), where thermal motion is practically frozen, and a nuclear resonance photon scattering (NRPS) technique to measure the out-of-plane tilt of the N_2 molecular axis relative to the graphite planes. The results obtained allow one to distinguish between a herringbone (HB) and pinwheel (PW) structures of nitrogen layers. The low temperature nitrogen molecular orientation is consistent with a 2-out HB ordering throughout a wide coverage range between sub-monolayer and up to bilayer formation where the tilt angles increase with increasing coverage.

Growth of organic films on a graphite surface at 80–200 K is studied by neutron diffraction during physical adsorption of methyl chloride [49], dichloromethane [50] and chloroform [51]. The film structure strongly depends on the geometry and polarity of adsorbed molecules. Solid methyl chloride film exhibits two rectangular structures in the sub-monolayer depending on the temperature and fractional coverage. For the less polar $CHCl_3$ molecules a bilayer film is formed from a mixture of two structures: one of them is hexagonal, the other exhibits a square unit cell. This behaviour is related mainly to the nucleation differences. Organic monolayers of saturated long-chain hydrocarbons were deposited on a Cu(1 1 1) surface [52] cooled down to 80–160 K. The alkane-chains ordering is shown by low energy electron diffraction (LEED) and high resolution elastic and inelastic He-atom scattering (HAS) to form at low temperatures highly ordered well-defined 2D lattices: a HB structure for octane–Cu(1 1 1) and a domain wall lattice with parallel arrangement of molecules for nonane–Cu(1 1 1). Whereas octane forms a commensurate lattice at a low temperature surface and undergoes a sharp first-order phase transition at higher temperatures, nonane forms a domain wall lattice and exhibits an unusual strongly broadened phase transition. This fact correlates with the odd–even effect known for bulk alkane structures: whereas even C_nH_{2n+2} alkanes have a monoclinic or triclinic unit cell, odd numbered alkanes have an orthorhombic unit cell. Organic–metal interfacial structure is of interest for molecule-based electronics and optoelectronics [53–55]. The interfacial electron transfer using two-photon photoemission is studied for two systems: hexafluorobenzene and a self-assembled monolayer of thiophenolate on Cu(1 1 1) [56]. The interaction between the organic molecule and surface reveals both energetics and dynamics of interfacial charge transfer and determines the characteristics of electronic and optoelectronic single-electron devices based on organic molecules.

4.2. Metal–semiconductor interfacial structures

Ultrathin gold films were obtained by deposition on a low temperature (173–300 K) semiconductor substrate surface [57]. The formation of an amorphous gold silicide layer, sandwiched between the Si crystal and pure metal film, and an Au layer, sandwiched between two Au–Si alloyed regions (13–28 A) are shown, one of them is located at the surface and the other under the Au layer [58].

For silver films (~0.3 nm) deposited on Ge(0 0 1) and Si(1 1 1, 0 0 1) 2D structures were found [59,60]. The first system is one of the prospective candidates for novel interface (surface) superconductors with $T_c = 2$–8 K due to weak Ag–Ge interactions and 2D Ag–Ge alloy islands formation [61]. With increasing average thickness of deposited silver the density of 3D islands increases at any temperature between 100 and 300 K. Second harmonic generation is shown for cold-deposited silver films [62]. The data obtained confirm the homogeneity of thin 15–20 nm films deposited on cooled surfaces and can be used for estimation of metal surface roughness.

Pb–Si(1 1 1) interface is chosen [63] as a good example of a metal–semiconductor interface uncomplicated by inter-diffusion or chemical reactivity. The Schottky barrier height of the Pb/Si diode is known to depend on the atomic reconstruction of the interface. The film is shown by surface X-ray diffraction to be initially highly disordered with irregular oscillations in the reflected intensity during growth. At a critical coverage of approximately five Pb layers at 100 K crystallised to form a well-ordered Pb(1 1 1) structure with subsequent growth being layer-by-layer and islanding being suppressed.

4.3. Bi- and triple hybrid metallic nanosystems

The surface properties of bimetallic systems have been widely investigated in recent years [64,65] due to the fact that a new class of mixed-metal compounds could be designed for possible applications in different areas such as electrochemistry, magnetism, material science and catalysis. While catalytic activity from bimetallic surfaces has been recognised for a long time [65], it is only in the last decade that the development of surface techniques has made systematic studies of these surfaces feasible [66]. The growth of metal ad-atomic films, deposited on metal surfaces at temperatures of 100–200 K, allows only surface diffusion to occur. The specific features of the nucleation and the kinetics of the formation of interface island monolayer during layer-by-layer or 3D film growth are dependent mostly on the similarity of lattice parameters for both metals, their ability to form bimetallic molecules and alloys, on the value of tensile stresses in the growing films, on the elastic energy relaxation mechanism, on the coverage degree and on the temperature of the supporting surface. Lifetimes of ad-atoms excited states and electronic spectral function of metal surfaces at low temperatures were studied by time-resolved two-photon photoemission technique [67].

Some recently studied [68] triple metallic hybrid nanostructured systems Cu/Co/Cu(0 0 1), Pd/Co/Pd(0 0 1), Fe/Cu/Fe(0 0 1) are found to have unusual magnetic properties due to interface magnetostriction and interface magnetic anisotropy. Truly new magnetic materials are expected to be manufactured by an artificial method based on nanoscale hybrid surface production. Magnetoresistance in spin valves and tunnelling magnetoresistance provide their applications as magnetic sensors and spin-valve magnetic reading and writing memories. Magnetic force microscopy (MFM) can be applied for the study of multilayer metallic systems [69].

Evaporation and deposition of metal atoms and clusters using oxide surfaces as supports is of increasing interest during last years, especially for clusters that are known to exhibit specific size-dependent properties due to prospective uses and modelling of highly effective and selective catalysts. The most intensively studied was Ag on MgO(1 0 0) single crystal [70]. In recent papers [71] the triple hybrid nanostructured systems Ag/MgO/Mo(1 0 0) and Ag/S/Mo(1 1 0) are studied by different spectroscopic techniques (electron energy loss, photoemission spectra and thermal desorption mass spectroscopy). In the last case sulfur multilayers supported on a Mo(1 1 0) single crystal surface at 110 K react with silver to form sulphide compounds. Upon annealing to higher temperature the silver sulphides promote the sulfidation of the Mo support leading to the MoS_x formation. At temperatures above 400 K Ag atoms diffuse into the bulk of molybdenum sulphide, forming $AgMoS_x$ compounds. AgS_y/MoS_x and $AgMoS_x$ systems were non-reactive towards molecular hydrogen under high vacuum conditions. However gas-phase atomic hydrogen reacted with the surfaces to form gaseous hydrogen sulphide.

Chemical vapour deposition (CVD) provides an attractive route for the controlled deposition of metal atoms and clusters on different surfaces [72] via thermal or photochemical decomposition of different chemical precursors. In recent papers [73,74] the deposition and decomposition of iron pentacarbonyl and other metal carbonyls, chromium hexacarbonyl and cyclopentadienyl rhodium dicarbonyl on single crystal surfaces Pt, Ag, Au and Cu(1 1 1) are studied at low temperatures 80–300 K using reflection absorption infra-red spectroscopy (RAIRS).

4.4. Reactions on single crystal metal surfaces

The other prospective LTSC method applied for surface hybrid nanostructures formation is exposing of metal single crystal surfaces to reactive gases at low temperatures. Chlorination of copper and silver single crystal surfaces of different crystallographic orientation was studied in [75]. Layers of silver chloride of 2 nm thick are formed by exposing the Ag(1 0 0) surface to gaseous chlorine at 150–250 K. Thin copper and silver chloride layers can be used also for formation of nanostructured materials—metallic clusters dispersed in the chloride phase: $Ag_x/AgCl$ and $Cu_x/CuCl$. Nanometer-scale local oxidation (LOCOS) of metal (Ti, Al,

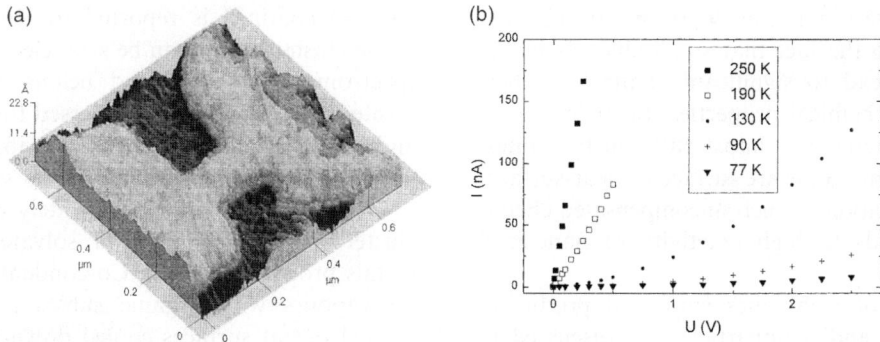

Fig. 6. Atomic force microscopy (AFM) image of titanium surface locally modified by a nanometer-size local oxidation (LOCOS) of metal (Ti) surface exposed to O_2 with the tip of a scanning probe microscope when applying a voltage between surface and tip enable the direct writing of metal oxide lines (10–20 nm): a Ti/TiO$_x$/Ti surface structure (a) and the current voltage characteristics of the junction at different temperatures (b) [76].

Cu) surfaces exposed to O_2 with the tip of a scanning probe microscope applying a voltage between surface and tip enables the direct writing of metal oxide lines (10–20 nm) (Fig. 6) [76]. For the first time the LOCOS technique using a scanning probe microscope was demonstrated for Si surface. High density nanometer-scale Si-pillars formation by sub-monolayer nitration in vacuum and subsequent oxidation was demonstrated in Ref. [77]. Efficient Si(1 0 0) etching by thermal-energy H-atoms has been achieved by exposure of Si(1 0 0) to disilane at 105 K [78]. Subsequent exposure to H atoms leads to the ejection of the etching product (SiH$_4$) for substrate surface temperature 105–480 K. O_2 molecules weakly chemisorbed on the Pt(1 1 1) surface can be displaced by incident CO molecules. The process is studied at surface temperatures between 95 and 110 K. The dynamic and kinetics of the process are found to behave distinctly differently above and below CO kinetic energy E_{CO} equal to 0.7 eV [79].

The photodissociation of neutral molecules on metal single crystal surfaces has been found in general to be altered from that of the same molecules in the gas phase, but generally having a substantially reduced cross-section (quenched) or going via low-energy dissociation pathways for low temperature photodissociation of CH_3J on Cu(1 1 0) [80] and *tert*-butyl nitrite on Ag(1 1 1) [81], low energy electrons induced dissociation of ammonia on Pt(1 1 1), Si(1 0 0), Ge(1 0 0) [82,83].

Silver surfaces are known to be a good catalyst for oxidation processes, however, organic molecules do not adsorb on silver surfaces at temperatures higher than 150 K. The situation modifies drastically in the presence of adsorbed oxygen [84]. The effective low temperature oxidation of cyanogen $(CN)_2$, CO, iodoethane, methanol, nitrobenzene and styrene molecules in the presence of adsorbed oxygen on the Ag(1 1 0), Pt(1 1 3) and Cu(1 1 0) surfaces has been reported in Refs. [85–87]. Dissociative sticking of metane on nickel single crystal surface is proposed as a model of the industrial reforming process takes place over a catalyst containing small nickel crystallites highly dispersed on a porous support material like aluminium/magnesium oxides [88]. Reactions of hydrogen atoms with acetone monolayers on graphite monolayer covered Pt(1 1 1) surface have been studied at 90 K by high resolution electron energy loss spectroscopy (HREELS) and thermal desorption spectroscopy techniques [89].

5. Low temperature synthesis and surface reactions of nanosized metal particles

Nanosized metal particles are assemblies of atoms with the size in the range of 1–100 nm composed of several hundreds of atoms. Metal particles with diameter less than 1 nm are usually referred to as clusters. Metallic clusters and particles in size ranging up to 10 nm are of most interest due to the fact that small changes in the particle size lead to significant changes of their physical and chemical properties. Such structures can be considered as almost fully surface ones, because all their atoms are surface or near-surface. High concentration of such uncompensated chemical bonds leads to high reactivity of nanosized particles.

A number of techniques employed producing metal clusters and nanoparticles are discussed in detail in a monograph [90]. Low and superlow temperatures give new possibilities to obtain small metal species [91].

Gas phase evaporation technique is based on the vapourisation of metals under a pressure of static inert gas. Different vapourisation techniques have been used including resistive heating, electron beam and laser vapourisation. The clusters and particles thus obtained deposit on the inner walls of reactive chamber and/or different substrates surfaces [92]. Fine Cu, Ag and Au particles (about 10 nm) are prepared by gas evaporating method combined with trapping of the formed particles in the cold liquids (ethanol, methanol and some alkanes). Mg particles of 2 nm in size were prepared in liquid He and larger particles about 100 nm in size in solid argon [93]. By varying the pressure and changing from He to argon intermediate size particles could be also obtained. Laser vapourisation is applicable to all metals. Thus, low energy neutral Fe, Co, Ni clusters beam deposited on different substrates produced nanostructured films with thickness up to 100 nm and determined their unusual structural and magnetic properties [94]. Laser ablation and electron beam evaporation are the most frequently used techniques for producing bimetallic species. Mass-spectroscopic study of the particles formed during supersonic nozzle expansion of Zn, Cd, Hg and Na evaporated together [96] shows, that most of particles consist of one heavy metal atom and up to 18 Na atoms surrounding.

The innovation of such techniques lies in the development of the ionised cluster beam source, it allows the users to control the kinetic energy of deposited clusters by the acceleration voltage applied. The formation of large surface clusters and metallic coatings is reported in Ref. [97]. Gas-phase cluster beams can be size selected by a mass spectrometer. Fractionated beams can be redeposited in a size-selected dispersed form. Thus Pt_n and Pd_n particles, where $n = 1$–15, have been prepared on Ag(1 0 0) single crystal surface [98].

Metal vapour synthesis is widely used for new cluster organometallics and solvated nanosized metals production [4,92]. Co-condensation of metal vapours with organic substances is usually carried out at surfaces cooled down to 77 K and below. At these temperatures metal atoms and small clusters are embedded into solid organic matrices. Upon warming, atoms agglomerate to higher clusters and particles up to 10–30 nm in size and stable colloidal solutions are formed at room temperatures. Further particle growth is prevented by solvating and electrostatic effects. Co-deposition of two metals simultaneously results in core-shell bimetallic particles. Fe-core particles (size controlled in the range from 3 to 16 nm) thus obtained by the Fe/Li co-condensate system in Ref. [94] are protected from surface oxidation by $Li_2O/Li_2 CO_3/Fe_2O_3$ coating layer. This approach is extended to Fe–Mg combination [96]. These results demonstrate for the first time that normally immiscible metals could be forced to form metastable clusters using this LTSC technique.

Alternatively, by allowing the co-condensate to melt and permeate a support oxide material (Al_2O_3, SiO_2, MgO) surface supported metal particles with crystallite sizes of 3–8 nm are obtained [99]. Bimetallic supported nanosized particles also could be prepared using such a technique [100]. Colloid metal particles in non-aqueous solvents have been used as precursors for the preparation of thin metallic films by solvent removal [101]. The films are found to be unique: they showed higher resistance than pure metallic films and were redissolvable. The alternative approach to small supported cluster production is the decomposition of organometallic compounds and cluster complexes [95], reduction of metal ions [102,103], template synthesis [104], high energy ball milling and sputtering [95] techniques.

Surface reduction/oxidation reactions on 'nanoelectrodes', which lead to deposition of a second metal such as lead, cadmium and thallium on colloidal gold particles, are investigated in [105]. The reactions are not initiated by applying an external potential, but by transfering electrons from free organic radicals generated radiolitically to nanometer-sized gold particles. Because of the extremely large surface of the particles, the surface reactions can readily be followed by spectrophotometric measurements. It is shown that second metal deposition and reoxidation substantially change the size and chemical nature of colloid nanoparticles. The electron properties of ad-atoms and electron exchange between metals of different work function strongly affect the chemical behaviour of the composite metal particles.

Chemical activity of metal nanoparticles at solid surfaces is considered in detail in a recent monograph [106]. It is demonstrated that chemical activity and selectivity of supported metal species can be explained by both electronic and structural properties of corresponding bulk metals and their variation with cluster size. Such an approach allows one to view these systems as a bridge, which connects surface reactions of metal clusters and nanoparticles with real catalysis.

The study of the reactivity of metal particles different in size in a broad temperature range allows us to find novel and unexpected chemical reactions. For instance, magnesium atoms and small clusters react with carbon tetrachloride at temperatures near 77 K, which is unusual for room temperatures and bulk metal. Poly-p-xylylene films with nanosize silver particles were used as catalytic system for methanol conversion to formaldehyde [107], treatment with metallic zinc at low temperatures caused an increase in the carbon tetrachloride de-chlorination rate and conversion into methane [108].

The preliminary results obtained show that conversion process could be carried out with higher yields at lower temperatures in comparison with the traditionally used catalysts. Silver organosols obtained by the CVD-method are used as catalytic system for ethylene epoxidation. The colloid of stabilised silver nanoparticles was deposited on alumina support. It is found that this process carried out using traditional catalytic systems leads to the same conversion values but allowed to use a smaller silver content [107].

6. Self-organised metal–organic nanostructures

The synthesis of metal–organic nanomaterials is of great interest from fundamental and applied viewpoints. Nanoscale hybrid (heterogeneous) materials, which are stable up to room temperature, offer an area of new intriguing technological applications [109,110]. For instance, a relatively large part of the volume of such particles consists of grain boundaries or interfaces. The interactions of such finely divided structures with photons and electrons can lead to unusual optical, magnetic and electric-physical behaviour [95,111]. The properties of nanoparticles and nanostructured films, the methods of their production, nanoparticles self-organisation processes, oriented films growth and templated synthesis are described in Refs. [95,109–112]. We present only some excellent examples of such materials produced by LTSC.

6.1. Polymer films with nanosize metal particles

Low temperature co-deposition of vapours of different metals such as Ag, Mg, Cd, Zn, Pb, Sm, Sn, Mn with the monomer p-xylylene reactive at 100–120 K has been used for production of nanosized metal particles encapsulated in a polymer matrix [113,114]. Polymerisation of the film system and formation of a polymer network just at low temperatures prevents in this case the higher aggregation of small metal species. Metal vapours are obtained by resistive heating of the bulk metal at 500–1200 °C depending on the volatility. The reactive monomer p-xylylene was prepared by pyrolysis of di-p-xylylene at 550–700 °C in the same vacuum set-up. After co-deposition of monomer with metal vapours on the surface cooled by liquid nitrogen during 3–30 min the co-condensate film was slowly heated up to the ambient temperature. The poly-p-xylylene films containing metal particles could be extracted from the vacuum cryostat for further investigations. The scheme of LTSC synthesis of poly-p-xylylene films with nanosize metals is presented in Fig. 7.

The metal-containing poly-p-xylylene films were examined by transmission electron microscopy (TEM). The metal content was determined by means of X-ray fluorescent analysis. It is found

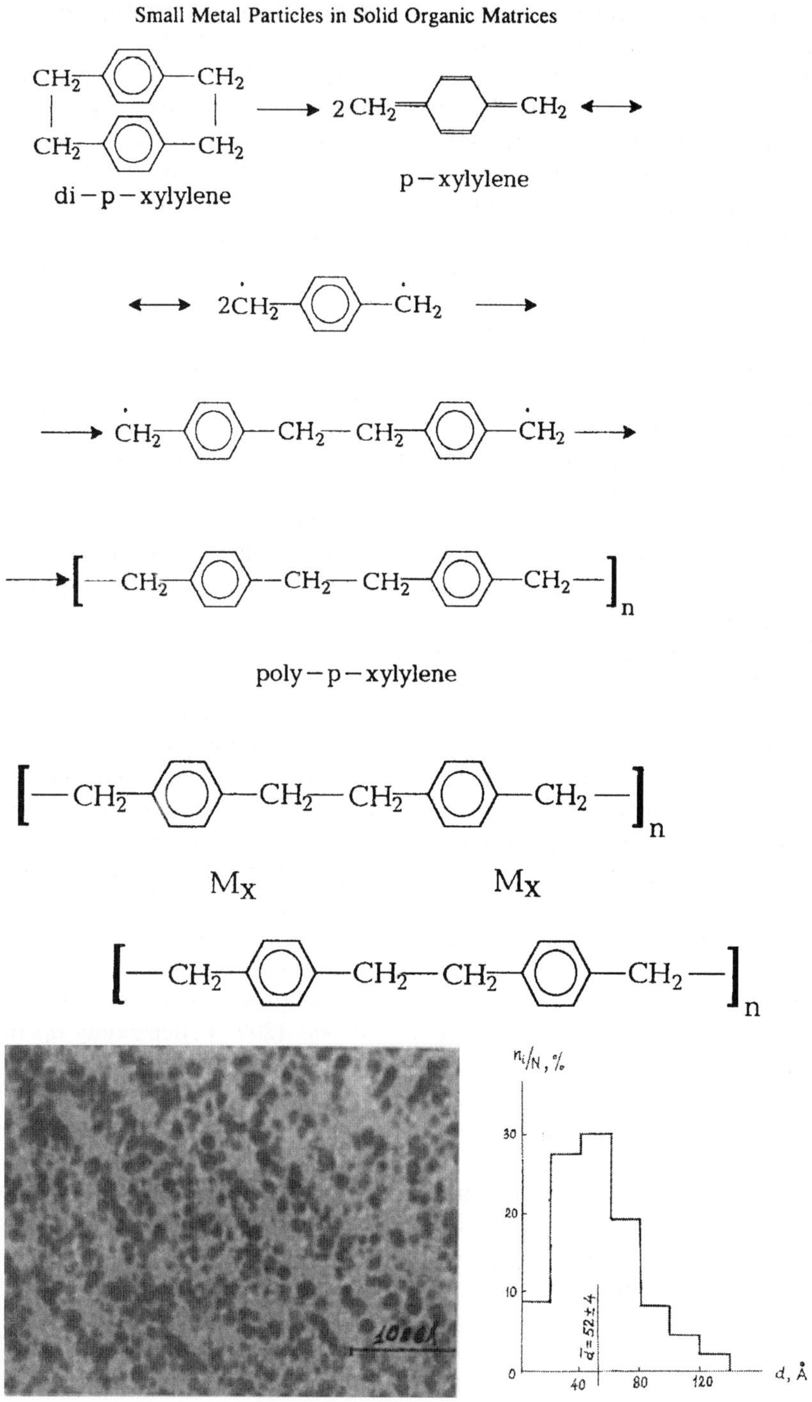

Fig. 7. The scheme of poly-*p*-xylylene films with nanosized metals low temperature synthesis and TEM picture and particles size distribution for poly-*p*-xylylene films with 6.5% Pb [113].

that globular Pb particles are rather homogeneously dispersed within poly-p-xylylene at room temperature (Fig. 7). The size of these particles was found to be 3–8 nm and is independent of the Pb content in the range 0.1–6.5 wt.%. The crystalline and metallic nature of the particles in the poly-p-xylylene films is shown by X-ray studies [113]. The lattice dimensions calculated from electron diffraction data for Zn and Cd particles formed in poly-p-xylylene are in a good agreement with those obtained in bulk metal. After keeping the Zn- and Cd- films for some days in the presence of oxygen, new phases with lattice parameters of ZnO and CdO were observed. Thus, metal particles encapsulated in poly-p-xylylene samples could be surrounded by surface oxide layers.

The poly-p-xylylene is a typical insulator. The addition of Pb up to 10 wt.% does not change the characteristics of the materials. The specific resistance of poly-p-xylylene films with metal particles is as high as 10^{16} Ω/cm^2, which has been measured by the electrical conductivity technique. The electrical conductivity of the metal–polymers films is studied during the sample formation and sample annealing in the temperature region 80–300 K [115].

Lead containing poly-p-xylylene material is found to be gas sensitive [116,117]. Measurements of the electrical resistivity for Pb–poly-p-xylylene materials in the presence of ammonia and water in the gas phase at room temperature show that the resistivity varies reversibly by a factor 4–6, it has been found that the resistance is a square-law function of ammonia concentration.

6.2. Nanometal–liquid crystals systems

The possibilities of stabilisation of silver nanoparticles using polar liquid crystals—long chain alkyl(alkoxy) cyanobiphenyls and poly-p-xylylene are studied in Ref. [118]. Silver containing cyanobiphenyl film samples ($l = 20$–50 μm) were obtained by reagent's vapour co-condensation on the cooled surfaces of special spectral cryostats under molecular beam conditions. The quartz, KBr or CaF$_2$ spectroscopic windows are used for transmission UVS–VIS and IR-spectroscopy and a polished copper cube for reflection IR-spectroscopy studies. Metal vapour is prepared by resistive heating of the bulk metal over the temperature range 1100–1200 K, cyanobiphenyl components were evaporated under resistive heating at 380–390 K.

The spectroscopic (IRS, UVS, ESR) study of silver-cyanobiphenyl co-condensates of different metal-to-ligand ratio (from 1:1 to 1:100) has been performed in the temperature range 80–300 K [119,121]. Most of the experiments were done with C_5H_{11}–C_6H_4–C_6H_4–CN (5CB) and 4-octyl-4′-cyanobiphenyl (8CB). The spectroscopic data obtained in combination with the results of quantum chemistry calculations of the model cyanophenyl system show the existence of sandwich-like Ag(CB)$_2$ complexes with cyanobiphenyl ligand molecules arranged by the "head-to-tail" principle [120]. The ESR-spectrum of silver-4-pentyl-4′-cyanobiphenyl (5CB) co-condensate film showed two doublet lines with $g(Ag^{107}) = 2.001$, $a(Ag^{107}) = 48.5$ mT and $g(Ag^{109}) = 2.003$, $a(Ag^{109}) = 55.7$ mT due to silver atoms ($J = 1/2$) stabilised in silver-cyanobiphenyl complexes [119]. The obtained value of the hyperfine structure constant allows us to estimate the degree of metal to ligand charge transfer. The partial charge on silver atoms in the proposed Ag(5CB)$_2$ π-complex structure is equal +0.23. The central anisotropic singlet line could be referred to silver nanoclusters. The estimation of silver cluster size stabilised in cyanobiphenyl matrix for the film sample with the ratio Ag/5CB = 1:10 at 90 K gives us the average value 25–30 atoms (3.6–4.2 nm). By heating from 90 to 200 K the relative intensities of the signals at lower and higher fields were decreased and the intensity of the central singlet line was increased. The effect of metal/ligand ratio in the film sample and components condensation rate on the relative part of silver atoms stabilized in π-complex structure and aggregated silver nanoclusters is shown. UVS–VIS spectra (Fig. 8) confirmed the formation of the complexes Ag/5CB, which are stable only at low temperatures and degrade upon heating up to 200 K. The complex absorption band at 340 nm disappeared at 200–300 K and a new wide band with maximum at 440 nm was detected [119,122]. The last one can be caused by surface plasmon absorption of silver nanoparticles, formed during Ag/CB complex degradation and followed by the aggregation of silver atoms/or small clusters.

The silver-cyanobiphenyl samples are encapsulated in poly-p-xylylene films under vacuum conditions using a special cryostat supplied with the inlet for xylylene monomer injecting into the system. The films are examined by TEM. The data of electron microscopy show two kinds of silver particles stabilised in liquid crystal matrix: globular silver particles of size 15–30 nm and anisotropic rod-like particles of more than 200 nm length (Fig. 8) [121]. The relative amount of the particles of definite shape and size depends upon the metal to ligand ratio in the co-condensate films and upon the conditions of thermal treatment of the samples. Thus, silver atoms or small clusters stabilised at low temperatures by interactions with the cyanobiphenyl molecule π-system can aggregate and self-assemble at a higher temperature in the anisotropic liquid-crystalline matrix. In this case the anisotropic matrix mediates the primary growth of anisotropic rod-like metal particles.

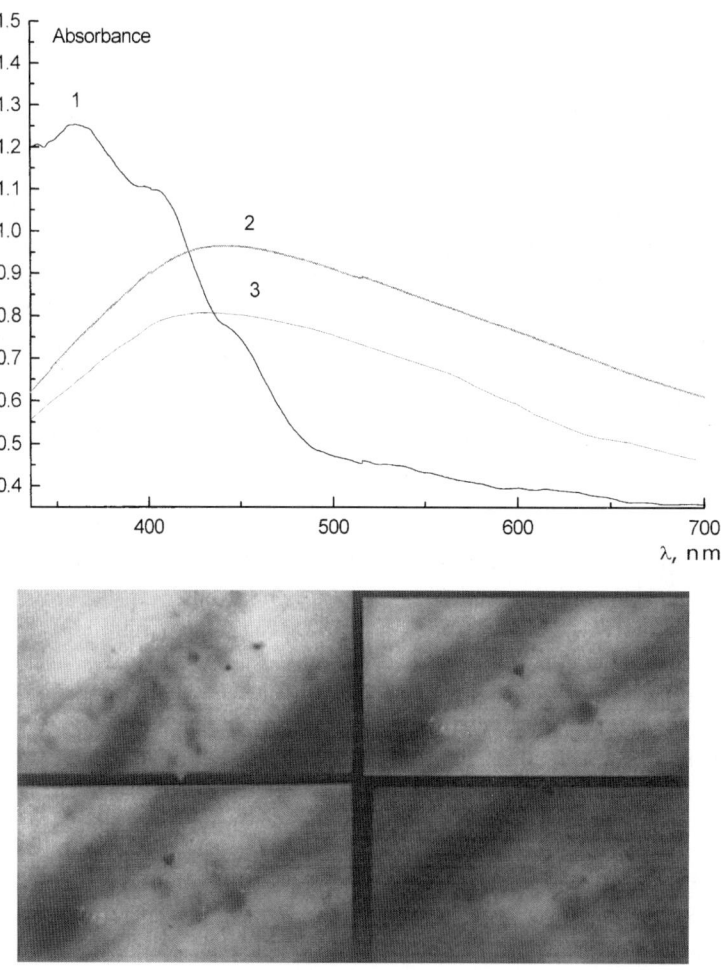

Fig. 8. Electronic (UVS–VIS) spectra of silver film co-condensate with mesogenic 4-pentyl-4′ cyanobiphenyl (Ag/ 5CB = 1/10) at: (1) $T = 90$ K, (2) $T = 200$ K, (3) $T = 300$ K ($l = 2$ μm) and TEM picture at 300 K. The figure shows two kinds of silver particles stabilised in a liquid crystal matrix at ambient temperature: globular silver particles of size 15–30 nm and anisotropic rod-like particles of more than 200 nm length [119].

6.3. Nanoparticles' assemblies and surface superlattices formation

Recently the main attention is given to formation of metal nanoparticles assemblies and their self-organizing on different support surfaces. Some examples can be found in Refs. [122–124]. Ordered structures of monolayers stabilised (protected) by organic ligand are considered for different applications in nanodevices construction. The study of the low temperature formation of metal dispersions in the different solvents shows that the solvents with functional groups C=O, OCH_3, CN, thiols, amines and etc. can stabilise metal particles.

Silver and gold nanoparticles are used as model object for the most of the studies. Nanocrystal superlattices are directly synthesised in gold colloid solutions with several thiol ligands [125] and using thiol and amine modified substrates [126]. Monodispersed silver nanocrystals stabilised by dodecanthiol are produced [127]. Detailed analysis of metal nanoparticles thiol stabilisation and their organisation in 1D, 2D and 3D structures is presented [128]. The formation of altered positive and negative charged gold and silver particles and their organisation in the presence of 4-aminothiophene and 4-carboxythiophene on glass substrate surface is described in Ref. [129]. 1D growth of gold particles on Si-substrates was achieved by gold particles' deposition on AFM-treated Si-surface.

Self-assembling of spherical nanoparticles of several nanometers (2–5 nm) in non-spherical aggregates is studied [130]. It is shown that the aggregation process and aggregates size, shape and morphology can be controlled by the substrate surface temperature: at 23 °C—spherical aggregates of smaller particles of (97 ± 17) nm diameter were obtained, at −23 °C—spherical particles of 0.5–1 μm, and at intermediate temperature 10 °C—1D chain aggregates of 50 nm length built from spherical sub-units. It is supposed that the formation of non-linear aggregates is the intermediate stage in the process of formation of large spherical aggregates from smaller particles and is controlled by the balance in enthalpy and entropy changes. The orientational ordering is achieved for gold nanorods [130]. The energy of femto- and nanosecond laser impulses was used in for reversed transformation of nanorods to nanospheres.

The effect of substrate surface is studied for the morphology of 2D and 3D superlattices of silver sulphide 5.8 nm in diameter, stabilised by dodecane thiol [131]. The authors used pyrolitic graphite and molibdenum disulphide as substrates. Controlled morphology of substrate crystal surfaces is important for pharmaceutical and food industry applications. Self-assembled thio-organic monolayers on gold crystal surfaces Au(1 1 1) are used as template mediated crystallisation of glizine [132]. Weak action of the tip for a scanning tunnelling microscope helped to obtain highly ordered sterene structures on the Si(1 0 0) surface. Modification of Si surfaces allowed authors to determine the orientation and length of molecular rows, duplicate parallel rows, to control the row direction and to change the angle of chain growth.

Self-assembling of alkanthiols monolayers on gold nanoparticles has been studied [133]. The introduction of fluor-hydrocarbon spacer chains increased the stability of ordered layers. Thermal stability increases by raising the number of methylene groups and for $n = 33$ the films are stable under air conditions during one hour at 150 °C. Stable films of gold nanoparticles (4 nm) were produced by polyethylene imine and poly-L-lyzine films with carboxylic acids stabilised gold nanoparticles [134]. Thin porous films of gold nanospheres on glass substrates are obtained using colloid crystals as templates [135,136].

Joint condensation of silver and methylacrylate vapours [137,138] onto a surface at 77 K followed by heating of the co-condensate up to room temperature leads to partial polymerisation of methylacrylate and to formation of organic colloids of silver nanoparticles stabilised by polymer layer. A typical absorption spectrum obtained after melting of the co-condensate film possesses a maximum at 400 nm, that is characteristic for spherical particles of colloidal silver <20 nm in diameter [137]. The TEM pictures of the samples indicate that silver particles are not exceeding 15 nm in size. Combined condensation of silver, lead and methylacrylate (MA) vapours in vacuum onto a surface cooled by liquid nitrogen followed by heating of the co-condensate film results in the formation of colloids consisting of bimetallic particles [138]. The size of the lead and silver/lead particles does not exceed 5 nm, i.e. they are smaller than the diameter of silver nanoparticles obtained previously under similar conditions. Absorption spectra of Ag/Pb/MA system show that the silver surface plasmon absorption band (416–420 nm) is shifted to the red ($\lambda_{max} = 438$ nm). The electron microscopy data allow us consider the aggregation of nanoparticles as the main reason of the red shift in an inert environment.

The process of metal nanoparticles formation has been studied both by optical spectroscopy and dynamic light scattering methods of the samples obtained by low temperature co-condensation of silver and organic monomer 2-(dimethylamino)-ethylmethacrylate (DMAEMA) [107]. The data obtained in combination with the results of TEM study show the possibility of the formation both of silver particles of 5–12 nm in diameter stabilised by surface polymer layer and their aggregates. Colloids of cryochemically synthesised silver nanoparticles in water, acetone, and toluene are found to be stabilised by macromolecular poly-DMAEMA layers formed at the surface of nanoparticles. The photoreduction of silver cations in the presence of the partially decarboxylated poly-(acrylic) acid allows the researchers to obtain stable particle aggregates of the rod-like type, and the length of these particles may achieve 500 nm [139].

Higly ordered organic monolayers formed on inorganic surfaces can be used for selective binding of different organic/inorganic building blocks of hybrid nanostructures. Biomolecules are some of the most useful and exciting building blocks available to material chemists. The molecular recognition properties of proteins and protein building blocks (peptides) and single stranded DNA molecules are unmatched by conventional synthetic analogies. Advances in molecular biology open up the strategies for designing peptides to bind selectively to metal and metal-oxide surfaces. Fig. 9 demonstrates two main possibilities of exploiting biomolecules as 'glue' for inorganic building blocks [140]. One is assembly of inorganic blocks by 'programmed' interaction between biomolecules with known recognition properties. In this case two inorganic materials modified with

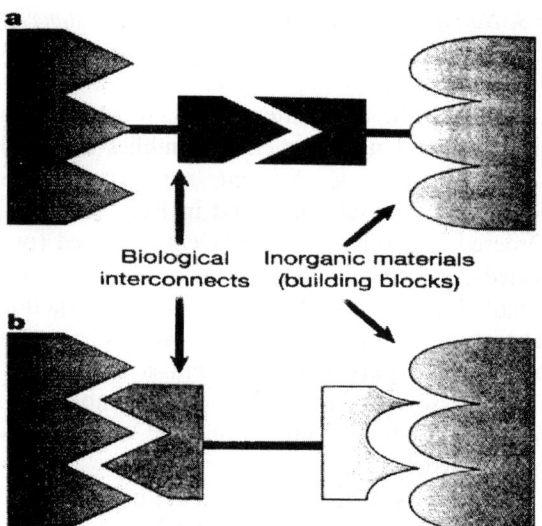

Fig. 9. Two main schemes of biomolecules used for binding inorganic building blocks: (a) assembly of inorganic blocks by 'programmed' interaction between biomolecules with known recognition properties; (b) assembly of novel inorganic materials by tailored evolution or interactions between biomolecules and inorganic surfaces [140].

proteins, antibodies, oligonucleotides or ligands which specifically recognise each other, allowing the formation of hybrid material. The second is assembly of novel inorganic materials by tailored evolution or interactions between biomolecules and inorganic surfaces. The last approach was realised in Ref. [141].

7. Conclusions and perspectives

Low and superlow temperatures allow researchers to study highly active metal species: atoms, small clusters and their nanosized aggregates. It was mentioned above that particles approximately 1 nm in size could be considered as the aggregates consisting of only surface atoms (the particles with 100% surface). Small metal species can be formed on different cooled surfaces and they can migrate on the surface by heating resulting in more large-sized particles. Such aggregation/stabilisation reactions are highly dependent upon surface temperature, deposition rate and are non-linear kinetically controlled processes.

Atoms on nanoparticles' surface are located in places with different co-ordination numbers, which are ordinary lower than for bulk substances. In some cases the initial spherical shape of particles deposited on the surface can be transformed and some irregularities appear. Such roughness is reflected in light absorption. The existence of receptor places on the nanoparticles surface and their number affect the migration kinetics, particles' aggregation and fine structure of their assemblies. For studying of such transformations further development of different surface probing techniques, and first of all different kinds of scanning microscopy at low temperatures is necessary.

The extended surface areas of nanoparticles and their high chemical activity open up the possibilities of surface synthesis of different chemical compounds. The nanoparticles' surfaces modified by different ligands can be consolidated in porous solid materials, or included in liquids or polymers. Low temperatures allow us to obtain small particles with size less than 1 nm. Their possible application is connected with the search of highly effective stabilisers. New catalysts and sorbents, new optic and corrosion protective coatings, new drugs and dyes, degassing agents for hazardous chemicals and biologically active substances can be produced by this way. For the successful solution of all mentioned tasks it is necessary to develop new ecological methods of nanosized substances synthesis, to develop highly accurate methods of structural and chemical analysis, to develop new precise physical properties measuring techniques and devices. The material presented in this review and the research experience of the authors allow us to formulate some recent and prospective areas of LTSC applications.

Thus, LTSC opens up the new prospects in 21st century. The application of low temperature condensed films gives us new possibilities for production of metal species of different size and to study their reactivity. It is a way to nano-technologies and new energy suppliers. The determination of the relation between the metal aggregate size (the number of metal atoms) and their reactivity is one of the most intriguing fundamental problems of the modern surface chemistry. The development of the new thermodynamic and kinetic models is of great importance for considering the reactivity of small reactive particles with size less than 1 nm. The reactive particle's size should be considered as a parameter, which achieves the thermodynamic function of temperature. High chemical activity of the nanosized particles, and metal nanoparticles, in particular, allow us to consider such systems as possible energy storage cells and special chemical nano-reactors. Using such systems it is possible to carry out the new reactions, which cannot be realised under ordinary conditions.

Using nanosized metal particles and metal-organic nanostructures stabilised in the films and/or on the surfaces of different supports, it is possible to produce systems with unusual practically important properties. When investigating nanoparticles and their surface properties the scientific problems are closely connected with technological and technical applications of such systems. The fundamental research is directed on establishing the relations between physical and chemical properties and the size and the shape of metal particles. Technical tasks are directed on utilisation of nanoparticles to produce new materials with unique optical, electrical, magnetic, mechanical, sensor and catalytic properties.

Nowadays the use of low temperatures is developed for the stabilisation of highly active small metal species at different surfaces, study of their reactions and production of small (milligram amounts) of nanostructured materials.

In the next 3–5 years the low temperature synthesis of new nanoparticulate substances and ecological nanomaterials in thousand gram amounts will be developed. The use of nanoparticles for surface functionalised modification; realisation of new chemical transformations, in particular using different types of electron tip scanning of the surface; the production of new nanoparticle sensing materials will be possible. To consider prospective applications the progress in the field of utilisation of nanomaterials should be noted. The first is nanoelectronics. Theoretical predictions show that in 1 cm^2 up to 10^{15} atoms can be placed, that will store all the information that humanity has

Table 3
Present and future applications of low temperature surface chemistry

At present time	In 3–5 years	Long term
Stabilisation and study of active species	Nanoparticles in chemical synthesis with controlling activity and selectivity	New paradigms in chemistry and physics of nanosized elements
Nanosized particles synthesis	Template synthesis and medium directed production of nanostructured materials	The discovery of new properties and important applications for nanosized species and their assemblies
Model reactions and size effects	Controlling synthesis and properties of nanosized surface heterostructures for catalysis	Direct measurement and control of nanosized particles surface energetics
Milligram amounts of new materials	Surface modification and reactions using electron tunnelling probe	Nanosized sensor and transducer devices
	Molecular and hybrid metal-organic structures for nanoelectronics and nanosensorics	Nanoparticles in biology, biochemistry and medicine

nowadays. These are sensors for toxic gases; high-selective and productive catalysts; membrane and high porous sorbents with high surface area and definite porous size; new superhard and superplastic materials and metal nano-alloys; protective coatings; materials for corrosion prevention; absorbing and reflecting ceramics; medicines with improved bio-accessibility and bio-transportation; new hybrid organic–inorganic materials and metal-bioceramic composites. Some bright examples of nanoscience applications are considered in detail in the first issue of the new Journal of Nanoparticle Research [142] and in recent papers [143,144]. The mentioned problems and possible prospects in LTSC are summarised in Table 3.

In long term prospects we expect the development and solution of the following problems. The low temperature chemistry transformations will be highly developed for all the elements of the periodic table and general features for low temperature chemistry at cooled surfaces will be established. The new properties of nanoparticle assemblies concerned with the number of their atoms and with the type of supramolecular organisation of the system will be discovered and described. The methods of controlling of nanoparticles' surface properties will be developed, the nanoparticles' surface energy will be measured and regulated. The theory of nanosized objects properties will be established and new methods of diagnostics in physics, chemistry, biology and medicine using nanoscience approaches will be proposed. New paradigms in physics, chemistry and biology can be expected.

Acknowledgements

The authors are most grateful to their colleagues from the Laboratory of Low Temperature Chemistry for help and fruitful discussions.

References

[1] M. Moskovits, G.A. Ozin (Eds.), Cryochemistry, Wiley-Interscience, NY, 1976, p. 594.
[2] G.B. Sergeev, V.A. Batyuk, Cryochemistry, Mir, Moscow, 1986, p. 321.
[3] G.B. Sergeev, V.V. Smirnov, Molecular Halogenation of Olefines, Moscow State University, Moscow, 1985, p. 240 (in Russian).
[4] K.J. Klabunde, Chemistry of Free Atoms and Particles, Academic Press, NY, 1980.
[5] G.B. Sergeev, Reactions in solid low temperature co-condensates, Mol. Cryst. Liq. Cryst. 313 (1998) 155–166.
[6] Yu.I. Petrov, Clusters and Small Particles, Nauka, Moscow, 1986.
[7] U. Kreibig, M. Volmer, Optical Properties of Metal Clusters. Springer, Berlin, 1995, p. 534.

[8] G.B. Sergeev, Cryochemistry of nanosize metal particles, Chemical Physics on the Border of XX1 Century: To the 100 Anniversary Academician Semenov, Nauka, Moscow, 1996, pp. 149–166 (in Russian).

[9] M.A. El-Sayed, Size dependence of gaseous cluster reactivity and evaporation dynamics as a mechanistic probe, J. Phys. Chem. 95 (1991) 3898–3906.

[10] G.B. Sergeev, M.Yu. Efremov, Size effects in reactions of solid organic compounds with metal particles, Mol. Cryst. Liq. Cryst. 278 (1996) 17–25.

[11] M.Yu. Efremov, A.F. Batsulin, G.B. Sergeev, Fast crystallization of amorphous water during the growth of a co-condensate film, Mendeleev Commun. (1999) 7–8.

[12] L. Andrews, M. Moskovits (Eds.), Chemistry and Physics of Matrix-Isolated Species, Elsevier, New York, 1989, p. 497.

[13] M.Ya. Melnikov, V.I. Pergushov, N.Yu. Osokina, Active intermediates onto silica surface. Alternative matrix isolation method and its application to study of radical reaction mechanisms, in: T. Miyazaki (Ed.), Proceedings of the Conference on Low temperature Chemistry, 1999, pp. 39–40.

[14] I.U. Goldschleger, A.V. Akimov, E.A. Misochko, C.A. Whight, Infrared and ESR spectroscopic studies of 2-C_2H_2F and 1-C_2H_2F radicals isolated in solid argon, J. Mol. Spectrosc. 205 (2001) 269–279.

[15] J.S. Ogden, A.J. Rest, R.L. Sweany, The infrared characterisation of interactions between dihydrogen and alkali halides in argon matrices: evidence for a dominant anion interaction, J. Phys. Chem. 99 (1995) 8485.

[16] M.A. Almond, A.J. Downs, Spectroscopy of matrix-isolated species, Advances in Spectroscopy, vol. 17, Wiley-Interscience, NY, 1987.

[17] J.A. Howard, R. Sutcliffe, J.S. Tse, H. Dahmane, B. Mile, Electron spin resonance spectra of the aluminium trimer in hydrocarbon matrices: a quartet $^4A_2$ state, J. Phys. Chem. 89 (1985) 3595.

[18] W.E. Klotzbuecher, M.A. Petrukhina, G.B. Sergeev, Holmium atoms and small clusters in inert matrices, J. Phys. Chem. A 101 (1997) 4548–4554.

[19] A.V. Nemukhin, A.Yu. Ermilov, M.A. Petrukhina, J. Smets, W.E. Klotzbuecher, Predicting lanthanide clusters properties: a comparison with the observed optical spectra of Ho_2, Spectrochim. Acta A 53 (1997) 1803.

[20] K.D. Bier, T.L. Haslett, A.D. Kirkwood, M. Moskovits, Transition metal diatomics: manganese dimer, manganese dimer ion (+1) and chromium–manganese (CrMn) molecule, J. Chem. Phys. 89 (1988) 6.

[21] S. Fedrigo, W. Harbich, J. Buttet, Optical response of Ag_2, Ag_3, Au_2, and Au_3 in argon matrices, J. Chem. Phys. 99 (1993) 5712.

[22] Z. Hu, B. Shen, Q. Zhou, S. Deosaran, J.R. Lombardi, D.M. Lindsay, W. Harbich, Raman spectra of mass-selected vanadium dimers in argon matrices, J. Chem. Phys. 95 (1991) 2206.

[23] J.C. Miller, L. Andrews, Laser photoluminescence of matrix-isolated CaMg, SrMg and SrCa in solid argon at 12 K, J. Am. Chem. Soc. 100 (1978) 6956.

[24] W.E. Klotzbuecher, G.A. Ozin, selective photoaggregation of metal atoms to small bimetallic clusters of known size. The chromium–molybednum system Cr_nMo_m, J. Am. Chem. Soc. 100 (1978) 2262.

[25] F. Galan, M. Fouassier, M. Tranquille, J. Mascetti, I. Papai, CO_2 coordination to nickel atoms: matrix isolation and density functional studies, J. Phys. Chem. 101 (1997) 2626–2633.

[26] Y.K. Lee, L. Manceron, An IR matrix isolation and DFT theoretical study of the first steps of the Ti(0) ethylene reaction: vinyl titanium hydride and titanacyclopropene, J. Chem. Phys. A 101 (1997) 9650.

[27] Y.K. Lee, Y. Hannachi, C. Xu, L. Andrews, L. Manceron, Infrared spectra of nickel–monoethylene. An infrared argon matrix and normal coordinate analysis study, J. Chem. Phys. 100 (1996) 2964–2970.

[28] K.J. Klabunde, G.H. Jeong, A.W. Olsen, in: J.A. Davies, P.L. Watson, A. Greenberg, J.F. Liebman (Eds.), Selective Hydrocarbon Activation: Principles and Progress, VCH, New York, 1990, p. 433 (Chapter 13).

[29] Z.H. Kafafi, R.H. Hauge, L. Fredin, W.E. Billups, J.L. Margrave, Activation of ethane, propane and cyclopropane by photoexcited iron atoms, J. Chem. Soc. Chem. Commun. (1983) 1230.

[30] S.F. Parker, C.H. Peden, P.H. Barrett, R.G. Pearson, Iron/alkene reactions: a matrix isolation infrared and Mossbauer investigation, Inorg. Chem. 22 (1983) 2813.

[31] F.G.N. Cloke, Zero oxidation state compounds of scandium, yttrium and the lanthanides, Chem. Soc. Rev. (1993) 17.

[32] P.L. Arnold, F.G.N. Cloke, P.B. Hitchcock, The first structurally authenticated zerovalent heteroarene complex of a lanthanide; synthesis and X-ray structure of bis(2,4,6-tri-tert-butyl-phosphorin)holmium (0), J. Chem. Soc. Chem. Commun. 5 (1997) 481–482.

[33] P.L. Arnold, F.G.N. Cloke, K. Khan, P. Scott, Bis(η-heteroarene)titanium complexes derived from 2,4, 6-tri-tert-butylpyridine and 2,4,6-tri-tert-butylphosphorin:conformational preference in solution and redox activity, J. Organomet. Chem. 528 (1997) 77–81.

[34] C. Elschenbroich, J. Hurley, B. Metz, W. Massa, G. Baum, Metal π-complexes of benzene derivatives. Tetraphenylsilane as a chelating ligand: synthesis, structural characterisation and reactivity of the tilted bis-(arene) metal complexes $[(C_6H_5)_2Si(\eta^6 - C_6H_5)_2]M$ (M = V, Cr), Organometallics 9 (1990) 889.

[35] C.G. Francis, S.I. Khan, P.R. Morton, Metal vapor routes to metal–isocianide complexes. Synthesis and molecular structure of tris–(μ-cyclohexyl isocianide)-tris(cyclohexyl isocyanide)triangulo-tripalladium, Inorg. Chem. 23 (1984) 3680.

[36] V.V. Zagorsky, G.B. Sergeev, Cryosynthesis of organometallic compounds in the solid phase, Mol. Cryst. Liq. Cryst. 186 (1990) 81.

[37] G.B. Sergeev, V.V. Zagorsky, F.Z. Badaev, Mechanism of the solid-phase reaction of magnesium with organic halides, J. Organomet. Chem. 243 (1983) 123.

[38] L. Tjurina, G. Barkovskii, A. Kashin, V. Smirnov, I. Beletsskaya, Cryosynthesis of indene–magnesium adduct, in: T. Miyazaki (Ed.), Proceedings of Third International Conference on Low Temperature Chemistry, 1999, pp. 139–140.

[39] K.J. Klabunde, A. Whetten, Metal cluster vs. atom reactivities. Calcium and magnesium vapor with alkyl halides and methane, J. Am. Chem. Soc. 108 (1986) 6529.

[40] V.N. Solov'ev, G.B. Sergeev, A.V. Nemukhin, S.K. Burt, I.A. Topol, IR matrix isolation and ab initio identification of products of the reactions of CH_3Cl and CH_3Br with Mg, J. Phys. Chem. A 101 (1997) 8625–8630.

[41] G.B. Sergeev, V.V. Zagorsky, M.V. Grishechkina, The reduction alcohols to hydrocarbons under their low temperature co-condensation with samarium, Metalorg. Chim. (Russ.) 1 (1988) 1187.

[42] G.B. Sergeev, V.S. Komarov, I.G. Tarkhanova, Catalytic activity of low temperature samarium co-condensates with hydrocarbons, Kinet. Catal. 31 (1990) 209.

[43] G.B. Sergeev, O.V. Zagorskaya, The interaction of carbon dioxide with ethylene and magnesium in presence of magnesium atoms, Vestnik Mosk. Univ. Ser. 2 Khimia (Russ.) 28 (1987) 362–364.

[44] V.N. Solov'ev, E.V. Polikarpov, A.V. Nemukhin, G.B. Sergeev, Matrix isolation and ab initio study of the reactions of magnesium atoms and clusters with CO_2, C_2H_4 and CO_2/C_2H_4 mixtures: formation of cyclic complexes, J. Phys. Chem. A 103 (1999) 6721–6725.

[45] G.B. Sergeev, T.I. Shabatina, V.N. Solov'ev, V.V. Zagorskii, Spectroscopic study of low temperature interactions in metal-organic co-condensates, Spectrochim. Acta 56 (2000) 2527.

[46] G.B. Sergeev, V.V. Smirnov, M.I. Shilina, T.N. Rostovshikova, The mechanism of olefine halogenation and hydrohalogenation in the solid phase, Mol. Cryst. Liq. Cryst. 161 (1988) 201.

[47] M.Yu. Efremov, V.S. Komarov, G.B. Sergeev, Mechanical evolution of the growing organic solid films of acetyl chloride and diethylamine and their specific interaction, Mol. Cryst. Liq. Cryst. 248 (1994) 111.

[48] Y. Finkelstein, R. Moreh, O. Shahal, Out-of-plane orientation of multilayer N_2 films adsorbed on Grafoil at 20 K, Surf. Sci. 437 (1999) 265–276.

[49] A.R.B. Shirazi, K. Knorr, CH_3Cl mono and submonolayers physisorbed on graphite, Surf. Sci. 243 (1991) 303.

[50] K. Madih-Ayadi, S. Rakotozafy, T. Ceva, B. Croset, N. Dupont-Pavlovsky, P. Convert, I. Mirebeau, E. Ressouche, Growth of dichlormethane phisisorbed on graphite. Thermodynamic and structural characterization, Surf. Sci. 436 (1999) 99–106.

[51] A. Bah, N. Dupont-Pavlovsky, X. Duval, Adsorption hysteresis and 2D phase nucleation of chloroform on graphite, Surf. Sci. 352 (1996) 518.

[52] D. Fuhrmann, R. Gerlach, H.G. Rubahn, Ch. Woll, Structure and phase transition of ultrathin films of alkanes adsorbed on Cu(1 1 1), Surf. Sci. 424 (1999) 145–154.

[53] M.A. Reed, C. Zhou, C.J. Muller, T.P. Burgin, J.M. Tour, Conductance of a molecular junction, Science 278 (1997) 252.

[54] A.A. Dhirani, P.H. Lin, P. Guyot-Sionnest, R.W. Zehner, L.R. Sita, Self-assembled molecular rectiniers, J. Chem. Phys. 106 (1997) 5249.

[55] V.K. Gupta, J.J. Skaife, T.B. Dubrovsky, N.L. Abbout, Optical Amplication of liquid receptor binding using liquid crystals, Science 279 (1998) 2077.

[56] X.Y. Zhu, T. Vondrak, H. Wang, C. Gahl, K. Ishioka, M. Wolf, Photo-induced electron transfer to molecular quantum structures on a metal surface, Surf. Sci. 451 (2000) 244–249.

[57] K. Oura, T. Hanawa, LEED–AES study of the gold–silicon system (1 0 0), Surf. Sci. 82 (1979) 202.

[58] G. Yang, J.H. Kim, S. Yang, A.H. Weiss, Depth profile analysis of surfaces produced by annealing ultra-thin films of Au deposited on Si(1 0 0), Surf. Sci. 367 (1996) 45–55.

[59] F. Komori, K. Kushida, K. Hottori, S. Arai, T. Limori, Grown of thin gold islands on Ge(0 0 1)-2*1 surfaces below room temperature, Surf. Sci. 438 (1999) 123–130.

[60] F. Moresco, M. Rocca, T. hildebrant, M. Henzler, Growth of ultrathin nanostructured Ag films on Si(1 1 1): a SPA-LEED study, Surf. Sci. 463 (2000) 22.

[61] K. Hattori, Y. Takahashi, T. Iivori, F. Komori, Low temperature STM/STS studies of Ag/Ge(1 0 0) surfaces: growth coulomb blockade and superconductivity, Surf. Sci. 357 (1996) 361.

[62] T. Bornemann, A. Otto, W. Heuer, H. Zacharias, Second harmonic generation by cold-desposited silver films, Surf. Sci. 420 (1999) 224–232.

[63] K.A. Edwards, P.B. Howes, J.E. Macdonald, T. Hibma, T. Bootsma, M.A. James, Observation of a structural transition during the low-temperature growth of the Si(1 1 1) 7*7—Pb interface, Surf. Sci. 424 (1999) 169–178.

[64] F. Bocquest, S. Robert, S. Gauthier, J.L. Duvault, J. Klein, On the low-temperature growth of Pb on Cu(1 0 0), Surf. Sci. 392 (1997) 86–102.

[65] L. Porte, M. Phaner-Goutorbe, J.M. Guigner, J.C. Bertolini, Structuring and catalytic activity of palladium thin layers deposited on the Ni(1 1 0) surface, Surf. Sci. 424 (1999) 262–270.

[66] D. Purdie, M. Hengsberger, M. Garnier, Y. Baer, An outlook for high-resolution spectroscopy of solid at low temperature, Surf. Sci. 407 (1999) L671–L675.

[67] H. Patek, M.J. Weida, H. Nagano, S. Ogawa, Electronic relaxation of alkali metal atoms on the Cu(1 1 1) surface, Surf. Sci. 451 (2000) 22–30.

[68] B. Heinrich, M. Kowalewski, J.F. Cochran, Engineering of magnetic properties using ultrathin metallic structures prepared by molecular beam epitaxy, Can. J. Chem. 76 (1998) 1595–1605.

[69] L. Abelmann, S. Porthun, M. Haast, C. Lodder, A. Moser, M.E. Best, P.J.A. Van Schendel, B. Steifel, H.J.

Hug, G.P. Heydon, A. Farley, S.R. Hoon, T. Pfaffelhuber, R. Proksch, K. Babcock, Comparing the resolution of magnetic force microscopes using the CAMST reference samples, J. Magn. Magn. Mater. 190 (1998) 135–147.

[70] E. Heifets, Y.F. Zhukovskii, E.A. Kotomin, M. Causa, The adhesion nature of the Ag/MgO(100) interface, Chem. Phys. Lett. 283 (1998) 395.

[71] M.H. Schaffner, F. Patthey, W.D. Schneider, Growth study of silver on MgO(100)/Mo(100), Surf. Sci. 417 (1998) 159–167.

[72] T. Kodas, M. Hampden-Smith, The Chemistry of Metal CVD, VCH, Weinheim, 1994, p. 530.

[73] S. Sato, Y. Ukisu, H. Ogawa, Y. Takasu, Infrared reflection absorption spectroscopy, X-ray photoelectron spectroscopy and temperature programmed desorption study on the adsorption and desorption of $Fe(CO)_5$ over silver surface, J. Chem. Soc. Faraday Trans. 89 (1993) 4387.

[74] L. Sun, E.M. McCash, Adsorbtion and decomposition of $Fe(CO)_5$ on Cu(111), Surf. Sci. 422 (1999) 77–86.

[75] B.V. Andryushechkin, K.N. Eltsov, V.M. Shelyuga, C. Tarducci, B. Cortigiani, U. Bardi, A. Atrei, Epitaxial growth of AgCl layers on the Ag(100) surface, Surf. Sci. 421 (1999) 27–32.

[76] R.J.M. Vullers, M. Ahlskog, C. Van Haesendock, Titanium nanostructures made by local oxidation with the atomic force microscope, Appl. Surf. Sci. 144–145 (1999) 584–588.

[77] M. Tabe, T. Yamamoto, Y. Terao, Nitridation and subsequent oxidation process of Si(111) and (100) surfaces for high-density Si pillar formation, Appl. Surf. Sci. 117 (1997) 131–135.

[78] S.K. Jo, B. Gong, G. Hess, J.M. White, J.G. Ekerdt, Low-temperature Si(100) etching: facile abstraction of $SiH_3(a)$ by thermal hydrogen atoms, Surf. Sci. 394 (1997) L162–L167.

[79] C. Akerlund, I. Zoric, B. Kasemo, Displacement of O_2 from Pt(111) by incident CO: dynamics and kinetics, Surf. Sci. 418 (1998) 543–554.

[80] C.C. Johnson, E.T. Jensen, Photodissociation of CH_3I on Cu(110)-I: surface effects on dissociation pathway, Surf. Sci. 451 (2000) 261–266.

[81] W. Zhao, C. Kim, J.M. White, Photodissociation of t-butyl nitrite on Ag(111): rotational and electronic-state distribution of the $NO(X^2\Pi)$ photofragments, Surf. Sci. 451 (2000) 267–275.

[82] Y.M. Sun, D. Sloan, H. Ihm, J.M. White, Electron induced surface chemistry: production and characterization of NH_2 and NH species on Pt(111), J. Vac. Sci. Technol. A 14 (1996) 1516.

[83] C. Bater, M. Sanders, J.H. Craig Jr., Electron-simulated dissociation of ammonia adsorbed on Ge(100), Surf. Sci. 451 (2000) 226–231.

[84] G. Wu, D. Staechiola, M. Kalchev, W.T. Tysoe, The adsorption and reaction of 2-iodethanol on Ag(111), Surf. Sci. 463 (2000) 81.

[85] S. Bougreois, R. Gouttebraon, M. Perderaeu, A static SIMS study of superficial reactions $(O_2,(CN)_2)$ on silver, Surf. Sci. 395 (1998) 356–362.

[86] G. Scott Jones, M.A. Barteau, J.M. Vohs, The formation of ether via the reaction of iodoethane with atomic oxygen on the Ag(110) surface, Surf. Sci. 420 (1999) 65–80.

[87] S.L. Silva, R.M. Lemor, F.M. Leibsle, STM studies of methanol oxidation to formate on Cu(110) surfaces: II. Codosing experiments, Surf. Sci. 421 (1999) 146–156.

[88] J.H. Larsen, I. Chorkendorff, From fundamental studies of reactivity on single crystals to design of catalysts, Surf. Sci. Rep. 35 (1999) 163–222.

[89] A. Dinger, C. Lutterloh, J. Biener, J. Kuppers, Reaction of hydrogen atoms with acetone monolayers adsorbed on graphite monolayer covered Pt(111) surfaces, Surf. Sci. 437 (1999) 116–124.

[90] K.J. Klabunde, Free Atoms, Clusters, and Nanoscale Particles, Academic Press, San Diego, 1994, p. 311.

[91] G.B. Sergeev, Modern Trends in Low Temperature Chemistry, Moscow State University, 1994.

[92] K. Klabunde, C. Gardenas-Trivino, Metal atom/vapor approaches to active metal cluster/particles, in: A. Furster (Ed.), Active Metals Preparation, Characterization, Applications, VCH, Weinheim, 1996, pp. 237–275.

[93] K. Kimura, Metal colloids produced by means of gas evaporation technique. IV. Size distribution of small Mg and In particles, Bull. Chem. Soc. Jpn. 60 (1987) 3093.

[94] G.N. Glavee, K. Eason, K.J. Klabunde, C.M. Sorensen, G.C. Hadjipanayis, Clusters of immisible metals. 2. Magnetic properties of iron–lithium bimetallic particles, Chem. Mater. 4 (1992) 1360.

[95] A.D. Pomogailo, A.S. Rosenberg, U.E. Uflyand, Nano-size Metals in Polymers, Khimia, Moscow, 2000, p. 672.

[96] K.J. Klabunde, D. Zhang, G.N. Glavee, C.M. Sorensen, Encapsulated nanoparticles in iron metal, Chem. Mater. 6 (1994) 784.

[97] H. Takaoka, I. Yamada, T. Takagi, Ionised-cluster beam deposition of optical-interference coatings, J. Vac. Sci. Technol. A36 (1985) 2665.

[98] H.-V. Roy, P. Fayet, F. Patthey, W.-D. Schneider, B. Delley, C. Massobrio, Evolution of electronic and geometric structure size-selected Pt and Pd clusters on Ag(110) observed by photoemission, Phys. Rev. B 49 (1994) 5612.

[99] N. Sun, K.J. Klabunde, High activity solid super base catalysts employing nanocrystals of metal oxides: isomerization and alkylation catalysis, including conversion of propylene–ethylene mixtures to pentenes and heptenes, J. Catal. 185 (1999) 506.

[100] G.N. Clavee, C.F. Kernizan, K.J. Klabunde, C.M. Sorensen, G.C. Hadjipanayis, Clusters of immiscible metals. Iron–lithium nanoscale bimetallic particle synthesis and behavior under thermal and oxidative treatments, Chem. Mater. 3 (1991) 967.

[101] K.J. Klabunde, Solvated atoms of platinum, palladium and gold. Precursors to colloids, films and catalysts, Plat. Metals Rev. 36 (1992) 80.

[102] E.M. Egorova, A.A. Revina, Synthesis of metallic nanoparticles in reverse micelles in the presence of quercitin, Colloids Surf., A: Physicochem. Eng. Aspects (1999) 2235.

[103] B.G. Ershov, A. Heinglein, Radiolytic formation of colloidal tin and tin–gold particles in aqueous solution, J. Phys. Chem. 97 (1993) 3436.

[104] C.A. Foss, G.L. Hornyak, J.A. Stockert, C.R. Martin, Optical properties of composite membranes containing arrays of nanoscopic gold cylinders, J. Phys. Chem. 96 (1992) 7497.

[105] F. Henglein, A. Henglein, P. Mulvaney, Surface chemistry of colloidal gold: deposition and reoxidation of Pb, Cd, and Tl, Ber. Bunsenges. Phys. Chem. 98 (1994) 180.

[106] V.I. Bukhtiyarov, Chemical reactivity of metal clusters, in: M.W. Roberts (Ed.), Interfacial Science, A Chemistry for the 21st Century' Monograph, 1999, p. 109.

[107] G.B. Sergeev, A.V. Nemukhin, B.M. Sergeev, T.I. Shabatina, V.V. Zagorskii, Cryoshytesis and properties of metal-organic nanomaterials, Nanostruct. Mater. 12 (1999) 1113–1116.

[108] T.N. Boronina, I. Lagadic, G.B. Sergeev, K.J. Klabunde, Activated and nonactivated zinc powder: reactivity towards chlorocarbons in water and AFM studies of surface morphology, Environ. Sci. Techn. 32 (1998) 2614.

[109] J. Fendler (Ed.), Nanoparticles and Nanostructured Films: Preparation Characterisation and Applications, Wiley-VCH, New York, 1998, p. 461.

[110] A.S. Edelstein, R.C. Cammarata (Eds.), Nanomaterials: Synthesis, Properties and Applications, Inst. of Physics Publ., Bristol-Philadelphia, 1998, p. 596.

[111] A.P. Alivisatos, Semiconductor clusters nanocrystals, and quantum dots, Science 271 (1996) 933.

[112] G.B. Sergeev, M.A. Petrukhina, Encapsulation of small metal particles in solid organic matrices, Progr. Solid State Chem. 24 (1996) 183–211.

[113] G.B. Sergeev, V.V. Zagorsky, M.A. Petrukhina, Nanosize metal particles in poly (p-xylylene) films obtained by low temperature codeposition, J. Mater. Chem. 5 (1995) 31.

[114] V.V. Zagorskii, V.E. Botchenkov, S.V. Ivashko, G.B. Sergeev, Cryochemical synthesis and physical–chemical properties of nano-dispersed metallopolymers, Nanostruct. Mater. 12 (1999) 863.

[115] V.V. Zagorskii, V.E. Botchenkov, S.V. Ivashko, G.B. Sergeev, Electric conductivity of organic films containing nanosize metal particles, Mater. Sci. Eng. 8–9 (1999) 329.

[116] G.B. Sergeev, V.V. Zagorsky, M.A. Petrukhina, S.A. Zav'yalov, E.I. Grigor'ev, L.I. Trakhtenberg, Preliminary study of the interaction of metal nanoparticle-containing poly-p-xylylene films with ammonia, Analyt. Commun. 34 (1997) 113.

[117] V.E. Bochenkov, N. Stephan, L. Brehmer, V.V. Zagorskii, G.B. Sergeev, Sensor activity of thin poly-p-xylylene films containing nanosize lead particles, Colloids Interf., in press.

[118] T.I. Shabatina, E.V. Vovk, N.V. Ozhegova, A.V. Nemukhin, Yu.N. Morosov, G.B. Sergeev, Synthesis and properties of metal-mesogenic nanostructures, Mater. Sci. Eng. 8–9 (1999) 53–56.

[119] T.I. Shabatina, E.V. Vovk, Y.N. Morosov, V.A. Timoshenko, G.B. Sergeev, Spectroscopic study of silver-containing mesogenic cyanobiphenyls in solid phase, Mol. Cryst. Liq. Cryst. 356 (2001) 143–148.

[120] N.V. Ozhegova, A.V. Nemukhin, T.I. Shabatina, G.B. Sergeev, Modeling structure and spectra of silver complexes in condensate films at polar liquid crystals, Mendeleev Commun. 23 (1998) 218.

[121] T.I. Shabatina, E.V. Vovk, Yu.N. Morosov, V.A. Timoshenko, G.B. Sergeev, Thermal behavior of silver-containing mesogenic cyanobiphenyl films, Colloids Interf., in press.

[122] L.M. Bronstein, D.M. Chernyshov, P.M. Valetsky, E.A. Wilder, R.J. Spontak, Metal nanoparticle grown in the nanostructured matrix of poly(octadecylsiloxane), Langmuir 22 (2000) 1.

[123] V.I. Roldugin, Quantum-size colloid metal systems, Russ. Chem. Rev. 69 (2000) 899.

[124] C.N.R. Rao, G.U. Kulkarni, P.J. Thomas, P.P. Edwards, Metal nanoparticles and their assemblies, Chem. Soc. Rev. 29 (1) (2000) 27–35.

[125] X.M. Lin, C.M. Sorensen, K.J. Klabunde, Ligand-induced gold nanocrystal superlattice formation in colloidal solutions, Chem. Mater. 11 (1999) 198.

[126] F. Tian, K.J. Klabunde, Nonaqueous gold colloids. Investigation of deposition and film growth on organically modified substrates and trapping of molecular gold clusters with an alkyl amine, New J. Chem. (1998) 1275.

[127] S. Fullam, S.N. Rao, D. Fitzmaurice, Noncovalent self-assembly of silver nanocrystal aggregates in solution, J. Phys. Chem. B 104 (26) (2000) 6164.

[128] A. Kumar, A.B. Mandale, M. Sastry, Sequential electrostatic assembly of amine-derivatized gold and carboxylic acid-derivatized silver colloidal particles on glass substrates, Langmuir 16 (17) (2000) 6921–6926.

[129] A.K. Boal, F. Ilhan, J.E. DeRouchey, T. Thurn-Albrecht, T.P. Russell, V.M. Rotello, Self-assembly of nanoparticles into structured spherical and network aggregates, Nature (London) 404 (6779) (2000) 746.

[130] B. Nikoobakht, Z.L. Wang, M.A. El-Sayed, Self-assembly of gold nanorods, J. Phys. Chem. B 104 (36) (2000) 8635.

[131] L. Motte, E. Lacaze, M. Maillard, M.P. Pileni, Self-assemblies of silver sulfide nanocrystals on various substrates, Langmuir 16 (8) (2000) 3803.

[132] J.F. Kang, J. Zaccaro, A. Ulman, A. Myerson, Nucleation and growth of glycine crystals on self-assembled monolayers on gold, Langmuir 16 (8) (2000) 3791.

[133] H. Fukushima, S. Seki, T. Nishikawa, H. Takiguchi, K. Tamada, K. Abe, R. Colorado Jr., M. Graupe, O.E. Shmakova, T.R. Lee, Microstructure, wettability, and thermal stability of semifluorinated self-assembled monolayers (SAMs) on gold, J. Phys. Chem. B 104 (31) (2000) 7417.

[134] L. Maya, E.A. Kenik, Polymer-mediated assembly of gold nanoclusters, Langmuir 16 (2000) 9151.

[135] P.M. Tessier, O.D. Velev, A.T. Kalambur, J.F. Rabolt, A.M. Lenhoff, E.W. Kaler, Assembly of gold nanostructured films templated by colloidal crystals and use in surface-enhanced raman spectroscopy, J. Am. Chem. Soc. 122 (39) (2000) 9554.

[136] J.J. Storhoff, A.A. Lazarides, R.C. Mucic, C.A. Mirkin, R.L. Letsinger, G.C. Schatz, What controls the optical properties of DNA-linked gold nanoparticle assemblies, J. Am. Chem. Soc. 122 (19) (2000) 4640.

[137] B.M. Sergeev, G.B. Sergeev, Y.J. Lee, A.N. Prusov, V.A. Polyakov, Cryochemical synthesis of silver organosols in methyl acrilate, Mendeleev Commun. (1997) 151.

[138] B.M. Sergeev, G.B. Sergeev, A.N. Prusov, Cryochemical synthesis of bimetallic nanoparticles in the silver–lead–methylacrylate system, Mendeleev Commun. (1998) 1.

[139] M.V. Kiryuhin, B.M. Sergeev, A.N. Prusov, V.G. Sergeev, Formation of nonspherical silver nanoparticles by photochemical reduction of silver cations in presence of a partially decarboxylated poly(acrylic) acid, Polym. Sci. Ser. B 42 (2000) 324.

[140] Ch.A. Mirkin, T.A. Taton, Semiconductors meet biology, Nature 405 (2000) 626.

[141] S.R. Whaley, D.S. English, E.L. Hu, P.F. Barbara, A.M. Belcher, Selection of peptides with semiconductor binding specificity for directed nanocrystal assembly, Nature (London) 405 (6787) (2000) 665.

[142] C.F. Quarte, Nanoscience and engineering: the next five years, J. Nanopart. Res. No. 1 (1999) 131.

[143] A.L. Buchachenko, Chemistry on the border of two centuries: achievements and prospects, Russ. Chem. Rev. 68 (2) (1999) 99.

[144] H. Gleiter, Nanostructured materials: basic concepts and microstructures, Acta Mater. 48 (N1) (2000) 1–29.

Biological surface science

Bengt Kasemo *

Department of Applied Physics, Chalmers University of Technology and Göteborg University, S-412 96 Göteborg, Sweden

Received 13 December 2000; accepted for publication 5 July 2001

Abstract

Biological surface science (BioSS), as defined here is the broad interdisciplinary area where properties and processes at interfaces between synthetic materials and biological environments are investigated and biofunctional surfaces are fabricated. Six examples are used to introduce and discuss the subject: Medical implants in the human body, biosensors and biochips for diagnostics, tissue engineering, bioelectronics, artificial photosynthesis, and biomimetic materials. They are areas of varying maturity, together constituting a strong driving force for the current rapid development of BioSS. The second driving force is the purely scientific challenges and opportunities to explore the mutual interaction between biological components and surfaces.

Model systems range from the unique water structures at solid surfaces and water shells around proteins and biomembranes, via amino and nucleic acids, proteins, DNA, phospholipid membranes, to cells and living tissue at surfaces. At one end of the spectrum the scientific challenge is to map out the structures, bonding, dynamics and kinetics of biomolecules at surfaces in a similar way as has been done for simple molecules during the past three decades in surface science. At the other end of the complexity spectrum one addresses how biofunctional surfaces participate in and can be designed to constructively participate in the total communication system of cells and tissue.

Biofunctional surfaces call for advanced design and preparation in order to match the sophisticated (bio) recognition ability of biological systems. Specifically this requires combined topographic, chemical and visco-elastic patterns on surfaces to match proteins at the nm scale and cells at the micrometer scale. Essentially all methods of surface science are useful. High-resolution (e.g. scanning probe) microscopies, spatially resolved and high sensitivity, non-invasive optical spectroscopies, self-organizing monolayers, and nano- and microfabrication are important for BioSS. However, there is also a need to adopt or develop new methods for studies of biointerfaces in the native, liquid state.

For the future it is likely that BioSS will have an even broader definition than above and include native interfaces, and that combinations of molecular (cell) biology and BioSS will contribute to the understanding of the "living state". © 2001 Elsevier Science B.V. All rights reserved.

Keywords: Adhesion; Biological compounds; Biological molecules – nucleic acids; Biological molecules – proteins; Solid–liquid interfaces

1. Introduction

Imagine the following six situations where the properties of solid surfaces are or may become important in practice:

---

* Tel.: +46-317723370; fax: +46-317723134.
 E-mail address: kasemo@fy.chalmers.se (B. Kasemo).

(i) A patient with severely degraded dental status is treated by a surgical procedure, where one or several dental implants, made from metal or ceramic, are implanted into the jawbone so that they after some healing in period can function as artificial teeth (Fig. 1). Other types of implants are shown in Fig. 2.

(ii) A blood or urine sample is distributed over a suitably prepared biosensor or biochip surface in order to diagnose a patient's health status or alternatively for a forensic identification (Fig. 3).

(iii) A small number of living cells of a particular kind, maybe a tissue culture or so-called stem cells for a certain type of human tissue, are placed in a scaffold made from some synthetic material, with the intention to make the cells grow in number and differentiate ex vivo into a new functional tissue, which later is placed in a patient, in order to repair a lost or degraded body function (Fig. 4).

(iv) Neural cells (neurons) are placed on a micropatterned surface, where they self-organize into a functioning neural network, which can be addressed chemically and/or electronically by in–out (I/O) connections so that a cell-based bioelectronic circuit is achieved.

(v) A particular kind of photosensitive, charge transfer proteins are attached to and organized on a specially designed material surface in order to harvest the energy of sun light with a high efficiency, thereby converting the light into chemical or electrical energy in a process mimicking the photosynthetic process of green plants or certain bacteria.

(vi) A surface is microfabricated with an array of specially architectured, soft protrusions in the micrometer size range, mimicking a shark or dolphin skin and thereby providing a dramatic reduction in hydro- or aerodynamic friction.

The *medical implant* example [1,2] is a clinical reality since many years and hundreds of thousands of patients have been treated with a major increase in life quality. There are many other examples of medical implants (Fig. 3), each one uniquely different from the others in the biological/clinical details e.g. artificial hip and knee joints, artificial blood vessels and heart valves, and synthetic intraocular lenses. Together they represent a large number of industries with total turnover approaching or even exceeding a hundred billion dollars per year.

The *biosensor* example [3–5] is representing an area, where commercial products already exist both for single sensors and array type sensors [6] (e.g. so-called DNA chips) which are used clinically or at the R&D level in biomedical industries and academia. Figs. 4 and 5 illustrate some of the current sensing principles and surface immobilization strategies. This area is in extremely rapid

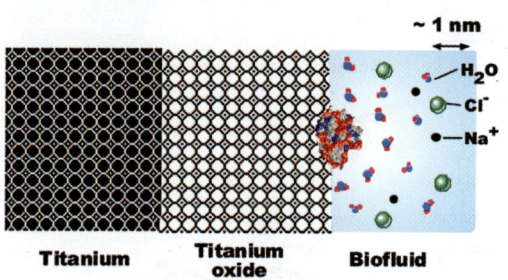

Fig. 1. Schematic illustration of the interface between a dental implant and the jawbone into which it is implanted, at different magnifications. After the surgical procedure the surface is first exposed to water, then to proteins, and eventually to cells (see Fig. 7). See Fig. 2 for other types of implants.

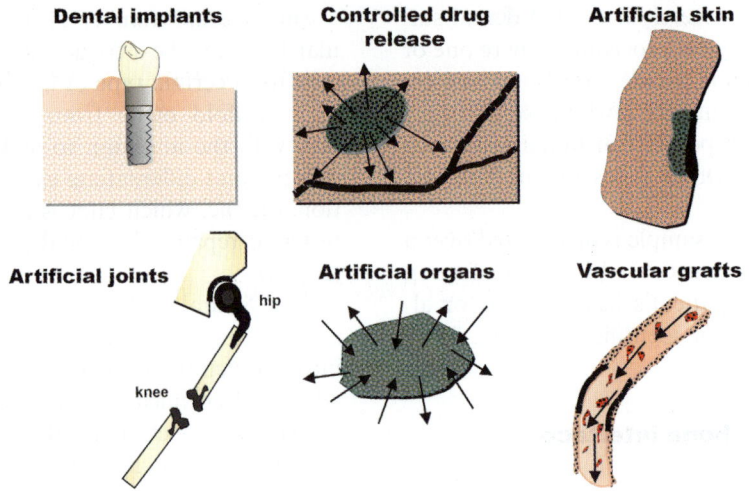

Fig. 2. A selected set of different medical implants.

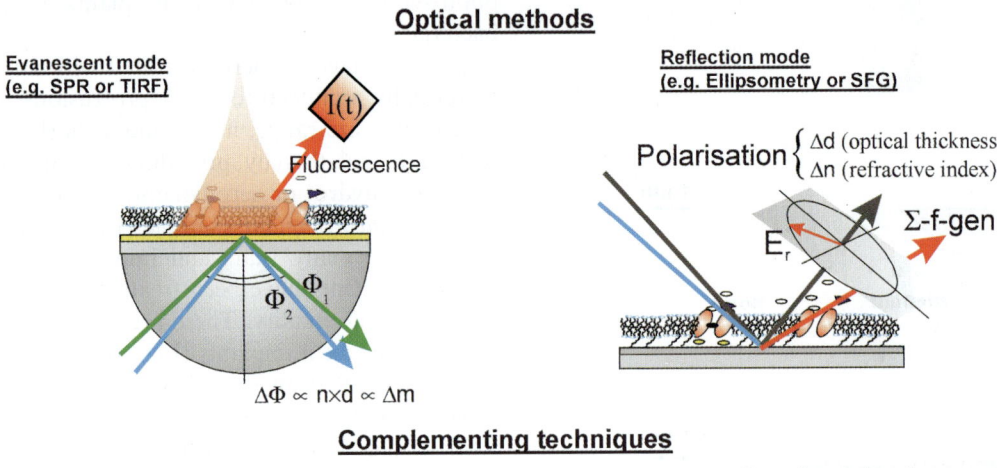

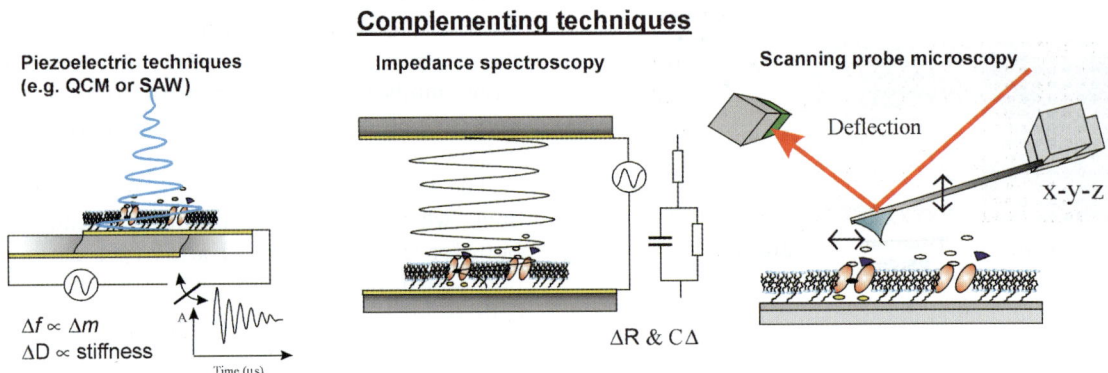

Fig. 3. Some common detection principles for biosensors and biochips.

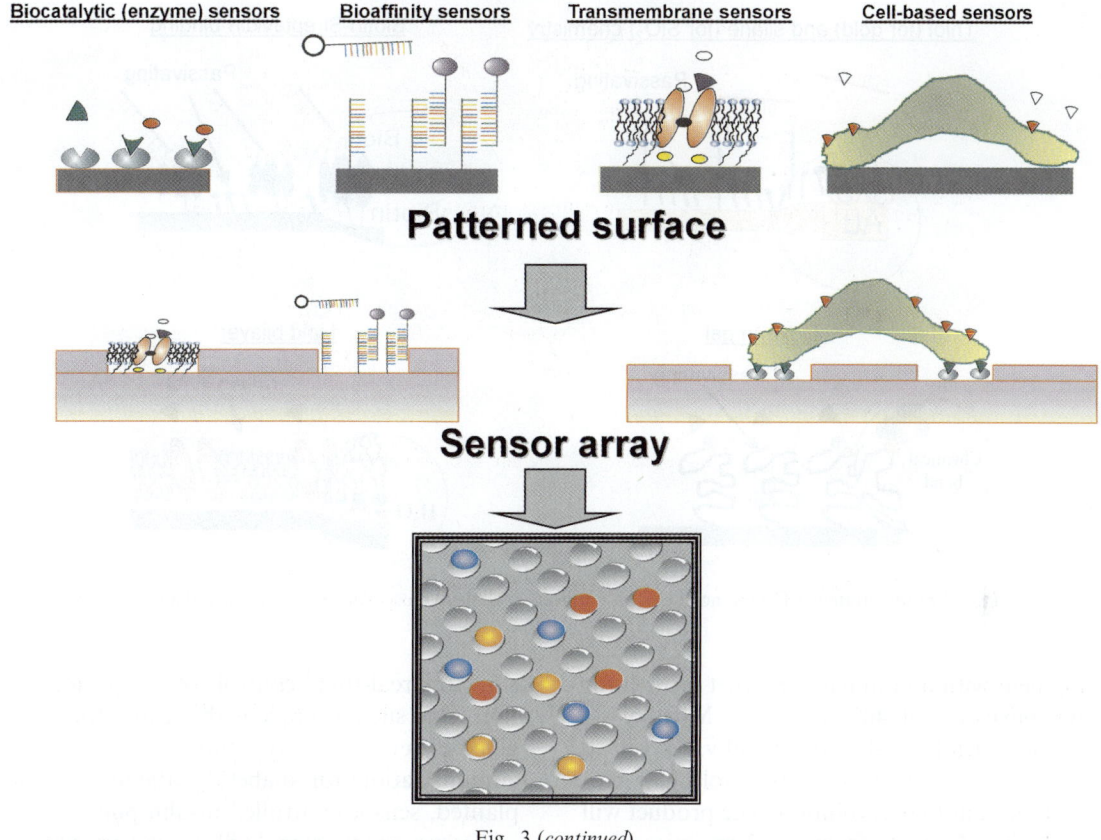

Fig. 3 (*continued*).

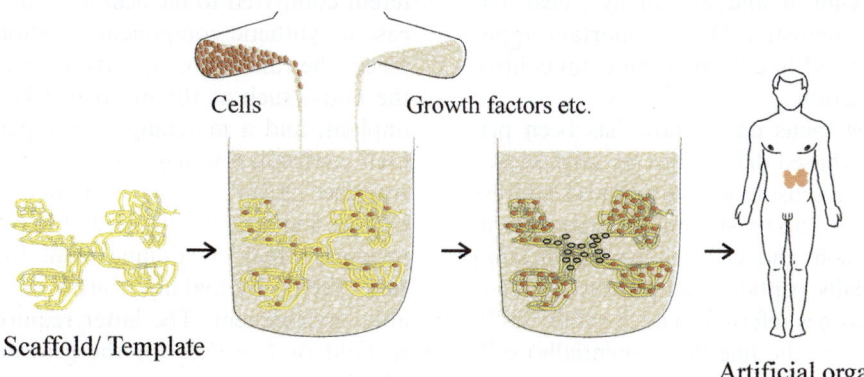

Fig. 4. Tissue engineering scaffold. A template of a synthetic material forms the "host" for a cell or tissue culture. The scaffold should promote the development of the culture into a functioning tissue.

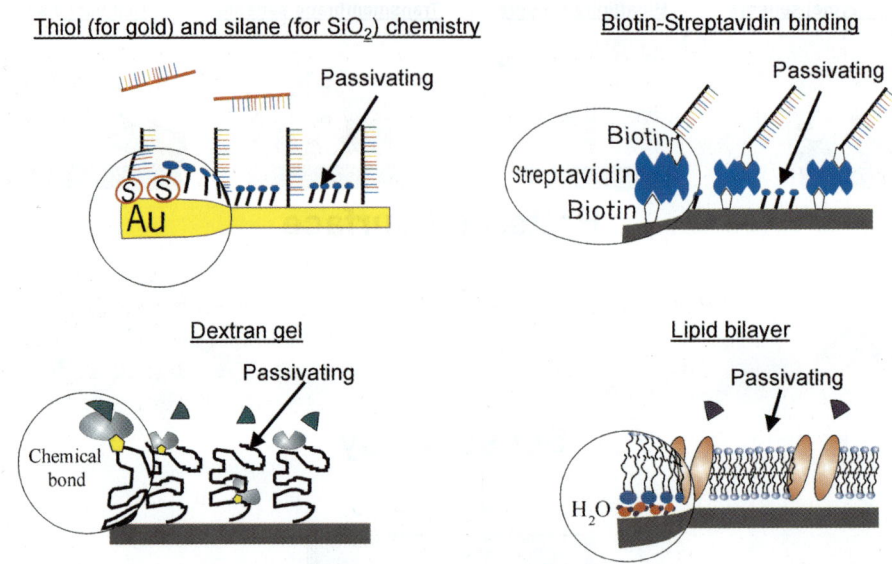

Fig. 5. Immobilization strategies for biomolecules on surfaces to achieve biosensor and biomedical diagnostic functions.

development with an estimated market growth for labchips/bioarrays of 40% per year. Most of its impact, technically and commercially, is yet to come in the wake of the so-called molecular biology or biotechnology revolution. The product will have important input from modern microelectronics, materials and surface science, advanced micro- and nanofabrication, and bioinformatics. The perspective for the future is unprecedented precision and speed in multiple (broad band), and individualized clinical and eventually "near patient" (home) diagnostics. Other important applications are expected in e.g. environmental control and food production.

Although the focus up till now has been primarily on DNA-based diagnostics by large sensor arrays, the future sensing arrays will also address biologically more "down stream" components in the cell expression and cell–cell communication processes, typically peptides and proteins (the latter area is sometimes referred to as "proteomics"). Even further down the line lies (potentially) cell-based diagnostics (cellomics).

Another type of sensors is the kind that can be implanted and used in feed back systems in vivo e.g. for real-time control of drug administration. One such example is the search for a reliable glucose sensor for real-time control of insulin administration for diabetes patients, via an implanted, sensor controlled insulin pump.

Tissue engineering [7,8] is an emerging area where most of the impact is also still to come, very much fueled by the molecular and cell biology revolution. The basic strategy for repair of lost body functions by way of tissue engineering is different compared to medical implants: In the latter case a synthetic component is fabricated that replaces the function of the lost or degraded part of the body, such as the metal ball head on the hip implant, and a matching counterpart made from high molecular weight polyethylene (HMWPE), constituting an artificial hip joint. In the tissue engineering strategy the idea (Fig. 4) is instead to produce a real functioning tissue from a cell culture "seed", by making it grow ex vivo in a suitable environment. The latter requires a synthetic scaffold or template for the growing cells/tissue, whose macroscopic geometry, microscopic topography and surface chemical properties cooperate with other stimuli such as extracellular signal substances and the composition of the physiological solution etc., to make the cell culture develop into a functioning tissue, ready for implantation. Practical tissue engineering is today realized only for a few "simple" tissues like skin and, to some extent, cartilage, while large research efforts are focussed on e.g. liver, pancreas, blood vessel, bladder, and heart valve tissues.

Bioelectronics [9–11] is still only at the very early, explorative research stage and it is too early to judge if it will or will not become a practical reality in the future. The basic idea is to take advantage of some unique aspects of nature's own way of information storage and processing. This can alternatively be done at the cellular level (sometimes referred to as "reverse engineering"; see below), where the primary building stones are networks of real cells, or by "forward engineering" at the biomolecular level, where the building stones are peptides, proteins, DNA strings etc., and combinations of them. Both in the cellular and molecular approach the biological components are likely to be steered into a suitable pattern in or on a microengineered template or substrate surface, providing chemical and/or electrical I/O communication. Bioelectronic components and circuits made by the molecular level, forward engineering approach are likely to resemble today's microelectronics circuits in some respects, but will also have new unique characteristics originating from the biomolecules' highly specialized functions, including their replication and recognition ability, and self-organization. The cell level, reverse engineering approach will result in entirely new circuit, component, and functional concepts. The idea is to "automatically" achieve signal- and information-storage pathways from the internal self-organization, architecture and multiple functions of living cells, and from the biosystem's inter- and intracellular communication pathways. The latter have evolved over billions of years, and thus one does not need to invent these functions and architectures again (the latter is the prime reason for calling this approach reverse engineering).

As in the bioelectronics case, biomimetic or *artificial photosynthesis* [12] is still an area at its infancy. The basic mechanism is known by principle (Fig. 6); it is based on electron excitation by incident, photons in suitable proteins or protein arrays, followed by charge separation (preventing charge recombination and concomitant energy loss), and finally redox reactions creating useful, energy rich (fuel) molecules. However, biomimetic structures that have the quantum efficiency, sustainability, and a self-repair ability approaching those of natural systems, and that can produce useful fuels (like hydrogen) are still just a vision and goal for future research. A strong driving

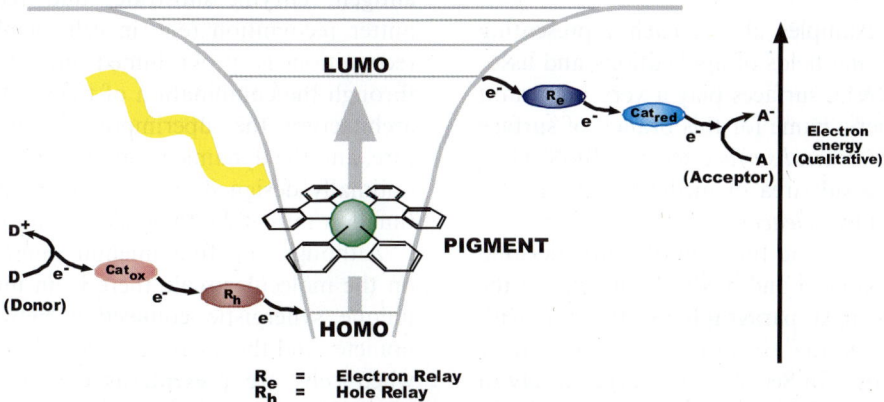

Fig. 6. The principle of photon to electron energy conversion in photosynthesis.

force is the rapidly growing need for sustainable energy systems. One strategy to realize such structures is to immobilize the photon-harvesting and redox protein arrays of real plants, or their synthetic counterparts, on surfaces with suitable chemical and electronic properties. These surfaces need to be functionilized both with respect to nano- and microtopography and with respect to surface chemistry. Another approach is to build entirely synthetic nanostructures mimicking the photosynthetic systems.

The low friction mimic of a sharkskin surface constitutes just one example of *biomimetic materials science* [13]. In biomimetic materials science the idea is to mimic the functional properties of biological materials/components or the processes by which they are manufactured in nature, in order to achieve new, superior materials properties or manufacturing advantages. A major driving force to attempt mimicking the sharkskin is the potentially much lower energy consumption of airplanes and sea vehicles that would be obtained. A more exotic example, already realized in real products, is swimsuits for faster swimming and low friction hull coatings on sailboats. Other biomimetic materials examples are antifreeze materials, utilizing the special features of some water holding proteins that can lower the freezing point of water in certain arctic fishes below 0 °C, and self-cleaning surfaces like some green leaves (the lotus leaf), and self-repairing surfaces.

1.1. Biological surface science

In all six examples above, each representing large and growing fields of applications and associated basic R&D, surfaces play a very important role. As a generic name for this branch of surface science I use *biological surface science* (BioSS) [14] (it includes the sub-area of medical applications; *biomedical surface science*).

The importance and function of synthetic solid surfaces in biological and medical contexts, is the subject of the next paragraph (Section 2), with special focus on the six examples or sub-areas mentioned above. In Section 3 the large variety of model systems available for BioSS studies, of different complexity (from water and amino acids to living cells) are briefly discussed. The same paragraph also discusses different types of surfaces and surface modifications, and analytical and preparation methods for biofunctional surfaces. However, these paragraphs are intentionally very short, since the focus is on general aspects and concepts rather than specific surfaces or methods. Finally some comments are attempted regarding the likely future development of BioSS.

The intention with the article is not to present a comprehensive review of the field, but rather to identify some important trends, directions, and concepts in BioSS and particularly to outline the opportunities for graduate students and surface scientists to contribute to this exciting field. References are therefore chosen primarily to provide starting points for further reading, and there is no ambition to cover all previous work in BioSS (i.e. the reference lists in the quoted references are just as important as the textual content).

2. The role of surfaces

2.1. Biorecognition

In most of the examples above, especially for (i)–(iv), *biorecognition* is a central component. Any attempt to make a sophisticated, functional surface for biointeractions must take into account the highly developed ability of biological systems to recognize specially designed features on the molecular scale. Famous examples are antibody–antigen, enzyme–substrate, and receptor–transmitter recognition (e.g. in cell membranes). The recognition is programmed into the molecules through the combination of their 3D topographic architecture, the superimposed chemical architecture, and the dynamic properties. Consequently an optimally designed surface for specific biological function must take these aspects into account.

Although the fundamental interactions occur on the molecular scale there is an interesting and unique synergistic connection between the nanometer and the micrometer length scales [1,15,16] when cells are present, as e.g. in the cases of medical implants, tissue engineering and cell-based bioelectronics. This is schematically illustrated in

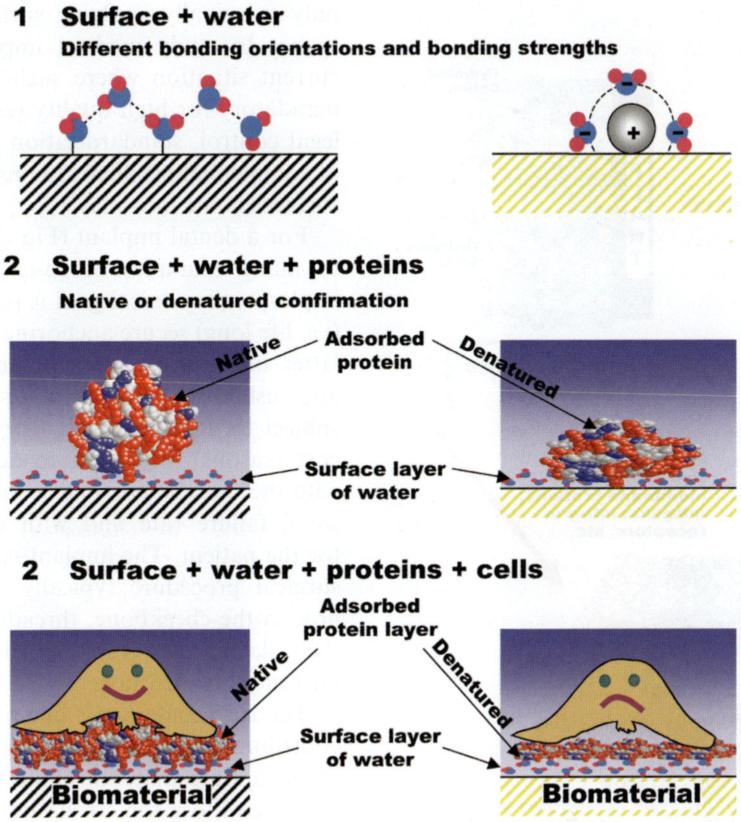

Fig. 7. Schematic illustration of the successive events following after implantation of a medical implant. The first molecules to reach the surface are water molecules (ns time scale). The water shell that is formed affect the protein interaction starting on the micro- to millisecond time scale, and continuing for much longer times. The water shell on the surface affects the protein interaction. Eventually cells reach the surface. Their surface interaction takes place via the protein coating whose properties is determined by the surface and water adlayer properties.

Figs. 7 and 8. The first picture (Fig. 7) shows the sequence of events after a surface has suddenly been placed in a biological milieu containing cells. The first molecules to reach the surface (time scale of order ns) are water molecules. Water is known to interact and bind very differently at surfaces depending on the surface properties (see Section 3). The properties of the surface water "shell" are an important factor influencing proteins and other molecules that arrive a little later. These water-soluble biomolecules also have hydration (water) shells and the interaction between the surface water shell with the biomolecular water shell influences the fundamental kinetic processes and the thermodynamics at the interface. For example, it may determine if proteins denature or not, their orientation, and coverage etc.

When cells arrive at the surface they "see" a protein-covered surface whose protein layer has properties that were initially determined by the preformed water shells. Thus, when we talk about cell–surface interactions, it is ultimately an interaction between cells and surface bound proteins (or other biomolecules). The latter is illustrated in Fig. 8, where the clean surface in the illustration is deliberately covered with a native or artificial biomembrane (supported bilayer), containing embedded receptors that can specifically interact (bind to and/or provide I/O signals etc.) with cells approaching the surface.

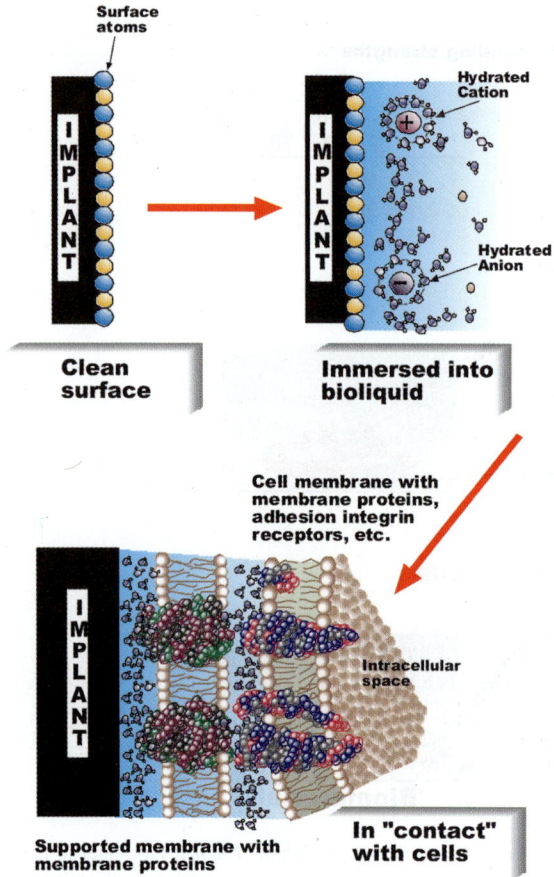

Fig. 8. A conceptual approach to convert a synthetic surface to a biomimetic surface by coating it with a supported biomembrane with built-in functional membrane-bound proteins.

Figs. 7 and 8 emphasize the above mentioned large size range that the functional units in biology cover, from water molecules and small proteins at sub-nanometer sizes, via cell membranes and supra-molecular complexes in the 10 nm range, to cells in the micrometer range and finally fully organized tissue and organs at the macroscopic size level.

2.2. Medical implants

In the case of medical implants (Figs. 1 and 3) the importance of surface science is quite obvious [1,2,14–16]. In 15–20 years time biomaterials research and development has transformed from only marginally applying surface analytical techniques to study medical implant surfaces, to a current situation where such use is the rule and mandatory for high quality research, and also for legal control, standardization and quality control of commercial products. Let us briefly inspect how surfaces come into play.

For a dental implant (Fig. 1), which is just one of many examples of so-called bone anchored implants, the clinical goal is to obtain a long term (i.e. life long) secure anchoring of the implant. The latter includes e.g. the implant's ability to carry and sustain the dynamic and static loads that it is subject to. It is obviously (for the patient and for cost reasons) important to achieve this function with the shortest possible healing time, with a very small failure rate and with minimal uncomfort for the patient. The implant is put in place after a surgical procedure typically involving drilling a hole in the cheekbone, threading of the hole, and then placing the screw-shaped end of the implant into the threaded hole.

The surface design is one of many components contributing to the clinical goal fulfillment. An inappropriate material choice or surface coating might lead to a too strong inflammatory response that adds on top of the already initiated inflammatory process by the surgical procedure. Then connective tissue rather than bone is formed around the implant, with concomitant loss of mechanical stability and eventual implant failure. A less perturbing surface may lead to a bone healing process, which is similar to the one in the absence of an implant i.e. as if the implant were a fairly passive spectator (sometimes this is referred to as inert materials or bone accepting materials, although no material is really inert in an absolute sense). Some surface coatings appear to have the ability to induce bone formation, for example the calcium phosphate materials usually referred to as hydroxyapatites.

More recently there has also been attempts to speed up the bone healing beyond the normal healing rate by adding so-called bone growth factors. The latter can e.g. be implemented by using a porous surface with a slow release of the desired biomolecules. Engineering of such surfaces requires a whole set of surface science tools. Examples of used surface science methods are coating technologies to deposit the desired surface material on a bulk carrier; surface spectroscopies like XPS, AES, and SIMS to control e.g. the Ca/P ratio; SEM or scanning probe techniques (SPMs) to control surface structure and chemistry, both of which influence the biological response; micro- and nanofabrication or other techniques to make porous surfaces; surface impregnation by molecules that are released at a pre-programmed rate to promote healing/bone formation, FTIR to measure adsorbed biological molecules etc.

Another challenging task for surface engineering of biomaterials is blood-compatible materials, such as the inner walls of artificial blood vessels, and the blood contacting surfaces of heart valves, dialyses equipment, and heart–lung machines. Normal blood is, naively speaking, a metastable state stabilized between two opposing driving forces—coagulation and anticoagulation. Almost any external perturbation in the form of an artificial surface tends to trigger the coagulation system, i.e. blood clotting may occur. The extent to which this happens, and is detrimental for the patient, depends on a number of factors such as where the surface is placed in the blood stream, the local hydrodynamics, the flow speed, contact time between the surface and blood and, of course, on the surface chemistry.

There is today clear evidence that the latter has a major influence on blood compatibility and that surfaces can be engineered to suppress blood coagulation, at least for considerable time. However, there is still a long way before surfaces that mimic the blood compatibility of the endothelial cells of normal blood vessel walls are achieved. For the surface scientist this task involves designing durable surface coatings that have the right chemical and micromechanical (visco-elastic) properties, which in turn may require a better understanding than today of how the native system works, so that a true biomimetic approach can be implemented.

2.3. Biosensors and biochips

A variety of surface based detection principles are employed for biosensors/biochips (Figs. 2 and 3). A common principle of a biosensor/biochip is that "detector molecules" are attached/immobilized on a solid surface in such a way that a specific signal is obtained from the sensor when the detector molecules *selectively* react with (bind to) the biomolecules they are designed to detect. In other words, the detection relies on the biorecognition between pre-designed native or synthetic reagent molecules and unknown sample molecules. The biorecognition event typically occurs when the pre-coated sensor is exposed to blood or urine samples or extracts thereof.

The detector read-out signals may be electrical, optical, or some other signal that is convertible into a measurable electrical signal. The detection may occur in real time, i.e. just when the recognition event occurs, or the sensor can be read-off at a later time. Two common examples are (i) antigens are pre-adsorbed on the sensor surface to detect the corresponding antibody(ies) in a clinical sample for diagnostics; (ii) a single strand DNA segment (i.e. a nucleic acid chain without cross-linking), representing a specific, known part of the genetic code, is attached to the surface so that all the dangling bonds of the non-hybridized, single strand are exposed and available for binding to a matching DNA strand. When the sensor is exposed to a DNA sample to be analyzed, one wants to obtain a specific response when there is perfect matching between the detector molecule and some sample molecules, while a different response should be obtained when the matching is imperfect. On a DNA chip many different such DNA segments are placed at different locations and thus many different DNA segments can be analyzed in one "shot" (bioarray, biochip). When the sensor array is combined with a sophisticated microfluidics system often utilizing functional hydrophobic–hydrophilic surfaces—we call the device a lab-on-a-chip or just labchip.

Other sensing principles are to e.g. employ enzymes that are immobilized on a surface with preserved function, and which can recognize specific biochemical substances, or the use of MALDI-MS to identify proteins and antibody–antigen complexes.

Surface science tools and expertise come into play in many ways in the preparation and use of biosensors and biochips. Immobilization of the detector molecules in such a way that they preserve their original recognition specificity, requires surface pre-coatings that simultaneously allow sufficiently strong binding of the detector molecules, and sufficiently weak and gentle perturbation that the molecules retain their functionality (Fig. 5). For example, the strong polarization forces at metal surfaces and the strong ionic or covalent interactions on many inorganic metal oxides and semiconductor materials may cause denaturation and loss of specificity of biomolecules. At the same time such materials may be desirable "platforms" for sensor arrays. A solution is then deposition of special spacer and/or linker molecules (typically some type of polymer or a native or synthetic biomembrane) on the original sensor substrate.

The preparation and characterization of such layers for useful deposition of detector molecules, obviously require sophisticated surface science input. The same holds for the preparation of the part of the sensor that provides a measurable signal. The most common detection approaches so far are based on optical detection in the visible or infrared. Optimization of the optical detection requires deposition of optically suitable layers e.g. regarding reflectivity, absorption properties, plasmon properties, field enhancements etc., which are all typical components in thin film design of optical coatings. Consequently advanced surface spectroscopy and microscopy come into play both for understanding of how the matrix and detector molecules bind to the sensor surfaces and for practical diagnostics. Regarding biochips, arrays of detector "spots" are required and here surface science plays a similar role as it does for the multiple microfabrication steps in semiconductor/microelectronics technology.

2.4. Tissue engineering

To reach the goal of tissue engineering—to grow a functional tissue ex vivo that can later be implanted into a human body—requires control over a large number of external parameters. Examples are the nutrients, extracellular signal substances, antibiotics, and the dynamic mechanical force fields etc. in the bioreactor where the tissue is grown. The surface of the template or scaffold material (Fig. 4), on which the original cell culture(s) are deposited and eventually develop into a tissue, must be specifically designed to promote, and not compromise, the tissue evolution. This calls first of all for a non-toxic surface for the cell culture, and for suitable protein adsorption properties, since the protein adsorption layer that always forms on the surface influences the cell–surface interactions. Since differentiated tissue is an ultimate goal, different parts of the scaffold may require different surface functionalities, i.e. different surface chemistries and topographies.

Cells in a tissue culture communicate with each other by sending out and receiving signaling molecules to/from the extracellular matrix. The total spectrum of such signals is influenced by the surfaces onto which cells attach. Thus, the surfaces of tissue engineering scaffolds are an integral part of the communicating and self-organizing tissue.

A common—but not the only one—concept in tissue engineering is a scaffold that eventually is dissolved as the tissue "matures". This calls for a sophisticated "programming" of the material so that it dissolves into harmless products at the right pace, and maybe also exposes different functional surfaces at different times. (The latter may also be a desired function of medical implants.) Many different surface technologies are potentially suitable for such surface functionalization, especially coating technologies that can be applied to 3D curved surfaces and porous surfaces, but also the standard surface characterization methods come into play. The 3D surface topography, which is very important for cell–surface interactions, may require micro- and nanofabrication methods.

2.5. Bioelectronics

Bioelectronics based on living cells (neurons etc.) call for patterned surfaces where the cells can be confined to certain geographic locations (in order to make a reproducible circuit), and where they can be kept alive and functional, which in turn requires suitable surface chemistries and topographies. Regarding the basic control of cells there is actually large similarities between cell-based electronics and tissue engineering. The surface patterns for bioelectronics require in addition adequate I/O connections for electrical or chemical communication. Similar requirements are at hand for bioelectronic devices based on biomolecules rather than whole cells, albeit at a generally smaller length scale (see also the text under artificial photosynthesis below). Consequently bioelectronic circuits require a combination of molecular cell biology with surface preparation/characterization and micro- or nanofabrication techniques.

2.6. Artificial photosynthesis

From a surface design point of view there are many similarities between biomolecule-based bioelectronics and artificial photosynthesis based on biomolecules. One common denominator is charge transfer within or between biomolecules, and between biomolecules and surfaces, either induced by external electrical fields (voltage bias) or by the electromagnetic field of light in the visible regime (e.g. photoexcitations of electron hole pairs). The relationship between the two areas becomes especially obvious when the bioelectronics concepts are extended towards opto(bio)electronic devices.

For the surface scientist a major contribution can be to produce chemical–topographic patterns on surfaces that adsorb and retain the biomolecular building blocks of the circuit elements at prescribed locations and in the right orientations (self-organization), without loss of functionality. In artificial photosynthesis with e.g. proteins (Fig. 6) there is also a major challenge to find and prepare surfaces that match the electronic states of the charge transfer proteins, e.g., as electron/hole acceptors or donors. Further requirements of the surface may be to make field enhancement structures that increase the quantum efficiency of "photon harvesting"; perhaps similar to the ones that produce the recently discovered giant surface enhanced Raman effect. Other ideas that come into play are to use optically active quantum dot arrays to catch and convert photons to energetic and "useful" electrons.

2.7. Biomimetic materials

This heading covers a large number of widely different scientific and technical applications. Examples involving surfaces as key elements are mimics of the lubrication and wear resistance of artificial joints, the low drag friction of shark and dolphin skin, the self-cleaning action of some green leaves, and threads made to mimic the strength and flexibility of spider web.

3. Model systems, research topics, and ongoing research

3.1. Biological model systems

In this section model systems are discussed and brief references are made to recent or ongoing research, illustrating the diversity of research areas and the challenging research problems in BioSS. The biologically relevant model systems available for surface science approaches cover such a broad range that any taste should find its topic of choice. Some of these areas—such as water and simple amino acids—are immediately approachable by existing UHV- and surface science preparative and analytical techniques. Others require entirely new or modified experimental approaches, such as those requiring wet preparation and in-liquid analyses or e.g. freeze-drying techniques, to maintain the biological properties and function of the studied systems. The model systems contained in the following list should be read with the accompanying label "... at surfaces":

- water
- amino acids, nucleic acids, lipids
- peptides, DNA segments
- proteins
- DNA
- lipids, biomembranes
- cells
- tissue, in vitro
- tissue, in vivo
- microorganisms

The list covers most but not all biointerfaces where there is an *interface between a synthetic (i.e. non-native) material surface and some component(s) of native biological systems*. I call them *hybrid biointerfaces*, in contrast to *native biointerfaces*, because they consist of one native biological component and one synthetic, non-native component. I will not treat native biointerfaces, such as cell–cell or microorganism–cell interfaces, although they are closely related to the hybrid interfaces; it is actually very likely that the knowledge and methods for these two types of interfaces will mutually interact and cross-fertilize in the future. For example, studies of properties and processes in real cell membranes and in so-called supported biomembranes (see below) are clearly benefiting from each other.

3.1.1. Water

Water at interfaces is an area where surface science studies will contribute significantly in the future, e.g. to the understanding of what is sometimes called "biological water". The latter refers to the special water structures that are formed near surfaces or at narrow interfaces between biological components, such as the hydration shells of proteins, DNA and biomembranes. Biological water at surfaces, or in small confined volumes, is in principle closely related to the special water structures that form at solid synthetic surfaces. They obtain their unique interfacial properties through a combination of the local physical–chemical interactions (the thermodynamics) at the interface, including the spatial constraints caused by the surface, and the kinetic constraints. They are also related to the hydration shells of solvated ions and to the Helmholtz layers at electrode surfaces in electrochemical cells, an area where surface science has contributed with significant new insight in the past two decades.

Since the presence of water has such a profound influence on both the thermodynamics and kinetics at biointerfaces it is a safe prediction that it will be a major topic in BioSS for a decade or more ahead. Recent conceptual and theoretical considerations can be found in the series of papers published by Vogler and collaborators (see Ref. [17] and references therein). Experimental work includes both direct studies of the force interactions at hybrid interfaces, as described by e.g. Israelachvili and collaborators [18], and kinetic and spectroscopic studies of interfacial water, using both biological and non-biological substances. There is also current progress in the modeling of water hydration shells on protein resistant self-assembled monolayers by Gruntze and co-workers [19], which is an important component both for understanding of what the requirements protein resistant surface is, and for understanding protein adsorption. It is also a central component for blood-compatible surfaces.

3.1.2. Amino acids and peptides

Amino acids are the building stones of one of the most elementary functional units in biology, namely proteins. Mapping their interactions with surfaces is therefore, apart from the inherent basic research interest, a precursor to the understanding of how peptides and proteins interact with surfaces. In the recent years a number of papers have been published [20–24] addressing how sub-monolayer to multilayer deposits of simple amino acids (glycine, alanine, ...) interact with single crystal surfaces or polycrystalline surfaces in vacuo. These studies include standard spectroscopic techniques, such as SIMS, HREELS, FTIR, and synchrotron light based spectroscopies applied to amino acid adsorbate layers. Some of these systems have also been characterized by LEED and TDS. There are also recent attempts to describe the amino acid–substrate interactions (including adsorbate–adsorbate interaction) by first principle calculations.

The amino acids—at least the simpler ones—are compatible with UHV evaporation techniques. For example glycine and several other amino acids can be evaporated from a simple Knudsen source at low temperature. More complex (fragile) amino acids may require deposition from the liquid phase, and then similar techniques are needed as have been developed for UHV studies of electrode double layers in electrochemistry. Important questions concern (i) how the amino acids bind to the surface, if the bonding is non-dissociative or dissociative, where and how the bonds are established (at the amine group, and/or at the carboxylic group, and/or at the R-side group etc.), (ii) strength and character of adsorbate–adsorbate interactions, (iii) molecular orientations, (iv) 2D crystal structure (when ordered layers are formed), and (v) the influence of water molecules on these properties. Another class of questions regards kinetic processes, e.g. to what extent can surfaces catalyze polymerization (e.g. forming peptide bonds) and other reactions between amino acids and co-adsorbates.

The same questions as above can be formulated for peptides, which are polymer chains of amino acids. They are important functional units in biology per se, and also constitute the amino-acid-chain sub-units of proteins. Studies of peptides are therefore one step up in complexity from amino acids, but also a step closer towards "real biology". (By analogy the step from single nucleic acids at surfaces to DNA segments at surfaces is a step both in complexity and towards biological relevance.) A natural extension of studies of amino acid monolayers is thus di-peptide, tri-peptide etc. adsorption studies. Up to date no systematic peptide studies, like the ones for amino acids, have to my knowledge been done in UHV, with state of the art surface science techniques. In contrast there is large ongoing activity focussing on liquid phase deposition of various peptide depositions and formulations on biomaterial surfaces in order to make them more functional for e.g. medical implant and tissue engineering applications. These depositions still lack the rigorous control that can be achieved in UHV, but surface spectroscopies like XPS and FTIR can and are applied, after deposition, to control the deposited layer composition.

3.1.3. Proteins

Proteins, which are (polymerized) peptide chains wrapped folded into special 3D structures (conformations), have since long been, and is currently the subject of extensive surface studies [25–42] due both to their central role in all biological processes and due to their importance in the context of bioengineering, e.g. for biosensors, biofouling and tissue engineering. The vast majority of past studies have focussed on the (macroscopic) kinetics, and to some extent on the energetics of adsorption on various surfaces, and under different conditions of liquid properties (protein concentration, pH, salinity,...). A typical experiment records the kinetics by which a surface is gradually covered by a single protein type or by proteins that compete for the surface sites. Common techniques are surface plasmon resonance, ellipsometry, fluorescence detection, and other optical methods, and gravimetric (QCM, QCM-D, SAW-devices) and radiolabelling methods. Protein adsorption is for most surfaces irreversible through additive contributions from e.g. van der Waal's interaction, polar and charged groups on the proteins, hydrogen bonding, and hydrophobic interactions.

The relative strengths of these contributions depend on protein type, on surface properties and on pH and salt concentration. Binding energies are difficult to measure for irreversible adsorption. A few surfaces bind proteins sufficiently weakly that reversible adsorption occurs, whereby heats of adsorption can be obtained from the kinetic measurements.

Very few surfaces are so-called protein resistant surfaces (see also under biomembranes below). Protein resistance means that the adsorbed amount, even at relatively high protein concentration in the bulk liquid, is below or near the detection limit. Theoretically it is still a major challenge to understand the mechanisms that contribute to protein resistance of surfaces, and even more so to understand the detailed contributions to strong(er) protein–surface bonds. Recent experimental work by Whitesides and co-workers [40–42] and by Gruntze and co-workers [19,39] constitute important steps towards understanding of protein adsorption, and protein resistance and ultimately of protein binding more generally.

Important challenges, not at all clarified at the molecular level, are to quantify the relative and absolute contributions of the various protein–surface interaction terms, and to ultimately understand how the relatively fragile protein structures (usually they denature well below 100 °C) are affected by the perturbation exerted by a solid surface. There is no doubt that the water shell of the surface and that of the (water soluble) proteins, and their mutual interaction, play an important role, which can be expressed in terms of hydrophobicity–hydrophilicity at the macroscopic scale, and in terms of water–water versus water–surface bonding strengths at the molecular level.

For example, a material surface of low polarizability in the bulk with a strongly bound water mono- or bilayer on the surface, is likely to only weakly perturb a protein with predominantly hydrophilic surface domains, while a hydrophobic surface is likely to strongly perturb a protein with mixed hydrophilic and hydrophobic domains, via hydrophobic interactions with the latter. These ideas have recently been quantified theoretically in some simple cases. Furthermore there are important entropy effects in the energetics of protein bonding at surfaces due to the highly organized and multiple state structures of proteins, and also due to the many different water structures that are possible.

Finally kinetic effects are of primary importance; the final structure of an adsorbed single protein, or protein adlayer, is almost always a long-lived metastable state, rather than the lowest energy state. In other words there is usually a sufficiently large activation barrier towards formation of the lowest energy states, so that further conformational changes are prevented. A common experimental manifestation of this is that the final structure of a protein adlayer often depends on the rate by which it is formed; at low bulk concentration of proteins the adsorption is slow and there is then longer time available for conformational changes to occur, in comparison with high bulk concentration and high adsorption rates, when neighboring molecules may sterically prevent slow conformational changes. The issue of protein conformation and denaturation at surfaces connects protein adsorption to the broad field of protein folding–unfolding in the bulk phase (see e.g. Refs. [38,43] and references therein).

A quantitative first principle description of protein bonding at surfaces lies far into the future. It will be preceded by corresponding descriptions of adsorbed amino acids and di- and tri-peptides (and the influence of water on such bonding). It will also require a quantitative first principles total energy description of proteins in the bulk phase, including the description of dynamics and multiple states. Spectroscopic techniques such as non-linear optical techniques and other laser based techniques (SFG, SHG,...), synchrotron based spectroscopies, neutron scattering, NMR, and scanning probe techniques are already addressing some of these questions. There are major challenges to adopt these techniques to the special "wet" requirements for proteins. The theoretical work ahead is equally demanding and important as the experimental work and will be absolutely necessary for progress in this area.

3.1.4. DNA

Studies of the DNA molecule at surfaces [44,45] is driven both by the inherent interest in understanding different aspects of this molecule, and by its importance for medical diagnostics using DNA biochip arrays. More recently this interest has been stirred by potential applications of DNA in (biomimetic) materials science and in molecular electronics. Generally, much less has been done regarding basic adsorption studies of nucleic acids and DNA compared to amino acid, peptide and protein adsorption, except in the sub-area related to DNA biochips. As a rule direct adsorption of DNA on such surfaces is of less interest than adsorption on pre-adsorbed spacer layers. The reason is that the surface perturbation from, say a metal, oxide or semiconductor surface, turns out or is suspected to be too strong to preserve the full function of the DNA molecule. For example, it is very likely that the dangling bonds of a non-hybridized DNA segment will bind to the surface and then be useless for detection of matching segments. The maintenance of the full recognition function of e.g. a single strand DNA segment is crucial for diagnostic applications as on biochips. Therefore various spacer layer and linker molecule strategies are currently developed to achieve both localization of the diagnostic DNA segments at different spots on a surface array, and preserved functionality.

The molecular electronics potential of DNA is currently a hot topic. Specifically there is intense research addressing the question whether the DNA molecule is a 1D conductor or not, and in that case, what the conductivity mechanisms are. Recent literature reports both high conductivity and insulating properties. Although some reports may be plagued by experimental difficulties there is another aspect; the question of conductivity or not must be more specific, and discriminate between different types of molecules, and take into account if the measurement is made with or without presence of the water shell of DNA etc. Microscopic (e.g. by SPMs) and spectroscopic characterization of the actual DNA configurations, preferably in the liquid phase, is thus vital for progress in this field.

3.1.5. Biomembranes

Biomembranes are and will be a future "hot" topic for BioSS and many other sub-disciplines, due to their enormous importance as functional units in living cells, and because of their potential role in many applications [46–53]. Biomembranes (Fig. 8 lower picture, Figs. 9 and 10) are self-organized structures of amphiphilic lipid molecules, the latter consisting of a hydrophobic (weakly interacting with water) lipid tail (linear hydrocarbon chain) and a hydrophilic (strongly interacting with water) terminal group such as a phosphate group (phospholipid molecules). In water solutions the thermodynamics drive these amphiphilic molecules to (self)assemble into structures that minimize the water interaction of the hydrophobic tails with surrounding water, and maximize the exposure of the hydrophilic end groups towards the water. This can result in many different self-assembled structures of which so-called unilamellar or bilayer membranes are the most interesting in the present context. They typically consist of a double layer of phospholipid molecules, with opposite orientations of the two layers i.e. with two consecutive layers of parallel hydrophobic tails in the interior, terminated on both sides by the hydrophilic end groups.

A specific structure occurs when the bilayer forms a closed spherical shell, with water both on the inside and the outside. This structure is named a vesicle or liposome, an entity that is a first crude approximation of a closed cell membrane, however, lacking the functional units of real cell membranes such as membrane-bound proteins, ion channels etc. One could refer to them as a primitive, "empty cell" that can be filled by both membrane-bound functional units (membrane proteins) and intracellular components, in order to

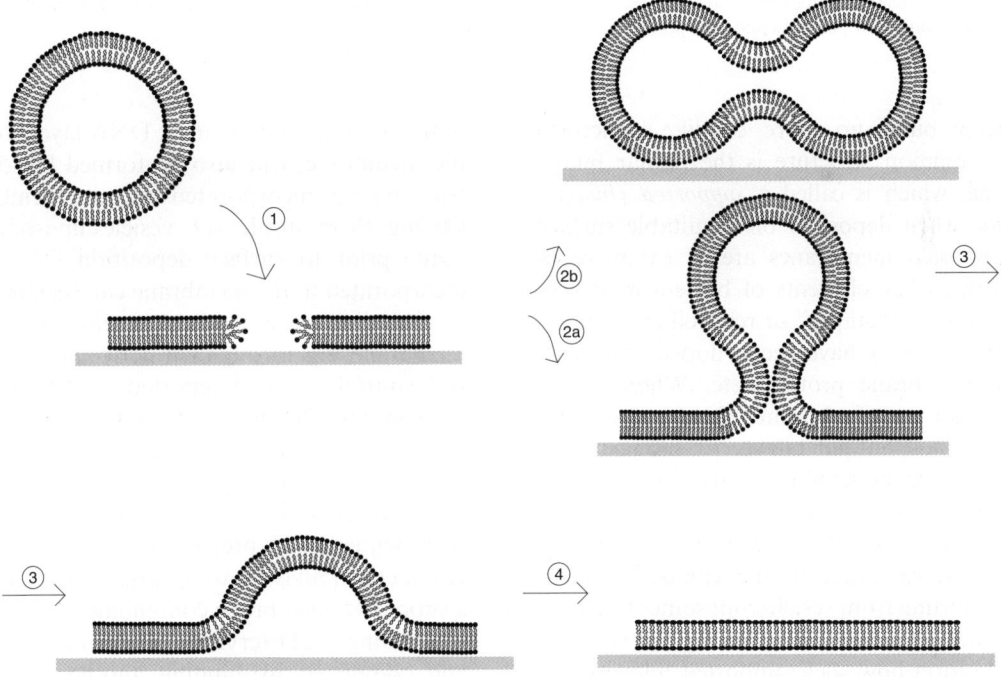

Fig. 9. Formation of SPB from vesicles (schematic; not proven by experiments).

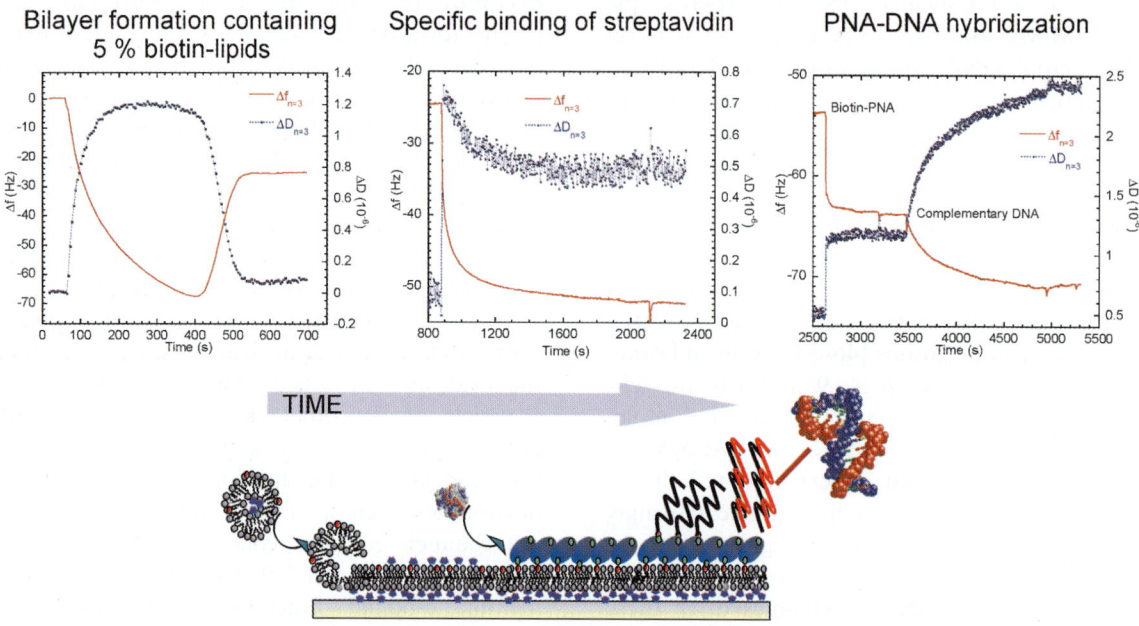

Fig. 10. A sequence of preparation steps that have been experimentally implemented [50] to build up a biofunctional surface for e.g. cell culturing or biosensing. The first step consists of formation of a biotin doped, phospholipid bilayer from vesicles, on top of which (2nd step) a 2D crystalline protein layer (streptavidin) is adsorbed. The third step is adsorption of DNA or PNA molecules, which are biotin coupled to the remaining free attachment sites on the streptavidin molecules. This platform can then be used e.g. for discriminative recognition of perfectly and non-perfectly matching DNA segments.

successively build up more cell-like structures. Another common structure is the planar bilayer membrane, which is called a *supported (bilayer) membrane* when deposited on a suitable surface. Such supported membranes are of extensive interest both as key elements of biosensor devices, and as mimics of "empty" or real cell membranes, the latter after they have been "doped" by incorporating membrane proteins etc. When the bilayer is made up of phospholipids, it is called a supported phospholipid bilayer or biomembrane (SPB). Such bilayer membranes are also of interest as functional coatings on medical implants, cell culture surfaces, and tissue engineering scaffolds.

Supported bialyer membranes can be formed on a surface starting from vesicles/liposomes (see Refs. [46–50], and references therein). Fig. 10 shows in schematic form how such supported bilayers may form. Supported bilayers with incorporated membrane proteins or protein or DNA layers on top of the membrane, can also be formed [48–53]. Proteins are e.g. incorporated into the membrane by placing them inside the vesicles/liposome membrane prior to surface deposition. The proteins incorporated in the membrane can then in turn act as a binding site for other proteins on top of the membrane, e.g. through antibody–antigen binding. A beautiful example reported by Reviakine and Brisson is the 2D protein crystals on top of a biotin doped bilayer imaged by AFM [49]. By suitable linker chemistry one can continue to add additional layers, to achieve a specific desired function. One such sequence of preparation steps executed by Höök et al. [50] is shown in Fig. 11, beginning with a supported membrane containing biotin, followed by binding a 2D (crystal) of streptavidin on top, and completed by binding another biotin layer linked to a single strand DNA or PNA segment in the outermost layer. This structure can then be used to detect recognition events between the single strand DNA segment and its matching counterpart, forming a double helix. For sensor applications it is important to find methods to pattern cells at surfaces. Such patterning has recently been demonstrated by several groups, see e.g. Ref. [53] and references therein.

3.1.6. Cells

Cells at surfaces is a fundamental ingredient both with regard to the cultivation of cell cultures in vitro, tissue engineering, and medical implants. In the future it may also be employed in cell-based biosensors and bioelectronics. The role of the surface chemistry and topography for the evolution and function of single cells and cell assemblies has been demonstrated in many experiments [54–63] but is still poorly understood. The recent rapid development in the area of so-called stem cells is particularly interesting in this context. Stem cells are cells that have not reached a high degree of specialization, but can "choose" to differentiate in different specialized directions depending on the local environment. Depending on the latter stimuli the cell can thus differentiate into different functions. It has also been shown that some stimuli can make a cell "go backwards" from a specialized to less specialized state.

The chemistry and topography of surfaces on which cells grow/divide are part of the total environment for the cells and can potentially influence their differentiation, i.e. they are part of the total cell–cell signaling/communication system. One key research area for the future is therefore how tailored surfaces should be designed to act as stimuli to guide cell differentiation, e.g. in tissue engineering, bioelectronics and cell-based sensors. For the surface scientist the chemical and topographic patterning of surfaces, and the associated characterization is a major opportunity and challenge for future research. Another is to use SPMs to probe or stimulate specific cell functions. Yet another one is non-invasive, spatially resolved spectroscopies and microscopies to study special fractions of living cells or cell membranes at high spatial resolution (bioimaging) and sensitivity (single molecule detection).

3.1.7. Tissue

Tissues at surfaces, in vitro and in vivo, have been demonstrated to be sensitive to the adjacent surfaces, but the mechanisms are understandably even less known than in the case of single cells and cell monocultures. Current research on tissue-material interfaces in vivo follows several parallel lines. One major direction is chemical modification of surfaces to enhance their biocompatibility and/or to provide controlled release of substances that promote the healing and sustained function of implants. Another one focuses on topographic modification of surfaces to promote desired cell–surface responses, and to some extent also protein–surface interaction. In some studies these surface modifications, i.e. chemical and topographic, are combined to achieve optimal performance of the surface. Additional functions of the surface may be to e.g. act as a reservoir that can release certain desired substances, such as growth hormones, in a time-programmed manner.

In the rapidly expanding tissue engineering field the questions and challenges are to some extent similar as for the in vivo situation with medical implants. There are however also entirely new challenges connected with the fact that tissue engineering demands that a given cell or tissue culture shall evolve into a desired tissue without the support from the living organism that is the host for in vivo implants. Thus tissue engineering demands scaffolds that cooperatively with other factors (nutrients, extracellular chemicals, force fields, temperature,...) steer the tissue culture in the required direction. In some sense one can say that the bioreactor must do for the tissue engineering scaffold + cell culture, what the living organism does for the medical implant during the healing-in period. However, the total demand is much higher, namely to make the orininal cell/tissue culture develop into a functioning organ. This may require special 3D architectures ranging from macroscopic dimensions to micrometer and nanometer dimensions, with superimposed chemical patterns and (bio)chemical signals, and probably with a time-programmed degradation rate of the templates as the tissue matures. The research opportunities for surface scientists in this area are similar to those for cell–surface interactions (see above).

3.1.8. Microorganisms

Microorganisms at surfaces are important e.g. in marine biofouling and bacterial infections on implant surfaces or in tissue cultures. They are also potentially of interest for growing simple organism cultures as model systems for a variety of biological experiments e.g. related to food production, development and evaluation of pharmaceuticals, and biomimetic energy production. Colonization of surfaces by e.g. bacteria (biofouling) is surface specific and can be influenced by surface chemistry, surface topography, and surface visco-elasticity. The settling down of microorganisms on a surface is—as in the case of cell adhesion—almost always preceded by protein adsorption. There is thus a causal relationship between protein adsorption and biofouling by microorganisms, in a similar way as sketched earlier for cell adhesion on a surface. Also the basic research opportunities and tools are similar as for cell culture growth and tissue engineering, while the more applied research may require much more specific and specially adapted tools, e.g. to study "real" marine biofouling or bacterial colonization on medical implants.

4. Surfaces

The material surfaces include almost all material types; metals, ceramics, carbon materials, polymers and composite materials. Although many native materials are currently used in practice, such as titanium for dental implants, stainless steel for orthopedic implants, PTFE for blood vessel replacements, silicones for internal drainage, PMMA for intraocular lenses, and so on, future more advanced materials and applications will often require the build up of sandwich type overlayers, with specific topographies and patterns, on the native surfaces, to obtain the desired function. This is actually already the case e.g. for intraocular lenses made from PMMA and almost always the case for advanced biosensors. In addition we know that surface patterns, topographically and chemically, on different length scales generate different responses of the biological system. This requirement of sophisticated surface structures and patterns derives from the very advanced recognition power of biological systems discussed earlier.

Native material surfaces are very unsophisticated compared to the biomolecular chemical + topographic architectures they are intended to interact with, and will in most cases at best cause a mild negative response in the biological system. In order to obtain a positive and selective response the molecular architecture should in general match some recognition sites of biomolecules on the biological side of the hybrid interface, so that it generates desired signal patterns to adjacent biomolecules or cells or sensing devices. This type of topographic + chemical microarchitectures are necessary both for applications and understanding-oriented basic research of biointerfacial processes.

Consequently the future development of biologically functional surfaces demands a very sophisticated machinery for surface preparation and characterization, including dry and wet chemical adlayer depositions, and nano- and microfabrication.

5. Methods

Since this article is more concept and idea oriented than method oriented this paragraph is deliberately very brief. A general statement about surface science methods and biointerfaces would be too trivial, since it would conclude that essentially all methods of surface science come into play in some contexts.

Well-established examples on the *analytical* side are the use of established surface spectroscopies (XPS, AES, SIMS,...) to characterize medical implants after manufacturing and sterilization, but before clinical insertion, and also after varying periods in vivo (retrieved implants). Synchrotron based spectroscopies will no doubt be important tools both for spectroscopic characterization, structure determination, and for imaging of biological components on surfaces. With regard to in situ characterization of surface bound biomolecules and larger entities, and especially for real-time recording of kinetic processes, the optical techniques are of prime importance. This includes both linear and non-linear (laser) techniques, and all spectral regions from UV and down in wavelength to the far IR. There is no doubt that scanning probe techniques will continue to increase in importance for studies of biomolecules at surfaces. This includes both STM and AFM [64,65] and spatially resolved optical techniques like SNOM. Complementary techniques that provide unique information in special cases, when sufficient surface sensitivity is achieved are e.g. NMR, neutron scattering and ESR.

Preparation by thin film deposition methods (PVD, CVD) and by glow discharge plasma methods are common in R&D on hybrid biointerfaces. Wet chemical methods such as self-assembled monolayers, colloidal chemistry, microemulsions, and micelles and vesicles come into play because they have the ability to produce interesting chemical and topographic patterns on surfaces. In addition they are usually fast and often inexpensive. Synthetic polymer, protein and oligonucleotide chemistry are increasingly important because they can produce ligands that can selectively anchor desired polymer chains, peptides, oligonucleitides or whole proteins to surfaces.

The whole range of nano- and microfabrication methods constitute important tools for the functional patterning of surfaces. They include electron beam and photolithography and the soft lithographies, like stamping and imprinting.

Interesting to note are the special demands on methods for BioSS, set by the fact that one deals with very fragile matter, often requiring a liquid environment for meaningful studies. This calls for new developments and/or to adaptations of experimental techniques to work non-invasively with e.g. cells under liquid environments, or perform sophisticated freeze drying to retain the features to be studied by methods that require a vacuum environment. For smaller entities like adsorbed proteins and biomembranes, both the microscopic inspection by SPMs and the spectroscopic characterization under water are key issues. In order to address the important and challenging task of dynamics of e.g. proteins and supported membranes, femtosecond techniques are required, in the aqueous environment. Consequently scanning probe techniques and time resolved laser spectroscopies will be central tools in the development of BioSS.

6. A look into the future

When surface science started to take off as a research field on its own, over thirty years ago, little was known even about the simplest model systems and the number and sophistication of the experimental and theoretical tools were very limited compared to today. The electronic structure of surfaces was largely unknown, we did not know e.g. how low energy electrons were scattered by surface atoms, hampering the use of LEED as a quantitative tool, the positions and orientation of CO on any surface were totally unknown, and we were very far from any quantitative potential energy surfaces for dissociation or reaction dynamics (even the 1D cuts were unknown). Total energies of chemical accuracy were impossible to calculate.

For simple systems the situation is today reversed; the total energies can be calculated, the electronic and atomic structures, and the lattice dynamics of surfaces are known in great detail. Potential energy surfaces can be calculated for dissociation and reaction of simple molecules, and the understanding is conceptually transparent even for much more complex systems. Important factors have been the development of new experimental probes, advanced preparation of well defined model systems, development of theoretical methods and simulation schemes to describe both the studied model systems and how the experimental probes (electrons, photons, ions, SPM tips etc.) interact with surfaces, and a vast increase in computational power. This development has allowed a broad and systematic, theoretical and experimental, exploration of model systems of varying complexity, which successively paved the way for a genuine, quantitative understanding today of the simpler model systems, and a semi-quantitative and/or conceptual understanding of quite complex systems.

This is the platform from which BioSS launches just now. We can foresee an exciting development of new knowledge about biomolecules at surfaces, initially mainly conceptual and qualitative rather than quantitative, except for the simplest model systems like water, and amino and nucleic acids. Successively we will map out and understand the structure, dynamics and biological functions of supported biomembranes with and without membrane-bound proteins, charge transfer in electron transfer proteins, DNA and ion channels attached to surfaces etc. Eventually (probably a decade or two) we will see the beginning of a detailed understanding of cell–surface interactions.

These basic research endeavors will be driven both by curiosity and the large number of applications of this knowledge; tissue engineering, biosensors, DNA and proteomic chips for drug development and unprecedented precision and individualization in medical diagnostics, and bioelctronics based on cell–cell self-organization and communication.

On the way there will be a dramatic improvement of existing experimental methods, and development of new ones, to prepare and characterize surfaces on the atomic and mesoscopic scale, including measurements of the dynamics and kinetics. These methods must, to a large extent, be adopted for real-time measurements at solid–liquid interfaces. The influence of the theoretical methods and simulations is likely to be even stronger than in the development of surface science referred to above.

The development will take place through strong interaction with adjacent fields to surface physics and chemistry, for example molecular biology, nanoscience and polymer science. Ultimately, it is likely that these studies will not only produce a deep understanding of material-biosystem interfaces, but also about the biosystems themselves.

Acknowledgements

I am very grateful to all members of the Chemical Physics Group at the Department of Applied Physics, Chalmers, who have so kindly helped me with invaluable reference searches and constructive comments on the manuscript.

References

[1] B. Kasemo, J. Gold, Advances in Dental Research 13 (1999) 8, and references therein.
[2] B.D. Ratner, Molecular Recognition 9 (1996) 617.
[3] W. Göpel, Biosensors and Bioelectronics 13 (1998) 723.
[4] A.F. Collings, F. Caruso, Reports on Progress in Physics 60 (1997) 165.
[5] H.A. Fishman, D.R. Greenwald, R.N. Zare, Annual Review of Biophysics and Biomolecular Structure 27 (1998) 165.
[6] D.J. Graves, Tibtech 17 (1999) 127.
[7] R. Langer, Accounts of Chemical Research 33 (2000) 94.
[8] J.A. Hubbell, Current Opinion in Solid State and Material Science 3 (1998) 246.
[9] R. Iyengar, Science 283 (1999) 381.
[10] P. Fromhertz, A. Offenhäusser, T. Vetter, J. Weis, Science 252 (1991) 1290.
[11] R.F. Service, Science 284 (1999) 578.
[12] A.J. Bard, M.A. Fox, Accounts of Chemical Research 28 (1995) 141.
[13] M. Sarikaya, Proceedings of the National Academy of Sciences 96 (1999) 14183.
[14] B. Kasemo, Current Opinion in Solid State and Materials Science 3 (1998) 451.
[15] B. Kasemo, J. Lausmaa, CRC Critical Review on Biocompatibility 2 (1986) 335.
[16] B. Kasemo, J. Lausmaa, Osseointegration in Oral Rehabilitation, 1993.
[17] E.A. Vogler, Advances in Colloid and Interface Science 74 (1998) 69.
[18] J. Israelachvili, H. Wennerström, Nature 379 (1996) 219.
[19] A.J. Pertsin, M. Grunze, Langmuir 16 (2000) 8829.
[20] S.M. Barlow, K.J. Kitching, S. Haq, N.V. Richardson, Surface Science 401 (1998) 322.
[21] M. Nyberg, J. Hasselström, O. Karis, N. Wassdahl, M. Weinelt, A. Nilsson, L.G.M. Pettersson, Journal of Chemical Physics 112 (2000) 5420.
[22] N.A. Booth, D.P. Woodruff, O. Schaff, T. Gießel, R. Lindsay, P. Baumgartel, A.M. Bradshaw, Surface Science 397 (1998) 258.
[23] X. Zhao, Z. Gai, R.G. Zhao, W.S. Yang, T. Sakurai, Surface Science 424 (1999) L348.
[24] P. Löfgren, A. Krozer, J. Lausmaa, B. Kasemo, Surface Science 370 (1997) 277.
[25] I. Lundström, Progress in Colloid and Polymer Science 70 (1985) 76.
[26] J.L. Brash, P.W. Wojciechowski, Interfacial Phenomena and Bioproducts (Edited by McGregor WC), Marcel Dekker, New York, 1996.
[27] T.A. Horbett, T.L. Brash, ACS Symposium Series 602, ACS, Washington, DC, 1995.
[28] V. Hlady, J. Buiks, Current Opinion in Biotechnology 7 (1996) 72.
[29] W. Norde, Macromolecular Symposium 103 (1996) 5.
[30] T. Zoungrana, G.H. Findenegg, W. Norde, Journal of Colloid and Interface Science 190 (1997) 437.

[31] H. Arwin, Thin Solid Films 313 (1998) 764.
[32] F. Höök, M. Rodahl, B. Kasemo, P. Brzezinski, Proceedings of the National Academy of Sciences 95 (1998) 12271.
[33] F. Höök, M. Rodahl, P. Brzezinski, B. Kasemo, Langmuir 14 (1998) 729.
[34] F. Caruso, D.N. Furlong, P. Kingshott, Journal of Colloid and Interface Science 186 (1997) 129.
[35] P. Tengvall, I. Lundström, B. Liedberg, Biomaterials 19 (1998) 407.
[36] H. Elwing, Biomaterials 19 (1998) 397.
[37] V. Hlady, J.N. Lin, J.D. Andrade, Biosensors Bioelectronics 5 (1990) 291.
[38] V.P. Zhdanov, B. Kasemo, Surface Review Letters 5 (1998) 615.
[39] P. Harder, M. Gruntze, J.H. Waite, Journal of Adhesion 73 (2000) 161.
[40] R.G. Chapman, E. Ostuni, S. Takayama, R.E. Holmlin, L. Yan, G.M. Whitesides, Journal of the American Chemical Society 122 (2000) 8303.
[41] R.G. Chapman, E. Ostuni, L. Yan, G.M. Whitesides, Langmuir 16 (2000) 6927.
[42] M. Mrksich, G.M. Whitesides, Annual Review of Biophysics and Biomolecular Structure 25 (1996) 55.
[43] V.P. Zhdanov, B. Kasemo, Proteins 29 (1997) 508.
[44] H.W. Fink, C. Schönenberger, Nature 398 (1999) 407.
[45] E. Braun, Y. Eichen, U. Sivan, G. Ben-Yoseph, Nature 391 (1998) 775.
[46] E. Sackman, Science 271 (1996) 43.
[47] J. Salafsky et al., Biochemistry 35 (1996) 14773.
[48] C.A. Keller, B. Kasemo, Biophysical Journal 75 (1998) 1397.
[49] I. Reviakine, A. Brisson, Langmuir 16 (2000) 1806.
[50] F. Höök et al., Langmuir, submitted for publication.
[51] W. Knoll, M. Zizlsperger, T. Liebermann, S. Arnold, A. Badia, M. Liley, D. Piscevic, F.J. Schmitt, Physiochemical and Engineering Aspects 161 (2000) 115.
[52] T. Stora, Z. Dienes, H. Vogel, C. Duschl, Langmuir 16 (2000) 547.
[53] J.S. Hovis, S.G. Boxer, Langmuir 16 (2000) 894.
[54] G.M. Whitesides, J.P. Mathias, C.T. Seto, Science 254 (1991) 1312.
[55] R.S. Kane, S. Takayama, E. Ostuni, D.E. Ingber, G.M. Whitesides, Biomaterials 20 (1999) 2363.
[56] C.S. Chen et al., Science 276 (1997) 1425.
[57] R.G. Flemming, C.J. Murphy, G.A. Abrams, S.L. Goodman, P.F. Nealey, Biomaterials 20 (1999) 573.
[58] Y. Xia, J.A. Rogers, K.E. Paul, G.M. Whitesides, Chemical Reviews 99 (1999) 1823.
[59] A.S. Curtis, C.D. Wilkinson, Journal of Biomaterials Science Polymer Edition 9 (1998) 1313.
[60] D. Brunette, B. Cheroudi, Journal of Biomechanical Engineering 121 (1999) 49.
[61] M. Mrksich, Cellular and Molecular Life Sciences 54 (1998) 653.
[62] R.S. Kane, S. Takayama, E. Ostuni, D.E. Ingber, G.M. Whitesides, Biomaterials 20 (1999) 2363.
[63] D. Qin, Y.N. Xia, B. Xu, H. Yang, C. Zhu, G.M. Whitesides, Advanced Materials 11 (1999) 1433.
[64] A. Engel, H.E. Gaub, D.J. Muller, Current Biology 9 (1999) R133.
[65] H.G. Hansma, K.J. Kirn, D.E. Laney, et al., Journal of Structural Biology 199 (1997) 99.

The surface science of enzymes

T.H. Rod, J.K. Nørskov *

Center for Atomic-scale Materials Physics, Department of Physics, Building 307, Technical University of Denmark, DK-2800 Lyngby, Denmark

Received 29 June 2000; accepted for publication 7 May 2001

Abstract

One of the largest challenges to science in the coming years is to find the relation between enzyme structure and function. Can we predict which reactions an enzyme catalyzes from knowledge of its structure—or from its amino acid sequence? Can we use that knowledge to modify enzyme function? To solve these problems we must understand in some detail how enzymes interact with reactants from its surroundings. These interactions take place at the surface of the enzyme and the question of enzyme function can be viewed as the surface science of enzymes. In this article we discuss how to describe catalysis by enzymes, and in particular the analogies between enzyme catalyzed reactions and surface catalyzed reactions. We do this by discussing two concrete examples of reactions catalyzed both in nature (by enzymes) and in industrial reactors (by inorganic materials), and show that although analogies exist and the two kinds of catalyst can be described by similar tools, nature and human effort have come up with different solutions. This on the other hand implies that new and improved catalysts may be made by learning from nature. © 2001 Elsevier Science B.V. All rights reserved.

Keywords: Biological molecules – proteins; Catalysis; Density functional calculations; Models of surface chemical reactions; Surface chemical reaction; Inorganic compounds

1. Introduction

Within a few years a large fraction of enzyme structures will be known from a number of sources. As the structures appear the question remains: What is the relationship between the structure of an enzyme and its function? Will we be able to understand why an enzyme with a given structure catalyzes a specific reaction? Will we be able to predict how changes in the structure of the enzyme affect its function?

These are some of the hardest and most important scientific problems of the next decades. If we can solve them we will have made an important step towards understanding some of the basic problems of life. In addition, we will have identified new possibilities for modifying enzymes and thereby for making new drugs, and new treatments, and for introducing new possibilities in biotechnological production.

The need for finding solutions for the above problems will become even more demanding as the

---

* Corresponding author. Tel.: +45-4525-3175; fax: +45-4593-2399.

  E-mail address: norskov@fysik.dtu.dk (J.K. Nørskov).

mapping of genomes continues to be more and more complete. Recently, the (almost) complete mapping of the fruit fly *Drosophila melangaster* was reported [1], and a draft of the human genome has been completed as well [2,3]. This mapping of the human genome will give us access to the amino acid sequence of all enzymes in the human body, but not the three-dimensional structure, nor the function of the enzymes.

Enzymes control essentially every aspect of the chemistry in living cells. They are catalysts helping reactions proceed that would not otherwise be possible under the mild conditions in a living cell, i.e. ambient temperature and pressure. Enzymes also help in selecting which molecules should react and which should not. In this way enzymes are the key to the regulation of the cell functions.

An enzyme catalyzed reaction must take place somewhere on the surface of the enzyme, which usually is folded up in a unique compact three-dimensional structure. Fig. 1 illustrates how an enzyme appears to an incoming reactant. Whether the enzyme interacts with part of another protein or with a smaller ligand, the reactant, or substrate, as it is called in biochemistry, it will have to find somewhere on the surface of the enzyme where it can adsorb and react further. The part of the surface where this happens is called the active site and often is only a small part of the total surface. In the present paper we discuss how to describe adsorption and further reaction of substrates on the surface of an enzyme. In particular we stress the analogies between the surface of an enzyme and the surface of inorganic catalysts, such as the metallic nano-particle shown in Fig. 2. Metallic nano-particles also interact with the surroundings via the surface and can act as catalysts just like the enzymes. The analogies go even further, since several of the most important industrial catalytic processes such as the ammonia synthesis or the methanol synthesis have analogies in enzymes, the nitrogenases and the methane mono-oxygenases (MMOs). One of the interesting questions one may ask is, why enzymes can catalyze such processes at room temperature and 1 atm pressure, while the inorganic counterparts usually require high temperatures and pressures. Perhaps we can also gain some new inspiration for developing new and

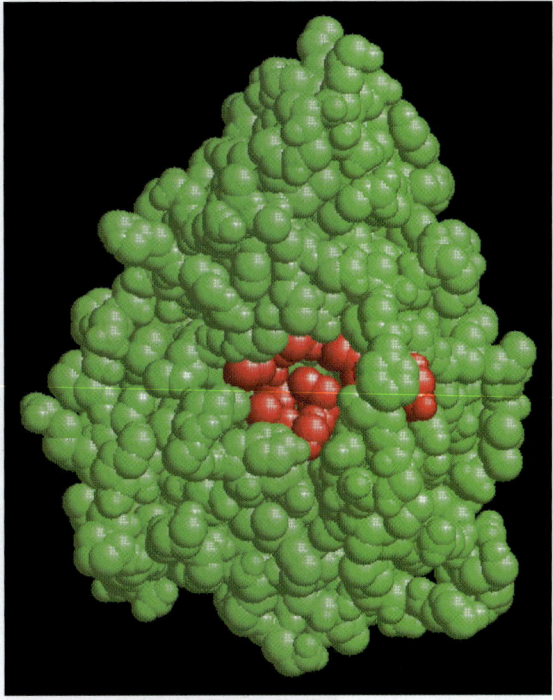

Fig. 1. Structure of the enzyme protein tyrosine phosphatase 1B (PTP1B), which catalyzes the abstraction of a phosphate group from other enzymes. The part of the enzyme where this reaction happens are colored red and is generally called the active site. The remaining atoms of the enzyme are colored green. Each atom is represented by a ball with radius equal to its van der Waals radius. Hydrogen atoms are, however, not shown for reasons of clarity. The structure is obtained from the Brookhaven Protein Data Base [70–72] where most of the known protein structures are stored and can be freely retrieved. The PDB ID for the above structure is 2HNP. The structure is based on X-ray diffraction spectroscopy by Barford et al. [73].

better industrial catalysts by studying the surface science of enzymes?

In the surface science of inorganic materials a number of experimental and theoretical techniques and a number of concepts have been developed which enable us to describe the interaction between molecules and surfaces at a very detailed level. This detail is usually not yet found in our understanding of enzyme processes. It is conceivable that some of the experimental and theoretical progress in inorganic surface science can be used

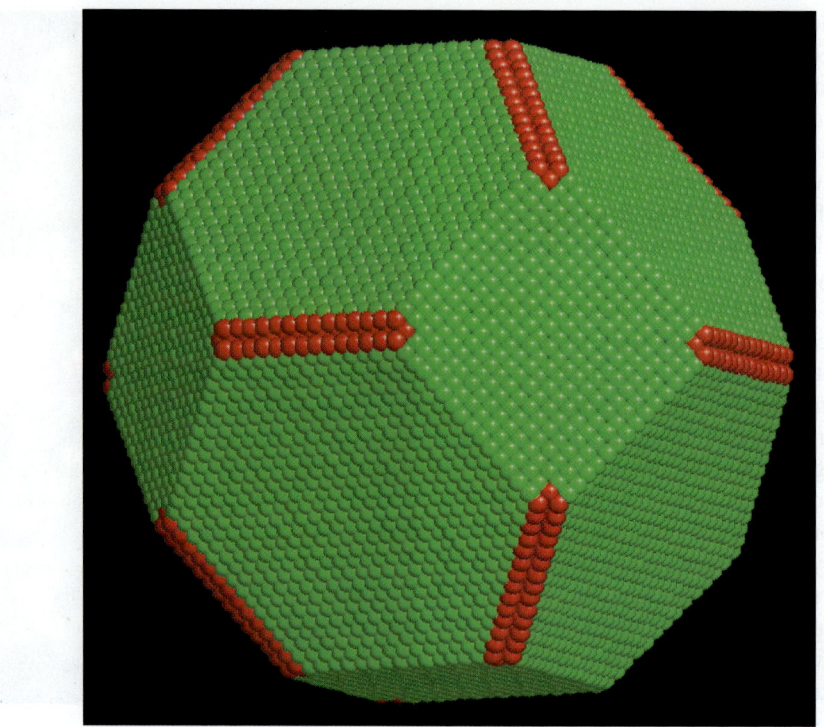

Fig. 2. Structure of a Cu nano-particle. Nano-particles like the one shown are often the active part of heterogenous catalysts used in the chemical industry or in the car exhaust. The preferable metal for the catalyst depends on the reaction catalyzed, Cu is for instance the preferred metal for the production of methanol (CH_3OH) from carbondioxide (CO_2) and hydrogen (H_2). As for enzymes, however, the catalyzed reaction occurs preferably on a small part of the particle. For the above Cu particle the atoms comprising this active part are the ones that sit on the corners. These atoms are colored red while the other Cu atoms are colored green. Comparing the above figure with Fig. 1 illustrates the analogue between the nature and human made catalysts, although they are quite different in composition.

to further develop our understanding of the way enzymes work.

In the following we first give a brief introduction to the structure and function of enzymes. Then we discuss two specific examples, the nitrogenase and the MMO. In doing so we try to illustrate how enzymes work and to contrast the enzyme function with what is known about the same kind of processes on metal surfaces. We conclude with a discussion of the possibilities of cross-fertilization between the surface science of inorganic materials and that of enzymes.

2. Enzyme structure and function

All enzymes are composed of proteins. The simplest enzymes consist of just a single protein, but many enzymes function by an interplay between several component proteins. Protein structure is typically described at different levels as the primary, secondary, and tertiary structure.

The structure of a protein can be understood as a folded chain where each link of the chain is composed of one out of twenty possible amino acids. The sequence of the amino acids comprising the chain is called the primary structure of the protein. The primary structure of a protein is determined by the genetic code(s) regulating its synthesis. This is illustrated in Fig. 3; the information encoded in the deoxyribonucleic acid molecule (DNA) (panel a of Fig. 3) is transcribed into the primary structure of the protein (panel b of Fig. 3).

We now continue down through Fig. 3. In panel c the secondary and tertiary structure have been formed by folding the chain into the three-dimensional shape of the protein. Secondary structure describes local structural motifs. Most common are α-helices and β-strands. α-Helices are regions where the chain is coiled into a kind of spiral, while the chain is extended in regions that form β-strands. These regular secondary structure elements are connected by more flexible loop regions. The tertiary structure describes how the secondary structure elements are folded into the native compact three-dimensional structure.

In panel c of Fig. 3 we indicate two representations of the structure of the enzyme protein tyrosine phosphatase 1B (PTP1B), previously indicated in Fig. 1. In the structure to the left of panel c the backbone atoms are traced in order to show how the chain is folded. The chain has been colored with gradually changing colors to make it easier to follow it from the beginning to the end. The right structure of panel c depicts the secondary structure elements along the chain, where α-helices are indicated by blue sheets and β-strands are indicated by yellow sheets with arrows. It is important to note that although the representations of PTP1B in part c of Fig. 3 show the molecule as rather open, it is actually a very compact molecule as illustrated in Fig. 1, where the same PTP1B enzyme is shown.

The protein folding process is studied intensely these years [4,5]. The three-dimensional structure is stabilized by the hydrogen bonds and disulfide bonds that can be made between different amino acids along the chain, and by the interaction with the environment (typical an aqueous salt solution). Due to the size and complexity of a protein it is difficult to describe all interactions accurately enough for predicting the protein structure from the genetic code. The hope is, that it will eventually be possible to predict the final three-dimensional structure directly from a knowledge of the amino acid sequence. This will save enormous amounts of experiments, since the determination of the full three-dimensional structure of a protein using e.g. X-ray diffraction is extremely demanding. Some theoretical approaches try to use information about the interactions between the different parts of the protein in direct dynamical simulations of the folding process using Monte Carlo, molecular dynamics or other techniques [6–8]. Other approaches are data driven. They use examples where both the amino acid sequence and the three-dimensional structure is known from experiments, to teach models what the relation is between sequence and structure, and then use these models to predict new structures [9,10].

Continuing with Fig. 3, the next and final level of complexity is related to the function of an enzyme. For our example enzyme PTP1B the function is to catalyze the reaction between water and enzymes with a phosphate group bound to a tyrosine (an amino acid) resulting in phosphate (HPO_4^{2-}) evolution. The removal of phosphate deactivates the enzyme and PTP1B thereby helps regulating cell activity [12]. The first part of this reaction, abstraction of PO_3^-, is illustrated in panel d of Fig. 3, where we zoom in to the active site of PTP1B. The left picture of panel d, indicates the active site prior to reaction. In the central picture of panel d a molecule (illustrating part of another enzyme) with a tyrosine bound phosphate has come into the active site, and the phosphate group now binds to both the tyrosine but also to a sulfur of a cysteine (yet another amino acid) in PTP1B. In the right picture the bond to tyrosine has been broken. Since PTP1B is a catalyst, the PO_3^- group has to leave again, which happen by reaction with water, in which case HPO_4^{2-} evolves [12].

An enzyme has active sites where the function of the enzyme takes place, but in many cases the function of the enzyme also depends critically on the structure of the whole protein [11]. If, for instance, the protein folds into another structure by being trapped in a local energy minimum, it may lose its activity or cause disease, as for instance in the mad cow disease [13,14]. The attachment of molecules or the interaction with other proteins can also change its activity strongly. Such effects can be very long range, over distances of 10–20 Å. This is a useful way of tuning the activity of an enzyme, and such effects are used extensively in the control of the complex environment that a living cell represents as exemplified above by phosphate abstraction by PTP1B. Another example are enzymes placed in cell membranes where a stimulus in the end of the enzyme sticking out of the cell can

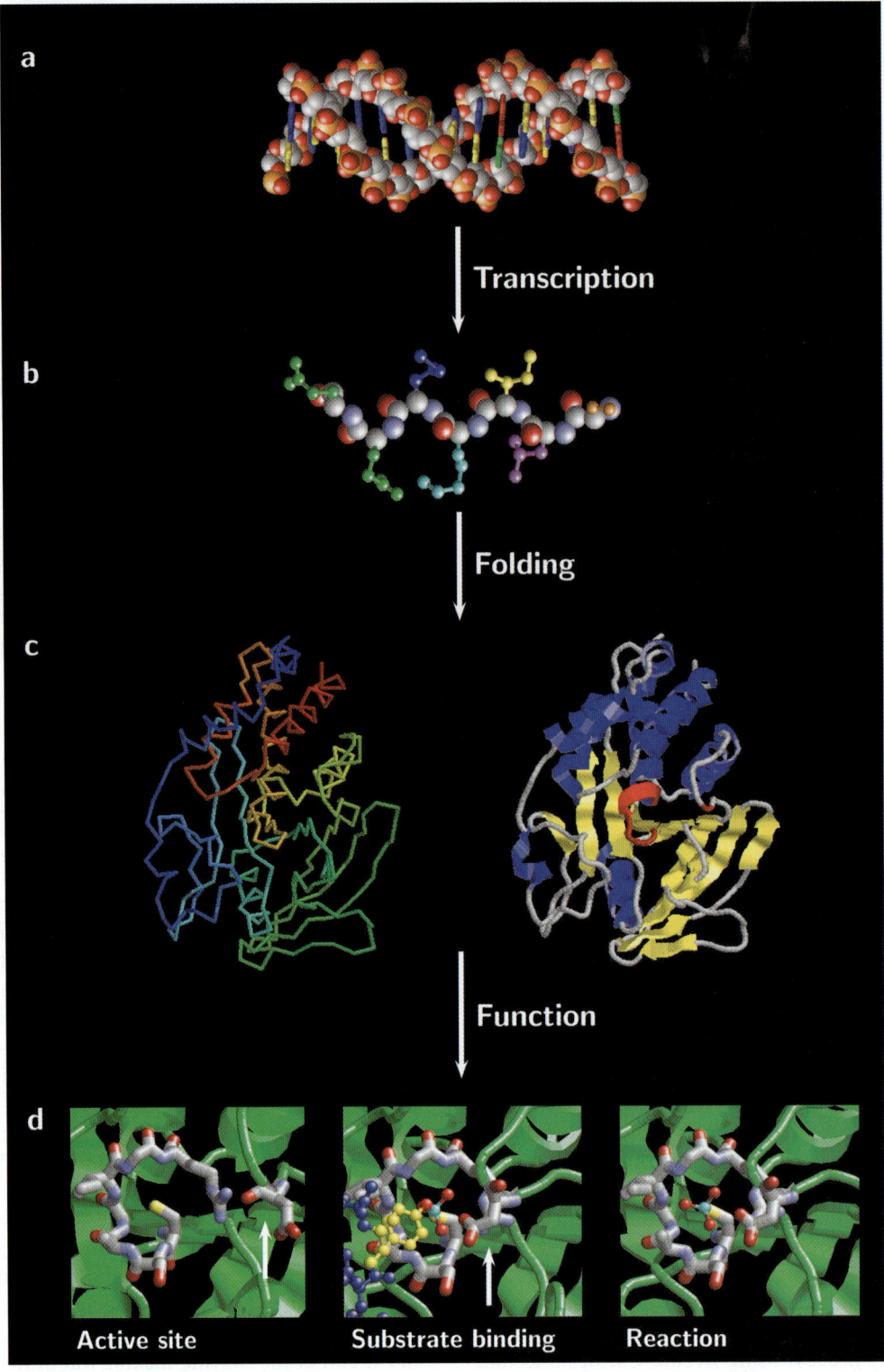

activate an action in the end of the enzyme sticking into the cell [15].

For PTP1B the active site is constructed of a region of the protein, but often the active site of an enzyme is a non-protein part bound by the protein. This non-protein part is called the cofactor and may be organic as well as inorganic. In case that the cofactor is organic it also is called a coenzyme [11].

Enzymes are characterized by their catalytic activity and selectivity. The selectivity is an important aspect of enzyme function. Often an enzyme is catalyzing simple processes, such as proton transfer or hydrolysis. This is not in itself very demanding, but enzymes are able to do it very selectively such that only one specific side chain of the substrate is affected. In this connection it is important that the enzyme has an active site that can recognize the ligand or site in another protein that should be modified. This recognition (or lock and key) effect is closely related to the design of the three-dimensional structure of the enzyme, because it requires a specific set of atoms around the active site [11]. The selectivity is demonstrated by our example enzyme PTP1B capable to remove a phosphate group from another enzyme only if that phosphate group is linked to a tyrosine. Hence, the name protein tyrosine phosphatase.

Catalytic activity is particularly important for more complex reactions. When stable inorganic molecules have to be activated or in the case of redox reactions involving electron transfer processes, the requirement for the active site in terms of ability to exchange electrons is large, and in such cases the active site is often a inorganic cofactor containing transition metals [16]. This is not unlike inorganic catalysts, where transition metals also are involved in most complex catalysts.

In the following we illustrate the function of an enzyme by two examples. They are chosen because they are both demanding reactions requiring the activation of stable molecules, N_2 and CH_4, respectively, and because they have inorganic

Fig. 3. Schematics of the relationship between genetic code, amino acid sequence, enzyme structure and catalytic function. A piece of DNA is illustrated in (a). The backbone atoms are shown with balls like the atoms in Fig. 1, but with phosphorus colored orange, oxygen red, and carbon gray. The side chains comprised of the purine bases adenine and guanine, and the pyrimidine bases cystosine and uracil are shown by sticks and the four different bases are given different colors. Each subsequent triple of paired bases along the chain, called a codon, is the code for one specific amino acid along the peptide illustrated in (b). The backbone atoms of this chain are space-filled balls with oxygen colored red, carbon gray, and nitrogen blue. The side chains of the amino acids are illustrated with smaller balls and sticks, and different side chains are given different colors. The mechanism behind the translation from DNA to the peptide is called transcription and is illustrated by the arrow from (a) to (b). The peptide in (b) is a segment of the complete chain of amino acids that comprises PTP1B. The sequence of amino acids is called the primary structure of the protein. The chain eventually folds into a three-dimensional structure indicated in (c), where two other representations are shown of the PTP1B shown in Fig. 1. In the left representation the backbone atoms (the large gray and blue balls in (b)) are traced in order to show the folding of the chain. The color changes gradually along the chain in the left picture starting with the blue color and ending with the red color. When the chain folds, regions of the chain is twisted into a specific kind of spirals called α-helixes, other regions are extended and are called β-strands. These structural motifs are secondary structure elements and are shown by the cartoon in the right picture in (c) with α-helixes colored blue and β-strands yellow. The active site where the function of the enzyme takes place is however red as in Fig. 1. The tertiary structure of the protein describes how the secondary structure elements are folded into the native structure which determines the overall three-dimensional structure of the protein. The three-dimensional structure determines the function of the enzyme, which is illustrated by the arrow pointing from (c) to (d). The function of the enzyme is to abstract a phosphate group linked to the amino acid tyrosine in other enzymes [74]. A closeup of the active site without substrate is shown in the left picture. The atoms comprising the active site are illustrated by thick sticks while the remaining parts are illustrated by a green cartoon. The yellow atom is a sulfur atom from the amino acid cysteine. In the central picture of (d) a substrate molecule with a phosphate group linked to tyrosine binds to the active site through the phosphate group. The substrate molecule is illustrated by blue balls and sticks, but with tyrosine colored yellow, the phosphorus atom cyan, and the oxygen atoms of the phosphate group red. The structure of PTP1B after it has abstracted the phosphate group from the tyrosine is shown in the right picture. In the left picture an aspartic acid is marked with a white arrow. Notice how that acid has swept over in the central picture to a position, where it can donate a proton to the tyrosine–phosphate oxygen. In this way the enzyme breaks the bond between tyrosine and the phosphate group. The last two structures have been published in Ref. [75] (PDB ID: 1PTU), and in Ref. [76] (PDB ID: 1A5Y), respectively.

counterparts so that we can compare directly the products of evolution with the products of human effort.

3. Examples of enzyme processes

The two enzymes we consider are nitrogenase and MMO. The first enzyme catalyzes the synthesis of ammonia (NH_3) from atmospheric N_2 while the latter catalyzes the synthesis of methanol (CH_3OH) from natural gas (methane, CH_4). Both are industrially important reactions. The industrial ammonia synthesis has been known since the early 1900s. Contrary to the enzyme process, however, it takes place under harsh conditions, ≈400 °C and 100 atm. One interesting question in this connection is why the enzyme can function under much milder conditions and whether we can learn about new possibilities for man made catalysts by studying the enzyme. The direct partial oxidation of methane to methanol has been a dream reaction for decades in chemical industry [17], but there is still no man made catalyst that can do this in one step. Both examples suggest that we might learn by studying the way nature has solved the problem.

3.1. Nitrogenase

The conversion of N_2 into a biologically accessible form of nitrogen is a difficult process. The formation of NH_3 requires the N–N bond to be broken. This is one of the strongest bonds in chemistry; the bond energy is about 1000 kJ/mol. The process therefore requires the participation of an effective catalyst to proceed at reasonable temperatures. Ammonia is synthesized from N_2 by two very different catalytic processes. The Haber–Bosch process takes place by passing N_2 and H_2 over Fe or Ru surfaces at high temperatures and pressures, about 400 °C and 100 atm [18,19]. In nature, on the other hand, the enzyme nitrogenase catalyzes the synthesis of ammonia from N_2, electrons, and protons at room temperature and atmospheric pressure [15,20].

In the following we first show what is known about the biological process; then compare them to the inorganic process; and finally discuss why the two processes require such different reaction conditions.

3.1.1. The enzyme process
The overall enzyme process can be expressed as

$$N_2 + 6e^- + 6H^+ \rightleftharpoons 2NH_3 \qquad (1)$$

in which the simultaneous H_2 evolution has been omitted. The enzyme nitrogenase consists of two proteins. The one is the MoFe protein where the above reaction takes place. The other is the Fe protein, which feeds electrons into the MoFe protein to be used for the reduction of N_2. Large amounts of energy is used in the process, both in the form of a high energy of the electrons and by using at least 16 adenosine triphosphate (ATP) per turn over of Eq. (1) [15,20]. ATP is the typical energy carrier in biological systems, and releases its energy by reacting with water in which case it forms adenosine diphosphate (ADP) and a phosphate group (HPO_4^{2-}).

The binding and hydrogenation of N_2 in nitrogenase is believed to take place on the FeMo cofactor (FeMoco), see Fig. 4. The geometrical structure of the cofactor has been established recently [21–24], and a considerable amount of insight has been obtained about the overall kinetics of the hydrogenation process [20,25]. The binding site of N_2 and the atomistic mechanism of N_2 hydrogenation to ammonia are, however, still largely unknown. One of the major problems is that adsorbed N_2 has never been observed. It is therefore not even known where on the FeMoco N_2 adsorbs and even less how it is hydrogenated.

Fortunately, CO does adsorb on the active site, and as in inorganic surface science CO has been used as a probe molecule to investigate the FeMoco. Infrared spectroscopy has, for instance, been used to characterize the adsorption site [26]. These experiments have shown that up to two CO molecules can be adsorbed per cluster, and the adsorption is associated with a considerable redshift in the C–O stretch frequency. Another interesting result is that CO adsorption on the active site in the protein shows the same characteristics as CO adsorption of the cluster when it is taken out of the protein environment [27]. This indicates that the adsorption properties are not dependent in

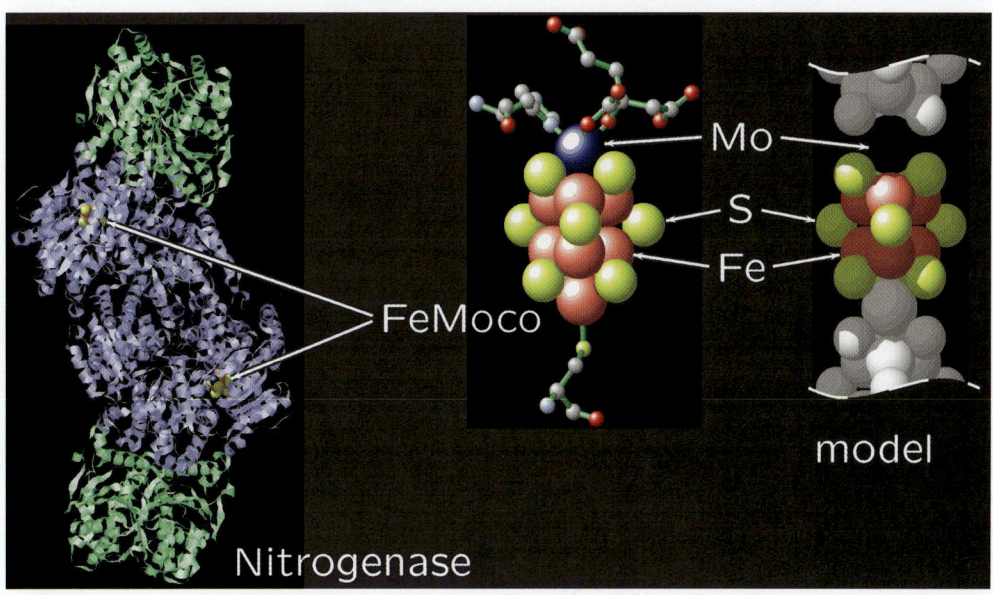

Fig. 4. The structure of nitrogenase as determined by X-ray crystallography published in Ref. [77] (PDB ID: 3MIN) (left). Also shown are the details of the active site, the FeMoco (center), and the model used to describe the FeMoco in the calculations (right).

detail on the environment and hence simple models including the cluster but not the details of the protein environment can be used to investigate the catalytic properties of the active site.

Since there is no direct experimental evidence for the interaction between N_2 and the active site, most of the hints about the atomistic mechanism comes from density functional theory (DFT) calculations [28–32]. These calculations can account for several aspects of the CO adsorption experiments [32].

DFT calculations using a $MoFe_6S_9$ complex to model the FeMoco, as indicated in the right-hand panel of Fig. 4 suggest that N_2 is adsorbed on the FeMoco without dissociating, and when electrons and protons are added to the N_2 molecule one by one, first one and then the second NH_3 molecule (or NH_4^+ ion) leaves the catalytic site [31,32]. In Fig. 5 we compare the energetics of adding H atoms (from H_2) one by one to an N_2 molecule in the gas phase and adsorbed on the $MoFe_6S_9$ complex. N_2 is adsorbed on the cofactor in going from panel a to b in Fig. 5. The first hydrogen is added to the adsorbed nitrogen in panel c. This cost energy for both the enzyme catalyzed reaction and for the gas phase reaction, but less so for the enzyme catalyzed reaction. From now on it is mainly downhill in energy. We add the next three hydrogen atoms one by one in panels d–f. Hydrazine (N_2H_4) is formed in panel f. Addition of a fifth hydrogen splits off the first ammonia as in panel g, and addition of yet another hydrogen forms the other and final ammonia as indicated in panel h. For the enzyme catalyzed reaction this ammonia is still adsorbed on the FeMoco but may eventually desorb. In that case the reaction is completed and we end up with the energy of panel i, which is the (calculated) energy of two ammonia molecules relative to N_2 and three H_2 molecules. The energy diagram in Fig. 5 shows that the critical step for the formation of ammonia is the addition of the first hydrogen atom to N_2, since this results in the least stable intermediate. In addition the energy diagram shows that the effect of the $MoFe_6S_9$ complex is significant since it stabilizes all the intermediates along the reaction path, particularly the least stable NNH intermediate by 110 kJ/mol, cf. panel c in Fig. 5.

The calculated energetics, Fig. 5, suggests that if the hydrogen entering the process comes directly

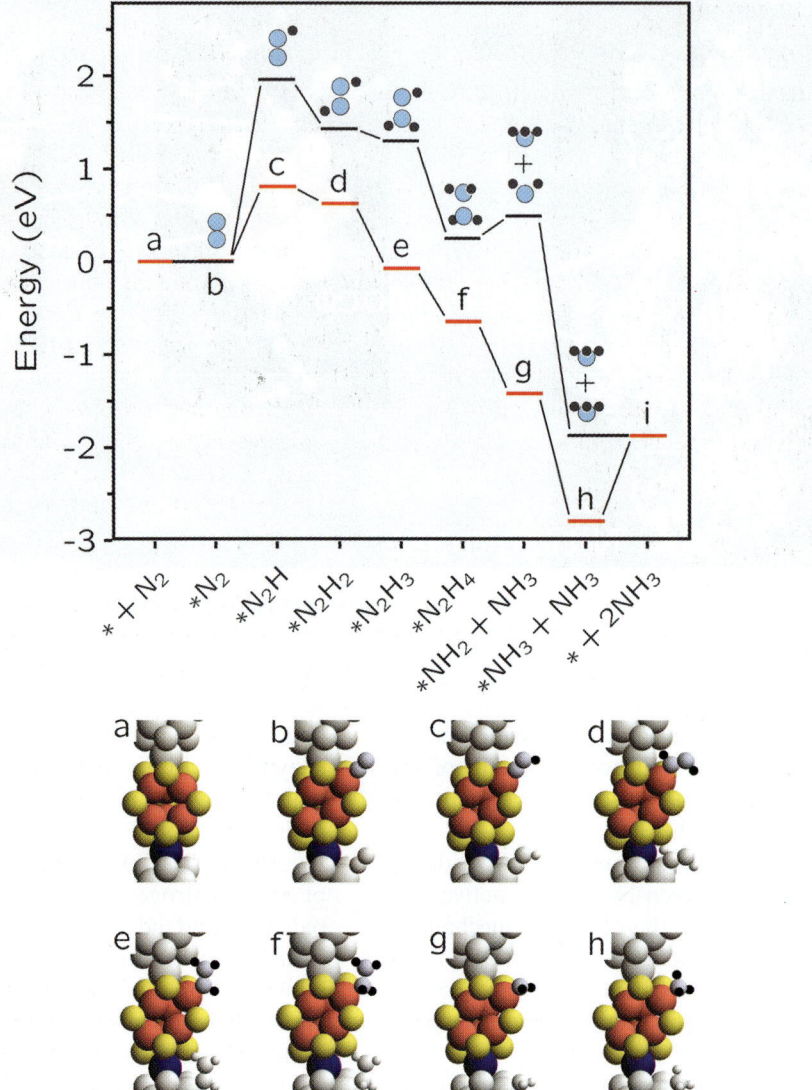

Fig. 5. The path of going from the reactants to the products of a reaction is called the reaction path. The above figure illustrates what happens along the reaction path for the hydrogenation of N_2 in the gas phase and on the $MoFe_6S_9$ complex modeling the FeMoco. In the top panel the calculated binding energies (1 eV = 100 kJ/mol) of the stable intermediate compounds are plotted as hydrogen is added one by one along the reaction path. The red bars indicate the energies of the intermediate compounds when N_2 is adsorbed on the $MoFe_6S_9$ complex and the black bars indicate the energies of the intermediate compounds when N_2 is in the gas phase. The energy is relative to the gas phase molecules $N_{2(g)}$ and $H_{2(g)}$, and zero-point energies are not included. The asterisk along the horizontal axis denotes either the $MoFe_6S_9$ complex in case of the red bars or vacuum in case of the black bars. The labels of the red bars refer to the corresponding intermediates of the complex reaction shown below the energy diagram. The gas phase structures are shown immediately above the corresponding black energy bars. Same color key as in Fig. 4, and with N being light blue, and H black. We now walk through the complex reaction. First N_2 adsorbs in going from (a) to (b). Then the first hydrogen is added to the adsorbed N_2 in going from (b) to (c). As seen from the energy diagram, this step is uphill in energy. Addition of three other hydrogen atoms through (d) and (e) results in hydrazine (N_2H_4) in (f). Addition of a fifth hydrogen atom splits off the first ammonia in (g), and addition of yet another hydrogen in (h) forms ammonia adsorbed on the complex, which eventually may desorb. All the steps after (c) are downhill in energy except for the final desorption of ammonia, indicating that the reaction proceeds easily as soon as the first hydrogen has been added (from Ref. [67]).

from H_2, the model enzyme still has a sizeable "barrier" associated with the NNH intermediate. However, in the enzyme the hydrogens are in the form of electrons and protons, and hence the energy of ("H" = $H^+ + e^-$) entering Eq. (1) is different from that of hydrogen in H_2. The effect of that is illustrated in Fig. 6 where the energetics of the enzyme catalyzed reaction is compared for the cases where the energy of the reacting hydrogens have the same energy as in H_2, and where the energy is 0.4 eV higher than that. It is seen that the effect of using high energetic electrons is to decrease the barrier by stabilizing the intermediates.

Thus, the enzyme lowers the reaction barrier by stabilizing the intermediates, and in addition it may use energetic electrons, which again lowers the barrier. These effects may be what makes the reaction go at room temperature [31]. If this is the case, it might be possible to produce ammonia by feeding electrons with a high energy into the isolated and solvated FeMoco. So far no one has been able to do that, but a first step in this direction has been taken by Pickett and coworkers [33]. They have reported that H_2 can be produced from protons from the solvent and electrons, provided the potential of the electrons is sufficiently negative. This demonstrates that it is possible to make the isolated cofactor catalyze a reaction where electrons and protons with a higher energy than that of H_2 are needed. Shilov and coworkers [34] have catalytically reduced acetylene by means of the FeMoco and various mercury compounds as reductants.

3.1.2. The surface catalyzed process

It is interesting to contrast the results in Fig. 5 with similar results for the surface reaction:

$$N_2 + 3H_2 \rightleftharpoons 2NH_3 \qquad (2)$$

This reaction on Fe and Ru surfaces has been the subject of a large number of experimental and theoretical studies. A detailed, molecular picture of the process has been developed [35–43]. The reaction proceeds via dissociation of N_2 and H_2 on the surface with subsequent hydrogenation of the adsorbed N atoms. DFT calculations of the reaction energetics on Ru surfaces illustrate the reaction mechanism, as indicated in Fig. 7. The dissociation of N_2 is the rate limiting step in this reaction. The barrier for dissociation is rather low; experiments find a dissociation barrier close to zero on Fe surfaces [36,37] and as low as 40 kJ/mol on stepped Ru [38]. Both values are consistent with the temperature dependence of the synthesis rate measured on an industrial catalyst [39–42]. They are also in excellent agreement with reaction barriers calculated using DFT [38,43]. The high temperatures and pressures are therefore not needed for N_2 dissociation to take place. H_2 dissociation is even more facile [39,44]. The problem in the surface process is that the bonding of the N and H atoms to the metal surfaces is so strong (cf. Fig. 7) that high temperatures are needed to have enough

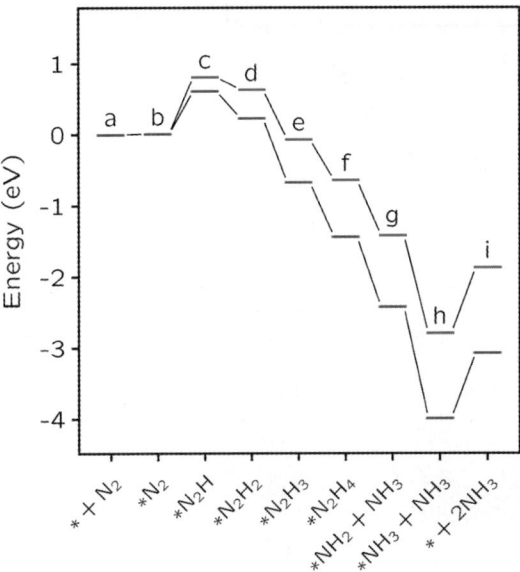

Fig. 6. Energetics along the reaction path for hydrogenation of N_2 adsorbed on the $MoFe_6S_9$ complex for two different energies of the entering electrons and protons. The reaction path and the intermediates are the same as shown in Fig. 5. The red bars are the binding energies of the intermediates when the energy of each hydrogen atom is equal to that of 1/2 H_2. These are the same as plotted in Fig. 5. The blue bars are the binding energies of the intermediates, when the energy of each hydrogen atom is 0.4 eV higher than the energy of 1/2 H_2. The result of shifting the energy of the hydrogens up is to shift each of the red bars down by the number of hydrogen atoms times the energy shift for each hydrogen atom, and hence the reaction is made more facile. The N_2 adsorption step and the NH_3 desorption step are unaffected by the shift in energy, since these steps involve no hydrogen atoms.

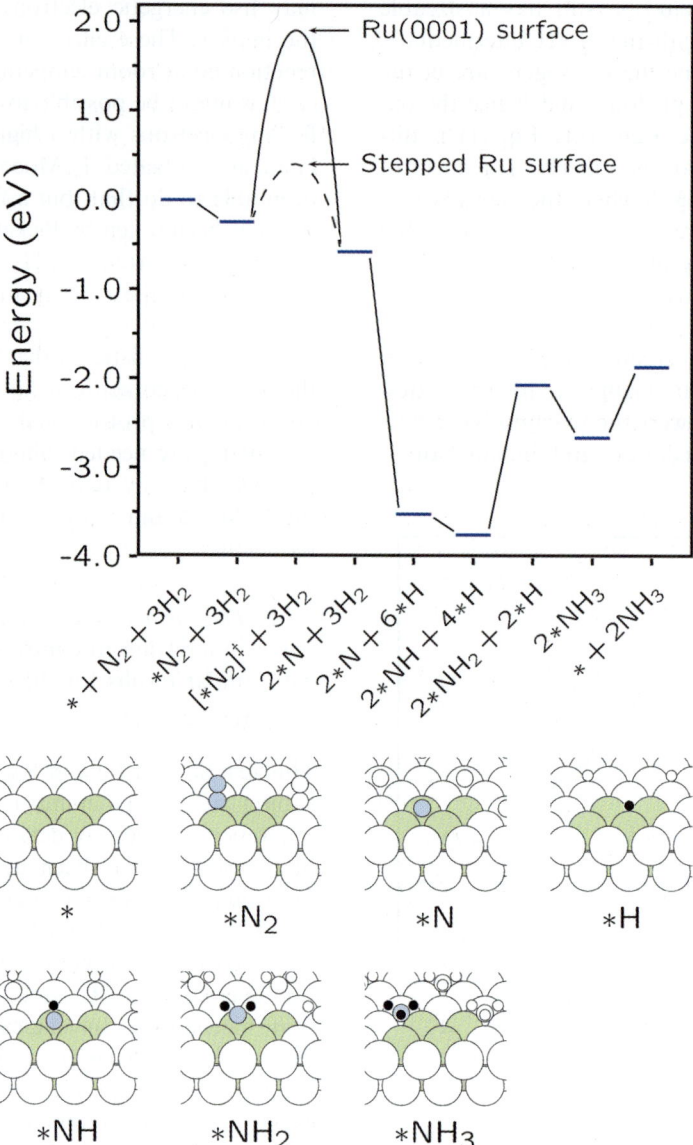

Fig. 7. Energy diagram for ammonia synthesis on Ru(0001). In the top panel the calculated energies for intermediates along the reaction path for the ammonia synthesis on Ru(0001) are shown, together with the calculated barriers for N_2 dissociation on a Ru(0001) surface and on stepped Ru surface. For the barriers see Ref. [38] for further details. The energy is defined relative to the energies of the clean Ru(0001) surface (*), and the gas phase molecules $N_{2(g)}$ and $H_{2(g)}$. Zero-point energies are not included. Structures of the most stable intermediates are shown below the energy diagram. One super cell is colored and the color code is the same as in Fig. 5, and with Ru being sand colored. N_2 is adsorbed in going from (a) to (b). After that N_2 dissociates in going from (b) to (c). This step is the hardest step for the reaction since there is a dissociation barrier. The calculated barriers for N_2 dissociation on a Ru(0001) surface and on stepped Ru surface is indicated in the figure. The barrier for dissociation on a step is sufficiently lower than for dissociation on a flat surface to make the step process completely dominant, even though the number of step sites may be significant lower than that of flat surface sites. In going from (c) to (d) we adsorb 6 hydrogen atoms, and in going from (d) to (e), two of these hydrogen atoms reacts with each a nitrogen atom to form two adsorbed NH. The intermediates in (d) and (e) are the most stable and from now on it goes uphill in energy as the remaining four hydrogen atoms react with the adsorbed NH species to form adsorbed NH_3 in (g) through (f). Eventually the two ammonia molecules desorb in going from (g) to (h) (from Ref. [67]).

clean, reactive surface available for dissociation and reaction [41]. The high temperatures has the side effect that the equilibrium in Eq. (2) is shifted to the left. This is not desirable, since no catalyst can produce more ammonia than the equilibrium amount. The high pressures are chosen to alleviate this problem, since that shifts the equilibrium back towards the products again.

A low temperature process based on N_2 and H_2 dissociation on the catalyst surface would therefore require a surface which does not bind the N and H atoms too strongly while keeping the barrier for N_2 dissociation low. Ru may be a slightly better catalyst than Fe because it does that to some extent, but a really good, low temperature catalyst based on the reaction Eq. (2) may be hard to find because the N–surface bond strength and the barrier for N_2 dissociation tends to be strongly coupled so that a weaker N–surface bond also results in a higher dissociation barrier [45,46].

We can now discuss the difference between the inorganic and the biological processes. As discussed above, the reason for the high temperature and pressure in the Haber–Bosch process is the fact that the N and H atoms are bound so strongly on the surface. The enzyme has solved this problem by not dissociating N_2 at all. The price is that the process of adding H to the adsorbed N_2 has a fairly high barrier. This is solved in the enzyme by feeding energy into the process in such a way that the H atoms are delivered as electrons and protons with a considerably higher energy than in the H_2 molecule.

3.2. Methane mono-oxygenase

Some organisms, the methanogenic bacteria produce considerable amounts of methane (CH_4) during their anaerobic metabolism. The amount of methane in the atmosphere is, however, kept down by methanotrophe bacteria, which use methane as their source for carbon and energy [47,48]. These bacteria either use the carbon directly for their biomass or oxidize the methane to CO_2 in order to obtain energy. The first step in the latter reaction is the partial oxidation of methane to methanol catalyzed by a group of enzymes named methane mono-oxygenases (MMOs). Some of these enzymes have active sites based on Fe, while other have active sites based on Cu [47,48]. The best characterized MMO is one with Fe in the active site, and we therefore focus on this enzyme here. We follow the same procedure as in the previous section by first showing what is known about the enzyme process, and then the surface process.

3.2.1. The enzyme process
The enzyme MMO catalyzes the reaction:

$$CH_4 + O_2 + 2H^+ + 2e^- \rightleftharpoons CH_3OH + H_2O \quad (3)$$

where the electrons are provided by reduced nicotineamide-adenine phosphate dinucleotide (NADPH) [47,48]. The NADPH is a nucleotide like ATP, but instead of being an energy carrier NADPH (and NADH) is the typical reduction agent (electron donor) in biochemical systems, delivering two electrons (e^-) and one proton (H^+) each time it is oxidized.

The X-ray structures of the protein containing the active site in its completely reduced and oxidized state have been determined [49–51,78]. The active site is an oxygen bridged iron dimer, depicted in Fig. 8, where the iron atoms are coordinated by nitrogen from histidine, and oxygen from glutamate, water and hydroxyl. Some of the ligands are quite mobile, and the Fe atoms go from being six-fold coordinated in the oxidized state as indicated in panel a of Fig. 8 to be five-fold coordinated in the reduced state as indicated in panel c of Fig. 8. Another structure of the active site in its oxidized state is shown in panel b of Fig. 8. In this structure a water molecule does not bridge the two iron atoms any more. Instead the Fe atoms are coordinated by an oxygen atom each from a solvent acetate.

The different crystal structures illustrate that the active site of the protein can change its structure. It is further known that complex formation with other protein components has a crucial effect on the rate of the reaction indicating that structural changes occur upon complex formation.

The reaction has been characterized in some detail by various spectroscopic and kinetic techniques. A sequence of five intermediates has been identified where O_2 adsorbs prior to CH_4 and after the two iron atoms have been reduced. The reaction sequence is illustrated in Fig. 9. First, the

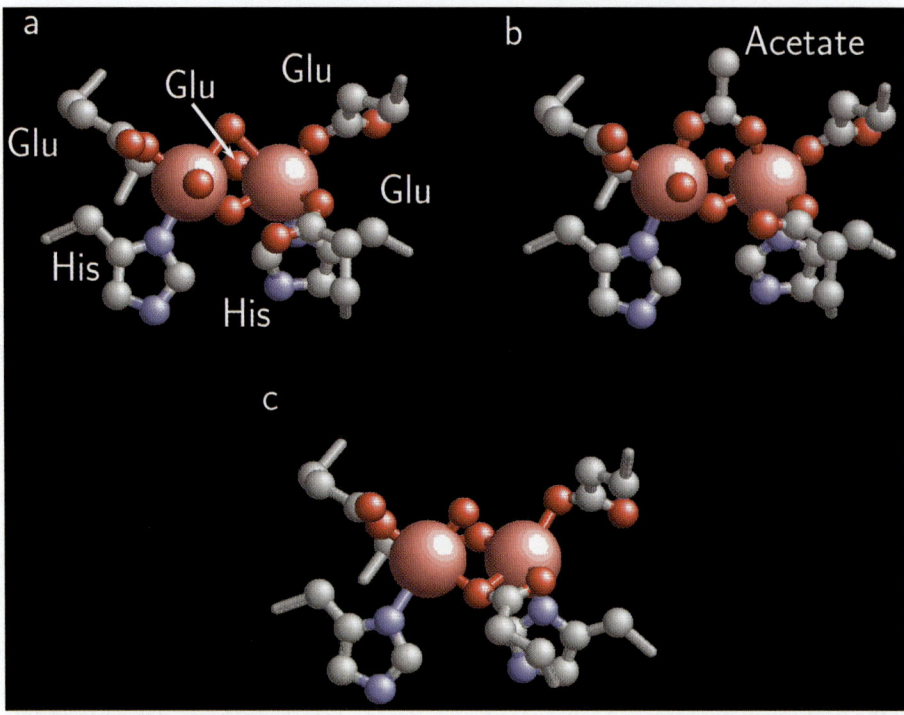

Fig. 8. Different crystal structures of the active site in MMO: (a) Oxidized resting state, denoted Ox in Fig. 9 (Ref. [50], PDB ID: 1MTY), (b) Oxidized state with acetate bridging the two Fe atoms (Ref. [49], PDB ID: 1MTY), (c) The reduced state, denoted Red in Fig. 9 (Ref. [78]). Gray balls indicate carbon, blue nitrogen, little red oxygen, and big brown iron. Hydrogen atoms are not shown for reasons of clarity. The iron atoms are coordinated by histidine (His) and glutamic acid (Glu) and water and hydroxyl groups (OH). They are six-fold coordinated in the oxidized states in (a) and (b) but are five-fold coordinated in the reduced state in (c). Both (a) and (b) represents oxidized states, but one of the bridging oxygens in (a) has been replaced with a bridging acetate in (b). The single oxygen atoms bound to the iron atoms are either from water or hydroxyl groups.

active site is fully reduced by two electrons to its ferrous state (Red in Fig. 9). After that O_2 is adsorbed to form the state O, which spontaneously converts to the state Q, through the transient state P* and the peroxo state P. The initial state and O are identified by their magnetic properties, while the states P and Q are identified by their optical properties. The state P* has been proposed because O decays faster than P increases. The resulting state Q activates the C–H bond of methane leading to the state T, which is an enzyme–product complex slowly dissociating into methanol and an oxidized ferric state (Ox in Fig. 9). The state T has been identified by using nitrobenzene instead of methane.

In spite of the knowledge accumulated so far, the details of the mechanism for oxidation of methane at the atomic level are largely unknown. Questions like what the structures of the intermediates are, where and how O_2 and CH_4 adsorb, and how the reaction between the reactants occurs remain to be answered. As in the case of nitrogenase, DFT calculations combined with spectroscopic results may help to provide answers.

DFT calculations have been reported by a number of groups [52–57]. They have used the X-ray crystal structures as a starting point for modeling of the active site. Different models have been employed by the different groups. There is qualitative agreement on how O_2 and CH_4 interacts with the core of the active site, although there are different suggestions as to how the Fe atoms are coordinated by the carboxylates and water in the intermediates P and Q. The DFT calculations

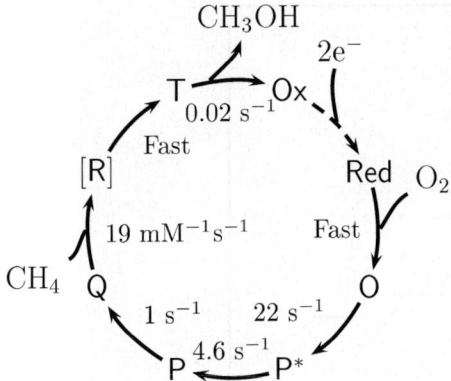

Fig. 9. Catalytic cycle of MMO which catalyzes the reaction: $CH_4 + O_2 + 2H^+ + 2e^- \rightleftharpoons CH_3OH + H_2O$ [47,48]. The reaction starts by adding two electrons to the oxidized resting state Ox, which results in the reduced state Red. After that we go through the states in alphabetical order starting with the state O, which is formed by O_2 adsorption on Red. This state decay through the states P* and P to the state Q. Upon methane adsorption the state T is formed through R. The state T finally dissociates into the Ox state and methanol. Protons and water should be added in order to balance the cycle stoichiometrically.

also have been able to reproduce the structural properties of P and Q, which have been revealed by spectroscopic methods, like EXAFS, Raman, IR, and Mössbauer spectroscopy [47,48].

Siegbahn and coworkers [30,52] and Siegbahn [54,55] are the only ones who have reported results for the whole reaction from the ferrous state to the ferric state by DFT calculations. We summarize the results from the most recent report [55].

Some of the structures obtained by Siegbahn are indicated in Fig. 10. His calculations suggest that the P structure is one where adsorbed O_2 bridge the two Fe atoms symmetrically. The O_2 then dissociates to form the Q structure. After that methane is activated by abstraction of H from methane by one of the bridging oxygen atoms. This results in the structure R and the transition state for this process is denoted by TS1 in Fig. 10. The resulting hydroxyl group recombines with the methyl group to form the structure T, where methanol bridges the two Fe atoms. The transition state for this latter process is denoted TS2 in Fig. 10.

The corresponding energetics also are shown in Fig. 10. It can be seen that the barriers in all cases are low enough to make the reaction go at room temperature.

In spite of the detailed DFT calculations, there remain important questions to be answered. For instance, the dependency of the formation rate of P and Q on pH [58] may suggest an alternative mechanism to the one suggested by the DFT calculations. The benefit (if any) of using electrons and protons instead of H_2 is also an interesting question.

3.2.2. The surface process

As mentioned above, there is no surface process whereby methane is converted directly into methanol. It would be extremely useful if such a process was to be found. It would make it possible to transform methane from the vast resources of natural gas into a product that was easy to transport and which could even be used directly as a substitute for gasoline in vehicles. Until such a catalyst is found, methanol is produced from methane by a two-step process combining several steps:

(1) The steam reforming process:

$$CH_4 + H_2O \rightleftharpoons CO + 3H_2 \quad (4)$$

which runs at very high temperatures (up to 1000 °C) over Ni based catalysts [59]

(2) The methanol synthesis process [60]:

$$CO_2 + 3H_2 \rightleftharpoons CH_3OH + H_2O \quad (5)$$

which is catalyzed by Cu based catalysts along with the reverse water-gas-shift reaction:

$$CO + H_2O \rightleftharpoons CO_2 + H_2 \quad (6)$$

The first Eq. (4) process is costly since the water has to be heated to high temperature. This makes it expensive at present to convert natural gas into methanol.

We are clearly in a situation where the human made technology is inferior to the enzyme "technology". There is no inorganic catalyst at present which can do what the enzyme can.

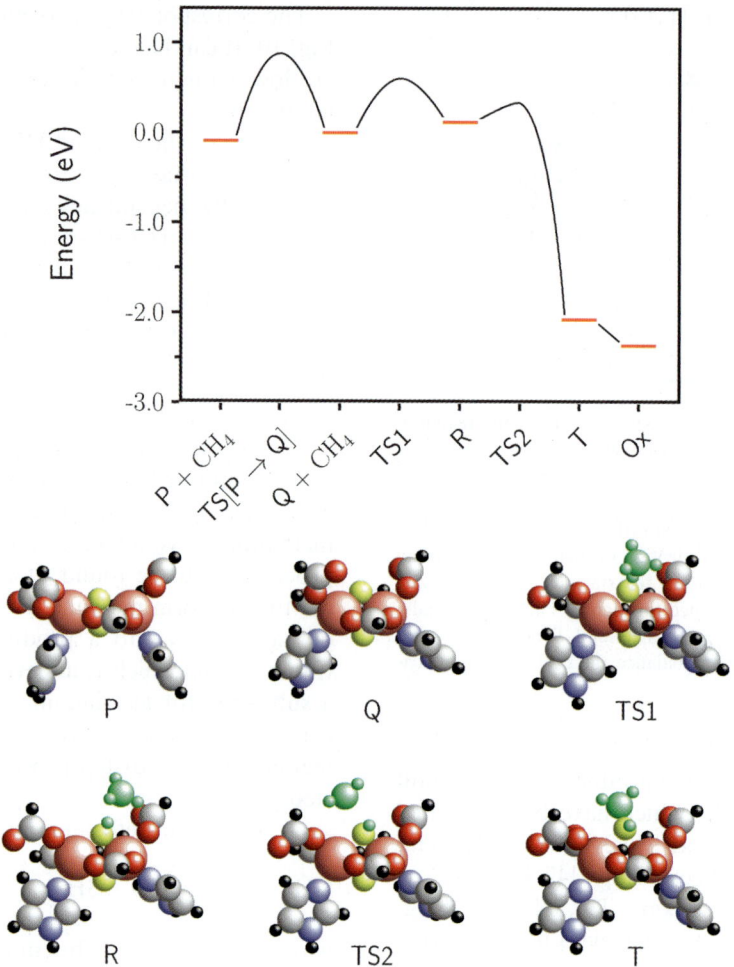

Fig. 10. Energy diagram for the catalytic cycle of MMO illustrated in Fig. 9. The energies are calculated by Siegbahn [55]. The energies of the intermediates are plotted in the top panel together with the calculated barriers between the intermediates. The energies are relative to the gas phase molecules CH_4 and O_2, and zero-point energies are not included. Some of the corresponding structures are shown below, where the yellow colored balls illustrate oxygen atoms from gas phase O_2, and where hydrogen atoms and the carbon atoms of methane are colored green. Other atoms are colored as in Fig. 8. The structures are rotated 180° relative to the structures in Fig. 8, and the letters P, Q, R and T refer to the states in Fig. 9. We start with the structure P to the left where oxygen is adsorbed. The structure Q is then formed by dissociation of the adsorbed dioxygen. The transition state for that process is TS (structure not shown). CH_4 then adsorbs and form the structure R where one of the oxygen atoms from the adsorbed dioxygen binds a hydrogen atom from CH_4, with a resulting elongated C–H bond. The transition state for this step is TS1 where the hydrogen atom is half between the carbon atom of CH_4 and the oxygen atom for the adsorbed dioxygen. The resulting OH group rotates such that the oxygen atom approaches and binds to the carbon atom to form the structure T, where methanol is adsorbed. The transition structure for that step is TS2. Eventually the structure T dissociates into CH_3OH and the Ox structure (not shown). To complete the cycle two electrons and protons should be added to (Ox) resulting in the Red state in Fig. 9 and release of H_2O. Addition of O_2 then results in the state P through O and P*, cf. Fig. 9.

4. Cross-fertilization

The possibilities for cross-fertilization between the surface science of solid surfaces and of enzymes are many. Here we focus on two avenues that seem promising.

First, there is a great need for a deeper understanding of reactions in enzymes. As indicated above, there is usually nothing like the detailed atomistic understanding of adsorption, bond breaking, and bond formation that is becoming standard in surface science. There are large possibilities for applying the vast array of experimental techniques developed to describe solid surfaces to investigations of enzymes. In a number of cases analogies already exist. A number of spectroscopic techniques are for instance applied to both solid surfaces and enzymes. Techniques as atomic force microscopy and scanning tunneling microscopy (STM) that have revolutionized surface science, also are beginning to be used to study enzymes or parts of enzymes, both in vacuum and under more realistic wet conditions [61,62]. There are also completely new possibilities that could be utilized in such studies. It would for instance be extremely interesting to use the STM tip to locate electron transfer paths and to transfer electrons directly into the enzyme. Electron transfer is not only a key process in the two examples discussed above, but more generally in many enzyme processes. Other possibilities is to study folding/unfolding and conformational changes for enzymes adsorbed on surfaces. Since the three-dimensional structure of enzymes to a large extent is determined by the environment, ultrahigh vacuum techniques developed for the study of surfaces are not directly applicable in the context of enzymes. However, one can imagine that it is possible to adsorb cofactors on surfaces and thereby study those by means of the ultrahigh vacuum techniques known in surface science. Some of the above examples have already been initiated, for other new technology is needed, but developments are rapid in this direction, see e.g. the contributions by Castner and Ratner [63], and Kasemo [64] in this volume.

The possibilities of applying the theoretical methods that have been used and tested extensively in surface science to enzyme processes are straightforward. Because the experimental detail is usually much larger in surface science the accuracy of the methods can be tested here and then the methods can be applied to enzyme problems with greater confidence. The main problem is to model the enzymes on a sufficiently detailed level. We are far from the goal, but new developments where molecular mechanics description of the protein surroundings is combined with a quantum mechanical description of the catalytic processes at the active site are being developed and offer new hope [65].

The combination of experimental and theoretical methods might change the way enzyme catalysis is described. The concepts of enzyme catalysis are still quite simple—the effect of the enzyme is usually described as a stabilization of the transition state relative to the transition state of the analogous uncatalyzed reaction [11,66]. It may well be that the whole reaction path changes in the presence of the enzyme such that a more advanced set of concepts is needed for a detailed description.

There are perhaps even larger possibilities for enzyme catalysis to teach surface scientists how to think of new, more clever catalytic reactions. It was shown above how nature has solved the problem of a direct conversion of methane into methanol while we still have no solid catalyst for a reaction that would profoundly change present petroleum technology. It is obvious that a more detailed understanding of the way nature has solved the problem might give the inspiration to a solution to the technical problem.

We conclude by illustrating this kind of inspiration from enzyme catalysis to surface catalysis in the case of the ammonia synthesis discussed above. An interesting question is whether an ammonia synthesis process like the biological one shown in Fig. 5 is possible directly on a metal surface, in particular if extra energy is fed into the reaction by using high energetic electrons. Again the DFT calculations can be used to study the question whether the FeMoco model system is unique in letting the process in Fig. 5 go so relatively easily, or whether a similarly facile process is possible at a metal surface.

Fig. 11 shows results for the direct hydrogenation of N_2 adsorbed on a Ru(0001) surface compared with the corresponding enzyme catalyzed reaction. As for the $MoFe_6S_9$ complex, the calculations for the surface have been performed by first adsorbing N_2 and then adding H atoms one by one. Each time a new H atom is added several binding sites on N_2 have been tried in order to find the most stable intermediate [31,67]. The energy diagram reveals that the energetics

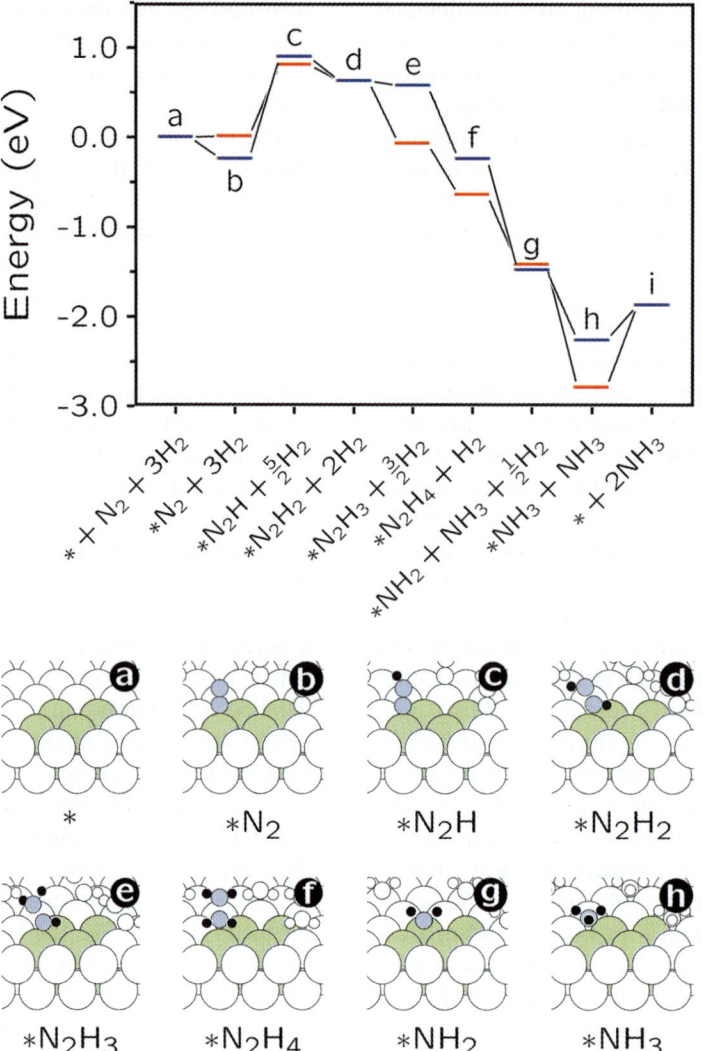

Fig. 11. Energy diagram for the hydrogenation of N_2 on the Ru(0 0 0 1) surface and on the $MoFe_6S_9$ complex modeling the FeMoco. In the top panel the energies of the most stable intermediates for the adsorption and hydrogenation of N_2 are plotted for Ru(0 0 0 1) (blue bars) and the $MoFe_6S_9$ complex (red bars). The energy is relative to the gas phase molecules $H_{2(g)}$ and $N_{2(g)}$ and zero-point energies are not included. The asterisk along the horizontal axis denotes either the $MoFe_6S_9$ complex in case of the red bars or the Ru(0 0 0 1) surface in case of the blue bars. The labels of the blue bars refer to the structures of the most stable intermediates for the Ru(0 0 0 1) surface shown below the energy diagram. Same color key as in Fig. 7. From Ref. [67]. By comparing the structures with the corresponding ones in Fig. 5 it is seen that the same intermediate structures are formed in the two reactions, and the above energy diagram demonstrates that the two reactions also are similar energetically. In both cases the least stable intermediate is the adsorbed NNH molecule, and hence the addition of the first hydrogen to N_2 is the critical step for both reactions.

for the hydrogenation of N_2 is very similar whether it is catalyzed by the $MoFe_6S_9$ complex or the Ru surface, in particular that the rate limiting step in both cases is the addition of the first hydrogen to N_2. The dissociation of N_2 is however still the most facile process on the Ru surface, since the barrier for that reaction is lower, cf. Fig. 7.

To make the hydrogenation process go on the surface instead of the dissociation process, we may use another metal with a higher barrier for N_2 dissociation. In fact, DFT calculations indicate that N_2 dissociation is impossible on the $MoFe_6S_9$ complex [67]. The barrier for hydrogenation may be decreased by using hydrogen in the form of electrons and protons as was illustrated in Fig. 6 for the $MoFe_6S_9$ complex [67]. The calculations do not have enough accuracy to predict accurately activation energies, but they indicate that the metal surface should in principle allow ammonia production at room temperature and atmospheric pressure if the electrons have a sufficiently high energy. This implies that it might be possible to form NH_3 at a metal surface, by using it as the cathode in a proton containing electrolyte in the presence of molecular nitrogen. There are two experimental reports that this may actually be possible [68,69], but much more experimental work is needed to investigate the possibility for achieving such a bioprocess in an inorganic environment.

5. Summary

Enzymes are crucial for life as they catalyze otherwise impossible reactions under physiological conditions. Enzymes catalyze a broad range of reactions from very simple hydrolysis to complicated redox processes, but usually a specific enzyme is very selective towards the reaction catalyzed. The activity and selectivity of an enzyme are determined by its three-dimensional structure. In principle the three-dimensional structure is given directly by the genetic code determining the amino acid sequence of the protein. Due to the complexity of the problem, this is however difficult in practice, but progress is fast in this direction. Several structures are now determined directly by experimental techniques, which continues to improve as well. The structure–function relation of enzymes is, however, still largely unknown. We suggest that surface science techniques to some extent can be adopted in enzyme research to improve the detailed description of enzymes and their function, but also that a detailed description of enzymes may give inspiration to new design principles for man-made catalysts.

We have exemplified the above points by giving two examples, the ammonia synthesis and the methanol synthesis. Ammonia is synthesized in nature from N_2 by the enzyme nitrogenase and methanol from methane by the enzyme MMO. The issue of finding the reaction mechanisms behind the enzyme catalyzed syntheses have in both cases been addressed by making DFT calculations in conjunction with spectroscopic and kinetic studies. DFT calculations have proven a valuable tool for finding mechanisms behind surface processes, and can for instance explain why the surface catalyzed ammonia synthesis need high temperature and pressure to go. We have shown here that DFT calculations suggest that the enzyme nitrogenase uses a different mechanism than the surface process.

In the enzyme process adsorbed N_2 is hydrogenated gradually by energetic electrons and protons, and this can explain why the enzyme can synthesize ammonia at ambient temperature and pressure. DFT calculations also suggest that a similar reaction can be catalyzed on a surface at low temperature if N_2 cannot dissociate and if the entering electrons and protons have a sufficiently high energy. We can also get inspiration from the enzyme MMO. Today we have no direct catalyzed pathway to synthesize methanol from methane. Instead methane is oxidized to CO and CO_2, which then is reduced to methanol. The enzyme illustrates that it is possible to synthesize methanol directly from methane, and the DFT calculations suggest a detailed reaction mechanism behind the synthesis.

Acknowledgements

We gratefully acknowledge Prof. P.E.M. Siegbahn for providing us with his calculated structures of the active site of MMO and for letting us read his manuscript before publication. We are also grateful to Prof. Lippard, Prof. Rosenzweig and D. Whittington for sending us the structure of the reduced MMO. Finally, we would like to thank Dr. M. Feig for helpful suggestions to improve the text.

The present work was financed in part by The Danish Research Councils through grant no. #9501775. The Center for Atomic-scale Materials Physics (CAMP) is sponsored by the Danish National Research Foundation.

References

[1] M.D. Adams et al., The genome sequence of Drosophila melanogaster, Science 287 (2000) 2185–2195.
[2] The human genome, Science 291 (2001). See also http://www.sciencemag.org/feature/data/genomes/landmark.shl.
[3] The human genome, Nature 409 (2001). See also http://www.nature.com/genomics/.
[4] V.S. Pande, A.Y. Grosberg, T. Tanaka, Heteropolymer freezing and design: towards physical models of protein folding, Rev. Mod. Phys. 72 (2000) 259–314.
[5] C.M. Dobson, A.R. Fersht (Eds.), Protein Folding, Philosophical Transactions of the Royal Society of London Series B, vol. 348, Royal Society, London, 1995.
[6] E.M. Boczko, C.L. Brooks III, First-principles calculation of the folding free-energy of a 3-helix bundle protein, Science 269 (1995) 393–396.
[7] D.A. Debe, M.J. Carlson, J. Sadanobu, S.I. Chan, W.A. Goddard III, Protein fold determination from sparse distance restraints: the restrained generic protein direct Monte Carlo method, J. Phys. Chem. B 103 (1999) 3001–3008.
[8] M. Gruebele, The fast protein folding problem, Annu. Rev. Phys. Chem. 50 (1999) 485–516.
[9] Q. Ning, T.J. Sejnowski, Predicting the secondary structure of globular-proteins using neural network models, J. Mol. Biol. 202 (1988) 865–884.
[10] P. Baldi, S. Brunak, Bioinformatics—the Machine Learning Approach, MIT Press, Cambridge, MA, 1998.
[11] T. Palmer, Understanding Enzymes, fourth ed., Prentice Hall, Ellis Horwood, London, 1995.
[12] D. Barford, A.K. Das, M. Egloff, The structure and mechanism of protein phosphatases: insight into catalysis and regulation, Annu. Rev. Biophys. Biomol. Struct. 27 (1998) 133–164.
[13] S.B. Prusiner, Molecular biology of prion diseases, Science 252 (1991) 1515–1522.
[14] Molecular Biology of Prion Diseases, Philosophical Transactions of the Royal Society of London Series B, vol. 343, 1994, p. 1306.
[15] L. Stryer, Biochemistry, fourth ed., W.H. Freeman, New York, 1995.
[16] W. Kaim, B. Schwederski, Bioinorganic Chemistry: inorganic Elements in the Chemistry of Life, Wiley, New York, 1991.
[17] J.M. Thomas, W.J. Thomas, Principles and Practise of Heterogeneous Catalysis, VCH, Weinheim, Germany, 1997.
[18] S.R. Tennison, in: J.R. Jennings (Ed.), Catalytic Ammonia Synthesis Fundamentals and Practice, Plenum Press, New York, 1991, p. 303.
[19] K.-I. Aika, K. Tamura, Ammonia synthesis over non-iron catalysts and related phenomena, in: A. Nielsen (Ed.), Ammonia: Catalysis and Manufacture, Springer, Berlin, 1995, p. 103.
[20] B.K. Burgess, D.J. Lowe, Mechanism of molybdenum nitrogenase, Chem. Rev. 96 (1996) 2983–3012.
[21] J. Kim, D.C. Rees, Structural models for the metal centers in the nitrogenase molybdenum–iron protein, Science 257 (1992) 1677–1682.
[22] M.K. Chan, J. Kim, D.C. Rees, The nitrogenase FeMo-cofactor and P-cluster pair: 2.2 Å resolution structures, Science 260 (1993) 792–794.
[23] J. Kim, D. Woo, D.C. Rees, X-ray crystal structure of the nitrogenase molybdenum–iron protein from Clostridium pasteurianum at 3.0 Å resolution, Biochemistry 32 (1993) 7104–7115.
[24] S.M. Mayer, D.M. Lawson, C.A. Gormal, S. Mark Roe, B.E. Smith, New insights into structure–function relationships in nitrogenase: a 1.6 Å resolution X-ray crystallographic study of Klebsiella pneumonia MoFe-protein, J. Mol. Biol. 292 (1999) 871–891.
[25] R.N.F. Thorneley, D.J. Lowe, Kinetics and mechanism of the nitrogenase enzyme system, in: T.G. Spiro (Ed.), Molybdenum Enzymes, Wiley, New York, 1985, pp. 221–285, Chapter 5.
[26] S.J. George, G.A. Ashby, C.W. Wharton, R.N.F. Thorneley, Time-resolved binding of carbon monoxide to nitrogenase monitored by stopped-flow infrared spectroscopy, J. Am. Chem. Soc. 119 (1997) 6450–6451.
[27] S.K. Ibrahim, K. Vincent, C.A. Gormal, B.E. Smith, S.P. Best, C.J. Pickett, The isolated iron–molybdenum cofactor of nitrogenase binds carbon monoxide upon electrochemically accessing reduced states, Chem. Commun. (1999) 1019–1020.
[28] I. Dance, Calculated details of a mechanism for conversion of N_2 to NH_3 at the FeMo cluster of nitrogenase, Chem. Commun. (1997) 165–166.
[29] I. Dance, Understanding structure and reactivity of new fundamental inorganic molecules: metal sulfides, metallocarbohedrenes, and nitrogenase, Chem. Commun. (1998) 523–530.
[30] P.E.M. Siegbahn, J. Westerberg, M. Svensson, R.H. Crabtree, Nitrogen fixation by nitrogenase: a quantum chemical study, Phys. Chem. B 102 (1998) 1615–1623.
[31] T.H. Rod, B. Hammer, J.K. Nørskov, Nitrogen adsorption and hydrogenation on a $MoFe_6S_9$ complex, Phys. Rev. Lett. 82 (1999) 4054–4057.
[32] T.H. Rod, J.K. Nørskov, Modeling the nitrogenase FeMo cofactor, J. Am. Chem. Soc. 122 (2000) 12751–12763.
[33] T. Le Gall, S.K. Ibrahim, C.A. Gormal, B.E. Smith, C.J. Pickett, The isolated iron–molybdenum cofactor of nitrogenase catalyses hydrogen evolution at high potential H^+, Chem. Commun. (1999) 773–774.

[34] T.A. Bazhenova, M.A. Bazhenova, G.N. Petrova, A.K. Shilova, A.E. Shilov, Catalytic reduction of acetylene and dinitrogen with the participation of the iron–molybdenum cofactor of nitrogenase and synthetic polynuclear molybdenum(III) complexes, Russ. Chem. Bull. 47 (1998) 861–867.

[35] N.D. Spencer, R.C. Schoonmaker, G.A. Somorjai, Iron single crystals as ammonia synthesis catalysts: effect of surface structure on catalytic activity, J. Catal. 74 (1982) 129.

[36] F. Bozso, G. Ertl, M. Grunze, M. Weiss, Interaction of nitrogen with iron surfaces, J. Catal. 49 (1977) 18–41.

[37] I. Alstrup, I. Chorkendorff, S. Ullmann, The dissociative chemisorption of nitrogen on iron (111) at elevated pressures, Z. Phys. Chem. 198 (1997) 123–134.

[38] S. Dahl, A. Logadottir, R.C. Egeberg, J.H. Larsen, I. Chorkendorff, E. Törnqvist, J.K. Nørskov, Role of steps in N_2 activation on Ru(0001), Phys. Rev. Lett. 83 (1999) 1814–1817.

[39] O. Hinrichsen, F. Rosowski, M. Muhler, G. Ertl, The microkinetics of ammonia synthesis catalyzed by cesium-promoted supported ruthenium, Chem. Eng. Sci. 51 (1996) 1683–1690.

[40] S. Dahl, P.A. Taylor, E. Törnqvist, I. Chorkendorff, The synthesis of ammonia over a ruthenium single crystal, J. Catal. 178 (1998) 679–686.

[41] P. Stoltze, J.K. Nørskov, Bridging the "pressure gap" between ultrahigh-vacuum surface physics and high-pressure catalysis, Phys. Rev. Lett. 55 (1985) 2502–2505.

[42] M. Bowker, I. Parker, K.C. Waugh, The application of surface kinetic data to the industrial synthesis of ammonia, Surf. Sci. 197 (1988) L223–L228.

[43] J.J. Mortensen, L.B. Hansen, B. Hammer, J.K. Nørskov, Nitrogen adsorption and dissociation on Fe(111), J. Catal. 182 (1999) 479–488.

[44] F. Bozso, G. Ertl, M. Grunze, M. Weiss, Chemisorption of hydrogen on iron surfaces, Appl. Surf. Sci. 1 (1977) 103–119.

[45] J.K. Nørskov, P. Stoltze, Theoretical aspects of surface reactions, Surf. Sci. 189/190 (1987) 91–105.

[46] M. Mavrikakis, L.B. Hansen, J.J. Mortensen, B. Hammer, J.K. Nørskov, Dissociation of N_2, NO, and CO on transition metal surfaces, in: D.G. Truhlar, K. Morokuma (Eds.), Transition State Modeling for Catalysis, vol. 721, ACS Symp. Ser, 1999, p. 245 (Chapter 19).

[47] B.J. Wallar, J.D. Lipscomb, Dioxygen activation by enzymes containing binuclear non-heme iron clusters, Chem. Rev. 96 (1996) 2625–2657.

[48] E.I. Solomon, T.C. Brunold, M.I. Davis, J.N. Kemsley, S.-K. Lee, N. Lehnert, F. Neese, A.J. Skulan, Y.-S. Yang, J. Zhou, Geometric and electronic structure/function correlations in non-heme iron enzymes, Chem. Rev. 100 (2000) 235–349.

[49] A.C. Rosenzweig, C.A. Frederick, S.J. Lippard, P. Nordlund, Crystal structure of a bacterial non-haem iron hydroxylase that catalyses the biological oxidation of methane, Nature 366 (1993) 537–543.

[50] A.C. Rosenzweig, H. Brandstetter, D.A. Whittington, P. Nordlund, S. Lippard, C.A. Frederick, Crystal structures of the methane monooxygenase hydroxylase from Methylococcus capsulatus (Bath): implications for substrate gating and component interactions, Proteins Struct. Funct. Genet. 29 (1997) 141–152.

[51] N. Elango, R. Radhakrishnan, W.A. Froland, B.J. Wallar, C.A. Earhart, J.D. Lipscomb, D.H. Ohlendorf, Crystal structure of the hydroxylase component of methane monooxygenase from Methylosinus trichosporium OB3b, Protein Sci. 6 (1997) 556–568.

[52] P.E.M. Siegbahn, R.H. Crabtree, Mechanism of C–H activiation by diiron methane monooxygenases: quantum chemical studies, J. Am. Chem. Soc. 119 (1997) 3103–3113.

[53] P.E.M. Siegbahn, R.H. Crabtree, P. Nordlund, Mechanism of methane monooxygenase a structural and quantum chemical perspective, J. Biol. Inorg. Chem. 3 (1998) 314–317.

[54] P.E.M. Siegbahn, Theoretical model studies of the iron dimer complex of MMO and RNR, Inorg. Chem. 38 (1999) 2880–2889.

[55] P.E.M. Siegbahn, O–O bond cleavage and alkane hydroxylation in methane monooxygenase, J. Biol. Inorg. Chem. 6 (2001) 27–45.

[56] H. Basch, K. Mogi, D.G. Musaev, K. Morokuma, Mechanism of the methane → methanol conversion reaction catalyzed by methane monooxygenase: a density functional study, J. Am. Chem. Soc. 121 (1999) 7249–7256.

[57] B.D. Dunietz, M.D. Beachy, Y. Cao, D.A. Whittington, S.J. Lippard, R.A. Friesner, Large scale ab initio quantum chemical calculations of the intermediates in the soluble methane monooxygenase catalytic cycle, J. Am. Chem. Soc. 122 (2000) 2828.

[58] S.-K. Lee, J.D. Lipscomb, Oxygen activation catalyzed by methane monooxygenase hydroxylase component: proton delivery during the O–O bond cleavage steps, Biochemistry 38 (1999) 4423–4432.

[59] J.R. Rostrup-Nielsen, in: J.R. Anderson, M. Boudart (Eds.), Catalysis—Science and Technology, Springer, Berlin, Germany, 1984 (Chapter 1).

[60] J.B. Hansen, Methanol synthesis, in: G. Ertl, H. Knözinger, J. Weitkamp (Eds.), Handbook of Heterogeneous Catalysis, vol. 4, VCH, Weinheim, Germany, 1997, pp. 1856–1876 (Chapter 3.5).

[61] C. Bai, Scanning Tunneling Microscopy and its Application, second ed., Springer, Berlin, Germany, 2000.

[62] V.J. Morris, A.R. Kirby, A.P. Gunning, Atomic Force Microscopy for Biologists, Imperial College Press, London, England, 1999.

[63] D.G. Castner, B.D. Ratner, Biomedical surface science: Foundations to frontiers, Surf. Sci. 500 (2002) 28.

[64] B. Kasemo, Biological surface science, Surf. Sci. 500 (2002) 656.

[65] P.D. Lyne, M. Hodoscek, M. Karplus, A hybrid QM–MM potential employing Hartree-Fock or density functional methods in the quantum region, J. Phys. Chem. A 103 (1999) 3462–3471.

[66] J.R. Knowles, Enzyme catalysis: not different just better, Nature 350 (1991) 121–124.
[67] T.H. Rod, A. Logadottir, J.K. Nørskov, Ammonia synthesis at low temperatures, J. Chem. Phys. 112 (2000) 5343–5347.
[68] G. Marnellos, M. Stoukides, Ammonia synthesis at atmospheric pressure, Science 282 (1998) 98–99.
[69] V. Kordali, G. Kyriacou, C. Lambrou, Electrochemical synthesis of ammonia at atmospheric pressure and low temperature in a solid polymer electrolyte cell, Chem. Commun. (2000) 1673–1674.
[70] F.C. Bernstein, T.F. Koetzle, G.J. Williams, E.E. Meyer Jr., M.D. Brice, J.R. Rodgers, O. Kennard, T. Shimanouchi, M. Tasumi, The protein data bank: a computer-based archival file for macromolecular structures, J. Mol. Biol. 112 (1977) 535.
[71] H.M. Berman, J. Westbrook, Z. Eng, G. Gilliland, T.N. Bhat, H. Weissig, I.N. Shindyalov, P.E. Bourne, The protein data bank, Nucleic Acids Res. 28 (2000) 235–242.
[72] http://www.rcsb.org/pdb/.
[73] D. Barford, A.J. Flint, N.K. Tonks, Crystal structure of human protein tyrosine phosphatase 1b, Science 263 (1994) 1397–1404.
[74] G.H. Peters, T.M. Frimurer, O.H. Olsen, Electrostatic evaluation of the signature motif (htv)cx_5r(s/t) in protein-tyrosine phosphatases, Biochemistry 37 (1998) 5383–5393.
[75] Z. Jia, D. Barford, A.J. Flint, N.K. Tonks, Structural basis for phosphotyrosine peptide recognition by protein tyrosine phosphatase 1b, Science 268 (1995) 1754–1757.
[76] A.D.B. Pannifer, A.J. Flint, N.K. Tonks, D. Barford, Visualization of the cysteinyl-phosphate intermediate of a protein-tyrosine phosphatase by X-ray crystallography, J. Biol. Chem. 273 (1998) 10454–10462.
[77] J.W. Peters, M.H.B. Stowell, M. Soltis, M.G. Finnegan, M.K. Johnson, D.C. Ress, Redox-dependent structural changes in the nitrogenase P-cluster, Biochemistry 36 (1997) 1181–1187.
[78] A.C. Rosenzweig, P. Nordlund, P.M. Takahara, C.A. Frederick, S.J. Lippard, Geometry of the soluble methane monooxygenase catalytic diiron center in two oxidation states, Chem. Biol. 2 (1995) 409–418.

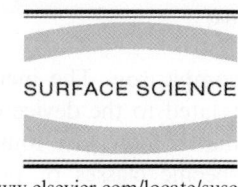

Computation with DNA on surfaces

Susan D. Gillmor, Paul P. Rugheimer, Max G. Lagally *

Department of Materials Science and Engineering, University of Wisconsin—Madison, 1509 University Avenue, Madison, WI 53706, USA

Received 31 July 2000; accepted for publication 10 April 2001

Abstract

DNA computation has the potential to tackle computationally difficult problems that have real-world implications. The parallel search capabilities of DNA make it a valuable tool to approach problems that have a large number of possible solutions, for which conventional computers have limited potential. Surface science can play a significant role in harnessing the parallel nature of DNA for computation. This article briefly reviews conventional computing architecture, discusses DNA computation, and describes the role of surface science in DNA computation. © 2001 Elsevier Science B.V. All rights reserved.

Keywords: Computer simulations; Surface chemical reaction; Biological molecules – nucleic acids

1. Introduction

Computer technology changes in the blink of an eye. In the beginning of the 1970s the first microcomputers were introduced with a clock frequency of about 100 kHz. Those systems were capable of performing about 60,000 instructions per second. By the early 1990s the fastest processor had a clock frequency of 100 MHz. Microprocessors running at this clock speed were able to perform as many as 150 million instructions per second (MIPS). The clock frequency has continued to increase, to a present value of 1 GHz, providing approximately 2.7 billion instructions per second. [1]

The most famous metric to gauge progress in computing was introduced by Gordon Moore. Although it has been slightly modified with time, "Moore's Law" [2] states that the number of transistors on a chip doubles every 18 months, a trend that has continued since Moore's observations in the mid-1960s. Moore's law is not a physical law, but instead is a consequence of human ingenuity, engineering prowess, and innovation.

Soon, the present numbers too will be surpassed, and today's computers will seem slow in

---

* Corresponding author. Tel.: +1-608-263-2078; fax: +1-608-265-4118.
E-mail address: lagally@engr.wisc.edu (M.G. Lagally).
[1] http://www.tomshardware.com/mainboard/00q2/000529/index-12.html. Benchmark performed using a 1 GHz Intel Pentium III and SiSoft Sandra software.

[2] In 1965, Gordon Moore made the observation that for the preceding three years, the number of components on an integrated chip had doubled each year. He predicted that the chips would follow this trend and increase in complexity until it reached 65,000 components per chip (~10 years time). The accuracy of his prediction astounded the computer industry. In fact, the density of transistors on a chip doubled yearly long after 1975 and has only recently slowed to doubling every 18 months. *Electronics* October 31, 1965.

0039-6028/01/$ - see front matter © 2001 Elsevier Science B.V. All rights reserved.
PII: S0039-6028(01)01524-2

comparison. The increase in speed is, of course, related to the device density on the chip. Yet, no matter how fast tomorrow's computers become, specific classes of problems overwhelm conventional computers, requiring them to calculate for months or even years. Many of these problems have considerable technological and societal as well as research importance, and so the development of computer technology to address them is of considerable interest.

In this article we discuss DNA computation as the basis for a class of possible future computers, and the potential role of surfaces in DNA computation. DNA computing holds vast potential for solving problems that are extremely difficult to solve by conventional computation methods. A deoxyribonucleic acid (DNA) molecule is a long string made up of four bases, adenine, thymine, guanine, and cytocine, arranged in a specific sequence determined by its function (see glossary and further discussion in Section 4.1). A DNA base is a ring-shaped molecule consisting of carbon, oxygen, nitrogen, and hydrogen. The arrangement of the bases on the DNA molecule defines a code. Within an organism, the DNA code represents the information necessary to assemble proteins, which in turn perform various functions throughout the body. Conventional computing uses a binary code, strings of 1s and 0s, representing switches that are on or off. In DNA computation, the bases can be considered bits of data, and a string of bases on a DNA molecule represents a quaternary code (a four base code system). Using the toolbox of biochemistry, various problems can be encoded in DNA. The DNA can be manipulated through a series of biomolecular reactions that represent the computation steps. After the computation is completed, the code of the remaining DNA is deciphered to reveal the solution to the problem. Because of the ability to perform biochemical operations simultaneously on many different DNA molecules, a problem that would take all the supercomputers in the world running in parallel years to solve could, in principle, be solved in a matter of minutes with a DNA computer.

The massive mapping of information onto DNA sequences is the basis for the ability of DNA to perform highly complex computations rapidly.

To provide a feeling for the magnitude of information storage in DNA, consider the following. On a written page, there are approximately 1500 characters, arranged in some order to represent words, sentences, and paragraphs. If 3 bases were to represent a character, 4500 bases could encode a page. If each of the 119 million volumes in the Library of Congress has 1000 pages, then approximately 2×10^{13} different DNA molecules each only 20 bases long could encode the entire contents of the library. A milliliter sample of 1 millimolar (1 mM) solution of different DNA molecules 20 bases long could represent the entire library as well as provide ~40,000 copies.

Why is DNA computation a potential area of future interest for surface scientists? DNA computation can be carried out in solution or with DNA bound to a surface. Performing the reactions necessary to compute with DNA is much easier on a surface than in solution. Surface science issues that need to be addressed for reliable DNA computation on a surface include development of the molecular surface chemistry to prepare the surface for DNA attachment, the subsequent attachment of the DNA, the development of reactions on the attached DNA to represent computation steps, the patterning of the attachment layer, and the detachment of DNA from the surface. We discuss aspects of each of these issues in this paper. Because a number of technical terms unique to biochemistry and DNA manipulation used in this article may be unfamiliar to the reader, we provide a glossary of terms at the end of the article.

2. Serial computation

As a baseline for computing power, let us consider the standard silicon-based microprocessors of today. Conventional silicon-based computing is primarily a serial process. With this serial architecture, today's fastest supercomputers are capable of performing hundreds of billions of operations per second. The most widely advertised feature of today's microprocessors is the clock speed, typically expressed in terms of MHz or GHz. The clock in a microprocessor supplies a repetitive signal that is used like a metronome to control the rate at which the various internal functions of the microprocessor are performed. The number of MIPS roughly scales with clock speed. The MIPS value is a universal metric that can be applied to numerous other computing architectures, including DNA computing.

Transistors and capacitors are the fundamental components of a microprocessor. Transistors are used to modify and route information, and capacitors are used to store information, in the form of 0s and 1s. Normally, a charged capacitor represents a 1 and an uncharged capacitor represents a 0. This on/off nature of the capacitors is a physical realization of the digital character of modern computers. The capacitor is either on (charged) or off (uncharged). A single bit is the smallest amount of information, either a single 1 or 0. By considering two or more bits together, larger and larger amounts of information can be represented. A single bit can only represent two possible numbers, 2 bits can represent four numbers, 3 bits can represent eight numbers, 4 bits can represent sixteen numbers, obeying the rule that the largest number that can be represented by a single N-bit string of data is 2^N. Although this digital storage of information has not changed since the first computers were built, initially computers could only process and store 2-bit and 4-bit numbers at a time. Larger numbers had to be reduced into 2-bit and 4-bit pieces. Today, 32-bit processors are becoming old-fashioned and 64-bit processors are becoming mainstream. By allowing larger numbers to be represented in one operation, modern 64-bit computers are able to perform more operations per second than the initial 2-bit computers, even when clock speed is accounted for. The number of bits dealt with in a single operation and the clock speed are both increasing with each new microprocessor generation. Although increasing either one of these properties by itself improves microprocessor performance, increasing both of them together yields the largest increase in the MIPS value.

A standard microprocessor for a personal computer operates by loading information from memory into one of the regions of the chip that performs the software instructions. The regions that perform the software instructions are referred to as the processor 'pipelines' because information flows from one end of the pipeline to the other like water through a pipe. More complex than a water pipe, a microprocessor pipeline consists of capacitors and transistors for storing and manipulating information, as well as interconnect wires for transferring the information from one place to another. A pipeline is typically 10–15 stages in length, which can be thought of as 10–15 positions in a production assembly line. Each stage is a place where the information is manipulated once according to the software instructions before being moved to the next stage. The length, or number of stages in a pipeline is a parameter that is determined when a new processor is designed. Designers must estimate the number of stages that would most benefit the overall speed of the processor, based upon what kind of computations they anticipate the computer is expected to perform. Except in special cases, a processor with fewer, longer pipelines is faster than a processor with many, shorter pipelines. Fig. 1 shows schematically a processor with three pipelines each containing five stages.

Since information flows down a pipeline from one end to the other, pipelined microprocessors process data serially. The pipeline design of the processor works as follows. The microprocessor loads information into the first stage of the pipeline. It then performs an instruction on the information in the first stage of the pipeline, moves the result to the second stage of the pipeline, and simultaneously loads new information into the first stage of the pipeline. When information is contained in both the first and second stages of the pipeline, the processor performs the necessary instructions on both stages. It then moves that information to the second and third stages and loads new information into the first stage. This process is repeated until the pipeline is full. At this point, the processor is performing calculations at 10–15 times the rate it would if it were to process one piece of information from start to finish.

Although the assembly-line-style stacking of information in pipelines greatly improves microprocessor performance, this architecture is serial in nature. Most calculations require that many operations be performed, and each operation must be applied separately from every other operation.

Because each operation and move requires several clock cycles to perform, the net result is that some computations can take a significant amount of time to complete. The long time required to perform some calculations is a disadvantage of this serial method of computation. To some extent, engineers are able to make the process parallel by building several pipelines into a single microprocessor (see Fig. 1), and by putting several independent processors to work simultaneously. 'Super computers' achieve high MIPS figures in this way. Inherently, though, the process remains serial, with at most a few hundred parallel pipelines in which information can be serially processed. A single microprocessor typically only has a handful of parallel pipelines. For example, the Advanced Micro Devices (AMD) Athlon chip consists of three fully parallel floating point unit pipelines, and three fully parallel integer pipelines [1].

2.1. Future of serial computing

Computers are able to run faster and perform more operations per second as their circuits get smaller and smaller. Most experts agree, however, that for silicon-based computing, this continued shrinkage reaches fundamental physical limits in approximately ten to fifteen years. At that point, transistors will be only a few hundred atoms across. Quantum mechanical effects, electrical noise, heat dissipation requirements, and crystal imperfections will require a change in paradigm in order to continue to increase computer speed. If single molecules could be used to transport current and function like today's silicon transistors, wires, and capacitors, then continuing Moore's law would be possible for many more years. Researchers already have successfully assembled molecular devices to create the elements of memory modules, logic elements, and wires. The successful assembly of 1000 molecular devices in an area a mere 30 nm in diameter, thousands of times denser than the devices produced in current silicon technology, has been demonstrated [2]. Although developments in molecular devices such as these will continue to make serial computation faster, practical problems that are considered computationally difficult still are not efficiently solvable on such computers, because of the serial nature of the processing architecture. For these problems alternative, non-serial computing architectures are being considered.

3. Parallel computation

The recent Internet boom has led to some massively parallel distributed computational schemes. The computational power of computers linked together by the internet and appropriate computer programs creates a large computational resource that can address computationally intensive problems that can be solved rapidly by a parallel process. Examples of this are the *Mersenne prime number search* [3] and the *SETI at home* project [4]. Mathematicians have developed terminology to describe the computational difficulty of problems. For example, the simple algorithm used by most people to multiply two *n*-digit numbers using pencil and paper takes approximately n^2 steps to complete. The reverse process, factoring, has a much higher degree of computational complexity.

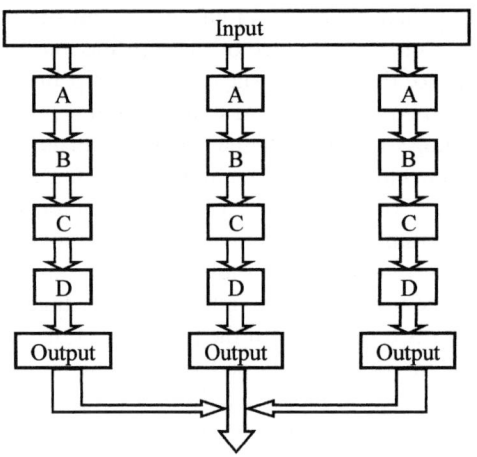

Fig. 1. A simplified schematic of the parallel pipeline architecture of modern desktop processors. A 5 stage pipeline architecture is shown, including the output stage as the fifth stage, with 3 pipelines working in parallel. Stages A through D could be operations such as loading data, loading instructions, aligning data, adding data, etc.

The best-known algorithm to factor an n-digit number with unknown factors into smaller integers requires at least $10^{\sqrt{n}}$ steps [5]. The number $10^{\sqrt{n}}$ gets extremely large for relatively small n. In fact, factoring a 200 digit number ($n = 200$) would require more computing power than any single computer possesses today. Problems that are solvable in a number of steps less than a polynomial in n (i.e., require no more than n^j operations, where j is a fixed number) are called tractable problems. Problems that scale as a super-polynomial, like $10^{\sqrt{n}}$ or n^n, are called intractable problems. Intractable does not mean unsolvable. It just means that the problem is not solvable efficiently. In practice this means that the problem can not be solved quickly.

Through the use of parallel computation, certain problems, even those that are considered intractable, can be solved relatively quickly. Distributed-computing schemes like SETI at home and the Mersenne prime number search are attempts to perform calculations with many computers that are simultaneously calculating different portions of the same problem. Through massive parallelization with ordinary computers, some intractable problems are solvable in a reasonable amount of time.

The *traveling salesman problem*, considered intractable by conventional computing schemes, is formulated as follows: Given a list of cities, a given starting city and a given final destination, construct the most efficient (shortest) route to visit all the cities exactly once. This type of problem can become very large very quickly. For example, visiting 18 cities has 18! ($=6.4 \times 10^{15}$) possible routes. If we assume that only one operation is required to evaluate each route, then a 1 GHz processor would take 2,371,000 s ($=27$ days) to search through the set of possible routes in a serial fashion to find the solution. In actuality, evaluating each route would take dozens or hundreds of operations, significantly increasing the time required to compute the correct answer. The best search algorithms avoid a purely serial approach, but they still increase in time exponentially with increasing number of possible solutions. In a truly parallel approach, possible routes are simultaneously compared to specific criteria determined by a computing algorithm and a large fraction of possible solutions is simultaneously rejected as incorrect.

It is worth considering just how far massively parallel computation could go. If every person on Earth (6 billion) had a computer capable of performing 1000 MIPS, 6×10^{18} operations per second could in theory be made available for a massively parallel computational effort. Compare this to the number of atoms on a surface. There are approximately 10^{15} atoms per square centimeter on the surface of most solids. Therefore, there are more than 10^{18} atoms on the surface of a solid that measures ~ 1 m^2. Through the use of any well chosen chemical reaction, all of these 10^{18} atoms can be affected (in one way or another) within one second or even in much less time. Spilling a glass of juice on a tabletop is not computation, but illustrates the potential of truly parallel computation.

4. DNA computation

To understand DNA computation, it is necessary to review the characteristics of the DNA molecule and our ability to tailor and to manipulate it for specific purposes. After a brief review of DNA, we discuss how it has been used to solve simple examples of intractable problems, its potential to solve non-trivial ones, and the advantages of surface-based DNA computation. DNA computation is still in its infancy, and its validity as a future methodology is by no means assured. Nevertheless, research stimulated by the desire to develop a DNA computer has implications for DNA manipulation, surface modification and functionalization, and the development of algorithms that transcend conventional computation.

4.1. An overview of DNA

A DNA molecule is composed of a string of bases (adenine, thymine, cytosine and guanine) bound together through a sugar-phosphate backbone (see Fig. 2). DNA can occur as a single strand that contains these four bases, or as hybridized double strands in the double-helix structure. In hybridized DNA, the bases on opposing

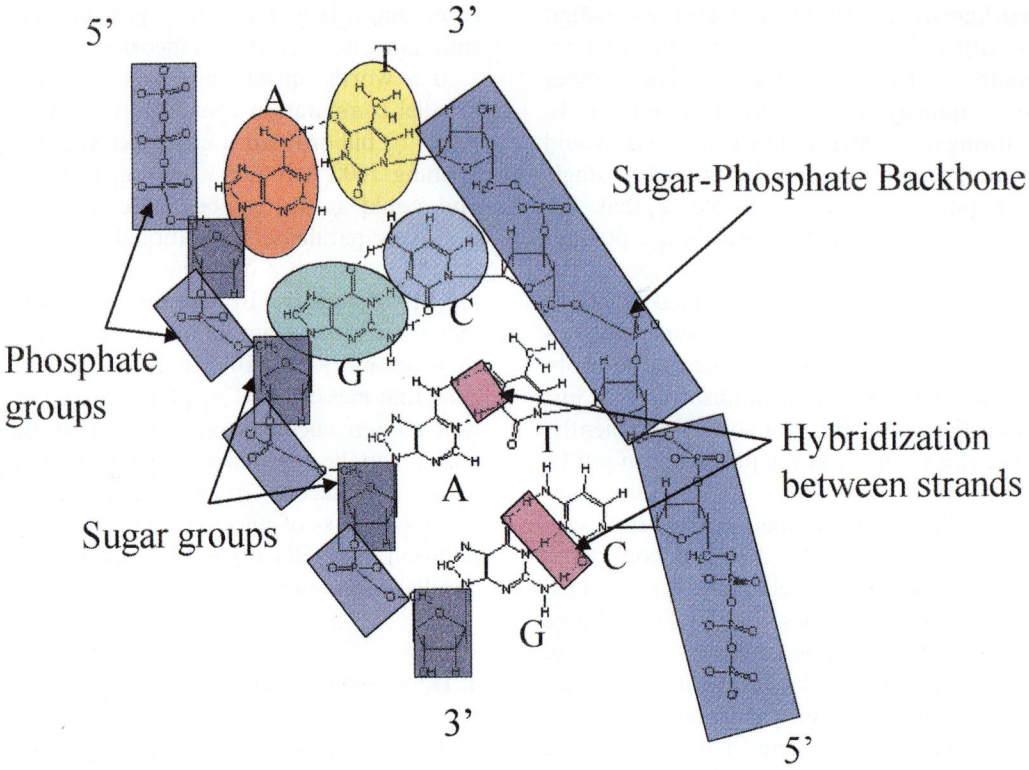

Fig. 2. Structure of double-stranded DNA. The four bases are arranged along the sugar-phosphate backbone in a particular order (the DNA sequence). Adenine (A) pairs with thymine (T), while cytosine (C) pairs with guanine (G). Weak hydrogen bonds between the bases hold the two DNA strands together. This pairing is called hybridization, and a particular single-stranded sequence has a corresponding complementary sequence. Reading from the top left the sequence is AGAG, with the complementary right strand reading TCTC. Image taken from Ref. [6].

single strands are paired, A–T and C–G, and connected through hydrogen bonds, so that the complementary string of bases mirrors the first string, in terms of having all the appropriate bases in the opposing position. The backbone of the DNA strands (both single and double-stranded DNA) has two distinct ends that have been dubbed "five prime" (5′) and "three prime" (3′). The 5′ end has an unconnected phosphate group while the 3′ end has an unconnected sugar. In the double-stranded configuration, a 3′ end is paired with a 5′ end (see Fig. 2 [6]). Enzymes (proteins that act as catalysts) key on these distinct end groups to orient themselves along the DNA strand for many operations such as transcription (the first step of protein synthesis), replication (the copying of DNA), or digestion (cutting a DNA strand) (see Glossary).

4.2. DNA as a computational tool

In a conventional computer, all data are represented in binary code, where each 0 or 1 is a bit and 8 of these integers, a byte, encode a character. For DNA, each base is the equivalent of a bit. In biology, 3 bases encode for each amino acid, (an amino acid is the biological character for protein synthesis) (see Glossary), and these 3 bases are therefore a biological byte. RNA (a biological molecule with a structure similar to DNA) and various enzymes access the information in the DNA and translate the code into proteins, which in turn carry out specific functions for the survival of the cell or organism. The genome in reality is a huge data storage site, which holds in its base sequences the information necessary to synthesize hundreds of thousands of proteins.

DNA can be used as a computational tool that can be highly parallel. We have already shown how DNA codes information. The parallel nature of DNA reactions can be illustrated simply. For example, one milliliter of a millimolar solution of DNA contains 10^{17} molecules of DNA, which we can synthesize to a specific length, say 20 bases long. Suppose we can make a solution of single-stranded DNA in which 10^{15} strands have different base sequences (on average, 100 copies of each sequence in the solution). We now introduce a milliliter of a different millimolar solution of single-stranded DNA, and among these 10^{17} molecules, there are 10^{14} different strands with approximately 1000 copies of each. Assume that there are 10^4 sequences that are complements to the original solution, and 10^{10} sequences that do not match any strands in the original solution. These strands all simultaneously seek to find and to hybridize with their exact complement, an extremely parallel process. After a nominal period of time to allow the reaction kinetics to proceed (1–24 h), we can analyze the solution to see which sequences have hybridized into a doubled-stranded structure and which ones are left in the single-stranded form. We should find 10^4 different sequences that have found complements, and we should find on the average 100 copies of each of them. The other molecules do not find a match and can be removed, as described later.

4.3. DNA computation—proof of principle

DNA can be used in computation in the following manner. First, information is encoded on single strands of DNA. Particular base sequences, e.g. AATCGATCG, can be easily and reliably synthesized with a DNA synthesizer. The processes of hybridization can be used to select specific sequences. Computation steps consist of selecting specific sequences based on some algorithm for solving a problem. Sequences not selected are removed. Each computation step further reduces the number of sequences until all computation steps are completed. The remaining sequences are then decoded. They represent the possible answers to the problem. Hybridization, de-hybridization, and the action of enzymes are employed to perform the computational steps and decipher the code.

Before we explain steps involved in DNA computation more fully, we describe an example of the types of problems that can efficiently be solved with DNA manipulation. In 1994, Leonard Adleman demonstrated the use of DNA in solving a traveling-salesman problem involving seven cities. The salesman must begin at city 0, end at city 4 (see Fig. 3a) and visit each city exactly once. Usually, traveling-salesman problems attempt to minimize distances as well, but to simplify the example, Adleman does not include distance in the encoding of the DNA, but rather designates specific inter-city routes as shown by the arrows in the graph of Fig. 3a. He uses the following algorithm to detect if a solution existed to the problem:

1. Generate random routes through the graph along the allowed city-to-city connections (Fig. 3a).
2. Keep only those routes that begin with city 0 and end with city 4.
3. Keep only those routes that enter exactly 7 cities.
4. Keep only those routes that enter all of the cities of the graph at least once.
5. If any routes remain, say "yes"; otherwise say "no".

He encodes the paths and the cities with particular single-stranded DNA sequences. For example, city 1 is encoded in a 20 base sequence, and the path from city X to city 1 is represented by a 20-base sequence, in which the first 10 bases are complementary to city X and the next 10 bases are complementary to city 1. When the three sequences of DNA that represent the path, for example, of city 3 to city 1, city 1, and the path of city 1 to city 6 hybridize, they form a potential travel route from city 3 to city 1 to city 6 (see Fig. 3b). Note that the path from city 6 to city 1 is represented by a different sequence than is the path from city 1 to city 6 (see Fig. 3c). The one-way and two-way streets are a specific characteristic of the problem postulated by Adleman. When Adleman combined the various strands that represented the cities and paths, they were in principle able to form

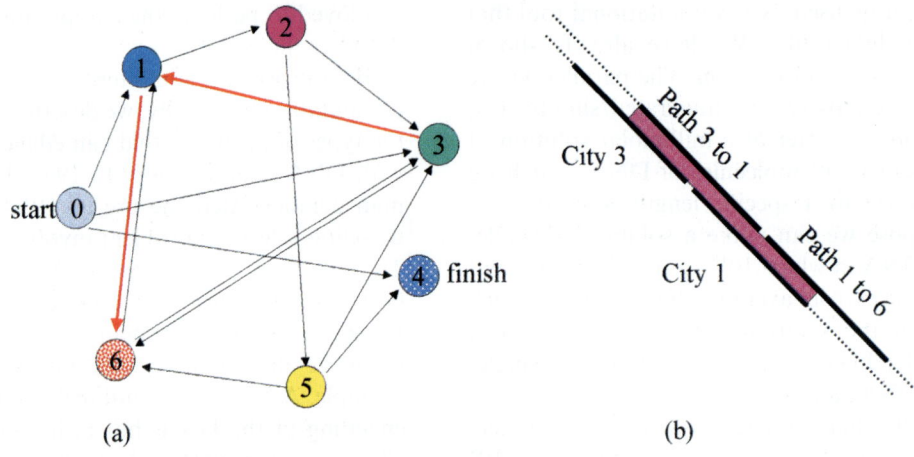

Fig. 3. Traveling salesman problem of Leonard Adleman [7]. The object is to find the path by which each city is visited exactly once. (a) Each circle represents a different city and each arrow represents a path with a specific direction; the paths shown are the only ones allowed. There are a several possible routes, but only one that represents the correct path. (b) Adleman represented each city (0–6) with a small sequence of DNA. Each path between the cities is represented with partial complementary sequences of two different cities, such as the path between city 3 and 1 as seen in (b). When the DNA strands interact, they hybridize to form "bridges" between the city strands and create possible routes for the traveling salesman. The correct solution to the problem is: 0 to 3 to 6 to 1 to 2 to 5 to 4. (c) The encoding of the cities and paths is specific and the encoding of the path from city 1 to city 6 is not the same as that of path from city 6 to city 1. The base sequences for the cities were chosen at random; the sequences for the paths were chosen to be complementary to the cities they connect.

all the possible combinations of routes for the travelling salesman (step 1 of his algorithm).

Afterwards, he "ligates" (essentially a weld, performed by an enzyme, between the ends of single strands abutting each other) the various fragments of DNA that had hybridized to form routes between cities. The ligation step produces two continuous complementary single strands of DNA, which, after de-hybridization, both encode for a particular route, even though one strand was originally "cities" and the other was "paths". Subsequently, he performs a polymerase chain reaction (PCR, a reaction that creates many copies of DNA strands in solution, see Glossary) and specifies that only those strands that started with city 0 or ended with city 4 were copied (step 2 of his algorithm). To find the correct solution to his problem, he then separates the DNA strands into specific size categories, through gel electrophoresis (a standard separation technique in molecular biology), and chooses only those strands that are 140 bases long (7 cities × 20 bases per city = 140 bases, step 3 of his algorithm). From this subset, he systematically checks that each DNA strand includes every city only once, eliminating those that skipped one city and visited another twice. Any route represented by the strands that skipped any city is eliminated (step 4 of his algorithm). The surviving DNA sequences encode for the route that visited each city exactly once (step 5 of his algorithm), which he then deciphers via DNA sequencing to find the correct answer. Adleman's paper [7] has become the basis of a new research field, DNA Computation [8]. The new field combines the expertise of molecular biologists, surface scientists, and material chemists with the interests of computer scientists to attempt to solve complex problems. Adleman's example in his demonstration of the utility of DNA as a computational tool can be easily solved using a conventional computer. For a traveling salesman problem involving 50 cities, DNA computation would be a viable, if not essential, approach.[3]

5. Surface-based DNA computation

As described so far, DNA computing would seem to be a molecular biology lab that consults with a computer scientist. In this solution-based, or "test-tube" approach, the molecular biologist synthesizes the sequences of the DNA strands to encode for variables of a specific problem to solve. The computer scientist provides the algorithms. The biochemical manipulation of the algorithm generally involves numerous unwieldy DNA reaction steps using very small amounts of liquids that are mixed, separated, and analyzed.

A potentially more reliable approach to DNA computation uses surfaces to support the DNA reactions. The principles are the same as described above, but there are numerous benefits of surface-based DNA computation. (1) With DNA bound to a surface, the computational operations become very simple. Computation steps are performed by introducing a solution to the surface and allowing it to react with the bound DNA on the surface. Later, DNA strands in the reaction solution are simply washed away. (2) When the DNA is attached to the surface, the loss of DNA between steps is minimal. In the solution-based approach, every transfer from one test tube to another leaves residual liquid behind, and in that liquid, some fraction of the DNA mixture, thus introducing errors. (3) Once immobilized on a surface, the single-stranded DNA strings (oligonucleotides) cannot interact with each other, while in solution, two complementary single-stranded sequences can hybridize. Thus a surface approach with tethered DNA strands would reduce interference and false signals. (4) Purification of DNA sequences, necessary at each step, is as simple as washing the surface with water to remove unwanted reaction products. Separation in a test-tube-based system is usually achieved through gel electrophoresis, and dialysis (a procedure that removes the majority of salts from a reaction mixture) is necessary sometimes to remove unwanted salts; both procedures are tedious and time consuming. (5) DNA computation on a surface can benefit from the wealth of tools available in silicon processing, designed for integrated-circuit fabrication, including patterning, etching, and transporting of silicon wafers, eliminating the need to design new materials-processing technology. Marrying the parallel processing of DNA to the controllability that surfaces allow gives a tangible methodology to implement DNA computation.

5.1. Process overview

In surface-based DNA computation, the information density is reduced to two dimensions from three dimensions in the test-tube approach. A 10×10 cm^2 surface, in which one DNA molecule occupies a 25 nm^2, can hold only about 10^{14} molecules of DNA [9]. This limitation is countered by a greater simplicity and reliability of the surface-based technique, and also the easy scalability to large sample areas and multiple surfaces.

The initial consideration in surface-based DNA computation is to prepare oligonucleotides with appropriate base sequences and to bind them to the surface for further manipulation to solve a specific problem. After determining a reliable surface attachment chemistry, an implementation of the DNA computation must be defined. A search technique to eliminate strands that are not answers to a given problem or step of a problem must be

---
[3] In a traveling salesman problem in which each city has two paths entering it and two paths exiting the city and if one examines only those routes that have traveled to n cities, possibly some more than once, there would be 2^n possible routes for n cities. For a 50-city tour, there would be 2^{50} ($\sim 1.1 \times 10^{15}$) possible routes.

defined. Finally, the strands that represent the answer to a given problem must be decoded. These steps are schematically illustrated in Fig. 4, with labels to identify a particular step. To explain these individual steps more explicitly, we demonstrate their application in a scheduling problem.

5.2. Implementation

Scheduling problems fall into the category of intractable problems, which are difficult to solve through conventional strategies, but would be manageable through a highly parallel approach. For example, a student wants to take a math, a literature, and a Latin class. The school offers a number of classes, including several in math, literature, and Latin, during the course of a day at three time slots (8 am, 10 am, and 1 pm). To find a schedule that satisfies the student's wishes, all of the possible schedules have to be searched and compared to the criterion of a math, a literature and a Latin class. Even in a trivial problem, the number of possible solutions can become large, as seen in Fig. 5. With only three time slots and seven classes available at each time slot, there are 343 different schedules. A parallel search method, such as DNA computing, to weed out the schedules that do not include the desired classes would make the problem less cumbersome. We now examine how DNA computation would work for our scheduling problem.

5.2.1. Design

Consider how to represent classes by encoding small DNA sequences. One strategy is to use a "word" design to store information in string of bases [10–12]. For a 5-base DNA word, there are 4^5 (1024) possible DNA sequences, and in an 8-base DNA word, 4^8 (64,536) possible sequences. A DNA word can have the following design:

$$5'-FFFFVVVVFFFF-3' \qquad (1)$$

where F is a fixed base (i.e. every strand has those bases at that location) and V is a variable base (i.e., the ones that differ from strand to strand). The variable-base region contains the encoded information; the fixed-base region is present simply to promote hybridization. Again, hybridization (the formation of double-stranded DNA) is essential for solving problems with DNA. In our problem, let each class be represented with a DNA word and let them be concatenated to form a "schedule:"

$$5'-F_1F_1F_1F_1XXXXXF_1F_1F_1F_1F_2F_2F_2F_2$$
$$\times YYYYYF_2F_2F_2F_2F_3F_3F_3F_3ZZZZZF_3F_3F_3F_3-3' \qquad (2)$$

where variables, X, Y, or Z represent the base sequences for the three desired classes, and where F_n are the fixed bases for each position. The fixed bases can also act as time divisions. For instance, the classes encoded with F_1 fixed bases would be the 8 am class; the classes with F_2 would be the 10 am class; and F_3, the 1 pm class.

5.2.2. Generation

Even for this simple scheduling problem, 343 different combinations are possible, and hence we need 343 different strands of DNA encoded to represent the different schedules. The generation of all these strands is best accomplished using a split-pool approach. For example, to make DNA strands consisting of three DNA words, DNA molecules representing math, literature, and Latin at 8 am are synthesized in three separate test tubes on support particles. They are subsequently mixed, and then redistributed into three new test tubes to add another word representing Latin, Japanese, and German at 10 am on the end of the strands, a different word in each test tube. In the final mixing, redistributing, and synthesis, chemistry, math and philosophy classes at 1 pm are added to create 27 different schedules of the three words as indicated Fig. 6 [13]. If such a strategy were repeated for 100 different words for 6 cycles, 10^{12} (100^6) different strands of DNA words would be generated in a combinatorial, not a serial, manner.

5.2.3. Attachment/Immobilization

Next, the DNA must be bound to the surface. A number of methods to tether DNA strands to a surface exist [14], many of them used in addressable DNA arrays (an array in which each DNA sequence is in a known, organized position, see glossary). DNA computation does not require an

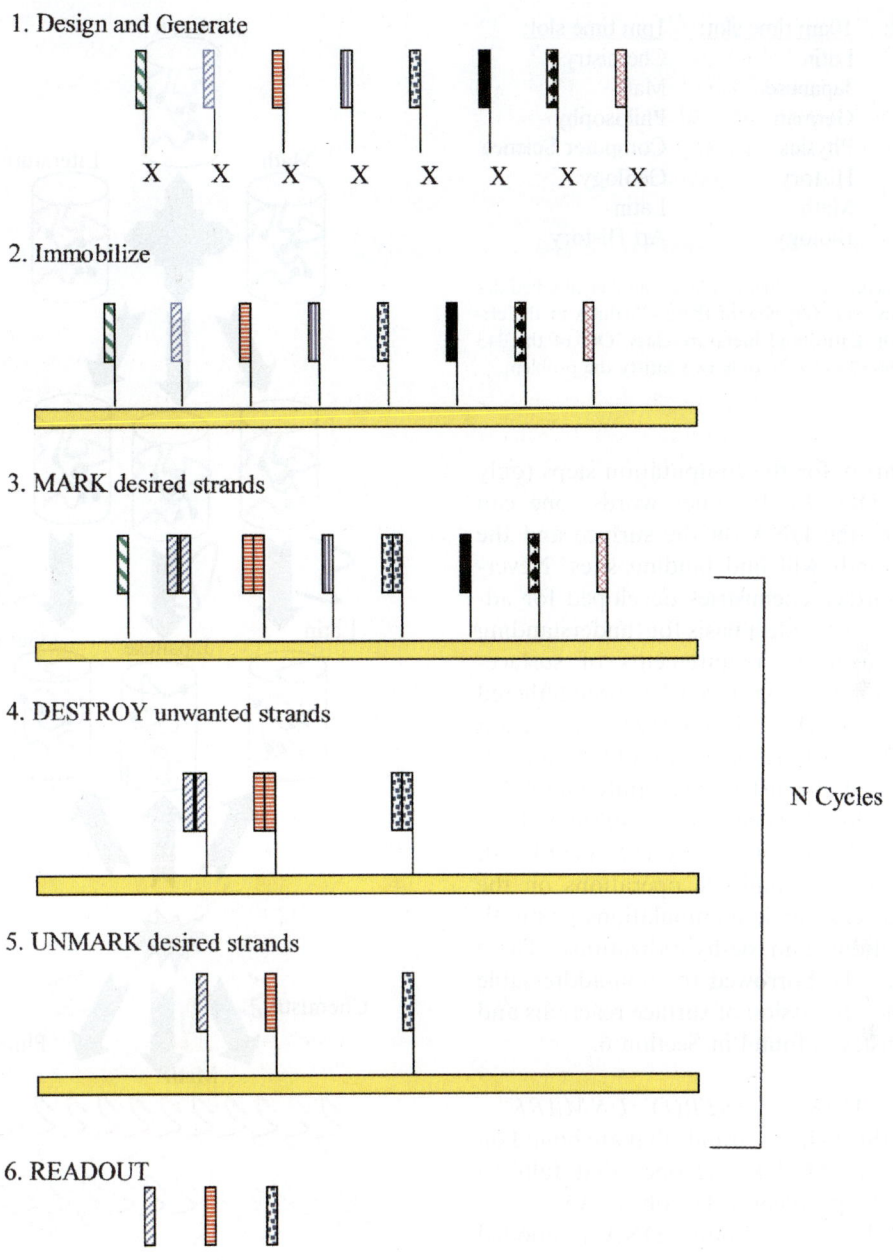

Fig. 4. A schematic outline of the DNA operations for surface-based DNA computation. (1) After designing an encoding scheme, generation involves creating a large pool of different DNA sequences to represent the different possible solutions to a problem. (2) We immobilize the strands on a surface, and (3) MARK, or hybridize, a specific subset of sequences with its complementary sequence. MARKed sequences correspond to a potentially correct answer in the computation problem. (4) Then we DESTROY the remaining single-stranded sequences on the surface with a judicious choice of enzymes that distinguish between the MARKed strands and the single strands. The single strands are cut off the surface through enzymatic digestion. (5) Afterwards, the complementary sequences are UNMARKED or separated from their complements to allow for further computation. Steps 3–5 can be repeated many times before READOUT is performed. (6) In READOUT, we determine the sequences of the strands that remain after the DNA computation operations [10,11].

8am time slot:	10am time slot:	1pm time slot:
Math	Latin	Chemistry
Literature	Japanese	Math
Latin	German	Philosophy
Spanish	Physics	Computer Science
Philosophy	History	Geology
Physics	Math	Latin
History	Biology	Art History

Fig. 5. Class Scheduling Problem. A large number of schedules are possible. However, very few of these schedules fit the criterion of a math, Latin, and literature class. Out of the 343 possible schedules (7 × 7 × 7), only two satisfy the problem.

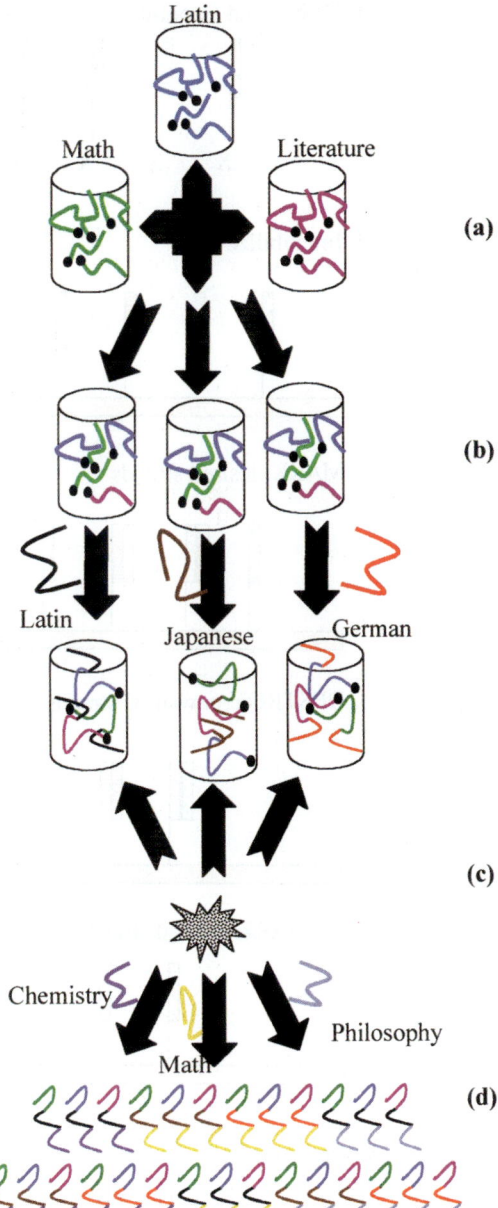

Fig. 6. Generation of the DNA strands. (a) Three DNA words, representing math, Latin, and literature classes encoded with the different colors, are synthesized separately on bead supports. (b) The DNA words are mixed together and redistributed into three new test tubes for the attachment of a second word, representing three new classes, onto the end of the original DNA strands. (c) The process can be repeated theoretically *ad infinitum*. (d) The set of DNA strands generated after three rounds is represented with the pattern scheme. There are 27 different DNA sequences with three words each.

addressable array for the computation steps (only for the READOUT). In other words, one can simply "pour" the DNA on the surface and the individual strands will find binding sites. Nevertheless, the surface chemistries developed for addressable arrays provide a basis for 'understanding the surface chemistry requirements of surface-based DNA computation. DNA has been tethered to plastic [15], glass [16–18], gold [19,9], and silicon surfaces [20,21]; various membranes [22]; and gels [23]. The surface chemistry is carefully chosen to create a stable link between the substrate and the DNA. This stable link, preferably a covalent bond, is desirable to allow multiple operations on the DNA, such as enzymatic manipulations or multiple hybridizations and de-hybridizations. These chemistries can be borrowed for non-addressable arrays. Further discussion of surface reactions and surface chemistry is found in Section 6.

5.2.4. Search (MARK, DESTROY, UNMARK)

To search through the strands that are bound on the surface, and MARK the ones that fulfill a condition of the problem to be solved, we use hybridization. The surface-bound DNA is flooded with the complement of a specific sequence. For example, in the class scheduling problem, after generating and randomly immobilizing on the surface the 343 different possible schedules in which three classes encoded as DNA words are linked together to form a "schedule" (see Eq. (2)), many copies of all the complementary sequences that encode for math class at 8 am, at 10 am, or at 1 pm, are introduced to the surface via a dilute solution.

The complements then search out all the strands that have a sequence that encode for a math class. Following Fig. 5, there are three math classes, one each at 8 am, 10 am, and 1 pm. All of the schedules with a math class at any of those times would be MARKed via hybridization of part of the strand. With an appropriate choice of enzyme, the MARKed strands are kept while the single-stranded DNA sequences are enzymatically digested, or cut from the surface. This is the DESTROY phase. All the DNA sequences that encode for schedules without a math class are DESTROYed. We are left with only those strands that have a math class somewhere in their sequence. These remaining DNA strands must be separated from their (partial) complements in order to search for those strands that also have a literature class in their sequence or schedule. We employ a highly concentrated salt solution and elevated temperatures to dissociate, i.e. to UNMARK, the strands, which does not, however, cause the tethered DNA to leave the surface, and then to wash the surface. To search for literature, we flood the surface with the complement for the literature classes, perform DESTROY using the proper enzyme, de-hybridize. We then repeat for Latin. The search strategy is illustrated schematically in Fig. 7.

5.2.5. READOUT

Among the strands remaining on the surface after this series of MARK, UNMARK, and DESTROY steps (arrived at by some optimized algorithm provided by the computer scientist) lie(s) the answer(s) to the problem to be solved by DNA computation. The location of any particular sequence on the surface is not known and does not need to be known. The operations of DNA computation sort the DNA into sequences that represent possible answers to the problem to be solved, and those that do not (the latter ones have been removed in the various computation steps), as seen in Fig. 7. But the process so far, although it has found all the sequences that can represent answer(s) to the problem, does not identify those sequences. To do so, readout arrays, in which DNA sequences are placed at specific locations in the array, are required [24]. For the schedule problem discussed above, we will have three separate arrays for simplicity. In the first array, we have bound the single strands of all the complementary sequences that encode for classes at the 8 am time slot, a different sequence or class in each array site. There are seven of them. The second array has all seven single strands of complementary sequences for the classes offered at the 10 am time slot, and the third array, the seven 1 pm-time-slot classes. We make a solution of the unknown strands by copying them from our computation surface, divide it into three parts and pour one part on each of the three arrays. In each array element, we include the complements to the fixed bases (see Fig. 8) that encode for the time slots. After hybridization occurs, we can detect which array site, and hence which sequence, has formed a double-stranded DNA structure. For the 8 am time slot, one class is indicated, literature. For the 10 am slot, two different classes are indicated, Latin and math. Finally, the READOUT for 1 pm reveals two different classes, Latin and math.

In this problem, there is more that one "correct" answer. With the classes indicated from the READOUT steps, we then need to revisit the original criterion and check that the schedules satisfy the problem. From our check, we find that there are two possible schedules, literature–Latin–math and literature–math–Latin, as seen in Fig. 9.

In our simple problem, finding the answer is trivial. One can easily imagine much more complex scheduling or sorting problems, however, in which there are both more initial classes and more "bins" (or time slots). In such cases, DNA computation, because of the extreme parallelism of the process, offers a significant advantage.

5.3. Surface science issues

The development of strategies for the attachment of DNA to surfaces is not solely the domain of DNA computation. In the past decade, both academic laboratories and the biotechnology industry have evolved a flourishing group of DNA-array makers, with applications in high-throughput analyses for, to name a few, gene expression [25], gene variation [26], toxicology [27], and drug development [28]. These efforts have led to the development of several different strategies for

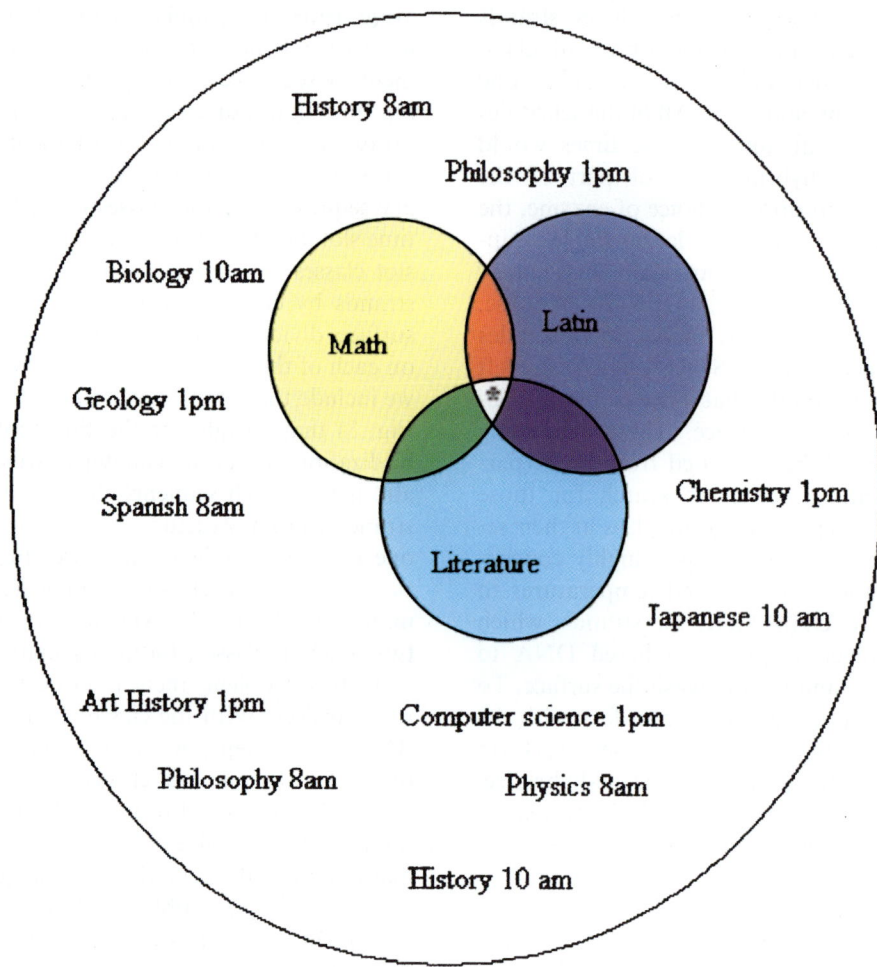

Fig. 7. Search strategy. Let the outer circle represent all the possible schedules for the classes (343). Let the yellow circle represent the DNA words that encode for schedules with a math class; the blue circle, the DNA words that encode for schedules with a Latin class; the light blue circle, a literature class. All the other schedules with the classes listed in the outer circle do not have any of the required classes. After the DNA computation operations are performed on the surface, only those sequences in the starred (*) region remain, as they are the only ones that satisfy the criterion of a math, a Latin, and a literature class. It is upon this subset that we perform READOUT to identify the sequences that remain from the large combinatorial set.

DNA attachment on the surface. Some laboratories use in situ synthesis on the surface to create the different DNA strands one base at a time. Others completely synthesize the entire strand before attaching it to the surface [29,30]. Several substrates have emerged as favorable choices, glass, silicon, and plastic to name a few. DNA computation requires many steps on the same surface, including hybridizations, enzymatic digestion, and PCR. Most of the DNA microarrays are designed for single use or maximally five reuses [31], and hence the corresponding DNA attachment schemes are not ideal for DNA computation.

A robust attachment chemistry that allows for multiple enzymatic reactions and for many hybridizations is the cornerstone of surface-based DNA computation. Such a system has been developed on gold films using alkane thiol chemistry [9,19,32]. Even though this chemistry is well understood, gold films are not ideal. A gold surface

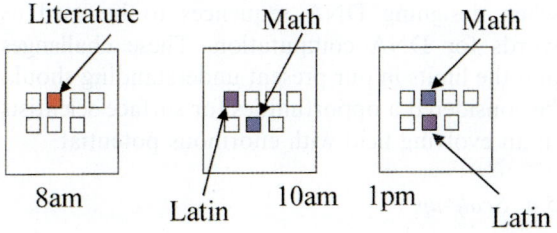

Fig. 8. READOUT with addressed arrays. From the three different addressed arrays for the 8 am, 10 am, and 1 pm time slots, we decipher the DNA sequences that contain DNA words representing classes that remain on the unaddressed DNA computation surface. In an addressed array, all the sites have known sequences at known locations. When an array element is indicated, we immediately know which DNA word is present and which class the word represents. For the 8 am classes, only the literature class is indicated; for the 10 am class, both math and Latin are present; for 1 pm, again math and Latin. The elements are in the same order in Fig. 5.

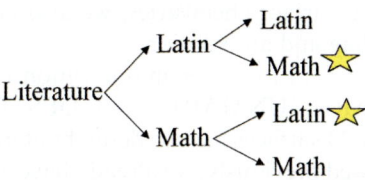

Fig. 9. Possible schedules. With three steps of MARK, DESTROY, UNMARK, the DNA computer reduces the 343 possible original schedules to four possibilities. When we examine those four and compare them to the original problem, we see that two of the four are real answers, and we have solved the problem. The two starred schedules satisfy the original scheduling problem: literature–Latin–math and literature–math–Latin.

(as well as glass, another common surface for DNA attachment) is rough with large hills and valleys invisible to the naked eye yet easily detectable with topographic profiling techniques. These undulations influence reactions on the surface and may well interfere with fluorescent-signal detection. Fluorescence is a common technique to detect biological molecules (see glossary). In addition, gold scratches easily, and these scratches lead to increased background noise after many DNA computation operations. Silicon, an extremely flat and less delicate surface, is well-characterized with a myriad of possible functionalization options [33]. Current investigations center on developing attachment techniques of DNA to silicon [20,21] and to plastics, which are less expensive.

Enzymatic reactions, such as DESTROY, are required in various DNA computation steps, but are not well understood on a surface. Many enzymes that react well in solution do not react at all on a surface. Each step of surface-based DNA computation that requires an enzymatic manipulation must be fully characterized to see if the enzyme will perform as well on the surface as in solution and to quantify the completeness of the surface reaction. For example, hundreds of enzymes cleave DNA, but of those tested for surface reactivity, only a few perform well. Of those few, the best cleaving enzyme might cut 95% of the strands on the surface [34,35]. For DNA computation, the ideal would be 100% function of each enzymatic operation; anything less leads to error in the computation.

On a surface, several variables influence the behavior of enzymes. The proximity of the enzyme to the surface has been found to be a decisive factor. When large spacer groups are inserted in the chemical attachment scheme to distance the DNA strands away from the surface, there is a significant increase in enzymatic efficiency [34,35]. The source of interference has not been determined. The surface morphology and chemical interaction with the enzymes may play a role [34] and are areas for further research. Other influences, such as steric interference between the enzyme and the surface and the use of short (<50 base pairs) strands, could also influence the enzymatic behavior.

The surface environment also affects hybridization [36] and fluorescence intensity signal from tagged strands [37]. Different types of chemical attachment of the DNA onto surfaces lead to various forms of accessibility of the individual DNA strands to hybridization. For example, in a common technique, fully synthesized DNA strands are placed onto a glass substrate coated with poly-lysine. The poly-lysine forms a net of positive charges uniformly over the surface. The DNA molecules carry an overall negative charge and are attracted to the positive substrate. Within the network of poly-lysine and single-stranded DNA, the hybridization efficiency of the DNA strands is reduced, as a portion of these DNA molecules can become entangled in the surface and inaccessible to their complements [16,36,38]. The interaction between the pre-synthesized single-stranded DNA and the surface is non-specific, i.e., an undefined non-covalent attachment. Background signal caused by non-specific binding of the unhybridized complements within the poly-lysine mesh limits the reuse of the microarray [31]. Non-specific binding from a variety of sources is a concern and an area for further research. In addition, the surface density of tethered DNA strands plays a role in the accessibility of the enzymes and strands complementary to the tethered strands, and hence the reaction kinetics [19,36,39].

The DNA strands present their own particular problems. First, a long single-stranded DNA molecule, either on the surface or in solution, can fold on itself to form secondary structures or internal hybridizations, as seen in Fig. 10. On a surface, DNA strands are less likely to interact with each other, but can still self-hybridize. The secondary structure, because it is a thermodynamically advantageous form for the unhybridized DNA strand, inhibits interaction between the strand and its complement [40]. Second, the temperature for hybridization and de-hybridization of DNA depends on the base content and length of the strands. The A–T pairing results in two hydrogen bonds between opposing strands, while the C–G pairing has three hydrogen bonds (see Fig. 2). Such a difference in the bonding between the base pairs leads to large differences in hybridization and de-hybridization temperatures for DNA strands that have many G–C pairings and to those that have few G–C pairings [41]. Both hairpin structures and G/C to A/T ratios need to be considered

(a) 5'—AACTGAGGTTTTCCTCAGTT—3'

(b) 5'—AACTGAGG
    3'—TTGACTCC

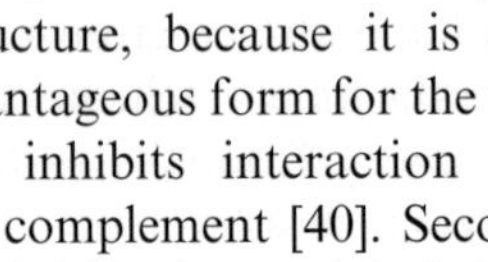

Fig. 10. Hairpin structure of single-stranded DNA. A hairpin, one possible secondary structure of single-stranded DNA molecules, can form if a strand contains complementary portions within its sequence. As the molecule twists and moves in solution or on a surface, it can self-hybridize along the complementary sections.

when designing DNA sequences to be used as words for DNA computation. These challenges and the limits in our present understanding should be considered a opportunities for surface scientists in an evolving field with enormous potential.

5.4. Scale-up

By the end of 2000, a proof of principle of surface-based DNA computation had been demonstrated with the solution of a 3-variable satisfiability probability (a type of intractable problem usually with a large pool of possible solutions, see Glossary), using addressed arrays [11]. If surface-based DNA computation is to be a viable tool for difficult problems, it must accommodate large problems, and must be accomplished in an unaddressed fashion. In addition to developing and optimizing surface chemistries, we also must consider scaling and automation.

DNA generation, its immobilization on a surface, MARK, UNMARK, and DESTROY on unaddressed surfaces have scale-up built into them. As discussed previously, we already have a strategy to generate 10^{12} different DNA strands efficiently using existing technology, and synthesizing machines exist that synthesize DNA in parallel. For MARK, the hybridization process for a surface covered with DNA is a highly parallel process. As the number of different single strands bound on the surface increases, the number of potential complements introduced to the surface in the MARK operation can also be raised. The de-hybridization of a surface covered with hybridized DNA, UNMARK, will not vary as the number of different strands on the surface increases, as UNMARK does not discriminate between different strands, but merely separates all complementary strands with a salt solution and heat. DESTROY, once again, is a highly parallel process. As the number of different strands increases, the step can be optimized to ensure that the enzymes interrogate each strand.

For READOUT, the method to encode the information greatly affects the efficiency of the operation. For example, with the word strategy in the class schedule problem discussed earlier, an addressed array is required for the DNA words that encode for the classes at each different time slot, as we discussed earlier. In the PCR reaction (the step in which we make many copies of the strands remaining on the unaddressed surface), we can tailor the amplification through the choice of a specific primer, or start sequence, which signals the enzymes to start copying at that specific sequence location. We can select the words one layer at a time for a multi-word sequence by selecting a primer specific to the fixed bases of each tier, and then applying the PCR products to the addressed array with the corresponding fixed bases of the words. Such repetition would lend itself well to automation. For other READOUT strategies, individually sequencing each DNA answer might be necessary. As illustrated by the human-genome project, large-scale sequencing of DNA is possible [42].

6. DNA arrays

After surface-based DNA computation operations or, for that matter, test-tube-based DNA computation, we must employ addressed arrays to decipher rapidly the base sequences to the remaining DNA strands that represent answers to our problem. To have confidence in the READOUT, the arrays must be reliable. As DNA computations increase in size, the READOUT arrays must accommodate the expansion, i.e., the information density in the array must increase. We describe the options available in DNA microarrays, also known as "gene chips" or "biochips", in the following subsections.

6.1. Present microarray techniques

The majority of DNA microarrays are fabricated on surfaces. A microarray is a pattern in which each element contains a unique DNA sequence. If one assumes a single layer of covalently bonded DNA molecules with a typical coverage of 10^{12} molecules per cm^2 [19,38,43] and a 100 μm diameter element size, each element would contain $\sim 10^8$ identical DNA molecules. The coverage of 10^{12} molecules per cm^2 comes from the fact that a DNA molecule is expected to occupy a 25 nm^2 area [9]. Methods to create DNA microarrays fall into two broad categories: (1) in situ synthesis, a base-by-base attachment to build up different DNA strands at different sites in the array or (2) ex situ synthesis, and subsequent attachment of different complete strands at individual array sites.

The former approach is exemplified by work of Fodor et al. [17] who have developed a photolithographic technique to create DNA arrays by means of a series of exposures through masks to control the addition of individual bases to specific sites on the surface (see Fig. 11, [44]). On a surface prepared for photolithographic synthesis, certain elements of the array are exposed through a mask to light. The light cleaves bonds of protecting groups on these elements (the de-protection step), as seen in Fig. 11. When the surface is subsequently flooded with a solution containing the base of choice, the reactive sites on the selected de-protected elements bind the bases, which on their opposite ends have another protecting group for the next cycle of light exposure and attachment. Unfortunately, the base attachment is not perfect, and some available sites on the selected elements do not interact with the chosen base.

The photolithographic in situ technique has been reported to have a repetitive yield of ~ 0.95

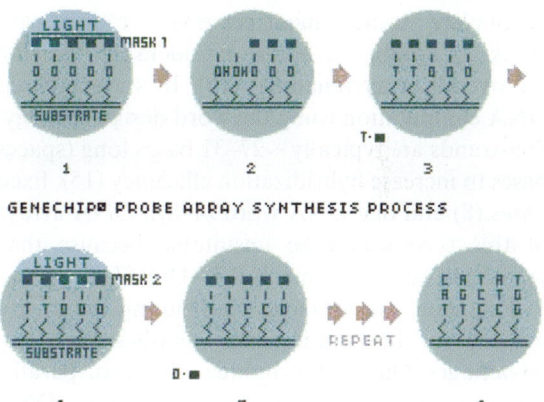

Fig. 11. Light-directed base-by-base synthesis of DNA arrays. A surface is covered with light sensitive protecting groups. When the substrate is exposed to light through a mask, regions of the surface react with the light and the protecting groups are removed. The de-protected regions are then able to interact with the DNA bases, which have been modified to terminate with a protecting group, and the bases become bound to the surface. After many such steps, many different DNA sequences are synthesized on the surface [44].

per base attachment [14,43,45], considerably limiting the length of the oligonucleotide that can be synthesized with fidelity. When a specific array element in the substrate is exposed again for the next base, unreacted binding sites from the previous exposure may interact with this next base and create an incorrect sequence [18,44,46]. The majority of the surface reaction inefficiency results in deletions in the given strand, e.g., the strand has only 19 instead of 20 bases [43]. Such strands do not have just a single-base mismatch but are incorrect from the deletion site on. If the deletion occurs at the tenth site of a 20-base sequence, the ten subsequent sites will be wrong. The sum total of these errors creates low overall fidelity. For example, if one chooses a 16-base-long sequence to synthesize on the surface, the overall fidelity of the strand would be 0.95^{16} or 44% correct sequences in the area allotted in a single array element. [4] Other base-by-base synthesis techniques, such as piezoelectric printing and ink jet methods, have similar yield and fidelity problems [47]. Elimination of these erroneous strands through purification might seem like the logical approach to improve the reliability. However, in situ methods of DNA strand synthesis on the surface do not allow for oligonucleotide purification afterwards to eliminate incomplete strands, incorrect ones, or ones with an excess number of bases, considerably reducing the overall array reliability [14]. In surface-based DNA computation using the word design strategy, the strands are typically ~27–31 bases long (spacer bases to increase hybridization efficiency (15), fixed bases (8) and one DNA word (4–8)). DNA arrays of this type would be unsuitable because they would introduce too much READOUT error.

Balancing the obstacles confronting the in situ method for DNA microarray synthesis are two advantages. One, the technique allows for parallel synthesis of a large number of different DNA strands. Two, arrays with relatively small, 20 μm × 20 μm, feature size and a pitch of 40 μm has been demonstrated via the photolithographic in situ method. On a standard 3 in. × 1 in. slide, the photolithographic method can create ~1.3 million different elements.

In the second method, completely synthesized strands are attached to elements on the surface. A pin tool, a tool similar to a fountain tip pen, is used to load the array mechanically. The oligonucleotides to be attached can be rigorously purified before surface attachment to allow for longer strands of high fidelity, something not possible with in situ synthesis. Shalon and Brown applied the ex situ technique [29] to create large arrays with long DNA strands. These arrays have proven to be a useful tool for investigating gene expression levels in diseases [48]. Their attachment chemistry of the DNA strands onto the surface, however, is not a specific covalent linkage. Instead, the strands are connected through the interaction between a poly-lysine coated slide and the DNA. As a polycation, the poly-lysine layer creates a positively charged surface, which attracts and binds the polyanion DNA. Unfortunately, the DNA strands can become enmeshed in the network of poly-lysine and then a portion of them are not accessible to hybridization [37,38,43]. If the attached DNA strands cannot hybridize with their complements, a lower hybridization signal will result. The non-specific binding of the DNA strands to the surface (any DNA that remains on the surface without a specific attachment is non-specifically bound, see glossary) with poly-lysine can trap complements that have not found their counterparts, leading to an increased background signal. Also, with a substrate that attracts DNA only with a weak electrostatic bond, droplets placed on a surface spread because of wetting and produce a variety of sizes for array elements. Droplets of different DNA solutions placed side-by-side can bleed into each other, resulting in cross-contamination. The technique therefore requires a features size on the order of 50–200 μm with a pitch, at minimum, of 200 μm [31]. The attachment of pre-synthesized DNA onto an unpatterned surface then comes at a cost of element or site density. A standard 3 in. × 1 in. microscope slide can accommodate ~20,000–40,000 sites for DNA sequences. However, synthesizing the DNA prior to creating the array and eliminating the incorrect strands greatly

---

[4] Assuming that all errors in the photochemistry are deletions, then 5% of the strands have the 16th site wrong; then 5% of the 95% remaining correct strands will have the 15th site wrong; 5% of the remaining 90.25% (0.95^2)...

reduces the possibility of error in the READOUT procedure and the DNA strands can be significantly longer via their synthesis through higher yield techniques such as controlled glass pore (CPG, see Glossary) and PCR methods.

6.2. Chemically patterned surfaces

Chemically patterning the substrate can overcome the limitations described above when complete single strands of DNA are attached to the surface to create an array [49–51]. A chemically patterned surface for DNA arrays is one that has an inert background, possibly hydrophobic (water repelling) in nature, and reactive array sites, possibly hydrophilic (water liking) in nature. When DNA solutions are placed onto the reactive sites of such a surface, the hydrophilic–hydrophobic interaction with the aqueous DNA solutions prevents DNA attachment between the elements of the array and thus acts as a barrier for diffusion of DNA between the elements, as seen in Fig. 12. Combined with a covalent-bonding attachment scheme for completely synthesized and purified DNA strands, high-purity arrays with high positional fidelity, excellent stability, small feature size, and minimal cross talk between elements can be fabricated [24]. With a chemically patterned surface, the array elements can be considerably smaller, 20 μm or less. The density of array sites can then be equal to that of base-by-base attachment methods. But, because pure pre-synthesized strands are attached, the reliability can be greater than in base-by-base fabricated arrays. In addition, the total information density of the array can be higher than arrays fabricated by the in situ method. Limited by yield in the base-by-base attachment, DNA sequences generally less than 40 bases long are synthesized in the in situ approach. In ex situ synthesis with the use of chemically patterned substrates, much longer, pre-purified strands of DNA can be attached. On the same 3 in. × 1 in. microscope slide, ∼200,000 50 μm square array sites would be available. If the strand is five times as long as what is possible with the in situ approach, the information density can be five times as high. Further research on new patterning techniques is necessary to expand the number of surfaces and to pattern in a manner compatible with existing DNA attachment chemistries.

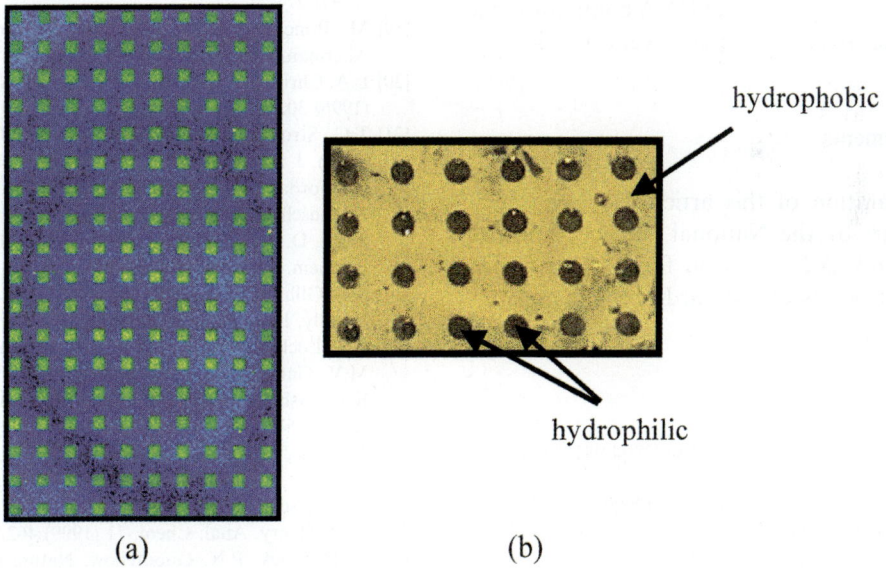

Fig. 12. DNA arrays using chemically patterned substrates. (a) An array with 50 μm square elements of DNA on a 100 μm pitch viewed in fluorescence. Single-stranded DNA is attached to the surface and its complement with a fluorescent tag is introduced to the sample. The strands hybridize and the areas containing DNA duplexes, the array sites, are detected fluorescently. (b) Illustration of chemically patterned substrate. Water condenses preferentially on the 1.5 mm diameter hydrophilic array sites.

7. Summary

In the few years since DNA computation was first proposed, great strides have been made in identifying the issues that must be addressed if we expect to develop a viable DNA computation technology. The most challenging problems are related to translating the algorithms into biochemical steps. These manipulations are not 100% efficient, hence fault tolerance must be built into any DNA computing scheme. The most viable DNA computing methodology is expected to use surfaces to support the DNA on which the biochemical operations must be made. The use of surfaces draws on the considerable background of knowledge in surface science. Yet, many surface science issues still need to be addressed. These include: the choice of surface and surface morphology, the development of methodologies for robust chemical attachment of DNA to the surface, the development of an understanding of enzymatic activity on molecules bound to a surface, the development of improved enzymes to carry out the biochemical reactions on molecules tethered to a surface, and the development of improved microarray fabrication loading techniques. The expertise of several disciplines is required to make DNA computation a reality. Surface science will play a central role.

Acknowledgements

The preparation of this article has been supported in part by the National Science Foundation. We thank A.E. Condon for her assistance with the finer points of NP-hard problems.

References

[1] http://www6.tomshardware.com/cpu/98q4/981015/sld004.html.
[2] M.A. Reed, J.M. Tour, Sci. Am. 282 (2000) 86.
[3] http://www.mersenne.org/.
[4] http://setiathome.ssl.berkeley.edu/.
[5] M. Mosca, R. Jozsa, A. Steane, A. Ekert, Phil. Trans. R. Soc. Lond. A 358 (2000) 261.
[6] http://esg-www.mit.edu:8001/esgbio/lm/nucleicacids/dna.html.
[7] L.M. Adleman, Science 266 (1994) 1021.
[8] DIMACs, Journal of Computational Biology, DNA7 Seventh International Meeting on DNA Based Computers June 10–13, 2001. http://www.cas.usf.edu/dna7/.
[9] L.M. Smith, R.M. Corn, A.E. Condon, M.G. Lagally, A.G. Frutos, Q. Liu, A.J. Thiel, J. Comput. Biol. 5 (1998) 255.
[10] A.G. Frutos, Q. Liu, A.J. Thiel, A.M.W. Sanner, A.E. Condon, L.M. Smith, R.M. Corn, Nucl. Acid Res. 25 (1997) 4748.
[11] Q. Liu, L. Wang, A.G. Frutos, A.E. Condon, R.M. Corn, L.M. Smith, Nature 403 (2000) 175.
[12] D.D. Shoemaker, D.A. Lashkari, D. Morris, M. Mittmann, R.W. Davis, Nature Genet. 14 (1996) 450.
[13] T.C. Strother, A.G. Frutos, Q. Liu, L. Wang, S.D. Gillmor, A.E. Condon, R.M. Corn, M.G. Lagally, L.M. Smith, Late Breaking Papers at the Genetics Programming 1998 Conference, 1998, p. 262.
[14] M.J. O'Donnell-Maloney, D.P. Little, Genet. Anal.: Biomolecular Eng. 13 (1996) 151;
A. Marshall, J. Hodgson, Nature Biotechnol. 16 (1998) 27.
[15] M.S. Shchepinov, S.C. Case-Green, E.M. Southern, Nucl. Acids Res. 25 (1997) 1155.
[16] N. Zammatteo, L. Jeanmart, S. Hamels, S. Courtios, P. Louette, L. Hevesi, J. Remacle, Anal. Biochem. 280 (2000) 143.
[17] S.P.A. Fodor, J.L. Read, M.C. Pirrung, L. Stryer, L.A. Tsai, D. Solas, Science 251 (1991) 767.
[18] S. Singh-Gasson, R.D. Green, Y. Yue, C. Nelson, F. Blattner, M.R. Sussman, F. Cerrina, Nature Biotechnol. 17 (1999) 974.
[19] M. Boncheva, L. Scheibler, P. Lincoln, H. Vogel, B. Åkerman, Langmuir 15 (1999) 4317.
[20] L.A. Chrisey, G.U. Lee, C.E. O'Ferrall, Nucl. Acid Res. 24 (1996) 3031.
[21] T.C. Strother, W. Cai, X.S. Zhao, R.J. Hamers, L.M. Smith, J. Am. Chem. Soc. 122 (2000) 1205.
[22] B.J. Jordan, J. Biochem. 124 (1998) 251.
[23] D. Guschin, G. Yershov, A. Zaslavsky, A. Gemmell, V. Shick, D. Proudnikov, P. Arenkov, A. Mirzabekov, Anal. Biochem. 250 (1997) 203.
[24] S.D. Gillmor, A.J. Thiel, T.C. Strother, L.M. Smith, M.G. Lagally, Langmuir 16 (2000) 7223.
[25] D.J. Lockhart, H.L. Dong, M.C. Byrne, M.T. Follettie, M.V. Gallo, M.S. Chee, M. Mittmann, C.W. Wang, M. Kobayashi, H. Horton, E.L. Brown, Nature Biotechnol. 14 (1996) 1675.
[26] M. Chee, R. Yang, E. Hubbell, A. Berno, X.C. Huang, D. Stern, J. Winkler, D.J. Lockhart, M.S. Morris, S.P.A. Fodor, Science 274 (1996) 610.
[27] C.M. Henry, Anal. Chem. 71 (1999) 462A.
[28] C. Debouck, P.N. Goodfellow, Nature Genet. 21 (1999) 48.
[29] D. Shalon, Genome Meth. 6 (1996) 639;
M. Schena, D. Shalon, R.W. Davis, P.O. Brown, Science 270 (1995) 467.

[30] T. Okamoto, T. Suzuki, N. Yamamoto, Nature Biotechnol. 18 (2000) 438.
[31] D.L. Bowtell, Nature Genet. 21 (1999) 33.
[32] R.G. Nuzzo, D.L. Allara, J. Am. Chem. Soc. 105 (1983) 4481.
[33] S.F. Bent, Surf. Sci. 500 (2002) 879.
[34] L. Wang, Q. Liu, R.M. Corn, A.E. Condon, L.M. Smith, J. Am. Chem. Soc. 122 (2000) 7435.
[35] A.G. Frutos, A.E. Condon, L.M. Smith, R.M. Corn, J. Am. Chem. Soc. 120 (1998) 10277.
[36] E. Southern, K. Mir, M. Shchepinov, Nature Genet. 21 (1999) 5.
[37] Z. Guo, R.A. Guilfoyle, A.J. Thiel, R. Wang, L.M. Smith, Nucl. Acids Res. 22 (1994) 5456.
[38] D.J. Duggan, M. Bittner, Y. Chen, P. Meltzer, J.M. Trent, Nature Genet. 21 (1999) 10.
[39] M.R. Henry, P.W. Stevens, J. Sun, D.M. Kelso, Anal. Biochem. 276 (1999) 204.
[40] Q. Liu, A.G. Frutos, A.J. Thiel, R.M. Corn, L.M. Smith, J. Computat. Biol. 5 (1998) 269.
[41] C.K. Mathews, K.E. van Holde, Biochemistry, second edition, New York, 1996.
[42] Press Release: Amersham Pharmacia Biotech, June 14, 1999. http://www.apbiotech.com/company/company_index.html.
[43] D.J. Graves, Trends Biotechnol. 17 (1999) 127.
[44] http://www.affymetrix.com.
[45] G.H. McGall, A.D. Barone, M. Diggelmann, S.P.A. Fodor, E. Gentalem, N. Ngo, J. Am. Chem. Soc. 119 (1997) 5081.
[46] R.J. Lipshutz, S.P.A. Fodor, T.R. Gingeras, Nature Genet. 21 (1999) 20.
[47] A. Marshall, J. Hodgson, Nature Biotechnol. 16 (1998) 27.
[48] A.A. Alizadeh, M.B. Eisen, R.E. Davis, C. Ma, I.S. Lossos, A. Rosenwald, J.G. Boldrick, H. Sabet, T. Tran, X. Yu, J.I. Powell, L.M. Yang, G.E. Marti, T. Moore, J. Hudson, L.S. Lu, D.B. Lewis, R. Tibshirani, G. Sherlock, W.C. Chan, T.C. Greiner, D.D. Weisenburger, J.O. Armitage, R. Warnke, R. Levy, W. Wilson, M.R. Grever, J.C. Byrd, D. Botstein, P.O. Brown, L.M. Staudt, Nature 403 (2000) 503.
[49] A.P. Blanchard, R.J. Kaiser, L.E. Hood, Biosens. Bioelectron. 11 (1996) 687.
[50] www.protogene.com.
[51] L.A. Chrisey, C.E. O'Ferrall, B.J. Spargo, C.S. Dulcey, J.M. Calvert, Nucl. Acid Res. 24 (1996) 3040.
[52] http://www.bs.jhmi.edu/proteinfacility/Oligos/oligochemistry.html
[53] D.A. Skoog, J.J. Leary, Principles of Instrumental Analysis, fourth edition, New York, 1992.

Glossary

Addressed/unaddressed arrays: An addressed array is an organized assembly of elements (DNA strands, proteins, legos, desserts...) principally on a surface, where each element has a designated location. In an unaddressed array, the location of a specific element (DNA strands, proteins, legos, desserts) is not known.

Amino acid: The building block of protein. It has the form of a central carbon (α-carbon) that has an amine group, a carboxylic acid, a side group and a hydrogen all attached to the central carbon. The side chains can be as simple as another hydrogen (glycine) and as complex as multi-ringed aromatic structures [41].

Controlled glass pore (CPG) synthesis: A standard DNA synthesis technique, it is a cycle of base attachment on a solid controlled-pore glass support. Through the use of a support, excess reagents can be removed via filtration, and the support eliminates the need for purification steps between base additions. The synthesis starts with the 3'-hydroxyl of a modified nucleotide attaching to a solid CPG support bead on a long spacer arm. The nucleotide has a dimethoxytrityl (DMT) group attached at the 5' location. Once it is removed, the base is ready to couple with another, which has been modified to have a DMT group to its 5' site. After the coupling, the unreacted bases undergo "capping" through an acetylation step that prevents coupling with any subsequent nucleotides. The synthesis cycle is repeated until the total sequence has been created. The strands are then removed from the beads and purified to remove the failed sequences [52].

Deoxyribonucleicacid (DNA): A molecule with the general form of five-member ring sugar with purine or pyrimidine base (adenine and guanine, cytosine and thymine respectively) at the 1' location, hydrogens at the 2' location, and a phosphate acid at the 4' location [41].

Dialysis: The diffusion through a semi-permeable membrane of low-molecular-weight reactants (i.e. salts) [41].

Digestion: The cutting of DNA, protein, or RNA molecules into their monomer forms (individual DNA and RNA bases, amino acids).

DNA base: A purine (adenine or guanine) or pyrimidine (cytosine or thymine) structure, which contains an amine base, attached to the sugar ring [41]. The bases have a particular characteristic manner of pairing, in which adenine forms two hydrogen bonds with thymine and guanine forms three hydrogen bond with cytosine. Given this characteristic pairing, each strand of DNA has a complementary strand in which all the bases form these hydrogen bonds and stabilize the structure in a double-stranded form. The interaction between a strand and its complement is known as hybridization.

DNA word: A method to encode information into DNA, similar to a byte.

Enzyme: A protein that catalyzes a reaction.

Fixed base: For the encoding described in Section 5.2.1, a fixed base is one that is identical for all the strands at that location in that set of DNA strands.

Fluorescence: The emission of light via the decay of excited atoms or molecules to the ground state [53]. In biological work, molecules with well defined excitation and emission spectra are routinely attached to DNA or RNA, i.e., the complement of a particular DNA sequence can have a fluorescent-molecule tag. Signals from these tags are recorded and analyzed.

Gel electrophoresis: A technique to separate electrically charged substances in a mixture. A solution of mixed DNA lengths (negatively charged due to its phosphoric backbone) is placed in a divot of the gel (a jello-like medium made of acrylamide or agarose) that is immersed in water. Electrodes are attached to the gel at opposite ends and power is applied. The DNA strands migrate towards the cathode, with the long DNA strands moving more slowly than the short strands [41].

Hybridization: The hydrogen-bond mediated interaction between a single DNA strand and its complement.

Ligate: To join, DNA bases or strands, amino acids... using enzymes.

MIPS: Million instructions per second. A widely used metric for comparing the performance of various computers. A 500 MIPS computer is capable of performing 500 million instructions per second.

Millimole: 6.022×10^{20} molecules.

Non-specific binding: The binding to the surface of molecules that do not have a specific covalent linkage. For example, the binding of DNA to a poly-lysine coated glass surface is a non-specific interaction.

Oligonucleotide: A string of DNA bases or RNA bases connected though a sugar-phosphate backbone.

Pin tool: A grooved tip that wicks small amounts of solution into its grooves and dispenses it onto a surface through capillary interaction. A pin tool operates similarly to a fountain tip pen.

Polymerase chain reaction (PCR): A technique to increase the number of copies of a specific DNA sequence through repeated cycles of denaturation and replication [41].

Primer: A short piece of DNA or RNA from which an enzyme can extend a DNA or RNA strand. It is commonly used to initiate PCR [41].

Protein: A string of amino acids in a sequence determined by a gene (primary structure) folded in a specific manner (secondary and tertiary structure).

Protein expression: The translation of a gene into a protein. The synthesis of an amino acid sequence into a polypeptide under the direction of a messenger RNA (mRNA) molecule, so that the nucleotide sequence of the mRNA is "translated" into the amino acid sequence of the protein [41].

Replication: The copying of a DNA strand.

Ribonucleic acid (RNA): A molecule with the general form of five member ring sugar with purine or pyrimidine base (adenine and guanine, cytosine and uracil respectively) at the 1' location, a hydroxyl group at the 2' location, and a phosphate acid at the 4' location [41].

Satisfiability (SAT) problem: A computation problem in the form: (a or b or c or d or...) and (e or f or g or h...) and (i or j or k or l...) and ...A correct solution will satisfy all of the clauses (for the above example, a solution would be {a, f, k ...}).

Single-base mismatch: For a given DNA strand, there exists a complement (see hybridization). A single-base mismatch is the perfect complement to the DNA strand except at one location.

Split-pool: A combinatorial method to create a large set of mixtures. For the large number of different DNA strands necessary for DNA computation, a short strand is synthesized on many glass beads in parallel. These beads are divided into many parts and mixed with other beads that have different short DNA sequences attached. For each part, a different new short strand of DNA is then attached to the pre-existing strands of DNA. The procedure is then repeated many times until a large combinatorial mixture is formed (see Fig. 7).

Transcription: The synthesis of an RNA molecule complementary to a DNA master for protein fabrication. The gene encoded in the base sequence of the DNA is "transcribed" into the RNA version of the same code [41].

Variable base: For the encoding described in Section 5.2.1, a variable base is one that is differs between strands at that location in that set of DNA strands.

An atomistic view of electrochemistry

D.M. Kolb *

Department of Electrochemistry, University of Ulm, 89069 Ulm, Germany

Received 2 January 2001; accepted for publication 21 June 2001

Abstract

One of the most important tasks of modern, physical electrochemistry is the development of an atomistic picture of the solid/liquid interface in order to provide the basis for a mechanistic understanding of electrochemical processes. Electrochemists seek answers to the same questions as their surface science colleagues (e.g., electronic and structure properties of surfaces and adlayers), but are faced with the fact that in electrochemistry the contact of the solid with a condensed phase, the electrolyte, makes life much more difficult. Nevertheless, electrochemists succeeded in the last 20 years to develop an *electrochemical surface science* by adopting experimental techniques and theoretical concepts from surface physicists.

This article describes the various routes electrochemists have used to obtain a detailed characterization of electrode surfaces in particular, and of the electrochemical interface in general. Success in physical electrochemistry is based on the development of non-traditional in situ methods to complement the classical, current- and voltage-based techniques. The former range from optical spectroscopies, linear and non-linear, to in situ X-ray diffraction and scanning tunneling microscopy. The current status of electrochemical surface science and its most important future goals are briefly addressed. © 2001 Elsevier Science B.V. All rights reserved.

Keywords: Scanning tunneling microscopy; Surface relaxation and reconstruction; Surface structure, morphology, roughness, and topography; Copper; Gold; Platinum; Single crystal surfaces; Solid–liquid interfaces

1. What is electrochemistry and why is it important?

Electrochemistry deals with chemical reactions that are connected with the transfer of electric charge between a chemical species and an electrode. The latter is usually a piece of metal immersed in an aqueous solution that contains the reaction partners. The transferred charge may be an electron or an ion. Consequently, we distinguish between electron and ion transfer reactions. Typical examples of both are the oxidation (reduction) of V^{2+} (V^{3+}) and the dissolution or deposition of a metal:

$$V^{2+}_{solv} \leftrightarrow V^{3+}_{solv} + e^-_{metal} \qquad (1)$$

$$Ag \leftrightarrow Ag^+_{solv} + e^-_{metal} \qquad (2)$$

Since an electron in the metal obviously is a reaction partner, electrochemical reactions can be driven and controlled by a voltage applied to the electrode (more precisely, applied between the working electrode and a second electrode), or—if the reactions occur spontaneously—can provide

---

* Tel.: +49-731-5025400; fax: +49-731-5025409.
  E-mail address: dieter.kolb@chemie.uni-ulm.de (D.M. Kolb).

electrical current. The generation of electricity is an intriguing aspect of electrochemistry, which caused the high rise of electrochemistry during the last century, when batteries and fuel cells were developed. The good old lead acid battery (invented in 1859 and still found in each car today) may serve as an example [1]. A second aspect of electrochemistry important in those days is related to the fact that many electrochemical reactions can be driven close to their equilibrium. Valuable thermodynamic data for chemical reactions can be conveniently and precisely determined from measurements of the voltage of a galvanic cell (Fig. 1c).

Electrochemical reactions play a pivotal role in many areas of our daily life, such as corrosion and corrosion inhibition, metal plating (there is hardly any piece of metal in one's vicinity that has not received an electrochemical surface finish, and there is no circuit board in our computers that has not experienced some electrochemical processing) or supplying electricity in portable devices. Life itself makes extensive use of electrochemistry to keep us going: Just think of ions in membranes and membrane potentials.

The interface between the electrode and an electrolyte is the heart of electrochemistry. It is the place where the charge transfer takes place, where gradients in electrical and chemical potentials constitute the driving forces for the electrochemical reactions. What does this interface look like? Early models of such a metal/electrolyte interface date back more than a century (Helmholtz 1874) and subsequently have been refined. The classical textbook model of the metal/solution interface is that of a plate condenser of molecular dimensions (Fig. 1). One plate is the metal proper with its surface excess charge, the other plate is built up by ions in solution.

This simple model of a plate condenser works surprisingly well in explaining most of the experimental observations. A whole branch of electrochemistry, called impedance spectroscopy [4], has developed, which describes and classifies electrochemical reactions by equivalent circuits out of capacitors and resistors (the latter characterizing the charge transfer during an electrochemical reaction). A characteristic feature of metal/aqueous solution interfaces is their remarkably high capacity, which ranges between 20 and 50 $\mu F\,cm^{-2}$. With the well-known equation for the capacity per unit area of a plate condenser $C = \varepsilon\varepsilon_0/d$, this can be rationalized by a "plate distance" d of 0.3 nm and a relative dielectric constant ε of 5–10 for the interfacial water, i.e., for water molecules in a strong electric field (including solvation water). By applying a potential drop across the electrochemical interface of up to 1 V (which for noble metal electrodes is indeed possible), high surface charges of up to about 20–50 $\mu C\,cm^{-2}$ can be achieved (corresponding to a surface charge of about 0.1–0.2 electrons per surface atom), and extremely high electric fields of about 3×10^7 $V\,cm^{-1}$ can be obtained. These are interesting parameters for any surface science study as we demonstrate shortly. The high capacity of an electrochemical interface is increasingly being used for an interesting energy storage device: the supercapacitor [5]. By employing high-surface-area material like carbon black with typically 1000 $m^2\,g^{-1}$ and 30 $\mu F\,cm^{-2}$ specific capacity, the total capacitance is 300 $F\,g^{-1}$. Operating the capacitor at 1 V, the energy stored is 150 $J\,g^{-1}$, or 150 $kJ\,kg^{-1}$. Remember that the high capacity value emerges in essence from the small "plate distance", which is achieved by dipping the solid into the solution. Electrochemical supercapacitors can, e.g., provide standby power for electronic devices and are used as energy source in case of short-term power break-downs.

Before we turn to electrochemical surface science with its use of modern surface-physics oriented methods, we briefly discuss the classical experimental approaches that provide the basis for every electrochemical experiment. The classic route to the study of electrochemical reactions rests on current and voltage measurements. Both quantities can be determined with high precision and are easily accessible, although they lack direct chemical or structural specificity. An electrochemist's daily bread consists of cyclic current–potential curves (cyclic voltammograms, CVs), in which the electric current flowing to the electrode in an external circuit is measured as a function of electrode potential E (in terms of voltage applied externally between the electrode under study and a reference electrode), as the latter is continuously changed at a constant sweep rate $v = dE/dt$. Such a CV is shown in Fig. 2 for a Pt(1 1 1) electrode in 0.1 M H$_2$SO$_4$. Four different processes can be seen to occur, which relate to a mere charging of the electrochemical interface, to sulfate or hydrogen adsorption, and to hydrogen evolution. How can one possibly be sure of such a detailed assignment? How do we know that certain electrons stem from sulfate adsorption, but others from hydrogen adsorption? Most electrochemical reactions can be studied under equilibrium conditions (like in Fig. 2 by choosing a low scan rate), which implies that the adsorbed species is in equilibrium with the corresponding solution species (e.g., sulfate ions or H$_3$O$^+$). Hence, by varying the concentration of a solution species, while keeping those of all others constant, the respective feature in the CV undergoes a potential shift associated with the concentration of the electroactive species in solution.

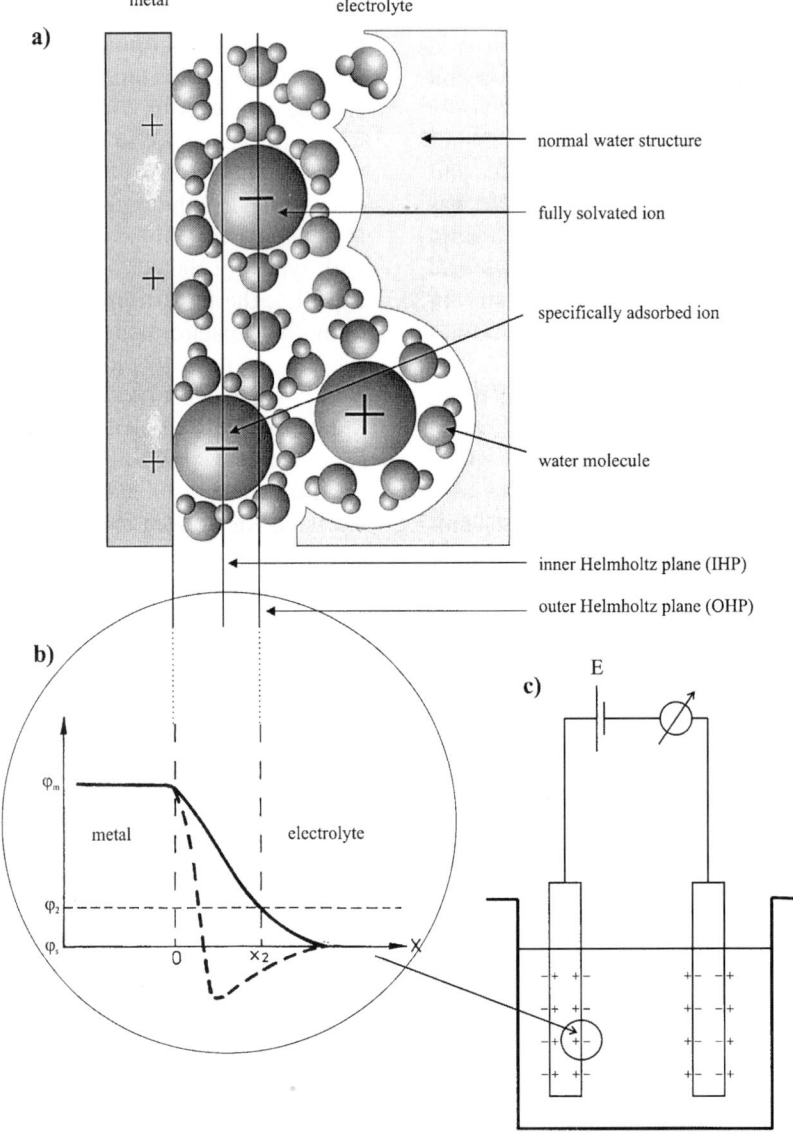

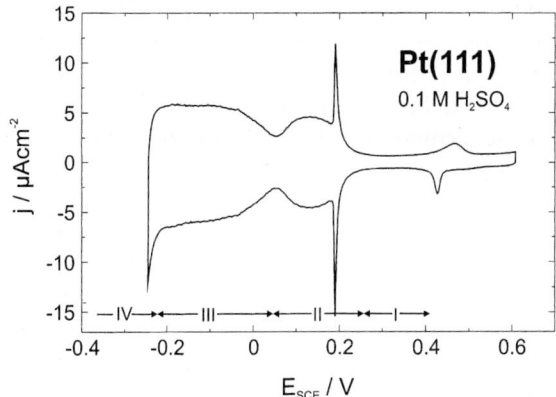

Fig. 2. CV for a Pt(1 1 1) electrode in 0.1 M H$_2$SO$_4$, recorded at a scan rate dE/dt = 10 mV s^{-1}. CVs in general show the electric current as a function of electrode potential while the latter is continuously changed at constant scan rate back and forth within a given potential range. They yield fingerprints of the processes occurring at the electrode–electrolyte interface. In this CV, four different processes can be envisaged, the corresponding potential ranges being marked by roman numbers. No. I denotes the so-called double-layer charging region, where no electrochemical reaction (i.e., no charge transfer through the interface) takes place and where the "Helmholtz capacitor" is simply charged or discharged by the potential variation. No. II shows the region where sulfate adsorption takes place on the positive potential scan (desorption on the negative scan), while no. III comprises the region of hydrogen adsorption (atomic hydrogen!) and desorption during the negative and the positive scan, respectively. In region no. IV, hydrogen evolution causes a steep rise of the cathodic current.

Thus, one can unravel the various reactions that contribute to the overall electric current.

CVs like the one in Fig. 2 have come a long, long way. Whenever our electrochemical ancestors studied reactions with solid electrodes, the results suffered badly from irreproducibility. Kinetic data were particularly difficult to obtain in a reliable and trustworthy form. This is understandable, since clean and structurally well-defined surfaces were not available. Imagine the extraordinary task to have a clean electrode surface in contact with an aqueous solution, when surface physicists had to retreat to ultrahigh vacuum (UHV) conditions in order to work with clean surfaces. Imagine the structurally ill-defined surfaces of the polycrystalline material used in those days, and remember that kinetic data sensitively depend on the defect structure of a surface. But electrochemists found an intriguing way out of that dilemma by using a liquid metal. The dropping mercury electrode (DME) opened the door to reliable electrochemical, particularly kinetic data: by letting liquid mercury flow through a glass capillary, the growing drops of Hg constituted the electrode. Hence, the surface was constantly renewed (typical drop times were in the seconds range) and it was defect free. The DME played an important role in electrochemistry, especially in electroanalytical chemistry, and Hg as electrode material enabled D.C.

Fig. 1. (a) Schematic diagram of the metal/electrolyte interface, showing fully and partially solvated ions. The idea behind this model is that of a plate condensor. One plate is the metal proper with its surface excess charge, the other plate is built up by solvated ions at closest approach, held in place by purely electrostatic forces (such ions are termed "non-specifically adsorbed"). This plane is called the outer Helmholtz plane (OHP). Its distance from the metal surface is estimated to be 0.3 nm, the radius of an solvated ion [2]. Ions with weakly bound solvation shells (e.g., anions, particularly halide ions like Cl$^-$, Br$^-$ and I$^-$) may give away part of those solvation shells to form a chemical bond with the electrode surface. These ions are called "specifically adsorbed". Their centers define the inner Helmholtz plane (IHP). Because the chemical interaction between (specifically adsorbed) ions and electrode surface causes more charge to be accumulated at the surface than required by electrostatics, counter charge (cations in the case of specifically adsorbed anions) is incorporated in the double layer for charge compensation. (b) Potential drop across the interface in case of non-specific (—) and specific (- - -) ion adsorption. ϕ_m, ϕ_s and ϕ_2 are potentials inside the metal, in the electrolyte, and at the outer Helmholtz plane x_2, respectively. A linear potential drop across the electrochemical interface is generally assumed for non-specifically adsorbing ions (like in a plate condensor), while in the case of specific adsorption a steeper potential gradient (close to the metal surface) and an "overshooting" of the potential with respect to the solution value is assumed [3]. (c) Schematic diagram of an electrochemical cell, showing two metal electrodes which are charged by an external voltage E. Because there are potential drops at the interfaces of both electrodes (at the working electrode and at the auxiliary electrode), the electrode potential of a working electrode would not be known in such a simple arrangement. Therefore, in practice a three-electrode arrangement is used, the third electrode being the reference electrode, which has a constant potential drop and against which the electrode potential of the working electrode is measured. A commonly employed reference is the saturated calomel electrode (SCE) and the corresponding electrode potentials are E_{SCE}.

Grahame to ingeniously design the thermodynamic basis for interfacial electrochemistry [6]. Of course, the chemical nature of the electrode material (e.g., Pt vs. Au) or the surface structure could not be varied, hence their role in electrochemical reactions could not be studied. This was clearly a deficiency. How has it overcome, how did CV's like the one in Fig. 2, which is characteristic of a clean and well-ordered single crystal surface, become reality?

2. The emergence of an electrochemical surface science

For a long time mercury was the dominating electrode material and electroanalytical questions (i.e., identification of ions in solution and concentration determinations) were a main issue in electrochemical research. Then came the heyday of (UHV-based) surface science, the beginning of which may be dated to the birth of this journal. Scientists began to characterize bare and adsorbate-covered surfaces (mostly single crystal) at an atomic or molecular level. By using a rapidly increasing number of highly surface-sensitive and structure-specific techniques information about surfaces was obtained, information that electrochemists always had longed for: the lateral arrangement of surface atoms, of atoms or molecules in an adlayer, binding sites and bond lengths, valence states and charge distributions, and much more. All this information was obtained for surfaces in UHV, with electrons as the most important probe (just think of low energy electron diffraction or photoelectron spectroscopy and their pivotal role in the sixties and seventies [7]). These experimental approaches were not directly applicable for electrochemical systems, because there the surfaces were in contact with a condensed phase, buried under a thick layer of electrolyte. Nevertheless, electrochemists gradually succeeded in establishing what I like to call *electrochemical surface science*, by learning from their surface science colleagues.

Developments in four different (though interconnected) directions laid the basis for an electrochemical surface science. All four of them are rooted in classical (i.e., UHV-based) surface science: (i) the development of new, non-electrochemical, structure- or molecule-specific methods that could be applied in situ, i.e., in an electrochemical environment; (ii) the use of single-crystal electrodes; (iii) the implementation of UHV-techniques in electrochemistry; and (iv) the adoption of theoretical concepts from solid state physics and statistical mechanics for the description of the electrochemical interface. Electrochemists began to establish contacts with surface science groups, and to invite surface scientists to their conferences. I recall an exciting conference in Snowmass, Colorado, in 1979 with the programmatic title "Non-traditional approaches to the study of solid/liquid interfaces", at which electrochemists discussed, among others, the implementation of spectroscopic methods in electrochemistry [8]. I remember that Charlie Duke (editor of this volume) was one of the invited speakers!

2.1. New experimental techniques

It was recognized quite early that classical electrochemical methods should be combined with techniques that are capable of providing information on a molecular basis and are applicable in situ. The obvious choice was optical spectroscopy, preferably in the visible and near-UV wave length range, where aqueous solutions are transparent. Reflectance spectroscopy, particularly electroreflectance (a modulation technique, where the change in reflectance with electrode potential is measured) was among the first non-traditional methods, with which questions of the optical properties of adsorbates, thin metal films and oxide overlayers were addressed [9]. Later when single crystal surfaces were employed, this method was used to derive surface structure information by determining rotational anisotropies at normal incidence or by involving surface states in the optical transitions [10]. The latter is demonstrated below. Then came surface enhanced Raman spectroscopy and despite intense discussions about the origin of the enhancement, vibronic information became accessible [11]. With the development of optical thin-layer cells for infrared studies, a major break-through was achieved. By reducing the thickness of the electrolyte film in front of the electrode surface to a few μm, in situ IR spectroscopy became possible, yielding important information on reaction intermediates [12]. Since such thin-layer cells also were applicable to X-ray studies, with different window material of course, the whole spectral region—from infrared to X-rays—became accessible to electrochemical studies. X-ray diffraction studies at grazing incidence began to provide structure information of astounding and hitherto unprecedented precision and detail [13]. A second major break-through came with the invention by Binnig and Rohrer of the scanning tunneling microscope that soon after its invention for surface studies under UHV conditions was adopted to electrochemical systems [14,15]. This technique advanced tremendously our knowledge about surface structure effects in electrochemistry, because imaging of electrode surfaces was now possible in-situ, in real space and with atomic-scale resolution.

2.2. The preparation of single crystal electrodes

For a detailed understanding of surface processes, the use of single crystal electrodes with structurally well-defined surfaces is mandatory. Because in surface science the classic way of preparing single crystal surfaces involves sputtering and annealing of the crystal in an UHV chamber, the first systematic studies with single crystal electrodes addressing the surface structure, were performed with UHV-prepared samples in an equipment that had a closed-transfer system between electrochemical cell and UHV chamber [16–18]. A major break-through (comparable in its impact on electrochemistry to the introduction of the scanning tunneling microscope, STM) was achieved by Clavilier et al. by demonstrating the preparation of high-quality single-crystal Pt surfaces simply with a Bunsen burner [19]. This flame-annealing, which subsequently had been shown to work also for Au and Ag, consisted of heating the crystal in the flame of a Bunsen burner, before it was quenched in ultrapure water and transferred to the electrochemical cell with a droplet of water adhering to the polished surface. The water droplet is crucial in protecting the surface from

1000 nm x 1000 nm

Fig. 3. STM image of a flame-annealed Pt(1 1 1) electrode in 0.1 M H_2SO_4 at +0.35 V vs. SCE. The image shows several atomically flat terraces, separated by monoatomic high steps. The latter are aligned along the main crystallographic directions of the substrate. Different terrace heights are represented by different shades of grey.

contaminants in the laboratory air. Fig. 3 shows the STM image of a flame-annealed Pt(1 1 1) electrode, revealing large, atomically flat terraces. The straight monoatomic-high steps indicate a high surface mobility during surface preparation. In the case of gold single crystals, all three low-index faces after flame-annealing and subsequent slow cooling leads to reconstructed surfaces, much like those obtained by sputtering and annealing under UHV conditions [20].

2.3. UHV techniques in electrochemistry: The ex situ approach

In the beginning, there were no structure-sensitive techniques available for in situ studies. Hence electrochemists adopted UHV techniques, particularly low energy electron diffraction (LEED), to check the quality of their UHV-prepared single crystal surfaces before and after electrochemical treatment, e.g., after some potential cycles. UHV studies were then extended to strongly bound adsorbates, such as iodide, metal or oxide overlayers, systems that were believed to survive with little or no changes the transfer into the UHV. An important step forward was achieved by W.N. Hansen, who showed that electrodes could be removed from solution with their electric double layer intact [21].

An important source of information became available to electrochemists when surface scientists started to investigate water adsorption on metal surfaces in UHV [22]. Although the studies had to be performed under conditions (e.g., low temperatures) totally different from those in electrochemistry, the information on metal–water interaction energies was extremely useful and helped to correct the classical double-layer model from the mid-fifties. In essence, the interaction of water with the commonly used electrode metals was much less than previously assumed. The growing interest of surface scientists in water signalled that the community became aware of electrochemistry as an interesting part of surface science.

2.4. New theoretical concepts

The fourth point that contributed in a significant way to the development of modern interfacial electrochemistry was the adoption of theoretical concepts and methods from solid state physics or statistical mechanics. This was particularly important for a better understanding of the electric double layer, or more generally speaking, of the solvent structure near a charged wall. Important results came from computer simulations of the electric double layer, which yielded new information about the spatial distribution of ions and water molecules near the electrode surface [23]. An example is given in Fig. 4, where the results of a simulation for two different Pt–water interfaces are reproduced [24]. The figure shows the oxygen density profiles for water near an uncharged Pt surface. The layering effect is clearly seen, together with a small influence of the surface crystallographic orientation on the water structure.

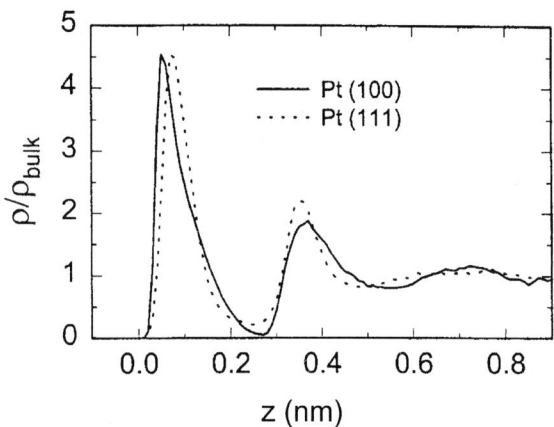

Fig. 4. Simulation of the oxygen density profile at the platinum–water interface for two different crystallographic orientations. The oxygen distribution reveals maxima at certain distances from the surface, which is indicative of a layering effect. From Ref. [24].

3. Metal surfaces in contact with an aqueous solution

Now that we are able to prepare well-ordered single crystal surfaces of metals like Cu, Ag, Au, Pd or Pt (surfaces that have been thoroughly studied in UHV by the surface science community), we can bring them in contact with an aqueous electrolyte, apply an external voltage and study their surface properties. We focus on phenomena that have been well-characterized in UHV and hence, admit a direct comparison between the two systems. These involve surface states, surface reconstruction, surface dynamics and phase transitions in adlayers.

3.1. Surface states at the metal/electrolyte interface

Surface states describe electrons which are strongly localized at the surface or interface of a solid. They arise as a consequence of lattice termination at the surface and are energetically situated in energy gaps for which delocalised bulk electronic states do not exist for motion normal to the surface. They are observed primarily for semiconductor surfaces where the existence of an energy gap for all spatial directions allows for easy detection of surface states. Although the bulk

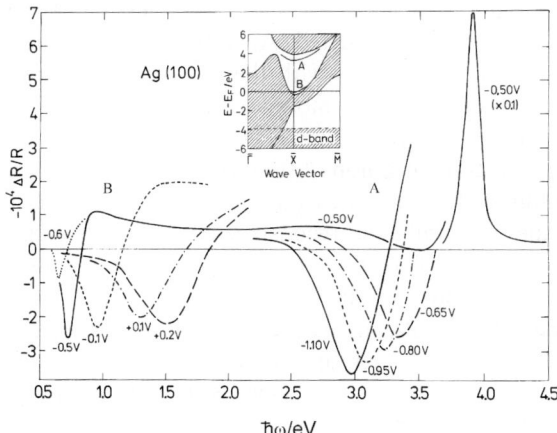

Fig. 5. Normal-incidence electroreflectance spectra of Ag(1 0 0) in 0.5 M NaF for various electrode potentials. The pronounced minima in $-\Delta R/R$ around 3 and 1 eV, which shift in energy with potential, arise from optical transitions from bulk states into the empty surface states A and B, respectively. The inset shows the surface band structure of Ag(1 0 0) with surface states A and B. From Ref. [10].

band structure of metals, by definition, shows no energy gap, the surface band structure can reveal such gaps in certain crystallographic directions of the surface Brillouin zone [25]. This is demonstrated in Fig. 5, where the surface band structure (i.e., the projection of the bulk band structure onto a plane parallel to the surface) for Ag(1 0 0) is shown as an inset, with surface states A and B in the energy gap. The energy of bulk as well as surface states depends on the local electrostatic potential the respective states feel. This fact actually leads to an interesting consequence for the electrochemical double layer with its strong potential gradient at the interface. Because surface states are so strongly localized at the surface, rapidly decaying along the surface normal in either direction, they see a different portion of the externally applied electrostatic potential than bulk states do. As a consequence, the energetic position of the surface states would vary with the electrode potential in a way which is distinctly different from that of the bulk states. From this it follows that optical transitions from bulk states into surface states should be influenced in their energy by the electrode potential [26]. Such a behavior is demonstrated in Fig. 5 for the ER spectrum of Ag(1 0 0) in 0.5 M NaF. A comparison of the experimentally observed transition energies and their dependence on the electrode potential with the calculations by Ho and Liu gave an almost perfect agreement, which supported our assignment of the above-mentioned spectral features to optical transitions into surface states [26].

3.2. Surface reconstruction of Au(100)

In order to derive a microscopic picture of surface processes such as adsorption or deposition, single crystals are generally used, for which the surface structure is believed to be precisely known. However, in general the atoms at a surface do not retain their bulk positions and hence the surface structure cannot be derived directly from the bulk structure which is known from X-ray crystallographic studies. The unsymmetrical environment of surface atoms due to missing neighbors leads to relaxations and rearrangements in the top few layers. Typically, atoms at metal surfaces tend to acquire a more densely packed structure because of an increased electron density between the surface atom cores. Such a rearrangement of surface atoms, which often involves a change in surface symmetry, is called *surface reconstruction* [27]. Typical examples of such phenomena are the bare low-index surfaces of gold which all reconstruct in order to lower the surface energy [28]. An eye-catching structure change due to surface reconstruction is seen with Au(1 0 0), which reconstructs into a hexagonal-close-packed surface ("hex-structure") when prepared under UHV conditions. While for the bare surface, the reconstructed state is energetically more favorable than the unreconstructed ("(1×1)") state, the opposite is normally the case for adsorbate-covered surfaces, because adsorption is energetically more favorable on the less dense, unreconstructed surface. Because of an activation barrier for reconstruction, removal of the adlayer that was responsible for lifting of the reconstruction, does not bring back the reconstructed surface. The latter process normally requires heating the sample.

For gold single crystal electrodes in contact with an aqueous solution, the same reconstruction phenomena as for surfaces in UHV are observed.

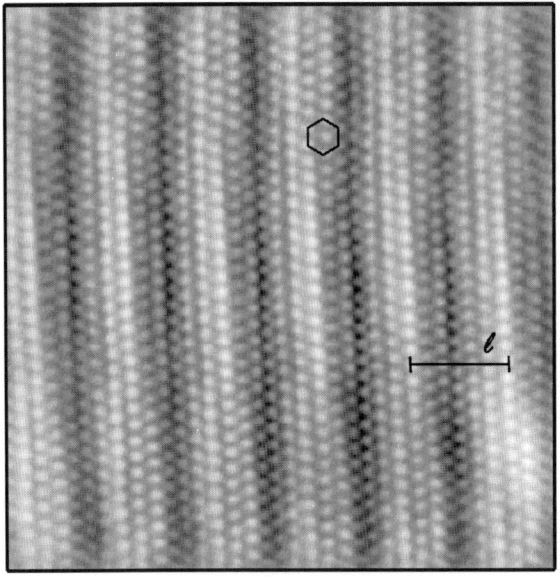

10nm x 10nm

$E_{SCE} = -0.2V$

Fig. 6. High-resolution STM image of a reconstructed Au(1 0 0) surface in 0.1 M H_2SO_4, showing the hexagonal arrangement of the surface atoms. Because the hexagonal top layer is no longer in registry with the underlying layers, the surface becomes buckled, leading to so-called reconstruction rows that run vertically in the figure. The corrugation length ℓ is 1.45 nm. The surface was prepared by flame-annealing. From Ref. [20].

Flame-annealed gold electrodes are reconstructed like in UHV, and the reconstruction can be retained upon immersion into the electrolyte, provided certain precautions are taken [29].

Fig. 6 shows the STM image of a freshly-prepared (flame-annealed), reconstructed Au(1 0 0) surface in 0.05 M H_2SO_4 at $E_{SCE} = -0.2$ V. The hexagonal arrangement of the surface atoms is clearly seen, together with the so-called reconstruction rows (they run vertically in Fig. 6), which reflect the fact that the (hex)-surface is slightly buckled, since it is no longer in registry with the underlying second layer. These rows are a convenient sign of a (1 0 0) surface to be reconstructed in all those cases in which atomic resolution is not obtained and hence, the hexagonal arrangement of the surface atoms cannot be seen directly. Well-prepared reconstructed Au(1 0 0) surfaces show large (hex)-domains with the reconstruction rows being aligned along monoatomic high steps.

Systematic studies have revealed that the specific adsorption of anions lifts reconstruction and the gold surface acquires the (1 × 1) structure. As a rule of thumb, the reconstructed surface is stable only at negative potentials. The transition potential depends on the type of anion and it shifts negatively with increasing strength of chemisorption: The stronger the interaction between anion and gold surface, the smaller is the stability range of the reconstructed surface.

A remarkable observation not possible for UHV can be made at the gold/electrolyte interface. Applying a negative potential to the unreconstructed Au(1 0 0) surface brings back the (hex)-reconstruction [30]. This *potential-induced* reconstruction has been observed for all three low-index faces of gold. Under UHV conditions the bare unreconstructed gold surfaces are in a metastable state, reconstruction of the uncharged surface being prevented at room temperature by an activation barrier. However, applying a sufficiently negative electrode potential to the unreconstructed Au-(1 0 0) surface not only frees the surface from anions, but also lowers the activation barrier to such an extent that reconstruction occurs within seconds or minutes, even at room temperature. This phenomenon has far-reaching consequences for electrochemical studies with gold single crystal electrodes: The surface structure of the electrode becomes a function of potential. Only at sufficiently positive potentials is the structure of the single crystal surface that expected from the bulk. At negative potentials a potential-induced reconstruction sets in transforming, e.g., a Au(1 0 0) surface into one that is Au(1 1 1)-like. Fig. 7 shows an STM image of a Au(1 0 0) electrode at negative potentials, after the initial, thermally induced (hex)-reconstruction had been lifted at positive potentials by sulfate adsorption. The potential-induced reconstruction is characterized by small domains that are rotated with respect to each other by 90°. Since the (hex)-surface has about a 25% higher packing density than the (1 × 1)-surface, lifting of the (hex)-reconstruction yields

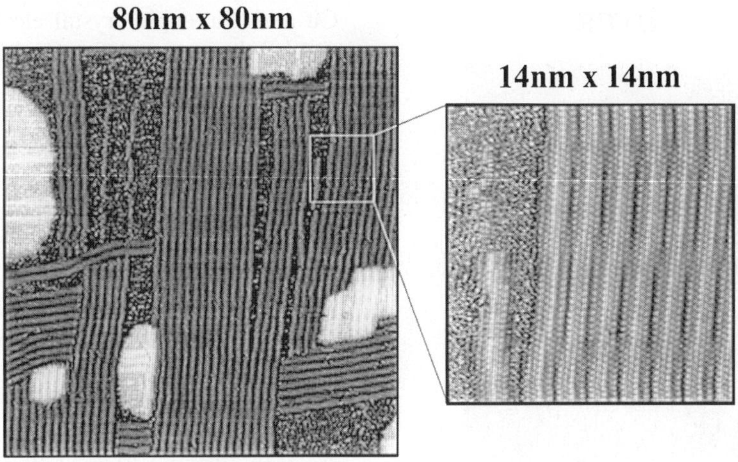

80nm x 80nm

14nm x 14nm

E_{SCE} = −250 mV

Fig. 7. STM image (80 × 80 nm^2) of a Au(1 0 0) electrode in 0.1 M H$_2$SO$_4$ at −0.25 V vs. SCE, where potential-induced reconstruction proceeds. The initially unreconstructed surface is being gradually transformed into the reconstructed form. The (14 × 14 nm^2) zoom shows a section of the surface, 3/4 of which has already been reconstructed; one single reconstruction row on the left hand side is seen to grow from bottom to the top of the image.

an unreconstructed surface that is littered with small monoatomic high gold islands (bright spots in the main image of Fig. 7) formed out of the excess atoms of the (hex)-surface. These islands partly act as gold supply for building up the (hex)-structure during potential-induced reconstruction. Such an event is seen in Fig. 7: the formerly round islands are eaten up by the reconstruction rows. An interesting detail is shown in the zoom of Fig. 7. The (1 × 1)-structure surrounding the reconstruction rows is in a highly mobile state that makes imaging of the individual surface atoms almost impossible.

3.3. Surface dynamics at metal/electrolyte interfaces

In this section we demonstrate how self diffusion on metal surfaces in contact with an aqueous electrolyte can be influenced by the electrode potential and by the solution composition. We also show how the mobility of surface atoms can be greatly enhanced by the electrochemical parameters, leading to interesting consequences like electrochemical annealing or surface faceting.

3.3.1. Frizzy steps

In UHV studies of stepped single crystal surfaces of Cu and Ag, so-called frizzy steps were observed. This term describes monoatomic high steps that appear irregular, i.e., "frazzled" in STM images. Ibach et al. have explained this intriguing observation by a rapid movement of kink sites, caused by adatoms that diffuse on the surface, preferentially along step edges [31]. Adatoms may be captured by the kink sites, leading to an expansion of the upper terrace; kink sites also can emit atoms causing a retreat of the upper terrace. The frizziness of a step thus reflects the constantly changing position of a monoatomic high step, monitored by the STM tip during each X-scan. Fig. 8 shows a typical example of a frizzy step on an electrode surface: Ag(1 1 1) in a sulfuric acid electrolyte [32]. The atomically resolved upper and lower terrace clearly indicates that the STM is working properly, so the noisy appearance of the step is not an instrumental artifact.

We expect a strong influence of the electrode potential on the dynamics of metal surfaces, particularly if it is close to the metal dissolution potential, and hence we may expect a pronounced

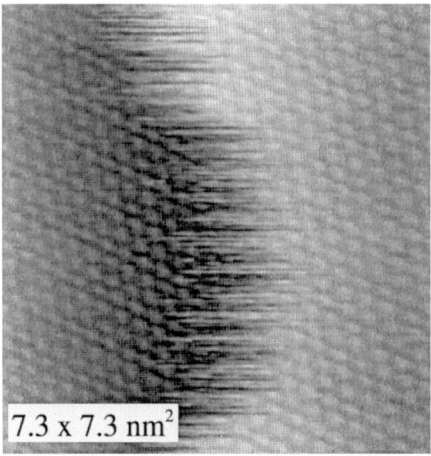

Fig. 8. High-resolution STM image of a Ag(1 1 1) electrode in sulfuric acid solution, showing the frazzled appearance of a monoatomic high step. This frizziness is caused by substrate atoms that either rapidly move along the step or become detached from the step and escape onto the lower terrace. From Ref. [32].

dependence of the frizziness of steps on the potential. This is indeed the case, as has been demonstrated for silver and gold electrodes. While for Ag(1 1 1) at negative potentials (e.g., −120 mV vs. SCE) the frizzy region extends roughly over 1 a.u. and is similar to that for Ag(1 1 1) in air, the frizzy region spans a width of about 8 a.u. at +0.08 V vs. SCE, which is quite close to the Ag dissolution potential [32]. In the latter case, the surface is already in a highly dynamic state.

The steps of a Au(1 1 1) electrode in a sulfuric acid solution are not frizzy. When halide ions such as chloride are added to the solution, however, the steps show frizziness [33]. Therefore the halide ions can switch on frizziness, i.e., can increase the surface mobility. This observation leads us to the next point.

3.3.2. Anion-enhanced surface mobility

In early work with polycrystalline Ag and Au electrodes, long-term changes in the double-layer capacity were often observed, which were tentatively explained by a slow smoothing of the rough surface. Later, Stickney and coworkers studied this electrochemical annealing in more detail with Cu and Au single crystal electrodes [34]. For example, the surface of a Cu(1 0 0) electrode which had been roughened by electrochemical reactions, could be restored by simply immersing the sample for some minutes in an HCl solution. STM studies with unreconstructed Au(1 0 0) and Au(1 1 1) electrodes revealed that the surface mobility of gold atoms could be significantly increased by applying positive potentials where anions are adsorbed on the surface, particularly at step edges. This was tested with the monoatomic high islands on the gold surface, which were created by lifting of the reconstruction. The rate at which the surface topography changed increased with the increasing strength of interaction between anion and gold surface (i.e., $HSO_4^- < Cl^- < Br^- < I^-$). One can visualize this anion-enhanced surface mobility by a local chemical bond between halide and gold atom at a step that weakens the bond of such a gold atom to its neighbors.

3.3.3. Potential-induced faceting

When a metal is in contact with its ions in solution (e.g., Ag in an aqueous solution of $AgNO_3$), an equilibrium potential is established with no external current flowing. This equilibrium can be dynamic, as the partial currents for dissolution and deposition, which are equal in magnitude, but opposite in sign, may be quite large. Silver is a good example of a dynamic equilibrium situation, as the exchange current density (the cathodic or anodic current density at the equilibrium potential) is extremely high, namely 13.4 A cm^{-2} for a 1 M Ag$^+$ solution. Hence it is not surprising that electrode surfaces, particularly polycrystalline ones, can undergo massive restructuring and faceting, when potentials close to their dissolution are applied. An intriguing effect, however, was reported by Arvia and coworkers some years ago, when they subjected polycrystalline platinum to fast potential cycles [35]. The cycles were run in the kHz range and the potential range span the hydrogen adsorption at the negative end and the surface oxide formation at the positive one. After cycling for an extended period of time (min to h) the Pt surface showed characteristic features of Pt(1 0 0) or Pt(1 1 1) in its electrochemical response. This indicated that depending on the exact limits for the potential range, (1 0 0) or (1 1 1) facets developed with time. Obviously, place exchange reactions during oxide formation and reduction causes a rearrangement of surface atoms and a faceting of the surface. We have to keep in mind that these processes happened millions of times due to the fast cycling. Hope has been expressed to use this simple method to tailor the surface structure of polycrystalline catalytic material for optimum conditions without having to utilize expensive single crystal surfaces.

3.4. Organic adlayers: structure and phase transitions

The adsorption of organic molecules at metal surfaces from aqueous solutions is a field of intensive research. This is partly due to the immense importance of organic molecules in applied areas such as additives in plating baths or as corrosion inhibitors, and partly because it is a source of many interesting surface phenomena such as two-dimensional phase transitions [36].

The main driving forces for the adsorption of organics are either attractive molecule-substrate interactions or the hydrophobic character of a molecule. A driving force for the two-dimensional condensation are strong attractive interactions between the molecules. For the latter interactions hydrogen-bridge bonding often plays an important role. And the substrate has a pronounced influence on the adlayer structure. In this respect, experiments with single crystal electrodes of different crystallographic orientation seem particularly rewarding. Among the many organic molecules that have been used in the study of adsorption on solid surfaces in general, and of the formation of ordered structures in particular, the two pyrimidine bases, uracil and thymine, received the highest attention, not the least because of the interesting question of spontaneous ordering of bio-molecules on surfaces, a question that relates in the widest sense to the origin of life.

Fig. 9 shows the CV for Au(1 1 1) in 0.05 M H_2SO_4 containing 12 mM uracil [37]. Despite the seemingly complex appearance of the CV, four regions of distinctly different adsorption behavior for uracil on Au(1 1 1) can be identified by a thermodynamic analysis of the electrochemical data. Phase transitions within the uracil adlayer at certain potentials are indicated by the appearance of current spikes. And from the packing density of the admolecules, clues about their orientation (upright or flat-lying) are derived. This example may demonstrate how much information one can obtain from a careful analysis of electrochemical data. Of course, all these conclusions had been checked afterwards by STM measurements, which in essence confirmed the above results about surface excesses and phase transitions, but added valuable structure information. While the physisorbed uracil layer

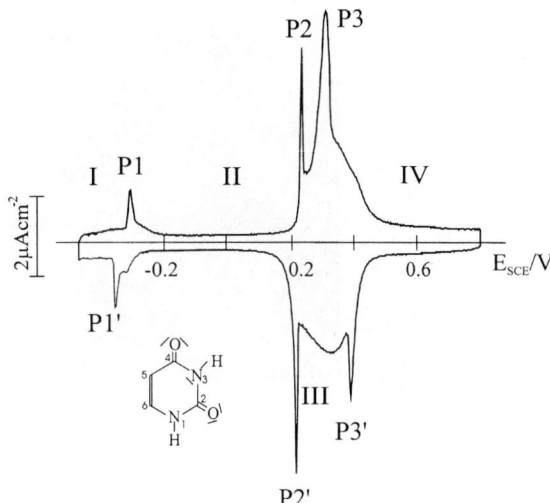

Fig. 9. Cyclic current-potential curve for Au(1 1 1) in 0.05 M H_2SO_4 + 12 mM uracil, recorded at a scan rate of 10 mV s^{-1}. The electrochemical processes connected with the current peaks P1 to P3 are in essence related to physisorption and chemisorption of the uracil molecule (the structure of which is given in the figure). Adsorption starts at negative potentials and consequently region I can be described by random (and mobile) adsorption. The current spike P1 indicates a phase transition to a condensed physisorbed state, which exists throughout region II, limited by spike P2. The low double-layer capacity of region II reveals that the adlayer is dense. From the surface excess of uracil determined by charge measurements, the area per molecule can be estimated to 42 Å^2, which implies flat-lying molecules. From the massive charge flowing in region III we can assume a reorientation of the molecules taking place that leads to a layer of chemisorbed molecules in region IV. The area per molecule in the latter region is much smaller (22 ± 4 Å^2), which hints at an upright position, where uracil binds to the gold surface through one of the two nitrogen atoms in the ring. From Ref. [37].

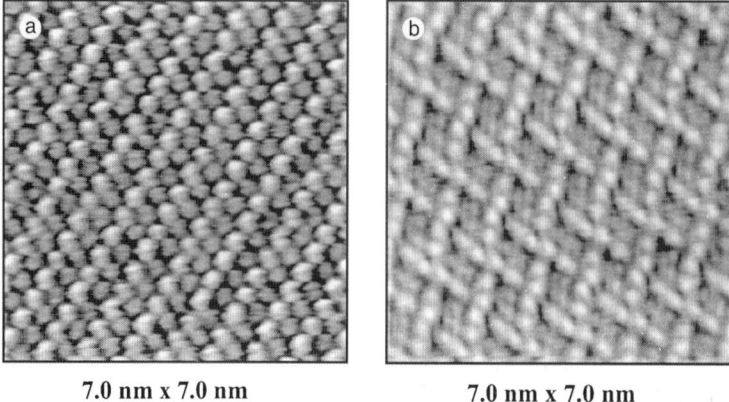

7.0 nm x 7.0 nm   7.0 nm x 7.0 nm

Fig. 10. High-resolution STM images of uracil molecules chemisorbed on Au(1 1 1) (panel a) and on Au(1 0 0)-(1 × 1) (panel b), in sulfuric acid solution. The influence of the surface symmetry on the adlayer structure is clearly seen. From Ref. [37].

does not lift reconstruction, the chemisorbed layer does. A high-resolution STM image of the latter is reproduced in panel(a) of Fig. 10, revealing a densely packed layer with an intermolecular distance of 4.9 ± 0.2 Å [37]. The STM-derived area per molecule, 21 ± 4 Å^2, agrees well with electrochemical data and suggests an upright position.

The structure of the physisorbed uracil layer has been found to be practically independent of the surface structure of the substrate, indicating that the lateral arrangement is determined by hydrogen bridge bonds. This is not the case for the chemisorbed uracil layer as one might expect as a consequence of a strong molecule–substrate bond. This is convincingly demonstrated by the two STM images in Fig. 10, which represent the chemisorbed phase on Au(1 1 1)-(1 × 1) and Au(1 0 0)-(1 × 1) [37]. The structural difference is intriguing; it is caused by a subtle balance between intermolecular forces, which in this case are highly directional, and a molecule–substrate interaction which searches for hollow sites. The short, four molecules long rows of the uracil adlayer on Au(1 0 0) has been tentatively explained by such a competition of forces [37].

4. Initial stages of metal plating

The study of metal deposition from aqueous solutions has a long tradition in electrochemistry. This reflects not only the tremendous importance of this process for the metal winning, refining and plating industries, but also the great interest of scientists in electrocrystallization phenomena, for which metal deposition provides an ideal case [38]. Metal deposition at the solid/solution interface also represents an interesting process to be compared with metal deposition from the vapor phase in an UHV chamber. The most significant difference between the two cases is that in the former equilibrium is easily established between the deposit and metal ions in solution and deposition can be conducted close to equilibrium conditions, while in the latter case deposition occurs far from equilibrium. (To match the electrochemical situation, Cu metal would have to be in thermodynamic equilibrium with Cu vapor!) Metal deposition and dissolution is governed by the electrode potential. When a metal is immersed into a solution containing its ion (e.g., a sheet of Cu metal in a 1 M CuSO$_4$ solution), an equilibrium potential is established, at which the rates of deposition and dissolution are equal, i.e., no net electric current is flowing. When we apply an external potential that is negative of this equilibrium potential E_r (the Nernst potential), to the electrode, metal ions are discharged and deposited onto the surface. At potentials positive of E_r metal is dissolved. The process continues in both cases, until the resulting metal ion concentration matches the new equilibrium potential. The equilibrium potential changes with metal ion concentration by $(59/z)$ mV per decade, z being the valence state of the ion, i.e., $z = 2$ for Cu^{++} [38].

4.1. The formation of metal monolayers (underpotential deposition)

Metal deposition onto a foreign metal substrate (e.g., Cu onto Au) often starts with the formation of a monolayer at underpotentials, i.e., at potentials positive to the Nernst potential for the respective bulk phase. This at first sight surprising observation (deposition should not be possible at potentials positive of E_r) can be easily understood by the fact that in such a case the metal–substrate interaction exceeds in strength that between the deposited metal atoms (e.g., Cu–Au as compared to Cu–Cu). This underpotential deposition (upd) is most conveniently studied by cyclic current–potential curves, where the current due to deposition (during the potential scan in negative direction) or dissolution (scan in positive direction) of the monolayer represents in essence the first derivative of the corresponding adsorption isotherm. Fig. 11 shows the CV for Cu upd on Au(1 1 1) in sulfuric acid solution, which demonstrates that the Cu monolayer is formed in two energetically distinctly different steps (at about +0.19 V and around +0.03 V vs. Cu/Cu^{++}). Bulk Cu deposition can only start negative of 0 V. The CV resembles a thermal desorption spectrum, frequently recorded in classical surface science to determine binding energies. In the latter case, the temperature of the substrate is increased at a constant rate dT/dt, and the rate of desorption, $d\theta/dt$, measured by mass spectrometry. Indeed, cyclic current–potential curves yield a similar type of information, the peak potential being a measure of the bond strength.

A multi-peak structure in the voltammogram of metal upd has been taken as a strong indication for the occurrence of ordered adsorption [38,39]. Indeed, LEED and RHEED studies with electrodes, which had been removed from the electrolyte and transferred into a UHV chamber, showed superstructures in the diffraction patterns which proved the existence of ordered adlayers in upd well before the use of STM in electrochemistry

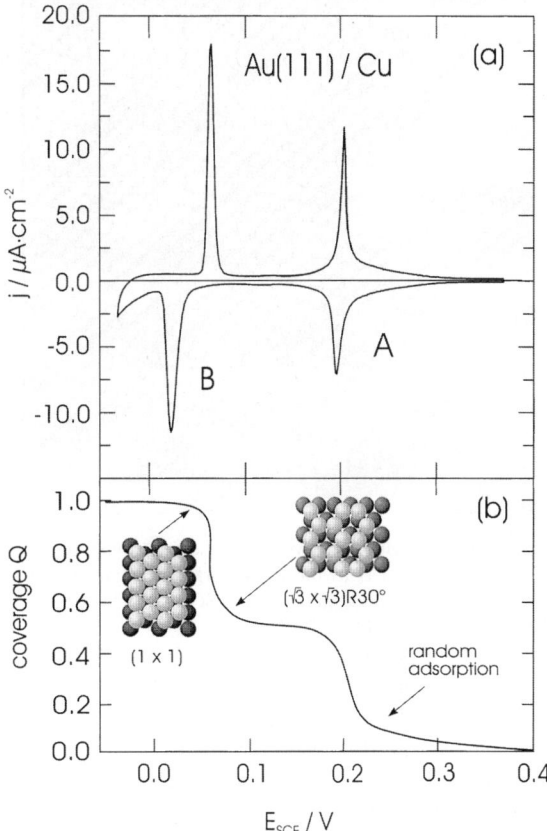

Fig. 11. (a) Cyclic current–potential curve and (b) adsorption isotherm $\theta(E)$ for the Cu monolayer formation on Au(1 1 1) in 0.05 M H$_2$SO$_4$ + 1 mM CuSO$_4$. Scan rate: 1 mV s^{-1}. The pronounced current peaks in the CV indicate that the Cu monolayer is formed in two steps: After the first peak at +0.2 V 2/3 of the Cu monolayer have been deposited; around 0 V the gold surface is covered by a full monolayer. The broad shoulder on the positive side of the +0.2 V peak shows that Cu deposition starts with random adsorption of Cu atoms. The adlayer structure of the Cu deposit at medium and high coverage is sketched in part (b).

[40]. For Cu on Au(1 1 1) in sulfuric acid solution systematic ex situ diffraction and AES measurements revealed that the Cu monolayer is formed in three steps: (i) random adsorption at coverages below 0.1; (ii) formation of a $(\sqrt{3} \times \sqrt{3})$ R30° superstructure at medium coverages, where the Cu adatoms are arranged in honey combs with a limiting coverage of 2/3; and (iii) formation of a full monolayer with a (1×1) structure (see structure models in Fig. 11b).

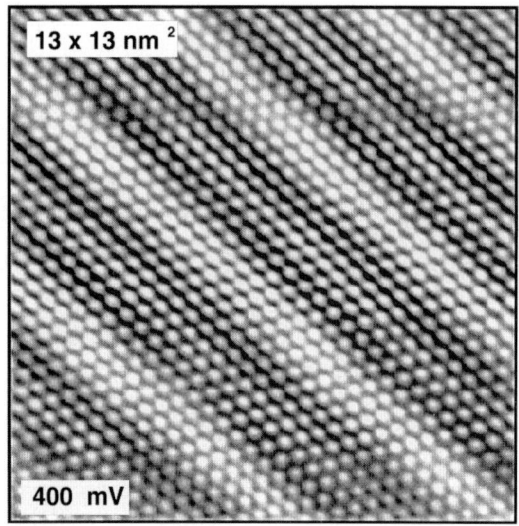

Fig. 12. High-resolution STM image of silver atoms, regularly adsorbed on Au(1 1 1) to form a distorted hexagon structure. The image was recorded in 0.05 M H_2SO_4 + 1 mM Ag_2SO_4 at +0.4 V vs. Ag/Ag^+. From Ref. [41].

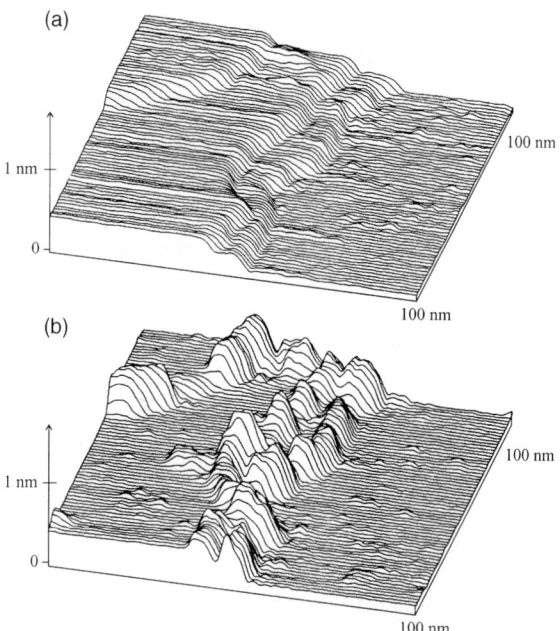

Fig. 13. STM images of Au(1 1 1) in 5 mM H_2SO_4 + 0.05 mM $CuSO_4$ before (panel a) and during (panel b) copper deposition. Image (b) demonstrates that nucleation and growth of Cu occurs preferentially at the steps of the substrate. From Ref. [42].

Metal deposits in submonolayer and monolayer amounts on foreign metal substrates obtained by upd have become a favorite playground for STM, as their manifold structures and structure transitions with changing coverage can be studied in detail and with relative ease by this technique. One example of a high-quality STM image with atomic resolution is given in Fig. 12, which shows the adlayer of underpotentially deposited Ag on Au-(1 1 1) in sulfuric acid solution at submonolayer coverage [41]. Because the Ag adatoms occupy sites on the gold surface, which are not equal, this results in a height modulation that is clearly reflected in the stripe pattern of Fig. 12. This image once more conveys the fascination of STM by being able to see individual atoms on surfaces, with an instrument that almost fits into the pocket of one's jacket (without controller!).

4.2. The initial stages of bulk deposition

When we apply an electrode potential negative of the equilibrium potential E_r, bulk deposition takes place, the rate of which strongly depends on the so-called overpotential, i.e., the difference between actual and equilibrium potential. Because metal deposition is a nucleation-and-growth process, surface defects such as steps or screw dislocations are known to play a crucial role as nucleation centers. Their density and local arrangement have a strong influence on the morphology of the deposit, particularly during the initial stages of growth. Because of the ability of STM to image surfaces in *real* space, this technique is especially suited to study nucleation at defects and the initial growth of metal clusters. Fig. 13 shows two STM images of a Au(1 1 1) surface before and during bulk Cu deposition [42]. The bare gold surface has atomically flat terraces separated by three monoatomic high steps. After a potential step to negative values deposition of bulk Cu occurs almost exclusively at the monoatomic high steps, the growing clusters decorating the surface defects. It is only after some time that Cu clusters start to grow on the terraces too.

Nucleation-and-growth behavior leads to inhomogeneous film thicknesses and hence, to rough metal overlayers. Scientists in the plating industry have known for a long time how to eliminate this problem: they add organic molecules to the plating baths, so-called levellers or brighteners, which modify either the nucleation or the growth behavior in such a way that smooth and uniform overlayers emerge. Although this is a rather secretive area, some typical additives are known: e.g., benzotriazol, thiourea or crystal violet. Addition of crystal violet induces a quasi-two-dimensional growth, which causes the substrate to be quickly buried by a rather uniform Cu film. Since the overall rate of Cu deposition is reduced significantly in the presence of crystal violet, we believe that the effect of the dye is to inhibit cluster growth normal to the surface, rather than to accelerate the growth parallel to the surface.

5. Electrochemical nanostructuring of surfaces

The past decade has witnessed a second career of the STM: a tool for positioning single atoms and molecules on surfaces. This was achieved by employing the tip–substrate interaction at close distance (say, of the order of an atomic diameter) to manipulate individual atoms or molecules with the tip, and directing them—one after the other—to predetermined positions. Impressive examples of this kind of nanostructuring of surfaces have been given by several groups working under UHV conditions and mainly at low temperatures [43–45]. Don Eigler's famous quantum corral, a circle out of 48 Fe atoms arranged with an STM tip on a Cu(1 1 1) surface, is probably the most frequently cited example of this kind [46].

Soon after surface scientists demonstrated how to use (or misuse) the STM for nanostructuring rather than mere imaging of surfaces, electrochemists started similar work. Their tip-generated entities were clearly larger than single atoms, because constructions from atoms would not have been stable enough to survive an electrochemical environment at room temperature. The most common approach to an electrochemical nanostructuring of surfaces was to use the tip to create surface defects, which then acted as nucleation centers for the metal deposition at preselected positions [47]. The defects were often produced by a mechanical contact between tip and sample ("tip crash"). A more recently developed strategy for positioning small metal clusters on electrode surfaces implied a two-step process, in which the metal was first deposited from solution onto the tip, followed by a burst-like dissolution and redeposition onto the sample right underneath the tip [48]. Such a procedure left the surface undamaged, but the cluster size was rather large, e.g., in the tens of nm range. In the following we briefly demonstrate a method, based on the so-called *jump-to-contact* between tip and sample [49], which also leaves the surface undamaged, but generates fast and reproducibly metal clusters containing of the order of only one hundred atoms.

The proposed mechanism for the tip-induced cluster formation is sketched in Fig. 14 [50], highlighting the two essentials of this method: first,

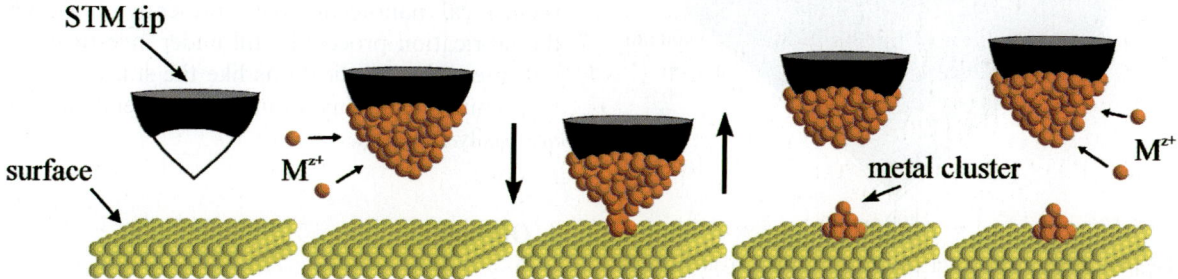

Fig. 14. Schematic representation of the mechanism that leads to tip-generated metal clusters. First of all, metal has to be deposited from solution onto the tip. The essential step is an approach of tip and substrate that leads to the so-called jump-to-contact (middle picture), where a connective neck between tip and substrate is formed. Upon retreat of the tip the neck breaks, leaving a metal cluster on the substrate. Continued metal deposition onto the tip supplies enough material for the next cluster generation (last picture).

metal is deposited onto the tip, by choosing a tip potential negative of the reversible deposition potential E_r (see Section 4). Secondly, the metal-loaded tip approaches the surface for a short period of time, during which the jump-to-contact occurs. The resulting connective neck breaks upon the subsequent retreat of the tip, leaving a small metal cluster on the surface. The direction of the material transfer depends on the cohesive energy of both sides, the situation depicted in Fig. 14 being that for, e.g., Cu and Au(1 1 1) substrate. (When nickel has been deposited onto the tip and the substrate is Au(1 1 1), gold will jump to the tip, leaving holes in the substrate.) Because of the negative tip potential, the tip is constantly "reloaded" with metal from solution and hence becomes ready for another cluster formation.

The jump-to-contact requires an approach down to about 0.3 nm tunnel gap. All three spatial coordinates of the tip can be externally controlled by a microprocessor, which makes the nanodecoration of an electrode surface with metal clusters a fully-automated process, allowing even complex patterns to be made rapidly and reproducibly. Two examples of tip-induced cluster formation are given in Figs. 15 and 16 [51], both referring to Cu on Au(1 1 1), as this system has turned out to work

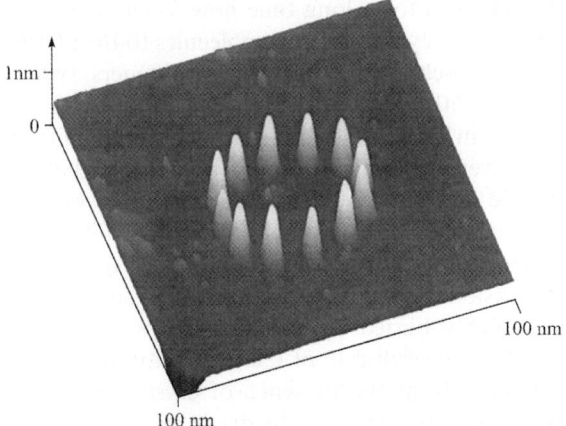

Fig. 16. STM image of 12 Cu clusters on a Au(1 1 1) electrode. The clusters are 0.8 nm high and were arranged in a circle of about 40 nm diameter. They were generated by the tip of an STM. From Ref. [51].

best. In Fig. 15 an array of 400 Cu clusters on Au(1 1 1) in 0.05 M H_2SO_4 + 1 mM $CuSO_4$ is shown. Fig. 15 demonstrates the amazing uniformity of the Cu clusters. In Fig. 16 the ability of surface patterning is demonstrated by a circle of 12 Cu clusters on Au(1 1 1), which is 40 nm in diameter. All Cu clusters are 0.8 nm high. It is important to point out that the electrode potential of the Au(1 1 1) during cluster fabrication and imaging always has been +10 mV vs. Cu/Cu^{++}. At such a value bulk Cu deposition from solution onto Au(1 1 1) is not possible. The cluster material originates exclusively from the Cu-covered tip. What now looks like a nice play ground for STM may well become the basis of an electrochemical nanotechnology. Presently, however, the fabrication process is still under investigation, but interesting applications like the study of electrochemical reactions on individual metal clusters are easily foreseen.

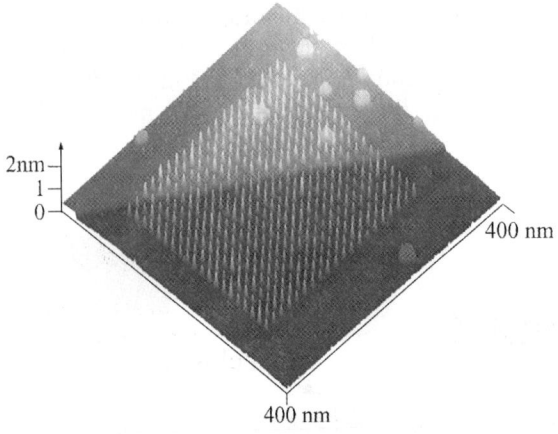

Fig. 15. Array of 400 Cu clusters on Au(1 1 1) in 0.05 M H_2SO_4 + 1 mM $CuSO_4$, generated by 400 single voltage pulses of 2 ms duration to the z-piezo of an STM. During each voltage pulse, tiny amounts of Cu are transferred from the STM tip to the gold substrate. The Cu clusters are about 0.6 nm in height. From Ref. [51].

6. Outlook

The task of characterizing on an atomic level the structure of bare or adsorbate-covered metal surfaces in contact with an aqueous electrolyte, which is of utmost importance for the mechanistic interpretation of many electrochemical processes, was such a demanding and exciting challenge that the main purpose of electrochemical studies, namely to measure reactions and turnovers, had to take a back seat. Now, as we know more about the surface structure of single-crystal electrodes and its dependence on the various electrochemical parameters, the establishment of structure–reactivity relations is the next challenge. This is crucial for the development of tailor-made electrocatalysts for fuel cells, just to mention one example which is currently receiving much attention from the scientific community as well as from the public. Fuel cell reactions tend to poison the surface of catalysts. Many researchers are currently engaged in the task of modifying the surface properties such that the desired reaction proceeds at a high pace without blocking the catalyst's surface by reaction intermediates.

The chemical modification of electrode surfaces is another promising area for future activities. The number of electrode materials suitable for certain electrochemical reactions is often limited either by their chemical stability (e.g., by corrosion) or by their costs (e.g., Pt as catalyst). In both respects, the chemical modification of surfaces—particularly by organic monolayers with their wide structural variety—appears promising. A molecular design of electrode surfaces and surface properties seems feasible.

Two more topics are at the verge of playing an important role in future surface science activities: bioelectrochemistry and electrochemical nanotechnology. The study of charge transfer reactions, which are pivotal for many biological systems, with structurally well-defined electrode surfaces, may well shed light on the complex electrochemical reactions occurring in living cells. The structuring of solid surfaces on a nanometer scale by electrochemical processes, e.g., by the use of an STM tip as has been demonstrated above, may be an interesting alternative to UHV-based techniques for the fabrication of quantum devices or of surface structures unique in their properties. It appears that the study of solid/liquid interfaces will remain an exciting area of surface science.

References

[1] J. Garche, On the historical development of the lead/acid battery, especially in Europe, J. Power Sources 31 (1990) 401.

[2] H. Gerischer, D.M. Kolb, J.K. Sass, The study of solid surfaces by electrochemical methods, Adv. Phys. 27 (1978) 437.

[3] J.O'M. Bockris, A.K.N. Reddy, Modern Electrochemistry, second ed., Plenum Press, New York, 1998.

[4] J.R. Macdonald, Impedance Spectroscopy, Wiley, New York, 1987.

[5] B.E. Conway, Electrochemical Supercapacitors, Kluwer, New York, 1999.

[6] D.C. Grahame, The electrical double layer and the theory of electrocapillarity, Chem. Rev. 41 (1947) 441.

[7] G. Ertl, J. Küppers, Low Energy Electrons and Surface Chemistry, VCH, Weinheim, 1985.

[8] Proceedings of the Snowmass Conference on: Non-traditional approaches to the study of the solid-electrolyte interface, Surf. Sci. 101 (1980).

[9] D.E. McIntyre, D.M. Kolb, Specular reflection spectroscopy of electrode surface films, Symp. Faraday. Soc. 4 (1970) 99.

[10] D.M. Kolb, UV-visible reflectance spectroscopy, in: R.J. Gale (Ed.), Spectroelectrochemistry: Theory and Practice, Plenum Press, New York, 1988, pp. 87–188.

[11] D.L. Jeanmaire, R.P. VanDuyne, Raman spectroelectrochemistry. Part. I: heterocyclic aromatic, and aliphatic amines adsorbed on the anodized silver electrode, J. Electroanal. Chem. 84 (1977) 1.

[12] A. Bewick, K. Kunimatsu, Infra red spectroscopy of the electrode–electrolyte interphase, Surf. Sci. 101 (1980) 131.

[13] M.F. Toney, O.R. Melroy, Surface X-ray scattering, in: H.D. Abruna (Ed.), Electrochemical Interfaces, Modern Techniques for In-Situ Interface Characterization, VCH, New York, 1991, p. 55.

[14] P. Lustenberger, H. Rohrer, R. Christoph, H. Siegenthaler, Scanning tunneling microscopy at potential-controlled electrode surfaces in electrolytic environment, J. Electroanal. Chem. 243 (1988) 225.

[15] J. Wiechers, T. Twomey, D.M. Kolb, R.J. Behm, An in situ scanning tunneling microscopy study of Au(1 1 1) with atomic scale resolution, J. Electroanal. Chem. 248 (1988) 451.

[16] A.T. Hubbard, J.L. Stickney, S.D. Rosasco, M.P. Soriaga, D. Song, Electrodeposition on a well-defined surface: Ag on Pt(1 1 1), J. Electroanal. Chem. 150 (1983) 165.

[17] A.S. Homa, E. Yeager, B.D. Cahan, LEED-AES Electrochemical Studies of H Adsorption on Pt(h k l), J. Electroanal. Chem. 150 (1983) 181.

[18] F.T. Wagner, P.N. Ross, LEED analysis of electrode surfaces, J. Electroanal. Chem. 150 (1983) 141.

[19] J. Clavilier, R. Faure, G. Guinet, R. Durand, Preparation of monocrystalline Pt microelectrodes and electrochemical study of the plane surfaces cut in the direction of

the (111) and (110) planes, J. Electroanal. Chem. 107 (1980) 205.

[20] A.S. Dakkouri, D.M. Kolb, Reconstruction of gold surfaces, in: A. Wieckowski (Ed.), Interfacial Electrochemistry, Marcel Dekker, New York, 1999, pp. 151–173.

[21] W.N. Hansen, The emersed double layer, J. Electroanal. Chem. 150 (1983) 133.

[22] P.A. Thiel, T.E. Madey, The interaction of water with solid surfaces: fundamental aspects, Surf. Sci. Rep. 7 (1987) 211.

[23] E. Spohr, Computer simulations of electrochemical interfaces, in: R.C. Alkire, D.M. Kolb (Eds.), Advances in Electrochem. Science and Engineering, vol. 6, Wiley-VCH, Weinheim, 1999, pp. 1–75.

[24] M.L. Berkowitz, I.C. Yeh, E. Spohr, Structure of water at the water–metal interface: molecular dynamics computer simulations, in: A. Wieckowski (Ed.), Interfacial Electrochemistry, Marcel Dekker, New York, 1999, pp. 33–45.

[25] A. Goldmann, V. Dose, G. Borstel, Empty electronic states at the (100), (110) and (111) surfaces of nickel, copper and silver, Phys. Rev. B 32 (1985) 1971.

[26] D.M. Kolb, W. Boeck, K.M. Ho, S.H. Liu, Observation of surface states on Ag(100) by infrared and visible electroreflectance spectroscopy, Phys. Rev. Lett. 47 (1981) 1921.

[27] G.A. Somorjai, M.A. VanHove, Adsorbate-induced restructuring of surfaces, Progr. Surf. Sci. 30 (1989) 201.

[28] M.A. VanHove, R.J. Koestner, P.C. Stair, J.P. Biberian, L.L. Kesmodel, I. Bartos, G.A. Somorjai, The surface reconstructions of the (100) crystal faces of iridium, platinum and gold, Surf. Sci. 103 (1981) 189.

[29] D.M. Kolb, Reconstruction phenomena at metal-electrolyte interfaces, Progr. Surf. Sci. 51 (1996) 109.

[30] J. Schneider, D.M. Kolb, Potential-induced surface reconstruction of Au(100), Surf. Sci. 193 (1988) 579.

[31] J.F. Wolf, B. Vicenzi, H. Ibach, Step roughness on vicinal Ag(111), Surf. Sci. 249 (1991) 233.

[32] M. Dietterle, T. Will, D.M. Kolb, Step dynamics at the Ag(111)-electrolyte interface, Surf. Sci. 327 (1995) L495.

[33] M. Giesen, D.M. Kolb, Influence of anion adsorption on the step dynamics on Au(111) electrodes, Surf. Sci. 468 (2000) 149.

[34] L.B. Goetting, B.M. Huang, T.E. Lister, J.L. Stickney, Preparation of Au single crystals for studies of the ECALE deposition of CdTe, Electrochim. Acta 40 (1995) 143.

[35] J.C. Canullo, W.E. Triaca, A.J. Arvia, Changes in the electrochemical response of polycrystalline Pt electrodes promoted by fast repetitive potential cycles, J. Electroanal. Chem. 175 (1984) 337.

[36] C. Buess-Herman, Phase transitions in adsorbed layers, in: A.F. Silva (Ed.), Trends in interfacial electrochemistry, NATO ASI, vol. C179, Reidel, Dordrecht, 1986, p. 205.

[37] Th. Dretschkow, Th. Wandlowski, Structural transitions in organic adlayers—a molecular view, in: K. Wandelt (Ed.), Solid/Liquid Interface Properties and Processes—A Surface Science Approach, Springer, Berlin, 2001, in press.

[38] E. Budevski, G. Staikov, W.J. Lorenz, Electrochemical Phase Formation and Growth, VCH, Weinheim, 1996.

[39] J.W. Schultze, D. Dickertmann, Potentiodynamic desorption spectra of metallic monolayers of Cu, Bi, Tl and Sb on Au(hkl) electrodes, Surf. Sci. 54 (1976) 489.

[40] Y. Nakai, M.S. Zei, D.M. Kolb, G. Lehmpfuhl, A LEED and RHEED investigation of Cu on Au(111) in the underpotential region, Ber. Bunsenges. Phys. Chem. 88 (1984) 340.

[41] M.J. Esplandiu, M.A. Schneeweiss, D.M. Kolb, An in situ scanning tunneling microscopy study of Ag electrodeposition on Au(111), Phys. Chem. Chem. Phys. 1 (1999) 4847.

[42] T. Will, M. Dietterle, D.M. Kolb, The initial stages of electrolytic Cu deposition: an atomistic view, in: A.A. Gewirth, H. Siegenthaler (Eds.), Nanoscale Probes of the Solid/Liquid Interface, Nato ASI, vol. E288, Kluwer, Dordrecht, 1995, p. 137.

[43] G. Meyer, S. Zöphel, K.H. Rieder, Manipulation of atoms and molecules with a low temperature scanning tunneling microscope, Appl. Phys. A 63 (1996) 557.

[44] I.-W. Lyo, Ph. Avouris, Field-induced nanometer to atomic-scale manipulation of silicon surfaces with the STM, Science 253 (1991) 173.

[45] M.T. Cuberes, R.R. Schlittler, J.K. Gimzewski, Room temperature supramolecular repositioning at molecular interfaces using a scanning tunneling microscope, Surf. Sci. 371 (1997) L231.

[46] M.F. Crommie, C.P. Lutz, D.M. Eigler, Confinement of electrons to quantum corrals on a metal surface, Science 262 (1992) 218.

[47] W. Li, J.A. Virtanen, R.M. Penner, Nanometer-scale electrochemical deposition of silver on graphite using a scanning tunneling microscope, Appl. Phys. Lett. 60 (1992) 1181.

[48] W. Schindler, D. Hofmann, J. Kirschner, Nanoscale electrodeposition: a new route to magnetic nanostructures? J. Appl. Phys. 87 (2000) 7007.

[49] U. Landman, W.D. Luedtke, N.A. Burnham, R.J. Colton, Atomistic mechanism and dynamics of adhesion, nanoindentation, and fracture, Science 248 (1990) 454.

[50] D.M. Kolb, G.E. Engelmann, J.C. Ziegler, Nanoscale decoration of electrode surfaces with an STM, Sol. State Ionics 131 (2000) 69.

[51] D.M. Kolb, R. Ullmann, T. Will, Nanofabrication of small copper clusters on gold(111) electrodes by a scanning tunneling microscope, Science 275 (1997) 1097.

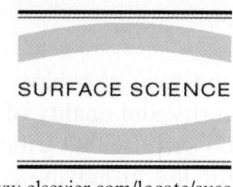

Surface science and the atomic-scale origins of friction: what once was old is new again

Jacqueline Krim *

Department of Physics, Box 8202, 104 Cox Hall—127 Stinson Drive, North Carolina State University, Raleigh, NC 27695-8202, USA

Received 4 August 2000; accepted for publication 12 April 2001

Abstract

Long neglected by physicists, the study of friction's atomic-level origins, or nanotribology, indicates that sliding friction stems from various unexpected sources, including sound energy, and static friction may arise from physisorbed molecules. Progress in this field will be discussed, with an emphasis on how the field of surface science has influenced our understanding of friction. © 2001 Elsevier Science B.V. All rights reserved.

Keywords: Friction; Physical adsorption; Phonons; Energy dissipation

1. Introduction

The fax from David Tabor came promptly, in response to my query about whether he and colleague Ken Johnson might be able to contribute to a special issue on "Fundamentals of Friction", which I was guest editing for the Bulletin of the Materials Research Society [1]. Tabor, whose renowned work on friction spanned well over forty years, unfortunately was not going to be able to complete a manuscript in the time frame allowed. But his letter did not end with that. He went on to describe how he had been thinking about my experiments and also about how lattice commensurability effects should impact sliding friction levels. Solid–solid phase transitions, for example, should produce changes in friction that my experimental techniques should be able to detect. As I thought about Tabor's research suggestions, I pondered whether forty years hence, at an age exceeding eighty, I too might still be investigating friction, and how fine it was that Tabor lived to participate in the current day renaissance in the topic of his life's work.

By most recent estimates, improved attention to friction and wear would save developed countries up to 1.6% of their gross national product, or over $100 billion annually in the US alone [2]. The magnitude of the financial loss associated with friction and wear arises from the fact that entire mechanical systems, be they coffee makers or automobiles, are frequently scrapped whenever only a few of their parts are badly worn. In the case of an automobile the energy consumed in its manufacture is equivalent to that consumed in 100,000 miles of operation [3]. More extreme examples include aircraft, which can be entirely destroyed for loss of one part. The consequences of friction

---

* Tel.: +1-9195132684; fax: +1-9195154496.
E-mail address: jkrim@unity.ncsu.edu (J. Krim).

and wear also have great impact on national security and quality of life issues, so it can come as no surprise that tribomaterials, materials designed for use in moving contact (sliding, rolling, abrasive, etc.), have for decades attracted the interest of materials scientists and mechanical and chemical engineers. What is surprising however, is how little is known even now about the fundamental origins of friction and wear.

As important as tribomaterials are to technology, their discovery has usually been serendipitous. To be sure, materials scientists have frequently been able to provide explanations for why tribomaterials perform as well as they do [4,5], and have also been able to substantially improve the performance of tribomaterials through the development of new alloys, composites and/or improved surface engineering methods. They have, however, been far less successful at a priori design of tribomaterials with improved performance, largely because friction and wear processes have not been understood at the molecular level. Why is so little known on the topic? The answer lies primarily in the fact that friction and wear are surface and interfacial phenomena which occur at a myriad of buried contacts which not only are extremely difficult to characterize, but are continuously evolving as the microscopic irregularities of the sliding surfaces touch and push into one another.

The late 1980's marked the advent of a renewed interest in fundamental areas of tribology, sparked by a number of new experimental and theoretical techniques capable of studying the force of friction in geometries which were well defined even at nanometer length scales [6]. These techniques benefited directly from advances in surface science throughout the 1970's, whereby improvements in ultra-high vacuum technology allowed scientists to prepare unprecedented, well-characterized crystalline surfaces. Experimental techniques such as the quartz crystal microbalance (QCM), the surface forces apparatus and the lateral force microscope, could now record friction in geometries involving a single contacting interface, a vastly simpler situation than that of contact between macroscopic objects. Faster computers meanwhile allowed large-scale simulations of condensed systems to be increasingly comparable to experiment in a very direct fashion [7].

The increasing overlap of the previously distinct areas of surface science and tribology has given rise to increased optimism that major breakthroughs will be achieved in upcoming decades. Issues of particular importance include: (1) Understanding the chemical and tribochemical reactions which occur in a sliding contact, and the energy dissipation mechanisms associated with such a contact. To fully characterize a system's behavior, one must not only estimate friction levels but also account for the effects of the heat that the friction generates. Are chemical reactions triggered? Do the contact points melt, etc.? (2) Characterization of the microstructural and mechanical properties of the contact regions between the sliding materials. (3) Merging and coordinating information gained on the atomic-scale with that observed at the macroscopic scale. Much of the current information is fragmented, with linkages between individual experimental results yet to have been established. (4) Development of realistic interaction potentials for computer simulations of materials of interest to tribological applications, and (5) Development of realistic laboratory test set-ups which are both well-controlled and relevant to operating machinery.

2. Working outside of a vacuum: tribology before 1970

The view of the Greek philosophers that vacuum was an impossibility hampered understanding of its basic principles until the mid-17th century (Fig. 1), greatly delaying progress in all vacuum-related fields [8]. The field of tribology by comparison dates well back to the construction of the Egyptian pyramids, if not hundreds of thousands of years earlier to the discovery of the use of flint stone for the sparking of fires [9]. Indeed, such tribological advances as Leonardo da Vinci's design of intricate gears and bearings [9] (some of which were not built until the industrial revolution provided sufficiently strong materials), and the landmark 18th century development of a timepiece allowing accurate longitudinal positioning of ships

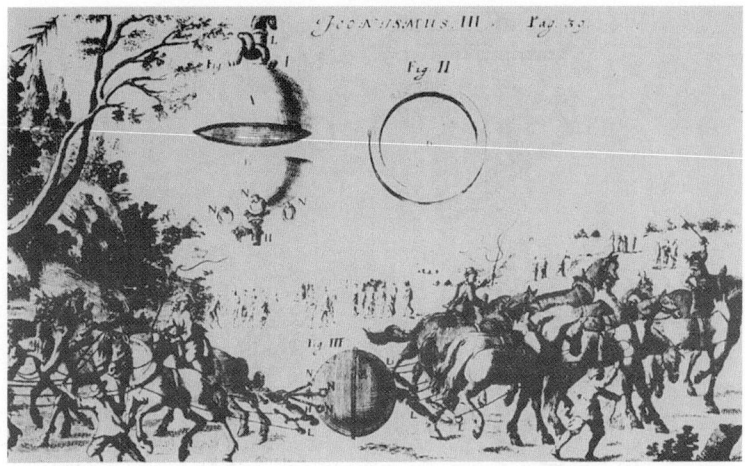

Fig. 1. A highly publicized demonstration by Otto Van Guericke, describing his 1672 invention of a vacuum pump (from Ref. [8]). The ability to produce vacuum in laboratory conditions would centuries later give rise to the discovery of the electron. This revolutionary advance in physics gave rise to the field of modern surface physics. Electrons travelling through ultra-high vacuum conditions and high potential differences were first demonstrated to be highly sensitive probes of surfaces in 1967 (Fig. 7), and provided the vast majority of information on surface structure and chemical composition up until the invention of the scanning probe microscope in the 1980's.

at sea (accomplished via a self-lubricating wooden gear) [10] could easily be termed "modern", given the overall longevity of the field. Modern study of friction began perhaps 500 years ago, when Leonardo da Vinci deduced the laws governing the motion of a rectangular block sliding over a planar surface (Fig. 2). Da Vinci's work had no historical influence, however, since his notebooks remained unpublished for hundreds of years. The French physicist Guillaume Amontons is credited with the first published account of the classic friction laws, which in 1699 described his observations of contacting solid surfaces [11].

Amontons observed that (1) the friction force that resists sliding at an interface is proportional to the "normal load", or force which presses the surfaces together, where the "coefficient of friction" is defined as the ratio of the frictional force to the load, and (2) the friction force is independent of the apparent area of contact: A small block experiences as much friction as does a large block of the same material so long as their weights are equal. A third law is frequently included with these, attributed to French physicist Charles-Augustin de Coulomb (better known for his work in electrostatics) in the 18th century: the friction force is independent of velocity for ordinary sliding speeds. Although the coefficient of friction is independent of the apparent contact area, loading force, and the sliding speed, it does in fact depend on whether the force is applied to an object at rest ("static friction") or already moving ("sliding friction"). Considering its simplicity, Amontons' law is amazingly well obeyed for a wide range of materials such as wood, ceramics, and metals.

Amontons' and Coulomb's classical friction laws have far outlived a variety of attempts to explain them on a fundamental basis. Surface roughness, which Coulomb unsuccessfully attempted to attribute friction to (Fig. 3), was ruled out definitively as a possible mechanism for most friction in the mid 1950's. Molecular adhesion, though, remained a strong possibility, a conclusion due in large part to Bowden, Tabor and coworkers at Cambridge University, England. Their group found that friction, although independent of apparent macroscopic contact area, is in fact proportional to the true contact area. That is, the microscopic irregularities of the surfaces touch and push into one another and the area of these contacting regions is directly proportional to the friction force. The Cambridge group subsequently

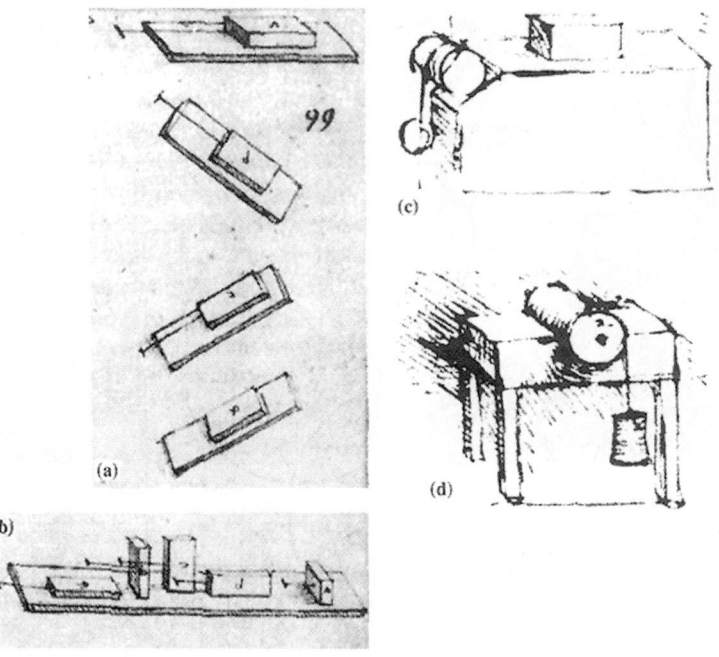

Fig. 2. Leonardo da Vinci's (A.D. 1452–1519) studies of friction. Sketches from the *Codex Atlanticus* and the *Codex Arundel* showing experiments to determine: (a) the force of friction between horizontal and inclined planes; (b) the influence of the apparent contact area upon the force of friction; (c) the force of friction on a horizontal plane by means of a pulley and (d) the friction torque on a roller and half bearing (from Ref. [9]).

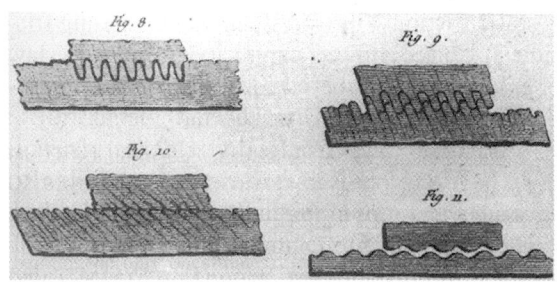

Fig. 3. Coulomb's representation of rough surfaces, published in 1785 (from Ref. [9]).

explored the possibility that friction arose from sufficiently strong bonding at the true contact points so as to produce continual tearing away of tiny fragments of material [12]. This explanation failed however to predict experimental observation. Indeed, it was proved incorrect by Tabor and one of his own graduate students, Jacob Israelachvili, who in the 1970's developed a "surface forces apparatus" (SFA) (Fig. 4) for atomic-scale friction measurements, which definitively exhibited friction in the total absence of wear.

For decades thereafter, Tabor would continue to ponder the fundamental dissipative mechanisms of friction. In a 1991 plenary lecture at a NATO sponsored conference on the Fundamentals of Friction [14] (the first meeting to bring together long-established tribologists with surface scientists new to the field, like myself), he would conclude that friction in the absence of wear must be due to strains building up in the sliding contact which were being released in the form of atomic vibrations [15]. Phonons, as such vibrations are called, were first suggested by Tomlinson in 1929 [16], with subsequent independent derivations by both Gary McClelland at IBM Almaden and Jeffrey Sokoloff at Northeastern University [17]. Friction arising from phonons occurs when atoms close to one surface are set into motion by the sliding action of atoms in the opposing surface (Fig. 5).

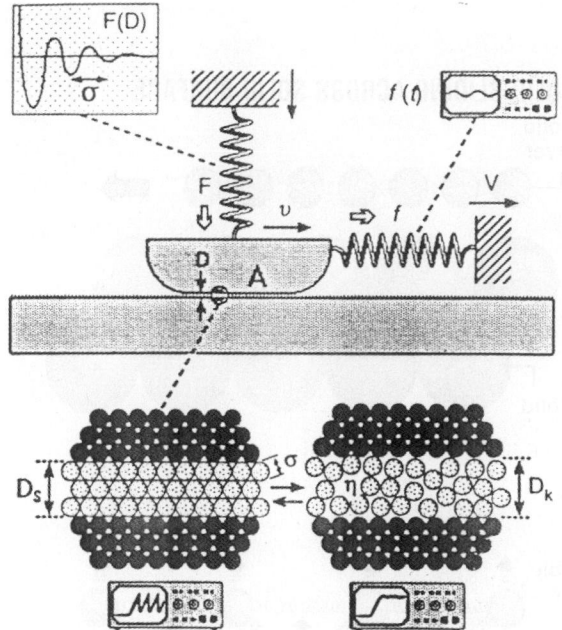

Fig. 4. Diagram of a SFA experiment (from Ref. [13]). The coiled springs are representative of any device used to measure the normal and shear forces between the samples. The SFA makes use of two cleaved mica surfaces, which are among the smoothest surfaces known. Investigators can place lubricant films, which can be as thin as a few molecules, between the mica surfaces and slide them about, to see how the films affect the friction.

The lattice vibrations are produced when mechanical energy needed to slide one surface over the other is converted to sound energy, which is eventually transformed into heat. Hence, to maintain the sliding motion, more mechanical energy must be added and one has to push harder.

In his presentation, Tabor could cite no experimental evidence, his own or otherwise, that phonons even existed, much less accounted for a major portion of the energy dissipation observed at the macroscopic scale. Nevertheless, by process of elimination he had decided that they must play an important role. Neither of us realized at the time that the sliding friction measurements of krypton monolayers which I had reported on a few months earlier [18] would prove to be the first definitive experimental evidence for the existence of phononic energy dissipation mechanisms in sliding friction. For one crucial missing piece in the puzzle would not be forthcoming until 1994 [19].

3. Energy dissipation in adsorbed films: shakers and movers

Tribologists were in fact not the only community at the time to be pondering energy dissipation mechanisms at moving interfaces. An entirely distinct "vibrations at surfaces" community, which emerged in the 1980's and 1990's, was also exploring energy dissipation within the context of the damping of small vibrational motions of atoms on surfaces [20].

Whenever atoms or molecules adsorb on surfaces, new vibrational modes emerge which are not present in either an isolated surface or the adsorbate alone. The modes that appear include "internal", stretching or torsional vibrations within a molecule, and "external" modes whereby the entire molecule or atom moves as a whole with respect to the surface. Before the mid 1980's, the only external vibrations to have been studied were the "frustrated" (i.e. damped) vibrations of physisorbed molecules on graphite in directions perpendicular to the surface. Their energies were determined using inelastic scattering of thermal energy neutrons, the standard technique for vibrational studies in bulk materials [21]. Only graphite surfaces could be studied at the time on account of high surface area sample requirements. (It is interesting to note that such samples were obtained from the carbon industry, which manufactured high surface area graphite lubricants.) Frustrated vibrations parallel to a surface, perhaps more directly relatable to sliding friction, were reported for the system nitrogen on graphite in 1990 [22], but no connections with sliding friction were made at the time. Indeed, even if the notion had been advanced, it would not have been obvious that the dissipation associated with small molecular vibrations of atoms adsorbed on a surface was comparable to frictional energy dissipation of atoms sliding several lattice spacings or more along a surface.

The first direct observation of external vibrations of a molecule adsorbed on a non-graphitic

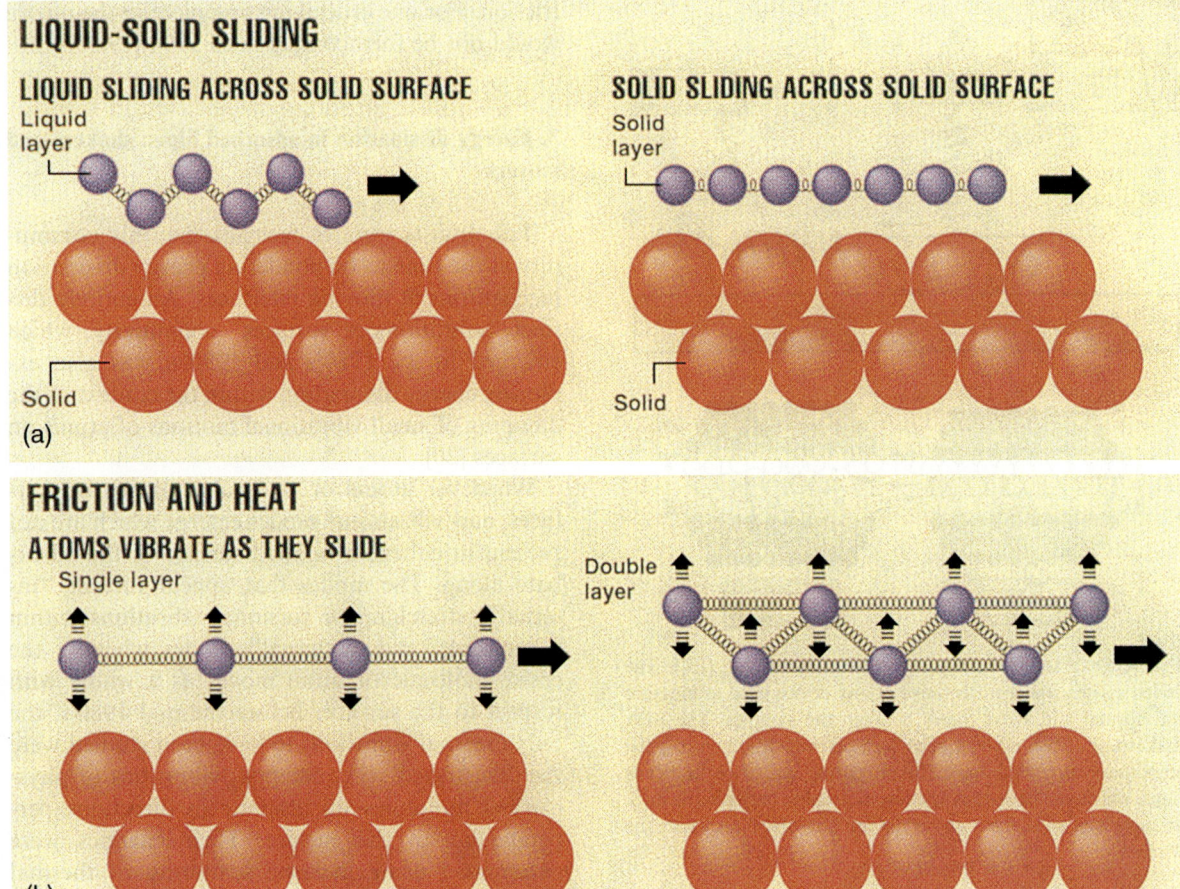

Fig. 5. Schematic of phonon friction (from Ref. [35]). Unlike matter in the visible world, a solid layer that's one atom thick (a) slides more easily over a solid surface than an equivalent liquid layer. That's because atoms in solids are more tightly bound together than atoms in liquids. Liquid atoms tend to fall between atoms of the surface beneath them, hindering their sideways movement. Solid atoms slip across the surface as a cohesive sheet. At the atomic level (b), a double layer of atoms is harder to slide over a solid surface than a single layer because the two layers of atoms jostle each other, producing extra heat. The extra heat creates more friction, and that means it takes more effort to slide the double layer of atoms.

substrate was reported in 1986 for CO molecules adsorbed on Pt(1 1 1) [23], obtained by means of the newly developed technique of inelastic scattering of thermal energy He atoms [24]. A large number of systems were subsequently explored by means of He scattering, and later advances in other fields also made possible the probing of external vibrational modes by means of electron energy loss spectroscopy [25], infrared spectroscopy [26], and Raman scattering [27]. Information on adsorbate vibration modes allows estimates to be made of the magnitude of a surface's corrugation, or how much the binding energy to a surface varies as an atom moves from one position to another along a surface. Such information is essential for correct modeling of sliding friction at the atomic scale. If there were no corrugation in the potential, there could be no phononic component of sliding friction, irrespective of how strongly bound the atoms are to the surface, for one primary manifestation of phononic friction is its hypersensitivity to commensurability and interfacial orientation effects.

Separated in both time and topic, the vibrations at surfaces community was not destined to intersect with the tribology community before the mid 1990's [28–30]. When the intersection did occur, it was within two contexts: (1) The surface corrugation information generated by the vibrations community was an essential ingredient in computer simulations of phononic friction, and (2) theorists in the vibrations community noticed that energy dissipation rates of electronic friction mechanisms (discussed below) [31] observed for vibrational modes of adatoms were close to the frictional energy dissipation rates which I had measured by means of QCM for adsorbed monolayers. Could it be that the two communities were in fact studying the same phenomena? For an incommensurate contact, it is increasingly apparent that the energy dissipation mechanisms are in fact the same.

The QCM (Fig. 6) is an instrument that operates on a time scale short enough to detect phonons, whose lifetimes are typically no longer than a few tens of nanoseconds. While for decades the QCM had been employed primarily for microweighing and time standard purposes, Allan Widom and I adapted it in the mid 1980's for sliding friction measurements of adsorbed layers on metal surfaces [32]. The basic component of a QCM is a single crystal of quartz that has very little internal dissipation (or friction). As a result it oscillates at an extremely sharp resonance frequency (usually 5–10 MHz) that is determined by its elastic constants and mass. The oscillations are driven by applying a voltage to thin metal electrodes that are deposited on the surface of the quartz in a manner that produces a crystalline texture, generally (1 1 1) in nature. Atomically thin films of a different material are then adsorbed onto the electrodes. The extra mass of the adsorbed layer lowers the resonance frequency of the microbalance, and the resonance is broadened by any frictional energy dissipation due to relative motion of the adsorbed layer and the microbalance. By simultaneously measuring the shift in frequency and the broadening of the resonance (as evidenced by a decrease in the amplitude of vibration of the microbalance), the sliding friction of the layer with respect to the metal substrate can be deduced. The friction can

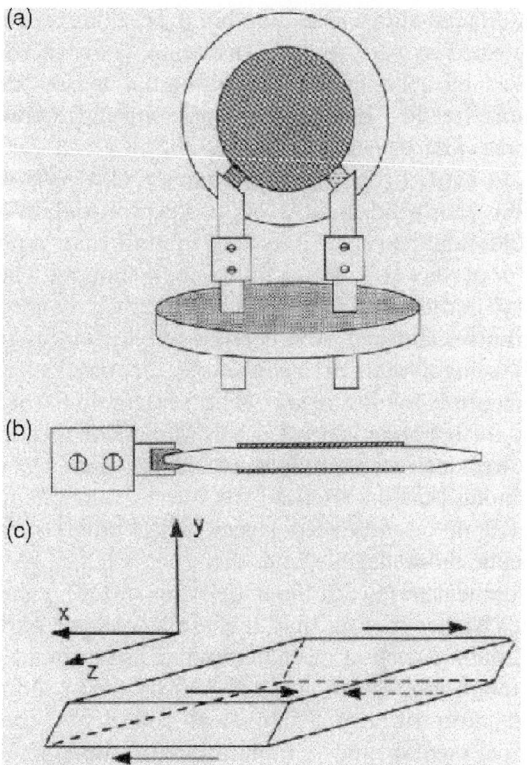

Fig. 6. Front (a) and side (b) views of a QCM. The shaded regions represent metal electrodes that are evaporated onto the major surfaces of the microbalance. Molecularly thin solid or liquid films adsorbed onto the surface of these electrodes (which are parallel to the x–z plane) depicted in (c) may exhibit measurable slippage at the electrode–film interface in response to the transverse shear oscillatory motion of the microbalance. The experiment is not unlike pulling a tablecloth out from under a table setting, whereby the degree of slippage is determined by the friction at the interface between the dishes (i.e. the adsorbed film material) and the tablecloth (i.e. the surface of the electrode).

be measured only if it is sufficiently low so as to result in significant sliding, which is accompanied by a measurable broadening of the resonance. For this reason, QCM measurements of sliding friction tend to be carried out on systems exhibiting very low friction, such as rare-gas solids adsorbed on noble metals. For the vast majority of other systems which exhibit higher friction (chemically bonded layers, etc.) the slippage of an adsorbed monolayer on the surface of the QCM is too small to produce a measurable broadening. In this case interfacial slippage and/or bond breaking can be detected by performing measurements on micron-sized particles, whose larger inertial masses can more readily overcome the stronger frictional forces [33].

In 1991, I reported my group's QCM observations that solid monolayers of krypton sliding on gold exhibited five times less friction than liquid monolayers [18]. I had found the "slippery when dry" nature of krypton to be extremely counter-intuitive and had even delayed publication of the data for a year, as I could offer no physical explanation for the results. The explanation would be forthcoming in 1994, when Mark Robbins and coworkers' successfully modeled the data by assuming that the friction was due to phonons excited in the adsorbed layers [34]. Liquid layers, being more flexible and therefore slightly more commensurate with the underlying surface, exhibited higher friction than their solid counter parts. Phononic friction mechanisms were also found to explain monolayers and bilayers of xenon sliding on silver surfaces [35,36], and collectively these experimental and computational studies have supplied definitive proof of the existence, if not dominance of phononic friction in the adsorbed monolayer systems.

A surprising aspect of the excellent agreement between the numerical simulation data and the experiments is the fact that friction arising from electronic mechanisms was totally neglected. Such friction is related to the resistance felt by mobile electrons in a conducting material as they are dragged along by forces exerted by the opposing surface [37]. Could the simulations have slightly overestimated the friction, masking electronic contributions? The answer is probably yes, since there is just enough uncertainty in the best estimates for corrugation levels of Au(1 1 1) and Ag(1 1 1) substrates to allow for the presence of a moderate level of electronic friction to have been concealed [38]. Indeed both electrical resistivity measurements by Schumacher et al. [39], as well as our own group's measurements of nitrogen sliding on lead in its normal and superconducting state [40], indicate that electronic contributions are non-negligible for the first adsorbed layer of atoms on conducting metallic substrates.

One of the more remarkable properties of the friction of the adsorbed layers studied so far is the total absence of static friction: An arbitrarily small applied force is sufficient to induce sliding [18,34]. While this is entirely consistent with atomic-scale theories of friction (both phononic and electronic) at clean crystalline interfaces, it is unheard of in the macroscopic world [41]. This gives rise to the obvious question of how fundamental dissipation mechanisms such as phononic and electronic effects are manifested in systems characterized by different length and timescales. Do they play a substantive role in wear-free friction at the macroscopic scale, as Tabor had suspected, or are they simply the primary energy dissipation mechanisms in molecularly thin films adsorbed on open surfaces due to the simplicity of the systems under study? The answers to these questions remain outstanding. But a growing body of literature, particularly that focussed on the role of commensurability effects in sliding friction, is helping to shed light on this issue.

4. Commensurability effects and static friction: a sticky topic

One major manifestation of phononic friction is its hypersensitivity to the relative commensurability of the two surfaces in sliding contact: A transition from commensurate to incommensurate sliding conditions theoretically can reduce the sliding friction levels by many orders of magnitude [42]. Another manifestation of phononic friction is the theoretical prediction that static friction should vanish for nearly every pair of clean surfaces which deform elastically: The force to overcome friction is simply proportional to the sliding speed times a friction coefficient characteristic of the two surfaces in contact. But one of the most common everyday experiences with friction at the macroscopic scale is the ever present occurrence of static friction: The force to initiate motion (which itself is quite variable, depending for example on how long the two surfaces have been in contact) is larger than that required to keep an object in motion. A closely associated phenomenon is that of "stick slip" friction [43], whereby for certain sliding speeds, the "velocity weakening" dependence of the transition from static to sliding friction leads to repetitive sticking and slipping at the interface, producing the all-too familiar screeching noises associated with brakes, fingernails, etc.

Stick–slip phenomena are frequently blamed for excessive interfacial wear rates, and as such have received much attention from the mainstream tribological community. The key to solving the mystery of static friction and stick–slip phenomenon appears to lie buried in the atomic-scale structure of the myriad of contacts formed between the two sliding surfaces, and the nature of molecules confined between them. These are not only extremely difficult to characterize, but are continuously evolving as the microscopic irregularities of the sliding surfaces touch and push into one another. The constant changing of the nature of the interfacial geometry of the contact areas (even in cases where the contact area is constant) gives rise to friction coefficients and stick slip event rates that are never exact numbers [43]. Moreover, the friction force at an individual asperity may or may not increase with applied load, depending on the structure of the contacting solids, and molecules confined within them [44,45].

Surface scientists new to the field of tribology have abandoned the approach of characterizing innumerable hidden interfaces. Instead, well-defined interfaces at nanometer length scales are prepared in advance of the measurements, usually involving contact at one, rather than multiple asperities [6]. The techniques which they employ benefit directly from progress in surface science in decades prior, whereby improvements in ultra-high vacuum technology allowed preparation of unprecedented, well-characterized crystalline surfaces. Three major advances in the period 1950–1970 gave rise to the ability to prepare well characterized surfaces: (1) The development and construction of routinely de-mountable vacuum-tight metallic enclosures in which residual pressures between 10^{-9} Torr and 10^{-10} Torr could be established and maintained without excessive difficulty, (2) The use of these enclosures to perform the elemental analysis of the constituents of a surface by means of Auger electron spectroscopy, and (3) The deployment of the diffraction of low-energy electrons (LEED) for structural studies of single crystal surfaces (Fig. 7). A great number of LEED/Auger studies were carried out starting in the 1970's on the structure of a wide range of single crystal surfaces, and the two-dimensional phases of atoms and molecules adsorbed on them [49–51] (Fig. 8).

Further advances in the structural characterization of surfaces came with the development of scanning probe microscopes in the 1980's, which were quickly adapted for non-conducting surfaces and for probes of microscopic-scale friction. Indeed, inspired by the concept of phonon friction, Gary McClelland collaborated with C. Mathew Mate at IBM Almaden in the mid 1980's to measure nanometer-scale friction. They did so by adapting a newly invented instrument: the atomic-force microscope, for measurements of lateral forces. With it, they published their first observations of friction, measured atom-by-atom, in a landmark 1987 paper. Their instrument revolutionized studies of friction at atomic-length scales [52].

An atomic-force microscope consists of a sharp tip mounted at the end of a compliant cantilever (Fig. 9). As the tip is scanned over a sample surface, forces that act on the tip deflect the cantilever. Various electrical and optical means (such as capacitance and interference) quantify the horizontal and vertical deflections. In the early 1990's, the IBM researchers succeeded in setting up their friction-force microscope in ultra-high vacuum, with a contact area estimated to be less than 20 atoms in extent. Their measurements yielded a friction force that exhibited no dependence on normal load [54]. According to the classical friction laws, this result would have implied zero friction. But not only was friction evident, the shear stress, or force per area required to maintain the sliding, was enormous: one billion newtons per square meter, or 150,000 pounds per square inch! That force is large enough to shear high-quality steel. Could there be frictional energy dissipative mechanisms as yet undetermined which are giving rise to such high levels?

Energy dissipation mechanisms and the fundamental origins of friction are the focus of ongoing efforts by Miquel Salmeron, University of

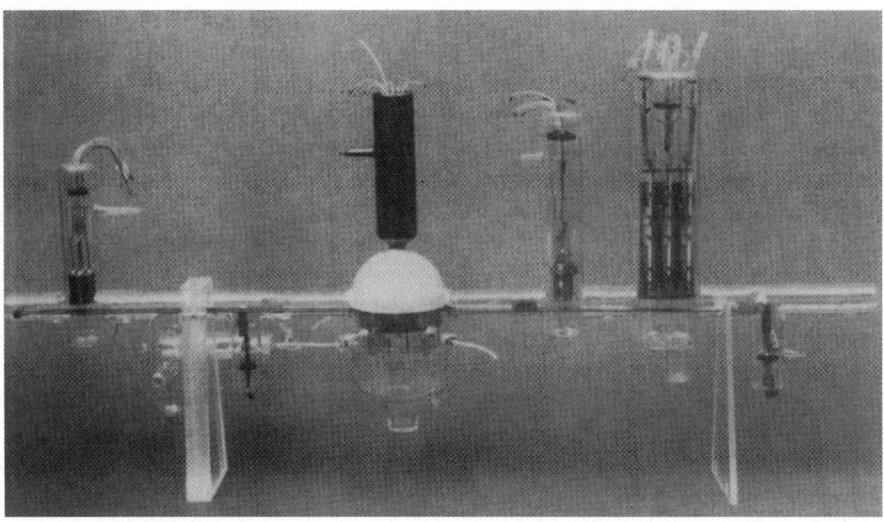

Fig. 7. This first ultra-high vacuum system became the first LEED/Auger Spectrometer when in 1967. R. Weber observed the differentiated Auger peaks of silicon and cesium [46] (from Ref. [8]). L.A. Harris made the first practical Auger Electron Spectroscopy measurements in 1968 using an electrostatic velocity analyzer [47], and Weber's demonstration of the use of the LEED apparatus for this purpose opened up a new era in surface physics. The system shown was originally built by P. Palmberg and used for his graduate research in W.R. Peria's Physical Electronics Laboratory [48]. Weber modified the system and also used it for his graduate research of alkali metal layers on semiconductor surfaces.

California, Lawrence Berkeley Laboratories, whose members are among an elite group worldwide who have succeeded at the difficult task of performing well-controlled atomic-force microscopy measurements in an ultra-high vacuum environment [45,55]. It is interesting to note that even in AFM geometry involving a single contacting asperity on a single crystal surface, static friction and stick–slip phenomenon are ever present. In the vast majority of cases, one stick–slip event is observed per unit cell of the substrate, even in cases where the atomic cell contains more than one species. Accounting for such phenomena in terms of energy dissipation, which is distributed among the tip, substrate, and cantilever, is an ongoing issue in the AFM community. The high-energy dissipation rates associated with AFM geometries may in fact be due to the creation of point defects and/or atom transfer to and from the tip. Carpick and Salmeron have published a comprehensive treatment of AFM measurement for studies of the fundamentals of friction [45]. It should be considered required reading material for scientists entering the field.

Among tribologists, AFM researchers are not the only ones to have adapted their techniques to UHV conditions. Many tribologists now routinely employ Auger Spectroscopy for surface chemical analysis of regions that have been exposed to sliding with and without the presence of a lubricant [56]. But the most direct merging of mainstream surface science with tribology has occurred in the laboratory directed by Andrew Gellman at Carnegie Mellon [57]. Gellman and coworkers have constructed a "UHV tribometer", which allows the tribological properties of two single crystal metal surfaces to be measured under the ultra-high vacuum conditions of a surface analysis apparatus (Fig. 10). The experiment allows measurements of both friction and adhesion between two single crystal surfaces brought into contact under a wide range of loads, and sheared with a wide range of sliding velocities. The experiments performed to date have systematically varied a number of surface characteristics in order to observe their effect on tribological properties, including the relative orientation of the single crystal lattices. Experiments with Ni(1 0 0) reveal varia-

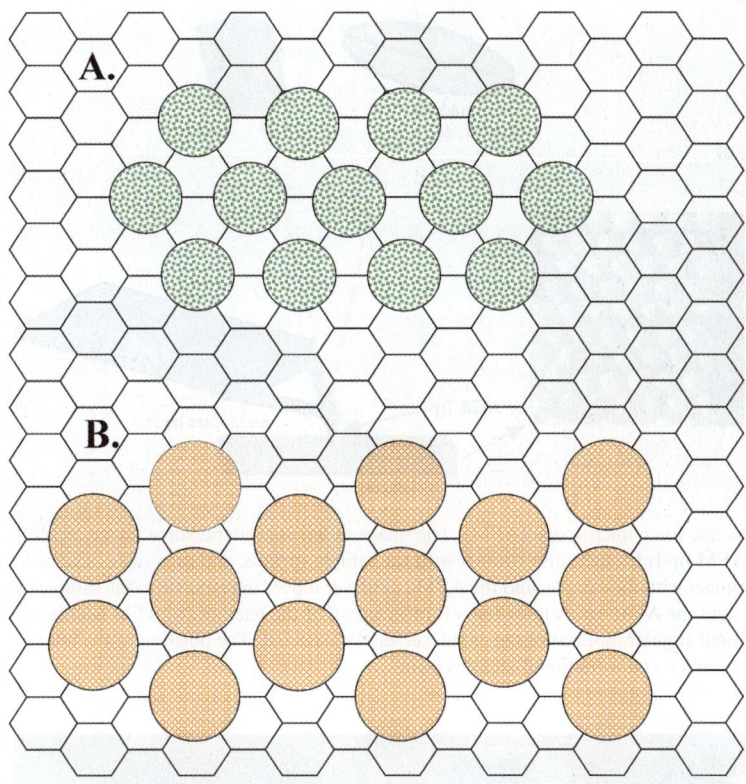

Fig. 8. Arrangement of molecules adsorbed in a commensurate fashion on the (0 0 0 1) face of graphite. (It is interesting to note that graphite samples were obtained from the carbon industry, which manufactured high surface area graphite lubricants.) (a) Arrangement of molecules of a diameter less than 4.26 Å, and (b) molecules of a diameter greater than 4.26 Å [49].

tions in friction (by about a factor of five) with relative orientation of the lattices [58]. While Gellman attributes the variation with orientation in this system to plastic deformation within the bulk solids, the result reveals how ubiquitous is the dependence of friction on interfacial commensurability, irrespective of the time or length scale under study. It is also routinely observed in surface forces apparatus [59] and atomic-force microscopy studies measurements [60,61].

Even though friction is routinely observed to depend on lattice commensurability, the variation is generally less than an order of magnitude, not the many orders-of-magnitude change (up to 14!) [42] predicted by theories of phononic dissipation mechanisms. Perhaps the discrepancy simply reflects the degree to which phononic dissipation mechanisms contribute to the total frictional observed. To shed light on the issue, our group is presently pursuing a detailed QCM study of friction as an adsorbed layer undergoes a transition from incommensurability to commensurability (Fig. 8) with the underlying substrate. In such a simple system we expect to observe changes in friction which are far closer to theory, just as adsorbed monolayers are to date the only systems to exhibit no static friction.

The question of course remains as to why static friction can be so ubiquitous when theoretically, two clean interfaces in sliding contact are not expected to exhibit it. The answer may ultimately prove to lie in "third body" effects, whereby additional adsorbed molecules act to initially pin the interface. Mark Robbins and coworkers have recently suggested that static friction is likely to be related to adhesive forces of thin adsorbed films (water, hydrocarbons, etc.) [41] which are known to be present on most surfaces: They have

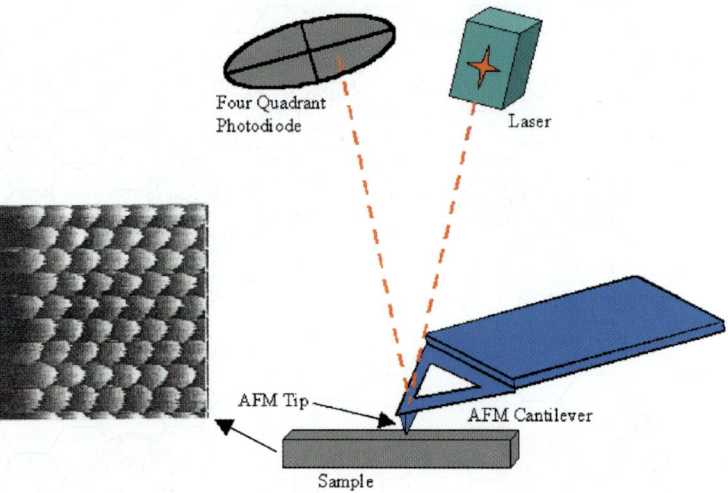

Fig. 9. Schematic of an atomic force microscope (AFM). The image at left is a lateral force image of MoS$_2$ taken from Ref. [53]. In contact AFM mode, the AFM tip is brought into contact with the sample surface, and then raster scanned across the sample surface. The AFM tip maintains contact with the sample, and the AFM cantilever moves in response to the sample's morphology and frictional forces between the sample and the AFM tip. A laser beam is reflected off of the back of the AFM cantilever and onto a four-quadrant photodiode. In this way, small angular motions of the cantilever can be detected. The position of the laser on the photodiode provides both a map of the sample surface and a frictional measurement.

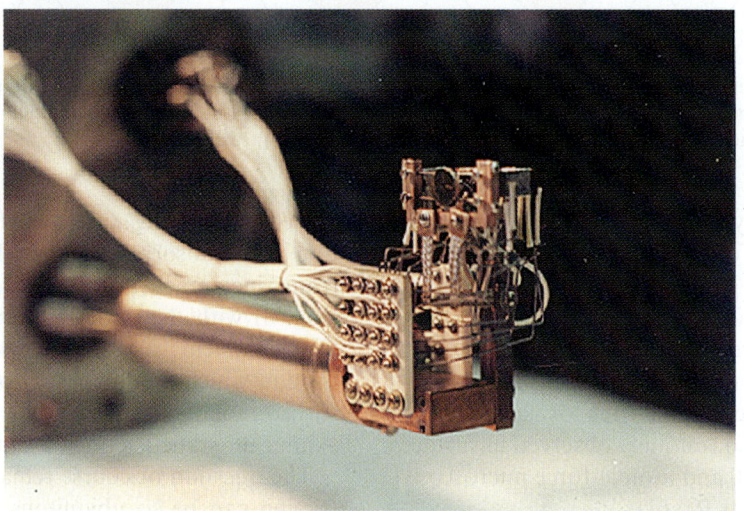

Fig. 10. Photograph of the UHV tribometer. One of the single crystal surfaces is the circular disk in the upper right of the photograph.

demonstrated this with computer simulations which indicate that these ubiquitous films behave like marbles which roll to the open niches of an incommensurate interface composed of, for example, ping–pong balls in contact with tennis balls. The marbles always find a local energy minimum (Fig. 11), so it always takes some energy to initiate sliding.

Experimental determination of the structure of films trapped between solid surfaces, and not just the structure of the surface contact points, thus remains one of the most important goals of the

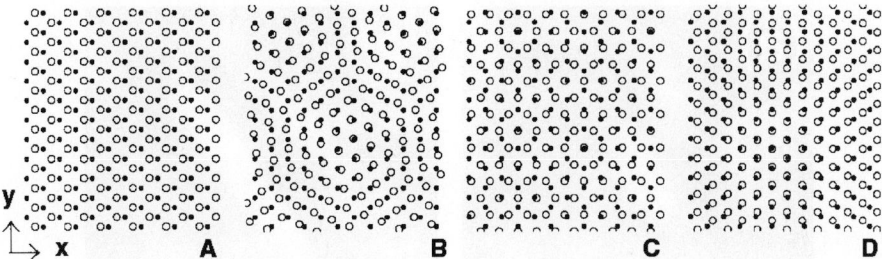

Fig. 11. Projections of atoms from the bottom (•) and top (○) surfaces into the plan of the walls. In (a)–(c) the two walls have the same structure and lattice constant, but the top wall has been rotated by 0, 11.6 or 90°, respectively. In (d) the walls are aligned, but the lattice constant of the top wall has been reduced by 12/13. Note that the atoms can only achieve perfect registry in the commensurate case (A) (from Ref. [41]).

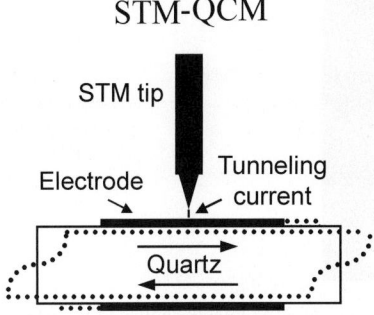

Fig. 12. Schematic of the STM-QCM apparatus. The STM tip scans the surface of a QCM which oscillates in transverse shear mode. This geometry provides a single asperity contact with nanometer scale contact area. The response of the QCM resonance to the applied normal force of the tip combined with STM images of the surface results in a unique view of a sliding contact and a powerful technique for studying the fundamentals of friction and lubrication [63].

nanotribology community. Current efforts to characterize the detailed atomic structure of films confined at an interface include combined synchrotron X-ray/Surface forces apparatus methods [62], and a combined STM-QCM apparatus. [63, 64] (Figs. 12 and 13).

5. Nanoscale machine lubrication: The squeaky wheels will get no grease

If the precise nature of the contacting asperities between macroscopic objects in sliding contact is determined, including the role of uncontrolled adsorbed species, the results of nanotribological studies can be directly implemented into mainstream tribological considerations. Meanwhile, the results of fundamental investigations of surfaces can be more readily applied to solid–vapor or solid–liquid interfaces, where the complicating factors associated with asperity contacts are less of an issue, and to micro-electro-mechanical systems (MEMS)/nano-electro-mechanical systems (NEMS) related issues, where machine components with incredibly small dimensions are rapidly approaching the length scales routinely probed by the nanotribological community. To name just one recent example, the design of a two-dimensional "car" complete with pulling power and an "engine" was recently published by Porto et al. [65]. It is "driven" by frictional push-off forces along a surface and bears a striking resemblance to the atoms and springs considered in the Tomlinson models. While the authors are quick to point out that the actual "cars" built according to their design need not be so small in scale (the general concept scales up in size to the mesoscale), the design illustrates how astoundingly small future machine components are envisioned.

Are such designs the way of the future, or just pure fancy? Consider, for a moment "ordinary" (i.e. macroscopic) motor vehicles, which in their early days were quite operational, but far from optimal mechanical devices. Ownership of an automobile in 1916 involved an overwhelming maintenance schedule requiring daily servicing of the lubricants, and major maintenance every 500 miles [3]. Without the major improvements in lubrication engineering that occurred subsequent to its

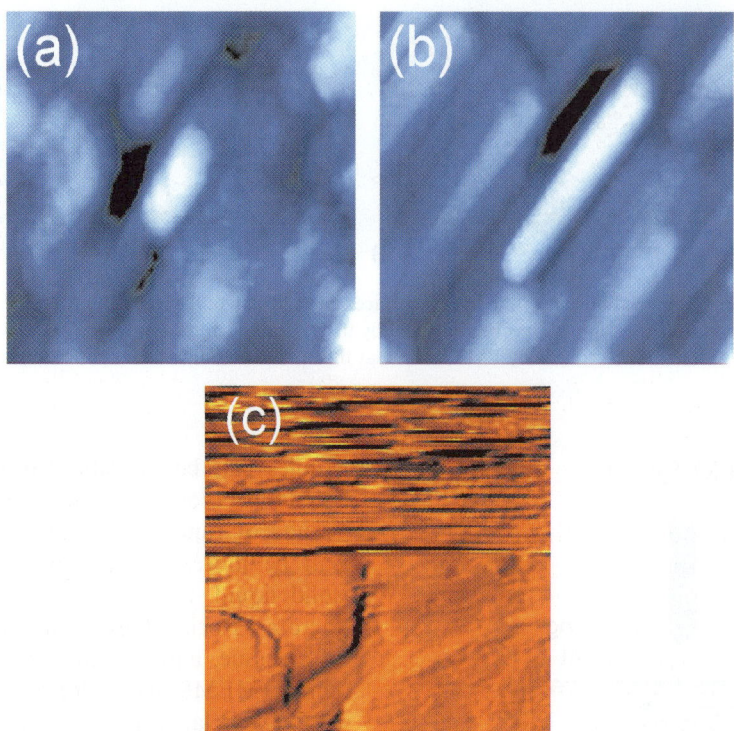

Fig. 13. Images recorded with the STM-QCM apparatus. Images (a) and (b) are a pair of STM images showing the surface of a QCM while stationary (a) and vibrating (b). The images are 177×177 nm^2, and the full vertical range is 15 nm. The quartz sample is a thin disk 8 mm in diameter. Both of its surfaces are coated with a thin (150 nm) metal film, which acts as an electrode for exciting the resonance of the quartz through piezoelectric coupling. This crystal has a 5 MHz fundamental frequency and vibrates in transverse shear mode, whereby vibration is in the plane of the surface. The direction and amplitude of vibration can be determined by comparing the stationary and vibrating images. Since the oscillation occurs very rapidly compared to the image acquisition time, vibrating features such as the prominent central mound appear smeared according to the extent of motion back and forth. Measurements of the amplitude and frequency reveal that the quartz surface readily achieves speeds over 1 m/s. Such speeds are typical for everyday objects in sliding contact. Therefore, the STM-QCM is capable of studying friction on the nanometer scale under realistic sliding conditions. In (c), an STM image (350×350 nm^2) of TBPP, a well-known lubricant additive, on a platinum QCM electrode at room temperature. The additive was vapor-deposited to form a 1 nm thick film on the surface. The underlying quartz crystal is vibrating in the bottom half of the image and stationary in the top half. Interestingly, the STM is unable to image the surface unless the quartz is vibrating. Most likely, the mobile and insulating lubricant is brushed away from the tip by the vibrating action of the surface, allowing the tip to image the conductive platinum surface [64].

invention, the motor vehicle would not have risen to its present status in society. Similarly, while there is widespread belief that the future will be revolutionized by MEMS and/or NEMS devices, the enormous promise will not materialize without substantial progress in overcoming the stiction, friction and wear associated with such devices [66]. Because MEMS devices must react to mechanical signals, many employ construction topologies that require physical motion. Suspended plates and beams that are fabricated a few microns away from their supporting substrates are in common use, and these structures typically have relatively large areas and very small stiffness. These combined characteristics makes MEMS devices highly susceptible to surface forces which can cause the suspended member to deflect towards the substrate, collapse and/or adhere permanently to the substrate [67]. With the current impetus towards device components extending well into the nanometer regime, the surface-related complications currently encountered with MEMS are expected to be even more severe: an atomistic understanding of the devices' material properties will be tantamount to their ability to function reliably, and it is clear that liquid and grease-like lubricants developed for macroscopic machinery will be of no use on such machines.

While the "MEMS train" has yet to leave the station, computer disk drive technology has soared forward with lightening speed [68]. The miniaturization of computer hard-disk storage devices is so advanced that many of the devices shipped today rely on a single layer of lubricant molecules adsorbed on a sub-micron thick hard protective coating of amorphous carbon to control the static friction and wear between the read/write head and the disk surface. The frictional properties of hard coating and lubricant molecules attached to solid surfaces are thus a matter of major interest to both fundamental and applied tribologists. (It is important make a distinction here between low frictional properties of an adsorbed film, and its overall ability to lubricate. More important, in many cases than low friction, is a lubricant's ability to remain on the surface.) Perry and co-workers at the University of Houston [69] have employed AFM with standard surface analytical methods to investigate the frictional properties of potential hard-coating material, vanadium carbide (VC), as a function of surface oxidation. In this study, a single-crystal VC sample was prepared under ultra-high vacuum conditions by sputtering and annealing, then characterized by LEED and Auger electron spectroscopy. Using a UHV-AFM, the frictional forces between a silicon nitride tip and the clean VC surface were measured as a function of the applied load. The surface was then exposed to a saturation coverage of molecular oxygen, known to react predominantly with the vanadium atoms of the VC surface, and then re-investigated in situ with AFM. The results of the frictional measurements appearing (Fig. 14) indicated a 38% reduction in the coefficient of friction upon surface oxidation. Oxidation is of interest for tribological considerations on account of its common occurrence, particularly in sliding contacts, under ambient conditions. The exact mechanism of the friction reduction observed is still being explored, within the context of phononic and electronic dissipation mechanisms, with electronic mechanisms appearing to be the most likely candidate.

AFM, as well as numerical simulations have also probed the frictional properties of model lubricant chain molecules. The average frictional forces of alkylsilane molecules containing two to eighteen carbon atoms adsorbed on silicon substrates decreases with chain length up to eight carbon atoms, and then remains relatively constant [70]. Salmeron and coworkers have proposed that the chain length dependence arises from the interplay between packing energy of the monolayer film and local deformations in the film [71], since below eight carbon atom chain lengths, the molecules are relatively disordered. Energy

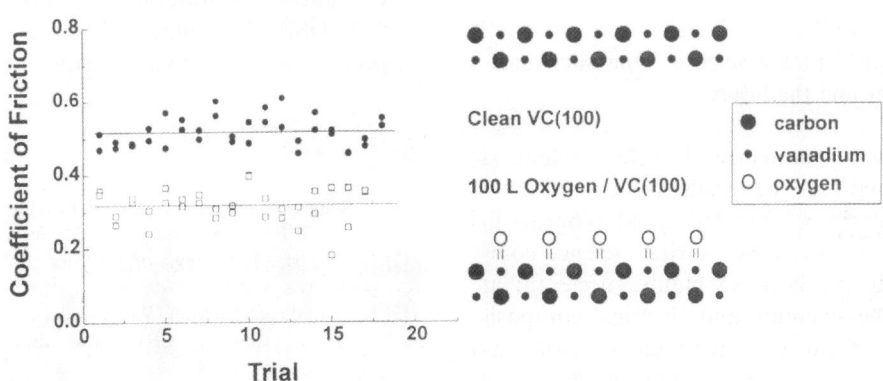

Fig. 14. Frictional forces measured using AFM under ultra-high vacuum conditions as a function of surface oxidation of single crystal VC (from Ref. [69]).

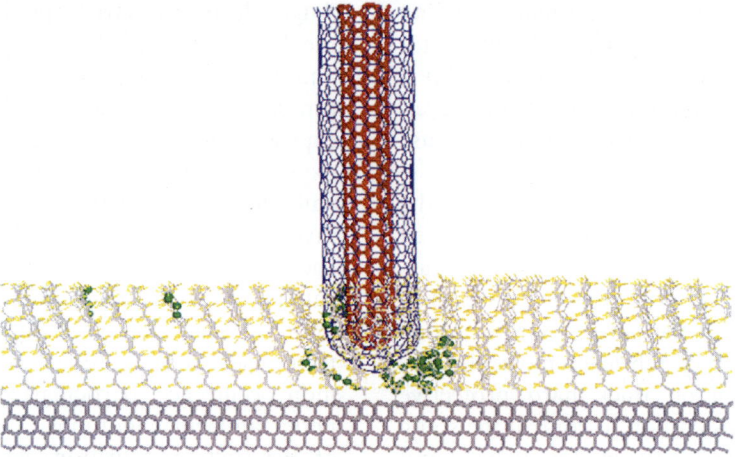

Fig. 15. Snapshot of a molecular dynamics simulation depicting a carbon, double-wall nanotube indenting a monolayer of *n*-alkane chains attached to diamond (1 1 1) (colored gray). The chains have 13 carbon atoms and are attached in a (2 × 2) arrangement (colored gray and yellow) to the diamond substrate. The nanotube is composed of a (5,5) capped nanotube (colored red) inside a (10,10) capped nanotube (colored blue). Gauche defects introduced into the monolayer by the indentation are depicted as large green spheres (from Ref. [72]).

dissipation mechanisms to be considered in such systems must go beyond simple electronic and phononic mechanisms, as now the vibrations within individual molecules (the "internal" vibrations as refereed to by the vibrations at surfaces community), as well as the creation of kinks and Gauche defects (deformation of extended chains) must be taken into account. Judith Harrison at the US Naval Academy is incorporating these effects into numerical simulations [72], and is rapidly converging on a solution to the problem, investigating all of the chain lengths which have been experimentally probed (Fig. 15).

6. Tribology and surface science: One slick partnership for now and the future

It is clear that fundamental surface scientists have now joined their materials science and engineering colleagues in efforts to speed progress in the area of tribology. Is the surface science community, which has been so highly successful at determining the structure and chemical composition of an open surface, up to the task of performing such characterizations at buried interfaces which are constantly moving? Only the future will tell. Meanwhile, pressed on by a confluence of forces ranging from environmental concerns over energy consumption to the ever shrinking nature of mechanical systems, the need to be able to a priori design tribomaterials is ever increasing. Whether for macroscopic applications in the automotive industry, or small-scale applications such as those involving MEMS, and ultimately NEMS, there is no doubt that the role of surface scientists will be critical.

Acknowledgements

The author is grateful for support from NSF and AFOSR. T. Coffey is thanked for assistance in preparation of the manuscript.

References

[1] J. Krim (Ed.), Fundamentals of Friction, Bulletin of the Materials Research Society, vol. 23, no. 6 (1998).
[2] H.P. Jost, Tribology: origin and future, Wear 136 (1990) 1.
[3] K.C. Ludema, Friction, Wear, and Lubrication: A Textbook in Tribology, CRC Press, Boca Raton, 1996, pp. 4–7.
[4] L. Pope, L. Fehrenbacher, W. Winer (Eds.), New Materials Approaches to Tribology: Theory and Applications, MRS, PA, 1989.

[5] I.M. Hutchings, Tribology: Friction and Wear of Engineering Materials, CRC Press, Boca Raton, 1992.
[6] J. Krim, The atomic-scale origins of friction, Sci. Am. (1996) 74.
[7] M.O. Robbins, M.H. Muser, Computer simulations of friction, lubrication and wear, in: B. Bhushan (Ed.), The Handbook of Modern Tribology, CRC Press, Boca Raton, 2001.
[8] T.E. Madey, Early applications of vacuum from, Aristotle to Langmuir, J. Vac. Sci. Technol. A 2 (1984) 110;
Otto von Guericke's Account of the Magdeburg Hemispheres Experiment, (translated by R. Sherman), in: T.E. Madey, W.C. Brown (Eds.), History of Vacuum Science and Technology, American Institute of Physics, New York, 1984, p. 67.
[9] D. Dowson, History of Tribology, Longmans, London, 1979.
[10] D. Sobel, Longitude: The True Story of a Lone Genius Who Solved the Greatest Scientific Problem of His Time, Walker and Company, New York, 1996.
[11] G. Amontons, De la Resistance causée dans les machines, Memoires de l'Academie Royale, A, Chez Gerard Kuyper, Amsterdam, 1706, pp. 257–282.
[12] Friction: Selected Reprints, A project of the AAPT Committee on Resource Letters, American Institute of Physics, New York, 1964.
[13] B. Bhushan, J.N. Israelachvili, U. Landman, Nanotribology: friction, wear and lubrication at the atomic scale, Nature 374 (1995) 607.
[14] I.L. Singer, H.M. Pollack (Eds.), Fundamentals of Friction: Macroscopic and Microscopic Processes, Kluwer, Dordrecht, 1992.
[15] D. Tabor, Friction as a Dissipative Process, in Ref. [14], p. 3.
[16] G.A. Tomlinson, A molecular theory of friction, Phil. Mag. 7 (1929) 905–939.
[17] G.M. McClelland and J.N. Glosli, Friction at the Atomic Scale, in Ref. [14], p. 405; J.B. Sokoloff, Theory of Atomic Level Sliding Friction, in Ref. [14], p. 423.
[18] J. Krim, D.H. Solina, R. Chiarello, Nanotribology of a Kr monolayer: a quartz crystal microbalance study of atomic-scale friction, Phys. Rev. Lett. 66 (1991) 181.
[19] M. Cieplak, E.D. Smith, M.O. Robbins, Molecular origins of friction: the force on adsorbed layers, Science 265 (1994) 1209.
[20] Ch. Woll, Low energy vibrations of molecular species adsorbed on metal surfaces, in: B.N.J. Persson, E. Tosatti (Eds.), Physics of Sliding Friction, Kluwer, Dordrecht, 1996.
[21] W. Press, Single Particle Rotations in Molecular Crystals, Springer, Heidelburg, 1981.
[22] F.Y. Hansen, V.L.P. Frank, H. Taub, L.W. Bruch, H.J. Lauter, J.R. Dennison, Corrugation in the nitrogen-graphite potential probed by inelastic neutron scattering, Phys. Rev. Lett. 64 (1990) 764.
[23] A.M. Lahee, J.P. Toennies, Ch. Woll, Low energy adsorbate vibrational modes observed with inelastic helium atom scattering, Surf. Sci. 177 (1986) 371.
[24] J.P. Toennies, in: W. Kress, F. de Wette (Eds.), Surface Phonons, Springer Series in Surface Science, Springer, Heidelberg, 1991.
[25] B. Voigtlander, D. Bruchmann, S. Lehwald, H. Ibach, Structure and adsorbate–adsorbate interactions of the compressed Ni(100)-(2*1) CO structure, Surf. Sci. 225 (1990) 151.
[26] C.J. Hirschmugl, G.P. Williams, F.M. Hoffmann, Y.J. Chabal, Adsorbate–substrate resonant interactions observed for CO on Cu(100) in the far infrared, Phys. Rev. Lett. 65 (1990) 480.
[27] W. Akemann, A. Ott, Vibrational modes of CO adsorbed on disordered copper films, J. Raman Spect. 22 (1991) 797.
[28] B.N.J. Persson, E. Tosatti (Eds.), Physics of Sliding Friction, Kluwer, Dordrecht, 1996.
[29] L.W. Bruch, F.Y. Hansen, Mode damping in a commensurate monolayer solid, Phys. Rev. B 55 (1997) 1782.
[30] B.N.J. Persson, E. Tosatti, D. Fuhrmann, G. Witte, Ch. Woll, Low-frequency adsorbate vibrational relaxation and sliding friction, Phys. Rev. B 59 (1999) 11777.
[31] B.N.J. Persson, Sliding Friction: Physical Principles and Applications, Springer, Berlin, 1998.
[32] J. Krim, A. Widom, Damping of a crystal oscillator by an adsorbed monolayer and its relation to interfacial viscosity, Phys. Rev. B 38 (1988) 12184.
[33] F.N. Dultsev, V.P. Ostanin, D. Klenerman, Hearing bond breakage measurement of bond rupture forces using a quartz crystal microbalance, Langmuir 16 (2000) 5036.
[34] M. Cieplak, E.D. Smith, M.O. Robbins, Molecular origins of friction, Science 265 (1994) 1209.
[35] C. Daly, J. Krim, Sliding friction of xenon monolayers and bilayers on Ag(111), Phys. Rev. Lett 76 (1996) 803, see also Alison Mack, There's the Rub, The Dallas Morning News, Discovery section, Monday, 14 August 1995.
[36] M.S. Tomassone et al., Dominance of phonon friction for a xenon film on a silver (111) surface, Phys. Rev. Lett. 79 (1997) 4798.
[37] B.N.J. Persson, Surface resistivity and vibrational damping in adsorbed layers, Phys. Rev. B 44 (1991) 3277.
[38] A. Liebsch et al., Electronic versus phononic friction of xenon on silver, Phys. Rev. B 60 (1999) 5034.
[39] D. Schumacher, in: Surface Scattering Experiments with Conduction Electrons, Springer Tracts Mod. Phys., vol. 128, Springer, Berlin, 1993, Secs. 4.2,3.
[40] A. Dayo, W. Alnasrallah, J. Krim, Superconductivity-dependent sliding friction, Phys. Rev. Lett. 80 (1998) 1690.
[41] G. He, M.M. Muser, M.O. Robbins, Adsorbed layers and the origin of static friction, Science 284 (1999) 1650.
[42] J.B. Sokoloff, Theory of energy dissipation in sliding crystal surfaces, Phys. Rev. B 42 (1990) 760, 6745.
[43] S. Granick, Soft matter in a tight spot, Physics Today 52 (July 1999) 26.

[44] J. Klein, E. Kumacheva, D. Mahalu, D. Perahia, L.J. Fetters, Reduction of frictional forces between solid-surfaces bearing polymer brushes, Nature 370 (1994) 634.

[45] R.W. Carpick, M. Salmeron, Scratching the surface: fundamental investigations of tribology with atomic force microscopy, Chem. Rev. 97 (1997) 1163.

[46] R.E. Weber, W.T. Peria, Use of LEED apparatus for the detection and identification of surface contaminants, J. Appl. Phys. 38 (1967) 4355.

[47] L.A. Harris, Analysis of materials by electron-excited Auger electrons, J. Appl. Phys. 39 (1968) 1419.

[48] P.W. Palmberg, W.T. Peria, Low energy electron diffraction studies on Ge and Na-covered Ge, Surf. Sci. 6 (1967) 57.

[49] M. Bienfait, Two-dimensional melting and polymorphism in adsorbed phases, Surf. Sci. 89 (1979) 13–27;
J. Benard et al., Adsorption on metal surfaces, an integrated approach, Studies Surf. Sci. Catal. 13 (1983).

[50] S.K. Sinha (Ed.), Ordering in Two-Dimensions, North Holland, New York, 1980.

[51] L.W. Bruch, M.W. Cole, E. Zaremba, Physical Adsorption: Forces and Phenomena, Clarendon Press, Oxford, 1997.

[52] C.M. Mate, G.M. McClelland, R. Erlandsson, S. Chang, Atomic-scale friction of a tungsten tip on a graphite surface, Phys. Rev. Lett. 59 (1987) 1942.

[53] S. Morita, S. Fujisawa, Y. Sugawara, Spatially quantized friction with lattice periodicity, Surf. Sci. Rep. 23 (1996) 1.

[54] G.J. Germann, S.R. Cohen, G. Neubauer, G.M. McClelland, H. Seki, D. Coulman, Atomic-scale friction of a diamond tip on diamond (1 0 0) and (1 1 1) surfaces, J. Appl. Phys. 73 (1993) 163.

[55] E. Meyer, R.M. Overney, L. Howald, R. Luthi, J. Frommer, H.-J. Guntherodt, Friction and wear of Langmuir–Blodgett films observed by friction force microscopy, Phy. Rev. Lett. 69 (1992) 1777.

[56] N. Forster, Rolling contact testing of vapor phase lubricants (III): surface analysis, Tribology Trans. 40 (1999) 1.

[57] C.F. McFadden, A.J. Gellman, Recent progress in ultra-high vacuum tribometry, Tribology Lett. 4 (1998) 155.

[58] A.J. Gellman, J.S. Ko, Tribological Properties of Single Crystalline Metal Surfaces, American Vacuum Society National Symposium, NS1-MoA7, October 1999, Seattle, WA, p. 35.

[59] M. Hirano, K. Shinjo, R. Kaneko, Y. Murata, Anisotropy of frictional forces in muscovite mica, Phys. Rev. Lett. 67 (1991) 2642.

[60] M. Hirano, K. Shinjo, R. Kanko, Y. Murata, Observation of superlubricity by scanning tunneling microscopy, Phys. Rev. Lett. 78 (1997) 1448.

[61] R.W. Carpick, D.Y. Sasaki, A.R. Burns, Large friction anisotropy of a polydiaceltylene monolayer, Tribology Lett. 7 (2000) 79.

[62] S.H.J. Izdiak et al., The X-ray surface forces apparatus: Structure of a thin smectic liquid crystal film under confinement, Science 264 (1994) 1915.

[63] B. Borovsky, B.L. Mason, J. Krim, Scanning tunneling microscope measurements, J. Appl. Phys. 88 (2000) 4017.

[64] B. Borovsky, M. Abdelmaksoud, J. Krim, STM-QCM studies of vapor-phase lubricants, in: S. Hsu (Ed.), Nanotribology: Critical Assesments and Research Needs, Kluwer, Dordrecht, 2000.

[65] M. Porto, M. Urbach, J. Klafter, Atomic-scale engines: Cars and wheels, Phys. Rev. Lett. 84 (2000) 1608.

[66] The little engine that couldn't, Science News 157 (July 22, 2000) 30.

[67] R. Maboudian, R.T. Howe, Stiction reduction processes for surface micromachines, Tribology Lett. 3 (1997) 215.

[68] B. Bhushan, Tribology and Mechanics of Magnetic Storage Devices, second ed., Springer, New York, 1996.

[69] P. Frantz, S. Didzuilis, P.B. Merrill, S.S. Perry, Spectroscopic and Scanning probe studies of oxygen and water on metal carbide surfaces, Tribology Lett. 4 (1998) 141.

[70] S.S. Perry et al., A correlation of molecular structure and interfacial friction, Abstr. Pap. Am. Chem. Soc. 217 (March 21, 1999) U586–U586, Part 1.

[71] X. Xiao, J. Hu, D.H. Charych, M. Salmeron, Chain length dependence of the frictional properties of alkylsilane molecules self-assembled on mica studied by atomic force microscopy, Langmuir 12 (1996) 235.

[72] J.A. Harrison et al., Phys. Rev. Lett., submitted for publication;
B. Bhushan (Ed.), The tribology of hydrocarbon surfaces investigated using molecular dynamics, Proc. Workshop on Tribology Issues and Opportunities in MEMS, Kluwer, Dordrecht, 1998, p. 285.

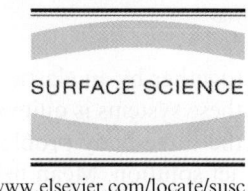

The surfaces of compact systems: from nuclei to stars

R.A. Broglia [a,b,*]

[a] *Department of Physics, University of Milan, and INFN Sez. di Milano, Via Celoria 16, 20133 Milan, Italy*
[b] *The Niels Bohr Institute, University of Copenhagen, Blegdamsveg 17, DK-2100 Copenhagen, Denmark*

Received 30 June 2000; accepted for publication 14 May 2001

Abstract

While providing information from worlds separated by five-to-six orders of magnitude in dimensions and in energy, the pairing properties (electrical resistance and viscosity), the electromagnetic response (spectrum of colours), the resilience to stress (elasticity), the ability to deform (plasticity), etc., associated with clusters of atoms and with atomic nuclei have surprisingly similar properties, once the proper scalings are done, and demonstrate the many analogies that can be drawn between different finite many-body systems. These analogies can be further extended to cosmic and to customer tailored nanometre materials. Femtometre materials, like the inner crust of a neutron star (pulsar), are made out of the same protons and neutrons which make infinite nuclear matter. However in pulsars, protons and neutrons are arranged in the form of finite nuclei immersed in a sea of free neutrons. This is the reason why these celestial objects rotate, conduct heat, emit neutrinos, etc., very differently from infinite nuclear matter. In fact, these phenomena reflect the properties of the corresponding atomic nuclei which form the pulsar. Among these properties, those associated with the nuclear surface are most important. Nanostructured materials are made out of atoms as their more common forms, but the atoms are arranged in nanometre or sub-nanometre-size clusters, which become the constituent grains, or building blocks, of new materials like, e.g., C_{60} fullerene. Because these tiny grains respond to light, mechanical stress and electricity quite differently from micron- or millimetre-sized grains, nanostructured materials display an array of novel attributes. At the basis of the new phenomena we find again the surface of the building blocks used to produce the new materials. A proper understanding of the interweaving of the single-particle motion with the static and dynamic deformations of the surface of finite many-body systems is likely to provide the key to open a whole new world of interdisciplinary research in such disparate fields as isolated atomic nuclei and clusters, new materials and compact stellar objects. The concepts and the experimental evidence needed to tool this key will be reviewed. Special emphasis will be set on the open questions still remaining to be answered to reach this goal. © 2001 Elsevier Science B.V. All rights reserved.

Keywords: Many body and quasi-particle theories; Superconductivity

1. Introduction

Finite many-particle systems can be viewed as drops of a liquid of interacting particles (protons and neutrons in the case of atomic nuclei, electrons and ions in the case of atomic aggregates, etc.) held

[*] Address: Dipartimento di Fisica, Instituto Nazionale di Fisica Nucleare (INFN), University of Milan, Via Celoria 16, 20133 Milan, Italy. Tel.: +39-02-583-57231; fax: +39-02-583-57487.
E-mail address: broglia@mi.infn.it (R.A. Broglia).

together by an elastic surface. To fully understand these systems is quite a formidable task, as already the three-body problem does not allow for an exact solution. Mean field theory is, in most cases, a very useful approximation to the problem. In it, one replaces the many-body equation describing the motion of all particles simultaneously, by the single-particle equation describing the motion of one particle at a time. Consequently particles move independently, in an average potential, which is generated by the particles themselves. The surface of this potential plays a paramount role not only in confining the motion of the particles but also in controlling their spill out. It can also act as an elastic membrane which can vibrate.

The vibrations of the surface of finite many-body systems dress the single-particle motion, renormalizing its properties and consequently, the properties of the entire system. In fact, in their trajectories particles bounce, most of the time, elastically off the surface. From time to time, however, they set the surface into vibration, vibration which can be reabsorbed at a later time by the same particle or by another particle. In the first case the particle carries around a vibration and becomes effectively heavier, which thus modifies, among other things, the specific heat of the system. In the second case, the vibration becomes a messenger between two particles, and thus acts as a glue. The resulting interaction is particularly efficient in producing pairs of particles. These pairs of particles have properties that are very different from those of single particles. In particular they may behave collectively as a liquid without viscosity, or, if charged, without resistance. That is, as a superfluid or as a superconductor.

The vibrational frequencies of the surface of finite many-body systems can be very low, becoming eventually zero in the case in which the surface acquires a permanent, static deformation. Both the single-particle motion and the collective surface vibrations of a deformed system can be quite different from the corresponding properties associated with neighbouring spherical systems.

In the universal language of field theory, first developed in connection with quantum electrodynamics, i.e., the theory describing the interaction of charged particles among themselves and with photons, these phenomena are associated with self-energy and vertex corrections and with induced interaction processes. These processes are intimately connected among themselves through very general relations, known as sum rules and reflecting the conservation of mass and charge. Furthermore, deformations and zero frequency vibrations are intimately connected with spontaneous symmetry breaking phenomena, and thus with the violation of the conservation of angular momentum, linear momentum, particle number, etc.

In what follows, we shall discuss some of the consequences the coupling of single-particle motion to static and dynamic deformations of the surface has on the physical properties of finite many-body systems. Special emphasis will be placed on the open questions and on the possible practical applications the subject has on the middle ground of physics [1]. The middle ground is an enormous domain, including everything intermediate in size between an atomic nucleus and a stellar object. It is the domain of everyday human experience. We shall start from the very small, the atomic nucleus, a paradigm of finite many-body systems whose properties are dominated by surface effects. This is because nuclei have a large ratio of surface to volume and, being very compact objects, a high surface tension. We shall provide experimental evidence and theoretical insight concerning the workings of the particle–vibration coupling mechanism in atomic nuclei. Pair formation and the phenomenon of superfluidity will be discussed in particular detail, also in connection with newly produced exotic forms of (nuclear) matter not previously observed in the Universe. Utilizing atomic nuclei as building blocks, we shall then concentrate our attention on very large systems, in particular on stellar compact objects. We shall again discuss the phenomenon of pair formation and of superfluidity in these systems, and its connection with the nuclear surface.

Armed with this experimental and theoretical insight we shall then proceed by analogy into the corresponding pair formation phenomenon and pairing in systems belonging to the middle ground of physics. Particular attention will be payed to superconductivity in nanometre organic materials made out of C_{60} fullerenes. The strength with which pairs of particles are held together in finite many-body systems due to the exchange of surface vibrations scales with surface tension. While this quantity is many orders of magnitude weaker in fullerenes than in atomic nuclei, the study of pairing in C_{60} is still a subject of particular interest. In fact, we shall emphasize that there is, potentially, a huge pay off in achieving a detailed understanding of these phenomena. Namely, the possibility of designing a real high T_c superconductor, that is a material which conducts electricity without resistance at temperatures not too different from room temperature.

After having studied, at the microscopic level, the consequences surface vibrations have on the interaction among particles and resulting in pair formation, we shall see how they affect the colour of finite many-body systems. We shall start the subject by discussing examples that have been around for centuries, namely that of small metallic clusters which give the colours to (stained) glasses. Making use of what one has learned during the last two decades on the "colour" of atomic nuclei, that is the response of nuclei to photons, it will be shown that it is likely that one already has the tools to design materials displaying costume tailored colours. Among other possibilities one can, by properly tuning the shape of the clusters, make them display more or less "metallic colours". From this knowledge it turns out that one can also design atomic wires which are of potential interest in microelectronics. This is again an open and quite promising subject of research.

We shall conclude the paper by emphasizing the very general physical concepts which explain why the surface of finite many-body systems is so important in determining their properties. Such concepts also explain why it is possible to work out an unified description of a great variety of objects ranging from the atomic nucleus to compact stellar objects, as well as materials which affect our daily life. The same concepts are, not surprisingly, also at the basis of the universal mathematical language used in this description: namely field theory of finite many-body systems.

2. Specific heat, superfluidity and superconductivity

The atom is the smallest unit into which matter can be divided that has the characteristic properties of a chemical element. Most of the atom is empty space. The rest consists of a cloud of negatively charged electrons whirling around a small, very dense, positively charged nucleus made out of protons and neutrons and contains essentially all the mass of the atom. The electric forces bind the electrons to the nucleus giving rise to a miniature planetary-like system. The strong and Coulomb forces acting among nucleons lead to a self-bound system. Inside it, nucleons move essentially independently of each other. Bouncing in their trajectories elastically off the nuclear surface, they describe closed orbits which resemble, at a length scale five orders of magnitude smaller, those of the electrons in the atom. This discovery won in 1963 the Nobel prize in physics to Maria Goeppert-Mayer and J. Hans D. Jensen [2].

2.1. The atomic nucleus

Nucleons, like electrons are fermions, i.e. have a half-integer spin $(1/2, 3/2, \ldots)$, and, according to quantum mechanics, obey the Pauli exclusion principle which allows only one particle to occupy each quantal state. Consequently, for the ground state of the system, the available orbitals of the mean field are filled from the bottom upwards to the Fermi energy. Under normal conditions, atomic nuclei are in their ground state, that is at zero temperature. This is because nuclei on earth, leaving aside those that arrive in the form of cosmic rays, are isolated. In fact, for two nuclei to interact, they need to have large kinetic energies, of the order of tens of MeV (1 MeV = 10^6 eV), so as to be able in a collision to overcome the Coulomb repulsion and reach within the range of the nuclear attraction. Energies of such magnitude contrast with the energies available at room temperature ($\approx 25 \times 10^{-3}$ eV). This is the reason why one needs large machines, so called atomic smashers, to make two atomic nuclei interact. In these machines, an atom containing a heavy atomic nucleus is stripped of most of its electrons becoming a heavy ion, which is subsequently accelerated and collimated. The resulting beam, aimed at a target of other heavy atomic nuclei, eventually lead to a heavy ion reaction (Fig. 1). In the event where the two nuclei fuse, the energy and angular momentum of relative motion becomes mostly excitation energy and angular momentum of the composite system. Typical values of these quantities are set, in (Fig. 2), in relation to temperatures and rotational frequencies observed in other physical systems. The thermalization of such a system depends, naturally, on the specific heat of the system, a quantity which is directly related to the density of levels around the Fermi energy and thus to the nuclear surface, as we shall explain in the next section.

2.1.1. Single-particle motion

To study the single-particle motion in nuclei the best probe one can use are transfer reactions. For example, aiming a beam of deuterons (the isotope $^{2}_{1}H_{1}$ ($\equiv d$)) of hydrogen containing one proton and one neutron) on a target nucleus, one can learn about the properties of the single-particle neutron levels lying above the Fermi energy, by studying

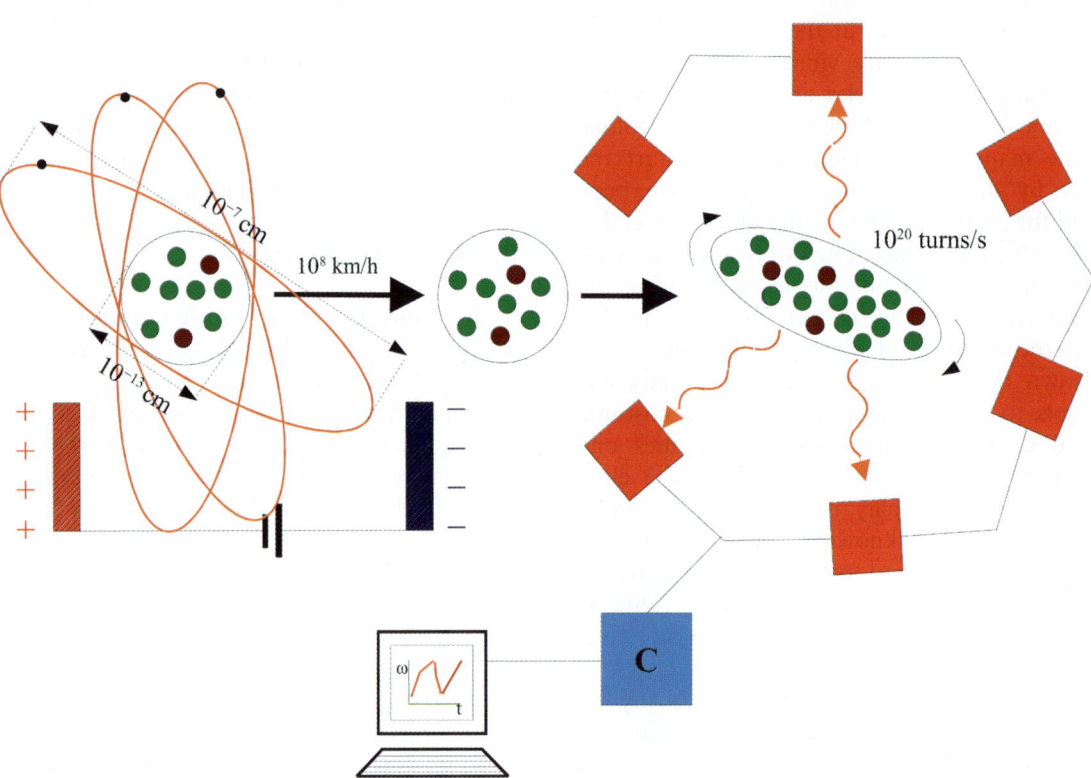

Fig. 1. Schematic representation of a heavy ion accelerator used to induce nuclear reactions. A neutral atom (left part of the figure), formed of a nucleus made out of protons (red dots) and neutrons (green dots), is stripped of the electrons (small black dots) moving around it. The system becomes positively charged and the resulting heavy ion is accelerated to velocities of the order of 100 millions of kilometers per hour in an electric field and smashed against the atomic nuclei of a target (middle). Out of the violent encounter between the two atomic nuclei a number of processes can take place: Coulomb excitation, particle transfer, etc. In some cases the two nuclei fuse, leading to a highly excited, rapidly rotating compound nucleus (right) which, in the process of cooling down, emits nucleons, alpha particles and γ-rays (orange wavy arrows). These γ-rays (photons) can be detected by use of a so called 4π-array, that is a set of detectors (red squares) covering a consistent fraction of the solid angle, an example of which is EUROBALL, the powerful array constructed by an European collaboration of nuclear physicists [3]. Once the signal is collected, it is analysed making use of specifically developed computer programs (software labeled C inside blue square) and the results eventually displayed on a screen.

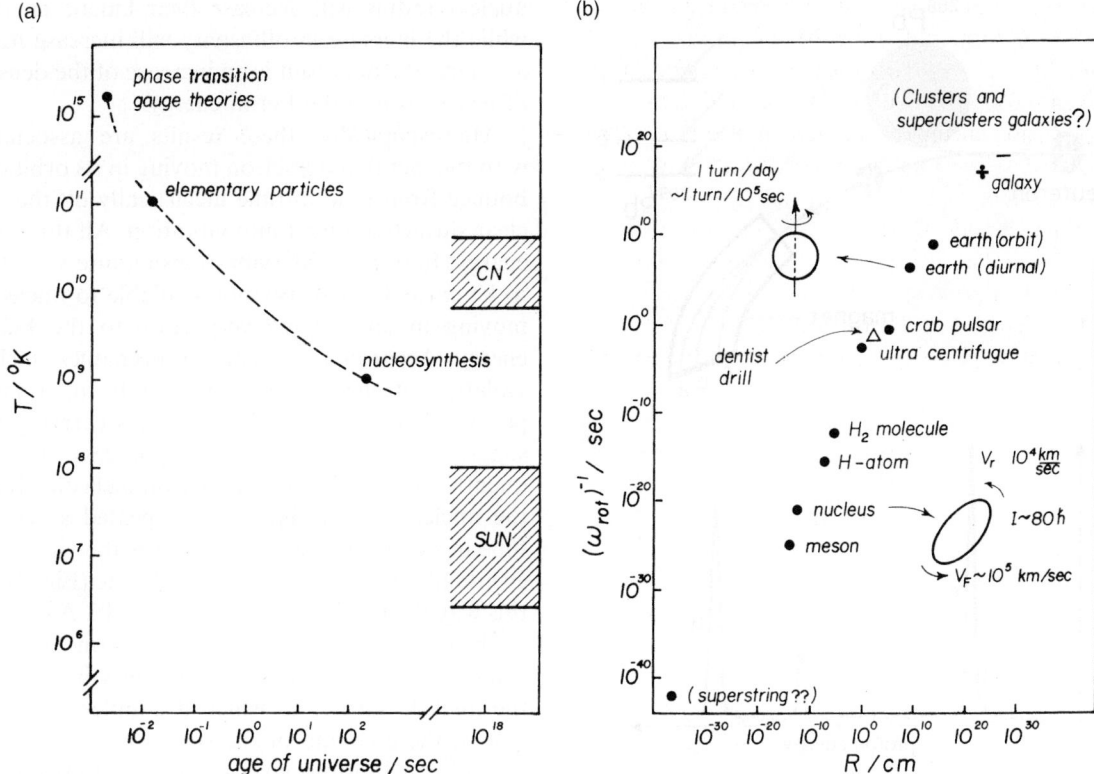

Fig. 2. Temperatures and rotational frequencies typical of reactions where two heavy ions fuse and eventually, after a time of the order of a few times 10^{-21} s equilibrate all their degrees of freedom, forming a single compound nucleus. (a) The scale of temperature has been set in connection with the temperature ascribed to the Universe at different times after the Big Bang through the first 3 min to our times [4]. (b) The inverse of the rotational period (in seconds) of a variety of rotating objects observed (or thought to exist, like superstrings which are the mathematical connotation of an elementary particle) in the Universe from the largest to the smallest, are quoted as a function of their radius. Both the inverse frequencies and the radii span 60 orders of magnitude, the relation between the two quantities being almost linear.

the spectrum and angular distribution of the emerging protons, resulting from the stripping of a neutron from the projectile (Fig. 3). The main outcome of these studies, in particular of those carried out on the closed shell nucleus ^{208}Pb, show that mean-field theory is able, in most cases, to correctly predict the sequence of single-particle levels. However, experiments indicate that the observed density of levels around the Fermi energy is higher than theoretically predicted. It is as if each of the nucleons described by mean-field theory had a mass that is ≈30–40% smaller than the free nucleon mass. This effective mass is known as the k-mass [5], indicating the momentum dependence associated with the non-locality in space of the mean-field (Fock-potential), a feature intimately connected with the fact that nucleons are fermions. Similar effects are found in the case of electrons in metals [6]. Furthermore, the levels lying close to the Fermi energy are less pure single-particle states than mean field theory posits. In fact, the occupation probability of the levels lying below but close to the Fermi energy is not 1 but more like 0.7–0.8. Similarly, levels lying above (and close) to the Fermi energy are not empty as mean field theory predicts, but display occupation factor of the order of 0.2–0.3. Finally, states lying a few MeV ($\geqslant 10$ MeV) away from the Fermi energy, display a width (finite lifetime). Mean field theory predicts these states to be sharp.

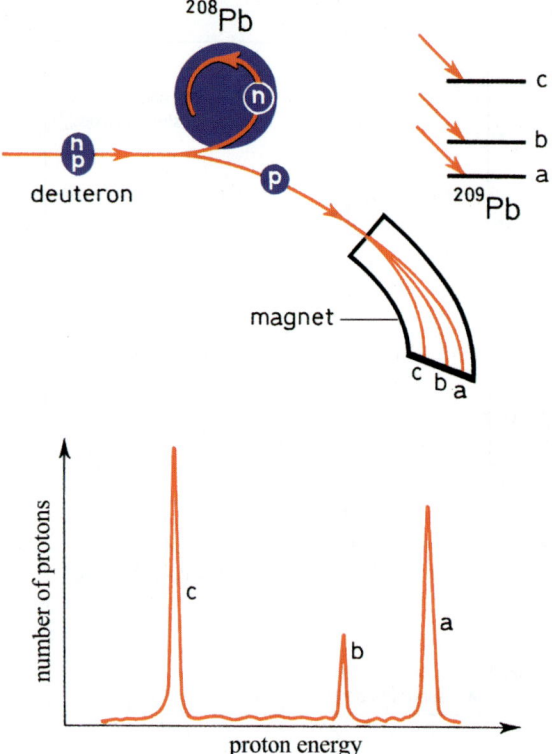

Fig. 3. Nuclear reaction in which a neutron is transferred from the projectile (deuteron) to the nucleus ^{208}Pb. The energy of the outgoing proton reflects both the Q-value of the reaction and the energy of the final state of the system.

nuclear radius will decrease their kinetic energy, while the increase in diffusivity will increase it. In any case, the net result is an increase of the density of levels around the Fermi energy.

Microscopically, these results are associated with the fact that a nucleon moving in its orbit can bounce from time to time inelastically off the nuclear surface setting it into vibration. All this costs energy. In fact, surface vibrations require 1–3 MeV to be excited, an energy not available to nucleons moving in levels lying very close to the Fermi energy. According to quantum mechanics, such a violation of energy can happen only for a short period of time, the resulting process carrying the suggestive name of "virtual" process. Consequently, the bouncing of a nucleon inelastically off the nuclear surface has to be repeated a second time in which the nucleon reabsorbs the vibration, eventually returning to its original state (Fig. 4). In this way the particle becomes "dressed". A dressed nucleon is heavier than a bare nucleon (40–80% heavier), as it has to carry around the vibrations of the nuclear surface to which it couples. In other words, the coupling of the single-particle motion to surface vibrations gives rise to an effective mass (called the ω-mass as it depends on the frequency of the vibrational mode), which is larger than the

The limitations mentioned above of mean field theory are connected, in the case of the atomic nucleus, with the fact that one is dealing with a static approximation to the many-particle problem. In fact, the presence of a mean field defines a surface which behaves like that of a liquid drop [7]. That is, as an elastic membrane which, not only confines the motion of the nucleons, but which can also vibrate at different frequencies determined by the surface tension ($\equiv 0.9$ MeV fm^{-2}) and by the associated inertia of the mode. These vibrations renormalize the properties of the mean field leading, among other things, to an increase in the nuclear radius and diffusivity. Consequently, levels lying somewhat above but close to the Fermi energy ϵ_F will decrease its energy coming closer to ϵ_F. Levels below (but close to) the Fermi energy will feel two contrasting effects. The increase in the

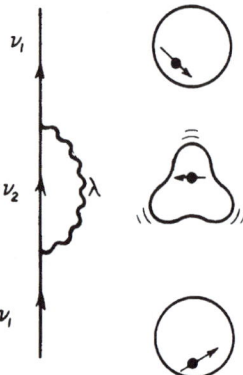

Fig. 4. Lowest-order process by which the single-particle motion is renormalized through the coupling to vibrations of the nuclear surface. The nucleon excites a surface vibration by bouncing inelastically off from the surface and absorbs it again at a later time. Particles are represented by an upwardgoing arrowed line (by a solid dot). The surface vibration is drawn as a wavy line.

bare nucleon mass [5]. The combined k-mass and ω-mass lead to an effective mass that is essentially equal to the bare mass of the nucleon, thus accounting for the observed density of levels at the Fermi energy and the associated occupation probabilities.

A number of nuclear properties, aside from the density of levels at the Fermi energy, depend on the ω-mass of the nucleons. In particular the symmetry energy. It is well known that the strong interaction has the same strength acting between all different nucleon species (protons and neutrons). Consequently, protons and neutrons like to be together, and it costs energy to separate them. The symmetry energy quantifies this cost. This energy is made out of two contributions of essentially the same magnitude: a potential and a kinetic energy contribution. The second one depends inversely on the ω-mass. Knowledge of the symmetry energy, and thus of the nucleon ω-mass is essential in describing such disparate phenomena as the stability and abundance of the nuclear species in the Universe, the occurrence of Supernovae (the fierce explosions second only to the Big Bang that mark the death of massive stars (cf. e.g. Refs. [8,9])), etc. These examples indicate that the study of the coupling of nucleons to vibrations of the nuclear surface is a subject lying at the forefront of nuclear research, of relevance for a variety of fields ranging from the very small to the very large.

2.1.2. Collective motion

Nuclear vibrations are excited by bombarding nuclei with high-energy photons, nucleons, electrons, etc. The vibrations are detected by observing how photons are absorbed by the nucleus, or how protons are scattered inelastically from the nuclear surface, etc. These experiments reveal that the nucleus display both elastic and plastic behaviour. In fact, the so called giant dipole resonance corresponding to a back and forth sloshing of protons against neutrons (Fig. 5(b)), excited in photoabsorption experiments has an energy centroid which scales with the inverse of the radius of the nucleus (a dependence observed for all nuclear

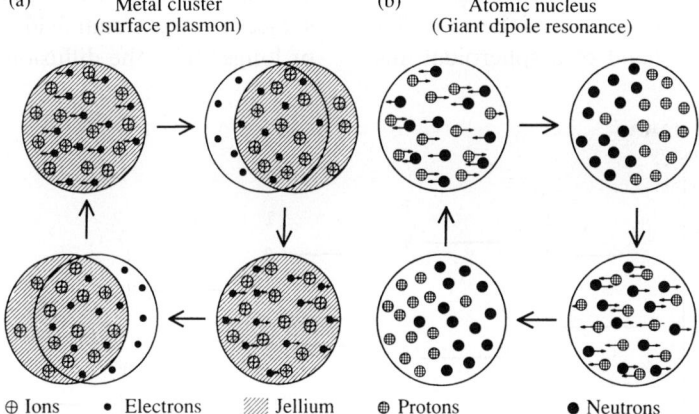

Fig. 5. Schematic representation of the giant dipole resonance in atomic nuclei and of the surface plasmon resonance in small metal clusters. The wavelength of the photon exciting these vibrations is large with respect to the diameter of the system. As a result the electric field associated with a passing gamma ray is nearly uniform across the system. (b) In the case of the excitation of the atomic nucleus, the field exerts a force on the positively charged protons, thus separating them from the neutrons. In fact, the neutrons act as having an (effective) negative charge, which oscillates out of phase with respect to the positively charged protons. (a) In the case of metal clusters, the electric field associated with the photon exerts a force on the positively charged ions and an identical force but with opposite direction on the electrons. Because the ions have a mass which is three orders of magnitude larger than that of the electron, the displacement of the electron cloud is much larger than that of the positive background. Simplified model called the jellium model smears the positive charges of the ions (atomic nuclei) uniformly over the whole volume of the cluster (after Ref. [10]).

resonances), i.e., with the nuclear momentum, a behaviour typical of elastic vibrations. Once excited into a giant dipole resonance the nucleus vibrates at an extremely high frequency, of the order of 5×10^{21} Hz (which corresponds to a vibrational energy of about 20 MeV). To make more vivid the qualification of "extremely" used above, it is useful to remember that the nominal range of human hearing extends from about 20 Hz to 2×10^4 Hz, while that of human vision ranges from 4×10^{14} to 8×10^{14} Hz. In order to excite giant resonances with suitable cross sections, use has to be made of fields that change quickly with time, with frequencies of the order of those associated with single-particle motion ($1/\omega \equiv 10^{23}$ Hz).

The lowest excited states of even–even nuclei display, with very few exceptions, quadrupole and octupole character, that is, they correspond to surface collective vibrations of multipolarity 2 and 3. The associated energies, 1–3 MeV, show a dependence with mass number inversely proportional to the area of the nuclear surface, typical of the vibrations of a liquid drop [7,10]. They reflect the plastic properties of atomic nuclei. In fact, away from closed shells, the energy of the lowest 2+ state can become particularly low in energy (Fig. 6). In these nuclei (^{152}Sm, ^{154}Sm or e.g. ^{238}U), the 2+ state is the lowest member of the so-called ground state rotational band of a spheroidal nucleus rotating around an axis perpendicular to the symmetry axis (Fig. 7). These states are excited with large cross sections by Coulomb and by nuclear fields which change slowly in time ($1/\omega \equiv 10^{20}$ Hz), as compared with typical frequencies associated with single-particle motion.

Systems displaying both elastic and plastic behaviour are well known from other fields of physics. In particular, the natural caoutchouc, that is, the rubber out of which the tires of cars are made. The basic ingredient of caoutchouc is a liquid, known as latex. It is made out of long polymer chains containing only carbon and hydrogen atoms. When this liquid is boiled together with sulphur, which fixes the polymers among them at certain points, it becomes caoutchouc. In this reaction, only one carbon atom out of 200 of them reacts with the sulphur. Consequently, over long distances the atoms of the polymers occupy reasonably fixed positions, as in a solid, while over short distances they are still quite free to move, as in a liquid. Under a stress, short in time as compared with the diffusion time of the atoms, the liquid part of the material has no time to change place allowing a change in position of the fixed points, and the system reacts elastically, displaying a high degree of elasticity. When the system is subject to an external field over a time of the order or longer than the diffusion time of the polymers,

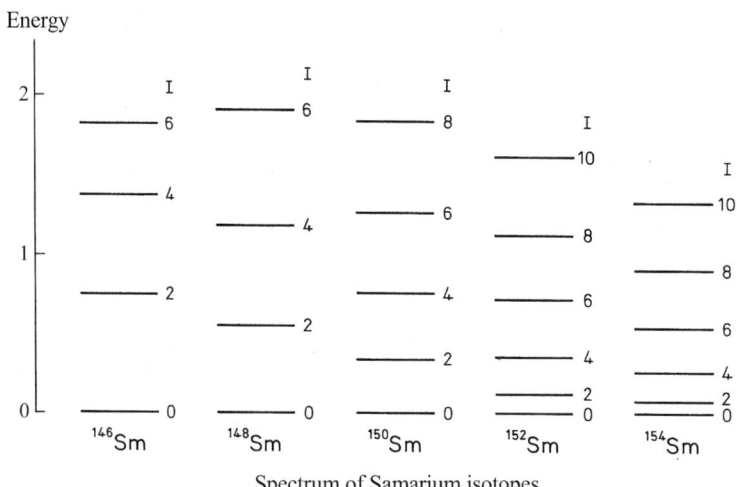

Fig. 6. Sequence of positive pairity states of the isotopes of $_{62}$Sm. The energy scale is in MeV.

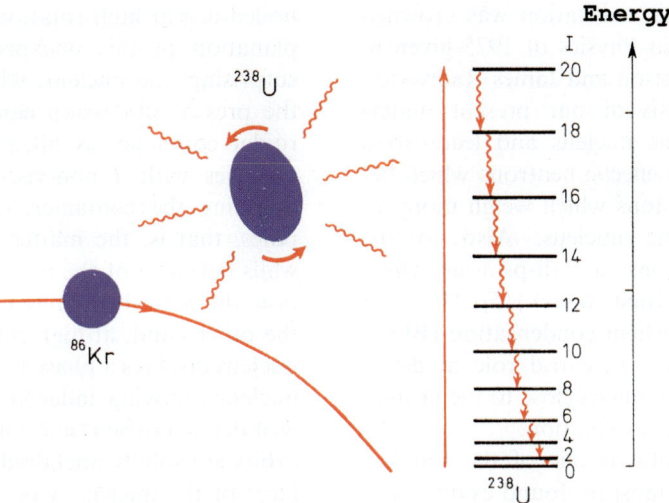

Fig. 7. Excitation of the rotational motion of a (target) nucleus of ^{238}U by a ^{86}Kr projectile through the Coulomb field (Coulomb excitation). The system cools down by emitting a number of γ-rays (red wavy arrowed lines) each carrying angular momentum two (in keeping with the quadrupole deformation of the nucleus ^{238}U) and different energy, connecting the members of the ground state rotational band of the system.

the fixed points have the possibility to change position in space and the caoutchouc deforms plastically. Evidence of this behaviour in the nuclear case is provided not only by low-lying vibrations and rotations, but also by nuclear fission. This process can be viewed as the division, similar to that of a cell, of a liquid drop into two smaller droplets as a result of deformation (Fig. 8). All these phenomena are essentially controlled by the nuclear surface.

2.1.3. Cooper pairs

The theory describing the coupling of the nucleons to vibrations and to static deformations of the nuclear surface provides an unified description of liquid drop and of independent-particle

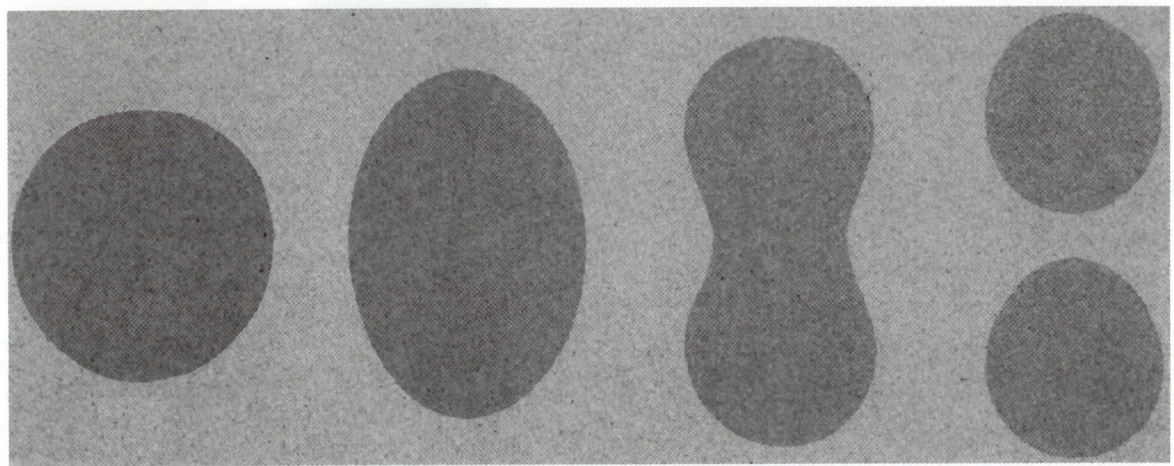

Fig. 8. A schematic sequence of events in the process of spontaneous nuclear fission.

behaviour [11–14]. This unification was crowned with the Nobel Prize in Physics of 1975 given to Aage Bohr, Ben Mottelson and James Rainwater. It constitutes the basis of our present understanding of the atomic nucleus and leads to a number of unexpected effects: neutrons which become charged and protons which weigh more inside than outside the nucleus. Also to the formation of di-neutrons and di-protons which behave like bosons (Cooper pairs) [15]. They can thus undergo Bose–Einstein condensation (BEC). This phenomenon plays a central role in determining the properties of nuclei close to the ground state, in particular nuclear rotation.

The rotational motion is one of the simplest and, at the same time, most profound examples of periodic motion (think only on the earth rotation around itself and around the Sun and on the succession between day and night and between summer and winter). A detailed study of the rotational frequency and of the variations of the rotational period as a function of time can provide important information on the physical properties of the system under study and on the forces acting inside it. Confronted with two eggs lying on the kitchen table, one of which we have been told has been hard boiled and the other not, one does not need to break them open to learn which of the two is cooked. It is sufficient to make them rotate by spinning them. One of the two will rotate for quite a while almost undisturbed, while the other one will stop rotating already after few revolutions. Obviously, the first of them is the hard boiled egg.

Strongly rotating nuclei (Fig. 2) can be produced in the laboratory either via Coulomb excitation (Fig. 7) or in fusion processes (Fig. 1). Because the system is charged, the rotating nucleus emits "light", that is gamma rays which provide detailed information on the properties of the system. A surprising result emerging from such measurements is that at low rotational frequencies, the moment of inertia of the system is considerably smaller, about half than the expected (rigid) moment of inertia, while at higher rotational frequencies the moment of inertia coincides more or less with the rigid moment of inertia of the system. It is as if at low rotational frequencies the nucleus behaves like a raw egg, while it spins like a hard-boiled egg at high rotational frequencies. The explanation of this unexpected behaviour is also surprising: the nucleus, which in connection with the present discussion can be viewed as a spheroidal container, is filled at low rotational frequencies with a non-viscous (superfluid) liquid. Spinning the container, only the matter at the poles, that is, the matter directly pushed by the walls (surface) of the container is set up into motion, the central core remaining at rest (Fig. 9). On the other hand, at high rotational frequencies the nucleus displays a phase transition into a system of nucleons moving independently of each other in well defined orbits (normal system). Each of these orbits are solidly anchored to the mean field (surface) of the nucleus. Consequently, each nucleon inside the nucleus reacts to a change in the rotational frequency with its full mass. The collective motion of the system thus coincides with that of a rigid body of the same mass, dimension, and deformation of the nucleus under study.

These phenomena are closely connected with the variety of effects belonging to the field of low-

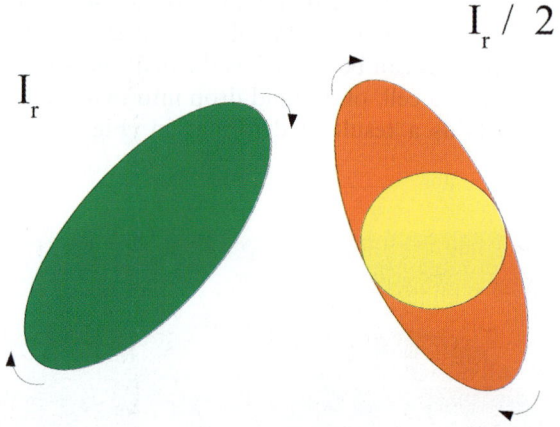

Fig. 9. Schematic representation of the reaction to rotation of a deformed nucleus. In the left part of the figure it is assumed that the system can be described in terms of the motion of independent nucleons (fermions). The associated moment of inertia is equal to the rigid moment of inertia I_r. In the right part of the figure it is assumed that the system is filled with a superfluid fluid (made out of pairs of fermions) displaying no friction. Consequently, under rotation the central core (yellow) remains essentially at rest and only the matter associated with the poles (orange) is involved in the rotation. The associated moment of inertia of this system is $\approx I_r/2$.

temperature physics, which go under the name of "superconductivity" [15,16] when they occur in an electric charged system such as electrons in a metal, and "superfluidity" when they occur in a neutral system such as an insulating liquid [17]. The behaviour of a superconducting metal is qualitatively different from that of a normal metal, in that it conducts electricity with zero resistance (hence the name). Similarly, a liquid that goes superfluid can flow through tiny capillars without apparent friction, and can even climb up over the rim of a vessel containing it and thereby gradually empty the vessel.

The importance of the subject has been acknowledged by the Swedish Royal Academy of Sciences with a host of Nobel Prizes in Physics (1913, 1962, 1972, 1973, 1978, 1987, 1996).

2.2. Bose–Einstein condensation

To sketch the general picture, let us concentrate on the common isotope of helium (^{4}He). Such an atom has zero total spin, and therefore, according to quantum mechanics should obey Bose–Einstein statistics. Consequently, when a large number of such particles are placed in a restricted volume and are cooled down below a certain temperature (which for helium turns out to be about 3 K), a phenomenon called BEC takes place: namely, a finite fraction of all the atoms begins to occupy a single quantum-mechanical state, the fraction increasing as the temperature decreases. Thus, in the same way that fermions are extremely individualistic, each of them occupying a different (quantal) state, bosons are extremely gregarious and love to be in the same (quantal) state of the system. The atoms in this special situation become locked together in their motion, like soldiers in a well drilled army, and can no longer behave independently. Thus, for example, if the liquid flows through a narrow capillary, the processes of scattering of individual atoms by roughness in the walls, which would produce so strong a viscous drag on any normal liquid as to effectively prevent it from flowing at all, is now quite ineffective, since all atoms (bosons) must scatter or none.

A similar picture applies to superconductivity in metals. However, in this case the "particles" that undergo BEC (and are therefore required to be bosons) are not individual electrons, which have spin 1/2 and are fermions. The "particles" are pairs of electrons (Cooper pairs) which form in the metal, and carry mass and charge double of that of a single electron aside from total spin zero. Cooper pairs condense when the metal is cooled down (e.g. below 7 K in the case of lead) and occupy a single quantal state. A similar phenomenon takes place in the nuclear case, where pairs of nucleons moving in time reversal states form Cooper pairs with total angular momentum zero (Fig. 10). Because the nucleus in its ground state or in the states belonging to the so called yrast band (that is the states of lowest energy for a given angular momentum, where all the energy of excitation is tied up in the collective, ordered rotational motion of the system as a whole) is at zero temperature, these Cooper pairs condense, giving rise to a superfluid non-viscous liquid.

The variation of the rotational frequency of deformed nuclei displays sharp discontinuities (Fig. 11). This phenomenon reflects the fact that rotation has opposite effect on the motion of the two members of each Cooper pair (Fig. 12). After a compound nucleus, with high angular momentum and excitation energy, has been formed in a fusion process, it cools down by evaporating particles. When the excited nucleus comes close to or eventually reaches the yrast line, it continues to slow down by emitting gamma rays each carrying two units of angular momentum. At the beginning of this process, the rotational frequency is so high that the associated energy is larger than the binding energy of each pair. Consequently no Cooper pair is present in the system, and the nucleus behaves as a normal system displaying a rigid moment of inertia. As the nucleus continues to decrease its rotational frequency it will eventually arrive at a (critical) frequency below which pair formation can take place, and the system can make a transition to the superfluid phase. Consequently, the moment of inertia of the system decreases by a factor of ≈ 2. The rotational frequency thus increases sharply. This is because the product of the moment of inertia and the rotational frequency is equal to the total angular momentum of the system, a quantity which is conserved. Thus, the

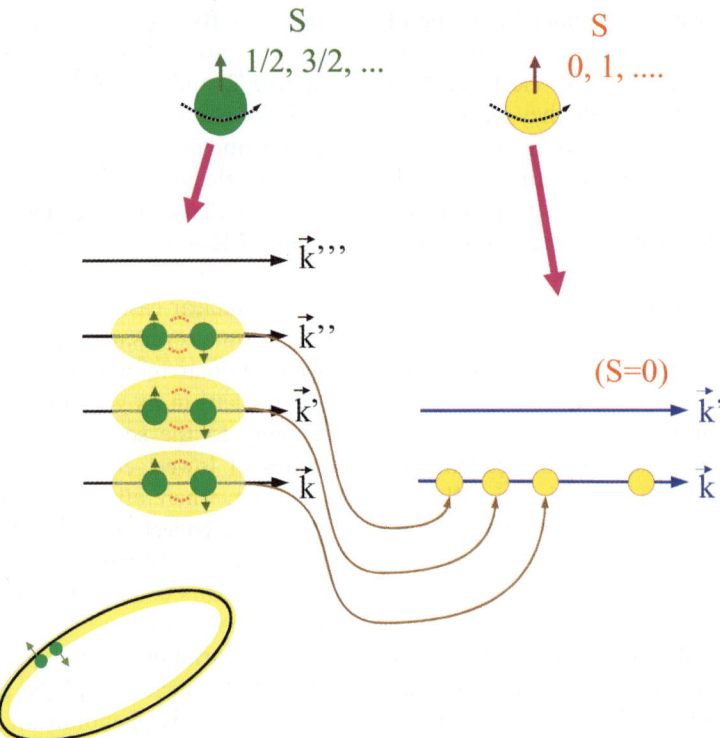

Fig. 10. Particles displaying half-integer values of the spin are called fermions, fulfill Fermi–Dirac statistics and according to the exclusion principles cannot occupy the same quantal state labeled in the figure by the quantum number $\vec{k}$ and by the intrinsic spin which can be in one of two states, namely spin up and spin down, associated with the two projections ($\pm 1/2$) of a spin $s = 1/2$ as that of the nucleon or of an electron, and explicitely shown in the figure in terms of an arrow pointing up or down respectively. Particles of integer spin are called bosons and fulfill Bose–Einstein statistics. They can all occupy the same quantal state. Coupling two fermions to integer spin, e.g. $S = 0$, one can form a boson (Cooper pair). If the temperature of the system is lower than the binding energy of the Cooper pair, a condensation phenomenon can also take place in this case. In nuclei the interaction responsible for the formation of Cooper pairs can be viewed as a surface phenomenon effect (cf. also Fig. 13).

observed discontinuities in the decrease of the rotational period is intimately related with the normal–superfluid phase transition taking place as a function of the rotational frequency. This transition is associated with a violent variation of the moment of inertia of the system, that is, to a nuclear quake.

In the case of low-temperature superconductivity the attraction among electrons is generated by the exchange of lattice phonons between them (and the energy needed to break the pair is of the order of few milli (10^{-3}) electron volts). In the nuclear case, roughly 50% of the pairing effects is due to a phenomenon similar to this one, namely the exchange of collective surface vibrations between nucleons [18]. The other 50% is due to the strong force acting between nucleons and arizing from the exchange of mesons (the carriers of the strong force) between nucleons (cf. Fig. 13). The effect this force has on nucleons moving in states connected by the operation of time reversal (Fig. 12) can be measured in scattering process between nucleons and expressed in terms of the $^1S_0$ phase shift (cf. Fig. 13 and corresponding caption). It is found that the exchange of mesons leads to a pairing force which is attractive at low relative momenta, that is in situations similar to that experienced by pairs of nucleons moving on the surface of the nucleus.

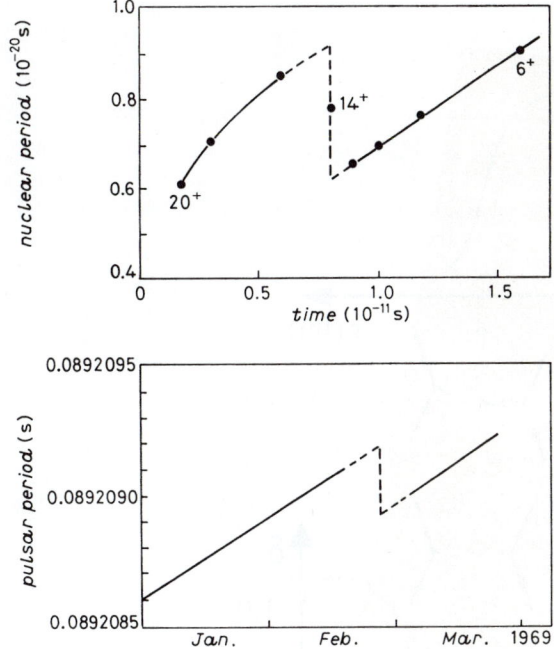

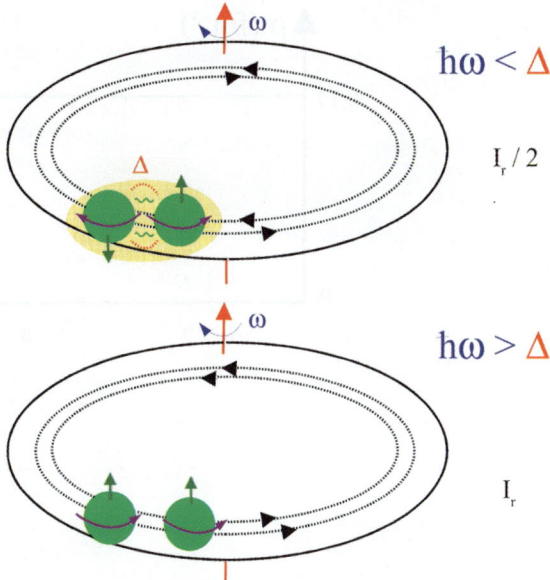

Fig. 11. Plot of the rotational period versus time for the nucleus ^{158}Er (top) and the pulsar Vela (bottom).

Fig. 12. Pair of nucleons moving in states of time-reversal (clockwise and anticlockwise, spin up and spin down) and forming a Cooper pair in a deformed nucleus which rotates as a whole with frequency ω. The binding energy of the pair is measured by the quantity Δ, known as the pairing gap [15], the interaction among the nucleons being represented by red dotted curves and green wavy lines (see also Fig. 13). For rotational frequencies such that the rotational energy $\hbar\omega$ is smaller than the pairing gap (top), the system behaves as a condensate of Cooper pairs (bosons), with the moment of inertia being one half the rigid moment of inertia (cf. Fig. 10). For rotational frequencies such that $\hbar\omega$ is larger than Δ, the nucleon moving clockwise in its trajectory is so much retarded in its revolution period with respect to the partner nucleon, that it cannot correlate efficiently any more with it and "align" its motion (and spin) with the rotational motion, becoming again a pair of fermions and not participating any more in the condensate (cf. Fig. 11).

The role the exchange of nuclear surface vibrations has on Cooper pair formation is larger for small rather than for large nuclei. This is because the surface–volume ratio, and thus the curvature of the system as well as the spill out of nucleons is larger for the light nuclei. Consequently, in these systems the collectivity of surface vibrations is larger than in the heavier systems. These effects become exacerbated in the case of very light exotic nuclei (cf. next subsection). We shall also see that effects similar to these open the possibility of creating real high T_c superconductors when applied to the case of fullerene based materials (cf. Section 2.4).

The central role played by the surface in the condensation process leading to nuclear superfluidity, where nucleons move without friction, is demonstrated by the fact that BEC essentially happens only at the surface of the nucleus. In other words, nucleons behave in the nuclear surface like a superfluid, while inside the nucleus they display normal properties, including friction. The pairing gap [15], the energy which provides a quantitative measure of the binding energy of the Cooper pairs and thus of the superfluidity of the system, is large ($\equiv 1$ MeV) at the nuclear surface, becoming essentially zero inside the nucleus.

2.3. Exotic forms of nuclear matter

Surface controlled pairing is also believed to be responsible for the existence of exotic forms of nuclear matter at very low densities. These so

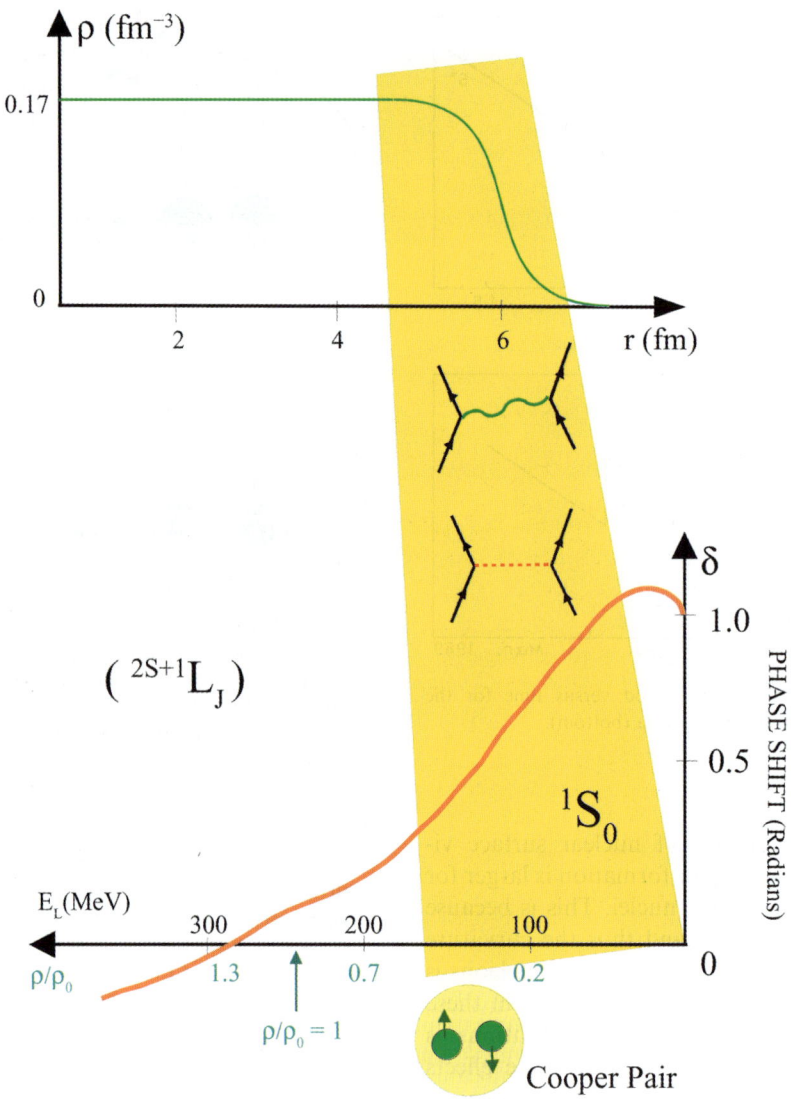

Fig. 13. (top) Nuclear density ρ in units of fm^{-3} (where $fm \equiv 10^{-13}$ cm), plotted as a function of the distance r (in units of fm) from the centre of the nucleus. Saturation density correspond to ≈ 0.17 fm^{-3}, equivalent to 2.8×10^{14} g/cm^3. Because of the short range of the nuclear force, the strong force, the nuclear density changes from 90% of saturation density to 10% within 0.65 fm, i.e. within the nuclear diffusivity. (bottom) Phase parameter associated with the elastic scattering of two nucleons moving in states of time reversal, so called $^1S_0$ phase shift, in keeping with the fact that the system is in a singlet state of spin zero. The solution of the Schrödinger equation describing the elastic scattering of a nucleon from a scattering centre (in this case another nucleon) is, at large distances from the scattering centre a superposition of the incoming wave and of the outgoing, scattering wave. The interaction of the incoming particle with the target particle changes only the amplitude of the outgoing wave. This amplitude can be written in terms of a real phase shift—or scattering phase—δ. Positive values of δ implies an attractive interaction, negative a repulsive one. For low relative velocities (kinetic energies E_L), i.e. around the nuclear surface where the density is low, the $^1S_0$ phase shift arizing from the exchange of mesons (like for example pions, represented by an horizontal dotted red line) between nucleons (represented by upward pointing arrowed lines) is attractive. This mechanism provides about half of the glue to nucleons moving in time reversal states to form Cooper pairs. These pairs behaves like boson and eventually condense in a single quantal state leading to nuclear superfluidity. Cooper pair formation is further assisted by the exchange of collective surface vibrations (green wavy curve in the scattering process) between the members of the pair.

called halo nuclei, which lie at the limits of stability of the chart of nuclides [19] (cf. Fig. 14), are composed of essentially all surface and display a very large spill out of nucleons.

2.3.1. Halo nuclei

These nuclei exist at the neutron drip line in the chart of nuclides (see below and Fig. 15(B)) and their study constitute one of the most important subjects lying at the forefront of nuclear research.

When neutrons are progressively added to a normal nucleus, the Pauli principle forces them into states of higher momentum. When the core becomes neutron-saturated, the nucleus expels most of the wavefunction of the last neutrons outside to form a halo, which because of its larger size can have lower momentum. In most cases neutrons essentially drip off from the nucleus, defining the lines of stability for neutron number in the chart of nuclides (neutron drip line, cf. Fig. 15(B), dashed lines marked $B_n = 0$). In some few, selected cases, the system becomes a halo nucleus. In halo nuclei, some of the constituents neutrons venture beyond the drop's surface and form a misty cloud or halo (similar effects are found for nuclei close to the proton drip line, cf. Fig. 15(B) dashed lines marked $B_p = 0$). Not surprisingly, these extended nuclei behave very differently from ordinary ("normal") nuclei lying along the stability valley in the chart of nuclides. In particular, they are larger than normal nuclei of the same mass number. For example, ^{11}Li is twice as large as ^{11}B, a stable nucleus containing the same number of nucleons and half the size of the lead nucleus ^{208}Pb, which holds 197 more particles (Fig. 14). In the case of ^{11}Li, the last two neutrons are very weakly bound. Consequently, these neutrons need very little energy to move away from the nucleus. There, they can remain in their "stratospheric" orbits, spreading out and forming a tenuous halo. If one neutron is taken away from ^{11}Li, a second neutron will come out immediately, leaving behind the core of the system, the ordinary nucleus ^{9}Li. This result indicates that pairing may play an important role in the stability of ^{11}Li. In fact, the properties of the exotic nucleus ^{11}Li can be understood, even quantitatively, in terms of the simplest scenario imaginable: the formation of a single di-neutron Cooper pair which is held together to the ^{9}Li core essentially by the exchange of surface vibrations among the halo nucleons (Fig. 15(A)) [21]. The importance of these vibrations is due to the high polarizability displayed by ^{11}Li [20].

Current research in nuclear physics and astrophysics focuses on exploring the behaviour of nucleonic matter under the extreme conditions associated with the drip lines (surface dominated nuclei). For this purpose, use is made of beams of new nuclear species created in modern accelerator laboratories [22]. To be noted that of all the atomic nuclei which are thought to be possible, only about half of them have been observed in the laboratory, and a small fraction exist naturally on Earth (cf. Fig. 15(B)). Of particular importance in this research are the mechanism for the creation of elements in the Universe. Unstable nuclei are involved in explosive burning in astrophysical environments. Prominent among these processes are explosive hydrogen burning (the rp-process) on the surface of accreting white dwarfs (novae) or neutron stars (X-ray bursts), as well as rapid neutron capture (the r-process) in very neutron-rich conditions of Supernova explosions (cf. Fig. 16). The r- and rp-process paths shown in Fig. 15(B), traverse regions of unstable nuclei of the chart of nuclides which are as yet unexplored. A third path, the s-process path, runs along the stable nuclei.

2.3.2. Neutron stars

The nuclear surface controls not only the properties of individual nuclei, but also the properties of "materials" whose building blocks are atomic nuclei. These femtometre materials display properties reflecting not only the ubiquitous role played by the interweaving of nucleons and surface vibrations, but also the marked dependence of the strong force with density. A textbook example is provided by neutron stars (pulsars) [23]. These remnants of fierce Supernova explosions are gigantic, rapidly rotating nuclei thirty kilometers across, held together by the gravitational force (Figs. 16 and 17), a discovery which led in 1974 to the Nobel Prize in physics for Anthony Hewish [26]. Neutron stars display, in the process of cooling down through radiation and particle

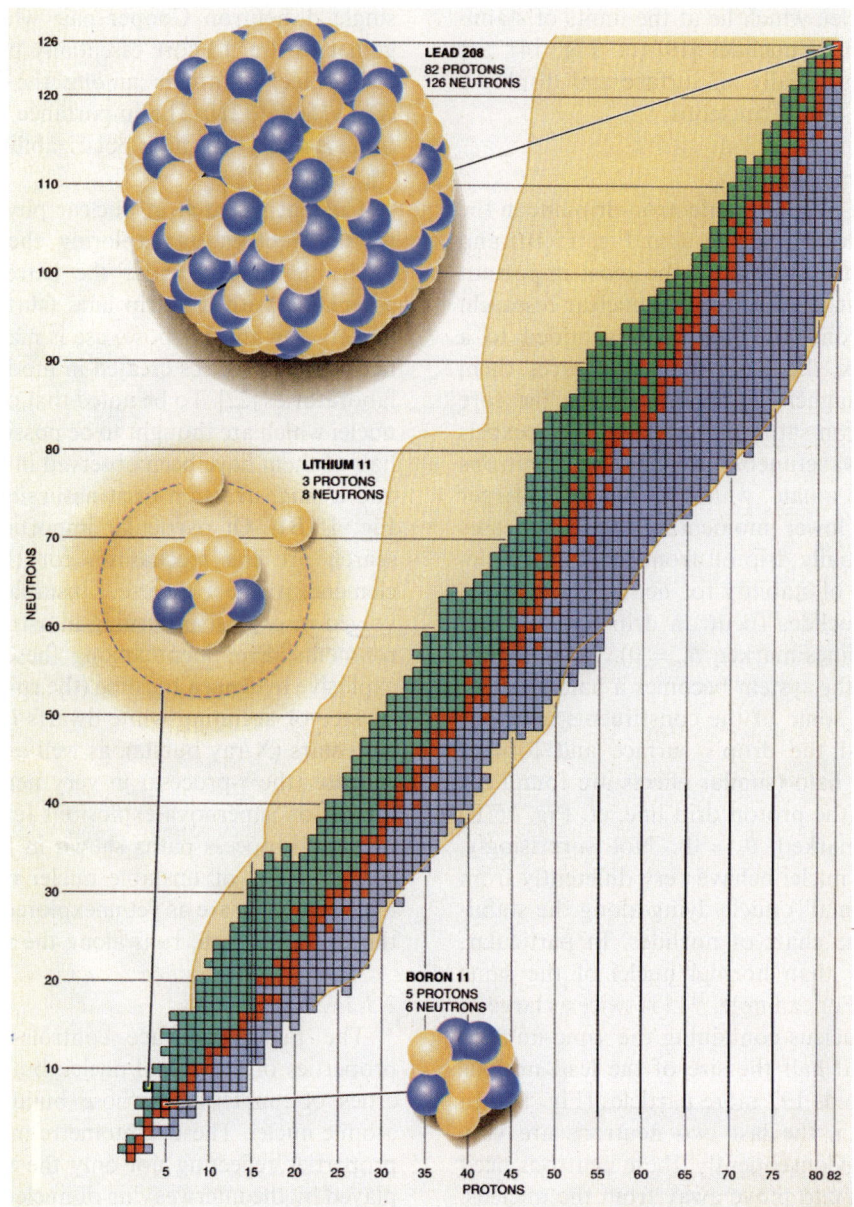

Fig. 14. Nuclei are represented here by squares, positioned horizontally according to the number of protons and vertically according to the number of neutrons they contain. Stable nuclei are shown as gold squares. Green squares indicate bound but unstable nuclei that have excess neutrons. Blue squares mark bound, proton-rich nuclei. The outer borders of these regions are called drip lines, along which large halo nuclei may be found. The extended halo of lithium 11, comprising two neutrons, makes this nucleus nearly twice as large as boron 11, a stable nucleus containing the same number of nucleons. ^{11}Li is half the size of a lead nucleus that holds 197 more particles, ^{208}Pb (after Ref. [19]).

emission, marked glitches (starquakes) (Fig. 11), which are likely to be connected with di-nucleon BEC [24,25]. In fact, neutron stars usually rotate with such precision that they are known as the best

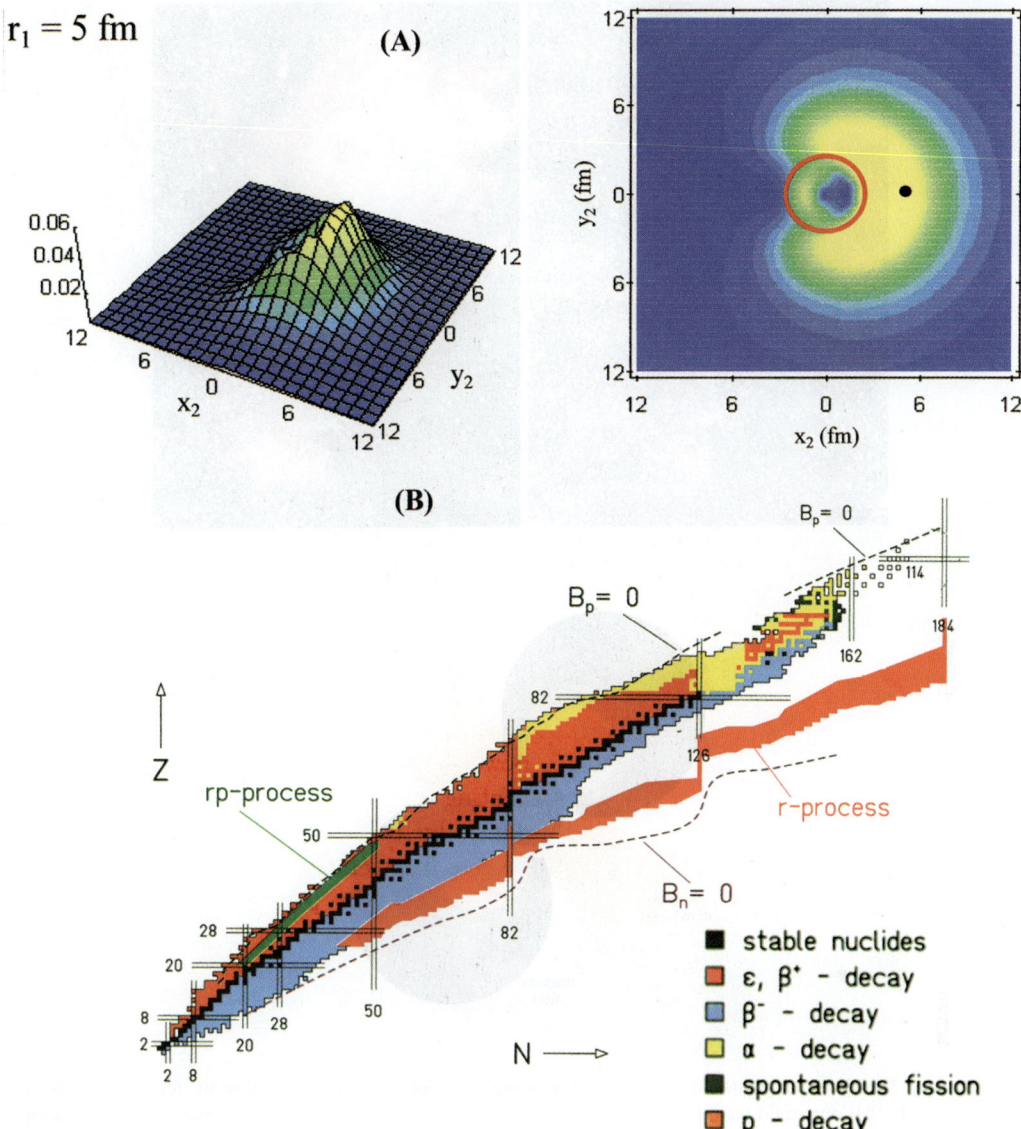

Fig. 15. (A) Spatial structure of the Cooper pair which describes the two halo neutrons of ^{11}Li [21]. Fixing one of the two neutrons at a distance of 5 fm from the origin of the nucleus (black dot) on the x-axis, the probability of finding the other neutron is displayed. The numbers appearing on the z-axis of the three-dimensional plot displayed to the left side are in units of fm^{-2}. The projection on the x–y plane is shown on the right panel, where the core ^{9}Li is shown as a red circle. (B) Map of the nuclear landscape, as in Fig. 14, where particular characteristics which makes it easier to appreciate the new area of research associated with exotic nuclei are emphasized. The nuclei existing naturally on Earth are marked by black squares. The magic numbers associated with increased stability (closed shells), are the convenient "landmarks" in the map and are indicated by double lines. Also shown are estimates of the borderline of stability, the so called drip lines, where the proton and neutron binding energies B_p and B_n become zero. Also shown are two paths astrophysicists identify as important in the creation of elements in the Universe: the r-process and the rp-process which transverse regions yet unexplored. A third path, the s-process path, runs along the stable nuclei.

time keepers in the Universe. But every so often their rotation rate increases. It is thought that these glitches are related to superfluidity inside the star, in particular superfluidity in the inner crust of

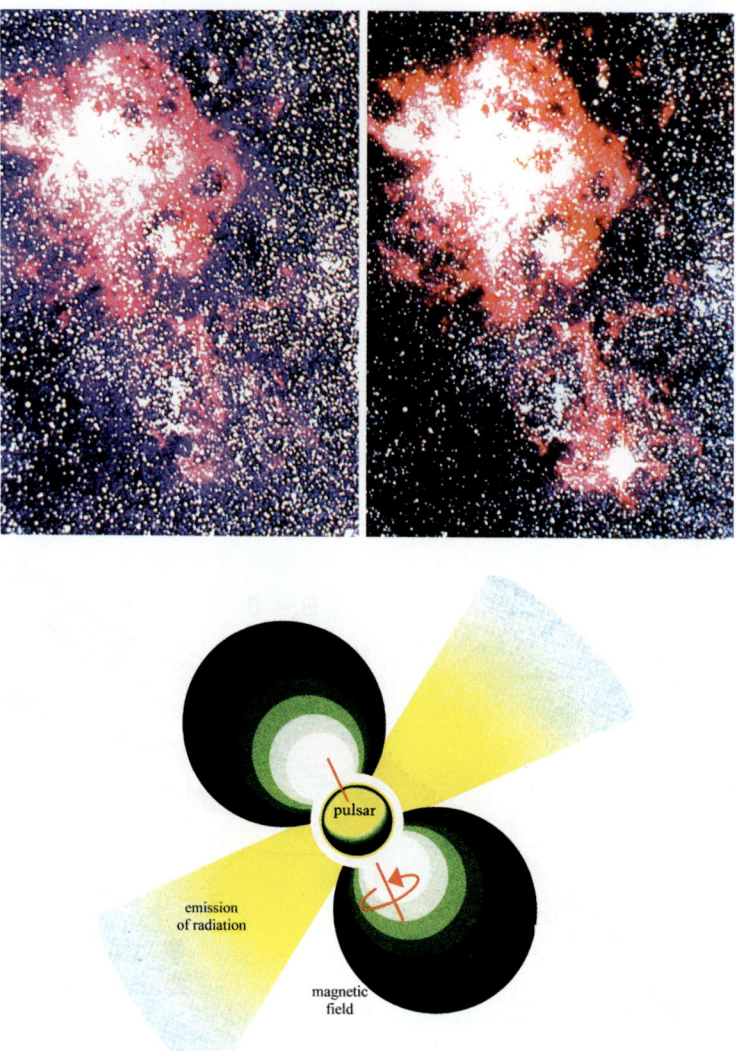

Fig. 16. On February 23, 1987, the great Magellean cloud was the scene of a supernova explosion. Supernova 1987A as it was christened was so bright that one could see it from the Southern hemisphere and during a period of about six months with the naked eye. (top) A photo of the Magellean cloud before (left) and after (right) the date indicated above. A bright spot on the lower part of the figure is clearly visible. (bottom) Neutron stars have an intense magnetic field which accelerates electrons and protons, inducing those particles to emit beams of radiation. These beams arrive to the earth with a frequency determined by the rotation of the star giving rise to a pulsating signal like light emitted by a lighthouse. Hence, the name pulsar.

the star, where nuclei forming a crystalline lattice are immersed in a sea of free, superfluid neutrons. Because a superfluid can rotate only by forming (quantized) vortices, the coexistence of superfluid neutrons and lattice nuclei in the crust leads to a special rotational dynamics. In it, vortex lines become pinned to the normal (bulk) nucleons of nuclei forming the lattice, where the pairing gap is small. Pulsar glitches are thought to be caused by the sudden release of pinned vortex lines [27] which eventually hit the outer surface of the star making it spin faster. The detailed understanding of these phenomena for the hundreds of pulsars known in the Universe, constitutes a subject lying at the forefront of astrophysical research.

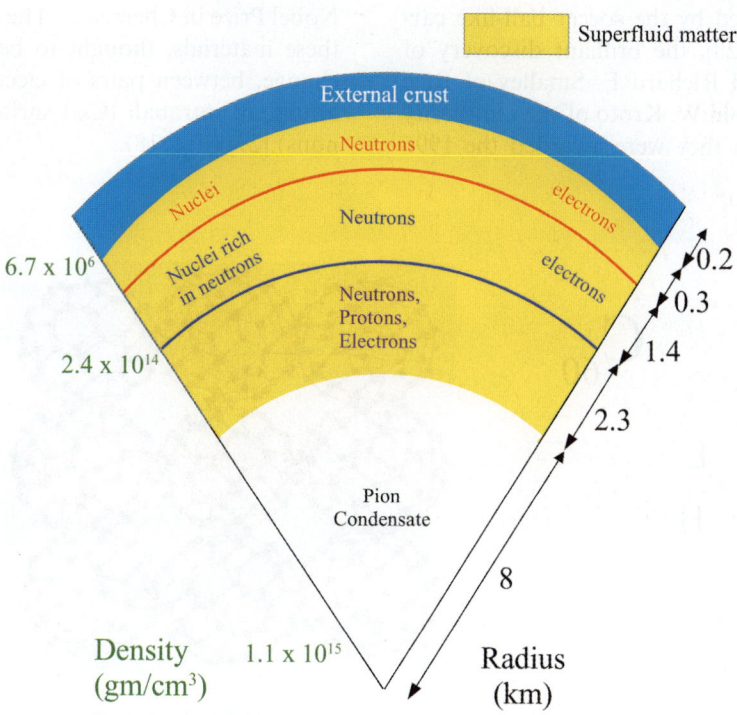

Fig. 17. Cross-section of a neutron star calculated making use of a so called stiff equation of state for neutron matter densities larger or equal than nuclear saturation density [23].

2.3.3. Femtometre materials

A Gedanken experiment designed to measure the electrical resistivity of a femtometre wire of the material filling the inner crust of a neutron star, but where neutrons are replaced by protons, will reveal a superconductor. In particular, let us think of a femtometre wire made out of $^{210}_{84}$Po$_{126}$, a nucleus with two protons moving outside the doubly closed shell system $^{208}_{82}$Pb$_{126}$ (with "magic numbers" of both protons (82) and neutrons (126), cf. Fig. 14). This nucleus displays a pairing gap (i.e. a binding energy of the Cooper pair) of the order of 1.2 MeV, arising from the nuclear pairing interactions (strong force and exchange of surface vibrational modes) strongly concentrated on the nuclear surface. If one could create a crystal where the average distance between each ^{210}Po nucleus was of the order of 15 fm, so that the distance between the surface of each nucleus was about 1 fm, one would produce a real high T_c superconductor. In fact, each nucleus will contribute a strongly correlated two proton Cooper pair to the conducting band. While the density of levels at the Fermi energy is expected to be somewhat lower than that associated with the single nucleus, one could still expect a pairing gap of the order of hundreds of keV.

2.4. Analogies: nanometre materials

While it is not possible to produce bulk samples of femtometre materials in the laboratory, one can recognize that the mechanism which is at the basis of this ideal material, namely superconductivity controlled by pairing interactions mediated by the surface of each single atomic nucleus and the density of levels of the whole crystal, is that which seems to be active in so called high T_c-superconductors made out of doped fullerenes. In these materials, the protons are replaced by electrons while the nuclei forming the Coulomb lattice of the inner neutron star crust, the building blocks of the material, are replaced by the soccer ball-like carbon molecule C_{60} [28], the brilliant discovery of Robert F. Curl and Richard E. Smalley of Rice University and Harold W. Kroto of the University of Sussex for which they were awarded the 1996 Nobel Prize in Chemistry. The Cooper pairs are, in these materials, thought to be formed by the exchange, between pairs of electrons of the doping atoms, of intraball (C_{60}) surface vibrations (phonons) [29] (Fig. 18).

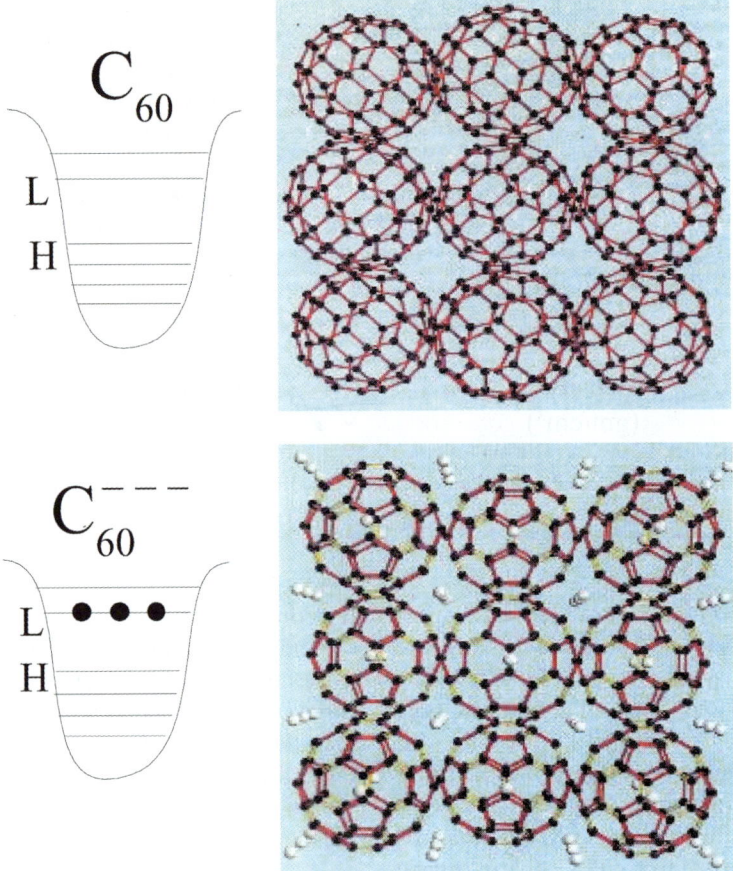

Fig. 18. (top) Model illustrating the packing of individual fullerenes clusters in the fcc lattice of solid C_{60}. (bottom) Model illustrating the packing of alkali metal ions into the tetrahedral and octahedral holes in the fcc $C_{60}K_3$ lattice (K stands for potassium but other alkali metals like Rb and Cs have been used to dope C_{60} solid leading to superconducting systems [29,75]). In the top, we schematically show the potential in which the electrons of C_{60} move leading to a "closed shell" system, that is, a system of fermions where all levels are completely filled up to the highest occupied molecular orbital (HOMO) and there is a "large" gap, of the order of a couple of electron volts between this level and the lowest unoccupied molecular orbital (LUMO). This is the reason why solid C_{60} is an insulating material. If one adds electrons to C_{60}, e.g. three electrons leading to C_{60}^{---} (bottom) they can correlate by exchanging vibrations of the molecule, and form Cooper pairs. However, because of the fact that the dimension of the Cooper pair is larger than the C_{60}^{---} system it cannot display a supercurrent. This can develop in e.g. a solid $C_{60}K_3$, where still the glue between the electrons is provided by the intramolecular vibrations, but where the size of the sample is much larger than the dimensions of the Cooper pair. To be noted that the glue between electrons in low-temperature (metallic) superconductors is provided by the intermolecular vibrations of the atoms of the crystal lattice. Due to the symmetries of the molecule C_{60}, its LUMO has degeneracy six, that is, it can accommodate six electrons, three with spin up and three with spin down. This is the reason why three metallic atoms (each providing on electron) are used to dope solid C_{60}, as the largest pairing correlations are obtained for a half filled (narrow) band.

These vibrations are much more effective in providing a glue between pairs of electrons than those associated with graphite. In fact, the critical temperature T_c, below which the system is superconductor is $T_c \equiv 5$ K for graphite compounds and $T_c \equiv 112$ K for hole doped C_{60} fullerite [30]. Because C_{60} can be viewed as made out of a wrapped carbon lattice plane by introducing defects in the honeycomb lattice so as to create 12 pentagons (C_{60} is made out of these 12 pentagons and of 20 hexagons), curvature effects and electron spill out seems to be at the basis of the increased electron–phonon coupling observed in going from graphite to C_{60}-based materials [31]. It is thus expected that materials made out of fullerenes with higher curvature and electron spill out than C_{60}, like C_{36}, C_{28} and C_{20} will display higher values of T_c [32,33]. This can be accomplished by reducing the number of hexagonal faces of the cage as one cannot reduce the number of pentagons. This is because according to Euler's theorem, 12 is the minimum number of pentagons a polyhedron can have. Therefore C_{20}, which is made out of exactly 12 pentagons, is the smallest of the family of fullerenes, displaying the largest surface to volume ratio. Consequently also the largest curvature and, potentially, the largest electron spill out. The associated phonons are thus likely to provide the strongest glue between pairs of electrons in fullerene based materials. While C_{20} fullerenes have recently been made using a gas-phase reaction method [34], no solid using this molecule as building block has yet been produced. Assuming that a C_{20} based solid can be made (which although covalent preserves the identity of the individual molecules) displaying a density of levels at the Fermi energy similar to that of the alkali doped C_{60} fullerite mentioned in connection with Fig. 18, one could expect a critical temperature T_c about eight times higher than that associated with A_3C_{60} [33]. In other words, one could expect essentially a room temperature superconductor.

To asses the validity of such possibilities one would need to understand in detail at least three main issues concerning fullerene based materials: (a) the electron–phonon coupling in a strong coupled situation (in particular for small fullerenes); (b) the screening of the Coulomb field which essentially amounts to understanding in detail the electron–plasmon coupling; and (c) the modifications of these properties in the solid. It is likely that such research may contribute not only to shed light on fullerene based materials, but also on high T_c superconductors at large [35].

Curvature, and thus surface driven superconductivity may also be encountered in nanotubes. These molecules are obtained by rolling a graphene sheet (a single layer from a three dimensional graphite crystal) into a cylinder. By capping each end of the cylinder with half a fullerene molecule, a "fullerene derived tubule" is formed [36]. Conducting nanotubes have been observed to exhibit a zero resistance and a high critical temperature [37]. Recently, the smallest possible carbon nanotube, with a diameter corresponding to a C_{20} dodecahedron [38,39] has been synthesized. Due to the very large curvature of the nanotube walls such an atomic wire may display an array of very interesting properties.

As it emerges from the discussion above, and from the very recent date of the (few) references existing on the subject, the study of the coupling of electrons to the surface of fullerenes and of nanotubes is a subject lying at the forefront of research in solid state physics and in the construction of nanostructured materials.

3. Absorption and emission of photons: the colour of finite systems

It is known that the properties of finite many-particle systems do not depend so much on the nature of the constituents particles or the forces acting among them, but on the fact that these are confined in their motion by a surface and that there are many [10]. Thus, the analogies which can be carried out between the atomic nuclei and atomic aggregates like fullerenes and metal clusters [40,41], can be used profitably in the selection of these building blocks needed to produce materials displaying customer designed characteristics. In particular colours and tonalities, which reflect the photoabsorption distribution of frequencies of finite systems [42]. From these analogies it emerges that the strategies devised during the last years in the design of nanophase materials [43], depend to a large extent on the surface of these clusters.

3.1. Metal clusters

A great deal of our basic insight of the world around us comes from the colours that we perceive. It is no accident that colours are used to describe emotions, and that they are an essential tool for artists to convey their feelings. An excellent example is provided by the stained glass used by Medieval architects in their cathedral windows. The architecture seems to defy gravity, whilst the warm colours of their *vitraux* produce a profoundly spiritual atmosphere when touched by the rays of the sun. Small silver and copper particles dissolved in the glass leads to the blues and the distinct ruby colours are produced by the presence of gold particles (Fig. 19). When trying to understand the "colour" of these particles, it is not enough to make reference to the optical response of either the individual atoms or bulk metal crystals. This is because we still do not know how the properties of a solid gradually evolve as atoms are brought together to form increasingly larger units. We understand reasonably well the end points of such evolution, but have still a poor knowledge of the intermediate situations. A sodium atom has a very simple absorption spectrum. It consists of essentially one line in the visible, the well-known yellow light. In quantum physics this phenomenon is well described as a one-electron transition from a quantal state, known as 3s, to an excited state, denoted by 3p. A sodium crystal, on the other hand, has a completely different spectrum. The absorption is strong in the infrared, goes through a minimum in the visible, and then rises again in the ultraviolet. The reason for this is that very low-energy photons can excite electrons from the continuum of states just below the energy of the last-occupied state (Fermi energy) to states just above the Fermi energy. The strong ultraviolet absorption is caused by interband transitions.

Suppose we chip off a small corner of a crystal, producing a microcrystal of only eight atoms. Again, the absorption spectrum changes completely. A relatively broad absorption maximum (resonance) appears in the visible (Fig. 20). This absorption is due to a collective excitation of outer electrons called a plasmon excitation (cf. Fig. 5). It can be viewed as a collective sloshing motion of the electrons from one side of the microcrystal to the other side of it, in a similar way to tea oscillating inside a cup which we carry in our hands. It is clear that such a motion has a strong dipole moment, behaving as an antenna and thus giving rise to a strong absorption band in the visible. The absorption of light by a sodium atom is well described within the picture of independent particle motion. This picture, however, cannot describe even qualitatively the resonance phenomenon (plasmon excitation) that dominates the optical response of a eight-atom cluster. Between one atom and eight we must completely change our way of looking at the optical response of the system. As we shall see this task is eased a good deal by making use of well-established results in nuclear physics, in particular those associated with the interplay between the giant dipole resonance (plasmon in the cluster case) and the surface of the system.

3.2. Damping of collective motion: the case of the atomic nucleus

As already mentioned, the surface of the nucleus, similar to that of a liquid drop, can be set into vibration. This implies, according to quantum mechanics, that the nuclear surface will fluctuate also when the nucleus is in its ground state due to the zero-point motion associated with the vibrations. The dipole vibration can then be viewed as taking place in an ensemble of nuclei displaying different shapes. The absorption of gamma rays will thus have a frequency distribution reflecting the differences in radii of the shapes measured along the three axis in a coordinate system fixed to the nucleus (Fig. 21). Each of the dipole vibrations along these axis carries one-third of the strength of the dipole mode (so called energy weighted sum rule (EWSR)). Because of symmetry relations, the most important of these fluctuations for the damping process are of quadrupole type, that is shapes that look like a spheroid, a cigar or a pancake being two particular realizations of it. A simple consequence of this picture is the fact that

Fig. 19. (a,b) Stained glass from the Middle Ages. (a) Adam, from the great west window of Canterbury Cathedral, England (*c.* 1180). The blue of the background arises through the presence of small silver and copper particles. (b) Three warriors (before 1166), from the church of St Patrokli in Soest, Germany. (c) An original Bohemian glass displaying its warm ruby colour arising from the presence of gold particles. (d) A glass produced *c.* 1700 in Berlin, where a glass-blower of this city reinvented the recipe of dissolving gold particles in glass.

in the case of nuclei displaying static prolate deformations one expects two peaks, one at lower energy associated with oscillations along the symmetry axis, and another at higher energy

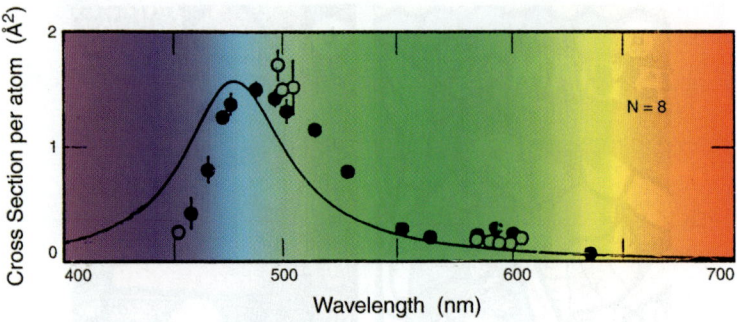

Fig. 20. Photoabsorption strength function associated with a cluster containing eight atoms of Na, as a function of the wavelength [44]. The frequency of the vibration of electrons against the ions (cf. Fig. 5) falls into the visible region of the spectrum, as indicated by the colours. The continuous curve corresponds to a theoretical calculation [45].

associated with vibrations in a plane perpendicular to the symmetry axis. Along the symmetry axis there is only one possible dipole mode, while there are two directions perpendicular to each other for dipolar vibrations in the plane perpendicular to the symmetry axis. Consequently, the lower peak carries one-third of the total intensity, while the higher peak carries two-thirds. Clear evidence for this effect has been observed (Fig. 22).

Nuclei cannot only become deformed, the ratio between the smallest and the largest radius being 1:1.2, they can also become superdeformed, when this ratio is 1:2. These are the largest deformations a nucleus can sustain without fissioning (Fig. 8), and it was a *tour de force* to detect the quadrupole transitions de-exciting systems rotating with such a shape [46]. With a new generation of γ-ray detectors, like EUROBALL [3], the brainchild of six European countries (Italy, France, Denmark, Germany, Sweden, England), which are able to cover, at a price tag of approximately 30 million dollars, a conspicuous fraction of the full solid angle, it has been possible to make even a more remarkable *tour de force*, namely that of measuring the γ-decay of a giant dipole resonance in coincidence with the γ-decay of the members of the superdeformed rotational band (Fig. 23). Thereby allowing the determination of the effect superdeformation of the nuclear surface has on the dipole vibration of the system. The results which have only recently became available [47], but which were predicted more than 15 years ago [48], are spectacular, implying a splitting between the energy associated with vibrations along the symmetry axis and in the plane perpendicular to it of about of 10 MeV as compared to a centroid energy of the order of 15 MeV. This result demonstrates the fact that large deformations make the atomic nucleus extremely polarizable [49].

3.2.1. Damping of plasmons: an analogy

Turning now to metal clusters, the calculated frequency of the photoabsorption peak (plasmon) turns out to be proportional to the frequency of the plasmons in bulk matter, a result that agrees well with the experimental findings. Consequently, it is largely independent of the dimensions of the cluster. This result, which can be traced back to the long range of the Coulomb force, is very different from that obtained in the nuclear case, where the centroid of the resonance was found to be inversely proportional to the dimension of the system. In spite of this difference, the method used in the nuclear case for calculating the spreading width of the vibrations can still be applied here [45]. One can make use of the fact that the back and forth sloshing of the electrons in a quadrupole-deformed cluster is associated with electric fields, and thus depolarizability factors, which differ greatly depending on whether the dipole vibration takes place along the largest or the smallest radius (Fig. 21, bottom).

A remarkable example of the central role played by the surface of the cluster in the damping of the plasmon oscillation is provided by the optical response of deformed clusters. From microscopic

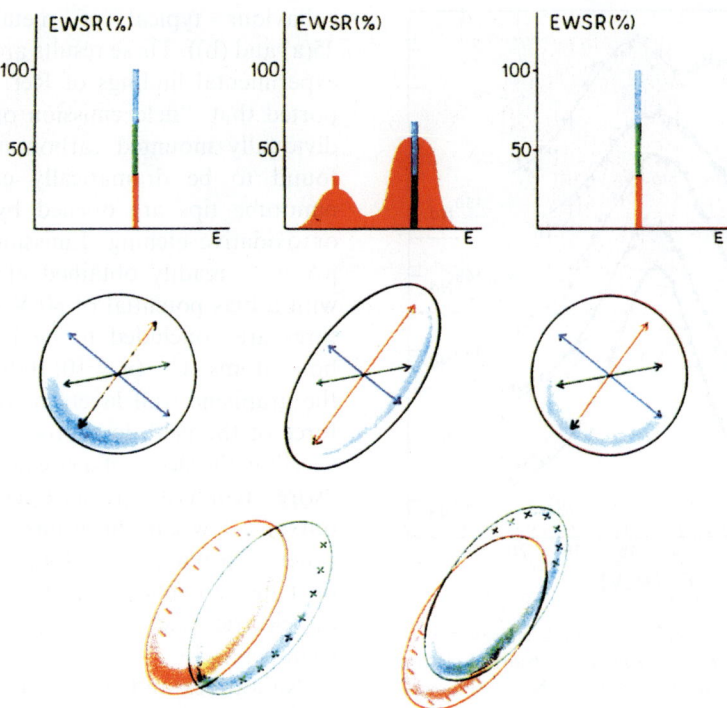

Fig. 21. Schematic representation of the mechanism that is at the basis for the damping of the giant dipole resonance in both atomic nuclei and metal clusters [10,49]. A finite system can vibrate as an antenna along three directions in space perpendicular to each other. The frequency of these vibrations in a nucleus are identical for a spherical configuration, in keeping with the fact that the frequency of the giant dipole resonance is inversely proportional to the radius of the system. For a quadrupole deformed (axially symmetric) nucleus, the single line splits into two, as there are two radii defining the system, one along the symmetry axis and one perpendicular to it. These situations are illustrated (top) where the EWSR of each of the contributions to the dipole vibration is shown as a function of excitation energy. Immediately below a schematic representation of the nuclear shapes and associated dipole vibrations is given. Because the nuclear shapes fluctuate as a function of time, the system will cover in a continuous fashion other deformations besides those explicitly shown, leading not only to a splitting of the strength, but also to a resonant peak with a distinct line shape (see middle top figure). (bottom) The dipole vibrations of electrons against the positive ions in a deformed metal cluster are shown. The positive and negative signs represent the excess of ions and of electrons respectively associated with the dynamical situation of departures from equilibrium during the vibration. The associated dipole fields are very different according to whether the vibration takes place along the symmetry axis or perpendicular to it.

considerations associated with the orbitals occupied by the electrons in the cluster ground state, sodium clusters containing 12 atoms are expected to display triaxial shapes. Consequently, one would expect a three peak structure for the associated surface plasmon. Experiments [44] have confirmed these expectations (Fig. 24). Microscopic calculations can explain these results in terms of dipolar vibrations along the x-, y- and z-axis of a coordinate system fixed to the cluster [50]. Although there are a number of nuclei that are also predicted to display triaxial equilibrium deformations, no giant dipole resonance with more than two peaks has been observed in the nuclear case. This is because the relatively large fluctuations and smaller static deformations displayed by the nuclear surface as compared with the cluster surface lead, in the nuclear case, to widths for each individual lines that blur the three peak structure. Summing up, the frequencies at which both atomic nuclei as well as metallic clusters absorb electromagnetic radiation, are strongly influenced by the shape of the surface of the system and by the spill-out of fermions from it.

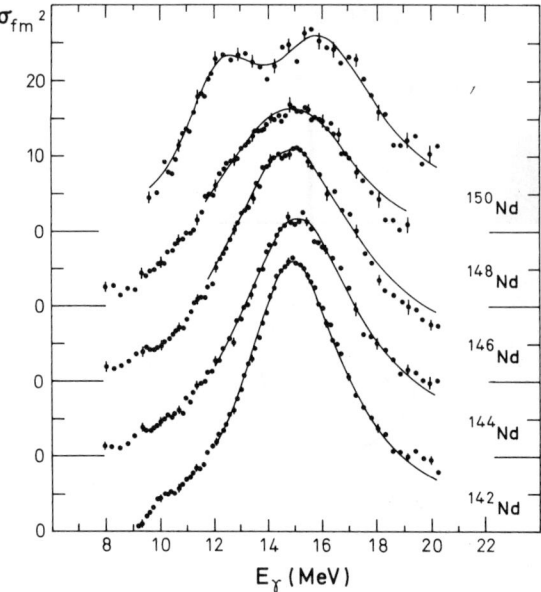

Fig. 22. Photoabsorption cross section for even isotopes of neodynium as a function of the energy of the absorbed photon (excitation energy of the system). While ^{142}Nd is spherical, ^{150}Nd is well deformed, the transition between spherical and deformed shapes been evidenced by a broadening of the resonance.

3.3. Atomic wires

Metals tend to be highly absorbing at long wavelengths (visible and infrared). Nanometre wires are thus expected to display this same property. Consequently, from the arguments developed in the previous sections, these wires are likely to be found among finite atomic aggregates displaying strong deformation and particle spill out. Among the systems satisfying these requirements, single-wall nanotubes and linear carbon chains seem to be particularly promising. In fact, theoretical [40,51,52] and experimental [53] studies have shown that they behave as metallic needles when subject to an electromagnetic field (cf. Fig. 25(c)). Consistent with this result, it has also been calculated [51] that, under standard bias conditions, i.e. subject to a moderate potential difference, linear chains made out of few carbon atoms are prolific emitters of electrons, the associated currents versus voltage curves displaying a behaviour typical of metallic systems (Fig. 25(a) and (b)). These results are consistent with the experimental findings of Ref. [54] where it is reported that: "field emission of electrons from individually mounted carbon nanotubes has been found to be dramatically enhanced when the nanotube tips are opened by laser evaporation or oxidative etching. Emission currents of 0.1–1 μA were readily obtained at room temperature, with a bias potential of 80 V. The emitting structures are concluded to be linear chains of carbon atoms C_n ($n = 10$–100), pulled out from the graphene wall layers of the nanotube by the force of the field, in a process that resembles unravelling the sleeve of a sweater". In other words, a "wire" (connected to an electron reservoir) made out of a few carbon atoms and of length of the order of a nanometre (Fig. 26), i.e. one billionth of a metre emits, under the influence of a modest electric field, a current as intense as a microampere.

Nanotubes and linear carbon chains are thus likely to constitute, among other things, the ultimate atomic-scale quantum wires and electron guns to make connections in nano-circuits and to produce extremely flat television screens [55] (Fig. 27).

It is interesting to note that while a chain of eight atoms is expected [51] to be able to emit, connected to an electron reservoir, currents of the order of microamperes, a linear chain of seven atoms is predicted to emit, under the same bias conditions, currents which are two orders of magnitude weaker. Making use of an analogy, it is as if by making a piece of wire just slightly shorter, one changed its electric properties completely. Within this contest it is worth remembering that Nobel Laureate Richard P. Feynman addressing the Annual Meeting of the American Physical Society on 29 December 1959 said: "...When we get to the very, very small world—say circuits of seven atoms—we have a lot of new things that would happen that represent completely new opportunities for design. Atoms on a small scale behave like nothing else on a large scale, for they satisfy the laws of quantum mechanics" [56]. In fact, it is precisely quantum mechanics which rules that C_7 is a much more stable system than C_8, and con-

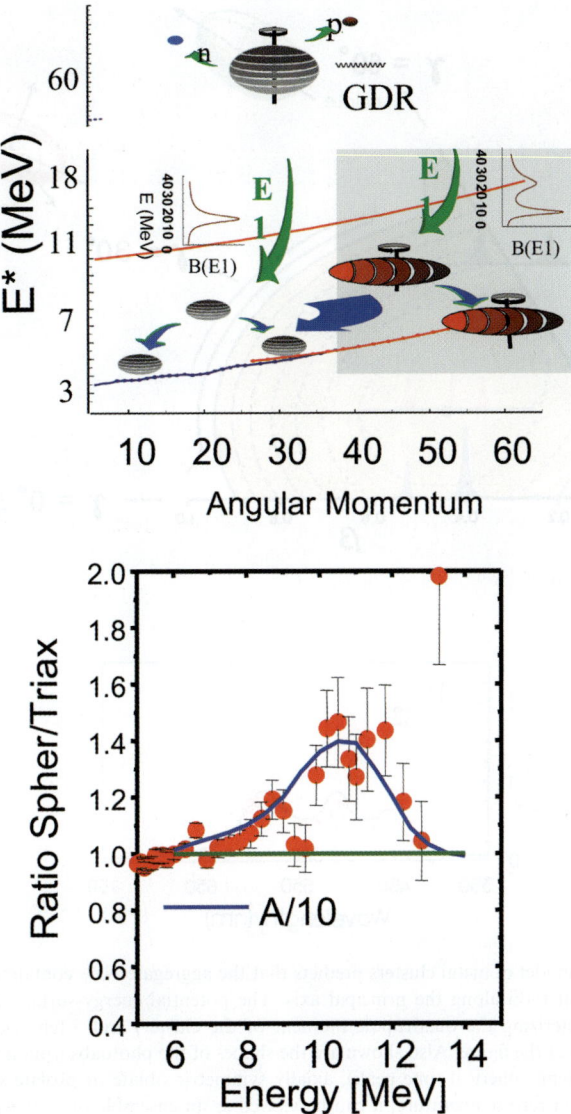

Fig. 23. Schematic representation of the formation and γ-decay of a superdeformed compound nucleus (upper part). In the bottom part we show the low-energy strength function associated with the decay of the giant dipole resonance of the nucleus ^{143}Eu in its superdeformed configuration (the ratio between the largest and the smallest radii being 2:1). The analysis of the data (blue curve going through the experimental points (red dots with error bars) above the background given by the green horizontal line) was carried out making use of a density of levels corresponding to a (density parameter) value of $A/10$, where A is the mass number of the nucleus. The measurements were carried out by recording hard γ-rays i.e. photons of energy in the range of energy from 5 to 20 MeV associated with the decay of the giant dipole resonance in coincidence with γ-rays corresponding to transitions connecting members of the superdeformed rotational band of ^{143}Eu. This was done to ensure that the dipole vibrations measured are based on the superdeformed minumum of the system.

sequently it is much more difficult to extract an electron out of C_7 than of C_8.

While C_7 may not be a too good a field emitter, it is interesting to note that the electromagnetic

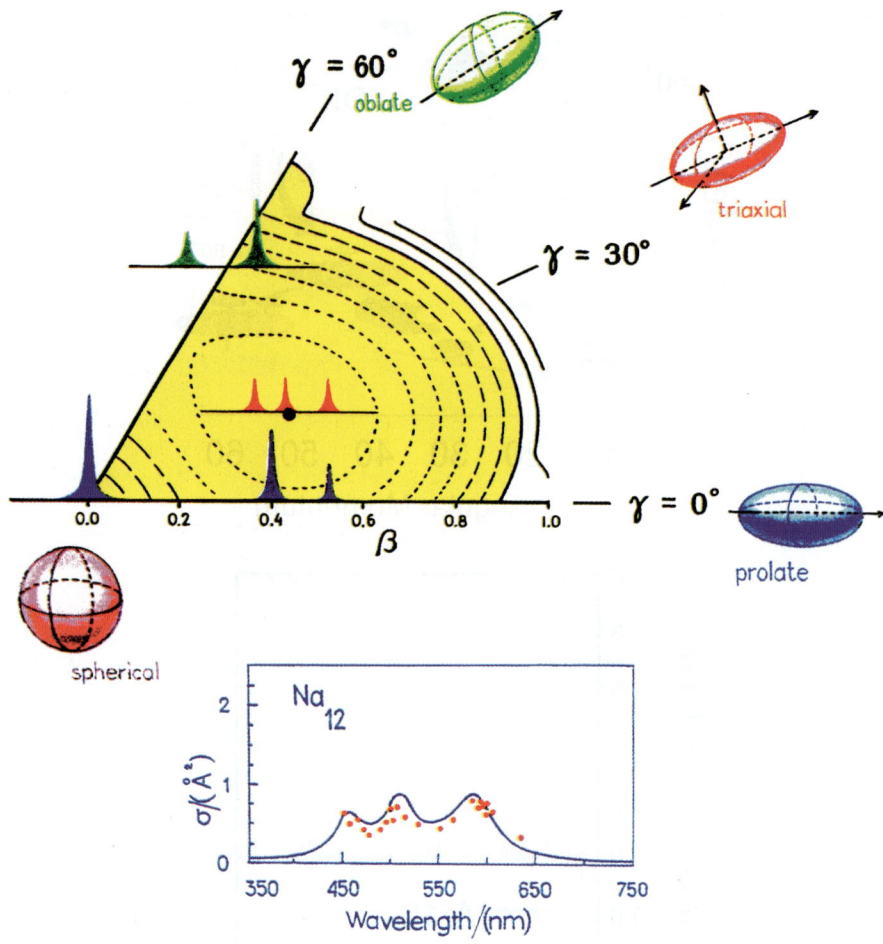

Fig. 24. The independent particle model of metal clusters predicts that the aggregate Na_{12} containing 12 atoms of sodium is strongly deformed, displaying three different radii along the principal axis. The potential energy surface of this cluster in the plane of the deformation parameters β (characterizing the quadrupole moment of the cluster) and γ (characterizing the departure from axial symmetry) is shown in the top part of the figure. Also shown are the shapes of the photoabsorption strength function (cf. also Fig. 21) for the different shapes of the system: spherical (one peak), axially symmetric oblate or prolate shapes (two peaks), triaxial (three peaks). Because the cluster is at non-zero temperature, it can be viewed as an ensemble of systems with different shapes, with probabilities fixed by the corresponding Boltzmann factors [49]. Calculations of the line-shape of the surface plasmon taking these effects into account have been carried out in Ref. [50]. They are shown (blue curve) in comparison with the data from [44] (red dots) in the lower part of the figure.

response of the anion C_7^- (among others) has important astrophysical implications, as we shall see below.

3.4. Diffuse interstellar bands

Just as dust particles in the earth atmosphere produce a red sunset, so starlight is reddened as it passes through clouds of interstellar dust. The visible–ultraviolet spectrum from distant stars is like the spectrum of the sun, our nearby star, crossed by a series of narrow, dark lines. But in the spectrum of distant stars, aside from these lines, which are associated with the absorption of light by atoms present in the outer layers of the sun, there are others that are much broader (or more

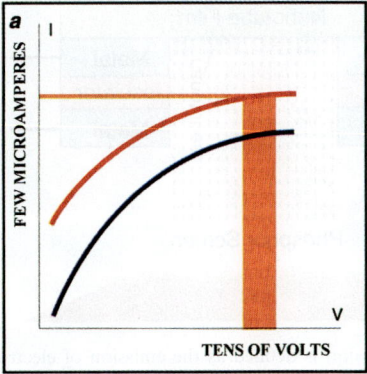 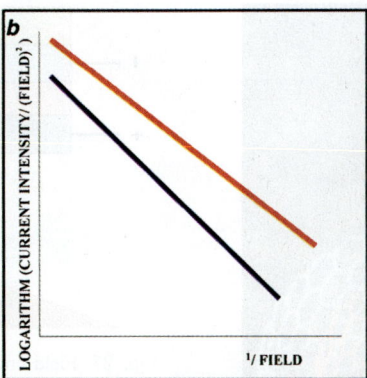 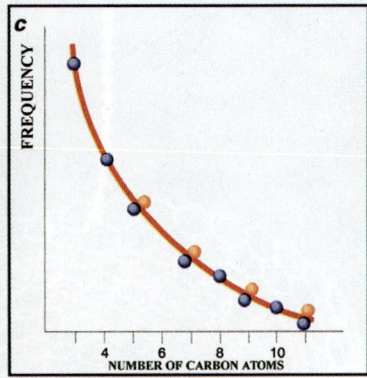

Fig. 25. (a) Applying a potential difference to linear carbon chains made out of seven (blue continuous curve) and of eight (red continuous curve) atoms leads, according to quantal calculations [51], to currents (known as field emission currents), with intensities differing by two orders of magnitudes. Chains with few and even number of carbon atoms can thus be very efficient atomic electron guns. (b) Plotting the logarithm of the current intensity I displayed on the ordinate of (a) divided by the square of the applied electric field as a function of the inverse of the field, corresponding to the abscissa of (a), a linear dependence is observed. Quantum mechanics predicts such a behaviour in the case of field emission from metallic particles. (c) Quantal calculations [51,52] predict that when a beam of photons, whose polarization plane contains the symmetry axis of the linear carbon chain C_n studied ($n = 3, 4, \ldots, 11$), is shone on the chain, the system absorbs energy at essentially a single frequency (blue dots). This frequency is quite close to that experimentally [53] observed (orange dots). The quantal results are well reproduced by a classical expression which depends on the ratio of the shortest to the longest radius of the chain, and which describes the behaviour of the macroscopic elongated metallic particles, of similar (scaled) shape as the linear carbon chain studied (red curve).

diffuse). For this reason they are referred to as bands rather than lines. The quest for the origin of these diffuse interstellar bands (DIB), dubbed by Harry Kroto "the last great problem in astronomy", is closely related to the presence of carbon in the Universe [57]. In particular, as it has been argued since 1977, of linear carbon chains [58–62]. One knows today that the answer to the origin of DIB is more articulated than that proposed in Ref. [58] (cf. e.g. Ref. [59] and Fig. 28). In any case, a number of optical gas phase electronic transitions (plasmons) of linear carbon chains anions (i.e. charged chains), recently observed in the laboratory [65], have been found to coincide with some of the DIB [63] (Fig. 28). In this connection, one can mention that first principal calculations of the spectrum of linear neutral carbon chains [64] as well as experimental data in inert gas matrices [53] (Fig. 25c), indicate also the coincidence of few of the corresponding lines with DIB. Because the spectrum of a quantal system is as revealing of its identity as fingerprints are of individuals, the findings of Ref. [63] reopens the case of linear carbon chains in connection to the question of DIB.

4. Resume and outlook

It is tempting to ask whether there is any fundamental reason or unifying concept, that makes it simple to understand and, in hindsight, "predict" the paramount role the surface of finite systems plays in determining their properties, and the properties of bulk materials they can form. The answer to this question is to be found in the principles lying at the foundations of quantum mechanics, the theoretical framework of the very small, where the symmetry concepts associated with the isotropy of space and time play a primary role [67]. The fact that space is isotropic implies that the quantal description of a system, in particular of a finite system, will not depend on its orientation. Thus, when a little nucleus, suspended in isotropic space and isolated from anything else deforms spontaneously in order to lower its energy, it has two choices for respecting the rules of the game: either to make the deformation static and then rotate collectively, averaging in this way the orientation of the system (collective rotations), or to make the deformation dynamic and vibrate

Fig. 26. Model of the tip of a multiwall nanotube (red and blue layers) showing a single C_n "atomic wire" extending out from the inner layer, held taut and straight by the electric field (after Ref. [54]). It is still an open question what the energy difference between the open-ended and the closed configurations of the multi wall nanotubes (MWN) used in Ref. [54] is under the biased applied. It has been shown that this difference decreases with increasing field [76], although it is unlikely it may be sufficient to lower the open end energy below that associated with the capped structure. How the situation is modified in MWN is an open question.

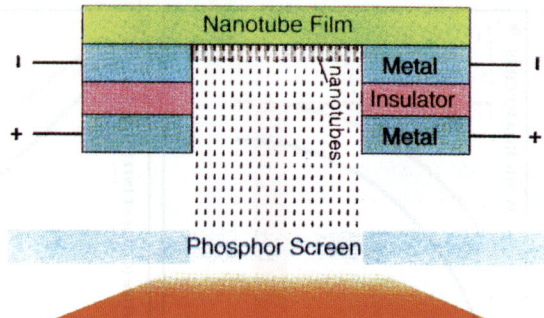

Fig. 27. Field emission is defined as the emission of electrons from the surface of a condensed phase into another phase, usually vacuum, under the action of an electrostatic field. The phenomenon consists in the tunneling of electrons through a potential barrier at the surface, in this case, of nanotubes (in the case of Fig. 25 of linear carbon chains). In the figure, a schematic diagram of the field emitting set up based on nanotubes is displayed [55]. Under a difference of potential it shoots electrons (—) at a phosphor display, triggering light emission.

(collective surface vibrations). Consequently, the phenomenon of spontaneous symmetry breaking makes the nuclear surface a source of collective motion. The motion of the nucleons will reflect, through the coupling of particles and vibrations, this privileged role of the nuclear surface. Because all the degrees of freedom determining the nuclear structure are exhausted by the degrees of freedom

Fig. 28. In 150 B.C. the Greek astronomer Hipparchus grouped the sky's stars into six brightness categories called magnitudes, first magnitudes representing the brightest stars he saw, sixth the faintest. The apparent magnitude (m) of a star—the magnitude as seen by Hipparchus and by us—depends on the stars real luminosity (i.e., on the amount of energy it radiates on watts) and on its distance. A star may seem bright to us either because it is nearby (even though faint) or because, though far away, it is very luminous. If we know the star's distance (d), we can find its luminosity from its apparent brightness. In astronomy, luminosity is expressed by a star's absolute magnitude (M), the value the apparent magnitude would have at a standard distance of 10 parsecs, that is, of 32.6 light years (the light year is the distance light travels in vacuum in a year). The two kinds of magnitude are related by the magnitude equation $M = m + 5 - 5\log d$, where d is in parsecs. To apply the method of spectroscopic distances we must correct the apparent magnitude of a star for the dimming by the interstellar dust. This means knowing the degree of extinction (A), the magnitude by which the apparent visual magnitude must be brightened (i.e., the amount by which its value must be decreased) before it is entered into the magnitude equation. One can find A because it scales with the degree of reddening. (a) The extinction function—the dependence of interstellar extinction to wavelength λ—has a rough λ^{-1} dependence in the optical and goes to zero at long wavelength (allowing correlated far-infrared and radio waves to penetrate interstellar space with penache). In the ultraviolet, interstellar space becomes remarkably opaque; the 2000 Å bumps have been found to be correlated with the laboratory spectrum of matrix-isolated nanometer-sized carbon particles (after Ref. [66]). (b) Two typical structures of carbon particles taken with high-resolution transmission electron microscope: (left) carbon particle with randomly oriented basic structural units, (right) onion-type carbon particle with several condensation seeds (cf. Ref. [56]). Theoretical calculations indicate the multishell fullerenes (carbon onions) to be the likely candidates for the 2200 Å bumps. (c) DIB observed as broad depressions in stellar spectra are seen from the ultraviolet through the near-infrared (after Ref. [66]). (d) Gas-phase electronic transitions of C_7^- (charged linear carbon chain containing seven atoms of carbonium) recorded in the laboratory [62] (upper trace). The bottom trace shows a Gaussian fit to the tabulated DIB. Those labeled with an asterisk coincide with the measured bands of C_7^- above.

of protons and neutrons, essentially all (low-energy) nuclear properties will depend on the surface of the system. This is also true for other physical systems such as metal clusters, fullerenes, nanotubes and linear carbon chains in particular, and finite many-body systems in general. Also for the materials constructed out of these systems.

It is interesting to think that the riddle connected with the surface of finite systems touches, through the phenomenon of spontaneous symmetry breaking, also on fundamental physical questions like the existence of Goldstone bosons [68–70]. That is, zero-mass particles like the photon, the carriers of the Coulomb force, and the pion, one of the carriers of the strong force. In the case of the nucleus, Goldstone bosons are intimately connected with zero-frequency vibrations and associated rotations [11,71,72]. The Higgs boson [73], the missing building block of the unified theory of elementary particles, the quest of which

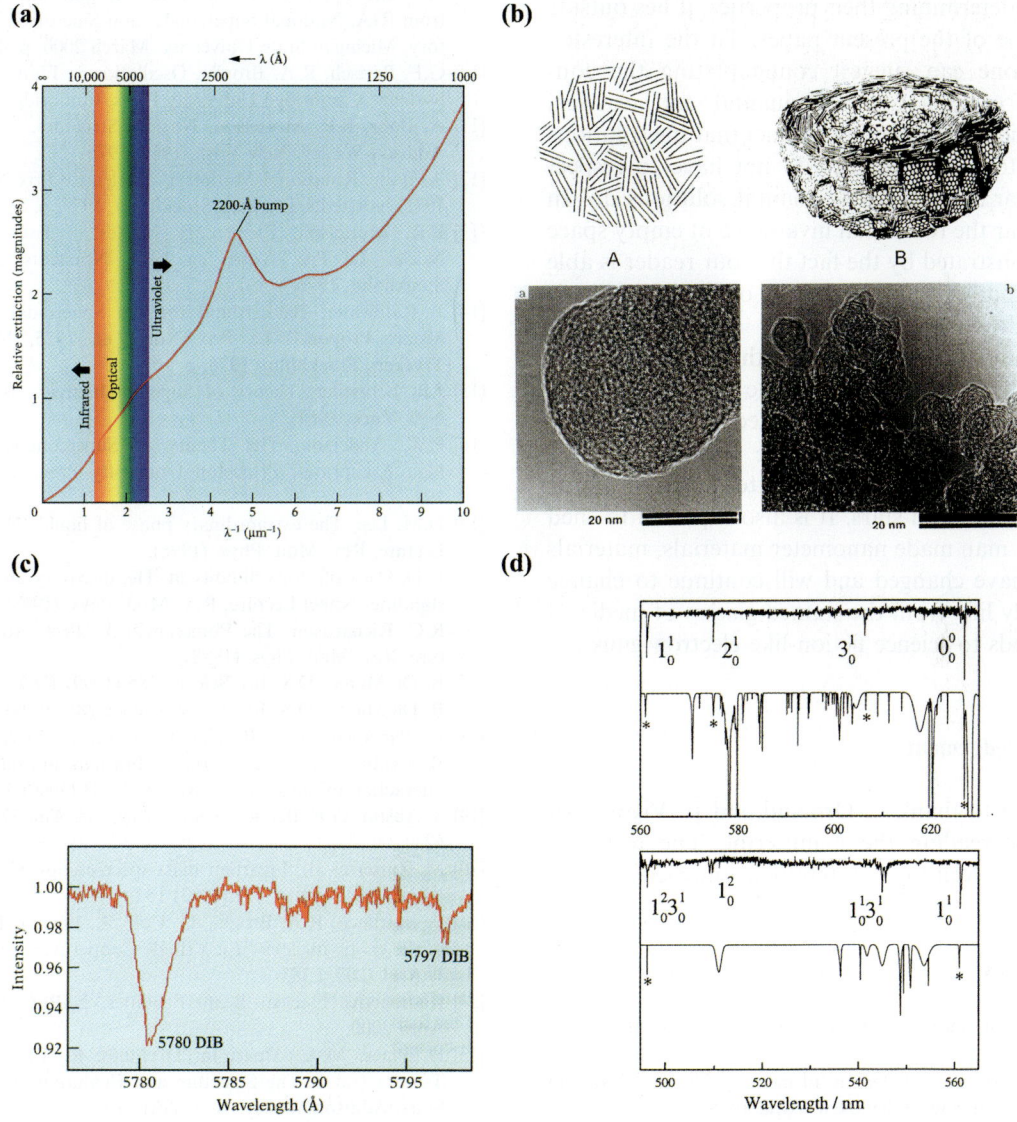

will be the main research activity of the largest accelerator ever built (Large Hadron Collider (LHC) of CERN in Switzerland), is a consequence of a "frustrated" phenomenon of spontaneous symmetry breaking as is the most mundane plasmon (electron vibrations) in condensed matter [74].

Once on the subject, one could ask the even deeper question: "why spontaneous symmetry breaking at all?". While this is the most basic and challenging question emerging in trying to understand the role played by the surface of finite systems in determining their properties, it lies outside the scope of the present paper. To the interested reader one can suggest contemplating the connection existing between quantal and classical mechanics, remembering the fact that the solutions of Newtonian mechanics do not have to respect particular symmetries as quantal solutions do. In particular the rotational invariance of empty space as demonstrated by the fact that our reader is able to find out whether the page he or she is reading is upside down or not.

A unified understanding of the surface of finite systems, of which the atomic nucleus can be viewed as a paradigm is expected to shed light, at the fundamental level, on the strategy used by nature in creating cosmic patterns, from exotic nuclei to neutron stars. It is also expected to shed light on man made nanometer materials, materials which have changed and will continue to change our daily life, from the stained glasses of medieval cathedrals to science fiction-like electron guns.

Acknowledgements

I wish to thank E. Ormand and E. Vigezzi for carefully reading the manuscript. The technical help of F. Marini is gratefully acknowledged.

References

[1] F. Dyson, Disturbing the Universe, Harper and Row, New York, 1974.
[2] M.G. Mayer, J.H. Jensen, Elementary Theory of Nuclear Shell Structure, Wiley, New York, 1955.
[3] J. Simpson, EUROBALL: present status and outlook, Heavy Ion Phys. 6 (1997) 253.
[4] S. Weinberg, The First Three Minutes, Bantam, New York, 1977.
[5] C. Mahaux, P.F. Bortignon, R.A. Broglia, C.H. Dasso, Dynamics of the shell model, Phys. Rep. 120 (1985) 1.
[6] A.L. Fetter, J.D. Walecka, Quantum Theory of Many-Particle Systems, McGraw-Hill, New York, 1971.
[7] N. Bohr, Neutron capture and nuclear constitution, Nature 137 (1936) 344.
[8] P. Donati, P. Pizzochero, P.F. Bortignon, R.A. Broglia, Temperature dependence of the nucleon effective mass and the physics of stellar collapse, Phys. Rev. Lett. 72 (1994) 2835.
[9] Scientific opportunities with fast fragmentation beams from RIA, National Superconduction Sincrotron Laboratory, Michigan State University, March 2000, p. 43.
[10] G.F. Bertsch, R.A. Broglia, Oscillation in Finite Quantal Systems, Cambridge University Press, Cambridge, 1994.
[11] A. Bohr, B.R. Mottelson, Nuclear Structure, Vols. I–II, Addison-Wesley, New York, 1969, 1975.
[12] A. Bohr, Rotational Motion in Nuclei, Le Prix Nobel en 1975, Norstedts Tryckeri, Stockholm, 1976, p. 56.
[13] B.R. Mottelson, Elementary Modes of Excitation in Nuclei, Le Prix Nobel en 1975, Norstedts Tryckeri, Stockholm, 1976, p. 80.
[14] J. Rainwater, Background for the Spheroidal Nuclear Model Proposal, Le Prix Nobel en 1975, Norstedts Tryckeri, Stockholm, 1976, p. 102.
[15] J.R. Schrieffer, Theory of Superconductivity, Benjamin, New York, 1964.
[16] P.W. Anderson, The Theory of Superconductivity in high T_c-cuprates, Princeton University Press, Princeton, 1997.
[17] D.M. Lee, The extraordinary phase of liquid ^{3}He, Nobel Lecture, Rev. Mod. Phys. (1998);
D.D. Osheroff, Superfluidity in ^{3}He: discovery and understanding, Nobel Lecture, Rev. Mod. Phys. (1998);
R.C. Richardson, The Pomeranchuck effect, Nobel Lecture, Rev. Mod. Phys. (1998);
B. De Marco, D.S. Jin, Science 285 (1999) 1703;
B. De Marco, D.S. Jin, News, Science 285 (1999) 1646.
[18] F. Barranco, R.A. Broglia, G. Gori, E. Vigezzi, P.F. Bortignon, J. Terasaki, Surface vibrations and the pairing interaction in nuclei, Phys. Rev. Lett. 83 (2000) 2147.
[19] S. Austin, G.F. Bertsch, Halo nuclei, Sci. Am. 272 (1995) 62.
[20] M. Zinser et al., Invariant mass-spectroscopy of ^{10}Li and ^{11}Li, Nucl. Phys. A 619 (1997) 157.
[21] F. Barranco, R.A. Broglia, G. Coló, E. Vigezzi, The halo of the exotic nucleus ^{11}Li: a single Cooper pair, Europhys. J. A 11 (2001) 385.
[22] Radioactive Nuclear Beam Facilities, NuPECC Report, April 2000.
[23] D. Pines, M.A. Alpar, in: D. Pines, R. Tamagaki, S. Tsuruta (Eds.), The Structure and Evolution of Neutron Stars, Addison-Wesley, New York, 1992.

[24] A. Alpar, Pulsars, glitches and superfluids, Phys. World 11 (1998) 25.

[25] P.M. Pizzochero, L. Viverit, R.A. Broglia, Vortex-nucleus and pinning forces in neutron stars, Phys. Rev. Lett. 79 (1997) 3347.

[26] A. Hewish, Pulsars and high density physics, Nobel Lecture, Rev. Mod. Phys. 47 (1975) 567.

[27] P. Anderson, M.A. Alper, D. Pines, J. Shaham, The rheology of neutron stars. Vortex-line pinning in the crust super-fluid, Philos. Mag. 45 (1982) 227.

[28] R.F. Curl, Dawn of the fullerenes: experiment and conjecture, Nobel Lecture, Rev. Mod. Phys. 69 (1997) 691;
H. Kroto, Symmetry, space, stars and C_{60}, Nobel Lecture, Rev. Mod. Phys. 69 (1997) 703;
R.E. Smalley, Discovering the fullerenes, Nobel Lecture, Rev. Mod. Phys. 69 (1997) 723.

[29] O. Gunnarsson, Superconductivity in fullerides, Rev. Mod. Phys. 69 (1997) 575.

[30] J.H. Schön, Ch. Kloc, B. Batlogg, High-temperature superconductivity in lattice-expanded C60, Nature 293 (2001) 2432.

[31] A. Devos, M. Lanoo, Electron–phonon coupling for aromatic molecular crystals: possible consequences for their superconductivity, Phys. Rev. B 58 (1998) 8236;
M. Lanoo, G.A. Baraff, M. Schlüter, D. Tomanek, Jahn–Teller effect for the negatively charged C_{60} molecule: analogy with the silicon vacancy, Phys. Rev. B 44 (1991) 12106;
K. Yabana, G.F. Bertsch, Comment on "Jahn–Teller effect for the negatively charged C_{60} molecule: Analogy with the silicon vacancy", Phys. Rev. B 46 (1992) 14263.

[32] M. Cotè, J.L. Grossman, M.L. Cohen, S. Louie, Electron–phonon interactions in solid C_{36}, Phys. Rev. Lett. 81 (1998) 679.

[33] N. Breda, R.A. Broglia, G. Coló, G. Onida, D. Provasi, E. Vigezzi, C_{28}: a possible room temperature organic superconductor, Phys. Rev. B 62 (2000) 130.

[34] H. Prinzbach, P. Landenberger, F. Wahl, J. Worth, L. Scott, M. Gelmont, D. Olevano, B. von Issendorf, Gas-phase production and photoelectron spectroscopy of the smallest fullerene, Nature 407 (2000) 60.

[35] A. Lanzara et al., Evidence for ubiquitous strong electron–phonon coupling in high temperature superconductors, Nature 412 (2001) 510.

[36] R. Saito, G. Dresselhaus, M.S. Dresselhaus, Physical Properties of Carbon Nanotubes, Imperial College Press, London, 1998.

[37] A. Yu et al., Supercurrents through single-walled carbon nanotubes, Science 284 (1999) 1508.

[38] L.-C. Qin, X. Zhao, K. Hirahara, Y. Miyamoto, Y. Ando, S. Iijima, The smallest carbon nanotube, Nature 408 (2000) 50.

[39] N. Wang, Z.K. Tang, G.D. Li, J.S. Chen, Single-Walled 4 Å carbon nanotubes arrays, Nature 408 (2000) 50.

[40] R.A. Broglia, F. Alasia, P. Arcagni, G. Coló, F. Ghielmetti, C. Mikani, H.E. Roman, Softening the long-wavelength electromagnetic response of finite quantal systems, Z. Phys. D 40 (1997) 240.

[41] W. de Heer, The physics of simple metal clusters: experimental aspects and simple models, Rev. Mod. Phys. 65 (1993) 611;
M. Brack, The physics of simple metal clusters: self-consistent jellium model and semiclassical approaches, Rev. Mod. Phys. 65 (1993) 677.

[42] G.F. Bertsch, Vibrations of the atomic nucleus, Sci. Am. 248 (1983) 40.

[43] R.W. Siegel, Creating nanophase materials, Sci. Am. 275 (1996) 42.

[44] K. Selby, M. Vollmer, J. Masui, V. Kresin, W.A. de Heer, W.D. Knight, Surface plasma resonances in free metal clusters, Phys. Rev. B 40 (1989) 5417.

[45] J.M. Pacheco, R.A. Broglia, Effect of surface fluctuations in the line shape of plasma resonances in small metal clusters, Phys. Rev. Lett. 62 (1989) 1400.

[46] P.J. Nolan, P.J. Twin, Superdeformed shapes at high angular momentum, Ann. Rev. Nucl. Part. Sci. 38 (1988) 533–562.

[47] A. Bracco, F. Camera, S. Leoni, B. Million, A. Maj, M. Kmiecik, The giant dipole resonance in superdeformed nuclei and the feeding of superdeformed bands, Nucl. Phys. A 687 (2001) 237c–244c.

[48] M. Diebel, M. Gallardo, T. Døssing, R.A. Broglia, Damping of the giant dipole resonance in hot, strongly rotating nuclei, Nucl. Phys. A 443 (1985) 415.

[49] P.F. Bortignon, A. Bracco, R.A. Broglia, Giant Resonances: Nuclear Structure at Finite Temperature, Harwood Academic Publishers, New York, 1998.

[50] M. Bernath, C. Yannouleas, R.A. Broglia, Deformation effects in the optical response of small metal clusters, Phys. Lett. A 156 (1991) 307–312.

[51] A. Lorenzoni, H.E. Roman, F. Alasia, R.A. Broglia, High-current field emission from an atomic quantum wire, Chem. Phys. Lett. 276 (1997) 237.

[52] K. Yabana, G.F. Bertsch, Optical response of small carbon clusters, Z. Phys. D 42 (1997) 219.

[53] D. Forney, P. Freivogel, M. Grutter, J.P. Maier, Electronic absorption spectra of linear carbon chains in neon matrices. IV. C_{2n+1} ($n = 2$–7), J. Chem. Phys. 104 (1996) 4954.

[54] A.G. Rinzler, J.H. Hafner, P. Nicolaev, L. Lou, S.G. Kim, D. Tomanek, P. Nordlander, D.T. Colbert, R.E. Smalley, Unraveling nanotubes: field emission from an atomic wire, Science 269 (1995) 1550.

[55] W. de Heer, A. Chatelain, D. Ugarte, A carbon nanotube field-emission electron source, Science 270 (1996) 1174.

[56] R.P. Feynman, There is a lot of space at the bottom, The Caltech Alumni Magazine (February 22, 1960); reprinted IEEE J. MEMS 1 (1992) 60.

[57] Th. Henning, F. Salama, Carbon in the Universe, Science 282 (1998) 2204.

[58] A.E. Douglas, Origin of diffuse interstellar lines, Nature 269 (1977) 130.

[59] P. Thaddeus, C.A. Gottlied, R. Mollaaghababa, J.M. Urtileh, Free carbenes in the instellar gas, J. Chem. Soc. Faraday Trans. 89 (1993) 2125.

[60] J. Fulara, D. Lessen, P. Freivogel, J.P. Maier, Laboratory evidence for highly unsaturated hydrocarbons as carriers of some of the diffuse interstellar bands, Nature 366 (1993) 439.

[61] D. Forney, M. Grutter, P. Freivogel, J.P. Maier, Electronic absorption spectra of carbon chain anions C_{2n+1} ($n = 2$–5) in neon matrices, J. Chem. Phys. A 101 (1997) 5292.

[62] P. Freivogel, M. Grutter, D. Formey, J.P. Maier, Electronic absorption spectra of carbon chain anions C_{2n}^- ($n = 4$–7) in neon matrices, J. Chem. Phys. 107 (1997) 22; P. Freivogel, M. Grutter, D. Forney, J.P. Maier, Electronic absorption spectra of C_4^- and C_6^- chains in neon matrices, J. Chem. Phys. 107 (1997) 4468.

[63] M. Tultj, D.A. Kirkwood, M. Pachkov, J.P. Maier, Gas-phase electronic transitions of carbon chain anions coinciding with diffuse interstellar bands, Astrophys. J. 506 (1998) L69.

[64] M. Bianchetti, P.F. Buonsante, F. Ginelli, H.E. Roman, R.A. Broglia, F. Alasia, Ab-initio study of the electromagnetic response and polarizability properties of carbon chains.

[65] L. Henrard, A.A. Lucas, Ph. Lambin, On the 2175 Å absorption band of hollow, onion-like carbon particles, Astrophys. J. 406 (1993) 92–96.

[66] J.B. Kaler, Cosmic Clouds, Scientific American Library, New York, 1997.

[67] A. Bohr, O. Ulfbeck, Primary manifestation of symmetry. Origin of quantal indeterminancy, Rev. Mod. Phys. 67 (1995) 1.

[68] G. Goldstone, Field theories with "superconductor" solutions, Nuovo Cimento 19 (1961) 154.

[69] P. Anderson, A Career in Theoretical Physics, World Scientific, Singapore, 1994.

[70] Y. Nambu, in: T. Eguchi, K. Nishijima (Eds.), Broken Symmetry, World Scientific, Singapore, 1995.

[71] P. Anderson, More is different, Science 177 (1972) 393.

[72] R.A. Broglia, J. Terasaki, N. Giovanardi, The Anderson–Goldstone–Nambu mode in finite and in infinite systems, Phys. Rep. 355 (2000) 1–18.

[73] M.J.G. Veltman, Scient. Am., The Higgs boson, November 1986, p. 88;
CERN collider homes in on Higgs boson, News, Nature 400 (1999) 601.

[74] P. Anderson, Use of solid state analogies in elementary particle physics, in: R. Arnowitt, P. Nath (Eds.), Gauge Theories and Modern Field Theory, Procs. of a Conference held at Northeastern University, Boston, 26 and 27 September 1975, MIT Press, Cambridge, MA, 1976, p. 311.

[75] C.M. Lieber, Z. Zhang, Physical properties of metal-doped fullerenes superconductors, Solid State Phys. 48 (1994) 349.

[76] L. Lou, P. Nordlander, R.E. Smalley, Fullerene nanotubes in electric field, Phys. Rev. B 52 (1995) 1429–1432.

Cosmic dust and our origins

J. Mayo Greenberg *

Raymond and Beverly Sackler Laboratory for Astrophysics, Leiden University, Leiden, The Netherlands

Received 20 November 2000; accepted for publication 28 June 2001

Abstract

The small solid particles in the space between the stars provide the surfaces for the production of many simple and complex molecules. Processes involving the effects of ultraviolet irradiation of the thin (hundredth micron) mantles are shown to produce a wide range of molecules and ions also seen in comets. Some of the more complex ones inferred from laboratory experiments are expected to play an important role in the origin of life. An outline of the chemical evolution of interstellar dust as observed and as studied in the laboratory is presented. Observations of comets are shown to provide substantial evidence for their being fluffy aggregates of interstellar dust as it was in the protosolar nebula, i.e. the interstellar cloud which collapsed to form the solar system. The theory that comets may have brought the progenitors of life to the earth is summarized. © 2001 Published by Elsevier Science B.V.

Keywords: Molecule–solid reactions; Photochemistry; Semi-empirical models and model calculations; Desorption induced by photon stimulation; Surface diffusion; Surface chemical reaction; Chemical vapor deposition; Porous solids

1. Introduction

Pervading all the space between the stars are clouds of gas and dust which often appear as dark nebulae such as that which appears in the shape of a horse's head in the region of the constellation Orion (see Fig. 1). Over 100 organic and inorganic molecules have been detected in the gas but it is the dust that contains the largest and most complex organic molecules; these dust grains are very, very small. Even the large ones are only tenths of microns in size and there are myriads of smaller ones which may be more like large molecules, such as polycyclic aromatic hydrocarbons (PAHs) and perhaps fullerenes. The role of interstellar dust in the cosmic scheme of things is fundamental to comets and other primitive solar system objects and is believed by many to provide the key to life's origins. The 'large' tenth micron sized particles are generally very cold with temperatures approaching as low as 5–6 K (−268 °C). Interactions on their surfaces with atoms, molecules, energetic photons and cosmic rays provide a major source of the chemical evolution in space. This is as much as we know today. But the subject of dark nebulae and what makes them dark has a patchy history. The earliest relevant ideas go back 100 years. Looking at the sky in the direction of the constellation Sagittarius, it is clear that there are tremendously dark lanes especially in the region toward the center of our Milky Way. This observation inspired Herschel in 1884 to say that there was undoubtedly a hole in the sky. Barnard started to

---

* Deceased.

Fig. 1. The Horsehead nebula, so called because in projection it looks like a horse's head, is a dark cloud of dust and gas in the Orion nebula in the region where much star formation is currently taking place. Its darkness is simply explained by the dust blocking of the starlight behind it.

take pictures and reported vast and wonderful cloud forms with their remarkable structure, lanes, holes and black gaps some of which are now called Bok globules. Clerke [1], in an astrophysics text she authored around the turn of the century, stated. "The fact is a general one, that in all the forest of the universe there are glades and clearings. How they come to be thus diversified we cannot pretend to say; but we can see that the peculiarity is structural that it is an outcome of the fundamental laws governing the distribution of cosmic matter. Hence the futility of trying to explain its origin, as a consequence, for instance, of the stoppage of light by the interposition of obscure bodies, or aggregations of bodies, invisibly thronging space". It was not until the work of Trumpler in 1930 [2] that the first evidence for interstellar reddening was found; which would imply that many stars were cooler than predicted. This reddening has nothing to do with the Doppler shift of receding objects. His observations indicated reddening even where he saw no clouds. In 1948 Whitford [3] published measurements of star colors versus spectral types over a wavelength range from about 350 nm (ultraviolet) to the near infrared. Spectral types characterize the size and temperature of stars. The relation was not the expected straight line, but rather, it showed curvature at the near ultraviolet and infrared regions. This was later updated in 1958 [4]. Things were beginning to make some physical sense from the point of view of small particle scattering. In 1935 Lindblad [5] published an article in Nature indicating that interstellar elemental abundances made it seem reasonable to grow particles in space. Eddington had long before hypothesized that it was so cold in space that anything that hits a small particle will stick. But the earliest attempts to explain interstellar extinction were not in terms of something that could grow in space but rather in terms of meteors [6,7]. Interstellar extinction is the apparent reduction of starlight.

It was in the 1940s that van de Hulst [8] broke with tradition and published the results of making particles out of atoms that were known to exist in space: H, O, C, and N. He assumed these atoms combined on a surface of some kind to form frozen saturated molecules. The surface was left as 'to be determined' (TBD). This is what later became known as the 'dirty ice' model and was the first application of surface physics to interstellar dust. The dirty ice model of dust was a logical follow up to the then existing information about the interstellar medium and contained the major idea of surface chemistry leading to the ices H_2O, CH_4 and NH_3. But it was not until the advent of infrared astronomical techniques, making it possible to observe silicate particles emitting at their characteristic 10 μm wavelength in the atmospheres of cool stars, that we had the cores (surfaces) on which the matter could form. TBD was now determined. It is interesting to note that their presence was predicted on theoretical grounds by Kamijo in 1963 [9]. As van de Hulst said, he chose to ignore the nucleation problem and just go ahead (where no one had gone before) with the assumption that 'something' would provide the seeds for the mantle to grow on. What a great and creative guess! So it was that by 1945 we had many of the theoretical basics to help us understand the sources of interstellar dust 'ices' but it was not until about 1970 that the silicates were ascertained. However, having a realistic dust model, van de Hulst developed the scattering tools to provide a good idea of dust properties.

After the extinction curve—how the light blockage increases from the red to the blue—and the inferred particle size had been well established, two investigators, Hall [10] and Hiltner [11], inspired by a prediction of Chandrasekhar on intrinsic stellar polarization, discovered instead, the general interstellar linear polarization. It turns out that the blockage of starlight by the dust produces a partial linear polarization of the starlight as if the clouds were polarizers. The implication of the linear polarization was that the extinction was caused by nonspherical particles aligned by magnetic fields.

The past 40 years have seen a revolution in the study of interstellar dust. This has been a threefold process. First of all, the observational access to the ultraviolet and the infrared clearly drove home the fact that there had to be a very wide range of particle sizes and types to account for the blocking of the starlight. Secondly, the infrared provided a probe of some of the chemical constituents of the dust. Thirdly, laboratory techniques were applied to the properties and evolution of possible grain materials. It was the advent of infrared techniques that made it possible to demonstrate conclusively that something like rocks (but very small, of course) constituted a large fraction of the interstellar dust. However the initial attempt to find the 3.1 μm feature of H_2O predicted by the dirty ice model was unsuccessful [12]. This was, at first, a total surprise to those of us who had accepted the dirty ice model. However, it supplied the incentive to perform the early experiments on the ultraviolet photoprocessing of low temperature mixtures of volatile molecules which would simulate the 'original' dirty ice grains [13]. By doing this we hoped to understand how and why the predicted H_2O was not clearly present. These experiments gave rise to predictions of a new component of interstellar dust in the form of complex organic molecules, as well as mantles of all sorts of ices on the silicates. The grain surfaces were becoming much more interesting.

Present studies of interstellar dust lead us in many directions; from chemical evolution of the space between the stars, to comets, to solar systems. I will try to present an overview of our current knowledge of the dust which brings many related astrophysical problems to the fore. Where and how small particles form and grow and evolve in the space between the stars, involve interactions between the solid dust and the gas atoms, molecules, the ultraviolet photons, and cosmic rays which drive the processes. The questions posed by our present information and the suggested needs for the future are discussed in the last section. The space age has ushered in some of the most dramatic developments and will ultimately give us the data we need to understand how solar systems are born out of collapsing clouds; how comets are made and whether comets are the carriers of the seeds of life's origins. The 'comet revolution' occurred when the Russian space probes VEGA 1/2 and the European space probe GIOTTO were sent out to intercept comet Halley and returned their remarkable results to earth [14]. Until 1986 no one had ever seen what a comet looked like close up and for the first time we were able to actually photograph the nucleus of a comet, which is always obscured by the gas and dust which it releases when it is heated by sunlight and produces the spectacular comet tail. Both the VEGA and the GIOTTO satellites passed between the comet and the sun so that the dust scattered light was minimized. Among other experiments on these satellites was a time-of-flight mass spectrometer which analyzed the dust which impacted the instrument at 70 km s^{-1}. From this it was discovered that a major fraction of the comet dust consisted of complex organic molecules as predicted from the interstellar dust—not just the minerals which many had anticipated. But it all starts with the interstellar dust.

2. The dust environment

Interstellar dust is coupled to and carried around the Milky Way by the gas clouds. These clouds come in a vast variety of shapes, sizes, densities and temperatures. They can, however, be qualitatively classified in two basic categories—diffuse clouds and molecular clouds. The diffuse clouds are not clearly distinguishable on a photograph and are limited in density to less than about 300 hydrogen atoms per cc and are at a temperature of 50–100 K. The molecular clouds can have kinetic temperatures as low as 20 K (even 10 K) and any density above 300 per cc including the densities at which clouds collapse to form stars. Diffuse clouds contain hydrogen in atomic form while molecular clouds are dominated by molecular hydrogen. They can be very dark, as commonly illustrated by the Horsehead nebula (see Fig. 1), and also by isolated blank regions in the sky commonly called Bok globules (see Fig. 2). These latter were suggested as concentrations of dust and gas which were collapsing to form stars.

The material environment is dominated by hydrogen which is the most abundant atom in the galaxy. But it is the condensable elements which are needed to make the solid dust. The most abundant ones are the 'organics' O, C, N which, all together, account for one atom for each 1000 H atoms. Down by another factor of 10 are the 'rockies' Si, Mg, Fe. Table 1 shows the abundances of these in the solar system. Now, if we add these all together the total mass of the dust is at most about 1% of the gas.

The energetic environment affecting the dust (other than that associated with phenomena like star formation and even more explosive supernovae) consists of the ultraviolet radiation from stars and the cosmic ray particles. One of the principal effects of the ultraviolet is that it heats the dust. As a result of a kinetic balance between absorption and emission of radiation the dust temperature in diffuse clouds is about 15 K but in molecular clouds where it is shielded, the temperature may be as low as 5–10 K [15,16]. The dust and gas temperature can be very different from each other, with the dust generally being at the lower temperature.

3. What do observations tell us about dust?

The main features which characterize interstellar solid particles are described by their size (sizes), shape, their chemical composition, and amount. All of these must be determined by remote observations using telescopes in space as well as on the ground, with a variety of optical wavelengths. It is only by combining all these observations that we can come to any valid description. Even then there remain disagreements on what dust really is. Nevertheless, there are some firm conclusions. The first of these has to do with size as deduced from the way the dust blocks the light of stars.

3.1. Interstellar extinction

The most obvious evidence for the dust is that it blocks the light of stars in a very special way. As illustrated in the average interstellar extinction curve, shown in Fig. 3, the light in the red is reduced less than the light in the blue and in the ultraviolet. It can be shown from the basic electromagnetic scattering properties of particles that the blockage is most effective when there is an approximate match of particle size to wavelength, whether the particles be interstellar dust or atmo-

Fig. 2. An isolated cloud of dust called a Bok globule. No obvious star formation is taking place in its vicinity.

Table 1
Solar system abundances obtained by studies of solar spectra and presumed to represent the relative proportions of the elements which make up the solar system

Element	Relative number
H	1
He	0.0098
C	4.44(−4)
N	0.93(−4)
O	7.44(−4)
Mg	0.38(−4)
Si	0.355(−4)
S	0.214(−4)
Fe	0.316(−4)

The carbon abundance is a key factor in the composition of interstellar dust and the consequences are discussed in Ref. [16]. The numbers in parentheses denote powers of 10. Note that the organics O, C, N are 10 times as abundant as the rockies Mg, Si, Fe.

spheric aerosols which, when seen as smog, make the sun look so red when low on the horizon. From the shape of the extinction curve we can deduce that the particles responsible for the visual extinction are about the size of smoke particles (from a cigarette) being about 0.1 μm in size (a mean radius) and those responsible for the hump and the far ultraviolet extinction are 10–100 times smaller [17].

3.2. Chemical composition

The principal molecular ingredients of the interstellar dust are deduced from the way they absorb or emit infrared radiation. Fortunately the so-called fingerprint region of the infrared

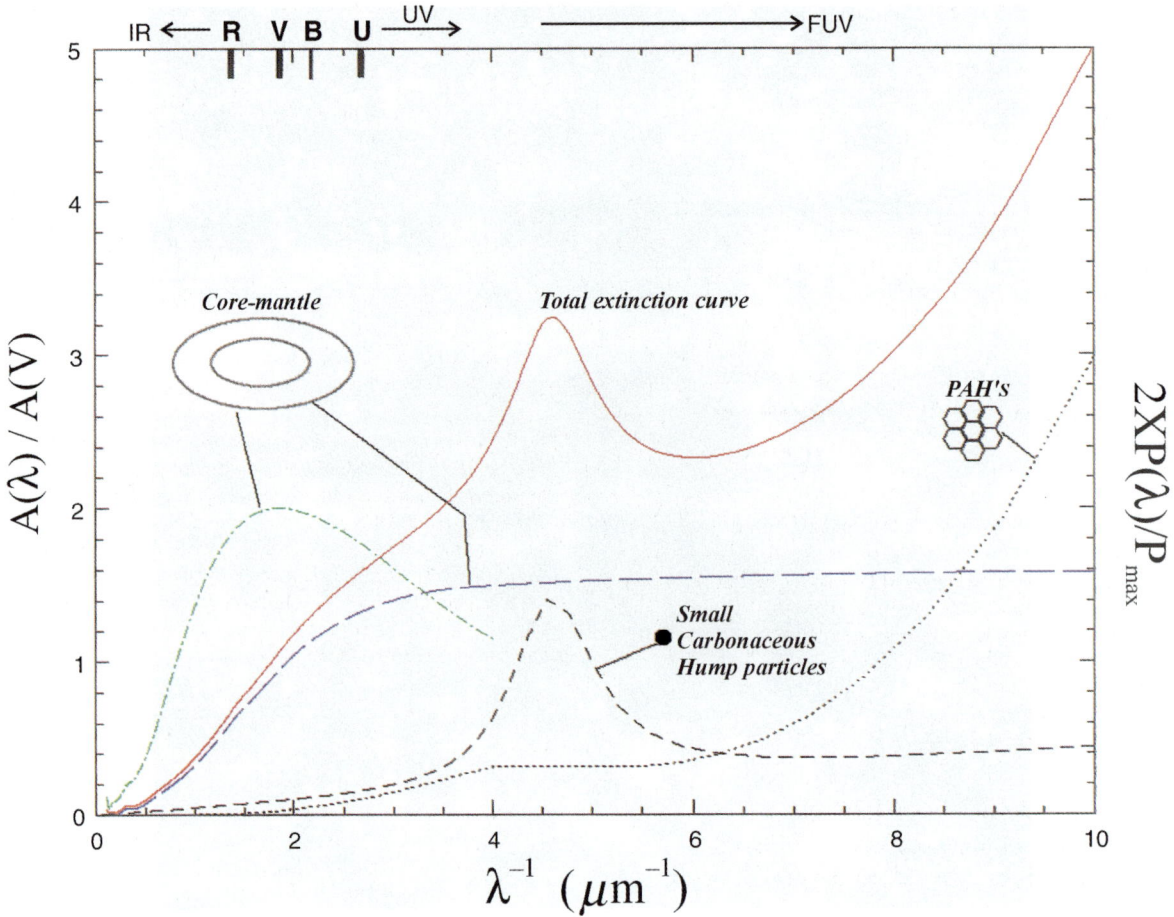

Fig. 3. The average extinction curve in the diffuse interstellar medium (solid line) and the mean polarization curve (dash-dot line) normalized to the maximum wavelength of polarization (P_{max}). The quantities $A(\lambda)$ and $A(V)$ are the amounts of blockage of starlight as a function of the wavelength (λ) and specifically in the visual (V) region of the spectrum at about 1.8 μm^{-1}. The various wavelength regions are indicated at the top: IR—infrared; R, V, B, U, —red, visual, blue, ultraviolet, FUV—far ultraviolet. If one compares the average extinction curve with the blocking of light by small solid particles, one finds that the long wavelength region (up to about 4 μm^{-1}) corresponds to the blocking of light by particles about 0.1 μm in mean radius. So-called Raleigh scattering which is proportional to λ^{-4} would be produced by particles much smaller than the wavelength. It turns out that the extinction of light by particles 0.1 μm in size levels off (saturates) when the wavelength is significantly shorter than the particle size no matter what the material properties of the particle are. This implies that the extra rise in the extinction curve at λ^{-1} about 4.6 μm^{-1} must be due to much smaller particles. The unified dust model [16] deconvolves the size distribution of the interstellar dust particles into three components: (1) the tenth micron silicate core-organic refractory mantle particles which provide all the polarization (dash-dot line) and a major fraction of the visible extinction (—); (2) very small (0.003 μm) carbonaceous particles responsible for the 216 nm hump in the extinction (–); (3) a population of PAH like particles (large molecules) responsible for ultraviolet extinction (···). The polarization of starlight is produced as radiation passes through clouds of aligned interstellar dust particles which act like a polarizer.

spectrum from 2.5 to 25 µm, where most vibrations of the molecular groups containing C, N, O and H occur, is accessible with modern telescopes. This region is indicated in Fig. 4. For example, the C–H stretching vibration is around 3.4 µm and the O–H stretch resonates at around 3.1 µm. Similarly the Si–O stretch of silicates (rocks) is at about 10 µm. Just knowing these features does not guarantee that we can identify the molecule in which the group exists. There are an infinite number of

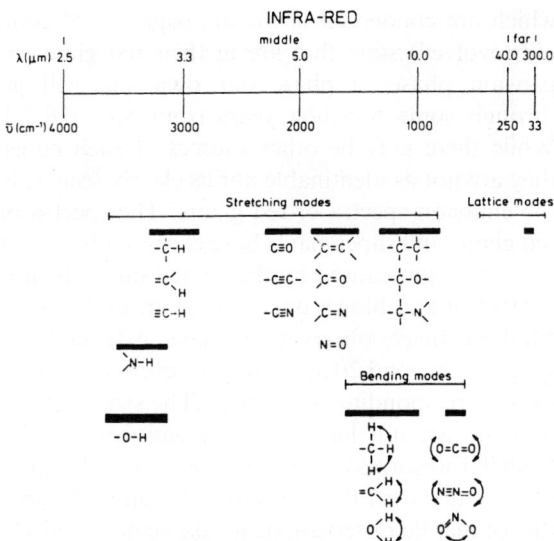

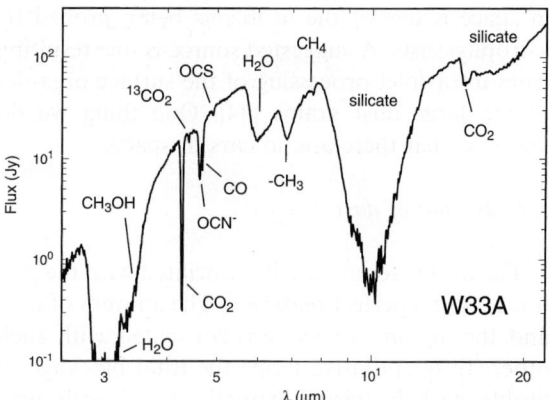

Fig. 4. Schematic of the infrared fingerprint region for vibrations of various molecular groups. Bending and stretching modes are indicated for a variety of molecular groups. The 3 µm region corresponds to stretching modes of C–H, N–H and O–H. For example, solid H_2O has a strong absorption at about 3.1 µm. Solid CO and CO_2 have strong C–O stretching absorptions at about 4.5 µm.

molecules that have C–H stretch features. By comparisons with laboratory spectra of simulations of the interstellar ices, it has been possible to identify about 15 molecules, radicals and ions. Fig. 5 shows observations made with the Infrared Space Observatory (ISO) of some of these molecules which are to be found in molecular cloud dust. In addition, a silicate absorption is clearly seen. This feature is always observed for the interstellar dust but it is not always accompanied by the ices. Even when we look far towards the center of our Milky Way, and looking mainly through diffuse clouds as early observed by Becklin et al. [18] (see Fig. 6), there is little or no ice (at 3.1 µm) to be seen [19–21]. What is seen in its place is the relatively insignificant absorption at 3.4 µm. This turns out to be the 'tip of the iceberg' indicating the presence of enormous amounts of complex organic molecules in the dust. What is equally interesting is that there is another component of complex organics which absorbs infrared poorly but is an efficient emitter of what are called the 'unidentified infrared features'. Although we can-

Fig. 5. Spectrum of the dust around the embedded protostar W33A in the wavelength region between about $\lambda = 2.5$ and 20 µm showing such identified molecules in the ice mantles as OCN^- [32,33] and OCS which are photoproduced in interstellar ices. The flux is given in janskys (jy). Note that the H_2O absorption is so large that it is completely saturated (little penetration). Identified absorption features are labeled according to the identified molecule or ion. Note the very broad feature at 10 µm which corresponds to the Si–O stretch of silicates.

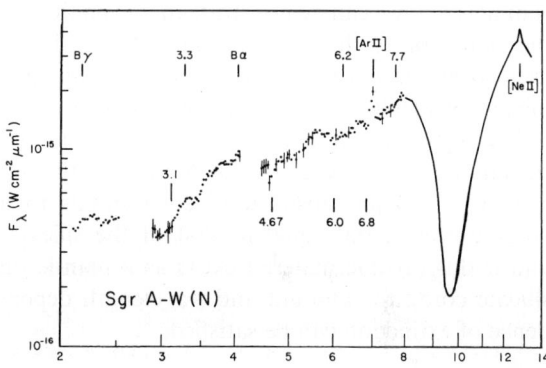

Fig. 6. Infrared spectrum of the galactic center [17,19,20] showing the lack of the ice band at 3.1 µm but the ubiquitous presence of the silicate band at 9.7 µm. The relatively inconspicuous feature at 3.4 µm is the 'tip of the iceberg' evidence for the major fraction of organic refractories in the interstellar dust. The abscissa is log λ. A few gaseous atomic lines such as the hydrogen Brackett γ, Brackett α and emission lines due to ionized argon and neon are noted. All the other features are due to dust components.

not specify precisely those molecules which emit these features, we are relatively certain that they are of the type called PAHs [22–24] made up of ensembles of benzene rings resembling the exhaust from automobiles. But exactly how they are made in space is one of the problems being probed by astrophysicists. A suggested source is one resulting from ultraviolet processing of the surface organics of the large dust grains [44]. One thing we do know, is that there are no cars in space.

3.3. Amount of dust

The dust is so intimately associated with the gas, it is almost a perfect marriage. The amount of dust and the amount of gas are correlated with each other. In quantitative terms the total blockage of visible starlight (visual extinction) is directly proportional to the number of atoms of hydrogen contained along the line of sight to the star [25]. The extinction is a measure of the total area of particles along the line of sight and the mean size of the particles is determined by the wavelength dependence. We can combine these to give the total volume of the dust and therefore have a measure of the amount of the dust for each hydrogen atom [26]. An important application of this line of reasoning is that we can demonstrate that even though the extinction to the galactic center shows almost exclusively the silicates, there are not enough cosmically available atoms of Si, Mg, and Fe (as silicates) to give the amount of dust needed to provide the amount of blockage of visible light. If the 'insignificant 3.4 µm absorption' corresponds to a volume of material equal to that of the silicates and if this organic material exists as a mantle on silicate cores, the amount and wavelength dependence of extinction can be satisfied.

4. Evolution of interstellar dust

The small particles in space have a complex physical and chemical history. This is summarized in Fig. 7. Although all the growth and variability of dust occurs in the space between the stars, the birthplace of the dust is in the atmospheres of stars and in supernovae.

4.1. The nuclei of interstellar dust

The nuclei or cores on which the solids can grow in space start out mostly as small silicate particles which are condensed in the atmospheres of cool, old (evolved) stars that are in their red giant expansion phase—a phase our own sun will go through some 5 billion years from now [26,27]. While there may be other sources of such cores, they are not as identifiable nor as clearly seen as in the emission spectra of red giants. The spectra of red giants and super giants have excess emissions at 10 and 20 µm caused by the heated dust which is formed in and blown out with gas from the stars. Modern space observations have extended the spectrum beyond 20 µm and show even more emissions corresponding to silicates. The sizes of these particles are not limited by the emission and interstellar absorption observations except that they must be no larger than the order of 1 µm, otherwise the ice mantle observations would be distorted. As a matter of fact, even as little as 2 or 3 µm would distort the 10 µm silicate band appreciably. This is a basic electromagnetic scattering phenomenon. The overall evidence convinces us that the silicates emitted by stars are probably in the range of 0.05 µm [28]. This is consistent with a wide range of astrophysical constraints [17].

4.2. Ice mantles

While many (more than 100) molecules are seen in the gas phase using mm and cm radio detection and about 15–20 molecules have been detected in the solid phase as mantles on the dust, both in a variety of space regions, the relative proportions of the identified molecules in the gas and solid phase are uncorrelated. A striking feature of this lack of correlation is that while the water molecule is generally the most abundant in the frozen solid ice mantles it is not very abundant in the gas. In fact both the observed and theoretically predicted gas phase abundances of water are generally much less than one hundredth that of carbon monoxide while it is observed that the water content in the solid ice mantles can be as high as 5–10 times as abundant as CO. The overabundance of H_2O cannot have resulted from accretion from the gas. Surface reactions are obviously needed to account for this seemingly anomalous result. First, an O atom hits the grain and sticks. One of many H atoms can then hit and while it does not remain

DUST IN SPACE

A Brief History of a Dust Grain

START
a) Small silicate particles condense in the atmospheres of cool stars and are ejected into space.

MOLECULAR CLOUDS

b) Silicate particles cool to T=5-15K and act as condensation nuclei for accretion of gas atoms and molecules as a mantle of frost.

c) Complex chemical reactions between gas and solids leads to an H_2O and CO dominated grain mantle.

d) Ultraviolet radiation of mantle breaks simple molecules leading to new combinations and to complex organic molecules (photoprocessing).

e) Star formation occurs leading to dissipation of portions of the collapsed cloud into surrounding low density (diffuse cloud) regions.

DIFFUSE CLOUDS

f) In the low density regions volatile icy mantles are either photoprocessed, evaporated, or destroyed leaving only the complex organic refractory mantles. These mantles are further subjected to photoprocessing and to destructive processes which can break off small pieces.

BACK TO b,c,d,e,f and repeat many times (~50) until the grain is consumed by star formation or becomes part of a comet.
Total mean life time of a dust grain = 5×10^9 years = Turnover time for the interstellar medium into and out of stars

Fig. 7. Brief history of a dust grain: the basic steps in its evolution from formation to destruction.

indefinitely on the surface it certainly remains long enough to encounter the sitting O and make OH. Another H atom comes along and does the same thing with the OH, thus making water. This is schematically illustrated in Fig. 8. The details of such processes are a subject of intensive laboratory and theoretical study by many groups looking for explanations, not only for the abundance of water but also for the abundance of a molecule such as methyl alcohol (the poisonous kind) in dust mantles. In any case, we have found ourselves coming back to van de Hulst's [8] idea of growing dust in space by surface reactions as well as by accretion. In Fig. 5 the relative over abundance of water relative to carbon monoxide is even more striking because it occurs in dust which has been heated causing a relative reduction of the CO which evaporates at a much lower temperature than H_2O.

But note that we have only come to the second step in grain evolution and we have only answered some of the questions raised by the observations. There are other molecules in the ices that are either not seen at all or are too under-abundant in the gas to have simply frozen out or even to have been formed by surface reactions. Such molecules as carbon dioxide and the cyanate ion (OCN^-) are candidates that may be created by the effects of ultraviolet on stirring up the ice mantles. This is discussed next.

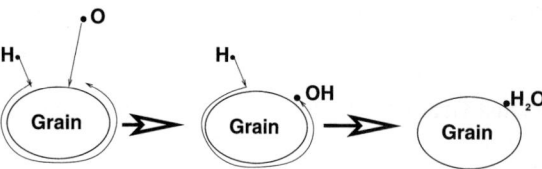

Fig. 8. A schematic illustration of the formation of hydrated molecules, and in particular H_2O, by sequential surface reactions between atoms and radicals with incoming H atoms.

4.3. Photoprocessing of ice mantles

A long standing puzzle is how to account for the presence of gas phase molecules at all, in view of the fact that their freeze out time on the dust is about 10–100 times shorter than the cloud lifetime. The answer to this puzzle and the reason for finding ice mantle molecules which come neither from the gas nor from surface reactions may be found in the effects of energetic processing of grain mantles.

During the lifetime of a molecular cloud the icy mantles are subjected to substantial energetic processing by the prevailing ultraviolet radiation and to the collisions by cosmic ray particles. The former seems to produce the dominant chemical modification while the latter may play an important role in bringing some of the solid molecules back into the gas.

Ultraviolet photons with energy greater than about 4 eV have sufficient energy to break chemical bonds in the ice mantle molecules. This leads to the creation of free (frozen) radicals and to detached hydrogen flying through and sometimes out of the mantle. Since the mantle thickness does not exceed about 0.02 μm all of the photons which impinge penetrate the mantle fully and are therefore effective. The internal chemical processing takes place as adjacent radicals combine and diffusing hydrogen is attached to another radical or molecule. For example, as schematically illustrated in Fig. 9, the breaking of molecules by photons

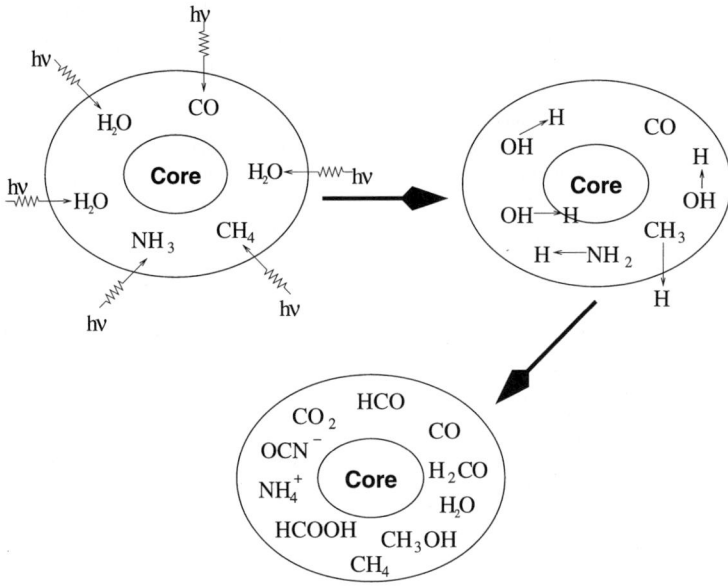

Fig. 9. Schematic of the sequence of bond breaking and recombination of radicals occurring in ultraviolet irradiated interstellar ice mantles. Initially the silicate core accretes (by surface formation or condensation) a mantle of water, methane, ammonia, and carbon monoxide. This mantle is subjected to bombardment by ultraviolet photons whose energy ($h\nu$) is sufficient to break molecular bonds so that, as depicted in the upper right panel, H atoms are detached and leave the radicals OH, etc. frozen (10 K) in the ice matrix. The lower panel depicts how the detached H atoms may combine with CO to produce the HCO radical and the CO may combine with the O from a doubly photoprocessed water to produce carbon dioxide. All the molecules in the lower panel are seen in both laboratory and interstellar spectra.

and subsequent formation of new ones occurs. Therefore, if a water molecule loses an H and an adjacent methane molecule loses an H the OH and CH$_3$ radicals would instantly combine to form methyl alcohol which is in fact observed in interstellar grains.

The chemistry and the infrared spectroscopy of the evolution of interstellar ice mantles is studied by comparing with results which are created in a laboratory simulation. The basic idea is a simple one and is illustrated by the cartoon (Fig. 10) of the interstellar dust chemical evolution simulator [29,30]. The equipment in current usage by a number of institutes consists of four basic parts: a closed cycle helium refrigerator, a vacuum system, a vacuum ultraviolet source and an infrared spectrometer. One creates a cold (10 K) surface in a vacuum and allows gas molecules to be accreted and subjected to irradiation by the ultraviolet. The laboratory ultraviolet flux is of course much higher than in space but it has been demonstrated that the laboratory results can be properly scaled to space so that 1 h of irradiation is equivalent to about 500 years in the diffuse interstellar space and about 5 million years in a molecular cloud. This ratio allows us to create changes in reasonable times to compare with the changes occurring in the lifetimes of molecular clouds. These changes can be studied by observing the changes using the infrared spectrometer. Fig. 11 shows the Leiden Lab setup. A typical example of the results of photoprocessing ices is shown in Fig. 12 with a comparison of the spectra of ices before and after ultraviolet. It cannot only be said that astronomical observations confirm the laboratory method but, in fact, that much of what has been discovered has required the data obtained in the laboratory [31]. The identification of molecules when mixed together in a frozen sample is not simply given in terms of the spectra of individual isolated species (for example see Fig. 13 with a comparison of CO in polar and nonpolar ices, [32]). Furthermore such reactive species as the cyanate ion which is observed in space results from ultraviolet radiation of mixtures including carbon monoxide and ammonia so that this provides evidence for the ultraviolet photoprocessing in space [33,34]. Another example is provided by the abundance of carbon dioxide in interstellar grains which is seen to be very efficiently and quickly produced in the laboratory upon irradiation of mixtures containing water and carbon monoxide which are two of the most abundant species in grain mantles. The sample laboratory spectra (Fig. 12) show some of these features. Laboratory studies are underway to test whether CO$_2$ can also be formed by pure surface reactions.

One of the critical consequences of the irradiation of mantles is the creation of a steady state abundance of free radicals. While some recombine to create new molecules there are always new ones being produced. The potential energy stored in the mantle by these radicals is substantial. It has been shown in the laboratory that if the sample can be warmed sufficiently to release these radicals from

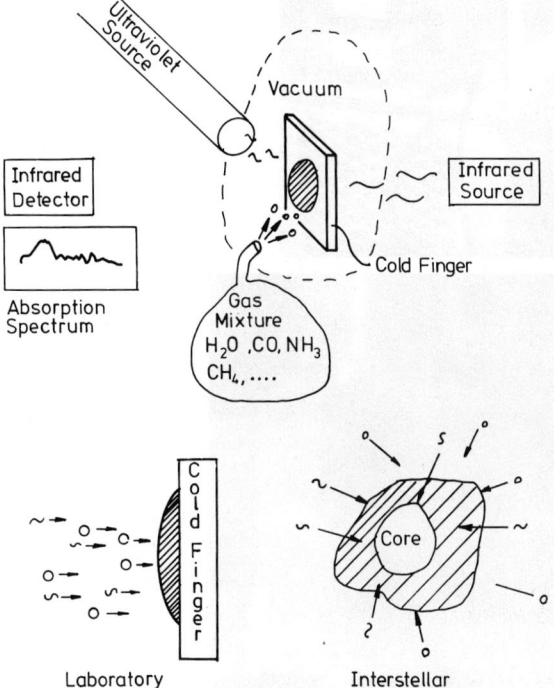

Fig. 10. A schematic of the laboratory analog method simulating the photoprocessing and the infrared spectroscopy of interstellar ice mantles. Gas mixture samples are slowly deposited on the cold finger at 10 K (dust surface analog) in a vacuum chamber (interstellar space). The material is irradiated by a vacuum UV hydrogen lamp (UV in clouds from stars and from cosmic rays) and the changes in infrared spectra are measured with an infrared spectrometer.

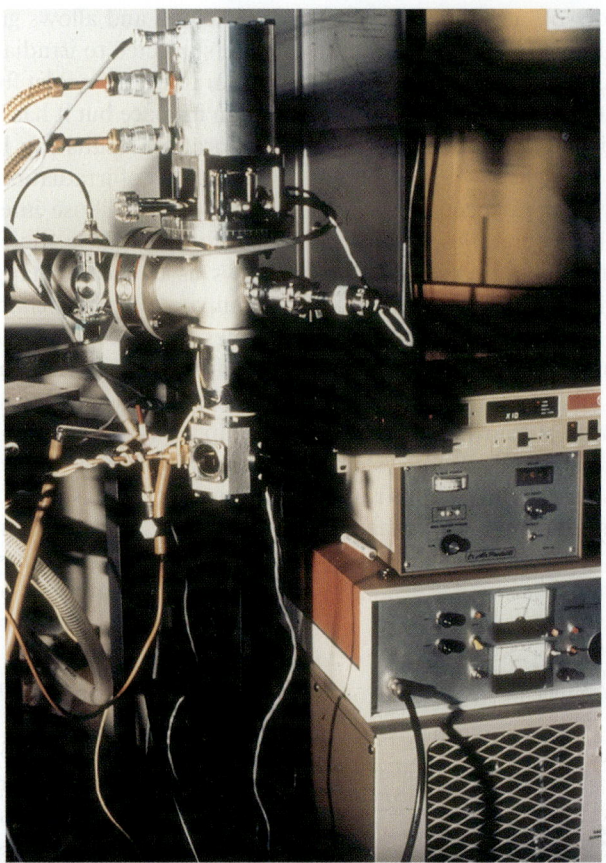
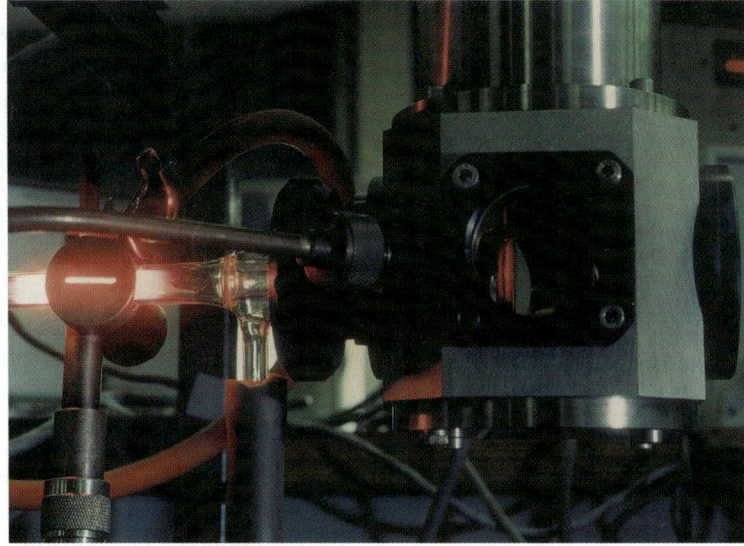

Fig. 11. A photograph of the Leiden Laboratory setup. The vacuum container of the cold finger is at the bottom of the cryocooler and the vacuum UV lamp at its left is seen close up in the other photo. The red color in the vacuum UV lamp is from hydrogen emission.

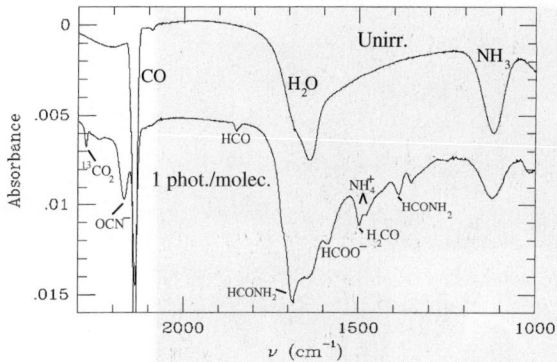

Fig. 12. A laboratory spectrum of a mixture of ices deposited at 10 K before (upper curve, showing the initial spectrum features corresponding to water, ammonia and carbon monoxide) and after irradiation by ultraviolet (lower curve). The presence of new molecules is indicated in the lower spectrum according to their absorptions. Note particularly the OCN$^-$ which is observed in both interstellar dust and in comets. The total carbon dioxide produced is very abundant but its absorption is at a shorter wavelength (higher wavenumber $\nu = 1/\lambda$) than is included. But one can see how abundant it is by noting the strength of the isotopic form and multiplying by the relative abundance of about $^{12}C/^{13}C = 80$.

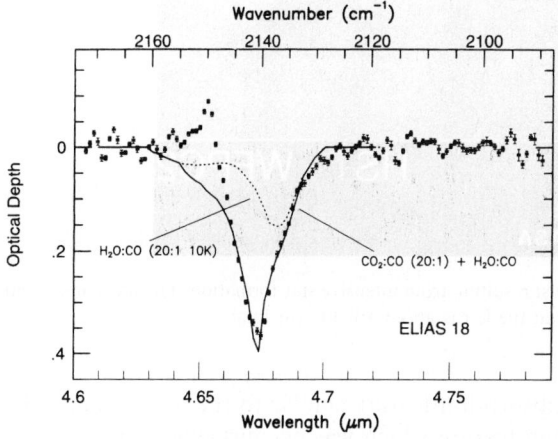

Fig. 13. Spectra showing the shape of the CO absorption and how it varies depending on whether it is totally imbedded in H$_2$O ($\cdots$) or partially on the outer surface (—) of the ice mantle [31]. The dots are observations of the absorption spectrum towards a star Elias 16.

their frozen sites enough energy is released by the rapid recombination of the stored radicals to lead to a runaway explosive reaction which desorbs a large amount of material to the gas. It has been proposed that in space such a trigger is provided by the heat generated when a heavy cosmic ray ion (principally Fe) plows through the mantle. It turns out that protons while relatively more abundant have little heating effect. The explosive desorption of mantle molecules triggered by cosmic ray iron nuclei in mantles with high radical concentrations appears to be the principal mechanism for maintaining a substantial concentration of gas phase molecules against freeze out on the grains. It turns out that a temperature increase of only about 15 K leads to sufficient diffusion of the free radicals in the initially 10 K ice mantle to trigger a chain (explosive) reaction [36]. The desorption problem in the interaction between dust and gas in molecular clouds is a basic one and is currently under intensive study. It plays an important role not only in keeping molecules in the gas but also by injecting different ones back into the gas phase; for example, CO$_2$ which is more abundantly made in the dust than in the gas [35–39].

The application of laboratory experiments to the study of evolution of interstellar grains may be one of the most important astrophysical fields at the present time and is a fertile field for many new ideas. There is a burgeoning of experiments using atomic beams as well as molecules on cold surfaces not only to explore how molecules like water, carbon dioxide, and formaldehyde are made but also to study how molecules are desorbed.

4.4. Organic refractory mantles

As noted earlier ice mantles are not to be seen in diffuse clouds. The transition from the molecular cloud to the diffuse cloud is a destructive one associated with the dramatic effects induced when the cloud collapses and forms stars [40]. This is exemplified by the classic photo (Fig. 14) taken with the Hubble telescope showing what look like cumulo nimbus clouds before a thunder storm but are really rapidly expanding interstellar clouds of gas and dust being blown apart by intensive star formation. In such an environment the volatile mantles are destroyed, evaporated or sputtered away. It appears, however, that not all the 'organics' in the ices are dissipated. In the laboratory when long term irradiated ices are warmed up they leave a

Fig. 14. The Eagle nebula exhibiting the expanding cloud of gas and dust resulting from intensive star formation. The ultraviolet light from hot stars causes the illumination at the colored edges. The pillar at the left is about a light year long.

residue of complex organic molecules which we call 'yellow stuff'. It had long been conjectured that such material would be created in interstellar space where it would provide the mantles on the silicates needed to account for the extinction (see Section 3.3 and Ref. [41]). When the C–H absorption feature was first observed in the direction of the galactic center (Fig. 6) it was thought to have confirmed this [42]. But what was made in the laboratory could only be what we might call first generation organic refractory mantle material. The 'yellow stuff' (see Fig. 15) possessed an infrared absorption feature similar to the astronomical 3.4 μm feature which was encouraging, but it did not duplicate the astronomical observations. This should not have been surprising because after the dust is ejected from its cocoon in the molecular cloud it finds itself in a much more hostile environment and the first generation yellow stuff material is now exposed to 10,000 times the ultraviolet flux compared with the molecular cloud. It was therefore to be expected that the organics which the actual interstellar dust accumulated were much more highly processed than the laboratory resi-

Fig. 15. A typical residue which remains after an ultraviolet irradiated laboratory ice sample is allowed to warm up to room temperature. Note the characteristic yellow color.

dues. The required amount of photoprocessing was difficult to produce in the lab. However, by placing laboratory samples on a space platform called the exobiology radiation assembly (ERA) which was part of the European EURECA satellite, it was possible to test the effect of long term ultraviolet radiation. A number of laboratory organic residues were put in a sealed sample carrier, and after a year in space (4 months actual solar ultraviolet radiation) they were returned. What went up yellow came back brown (see Fig. 16). The color change indicates that the material has become relatively richer in carbon. When the 'yellow, now turned brown stuff' was studied in the infrared, the astronomical 3.4 µm feature was exactly duplicated as shown in Fig. 17 [43]. This is an indication that the extra radiation leads to a closer representation of the interstellar organic refractory mantles which exist in space.

4.5. The interstellar dust evolution cycle

A theoretical study of the principal constituents of interstellar dust must take into account all the observational constraints. Not all astronomers agree on the same model of dust. One thing which is now generally accepted is that the solid particles are sequentially and repeatedly being cycled into and out of diffuse and molecular clouds. In consequence one can start with how the dust appears in, say diffuse clouds, and try to follow what happens to it when it is swept up into a molecular cloud and finally how, after ejection it recovers its diffuse cloud character. This is what is schematically followed in Fig. 18. The processing which occurs in the molecular cloud phase consists of both accretion and photoprocessing [45,46] while that which occurs in the diffuse cloud phase is discussed in Ref. [44]. Fig. 19 depicts an average diffuse cloud core–mantle grain and an average precometary dust grain; i.e. a grain which is at its final stage in the collapsing cloud having accreted all condensable gases and small particles before the star forms. The elongated shape is needed to account for the interstellar polarization produced by aligned dust grains.

A single cycle lasts about 10^8 years and as many as 50 may occur before the dust is consumed in

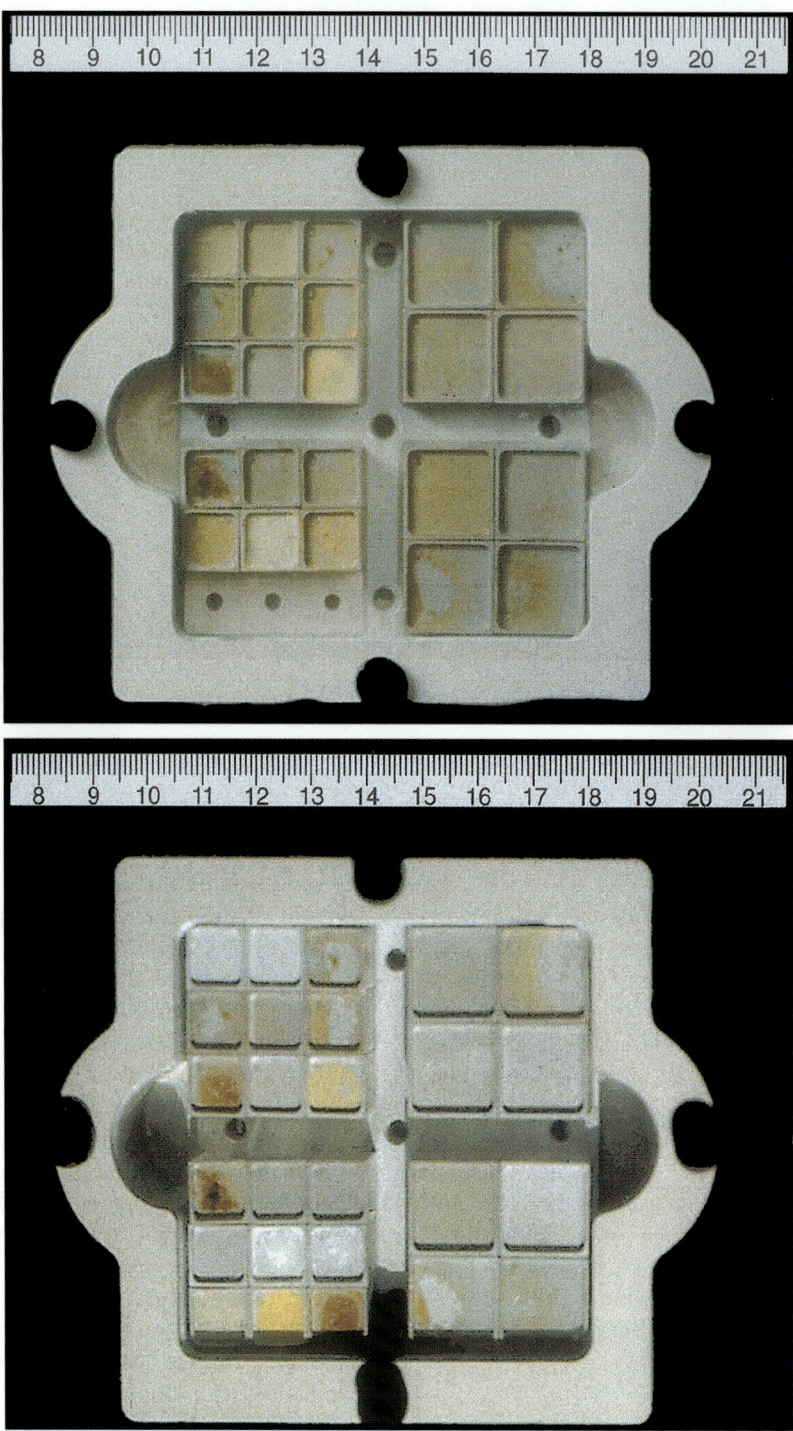

Fig. 16. Photos of the samples of various organic materials before and after ultraviolet radiation by the sun while on the ERA platform of the EURECA satellite above the earth's atmosphere. Note particularly the four (large) samples in the upper right of the sample carrier which went up as yellow organic residues and were returned brown. They are called A, B, C, D in clockwise sequence.

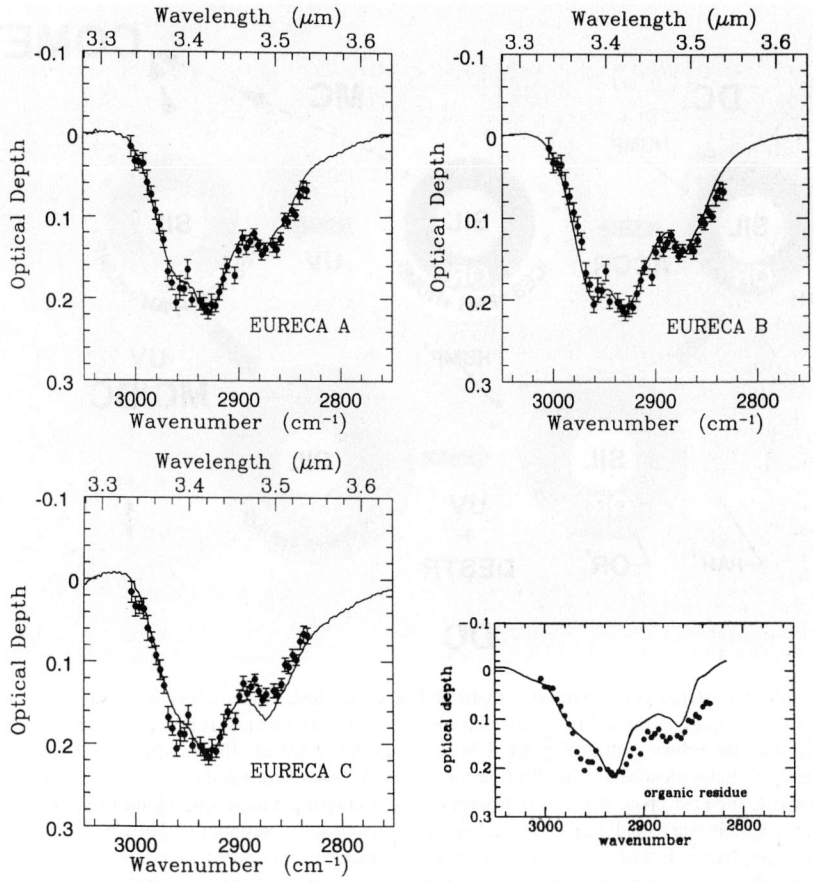

Fig. 17. Infrared absorption spectra of samples A, B, C compared with the 3.4 μm absorption towards the galactic center. For comparison is shown the same region of the spectra for a sample of the yellow stuff (lower right).

star formation as noted in Fig. 7. An interesting consequence of this concept is that if one assumes a mean lifetime of a large dust grain to be 5 billion years then those which aggregated to form the solar system comets 4.5 billion years ago carry a chemical history back almost 10 billion years—close to the beginning of the universe. What the first dust looked like is still open to conjecture but it is unlikely to have been the result of the evolutionary picture above because in the earliest galaxies seen at high red shift the energetics of the large hot stars and the high rates of formation of supernovae would not have permitted the growth of mantles and the concomitant production of PAH and hump particles [47,48].

5. What is a comet?

Comets have long been thought to be the most pristine bodies in the solar system. They are relics of the collapsing cloud which went into making the sun and the planets. They may appear as spectacular bright objects spanning distances of hundreds of thousands of kilometers and covering large parts of the sky as shown in the color photo of Hale–Bopp (Fig. 20). But the source of this display is a rather small object—the nucleus—which is only kilometers across as was first seen in the photographs of Halley taken by GIOTTO (Fig. 21). The peacock-like display of a feathery tail and coma is made up of the gas and dust which boils off as the nucleus comes close to the sun. It

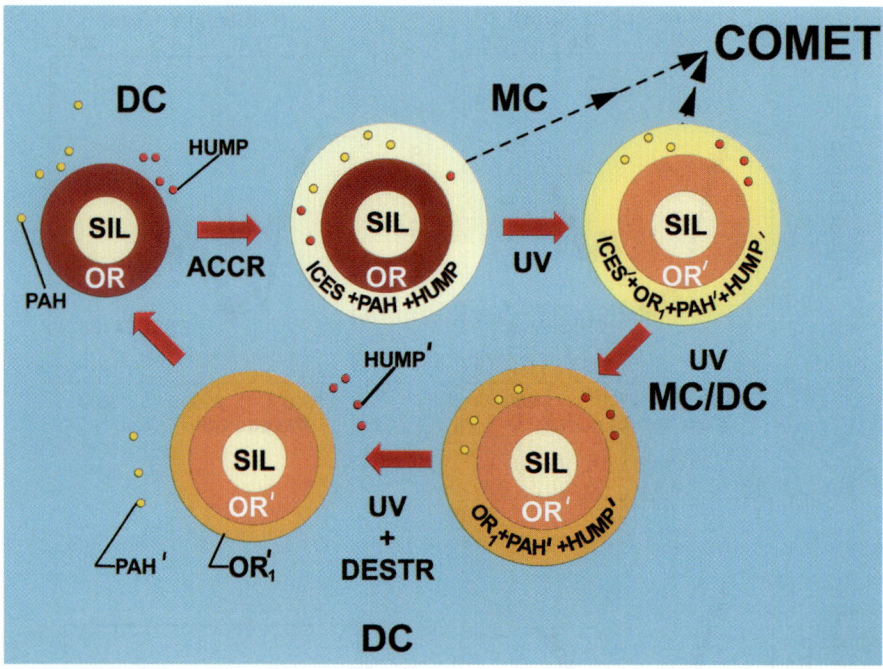

Fig. 18. Cyclic evolution of interstellar grains. One starts with diffuse cloud dust. Upper left is an average tenth micron radius silicate core–organic refractory mantle grain in the diffuse interstellar medium (DC). The mantle is heavily processed organic materials—even more heavily processed than the brown stuff in Fig. 14. Schematically illustrated are the hundreds of thousands to millions of very small carbonaceous hump particles (denoting their effect on the interstellar extinction) and even smaller PAH particles as described in Fig. 2. Following clockwise, the next phase depicts what happens after entering a molecular cloud (MC) showing the accretion of a complex ice mantle along with the very small particles and, simultaneously with accretion, the ultraviolet photoprocessing of the outer mantle and the organic inner mantle as well as the PAH and small carbonaceous particles (primes denote modified material). The next phase corresponds to the emergence out of the molecular cloud (MC/DC) after star formation when the ices are evaporated/destroyed leaving first generation organics (OR_1). And finally, the ultraviolet processing and partial destruction of the newly added first generation organic material ($OR_1 \to OR_1'$) as well as re-emergence and reforming of PAH and hump particles leading back to the 'original' diffuse cloud (DC) dust. The arrows leading upward depict the kinds of dust, which would make up the protostellar material (material in the collapsing cloud in which the star and planets were born) aggregating to form comets. In this representation it is assumed that this occurs with little or no evaporation and reforming of ices. A single cycle lasts about 10^8 years and as many as 50 may occur before the dust is consumed in star formation.

was the pioneering work of Whipple [49,50] which suggested that a comet should be an icy aggregate containing meteoritic type rocky materials. The more recent theories which provide a fundamental basis for this will be discussed in Section 6. The source of comets as having been made in the early solar system and thrown out to distances of 50,000 times the distance from the earth to the sun (astronomical unit (AU) 1.5×10^{13} cm) there to reside for 4.5 billion years along with hundreds of billions of companions in what is called the Oort cloud (of comets) was the contribution of Oort [51]. These 'nascent' comets which are too distant to be seen are almost in interstellar space and remain very cold. It is not until such a comet which has been in cold storage in the Oort cloud since the birth of the solar system receives a small gravitational perturbation from some object far removed from our solar system that it leaves its home to join with the planets closer to the sun. As it approaches the sun it is heated and the frozen gases (mostly water ice and carbon monoxide) evaporate and like a wind carry the dust along with them. The illumination of the dust by sunlight and the fluorescent emission of the outflowing emitted gases induced by the solar ultraviolet which lead to

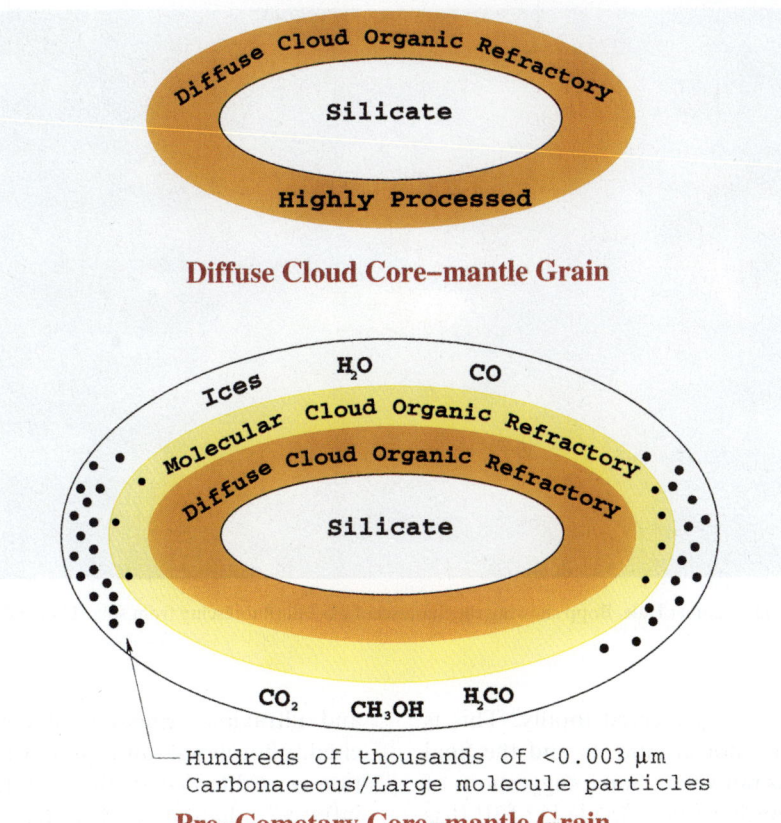

Fig. 19. Upper figure depicts a diffuse cloud silicate core–organic refractory mantle particle. It has to be nonspherical (here elongated) in order to provide for interstellar optical polarization. Lower figure depicts a fully accreted grain in the protosolar nebula; i.e. an average large grain on which all available gas phase molecules have accreted along with all the hundreds of thousands of very small grains/large molecules remaining in the dense collapsing cloud out of which the sun and planets were born.

the extended region of visible light extending over thousands of kilometers in the coma and over hundreds of thousands of kilometers in the tail make the appearance of the comet so dramatic. The source of all this is the small dark nucleus shown in Fig. 21. We are now aware of another system of comets which is much closer (perhaps only hundreds of AU from the sun) known as the Kuyper belt although it was likely to have originally formed even farther out than the Oort cloud comets. These comets started from the same or similar material in the protosolar nebula (the dense cloud of gas and dust out of which the sun and planets formed) as the Oort cloud variety. Comets are the best probe we have of the origin of the solar system.

6. The composition of comets

A matter of common belief is that comets have preserved the elemental composition of the presolar nebula with the exception of helium which could not condense and hydrogen which have remained, as in the collapsing cloud, in almost totally gaseous form. Whatever hydrogen was maintained would have been in molecular combinations with other elements. The major question is the degree to which all the volatiles might have frozen onto the dust in the final cloud collapse phase and maintained their identity during aggregation into comets. Did some of the icy mantles of interstellar dust evaporate and recondense after partaking of chemical processing in the presolar

Fig. 20. A photo of the large comet Hale–Bopp showing the fluorescent gas tail (fluorescing from solar UV) and the dust tail scattering the sun light.

nebula or was the dust preserved totally? This is the subject of many hot arguments and the final agreed on answer is not yet in. Some suggestions of particle evaporation have been made [52,53]. It is however well documented by many modern observations that most if not all of the interstellar dust icy mantles are preserved.

Classically, the composition of comet nuclei was derived primarily from the coma (the cloud of gas and dust closest to the nucleus) volatile molecules dominated by water or OH. The dust was considered mostly in terms of its scattering properties. The discovery of silicates in comets as an infrared emission feature [54] confirmed the existence of refractory material in comets along with the ices (volatiles). The idea of organic refractories as a major comet nucleus constituent was first introduced in the interstellar dust model of comets by Greenberg [55]. In fact it was predicted to be as important as the silicates. But it was the mass spectrometric evidence of the GIOTTO/VEGA space probes which provided the first proof that the refractory material in comet dust consisted of both the organic elements (O, C, N) as well as the rocky elements (Mg, Si, Fe) [56–59] in approximately equal mass proportions. While the visual and ultraviolet emission of coma molecules excited, photolyzed, and ionized by the solar radiation is used to deduce the volatile components, the infrared radiation by the heated dust tells us about the refractory components.

6.1. Interstellar ices and comet coma volatiles

A recent compilation of the molecules observed in a variety of comet comae and in interstellar dust mantles reveals a remarkable similarity (see Ref. [60]). Water is of course always the dominant volatile species with CO next and CO_2 and CH_3OH running a close third and even such relatively exotic species as OCS and the cyanate ion OCN^- being represented. It is interesting that as more and more comet molecules are identified the similarity in the comparison between interstellar dust and comet molecules is enhanced. The remarkably bright comet Hale–Bopp provided the vehicle to observe many new molecules [60]. These new results prompted the remark that "It supports models (of comet nuclei) in which cometary volatiles formed in the interstellar medium and suffered little processing in the solar nebula" [61]. Similar statements have been made by a number of comet

Fig. 21. A composite photo of the nucleus of comet Halley taken with the camera on board the GIOTTO space vehicle. The sun is 30° above the horizontal on the left and between 15° and 12° below the image plane. Note that the nucleus is very dark and that the emission of dust and gas occurs locally on those parts facing the sun while those in the shadow do not emit. The small hill is lit only on one side where it is evaporating (image courtesy Max-Planck-Institute für Aeronomy).

Table 2
Comparison of the abundances of ices in the interstellar medium (towards IRS9, a high mass protostar) and of cometary volatiles (at ~1 AU)

Species	Interstellar ices	Cometary volatiles
H_2O	100	100
CO	15	2–20
CH_3OH	6.3	1–7
CO_2	12	2–6
H_2CO	<3	0.05–4
HCOOH	3	~0.1
CH_4	1.6	0.7
Other hydrocarbons	?	~1 (C_2H_2, C_2H_6)
NH_3	<6	0.5
O_3	⩽2	?
OCN^-	0.5	0.2 (nitriles + HNCO)
OCS	0.2	0.4 (OCS + CS)
SO_2	?	~0.1
H_2	∃1	?
N_2	?	?
O_2	?	?

Cometary abundances are from Table 7 in Ref. [60]. ISM ice abundances are adopted largely from a compilation of Schutte [31] from various sources. The values are numbers of molecules relative to H_2O taken as 100. Although they generally follow the cometary distribution it must be noted that the sun was a low mass protostar when comets formed and the interstellar ices were probably somewhat different in detail.

specialists [60,62,63]. How cold the water was when it originally condensed (or formed) was inferred from the state in which the water molecule is observed in the coma; i.e. whether the two symmetric H atoms exist with their nuclear spins parallel (ortho) or anti-parallel (para). Because the ortho state has spin 1 it can exist in three projections while the para state with spin 0 has only the one, so that they are normally (in thermodynamic equilibrium) found to be in the ratio 3:1. But, in comet Halley this ratio was found to be substantially smaller, corresponding to a temperature of condensation of 23 K (interstellar) thus excluding formation in a warm solar nebula [64] and any long time heating in the solid thereafter. It is worth noting that such molecules as S_2, SO_2, H_2S and OCS and OCN^- owe their consideration and identification in comets as of interstellar dust origin (see Table 2) largely if not exclusively because of experiments done with photoprocessing of ice mixtures in the laboratory [65–68] at temperatures corresponding to the interstellar dust. Comets appear to have been born cold and remained cold and relatively unmodified since they were born along with the earth and other planets.

6.2. Amorphous versus crystalline ice: further evidence for interstellar dust in comets

An important and useful example of the surface physics of ice is the study of the conditions of water molecule condensation which determine whether the resulting ice is to be found in an amorphous or crystalline state in comets and elsewhere. The conditions for formation and preservation of amorphous ice have been studied both theoretically and experimentally [69]. When a water molecule hits a surface it may diffuse to neighboring sites. If the diffusion rate is slow compared with the rate of deposition the molecule is essentially frozen in the vicinity of its adsorption site and does not have time to find and settle in a crystalline site before the next molecule hits; i.e. if the flux of impinging molecules is lower than a critical amount the crystalline form is deposited. Thus amorphous ice may be formed under the condition that the diffusion distance during the time of coverage of the surface by adatoms (molecules) is smaller than the lattice constant of the substrate ice. It may be shown that on the interstellar dust in the parent molecular cloud the ice is both deposited and preserved in its amorphous form. If the interstellar ice is evaporated and recondensed in the solar nebula the only region where ice could reform is limited to the region beyond the asteroid belt which is where comets may have formed. The temperature in this region is such that diffusion rate is high enough and the flux of molecules is low enough (lower than the critical flux) that the ice is formed crystalline. Thus whether comets formed directly from interstellar dust or from dust from which the ice was evaporated and recondensed determines whether the ice in comets was initially amorphous or crystalline respectively. Furthermore, since the ice condensation region of the solar nebula encompasses temperatures significantly higher than that corresponding to the condensation temperature derived from the ortho/para ratio observed in the water evaporated from comets we may conclude that the ice in comets is indeed amorphous. This has been a classical assumption which we can now claim is justified by the solution of a problem in surface physics.

6.3. Interstellar and comet organic refractories

When pieces of the comet are ejected and blown away by the expanding heated gas the volatile constituents are removed leaving only the more refractory components called comet dust. One of the remarkable results established by the space missions GIOTTO and VEGA to Comet Halley (see Introduction) was that its dust was not, as some people expected, just of rocky composition (Mg, Si, Fe) but consisted of about equal amounts of the organics (O, C, N) as indicated in Fig. 22. This was one of the principal predictions of the interstellar dust model. The mass spectroscopic results not only confirmed the presence of an or-

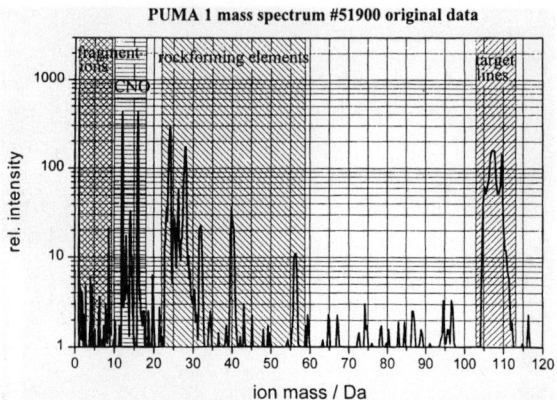

Fig. 22. An example of the kinds of mass spectra typical of comet Halley dust. The dust hit at ~ 70 km s^1 and was reduced to its ionized elemental components. Note that all particles had a mixture of the organic elements along with the rockies but the ratio varies. The peaks labelled O, C, and N primarily come from the organic mantles of the interstellar dust in the comet dust while 'rock forming elements' beyond mass 24 correspond to the elements Mg, Si, Fe, etc.

ganic refractory component of the dust but gave indication that it occurs as a mantle on the silicates [58] further substantiating the interstellar dust concept [17]. The organic part of comet dust revealed a large abundance of molecular species that are recognized to be of relevance to prebiotic chemistry; i.e. molecules which are believed to have been building blocks of the first living organisms. Mass spectroscopy and gas chromatography of laboratory analogue organics—the yellow stuff and brown stuff discussed in Section 4.4—also reveal the presence of a number of important prebiotic species. Amino acids are present in significant amounts but also molecules like hexamethylene tetramine which can also be sources of amino acids in the right environment [45,87,88]. The brown stuff is rich in PAHs [44] and one of the molecules observed in comets was phenanthrene—an intermediate sized PAH [89].

7. The structure of comets

Even if we know that presolar interstellar dust maintains its composition when aggregated to form a comet nucleus there would still be questions remaining about the end product of the way it aggregated. And furthermore what does the interior of a nucleus look like after 4.5 billion years? A good deal of theoretical work is going into looking for a solution to this problem (thermal processing, collisions etc.). We cannot hope to have the final answer to these questions until we can sample a comet directly or, better yet, bring back a piece. But in the meanwhile we have been able to use what comes off the comet to arrive at some pretty clear notions about the density of comets.

7.1. Comet dust density

One of the pieces needed to solve the puzzle of how a comet nucleus is constructed is to look at the pieces of a comet—its dust. GIOTTO and VEGA provided a major clue in establishing the mass (size) distribution of the dust fragments of the Halley nucleus [70,71] shown in Fig. 23. These are the only real (direct) data on comet dust masses less than 10^{-6} g. Because of the concerted effort between space and ground based observers a great deal of additional information was made available. Among these was the infrared observation of the comet dust. As had been observed in other comets before, there was the presence of a 10 μm emission feature produced by heated silicates. Now it is well known that only silicate particles much smaller than 10 μm in size (those indicated in Fig. 23 with a mass of $\sim 10^{-11}$ g) can emit such a feature, and when one looked at the Halley dust mass distribution there just were not enough particles small enough to be the source of the emission if they were compact. The answer to this question already existed in the prediction that the dust consists not only of aggregates but these aggregates are not compact [72,73]. Simply stated, if one considers a compact particle as a collection of fine pieces and separates them, the ensemble acts like a sum of the small particles. As indicated in Fig. 23 a 10^{-9} g fluffy ensemble (porosity $P = 0.99$) of 100 particles of 10^{-11} g acts like a sum of the small units so that by considering the Halley mass distribution as for very porous particles there are enough because we can include all those with masses ~ 100 times larger than the maximum if compact. It is beyond the scope of this article to go into the details of the calculations or the arguments but the end result is that one can show that the constraint imposed by the observed Halley dust mass distribution is that the dust consists of very fluffy aggregates of silicate core–organic refractory particles whose individual mean masses correspond to the hashed region of masses ($\sim 10^{-14}$ g) indicated in Fig. 23. The porosity had to be about 95% [74] corresponding to 95% vacuum.

7.2. Comet nucleus

As seen in the photograph of comet Halley (Fig. 21) the comet nucleus is a relatively small solar system body and even the largest one seen recently (Hale–Bopp) is only the order of 20 km in size, while they may be as small as or smaller than 1 km in size. Within the framework of the interstellar dust model one may reconstruct the comet nucleus from its dust if one assumes that the ejected pieces of the nucleus, heated by the sun, maintain their basic structure while evaporating all volatiles and leave only the skeletal remains (Fig. 24). This amounts to reconstructing each of the interstellar

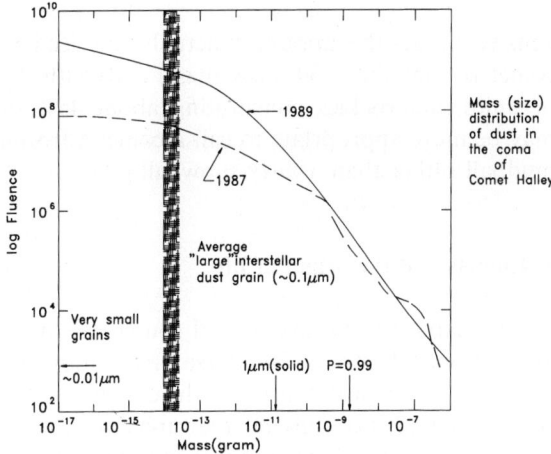

Fig. 23. The measured distribution of dust masses coming off of comet Halley [70,71]. The mean mass corresponding to the core–mantle interstellar dust distribution is indicated by the hatched region. Note that there are particles whose sizes are much smaller which may correspond to the hump and for ultraviolet particles in the unified dust model size spectrum. The larger dust grains are aggregates of the core–mantle grains.

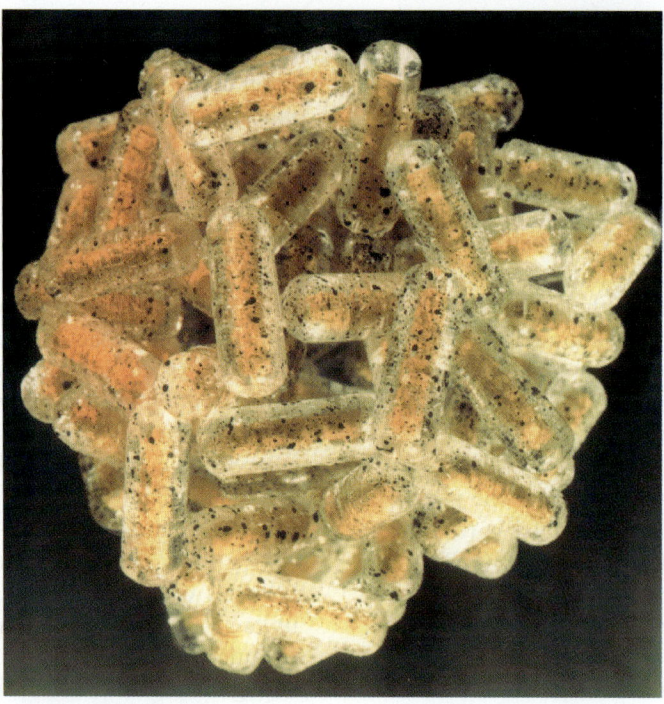

Fig. 24. A schematic model of a piece of a comet consisting of an aggregate of 100 average interstellar grains. Each grain consists of a silicate-core organic refractory mantle particle with an outer mantle of ices in which are imbedded the very finely divided carbonaceous and PAH particles as accreted in the final stages of cloud contraction to the protosolar nebula. The model is designed to be 3 µm in diameter and to have a porosity of $P = 0.8$. The corresponding density of a comet made of such material would be $\rho = 0.28$ g cm^{-3}.

grains in the aggregate by putting the less refractory organics (those that evaporate at temperatures of >450 K and are not seen by the GIOTTO/VEGA mass spectrometer in the dust) back on each core along with the icy mantle molecules. The result of this is pictured as a model of 'A Piece of a Comet' which was actually constructed in 1986 [75]. The model consists of 100 average core–mantle interstellar grains which have fully accreted all condensibles in the presolar cloud along with the hundreds of thousands of very small carbonaceous particles and PAH particles. The overall dimension is about 3 µm. The porosity as shown is about 80% meaning that a comet made this way would be 80% empty space. It is now generally accepted that comets are indeed low density objects but exactly how low is still an open question. The model would predict a comet nucleus density of about three tenths that of ice but recall that a large part of the stuff in comets has a much higher density. In fact the amount of actual water ice in a comet is only about 30% (by mass) and, with the organics and rockies constituting about 45%, it may be more appropriate to call a comet a frozen mudball rather than a dirty snowball [76].

8. Comets and the origin of life

The search for extraterrestrial sources of life is an old one. There is of course the concept of Panspermia which is defined as the case where life is transported throughout the universe having originated on one planet and is then transported to another. While this theory may not be excluded because of the evidence of its possibility in Martian meteorites being transported to the earth (for a recent review see Ref. [77]) and vice versa we shall not discuss this because it begs the question of how life started in the first place. The concept that prebiotic molecules might be created right here on earth was the incentive for the Miller–Urey experiment [78] and was probably inspired by Oparin's theories of life's origins from basic chemical processes. By passing an electrical discharge through a vapor mixture including water, methane and ammonia as a simulation of lightning in an early earth atmosphere, Miller showed with careful chemical analysis that many molecules such as amino acids, which are building blocks of life, were produced. However we now know that the earliest atmosphere which exited at the time when life emerged was almost totally carbon dioxide and in such a gas the production of prebiotic building blocks by such a process is extremely limited. This has led to the search for initiating prebiotic chemistry by extraterrestrial sources.

There are many objects hitting the earth which contain complex organic matter—interplanetary dust particles, micrometeorites, etc. as well as comets. In fact, the overall input from comets is not even the predominant source being outweighed by IDPs and meteorites. It appears most likely that all the objects which bring organics have started with interstellar dust at some stage so that one could say that the principal source of extraterrestrial organics is the interstellar dust. There are a number of cogent reasons to believe that comets are (were) the chosen vehicle to bring interstellar created building blocks to initiate life on earth—the right stuff at the right time.

8.1. Timeliness

There is an interesting coincidence between when life first appeared and the massive bombardment of the earth by comets and other objects was just beginning to tail off. It can be seen from the cratering record on other solar system objects, where the record has not been erased by weathering and and crustal evolution (see Ref. [79] for a review), that during the first 700 million years after the earth formed it had undergone about 20 million impacts by comets—at least one 5 km comet every 10 years and that it was not until 3.8 billion years ago that this rate dropped quickly. It appears from the fossil evidence that life was already extensive 3.6 billion years ago and possibly even 200 million years before that based on carbon isotopic evidence [97]. We may even speculate that the evidence for the earliest living organisms could be extended further back in time by noting that the lack of evidence is possibly because the amount of datable rocks decreases with time similarly to the amount of identifiable fossils so that the percentage of fossil material may not be zero 3.8 billion years ago. In any case if we assume that all that heavy bombardment was delivering millions of death dealing blows to the earth similar to the single one which took care of the dinosaurs 65 million years ago there was not a great window in time for life to have emerged. The question is whether comets were the source of life as well as death.

8.2. The right stuff

The question is what constitutes the 'right stuff' for most rapidly initiating the prebiotic chemical evolution. There is certainly a great abundance of complex organics in each comet Halley equal in mass to about 1% of the current 'living mass' on the earth. Not only that but if we can believe that comets have preserved the interstellar dust better than meteorites then these organics already resemble those of living organisms in that they have a specific mirror symmetry—the amino acids are left handed [80,81]. However the packaging may be even more important than the contents as has been emphasized by Krueger and Kissel [82,83]. It involves the physical structure of comet dust as well as its chemistry. If comets are indeed very fluffy aggregates of the interstellar dust as postulated by Greenberg [84], the early bombardment of the earth by comets brought an enormous amount of complex organics to the earth in the form of very low density comet dust particles. Small fluffy aggregates of interstellar dust would have not been heated sufficiently to modify the organics [85] and could have floated to the earth preserving their character except for the evaporation of the volatile components which actually could cool the remaining material. Favorable factors provided by comet dust to the initiation of life involve three factors: (1) high porosity, (2) inorganic (mineral) surfaces for significant catalysis, and of course, (3) the presence of preformed chiral organics. We envisage that these comet dust fragments may fulfill the requirements for chemical thermodynamics to start molecular self-organization [82]. Each comet dust particle, consisting of a porous aggregate of submicron silicate cores with organic refractory mantles sitting within a water bath (the local "ocean" containing nutrients much of which were also brought by comets), satisfies the conditions for non-equilibrium thermodynamics [86]. A very porous comet dust grain pictured as a piece of a comet minus the volatiles may be thought of as having an inside and an outside in the sense that the diffusion of small nutrient molecules into it from the bath occurs more readily than the diffusion of large molecules outward. There is already a substantial concentration gradient of preformed complex organic molecules in the initial structure. Therefore reactions within are unidirectional towards higher complexity and organization. An optimum size of the porous structure depends on the combined diffusion-reaction equations. If the aggregate is too small the small molecules pass through without interacting and the large ones may even diffuse out. If it is too large the small molecules interact only on the outside and do not penetrate fully. For example an aggregate of 100 particles as pictured in Fig. 24 with a diameter of 3 μm (but with the ices gone) may allow many small molecules to pass through without a collision. A rather simple calculation suggests the order of 5–10 μm diameter as an optimal aggregate size. The first discussion of this mechanism by Krueger and Kissel [82] pointed out particularly to the chemical composition of the organics predicted by Greenberg [87,88] provide excellent precursors for the making of biological molecules. An additional advantage of the comet dust is that if one tenth of a percent of the comet is shed as 5 μm dust particles there would be available 10^{25} (!) of them as possible seeds. Laid end to end they would stretch 100 times the distance to the nearest star. One can always argue that if the chance of any one being a life precursor is zero then the vast multiplying factor is not significant. But the value of large numbers of concentrated 'cell-like' chemical factories compared with a dilute soup of prebiotic molecules is difficult to counter. Kissel and Krueger estimate that 1 in 10^4 could make it [83]!

8.3. Delivery

The safe delivery of these dust particles with their organics is not a trivial problem when one considers the energy released in an impact occurring at say 20 km s^{-1}. However there are very special characteristics of the comet nucleus which could have provided the conditions needed for bringing a significant fraction of its organics unheated (or unpyrolized) to the earth's surface. First of all one should bear in mind that not every collision will be head-on. Secondly, comets are rather fragile objects and may break up on approaching the earth and shed a great deal of dust [90,91] as was recently demonstrated by the break up of comet Shoemaker–Levy in the vicinity of Jupiter. Thirdly the earth's early atmosphere was as much as 10 times denser than now providing a much thicker cushion for small particles to be slowed down [92]. Fourthly, comet dust is very underdense and, being made up of tenth micron units, may (according to Basiuk et al. [85]) undergo insufficient heating to pyrolize the organics. The fluffiness of comet dust and the tenth micron size of its elementary units is therefore highly favorable to its low entry heating and survival to float 'gently' to the earth. Another possibility for controlling the overheating is the cooling by evaporation of volatiles initially contained in the dust [93–96]. Small fluffy particles have a great chance of surviving with their organics intact.

9. Space missions to comets

The future prospects in our understanding of interstellar dust, comets and our origins are exciting. Already from the first space missions to comet Halley we became aware of their complex chemistry. Both the US NASA and the European ESA have planned missions to comets and asteroids which will reveal even more details about their composition. It is the goal of the ESA Rosetta mission to explore the nucleus of a comet (P/Wirtanen) in detail from close distance over an extended period of time not only remotely from the Orbiter, but also by means of a Lander on the surface of the nucleus. Laboratory analog samples in Leiden will be used to test the validity of the origins of prebiotic organics in interstellar space. The molecular composition of the organic phase of the solid cometary particles will be used to establish its exobiological relevance as possible organic precursors to the origin of life.

The Rosetta orbiter will both see (photograph) the comet and will measure its mass from its orbit just as the mass of the moon is determined by its orbit around the earth. Thus by knowing simultaneously the size and mass, we will for the first time know the density of a comet—how fluffy it really is! What might we anticipate being accomplished in the next hundred years? When we retrieve bulk material from a comet intact and bring it back to earth one thing to try is to distribute 5–10 µm pieces in appropriate nutrient water baths and follow their chemical evolution. We can expect that new experimental techniques will have developed to the point at which this will be possible. In conclusion, it is fascinating to think that ongoing research may well reveal more information about the directed processes of chemical evolution and a greater insight into the mechanisms by which these life-forming processes reached the earth. And in any case we may have in our hands the interstellar dust out of which the solar system was born.

Acknowledgements

I would like to express my appreciation to C.-J. Shen and to Dr. W. Schutte for help in preparing this manuscript. I thank Dr. Schutte also for allowing me to use some of his recent laboratory spectra.

References

[1] A.M. Clerke, in: Problems in Astrophysics, Black, London, 1903, p. 567.
[2] R.J. Trumpler, Preliminary results on the distances dimensions and space distribution of open star clusters, Lick Obs. Bull. XIV (1930) 154.
[3] A.E. Whitford, An extension of the interstellar absorption curve, Astrophys. J. 107 (1948) 102.
[4] A.E. Whitford, The law of interstellar reddening, Astron. J. 63 (1958) 201.
[5] B. Lindblad, A condensation theory of meteoric matter and its cosmological significance, Nature 135 (1935) 133.
[6] J.L. Greenstein, A determination of selective absorption based on the spectrophotometry of reddened B stars, Pub. Washburn Obs. 15 (5) (1934) 151.
[7] C. Schalén, Zur Frage eines allgemeinen absorption des lichtes im Weltraum, Astron. Nachr. 236 (1929) 249–258.
[8] H.C. Van de Hulst, The solid particles in interstellar space, Rech. Astron. Obs. Utrecht 11 (pt. 2) (1949).
[9] F. Kamijo, A theoretical study on the long period variable star III, Pub. Astr. Soc. Jpn. 15 (1963) 440–448.
[10] J.S. Hall, Observations of the polarized light from stars, Science 109 (1949) 166.
[11] W.A. Hiltner, Polarization of light from distant stars by interstellar medium, Science 109 (1949) 165.
[12] R.E. Danielson, N.J. Woolf, J.E. Gaustad, Search for interstellar ice absorption in the infrared spectrum of mu Cephei, Astrophys. J. 141 (1965) 116.
[13] J.M. Greenberg, A.J. Yencha, J.W. Corbett, H.L. Frisch, Ultraviolet effects on the chemical composition and optical properties of interstellar grains, Mém. Soc. Roy. Sci. Liège tome III (6e série) (1972) 425–436.
[14] Encounters with comet Halley: the first results, Nature 321 (1986) 259–366.
[15] J.M. Greenberg, Interstellar grain temperatures effects of grain materials and radiation fields, Astron. Astrophys. 12 (1971) 240–249.
[16] J.M. Greenberg, A. Li, Evolution and emission of cold warm and hot dust populations in diffuse and molecular clouds, in: D.L. Block, J.M. Greenberg (Eds.), New Extragalactic Perspectives in the New South Africa, Kluwer, Dodrecht, 1996, pp. 118–134.
[17] A. Li, J.M. Greenberg, A unified model of interstellar dust, Astron. Astrophys. 323 (1997) 566–584.
[18] E.E. Becklin, G. Neugebauer, S.P. Willner, et al., Infrared observations of the galactic center IV, Astrophys. J. 220 (1978) 831–835.
[19] I. Butchart, A.D. McFadzean, D.C.B. Whittet, T.R. Geballe, J.M. Greenberg, Three micron spectroscopy of the galactic centre source IRS 7, Astron. Astrophys. 154 (1986) L5–L7.
[20] S.P. Willner, R.W. Russell, R.C. Puetter, et al., The 4–8 µm spectrum of the galactic center, Astrophys. J. 229 (1979) L65–L68.
[21] S.P. Willner, R.C. Puetter, R.W. Russell, et al., Unidentified infrared spectral features, Astrophys. Space Sci. 65 (1979) 95–101.
[22] L.J. Allamandola, A.G.G.M. Tielens, J.R. Barker, Polycyclic aromatic hydrocarbons and the unidentified infrared emission bands: auto exhaust along the Milky Way, Astrophys. J. 290 (1985) L25–L28.
[23] A. Leger, J.L. Puget, Identification of the unidentified IR emission features of interstellar dust? Astron. Astrophys. Lett. 137 (1984) L5–L8.
[24] A. Leger, L. d'Hendecourt, N. Boccara (Eds.), Polycyclic Aromatic Hydrocarbons and Astrophysics, NATO ASI series C191, Reidel, Dordrecht, 1987, p. 402.

[25] L. Spitzer Jr., Physical Processes in the Interstellar Medium, Wiley, New York, 1978.
[26] J.M. Greenberg, Interstellar dust, in: J.A.M. McDonnell (Ed.), Cosmic Dust, Wiley, New York, 1978, pp. 187–294.
[27] R.D. Gehrz, Sources of stardust in the galaxy, in: L.J. Allamandola, A.G.G.M. Tielens (Eds.), Interstellar Dust (IAU symposium 135), Kluwer, Dordrecht, 1989, pp. 445–453.
[28] J. Tinbergen, J.M. Greenberg, C. De Jager, Multicolor linear polarimetry of Betelgeuse and Antares, Astron. Astrophys. 95 (1981) 215–220.
[29] W. Hagen, L.J. Allamandola, J.M. Greenberg, Interstellar molecule formation in grain mantles: the laboratory analog experiments results and implications, Astrophys. Space Sci. 65 (1979) 215–240.
[30] W. Hagen, L.J. Allamandola, J.M. Greenberg, Infra-red absorption lines by molecules in grain mantles, Astron. Astrophys. 86 (1980) L1, L4–L6.
[31] W.A. Schutte, Ices in the interstellar medium, in: P. Ehrenfreund, et al. (Eds.), Laboratory Astrophysics and Space Research, Kluwer, Dordrecht, 1999, pp. 69–103.
[32] J.E. Chiar, A.J. Adamson, T.H. Kerr, et al., High-resolution studies of solid CO in the Taurus dark cloud: characterizing the ices in quiescent clouds, Astrophys. J. 455 (1995) 234–243.
[33] R.J.A. Grim, J.M. Greenberg, Ions in grain mantles: the 4.62 micron absorption by OCN-in W33A, Astrophys. J. 321 (1987) L91–L96.
[34] W.A. Schutte, J.M. Greenberg, Further evidence for the OCN-asssignment to the XCN band in astrophysical analogs, Astron. Astrophys. 317 (1997) L43–L46.
[35] J.M. Greenberg, A.J. Yencha, Exploding interstellar grains and complex molecules, in: J.M. Greenberg, H.C. Van de Hulst (Eds.), Interstellar dust and related topics, IAU symp. 52, Reidel, Dordrecht, 1973, pp. 363–373.
[36] L.B. d'Hendecourt, L.J. Allamandola, F. Baas, J.M. Greenberg, Interstellar grain explosions: molecule cycling between gas and dust, Astron. Astrophys. 109 (1982) L12–L14.
[37] W.A. Schutte, J.M. Greenberg, Explosive desorption of icy grain mantles in dense clouds, Astron. Astrophys. 244 (1991) 190–204.
[38] J.M. Greenberg, L.B. d'Hendecourt, Evolution of ices from interstellar space to the solar system, in: J. Klinger, et al. (Eds.), Ices in the Solar System, Reidel, Dordrecht, 1985, pp. 185–204.
[39] J.M. Greenberg, C.X. Mendoza-Gómez, M.S. De Groot, R. Breukers, Laboratory dust studies and gas-grain chemistry, in: T.J. Millar, D.A. Williams (Eds.), Dust and Chemistry in Astronomy, Proc. Conf. Manchester, IOP Publ. Ltd., January 1992, pp. 265–288.
[40] J.M. Greenberg, Dust in dense clouds, in: J.E. Beckman, J.P. Phillips (Eds.), Submillimetre Wave Astronomy, Cambridge University Press, Cambridge, 1982, pp. 261–306.
[41] J.M. Greenberg, Interstellar dust, in: J.A.M. McDonnell (Ed.), Cosmic Dust, Wiley, New York, 1978, pp. 187–294.

[42] D.A. Allen, D.T. Wickramasinghe, Diffuse interstellar absorption band between 2.9 and 4.0 µm, Nature 294 (1981) 239–240.
[43] J.M. Greenberg, A. Li, C.X. Mendoza-Gómez, W.A. Schutte, P.A. Gerakines, M. De Groot, Approaching the interstellar grain organic refractory component, Astrophys. J. Lett. 455 (1995) L177–L180.
[44] J.M. Greenberg, J.S. Gillette, G.M. Munoz-Caro, T.B. Mahajan, R.N. Zare, A. Li, W.A. Schutte, M. De Groot, C. Mendoza-Gomez, Ultraviolet photoprocessing of interstellar dust mantles as a source of polycyclic aromatic hydrocarbons and other conjugated molecules, Astrophys. J. Lett. 531 (2000) L71–L73.
[45] M.P. Bernstein, S.A. Sandford, L.J. Allamandola, et al., UV irradiation of polycyclic aromatic hydrocarbons in ices: production of alcohols, quinones, and ethers, Science 283 (1999) 1135–1138.
[46] J.M. Greenberg, A. Li, Tracking the organic refractory component from interstellar dust to comets, Adv. Space Res. 24 (1999) 497–504.
[47] A. Dey, W. Van Breugel, W. Vacca, R. Antonucci, Triggered star formation in a massive galaxy at $z = 3.8$:4C 41.17, Astrophys. J. 490 (1997) 698.
[48] H.J.A. Röttgering, P.N. Best, M.D. Lehnert (Eds.), The Most Distant Radio Galaxies, Royal Netherlands Academy of Arts and Sciences, 1999.
[49] F.L. Whipple, A comet model. I. The acceleration of comet Encke, Astrophys. J. 111 (1950) 375–394.
[50] F.L. Whipple, A comet model, II. Physical relations for comets and meteors, Astrophys. J. 113 (1951) 464–474.
[51] J.H. Oort, The structure of the cloud of comets surrounding the solar system and a hypothesis concerning its origin, B.A.N. 11 (1950) 91–110.
[52] B. Fegley, Chemical and physical processing of presolar materials in the solar nebula and the implications for preservation of presolar materials in comets, Space Sci. Rev. 90 (1999) 239–252.
[53] B. Fegley, Chemistry of the solar nebula, in: J.M. Greenberg, C.X. Mendoza-Gómez, V. Pirronello (Eds.), The Chemistry of Life's Origins, Kluwer, Dordrecht, 1993, pp. 75–147.
[54] R.W. Maas, E.P. Ney, N.J. Woolf, The 10-micron emission peak of comet Bennett 1969i, Astrophys. J. 160 (1970) L101–L104.
[55] J.M. Greenberg, What are comets made of—a model based on interstellar dust, in: L.L. Wilkening (Ed.), Comets, Univ. of Arizona Press, Tucson, 1982, pp. 131–163.
[56] J. Kissel, D.E. Brownlee, K. Büchler, et al., Composition of comet Halley dust particles from Giotto observations, Nature 321 (1986) 336–337.
[57] J. Kissel, R.Z. Sagdeev, J.L. Bertaux, et al., Composition of comet Halley dust particles from Vega observations, Nature 321 (1986) 280–282.
[58] E.K. Jessberger, J. Kissel, Chemical properties of cometary dust and a note on carbon isotopes, in: R. Newburn, M. Neugebauer, J. Rahe (Eds.), Comets in the Post-Halley Era, Kluwer, Dordrecht, 1991, pp. 1075–1092.

[59] J. Kissel, In situ measurements of evolved solids in space with emphasis on cometary particles, in: J.M. Greenberg, A. Li (Eds.), Formation and Evolution of Solids in Space, Kluwer, Dordrecht, 1999, p. 427.

[60] J. Crovisier, Solids and volatiles in comets: from nuclei to cometary atmospheres, in: J.M. Greenberg, A. Li (Eds.), Formation and Evolution of Solids in Space, Kluwer, Dordrecht, 1999, pp. 389–426.

[61] D. Bockelee-Morvan, D.C. Lis, J.E. Wink, et al., New molecules found in comet C/1995 01 (Hale-Bopp); investigating the link between cometary and interstellar material, Astron. Astrophys. 353 (2000) 1101–1114.

[62] M.J. Mumma, P.R. Weissman, S.A. Stern, Comets and the origin of the solar system: Reading the Rosetta stone, in: E.H. Levy, J.I. Lunine, M.S. Matthews (Eds.), Planets and Protostars III, Univ. of Arizona Press, Tucson, 1993, pp. 1177–1252.

[63] M.J. Mumma, M.A. DiSanti, N. Dello Russo, et al., Detection of abundant ethane and methane, Science 272 (1996) 1310–1314.

[64] M.J. Mumma, W.A. Weaver, H.P. Larson, The ortho–para ratio of water in comet P/Halley, Astron. Astrophys. 187 (1987) 419–424.

[65] J.M. Greenberg, R. Grim, L.J. Van Ijzendoorn, Interstellar S_2 in comets, in: C.-I. Lagerkvist, B.A. Lindblad, et al. (Eds.), Asteroids, Comets, Meteors II, Reprocentrum HSC, Uppsala, 1986, pp. 218–220.

[66] R.J.A. Grim, J.M. Greenberg, Photoprocessing of H_2S in interstellar grain mantles as an explanation for S_2 in comets, Astron. Astrophys. 181 (1987) 155–168.

[67] R.J.A. Grim, J.M. Greenberg, Photochemical studies of S_2 formation and their implications on the source and evolution of comets, The comet nucleus sample return mission, Proc. Workshop Canterbury, UK, 15–17 July, 1986, SP-249, pp. 143–151.

[68] W.A. Schutte, J.M. Greenberg, Further evidence for the OCN^- assignment to the XCN band in astrophysical analogs, Astron. Astrophys. 317 (1997) L43–L46.

[69] A. Kouchi, T. Yamamoto, T. Kozasa, T. Kuroda, J.M. Greenberg, Conditions for condensation and preservation of amorphous ice and crystallinity of astrophysical ices, Astron. Astrophys. 290 (1994) 1009–1018.

[70] J.A.M. McDonnell, W.M. Alexander, W.M. Burton, et al., The dust distribution within the inner coma of comet P/Halley 1982i: encounter by Giotto's impact detectors, Astron. Astrophys. 187 (1987) 719–741.

[71] J.A.M. McDonnell, J. Kissel, E. Grün, et al., In situ exploration of the dusty coma of comet P/Halley at Giotto's encounter—flux rates and time profiles from 10^{-19} kg to 10^{-5}, Adv. Space Res. 9 (3) (1989) 277–280.

[72] J.M. Greenberg, Fluffy comets, in: C.-I. Lagerkvist, B.A. Lindblad, et al. (Eds.), Asteroids, Comets, Meteors II, Reprocentrum HSC, Uppsala, 1986, pp. 221–223.

[73] B. Donn, Comet nucleus: some characteristics and a hypothesis on origin and structure, in: C. Ponnamperuma (Ed.), Comets and the Origin of Life, Kluwer, Dodrecht, 1981, pp. 21–29.

[74] J.M. Greenberg, J.I. Hage, From interstellar dust to comets: a unification of observational constraints, Astrophys. J. 361 (1990) 260–274.

[75] J.M. Greenberg, Evidence for the pristine nature of comet Halley, The comet nucleus sample return mission, Proc. Workshop Canterbury, UK, 15–17 July, 1986, SP-249, pp. 47–55.

[76] J.M. Greenberg, Making a comet nucleus, Astron. Astrophys. 330 (1998) 375–380.

[77] G. Horneck, The microbial world and the case for Mars, planetary and space, Science 48 (2000) 1053–1063.

[78] S.L. Miller, A production of amino acids under possible primitive Earth conditions, Science 117 (1953) 528–529.

[79] C.F. Chyba, Impact delivery and erosion of planetary oceans in the early inner Solar System, Nature 343 (1990) 129–133.

[80] J.R. Cronin, S. Pizzarello, Enantiomeric analysis of chiral organic compounds from the Murchison meteorite, Science 275 (1997) 951.

[81] M.H. Engel, S.A. Macko, Isotopic evidence for extraterrestrial non-racemic amino acids in the Murchison meteorite, Nature 389 (1997) 265–268.

[82] F.R. Krueger, J. Kissel, Biogenesis of cometary grains—thermodynamic aspects of self-organization, Origins Life Evol. Biosphere 19 (1989) 87–93.

[83] J. Kissel, F.R. Krueger, Urzeugung aus kometenstaub? Spektrum der Wissenschaft Mai (2000) 65–71.

[84] J.M. Greenberg, Predicting that comet Halley is dark, Nature 321 (1986) 385.

[85] V.A. Basiuk, J. Douda, R. Navarro-Gonzalez, Transport of extraterrestrial biomolecules to the earth: problem of thermal stability, Adv. Space Res. 24 (1999) 505–514.

[86] G. Nicolis, I. Prigogine, Self-organization in nonequilibrium systems, Wiley, New York, 1977.

[87] R. Briggs, G. Ertem, J.P. Ferris, J.M. Greenberg, P.J. McCain, C.X. Mendoza-Gómez, W. Schutte, Comet Halley as an aggregate of interstellar dust and further evidence for the photochemical formation of organics in the interstellar medium, Origins Life Evol. Biosphere 22 (1992) 287–307.

[88] V.K. Agarwal, W. Schutte, J.M. Greenberg, J.P. Ferris, R. Briggs, S. Connor, C.P.E.M. Van de Bult, F. Baas, Photochemical reactions in interstellar grains, photolysis of CO, NH_3 and H_2O, Origins Life Evol. Biosphere 16 (1985) 21–40.

[89] G. Moreels, J. Clairemidi, P. Hermine, P. Brechignac, P. Rousselot, Detection of a polycyclic aromatic molecule in comet P/Halley, Astron. Astrophys. 282 (1994) 643–656.

[90] H. Rickman, J.M. Greenberg, Tidal splitting of comets in earth's vicinity, in: J. Andersen (Ed.), Highlights of Astronomy, vol. 11A, 1998, pp. 262–265.

[91] B.J.R. Davidsson, Tidal splitting and rotational breakup of solid spheres, Icarus 142 (1999) 525–535.

[92] J.F. Kasting, Early evolution of the atmosphere and ocean, in: J.M. Greenberg, C.X. Mendoza-Gómez, V. Pirronello (Eds.), The Chemistry of Life's Origins, Kluwer, Dordrecht, 1993, pp. 149–176.

[93] J.M. Greenberg, The chemical composition of comets and possible contribution to planet composition and evolution, in: R. Smoluchowski, J.N. Bahcall, M.S. Matthews (Eds.), The Galaxy and the Solar System, Univ. of Arizona Press, Tucson, 1986, pp. 103–115.

[94] J.M. Greenberg, C.X. Mendoza-Gómez, The seeding of life by comets, Adv. Space Res. 12 (4) (1992) 169–180.

[95] C. Chyba, C. Sagan, Endogenous production, exogenous delivery and impact-shock synthesis of organic molecules: an inventory for the origins of life, Nature 355 (1992) 125–132.

[96] C.F. Chyba, P.J. Thomas, L. Brookshaw, C. Sagan, Cometary delivery of organic molecules to the early earth, Science 249 (1990) 366–373.

[97] M. Schidlowski, The initiation of biological processes on earth: summary of empirical evidence, in: M.H. Engel, S.A. Macko (Eds.), Organic Geochemistry, Plenum, New York, 1993, pp. 639–655.

It's a dusty Universe: surface science in space

David A. Williams [a], Eric Herbst [b],*

[a] *Department of Physics and Astronomy, University College London, Gower Street, London WC1E 6BT, UK*
[b] *Departments of Physics and Astronomy, The Ohio State University, Columbus OH 43210-1106, USA*

Received 27 June 2000; accepted for publication 24 April 2001

Abstract

We live in a dusty Universe! Dust is not only found in our solar system among the planets but is found in a wide variety of objects throughout the Universe, mainly in those regions between the stars known as interstellar clouds. Interstellar dust particles, which consist of cores of silicates and carbonaceous material often surrounded by icy mantles, are most probably highly irregular in shape with a size distribution from micro- to nanometers. Interstellar dust is important for many reasons, including the template it provides for surface chemical reactions that form, among other species, the most important interstellar molecule—H_2. In this article, we discuss the evidence for interstellar dust, its physical and chemical properties, its role in interstellar surface chemistry, and what remains to be learned. © 2001 Elsevier Science B.V. All rights reserved.

Keywords: Models of surface chemical reactions; Models of non-equilibrium phenomena; Atom–solid interactions; Diffusion and migration; Silicon oxides; Amorphous surfaces

1. The Universe is dusty!

1.1. Introduction

Around 1784, William Herschel, the musician turned astronomer at the court of King George III, noted the existence of small regions in the sky where there appeared to be a complete absence of stars. These regions were most easily seen against the rich star-fields of the Milky Way. Herschel was struck by these apparent voids in the distribution of stars, and called them "holes in the sky". From the evidence at the time, it was unclear whether these "holes" were true absences of stars, as Herschel proposed, or whether—as we now know—something was obscuring the radiation from a distant star field. In fact, although Herschel's observation was not interpreted in that way, this was probably the first indication that there is material in the space between the stars, the interstellar medium, and that this material is capable of absorbing starlight. Fig. 1 shows an apparent void in space, now known to be a rather dense interstellar cloud that is blocking the light of background stars.

The interstellar medium has been actively studied by astronomers since that time. At first, the obscuration was regarded merely as a hindrance to the observation of objects of proper astronomical concern, the stars and galaxies. Therefore, the efforts that astronomers made were simply to find

---

* Corresponding author. Tel.: +1-614-292-6951/5713; fax: +1-614-292-7557.
E-mail address: herbst@mps.ohio-state.edu (E. Herbst).

Fig. 1. The "Black Cloud" B68. The dust particles in this dense interstellar cloud extinguish most of the background visible starlight. Dense interstellar clouds are studied at longer wavelengths, which can penetrate the dust. Permission to reproduce the figure here has been granted by ESO, which holds the copyright. Taken from web site http://antwrp.gsfc.nasa.gov/apod/ap990511.html.

ways of accounting for the extinction as accurately as possible, so that the properties of stars and galaxies might be accurately known. Gradually, however, it became clear that the obscuring medium was interesting in its own right [1]. A variety of observations showed that the extinguishing agents in interstellar space are dust grains of a size comparable with the wavelength of visible light. These particles not only absorb starlight quite efficiently, but scatter it, too. The absorbed radiation heats the dust grains, which then re-radiate in the infrared to the millimeter-wave (depending on temperature): what goes in as optical radiation must come out as thermal radiation. It is now known that dust grains contribute approximately 1% of the mass of material in interstellar clouds, with a number density of dust roughly 1.3×10^{-12} that of the gas phase, if one assumes the "standard" grain radius of 0.1 μm. The gas is dominated by either atomic or molecular hydrogen, depending on the overall density of the gas, which can range from 10^2 cm^{-3} in the more diffuse regions to upwards of 10^4 cm^{-3} in denser sources.

The study of interstellar dust has revealed much about its properties and proved to be an exciting study of small particles in situ. More surprisingly still, it is now clear that dust grains are a critical and active component of the Galaxy, and indeed of other galaxies. Dust grains play a part in establishing the thermal and chemical structure of interstellar clouds, in the collapse of these clouds in the process of star formation, and—of course—in the formation of planets, for which they are the raw material. We shall describe some aspects of the roles of dust in this article.

1.2. Evidence for dust

Box 1 lists the various ways in which astronomers obtain information about interstellar dust. The traditional approach has been through a study of interstellar extinction and how it varies with wavelength, from the near infrared to the vacuum ultraviolet. This curve has rather little structure except for a wide bump at a wavelength near 220 nm, and can be readily modeled by a distribution of dust grains of sizes ranging from a few nanometers to about 1 μm. The grains are often assumed to be dielectric materials, such as silicates (for which there is infrared spectral evidence, see below), and carbonaceous material of various kinds is often included in such computations. In fact, these exercises do not give a unique solution because there are too many free parameters available to distinguish among different grain models.

Although these calculations are often carried out for grains that are assumed to be spherical, it is clear that some of them, at least, cannot be

---

Box 1. Methods of studying interstellar dust

1. General extinction curve of starlight in visible and ultraviolet.
2. Scattering and polarization of starlight.
3. Infrared absorption and emission features.
4. Re-radiation of absorbed starlight in infrared and millimeter-wave.
5. Depletions of elements from the gas phase.
6. Collection/analysis of interplanetary dust particles, especially the unprocessed component.

spherical. Starlight is often a few percent linearly polarized, and this is interpreted as being due to a selective extinction caused by dust grains that are partially aligned by the interstellar magnetic field. Thus, the interstellar medium acts rather like polaroid sunglasses, preferentially extinguishing one plane of polarization.

Starlight scattered by the dust can in some circumstances be detectable. These situations generally occur near to bright stars, and are called reflection nebulae. The variation of the scattered light with wavelength and with position in the nebula places a valuable extra constraint on the dust grain composition and size distribution, independently of the information contained in the extinction curve. However, accurate measurements of scattered light are rather difficult to make.

Atoms contained in dust grains are obviously not present in the gas! Some detailed information about the chemical (not physical) nature of the dust can be obtained by asking: what is missing from the interstellar gas? These "depletions" from an assumed standard stellar composition are very instructive indeed. Box 2 shows typical stellar abundances of selected atoms (with respect to hydrogen) along with the fraction of these atoms remaining in the gas phase in one particular cloud towards a bright star called Zeta Ophiuchi [2]. Some elements like iron, calcium, and titanium are very heavily depleted in this cloud; for example, less than one calcium atom in a thousand remains in the gas. Carbon is modestly depleted, but its total abundance is in fact very much greater than that of the metals, so the amount of carbon in the dust is quite large.

The most informative approach to determine the nature of the dust lies in infrared spectroscopy. This can only with great difficulty be carried out from the ground, because of the complications introduced by the Earth's atmosphere. The most accurate recent measurements have been made by a European satellite—the Infrared Space Observatory (ISO). The composition of cold and warm dust has been determined by ISO observations. Absorption by amorphous and crystalline silicates has been clearly identified, as has absorption by molecular ices including water, carbon dioxide, carbon monoxide, and methanol [3]. Fig. 2 shows an absorption spectrum of cold interstellar dust taken against continuum infrared radiation from an embedded stellar object within a dense cloud. The absorption features have been assigned in terms of molecular ice mantles and silicate dust cores. Warm dust shows spectral emissions of a range of features in the near- to mid-infrared that are characteristic of polycyclic aromatic hydrocarbons (PAHs) [4]. A typical emission spectrum is shown in Fig. 3. However, no identifications of individual PAHs have yet been made, nor is it clear whether the emitters are individual large molecules or assemblies of smaller molecules in a solid matrix.

The continuum re-emission of starlight absorbed by dust grains of a size comparable with the wavelength of visible light is detected by infrared observatories like ISO and by ground-based telescopes operating at millimeter wavelengths. This continuum emission is proving to be a very effective way of detecting the presence of interstellar matter, not only in our own Galaxy, but quite distant galaxies, too [5]. Dust has therefore been present in the Universe for a very long time, and the elements required for dust must have been formed rapidly in the early stages of the existence of the galaxies.

There is one further strand of information that helps us to determine the nature of interstellar dust. We can collect interplanetary dust as it falls

Box 2. Elemental depletions from the gas towards Zeta Ophiuchi					
Element	Stellar	Depletion	Element	Stellar	Depletion
C	2.1×10^{-4}	0.63	Cr	3.2×10^{-7}	1.1×10^{-2}
O	4.6×10^{-4}	0.72	Fe	2.7×10^{-5}	7.1×10^{-3}
Mg	2.5×10^{-5}	0.23	Ti	6.5×10^{-8}	4.2×10^{-3}
Si	1.9×10^{-5}	0.09	Ca	1.6×10^{-6}	3.2×10^{-4}

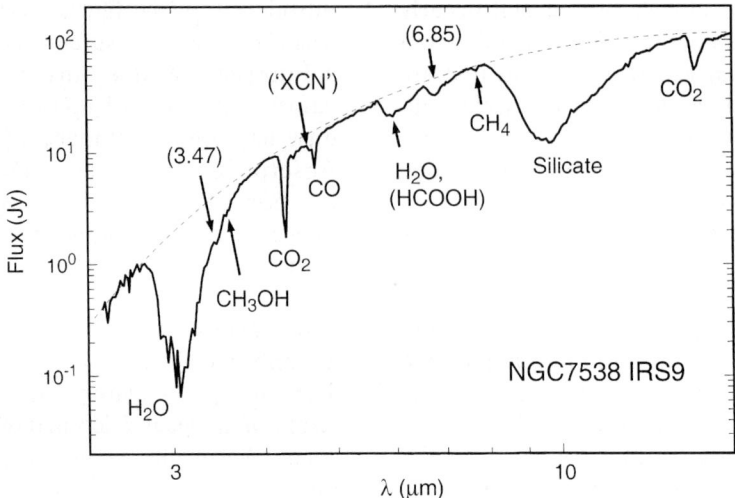

Fig. 2. The ISO spectrum of the embedded stellar object NGC7538 IRS9. The flux of radiation (in Janskys; 1 Jy = 10^{-26} (W/m^2)/Hz is plotted against the wavelength in μm in the wavelength range 2.4–17 μm. The spectrum shows absorption features arising from vibrations of various cold molecular species in the ice mantles and cores of dust grains. The dashed line is an estimate of the continuum. Uncertain (viz. 3.47 and 6.85 μm) or ambiguous assignments (viz. XCN and HCOOH) are in brackets. Taken, with permission, from volume 122 of the Astronomical Society of the Pacific Conference Series, 390 Ashton Avenue, San Francisco, CA.

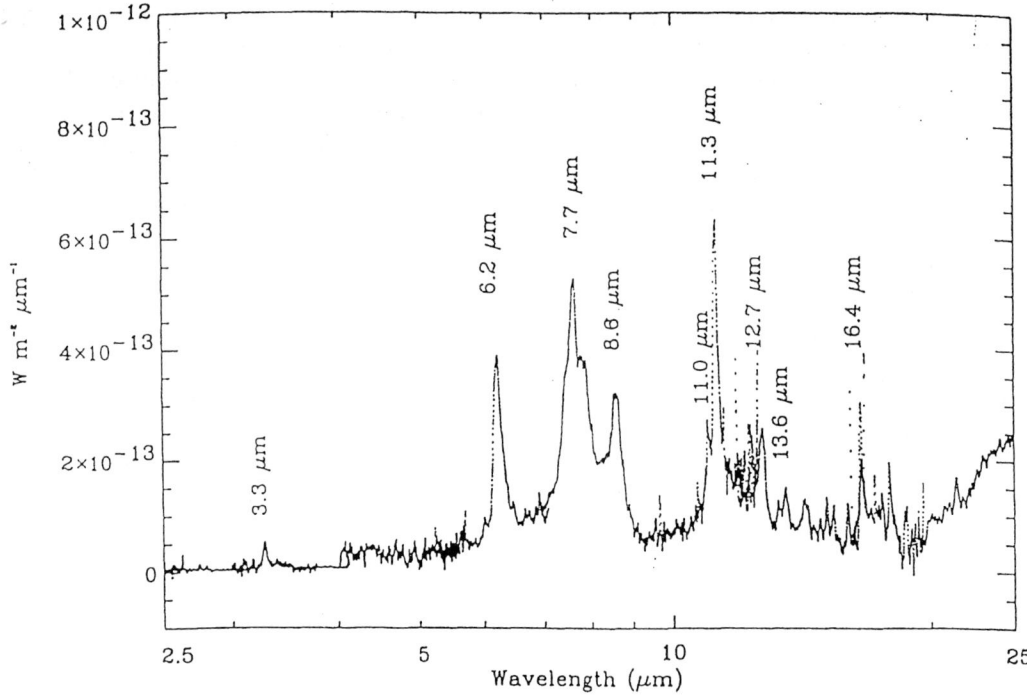

Fig. 3. The ISO spectrum of the warm reflection nebula NGC7023 plotted in terms of watts per square meter per μm vs. wavelength in μm. The spectrum shows emission features at wavelengths from 3.3 to 11.3 μm normally interpreted in terms of the vibrations of PAHs. Taken, with permission, from C. Moutou et al. [23].

onto the Earth. In general, interplanetary dust comes from a variety of sources, including the erosion of asteroids and the impact of asteroids on planets. Such dust has been heavily affected by the physical processing the material has received. However, interplanetary dust also contains a component that seems to be unmodified; it shows no evidence of a high temperature phase, nor of shocks, and this component may be unmodified interstellar dust. Interstellar dust is expected to enter the solar system as it moves through interstellar space. Such dust was apparently identified by the Ulysses spacecraft, on its journey to the outer parts of the solar system [6]. Recent work by Bradley and co-workers has identified glasses with embedded metals and sulfides, or "GEMS", that have a remarkable similarity in terms of physical and chemical composition and sizes with what is expected for interstellar dust [7]. Specifically, GEMS are typically 100 nm in size, and are mainly glassy with 10 nm-size globules of iron and nickel sulfides. Fig. 4 shows a picture of recovered GEMS. It is also interesting to look at the morphology of collected interplanetary particles larger than GEMS. Fig. 5 shows a rather large (10 µm) piece of dust obtained in the stratosphere with

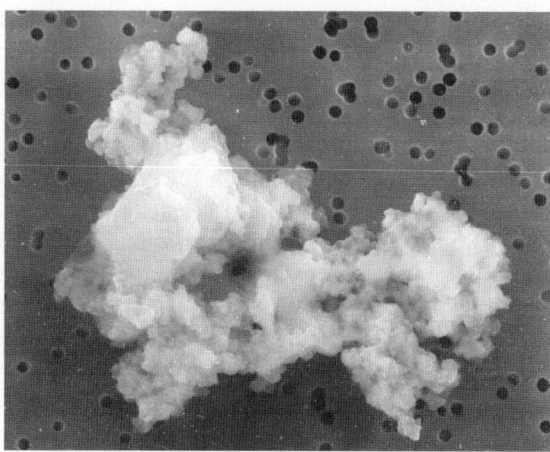

Fig. 5. A 10-µm interplanetary dust particle known as a Brownlee particle. Collected in the stratosphere, it is composed of glass, carbon, and mineral silicates. Acknowledgment is made to NASA for allowing reproduction of this picture from web site http://stardust.jpl.nasa.gov/science/sci2.html.

portions containing carbon and different types of silicate minerals. This image may suggest what some of the larger interstellar grains may look like.

Information from all these strands leads to some uniformity of view as to the nature of interstellar dust, which is listed in Box 3.

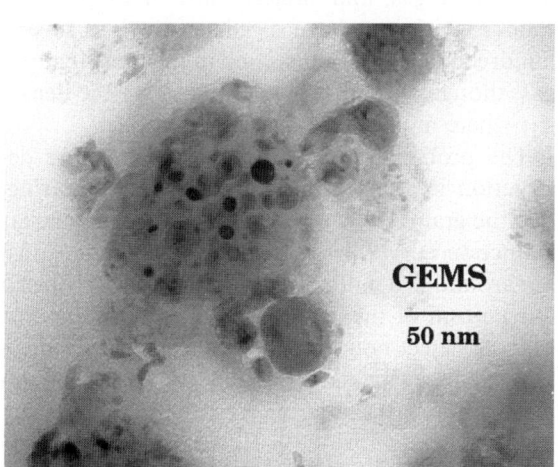

Fig. 4. Particles known as GEMS (glasses with embedded metal and sulfides). Usually a hundred nm in size, these primitive interplanetary dust particles are glassy with large numbers of smaller grains. Acknowledgment is made to NASA for allowing reproduction of this picture from web site http://stardust.jpl.nasa.gov/science/sci2.html.

Box 3. The nature of interstellar dust
1. Ranges in size from about 1 nm to 10 µm, with many more small grains than large.
2. The larger grains are likely to be non-spherical, and perhaps porous, fluffy, and even fractal.
3. Composition is likely to contain cores of metallic silicates, carbonaceous material, and GEMS, probably in different populations rather than mixed in individual grains.
4. Mantles of ices (H_2O, CO, CO_2, CH_3OH) are found in cold dense regions.

2. Sources and sinks of interstellar dust

2.1. Where does the dust come from?

Hold a glass above a candle flame and it becomes covered in soot. The flame of the candle is quite evidently smoky, and sooty particles formed somehow in the hot gas of the flame are expelled as the gas rises upwards against gravity. A similar process seems to be working in the atmospheres of hot stars. Towards the end of their lives, some stars develop rather extended atmospheres that may be cool and dense enough for solid nuclei to form. Then any atoms and molecules that are supersaturated will be rapidly deposited on these nuclei, forming solid particles. Radiation pressure from the central star is then capable of accelerating the dust and establishing an outflow in the entire dusty envelope, which travels out from the star until it mixes with the interstellar gas.

We do not fully understand the details of the nucleation process, whether in a stellar atmosphere or in a candle flame. However, this general picture for the formation of interstellar dust seems to be a valid one, for we can actually see in real time the formation of dust in a cyclic pulsation process in some stellar atmospheres. First, we see the rapid darkening of the underlying star as the dust forms abruptly in a runaway event; this is accompanied by a rise in the infrared emission from the star, as starlight is converted to heat in the dust. Then as the gas and dust drift away from the star, the effects of extinction are diluted, and the star shines brightly once again. Such outflows from stars are a fairly common phenomenon.

Outflows can be carbon rich or oxygen rich. This means that, after hydrogen and helium, the most abundant element is either carbon or oxygen, respectively. The stellar nuclear physics and the convective processes in the star determine the category to which the outflow belongs. The two types determine the nature of the dust that is produced. Where carbon is more abundant than oxygen, the oxygen is almost entirely locked up in carbon monoxide, and the excess carbon produces a rich chemistry of carbon-based molecules and of carbonaceous dust, i.e. sooty grains. On the other hand, where oxygen in an envelope is more abundant than carbon, the excess oxygen (after CO is formed) makes metallic oxides that nucleate to form solid oxides and silicates.

Cool stellar envelopes are not the only source of dust. Stellar explosions can also make a substantial contribution. Novae and supernovae are eruptions of stars near the end of their lives. An eruption can eject a parcel of gas in which conditions may evolve in such a way that nucleation occurs and dust forms. This is often seen in a nova: first, the star brightens enormously; then it fades and the infrared emission rises as the dust is formed, and finally the star becomes optically bright once more as the dusty ejecta are dissipated.

In total, these sources inject into the interstellar space of our Galaxy about 30 000 times the mass of our Sun, in the form of dust, in each year. Similar processes occur in other galaxies, too. Will the Galaxy get more and more dusty, as time goes on?

2.2. What happens to the dust?

There is a limit to the amount of dirt (or dust) that the Galaxy contains, because there are "cleansing" processes in operation, too. The dust ejected from a star does not remain localized around that source, like a pall of smoke over a bonfire on a windless day. The interstellar medium is rather violently stirred by explosions of supernovae; these occur rather frequently in the Galaxy (perhaps more than once a century) and have Galaxy-wide effects. The explosions maintain a complicated density–temperature structure in the interstellar gas, and subject most of it to strong shocks which may have speeds as much as several hundred kilometers per second in the most tenuous gas, though slower, around 10 km/s, in the denser gas where most of the dust grains reside.

The probable fate of the dust is therefore destruction in shocks, via sputtering or shattering. For the grains of sizes that generate the interstellar extinction curve, the consequent lifetimes of dust are around 500 million years. This lifetime is just a few percent of the age of the Galaxy. So the dust that we detect now is really rather recent, in these terms! Somewhat larger grains, however, are affected more by shattering and their lifetimes are smaller by a factor of 10. However, in spite of all the supernovae and their associated shocks, about one-tenth of the dust in the interstellar medium at any time has not been shocked.

Starlight can also alter the chemical and physical nature of carbonaceous dust; for example, H-rich tetrahedrally bonded amorphous carbon is converted under UV irradiation to an H-poor triply bonded form. The two forms have quite different optical and chemical properties. Similarly, crystalline silicates that may be a component emerging from O-rich stellar atmospheres can be significantly affected by the flux of energetic protons and electrons (the cosmic ray particles). The long-range order in the lattice is removed and the material becomes amorphous. Around young low-mass stars, dust particles coagulate into larger bodies, eventually forming primitive objects such as comets, and into the planetesimals that are the building blocks of planets.

3. Surface chemistry: the formation of molecular hydrogen

In dense interstellar clouds, a large number of molecular species have been identified in the gas phase via high resolution spectral techniques, chiefly rotational spectroscopy in the millimeter-wave region of the electromagnetic spectrum. These molecules range in complexity from H_2 to a 13-atom linear nitrile, $HC_{11}N$, and are mainly organic in nature [8]. Many of the molecules are quite unusual by terrestrial standards and serve to show that our standard ideas on what molecules are stable and what molecules are not stable are biased by terrestrial conditions. Since hydrogen is the dominant element in the Universe, H_2 is the most abundant molecule by far, with CO in second place, four orders of magnitude lower in concentration. More complex organic species are even less abundant, although the amount of material tied up in these species is very large compared with the analogous amount on our planet. How and where are the molecules formed?

Although the low density ($n \approx 10^4$ cm^{-3}), low temperature ($T \approx 10$ K) conditions in "dense" interstellar clouds do not appear to be favorable for chemistry, there is no real alternative to this local hypothesis. Interstellar clouds are formed from the detritus of previous generations of stars that blow out matter in the form of dust particles and gas. Despite the fact that molecules can indeed be formed in the ejecta of old stars, these molecules are "soon" (within about 100 years) torn apart by the harsh ultraviolet radiation field found in interstellar space. So the gas-phase material forming interstellar clouds is essentially atomic in nature. Molecular synthesis then takes place over the eons despite the low densities and temperatures; the molecules formed by the in situ chemistry are protected from radiation to a great extent by the dust particles.

For more than 30 years, chemists and astronomers have investigated gas-phase formation and destruction routes for most of the molecules detected in the gas phase. The major reactions are exothermic ion–molecule reactions, since these are rapid and are known to occur even at very low temperatures. Atomic and molecular ions are produced in dense interstellar clouds via collisions with cosmic rays, which are particles (mainly protons) formed in supernovae and moving at relativistic speeds throughout the Universe. Simulations that include only gas-phase chemical reactions are successful in producing most ($\sim 80\%$) of the molecules observed in the gas phase [8,9]. But there are some glaring exemptions, among which the dominant molecule H_2 is clearly the most dramatic. This species cannot be formed when two hydrogen atoms collide in the gas, because the two atoms cannot stick together with high efficiency. Instead, the only feasible synthetic pathway involves the surfaces of interstellar dust particles.

Consider a hydrogen atom striking a low temperature dust grain. The sticking probability is known to be high for a variety of surfaces representative of the interstellar medium. An atom of hydrogen can stick to a grain, if not to another hydrogen atom, because the grain is a thermodynamic entity and converts the energy of collision into a rise in temperature. The surface formation of H_2 can occur via two different routes. First, let us imagine a situation in which weak binding occurs between hydrogen atoms and the grain. An adsorbed H atom is then relatively free to diffuse over the grain either by tunneling from binding site to site or by thermally hopping over the barriers between such sites. The detailed nature of the diffusion has been investigated by many people, but appears to be very sensitive to the composition of the surface, including impurities. Since the binding is weak, re-evaporation (desorption) into the gas phase is also possible, even if the temperature is very low. The formation of molecular hydrogen occurs when two diffusing H atoms approach one another and transfer enough of the energy of molecular formation to the grain that the H_2 species is stabilized. The newly formed molecule can desorb rapidly from the grain if some of the exothermicity of reaction can be channeled into translational motion perpendicular to the surface. Evaporation at a later time is also possible. The diffusive process is known as the Langmuir–Hinshelwood mechanism, and is probably the dominant interstellar process on surfaces such as silicates and amorphous ice. Following work of Salpeter and co-workers, astronomers had long thought that H atoms landing on grains move rapidly over the entire grain by tunneling from site to site; such rapid motions ensure that there is a high efficiency of forming H_2 despite the possibility of evaporation [10]. Within the last several years, however, detailed experimental measurements on the diffusive formation of H_2 have been carried out at low relevant temperatures [11] on olivine and amorphous carbon. It was found that diffusion of H atoms occurs much more slowly than envisaged by astronomers and that it occurs mainly by thermal hopping rather than tunneling. It was also found that the newly formed H_2 is desorbed from the cold surface a significant percentage of the time. Taking into account the new laboratory results, it still appears that, under interstellar conditions, H_2 can be formed by a diffusive mechanism on a surface such as olivine, but the temperature range over which this can happen is much smaller than previously assumed. At lower temperatures, diffusion does not occur efficiently, while at higher temperatures, evaporation occurs before reaction.

An alternative possibility for surface H_2 formation occurs if the H atoms are bound sufficiently strongly to the surface that diffusion is not competitive. In this case, evaporation is also unlikely to be important, and interstellar grains will "rapidly" form monolayers of atomic hydrogen. The formation of molecular hydrogen can then occur via an Eley–Rideal mechanism, in which a gas-phase hydrogen atom lands atop a surface hydrogen atom, forming a molecule which leaves the grain surface. Although experiments on such a mechanism have yet to be undertaken, theoreticians have been quick to join the fray. Fully quantum calculations on a material with the properties of graphite show that H_2 formation can occur via an Eley–Rideal mechanism [12].

Although the exact mechanism of H_2 formation on grains is still uncertain, the total conversion of atomic to molecular hydrogen can occur on grain surfaces in a time of $\sim 10^5$ years if it is assumed that nearly all H atoms that land on grains eventually form H_2. The time of 10^5 years is relatively short in an astronomical sense, but this time scale could be much longer if the formation of H_2 on grains is much less than 100% efficient. The actual astronomical efficiency is, of course, highly uncertain. Clearly, astronomers need help from scientists more knowledgeable in surface science!

4. Surface chemistry: the formation of other species

Molecular hydrogen is not the only species that can be formed at low temperatures on the surfaces of interstellar grains. If hydrogen atoms diffuse rapidly over a grain surface, they can interact not only with other hydrogen atoms, but with heavier species as well. Although more slowly moving than H atoms, heavier species can also react with one another. In a low temperature medium, the reactive species are likely to be limited in the main to atoms and radicals since they can react with zero or at most small activation energy. Positively charged species, which form in the interstellar gas, are likely to be neutralized when they strike granular surfaces because the surfaces are thought to be negatively charged given the greater thermal velocity of electrons compared with positive atomic and molecular ions. There is the usual astronomical uncertainty in this argument. Neutralization can also occur on neutral grains via resonant or Auger processes.

Unlike the gas phase, most reactions occurring on the grain surfaces are thought to be associative in nature, like the formation of H_2. If we consider two reactive species A and B, an association reaction is one in which only one product, labeled C, is formed:

$$A + B \rightarrow C \quad (1)$$

while a "normal" reaction produces more than one product. The difficulty in producing only one product is that the species A and B must literally stick together. In the gas, this sticking can occur by the emission of excess energy via radiation ("radiative association") or, at sufficiently high densities, via an inelastic collision with a third body. On a grain surface, the grain acts as the third body so that energy is always being taken out of the reactive system. Thus, it is expected that "normal" reactions on grains will occur only if there is no association product with a strong chemical bond between A and B. The competition between associative and normal reactions is not fully understood, but some work has recently been undertaken on this subject [13].

Some of the types of (exothermic) surface reactions that can occur under interstellar conditions are shown in Box 4.

Box 4. Types of grain surface reactions in interstellar clouds

(i) $H + X \rightarrow HX$
(ii) $X + Y \rightarrow XY$
(iii) $H + HX \rightarrow H_2 + X$
(iv) $H_2 + X \rightarrow H + HX$

The first type is obviously analogous to molecular hydrogen formation. Among the salient reaction pathways occurring via sequences of reactions of type (i) are the hydrogenation of O, C, N, and S into the saturated forms H_2O, CH_4, NH_3, and H_2S, all formed much more efficiently than in the gas phase. An example of reaction type (ii) is the formation of CO_2 from CO and O. Reaction class (iii) contains processes such as the abstraction of a hydrogen atom when H reacts with H_2S to form H_2 and HS with a small activation energy. Finally, an example of reaction type (iv), which always occurs with some activation energy, is the hydrogenation of OH via H_2: $OH + H_2 \rightarrow H_2O + H$. Although as dense clouds mature, H_2 becomes far more abundant than H in the gas phase ($H/H_2 \sim 10^{-3}$–10^{-4}), the reactivity of atomic hydrogen on grains is such that it is still the dominant reactive partner.

One very important sequence of reactions in class (i) is the slow hydrogenation of surface CO into methanol via successive reactions with atomic hydrogen: $CO + H \rightarrow HCO$; $HCO + H \rightarrow H_2CO$; $H_2CO + H \rightarrow H_3CO$; $H_3CO + H \rightarrow CH_3OH$. There are competitive normal channels and there are activation energy barriers for several of the reactions along the pathway. Yet low temperature laboratory work and theory indicate that the production of methanol may take place [13].

All of the above reaction types occur in interstellar clouds if the mechanism for surface chemistry is indeed diffusive in nature. If, however, the Eley–Rideal mechanism dominates, which reactions are important has yet to be even guessed at. One class of dust particles for which this latter mechanism is likely to be operative are those particles considerably smaller than the standard size of 0.1 μm yet too large to be thought of as gas-phase molecules. For such small particles, weakly bound adsorbates are likely to be desorbed efficiently by an assortment of thermal pulses, and chemical bonds between adsorbate and particle are more likely to be formed, so that strongly bound species may dominate [14].

5. Interstellar surface chemistry: simulation

In order to model the chemistry of interstellar clouds, astronomers have to follow the chemistry occurring both in the gas-phase and on the surfaces of the dust particles. The gas-phase chemistry is typically treated via rate equations. Consider a gas-phase species C that is one product of a reaction between A and B and is destroyed by reaction with D. Its rate equation is given by the expression

$$d[C]/dt = k_{A-B}[A][B] - k_{C-D}[C][D] \quad (2)$$

where the symbol [ ] refers to concentration, and the k's are so-called rate coefficients. If surface chemistry could be ignored the solution of the gas-phase chemistry could be undertaken by simultaneous integration of rate equations for all species in the model given initial concentrations and either fixed or varying physical conditions. Such simulations have been done, although they are not 100% gas-phase in nature since they contain high abundances of H$_2$ as an initial condition. The result is a series of time-dependent concentrations which for many molecules at selected times are in reasonable agreement with what is observed. But the dust cannot be ignored! The existence of dust particles leads to two new terms on the right-hand side of Eq. (2)—one for loss of C onto the surface (adsorption) and one for desorption of surface C back into the gas phase. At the low temperatures of most regions in interstellar clouds, thermal evaporation of adsorbates within even astronomical time scales is limited to light species such as H, H$_2$, and He. Several intermittent sources of energy have been suggested to cause non-thermal desorption, including grain–grain collisions, cosmic ray bombardment, the exoergicity of chemical reactions, and photons. The efficiency of these mechanisms does not appear to be high, so that in the absence of star formation (which raises the temperature or causes shock waves) adsorption eventually triumphs for heavy species, the gas phase molecules decline in concentration, and large mantles inexorably build up on the interstellar grains. Of course, "eventually" is a long time even in astronomy, and gas-phase abundances can remain high for quite some time!

The mathematical treatment of surface chemistry depends on the operative mechanism, which we here assume to be diffusive. Consider the surface of a grain with large numbers of diffusive and reactive species. If we once again consider a species C that is formed by reaction of A and B and destroyed by reaction with species D, we obtain an analogous rate formula of the type

$$dN_C/dt = K_{A-B} N_A N_B - K_{C-D} N_C N_D \quad (3)$$

where the symbol N stands for the number of molecules of a particular species on the surface. The rate coefficients K are quite different from their gas-phase counterparts since they depend on diffusion rather than simple straight-line motions, as in the gas. If we assume that diffusion occurs via thermal hopping from site to site, the rate coefficient K_{A-B} can be expressed via the relation [15]

$$K_{A-B} = \kappa\{v_A \exp(-E_A/kT) + v_B(-E_B/kT)\}/N_{tot} \quad (4)$$

where v_A and v_B are surface vibrational frequencies for A and B, respectively, E_A and E_B are energy barriers against diffusion from one binding site to another for A and B, respectively, k is the Boltzmann constant, T is the temperature, N_{tot} is the number of binding sites on the surface, and κ is a coefficient expressing the likelihood of reaction if species A and B approach one another closely. The parameter κ is assumed to be unity for exothermic reactions unless there is an activation energy barrier against reaction, in which case $\kappa < 1$ is determined by the greater of the tunneling or hopping probability. The frequencies multiplied by Boltzmann factors give the rate per second of hopping to an adjacent binding site. Upon substitution of the expression for K_{A-B} into Eq. (3), one can see that the overall rate of formation of C can be thought of as the sum of two terms—one involving the diffusion of A and the other that of B. For A, the term consists of the probability per unit time that species A moves from one binding site to another multiplied by the probability (N_B/N_{tot}) that the neighboring site is occupied by B, multiplied by the number of species A on the surface, multiplied by the probability of reaction upon a close encounter of A and B. A similar term, with equally many factors, exists for B. The destruction of species C by species D can be similarly formulated. The rate law for surface species C in Eq. (3) has to be modified, if gas-phase species are included, by appropriate terms for adsorption onto and desorption from the surface.

Do these rate laws for surface chemistry pertain to diffusive surface chemistry on interstellar grains? Not with any certainty! The basic problem is that grains are small, and large numbers of reactive species may not be present at any given time. If, on average, the number of species A on a grain is significantly less than unity, it is not at all clear that the abundance of A is determined by its rate of diffusion on the grain. Consider, instead, the so-called accretion limit: species A lands on a grain, species B lands thereafter, and before a third species can land, A and B react to form C. The rate of production of C in a given time interval is not dependent on the diffusion rates of A and B as long as they are fast enough to cause reaction before the third species lands on the surface. In this instance, it is closer to reality to avoid the use of rate equations and to utilize a Monte Carlo approach, in which the order of successive landings on a particular grain is calculated probabilistically, and reaction is assumed to occur before another landing if the species are (a) reactive and (b) capable of diffusing rapidly. One problem with this approach, however, is that the accretion limit is not always applicable.

Because the Monte Carlo approach has not yet been successfully introduced into a true time-dependent calculation, simulations of grain chemistry (often coupled with gas-phase chemistry) using rate equations are much more common, even if of uncertain value. For truly simple systems in which the gas-phase abundances of a small number of species are held fixed although the species are allowed to accrete onto grains, the rate equation method can be tested against the Monte Carlo approach in the accretion limit. Such a comparison shows that when a cloud evolves to the point that much of the gas-phase atomic hydrogen has been converted into molecular hydrogen, the remaining rapid processes occurring on dust surfaces may not be handled well by the rate equation approach. In particular, for grains at a temperature of 10 K, it is found that for a limited system of O and H atoms adsorbing onto grains, the rate equations tend to produce too much H_2 and not enough O_2 under some conditions [16].

Some semi-empirical modifications of the rate equation method for grain surface chemistry have been formulated that allow the results of this approach to agree with the results of the Monte Carlo method in the limit that the latter approach is correct (the accretion limit) [16]. The original formulation has since been modified. For the simple O, H system, one must limit the diffusion rates of H and O so that they do not exceed the larger of the evaporation and accretion rates for each atom. The correction is much more important for the faster-moving H. These suggested modifications have not met with universal praise, but providence has been merciful to their authors. The recent experimental work concerning the formation of H_2 on olivine indicates that the diffusion rate of H is sufficiently slow that the proposed modification is barely necessary at 10 K, so that the simple rate method may be a reasonable approximation after all.

What do the admittedly imperfect simulations of surface chemistry tell us? This depends on the physical conditions of the astronomical source and how they change with time. As long as a region remains cold and free of shock waves, mantles containing an assortment of ices grow due both to surface chemistry and to adsorption from the gas. Other regions experience a sufficient warming up due to star formation that the mantles are lost. The molecules ejected into the gas then strongly affect the gas-phase chemistry, as we shall see.

6. Interstellar ices

Stars and star-forming regions often emit enough infrared radiation to act as spectroscopic light sources. Their emission can be partially absorbed by cold material in dense clouds between them and us. The absorption is in the form of sharp features, attributable to gaseous molecules, and, as mentioned earlier, broad features, attributable to molecules both in the cores of dust particles and on their surfaces. Some broad features and their assignments are shown in Fig. 2. The broad absorption spectra, when coupled with laboratory simulations, tell us the species responsible for the absorption, the relative abundances of the species, and even the environments in which the molecules are located. Since the features are broad, one must be careful about definitive assignments, but it must be realized that strong features can only be carried by molecular species that contain the more abundant elements present in the Universe.

One of the best-studied objects lies in the direction of the so-called "field star" Elias 16, for which the line of sight passes through dense interstellar material in the constellation Taurus [17]. The strongest mantle features seen in the spectrum belong to water ice, which is estimated to contain much of the elemental oxygen available (roughly 10^{-4} of the abundance of hydrogen) and to comprise up to 100 monolayers of material per grain. Next in abundance come CO and CO_2 (dry ice), which constitute 25% and 22% of the water ice, respectively. The CO is divided into two environments—an apolar environment (22%) and a polar one (3%).

More commonly, infrared absorption spectra arise from cool gas and dust in front of IR-emitting young stellar objects ("protostars") within dense interstellar clouds. The spectrum in Fig. 2 is of this type. These protostars tend to warm up the area around them so that the physical conditions in the foreground absorbing regions are neither as well characterized nor as homogeneous as when the source of the infrared radiation is a background field star. Still, the existence of ices in these regions implies that temperatures are quite low ($\ll 100$ K). In addition to the H_2O, CO, and CO_2 seen towards Elias 16, significant abundances of methanol, methane, formaldehyde, OCS, and formic acid are present in grain mantles.

Whatever the source of the radiation, the derived abundances of at least some of these surface species, especially water, CO_2, and CH_3OH, are much too high to be explained by gas-phase processes followed by adsorption within reasonable time periods. Consequently, it would appear that surface chemistry is needed to explain their abundances, and excellent laboratory work supports this interpretation [18]. Currently, detailed simulations of gas-phase and diffusive grain chemistry in dense cold regions are able to produce large abundances of water ice, carbon monoxide, and methanol, but do not appear able to explain the large abundance of dry ice (carbon dioxide) [19].

Surface chemistry can be activated by photochemical energy or the bombardment of high-energy particles such as cosmic rays. Laboratory groups have shown that high doses of radiation or particulate bombardment can indeed enhance the production rate of CO_2 and many other molecules [18]. In addition, with suitable starting materials, large abundances of refractory organic species can be produced. The question remains as to whether these results are pertinent to the much lower photon and particulate fluxes present in interstellar clouds, even in the vicinity of newly formed stars. Simulations of both laboratory and interstellar photochemistry are clearly needed.

Finally, when many monolayers of ices are present, reactants need not remain on the surface but can diffuse vertically into the mantle. Chemistry can then occur underneath the surface.

7. Protoplanetary disks, comets, and planets

In recent years, astronomers have spent much time studying the formation of low-mass stars analogous to the sun. As dense cores within large interstellar clouds collapse to form such stars, they develop so-called "accretion" disks of gas and dust that revolve around a central object. The circular disks, which are quite planar in nature close to the central star, are sometimes referred to as protoplanetary, since it is thought that they will evolve eventually into solar-type planetary systems. Protoplanetary disks are typically much larger than our solar system, but the material in the inner portions is gradually accreting inwards as it revolves around the central object. The densities are much higher than in normal interstellar clouds, especially in those regions close to the newly forming star. In the mid-plane of a typical disk, much of the gas-phase molecular material has already been adsorbed onto the dust particles.

Simulations of the chemistry occurring in protoplanetary disks must start with a portion of a dense interstellar cloud, follow the gas-phase and surface chemistry during collapse into the disk phase, and then follow the chemistry occurring in the disk. A variety of groups have carried out such calculations (e.g. Ref. [20]) but only recently has surface chemistry been included in the models. The simulations show the inventory of both gas-phase and surface molecules to be strongly dependent on the age of the protoplanetary disk and the position in the object. The results for the outer regions can be compared with radioastronomical observations of the gas-phase molecules, while the results for the inner regions, where the gas phase is heavily depleted, can be compared with molecules found in primitive solar system bodies such as comets.

One of the most interesting recent cometary observations concerns the deuterium-to-hydrogen elemental abundance ratio in the cometary molecules HCN and H_2O. The ratios DCN/HCN and HDO/H_2O are several orders of magnitude higher than the elemental D/H ratio seen in most bodies of the Universe, an effect known as fractionation. The fractionation seen in molecules in cold interstellar clouds is even stronger than in comets due to a variety of gas-phase and surface chemical processes. Models of protoplanetary disks show that much of the fractionation detected in interstellar clouds is preserved on the surfaces of dust particles in the inner regions of disks, which are presumably the forebears of comets [21].

8. Hot cores

The formation of high-mass stars leads to some spectacular events. It also leads to some more prosaic happenings, among which are so-called "hot cores", named imaginatively after the first such object, seen in the Orion dense interstellar cloud. Hot cores are small (≈ 0.3 light years), dense ($n \approx 10^6$–10^7 cm^{-3}), and warm ($T \approx 100$–200 K) clumps of material, in which the molecular material in the gas phase is noticeably different from that in cooler regions of dense interstellar clouds. In particular, small saturated molecules such as NH_3, H_2O and methanol (CH_3OH) are far more abundant than in the ambient medium, while larger organic molecules such as methyl formate ($HCOOCH_3$), dimethyl ether (CH_3OCH_3), ethanol (C_2H_5OH), and ethyl cyanide (C_2H_5CN) are seen only in such regions. The detection of large abundances of ethanol in hot cores was particularly noticed by the press, with headlines such as "pub at the end of the Universe" and "Scotch on the rocks." The detection of somewhat larger species than currently seen in hot cores (methyl-ethyl ether, diethyl ether, ethyl formate, etc.) is not yet feasible because the laboratory spectra in the relevant spectral ranges used by radioastronomers to detect molecules have not been studied sufficiently.

The dominant, perhaps even correct, school of thought is that the molecules detected in hot cores arise from chemical processes occurring both before star formation proceeds and during this process. Before the onset of the collapse leading to star formation, gas-phase and surface chemistry proceed under typical interstellar conditions ($n \approx 10^4$ cm^{-3}, $T \approx 10$ K). Mantles containing hydrogen-rich ices are synthesized on grains, with atomic hydrogen playing the main role in the hydrogenation leading to H_2O, NH_3, CH_3OH and other species. During star formation, the temperature gradually rises to 100–200 K and the grain mantles evaporate into the gas. Some grain chemistry may also occur during the heating up period, especially involving species in the inner mantle, which desorb relatively slowly. There follows a "high" temperature gas-phase chemistry that takes roughly 10^5 years to expunge the initial conditions and reach a new steady state. During this period, gas-phase reactions are thought to convert a significant proportion of the liberated methanol into more complex organic alcohols (e.g. ethanol), ethers (e.g. dimethyl ether), and other classes of molecules [22]. Thus, what one detects is the result of both a previous era of low temperature surface chemistry and a more recent era of gas-phase chemistry. Selected molecules such as H_2S can even be used as clocks of the ages of the hot core regions.

Hot cores are not identical in chemical composition. In addition to the effect of age, there are also much greater differences that must reflect different surface processes from the previous cold era. For example, in the Orion molecular cloud, there are two closely spaced hot cores, one of which (The "Compact Ridge") shows higher abundances of oxygen-containing organics while the other (The "Hot Core") shows higher abundances of nitrogen-containing species. One possibility for explaining these chemical differences is that during the cold era the Hot Core was somewhat warmer than the Compact Ridge so that gas-phase CO could not reside on grain surfaces long enough to be hydrogenated into methanol, the precursor of oxygen-containing organics.

One molecule of extreme interest because of its association with life is the simplest amino acid—glycine. Despite much effort, however, glycine has not been detected unambiguously in hot cores. Of course, looked at more rationally, regions around young high-mass stars are not likely to form life-sustaining or even giant planets since high-mass stars live for only relatively short periods of time. Thus, the idea of an association between biology and hot cores is unlikely.

9. Unsolved problems and future prospects

Goethe once wrote that "We know accurately only when we know little; with knowledge doubt increases." For many years, many if not most astronomers assumed with little experimental basis to support them that molecular hydrogen was formed on the cold surfaces of interstellar dust particles but that the formation and depletion of almost all other molecules could be understood without reference to dust particles. This view is now untenable as the importance of dust particles and their surface chemistry becomes better appreciated if not yet totally understood. Yet, at the same time, what we thought we knew about the formation of H_2 turns out to be an unrealistic simplification of a process which depends intimately on details we do not yet know accurately, such as the shape of interstellar dust particles, and their detailed chemical nature.

Given the current intense interest in H_2 formation on cold surfaces and the number of experiments now existing or being set up to understand the chemistry, we can expect that in the next few years, we will have a detailed understanding of the various ways in which H_2 can be made on assorted cold surfaces in the laboratory. We may even have experiments involving small particles rather than larger surfaces. At the same time, our understanding of the physical and chemical nature of dust particles is likely to improve, as new techniques and satellites for observing dust are brought into play. Perhaps within the next decade, the interstellar H_2 problem will be solved.

Of course, the chemistry of H_2 is not the only chemistry that occurs on the surfaces of interstellar dust particles. Many other reactions occur as well, and experiments on some of these are being undertaken. Our knowledge of types of reactions, especially hydrogenation processes, will soon be much greater than it has been up to now, and this increasing knowledge will enable astronomers to understand the chemistry of interstellar clouds in far more detail.

We now know that surface chemistry is not confined to the Earth, but is a widespread phenomenon throughout the Universe. The surface is provided by interstellar dust particles, which are very small and highly varied. Despite their distance from us, we have learned much about them. We also know that some very abundant molecules—especially molecular hydrogen—are formed on their surfaces, although the exact nature of how the formation occurs still eludes us. It is indeed a dusty Universe, and the ubiquitous presence of the dust means that it is also a Universe filled with surface science.

Acknowledgements

DAW and EH acknowledge the support of NATO for providing the funds to enable them to visit each other's laboratories and collaborate effectively. We thank Dr. J. Bowey for useful discussions. DAW also acknowledges with thanks the support of his research by the (UK) Particle Physics and Astronomy Research Council. EH is grateful to the National Science Foundation for support of his research in astrochemistry.

References

[1] T.J. Millar, D.A. Williams (Eds.), Dust and Chemistry in Astronomy, Institute of Physics, Philadelphia, 1993.
[2] T.P. Snow, A.N. Witt, Interstellar depletions updated: where all the atoms went, Astrophys. J. 468 (1996) L65.
[3] D.C.B. Whittet, W.A. Schutte, A.G.G.M. Tielens, A.C.A. Boogert, T. de Graauw, P. Ehrenfreund, P.A. Gerakines, F.P. Helmich, T. Prusti, E.F. van Dishoeck, An ISO SWS view of interstellar ices: first results, Astron. Astrophys. 315 (1996) L357.
[4] L.J. Allamandola, S.A. Sandford, J.R. Barker, Polycyclic aromatic hydrocarbons and the unidentified infrared emission bands—auto exhaust along the Milky Way, Astrophys. J. 290 (1985) L25;
A. Léger, J.L. Puget, Identification of the 'unidentified' IR emission features of interstellar dust?, Astron. Astrophys. 137 (1984) L5.
[5] K. Ohta, T. Yamada, K. Nakanishi, K. Kohno, M. Akiyama, R. Kawabe, Detection of molecular gas in the quasar BR1020-0725 at redshift $z = 4.69$, Nature 382 (1996) 426;

A. Omont, P. Petitjean, S. Guilloteau, R.G. McMahon, P.M. Solomon, E. Pontal, Molecular gas and dust around a radio-quiet quasar at redshift 4.69, Nature 382 (1996) 428.

[6] E. Grün, B. Gustafson, I. Mann, M. Baguhl, R. Riemann, M. Horanyi, C. Polanskey, Dust streams from comet Shoemaker-Levy 9? Astron. Astrophys. 286 (1994) 915.

[7] J.P. Bradley, L.P. Keller, T.P. Snow, M.S. Hanner, G.J. Flynn, J.C. Gezo, S.J. Clemett, D.E. Brownlee, J.E. Bowey, An infrared spectral match between GEMS and interstellar grains, Science 285 (1999) 1716.

[8] T.J. Millar, P.R.A. Farquhar, K. Willacy, The UMIST database for astrochemistry 1995, Astron. Astrophys. Suppl. Ser. 121 (1997) 139.

[9] R. Terzieva, E. Herbst, The sensitivity of gas-phase chemical models of interstellar clouds to C and O elemental abundances and to a new formation mechanism for ammonia, Astrophys. J. 501 (1998) 207.

[10] D. Hollenbach, E.E. Salpeter, Surface adsorption of light gas atoms, J. Chem. Phys. 53 (1970) 79;
D. Hollenbach, E.E. Salpeter, Surface recombination of hydrogen molecules, Astrophys. J. 163 (1971) 155.

[11] N. Katz, I. Furman, O. Biham, V. Pirronello, G. Vidali, Molecular hydrogen formation on astrophysically relevant surfaces, Astrophys. J. 522 (1999) 305.

[12] A.J. Farebrother, A.J.H.M. Meijer, D.C. Clary, A.J. Fisher, Formation of molecular hydrogen on a graphite surface via an Eley–Rideal mechanism, Chem. Phys. Lett. 319 (2000) 303.

[13] K. Hiraoka, N. Ohashi, Y. Kihara, K. Yamamoto, T. Sato, A. Yamashita, Formation of formaldehyde and methanol from the reaction of H atoms with solid CO at 10–20 K, Chem. Phys. Lett. 229 (1994) 408;
See also K. Hiraoka, T. Miyagoshi, T. Takayama, K. Yamamoto, Y. Kihara, Gas-grain processes for the formation of CH_4 and H_2O: reactions of H atoms with C, O, and CO in the solid phase at 12 K, Astrophys. J. 498 (1998) 710.

[14] A.G.G.M. Tielens, PAHs and astrochemistry, in: T.J. Millar, D.A. Williams (Eds.), Dust and Chemistry in Astronomy, Institute of Physics, Philadelphia, 1993, p. 103.

[15] A.G.G.M. Tielens, L.J. Allamandola, Composition, structure, and chemistry of interstellar dust, in: D.J. Hollenbach, H.A. Thronson Jr. (Eds.), Interstellar Processes, Reidel, Dordrecht, 1987, p. 397.

[16] P. Caselli, T.I. Hasegawa, E. Herbst, A proposed modification of the rate equations for reactions on grain surfaces, Astrophys. J. 495 (1998) 309.

[17] W.A. Schutte, Ices in the interstellar medium, in: P. Ehrenfreund, et al. (Eds.), Laboratory Astrophysics and Space Research, Kluwer, Dordrecht, 1998, p. 69.

[18] P.A. Gerakines, W.A. Schutte, P. Ehrenfreund, Ultraviolet processing of interstellar ice analogs. I. Pure ices, Astron. Astrophys. 312 (1996) 289.

[19] E.F. van Dishoeck, G.A. Blake, Chemical evolution of star-forming regions, Annu. Rev. Astron. Astrophys. 36 (1998) 317.

[20] K. Willacy, H.H. Klahr, T.J. Millar, Th. Henning, Gas and grain chemistry in a protoplanetary disk, Astron. Astrophys. 338 (1998) 995.

[21] Y. Aikawa, E. Herbst, Deuterium fractionation in protoplanetary disks, Astrophys. J. 526 (1999) 314.

[22] S.B. Charnley, Sulfuretted molecules in hot cores, Astrophys. J. 481 (1997) 396.

[23] C. Moutou et al., in: L. Hendecourt, C. Jones (Eds.), Solid Interstellar Matter: the ISO Revolution. EDP Sciences and Springer, Berlin, 1999.

 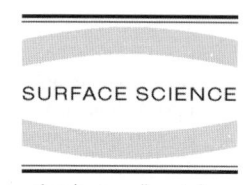

Far-out surface science: radiation-induced surface processes in the solar system

Theodore E. Madey [a,*], Robert E. Johnson [b], Thom M. Orlando [c]

[a] *Laboratory for Surface Modification and Department of Physics and Astronomy, Rutgers, The State University of New Jersey, Piscataway, NJ 08854-8019, USA*
[b] *Department of Engineering Physics, University of Virginia, Charlottesville, VA 22903, USA*
[c] *School of Chemistry and Biochemistry and School of Physics, Georgia Institute of Technology, Atlanta, GA 30332-0400, USA*

Received 10 August 2000; accepted for publication 14 May 2001

Abstract

Interplanetary space is a cosmic laboratory for surface scientists. Energetic photons, ions and electrons from the solar wind, together with galactic and extragalactic cosmic rays, constantly bombard surfaces of planets, planetary satellites, dust particles, comets and asteroids. Many of these bodies exist in ultrahigh vacuum environments, so that direct particle–surface collisions dominate the interactions. In this article, we discuss the origins of the very tenuous planetary atmospheres observed on a number of bodies, space weathering of the surface of asteroids and comets, and magnetospheric processing of the surfaces of Jupiter's icy satellites. We emphasize non-thermal processes and the important relationships between surface composition and the gas phase species observed. We also discuss what laboratory and computational modeling should be done to support the current and future space missions—e.g. the Genesis mission to recover solar wind particles, the Cassini mission to probe Saturn, the Europa Lander mission to explore the subsurface ocean hypothesis, and the Pluto/Kuiper Express to sample the outer reaches of the solar system. © 2001 Elsevier Science B.V. All rights reserved.

Keywords: Models of non-equilibrium phenomena; Sputtering; Electron stimulated desorption (ESD); Photon stimulated desorption (PSD); Ion bombardment; Amorphous surfaces; Silicon oxides; Water; Alkali metals

1. Introduction

1.1. We are constantly being bombarded!

Interplanetary space is a cosmic, ultrahigh vacuum laboratory for surface scientists. The "space" between planetary bodies, and even the rarified atmospheres above a number of large bodies, correspond to much better vacuum than that generally attainable in laboratories. Although interplanetary space conjures up a mental image of a vast, peaceful void of total emptiness, in reality, it is a stormy and sometimes very violent environment permeated by energetic particles and radiation constantly emanating from the Sun. This outpouring from the Sun is called the *solar wind*, and is an expanding mixture of fully ionized gaseous material; i.e., an electrically neutral plasma containing electrons and ions that

---

[*] Corresponding author. Fax: +1-732-445-4991.
E-mail address: madey@physics.rutgers.edu (T.E. Madey).

carries a magnetic field and constantly streams outward from the inner solar corona. In addition to the solar wind, photons, solar flare ions and cosmic rays bath the solar system, so that *most planetary and interplanetary surfaces are, in fact, continuously bombarded by radiation*. Mankind is rather fortunate since the magnetic field and atmosphere of Earth shield us surface inhabitants from the harmful effects of this continuous unfriendly assault!

The magnetic field of the Earth creates a magnetosphere that "funnels" most of the charged particles near the poles creating the great northern and southern light shows, i.e., the beautiful optical displays of the aurora borealis and aurora australis, respectively. A little farther from home, the photons from the Sun can cause our Moon to "glow" even during the daytime by stimulating the removal of alkali metal atoms from its surface. As we proceed beyond the outer reaches of the so-called traditional "habitable zone", we encounter the asteriod belt which is composed of a multitude of "minor" planets or, more simply, planetary fragments. On an unfortunate occasion, an asteriod can plummet towards the Earth (a scenario that Hollywood box offices find attractive and one that has been suggested to explain the extinction of dinosaurs). Asteriods that do not annihilate themselves by slamming into larger planetary objects are space weathered by radiation. The small interstellar ice grains and dust particles, which collectively account for much of the measurable mass of the universe, are also constantly "processed" by radiation. In the outer solar system, the magnetospheres of Saturn and Jupiter trap energetic ions and electrons that irradiate the low temperature ice present in the rings of Saturn and the rocky and icy material of Jupiter's satellites. The four major satellites of Jupiter are Io, Ganymede, Europa and Callisto. These spectacular objects were first observed by Galileo Galilei in 1610 and described in Sidereus Nuncius (Starry Messenger): *Magna longeque admirabilia spectacula pandens suspiciendaque proponens unicuique.* (Revealing the great and most spectacular display to everyone to contemplate.) The surfaces of these heavenly bodies are significantly altered by radiation bombardment!

1.2. The surfaces of interest

In this paper, we focus on bodies in the solar system which have tenuous atmospheres, comparable to ultrahigh vacuum (UHV) in the laboratory ($<1 \times 10^{-8}$ Pa). These include the planet Mercury, the Moon, the icy satellites (moons) of Jupiter and other satellites, asteroids, etc. In a recent article in Newsweek [1] entitled "Shoot the Moon", a headline trumpeted "with 66 Moons in the solar system, it's sheer lunacy out there—Moons with water, atmospheres or both are good places to discover how life on earth got started". For most of these bodies, thermal processes and direct particle–surface collisions dominate the surface interactions and lead to fascinating surface chemistry and physics, including the formation of tenuous, ballistic atmospheres (i.e., UHV atmospheres in which gravity is the dominant force affecting the trajectories of atmospheric atoms and molecules).

There is a wide and complex spectrum of materials whose surfaces are exposed to the space environment. Some of our earliest and most direct information about materials in space has literally fallen from the sky in the form of meteorites. Most are chondrites, iron-rich bodies containing silicates and ppm traces of most of the elements in the periodic table [2]!

The Moon is the only planetary body (besides Earth) for which rock samples have been collected from known locations, thanks to the Apollo Moon-lander missions. The lunar samples are predominately silicates, with SiO_2 as the dominant constituent. Other oxide components vary from location to location, and include Al_2O_3, MgO, FeO, CaO and TiO_2, with traces of Na_2O and K_2O [2]. Lunar Prospector maps of the Moon show the surface distribution of Fe, Ti, Th and K [3]. Optical reflectance studies of Mercury provide evidence for Mg silicates.

The cold planets (Jupiter and beyond) and the icy satellites of Jupiter are largely covered by layers of condensed molecules (e.g., NH_3, CH_4, H_2O, CO_2, N_2,...) [2]. The icy satellites of Jupiter also contain large amounts of water ice and non-ice regions that may contain hydrated minerals (possibly from a subsurface ocean) and radiation-processed materials, including sulfuric acid.

Thus, the challenge to surface scientists is in the study of oxide surfaces, including complex multi-component minerals, and the surfaces of condensed molecular solids containing dissolved minerals. These are far more complex than the elemental solids and binary oxides typically studied in surface science laboratories!

2. The radiation environment

2.1. Solar photons and the solar wind

Solar photon radiation is by far the largest source of radiant energy for surface processing, and most of this energy is in the form of infrared and visible photons. Some of this energy is absorbed by objects in the solar system, and causes heating of the surface regions. In addition to possibly inducing equilibrium chemistry on exposed surfaces, the heating leads to thermal desorption of gases from a number of 'airless' bodies (see Section 3.1, below). Here we will primarily describe processes induced by more energetic radiation. For instance, solar photons with energies above ~4 eV can induce bond breaking causing stimulated desorption of absorbates. Higher energy photons and energetic charged particles can induce chemistry in surfaces causing, for instance, alterations in the reflectance properties, as well as desorption of bulk species. Finally, low energy electrons, in addition to being important for surface charging, can cause electronic excitations that lead to desorption These processes will be discussed in later sections. Table 1 lists the particle fluxes and the energy fluxes in a number of environments. These are elaborated on below.

The flux of energetic particles available for surface processing spans an enormous range of energies [4,5]. Surfaces in the interplanetary medium are exposed to the continuous expansion of the solar corona referred to as the solar wind (see Fig. 1), primarily consisting of ions representing the solar abundance of elements with flow energies ~1 keV/amu and a thermal distribution ~10 eV. A corresponding flow of electrons occurs also with similar thermal energies. This flux, given in Table 1, decreases like the solar photon flux as $\sim 1/R^2$,

where R is usually given relative to the distance from the Sun to the Earth (1 AU, Table 1). Superimposed on this are more energetic particles associated with active regions on the Sun and solar flares (10's of keV to 10's of MeV) and the cosmic-ray background flux given in Fig. 2.

2.2. Magnetospheres of giant planets

Of current topical interest is the large flux of energetic ions and electrons trapped in the magnetospheres of the giant planets, roughly equivalent to Earth's van Allen belts. Our Moon lies outside the earth's magnetosphere for most of its orbit, and receives the full blast of the solar wind

Table 1
Radiation flux

Radiation	Flux (cm^{-2} s^{-1})	Energy flux (eV cm^{-2} s^{-1})
Solar photons (1 AU)[a]		(8.5×10^{17})
IR (<1.8 eV)	2×10^{17}	4.0×10^{17}
Visible (1.8–3.1 eV)	1×10^{17}	3.7×10^{17}
UV-A (3.1–3.9 eV)	2×10^{16}	7.2×10^{16}
UV-B (3.9–4.4 eV)	2.5×10^{15}	1.0×10^{16}
UV-C (4.4–12.4 eV)	1×10^{14}	4.3×10^{15}
EUV (>12.4 eV)	1×10^{10}	1.3×10^{11}
Lyman-α	3×10^{11}	3×10^{12}
Solar wind (1 AU)[a]		
1 keV/u H$^+$(96%), He^{++}(4%)[b]	4×10^{8}	4×10^{11}
electrons (~12 eV)	4×10^{8}	
Cosmic ray ions (see Fig. 2.1)	2	6×10^{9}
Magnetospheric plasmas		
Europa (9.4 R_J)[c] (see Fig. 2.2)	3×10^{8}	(8×10^{13})[c]
Dione (6.3 R_S)[d]	1×10^{12}	3×10^{12}

[a] Average solar conditions at 1 AU (distance from Sun to Earth) [83]; decay as R^{-2} with R in AU (e.g., Jupiter is 5.2 AU, Saturn is 9.54 AU).
[b] Other ions are approximately solar [83] or cosmic [82] abundance.
[c] Orbital distance in Jupiter radii (R_J). Flux is for ions; total energy flux includes energetic electrons [5]. Low energy plasma determines flux, high energy plasma determines the energy flux.
[d] Orbital distance in Saturnian radii (R_S). Flux and energy flux are for ions. Low energy plasma determines flux, high energy plasma determines the energy flux [69].

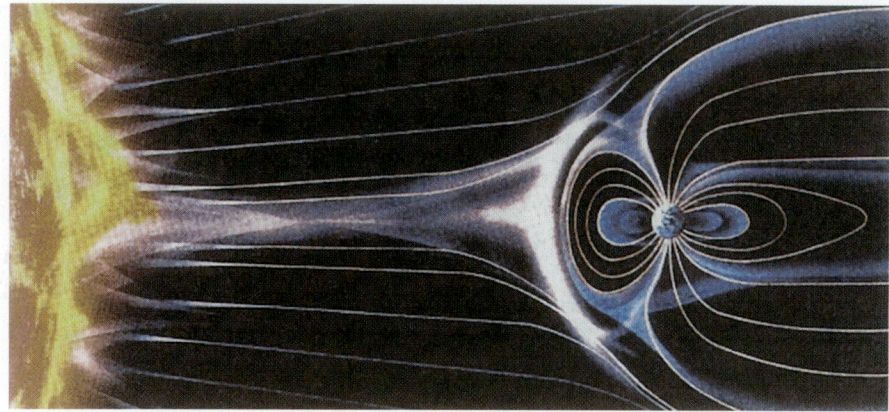

Fig. 1. Artist's conception of solar wind interacting with magnetosphere of Earth (not to scale!). From Rice University with permission.

during that time. However, for a few days around the time of full moon each month, the Moon passes through the earth's magnetotail and is largely shielded from solar wind particles. During the full moon, the Moon is exposed primarily to solar photons and cosmic-ray particles. In contrast, many of the moons and ring particles of the giant planets lie within their magnetospheres. Therefore, they are exposed to the flux of trapped ions and electrons. The charged particle energy flux to the surface given in Table 1 is larger than the chemically interesting UV flux (photons $> \sim 6$ eV) when scaled to Jupiter's distance from the Sun. In fact, on most of the large moons in the outer solar system, (with the exception of Titan, which has an atmosphere larger than the Earth's), the surfaces are altered by the trapped particle radiation. This radiation environment has been measured by the Pioneer, Voyager and Galileo spacecrafts.

Whereas the trapped plasma data for Saturn, Neptune and Uranus are sparse, the Jovian environment is well described. In Fig. 3 is given a recent description of the energetic particle flux at Europa [6]. It is seen that the energetic electron and proton fluxes dominate, but there is also a large flux of multiply-charged oxygen and sulfur ions. The mass spectrum of trapped magnetospheric ions at Europa is seen to be very different from that for the solar particles and the cosmic rays (e.g., Fig. 2). This indicates that the sulfur and oxygen ions, and probably also the protons, are produced and accelerated within Jupiter's magnetosphere. In fact, principal sources of the ions in the giant planet magnetospheres are the surfaces of the imbedded moons and ring particles, indicating the surface processing of these bodies is self sustained. In addition to the energetic ions, a flux of lower energy ions (referred to as thermal plasma) can flow onto the surfaces of the Moons and the ring particles. Because of the rapid rotation of the Jupiter's magnetosphere and the distance to the Galilean moons these particles (mainly H^+, O^{n+}, and S^{n+}) have energies of 100's of eV to ~ 10 keV and are important for sputtering but constitute a smaller energy flux than for the energy range in Fig. 3. Similar energetic plasmas are trapped in the magnetospheres of the other outer planets [4] as indicated in Table 1 for Saturn.

Whereas the cosmic-ray particle bombardment is isotropic, the solar wind and the trapped magnetospheric particle radiations are not. The geometry of the bombardment is shown schematically in Fig. 4 for Europa. Jupiter's icy moons, like our Moon, are phase-locked to the parent planet. Since they rotate around Jupiter slower than Jupiter's field rotates, the trapped plasma flows preferentially onto the hemisphere trailing the motion [4]. In addition, the moon Ganymede has an intrinsic magnetic field, one of the great discoveries by the Galileo spacecraft. This filters

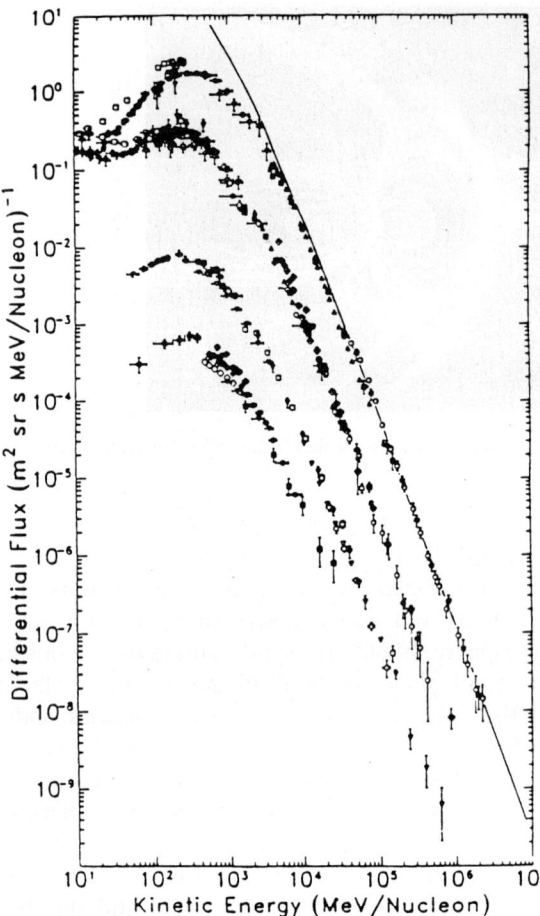

Fig. 2. Summary of energy spectra data measured at Earth for the cosmic-ray ion flux versus ion energy, expressed as MeV per nucleon. The four sets of compiled data are for the elements, hydrogen, helium, carbon and iron in order of decreasing flux. The solid line indicates the total proton energy spectrum extrapolated to the interstellar medium after correcting for solar modulation [82]. That is, the Sun's magnetic field deflects the low energy galactic protons, so the corrected spectrum corresponds to protons incident on objects in the distant reaches of the solar system (e.g, the Oort cloud).

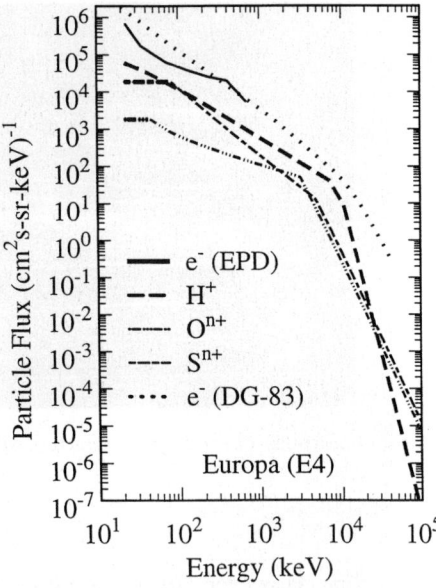

Fig. 3. Energy and mass spectra of energetic charged particles, H^+, O^{n+} and S^{n+} incident on Europa, measured by Galileo (from Ref. [6] with permission).

the particles that have access to the equatorial regions leading to preferential bombardment of the poles. Indeed, weak polar auroras are seen due to surface sputtering. Therefore, there are important spatial differences in the radiation flux to the surface that can be useful in trying to understand the surface processing. For example, the reflectance spectrum of the leading and trailing hemispheres are seen to be different, which has long been a rationale for studying the radiation processing of these surfaces [4].

Model calculations have been made of energy deposition into the surface layers of Europa. Weakly interacting protons and electrons affect the surface well below the typical optical depth on time scales are much shorter than typical geological time scales ($\sim 10^7$ yrs on Europa). Further, the heavy ions (O^{n+}, S^{n+}) interact strongly with the surface (low penetration) depositing very large amounts of energy in a small layer (<0.1 μm) in very short times. Adding to this the energy deposited by the 'thermal' plasma flux suggests that near surface materials could be fully decomposed on a stable surface. Whereas geological processes are very slow, surface fragmentation and stirring by micrometeorites occurs on comparable time scales but preferentially on the hemisphere leading the motion. These spatial differences allow the possibility of separating surface weathering effects. Below we describe the effects of thermal heating and particle bombardment on the surfaces of solar system materials.

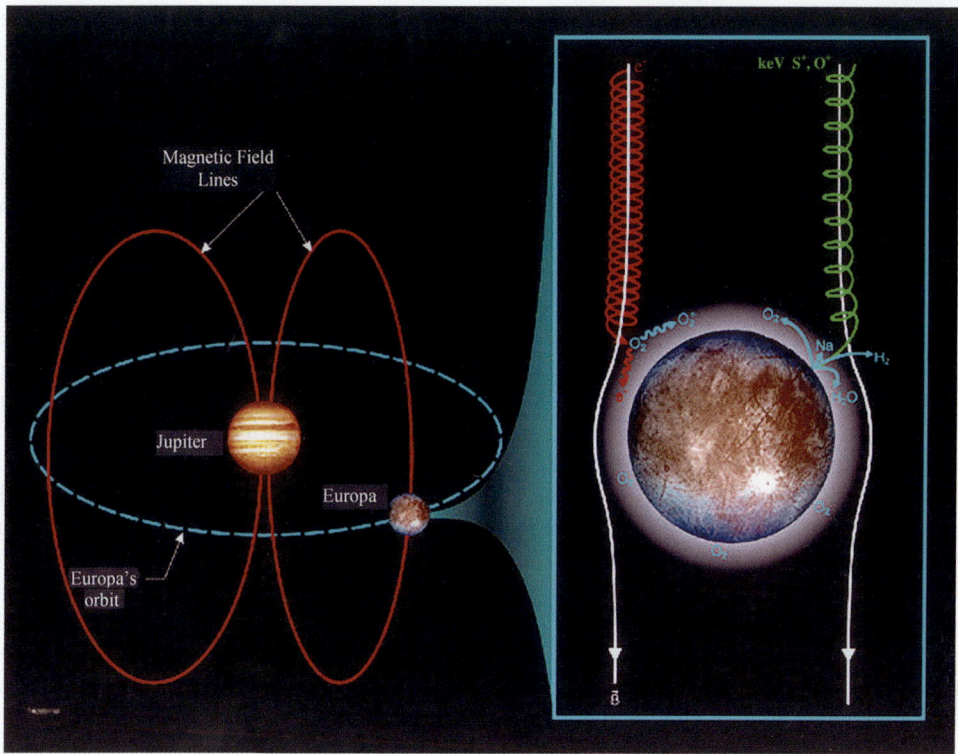

Fig. 4. Schematic of interaction with the magnetosphere. Left hand side, Europa and magnetic field lines rotate around Jupiter at different speeds. Right hand side, energetic ions and electrons moving along the field lines can sputter and decompose the surface, producing a tenuous atmosphere and ionosphere.

3. Relevant surface phenomena

3.1. Thermal processes and desorption from surfaces

The temperature extremes due to weather on Earth, from the coldest night in Antarctica to the hottest day in the Sahara, range from roughly 200 to 330 K. Much greater temperature extremes are experienced in the solar system due to the presence (or absence) of a flux of solar infrared photons, especially on surfaces that are not protected by an atmosphere. Solar heating at "noontime" at Mercury's equator causes surface temperatures of ~ 700 K, while the side of Mercury away from the Sun cools to ~ 100 K (due to radiation losses) during the long night. The lunar subsolar point (i.e., the equator at high noon) reaches ~ 400 K, while the darkside of the Moon is ~ 100 K. The extremes of temperature (and the swings in temperature) mean that thermally-induced surface chemistry, especially thermal desorption of surface species, is constantly occurring. Thermal desorption refers to heating of a substrate and the accompanying desorption or evaporation of surface species [7]. The temperature at which an atom or molecule desorbs from a surface is directly related to its bond strength or adsorption energy to the surface [8].

Surface science studies of thermal desorption of atoms and molecules from oxide surfaces, and of desorption from condensed gases, can provide insights into thermal processes that affect the weathering and atmospheres of solar system bodies. At 700 K, for example, the desorption rate of alkali atoms (Na and K) from a silicon dioxide surface is considerable; this suggests that thermal processes may contribute to the Na and K in the tenuous atmosphere of Mercury [7]. The Na and K may then adsorb on the cold side, and be released again as the planet rotates to face the Sun. Whereas water ice has a very low vapor pressure at 150 K ($\sim 10^{-4}$ Pa), nonetheless, this corresponds to a vaporization or desorption rate of ~ 1 monolayer/sec. On geological time scales, H_2O vapor can be transported from "warm" to "cold" regions on icy satellites, and may be responsible for changes in observable surface features on Ganymede. Thermally-driven sublimation and condensation of small molecules must occur constantly at the surfaces of all the outer planets and their icy satellites.

The role of thermal desorption phenomena on interstellar "space dust" is considered in the article by Williams and Herbst, also in this volume [9].

Despite the importance of thermal processes in explaining some aspects of weathering, tenuous atmospheres, etc., there are many phenomena that cannot be explained by thermal effects alone. As indicated above, non-thermal radiation effects (e.g., bombardment of surfaces by energetic solar photons, the solar wind, and planetary magnetospheres) play a dominant role in interplanetary space.

3.2. Energetic particle/solid interactions

For eons, the surfaces of lunar soils, interplanetary dust particles, asteroids, comets and icy surfaces have been bombarded by macro- and microscopic meteorites, galactic and extragalactic cosmic rays, solar ultraviolet photons, and solar wind and flare ions as discussed above. These impact events can result in physical (structural) and chemical (compositional) changes which are observable using ground based and orbital telescopes and spacecraft (i.e. Voyager, Hubble Space Telescope, International Ultraviolet Explorer, Galileo, etc). The process of bombardment-induced surface alteration or damage has been coined "space weathering"—a term that has been borrowed from the geological processes of erosion and degradation on earth caused by air and water.

Though much of the material removal and the chemical alterations due to space weathering happen over geological time frames and distances, *the physical and chemical basis for important material removal and alteration processes involves atomic and molecular collisions and electronic excitations in the surface material.* It is well known that the slowing down of energetic particles is due to the interaction of the colliding entity with the atomic nuclei and electrons in the target material. For the relevant incident ions and photons described above, the production of excitations and ionization are the principal mode of interaction. In fact, the surface science community has studied these type of interactions for decades and a detailed understanding of the principles of collisional sputtering as well as electronically-induced sputtering or desorption induced by electronic transitions (DIET) [10] has emerged.

Some definitions of experimental methods used by surface scientists to study radiation-induced desorption of atoms, molecules and ions from surfaces are appropriate here. DIET includes photon-stimulated desorption (PSD) as well as electron-stimulated desorption (ESD). PSD refers to the desorption of atoms or ions as a result of a direct photon-induced electronic excitation of a surface species, such as a bandgap excitation, a valence electron excitation, or a core excitation [11]. Solar photons can be responsible for PSD. ESD refers to desorption initiated by an electronic excitation during electron bombardment of a surface [12,13]. For surfaces in space, the electrons may originate in the solar wind, or the magnetosphere plasma, or they can include secondary electrons released when the substrate is bombarded by photons or ions. The manifold of electronic excitations that lead to desorption in ESD and PSD is nearly the same in most cases, although there are differences in cross-sections and energy dependencies. Electron and photon bombardment can also produce alterations in surface chemistry.

Secondary ion mass spectrometry (SIMS) refers to bombardment of surfaces by ions (e.g., H^+, He^{++}) and the accompanying collision-induced sputtering of secondary ions from the target surfaces. The mass distribution of sputtered ions (as measured using a sensitive mass spectrometer) is a diagnostic of the surface chemical composition. For surfaces in space, the bombarding primary ions can include energetic ions in the solar wind, cosmic rays, or magnetospheric plasmas (cf. Table 1 and Fig. 2).

In one interesting collisional-sputtering experiment, Elphic et al. [14] simulated the influence of solar wind ions on the lunar surface in a laboratory SIMS study. They bombarded lunar soil simulants with 1.5 keV H^+ and He^{++}. While these ions are not efficient sputterers, significant fluxes of secondary lunar ions were seen, including Na^+, Mg^+, Al^+, Si^+, K^+, Ca^+, Ti^+, Mn^+, and Fe^+. The authors predict that fluxes of ions should be measurable from a detector in a 100 km lunar orbit.

In this article, we describe some advances and recent joint studies between planetary scientists and surface scientists which not only explain some remarkable observations, but which also reveal important aspects of radiation-induced surface processes on targets such as lunar soil surrogates, minerals, silicate particles, and ice—the "rock" of the outer solar system.

4. Some radiation-induced surface processes in the solar system

4.1. The origin of the Moon and Mercury glow

It has been shown that space weathering on the Moon involves high velocity and large momentum impact events (e.g., bombardment of the surface by micrometeorites—a form of "space dust" in the solar system). These result in the production of a melt and vaporization. Subsequent re-deposition of material produces "optical" coatings containing submicron sized particles of reduced iron. Photon and charged particle interactions with the lunar surface do not appear to control the lunar optical properties. However, non-thermal desorption of alkali atoms has been shown to contribute to the "glow" observed from our Moon and Mercury.

In the mid to late 1980's, neutral sodium and potassium vapor were discovered in the ultrahigh vacuum atmospheres of Mercury and the Moon [15–17] (Fig. 5). This was a surprise: Na and K are both very reactive atoms, and they are not observed in dense atmospheres. Both Mercury and the Moon have tenuous ballistic atmospheres (density too low for significant numbers of atom–atom collisions), and neither can retain Na or K for more than a few hours: the neutral Na and K are lost via photoionization by solar photons, and by readsorption. The atmospheric Na and K must be continuously resupplied, and the mechanisms proposed all involve surface processes [7,18–20]. Suggestions include collisional sputtering by the solar wind ions, thermal desorption, PSD, ESD, and micrometeorite impacts. Until recently, there were few data and no general agreement about which processes dominate. New results have shed light on this phenomenon: in laboratory UHV measurements, the photon-stimulated desorption of Na atoms from SiO_2 surfaces that simulate lunar silicates has been reported [21]. Bombardment of such surfaces at temperatures of ~ 250 K by ultraviolet photons (energy > 4 eV, wavelength $\leqslant 300$ nm) is found to cause very efficient desorption of Na atoms. Desorption is induced by electronic excitations (i.e., a photo-induced charge transfer) rather than by thermal processes (Fig. 6). The flux at the lunar surface of ultraviolet photons from the Sun is adequate to insure that PSD of sodium contributes substantially to the Moon's atmosphere [22]. On Mercury (as indicated in Section 3.1) solar heating of the surface is high enough that thermal desorption will also be a potential source of atmospheric sodium.

Where does the Na (K) come from? The small natural abundance is depleted via PSD at an approximate rate less than 1 monolayer per month but much of it is readsorbed (recycled); the surface Na(K) may also be restored by diffusion from the bulk, as well as by micrometeorite impact [23,24].

4.2. Surface charging in space

There is nothing neutral about objects isolated in a space environment: they invariably charge to a floating electrostatic potential that is determined by the balance between incident and emitted charged particles. Solar photons together with electrons and ions in the solar wind are the dominant bombarding species outside the planetary magnetospheres; photoemitted electrons and electrons created by secondary emission are the dominant charged species emitted from the surface

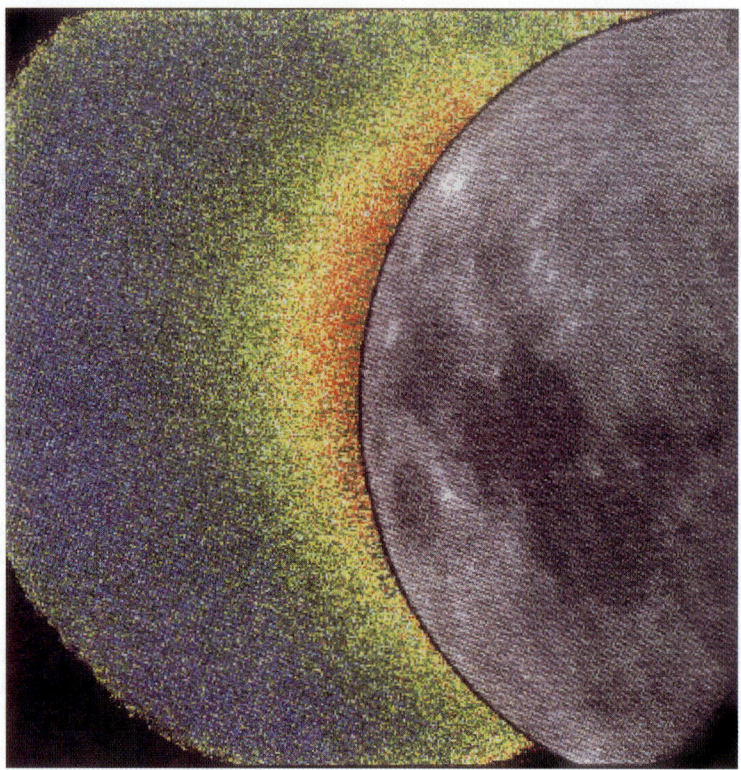

Fig. 5. Image illustrating the sodium vapor atmosphere of the Moon. The solar radiation is incident from the left, and the false color "haze" represents a high resolution spectrographic image of Na D-line emission (589 nm) due to sunlight scattered from Na atoms. The bright region just above the Moon's surface on the left (the "equator") is the region of highest Na vapor concentration. The Na emission is detected out to >10 lunar radii [20,23]. This image is from A.E. Potter, with permission [17].

[25,26]. At the Moon the sunlit hemisphere charges up to 20 V positive, mainly due to photoemission by ultraviolet photons [27]. The lunar terminator and nightside can have large negative potentials, −100 and −1000 V, respectively [28]. Small objects such as dust grains on the surface of larger objects, such as the Moon and Mercury, may become positively charged. These dust grains are repelled by the positively-charged Moon and can become levitated above the surface; there have been multiple reports of dust grains suspended above the lunar surface [29]. Charging of grains in Saturn's magnetosphere [30] is sufficient to affect the orbital dynamics of small grains and hence, the spatial morphology of Saturn's E-ring [31].

The photoelectric charging of dust particles in vacuum was recently tested experimentally using a simple apparatus [26]. The authors measured the ultraviolet light-induced charging of individual 100 μm—diameter grains of various particles, including glass and several metals. Insulating glass particles were clearly shown to be positively charged due to ultraviolet photoemission. The sign and magnitude of the net charge on the particles could be affected by their falling through a photoelectron "sheath" above a low work-function surface.

4.3. Space weathering of asteroids and comets

Most planetary scientists now believe that asteroids are remnants of a planet that was fragmented, and that asteroids are the "parents" of meteorites. Though it has been long suspected that a spectral signature could provide a link between asteroids and meteorites, the evidence for surface

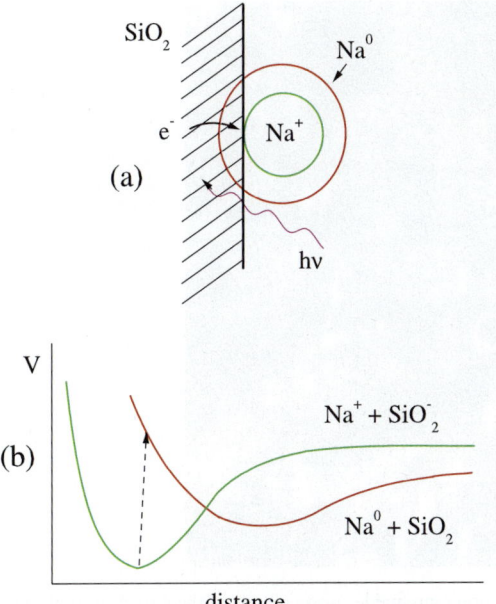

Fig. 6. Schematic mechanism for PSD of sodium from SiO_2, a model mineral for the lunar surface: (a) The sodium at the surface is ionic Na^+; a solar ultraviolet photon excites an electron in the substrate which attaches to the Na^+. This charge transfer converts Na^+ to neutral Na^0. Because Na^0 has a larger radius than Na^+, the atom is in a repulsive state and can desorb (see (b) schematic of interaction potential V as a function of distance from the surface). This process is described in Ref. [21].

alterations and space weathering of asteroids (as seen in reflectance spectra and Galileo images of asteroids) may make this connection somewhat tenuous [32,33]. To test these, recent laboratory simulations of micrometeorites and solar wind irradiation of olivine, a major constituent of ordinary chondritic meteorites and S-type asteroids [2] show rather dramatic chemical alterations of the surface [34]. These alterations seem to change the optical reflectance in the near infrared and visible regions and may account for the optical differences observed in telescopic and remote observations of S-type asteroids and the bulk properties of collected ordinary chondritic meterorites.

Comets are composed of material left over from the formation of the Sun and planets and both observations and models indicate the presence of a large number of comets in the Kuiper belt and Oort Cloud. The Kuiper belt is located several thousand astronomical units away from the Sun and is considered to be the source of many short period (<200 years) comets, whereas the Oort Cloud at 10^4–10^5 AU is the source of the longer period comets. A limitation on the direct observations of comets in the outer solar system is the low optical albedo for reflection. It has been suggested that bombardment with galactic cosmic-ray particles, ions, and electrons and radiation-induced processing of hydrocarbons on the comet mantle surface leads to the general darkening. Though this mantle material is darkened, the subsurface and core (nucleus) is ice-rich and porous. Far ultraviolet images of comets like Hale–Bopp taken by the Midcourse Space Experiment spacecraft reveal huge clouds of atomic hydrogen and oxygen around the comets. These clouds can span many millions of kilometers and the strength of the Lyman alpha emission at 121.6 nm reveals a source term of 10^{29} hydrogen atoms per sec. The velocity of this gas is 8 km/s, about 10 times faster than the velocity expected for sublimation. Although radiation-induced surfaces processes have not been implicated as the hydrogen and oxygen source term, surface science studies have shown that electronic excitation of ice will produce hydrogen and oxygen atoms with non-thermal velocity distributions [35–37]. However, it has been suggested that "weathering" of comets at large distances from the Sun and cometary grain evolution are enhanced by "surface radiolysis" (i.e., radiation-induced changes in the surface chemistry) [38] and that many of the large organic molecules seen in the gas phase are produced by stimulated desorption [39].

4.4. Radiation processing of icy satellites: the Jovian system

4.4.1. Background; evidence for oxygen

Information on the icy Galilean satellite surfaces comes primarily from ground based and Earth-orbital telescopes, two Voyager flybys and the recent Galileo Jupiter mission flybys. As a result of these successful activities, one of the most exciting current research areas in planetary science is the study of the composition and chemistry of the Galilean satellites, Io, Europa, Ganymede and

Fig. 7. This view of the Galileo Regio region of the surface of Jupiter's moon Ganymede shows details of the dark terrain that makes up roughly half of the surface. Dark areas may have originated from dark material released during meteorite impact. One of the many ancient impact craters is shown on the middle of the left. The image was taken on June 27, 1996 at a range of 7.625 km. Photo is courtesy of NASA/JPL/Caltech, from web site: http://www.jpl.nsa.gov/galileo/ganymede/p47066.html.

Callisto (Jupiter has 17 satellites, and these are the largest). Io, the first large satellite, is the most volcanically active body in our solar system. Europa is the second large satellite and has a dense core and an ice crust which may be "floating" on top of a liquid subsurface ocean. Ganymede, the third body, is also differentiated with scars from multiple meteorite impacts (Fig. 7) and Callisto, the fourth large satellite, seems to have also suffered many impact events and was recently suggested to have a subsurface ocean.

These satellites are located in the region of the solar system where low temperature water and other "volatile" species either stick to the surface or are trapped in the ice and mineral matrices for extremely long periods—nearly the age of the solar system. Therefore, a chemically interesting and rich mixture exists which may provide a unique window concerning the evolutionary processes that have occurred over eons of time. In addition, Jupiter's magnetosphere supplies a large dose of energetic electrons and ions to the surfaces of these moons as described above, which can lead to sputtering, material implantation, and radiolysis.

Motivated by the Voyager data, extensive early experiments by Brown, Lanzerotti and co-workers [40,41] showed that the sputtering rates of ices by energetic ion and electron bombardment are comparable to the sublimation rates [4]. More recently, UV photon-stimulated desorption was shown to be efficient [42]. This bombardment can also lead to the formation of molecules, some of which may be of biological relevance. For example, the signature of condensed O_2 has been reported in the optical reflectance measurements of Ganymede [43] and oxygen atmospheres have been observed at Europa and Ganymede [44]. This was predicted based on laboratory studies which showed that low temperature water ice is efficiently decomposed and sputtered by ionizing radiation [45]. Whereas the excitations and ionizations lead primarily to radicals, the mass spectrum of ejecta was shown to be dominated by the most weakly bound species, H_2 and O_2, and well as H_2O [41,46]. Since the H_2 readily escapes to space and the H_2O recondenses, the O_2 is seen to be the primary species in the ambient gas. Whereas the early experiments demonstrated that O_2 is produced in

DIET processes by fast ions, it was also shown to be produced very efficiently under normal sputtering by lower energy ions [47].

Recently, a group of surface scientists examined the role of low energy electrons in the production of O_2 from low-temperature (<150 K) water-ice [48]. An "image" of the O_2 yield as a function of electron-beam spot position on a thin ice sample is shown in Fig. 8. Regions selected to spell out "O_2" were exposed to a 10^{15} cm^{-2} dose of 50 eV electrons at a sample temperature of 120 K. After this exposure, the electron beam was then rastered over the sample, and the O_2 yield was measured as a function of the electron beam spot position. The bright and dark areas correspond to high and low O_2 yields, respectively. This electron "write" and "read" experiment demonstrated a two-step "precursor" dissociation mechanism which is schematically illustrated in Fig. 9. Briefly, a water molecule is first excited by an electron, ion or photon impact. Since small changes in the ice structure or dynamic relaxation can cause the excited state lifetime to increase with temperature, the dissociation probability might also be expected to increase. The dissociation fragments of water (H or OH) can reactively scatter to produce a stable precursor molecule, possibly HO_2 or H_2O_2. A second electronic excitation then directly dissociates this precursor to form O_2, which can desorb.

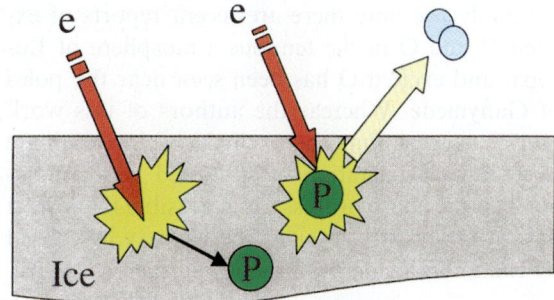

Fig. 9. Proposed scheme for the production of molecular oxygen on icy satellite surfaces which involves radiation-induced production of a precursor species, P, presumably HO_2 or H_2O_2, followed by radiation-induced production of O_2. The temperature dependence of the yield of O_2 is in the first step.

The mechanism proposed above is applicable to electronic excitations caused by essentially any radiation source and evidence for oxygen has been seen on a number of icy moons. One possible piece of evidence is the set of observations which suggests there is an absorption band close to that of O_3 on Ganymede and the icy Saturnian satellites [49]. Although O_3 is not firmly identified and is not a direct product of irradiation of ice, it is possibly formed after the accumulation of O_2 in some condensed form [50,51]. However, the strongest piece of evidence that the icy surfaces are being processed is the recent observation of H_2O_2 in the ice of Europa [52] and of other icy satellites. These observations are all consistent with the production and desorption of H_2 leaving behind an oxidizing surface [4,6]. Whereas the atmospheres of Europa and Ganymede contain radiolytically produced O_2, that of Callisto has predominantly CO_2 [53]. This may be due to slow outgassing of a primordial gas, but has also been suggested as a radiation decomposition product [54].

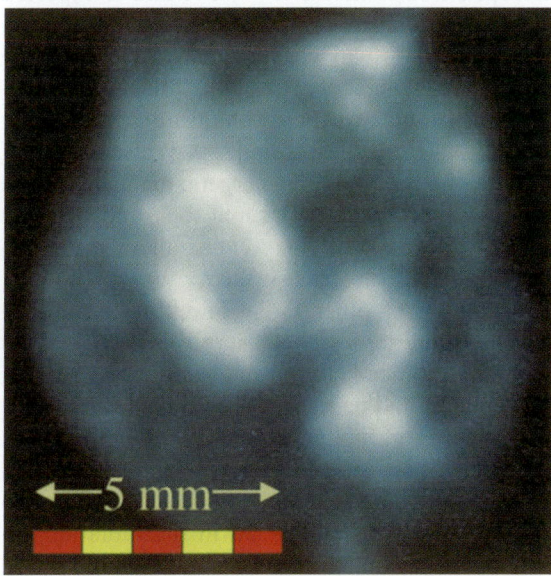

Fig. 8. An "image" of the O_2 yield as a function of electron-beam spot position on a thin ice sample. Regions selected to spell out "O_2" were exposed to 10^{15} cm^{-2} dose at 50 eV at a sample temperature of 120 K. After this exposure, the electron beam was then rastered over the sample, and the O_2 yield was measured as a function of the electron beam spot position. The bright and dark areas correspond to high and low O_2 yields, respectively. The circular profile of the sample is evident. (Copyright release from Ref. [48]).

Finally, we note there are recent reports of excited H and O in the tenuous atmosphere of Europa, and excited O has been seen near the poles of Ganymede. Whereas the authors of this work [44,55] suggest that the excited species originate from impact excitation of sputtered water molecules (a gas phase process), it is possible that direct desorption of these excited species may be caused by radiation incident on the icy surface.

4.4.2. NIMS observations and implications for surface composition

Some of the most spectacular images of the satellite surfaces have been obtained using the Near Infrared Mapping Spectrometer (NIMS) onboard the Galileo spacecraft. The NIMS measures reflected light from surface zones ranging from about 1 to several 100's of kilometers in size. (Slightly larger than typical surfaces studied by conventional surface science techniques!) The spectrograph records 408 spectral channels between 0.7 and 5.2 µm and emphasizes a spectral regime that is dominated by vibrational overtone and combination bands of water ice. Distortions in certain bands have been correlated with the disrupted regions on the surface of Europa (see Fig. 10) and it has been suggested that the material associated with these disrupted regions are hydrated salt minerals such as epsomite ($MgSO_4 \cdot 7H_2O$), mirabilite ($Na_2SO_4 \cdot 10H_2O$) and Natron ($Na_2CO_3 \cdot 10H_2O$). Such materials were predicted from models for formation of these objects and are consistent with possible cryo-volcanism and injection from a subsurface briny ocean [56,57]. These materials have been shown to be stable under Europa's environmental conditions [58]. The possibility of a briny ocean, also suggested also by the Galileo magnetometer measurements [59], is very intriguing since it is an environment which could harbor very primitive life forms.

Recent work has suggested that these NIMS bands can instead be interpreted in terms of the presence of hydrated sulfuric acid [60]. This material could be produced by radiolysis of water and sulfur bearing species, or, less easily, by the decomposition of the sulfate salts discussed above [4]. The presence of hydrated sulfate ions and sulfuric acid is also consistent with the spectral signatures in the UV and visible of small amounts of SO_2 and S_x. Though the radiation chemistry leading to the production of sulfuric acid from one of the metal sulfates above is complicated and the yields are low, a reasonable buildup could occur over geological times [60]. Although a surface containing hydrated sulfuric acid might initially appear to be less-than-inviting for life, it along with the other surface oxidants can be a source of energy if subducted into the proposed ocean [6,61]!

4.4.3. Related observations

The discovery of gas-phase sodium atoms in the tenuous atmosphere of Europa [62] provides evidence that radiation-induced decomposition of the surface is ongoing [54]. Desorbed Na atoms that do not escape can be readsorbed in the predominantly icy regions of Europa. There are recent laboratory reports that ESD and PSD can efficiently desorb neutral alkali atoms from the surfaces of alkali-covered thin H_2O ice films (simulating the icy surface of Europa) [63]. The appearance threshold for desorption of K atoms is ~4 eV, and is expected to be similar for Na. The proposed mechanism of desorption is an electronically-excited charge transfer from ice to alkali ion, followed by desorption (similar to the mechanism suggested for PSD of Na from the Moon [21]). We suggest that along with sputtering by energetic magnetospheric ions [54], ultraviolet solar photons and electron fluxes with $E > 4$ eV may cause alkali metal desorption from Europa's surface.

A number of other radiation effects have been proposed but are less well established. Differences in the reflectance between the leading and trailing hemispheres [64] have been seen on all of the large icy satellites of Jupiter and Saturn. The observed difference in reflectance between the poles of Ganymede and its surface at low latitudes has also been attributed to differences in the radiation environment [6] as discussed earlier. In addition, the presence of trapped CO_2 and SO_2 on on Europa, Ganymede and Callisto and a CO_2 atmosphere on Callisto may be due to radiation induced decomposition and sputtering [4,56,60]. In fact, based on the energy flux to the surface described earlier the ambient gas over the surface of Europa should contain additional sputtered material. If organics are present on the surface this could include large

Fig. 10. False color Galileo image of the icy plains and disrupted terrain on Europa. The plains are represented by the gray and blue regions and the mottled terrain is represented by patchy brown regions toward the left. Also visible are numerous linea which exhibit the same reddish-brown color as the disrupted mottled terrain. This brown color represents the presence of non-ice materials. Photo is courtesy of NASA/JPL/Caltech, from web site:http://www.jpl.nasa.gov/galileo/ganymede/p47906.html.

organic molecules as seen in SIMS experiments on organics in the matrices. On volcanic Io the surface materials are continuously replenished and are primarily coated in frozen volcanic SO_2 [65]. Color variations between the equator and the poles have been suggested as due to radiation-induced surface processing of the SO_2 [66] and the loss of Na from the surface has also been suggested as due to decomposition and sputtering of the surface [67,68].

It must be remembered that the NIMS sampling depths are only on the order of sub-micrometers to millimeters so the reflectivity measurements only sample the surface. Also, many of the bands are broad, the spectra are affected by the local chemical environment, and the surfaces are typically porous regoliths. Thus, the answer or "resolution" to the questions concerning the identification of the materials and chemical compositions present on planetary surfaces and interstellar media will rely heavily upon collaborations with the surface science community!

4.5. Radiation processing of the icy satellites: Saturn and beyond

4.5.1. Saturn's icy satellites and ring particles

Like the large moons of Jupiter, the surfaces of the icy satellites of Saturn as well as Saturn's ring particles are eroded and modified by the plasma trapped in Saturn's magnetosphere. This is a self-sustained process, as the predominant source of this plasma is the surfaces of the Moons and ring particles that orbit within the magnetosphere [69]. Whereas the density of particles in Saturn's main rings is high and the plasma formed is rapidly quenched, the spatially largest ring in the solar system (Saturn's E-ring) co-orbits with the icy satellites and is imbedded in the hot trapped plasma. This is composed of ~1 μm grains whose orbital motion is determined also by their surface charge in the ambient plasma. It is this charging that is thought to cause the large ring's spatial extent [31].

The surfaces of these ring particles and of the icy satellites serve as sputtering sources that produce a giant torus of neutral gas and plasma [4,69]. Quite remarkably, the OH component of the sputtered neutrals has been imaged by the Hubble space telescope and shows that, unlike Jupiter's magnetosphere, the ambient density of neutrals is larger than the plasma density [70].

The direct observation of the principal sputter product (OH from H_2O dissociated in the gas phase) along with the observations of trapped O_3 like [49] and H_2O_2 like [52] features in UV reflectance of the moons indicate that radiation processing of the surface materials is occurring. Since these satellites are phase-locked to Saturn there are also leading–trailing hemisphere asymmetries in the reflectance, with the more heavily bombarded side being redder in the visible [71].

4.5.2. Uranus, Neptune and Pluto

With increasing distance from the Sun more volatile species condense out and form components of the surface [65]. In fact, because of its relatively large cohesive energy, water ice is often thought of as the "rock" of the outer solar system with CO, CO_2, CH_4, N_2, and NH_3 forming the possible gases and condensates.

Uranus and Neptune each have a number of icy Moons but of interest typically are the largest ones, (Miranda, Ariel, Umbriel, Titania and Oberon at Uranus and Triton and Nereid at Neptune). In addition, Pluto has a relatively large Moon (Charon) which is about half its size. Whereas the surfaces of the Moons of Uranus and Neptune are weathered principally by the magnetospheric plasmas [72], the surface of Charon is exposed only to the solar wind and cosmic rays. This is also the case for Pluto's surface when it is far from the Sun on its eccentric orbit and the atmosphere freezes out [73]. Pluto and Triton are thought to be related to the Kuiper belt objects discussed earlier and have methane and nitrogen atmospheres over an icy subsurface.

Uranus is unique in that its axis is tipped ~90° to the ecliptic plane, due presumably to a collision with a large object; there is a large angle between the magnetic and rotation axis, so the plasma trapping is not efficient. Therefore, the main plasma component is probably energetic protons from hydrogen in the atmosphere of Uranus, or the penetrating solar wind. It appears that the surfaces are radiation-weathered dark ice that contain some carbon species, probably CO or CO_2 [65]. It also has an unusual ring composed of dark particles, probably a radiation-carbonized material.

5. Upcoming planetary missions and challenges for surface scientists

5.1. Upcoming missions

There are a number of exciting experiments that will be flown on upcoming NASA space missions, and these will present challenging opportunities to the surface science community. We describe a few of these below.

5.1.1. Capturing the Sun: the Genesis Discovery mission [74]

The Genesis mission has an ambitious goal, to collect and return solar matter for analysis in terrestrial laboratories. Collection and analysis of the solar wind ions will provide unprecedented insights into the composition of the original solar nebula, the matter that condensed to form our solar system. Scheduled to be launched in late 2000, the Genesis spacecraft is built around a sample return canister that houses collector arrays. These arrays are high purity materials that will be exposed to the solar wind and will collect ions by implantation. The arrays will be located a million miles from Earth at a point in space that every first-year physics student has computed, where Earth and Sun's gravity balance each other. After two years, the collector arrays will be restored in the contamination-tight canister and returned to Earth for recovery and testing.

Analysis of the precious samples will focus on establishing the isotopic and elemental abundances of the solar wind. A number of sensitive and accurate analytical methods are expected to be applied, including resonance ionization mass spectrometry (RIMS). RIMS is a surface analytical method that has a unique combination of capabilities no other single analytical instrument can match, i.e., unambiguous identification of elemental impurities in the surface regions of solids at trace levels with single monolayer resolution [75]. The capabilities of RIMS are well matched to the needs of the Genesis mission, where solar wind constituents will be implanted at levels down to 0.1 ppt (parts per trillion) at depths $\leqslant 100$ nm. The specific RIMS instrument proposed for this work is referred to as SARISA (surface analysis by resonance ionization of sputtered atoms), and it employs resonantly-enhanced multi-photon ionization (REMPI) to optimize selectivity and sensitivity [76] in the detection of trace atoms sputtered from the target. REMPI can photoionize ground state atoms with an efficiency near 100%, so that a substantial fraction of the sputtered atoms can be detected. This is indeed an ideal tool for these important studies!

5.1.2. Exploring Saturn: the CASSINI mission

Cassini is likely to be the last of the large spacecraft sent to the outer solar system in the foreseeable future. Cassini will launch a probe into the atmosphere of Titan, the large Moon of Saturn which has an atmosphere thicker than that of Earth. Like the Galileo spacecraft, which is now completing its study of the Jovian system, the Cassini orbiter has a full complement of instruments to study Saturn and Titan, including a UV spectrometer, the ability to image through a number of optical filters, and the visual and infrared mapping spectrometer, the equivalent of Galileo's NIMS discussed above but with a somewhat larger spectral range [73]. In addition, there is a full complement of instruments to measure the plasma and magnetic fields and to detect "dust" impacts. The surface studies, therefore, are done remotely in orbit.

Although a principal goal of the Cassini mission is to understand the atmospheres of Saturn and Titan, another main goal is to understand the origins and evolution of the icy satellites and ring particles that are imbedded in the hot plasma trapped in Saturn's magnetosphere. Again surface physics is a dominant concern since the trapped plasma is self-sustained by sputtering and decomposition of the surface materials [4]. Because the icy Moons of Saturn are smaller than the Galilean satellites, there is no gravity filter and most of the sputtered material escapes directly from the satellites and orbits as an ambient neutral gas within Saturn's magnetosphere. Quite remarkably the OH component of this toroidal neutral atmosphere has already been imaged [70].

Since the neutrals ejected from the surface are eventually ionized by the electrons in the plasma, the plasma-ion composition is a *direct measure of the satellite's surface composition* [3]. This is an exciting prospect for surface scientists. That is, measuring the accumulated plasma is equivalent to performing a post-ionization SIMS study of the surfaces of the Moons and ring particles. Since it is also well known to the surface science community that, although the most volatile species are removed first by sputtering, after long-term bombardment the ejected material is a reflection of the *bulk* composition [77]. Therefore, whereas the spectral instrument only give a description of the surface layer depleted in the most volatile species, plasma mass spectrometry can indicate the subsurface materials in the absence of rapid re-coating.

Because of the above, a critical instrument for surface science at Saturn is CAPS (the Cassini Plasma Science Instrument [78]). This is the first time-of-flight mass spectrometer sent to the outer solar system. Whereas the Galileo plasma science studies were not able to give reasonable mass analysis of the low energy plasma and no information on molecular ions, CAPS will determine both the atomic and molecular composition of the low energy (<40 keV) component of the sputter-produced plasma. Cassini will also have detectors to measure the flux and masses of the very energetic ions (e.g., Fig. 3 for Galileo), but these highly accelerated ions will mostly be atomic. Though an indication of the atomic composition, these ions will also have experienced considerable transport during the acceleration process, so that the source region is not as easily identified. CAPS also can distinguish freshly formed ions from equilibrated ions based on the energy spectra of the ions. Therefore, the ability of CAPS to measure the molecular composition of the ions freshly formed from the sputter ejecta, combined with the spectral mapping of the satellite surfaces, will allow planetary scientists working with surface scientists to determine the composition of Saturn's Moons and ring particles. This is an exciting and challenging prospect which begins in 2004 when Cassini arrives at Saturn.

5.1.3. Europa Orbiter

Europa is of one of the highest priority targets for an outer solar system exploration mission. If subsurface liquid water or even warm subsurface ice were to exist on Europa, there is the possibility of prebiotic processes or even the possibility of life, perhaps forming near undersea volcanic vents. As a first step the Europa Orbiter is proposed to determine the presence or absence of a subsurface ocean, to characterize the three dimensional distribution of any subsurface liquid water and its overlying ice layers, and to understand the formation of surface features (e.g., Fig. 10), including sites of recent or current activity. Although the final payload has not been decided this will include instruments for precise determination of gravitational field, laser altimetry, ice-penetrating radar, and, more relevant to our discussions, imaging using either filters or a spectrometer and, possibly, a mass spectrometer. The proposed schedule is: following launch in 2003, Europa Orbiter will arrive at Jupiter's system in 2007 and enter orbit around Europa in 2009.

5.1.4. Pluto/Kuiper express

Pluto, the smallest planet, has remained enigmatic since its discovery in 1930 and is the only planet in our solar system not yet viewed close-up by spacecraft. Given its distance and size, it is a challenge for planetary astronomers. Although there has been recent progress in determining the atmosphere and surface composition, many of the key questions about Pluto and its satellite Charon await the close-up observation of a space flight mission, the Pluto/Kuiper Express [71]. The goals of the mission are to characterize surface geology and morphology and map the surface composition of Pluto and Charon, and to characterize the neutral atmosphere of Pluto and its escape rate. A possible extension is to fly by one or more Kuiper Disk objects. The strawman payload contains imaging/mapping spectrometers in the visible and IR, and a UV spectrometer. The study of the morphology of these objects will use the radio link to earth and be based on orbit analysis and occultations of the spacecraft. Although it is called a planet, Pluto may have been a member of the Kuiper Belt of icy objects mentioned above. It also is thought to be similar in many ways to Triton, the large moon of Neptune which is already known to have an active surface. On both objects and on the Kuiper Belt objects the surface properties are thought to be affected by the incident radiation as discussed. The launch is planned for around 2010 or soon after with the arrival 8–12 years later.

5.2. Challenges to surface scientists in model studies

5.2.1. Materials

To enhance their impact and credibility in the planetary sciences community, surface scientists will need to move away from crystalline surfaces of elemental solids and binary oxides and study more complex materials characteristic of the solar system. Many interesting crystalline materials exist that are good models for the Moon's surface, e.g., surfaces of feldspars ($NaAlSi_3O_3$, $KAlSi_3O_8$), or olivine, or plagioclase. Iron silicates are ubiquitous in meteorites and (presumably) asteroids. Due to space weathering, amorphous non-stoichiometric surfaces are ever present on the airless bodies of the solar system. Ices of condensed gases are a fertile areas as models of icy satellites and planets, particularly mixtures of gases and briny solutions. Whereas the thermal and non-thermal chemistry of water ice surfaces have been extremely well studied via TPD, ESD, PSD, etc., there is little or nothing known about ESD/PSD of frozen salt solutions or hydrated salt minerals. Porous materials—low temperature ices as well as powdered solids—present new challenges.

5.2.2. Beam damage and space weathering

The effects of all forms of radiation on the space weathering of asteroid-like surfaces need to be examined. Ion, electron and even laser beams (to simulate melting that occurs in micrometeorite impact) can be used to bombard mineral surfaces. What are the structural/chemical/electronic changes that lead to alterations in optical properties?

5.2.3. Role of ion sputtering and DIET

It appears to be clear that PSD can contribute to the lunar Na atmosphere and that DIET leads to the production of atomic and molecular oxygen at/near Europa and Ganymede, but the detailed role of DIET processes (ESD and PSD) in desorption from more complex surfaces is not known. Thresholds, absolute cross-sections and yields, and velocity distributions of products are all necessary information for planetary modelers. Cross-sections and sputtering yields for ion/solid collisions are generally not available for many magnetosphere ions colliding with complex planetary surfaces. The possible effects of DIET and sputtering on unusual surface isotopic abundances, and the huge disparity between surface composition and atmospheric abundances (in tenuous atmospheres) require further investigation. Ion-induced chemical reactions (e.g., formation of volatile H-containing molecules due to bombardment of surfaces by protons in the solar wind) are virtually unexplored. Theoretical calculations of thresholds, cross-sections and desorption yields are non-existent.

5.2.4. Model calculations

Because experiments cannot be carried out for the full range of ion and electron energies and ion types indicated in Figs. 2, 3 and Table 1, model descriptions of the radiolysis and DIET processes are absolutely essential. The model calculations that are required range from simple analytical analysis of laboratory data to detailed calculations of the effects of the interaction of radiation on surfaces in a vacuum. For ion–solid collisions, the energy deposition and initial transport are reasonably well in hand, and considerable progress has been made in calculating desorption of adsorbed species for many special cases; however, there are huge gaps in our understanding that do not allow us to extrapolate ideas learned to new materials or to other radiation types and other energy regimes. In addition, many of the simple models used for sputtering, such as the collision cascade and thermal spike models, either fail totally or are inapplicable to the processes described above. Therefore, in many instances, detailed classical or quantum molecular dynamics (MD) calculations may be required. For instance, thermal spike models of sputtering from a cylindrically energized region about an ion track appeared to well fit and allow extrapolation of the data on the electronic sputtering of ices by fast ions [79]. However, recent MD calculations have shown that apparent agreement to be completely fortuitous [80] at the high excitation densities of interest.

Moreover, lack of understanding exists at a more fundamental level. At present, although the database for electronic sputtering of ices by energetic ions is relatively large [4,81], a detailed understanding of the radiative and non-radiative decay processes leading to desorption exists only for the rare-gas solids. The electronic processes in molecular solids are only understood phenomenologically, as indicated in some of the discussion above. This is not surprising as these processes are strongly affected by local temperature, structure and composition, and by the complex chemistry induced in the surfaces and within the solids. However, although the description of such processes can be difficult, the motivation for understanding these processes is rather high. That is, a detailed understanding of the relevant surface processes can lead to an understanding of the origins and evolution of interesting objects in our solar system and, possibly, the origins of potential biological process on objects other than the Earth.

Acknowledgements

T.E. Madey acknowledges fruitful collaborations with Dr. Boris Yakshinskiy and Prof. V.N. Ageev, and partial support from the National

Science Foundation under grant no. 0075995, and by NASA Planetary Atmospheres Division. R.E. Johnson thanks W.L. Brown and L.J. Lanzerotti for introducing him to this subject and for many fruitful collaborations, and acknowledges long-term support from NSF and NASA. Thom M. Orlando thanks Prof. Tom McCord of the Hawaii Institute of Geophysics and Planetology, University of Hawaii for an introduction to planetary science and a collaboration concerning the issue of salts on Europa. Thom M. Orlando also acknowledges fruitful collaborations with Drs. Glenn Teeter, Greg Kimmel and Matthew Sieger (Pacific Northwest National Laboratory) and thanks the United States Department of Energy, Office of Science for support.

References

[1] Shoot the Moon, Newsweek 26 (1999) 64.
[2] K. Lodders, B. Fegley Jr., The Planetary Scientist's Companion, Oxford University Press, New York, 1998.
[3] Science (special issue on the lunar prospector) 281 (1998) 1405–1560.
[4] R.E. Johnson, in: J. Klinger, D. Reidel (Eds.), Ices in the Solar System, 1985, p. 337.
[5] R. Lundin, Erosion by the solar wind, Science 291 (2001) 1909.
[6] J.F. Cooper, R.E. Johnson, B.H. Mauk, H.B. Garrett, N. Gehrels, Energetic ion and electron irradiation of the icy Galilean Satellites, Icarus, 149 (2001) 133.
[7] T.E. Madey, B.V. Yakshinskiy, V.N. Ageev, R.E. Johnson, Desorption of alkali atoms and ions from oxide surfaces: relevance to origins of Na, K in atmospheres of Mercury and the Moon, J. Geophys. Res. 103 (1998) 5873.
[8] D.P. Woodruff, T.A. Delchar, Modern Techniques of Surface Science, second ed., Cambridge University Press, Cambridge, 1994.
[9] D.A. Williams, E. Herbst, It's a dusty Universe: surface science in space, Surf. Sci. 500 (2002) 823.
[10] T.E. Madey, F.M. Zimmermann, R.A. Bartynski (Eds.), Proc. of Eighth International Workshop on Desorption Induced by Electronic Transitions, Surf. Sci. 451 (2000) 1–275.
[11] X.L. Zhou, X.Y. Zhu, J.M. White, Photochemistry at adsorbate/metal interfaces, Surf. Sci. Rpts. 13 (1991) 73.
[12] T.E. Madey, Electron and photon-stimulated desorption: probes of structure and bonding at surfaces, Science 234 (1986) 316–322.
[13] V.N. Ageev, Y.A. Kuznetsov, B.V. Yakshinskii, T.E. Madey, Electron stimulated desorption of alkali metal ions and atoms: local surface field relaxation, Nucl. Instr. Meth. B 101 (1995) 69–72.
[14] R.C. Elphic, H.O. Funsten III, B.L. Barraclough, D.J. McComas, M.T. Paffett, D.T. Vaniman, G. Heiken, Lunar surface composition and solar wind-induced secondary ion mass spectrometry, Geophys. Res. Lett. 18 (1991) 2165–2168.
[15] A.E. Potter Jr., T.H. Morgan, Discovery of sodium in the atmosphere of Mercury, Science 229 (1985) 651–653.
[16] A.E. Potter Jr., T.H. Morgan, Discovery of sodium and potassium vapor in the atmosphere of the Moon, Science 241 (1988) 675–680.
[17] A.E. Potter Jr., T.H. Morgan, Coronagraphic observation of the lunar sodium surface, J. Geophys. Res. 103 (1998) 8581–8586.
[18] M.A. McGrath, R.E. Johnson, L.J. Lanzerotti, Sputtering of sodium on the planet Mercury, Nature 323 (1986) 694–696.
[19] D.M. Hunten, A.L. Sprague, Origin and character of the lunar and Mercurian atmospheres, Adv. Space Res. 19 (1997) 1551–1560.
[20] M. Mendillo, J. Baumgardner, Constraints on the origin of the Moon's atmosphere from observations during lunar eclipse, Nature 377 (1995) 404–406.
[21] B.V. Yakshinskiy, T.E. Madey, Evidence for role of photon stimulated desorption as a source of Na in the lunar atmosphere, Nature 400 (1999) 642.
[22] D.F. Heath, M.P. Thekaekara, in: O.R. White (Ed.), The Solar Output and its Variation, Colorado Assocation University Press, Boulder, 1977, p. 193.
[23] M. Mendillo, J. Baumgardner, J. Wilson, Observational test for the solar wind sputtering origin of the Moon's extended sodium atmosphere, Icarus 137 (1999) 13–23.
[24] S. Verami, C. Barbieri, C. Benn, G. Cremonese, Possible detection of meteor stream effects on the lunar sodium atmosphere, Planet. Space Sci. 46 (1998) 1003–1006.
[25] E.D. Whipple, Potentials of surface in space, Rep. Prog. Phys. 44 (1981) 1197–1250.
[26] A.A. Sickafoose, J.E. Colwell, M. Horanyi, S. Robertson, Photoelectric charging of dust particles in vacuum, Phys. Rev. Lett. 84 (2000) 6034.
[27] R.H. Manke, Plasma and potential at the lunar surface, in: R.J.L. Grace (Ed.), Photon and Particle Interactions with Surfaces in Space, D. Reidel, Norwell, MA, 1973, pp. 347–361.
[28] J. Benson, Direct measurement of the plasma screening length and surface potential near the lunar terminator, J. Geophys. Res. 82 (1977) 1917–1920.
[29] H.A. Zook, J.E. McCoy, Large scale lunar horizon glow and a high altitude lunar dust exosphere, Geophys. Res. Lett. 18 (1991) 2117–2120.
[30] S. Jurac, R.E. Johnson, R.A. Baragiola, E.C. Sittler, Charging of E-ring grains, J. Geophys. Res. 100 (1995) 14821–14831.
[31] M.J. Horanyi, J.A. Burns, D.P. Hamilton, The dynamics of Saturn's E-ring particles, Icarus 97 (1992) 248–259.

[32] B.E. Clark, R.E. Johnson, Interplanetary weathering: surface erosion in outer space, EOS Trans. 77 (1996) 141–145.

[33] R.J. Sullivan, R. Greeley, R.T. Pappalardo, E. Asphaug, J.M. Moore, D. Morrison, M.J. Belton, M. Carr, C.R. Chapman, P. Geissler, R. Greenberg, J. Granahan, J.W. Head, R. Kird, A. McEwen, P. Lee, P.C. Thomas, J. Veverka, Geology of 243 Lda, Icarus 120 (1996) 119–139.

[34] C.A. Dukes, R.A. Baragiola, L.A. McFadden, Surface modification of olivine by H^+ and He^+ bombardment, J. Geophys. Res. 104 (1999) 1865–1872.

[35] M.T. Sieger, W.C. Simpson, T.M. Orlando, Electron-stimulated desorption of D^+ from D_2O ice: surface structure and electronic excitations, Phys. Rev. B 56 (1997) 4925–4937.

[36] G.A. Kimmel, T.M. Orlando, Low-energy electron-stimulated dissociation of amorphous D_2O Ice: D(2S), $O(^3P_{2,1,0})$ and $O(^1D_2)$ yields and velocity distributions, Phys. Rev. Lett. 75 (1995) 2606.

[37] T.M. Orlando, M.T. Sieger, Electronic excitations and dissociation processes on low-temperature ice surfaces, J. Geophys. Res., in preparation.

[38] G. Strazzulla, G.A. Baratta, R.E. Johnson, B. Donn, The primordial comet mantle: irradiation production of a stable organic crust, Icarus 91 (1991) 101–104.

[39] M.H. Moore, T. Tanabe, Mass spectra of sputtered polyoxymethylene: implications for comets, J. Astrophys. 365 (1990) L39–L42.

[40] W.L. Brown, L.J. Lanzerotti, R.E. Johnson, Fast ion bombardment of ices and its astrophysical implications, Science 218 (1982) 525–531.

[41] W.L. Brown, W.M. Augustyniak, E. Simmons, K.J. Marcantonio, L.J. Lanzerotti, R.E. Johnson, J.W. Boring, C.T. Reimann, G. Foti, V. Pirronello, Erosion and molecule formation in condensed gas films by electronic energy loss of fast ions, Nucl. Instr. Meth. Phys. Res. 198 (1982) 1–8.

[42] M. Westley, R.A. Baragiola, R.E. Johnson, G. Baratta, Photo desorption from low temperature water ice: astrophysical implications, Nature 373 (1995) 405–407.

[43] J.R. Spencer, W.M. Calvin, M.J. Person, Charge-coupled device spectra of the Galilean satellites: molecular oxygen on Ganymede, J. Geophys. Res. 100 (1995) 19049–19056.

[44] D.T. Hall, D.F. Strobel, P.D. Feldman, M.A. McGrath, H.A. Weaver, Detection of an oxygen atmosphere on Jupiter's moon Europa, Nature 373 (1995) 677–679.

[45] R.E. Johnson, L.J. Lanzerotti, W.L. Brown, Planetary applications of ion induced erosion of condensed-gas frosts, Nucl. Instru. Meth. Phys. Res. 198 (1982) 147–157.

[46] C.T. Reimann, J.W. Boring, R.E. Johnson, J.W. Garrett, K.R. Farmer, W.L. Brown, K.J. Marcantonio, W.M. Augustyniak, Ion-induced molecular ejection from D_2O ice, Surf. Sci. 147 (1984) 227–240.

[47] A. Bar-Nun, G. Hermann, M.L. Rappaport, Yu. Mekler, Ejection of H_2O, O_2, H_2 and H from water ice by 0.5–6 keV H^+ and Ne^+ ion bombardment, Surf. Sci. 150 (1985) 143–156.

[48] M.T. Sieger, W.C. Simpson, T.M. Orlando, Production of O_2 on icy satellites by electronic excitations of low-temperature water ice, Nature 394 (1998) 554–556.

[49] K.S. Noll, T.L. Rousch, D.P. Cruikshank, R.E. Johnson, Y.J. Pendleton, Detection of ozone on Saturn's satellites Rhea and Dione, Nature 388 (1997) 45–47.

[50] P.A. Gerakines, W.A. Schutte, P. Ehrenfreund, Ultraviolet processing of interstellar ice analogs. I. Pure ices, Astron. Astrophys. 312 (1996) 289–305.

[51] R.E. Johnson, W.A. Jesser, O_2/O_3 micro-atmospheres in the surface of Ganymede, Astrophys. J. Lett. 480 (1997) L79–L82.

[52] R.W. Carlson, M.S. Anderson, R.E. Johnson, W.D. Smythe, A.R. Hendrix, C.A. Barth, L.A. Soderblom, G.B. Hansen, T.B. McCord, J.B. Dalton, R.N. Clark, A.C. Ocampo, J.H. Shirley, D.L. Matson, Hydrogen peroxide on the surface of Europa, Science 283 (1999) 2062–2064.

[53] R. Carlson, A tenuous carbon dioxide atmosphere on Jupiter's moon Callisto, Science 283 (1999) 820–821.

[54] R.E. Johnson, Sodium at Europa, Icarus 143 (2000) 429–433.

[55] C.A. Barth, C.W. Hord, A.I.F. Stewart, W.R. Pryor, K.E. Simmons, W.E. McClintock, Galileo ultraviolet spectrometer observations of atomic hydrogen in the atmosphere of Ganymede, Geophys. Res. Lett. 24 (1997) 2147.

[56] T.B. McCord et al., Salts on Europa's surface detected by Galileo's new infrared mapping spectrometer, Science 280 (1998) 1242–1245.

[57] J.S. Kargel, Brine volcanism and the interior structures of asteroids and icy satellites, Icarus 94 (1991) 368–390.

[58] T.B. McCord, T.M. Orlando, G. Teeter, G.B. Hansen, M.T. Sieger, N. Petrik, L. Van Keulen, Thermal and radiation stability of hydrated salt minerals epsomite, mirabilite, and natron under Europa's environmental conditions, J. Geophys. Res., 106 (2001) 3311.

[59] K.K. Khurana et al., Induced magnetic fields as evidence for subsurface oceans in Europa and Callisto, Nature 395 (1998) 777–780.

[60] R.W. Carlson, R.E. Johnson, M.S. Anderson, Sulfuric acid on Europa and the radiolytic sulfur cycle, Science 286 (1999) 97–99.

[61] C. Chyba, Energy for microbial life on Europa, Nature 403 (2000) 381–382.

[62] M.E. Brown, R.E. Hill, Discovery of an extended sodium atmosphere around Europa, Nature 380 (1996) 229–231.

[63] B.V. Yakshinskiy, T.E. Madey, Electron and photon-stimulated desorption of K from ice surfaces, J. Geophys. Res., in press.

[64] M.K. Pospieszalska, R.E. Johnson, Magnetospheric ion bombardment profiles of satellites: Europa and Dione, Icarus 78 (1989) 1–13.

[65] J.A. Burns, M.S. Matthews, Satellites, University of Arizona Press, Tucson, 1986.

[66] R.E. Johnson, R.M. Killen, J.H. Waite, W.S. Lewis, Europa's surface and sputter-produced ionosphere, Geophys. Res. Lett. 25 (1998) 3257–3260.

[67] D.B. Chrisey, R.E. Johnson, J.W. Boring, J.A. Phipps, Ejection of sodium from sodium sulfide by the sputtering of the surface of Io, Icarus 75 (1988) 233–244.

[68] R.C. Wiens, D.S. Burnett, W.F. Calaway, C.S. Hansen, K.R. Lykke, M. Pellin, Sputtering products of sodium sulfate: Implications for Io's surface and for sodium-bearing molecules in the Io torus, Icarus 128 (1997) 386–397.

[69] S. Jurac, R.E. Johnson, J.D. Richardson, Saturn's E ring and production of the neutral torus, Icarus, 149 (2001) 384–396.

[70] J.D. Richardson, A. Eviatar, M.A. McGrath, V.M. Vasyliunas, OH in Saturn's magnetosphere: observation and implications, J. Geophys. Res. 103 (1998) 20245–20255.

[71] Pluto/Kuiper Express, http:/spacescience.nasa.gov/ao/99-oss-04/.

[72] L.J. Lanzerotti, W.L. Brown, C.G. Maclennan, A.F. Cheng, S.M. Drimigis, R.E. Johnson, Effects of charged particles on the surfaces of the Moons and Rings of Uranus, J. Geophys. Res. 92 (1987) 14949–14956.

[73] R.E. Johnson, Effect of irradiation on the surface of Pluto, Geophys. Res. Lett. 16 (1989) 1233–1236.

[74] http://www.gps.caltech.edu/genesis/genesis3.html.

[75] W.F. Calaway, M.P. McCann, M.J. Pellin, Impurity characterization of solar wind collectors for the Genesis Discovery mission by resonance ionization mass spectrometry, Proc. of 1998 MRS Symposium, Materials in Space-Science Technology and Exploration.

[76] Z. Ma, R.N. Thompson, K.R. Lykke, M.J. Pellin, A.M. Davis, New instrument for microbeam analysis incorporating submicron imaging and resonance ionization mass spectrometry, Rev. Sci. Instrum. 66 (1995) 3168–3176.

[77] R.E. Johnson, E.C. Sittler, Sputter-produced plasma as a measure of satellite surface composition: Cassini mission, Geophys. Res. Lett. 17 (1990) 1629–1632.

[78] CAPS2000, web page: http://www.jpl.nasa.gov/cassini/Science/MAPS/CAPS.html.

[79] R.E. Johnson, M.K. Pospieszalska, W.L. Brown, Linear to quadratic transition in electronic sputtering of solid N_2 and O_2, Phys. Rev. B. 44 (1991) 7263–7272.

[80] E.M. Bringa, R.E. Johnson, M. Jakas, Molecular dynamics simulation of electron sputtering from a cylindrical track, Phys. Rev. B 60 (1999) 15107–15116.

[81] R.E. Johnson, T.I. Quickenden, Photolysis and radiolysis of water ice on outer solar system bodies, J. Geophys. Res. 102 (1997) 10985–10996.

[82] J.A. Simpson, Elemental and isotopic composition of the galactic cosmic rays, Ann. Rev. Nucl. Part. Sci. 33 (1983) 323–381.

[83] A.S. Jurac, Handbook of Geophysics and Space Environment, Air Force Geophysics Laboratory, Hanscom AFB, MA, USA, 1985, pp. 2-1–2-21.

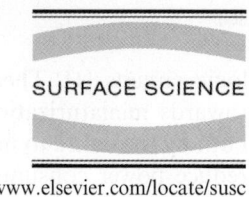

The surface science of semiconductor processing: gate oxides in the ever-shrinking transistor

Marcus K. Weldon [a,*], K.T. Queeney [b], Joseph Eng Jr. [c], Krishnan Raghavachari [c], Yves J. Chabal [c]

[a] *Bell Laboratories, Lucent Technologies, 600-700 Mountain Avenue, Murray Hill, NJ 07974, USA*
[b] *Department of Chemistry, Smith College, Northampton, MA 01063, USA*
[c] *Agere Systems, 600-700 Mountain Avenue, Murray Hill, NJ 07974, USA*

Received 1 August 2000; accepted for publication 17 August 2001

Abstract

Due to the extreme dimensional scaling required by Moore's law, Si device technology is increasingly subject to the limitations imposed by the intrinsic physics and chemistry of surfaces and interfaces. In this review we outline ways in which fundamental surface science has contributed an understanding to the microelectronics community and discuss areas where surface science may impact future development. We focus on the example of silicon dioxide (SiO_2) on silicon, since this interface lies at the heart of modern transistor technology and has therefore received a great deal of attention in recent years. We highlight a number of experimental and theoretical approaches that have elucidated the fundamental phenomena associated with the formation and evolution of this critical technological interface, revealing the remarkable interdependence of science and technology that now characterizes this rapidly evolving industry. © 2001 Published by Elsevier Science B.V.

Keywords: Silicon; Silicon oxides; Oxidation; Vibrations of adsorbed molecules; Ab initio quantum chemical methods and calculations; Low index single crystal surfaces

1. Introduction

At the dawn of the 21st century, we find ourselves firmly entrenched in the information age, an era made possible by advances in microelectronics during the last half of the 20th century. Computers and microelectronic devices, once exclusively the tools of scientists and engineers, are now inseparably woven into the fabric of mainstream society, causing a revolution in communications, commerce, information management, and automation. Just as the industrial revolution began with the invention of the steam engine, the information age began with the invention of the transistor.

Originally designed as a solid-state amplifier to replace vacuum tubes, the transistor found a far more important use when researchers at Bell Laboratories realized that the transistor's small size and low power consumption make it ideal for

---

[*] Corresponding author. Tel.: +1-908-582-5645; fax: +1-908-582-3958.
E-mail address: marcus@bell-labs.com (M.K. Weldon).

logic circuits [1]. There is an ever-present trend towards miniaturization of transistor size, driven both by the desire to increase switching speeds and reduce power consumption as well as the economies afforded by increased throughput (fabrication of more transistors per wafer). The breathtaking rate of transistor size scaling is captured in a concise statement known as Moore's law, named after Gordon Moore of Intel. In 1965, Moore published a semilog plot of the number of transistors on a silicon chip as a function of the date that the chip would become available, based on an extrapolation from the progress in miniaturization leading up to that time. The plot took the form of a straight line and predicted that the number of transistors per chip would double every year. Amazingly, the semiconductor industry has kept pace with Moore's law for many years, although more recently the rate of doubling has been relaxed to every 18–24 months (Fig. 1) [2,3]. The success of this push toward transistor miniaturization is illustrated by comparison of the first transistor (built in 1947) with a transistor fabricated using current state-of-the-art technology (Fig. 2).

A drastic reduction in the cost per transistor has accompanied this aggressive transistor size scaling. For example, the 1999 international technology roadmap for semiconductors (published by SIA, the Semiconductor Industry Association) projects that the cost per transistor will decrease by nearly a factor of 8 in the six-year period from 1999 to 2005, from 1735 to 217 microcents/transistor [4]. (To put these costs in perspective, it is currently more expensive to print one character in a newspaper than it is to fabricate one transistor.) This combination of low absolute cost per transistor and continuing cost reduction has made widespread use of microelectronic devices economically feasible.

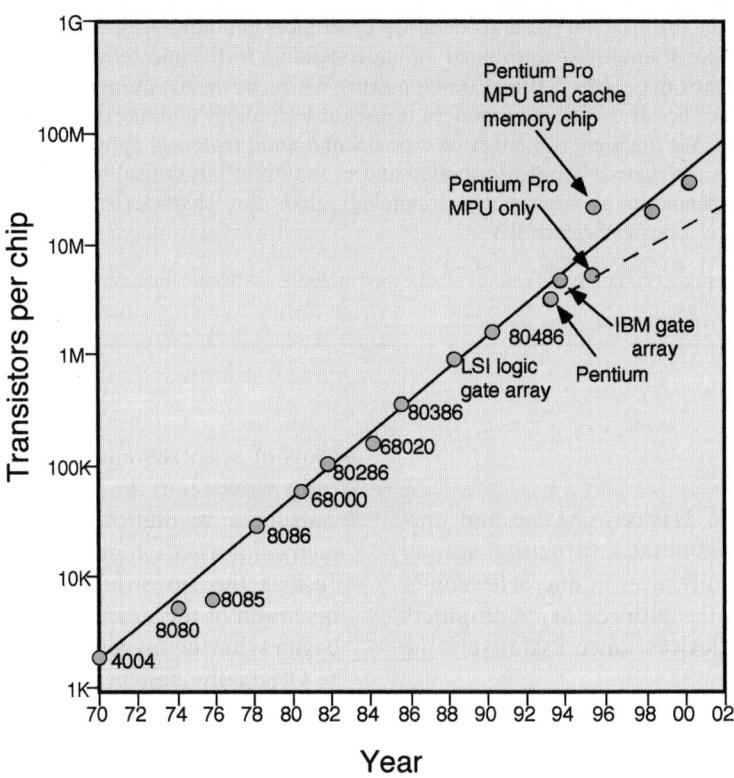

Fig. 1. Moore's law plot showing the progress in transistor scaling since 1970.

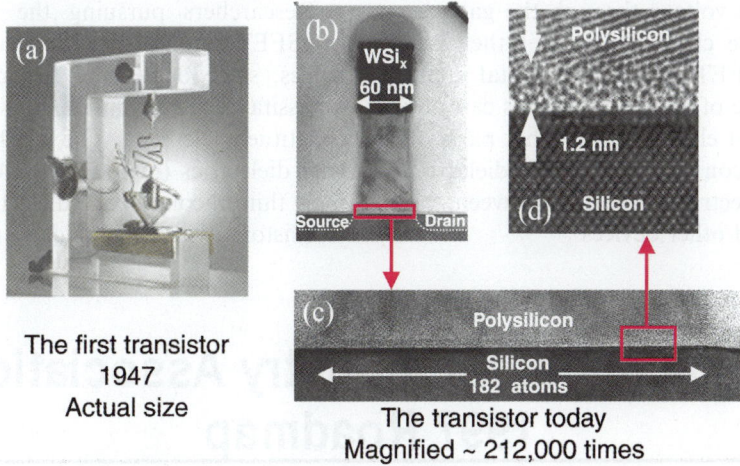

Fig. 2. (a) The first transistor; (b) a transmission electron micrograph of a current state-of-the-art transistor; (c) a closeup of the channel region between the source and drain; and (d) a closeup of the silicon oxide gate dielectric layer. TEM courtesy of Frieder Baumann.

Despite the success story outlined above, it is now widely recognized that the miniaturization scheme predicted by Moore's law cannot continue for much longer with the materials conventionally used for fabricating transistors. However, a substantial deviation from the pace dictated by Moore's law would be disastrous, as it would inhibit the technology-driven economic expansion that we now enjoy. Consequently, researchers in the semiconductor industry are searching intensively for viable new materials and processing strategies that will allow continued progress in transistor miniaturization. The physical characteristics of the transistor—discussed in more detail below—are such that the intrinsic physics and chemistry of ultrathin films and solid interfaces are of paramount importance in this search. It is therefore in this critical arena in which surface science promises to make significant contributions.

To put the current materials challenges into proper perspective, it is useful to take a brief look both at how transistors work and at the materials used to make them. Fig. 3 shows a schematic diagram of a transistor (design circa 1985), complete with metal interconnects and dielectric isolation layers [5]. This type of transistor, known as a metal-oxide semiconductor field effect transistor (MOSFET), continues to be the predominant de-

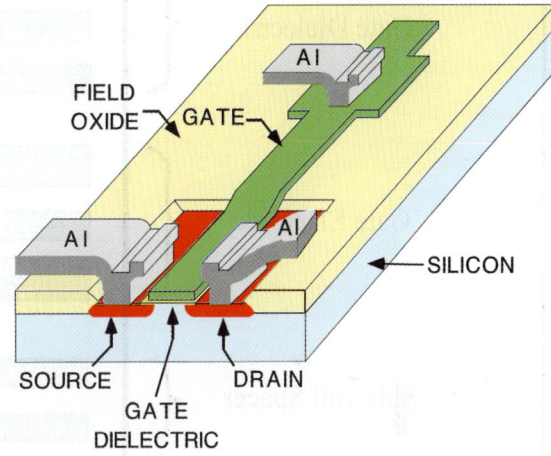

Fig. 3. Schematic diagram of a transistor (design circa 1985).

vice in ultralarge scale integrated circuits because it is considerably simpler to scale down compared to other types of transistors [6]. The central core of a MOSFET is comprised of a source and a drain (regions of doped silicon), a gate dielectric (an extremely thin silicon dioxide layer), and a gate electrode—a layer of poly-crystalline silicon, rendered conductive by heavy doping, to bias the gate dielectric. Essentially, the MOSFET acts as a switch, allowing current to flow from the source to the drain only when the gate electrode supplies the appropriate bias voltage through the gate dielectric to the active channel region. Other key elements of a MOSFET include thin metal interconnect layers (made of aluminum in this case) to connect the transistor electrically to other parts of the circuit and silicon dioxide (SiO_2) dielectric layers to provide electrical isolation between the Al interconnects and other devices.

Researchers pursuing the miniaturization of MOSFETs are faced with several daunting challenges, since decreasing the size of a transistor necessitates reduction in the scale of each of its constituent elements The 1999 SIA roadmap for gate dielectrics (table in Fig. 4) demands exceedingly thin silicon dioxide dielectric layers for future transistors, with SiO_2 thicknesses of 1.0 nm (cor-

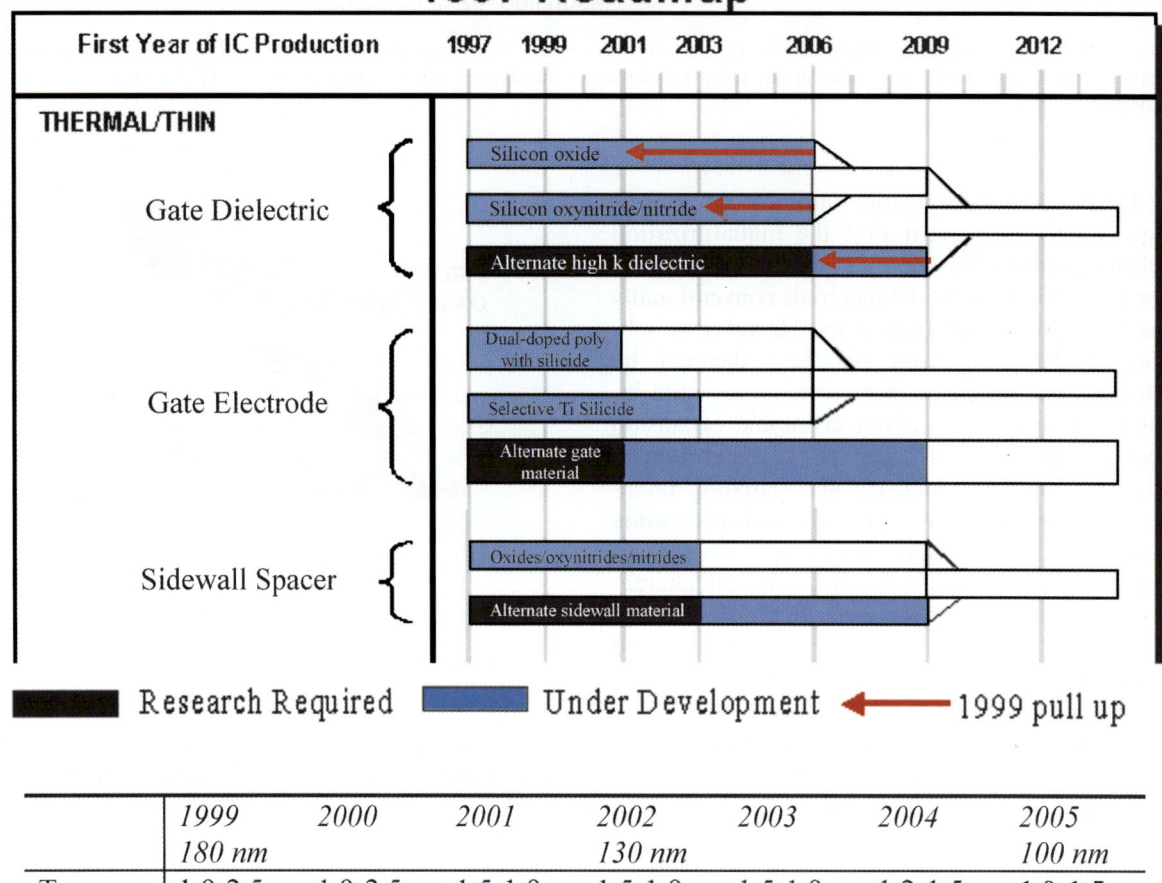

Fig. 4. Section of the 1997 SIA roadmap for silicon device technology showing some of the critical areas for research and development in interfacial materials. The red arrows indicate the amended scheduling agreed upon in 1999. The table indicates the specific schedule for gate oxide thickness reduction. The 180, 130, and 100 nm headings refer to the distance between source and drain in the transistor. No known method of making electrically reliable 1.0–1.5 nm gate oxides currently exists.

responding to merely three layers of SiO_2) by 2005 (Fig. 2) [4]. Electron tunneling through these ultrathin SiO_2 films is a serious problem during gate biasing, furthermore, dopants in the poly-crystalline Si gate electrode (in particular boron) tend to diffuse through the gate oxide during biasing, eventually winding up in the channel region between the source and the drain and causing significant, irreversible variations in the electrical performance of the MOSFET [7,8]. As will be addressed in more detail below, these fundamental limitations in the utility of SiO_2 as a gate dielectric are driving the current search to identify an alternative material (or materials) to replace SiO_2 and permit further miniaturization [9].

In addition to the failure of SiO_2 as a gate dielectric material at the transistor dimensions predicted by Moore's law within the next 10 years, several other problems arise from continued transistor scaling. For example, compared to circuits from a decade ago, today's circuits have metal interconnects with considerably smaller cross-sections. As a result, current densities in the interconnects are high enough to cause diffusion of metal atoms, a process known as electromigration [10]. Although mechanisms of electromigration are not completely understood, it is recognized that this problem can be circumvented by replacing aluminum, the metal traditionally used for interconnects, with a metal that has a lower resistivity. Copper, which is relatively abundant and second only to silver in conductivity, has emerged as the top candidate for the interconnect material in future generations of devices [11,12]. However, poor metal adhesion and unwanted diffusion into underlying layers make copper processing problematic [10].

Another, closely associated, materials problem is to find an alternative to SiO_2 for dielectric isolation of the copper interconnection layers. This need arises because the dielectric constant, k, of SiO_2 is large enough that it causes parasitic capacitances that slow down the overall speed of the circuit. Alternative low-k dielectric materials, both organic and inorganic, have been examined as possible replacements. However, a wide range of issues must be considered for successful implementation, including uniformity, adhesion, chemical resistance, thermal stability, density, coefficient of thermal expansion, and copper diffusion coefficient [10,13–15].

The materials issues listed above do not constitute an exhaustive list of those facing the semiconductor industry and are intended only as a sampling of the types of problems faced by researchers in this field. What is notable in each case described is that, throughout the transistor architecture, the success of aggressive scaling has brought the materials problems of microelectronics into the realm of ultrathin films and interfaces. In this domain surface science is poised to make substantial contributions, since the length scales probed by surface analytical techniques are comparable to the dimensions of microelectronic devices. To illustrate the impact of surface science in more concrete terms, we will focus below on how surface analytical techniques have been used to study the physical characteristics of ultrathin silicon oxide gate dielectrics.

2. The Si/SiO_2 interface

As outlined above current device technology utilizes a thin layer of SiO_2, grown on the Si substrate, as the gate dielectric. The choice of SiO_2 was based on both its thermodynamically driven formation as the native oxide on Si and on its excellent electrical characteristics. Specifically, silicon dioxide growth gives rise to low interfacial defect densities ($\sim 10^{12}$ cm^{-2}, corresponding to less than 1/100th of a monolayer) and exhibits high resistivity ($\geqslant 10^{15}$ Ω cm), a large band gap (9 eV) and good dielectric strength (10^7 V/cm). However, device scaling has now progressed to the point where gate oxides of ~ 1.3 nm will soon be needed to provide the requisite drain current response to the (decreasing) applied voltage. Such gate oxide films are ~ 5 SiO_2 layers across, with at least two of these layers occurring at the Si/SiO_2 interfaces between the gate oxide and the $Si(1 0 0)$ substrate at the bottom, and the poly-crystalline Si gate at the top [16]. Moreover, depending on the degree of interfacial roughness, of course, the number of "interfacial" Si atoms may be higher. Consequently, Si technology has progressed to the point where it has become critical to understand and control this Si/SiO$_2$ interface at the monolayer level—the defining thickness regime of surface science.

Despite the intrinsic near perfection of the Si/SiO$_2$ interface that can be loosely ascribed to the natural thermodynamics of the system, at the monolayer level there necessarily exists an inability to abruptly change from one material (Si) to another (SiO$_2$) without the formation of some transition region. For example, even in the limit of a perfect interface, the last layer of Si substrate atoms *cannot* be coordinated to four oxygen atoms (as required for stoichiometric silicon dioxide). This alone will degrade device performance by the creation of non-bulk electronic states [17]. While this fundamental limit is not yet upon us, it looms large on the horizon, as the projected miniaturization of transistor technology requires an oxide thickness of less than 1 nm by 2012. The prevailing issues, therefore, pertain to the extent to which actual film growth results in thicker transition layers and chemical defects that further modify the electronic response from this "ideal" limit.

The first reports of significant additional interfacial material occurred in the late 1980s, when the state-of-the-art research devices displayed gate oxide thicknesses of ~100 Å. X-ray photoelectron spectroscopy was the first surface science technique to be widely applied to the study of these systems. The attachment of electronegative O atoms to Si results in the appearance of additional Si core-level peaks; analysis of the thickness dependence led to the conclusion that these features arose from a 10 to 30 Å thick interfacial transition layer (Fig. 5) [18–20]. This early work marked the onset of a veritable stampede of research aimed at investigating the physical and chemical structure of this interface, as well as the impact of such interfacial layers on device performance, as a function of processing conditions. While these investigations led to some advances in understanding, only in the past few years have the sensitivity, sophistication and availability of a number of key experimental techniques and theoretical methodologies been sufficient to allow a truly coherent, detailed picture to emerge. Therefore, our focus here will be on reviewing a few recent quintessential studies, rather than on attempting to provide a full historical overview of the literature.

2.1. The Si/SiO$_2$ interface: a core-level view

X-ray photoelectron spectroscopy, the first surface science technique to yield valuable information on the Si/SiO$_2$ interface, remains a crucial technique for such studies. We therefore begin our overview of recent surface studies of silicon oxidation with a discussion of the recent controversy regarding the assignment of XPS core-level shifts of Si, since this issue lies at the heart of the utility of XPS for studying this (and other) interfaces. The controversy centers around the role that second-nearest-neighbor atoms play in determining the binding energy of a given (Si) atom. The traditional interpretation of core-level spectra is based on a formal oxidation state model, wherein the direct attachment of electronegative O atoms to Si shifts the relevant Si core-level peaks by ~1 eV per nearest neighbor O atom (Fig. 5) [21]. The relative number of silicon atoms in different oxidation states can then be used to develop structural models of the Si/SiO$_2$ interface. This simple (and widely accepted) model assumes implicitly that the effect of second-nearest-neighbor O atoms on the core-level shifts of Si atoms is small. While such second-nearest-neighbor effects are discussed extensively in the XPS literature, until recently no definitive study has been undertaken to justify this assumption.

Banaszak-Holl, McFeely and coworkers investigated a model system designed to address this issue in the case of the Si/SiO$_2$ interface and therefore to allow definitive assignment of the transition region species identified with XPS [22,23]. The model system was comprised of a spherosiloxane molecule, essentially a cube with sides spanned by Si–O–Si linkages and vertexes comprised of tetrahedral H–Si–O$_3$ units. They argued that exposure of the clean Si(1 0 0) surface to this molecule resulted in attachment of the intact cube to one side of a Si dimer on the (2 × 1) reconstructed surface (so-called "single vertex attachment"), along with formation of an Si–H bond on the other end of the dimer (Fig. 6, bottom

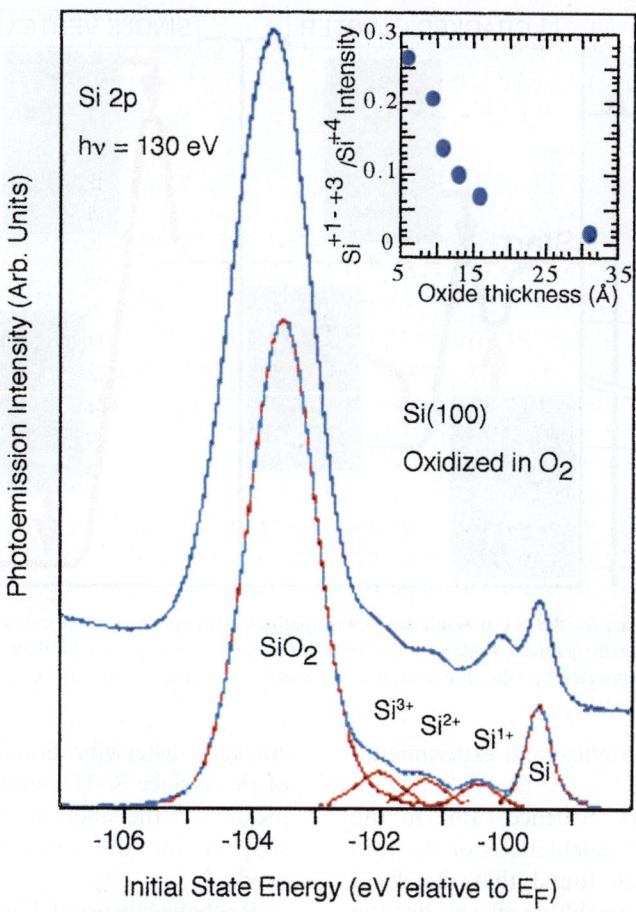

Fig. 5. X-ray photoelectron spectra of an oxide grown on Si(1 0 0). The raw data are shown in the upper curve, the baseline corrected data are below, together with the peak fit showing the individual sub-oxide features (from Ref. [18]). Inset: Typical relative sub-oxide intensity as a function of thickness (from Ref. [28]).

right). Indeed, such a reaction scheme is in accord with the known reactivity of this surface for a variety of other reagents: H_2O, CH_3CH_2OH and SiH_4 each dissociate across the dimers to produce HO–Si–Si–H, CH_3CH_2O–Si–Si–H and H_3Si–Si–Si–H, respectively. Core-level spectra of the resulting surface seemed to demonstrate that second-nearest-neighbor effects were indeed active in this system, since the resulting core-level shifts could only be assigned within the single vertex attachment model by incorporating such effects. Specifically, for such a structural model, the eight Si atoms of the cube are all directly coordinated to three oxygens (denoted +3 oxidation state), so that only one significantly shifted Si core-level peak should be observed (shifted by 3.5 eV) according to the formal oxidation state model. However, the spectrum actually observed contains two additional features shifted by 1.0 and 2.2 eV relative to bulk Si. This latter feature was controversially attributed to the Si of the cube that was attached to the surface (denoted 2 in Fig. 6); the 1.0 eV feature was then attributed to the surface Si (denoted 1) under the influence of the second-nearest-neighbor oxygen atoms of the cube.

The implications of this new assignment scheme were profound, since the composition of the Si/SiO_2 interface deduced from all prior XPS studies was now brought into question. However, before widespread acceptance of this new model could be

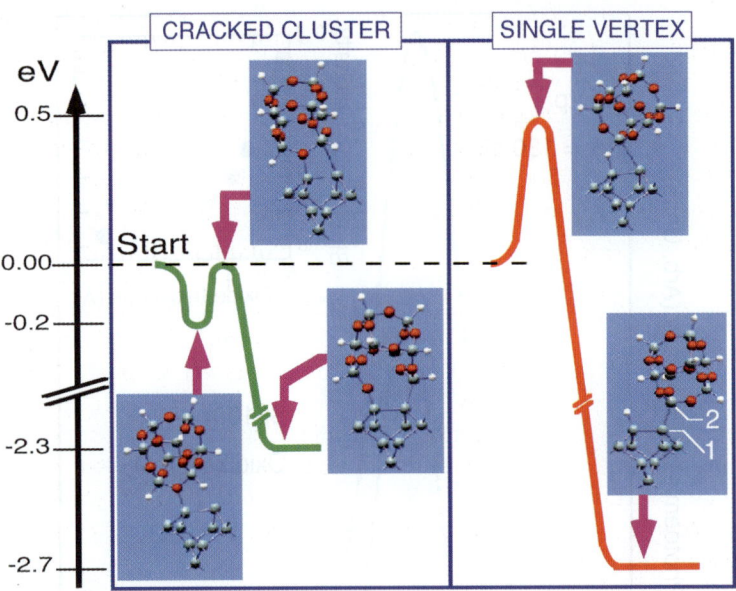

Fig. 6. Calculated energy diagram for the two possible adsorption pathways for spherosiloxane cubes on Si(1 0 0)-(2 × 1). The right panel shows the (thermodynamically favored) single vertex attachment proposed by Banaszak-Holl and McFeely; the left panel shows the cracked cluster geometry supported by vibrational studies and kinetic considerations (from Ref. [26]).

expected, further key theoretical and experimental studies were warranted.

The first contradictory evidence came in the form of the ab initio slab calculations of Pasquarello et al., [21,24] which found that the effect of such second-nearest-neighbor effects on the core-level spectrum of silicon was *negligible*. The theoretical results suggested that perhaps the adsorption model cited in the initial work was incorrect; the counter hypothesis was that the discrepancy might lie in the failure of the ab initio calculations to accurately reproduce core-level shifts. This question was subsequently addressed in a study by Raghavachari and Eng [25] that employed quantum chemical cluster calculations, in combination with surface infrared spectroscopic measurements, of the adsorbate-covered surface. This study found that, although the single vertex attachment initially proposed by Banaszak-Holl and McFeely is indeed thermodynamically favored, kinetic factors favor decomposition of the cube structure via Si–O bond scission ("cracked cluster" model) to form the adsorbate species shown in Fig. 6 (left) [26]. Not only was there excellent agreement between experiment and theory for the cracked cluster vibrational modes, but the absence of the surface Si–H species that is a necessary by-product of the single vertex model provided clear support for the predominance of the kinetic product.

Raghavachari et al. [26] have recently calculated the core-level spectrum of this surface species and found it to be in excellent agreement with the XPS data without requiring second-nearest-neighbor interactions. Specifically, seven Si atoms of the cracked cube are coordinated to three oxygens (+3 oxidation state), one is bound to two oxygens (+2) and the surface Si is bound to one oxygen (+1), producing core-level peaks at 3.6, 2.2 and 1.0 eV with a 7:1:1 intensity ratio, just as is observed experimentally.

Therefore, although there remains some residual contention in the literature—in particular, the need to reconcile work of Raghavachari and Eng with recent scanning tunneling microscopy (STM) results on this same system—a convincing body of work now exists in the surface physics literature that supports the original formal oxidation state model. It is important to point out that although the net effect of these surface studies was to support the original hypothesis, the importance of these recent contributions should not be underestimated. Not only do such studies exemplify the level of detail with which Si surface reactions can be understood (see the chapter by S.F. Bent for further examples), but they also underscore the critical technological need to definitively characterize these silicon–oxide interfaces.

2.2. The Si/SiO₂ interface: a vibrational view from the top down

The conclusions described above have notable implications for researchers seeking to elucidate the microscopic details of the growth of genuine ultrathin oxide films. As mentioned earlier, XPS was the first surface technique with sufficiently high sensitivity and selectivity to be used to probe the ultrathin oxide regime [18,20]. A large number of such studies have been published over the past decade, with the essential finding being that in all state-of-the-art films today there is a 3–5 Å sub-stoichiometric silicon oxide layer at the interface, comprised of Si^{+1}, Si^{+2} and Si^{+3} species [27]. However, due to the relative insensitivity to the chemical environment beyond the nearest coordination shell, XPS has little to say regarding the local or extended structure of the film, nor the microscopic mechanism by which Si oxidizes. Therefore, infrared spectroscopy has become increasingly used as a complementary technique with both monolayer sensitivity and the ability to probe buried interfaces (below the surface). Since infrared spectroscopy probes the vibrations of structural—units which are sensitive to the precise details of the chemical coordination, i.e. the valence-level electronic structure—it can provide unique insight into the microscopic structure of the interfacial region.

Despite the intrinsic appeal of infrared spectroscopy as a technique to probe these films, the interpretation of spectra is a complex problem. This complexity is a consequence of the sensitivity of vibrational excitations to many factors: for example, bond type, bond length and angle, and through-space dipole interactions both between oscillators and with the high dielectric substrate [28]. Fortunately, by careful modeling of the effect of each type of interaction on the observed vibrational spectrum, the essential film character can be deduced.

Perhaps the best approach for characterizing the Si/SiO₂ interface region in genuine transistor gate oxide films is to acquire spectra as a function of oxide film thickness and to follow the changes apparent as one moves from the bulk film limit to the thinnest interfacial oxide layers. Such a study has recently been reported, with infrared spectra recorded as a 31 Å thermal oxide film is successively thinned in dilute hydrofluoric acid down to ∼6 Å [28]. The principle change observed over this thickness regime is the continuous shift of the characteristic transverse optical (TO) and longitudinal optical (LO) phonon modes to lower frequency as the thickness is decreased (Fig. 7). For pedagogical convenience the TO mode can be thought of as an extended mode version of the T_2 vibration of the constituent tetrahedra in which two oxygens move towards the central Si and two move away, with the Si moving to maintain the center of mass position [29]. The LO is then derived from the TO mode by addition of a coulombic coupling term.

Queeney et al. [28] have shown by a quantitative analysis of the TO/LO mode behavior that only the presence of a sub-stoichiometric layer at the interface can explain both the direction *and* differing magnitudes of the respective shifts observed. Specifically, they used the effective medium approximation to deduce the effective dielectric response (ε_{eff}) of a non-stoichiometric layer mixed with bulk SiO₂; the resultant infrared spectrum is then calculated from TO = Im(ε_{eff}) and LO = −1/Im(ε_{eff}). SiO was used as a prototypical non-stoichiometric oxide layer (as ε_{SiO} has been tabulated in the infrared region of interest); the agreement between the spectrum calculated for a theoretical layer composed of 35% SiO₂ and 65% SiO and experimental data for a 4.5 Å film is quite remarkable (Fig. 8). However, at this level of analysis, the agreement must be regarded as indicating only the *qualitative* correctness of this picture of the interface, since such studies cannot answer essential questions regarding how and when this sub-stoichiometric interface was formed during the film growth process. To obtain such microscopic

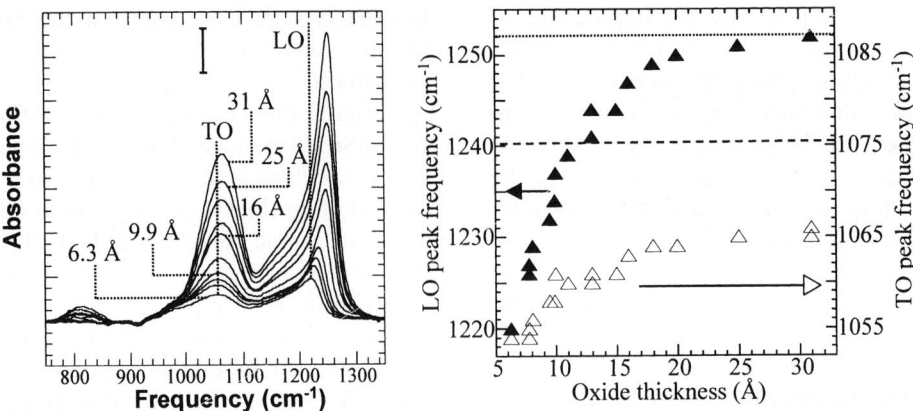

Fig. 7. Left: Infrared spectra of the Si–O stretching vibrations of a thermally oxidized Si(1 0 0) wafer as a function of thinning in dilute hydrofluoric acid. Right: The TO and LO mode frequencies plotted over the same thickness regime (from Ref. [28]).

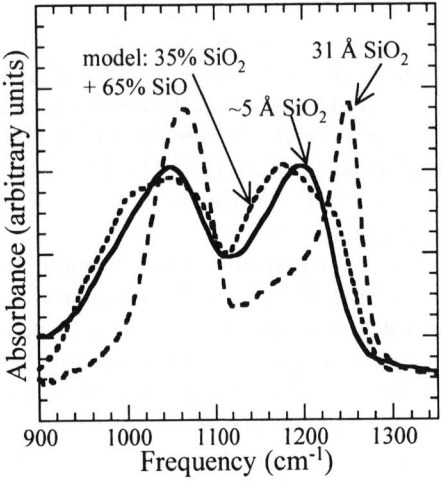

Fig. 8. Comparison of experimental infrared spectra with the calculated spectrum for a layer comprised of 65% SiO and 35% SiO_2 (from Ref. [28]).

mechanistic insight one needs the full control and analytical capabilities of an ultrahigh vacuum environment, allowing the study of the sub-monolayer growth regime in a step-by-step manner, as described in the following.

2.3. The Si/SiO$_2$ interface: a vibrational view from the bottom up

The initial oxidation of Si has recently been studied in great detail using a combination of surface infrared spectroscopy and ab initio quantum cluster calculations. The cluster calculations employed in these studies utilize gradient-corrected density functional techniques and yield accurate energetics for the different possible reaction products, together with their associated vibrational frequencies. Careful comparison between the theoretical and experimental spectra enables unambiguous structural assignments to be made in most cases.

The majority of the work to date has focussed on the water-induced oxidation of the clean Si(1 0 0)-(2 × 1) reconstructed surface [30–33], exposing this surface to sub-monolayer quantities of oxygen (in the form of H_2O) and then annealing to elevated temperatures to simulate growth conditions. Using this classical surface science approach it has been possible to follow the conversion of oxygen from one chemical form to another, as well as to delineate the lateral migration of O to form oxygen agglomerates prior to completion of the first monolayer. The essential findings can be summarized as follows:

The initial uniform H–Si–Si–OH dimer structure (Fig. 9, left) first decomposes at ~475 K to form a mixed overlayer of one oxygen-containing dimers, H–Si–O–Si–H, and two oxygen-containing dimers, H–(O)Si–O–Si–H, as well as oxygen-free dimers, H–Si–Si–H (Fig. 9, middle) [30]. Importantly, the quantitative agreement between the theoretical and experimental vibrational fre-

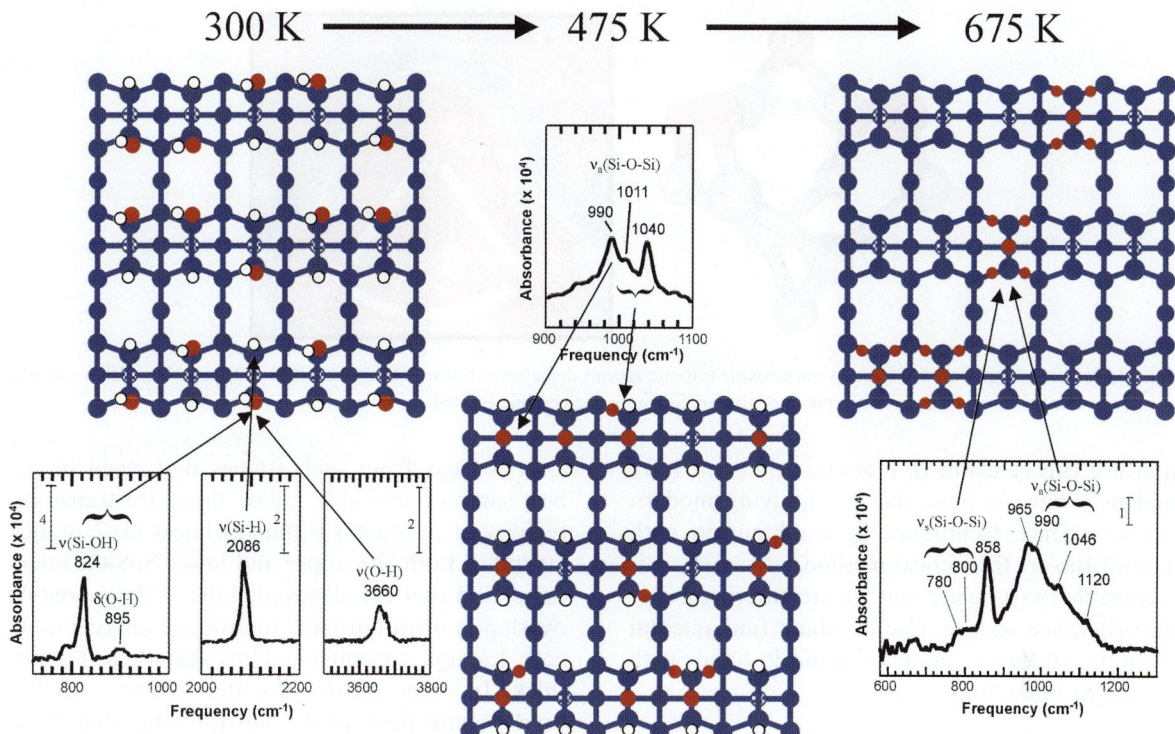

Fig. 9. Composite annealing behavior of a water-exposed Si(1 0 0)-(2 × 1) surface in ultrahigh vacuum. Infrared spectra of the relevant Si–O stretching vibrations are shown, together with schematic illustrations of the top view of the surface at each stage deduced from the data. Atomic species depicted as follows: Blue—Si, White—H, Red—O.

quencies associated with these structures made such a definitive assignment possible. Furthermore, this tendency for oxygen to agglomerate is found to be thermodynamically driven: e.g. in all cases it is energetically more favorable to have one dimer with n oxygens, for $n = 2$–5, and $(n - 1)$ oxygen-free dimers, than it is to have n dimers each with one oxygen [34]. Among the agglomerated structures, dimers containing three and five oxygens were found to be particularly stable.

To this point in the annealing sequence the dimer structures have remained H-terminated; at higher temperatures (675 K) H desorption leads to the presence of surface Si dangling bonds (unsaturated valence orbitals) which facilitate the subsequent diffusion of oxygen [35]. As a result, rapid agglomeration of oxygen is observed, with the resultant surface almost entirely comprised of five oxygen-containing dimer structures (Fig. 9, right) [32]. Interestingly, the dangling bonds on these oxygenated structures recombine to form a new Si–Si surface bond (beneath the oxygen-inserted dimer bond) in a triangular configuration termed an "epoxide" linkage (Fig. 10). Again ab initio calculations both provide the basis for interpreting the observed spectra and confirm the stability of such species relative to their dangling bond analogs when multiple oxygens are present in the Si backbonds: the flexibility of the backbond Si–O–Si linkages allows the formation of such a relatively strained "epoxide" structure. Further studies of this surface using repeated dosing and annealing cycles to complete the first monolayer of oxidation [33] have provided the first tantalizing insight into the formation of more extended vibrational modes [36] that are the genesis of the TO and LO modes described previously.

In summary, although much remains to be done before a comprehensive mechanistic understanding of the formation and evolution of the Si/SiO$_2$

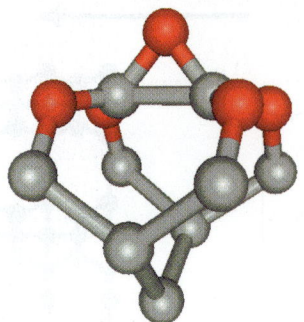 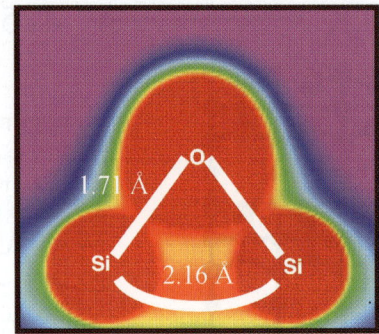

Fig. 10. Calculated structure for five-oxygen epoxide (atomic species depicted as follows: Grey—Si, Red—O), together with a contour plot of the total electron density in the plane of the SiÔSi triangular epoxide linkage.

interface can be claimed, it is clear that this lofty goal may now be realizable by applying modern surface science techniques in combination with state-of-the-art theoretical methods. As such, it is reasonable to speculate that we are entering a new era of surface science, one in which fundamental scientific endeavor marches hand-in-hand with current Si technology.

2.4. The Si/SiO₂ interface: an electronic side view

A prime illustration of the coupling of fundamental surface science with cutting-edge technological advances is provided by a recent scanning transmission electron microscopy (STEM) study of the atomic-scale electronic structure of ultrathin gate oxides [16,37]. The essence of this measurement is to cross-section the sample of interest and thin this cross-sectional slice down to ∼100 atoms thick. A highly focussed (∼2.7 Å spot size) 100 keV electron beam is then used to view individual atom columns by the Rutherford scattering of the incident electron beam to large angles, resulting in the so-called annular dark field (ADF) image shown in Fig. 11. However, there is also a vital second component to the measurement: atomic-scale electron energy loss spectroscopy is performed at each point so that a map of chemical composition is obtained as the interface is traversed (Figs. 11 and 12). The latter measurement provides a spatially resolved electronic density of states akin to that attainable using XPS [37,38], with the proviso that EELS yields higher sensitivity to the next-nearest neighbors of a given atom than does XPS.

It is clear from such studies that, even in the best samples available today, there are transition regions of at least 1.6 nm thickness (1–2 monolayers) at both the upper and lower Si/SiO₂ interfaces, and that the electronic tails of these regions overlap in ultrathin films to produce unacceptable gate leakage currents in films less than 1.3 nm thick. In other words, in order to approach the fundamental limit of 0.7 nm for SiO₂ that these studies suggest, the interfacial thickness must be reduced to the monolayer level or below. Indeed, STEM studies of ultrathin SiO₂ grown on Si(1 0 0) substrates prepared via different methods demonstrated that the electronic structure of these overlap regions is highly sensitive to surface preparation and the resulting roughness of the Si substrate. Controlling the evolution of Si surface morphology is therefore of paramount importance in the growth of Si/SiO₂ in the demanding ultrathin regime.

3. The passivation and etching of silicon

To conclude this overview we therefore step back to consider the issue of how to chemically engineer a "perfect" Si surface under ambient conditions, with the goal being to prepare an ideal interface for subsequent device processing (e.g. oxidation). The etching of silicon in fluoride solutions is the predominant [39] methodology employed commercially for this purpose. Hydrofluoric acid (HF$_{aq}$) is well known to dissolve silicon dioxide but it is only in the last decade that two essential

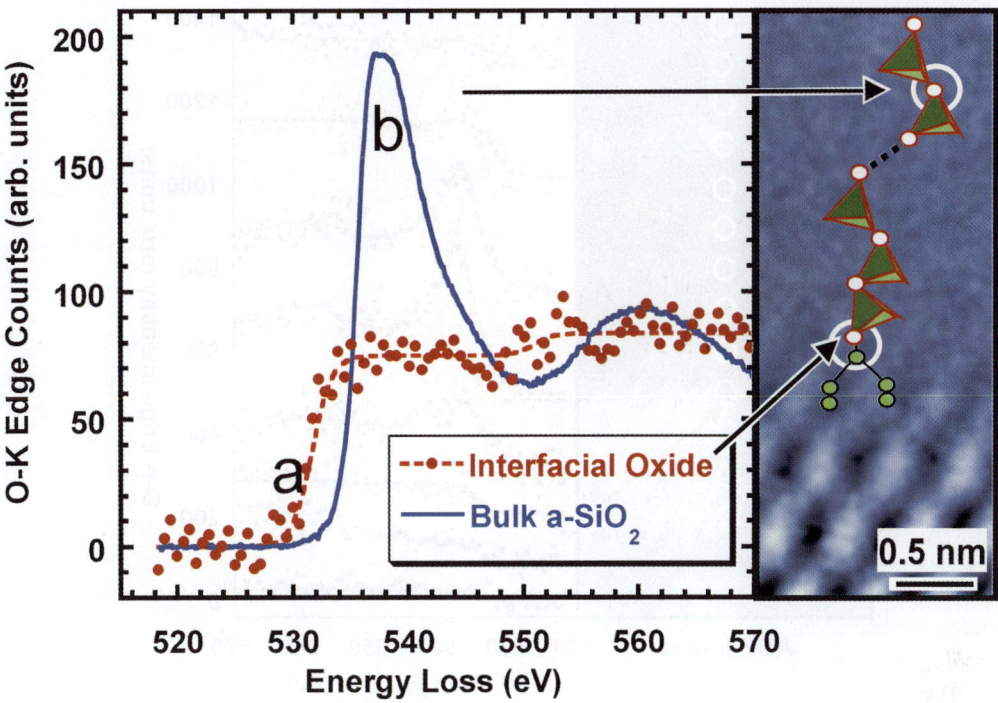

Fig. 11. Left panel: The measured EELS O K edges of bulk SiO_2 and for O atoms at an atomically smooth Si/SiO_2 interface. Right panel: The corresponding ADF image of this interface (from Ref. [16]).

aspects of HF etching of silicon surfaces have been uncovered, as a direct result of surface science investigations.

3.1. Passivation

The first discovery was that *hydrogen* passivates the underlying Si surface after removal of the native oxide layer. Although the unique stability of HF-etched surfaces had been pointed out as early as 1958 [40,41], the realization that the surface is fully hydrogen terminated was relatively recent [42–44]. It was first believed that the hydrophobic nature of HF etched surfaces was due to F termination (i.e. Teflon-like), since fluorine had been detected on the surface of HF-etched silicon [45,46]. Infrared absorption spectroscopy was employed to show that, in fact, the surfaces are H-terminated (more 'paraffin-like', due to the nonpolar nature of the Si–H bond) [47]. The tendency to assume fluorine termination of silicon had arisen not only because of the high stability of the Si–F bond (6 eV versus 3.5 eV for Si–H), but also because of the accepted explanation that SiO_2 dissolution leads automatically to F-terminated Si (via the reaction $SiO_2 + HF \rightarrow SiF_x + H_2O$). The breakthrough came with the realization that the Si–F bond is so highly polar that it in turn causes polarization of the Si–Si backbonds, allowing HF attack of these backbonds [42], followed by removal of the surface Si atom as SiF_4 and passivation of the *underlying* layer by hydrogen. The process is illustrated in Fig. 13a and has been quantitatively confirmed by first-principles quantum chemical cluster calculations [48].

3.2. Etching

The second key aspect of surface etching is that, subsequent to oxide removal, anisotropic etching of the silicon itself can take place, particularly in more basic fluoride solutions. This inherent ability to structurally modify silicon surfaces while

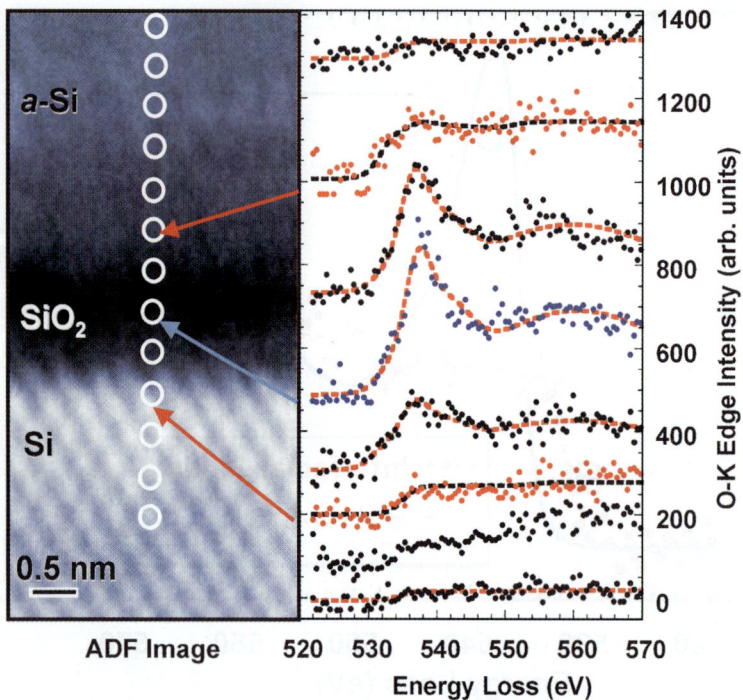

Fig. 12. Right panel: Background-corrected EELS O K edge spectra recorded point-by-point across a gate stack (c-Si/SiO$_2$/a-Si) containing a thin gate oxide. The smooth curves are best fits obtained using a mixture of bulk and interfacial oxide spectra from Fig. 11. Left panel: The ADF image showing where each spectrum was taken (from Ref. [16]).

maintaining a passivation layer of Si–H to prevent further reaction (e.g. oxidation or reaction with ambient contaminants) is critical to controlling interface roughness and therefore of great interest to the microelectronics industry. The full impact of this discovery was first appreciated for the chemical etching of Si(1 1 1), for which large atomically flat and unreconstructed Si(1 1 1) terraces (∼1000 Å wide), with ideal monohydride termination and separated by atomically straight steps, could be produced by immersion in ammonium fluoride solution [49]. Fig. 14 shows the contrast between a surface etched in pure HF, where the initial Si/SiO$_2$ interfacial roughness is preserved, and the same surface atomically smoothed by ammonium fluoride etching. The latter surface is characterized by a single, sharp Si–H stretching vibration associated with the ideal H termination on the terraces and more complex, but well-defined features for hydrogen at the step edges (Fig. 15).

The mechanism responsible for this perfection involves sequential oxidation and etching (Fig. 13b). As in most silicon aqueous etchants, oxidation is the rate-limiting step, consistent with the fact that the surfaces remain unreconstructed and H-terminated. Thus, the surface morphology is determined by the site specificity of the oxidation reaction. Using a combination of STM images and kinetic Monte Carlo (KMC) simulations of the incipient etched surfaces, Hines and coworkers have been able to extract the relative reactivities of all the relevant Si(1 1 1) surface sites (terrace, step, kinks, etc.), as summarized in Fig. 16 [50]. For instance, the distorted kink site is 10^7 times more reactive than the flat terrace site, even though both sites are nominally terminated by a single hydrogen atom (monohydride)! Importantly, by correlating the structure with the observed reactivity, they have been able to propose a mechanism for the rate limiting oxidation, involving a pentavalent transition state to H displacement (Fig. 16) [51]. In

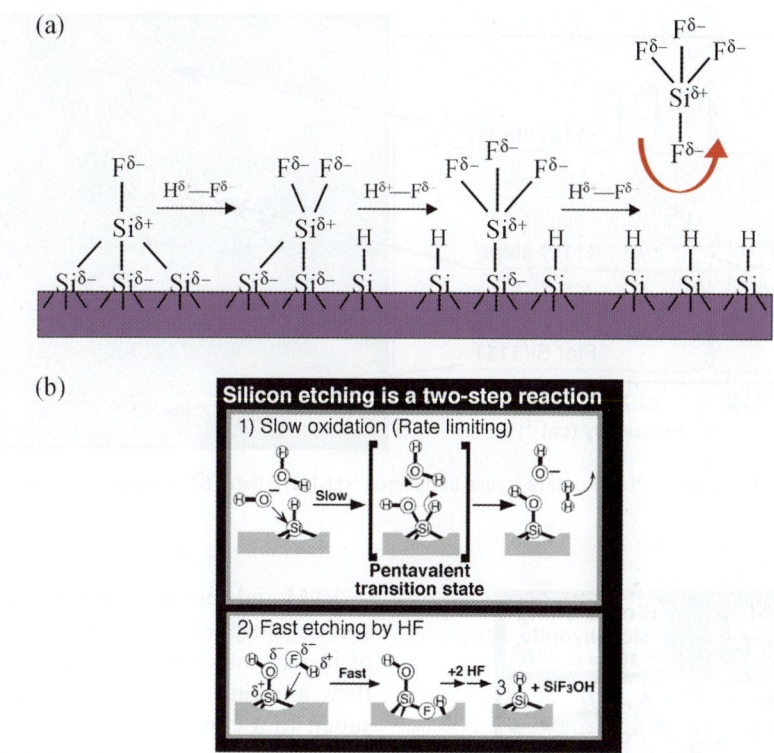

Fig. 13. Mechanism of hydrofluoric acid passivation (top) and etching (bottom) of silicon (from Refs. [48] and [51], respectively).

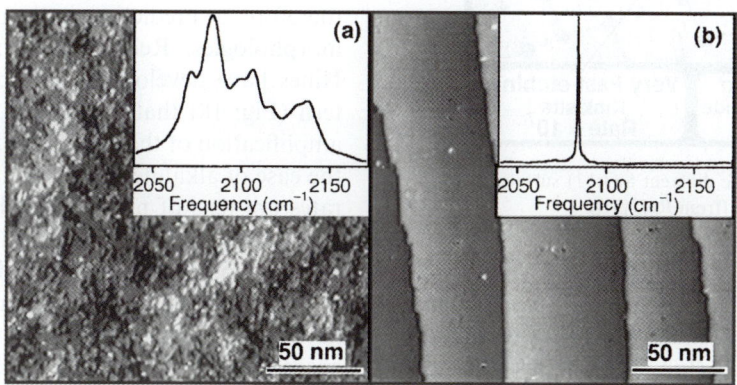

Fig. 14. STM images of Si(1 1 1) surfaces etched in concentrated HF (left) and buffered NH$_4$F:HF (right). Infrared spectra of equivalent surfaces are displayed as insets in each panel.

this model Si centers held in a rigid, tetrahedral environment react much more slowly than those held in less rigid or distorted environments. The key finding of the combined STM data and KMC simulations is that the morphological features observed in the fluoride etching of Si(1 1 1) can all be explained by *site-specific* kinetics. This important understanding has brought new insight into the role of additives that strongly modify etching anisotropy. For instance, the addition of 1%

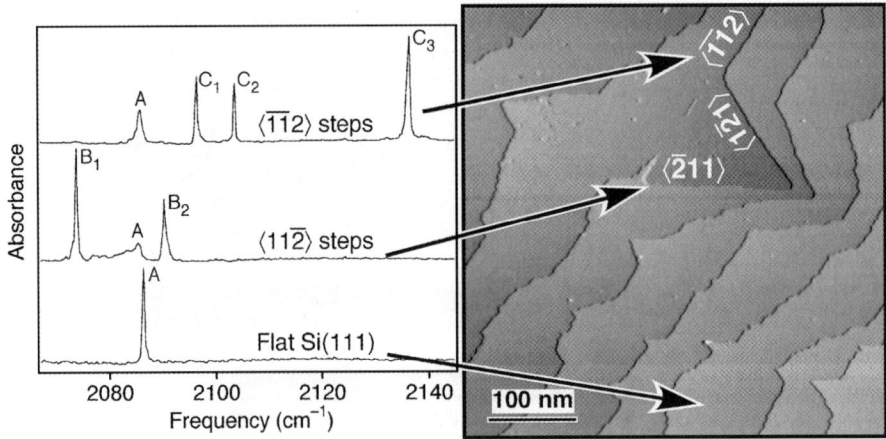

Fig. 15. Infrared spectra of a stepped Si(1 1 1) surface etched in buffered NH$_4$F:HF (left). STM image of an equivalent surface is shown in the right panel.

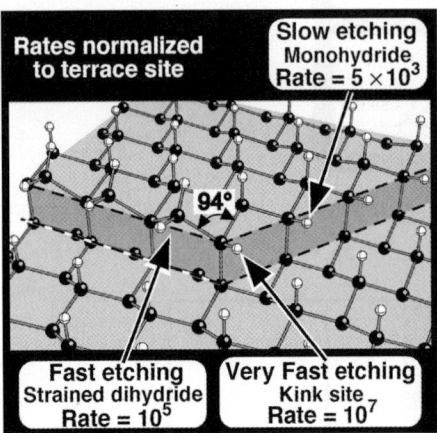

Fig. 16. Illustration of the different Si(1 1 1) surface sites with their respective etch rates (from Ref. [51]).

isopropyl alcohol, (IPA, a reputedly inactive chemical) to NH$_4$F dramatically alters the Si surface morphology [52] (Fig. 17). In this case, the simulations show that IPA does not need to be involved in the etching process itself to yield the resultant morphology, but can simply block the most reactive sites, reducing their reaction rate by up to 5 orders of magnitude and thus dramatically altering the final morphology. The concentration and location of these blocking species is now being investigated by other surface spectroscopic techniques, including IR and XPS.

STM and spectroscopic investigations of etched surface morphologies provide an impressive degree of insight into surface defect reactivity. However, they are time consuming and therefore poorly suited to a comprehensive study of potential oxidants, such as etchants to control the technologically important Si(1 0 0) surface. Fortunately, measuring macroscopic etch rates can provide important clues to the underlying reactions and is therefore a predictive indicator of atomic-scale morphologies. Recognizing this fact, Wind and Hines have developed a micromachined test pattern (Fig. 18) that provides a 100-fold geometric amplification of the etch rates [53]. For instance, in the case of alkaline silicon etchants, for which etch rates are tens of microns per hour, the retraction rate of 1° wedges is a few millimeters per hour—a distance easily seen by eye and quantified with an optical microscope. Consequently, the etch rates of all lattice planes represented in this 360° angular space can be quantified by inspection. Furthermore, one finds that the observed etch rates can be decomposed into contributions from the step and terrace sites with a single set of kinetic parameters which are themselves derived from the prior STM work described above—another elegant demonstration of the power of surface science when applied to well-defined real systems.

What has emerged from studies of wet chemical silicon etching is that amazingly simple and local

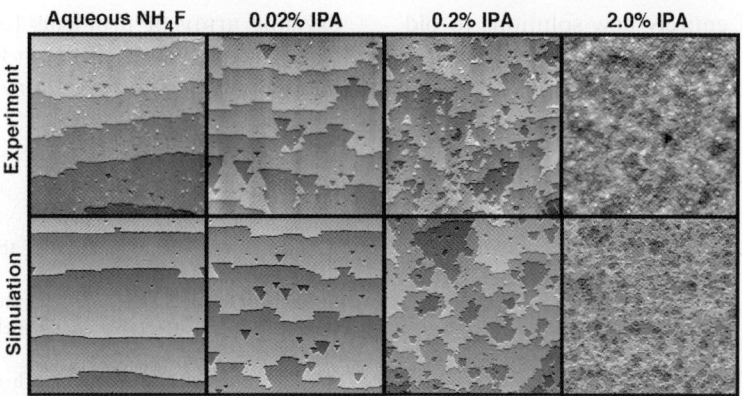

Fig. 17. Comparison of STM images and KMC simulations of Si(1 1 1) surfaces etched in buffered NH_4F:HF solutions containing varying amounts of IPA (from Ref. [52]).

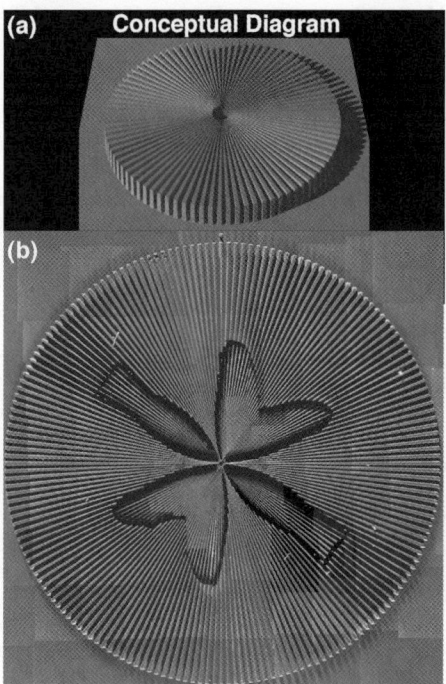

Fig. 18. Top: Conceptual illustration of a Si wafer micromachined to form a 360° array of lattice planes, each differing from the next by 2° in orientation. Bottom: Composite optical micrograph of the resultant pattern produced by immersing such a test wafer in an alkaline etching solution. The (1 1 1) direction is at ~45° to the vertical in this image (from Ref. [53]).

chemistry can lead to a rich tapestry of macroscopic morphologies. Traditional spectroscopies such as infrared absorption spectroscopy have been particularly powerful tools to identify and characterize surface species, mostly because of the large body of work performed in ultrahigh vacuum in the 1980s [54–56]. On the basis of our current understanding, chemical control of atomic-scale surface morphology can be envisioned, but a database of oxidant reactivities is necessary to make this vision a reality. New approaches—such as that the micromachined test pattern—clearly make such a broad survey of a host of new etchants now possible.

4. Conclusions and outlook

The examples presented in this short review demonstrate how surface science has made significant contributions to our understanding of the fundamental physics and chemistry that underlies complex microelectronic devices. Such understanding is crucial for the continued optimization of such devices as they approach smaller and smaller length scales, where the molecular-level interactions that are the "bread and butter" of the surface science community become the relevant length scales for device performance. In addition to allowing researchers to scale current technologies, surface science can play perhaps an even more essential role by paving the way for new technologies. As described in the introduction, several areas of development in microelectronics are in great need of entirely new solutions to old problems.

The system described in detail herein, the gate dielectric, is perhaps the most striking example of such a need. Regardless of how well researchers manage to fine-tune silicon oxidation to reach the fundamental limit of SiO_2 scaling, no amount of engineering can push this material beyond its fundamental limits. Therefore, the microelectronics community is currently engaged in a race to identify the next material for a gate dielectric, one with a dielectric constant significantly higher than that of SiO_2 to provide smaller effective gate dielectric thicknesses while allowing the actually physical thickness of the gate dielectric to remain within realizable lithographic limits.

Early in the search for such a material, researchers recognized that the thermodynamic stability of the Si/SiO_2 interface is fairly unique, and furthermore that the strong driving force for silicon oxidation makes it difficult to deposit other oxide materials on silicon without them decomposing at the elevated temperatures require for subsequent processing steps [57]. Recent XPS and medium energy ion scattering studies have shown, for instance, that deposition of tantalum pentoxide (Ta_2O_5), a potentially promising candidate, results in the eventual formation of undesirable interfacial layers presumed to correspond to SiO_2 and reduced forms of Ta [58–61]. A group of researchers at Oak Ridge National Labs has circumvented this problem by careful deposition of an interfacial barrier layer and subsequent epitaxial growth of strontium barium titanate [62]. In that case, once again the technique of combined STEM and EELS proved critical in identifying the formation of the desired interfacial characteristics.

The above are but two examples of the dynamic interplay between surface science and technology at the forefront of new materials identification and characterization. Such interplay need not be limited to advances in traditional microelectronic devices, of course, but also arises naturally in areas such as the integration of optically active elements with silicon-based devices for optoelectronic applications. In short, the intrinsic importance of surfaces and interfaces in both current and future technologies presents surface scientists with a dazzling array of problems to solve, providing a rare natural synergy between fundamental science and technological application.

References

[1] I.M. Ross, The foundation of the silicon age, Bell Labs Tech. J. 2 (1997) 3.

[2] C.M. Melliar-Smith, D.E. Haggan, W.W. Troutman, Key steps to the integrated circuit, Bell Labs Tech. J. 2 (1997) 15.

[3] P. Burggraaf, 2000 Begins with a revised industry roadmap, Sol. Stat. Tech. 43 (2000) 31.

[4] S.I.A., The National Technology Roadmap for Semiconductors, Sematech, Austin, 1999.

[5] L.C. Parillo, in: S.M. Sze (Ed.), VLSI Technology, McGraw Hill, New York, 1983, p. 445.

[6] S.M. Sze, Semiconductor Devices: Physics and Technology, Wiley, New York, 1985.

[7] M.L. Green, D. Brasen, K.W. Evans-Lutterodt, L.C. Feldman, K. Krisch, Rapid thermal oxidation of silicon with N_2O between 800 and 1200 °C: incorporated nitrogen and interfacial roughness, Appl. Phys. Lett. 65 (1994) 848.

[8] D.A. Buchanan, Scaling the gate dielectric: materials, integration and reliability, IBM J. Res. Dev. 43 (1999) 245.

[9] J.C. Demmin, The search continues for high-k gate dielectrics, Solid State Technol. 44 (2001) 46.

[10] A.E. Braun, Aluminum persists as the copper age dawns, Semicond. Int. 9 (1999) 58.

[11] S.P. Murarka, S.W. Hymes, Copper metalization for ULSI and beyond, Crit. Rev. Solid State Mater. Sci. 20 (1995) 87.

[12] J. Li, T.E. Seidel, J.W. Mayer, Copper-based metallization in ULSI structure. Part II: Is Cu ahead of its time as an on-chip interconnect material? MRS Bull. 19 (1994) 15.

[13] L. Peters, Low-k dielectrics: will spin-on or low-k prevail? Semicond. Int. 6 (2000) 108.

[14] R.S. List, A. Singh, A. Ralston, G. Dixit, Integration of low-k dielectric materials into sub-0.25-µm interconnects, MRS Bull. 22 (1997) 61.

[15] E.T. Ryan, A.J. Mckerrow, J. Leu, P.S. Ho, Materials issues and characterization of low-k dielectric materials, MRS Bull. 22 (1997) 49.

[16] D.A. Muller, T. Sorsch, S. Moccio, F.H. Baumann, K. Evans-Lutterodt, G. Timp, The electronic structure at the atomic scale of ultrathin gate oxides, Nature 399 (1999) 758.

[17] D.R. Hamann, Energetics of silicon suboxides, Phys. Rev. B 61 (2000) 9899.

[18] F. Himpsel, F.R. McFeely, A. Taleb-Ibrahimi, J.A. Yarmoff, G. Hollinger, Microscopic structure of the SiO_2/Si interface, Phys. Rev. B 38 (1988) 6084.

[19] A. Ourmazd, D.W. Taylor, J.A. Rentschler, J. Bevk, Si to SiO_2 transformation: interfacial structure and mechanism, Phys. Rev. Lett. 59 (1987) 213.

[20] F.J. Grunthaner, P.J. Grunthaner, Chemical and electronic structure of the Si/SiO$_2$ interface, Mater. Sci. Rep. 1 (1986) 65.
[21] A. Pasquarello, M.S. Hybertsen, R. Car, Theory of Si 2p core-level shifts at the Si(0 0 1)–SiO$_2$ interface, Phys. Rev. B 53 (1996) 10942.
[22] M.M. Banaszak-Holl, F.R. McFeely, Si/SiO$_2$ interface: new structures and well-defined model systems, Phys. Rev. Lett. 71 (1993) 2441.
[23] K.Z. Zhang, K.E. Litz, M.M. Banaszak-Holl, F.R. McFeely, The role of second-nearest neighbor effects in photoemission: are silicon surfaces and interfaces special? Appl. Phys. Lett. 72 (1998) 46.
[24] A. Pasquarello, M.S. Hybertsen, R. Car, Spherosiloxane H$_8$Si$_8$O$_{12}$ clusters on Si(0 0 1): first-principles calculation of Si 2p core-level shifts, Phys. Rev. B 54 (1996) R2339.
[25] K. Raghavachari, J. Eng Jr., New structural model of Si/SiO$_2$ interfaces derived from spherosiloxane clusters: implications for Si 2p photoemission spectroscopy, Phys. Rev. Lett. 84 (2000) 935.
[26] K. Raghavachari, A. Pasquarello, J. Eng Jr., M.S. Hybertsen, Chemisorption pathways and Si 2p core level shifts for the interaction of spherosiloxane clusters with Si(1 0 0): implications for photoemission in Si/SiO$_2$ systems, Appl. Phys. Lett. 76 (2000) 3873.
[27] H.Z. Massoud, E.H. Poindexter, C.R. Helms, Proceedings of the 1st–3rd International Symposium on the Physics and Chemistry of SiO$_2$ and the Si/SiO$_2$ interface, The Electrochemical Society, Pennington, NJ, 1988–1996.
[28] K.T. Queeney, M.K. Weldon, J.P. Chang, Y.J. Chabal, A.B. Gurevich, J. Sapjeta, R.L. Opila, Infrared spectroscopic analysis of the Si/SiO$_2$ interface of thermally oxidized silicon, J. Appl. Phys. 87 (2000) 1322.
[29] A. Pasquarello, R. Car, Dynamic charge tensors and infrared spectrum of amorphous SiO$_2$, Phys. Rev. Lett. 79 (1997) 1766.
[30] M.K. Weldon, B.B. Stefanov, K. Raghavachari, Y.J. Chabal, Initial H$_2$O-induced oxidation of Si(0 0 1)-(2 × 1), Phys. Rev. Lett. 79 (1997) 2851.
[31] A.B. Gurevich, B.B. Stefanov, M.K. Weldon, Y.J. Chabal, K. Raghavachari, Heterogeneous nucleation of oxygen on silicon: hydroxyl-mediated inter-dimer coupling on Si(0 0 1)-2 × 1, Phys. Rev. B 58 (1998) R13434.
[32] B.B. Stefanov, A.B. Gurevich, M.K. Weldon, Y.J. Chabal, K. Raghavachari, Silicon epoxide: unexpected intermediate during silicon oxide formation, Phys. Rev. Lett. 81 (1998) 3908.
[33] M.K. Weldon, K.T. Queeney, A.B. Gurevich, Y.J. Chabal, B.B. Stefanov, K. Raghavachari, Mechanistic studies of silicon oxidation, J. Vac. Sci. Technol. B 17 (1999) 1795.
[34] B.B. Stefanov, K. Raghavachari, Oxidation of Si(0 0 1)-2 × 1: thermodynamics of oxygen insertion and migration, Surf. Sci. 389 (1997) L1159.
[35] B.B. Stefanov, K. Raghavachari, Pathways for initial water-induced oxidation of Si(0 0 1), Appl. Phys. Lett. 73 (1998) 824.
[36] K.T. Queeney, M.K. Weldon, K. Raghavachari, Y.J. Chabal, The microscopic origins of the TO and LO mode of a substoichiometric silicon oxide film, in preparation.
[37] P.E. Batson, Simultaneous STEM imaging and electron energy loss spectroscopy with atomic column sensitivity, Nature 366 (1993) 727.
[38] J.B. Neaton, D.A. Muller, N.W. Ashcroft, Electronic properties of the Si/SiO$_2$ interface from first principles, Phys. Rev. Lett. 85 (2000) 1298.
[39] W. Kern, Handbook of Silicon Wafer Cleaning Technology: Science, Technology and Applications, Noyes Publications, Park Ridge, NJ, 1993.
[40] T.M. Buck, F.S. McKim, Effects of certain chemical treatments and ambient atmospheres on surface properties of silicon, J. Electrochem. Soc. 105 (1958) 709.
[41] M.M. Atalla, E. Tannenbaum, E.J. Scheibner, Bell Syst. Tech. J. 38 (1959) 749.
[42] H. Ubara, T. Imura, A. Hiraki, Formation of silicon-hydrogen bonds on the surface of microcrystalline silicon covered with silicon oxides by HF treatment, Solid State Commun. 50 (1984) 673.
[43] E. Yablonovitch, D.L. Allara, C.C. Chang, T. Gmitter, T.B. Bright, Unusually low surface-recombination velocity on silicon and germanium surfaces, Phys. Rev. Lett. 56 (1988) 249.
[44] Y.J. Chabal, V.A. Burrows, G.S. Higashi, K. Raghavachari, Infrared absorption spectroscopy of Si(1 1 1) surfaces after HF treatment: hydrogen termination and surface morphology, J. Appl. Phys. 53 (1988) 998.
[45] B.R. Weinberger, H.W. Deckman, E. Yablonovitch, T. Gmitter, W. Kobasz, S. Garoff, The passivation of electrically active sites on the surface of crystalline silicon by fluorination, J. Vac. Sci. Technol. A 3 (1985) 887.
[46] B.R. Weinberger, G.G. Peterson, T.C. Eschrich, H.A. Drasinski, Surface chemistry of hydrogen fluoride passivated silicon: X-ray photoelectron and ion scattering spectroscopy results, J. Appl. Phys. 60 (1986) 3232.
[47] Y.J. Chabal, in: M. Balkanski (Ed.), Optical Properties, vol. 2, Elsevier, Amsterdam, 1994, p. 187.
[48] G.W. Trucks, K. Raghavachari, G.S. Higashi, Y.J. Chabal, Mechanism of HF etching of silicon surfaces: a theoretical understanding of hydrogen passivation, Phys. Rev. Lett. 65 (1990) 504.
[49] G.S. Higashi, Y.J. Chabal, G.W. Trucks, K. Raghavachari, Ideal hydrogen termination of the Si(1 1 1) surface, Appl. Phys. Lett. 56 (1990) 656.
[50] J. Flidr, Y.-C. Huang, T.A. Newton, M.A. Hines, The formation of etch hillocks during step-flow etching of Si(1 1 1), Chem. Phys. Lett. 302 (1999) 85.
[51] J. Flidr, Y.-C. Huang, T.A. Newton, M.A. Hines, Extracting site-specific reaction rates from steady state surface morphologies: kinetic Monte Carlo simulations of aqueous Si(1 1 1) etching, J. Chem. Phys. 108 (1998) 5542.
[52] T.A. Newton, Y.-C. Huang, L.A. Lepak, M.A. Hines, The site-specific reactivity of isopropanol on aqueous silicon etching: controlling morphology with surface chemistry, J. Chem. Phys. 111 (1999) 9125.

[53] R.A. Wind, M.A. Hines, Micromachined test pattern approach to determining Si etch rates, Surf. Sci. 460 (2000) 21.

[54] Y.J. Chabal, Hydride formation on the Si(100):H_2O surface, Phys. Rev. B 29 (1984) 3677.

[55] Y.J. Chabal, K. Raghavachari, Surface infrared study of Si(100)-(2 × 1) H, Phys. Rev. Lett. 53 (1984) 282.

[56] Y.J. Chabal, S.B. Christman, Evidence of dissociation of water on the Si(100)-2 × 1 surface, Phys. Rev. B 29 (1984) 6974.

[57] K.J. Hubbard, D.G. Schlom, Thermodynamic stability of binary oxides in contact with silicon, J. Mater. Res. 11 (1996) 2757.

[58] H.C. Lu, N. Yasuda, E. Garfunkel, T. Gustafsson, J.P. Chang, R.L. Opila, G. Alers, Structural properties of thin films of high dielectric constant materials on silicon, Microelec. Eng. 48 (1999) 287.

[59] J.P. Chang, M.L. Steigerwald, R.M. Fleming, R.L. Opila, G.B. Alers, Thermal stability of Ta_2O_5 in metal-oxide-metal capacitor structures, Appl. Phys. Lett. 74 (1999) 3705.

[60] G. Lucovsky, B. Rayner, H. Niimi, R. Therrien, R. Johnson, J.G. Hong, Local atomic structure and chemical stability of high-k gate dielectrics for advanced silicon metal oxide semiconductor devices: (Proceedings of the 25th International Conference on the Physics of Semiconductors, 200, Part II), Springer Proc. Phys. 87 (2001) 1759.

[61] A.Y. Mao, K.A. Son, J.M. White, D.L. Kwong, D.A. Roberts, R.N. Vrtis, Study of thermal stability of CVD Ta_2O_5/Si interface: (Ultrathin SiO_2 and High-K Materials for ULSI Gate Dielectrics), Mater. Res. Soc. Symp. Proc. 567 (1999) 473.

[62] R.A. McKee, F.J. Walker, M.F. Chisholm, Crystalline oxides on silicon: the first five monolayers, Phys. Rev. Lett. 81 (1998) 3014.

Organic functionalization of group IV semiconductor surfaces: principles, examples, applications, and prospects

Stacey F. Bent *

Department of Chemical Engineering, Stanford University, Stanford, CA 94305-5025, USA

Received 5 July 2000; accepted for publication 22 June 2001

Abstract

Organic functionalization is emerging as an important area in the development of new semiconductor-based materials and devices. Direct, covalent attachment of organic layers to a semiconductor interface provides for the incorporation of many new properties, including lubrication, optical response, chemical sensing, or biocompatibility. Methods by which to incorporate organic functionality to the surfaces of semiconductors have seen immense progress in recent years, and in this article several of these approaches are reviewed. Examples are included from both dry and wet processing environments. The focus of the article is on attachment strategies that demonstrate the molecular nature of the semiconductor surface. In many cases, the surfaces mimic the reactivity of their molecular carbon or organosilane counterparts, and examples of functionalization reactions are described in which direct analogies to textbook organic and inorganic chemistry can be applied. This article addresses the expected impact of these functionalization strategies on emerging technologies in nanotechnology, sensing, and bioengineering. © 2001 Elsevier Science B.V. All rights reserved.

Keywords: Single crystal surfaces; Silicon; Germanium; Diamond; Chemisorption; Alkenes; Infrared absorption spectroscopy; Scanning tunneling microscopy

1. Introduction

The microelectronics revolution that was started fifty years ago has grown into an industry that drives much of the world's technology today. It is responsible for a nearly $1 trillion worldwide electronics market. Microelectronics use semiconductor materials as their basic building block. These semiconducting solids—such as silicon, gallium arsenide, and germanium—have become ubiquitous. Semiconductor-based devices can now be found in everything from automobiles and home appliances, to means of communication and equipment in our doctor's office.

The rapidity with which microelectronics technologies have advanced is unprecedented. For example, the number of electronic components on a single silicon chip doubles every 18 months, and computer memory that used to cost thousands of dollars can now be fabricated for only pennies. The widespread availability of microelectronic-controlled machines depends upon these revolutionary advances. Many of these advances have resulted from miniaturization of the most

---

* Tel.: +1-650-723-0385; fax: +1-650-723-9780.
  E-mail address: stacey.bent@stanford.edu (S.F. Bent).

fundamental building blocks of electronic devices, in which more and more active units are packed into smaller and smaller areas on the semiconductor chip. Integrated circuits now contain nearly one billion transistors on a chip less than a square inch in size [1,2].

As we enter a new century, the sizes of microelectronic devices have gotten so small that the active elements now approach molecular dimensions. For example, it is predicted by an industry guideline known as the Technology Roadmap for semiconductors [3] that by the year 2014, lateral device dimensions will reach 35 nm (350 Å); vertical dimensions will be even smaller. Because molecules themselves are on the order of a few tenths of nanometers in size, it follows that future technologies increasingly will depend upon the functionality of *individual molecules and atoms*.

Surface phenomena have always been a cornerstone of the microelectronics industry. Processes such as epitaxy [4], chemical vapor deposition, etching, oxidation and passivation [5], which are used routinely in the industry, involve chemical or physical processes occurring *at the surface* of the semiconductor wafer. However, with the rapid miniaturization of devices, understanding of the atomic-level phenomena underlying these processes becomes even more critical. The sizes that microelectronics are projected to reach in the near future means that a given material in a device might be just a few layers thick. As a result, much of the functionality of a device will be due to physical processes that occur within a few molecular layers of some interface, such as a buried interface, e.g. between Si/SiO_2 or Si/metal, or an outer surface exposed to ambient. For this reason, research on the interfacial chemistry at semiconductor surfaces has seen a tremendous amount of growth over the past decade, and promises to be an increasingly important field in the future.

The importance of atomic level surface chemistry is highlighted in the growing field of organic functionalization of semiconductors. An area that has been researched extensively over the past few years, surface functionalization or organic modification is the process of depositing layers of organic molecules (i.e. those that contain carbon) at semiconductor surfaces. Fig. 1 illustrates a semi-

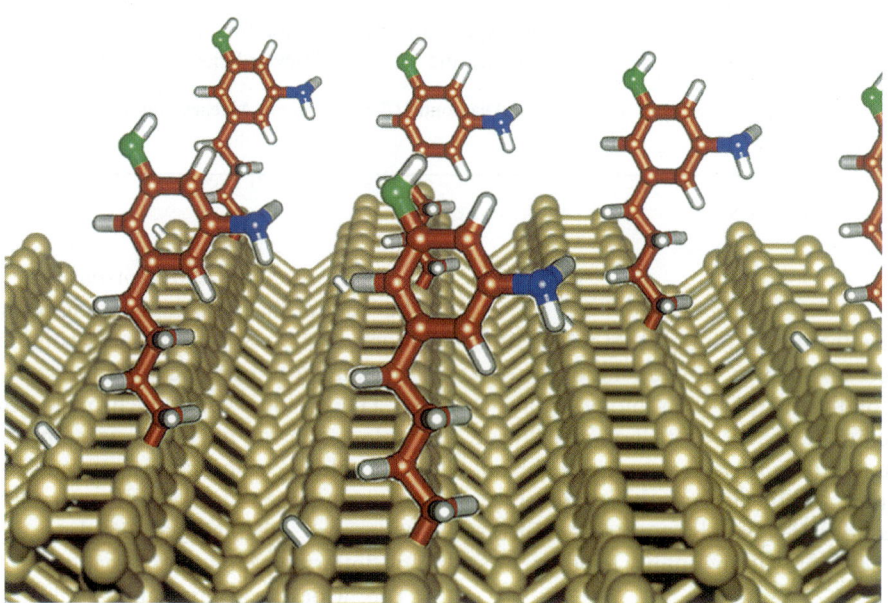

Fig. 1. Illustration of organic molecules attached to a silicon substrate in the initial stages of film deposition. Layered organic/semiconductor materials are being explored for possible application in sensor development, molecular electronics, and biocompatible implants.

conductor surface upon which organic molecules have been attached. The motivation to deposit organic layers on a semiconductor stems from a desire to impart some property of the organic material to the semiconductor device. Over the same time period during which semiconductor (typically Si, Ge, GaAs) based devices were undergoing rapid size scaling, revolutionary advances also were made in the development of new solid state organic materials. Organic molecules comprise more than 95% of all known chemical compounds. Carbon can form a myriad of molecules differing in shape, size and composition [6]. For example, the organic molecules attached to the surface in Fig. 1 will exhibit certain properties based on the types and specific groupings of atoms in the molecules. As a result, organic materials offer engineers great flexibility in designing and creating unique molecular properties, that can then be exploited to provide new capabilities in optical, electronic, and mechanical function as well as in chemical and biological activity.

An especially promising approach is to *combine* organic materials with conventional semiconductor materials [7]. Microelectronics technology provides an opportunity to create hybrid devices exploiting the best properties and features of both organic and inorganic materials. As an example, consider a chemical or biological sensor built with a mixed organic/semiconductor design. Organic molecules such as those shown in Fig. 1 can be deposited on a silicon substrate. The organic layer may be terminated with a variety of end groups that respond to different chemical or biological stimuli. A "sensing" response occurs if a species of interest binds to the end group, causing transduction of signal within the organic layer. If this signal can be coupled into the silicon substrate, then all of the capabilities of silicon-based microelectronic circuitry—such as signal amplification, processing and storage—become available. The end product is a chip-based chemical or biological sensor.

Hybrid organic/semiconductor materials also are being explored for use in molecular electronics and computing and for imparting biocompatibility to semiconductor devices for implantation. Organic materials are also finding use in more conventional microelectronics processing, such as for new-generation dielectric materials for metal interconnect isolation, or for surface passivation and protection.

Fortunately, the surfaces of most semiconductors of interest for microelectronics have characteristics that enable organic molecules to be attached by a number of different chemical reactions. These attachment chemistries provide an insightful view of the reactivity of semiconductor surfaces. They also demonstrate how the wealth of chemical understanding that has been developed over the past century within the field of organic chemistry can now be used to manipulate semiconductor surfaces to impart properties useful for a variety of applications. The focus of this article is on several different approaches that have been taken to functionalize semiconductor surfaces with organic molecules. The list of attachment reactions covered in this article is not intended to be comprehensive, but rather a selected group of systems has been chosen for the level of insight they provide into the molecular nature of semiconductor surfaces. The article will use these examples to illustrate some general principles describing the chemistry at semiconductor surfaces. For more complete technical reviews on the subject of attachment reactions at semiconductor surfaces, the reader is referred to Refs. [8–11].

2. The silicon surface

We will begin with an investigation into the structure of the silicon surface. Silicon is the predominant semiconductor material used in the microelectronics industry, due in part to several important properties. Silicon can be produced in single crystalline form at better than 99.999999999% purity, it forms an excellent oxide at its surface, and its electronic properties can be tuned dramatically by substituting only a small fraction of silicon atoms in the lattice with another element in a process called "doping". Silicon is a covalent solid that crystallizes into a diamond cubic lattice structure, illustrated in Fig. 2. In a covalent solid, bonding is produced by overlap of the highly directional electronic orbitals of the

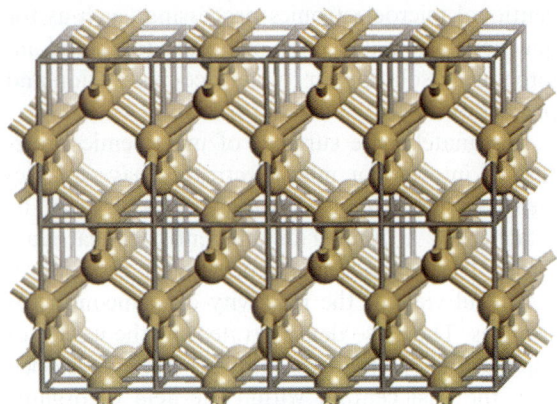

Fig. 2. The diamond cubic lattice is the crystal structure found for diamond, silicon and germanium. Each atom is bonded to four neighboring atoms in a tetrahedral geometry.

atoms. Like carbon, its group IV homolog, silicon atoms hybridize into a tetrahedral bonding configuration. In the bulk solid, the diamond cubic lattice allows silicon atoms to achieve this tetrahedral configuration. At the surface of the material, however, the bulk is truncated, so the stable bulk tetrahedral bonding is disturbed.

The reactivity of the silicon surface is controlled in part by the unsatisfied bonding orbitals, or so-called dangling bonds, that remain upon truncating the bulk. A surface silicon atom that is bonded to only two other silicon atoms instead of four, as shown at the top surface of the crystal illustrated in Fig. 2 for example, would have two dangling bonds. Dangling bonds contain single electrons, whereas normal covalent bonds contain two spin-paired electrons. At the surface, atoms can readjust to minimize the total free energy of the system and eliminate the dangling bonds. This process is referred to as surface "reconstruction". The energy minimization is a trade-off between energy gained by forming new local bonds (to eliminate the dangling bonds) with energy lost because of bond strain that results from its new configuration. This energy trade-off often leads to complex surface reconstructions [12–15].

The crystallographic faces of silicon that are most important industrially are the Si(1 0 0) and the Si(1 1 1) surfaces. The numbers in parentheses refer to a crystallographic notation known as Miller indices, which designate the particular orientation at which the bulk is terminated. The top surface in Fig. 2, e.g., has the (1 0 0) orientation. Both the (1 0 0) and (1 1 1) surfaces undergo extensive reconstructions, i.e. their surface atomic geometry differs significantly from that of the bulk. Moreover, the two surfaces have markedly different surface structures [14,15]. The Si(1 0 0) surface reconstructs into a (2×1) structure, where (2×1) designates the new periodicity of the surface atoms. The Si(1 0 0)-2×1 surface, illustrated in Fig. 3(a), consists of pairs of silicon atoms in adjacent rows that have bonded to each other, thereby reducing the number of dangling bonds. These pairs of silicon atoms are called dimers. Si(1 1 1), on the other hand, reconstructs into a complex (7×7) structure that contains 49 surface atoms in the new unit cell.

It is useful to consider in more detail the bonding of the dimers on the Si(1 0 0)-2×1 surface, because they play such an important role in

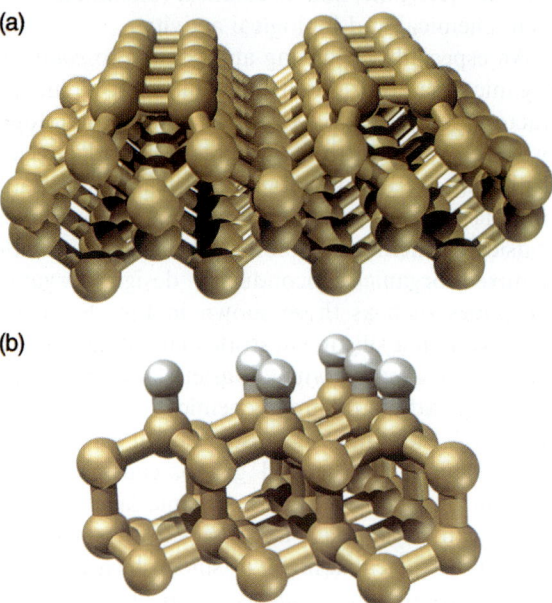

Fig. 3. Two surfaces of silicon that are important in organic functionalization chemistry. (a) The clean Si(1 0 0)-2×1 surface, with rows of silicon dimers lining the surface. These dimers play an important role in the chemistry of organic molecules at this surface. (b) The hydride-terminated Si(1 1 1) surface. The golden spheres represent silicon atoms; hydrogen atoms are shown in white. The hydrogen–silicon bonds are the important molecular group in hydrosilylation attachment chemistry, discussed in the text.

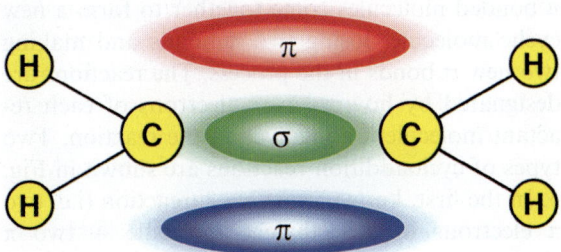

Fig. 4. Illustration of a σ bond and a π bond in a double-bonded molecule. The σ bond (green) is cylindrically symmetric around the molecular axis, while the two lobes of the π bond (red and blue) are separated by a nodal plane. The molecule shown is ethylene.

the reactivity of this surface toward organic molecules. The Si(1 0 0)-2 × 1 dimers are attached by something akin to a double bond [16]. In the language of chemical bonding, a double bond consists of two types of bonds: a σ bond (one that has symmetry around the axis connecting the two atoms) and a π bond (one with a nodal plane along the axis) [17]. Fig. 4 provides an illustration of a σ and a π bond in a molecule; note that the π bond has two lobes.

However, in contrast to the π bond in the molecule shown in Fig. 4 (ethylene), the π bond between the silicon atoms is sufficiently weak that the dimer is not held in a symmetric configuration. Instead, the energy of the Si(1 0 0)-2 × 1 surface can be reduced even further if the dimer moves from a symmetric configuration to one in which the dimer is tilted. Hence, at low temperatures, the dimers of the Si(1 0 0)-2 × 1 surface are observed to be tilted [14]. At higher temperatures, thermal energy induces a rapid change in the direction of the tilt which causes the dimers to appear symmetric, as represented in Fig. 3(a). The presence of the π bond in the silicon dimer and the dimer tilting will both be important in understanding the reactivity of the surface.

Even after reconstruction, both the Si(1 0 0)-2 × 1 and Si(1 1 1)-7 × 7 surfaces are highly reactive. Both surfaces are typically cleaned and prepared in an ultrahigh vacuum environment ($p = 10^{-10}$ Torr) by bombarding the surface with argon ions to sputter away surface layers and/or heating to temperatures near 900 °C. The resulting surfaces are not stable in air, instead quickly oxidizing to form an SiO_2 layer. Hence, experiments on the clean Si(1 0 0)-2 × 1 and Si(1 1 1)-7 × 7 surfaces must be performed in vacuum, and the organic reactants for functionalization are introduced in vapor form into the vacuum chamber.

On the other hand, silicon surfaces can be rendered relatively stable in air (i.e. relatively resistant to oxidation) by coating the surface with hydrogen. A number of recipes have been developed to terminate the silicon surface with Si–H (silicon hydride) groups. An excellent overview of this process is given by Weldon et al. [5]. The most common methods involve exposure to dilute, aqueous hydrofluoric acid solutions, and these wet etch procedures are used both at bench-top research laboratories and large-scale microfabrication facilities alike. Not only does terminating a silicon surface with hydrogen atoms render it passivated toward many reactions, but it also changes the surface structure. Because the reconstructed surface contains some weak Si–Si bonds, replacement of these with strong Si–H bonds can relax the surface reconstructions. One important surface for organic functionalization reactions is that of hydride-terminated Si(1 1 1). Due to the strong bonding and passivating nature of hydrogen at the interface, Si(1 1 1) that is hydride terminated does not reconstruct, revealing instead a nearly bulk-like periodicity of the surface silicon atoms [18]. The structure of the hydride-terminated Si(1 1 1) surface is shown in Fig. 3(b).

Both the bare and hydrogen-terminated silicon surfaces have been used as substrates for organic functionalization. Because of their different relative reactivities, most of the reactivity studies of bare silicon are performed under ultrahigh vacuum conditions while reactions with hydride-terminated silicon involve wet chemical methods. Despite the different approaches and surface structures, however, a unifying feature in all of these attachment chemistries is the molecular nature of the chemistry, which in turn allows some common principles to be formulated for functionalization chemistry at semiconductor surfaces.

3. Principles of cycloaddition chemistry at semiconductor surfaces

The covalent nature of the semiconductor surface permits its reactivity to be described within a molecular framework. This covalency leads to a reactivity that is often quite distinct from that of metals. On metals, bonding may not be strongly site specific. This characteristic is markedly different from that of a surface such as silicon, where bonding is both localized and directional. To a significant degree, the reactivity of semiconductor surfaces can be understood by drawing analogies to known molecular systems, and in particular to systems in organic chemistry. In this section, this molecular analogy is developed to understand and predict one class of organic reactions that can be applied to surface functionalization of the Si(1 0 0)-2 × 1 surface. These are known as cycloaddition reactions. In Section 4, examples of cycloaddition chemistry will be described. Then in Section 5, some of the same principles developed for surface cycloaddition reactions will be extended to other classes of organic reactions used to functionalize silicon.

The Si(1 0 0)-2 × 1 reconstructed surface, shown in Fig. 3(a), to some extent mimics an organic reagent. As described above, the dimers of the Si(1 0 0)-2 × 1 surface are attached by two bonds: a σ bond and a π bond (Fig. 4) [16]. The π bond across the dimer is quite weak, but nevertheless its presence provides a nice way of describing the dimer [17]. Because of these two bonds, it is useful to draw analogies between the Si dimer and a carbon–carbon double bond, since C and Si are both in the same group (group IV) of the periodic table. Carbon–carbon double bonds, known as alkenes, are a very well studied functional group in organic chemistry. The simplest alkene molecule, ethylene, is the molecule displayed in Fig. 4.

A number of reactions are known which can form bonds with an alkene group. One synthetically useful class of these reactions are "cycloaddition" reactions. Cycloadditions are widely used in organic synthesis as a means of forming new carbon–carbon bonds and new carbon rings because of their versatility and high stereoselectivity [19–21]. Cycloadditions are reactions in which two π bonded molecules come together to form a new cyclic molecule, losing two π bonds and making two new σ bonds in the process. The reactions are designated by how many π electrons of each reactant molecule are involved in the reaction. Two types of cycloaddition reactions are shown in Fig. 5. In the first, known as a [2 + 2] reaction (i.e. two π electrons in the ethylene molecule + two π electrons in the other ethylene molecule), two alkenes come together to form a new four-membered ring. The second type of cycloaddition in Fig. 5 is the [4 + 2] reaction, also known as the Diels–Alder reaction [19]. This important reaction is named after Otto Diels and Kurt Alder, who received the Nobel Prize in Chemistry in 1950 for their studies of this reaction. Here, a "diene" molecule with two neighboring (conjugated) π

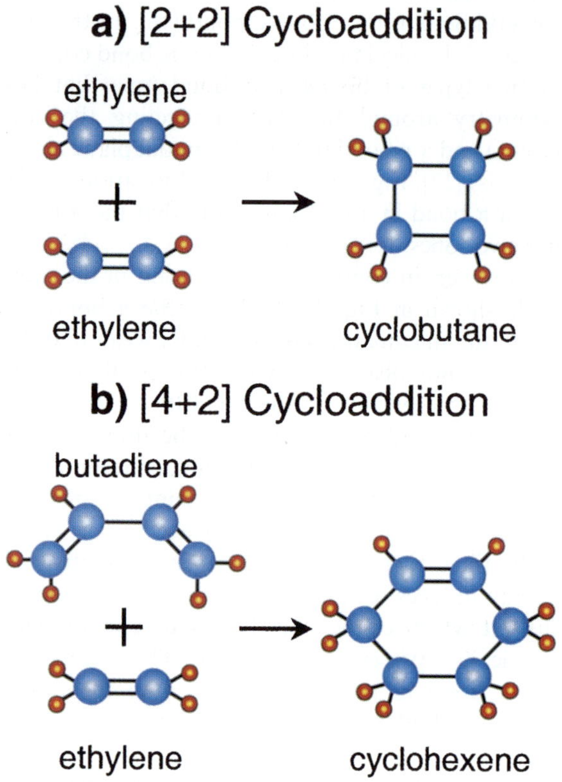

Fig. 5. Cycloaddition reactions are an important class of reactions for forming new ring compounds in organic chemistry. The [2 + 2] cycloaddition between two alkenes forms a four-membered ring, while the [4 + 2] cycloaddition between an alkene and a diene forms a six-membered ring. The designations refer to the number of π electrons involved in the reaction.

bonds such as butadiene reacts with an alkene (ethylene) to form a new six-membered ring. In Fig. 5, the π bond and the σ bond are each represented by a line, so that the double bond of ethylene is shown with two lines while the product cyclobutane, which contains only σ bonds, is shown with single lines.

The cycloaddition reactions in Fig. 5 are subject to what are know as the Woodward–Hoffman selection rules. These selection rules stem from an analysis based on frontier orbital theory which examines the symmetries of the highest occupied and lowest unoccupied molecular orbitals (HOMO and LUMO, respectively) of the reactants as they come together to form the reaction product. Frontier orbital theory was formulated by Woodward and Hoffman [22]. Professor Hoffman was awarded the Nobel Prize in Chemistry in 1981, together with Professor Fukui of Kyoto University, for his contributions in this area. The resulting Woodward–Hoffman selection rules are widely used for predicting how readily an organic reaction will occur. Such an analysis is illustrated in Fig. 6 for [4 + 2] and [2 + 2] cycloaddition reactions. In cycloaddition reactions, the frontier π orbitals (characterized by positive and negative lobes, shown in different colors in Fig. 6) must overlap "in phase" for the reaction to be symmetry allowed.

Because of the particular properties of these π orbitals in the [2 + 2] reaction, [2 + 2] cycloadditions are found to be "symmetry forbidden". That is, the Woodward–Hoffman symmetry rules dictate that this reaction should not occur without significant energy activation. In fact, in organic chemistry this reaction is largely limited to synthesis involving photochemical activation. The [4 + 2] reaction, in contrast, has frontier orbital properties (shown in Fig. 6) that make it "allowed" by symmetry considerations. Indeed, Diels–Alder reactions are commonly used in organic synthesis as a means of forming new C–C bonds and ring structures.

If the analogy that we have drawn between the Si=Si dimer on the Si(1 0 0)-2 × 1 surface and an alkene group is reasonable, then certain parallels might be expected to exist between cycloaddition reactions in organic chemistry and reactions that occur between alkenes or dienes and the silicon

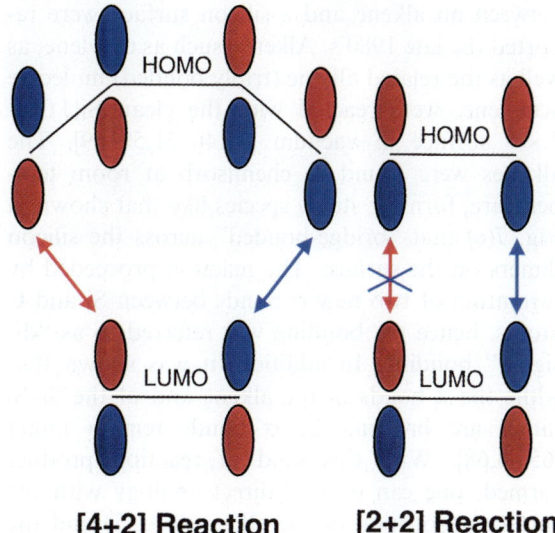

Fig. 6. Examination of two cycloaddition reactions by frontier orbital analysis. The frontier orbitals under consideration are the highest occupied molecular orbital (HOMO) and the lowest unoccupied molecular orbital (LUMO). The blue and red lobes represent the mathematical sign of the molecular orbital, such that blue combines with blue and red combines with red. Mixing the orbitals leads to a symmetry forbidden reaction, as in the case of the [2 + 2] reaction.

surface. We can therefore formulate the following guiding principle. To the extent that the simple picture of surface Si=Si double bonds is valid, similar symmetry selection rules as developed for organic reactions should apply to the surface. On the other hand, as the analogy of the dimer as a true double bond becomes less accurate, deviations from the symmetry selection rules will be expected. Indeed, this prediction has been borne out in a number of recent studies of cycloaddition reactions on Si(1 0 0)-2 × 1 [23–51], as well as on the related surfaces of Ge(1 0 0)-2 × 1 [52–55] and C(1 0 0)-2 × 1 [56–58]. These studies are described in Section 4.

4. Examples

4.1. Examples of [2 + 2] cycloaddition reactions on silicon

The first examples of what can be categorized as [2 + 2] type cycloaddition reactions occurring between an alkene and a silicon surface were reported the late 1980's. Alkenes such as ethylene, as well as the related alkyne (triply bonded) molecule acetylene, were reacted with the clean Si(1 0 0)-2 × 1 surface in vacuum [31,46–51,59–69]. The alkenes were found to chemisorb at room temperature, forming stable species like that shown in Fig. 7(c) that "bridge-bonded" across the silicon dimers on the surface. The reaction proceeded by formation of two new σ bonds between Si and C atoms, hence the bonding was referred to as "di-sigma" bonding. In addition, it was shown that while the π bonds of the alkene and of the Si–Si dimer are broken, the σ bonds remain intact [65,66,68]. With this kind of reaction product formed, one can draw a direct analogy with the surface adsorption of the alkene (Fig. 7) and the [2 + 2] cycloaddition reaction shown in Fig. 5(a).

Just as two ethylene molecules ($CH_2=CH_2$) react to form a four-membered cyclobutane ring, ethylene reacts with the silicon dimer ($Si=Si$) on the Si(1 0 0)-2 × 1 surface to form a four-membered $Si_2(CH_2)_2$ ring with the surface.

Interestingly, the surface [2 + 2] cycloaddition is a relatively fast reaction, occurring readily with most alkenes at room temperature [9]. This high reactivity would not be expected for a true, symmetric 2 + 2 reaction, which formally is symmetry forbidden and is slow for the analogous homogeneous reaction [21,22]. The symmetry analysis shown in Fig. 6, however, applies to a system in which the Si dimer is symmetric and closely mimics an alkene. This is the pathway illustrated in Fig. 7(a). However, the Si dimer is not symmetric in its lowest energy configuration. The lowest energy state of the Si(1 0 0)-2 × 1 surface is actually one in

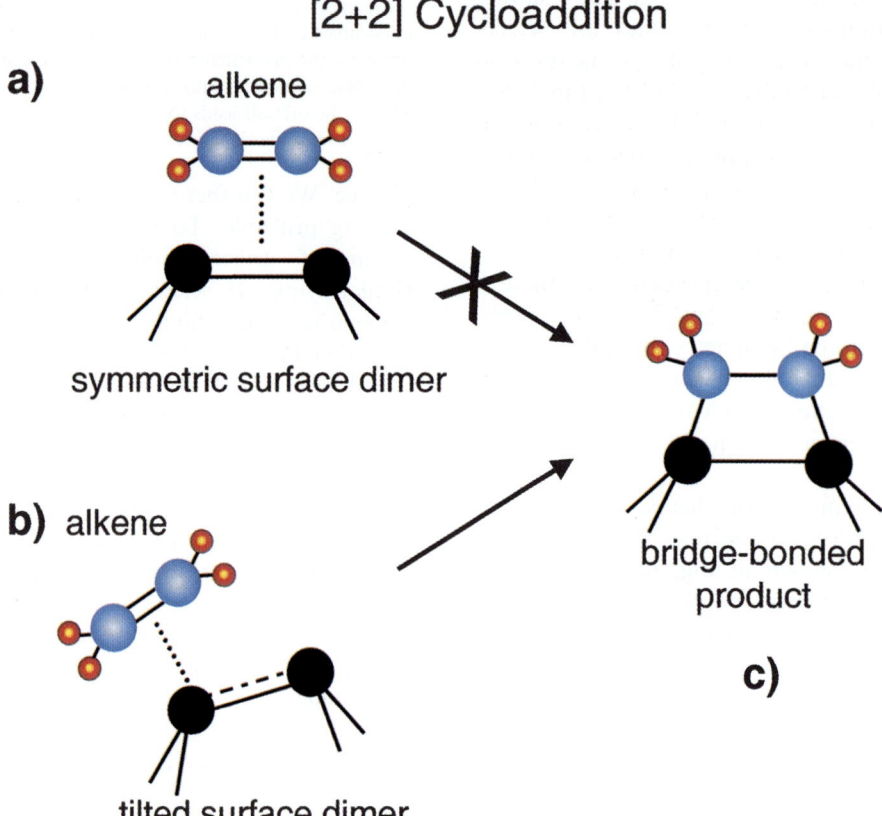

Fig. 7. A [2 + 2] reaction at the silicon surface occurs between an alkene molecule and the silicon dimer. The product is bridge-bonded across the dimer.

which the silicon dimers are tilted. It has been proposed that this low symmetry (tilted) geometry allows the reaction to proceed through an asymmetric pathway in which the alkene approaches the dimer from one side, as shown in Fig. 7(b) [10,31,64,69]. This asymmetric approach is of a lower symmetry and can occur with little or no energetic barrier. This example serves as an interesting case of a solid state effect in the silicon (tilting of the dimers) controlling a surface reaction (via relaxation of the symmetry constraints).

The result of the [2 + 2]-like reaction is a tightly bonded organic group, directly attached to the silicon surface through strong Si–C bonds (~82 kcal/mol each [25]), as illustrated in Fig. 7(c). Furthermore, as reported above, the reaction occurs readily for many alkenes. Consequently, [2 + 2] cycloaddition has attracted much attention as a means to functionalize the Si(1 0 0)-2 × 1 surface. Many beautiful examples have been reported of silicon surfaces functionalized by means of [2 + 2] chemistry using alkene or dialkene precursors.

One example is the work by Hamers and coworkers. In 1997, this group began a series of studies showing that this type of attachment chemistry can be used to form well-ordered monolayer organic films across the surface of silicon [31–36]. By adsorbing cyclopentene (the molecule shown in Fig. 8) at room temperature in vacuum, layers that are ordered in registry with the underlying silicon surface lattice could be generated [32,35].

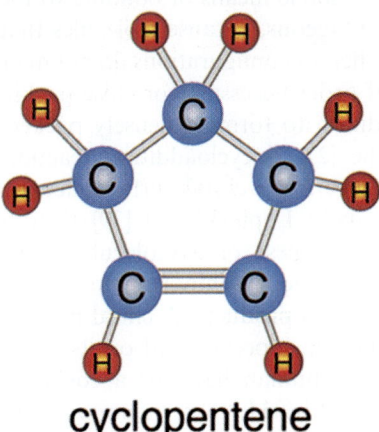

Fig. 8. The cyclopentene molecule.

The experiments by Hamers and coworkers were performed on a Si surface that was intentionally cut slightly (4°) off-axis from the (1 0 0) crystal face. This miscut has the effect of producing a stepped surface, where between each step resides a terrace ≈8 atoms wide. The orientation of the dimer rows on the terraces is indicated schematically in Fig. 9(a). Fig. 9(b) shows a scanning tunneling microscope (STM) image of the clean 4° miscut Si(1 0 0) surface [35]. In the picture, each oblong white feature is an image of a Si–Si dimer. It is readily seen that on the flat terrace regions, the Si=Si dimers all maintain the same orientation with respect to the step. If cyclopentene molecules are now adsorbed on this surface, and if they bond across the dimer, as expected for a [2 + 2] cycloaddition, we would expect to see rows of bound cyclopentene that follow the underlying silicon rows. In fact, the STM images obtained after cyclopentene exposure clearly show ordered molecular adsorption of cyclopentene at this surface. In Fig. 9(c), the white features (larger than those in Fig. 9(b)) are images of individual adsorbed cyclopentene molecules. These images offer beautiful evidence for the degree of molecular order that can be achieved in this reactive system. Long range order on a length scale of millimeters was confirmed by measuring the retention of orientational anisotropy across the length of the silicon sample using polarized vibrational spectroscopy [32]. These studies provide support for the use of cycloaddition chemistry at the Si(1 0 0)-2 × 1 surface as a method to prepare ordered monolayer organic films.

The success in forming ordered overlayers is especially significant in light of the following characteristics of this system. As with ethylene, cyclopentene was found to react at the Si(1 0 0)-2 × 1 surface with near unity reactivity, i.e., each gas molecule that collides with the surface has an almost perfect chance of sticking and bonding at the surface. Furthermore, diffusion is expected to be very slow in these systems at room temperature, suggesting that once the molecules have bonded to the surface, they remain at or near the same site on the surface [9]. The formation of ordered overlayers in the absence of significant surface mobility is in contrast to the behavior of other growth

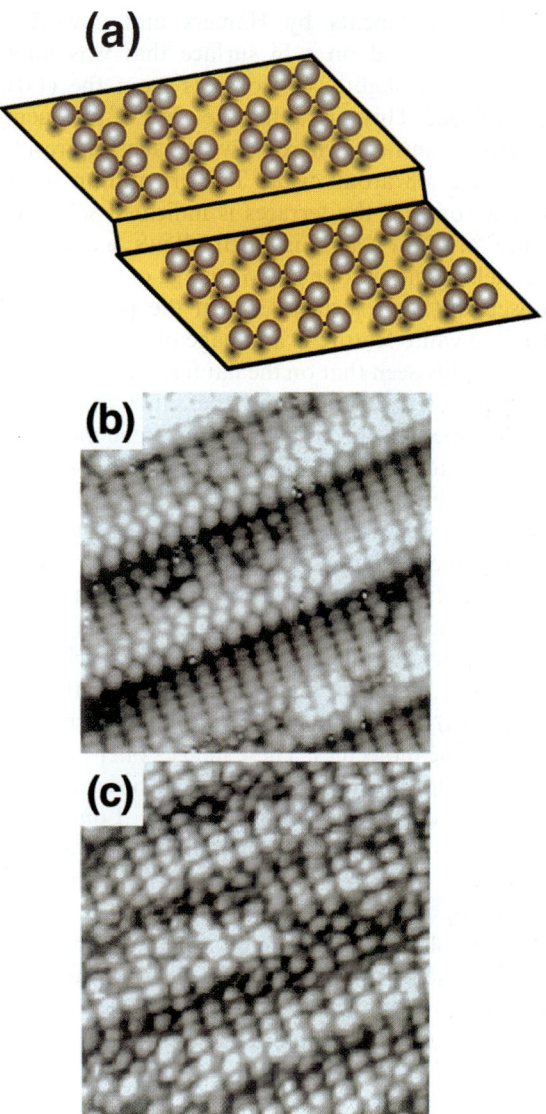

Fig. 9. (a) Schematic illustration of the oriented rows of silicon dimers present on a clean Si(1 0 0) surface that has been intentionally miscut by 4°. (b) STM images of this surface, and (c) the same surface after adsorption of cyclopentene. Figure reprinted from Ref. [35] with permission from Elsevier Science.

coverage. This is usually achieved by holding the substrate at an elevated temperature. These cyclopentene films, therefore, do not grow like ordinary metal or semiconductor crystals. Rather, the strong and directional bonding at the interface is sufficient to direct the molecules into the necessary sites, so that the semiconductor surface acts as a template for the placement of the organic molecules. The effect is one in which the structure of the underlying surface is propagated into ordered overlayers of organic molecules.

Ordered overlayers are not achieved in all systems, however, and it appears that a certain rigidity and molecular size is necessary to form well-ordered films, as evident in the following example. Although Hamers et al. have shown that other alkenes and dialkenes (such as 1,5-cyclooctadiene, displayed in Fig. 10) could be used to produce ordered monolayer films through the same [2 + 2] attachment [33], a number of other alkenes investigated did not adsorb with a high degree of order. For example, cyclohexadiene and cyclooctatetraene, both illustrated in Fig. 10, while reacting readily with the Si(1 0 0)-2 × 1 surface, did not form ordered films on Si(1 0 0)-2 × 1 [34,36]. This reduced packing may stem from either disorder in the configurations of individual adsorbates or from too low a final (saturation) coverage of molecules at the surface. Rigidity of the organic molecule, generally provided by a ring structure, was found to help in the formation of ordered layers. In addition, having a single type of functional group or a single means of bonding to the surface was advantageous, because molecules that bind in several different configurations do not maintain the degree of order necessary for close packing [34].

In addition to forming densely packed organic layers, the [2 + 2] cycloaddition reaction can be used to generate a chiral surface, according to a recent study by Lopinski et al. [39]. Chirality refers to the handedness of a molecule, and chirality plays an enormously important role in the chemistry of natural products. A chiral molecule is one that cannot be superimposed on its mirror image similar to a human hand or a conch shell. For example, in Fig. 11, there is no way to take the image of either the molecule or the hand on the left side of the picture and superimpose it with the

systems. In most common methods of semiconductor film growth, such as molecular beam epitaxy (MBE) or chemical vapor deposition (CVD), diffusion of the depositing species at the surface is a requisite for the formation of good overlayers [1]. Motion of atoms or molecules to find the optimal binding sites is necessary for conformal film

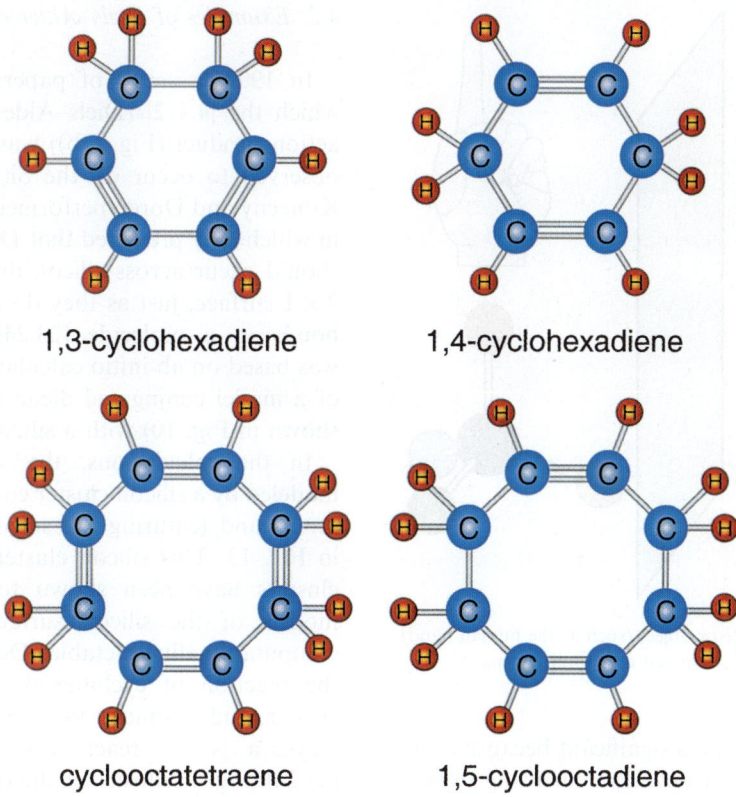

Fig. 10. The 1,3-cyclohexadiene, 1,4-cyclohexadiene, 1,5-cyclooctadiene, and cyclooctatetraene molecules.

corresponding image on the right (the mirror image). Both objects are chiral. Chiral molecules that share the same stoichiometry and bonding sequence but are non-superimposable are known as enantiomers. An example is the molecule pictured in Fig. 11. Nature makes broad use of enantiomeric species, and most biochemical processes select for one enantiomer over another. For example, biological enzymes distinguish between enantiomers.

Lopinski et al. have demonstrated the formation of a synthetic chiral surface by adsorbing the molecule 1S(+)-3-carene on the Si(1 0 0)-2 × 1 surface [39]. Carene, illustrated in Fig. 12(a), is a chiral, two-ringed molecule containing a C=C double bond and a three-carbon-atom ring (cyclopropane). Carene initially adsorbs by [2 + 2] cycloaddition between the alkene group and the silicon dimer. The final adsorbed state is believed to bridge across two dimers by reacting also at the cyclopropane group. Calculated structures for carene molecule adsorbed at a Si(1 0 0)-2 × 1 surface are indicated in Fig. 12. The structure in Fig. 12(b) is of the [2 + 2] cycloaddition product; that in Fig. 12(c) is of carene bridging across two dimers.

The carene molecule was chosen in the study because it contains a bulky dimethyl group that geometrically hinders adsorption at one face of the molecule. This so-called steric hindrance forces the [2 + 2] reaction to take place with only one facial orientation of the molecule. Lopinski et al. verified that the reaction was enantiospecific, i.e. produced only one of the possible chiral products, using scanning tunneling microscopy. The STM image allows the determination of absolute chirality of adsorbed molecules, as shown in earlier work from the same group [38]. The resulting carene adsorbate contains chiral centers and produces a chiral

4.2. Examples of Diels–Alder reactions on silicon

In 1997, a series of papers was published in which the [4 + 2] (Diels–Alder) cycloaddition reaction product (Fig. 5(b)) was first predicted and observed to occur on the Si(1 0 0)-2 × 1 surface. Konecny and Doren performed a theoretical study in which they predicted that Diels–Alder reactions should occur across silicon dimers on a Si(1 0 0)-2 × 1 surface, just as they do across C=C double bonds in a molecule [23,24]. Their prediction was based on ab initio calculations of the reaction of a model conjugated diene (1,3-cyclohexadiene, shown in Fig. 10) with a silicon surface.

In the calculations, the silicon surface was modeled by a silicon cluster containing nine silicon atoms and featuring the silicon dimer, as shown in Fig. 13. This silicon cluster and related larger clusters have been shown to provide adequate models of the silicon surface while remaining computationally tractable. Doren et al. followed the reaction of cyclohexadiene with the silicon cluster, and examined two possible reaction pathways: a [4 + 2] reaction or a [2 + 2] reaction [23,24]. Fig. 13 illustrates the two possible reaction products for a related molecule, 1,3-butadiene (see Fig. 5). The calculations showed that there is a substantial thermodynamic driving force (54 kcal/mol) to form the product of a [4 + 2] reaction between cyclohexadiene and the silicon dimer. This is significantly more energy than is gained in

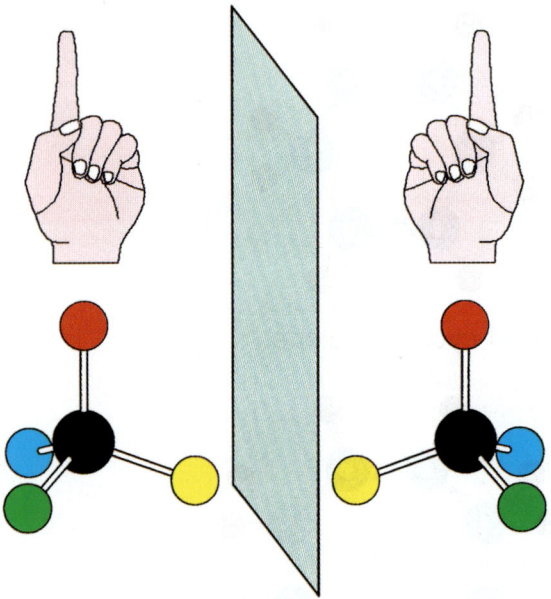

Fig. 11. A chiral molecule or object (such as the human hand) is one that cannot be superimposed on its mirror image.

surface [39]. This result is significant because of its implications for sensor development using organically modified surfaces. In molecular recognition, distinguishing between enantiomers of molecules requires a chiral probe. Hence, if organic layers on silicon are to be used to achieve molecular recognition, particularly in biological systems, chiral surfaces such as this one must be utilized.

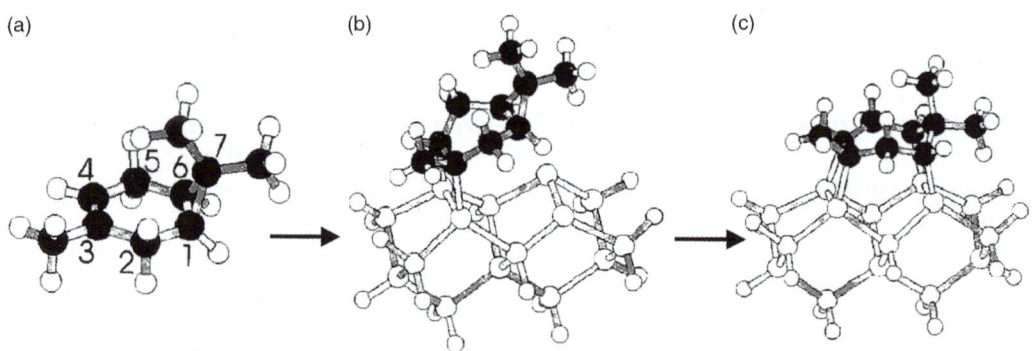

Fig. 12. Structures of (a) the 1S(+)-3-carene molecule; (b) carene molecule bonded by [2 + 2] cycloaddition across the silicon dimer; (c) carene molecule bonded in a bridging configuration across two silicon dimers. Both adsorbed structures are chiral. Figure adapted with permission from Ref. [39]. Copyright 1999 American Chemical Society.

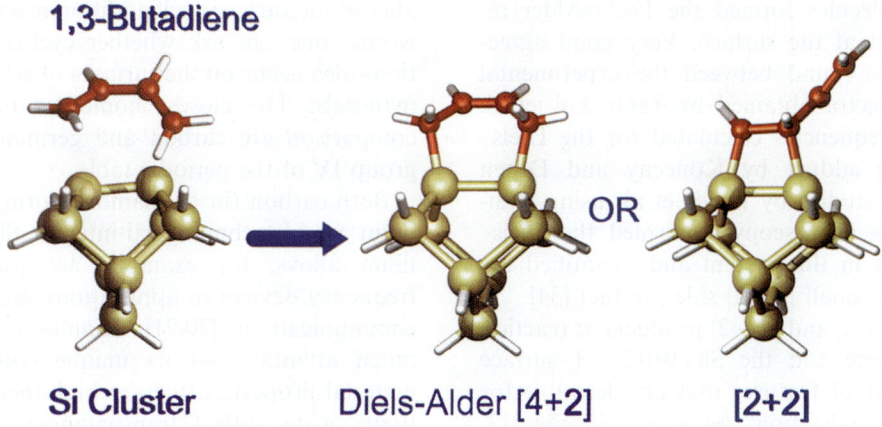

Fig. 13. In ab initio theoretical studies of silicon surface reactivity, the Si(1 0 0)-2 × 1 surface is modeled using a nine silicon atom cluster. Two possible outcomes for the reaction of the Si cluster with 1,3-butadiene are shown: the product of the Diels–Alder reaction and the product of the [2 + 2] reaction.

the competing [2 + 2] reaction (39 kcal/mol). In addition, the calculations predicted that the reaction should occur with little or no energetic barrier across a symmetric silicon dimer, consistent with the Woodward–Hoffman rules for a [4 + 2] reaction [23,24].

Studies by Teplyakov et al. provided the experimental evidence for the formation of the Diels–Alder reaction product at the Si(1 0 0)-2 × 1 surface [26,27]. A combination of surface-sensitive techniques was applied to make the assignment, including surface vibrational spectroscopy, thermal desorption studies, and synchrotron-based X-ray absorption spectroscopy. Vibrational spectroscopy in particular provides a molecular "fingerprint" and is useful in identifying bonding and structure in the adsorbed molecules. Fig. 14 provides an example of a vibrational spectrum used to assign the bonding of the butadiene molecule at a surface. The peaks shown in the spectrum correspond to absorption of light at the energies of vibration (in a unit of reciprocal centimeters, or cm^{-1}, called wavenumber) for carbon–hydrogen bonds in the molecules. Because there are multiple C–H bonds present in the molecule, the spectrum contains multiple peaks. By comparing the spectrum for the butadiene molecule (Fig. 14(a)) with the spectrum for reacted butadiene (Fig. 14(b)), the bonding configuration of the reacted molecule

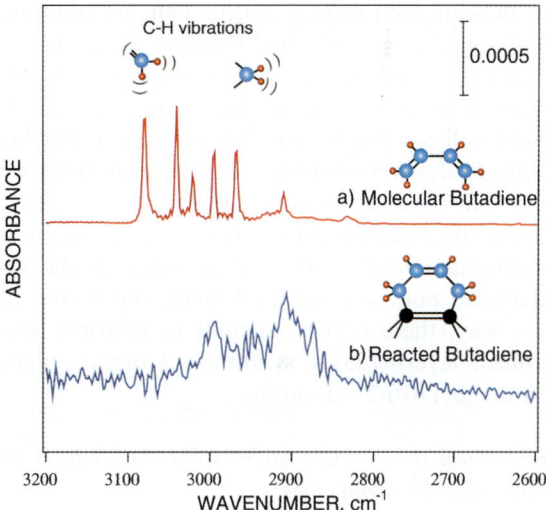

Fig. 14. Measured vibrational spectra giving information on the bonding of butadiene at the Si(1 0 0)-2 × 1 surface. The peaks shown in the spectrum correspond to absorption of light at the energies of vibration over the spectral region for carbon–hydrogen bond stretches. The units of energy are in wavenumber, or cm^{-1}.

can be assigned. An analysis of the vibrational spectra of adsorbed butadiene on Si(1 0 0)-2 × 1 in which several isotopic forms of butadiene (i.e., some of the H atoms were substituted with D atoms) were compared showed that the majority of butadiene molecules formed the Diels–Alder reaction product at the surface. Very good agreement was also found between the experimental vibrational spectra obtained by Teplyakov et al. [26,27] and frequencies calculated for the Diels–Alder surface adduct by Konecny and Doren [23,24]. Later studies by Hovis et al. using scanning tunneling microscopy confirmed the Diels–Alder product in this system and quantified the formation of a small [2 + 2] side product [34].

Both the [2 + 2] and [4 + 2] products of reaction between a diene and the Si(1 0 0)-2 × 1 surface have a number of features that are desirable for surface functionalization. As shown in Fig. 13, both reactions produce organic molecules that are strongly bound to the silicon surface through two covalent Si–C bonds. The high strength of these bonds means that the organic layer at the surface is typically stable to temperatures in excess of one to two hundred degrees celsius [30]. In addition, both of the cycloaddition reaction products between the diene and the surface retain one double bond in the organic layer. This remaining double bond (alkene group) can be used as a starting point for further derivatization of the surface. For example, the alkene group in the first layer can act as the attachment point for the second layer, a new functional group in the second layer as the attachment point for the third layer, and so on. In this way, there is the potential to perform controlled, layer-by-layer synthesis of organic films on the Si(1 0 0)-2 × 1 surface.

4.3. Down the periodic table: solid state effects on cycloaddition chemistry

It is intriguing that the analogs of key reactions in organic chemistry—such as the Diels–Alder and [2 + 2] cycloadditions—occur across the silicon dimer on the Si(1 0 0)-2 × 1 surface. These results lend support to the analogy that was drawn between a dimer on Si(1 0 0)-2 × 1 and the C=C double bond. The success of this analogy, in turn, advances the concept of describing the reactivity of a covalent surface within a molecular framework. It has therefore been a challenge to both experimentalists and theoreticians to test the limits of this molecular framework by probing the generality of the surface cycloaddition reaction. In other words, one can ask whether cycloaddition reactions also occur on the surfaces of related covalent materials. The closest homologs to silicon for comparison are carbon and germanium, also in group IV of the periodic table.

Both carbon (in its diamond form) and germanium are of technological interest. Silicon germanium alloys, for example, are used for high frequency devices in applications such as wireless communications [70,71]. Diamond has received much attention for its unique combination of material properties, such as high thermal conductivity, wide optical transparency, hardness and chemical inertness [72].

Like Si(1 0 0), the (1 0 0) crystallographic face of both Ge [73] and diamond [74] undergo a (2 × 1) reconstruction to form rows of surface dimers. These dimers each are believed to have some degree of double-bond character, with a strong σ bond and a partial π bond [73,74]. However, the surfaces differ in the strength of the π bond and the degree of tilt that is observed in the dimers. The dimers on both the Si and Ge surfaces tilt due to the weakness of the π bond. In the case of Si(1 0 0)-2 × 1, the dimers are observed in their tilted form only at low temperature, as discussed previously, whereas on Ge(1 0 0)-2 × 1, the dimers are tilted at both low and moderate temperatures [14]. An appropriate picture of the dimers on these surfaces is that shown in Fig. 7(b). In contrast, the diamond dimer is essentially different. Due to good π bond formation between C atoms, the diamond surface forms symmetric dimers, best described by the picture in Fig. 7(a), over a large temperature range [74].

Over the past few years, researchers have begun to probe for detailed parallels between the surface chemistry of Si(1 0 0) with that of Ge(1 0 0) and C(1 0 0), with special consideration paid to the role of the surface dimers. The first published studies focused on Ge surfaces. Adsorption studies of alkene-containing molecules such as ethylene [52], butadiene [53], cyclopentene [54], and cyclohexadiene [55] at the Ge(1 0 0)-2 × 1 surface in ultrahigh vacuum revealed that like Si, Ge(1 0 0) can participate in cycloaddition chemistry. Both [2 + 2] and [4 + 2] reaction products have been observed experimentally. Theoretical investigations by Musgrave and coworkers have also revealed close parallels in energetics between cycloaddition reactions on Si(1 0 0) vs Ge(1 0 0). Namely, on both surfaces, the [4 + 2] cycloaddition product is more stable by ≈15–20 kcal/mol than the [2 + 2] cycloaddition product [25]. Indeed, experimental studies of Diels–Alder reactions on Ge(1 0 0)-2 × 1 show close similarities in bonding (based on the vibrational spectra) with the Si-based product [53]. Although the formation of well-ordered monolayers, as in the case of silicon, has not yet been reported, it might be expected that the proper choice of molecular precursor will lead to similarly packed monolayers on Ge.

Unlike on Si, however, the Diels–Alder reaction was found to be reversible on Ge(1 0 0)-2 × 1, leading to loss (by desorption) of the organic layer upon heating the surface to 300 °C [53]. The reverse of the Diels–Alder reaction is called the retro-Diels–Alder reaction, and it also finds close parallels in molecular organic chemistry. This difference between Si and Ge surfaces in the behavior of the organic layer upon heating is a significant one. It is the result of a fundamental difference in the strength of the bonds formed between carbon and silicon atoms versus carbon and germanium atoms. A C–Ge bond is weaker than a C–Si bond by almost 10 kcal/mol [25] and bond for bond, an organic molecule bound to Ge will be held less tightly than one bound to Si. Exploiting the reversibility of the attachment reaction on Ge due to its weaker bonding is being considered as a way to spatially pattern organic layers on the surface [53].

Recently, there have been reports of both [4 + 2] and [2 + 2] cycloaddition reactions occurring on the diamond(1 0 0)-2 × 1 surface [56–58]. Unlike germanium or silicon, the diamond surface allows *direct comparison* to be made between chemistry in a molecular system (i.e. organic chemistry) versus chemistry on the surface of an extended solid (i.e. surface chemistry). First, both reactions directly involve carbon atoms. Second, like the corresponding alkene, the carbon–carbon dimer on the diamond(1 0 0)-2 × 1 surface is symmetric. In agreement with this analysis, studies by Russell and coworkers [56,57] and by Hossain et al. [58] have shown that the carbon dimers on the diamond(1 0 0) surface behave like C=C double bonds toward organic cycloaddition reactions. For example, the reaction between butadiene and the C(1 0 0)-2 × 1 surface was found to form predominantly the Diels–Alder product. The resulting six-membered ring tethered to the surface exhibited remarkable similarities to cyclohexane, which is the Diels–Alder product of the related reaction between butadiene and ethylene, shown in Fig. 5 [56]. Recent theoretical calculations by Fitzgerald and Doren have investigated the butadiene/diamond(1 0 0) reaction system, and their results are consistent with formation of a majority [4 + 2] product with a minor [2 + 2] product [75].

Studies of [2 + 2] reactions on diamond provide further insight into the reactivity of covalent solids [57] and further evidence for a solid state effect on surface reactivity in these systems. Recall that [2 + 2] cycloadditions are not generally observed in molecular systems because they are symmetry forbidden according to the Woodward–Hoffman rules. At the surfaces of Si and Ge solids, [2 + 2] reactions are believed to occur because the surface dimers tilt, providing an asymmetric reaction approach (transition state) that relaxes the symmetry restrictions [31,69]. Because on diamond the dimers do not tilt [74], it would be expected that the diamond surface, of all the group IV surfaces, would behave most like its C=C analog. Following the Woodward–Hoffman selection rules, a [2 + 2] reaction along a symmetric reaction pathway would be unlikely to occur. Consistent with this expectation, it was found that while cyclopentene does undergo [2 + 2] addition on C(1 0 0), the reaction probability is three orders of magnitude lower than that on Si(1 0 0) or Ge(1 0 0) [57]. Hence the structure of the surface exerts a strong influence on the surface reactivity.

5. Further applications: functionalization of hydride-terminated Si surfaces

Both [2 + 2] and [4 + 2] cycloaddition reactions have now been demonstrated as a means of functionalizing the surfaces of three covalent semiconductor solids: C, Si, and Ge. Although work is ongoing, the results already seem to be quite general for alkenes and dienes reacting on these surfaces. Cycloaddition chemistry is a good method for forming organic layers on semiconductor substrates. By proper choice of organic reagent, a variety of useful surface layers can be engineered, including molecular monolayers exhibiting good stability, chirality, and a high degree of order. The understanding of cycloaddition reactions at semiconductor surfaces is now reaching a level where scientists can begin to apply these reactions to the rational design of organic/semiconductor interfaces.

Conceptually, the framework for describing cycloaddition reactions in organic chemistry appears to apply moderately well to reactions at these surfaces. These parallels allow one to apply the broad knowledge base available from organic chemistry to tailor the reactions at the surface in order to obtain targeted products. Furthermore, the general principles that have can be developed for the cycloaddition systems can be extended to understand an even broader range of reactions at semiconductors. For example, it is evident that the surface structure plays a role in influencing surface selection rules and reaction pathways. It has also been shown that at these covalent surfaces, the formation of strong directional bonds with rigid organic molecules allows the production of ordered overlayers, even in the absence of diffusion.

There are many additional applications in which these guiding principles and a molecular approach to designing surface reactions can be used to functionalize semiconductor surfaces. One area of application that exhibits great promise is the solution-based functionalization of hydrogen-terminated silicon surfaces. Even before vacuum-based cycloaddition methods began to be exploited for attaching organic groups to semiconductor surfaces (see Section 4), remarkable advances were made in solution-based functionalization chemistry. In these so-called "wet-chemical" methods, monolayers of organic films could be obtained on silicon that imparted outstanding resistance to oxidation and provided superb protection and passivation to the underlying substrate.

Solution-based functionalization chemistry exhibits a number of advantages over the vacuum approaches. These include a much simpler apparatus and the potential for significantly higher reaction rates due to higher reactant fluxes present in the liquid. In addition, for certain applications, the solution-based method may be easier to integrate into an existing processing protocol. On the other hand, the degree of control over surface cleanliness and structure, and the ability to perform in situ diagnostics, may be reduced for this method compared to a vacuum approach. For some applications, the surface control and range of attachment configurations that are possible by the vacuum-based approach may prove critical. For example, in the development of molecular electronics, the electrical properties will depend upon the structural properties of the surface and the specific interfacial bonding, which can be well controlled in a vacuum environment. In other applications, the substrate being modified may not be able to withstand the harsher treatment used in solution processing. The choice of using wet chemistry (solution based) vs dry chemistry (vacuum based) ultimately will depend upon the needs of the particular application, and both methods continue to be actively pursued.

Most solution based functionalization approaches use the hydride-terminated silicon surface as a starting point. Hydride-terminated silicon, particularly the H:Si(1 1 1) surface, can be generated with a high degree of order and atomic flatness, and can be rendered quite passive against oxidation and contamination [5]. The structure of the H:Si(1 1 1) surface is shown in Fig. 3(b). The hydride-terminated surface does not require vacuum conditions either for its preparation or subsequent reaction [5].

Recall that for the Si(1 0 0)-2 × 1 surface, the important functional group was the Si=Si dimer, for which analogies were drawn to alkene chemistry. On the H:Si(1 1 1) surface, the group of interest for functionalization is typically the Si–H bond. The literature on organosilicon chemistry (i.e., the chemistry of molecules with C–Si bonds) includes many known reactions for Si–H bonds. One major class of these reactions is "hydrosilylation", whereby the silicon-hydrogen bond adds across an alkene (C=C) or alkyne (C≡C) group. Examples in which this chemistry is used to functionalize silicon surfaces are described in the following sections.

5.1. Hydrosilylation chemistry at the H:Si(111) surface

The first report of a densely packed organic monolayer bound directly to silicon through Si–C bonds was based upon the hydrosilylation reaction on the H:Si(1 1 1) surface [76,77]. In pioneering work, Chidsey and coworkers demonstrated that long-chain 1-alkenes (those with the double bond at one end of the chain) could be attached to hydrogen-terminated silicon surfaces at 100 °C in the presence of a molecule called a radical initiator [76,77]. A radical initiator is a species used to create free radicals in reactions. Chidsey et al. proposed that the surface attachment mechanism was analogous to the hydrosilylation reaction in solution phase chemistry. In the proposed surface mechanism (illustrated in Fig. 15), the radical initiator starts the reaction by abstracting H from a surface Si–H bond, producing a silicon dangling bond (which is also referred to as a silicon radical). In Fig. 15, a surface dangling bond is designated by an orbital with a dot, where the dot represents a single, unpaired electron. The silicon dangling bond then reacts with the alkene molecule, forming a new Si–C bond at the surface and leading to the attachment of the organic molecule, which itself is left with a carbon radical as shown in Fig. 15. This surface bound organic radical is thought to abstract a hydrogen atom either from an unreacted alkene molecule or from a neighboring Si–H group on the surface. The net reaction sequence produces a stable, closed-shell organic group bound directly to the Si(1 1 1) surface.

Studies by Chidsey and coworkers [76,77] showed a remarkable stability and robustness to the organic monolayers formed by this process. A number of diagnostic techniques were used to characterize the monolayers, including vibrational spectroscopy, X-ray spectroscopy and reflectivity, ellipsometry and wetting measurements. These showed that the monolayers were tightly packed, with a density close to 90% that of crystalline hy-

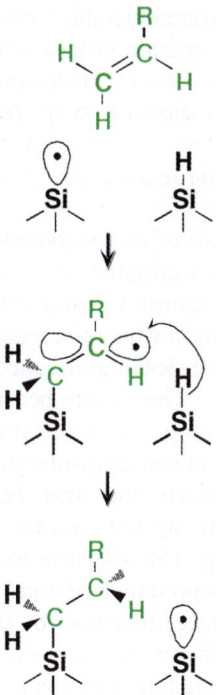

Fig. 15. Proposed mechanism for surface hydrosilylation. Initial loss of silicon hydride generates a silicon dangling bond. Reaction between the silicon and an alkene molecule leads to an attached alkyl radical, which may abstract a hydrogen atom from a neighboring silicon [84].

drocarbons. More importantly, the organic layers formed by hydrosilylation were stable in a number of environments, including boiling water, organic solvents, acids and bases, and air. The silicon substrate was found to be protected by the monolayer against oxidation after many weeks of exposure to air [76,77].

After the reports of the radical-initiated hydrosilylation process, a number of different approaches for functionalization based on this reaction were developed [8,77–79]. It was shown that the reaction could proceed not just with a radical initiator, but also with other methods for generating the initial silicon radical (dangling bond). The use of heat (>150 °C) succeeded by thermally breaking the Si–H bond, although the monolayers produced by this method were of a lower quality [77]. Another method involved use of acidic molecules (called Lewis acids) to catalyze the reaction [78,79]. Related approaches using reactions of organometallic species (molecules involving both organic groups and metal atoms) to functionalize silicon surfaces through the Si–H bonds were also shown to work [80–83].

5.2. Radical chain reactions

A recent revisit of the proposed reaction mechanism for radical-initiated hydrosilylation [76,77] has led to two beautiful demonstrations of silicon functionalization in which the reaction propagates along the surface, leaving attached organic molecules in its wake. This occurs because the reaction is "self-propagating". In a self-propagating reaction, one reaction causes a subsequent reaction to occur, which starts the next reaction, and so on. Recall that in the surface hydrosilylation mechanism (Fig. 15), the process is initiated by generating a silicon dangling bond on the hydride-terminated surface. Reaction of this silicon dangling bond with an alkene molecule leads to a surface-bound organic group containing a carbon radical, which must abstract a hydrogen atom in order to become a stable, closed shell species. If the adsorbate abstracts a hydrogen atom from a neighboring Si–H group, a new silicon dangling bond is produced, which can begin the reaction a new. This process is also known as a radical chain reaction.

This proposed radical-chain reaction mechanism was recently confirmed in hydrosilylation experiments performed in vacuum. In one report, Cicero et al. prepared hydride-terminated Si(1 1 1) surfaces using standard wet-etch methods, then introduced the sample into an ultrahigh vacuum chamber [84]. In vacuum, the tip of a STM was used to generate isolated silicon dangling bond sites surrounded by silicon hydride groups [85,86]. This surface was then exposed to styrene, a molecule which consists of an alkene group attached to a benzene ring, as shown in Fig. 16. In the STM image of Fig. 17(A), islands of adsorbed styrene are observed on the silicon surface in areas surrounding the locations of the initial dangling bonds, which are designated by black dots [84]. In other words, one dangling bond on the surface led to the deposition of many styrene molecules. This result supports the presence of a radical-chain

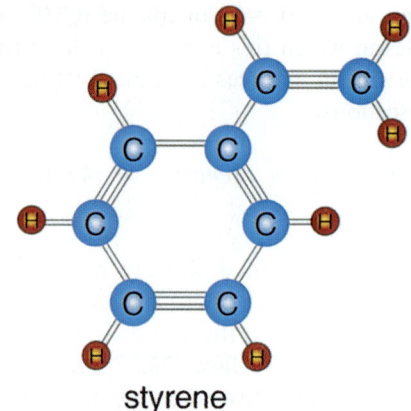

Fig. 16. The styrene molecule.

reaction during which hydrogen atoms are abstracted from neighboring silicon atoms. The observation of bunched-up islands of styrene rather than one-dimensional lines is attributed to a "random walk" propagation of the reaction. Cicero et al. suggest that the reaction stops at a silicon atom that no longer has any neighboring hydrogen atoms [84]. Note that this system provides another clear example of the localization of the reaction and limited mobility of the organic molecules at silicon surfaces. If the styrene molecules were mobile, the bunching effect would not be observed.

In another report, Lopinski and coworkers have used the same concept of the radical-initiated hydrosilylation reaction on the Si(1 0 0)-2 × 1 surface to induce self-directed growth of molecular "wires" on the surface [87]. On the Si(1 0 0)-2 × 1 surface, the radical chain reaction propagates primarily along the direction of the dimer row, leading to "lines" of organic adsorbates, as shown in Fig. 17(B). Molecular assemblies as long as 130 Å (corresponding to 34 styrene molecules) were observed. Lopinski et al. suggested that by using different reactant alkenes and seed conditions at the surface, this reaction may be used to engineer more complex molecular nanostructures on silicon.

5.3. Photoinitiation and photopatterning

Another interesting extension of the hydrosilylation reaction involves photoinitiation. Chidsey and coworkers have shown that shining ultraviolet

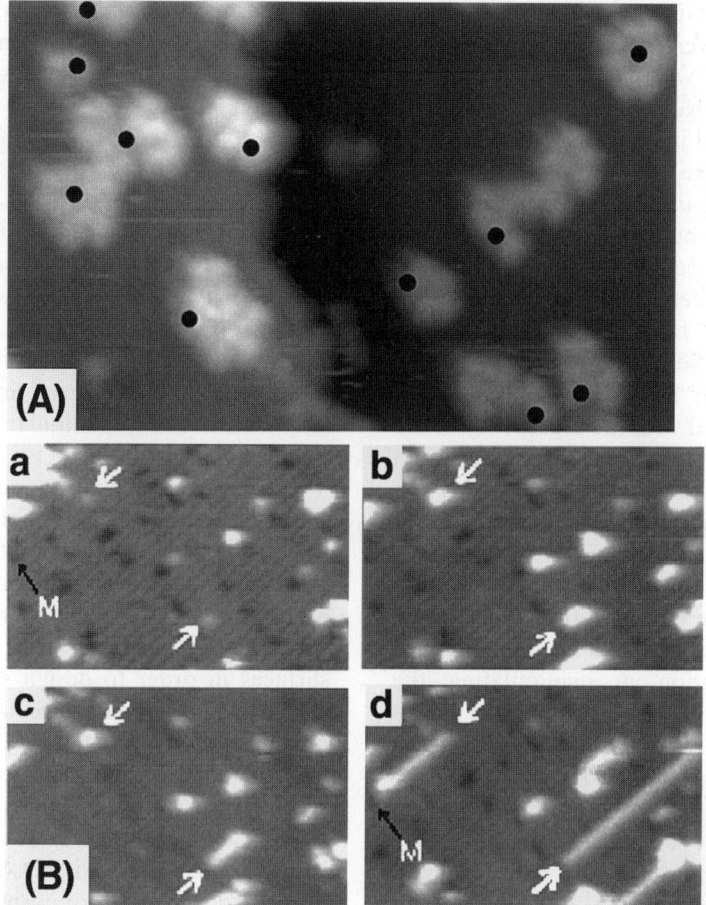

Fig. 17. (A) STM image of hydrocarbon islands formed by radical-initiated hydrosilylation chemistry of styrene on H-terminated Si(1 1 1) [84] (reprinted with permission from Ref. [84]. Copyright 2001 American Chemical Society). (B) STM image of hydrocarbon lines formed by the same mechanisms on H-terminated Si(1 0 0)-2 × 1. The four images show the propagation of the styrene nanostructures with increasing styrene exposure [87] (reprinted from Ref. [87] with permission from nature).

light on the H:Si(1 1 1) surface while exposing the surface to a 1-alkene also led to the formation of organic monolayers [88,89]. Buriak and coworkers have used the photoinitiation of hydrosilylation in an exciting application of this process [90]. They have examined functionalization reactions on *porous* silicon, which is a cousin of single-crystalline silicon. Porous silicon contains a complex array of pores at the surface, generating a surface structure of silicon reminiscent of stalagmites [8]. Porous silicon is generated by specialized etching of crystalline silicon, and like crystalline silicon, can be hydride terminated. In contrast to crystalline silicon, though, porous silicon is a strong emitter of light (called "luminescence"). In fact, it luminesces in response to electrical, optical, and chemical stimuli, making it of great interest for use in light emitting devices.

Buriak et al. demonstrated that the hydrosilylation reaction of alkenes could be induced by visible white light on photoluminescent samples of hydride-terminated porous silicon [8,90]. Using a simple optical setup in which light was only shined on certain regions of the silicon, they showed that using the photoreaction, they could form patterns of organic layers on the surface. Regions that were exposed to white light in the presence of alkene molecules became covered with organic groups, whereas regions that were kept dark remained terminated with hydrogen. Subsequently, the porous silicon was etched in basic solution. Areas of the substrate that were functionalized remained intact because the organic layer acted to protect the underlying silicon, but areas that were hydride terminated were destroyed during the etch process.

A sample that had been subjected to this treatment is shown in Fig. 18 while being illuminated with ultraviolet light. Some regions of the sample appear orange. These are the areas that were photochemically coated with the organic layer; the orange light is the visible luminescence that occurs when alkyl-functionalized porous silicon is exposed to ultraviolet light. The dark regions of the sample reflect areas that do not photoluminescence. These are areas that were not hydrosilylated and were etched in the basic solution. This simple example serves to demonstrate the potential of organic functionalization in manipulating the properties of a silicon substrate. Adding the ability to pattern these attached organics, as shown in the work by Buriak and coworkers [90], only expands the possibilities for future device applications.

6. Prospects

The microelectronics industry is headed toward increasingly miniaturized devices. Within the next decade or two, the industry will be facing the necessity to fabricate atomic-scale devices. As semiconductor devices continue their downward trend in size, the importance of interfacial phenomena and atomic scale manipulation will continue to rise.

The tools available to surface chemists and surface physicists already enable an unprecedented level of control in the ability to modify surfaces at the atomic level. New capabilities in both the experimental and theoretical arenas will give scientists and engineers increasing power to manipulate surfaces in order to design specific functionality. The benefits of this molecular-level surface engineering will be felt in many areas of microelectronic fabrication—in atomic layer deposition, in surface passivation, in deposition of ultrathin dielectrics, for example. It will then go beyond existing processing and into the realm of new uncharted territory. For example, signs are already appearing of molecular electronics and molecular computing [91].

One of the emerging areas in interface science is organic functionalization of semiconductors. This technology provides the opportunity to create hybrid devices exploiting the properties and features of both organic and inorganic materials. The addition of organic layers at the interface provides for myriad possibilities, including the incorporation of lubrication, optical response, chemical sensing, or biocompatibility properties to semiconductor surfaces. Thus, the field of organic modification ties directly into a number of technological areas that will become of increasing importance.

One area of impact is nanotechnology. Nanotechnology refers to the development of materials, systems and devices with a characteristic length scale of 1–100 nm. For comparison, a H atom is

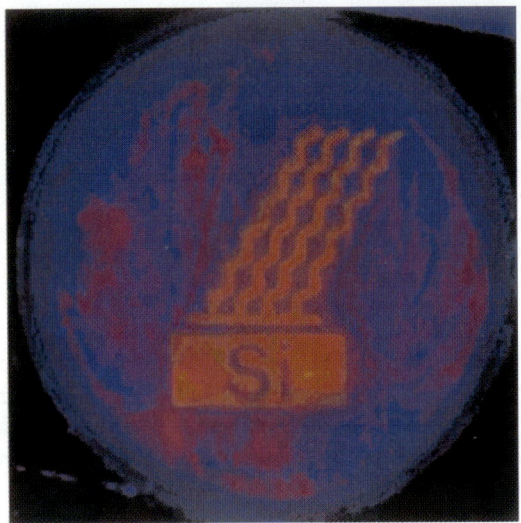

Fig. 18. Photograph of a porous silicon sample that was photopatterned using the light-promoted hydrosilylation reaction. The orange regions of the pattern are coated with an organic layer that protected the surface against etching, allowing the underlying silicon to luminescence. The dark blue regions were unprotected, and after etching they no longer luminescence (reproduced with permission of Wiley-VCH from Ref. [90]).

roughly 0.1 nm in diameter. Such molecular-scale materials are being explored for many applications in which novel electronic, optical or mechanical properties are desired. One example is a so-called molecular computer, in which switches and memory components will be constructed out of molecules instead of patterned out of silicon. Nanotechnology in effect relies upon manipulating atoms into specific locations in order to construct such novel devices. It is evident that methods being developed for organic functionalization can be used to produce films of molecules on semiconductor surfaces with nanometer scale thicknesses, i.e. one molecule high. Achieving nanometer scale control in the lateral direction is also possible. One example is the use of radical-initiated hydrosilylation of alkenes to form self-limiting islands or lines of organic material several approximately 10–100 nm in size [84,87]. Other possibilities include writing complex patterns in attached organic layers with the tip of a STM [45], or utilizing polymerization in the organic layer. The opportunities to make contributions in the area of nanotechnology by creative use of functionalization strategies are vast.

Another area closely tied to organic functionalization is molecular recognition and chemical or biological sensing. Organic molecules are key players in molecular recognition. Coupling organic molecules to semiconductor solids provides opportunity to fabricate integrated chip-based sensors. In such a scheme, the organic layer provides the molecular recognition function, while the semiconductor provides capabilities for signal processing, data storage, logic, and even wireless communication. Using this approach, the fabrication of compact sensor that can be used out in the field and communicate data remotely can be envisioned. Key to such a device is the attachment of an organic layer on the semiconductor substrate, where the organic layer can be engineered with a variety of end groups designed for molecular recognition. It remains a challenge to develop strategies for incorporating molecular receptors of a general nature onto the semiconductor surface.

Even bioengineering stands to gain significantly from advances made in organic surface modification. Surface modification is known to play a decisive role in influencing cell growth [92,93]. For example, interfacial properties can be used either to inhibit or to accelerate cell adhesion or cell growth. Strategies utilizing patterned organic layers that combine adhesion proteins, cell growth factors, and inhibiting molecules can be used to direct two-dimensional cell growth. It has been demonstrated, for example, that the direction of axon extension from nerve cells can be controlled by printing μm-scale lines of proteins on glass surfaces [94,95]. Directed cell growth, in turn, can be used for tissue engineering. Organic modification can also be used to impart biocompatibility to semiconductor-based devices designed to be implanted into living organisms.

There are clearly a large number of applications for which organic functionalization of semiconductor surfaces can have a great impact. Each of these applications requires a deep understanding of how to modify the semiconductor surfaces controllably. Methods to incorporate organic functionality to the surfaces of semiconductors have seen immense progress in recent years. We have shown in this review how direct, covalent attachment of organic molecules at the surface is being used to bind organic layers onto semiconductors. Methods developed for both dry and wet environments can impart excellent control of the interfacial properties. For example, organic layers can be deposited which protect the underlying substrate from attack by etchants, impart chiral properties to the surface, or add new optical properties to silicon.

A number of challenges remain, however, before many of the futuristic applications—in nanotechnology, sensing, and others areas—can be realized using this technology. One future need is to increase the degree of control over both the selectivity (i.e., in producing one product over competing products) and the degree of order in the organic layer. This in turn requires a deeper understanding of the driving forces that control the attachment reactions. Evidence points to the importance of both the reaction rates (kinetics) and the energetic stability of the products (thermodynamics) in controlling these reaction systems. Therefore, it is important to understand how to design the system (by choice of reaction conditions or molecular precursors) in order to manipulate the product distribution. More work is also needed in determining the detailed reaction mechanisms.

One of the key challenges is the ability to deposit multiple organic layers in a controllable fashion. So far, most of the work in organic functionalization by covalent Si–C attachment has focused on the initial monolayer. There are some nice exceptions, however, such as the work by Bitzer and Richardson [96]. It is clear that many applications require flexibility in the deposition of multiple layers. In order to extend the organic modification to allow for next-layer attachment, a bifunctional or polyfunctional organic molecule must be used for the first layer. This layer, in turn, will retain a reactive functional group for further attachment. This review, and indeed most of the work on organic functionalization by a covalent Si–C linkage, has focused on the alkene functional group. Research on other functional groups is in its infancy. There is clearly a great need not only for understanding how different functional groups react at the semiconductor surface, but also for learning how to direct the reaction of a polyfunctional molecule in a stepwise fashion.

The application of organic chemistry at semiconductor surfaces for forming organic monolayers and thin films is becoming an important and growing area. By focusing on the molecular nature of the surface, the functionalization methods ultimately will provide the potential for control over the monolayer bonding and order at the surface. The impact of these studies on emerging technologies in microelectronics, sensing, and photonics is expected to be significant in the coming decades.

Acknowledgements

The author expresses appreciation for the many fruitful interactions with students and colleagues that contributed immensely to this review. Thanks are also due to Prof. Charles Musgrave, Collin Mui, George Wang, Michael Filler, and Michael Manzo for their contribution to the figures, and to Prof. Jillian Buriak, Prof. Christopher Chidsey, Prof. Robert Hamers and Dr. Robert Wolkow for graciously providing permission to reprint their figures. The National Science Foundation (Grants nos. DMR-9896333 and CHE-9900041) and the Beckman Foundation are gratefully acknowledged for financial support which made this review possible. The author also thanks the Camille and Henry Dreyfus Foundation for support through a Camille Dreyfus Teacher–Scholar award.

References

[1] S.A. Campbell, The Science and Engineering of Microelectronic Fabrication, Oxford, New York, 1996.
[2] C.Y. Chang, S.M. Sze (Eds.), ULSI Technology, McGraw-Hill, New York, 1996.
[3] International Technology Roadmap for Semiconductors, Semiconductor Industry Association, 1999.
[4] See, for example: (a) J.R. Arthur, Molecular beam epitaxy, Surf. Sci. 500 (2002) 189;
(b) P. Finnie, Y. Homma, Epitaxy: the motion picture, Surf. Sci. 500 (2002) 437.
[5] M.K. Weldon, K.T. Queeney, J. Eng Jr., K. Raghavachari, Y.J. Chabal, The surface science of semiconductor processing, Surf. Sci. 500 (2002) 859.
[6] J.W. Hill, D.K. Kolb, Chemistry for Changing Times, Prentice Hall, Englewood Cliffs, NJ, 1995.
[7] J.T. Yates Jr., A new opportunity in silicon-based microelectronics, Science 279 (1998) 335.
[8] J.M. Buriak, Organometallic chemistry on silicon surfaces: formation of functional monolayers bound through Si–C bonds, Chem. Commun. 12 (1999) 1051.
[9] R.A. Wolkow, Controlled molecular adsorption on silicon: laying a foundation for molecular devices, Ann. Rev. Phys. Chem. 50 (1999) 413.
[10] R.J. Hamers, S.K. Coulter, M.D. Ellison, J.S. Hovis, D.F. Padowitz, M.P. Schwartz, C.M. Greenlief, J.N. Russell Jr., Cycloaddition chemistry of organic molecules with semiconductor surfaces, Acc. Chem. Res. 33 (2000) 617.
[11] S.F. Bent, Attaching organic layers to semiconductor surfaces, J. Phys. Chem. B, submitted.
[12] J.J. Boland, The importance of structure and bonding in semiconductor surface chemistry: hydrogen on the Si(1 1 1)-7×7 surface, Surf. Sci. 244 (1991) 1.
[13] A. Zangwill, Physics at Surfaces, Cambridge University Press, New York, 1988.
[14] C.B. Duke, Semiconductor surface reconstruction: the structural chemistry of two-dimensional surface compounds, Chem. Rev. 96 (1996) 1237.
[15] H.N. Waltenberg, J.T. Yates Jr., Surface chemistry of silicon, Chem. Rev. 95 (1995) 1589.
[16] R.J. Hamers, R.M. Tromp, J.E. Demuth, Electronic and geometric structure of Si(1 1 1)-(7×7) and Si(0 0 1) surfaces, Surf. Sci. 181 (1987) 346.
[17] J.J. Boland, Scanning tunneling microscopy of the interaction of hydrogen with silicon surfaces, Adv. Phys. 42 (1993) 129.

[18] G.S. Higashi, Y.J. Chabal, G.W. Trucks, K. Raghavachari, Ideal hydrogen termination of the Si(111) surface, Appl. Phys. Lett. 56 (1990) 656.
[19] A. Wassermann, Diels–Alder Reactions: Organic Background and Physico-Chemical Aspect, Elsevier, New York, 1965.
[20] G.B. Gill, M.R. Willis, Pericyclic Reactions, Chapman and Hall, London, 1974.
[21] W. Carruthers, Cycloaddition Reactions in Organic Synthesis, Pergamon Press, New York, 1990.
[22] R.B. Woodward, R. Hoffmann, The Conservation of Orbital Symmetry, Academic Press, New York, 1970.
[23] R. Konecny, D. Doren, Theoretical prediction of a facile Diels–Alder reaction on the Si(100)-2 × 1 surface, J. Am. Chem. Soc. 119 (1997) 11098.
[24] R. Konecny, D.J. Doren, Cycloaddition reactions of unsaturated hydrocarbons on the Si(100)-(2 × 1) surface: theoretical predictions, Surf. Sci. 417 (1998) 169.
[25] C. Mui, S.F. Bent, C.B. Musgrave, A theoretical study of the structure and thermochemistry of 1,3-butadiene on the Ge/Si(100)-2 × 1 surface, J. Phys. Chem. 104 (2000) 2457.
[26] A.V. Teplyakov, M.J. Kong, S.F. Bent, Vibrational spectroscopic studies of Diels–Alder reactions with the Si(100)-2 × 1 surface as a dienophile, J. Am. Chem. Soc. 119 (1997) 11100.
[27] A.V. Teplyakov, M.J. Kong, S.F. Bent, Diels–Alder reactions of butadienes with the Si(100)-2 × 1 surface as a dienophile: vibrational spectroscopy, thermal desorption and NEXAFS studies, J. Chem. Phys. 108 (1998) 4599.
[28] M.J. Kong, A.V. Teplyakov, J.G. Lyubovitsky, S.F. Bent, NEXAFS studies of adsorption and reaction of benzene on Si(100)-2 × 1, Surf. Sci. 411 (1998) 286.
[29] G.T. Wang, C. Mui, C.B. Musgrave, S.F. Bent, Cycloaddition of cyclopentadiene and dicyclopentadiene on Si(100)-2 × 1: comparison of monomer and dimer adsorption, J. Phys. Chem. 103 (1999) 6803.
[30] M.J. Kong, A.V. Teplyakov, J. Lyubovitsky, J. Jagmohan, S.F. Bent, Interaction of C6 cyclic hydrocarbons with a Si(100)-2 × 1 surface: adsorption and hydrogenation reactions, J. Phys. Chem. 104 (2000) 3000.
[31] H. Liu, R.J. Hamers, Stereoselectivity in molecule–surface reactions: adsorption of ethylene on the Si(001) surface, J. Am. Chem. Soc. 119 (1997) 7593.
[32] R.J. Hamers, J.S. Hovis, S. Lee, H. Liu, J. Shan, Formation of ordered, anisotropic organic monolayers on the Si(001) surface, J. Phys. Chem. B 101 (1997) 1489.
[33] J.S. Hovis, R.J. Hamers, Structure and bonding of ordered organic monolayers of 1,5-cyclooctadiene on the Si(100) surface, J. Phys. Chem. 101 (1997) 9581.
[34] J.S. Hovis, H. Liu, R.J. Hamers, Cycloaddition chemistry of 1,3-dienes on the silicon(001) surface: competition between [4 + 2] and [2 + 2] reactions, J. Phys. Chem. 102 (1998) 6873.
[35] J.S. Hovis, H. Liu, R.J. Hamers, Cycloaddition chemistry and formation of ordered organic monolayers on silicon(001) surfaces, Surf. Sci. 402 (1998) 1.
[36] J.S. Hovis, R.J. Hamers, Structure and bonding of ordered organic monolayers of 1,3,5,7-cyclooctatetraene on the Si(001) surface: surface cycloaddition chemistry of an antiaromatic molecule, J. Phys. Chem. 102 (1998) 687.
[37] G.P. Lopinski, T.M. Fortier, D.J. Moffatt, R.A. Wolkow, Multiple bonding geometries and binding state conversion of benzene/Si(100), J. Vac. Sci. Technol. A 16 (1998) 1037.
[38] G.P. Lopinski, D.J. Moffatt, D.D.M. Wayner, R.A. Wolkow, Determination of the absolute chirality of individual molecules using the scanning tunneling microscope, Nature 392 (1998) 909.
[39] G.P. Lopinski, D.J. Moffatt, D.D.M. Wayner, M.Z. Zgierski, R.A. Wolkow, Asymmetric induction at a silicon surface, J. Am. Chem. Soc. 121 (1999) 4532.
[40] S. Gokhale, P. Trischberger, D. Menzel, W. Widdra, H. Dröge, H.P. Steinrück, U. Birkenheuer, U. Gutdeutsch, N. Rösch, Electronic structure of benzene adsorbed on single-domain Si(001)-2 × 1: a combined experimental and theoretical study, J. Chem. Phys. 108 (1998) 5554.
[41] B. Borovsky, M. Krueger, E. Ganz, Metastable adsorption of benzene on the Si(001) surface, Phys. Rev. B 57 (1998) 4269.
[42] U. Birkenheuer, U. Gutdeutsch, N. Rosch, Geometrical structure of benzene absorbed on Si(001), Surf. Sci. 409 (1998) 213.
[43] K.W. Self, R.I. Pelzel, J.H.G. Owen, C. Yan, W. Widdra, W.H. Weinberg, Scanning tunneling microscopy study of benzene adsorption on Si(100)-2 × 1, J. Vac. Sci. Technol. A 16 (1998) 1031.
[44] J.W. Lyding, T.C. Shen, G.C. Abeln, C. Wang, P.A. Scott, J.R. Tucker, P. Avouris, R.E. Walkup, Ultrahigh-vacuum scanning tunneling microscope-based nanolithography and selective chemistry on silicon surfaces, Isr. J. Chem. 36 (1996) 3.
[45] G.C. Abeln, S.Y. Lee, J.W. Lyding, D.S. Thompson, J.S. Moore, Nanopatterning organic monolayers on Si(100) by selective chemisorption of norbornadiene, Appl. Phys. Lett. 70 (1997) 2747.
[46] A.J. Mayne, A.R. Avery, J. Knall, T.S. Jones, G.A.D. Briggs, W.H. Weinberg, An STM study of the chemisorption of C_2H_4 on Si(100)(2 × 1), Surf. Sci. 284 (1993) 247.
[47] C. Huang, W. Widdra, W.H. Weinberg, Adsorption of ethylene on the Si(100)-2 × 1 surface, Surf. Sci. 315 (1994) L953.
[48] L. Clemen, R.M. Wallace, P.A. Taylor, M.J. Dresser, W.J. Choyke, W.H. Weinberg, J.T. Yates Jr., Adsorption and thermal behavior of ethylene on Si(100)-(2 × 1), Surf. Sci. 268 (1992) 205.
[49] W. Widdra, C. Huang, S.I. Yi, W.H. Weinberg, Coadsorption of hydrogen with ethylene and acetylene on Si(100)-(2 × 1), J. Chem. Phys. 105 (1996) 5605.
[50] S.H. Xu, M. Keeffe, Y. Yang, C. Chen, M. Yu, G.J. Lapeyre, F. Rotenberg, J. Denlinger, J.T. Yates Jr., Photoelectron diffraction imaging for C_2H_2 and C_2H_4 chemisorbed on Si(100) reveals a new bonding configuration, Phys. Rev. Lett. 84 (2000) 939.

[51] D.C. Sorescu, K.D. Jordan, Theoretical study of the adsorption of acetylene on the Si(100) surface, J. Phys. Chem. B 104 (2000) 8259.

[52] P. Lal, A.V. Teplyakov, Y. Noah, M.J. Kong, G.T. Wang, S.F. Bent, Adsorption of ethylene on the Ge(100)-2 × 1 surface: coverage and time-dependent behavior, J. Chem. Phys. 110 (1999) 10545.

[53] A.V. Teplyakov, P. Lal, Y.A. Noah, S.F. Bent, Evidence for a retro-Diels–Alder reaction on a single crystalline surface: butadienes on Ge(100), J. Am. Chem. Soc. 120 (1998) 7377.

[54] R.J. Hamer, J.S. Hovis, C.M. Greenlief, D.F. Padowitz, Scanning tunneling microscopy of organic molecules and monolayers on silicon and germanium (001) surfaces, Jpn. J. Appl. Phys. 38 (1999) 3879.

[55] S.W. Lee, L.N. Nelen, H. Ihm, T. Scoggins, C.M. Greenlief, Reaction of 1,3-cyclohexadiene with the Ge(100) surface, Surf. Sci. 410 (1998) L773.

[56] G.T. Wang, S.F. Bent, J.N. Russell Jr., J.E. Butler, M.P. D'Evelyn, Functionalization of diamond(100) by Diels–Alder chemistry, J. Am. Chem. Soc. 122 (2000) 744.

[57] J.S. Hovis, S.K. Coulter, R.J. Hamers, M.P. D'Evelyn, J.N. Russell, J.E. Butler, Cycloaddition chemistry at surfaces: reaction of alkenes with the diamond(001)-2 × 1 surface, J. Am. Chem. Soc. 122 (2000) 732.

[58] M.Z. Hossain, T. Aruga, N. Takagi, T. Tsuno, N. Fujimori, T. Ando, M. Nishijima, Diels–Alder reaction on the clean diamond(100) 2 × 1 surface, Jpn. J. Appl. Phys. 38 (1999) L1496.

[59] M.J. Bozack, W.J. Choyke, L. Muehlhoff, J.T. Yates Jr., Reaction chemistry at the Si(100) surface-control through active-site manipulation, J. Appl. Phys. 60 (1986) 3750.

[60] M. Nishijima, J. Yoshinobu, H. Tsuda, M. Onchi, The adsorption and thermal-decomposition of acetylene on Si(100) and vicinal Si(100)9-degrees, Surf. Sci. 192 (1987) 383.

[61] J. Yoshinobu, H. Tsuda, M. Onchi, M. Nishijima, The adsorbed states of ethylene on Si(100)c(4 × 2), Si(100)-(2 × 1), and vicinal Si(100): electron energy loss spectroscopy and low-energy electron diffraction studies, J. Chem. Phys. 87 (1987) 7332.

[62] C.C. Cheng, R.M. Wallace, P.A. Taylor, W.J. Choyke, J.T. Yates Jr., Direct determination of absolute monolayer coverages of chemisorbed C_2H_2 and C_2H_4 on Si(100), J. Appl. Phys. 67 (1990) 3693.

[63] P.A. Taylor, R.M. Wallace, C.C. Cheng, W.H. Weinberg, M.J. Dresser, W.J. Choyke, J.T. Yates Jr., Adsorption and decomposition of acetylene on Si(100)-(2 × 1), J. Am. Chem. Soc. 114 (1992) 6754.

[64] Q. Liu, R. Hoffmann, The bare and acetylene chemisorbed Si(100) surface, and the mechanism of acetylene chemisorption, J. Am. Chem. Soc. 117 (1995) 4082.

[65] B.I. Craig, P.V. Smith, Structures of small hydrocarbons adsorbed on Si(001) and Si terminated Beta-SiC(001), Surf. Sci. 276 (1992) 174.

[66] A.J. Fisher, P.E. Blchl, G.A.D. Briggs, Hydrocarbon Adsorption on Si(001): when does the Si dimer bond break?, Surf. Sci. 374 (1997) 298.

[67] W. Pan, T. Zhu, W. Yang, First-principles study of the structural and electronic properties of ethylene adsorption on Si(100)-(2 × 1) surface, J. Chem. Phys. 107 (1997) 3981.

[68] B.Q. Meng, D. Maroudas, W.H. Weinberg, Structure of chemisorbed acetylene on the Si(001)-(2 × 1) surface and the effect of coadsorbed atomic hydrogen, Chem. Phys. Lett. 278 (1997) 97.

[69] G.P. Lopinski, D.J. Moffatt, D.D.M. Wayner, R.A. Wolkow, How stereoselective are alkene addition reactions on Si(100)?, J. Am. Chem. Soc. 122 (2000) 3548.

[70] L.E. Larson, High-speed Si/SiGe technology for next generation wireless system applications, J. Vac. Sci. Technol. B 16 (1998) 1541.

[71] B.S. Meyerson, Silicon:germanium-based mixed-signal technology for technology for optimization of wired and wireless telecommunications, IBM J. Res. Develop. 44 (2000) 391.

[72] J.E. Field, The Properties of Natural and Synthetic Diamond, Academic Press, London, 1992.

[73] (a) J.A. Kubby, J.E. Griffith, R.S. Becker, J.S. Vickers, Tunneling microscopy of Ge(001), Phys. Rev. B 36 (1987) 6079;
(b) Y.J. Chabal, High resolution infrared spectroscopy of adsorbates on semiconductor surfaces: hydrogen on Si(100) and Ge(100), Surf. Sci. 168 (1986) 594;
(c) X. Torrelles, H.A. van der Vegt, V.H. Etgens, P. Fajardo, J. Alvarez, S. Ferrer, The structure of the Ge(001)-(2 × 1) reconstruction investigated with X-ray diffraction, Surf. Sci. 364 (1996) 242.

[74] (a) B.D. Thoms, J.E. Butler, HREELS and LEED of H/C(100): the 2 × 1 monohydride dimer row reconstruction, Surf. Sci. 328 (1995) 291;
(b) T.W. Mercer, P.E. Pehrsson, Surface state transitions on the reconstructed diamond C(100) surface, Surf. Sci. 399 (1998) L327;
(c) M.P. D'Evelyn, in: Handbook of Industrial Diamonds and Diamond Films, Dekker, New York, 1998.

[75] D.R. Fitzgerald, D.J. Doren, Functionalization of diamond(100) by cycloaddition of butadiene: first-principles theory, J. Am. Chem. Soc. 122 (2000) 12334.

[76] M.R. Linford, C.E.D. Chidsey, Alkyl monolayers covalently bonded to silicon surfaces, J. Am. Chem. Soc. 115 (1993) 12631.

[77] M.R. Linford, P. Fenter, P.M. Eisenberger, C.E.D. Chidsey, Alkyl monolayers on silicon prepared from 1-alkenes and hydrogen-terminated silicon, J. Am. Chem. Soc. 117 (1995) 3145.

[78] J.M. Buriak, M.L. Allen, Lewis acid mediated functionalization of porous silicon with substituted alkenes and alkynes, J. Am. Chem. Soc. 120 (1998) 1339.

[79] J.M. Buriak, M.P. Stewart, T.W. Geders, M.J. Allen, H.C. Choi, J. Smith, D. Raftery, L.T. Canham, Lewis acid mediated hydrosilylation on porous silicon surfaces, J. Am. Chem. Soc. 121 (1999) 11491.

[80] A. Bansal, X. Li, I. Lauermann, N.S. Lewis, S.I. Yi, W.H. Weinberg, Alkylation of Si surfaces using a two-step halogenation/Grignard route, J. Am. Chem. Soc. 118 (1996) 7225.

[81] J.H. Song, M.J. Sailor, Functionalization of nanocrystalline porous silicon surfaces with aryllithium reagents: formation of silicon–carbon bonds by cleavage of silicon–silicon bonds, J. Am. Chem. Soc. 120 (1998) 2376.

[82] N.Y. Kim, P.E. Laibinis, Derivatization of porous silicon by grignard reagents at room temperature, J. Am. Chem. Soc. 120 (1998) 4516.

[83] N.Y. Kim, P.E. Laibinis, Improved polypyrrole/silicon junctions by surfacial modification of hydrogen-terminated silicon using organolithium reagents, J. Am. Chem. Soc. 121 (1999) 7162.

[84] R.L. Cicero, C.E.D. Chidsey, G.P. Lopinski, D.D.M. Wayner, R.A. Wolkow, Olefin additions on H-Si(1 1 1): evidence for a surface chain reaction initiated at isolated dangling bonds, Langmuir, in press.

[85] T.C. Shen, C. Wang, G.C. Abeln, J.R. Tucker, J.W. Lyding, P. Avouris, R.E. Walkup, Atomic-scale desorption through electronic and vibrational excitation mechanisms, Science 268 (1995) 1590.

[86] P. Avouris, R.E. Walkup, A.R. Rossi, H.C. Akpati, P. Nordlander, T.C. Shen, G.C. Abeln, J.W. Lyding, Breaking individual chemical bonds via STM-induced excitations, Surf. Sci. 363 (1996) 368.

[87] G.P. Lopinski, D.D.M. Wayner, R.A. Wolkow, Self-directed growth of molecular nanostructures on Si(1 0 0), Nature 406 (2000) 48.

[88] J. Terry, M.R. Linford, C. Wigren, R. Cao, P. Pianetta, C.E.D. Chidsey, Determination of the bonding of alkyl monolayers to the Si(1 1 1) surface using chemical-shift, scanned-energy photoelectron diffraction, Appl. Phys. Lett. 71 (1997) 1056.

[89] J. Terry, M.R. Linford, C. Wigren, R.Y. Cao, P. Pianetta, C.E.D. Chidsey, Alkyl-terminated Si(1 1 1) surfaces: a high-resolution core-level photoelectron spectroscopy study, J. Appl. Phys. 85 (1999) 213.

[90] M.P. Stewart, J.M. Buriak, Photopatterned hydrosilylation on porous silicon, Angew. Chem. Int. Ed. 37 (1998) 3257.

[91] C.P. Collier, E.W. Wong, M. Belohradsky, F.M. Raymo, J.F. Stoddart, P.J. Kuekes, R.S. Williams, J.R. Heath, Electronically configurable molecular-based logic gates, Science 285 (1999) 391.

[92] S. Zhang, L. Yan, M. Altman, M. Lssle, H. Nugent, F. Frankel, D.A. Lauffenburger, G.M. Whitesides, A. Rich, Biological surface engineering: a simple system for cell pattern formation, Biomaterials 20 (1999) 1213.

[93] Y. Ito, Surface micropatterning to regulate cell functions, Biomaterials 20 (1999) 2333.

[94] P. Clark, S. Britland, P. Connolly, Growth cone guidance and neuron morphology on micropatterned laminim surfaces, J. Cell Sci. 105 (1993) 203.

[95] C.D. James, R. Davis, M. Meyer, A. Turner, S. Turner, G. Withers, L. Kam, G. Banker, H. Craighead, M. Isaacson, J. Turner, W. Shain, Aligned microcontact printing of micrometer-scale poly-L-lysine structures for controlled growth of cultured neurons on planar microelectrode arrays, IEEE Trans. Biomed. Eng. 47 (2000) 17.

[96] T. Bitzer, N.V. Richardson, Route for controlled growth of ultrathin polyimide films with Si–C bonding to Si(1 0 0)-2 × 1, Surf. Sci. 144 (1999) 339.

Surfaces and interfaces in polymer-based electronics

M. Fahlman [a,*], W.R. Salaneck [b]

[a] *Department of Science and Technology, Linköping University, Campus Norrköping, SE-60174 Norrköping, Sweden*
[b] *Department of Physics, Linköping University, SE-58183 Linköping, Sweden*

Received 19 July 2000; accepted for publication 28 June 2001

Abstract

Research on electronics applications such as light-emitting devices for flat-panel displays, transistors, sensors and even solid state lasers based on conducting polymers is presently under way and in some cases has reached the stage of prototype production. The mechanisms for charge injection and conduction in these materials are being studied, as are the physics of luminescence and its quenching. Lately, research into controlling film morphology through self-organizing techniques also has gained interest. Though the present interest in conducting polymers mainly concerns the pristine semiconducting state, doped conducting polymers are also studied for potential use in many applications.

In this paper, we present an overview of some of the central issues in surface and interface science in the field, as well as provide our view on what may lie ahead in the future. Specifically, the importance of metal/polymer, polymer/metal and polymer/polymer interfaces is addressed. We illustrate these using polymer-based light-emitting devices, though the same type of issues appear in other polymer-based applications such as transistors and solar cells. © 2001 Elsevier Science B.V. All rights reserved.

Keywords: Photoelectron spectroscopy; Electroluminescence; Photoluminescence; Heterojunctions; Conductivity; Surface chemical reaction

1. Introduction

The concept of polymers did not achieve general acceptance until the early years of the 20th century. Popularly known as "plastics", polymers are large molecules constructed from smaller structural units (monomers) covalently bonded together in any conceivable pattern. By modifying the monomer building blocks and the bonding scheme, the mechanical and thermal properties of the polymers can be controlled. Polymers can be made 'rubbery' or brittle, soft or hard, soluble or non-soluble, and in practically any color by blending with suitable chromophores. The special properties of polymers allow for a variety of convenient processing, such as injection molding, spin coating and 'spray' painting. This is in striking contrast to the processing of metals and even more so, inorganic semiconductors like silicon. Hence, polymers can be found in almost all products in our present day society: clothes, furniture, home appliances, cars, airplanes, electronics etc. There exist naturally occurring polymers like proteins and rubber, but most polymers in use today are

*Corresponding author. Tel.: +46-11363322; fax: +46-363070.
E-mail address: matfa@itn.liu.se (M. Fahlman).

synthesized, and hence are called synthetic polymers.

Polymers are mainly used for their structural properties, be it the elasticity of plastic food wrap or the hard yet light and easily moldable plastic parts in a car. The idea of utilizing the electrically conducting properties of polymers was not proposed until the 1960's [1], but since then polymers have been used as active components in a variety of electronic applications. For instance, carbon and metal filled polymers are used as moldable semiconductors in the electronic industry and polymers and organic molecules long have served as photoreceptors in electrophotographic copying machines. In this paper, however, we will concentrate on a particular class of electronic polymers, the so-called conductive polymers.

In 1977, it was discovered that an alternating-bond conjugated polymer, *trans*-polyacetylene, could be doped and thereby transformed into a good electrical conductor [2]. This discovery led to the Nobel Prize in Chemistry 2000 being awarded to Profs. A.J. Heeger, A.G. MacDiarmid and H. Shirakawa. Early on, most of the research effort was concentrated to the conducting properties of the doped alternating-bond conjugated polymers [3], and alternating-bond conjugated polymers were often referred to as "conducting polymers". We use this notation throughout the text to describe this class of polymers both when in their pristine and doped forms. Using oriented samples of doped *trans*-polyacetylene, conductivities almost comparable to that of single crystal of copper was reached [4,5]. Interest centered on the prospect of replacing existing conducting materials with the polymers, utilizing the various types of polymer processing to create new types of applications or to make existing ones less expensive. Conducting polymers are used as transparent conductors [6], in conducting textiles [7], and in various anti-static coatings [8]. The mechanism for electron conduction in these materials differs from that of 'normal' metals and is still under intense study [9].

The interest in conducting polymers has gradually expanded to include semiconducting properties of their pristine state. Conducting polymers and oligomers [10] now are used in the development of a variety of organic-based applications: light-emitting devices [11], anti-corrosion coatings [12], photo-voltaic devices [13,14], rechargeable batteries [15], transistors and integrated circuits [16–19], and lasers [20–23]. Here too, the hope is to use the unique processing capabilities of the polymer materials to achieve cost reduction. One can envision manufacturing displays or cheap throw-away electronics by roll-to-roll processing (see Fig. 1) in which the patterning of the circuits/pixels is obtained by ink-jet printing or off-set printing. Compare this with the highly complex and expensive inorganic semiconductor electronics processing and it is easy to see the advantages that the polymer-based technology offers.

Perhaps the most promising application to date is the polymer-based light-emitting device (LED) given the current high interest in developing ultra thin flat-panel displays for computer monitors and television sets. The use of polymers as light-emitting materials was proposed as early as 1969 [24]. Electroluminescence by various organic materials was demonstrated in the following decades [25,26].

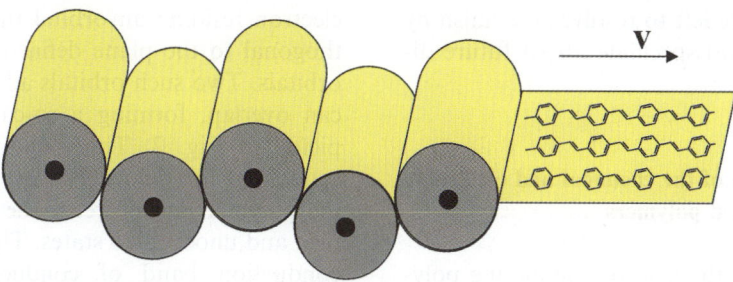

Fig. 1. An idealized picture of roll-to-roll processing of polymer-based electronics. As the flexible substrate (yellow) passes between the rotating cylinders, patterned layers of polymers are printed onto the surface.

In 1990, results on a light-emitting diode with poly(*p*-phenylenevinylene) (PPV), as the emitting layer were published [27]. The field of polymer-based LEDs has been rapidly expanding ever since. Only ten years after the first proof of concept, a number of companies have already developed prototypes of both monochromatic and full color displays that may see market introduction within a year or two from now. In the future, polymer-based emissive displays may well be found in your cell-phone, computer, car and in various home electronics as well as in large screen advertising. Displays that use the reflective properties of these materials may also see use in cheap large area displays for advertising or 'rewritable' newspapers: 'paper' electronics. The electronic and chemical structure of the polymers as well as of polymer/metal interfaces occurring at the electrode contacts will play key roles throughout the development of these technologies. Using self-organizing techniques to define the structure of devices also is likely if cheap and simple production is to be realized. Hence, the interfacial properties of the individual layers in a device, as well as polymer–polymer interface interactions, need to be studied.

In this article we present an overview of some results that already have proven to be important and speculate about the future. First, however, we give a brief introduction to the chemical and electronic properties of conducting polymers, i.e., alternating-bond conjugated polymers. Second, we describe polymer-based light-emitting devices illustrate ways in which surface science has helped solve 'real' problems in the development of this particular application. We then discuss two new applications, organic lasers and so-called electronic ink, and some of their respective surface science issues that are left to resolve. We finish by giving a summary and speculate about future directions in the field.

2. A brief description of the chemical and electronic structure of conjugated polymers

The emergence of the field of conducting polymers came about in a rather inauspicious way. In the early 1970's, a graduate student at the Tokyo Institute of Technology accidentally discovered a new way to synthesize *trans*-polyacetylene, producing free-standing silvery films instead of the usual dark powder. The films were still found to be insulating [28]. However, when exposed to dopants, an increase in conductivity of up to nine orders of magnitude was achieved [2] and a new era in the field of conjugated polymers was born.

What is then the difference between the conducting polymers and the polymers we usually refer to as plastics? Polymers are macromolecules consisting of a great number of repeating units, which are coupled to each other forming a chain. Such (linear) chains can also cross-link and form more complex, three-dimensional structures, which is the case in many commercial polymers, since this improves the mechanical properties of the materials. The repeating units can of course include any group of atoms, but for the vast majority of polymers, including conducting polymers, the repeat units primarily consist of carbons. So, conducting polymers are constructed from the same type of elements as other polymers and are mostly linear (not cross-linked). The conducting polymers discussed in this paper have a bonding pattern consisting of alternating single and double carbon bonds along the backbone of the polymer chain. Conventional organic polymers have four-fold saturated carbons along the backbone of the chain. The four valence electrons available for bond formation reside in orbitals having their major lobes directed in a tetrahedral geometry and form bonds with neighboring atoms in that geometry. In contrast, conducting polymers the carbon atoms have three strongly bonding "σ" orbital lobes residing in a plane at a 120° angle to each other with the fourth remaining valence electron residing an orbital that has its lobes orthogonal to the plane defined by the three other orbitals. Two such orbitals adjacent to each other can overlap, forming a so-called π-bond as depicted in Fig. 2. This bonding scheme induces states that are delocalized over large segments of the polymer and decrease the gap between occupied and unoccupied states. The valence band and conduction band of conducting polymers are generally derived from such π-bonds. The band gap of these polymers tends to lie between 1.5–3 eV, in

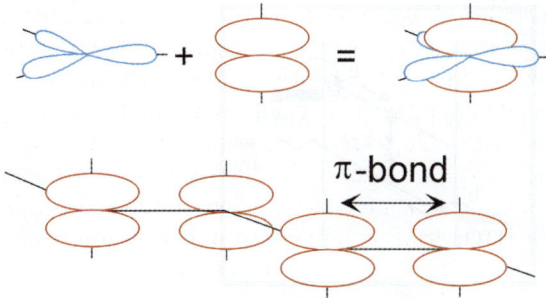

Fig. 2. The atomic orbitals of the four carbon valence electrons in conducting polymers. The red orbital is orthogonal to the plane defined by the three blue orbitals. The overlap of the red orbitals along the backbone create the π-bonds.

the same range as inorganic semiconductors. Since the bonds formed by the σ orbitals are significantly stronger than the π-bonds, the polymers do not break apart when excited states are created in the π-electron levels. Also, in most films, there is only a weak overlap between the π-orbitals between neighboring polymer chains, so electrons and holes tend to be delocalized on individual polymer chains, although they can hop between chains. Thus, these films behave much like inorganic semiconductors.

Like inorganic semiconductors, alternating-bond conjugated polymers can be made conducting by doping. Doping means charge transfer, be it by oxidation (p-type doping) or reduction (n-type doping), as well as the insertion of a counter ion into the vicinity of the polymer. This counter ion helps screen the charge on the polymer chain and ensures charge neutrality for the system as a whole. In conducting polymers, charge transfer doping and optical excitations induce large (local) modifications of both the electronic and the geometric structure of the polymer. This local modification introduces new energy state(s) in the band gap that facilitate the conduction of charge.

Charge transport in conducting polymers differs from that of crystalline semiconductors where electrons move in a periodic potential from the crystal lattice. For a perfect crystal, the electronic states become delocalized over the whole sample and the electron (and hole) transport occurs in energy bands: a concept familiar from solid state physics [29]. In reality, the lattice always have defects, but if the defect density is low, they can be treated as scattering centers that increase the resistivity. In contrast, conducting polymer films are usually highly amorphous, and the conducting polymers themselves tend to include a number of defects as well. Hence, the concept of three-dimensional delocalized electrons and band transport no longer holds since the potential from the lattice is now random, not periodic, which results in localized electronic states. Instead electrons (and holes) are transported through so-called variable range hopping (at least for moderate doping levels), i.e., tunneling between localized states [30].

3. Polymer-based light-emitting devices

The first light-emitting diode (LED) fabricated based on an alternating-bond conjugated polymer was a simple single-layer device consisting of an indium tin oxide (ITO) covered glass substrate as the hole injector, a thin PPV film spun on top as the light-emitting layer, with aluminum as the cathode electron-injecting contact evaporated onto the polymer film [27]. A simple schematic picture of such a device is depicted in Fig. 3. Under an appropriate applied voltage, electrons are injected into the conduction band of the semiconducting polymer from the low work function (aluminum) metal electrode, and holes are injected into the valence band from the high work function (ITO) electrode. Neither the cathode nor the anode contacts are generally ohmic, so the charge injection occurs by tunneling through the barriers at the interfaces. The barriers are not dependent solely on the work functions of the electrodes and the position of the valence and conduction band in the pristine conducting polymers, as discussed later. In conducting polymers, injected charges produce localized bond order deformations, that is, self-localized charge carrying species (known as polarons in non-degenerate ground state polymers). These charges move through the device under the influence of the external field (bias). Polaron-exciton states are generated in the emissive layer when oppositely charged polarons meet and combine. An exciton thus formed may be either singlet state ($S = 0$) or triplet state ($S = 1$). The

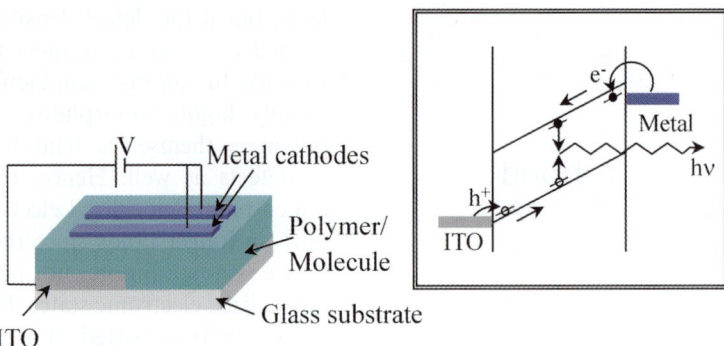

Fig. 3. Schematic picture of a single layer polymer light emitting diode. Injection under bias (V) of electrons (e⁻) and holes (h⁺) is illustrated in the panel to the right. The Fermi levels of the metal cathode and ITO anode are depicted, as are the conduction and valence bands of the polymer. The electron and hole polarons migrate through the film due to the electric field. If electron and hole polarons meet, recombination can occur producing photons ($h\nu$).

spin-allowed radiative emission is from the singlet excitons only. The excitons are localized with the wave function of the singlet exciton extending ∼30 Å in PPV [31]. Due to the localization of the singlet exciton, the resulting light emitted is red shifted in energy from the band gap by the exciton binding energy, calculated to be ∼0.2 eV [31].

The electroluminescence efficiency of early devices based on aluminum electron-injecting electrodes was low. This was due to a wide variety of problems, including impurities in the polymer film and at the contact interfaces, and poor matching of the electrode work functions with the valence and conduction bands of the polymer resulting in high barriers towards charge injection. The efficiency of these devices has improved, however, to more than 20 Lumens/Watt today. Surface science techniques have played an important role in this remarkable improvement, which took ∼30 years for inorganic LEDs. Some highlights of the research involved are noted below.

One of the first issues tackled in the development of polymer LEDs was improving the quality of the polymer material, i.e., improving the photoluminescence efficiencies of the material itself in the hope that this would carry over into improved electroluminescence efficiency in the actual devices. Time-resolved photoluminescence studies showed that upon creation, the light-producing singlet excitons move to the segments on the chains that have the lowest energy, i.e., longest conjugation length, where they relax and emit light [32–34]. The excitons also could dissociate, however, i.e., the electron and hole are separated resulting in non-radiative relaxation. Exciton dissociation usually occurs due to defects on the polymer chain or inter-chain interactions. Hence, to improve the luminescence efficiency of the polymers one must reduce the number of defects on the chains or introduce 'blocking' (non-conjugated) segments on the chains that would prevent the excitons from migrating, reducing the chance that they wind up at a defect site and dissociate. Both approaches have been used.

PPV like most conducting polymers is not soluble in common organic solvents, so substituents that improve solubility must be introduced. In order to make full color displays, the band gap of the conducting polymers must be tuned. By controlling the positions of the valence and conduction band edges, the injection properties also can be improved. So, by using ring substituents in PPV, both the band gap and the positions of the band edges relative to the vacuum level can be controlled while at the same time rendering the polymers soluble. The effects of a variety of ring and backbone substituents have been studied using ultraviolet photoelectron spectroscopy (UPS) combined with quantum chemical calculations [35]. The position of the valence band and its dispersion was determined by UPS and the ring substituents were found to affect the electronic properties in

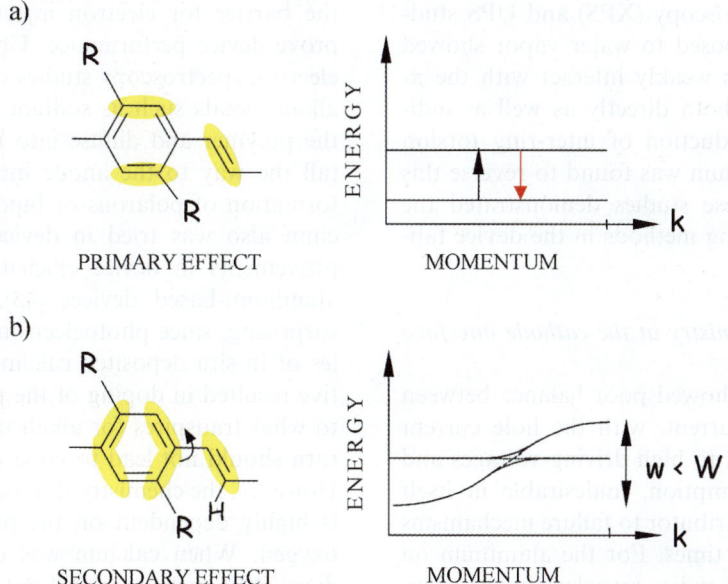

Fig. 4. The (a) primary (substituent) and (b) secondary (steric torsion) effects on the two highest lying π-bands in PPV. The molecular orbitals of the benzene ring and vinyl groups corresponding to the second highest occupied band (a) and highest occupied band (b) are depicted in yellow. R represents a substituent group. When the substituent groups are π-donators, the frontier energy bands are destabilized and move up (black arrow) relative to the vacuum level. If the substituent groups are π-acceptors, the frontier energy bands are stabilized and move down in energy (red arrow). The bandwidth (W) of the can be decreased due to torsion induced by steric hindrance between the substituent groups and the vinyl hydrogen.

two ways: *directly* by either stabilizing (withdrawing charge, red arrow) or destabilizing (donating charge, black arrow) the valence band (see panel (a) of Fig. 4), or *indirectly* by introducing an inter-ring torsion through steric interactions between the ring substituents and the vinyl hydrogen as depicted in panel (b) of Fig. 4. Inter-ring torsion decreases the overlap of the orbitals that are positioned orthogonal to the plane defined by the polymer backbone as discussed earlier (red orbitals in Fig. 2). Hence it decreases the dispersion of the valence and conduction bands. For alkoxy-substituents where the oxygen atom of the chain bonds to the ring, a weak hydrogen bond is formed with the vinyl hydrogen inducing a coplanar conformation instead of inter-ring torsion. This effect is independent on the subsequent chain length of the saturated carbon part of the alkoxy substituents. Hence, it is possible to use long substituent chains to obtain good solubility and film-forming properties while having a near coplanar conformation thanks to the hydrogen bond!

Optical absorption and photoluminescence measurements were performed to study the effects of the modifications on the band gap and luminescence efficiency [36]. The position of the valence and conduction bands also can be controlled through introduction of different conjugated or non-conjugated segments along the backbone [37].

Having achieved high luminescence polymers with good solubility, there were still concerns about the stability of the polymers. Would they degrade during extended use or after long storage? Hence the environmental stability of the polymers was studied. The presence of water and oxygen molecules was found to affect the chemical and electronic properties. Photoluminescence studies showed that photo-oxidation of the polymers causes the formation of carbonyl groups along the backbone, disrupting the conjugation and acting as luminescence quenching sites [38,39]. These carbonyl groups are stable and cannot be removed by heat treatments. Water also has an effect on the electronic structure of the polymers. X-ray photoelectron spectroscopy (XPS) and UPS studies of PPV films exposed to water vapor showed that water molecules weakly interact with the π-electronic structure both directly as well as indirectly through introduction of inter-ring torsion [40]. Heating in vacuum was found to reverse this effect, however. These studies demonstrated the need for encapsulating methods in the device fabrication.

3.1. Physics and chemistry at the cathode interface

The first devices showed poor balance between electron and hole current, with the hole current dominating. This led to high driving voltages and higher power consumption, undesirable in itself and also a likely contributor to failure mechanisms leading to short life times. For the aluminum on PPV interface, XPS studies revealed that the aluminum forms covalent bonds primarily with the vinyl groups, disrupting the conjugation and making the polymer insulating. Since the aluminum atoms diffuse roughly 30 Å into the film, the 'real' interface is a 30 Å insulating aluminum-PPV layer sandwiched between the metallic aluminum and the conducting polymer film, as indicated in Fig. 5, which explains the poor electron injection. In order to improve electron injection, different cathode materials were tried. Photoelectron spectroscopy combined with quantum chemical theory provided some of the guidelines. Low work function metals such as alkali metals should reduce the barrier for electron injection and hence improve device performance. Unfortunately, photoelectron spectroscopy studies on PPV showed that alkali metals such as sodium and rubidium dope the polymer and diffuse into the bulk of the film (all the way to the anode interface), causing the formation of polarons or bipolarons [41,42]. Calcium also was tried in devices and showed improvements in device efficiency as compared to aluminum-based devices [43]. This was at first surprising, since photoelectron spectroscopy studies of in situ deposited calcium on a PPV-derivative resulted in doping of the polymer film similar to what transpires for alkali metals [44], which in turn should not lead to good device performance. However, the chemistry that occurs at the interface is highly dependent on the presence of water or oxygen. When calcium was deposited onto oxidized PPV surfaces, it did not diffuse into the film and dope the polymer chains, but instead formed a thin 'oxide'-layer between the metallic calcium and the PPV surface, resulting in a Schottky-barrier type contact with the polymer film [45]. By preparing a large number of devices at different background pressures of oxygen during deposition, it was found that without the presence of oxygen, the devices would short circuit due to the diffusion of calcium and subsequent doping [46]. Too much oxygen present on the other hand would lead to a too thick oxide layer at the interface reducing the chance of electrons tunneling through it. The 'ideal' vacuum turned out to be roughly equivalent of 10^{-7} mbar, which is typical of the metal evaporation chambers used for device fabrication. This demonstrated that oxides at the cathode contact can stabilize the interface and that it is not necessary to use vacuums of the 10^{-9} to 10^{-10} mbar range, which would make large scale production of the LEDs much more expensive. Studies on other metal/polymer interfaces have shown that depending upon the work function of the metal and the ambient pressure, either doping of the polymer or covalent bonding occurs at the metal-on-polymer interface. The cathode contacts are never abrupt interfaces between two 'pristine' materials. Rather, chemistry always occurs at this interface which has to be taken into account when designing and fabricating the devices.

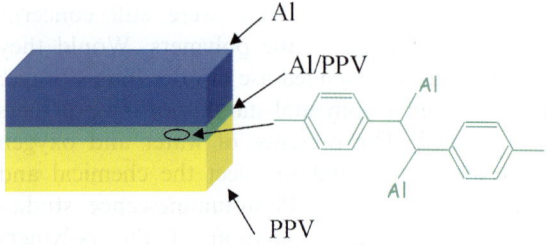

Fig. 5. The Al/PPV interface with a ∼30 Å sandwich layer (green) of insulating/non-conjugated Al-PPV complexes between the metallic aluminum (blue) and the pristine (semiconducting) PPV (yellow). The poor electron mobility in the insulating region will limit the injection current and defects, acting as electron traps, will further reduce the number of charge carriers.

Though calcium cathodes work well in the lab., the fact that they easily oxidize makes them less suitable for commercial devices that have to demonstrate long lifetimes. Aluminum has better environmental stability and other advantages in terms of large scale processing, so it is a better candidate for cathode material in those regards. In order to use aluminum contacts, however, one first must solve the problems of aluminum-induced degradation of the polymer films as well as the high barrier for electron injection. One promising method is the introduction of sandwich structures at the cathode interface. By depositing a thin (5–10 Å) insulating layer on top of the emissive organic film, one can significantly decrease the aluminum-induced degradation as well as lower the energy barrier for electron injection by improving the alignment of the cathode Fermi level with the polymer conduction band, as indicated in Fig. 6. Such a device using LiF as the insulating layer demonstrated remarkable improvements over the normal aluminum cathode LEDs [47]. Other forms of insulating 'tunneling' barriers have been introduced successfully, such as Al_2O_3 [48] and Langmuir–Blodgett films [49]. It was discovered however that even sub-monolayers of LiF improved device performance, which could not be explained by a tunneling barrier mechanism. XPS and UPS studies showed that depending on the emissive material, LiF may dissociate at the interface when sandwiched between the aluminum and the organic film, with the lithium doping the near interface region. The thin n-doped region enhances the electron injection by providing new states closer to the aluminum Fermi level as well as improving electron transport away from the interface due to the improved conductivity, all of which improves the device performance. The lithium does not seem to diffuse into and through the emissive layer unlike the case of alkali metal deposition. This could be due to that when Li atoms are deposited through evaporation, they come in 'hot' (high kinetic energy) when impacting on the polymer/molecular surface, whereas the energy of the dissociated Li is significantly lower. For other polymer materials, LiF does not dissociate in the Al/LiF/polymer structure but serves to protect the polymer from interacting with aluminum. Aluminum chemically bonds to some polymers creating defect sites (disrupting the alternating-bond conjugation) at the interface that can 'trap' injected charge carriers [50]. Other forms of insulating materials are being tried and surface science techniques, photoelectron spectroscopy in particular, are playing a key role in characterizing and optimizing the thin sandwich layers, determining both the chemical compositions of the interface as well as the shifts in the energy band positions and shapes.

What are the future issues of interest concerning electron injection (LEDs, transistors) and electron extraction (solar cells) in polymer-based

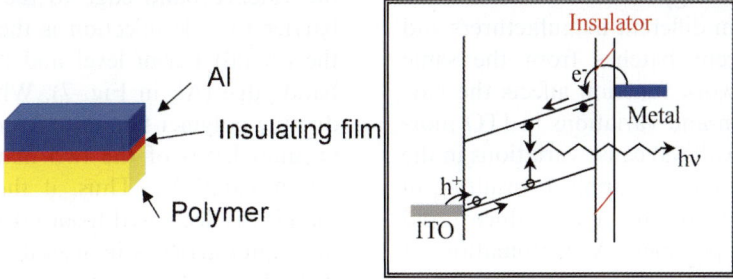

Fig. 6. The effect of a thin (1–10 Å) insulating sandwich layer between a metallic cathode and the emissive polymer film. Injection under bias (V) of electrons (e^-) and holes (h^+) is illustrated in the panel to the right. The large voltage drop over the insulating film under forward bias lowers the conduction band of the polymer so that energy difference (ΔE) between the Fermi level of the metal cathode and the conduction band is decreased. This increases the injection current, which depends on $\sim\exp(-\Delta E/k_B T)$, as long as the insulating layer is thin enough for the electrons to 'tunnel' through. (Quantum mechanics define a probability for an electron to tunnel through a 'forbidden' region (high energy) in order to reach an allowed region.)

electronics? One of the main driving forces in the field is to facilitate roll-to-roll processing of the electronic components, eliminating vacuum and high temperature steps whenever possible. Using surfactant molecules instead of inorganic materials such as alkali halides at the interface between the electrode metal and the emissive polymer to create a dipole improving the alignment of the metal Fermi level with the polymer conduction band and/or serving as a protective barrier is one exiting possibility that would be compatible with the envisioned processing techniques. Two issues would be to get the right type of order of the surfactants on the polymer surface to get the desired dipole and the right thickness to obtain optimal charge injection (too thick and the electrons from the metal contact may not tunnel through, too thin and the metal atoms deposited into the cathode layer may diffuse into the polymer film). Laminating a thin aluminum layer onto an electronic device instead of carrying out a vacuum deposition of the aluminum cathode contact would be another interesting problem to tackle, which would require intense study of the interface.

3.2. Physics and chemistry at the anode interface

The interface at the anode side of polymer-based electroluminescent devices also has been studied intensely. ITO, which traditionally has been used as the transparent hole-injecting contact in polymer light-emitting devices, is not well understood or well controlled. In particular, large variations occur in both morphology and work function between substrates from different manufacturers and even between different batches from the same manufacturer. The work function affects the barrier for hole injection and variations in ITO morphology (peaks and valleys) cause variations in the electric field when voltage is applied, resulting in local areas of high fields that may induce rapid degradation of the polymer. A combination of atomic force microscopy and photoelectron spectroscopy studies showed that the morphology and work function of the ITO surfaces can be controlled to a degree by surface treatments such as ozone cleaning and acid etchings [51]. Device studies on such treated ITO anodes showed that device performance can be significantly improved [52]. Unfortunately, the stability of the polymer-on-ITO interface also was questioned because studies showed that indium diffuses into the polymer film [53,54]. The indium present in the polymer film acts as trap sites for the charge carriers, reducing the luminescence yield. Some studies suggest that in addition oxygen-containing species migrate from the ITO into the polymer film during use [55], forming carbonyl defects in the polymer, also reducing the luminescence.

In order to stabilize the anode interface and prevent/decrease indium and oxygen diffusion into the emissive polymer layer, interfacial layers of (conducting) polymers between the ITO surface and the light-emitting polymer medium have been introduced [56–58]. The conducting layers were expected to act as hole transport layers while 'smoothing' out the rough surface and decreasing indium and oxygen diffusion by acting as a barrier. By studying how the energy levels of the polymer(s) line up at the ITO/conducting-polymer and conducting-polymer/emissive-polymer interfaces, however, it was found that the effect of the interfacial layers were not limited to smoothing the surface and indium/oxygen diffusion reduction. An important consequence of a conducting polymer sandwich layer is that it determines the barrier for hole injection [59]. An energy level diagram depicting this effect is shown in Fig. 7. Any polymer has a specific ionization potential (IP) which is the minimum energy needed to remove an electron from the polymer, i.e., to bring an electron from the valence band edge to the vacuum level. The barrier to hole injection is the difference between the (metal) Fermi level and the polymer valence band edge (ΔE in Fig. 7). When spinning a conducting polymer film onto a metallic substrate, the vacuum levels of the two materials usually align (offset <0.2 eV). Thus, if the work function of the ITO is decreased from ~4.8 to 4.4 eV, the hole injection barrier is increased by the same amount [59]. Hence, the variations in the ITO work function that exist not only from manufacturer to manufacturer, but also between batches of ITO from the same manufacturer, and in the same batch over time, affect the hole injection properties and cause a large variation in device performance.

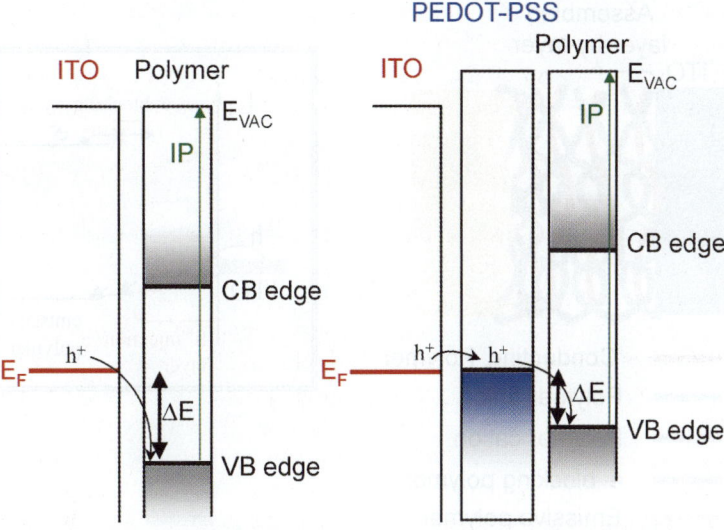

Fig. 7. Energy band diagram for an emissive polymer spun onto ITO (left) and a pristine emissive polymer spun onto a doped conductive polymer (PEDOT-PSS) spun onto ITO (right). The Fermi energy of ITO and the doped polymer, as well as the energy of the valence band (VB edge) and conduction band (CB edge) of the emissive polymer, is depicted in the panels. Holes (h+) are shown being injected into the emissive polymer film. IP is the ionization potential of the emissive polymer. ΔE is the energy difference between the metal Fermi level and the valence band edge of the emissive polymer (i.e., the hole injection barrier). If the doped conducting polymer has a higher work function, i.e., distance between the vacuum level (E_{VAC}) and the Fermi energy (E_F), than ITO, the hole injection barrier is decreased as depicted in the right panel.

Since the conducting polymer film is not affected by a change in the ITO work function (metal/metal interface), however, the emissive polymer spun on top is not affected either since it merely follows the vacuum level of the conducting polymer. The barrier towards hole injection remains unchanged, which in turn permits a better control of device performance, crucial to commercialization.

Another way to improve device performance by modifying the interface at the hole-injecting contact is by using alternating polycation/polyanion (hole transporting) sandwich layers between the ITO and the emissive polymer, as indicated in Fig. 8. Such a sandwich structure is assembled (mono) layer by (mono) layer, which allows for control of the electronic properties of each layer [60,61]. By decreasing the doping level of each individual layer going away from the ITO surface, the IP for each layer was increased creating a graded slope from the ITO Fermi level to the emissive layer's valence band edge [62]. The doping level for each sub-layer was characterized by optical absorption and XPS. Due to the sandwich structure, instead of having one large hole injection barrier to overcome, there are many small barriers, which was found to improve hole injection. In order to enhance the electron–hole recombination, an electron-blocking interface was added between the graded hole injection sandwich structure and the emissive polymer, further improving device performance [62]. This is yet another example on how device performance can be improved and controlled by modifying interfaces and their properties at the nanometric level, i.e., via 'nano-engineering'.

Much of the future research concerning hole injection (or hole extraction in solar cells) will focus on improving the hole-transport layers both in terms of their energy level matching but also their morphology. Even more interesting is the research on how to replace ITO with a conducting

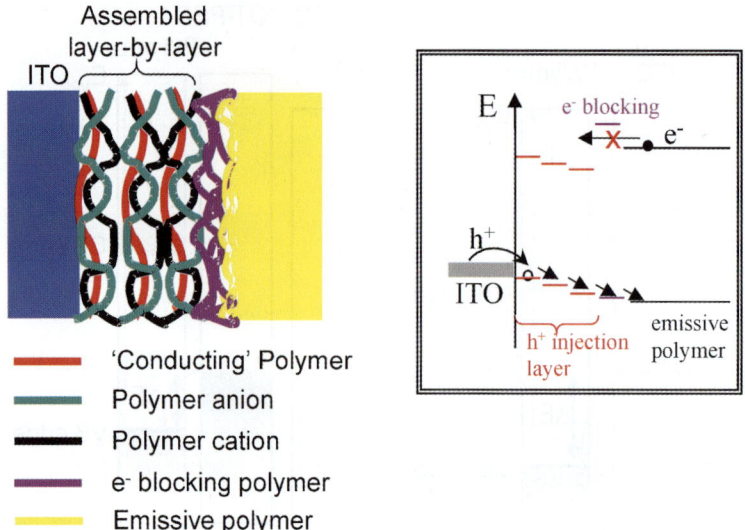

Fig. 8. Illustration of a (graded) doped region (red, green, black) assembled layer-by-layer at the ITO interface with an added electron-blocking layer (purple) on top. The highest level of doping is in the monolayer at the ITO surface, decreasing monolayer by monolayer moving away from the interface. This results in a progressively increasing ionization potential of the individual monolayers. The electron-blocking layer presents a huge barrier to overcome for electrons (e$^-$) coming from the emissive layer due to its low electron affinity, where as the ionization potential of the blocking layer is well matched to the valence band of the emissive polymer, enabling efficient hole (h$^+$) transport.

polymer, in order to facilitate roll-to-roll processing which requires flexible layers. An advantage could be obtained by using the liquid properties of a conducting polymer solution to print the anode contact directly in the desired circuit pattern by for instance ink-jet [63] or screen printing [64] techniques! Understanding the surface interactions between the liquid drops and the substrate to be printed will be crucial to control both line width and resolution of such printed circuits.

Another simple yet powerful patterning technique is soft lithography [65], illustrated in Fig. 9. A stamp 'negative' is created by normal lithography and etching of a silicon substrate. Using this as a template, a polysiloxane rubber is poured into it and cured. By lifting off the rubber from the silicon template a stamp is obtained that can be repeatedly used to create patterns on substrates without the harsh chemical etching or heat treatments that occur when patterning by photolithography! In practice, the rubber stamp is coated with a monolayer of self-assembling molecules. A pattern can then be stamped onto a substrate as depicted in Fig. 9 and by choosing suitable end groups (hydrophobic/hydrophilic) the stamped pattern can determine the organization of a following coating obtained from solution. This technique is not only relatively cheap as compared with photolithography, but its 'mildness' and the fact that it can be carried out in air makes it suitable for patterning of biological materials as well as conducting polymers. Many interesting research topics are found in this patterning process concerning the physical organization of the self assembled monolayer (SAM) on the substrate surface and the organization of the (conducting) polymer layer on the (SAM)/substrate pattern. The chemical interactions that may occur at these interfaces and the charge injection (extraction) properties of the interfaces also must be understood and controlled as well.

The sheet resistance ($\Omega/\square$) of the conducting polymer coatings suitable as anode electrodes is too high at the moment, so new ways of decreasing and controlling the (long term) sheet resistance of the polymer films are needed, defining added challenges for surface scientists in this field. (A transparent conducting polymer that could be

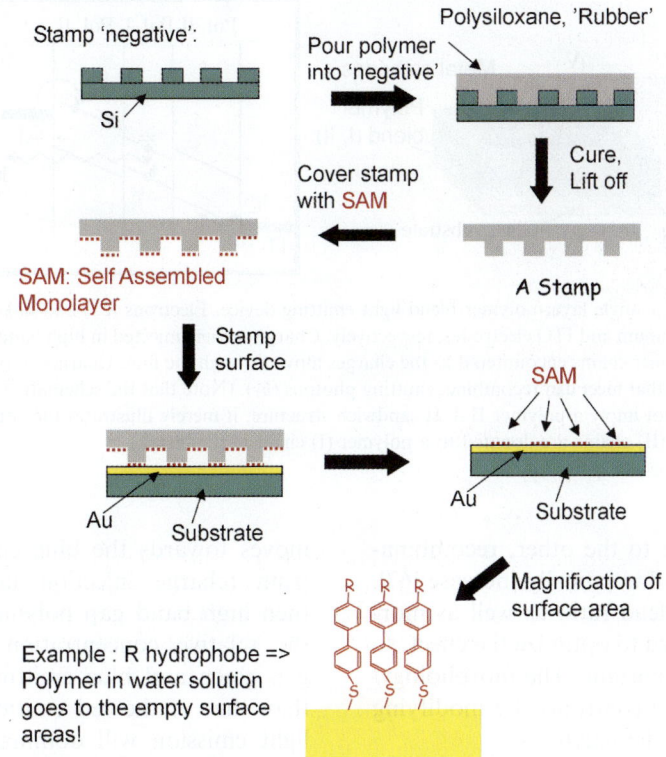

Fig. 9. Schematic illustration of the soft lithography process from the creation of a stamp to the patterning of a substrate. A stamp negative of Si is created by normal lithography and etching techniques. A rubber polymer is poured into the stamp form and cured by heating, where upon it is lifted off. The now hard rubber stamp is then coated with a SAM, and is then pressed onto a gold surface. The resulting pattern defines where a subsequent coating can and cannot contact with the gold surface.

used as the electron-injecting cathode would be desirable for roll-to-roll processing purposes creating an 'all plastic' device, but no such candidate is available at present.)

3.3. Multi-layer devices: polymer/polymer interfaces

An obvious advantage of conducting polymers compared to inorganic semiconductors is that polymers are soluble and can be spin-cast as thin films from solution. Instead of building up thin multi-layers in ultrahigh vacuum, one can make a blend of two or more polymers, spin-cast them as a film in air and have the interfaces already built into that single layer [66]. As mentioned earlier, one of the main problems with polymer LEDs is that charge carriers (holes in particular) may go through the device without recombining and creating a photon. That creates wasted current and hence wasted power. The recombination probability can be enhanced by inserting electron (or hole) blocking layers as discussed earlier, but that increases the thickness of the devices and therefore the driving voltage, plus it adds complexity and cost to the device production. The problem of non-recombining charge carriers can be addressed using blends. By blending two conducting polymers, where one of the polymers (I) has its valence band and conduction band within the other polymers (II) band gap, as illustrated in Fig. 10, the electrons and holes that are transported by polymer (II) will be donated to polymer (I). This occurs because the electron and hole charge carrier states will be lower in energy on polymer (I). Since the charge carriers are collected on the polymer (I) chains in the film instead of running through the

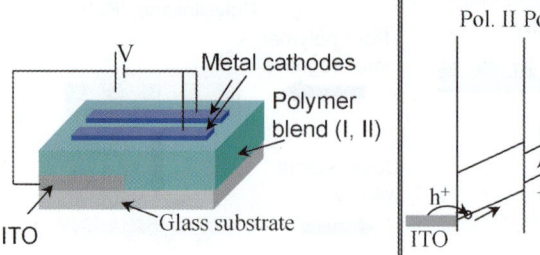

Fig. 10. Schematic picture of a single layer polymer blend light emitting device. Electrons (e⁻) and holes (h⁺) are injected under an applied bias (V) from the aluminum and ITO electrodes, respectively. Charge carriers injected in high band gap polymer chains (II) will transfer to low band gap polymer chains encountered as the charges move through the film. Charges of opposite sign collected on the low band gap polymer chains that meet can recombine, emitting photons (hv). (Note that the schematic valence and conduction band diagram in right panel does not imply a polymer II–I–II sandwich structure, it merely illustrates the situation where holes and electronics injected into polymer (II) chains are donated to a polymer (I) chain in the blend.)

film from one electrode to the other, recombination and light emission dramatically increase [67]. Of course the proper blend ratio as well as morphology must be achieved to optimize the effect, so choice of solvents is important. The morphology/domain size may also be controlled by modifying the ring substituents of the polymers.

In order to make a full color emissive display using inorganic LEDs, one LED for each color is needed for each pixel (the individual points that define a picture on the screen). It would be more desirable to just have one LED for each pixel capable of emitting all necessary colors. By blending conducting polymers with different emission wavelengths and different charge transport characteristics, LEDs can be fabricated with the emission color being controlled by the applied voltage over the device [68]. Phase separation of the different polymers occurs in the blends, causing the formation of domains with different sizes and emission characteristics. For the polymers chosen, the low band gap polymers required lower electric fields for charge injection the high band gap polymers. Also, some of the charge carriers injected onto a high band gap polymers may be transferred to a low band gap polymer if they are in the vicinity of each other as illustrated in Fig. 10. So devices can be fabricated that for low applied voltage emit in the red (charge injection into the low band gap polymers dominates). For higher voltages, the emission progressively moves towards the blue end of the visible spectrum (charge injection into the medium and then high band gap polymers starts to occur). If the relative concentration of the higher band gap green and blue emitting polymers is greater than that of the red emitter, the green and blue light emission will dominate even though emission still occurs from the low band gap polymer, thus allowing for a voltage controlled color emission. Polymer blends also can be used to fabricate sub-micron sized LEDs (by controlling the domain size of the polymer that emits at a certain voltage) as well as white light emitters [69].

The use of polymer blends [14,70] also is of importance for polymer solar cells. Here the object is to separate the charges of an exciton formed by light absorption, into two different polymers. Electrons are then transported to one electrode, holes to the other, thus creating a current. The film morphology, i.e., the mixing of the two polymer/molecular species, dramatically affect device performance. On one hand the interface area between the two components should be as large as possible to increase the chance of charge separation of the excitons. However, in order to facilitate charge transfer through the device, the interactions of polymer chains of the *same* type must also be enhanced. Often a conducting polymer is blended with a fullerene derivative where the fullerene acts as the electron acceptor [71].

Hence, future research into polymer blends probably will continue to focus on the polymer–polymer and polymer–molecular dye interactions that control phase segregation and also surface segregation. For example, even if a proper blend morphology is obtained in the bulk, surface phase segregation leading to an enrichment of the hole-transporting (donor) material at the electron extracting electrode would severely reduce device efficiency.

4. Solid state polymer lasers and electronic ink

Just as having 'plastic' light-emitting diodes has its obvious advantages, having plastic lasers also would open up new possibilities. The benefits of conducting polymer lasers would be that the emission color of the polymers are relatively easy to tune and the fabrication of such lasers should be simpler and cheaper than the inorganic ones if roll-to-roll processing could be realized. Lasing in optically pumped conducting polymer thin films already have been achieved [20–22]. The main goal at present is to develop charge injection laser based on polymers. An organic-molecule-based (tetracene) charge injection laser has been fabricated [72], though this design requires single-crystal films that are not compatible with solution processing, much less roll-to-roll.

The photoluminescence of conducting polymers tend to be shifted relative to the corresponding absorption spectra so that there is only weak overlap with the ground state absorption as indicated in panel (a) of Fig. 11. It may still be a problem with luminescence from the material being reabsorbed thus preventing light from escaping the material. Hence, the use of polymer blends where the optically pumped absorption takes place in a high band gap material with the excitons transferred into a low band gap material where the emission occurs (similar to the polymer blend LED) has been presented [73]. This shifts the luminescence energy further away from the absorption, decreasing the absorption losses. A simple diagram of the absorption and emission processes in a conducting polymer is depicted in the right panel (b) of Fig. 11. The I → II transition and

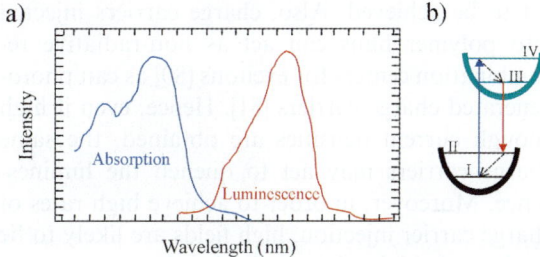

Fig. 11. (a, left panel) Generalized absorption and photoluminescence spectra for conducting polymers and (b, right panel) energy diagram of absorption and luminescence processes in conducting polymers. Overlap between the luminescence and absorption spectrum can cause absorption of the laser photons in the material and hence reduce the light output. The I → II transition and III → II transition represent absorption and emission. II → I and IV → III are phonon-assisted relaxation transitions.

III → II transition represent absorption and emission. If the II → I and IV → III phonon-assisted relaxation transitions are fast as compared to the life time of III, population inversion between levels III and IV can be achieved and this is the case for some conducting polymers that also have very high luminescence yields. Hence, the main requirement for lasing in these materials is satisfied.

Various types of (micro) cavity structures have been used: sandwich structures of a polymer film between a dielectric mirror and a metal [20], between two dielectric mirrors and using dielectric Bragg reflectors [74]. Spherically [75] and cylindrically shaped [76] microcavities also have been used showing high quality factors, i.e., near lossless internal reflections. Cylindrically shaped cavity structures also are easily fabricated. For example, a 'microring' cavity can be obtained simply by dipping an optical fiber in a conducting polymer solution and after a quick drying process, a uniform coating of suitable thickness is produced [77,78].

Achieving charge injection polymer lasers has many challenges. The threshold transient current density necessary for lasing is thought to be in the order of 10^4 A/cm^2. Current densities as high as 10^6 A/cm^2 has been demonstrated for electrically pulsed polymer LEDs [79], but similar values for a polymer and device structure suitable to lasing has yet to be achieved. Also, charge carriers injected into polymer films can act as non-radiative recombination centers for excitons [80] as can photo-generated charge carriers [81]. Hence, even if high enough current densities are obtained, the same charge carriers may act to quench the luminescence. Moreover, in order to achieve high rates of charge carrier injection, high fields are likely to be used, and high electric fields also may cause quenching of the luminescence by dissociating the excitons [82]. Also, due to the low values of mobility of the polymer materials and the often poor balance between the number of electron and hole charge carriers in the polymer films, the highest density of electron–hole pairs are likely to be at the cathode interface, where the probability for non-radiative recombination is highest.

As should be evident from the various problems listed, this particular area is ripe for significant research regarding interface properties. It is likely that many breakthroughs in nano-engineering the interfaces remain to be made before charge injection polymer lasers can be fabricated.

Another exciting application deals with so-called electronic ink. The ink can be printed to a suitable surface such as paper or plastics using normal printing techniques. Displays using electronic inks do not emit light but work in a reflective mode, i.e., just like normal ink on printed poster or in a newspaper. Electronic ink differs from normal ink in that it can be switched off and on using an electric field. This could be switching from white to black or from white to some particular color. The electronic ink should be bi-stable, i.e., once it is been switched from on or off, it should retain its status even after the field is turned off. Using electronic ink, non-emissive displays for electronic newspapers or advertisement boards could be fabricated, more examples of cheap, throw-away electronics. Here too, the surface interactions between the ink and the to-be-printed surface needs to be further researched as well as the interactions between the pigments in the ink. One version of an electronic ink is depicted in Fig. 12. This particular electronic ink is a liquid that contains millions of tiny 'microcapsules', each one containing white particles suspended in a dark dye [83]. When an electric field is applied, the white

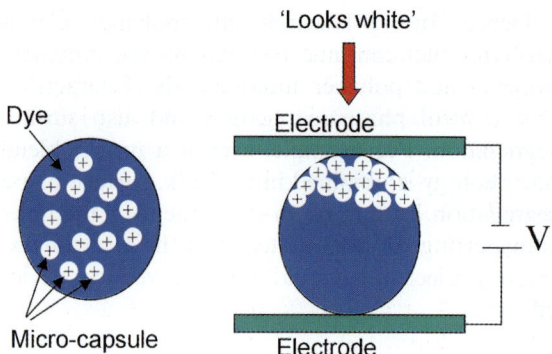

Fig. 12. A microcapsule (blue dye and white, positively charged particles) in the left panel. The right panel depicts a schematic of a device. When a bias (V) is applied, the white particles move to the surface of the microcapsule facing the negatively charged electrode, creating a white spot there.

particles move to the surface of one the sides of the microcapsule (depending on the direction of the field), as indicated in Fig. 12. For an applied field that draws the white particles to the side facing a viewer, the printed surface appears white. An electric field in the opposite direction pulls the particles to the other end of the microcapsule where they are hidden by the dye, making the surface appear dark.

5. Prospects for the future

Plastic electronic devices already are being introduced into the market. Almost all products in the future may have electronic components, and just as we today use electronics such as cell phones to communicate person to person, product to product communication surely will be realized in the not too distant future. Scientists and engineers are already envisioning electronic components such as keyboards and cell phones built into clothes. In the kitchens of tomorrow the produce containers may be able to tell the fridge if the contents are about to be spoiled or if the supply of milk is running short. Family members could call up the fridge and find out if they need to pick something up on their way home from work. One also envisions cheap 'throw-away' electronics on paper or plastics if manufacturing costs could be reduced. For instance, 'smart' barcode stickers that communicate what you bought to a computer as you exit the store, automatically charging the cost to your credit card: no more waiting in lines.

As mentioned earlier, a way to reduce production costs using polymer-based electronics would be roll-to-roll processing where the electronic circuitry is printed onto paper or plastic substrates using techniques such as offset printing or ink jet printing. At paper mills, paper is routinely produced on to rolls at a rate of ~120 km/h, though similar rates are unlikely to be achieved for paper electronics. At any rate, if a 'second' electronic revolution is to take place, a cheaper alternative to silicon must be found, and the alternative semiconductor material has to be easy to recycle or we will wind up with mountains of electronic 'garbage'. Specific research topics of future interest have already been listed previously in the article and below we will summarize some of the more important problems that remain to be solved.

The most mature conducting polymer-based technology is probably light-emitting devices. However, before they can be incorporated into high quality displays that require long life times, the environment-induced degradation of the devices must be reduced. Though the exact mechanism remains to be determined, many devices fail over time probably due to oxidation/corrosion of the cathode metal. Water and/or oxygen are suspected to be the main cause for this. Hence, clever new ways to encapsulate the devices preventing water and oxygen from diffusing to the cathode contact, or new cathode interface structures capable of achieving the required injection properties while using cathode metals that are less susceptible to oxidation, must be developed. The use of inorganic blocking/sealing layers to encapsulate the devices have so far proven to be incompatible with the roll-to-roll processing that is desired (flexing the devices cause cracks in the inorganic barriers layers).

Another area where more research is needed deals with the wetting of polymer solutions on surfaces. The interfaces in commercial devices must not only have the correct electronic properties in terms of energy level alignments, turn on voltages and charge injection rates, but also the correct physical properties to ensure precise control of film thickness as well as narrow line widths of the printed patterns. The latter is especially important for all-polymer transistors and integrated circuits. For instance, in order for offset printing to work at the high printing speeds desired, exact control of the wetting/adhesion on both the surface to be printed as well as the 'press' is needed. The interfacial properties between the ink jet drops and the printed surface also need to be researched to further develop the printing technique. Hence, with both surface treatments of the to-be-printed substrates and polymer solution formulation likely being of crucial importance, there is a huge variety of polymer/polymer and metal/polymer interfaces for the surface scientists to probe and control.

Of course, in almost any electronic application, transistors are needed, and organic transistors mass fabricated by roll-to-roll processing or similar techniques are being developed. Conducting polymers as well as shorter segments of polymers, oligomers, are used. Due to the low mobility, typically $\sim 10^{-2}$ cm^2/Vs or less in non-ordered films, polymer transistors are unlikely to ever compete with inorganic semiconductors in applications were speed is a premium. However, if roll-to-roll processing or a similar cheap fabrication method can be perfected, organic transistors are likely to find their way into a lot of cheap 'throw-away' electronics. For the organic transistors, similar type of problems as for the polymer LEDs remain to be solved. As mentioned, control of film morphology may even be of higher importance for the transistors, since the electron/hole mobility of films in which the polymer chains are aligned parallel to each other is significantly higher than for amorphous films. Narrow line widths for the channels are of course desired as well.

There is a great need for new surface science insights in order to realize charge injection polymer-based lasers. New types of charge-injecting interfaces must be developed so that the threshold current density necessary for lasing is achieved while avoiding quenching of the luminescence. This provides a formidable challenge to the field, a challenge that may not be met, though it would be a major boost to the opto-electronics industry if charge injection polymer thin film lasers were to be realized.

Conductive polymers are of course just one class of organic materials that can be used as active components in electronic devices. Displays based on light-emitting organic molecules have already reached the market in car radios and cell phones, and charge injection lasers have been fabricated in laboratories. C_{60} derivatives are being combined with conducting polymers in photovoltaic cells and one can envision building 'pancake' stacks of graphite-like molecules forming nano-wires. Research into so-called nano-tubes also is likely to continue to be a fruitful field. Magnetic molecules and polymer magnets also have been realized, which could open up whole new areas of applications, such as spintronics. This field is fairly new, however so much fundamental research on the properties of the materials and their interfaces with metal contacts remains to be done. As the technology moves towards nano-structured electronic devices, organic electronic materials are likely to be included in many applications, and the properties of organic films and interfaces should continue to be rich areas for research.

Acknowledgements

Research in Norrköping on interfaces in polymer electronics is supported by grants from the Swedish Research Council for Engineering Sciences (TFR), and the Swedish Natural Science Research Council (NFR). MF is supported by the Swedish Foundation for Strategic Research (SSF) financed Center for Advanced Molecular Materials (CAMM). Research on condensed molecular solids and polymers in Linköping is supported in general by grants from NFR, TFR, the Carl Tryggers Foundation, SSF and SISITOMAS (project no. 0261).

References

[1] H. Naarmann, vol. 1179715, 1197228, 1179716, BASF Corp., Germany, 1963.
[2] C.K. Chiang, C.R. Fincher, Y.W. Park, A.J. Heeger, H. Shirakawa, E.J. Louis, S.C. Gua, A.G. MacDiarmid, Electrical conductivity in doped polyacetylene, Phys. Rev. Lett. 39 (1977) 1098.
[3] T.A. Skotheim, Handbook of Conducting Polymers, vol. I and II, Marcel Dekker, New York, 1986.
[4] H. Naarmann, Synthesis of new conductive polymers, Synth. Met. 17 (1987) 223.
[5] H. Naarmann, N. Theophilou, New process for the production of metal-like, stable polyacetylene, Synth. Met. 22 (1987) 1.
[6] S. Roth, W. Graupner, P. McNeillis, Survey of industrial applications of conducting polymers, Acta Phys. Pol. 87 (1995) 699.
[7] H.H. Kuhn, in: M. Aldissi (Ed.), Intrinsically Conducting Polymers: An emerging technology, vol. 246, Kluwer, Dordrecht, 1993, pp. 25–34.
[8] www.zipperling.de/.
[9] R.S. Kohlman, A.J. Epstein, in: T. Skotheim, J.R. Reynolds, R.L. Elsenbaumer (Eds.), Handbook of Conducting Polymers, Marcel Dekker, New York, 1997, pp. 85–121.
[10] K. Müllen, G. Wegner (Eds.), Electronic Materials: The Oligomer Approach, Wiley-VCH, Weinheim, 1998.
[11] R.H. Friend, R.W. Gymer, A.B. Holmes, E.G.J. Staring, R.N. Marks, C. Taliani, D.D.C. Bradley, D.A. dos Santos, J.L. Brédas, M. Lögdlund, W.R. Salaneck, Conjugated polymer electroluminescence, Nature 397 (1998) 121.
[12] B. Wessling, Corrosion prevention with an organic metal (polyaniline): surface ennobling, passivation, corrosion test results, Mater. Corros. 47 (1996) 439.
[13] N.S. Sariciftci, L. Smilowitz, A.J. Heeger, F. Wudl, Photoinduced electron transfer from a conducting polymer to buckminsterfullerene, Science 258 (1992) 1474.
[14] J.J.M. Halls, C.A. Walsh, N.C. Greenham, E.A. Marseglia, R.H. Friend, S.C. Moratti, A.B. Holmes, Efficient photodiodes from interpenetrating polymer networks, Nature 376 (1995) 498–500.
[15] T. Tatsuma, T. Sotumura, T. Sato, D.A. Buttry, N. Oyama, Dimercaptan-polyaniline cathodes for lithium batteries, J. Electrochem. Soc. 142 (1995) 182–184.
[16] F. Garnier, R. Hajlaoui, A. Yassar, P. Srivastava, All-polymer field-effect transistor realized by printing techniques, Science 265 (1994) 684.
[17] X.C. Li, H. Sirringhaus, F. Garnier, A.B. Holmes, S.C. Moratti, N. Feeder, W. Clegg, S.J. Teat, R.H. Friend, A highly pi-stacked organic semiconductor for thin film transistors based on fused thiophenes, J. Am. Soc. 120 (1998) 2206.
[18] H. Sirringhaus, N. Tessler, R.H. Friend, Integrated optoelectronic devices based on conjugated polymers, Science 280 (1998) 1741.
[19] G.H. Gelinck, T.C.T. Geuns, D.M.d. Leeuw, High-performance all-polymer integrated circuits, Appl. Phys. Lett. 77 (2000) 1487–1489.
[20] N. Tessler, G.J. Denton, R.H. Friend, Lasing of conjugated polymer microcavities, Nature 382 (1996) 695.

[21] S.V. Frolov, M. Ozaki, W. Gellerman, Z.V. Vardeny, K. Yoshino, Mirrorless lasing in conducting polymer poly(2,5-dooctyloxy-p-phenylenevinylene) films, Jpn. J. Appl. Phys. 35 (1996) 1371–1373.

[22] F. Hide, M.A. Diaz-Garcia, B. Schwartz, M.R. Andersson, Q. Pei, A.J. Heeger, Semiconducting polymers: a new class of solid-state laser materials, Science 273 (1996) 1833.

[23] M.A. Diaz-Garcia, F. Hide, B.J. Schwartz, M.R. Andersson, Q.B. Pei, A.J. Heeger, Plastic lasers: semiconducting polymers as a new class of solid-state laser materials, Synth. Met. 84 (1997) 455–462.

[24] J. Dresner, Double injection electroluminescence in anthracene, RCA Rev. 30 (1969) 322.

[25] K. Kaneto, K. Yoshino, K. Kao, Y. Inuishi, Electroluminescence in polyethylene terephthalate, Jpn. J. Appl. Phys. 13 (1974) 1023.

[26] R.H. Partridge, Electroluminescence from polyvinylcarbazole films. I. Carbazole cations, Polymer 24 (1983) 733.

[27] J.H. Burroughes, D.D.C. Bradley, A.R. Brown, R.N. Marks, K. Mackay, R.H. Friend, P.L. Burn, A.B. Holmes, Light emitting diodes based upon conjugated polymers, Nature 347 (1990) 539–541.

[28] T. Ito, H. Shirakawa, S. Ikeda, J. Polym. Sci. Polym. Chem. Ed. 12 (1974) 11–20.

[29] N.W. Ashcroft, N.D. Mermin, Solid State Physics, W.B. Saunders, Philadelphia, 1976.

[30] N.F. Mott, Metal–insulator transitions, second ed., Taylor and Francis, London, 1990.

[31] J. Cornil, D. Beljonne, C.M. Heller, I.H. Campbell, B.K. Laurich, D.L. Smith, J.L. Brédas, Photoluminescence spectra of oligo-paraphenylenevinylenes: a joint theoretical and experimental characterization, Chem. Phys. Lett. 278 (1997) 139–145.

[32] R. Kersting, U. Lemmer, R.F. Mahrt, K. Leo, H. Kurz, H. Bässler, E.O. Göbel, Femtosecond energy relaxation in pi-conjugated polymers, Phys. Rev. Lett. 70 (1993) 3820.

[33] G.R. Hayes, I.D.W. Samuel, R.T. Philipps, Exciton dynamics in electroluminescent polymers studied by femtosecond time-resolved photoluminescence spectroscopy, Phys. Rev. B 52 (1995) 11569.

[34] W.R. Salaneck, S. Stafström, J.L. Brédas, Conjugated Polymer Surfaces and Interfaces, Cambridge University Press, Cambridge, 1996.

[35] M. Fahlman, O. Lhost, F. Meyers, J.L. Brédas, S.C. Graham, R.H. Friend, P.L. Burn, A.B. Holmes, K. Kaeriyama, Y. Sonoda, M. Lögdlund, S. Stafström, W.R. Salaneck, Experimental and theoretical studies of the electronic structure of substituted and unsubstituted poly(p-phenylenevinylene) (PPV), Macromolecules 28 (1995) 1959.

[36] H.S. Woo, S.C. Graham, D.A. Halliday, D.D.C. Bradley, R.H. Friend, P.L. Burn, A.B. Holmes, Photoinduced absorption and photoluminescence in poly(2,5-dimethoxy-p-phenylene vinylene), Phys. Rev. B 46 (1992) 7379.

[37] F. Meyers, A.J. Heeger, J.L. Brédas, Fine tuning of the band gap in conjugated polymers via control of block copolymer sequences, J. Chem. Phys. 97 (1992) 2750.

[38] H. Antoniadis, L.J. Rothberg, F. Papadimitrakopoulos, M. Yan, M.E. Galvin, M.A. Abkowitz, Enhanced carrier photogeneration by defects in conjugated polymers and its mechanism, Phys. Rev. B 50 (1994) 14911.

[39] N.T. Harrison, G.R. Hayes, R.T. Phillips, R.H. Friend, Singlet intrachain exciton generation and decay in poly-(p-phenylene vinylene), Phys. Rev. Lett. 77 (1996) 1881–1884.

[40] K. Xing, M. Fahlman, M. Lögdlund, D.A. dos Santos, V. Parente, R. Lazzaroni, J.L. Brédas, R. Gymer, W.R. Salaneck, The interaction of poly(p-phenylenevinylene) with air, Adv. Mater. 8 (1996) 971.

[41] M. Fahlman, D. Beljonne, M. Lögdlund, P.L. Burn, A.B. Holmes, R.H. Friend, J.L. Brédas, W.R. Salaneck, Experimental and theoretical studies of the electronic structure of Na-doped poly(p-phenylenevinylene), Chem. Phys. Lett. 214 (1993) 327.

[42] G. Iucci, K. Xing, M. Lögdlund, M. Fahlman, W.R. Salaneck, Polaron to bipolaron transition in a conjugated polymer: Rubidium-doped poly(p-phenylenevinylene), Chem. Phys. Lett. 244 (1995) 139.

[43] D. Braun, A.J. Heeger, Visible light emission from semiconducting polymer diodes, Appl. Phys. Lett. 58 (1991) 1982.

[44] P. Dannetun, M. Fahlman, C. Fauquet, K. Kaerijama, Y. Sonoda, R. Lazzaroni, J.L. Brédas, W.R. Salaneck, in: J.L. Brédas, W.R. Salaneck, G. Wegner (Eds.), Organic Materials, North Holland, Amsterdam, 1994, p. 113.

[45] E. Ettedgui, H. Razafitrimo, Y. Gao, B.R. Hsieh, Schottky barrier formation at the Ca/Poly(p-phenylenevinylene) interface and its role in tunneling at the interface, Synth. Met. 78 (1996) 247.

[46] P. Bröms, J. Birgersson, N. Johnsson, M. Lögdlund, W.R. Salaneck, Calcium electrodes in polymer LED's, Synth. Met. 74 (1995) 179.

[47] L.S. Hung, C.W. Tang, M.G. Mason, Enhanced electron injection in organic electroluminescence, Appl. Phys. Lett. 70 (1997) 152.

[48] H. Tang, J. Shinar, Bright high efficiency blue organic LEDs with Al2O3/Al cathodes, Appl. Phys. Lett. 71 (1997) 2560.

[49] Y.-E. Kim, H. Park, J.-J. Kim, Enhanced quantum efficiency in polymer EL devices by inserting a tunneling barrier formed by L-B films, Appl. Phys. Lett. 69 (1996) 599.

[50] G. Greczynski, M. Fahlman, W.R. Salaneck, An experimental study of poly(9,9-dioctyl-fluorene) and its interfaces with Li, Al and LiF, J. Chem. Phys. 113 (2000) 2407–2412.

[51] P. Bröms, Th. Kugler, W.R. Salaneck, in: P. Bröms, Polymer Light Emitting Devices, Ph.D. dissertation No. 570, Linköping University, Linköping, 1999, pp. 89–105.

[52] J.S. Kim, M. Granstrom, R.H. Friend, N. Johansson, W.R. Salaneck, R. Daik, W.J. Feast, F. Cacialli, Indium-tin oxide treatments for single-and double-layer polymeric light-emitting diodes: The relation between the anode physical, chemical, and morphological properties and the device performance, J. Appl. Phys. 84 (1998) 6859.

[53] G. Sauer, M. Kilo, M. Hund, A. Wokaun, S. Karg, M. Meier, W. Riess, M. Schwoerer, H. Suzuki, J. Simmerer, H. Meyer, D. Haarer, Characterization of polymeric light emitting diodes by SIMS depth profiling analysis, Fresenius J. Anal. Chem. 353 (1995) 5–8.

[54] A.R. Schlatmann, D.W. Floet, A. Hilberer, F. Garten, P.J.M. Smulders, T.M. Klapwijk, G. Hadziioannou, Indium contamination from the indium-tin-oxide electrode in polymer light-emitting diodes, Appl. Phys. Lett. 69 (1996) 1764–1766.

[55] J.C. Scott, J.H. Kaufman, P.J. Brock, R. DiPietro, J. Salem, J.A. Gotia, Degradation and failure of MEH-PPV light-emitting diodes, J. Appl. Phys. 79 (1996) 2745.

[56] S.A. Carter, M. Angelopoulos, S. Karg, P.J. Brock, J.C. Scott, Polymeric anodes for improved polymer light-emitting diode performance, Appl. Phys. Lett. 70 (1997) 2067.

[57] Y. Cao, G. Yu, C. Zhang, R. Menon, A.J. Heeger, Polymer light-emitting diodes with polyethylene dioxythiophene-polystyrene sulfonate as the transparent material, Synth. Met. 87 (1997) 171.

[58] J.C. Carter, I. Grizzi, S.K. Heeks, D.J. Lacey, S.G. Latham, P.G. May, O.R.d.l. Panos, K. Pichler, C.R. Towns, H.F. Wittman, Operating stability of light-emitting polymer diodes based on poly(p-phenylene vinylene), Appl. Phys. Lett. 71 (1997) 34–36.

[59] T. Kugler, W.R. Salaneck, H. Rost, A.B. Holmes, Polymer band alignment at the interface with indium tin oxide: Consequences for light emitting devices, Chem. Phys. Lett. 310 (1999) 391.

[60] G. Decher, Fuzzy nanoassemblies: toward layered polymeric multicomposites, Science 277 (1997) 1232.

[61] O. Onitsuka, A.C. Fou, M. Ferreira, B.R. Hsieh, M.F. Rubner, Enhancement of light emitting diodes based on self-assembled heterostructures of poly(p-phenylene vinylene), J. Appl. Phys. 80 (1996) 4067.

[62] P.K.H. Ho, J.-S. Kim, J.H. Burroughes, H. Becker, S.F.Y. Li, T.M. Brown, F. Cacialli, R.H. Friend, Molecular-scale interface engineering for polymer light-emitting diodes, Nature 404 (2000) 481.

[63] H. Sirringhaus, T. Kawase, R.H. Friend, T. Shimoda, M. Inbasekaran, W. Wu, E.P. Woo, High resolution inkjet printing of all-polymer transistor circuits, Science 290 (2000) 2123.

[64] D.A. Pardo, G.E. Jabbour, N. Peyghambarian, Application of screen printing in the fabrication, Adv. Mat. 12 (2000) 1249–1252.

[65] S. Brittain, K. Paul, M.Z. Xiao, G. Whitesides, Soft lithography and microfabrication, Phys. World 11 (1998) 31–36.

[66] C. Zhang, H. von Seggern, Pakbaz-K, B. Kraabel, H.-W. Schmidt, A.J. Heeger, Blue electroluminescent diodes using blends of poly(p-phenylene vinylene) in poly(9-vinylcarbazole), Synth. Met. 62 (1994) 35.

[67] J. Birgerson, K. Kaeriyama, P. Barta, P. Bröms, M. Fahlman, T. Granlund, W.R. Salaneck, Efficient blue-light emitting devices from conjugated polymer blends, Adv. Mat. 8 (1996) 982.

[68] M. Berggren, O. Inganäs, C. Gustafsson, J. Rasmusson, M.R. Andersson, T. Hjertberg, O. Wennerström, Light-emitting diodes with variable colours from polymer blends, Nature 372 (1994) 444.

[69] M. Granström, M. Berggren, O. Inganäs, M.R. Andersson, T. Hjertberg, O. Wennerström, Phase separation of conjugated polymers—tools for new functions in polymer LEDs, Synth. Met. 85 (1997) 1193.

[70] J.J. Dittmer, E.A. Marseglia, R.H. Friend, Electron trapping in dye/polymer blend photovoltaic cells, Adv. Mat. 12 (2000) 1270–1274.

[71] S.E. Shaheen, C.J. Brabec, N.S. Sariciftci, F. Padinger, T. Fromherz, J.C. Hummelen, 2.5% efficient organic plastic solar cells, Appl. Phys. Lett. 78 (2001) 841–843.

[72] H. Schon, C. Kloc, A. Dodabalapur, B. Batlogg, An organic solid state injection laser, Science 289 (2000) 599–601.

[73] M. Berggren, A. Dodabalapur, R.E. Slusher, Z. Bao, Organic lasers based on Förster transfer, Synth. Met. 91 (1997) 65.

[74] M. Berggren, A. Dodabalapur, R. Slusher, Z. Bao, Light amplification in organic thin films using cascade energy transfer, Nature 389 (1997) 466.

[75] M. Berggren, A. Dodabalapur, R. Slusher, Z. Bao, Solid-state droplet laser made from an organic blend with a conjugated polymer emitter, Adv. Mat. 9 (1997) 968.

[76] S.V. Frolov, M. Shkunov, A. Fujii, K. Yoshino, Z.V. Vardeny, Lasing and stimulated emission in pi-conjugated polymers, IEEE J. Quant. Elec. 36 (2000) 2.

[77] S.V. Frolov, Z.V. Vardeny, K. Yoshino, Plastic microring lasers on fibers and wires, Appl. Phys. Lett. 72 (1998) 1802.

[78] S.V. Frolov, A. Fujii, Z. Chinn, Z.V. Vardeny, K. Yoshino, R.V. Gregory, Cylindrical microlasers and light-emitting devices from conducting polymers, Appl. Phys. Lett. 72 (1998) 2811.

[79] D.G. Lidzey, D.D.C. Bradley, S.F. Alvarado, P.F. Seidler, Electroluminescence in polymer films, Nature 386 (1997) 135.

[80] P. Dyreklev, O. Inganäs, J. Paloheimo, H. Stubb, Photoluminescence quenching in a polymer thin-film field-effect luministor, J. Appl. Phys. 71 (1992) 2816.

[81] D.D.C. Bradley, R.H. Friend, Light-induced luminescence quenching in precursor-route poly(p-phenylene vinylene), J. Phys. Condens. Matter 1 (1989) 3671.

[82] D. Moses, H. Okumoto, C.H. Lee, A.J. Heeger, T. Ohnishi, T. Noguchi, Mechanism of carrier generation in poly(phenylene vinylene): Transient photoconductivity and photoluminescence at high electric fields, Phys. Rev. B 54 (1996) 4748.

[83] http:www.eink.com.

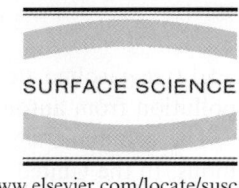

Role of surface science in catalysis

John H. Sinfelt *

Senior Scientific Advisor Emeritus, ExxonMobil Research and Engineering Company, P.O. Box 364, Oldwick, NJ 08858, USA

Received 30 June 2000; accepted for publication 16 April 2001

Abstract

Around the time of World War I, Langmuir advanced a simple theory of chemisorption and showed how it could be used to formulate rate laws for reactions occurring on surfaces. From that time on, surface science has played an important role in heterogeneous catalysis. Between the two world wars, simple studies of extents of adsorption by catalyst surfaces led to the concept of activated adsorption and to a universally used method for determining the high surface areas associated with the pore structures of catalytic materials. After World War II, the application of various spectroscopic and structural probes made it possible to investigate catalyst surfaces at a more microscopic level. Studies with idealized surfaces such as the faces of single crystals in ultra-high vacuum apparatus also made their appearance. By the end of the twentieth century, direct information was being obtained on the rates of elementary reactions of well-defined surface species. The results of such work are beginning to put "finishing touches" on the great insight of early pioneers in surface science and heterogeneous catalysis. Much has been accomplished, but exciting opportunities still remain. © 2001 Elsevier Science B.V. All rights reserved.

Keywords: Catalysis; Chemisorption; Adsorption isotherms; Adsorption kinetics; Physical adsorption; Surface chemical reaction; Clusters; Single crystal surfaces

1. Introduction

A catalyst is an entity which accelerates a chemical reaction without being consumed itself in the process. Without catalysts, various chemical reactions of great importance would proceed so slowly that they could not even be detected, although the reaction conditions (temperature and pressure) are thermodynamically favorable for the occurrence of the reactions [1–5]. Suitable catalysts provide a solution to this problem. They make it possible for the reactions to proceed at rates high enough to permit their commercial exploitation on a large scale, with resulting economic benefits for everyone.

Catalytic phenomena affect virtually all aspects of our lives. They are crucial in many processes occurring in living things, where enzymes are the catalysts. They are important in the processing of foods and the production of medicines, in the refining of petroleum and the manufacture of synthetic fibers and plastics, and in the production of many different chemicals with all kinds of uses [6–8]. Moreover, catalysts play an important role in the preservation of our environment, as demonstrated by their vital contributions in making

---

* Tel.: +1-908-439-3603; fax: +1-908-730-3301.

lead-free gasoline a reality [9,10] and in decreasing pollution from automobile exhaust gases [11].

The economic significance of catalysis is enormous. In the United States alone, the annual value of products manufactured with the use of catalysts is roughly in the vicinity of one trillion dollars [12]. This amazing success story developed steadily during the course of the twentieth century [13]. Although technological progress has generally been ahead of our knowledge of the underlying science during this entire period, it would be a mistake to conclude that science has played little or no role in the development of catalysis. Major technological advances in catalysis have generally stimulated efforts to understand the scientific issues underlying the technology. As the science has developed, it has frequently been important in shaping further advances of the technology. Sometimes the basic science efforts have led to the discovery of new catalysts that have improved the technology dramatically [14].

Surface science has been associated with catalysis for a long time. The association is not with catalysis in general, but rather with the particular branch of the subject known as heterogeneous catalysis. The adjective heterogeneous refers to the fact that the reactants are present in one phase and the catalyst in another, with the catalytic action occurring at the interface or surface between them. In the branch of the subject known as homogeneous catalysis, by contrast, the reactant molecules and the catalyst are present in a single phase, as in a liquid solution. Since the reaction of interest in this case occurs within the phase, rather than being confined to the surface, homogeneous catalysis is outside the scope of the present paper.

In examining the role of surface science in heterogeneous catalysis, we divide the discussion into three main parts. First we attempt to impart some feeling for why the topic is important. We do this by reviewing a few of the technological highlights in heterogeneous catalysis, along with key aspects of the science underlying some of them. Second, we present a number of examples illustrating the connection between surface science and heterogeneous catalysis. Finally, we attempt to make some projections about the possible role of surface science in catalysis in the future.

2. Technological highlights in heterogeneous catalysis

The large-scale industrial application of heterogeneous catalysis had its beginning in the latter part of the nineteenth century, although the awareness of the phenomenon of catalysis was already very old at that time. It is of interest to note here that the science of physical chemistry, including the earliest developments in the areas of chemical thermodynamics and chemical kinetics, also had its beginning in the latter part of the nineteenth century. Scientists such as van't Hoff, Arrhenius, Nernst, and Ostwald were prominently identified with the emergence of physical chemistry at that time. The new discipline played an important role in some of the early technological applications of catalysis.

2.1. Chemical industry

In the 1880s and 1890s a German chemical company, Badische Anilin und Soda-Fabrik, developed a heterogeneous catalytic process (the so-called "contact process") for oxidizing sulfur dioxide to sulfur trioxide [15]:

$$2SO_2 + O_2 \rightleftarrows 2SO_3$$

This reaction is the key step in the commercial manufacture of sulfuric acid (H_2SO_4). The sulfur dioxide used for the reaction is obtained by burning elemental sulfur or iron pyrites (FeS_2) mined from naturally occurring deposits in the earth's crust. In the original Badische process, the reaction was carried out over a platinum catalyst. However, the susceptibility of the catalyst to poisoning by impurities in the reaction gases led eventually to its replacement by a vanadium pentoxide catalyst. The sulfur trioxide formed in the reaction is converted to sulfuric acid by reaction with water:

$$SO_3 + H_2O \rightarrow H_2SO_4$$

Sulfuric acid is one of the most important industrial chemicals. It is widely used within the chemical industry, and in other industries as well. For example, it finds extensive application in the metallurgical industry, in batteries, and in the manufacture of fertilizers, dyes, drugs, pigments, explosives, and textiles [16]. The uses of sulfuric acid are so wide ranging that its annual production has frequently been considered to be a reasonable index of industrial prosperity. In a recent year the world-wide production of sulfuric acid was approximately 145,000,000 metric tons [17]. The role of heterogeneous catalysis in the production of this critically important basic chemical has been a decisive one for more than a century.

Not long after the development of the contact process for sulfuric acid manufacture, two other enormously important processes based on heterogeneous catalysis were introduced into the chemical industry: (1) the process resulting from the pioneering research of Fritz Haber for the production of ammonia from elemental nitrogen and hydrogen [18], and (2) the Ostwald process for the oxidation of ammonia to nitric oxide in the manufacture of nitric acid [19]. The second of these processes bears the name of the scientist most closely associated with the recognition of catalysis as a phenomenon in chemical kinetics [2], the crucial insight that provided a foundation for systematic scientific inquiry into the subject.

The heterogeneous catalytic processes for the synthesis and oxidation of ammonia are the important steps in the production of various fertilizers, which in turn increase the output of agriculture and thereby enhance the world's food supply [20]. The nitric acid obtained via ammonia oxidation, in addition to being used to produce nitrates for fertilizers, also finds use in the manufacture of explosives such as trinitrotoluene and nitroglycerine [21]. In a very different kind of application of nitroglycerine, it is used as a vasodilator in the treatment of heart disease. Total world-wide production of ammonia in a recent year was about 120,000,000 metric tons [22]. Nitric acid is produced on a similar large scale.

The chemical reaction in the Haber process is represented by the equation:

$$N_2 + 3H_2 \rightleftarrows 2NH_3$$

The elemental nitrogen and hydrogen for the reaction are obtained from air, steam, and natural gas. The technology for accomplishing this is described in more detailed treatments of ammonia synthesis [20]. To obtain a sufficiently high rate of formation of ammonia in the synthesis reaction, it is common practice to use an iron catalyst at temperatures in the vicinity of 700 K. However, since the reaction is exothermic, the high temperature is undesirable from the standpoint of thermodynamics. To compensate for the adverse effect of the high temperature on the equilibrium yield of ammonia, pressures of 200 atm or higher are employed to drive the reaction further in the direction of increased ammonia formation.

The implementation of ammonia synthesis on a commercial scale was due to Badische Anilin und Soda Fabrik [20], the same company that developed the contact process for sulfuric acid manufacture. Two scientists at that company, Carl Bosch and Alwin Mittasch, played important roles in the development of the commercial process based on Haber's initial work. The work of Mittasch led to a commercially satisfactory iron catalyst that contained small amounts of potassium oxide and aluminum oxide as promoters [23]. The promoters increased the catalytic activity of the iron and rendered it more resistant to loss of surface area by sintering. Bosch introduced new ideas in the design of suitable equipment for conducting the reaction at high pressure on an industrial scale [20]. Important contributions were also made in the development of satisfactory procedures for the preparation and purification of the reactant gases. The first industrial plant began operations in Germany shortly before the beginning of World War I [20]. The success of this pioneering venture led to the widespread adoption of the Haber process in chemical plants throughout the world. From the standpoint of both science and technology, ammonia synthesis ranks among the all-time great achievements in the history of catalysis.

In the Ostwald process for the production of nitric acid [19], the ammonia oxidation step is represented by the stoichiometric equation:

$$4NH_3 + 5O_2 \rightarrow 4NO + 6H_2O$$

Platinum or a platinum–rhodium alloy in the form of a gauze is used as the catalyst for the reaction, which is generally conducted with oxygen in excess at temperatures of approximately 1100–1200 K [19]. The exit gas from the reactor is cooled and the nitric oxide undergoes oxidation to nitrogen dioxide:

$2NO + O_2 \rightleftarrows 2NO_2$

Contact of the nitrogen dioxide with water in an absorption tower produces the desired nitric acid:

$3NO_2 + H_2O \rightarrow 2HNO_3 + NO$

The nitric oxide formed as a side product in the reaction is recycled in the process so that it can be reoxidized to the dioxide.

The large-scale industrial application of ammonia oxidation was initiated in Germany in the early stages of World War I to provide a source of nitric acid for explosives and munitions. The country was cut off from an external source of nitrates by a blockade, and therefore had to develop the technology to maintain its war effort. By the end of the war, ammonia oxidation plants were also in operation in the United States and a number of other countries.

Thus, over a period of about three decades extending from the latter years of the nineteenth century through the first two decades of the twentieth century, heterogeneous catalysis became firmly entrenched in the chemical industry through the development of processes for the manufacture of sulfuric acid, ammonia, and nitric acid on a large scale. These processes are still widely employed in the industry. During the twentieth century, many other processes based upon heterogeneous catalysis also evolved in the chemical industry. A few of these include: (a) the high pressure synthesis of methanol from carbon monoxide and hydrogen [24], (b) the oxidation of ethylene to ethylene oxide [25] for manufacture of ethylene glycol anti-freeze, (c) the dehydrogenation of butene to butadiene [26] for production of synthetic rubber, (d) the reaction of propylene with oxygen and ammonia to yield acrylonitrile [27] for production of polymeric materials such as Orlon, and (e) the polymerization of ethylene and propylene with Ziegler-type catalysts to produce highly crystalline polyethylene and polypropylene [28]. These examples of applications of heterogeneous catalysis in the chemical industry constitute only a very small number of the many that could be cited.

2.2. Petroleum industry

The first large-scale applications of heterogeneous catalysis in the petroleum industry lagged significantly behind those in the chemical industry. Thus, when the first catalytic cracking process in petroleum refining was introduced by Houdry in 1936 [29], the contact process for manufacturing sulfuric acid had already been used commercially for about four decades.

In catalytic cracking, large hydrocarbon molecules are decomposed into smaller ones by scission of carbon–carbon bonds at temperatures of 700–800 K with a suitable catalyst. A primary objective is the production of hydrocarbons in the gasoline boiling range, i.e., hydrocarbon molecules containing five to ten carbon atoms. An example of a reaction that could occur is the decomposition of diphenylethane to benzene and styrene:

$C_6H_5-CH_2-CH_2-C_6H_5$
$\rightarrow C_6H_6 + CH_2=CH-C_6H_5$

Another is the decomposition of hexadecane molecules by multiple rupture of C–C bonds. The products from the cracking of any given hexadecane molecule depend on which C–C bonds are broken, but there is a strong preference for "cracking patterns" yielding primary products containing three to six carbon atoms [29]:

$C_{16}H_{34} \rightarrow C_3, C_4, C_5, C_6$ hydrocarbons

Catalytic cracking increases greatly the amount of gasoline that can be obtained from a barrel of crude oil. Hydrocarbons in the gasoline boiling range occurring naturally in crude oil, constituting so-called "straight-run gasoline", amount to only about $20 \pm 5\%$ of the total. But, apart from increasing the amount of gasoline obtained from crude oil, catalytic cracking yields gasoline of better quality than straight-run gasoline. The quality is also better than that of gasoline obtained by simply cracking large hydrocarbon molecules thermally. In particular, the gasolines produced in catalytic cracking have higher "anti-knock" ratings, which are commonly represented by octane numbers. The higher octane numbers are due to higher ratios of branched to straight-chain alkanes and to the presence of more aromatic hydrocarbons [30].

The catalysts used in commercial cracking units are aluminosilicates. These materials have acidic properties, as evidenced by the affinities of their surfaces for basic molecules like ammonia, pyridine, and quinoline [31]. The acidic nature of a silica–alumina catalyst of the type used in the early years of catalytic cracking was considered in 1949 by C.L. Thomas, who proposed a structure consisting of SiO_4 and AlO_4 tetrahedral units linked through the sharing of oxygen atoms at the corners of these units [31]. Since aluminum is trivalent and silicon is tetravalent, there is an imbalance of one unit of negative charge associated with each aluminum-containing tetrahedron in the structure. To preserve electroneutrality, Thomas suggested that a proton (H^+) was associated with each tetrahedral unit containing an aluminum atom. With regard to the source of the protons constituting the acid sites, it is known that a small amount of chemically bound water is present in the structure of a silica–alumina catalyst that has been heated to a temperature of about 800 K in its preparation [31]. Hydroxyl groups present at the surface are clearly revealed by infrared spectroscopy [32]. A proton associated with an aluminum-containing tetrahedral unit at the surface is depicted in Fig. 1.

Despite the complexity of the catalytic cracking process, the chemistry can be rationalized satisfactorily by the same mechanistic arguments that have long been invoked for simpler hydrocarbon reactions in liquid acid media. There is considerable evidence that the liquid phase reactions involve carbocation (ionic) intermediates [33]. Therefore, it has been useful to postulate that similar intermediates are involved in the surface reactions occurring in catalytic cracking [29]. The justification for doing this is simply that it permits one to account satisfactorily for a large body of catalytic cracking data, especially data on distributions of reaction products. It also provides one with a reasonable way to account for the striking differences between catalytic and thermal cracking. The reactions in thermal cracking involve free radicals [34], rather than carbocations, as intermediates.

Dramatic improvements in the catalytic cracking process were achieved as a result of the replacement of amorphous silica–alumina catalysts by crystalline aluminosilicates, commonly known as zeolites. This began during the 1960s [35]. The amounts of gasoline that can be obtained in catalytic cracking operations with zeolite catalysts are substantially higher than the amounts produced with the former catalysts, and the yields of gaseous products are correspondingly lower. Moreover, the gasoline has better anti-knock properties because it contains more branched-chain alkanes and aromatic hydrocarbons.

The zeolites of interest for catalytic cracking have the faujasite structure shown in Fig. 2. In this structure, SiO_4 and AlO_4 tetrahedra are the

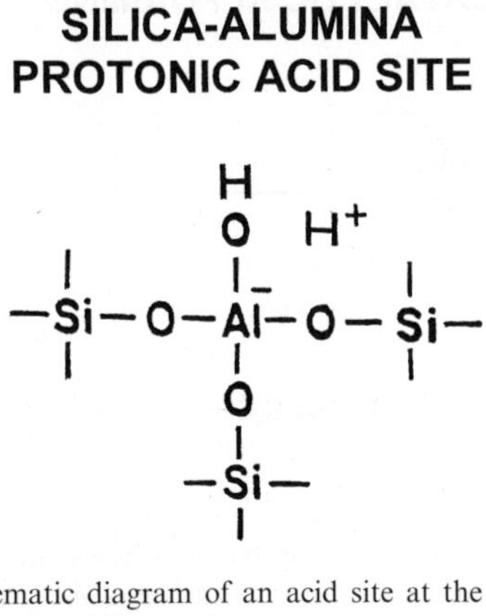

Fig. 1. Schematic diagram of an acid site at the surface of a silica–alumina cracking catalyst.

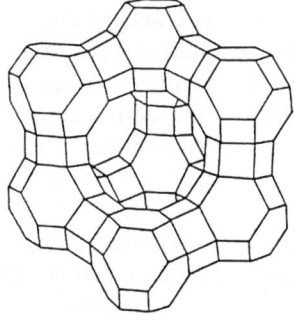

Fig. 2. Schematic diagram of the structure of a faujasite-type zeolite. The lines represent oxygen anions, and the points of intersection of the lines represent silicon or aluminum cations.

primary structural units. The silicon and aluminum ions are represented by the points of intersection in the drawing [36]. A line represents an oxygen anion bridging a silicon cation to an aluminum cation or another silicon cation. In addition to the primary structural units, there are secondary structural units in the form of cubooctahedra, which are truncated octahedra. These units are commonly known as sodalite units. At either the left- or right-hand end of the drawing in Fig. 2, for example, the reader can see a pair of sodalite units in which a hexagonal face of one is bonded to a similar face of the other through oxygen anions located between the faces. The two hexagonal faces and the bridging oxygen anions form a unit frequently called a hexagonal prism, which is considered to be another secondary structural unit.

In the extended faujasite structure, each sodalite unit is connected to four others in a diamond-like configuration. In Fig. 2 we see that the sodalite units and connecting hexagonal prisms surround a cavity that is called a supercage. The supercage has a diameter of about 12 Å and is connected to four other similar cavities through openings such as the one shown in the front of the drawing in Fig. 2. The opening is bounded by 12 faces, half of them from six sodalite units and the other half from six hexagonal prisms. The diameter of the opening is 7.4 Å. We see that the supercages connected to one another constitute an intracrystalline pore structure. The acid sites for cracking reactions are protons located inside the supercages. The protons are bound to oxygen anions of the surrounding hexagonal prisms. Thus, when faujasite-type zeolites are employed as cracking catalysts, the reactions occur on the interior surfaces of the primary zeolite crystals. In amorphous silica–alumina catalysts, by contrast, the active cracking sites are present on the external surfaces of primary solid particles packed together in non-crystalline aggregates. The pore structure of an amorphous catalyst is therefore associated with spaces between the primary particles, thereby differing from the intracrystalline pore structure associated with zeolite catalysts.

Zeolite catalysts are now used in catalytic cracking units in refineries throughout the world.

Their introduction in the cracking process ranks with the original pioneering work of Houdry and the early development of the fluidized-bed version of the process [35] as the major events in the field of catalytic cracking since its inception nearly 65 years ago.

Only a few years after the first Houdry catalytic cracking unit went on stream, another major new process called hydroforming was introduced in petroleum refining. At a later date, the process became known as catalytic reforming. A joint effort of three companies, the Standard Oil Development Company (Exxon), the M.W. Kellogg Company, and the Standard Oil Company of Indiana (Amoco), led to the first commercial application of the process in 1940 [37]. The reactions of primary interest in the process are the conversions of alkanes and cycloalkanes to high-octane-number aromatic hydrocarbons. Large amounts of hydrogen are formed in these reactions. Typical examples of the reactions are shown in Fig. 3.

The process was originally conducted with a molybdena on alumina catalyst at temperatures in the range of 750–800 K and a pressure of approximately 15 atm. The hydrocarbon feed stream was combined with a hydrogen-rich recycle stream before entering the reaction zone. The hydrogen to hydrocarbon mole ratio was typically about three to one, so that the hydrogen partial pressure represented a substantial fraction of the total pressure throughout the reaction zone. At first glance, one

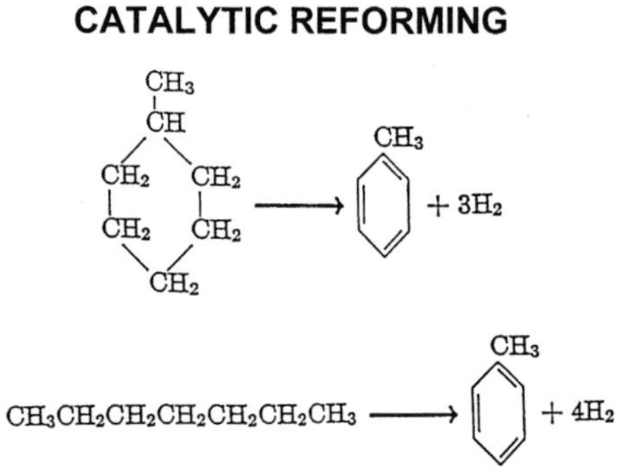

CATALYTIC REFORMING

methylcyclohexane → toluene + $3H_2$

$CH_3CH_2CH_2CH_2CH_2CH_2CH_3$ → toluene + $4H_2$

Fig. 3. Two important kinds of reactions occurring in catalytic reforming, as represented by the dehydrogenation of methylcyclohexane and the dehydrocyclization of n-heptane, both of which yield toluene and hydrogen as products.

might be skeptical about operating at an elevated hydrogen partial pressure, since the key reactions involve removal of hydrogen from the hydrocarbon molecules. However, the hydrogen pressure was not high enough to decrease the yield of aromatic hydrocarbons significantly at the high temperatures used in the process. But, it was high enough to suppress extreme types of surface dehydrogenation processes that produce carbonaceous residues on the catalyst. This is a very important consideration. When the hydrogen partial pressure is too low, the rate of formation of residues on the catalyst can be so great that the catalyst deactivates catastrophically. The elevated hydrogen partial pressure does not eliminate this cause of deactivation, but it moderates it to an extent that makes it possible to operate the process successfully.

In 1949 the Universal Oil Products Company started up a commercial reforming operation with a platinum catalyst [38]. The venture was very successful and generated much interest throughout the petroleum industry. Various other companies developed their own processes based on platinum catalysts [37]. The platinum in these catalysts was present in a very small amount, typically about 0.5 wt.%, and was dispersed throughout the pores of a high surface area alumina. The average diameter of the pores was about 10 nm. The platinum was present in the form of extremely small crystallites (later called clusters) of the order of 1 nm in size, and was therefore one of the early nanoscale materials used widely in industry. A diagram representing one of these platinum entities as a cubooctahedron is shown in Fig. 4. An actual catalyst particle with its network of pores is de-

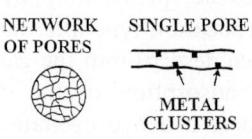

Fig. 5. Schematic diagram of a typical particle of catalyst of the kind used in catalytic reforming. The network of pores is illustrated on the left, and the metal clusters in a single pore are depicted on the right.

picted schematically in the left-hand side of Fig. 5. On the right-hand side of Fig. 5, a single pore is magnified to show the platinum entities on the pore wall [10]. Reactant molecules must diffuse into the pores to contact the platinum.

The surface of the alumina in the reforming catalysts exhibits acidic properties, largely due to the presence of chlorine ions (Cl^-) on the surface. The chlorine ions interact with hydroxyl (OH) groups to enhance the protonic acidity arising therefrom [39]. Thus, a reforming catalyst possesses two different types of sites, platinum metal sites and acidic sites. Some of the important reforming reactions are multistep processes in which certain steps occur on platinum sites and other steps on acidic sites. Evidence for this bifunctional catalysis was reported in an important paper in 1953 by Mills, Heinemann, Milliken, and Oblad of the Houdry Process Corporation [40].

For example, in the conversion of methylcyclopentane to benzene (Fig. 6), the first step in the sequence is the dehydrogenation of methylcyclopentane to methylcyclopentenes on platinum sites. The methylcyclopentenes desorb into the vapor space within the pores and readsorb on acidic sites on the alumina. Here, they rearrange to

PLATINUM CLUSTER

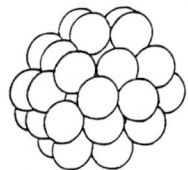

Fig. 4. Arrangement of platinum atoms in a cluster about 1 nm in size.

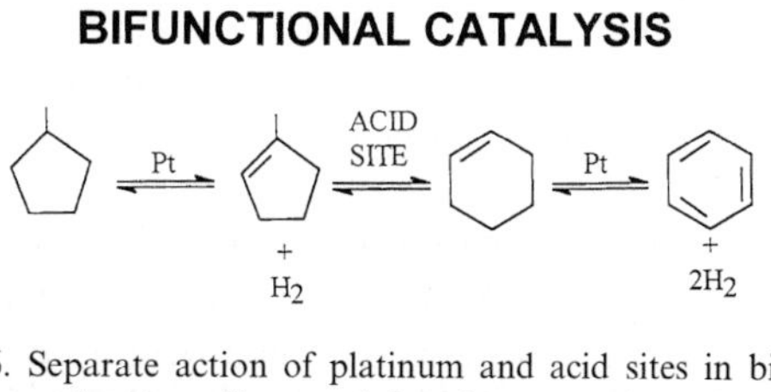

BIFUNCTIONAL CATALYSIS

Fig. 6. Separate action of platinum and acid sites in bifunctional catalysts, as illustrated for the conversion of methylcyclopentane to benzene.

form cyclohexene, presumably by a mechanism involving carbocation-type intermediates. The cyclohexene then desorbs into the gas phase. This is followed by adsorption of the cyclohexene on platinum, where it dehydrogenates to benzene to complete the sequence.

Quantitative evidence for the applicability of such mechanisms was obtained in kinetic studies of several of the major reforming reactions by Sinfelt et al. in the late 1950s and early 1960s [41–43]. The reactions included the isomerization of alkanes as well as the conversion of methylcyclopentane to benzene. Over the range of conditions normally employed in catalytic reforming, it was found that the transformations on the acid sites were the rate-controlling steps.

Thus, bifunctional catalysis plays a crucial role in reforming, but it is not the whole story. One of the major reactions illustrated in Fig. 3, the dehydrogenation of a cyclohexane homolog to an aromatic hydrocarbon, involves only the platinum metal sites of the catalyst [39,44]. The other reaction illustrated in the same figure, the dehydrocyclization of a straight-chain alkane to an aromatic hydrocarbon, appears to proceed by several reaction paths, monofunctional as well as bifunctional [39].

Throughout the 1950s and 1960s reforming with platinum–alumina catalysts became increasingly important as a source of high-octane-number components (aromatic hydrocarbons) for gasoline. Moreover, as the 1970s approached, it was clear that reforming would have to play an even greater role, since the removal of lead from gasoline would be required for environmental reasons. Since lead had contributed very significantly to the anti-knock properties of gasoline, its removal meant that gasoline would have to be enriched with other very high-octane-number components (e.g., aromatic hydrocarbons) to maintain the same anti-knock rating. This provided a stimulus for research to improve the reforming process.

Fortunately, research on bimetallic catalysts initiated in the 1960s showed promise for meeting this need. In a long-term fundamental research program in this area, Sinfelt developed the idea of a "bimetallic cluster" [9,10,45–47]. Such an entity

BIMETALLIC CLUSTER

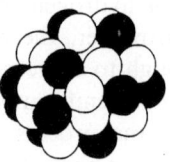

Fig. 7. Arrangement of atoms in a bimetallic cluster about 1 nm in size. The two kinds of atoms are represented by the black and white spheres.

might be visualized to have a structure like that illustrated in Fig. 7, i.e., a cubooctahedron composed of metal atoms of two different kinds [8]. It contains the same number of metal atoms as the cubooctahedron in Fig. 4, which we have already used to depict a monometallic cluster of platinum atoms. In a typical catalyst, the bimetallic clusters are dispersed throughout the pores of a material such as alumina, in the same manner as the platinum clusters illustrated in Fig. 5. Studies of the structures of bimetallic clusters present in a variety of catalysts, using extended X-ray absorption fine structure (EXAFS) as a probe, have generally indicated that one of the two kinds of metal atoms tends to concentrate at the surface or boundary of the cluster [48,49]. Thus, a homogeneous distribution of the two kinds of metal atoms within a bimetallic cluster appears to be the exception rather than the rule.

A platinum–iridium reforming catalyst emerged from the research program on bimetallic clusters [9,10]. It was first used in an Exxon refinery in 1971, and subsequently found wide application in the 1970s in reforming units throughout the world. In roughly the same time frame, a platinum–rhenium catalyst developed by Chevron [50] also found wide application in reforming. Not long after the introduction of the platinum–iridium and platinum–rhenium catalysts, the Universal Oil Products Company announced a modified reforming process featuring much lower pressure operation with a platinum–tin catalyst [51]. All of these developments represented major improvements in catalytic reforming, as did the zeolite catalysts in catalytic cracking.

3. How has heterogeneous catalysis been influenced by surface science?

The technological highlights considered in the previous section were intended to give the reader a feeling for some of the great accomplishments in heterogeneous catalysis that have established it as a major force in our civilization during the twentieth century. In parallel with these accomplishments, the science of heterogeneous catalysis has progressed steadily. The progress has been due to improved understanding of the kinetic and mechanistic features of catalytic reactions, and to a better understanding of the catalyst surfaces themselves. In this section of the paper, we present a series of examples illustrating the role that surface science has played in this progress.

3.1. Langmuir chemisorption and rate laws for surface reactions

A good place to begin a discussion of the role of surface science in heterogeneous catalysis is with the work of Irving Langmuir around the time of World War I. Langmuir derived a very simple relation for the chemisorption of a gas on the surface of a solid, in which the amount adsorbed was related to the pressure of the gas at equilibrium [52]. The relation is illustrated very simply by the chemisorption isotherm shown in Fig. 8. The extent of adsorption A increases with increasing pressure p, but eventually approaches a saturation limit when the pressure becomes sufficiently high. This means that the chemisorption process is complete when the surface is covered by a single layer of molecules. Langmuir's theory was based on elementary kinetic arguments that can be found in any textbook on physical chemistry. Its importance for heterogeneous catalysis resided in the fact that reaction rate expressions based on unknown surface concentrations could be converted to expressions relating rates to known partial pressures or concentrations of species in the gas phase [52].

Thus, in the case of a single gas-phase species undergoing a unimolecular reaction on the surface, the pressure dependence of the reaction rate is determined simply by the pressure dependence of A in Fig. 8, with adsorption and desorption of the reactant occurring at rates much higher than the rate of the surface reaction. In this situation, an adsorption equilibrium is established between the gas-phase species and the surface species undergoing reaction. The dependence of the reaction rate on the pressure corresponds to first-order kinetics at pressures low enough for A to be directly proportional to the pressure. On the other hand, the dependence corresponds to zero-order kinetics at pressures high enough for A to approach its saturation limit. For an intermediate range of pressure, where the variation of A with pressure can be approximated by a power law with a fractional exponent, the kinetics exhibit a fractional order. Thus, the Langmuir adsorption theory provided a basis for understanding the kinetics of a unimolecular surface reaction over a wide range of reactant pressures.

For a bimolecular surface reaction, the Langmuir adsorption theory provided an explanation for a number of interesting features in rate equations. Apart from the fractional reaction orders often observed, retardation of the reaction by a product, and even by a reactant, were explainable. As an example of a surface catalyzed reaction exhibiting fractional orders with respect to one or both reactants, and substantial retardation by the product even at low conversions, we have the oxidation of sulfur dioxide to sulfur trioxide considered earlier in this paper [53]. A reaction exhibiting

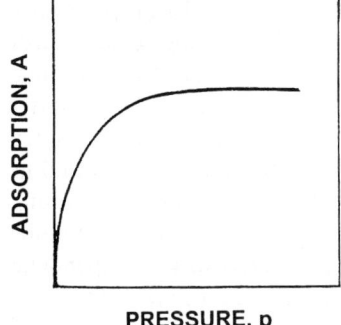

Fig. 8. Langmuir adsorption isotherm showing the extent of chemisorption of a gas as a function of pressure.

retardation by a reactant is the oxidation of carbon monoxide to carbon dioxide on platinum, in which the rate is inversely proportional to the carbon monoxide pressure over a range of temperatures [54]. The retardation effects are consequences of competition between various reactant and product species in the formation of an adsorption layer on the catalyst surface.

3.2. Activated adsorption

In applying his adsorption theory to surface catalyzed reactions, Langmuir suggested that the adsorption and desorption steps in the reaction would frequently be fast relative to a surface reaction step. Adsorption equilibria would then be effectively established between species in the gas phase and those on the surface. Hinshelwood adopted this viewpoint in the interpretation of kinetic data for a number of surface catalyzed reactions in the 1920s [54]. Consequently, this mechanism of catalytic action became known as the Langmuir–Hinshelwood mechanism.

The possibility that chemisorption could in some cases be a slow process with an appreciable activation energy was considered by Taylor in a 1931 paper [55]. He introduced the term "activated adsorption" in referring to the phenomenon. A particular type of adsorption experiment in which the amount of gas adsorbed on a surface at a given pressure is determined as a function of temperature is useful in providing evidence for activated adsorption. A plot of such data is called an adsorption isobar. In some investigations, isobars of the kind illustrated in Fig. 9 were obtained. Beginning at low temperature, one observes an initial decline in the amount of gas adsorbed as the temperature is increased. In this region the adsorption is fast and readily equilibrated. The exothermic nature of the process accounts for the negative temperature dependence. As the temperature is increased further, a new kind of adsorption comes into play. It is a slow process that is not equilibrated in the experiment. Since a rate phenomenon is being observed in this temperature region, the amount adsorbed actually increases with temperature, an indication of an activated process. Eventually, a temperature is reached where

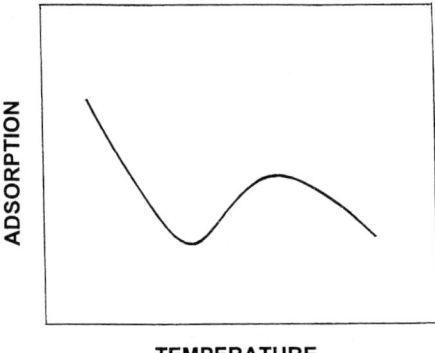

Fig. 9. Adsorption isobar showing the onset of an activated adsorption process immediately beyond the minimum in the curve relating extent of adsorption to temperature.

the activated process is fast enough to be equilibrated, and one then observes the decrease in adsorption with increasing temperature that is normally expected.

The observation of activated adsorption called attention to the possibility that the chemisorption of a reactant could be the rate-controlling step in a catalytic reaction. Evidence that this could be the case for a very important reaction in heterogeneous catalysis was obtained only a few years after Taylor first introduced his idea of activated adsorption.

3.3. The rate of nitrogen chemisorption on iron—its relevance in ammonia synthesis

In 1934 Emmett and Brunauer measured rates of chemisorption of nitrogen on an iron catalyst representative of that used for ammonia synthesis [56]. It contained 1.3 wt.% Al_2O_3 and 1.59 wt.% K_2O as promoters. In the range of temperature from 497 to 725 K data were reported in the form of curves showing the amount of nitrogen adsorbed as a function of time at a number of temperatures. The temperature dependence of the nitrogen chemisorption rates obtained from the curves indicated an activation energy of approximately 16 kcal/mol for the chemisorption process. A more detailed analysis of the data at a later date indicated that the activation energy increased as the adsorption proceeded, from an initial value of 10 to a final value of 21.5 kcal/mol [57]. The heat of adsorption decreased correspondingly from 44 to 30 kcal/mol. The activation energy therefore increased as the strength of binding of the nitrogen to the surface decreased, a common observation in chemisorption.

Where comparisons of rates of nitrogen chemisorption and of ammonia synthesis from nitrogen and hydrogen could be made on the same catalyst at the same temperature, Emmett and Brunauer found reasonable agreement between them. This provided the basis for their conclusion that nitrogen chemisorption is the rate limiting step in ammonia synthesis. Thus, the main features of ammonia synthesis can be described by representing it as a two-step process. In the first step, nitrogen molecules are dissociatively chemisorbed on the iron catalyst to form chemisorbed nitrogen atoms:

$$N_2 + 2* \rightarrow 2N*$$

The asterisk represents a surface site, and $N*$ represents a chemisorbed nitrogen atom. The latter reacts readily with hydrogen to form ammonia in the second step:

$$2N* + 3H_2 \rightleftarrows 2NH_3 + 2*$$

Equilibrium is established in this step, and hence the concentration of unoccupied surface sites is determined by the concentrations of ammonia and hydrogen over the catalyst. Because of the equilibrated nature of the step, its further dissection into elementary steps is not important for an understanding of the reaction kinetics. The reaction rate is simply the rate of the first step, which depends only on the nitrogen pressure or concentration in the gas phase and on the concentration of unoccupied sites available for the chemisorption of the nitrogen.

Thus, surface science studies of activated adsorption played a crucial role in elucidating the kinetics of ammonia synthesis, a reaction we considered earlier in connection with the important role it has played in advancing heterogeneous catalysis in the chemical industry. The evidence that the chemisorption step in a catalytic reaction could be the slow step broadened the view of catalytic kinetics embraced earlier in Langmuir–Hinshelwood mechanisms.

3.4. BET method for determining surface areas

Other surface science investigations of Emmett and Brunauer in the 1930s led to a generally applicable method for the determination of the total surface area of a solid catalyst [58]. The surface area we are considering is not simply the area associated with the boundary surfaces of the catalyst particles charged to a reactor. We refer to the latter area as the external, or geometric, surface area of the particles, which in the case of catalysts generally constitutes only a very small fraction of the surface area of interest. Catalysts of practical importance are usually porous materials, with average pore diameters of 100 Å (10 nm) being very common. There is an enormous amount of surface associated with these pores, the so-called "internal surface". It can easily be four to five orders of magnitude greater than the external surface area of the catalyst particles. Since this internal surface is accessible to reactant molecules, it is the important surface determining the catalytic activity of a material. Consequently, it is necessary to have a reliable measure of the area associated with this surface to put studies of the catalytic properties of materials on a sound basis. This was the objective of Emmett and Brunauer when they conducted their research on the problem in the 1930s.

These investigators felt that a gas adsorption method could be the answer to the problem. To develop a method of general applicability, they concentrated on physical adsorption rather than chemisorption, since the latter depends on the chemical nature of the surface. However, physical adsorption measurements are made at temperatures near the boiling point of the adsorbate, and there is the complicating feature that the process does not stop with the formation of a single layer of adsorbed molecules. One is dealing with multilayer adsorption, and it is necessary to have a way to deduce from an adsorption isotherm the number of molecules of gas adsorbed when a monolayer is completed. Once this number is known,

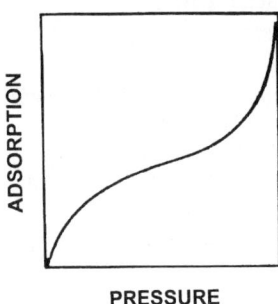

Fig. 10. Isotherm for the physical adsorption of a gas at a temperature near its normal boiling point.

the surface area is readily calculated by multiplying the number by an appropriate value of the area occupied per adsorbed molecule (16.2 Å^2 in the case of a nitrogen molecule, for example).

An adsorption isotherm for nitrogen at its boiling point (77 K) typically looks like the curve sketched in Fig. 10. There is a middle region where the isotherm is roughly linear. At higher pressures approaching the saturation vapor pressure (1 atm, since the temperature is the boiling point of the adsorbate), the curve begins to slope upward due to condensation in the pores. It appeared to Emmett and Brunauer that the beginning of the roughly linear region of the isotherm corresponded to the completion of a monolayer and the beginning of a second adsorption layer. Although this provided a way to make a crude estimate of monolayer adsorption for nitrogen and certain other gases on a variety of catalytic materials, there were cases where it was impossible to make such estimates by simple inspection of the adsorption isotherms.

Emmett and Brunauer therefore concluded that they needed to develop a theory of multilayer adsorption that would provide them with an equation which could be used to calculate the amount of adsorption corresponding to monolayer coverage. This led them into a collaboration with the physicist Edward Teller which culminated in 1938 in a theory known widely as the BET theory [58]. The theory led to an equation that could conveniently be used to present the data for a substantial region of an adsorption isotherm in the form of a linear plot, the slope and intercept of which could be used to obtain a value for the monolayer adsorption.

The BET method for determination of surface area has proved to be an extremely important product of surface science, not only for heterogeneous catalysis but also for important gas separation and purification processes based on adsorption by solid surfaces. It provided a reliable, quantitative method of characterizing a crucial property of a catalyst or adsorbent, making it possible to account for differences in surface areas in comparisons of the performance of such materials in practice.

3.5. Selective chemisorption for characterization of multicomponent catalysts

Most solid catalysts do not consist of a single element or compound. If one wishes to employ a gas adsorption method to obtain information on a particular component in the surface of a multicomponent catalyst, it is necessary to use chemisorption rather than physisorption, since chemical specificity is needed in this situation. Physisorption can be excluded by determining the amount of adsorption on the surface at temperatures far higher than the boiling point of the adsorbate. In this kind of application of gas adsorption as a surface probe, Emmett and Brunauer were again the pioneers [59]. They were interested in determining the extent to which small amounts of alumina and potassia (K_2O) promoters in an iron catalyst concentrate at the surface. As noted earlier, this type of catalyst is widely used in ammonia synthesis.

Carbon monoxide chemisorption was used to determine how much of the surface consisted of exposed iron, since the chemisorption occurred very selectively on the iron. Emmett and Brunauer found, by comparing iron catalysts with and without the promoters, that the promoters decreased the chemisorption capacity by a far greater amount than would have been expected if their concentration in the surface was the same as their overall concentration in the bulk. For an iron catalyst containing fewer than two moles of promoter per one hundred moles of iron, these investigators concluded that the promoters constituted approximately 60% of the surface. Moreover, with the use of carbon dioxide chemisorption, which was specific to the alkali (K_2O), they were able to determine the fraction of the surface taken up by this promoter alone. Thus, through an ingenious selection of adsorbates and conditions for their measurements, Emmett and Brunauer were able to make a very reasonable assessment of the composition of the surface of the complex ammonia synthesis catalyst.

This pioneering work paved the way for other important applications of selective chemisorption in the characterization of catalyst surfaces. Its use in the characterization of the extent of dispersion of a metal on a support is particularly noteworthy, since supported metals constitute a very widely used class of catalysts [60]. In this type of application one utilizes a gas that chemisorbs on the metal, but not on the support. The use of hydrogen as an adsorbate for this purpose was reported in the 1950s for platinum, nickel and other Group VIII metals supported on silica [61,62], and in 1960 for platinum dispersed on alumina [63,64] in reforming catalysts of the kind described earlier. The latter application received a great deal of attention because of the importance of the catalytic reforming process. The measurements revealed a very high chemisorption of hydrogen by the platinum, approximately one atom of hydrogen per atom of the metal. In the paper of Spenadel and Boudart, a particularly lucid discussion of the nature of the highly dispersed platinum entities was presented. It was concluded that such entities, if they were present as three-dimensional structures with approximate spherical symmetry, would have diameters of about 1 nm, as represented earlier in Fig. 4.

The importance of selective chemisorption for determining the dispersion of a supported metal lies in the fact that the number of surface metal atoms in the catalyst can then be determined. This makes it possible to use supported metals as samples for fundamental studies of the catalytic activities of metals, since one can then determine the catalytic activity per metal atom in the surface or per unit surface area of metal. Only a few months after Spenadel and Boudart reported their hydrogen chemisorption results for a platinum on alumina reforming catalyst, Sinfelt et al. reported a value for the rate of dehydrogenation of methylcyclohexane to toluene per square centimeter of platinum in such a catalyst [44]. It was the first time that such a determination of catalytic activity per unit metal surface area had been reported for a platinum reforming catalyst. This paper, and many others from the Exxon laboratories in the 1960s [65–71] for a variety of reactions and supported metals, placed needed emphasis on the desirability of taking metal dispersion into account when data on the catalytic activities of supported metals are reported in the literature. Papers of Boudart et al. provided additional emphasis [72, 73]. Accounting for metal dispersion has been a crucial first step in placing studies of the catalytic properties of supported metals on a firmer foundation.

Another important application of selective chemisorption in heterogeneous catalysis is its use in demonstrating surface segregation in bimetallic catalysts, either alloys or supported bimetallic clusters. The use of hydrogen chemisorption to distinguish nickel from copper in the surface of a nickel–copper alloy provides a good example [74,75]. If we define strongly chemisorbed hydrogen as the amount that is retained by a surface after evacuation to a pressure of 10^{-6} Torr subsequent to the measurement of an adsorption isotherm at room temperature, we find that nickel exhibits such chemisorption, whereas copper does not [75]. Therefore, for a nickel–copper alloy, one expects the amount of strongly chemisorbed hydrogen per unit surface area to be lower than it is for pure nickel. However, for samples with surface areas of about 1 m^2/g, the amount of strong chemisorption on an alloy containing only a few percent copper is enormously lower than one would expect if the composition of the surface were the same as that of the alloy overall. The low chemisorption is due to extreme segregation of copper in the surface. For nickel–copper alloys containing as little as 5% copper, it appears that copper is the dominant component in the surface.

3.6. Probing catalyst surfaces by the methods of chemical physics

Our previous examples of advances in surface science that have been important for heterogeneous catalysis were all based on extents of adsorption of gases by catalyst surfaces. We have seen how simple considerations of adsorption isotherms and rates of adsorption have been vitally important for advancing our ideas about catalyst surfaces and the reactions that occur on such surfaces. They are rooted in classical physical chemistry and have had an enormous impact in shaping the subject of heterogeneous catalysis.

After World War II, some new approaches rooted more in chemical physics were introduced for the investigation of catalyst surfaces and chemisorbed molecules. One of these involved studies of the magnetic properties of catalysts, an area pioneered by the late Pierce Selwood [76]. A noteworthy example from Selwood's laboratory was concerned with the effect of chemisorbed hydrogen on the magnetic properties of the nickel crystallites present in a nickel–silica catalyst [77]. It was observed that the magnetization of the nickel decreased when hydrogen was chemisorbed on its surface. In the chemisorption process hydrogen molecules are dissociated into atoms, and Selwood's work indicated that each chemisorbed hydrogen atom erased the contribution of one nickel atom to the magnetization. Since the magnetic properties are associated with the d electron structure of the nickel, the result indicated that this structure is strongly affected, at least for the surface nickel atoms, when nickel–hydrogen bonds are formed in chemisorption. This provided support for ideas relating the chemisorption and catalytic properties of transition metals to their d-electron structures.

Another new approach was pioneered by Eischens. It involved the development of experimental methodology for observing infrared spectra of adsorbed molecules [32]. When carbon monoxide was chemisorbed on supported platinum or palladium crystallites, infrared absorption bands were observed for two different structures, a linear one and a bridged one. These structures are illustrated

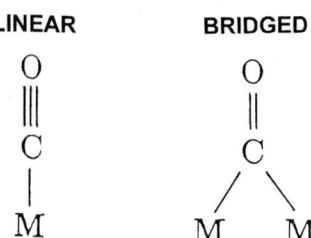

Fig. 11. Two different modes of chemisorption of carbon monoxide on metal surfaces, identified as the linear and bridged structures. The letter M represents a surface metal atom.

in Fig. 11, where M represents a surface metal atom.

In the chemisorption of ammonia on a silica–alumina cracking catalyst, infrared absorption bands for two different surface species were again observed. One of the bands was observed at a wave length indicative of an ammonium ion, presumably formed by the interaction of an ammonia molecule with a surface proton such as that depicted earlier in Fig. 1:

$$NH_3 + H^+ \rightarrow NH_4^+$$

The other band occurred at a wave length attributable to ammonia molecules chemisorbed without a change in configuration, i.e., as NH_3. The surface site on which an ammonia molecule is chemisorbed in this manner is known as a Lewis acid site, named after the noted physical chemist G.N. Lewis. A site of this kind can be formed by the elimination of a molecule of H_2O from the protonic acid site depicted earlier in Fig. 1. As a consequence of such a dehydration process, the aluminum atom in the structure is exposed with an unoccupied p orbital, which can accept the lone-electron pair of a nitrogen atom in a molecule of ammonia. The interaction is illustrated in Fig. 12. The relative concentrations of Lewis and protonic acid sites on the surface of a silica–alumina catalyst can be determined from the intensities of the absorption bands associated with the surface species NH_3 and NH_4^+, respectively. Thus, infrared spectroscopic studies have provided valuable information on the sites responsible for the impor-

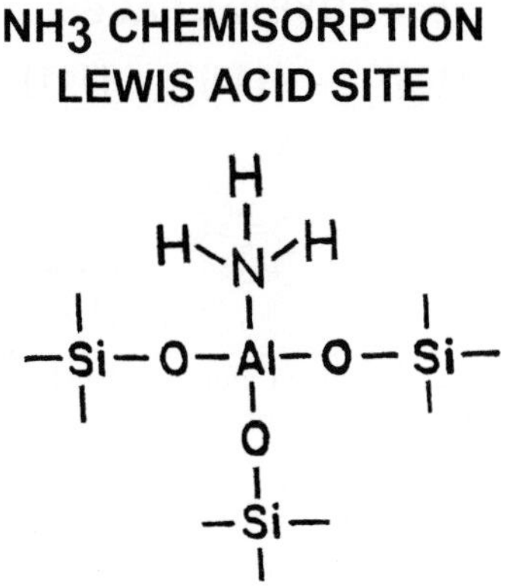

**NH₃ CHEMISORPTION
LEWIS ACID SITE**

```
        H
        |
     H~ N ~H
     |  |  |
  —Si—O—Al—O—Si—
     |  |  |
        O
        |
       —Si—
        |
```

Fig. 12. An ammonia molecule chemisorbed on a Lewis acid site of a silica–alumina cracking catalyst.

tant reactions occurring in the catalytic cracking of petroleum fractions.

Electron spectroscopy, pioneered by Siegbahn et al. in Sweden [78], has also found important applications as a probe of catalyst surfaces. In this type of spectroscopy, electrons are emitted from a substance as a result of some form of excitation. When X-rays provide the excitation, one speaks of X-ray photoelectron spectroscopy (XPS). From a measurement of the kinetic energy of the emitted electrons, one can determine the binding energy of the electrons in some level within the emitting atom, since the energy of the exciting X-rays is known. Since an electron binding energy is a characteristic feature of the atoms of a given element, the determination of binding energies by XPS makes it possible to identify the kinds of elements present in a material. A line in an XPS spectrum is associated with electrons having a particular binding energy. The intensity of the line provides information on the amount of the element present. It is critically important for surface science that XPS is a surface sensitive spectroscopy. This arises because electrons emitted by atoms in subsurface layers are subject to inelastic scattering processes. These processes strongly decrease the probability that such electrons will emerge from the material with the expected kinetic energy.

An example of the use of XPS in probing catalyst surfaces is its application to ruthenium–copper catalysts [79]. Although ruthenium and copper are immiscible in the bulk, chemisorption and catalysis studies had suggested very strongly that the copper was present on the surface of the ruthenium, much like a chemisorption layer [46,80]. In agreement with this picture, XPS studies showed that the ratio of the intensity of a copper line in the spectrum to that of a ruthenium line was an order of magnitude greater than would have been expected on the basis of the overall composition.

Another form of spectroscopy which has proved to be useful for obtaining information on catalytic materials is X-ray absorption spectroscopy. A particular aspect of this kind of spectroscopy known as EXAFS attracted the attention of Farrel Lytle in the early 1960s. The extended fine structure refers to the oscillations in absorption coefficient observed on the high energy side of an absorption edge. As mentioned earlier in the paper, EXAFS studies have proved to be very useful for probing the structures of bimetallic clusters, including those present in petroleum reforming catalysts used in the production of high-octane number, lead-free gasoline [81,82]. Although EXAFS is generally a probe of bulk rather than surface structure, it becomes a surface probe in studies of the bimetallic clusters in a reforming catalyst, since very high fractions of the atoms in these nanometer-size clusters are surface atoms.

Finally, an area of spectroscopy that has contributed much to the surface science of catalysts in recent years is nuclear magnetic resonance (NMR). The writer has been very fortunate to collaborate with Professor Charles Slichter and his students in the use of NMR in surface science investigations of supported metal catalysts. One part of the work was concerned with molecules chemisorbed on the metal clusters in the catalysts [83]. Various molecules, including carbon monoxide, ethylene, and acetylene, were chemisorbed on a number of different Group VIII metals in these investigations. The structures of the chemisorbed species were determined, and, in the case of the hydrocarbon molecules, kinetic data were obtained for simple surface reaction processes such as the scission of carbon–carbon bonds and the formation and dissociation of carbon–hydrogen bonds [83–85]. Kinetic data for such elementary surface reactions contribute significantly to our understanding of the kinetics of more complex catalytic reactions of hydrocarbons.

3.7. Studies with idealized surfaces

Shortly before and after World War II, investigations with idealized catalyst surfaces were initiated by a circle of scientists who felt that such an approach would lead to a deeper understanding of fundamental issues in heterogeneous catalysis. One of these scientists was Otto Beeck, who employed evaporated metal films in classical studies of the chemisorption and catalytic properties of metals [86]. Beeck's work was followed by other classical research on metal films by the late Charles Kemball [87], in which the mechanisms of exchange reactions of deuterium with hydrocarbons were investigated in detail. Evaporated metal films were also employed by Trapnell in a broad exploration of the specificity of metal surfaces for chemisorption [88].

By the late 1950s and early 1960s, idealized surfaces were being identified increasingly with specific lattice planes of single crystals under ultra-high vacuum conditions (pressures of the order of 10^{-10} Torr). Germer [89] and Farnsworth [90] were pioneers in surface science investigations employing single crystals. In the 1960s, they were joined by Gabor Somorjai, Gerhard Ertl, Lanny Schmidt, John Yates, and Mike White, among others, in developing this modern approach to surface science. The new era of surface science was stimulated greatly by advances in ultra-high vacuum technology that originated in the massive research effort devoted to the exploration of outer space. Also, it owed much to the development of probes such as low energy electron diffraction and electron spectroscopy for the detailed examination of surfaces.

Figs. 13 and 14 illustrate the arrangements of atoms in the (1 0 0), (1 1 0), and (1 1 1) faces of body-centered cubic (bcc) and face-centered cubic (fcc) single crystals, respectively. In the bcc crystal the atoms are packed most densely in the (1 1 0) face; in the (fcc) crystal the packing density is highest for the (1 1 1) face. The metals platinum, palladium, and nickel exhibit the fcc structure,

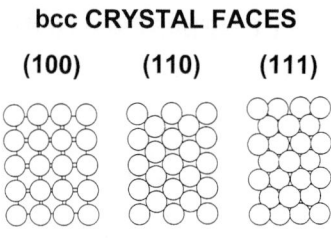

Fig. 13. The arrangements of atoms in the (1 0 0), (1 1 0), and (1 1 1) faces of single crystals of metals with bcc structures.

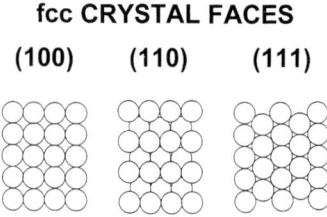

Fig. 14. The arrangements of atoms in the (1 0 0), (1 1 0), and (1 1 1) faces of single crystals of metals with fcc structures.

while iron has the bcc structure. For studies of catalytic reactions with single crystal surfaces having areas of about 1 cm^2, Somorjai and co-workers developed an experimental apparatus which permitted isolation of the single crystal from the ultra-high vacuum system after the surface had been characterized by electron diffraction and electron spectroscopy techniques [91]. The isolated crystal was then exposed to reactant gases at higher pressures.

For measurements of rates of synthesis of ammonia from a 3/1 hydrogen–nitrogen mixture on the (1 1 1), (1 0 0), and (1 1 0) faces of iron single crystals at a temperature of 798 K, a pressure of 20 atm was employed [91]. It was observed that the reaction rate was highest on the (1 1 1) face, over 400 times higher than the rate on the (1 1 0) face and roughly 15 times higher than the rate on the (1 0 0) face. This result was important because it provided direct evidence for the long-held view that the (1 1 1) plane of α-iron is the most active plane for ammonia synthesis [92]. As early as 1909, Mittasch had shown that iron catalysts prepared by careful reduction of magnetite (Fe_3O_4) were the ones with the highest activity [23], and subsequent structural studies showed that catalysts prepared in this manner preferentially exposed (1 1 1) planes at the surface [92].

In a comparison of rates of chemisorption of nitrogen on the (1 1 1), (1 0 0), and (1 1 0) crystal faces of iron, Ertl [93] found the order of variation of rates to be the same as that observed by the Somorjai group for ammonia synthesis. The rate on the (1 1 1) face was two orders of magnitude higher than the rate on the least active face. It must be appreciated that the results of the nitrogen chemisorption studies were obtained at a pressure many orders of magnitude lower than that used in the catalytic studies. In the chemisorption measurements, the experimental methodology associated with the ultra-high vacuum apparatus limited the investigations to pressures lower than about 10^{-4} Torr. Ertl [93] has compared sticking coefficients obtained in the low pressure nitrogen chemisorption experiments with reaction probabilities obtained in the higher pressure ammonia synthesis experiments. These quantities, defined as the fractions of nitrogen molecules colliding with the surface that were chemisorbed in the one investigation or converted to ammonia in the other, were of the same magnitude, in the range of 10^{-7}–10^{-8}.

The results of the Ertl and Somorjai groups on iron single crystals lend strong support to the original view of Emmett and Brunauer that nitrogen chemisorption is the rate-controlling step in ammonia synthesis. The chemisorption and catalytic results of the Ertl and Somorjai groups hold together very nicely in view of the enormous difference in the pressures of the experiments. Moreover, the extremely low values of the sticking coefficients and reaction probabilities are in excellent accord with values of sticking coefficients obtained from data on rates of nitrogen chemisorption reported by Emmett and Brunauer [56] for a "real" ammonia synthesis catalyst in the early 1930s.

Since the rate of ammonia synthesis on an iron single crystal is strongly dependent on the particular crystal face that is exposed at the surface, the reaction is said to be structure sensitive [94]. For supported iron catalysts the structure sensitivity is manifested in a pronounced dependence of the turnover frequency of the reaction on the size of the iron clusters or crystallites present in the catalysts [94]. The turnover frequency is simply the reaction rate per surface metal atom. Comparisons of turnover frequencies must be made for a given set of reaction conditions (temperature, pressure, and reactant mole fractions). Although major features of nitrogen chemisorption and ammonia synthesis studies on iron single crystals conform very well with observations made in similar studies with real catalysts, it must be recognized that structure sensitivity can be a complicating factor when one compares the turnover frequency on an actual metal catalyst with that obtained on a metal single crystal.

However, there are a number of examples of metal catalyzed reactions that do not exhibit structure sensitivity. In such cases the turnover frequency on the small metal clusters present in a supported metal catalyst is essentially the same as it is on a metal single crystal. Boudart and Djéga-Mariadassou have listed a number of examples of structure-insensitive reactions [94]. They include the hydrogenation of cyclopropane and cyclohexene on platinum, the hydrogenation of ethylene, benzene, and carbon monoxide on nickel, and the oxidation of carbon monoxide on palladium. Thus, for structure-insensitive reactions, it would appear that rate data on metal single crystals can be used directly by catalytic chemists in their work with simple metal catalysts. This provides confidence that surface science studies with single crystals can have a relevance for heterogeneous catalysis that is very direct.

This direct kind of relevance is particularly important, and is illustrated very impressively by recent work of Lanny Schmidt and his students [95]. These workers have pursued surface science investigations on the kinetics of elementary reaction steps that are likely to be important in the catalytic oxidation of hydrocarbons, and they have made direct use of the results. They have found a way to obtain very high yields of carbon monoxide and hydrogen from methane–air or methane–oxygen mixtures with the aid of a solid catalyst surface. The overall reaction is represented by the stoichiometric equation:

$$2CH_4 + O_2 \rightarrow 2CO + 4H_2$$

The surprising result was achieved by passing a methane–air mixture ($CH_4/O_2 = 1.7$) preheated to 735 K through a rhodium-coated ceramic monolith at a flow rate giving a residence time of the order of a millisecond in the reaction zone. The reactor is operated nearly adiabatically so that the heat released by the exothermic oxidation reactions raises the temperature to a level of 1200 K. This kind of operation is reminiscent of the Ostwald process for ammonia oxidation in the commercial manufacture of nitric acid discussed earlier in this paper. The monolith form of catalyst also reminds one of the catalytic converters used for pollution control in automobile exhaust systems, although the end result of the reactions is different.

The Schmidt group also investigated platinum-coated monoliths for methane oxidation. They work, but the results are less impressive. The surface science studies of the group provided them with an explanation of why the rhodium-coated monolith gives better results [96,97]. Their explanation is based on a consideration of the rate of formation of water by the sequence of elementary steps:

$$H(s) + O(s) \rightleftarrows OH(s)$$

$$H(s) + OH(s) \rightleftarrows H_2O(s)$$

$$H_2O(s) \rightleftarrows H_2O(g)$$

The symbols (s) and (g) signify surface and gas species, respectively. On platinum the overall rate of formation of water in the sequence is greater than it is on rhodium. The other step of interest, of course, is the recombination of surface H atoms to yield hydrogen in the gas:

$$H(s) + H(s) \rightleftarrows H_2(g)$$

On rhodium this step competes with the sequence of steps yielding water much more effectively than it does on platinum. This accounts for the much higher selectivity to hydrogen observed in the methane oxidation experiments on the rhodium-coated monolith.

This work has important technological possibilities, since it provides an alternative way to produce synthesis gas, i.e., mixtures of carbon monoxide and hydrogen. Such mixtures are of interest for the production of methanol or liquid hydrocarbons. Currently, synthesis gas is produced by the reaction of methane with steam at a temperature of about 1100 K over a nickel catalyst. The product composition approaches that expected when equilibria are established for the two reactions:

$$CH_4 + H_2O \rightleftarrows CO + 3H_2$$

$$CO + H_2O \rightleftarrows CO_2 + H_2$$

The second of these is commonly known as the water gas shift reaction. The feed stream to the reactor contains excess steam, and the exit gas consists of a mixture of CO, H_2, CO_2, and H_2O. The yield of hydrogen is not high enough at this point in the process. It is increased by admixture of the gas with steam prior to contact with an iron catalyst in a second reactor at a lower temperature. This works because the reaction occurring in the second reactor is the water gas shift reaction. Since it is exothermic, it occurs more extensively as the temperature is decreased, thereby yielding additional hydrogen. The complexity of a steam reforming plant, in the opinion of the Schmidt group, provides significant incentive to produce synthesis gas by the reaction of methane with oxygen. The choice of one process over the other depends ultimately on the overall economics, which can be a complex matter.

In their work on methane oxidation, Schmidt and coworkers concluded that the results could be interpreted entirely in terms of surface catalyzed processes. In an extension of the research to the oxidation of ethane, in which ethylene is the important product, they reached a similar conclusion [98]. However, in later work they seem to be less certain that homogeneous gas phase reactions do not play a role [99], and recent work by another group [100] indicates that such reactions make an important contribution to the overall results. In the latter work, the investigators conclude that the major role of the catalyst is to accelerate exothermic reactions at the front of the reactor to increase the downstream temperature to the point where the gas-phase reactions occur readily.

4. Future outlook

Beginning with the work of Langmuir around the time of World War I, heterogeneous catalysis has clearly been influenced by surface science. During the period between World War I and World War II, the surface science of interest emerged for the most part from studies of the extent of adsorption of gases on catalyst surfaces, i.e., from measurements of adsorption isotherms, adsorption isobars, and rates of adsorption. Studies of this type have been at the heart of some of the most important contributions of surface science to heterogeneous catalysis. They still provide essential information in catalysis research programs.

After World War II, many new approaches for the experimental investigation of catalyst surfaces were developed. They involved the application of various spectroscopic and structural probes which yielded information at a microscopic level to supplement the information derived from classical adsorption studies. Studies employing single crystal surfaces in ultra-high vacuum apparatus also made their appearance. Overall, the progress in surface science during the second half of the twentieth century was associated primarily with the application of new probes from the arsenal of the physicist and the chemical physicist and with advances in experimental capabilities in general.

As we move into the twenty-first century, it is becoming increasingly clear that modern surface science can yield information of value in heterogeneous catalysis. This is true whether the surface science is done with actual catalysts or with idealized surfaces. For example, information on the kinetics of elementary reactions of well-defined surface species can be obtained in experiments on metal single crystals, on polycrystalline metal foils, and on supported metal crystallites or clusters. In the latter case this has been done with the aid of NMR in defining the structures of the species involved in the elementary surface reactions [83–85]. The acquisition of data on kinetic parameters for elementary surface steps is important for a detailed understanding of heterogeneous catalysis. However, it can also find valuable application in the modeling of complex catalytic reactions, an activity that provides important guidance in the development of a new catalytic process.

Some examples of this are found in recent papers on hydrocarbon oxidation [95,100] that we cited earlier. The papers describe the overall reactions in terms of plausible elementary steps, each with its own kinetic parameters. In one case [100] the contribution of homogeneous gas-phase reactions is included. The approach employed in the modeling has been called "microkinetic analysis" by Dumesic et al. [101]. The many kinetic parameters of elementary reactions required for implementation of the approach happened to be available in the hydrocarbon oxidation examples. This is understandable, since the important surface species were all relatively simple entities, the kinds of species that one would expect to be the subjects of surface science investigations. It is fortunate that results of early investigations of the kinetics of elementary surface reactions of such species have found almost immediate application in catalytic processes of great interest. There is a clear connection in this instance between research conducted with the objective of basic understanding and research conducted with the objective of practical application. It would be nice if the connection were always this clear.

Experimental determinations of kinetic parameters for elementary surface reactions are an important part of long-term fundamental research in heterogeneous catalysis. Investigations of this type for elementary steps in the gas phase have long been of interest to chemical kineticists. The studies have provided much understanding of homogeneous gas-phase reactions, and the results have found many practical applications. The achievements in the area provide additional inspiration for surface scientists to continue their investigations of elementary reactions on surfaces. Such research requires a great deal of effort, but on the basis of what we have seen so far it should reap dividends in the long run.

In the area of high-temperature oxidation of hydrocarbons on metal surfaces, the experimental information on elementary reactions of the surface species H, O, OH, H_2O, CO, and C provides excellent material for the theoreticians to be thinking about, in view of the simplicity of the species involved. There is an opportunity here for experiment and theory to come together in an area that has the potential for major technological innovations.

The great advances in computers have focused much attention on the ability of basic theory to describe complicated phenomena in virtually all areas of science. In chemistry we recognize this by referring to "computational chemistry", which in effect sets it off as another discipline within the field. We can expect computational chemistry to play an increasingly important role in surface science and catalysis in the coming years. It will be evident over the whole gamut of activities from basic research to final application.

In this context, another challenging problem for theoreticians to consider is the conversion of nitric oxide on a surface. The practical interest here, of course, is the elimination of nitric oxide as a pollutant in exhaust gases from automobiles and from the smokestacks of power plants [11,102]. Reactions to consider are: (1) the direct decomposition to yield nitrogen and oxygen, and (2) reduction of the nitric oxide by carbon monoxide, hydrogen, or hydrocarbons. The direct decomposition reaction has the problem that the oxygen formed in the rupture of the nitrogen–oxygen bond is very strongly bound to a surface such as that of platinum, thereby inhibiting the reaction. This suggests a surface that would bind the oxygen less tenaciously, but would such a hypothetical surface be able to break the nitrogen–oxygen bond? Perhaps basic theory might provide some guidance here.

Reduction reactions of NO are known to occur in the platinum–palladium–rhodium catalysts used in present exhaust gas converters largely because of the presence of the rhodium [11]. However, the full effectiveness of the rhodium can be realized only if the air to fuel ratio in the vapor admitted to the engine is controlled within a very narrow range. A catalyst that would operate effectively over a wider range of air to fuel ratios would decrease the demands on the control system. Interestingly, iridium exhibits the ability to catalyze nitric oxide reduction even when there is a net oxidizing environment. However, the use of iridium in exhaust gas catalysts is precluded by its limited availability. Thus, the matter of improving exhaust gas catalysts is complex, involving a variety of different considerations. While current systems are performing amazingly well and represent a huge triumph for catalysis, the ever increasing demands on them will require a great deal of dedication on the part of people working in this area. The problems become increasingly difficult as the demands increase, but there is still a lot of opportunity for surface scientists to have an impact in this challenging field.

Most of the work in modern surface science related to catalysis has been done with metals. However, metals are by no means the only catalysts of interest. Non-metallic solids also play a very important role in heterogeneous catalysis. In the future, we can anticipate that the surface science community will devote more attention to these solids. From the viewpoint of catalysis, crystalline aluminosilicates (zeolites) are particularly important members of this broad class of materials.

As noted earlier, certain types of zeolites have found very extensive application in the catalytic cracking of petroleum in recent decades. The nature of the diffusional process involved in the transport of molecules into and out of the intracrystalline pore systems in zeolites is an intriguing scientific issue, since the pores have dimensions not much larger than those of ordinary molecules. Weisz has introduced the term "configurational diffusion" in referring to the phenomenon [103]. What is the nature of the phenomenon? Is the process strictly one of surface diffusion, or is there a possibility of parallel contributions of surface diffusion and a "gas-like" diffusion process? It is reasonable to expect the diffusional properties to depend on the affinity of the surface for the diffusing molecules. In view of the fact that the pore sizes approach the dimensions of the molecules, the pore surfaces will undoubtedly influence the diffusion process, whether the diffusing species is adsorbed or retains in large part the character of a gas-like entity [104]. Consequently, the diffusion process is an issue within the scope of surface science. Although the dimensions of interest in zeolite pore systems are typically about 1 nm and smaller, the variation among different zeolites within the limited range of dimensions possible, coupled with a comparable variation in the sizes of the diffusing molecules, can easily have the effect of altering diffusion coefficients by orders of magnitude. Consequently, one is dealing with a sensitive phenomenon and can anticipate a variety of interesting catalytic consequences. Indeed, such consequences are already well known as examples of the phenomenon of size and shape selectivity in catalysis by zeolites [103]. Thus, there would appear to be much incentive to obtain a clearer understanding of the processes occurring in the pores of zeolites.

The consideration of effects of size and shape of reactant molecules and of zeolite pores is reminiscent of enzyme catalysis, where geometry plays an important role in the specificity that is observed in the action of an enzyme on a substrate molecule. Thus, both zeolite catalysis and enzyme catalysis have the common feature that a chemical transformation occurs in a confined space of molecular dimensions. There would appear to be an opportunity here for a profitable interaction between people interested in the surface science issues in zeolite pores and the people investigating fundamental aspects of enzyme catalysis.

Zeolites, and amorphous aluminosilicates as well, are solid acids, as we have seen earlier in this article. Their abilities to catalyze hydrocarbon transformations are associated with the acidic properties. It has been postulated that the elementary reaction steps in the transformations involve ionic-type species known as carbocations. The postulate was based originally on the observation that many features of reactions catalyzed by the materials could be interpreted satisfactorily by mechanisms similar to those that had been successfully advanced for reactions occurring in the presence of acidic substances such as sulfuric acid and aluminum chloride [29]. The participation of carbocations as intermediates in the latter has been accepted for a long time, and there is spectroscopic evidence of their existence in liquid acid systems [33].

The success of mechanistic arguments involving carbocations in accounting for the distribution of reaction products in catalytic cracking on a solid acid surface has been very impressive, especially when one considers the complexity of the cracking process. The hypothetical carbocations on the surface of a cracking catalyst should not be viewed as free ions, although for simplicity it is common practice to do so in discussions of the role played by such species in cracking reactions. Realistically such species should be associated with complementary anionic sites in the catalyst surface. The species are preferably viewed as polarized surface entities rather than free ions.

In view of the great importance of solid acid surfaces in catalytic cracking and related hydrocarbon conversion processes, it would appear to be worthwhile to attempt a more direct investigation of the hypothetical carbocations. In such an investigation, it would be desirable to study hydrocarbon species chemisorbed on the catalyst surface at temperatures as close as possible to those utilized for the reactions.

In the 1960s, shortly before his death, Professor P.J.W. Debye at Cornell University had begun investigations of the dielectric properties of adsorbed molecules. At the time, Professor Debye was a consultant to Esso (Exxon) Research and Engineering Company, which accounts for the writer's awareness of the work then in progress. Unfortunately, the work was terminated by Professor Debye's death in 1966.

Professor Debye's approach involved the use of a condenser consisting of two concentric cylinders. The catalyst was packed in the annulus between the cylinders. The condenser was part of a system in which a gas could be adsorbed on the catalyst under different conditions of temperature and pressure. From measurements of the capacitance with and without the adsorbed gas, the capacitance due to the adsorbed molecules could be determined. By comparing this with the capacitance of the same number of gas molecules in the absence of the catalyst, one could presumably determine the dielectric changes in the molecules as a result of adsorption.

It appears to the writer that Professor Debye's approach could find use in the characterization of polarized hydrocarbon species chemisorbed on the acidic sites in the surfaces of aluminosilicate cracking catalysts. The writer does not know what plans, if any, that Professor Debye had to address this matter. As the writer recalls (after 34 years), the plan called for investigating first the adsorption of a simple molecule such as nitrogen on alumina at low temperature (physical adsorption). Investigations of the chemisorption of simple molecules would have come later. Other investigators might want to consider reopening research on the dielectric properties of adsorbed molecules following Debye's approach.

In closing this section on the future outlook for surface science in catalysis, the writer concludes that the opportunities are great. The past has demonstrated clearly that progress in heterogeneous catalysis has been closely linked to progress in surface science, from the time of Langmuir to the present. Why should the future be any different?

References

[1] E. Rideal, H.S. Taylor, Catalysis in Theory and Practice, Macmillan Company Limited, London, 1926.
[2] G.M. Schwab, Catalysis: From the Standpoint of Chemical Kinetics, D. Van Nostrand, New York, 1937, p. 8 (translation by H.S. Taylor, R. Spence of the first (1931) German edition).
[3] S. Glasstone, K.J. Laidler, H. Eyring, The Theory of Rate Processes, first ed., McGraw-Hill Book Company, New York and London, 1941 (pp. 339–399).
[4] L. Pauling, General Chemistry, second ed., W.H. Freeman and Company, San Francisco and London, 1953 (p. 409).
[5] J.H. Sinfelt, The evolution of catalytic science and technology (Gold Medal Address, American Institute of Chemists), The Chemist 61 (10) (1984) 6–10.
[6] J.H. Sinfelt, Some reflections on catalysis, Chemistry and Industry (London) June 4 (11) (1984) 403–406 (Perkin Medal Address).
[7] J.H. Sinfelt, Structure of metal catalysts, Rev. Mod. Phys. 51 (1979) 569–589.
[8] J.H. Sinfelt, Catalysis: an old but continuing theme in chemistry, Proc. Am. Philos. Soc. 143 (1999) 388–399.
[9] J.H. Sinfelt, Bimetallic Catalysts: Discoveries, Concepts and Applications, Wiley, New York, 1983.
[10] J.H. Sinfelt, Bimetallic catalysts, Scientific Am. 253 (3) (1985) 90–98.
[11] K.C. Taylor, Automobile catalytic converters, in: J.R. Anderson, M. Boudart (Eds.), Catalysis—Science and Technology, vol. 5, Springer, Berlin, 1984, pp. 119–170.
[12] G. Bylinsky, The magic of designer catalysts, Fortune May 27 (1985) 82–88.
[13] M. Prettre, Catalysis and Catalysts, Dover Publications, New York, 1963 (translation by David Antin of the third (1961) edition of Catalyse et Catalyseurs, published by the Presses Universitaires de France).
[14] J.H. Sinfelt, Influence of technology on catalytic science, Ind. Eng. Chem. Fundam. 25 (1986) 2–9.
[15] E.B. Maxted, Catalysis and its Industrial Applications, J.&A. Churchill, London, 1933 (p. 482).
[16] H.T. Briscoe, General Chemistry for Colleges, third ed., The Riverside Press, Cambridge, MA, 1943 (pp. 626 and 627).
[17] N.N. Greenwood, A. Earnshaw, Chemistry of the Elements, second ed., Butterworth-Heinemann, Oxford, 1997 (p. 710).
[18] W.G. Frankenburg, The catalytic synthesis of ammonia from nitrogen and hydrogen, in: P.H. Emmett (Ed.), Catalysis, vol. III, Reinhold, New York, 1955, pp. 171–263.
[19] G. Chinchen, P. Davies, R.J. Sampson, The historical development of catalytic oxidation processes, in: J.R. Anderson, M. Boudart (Eds.), Catalysis—Science and Technology, vol. 8, Springer, Berlin, 1987, pp. 1–67.
[20] S.A. Topham, The history of the catalytic synthesis of ammonia, in: J.R. Anderson, M. Boudart (Eds.), Catalysis—Science and Technology, vol. 7, Springer, Berlin, 1985, pp. 1–50.
[21] R.C. Fuson, H.R. Snyder, Organic Chemistry, Wiley, New York, 1942 (pp. 39 and 58).
[22] Ref. [17], p. 422.
[23] A. Mittasch, Early studies of multicomponent catalysts, Adv. Catal. 2 (1950) 81–104.
[24] G. Natta, Synthesis of methanol, in: P.H. Emmett (Ed.), Catalysis, vol. III, Reinhold, New York, 1955, pp. 349–411.
[25] H.H. Voge, C.R. Adams, Catalytic oxidation of olefins, Adv. Catal. 17 (1967) 151–221.
[26] K.K. Kearby, Catalytic dehydrogenation, in: P.H. Emmett (Ed.), Catalysis, vol. III, Reinhold, New York, 1955, pp. 453–491.
[27] B.C. Gates, J.R. Katzer, G.C.A. Schuit, Chemistry of Catalytic Processes, McGraw-Hill, New York, 1979 (pp. 349–389).
[28] Ref. [27], pp. 150–176.
[29] H.H. Voge, Catalytic cracking, in: P.H. Emmett (Ed.), Catalysis, vol. VI, Reinhold, New York, 1958, pp. 407–493.
[30] A.G. Oblad, T.H. Milliken, G.A. Mills, The effects of the variables in catalytic cracking, in: B.T. Brooks, C.E. Boord, S.S. Kurtz, L. Schmerling (Eds.), The Chemistry of Petroleum Hydrocarbons, vol. II, Reinhold, New York, 1955, pp. 165–188.
[31] L.B. Ryland, M.W. Tamele, J.N. Wilson, Cracking catalysts, in: P.H. Emmett (Ed.), Catalysis, vol. VII, Reinhold, New York, 1960, pp. 1–91.
[32] R.P. Eischens, W.A. Pliskin, The infrared spectra of adsorbed molecules, Adv. Catal. 10 (1958) 1–56.
[33] R.T. Morrison, R.N. Boyd, Organic Chemistry, second ed., Allyn and Bacon, Boston, 1966 (pp. 199 and 200).
[34] F.O. Rice, K.K. Rice, The Aliphatic Free Radicals, Johns Hopkins Press, Baltimore, 1935.

[35] H. Heinemann, A brief history of industrial catalysis, in: J.R. Anderson, M. Boudart (Eds.), Catalysis—Science and Technology, vol. 1, Springer, Berlin, 1981, pp. 1–41.
[36] D.W. Breck, Zeolite Molecular Sieves, Wiley, New York, 1974.
[37] F.G. Ciapetta, R.M. Dobres, R.W. Baker, Catalytic reforming of pure hydrocarbons and petroleum naphthas, in: P.H. Emmett (Ed.), Catalysis, vol. VI, Reinhold, New York, 1958, pp. 495–692.
[38] V. Haensel, Aromatization, hydroforming and platforming, in: B.T. Brooks, C.E. Boord, S.S. Kurtz, L. Schmerling (Eds.), The Chemistry of Petroleum Hydrocarbons, vol. II, Reinhold, New York, 1955, pp. 189–219.
[39] J.H. Sinfelt, Catalytic reforming of hydrocarbons, in: J.R. Anderson, M. Boudart (Eds.), Catalysis—Science and Technology, vol. 1, Springer, Berlin, 1981, pp. 257–300.
[40] G.A. Mills, H. Heinemann, T.H. Milliken, A.G. Oblad, Houdriforming reactions. Catalytic mechanism, Ind. Eng. Chem. 45 (1953) 134–137.
[41] J.H. Sinfelt, H. Hurwitz, J.C. Rohrer, Kinetics of n-pentane isomerization over Pt–Al_2O_3 catalyst, J. Phys. Chem. 64 (1960) 892–894.
[42] J.H. Sinfelt, J.C. Rohrer, Kinetics of the catalytic isomerization–dehydroisomerization of methylcyclopentane, J. Phys. Chem. 65 (1961) 978–981.
[43] J.H. Sinfelt, H. Hurwitz, J.C. Rohrer, Role of dehydrogenation activity in the catalytic isomerization and dehydrocyclization of hydrocarbons, J. Catal. 1 (1962) 481–483.
[44] J.H. Sinfelt, H. Hurwitz, R.A. Shulman, Kinetics of methylcyclohexane dehydrogenation over Pt–Al_2O_3, J. Phys. Chem. 64 (1960) 1559–1562.
[45] J.H. Sinfelt, Supported bimetallic cluster catalysts, J. Catal. 29 (1973) 308–315.
[46] J.H. Sinfelt, Catalysis by alloys and bimetallic clusters, Acc. Chem. Res. 10 (1977) 15–20.
[47] J.H. Sinfelt, Structure of bimetallic clusters, Acc. Chem. Res. 20 (1987) 134–139.
[48] J.H. Sinfelt, G.H. Via, F.W. Lytle, Application of EXAFS in catalysis. Structure of bimetallic cluster catalysts, Catal. Rev. 26 (1) (1984) 81–140.
[49] G.H. Via, K.F. Drake Jr., G. Meitzner, F.W. Lytle, J.H. Sinfelt, Analysis of EXAFS data on bimetallic clusters, Catal. Lett. 5 (1990) 25–34.
[50] H.E. Kluksdahl, Reforming a sulfur-free naphtha with a platinum–rhenium catalyst, US Patent 3,415,737, 1968.
[51] E.A. Sutton, A.R. Greenwood, F.H. Adams, A new processing concept: continuous platforming, Oil Gas J. 70 (May 22) (1972) 52–56.
[52] I. Langmuir, The constitution and fundamental properties of solids and liquids. Part I. Solids, J. Am. Chem. Soc. 38 (1916) 2221–2295.
[53] J.K. Dixon, J.E. Longfield, Oxidation of ammonia, ammonia and methane, carbon monoxide and sulfur dioxide, in: P.H. Emmett (Ed.), Catalysis, vol. VII, Reinhold, New York, 1960, pp. 281–345.

[54] C.N. Hinshelwood, The Kinetics of Chemical Change, Clarendon Press, Oxford, 1940 (pp. 178–233).
[55] H.S. Taylor, The activation energy of adsorption processes, J. Am. Chem. Soc. 53 (1931) 578–597.
[56] P.H. Emmett, S. Brunauer, The adsorption of nitrogen by iron synthetic ammonia catalysts, J. Am. Chem. Soc. 56 (1934) 35–41.
[57] S. Brunauer, K.S. Love, R.G. Keenan, Adsorption of nitrogen and the mechanism of ammonia decomposition over iron catalysts, J. Am. Chem. Soc. 64 (1942) 751–758.
[58] S. Brunauer, P.H. Emmett, E. Teller, Adsorption of gases in multimolecular layers, J. Am. Chem. Soc. 60 (1938) 309–319.
[59] P.H. Emmett, S. Brunauer, Accumulation of alkali promoters on surfaces of iron synthetic ammonia catalysts, J. Am. Chem. Soc. 59 (1937) 310–315.
[60] J.H. Sinfelt, Highly dispersed catalytic materials, Ann. Rev. Mater. Sci. 2 (1972) 641–662.
[61] G.K. Boreskov, Catalytic activation of dioxygen, in: J.R. Anderson, M. Boudart (Eds.), Catalysis—Science and Technology, vol. 3, Springer, Berlin, 1982, pp. 39–137.
[62] G.C.A. Schuit, L.L. van Reijen, The structure and activity of metal-on-silica catalysts, Adv. Catal. 10 (1958) 242–317.
[63] L. Spenadel, M. Boudart, Dispersion of platinum on supported catalysts, J. Phys. Chem. 64 (1960) 204–207.
[64] J.J. Keavney, S.F. Adler, The physical nature of supported platinum, J. Phys. Chem. 64 (1960) 208–212.
[65] D.J.C. Yates, W.F. Taylor, J.H. Sinfelt, Catalysis over supported metals. I. Kinetics of ethane hydrogenolysis over nickel surfaces of known area, J. Am. Chem. Soc. 86 (1964) 2996–3001.
[66] J.H. Sinfelt, W.F. Taylor, D.J.C. Yates, Catalysis over supported metals. III. Comparison of metals of known surface area for ethane hydrogenolysis, J. Phys. Chem. 69 (1965) 95–101.
[67] J.A. Cusumano, G.W. Dembinski, J.H. Sinfelt, Chemisorption and catalytic properties of supported platinum, J. Catal. 5 (1966) 471–475.
[68] J.H. Sinfelt, D.J.C. Yates, Catalytic hydrogenolysis of ethane over the noble metals of group VIII, J. Catal. 8 (1967) 82–90.
[69] J.H. Sinfelt, D.J.C. Yates, Studies of ethane hydrogenolysis over group VIII metals: supported osmium and iron, J. Catal. 10 (1968) 362–367.
[70] D.J.C. Yates, J.H. Sinfelt, An investigation of the dispersion and catalytic properties of supported rhenium, J. Catal. 14 (1969) 182–186.
[71] R.A. Della Betta, J.A. Cusumano, J.H. Sinfelt, Cyclopropane–hydrogen reaction over the group VIII noble metals, J. Catal. 19 (1970) 343–349.
[72] M. Boudart, A. Aldag, J.E. Benson, N.A. Dougharty, C.G. Harkins, On the specific activity of platinum catalysts, J. Catal. 6 (1966) 92–102.
[73] M. Boudart, L.D. Ptak, Reactions of neopentane on transition metals, J. Catal. 16 (1970) 90–96.

[74] P. van der Plank, W.M.H. Sachtler, Surface composition of equilibrated copper–nickel alloy films, J. Catal. 7 (1967) 300–303.

[75] J.H. Sinfelt, J.L. Carter, D.J.C. Yates, Catalytic hydrogenolysis and dehydrogenation over copper–nickel alloys, J. Catal. 24 (1972) 283–296.

[76] P.W. Selwood, Magnetochemistry, second ed., Interscience, New York, 1956 (pp. 374–408).

[77] P.W. Selwood, Adsorption and Collective Paramagnetism, Academic Press, New York, 1962.

[78] K. Siegbahn, C. Nordling, A. Fahlman, R. Nordberg, K. Hamrin, J. Hedman, G. Johansson, T. Bergmark, S.-E. Karlsson, I. Lindgren, B. Lindberg, ESCA-Electron Spectroscopy for Chemical Analysis, Almqvist & Wiksells Boktryckeri AB, Uppsala, 1967.

[79] C.R. Helms, J.H. Sinfelt, Electron spectroscopy (ESCA) studies of Ru–Cu catalysts, Surf. Sci. 72 (1978) 229–242.

[80] J.H. Sinfelt, Y.L. Lam, J.A. Cusumano, A.E. Barnett, Nature of ruthenium–copper catalysts, J. Catal. 42 (1976) 227–237.

[81] J.H. Sinfelt, G.H. Via, F.W. Lytle, Structure of bimetallic clusters. Extended X-ray absorption fine structure (EXAFS) studies of Pt–Ir clusters, J. Chem. Phys. 76 (1982) 2779–2789.

[82] G. Meitzner, G.H. Via, F.W. Lytle, J.H. Sinfelt, Structure of bimetallic clusters. Extended X-ray absorption fine structure (EXAFS) of Pt–Re and Pd–Re clusters, J. Chem. Phys. 87 (1987) 6354–6363.

[83] P.-K. Wang, J.-P. Ansermet, S.L. Rudaz, Z. Wang, S. Shore, C.P. Slichter, J.H. Sinfelt, NMR studies of simple molecules on metal surfaces, Science 234 (1986) 35–41.

[84] D.B. Zax, C.A. Klug, C.P. Slichter, J.H. Sinfelt, Reaction kinetics of ethylene conversion and hydrogen exchange on Pt by ^{2}H NMR, J. Phys. Chem. 93 (1989) 5009–5012.

[85] C.A. Klug, C.P. Slichter, J.H. Sinfelt, A ^{2}H NMR study of the formation of ethylidyne from acetylene and hydrogen coadsorbed on Pt, J. Phys. Chem. 95 (1991) 2119–2121.

[86] O. Beeck, Catalysis—A challenge to the physicist, Rev. Mod. Phys. 17 (1945) 61–71.

[87] C. Kemball, The catalytic exchange of hydrocarbons with deuterium, Adv. Catal. 11 (1959) 223–262.

[88] B.M.W. Trapnell, Chemisorption, first ed., Butterworth, London, 1955 (p. 173).

[89] L.H. Germer, A new electron diffraction technique, potentially applicable to research in catalysis, Adv. Catal. 13 (1962) 191–201.

[90] H.E. Farnsworth, The clean single-crystal-surface approach to surface reactions, Adv. Catal. 15 (1964) 31–63.

[91] G.A. Somorjai, The surface science of heterogeneous catalysis, Proceedings of the Robert A. Welch Foundation Conferences on Chemical Research, Houston, Texas, November 9–11, 1981, Heterogeneous Catalysis, vol. XXV, pp. 83–127.

[92] C. Bokhoven, C. van Heerden, R. Westrik, P. Zwietering, Research on ammonia synthesis since 1940, in: P.H. Emmett (Ed.), Catalysis, vol. III, Reinhold, New York, 1955, pp. 265–348.

[93] G. Ertl, Primary steps in ammonia synthesis, Proceedings of the Robert A. Welch Foundation Conferences on Chemical Research, Houston, Texas, November 9–11, 1981, Heterogeneous Catalysis, vol. XXV, pp. 179–207.

[94] M. Boudart, G. Djéga-Mariadassou, Kinetics of Heterogeneous Catalytic Reactions, Princeton University Press, Princeton, NJ, 1984 (pp. 155–193).

[95] D.A. Hickman, L.D. Schmidt, Production of syngas by direct catalytic oxidation of methane, Science 259 (1993) 343–346.

[96] W.R. Williams, C.M. Marks, L.D. Schmidt, Steps in the reaction $H_2 + O_2 \rightleftharpoons H_2O$ on Pt: OH desorption at high temperatures, J. Phys. Chem. 96 (1992) 5922–5931.

[97] M.P. Zum Mallen, W.R. Williams, L.D. Schmidt, Steps in H_2 oxidation on Rh: OH desorption at high temperatures, J. Phys. Chem. 97 (1993) 625–632.

[98] M. Huff, L.D. Schmidt, Ethylene formation by oxidative dehydrogenation of ethane over monoliths at very short contact times, J. Phys. Chem. 97 (1993) 11815–11822.

[99] A.S. Bodke, D.A. Olschki, L.D. Schmidt, E. Ranzi, High selectivities to ethylene by partial oxidation of ethane, Science 285 (1999) 712–715.

[100] M.C. Huff, I.P. Androulakis, J.H. Sinfelt, S.C. Reyes, The contribution of gas-phase reactions in the Pt-catalyzed conversion of ethane–oxygen mixtures, J. Catal. 191 (2000) 46–54.

[101] J.A. Dumesic, D.F. Rudd, L.M. Aparicio, J.E. Rekoske, A.A. Trevino, The microkinetics of heterogeneous catalysis, American Chemical Society, Washington, DC, 1993.

[102] M. Iwamoto, H. Hamada, Removal of nitrogen monoxide from exhaust gases through novel catalytic processes, Catal. Today 10 (1991) 57–71.

[103] P.B. Weisz, Zeolites—New horizons in catalysis, Chem. Technol. 3 (1973) 498–505.

[104] S.C. Reyes, J.H. Sinfelt, G.J. DeMartin, Diffusion in porous solids: The parallel contribution of gas and surface diffusion processes in pores extending from the mesoporous region into the microporous region, J. Phys. Chem. B 104 (2000) 5750–5761.

[35] H. Heinemann, A brief history of industrial catalysis, in: J.R. Anderson, M. Boudart (Eds.), Catalysis—Science and Technology, vol. 1, Springer, Berlin, 1981, pp. 1–41.
[36] D.W. Breck, Zeolite Molecular Sieves, Wiley, New York, 1974.
[37] F.G. Ciapetta, R.M. Dobres, R.W. Baker, Catalytic reforming of pure hydrocarbons and petroleum naphthas, in: P.H. Emmett (Ed.), Catalysis, vol. VI, Reinhold, New York, 1958, pp. 495–692.
[38] V. Haensel, Aromatization, hydroforming and platforming, in: B.T. Brooks, C.E. Boord, S.S. Kurtz, L. Schmerling (Eds.), The Chemistry of Petroleum Hydrocarbons, vol. II, Reinhold, New York, 1955, pp. 189–219.
[39] J.H. Sinfelt, Catalytic reforming of hydrocarbons, in: J.R. Anderson, M. Boudart (Eds.), Catalysis—Science and Technology, vol. 1, Springer, Berlin, 1981, pp. 257–300.
[40] G.A. Mills, H. Heinemann, T.H. Milliken, A.G. Oblad, Houdriforming reactions. Catalytic mechanism, Ind. Eng. Chem. 45 (1953) 134–137.
[41] J.H. Sinfelt, H. Hurwitz, J.C. Rohrer, Kinetics of n-pentane isomerization over Pt–Al_2O_3 catalyst, J. Phys. Chem. 64 (1960) 892–894.
[42] J.H. Sinfelt, J.C. Rohrer, Kinetics of the catalytic isomerization–dehydroisomerization of methylcyclopentane, J. Phys. Chem. 65 (1961) 978–981.
[43] J.H. Sinfelt, H. Hurwitz, J.C. Rohrer, Role of dehydrogenation activity in the catalytic isomerization and dehydrocyclization of hydrocarbons, J. Catal. 1 (1962) 481–483.
[44] J.H. Sinfelt, H. Hurwitz, R.A. Shulman, Kinetics of methylcyclohexane dehydrogenation over Pt–Al_2O_3, J. Phys. Chem. 64 (1960) 1559–1562.
[45] J.H. Sinfelt, Supported bimetallic cluster catalysts, J. Catal. 29 (1973) 308–315.
[46] J.H. Sinfelt, Catalysis by alloys and bimetallic clusters, Acc. Chem. Res. 10 (1977) 15–20.
[47] J.H. Sinfelt, Structure of bimetallic clusters, Acc. Chem. Res. 20 (1987) 134–139.
[48] J.H. Sinfelt, G.H. Via, F.W. Lytle, Application of EXAFS in catalysis. Structure of bimetallic cluster catalysts, Catal. Rev. 26 (1) (1984) 81–140.
[49] G.H. Via, K.F. Drake Jr., G. Meitzner, F.W. Lytle, J.H. Sinfelt, Analysis of EXAFS data on bimetallic clusters, Catal. Lett. 5 (1990) 25–34.
[50] H.E. Kluksdahl, Reforming a sulfur-free naphtha with a platinum–rhenium catalyst, US Patent 3,415,737, 1968.
[51] E.A. Sutton, A.R. Greenwood, F.H. Adams, A new processing concept: continuous platforming, Oil Gas J. 70 (May 22) (1972) 52–56.
[52] I. Langmuir, The constitution and fundamental properties of solids and liquids. Part I. Solids, J. Am. Chem. Soc. 38 (1916) 2221–2295.
[53] J.K. Dixon, J.E. Longfield, Oxidation of ammonia, ammonia and methane, carbon monoxide and sulfur dioxide, in: P.H. Emmett (Ed.), Catalysis, vol. VII, Reinhold, New York, 1960, pp. 281–345.
[54] C.N. Hinshelwood, The Kinetics of Chemical Change, Clarendon Press, Oxford, 1940 (pp. 178–233).
[55] H.S. Taylor, The activation energy of adsorption processes, J. Am. Chem. Soc. 53 (1931) 578–597.
[56] P.H. Emmett, S. Brunauer, The adsorption of nitrogen by iron synthetic ammonia catalysts, J. Am. Chem. Soc. 56 (1934) 35–41.
[57] S. Brunauer, K.S. Love, R.G. Keenan, Adsorption of nitrogen and the mechanism of ammonia decomposition over iron catalysts, J. Am. Chem. Soc. 64 (1942) 751–758.
[58] S. Brunauer, P.H. Emmett, E. Teller, Adsorption of gases in multimolecular layers, J. Am. Chem. Soc. 60 (1938) 309–319.
[59] P.H. Emmett, S. Brunauer, Accumulation of alkali promoters on surfaces of iron synthetic ammonia catalysts, J. Am. Chem. Soc. 59 (1937) 310–315.
[60] J.H. Sinfelt, Highly dispersed catalytic materials, Ann. Rev. Mater. Sci. 2 (1972) 641–662.
[61] G.K. Boreskov, Catalytic activation of dioxygen, in: J.R. Anderson, M. Boudart (Eds.), Catalysis—Science and Technology, vol. 3, Springer, Berlin, 1982, pp. 39–137.
[62] G.C.A. Schuit, L.L. van Reijen, The structure and activity of metal-on-silica catalysts, Adv. Catal. 10 (1958) 242–317.
[63] L. Spenadel, M. Boudart, Dispersion of platinum on supported catalysts, J. Phys. Chem. 64 (1960) 204–207.
[64] J.J. Keavney, S.F. Adler, The physical nature of supported platinum, J. Phys. Chem. 64 (1960) 208–212.
[65] D.J.C. Yates, W.F. Taylor, J.H. Sinfelt, Catalysis over supported metals. I. Kinetics of ethane hydrogenolysis over nickel surfaces of known area, J. Am. Chem. Soc. 86 (1964) 2996–3001.
[66] J.H. Sinfelt, W.F. Taylor, D.J.C. Yates, Catalysis over supported metals. III. Comparison of metals of known surface area for ethane hydrogenolysis, J. Phys. Chem. 69 (1965) 95–101.
[67] J.A. Cusumano, G.W. Dembinski, J.H. Sinfelt, Chemisorption and catalytic properties of supported platinum, J. Catal. 5 (1966) 471–475.
[68] J.H. Sinfelt, D.J.C. Yates, Catalytic hydrogenolysis of ethane over the noble metals of group VIII, J. Catal. 8 (1967) 82–90.
[69] J.H. Sinfelt, D.J.C. Yates, Studies of ethane hydrogenolysis over group VIII metals: supported osmium and iron, J. Catal. 10 (1968) 362–367.
[70] D.J.C. Yates, J.H. Sinfelt, An investigation of the dispersion and catalytic properties of supported rhenium, J. Catal. 14 (1969) 182–186.
[71] R.A. Della Betta, J.A. Cusumano, J.H. Sinfelt, Cyclopropane–hydrogen reaction over the group VIII noble metals, J. Catal. 19 (1970) 343–349.
[72] M. Boudart, A. Aldag, J.E. Benson, N.A. Dougharty, C.G. Harkins, On the specific activity of platinum catalysts, J. Catal. 6 (1966) 92–102.
[73] M. Boudart, L.D. Ptak, Reactions of neopentane on transition metals, J. Catal. 16 (1970) 90–96.

[74] P. van der Plank, W.M.H. Sachtler, Surface composition of equilibrated copper–nickel alloy films, J. Catal. 7 (1967) 300–303.

[75] J.H. Sinfelt, J.L. Carter, D.J.C. Yates, Catalytic hydrogenolysis and dehydrogenation over copper–nickel alloys, J. Catal. 24 (1972) 283–296.

[76] P.W. Selwood, Magnetochemistry, second ed., Interscience, New York, 1956 (pp. 374–408).

[77] P.W. Selwood, Adsorption and Collective Paramagnetism, Academic Press, New York, 1962.

[78] K. Siegbahn, C. Nordling, A. Fahlman, R. Nordberg, K. Hamrin, J. Hedman, G. Johansson, T. Bergmark, S.-E. Karlsson, I. Lindgren, B. Lindberg, ESCA-Electron Spectroscopy for Chemical Analysis, Almqvist & Wiksells Boktryckeri AB, Uppsala, 1967.

[79] C.R. Helms, J.H. Sinfelt, Electron spectroscopy (ESCA) studies of Ru–Cu catalysts, Surf. Sci. 72 (1978) 229–242.

[80] J.H. Sinfelt, Y.L. Lam, J.A. Cusumano, A.E. Barnett, Nature of ruthenium–copper catalysts, J. Catal. 42 (1976) 227–237.

[81] J.H. Sinfelt, G.H. Via, F.W. Lytle, Structure of bimetallic clusters. Extended X-ray absorption fine structure (EXAFS) studies of Pt–Ir clusters, J. Chem. Phys. 76 (1982) 2779–2789.

[82] G. Meitzner, G.H. Via, F.W. Lytle, J.H. Sinfelt, Structure of bimetallic clusters. Extended X-ray absorption fine structure (EXAFS) of Pt–Re and Pd–Re clusters, J. Chem. Phys. 87 (1987) 6354–6363.

[83] P.-K. Wang, J.-P. Ansermet, S.L. Rudaz, Z. Wang, S. Shore, C.P. Slichter, J.H. Sinfelt, NMR studies of simple molecules on metal surfaces, Science 234 (1986) 35–41.

[84] D.B. Zax, C.A. Klug, C.P. Slichter, J.H. Sinfelt, Reaction kinetics of ethylene conversion and hydrogen exchange on Pt by ^{2}H NMR, J. Phys. Chem. 93 (1989) 5009–5012.

[85] C.A. Klug, C.P. Slichter, J.H. Sinfelt, A ^{2}H NMR study of the formation of ethylidyne from acetylene and hydrogen coadsorbed on Pt, J. Phys. Chem. 95 (1991) 2119–2121.

[86] O. Beeck, Catalysis—A challenge to the physicist, Rev. Mod. Phys. 17 (1945) 61–71.

[87] C. Kemball, The catalytic exchange of hydrocarbons with deuterium, Adv. Catal. 11 (1959) 223–262.

[88] B.M.W. Trapnell, Chemisorption, first ed., Butterworth, London, 1955 (p. 173).

[89] L.H. Germer, A new electron diffraction technique, potentially applicable to research in catalysis, Adv. Catal. 13 (1962) 191–201.

[90] H.E. Farnsworth, The clean single-crystal-surface approach to surface reactions, Adv. Catal. 15 (1964) 31–63.

[91] G.A. Somorjai, The surface science of heterogeneous catalysis, Proceedings of the Robert A. Welch Foundation Conferences on Chemical Research, Houston, Texas, November 9–11, 1981, Heterogeneous Catalysis, vol. XXV, pp. 83–127.

[92] C. Bokhoven, C. van Heerden, R. Westrik, P. Zwietering, Research on ammonia synthesis since 1940, in: P.H. Emmett (Ed.), Catalysis, vol. III, Reinhold, New York, 1955, pp. 265–348.

[93] G. Ertl, Primary steps in ammonia synthesis, Proceedings of the Robert A. Welch Foundation Conferences on Chemical Research, Houston, Texas, November 9–11, 1981, Heterogeneous Catalysis, vol. XXV, pp. 179–207.

[94] M. Boudart, G. Djéga-Mariadassou, Kinetics of Heterogeneous Catalytic Reactions, Princeton University Press, Princeton, NJ, 1984 (pp. 155–193).

[95] D.A. Hickman, L.D. Schmidt, Production of syngas by direct catalytic oxidation of methane, Science 259 (1993) 343–346.

[96] W.R. Williams, C.M. Marks, L.D. Schmidt, Steps in the reaction $H_2 + O_2 \rightleftharpoons H_2O$ on Pt: OH desorption at high temperatures, J. Phys. Chem. 96 (1992) 5922–5931.

[97] M.P. Zum Mallen, W.R. Williams, L.D. Schmidt, Steps in H_2 oxidation on Rh: OH desorption at high temperatures, J. Phys. Chem. 97 (1993) 625–632.

[98] M. Huff, L.D. Schmidt, Ethylene formation by oxidative dehydrogenation of ethane over monoliths at very short contact times, J. Phys. Chem. 97 (1993) 11815–11822.

[99] A.S. Bodke, D.A. Olschki, L.D. Schmidt, E. Ranzi, High selectivities to ethylene by partial oxidation of ethane, Science 285 (1999) 712–715.

[100] M.C. Huff, I.P. Androulakis, J.H. Sinfelt, S.C. Reyes, The contribution of gas-phase reactions in the Pt-catalyzed conversion of ethane–oxygen mixtures, J. Catal. 191 (2000) 46–54.

[101] J.A. Dumesic, D.F. Rudd, L.M. Aparicio, J.E. Rekoske, A.A. Trevino, The microkinetics of heterogeneous catalysis, American Chemical Society, Washington, DC, 1993.

[102] M. Iwamoto, H. Hamada, Removal of nitrogen monoxide from exhaust gases through novel catalytic processes, Catal. Today 10 (1991) 57–71.

[103] P.B. Weisz, Zeolites—New horizons in catalysis, Chem. Technol. 3 (1973) 498–505.

[104] S.C. Reyes, J.H. Sinfelt, G.J. DeMartin, Diffusion in porous solids: The parallel contribution of gas and surface diffusion processes in pores extending from the mesoporous region into the microporous region, J. Phys. Chem. B 104 (2000) 5750–5761.

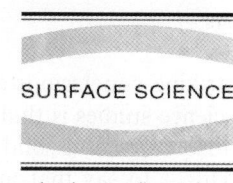

The surface chemistry of catalysis: new challenges ahead

Francisco Zaera *

Department of Chemistry, University of California, Riverside, CA 92521, USA

Received 27 June 2000; accepted for publication 16 April 2001

Abstract

Surface scientists have over the last few decades greatly advanced the atomic level understanding of the surface chemical reactions associated with heterogeneous catalysis. Nonetheless, many fundamental questions still remain unanswered. In this review, a critical analysis of the state of the art of this field is provided, and a number of future research directions are suggested. © 2001 Elsevier Science B.V. All rights reserved.

Keywords: Catalysis; Surface chemical reaction; Chemisorption; Surface energy; Single crystal surfaces

1. Introduction

Heterogeneous catalysis, by which a chemical reaction is accelerated with the aid of a solid surface, is a mature field that profoundly affects our everyday lives. Indeed, more than 80% of the industrial chemical processes in use nowadays rely on one or more catalytic reactions [1]. A number of those, including oil refining, petrochemical processing and the manufacturing of commodity chemicals (olefins, methanol, ethylene glycol, etc.), are already well established. But many others, such as the clean manufacturing of specific pharmaceuticals and the abatement of pollutants, represent industrial challenges requiring the development of entirely new chemistry.

Progress in catalysis has in many instances been serendipitous. Initial industrial designs required exhaustive trial and error studies, and only the results from those trials provided the body of information that allowed for the development of some empirical intuition. With the advent of new surface-sensitive analytical techniques, however, a more basic understanding of the surface chemistry underlying heterogeneous catalytic processes developed. Much progress has been made in this respect, but research on the microscopic details of catalysis is still a young field. In this review we highlight a number of what we consider key directions in surface science for the advancement of the field of catalysis in the next century.

2. The surface chemical bond

Heterogeneous catalysis relies on the opening of new, fast reaction channels involving the adsorption and conversion of reactants on the surface of the catalyst. Consequently, the key to the success of catalytic processes lies in the details of the interactions between the surface and the adsorbed molecules—the adsorbates. One of the most

---

* Tel.: +1-909-787-5498; fax: +1-909-787-3962.
E-mail address: francisco.zaera@ucr.edu (F. Zaera).

striking conclusions arising from modern surface-science studies is that the chemical bonds between adsorbates and solid surfaces are quite localized. This is to say that most of the electronic interactions of individual adsorbates occur with only a small ensemble of surface atoms. The localized nature of the chemistry that takes place on solid surfaces contrasts with the delocalized aspects of many physical properties of solids such as light reflectivity and electrical conductivity. Because of this localization of the surface bonds, a large number of catalytic systems can be described in terms of analogous molecular chemistry [2]. Specifically, clear parallels have been identified between systems involving solid surfaces and those associated with the discrete molecules containing organic fragments bonded to metal centers known as organometallic compounds [3,4].

One clear manifestation of the localization of surface bonds is the specific geometry that the atoms of the adsorbed molecules adopt with respect to those of the surface of the solid. Again, the geometry of adsorbed systems often parallels that of similar small molecules. An early example of such similarity is provided by the coordination of carbon monoxide to transition metal centers. The most common adsorption geometries of that molecule on surfaces, namely, on-top of one metal atom (in a linear arrangement), in a twofold bridge between two metal atoms, and nested in three or fourfold hollow sites, follow closely those in organometallic complexes [5]. Even the unique tilted CO adsorption arrangement observed on a few metal surfaces has been modeled with isolated molecules [6]. Fig. 1 provides another example of this organometallic-surface similarity, in this case for the coordination of an organic fragment to a kernel of metal centers [5].

The analogy between organometallic and surface chemistry has proven quite powerful, especially in the area of hydrocarbon chemistry. Proposed bonding models for olefins, alkyls, and other relevant surface intermediates based on crystallographic studies with organometallic models have often been corroborated by surface structural studies. Because it is significantly more difficult to determine the structure of surface species than that of discrete molecules, it is now routine to start with the information available in the literature on individual organometallic compounds in order to infer the structure and reactivity of adsorbates. It is nevertheless possible to obtain structural information on surface species directly. Initial surface structural studies relied on

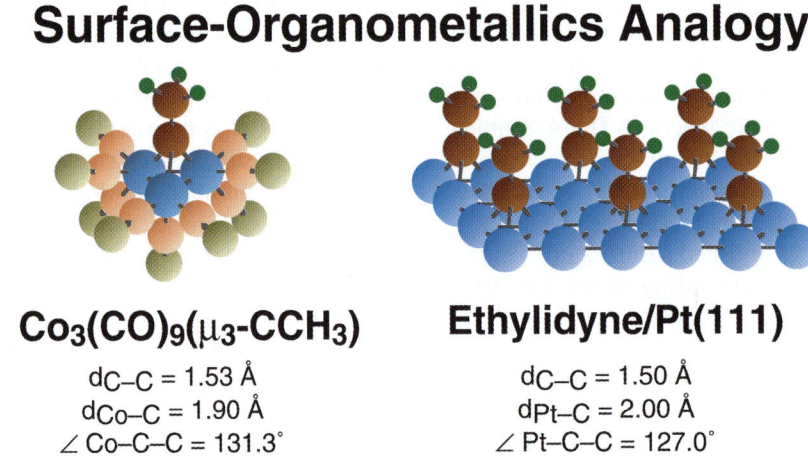

Fig. 1. Example of the bonding similarities between surface species and organometallic compounds. This figure contrasts the structures of ethylidyne molecular fragments in $Co_3(CO)_9(\mu_3\text{-}CCH_3)$, as determined by X-ray crystallography, and on Pt(1 1 1) surfaces, as calculated from low energy electron diffraction (LEED) data [5]. Notice in particular the almost identical local geometry of the molecular fragments inside the metal triangles on the surface and within the discrete molecule. d_{C-C}, d_{Co-C}, and d_{Pt-C} refer to the corresponding bond distances between the two stated atoms, while $\angle Co\text{-}C\text{-}C$ and $\angle Pt\text{-}C\text{-}C$ refer to the appropriate bond angles.

general diffraction techniques, and were limited by the required long-range order of surfaces. Fortunately, a number of additional methodologies capable of probing local geometries, including some new scanning microscopies [7], have been recently added to our toolbox [8–11] The development of novel ways of employing optical probes such as infrared and Raman spectroscopies for surface structural determinations have helped in this endeavor as well, in particular in the study of interfaces under non-vacuum environments [12]. Nevertheless, much more work is still required to develop a systematic picture of the structural and electronic details of adsorbates on surfaces.

3. Surface energetics

In addition to the geometrical considerations mentioned above, the viability of a catalytic process also depends on the strength (energy) of the adsorbate–surface bonds. In that respect, the analogy with organometallic chemistry can only be taken so far, because the energetics of the bonds formed between reaction intermediates and catalytic surfaces are altered in a significant manner by the presence of other nearby surface species. Since the energy required for a reaction to proceed (the activation energy) is closely related to the energies of the chemical bonds involved, the presence of such "spectator" adsorbates on the surface of a catalyst leads to appreciable changes in catalytic reaction rates. These effects need to be included in any model aimed at describing the rates of complex heterogeneous processes.

Initial studies on the kinetics (rates) of simple adsorption and desorption steps at a molecular level addressed this problem by assuming that activation energies depend on the coverage, that is, the surface concentration, of the adsorbates [13–15]. In addition, more recent and accurate kinetic experiments have indicated that reaction rates also depend on the local spatial arrangement of the adsorbates on the surface. An example of this is provided in Fig. 2 for the case of the oxidation of carbon monoxide on Pt(1 1 1). The results in that figure clearly show that the clustering of adsorbed oxygen atoms can significantly enhance the rate of carbon dioxide production [16]. A large body of fundamental experimental [17] and theoretical [18,19] work has focused on explaining the nature of the interactions among adsorbates, but little has been done to incorporate that knowledge into kinetic models to describe reaction rates in catalytic processes [20,21]. This remains another challenge for the future.

Part of the difficulty with the development of an appropriate description of surface reaction kinetics resides in the lack of good-quality data to test theoretical models. Extensive kinetic measurements have been carried out on real catalysts over the years [22–24]. These involve the use of ill-characterized, highly heterogeneous surfaces, however, and have provided poorly defined average rates with almost no connection to specific elementary reaction steps. Better designed desorption kinetic studies have been performed for a number of adsorbates on model surfaces under controlled vacuum environments [25,26]. Unfortunately, those experiments usually involve the simultaneous and correlated change of surface temperature and surface coverages, a connection that severely limits the usefulness of the data. Finally, work using gas beams has focused mainly on the determination of the initial probabilities for molecules to stick to surfaces, and has typically involved gas-phase molecules traveling at speeds much faster than those encountered in realistic catalytic systems [27,28].

A new generation of molecular beam work has emerged in recent years specifically aimed at the emulation of catalytic reactions with well-characterized surfaces under controlled environments [15,16,29,30]. The components of a typical set-up used in this research are schematically depicted in Fig. 3. The new focus has been to follow not only the adsorption kinetics of the reactants, but also the rate of formation of the products. In some cases the probability for reactants to stick on the surface under catalytic conditions, where two or more types of molecules compete for the same adsorption sites on the surface, has also been determined. Beams with molecular velocities similar to those associated with catalysis have also been used. Lastly, pulsing of the beams has allowed for the study of kinetic transients, in particular those

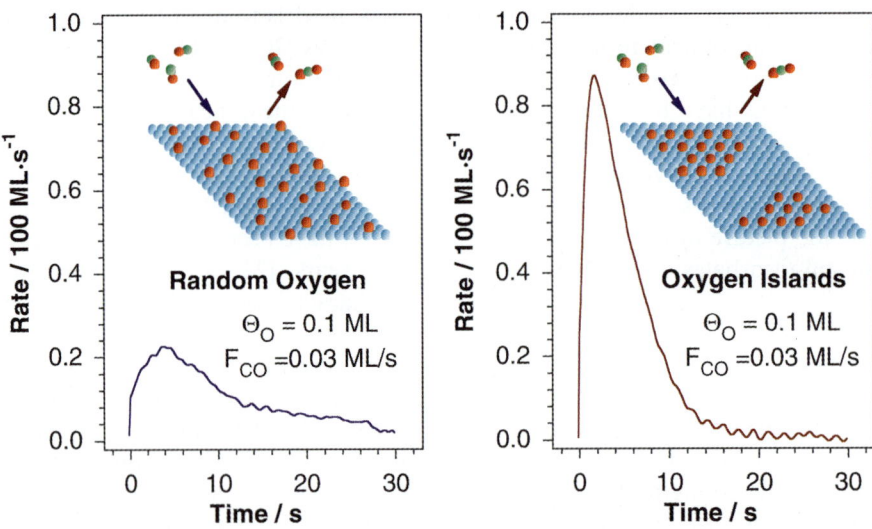

Fig. 2. Oxidation kinetics for carbon monoxide on oxygen-covered platinum single-crystal surfaces illustrating the importance that the spatial distribution of the adsorbates has in determining reaction rates [16]. Two different sample preparation methods were used to obtain a random distribution of O atoms on the surface in one case (left) and a clustered arrangement in ordered surface islands in the other (right). Large differences in rates were observed between the two cases, in spite of the fact that both kinetic runs started with the same average CO fluxes ($F_{CO} = 0.03$ ML/s) and oxygen concentrations on the surface ($\Theta_O = 0.1$ ML). This example illustrates the role that neighboring adsorbates play in modifying both the energetics of the reacting species and the rates of reaction.

of weakly adsorbed intermediates. Examples on the kind of detailed kinetic data obtainable with this approach include the identification of coverage and surface ensemble effects on surface reaction rates for the cases of CO oxidation [16], NO reduction [31], and ethylene hydrogenation [32] reactions. Again, only by obtaining high-quality isothermal kinetic data for catalytic reactions systematically as a function of temperature and surface coverages (or, more appropriately, as a function of impinging frequencies of reactants on the surface) will it be possible to establish kinetic models to fully describe reaction rates in catalysis.

4. The transition state

The success of catalysis lies on the ability of a given catalyst to promote a specific reaction by accelerating its rate relative to those of other undesirable alternatives. A good understanding of the specific manner by which surface reactions take place at an atomic level is therefore needed to design appropriate catalytic processes. For one, the conversion of reactants is controlled by the changes in energy induced by the specific atomic motions involved, and normally follows the path with the lowest energy of activation. It is therefore highly desirable to map out this energy dependence systematically as a function of atomic coordinates, that is, to develop an accurate potential energy surface. It is also quite useful to characterize the molecular conformation corresponding to the highest energy point the system needs to cross to go from reactants to products—the transition state. This knowledge can be used to envision the changes needed in the catalyst in order to optimize its performance.

Transient Kinetics with Molecular Beams

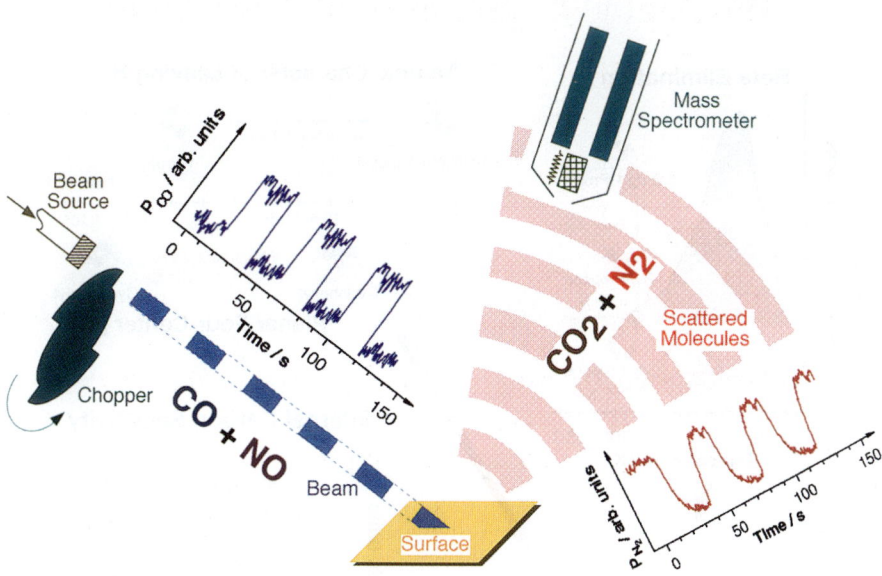

Fig. 3. Schematic depiction of a molecular beam set-up designed to study the kinetics of catalytic reactions under controlled conditions. A collimated beam with a realistic energy distribution is generated under vacuum (left side) and directed towards the catalytic surface (bottom square) while the time evolution of the partial pressures of both reactants and products is followed by mass spectrometry (upper right). The partial pressures measured this way can then be directly related to reaction rates. Chopping of the beam (by the chopper placed right after the beam source in the upper right corner) is used to probe kinetic transients around the steady-state catalytic conditions. Typical data obtained with this experimental arrangement are provided here in the two graphs of P_{CO} and P_{N_2} (partial pressures of CO and N_2, respectively) versus time for the reduction of NO by CO on Rh(1 1 1) [98].

In probing transition states, surface science is advised to borrow some well-established methods from other fields of chemistry. For instance, the labeling of specific atoms via their substitution with isotopes of different masses is ideal for the identification of the reactive molecular fragments in adsorbates. This has been particularly useful in the determination of the reaction atomic details in hydrocarbon conversion processes [33]. Isotope substitution also allows for the selective identification of reactive adsorbates, in particular when special surface ensembles are involved. For instance, ^{15}N labeling has been recently used to demonstrate that the reduction of NO on rhodium occurs at the periphery of nitrogen surface islands and involves the formation of a N–NO intermediate [34]. Changes in reaction rates upon isotope substitution provide additional information on the details of reaction pathways [35]. Other substitutions can be employed to identify the effects of changes in electronic distributions within molecules on reactivity [36], and reactants with restricted structures such as strained cyclic molecules can be used to infer transition state geometries [37]. Fig. 4 illustrates how these ideas have been used to determine the four-center geometry and negative ion nature of the leaving hydrogen atom in the dehydrogenation of adsorbed alkyl groups from their beta (second from the attached end) position. Even though these systematic substitutions are used routinely in other areas of chemistry, the power of this approach has seldom been exploited in surface studies.

5. The catalytic site

Another aspect unique to solid surfaces is the fact that their complex structures include atomic ensembles for catalysis—catalytic sites—difficult

β-Hydride Elimination Mechanism

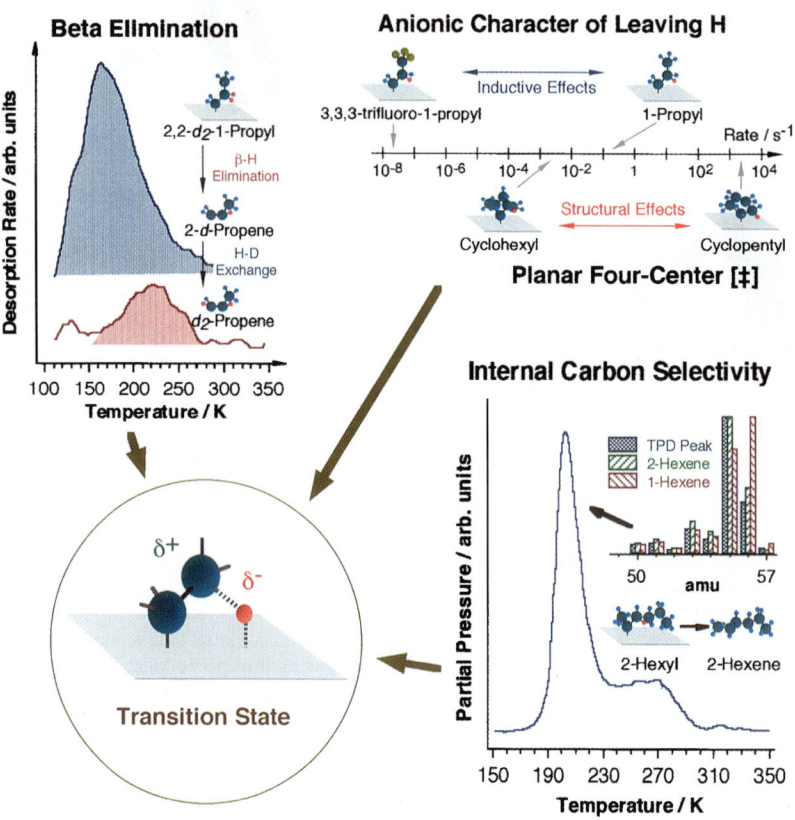

Fig. 4. Representative data from substitution studies used to determine the nature of the transient state in β-hydride elimination from adsorbed alkyl groups. Top, left: the preferential desorption of 2-d-propene (top trace with maximum around 160 K) from thermal decomposition of 2,2-d_2-1-propyl groups indicates that hydrogen (deuterium) removal occurs selectively at the beta carbon [99]. Only at higher (∼220 K) temperatures some d_2-propene is also produced by subsequent exchange reactions (bottom trace). Top, right: the large decrease in dehydrogenation rate—from about 0.1 to $< 2 \times 10^8$ s^{-1}—induced by substitution of the γ-hydrogens in 1-propyl groups with fluorine atoms (in 3,3,3-trifluoro-1-propyl) is due to a change in the electronic distribution within the molecule known as an inductive effect, and points to the negatively-charged (anionic) character of the leaving β-hydrogen [36]. In addition, the rate changes resulting from the constraints imposed by the structure of cyclic alkyls suggest a planar four-center [Pt–C–CH] transition state. In particular, note that the conversion of the planar cyclopentyl fragment is about six orders of magnitude faster than that of the buckled cyclohexyl species [37]. Bottom, right: Hydrogen atoms are extracted from inner (more substituted) carbons, hence the selectivity for 2-hexene over 1-hexene formation from 2-hexyl adsorbates [54]. All this can be summarized in the transition state structure shown inside the lower left circle.

to reproduce with discrete molecules. In fact, it is the strong dependence of the rates of some reactions on the structure of the catalyst what makes heterogeneous catalysis so useful. This correlation between surface structure and reactivity has been explored in the past via controlled surface-science studies on single crystals with different exposed planes [38,39]. One pristine example of the success of this approach comes from the study of ammonia synthesis from molecular nitrogen (the Haber process). Fig. 5 reveals that dramatic changes in reaction rates are observed for this reaction when

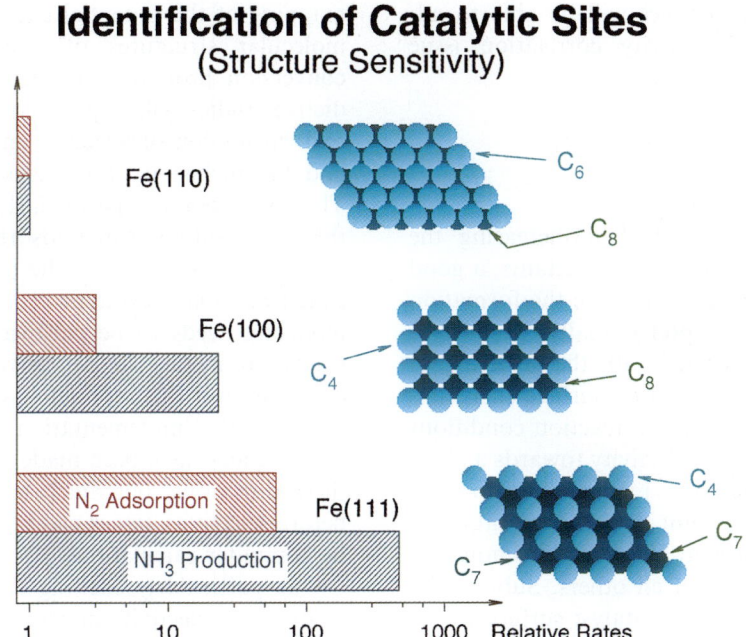

Fig. 5. Surface-science studies on the dependence of reaction rates on surface structure can lead to the identification of catalytic sites. Shown here are the relative rates measured for the dissociative adsorption of N_2 [40] and for the catalytic conversion of $N_2 + H_2$ to ammonia [41] on three different iron single crystals. The similar rate trends seen as a function of surface structure in the two cases indicate that molecular nitrogen dissociation is the step that limits the overall rate of ammonia synthesis. The correlation between surface structure and reactivity is likely explained by the different number of surrounding neighbors around a given surface atoms on the different surface planes. In the schematic representation of the three surfaces used here and shown on the right of the figure, atoms with 4, 6, 7, and 8 nearest-neighbors (C_4, C_6, C_7, and C_8, respectively) are indicated. In this context, the particularly high activity of the (1 1 1) plane points to the importance of the so-called C_7 site.

iron surfaces with different structures are used [40,41]. Two main conclusions were reached from that work: (1) the rate of ammonia production is limited by the initial adsorption and dissociation of the nitrogen molecule, N_2, into adsorbed nitrogen atoms (notice in particular the nice correlation between the rates of the two processes as a function of surface structure); and (2) specific so-called C_7 sites are needed for N_2 dissociation. The latter conclusion is of great significance for the design of ammonia synthesis catalysts.

Understanding the characteristics of the catalytic site is particularly important in complex materials. For example, the catalytic activity of most oxides has traditionally been ascribed to the presence of defects on the surface. Unfortunately, the exact nature of those defects is not yet fully known. A number of different types of sites, including oxygen vacancies, structural dislocations, reduced or oxidized metal atoms, and hydroxide, peroxides and other oxygen-containing species, have all been invoked to explain catalytic processes. Nevertheless, the establishment of a direct correlation between the existence of those sites on the surface and the activity of the catalyst awaits a better methodology. Here is where surface chemists can make a difference. The extension of traditional surface-science studies to multi-component surfaces such as alloys [42], oxides [43], and carbides [44] has already started. A combination of chemical and physical methods can also be used to prepare and to characterize specific defect sites in situ on well-defined substrates, as illustrated by our recent report on the titration of low-coordination nickel sites in nickel oxide [45]. Reactivity studies on surfaces with high concentrations of specific surface ensembles will be able to address this structure–reactivity correlation issue directly.

6. Selectivity

Perhaps more important than increasing the overall rate of conversion of the reactants, a good catalyst should be able to promote the formation of the desired products preferentially. Because of the complexities associated with the structure of catalytic sites and their effect on surface energetics and kinetics, small changes in reaction conditions can drastically alter this selectivity towards a given process. Again, it is worth remembering that kinetics (reaction rates) lie at the heart of catalysis: the role of a catalyst is to selectively promote one reaction at the expense of all others. Subtle variations in the nature of the catalyst surface (in its composition or its structure) or in the relative coverages or spatial distribution of the adsorbates may induce large relative kinetic changes, and result in a drastic switch in the preferred conversion route of the reactants. This is well exemplified by the vast number of chemicals manufacturing processes that depend on fine tuning of the procedures for catalyst preparation or the reaction conditions [46]. Particularly puzzling there is the fact that catalysts based on seemingly similar transition metals lead to widely diverse product distributions. Another example is that of hydrocarbon partial oxidations, where it is often desirable to produce oxygenated hydrocarbons such as alcohols, aldehydes, ketones, and epoxides and to avoid the formation of the more energetically stable but undesirable carbon dioxide total oxidation product [47,48]. A complete understanding of the parameters that affect selectivities is essential for the controlled design of catalytic processes.

The problem of selectivity represents a challenge not only to experimentalists, but also to theoreticians. Quality kinetic data need to be complemented with accurate calculations on adsorbate–surface interactions [49]. The difficulties to surmount here are many. For one, good estimates of the energy released by adsorption of surface intermediates need to be complemented with a mapping of the potential energy surface over the molecular structures of relevance to catalytic conversion processes, as mentioned before. Predictive studies will require both the identification and calculation of accurate energies of activation and the inclusion of the effect of coadsorbates. This latter point is particularly critical in view of the significant effect that adsorbed species exert on each other, especially at the high coverages associated with many catalytic processes. Finally, more modeling needs to be performed with the heavy transition metals so pervasive in catalytic processes, even if this represents a significant complication in the implementation of the calculations. Some strides have been made in recent years in all these directions, but much more work is needed before selectivity issues can be properly addressed by computer calculations.

One particularly exciting direction in terms of controlling selectivity in catalysis is that related to the manufacturing of chiral compounds. Chirality, the result of four different groups coordinating to a given carbon atom, has long been recognized as a crucial property in biology. Molecules containing a chiral carbon atom can exist in two identical conformations, the only difference being that they are non-superimposable mirror images of each other. Typically, only one of those so-called enantioisomers is biologically active, while the other is inert or even poisonous. The production of pure chiral pharmaceuticals is a growing industry in which heterogeneous catalysis is likely to play a prominent role. Unfortunately, only a handful of examples of catalytic processes for the selective production of a single enantioisomer are available to date [50], and no good understanding on how those work exists yet. Clearly, the selectivity of such catalysts must be driven by subtle energy differences between the two slightly different transition states for the two enantioisomers of a given chiral molecule. Some interesting preliminary data on the manifestation of those differences have been reported in model asymmetric surfaces [51,100], but much is still to be done in this direction of research. In the long run, it should be possible to design industrial processes for the manufacturing of chiral compounds via the controlled manipulation of catalytic sites.

7. The pressure gap

As mentioned before, surface scientists have made a great deal of progress towards understanding the fundamentals of bonding and reactivity of adsorbates on surfaces. The relevance of those studies to realistic catalytic systems is, however, still highly debated. The problem is that while most of the surface-science studies reported to date have been carried under well-controlled vacuum conditions, catalytic processes are typically performed at much higher (1–100 atm) pressures. It is not yet known to what extent the intermediates characterized by the former studies are involved in the reactions that take place in the latter systems.

This so-called "pressure gap" has been addressed by a number of research groups with mixed success. One difficulty here is the fact that the complexity of catalytic systems often makes traditional kinetics studies with real catalysts difficult to generalize. Results from such research often vary widely depending on how the catalysts are prepared. According to the severity of the changes in reactivity observed when changing catalyst preparation procedures, catalytic reactions have been divided into two broad groups. On one end, demanding reactions such as ammonia synthesis and CO, nitrogen oxides and hydrocarbon oxidation typically require high pressures and temperatures. At the other extreme, non-demanding reactions, hydrogenation of unsaturated hydrocarbons in particular, take place under mild conditions (atmospheric pressures, room temperature). The two types of reactions may follow significantly different kinds of mechanisms.

It is generally believed that during the course of demanding catalytic reactions the surface of the catalyst is covered with small concentrations of strongly bonded intermediates similar to those characterized by vacuum surface analytical techniques. Because of this, the knowledge acquired by surface-science studies is likely to transfer directly to catalysis. In the case of the conversion of methane, for example, catalytic kinetic runs on single crystals have been successfully used to reproduce data obtained using real supported catalysts [52]. Much has been learned about the initial adsorption of methane [28] as well as about the chemistry of the resulting methyl surface groups [53,54] by studying appropriate controlled systems under vacuum. That knowledge is being used to design viable commercial processes for the conversion of natural gas to more useful chemicals.

Non-demanding catalytic processes, on the other hand, present a much bigger challenge for the surface-science community. It has become clear that during non-demanding catalytic reactions the surface of the catalysts is covered with large amounts of strongly-held surface species, but the role of those molecular fragments has not yet been elucidated [46,55]. What has been shown so far is that they are not direct participants in the conversion of reactants to products, but that they affect the reaction in an indirect way instead. In the case of hydrocarbon conversions, for instance, the carbon-containing deposits that grow on the surface are believed to store hydrogen [56], and to perhaps passivate the metal to avoid deep dehydrogenation and decomposition of the reactants [46,55]. Unraveling the role of strongly bonded surface adsorbates in non-demanding reactions will require a great deal of creativity.

8. The materials gap

Another source of divergence between the chemistry of simple model systems and that which takes place on real catalysts is often referred to as the "materials gap". This gap has to do with the fact that most practical catalysts consist of small particles finely dispersed on high surface-area porous solids, and that the size and shape of those particles influence the catalytic chemistry in ways not easy to reproduce with single crystals, foils, or films. For one, the electronic properties of many solid particles, of metals in particular, change dramatically as their diameter becomes smaller than 10–100 Å; most supported catalysts involve particles in that size range. Moreover, catalytic particles often interact in complex ways with their supporting surfaces. In fact, the performance of some catalysts depends in part on the activity of the support itself. Reforming catalysts, for instance, rely as much on the acid properties of the silico–alumina support as on the hydrogenation–dehydrogenation activity of the active transition metals [24]. It is also possible for adsorbates to migrate from one phase, where they are produced, to another, where the final reactions take place [57]. Finally, in extreme cases, a strong interaction between the metal and the support makes the catalyst perform in ways not accountable by either phase alone [58]. None of these phenomena can be easily studied by using single crystals.

A number of approaches have been reported for the production of more elaborate model systems to better reproduce the complexities originating from interactions among the different components of practical catalysts. In particular, metal catalysts deposited on oxide supports have been emulated by the sequential physical deposition of thin oxide films and metal particles on well-defined substrates [59–61]. The example depicted in Fig. 6 nicely illustrates the series of steps associated with this type of work. First, model catalysts are prepared in situ under vacuum. In the case shown here, the initial deposition of a ∼10 nm thick silica film on Mo(1 0 0) was followed by evaporation of palladium particles. Next, the resulting surface is characterized with the aid of surface-sensitive spectroscopies. Scanning tunneling microscopy (STM, a technique used to map surface topography at the atomic level by rastering a sharp needle) was employed in our example to check that the resulting metal particles display a reasonably narrow distribution of sizes, averaging about

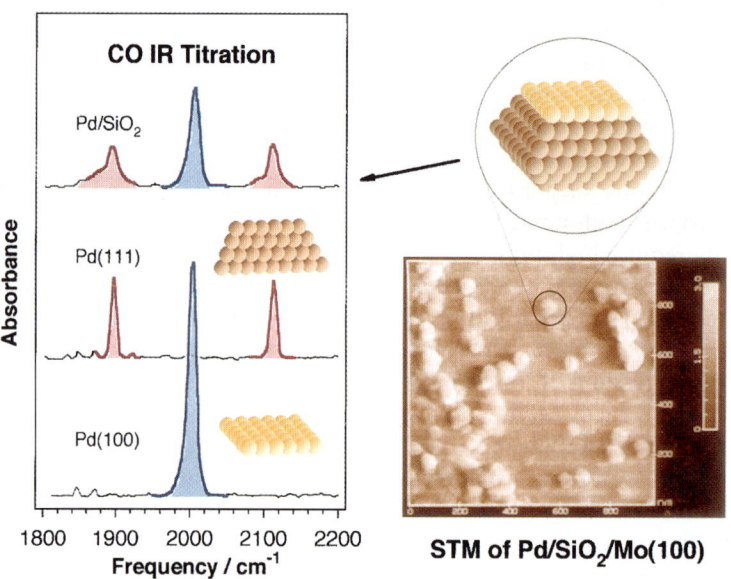

Fig. 6. Set of experimental data illustrating the steps typically followed during the development of model surfaces for the emulation of supported catalysts [59]. In this case, palladium particles were first deposited on a ∼10 nm SiO_2 thick film (prepared by evaporation of silicon on a Mo(1 0 0) single crystal in the presence of gas-phase oxygen). STM was then used to determine the particle size distribution (bottom, right). It was found that this sample contained approximately 50 Å diameter round metal particles. The surface was subsequently characterized by investigation of the vibrational frequencies of CO adsorbed on those particles, that is, by CO titration experiments such as those illustrated in the left panel. By comparing the infrared (IR) spectra of the adsorbed CO on the model catalyst with those from different palladium single crystals (the different traces of signal intensity—absorbance—versus frequency shown in the left panel), the exposed surface of the metallic particles were determined to contain an approximately 1:1 mixture of (1 1 1) and (1 0 0) planes. Finally, catalytic experiments were performed on the supported catalyst model (not shown).

50 Å in diameter. Independent studies on the adsorption of CO by infrared spectroscopy were also used to establish the nature of the surface of the metal particles, which were determined to contain mostly (1 0 0) and (1 1 1) planes. Finally, the catalytic activity of the model systems is checked. The oxidation of CO was tested in this example (results not shown). Studies applying this methodology on more demanding reactions are still in the future.

A more subtle aspect of the materials gap is the fact that the initial structure and composition of the catalyst may not be retained under the conditions of the reaction. Indeed, the fact that the atoms in solid surfaces rearrange to minimize their energy has been recognized for some time [5]. Adsorbates in particular are known to not only induce surface structural changes, but also facilitate the segregation of specific components. It is therefore imperative to study catalytic surfaces in situ during the course of the reactions in order to characterize their active phase. This problem has barely been addressed to date. Changes in surface composition can sometimes be assessed by post-mortem analysis, but caution must be exerted when extrapolating those results to the description of the active catalyst because of the potential reversibility of the changes induced by the catalytic environment. Surface structures can also change dramatically when in the presence of high pressures of reactants, as recently documented by in situ STM studies [62]. Again, work to answer these questions is yet to be done.

9. New experimental approaches

Addressing the issues raised in the previous discussion requires the development of new experimental methodologies. In terms of the study of chemical reactions at the atomic level, significant and exciting advances have recently come from the development of STM and its close relative, atomic force microscopy (AFM). These scanning microscopies provide the unique ability to image individual surface atoms, and to follow their fate in real time as they react. STM studies have already began to provide new information on the role of steps, kinks and other defects on the overall chemistry of surfaces. It is even possible to manipulate individual adsorbates with the tip of the scanning probes, to the point of inducing specific chemical reactions on the surface [63,64]. Fig. 7 illustrates how this has been accomplished in the case of the dimerization of phenyl groups on copper surfaces [65]. The use of scanning microscopy for the mapping and manipulation of individual surface sites constitutes a particularly promising new direction in surface science.

It is also crucial to be able to identify and characterize active reaction intermediates. As mentioned before, many intermediates adsorb only weakly on the surface, are present in small concentrations during reactions, and desorb readily upon removal of the reactants from the gas phase. All this makes their isolation under the controlled vacuum conditions typically needed for surface characterization almost impossible. Moreover, their detection needs to be achieved in the presence of large surface coverages of other adsorbed species.

The characterization of weakly bonded catalytic intermediates in situ during reactions has led to a departure from the traditional thinking in surface science. Originally, as vacuum technology was developed, a number of electron, ion, and other particle-based techniques were designed to probe surfaces selectively. While the penetrating power of photons renders most optical techniques surface insensitive, the strong interaction of electrons and ions with condensed matter makes them incapable of traveling deeply into solids, and therefore ideal for providing information on the chemical and physical characteristics of the few outer layers of those solids. In view of the need to study catalytic reactions in situ under non-vacuum environments, however, this idea has been recently revised. In particular, there has been a renewed interest in the use of optical spectroscopies for surface studies [12].

In order to use optical techniques for the characterization of surface reactions, however, the discrimination of information pertaining to surfaces from that due to the bulk of the material has to be achieved indirectly. For instance, the identification of adsorbates can be accomplished by their unique spectroscopic features, those that set

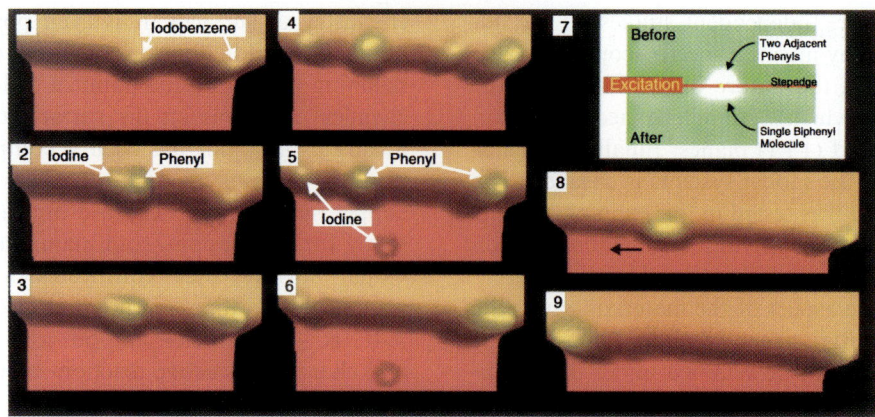

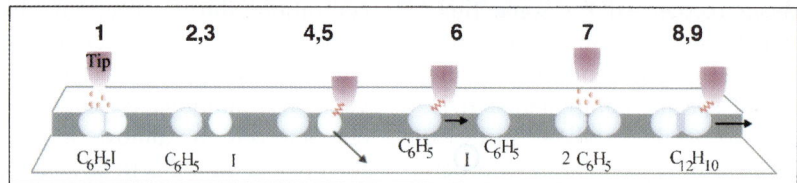

Fig. 7. Example of atomic-level manipulation of surface reactions by STM. This figure shows a series of STM images illustrating the steps involved in a tip-induced reaction where two phenyl groups are coupled to form a biphenyl molecule [65]. (1) Two iodobenzene (C_6H_5I) molecules are adsorbed at a Cu(1 1 1) step edge. (2, 3) The iodine atom (I) is abstracted from both molecules by using a voltage pulse. (4) Iodine atoms (small protrusions, I) and phenyl fragments (large balls, C_6H_5) are further separated by lateral manipulation. (5) The iodine atom located between the two phenyls is removed onto the lower terrace to clear the path between the two phenyls. (6) The left phenyl fragment is moved by the STM tip close to the right phenyl moiety to prepare for their association. (7) Electron excitation is used to fuse the two phenyl moieties into one biphenyl molecule (C_6H_5–C_6H_5). This panel shows a background subtracted STM image with a phenyl couple in its center. The upper and lower parts correspond to the stages before and after the chemical association. (8, 9) Verification of successful chemical association of two phenyls. The synthesized biphenyl molecule is pulled by the STM tip across the image, from left to right. The bottom diagram offers a schematic representation of the whole process.

them apart from the other elements already present either on the surface or in the bulk of the solid. Alternatively, optical techniques can be made to specifically probe surfaces by taking advantage of the reduced symmetry of interfaces, as in the recent studies of catalytic processes by reflection–absorption infrared spectroscopy [66–69], non-linear optical spectroscopies such as second-harmonic and sum-frequency generation [70–72], and nuclear magnetic resonance [73]. Much more is expected from these approaches.

The use of in situ spectroscopic characterization techniques in the study of catalytic processes addresses the first of the problems mentioned above, that associated with the difficulties encountered in isolating weakly bonded, short-lived catalytic intermediates. It does not, however, offer an answer to the second problem, the one related with the discrimination of reaction intermediates from other strongly bonded surface molecular fragments. For that, it is highly desirable to perform surface spectroscopic characterization experiments in a differential mode, via the induction of transients. The example illustrated in Fig. 8 is that of the identification of the weakly π-bonded olefin adsorbed on top of an alkylidyne-covered platinum single crystal surface believed to be key in hydrogenation–dehydrogenation surface steps [46,74,75]. Even though it is clear that the use of differential spectroscopy should help unlock the mysteries of the intermediates involved in catalysis, progress in this direction has been hindered by the associated experimental difficulties. Surmounting those difficulties represents one of the

Characterization of Weakly-Adsorbed Intermediates with Infrared Spectroscopy

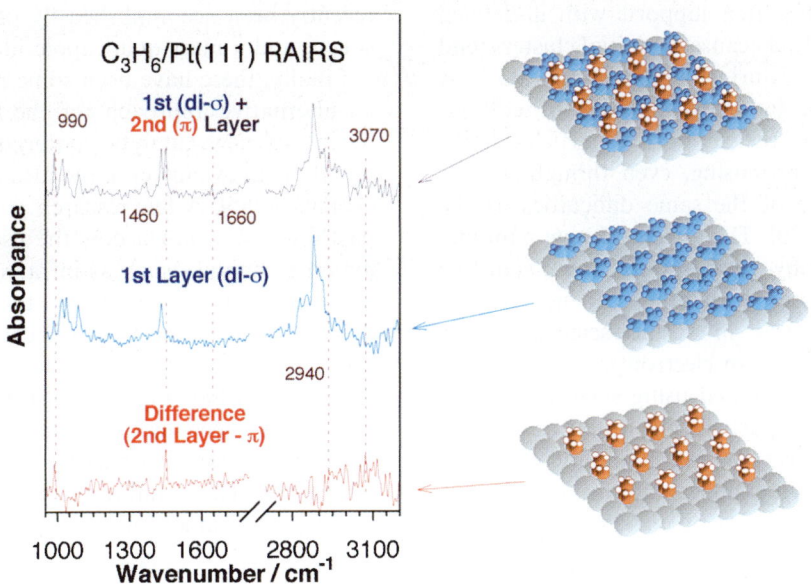

Fig. 8. Infrared spectroscopy can be used in a differential mode to detect weakly adsorbed intermediates in situ during catalytic reactions. The data shown here are for the adsorption of propylene on Pt(1 1 1) single-crystal surfaces [74]. The middle trace corresponds to the absorbance (signal intensity) versus frequency spectrum of the propylene molecules strongly adsorbed in the first layer, known as di-σ bonded species. The top spectrum corresponds to the data obtained once two layers have been deposited on the surface. The bottom trace, obtained by subtracting the middle spectrum from the top one, highlights the features associated with the second, weakly bonded, so-called π layer. Those π species are difficult to isolate under the vacuum conditions normally used in surface-science experiments, but are often critical participants in catalytic processes.

fundamental challenges in the surface science of catalysis.

Several new directions are also developing in connection with the quest for accurate kinetic measurements in catalysis. In terms of the use of molecular beams, one of the main limitations at this point is that there are currently no designs available for the generation of the high-flux beams needed to reproduce non-demanding reactions under vacuum [76]. Recall that the weakly adsorbed species believed to intervene in those processes are short-lived and likely to build-up on the surface in detectable amounts only under the high impinging frequency condition that prevail in atmospheric pressure environments. It is also conceivable to combine pulsed molecular beams with in situ optical spectroscopies for the characterization of the chemistry of weakly-adsorbed transient intermediates. In terms of the measurements of bond energies on surfaces, the development of a new microcalorimeter by King et al. provides a direct way to evaluate coverage-dependent adsorption energetics on well-defined (single-crystal) surfaces [77]. And from the computational point of view, Monte-Carlo simulations promise to help unravel the contributions of the different aspects of the local energetics of surface reactions to their overall kinetic behavior [21,78].

In reference to the materials gap, a number of novel synthetic methods are being tried to produce model supported catalysts with narrow and reproducible particle size distributions. One of the early ideas in this direction involved the decomposition of adsorbed organometallic clusters [79]. Unfortunately, that approach has proven only partially successful, because it still relies on the use of high surface-area supports with ill-defined structures, and also because the initial clusters tend to sinter (coalesce) during catalyst preparation. A newer procedure involving the use of colloids as precursors for the supported particles looks somewhat more promising, even though it does suffer from some of the same difficulties as the organometallics [80]. The advent of new nanometer-size lithography promises to circumvent the problems associated with other chemical manufacturing procedures in surface-science studies [81,82]. Fig. 9 shows an electron microscopy picture of a surface prepared using such method, in this case platinum particles deposited on a silica support. At present, nanolithography can only carve features down to a few tenths of nanometers, still too big a dimension to produce particles of relevance for most catalytic processes, but this limitation is likely to be surpassed in the near future. Lastly, the deposition of mass-selected metal clusters with one to thirty atoms each has recently been accomplished by using a sputtering source and a mass quadrupole filter [83,84].

Finally, there have been some recent reports on an alternative approach for the formulation and testing of new catalytic materials and processes based on an extension of the idea of combinatorial chemistry [85]. A few research groups have in the past few years shown how the use of clever automated techniques such as infrared thermography, resonance-enhanced multi-photon ionization, and differential mass spectroscopy can be used for the rapid screening of large quantities of potential catalysts [86–88]. Fig. 10 illustrates how this approach has been used for the screening of Mo–V–Nb alloys for the oxidative dehydrogenation of ethane [89]. The initial results in this area are quite exciting, and much more work is expected. It is

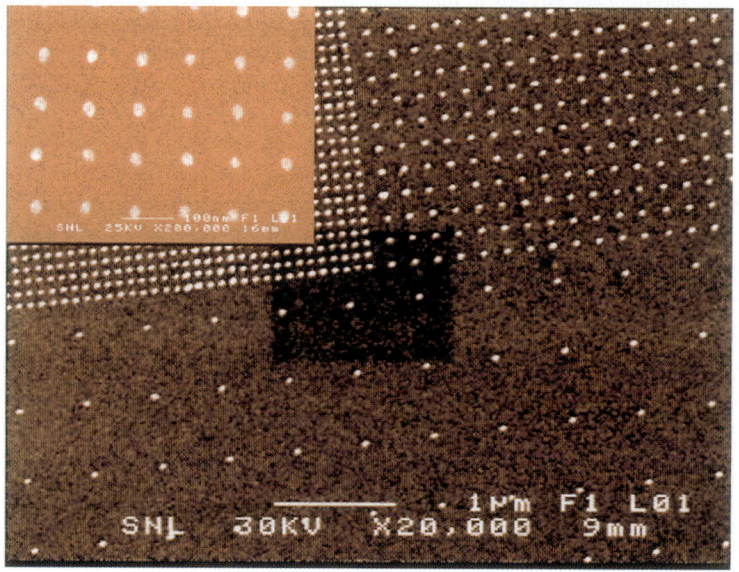

Fig. 9. Electron microscopy photograph of a sample prepared using nanolithography showing three different sets of arrays of platinum particles on a silica support with different inter-particle spacings [82]. The inset displays the details of the most densely packed area, where the Pt particles have an average diameter of ∼30 nm and a separation of 100 nm. Nanolithography promises to provide a way to prepare well-defined models for the study of supported catalysts.

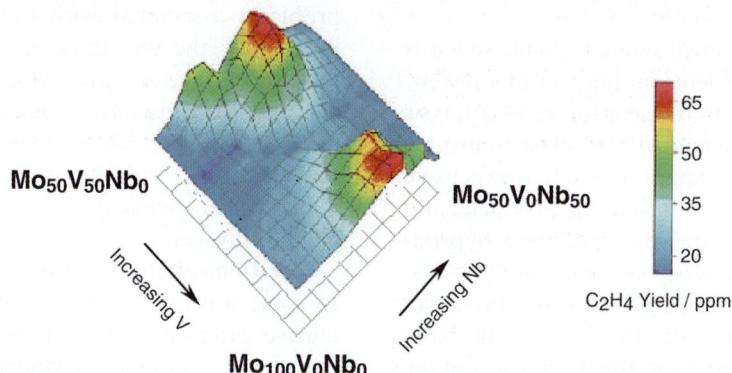

Fig. 10. Example of combinatorial screening of potential catalytic systems. Shown here is a plot of the reaction rates obtained for the oxidative dehydrogenation of ethane on Mo–V–Nb surfaces as a function of alloy composition [89]. An array of alloys was prepared by systematically varying the relative amounts of the three metals as a function of position on the underlying wafer, and the reactivity of the resulting samples was tested by an automated rasterable probe with concentric capillaries for gas feeding and mass-spectrometry detection. This "multiple sample" approach promises to expedite the search for viable catalysts, but still requires the chemical insight provided by the fundamental surface-science studies discussed in this review.

important to note, however, that the optimum use of this "multiple sample" approach still requires the good basic chemical understanding developed by the more standard studies discussed here.

10. Designing new catalytic processes

The holly grail in catalysis has been to be able to design catalytic processes from first principles. This is particularly desirable because of the complexity of the systems involved: the random development of catalytic processes is a hopeless task that requires the identification of an optimized system from innumerable permutations of materials and reaction conditions. Ideally, it should be possible to accomplish the systematic design of active sites, the targeted control of reaction energetics, and the manipulation of transition states to achieve high selectivities towards the products of interests. In this sense, the surface-science studies carried out to date have enhanced our atomic level view of catalytic systems, and that knowledge has been incorporated into a semi-empirical and intuitive approach to catalyst design. However, a complete understanding of the basics behind catalysis is still quite far off. Several attempts to formalize the reasoning process behind catalyst designing have shown only limited success [90,91], and the availability of better algorithms for this await future advances in the basic science of surface chemistry.

As stated above, one of the main roles of surface science in this area is to expand our basic knowledge about the surface reactions relevant to catalysis. The more we understand the microscopic details of surface reactions, the less guessing will be required for the design of new processes. The gradual nature of this evolution often makes it easy to underestimate the synergy between catalysis and surface science. When taking a historic perspective, however, the contributions of surface science to catalysis become quite evident. In particular, a better-developed picture of the surface-adsorbate chemical bond has led to a clearer definition of the catalytic problems. Kinetic and energetic measurements in model systems have greatly advanced our ideas on how to improve selectivities towards specific reactions. A more detailed understanding of the structure and electronic properties of solid surfaces has focused the search for catalytic materials. The field is still growing; we can look forward to more advances in these directions in the next decades.

Additional progress will surely be made towards solving the basic problems outlined in this review. Perhaps the key issue in the near future of catalysis is selectivity. Selective catalytic processes consume less reactants, do not require subsequent expensive separation procedures to isolate the products, and do not produce undesirable or polluting byproducts. Given the complexity of most catalytic systems, however, designing highly selective processes requires a better grip on the underlying basic chemistry involved. For that, the description of the surface–adsorbate chemical bond needs to be taken to the next level, one that better incorporates the role of neighboring surface atoms. Appropriate models for the kinetics of reactions in condensed matter systems, on surfaces in particular, need to be developed. The detection and characterization of direct catalytic surface intermediates have proven quite elusive, and needs to be pursued with renewed vigor. The preparation of complex surface models to better represent the processes that occur in real catalytic systems needs to be improved.

Underlying the search for answers to the fundamental questions mentioned above, there is also a need for the development of new technical methodology. Here the future looks quite bright. For instance, the mapping of individual atoms on surfaces, unthinkable a few decades go, is already a reality. Scanning probes are just starting to blossom, and will soon aid in the characterization of individual molecular bond-forming and bond-breaking steps and in the manipulation of individual atoms and molecules on surfaces. Additional advances in the use of optical spectroscopies for the study of non-vacuum interfaces, for the characterization of solid–liquid boundaries in particular, will unravel a new world of surface chemistry. The use of molecular beams specifically designed for the kinetic study of catalytic reactions should close the pressure gap between the model systems used in vacuum studies and more realistic catalytic processes under atmospheric pressures. The advent of new sample preparation methods such as nanolithography and colloid, chemical, and physical depositions will empower the surface-science community with the tools to address the problems associated with the materials gap, allowing for the emulation of more complex and realistic surface systems. More powerful computers and new quantum mechanics and kinetics methods will provide a more detailed account of the electronic and structural factors associated with surface chemical bonds and with specific kinetic pathways.

Questions about catalysis are not just academic. In some ways, catalysis is a mature field with extensive practical applications in today's society. Petroleum refining, chemical processing, and emission control amount to a business of over $10 billion worldwide which produces goods worth more than a hundred times that amount. Future improvement of the processes associated with those established industries is likely to be incremental. On the other hand, the continuous demand for new products requires new innovative ideas. For one, many stoichiometric organic synthesis are expensive, highly polluting, and in need of updating [92]. Heterogeneous catalysis is preferred to the homogeneous counterpart for this, because the latter is often expensive and requires the difficult separation of the catalyst from the products [93]. Pollution prevention and abatement is imposing additional new demands on catalysis, not only in emission control, but also in remediation processes such as the cleaning of underground waters [94]. Alternative energy sources require creative catalytic processes, among others clean and small reforming reactors [95] and efficient fuel cells [96]. New selective chemical sensors are in continuous demand [97]. The number of problems where catalysis can have a big impact is constantly growing. Basic surface chemistry knowledge will be indispensable if this progress is to materialize.

Acknowledgements

Funding for the preparation of this manuscript has been provided by grants from the National Science Foundation and the Department of Energy.

References

[1] A.M. Thayer, Catalyst suppliers face changing industry, Chem. Eng. News March 9 (1992) 27.
[2] R. Hoffmann, Solids and Surfaces: A Chemist's View of Bonding in Extended Structures, VCH, New York, 1988.
[3] M.R. Albert, J.T. Yates Jr., The Surface Scientist's Guide to Organometallic Chemistry, American Chemical Society, Washington, DC, 1987.
[4] F. Zaera, An organometallic guide to the chemistry of hydrocarbon moieties on transition metal surfaces, Chem. Rev. 95 (1995) 2651.
[5] G.A. Somorjai, Introduction to Surface Chemistry and Catalysis, Wiley, New York, 1994.
[6] J.P. Fulmer, F. Zaera, W.T. Tysoe, The orientation and bonding of CO on Mo(100) using angle-resolved photoelectron spectroscopy and near-edge X-ray absorption fine structure, J. Chem. Phys. 87 (1987) 7265.
[7] M.A. Van Hove, J. Cerda, P. Sautet, M.L. Bocquet, M. Salmeron, Surface structure determination by STM vs. LEED, Prog. Surf. Sci. 54 (1997) 315.
[8] D.P. Woodruff, T.A. Delchar, Modern Techniques of Surface Science, second ed., Cambridge University Press, Cambridge, 1994.
[9] U. Starke, J.B. Pendry, K. Heinz, Diffuse low-energy electron diffraction, Prog. Surf. Sci. 52 (1996) 53.
[10] S. Stöhr, NEXAFS spectroscopy, in: G. Ertl, R. Gomer, D.L. Mills (Eds.), Springer Series in Surface Science, vol. 25, Springer, Berlin, 1992.
[11] F. Zaera, In situ NEXAFS characterization of surface intermediates, in: Y. Iwasawa (Ed.), X-Ray Absorption Fine Structure for Catalysis and Surfaces, World Scientific, Singapore, 1996, p. 362.
[12] F. Zaera, Surface structural determinations: optical methods, in: J.H. Moore, N.D. Spencer (Eds.), Encyclopedia of Chemical Physics and Physical Chemistry, IOP Publishing, Philadelphia, 2000, in press.
[13] K.A. Peterlinz, T.J. Curtiss, S.J. Sibener, Coverage dependent desorption kinetics of CO from Rh(111) using time-resolved specular helium scattering, J. Chem. Phys. 95 (1991) 6972.
[14] D.H. Wei, D.C. Skelton, S.D. Kevan, Molecular interactions on surfaces, J. Vac. Sci. Technol. A 12 (1994) 2029.
[15] G. Ertl, Reactions at well-defined surfaces, Surf. Sci. 299/300 (1994) 742.
[16] F. Zaera, J. Liu, M. Xu, Isothermal study of the kinetics of carbon monoxide oxidation on Pt(111): rate dependence on surface coverages, J. Chem. Phys. 106 (1997) 4204.
[17] L. Österlund, M.Ø. Pedersen, I. Stensgaard, E. Lægsgaard, F. Besenbacher, Quantitative determination of adsorbate–adsorbate interactions, Phys. Rev. Lett. 83 (1999) 4812.
[18] J.J. Mortensen, B. Hammer, J.K. Norskov, A theoretical study of adsorbate–adsorbate interactions on Ru(0001), Surf. Sci. 414 (1998) 315.
[19] D. Loffreda, D. Simon, P. Sautet, Dependence of stretching frequencies on surface coverage and adsorbate–adsorbate interactions: a density-functional theory approach of CO on Pd(111), Surf. Sci. 425 (1999) 68.
[20] S.J. Lombardo, A.T. Bell, A review of theoretical models of adsorption diffusion, desorption, and reaction of gases on metal surfaces, Surf. Sci. Rep. 13 (1991) 1.
[21] E.W. Hansen, M. Neurock, Modeling surface kinetics with first-principles-based molecular simulation, Chem. Eng. Sci. 54 (1999) 3411.
[22] G.C. Bond, Catalysis by Metals, Academic Press, London, 1962.
[23] J.M. Thomas, W.J. Thomas, Introduction to the Principles of Heterogeneous Catalysis, Academic Press, London, 1967.
[24] B.C. Gates, J.R. Katzer, G.C.A. Schuit, Chemistry of Catalytic Processes, McGraw-Hill, New York, 1979.
[25] D. Menzel, Desorption phenomena, in: R. Gomer (Ed.), Interactions on Metal Surfaces, Springer, New York, 1975, p. 102.
[26] R.J. Madix, Surface reactivity: heterogeneous reactions on single crystal surfaces, Acc. Chem. Res. 12 (1979) 265.
[27] J.A. Barker, D.J. Auerbach, Gas–surface interactions and dynamics: thermal energy atomic and molecular beam studies, Surf. Sci. Rep. 4 (1985) 1.
[28] S.T. Ceyer, Dissociative chemisorption: dynamics and mechanisms, Annu. Rev. Phys. Chem. 39 (1988) 479.
[29] D.A. King, M.G. Wells, Molecular beam investigation of adsorption kinetics on bulk metal targets: nitrogen on tungsten, Surf. Sci. 29 (1972) 454.
[30] C.S. Gopinath, F. Zaera, Transient behavior during the isothermal reduction of NO by CO on Rh(111) as studied with effusive collimated molecular beams, J. Phys. Chem. B 104 (2000) 3194.
[31] F. Zaera, C.S. Gopinath, Role of adsorbed nitrogen atoms during the catalytic reduction of NO on rhodium surfaces, J. Chem. Phys. 111 (1999) 8088.
[32] H. Öfner, F. Zaera, Isothermal kinetic measurements for the hydrogenation of ethylene on Pt(111) under vacuum: significance of weakly bound species in the reaction mechanism, J. Phys. Chem. 101 (1997) 396.
[33] F. Zaera, D. Chrysostomou, Propylene on Pt(111). II. Hydrogenation, dehydrogenation, and H–D exchange, Surf. Sci. 457 (2000) 89.
[34] F. Zaera, C.S. Gopinath, Evidence for a N_2O Intermediate in the catalytic reduction of NO to N_2 on rhodium surfaces, Chem. Phys. Lett. 332 (2000) 209.
[35] F. Zaera, S. Tjandra, T.V.W. Janssens, Selectivity among dehydrogenation steps for alkyl groups on metal surfaces: comparison between nickel and platinum, Langmuir 14 (1998) 1320.
[36] A.J. Gellman, Transition states for surface-catalyzed chemistry, Acc. Chem. Res. 33 (2000) 19.
[37] B.E. Bent, Mimicking aspects of heterogeneous catalysis: generating, isolating and reacting proposed surface intermediates on single crystals in vacuum, Chem. Rev. 96 (1996) 1361.

[38] S.M. Davis, F. Zaera, G.A. Somorjai, Surface structure and temperature dependence of n-hexane skeletal rearrangement reactions catalyzed over platinum single crystal surfaces: marked structure sensitivity of aromatization, J. Catal. 85 (1984) 206.

[39] G.A. Somorjai, F. Zaera, Heterogeneous catalysis on the molecular scale, J. Phys. Chem. 86 (1982) 3070.

[40] G. Ertl, Surface science and catalysis—studies on the mechanism of ammonia synthesis: The P.H. Emmett Award Address, Catal. Rev. Sci. Eng. 21 (1980) 201.

[41] F. Zaera, A.J. Gellman, G.A. Somorjai, surface science studies of catalysis: classification of reactions, Acc. Chem. Res. 19 (1986) 24.

[42] J.A. Rodriguez, Physical and chemical properties of bimetallic surfaces, Surf. Sci. Rep. 24 (1996) 225.

[43] V.E. Henrich, The surface of metal oxides, Rep. Prog. Phys. 48 (1985) 1481.

[44] J.G. Chen, NEXAFS Investigations of transition metal oxides nitrides, carbides, sulfides and other interstitial compounds, Surf. Sci. Rep. 30 (1997) 1.

[45] H. Öfner, F. Zaera, Surface defect characterization in oxygen-dosed nickel surfaces and in NiO thin films by CO adsorption–desorption experiments, J. Phys. Chem. B 101 (1997) 9069.

[46] F. Zaera, The surface chemistry of hydrocarbons on transition metal surfaces: a critical review, Isr. J. Chem. 38 (1998) 293.

[47] J.G. Serafin, A.C. Liu, S.R. Seyedmonir, Surface science and the silver-catalyzed epoxidation of ethylene: an industrial perspective., J. Mol. Catal. A 131 (1998) 157.

[48] F. Zaera, N.R. Gleason, B. Klingenberg, A.H. Ali, Partial oxidation of hydrocarbons on nickel: from surface science mechanistic studies to catalysis, J. Mol. Catal. A 146 (1999) 13.

[49] M. Neurock, V. Pallassana, R.A. van Santen, The importance of transient states at higher coverages in catalytic reactions, J. Am. Chem. Soc. 122 (2000) 1150.

[50] P.B. Wells, A.G. Wilkinson, Platinum group metals as heterogeneous enantioselective catalysts, Top. Catal. 5 (1998) 39.

[51] C.S. McFadden, P.S. Cremer, A.J. Gellman, Adsorption of chiral alcohols on chiral metal surfaces, Langmuir 12 (1996) 2483.

[52] D.W. Goodman, Model catalytic studies over metal single crystals, Acc. Chem. Res. 17 (1984) 194.

[53] F. Zaera, Preparation and reactivity of alkyl groups adsorbed on metal surfaces, Acc. Chem. Res. 25 (1992) 260.

[54] C.J. Jenks, B.E. Bent, F. Zaera, The chemistry of alkyl iodides on copper surfaces. 2. influence of surface structure on reactivity, J. Phys. Chem. B 104 (2000) 3017.

[55] F. Zaera, G.A. Somorjai, Role of hydrogen in low and high pressure hydrocarbon reactions, in: Z. Paál, P.G. Menon (Eds.), Hydrogen Effects in Catalysis: Fundamentals and Practical Applications, Marcel Dekker, New York, 1988, p. 425.

[56] S.M. Davis, F. Zaera, G.A. Somorjai, The reactivity and composition of strongly adsorbed carbonaceous deposits on platinum model of the working hydrocarbon conversion catalyst, J. Catal. 77 (1982) 439.

[57] G.M. Pajonk, Spillover effects, in: G. Ertl, H. Knözinger, J. Weitkamp (Eds.), Handbook of Heterogeneous Catalysis, vol. 3, Wiley-VCH, Weinheim, 1997, p. 1064.

[58] S.A. Stevenson, J.A. Dumesic, R.T.K. Baker, E. Ruckenstein, Metal–Support Interactions in Catalysis, Sintering and Redispersion, Van Nostrand Reinhold, New York, 1987.

[59] D.W. Goodman, Model studies in catalysis using surface science probes, Chem. Rev. 95 (1995) 523.

[60] M. Bäumer, J. Libuda, H.-J. Freund, Metal deposits on thin well ordered oxide films: morphology, adsorption and reactivity, in: R.M. Lambert, G. Pacchioni (Eds.), Chemisorption and Reactivity on Supported Clusters and Thin Films, Kluwer, Amsterdam, 1997, p. 61.

[61] C.R. Henry, Surface studies of supported model catalysts, Surf. Sci. Rep. 31 (1998) 235.

[62] J.A. Jensen, K.B. Rider, Y. Chen, M. Salmeron, G.A. Somorjai, High pressure, high temperature scanning tunneling microscopy, J. Vac. Sci. Technol. B 17 (1999) 1080.

[63] P.S. Weiss, M.M. Kamma, T.M. Graham, S.J. Stranick, Imaging benzene molecules and phenyl radicals on Cu(1 1 1), Langmuir 14 (1998) 1284.

[64] L.J. Lauhon, W. Ho, Electronic and vibrational excitation of single molecules with a scanning tunneling microscope, Surf. Sci. 451 (2000) 219.

[65] S.-W. Hla, L. Bartels, G. Meyer, K.-H. Rieder, Inducing all steps of a chemical reaction with the scanning tunneling microscope tip: towards single molecule engineering, Phys. Rev. Lett. 85 (2000) 2777.

[66] F.M. Hoffmann, Infrared reflection–absorption spectroscopy of adsorbed molecules, Surf. Sci. Rep. 3 (1983) 107.

[67] H. Hoffmann, N.A. Wright, F. Zaera, P.R. Griffiths, differential-polarization dual-beam FT-IR spectrometer for surface analysis, Talanta 36 (1989) 125.

[68] A. Bandara, J. Kubota, A. Wada, K. Domen, C. Hirose, Adsorption and reactions of formic acid on (2×2)-NiO(1 1 1)/Ni(1 1 1) surface. 2. IRAS study under catalytic steady-state conditions, J. Phys. Chem. B 101 (1997) 361.

[69] C.J. Hirschmugl, Frontiers in infrared spectroscopy at surfaces and interfaces, Surf. Sci. 500 (2002) 577.

[70] Y.R. Shen, Surfaces probed by nonlinear optics, Surf. Sci. 300 (1994) 551.

[71] G.A. Somorjai, X.C. Su, K.R. McCrea, K.B. Rider, Molecular surface studies of adsorption and catalytic reaction on crystal (Pt, Rh) surfaces under high pressure conditions (atmospheres) using sum frequency generation (SFG)—surface vibrational spectroscopy and scanning tunneling microscopy (STM), Top. Catal. 8 (1999) 23.

[72] C.T. Williams, D.A. Beattie, Probing buried interfaces with nonlinear optical spectroscopy, Surf. Sci. 500 (2002) 545.

[73] J.A. Norcross, C.P. Slichter, J.H. Sinfelt, Nuclear magnetic resonance study of platinum catalysts containing alkali atoms, Catal. Today 53 (1999) 343.

[74] F. Zaera, D. Chrysostomou, Propylene on Pt(111). I. Characterization of surface species by infrared spectroscopy, Surf. Sci. 457 (2000) 71.

[75] F. Zaera, On the mechanism for the hydrogenation of olefins on transition-metal surfaces: the chemistry of ethylene on Pt(111), Langmuir 12 (1996) 88.

[76] J.M. Guevremont, S. Sheldon, F. Zaera, Design and characterization of collimated effusive gas beam sources: effect of source dimensions and backing pressure on total flow and beam profile, Rev. Sci. Instrum., in press.

[77] W.A. Brown, R. Kose, D.A. King, Femtomole adsorption calorimetry on single-crystal surfaces, Chem. Rev. 98 (1998) 797.

[78] V.P. Zhdanov, B. Kasemo, Kinetic phase transitions in simple reactions on solid surfaces, Surf. Sci. Rep. 20 (1994) 111.

[79] B.C. Gates, Supported metal clusters: synthesis, structure and catalysis, Chem. Rev. 95 (1995) 511.

[80] J.H. Fendler (Ed.), Nanoparticles and Nanostructured Films, Wiley-VCH, Weinheim, 1998.

[81] M.X. Yang, D.H. Gracias, P.W. Jacobs, G.A. Somorjai, Lithographic fabrication of model systems in heterogeneous catalysis and surface science studies, Langmuir 14 (1998) 1458.

[82] S. Johansson, K. Wong, V.P. Zhdanov, B. Kasemo, Nanofabrication of model catalysts and simulations of their reaction kinetics, J. Vac. Sci. Technol. A 17 (1999) 297.

[83] M.-H. Schaffner, J.-F. Jeanneret, F. Patthey, W.-D. Schneider, An ultrahigh vacuum sputter source for in situ deposition of size-selected clusters: Ag on graphite, J. Phys. D 31 (1988) 3177.

[84] U. Heiz, A. Sanchez, S. Abbet, W.-D. Schneider, Catalytic oxidation of carbon monoxide on monodispersed platinum clusters: each atom counts, J. Am. Chem. Soc. 121 (1999) 3214.

[85] R. Dagani, A faster route to new materials, Chem. Eng. News 8 (1999) 51.

[86] S.J. Taylor, J.P. Morken, Thermographic selection of effective catalysts from an encoded polymer-bound library, Science 280 (1998) 267.

[87] M. Orschel, J. Klein, H.W. Schmidt, W.F. Maier, Detection of reaction selectivity on catalyst libraries by spatially resolved mass spectrometry, Angew. Chem. Int. Ed. Engl. 38 (1999) 2791.

[88] T. Bein, Efficient assays for combinatorial methods for the discovery of catalysts, Angew. Chem. Int. Ed. Engl. 38 (1999) 323.

[89] P. Cong, A. Dehestani, R. Doolen, D.M. Giaquinta, S. Guan, V. Markov, D. Poojary, K. Self, H. Turner, W.H. Weinberg, Combinatorial discovery of oxidative dehydrogenation catalysts within the Mo–V–Nb–O system, Proc. Natl. Acad. Sci. 96 (1999) 11077.

[90] J.B. Butt, Progress towards the a priori determination of catalytic properties, AIChE J. 22 (1976) 1.

[91] L.L. Hegedus (Ed.), Catalyst Design: Progress and Perspectives, Wiley, New York, 1987.

[92] K. Tanabe, W.F. Hölderich, Industrial application of solid acid-base catalysts, Appl. Catal. A 181 (1999) 399.

[93] B. Cornils, W.A. Herrmann (Eds.), Applied Homogeneous Catalysis with Organometallic Compounds: A Comprehensive Handbook in Two Volumes, Weinheim-VCH, New York, 1996.

[94] R.W. Puls, C.J. Paul, R.M. Powell, The application of in situ permeable reactive (zero-valent iron) barrier technology for the remediation of chromate-contaminated groundwater: a field test, Appl. Geochem. 14 (1999) 989.

[95] R.J. Farrauto, R.M. Heck, Environmental catalysis into the 21st century, Catal. Today 55 (2000) 179.

[96] G.J.K. Acres, J.C. Frost, G.A. Hards, R.J. Potter, T.R. Ralph, D. Thompsett, G.T. Burstein, G.J. Hutchings, Electrocatalysts for fuel cells, Catal. Today 38 (1997) 393.

[97] J. Janata, Environmental chemical sensors—new challenge and opportunity, Crit. Rev. Anal. Chem. 28 (1998) 27.

[98] C.S. Gopinath, F. Zaera, A molecular beam study of the kinetics of the catalytic reduction of NO by CO on Rh(111) single crystal surfaces, J. Catal. 186 (1999) 387.

[99] S. Tjandra, F. Zaera, Adsorption and thermal decomposition of propyl iodides on Ni(100) surfaces, Langmuir 10 (1994) 2640.

[100] J. Kubota, F. Zaera, Adsorption of modifiers as key in imparting chirality to platinum catalysts, J. Am. Chem. Soc., submitted.

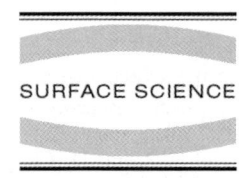

Impact of surface science on the understanding of kinetics of heterogeneous catalytic reactions

Vladimir P. Zhdanov *

Boreskov Institute of Catalysis, Russian Academy of Sciences, Novosibirsk 630090, Russia

Received 18 April 2000; accepted for publication 24 May 2001

Abstract

The kinetics of chemical reactions in gas and liquid phases are usually described by employing the conventional mass-action law equations. The laws governing the kinetics of heterogeneous catalytic reactions are as a rule much more complex due to adsorbate–adsorbate lateral interactions, surface heterogeneity, spontaneous and adsorbate-induced changes in a surface, and/or limited mobility of reactants. The importance of these factors was recognized by the heterogeneous catalysis community far before the surface-science era. Only with the development of surface science, however, has it become possible to study in detail the non-ideality of rate processes on solid surfaces. In the present paper, we survey the main conceptual results currently available in this field and illustrate the impact of surface science on its development. Specifically, we outline the approaches used to describe elementary reaction steps and the whole reaction kinetics near and far from equilibrium, including such topics as kinetic phase transitions, pattern formation, kinetic oscillations and chaos, and pressure- and structure-gap problems. All these phenomena and problems are demonstrated to provide promising opportunities for further experimental and theoretical studies. © 2001 Elsevier Science B.V. All rights reserved.

Keywords: Catalysis; Chemisorption; Models of surface chemical reactions

1. Introduction

This essay written for a general readership started at the end of the 20th century and finished in the new millennium. This has been a particularly good time to look at the big picture and ask such questions as "how did we get here?" and "where do we go from here?". With respect to physics in general, these questions were formulated in a more explicit form and addressed in an interesting issue of Physics World [1] published in December 1999. Reviewing the responses of physicists, Peter Rodgerts and Martin Durrani, the Editors of Physics World, write: "The picture that emerges is of a dynamic and vibrant community that is eager to continue the work of Galileo, Newton, Maxwell, Einstein et al. on the grand challenges while, at the same time, continuing to push physics into new areas ...". The latter is directly applicable to surface science because this flourishing subfield of physics has exerted tremendous impact on many other disciplines, including heterogeneous catalysis in general and on

---

* Fax: +7-3832-344687.
  E-mail address: zhdanov@catalysis.nsk.su (V.P. Zhdanov).

the understanding of the kinetics of heterogeneous catalytic reactions in particular.

By definition, catalysis is a phenomenon of acceleration of the rate of a chemical reaction by a substance, catalyst, not consumed in the reaction. Broadly speaking, catalysis is a branch of chemistry, aimed at acceleration of chemical reactions. Heterogeneous catalysis employing solid surfaces as catalysts is formally a subfield of catalysis [2]. In fact, however, it is a fascinating interdisciplinary branch of natural sciences. In particular, the experimental methods used now in heterogeneous catalysis are primarily those developed in surface science. The effect of surface science on heterogeneous catalysis is discussed in this volume of Surface Science by several authors including Hans-Joachim Freund, John Sinfelt and Francisco Zaera (for more extensive reviews, see Ref. [3]). Complementing their presentation, we focus our attention on the kinetics of heterogeneous catalytic reactions.

2. Complexity and practical importance

For scientists working in chemistry or in the related subfields of physics, the understanding of the kinetics of heterogeneous catalytic reactions is recognized to be important from many different perspectives. For those whose field of expertise is distant from chemistry or for undergraduate students choosing a research field and trying to have fun, to match their talents and skills with the problems available in that field, and to be aware of the career opportunities outside the University walls, an explicit discussion of these perspectives might be of interest.

In the pursuit of fun and in the interest of curiosity, it is instructive to look through tabulations of opinions about the biggest unsolved problem in physics. Along with such fundamental problems as the nature of observation in quantum mechanics or the possibility of explanation of life via the physical laws we know, complexity is a common theme [1]. Kinetics of heterogeneous catalytic reactions are encompassed by this theme. Indeed, all types of the kinetic processes occurring in the bulk may be observed at the interface as well. Moreover, the kinetics of heterogeneous catalytic reactions are actually much richer and more interesting compared to those in the bulk.

From a practical point of view, the understanding of the kinetics of catalytic reactions is important, because heterogeneous catalysis is the mainstay of the chemical industry. More than 90% of the chemical manufacturing processes in use throughout the world utilize catalysis and primarily heterogeneous catalysis [4]. To predict the performance of catalytic reactors, one needs to know the reaction kinetics and employ it in simulations. This step is standard for projecting common reactors operating in the steady-state regime.

New trends in chemical engineering are based on using the benefits of unsteady-state conditions. A classical theoretically predicted and practically realized example is that of reverse-flow operation in a reactor during exothermic reactions, e.g., for SO_2 oxidation [5,6]. In this case, the catalyst not only accelerates the reaction but also serves as a heat generator. The latter makes it possible to exclude the external recuperative heat exchange and to simplify the reactor construction.

Sometimes, the unsteady-state conditions are inevitable. For example, the catalytic cleaning of gasoline motor vehicle exhaust gas is based on the so-called three-way Pt–Rh catalysts (TWC), converting simultaneously the two reducing pollutants, CO and uncombusted hydrocarbons, and the oxidizing pollutants, NO, NO_2 and N_2O (called NO_x), to non-toxic products [7]. The main idea of the TWC is that complete conversion of the reducing pollutants can be achieved if the concentrations of these species are exactly matched by NO_x plus oxygen. Technologically, this condition is met by using an electronic engine control module, which employs a signal from a solid-state exhaust oxygen sensor for maintaining the air/fuel (A/F) mixture in a narrow range around stoichiometry. With such sensors, the A/F ratio typically oscillates (about 1 cycle per second) within ±0.2 A/F of the balanced composition due to inherent time delay in the A/F correction. The oscillations can be compensated (at least partly) by introducing ceria into the TWC. This function of ceria is based on its ability to take up, store and release oxygen.

In the future, the role of "smart" catalytic technologies employing complex catalysts, aimed at high conversion at low temperatures, and/or unsteady-state regimes, will increase. The role of kinetic simulations will increase as well, because without such simulations the optimization of the behaviour of catalytic systems is hardly possible.

3. Reaction kinetics on different size scales

Looking through Sections 1 and 2, researchers, whose knowledge base is far from chemistry, or undergraduate students might ask: Where do heterogeneous catalytic reactions occur? By definition, heterogeneous reactions occur on interfaces between different phases. Below we discuss some of the most interesting cases of reactions at the gas–metal interface. To illustrate the type of kinetic problems here, it is instructive to recall size scales characterising catalytic processes. Typically (Fig. 1), a catalytic reactor is filled by porous pellets with nm catalyst particles deposited on the walls of pores. Thus, one can study reaction kinetics on (i) the surface of nm catalyst particles, (ii) in pellets, and/or (iii) the whole reactor. To describe cases (ii) and (iii), the results obtained in case (i) should be combined with equations for mass and heat transport in pellets or a reactor, respectively. Additional complications arising on steps (ii) and (iii) are traditionally explored in chemical engineering. Surface science on these steps is irrelevant. For this reason, we do not discuss below the kinetics of type (ii) and (iii).

Although nm supported catalyst particles have been in practical use for several decades already, the understanding of what is on their surface under the reactive conditions is still limited, because, due to high (atmospheric) pressure and difficulties with accessibility, only a few of the physical methods traditionally used to explore surface processes are applicable. To prevent surface contamination and employ the full advantage of the surface science physical methods, academic studies of catalytic reactions have usually been performed on single- or poly-crystal samples (with the size from a few mm to 1 cm) under ultrahigh vacuum (UHV) conditions.

To bridge the size gap, there has recently been a rapid development of surface-science based techniques aimed at preparation of well controlled arrays of supported particles on planar surfaces, that mimic supported catalysts. For small (a few nm) particles, these methods, e.g., are carefully controlled evaporation and annealing of condensed particles, or deposition of clusters by cluster beams [9]. For larger sizes ($\geqslant 10$ nm), electron beam lithography [10,11] can be employed, to make nearly perfect arrays of supported particles with fairly uniform size and shape distributions, which can be varied systematically.

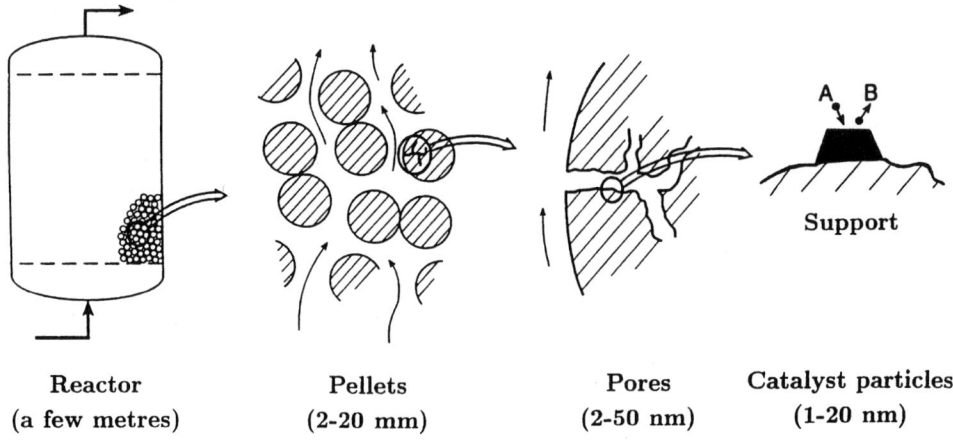

Fig. 1. Size scales in a packed bed catalytic reactor. (Adapted from Ref. [8].)

Thus, surface science makes it possible to study mechanism and kinetics of heterogeneous catalytic reactions on all the size scales.

4. How heterogeneous reactions occur

Chemical reactions occur via elementary reaction steps. For heterogeneous catalytic reactions, these are monomolecular or dissociative adsorption, $A_{gas} \to A_{ads}$ and $(B_2)_{gas} \to 2B_{ads}$, monomolecular or associative desorption, $A_{ads} \to A_{gas}$ and $2B_{ads} \to (B_2)_{gas}$, and bimolecular steps including adsorbed species,

$$A_{ads} + B_{ads} \to Product, \qquad (1)$$

or adsorbed and gas-phase particles,

$$A_{ads} + B_{gas} \to Product. \qquad (2)$$

In the former case, "Product" means $(AB)_{gas}$, $(AB)_{ads}$, or $C_{ads} + D_{ads}$. In the latter case, "Product" is $(AB)_{gas}$ or $C_{ads} + D_{ads}$.

Steps (1) and (2) are now associated with Langmuir and Hinshelwood and Eley and Rideal, respectively, in honour of their pioneering studies performed before World War II.

Mechanisms of practically important reactions usually include adsorption, desorption, and Langmuir–Hinshelwood (LH) steps. The cases when the Eley–Rideal step plays a significant role are rare.

5. Why are the reaction kinetics complex?

5.1. Ideal adsorbed layer

Such words as "complicated" or "complexity" (Section 2) are, of course, relative. To use them effectively one should first define the specific meaning of the words "simple" and "simplicity". For the scientists studying heterogeneous reactions, the latter words are usually associated with the kinetic equations corresponding to an ideal (or Langmuir) adsorbed layer, when the surface is uniform and stable and there are no adsorbate–adsorbate lateral interactions. In this case, the kinetic equations for various surface rate processes can easily be derived provided that each adsorbed particle occupies one site. For example, the rates of desorption, LH and Eley–Rideal steps ((1) and (2)) are proportional to coverage, the product of A and B coverages, and the product of A coverage and B pressure, respectively (for a given species, coverage is defined as a fraction of adsorbed sites occupied by this species). Using such equations, one can calculate the rate of heterogeneous reactions occurring via several steps under steady-state conditions. For typical reaction mechanisms, the corresponding expressions were derived by Hougen and Watson [12] (see also Ref. [13]). Since then, the Hougen–Watson-type equations are widely used in academic and applied studies of catalytic reactions [14].

In real overlayers, adsorbed on single- or poly-crystal samples, the reaction kinetics are complicated due to (i) surface heterogeneity, (ii) adsorbate–adsorbate lateral interactions, (iii) spontaneous and adsorbate-induced surface restructuring, and/or (iv) limited mobility of adsorbed species. For the nm catalyst particles, there are additional specific factors resulting in the complexity of reaction kinetics (Section 10).

5.2. Surface heterogeneity

The structure of heterogeneous surfaces may in general be very complex. The reaction kinetics on such surfaces may be far from those derived for the ideal Langmuir layer. This was realized already in the 1920s when Taylor formulated [15] (see the review [16]) his principle of active sites or active centres at which reactions occur. Such sites are usually believed to be associated with various defects. For example, single-crystal surfaces contain terraces separated by steps. In this case, active sites may be associated with step or kink sites (kinks are defects of steps). The surface-science based studies indicate, for example, that in H_2 oxidation on Ni(1 1 0) [17] or CO oxidation [18] and isobutane conversion (into C_1–C_3 gases) [19] on Pt, the step sites are really more active compared to terrace sites (see also Monte Carlo simulations [20]). This is however not always the case, because there are also reports [21] that in some other reactions (with participation of ethylene, nitric oxide and methanol) on Pt there is no correlation between catalytic activity and step sites.

The first models describing adsorption on heterogeneous surfaces were proposed in the 1930s by Zeldovich [22] and Roginsky [23]. The main idea was to introduce an adsorption energy distribution for adsorption sites. The correlations in the arrangement of adsorption sites with different adsorption energies were neglected. This approach can be used to fit the adsorption isotherms (i.e., the dependence of adsorbate coverage on pressure under the adsorption–desorption equilibrium) and reaction kinetics [24]. A classical example here is the kinetic model proposed by Temkin and Pyzhov [25] for ammonia syntheses on the Fe-based catalyst.

Traditionally, the concept of heterogeneity is related to "real" surfaces. Numerous surface-science based studies indicate that the single-crystal surfaces are in fact often heterogeneous as well. On Pt(1 1 1), for example, CO adsorption occurs primarily on top sites at coverages below one third monolayer, $\theta_{CO} \leqslant 1/3$ ML, while at higher coverages both top and bridge sites are occupied [26]. In contrast, oxygen adsorbs on fcc-hollow sites. On more open faces, the situation is often more complex.

5.3. Adsorbate–adsorbate lateral interactions

In theoretical chemistry, rate processes are usually treated in terms of nuclear motion along the potential-energy surface formed during interaction of reactants. This surface contains potential wells corresponding to initial and final species. The wells are separated by barriers. A rate process occurs when a system crosses a barrier. At thermal conditions, according to the Arrhenius law, the probability of such events is exponentially low, $\propto \exp(-\Delta E/k_B T)$ (ΔE is the energy difference between the activated state (near the barrier) and initial (ground) state).

In adsorbed layers, at appreciable coverages, the potential-energy surface usually depends not only on the coordinates of particles directly participating in a rate process but also on the arrangement of other particles located in adjacent sites. Lateral–lateral interaction of reactants with

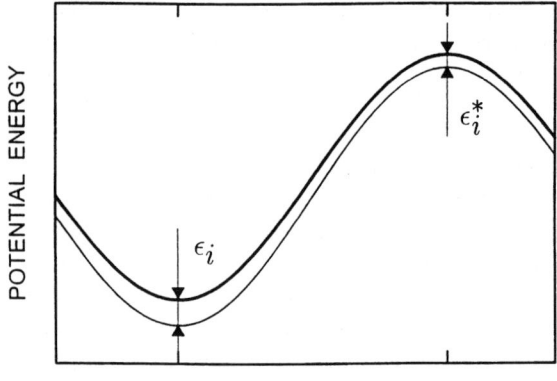

Fig. 2. Schematic cross-section of the potential-energy surface along the coordinate corresponding to the transition from the initial state to the activated state. The thin and thick solid lines are for the cases when the reacting particles have no and have neighbours, respectively. ϵ_i and ϵ_i^* are the lateral interactions (with neighbours) in the initial and activated states. The subscript i characterizes the arrangement of particles.

the latter particles slightly shifts potential-energy (Fig. 2). The resultant change in ΔE is given by $\epsilon_i^* - \epsilon_i$, where ϵ_i and ϵ_i^* are the lateral adsorbate–adsorbate interactions in the initial and activated states. The transition probability is accordingly proportional to $\exp[-(\epsilon_i^* - \epsilon_i)/k_B T]$. From the point of view of chemical kinetics, the interactions ϵ_i and ϵ_i^* are equally important. The energetics of adsorption depends however only on the interactions in the ground state. Practically, this means that for rapid adsorbate diffusion the probabilities of different arrangements of adsorbed particles depend only on the latter interactions and can be calculated by using canonical or grand canonical distribution, because in this case the adsorbed overlayer is close to equilibrium. The lateral interaction in the activated state may explicitly manifest itself, e.g., in the coverage dependence of the energy distribution of the reaction products directly desorbing to the gas phase (this effect analysed in Ref. [27] was observed in several reactions [28]).

Lateral interactions (in the ground state) result in ordering of adsorbed particles. With the development of the low-energy electron diffraction (LEED) technique, this phenomenon was observed in thousands of adsorbate/substrate systems (see, e.g., Ref. [19]). Analysing LEED data obtained at different coverages and temperatures makes it possible to construct adsorbate–substrate phase diagrams [29]. Comparing measured and calculated phase diagrams, one can obtain values of lateral interactions.

Another common way to evaluate lateral interactions is based on the analysis of desorption kinetics. Due to adsorbate–adsorbate lateral interactions or surface heterogeneity, the desorption rate usually rapidly decreases with decreasing coverage. If temperature is constant, this effect often complicates experimental studies of desorption kinetics. To overcome this problem, experimenters increase temperature during desorption (often in the linear fashion, i.e., $T(t) = T(0) + \beta t$, where β is the heating rate). The temperature dependence of the desorption rate in such temperature-programmed desorption (TPD) regimes is called the TPD spectrum. Lateral interactions result in broadening or splitting of TPD spectra and accordingly can be estimated by comparing the measured and calculated TPD kinetics as was first shown by Goymour and King [30] and Adams [31] (see also the reviews [32,33] and recent simulations [34,35]). The bulk of TPD data obtained before 1984 was collected by Morris et al. [32]. More recent TPD measurements can be found in an almost every issue of Surface Science.

Quantitative information on lateral interactions can also be extracted from high-resolution scanning tunneling microscopy (STM) data by analysing distribution of adsorbed particles [36].

The values of lateral interactions between two nearest-neighbour (nn) particles are often in the range 1–4 kcal/mol. This means that the lateral interactions are relatively weak compared to the adsorption energy and accordingly can hardly be accurately calculated even with current computer facilities. On the other hand, their qualitative behaviour is well known [37,38]. In particular, dipole–dipole and elastic interactions are usually repulsive, $\epsilon(r) \propto 1/r^3$ (r is the adsorbate–adsorbate distance). Indirect substrate-mediated interaction behaves as $\epsilon(r) \propto \sin(2k_F r)/r^5$ (k_F is the Fermi momentum). Attractive Van der Waals interaction, $\epsilon(r) \propto -1/r^6$, is usually important only for adjacent particles.

At low coverage, adsorbed particles have no neighbours. At high coverages, all the nn sites are often occupied. This means that the change of the adsorption energy with increasing coverage is about $z\epsilon_1 \sim$ 5–30 kcal/mol, where ϵ_1 is the nn interaction, and z the number of nn sites. This difference may result in dramatic changes in the reaction kinetics with increasing coverage. This was clear already in the 1930s [25]. General equations describing the effect of lateral interactions on the kinetics of simplest reactions were derived in the early 1980s [39] (see also Ref. [33]).

5.4. Surface restructuring

The fact that adsorption often results in considerable changes in the surface structure and that such changes may influence the reaction kinetics was not properly appreciated by the catalysis community before the surface-science era, even though the possibility of this phenomenon was anticipated already in the 1920s. For example, some of the statements in the cited paper by Taylor [15] can be interpreted in this meaning (see the review [16]). Explicitly, the importance of adsorbate-induced changes in catalysts was emphasized by Boreskov since the beginning of the 1960s. Historically, this was connected with his studies of SO_2 oxidation on the V-based catalyst (the changes of the structure of this catalyst were evident).

In the surface-science community, the significance of spontaneous and adsorbate-induced changes in the surface was first (before the beginning of the 1980s) underestimated as well. Since then, it has however become clear that this phenomenon is pervasive [40–42]. It lies behind a variety of interesting processes including oscillations and chaos in the kinetics of various catalytic reactions [40,41]. The latter two phenomena are often observed on Pt(1 0 0) and Pt(1 1 0). In both cases, the bulk (1 × 1) periodicity of clean surfaces (Fig. 3) is higher in energy compared to a close-packed quasi-hexagonal or (2 × 1) missing-row arrangement, respectively. Adsorption may stabilize the (1 × 1) phase. The interplay between such surface changes and elementary reaction steps is

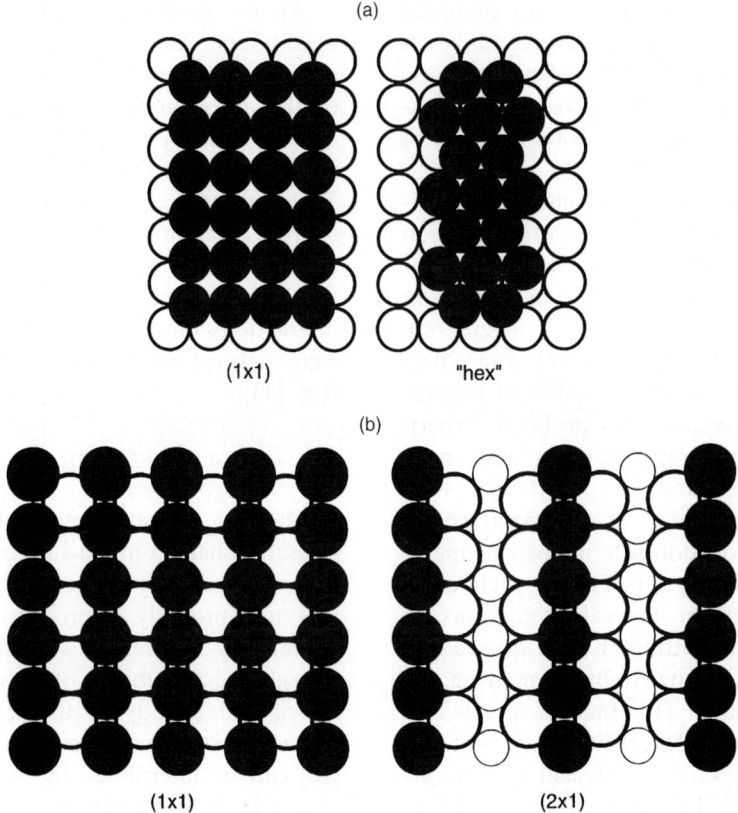

Fig. 3. Truncated-bulk and reconstructed (1 0 0) and (1 1 0) surfaces ((a) and (b)) of Pt. Filled and open circles show Pt atoms in the first and second layers. Small open circles exhibit the third layer.

often the main reason (at least at the UHV conditions) of the complexity of reaction kinetics on Pt.

The statistical models of spontaneous and adsorbate-induced surface restructuring or roughening are still not well developed (for the state of the art in this field, see Refs. [42,43]). For this reason, the understanding of the effect of surface restructuring on elementary rate processes in adsorbed overlayer is limited as well (for a review of the corresponding models, see Ref. [44]).

5.5. Limited mobility of adsorbed species

In chemically reactive adsorbed overlayers, the probabilities of arrangements of adsorbed particles and accordingly the reaction rate are defined by the interplay between adsorption, reaction, and adsorbate diffusion. Surface-science studies [45–47] indicate that the activation energy for surface diffusion is often relatively low (about 25% of the adsorption energy). For this reason, the situations when the rates of diffusion of all the adsorbed species involved into reaction are fast compared to other elementary steps are frequent. In this case, the adsorption overlayer is close to equilibrium in the sense that the arrangements of adsorbed particles can be calculated by using the prescriptions of statistical physics. The situations when diffusion of some of the reactants is slow compared to other step are however also possible. In the latter case, the arrangements of adsorbed particles is often far from equilibrium. In particular, immobile reactant may form islands or domains (especially during the transient reaction regimes as illustrated in Fig. 4).

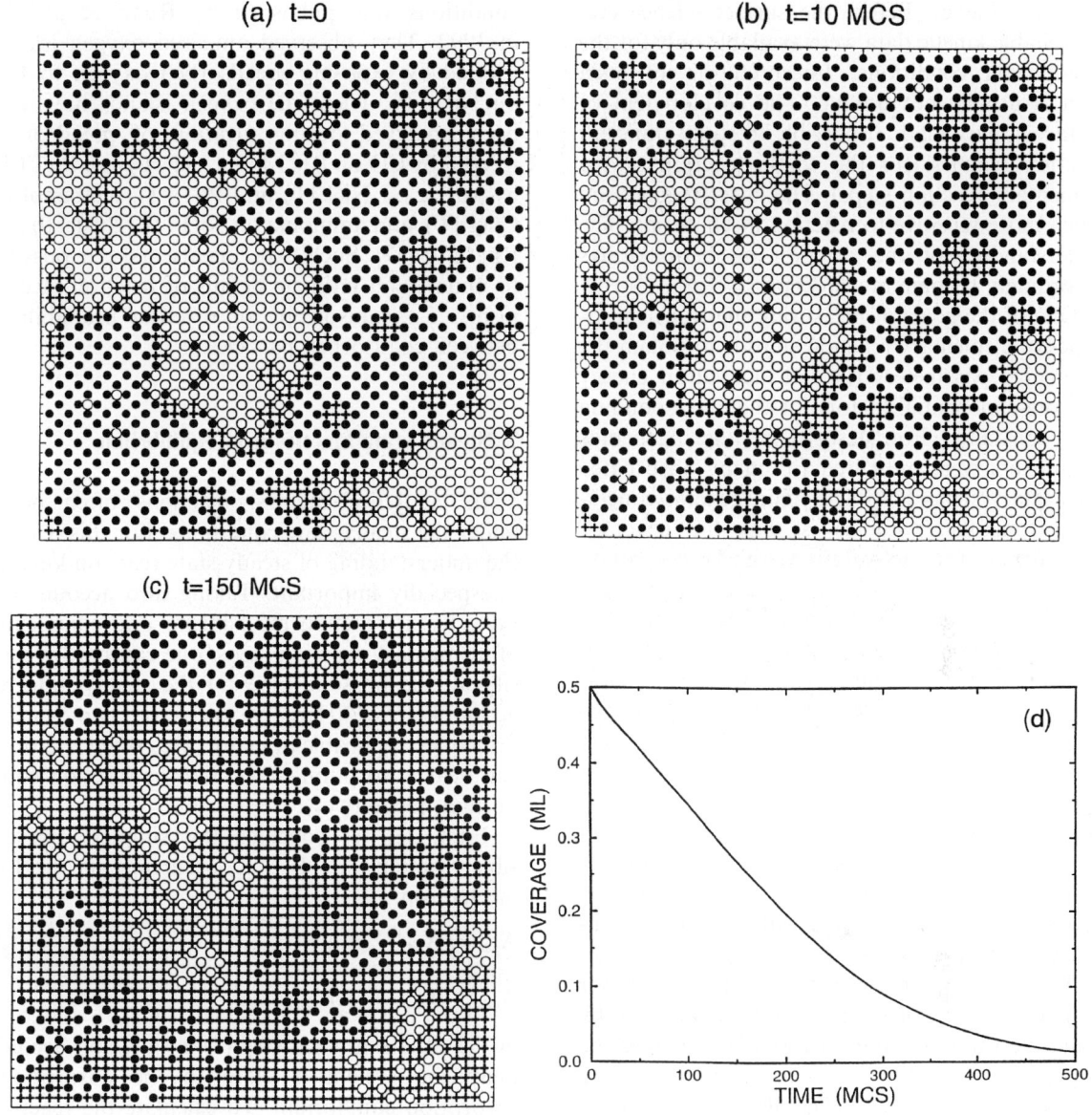

Fig. 4. A 50 × 50 fragment of the 500 × 500 lattice during the fast Monte Carlo $A_2 + B$ reaction in the case when preadsorbed B particles form $c(2 \times 2)$ domains [48]. B diffusion is considered to be negligibly slow compared to reaction. Filled and open circles indicate B particles located on two $c(2 \times 2)$ sublattices. Plus signs show sites occupied by A particles (due to rapid dissociative adsorption of A_2 molecules). Vacant sites are not shown. Panel (a) exhibits the initial distribution of reactants. The initial B coverage is 0.5 ML. Panels (b) and (c) display the surface after 10 and 150 Monte Carlo steps (MCS) for reaction. In the latter cases, the B coverage is 0.48 and 0.27 ML, respectively. Panel (d) exhibits B coverage as a function of time.

6. Transient kinetics

To describe reaction kinetics, one needs to know a reaction mechanism and the rate constants of elementary reaction steps. Another version of this statement is that to establish a reaction mechanism and to get the rate constants of elementary reaction steps, one should scrutinize the reaction kinetics. Before the surface-science era, the reliable kinetic data were available only for the steady-state conditions. Interpretation of such data is usually not unique. The development of surface science made it possible to study transient reaction regimes (for the whole reaction or for elementary reaction steps). The latter regimes are much more sensitive to reaction mechanisms. The understanding of the reaction mechanisms has accordingly become much deeper.

One of the goals of measuring the transient kinetics of elementary reaction steps is to obtain the Arrhenius parameters (i.e., an activation energy and pre-exponential factor) for the corresponding rate constants. For desorption, the Arrhenius parameters were measured for many adsorbate–substrate systems [49]. The main general conclusion from these studies is that the coverage dependence of the activation energy for gesorption is usually strong. As we already noted in Section 5.2, the changes in the activation energy are often 10–30 kcal/mol. For elementary reactions, the reliable experimental data on the Arrhenius parameters are in fact lacking. To get such data, one needs to measure the reaction rate in a relatively wide range of temperatures at constant reactant coverages. The latter condition is important, because the kinetics are usually far from ideal and accordingly the Arrhenius procedure may result in large errors if the coverages are not constant. To maintain constant coverage with increasing temperature however is difficult.

Theoretically, the pre-exponential factor for elementary reaction steps can be estimated by using the transition state theory which relates the populations of the transition (activated) and ground states (Fig. 2). The list of "standard" values of pre-exponential factors was tabulated [49] (see also Ref. [33]). An activation energy can in principle be calculated by employing the methods of quantum chemistry (see, e.g., recent ab initio calculations [50] of the activation energy for CO oxidation on Pt(1 1 1)). This approach is now becoming fashionable [51].

With the development of STM, one can directly monitor the arrangement of adsorbed particles during transient reaction regimes. The first report showing STM snapshots obtained under reactive conditions was published by Ruan et al. [52] in 1992. They observed nm-sized oxygen islands formed by exposing preadsorbed oxygen to H_2S on Ni(1 1 0). Then, the oxygen islands have also been observed during titration of preadsorbed oxygen by CO on Rh(1 1 0), Cu(1 1 0) and Pt(1 1 1), hydrogen on Ni(1 1 0) and Pt(1 1 1), methanol on Cu(1 1 0), and ammonia on Ni(1 1 0), Cu(1 1 0) and Rh(1 1 0) [47]. Such experiments make it possible to understand intimate details of reaction mechanisms, e.g., anisotropy of reaction probabilities between coadsorbed particles.

7. Steady-state kinetics near equilibrium

In catalytic reactors, reactions often occur under the steady-state conditions. For this reason, the understanding of steady-state reaction kinetics is especially important. Taking into account the diversity of reaction mechanisms and complexity of elementary reaction steps, one can hardly classify all the possible types of the steady-state reaction kinetics. Here we focus on two general principles applicable in many cases when reactions run near adsorption–desorption or reaction equilibrium, respectively.

First, let us consider the simplest irreversible reaction, $A + B \rightarrow C$, occurring via the LH mechanism,

$$A_{gas} \rightleftharpoons A_{ads}, \quad B_{gas} \rightleftharpoons B_{ads}, \qquad (3)$$

$$A_{ads} + B_{ads} \rightarrow (AB)_{gas}, \qquad (4)$$

in the situation when step (4) is slow and accordingly the system is close to the adsorption–desorption equilibrium. To calculate the reaction rate, W, one can use in this case the reactant coverages corresponding to adsorption equilibrium. For the Langmuir layer, an elementary analysis based on this approximation yields

$$W = k_r K_A P_A K_B P_B / (K_A P_A + K_B P_B)^2, \qquad (5)$$

where k_r is the rate constant of the Langmuir–Hinshelwood step, K_A and K_B are the adsorption equilibrium constants, and P_A and P_B are the reactant pressures. For fixed B pressure, the reaction rate first linearly increases with increasing P_A and

tween nn adsorbed particles. Comparing the ideal and non-ideal kinetics indicates that the asymptotic dependences of the reaction rate on P_A are the same in both cases, but the transition region is much wider in the non-ideal case.

The second generic case is when a reaction with reversible steps runs near the reaction equilibrium. According to thermodynamics, the reaction rate should be equal to zero at equilibrium. Then, one may ask: "what should we have near equilibrium?". The response to this question is trivial for a reaction occurring in the bulk via two (forward and inverse) steps. For the $A + B \rightleftharpoons C$ reaction, for example, the reaction rates in the two directions are $W_+ = k_1 P_A P_B$ and $W_- = k_{-1} P_C$, where k_{-1} and k_1 are the corresponding rate constants. Thus, the total reaction rate is given by $W = W_+ - W_- = W_+(1 - KP_C/P_A P_B)$, where $K = k_{-1}/k_1$ is the equilibrium constant. Catalytic reactions run however via a few steps. For example, one of the mechanisms of ammonia synthesis on Fe-based catalysts is as follows

$$3\ (H_2)_{gas} \rightleftharpoons 2H_{ads}, \quad (6)$$

$$1\ (N_2)_{gas} \rightleftharpoons 2N_{ads}, \quad (7)$$

$$2\ N_{ads} + H_{ads} \rightleftharpoons NH_{ads}, \quad (8)$$

$$2\ NH_{ads} + H_{ads} \rightleftharpoons (NH_2)_{ads}, \quad (9)$$

$$2\ (NH_2)_{ads} + H_{ads} \rightleftharpoons (NH_3)_{gas}. \quad (10)$$

The corresponding balance equation, obtained by multiplying the steps above by the stoichiometric numbers indicated on the left-hand side and then summarizing all the left- and right-hand terms, is

$$3H_2 + N_2 \rightleftharpoons 2NH_3. \quad (11)$$

What should we have here near equilibrium? The response to this equation, given by Boreskov [53], is

$$W = W_+ - W_- = W_+[1 - (KP_{NH_3}^2/P_{H_2}^3 P_{N_2})^{1/n}], \quad (12)$$

where n is the stoichiometric number corresponding to the rate-limiting step (e.g., $n = 1$ for step (7)). Thus, if a reaction occurs via one-route mechanism and there is a rate-limiting step, the

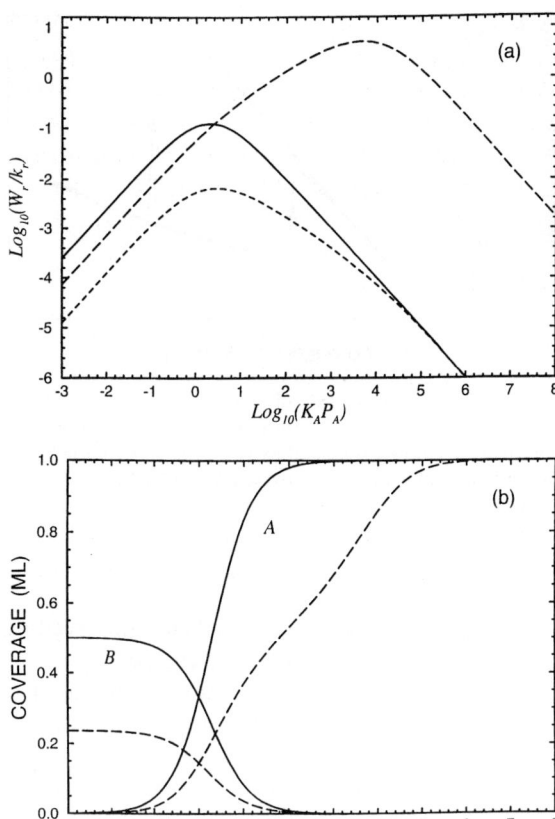

Fig. 5. Steady-state kinetics of the $A + B \rightarrow AB$ reaction occurring on a square lattice via the LH mechanism (steps (3) and (4)) near the adsorption–desorption equilibrium. Panels (a) and (b) show the reaction rate and reactant coverages as a function of A pressure for fixed B pressure. The solid lines are for the ideal Langmuir layer. The dashed lines are for the model taking into account (in the quasi-chemical approximation [33]) repulsive lateral interactions, $\epsilon_{AA} = \epsilon_{BB} = \epsilon_{AB} = 2k_B T$, between nn adsorbed particles. In the latter case, the reaction rate has been calculated for the two situations when (i) there are no lateral interactions in the activated state for the LH step (long dashed curves) and (ii) the lateral interactions in the activated state are the same as in the ground state (short dashed curves), respectively. k_r is the rate constant of the LH step, and K_A is the adsorption equilibrium constants (for the kinetics with lateral interactions, k_r and K_A correspond to the low-coverage limit).

then after reaching a maximum becomes inversely proportional to P_A. To illustrate, the effect of the non-ideality of the reaction steps on the steady-state reaction kinetics, we show (Fig. 5) the corresponding kinetics calculated with repulsive lateral interactions, $\epsilon_{AA} = \epsilon_{BB} = \epsilon_{AB} = 2k_B T$, be-

reaction rates in the forward and inverse directions are interconnected via the ratio of the reactant pressures (in analogy with the balance equation) and the stoichiometric number of the rate-limiting step.

Knowledge accumulated in heterogeneous catalysis indicates that interpretation of the steady-state reaction kinetics based only on the dependence of the reaction rate on reactant pressures is usually not unique. If the kinetic measurements are combined with surface-science based monitoring of adsorbate coverages however, the understanding of a reaction mechanism becomes much deeper and the interpretation of reaction kinetics more reliable.

8. Far from equilibrium

Under the steady-state conditions, the kinetics of reactions running near adsorption–desorption or reaction equilibrium usually have no singularities (i.e., the dependence of the reaction rate or reactant pressures is smooth). Often, however, catalytic reactions occur far from equilibrium. In this case, with changing governing parameters (e.g., pressure), one can observe kinetic phase transitions, oscillations, chaos, and pattern formation.

8.1. Kinetic phase transitions

Kinetic phase transitions, observed since the early 1970s in such catalytic reactions CO, hydrogen and hydrocarbon oxidation and NO reduction on Pt, Rh and Pd, belong to the first-order class [54–56]. This means that with increasing pressure one can monitor a stepwise increase (or decrease) in the reaction rate. On the way back, there is usually hysteresis as indicated in Fig. 6, just as in real first-order phase transitions. Mathematically, such behaviour corresponds to the situation when a system of non-linear kinetic equations have three solutions for adsorbate coverages. Two of them are stable and the third (intermediate) one is unstable. The transition from one stable state to another occurs via propagation of chemical waves [55]. In the most recent experi-

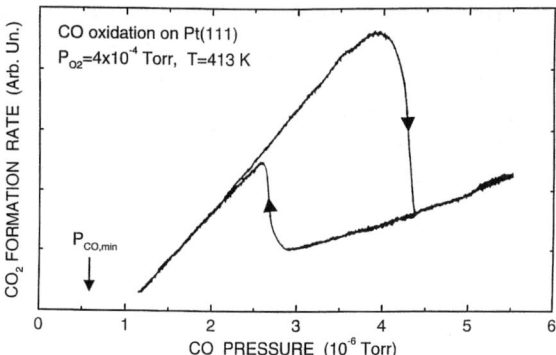

Fig. 6. Hysteresis in the rate of CO oxidation on Pt(1 1 1). In the low reactive state, the surface is covered primarily by CO and the reaction rate is limited by O_2 adsorption. In the highly reactive state, the surface is covered primarily by oxygen and the CO and O_2 adsorption match each other. (Redrawn from Ref. [57].)

ments [57], such waves were explicitly observed by using photoelectron emission microscopy.

A *continuous* kinetic phase transition was first found [58] in Monte Carlo simulations of the $2A + B_2 \rightarrow 2AB$ with irreversible A and B_2 adsorption in the case when there is no diffusion of adsorbed species. For this reaction scheme, the lattice is completely covered by B if the A impingement rate is low. With increasing the A impingement rate, the model predicts first a continuous kinetic phase transition to the reactive state with the surface covered primarily by B and then a first-order transition to the unreactive state with the lattice completely covered by A. Experimentally, to our knowledge, continuous kinetic phase transitions have not been observed yet.

8.2. Kinetic oscillations and chaos

The first reports on kinetic oscillations in heterogeneous catalytic reactions were published in the early 1970s (see the reviews [59,60]). Later on, regular and chaotic oscillations were found in about 30 surface reactions on practically all types of catalysts including single crystals, poly-crystalline samples (foils, ribbons and wires) and supported catalysts at a pressure range from 10^{-12} bar to atmospheric pressure [40,41,59,60]. For a given temperature and total pressure, kinetic oscillations are usually observed in a relatively narrow range of reactant pressures. Often oscillations are periodic. Sometimes, changing a governing parameter (e.g., pressure of one of the reactants), one can observe the transition from periodic oscillations to chaos. Usually, this transition occurs via period doubling as shown, e.g., in Fig. 7. In general theory of non-linear kinetic processes, this scenario proposed originally by Feigenbaum is now considered to be universal [61].

The contribution of surface-science based studies to scrutinizing the physics behind oscillations was crucial. The experience accumulated in this field indicates that oscillations are often observed in the systems where a rapid bistable catalytic cycle (Section 8.1) is combined with a relatively slow "side" process, e.g., with adsorbate-induced surface restructuring [62], oxide formation [63], or carbon deposition [64]. The understanding of the corresponding oscillatory kinetics is still far from complete, because elementary reaction steps and side processes are usually far from ideal (Section 5).

The bulk of simulations of oscillations in reactions on solid surfaces [60] is based on the mean-field (MF) equations ignoring the non-ideality of surface rate processes. This approach is reasonable in many cases. For example, the properly modified MF equations can be used to describe, at least phenomenologically, oscillations resulting from

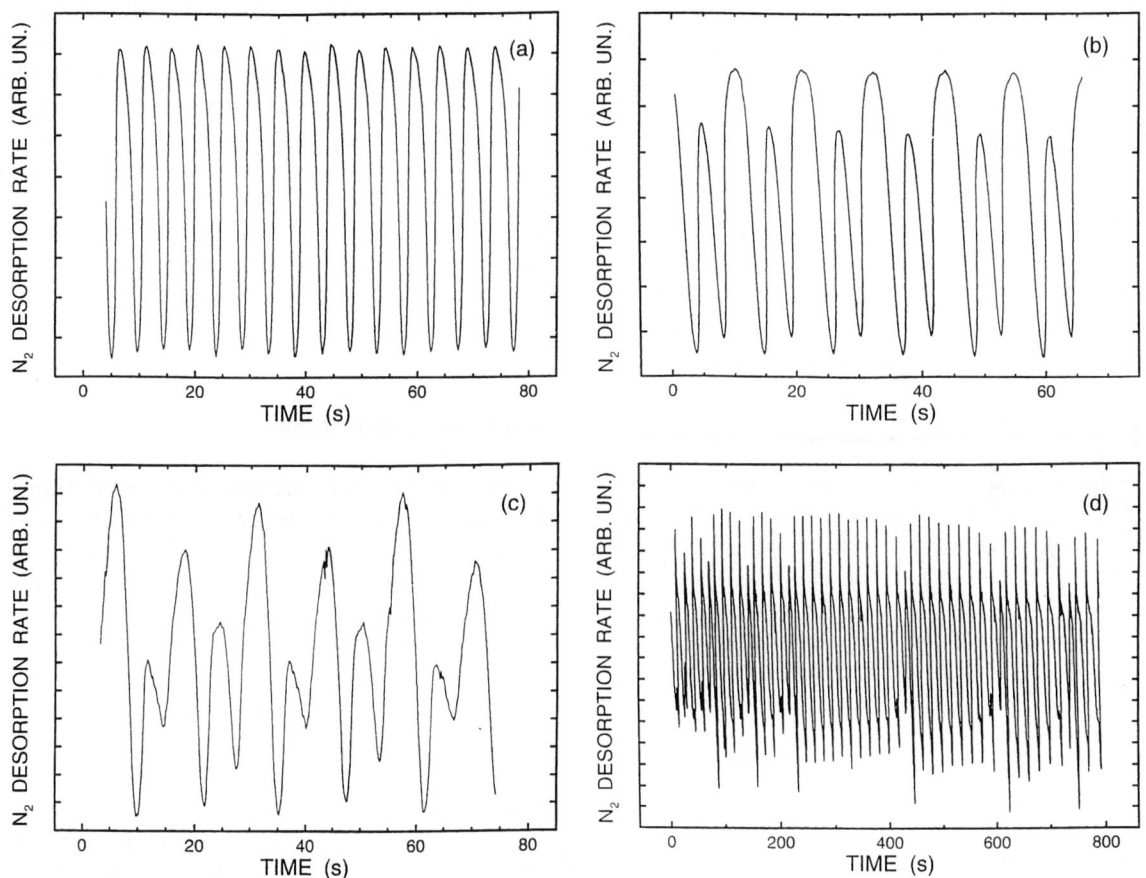

Fig. 7. Rate of N_2 desorption as a function of time during the NO–H_2 reaction on Pt(1 0 0) at $P_{NO} = 3 \times 10^{-9}$ bar and $T = 460$ K: (a) period-1 oscillations at $P_{NO}/P_{H_2} = 1$; (b) period-2, (c) period-4, and (d) and aperiodic oscillations at $P_{NO}/P_{H_2} \simeq 1.4$. (Redrawn from Ref. [65].)

the interplay of chemical steps and surface restructuring [40]. Alternatively, one can employ the Monte Carlo method to simulate oscillations. For simulations of patterns on the nm scale, this method is superior, because it easily allows taking into account and exploring in detail various complicating factors which can hardly be treated analytically. For example, we show (Fig. 8) oscillatory Monte Carlo kinetics [66] of NO reduction on Pt(1 0 0). The model employed takes into account NO-induced surface restructuring, described in terms of the theory of first-order phase transitions. NO molecules are adsorbed (Fig. 9) primarily on the NO-induced (1×1) islands. During oscillations, the islands periodically grow and shrink.

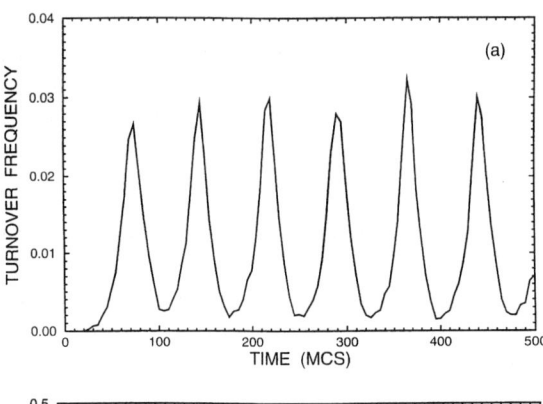

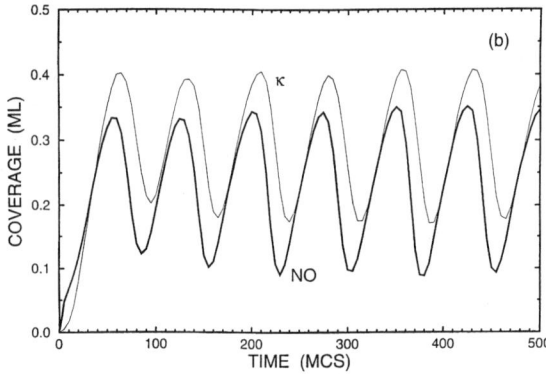

Fig. 8. Oscillatory Monte Carlo kinetics [66] of NO reduction by H_2 on Pt(1 0 0): (a) NO decomposition rate (per site per MCS) and (b) NO coverage and fraction, κ, of Pt atoms in the state corresponding to the (1×1) phase (Fig. 3(a)). The reaction occurs via NO decomposition on the (1×1) islands. (Redrawn from Ref. [66].)

8.3. Pattern formation

Pattern formation in chemical reactions was first analysed by Turing [67]. He predicted that spontaneous spatial organization is possible if reactant diffusion is coupled with non-linear chemical kinetics allowing (without spatial perturbations) oscillations. In heterogeneous catalytic reactions, pattern formation was observed on all the size scales from a few nm, e.g., in H_2 oxidation on a Pt tip of a field ion microscope [68], to a few mm, e.g., in CO oxidation on Pt(1 0 0) [70] or in H_2 oxidation on supported Pt catalysts [71]. Often, the pattern formation in catalytic reactions is directly connected with kinetic bistability and oscillations (see, e.g., Fig. 10 illustrating "out-of-phase" oscillations on the (1 0 0) facets of a μm-sized Ir tip during the NO–H_2 reaction [72] and Fig. 11 showing target patterns observed on the 100-μm scale during CO oxidation on Pt(1 1 0) [73]). The latter is not however a necessary condition for pattern formation. In general, this phenomenon may also occur due to adsorbate–adsorbate interactions. For example, we refer to the first STM study [69] of the island formation on the nm scale under the steady-state conditions (during CO oxidation on Cu(1 1 0)) or to general Monte Carlo simulations [20,74].

9. Pressure-gap problem

One of the topics actively discussed during the formative years of surface science ($\sim$1965–1990 [75]) was whether one can use the kinetic data obtained at UHV conditions to describe the reaction kinetics at practical (e.g., atmospheric) pressures. The first attempts to solve this so-called pressure-gap problem were aimed at ammonia syntheses on the K-promoted Fe catalyst [76] and H_2 oxidation on Pt(1 1 1) [77]. In the first study, Stoltze and Norskov [76] demonstrated (Fig. 12) that the agreement between the UHV-based and real kinetics might be within a factor of 2 (see, however, comments [78,79]). With our present knowledge, this almost perfect agreement, connected partly with the fact that at high pressures the measured outcome NH_3 mole fraction is lim-

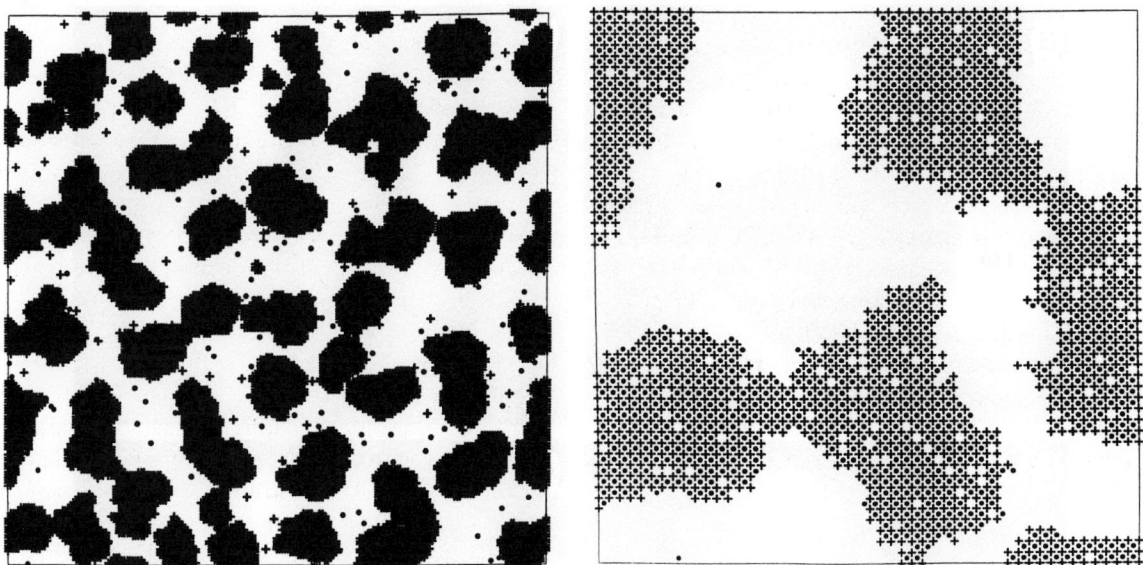

Fig. 9. Typical snapshots of a (200 × 200) lattice for the run, shown in Fig. 8, at the stages when the fraction of Pt atoms in the metastable state is maximum. The left panel shows the whole lattice. The right panel, representing a (50 × 50) fragment of the lattice, exhibits the structure of the restructured islands in detail. Filled circles and plus signs indicate, respectively, NO molecules and substrate atoms in the state corresponding to the (1 × 1) phase (cf. Fig. 3(a)). Metal atoms in the stable state corresponding to the "hex" phase are not shown. (Redrawn from Ref. [66].)

ited by thermodynamics, can unfortunately hardly be considered as typical. In the second study [77], the reaction rate, extrapolated from low to moderate pressures, was found to exceed the measured rate by a factor of about 10 (Fig. 13). Further simulations were focused on CO oxidation on Pt [80], methanol synthesis on Cu [81], NO–CO reaction on Rh [82] and some other reactions [83]. The experience accumulated in this field indicates that extrapolation of kinetic data obtained at UHV to practical conditions is hardly possible if one ignores the non-ideality of the reaction kinetics. Incorporation of the latter type of features into the reaction kinetic models is still a challenging problem, because the adsorbate coverages corresponding to practical pressures are often much higher than those typical for UHV conditions.

The simulations mentioned above have resulted in a deeper insight into the mechanism of the reactions under consideration, but did not result in the appearance of new catalysts for the corresponding processes. One of the few examples where it actually had been possible by combining fundamental structural, theoretical, and reactivity surface-science based studies to design a new catalyst which was demonstrated to work under realistic conditions has recently been described in detail by Larsen and Chorkendorff [84]. Their review is focused on methane conversion in the steam reforming process. For this important reaction, a new gold-nickel alloy catalyst, which is much more resistant to carbon formation compared to the conventional catalysts, was invented after realizing that such alloys could be synthesized from otherwise immiscible metals.

10. Structure-gap problem

A common class of real catalysts consists (Fig. 1) of small metal particles deposited on the internal surface of a more or less inactive porous support (this design is used to maximize the active surface area per unit weight and volume of the catalyst). In practice, such catalysts are shaped either as porous pellets (2–20 mm in diameter)

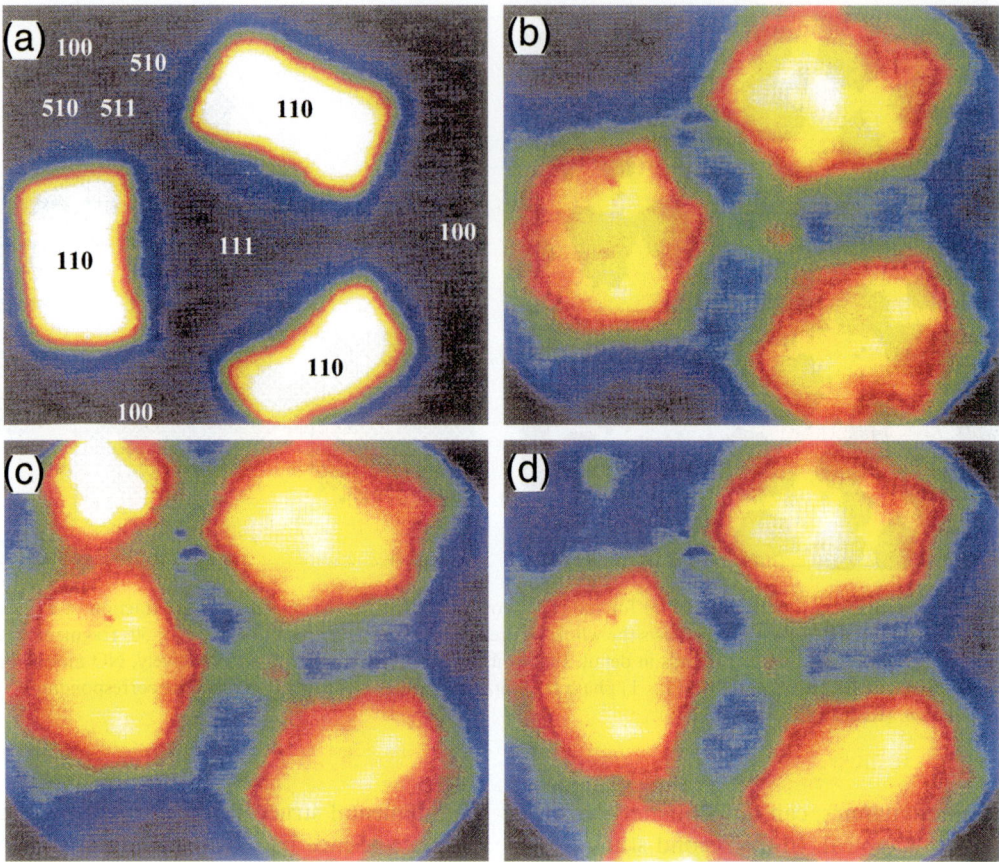

Fig. 10. Images of a μm-sized Ir tip of a field emission microscope. Panel (a) shows a clean emitter with a central (1 1 1) facet and peripheral (1 1 0) and (1 0 0) facets (white colour corresponds to the maximum current). The other three panels exhibit current distribution observed during the NO–H$_2$ reaction at $P_{NO} = 1.5 \times 10^{-6}$ Torr, $P_{H_2} = 4.6 \times 10^{-5}$ Torr and $T = 341$ K for $t = 0$ s (b), 4 s (c) and 100 s (d). Irregular kinetic oscillations occur only on the (1 0 0) facets. Initially (b), these facets are dark. At $t = 4$ s (c), the left (1 0 0) facet is bright. At $t = 100$ s (d), the bottom (1 0 0) facet is bright. The size of facet is expected to be about 0.1 μm. (Redrawn from Ref. [72].)

or so-called monoliths with a porous "washcoat". The size of pores varies in a large range from 1–2 nm (micropores) to ⩾ 50 nm (macropores). Often the pores are mesoscopic (2–50 nm). The size of metal particles, d, may vary in a wide range as well. In zeolites (with pores of about 1 nm), metal particles often contain only a few atoms (such particles are called clusters). More typical size is 1–20 nm. Sometimes, however, supported particles are much larger (e.g., the silver particles employed for selective oxidation of ethylene are roughly 10^3 nm in size, and Pt and Rh particles of car exhaust cleaning catalysts are typically 10–100 nm after some time of use).

The first systematic experimental investigations of the relationship between the particle size and catalytic activity were done in the middle of the fifties [85,86] after the development of a technique for measuring the surface area of noble-metal catalysts by means of the selective chemisorption of hydrogen. Since then, specific catalytic activities of different catalysts in different reactions have been measured in thousands of studies. Although the early results were often contradictory [86], it was firmly established that catalytic reactions could be divided into four classes. (i) The turnover rate (i.e., the rate per surface metal atom or per adsorption site) of *structure-insensitive* reactions is

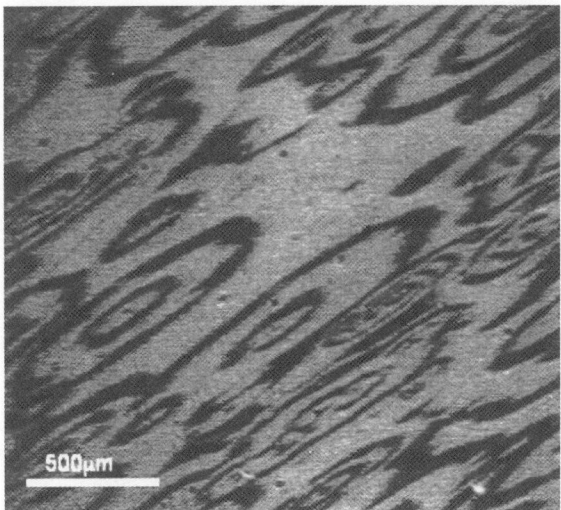

Fig. 11. Target patterns obtained by using reflection anisotropy microscopy during CO oxidation on Pt(1 1 0) at $P_{CO} = 6 \times 10^{-5}$ mbar, $P_{O_2} = 3 \times 10^{-4}$ mbar and $T = 473$ K. Bright and dark patches are covered primarily by oxygen and CO, respectively. (Redrawn from Ref. [73].)

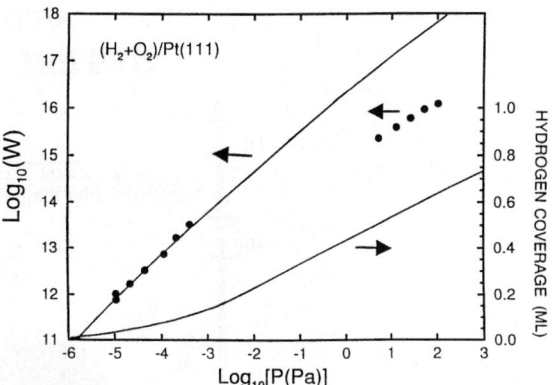

Fig. 13. Steady-state rate of the H_2–O_2 reaction (in units of O_2 molecules per cm^2 s) and hydrogen coverage of Pt(1 1 1) as a function of pressure of the stoichiometric mixture of reactants at 360 K. Circles and solid lines correspond to experiment and theory, respectively. The model parameters were obtained on the basis of the transient kinetics measured at UHV conditions. Note that the model predicts rather slow increase in the reaction rate and hydrogen coverage with increasing pressure. This is a direct consequence of the hydrogen–hydrogen lateral interactions taken into account in the calculations. (Redrawn from Ref. [77].)

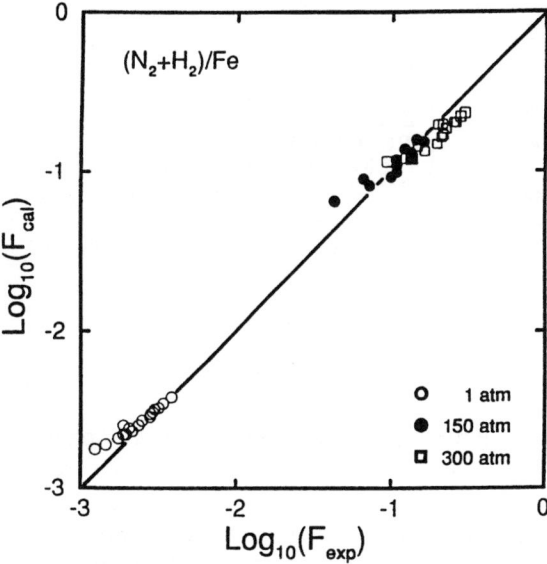

Fig. 12. Comparison of calculated and measured NH_3 mole fraction out of a catalytic reactor containing K-promoted Fe catalyst. The data set includes a broad range of the ammonia syntheses conditions. Extrapolation to the high pressure region was executed with the parameters corresponding to the low pressure region. The deviations between experiment and theory is within a factor of approximately 2. (Redrawn from Ref. [76].)

nearly independent of the particle size (within a factor of 2). Such reactions can however be considered as exceptions at $d \sim 10$ nm. As a rule, catalytic reactions are *structure-sensitive* on this scale. In this case, the turnover rates may (ii) increase, (iii) pass through a maximum, or (iv) decrease with increasing the particle size.

The fact that the turnover rate of catalytic reactions often depends on d indicates that the relationship between the reaction kinetics on nm particles and those measured on single-crystal surfaces is not straightforward. This structure gap between the kinetics may have different origins as indicated schematically in Fig. 14. At the smallest length scales of the order of 1 nm, metal particles have 2D or 3D sizes comparable to the electronic screening length in metals, and the electron structure is significantly different from that of bulk metals. Consequently their catalytic activities are also different. At somewhat larger sizes, 3–4 nm, the electronic properties of the particles are already close to those of the bulk metal, except at the atoms contacting or very near the support where

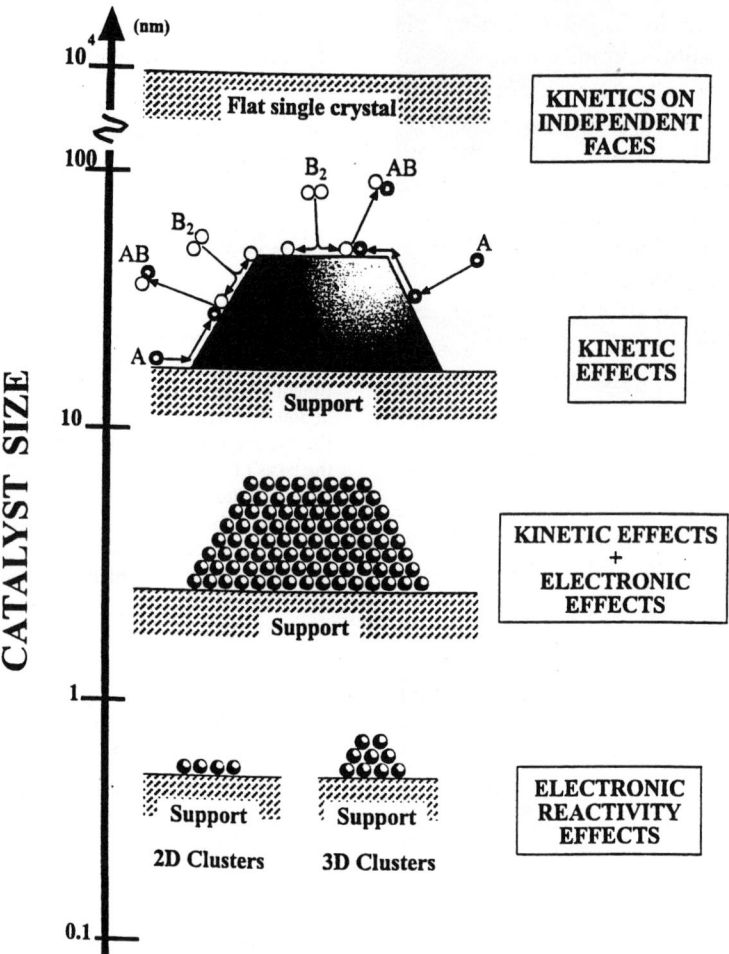

Fig. 14. Electronic and kinetic effects as ingredients of the structure gap in heterogeneous catalysis.

metal support interaction may modify the catalytic properties. The latter may be significantly perturbed, electronically, and thereby also have different catalytic activities. Since the particles are small, these sites may be important or even dominant for the overall catalytic kinetics. At still larger sizes, above about 10 nm, the particles are electronically essentially identical to bulk metal but may still exhibit remarkably different kinetics, compared to single crystals. In this case, the reasons of structure sensitivity are purely kinetic. The basic underlying mechanisms for kinetic phenomena which do not occur on single crystals are [87]: (i) the different catalytic activities on different facets of a small supported crystalline particle become coupled in a strongly non-linear fashion due to diffusion occurring over facet boundaries, (ii) equilibrium-shape changes of small particles, caused by adsorbates, induce changes in catalytic behaviour, (iii) different kinetic rate constants at the facet boundaries of a supported particle compared to those for the perfect facets give rise to new kinetics, and (iv) spillover by diffusion of reactants, between the particle and its support, also create new kinetics. All these factors are especially important for catalytic reactions occurring far from adsorption–desorption equilibrium.

The role of surface science in clarifying the reasons of structure sensitivity of catalytic reactions was decisive, because numerous surface-science based studies have explicitly shown that the energetics of adsorption and reaction kinetics are often quite different on different faces of single crystals. Only with the development of surface science, it has become possible to scrutinize the physical factors behind structure sensitivity.

11. Conclusion

In this brief review, I have outlined general concepts used to rationalize the kinetics of heterogeneous catalytic reactions. Although most of these concepts appeared far before the surface-science era, only with application of surface-science methods has it become clear what we really have in chemically reactive adsorbed overlayers and how the concepts practically work. In the most interesting and intriguing areas, such as oscillations, chaos and 2D pattern formation, the role of surface science was crucial. Before the development of surface science, these branches of the 2D kinetics simply did not exist.

During the formative years of surface science, academic studies of catalytic reactions were primarily executed on single-crystal surfaces. The most recent surface-science based academic studies [9,10,87,88] are aimed at fabrication of model nm-sized supported catalysts under well controlled conditions. This branch of surface science is expected to extend our understanding of the reaction kinetics on the nm scale.

In the future, the role of surface science in heterogeneous catalysis in general and in kinetic experiments and simulations in particular will only increase. Taking into account the complexity and practical importance of heterogeneous catalytic reactions, we may conclude that the surface-science based studies of the kinetics of such reactions merit attention of a new generation of scientists.

Acknowledgements

Preparing this review, I bore in mind my talks with the late Georgii K. Boreskov and Kirill I. Zamaraev and also with Bengt Kasemo.

References

[1] Past, Present and Future, Physics Prepares for the 21st Century, Phys. World 12 (12) (1999).
[2] A.P. Kieboom, J. Moulijn, P.W.N.M. van-Leeuwen, R.A. van Santen, History of catalysis, in: R.A. van-Santen et al. (Eds.), Catalysis: An Integrated Approach, Studies on Surface Science Catalysis, vol. 123, Elsevier, Amsterdam, 1999, p. 3.
[3] B.C. Gates, H. Knözinger (Eds.), Impact of Surface Science on Catalysis, Academic Press, San Diego, 2000.
[4] J.M. Thomas, W.J. Thomas, Principles and Practice of Heterogeneous Catalysis, VCH, Weinheim, 1997.
[5] G.K. Boreskov, Y.S. Matros, O.V. Kiselev, G.A. Bunimovich, Performance of heterogeneous catalytic reactions in the unsteady-state regime, Proc. Acad. Sci. USSR 23 (1977) 160 (in Russian).
[6] Y.S. Matros, G.A. Bunimovich, Reverse-flow operation in fixed bed catalytic reactors, Catal. Rev. Sci. Engng. 38 (1996) 1.
[7] K.C. Taylor, Nitric-oxide catalysis in automotive exhaust systems, Catal. Rev. Sci. Engng. 35 (1993) 457.
[8] L.D. Schmidt, The Engineering of Chemical Reactions, Oxford, New York, 1998.
[9] C.R. Henry, Surface studies of supported model catalysts, Surf. Sci. Rep. 31 (1998) 231.
[10] P.L.J. Gunter, J.W. Niemantsverdriet, F.H. Ribeiro, G.A. Somorjai, Surface science approach to modeling supported catalysts, Catal. Rev. Sci. Engng. 39 (1997) 77.
[11] K. Wong, S. Johansson, B. Kasemo, Nanofabricated model catalysts: manufacturing and model studies, Faraday Discuss. 105 (1997) 237.
[12] A.O. Hougen, K.M. Watson, Chemical Process Principles, Wiley, New York, 1947.
[13] R.I. Masel, Principles of Adsorption and Reaction on Solid Surfaces, Wiley, New York, 1996.
[14] R. Mezaki, H. Inoue, Rate Equations of Solid-Catalyzed Reactions, University of Tokyo, Tokyo, 1991.
[15] H.S. Taylor, A theory of the catalytic surface, Proc. Roy. Soc. (London) A 108 (1925) 105.
[16] M. Boudart, Principles of heterogeneous reactions, in: G. Ertl, H. Knözinger, J. Weitkamp (Eds.), Handbook of Heterogeneous Catalysis, vol. 1, VCH, Weinheim, 1997, p. 1.
[17] P.T. Sprunger, Y. Okawa, F. Besenbacher, I. Stensgaard, K. Tanaka, STM investigation of the coadsorption and reaction of oxygen and hydrogen on Ni(1 1 0), Surf. Sci. 344 (1995) 98.

[18] J.T. Yates, Surface chemistry at metallic step sites, J. Vac. Sci. Technol. A 13 (1995) 1359.
[19] G.A. Somorjai, Introduction to Surface Chemistry and Catalysis, Wiley, New York, 1994.
[20] V.P. Zhdanov, B. Kasemo, Kinetics of the $2A + B_2 \to 2AB$ reaction complicated by spatial constraints for adsorption, formation of islands, and steps, Surf. Sci. 412 (1998) 527.
[21] L.P. Ford, P. Blowers, R.I. Masel, Role of steps and kinks in catalytic activity, J. Vac. Sci. Technol. A 17 (1999) 1705.
[22] B.Y. Zeldovich, On the theory of the Freundlich adsorption isotherms, Acta Physicochimica URSS 3 (1935) 791.
[23] S.Z. Roginsky, Adsorption and catalysis, in: S.Z. Roginsky (Ed.), Problems of Kinetics and Catalysis, vol. III, ONTI, Leningrad, 1937, p. 356.
[24] D.J. Dooling, J.E. Rekoske, L.J. Broadbelt, Microkinetic models of catalytic reactions on nonuniform surfaces: application to model and real systems, Langmuir 15 (1999) 5846.
[25] M. Temkin, V. Pyzhov, Kinetics of ammonia synthesis on the promoted Fe catalyst, J. Phys. Chem. 13 (1939) 851 (in Russian).
[26] B.N.J. Persson, Ordered structures and phase transitions in adsorbed layers, Surf. Sci. Rep. 15 (1992) 1.
[27] V.P. Zhdanov, Effect of lateral interaction of adsorbed particles on the average energy of desorption or reaction products, Surf. Sci. 165 (1986) L31.
[28] D.J. Bald, S.L. Bernasek, The internal energy of CO_2 produced from catalytic oxidation of CO by NO, J. Chem. Phys. 109 (1998) 746.
[29] M.A. Van Hove, W.H. Weinberg, C.-M. Chan, Low-Energy Electron Diffraction, Springer, Berlin, 1986.
[30] C.G. Goymour, D.A. King, Lateral interaction model for desorption kinetics, J. Chem. Soc., Faraday Trans. 1 69 (1973) 749.
[31] D.L. Adams, Consequence of adsorbate–adsorbate interactions for thermal desorption and LEED measurements, Surf. Sci. 42 (1974) 12.
[32] M.A. Morris, M. Bowker, D.A. King, in: C.H. Bamford, C.F.H. Tipper, R.G. Compton (Eds.), Simple Processes at the Gas-Solid Interface, vol. 19, Comprehensive Chemical Kinetics, Elsevier, Amsterdam, 1984, p. 163.
[33] V.P. Zhdanov, Elementary Physicochemical Processes on Solid Surfaces, Plenum Press, New York, 1991.
[34] V.P. Zhdanov, B. Kasemo, Simulation of oxygen desorption from Pt(1 1 1), Surf. Sci. 415 (1998) 403.
[35] C. Stampfl, H.J. Kreuzer, S.H. Payne, H. Pfnür, M. Scheffler, First-principles theory of surface thermodynamics and kinetics, Phys. Rev. Lett. 83 (1999) 2993.
[36] L. Österlund, M.O. Pedersen, I. Stensgaard, E. Legsgaard, F. Besenbacher, Quantitative determination of adsorbate–adsorbate interactions, Phys. Rev. Lett. 83 (1999) 4812.
[37] T.L. Einstein, Theory of indirect interaction between chemisorbed atoms, Crit. Rev. Sol. State Mater. Sci. 7 (1978) 261.
[38] J.K. Norskov, in: D.A. King, D.P. Woodruff (Eds.), Coadsorption, Promoters and Poisons, Elsevier, Amsterdam, 1993, p. 1.
[39] V.P. Zhdanov, Lattice-gas model for description of adsorbed molecules, Surf. Sci. 111 (1981) 63.
[40] M. Gruyters, D.A. King, Effects of restructuring in adsorption and reaction dynamics at metal surfaces, J. Chem. Soc., Faraday Trans. 93 (1997) 2947.
[41] R. Imbihl, G. Ertl, Oscillatory kinetics in heterogeneous catalysis, Chem. Rev. 95 (1995) 697.
[42] D.A. King, D.P. Woodruff (Eds.), Phase Transitions and Adsorbate Restructuring at Metal Surfaces, Elsevier, Amsterdam, 1994.
[43] V.P. Zhdanov, B. Kasemo, Surface roughening: kinetics, adsorbate-induced effects and manifestation in catalytic reactions, J. Chem. Phys. 108 (1998) 4582.
[44] V.P. Zhdanov, P.R. Norton, Surface reconstruction and rate processes in adsorbed overlayers, Langmuir 12 (1996) 101.
[45] A.G. Naumovets, Yu.S. Vedula, Surface diffusion of adsorbates, Surf. Sci. Rep. 4 (1985) 365.
[46] R. Gomer, Diffusion of adsorbates on metal surfaces, Rep. Rrogr. Phys. 53 (1990) 917.
[47] J. Wintterlin, Scanning tunneling microscopy studies of catalytic reactions, in: B.C. Gates, H. Knözinger (Eds.), Impact of Surface Science on Catalysis, Academic Press, San Diego, 2000, p. 131.
[48] V.P. Zhdanov, Domain growth and reaction kinetics in adsorbed overlayers, Phys. Rev. Lett. 77 (1996) 2109.
[49] V.P. Zhdanov, Arrhenius parameters for rate processes on solid surfaces, Surf. Sci. Rep. 12 (1991) 183.
[50] A. Alavi, P. Hu, T. Deutsch, P.L. Silvestrelli, J. Hutter, CO oxidation on Pt(1 1 1): an ab initio density functional study, Phys. Rev. Lett. 80 (1998) 3650.
[51] Q.F. Ge, R. Kose, D.A. King, Adsorption energetics and bonding from femtomole calorimetry and from first principles theory, in: B.C. Gates, H. Knözinger (Eds.), Impact of Surface Science on Catalysis, Academic Press, San Diego, 2000, p. 207.
[52] L. Ruan, F. Besenbacher, I. Stensgaard, E. Lagsgaard, Atom-resolved studies of the reaction between H_2S and O on Ni(1 1 0), Phys. Rev. Lett. 69 (1992) 3523.
[53] G.K. Boreskov, Relationship between the reaction orders and activation energies for the forward and reverse directions, J. Phys. Chem. 19 (1945) 92 (in Russian).
[54] L.F. Razon, R.A. Schmitz, Intrinsically instable behaviour during the oxidation of carbon monoxide on platinum, Catal. Rev. Sci. Engng. 28 (1986) 89.
[55] V.P. Zhdanov, B. Kasemo, Kinetic phase transitions in simple reactions on solid surfaces, Surf. Sci. Rep. 20 (1994) 111.
[56] H.C. Kang, W.H. Weinberg, Modeling the kinetics of heterogeneous catalysis, Chem. Rev. 95 (1995) 667.
[57] M. Berdau, G.G. Yelenin, A. Karpowicz, M. Ehsasi, K. Christmann, J.H. Block, Macroscopic and mesoscopic characterization of a bistable reaction system: CO oxidation on Pt(1 1 1) surface, J. Chem. Phys. 110 (1999) 11551.
[58] R.M. Ziff, E. Gulari, Y. Barshad, Kinetic phase transitions in irreversible surface-reaction model, Phys. Rev. Lett. 56 (1986) 2553.

[59] F. Schüth, B.E. Henry, L.D. Schmidt, Oscillatory reactions in heterogeneous catalysis, Adv. Catal. 39 (1993) 51.
[60] M.M. Slinko, N.I. Jaeger, Oscillatory Heterogeneous Catalytic Systems, Elsevier, Amsterdam, 1994.
[61] S.K. Scott, Chemical Chaos, Clarendon Press, Oxford, 1991.
[62] G. Ertl, P.R. Norton, J. Rüstig, Kinetic oscillations in the platinum-catalyzed oxidation of CO, Phys. Rev. Lett. 49 (1982) 177.
[63] B.C. Sales, J.E. Turner, M.B. Maple, Oscillatory oxidation of over Pt, Pd and Ir catalysts: theory, Surf. Sci. 114 (1982) 381.
[64] N.A. Collins, S. Sundaresan, Y.J. Chabal, Studies of self-sustained reaction-rate oscillations: the carbon model, Surf. Sci. 180 (1987) 136.
[65] P. Cobden, J. Siera, K. Tanaka, B.E. Nieuwenhuys, Oscillatory reduction of nitric-oxide with hydrogen over Pt(100), J. Vac. Sci. Technol. A 10 (1992) 2487.
[66] V.P. Zhdanov, Surface restructuring, kinetic oscillations, and chaos in heterogeneous catalytic reactions, Phys. Rev. E 60 (1999) 7554.
[67] A.M. Turing, The chemical basis of morphogenesis, Philos. Trans. Roy. Soc. B 327 (1952) 37.
[68] V. Gorodetskii, J. Lauterbach, H.H. Rotermund, J.H. Block, G. Ertl, Coupling between adjacent crystal planes in heterogeneous catalysis by propagating reaction-diffusion waves, Nature 370 (1994) 276.
[69] W.W. Crew, R.J. Madix, A scanning tunneling microscopy study of the oxidation of CO on Cu(110) at 400 K: site specificity and reaction kinetics, Surf. Sci. 349 (1996) 275.
[70] H.H. Rotermund, Imaging of dynamic processes on surfaces by light, Surf. Sci. Rep. 29 (1997) 265.
[71] P.C. Pawlicki, R.A. Schmitz, Spatial effects on supported catalysts, Chem. Engng. Progr. 83 (1987) 40.
[72] P.D. Cobden, Y. van Breugel, B.E. Nieuwenhuys, Local independent oscillation centers present on an iridium field emitter during NO–H_2 reaction, Surf. Sci. 402 (1998) 155.
[73] J. Dicke, P. Erichesen, J. Wolff, H.H. Rotermund, Reflection anisotropy microscopy: improved set-up and application to CO oxidation on platinum, Surf. Sci. 462 (2000) 90.
[74] V.P. Zhdanov, Pattern formation in catalytic reactions due to lateral adsorbate–adsorbate interactions, Langmuir 17 (2001) 1793–1799.
[75] C.B. Duke, Foreword to surface science: the first thirty years, Surf. Sci. 299/300 (1993).
[76] P. Stoltze, J.K. Norskov, Bridging the pressure gap between ultrahigh-vacuum surface physics and high-pressure catalysis, Phys. Rev. Lett. 55 (1985) 2502.
[77] V.P. Zhdanov, V.I. Sobolev, V.A. Sobyanin, Hydrogen-oxygen reaction on Pt(111), Surf. Sci. 175 (1986) L747.
[78] M. Bowker, I. Parker, K.C. Waugh, The application of surface kinetic data to the industrial synthesis of ammonia, Surf. Sci. 197 (1988) L223.
[79] B. Fastrup, Temperature-programmed adsorption and desorption of nitrogen on iron ammonia-synthesis catalysts, J. Catal. 150 (1994) 345.
[80] V.P. Zhdanov, B. Kasemo, Steady-state kinetics of CO oxidation on Pt: extrapolation from 10^{-10} to 1 bar, Appl. Surf. Sci. 74 (1994) 147.
[81] C.V. Ovesen, B.S. Clausen, J. Schiotz, P. Stoltze, H. Topsoe, J.K. Norskov, Kinetic implications of dynamic changes in catalytic morphology during methanol synthesis over Cu/ZnO catalysts, J. Catal. 168 (1997) 133.
[82] V.P. Zhdanov, B. Kasemo, Mechanism and kinetics of the NO–CO reaction on Rh, Surf. Sci. Rep. 29 (1997) 31.
[83] P. Stoltze, Microkinetic simulation of catalytic reactions, Progr. Surf. Sci. 65 (2000) 65.
[84] J.H. Larsen, I. Chorkendorff, From fundamental studies of reactivity on single crystals to the design of catalysts, Surf. Sci. Rep. 35 (1999) 163.
[85] G.K. Boreskov, V.S. Chesalova, Specific catalytic activity of metals, J. Phys. Chem. 30 (1956) 2560 (in Russian).
[86] M. Che, C.O. Bennett, The influence of particle-size on the catalytic properties of supported metals, Adv. Catal. 36 (1989) 55.
[87] V.P. Zhdanov, B. Kasemo, Simulations of the reaction kinetics on nonometer supported catalyst particles, Surf. Sci. Rep. 39 (2000) 25.
[88] H.-J. Freund, M. Bäumer, H. Kuhlenbeck, Catalysis and surface science: what do we learn from studies of oxide-supported cluster model systems? in: B.C. Gates, H. Knözinger (Eds.), Impact of Surface Science on Catalysis, Academic Press, San Diego, 2000, p. 131.

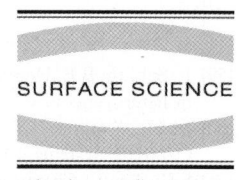

Role of surface and interface science in chemical vapor deposition diamond technology

L.K. Bigelow [a], M.P. D'Evelyn [b],*

[a] *Bigelow Consulting Corporation, Boylston, MA 01505, USA, formerly with Norton Diamond Film of Saint-Gobain Industrial Ceramics, USA*
[b] *General Electric Corporate Research and Development, P.O. Box 8, Schenectady, NY 12301, USA*

Received 15 August 2000; accepted for publication 3 May 2001

Abstract

Diamond is well known as the hardest material in nature. It also has other unique bulk physical and mechanical properties, such as very high thermal conductivity and broad optical transparency, which enable a number of new applications now that large areas of diamond can be fabricated by the new diamond plasma chemical vapor deposition (CVD) technologies. However, some of the most interesting properties of diamond, including the ability to be grown over large areas by CVD processes, result not from its bulk properties but from its special and unique surface chemistry. The surface chemistry derived properties are as remarkable as the bulk properties, and in the end may enable the development of new applications, technologies, and industries which are at least as important as those based on the bulk properties. Some of these surface properties are extreme chemical inertness, low surface energy, low friction coefficients, negative electron affinity, biological inertness, and high over-voltage electrode behavior. The surface science and some of the interesting ongoing research in these areas are explored and illustrated, and unresolved questions are highlighted. © 2001 Elsevier Science B.V. All rights reserved.

Keywords: Models of surface kinetics; Chemical vapor deposition; Growth; Surface chemical reaction; Surface energy; Diamond; Polycrystalline surfaces

1. Introduction

Diamond is best known for its qualities as a gemstone and for its use as an industrial abrasive. As a gem, diamond is highly valued because it has a crystal structure that can be cleaved to bring out the brilliance and fire created by its high refractive index and optical transparency. As an industrial abrasive, it is unique because of its unmatched hardness, which makes it the abrasive of choice to grind or cut any other hard material, and the only one that can be used to mechanically polish diamond itself.

The development of large area coatings of diamond by plasma chemical vapor deposition (CVD) processes has made it possible to utilize of some of diamond's unique bulk and surface properties in new applications that were not possible with the small stones found in nature, or with high-pressure

---

* Corresponding author. Tel.: +1-518-3877133; fax: +1-518-3877563.
E-mail addresses: lkbigelow@aol.com (L.K. Bigelow), develyn@crd.ge.com (M.P. D'Evelyn).

0039-6028/01/$ - see front matter © 2001 Elsevier Science B.V. All rights reserved.
PII: S0039-6028(01)01545-X

synthetic diamond. Many of these new potential opportunities stem not from the bulk properties important in gemstones and abrasive applications, but from the surface chemistry and surface properties of diamond. It is the objective of this review article to examine a few of the most interesting of these new opportunities, including speculative ones, and to discuss the surface science issues which must be explored to understand and develop them into commercially successful new technologies.

2. Diamond growth overview

Many of the most interesting features of diamond, including the CVD growth process itself, derive from the special surface chemistry of diamond. These surface chemistry related properties and processes in the end may enable the development of new applications, technologies, and industries which are at least as important as those based on diamond's strength and hardness. Some of these special surface properties are extreme chemical inertness, low surface energy, low friction coefficients, negative electron affinity, biological inertness, and unique behavior when used as an electrode in aqueous solutions.

Diamond is one of the most stable chemical structures in nature, but graphite has an even more favorable free energy of formation than diamond under normal ambient pressures and temperatures. The Gibbs free energy of formation at room temperature and 1 atm is 2.9 kJ/mol more negative for graphite than for diamond [1]. When carbon condenses under other than extreme pressures, the formation of graphite is thermodynamically favored instead of the growth of diamond. If one were to try to transform graphite to diamond at room temperature, a pressure of at least 16,000 atm would be required, although the kinetics of transformation would of course be vanishingly slow at this temperature. However, at higher temperatures the pressure stability region of graphite expands even further, and even higher pressures are required to favor the growth of diamond. At 1700 K in excess of 50 kbar are required to reach the diamond stability region, which creates a serious technical barrier to the growth of synthetic diamond by high pressure processes.

Despite these barriers, synthetic diamond is now grown on a large scale in high-pressure presses under diamond-stable conditions by dissolution of graphite in metal "solvent-catalysts" and precipitation of diamond. This process was developed in the mid 1950s [2] and commercialized by General Electric, followed soon after by De Beers and others. The worldwide market for synthetic superabrasives, including diamond and cubic boron nitride grit, polycrystalline diamond and cubic boron nitride (manufactured under high pressures and high temperatures), and diamond and cubic boron nitride powder, is approximately one billion US dollars per year and continues to expand significantly.

Questions remain concerning how diamond forms in nature. Conventional models have diamond growing over geological eons, 200 or more kilometers below the surface of the earth under high temperatures and pressures, and subsequently being transported to the surface by primary volcanic intrusions. In South Africa, the type of volcanic rock commonly associated with diamonds is called kimberlite, after the area where it was first discovered. In the 1970s, in Australia, diamonds were also discovered in a significantly different type of volcanic intrusion known as lamproite [3]. More recently, it has also been proposed that diamond can form in the interior of large planets such as Uranus and Neptune, whose atmospheres contain CH_4, water, and ammonia, through the conversion of methane to diamond at pressures in the range of 10–50 GPa and 2000–3000 K [4]. The authors propose that this process of conversion to diamond even produces enough energy to contribute to the observed energy budget of these planets!

3. Surface chemistry of chemical vapor deposition diamond growth

Growth of diamond by CVD [5,6] is driven by the chemistry and kinetics of gas-phase and surface reactions, rather than thermodynamics. Diamond is typically grown using a dilute mixture of a hydrocarbon such as methane in hydrogen.

Additional species, such as oxygen, nitrogen, fluorine, and chlorine, are sometimes used, but consideration of the mechanistic role of the latter species is beyond the scope of this article. Atomic hydrogen is generated by inputting power into the gas phase, for example, by means of a white-hot filament, a microwave discharge, or an arc discharge, which in turns reacts with the hydrocarbon to form reactive species such as CH_3 and C_2H_2. These species are then transported to the substrate surface, where the reactive precursors adsorb and are subsequently converted to diamond by a series of reactions that are still not completely understood. Surface chemistry plays a crucial role in determining the growth rate and quality of diamond films, and surface science methods have contributed significantly to the current level of understanding of the growth mechanism.

Virtually all commercial CVD diamond is polycrystalline, but on the atomic scale growth takes place on single-crystal grains. The symmetry of individual crystallites can be best understood by reference to the crystal structure of diamond, shown in Fig. 1. Each carbon atom is bonded to four neighbors, and these tetrahedra are arranged in such a way that eight atoms fit in a cube whose edge is the lattice constant (3.567 Å). Facets on crystals always occur perpendicular to the slowest-growing crystallographic directions, which in CVD diamond normally correspond to the so-called [1 0 0] and [1 1 1] crystallographic directions [7]. Referring to Fig. 1, the ideal-structure (1 0 0) surface (perpendicular to the [1 0 0] direction) comprises vertically oriented zig-zag chains of atoms; as each new atomic layer is added the direction of the chains rotates by 90°. On ideal-structure (1 1 1) surfaces each surface atom is bonded to three atoms below to form an array of puckered hexagons. Square (1 0 0) and hexagonal or triangular (1 1 1) facets characteristic of individual diamond crystals (Fig. 1(d)) are also present in polycrystalline CVD diamond films, as shown in Fig. 2. The properties of the film, including strength, abrasion resistance, fracture toughness, transparency, surface roughness, etc., are determined by the microstructure, including grain size and orientation, stresses, dislocations, and incorporation of impurities. These impurities can include elements such as nitrogen, or metals, or non-diamond phases such as carbon atoms in a graphitic bonding configuration, known as sp^2 carbon. The microstructure is in turn determined by the detailed surface growth mechanism, including the relative growth rates in the [1 0 0] and [1 1 1] directions, v_{100} and v_{111}, respectively, and the relative rates of generation of new crystallographic directions (twinning or re-nucleation) versus continued growth in a given direction (homoepitaxy) [8]. Examples of CVD diamond surface structures that are produced by different growth chemistries and growth conditions are shown in Figs. 2 and 3. The polycrystalline structure of the diamond in Fig. 2(a) is typical for many commercial coatings on cutting tool inserts or rotary tools. The square (1 0 0) diamond crystal faces shown in a surface view in Fig. 2(b) and in a fracture cross-section in Fig. 3 were grown within a transitional parameter space between nearly pure (1 0 0) facet growth and a mode which generates a microcrystalline diamond morphology. The microcrystalline morphology can be seen between the (1 0 0) faces in Fig. 2, and the (1 0 0) growth columns in Fig. 3.

Our current level of understanding of the atomic-scale mechanism of diamond CVD results from a close interplay of experiment and modeling, although there are a number of basic questions that remain unanswered. Under typical diamond growth conditions there is good evidence that most of the carbon atoms originate from methyl radicals, CH_3 [6,9–11]. However, diamond growth from acetylene, C_2H_2, occurs with a lower reaction probability [12,13], and under more extreme conditions atomic carbon, C_2, and alkyl halides can act as diamond precursors, although the latter are beyond the scope of this article.

What is the molecular-level mechanism by which CH_3 and/or C_2H_2 become incorporated into (1 0 0) and (1 1 1) diamond surfaces? To provide a starting point for kinetic growth models, several groups have examined hydrogenated (1 0 0) and (1 1 1) diamond surfaces or facets after quenching from growth conditions by scanning tunneling microscopy (STM). On the (1 0 0) surface, a number of authors have observed a (2×1) reconstruction [14–16], that is, a doubling of the surface unit cell size in one direction, as shown in

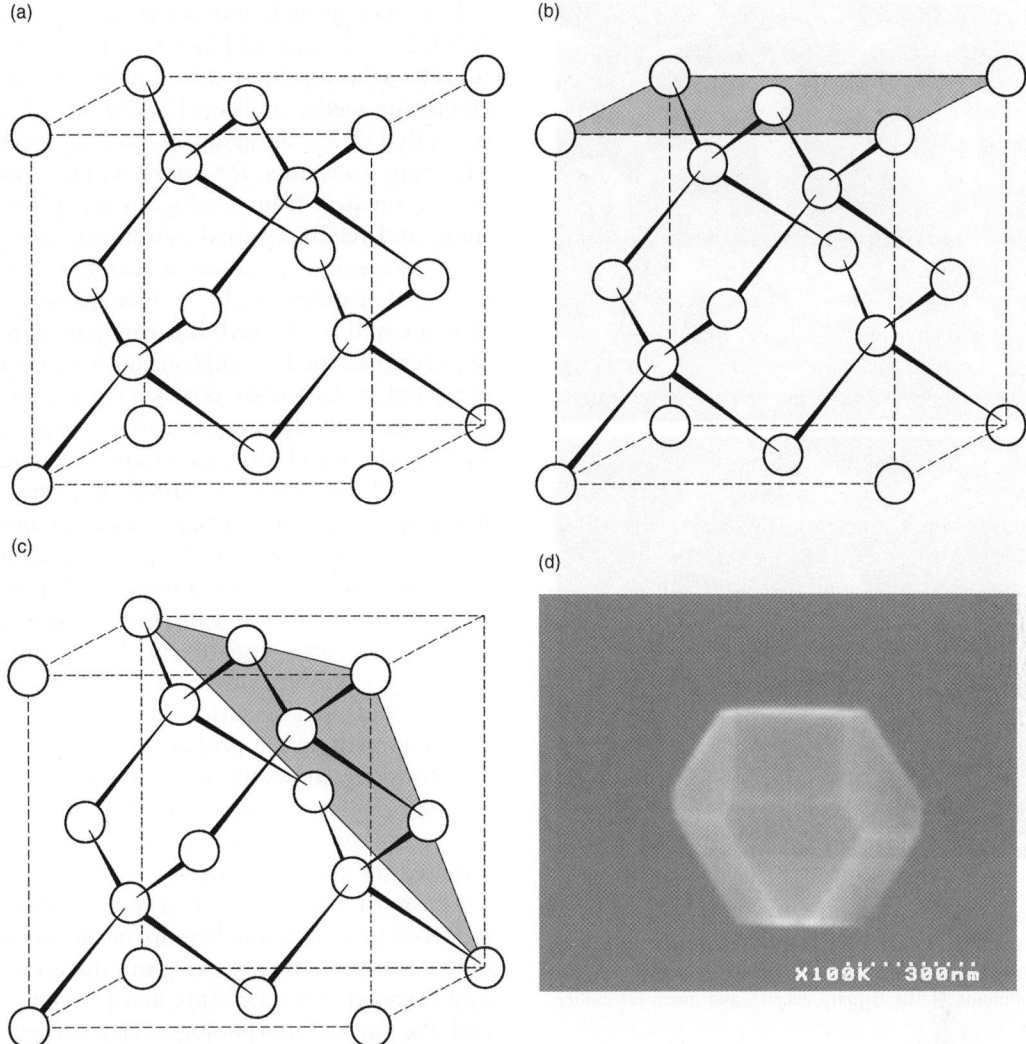

Fig. 1. Lattice structure of diamond. (a) Cubic unit cell, showing fourfold bonding to each carbon atom. (b) (1 0 0) surface (ideal structure), comprising vertically oriented zig-zag chains of first- and second-layer atoms. (c) (1 1 1) surface (ideal structure), comprising puckered hexagons of first- and second-layer atoms. (d) Isolated diamond crystal, with square (1 0 0) and hexagonal (1 1 1) facets.

Fig. 4(a). (1 1 1) surfaces are often highly defective on the micron scale, but unreconstructed (1 × 1) surface domains, that is, with a surface unit cell that is indistinguishable from the ideal structure (Fig. 1(c)), have been observed by several groups [15,17], as shown in Fig. 4(b). These surface structures are well understood. The linear features on the (1 0 0)-2 × 1:H surface comprise rows of C–C dimers, in which each surface carbon atom is bonded to one hydrogen atom and has two backbonds to the diamond lattice, as shown schematically in Fig. 5(a). The orientation of the dimer rows in adjacent atomic layers is rotated by 90° due to the symmetry of the lattice. The length of the monohydride dimer bond has been measured as 1.60 Å by dynamical low energy electron diffraction (LEED) [18], in excellent agreement with theory [19–21]. The vibrational modes of the C–H

Fig. 2. (a) Typical polycrystalline diamond coating on a cutting tool. Growth steps, twinning, and apparent re-nucleation at crystal plane intersections are visible. (b) Surface morphology of CVD diamond grown in a transitional parameter space between dominant (1 0 0) (square facets) and microcrystalline structures.

Diamond growth models beginning from the (1 0 0)-2 × 1:H and (1 1 1)-1 × 1:H surface structures have been proposed by a number of authors, employing methyl radicals [27–29], acetylene [30], or both [31–33] as assumed growth precursors. The rate constants for most surface reactions were estimated from analogous gas-phase reactions, and relatively good agreement with experimental growth rates has been achieved. A number of global features of the surface growth mechanism are well established. Incident hydrogen atoms abstract (i.e., an Eley–Rideal mechanism) hydrogen from surface C–H bonds, creating dangling-bond sites at which hydrocarbons can adsorb. Further surface H-abstraction and rearrangement reactions form new C–C bonds and remove the remaining H atoms, incorporating the adsorbed species into the diamond lattice. However, beyond this global understanding a number of basic questions remain under active discussion. What are the roles of different growth precursors, especially CH_3 and C_2H_2? Are etching (removal of adsorbed CH_x species) and/or surface migration important during growth and in producing smooth surfaces? We consider these questions briefly in the remainder of this section but defer others, such as: why are $\sim 10^4$ hydrogen-atom surface reactions required for each atom of diamond deposited [6]?

The most definitive test of a diamond growth model is to test its predictions on all experimentally observed surfaces, including the growth rate as a function of temperature and gas composition and the surface morphology. This program represents a challenge to theory based on kinetics of elementary surface reactions because of the large range of time scales involved and the large number of atoms necessary to adequately address questions of surface morphology. To date this test has been performed only by Battaile et al. [32,33], who considered growth from both CH_3 and C_2H_2 and employed a sophisticated computer simulation method capable of sampling a wide range of time scales. These authors achieved reasonable agreement with experimental homoepitaxial growth kinetics data in the [1 0 0] and [1 1 1] directions [7,34] with two key findings. First, reasonable growth rates in the [1 1 1] direction could only be achieved if a significant level of C_2H_2 was present in the gas

dimer species are coupled, producing symmetric and asymmetric stretch modes at 2919 and 2899 cm^{-1}, respectively, as observed by surface infrared spectroscopy [22], again in excellent agreement with theory. The (1 1 1)-1 × 1:H surface comprises a simple termination of the bulk lattice, with one hydrogen atom bonded to each surface carbon atom, as shown schematically in Fig. 5(b). The C–C bond lengths are essentially unperturbed, as established by dynamical LEED [23] and X-ray diffraction [24], and the C–H bond produces a sharp vibrational mode at 2832 cm^{-1}, as observed by infrared–visible sum frequency generation [25] and infrared spectroscopy [26].

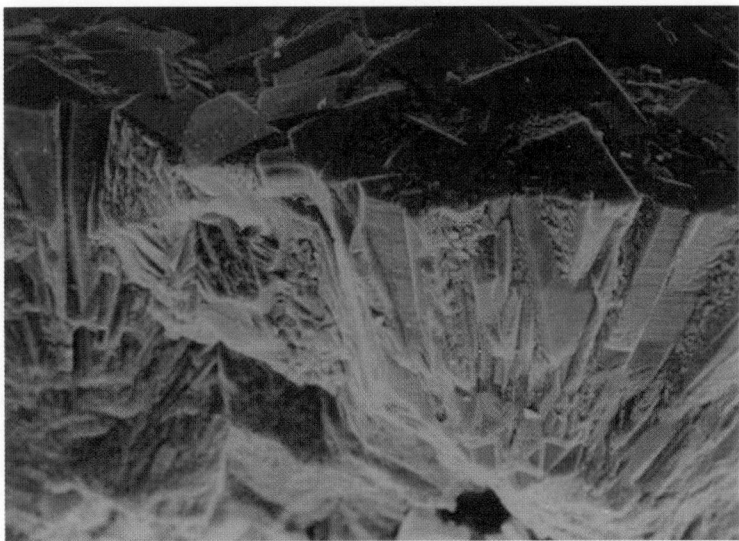

Fig. 3. Fracture cross-section of the film in Fig. 2(b), showing the columnar growth of the (1 0 0) crystals, and the interspersed microcrystalline diamond. The nucleating substrate surface is at the bottom of the micrograph.

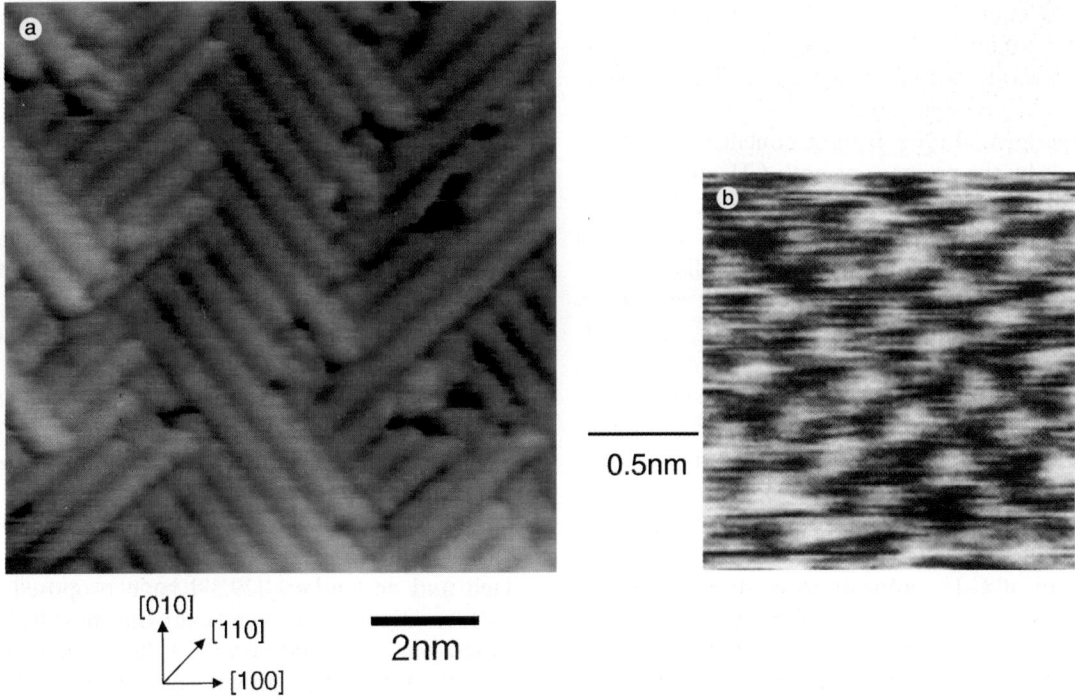

Fig. 4. STM images of (a) (1 0 0)-2 × 1 domains and (b) (1 1 1)-1 × 1 domains on CVD-grown diamond surfaces (from Ref. [15], with permission). In (a) the surface unit cell is doubled in size in the direction perpendicular to the rows—hence the (2 × 1) designation—whereas in (b) the surface unit cell is identical to that of an ideal surface where the atoms are frozen in place at their bulk positions (cf. Fig. 1(c)).

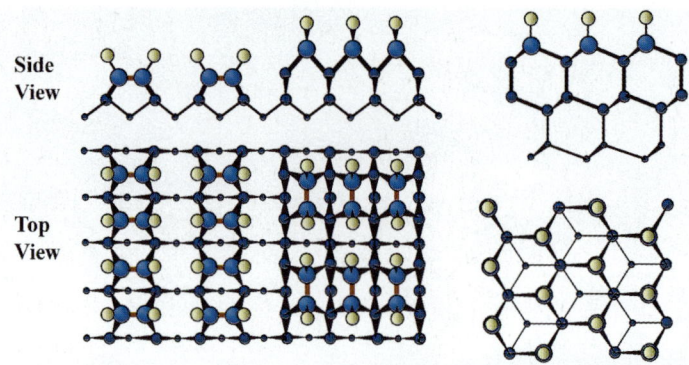

Fig. 5. Atomic structure models of (a) (1 0 0)-2 × 1 and (b) (1 1 1)-1 × 1 diamond surfaces. Light circles: H atoms. Larger, smaller dark circles: top-layer, lower-layer C atoms. The 90° rotation in the direction of the dimer rows in (a) as the surface height is raised by one atomic layer is a simple consequence of the symmetry of the diamond lattice (Fig. 1(b)).

phase in order to nucleate new layers, even though most of the growth resulted from CH_3. Second, [1 0 0] growth was significantly faster than experiment unless etching of CH_x species, by a combination of H-induced reactions and simple thermal desorption, was assumed to be relatively facile. The latter conclusion was supported by high-level ab initio calculations of the energetics of several key surface intermediates.

Experimental tests of these conclusions is made difficult by the presence of both CH_3 and C_2H_2 in virtually all diamond CVD reactors. However, Martin and co-workers [11,12,35] have shown that it is possible to grow high-quality diamond in a flow tube, where the gas velocity is sufficiently high that reactions interconverting CH_3 and C_2H_2 are negligible in the region of diamond growth. Recently, D'Evelyn et al. [35] have shown that diamond growth from H and CH_3 can produce well-faceted cubo-octahedra, as shown in Fig. 6(a). The observation of comparable growth rates in the [1 0 0] and [1 1 1] directions—implied by the shapes of the crystals—despite a very low partial pressure of C_2H_2, appears to contradict the first conclusion of Battaile et al. Growth from H and C_2H_2 produced octahedra with a high concentration of contact twins, as shown in Fig. 6(b). The latter results demonstrate that CH_3 and C_2H_2 are not interchangeable in the growth mechanism and, in particular, that defect-free incorporation into the lattice is less likely with C_2H_2, at least on the

(1 1 1) surface. The apparent absence of a nucleation requirement by 2-carbon species during [1 1 1] growth suggests that the steady-state growth surface has a significant coverage of adsorbed species, i.e., does not closely resemble the (1 1 1)-(1 × 1) surface, which in turn suggests that etching reactions may not as facile as modeled by Battaile et al. [33]. Further support for a more complex steady-state growth surface comes from recent modeling work [36] and the dependence of STM surface images on post-growth annealing/etching in atomic hydrogen [37].

Most diamond growth models assume that hydrocarbon precursors either become incorporated into the lattice where they stick initially or else desorb, i.e., surface migration is negligible. Yet the facets of as-grown CVD diamond surfaces are normally relatively smooth on the nanometer-to-micron scale, and surface diffusion is known to be important in the production of smooth surfaces of many materials. Random growth on arbitrary surface sites should produce atomically rough surfaces, in contradiction to experiment. Frenklach and co-workers [29,38] have proposed that surface migration is significant during [1 0 0] diamond growth, at least on the 5–20 Å length scale, and facilitates the growth of dimer rows and smooth surfaces. These predictions are supported by semi-empirical and high-level quantum chemical calculations of the structures and energetics of various surface intermediates and reactions

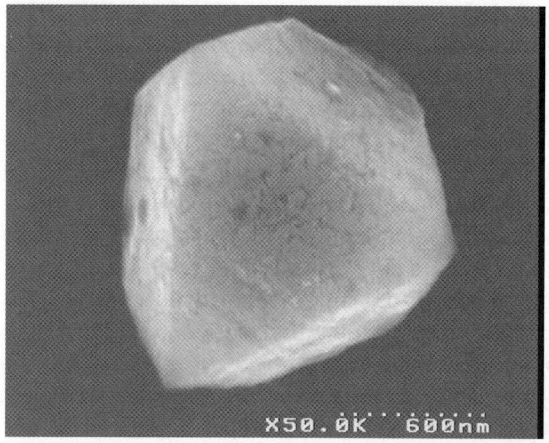

Fig. 6. Diamond crystals grown in a flow-tube at 800 °C from (a) $H + CH_3$; (b) $H + C_2H_2$. The crystal in (a) is well faceted, with (1 0 0) and (1 1 1) facets of similar area, which implies that methyl radicals alone are capable of growing diamond in [1 0 0] and [1 1 1] directions at similar rates. The crystal in (b) is terminated by highly defective (1 1 1) surfaces only, which indicates that [1 1 1] growth from acetylene leads to many defects and is considerably slower than growth in the [1 0 0] direction.

that etching and surface migration can occur under growth conditions, but this remains to be firmly established. Hydrogen plasma treatments can produce smooth (1 0 0) and (1 1 1) surfaces under some conditions [39–43], and can roughen [42–44] or produce faceted pits [45,46] under other conditions. Most authors have attributed the smoothing phenomenon to etching. However, Rawles et al. [42,43] found that the size of diamond particles was unchanged during H-atom-induced facet formation and was independent of particle packing density and gas flow rates, and argued that this observation supported surface migration but was inconsistent with etching and etching/regrowth mechanisms.

Definitive answers to the questions posed above require further experimental and theoretical surface science investigation. An improved fundamental understanding of the CVD growth mechanism should enable improvements in the quality and decreases in the cost of CVD diamond. Particularly valuable would be in situ measurement of the coverage and type of surface hydrogen during growth; quantification of the evolution of nanometer-scale surface roughness; and the kinetics of etching and surface migration induced by hydrogen atoms.

4. Properties

The chemical composition of the diamond surface is important in determining certain physical properties and for commercial applications. These properties include friction coefficient, surface electrical conductivity, surface energy for wetting and bonding, negative electron affinity, and plasma erosion resistance. As discussed above, the surface bond structure and tendency to reconstruct depends sensitively on the crystal orientation, growth mode and surface treatment process.

4.1. Surface bonding

In this section we discuss the bonding and energetics of several diamond surface structures that are or might be important for applications. More comprehensive reviews can be found elsewhere [19,47–49].

[29,38]. Direct evidence for surface migration during growth is so far lacking, but it is not clear that the degree of smoothness observed on CVD-grown facets can be accounted for in the absence of migration.

Separate experiments in hydrogen plasmas (etching, rather than growth conditions) suggest

Hydrogenated diamond surfaces are important for applications as well as in CVD growth. Several types of studies have demonstrated that the surface C–H bond is essentially identical to that in alkanes, with a bond energy of about 96 kcal/mol [50]. As noted above, the vibrational frequencies of adsorbed hydrogen agree closely with theory. We also saw in the previous section that abstraction of surface hydrogen by gas-phase atomic hydrogen plays a crucial role in the growth mechanism. CVD modeling studies have assumed an activation energy of about 7 kcal/mol [30,51], similar to that in analogous gas phase abstraction reactions. Krasnoperov et al. [52] measured an identical activation energy for the surface abstraction reaction on polycrystalline CVD diamond at pressures that were high enough to reach thermal equilibrium between the surface and the impinging gas-phase species. A much smaller apparent activation energy has been observed under ultrahigh vacuum conditions [53,54], which indicates that translational energy in the incoming H atom is more effective in surmounting the activation barrier than is thermal motion of the surface atoms. The sticking coefficient for atomic hydrogen on bare surface sites appears to be approximately one.

The rates of desorption of H_2 from the (1 0 0)-2×1:H and (1 1 1)-1×1:H monohydride surfaces are approximately proportional to the surface concentration of hydrogen atoms. This behavior is unusual on surfaces—hydrogen desorption rates are more typically proportional to the square of the surface concentration. The corresponding activation energies for hydrogen desorption are 80–88 kcal/mol [55,56] and 85–92 kcal/mol [25,57], respectively, consistent with a C–H bond energy near 96 kcal/mol. Following desorption the surfaces adopt a 2×1 reconstruction, comprising C=C dimers linked by a σ bond and a weak π bond on (1 0 0) or π-bonded –C–C– chains on (1 1 1) [19,47–49]. Naively, the simplest desorption mechanism on the (1 0 0) surface would involve H atoms on a dimer moving together and forming an H–H bond while breaking two C–H bonds and forming a surface π bond. However, such a mechanism is symmetry forbidden [58] and, by analogy to hydrocarbons, a much higher activation energy (120 kcal/mol for H_3C–CH_3 →

H_2C=CH_2 + H_2 [59]) might be expected. The dynamical nature of the H_2 desorption mechanism is as yet unknown. The first-order kinetics imply that hydrogen tends to cluster together on the surface rather than being randomly distributed, but the roles of π-bonding on the clean surface, bond strain, surface diffusion, and perhaps other factors remain to be elucidated.

The existence of a π bond (albeit highly strained) on the clean (1 0 0)-2×1 surface suggests that so-called Diels–Alder chemistry—reaction between one molecule with a C=C σ + π bond and a second molecule with two C=C bonds separated by a C–C σ bond to form a cyclic structure—might be possible, creating new possibilities for functionalization of diamond surfaces. Recent studies have shown this to be the case. Adsorption of 1,3-butadiene (H_2C=CH–CH=CH_2) occurs readily and produces the expected –CH_2–CH=CH–CH_2– product, as shown by infrared spectroscopy [60] and high-resolution electron energy loss spectroscopy [61]. There is a twist, however, which apparently results from the highly strained nature of the surface π bond. Adsorption of a simple alkene (cyclopentene), which is nominally symmetry forbidden (the analogous molecular reaction does not take place) also occurs on the diamond surface [62], albeit with a low sticking coefficient.

Hydrogenated diamond surfaces are stable in oxygen or air up to a temperature of about 300 °C, above which the surface hydrogen is progressively replaced by surface oxygen species [63].

The chemistry of oxygen-containing species on diamond surfaces is important and, for the most part, analogous to that in organic molecules. Unfortunately, perhaps, the details are fairly complex. Oxygenated diamond surfaces typically comprise a combination of bridge-bonded –O– and linear >C=O species and, if some surface hydrogen is also present, –OH and –COOH groups [63–66]. Upon heating, adsorbed oxygen desorbs only as carbon containing gas products, including CO and CO_2, starting at desorption temperatures near 600 °C [67]. Desorption occurs over a wider range of temperatures for oxygen than for hydrogen, reflecting a range of binding energies as would be expected from the range of structures. Frenklach et al. have determined the activation energy for desorption to be 45 kcal/mol by β-scission of a C–C=O backbond [68].

Halogen termination of diamond might be expected to be very stable and to have useful properties, by analogy to alkyl halides and Teflon™. However, in contrast to the alkane-like behavior described above, the chemistry of halogens on diamond surfaces shows large departures from the behavior of analogous alkyl halides, which have C–X bond energies of about 116 or 82 kcal/mol for fluorine or chlorine, respectively. Experimental evidence indicates that desorption of F and Cl is atomic rather than recombinative [69]. Based on bond energies calculated with a cluster analogue of diamond and assuming a normal pre-exponential factor, one would predict that less than 10% of a monolayer would desorb upon heating to temperatures of 1200 or 700 °C for F or Cl, respectively [70]. In stark contrast, F and Cl begin "oozing" (desorbing) from the surface at or below room temperature. Fluorine begins desorbing near room temperature but residual F has been found by some authors to remain at temperatures as high as 1200 °C [69,71,72]. For chlorine on diamond (1 0 0) and (1 1 1) desorption was observed over the range −50 to 300 °C [69], while on diamond powder a significantly higher desorption temperature range was observed, ≈300–1100 °C [73]. The extremely broad range of desorption temperatures imply a strongly coverage-dependent bond energy, as low as 20 kcal/mol at full coverage. The bond energies for F and Cl have been calculated as 103 and 48 kcal/mol at monolayer coverage [74], somewhat reduced from the low-coverage results [70]. However, it is clear that further contributions to the destabilization of C–X bonds at high coverage remain to be understood, whether from steric repulsion, dipole repulsion, a decreasing ability of the surface to donate charge to the more electronegative halogen atoms, or other effects.

4.2. Surface energy

Surface termination of the diamond has important implications for many industrially important processes such as metallization and bonding to diamond. If the diamond is annealed at 1000 °C in ultrahigh vacuum, the adsorbed hydrocarbons, hydrogen and oxide species are removed, and the surface reconstructs. A clean surface of C–C saturated bonds is actually quite stable and unreactive, at least by comparison to other materials such as silicon. Fabis has reported on the effect of plasma treatment of cleaned diamond surfaces with H, O, and CF_4 plasmas on the surface energy of the diamond, and the impact that has on polymer adhesion for electronics packaging [75].

The surface energy and the availability of electrons for further bonding to materials on the surface is strongly affected by these species terminating the surface, as illustrated by the coverage dependence of C–X bond energies. Fabis measured the surface energy of diamond after various surface treatments, using a series of liquids of different surface tensions, following the work of Kaifu and Komai [76]. As shown in Table 1, the O terminated surface is the most wettable, and is also hydrophilic.

Processing treatments which create polar oxygen functionalities such as carbonyl (C=O) and ether (C–O) are a key requirement for strong bonding of polymers to diamond. In contrast, clean high temperature annealed (1000 °C in 10^{-10} Torr) surfaces with C=C saturated bonds, hydrogen terminated surfaces, and even fluorine terminated surfaces are hydrophobic and are resistant to wetting or bonding. XPS analyses reported by Fabis indicate C–O and C=O surface functional groups, and C–F (with some possible C–O residuals) are present on the surfaces treated with CF_4 plasmas.

Table 1
Surface energy measurements of CVD diamond as a function of surface treatment of polished samples (Ra = 10 nm), from Fabis [75]

Surface treatment	XPS functional group	Surface energy (dynes/cm)
1000 °C: 1200 s, at 10^{-10} Torr	C–C	22
H-plasma: 1800 s, 1 KW, 2.45 GHz	C–H	28
CF_4-plasma: 1800 s, 1 KW, 13.56 MHz	C–F (C–O)	38
O-plasma: 1800 s, 1 KW, 13.56 MHz	C–O, C=O	50

The surface activity of the polymer or metal to be bonded to diamond is of course the other half of the equation which determines the actual strength of the bond to a treated diamond surface. The adherence of an epoxy polymer used in electronics plastics packaging under a series of boil/peel cycles was found by Fabis to directly correlate with the surface energy measurement in Table 1. Surface treatment of the diamond after deposition to increase its surface energy is also critical to the adherence of blanket or patterned metallizations for the fabrication of thermal management substrates for electronic devices.

5. Applications

5.1. Surface factors in diamond friction

CVD diamond is known to have a low friction coefficient under appropriate conditions. This is a function of its surface termination state, and is affected by prior processing and by the environment and conditions of frictional contact, as well as the surface roughness and crystal structure of the diamond faces. In vacuum, diamond-on-diamond sliding contact eventually leads to an increase in friction from initial values of 0.1 to 1.4, with significant surface damage occurring in the process [77]. This can be attributed first to shearing and fracture of local high spots, and then to the removal under high local forces of the critical surface terminating species, such as O or H. Those species, when present reduce the surface energy, and prevent local atomic attraction or actual bonding across the interface of carbon atoms at contact asperities.

Those surface chemistries and structures of the diamond surface that promote low frictional properties are of major importance in bearing and seal applications. These are presently under development. They take advantage of the strong bonding of diamond to its surface terminating species, which reduces local bonding with the contacting frictional face and reduces frictional forces. In many applications, the chemical inertness and low surface energy of diamond can also prevent the build-up of deposits that would normally reduce seal life.

5.2. Diamond electrodes

The use of doped diamond electrodes in electrochemical analysis and in electrochemical synthesis and destruction is an area of increasing research interest and potential industrial importance. CVD diamond has three advantages over conventional graphite electrodes used in electrochemistry, according to Angus et al. [78]. These are chemical stability and robustness, large potential window for water stability, and low background current densities. Boron doping at low concentrations produces p-type semi-conductivity at an acceptor level 0.37 eV above the valence band [79]. At high B concentrations, in the range of 10^{20}–10^{21} cm^{-3}, diamond becomes a semi-metal with conductivities as high as 10^{-3} Ω cm. These conductivities can be reached with B/C ratios in the gas greater than about 500 ppm [78].

Conductive CVD diamond can be grown on substrates such as Si, W, and Mo by doping with boron in the gas phase. Angus et al. [78] has surveyed the literature relating the ratio B/C in the gas phase to B/C in the grown diamond. He found a wide range of rates of incorporation for different deposition processes, but the amount of B incorporated in the diamond was usually proportional to the ratio in the gas for a given deposition experiment. In general, boron incorporation was greater in hot filament depositions than in microwave growth, and was greater on (1 1 1) surfaces than (1 0 0) surfaces. It is speculated that there may be more activated species on the surface of the diamond in microwave deposition, which reduces the concentration of adsorbed boron on the surface. The presence of oxygen also greatly reduces the incorporation ratio of boron, apparently by oxidation of boron to B_2O_3 [80].

Highly conductive diamond is desirable for electrosynthesis and electrodestructive applications, while electrochemistry can utilize diamond of lower conductivity. Further, higher quality diamond with less incorporation of non-diamond carbon produces a broader window of water stability for electrochemistry (Fig. 7).

This figure shows the difference in range of water stability in electrolysis experiments in 0.5 M H_2SO_4 for high quality diamond, low quality

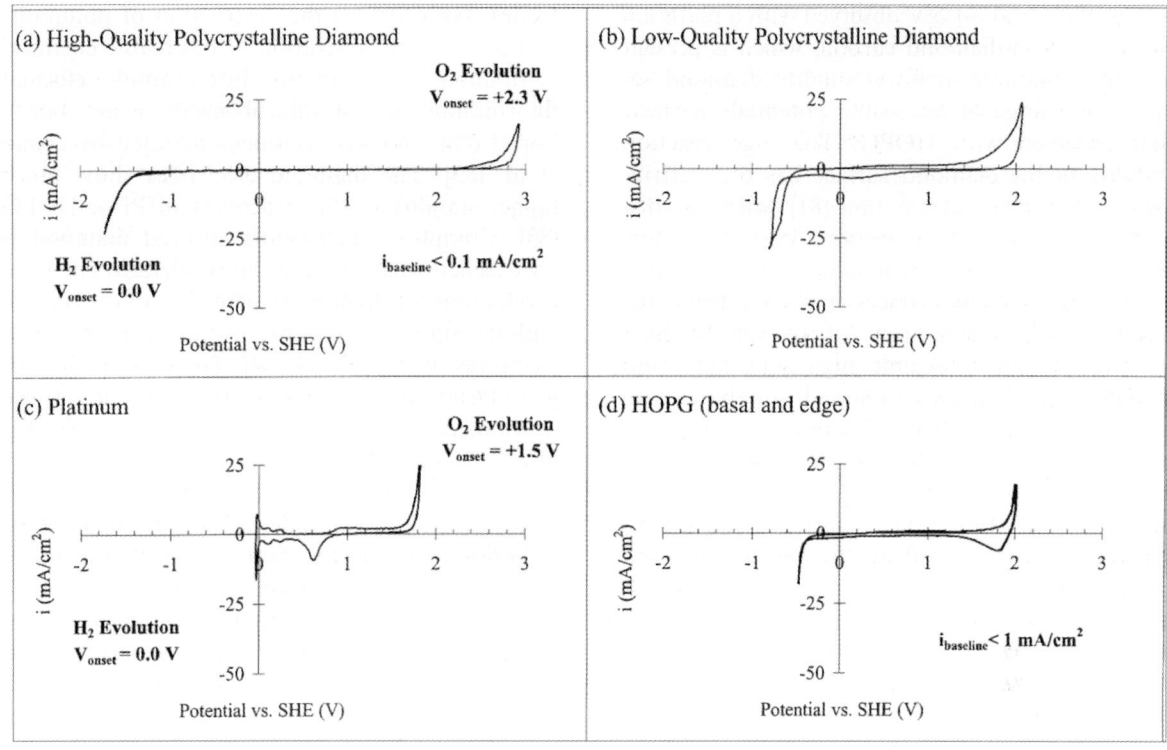

Fig. 7. Effect of type of electrode (diamond/platinum/HOPG), and of diamond quality on electrode potential range for water decomposition. (from Angus et al. [78], reproduced by permission of the publisher of New Diamond and Frontier Carbon Technology, MYU K.K.). The horizontal axes show the electrochemical potential applied to the diamond electrodes relative to a standard hydrogen reference electrode (SHE). The current flow (vertical axes) increases sharply at the potential at which breakdown of the aqueous electrolyte into hydrogen (negative potential) or oxygen (positive potential) takes place. Note that the high quality diamond electrode (a), has the widest range of negative and positive potential before breakdown of the aqueous solution (-1.25 to $+2.3$ V). The low surface energy of the hydrogen-terminated surface of diamond may contribute to this behavior. The potential range observed with lower quality diamond (b) is less than with high quality diamond, probably because the defects on the surface are graphitic in behavior, and are more reactive. The platinum (c) and HOPG (d) electrodes result in the breakdown of the electrolyte at lower potential limits than either of the diamond types. The importance of this behavior of diamond electrodes is that electrochemical methods can be used to react and destroy many chemicals, such as contaminants in waste water, before actually decomposing the water itself. Electrochemistry with CVD diamond electrodes therefore has potential as a method of waste water treatment. CVD diamond electrodes can also precipitate heavy metal from waste water, and can power a number of chemical reactions in water, other solutions, and molten salts, while often exhibiting much greater resistance to attack and corrosion than other materials. Diamond electrodes can be used to detect and measure a much larger range of dissolved substances than electrodes of other materials by measuring the breakdown characteristics of the compounds dissolved in the aqueous electrolyte. Examples of substances that have been detected via diamond electrodes include polyalkylene glycol, dopamine, azide, sulfa drugs, and histamines.

diamond, platinum, and highly oriented pyrolytic graphite (HOPG). In this type of experiment, one varies the electrical potential and measures the current that flows as substances dissolved in an aqueous solution break down or dissociate or as the water itself is dissociated into hydrogen and oxygen. Electrochemical reactions of dissolved species can only be observed usefully if the water itself is not decomposing! For high quality diamond, an aqueous solution evolves hydrogen at a potential of -1.25 V and oxygen at a potential of $+2.3$ V, a much wider range than the corresponding values of 0.0 and +1.5 V observed with a platinum electrode. Non-diamond carbon, which is present in larger amounts in lower quality diamond reduces the range of accessible potentials to near that observed with HOPG. The high reaction stability of the diamond surface has been attributed to hydrogen termination [81] which, as discussed in the previous section, leads to a low surface energy and hydrophobic behavior. However, even diamond surfaces that have been oxidized by local generation of oxygen at high positive potentials exhibit high sensitivity and stability, which makes diamond suitable as an electrode for electrochemical determinations under oxidizing conditions, for example, dopamine in ascorbic acid [81]. The importance of diamond quality in resisting attack during electrochemical processes is pointed out by DeClements and others, following the results of their research on anodic polarization in concentrated KOH [82]. Low quality diamond films are attacked at the grain boundaries, but separate high quality film samples showed no signs of corrosion or morphological damage under the same experimental conditions.

Diamond electrodes can be used for analyzing dissolved compounds that cannot be detected with other electrodes. This is possible because the water is dissociated at a lower potential with these other electrodes than are the dissolved compounds. For example, the presence of polyalkylene glycol can be analyzed using diamond electrodes, whereas platinum electrodes cause electrolysis of the water before oxidation of the polyalkylene glycol, and the reaction potential for the glycol is not reached [78]. With diamond, the peak heights are linear with the glycol concentration. The wide potential window and low baseline current of diamond make this possible.

Other electroanalytical applications have been demonstrated on diamond that are unreliable or impossible on glassy carbon (GC). Some of these are detection or measurement of azide, polyamines, sulfa drugs, histamines, and NADH, a coenzyme used in several dehydrogenase-based biosensors [81].

A great deal of interest and research activity is focussed on commercially developing the electrochemical processing capability of diamond electrodes, both for, (1) the destruction of pollutants and for, (2) the generation of desireable chemical compounds and elements. For example, efficient fluorination of 1,4-diflurobenzene using boron doped diamond has been demonstrated by Okino et al. [83]. The diamond electrodes show much higher stability in this process than Pt or HOPG [83]. Compton has recently showed diamond is an excellent material for the sonoelectrochemical production of hydrogen peroxide from oxygen with its combination of mechanical properties and electrode characteristics [84]. The effect of hydrogen plasma pre-treatment of the diamond surface in sonoelectrochemical processing of dioxygen has also been studied [85].

Industrial applications of the diamond electrode technology are under active development for pollution control, for example in treatment of waste water effluent from industrial plants. The oxidation of organic compounds to CO_2 at the surface of a diamond electrode has been demonstrated with a current efficiency greater than 85% [86]. Experiments on the oxidation of phenol (an example of an aromatic compound), and acetic acid (an example of an aliphatic compound) were successful in reducing these contaminants to low levels (less than 3 ppm in the case of phenol). The electrodes were not damaged or poisoned after weeks of operation. Cyanide can also be eliminated, especially in the presence of chloride ions, which help to catalyse the oxidation reaction.

Large area (50×60 cm^2) boron doped diamond electrodes on substrate materials such as titanium, zirconium, niobium, tantalum, tungsten, and graphite have been developed [87]. Examples of boron doped diamond electrodes produced on these materials by the Fraunhofer Institute are shown in Fig. 8. Further research to understand the details of the oxidation and reduction reactions at the diamond surface will lead to electrodes optimized for industrial applications. The reduction of cadmium and copper from aqueous solutions has also been demonstrated [88]. These toxic metals precipitate out as fine particulates which do not adhere to the diamond electrode, thus permitting easy disposal of the precipitated metals, and re-use of the electrodes. The clean surface that

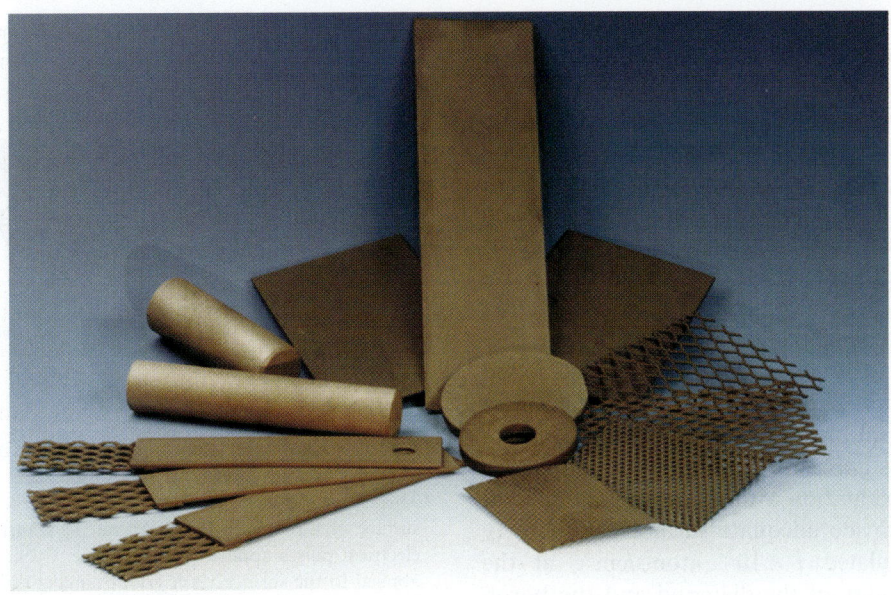

Fig. 8. Boron doped diamond electrodes on titanium, tantalum, niobium and graphite base material produced by the Fraunhofer Institute, Braunschweig. (Reproduced by permission of New Diamond and Frontier Carbon Technology—from vol. 9, pp. 229–240.)

remains is a direct result of the low surface energy and chemical stability of diamond.

6. Future research directions

Some opportunities for improved fundamental understanding of diamond surface and growth chemistry have already been given. In this section we enumerate some novel applications.

6.1. Negative electron affinity

Diamond has the property that with certain types of surface termination, such as clean or hydrogen terminated surfaces, the conduction band minimum lies above the vacuum level at the surface, and the diamond exhibits negative electron affinity. This means there is no activation barrier to ejection of an electron from the surface, in contrast to most other materials, which require an applied potential at the surface or a high temperature to initiate electron emission into space. This behavior has created considerable research interest [89–91] because it could potentially lead to highly efficient cold cathode emission devices, and products such as large area flat panel displays. The market for field emission displays is potentially very large market because the devices would be bright ("sun visible") and highly energy efficient. However, the performance of diamond in field emission display development to date has been disappointing because the density of emission sites has been too low, and the emission current has also been too low. A factor in this is the lack so far of an effective way to n-type dope the diamond to support significant emission from NEA surfaces or from impurity based conduction sites. Nanotubes appear at present to be the most promising approach to field emission sources for flat panel displays.

6.2. Bio-sensors and detectors

The surface and electronic properties of diamond also make it a candidate for the development of new bio-sensors and detectors, and advanced diagnostic equipment such as in vivo blood chemistry monitors. First of all, the extreme inertness, especially of the hydrogen terminated diamond surface, means that diamond is the most inert and least likely to be rejected of all materials for both structural and sensor applications in the body. This bio-inertness and low tendency to induce blood clots has been reported many times. Significant in vivo testing of products has not yet begun, but there is growing activity to develop diamond based body replacement parts, such as permanent hip joint replacements and heart valve components. This is an area for important future research and commercial competition.

Troupe et al. have reported on diamond based glucose sensors. The diamond electrodes have to be boron doped to provide adequate conductivity for use in this application [92]. The diamond was doped by implantation with boron and annealed at 850 °C to provide adequate surface conductivity (with a calculated 1×10^{20} atoms/cm^{-3} at the surface). Stability of the diamond and the boron in the lattice was measured by immersion in deionized water for four weeks at 37 °C. The concentration of boron in the water reached only 4 ppb, indicating virtually no leaching of boron from the diamond surface. Several designs were tried. A device using an oxidized, activated diamond surface, with specially selected adsorbed bio-molecules to assist in the surface reactions provided the best sensitivity for glucose detection.

A range of new bio-agent and chemical detectors have been proposed, based on the wide band gap of diamond, in combination with its inert and stable surface. A recently demonstrated technology with the potential to identify a wide range of absorbed species on the diamond surface and thereby detect dangerous bio-agents or process-critical chemical constituents is called charged-coupled deep level transient spectroscopy (Q-DLTS). The principle is that adsorption of any compound on the surface of the diamond will result in a change in the electronic states at the diamond surface, which can be characterized by its effect on the transient response to imposed voltage pulses. The voltage pulses may be applied by an electrode grid on the surface of the diamond. The surface potential energy and the density of trapping centers will be characteristically different with the adsorption of different species on the surface of the diamond, and thus can be detected and

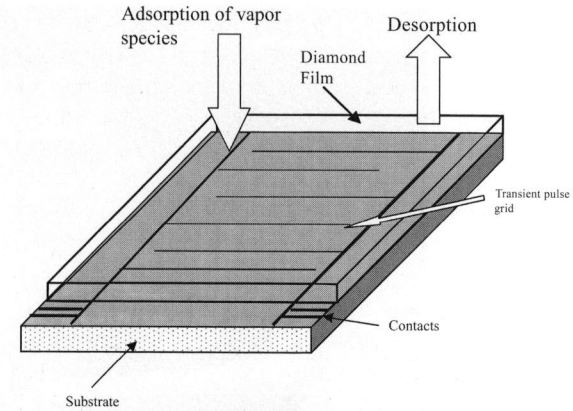

Fig. 9. Schematic of detector grid and design for proposed Q-DLTS chemical and bio-agent sensor using CVD diamond. This detector is based on the effect that adsorbed species on the surface of a diamond film have on the transient response to electrical pulses applied to the surface. Cyclic bias pulses are applied to the surface via a grid, changing the charge state of the surface and of subsurface electronic trapping centers. The surface charge is measured by integrating the current that flows as the surface charge decays when the bias voltage is turned off. The charging and decaying behavior is affected by adsorbed molecules on the surface, which can be differentiated and identified by their characteristic effects and decay signatures. The frequency of the pulses is varied to produce curves of charge versus pulse duration that are characteristic of the system and the adsorbed species.

quantified by analyzing the response to the voltage pulses. This surface analysis technique has been demonstrated in a detector using DLC on quartz. Adsorbed species (isopropyl alcohol and water) can be differentiated by the transient response to pulses from the array of inter-digital electrodes on the surface [93]. A schematic of one design concept using CVD diamond is shown in Fig. 9.

It is expected in this method that the effect of different adsorbed species on the surface will be independent of each other, so that many agents can be detected by a single detector. Much work remains to characterize and catalog the effect of each adsorbate on diamond made with different surface textures, surface treatments, and surface terminations. However, the Q-DLTS method is sensitive enough to detect a charge of 2000 electrons on the surface, and may be able to detect airborne contaminants in the parts per trillion range. Bio-agents are typically transmitted by aerosols. The Q-DLTS sensor is sensitive to particles. It is hoped that bio-agents such as viral agents, toxins and spores may be detectable directly or indirectly by developing a database of characteristic and differentiable effects on the transient response of the diamond. Such systems, designed with compact diamond detector substrates and with an analytical database on-board for each species of concern, should be easily portable. This would make the devices ideal for individual use for working near hazardous chemicals, or for field use to monitor and protect military and civilian personnel against chemical or biological agents.

The potential uses for disease control, hospital and industrial air quality monitoring, and chemical process monitoring and control are staggering, if the technology proves to be as sensitive as projected and also manufacturable and cost effective. Surface science will play a key role in identifying the important surface variables and treatments to optimize sensitivity for different types of chemicals and agents of interest. An added feature, which is a benefit of the bulk properties of diamond is the possibility to heat the detector, using a resistive grid beneath the substrate or by another method. This would desorb the adsorbed species from the surface so the device could continuously monitor and update the environment. It would become reuseable over long periods in service.

7. Conclusion

A detailed understanding and application of surface science is essential to the present industrial use of diamond and to the development of the exciting new products diamond will make possible. An example of an important practical problem overcome by surface studies was reviewed earlier, the problem of metallization of diamond. Until the factors that affect the surface energy of diamond were understood and the solutions applied, a great deal of difficulty and inconsistency was experienced by those trying to bond patterned or blanket metallization to diamond. However, with appropriate pre-treatment, excellent adherence and reliable product performance has been achieved. It is certain that the same principle will apply to the development of many other new diamond products, including the selected ones we have reviewed here.

References

[1] M.W. Chase Jr., C.A. Davies, J.R. Downey Jr., D.J. Frurip, R.A. McDonald, A.N. Syverud, JANAF Themochemical Tables, third ed., Part I, Al–Co, J. Phys. Chem. Ref. Data (USA) 14 (1985) 1.

[2] F.P. Bundy, H.T. Hall, H.M. Strong, H.G. Wentorf, Manmade diamonds, Nature 176 (1955) 51–55.

[3] H.O.A. Meyer, M. Seal, Natural diamond, in: M.A. Prelas, G. Popovici, L.K. Bigelow (Eds.), Handbook of Industrial Diamonds and Diamond Films, Marcel Dekker, New York, 1998, pp. 481–526.

[4] L.R. Bennedetti, J.H. Nguyen, W.A. Caldwell, H. Liu, M. Kruger, R. Jeanloz, Dissociation of CH_4 at high pressures and temperatures: diamond formation in giant planet interiors, Science 286 (1999) 100.

[5] J.E. Butler, R.L. Woodin, Thin film diamond growth mechanisms, Phil. Trans. R. Soc. Lond. A. 342 (1993) 209–224.

[6] D.G. Goodwin, J.E. Butler, Theory of diamond chemical vapor deposition, in: M.A. Prelas, G. Popovici, L.K. Bigelow (Eds.), Handbook of Industrial Diamonds and Diamond Films, Marcel Dekker, New York, 1998, pp. 527–582.

[7] C.J. Chu, R.H. Hauge, J.L. Margrave, M.P. D'Evelyn, Growth kinetics of (100), (110), and (111) homoepitaxial diamond films, Appl. Phys. Lett. 61 (1992) 1393–1395.

[8] R.E. Clausing, Diamond morphology, in: M.A. Prelas, G. Popovici, L.K. Bigelow (Eds.), Handboook of Industrial Diamonds and Diamond Films, Marcel Dekker, New York, 1998, pp. 19–47.

[9] M.P. D'Evelyn, C.J. Chu, R.H. Hauge, J.L. Margrave, Mechanism of diamond growth by chemical vapor deposition: carbon-13 studies, J. Appl. Phys. 71 (1992) 1528–1530.

[10] C.E. Johnson, W.A. Weimer, F.M. Cerio, Efficiency of methane and acetylene in forming diamond by microwave plasma assisted chemical vapor deposition, J. Mater. Res. 7 (1992) 1427.

[11] S.J. Harris, L.R. Martin, Methyl versus acetylene as diamond growth species, J. Mater. Res. 5 (1990) 2313.

[12] L.R. Martin, High-quality diamonds from an acetylene mechanism, J. Mater. Sci. Lett. 12 (1993) 246.

[13] M.A. Cappelli, M.H. Loh, In situ mass spectrometric sampling during supersonic arcjet synthesis of diamond, Diamond Relat. Mater. 3 (1994) 417–421;
M.H. Loh, M.A. Cappelli, CH_3 detection in a low-density supersonic arcjet plasma during diamond synthesis, Appl. Phys. Lett. 70 (1997) 1052.

[14] T. Tsuno, T. Imai, Y. Nishibayashi, K. Hamada, N. Fujimori, Epitaxially grown diamond (001) $2 \times 1/1 \times 2$ surface investigated by scanning tunneling microscopy in air, Jpn. J. Appl. Phys. 30 (1991) 1063–1066.

[15] T. Tsuno, T. Imai, S. Shikata, N. Fujimori, Surface structure and morphology of diamond epitaxial films, in: S. Saito et al. (Eds.), Advances in New Diamond Science and Technology, MYU, Tokyo, 1994, pp. 241–246.

[16] H. Kawarada, H. Sasaki, A. Sato, Scanning-tunneling-microscope observation of the homoepitaxial diamond (001) 2×1 reconstruction observed under atmospheric pressure, Phys. Rev. B 52 (1995) 11351–11358.

[17] H. Sasaki, H. Kawarada, Structure of chemical vapor deposited diamond (111) surfaces by scanning tunneling microscopy, Jpn. J. Appl. Phys. 32 (1993) L1771–L1774.

[18] Y.M. Yang, K.W. Wong, S.T. Lee, M. Nishitani-Gamo, I. Sakaguchi, K.P. Loh, T. Ando, Surface structure of C(100)-(2×1)–H studied by a quantitative LEED analysis, Phys. Rev. B 59 (1999) 10347–10350.

[19] M.P. D'Evelyn, Surface properties of diamond, in: M.A. Prelas, G. Popovici, L.K. Bigelow (Eds.), Handbook of Industrial Diamonds and Diamond Films, Marcel Dekker, New York, 1998, pp. 89–146.

[20] T.I. Hukka, T.A. Pakkanen, M.P. D'Evelyn, Chemisorption of hydrogen on the diamond (100)2×1 surface—an ab initio study, J. Phys. Chem. 98 (1994) 12420.

[21] J. Furthmüller, J. Hafner, G. Kresse, Structural and electronic properties of clean and hydrogenated diamond (100) surfaces, Europhys. Lett. 28 (1994) 659.

[22] J.N. Russell Jr., J.E. Butler, G.T. Wang, S.F. Bent, J.S. Hovis, R.J. Hamers, M.P. D'Evelyn, in press.

[23] W.S. Yang, J. Sokolov, F. Jona, P.M. Marcus, Bulk-like structure of diamond (111), Solid State Commun. 41 (1982) 191–193.

[24] W.J. Huisman, J.F. Peters, S.A. de Vries, E. Vlieg, W.-S. Yang, T.E. Derry, J.F. van der Ween, Structure and morphology of the as-polished diamond (111)-1×1 surface, Surf. Sci. 387 (1997) 342.

[25] R.P. Chin, J.Y. Huang, Y.R. Shen, T.J. Chuang, H. Seki, Interaction of atomic hydrogen with the diamond C(111) surface studied by infrared–visible sum-frequency-generation spectroscopy, Phys. Rev. B 52 (1995) 5985–5995.

[26] C.-L. Cheng, J.-C. Lin, H.-C. Chang, The absolute absorption strength and vibrational coupling of CH stretching on diamond (111), J. Chem. Phys. 106 (1997) 7411–7421.

[27] B.J. Garrison, E.J. Dawnkaski, D. Srivastava, D.W. Brenner, Molecular dynamic simulations of dimer opening on a diamond {001} (2×1) surface, Science 255 (1992) 835.

[28] S.J. Harris, D.G. Goodwin, Growth on the reconstructed diamond (100) surface, J. Phys. Chem. 97 (1993) 23.

[29] S. Skokov, B. Weiner, M. Frenklach, Elementary reaction mechanism for growth of diamond (100) surfaces from methyl radicals, J. Phys. Chem. 98 (1994) 7073.

[30] M. Frenklach, K.E. Spear, Growth mechanism of vapor-deposited diamond, J. Mater. Res. 3 (1988) 133.

[31] M.E. Coltrin, D.S. Dandy, An elementary reaction mechanism for the growth of diamond, J. Appl. Phys. 74 (1993) 5803.

[32] C.C. Battaile, D.J. Srolovitz, J.E. Butler, Atomic-scale simulations of chemical vapor deposition on flat and vicinal diamond surfaces, J. Cryst. Growth 194 (1998) 353.

[33] C.C. Battaile, D.J. Srolovitz, I.I. Oleinik, D.G. Pettifor, A.P. Sutton, S.J. Harris, J.E. Butler, Etching effects during the chemical vapor deposition of (100) diamond, J. Chem. Phys. 111 (1999) 4291.

[34] R.E. Rawles, W.G. Morris, M.P. D'Evelyn, Kinetics and morphology of homoepitaxial CVD growth on diamond (100) and (111), Mater. Res. Soc. Symp. Proc. 416 (1996) 13.

[35] M.P. D'Evelyn, J.D. Graham, L.R. Martin, [100] versus [111] diamond growth from methyl radicals and/or acetylene, J. Cryst. Growth 231 (2001) 506.

[36] A. Netto, M. Frenklach, Kinetic Monte Carlo simulation of diamond film growth with the inclusion of surface migration, Mater. Res. Soc. Symp. Proc. 527 (1998) 383.

[37] R.E. Stallcup II, L. Villareal, S.C. Lim, I. Akwani, A.F. Aviles, J.M. Perez, Atomic structure of the diamond (100) surface studied using scanning tunnelling microscopy, J. Vac. Sci. Technol. B 14 (1996) 929–932.

[38] M. Frenklach, S. Skokov, Surface migration in diamond growth, J. Phys. Chem. B 101 (1997) 3025.

[39] B.D. Thoms, M.S. Owens, J.E. Butler, C. Spiro, Production and characterization of smooth, hydrogen-terminated diamond C(100), Appl. Phys. Lett. 65 (1994) 2957.

[40] O.M. Küttel, L. Diederich, E. Schaller, O. Carnal, L. Schlapbach, The preparation and characterization of low surface roughness (111) and (100) natural diamonds by hydrogen plasma, Surf. Sci. 337 (1995) L812–L818.

[41] K. Hayashi, S. Yamanaka, H. Okushi, K. Kajimura, Stepped growth and etching of (001) diamond, Diamond Relat. Mater. 5 (1996) 1002–1005.

[42] R.E. Rawles, R. Gat, W.G. Morris, M.P. D'Evelyn, Hydrogen plasma treatment of natural and homoepitaxial diamond, Mater. Res. Soc. Symp. Proc. 416 (1996) 299–304.

[43] R.E. Rawles, S.F. Komarov, R. Gat, W.G. Morris, J.B. Hudson, M.P. D'Evelyn, Mechanism of surface smoothing of diamond by a hydrogen plasma, Diamond Relat. Mater. 6 (1997) 791–795.

[44] B. Koslowski, S. Strobel, M.J. Wenig, R. Martschat, P. Ziemann, On the roughness of hydrogen-plasma treated diamond (100) surfaces, Diamond Relat. Mater. 7 (1998) 322.

[45] C.-L. Cheng, H.-C. Chang, J.-C. Lin, K.-J. Song, J.-K. Wang, Direct observation of hydrogen etching anisotropy on diamond single crystal surfaces, Phys. Rev. Lett. 78 (1997) 3713.

[46] T.W. Mercer, J.N. Russell Jr., P.E. Pehrsson, The effect of a hydrogen plasma on the diamond (110) surface, Surf. Sci. 392 (1997) L21.

[47] J. Wei, J.T. Yates Jr., Diamond surface chemistry I—a review, Crit. Rev. Surf. Chem. 5 (1995) 1.

[48] J.E. Butler, B.D. Thoms, M. McGonigal, J.N. Russell Jr., P.E. Pehrsson, Hydrogen chemistry on diamond surfaces, in: M.A. Prelas et al. (Eds.), Wide Band Gap Electronic Materials, Kluwer, Amsterdam, 1995, pp. 105–114.

[49] H. Kawarada, Hydrogen-terminated diamond surfaces and interfaces, Surf. Sci. Rep. 26 (1996) 205–259.

[50] D. Gutman, The controversial heat of formation of the $t-C_4H_9$ radical and the tertiary C–H bond energy, Acc. Chem. Res. 23 (1990) 375.

[51] S.J. Harris, Mechanism for diamond growth from methyl radicals, Appl. Phys. Lett. 56 (1990) 2298.

[52] L.N. Krasnoperov, I.J. Kalinovski, H.-N. Chu, D. Gutman, Heterogeneous reactions of H atoms and CH_3 radicals with a diamond surface in the 300–1173 K temperature range, J. Phys. Chem. 97 (1993) 11787.

[53] B.D. Thoms, J.N. Russell Jr., P.E. Pehrsson, J.E. Butler, Adsorption and abstraction of hydrogen on polycrystalline diamond, J. Chem. Phys. 100 (1994) 8425.

[54] D.D. Koleske, S.M. Gates, B.D. Thoms, J.N. Russell Jr., J.E. Butler, Hydrogen on polycrystalline diamond films: studies of isothermal desorption and atomic deuterium abstraction, J. Chem. Phys. 102 (1995) 992.

[55] C. Su, J.-C. Lin, Thermal desorption of hydrogen from the diamond C(100) surface, Surf. Sci. 406 (1998) 149.

[56] K. Bobrov, H. Shechter, M. Folman, A. Hoffman, Deuterium adsorption–desorption from diamond (100) single crystal surfaces studied by TPD, Diamond Relat. Mater. 7 (1998) 170.

[57] C. Su, K.-J. Song, Y.L. Wang, H.-L. Lu, T.J. Chuang, J.-C. Lin, Hydrogen chemisorption and thermal desorption on the diamond C(111) surface, J. Chem. Phys. 107 (1997) 7543.

[58] R.B. Woodward, R. Hoffman, The Conservation of Orbital Symmetry, Verlag Chemie, Weinheim, 1970.

[59] M.S. Gordon, T.N. Truong, J.A. Pople, Thermal decomposition pathways of ethane, Chem. Phys. Lett. 130 (1986) 245.

[60] G.T. Wang, S.F. Bent, J.N. Russell Jr., J.E. Butler, M.P. D'Evelyn, Functionalization of diamond (100) by Diels-Alder chemistry, J. Am. Chem. Soc. 122 (2000) 744.

[61] M.Z. Hossain, T. Aruga, N. Takagi, T. Tsuno, N. Fujimori, T. Ando, M. Nishijima, Diels-Alder reaction on the clean diamond $(100)2 \times 1$ surface, Jpn. J. Appl. Phys. 38 (1999) L1496.

[62] J.S. Hovis, S.K. Coulter, R.J. Hamers, M.P. D'Evelyn, J.N. Russell Jr., J.E. Butler, Cycloaddition chemistry at surfaces: reaction of alkenes with the diamond (001)-2×1 surface, J. Am. Chem. Soc. 122 (2000) 732.

[63] T. Ando, K. Yamamoto, M. Ishii, M. Kamo, Y. Sato, Vapour-phase oxidation of diamond surfaces in O_2 studied by diffuse reflectance Fourier-transform infrared and temperature-programmed desorption spectroscopy, J. Chem. Soc. Faraday Trans. 89 (1993) 3635.

[64] L.M. Struck, M.P. D'Evelyn, Interaction of hydrogen and water with diamond (100): infrared spectroscopy, J. Vac. Sci. Technol. A 11 (1993) 1992.

[65] M.Z. Hossain, T. Kubo, T. Aruga, N. Takagi, T. Tsuno, N. Fujimori, M. Nishijima, Chemisorbed states of atomic oxygen and its replacement by atomic hydrogen on the diamond (100)-(2×1) surface, Surf. Sci. 436 (1999) 63.

[66] P.E. Pehrsson, T.W. Mercer, Oxidation of the hydrogenated diamond (100) surface, Surf. Sci. 460 (2000) 49; P.E. Pehrsson, T.W. Mercer, Oxidation of heated diamond (100):H surfaces, Surf. Sci. 460 (2000) 74.

[67] R.E. Thomas, R.A. Rudder, R.J. Markunas, Thermal desorption from hydrogenated and oxygenated diamond (100) surfaces, J. Vac. Sci. Technol. A 10 (1992) 2451.

[68] M. Frenklach, D. Huang, R.E. Thomas, R.A. Rudder, R.J. Markunas, Activation energy and mechanism of CO desorption from (100) diamond surface, Appl. Phys. Lett. 63 (1993) 3090.

[69] A. Freedman, Halogenation of diamond (100) and (111) surfaces by atomic beams, J. Appl. Phys. 75 (1994) 3112.

[70] T.I. Hukka, T.A. Pakkanen, M.P. D'Evelyn, Chemisorption of fluorine, chlorine, HF, and HCl on the diamond (100) 2×1 surface: an ab initio study, J. Phys. Chem. 99 (1995) 4710.

[71] T. Yamada, T.J. Chuang, H. Seki, Chemisorption of fluorine, hydrogen and hydrocarbon species on the diamond C(111) surface, Mol. Phys. 76 (1992) 887.

[72] V.S. Smentkowski, J.T. Yates Jr., X. Chen, W.A. Goddard III, Fluorination of diamond—C_4H_9I and CF_3I photochemistry on diamond (100), Surf. Sci. 370 (1997) 209.

[73] T. Ando, K. Yamamoto, S. Suehara, M. Kamo, Y. Sato, S. Shimosaki, M. Nishitani-Gamo, Interaction of chlorine with hydrogenated diamond surface, J. Chin. Chem. Soc. 42 (1995) 285.

[74] K. Larsson, S. Lunell, Stability of halogen-terminated diamond (11,1) surfaces, J. Phys. Chem. A 101 (1997) 76.

[75] P. Fabis, Material property, compatibility, and reliability issues in diamond-enhanced, Ga As-based plastic packages, Microelectron. Reliab. 1344(C) (1999) 1–17.

[76] K. Kaifu, T.J. Komai, Wetting characteristics and blood clotting on surfaces of acylated chitins, Biomed. Mater. Res. 16 (1982) 757.

[77] M.L. Languell, M.A. George, J.J. Wert, J.L. Davidson, Morphological response of diamond films in dry sliding contact, J. Met. 46 (7) (1994) 66–70.

[78] J.C. Angus, H.B. Martin, U. Landau, Y.E. Evstefeeva, B. Miller, N. Vinokur, Conducting diamond electrodes: applications in electrochemistry, New Diamond Front. Carbon Technol. 9 (1999) 175–187.

[79] A.T. Collins, The optical and electronic properties of semiconducting diamond, Phil. Trans. Roy. Soc. A 342 (1993) 233.

[80] H. Kawarada, H. Matsuyama, Y. Yokota, T. Sogi, A. Yamaguchi, A. Hiraki, Excitonic recombination radiation in undoped and boron-doped chemical-vapor-deposited diamonds, Phys. Rev. B 47 (1993) 3633.

[81] T.N. Rao, A. Fujishima, Recent advances in electrochemistry of diamond, Diamond Relat. Mater. 9 (2000) 384–389.

[82] R. DeClements, G.M. Swain, The formation and electrochemical activity of microporous diamond thin film electrodes in concentrated KOH, J. Electrochem. Soc. 144 (3) (1997) 856–866.

[83] F. Okino, H. Shibata, S. Kawasaki, H. Touhara, K. Momota, M. Nishitani-Gamo, L. Sakaguchi, T. Ando, Electrochemical fluorination of 1,4-difluorobenzene using boron-doped diamond thin-film electrodes, Electrochem. Solid State Lett. 2 (1999) 382.

[84] R.G. Compton, F. Marken, C.H. Goeting, R.A.J. McKeown, J.S. Foord, G. Scarsbrook, R.S. Sussman, A.J. Whitehead, Sonoelectrochemical production of hydrogen peroxide at polished boron-doped diamond electrodes, Chem. Commun. (1998) 1921.

[85] C.H. Goeting, F. Marken, A. Gutierrez-Sosa, R.G. Compton, J.S. Foord, Electrochemically induced surface modifications of boron-doped diamond electrodes: an X-ray photoelectron spectroscopy study, Diamond Relat. Mater. 9 (2000) 390–396.

[86] M. Fryda, D. Herrmann, L. Schaefer, C.-P. Klages, A. Perret, W. Haenni, C. Comninellis, D. Gandini, Properties of diamond electrodes for wastewater treatment, New Diamond Front. Carbon Technol. 9 (1999) 229–240.

[87] F. Beck, H. Krohn, W. Kaiser, M. Fryda, C.P. Klages, L. Schaefer, Boron doped diamond/titanium composite electrodes for electrochemical gas generation from aqueous electrolytes, Electrochimica Acta 44 (1998) 525–532.

[88] A. Perret, W. Haenni, N. Skinner, X.-M. Tang, D. Gandini, C. Comninellis, B. Correa, G. Foti, Electrochemical behavior of synthetic diamond thin film electrodes, Diamond Relat. Mater. 8 (1999) 820–823.

[89] P.W. May, M.-T. Kuo, M.N.R. Ashfold, Field emission conduction mechanisms in chemical-vapour-deposited diamond and diamond-like carbon films, Diamond Related Mater. 8 (1999) 1490–1495.

[90] P.W. May, J.C. Stone, M.N.R. Ashfold, K.R. Hallam, W.N. Wang, N.B. Fox, The effect of diamond surface termination species upon field emission properties, Diamond Relat. Mater. 7 (1988) 671–676.

[91] P.K. Baumann, R.J. Nemanich, Electron emission for metal–diamond interfaces (100), (111) and (110) interfaces, Diamond Relat. Mater. 7 (1998) 612–619.

[92] C.E. Troupe, I.C. Drummond, C. Graham, J. Grice, P. Hohn, J.I.B. Wilson, M.G. Jubber, N.A. Morrison, Diamond-based glucose sensors, Diamond Relat. Mater. 7 (1998) 575–580.

[93] V.I. Polyakov, A.I. Rukovishnikov, A.V. Khomich, B.L. Druz, A. Dania, A. Hayes, M.A. Prelas, B.V. Tompson, T.K. Ghosh, S.K. Loyalka, Surface phenomena of the thin diamond-like carbon films, Mat. Res. Soc. Symp. Proc. 555 (1999) 345–350.

Surface Science 500 (2002) 1005–1023

www.elsevier.com/locate/susc

The surface science of xerography

Charles B. Duke [a,*], Jaan Noolandi [b], Tracy Thieret [c]

[a] *Xerox Corporation, Wilson Center for Research and Technology, 800 Phillips Road 114-38D, Webster, NY 14580, USA*
[b] *Xerox Corporation, Xerox Research Center Canada, 2660 Speakman Drive, Mississauga, Ont., Canada L5K 2L1*
[c] *Xerox Corporation, Wilson Center for Research and Technology, 800 Phillips Road 114-41D, Webster, NY 14580, USA*

Received 22 May 2000; accepted for publication 19 April 2001

Abstract

Over the past four decades xerography, the dry ink marking process developed by the photocopy industry, has grown from nothing into a $170 billion industry worldwide. This amazing commercial success is due to the fact that during this period, xerographic technology experienced constant and often-dramatic improvement created by sustained industrywide research and development. Indeed, the development of the xerographic copying and printing industry is one of the great applied surface science successes of all time. In this article we outline the story of the advances in xerographic technology during the past four decades, describe the profound dependence on these advances of the control of surface and interface properties of increasingly sophisticated multi-component materials systems, and indicate the potential impact on the industry of the continuing development of the surface and interface science of the multi-component materials packages used in xerographic technology. © 2001 Elsevier Science B.V. All rights reserved.

Keywords: Adhesion; Dielectric phenomena; Electrical transport (conductivity, resistivity, mobility, etc.); Photoconductivity; Surface electronic phenomena (work function, surface potential, surface states, etc.); Surface energy; Surface melting

1. Introduction

The practice of xerography began in 1938 with Chester Carlson's first xerographic print [1,2]. That humble beginning in an apartment in New York City spawned the biggest change in work practices in the history of the office. The first demonstration of the system contained the essential elements of the modern electrophotographic copier or printer. The xerographic process consists of creating an electrostatic image on a photoconducting drum or belt, developing that image with a pigmented charged powder called toner, transferring that image to a substrate, typically paper, and then melting the toner to "fuse" it onto the substrate. When in 1959 this process was embedded into the legendary 914 copier, the world changed forever. Hand copying and carbon paper were banished; access to personal information was democratized; and the era of personal publishing began. Most readers of this article have never known a world without the convenience of personal copying and printing. 1959 marked the dawn of the modern information era.

From these modest beginnings, printing and copying using the xerographic process is an industry that has been generated by over four decades

* Corresponding author. Tel.: +1-716-4222106; fax: +1-716-2655080.
E-mail address: cduke@crt.xerox.com (C.B. Duke).

0039-6028/01/$ - see front matter © 2001 Elsevier Science B.V. All rights reserved.
PII: S0039-6028(01)01527-8

of continual improvement of the implementation of the xerographic process steps. Many of the improvements are the result of sophisticated materials packages whose design and operation rely heavily on the understanding and control of surface and interface properties. Thus, xerography is a technology that has been generated and advanced by the control and utilization of interfacial phenomena. The story of the development of the xerographic copier/printer industry as a case history in value of basic research in US industry has been summarized by the Committee for Economic Development [3]. In this article we describe the xerographic process, indicate some of the major material challenges in designing a modern, commercial xerographic marking engine, and develop several illustrative examples of the role of surface and interface science in overcoming these challenges.

This might be a footnote in history of minor interest to the intended audience for this volume, first year graduate students in a wide variety of scientific and engineering disciplines, but for one thing: During the past decade the world has changed profoundly and with it the practice of both science and engineering. The world has entered an era of unprecedented pace in the generation of new knowledge and its commercial application in global markets. The search for new knowledge is increasingly pursued in a context in which its application to generate economic value is valued even more highly than the novelty and impact of the knowledge itself. Few of today's first year graduate students will pursue their careers without repeatedly feeling the need to solve practical problems under tight time constraints. The story of the development of xerographic technology by solving difficult, practical surface science problems is a harbinger of the future. By appreciating the whys and hows of the development of xerography, you set forth the elements of a blueprint for your own professional success in this new world.

2. Industry overview

Xerographic printing and copying is big business. The scope of the industry is monitored by several consulting firms, one of which is CAP ventures from whose services we take the data noted below [4]. From the humble beginnings of Chester Carlson's kitchen experiments in 1938 and the first automatic copier in 1959, the worldwide market served by xerographic marking engines and the services built around them has exploded to become $167 B in 1998 growing at 12% per year. This corresponds to 1.2 trillion pages made by xerographic printers, copiers and fax machines. For comparison, the corresponding number of pages made on desktop ink jet marking engines is 160 billion, i.e., nearly 10 times smaller. As indicated in Ref. [3], this entire industry and its amazing growth is an outcome of an aggressive research and development (R&D) program carried out over six decades. Much of this R&D was devoted to solving surface and interface technical problems needed to produce better image quality, improved reliability and lower cost of xerographic products.

An important aspect of the industry is that its business model derives an important portion of its profits from selling "consumables", specifically the dry ink and replacements for some of the parts that wear out over the life of the engine (like tires on an automobile). Dry ink is the black powder that you get all over you when it is not adequately fused to the paper or when you try to change the bottle of dry ink and spill some. In 1998 the dry ink market alone was over $8 B growing at 10% per year, and shared by over 12 multi-national firms, with the largest shares being held by Xerox and Canon. One of the important replacement parts is the photoreceptor, the worldwide market for which in 1998 was approximately $10 B. As we see below, both photoreceptors and the dry ink are complex, multi-component organic composites designed and tested using sophisticated surface science concepts and instrumentation. Thus, through the vehicle of xerographic consumables, surface and interface science and technology are exerting a decisive influence on over $20 B of cash flow every year: a number that is comparable to the entire US annual expenditure for basic research in recent years [5].

3. The xerographic process: description of process steps

As indicated above, the xerographic process consists of five process steps that begin by charging a photoconductive belt or drum; generating a latent image on this photoconductor by imagewise exposure to light; development of this latent image with charged toner; transfer of the toner image to a substrate; and fusing of the image to that substrate. A sixth process step, the cleaning of residual charge off the photoreceptor, is needed in commercial devices to reuse the same photoreceptor for subsequent imaging. We begin this section with an explanation of the nature and operation of the subsystems that together comprise a monochrome xerographic marking engine, as illustrated in Fig. 1 [1,2,6–10].

3.1. Photoreceptor

The photoreceptor is the central element of the xerographic process. This device's fundamental responsibility is the transport of the page images of a document in their various forms through the steps of the process. While performing this function the photoreceptor also serves as the image-generating surface. Specifically, charge deposited on the photoreceptor is discharged imagewise by the photogeneration in a charge-generator layer of holes that migrate to the negatively charged upper surface to discharge this surface as shown in Fig. 2 [1,2]. The device is a loop that recycles through the process steps. It may be realized either as a rigid drum or as a belt, as shown in Fig. 1. A detailed description of the operation of multi-layer photoreceptors is given in Refs. [2,6]. Designing and fabricating a multi-layer photoreceptor that is stable for hundreds of thousands of impressions and supports resolutions of up to 1200 spots per inch uniformly across its page-size surface is a challenging technical problem which involves the solution of multiple interface science problems as indicated in the caption to Fig. 2. For contributions to this feat for the belt photoreceptors used in the Xerox 10 series and subsequent products Damodar Pai, Jack Yanus, and Milan Stolka of

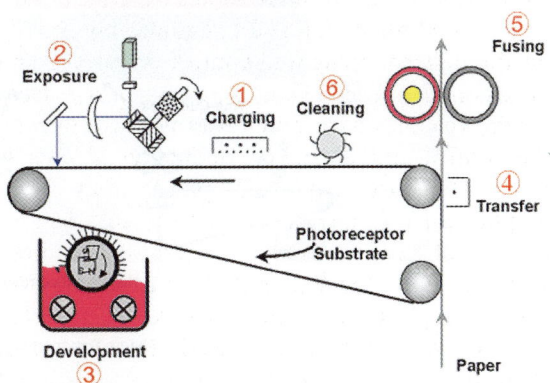

Fig. 1. Schematic diagram of the monochrome xerographic process. A photoreceptor belt is uniformly charged in step (1). An image is written on this belt by a laser in step (2), thereby generating a charge image on the belt. This charge image is converted into a powder image of toner on the belt in the development step (3). This powder image is transferred to a sheet of paper in the transfer step (4) and subsequently fused to the paper in step (5). Residual toner on the photoreceptor is cleaned off in step (6) and the process repeats.

Xerox received the "Heroes of Chemistry 2000" award sponsored by the Industry Relations Office of the American Chemical Society. This illustrates a first lesson for prospective surface scientists: Outstanding applied surface science is recognized not only by one's sponsors, but also by one's peers.

3.2. Charging

In most xerographic architectures the charging subsystem is presented with a clean (of toner) and erased (of static electric charge) photoreceptor as indicated in Fig. 3 [1]. Its output is a uniformly charged photoreceptor surface at a defined voltage. The charging device is typically a fine tungsten wire operated at a few hundred volts DC and a few thousand volts AC. This current produces an ionization of the air in its vicinity. These ions are the source of a "corona wind" that charges the photoreceptor. A corotron-charging device is built by surrounding the wire with a grounded aluminum housing on three sides [1]. In modern charging

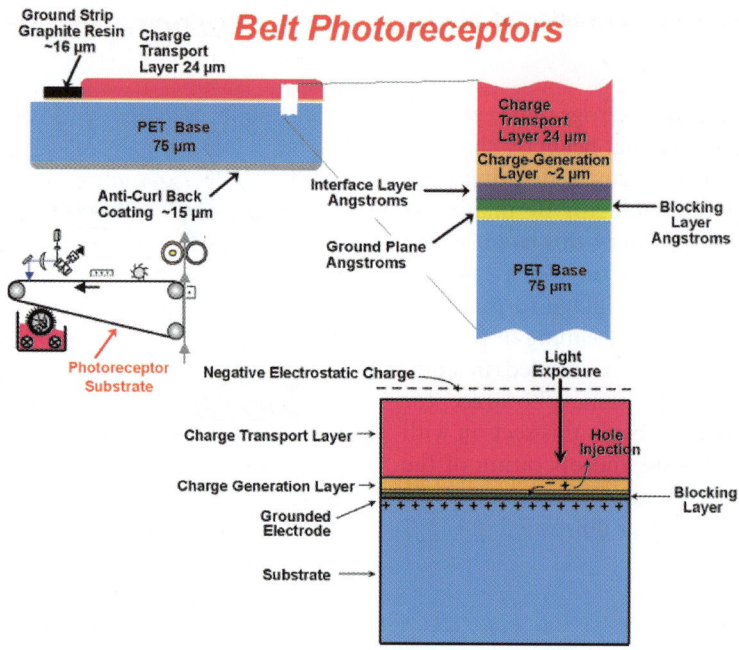

Fig. 2. Successive blow-up diagrams of the structure and operation of a multi-layer photoreceptor belt. The multiple interface, photogenerator and transport layers are coated on a polyethylene terephthalate (PET) belt. The operation of the photoreceptor in a xerographic marking engine is shown in the diagram in the left-hand panel. The physical processes that occur to discharge the photoreceptor upon exposure to light are indicated in the diagram in the lower right-hand corner. The light penetrates through the charge transport layer and is absorbed in the charge generation layer where it generates an electron–hole pair. The hole migrates to the front of the photoreceptor where it discharges the charge pattern on the surface. The electron migrates to the grounded electrode and flows to ground. Designing multi-layer photoreceptors involves the solution of sophisticated interface science problems associated with charge injection across the various interfaces, assuring that charge does not get trapped either in the layers of the photoreceptor or their interfaces, and guaranteeing that charge does not spread as it is generated and transported across photoreceptor so that resolutions of up to 1200 dpi may be achieved uniformly over page-size images.

devices, called scorotrons, a biased mesh grid is interposed between the wire and the photoreceptor [1]. This grid enforces a charge limit. As the device charges toward the grid potential, the electric field between the grid and the photoreceptor surface drops to zero, leading to a known uniform voltage at the exit from the charging station. This is a good example of the use of passive elements operated using physical principles to stabilize the xerographic process.

3.3. Exposure/illuminator

The exposure station receives a photoreceptor with a uniform charge spread two dimensionally across its surface. In modern digital xerographic marking engines the photoreceptor is discharged imagewise by a laser beam so that the output is a latent image in which the charged and discharged areas of the photoreceptor are a reflection of the image that ultimately appears on the output medium. Such a station is indicated schematically in Fig. 4 [1,2]. Usually the discharged areas correspond to the black areas in a monochrome image. Gray areas are described by halftone dot patterns [11]. Typical resolutions are 600–1200 dots per inch (dpi).

The desired imaging property of the photoreceptor is that in the dark, the device functions as a capacitor, maintaining an electrostatic voltage between the upper surface and the ground plane on the lower surface. When exposed to light, however, the device becomes a conductor. The voltage drops to near zero. Because there is little lateral

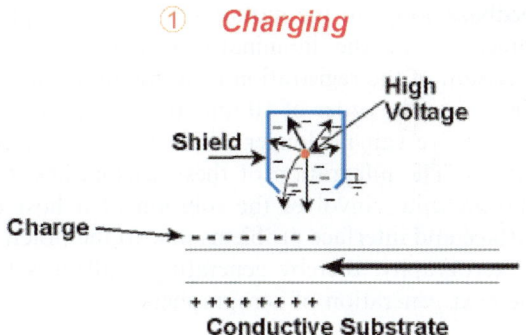

Fig. 3. Schematic diagram of a corotron charging subsystem. A corotron wire at high voltage generates negative ions that flow onto the photoreceptor belt to give it a uniform charge as it passes below an orifice in the grounded shield surrounding the wire. Sometime a grid is added below the orifice to better control the potential on the photoreceptor. In this case the charging subsystem is called a scorotron. In recent years completely solid-state versions of these devices have been built and tested, but have seen limited deployment in the industry. The perfection of such "charging bars" remains a challenge for the next generation.

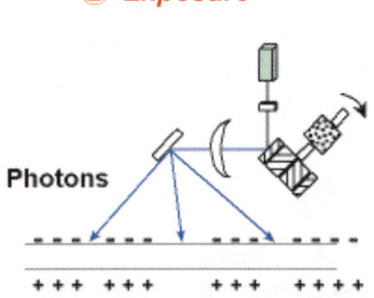

Fig. 4. Schematic diagram of the exposure ("illuminator") subsystem. A laser beam is swept across the surface of the photoreceptor by a rotating polygon of mirrors. The beam is turned on and off imagewise either by a modulator or by the current through a laser diode. Everywhere that the beam strikes the photoreceptor, it is discharged as indicated by the absence of a charge pattern. Many technical challenges face the designer of a commercially successful diode-laser-based illuminator including selecting a combination of laser life, laser power, laser spot size, laser pulse shape, photoconductor sensitivity, optical and polygon design that satisfy the cost, stability and compactness requirements of a modern xerographic marking engine.

conduction, a small spot of light produces a localized area of voltage difference. These properties are achieved by the multi-layer photoreceptor architecture indicated in Fig. 2. Residual, trapped charges prohibit the exposed voltage from achieving the ground potential. Aging of the photoreceptor includes increases in both the "dark decay", where the device voltage decreases with time even in the dark, and the "residual" potential, the minimum voltage achievable upon exposure. Reducing aging and dark decay to allowable limits is a challenging problem involving considerable use of surface science techniques and tools to identify and mitigate the root causes of these phenomena.

A modulated or diode laser provides a convenient source of light to expose the photoreceptor in an imagewise way. The light is scanned laterally using a polygon mirror rotated at a constant angular velocity. The photoreceptor motion provides the second dimension linear variation so that a raster scan is produced on the photoreceptor surface much in the same way that a television picture is generated. The resolution is typically limited by practical design considerations (e.g., laser power, photoreceptor sensitivity, cost of the optics) rather than fundamental physical phenomena (e.g., charge spreading on the surface of the photoreceptor). More detailed descriptions of the "illuminator" subsystem that exposes the photoreceptor may be found in Refs. [1,2]. The design and fabrication of the diode lasers for use in illuminators is a challenging technical problem in its own right. Don Scifres, Robert Burhnam and William Streifer developed the pioneering technology for one of the early generations of semiconductor diode lasers at the Xerox Palo Alto Research Center (PARC). Scifres spun out the design and manufacturing technology from PARC as the firm SDL to manufacture these lasers for Xerox. Subsequently SDL has become a major supplier of lasers to the entire optical communications industry, and Scifres has remained its chief executive officer (CEO). Scifres and Burnham were elected to the US National Academy of Engineering for their contributions to this technology. Scifres also was awarded the 1997 George E. Pake Prize of the American Physical Society for his role in commercializing the technology and thereby generating jobs for young scientists and engineers. This illustrates a second lesson for prospective surface scientists: Such are the rewards awaiting successful young scientists and engineers in the new world of global commerce.

3.4. Registration

In color-printing systems the photoreceptor must be imaged separately for each of four (cyan, magenta, yellow and black or "CMYK") color separations. One way to do this, the multi-pass cyclic architecture, is shown in Fig. 5. The four color panels must be registered relative to one another. Mis-registration results in undesirable color shifts, image blur, unimaged areas, and other defects. The timing of the individual exposure steps is regulated such that the images overlay one another within a few 10s of microns. In the color-printing architecture shown in Fig. 5, there is an additional registration requirement. When the individual color separations are built up on the output media, these transfer steps must occur with the same level of precision. A description of the various architectures used to implement color xerography may be found in Refs. [1,6]. The solution of the color registration problems is an active area of research in modern control system design. The use of novel sensors on the photoreceptor and feedback loops to the motion control of the photoreceptor or the illuminator is enhancing the precision of the registration from the order of 80–100 μm to the order of 10 μm, at which point the human eye can no longer detect mis-registration defects. The integration of these sensors into the photoreceptor involves the solution of a host of surface and interface problems new to the practice of xerography, thereby generating challenges for the next generation of xerographers.

3.5. Development

The development subsystem is presented with a latent image on the photoreceptor. Upon exit the toned image is visible on its surface. A schematic diagram of a magnetic brush development subsystem is shown in panel (a) of Fig. 6 [1,6], although many other types of subsystems are used commercially [1,6]. Within the magnetic brush development subsystem there are two major materials components collectively called "developer" and illustrated in panel (b) of Fig. 6. The first is the toner, or dry ink. Toners are complex multi-component composite materials packages the composition of which is indicated in panel (c) of Fig. 6. Toner is the most challenging materials package in the system

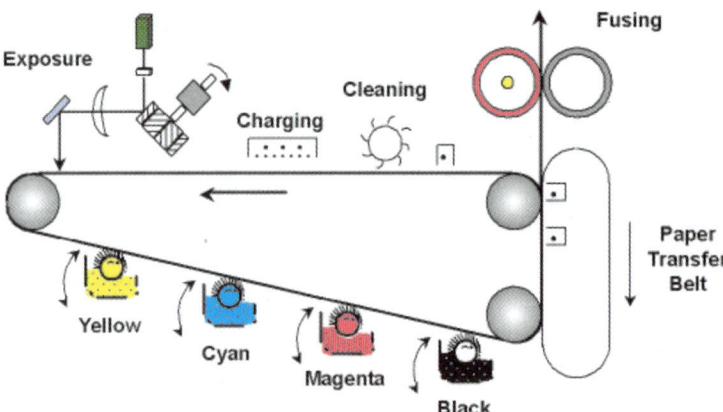

Fig. 5. Schematic diagram of the cyclic color xerographic process. In this process one color is imaged, developed and transferred in each pass of the photoreceptor belt. The developers for the other colors are cammed out as indicated by the arrows in the diagram. The four-color image is built up on the paper as each of the four developed powder images is transferred sequentially. Then, the final four-color image is fused as it exits the transfer zone.

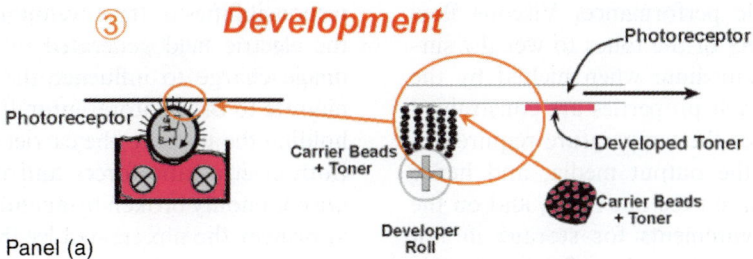

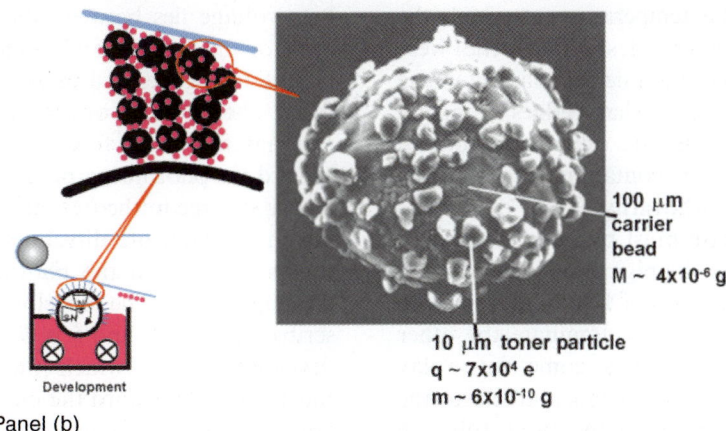

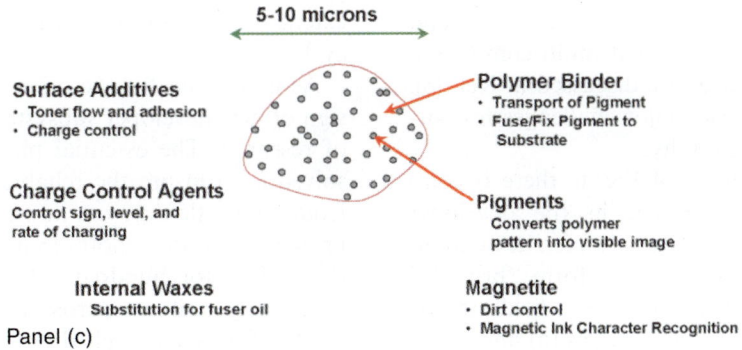

Fig. 6. Successive blow-ups of a development subsystem (panel (a)), development zone and carrier bead (panel (b)), and toner particle (panel (c)). Key parameters and features of each of these are indicated in the figure.

because there are so many properties it must exhibit. These include powder and viscous flow, charging, melting properties, color, toxicity, size, and adhesion to the various surfaces with which it comes in contact during the xerographic process.

Many of these properties deal with how the toner interacts with surfaces. Powder flow governs the capability of the toner to be dispensed from the hopper to the developer housing and once there, to mix well with the contents. Additives are required to generate suitable performance. Viscous flow determines the ability of the toner to wet the surface of the output medium when melted by the fusing subsystem. Melt properties are constrained by a desire to reduce the temperature required to fuse the toner to the output media, and hence energy consumption, on the one hand, and on the other hand by requirements for storage in hot environments. There are stories of toner bottles turning into big crayon containers after transport in trucks across the desert because their blocking temperatures (i.e., the temperature at which the toner spontaneously aggregates in the bottle) were too low. Size distribution is a determinant in image quality. Large toners make sharp edges in images look blurry. Small toners are difficult to contain and can become airborne contaminants. Pigments may not be chosen arbitrarily. They must pass toxicity testing and also must produce the desired colors. The loading of the pigments into the toner polymer base (see panel (c) of Fig. 6) must not be excessive or the pigment will dominate the other properties. Release properties come into play when, during fusing, the toner must adhere to the output media rather than to the fuser roll (see Fig. 8) to be deposited on a subsequent sheet in a process called hot offset. With all these constraints, the challenge of constructing a material that satisfies them is quite a challenge. The design and fabrication of the sophisticated multi-component toner materials packages are discussed in Ref. [8]. They remain one of the major challenges for surface scientists in xerography.

As shown in panel (b) of Fig. 6, there is commonly a second component in the developer housing. The "carrier" is typically a ferrite of about 100 μm diameter. These carriers form "brushes" which are rotated by the magnetic fields in the developer roll as indicated in panels (a) and (b) of Fig. 6. The toner, when agitated against the carrier, develops a triboelectric charge and adheres to it. Typically this requires the ferrite "core" of the carrier to be coated with an organic polymer that yields the desired charge exchange. Together this mixture is referred to as "developer." Developer "aging" occurs when the toners impact the surfaces of the carrier particles, diminishing the carrier surface area available for tribocharging. One of the responsibilities of the development system is to use the electric field generated by the photoreceptor image charge to influence the charged toners to migrate to the photoreceptor. Thus, the two forces holding the toner to the carrier must be overcome. Both electrostatic forces and the adhesion forces are commonly broken by agitation of the developer in or near the nip created by the development roll and the photoreceptor surface.

A bias is applied to the housing to generate the electric fields for development. The value of this bias voltage lies between the photoreceptor voltages for the charged and discharged areas. Thus, a potential is generated in the nip that produces a force, the product of the toner charge and the external field that for the toned regions points toward the photoreceptor. The same bias produces a similar force in the untoned regions of the image but in the opposite direction, thereby driving the toners away from the photoreceptor toward the developer roll. The development subsystem described in Figs. 1 and 5 is called "discharged area development" in which the negatively charged toners migrate toward the uncharged (i.e. exposed) areas of the photoreceptor [1,2]. Unlike light–lens copiers, in which the charged areas are developed, this is the common mode of exposure for digital printers and copiers, selected primarily in order to extend the life of the laser by reducing its duty cycle.

As is the case for the other subsystems, the design of new developer subsystems is an active area of research. The essential physics problem to be solved is to insure the reliable transport of toner from the bottle that you insert into your copier or printer onto the photoreceptor at precisely the right place for hundreds of thousands of copies completely uniform across the page. One might think of this as applied soft condensed matter physics. In the early 1980s a new system design called highly agitated zone (HAZE) development was invented which provided both excellent broad area and image detail development with hardware that was smaller and lower cost than its predecessors. This design was first incorporated in a 62-copies-per-minute Xerox product, the Xerox 1065 Marathon copier, introduced in 1987. Over the past 13 years, this product family has produced more than 100 000 machines in the US. Furthermore, the HAZE design was subsequently incorporated in several additional major Xerox product platforms for copiers and printers. The combined revenue from all of the products embodying this development system was nearly half of Xerox' total revenue in 1996. In recognition of the contribution of the HAZE development process to the success of three generations of Xerox' current products, the inventor of the process, Dan Hays of the Xerox Wilson Center for Research and Technology, was awarded the 1997 American Institute of Physics Prize for Industrial Applications of Physics. More innovations like this are in the pipeline, especially with the decreasing size of marking engines and the move to color. Lots of surface and interface problems must be solved to move little 5 µm toner particles around over 10 in. swaths, so there is ample opportunity for surface and interface scientists in this area. This illustrates a third lesson for prospective surface scientists: Opportunity abounds, and fame as well as fortune awaits those who generate important new inventions.

3.6. Transfer

The transfer step begins with a toned photoreceptor image and a sheet of media. The toner is transferred from the photoreceptor to the media with a minimum of residual mass remaining on the photoreceptor as indicated in Fig. 7. This process uses electrostatic forces on the toner charge to encourage it to move to the paper. Corotrons or biased rolls are used to manipulate the toners electrostatically. When printing color in the normal way by mixing together a set of primaries, variations in the transfer efficiency alter the amount of transferred toner from each constituent and result in noticeable differences in color reproduction. Transfer efficiencies in excess of 98% are typically achieved in modern xerographic marking engines, but even at this figure efficient, stable transfer remains one of the major research problems limiting the performance of modern xerographic marking engines. This is a promising area for a prospective surface scientist to establish her or his fame and fortune.

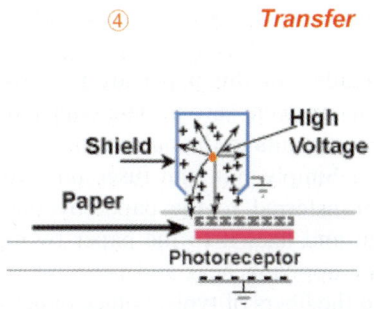

Fig. 7. Schematic diagram of a transfer subsystem. A corotron biases the incoming paper positively so that it attracts the negatively charged toner to transfer from the photoreceptor. The geometries shown in this figure are highly idealized.

3.7. Media handling

In order to position the media to receive the xerographic image at the transfer station, additional processes must be considered. Sheets of paper (or vugraphs) must be removed from a stack, transported to the transfer station, and presented to it at the precise time such that the arrival of the photoreceptor image and the media may be synchronized within an accuracy of a few tenths of millimeters. The most common type of media is cut sheets, although the use of continuous webs is sometimes employed in high-throughput printing systems. The properties of media vary greatly from one type to another and sometimes even from sheet to sheet (e.g., the use of different weights and finishes of paper in a single document). In some printing systems the media are escorted through the system with gripper bars or tacked to an electrostatically charged transport medium. In most electrophotographic systems, however, the media are passed between subsystems and, after marking, sent to the finishing station, by mechanical devices that rely on frictional forces. The development of gentle media handling systems based on distributed sensors and actuators is a major frontier of modern printing systems research, due to the high importance of handling a wide range of media in color printing. This is a huge opportunity for the development and application of unique microelectromechanical (MEMS) devices and technology to the printing business by the next generation of applied physicists.

3.8. Fusing

Most readers of this paper are probably aware of the fuser in their xerographic copier or printer because documents in the output tray are warm to the touch. Simply put, the fuser takes the toner powder transferred to the paper by the transfer subsystem and fixes it to the paper by melting it, usually by applying heat and pressure, so that it flows into the fibers of typical office paper or into a specially prepared surface layer on coated stock or transparencies. Once the toner is on the media as it exits the transfer subsystem prior to fusing, it may be readily smudged (or even blown off) because the only forces holding it are the electrostatic forces remaining from the transfer step. Fusing melts the layers of toner on the media to achieve fixing the image as indicated in Fig. 8 [8,9]. The heating may take place using a radiant source or more commonly a heated roll, which assists the fixing of the toner to the media by pushing it into the media. These rolls are sometimes coated with a functional oil (called a "release agent") that forms covalent bonds with the roll surface in order to keep the toner from adhering to it. Roll fusers are complicated multi-component materials systems, embodying multiple layers of composite materials.

Their design and fabrication are discussed in Refs. [9,10]. When an electrophotographic printer/copier is first turned on, the "warm up" time is consumed in heating the fuser. Indeed, during warm-up, all the power available to the device is used to heat the fuser as rapidly as possible. As xerographic devices become faster, image quality requirements become higher, and power constraints become tighter, research is moving toward belt fusers, that can warm up much faster and use a lot less power than roll fusers. This is an arena ripe with opportunity for the applied surface scientist and the control engineer to collaborate on designing smart fusers and associated toners that precisely fuse the images on each sheet, at minimal power, image by image, without a smudge or smear to be seen.

3.9. Cleaning and erase

The photoreceptor cycle is completed in the cleaning and erase stations. Input to these subsystems is a surface that has untransferred residual toner and some remaining charge. The erase station, not shown in Figs. 1 and 5, consists of a "flood" exposure of the photoreceptor performed by a suitable light source. This step is designed to discharge the photoreceptor and residual toner as completely as possible so that the toner may be removed more easily from the photoreceptor and a consistent initial state may be prepared for the charging subsystem.

Cleaning, illustrated schematically in Fig. 9, is performed using one or more of a number of technologies. It is common in many printers to use an elastomer blade to scrape the toners off without damaging the surface of the photoreceptor. Biased cleaning brushes rotating in contact with the photoreceptor belt also are used frequently. The collected toner is placed into a waste toner bottle, to be recycled later. Perhaps the highest goal for cleaning subsystems is to eliminate them entirely by designing transfer subsystems that are sufficiently efficient, thus enabling a lower machine price by reducing both the cost and number of required subsystems. This is currently one of the greatest technical challenges facing a prospective surface scientist working on xerography.

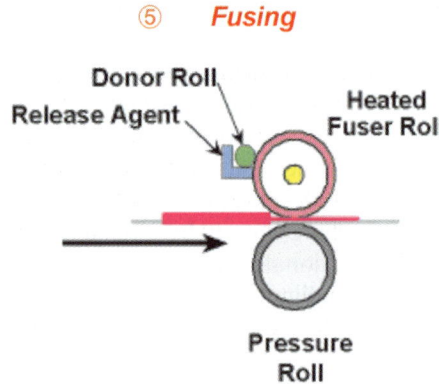

Fig. 8. Schematic diagram of a fusing subsystem based on using a release agent. The release agent coats the heated fuser roll so that the toner does not stick to the roll. Under the influence of heat and pressure in the fuser, the toner particles are "fused" together (i.e., the toner powder becomes a thin polymer film) and fixed to the paper as described in Fig. 11.

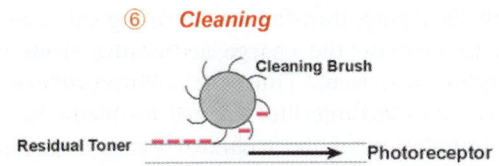

Fig. 9. Schematic diagram of a brush cleaning subsystem. By either mechanical or electrostatic forces, the residual toner is stripped of the photoreceptor and collected in a waste bottle (not shown).

4. Material challenges in xerography: the surface connection

As noted above, much of the functional performance of the xerographic process is generated by the use of sophisticated materials packages, typically multi-component organic composites, which must satisfy many constraints. Due to their composite nature, interfacial properties play a dominant role in their performance and stability. The specifics vary with the nature of the materials package, with multi-layer systems, film composites, and multi-component powders being three broad material classes exhibiting distinct design rules and criteria.

Photoreceptors (Fig. 2) and fuser rolls (Fig. 8) are two prominent examples of multi-layer mostly organic composites. Both embody sophisticated boundary layer treatments to ensure the mechanical integrity of the device under hostile operating conditions (e.g., heat, corona charging, media contact and release and repeated flexion and tension). Wear is a central concern for both—due primarily to media contact for the fuser rolls but also due to the transfer and cleaning subsystems for the photoreceptor. Moreover both must exhibit good release properties: the photoreceptor to release toner powder in the transfer step and the fuser roll to release melted toner on the media upon exit from the fuser. Finally, the electrical, optical, and thermal properties of these devices must be designed carefully. Fuser rolls need adequate thermal conductivity. Charges generated in the charge generation layer (CGL) of a photoreceptor must cross multiple interfaces and move through the device to discharge the receptor upon exposure. All of these functional requirements require exquisitely designed interfacial properties both between the various layers of the photoreceptors and fuser rolls and between the base polymer and its loading material for the CGL.

Developer and toner powders (Fig. 6) have different design criteria. The toner and carrier coating must exhibit the correct magnitude and kinetics of contact charge exchange: very delicate surface properties. They must flow well in powder form for the developer subsystem to work. The toner must exhibit viscoelastic flow when subjected to heat and pressure in order to get a good fix to the paper, a suitable gloss on the image, and the reduction of light scattering within the fused toner layer to acceptable levels for transparencies. The humidity sensitivity (i.e., dependence on adsorbed water) of the toner powder flow and contact charge exchange must be controlled. All of these characteristics result from controlling the surface and interface properties of the external and internal surfaces in the toners and carriers. We develop this theme in the following section by considering three specific examples in more detail.

5. Microscopic control of organic composites

5.1. Multi-layer photoreceptors

The dual-layer photoreceptor design shown in panel (a) of Fig. 2 is based on separating the functions of photogeneration and charge carrier transport. This allows the flexibility to select photogenerator materials optimized for different wavelengths of the light sources used in xerography (e.g. visible LED image bars, IR diode lasers) while designing the rest of the photoreceptor to satisfy other constraints on its behavior (e.g., wear, speed, stability). The current trend is toward the use of organic photoconductors. The substrate can be a metal drum or a flexible metalized polyester belt, as shown in Fig. 2. The substrate is coated with an undercoat (blocking) layer which serves to prevent the injection of charge carriers in the dark from the grounded electrode. The CGL consists of pigment particles dispersed in a polymer binder. The pigment is selected and optimized for the specific application. Some of the pigments used include perylenes, phthalocyanines and azo compounds [2,6]. The function of the thin CGL (about 0.2–2 µm) is to absorb the incident light and photogenerate charge carriers efficiently. The photogenerated charge carrier must be injected and transported through the next layer to reach the surface of the photoreceptor. The charge transport layer (CTL) is coated on top of the CGL and consists of a film (15–30 µm thick) of insulating polymer doped with charge-transporting molecules. Most of the active molecules, which include aryl amines and hydrazone, transport only holes and the photoreceptor surface is charged negative. The CTL should be transparent to the incident light and the layer is designed to be clear and amorphous. Additional layers are included in practical photoreceptors, e.g., an adhesive layer on the metalized polyester to improve adhesion of belt photoreceptors and polymer overcoats on top of the CTL to extend the life of the photoreceptor. The basic phenomena in the operation of the organic photoreceptors are the photogeneration, injection and transport of charge carriers. Detailed descriptions of the design and operation of these devices have been given in Ref. [2].

Surface and interface science plays a key role in the design of practical photoreceptors. The CGL is typically a pigment dispersed in a binder, which requires control of the pigment/binder interfaces for effective photogeneration. The injection of holes into the CTL and electrons into the ground electrode requires engineering of the electronic properties of the CGL interfaces. The adhesion of the CTL, CGL, and backplane layers typically is achieved by special interfacial chemical treatments. Polyesters, polyamides, poly(vinyl butyral), poly(vinyl alcohol), polyurethane, and polyacrylonitrile have been used as adhesives in contact with the supporting substrate. The adhesive layer is of a thickness from about 0.001 µm to about 1 µm. This layer may also contain conductive and nonconductive particles, such as zinc oxide, titanium dioxide, silicon nitride, and carbon black to provide desirable electrical and optical properties [12]. Often the top surface of the photoreceptor is coated with an overcoat designed to reduce the wear on the photoreceptor by the charging, transfer, and cleaning subsystems and to increase the charge acceptance from the charging subsystem. Thus, the solution of practical surface and interface science problems lies at the foundation of the technical issues associated with fabricating cost-effective, long-life commercial multi-layer photoreceptors.

5.2. Dry ink: conventional and chemical toner

Conventional toners are fabricated by dispersing pigments in a polymer base, grinding and classifying. Descriptions of their design and processing are available in the literature [7]. Chemical toner represents a major advance over conventional processing. This new chemical technology involves molecular design and micro-fabrication allowing precision manufacturing of custom toner particles on the micrometer scale. Particle size distribution and surface characteristics are strong drivers of image quality. Narrow size distributions and spherical particles have more uniform charge per unit mass (Q/m) and hence, more predictable behavior in the development subsystem. Conventional toner manufacture yields a wide distribution of particle sizes and a particle morphology characteristic of broken glass with jagged edges. Chemical toner technology permits the manufacturer to specify a narrow distribution of nearly spherical particles produced by the process, as well as a wide range of materials compositions.

A number of chemical toner processes have been commercialized. Among these are suspension polymerization, emulsion aggregation (EA), solvent dispersion, and encapsulation. We discuss here EA as practiced by the Xerox group, Nippon Carbide and Konica [13–15]. The EA process is a powerful technology to produce monodispersed composite particles from the submicron to the micrometer range. Controlled aggregation of submicron polymer latex particles, together with pigment particles, is an important process in the development and production of toners as indicated in Fig. 10. Starting from a latex suspension, larger aggregates of well defined size and narrow size distribution form under moderate shear with various additives. The final aggregates are heated above their glass transition temperature to coalesce, followed by

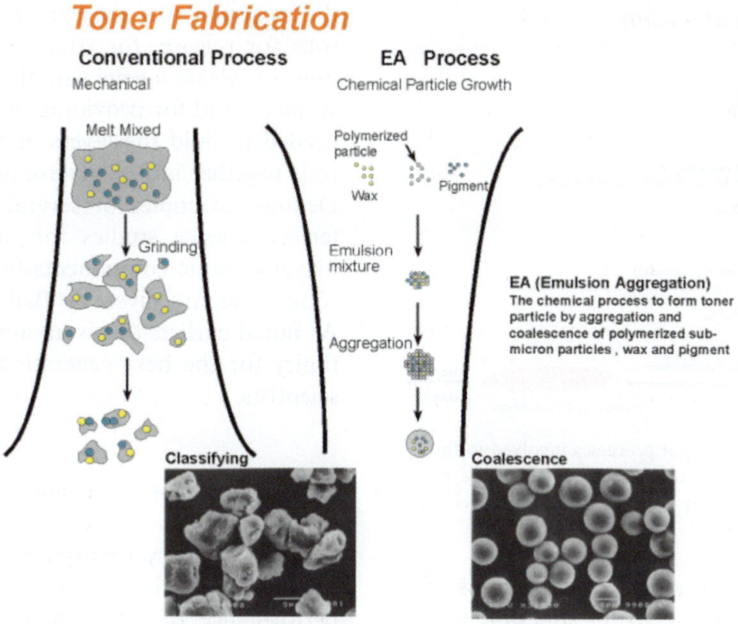

Fig. 10. Schematic diagram comparing fabrication of conventional toner to that of chemical toner. The various steps of both fabrication processes are indicated in the figure. The differences in the morphologies of the resulting toner particles are evident in the pictures at the bottom of the figure.

washing and drying for use as toners. The particle properties, as well as the sizes and size distributions, depend critically on the aggregation process, as controlled by surfactants and additives that modify their interfacial properties.

The observed dependencies of the final aggregate size on process parameters such as initial latex particle size, ionic strength, and ratio of cationic to anionic surfactants can be explained by taking into account the kinetics of aggregation, and using a charged liquid droplet model for the aggregate energetics [16]. The morphology of EA toner particles depends in a complex way on the conditions used during the coalescence of the latex aggregates. The coalescence portion of the EA toner making process involves converting the tightly bound aggregates of latex and pigments into particles that can be employed in the xerographic process. The coalescence of latex particles in a drying paint film is perhaps the phenomenon having most in common with the EA latex coalescence process. Primary aggregates approximately 1 μm in diameter can contain between 100 and 200 latex particles for a latex 200 nm in diameter. A toner-sized secondary aggregate of diameter 5 μm would contain about 100 primary aggregates or 15 000 latex particles.

The advantages of the EA process to make chemical toner are small toner size, narrow particle size distributions, tunable morphology, wide materials design latitude, and low cost. The particles are finer and more uniform in size and shape compared with the crushed multi-component toner particles. These qualities are important for obtaining more effective development and image transfer. Therefore they yield superior image quality in the xerographic process. The controlled and adjustable modification of the interfacial properties of these specially designed materials using surfactants and additives makes this quality improvement possible.

5.3. Fuser components

Roll fusing involves the melting, coalescence, and spreading of toner particles, as well as the

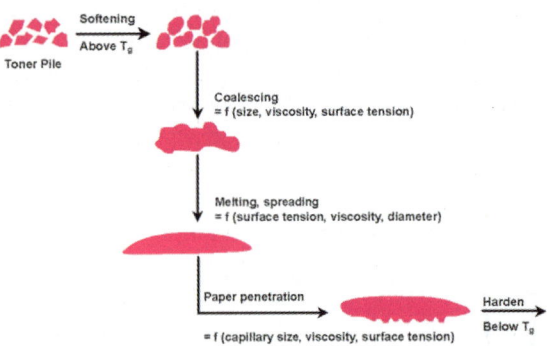

Fig. 11. Diagram of the physical processes involved in fusing. These are shown as they occur after the toner on the paper enters the fuser as indicated in Fig. 8 (adapted from Ref. [2]).

adhesion of toner to paper and the cohesive strength of toner, relative to its adhesion to the fuser roll as indicated in Fig. 11 [2]. The important variables in the melting, coalescence, and spreading of toners are temperature, time, pressure, viscosity, particle size, and the surface tension of toner [2,9,10,17]. In particular, adhesion of an image to a substrate depends on the surface energies, the contact area between the surfaces, the toner modulus, the hardness and the brittleness of the toner. In general, coalescence and toner fusion is increased proportional to the pressure, surface tension and dwell time and inversely proportional to the viscosity and particle size. One important variable affecting coalescence is the melt viscoelasticity. External additives on the fuser roll, such as silicone oil, assist in releasing the toner. These additives can, however, increase the coalescence temperature in some toners, which can cause toner aggregation and hence grainy images: not a desirable outcome. The effect of additives such as waxes in the toner can also enhance toner release from the fuser roll and reduce or eliminate the necessity of using oil in the fusing subsystem. In general, materials screening for fusing involves initially selecting the viscoelastic properties of toner resins for optimum fusing performance, and then studying the effects of various additives such as surface charge control agents on fusing. Interface science provides the foundation for designing the treatments of the surface layers of the fuser rolls for release, for studying the chemical reactions of release agents with the fuser roll and toner surfaces, and for providing the specialty adhesives needed to hold the layers of the fuser and donor rolls together under adverse operating conditions. Detailed examples of several applications of interface science studies for the design of multiplayer organic components for fuser and transfer subsystems are given by Badesha and Swift [18]. As noted earlier, this is an arena of major opportunity for the next generation of applied surface scientists.

6. Semiconductor processing connections

While most xerographic components involve organic composites, the revolution in the size and performance of semiconductors also is affecting xerographic marking engines. The most important of these influences are the use of laser diodes in the exposure subsystem, the use of ink jet arrays to mark directly on the output media, and the increased use of active feedback in xerographic process control. In this section we explore some of the impacts of the semiconductor process revolution on xerographic marking engines.

6.1. Semiconductor diode lasers

In high-speed printing tradeoffs are required between high resolution, high speed, and manufacturing tolerances. Fig. 4 shows a raster output scanning illuminator subsystem with a polygon scanner photoreceptor and laser system along with scanning and correction optics. High resolution requires large polygon facets, on the other hand, whereas high speed is advantaged by having many facets. The use of multiple beam diode lasers is able to circumvent part of this tradeoff by allowing higher speeds to be attained with fewer facets. In addition, as higher resolution becomes more important, it is advantageous to go to shorter wavelengths, so that higher resolution can be achieved with small polygon facets and good depth of focus. Alternatively, short wavelengths can be used at today's resolution of 600 dpi to increase the speed of a printing system by using a large number of smaller polygon facets. A Moore's law type assessment reveals that print resolution for office printers is doubling about every 12 years and that 1200 dpi is upon us. A blue laser print engine operating at a resolution 1200 dpi enables offset quality xerographic printing at high speeds provided that a photoreceptor with suitable spectral sensitivity becomes available commercially.

III–V compound semiconductor materials for semiconductor diode lasers are prepared by metal organic chemical vapor deposition (MOCVD). In this technique the group 3 species (aluminum, gallium, indium) are transported into a growth chamber as the vapors of organo-metallic precursors e.g. trimethylgallium. The group 5 species, however, are transported as hydrides for example arsine, phosphine, or ammonia. The development of this technique was a triumph of surface science in the 1970s [19,20].

Laser diodes are fabricated from materials of exceptional structural and optoelectronic perfection. Structural defects (such as dislocations) or impurities can seriously degrade the luminescence efficiency of semiconductor materials. Thus, the materials from which laser diodes are made start with single crystal substrates to ensure that the deposited films are as structurally perfect as possible. As indicated in Fig. 12, red and near-infrared lasers are grown on gallium arsenide single crystal substrate wafers, fiber optic devices are deposited over indium phosphate substrates, and nitride lasers are deposited on sapphire substrates. Nitride laser structures differ from infrared or red laser diodes in a number of important respects. IR and red lasers, used today in xerographic marking engines, are grown perfectly lattice matched, on gallium arsenide substrates [21]. Therefore the dislocation density is very small, representing the defect density of the substrate seed crystal. On the other hand, nitride laser structures are most often deposited on sapphire substrates, corresponding to a 14% lattice mis-match, which produces an enormous defect density [22]. Such a large concentration of defects would render IR or red emitting materials optically inactive. The nitrides apparently do not suffer from these dislocations, however, a surprising fact that is still under investigation. The ongoing development of nitride lasers is a frontier in modern materials and interface science, which directly impacts xerographic marking technology [23].

6.2. *Microelectromechanical systems (MEMS) markers*

The field of MEMS, deals with micron scale machining, mechanical functionality, and low

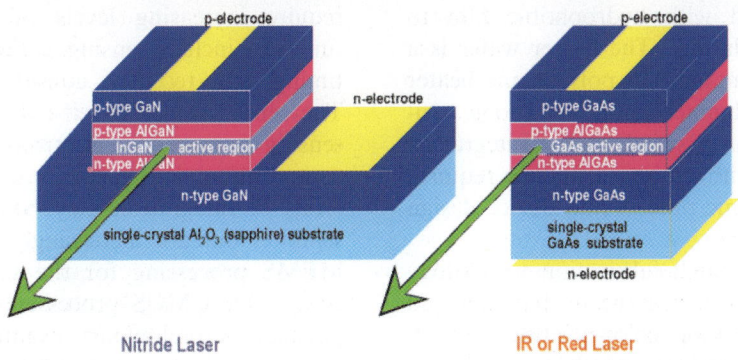

Fig. 12. Schematic diagram of the MOCVD fabricated blue (left-hand panel) and red (right-hand panel) laser diodes for xerographic exposure systems. The emission of light from the active region is indicated by a green arrow. The different layers of these diode lasers are indicated. The geometry is simplified relative to that of a production laser. The structures are comprized of epitaxial layers of single-crystal material. Different active layers are required to generate light in different spectral regions: GaAs for IR and InGaN for blue. Different substrates are required in order to get sufficiently accurate lattice matching for adequate epitaxial growth.

manufacturing cost for a large number of identical parts, as well as enabling high-bandwidth mechanical behavior [24]. An example is a thermal ink jet (TIJ) head that integrates electronics and sensor actuator functionality, and uses polymers in a large area application. MEMS technology already has exerted a profound impact on the printing industry by virtue of its use to generate an unprecedented cost/performance ratio for TIJ products that currently dominate the desktop color-printing market [25].

The two main performance qualifiers for printing products are image quality and print speed. One approach to increasing print speed is through an increasing number ejectors per print head, economically enabled through a high degree of modularity and integration, including on-chip addressing logic. In terms of image quality, a high integration level enables ever-increasing image resolution. The implementation of this strategy relies on MEMS technology to combine fluid flow pathways, ejector nozzles, fluid reservoirs, heater elements, MOS power drivers and addressing logic onto one silicon die. In the Xerox process [26] TIJ die are fabricated by wafer level bonding of a bulk micromachined silicon "channel wafer" to an MOS "heater wafer" with an intermediate polyimide spacer layer. The wafer bonding is based on a thermal cure adhesive process. The channel wafer contains the fluid flow channels and local link reservoirs, micromachined using wet anisotropic etching of crystalline silicon. The front face is subsequently coated with hydrophobic film to avoid flooding by the ink. The heater wafer is a MOS wafer that contains 128 polysilicone heater elements (one per channel), addressing logic, driver circuitry and power switches. The integration of these elements with on-board fluidics requires several microelectronic process and device design tradeoffs.

The adoption of batch fabrication technology to produce disposable print heads has changed the rules of the desktop color-printing market, launching low end printing products onto cost and performance trajectories that are analogous to semiconductor industry trajectories. The technology consists of a unique blend of large-scale integration (LSI), microfluidics and thermodynamics. This integration requires design compromises to be made across various energy domains (e.g., electrical, thermal, mechanical, chemical), and various technologies (e.g., MOS, micromachining, ink, and paper). Yet the efficiency of the batch fabrication techniques is sufficiently high for disposable print heads to be economically viable. The design, manufacture and commercial introduction of such disposable heads illustrate well the complexity of the processes and devices to which modern applied surface scientists are making major contributions.

6.3. Sensing and control

Even with all the emphasis on reproducibility of the process and the materials components, external disturbances still cause the quality of the printed output to vary. Variations in temperature, humidity, toner consumption, and media composition all lead to corresponding variations in the appearance of the output prints. For monochrome printing, image quality stabilization was resolved using simple algorithms and a few sensors. In color printing the subsystem latitudes that produce predictable color images are more restricted, and output variations much less tolerable. The products of semiconductor processing have been utilized to mitigate this problem in two specific ways.

Firstly, the sensors that are used to detect variations in the intermediate and final process outputs require increasing levels of integration. Their functions include sensing, calibration, networking, timing, and recently, considerable computation. The sensor device is self-contained including the sensing element, drive electronics, communications, computation, and memory all coresident. These levels of integration may be obtained using assemblies but are increasingly implemented using MEMS processing for the sensing element and compatible CMOS processes for the other components. A preliminary example of this type of integration is the spectrophotometer produced by MicroParts [27]. It has an optical input and integrates a MEMS fabricated diffraction grating coupled to a CCD array for spectral discrimination and detection. The output of the sensor is an array of numbers corresponding to the visible spectrum of the light at the optical input.

Secondly, with inexpensive sensors providing frequent, accurate, and calibrated readings and the performance of semiconductor chips constantly increasing, sophisticated process control algorithms running in standard commercial semiconductor processors become feasible. Ten years ago with 8 bit devices even elementary arithmetic operations (e.g. floating point division) occupied 10s of milliseconds (ms). The dwell time for sensed features on the photoreceptor is less than 5 ms for even moderate speed printers. Thus, the feature could pass under the sensor without being detected because the processor was busy doing division. Real-time sensing and computation was impossible. Now matrix inversion, Fast Fourier Transform, and other complex computations may be performed in less than 1 ms and may be interrupted by real-time events in multi-tasking environments. This capability, thanks to the results of Moore's law, has been combined with the last three decades of advances in automatic control theory to provide both the algorithms and the computational environment to compute process adjustments in real time. These adjustments alter the behavior of the xerographic system to provide disturbance rejection resulting in consistent, predictable image quality from a very complex color-printing system [28].

This trend of using MEMS devices as sensors, actuators, and mechanically functional units plus the application of the computational consequences of Moore's law will continue with the result that more commercial semiconductor devices will be integrated into future photocopy machines and printers.

7. Direct marking and the future of xerography

Perhaps the greatest competitive challenge to xerography in the digital era is its replacement by smaller, simpler and hence less expensive direct marking devices. Direct marking involves replacing the charging, exposure and development subsystems shown in Fig. 1 with a direct marking head that writes on a consumable such as paper, or an intermediate belt or drum from which the image is transferred to the consumable medium. Commercially successful direct marking engines have fewer parts, less weight, smaller footprints, and simpler architectural design than their xerographic counterparts. This technology also offers a potential cost advantage of direct marking inks with respect to xerographic developer, one for achieving liquid ink/lithographic document appearance and image quality, and the prospect of easier maintenance with fewer replacement parts. For the time being, however, they operate at slower speeds and with generally lower image quality than xerographic printers. One approach to improving the image quality and extending the range of media for both xerographic and direct marking is by exploring marking engines in which the marker produces an image on an intermediate belt or drum from which it is subsequently transferred and fused to the final medium in a single "transfuse" step. In the Oce printing scheme [29], an intermediate transfer medium can be preheated to the toner melting temperature at the print fuser, and in one belt revolution the page panel scrolls through the high-pressure fuser, handing the image off to the consumable media (e.g., paper or transparency). In this technology, the imaging of the toner exiting the direct marking head is always carried out on the intermediate transfer medium eliminating substrate variability (e.g., different papers, transparencies). After the toner is imaged on the intermediate transfer medium, it is fused and transferred at the same time to the substrate. Hence the substrates can encompass a wide range of paper weights, finishes, and coatings. This type of subsystem, in which transfer and fusing are thus combined, is a current frontier in xerographic marking. The belt can be heated because it does not come into contact with a temperature-sensitive element like a photoreceptor. An addition, the subsystem can deliver a very thin layer of a release agent to the belt [30] before it comes in contact with the imaging material (toner), enabling efficient release of the toner from the intermediate transfer medium, which can be in the form of an endless belt [29]. Obviously, the design of the toner, the intermediate medium and the transfuse subsystem process depend sensitively on the surface and interface properties of the toner and medium, and their dependence on temperature. It is evident, therefore, that the solution of surface and interface science problems plays a major role in the design and implementation of such transfuse subsystems.

Historically, xerography rose to commercial prominence because of its unique suitability to make black and white (monochrome) copies via light-lens exposure. The digital era has fundamentally changed its competitive positioning. First, color, unachievable at adequate image quality with light–lens exposure, is enabled by the use of digital scanners and color-correction software. Second, the use of digital "originals" in printers and copiers has obviated the competitive advantages of light-lens xerography and rendered direct marking engines competitive with xerographic marking engines. At the desktop, where acquisition cost is king, slow speeds are tolerable, and image quality can be "good enough", color TIJ marking directly on paper have captured the market. In the high volume, multiple-copy markets (e.g., magazines, newspapers) lithography has been king for many years. Thus, color xerography is sandwiched in the middle, advantaged relative to each by special features (e.g., speed and image quality relative to TIJ and short-run cost and variable data capability relative to lithography) but fighting for its share of the short-run digital color-printer/copier market.

The winners and losers in this competition will be determined largely by the speed and efficacy of the solution of sophisticated surface and interface problems associated with the toners or inks and their interactions with both the image bearing media and the various surfaces with which they come in contact during the marking process. Long life, low cost and high image quality require sophisticated consumables that reduce the complexity of the marking engine by assuming added complexity in the materials themselves. In each market space the race will be won by the marking technology that yields solutions most quickly to the shortfalls in areas of high customer value. For direct marking devices, key issues include image quality, image permanence, and range of output media. For xerography they are cost, media latitude, and in some markets image appearance relative to lithography. For lithography they are short-run cost, automation of setup and maintenance of high image quality, and suitability for office environments. Progress on all of these rests on surface and interface science because in all cases the results are dominated by the consequences of the interactions of the ink or toner with the surfaces that they touch during the course of the marking process. For direct marking, the ink–media interaction is the key to image quality and permanence. For xerography, lowering the cost and improving the reliability of the process requires ever more sophisticated multi-component photoreceptor, toner, carrier and fuser materials packages. For lithography, process control via sensors and electronics and effluent control are key issues. All of these are dominated by media-materials interactions driven by surface and interface phenomena. Thus, we can confidently anticipate that in the future, as in the past [1,2, 6–10], the solution of challenging applied problems in surface and interface science by talented and creative surface scientists, will enable the inexorable march of color-marking engines, xerographic as well as others, to better, faster, cheaper.

8. Synopsis

The amazing development of the xerographic copier and printer industries since the 1950s has been a direct consequence of sustained investment in basic research underlying the Xerographic process and associated materials [3]. It resulted from the diligent, systematic application of physics, chemistry and surface science, leading to wealth for investors and excellent careers, sometimes even recognition and fame, for the scientists and engineers who pioneered the major advances. The advent of the digital, networked era has changed the calculus relative to the stand-alone light–lens copier era (1955–85), rendering other marking technologies more competitive in certain market segments. Commercial success in this new era depends even more upon exploitation of the fruits of research in surface and interface science, than in the copier era. There is a continuing big payoff for applications of surface and interface science in the printer/copier industry and a generous supply of as yet unsolved problems to challenge the next generation of creative applied scientists and engineers.

References

[1] D.A. Hays, Xerography, in: G.L. Trigg (Ed.), Encyclopedia of Applied Physics, vol. 23, Wiley–VCH, Weinheim, 1998, p. 541.
[2] D.M. Pai, B.E. Springett, Physics of electrophotography, Rev. Mod. Phys. 65 (1993) 163.
[3] J.S. Weston (Ed.), Xerographic Process Research, America's Basic Research: Prosperity Through Discovery, Committee for Economic Development, New York, 1998, p. 73.
[4] J.E. Shane, USA and European Marking Materials Forecast and Trends, CAP Ventures, Norwell, 2000.
[5] Science and Engineering Indicators, National Science Board, Washington, DC, 1998, pp. 4–10.
[6] L.B. Schein, Electrophotography and Development Physics, second ed., Springer, Berlin, 1992.
[7] M.E. Scharfe, D.M. Pai, R.J. Gruber, Electrophotography, in: J. Sturge, V. Walworth, A. Shepp (Eds.), Imaging Processes and Materials, eighth ed., Van Nostrand Reinhold, New York, 1989, p. 135.
[8] R.J. Gruber, P.C. Julien, Dry toner technology, in: A.S. Diamond (Ed.), Handbook of Imaging Materials, second ed., Marcel-Dekker, New York, 1991, p. 159.
[9] D. Battat, Soft roll fusing process elements, in: J. Bares (Ed.), Color Hard Copy and Graphic Arts III, Proc. SPIE, vol. 2171, 1994, p. 157.
[10] D.A. Seanor, The role of materials science in the fixing of xerographic images, in: A. De Clerg, F. Frey, S. Ohno (Eds.), Proceedings of the International Congress on Imaging Science, vol. 2, International Committee on the Science of Photography, Antwerp, 1998, p. 149.
[11] H.R. Kang, Digital Color Halftoning, Society of Photo-optical Instrumentation Engineers, Bellingham, 1999.
[12] B. Keoshkerian, G. Liebermann, C.-K. Hsiao, J.D. Mayo, D. Murti, S.J. Gardner, Hydroxygallium phthalocyanine photoconductive imaging members, US Patent 5,563,261, Oct. 8, 1996.
[13] R.D. Patel, M.A. Hopper, P.F. Smith, B.S. Ong, A.K. Chen, Wax containing colorants, US Patent 5,994,020, Nov. 30, 1999.
[14] H. Shimomura, Y. Hasegawa, H. Serizawa, M. Masatoshi, Toners for developing electrostatic image, US Patent 5,591,556, Jan. 7, 1997.
[15] M. Koyama, K. Hayashi, T. Kikuchi, H. Yamazaki, Toner for developing an electrostatic latent image, Developer and a method of producing an image using the toner, US Patent 5,830,617, Nov. 3, 1998.
[16] N. Bohr, J.A. Wheeler, Mechanism of nuclear fission, Phys. Rev. 56 (1939) 426.
[17] D.A. Seanor, Fusing against bare metal surfaces, Photo. Sci. Eng. 22 (1978) 240.
[18] S.S. Badesha, J.A. Swift, Practical surfaces: beyond the wheel, Surf. Sci. 500 (2002) 1024.
[19] A.Y. Cho, Preparation and properties of GaAs devices by molecular-beam epitaxy, J. Electrochem. Soc. 122 (1975) C262.
[20] J.R. Arthur, Molecular beam epitaxy, Surf. Sci. 500 (2002) 189.
[21] R.S. Geels, D.F. Welch, D.R. Scifres, D.P. Bour, D.W. Treat, R.D. Bringans, Dual spot visible laser diodes, Electron. Lett. 28 (1992) 1460.
[22] S.D. Lester, F.A. Ponce, M.G. Craford, D.A. Steigerwald, High dislocation densities in high efficiency GaN-based light-emitting diodes, Appl. Phys. Lett. 66 (1995) 1249.
[23] R.D. Bringans, Application of blue diode lasers to printing, Mat. Res. Soc. Symp. Proc. 482 (1998) 1203.
[24] E. Peeters, Micro electro mechanical systems: pyrite or pure gold?, APS News 8 (1999) 5.
[25] M.P. O'Horo, N.V. Deshpande, D.J. Drake, Drop generation processes in TIJ printheads, in: Proceedings of the 10th International Congress on Advances in Non-Impact Printing Technologies XXX, Society for Imaging Science and Technology, 1994, p. 418.
[26] E. Peeters, S. Verdonckt-Vandebroek, Thermal ink jet technology, IEEE Circ. Dev. 13 (1997) 19.
[27] VIS-LIGA-Spectrophotometer for Analysis and Color Measurement, STEAG MicroParts GmbH, Havert 7, Dortmund D-44227, Germany, 2000.
[28] T.E. Thieret, T.A. Henderson, M.A. Butler, Method and control system architecture for controlling tone reproduction in a printing device, US Patent 5,471,313, Nov. 28, 1995.
[29] H.A.M. Loonen, M. Miedema, B. Schoustra, J.A. Verbundt, E.H.A.M. Smit, Toner image transfer apparatus including intermediate transfer medium, US Patent 5,361,126, Nov. 1, 1994.
[30] J.S. Berkes, J.S. Chambers, Brush for applying release agent to intermediate transfer member, US Patent 5,434,657, Jul. 18, 1995.

 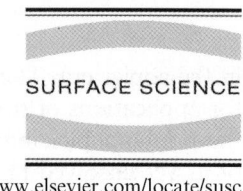

Practical surfaces: beyond the wheel

Santokh S. Badesha, Joseph A. Swift *

Xerox Corporation, Wilson Center for Research and Technology, 800 Phillips Road, (0114-39D), Webster, NY 14580, USA

Received 13 September 2000; accepted for publication 12 April 2001

Abstract

In this paper, we explore the challenges of designing practical surfaces for the 21st century. By building upon examples from different industries, we derive a number of useful tenets intended to guide those who face these challenges through the maze of technical and economic issues that they may confront. Given that future opportunities abound for surfaces having new and better properties, we describe some emerging materials systems that are likely to yield these improvements. © 2001 Elsevier Science B.V. All rights reserved.

Keywords: Surface structure, morphology, roughness, and topography

1. Introduction

As the dawn of the new millennium breaks, we find the contemporary surface scientist confronted, if not overwhelmed, by a complex array of challenges. Due to the ever-increasing interdependency amongst science and technology (S&T) and economics, these challenges involve economic and social as much as technical issues. Thus, we are witnessing an exciting period, in which the escalating intensity of global competition creates enormous pressure upon firms, universities, S&T institutes, and developed nations to bring to fruition profitable results from their R&D investments. These factors are stimulating R&D activity to historically high levels, which may eventually benefit all of mankind. In spite of the generally favorable climate for S&T, not every technology-intensive establishment will flourish in this millennium. For such organizations to succeed in global competition, they need to develop the capability to deploy technology effectively to maintain a competitive edge through product differentiation, production efficiency, and responsiveness to market interests [1].

Many firms are recognizing this, although some have done so too late [2]. The list of firms that have gone out of business, due to their inability to compete effectively, or that have been acquired under "fire sale" conditions, is tragically long. In the air transportation sector, for example, we witnessed the demise of many regional carriers, such as Eastern, Mohawk, and Peoples airlines. During the same period, start ups such as Southwest and JetBlue are finding innovative ways to fill the voids created by disruptions in this industry. A disturbing new paradigm is being forged in this country's bankruptcy courts involving new dot.com companies. The courts are ruling against

---

* Corresponding author. Tel.: +1-716-422-4910; fax: +1-716-265-5837.
  E-mail address: jswift@crt.xerox.com (J.A. Swift).

a growing number of the under-funded and asset-poor dot.coms, which were born during the past few years and are finding that they cannot survive into adolescence. The courts are forcing them to go directly into chapter 7, namely dissolution, and not permitting them to reorganize (i.e. under chapter 11) because the courts believe there is no hope for them. In other sectors, well-founded and well-managed firms have had or are presently experiencing setbacks. For example, Sears and Montgomery Ward, once front runners in retail sales have lost substantial market share to discount retailers, such as Wal Mart and Target that have employed aggressive marketing, supplemented by modern technology, to win market share. Bell labs, once the pinnacle of US research establishments, has been spun-off into a now struggling commercial venture, Lucent Technologies. The pre-eminence of US steel producers, Bethlehem Steel Corp. and US Steel Corp., has been upset by firms such as Nucor, which use a more efficient minimill technology. Firestone is being financially threatened by a major product failure whose recall may exceed $1 billion. Xerox, the world's leader in plain paper printing and copying, has had difficulties in bringing its best ideas to market and has been accused of "fumbling its future" [3]. Each of these distressing examples can be attributed, in one way or another, to the ineffective adaptation of S&T. Each indicates that maximal return has not been achieved from the R&D investments made by the establishment.

Moreover, universities as well as government laboratories are becoming more actively involved in driving products of those sectors' R&D spending beyond promising raw ideas and into commercial reality. Zacks [4] reports that university inventions in 1998 supported 280,000 jobs and generated an estimated $33 billion in economic activity. This behavior is a new source of competition for, and thereby a new threat to, the traditional commercial firm. Alternately, depending upon the type of linkages between the commercial and non-commercial labs, the trend may represent a wellspring of new ideas and opportunities.

Thus, we are observing an increasingly tense, but symbiotic, interdependency amongst science (viz. the generation of knowledge), technology (viz. the practical application of that knowledge), and innovation (viz. the productive use and profitable adaptation of that knowledge). This suggests that the technology intensive organizations that survive in the next millennium will be those that can effectively innovate in spite of the ever-changing competitive climate. Those to survive will be amongst the best at moving technology from the laboratory and into the market place.

One of the world's greatest inventors, Albert Einstein, is credited with saying that we can never solve our significant problems from the same level of thinking we were at when we created the problem. The objectives of this chapter are to help elevate the level of thinking of those confronting these modern challenges and to provide the reader with some useful tenets for developing the practical surfaces of the future.

2. Functional surfaces

Surfaces capable of performing a function are considered mono-functional. Surfaces that perform more than one function are described as multi-functional. At first glance, these statements appear to be obvious and perhaps trivial. Nevertheless, the distinctions made between mono- and multi-functional set the stage for discussions of how surfaces can be designed to achieve a variety of different and often conflicting functions.

2.1. Functionality

The degree of functionality characteristic of any surface depends upon the level or degree to which the function exists or upon the number of different functions that surface performs. The aluminum oxide and silicon dioxide mixture used historically as the catalyst cracker in gasoline production [5] and the metal fluoride antireflective coatings used on eye glasses [6] are good examples of single function surfaces. The functional capability, or functionality, of these surfaces is measured by how effectively the surface performs the cited function.

In contrast, the modern automobile tire is a classic example of a multi-functional surface, which accounts for a large portion of the overall performance. The tire's surface provides traction and load bearing, heat and static energy dissipation, as well as tear and puncture resistance. It also contributes to aesthetics, tolerates contamination, endures UV exposure, adapts to high and low temperatures as well as dry and wet environments, carries warning, brand identity, and other labels, self-renews, and can even be run flat in some cases. The degree of functionality of a multi-functional surface can be measured by the number of functions the surface performs and how effectively those functions are performed. In consideration of the commercial importance and societal impacts of the often ignored but frequently kicked tire, it is hard to conceive of a more ideal product, whose surface must contend with the aforementioned range of mechanical, thermal, electric, chemical, and environmental stresses. Its ability to thrive and to function, nearly flawlessly, in virtually every earthly environment are factors that account for the tire representing one of technology's most amazing multi-functional practical surfaces. Estimates shown in Fig. 1 reveal that the total R&D investment over the past five years to develop the modern auto tire is in excess of $9 billion. The tire ranks high on any listing of expensive, albeit practical surfaces.

2.2. Complexity

A surface can be considered complex if it is made up of two or more parts. Complexity can be quantified by the number of component ingredients, by the number of constituent phases, regions, layers, or domains, by the number, or intricacy, of manufacturing processes used to make that surface, or, by a combination of these factors. The tire also is an excellent example of a highly complex composite. Its surface is comprised of natural and synthetic rubbers, reinforcing fillers, antidegradants, curatives, processing aides, and adhesion promoters [7]. The bulk of a contemporary, all-season, radial tire can contain 4 lb of eight types of natural rubber, 5 lb of eight types of carbon black, 1 lb of steel cord, 1 lb of polyester and nylon, and 3 lb of 40 different chemicals, waxes, pigments, oils, and the like. Much of these are present within the tire's surfaces. Below the surfaces reside a large number of layers and interlayer interfaces. Further adding to the complexity, there can be five, or more, manufacturing processes used to make a single tire. These typically involve mixing and blending, layer assembly, mold forming, curing, and inspecting. The number of potential interactions amongst the manufacturing processes and the compositions that must be addressed in order for the tire to deliver all of the end functions further complicates things. The tire epitomizes high technological complexity.

2.3. Complexity—the dilemma

The selection and development of technologies to be used in future tires present a practical, but potentially serious, challenge to tire manufacturers. Such challenges extend to every manufacturer, but particularly to those who manufacture technology-rich products. Specifically, the embedded technologies must not only assure product competitiveness through features and pricing, but also must render the product intrinsically safe throughout its life. For example, pre-mature failure of one or more of the constituents of a tire can cause catastrophic failure, which in the case of automobile tires results in life threatening situations. When failure of this sort occurs within a large population of tires (as was the case that led to recall of more than 14 million of Firestone's 500 radials in 1978 and more recently with 6.5 million of the ATX and ATXII lines of tires in 2000), it

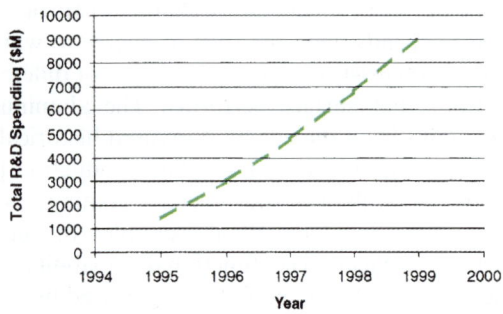

Fig. 1. Graph of aggregate R&D spending by tire firms during the period from 1995 to 1999 indicating that approximately $9 billion has been invested in developing the present generation of automobile tire.

can not only threaten the safety of the user but also the reputation, if not the survival, of the manufacturer. The challenge, then, is to choose product technologies that enable competitive performance while resulting in adequate reliability and safety.

Since a high degree of product complexity may lead to, but not necessarily cause, a higher probability of failure, the dilemma facing the new product technologist is to judge when adding complexity to achieve higher functionality is warranted and the technical risk is acceptable. Most high tech commercial firms utilize formal technology management processes that embody a progression of checks and balances to assure that immature technologies do not get to the market too soon and risk falling short of the expected quality, or are cause of a liability problem.

3. The functionality–complexity model

The functionality and complexity factors can be linked to form the framework indicated in Fig. 2. Here the model is used to illustrate the progress of the tire from the earliest era, where tires were made of solid rubber (essentially, natural rubber and carbon black) and functioned solely to provide load bearing, drive and breaking traction, and little else, to the present. This early era auto tire had low functionality and low complexity and therefore can be positioned in the lower left corner within the framework.

Several decades later, the tire's appearances and number of functions had changed significantly. In addition to the basic capabilities, the surface of the pneumatic ("balloon") tire of the late 1940 and 1950s wore a distinctive whitewall, carried information labels, cushioned the ride better, and provided improved traction in adverse weather. To achieve this functional increase, the design, composition, and methods for manufacture became more complex. Perhaps, the most renowned difference was incorporation of an inner tube, which allowed a layer of air to support the outer rubber shell of this two-layer tire. Typical life of the tube tire was about 10,000–15,000 miles. This tube-type design enabled safe and comfortable high-speed auto travel and was a substantial factor in the growth of the auto industry. Tires of this generation were more complex and served a greater number of functions than the earliest tires. Therefore, they appear in a central position in our framework.

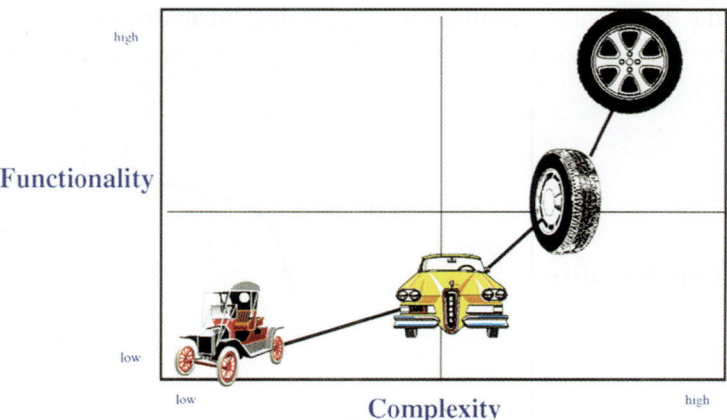

Fig. 2. The functionality–complexity framework: mapping the progress of the tire over its history. Functionality is quantified in terms of the level or number of functions the surface performs. Complexity can be measured in terms of the number of ingredients or components used to enable the function(s). The earliest tires as shown on the automobile symbol are positioned in the lower left quadrant since they possessed only a few features and were made of simple construction. Each of the three follow-on generations of tires was more complex and richer in features. To reflect this trend, each successive generation is positioned higher and towards the right in this framework. The auto and tire symbols are arbitrary and are intended only to indicate the era of their origin.

The more complex, biased-ply tire was developed to respond to the explosive industry growth and the then-contemporary requirements of the 1960s autos. Featuring a single tube-less design, longer life, lower rolling friction and a wider range of available sizes, these attributes of this generation tire eventually made the tube tires that were used on passenger cars obsolete. The bias-ply tire can be appropriately positioned slightly to the right and above the tube tire in the framework.

The tire design representing today's leading-edge technology is best exemplified, perhaps, by Dayton's Premium GT Touring™ tire. It is a complex, multi-belt radial design with outstanding long life, employing leading-edge antinoise technology. Its functional surface features wide channeling grooves that force water away from tread elements to assure positive contact with pavement under a wide range of environmental conditions. Many of the component materials, such as Kevlar™ and steel fiber reinforcements as well as extreme-stress resistant composites to enable run-flat capabilities can be easily regarded as "leading edge". This tire, along with all similarly featured ones that comprise this generation, is positioned within the upper right quadrant of the functionality–complexity framework.

At this point, we have mapped major developments in the history of the tire through three of the four quadrants of the functionality–complexity framework. We observe progress through advancement of the various technology states from the lower left region of the framework (i.e. earliest state, low in complexity and functionality) across the center-lower region and into the upper right (i.e. latest technology, most complex and highest functionality). This movement is stereotypical of many modern technologies. For example, the personal computer (PC), digital communications, flat screen television, and xerography can each be characterized by a continuous escalation in functionality and complexity. The bounds on functionality and complexity are limited only by future advancements. The framework can be viewed to be evergreen since it is not constrained by any obvious boundaries or any one technology.

A common characteristic of many technologies is that progress towards achieving given levels of performance often begins slowly, gaining momentum to eventually reach saturation. This S-curve behavior, as depicted in Fig. 3, is a well-recognized characteristic [1,2,8]. Various embodiments of it have been used in different ways for temporal mapping of technology progress, maturity, and/or diffusion. It has also been adapted as a technology forecasting methodology [9].

Tire features discussed within the context of the functionality–complexity framework can be viewed in S-curve methodology. One or more individual attributes, such as handling, ride comfort,

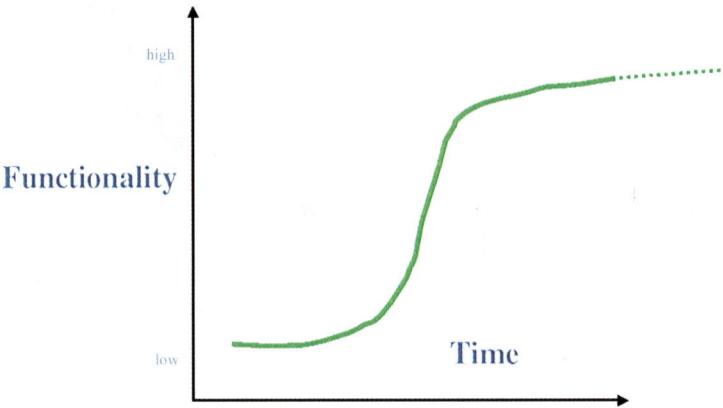

Fig. 3. Diagram of the technology S-curve illustrating the temporal growth of a functional capability. The rate of improvement of a technological feature (such as a tire's ride comfort) is slow at first, then transitions quite rapidly to a higher level before establishing a new performance level.

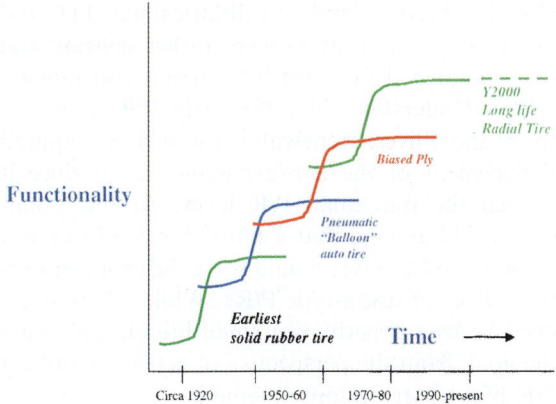

Fig. 4. Integration of the major tire technologies into a series of S-curves depicting the long life span of advancements in this industry. Much, if not all, of the functional features found in an early generation of tire establish the basis for the higher degree of functionality developed in subsequent generations.

or life, or, the sum of all of the attributes together can be expressed as one or more S-curve that span a period. Fig. 4 illustrates a series of technology S-curves for the four generations of automobile tire where a functional feature, such as handling ability or wet handling, represents the functionality of interest. The 1920s era tires constructed from narrow, solid designs contributed minimally to the handling of the car (note that this level of functionality is illustrated by the lower left S-curve of Fig. 4). By Y2K, high performance radial tires contributed significantly to overall handling performance and enabled safe, high-speed travel under all weather conditions. This higher level of functionality is expressed by the uppermost S-curve in the figure. From Fig. 4, we see the tendency of successive generations of technology to render obsolete the earlier generation, and, that the life cycle of the underlying technologies developed in this "mature" industry appears to last for nearly a decade. Since replacement of the older generation with the newer one does not occur instantaneously, we see overlap between any two successive generations, which can last for several years.

Electronics and software used in PCs and other computers fall at the other end of the technology life spectrum where it is common for competitors to introduce new products at a high frequency thereby driving up the rate at which these products become obsolete. Regardless of the expected technology life, a principle-based approach can be used to develop the next generation technologies that eventually make obsolete the present generation. In the following sections, we examine the genesis of, and rationale behind, some new surface technologies and provide case examples from the reprographics industry to illustrate this theme.

4. Crafting of the functional surface

The discussion will now focus upon polymeric functional surfaces where design of the practical surface begins with selecting a suitable polymer phase, which in the case of the least complex system may comprise the entire structure. In progressively more complex systems, such as mixtures, blends, and composites/nanocomposites, the polymer serves as the host, matrix, or binder phase and may, or may not, be the surface-dominate component. However, in each of these multiple-component systems, the host polymer plays a critically important role in the resultant surface's functionality and consequently requires particular attention.

4.1. Knowing the problem

The first step in solving any problem involving change of, or improvement to, a surface is defining the problem precisely. Often, this is not a trivial matter.

Analogous to the link between the auto industry and the tire, the reprographics industry is dependent upon many unique surfaces that have been crafted to perform the highly specialized and often-complex functions used to produce xerographic copies. The books by Williams [10], Scharfe [11], and Hays [12] give not only concise overviews of this fascinating technology, but also summary accounts of the relevant literature. The reader is also encouraged to refer to the accompanying paper by Duke et al. (titled "The surface science of xerography") and associated figures, which provides more detailed descriptions of the devices mentioned here.

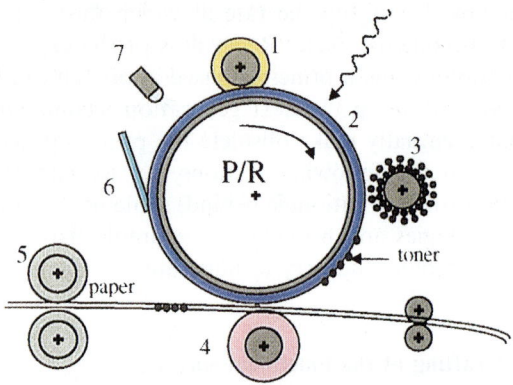

Fig. 5. The major elements of the xerographic process—the P/R receives charge from the BCR (1) and upon exposure to light (2) generates and transports electron holes to selectively discharge corresponding to the image being reproduced (not shown). Toner is applied to develop (3) the latent image and is then transferred to paper by the electrostatic field supplied by the BTR (4). The paper and toner move through the fuser (5) where heat and pressure solidify the toner. A cleaner (6) removes any toner residing on the P/R and an erase lamp (7) restores the P/R surface to its original state.

At the center of all electrophotographic processes is a photoreceptor (P/R) surface as shown in Fig. 5. The P/R receives an electrostatic surface charge from a high voltage source, such as a biased charging roll (BCR), and upon subsequent exposure to light, generates and transports electron–holes to selectively discharge regions of the P/R surface in a pattern corresponding to the image being reproduced. Fine, charged powder known as toner is then applied to develop the latent electrostatic pattern on the P/R surface. A biased transfer roll (BTR), which contacts the backside of the copy sheet to form a rolling sandwich between the BTR surface, the paper, the toner, and the P/R is then used to electrostatically transfer the toner from the P/R surface onto paper. The toner is then permanently affixed to paper by rolling contact with a heated pair of rolls in a process referred to as fusing. These individual processes are linked into a synchronized, continuous process that produces electrophotographic copies at high rates (e.g. up to 200 ppm). This process is known as the "xerographic process".

The early generations of P/Rs were usually made from thin (10–50 μm) selenium or selenium alloy layers coated onto a cylindrical metal ground plane. These thin layers were rather delicate and easily damaged or worn by contact with moving paper. Understandably, the early P/R had relatively short lives. Renewal of the surface required replacement of the entire metallic drum since it carried the functional P/R layer. At this point, one would assume that a more durable or even a thicker, surface layer would solve the problem with short lives of drum-type P/Rs. While this may indeed be true if perhaps a small life increase were the goal, from the perspective of achieving utmost P/R life, better options emerged.

Synthetic polymers have emerged to offer important benefits for use as P/Rs. Several factors made this possible. First, movement within the industry towards belt-based, machine architectures opened the door for large area, belt-type P/Rs. Second, it was discovered that polymer P/Rs could be made into composites where charge generation and transport were separate mechanisms and these composites could be designed into low charge trapping transport layers that could easily be coated as durable and flexible thin layers onto thin flexible substrates such as Mylar™ (see article by Duke, and particularly Figs. 1 and 2 in that article). The multi-layered films could then easily and inexpensively be joined into long, endless belt P/Rs. Later, it was discovered that organic P/Rs could be made to be infrared sensitive, which turned out to be a feature needed to enable the use of fast, low cost, IR lasers for high-speed full color xerography. Given this progression of events, we can ask why did polymer P/Rs emerge to replace the metallic-substrate P/Rs, even though they were not substantially more durable nor wear resistant than selenium alloys?

The answer, of course, lies in knowing what problem needed to be solved. In this case, new machine architectures allowed new materials and design options to come forward. The use of long P/R belts was not only enabled, but was required by the new architectures. Once this occurred, the short lives associated with selenium drum P/Rs that were caused by wear could be spread over a larger area thereby effectively increasing the operational life. These organic P/Rs (OP/Rs) are now an industry standard and represent a class of complex, multi-functional devices, whose surfaces are vitally important to today's high quality, high-speed photocopier performance.

4.2. Knowing the functional requirements

Another example of functionally tailored surfaces vital to modern xerography is found on the BCRs and the BTRs that were illustrated in Fig. 5. They are similar in construction and function. Typically, they are comprised of one, or more, electrically conductive layers of filled synthetic rubber mounted on a rigid metal core. Often a surface treatment and/or primer is used to promote adhesion between the rubber composite and core. The rubber layers are assembled such that the innermost layer is relatively thick and sufficiently soft to provide a wide contact area with either the P/R or the paper. The topmost layer is relatively hard to achieve the required surface wear and dirt resistant properties. The rolls function by making rolling contact with either the P/R surface (BCR) or backside of the paper (BTR) and with application of the appropriate bias voltages (i.e. $\approx$600–2500 V) perform the xerographic charging and transfer functions. Because small amounts of ozone may be generated at the high voltages, the exposed materials must be oxidization resistant. The proper selection of electrical, chemical and mechanical properties of the constituent materials enable these rolls to perform all of the functions involving motion, pressure, and electric biasing.

Since the rolls perform nearly identical functions, one might ask if a common part could be used to do these functions. In other words, can the exact same part, made from the same materials, in the same configuration, be used to perform the charging and transfer functions? If two identical parts are adequate, the R&D investment needed to develop a solution to satisfy both requirements can be halved! Thus, knowing the functional requirements for the technology being developed is as important as knowing what problem is to be solved.

Another way to view this situation is to assume that one has successfully developed and commercialized a new surface technology and then explore the question, where else can it be used? Such technology diffusion is known as "leverage". If the technology can be "dropped into" or easily transformed to meet the requirements of other applications, the return for the initial sunk R&D investment is multiplied. The ability to identify such opportunities for leverage of new technologies becomes a powerful argument when one is trying to generate new or additional R&D funding.

4.3. Knowing what can go wrong

Several other surfaces important to the high-speed xerography appear on the four components contained in the fuser subsystem as illustrated in Fig. 6. These include (1) the fuser roll, (2) pressure roll, (3) oil-metering roll and (4)-donor roll. In order to make the toned image that has been transferred to paper permanent, the combined action of heat and pressure is used to melt the thermal plastic toner into the structure of the paper. This process is referred to as "fusing". To accomplish this, the image-bearing paper is fed through the high-pressure intersection of a pair of nip-forming rollers, where at least one, the fuser, is heated to high temperatures (e.g. 175–200 °C).

As in the previous examples, the functions of these rolls are very complex. For a better

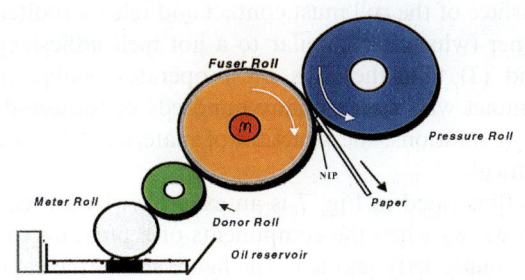

Fig. 6. Schematic diagram of a modern xerographic fuser. The heated, high-pressure nip formed by contact between the pressure and fuser rolls melts the toner into the paper and thereby makes the image permanent. Silicone oil, used as a liquid release layer on the fuser roll's surface to minimize toner adhesion and help keep the surfaces clean, is delivered from the oil reservoir by the metering and donor rolls. As discussed in the article, adverse interactions between the oil and the rolls can cause premature failure of the rolls.

understanding, the multiplicity of functions can be segregated into logical groupings, such as; heat transfer (which requires optimization of the thermal properties of the materials), toner release (requiring chemical properties and mass transport optimization), and dynamic pressure (involving mechanical properties). This classification methodology can provide useful insights to guide in the selection of design and materials options for developing the next generation of these components.

In the design of the modern fuser system, at least one of the rollers must be soft and sufficiently conformable to create an interface, known as the "nip" which exists at the intersection between the fuser and pressure rolls (see Fig. 6). In a pressure roll-forming nip, the pressure roll is soft and conformable. Alternately, in a soft fuser-hard pressure configuration, the fuser roll is used to create the nip. In either case, a suitably chosen elastomer comprises the major portion of the conformable layer of the soft roll. Often, a covering layer on the soft roll is used to provide the low surface energy properties required for efficient toner release. In addition, a silicone-based release fluid known as fuser oil is continually applied in a thin layer to the roll surfaces to assist in toner release and to minimize the unwanted build of toner on the rolls. Knowing that: (1) these rolls must function at high temperatures and under rolling loads that approach 50–300 psi, (2) the surface of the roll must contact and release molten toner (which acts similar to a hot melt adhesive), and (3) that the rolls must operate reliably in contact with oil for many hundreds of thousands of revolutions, the selection of materials becomes critical.

Illustrated in Fig. 7 is an example of what can go wrong when the components of a pressure roll are improperly selected. The figure shows the catastrophic destruction that occurred during a functional life test. This particular sleeved pressure roll was designed with three elements; a rigid metal shaft, a thick layer of thermally stable composite rubber, and a thin, outer sleeve made from a low surface energy, heat resistant polymer. It has dimensions of approximately 2.5 in. in diameter and 16 in. in length and weight of nearly 3

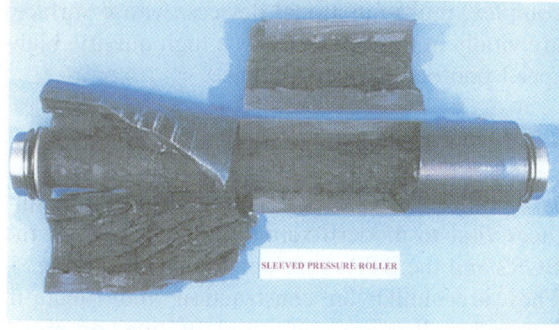

Fig. 7. Picture of a pressure roll that failed catastrophically during a life test. Failure was initiated by delamination of the innermost rubber layer close to the metal core.

lb. Appropriate materials and processes were used to promote adhesion at the inter-layer interfaces. A cutout was removed from the roll (shown centered in Fig. 7) reveals details of the innermost thick layer and the mode of failure. It provides evidence of extensive delamination of the layer near the metal core, which extends across about two-thirds of the roll. Post-mortem analysis of the mechanical and thermal loads acting on the rubber layer confirmed the presence of high mechanical stresses coupled with a strong thermal gradient at the sleeve-to-rubber interface and at the polymer-to-metal interfaces. Long term effects of these stresses were found to cause slow deterioration of small regions of the interfaces that nucleated the failure. If left in this state, this problem could have led to a substantial component recall, estimated to be in excess of several millions of dollars. Although changes to the rubber formulation, the curing process, and the interface chemistry were all shown to be solutions to this problem, it was the complexity and affiliated unreliability of this multilayer design that turned out to be the key factor in its rapid obsolescence.

Clearly, knowledge of potential failures and their underpinning mechanism(s) can be highly useful. A detailed understanding of what can, and does go wrong, can provide important insights that can lead to more informed decisions about the viability of any new technology and/or design options. This knowledge can also be an enabler of shorter technology development times.

4.4. Selecting the design and defining the surface

As suggested by the foregoing examples, selecting the proper design is a critical step in crafting any new functional surface. The performance of the entire component often hinges upon the strength, or weakness, of the substructure(s).

Up to this point, there has been much inferred about the "functional surface". We can formalize the definition: as the surface that establishes the outermost boundary of a component and serves as the working interface between the object and its operational environment. As noted earlier, the functional surface provides much of the observable performance, and, can be a contributor to failure. Although the functional surface may be a gas (such as a monolayer of gas absorbed on a reactive substrate), a liquid (such as the oil layer that provides toner release in a xerographic fuser), a solid (such as the tire's tread), or a combination of these, it is the solid surfaces that are the focus of this chapter. In the simplest designs, the surface and the bulk are compositionally identical. In arrangements that are more complex however, the surface is most likely to be compositionally different, because it must perform one, or more, functions that are different from those performed by the rest of the component. In other complex designs, a great number of layers of compositionally different materials may comprise the substructure. To add further to the complexity, the orientation of the layers may be different. For example, in the plied layers found within a tire, some of the sublaminate layers are formed in an "in-surface-plane" orientation, while others are rotated in an orientation that is off-angle. This crisscrossing combination provides inter-reinforcing layers that exhibit great strength.

5. Designing the materials

Assuming that a relevant problem has been selected, that something is known about the functional requirements and design options, and that insights into failure modes exist, the next step is to identify, select, or design the material or materials to be used in the application. At times, this sequence of events may be inverted. New material are constantly being discovered or invented that demonstrate new and interesting properties. These properties suggest inserting this new material into applications where performance shortfalls exist. In this case, where a problem exists waiting for a solution, the new material becomes that "ideal" problem solving material. Adoption and diffusion of the new technology usually occur rapidly in these cases because the economic impacts can easily be determined.

5.1. Separating the controlling variables

One approach to tackling what can otherwise be an overwhelming materials design challenge involves the ancient principle of divide and conquer.

When we, at Xerox, design a material for an application involving a multitude of functions, we prefer to have, or to create if possible, the situation where each controlling variable such as, the composition or the processing, affect one, and only one, resultant property of the material. In this way, the complex multi-dimensional, material optimization problem can be decomposed into a series of one-dimensional problems.

An example of where the principle of separation of variables may be applied relates to the new machine architectures as illustrated in Fig. 8. Here, we observe two belts, the intermediate transfer belt (ITB) and the transfuse belt working together, in tandem, to make full color copies. Today's color print engines transfer a total of four, single color, toner images to an intermediate belt in precise registry. Then, once the final of the four colors is delivered to the ITB, the multiple toner layers are transferred from the ITB surface as a single unit onto a second belt, the transfuse belt, before final transfer to paper.

In this sequence, each initial toner transfer from the P/R to the ITB is electrostatic. The first biased roll (BTR), as described earlier, is positioned inside the ITB and electrically biases the back of the belt opposite the P/R. This produces a strong electrostatic transfer field that attracts the toner from the P/R to the surface of the ITB. Effective transfer in this case requires control of both the

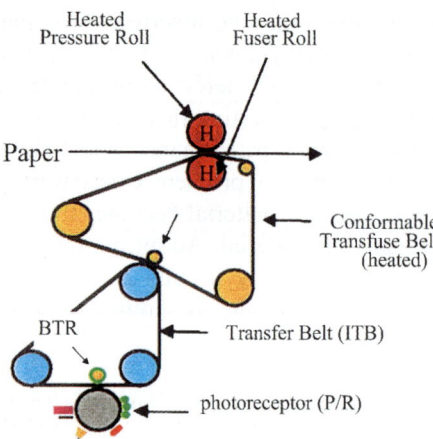

Fig. 8. Schematic of a contemporary two belt, "tandem" architecture illustrating an ITB and a transfuse belt working together to make full color images. The process begins when a single color, toner image is developed on a P/R surface. Up to three other colors of toners are developed on each of three other P/Rs (not shown). The single color images are then transferred electrostatically in registration from the P/Rs by the first BTRs onto the ITB. The full color image is then transferred from the ITB to the heated surface of the transfuse belt and again onto the paper where it is fused.

ITBs volume resistivity and its lateral resistivity. Volume resistivity is the parameter that determines the voltage drop across the ITB and therefore controls the transfer field between the ITB and the P/R. Lateral or surface resistivity, on the other hand, determines the variation of the transfer field entering and leaving the nip. The areas immediately before, and immediately after, the BTR-to-P/R contact zone are referred to as "pre-nip" and "post-nip" regions. Fletcher provides an excellent discussion about this aspect of the xerographic process [13] and describes the importance of "field tailoring" to avoid high voltage breakdown of the air in the critical pre- and post-nip regions. Pre-nip air breakdown has been shown to be a major cause of image disturbances during the transfer processes. To avoid or minimize the impacts of this unwanted breakdown, both the belt material's surface and its volume resistivity have to be controlled within pre-determined ranges.

These electrical properties of the ITB illustrate an example of where the critical controlling variables are (nearly) separable. Unfortunately, the transfer process we have described requires resistivities that are in a range that is traditionally difficult to achieve reliably. There are several materials system options to choose from: (1) conductive particle filled homopolymers or copolymers similar to those used in BTRs, (2) intrinsically conducting homopolymers, and (3) mixtures of 1 and 2.

A well-known means to alter the electrical resistivity of a polymer is to add fine, electrically conductive particles to the polymer to create a conductive composite. The final resistivity of the composite can be selected by adjusting the volume fraction of conductive particles carried within the polymer. Being commercially available and low in cost, carbon blacks (CB) are often used as the conductive filler in these composites. The steady state resistivity of a CB filled composite at room temperature is determined primarily by the components and their ratio. At low carbon black concentrations, the resistivity emulates that of the insulating polymer. At high carbon black concentrations, the resistivity approaches that of the conductive carbon black. For intermediate carbon black concentrations, the resistivity decreases strongly with minor changes in carbon black content. A percolation threshold phenomenon as illustrated in Fig. 9, occurs at a critical carbon

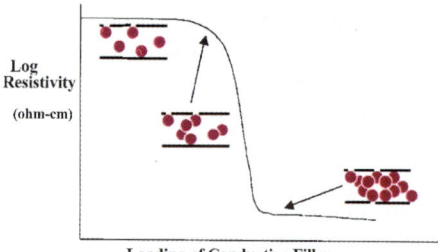

Fig. 9. Graph of the log of electrical resistivity of a conductive particle filled polymer as a function of carbon black loading. Illustrated is the sudden change of resistance that corresponds to a small change in filler loading where the resistivity changes from insulating to conducting. This phenomenon is referred to as a volume percolation effect and the point where the change begins to occur is known as the percolation threshold. The figure also illustrates the particle spacing for each domain. Note that the onset of percolation coincides with formation of the first continuous chains of filler particles. No further change in resistivity occurs once a large number of particle chains is established.

black concentration at which point the resistivity drops precipitously and formation of particle-to-particle chains occurs. This critical carbon black concentration is an example of multiple interacting, control variables, including, but not necessarily limited to, the type of CB, its particle size and shape, its resistivity, the surface chemistry, as well as the processing parameters for dispersing the carbon black in the polymer. The figure also shows, schematically, the typical change in resistivity as carbon black particles go from not touching, to forming networks that extend for macroscopic distances. This large resistivity change, which corresponds to small variations in carbon black concentration, makes the resistivity at intermediate concentrations hypersensitive to variations in particle size, morphology, surface chemistry, and processing.

The resultant resistivities obtained from these traditional percolation-controlled systems are typically binary. That is, they are either insulator-like or conductor-like. In between resistivities are difficult to obtain and hard to control reliably. Until recently, there have been no general solutions to this problem with conductive particle filled polymer systems. Therefore, their use was typically limited to the conductor-like end of the spectrum.

A particularly innovative solution to this problem that was brought to light by Tarnawskyj and coworkers [14] is based upon the unique chemistry that occurs when Accufluor® CB is used in combination with VITON™, a fluorine-containing polymer manufactured by E.I. DuPont®. Accufluor® is the trade name given by Allied Signal® to certain CB with fluorinated surfaces. The reaction that occurs between the fluorine of the CB and the reactants used to crosslink the VITON™ leads to a novel way to tune the filler–polymer conductivity. Moreover, changes to the surface properties of the fluorine containing carbon black result in changes to the surface energy, which can affect, at least, the dispersion properties of the filler in the host polymer. In this system, better and more consistent dispersion is a determinant of more consistent composite properties. This softens the resistivity–concentration transition in a way that enables reliable manufacture of composites with predictable and tightly controlled

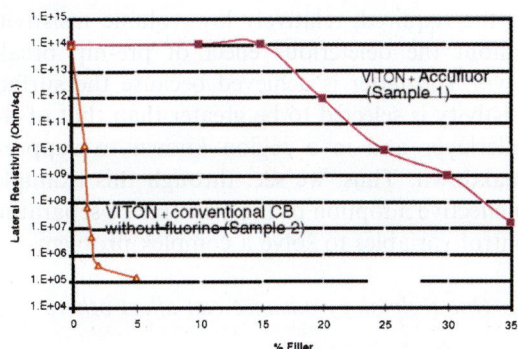

Fig. 10. Graph of lateral resistivity for an Accufluor–VITON™ system (sample 1) compared to a conventional CB filled VITON™ (sample 2). The shallow slope of sample 1 suggests that precise resistivities from within a seven-decade range can be selected.

intermediate resistivities. In Fig. 10, we contrast the resistivity behavior of two VITON™-based systems; one having conventional carbon black and the other filled with the new Accufluor® CB. Two noteworthy differences occur with the new filler system. First, the transition from insulating to "controlled conductivity" occurs at a substantially higher CB loading. This implies that the transition may be easier to manage. Second, the near-linear response of resistivity to CB loading enables design of composites with precise resistivity targets.

Returning to the ITB example that requires independent control of the volume and lateral resistivities, the introduction of a filler based system with "tailored electrical resistivity" can be used to fulfill both of these needs. For example, a two-layer design can be selected where the bottom layer can be constructed by melt extrusion of a composite of polymer and Accufluor CB that is designed to have a bulk resistivity in the range of, for example 1×10^7–7×10^7 Ω cm. Over this, a top layer coating of a similar polymer but having slightly less filler results in an effective surface resistivity of about 2–2.5 orders of magnitude greater than the substrate layer. This simultaneous achievement of dual objectives, viz. a partially resistive layer on a more conducting base, enables the ITB to function as intended by delivering acceptable bias voltages to the backside of the copy via the required, relatively low volume resistivity without the deleterious effect of pre-nip breakdown. The latter is achieved because the surface resistivity is selected to be greater than that of the underlayer and in a region known to suppress breakdown. Thus, we see, through this example, an effective adoption of the principle of separating control variables to solve a complex problem.

5.2. One technology advance can affect others

The movement to digital color in the xerographic industry set the stage for adoption of the tandem belt architecture as shown in Fig. 8. Instead of building a full color image directly on a single P/R surface, four P/Rs are used in series. Each creates one of the four fundamental colors of the final image, which is transferred continuously onto the ITB and then onto paper. Migration of the function requiring contact with paper from the P/R to the ITB eliminated the requirement for the P/R to accommodate paper abrasion. Therefore, the less than acceptable durability and short life of the early generation drum-type P/R suddenly became a non-issue. Further, since four P/Rs would be needed in the new color processes, the small size of drum-type P/Rs enabled more compact machine architectures. Interestingly, features that were disadvantages in earlier black-only processes became advantages in the new full color processes. Technology shifts in one application can often affect others and assumptions based on prior constraints may not necessarily persist during technology shifts.

5.3. Add a little heat to the mix

Referring again to Fig. 8, we see that the second transfer of toner from the ITB surface is performed at elevated temperatures. In this case, toner transfer occurs by a combination of electric field and heat activated "tacky" transfer and the belt that performs both the transfer and fusing functions is referred to as a transfuse belt. For this belt, the new functional requirement is control of the belt's electrical resistivity at high temperature. This may be accomplished by noting that a correlation exists between the high temperature resistivity of a composite and its thermal expansion. Heaney [15], for example, reports of the changes to the resistivity of a high-density polyethylene (HDPE) composite upon heating. In this carbon black filled HDPE system, the resistivity response over the temperature range of 25–180 °C showed resistivity to move from 10^{-1} to 10^{-8} Ω cm. A five order of magnitude increase occurred between 120 and 133 °C, which, of course, is the temperature at which polyethylene melts. By simultaneously measuring the thermal expansion of the composite and its resistivity, Heany was able to show that the resistivity is a smooth function when adjusted for temperature. This correlation confirms that the volume fraction of the filler within the polymer is effectively reduced when the polymer expands and the filler does not. This behavior suggests that thermal expansion is a way to affect electrical behavior. For applications employing heat, one may be able to take advantage of this phenomenon to design an appropriately functioning material.

These examples shows what one can expect to encounter in attempting to customize the performance of a functional material. It is possible with some properties to achieve complete separation of the functions and thereby independently adjust each to the desired final balance without complication. On the other hand, the presence of interactions with heat for example, may render it impossible to accomplish separation of the controlling variables. In this event, the higher level of complexity suggests that optimization requires an intensive empirical approach. Insights into whether the development effort will be straightforward or complex and into what level of activity will be required to craft the target system to achieve the desired mix of performance can be important during the planning stages of a research project.

6. Novel materials and structures

Since novel materials having innovative surfaces are emerging to be a determinant of future growth of many high tech establishments, we now turn to some recent and more interesting materials options and where they may lead us.

6.1. Contemporary copolymers

Copolymers are polymers formed from two or more monomers, which represent a class of polymers that enable modification of one or more secondary properties without significant alteration to the primary properties. As mentioned earlier, it is a useful design objective to be able to alter one property independent of the others. In the Accu-fluor® Carbon Black example above, it is the modification of the surface properties, independent of the electrical properties of the filler that gives rise to better dispersion of the filler in the polymer. This leads to more precise control of the resistivity. In many polymer systems, modification of the surface properties such as surface energy, independent of the other properties is a similarly desirable objective. Ober et al.'s [21] work with block and branched copolymers provides insights into how copolymers may achieve this. Ober et al. showed that it is possible to synthesize novel copolymers with branched side groups, identified as monodendrons, where the number of branches on the side groups can be controlled and that the functional end groups on the branched "arms" can be pre-determined. In this work, the use of semi-fluorinated branches created stable, low surface-energy copolymers that effectively masked strong polar groups residing elsewhere on the copolymer. This methodology may be used to achieve those features previously difficult to obtain. For example, a copolymer could be designed to have a functional surface exhibiting what would appear to be conflicting attributes, such as the combination of low surface energy and self-adhesion.

6.2. Nanocomposites

Nanocomposite is the term for a polymer composite that contains ultra-fine solid phases where the sizes of all of the solids approach molecular dimensions [16]. The use of nanometer size fillers provides another path for modifying the properties of a surface and enables new combinations of properties that are otherwise difficult, if not impossible, to obtain. Some new nanocomposites are showing dramatic enhancements in important properties such as elastic modulus, wear resistance, thermal conductivity, and fluid permeability.

Giannelis [17] was one of the first to show that layered, inorganic silicates can be made compatible with and readily dispersed in a wide range of polymers to form polymer–ceramic nanocomposites. In this work, an ion exchange technique was used to add an organic cation (i.e. alkyl ammonium cation) to the inorganic filler, which lowered the surface energy of the silicate and improved wetting by the polymers. Giannelis then melt blended the treated, ultra-fine filler into the polymer at a temperature just above its melting point. This technique was extended to reactive silicones where crosslinking of a well dispersed, treated silica with a silicone revealed that the original silicate dispersion can be preserved through the liquid to solid transformation. The resultant solid nanocomposite was found to exhibit dramatic enhancements in thermal stability, mechanical modulus, and solvent swelling. The decrease in solvent swelling behavior in these nanocomposites indicates that the layered-silicate fillers act as a good reinforcing agent for siloxane systems. This is in contrast to the more conventional fillers such as, talc and kaolin, which do not affect swelling. This new behavior is related to the large surface area of the layered silicate and to the formation of filler-bound polymer, which is unable to swell to the same degree as the unfilled polymer. In addition, the thermal stability of the nanocomposite is enhanced relative to unfilled siloxanes. Furthermore, Giannelis observed that a 300–400% increase in modulus could be achieved in these nanocomposite systems.

Similar results, obtained in fluoroelastomeric-based nanocomposites by Badesha et al. [18], demonstrate that low molecular weight fluoroelastomers can be intercalated (dispersed) into a similar, commercially available, layered silicate. These finding suggest that nanoscale silicates are a good reinforcement for the earlier-cited VITON™-based systems, which can lead to tougher, more durable surfaces.

6.3. Sol gels: a solution for future generation fusers

Returning to the fuser example, the fuser subsystem in Fig. 6 also uses an oil-metering roll

(MR) and a donor roll (DR), which serve to precisely supply a release-promoting silicone fluid to the surface of the heated fuser roll. The fluid serves as a low surface tension, shearable layer that prevents adhesive bonding of the molten toner to the hot surface and enables the fuser surface to stay clean. Early attempts to identify a suitable material for the donor roll were frustrated by a failure mechanism that was traced to hot oil swelling of the soft outer layer. Compare the surface of the new donor roll in Fig. 11 (top) to that of one having been exposed for only a few operational hours to heat and silicone oil, Fig. 11 (center). Here, the surface shows early indications of degradation as can be seen by the stained appearing regions at the center and to the left of the centermost roll. Within a few hundred thousand cycles, a catastrophic failure occurred resulting in loss of large chucks of silicone from the surface (not shown).

The group of scientists at Xerox who set out to solve this problem of swell induced failure recognized that silicones and fluoropolymers were the two polymers routinely used for the rolls' surface. Each was known to have different advantages and disadvantages which must be understood and traded off in the final application. For example, silicones are easily wet by release oils and thereby can minimize toner adhesion. Unfortunately, silicones have poor abrasion characteristics, especially at high temperatures. Fluoropolymers, on the other hand, are tough and exhibit long wear lives, but are poorly wet by release agents. The natural thought would be to identify a way to combine these two dissimilar materials and through a new polymer–polymer composite obtain the most desirable properties of each. This thinking led to a sol–gel process to enable a series of new, high temperature materials [19]. First, it was discovered that sol–gel procedures could be used to graft a thin layer of siloxanes on the surface of a fluoropolymer. The work led to the discovery that the addition of silica and titanium within the backbone of the base polymer would produce more durable, high temperature performance.

The surfaces of fuser rollers prior to the introduction of silicone graft surfaces typically exhibited functional lives in the range of a few hundred thousand cycles. Moreover, as noted by the earlier pressure roll example, the design of other surfaces required costly and complex, multi-layer processing in their manufacture. The use of innovative Si-VITON™ graft polymers in these applications has not only enabled simpler designs, but also has improved reliability to the point where lives of these components now exceeds 10 million cycles. The bottommost roll shown in Fig. 11 illustrates the surface of a new silicone-VITON™ grafted DR after a life test of nearly 5 million cycles. Note that with the exception of a small cut-out that was intentionally removed from the surface, no change has occurred anywhere on the surface.

The conclusion is that significant changes that are brought about by new technology can be made to impact upon what may appear to be small, or even insignificant, parts of larger systems. However, by doing so (for example, by use of this cleaver sol–gel chemistry to improve a simple roller), it is possible to generate millions of dollars in savings.

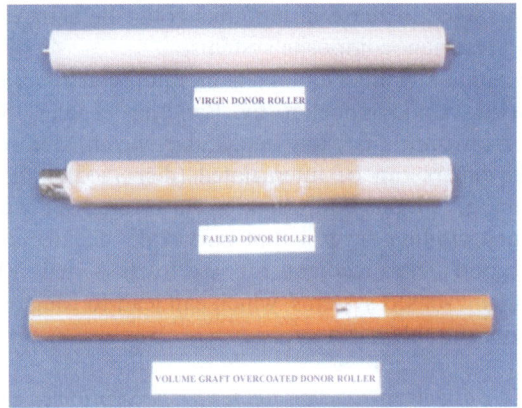

Fig. 11. Photograph of Donor Roll surfaces depicting various periods of use: new (top), slightly used and failed (center), and used for 5 million cycles without failure (bottom).

6.4. The role of novel processes

The functional properties of a surface can be influenced not only by the selection of constituent materials but also by the processing used to manufacture that surface. It has been a longstanding objective of modern commerce to migrate away from organic solvent-based processes to environmentally friendly ones. In this regard, supercritical fluid (SCF) technology has received much attention and has recently been adapted for polymer synthesis and composite processing. The recent work of Ober et al. [22] shows that supercritical carbon dioxide (SCF-CO_2) can be used as an environmentally friendly solvent for synthesizing certain fluoropolymers. The resulting polymers have been shown to have remarkable properties. For example, some have revealed small-feature, image resolution capability and therefore show great potential for use as the thin-layer photoresist in the manufacture of microelectronic circuits. Processes can be as important to the achievement of new surface properties as the selection and design of the material itself since new synthetic design pathways are created.

7. Economics, science and technology

In this section, we explore the question, "how does one go about assigning "value" to technology"? Similarly, the question asked frequently by corporate managers is "how do we know that we are getting *value* from the firm's R&D spending"?

Economics is as much of a challenge to the surface scientist as the S&T itself is. Many of the best of the world's inventors have been frustrated by the economics of S&T. Chester Carlson, the inventory of xerography, for example, was not the first to discover that, without appropriate financial sponsorship, even one of the best ideas in the world could end up as "the billions that nobody wanted" [20]. The economics of S&T can be examined from many perspectives. Business investment decisions are often multi-faceted and thereby complex. While it is not always clear why some ideas are funded and others are not, the persistence of a strong proponent, such as Chester Carlson, coupled with the existence of recognizable value often prove to be the factors that are critical to the outcome of these decisions.

From the technology perspective, the determinants of value are perhaps not so complicated. The earlier-cited technology S-curve can be modified as shown in Fig. 12a to include value as the dependent variable, where *value* is defined as the ratio of functionality to cost. From this, we observe that functionality and cost contribute to value in inverse ways. If a new technology enables a product to perform more or better functions, value increases because these improvements enable product differentiation in the marketplace. Similarly, if a technology enables cost reductions without changing the functionality of a product, then value increases. Advances in manufacturing process technology often enable this type of cost reduction. The most valuable technologies are those that deliver functional advancements and enable lower cast. Envision, for example, the ultimate generation tire that lasts forever and costs substantially less than its predecessor does!

The rational, competitive firm invests in those value-based technologies that enable it to compete effectively and earn a profit. Since it may take a long time for any one R&D investment to pay off, sustained growth in the high tech firm is often linked to the firm's ability to deploy a continuous stream of technologies, which are used to assure product differentiation, production efficiency, and responsiveness to changing markets. In this light and as noted earlier, technology leverage is a potentially valuable concept. Fig. 12b illustrates the concept of technology leverage, which is shown to produce value in a series of related applications. The earlier described OP/R technology is a good example of leveraging a technology into many applications. Due to their variety, low cost, and ability to function over a broad range of process speeds and conditions, OP/Rs have emerged over a 20-year period to become an industry standard. The first OP/Rs were introduced in 1982 in the Xerox 1075®, a 70 ppm black and white (B/W) copier/duplicator. Shortly thereafter, they were adapted for use in the 1090® machine, which was a faster (90 ppm) version of the 1075 and in 1988 were advanced into the 5090 product @ 135 ppm. They now are used in an almost countless list of B/W and color products that are supplied by numerous manufacturers worldwide.

Upon its introduction, the 1075 was a radically new product design, which quickly established

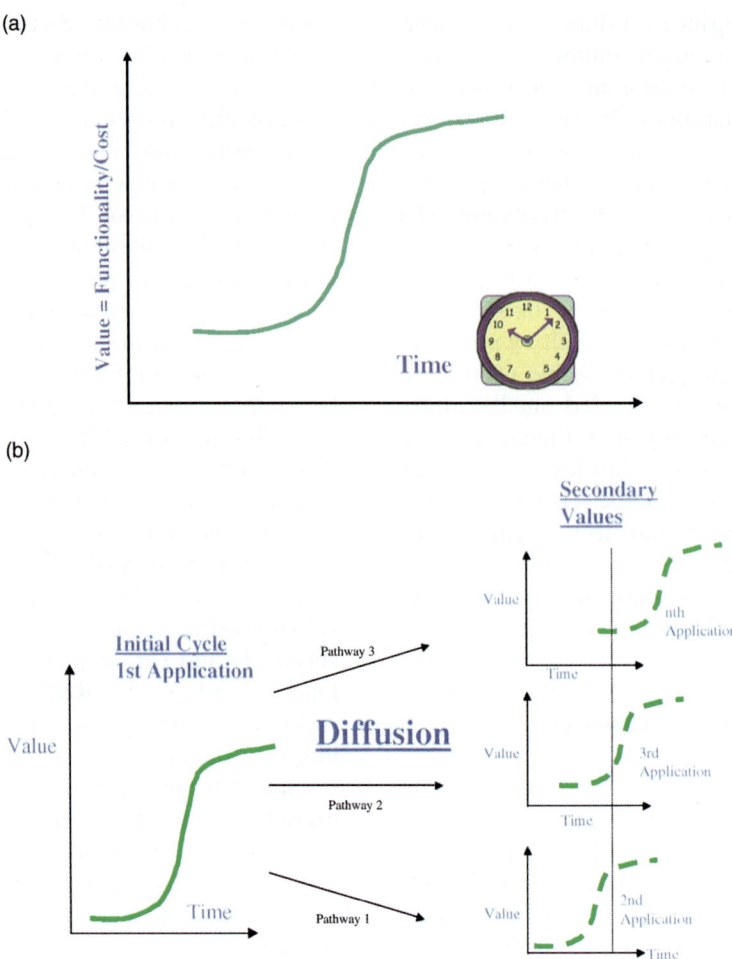

Fig. 12. (a) Illustration of technology value in terms of an S-curve behaviour where value is the level of functionality normalized by cost. (b) Schematic illustration of the technology leverage concept. Higher value is obtained by driving the diffusion of the parent technology into subsequent applications. In the case of the OP/R technology as discussed in the article, diffusion occurred over a period of more than 20 years and impacted dozens of products.

industry benchmarks for size, speed, quality, and reliability. Many of these features were derived from use of OP/R technology, which unquestionably contributed to the success of this lead product. It is quite remarkable to note that even today, the follow-on adaptations of OP/R technology are not only continuing to accrue value, but will likely do so into the foreseeable future. From this example, we see that the purposeful diffusion of a new technology can produce a near-limitless wellspring of value.

8. Synopsis

In this paper, we present a new way to view the functionality of surfaces. We explore technological complexity and examine the challenges of designing practical surfaces with multi-functionality. We describe principles and illustrative examples relating to the separation of variables as a means of simplifying the otherwise overwhelming challenges that one may face in crafting new surfaces. We develop the functionality–complexity framework and use it to map major evolution of some interesting technologies.

Good science (i.e. discovery) leads to good technology (i.e. development) and both set the stage for innovation (use). Innovation is a cyclic process, which is not complete until the final stage (use) is achieved, at least once. Technological value stems from achieving advancements in performance and in lowering costs. Value also occurs by leveraging new technology into a series of applications.

The challenge to the reader who is interested in this theme is to look beyond scientific discovery and beyond invention towards implementation and commercialization as being the end point for deriving value of a new technology whether in the context of the established firm or in a more entrepreneurial mode. Look beyond the wheel because it is the number and quality of the turns of the innovation cycle that determines the eventual value of the underlying discovery. The challenge to the student of surface science is to recognize and respond to the limitless opportunities for practical surfaces that lie ahead in this new millennium.

Acknowledgements

The authors wish to extend their gratitude to the following individuals; L. Dunlap, J. Mort, I. Tarnawskyj, B. Springett, and C. Duke.

References

[1] P.A. Roussel, K.N. Saad, T.J. Erickson, Managing the Link to Corporate Strategy, Third Generation R&D, Arthur D. Little, 1991.
[2] C.M. Christensen, The Innovator's Dilemma, When New Technologies Cause Great Firms to Fail, Harvard Business School Press, 1997.
[3] D.K. Smith, R.C. Alexander, Fumbling the Future—How Xerox Invented, Then Ignored the First Personal Computer, William Morrow and Co, 1988.
[4] R. Zacks, The University Research Scorecard, Technology review, MIT Press, Cambridge, MA, 2000, pp. 88–90.
[5] L.B. Andersen, L.A. Wenzel, Introduction to Chemical Engineering, McGraw-Hill, New York, 1961, p. 314ff.
[6] A.K. Varshneya, Fundamentals of Inorganic Glasses, Academic Press, Harcourt Brace & Company, London, 1994, p. 503ff.
[7] www.goodyear.com/us/tire school/index.html, Basic Ingredients to Make a Tire.
[8] W.E. Souder, Managing New Product Innovations, D.C. Health and Company, 1987, pp. 39–42.
[9] A.L. Porter et al., Forecasting and Management of Technology, Wiley, New York, 1991, p. 169ff.
[10] E.M. Williams, The Physics and Technology of Xerographic Processes, Wiley, New York, 1984.
[11] M. Scharfe, Electrophotography Principles and Optimization, Research Studies Press, 1984.
[12] D.A. Hays, in: Xerography, Encyclopedia of Applied Physics, vol. 23, Wiley-VCH, New York, 1998.
[13] C.A. DiRubio, G.M. Fletcher, Proceedings of the IS&T NIP12 International Conference on Digital Printing Technology, 1996.
[14] M.A. Abkowitz, F.E. Knier Jr., I.W. Tarnawskyj, M. Stolka, J. Mammino, K.-Y. Law, US Patent 5,856,013, Ohmic Contact-Providing Compositions, issued 1/5/99.
[15] M.B. Heaney, Resistance-expansion-temperature behavior of a disordered conductor–insulator composite, Appl. Phys. Lett. 69 (1996) 2602–2604.
[16] S. Komarneni, Nanocomposites, J. Mater. Chem. 2 (12) (1992) 1218–1230.
[17] E.P. Giannelis, Polymer layered silicate nanocomposites, Adv. Mater. 8 (1) (1996) 29–35.
[18] S.S. Badesha, A. Henry, C. Eddy, J. Maliborski, Polymer nanocomposites, US Patent 5,840,796, 1998.
[19] S.S. Badesha et al., Fuser member, US Patent 5,141,788 (8/25/92);
S.S. Badesha et al., Sol–gel for the preparation of volume graft, USP 5,366,772 (11/22/1994).
[20] J.H. Dessauer, My Years with Xerox, The Billions Nobody Wanted, Mannor Books, 1975.
[21] C.K. Ober, M. Xiang, X. Li, K. Char, J. Genzer, E. Sivaniah, E.J. Kramer, D.A. Fischer, Surface stability in liquid-crystalline block copolymers with semifluorinated monodendron side groups, Macromolecules, August 2000, pp. 6106–6119.
[22] C. Ober et al., Supercritical CO_2 processing for submicron imaging of fluoropolymers, Chem. Mater. 12 (1) (2000) 41–48.

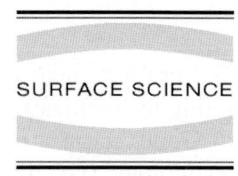

Author index

Arthur, J.R., Molecular beam epitaxy 500 (2002) 189

Bader, S.D., Magnetism in low dimensionality 500 (2002) 172
Badesha, S.S. and J.A. Swift, Practical surfaces: beyond the wheel 500 (2002) 1024
Beattie, D.A., see Williams 500 (2002) 545
Bent, S.F., Organic functionalization of group IV semiconductor surfaces: principles, examples, applications, and prospects 500 (2002) 879
Biesalski, M., see Tirrell 500 (2002) 61
Bigelow, L.K. and M.P. D'Evelyn, Role of surface and interface science in chemical vapor deposition diamond technology 500 (2002) 986
Bonn, M., A.W. Kleyn and G.J. Kroes, Real time chemical dynamics at surfaces 500 (2002) 475
Broglia, R.A., The surfaces of compact systems: from nuclei to stars 500 (2002) 759

Carter, E.A., see Starrost 500 (2002) 323
Castner, D.G. and B.D. Ratner, Biomedical surface science: Foundations to frontiers 500 (2002) 28
Chabal, Y.J., see Weldon 500 (2002) 859

Dai, H., Carbon nanotubes: opportunities and challenges 500 (2002) 218
D'Evelyn, M.P., see Bigelow 500 (2002) 986
Diño, W.A., H. Kasai and A. Okiji, Dynamical phenomena including many body effects at metal surfaces 500 (2002) 105
Duke, C.B., J. Noolandi and T. Thieret, The surface science of xerography 500 (2002) 1005

Eberhardt, W., Clusters as new materials 500 (2002) 242
Eng Jr., J., see Weldon 500 (2002) 859

Fahlman, M. and W.R. Salaneck, Surfaces and interfaces in polymer-based electronics 500 (2002) 904
Ferrer, S. and Y. Petroff, Surface science done at third generation synchrotron radiation facilities 500 (2002) 605
Finnie, P. and Y. Homma, Epitaxy: the motion picture 500 (2002) 437
Freund, H.-J., Clusters and islands on oxides: from catalysis via electronics and magnetism to optics 500 (2002) 271

Ganduglia-Pirovano, M.V., see Stampfl 500 (2002) 368
Gillmor, S.D., P.P. Rugheimer and M.G. Lagally, Computation with DNA on surfaces 500 (2002) 699
Greenberg, J.M., Cosmic dust and our origins 500 (2002) 793
Grey, F., see Hasegawa 500 (2002) 84
Groß, A., The virtual chemistry lab for reactions at surfaces: Is it possible? Will it be useful? 500 (2002) 347

Hanley, L. and S.B. Sinnott, The growth and modification of materials via ion–surface processing 500 (2002) 500
Hasegawa, S. and F. Grey, Electronic transport at semiconductor surfaces—from point-contact transistor to micro-four-point probes 500 (2002) 84
Herbst, E., see Williams 500 (2002) 823
Hirschmugl, C.J., Frontiers in infrared spectroscopy at surfaces and interfaces 500 (2002) 577
Homma, Y., see Finnie 500 (2002) 437

Ismail, see Plummer 500 (2002) 1

Johnson, R.E., see Madey 500 (2002) 838

Kasai, H., see Diño 500 (2002) 105
Kasemo, B., Biological surface science 500 (2002) 656
Kirschner, J., see Shen 500 (2002) 300
Kleyn, A.W., see Bonn 500 (2002) 475
Kokkoli, E., see Tirrell 500 (2002) 61
Kolb, D.M., An atomistic view of electrochemistry 500 (2002) 722
Krim, J., Surface science and the atomic-scale origins of friction: what once was old is new again 500 (2002) 741
Kroes, G.J., see Bonn 500 (2002) 475

Lagally, M.G., see Gillmor 500 (2002) 699

Madey, T.E., R.E. Johnson and T.M. Orlando, Far-out surface science: radiation-induced surface processes in the solar system 500 (2002) 838
Matzdorf, R., see Plummer 500 (2002) 1
Melechko, A.V., see Plummer 500 (2002) 1

Naumovets, A.G. and Z. Zhang, Fidgety particles on surfaces: how do they jump, walk, group, and settle in virgin areas? 500 (2002) 414

Noolandi, J., see Duke 500 (2002) 1005
Nørskov, J.K., see Rod 500 (2002) 678

Okiji, A., see Diño 500 (2002) 105
Orlando, T.M., see Madey 500 (2002) 838

Petroff, Y., see Ferrer 500 (2002) 605
Pierce, J.P., see Plummer 500 (2002) 1
Plummer, E.W., Ismail, R. Matzdorf, A.V. Melechko, J.P. Pierce and J. Zhang, Surfaces: a playground for physics with broken symmetry in reduced dimensionality 500 (2002) 1

Queeney, K.T., see Weldon 500 (2002) 859

Raghavachari, K., see Weldon 500 (2002) 859
Ramana Murty, M.V., Sputtering: the material erosion tool 500 (2002) 523
Ratner, B.D., see Castner 500 (2002) 28
Reuter, K., see Stampfl 500 (2002) 368
Rod, T.H. and J.K. Nørskov, The surface science of enzymes 500 (2002) 678
Rosei, F. and R. Rosei, Atomic description of elementary surface processes: diffusion and dynamics 500 (2002) 395
Rosei, R., see Rosei 500 (2002) 395
Rugheimer, P.P., see Gillmor 500 (2002) 699

Safran, S.A., Statistical thermodynamics of soft surfaces 500 (2002) 127
Salaneck, W.R., see Fahlman 500 (2002) 904
Scheffler, M., see Stampfl 500 (2002) 368
Sergeev, G.B. and T.I. Shabatina, Low temperature surface chemistry and nanostructures 500 (2002) 628
Shabatina, T.I., see Sergeev 500 (2002) 628
Shen, J. and J. Kirschner, Tailoring magnetism in artificially structured materials: the new frontier 500 (2002) 300

Sinfelt, J.H., Role of surface science in catalysis 500 (2002) 923
Sinnott, S.B., see Hanley 500 (2002) 500
Stampfl, C., M.V. Ganduglia-Pirovano, K. Reuter and M. Scheffler, Catalysis and corrosion: the theoretical surface-science context 500 (2002) 368
Starrost, F. and E.A. Carter, Modeling the full monty: baring the nature of surfaces across time and space 500 (2002) 323
Swift, J.A., see Badesha 500 (2002) 1024

Thieret, T., see Duke 500 (2002) 1005
Tirrell, M., E. Kokkoli and M. Biesalski, The role of surface science in bioengineered materials 500 (2002) 61

Vlieg, E., Understanding crystal growth in vacuum and beyond 500 (2002) 458

Weldon, M.K., K.T. Queeney, J. Eng Jr., K. Raghavachari and Y.J. Chabal, The surface science of semiconductor processing: gate oxides in the ever-shrinking transistor 500 (2002) 859
Williams, C.T. and D.A. Beattie, Probing buried interfaces with non-linear optical spectroscopy 500 (2002) 545
Williams, D.A. and E. Herbst, It's a dusty Universe: surface science in space 500 (2002) 823
Woodruff, D.P., Solved and unsolved problems in surface structure determination 500 (2002) 147

Zaera, F., The surface chemistry of catalysis: new challenges ahead 500 (2002) 947
Zhang, J., see Plummer 500 (2002) 1
Zhang, Z., see Naumovets 500 (2002) 414
Zhdanov, V.P., Impact of surface science on the understanding of kinetics of heterogeneous catalytic reactions 500 (2002) 966

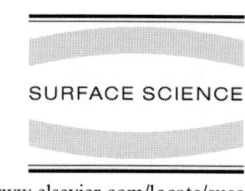

Subject index

THEORETICAL METHODS, MODELS AND TECHNIQUES

Ab initio quantum chemical methods and calculations

F. Starrost and E.A. Carter, Modeling the full monty: baring the nature of surfaces across time and space
500 (2002) 323

A. Groß, The virtual chemistry lab for reactions at surfaces: Is it possible? Will it be useful? 500 (2002) 347

M.K. Weldon, K.T. Queeney, J. Eng Jr., K. Raghavachari and Y.J. Chabal, The surface science of semiconductor processing: gate oxides in the ever-shrinking transistor
500 (2002) 859

Atomistic dynamics

A. Groß, The virtual chemistry lab for reactions at surfaces: Is it possible? Will it be useful? 500 (2002) 347

M. Bonn, A.W. Kleyn and G.J. Kroes, Real time chemical dynamics at surfaces 500 (2002) 475

Computer simulations

A. Groß, The virtual chemistry lab for reactions at surfaces: Is it possible? Will it be useful? 500 (2002) 347

L. Hanley and S.B. Sinnott, The growth and modification of materials via ion–surface processing 500 (2002) 500

S.D. Gillmor, P.P. Rugheimer and M.G. Lagally, Computation with DNA on surfaces 500 (2002) 699

Density functional calculations

F. Starrost and E.A. Carter, Modeling the full monty: baring the nature of surfaces across time and space
500 (2002) 323

A. Groß, The virtual chemistry lab for reactions at surfaces: Is it possible? Will it be useful? 500 (2002) 347

C. Stampfl, M.V. Ganduglia-Pirovano, K. Reuter and M. Scheffler, Catalysis and corrosion: the theoretical surface-science context 500 (2002) 368

T.H. Rod and J.K. Nørskov, The surface science of enzymes
500 (2002) 678

Electron density, excitation spectra calculations

W.A. Diño, H. Kasai and A. Okiji, Dynamical phenomena including many body effects at metal surfaces
500 (2002) 105

Equilibrium thermodynamics and statistical mechanics

S.A. Safran, Statistical thermodynamics of soft surfaces
500 (2002) 127

Many body and quasi-particle theories

W.A. Diño, H. Kasai and A. Okiji, Dynamical phenomena including many body effects at metal surfaces
500 (2002) 105

R.A. Broglia, The surfaces of compact systems: from nuclei to stars 500 (2002) 759

Models of non-equilibrium phenomena

D.A. Williams and E. Herbst, It's a dusty Universe: surface science in space 500 (2002) 823

T.E. Madey, R.E. Johnson and T.M. Orlando, Far-out surface science: radiation-induced surface processes in the solar system 500 (2002) 838

Models of surface chemical reactions

H.-J. Freund, Clusters and islands on oxides: from catalysis via electronics and magnetism to optics 500 (2002) 271

A. Groß, The virtual chemistry lab for reactions at surfaces: Is it possible? Will it be useful? 500 (2002) 347

T.H. Rod and J.K. Nørskov, The surface science of enzymes
500 (2002) 678

D.A. Williams and E. Herbst, It's a dusty Universe: surface science in space 500 (2002) 823

V.P. Zhdanov, Impact of surface science on the understanding of kinetics of heterogeneous catalytic reactions
500 (2002) 966

Models of surface kinetics

J.R. Arthur, Molecular beam epitaxy 500 (2002) 189

A.G. Naumovets and Z. Zhang, Fidgety particles on surfaces: how do they jump, walk, group, and settle in virgin areas?
500 (2002) 414

L.K. Bigelow and M.P. D'Evelyn, Role of surface and interface science in chemical vapor deposition diamond technology
500 (2002) 986

Molecular dynamics

F. Starrost and E.A. Carter, Modeling the full monty: baring the nature of surfaces across time and space
500 (2002) 323

L. Hanley and S.B. Sinnott, The growth and modification of materials via ion–surface processing 500 (2002) 500

Monte Carlo simulations

F. Starrost and E.A. Carter, Modeling the full monty: baring the nature of surfaces across time and space 500 (2002) 323

Non-equilibrium thermodynamics and statistical mechanics

C. Stampfl, M.V. Ganduglia-Pirovano, K. Reuter and M. Scheffler, Catalysis and corrosion: the theoretical surface-science context 500 (2002) 368

Semi-empirical models and model calculations

J.M. Greenberg, Cosmic dust and our origins 500 (2002) 793

EXPERIMENTAL SAMPLE PREPARATION AND CHARACTERIZATION METHODS

Adsorption isotherms

J.H. Sinfelt, Role of surface science in catalysis 500 (2002) 923

Atomic force microscopy

D.G. Castner and B.D. Ratner, Biomedical surface science: Foundations to frontiers 500 (2002) 28

Chemical vapor deposition

H. Dai, Carbon nanotubes: opportunities and challenges 500 (2002) 218

J.M. Greenberg, Cosmic dust and our origins 500 (2002) 793

L.K. Bigelow and M.P. D'Evelyn, Role of surface and interface science in chemical vapor deposition diamond technology 500 (2002) 986

Desorption induced by electronic transitions (DIET)

W.A. Diño, H. Kasai and A. Okiji, Dynamical phenomena including many body effects at metal surfaces 500 (2002) 105

Electron stimulated desorption (ESD)

T.E. Madey, R.E. Johnson and T.M. Orlando, Far-out surface science: radiation-induced surface processes in the solar system 500 (2002) 838

Photon stimulated desorption (PSD)

T.E. Madey, R.E. Johnson and T.M. Orlando, Far-out surface science: radiation-induced surface processes in the solar system 500 (2002) 838

Electrical transport measurements

S. Hasegawa and F. Grey, Electronic transport at semiconductor surfaces—from point-contact transistor to micro-four-point probes 500 (2002) 84

H. Dai, Carbon nanotubes: opportunities and challenges 500 (2002) 218

Electron microscopy

P. Finnie and Y. Homma, Epitaxy: the motion picture 500 (2002) 437

Electron–solid diffraction

Reflection high-energy electron diffraction (RHEED)

J.R. Arthur, Molecular beam epitaxy 500 (2002) 189

Ellipsometry

C.J. Hirschmugl, Frontiers in infrared spectroscopy at surfaces and interfaces 500 (2002) 577

Field ion microscopy

A.G. Naumovets and Z. Zhang, Fidgety particles on surfaces: how do they jump, walk, group, and settle in virgin areas? 500 (2002) 414

Ion etching

L. Hanley and S.B. Sinnott, The growth and modification of materials via ion–surface processing 500 (2002) 500

M.V. Ramana Murty, Sputtering: the material erosion tool 500 (2002) 523

Laser methods

M. Bonn, A.W. Kleyn and G.J. Kroes, Real time chemical dynamics at surfaces 500 (2002) 475

Molecular beam epitaxy

J.R. Arthur, Molecular beam epitaxy 500 (2002) 189

P. Finnie and Y. Homma, Epitaxy: the motion picture 500 (2002) 437

E. Vlieg, Understanding crystal growth in vacuum and beyond 500 (2002) 458

Molecule–solid reactions

M. Bonn, A.W. Kleyn and G.J. Kroes, Real time chemical dynamics at surfaces 500 (2002) 475

J.M. Greenberg, Cosmic dust and our origins 500 (2002) 793

Non-linear optical methods

C.T. Williams and D.A. Beattie, Probing buried interfaces with non-linear optical spectroscopy 500 (2002) 545

Sum frequency generation

C.T. Williams and D.A. Beattie, Probing buried interfaces with non-linear optical spectroscopy 500 (2002) 545

Photoelectron spectroscopy

D.G. Castner and B.D. Ratner, Biomedical surface science: Foundations to frontiers 500 (2002) 28
W. Eberhardt, Clusters as new materials 500 (2002) 242
M. Fahlman and W.R. Salaneck, Surfaces and interfaces in polymer-based electronics 500 (2002) 904

Synchrotron radiation photoelectron spectroscopy

S. Ferrer and Y. Petroff, Surface science done at third generation synchrotron radiation facilities 500 (2002) 605

Photon absorption spectroscopy

C.J. Hirschmugl, Frontiers in infrared spectroscopy at surfaces and interfaces 500 (2002) 577

Infrared absorption spectroscopy

C.J. Hirschmugl, Frontiers in infrared spectroscopy at surfaces and interfaces 500 (2002) 577
S.F. Bent, Organic functionalization of group IV semiconductor surfaces: principles, examples, applications, and prospects 500 (2002) 879

X-ray absorption spectroscopy

S. Ferrer and Y. Petroff, Surface science done at third generation synchrotron radiation facilities 500 (2002) 605

Scanning tunneling microscopy

S. Hasegawa and F. Grey, Electronic transport at semiconductor surfaces—from point-contact transistor to micro-four-point probes 500 (2002) 84
F. Rosei and R. Rosei, Atomic description of elementary surface processes: diffusion and dynamics 500 (2002) 395
A.G. Naumovets and Z. Zhang, Fidgety particles on surfaces: how do they jump, walk, group, and settle in virgin areas? 500 (2002) 414
P. Finnie and Y. Homma, Epitaxy: the motion picture 500 (2002) 437
D.M. Kolb, An atomistic view of electrochemistry 500 (2002) 722
S.F. Bent, Organic functionalization of group IV semiconductor surfaces: principles, examples, applications, and prospects 500 (2002) 879

Secondary ion mass spectroscopy

D.G. Castner and B.D. Ratner, Biomedical surface science: Foundations to frontiers 500 (2002) 28

M.V. Ramana Murty, Sputtering: the material erosion tool 500 (2002) 523

Sputter deposition

M.V. Ramana Murty, Sputtering: the material erosion tool 500 (2002) 523

X-ray scattering, diffraction, and reflection

E. Vlieg, Understanding crystal growth in vacuum and beyond 500 (2002) 458
S. Ferrer and Y. Petroff, Surface science done at third generation synchrotron radiation facilities 500 (2002) 605

PHENOMENA

Adhesion

M. Tirrell, E. Kokkoli and M. Biesalski, The role of surface science in bioengineered materials 500 (2002) 61
B. Kasemo, Biological surface science 500 (2002) 656
C.B. Duke, J. Noolandi and T. Thieret, The surface science of xerography 500 (2002) 1005

Adsorption kinetics

C.T. Williams and D.A. Beattie, Probing buried interfaces with non-linear optical spectroscopy 500 (2002) 545
J.H. Sinfelt, Role of surface science in catalysis 500 (2002) 923

Atom–solid interactions

D.A. Williams and E. Herbst, It's a dusty Universe: surface science in space 500 (2002) 823

Bending of surfaces

S.A. Safran, Statistical thermodynamics of soft surfaces 500 (2002) 127

Catalysis

W. Eberhardt, Clusters as new materials 500 (2002) 242
H.-J. Freund, Clusters and islands on oxides: from catalysis via electronics and magnetism to optics 500 (2002) 271
F. Starrost and E.A. Carter, Modeling the full monty: baring the nature of surfaces across time and space 500 (2002) 323
A. Groß, The virtual chemistry lab for reactions at surfaces: Is it possible? Will it be useful? 500 (2002) 347
C. Stampfl, M.V. Ganduglia-Pirovano, K. Reuter and M. Scheffler, Catalysis and corrosion: the theoretical surface-science context 500 (2002) 368
C.T. Williams and D.A. Beattie, Probing buried interfaces with non-linear optical spectroscopy 500 (2002) 545
C.J. Hirschmugl, Frontiers in infrared spectroscopy at surfaces and interfaces 500 (2002) 577

S. Ferrer and Y. Petroff, Surface science done at third generation synchrotron radiation facilities 500 (2002) 605

T.H. Rod and J.K. Nørskov, The surface science of enzymes 500 (2002) 678

J.H. Sinfelt, Role of surface science in catalysis 500 (2002) 923

F. Zaera, The surface chemistry of catalysis: new challenges ahead 500 (2002) 947

V.P. Zhdanov, Impact of surface science on the understanding of kinetics of heterogeneous catalytic reactions 500 (2002) 966

Chemisorption

M. Tirrell, E. Kokkoli and M. Biesalski, The role of surface science in bioengineered materials 500 (2002) 61

D.P. Woodruff, Solved and unsolved problems in surface structure determination 500 (2002) 147

H. Dai, Carbon nanotubes: opportunities and challenges 500 (2002) 218

W. Eberhardt, Clusters as new materials 500 (2002) 242

F. Rosei and R. Rosei, Atomic description of elementary surface processes: diffusion and dynamics 500 (2002) 395

M. Bonn, A.W. Kleyn and G.J. Kroes, Real time chemical dynamics at surfaces 500 (2002) 475

S. Ferrer and Y. Petroff, Surface science done at third generation synchrotron radiation facilities 500 (2002) 605

S.F. Bent, Organic functionalization of group IV semiconductor surfaces: principles, examples, applications, and prospects 500 (2002) 879

J.H. Sinfelt, Role of surface science in catalysis 500 (2002) 923

F. Zaera, The surface chemistry of catalysis: new challenges ahead 500 (2002) 947

V.P. Zhdanov, Impact of surface science on the understanding of kinetics of heterogeneous catalytic reactions 500 (2002) 966

Conductivity

W.A. Diño, H. Kasai and A. Okiji, Dynamical phenomena including many body effects at metal surfaces 500 (2002) 105

M. Fahlman and W.R. Salaneck, Surfaces and interfaces in polymer-based electronics 500 (2002) 904

Corrosion

C. Stampfl, M.V. Ganduglia-Pirovano, K. Reuter and M. Scheffler, Catalysis and corrosion: the theoretical surface-science context 500 (2002) 368

Desorption induced by photon stimulation

M. Bonn, A.W. Kleyn and G.J. Kroes, Real time chemical dynamics at surfaces 500 (2002) 475

J.M. Greenberg, Cosmic dust and our origins 500 (2002) 793

Dielectric phenomena

C.B. Duke, J. Noolandi and T. Thieret, The surface science of xerography 500 (2002) 1005

Diffusion and migration

D.A. Williams and E. Herbst, It's a dusty Universe: surface science in space 500 (2002) 823

Electrical transport (conductivity, resistivity, mobility, etc.)

J. Shen and J. Kirschner, Tailoring magnetism in artificially structured materials: the new frontier 500 (2002) 300

C.J. Hirschmugl, Frontiers in infrared spectroscopy at surfaces and interfaces 500 (2002) 577

C.B. Duke, J. Noolandi and T. Thieret, The surface science of xerography 500 (2002) 1005

Energy dissipation

C.J. Hirschmugl, Frontiers in infrared spectroscopy at surfaces and interfaces 500 (2002) 577

J. Krim, Surface science and the atomic-scale origins of friction: what once was old is new again 500 (2002) 741

Epitaxy

J.R. Arthur, Molecular beam epitaxy 500 (2002) 189

J. Shen and J. Kirschner, Tailoring magnetism in artificially structured materials: the new frontier 500 (2002) 300

F. Rosei and R. Rosei, Atomic description of elementary surface processes: diffusion and dynamics 500 (2002) 395

P. Finnie and Y. Homma, Epitaxy: the motion picture 500 (2002) 437

Etching

F. Starrost and E.A. Carter, Modeling the full monty: baring the nature of surfaces across time and space 500 (2002) 323

Friction

J. Krim, Surface science and the atomic-scale origins of friction: what once was old is new again 500 (2002) 741

Growth

F. Starrost and E.A. Carter, Modeling the full monty: baring the nature of surfaces across time and space 500 (2002) 323

F. Rosei and R. Rosei, Atomic description of elementary surface processes: diffusion and dynamics 500 (2002) 395

P. Finnie and Y. Homma, Epitaxy: the motion picture 500 (2002) 437

E. Vlieg, Understanding crystal growth in vacuum and beyond 500 (2002) 458

L. Hanley and S.B. Sinnott, The growth and modification of materials via ion–surface processing 500 (2002) 500

G.B. Sergeev and T.I. Shabatina, Low temperature surface chemistry and nanostructures 500 (2002) 628

L.K. Bigelow and M.P. D'Evelyn, Role of surface and interface science in chemical vapor deposition diamond technology 500 (2002) 986

Ion bombardment

L. Hanley and S.B. Sinnott, The growth and modification of materials via ion–surface processing 500 (2002) 500

T.E. Madey, R.E. Johnson and T.M. Orlando, Far-out surface science: radiation-induced surface processes in the solar system 500 (2002) 838

Ion implantation

L. Hanley and S.B. Sinnott, The growth and modification of materials via ion–surface processing 500 (2002) 500

Ion–solid interactions

L. Hanley and S.B. Sinnott, The growth and modification of materials via ion–surface processing 500 (2002) 500

Luminescence

Electroluminescence

M. Fahlman and W.R. Salaneck, Surfaces and interfaces in polymer-based electronics 500 (2002) 904

Photoluminescence

M. Fahlman and W.R. Salaneck, Surfaces and interfaces in polymer-based electronics 500 (2002) 904

Magnetic phenomena (cyclotron resonance, phase transitions, etc.)

E.W. Plummer, Ismail, R. Matzdorf, A.V. Melechko, J.P. Pierce and J. Zhang, Surfaces: a playground for physics with broken symmetry in reduced dimensionality 500 (2002) 1

S.D. Bader, Magnetism in low dimensionality 500 (2002) 172

W. Eberhardt, Clusters as new materials 500 (2002) 242

J. Shen and J. Kirschner, Tailoring magnetism in artificially structured materials: the new frontier 500 (2002) 300

C.T. Williams and D.A. Beattie, Probing buried interfaces with non-linear optical spectroscopy 500 (2002) 545

Oxidation

C. Stampfl, M.V. Ganduglia-Pirovano, K. Reuter and M. Scheffler, Catalysis and corrosion: the theoretical surface-science context 500 (2002) 368

M.K. Weldon, K.T. Queeney, J. Eng Jr., K. Raghavachari and Y.J. Chabal, The surface science of semiconductor processing: gate oxides in the ever-shrinking transistor 500 (2002) 859

Photochemistry

M. Bonn, A.W. Kleyn and G.J. Kroes, Real time chemical dynamics at surfaces 500 (2002) 475

J.M. Greenberg, Cosmic dust and our origins 500 (2002) 793

Photoconductivity

C.B. Duke, J. Noolandi and T. Thieret, The surface science of xerography 500 (2002) 1005

Physical adsorption

J. Krim, Surface science and the atomic-scale origins of friction: what once was old is new again 500 (2002) 741

J.H. Sinfelt, Role of surface science in catalysis 500 (2002) 923

Quantum effects

H. Dai, Carbon nanotubes: opportunities and challenges 500 (2002) 218

W. Eberhardt, Clusters as new materials 500 (2002) 242

Second harmonic generation

M. Bonn, A.W. Kleyn and G.J. Kroes, Real time chemical dynamics at surfaces 500 (2002) 475

C.T. Williams and D.A. Beattie, Probing buried interfaces with non-linear optical spectroscopy 500 (2002) 545

Self-assembly

D.G. Castner and B.D. Ratner, Biomedical surface science: Foundations to frontiers 500 (2002) 28

M. Tirrell, E. Kokkoli and M. Biesalski, The role of surface science in bioengineered materials 500 (2002) 61

S.A. Safran, Statistical thermodynamics of soft surfaces 500 (2002) 127

C.T. Williams and D.A. Beattie, Probing buried interfaces with non-linear optical spectroscopy 500 (2002) 545

Sputtering

L. Hanley and S.B. Sinnott, The growth and modification of materials via ion–surface processing 500 (2002) 500

M.V. Ramana Murty, Sputtering: the material erosion tool 500 (2002) 523

T.E. Madey, R.E. Johnson and T.M. Orlando, Far-out surface science: radiation-induced surface processes in the solar system 500 (2002) 838

Sticking

A. Groß, The virtual chemistry lab for reactions at surfaces: Is it possible? Will it be useful? 500 (2002) 347

Superconductivity

R.A. Broglia, The surfaces of compact systems: from nuclei to stars 500 (2002) 759

Surface chemical reaction

M. Tirrell, E. Kokkoli and M. Biesalski, The role of surface science in bioengineered materials 500 (2002) 61

A. Groß, The virtual chemistry lab for reactions at surfaces: Is it possible? Will it be useful? 500 (2002) 347

C. Stampfl, M.V. Ganduglia-Pirovano, K. Reuter and M. Scheffler, Catalysis and corrosion: the theoretical surface-science context 500 (2002) 368

F. Rosei and R. Rosei, Atomic description of elementary surface processes: diffusion and dynamics 500 (2002) 395

M. Bonn, A.W. Kleyn and G.J. Kroes, Real time chemical dynamics at surfaces 500 (2002) 475

S. Ferrer and Y. Petroff, Surface science done at third generation synchrotron radiation facilities 500 (2002) 605

G.B. Sergeev and T.I. Shabatina, Low temperature surface chemistry and nanostructures 500 (2002) 628

T.H. Rod and J.K. Nørskov, The surface science of enzymes 500 (2002) 678

S.D. Gillmor, P.P. Rugheimer and M.G. Lagally, Computation with DNA on surfaces 500 (2002) 699

J.M. Greenberg, Cosmic dust and our origins 500 (2002) 793

M. Fahlman and W.R. Salaneck, Surfaces and interfaces in polymer-based electronics 500 (2002) 904

J.H. Sinfelt, Role of surface science in catalysis 500 (2002) 923

F. Zaera, The surface chemistry of catalysis: new challenges ahead 500 (2002) 947

L.K. Bigelow and M.P. D'Evelyn, Role of surface and interface science in chemical vapor deposition diamond technology 500 (2002) 986

Surface diffusion

H.-J. Freund, Clusters and islands on oxides: from catalysis via electronics and magnetism to optics 500 (2002) 271

F. Rosei and R. Rosei, Atomic description of elementary surface processes: diffusion and dynamics 500 (2002) 395

A.G. Naumovets and Z. Zhang, Fidgety particles on surfaces: how do they jump, walk, group, and settle in virgin areas? 500 (2002) 414

J.M. Greenberg, Cosmic dust and our origins 500 (2002) 793

Surface electrical transport (surface conductivity, surface recombination, etc.)

S. Hasegawa and F. Grey, Electronic transport at semiconductor surfaces—from point-contact transistor to micro-four-point probes 500 (2002) 84

W.A. Diño, H. Kasai and A. Okiji, Dynamical phenomena including many body effects at metal surfaces 500 (2002) 105

Surface electronic phenomena (work function, surface potential, surface states, etc.)

S. Hasegawa and F. Grey, Electronic transport at semiconductor surfaces—from point-contact transistor to micro-four-point probes 500 (2002) 84

C.B. Duke, J. Noolandi and T. Thieret, The surface science of xerography 500 (2002) 1005

Surface energy

G.B. Sergeev and T.I. Shabatina, Low temperature surface chemistry and nanostructures 500 (2002) 628

F. Zaera, The surface chemistry of catalysis: new challenges ahead 500 (2002) 947

L.K. Bigelow and M.P. D'Evelyn, Role of surface and interface science in chemical vapor deposition diamond technology 500 (2002) 986

C.B. Duke, J. Noolandi and T. Thieret, The surface science of xerography 500 (2002) 1005

Surface melting

C.B. Duke, J. Noolandi and T. Thieret, The surface science of xerography 500 (2002) 1005

Surface relaxation and reconstruction

D.P. Woodruff, Solved and unsolved problems in surface structure determination 500 (2002) 147

E. Vlieg, Understanding crystal growth in vacuum and beyond 500 (2002) 458

D.M. Kolb, An atomistic view of electrochemistry 500 (2002) 722

Surface structure, morphology, roughness, and topography

E.W. Plummer, Ismail, R. Matzdorf, A.V. Melechko, J.P. Pierce and J. Zhang, Surfaces: a playground for physics with broken symmetry in reduced dimensionality 500 (2002) 1

S.A. Safran, Statistical thermodynamics of soft surfaces 500 (2002) 127

D.P. Woodruff, Solved and unsolved problems in surface structure determination 500 (2002) 147

E. Vlieg, Understanding crystal growth in vacuum and beyond 500 (2002) 458

M.V. Ramana Murty, Sputtering: the material erosion tool 500 (2002) 523

D.M. Kolb, An atomistic view of electrochemistry 500 (2002) 722

S.S. Badesha and J.A. Swift, Practical surfaces: beyond the wheel 500 (2002) 1024

Surface tension

S.A. Safran, Statistical thermodynamics of soft surfaces 500 (2002) 127

Surface thermodynamics (including phase transitions)

S.A. Safran, Statistical thermodynamics of soft surfaces 500 (2002) 127

C. Stampfl, M.V. Ganduglia-Pirovano, K. Reuter and M. Scheffler, Catalysis and corrosion: the theoretical surface-science context
500 (2002) 368

A.G. Naumovets and Z. Zhang, Fidgety particles on surfaces: how do they jump, walk, group, and settle in virgin areas?
500 (2002) 414

Surface waves

Phonons

J. Krim, Surface science and the atomic-scale origins of friction: what once was old is new again
500 (2002) 741

Tribology

C.T. Williams and D.A. Beattie, Probing buried interfaces with non-linear optical spectroscopy
500 (2002) 545

Vibrations of adsorbed molecules

C.J. Hirschmugl, Frontiers in infrared spectroscopy at surfaces and interfaces
500 (2002) 577

M.K. Weldon, K.T. Queeney, J. Eng Jr., K. Raghavachari and Y.J. Chabal, The surface science of semiconductor processing: gate oxides in the ever-shrinking transistor
500 (2002) 859

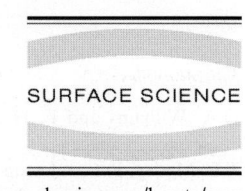

Materials index

ELEMENTAL AND CHEMICAL IDENTITY

Elements

Alkali metals

T.E. Madey, R.E. Johnson and T.M. Orlando, Far-out surface science: radiation-induced surface processes in the solar system 500 (2002) 838

Carbon

H. Dai, Carbon nanotubes: opportunities and challenges 500 (2002) 218

Cobalt

S.D. Bader, Magnetism in low dimensionality 500 (2002) 172

Copper

D.M. Kolb, An atomistic view of electrochemistry 500 (2002) 722

Diamond

S.F. Bent, Organic functionalization of group IV semiconductor surfaces: principles, examples, applications, and prospects 500 (2002) 879

L.K. Bigelow and M.P. D'Evelyn, Role of surface and interface science in chemical vapor deposition diamond technology 500 (2002) 986

Germanium

S.F. Bent, Organic functionalization of group IV semiconductor surfaces: principles, examples, applications, and prospects 500 (2002) 879

Gold

D.M. Kolb, An atomistic view of electrochemistry 500 (2002) 722

Iron

S.D. Bader, Magnetism in low dimensionality 500 (2002) 172

Platinum

D.M. Kolb, An atomistic view of electrochemistry 500 (2002) 722

Ruthenium

C. Stampfl, M.V. Ganduglia-Pirovano, K. Reuter and M. Scheffler, Catalysis and corrosion: the theoretical surface-science context 500 (2002) 368

Silicon

S. Hasegawa and F. Grey, Electronic transport at semiconductor surfaces—from point-contact transistor to micro-four-point probes 500 (2002) 84

M.K. Weldon, K.T. Queeney, J. Eng Jr., K. Raghavachari and Y.J. Chabal, The surface science of semiconductor processing: gate oxides in the ever-shrinking transistor 500 (2002) 859

S.F. Bent, Organic functionalization of group IV semiconductor surfaces: principles, examples, applications, and prospects 500 (2002) 879

Silver

S. Hasegawa and F. Grey, Electronic transport at semiconductor surfaces—from point-contact transistor to micro-four-point probes 500 (2002) 84

Compounds (non-molecular solids)

Alloys

W.A. Diño, H. Kasai and A. Okiji, Dynamical phenomena including many body effects at metal surfaces 500 (2002) 105

Aluminum oxide

H.-J. Freund, Clusters and islands on oxides: from catalysis via electronics and magnetism to optics 500 (2002) 271

Biological compounds

D.G. Castner and B.D. Ratner, Biomedical surface science: Foundations to frontiers 500 (2002) 28

M. Tirrell, E. Kokkoli and M. Biesalski, The role of surface science in bioengineered materials 500 (2002) 61

C.J. Hirschmugl, Frontiers in infrared spectroscopy at surfaces and interfaces 500 (2002) 577

B. Kasemo, Biological surface science 500 (2002) 656

Inorganic compounds

T.H. Rod and J.K. Nørskov, The surface science of enzymes 500 (2002) 678

Silicon oxides

D.A. Williams and E. Herbst, It's a dusty Universe: surface science in space 500 (2002) 823

T.E. Madey, R.E. Johnson and T.M. Orlando, Far-out surface science: radiation-induced surface processes in the solar system 500 (2002) 838

M.K. Weldon, K.T. Queeney, J. Eng Jr., K. Raghavachari and Y.J. Chabal, The surface science of semiconductor processing: gate oxides in the ever-shrinking transistor 500 (2002) 859

Molecules

Alkenes

S.F. Bent, Organic functionalization of group IV semiconductor surfaces: principles, examples, applications, and prospects 500 (2002) 879

Biological molecules – nucleic acids

B. Kasemo, Biological surface science 500 (2002) 656

S.D. Gillmor, P.P. Rugheimer and M.G. Lagally, Computation with DNA on surfaces 500 (2002) 699

Biological molecules – proteins

D.G. Castner and B.D. Ratner, Biomedical surface science: Foundations to frontiers 500 (2002) 28

B. Kasemo, Biological surface science 500 (2002) 656

T.H. Rod and J.K. Nørskov, The surface science of enzymes 500 (2002) 678

Fullerenes

W. Eberhardt, Clusters as new materials 500 (2002) 242

Water

T.E. Madey, R.E. Johnson and T.M. Orlando, Far-out surface science: radiation-induced surface processes in the solar system 500 (2002) 838

PHYSICAL STATE

Surfaces

Amorphous surfaces

D.A. Williams and E. Herbst, It's a dusty Universe: surface science in space 500 (2002) 823

T.E. Madey, R.E. Johnson and T.M. Orlando, Far-out surface science: radiation-induced surface processes in the solar system 500 (2002) 838

Polycrystalline surfaces

L.K. Bigelow and M.P. D'Evelyn, Role of surface and interface science in chemical vapor deposition diamond technology 500 (2002) 986

Single crystal surfaces

J.R. Arthur, Molecular beam epitaxy 500 (2002) 189

A.G. Naumovets and Z. Zhang, Fidgety particles on surfaces: how do they jump, walk, group, and settle in virgin areas? 500 (2002) 414

D.M. Kolb, An atomistic view of electrochemistry 500 (2002) 722

S.F. Bent, Organic functionalization of group IV semiconductor surfaces: principles, examples, applications, and prospects 500 (2002) 879

J.H. Sinfelt, Role of surface science in catalysis 500 (2002) 923

F. Zaera, The surface chemistry of catalysis: new challenges ahead 500 (2002) 947

Low index single crystal surfaces

M.K. Weldon, K.T. Queeney, J. Eng Jr., K. Raghavachari and Y.J. Chabal, The surface science of semiconductor processing: gate oxides in the ever-shrinking transistor 500 (2002) 859

Surface defects

E.W. Plummer, Ismail, R. Matzdorf, A.V. Melechko, J.P. Pierce and J. Zhang, Surfaces: a playground for physics with broken symmetry in reduced dimensionality 500 (2002) 1

Interfaces

Solid–liquid interfaces

D.G. Castner and B.D. Ratner, Biomedical surface science: Foundations to frontiers 500 (2002) 28

E. Vlieg, Understanding crystal growth in vacuum and beyond 500 (2002) 458

S. Ferrer and Y. Petroff, Surface science done at third generation synchrotron radiation facilities 500 (2002) 605

B. Kasemo, Biological surface science 500 (2002) 656

D.M. Kolb, An atomistic view of electrochemistry 500 (2002) 722

Solid–gas interfaces

F. Starrost and E.A. Carter, Modeling the full monty: baring the nature of surfaces across time and space 500 (2002) 323

Heterostructures

Heterojunctions

J.R. Arthur, Molecular beam epitaxy 500 (2002) 189

M. Fahlman and W.R. Salaneck, Surfaces and interfaces in polymer-based electronics 500 (2002) 904

Quantum wells

J.R. Arthur, Molecular beam epitaxy 500 (2002) 189

Superlattices

J.R. Arthur, Molecular beam epitaxy 500 (2002) 189

Adatoms

A.G. Naumovets and Z. Zhang, Fidgety particles on surfaces: how do they jump, walk, group, and settle in virgin areas? 500 (2002) 414

Clusters

W. Eberhardt, Clusters as new materials 500 (2002) 242
H.-J. Freund, Clusters and islands on oxides: from catalysis via electronics and magnetism to optics 500 (2002) 271
A.G. Naumovets and Z. Zhang, Fidgety particles on surfaces: how do they jump, walk, group, and settle in virgin areas? 500 (2002) 414
G.B. Sergeev and T.I. Shabatina, Low temperature surface chemistry and nanostructures 500 (2002) 628
J.H. Sinfelt, Role of surface science in catalysis 500 (2002) 923

Porous solids

J.M. Greenberg, Cosmic dust and our origins 500 (2002) 793

ELECTRICAL AND MAGNETIC BEHAVIOR

Surfaces and/or films

Insulating films

H.-J. Freund, Clusters and islands on oxides: from catalysis via electronics and magnetism to optics 500 (2002) 271

Magnetic films

S.D. Bader, Magnetism in low dimensionality 500 (2002) 172
J. Shen and J. Kirschner, Tailoring magnetism in artificially structured materials: the new frontier 500 (2002) 300

Magnetic surfaces

S.D. Bader, Magnetism in low dimensionality 500 (2002) 172
S. Ferrer and Y. Petroff, Surface science done at third generation synchrotron radiation facilities 500 (2002) 605

Metallic surfaces

W.A. Diño, H. Kasai and A. Okiji, Dynamical phenomena including many body effects at metal surfaces 500 (2002) 105

Interfaces

Magnetic interfaces

S.D. Bader, Magnetism in low dimensionality 500 (2002) 172
J. Shen and J. Kirschner, Tailoring magnetism in artificially structured materials: the new frontier 500 (2002) 300

Heterostructures and thin film structures

Metal–metal magnetic heterostructures

J. Shen and J. Kirschner, Tailoring magnetism in artificially structured materials: the new frontier 500 (2002) 300